The Elements

Name	Symbol	Atomic Number	Relative Atomic Mass	Name	Symbol	Atomic Number	Relative Atomic Mass
Actinium	Ac	89	227.028	Mendelevium	Md	101	(258)
Aluminum	Al	13	26.9815	Mercury	Hg	80	200.59
Americium	Am	95	(243)	Molybdenum	Mo	42	95.94
Antimony	Sb	51	121.760	Neodymium	Nd	60	144.24
Argon	Ar	18	39.948	Neon	Ne	10	20.1797
Arsenic	As	33	74.9216	Neptunium	Np	93	237.048
Astatine	At	85	(210)	Nickel	Ni	28	58.6934
Barium	Ba	56	137.327	Niobium	Nb	41	92.9064
Berkelium	Bk	97	(247)	Nitrogen	N	7	14.0067
Beryllium	Be	4	9.01218	Nobelium	No	102	(259)
Bismuth	Bi	83	208.980	Osmium	Os	76	190.23
Bohrium	Bh	107	(262)	Oxygen	O	8	15.9994
Boron	B	5	10.811	Palladium	Pd	46	106.42
Bromine	Br	35	79.904	Phosphorus	P	15	30.9738
Cadmium	Cd	48	112.411	Platinum	Pt	78	195.084
Calcium	Ca	20	40.078	Plutonium	Pu	94	(244)
Californium	Cf	98	(251)	Polonium	Po	84	(209)
Carbon	C	6	12.0107	Potassium	K	19	39.0983
Cerium	Ce	58	140.115	Praseodymium	Pr	59	140.908
Cesium	Cs	55	132.905	Promethium	Pm	61	(145)
Chlorine	Cl	17	35.453	Protactinium	Pa	91	231.036
Chromium	Cr	24	51.9961	Radium	Ra	88	(226)
Cobalt	Co	27	58.9332	Radon	Rn	86	(222)
Copper	Cu	29	63.546	Rhenium	Re	75	186.207
Curium	Cm	96	(247)	Rhodium	Rh	45	102.906
Darmstadtium	Ds	110	(271)	Roentgenium	Rg	111	(272)
Dubnium	Db	105	(262)	Rubidium	Rb	37	85.4678
Dysprosium	Dy	66	162.50	Ruthenium	Ru	44	101.07
Einsteinium	Es	99	(252)	Rutherfordium	Rf	104	(261)
Erbium	Er	68	167.26	Samarium	Sm	62	150.36
Europium	Eu	63	151.965	Scandium	Sc	21	44.9559
Fermium	Fm	100	(257)	Seaborgium	Sg	106	(263)
Fluorine	F	9	18.9984	Selenium	Se	34	78.96
Francium	Fr	87	(223)	Silicon	Si	14	28.0855
Gadolinium	Gd	64	157.25	Silver	Ag	47	107.868
Gallium	Ga	31	69.723	Sodium	Na	11	22.9898
Germanium	Ge	32	72.64	Strontium	Sr	38	87.62
Gold	Au	79	196.967	Sulfur	S	16	32.065
Hafnium	Hf	72	178.49	Tantalum	Ta	73	180.948
Hassium	Hs	108	(265)	Technetium	Tc	43	(98)
Helium	He	2	4.00260	Tellurium	Te	52	127.60
Holmium	Ho	67	164.930	Terbium	Tb	65	158.925
Hydrogen	H	1	1.00794	Thallium	Tl	81	204.383
Indium	In	49	114.818	Thorium	Th	90	232.038
Iodine	I	53	126.904	Thulium	Tm	69	168.934
Iridium	Ir	77	192.217	Tin	Sn	50	118.710
Iron	Fe	26	55.845	Titanium	Ti	22	47.867
Krypton	Kr	36	83.798	Tungsten	W	74	183.84
Lanthanum	La	57	138.905	Uranium	U	92	238.029
Lawrencium	Lr	103	(260)	Vanadium	V	23	50.9415
Lead	Pb	82	207.2	Xenon	Xe	54	131.293
Lithium	Li	3	6.941	Ytterbium	Yb	70	173.04
Lutetium	Lu	71	174.967	Yttrium	Y	39	88.9059
Magnesium	Mg	12	24.3050	Zinc	Zn	30	65.409
Manganese	Mn	25	54.9380	Zirconium	Zr	40	91.224
Meitnerium	Mt	109	(266)				

Atomic masses in this table are relative to carbon-12 and are the values recommended by the International Union of Pure and Applied Chemistry (IUPAC) (22 June 2007). For certain radioactive elements the numbers listed (in parentheses) are the mass numbers of the most stable isotopes.

Mastering CHEMISTRY™

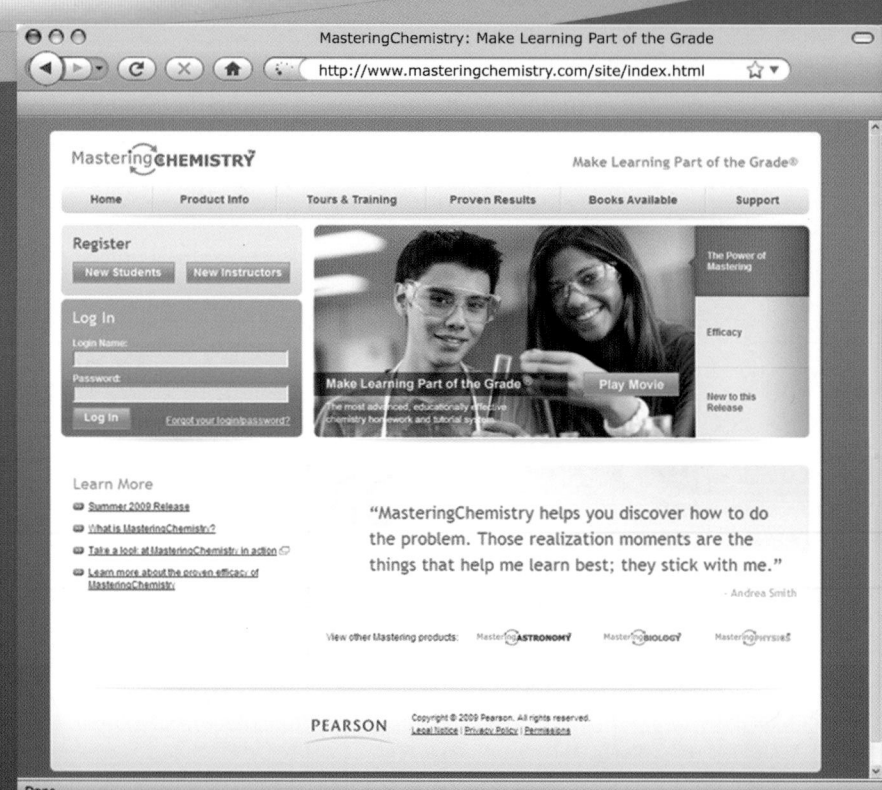

Save Time.

. . . .

Improve Results.

More than 1 million students have used a Pearson Mastering product to get a better grade.

MasteringChemistry is an online tutorial and homework system that is designed to provide students with customized coaching and individualized feedback to help improve problem-solving skills. *Mastering Chemistry* helps you complete homework efficiently and effectively and includes tutorials that provide targeted help.

- Hints that you can choose to open up to provide extra help when you get stuck on a problem
- Feedback specific to common errors to help guide you to the correct final answer
- Partial credit for the parts of an exercise that you get right
- 24/7 access—you decide when you want to study
- A full interactive eText
- A set of self-study quizzes

To take advantage of all that *MasteringChemistry* has to offer, you will need an access code. If you do not already have an access code, you can buy one online at **www.masteringchemistry.com**

Personalized Learning

In MasteringChemistry you are treated as an individual with specific learning needs.

The study and assessment resources that come with your textbook allow you to review content and develop what you need to know, on your own time, and at your own pace.

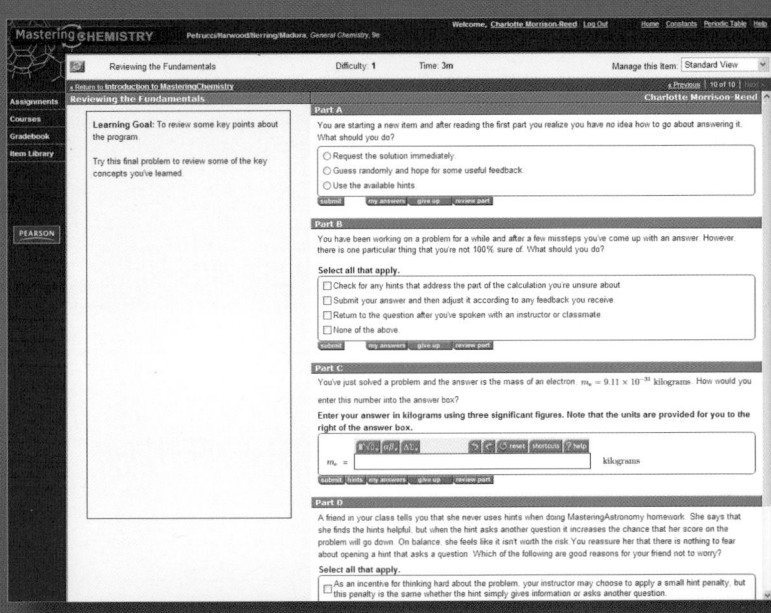

MasteringChemistry Benefits for Students:

- **Online homework questions** with immediate grading so you know right away how well you did and can learn from your mistakes.

- **Immediate and specific feedback** on wrong answers provides you with personal coaching. Specific feedback on common errors helps explain why a particular answer is not correct.

- **Hints provide individualized coaching**. Skip the hints you don't need and access only the ones that you need, for the most efficient path to the correct solution.

- An **interactive eText** so you can access the textbook anywhere.

- **Self-quizzing** to test yourself on each chapter.

- **Review your assignments anytime** (including all the hints for each problem) to study for midterms and exams.

"Textbooks give the answers to certain problems, but they don't help with the method or in why you got an answer wrong. By helping with these aspects, *MasteringChemistry* lets me learn from my mistakes rather than staying confused."
Student Brittany Schieron, University of California, Davis

Save Time. Improve Results. www.masteringchemistry.com

Pearson eText

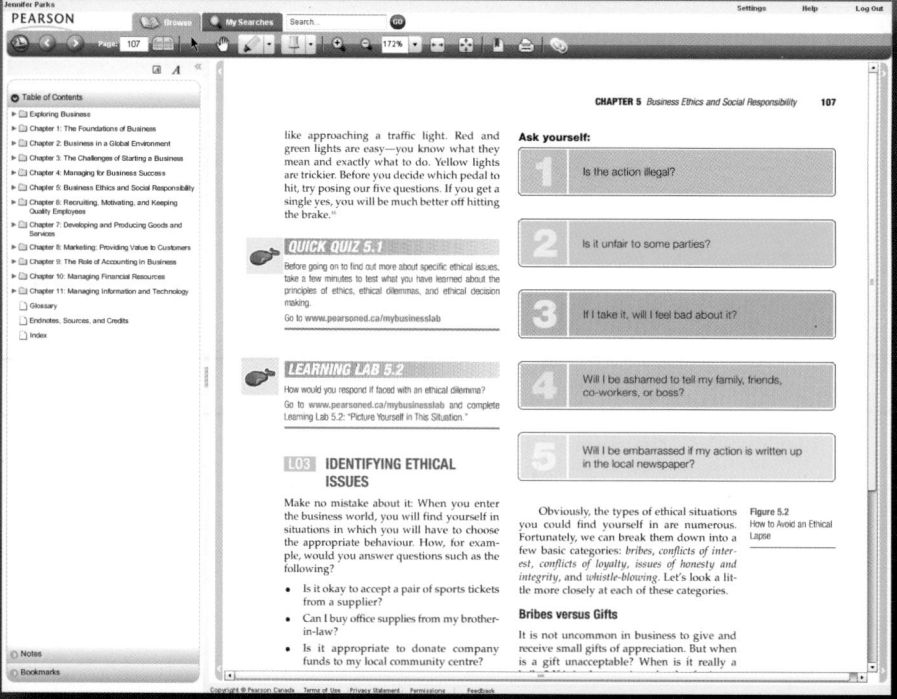

Pearson eText gives students access to the text whenever and wherever they have access to the internet. eText pages look exactly like the printed text, offering powerful new functionality for students and instructors.

Users can create notes, highlight text in different colours, create bookmarks, zoom, click hyperlinked words and phrases to view definitions, and choose single-page or two-page view.

Pearson eText allows for quick navigation using a table of contents and provides full-text search. The eText may also offer links to associated media files, enabling users to access videos, animations, or other activities as they read the text.

Save Time. Improve Results. www.masteringchemistry.com

GENERAL CHEMISTRY

Principles and Modern Applications TENTH EDITION

RALPH H. PETRUCCI
CALIFORNIA STATE UNIVERSITY, SAN BERNARDINO

F. GEOFFREY HERRING
UNIVERSITY OF BRITISH COLUMBIA

JEFFRY D. MADURA
DUQUESNE UNIVERSITY

CAREY BISSONNETTE
UNIVERSITY OF WATERLOO

Pearson Canada
Toronto

Library and Archives Canada Cataloguing in Publication

General Chemistry: Principles and Modern Applications / Ralph H. Petrucci ...
[et al.].—10th ed.

Includes index.

ISBN 978-0-13-206452-1
1. Chemistry—Textbooks.   I. Petrucci, Ralph H.   II. Title.

QD31.3.P47 2010   540   C2009-902505-1

ISBN: 978-0-13-206452-1

Vice-President, Editorial Director: Gary Bennett
Acquisitions Editor: Cathleen Sullivan
Marketing Manager: Kimberly Ukrainec
Supervising Developmental Editor: Maurice Esses
Production Editor: Lila Campbell
Copy Editor: Dawn Hunter
Proofreaders: Rosemary Tanner, Heather Sangster of Strong Finish
Production Coordinators: Lynn O'Rourke, Patricia Ciardullo
Compositor: GGS Higher Education Resources, a Division of PreMedia Global, Inc.
Photo Research: Heather Jackson
Art Director: Julia Hall
Cover Designer: Miguel Acevedo
Interior Designer: Quinn Banting
Cover Image: GGS Higher Education Resources, a Division of PreMedia Global, Inc.

For permission to reproduce copyrighted material, the publisher gratefully acknowledges the copyright holders listed on pages PC1–PC2, which are considered an extension of this copyright page.

2 3 4 5   14 13 12 11 10

Printed and bound in the United States of America.

WARNING: Many of the compounds and chemical reactions described or pictured in this book are hazardous. Do not attempt any experiment pictured or implied in the text except with permission in an authorized laboratory setting and under adequate supervision.

Brief Table of Contents

Contents

Focus On Discussions on MasteringChemistry™ (www.masteringchemistry.com)

About the Authors

Ralph H. Petrucci

Ralph Petrucci received his B.S. in Chemistry from Union College, Schenectady, NY, and his Ph.D. from the University of Wisconsin–Madison. Following ten years of teaching, research, consulting, and directing the NSF Institutes for Secondary School Science Teachers at Case Western Reserve University, Cleveland, OH, Dr. Petrucci joined the planning staff of the new California State University campus at San Bernardino in 1964. There, in addition to his faculty appointment, he served as Chairman of the Natural Sciences Division and Dean of Academic Planning. Professor Petrucci, now retired from teaching, is also a coauthor of *General Chemistry* with John W. Hill, Terry W. McCreary, and Scott S. Perry.

F. Geoffrey Herring

Geoff Herring received both his B.Sc. and his Ph.D. in Physical Chemistry, from the University of London. He is currently a Professor Emeritus in the Department of Chemistry of the University of British Columbia, Vancouver. Dr. Herring has research interests in biophysical chemistry and has published more than 100 papers in physical chemistry and chemical physics. Recently, Dr. Herring has undertaken studies in the use of information technology and interactive engagement methods in teaching general chemistry with a view to improving student comprehension and learning. Dr. Herring has taught chemistry from undergraduate to graduate levels for 30 years and has twice been the recipient of the Killam Prize for Excellence in Teaching.

Jeffry D. Madura

Jeffry D. Madura is a Professor in the Department of Chemistry and Biochemistry at Duquesne University located in Pittsburgh, PA. He earned a B.A. from Thiel College in 1980 and a Ph.D. in Physical Chemistry from Purdue University in 1985. The Ph.D. was followed by a postdoctoral fellowship in biophysics with Professor J. Andrew McCammon at the University of Houston. Dr. Madura's research interests are in computational chemistry and biophysics. He has published more than 80 papers in physical chemistry and chemical physics. Dr. Madura has taught chemistry from undergraduate to graduate levels for 20 years and was the recipient of a Dreyfus Teacher-Scholar Award. He also received the Bayer School of Natural and Environmental Sciences and the Duquesne University Presidential Award for Excellence in Scholarship in 2007.

Carey Bissonnette

Carey Bissonnette is Continuing Lecturer in the Department of Chemistry at the University of Waterloo, Ontario. He received his B.Sc. from the University of Waterloo in 1989 and his Ph.D. in 1993 from the University of Cambridge in England. His research interests are in the development of methods for modeling dynamical processes of polyatomic molecules in the gas phase. He has won awards for excellence in teaching, including the University of Waterloo's Distinguished Teacher Award in 2005. Dr. Bissonnette has made extensive use of technology in both the classroom and the laboratory to create an interactive environment for his students to learn and explore. For the past several years, he has been actively engaged in undergraduate curriculum development, high-school liaison activities, and the coordination of the university's high-school chemistry contests, which are written each year by students around the world.

Preface

"Know your audience." For this new edition, we have tried to follow this important advice by attending even more to the needs of students who are taking a serious journey through this material. We also know that most general chemistry students have career interests not in chemistry but in other areas such as biology, medicine, engineering, environmental science, and agricultural sciences. And we understand that general chemistry will be the only university or college chemistry course for some students, and thus their only opportunity to learn some practical applications of chemistry. We have designed this book for all these students.

Students of this text should have already studied some chemistry. But those with no prior background and those who could use a refresher will find that the early chapters develop fundamental concepts from the most elementary ideas. Students who do plan to become professional chemists will also find opportunities in the text to pursue their own special interests.

The typical student may need help identifying and applying principles and visualizing their physical significance. The pedagogical features of this text are designed to provide this help. At the same time, we hope the text serves to sharpen student skills in problem solving and critical thinking. Thus, we have tried to strike the proper balances between principles and applications, qualitative and quantitative discussions, and rigor and simplification.

Throughout the text and on the Mastering Chemistry site (www.mastering chemistry.com) we provide real-world examples to enhance the discussion. Examples relevant to the biological sciences, engineering, and the environmental sciences will be found in numerous places. This should help to bring the chemistry alive for these students and help them understand its relevance to their career interests. It also, in most cases, should help them master core concepts.

ORGANIZATION

In this edition we retain the core organization of the ninth edition of this text, but with additional depth and breadth of coverage of material in several areas. After a brief overview of core concepts in Chapter 1, we introduce atomic theory, including the periodic table, in Chapter 2. The periodic table is an extraordinarily useful tool, and presenting it early allows us to use the periodic table in new ways throughout the early chapters of the text. In Chapter 3 we introduce chemical compounds and their stoichiometry. Organic compounds are included in this presentation. The early introduction of organic compounds allows us to use organic examples throughout the book. Chapters 4 and 5 introduce chemical reactions. We discuss gases in Chapter 6, partly because they are familiar to students (which helps them build confidence), but also because some instructors prefer to cover this material early to better integrate their lecture and lab programs. Note that Chapter 6 can easily be deferred for coverage with the other states of matter, in Chapter 12. In Chapter 8 we delve more deeply into wave mechanics, although we do so in a way that allows omission of this material at the instructor's discretion. As with previous editions, we have emphasized real-world chemistry in the final chapters that cover descriptive chemistry (Chapters 21–24), and we have tried to make this material easy to bring forward into earlier parts of the text. Moreover, many topics in these chapters can be covered selectively, without requiring the study of entire chapters.

The text ends with comprehensive chapters on organic chemistry (Chapters 26 and 27) and biochemistry (Chapter 28).

CHANGES TO THIS EDITION

For this edition, we have strengthened the pedagogical apparatus and increased the depth of coverage in selected areas—all in accordance with contemporary thoughts about how best to teach general chemistry. We have also made a number of smaller organizational changes to improve the flow of information. The following summarizes the major improvements made throughout the book.

- *Logical approach to solving problems.* All worked examples are presented consistently throughout the text by using a tripartite structure of Analyze-Solve-Assess. This presentation not only encourages students to use a logical approach in solving problems but also provides them with a way to start when they are trying to solve a problem that may seem, at first, impossibly difficult. The approach is used implicitly by those who have had plenty of practice solving problems; but for those who are just starting out, the Analyze-Solve-Assess structure will serve to remind students to (1) analyze the information and plan a strategy, (2) implement the strategy, and (3) check or assess their answer to ensure that it is a reasonable one.

- *Integrative Practice Examples and End of Chapter Exercises.* Users of previous editions have given us very positive feedback about the quality of the integrative examples at the end of each chapter and the variety of the end-of-chapter exercises. We have added two practice examples (Practice Example A and Practice Example B) to every Integrative Example in the text. Rather than replace end-of-chapter exercises with new exercises, we have opted in most chapters to increase the number of exercises. In most chapters, at least 10 new exercises have been added; and in many chapters, 20 or more exercises have been added.

- *Use of IUPAC recommendations.* We are pleased that our book serves the needs of instructors and students around the globe. Because communication among scientists in general, and chemists in particular, is made easier when we agree to use the same terms and notations, we have decided to follow—with relatively few exceptions—recommendations made by the International Union of Pure and Applied Chemistry (IUPAC). In particular, the version of the periodic table that now appears throughout the text is based on the one currently endorsed by IUPAC. The IUPAC-endorsed version places the elements lanthanum (La) and actinium (Ac) in the lanthanides and actinides series, respectively, rather than in group 3. Interestingly, almost every other chemistry book still uses the old version of the periodic table, even though the proper placement of La and Ac has been known for more than 20 years!

We have also made the following important changes in specific chapters and appendices:

- In Chapters 1 to 6, many problems are solved by using both a stepwise approach and a conversion pathway approach. Students with no chemistry background may be intimidated by the conversion pathway approach and may prefer a stepwise approach. Those who require only a refresher will likely prefer and use the conversion pathway approach. We hope that the needs of both instructors and students will be well served by showing both approaches in the early chapters.

- In Chapter 6 (Gases), we have changed the definition of standard temperature and pressure (STP) to conform to the IUPAC recommendations.

Also, we have added discussion about the molar volumes of gases and the distribution of molecular speeds. The discussion about the distribution of molecular speeds may be used as a springboard for justifying the Arrhenius form of the rate constant in Chapter 14 Chemical Kinetics.

- In Chapter 8 (Electrons in Atoms), we have put the material on the particle-in-a box into a separate section that can be used, or excluded, at the instructor's discretion. The discussion has been expanded slightly to illustrate how wave functions are used to make probability statements for an electron in a particular state.

- In Chapter 10 (Chemical Bonding I: Basic Concepts), we have introduced the dash-and-wedge symbolism for representing three-dimensional structures of molecules, and this symbolism is used throughout the remainder of the text. Also, we have added a new Are You Wondering? box comparing oxidation states and formal charges.

- Chapter 12 (Intermolecular Forces: Liquids and Solids), has been reorganized so that intermolecular forces are discussed first. Trends in the properties of liquids and solids are then discussed in terms of the intermolecular forces contributing to the attraction among the entities making up the substance.

- In Chapter 14 (Chemical Kinetics), we have adopted the IUPAC recommendation for defining reaction rates, which takes into account the stoichiometric coefficients of the balanced chemical equation. We have also included a new Are You Wondering? box that provides a molecular interpretation of reaction progress.

- In Chapter 15 (Principles of Chemical Equilibrium), we have expanded the discussion of the relationships among activities, pressure and concentrations, and also among K, K_p and K_c.

- In Chapter 16 (Acids and Bases), we have used curved arrows in a manner that is consistent with their use in organic chemistry, that is, to emphasize the movement of electron pairs in acid–base reactions. Also, we present an improved and modernized discussion of the connection between molecular structure and acid strength.

- In Chapter 20 (Electrochemistry), we have made some changes in notation that are recommended by IUPAC. Most importantly, we introduce the concept of *electron number*, z, and use it in place of n in the Nernst equation and other equations.

- Chapter 21 (Chemistry of the Main-Group Elements I: Groups 1, 2, 13, and 14); and Chapter 22 (Chemistry of the Main-Group Elements II: Groups 18, 17, 16, 15, and Hydrogen) have been updated to include discussion of interesting and important materials, such as crown ether complexes, zeolites, and graphene. In Chapter 21, we introduce the concept of charge density and use it throughout these two chapters to rationalize similarities and differences in properties of elements.

- We have devoted two chapters to organic chemistry. Chapter 26 (Structures of Organic Compounds) focuses on the structures, conformations, preparation, and uses of organic compounds. Chapter 27 (Reactions of Organic Compounds) focuses on a few important types of reactions and their mechanisms. In examining these reactions, emphasis is placed on concepts introduced earlier in the text, such as acid or base strength, electronegativity, and polarizability.

- In Appendix D, we have added the molar heat capacity for each of the substances listed in Table D.2 (Thermodynamic Properties of Substances at 298.15 K). We have also provided a new Table D.5 of Isotopic Masses and Their Abundance.

FEATURES OF THIS EDITION

We have made a careful effort with this edition to incorporate features that will facilitate the teaching and learning of chemistry.

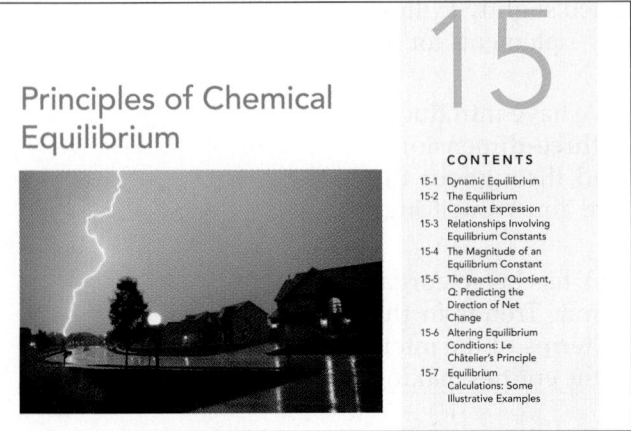

Principles of Chemical Equilibrium

15

CONTENTS

Chapter Opener
Each chapter opens with listing of the main headings to provide a convenient overview of the chapter's **Contents**.

Key Terms
Key terms are boldfaced where they are defined in the text. A **Glossary** of key terms with their definitions is presented in Appendix F.

If two elements form more than a single compound, the masses of one element combined with a fixed mass of the second are in the ratio of small whole numbers.

Highlighted Boxes
Significant equations, concepts, and rules are highlighted against a color background for easy reference.

🔍 13-1   CONCEPT ASSESSMENT

In one mole of a solution with a mole fraction of 0.5 water, how many water molecules would there be?

Concept Assessment
Concept Assessment Questions (many of which are qualitative) are distributed throughout the body of the chapters. They enable students to test their understanding of basic concepts before proceeding further. Full solutions are provided near the back of the book in Appendix G.

EXAMPLE 13-5   Using Henry's Law.

At 0 °C and an O_2 pressure of 1.00 atm, the aqueous solubility of $O_2(g)$ is 48.9 mL O_2 per liter. What is the molarity of O_2 in a saturated water solution when the O_2 is under its normal partial pressure in air, 0.2095 atm?

Analyze

Think of this as a two-part problem. (1) Determine the molarity of the saturated O_2 solution at 0 °C and 1 atm. (2) Use Henry's law in the manner just outlined.

Solve

Determine the molarity of O_2 at 0 °C when P_{O_2} = 1 atm.

$$\text{molarity} = \frac{0.0489 \text{ L } O_2 \times \dfrac{1 \text{ mol } O_2}{22.4 \text{ L } O_2 \text{ (STP)}}}{1 \text{ L soln}} = 2.18 \times 10^{-3} \text{ M } O_2$$

Evaluate the Henry's law constant.

$$k = \frac{C}{P_{gas}} = \frac{2.18 \times 10^{-3} \text{ M } O_2}{1.00 \text{ atm}}$$

Apply Henry's law.

$$C = k \times P_{gas} = \frac{2.18 \times 10^{-3} \text{ M } O_2}{1.00 \text{ atm}} \times 0.2095 \text{ atm} = 4.57 \times 10^{-4} \text{ M } O_2$$

Assess

When working problems involving gaseous solutes in a solution in which the solute is at very low concentration, use Henry's law.

PRACTICE EXAMPLE A: Use data from Example 13-5 to determine the partial pressure of O_2 above an aqueous solution at 0 °C known to contain 5.00 mg O_2 per 100.0 ml of solution.

PRACTICE EXAMPLE B: A handbook lists the solubility of carbon monoxide in water at 0 °C and 1 atm pressure as 0.0354 mL CO per milliliter of H_2O. What pressure of CO(g) must be maintained above the solution to obtain 0.0100 M CO?

Examples with Practice Examples A and B
Worked-Out Examples throughout the text illustrate how to apply the concepts. In many instances, a drawing or photograph is included to help students visualize what is going on in the problem. More importantly, all worked-out Examples now follow a tripartite structure of **Analyze-Solve-Assess** to encourage students to adopt a logical approach to problem solving.

Two **Practice Examples** are provided for each worked-out Example. The first, **Practice Example A**, provides immediate practice in a problem very similar to the given Example. The second, **Practice Example B**, often takes the student one step further than the given Example and is similar to the end-of-chapter problems in terms of level of difficulty. Answers to all the Practice Examples are given on the **MasteringChemistry™** site (www.masteringchemistry.com).

◄ Vapor-pressure lowering, as expressed through Raoult's law for ideal solutions, is also a colligative property.

Marginal Glosses
Marginal Glosses help clarify important points.

Keep In Mind Notes

Keep In Mind margin notes remind students about ideas introduced earlier in the text that are important to an understanding of the topic under discussion. In some instances they also warn students about common pitfalls.

Are You Wondering?

Are You Wondering? boxes pose and answer good questions that students often ask. Some are designed to help students avoid common misconceptions; others provide analogies or alternate explanations of a concept; and still others address apparent inconsistencies in the material that the students are learning. These topics can be assigned or omitted at the instructor's discretion.

Focus On Discussions

References are given near the end of chapter to a **Focus On** essay which is found on the **Mastering-Chemistry**™ site (www.masteringchemistry.com). These essays describe interesting and significant applications of the chemistry discussed in the chapter. They help show the importance of chemistry in all aspects of daily life.

Additional Material on Organic Chemistry Chapter

Chapter 27 includes references to discussions of Organic Acids and Bases; A Closer Look at the E2 Mechanism; and Carboxylic Acids and Their Derivatives: The Addition-Elimination Mechanism that are found on the MasteringChemistry™ site (www.masteringchemistry.com).

Summary

A prose **Summary** is provided for each chapter. The Summary is organized by the main headings in the chapter and incorporates the key terms in boldfaced type.

Integrative Example

An **Integrative Example** is provided near the end of chapter. These challenging examples show students how to link various concepts from the chapter and earlier chapters to solve complex problems. Each Integrative Example is now accompanied by a **Practice Example A** and **Practice Example B**. Answers to these Practice Examples are given on the **MasteringChemistry**™ site (www.mastering chemistry.com).

KEEP IN MIND

that the placement of the two curves in liquid–vapor equilibrium diagrams is such that the vapor is richer in the more volatile component than is the liquid. The more volatile component is the one with the higher vapor pressure or lower boiling point.

13-1 ARE YOU WONDERING...

What is the nature of the intermolecular forces in a mixture of carbon disulfide and acetone?

Carbon disulfide is a nonpolar molecule, and so in the pure substance the only intermolecular forces are weak London dispersion forces; carbon disulfide is a volatile liquid. Acetone is a polar molecule, and in the pure substance dipole–dipole forces are strong. Acetone is somewhat less volatile than carbon disulfide. In a solution of acetone in carbon disulfide (case 3 on page 564), the dipoles of acetone molecules polarize carbon disulfide molecules, giving rise to *dipole–induced dipole* interactions.

Mastering CHEMISTRY   www.masteringchemistry.com

What is the most abundant element? This seemingly simple question does not have a simple answer. To learn more about the abundances of elements in the universe and in the Earth's crust, go to the Focus On feature for Chapter 2, entitled Occurrence and Abundances of the Elements, on the MasteringChemistry site.

Summary

13-1 Types of Solutions: Some Terminology—In a solution, the **solvent**—usually the component present in greatest amount—determines the state of matter in which the solution exists (Table 13.1). **A solute** is a solution component dissolved in the solvent. Dilute solutions contain relatively small amounts of solute and concentrated solutions, large amounts.

13-2 Solution Concentration—Any description of the composition of a solution must indicate the quantities of solute and solvent (or solution) present. Solution concentrations expressed as mass percent, volume percent, and mass/volume percent all have practical importance, as do the units, parts per million (**ppm**), parts per billion (**ppb**), and parts per trillion (**ppt**). However, the more fundamental concentration units are mole fraction, molarity, and molality. Molarity (moles of solute per liter of solution) is temperature dependent, but mole fraction and **molality** (moles of solute per kilogram of solvent) are not.

help us to visualize **fractional distillation**, a common method of separating the volatile components of a solution. Such curves also illustrate the formation of azeotropes in some nonideal solutions. **Azeotropes** are solutions that boil at a constant temperature and produce vapor of the same composition as the liquid; they have boiling points that in some cases are greater than the boiling points of the pure components and in some cases, less (Fig. 13-15).

13-7 Osmotic Pressure—**Osmosis** is the spontaneous flow of solvent through a semipermeable membrane separating two solutions of different concentration. The net flow is from the less to the more concentrated solution (Fig. 13-17). Osmotic flow can be stopped by applying a pressure, called the **osmotic pressure**, to the more concentrated solution. In **reverse osmosis**, the direction of flow is reversed by applying a pressure that exceeds the osmotic pressure to the more concentrated solution. Both osmosis and reverse osmosis have important practical applications. Osmotic pressure can be calculated with a simple

---

Integrative Example

Peroxyacetyl nitrate (PAN) is an air pollutant produced in photochemical smog by the reaction of hydrocarbons, oxides of nitrogen, and sunlight. PAN is unstable and dissociates into peroxyacetyl radicals and $NO_2(g)$. Its presence in polluted air is like a reservoir for NO_2 storage.

$$CH_3COONO_2 \longrightarrow CH_3COO\cdot + NO_2$$
PAN   Peroxyacetyl radical

The first-order decomposition of PAN has a half-life of 35 h at 0 °C and 30.0 min at 25 °C. At what temperature will a sample of air containing 5.0×10^{14} PAN molecules per liter decompose at the rate of 1.0×10^{12} PAN molecules per liter per minute?

Analyze

This problem, which requires four principal tasks, is centered on the relationship between rate constants and temperature (equation 14.22) and between a rate constant and the rate of a reaction (equation 14.6). Specifically, we will need to (1) convert the two half-lives to values of k; (2) use those values of k and their related temperatures to determine the activation energy of the reaction; (3) find the value of k corresponding to the decomposition rate specified; and (4) calculate the temperature at which k has the value determined in (3).

Solve

Determine the value of k at 0 °C for the first-order reaction.

$$k = 0.693/t_{1/2}$$

$$k = \frac{0.693}{35\,\text{h}} \times \frac{1\,\text{h}}{60\,\text{min}} = 3.3 \times 10^{-4}\,\text{min}^{-1}$$

PRACTICE EXAMPLE A: At room temperature (20 °C), milk turns sour in about 64 hours. In a refrigerator at 3 °C, milk can be stored three times as long before it sours. **(a)** Estimate the activation energy of the reaction that causes the souring of milk. **(b)** How long should it take milk to sour at 40 °C?

PRACTICE EXAMPLE B: The following mechanism can be used to account for the change in apparent order of unimolecular reactions, such as the conversion of cyclopropane (A) into propene (P), where A* is an energetic form of cyclopropane that can either react or return to unreacted cyclopropane.

$$A + A \underset{k_{-1}}{\overset{k_1}{\rightleftharpoons}} A^* + A$$

$$A^* \xrightarrow{k_2} P$$

Show that at low pressures of cyclopropane, the rate law is second order in A and at high pressures, it is first order in A.

Exercises

Homogeneous and Heterogeneous Mixtures

1. Which of the following do you expect to be most water soluble, and why? $C_{10}H_8(s)$, $NH_2OH(s)$, $C_6H_6(l)$, $CaCO_3(s)$.
2. Which one of the following do you expect to be moderately soluble both in water and in benzene $[C_6H_6(l)]$, and why? **(a)** butyl alcohol, C_4H_9OH; **(b)** naphthalene, $C_{10}H_8$; **(c)** hexane, C_6H_{14}; **(d)** $NaCl(s)$.
3. Substances that dissolve in water generally do not dissolve in benzene. Some substances are moderately soluble in both solvents, however. One of the following is such a substance. Which do you think it is and why?

(a) *para*-Dichlorobenzene (a moth repellent)

(b) Salicyl alcohol (a local anesthetic)

Vitamin C

Vitamin E

End-of-Chapter Questions and Exercises

Each chapter ends with four categories of questions:

Exercises are organized by topic subheads and are presented in pairs. Answers to selected questions (i.e., those numbered in red) are given on the **MasteringChemistry**™ site (www.masteringchemistry.com).

Integrative and Advanced Exercises

87. A typical root beer contains 0.13% of a 75% H_3PO_4 solution by mass. How many milligrams of phosphorus are contained in a 12 oz can of this root beer? Assume a solution density of 1.00 g/mL; also, 1 oz = 29.6 mL.
88. An aqueous solution has 109.2 g KOH/L solution. The solution density is 1.09 g/mL. Your task is to use 100.0 mL of this solution to prepare 0.250 *m* KOH. What mass of which component, KOH or H_2O, would you add to the 100.0 mL of solution?
89. The term "proof," still used to describe the ethanol content of alcoholic beverages, originated in seventeenth-century England. A sample of whiskey was poured on gunpowder and set afire. If the gunpowder ignited after the whiskey had burned off, this "proved" that the whiskey had not been watered down. The minimum ethanol content for a positive test was about 50%, by volume. The 50% ethanol solution became known as "100 proof." Thus, an 80-proof whiskey would be 40% CH_3CH_2OH by volume. Listed in the table below are some data for several aqueous solutions of ethanol. *With a minimum amount of calculation*, determine which of the solutions are more than 100 proof. Assume that the density of pure ethanol is 0.79 g/mL.
90. Four aqueous solutions of acetone, CH_3COCH_3, are prepared at different concentrations: **(a)** 0.100% CH_3COCH_3, by mass; **(b)** 0.100 M CH_3COCH_3; **(c)** 0.100 *m* CH_3COCH_3; and **(d)** $\chi_{acetone} = 0.100$. Estimate the highest partial pressure of water at 25 °C to be found in the equilibrium vapor above these solutions. Also, estimate the lowest freezing point to be found among these solutions.
91. A solid mixture consists of 85.0% KNO_3 and 15.0% K_2SO_4, by mass. A 60.0 g sample of this solid is added to 130.0 g of water at 60 °C. Refer to Figure 13-8.
    (a) Will all the solid dissolve at 60 °C?
    (b) If the resulting solution is cooled to 0 °C, what mass of KNO_3 should crystallize?
    (c) Will K_2SO_4 also crystallize at 0 °C?
92. Suppose you have available 2.50 L of a solution ($d = 0.9767$ g/mL) that is 13.8% ethanol (C_2H_5OH), by mass. From this solution you would like to make the *maximum* quantity of ethanol-water antifreeze solution that will offer protection to −2.0 °C. Would you add more ethanol or more water to the solution? What mass of liquid would you add?
93. Hydrogen chloride is a colorless gas, yet when a bottle of concentrated hydrochloric acid [HCl(conc aq)] is

Integrative and Advanced Exercises are more advanced than the preceding *Exercises*. They are not grouped by topic or type. They integrate material from sections of the chapter and sometimes from multiple chapters. In some instances, they introduce new ideas or pursue specific ideas further than is done in the chapter. Answers to selected questions (i.e., those numbered in red) are given on the **MasteringChemistry**™ site (www.masteringchemistry.com).

Feature Problems

113. Cinnamaldehyde is the chief constituent of cinnamon oil, which is obtained from the twigs and leaves of cinnamon trees grown in tropical regions. Cinnamon oil is used in the manufacture of food flavorings, perfumes, and cosmetics. The normal boiling point of cinnamaldehyde, C_9H_8O, is 246.0 °C, but at this temperature it begins to decompose. As a result, cinnamaldehyde cannot be easily purified by ordinary distillation. A method that can be used instead is *steam distillation*. A heterogeneous mixture of cinnamaldehyde and water is heated until the sum of the vapor pressures of the two liquids is equal to barometric pressure. At this point, the temperature remains constant as the liquids vaporize. The mixed vapor condenses to produce two immiscible liquids; one liquid is essentially pure water and the other, pure cinnamaldehyde. The following vapor pressures of cinnamaldehyde are given: 1 mmHg at 76.1 °C; 5 mmHg at 105.8 °C; and 10 mmHg at 120.0 °C. Vapor pressures of water are given in Table 13.2.
     (a) What is the approximate temperature at which the steam distillation occurs?

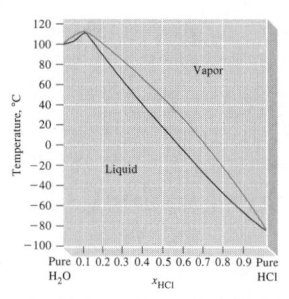

boils in an open container, the composition changes. Explain why this is so.

Feature Problems require the highest level of skill to solve. Some deal with classic experiments; some require students to interpret data or graphs; some suggest alternative techniques for problem solving; some are comprehensive in their scope; and some introduce new material. These problems are a resource that can be used in several ways: for discussion in class, for individually assigned homework, or for collaborative group work. Answers to selected questions (i.e., those numbered in red) are given on the **MasteringChemistry**™ site (www.masteringchemistry.com).

Self-Assessment Exercises

117. In your own words, define or explain the following terms or symbols: **(a)** x_B; **(b)** P_A°; **(c)** K_f; **(d)** i; **(e)** activity.
118. Briefly describe each of the following ideas or phenomena: **(a)** Henry's law; **(b)** freezing-point depression; **(c)** recrystallization; **(d)** hydrated ion; **(e)** deliquescence.
119. Explain the important distinctions between each pair of terms: **(a)** molality and molarity; **(b)** ideal and nonideal solution; **(c)** unsaturated and supersaturated solution; **(d)** fractional crystallization and fractional distillation; **(e)** osmosis and reverse osmosis.
120. An aqueous solution is 0.010 M CH_3OH. The concentration of this solution is also very nearly **(a)** 0.010% CH_3OH (mass/volume); **(b)** 0.010 *m* CH_3OH; **(c)** $x_{CH_3OH} = 0.010$; **(d)** 0.990 M H_2O.
121. The most likely of the following mixtures to be an ideal solution is **(a)** NaCl–H_2O; **(b)** $CH_3CH_2OH–C_6H_6$; **(c)** $C_7H_{16}–H_2O$; **(d)** $C_7H_{16}–C_8H_{18}$.
122. The solubility of a nonreactive gas in water increases with **(a)** an increase in gas pressure; **(b)** an increase in

likely to be equal; **(d)** 1.00 for the solvent and 0.00 for the solute.
125. A solution prepared by dissolving 1.12 mol NH_4Cl in 150.0 g H_2O is brought to a temperature of 30 °C. Use Figure 13-8 to determine whether the solution is unsaturated or whether excess solute will crystallize.
126. NaCl(aq) isotonic with blood is 0.92% NaCl (mass/volume). For this solution, what is **(a)** $[Na^+]$; **(b)** the total molarity of ions; **(c)** the osmotic pressure at 37 °C; **(d)** the approximate freezing point? (Assume that the solution has a density of 1.005 g/mL.)
127. A solution ($d = 1.159$ g/mL) is 62.0% glycerol, $C_3H_8O_3$, and 38.0% H_2O, by mass. Determine **(a)** the molarity of $C_3H_8O_3$ with H_2O as the solvent; **(b)** the molarity of H_2O with $C_3H_8O_3$ as the solvent; **(c)** the molality of H_2O in $C_3H_8O_3$; **(d)** the mole fraction of $C_3H_8O_3$; **(e)** the mole percent of water.
128. Which aqueous solution from the column on the right has the property listed on the left? Explain your choices.

Self-Assessment Exercises are designed to help students review and prepare for some of the types of questions that often appear on quizzes and exams. Students can use these questions to decide whether they are ready to move on to the next chapter or first spend more time working with the concepts in the current chapter. Answers with explanations to selected questions (i.e., those numbered in red) are given on the **MasteringChemistry**™ site (www.mastering chemistry.com).

Appendices

Five Appendices at the back of the book provide important information:

Appendix A succinctly reviews of some basic **Mathematical Operations**.

Appendix B concisely describes some basic **Physical Concepts**.

Appendix C summarizes the conventions of **SI Units**.

Appendix D provides 5 useful **Data Tables**, including the new Table D.5 Isotopic Masses and Their Abundance.

Appendix E provides guidelines, along with an example, for constructing **Concept Maps**.

Appendix F consists of a **Glossary** of all the key terms in the book.

Appendix G provides **Answers to the Concept Assessment Questions**.

- For easy reference, the **Periodic Table** and a **Tabular Listing of Elements** are presented on the inside of the front cover.

- For convenience, listings of **Selected Physical Constants, Some Conversion Factors, Some Useful Geometric Formulas**, and **Locations of Important Data and Other Useful Information** are presented on the inside of the back cover.

SUPPLEMENTS

For the Instructor and the Student

- The **MasteringChemistry™ website** (http://www.masteringchemistry.com) is a comprehensive site that offers many learning and teaching tools. For instructor-assigned homework, MasteringChemistry™ provides the first adaptive-learning online tutorial and assessment system. Based on extensive research of precise concepts students struggle with, the system is able to coach students with feedback specific to their needs and with simpler problems upon request. The result is a large set of targeted tutorials that help optimize study time and maximize learning for students. In addition, the MasteringChemistry™ site also includes a Pearson eText. This robust eText platform enables students and instructors to highlight sections, add notes, share notes, and magnify any of the images or pages without distortion. The MasteringChemistry™ site also contains a Study Area which contains a new Self Quizzing feature for students and an electronic version of the Math Review Toolkit; students can access the content in the Study Area without instructor involvement.

For the Instructor

- An **Instructor's Resource CD-ROM** (978-013-509778-6) provides instructors with the following 10 supplements designed to facilitate lecture presentations, encourage class discussions, aid in creating tests, and foster learning:

  - An **Instructor's Resource Manual**, organized by chapter, provides detailed lecture outlines, describes some common student misconceptions, and demonstrates how to integrate the various instructor resources into the course.

  - The **Complete Solutions Manual** contains full solutions to all the end-of chapter exercises and problems (including those Self-Assessment Exercises that are not discussion questions), as well as full solutions to all the Practice Examples A and B in the book.

  - A **Testbank (Test Item File)** in Word provides more than 2700 questions. Many of the questions are in multiple-choice form, but there are also true/false and short-answer questions. Each question is accompanied by the correct answer, the relevant chapter section in the textbook, and a level of difficulty (i.e., 1 for Easy, 2 for Moderate, and 3 for Challenging).

  - The **Computerized Testbank (Pearson TestGen)** presents the testbank in a powerful program that enables instructors to view and edit existing questions, create new questions, and generate quizzes, tests, exams, or homework. TestGen also allows instructors to administer tests on a local area network, have the tests graded electronically, and have the results prepared in electronic or printed reports.

- **PowerPoints Set 1** consists of all the figures and photos in the textbook in PowerPoint format.

- **PowerPoints Set 2** provides lecture outlines for each chapter of the textbook.

- **PowerPoints Set 3** provides questions for Personal Response Systems (i.e., clickers) that can be used to engage students in lectures and to obtain immediate feedback about their understanding of the concepts being presented.

- **PowerPoints Set 4** consists of the all worked Examples from the textbook in PowerPoint format.

- **Focus On Discussions** consist of all the Focus On Essays referenced in the textbook which students can find on the **MasteringChemistry**™ site (www.masteringchemistry.com).

- **Additional Material on Organic Chemistry** consists of discussions of Organic Acids and Bases; A Closer Look at the E2 Mechanism; and Carboxylic Acids and Their Derivatives: The Addition-Elimination Mechanism that are referenced in Chapter 27. Students can find this material on the MasteringChemistry™ site (www.mastering chemistry.com).

- **Answers to Practice Examples and to selected End-of-Chapter Exercises and Problems** (i.e., those numbered in red in the textbook) are provided here for the convenience of instructors. This same material is also available to students on the **MasteringChemistry**™ site (www.masteringchemistry.com).

- The **Complete Solutions Manual** is also available in printed form (978-013-504293-9). With instructor approval, arrangements can be made with the publisher to make this manual available to students.

- **Transparency Package** (978-013-703215-0) provides colour acetates of selected figures, tables, and photos from the textbook.

- A prebuilt **WebCT**® **Course** (978-013-703208-2) has been prepared to accompany the book.

- **Pearson's Technology Specialists** work with faculty and campus course designers to ensure that Pearson technology products, assessment tools, and online course materials are tailored to meet your specific needs. This highly qualified team is dedicated to helping students take full advantage of a wide range of educational resources, by assisting in the integration of a variety of instructional materials and media formats. Your local Pearson Education sales representative can provide you with more details about this service program.

- **The CourseSmart eTextbook** (978-013-509775-5) goes beyond traditional expectations—providing instant, online access to the textbooks and course materials you need at a lower cost for students. And even as students save money, you can save time and energy with a digital eTextbook that allows you to search for the most relevant content at the very moment you need it. Whether it's evaluating textbooks or creating lecture notes to help students with difficult concepts, CourseSmart can make life a little easier. See how when you visit www.coursesmart.com/instructors.

For the Student

- Along with an **Access Code Card for Mastering Chemistry**™, each new copy of the book is accompanied by a 10-page **Study Card**

(978-013-703212-9). This card provides a convenient concise review of some of the key concepts and topics discussed in each chapter of the textbook.

- The **Selected Solutions Manual** (978-013-504292-2) provides full solutions to all the end-of chapter exercises and problems that are numbered in red.
- The **Math Review Toolkit** (978-013-612039-1) contains a review of the essential math skills required for each chapter of the textbook.
- **The CourseSmart eTextbook** (978-013-509775-5) goes beyond traditional expectations—providing instant, online access to the textbooks and course materials you need at an average savings of 50%. With instant access from any computer and the ability to search your text, you'll find the content you need quickly, no matter where you are. And with online tools like highlighting and note-taking, you can save time and study efficiently. See all the benefits at www.coursesmart.com/students.

ACKNOWLEDGMENTS

We are grateful to the following instructors who provided formal reviews of parts of the manuscript.

Brian M. Baker *University of Notre Dame*

Robert J. Balahura *University of Guelph*

John Carran *Queen's University*

Chin Li Cheung *University of Nebraska, Lincoln*

Savitri Chandrasekhar *University of Toronto – Scarborough*

H. Floyd Davis *Cornell University*

David Dick *College of the Rockies*

Randall S. Dumont *McMaster University*

Philip Dutton *University of Windsor*

Lucio Gelmini *Grant MacEwan College*

Kevin Grundy *Dalhousie University*

P. Shiv Halasyamani *University of Houston*

C. Alton Hassell *Baylor University*

Sheryl Hemkin *Kenyon College*

Michael Hempstead *York University*

Hugh Horton *Queen's University*

Robert C. Kerber *Stony Brook University*

Pippa Lock *McMaster University*

J. Scott McIndoe *University of Victoria*

Umesh Parshotam *University of Northern British Columbia*

Darrin Richeson *University of Ottawa*

Lawton Shaw *Athabasca University*

Roberta Šilerová *John Abbot College*

Andreas Toupadakis *University of California, Davis*

A. van der Est *Brock University*

Rashmi Venkateswaren *University of Ottawa*

Deborah Walker *University of Texas at Austin*

Todd Whitcombe *University of Northern British Columbia*

Milton J. Wieder *Metropolitan State College of Denver*

Vance Williams *Simon Fraser University*

We would like to especially acknowledge the valuable assistance that Stephen Forsey (University of Waterloo) provided in the crafting of the new Chapter 27.

We would like to thank the following instructors for technically checking selected chapters of the new edition during production.

Chin Li Cheung *University of Nebraska, Lincoln*

David Dick *College of the Rockies*

Philip Dutton *University of Windsor*

J. Scott McIndoe *University of Victoria*

Todd Whitcombe *University of Northern British Columbia*

Milton J. Wieder *Metropolitan State College of Denver*

We are most grateful to our coauthor Ralph Petrucci for taking the extraordinary additional step of carefully checking every single page before we went to press.

Responding to feedback from our colleagues and students is the most important element in improving this book from one edition to the next. Please do not hesitate to email us. Your observations and suggestions are most welcome.

CAREY BISSONNETTE
cbissonn@uwaterloo.ca

JEFFRY D. MADURA
madura@duq.edu

F. GEOFFREY HERRING
fgh@chem.ubc.ca

WARNING: Many of the compounds and chemical reactions described or pictured in this book are hazardous. Do not attempt any experiment pictured or implied in the text except with permission in an authorized laboratory setting and under adequate supervision.

Matter: Its Properties and Measurement

CONTENTS

A Hubble Space Telescope image of a cloud of hydrogen gas and dust (lower right half of the image) that is part of the Swan Nebula (M17). The colors correspond to light emitted by hydrogen (green), sulfur (red), and oxygen (blue). The chemical elements discussed in this text are those found on Earth and, presumably, throughout the universe.

From the clinic that treats "chemical dependency" to a theatrical performance with "good chemistry" to the food label stating "no chemicals added," chemistry and chemicals seem an integral part of life, even if everyday references to them are often misleading. A label implying the absence of chemicals in a food makes no sense. All foods consist entirely of chemicals, even if "organically grown." In fact, all material objects—whether living or inanimate—are made up only of chemicals, and we should begin our study with that thought clearly in mind.

By manipulating materials in their environment, people have always practiced chemistry. Among the earliest applications were the glazing of pottery, the smelting of ores to produce metals, the tanning of hides, the dyeing of fabrics, and the making of cheese, wine, beer, and soap. With modern knowledge, though, chemists can decompose matter into its smallest components (atoms) and reassemble those components into materials that do not exist naturally and that often exhibit unusual properties.

Thus, motor fuels and thousands of chemicals used in the manufacture of plastics, synthetic fabrics, pharmaceuticals, and pesticides can all be made from petroleum. Modern chemical knowledge is also needed to understand the processes that sustain life and to understand and control processes that are detrimental to the environment, such as the formation of smog and the destruction of stratospheric ozone. Because it relates to so many areas of human endeavor, chemistry is sometimes called the central science.

Early chemical knowledge consisted of the "how to" of chemistry, discovered through trial and error. Modern chemical knowledge answers the "why" as well as the "how to" of chemical change. It is grounded in principles and theory, and mastering the principles of chemistry requires a systematic approach to the subject. Scientific progress depends on the way scientists do their work—asking the right questions, designing the right experiments to supply the answers, and formulating plausible explanations of their findings. We begin with a closer look into the scientific method.

1-1   The Scientific Method

Science differs from other fields of study in the *method* that scientists use to acquire knowledge and the special significance of this knowledge. Scientific knowledge can be used to explain natural phenomena and, at times, to *predict* future events.

The ancient Greeks developed some powerful methods of acquiring knowledge, particularly in mathematics. The Greek approach was to start with certain basic assumptions, or premises. Then, by the method known as *deduction*, certain conclusions must logically follow. For example, if $a = b$ and $b = c$, then $a = c$. Deduction alone is not enough for obtaining scientific knowledge, however. The Greek philosopher Aristotle *assumed* four fundamental substances: air, earth, water, and fire. All other materials, he believed, were formed by combinations of these four elements. Chemists of several centuries ago (more commonly referred to as alchemists) tried, in vain, to apply the four-element idea to turn lead into gold. They failed for many reasons, one being that the four-element assumption is false.

The scientific method originated in the seventeenth century with such people as Galileo, Francis Bacon, Robert Boyle, and Isaac Newton. The key to the method is to make no initial assumptions, but rather to make careful observations of natural phenomena. When enough observations have been made so that a pattern begins to emerge, a generalization or natural law can be formulated describing the phenomenon. **Natural laws** are concise statements, often in mathematical form, about natural phenomena. The form of reasoning in which a general statement or natural law is inferred from a set of observations is called *induction*. For example, early in the sixteenth century, the Polish astronomer Nicolas Copernicus (1473–1543), through careful study of astronomical observations, concluded that Earth revolves around the sun in a circular orbit, although the general teaching of the time, not based on scientific study, was that the sun and other heavenly bodies revolved around Earth. We can think of Copernicus's statement as a natural law. Another example of a natural law is the radioactive decay law, which dictates how long it takes for a radioactive substance to lose its radioactivity.

The success of a natural law depends on its ability to explain, or account for, observations and to predict new phenomena. Copernicus's work was a great success because he was able to predict future positions of the planets more accurately than his contemporaries. We should not think of a natural law as an *absolute* truth, however. Future experiments may require us to modify the law. For example, Copernicus's ideas were refined a half-century later by Johannes Kepler, who showed that planets travel in elliptical, not circular, orbits. To verify a natural law, a scientist designs *experiments* that show whether the conclusions deduced from the natural law are supported by experimental results.

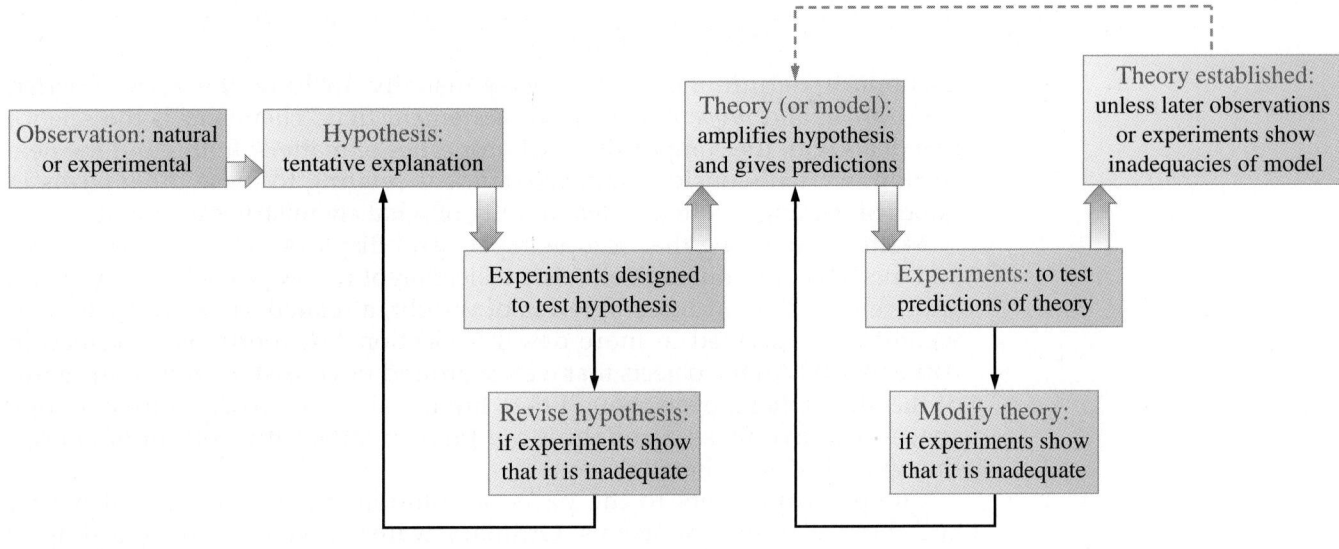

▲ FIGURE 1-1
The scientific method illustrated

A **hypothesis** is a tentative explanation of a natural law. If a hypothesis survives testing by experiments, it is often referred to as a theory. In a broader sense, a **theory** is a model or way of looking at nature that can be used to explain natural laws and make further predictions about natural phenomena. When differing or conflicting theories are proposed, the one that is most successful in its predictions is generally chosen. Also, the theory that involves the smallest number of assumptions—the simplest theory—is preferred. Over time, as new evidence accumulates, most scientific theories undergo modification, and some are discarded.

The **scientific method** is the combination of observation, experimentation, and the formulation of laws, hypotheses, and theories. The method is illustrated by the flow diagram in Figure 1-1. Scientists may develop a pattern of thinking about their field, known as a *paradigm*. Some paradigms may be successful at first but then become less so. When that happens, a new paradigm may be needed or, as is sometimes said, a "paradigm shift" occurs. In a way, the method of inquiry that we call the scientific method is itself a paradigm, and some people feel that it, too, is in need of change. That is, the varied activities of modern scientists are more complex than the simplified description of the scientific method presented here.* In any case, merely following a set of procedures, rather like using a cookbook, will not guarantee scientific success.

Another factor in scientific discovery is chance, or serendipity. Many discoveries have been made by accident. For example, in 1839, the American inventor Charles Goodyear was searching for a treatment for natural rubber that would make it less brittle when cold and less tacky when warm. In the course of this work, he accidentally spilled a rubber–sulfur mixture on a hot stove and found that the resulting product had exactly the properties he was seeking. Other chance discoveries include X-rays, radioactivity, and penicillin. So scientists and inventors always need to be alert to unexpected observations. Perhaps no one was more aware of this than Louis Pasteur, who wrote, "Chance favors the prepared mind."

▲ Louis Pasteur (1822–1895). This great practitioner of the scientific method was the developer of the germ theory of disease, the sterilization of milk by pasteurization, and vaccination against rabies. He has been called the greatest physician of all time by some. He was, in fact, not a physician at all, but a chemist—by training and by profession.

🔍 **1-1   CONCEPT ASSESSMENT**

Is the common saying "The exception proves the rule" a good statement of the scientific method? Explain.

◄ Answers to Concept Assessment questions are given in Appendix G.

*W. Harwood, *JCST*, **33**, 29 (2004). *JCST* is an abbreviation for *Journal of College Science Teaching*.

1-2 Properties of Matter

Dictionary definitions of chemistry usually include the terms *matter*, *composition*, and *properties*, as in the statement that "chemistry is the science that deals with the composition and properties of matter." In this and the next section, we will consider some basic ideas relating to these three terms in hopes of gaining a better understanding of what chemistry is all about.

Matter is anything that occupies space and displays the properties of *mass* and inertia. Every human being is a collection of matter. We all occupy space, and we describe our mass in terms of weight, a related property. (Mass and weight are described in more detail in Section 1-4. Inertia is described in Appendix B.) All the objects that we see around us consist of matter. The gases of the atmosphere, even though they are invisible, are matter—they occupy space and have mass. Sunlight is *not* matter; rather, it is a form of energy. Energy is discussed in later chapters.

Composition refers to the parts or components of a sample of matter and their relative proportions. Ordinary water is made up of two simpler substances—hydrogen and oxygen—present in certain fixed proportions. A chemist would say that the composition of water is 11.19% hydrogen and 88.81% oxygen by mass. Hydrogen peroxide, a substance used in bleaches and antiseptics, is also made up of hydrogen and oxygen, but it has a different composition. Hydrogen peroxide is 5.93% hydrogen and 94.07% oxygen by mass.

Properties are those qualities or attributes that we can use to distinguish one sample of matter from others; and, as we consider next, the properties of matter are generally grouped into two broad categories: physical and chemical.

Physical Properties and Physical Changes

A **physical property** is one that a sample of matter displays without changing its composition. Thus, we can distinguish between the reddish brown solid, copper, and the yellow solid, sulfur, by the physical property of *color* (Fig. 1-2).

Another physical property of copper is that it can be hammered into a thin sheet of foil (see Figure 1-2). Solids having this ability are said to be *malleable*. Sulfur is not malleable. If we strike a chunk of sulfur with a hammer, it crumbles into a powder. Sulfur is *brittle*. Another physical property of copper that sulfur does not share is the ability to be drawn into a fine wire (ductility). Also, sulfur is a far poorer conductor of heat and electricity than is copper.

Sometimes a sample of matter undergoes a change in its physical appearance. In such a **physical change**, some of the physical properties of the sample may change, but its composition remains unchanged. When liquid water freezes into solid water (ice), it certainly looks different and, in many ways, it is different. Yet, the water remains 11.19% hydrogen and 88.81% oxygen by mass.

▶ FIGURE 1-2
Physical properties of sulfur and copper
A lump of sulfur (left) crumbles into a yellow powder when hammered. Copper (right) can be obtained as large lumps of native copper, formed into pellets, hammered into a thin foil, or drawn into a wire.

Chemical Properties and Chemical Changes

In a **chemical change**, or **chemical reaction**, one or more kinds of matter are converted to new kinds of matter with different compositions. The key to identifying chemical change, then, comes in observing a *change in composition*. The burning of paper involves a chemical change. Paper is a complex material, but its principal constituents are carbon, hydrogen, and oxygen. The chief products of the combustion are two gases, one consisting of carbon and oxygen (carbon dioxide) and the other consisting of hydrogen and oxygen (water, as steam). The ability of paper to burn is an example of a chemical property. A **chemical property** is the ability (or inability) of a sample of matter to undergo a change in composition under stated conditions.

Zinc reacts with hydrochloric acid solution to produce hydrogen gas and a solution of zinc chloride in water (Fig. 1-3). This reaction is one of zinc's distinctive chemical properties, just as the inability of gold to react with hydrochloric acid is one of gold's chemical properties. Sodium reacts not only with hydrochloric acid but also with water. In some of their physical properties, zinc, gold, and sodium are similar. For example, each is malleable and a good conductor of heat and electricity. In most of their chemical properties, though, zinc, gold, and sodium are quite different. Knowing these differences helps us to understand why zinc, which does not react with water, is used in roofing nails, roof flashings, and rain gutters, and sodium is not. Also, we can appreciate why gold, because of its chemical inertness, is prized for jewelry and coins: It does not tarnish or rust. In our study of chemistry, we will see why substances differ in properties and how these differences determine the ways in which we use them.

1-3 Classification of Matter

Matter is made up of very tiny units called **atoms**. Each different type of atom is the building block of a different chemical **element**. Presently, the International Union of Pure and Applied Chemistry (IUPAC) recognizes 112 elements, and *all* matter is made up of just these types! The known elements range from common substances, such as carbon, iron, and silver, to uncommon ones, such as lutetium and thulium. About 90 of the elements can be obtained from natural sources. The remainder do not occur naturally and have been created only in laboratories. On the inside front cover you will find a complete listing of the elements and also a special tabular arrangement of the elements known as the *periodic table*. The periodic table is the chemist's directory of the elements. We will describe it in Chapter 2 and use it throughout most of the text.

Chemical **compounds** are substances comprising atoms of two or more elements joined together. Scientists have identified millions of different chemical compounds. In some cases, we can isolate a molecule of a compound. A **molecule** is the smallest entity having the same proportions of the constituent atoms as does the compound as a whole. A molecule of water consists of three atoms: two hydrogen atoms joined to a single oxygen atom. A molecule of hydrogen peroxide has two hydrogen atoms and two oxygen atoms; the two oxygen atoms are joined together and one hydrogen atom is attached to each oxygen atom. By contrast, a molecule of the blood protein gamma globulin is made up of 19,996 atoms, but they are of just four types: carbon, hydrogen, oxygen, and nitrogen.

▲ FIGURE 1-3
A chemical property of zinc and gold: reaction with hydrochloric acid
The zinc-plated (galvanized) nail reacts with hydrochloric acid, producing the bubbles of hydrogen gas seen on its surface. The gold bracelet is unaffected by hydrochloric acid. In this photograph, the zinc plating has been consumed, exposing the underlying iron nail. The reaction of iron with hydrochloric acid imparts some color to the acid solution.

◀ The International Union of Pure and Applied Chemistry (IUPAC) is recognized as the world authority on chemical nomenclature, terminology, standardized methods for measurement, atomic mass, and more. Along with many other activities, IUPAC publishes journals, technical reports, and chemical databases, most of which are available at www.iupac.org.

◀ The identity of an atom is established by a feature called its atomic number (see Section 2-3). Recent report of other new elements, such as elements 113 to 116 and 118, await confirmation. Characterizing "superheavy" elements is a daunting challenge; they are produced only a few atoms at a time and the atoms disintegrate almost instantaneously.

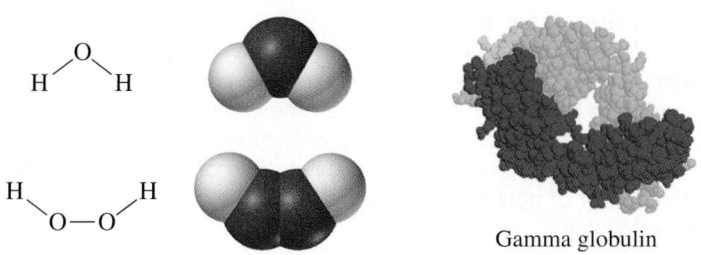

Gamma globulin

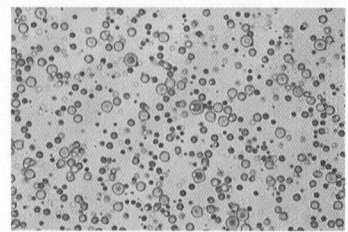

▲ Is it homogeneous or heterogeneous? When viewed through a microscope, homogenized milk is seen to consist of globules of fat dispersed in a watery medium. Homogenized milk is a *heterogeneous* mixture.

The composition and properties of an element or a compound are uniform throughout a given sample and from one sample to another. Elements and compounds are called **substances**. (In the chemical sense, the term *substance* should be used only for elements and compounds.) A *mixture* of substances can vary in composition and properties from one sample to another. One that is uniform in composition and properties throughout is said to be a **homogeneous mixture** or a *solution*. A given solution of sucrose (cane sugar) in water is uniformly sweet throughout the solution, but the sweetness of another sucrose solution may be rather different if the sugar and water are present in different proportions. Ordinary air is a homogeneous mixture of several gases, principally the *elements* nitrogen and oxygen. Seawater is a solution of the *compounds* water, sodium chloride (salt), and a host of others. Gasoline is a homogeneous mixture or solution of dozens of compounds.

In **heterogeneous mixtures**—sand and water, for example—the components separate into distinct regions. Thus, the composition and physical properties vary from one part of the mixture to another. Salad dressing, a slab of concrete, and the leaf of a plant are all heterogeneous. It is usually easy to distinguish heterogeneous from homogeneous mixtures. A scheme for classifying matter into elements and compounds and homogeneous and heterogeneous mixtures is summarized in Figure 1-4.

Separating Mixtures

▶ It is composition, particularly its variability, that helps us distinguish the several classifications of matter.

▶ Solutions can be gaseous and liquids as described here, but they can also be solids. Some alloys are examples of solid solutions.

A mixture can be separated into its components by appropriate physical means. Consider again the heterogeneous mixture of sand in water. When we pour this mixture into a funnel lined with porous filter paper, the water passes through and sand is retained on the paper. This process of separating a solid from the liquid in which it is suspended is called *filtration* (Fig. 1-5a). You will probably use this procedure in the laboratory. Conversely, we cannot separate a homogeneous mixture (solution) of copper(II) sulfate in water by filtration because all components pass through the paper. We can, however, boil the solution of copper(II) sulfate and water. In the process of *distillation*, a pure liquid is condensed from the vapor given off by a boiling solution. When all

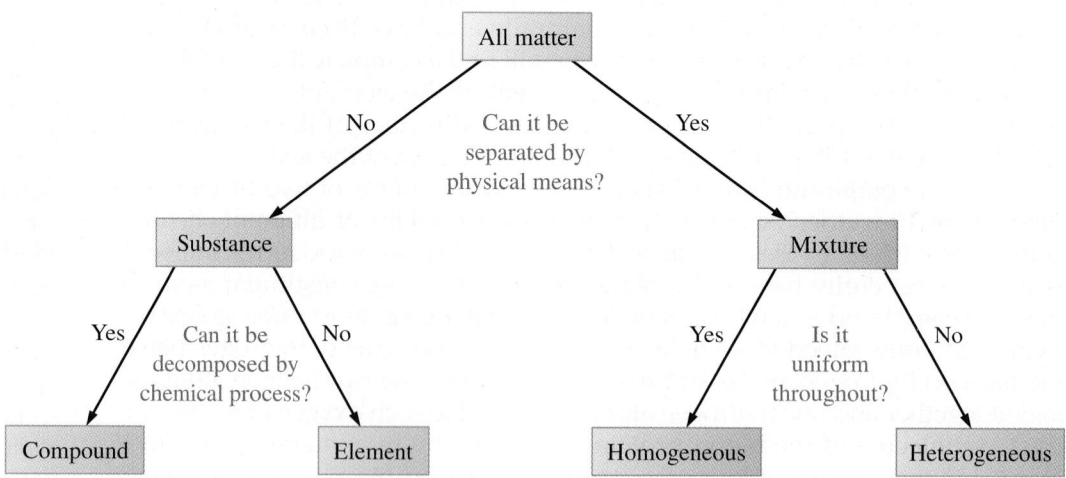

▲ FIGURE 1-4
A classification scheme for matter
Every sample of matter is either a single substance (an element or compound) or a mixture of substances. At the molecular level, an element consists of atoms of a single type and a compound consists of two or more different types of atoms, usually joined into molecules. In a homogeneous mixture, atoms or molecules are randomly mixed at the molecular level. In heterogeneous mixtures, the components are physically separated, as in a layer of octane molecules (a constituent of gasoline) floating on a layer of water molecules.

(a)

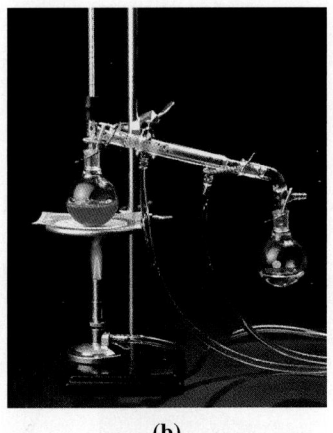
(b)

(c)

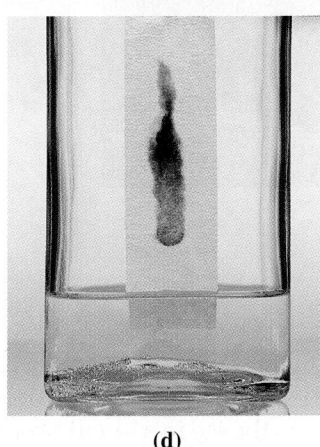
(d)

◄ FIGURE 1-5
Separating mixtures: a physical process
(a) Separation of a heterogeneous mixture by filtration: Solid copper(II) sulfate is retained on the filter paper, while liquid hexane passes through. **(b)** Separation of a homogeneous mixture by distillation: Copper(II) sulfate remains in the flask on the left as water passes to the flask on the right, by first evaporating and then condensing back to a liquid. **(c)** Separation of the components of ink using chromatography: A dark spot of black ink can be seen just above the water line as water moves up the paper. **(d)** Water has dissolved the colored components of the ink, and these components are retained in different regions on the paper according to their differing tendencies to adhere to the paper.

the water has been removed by boiling a solution of copper(II) sulfate in water, solid copper(II) sulfate remains behind (Fig. 1-5b).

Another method of separation available to modern chemists depends on the differing abilities of compounds to adhere to the surfaces of various solid substances, such as paper and starch. The technique of *chromatography* relies on this principle. The dramatic results that can be obtained with chromatography are illustrated by the separation of ink on a filter paper (Fig. 1-5c–d).

Decomposing Compounds

A chemical compound retains its identity during physical changes, but it can be decomposed into its constituent elements by *chemical changes*. The decomposition of compounds into their constituent elements is a more difficult matter than the mere physical separation of mixtures. The extraction of iron from iron oxide ores requires a blast furnace. The industrial production of pure magnesium from magnesium chloride requires electricity. It is generally easier to convert a compound into other compounds by a chemical reaction than it is to separate a compound into its constituent elements. For example, when heated, ammonium dichromate decomposes into the substances chromium(III) oxide, nitrogen, and water. This reaction, once used in movies to simulate a volcano, is illustrated in Figure 1-6.

States of Matter

Matter is generally found in one of three *states*: solid, liquid, or gas. In a **solid**, atoms or molecules are in close contact, sometimes in a highly organized arrangement called a *crystal*. A solid has a definite shape. In a **liquid**, the atoms or molecules are usually separated by somewhat greater distances than in a solid. Movement of these atoms or molecules gives a liquid its most distinctive property—the ability to flow, covering the bottom and assuming the shape of its container. In a **gas**, distances between atoms or molecules are much greater

▲ FIGURE 1-6
A chemical change: decomposition of ammonium dichromate

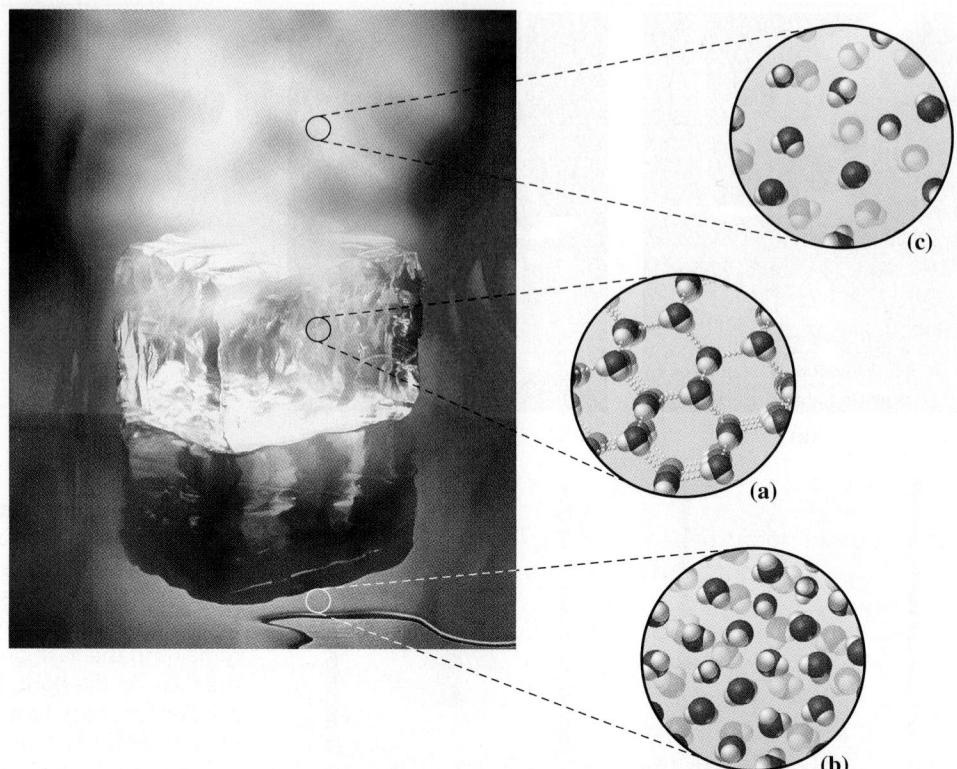

► FIGURE 1-7
Macroscopic and microscopic views of matter
The picture shows a block of ice on a heated surface and the three states of water. The circular insets show how chemists conceive of these states microscopically, in terms of molecules with two hydrogen atoms joined to one of oxygen. In ice **(a)**, the molecules are arranged in a regular pattern in a rigid framework. In liquid water **(b)**, the molecules are rather closely packed but move freely. In gaseous water **(c)**, the molecules are widely separated.

than in a liquid. A gas always expands to fill its container. Depending on conditions, a substance may exist in only one state of matter, or it may be present in two or three states. Thus, as the ice in a small pond begins to melt in the spring, water is in two states: solid and liquid (actually, three states if we also consider water vapor in the air above the pond). The three states of water are illustrated at two levels in Figure 1-7.

The *macroscopic level* refers to how we perceive matter with our eyes, through the outward appearance of objects. The *microscopic level* describes matter as chemists conceive of it—in terms of atoms and molecules and their behavior. In this text, we will describe many macroscopic, observable properties of matter, but to explain these properties, we will often shift our view to the atomic or molecular level—the microscopic level.

1-4 Measurement of Matter: SI (Metric) Units

► Nonnumerical information is *qualitative*, such as the color blue.

Chemistry is a *quantitative* science, which means that in many cases we can measure a property of a substance and compare it with a standard having a known value of the property. We express the measurement as the product of a *number* and a *unit*. The unit indicates the standard against which the measured quantity is being compared. When we say that the length of the playing field in football is 100 yd, we mean that the field is 100 times longer than a standard of length called the yard (yd). In this section, we will introduce some basic units of measurement that are important to chemists.

The scientific system of measurement is called the *Système Internationale d'Unités* (International System of Units) and is abbreviated **SI**. It is a modern version of the metric system, a system based on the unit of length called a *meter* (m). The meter was originally defined as 1/10,000,000 of the distance from the equator to the North Pole and translated into the length of a metal bar kept in Paris. Unfortunately, this length is subject to change with temperature, and it cannot be exactly reproduced. The SI system substitutes for the standard meter bar an unchanging, reproducible quantity: 1 meter is the distance traveled by light in a vacuum in 1/299,792,458 of a second. Length is one of the seven

► The *definition of the meter*, formerly based on the atomic spectrum of ^{86}Kr, was changed to the speed of light in 1983. Effectively, the speed of light is now defined as 2.99792458×10^8 m/s.

TABLE 1.1 SI Base Quantities		
Physical Quantity	Unit	Symbol
Length	meter[a]	m
Mass	kilogram	kg
Time	second	s
Temperature	kelvin	K
Amount of substance[b]	mole	mol
Electric current[c]	ampere	A
Luminous intensity[d]	candela	cd

[a]The official spelling of this unit is "metre," but we will use the American spelling.
[b]The mole is introduced in Section 2-7.
[c]Electric current is described in Appendix B and in Chapter 20.
[d]Luminous intensity is not discussed in this text.

TABLE 1.2 SI Prefixes	
Multiple	Prefix
10^{18}	exa (E)
10^{15}	peta (P)
10^{12}	tera (T)
10^{9}	giga (G)
10^{6}	mega (M)
10^{3}	kilo (k)
10^{2}	hecto (h)
10^{1}	deka (da)
10^{-1}	deci (d)
10^{-2}	centi (c)
10^{-3}	milli (m)
10^{-6}	micro (μ)[a]
10^{-9}	nano (n)
10^{-12}	pico (p)
10^{-15}	femto (f)
10^{-18}	atto (a)
10^{-21}	zepto (z)
10^{-24}	yocto (y)

[a]The Greek letter μ (pronounced "mew").

fundamental quantities in the SI system (see Table 1.1). All other physical quantities have units that can be derived from these seven. SI is a *decimal* system. Quantities differing from the base unit by powers of ten are noted by the use of prefixes. For example, the prefix *kilo* means *one thousand* (10^3) times the base unit; it is abbreviated as k. Thus 1 *kilo*meter = 1000 meters, or 1 km = 1000 m. The SI prefixes are listed in Table 1.2.

Most measurements in chemistry are made in SI units. Sometimes we must convert between SI units, as when converting kilometers to meters. At other times we must convert measurements expressed in non-SI units into SI units, or from SI units into non-SI units. In all of these cases we can use a *conversion factor* or a series of conversion factors in a scheme called a conversion pathway. Later in this chapter, we will apply conversion pathways in a method of problem solving known as *dimensional analysis*. The method itself is described in some detail in Appendix A.

◀ It is a good idea to *memorize the most common SI prefixes* (such as G, M, k, d, c, m, μ, n, and p) because you can't survive in a world of science without knowing the SI prefixes.

Mass

Mass describes the quantity of matter in an object. In SI the standard of mass is 1 *kilogram* (kg), which is a fairly large unit for most applications in chemistry. More commonly we use the unit *gram* (g) (about the mass of three aspirin tablets).

Weight is the force of gravity on an object. It is directly proportional to mass, as shown in the following mathematical expressions.

$$W \propto m \quad \text{and} \quad W = g \times m \qquad (1.1)$$

An object has a fixed mass (m), which is independent of where or how the mass is measured. Its weight (W), however, may vary because the acceleration due to gravity (g) varies slightly from one point on Earth to another. Thus, an object that weighs 100.0 kg in St. Petersburg, Russia, weighs only 99.6 kg in Panama (about 0.4% less). The same object would weigh only about 17 kg on the moon. Although the weight of an object varies from place to place, its mass is the same in all locations. The terms *weight* and *mass* are often used interchangeably, but only mass is a measure of the quantity of matter. A common laboratory device for measuring mass is called a balance. A balance is often called, incorrectly, a scale.

The principle used in a balance is that of counteracting the force of gravity on an unknown mass with a force of equal magnitude that can be precisely measured. In older two-pan beam balances, the object whose mass is being determined is placed on one pan and counterbalancing is achieved through the force of gravity acting on *weights*, objects of precisely known mass, placed on the other pan. In the type of balance most commonly seen in laboratories today—the electronic balance—the counterbalancing force is a magnetic force produced by passing an electric current through an electromagnet. First, an initial balance condition is achieved when no object is present on the balance pan. When the

◀ The symbol $\propto$ means "proportional to." It can be replaced by an equality sign and a proportionality constant. In expression (1.1), the constant is the acceleration due to gravity, g. (See Appendix B.)

▲ An electronic balance.

object to be weighed is placed on the pan, the initial balance condition is upset. To restore the balance condition, additional electric current must be passed through the electromagnet. The magnitude of this additional current is proportional to the mass of the object being weighed and is translated into a mass reading that is displayed on the balance. An electronic balance is shown in the margin.

🔍 1-2 CONCEPT ASSESSMENT

Would either the two-pan beam balance or the electronic balance yield the same result for the mass of an object measured on the moon as that measured for the same object on Earth? Explain.

Time

In daily use we measure time in seconds, minutes, hours, and years, depending on whether we are dealing with short intervals (such as the time for a 100 m race) or long ones (such as the time before the next appearance of Halley's comet in 2062). We can use all these units in scientific work also, although in SI the standard of time is the *second* (s). A time interval of 1 second is not easily established. At one time it was based on the length of a day, but this is not constant because the rate of Earth's rotation undergoes slight variations. In 1956, the second was defined as 1/31,556,925.9747 of the length of the year 1900. With the advent of atomic clocks, a more precise definition became possible. The second is now defined as the duration of 9,192,631,770 cycles of a particular radiation emitted by certain atoms of the element cesium (cesium-133).

▶ Electromagnetic radiation is discussed in Section 8-1.

Temperature

To establish a temperature scale, we arbitrarily set certain fixed points and temperature increments called degrees. Two commonly used fixed points are the temperature at which ice melts and the temperature at which water boils, both at standard atmospheric pressure.*

On the **Celsius** scale, the melting point of ice is 0 °C, the boiling point of water is 100 °C, and the interval between is divided into 100 equal parts called Celsius degrees. On the **Fahrenheit** temperature scale, the melting point of ice is 32 °F, the boiling point of water is 212 °F, and the interval between is divided into 180 equal parts called Fahrenheit degrees. Figure 1-8 compares the Fahrenheit and Celsius temperature scales.

The SI temperature scale, called the **Kelvin** scale, assigns a value of zero to the lowest possible temperature. The zero on the Kelvin scale is denoted 0 K and it comes at –273.15 °C. We will discuss the Kelvin temperature scale in detail in Chapter 6. For now, it is enough to know the following:

- The interval on the Kelvin scale, called a *kelvin*, is the same size as the Celsius degree.
- When writing a Kelvin temperature, we do not use a degree symbol. That is, we write 0 K or 300 K, not 0 °K or 300 °K.
- The Kelvin scale is an absolute temperature scale; there are no negative Kelvin temperatures.

▶ The SI symbol for Kelvin temperature is T and that for Celsius temperature is t but shown here as $t(°C)$. The Fahrenheit temperature, shown here as $t(°F)$, is not recognized in SI.

In the laboratory, temperature is most commonly measured in Celsius degrees; however, these temperatures must often be converted to the Kelvin scale (in describing the behavior of gases, for example). Occasionally, particularly in some engineering applications, temperatures must be converted

---

*Standard atmospheric pressure is defined in Section 6-1. The effect of pressure on melting and boiling points is described in Chapter 12.

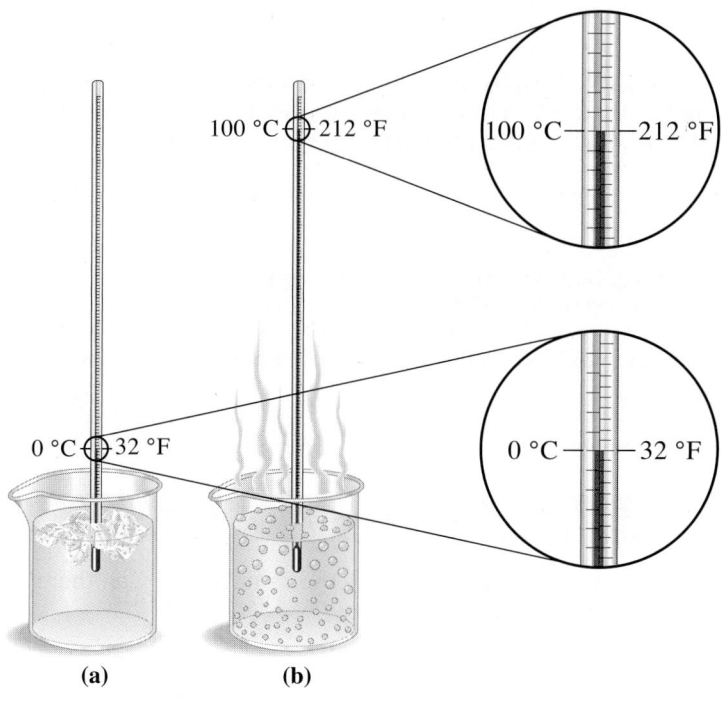

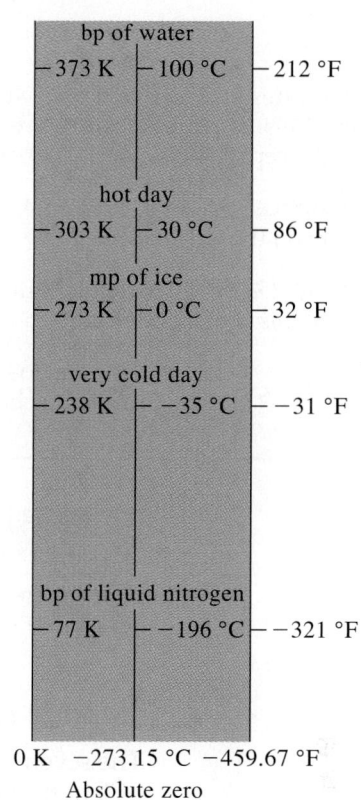

▲ FIGURE 1-8
A comparison of temperature scales
(a) The melting point (mp) of ice. (b) The boiling point (bp) of water.

between the Celsius and Fahrenheit scales. Temperature conversions can be made in a straightforward way by using the algebraic equations shown below.

$$\text{Kelvin from Celsius} \quad T(\text{K}) = t(^\circ\text{C}) + 273.15$$

$$\text{Fahrenheit from Celsius} \quad t(^\circ\text{F}) = \frac{9}{5}t(^\circ\text{C}) + 32$$

$$\text{Celsius from Fahrenheit} \quad t(^\circ\text{C}) = \frac{5}{9}\left[t(^\circ\text{F}) - 32\right]$$

The factors $\frac{9}{5}$ and $\frac{5}{9}$ arise because the Celsius scale uses 100 degrees between the two chosen reference points and the Fahrenheit scale uses 180 degrees: $180/100 = \frac{9}{5}$ and $100/180 = \frac{5}{9}$. The diagram in Figure 1-8 illustrates the relationship among the three scales for several temperatures.

EXAMPLE 1-1   Converting Between Fahrenheit and Celsius Temperatures

The predicted high temperature for New Delhi on a given day is 41 °C. Is this temperature higher or lower than the predicted daytime high of 103 °F for the same day in Phoenix, Arizona, reported by a newscaster?

Analyze

We are given a Celsius temperature and seek a comparison with a Fahrenheit temperature. To convert the given Celsius temperature to a Fahrenheit temperature, we use the equation given previously that expresses $t(^\circ\text{F})$ as a function of $t(^\circ\text{C})$.

Solve

$$t(^\circ\text{F}) = \frac{9}{5}t(^\circ\text{C}) + 32 = \frac{9}{5}(41) + 32 = 106 \ ^\circ\text{F}$$

The predicted temperature for New Delhi, 106 °F, is 3 °F higher than for Phoenix, 103 °F.

(continued)

Assess

For temperatures at which $t(°C) > -40\ °C$, the Fahrenheit temperature is greater than the Celsius temperature. If the Celsius temperature is lower than $-40\ °C$, then $t(°F)$ is lower than (more negative than) $t(°C)$ (Fig. 1-8). Concept Assessment 1-3 asks you to think further about the relationship between $t(°C)$ and $t(°F)$.

PRACTICE EXAMPLE A: A recipe in an American cookbook calls for roasting a cut of meat at 350 °F. What is this temperature on the Celsius scale?

PRACTICE EXAMPLE B: A particular automobile engine coolant has antifreeze protection to a temperature of $-22\ °C$. Will this coolant offer protection at temperatures as low as $-15\ °F$?

Mastering **CHEMISTRY**   Answers to Practice Examples are given on the Mastering Chemistry site: www.masteringchemistry.com.

1-3   CONCEPT ASSESSMENT

Can there be a temperature at which °C and °F have the same value? Can there be more than one such temperature? Explain.

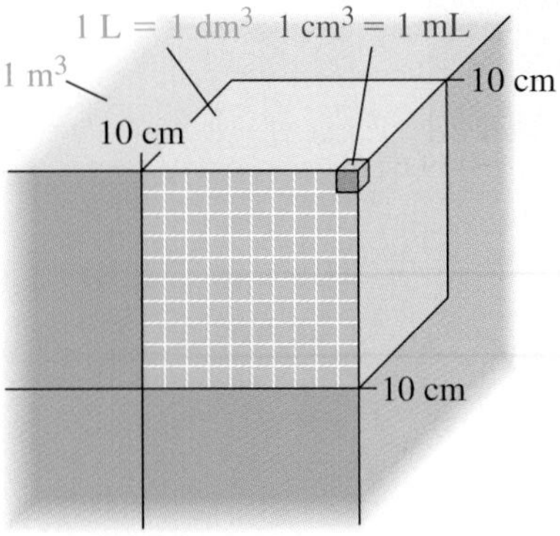

1 L = 1 dm³   1 cm³ = 1 mL
1 m³
10 cm
10 cm
10 cm

▲ FIGURE 1-9
Some metric volume units compared
The largest volume, shown in part, is the SI standard of 1 cubic meter (m³). A cube with a length of 10 cm (1 dm) on edge (in blue) has a volume of 1000 cm³ (1 dm³) and is called 1 liter (1 L). The smallest cube is 1 cm on edge (red) and has a volume of 1 cm³ = 1 mL.

▶ The official spelling is *litre*, but we will use the American spelling, *liter*.

Derived Units

The seven units listed in Table 1.1 are the SI units for the fundamental quantities of length, mass, time, and so on. Many measured properties are expressed as combinations of these fundamental, or base, quantities. We refer to the units of such properties as *derived units*. For example, velocity is a distance divided by the time required to travel that distance. The unit of velocity is length divided by time, such as m/s or m s⁻¹. Some derived units have special names. For example, the combination kg m⁻¹ s⁻² is called the *pascal* (Chapter 6) and the combination kg m² s⁻² is called the *joule* (Chapter 7). Other examples are given in Appendix C.

An important measurement that uses derived units is *volume*. Volume has the unit (length)³, and the SI standard unit of volume is the *cubic meter* (m³). More commonly used volume units are the *cubic centimeter* (cm³) and the *liter* (L). One liter is defined as a volume of 1000 cm³, which means that one *milliliter* (1 mL) is equal to 1 cm³. The liter is also equal to one *cubic decimeter* (1 dm³). Several volume units are depicted in Figure 1-9.

Non-SI Units

Although its citizens are growing more accustomed to expressing distances in kilometers and volumes in liters, the United States is one of the few countries where most units used in everyday life are still non-SI. Masses are given in pounds, room dimensions in feet, and so on. In this book, we will not routinely use these non-SI units, but we will occasionally introduce them in examples and end-of-chapter exercises. In such cases, any necessary relationships between non-SI and SI units will be given or can be found on the inside back cover.

1-1   ARE YOU WONDERING...

Why attaching the units to a number is so important?

In 1993, NASA started the Mars Surveyor program to conduct an ongoing series of missions to explore Mars. In 1995, two missions were scheduled that would be launched in late 1998 and early 1999. The missions were the Mars Climate Orbiter (MCO) and the Mars Polar Lander (MPL). The MCO was launched December 11, 1998, and the MPL, January 3, 1999.

Nine and a half months after launch, the MCO was to fire its main engine to achieve an elliptical orbit around Mars. The MCO engine start occurred on September 23, 1999, but the MCO mission was lost when the orbiter entered the Martian atmosphere on a lower-than-expected trajectory. The MCO entered the low orbit because the computer on Earth used British Engineering units, whereas the MCO computer used SI units.

This error in units brought the MCO 56 km above the surface of Mars instead of the desired 250 km. At 250 km, the MCO would have successfully entered the desired elliptical orbit, and the $168 million orbiter would probably not have been lost.

◀ The development of science requires careful *quantitative measurement.* Theories have stood or fallen based on their agreement or otherwise with experiments in the fourth significant figure or beyond. Problem solving, the handling of units, and the use of significant figures (Section 1-7) are important in all areas of science.

1-5   Density and Percent Composition: Their Use in Problem Solving

Throughout this text, we will encounter new concepts about the structure and behavior of matter. One means of firming up our understanding of these new concepts is to work problems that relate concepts that we already know to those we are trying to understand. In this section, we will introduce two quantities frequently required in problem solving: density and percent composition.

Density

Here is an old riddle: "What weighs more, a ton of bricks or a ton of cotton?" If you answer that they weigh the same, you demonstrate a clear understanding of the meaning of weight and, indirectly, of the quantity of matter to which weight is proportional, that is, mass. Anyone who answers that the bricks weigh more than the cotton has confused the concepts of weight and density. Matter in a brick is more concentrated than in cotton—that is, the matter in a brick is confined to a smaller volume. Bricks are more dense than cotton. **Density** is the ratio of mass to volume.

$$\text{density } (d) = \frac{\text{mass } (m)}{\text{volume } (V)} \qquad \textbf{(1.2)}$$

Mass and volume are both extensive properties. An **extensive property** is *dependent* on the quantity of matter observed. However, if we divide the mass of a substance by its volume, we obtain density, an intensive property. An **intensive property** is *independent* of the amount of matter observed. Thus, the density of pure water at 25 °C has a unique value, whether the sample fills a small beaker (small mass/small volume) or a swimming pool (large mass/large volume). Intensive properties are especially useful in chemical studies because they can often be used to identify substances.

The SI base units of mass and volume are kilograms and cubic meters, respectively, but chemists generally express mass in grams and volume in cubic centimeters or milliliters. Thus, the most commonly encountered density unit is grams per cubic centimeter (g/cm^3) or the identical unit grams per milliliter (g/mL).

The mass of 1.000 L of water at 4 °C is 1.000 kg. The density of water at 4 °C is 1000 g/1000 mL, or 1.000 g/mL. At 20 °C, the density of water is 0.9982 g/mL. Density is a function of temperature because volume varies with temperature, whereas mass remains constant. One reason that climate change is a concern is because as the average temperature of seawater increases, the seawater will become less dense, its volume will increase, and sea level will rise—*even if no continental ice melts.*

Like temperature, the state of matter affects the density of a substance. In general, solids are denser than liquids and both are denser than gases, but there are notable overlaps in densities between solids and liquids. Following are the ranges of values generally observed for densities; this information should prove useful in solving problems.

- Solid densities: from about 0.2 g/cm^3 to 20 g/cm^3
- Liquid densities: from about 0.5 g/mL to 3–4 g/mL
- Gas densities: mostly in the range of a few grams per liter

In general, densities of liquids are known more precisely than those of solids (which may have imperfections in their microscopic structures). Also, densities of elements and compounds are known more precisely than densities of materials with variable compositions (such as wood or rubber).

There are several important consequences of the different densities of solids and liquids. A solid that is insoluble and floats on a liquid is *less* dense than the liquid, and it displaces a *mass* of liquid equal to its own mass. An insoluble solid that sinks to the bottom of a liquid is *more* dense than the liquid and displaces a *volume* of liquid equal to its own volume. Liquids that are immiscible in each other separate into distinct layers, with the most dense liquid at the bottom and the least dense liquid at the top.

🔍 **1-4   CONCEPT ASSESSMENT**

Approximately what fraction of its volume is submerged when a 1.00 kg block of wood (d = 0.68 g/cm^3) floats on water?

Density in Conversion Pathways

If we measure the mass of an object and its volume, simple division gives us its density. Conversely, if we know the density of an object, we can use density as a conversion factor to determine the object's mass or volume. For example, a cube of osmium 1.000 cm on edge weighs 22.59 g. The density of osmium (the densest of the elements) is 22.59 g/cm^3. What would be the mass of a cube of osmium that is 1.25 in. on edge (1 in. = 2.54 cm)? To solve this problem, we begin by relating the volume of a cube to its length, that is, $V = l^3$. Then we can map out the *conversion pathway*:

$$\text{in. osmium} \longrightarrow \text{cm osmium} \longrightarrow \text{cm}^3 \text{ osmium} \longrightarrow \text{g osmium}$$

(converts in. to cm)   (converts cm to cm^3)   (converts cm^3 to g osmium)

$$? \text{ g osmium} = \left[1.25 \text{ in.} \times \frac{2.54 \text{ cm}}{1 \text{ in.}}\right]^3 \times \frac{22.59 \text{ g osmium}}{1 \text{ cm}^3} = 723 \text{ g osmium}$$

At 25 °C the density of mercury, the only metal that is liquid at this temperature, is 13.5 g/mL. Suppose we want to know the volume, in mL, of 1.000 kg of mercury at 25 °C. We proceed by (1) identifying the known information: 1.000 kg of mercury and d = 13.5 g/mL (at 25 °C); (2) noting what we are trying to determine—a volume in milliliters (which we designate mL mercury); and (3) looking for the relevant conversion factors. Outlining the conversion pathway will help us find these conversion factors:

$$\text{kg mercury} \longrightarrow \text{g mercury} \longrightarrow \text{mL mercury}$$

We need the factor 1000 g/kg to convert from kilograms to grams. Density provides the factor to convert from mass to volume. But in this instance, we need to use density in the *inverted* form. That is,

$$? \text{ mL mercury} = 1.000 \text{ kg} \times \frac{1000 \text{ g}}{1 \text{ kg}} \times \frac{1 \text{ mL mercury}}{13.5 \text{ g}} = 74.1 \text{ mL mercury}$$

Examples 1-2 and 1-3 further illustrate that numerical calculations involving density are generally of two types: determining density from mass and volume measurements and using density as a conversion factor to relate mass and volume.

EXAMPLE 1-2 Relating Mass, Volume, and Density

The stainless steel in the solid cylindrical rod pictured below has a density of 7.75 g/cm^3. If we want a 1.00 kg mass of this rod, how long a section must we cut off? Refer to the inside back cover for the formula to calculate the volume of a cylinder.

1.000 in.

Analyze

We are given the density, d, and the desired mass, m. Because $d = m/V$, we can solve for V and then use the formula for the volume of a cylinder, $V = \pi r^2 h$, to calculate h, the length of rod we seek. Two different mass units (g and kg) and two different length units (centimeters and inches) appear in the information given in this problem, so we anticipate having to make at least two unit conversions. To avoid errors, we include units in all steps.

Solve

Solve equation (1.2) for V. The reciprocal of density, $1/d$, is a conversion factor for converting from mass to volume.

$$V = \frac{m}{d} = m \times \frac{1}{d}$$

Calculate the volume of the rod that will have a mass of 1.00 kg. A conversion from kg to g is required in this step.

$$V = 1.00 \text{ kg} \times \frac{1000 \text{ g}}{1 \text{ kg}} \times \frac{1 \text{ cm}^3}{7.75 \text{ g}} = 129 \text{ cm}^3$$

Solve $V = \pi r^2 h$ for h and then calculate h. We must be certain to use the radius of the rod (one-half the diameter) and to express the radius in centimeters.

$$h = \frac{V}{\pi r^2} = \frac{129 \text{ cm}^3}{3.1416 \times (0.500 \text{ in.} \times 2.54 \text{ cm}/1 \text{ in.})^2} = 25.5 \text{ cm}$$

Assess

One way to check whether our answer is correct is to work the problem in reverse. For example, we calculate $d = 1.00 \times 10^3 \text{ g}/[3.1416 \times (1.27 \text{ cm})^2 \times 25.5 \text{ cm}] = 7.74 \text{ g/cm}^3$, which is very close to the given density. We are confident that our answer, $h = 25.5$ cm, is correct.

PRACTICE EXAMPLE A: To determine the density of trichloroethylene, a liquid used to degrease electronic components, a flask is first weighed empty (108.6 g). It is then filled with 125 mL of the trichloroethylene to give a total mass of 291.4 g. What is the density of trichloroethylene in grams per milliliter?

PRACTICE EXAMPLE B: Suppose that instead of using the cylindrical rod of Example 1-2 to prepare a 1.000 kg mass we were to use a solid spherical ball of copper ($d = 8.96 \text{ g/cm}^3$). What must be the radius of this ball?

EXAMPLE 1-3 Determining the Density of an Irregularly Shaped Solid

A chunk of coal is weighed twice while suspended from a spring scale (see Figure 1-10). When the coal is suspended in air, the scale registers 156 g; when the coal is suspended underwater at 20 °C, the scale registers 59 g. What is the density of the coal? The density of water at 20 °C is 0.9982 g cm^{-3}.

Analyze

We need the ratio of mass to volume of the chunk of coal. The mass of the coal is easily obtained; it is what registers on the scale when the coal is suspended in air: 156 g. But what is the volume of this chunk of coal? The key to this calculation is the weight measurement under water. The coal weighs less than 156 g when submerged in water because the water exerts a buoyant force on the coal. The buoyant force is the difference

(continued)

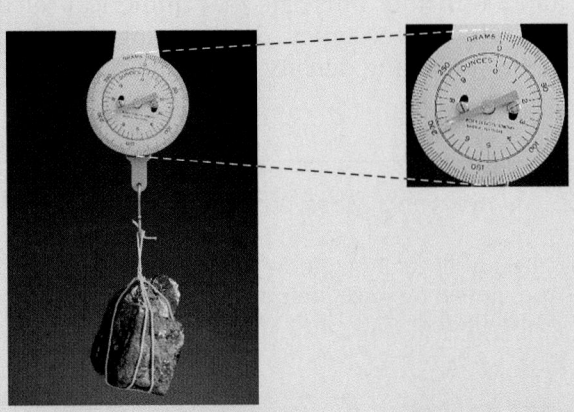

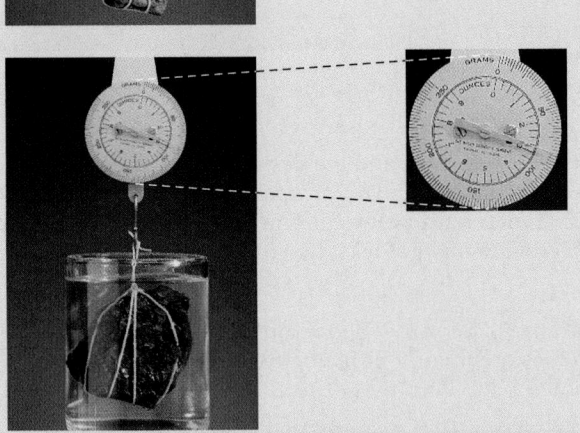

▶ FIGURE 1-10
Measuring the volume of an irregularly shaped solid
When submerged in a liquid, an irregularly shaped solid displaces a volume of liquid equal to its own. The necessary data can be obtained by two mass measurements of the type illustrated here; the required calculations are like those in Example 1-3.

between the two weight measurements: 156 g − 59 g = 97 g. Recall the statement on page 14 that a submerged solid displaces a volume of water equal to its own volume. We don't know this volume of water directly, but we can use the mass of displaced water, 97 g, and its density, 0.9982 g/cm³, to calculate the volume of displaced water. The volume of the coal is equal to the volume of displaced water.

Solve

The mass of the chunk of coal is 156 g. If we use m_{water} to denote the mass of displaced water, then the volume of the displaced water is calculated as follows:

$$V = \frac{m_{water}}{d} = \frac{156\ g - 59\ g}{0.9982\ g/cm^3} = 97\ cm^3$$

The volume of the chunk of coal is the same as the volume of displaced water. Therefore, the density of the coal is

$$d = \frac{156\ g}{97\ cm^3} = 1.6\ g/cm^3$$

Assess

To determine the density of an object, we might think it is necessary to make measurements of both the mass and volume of the object. Example 1-3 shows that a volume measurement is not necessary. The steps in our calculation can be combined to give the following expression:

(density of object)/(density of water) = (weight in water)/(weight in air − weight in water).

The expression above clearly shows that the density of an object can be determined by making two weight measurements: one in air, and the other in a fluid (such as water) of known density.

PRACTICE EXAMPLE A: A graduated cylinder contains 33.8 mL of water. A stone with a mass of 28.4 g is placed in the cylinder and the water level rises to 44.1 mL. What is the density of the stone?

PRACTICE EXAMPLE B: In the situation shown in the photograph, when the ice cube melts completely, will the water overflow the container, will the water level in the container drop, or will the water level remain unchanged? Explain.

Percent Composition as a Conversion Factor

In Section 1-2, we described composition as an identifying characteristic of a sample of matter. A common way of referring to composition is through percentages. Percent (*per centum*) is the Latin for *per* (meaning "for each") and *centum* (meaning "100"). Thus, percent is the number of parts of a constituent in 100 parts of the whole. To say that a seawater sample contains 3.5% sodium chloride by mass means that there are 3.5 g of sodium chloride in every 100 g of the seawater. We make the statement in terms of grams because we are talking about percent *by mass*. We can express this percent by writing the following ratios:

$$\frac{3.5 \text{ g sodium chloride}}{100 \text{ g seawater}} \quad \text{and} \quad \frac{100 \text{ g seawater}}{3.5 \text{ g sodium chloride}} \qquad \textbf{(1.3)}$$

In Example 1-4, we will use one of these ratios as a conversion factor.

EXAMPLE 1-4   Using Percent Composition as a Conversion Factor

A 75 g sample of sodium chloride (table salt) is to be produced by evaporating to dryness a quantity of seawater containing 3.5% sodium chloride by mass. What volume of seawater, in liters, must be taken for this purpose? Assume a density of 1.03 g/mL for seawater.

Analyze

The conversion pathway is g sodium chloride → g seawater → mL seawater → L seawater. To convert from g sodium chloride to g seawater, we need the conversion factor in expression (1.3), with g seawater in the numerator and g sodium chloride in the denominator. To convert from g seawater to mL of seawater, we use the reciprocal of the density of seawater as the conversion factor. To make the final conversion, from mL seawater to L of seawater, we use the fact that 1 L = 1000 mL.

Solve

Following the conversion pathway described above, we obtain

$$? \text{ L seawater} = 75 \text{ g sodium chloride} \times \frac{100 \text{ g seawater}}{3.5 \text{ g sodium chloride}}$$

$$\times \frac{1 \text{ mL seawater}}{1.03 \text{ g seawater}} \times \frac{1 \text{ L seawater}}{1000 \text{ mL seawater}}$$

$$= 2.1 \text{ L seawater}$$

Assess

In solving this problem, we set up a conversion pathway, and then we thought about the conversion factors that were required. We will make use of this approach throughout the text.

PRACTICE EXAMPLE A: How many kilograms of ethanol are present in 25 L of a gasohol solution that is 90% gasoline to 10% ethanol by mass? The density of gasohol is 0.71 g/mL.

PRACTICE EXAMPLE B: Common rubbing alcohol is a solution of 70.0% isopropyl alcohol by mass in water. If a 25.0 mL sample of rubbing alcohol contains 15.0 g of isopropyl alcohol, what is the density of the rubbing alcohol?

1-2   ARE YOU WONDERING...

When to multiply and when to divide in doing problems with percentages?

A common way of dealing with a percentage is to convert it to decimal form (3.5% becomes 0.035) and then to multiply or divide by this decimal, but students sometimes can't decide which to do. Expressing percentage as a conversion factor and

(continued)

using it to produce a necessary cancellation of units gets around this difficulty. Also, remember that

The quantity of a *component* must always be less than the quantity of the entire mixture. (*Multiply* by percentage.)

The quantity of the entire *mixture* must always be greater than the quantity of any of the components. (*Divide* by percentage.)

Component

MIXTURE

If, in Example 1-4, we had not been careful about the cancellation of units and had multiplied by percentage (3.5/100) instead of dividing by it (100/3.5), we would have obtained the numerical answer 2.5×10^{-3}. This would be a 2.5 mL sample of seawater, weighing about 2.5 g. Clearly, a sample of seawater that *contains* 75 g of sodium chloride must have a mass *greater than* 75 g.

1-6 Uncertainties in Scientific Measurements

All measurements are subject to error. To some extent, measuring instruments have built-in, or inherent, errors, called **systematic errors**. (For example, a kitchen scale might consistently yield results that are 25 g too high or a thermometer a reading that is 2°C too low.) Limitations in an experimenter's skill or ability to read a scientific instrument also lead to errors and give results that may be either too high or too low. Such errors are called **random errors**.

▶ *Random errors* are observed by scatter in the data and can be dealt with effectively by taking the average of many measurements. *Systematic errors*, conversely, are the bane of the experimental scientist. They are not readily apparent and must be avoided by carefully calibrating a method against a known sample or result. Systematic errors influence the *accuracy* of a measurement, whereas random errors are linked to the *precision* of measurements.

Precision refers to the degree of reproducibility of a measured quantity—that is, the closeness of agreement when the same quantity is measured several times. The precision of a series of measurements is *high* (or good) if each of a series of measurements deviates by only a small amount from the average. Conversely, if there is wide deviation among the measurements, the precision is *poor* (or low). **Accuracy** refers to how close a measured value is to the accepted, or actual, value. High-precision measurements are not always accurate—a large systematic error could be present. (A tight cluster of three darts near the edge of a dart board can be considered precise but not very accurate if the intention was to strike the center of the board.) Still, scientists generally strive for high precision in measurements.

To illustrate these ideas, consider measuring the mass of an object by using the two balances shown on page 19. One of the balances is a single-pan balance that gives the mass in grams with only one decimal place. The other balance is a sophisticated analytical balance that gives the mass in grams with four decimal places. The accompanying table gives results obtained when the object is weighed three times on each balance. For the single-pan balance, the average of the measurements is 10.5 g, with measurements ranging from 10.4 g to 10.6 g. For the analytical balance, the average of the measurements is 10.4978 g, with measurements ranging from 10.4977 g to 10.4979 g. The scatter in the data obtained with the single-pan balance (±0.1 g) is greater than that obtained with the analytical balance (±0.0001 g). Thus, the results obtained by using the single-pan balance have lower (or poorer) precision than those obtained by using the analytical balance.

	Pan Balance	Analytical Balance
Three measurements	10.5, 10.4, 10.6 g	10.4978, 10.4979, 10.4977 g
Their average	10.5 g	10.4978 g
Reproducibility	±0.1 g	±0.0001 g
Precision	low or poor	high or good

1-5   CONCEPT ASSESSMENT

Can a set of measurements be precise without being accurate? Can the average of a set of measurements be accurate and the individual measurements be imprecise? Explain.

1-7   Significant Figures

Consider these measurements made on a low-precision balance: 10.4, 10.2, and 10.3 g. The reported result is best expressed as their average, that is, 10.3 g.

A scientist would interpret these results to mean that the first two digits—10—are known with certainty and the last digit—3—is uncertain because it was estimated. That is, the mass is known only to the nearest 0.1 g, a fact that we could also express by writing 10.3 ± 0.1 g. To a scientist, the measurement 10.3 g is said to have *three* **significant figures**. If this mass is reported in kilograms rather than in grams, 10.3 g = 0.0103 kg, the measurement is still expressed to *three* significant figures even though more than three digits are shown. When measured on an analytical balance, the corresponding reported value might be 10.3107 g—a value with *six* significant figures. The number of significant figures in a measured quantity gives an indication of the capabilities of the measuring device and the precision of the measurements.

We will frequently need to determine the number of significant figures in a numerical quantity. The rules for doing this, outlined in Figure 1-11, are as follows:

- All nonzero digits are significant.
- Zeros are also significant, but with two important *exceptions* for quantities less than one. Any zeros (1) preceding the decimal point, or (2) following the decimal point and preceding the first nonzero digit, are *not* significant.
- The case of terminal zeros that precede the decimal point in quantities greater than one is *ambiguous*.

The quantity 7500 m is an example of an ambiguous case.

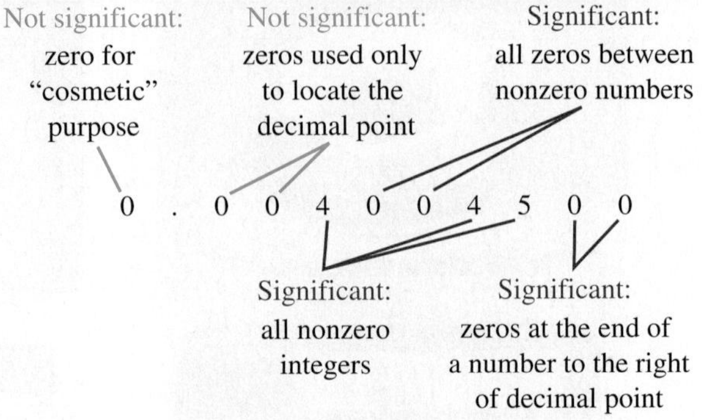

▶ FIGURE 1-11
Determining the number of significant figures in a quantity
The quantity shown here, 0.004004500, has *seven* significant figures. All nonzero digits are significant, as are the indicated zeros.

Not significant: zero for "cosmetic" purpose

Not significant: zeros used only to locate the decimal point

Significant: all zeros between nonzero numbers

$$0 \, . \, 0 \; 0 \; 4 \; 0 \; 0 \; 4 \; 5 \; 0 \; 0$$

Significant: all nonzero integers

Significant: zeros at the end of a number to the right of decimal point

Do we mean 7500 m, measured to the nearest meter? Nearest 10 meters? If all the zeros are significant—if the value has *four* significant figures—we can write 7500. m. That is, by writing a decimal point that is not otherwise needed, we show that all zeros preceding the decimal point are significant. This technique does not help if only one of the zeros, or if neither zero, is significant. The best approach here is to use exponential notation. (Review Appendix A if necessary.) The coefficient establishes the number of significant figures, and the power of ten locates the decimal point.

2 significant figures	*3 significant figures*	*4 significant figures*
7.5×10^3 m	7.50×10^3 m	7.500×10^3 m

Significant Figures in Numerical Calculations

Precision must neither be gained nor be lost in calculations involving measured quantities. There are several methods for determining how precisely to express the result of a calculation, but it is usually sufficient just to observe some simple rules involving significant figures.

▶ A more exact rule on multiplication/division is that the result should have about the same relative error—for example, expressed as parts per hundred (percent) or parts per thousand—as the least precisely known quantity. Usually the significant figure rule conforms to this requirement; occasionally, it does not (see Exercise 67).

> The result of multiplication or division may contain only as many significant figures as the *least* precisely known quantity in the calculation.

In the following chain multiplication to determine the volume of a rectangular block of wood, we should round off the result to *three* significant figures. Figure 1-12 may help you to understand this.

$$14.79 \text{ cm} \times 12.11 \text{ cm} \times 5.05 \text{ cm} = 904 \text{ cm}^3$$
$$\text{(4 sig. fig.)} \quad \text{(4 sig. fig.)} \quad \text{(3 sig. fig.)} \quad \text{(3 sig. fig.)}$$

In adding and subtracting numbers, the applicable rule is as follows:

▶ In addition and subtraction the *absolute* error in the result can be no less than the absolute error in the least precisely known quantity. In the summation at the right, the absolute error in one quantity is ±0.1 g; in another, ±0.01 g; and in the third, ±0.001 g. The sum *must* be expressed with an absolute error of ±0.1 g.

> The result of addition or subtraction must be expressed with the same number of digits beyond the decimal point as the quantity carrying the *smallest* number of such digits.

Consider the following sum of masses.

$$
\begin{array}{r}
15.02 \text{ g} \\
9986.0 \text{ g} \\
\underline{3.518 \text{ g}} \\
10{,}004.53\,8 \text{ g}
\end{array}
$$

▲ FIGURE 1-12
Significant-figure rule in multiplication
In forming the product 14.79 cm × 12.11 cm × 5.05 cm, the least precisely known quantity is 5.05 cm. Shown on the calculators are the products of 14.79 and 12.11 with 5.04, 5.05, and 5.06, respectively. In the three results, only the first two digits, 90..., are identical; variations begin in the third digit. We are certainly not justified in carrying digits beyond the third. We express the volume as 904 cm^3. Usually, instead of a detailed analysis of the type done here, we can use a simpler idea: *The result of a multiplication may contain only as many significant figures as does the least precisely known quantity.*

The sum has the same uncertainty, ±0.1 g, as does the term with the *smallest* number of digits beyond the decimal point, 9986.0 g. Note that this calculation is *not* limited by significant figures. In fact, the sum has more significant figures (six) than do any of the terms in the addition.

There are two situations when a quantity appearing in a calculation may be *exact*, that is, not subject to errors in measurement. This may occur

- by definition (such as 1 min = 60 s, or 1 in. = 2.54 cm)
- as a result of counting (such as *six* faces on a cube, or *two* hydrogen atoms in a water molecule)

◀ Later in the text, we will need to apply ideas about significant figures to logarithms. This concept is discussed in Appendix A.

> *Exact* numbers can be considered to have an unlimited number of significant figures.

◀ As added practice in working with significant figures, review the calculations in Section 1-6. You will note that they conform to the significant figure rules presented here.

🔍 **1-6   CONCEPT ASSESSMENT**

Which of the following is a more precise statement of the length 1 inch: 1 in. = 2.54 cm or 1 m = 39.37 in.? Explain.

Rounding Off Numerical Results

To three significant figures, we should express 15.453 as 15.5 and 14,775 as 1.48 × 10^4. If we need to drop just one digit, that is, to round off a number, the rule that we will follow is to increase the final digit by one unit if the digit dropped is 5, 6, 7, 8, or 9 and to leave the final digit unchanged if the digit dropped is 0, 1, 2, 3, or 4.* To three significant figures, 15.44 rounds off to 15.4, and 15.45 rounds off to 15.5.

◀ Some people prefer the *"round 5 to even"* rule. Thus 15.55 rounds to 15.6, and 17.65 rounds to 17.6. In banking and with large data sets, rounding needs to be unbiased. With a small number of data, this is less important.

*C. J. Guare, *J. Chem. Educ.*, **68**, 818 (1991).

EXAMPLE 1-5 Applying Significant Figure Rules: Multiplication/Division

Express the result of the following calculation with the correct number of significant figures.

$$\frac{0.225 \times 0.0035}{2.16 \times 10^{-2}} = ?$$

Analyze

By inspecting the three quantities, we see that the least precisely known quantity, 0.0035, has *two* significant figures. Our result must also contain only *two* significant figures.

Solve

When we carry out the calculation above by using an electronic calculator, the result is displayed as 0.0364583. In our analysis of this problem, we determined that the result must be rounded off to two significant figures, and so the result is properly expressed as 0.036 or as 3.6×10^{-2}.

Assess

To check for any possible calculation error, we can estimate the correct answer through a quick mental calculation by using exponential numbers. The answer should be $(2 \times 10^{-1})(4 \times 10^{-3})/(2 \times 10^{-2}) \approx 4 \times 10^{-2}$, and it is. Expressing numbers in exponential notation can often help us quickly estimate what the result of a calculation should be.

PRACTICE EXAMPLE A: Perform the following calculation, and express the result with the appropriate number of significant figures.

$$\frac{62.356}{0.000456 \times 6.422 \times 10^3} = ?$$

PRACTICE EXAMPLE B: Perform the following calculation, and express the result with the appropriate number of significant figures.

$$\frac{8.21 \times 10^4 \times 1.3 \times 10^{-3}}{0.00236 \times 4.071 \times 10^{-2}} = ?$$

EXAMPLE 1-6 Applying Significant Figure Rules: Addition/Subtraction

Express the result of the following calculation with the correct number of significant figures.

$$(2.06 \times 10^2) + (1.32 \times 10^4) - (1.26 \times 10^3) = ?$$

Analyze

If the calculation is performed with an electronic calculator, the quantities can be entered just as they are written, and the answer obtained can be adjusted to the correct number of significant figures. To determine the correct number of significant figures, identify the largest quantity, and then write the other quantities with the same power of ten as appears in the largest quantity. The answer can have no more digits beyond the decimal point than the quantity having the smallest number of such digits.

Solve

The largest quantity is 1.32×10^4 and thus, we write the other two quantities as 0.0206×10^4 and 0.126×10^4. The result of the required calculation must be rounded off to two decimal places.

$$(2.06 \times 10^2) + (1.32 \times 10^4) - (1.26 \times 10^3)$$
$$= (0.0206 \times 10^4) + (1.32 \times 10^4) - (0.126 \times 10^4)$$
$$= (0.0206 + 1.32 - 0.126) \times 10^4$$
$$= 1.2146 \times 10^4$$
$$= 1.21 \times 10^4$$

Assess

If you refer back to the margin note on page 20, you will see that there is another way to approach this problem. To determine the absolute error in the least precisely known quantity, we write the three quantities as $(2.06 \pm 0.01) \times 10^2$, $(1.32 \pm 0.01) \times 10^4$ and $(1.26 \pm 0.01) \times 10^3$. We conclude that 1.32×10^4 has the largest absolute error ($\pm 0.01 \times 10^4$) and so, the absolute error in the result of the calculation above is also $\pm 0.01 \times 10^4$. Thus, 1.2146×10^4 is rounded to 1.21×10^4.

PRACTICE EXAMPLE A: Express the result of the following calculation with the appropriate number of significant figures.

$$0.236 + 128.55 - 102.1 = ?$$

PRACTICE EXAMPLE B: Perform the following calculation, and express the result with the appropriate number of significant figures.

$$\frac{(1.302 \times 10^3) + 952.7}{(1.57 \times 10^2) - 12.22} = ?$$

In working through the preceding examples, you likely used an electronic calculator. What's nice about using electronic calculators is that we don't have to write down intermediate results. In general, disregard occasional situations where intermediate rounding may be justified and store all intermediate results in your electronic calculator without regard to significant figures. Then, round off to the correct number of significant figures only in the final answer.

KEEP IN MIND

that addition and subtraction are governed by one significant-figure rule, and multiplication and division are governed by a different rule.

 Mastering**CHEMISTRY**   **www.masteringchemistry.com**

In the late 1960s, scientists heatedly debated the reported discovery of a new form of water called polywater. For a discussion of the polywater debate and the importance of the scientific method in helping the scientific community reach a consensus, go to the Focus On feature for Chapter 1 (The Scientific Method at Work: Polywater) on the MasteringChemistry site.

Summary

1-1 The Scientific Method—The **scientific method** is a set of procedures used to develop explanations of natural phenomena and possibly to predict additional phenomena. The four basic stages of the scientific method are (1) gathering data through observations and experiments; (2) reducing the data to simple verbal or mathematical expressions known as **natural laws**; (3) offering a plausible explanation of the data through a **hypothesis**; (4) testing the hypothesis through predictions and further experimentation, leading ultimately to a conceptual model called a **theory** that explains the hypothesis, often together with other related hypotheses.

1-2 Properties of Matter—**Matter** is defined as anything that occupies space, possesses mass, and displays inertia. **Composition** refers to the component parts of a sample of matter and their relative proportions. **Properties** are the qualities or attributes that distinguish one sample of matter from another. Properties of matter can be grouped into two broad categories: **physical** and **chemical**.

Matter can undergo two types of changes: **chemical changes** or **reactions** are changes in composition; **physical changes** are changes in state or physical form and do not affect composition.

1-3 Classification of Matter—The basic building blocks of matter are called **atoms**. Matter that is composed of a collection of a single type of atom is known as an **element**. A sample of matter composed of two or more elements is known as a **compound**. A **molecule** is the smallest entity of a compound having the same proportions of the constituent atoms as does the compound as a whole. Collectively, elements and compounds compose the types of matter called **substances**. **Mixtures** of substances can be classified as **homogeneous** or **heterogeneous** (Fig. 1-4). The three states of matter are **solid**, **liquid**, and **gas**.

1-4 Measurement of Matter: SI (Metric) Units—Chemistry is a quantitative science, meaning that chemical measurements are usually expressed in terms of a number and an accompanying unit. The scientific system of measurement, called the *Système Internationale d'Unités* (abbreviated **SI**), involves seven base quantities (Table 1.1). **Mass** describes a quantity of matter. Weight measures the force of gravity on an object; weight is related to, but different from, mass. The temperature scales used by chemists are the **Celsius** and **Kelvin** scales. The **Fahrenheit** temperature scale, commonly used in daily life in the United States, is also used in some industrial settings. The three temperature scales can be related algebraically (Fig. 1-8).

1-5 Density and Percent Composition: Their Use in Problem Solving—Mass and volume are **extensive properties**; they *depend* on the amount of matter in a sample. **Density**, the ratio of the mass of a sample to its volume, is an **intensive property**, a property *independent* of the amount of matter sampled. Density is used as a conversion factor in a variety of calculations.

1-6 Uncertainties in Scientific Measurements—Measurements are subject to **systematic** and **random** errors. In making a series of measurements, the degree to which the measurements agree with one another is known as the **precision** of the measurement, while the degree to which the measurement agrees with the actual value is referred to as the **accuracy** of the measurement.

1-7 Significant Figures—The proper use of **significant figures** is important in that it prevents the suggestion of a higher degree of precision in a calculated quantity than is warranted by the precision of the measured quantities used in the calculation. The precision of an answer cannot be greater than the precision of the numbers used in the calculation. In addition to reporting the correct number of significant figures in a calculated quantity, it is important to know the rules for rounding off numerical results.

Integrative Example

Consider a 58.35 g hexagonal block of wood that is 5.00 cm on edge and 1.25 cm thick, with a 2.50 cm diameter hole drilled through its center. Also given are the densities of the liquids hexane ($d = 0.667$ g/mL) and decane ($d = 0.845$ g/mL). Assume that the density of a mixture of the two liquids is a linear function of the volume percent composition of the solution. Determine the volume percent of hexane required in the solution so that the hexagonal block of wood will just barely float on the solution.

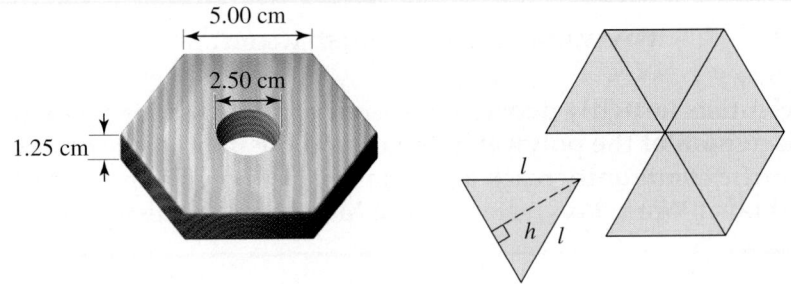

Analyze
The first task is to determine the density of the wood block, $d = m/V$. The mass is given, so the critical calculation is that of the volume. The key to the volume calculation is recognizing that the volume of the block is the *difference* between two volumes: The volume of the block if there were no hole *minus* the volume of the cylindrical hole. The second task is to write a simple equation relating density to the volume percent composition of the liquid solution, and then to solve that equation for the volume percent hexane that yields a solution density equal to the density calculated for the wood.

Solve
The solid hexagonal block can be divided into six smaller blocks, each an equilateral triangle of length, l, and height, h. The area of a triangle is given by the formula

$$A = \frac{1}{2}(\text{base} \times \text{height}) = \frac{1}{2} \times l \times h$$

Only the base, l, is given (5.00 cm). To express h in terms of l, we use the Pythagorean theorem for the right triangle pictured, that is, $a^2 + b^2 = c^2$, rearranged to the form $a^2 = c^2 - b^2$.

$$h^2 = l^2 - \left(\frac{l}{2}\right)^2 = l^2 - \frac{l^2}{4} = \frac{3l^2}{4} \text{ and}$$

$$h = \frac{\sqrt{3}}{2} \times l$$

Now, for the area of one of the six triangles we have

$$A = \frac{1}{2} \times l \times \frac{\sqrt{3}}{2} \times l = \frac{\sqrt{3}}{4} \times l^2$$

Substituting $l = 5.00$ cm and multiplying the cross-sectional area by the thickness, 1.25 cm, we obtain a volume of

$$V = \frac{\sqrt{3}}{4} \times (5.00 \text{ cm})^2 \times 1.25 \text{ cm} = 13.5 \text{ cm}^3$$

The volume of the hexagonal block of wood, without the cylindrical hole, is that of six triangular blocks.

$$V = 6 \times 13.5 \text{ cm}^3 = 81.0 \text{ cm}^3$$

The volume of the cylindrical hole with a radius of 1.25 cm (one-half the 2.50 cm diameter) and a height of 1.25 cm is

$$V = \pi r^2 h = 3.1416 \times (1.25 \text{ cm})^2 \times 1.25 \text{ cm}$$
$$= 6.14 \text{ cm}^3$$

The volume of the block of wood is the difference

$$V = 81.0 \text{ cm}^3 - 6.14 \text{ cm}^3 = 74.9 \text{ cm}^3$$

The density of the wood is

$$d = \frac{m}{V} = \frac{58.35 \text{ g}}{74.9 \text{ cm}^3} = 0.779 \text{ g/cm}^3 \text{ or } 0.779 \text{ g/mL}$$

The general formula for a linear (straight-line) relationship is

$$y = mx + b$$

In the present case, let y represent the density, d, of a solution (in g/mL) and x, the volume fraction of hexane (volume percent/100). By substituting the density of pure decane into the equation, we note that $x = 0$ and find that $b = 0.845$.

$$d = 0.845 = (m \times 0) + b$$

Now, using the density of pure hexane, $d = 0.667$, and the values $x = 1.00$ and $b = 0.845$, we obtain the value of m.

$$d = 0.667 = (m \times 1.00) + 0.845$$
$$m = 0.667 - 0.845 = -0.178$$

Our final step is to find the value of x for a solution having the same density as the wood: 0.779 g/mL.

$$d = 0.779 = -0.178x + 0.845$$
$$x = \frac{0.845 - 0.779}{0.178} = 0.37$$

The volume fraction of hexane is 0.37, and the volume percent composition of the solution is 37% hexane and 63% decane.

Assess

There is an early point in this calculation where we can check the correctness of our work. We have two expectations for the density of the block of wood: (1) it should be less than 1 g/cm^3 (practically all wood floats on water), and (2) it must fall between 0.667 g/cm^3 and 0.845 g/cm^3. If the calculated density of the wood were outside this range, the block of wood would either float on both liquids or sink in both of them, making the rest of the calculation impossible.

Another point to notice in this calculation is that we are justified in carrying three significant figures throughout the calculation up to the last step. There, because we must take the difference between two numbers of similar magnitudes, the number of significant figures drops from three to two.

PRACTICE EXAMPLE A: Magnalium is a solid mixture (an alloy) of aluminum metal and magnesium metal. An irregularly shaped chunk of a sample of magnalium is weighed twice, once in air and once in vegetable oil, by using a spring scale (see Figure 1-10). The weight in air is 211.5 g and the weight in oil is 135.3 g. If the densities of pure aluminum, pure magnesium, and vegetable oil are 2.70 g/cm^3, 1.74 g/cm^3, and 0.926 g/cm^3, respectively, then what is the mass percent of magnesium in this chunk of magnalium? Assume that the density of a mixture of the two metals is a linear function of the mass percent composition.

PRACTICE EXAMPLE B: A particular sample of seawater has a density of 1.027 g/cm^3 at 10 °C and is 2.67% sodium chloride by mass. Given that sodium chloride is 39.34% sodium by mass and that the mass of a single sodium atom is 3.817×10^{-26} kg, calculate the maximum mass of sodium and the maximum number of sodium atoms that can be extracted from a 1.5 L sample of this seawater.

Mastering**CHEMISTRY**

You'll find a link to additional self study questions in the study area on www.masteringchemistry.com

---

Exercises

(see also Appendices A-1 and A-5)

The Scientific Method

1. What are the principal reasons that one theory might be adopted over a conflicting one?
2. Can one predict how many experiments are required to verify a natural law? Explain.
3. A common belief among scientists is that there exists an underlying order to nature. Einstein described this belief in the words "God is subtle, but He is not malicious." What do you think Einstein meant by this remark?
4. Describe several ways in which a scientific law differs from a legislative law.
5. Describe the necessary characteristics of an experiment that is suitable to test a theory.
6. Describe the necessary characteristics of a scientific theory.

Properties and Classification of Matter

7. State whether the following properties of matter are physical or chemical.
   (a) An iron nail is attracted to a magnet.
   (b) A piece of paper spontaneously ignites when its temperature reaches 451 °F.
   (c) A bronze statue develops a green coating (patina) over time.
   (d) A block of wood floats on water.
8. State whether the following properties are physical or chemical.
   (a) A piece of sliced apple turns brown.
   (b) A slab of marble feels cool to the touch.
   (c) A sapphire is blue.
   (d) A clay pot fired in a kiln becomes hard and covered by a glaze.
9. Indicate whether each sample of matter listed is a substance or a mixture; if it is a mixture, indicate whether it is homogeneous or heterogeneous.
   (a) clean fresh air
   (b) a silver-plated spoon
   (c) garlic salt
   (d) ice
10. Indicate whether each sample of matter listed is a substance or a mixture; if it is a mixture, indicate whether it is homogeneous or heterogeneous.
    (a) a wooden beam
    (b) red ink
    (c) distilled water
    (d) freshly squeezed orange juice
11. Suggest physical changes by which the following mixtures can be separated.
    (a) iron filings and wood chips
    (b) ground glass and sucrose (cane sugar)
    (c) water and olive oil
    (d) gold flakes and water
12. What type of change—physical or chemical—is necessary to separate the following?
    [*Hint:* Refer to a listing of the elements.]
    (a) sugar from a sand/sugar mixture
    (b) iron from iron oxide (rust)
    (c) pure water from seawater
    (d) water from a slurry of sand in water

Exponential Arithmetic

13. Express each number in exponential notation. (a) 8950.; (b) 10,700.; (c) 0.0240; (d) 0.0047; (e) 938.3; (f) 275,482.
14. Express each number in common decimal form. (a) 3.21×10^{-2}; (b) 5.08×10^{-4}; (c) 121.9×10^{-5}; (d) 16.2×10^{-2}.
15. Express each value in exponential form. Where appropriate, include units in your answer.
    (a) speed of sound (sea level): 34,000 centimeters per second
    (b) equatorial radius of Earth: 6378 kilometers
    (c) the distance between the two hydrogen atoms in the hydrogen molecule: 74 trillionths of a meter
    (d) $\dfrac{(2.2 \times 10^3) + (4.7 \times 10^2)}{5.8 \times 10^{-3}} =$
16. Express each value in exponential form. Where appropriate, include units in your answer.
    (a) solar radiation received by Earth: 173 thousand trillion watts
    (b) average human cell diameter: 1 ten-millionth of a meter
    (c) the distance between the centers of the atoms in silver metal: 142 trillionths of a meter
    (d) $\dfrac{(5.07 \times 10^4) \times (1.8 \times 10^{-3})^2}{0.065 + (3.3 \times 10^{-2})} =$

Significant Figures

17. Indicate whether each of the following is an exact number or a measured quantity subject to uncertainty.
   (a) the number of sheets of paper in a ream of paper
   (b) the volume of milk in a liter bottle
   (c) the distance between Earth and the sun
   (d) the distance between the centers of the two oxygen atoms in the oxygen molecule

18. Indicate whether each of the following is an exact number or a measured quantity subject to uncertainty.
   (a) the number of pages in this text
   (b) the number of days in the month of January
   (c) the area of a city lot
   (d) the distance between the centers of the atoms in a gold medal

19. Express each of the following to *four* significant figures. **(a)** 3984.6; **(b)** 422.04; **(c)** 186,000; **(d)** 33,900; **(e)** 6.321×10^4; **(f)** 5.0472×10^{-4}.

20. How many significant figures are shown in each of the following? If this is indeterminate, explain why.
   (a) 450; **(b)** 98.6; **(c)** 0.0033; **(d)** 902.10; **(e)** 0.02173; **(f)** 7000; **(g)** 7.02; **(h)** 67,000,000

21. Perform the following calculations; express each answer in exponential form and with the appropriate number of significant figures.
   (a) $0.406 \times 0.0023 =$
   (b) $0.1357 \times 16.80 \times 0.096 =$
   (c) $0.458 + 0.12 - 0.037 =$
   (d) $32.18 + 0.055 - 1.652 =$

22. Perform the following calculations; express each number and the answer in exponential form and with the appropriate number of significant figures.
   (a) $\dfrac{320 \times 24.9}{0.080} =$
   (b) $\dfrac{432.7 \times 6.5 \times 0.002300}{62 \times 0.103} =$
   (c) $\dfrac{32.44 + 4.9 - 0.304}{82.94} =$
   (d) $\dfrac{8.002 + 0.3040}{13.4 - 0.066 + 1.02} =$

23. Perform the following calculations and retain the appropriate number of significant figures in each result.
   (a) $(38.4 \times 10^{-3}) \times (6.36 \times 10^5) =$
   (b) $\dfrac{(1.45 \times 10^2) \times (8.76 \times 10^{-4})}{(9.2 \times 10^{-3})^2} =$

(c) $24.6 + 18.35 - 2.98 =$
(d) $(1.646 \times 10^3) - (2.18 \times 10^2) + [(1.36 \times 10^4) \times (5.17 \times 10^{-2})] =$
(e)
$$\dfrac{-7.29 \times 10^{-4} + \sqrt{(7.29 \times 10^{-4})^2 + 4(1.00)(2.7 \times 10^{-5})}}{2 \times (1.00)}$$
[*Hint:* The significant figure rule for the extraction of a root is the same as for multiplication.]

24. Express the result of each of the following calculations in exponential form and with the appropriate number of significant figures.
   (a)
   $(4.65 \times 10^4) \times (2.95 \times 10^{-2}) \times (6.663 \times 10^{-3}) \times 8.2 =$
   (b) $\dfrac{1912 \times (0.0077 \times 10^4) \times (3.12 \times 10^{-3})}{(4.18 \times 10^{-4})^3} =$
   (c) $(3.46 \times 10^3) \times 0.087 \times 15.26 \times 1.0023 =$
   (d) $\dfrac{(4.505 \times 10^{-2})^2 \times 1.080 \times 1545.9}{0.03203 \times 10^3} =$
   (e)
   $$\dfrac{(-3.61 \times 10^{-4}) + \sqrt{(3.61 \times 10^{-4})^2 + 4(1.00)(1.9 \times 10^{-5})}}{2 \times (1.00)}$$
   [*Hint:* The significant figure rule for the extraction of a root is the same as for multiplication.]

25. An American press release describing the 1986 nonstop, round-the-world trip by the ultra-lightweight aircraft *Voyager* included the following data:
   flight distance: 25,012 mi
   flight time: 9 days, 3 minutes, 44 seconds
   fuel capacity: nearly 9000 lb
   fuel remaining at end of flight: 14 gal
   To the maximum number of significant figures permitted, calculate
   (a) the average speed of the aircraft in kilometers per hour
   (b) the fuel consumption in kilometers per kilogram of fuel (assume a density of 0.70 g/mL for the fuel)

26. Use the concept of significant figures to criticize the way in which the following information was presented. "The estimated proved reserve of natural gas as of January 1, 1982, was 2,911,346 trillion cubic feet."

Units of Measurement

27. Perform the following conversions.
   (a) $0.127 \text{ L} =$ _____ mL
   (b) $15.8 \text{ mL} =$ _____ L
   (c) $981 \text{ cm}^3 =$ _____ L
   (d) $2.65 \text{ m}^3 =$ _____ cm^3

28. Perform the following conversions.
   (a) $1.55 \text{ kg} =$ _____ g
   (b) $642 \text{ g} =$ _____ kg
   (c) $2896 \text{ mm} =$ _____ cm
   (d) $0.086 \text{ cm} =$ _____ mm

29. Perform the following conversions from non-SI to SI units. (Use information from the inside back cover, as needed.)

(a) $68.4 \text{ in.} =$ _____ cm
(b) $94 \text{ ft} =$ _____ m
(c) $1.42 \text{ lb} =$ _____ g
(d) $248 \text{ lb} =$ _____ kg
(e) $1.85 \text{ gal} =$ _____ dm^3
(f) $3.72 \text{ qt} =$ _____ mL

30. Determine the number of the following:
   (a) square meters (m^2) in 1 square kilometer (km^2)
   (b) cubic centimeters (cm^3) in 1 cubic meter (m^3)
   (c) square meters (m^2) in 1 square mile (mi^2) (1 mi = 5280 ft)

31. Which is the greater mass, 3245 μg or 0.00515 mg? Explain.
32. Which is the greater mass, 3257 mg or 0.000475 kg? Explain.
33. The non-SI unit, the hand (used by equestrians), is 4 inches. What is the height, in meters, of a horse that stands 15 hands high?
34. The unit *furlong* is used in horse racing. The units *chain* and *link* are used in surveying. There are exactly 8 furlongs in 1 mi, 10 chains in 1 furlong, and 100 links in 1 chain. To three significant figures, what is the length of 1 link in centimeters?
35. A sprinter runs the 100 yd dash in 9.3 s. At this same rate,
    (a) how long would it take the sprinter to run 100.0 m?
    (b) what is the sprinter's speed in meters per second?
    (c) how long would it take the sprinter to run a distance of 1.45 km?
36. A non-SI unit of mass used in pharmaceutical work is the grain (gr) (15 gr = 1.0 g). An aspirin tablet contains 5.0 gr of aspirin. A 155 lb arthritic individual takes two aspirin tablets per day.
    (a) What is the quantity of aspirin in two tablets, expressed in milligrams?

(b) What is the dosage rate of aspirin, expressed in milligrams of aspirin per kilogram of body mass?
(c) At the given rate of consumption of aspirin tablets, how many days would it take to consume 1.0 kg of aspirin?

37. In SI units, land area is measured in *hectares* (1 hectare = 1 hm^2). The commonly used unit for land area in the United States is the *acre*. How many acres correspond to 1 hectare? (1 mi^2 = 640 acres, 1 mi = 5280 ft, 1 ft = 12 in.).
38. In an engineering reference book, you find that the density of iron is 0.284 lb/in.^{3}. What is the density in g/cm^3?
39. In a user's manual accompanying an American-made automobile, a typical gauge pressure for optimal performance of automobile tires is 32 lb/in.^{2}. What is this pressure in grams per square centimeter and kilograms per square meter?
40. The volume of a red blood cell is about 90.0×10^{-12} cm^3. Assuming that red blood cells are spherical, what is the diameter of a red blood cell in millimeters?

Temperature Scales

41. We want to mark off a thermometer in both Celsius and Fahrenheit temperatures. On the Celsius scale, the lowest temperature mark is at -10 °C, and the highest temperature mark is at 50 °C. What are the equivalent Fahrenheit temperatures?
42. The highest and lowest temperatures on record for San Bernardino, California, are 118 °F and 17 °F, respectively. What are these temperatures on the Celsius scale?
43. The absolute zero of temperature is -273.15 °C. Should it be possible to achieve a temperature of -465 °F? Explain.
44. A family/consumer science class is given an assignment in candy-making that requires a sugar mixture to be brought to a "soft-ball" stage (234–240 °F).

A student borrows a thermometer having a range from -10 °C to 110 °C from the chemistry laboratory to do this assignment. Will this thermometer serve the purpose? Explain.

45. You decide to establish a new temperature scale on which the melting point of mercury (-38.9 °C) is 0 °M, and the boiling point of mercury (356.9 °C) is 100 °M. What would be (a) the boiling point of water in °M; and (b) the temperature of absolute zero in °M?
46. You decide to establish a new temperature scale on which the melting point of ammonia (-77.75 °C) is 0 °A and the boiling point of ammonia (-33.35 °C) is 100 °A. What would be (a) the boiling point of water in °A; and (b) the temperature of absolute zero in °A?

Density

47. A 2.18 L sample of butyric acid, a substance present in rancid butter, has a mass of 2088 g. What is the density of butyric acid in grams per milliliter?
48. A 15.2 L sample of chloroform at 20 °C has a mass of 22.54 kg. What is the density of chloroform at 20 °C, in grams per milliliter?
49. To determine the density of acetone, a 55.0 gal drum is weighed twice. The drum weighs 75.0 lb when empty and 437.5 lb when filled with acetone. What is the density of acetone expressed in grams per milliliter?
50. To determine the volume of an irregularly shaped glass vessel, the vessel is weighed empty (121.3 g) and when filled with carbon tetrachloride (283.2 g). What is the volume capacity of the vessel, in milliliters, given that the density of carbon tetrachloride is 1.59 g/mL?
51. A solution consisting of 8.50% acetone and 91.5% water by mass has a density of 0.9867 g/mL. What

mass of acetone, in kilograms, is present in 7.50 L of the solution?

52. A solution contains 10.05% sucrose (cane sugar) by mass. What mass of the solution, in grams, is needed for an application that requires 1.00 kg sucrose?
53. A fertilizer contains 21% nitrogen by mass. What mass of this fertilizer, in kilograms, is required for an application requiring 225 g of nitrogen?
54. A vinegar sample is found to have a density of 1.006 g/mL and to contain 5.4% acetic acid by mass. How many grams of acetic acid are present in 1.00 L of this vinegar?
55. Calculate the mass of a block of iron ($d = 7.86$ g/cm^3) with dimensions of 52.8 cm $\times$ 6.74 cm $\times$ 3.73 cm.
56. Calculate the mass of a cylinder of stainless steel ($d = 7.75$ g/cm^3) with a height of 18.35 cm and a radius of 1.88 cm.

57. The following densities are given at 20 °C: water, 0.998 g/cm³; iron, 7.86 g/cm³; aluminum, 2.70 g/cm³. Arrange the following items in terms of *increasing* mass.
    (a) a rectangular bar of iron,
    $$81.5 \text{ cm} \times 2.1 \text{ cm} \times 1.6 \text{ cm}$$
    (b) a sheet of aluminum foil,
    $$12.12 \text{ m} \times 3.62 \text{ m} \times 0.003 \text{ cm}$$
    (c) 4.051 L of water

58. To determine the approximate mass of a small spherical shot of copper, the following experiment is performed. When 125 pieces of the shot are counted out and added to 8.4 mL of water in a graduated cylinder, the total volume becomes 8.9 mL. The density of copper is 8.92 g/cm³. Determine the approximate mass of a single piece of shot, assuming that all of the pieces are of the same dimensions.

59. The density of aluminum is 2.70 g/cm³. A square piece of aluminum foil, 22.86 cm on a side is found to weigh 2.568 g. What is the thickness of the foil, in millimeters?

60. The angle iron pictured here is made of steel with a density of 7.78 g/cm³. What is the mass, in grams, of this object?

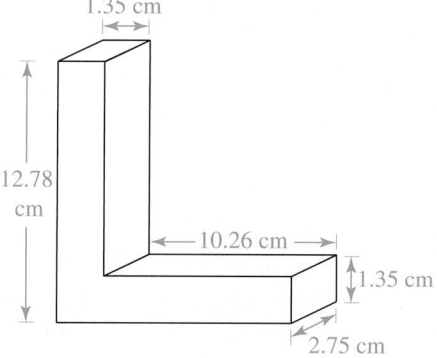

61. In normal blood, there are about 5.4×10^9 red blood cells per milliliter. The volume of a red blood cell is about 90.0×10^{-12} cm³, and its density is 1.096 g/mL. How many liters of whole blood would be needed to collect 0.5 kg of red blood cells?

62. A technique once used by geologists to measure the density of a mineral is to mix two dense liquids in such proportions that the mineral grains just float. When a sample of the mixture in which the mineral calcite just floats is put in a special density bottle, the weight is 15.4448 g. When empty, the bottle weighs 12.4631 g, and when filled with water, it weighs 13.5441 g. What is the density of the calcite sample? (All measurements were carried out at 25 °C, and the density of water at 25 °C is 0.9970 g/mL).

▲ At the left, grains of the mineral *calcite* float on the surface of the liquid bromoform ($d = 2.890$ g/mL). At the right, the grains sink to the bottom of liquid chloroform ($d = 1.444$ g/mL). By mixing bromoform and chloroform in just the proportions required so that the grains barely float, the density of the calcite can be determined (Exercise 62).

Percent Composition

63. In a class of 76 students, the results of a particular examination were 7 A's, 22 B's, 37 C's, 8 D's, 2 F's. What was the percent distribution of grades, that is, % A's, % B's, and so on?

64. A class of 84 students had a final grade distribution of 18% A's, 25% B's, 32% C's, 13% D's, 12% F's. How many students received each grade?

65. A solution of sucrose in water is 28.0% sucrose by mass and has a density of 1.118 g/mL. What mass of sucrose, in grams, is contained in 3.50 L of this solution?

66. A solution containing 12.0% sodium hydroxide by mass in water has a density of 1.131 g/mL. What volume of this solution, in liters, must be used in an application requiring 2.25 kg of sodium hydroxide?

Integrative and Advanced Exercises

67. According to the rules on significant figures, the product of the measured quantities 99.9 m and 1.008 m should be expressed to three significant figures—101 m². Yet, in this case, it would be more appropriate to express the result to *four* significant figures—100.7 m². Explain why.

68. For a solution containing 6.38% para-diclorobenzene by mass in benzene, the density of the solution as a function of temperature (t) in the temperature range 15 to 65 °C is given by the equation

    $$d(\text{g/mL}) = 1.5794 - 1.836 \times 10^{-3} (t - 15)$$

    At what temperature will the solution have a density of 1.543 g/mL?

69. A solution used to chlorinate a home swimming pool contains 7% chlorine by mass. An ideal chlorine level for the pool is one part per million (1 ppm). (Think of 1 ppm as being 1 g chlorine per million grams of water.) If you assume densities of 1.10 g/mL for the chlorine solution and 1.00 g/mL for the swimming pool water, what volume of the chlorine solution, in liters, is required to produce a chlorine level of 1 ppm in an 18,000-gallon swimming pool?

70. A standard 1.000 kg mass is to be cut from a bar of steel having an equilateral triangular cross section with sides equal to 2.50 in. The density of the steel is 7.70 g/cm^3. How many inches long must the section of bar be?

71. The volume of seawater on Earth is about 330,000,000 mi^3. If seawater is 3.5% sodium chloride by mass and has a density of 1.03 g/mL, what is the approximate mass of sodium chloride, in tons, dissolved in the seawater on Earth (1 ton = 2000 lb)?

72. The diameter of metal wire is often referred to by its American wire-gauge number. A 16-gauge wire has a diameter of 0.05082 in. What length of wire, in meters, is found in a 1.00 lb spool of 16-gauge copper wire? The density of copper is 8.92 g/cm^3.

73. Magnesium occurs in seawater to the extent of 1.4 g magnesium per kilogram of seawater. What volume of seawater, in cubic meters, would have to be processed to produce 1.00×10^5 tons of magnesium (1 ton = 2000 lb)? Assume a density of 1.025 g/mL for seawater.

74. A typical rate of deposit of dust ("dustfall") from unpolluted air was reported as 10 tons per square mile per month. **(a)** Express this dustfall in milligrams per square meter per hour. **(b)** If the dust has an average density of 2 g/cm^3, how long would it take to accumulate a layer of dust 1 mm thick?

75. In the United States, volume of irrigation water is usually expressed in acre-feet. One acre-foot is a volume of water sufficient to cover 1 acre of land to a depth of 1 ft (640 acres = 1 mi^2; 1 mi = 5280 ft). The principal lake in the California Water Project is Lake Oroville, whose water storage capacity is listed as 3.54×10^6 acre-feet. Express the volume of Lake Oroville in **(a)** cubic feet; **(b)** cubic meters; **(c)** U.S. gallons.

76. A Fahrenheit and a Celsius thermometer are immersed in the same medium. At what Celsius temperature will the numerical reading on the Fahrenheit thermometer be
    (a) 49° less than that on the Celsius thermometer;
    (b) twice that on the Celsius thermometer;
    (c) one-eighth that on the Celsius thermometer;
    (d) 300° more than that on the Celsius thermometer?

77. The accompanying illustration shows a 100.0 mL graduated cylinder half-filled with 8.0 g of diatomaceous earth, a material consisting mostly of silica and used as a filtering medium in swimming pools. How many milliliters of water are required to fill the cylinder to the 100.0 mL mark? The diatomaceous earth is insoluble in water and has a density of 2.2 g/cm^3.

78. The simple device pictured here, a pycnometer, is used for precise density determinations. From the data presented, together with the fact that the density of water at 20 °C is 0.99821 g/mL, determine the density of methanol, in grams per milliliter.

Empty
25.601 g

Filled with water
at 20 °C: 35.552 g

Filled with methanol
at 20 °C: 33.490 g

79. If the pycnometer of Exercise 78 is filled with ethanol at 20 °C instead of methanol, the observed mass is 33.470 g. What is the density of ethanol? How precisely could you determine the composition of an ethanol–methanol solution by measuring its density with a pycnometer? Assume that the density of the solution is a linear function of the volume percent composition.

80. A pycnometer (see Exercise 78) weighs 25.60 g empty and 35.55 g when filled with water at 20 °C. The density of water at 20 °C is 0.9982 g/mL. When 10.20 g lead is placed in the pycnometer and the pycnometer is again filled with water at 20 °C, the total mass is 44.83 g. What is the density of the lead in grams per cubic centimeter?

81. The Greater Vancouver Regional District (GVRD) chlorinates the water supply of the region at the rate of 1 ppm, that is, 1 kilogram of chlorine per million kilograms of water. The chlorine is introduced in the form of sodium hypochlorite, which is 47.62% chlorine. The population of the GVRD is 1.8 million persons. If each person uses 750 L of water per day, how many kilograms of sodium hypochlorite must be added to the water supply each week to produce the required chlorine level of 1 ppm?

82. A Boeing 767 due to fly from Montreal to Edmonton required refueling. Because the fuel gauge on the aircraft was not working, a mechanic used a dipstick to determine that 7682 L of fuel were left on the plane. The plane required 22,300 kg of fuel to make the trip. In order to determine the volume of fuel required, the pilot asked for the conversion factor needed to convert a volume of fuel to a mass of fuel. The mechanic gave the factor as 1.77. Assuming that this factor was in metric units (kg/L), the pilot calculated the volume to be added as 4916 L. This volume of fuel was added and the 767 subsequently ran out the fuel, but landed safely by gliding into Gimli Airport near Winnipeg. The error arose because the factor 1.77 was in units of pounds per liter. What volume of fuel should have been added?

83. The following equation can be used to relate the density of liquid water to Celsius temperature in the range from 0 °C to about 20 °C:

$$d(\text{g/cm}^3) = \frac{0.99984 + (1.6945 \times 10^{-2}t) - (7.987 \times 10^{-6}t^2)}{1 + (1.6880 \times 10^{-2}t)}$$

(a) To four significant figures, determine the density of water at 10 °C.
(b) At what temperature does water have a density of 0.99860 g/cm^3?
(c) In the following ways, show that the density passes through a maximum somewhere in the temperature range to which the equation applies.
   (i) by estimation
   (ii) by a graphical method
   (iii) by a method based on differential calculus

84. A piece of high-density Styrofoam measuring 24.0 cm by 36.0 cm by 5.0 cm floats when placed in a tub of water. When a 1.5 kg book is placed on top of the Styrofoam, the Styrofoam partially sinks, as illustrated in the diagram below. Assuming that the density of water is 1.00 g/mL, what is the density of Styrofoam?

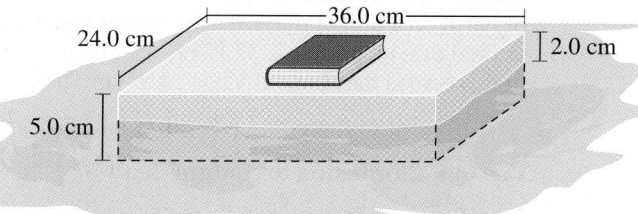

85. A tabulation of data lists the following equation for calculating the densities (d) of solutions of naphthalene in benzene at 30 °C as a function of the mass percent of naphthalene.

$$d(\text{g/cm}^3) = \frac{1}{1.153 - 1.82 \times 10^{-3}(\%N) + 1.08 \times 10^{-6}(\%N)^2}$$

Use the equation above to calculate (a) the density of pure benzene at 30 °C; (b) the density of pure naphthalene at 30 °C; (c) the density of solution at 30 °C that is 1.15% naphthalene; (d) the mass percent of naphthalene in a solution that has a density of 0.952 g/cm^3 at 30 °C. [*Hint*: For (d), you need to use the quadratic formula. See Section A-3 of Appendix A.]

86. The total volume of ice in the Antarctic is about 3.01×10^7 km^3. If all the ice in the Antarctic were to melt completely, estimate the rise, h, in sea level that would result from the additional liquid water entering the oceans. The densities of ice and fresh water are 0.92 g/cm^3 and 1.0 g/cm^3, respectively. Assume that the oceans of the world cover an area, A, of about 3.62×10^8 km^2 and that the increase in volume of the oceans can be calculated as $A \times h$.

87. An empty 3.00 L bottle weighs 1.70 kg. Filled with a certain wine, it weighs 4.72 kg. The wine contains 11.5% ethyl alcohol by mass. How many grams of ethyl alcohol are there in 250.0 mL of this wine?

88. The filament in an incandescent light bulb is made from tungsten metal (d = 19.3 g/cm^3) that has been drawn into a very thin wire. The diameter of the wire is difficult to measure directly, so it is sometimes estimated by measuring the mass of a fixed length of wire. If a 0.200 m length of tungsten wire weighs 42.9 mg, then what is the diameter of the wire? Express your answer in millimeters.

89. Blood alcohol content (BAC) is sometimes reported in weight-volume percent and, when it is, a BAC of 0.10% corresponds to 0.10 g ethyl alcohol per 100 mL of blood. In many jurisdictions, a person is considered legally intoxicated if his or her BAC is 0.10%. Suppose that a 68 kg person has a total blood volume of 5.4 L and breaks down ethyl alcohol at a rate of 10.0 grams per hour.* How many 145 mL glasses of wine, consumed over three hours, will produce a BAC of 0.10% in this 68 kg person? Assume the wine has a density of 1.01 g/mL and is 11.5% ethyl alcohol by mass. (*The rate at which ethyl alcohol is broken down varies dramatically from person to person. The value given here for the rate is a realistic, but not necessarily accurate, value.)

Feature Problems

90. In an attempt to determine any possible relationship between the year in which a U.S. penny was minted and its current mass (in grams), students weighed an assortment of pennies and obtained the following data.

1968	1973	1977	1980	1982	1983	1985
3.11	3.14	3.13	3.12	3.12	2.51	2.54
3.08	3.06	3.10	3.11	2.53	2.49	2.53
3.09	3.07	3.06	3.08	2.54	2.47	2.53

What valid conclusion(s) might they have drawn about the relationship between the masses of the pennies within a given year and from year to year?

91. In the third century BC, the Greek mathematician Archimedes is said to have discovered an important principle that is useful in density determinations. The story told is that King Hiero of Syracuse (in Sicily) asked Archimedes to verify that an ornate crown made for him by a goldsmith consisted of pure gold and not a gold–silver alloy. Archimedes

had to do this, of course, without damaging the crown in any way. Describe how Archimedes did this, or if you don't know the rest of the story, rediscover Archimedes's principle and explain how it can be used to settle the question.

92. The Galileo thermometer shown in the photograph is based on the dependence of density on temperature. The liquid in the outer cylinder and the liquid in the partially filled floating glass balls are the same except that a colored dye has been added to the liquid in the balls. Explain how the Galileo thermometer works.

93. The canoe gliding gracefully along the water in the photograph is made of concrete, which has a density of about 2.4 g/cm³. Explain why the canoe does not sink.

94. The accompanying sketches suggest four observations made on a small block of plastic material. Tell what conclusions can be drawn from each sketch, and conclude by giving your best estimate of the density of the plastic.

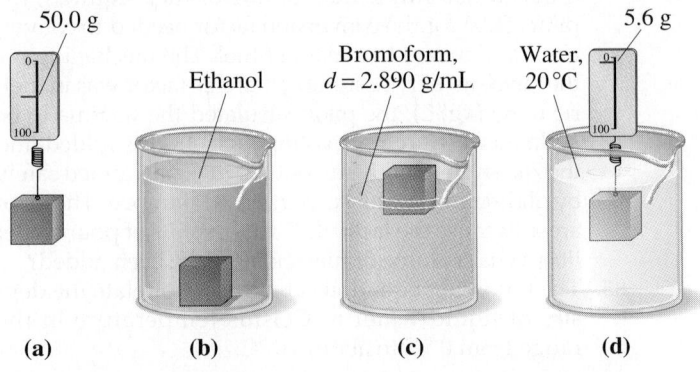

(a)   (b)   (c)   (d)

95. As mentioned on page 13, the MCO was lost because of a mix-up in the units used to calculate the force needed to correct its trajectory. Ground-based computers generated the force correction file. On September 29, 1999, it was discovered that the forces reported by the ground-based computer for use in MCO navigation software were low by a factor of 4.45. The erroneous trajectory brought the MCO 56 km above the surface of Mars; the correct trajectory would have brought the MCO approximately 250 km above the surface. At 250 km, the MCO would have successfully entered the desired elliptic orbit. The data contained in the force correction file were delivered in lb-sec instead of the required SI units of newton-sec for the MCO navigation software. The newton is the SI unit of force and is described in Appendix B. The British Engineering (gravitational) system uses a pound (lb) as a unit of force and ft/s² as a unit of acceleration. In turn, the pound is defined as the pull of Earth on a unit of mass at a location where the acceleration due to gravity is 32.174 ft/s². The unit of mass in this case is the slug, which is 14.59 kg. Thus,

$$\text{BE unit of force} = 1 \text{ pound} = (\text{slug})(\text{ft/s}^2)$$

Use this information to confirm that

$$\text{BE unit of force} = 4.45 \times \text{SI unit of force}$$
$$1 \text{ pound} = 4.45 \text{ newton}$$

Self-Assessment Exercises

96. In your own words, define or explain the following terms or symbols: (a) mL; (b) % by mass; (c) °C; (d) density; (e) element.

97. Briefly describe each of the following ideas: (a) SI base units; (b) significant figures; (c) natural law; (d) exponential notation.

98. Explain the important distinctions between each pair of terms: (a) mass and weight; (b) intensive and extensive properties; (c) substance and mixture; (d) systematic and random errors; (e) hypothesis and theory.

99. The fact that the volume of a fixed amount of gas at a fixed temperature is inversely proportional to the gas pressure is an example of (a) a hypothesis; (b) a theory; (c) a paradigm; (d) the absolute truth; (e) a natural law.

100. A good example of a homogeneous mixture is
    (a) a cola drink in a tightly capped bottle
    (b) distilled water leaving a distillation apparatus
    (c) oxygen gas in a cylinder used in welding
    (d) the material produced in a kitchen blender

101. Compared with its mass on Earth, the mass of the same object on the moon should be **(a)** less; **(b)** more; **(c)** the same; **(d)** nearly the same, but somewhat less.

102. Of the following masses, two are expressed to the nearest milligram. The two are **(a)** 32.7 g; **(b)** 0.03271 kg; **(c)** 32.7068 g; **(d)** 32.707 g; **(e)** 30.7 mg; **(f)** 3×10^3 μg.

103. The highest temperature of the following group is **(a)** 217 K; **(b)** 273 K; **(c)** 217 °F; **(d)** 105 °C; **(e)** 373 K.

104. Which of the following quantities has the greatest mass?
(a) 752 mL of water at 20 °C
(b) 1.05 L of ethanol at 20 °C ($d = 0.789$ g/mL)
(c) 750 g of chloroform at 20 °C ($d = 1.483$ g/mL)
(d) a cube of balsa wood ($d = 0.11$ g/cm^3) that is 19.20 cm on edge

105. The density of water is 0.9982 g/cm^3 at 20 °C. Express the density of water at 20 °C in the following units: **(a)** g/L; **(b)** kg/m^3; **(c)** kg/km^3.

106. Two students each made four measurements of the mass of an object. Their results are shown in the table below.

	Student A	Student B
Four measurements:	51.6, 50.8, 52.2, 50.2 g	50.1, 49.6, 51.0, 49.4 g
Their average:	51.3 g	50.0 g

The exact mass of the object is 51.0 g. Whose results are more precise, Student A's or Student B's? Whose results are more accurate?

107. The reported value for the volume of a rectangular piece of cardboard with the dimensions 36 cm × 20.2 cm × 9 mm should be **(a)** 6.5×10^3 cm^3; **(b)** 7×10^2 cm^3; **(c)** 655 cm^3; **(d)** 6.5×10^2 cm^3.

108. List the following in the order of increasing precision, indicating any quantities about which the precision is uncertain: **(a)** 1400 km; **(b)** 1516 kg; **(c)** 0.00304 g; **(d)** 125.34 cm; **(e)** 2000 mg.

109. *Without doing detailed calculations*, explain which of the following objects contains the greatest mass of the element iron.
(a) A 1.00 kg pile of pure iron filings.
(b) A cube of wrought iron, 5.0 cm on edge. Wrought iron contains 98.5% iron by mass and has a density of 7.7 g/cm^3.
(c) A square sheet of stainless steel 0.30 m on edge and 1.0 mm thick. The stainless steel is an alloy (mixture) containing iron, together with 18% chromium, 8% nickel, and 0.18% carbon by mass. Its density is 7.7 g/cm^3.
(d) 10.0 L of a solution characterized as follows: $d = 1.295$ g/mL. This solution is 70.0% water and 30.0% of a compound of iron, by mass. The iron compound consists of 34.4% iron by mass.

110. A lump of pure copper weighs 25.305 g in air and 22.486 g when submerged in water ($d = 0.9982$ g/mL) at 20.0 °C. Suppose the copper is then rolled into a 248 cm^2 foil of uniform thickness. What will this thickness be, in millimeters?

111. Water, a compound, is a substance. Is there any circumstance under which a sample of pure water can exist as a heterogeneous mixture? Explain.

112. Appendix E describes a useful study aid known as concept mapping. Using the method presented in Appendix E, construct a concept map illustrating the different concepts presented in Sections 1-2, 1-3, and 1-4.

2

Atoms and the Atomic Theory

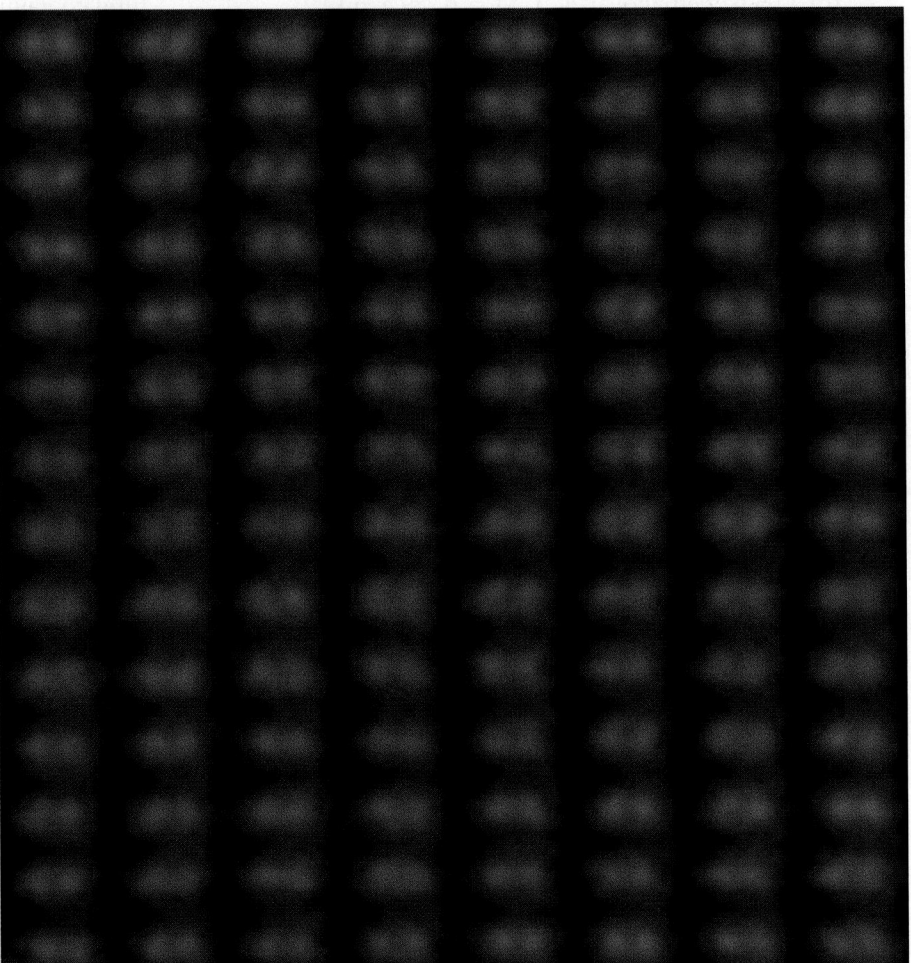

Image of silicon atoms that are only 78 pm apart; image produced by using a scanning transmission electron microscope (STEM). The hypothesis that all matter is made up of atoms has existed for more than 2000 years. It is only within the last few decades, however, that techniques have been developed that can render individual atoms visible.

We begin this chapter with a brief survey of early chemical discoveries, culminating in Dalton's atomic theory. This is followed by a description of the physical evidence leading to the modern picture of the *nuclear atom*, in which protons and neutrons are combined into a nucleus with electrons in space surrounding the nucleus. We will also introduce the periodic table as the primary means of organizing elements into groups with similar properties. Finally, we will introduce the concept of the mole and the Avogadro constant, which are the principal tools for counting atoms and molecules and measuring amounts of substances. We will use these tools throughout the text.

2-1   Early Chemical Discoveries and the Atomic Theory

Chemistry has been practiced for a very long time, even if its practitioners were much more interested in its applications than in its underlying principles. The blast furnace for extracting iron from iron ore appeared as early as A.D. 1300, and such important chemicals as sulfuric acid (oil of vitriol), nitric acid (aqua fortis), and sodium sulfate (Glauber's salt) were all well known and used several hundred years ago. Before the end of the eighteenth century, the principal gases of the atmosphere—nitrogen and oxygen—had been isolated, and natural laws had been proposed describing the physical behavior of gases. Yet chemistry cannot be said to have entered the modern age until the process of combustion was explained. In this section, we explore the direct link between the explanation of combustion and Dalton's atomic theory.

Law of Conservation of Mass

The process of combustion—burning—is so familiar that it is hard to realize what a difficult riddle it posed for early scientists. Some of the difficult-to-explain observations are described in Figure 2-1.

In 1774, Antoine Lavoisier (1743–1794) performed an experiment in which he heated a sealed glass vessel containing a sample of tin and some air. He found that the mass before heating (glass vessel + tin + air) and after heating (glass vessel + "tin calx" + remaining air) were the same. Through further experiments, he showed that the product of the reaction, tin calx (tin oxide), consisted of the original tin together with a portion of the air. Experiments like this proved to Lavoisier that oxygen from air is essential to combustion, and also led him to formulate the **law of conservation of mass**:

> The total mass of substances present after a chemical reaction is the same as the total mass of substances before the reaction.

This law is illustrated in Figure 2-2, where the reaction between silver nitrate and potassium chromate to give a red solid (silver chromate) is monitored by placing the reactants on a single-pan balance—the total mass does not change. Stated another way, the law of conservation of mass says that matter is neither created nor destroyed in a chemical reaction.

▲ FIGURE 2-1
Two combustion reactions
The apparent product of the combustion of the match—the ash—weighs *less* than the match. The product of the combustion of the magnesium ribbon (the "smoke") weighs *more* than the ribbon. Actually, in each case, the total mass remains unchanged. To understand this, you have to know that oxygen gas enters into both combustions and that water and carbon dioxide are also products of the combustion of the match.

(a)   (b)

◀ FIGURE 2-2
Mass is conserved during a chemical reaction
(a) Before the reaction, a beaker with a silver nitrate solution and a graduated cylinder with a potassium chromate solution are placed on a single-pan balance, which displays their combined mass—104.50 g. **(b)** When the solutions are mixed, a chemical reaction occurs that forms silver chromate (red precipitate) in a potassium nitrate solution. Note that the total mass—104.50 g—remains unchanged.

EXAMPLE 2-1   Applying the Law of Conservation of Mass

A 0.455 g sample of magnesium is allowed to burn in 2.315 g of oxygen gas. The sole product is magnesium oxide. After the reaction, no magnesium remains and the mass of unreacted oxygen is 2.015 g. What mass of magnesium oxide is produced?

Analyze

The total mass is unchanged. The total mass is the sum of the masses of the substances present initially. The mass of magnesium oxide is the total mass minus the mass of unreacted oxygen.

Solve

First, determine the total mass before the reaction.

mass before reaction = 0.455 g magnesium + 2.315 g oxygen
= 2.770 g mass before reaction

The total mass after the reaction is the same as before the reaction.

2.770 g mass after reaction = ? g magnesium oxide after reaction
+ 2.015 g oxygen after reaction

Solve for the mass of magnesium oxide.

? g magnesium oxide after reaction = 2.770 g mass after reaction
− 2.015 g oxygen after reaction
= 0.755 g magnesium oxide after reaction

Assess

Here is another approach. The mass of oxygen that reacted is 2.315 g − 2.015 g = 0.300 g. Thus, 0.300 g oxygen combined with 0.455 g magnesium to give 0.300 g + 0.455 g = 0.755 g magnesium oxide.

PRACTICE EXAMPLE A: A 0.382 g sample of magnesium is allowed to react with 2.652 g of nitrogen gas. The sole product is magnesium nitride. After the reaction, the mass of unreacted nitrogen is 2.505 g. What mass of magnesium nitride is produced?

PRACTICE EXAMPLE B: A 7.12 g sample of magnesium is heated with 1.80 g of bromine. All the bromine is used up, and 2.07 g of magnesium bromide is the only product. What mass of magnesium remains *unreacted*?

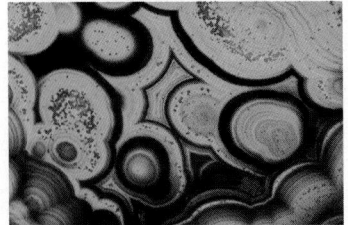
(a)

(b)

▲ The mineral malachite **(a)** and the green patina on a copper roof **(b)** are both basic copper carbonate, just like the basic copper carbonate prepared by Proust in 1799.

🔍 2-1   CONCEPT ASSESSMENT

Jan Baptista van Helmont (1579–1644) weighed a young willow tree and the soil in which the tree was planted. Five years later he found that the mass of soil had decreased by only 0.057 kg, while that of the tree had increased by 75 kg. During that period he had added only water to the bucket in which the tree was planted. Helmont concluded that essentially all the mass gained by the tree had come from the water. Was this a valid conclusion? Explain.

Law of Constant Composition

In 1799, Joseph Proust (1754–1826) reported, "One hundred pounds of copper, dissolved in sulfuric or nitric acids and precipitated by the carbonates of soda or potash, invariably gives 180 pounds of green carbonate."* This and similar observations became the basis of the **law of constant composition**, or the **law of definite proportions**:

> All samples of a compound have the same composition—the same proportions by mass of the constituent elements.

To see how the law of constant composition works, consider the compound water. Water is made up of two atoms of hydrogen (H) for every atom of oxygen (O), a fact that can be represented symbolically by a *chemical formula*, the familiar H_2O.

*The substance Proust produced is actually a more complex substance called *basic* copper carbonate. Proust's results were valid because, like all compounds, basic copper carbonate has a constant composition.

The two samples described below have the same proportions of the two elements, expressed as percentages by mass. To determine the percent by mass of hydrogen, for example, simply divide the mass of hydrogen by the sample mass and multiply by 100%. For each sample, you will obtain the same result: 11.19% H.

Sample A and Its Composition		Sample B and Its Composition	
10.000 g		27.000 g	
1.119 g H	% H = 11.19	3.021 g H	% H = 11.19
8.881 g O	% O = 88.81	23.979 g O	% O = 88.81

EXAMPLE 2-2   Using the Law of Constant Composition

In Example 2-1 we found that when 0.455 g of magnesium reacted with 2.315 g of oxygen, 0.755 g of magnesium oxide was obtained. Determine the mass of magnesium contained in a 0.500 g sample of magnesium oxide.

Analyze

We know that 0.755 g of magnesium oxide contains 0.455 g of magnesium. According to the law of constant composition, the mass ratio 0.455 g magnesium/0.755 g magnesium oxide should exist in all samples of magnesium oxide.

Solve

Application of the law of constant composition gives

$$\frac{0.455 \text{ g magnesium}}{0.755 \text{ g magnesium oxide}} = \frac{? \text{ g magnesium}}{0.500 \text{ g magnesium oxide}}$$

Solving the expression above, we obtain

$$? \text{ g magnesium} = 0.500 \text{ g magnesium oxide} \times \frac{0.455 \text{ g magnesium}}{0.755 \text{ g magnesium oxide}}$$

$$= 0.301 \text{ g magnesium}$$

Assess

You can also work this problem by using mass percentages. If 0.755 g of magnesium oxide contains 0.455 g of magnesium, then magnesium oxide is (0.455 g/0.755 g) × 100% = 60.3% magnesium by mass and (100% − 60.3%) = 39.7% oxygen by mass. Thus, a 0.500 g sample of magnesium oxide must contain 0.500 g × 60.3% = 0.301 g of magnesium and 0.500 g × 39.7% = 0.199 g of oxygen.

PRACTICE EXAMPLE A: What masses of magnesium and oxygen must be combined to make exactly 2.000 g of magnesium oxide?

PRACTICE EXAMPLE B: What substances are present, and what are their masses, after the reaction of 10.00 g of magnesium and 10.00 g of oxygen?

2-2   CONCEPT ASSESSMENT

When 4.15 g magnesium and 82.6 g bromine react, (1) all the magnesium is used up, (2) some bromine remains unreacted, and (3) magnesium bromide is the only product. With this information alone, is it possible to deduce the mass of magnesium bromide produced? Explain.

Dalton's Atomic Theory

From 1803 to 1808, John Dalton, an English schoolteacher, used the two fundamental laws of chemical combination just described as the basis of an atomic theory. His theory involved three assumptions:

1. Each chemical element is composed of minute, indivisible particles called atoms. Atoms can be neither created nor destroyed during a chemical change.

▲ John Dalton (1766–1844), developer of the atomic theory. Dalton has not been considered a particularly good experimenter, perhaps because of his color blindness (a condition sometimes called daltonism). However, he did skillfully use the data of others in formulating his atomic theory. (The Granger Collection)

KEEP IN MIND

that all we know is that the second oxide is twice as rich in oxygen as the first. If the first is CO, the possibilities for the second are CO_2, C_2O_4, C_3O_6, and so on. (See also Exercise 18.)

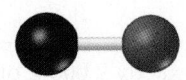

▲ FIGURE 2-3
Molecules CO and CO_2 illustrating the law of multiple proportions
The mass of carbon is the same in the two molecules, but the mass of oxygen in CO_2 is twice the mass of oxygen in CO. Thus, in accordance with the law of multiple proportions, the masses of oxygen in the two compounds, relative to a fixed mass of carbon, are in a ratio of small whole numbers, 2:1.

2. All atoms of an element are alike in mass (weight) and other properties, but the atoms of one element are different from those of all other elements.

3. In each of their compounds, different elements combine in a simple numerical ratio, for example, one atom of A to one of B (AB), or one atom of A to two of B (AB_2).

If atoms of an element are indestructible (assumption 1), then the *same* atoms must be present after a chemical reaction as before. The total mass remains unchanged. Dalton's theory explains the law of conservation of mass. If all atoms of an element are alike in mass (assumption 2) and if atoms unite in *fixed* numerical ratios (assumption 3), the percent composition of a compound must have a unique value, regardless of the origin of the sample analyzed. Dalton's theory also explains the law of constant composition.

Like all good theories, Dalton's atomic theory led to a prediction—the **law of multiple proportions**.

> If two elements form more than a single compound, the masses of one element combined with a fixed mass of the second are in the ratio of small whole numbers.

To illustrate, consider two oxides of carbon (an oxide is a combination of an element with oxygen). In one oxide, 1.000 g of carbon is combined with 1.333 g of oxygen, and in the other, with 2.667 g of oxygen. We see that the second oxide is richer in oxygen; in fact, it contains twice as much oxygen as the first, $2.667 \text{ g}/1.333 \text{ g} = 2.00$. We now know that the first oxide corresponds to the formula CO and the second, CO_2 (Fig. 2-3).

The characteristic relative masses of the atoms of the various elements became known as atomic weights, and throughout the nineteenth century, chemists worked at establishing reliable values of relative atomic weights. Mostly, however, chemists directed their attention to discovering new elements, synthesizing new compounds, developing techniques for analyzing materials, and in general, building up a vast body of chemical knowledge. Efforts to unravel the structure of the atom became the focus of physicists, as we see in the next several sections.

2-2 Electrons and Other Discoveries in Atomic Physics

Fortunately, we can acquire a qualitative understanding of atomic structure without having to retrace all the discoveries that preceded atomic physics. We do, however, need a few key ideas about the interrelated phenomena of electricity and magnetism, which we briefly discuss here. Electricity and magnetism were used in the experiments that led to the current theory of atomic structure.

Certain objects display a property called electric charge, which can be either positive (+) or negative (−). Positive and negative charges attract each other, while two positive or two negative charges repel each other. As we learn in this section, all objects of matter are made up of charged particles. An object having equal numbers of positively and negatively charged particles carries no net charge and is electrically neutral. If the number of positive charges exceeds the number of negative charges, the object has a net positive charge. If negative charges exceed positive charges, the object has a net negative charge. Sometimes when one substance is rubbed against another, as in combing hair, net electric charges build up on the objects, implying that rubbing separates

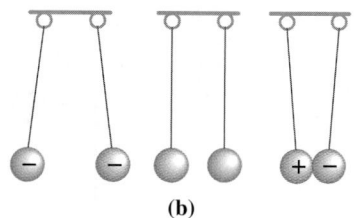

(a)   **(b)**

◀ We will use *electrostatics* (charge attractions and repulsions) to explain and understand many chemical properties.

▲ FIGURE 2-4
Forces between electrically charged objects
(a) Electrostatically charged comb. If you comb your hair on a dry day, a static charge develops on the comb and causes bits of paper to be attracted to the comb. **(b)** Both objects on the left carry a negative electric charge. Objects with like charge repel each other. The objects in the center lack any electric charge and exert no forces on each other. The objects on the right carry opposite charges—one positive and one negative—and attract each other.

some positive and negative charges (Fig. 2-4). Moreover, when a stationary (static) positive charge builds up in one place, a negative charge of equal size appears somewhere else; charge is balanced.

Figure 2-5 shows how charged particles behave when they move through the field of a magnet. They are deflected from their straight-line path into a curved path in a plane perpendicular to the field. Think of the field or region of influence of the magnet as represented by a series of invisible "lines of force" running from the north pole to the south pole of the magnet.

The Discovery of Electrons

CRT, the abbreviation for cathode-ray tube, was once a familiar acronym. Before liquid crystal display (LCD) was available, the CRT was the heart of computer monitors and TV sets. The first cathode-ray tube was made by Michael Faraday (1791–1867) about 150 years ago. When he passed electricity through glass tubes from which most of the air had been evacuated, Faraday discovered **cathode rays**, a type of radiation emitted by the negative terminal, or *cathode*. The radiation crossed the evacuated tube to the positive terminal, or *anode*. Later scientists found that cathode rays travel in straight lines and have properties that are independent of the cathode material (that is, whether it is iron, platinum, and so on). The construction of a CRT is shown in Figure 2-6. The cathode rays produced in the CRT are invisible, and they can be detected only by the light emitted by materials that they strike. These materials, called *phosphors*, are painted on the end of the CRT so that the path of the cathode rays can be revealed. (*Fluorescence* is the term used to describe the emission of light by a phosphor when it is struck by

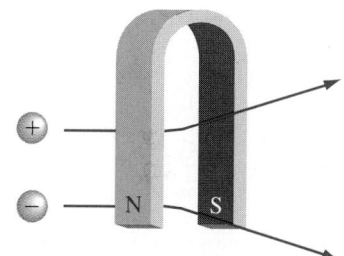

▲ FIGURE 2-5
Effect of a magnetic field on charged particles
When charged particles travel through a magnetic field so that their path is perpendicular to the field, they are deflected by the field. Negatively charged particles are deflected in one direction, and positively charged particles in the opposite direction. Several phenomena described in this section depend on this behavior.

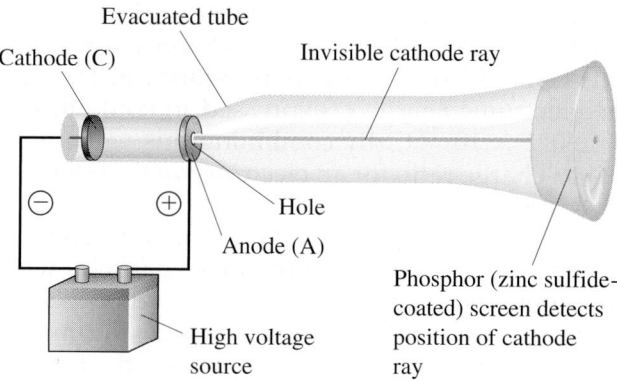

Cathode (C)
Evacuated tube
Invisible cathode ray
Hole
Anode (A)
High voltage source
Phosphor (zinc sulfide-coated) screen detects position of cathode ray

◀ FIGURE 2-6
A cathode-ray tube
The high-voltage source of electricity creates a negative charge on the electrode at the left (cathode) and a positive charge on the electrode at the right (anode). Cathode rays pass from the cathode (C) to the anode (A), which is perforated to allow the passage of a narrow beam of cathode rays. The rays are visible only through the green fluorescence that they produce on the zinc sulfide–coated screen at the end of the tube. They are invisible in other parts of the tube.

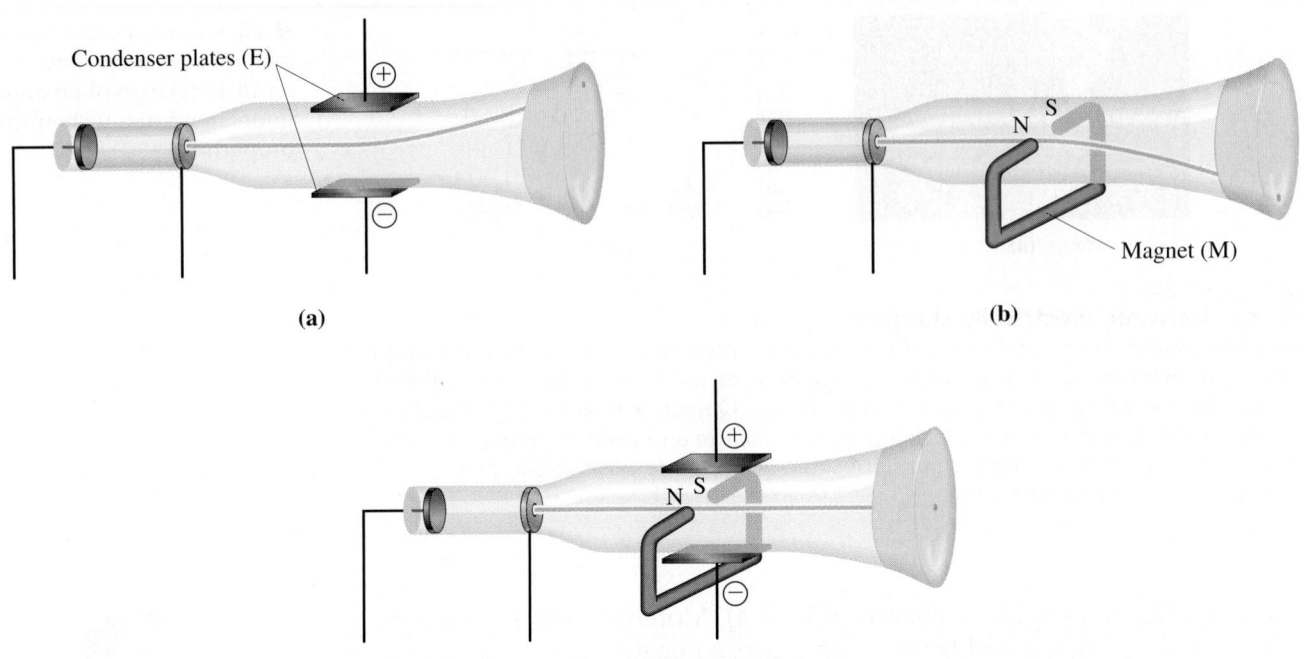

Condenser plates (E)

(a)

Magnet (M)

(b)

(c)

▲ FIGURE 2-7
Cathode rays and their properties
(a) Deflection of cathode rays in an electric field. The beam of cathode rays is deflected as it travels from left to right in the field of the electrically charged condenser plates (E). The deflection corresponds to that expected of negatively charged particles.
(b) Deflection of cathode rays in a magnetic field. The beam of cathode rays is deflected as it travels from left to right in the field of the magnet (M). The deflection corresponds to that expected of negatively charged particles. **(c)** Determining the mass-to-charge ratio, m/e, for cathode rays. The cathode-ray beam strikes the end screen undeflected if the forces exerted on it by the electric and magnetic fields are counterbalanced. By knowing the strengths of the electric and magnetic fields, together with other data, a value of m/e can be obtained. Precise measurements yield a value of -5.6857×10^{-9} g per coulomb. (Because cathode rays carry a negative charge, the sign of the mass-to-charge ratio is also negative.)

energetic radiation.) Another significant observation about cathode rays is that they are deflected by electric and magnetic fields in the manner expected for negatively charged particles (Fig. 2-7a, b).

In 1897, by the method outlined in Figure 2-7(c), J. J. Thomson (1856–1940) established the ratio of mass (m) to electric charge (e) for cathode rays, that is, m/e. Also, Thomson concluded that cathode rays are negatively charged *fundamental* particles of matter found in all atoms. (The properties of cathode rays are *independent* of the composition of the cathode.) Cathode rays subsequently became known as **electrons**, a term first proposed by George Stoney in 1874.

Robert Millikan (1868–1953) determined the electronic charge e through a series of oil-drop experiments (1906–1914), described in Figure 2-8. The currently accepted value of the electronic charge e, expressed in coulombs to five significant figures, is -1.6022×10^{-19} C. By combining this value with an accurate value of the mass-to-charge ratio for an electron, we find that the mass of an electron is 9.1094×10^{-28} g.

Once the electron was seen to be a fundamental particle of matter found in all atoms, atomic physicists began to speculate on how these particles were incorporated into atoms. The commonly accepted model was that proposed by J. J. Thomson. Thomson thought that the positive charge necessary to counterbalance the negative charges of electrons in a neutral atom was in the form of

▶ The coulomb (C) is the SI unit of electric charge (see also Appendix B).

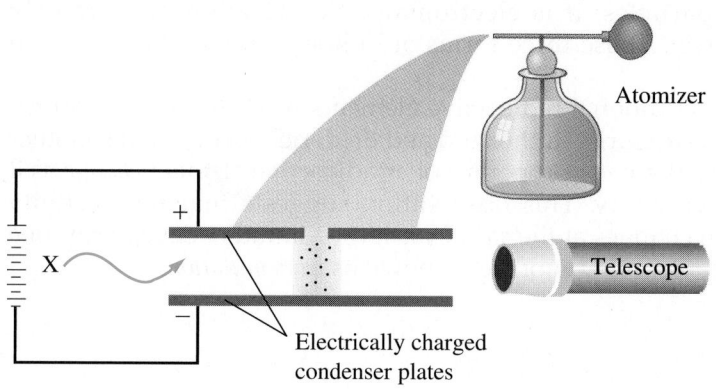

Millikan's oil-drop experiment
Ions (charged atoms or molecules) are produced by energetic radiation, such as X-rays (X). Some of these ions become attached to oil droplets, giving them a net charge. The fall of a droplet in the electric field between the condenser plates is speeded up or slowed down, depending on the magnitude and sign of the charge on the droplet. By analyzing data from a large number of droplets, Millikan concluded that the magnitude of the charge, q, on a droplet is an *integral* multiple of the electric charge, e. That is, $q = ne$ (where $n = 1, 2, 3, \ldots$).

a nebulous cloud. Electrons, he suggested, floated in a diffuse cloud of positive charge (rather like a lump of gelatin with electron "fruit" embedded in it). This model became known as the plum-pudding model because of its similarity to a popular English dessert. The plum-pudding model is illustrated in Figure 2-9 for a neutral atom and for atomic species, called *ions*, which carry a net charge.

X-Rays and Radioactivity

Cathode-ray research had many important spin-offs. In particular, two natural phenomena of immense theoretical and practical significance were discovered in the course of other investigations.

In 1895, Wilhelm Roentgen (1845–1923) noticed that when cathode-ray tubes were operating, certain materials *outside* the tubes glowed or fluoresced. He showed that this fluorescence was caused by radiation emitted by the cathode-ray tubes. Because of the unknown nature of this radiation, Roentgen coined the term *X-ray*. We now recognize the X-ray as a form of high-energy electromagnetic radiation, which is discussed in Chapter 8.

Antoine Henri Becquerel (1852–1908) associated X-rays with fluorescence and wondered if naturally fluorescent materials produce X-rays. To test this idea, he wrapped a photographic plate with black paper, placed a coin on the paper, covered the coin with a uranium-containing fluorescent material, and exposed the entire assembly to sunlight. When he developed the film, a clear image of the coin could be seen. The fluorescent material had emitted radiation (presumably X-rays) that penetrated the paper and exposed the film. On one occasion, because the sky was overcast, Becquerel placed the experimental assembly inside a desk drawer for a few days while waiting for the weather to clear. On resuming the experiment, Becquerel decided to replace the original photographic film, expecting that it may have become slightly exposed. He developed the original film and found that instead of the expected feeble image, there was a very sharp one. The film had become strongly exposed because the uranium-containing material had emitted radiation continuously, even when it was not fluorescing. Becquerel had discovered **radioactivity**.

Ernest Rutherford (1871–1937) identified two types of radiation from radioactive materials, alpha (α) and beta (β). **Alpha particles** carry two fundamental units of positive charge and have essentially the same mass as helium atoms. In fact, alpha particles are identical to He^{2+} ions. **Beta particles** are negatively charged particles produced by changes occurring within the nuclei of radioactive atoms and have the same properties as electrons. A third form of radiation, which is not affected by electric or magnetic fields, was discovered in 1900 by Paul Villard. This radiation, called **gamma rays** (γ),

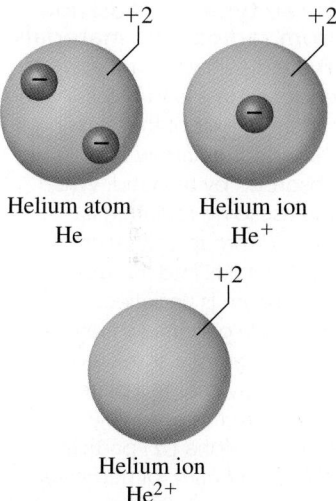

▲ FIGURE 2-9
The plum-pudding atomic model
According to this model, a helium atom would have a +2 cloud of positive charge and two electrons (−2). If a helium atom loses one electron, it becomes charged and is called an *ion*. This ion, referred to as He^+, has a net charge of 1+. If the helium atom loses both electrons, the He^{2+} ion forms.

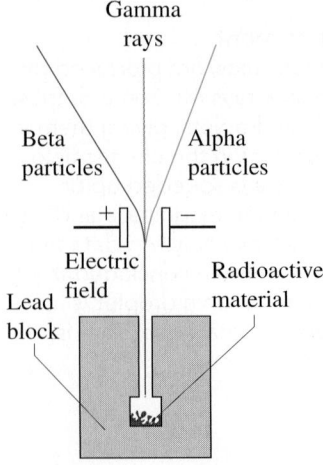

▲ FIGURE 2-10
Three types of radiation from radioactive materials
The radioactive material is enclosed in a lead block. All the radiation except that passing through the narrow opening is absorbed by the lead. When the escaping radiation is passed through an electric field, it splits into three beams. One beam is undeflected—these are gamma (γ) rays. A second beam is attracted to the negatively charged plate. These are the positively charged alpha (α) particles. The third beam, of negatively charged beta (β) particles, is deflected toward the positive plate.

is not made up of particles; it is electromagnetic radiation of extremely high penetrating power. These three forms of radioactivity are illustrated in Figure 2-10.

By the early 1900s, additional radioactive elements were discovered, principally by Marie and Pierre Curie. Rutherford and Frederick Soddy made another profound finding: The chemical properties of a radioactive element *change* as it undergoes radioactive decay. This observation suggests that radioactivity involves fundamental changes at the *subatomic* level—in radioactive decay, one element is changed into another, a process known as *transmutation*.

2-3 The Nuclear Atom

In 1909, Rutherford, with his assistant Hans Geiger, began a line of research using α particles as probes to study the inner structure of atoms. Based on Thomson's plum-pudding model, Rutherford expected that most particles in a beam of α particles would pass through thin sections of matter largely undeflected, but that some α particles would be slightly scattered or deflected as they encountered electrons. By studying these scattering patterns, he hoped to deduce something about the distribution of electrons in atoms.

The apparatus used for these studies is pictured in Figure 2-11. Alpha particles were detected by the flashes of light they produced when they struck a zinc sulfide screen mounted on the end of a telescope. When Geiger and Ernst Marsden, a student, bombarded very thin foils of gold with α particles, they observed the following:

- The majority of α particles penetrated the foil undeflected.
- Some α particles experienced slight deflections.
- A few (about 1 in every 20,000) suffered rather serious deflections as they penetrated the foil.
- A similar number did not pass through the foil at all, but bounced back in the direction from which they had come.

The large-angle scattering greatly puzzled Rutherford. As he commented some years later, this observation was "about as credible as if you had fired a 15-inch shell at a piece of tissue paper and it came back and hit you." By 1911, though, Rutherford had an explanation. He based his explanation on a model of the atom known as the *nuclear atom* and having these features:

1. Most of the mass and all of the positive charge of an atom are centered in a very small region called the *nucleus*. The remainder of the atom is mostly *empty space*.
2. The magnitude of the positive charge is different for different atoms and is approximately one-half the atomic weight of the element.
3. There are as many electrons outside the nucleus as there are units of positive charge on the nucleus. The atom as a whole is electrically neutral.

▶ Perhaps because he found it tedious to sit in the dark and count spots of light on a zinc sulfide screen, Geiger was motivated to develop an automatic radiation detector. The result was the well-known Geiger counter.

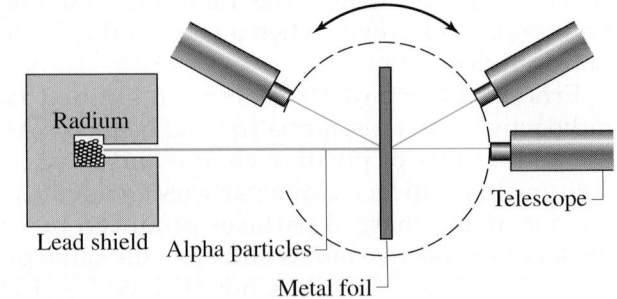

▶ FIGURE 2-11
The scattering of α particles by metal foil
The telescope travels in a circular track around an evacuated chamber containing the metal foil. Most α particles pass through the metal foil undeflected, but some are deflected through large angles.

Radium

Lead shield   Alpha particles

Metal foil

Telescope

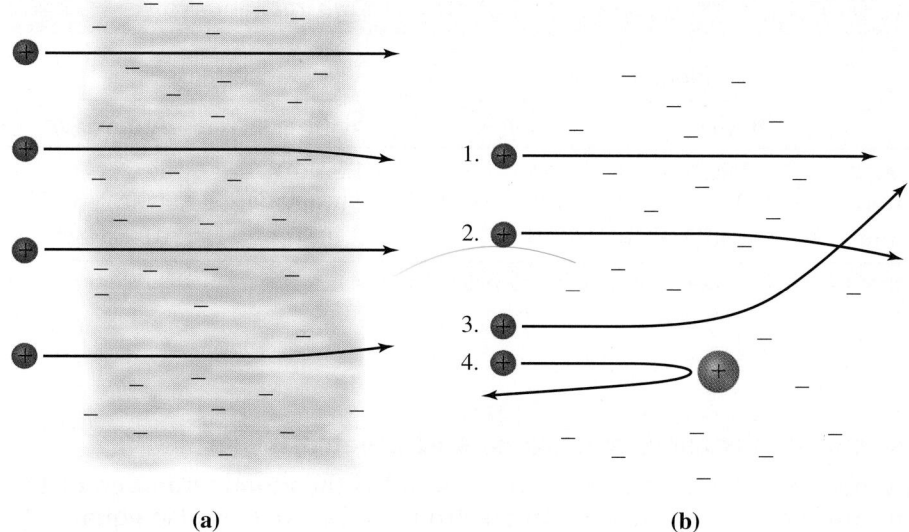

(a)   **(b)**

▲ FIGURE 2-12
Explaining the results of α-particle scattering experiments
(a) Rutherford's expectation was that small, positively charged α particles should pass through the nebulous, positively charged cloud of the Thomson plum-pudding model largely undeflected. Some would be slightly deflected by passing near electrons (present to neutralize the positive charge of the cloud). **(b)** Rutherford's explanation was based on a nuclear atom. With an atomic model having a small, dense, positively charged nucleus and extranuclear electrons, we would expect the four different types of paths actually observed:

   1. undeflected straight-line paths exhibited by most of the α particles
   2. slight deflections of α particles passing close to electrons
   3. severe deflections of α particles passing close to a nucleus
   4. reflections from the foil of α particles approaching a nucleus head-on

Rutherford's initial expectation and his explanation of the α-particle experiments are described in Figure 2-12.

Discovery of Protons and Neutrons

Rutherford's nuclear atom suggested the existence of positively charged fundamental particles of matter in the nuclei of atoms. Rutherford himself discovered these particles, called **protons**, in 1919 in studies involving the scattering of α particles by nitrogen atoms in air. The protons were freed as a result of collisions between α particles and the nuclei of nitrogen atoms. At about this same time, Rutherford predicted the existence in the nucleus of electrically neutral fundamental particles. In 1932, James Chadwick showed that a newly discovered penetrating radiation consisted of beams of *neutral* particles. These particles, called **neutrons**, originated from the nuclei of atoms. Thus, it has been only for about the past 100 years that we have had the atomic model suggested by Figure 2-13.

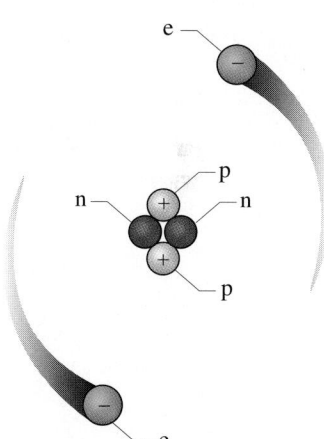

▲ FIGURE 2-13
The nuclear atom—illustrated by the helium atom
In this drawing, electrons are shown much closer to the nucleus than is the case. The actual situation is more like this: If the entire atom were represented by a room, 5 m × 5 m × 5 m, the nucleus would occupy only about as much space as the period at the end of this sentence.

🔍 **2-3   CONCEPT ASSESSMENT**

In light of information presented to this point in the text, explain which of the three assumptions of Dalton's atomic theory (page 37) can still be considered correct and which cannot.

TABLE 2.1   Properties of Three Fundamental Particles

	Electric Charge		Mass	
	SI (C)	Atomic	SI (g)	Atomic (u)[a]
Proton	$+1.6022 \times 10^{-19}$	$+1$	1.6726×10^{-24}	1.0073
Neutron	0	0	1.6749×10^{-24}	1.0087
Electron	-1.6022×10^{-19}	-1	9.1094×10^{-28}	0.00054858

[a]u is the SI symbol for atomic mass unit (abbreviated as amu).

▶ The masses of the proton and neutron are different in the fourth significant figure. The charges of the proton and electron, however, are believed to be exactly equal in magnitude (but opposite in sign). The charges and masses are known much more precisely than suggested here. More precise values are given on the inside back cover.

Properties of Protons, Neutrons, and Electrons

The number of protons in a given atom is called the **atomic number**, or the **proton number, Z**. The number of electrons in the atom is also equal to Z because the atom is electrically neutral. The total number of protons and neutrons in an atom is called the **mass number, A**. The number of neutrons, the **neutron number**, is $A - Z$. An electron carries an atomic unit of negative charge, a proton carries an atomic unit of positive charge, and a neutron is electrically neutral. Table 2.1 presents the charges and masses of protons, neutrons, and electrons in two ways.

The **atomic mass unit** (described more fully on page 46) is defined as exactly 1/12 of the mass of the atom known as carbon-12 (read as carbon twelve). An atomic mass unit is abbreviated as amu and denoted by the symbol u. As we see from Table 2.1, the proton and neutron masses are just slightly greater than 1 u. By comparison, the mass of an electron is only about 1/2000th the mass of the proton or neutron.

The three subatomic particles considered in this section are the only ones involved in the phenomena of interest to us in this text. You should be aware, however, that a study of matter at its most fundamental level must consider many additional subatomic particles. The electron is believed to be a truly fundamental particle. However, modern particle physics now considers the neutron and proton to be composed of other, more fundamental particles.

2-4   Chemical Elements

Now that we have acquired some fundamental ideas about atomic structure, we can more thoroughly discuss the concept of chemical elements.

All atoms of a particular element have the same atomic number, Z, and, conversely, all atoms with the same number of protons are atoms of the same element. The elements shown on the inside front cover have atomic numbers from $Z = 1$ to $Z = 112$. Each element has a name and a distinctive symbol. **Chemical symbols** are one- or two-letter abbreviations of the name (usually the English name). The first (but never the second) letter of the symbol is capitalized; for example: carbon, C; oxygen, O; neon, Ne; and silicon, Si. Some elements known since ancient times have symbols based on their Latin names, such as Fe for iron (*ferrum*) and Pb for lead (*plumbum*). The element sodium has the symbol Na, based on the Latin *natrium* for sodium carbonate. Potassium has the symbol K, based on the Latin *kalium* for potassium carbonate. The symbol for tungsten, W, is based on the German *wolfram*.

Elements beyond uranium ($Z = 92$) do not occur naturally and must be synthesized in particle accelerators (described in Chapter 25). Elements of the very highest atomic numbers have been produced only on a limited number of occasions, a few atoms at a time. Inevitably, controversies have arisen about

▶ The discovery of element 112 has recently been authenticated by IUPAC. However, element 112 has not yet been given a name or symbol.

▶ Other atomic symbols not based on English names include Cu, Ag, Sn, Sb, Au, and Hg.

which research team discovered a new element and, in fact, whether a discovery was made at all. However, international agreement has been reached on the first 112 elements; each one, except element 112, has an official name and symbol.

Isotopes

To represent the composition of any particular atom, we need to specify its number of protons (p), neutrons (n), and electrons (e). We can do this with the symbolism

$$\text{number p + number n} \longrightarrow {}^{A}_{Z}E \longleftarrow \text{symbol of element} \qquad (2.1)$$
$$\text{number p} \longrightarrow$$

This symbolism indicates that the atom is element E and that it has atomic number Z and mass number A. For example, an atom of aluminum represented as ${}^{27}_{13}Al$ has 13 protons and 14 neutrons in its nucleus and 13 electrons outside the nucleus. (Recall that an atom has the same number of electrons as protons.)

Contrary to what Dalton thought, we now know that atoms of an element do not necessarily all have the same mass. In 1912, J. J. Thomson measured the mass-to-charge ratios of positive ions formed from neon atoms. From these ratios he deduced that about 91% of the atoms had one mass and that the remaining atoms were about 10% heavier. All neon atoms have 10 protons in their nuclei, and most have 10 neutrons as well. A very few neon atoms, however, have 11 neutrons and some have 12. We can represent these three different types of neon atoms as

$$ {}^{20}_{10}Ne \qquad {}^{21}_{10}Ne \qquad {}^{22}_{10}Ne $$

Atoms that have the *same* atomic number (Z) but *different* mass numbers (A) are called **isotopes**. Of all Ne atoms on Earth, 90.51% are ${}^{20}_{10}Ne$. The percentages of ${}^{21}_{10}Ne$ and ${}^{22}_{10}Ne$ are 0.27% and 9.22%, respectively. These percentages— 90.51%, 0.27%, 9.22%—are the **percent natural abundances** of the three neon isotopes. Sometimes the mass numbers of isotopes are incorporated into the names of elements, such as neon-20 (neon twenty). Percent natural abundances are always based on *numbers*, not masses. Thus, 9051 of every 10,000 neon atoms are neon-20 atoms. Some elements, as they exist in nature, consist of just a single type of atom and therefore do not have naturally occurring isotopes.* Aluminum, for example, consists only of aluminum-27 atoms.

Ions

When atoms lose or gain electrons, for example, in the course of a chemical reaction, the species formed are called **ions** and carry net charges. Because an electron is negatively charged, adding electrons to an electrically neutral atom produces a negatively charged ion. Removing electrons results in a positively charged ion. The number of protons does not change when an atom becomes an ion. For example, ${}^{20}Ne^{+}$ and ${}^{22}Ne^{2+}$ are ions. The first one has 10 protons, 10 neutrons, and *9* electrons. The second one also has 10 protons, but 12 neutrons and *8* electrons. The charge on an ion is equal to the number of protons *minus* the number of electrons. That is

$$\text{number p + number n} \longrightarrow {}^{A}_{Z}E^{\#\pm} \qquad \text{number p − number e} \qquad (2.2)$$
$$\text{number p} \longrightarrow$$

Another example is the ${}^{16}O^{2-}$ ion. In this ion, there are 8 protons (atomic number 8), 8 neutrons (mass number − atomic number), and *10* electrons $(8 - 10 = -2)$.

▶ Because neon is the only element with $Z = 10$, the symbols ${}^{20}Ne$, ${}^{21}Ne$, and ${}^{22}Ne$ convey the same meaning as ${}^{20}_{10}Ne$, ${}^{21}_{10}Ne$, and ${}^{22}_{10}Ne$.

▶ Odd-numbered elements tend to have fewer isotopes than do even-numbered elements. Section 25-7 will explain why.

▶ Usually all the isotopes of an element share the same name and atomic symbol. The exception is hydrogen. Isotope ${}^{2}_{1}H$ is called deuterium (symbol D), and ${}^{3}_{1}H$ is tritium (T).

▶ In this expression, #± indicates that the charge is written with the number (#) *before* the + or − sign. However, when the charge is 1+ or 1−, the number 1 is not included.

*****Nuclide** is the general term used to describe an atom with a particular atomic number and mass number. Although there are several elements with only one naturally occurring nuclide, it is possible to produce additional nuclides of these elements—isotopes—by artificial means (Section 25-3). The artificial isotopes are radioactive, however. In all, the number of synthetic isotopes exceeds the number of naturally occurring ones by several fold.

EXAMPLE 2-3   Relating the Numbers of Protons, Neutrons, and Electrons in Atoms and Ions

Through an appropriate symbol, indicate the number of protons, neutrons, and electrons in **(a)** an atom of barium-135 and **(b)** the double negatively charged ion of selenium-80.

Analyze

Given the name of an element, we can find the symbol and the atomic number, Z, for that element from a list of elements or a periodic table. To determine the number of protons, neutrons, and electrons, we make use of the following relationships:

$$Z = \text{number p} \qquad A = \text{number p} + \text{number n} \qquad \text{charge} = \text{number p} - \text{number e}$$

The relationships above are summarized in expression (2.2).

Solve

(a) We are given the name (barium) and the mass number of the atom (135). From a list of the elements or a periodic table we obtain the symbol (Ba) and the atomic number ($Z = 56$), leading to the symbolic representation

$$^{135}_{56}\text{Ba}$$

From this symbol one can deduce that the neutral atom has 56 protons; a neutron number of $A - Z = 135 - 56 = 79$ neutrons; and a number of electrons equal to Z, that is, 56 electrons.

(b) We are given the name (selenium) and the mass number of the ion (80). From a list of the elements or a periodic table we obtain the symbol (Se) and the atomic number (34). Together with the fact that the ion carries a charge of 2−, we have the data required to write the symbol

$$^{80}_{34}\text{Se}^{2-}$$

From this symbol, we can deduce that the ion has 34 protons; a neutron number of $A - Z = 80 - 34 = 46$ neutrons; and 36 electrons, leading to a net charge of $+34 - 36 = -2$.

Assess

When writing the symbol for a particular atom or ion, we often omit the atomic number. For example, for $^{135}_{56}\text{Ba}$ and $^{80}_{34}\text{Se}^{2-}$, we often use the simpler representations ^{135}Ba and $^{80}\text{Se}^{2-}$.

PRACTICE EXAMPLE A: Use the notation $^{A}_{Z}\text{E}$ to represent the isotope of silver having a neutron number of 62.

PRACTICE EXAMPLE B: Use the notation $^{A}_{Z}\text{E}$ to represent a tin ion having the same number of electrons as an atom of the isotope cadmium-112. Explain why there can be more than one answer.

🔍 **2-4   CONCEPT ASSESSMENT**

What is the single exception to the statement that all atoms comprise protons, neutrons, and electrons?

▶ Ordinarily we expect like-charged objects (such as protons) to repel each other. The forces holding protons and neutrons together in the nucleus are very much stronger than ordinary electrical forces (Section 25-6).

▶ This definition also establishes that one atomic mass unit (1 u) is *exactly* 1/12 the mass of a carbon-12 atom.

Isotopic Masses

We cannot determine the mass of an individual atom just by adding up the masses of its fundamental particles. When protons and neutrons combine to form a nucleus, a very small portion of their original mass is converted to energy and released. However, we cannot predict exactly how much this so-called nuclear binding energy will be. Determining the masses of individual atoms, then, is something that must be done by experiment, in the following way. By international agreement, one type of atom has been chosen and assigned a specific mass. This standard is an atom of the isotope carbon-12, which is assigned a mass of exactly 12 atomic mass units, that is, 12 u. Next, the masses of other atoms relative to carbon-12 are determined with a **mass spectrometer**. In this device, a beam of gaseous ions passing through electric and magnetic fields separates into components of differing masses. The

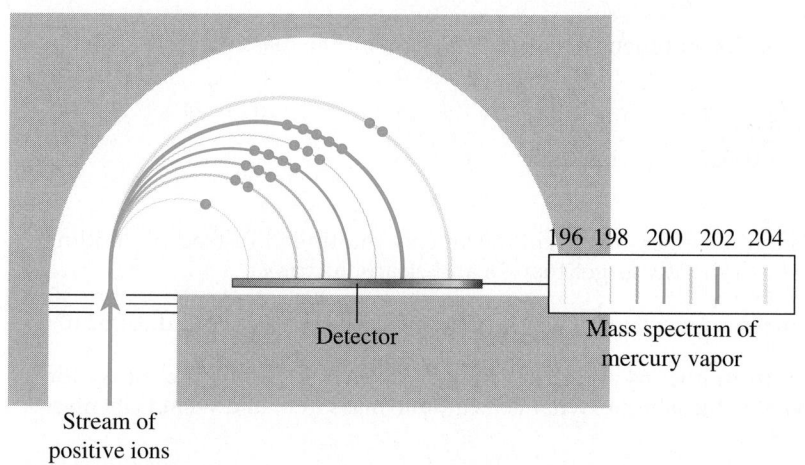

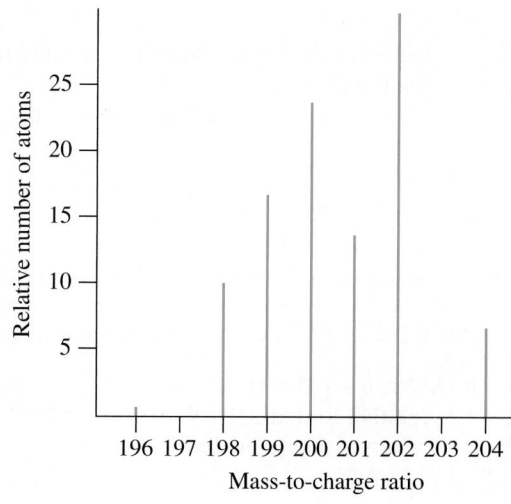

Mass spectrum of mercury vapor

▲ FIGURE 2-14
A mass spectrometer and mass spectrum
In this mass spectrometer, a gaseous sample is ionized by bombardment with electrons in the lower part of the apparatus (not shown). The positive ions thus formed are subjected to an electrical force by the electrically charged velocity selector plates and a magnetic force by a magnetic field perpendicular to the page. Only ions with a particular velocity pass through and are deflected into circular paths by the magnetic field. Ions with different masses strike the detector (here a photographic plate) in different regions. The more ions of a given type, the greater the response of the detector (intensity of line on the photographic plate). In the mass spectrum shown for mercury, the response of the ion detector (intensity of lines on photographic plate) has been converted to a scale of relative numbers of atoms. The percent natural abundances of the mercury isotopes are ^{196}Hg, 0.146%; ^{198}Hg, 10.02%; ^{199}Hg, 16.84%; ^{200}Hg, 23.13%; ^{201}Hg, 13.22%; ^{202}Hg, 29.80%; and ^{204}Hg, 6.85%.

◀ The primary standard for atomic masses has evolved over time. For example, Dalton originally assigned H a mass of 1 u. Later, chemists took naturally occurring oxygen at 16 u to be the definition of the atomic-weight scale. Concurrently, physicists defined the oxygen-16 isotope as 16 u. This resulted in conflicting values. In 1971 the adoption of carbon-12 as the universal standard resolved this disparity.

separated ions are focused on a measuring instrument, which records their presence and amounts. Figure 2-14 illustrates mass spectrometry and a typical mass spectrum.

Although mass numbers are whole numbers, the actual masses of individual atoms (in atomic mass units, u) are never whole numbers, except for carbon-12. However, they are very close in value to the corresponding mass numbers, as we can see for the isotope oxygen-16. From mass spectral data the ratio of the mass of ^{16}O to ^{12}C is found to be 1.33291. Thus, the mass of the oxygen-16 atom is

$$1.33291 \times 12 \text{ u} = 15.9949 \text{ u}$$

which is very nearly equal to the mass number of 16.

EXAMPLE 2-4   Establishing Isotopic Masses by Mass Spectrometry

With mass spectral data, the mass of an oxygen-16 atom is found to be 1.06632 times that of a nitrogen-15 atom. Given that ^{16}O has a mass of 15.9949 u (see above), what is the mass of a nitrogen-15 atom, in u?

Analyze

Given the ratio (mass of ^{16}O)/(mass of ^{15}N) = 1.06632 and the mass of ^{16}O, 15.9949 u, we solve for the mass of ^{15}N.

Solve

We know that

$$\frac{\text{mass of } ^{16}\text{O}}{\text{mass of } ^{15}\text{N}} = 1.06632$$

(continued)

We solve the expression above for the mass of ^{15}N and then substitute 15.9949 u for the mass of ^{16}O. We obtain the result

$$\text{mass of }^{15}\text{N} = \frac{\text{mass of }^{16}\text{O}}{1.06632} = \frac{15.9949 \text{ u}}{1.06632} = 15.0001 \text{ u}$$

Assess

The mass of ^{15}N is very nearly 15, as we should expect. If we had mistakenly multiplied instead of dividing by the ratio 1.06632, the result would have been slightly larger than 16 and clearly incorrect.

PRACTICE EXAMPLE A: What is the ratio of masses for $^{202}Hg/^{12}C$, if the isotopic mass for ^{202}Hg is 201.97062 u?

PRACTICE EXAMPLE B: An isotope with atomic number 64 and mass number 158 is found to have a mass ratio relative to that of carbon-12 of 13.16034. What is the isotope, what is its atomic mass in u, and what is its mass relative to oxygen-16?

2-5   Atomic Mass

▶ Carbon-14, used for radiocarbon dating, is formed in the upper atmosphere. The amount of carbon-14 on Earth is too small to affect the atomic mass of carbon.

KEEP IN MIND

that the fractional abundance is the percent abundance divided by 100%. Thus, a 98.93% abundance is a 0.9893 abundance.

In a table of atomic masses, the value listed for carbon is 12.0107, yet the atomic mass standard is *exactly* 12. Why the difference? The atomic mass standard is based on a sample of carbon containing only atoms of carbon-12, whereas naturally occurring carbon contains some carbon-13 atoms as well. The existence of these two isotopes causes the observed atomic mass to be greater than 12. The **atomic mass (weight)*** of an element is the average of the isotopic masses, *weighted* according to the naturally occurring abundances of the isotopes of the element. In a weighted average, we must assign greater importance—give greater weight—to the quantity that occurs more frequently. Since carbon-12 atoms are much more abundant than carbon-13, the weighted average must lie much closer to 12 than to 13. This is the result that we get by applying the following general equation, where the right-hand side of the equation includes one term for each naturally occurring isotope.

$$\begin{array}{l}\text{at. mass} \\ \text{of an} \\ \text{element}\end{array} = \left(\begin{array}{l}\text{fractional} \\ \text{abundance of} \times \text{isotope 1} \\ \text{isotope 1}\end{array}\right) + \left(\begin{array}{l}\text{fractional} \\ \text{abundance of} \times \text{isotope 2} \\ \text{isotope 2}\end{array}\right) + \ldots \quad \textbf{(2.3)}$$

The first term on the right side of equation (2.3) represents the contribution from isotope 1; the second term represents the contribution from isotope 2; and so on.

We will use equation (2.3), with appropriate data, in Example 2-6, but first let us illustrate the ideas of fractional abundance and a weighted average in a different way in establishing the atomic mass of naturally occurring carbon. The mass spectrum of carbon shows that 98.93% of carbon atoms are carbon-12 with a mass of exactly 12 u; the rest are carbon-13 atoms with a mass of 13.0033548378 u. Therefore:

$$\begin{aligned}\text{at. mass of naturally} \atop \text{occurring carbon} &= 0.9893 \times 12 \text{ u} + (1 - 0.9893) \times 13.0033548378 \text{ u} \\ &= 13.0033548378 \text{ u} - 0.9893 \times (13.0033548378 \text{ u} - 12 \text{ u}) \\ &= 13.0033548378 \text{ u} - 0.9893 \times (1.0033548378 \text{ u}) \\ &= 13.0033548378 \text{ u} - 0.9893 \\ &= 12.0107 \text{ u}\end{aligned}$$

*Since Dalton's time, atomic masses have been called atomic weights. They still are by most chemists, yet what we are describing here is mass, not weight. Old habits die hard.

It is important to note that, in the setup above, 12 u and the "1" appearing in the factor (1 − 0.9893) are exact numbers. Thus, by applying the rules for significant figures (see Chapter 1), the atomic mass of carbon can be reported with four decimal places.

To determine the atomic mass of an element having three naturally occurring isotopes, such as potassium, we would have to include three contributions in the weighted average, and so on.

The percent natural abundances of most of the elements remain very nearly constant from one sample of matter to another. For example, the proportions of ^{12}C and ^{13}C atoms are the same in samples of pure carbon (diamond), carbon dioxide gas, and a mineral form of calcium carbonate (calcite). We can treat all natural carbon-containing materials as if there were a single *hypothetical* type of carbon atom with a mass of 12.0107 u. This means that once weighted-average atomic masses have been determined and tabulated, we can simply use these values in calculations requiring atomic masses.

Sometimes a qualitative understanding of the relationship between isotopic masses, percent natural abundances, and weighted-average atomic mass is all that we need, and no calculation is necessary, as illustrated in Example 2-5. Example 2-6 and the accompanying Practice Examples provide additional applications of equation (2.3).

The table of atomic masses (inside the front cover) shows that some atomic masses are stated more precisely than others. For example, the atomic mass of F is given as 18.9984 u and that of Kr is given as 83.798 u. In fact, the atomic mass of fluorine is known even more precisely (18.9984032 u); the value of 18.9984 u has been rounded off to six significant figures. Why is the atomic mass of F known so much more precisely than that of Kr? Only one type of fluorine atom occurs naturally: fluorine-19. Determining the atomic mass of

---

EXAMPLE 2-5   Understanding the Meaning of a Weighted-Average Atomic Mass

The two naturally occurring isotopes of lithium, lithium-6 and lithium-7, have masses of 6.01512 u and 7.01600 u, respectively. Which of these two occurs in greater abundance?

Analyze

Look up the atomic mass of Li and compare it with the masses of ^6Li and ^7Li. If the atomic mass of Li is closer to that of ^6Li, then ^6Li is the more abundant isotope. If the atomic mass of Li is closer to that of ^7Li, then ^7Li is the more abundant isotope.

Solve

From a table of atomic masses (inside the front cover), we see that the atomic mass of lithium is 6.941 u. Because this value—a weighted-average atomic mass—is much closer to 7.01600 u than to 6.01512 u, lithium-7 must be the more abundant isotope.

Assess

Atomic masses of specific isotopes can be determined very precisely. The values given above for ^6Li and ^7Li have been rounded to five decimal places. The precise values are 6.015122795 u and 7.01600455 u.

PRACTICE EXAMPLE A: The two naturally occurring isotopes of boron, boron-10 and boron-11, have masses of 10.0129370 u and 11.0093054 u, respectively. Which of these two occurs in greater abundance?

PRACTICE EXAMPLE B: Indium has two naturally occurring isotopes and a weighted atomic mass of 114.818 u. One of the isotopes has a mass of 112.904058 u. Which of the following must be the second isotope: ^{111}In, ^{112}In, ^{114}In, or ^{115}In? Which of the two naturally occurring isotopes must be the more abundant?

EXAMPLE 2-6   **Relating the Masses and Natural Abundances of Isotopes to the Atomic Mass of an Element**

Bromine has two naturally occurring isotopes. One of them, bromine-79, has a mass of 78.9183 u and an abundance of 50.69%. What must be the mass and percent natural abundance of the other, bromine-81?

Analyze

Although the atomic mass of Br is not given explicitly, it is a known quantity. From the inside front cover, we find that the atomic mass of Br is 79.904 u. We need to apply two key concepts: (1) the atomic mass of Br is a weighted average of the masses of ^{79}Br and ^{81}Br, and (2) the percent natural abundances of ^{79}Br and ^{81}Br must add up to 100%.

Solve

The atomic mass of Br is a weighted average of the masses of ^{79}Br and ^{81}Br:

$$\text{atomic mass} = \begin{pmatrix} \text{fraction of atoms} \\ \text{that are } ^{79}\text{Br} \times \\ \text{mass of } ^{79}\text{Br} \end{pmatrix} + \begin{pmatrix} \text{fraction of atoms} \\ \text{that are } ^{81}\text{Br} \times \\ \text{mass of } ^{81}\text{Br} \end{pmatrix}$$

Because the percent natural abundances must total 100%, the percent natural abundance of ^{81}Br is $100\% - 50.69\% = 49.31\%$. Substituting 79.904 u for the atomic mass, 78.9183 u for the mass of ^{79}Br, and the fractional abundances of the two isotopes, we obtain

$$79.904 \text{ u} = (0.5069 \times 78.9183 \text{ u}) + (0.4931 \times \text{mass of } ^{81}\text{Br})$$

$$= 40.00 \text{ u} + (0.4931 \times \text{mass of } ^{81}\text{Br})$$

$$\text{mass of } ^{81}\text{Br} = \frac{79.904 \text{ u} - 40.00 \text{ u}}{0.4931} = 80.92 \text{ u}$$

To four significant figures, the natural abundance of the bromine-81 isotope is 49.31% and its mass is 80.92 u.

Assess

We can check the final result by working the problem in reverse and using numbers that are slightly rounded. The atomic mass of Br is $50.69\% \times 78.92 \text{ u} + 49.31\% \times 80.92 \text{ u} \approx \frac{1}{2}(79 \text{ u} + 81 \text{ u}) = 80 \text{ u}$. The estimated atomic mass (80 u) is close to the actual atomic mass of 79.904 u.

PRACTICE EXAMPLE A: The masses and percent natural abundances of the three naturally occurring isotopes of silicon are ^{28}Si, 27.9769265325 u, 92.223%; ^{29}Si, 28.976494700 u, 4.685%; ^{30}Si, 29.973377017 u, 3.092%. Calculate the weighted-average atomic mass of silicon.

PRACTICE EXAMPLE B: Use data from Example 2-5 to determine the percent natural abundances of lithium-6 and lithium-7.

fluorine means establishing the mass of this type of atom as precisely as possible. The atomic mass of krypton is known less precisely because krypton has six naturally occurring isotopes. Because the percent distribution of the isotopes of krypton differs very slightly from one sample to another, the weighted-average atomic mass of krypton cannot be stated with high precision.

2-5   CONCEPT ASSESSMENT

The value listed for chromium in the table of atomic masses inside the front cover is 51.9961 u. Should we conclude that naturally occurring chromium atoms are all of the type $^{52}_{24}$Cr? The same table lists a value of 65.409 u for zinc. Should we conclude that zinc occurs as a mixture of isotopes? Explain.

2-6 Introduction to the Periodic Table

Scientists spend a lot of time organizing information into useful patterns. Before they can organize information, however, they must possess it, and it must be correct. Botanists had enough information about plants to organize their field in the eighteenth century. Because of uncertainties in atomic masses and because many elements remained undiscovered, chemists were not able to organize the elements until a century later.

We can distinguish one element from all others by its particular set of observable physical properties. For example, sodium has a low density of 0.971 g/cm^3 and a low melting point of 97.81 °C. No other element has this same combination of density and melting point. Potassium, though, also has a low density (0.862 g/cm^3) and low melting point (63.65 °C), much like sodium. Sodium and potassium further resemble each other in that both are good conductors of heat and electricity, and both react vigorously with water to liberate hydrogen gas. Gold, conversely, has a density (19.32 g/cm^3) and melting point (1064 °C) that are very much higher than those of sodium or potassium, and gold does not react with water or even with ordinary acids. It does resemble sodium and potassium in its ability to conduct heat and electricity, however. Chlorine is very different still from sodium, potassium, and gold. It is a gas under ordinary conditions, which means that the melting point of solid chlorine (−101 °C) is far below room temperature. Also, chlorine is a nonconductor of heat and electricity.

Even from these very limited data, we get an inkling of a useful classification scheme of the elements. If the scheme is to group together elements with similar properties, then sodium and potassium should appear in the same group. And if the classification scheme is in some way to distinguish between elements that are good conductors of heat and electricity and those that are not, chlorine should be set apart from sodium, potassium, and gold. The classification system we need is the one shown in Figure 2-15 (and inside the front cover), known as the **periodic table** of the elements. In

1 1A	2 2A	3 3B	4 4B	5 5B	6 6B	7 7B	8 8B	9 8B	10 8B	11 1B	12 2B	13 3A	14 4A	15 5A	16 6A	17 7A	18 8A
1 H 1.00794																	2 He 4.00260
3 Li 6.941	4 Be 9.01218											5 B 10.811	6 C 12.0107	7 N 14.0067	8 O 15.9994	9 F 18.9984	10 Ne 20.1797
11 Na 22.9898	12 Mg 24.3050											13 Al 26.9815	14 Si 28.0855	15 P 30.9738	16 S 32.065	17 Cl 35.453	18 Ar 39.948
19 K 39.0983	20 Ca 40.078	21 Sc 44.9559	22 Ti 47.867	23 V 50.9415	24 Cr 51.9961	25 Mn 54.9380	26 Fe 55.845	27 Co 58.9332	28 Ni 58.6934	29 Cu 63.546	30 Zn 65.409	31 Ga 69.723	32 Ge 72.64	33 As 74.9216	34 Se 78.96	35 Br 79.904	36 Kr 83.798
37 Rb 85.4678	38 Sr 87.62	39 Y 88.9059	40 Zr 91.224	41 Nb 92.9064	42 Mo 95.94	43 Tc (98)	44 Ru 101.07	45 Rh 102.906	46 Pd 106.42	47 Ag 107.868	48 Cd 112.411	49 In 114.818	50 Sn 118.710	51 Sb 121.760	52 Te 127.60	53 I 126.904	54 Xe 131.293
55 Cs 132.905	56 Ba 137.327	57–71 La–Lu	72 Hf 178.49	73 Ta 180.948	74 W 183.84	75 Re 186.207	76 Os 190.23	77 Ir 192.217	78 Pt 195.084	79 Au 196.967	80 Hg 200.59	81 Tl 204.383	82 Pb 207.2	83 Bi 208.980	84 Po (209)	85 At (210)	86 Rn (222)
87 Fr (223)	88 Ra (226)	89–103 Ac–Lr	104 Rf (261)	105 Db (262)	106 Sg (266)	107 Bh (264)	108 Hs (277)	109 Mt (268)	110 Ds (271)	111 Rg (272)							

*Lanthanide series	57 La 138.905	58 Ce 140.116	59 Pr 140.908	60 Nd 144.242	61 Pm (145)	62 Sm 150.36	63 Eu 151.964	64 Gd 157.25	65 Tb 158.925	66 Dy 162.500	67 Ho 164.930	68 Er 167.259	69 Tm 168.934	70 Yb 173.04	71 Lu 174.967
†Actinide series	89 Ac (227)	90 Th 232.038	91 Pa 231.036	92 U 238.029	93 Np (237)	94 Pu (244)	95 Am (243)	96 Cm (247)	97 Bk (247)	98 Cf (251)	99 Es (252)	100 Fm (257)	101 Md (258)	102 No (259)	103 Lr (262)

▲ FIGURE 2-15
Periodic table of the elements
Atomic masses are relative to carbon-12. For certain radioactive elements, the numbers listed in parentheses are the mass numbers of the most stable isotopes. Metals are shown in tan, nonmetals in blue, and metalloids in green. The noble gases (also nonmetals) are shown in pink.

KEEP IN MIND

that the periodic table shown in Figure 2-15 is the one currently recommended by IUPAC. The discovery of element 112 has recently been authenticated by IUPAC (in May 2009) but the element has not yet been named. Elements with atomic numbers greater than 112 have been reported but not fully authenticated. In Figure 2-15, lutetium (Lu) and lawrencium (Lr) are the last members of the lanthanide and actinide series, respectively. A strong argument* has been made for placing Lu and Lr in group 3, meaning the lanthanide series would end with ytterbium (Yb) and the actinide series would end with nobelium (Nb). To date, IUPAC has not endorsed placing Lu and Lr in group 3.

* See W. B. Jensen, *J. Chem. Educ.*, **59**, 634 (1982).

Chapter 9, we will describe how the periodic table was formulated, and we will also learn its theoretical basis. For the present, we will consider only a few features of the table.

Features of the Periodic Table In the periodic table, elements are listed according to increasing atomic number starting at the upper left and arranged in a series of horizontal rows. This arrangement places similar elements in *vertical* **groups**, or **families**. For example, sodium and potassium are found together in a group labeled 1 (called the *alkali metals*). We should expect other members of the group, such as cesium and rubidium, to have properties similar to sodium and potassium. Chlorine is found at the other end of the table in a group labeled 17. Some of the groups are given distinctive names, mostly related to an important property of the elements in the group. For example, the group 17 elements are called the *halogens*, a term derived from Greek, meaning "salt former."

▶ That elements in one group have similar properties is perhaps the most useful simplifying feature of atomic properties. Significant differences within a group do occur. The manner and reason for such differences is much of what we try to discover in studying chemistry.

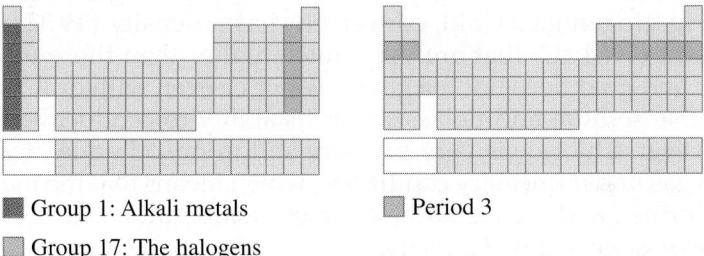

■ Group 1: Alkali metals   ■ Period 3

■ Group 17: The halogens

Each element is listed in the periodic table by placing its symbol in the middle of a box in the table. The atomic number (Z) of the element is shown above the symbol, and the weighted-average atomic mass of the element is shown below its symbol. Some periodic tables provide other information, such as density and melting point, but the atomic number and atomic mass are generally sufficient for our needs. Elements with atomic masses in parentheses, such as plutonium, Pu (244), are produced synthetically, and the number shown is the mass number of the most stable isotope.

It is customary also to divide the elements into two broad categories—**metals** and **nonmetals**. In Figure 2-15, colored backgrounds are used to distinguish the metals (tan) from the nonmetals (blue and pink). Except for mercury, a liquid, metals are solids at room temperature. They are generally malleable (capable of being flattened into thin sheets), ductile (capable of being drawn into fine wires), and good conductors of heat and electricity, and have a lustrous or shiny appearance. The properties of nonmetals are generally opposite those of metals; for example, nonmetals are poor conductors of heat and electricity. Several of the nonmetals, such as nitrogen, oxygen, and chlorine, are gases at room temperature. Some, such as silicon and sulfur, are brittle solids. One—bromine—is a liquid.

▶ There is lack of agreement on just which elements to label as metalloids. However, they are generally considered either to lie adjacent to the stair-step line or to be close by.

Two other highlighted categories in Figure 2-15 are a special group of nonmetals known as the **noble gases** (pink), and a small group of elements, often called **metalloids** (green), that have some metallic and some nonmetallic properties.

The *horizontal* rows of the table are called **periods**. (The periods are numbered at the extreme left in the periodic table inside the front cover.) The first period of the table consists of just two elements, hydrogen and helium. This is followed by two periods of eight elements each, lithium through neon and sodium through argon. The fourth and fifth periods contain 18 elements each, ranging from potassium through krypton and from rubidium through xenon. The sixth period is a long one of 32 members. To fit this period in a table that is held to a maximum width of 18 members, 15 members of the period are placed at the bottom of the periodic table. This series of 15 elements start with lanthanum ($Z = 57$), and these elements are called the **lanthanides**. The seventh and final period is incomplete (some members are yet to be discovered), but it is known to be a long one. A 15-member series is also extracted from the seventh period and placed at the bottom of the table. Because the elements in this series start with actinium ($Z = 89$), they are called the **actinides**.

The labeling of the groups of the periodic table has been a matter of some debate among chemists. The 1-18 numbering system used in Figure 2-15 is the one most recently adopted. Group labels previously used in the United States consisted of a letter and a number, closely following the method adopted by Mendeleev, the developer of the periodic table. As seen in Figure 2-15, the A groups 1 and 2 are separated from the remaining A groups (3 to 8) by B groups 1 through 8. The International Union of Pure and Applied Chemistry (IUPAC) recommended the simple 1 to 18 numbering scheme in order to avoid confusion between the American number and letter system and that used in Europe, where some of the A and B designations were switched! Currently, the IUPAC system is officially recommended by the American Chemical Society (ACS) and chemical societies in other nations. Because both numbering systems are in use, we show both in Figure 2-15 and in the periodic table inside the front cover. However, except for an occasional reminder of the earlier system, we will use the IUPAC numbering system in this text.

◄ Mendeleev's arrangement of the elements in the original periodic table was based on observed chemical and physical properties of the elements and their compounds. The arrangement of the elements in the modern periodic table is based on atomic properties–atomic number and electron configuration.

Useful Relationships from the Periodic Table

The periodic table helps chemists describe and predict the properties of chemical compounds and the outcomes of chemical reactions. Throughout this text, we will use it as an aid to understanding chemical concepts. One application of the table worth mentioning here is how it can be used to predict likely charges on simple monatomic ions.

Main-group elements are those in groups 1, 2, and 13 to 18. When main-group metal atoms in groups 1 and 2 form ions, they lose the same number of electrons as the IUPAC group number. Thus, Na atoms (group 1) lose one electron to become Na^+, and Ca atoms (group 2) lose two electrons to

---

EXAMPLE 2-7   Describing Relationships Based on the Periodic Table

Refer to the periodic table on the inside front cover, and indicate

    (a) the element that is in group 14 and the fourth period;
    (b) two elements with properties similar to those of molybdenum (Mo);
    (c) the ion most likely formed from a strontium atom.

Analyze

For **(a)**, the key concept is that the rows (periods) are numbered 1 through 7, starting from the top of the periodic table, and the groups are numbered 1 through 18, starting from the left side. For **(b)**, the key concept is that elements in the same group have similar properties. For **(c)**, the key concept is that main-group metal atoms in groups 1 and 2 form positive ions with charges of +1 and +2, respectively.

Solve

    (a) The elements in the fourth period range from K ($Z = 19$) to Kr ($Z = 36$). Those in group 14 are C, Si, Ge, Sn, and Pb. The only element that is common to both of these groupings is Ge ($Z = 32$).
    (b) Molybdenum is in group 6. Two other members of this group that should resemble it are chromium (Cr) and tungsten (W).
    (c) Strontium (Sr) is in group 2. It should form the ion Sr^{2+}.

Assess

In Chapter 8, we will examine in greater detail reasons for the arrangement of the periodic table.

PRACTICE EXAMPLE A: Write a symbol for the ion most likely formed by an atom of each of the following: Li, S, Ra, F, I, and Al.

PRACTICE EXAMPLE B: Classify each of the following elements as a main-group or transition element. Also, specify whether they are metals, metalloids, or nonmetals: Na, Re, S, I, Kr, Mg, U, Si, B, Al, As, H.

become Ca^{2+}. Aluminum, in group 13, loses three electrons to form Al^{3+} (here the charge is "group number minus 10"). The few other metals in groups 13 and higher form more than one possible ion, a matter that we deal with in Chapter 9.

When nonmetal atoms form ions, they gain electrons. The number of electrons gained is normally 18 minus the IUPAC group number. Thus, an O atom gains $18 - 16 = 2$ electrons to become O^{2-}, and a Cl atom gains $18 - 17 = 1$ electron to become Cl^-. The "18 minus group number" rule suggests that an atom of Ne in group 18 gains no electrons: $18 - 18 = 0$. The very limited tendency of the noble gas atoms to form ions is one of several characteristics of this family of elements.

The elements in groups 3 to 12 are the **transition elements**, and because all of them are metals, they are also called the **transition metals**. Like the main-group metals, the transition metals form positive ions, but the number of electrons lost is not related in any simple way to the group number, mostly because transition metals can form two or more ions of differing charge.

2-7   The Concept of the Mole and the Avogadro Constant

Starting with Dalton, chemists have recognized the importance of relative numbers of atoms, as in the statement that *two* hydrogen atoms and *one* oxygen atom combine to form *one* molecule of water. Yet it is physically impossible to count every atom in a macroscopic sample of matter. Instead, some other measurement must be employed, which requires a relationship between the measured quantity, usually mass, and some known, but uncountable, number of atoms. Consider a practical example of mass substituting for a desired number of items. Suppose you want to nail down new floorboards on the deck of a mountain cabin, and you have calculated how many nails you will need. If you have an idea of how many nails there are in a pound, then you can buy the nails by the pound.

▶ Because the value of Avogadro's number depends, in part, on a measurement, the value has changed slightly over the years. The values recommended since 1986 by the Committee on Data for Science and Technology (CODATA) are listed below.

Year	Avogadro's Number
1986	6.0221367×10^{23}
1998	$6.02214199 \times 10^{23}$
2002	6.0221415×10^{23}
2006	$6.02214179 \times 10^{23}$

The SI quantity that describes an amount of substance by relating it to a number of particles of that substance is called the *mole* (abbreviated *mol*). A **mole** is the amount of a substance that contains the same number of elementary entities as there are atoms in exactly 12 g of pure carbon-12. The "number of elementary entities (atoms, molecules, and so on)" in a mole is the **Avogadro constant, N_A.**

$$N_A = 6.02214179 \times 10^{23} \text{ mol}^{-1} \qquad \textbf{(2.4)}$$

The Avogadro constant consists of a number, $6.02214179 \times 10^{23}$, known as Avogadro's *number*, and a unit, mol^{-1}. The unit mol^{-1} signifies that the entities being counted are those present in 1 mole.

The value of Avogadro's number is based on both a definition and a measurement. A mole of carbon-12 is *defined* to be 12 g. If the mass of one carbon-12 atom is *measured* by using a mass spectrometer (see Figure 2-14), the mass would be about 1.9926×10^{-23} g. The ratio of these two masses provides an estimate of Avogadro's number. In actual fact, accurate determinations of Avogadro's number make use of other measurements, not the measurement of the mass of a single atom of carbon-12.

Often the value of N_A is rounded off to $6.022 \times 10^{23} \text{ mol}^{-1}$, or even to $6.02 \times 10^{23} \text{ mol}^{-1}$.

▶ When rounding Avogadro's number or any other accurately known value, keep one more significant figure than that of the least accurate number in the calculation to avoid rounding errors.

If a substance contains atoms of only a single isotope, then

$$1 \text{ mol } ^{12}\text{C} = 6.02214 \times 10^{23} \text{ }^{12}\text{C atoms} = 12.0000 \text{ g}$$
$$1 \text{ mol } ^{16}\text{O} = 6.02214 \times 10^{23} \text{ }^{16}\text{O atoms} = 15.9949 \text{ g (and so on)}$$

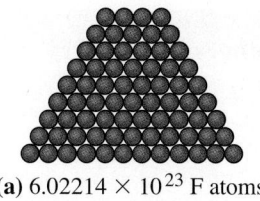

(a) 6.02214×10^{23} F atoms
= 18.9984 g

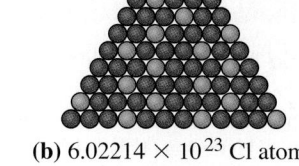

(b) 6.02214×10^{23} Cl atoms
= 35.453 g

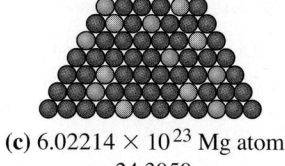

(c) 6.02214×10^{23} Mg atoms
= 24.3050 g

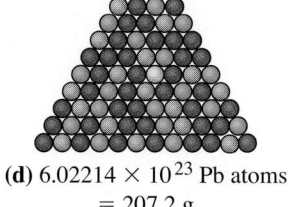

(d) 6.02214×10^{23} Pb atoms
= 207.2 g

▲ FIGURE 2-16
Distribution of isotopes in four elements
(a) There is only one type of fluorine atom, ^{19}F (shown in red). (b) In chlorine, 75.77%
of the atoms are ^{35}Cl (red) and the remainder are ^{37}Cl (blue). (c) Magnesium has one
principal isotope, ^{24}Mg (red), and two minor ones, ^{25}Mg (gray) and ^{26}Mg (blue).
(d) Lead has four naturally occurring isotopes: 1.4% ^{204}Pb (yellow), 24.1% ^{206}Pb (blue),
22.1% ^{207}Pb (gray), and 52.4% ^{208}Pb (red).

Most elements are composed of mixtures of two or more isotopes so that the
atoms in a sample of the element are not all of the same mass but are present in
their naturally occurring proportions. Thus, in one mole of carbon, most of the
atoms are carbon-12, but some are carbon-13. In one mole of oxygen, most of
the atoms are oxygen-16, but some are oxygen-17 and some are oxygen-18. As
a result,

$$1 \text{ mol of C} = 6.02214 \times 10^{23} \text{ C atoms} = 12.0107 \text{ g}$$

$$1 \text{ mol of O} = 6.02214 \times 10^{23} \text{ O atoms} = 15.9994 \text{ g, and so on.}$$

The Avogadro constant was purposely chosen so that the mass of one mole of
carbon-12 atoms—exactly 12 g—would have the same *numeric* value as the
mass of a single carbon-12 atom—exactly 12 u. As a result, for all other elements
the numeric value of the mass in grams of one mole of atoms and the weighted-
average atomic mass in atomic mass units are equal. For example, the weighted
average atomic mass of lithium is 6.941 u and the mass of one mole of lithium
atoms is 6.941 g. Thus, we can easily establish the mass of one mole of atoms,
called the **molar mass, M**, from a table of atomic masses.* For example, the
molar mass of lithium is 6.941 g Li/mol Li. Figure 2-16 attempts to portray the
distribution of isotopes of an element, and Figure 2-17 pictures one mole each of
four common elements.

◀ The weighted-average
atomic mass of carbon was
calculated on page 48.

KEEP IN MIND

that molar mass has the unit
g/mol.

◀ FIGURE 2-17
One mole of an element
The watch glasses contain one mole of copper atoms (left)
and one mole of sulfur atoms (right). The beaker contains
one mole of mercury atoms as liquid mercury, and the
balloon, of which only a small portion is visible here,
contains one mole of helium atoms in the gaseous state.

*Atomic mass (atomic weight) values in tables are often written without units, especially if they
are referred to as *relative* atomic masses. This simply means that the values listed are in relation
to *exactly* 12 (rather than 12 u) for carbon-12. We will use the atomic mass unit (u) when referring
to atomic masses (atomic weights). Most chemists do.

🔍 **2-6** CONCEPT ASSESSMENT

Dividing the molar mass of gold by the Avogadro constant yields the mass of any individual atom of naturally occurring gold. In contrast, no naturally occurring atom of silver has the mass obtained by dividing the molar mass of silver by the Avogadro constant. How can this be?

Thinking About Avogadro's Number

Avogadro's number (6.02214×10^{23}) is an enormously large number and practically inconceivable in terms of ordinary experience. Suppose we were counting garden peas instead of atoms. If the typical pea had a volume of about 0.1 cm^3, the required pile of peas would cover the United States to a depth of about 6 km (4 mi). Or imagine that grains of wheat could be counted at the rate of 100 per minute. A given individual might be able to count out about 4 billion grains in a lifetime. Even so, if all the people currently on Earth were to spend their lives counting grains of wheat, they could not reach Avogadro's number. In fact, if all the people who ever lived on Earth had spent their lifetimes counting grains of wheat, the total would still be far less than Avogadro's number. (And Avogadro's number of wheat grains is far more wheat than has been produced in human history.) Now consider a much more efficient counting device, a modern personal computer; it is capable of counting at a rate of about 1 billion units per second. The task of counting out Avogadro's number would still take about 20 million years!

Avogadro's number is clearly not a useful number for counting ordinary objects. However, when this inconceivably large number is used to count inconceivably small objects, such as atoms and molecules, the result is a quantity of material that is easily within our grasp, essentially a "handful."

▶ You will have an opportunity to calculate a value of N_A at several points in the text, starting with Exercise 113 in Chapter 3.

2-8 Using the Mole Concept in Calculations

Throughout the text, the mole concept will provide conversion factors for problem-solving situations. With each new situation, we will explore how the mole concept applies. For now, we will deal with the relationship between numbers of atoms and the mole. Consider the statement: 1 mol S = 6.022×10^{23} S atoms = 32.065 g S. This allows us to write the conversion factors

$$\frac{1 \text{ mol S}}{6.022 \times 10^{23} \text{ S atoms}} \quad \text{and} \quad \frac{32.065 \text{ g S}}{1 \text{ mol S}}$$

In calculations requiring the Avogadro constant, students often ask when to multiply and when to divide by N_A. One answer is always to use the constant in a way that gives the proper cancellation of units. Another answer is to think in terms of the expected result. In calculating a number of atoms, we expect the answer to be a very large number and certainly *never* smaller than one. The number of moles of atoms, conversely, is generally a number of more modest size and will often be less than one.

In the following examples, we use atomic masses and the Avogadro constant in calculations to determine the number of atoms present in a given sample. Atomic masses and the Avogadro constant are known rather precisely, and students often wonder how many significant figures to carry in atomic masses or the Avogadro constant when performing calculations. Here is a useful rule of thumb.

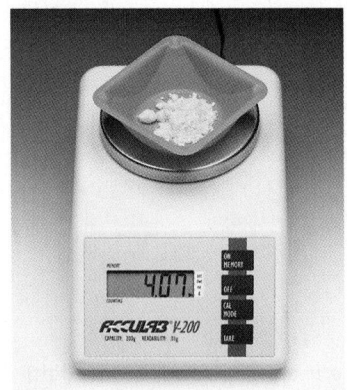

▲ FIGURE 2-18
Measurement of 7.64×10^{22} S atoms (0.127 mol S)—Example 2-8 illustrated
The balance was set to zero (tared) when just the weighing boat was present. The sample of sulfur weighs 4.07 g.

To ensure the maximum precision allowable, carry at least one more significant figure in well-known physical constants than in other measured quantities.

For example, in calculating the mass of 0.600 mol of sulfur, we should use the atomic mass of S with *at least* four significant figures. The answer 0.600 mol S × 32.07 g S/mol S = 19.2 g S is a more precise response than 0.600 mol S × 32.1 g S/mol S = 19.3 g S.

EXAMPLE 2-8 Relating Number of Atoms, Amount in Moles, and Mass in Grams

In the sample of sulfur weighing 4.07 g pictured in Figure 2-18, **(a)** how many moles of sulfur are present, and **(b)** what is the total number of sulfur atoms in the sample?

Analyze

For **(a)**, the conversion pathway is g S → mol S. To carry out this conversion, we multiply 4.07 g S by the conversion factor (1 mol S/32.07 g S). The conversion factor is the molar mass inverted. For **(b)**, the conversion pathway is mol S → atoms S. To carry out this conversion, we multiply the quantity in moles from part **(a)** by the conversion factor (6.022×10^{23} atoms S/1 mol S).

Solve

(a) For the conversion g S → mol S, using ($1/M$) as a conversion factor achieves the proper cancellation of units. The result of this calculation should be stored without rounding it off because it is required in part (b).

$$? \text{ mol S} = 4.07 \text{ g S} \times \frac{1 \text{ mol S}}{32.07 \text{ g S}} = 0.127 \text{ mol S}$$

(b) The conversion mol S → atoms S is carried out using the Avogadro constant as a conversion factor.

$$? \text{ atoms S} = 0.127 \text{ mol S} \times \frac{6.022 \times 10^{23} \text{ atoms S}}{1 \text{ mol S}} = 7.64 \times 10^{22} \text{ atoms S}$$

Assess

By including units in our calculations, we can check that proper cancellation of units occurs. Also, if our only concern is to calculate the number of sulfur atoms in the sample, the calculations carried out in parts **(a)** and **(b)** could be combined into a single calculation, as shown below.

$$? \text{ atoms S} = 4.07 \text{ g S} \times \frac{1 \text{ mol S}}{32.07 \text{ g S}} \times \frac{6.022 \times 10^{23} \text{ atoms S}}{1 \text{ mol S}} = 7.64 \times 10^{22} \text{ atoms S}$$

Had we rounded 4.07 g S × (1 mol S/32.07 g S) to 0.127 mol S and used the rounded result in part (b), we would have obtained a final answer of 7.65×10^{22} atoms S. With a single line calculation, we do not have to write down an intermediate result and we avoid round-off errors.

PRACTICE EXAMPLE A: What is the mass of 2.35×10^{24} atoms of Cu?

PRACTICE EXAMPLE B: How many lead-206 atoms are present in a 22.6 g sample of lead metal? [*Hint:* See Figure 2-16.]

Example 2-9 is perhaps the most representative use of the mole concept. Here it is part of a larger problem that requires other unrelated conversion factors as well. One approach is to outline a conversion pathway to get from the given to the desired information.

EXAMPLE 2-9 Combining Several Factors in a Calculation—Molar Mass, the Avogadro Constant, Percent Abundances

Potassium-40 is one of the few naturally occurring radioactive isotopes of elements of low atomic number. Its percent natural abundance among K isotopes is 0.012%. How many ^{40}K atoms are present in 225 mL of whole milk containing 1.65 mg K/mL?

(continued)

Analyze

Ultimately we need to complete the conversion mL milk $\rightarrow$ atoms ^{40}K. There is no single conversion factor that allows us to complete this conversion in one step, so we anticipate having to complete several steps or conversions. We are told the milk contains 1.65 mg K/mL = 1.65×10^{-3} g K/mL, and this information can be used to carry out the conversion mL milk $\rightarrow$ g K. We can carry out the conversions g K $\rightarrow$ mol K $\rightarrow$ atoms K by using conversion factors based on the molar mass of K and the Avogadro constant. The final conversion, atoms K $\rightarrow$ atoms ^{40}K, can be carried out by using a conversion factor based on the percent natural abundance of ^{40}K. A complete conversion pathway is shown below:

$$\text{mL milk} \rightarrow \text{mg K} \rightarrow \text{g K} \rightarrow \text{mol K} \rightarrow \text{atoms K} \rightarrow \text{atoms } ^{40}K$$

Solve

The required conversions can be carried out in a stepwise fashion, or they can be combined into a single line calculation. Let's use a stepwise approach. First, we convert from mL milk to g K.

$$? \text{ g K} = 225 \text{ mL milk} \times \frac{1.65 \text{ mg K}}{1 \text{ mL milk}} \times \frac{1 \text{ g K}}{1000 \text{ mg K}} = 0.371 \text{ g K}$$

Next, we convert from g K to mol K,

$$? \text{ mol K} = 0.371 \text{ g K} \times \frac{1 \text{ mol K}}{39.10 \text{ g K}} = 9.49 \times 10^{-3} \text{ mol K}$$

and then we convert from mol K to atoms K.

$$? \text{ atoms K} = 9.49 \times 10^{-3} \text{ mol K} \times \frac{6.022 \times 10^{23} \text{ atoms K}}{1 \text{ mol K}} = 5.71 \times 10^{21} \text{ atoms K}$$

Finally, we convert from atoms K to atoms ^{40}K.

$$? \text{ atoms } ^{40}K = 5.71 \times 10^{21} \text{ atoms K} \times \frac{0.012 \text{ atoms } ^{40}K}{100 \text{ atoms K}} = 6.9 \times 10^{17} \text{ atoms } ^{40}K$$

Assess

The final answer is rounded to two significant figures because the least precisely known quantity in the calculation, the percent natural abundance of ^{40}K, has two significant figures. It is possible to combine the steps above into a single line calculation.

$$? \text{ atoms } ^{40}K = 225 \text{ mL milk} \times \frac{1.65 \text{ mg K}}{1 \text{ mL milk}} \times \frac{1 \text{ g K}}{1000 \text{ mg K}} \times \frac{1 \text{ mol K}}{39.10 \text{ g K}}$$

$$\times \frac{6.022 \times 10^{23} \text{ atoms K}}{1 \text{ mol K}} \times \frac{0.012 \text{ atoms } ^{40}K}{100 \text{ atoms K}}$$

$$= 6.9 \times 10^{17} \text{ atoms } ^{40}K$$

PRACTICE EXAMPLE A: How many Pb atoms are present in a small piece of lead with a volume of 0.105 cm³? The density of Pb = 11.34 g/cm³.

PRACTICE EXAMPLE B: Rhenium-187 is a radioactive isotope that can be used to determine the age of meteorites. A 0.100 mg sample of Re contains 2.02×10^{17} atoms of ^{187}Re. What is the percent abundance of rhenium-187 in the sample?

Mastering**CHEMISTRY**   www.masteringchemistry.com

What is the most abundant element? This seemingly simple question does not have a simple answer. To learn more about the abundances of elements in the universe and in the Earth's crust, go to the Focus On feature for Chapter 2, entitled Occurrence and Abundances of the Elements, on the MasteringChemistry site.

Summary

2-1 Early Chemical Discoveries and the Atomic Theory—Modern chemistry began with eighteenth-century discoveries leading to the formulation of two basic laws of chemical combination, the **law of conservation of mass** and the **law of constant composition (definite proportions)**. These discoveries led to Dalton's atomic theory—that matter is composed of indestructible particles called atoms, that the atoms of an element are identical to one another but different from atoms of all other elements, and that chemical compounds are combinations of atoms of different elements. Based on this theory, Dalton proposed still another law of chemical combination, the **law of multiple proportions**.

2-2 Electrons and Other Discoveries in Atomic Physics—The first clues to the structures of atoms came through the discovery and characterization of **cathode rays (electrons)**. Key experiments were those that established the mass-to-charge ratio (Fig. 2-7) and then the charge on an electron (Fig. 2-8). Two important accidental discoveries made in the course of cathode-ray research were of **X-rays** and **radioactivity**. The principal types of radiation emitted by radioactive substances are **alpha (α) particles, beta (β) particles**, and **gamma (γ) rays** (Fig. 2-10).

2-3 The Nuclear Atom—Studies on the scattering of α particles by thin metal foils (Fig. 2-11) led to the concept of the nuclear atom—a tiny, but massive, positively charged nucleus surrounded by lightweight, negatively charged electrons (Fig. 2-12). A more complete description of the nucleus was made possible by the discovery of **protons** and **neutrons**. An individual atom is characterized in terms of its **atomic number (proton number) Z** and **mass number, A**. The difference, $A - Z$, is the **neutron number**. The masses of individual atoms and their component parts are expressed in **atomic mass units (u)**.

2-4 Chemical Elements—All elements from $Z = 1$ to $Z = 112$ have been characterized and all but element 112 have been given a name and **chemical symbol**. Knowledge of the several elements following $Z = 112$ is more tenuous. **Nuclide** is the term used to describe an atom with a particular atomic number and a particular mass number. Atoms of the same element that differ in mass number are called **isotopes**. The **percent natural abundance** of an isotope and the precise mass of its atoms can be established with a **mass spectrometer** (Fig. 2-14). A special symbolism (expression 2.2) is used to represent the composition of an atom or an **ion** derived from the atom.

2-5 Atomic Mass—The **atomic mass (weight)** of an element is a weighted average based on an assigned value of exactly 12 u for the isotope carbon-12. This weighted average is calculated from the experimentally determined atomic masses and percent abundances of the naturally occurring isotopes of the element through expression (2.3).

2-6 Introduction to the Periodic Table—The **periodic table** (Fig. 2-15) is an arrangement of the elements in horizontal rows called **periods** and vertical columns called **groups** or **families**. Each group consists of elements with similar physical and chemical properties. The elements can also be subdivided into broad categories. One categorization is that of **metals, nonmetals, metalloids**, and **noble gases**. Another is that of **main-group elements** and **transition elements (transition metals)**. Included among the transition elements are the two subcategories **lanthanides** and **actinides**. The table has many uses, as will be seen throughout the text. Emphasis in this chapter is on the periodic table as an aid in writing symbols for simple ions.

2-7 The Concept of the Mole and Avogadro Constant—The **Avogadro constant**, $N_A = 6.02214 \times 10^{23}$ mol^{-1}, represents the number of carbon-12 atoms in *exactly* 12 g of carbon-12. More generally, it is the number of elementary entities (for example, atoms or molecules) present in an amount known as one **mole** of substance. The mass of one mole of atoms of an element is called its **molar mass, M**.

2-8 Using the Mole Concept in Calculations—Molar mass and the Avogadro constant are used in a variety of calculations involving the mass, amount (in moles), and number of atoms in a sample of an element. Other conversion factors may also be involved in these calculations. The mole concept is encountered in ever broader contexts throughout the text.

Integrative Example

A stainless steel ball bearing has a radius of 6.35 mm and a density of 7.75 g/cm^3. Iron is the principal element in steel. Carbon is a key minor element. The ball bearing contains 0.25% carbon, by mass. Given that the percent natural abundance of ^{13}C is 1.108%, how many ^{13}C atoms are present in the ball bearing?

Analyze

The goal is to determine the number of carbon-13 atoms found in a ball bearing with a particular composition. The critical point in this problem is recognizing that we can relate number of atoms to mass by using molar mass and Avogadro's constant. The first step is to use the radius of the ball bearing to determine its volume. The second step is to determine the mass of carbon present by using the density of steel along with the percent composition. The third step uses the molar mass of carbon to convert grams of carbon to moles of carbon; Avogadro's constant is then used to convert moles of carbon to the number of carbon atoms. In the final step, the natural abundance of carbon-13 atoms is used to find the number of carbon-13 atoms in the total number of carbon atoms in the ball bearing.

Solve

The ball-bearing volume in cubic centimeters is found by applying the formula for the volume of a sphere, $V = 4/3\,\pi r^3$. Remember to convert the given radius from millimeters to centimeters, so that the volume will be in cubic centimeters.

$$V = \frac{4\pi}{3}\left[6.35\text{ mm} \times \frac{1\text{ cm}}{10\text{ mm}}\right]^3 = 1.07\text{ cm}^3$$

The product of the volume of the ball bearing and the density of steel equals the mass. The mass of the ball bearing multiplied by the percent carbon in the steel gives the mass of carbon present.

$$?\text{ g C} = 1.07\text{ cm}^3 \times \frac{7.75\text{ g steel}}{1\text{ cm}^3\text{ steel}} \times \frac{0.25\text{ g C}}{100\text{ g steel}} = 0.021\text{ g C}$$

The mass of carbon is first converted to moles of carbon by using the inverse of the molar mass of carbon. Avogadro's constant is then used to convert moles of carbon to atoms of carbon.

$$?\text{ C atoms} = 0.021\text{ g C} \times \frac{1\text{ mol C}}{12.011\text{ g C}} \times \frac{6.022 \times 10^{23}\text{ C atoms}}{1\text{ mol C}}$$
$$= 1.1 \times 10^{21}\text{ C atoms}$$

The number of ^{13}C atoms is determined by using the percent natural abundance of carbon-13.

$$?\ ^{13}\text{C atoms} = 1.1 \times 10^{21}\text{ C atoms} \times \frac{1.108\ ^{13}\text{C atoms}}{100\text{ C atoms}}$$
$$= 1.2 \times 10^{19}\ ^{13}\text{C atoms}$$

Assess

The number of carbon-13 atoms is smaller than the number of carbon atoms, which it should be, given that the natural abundance of carbon-13 is just 1.108%. To avoid mistakes, every quantity should be clearly labeled with its appropriate unit so that units cancel properly. Two points made by this problem are, first, that the relatively small ball bearing contains a large number of carbon-13 atoms even though carbon-13's abundance is only 1.108% of all carbon atoms. Second, the size of any atom must be very small.

PRACTICE EXAMPLE A: Calculate the number of ^{63}Cu atoms in a cubic crystal of copper that measures 25 nm on edge. The density of copper is 8.92 g/cm^3 and the percent natural abundance of ^{63}Cu is 69.17%.

PRACTICE EXAMPLE B: The United States Food and Drug Administration (USFDA) suggests a daily value of 18 mg Fe for adults and for children over four years of age. The label on a particular brand of cereal states that one serving (55 g) of dry cereal contains 45% of the daily value of Fe. Given that the percent natural abundance of ^{58}Fe is 0.282%, how many full servings of dry cereal must be eaten to consume exactly one mole of ^{58}Fe? The atomic weight of ^{58}Fe is 57.9333 u. Is it possible for a person to consume this much cereal in a lifetime, assuming that one full serving of cereal is eaten every day?

Mastering**CHEMISTRY**

You'll find a link to additional self study questions in the study area on www.masteringchemistry.com

Exercises

Law of Conservation of Mass

1. When an iron object rusts, its mass increases. When a match burns, its mass decreases. Do these observations violate the law of conservation of mass? Explain.

2. When a strip of magnesium metal is burned in air (recall Figure 2-1), it produces a white powder that weighs more than the original metal. When a strip of magnesium is burned in a flashbulb, the bulb weighs the same before and after it is flashed. Explain the difference in these observations.

3. A 0.406 g sample of magnesium reacts with oxygen, producing 0.674 g of magnesium oxide as the only product. What mass of oxygen was consumed in the reaction?

4. A 1.446 g sample of potassium reacts with 8.178 g of chlorine to produce potassium chloride as the only product. After the reaction, 6.867 g of chlorine remains unreacted. What mass of potassium chloride was formed?

5. When a solid mixture consisting of 10.500 g calcium hydroxide and 11.125 g ammonium chloride is strongly heated, gaseous products are evolved and 14.336 g of a solid residue remains. The gases are passed into 62.316 g water, and the mass of the resulting solution is 69.605 g. Within the limits of experimental error, show that these data conform to the law of conservation of mass.

6. Within the limits of experimental error, show that the law of conservation of mass was obeyed in the following experiment: 10.00 g calcium carbonate (found in limestone) was dissolved in 100.0 mL hydrochloric acid ($d = 1.148$ g/mL). The products were 120.40 g solution (a mixture of hydrochloric acid and calcium chloride) and 2.22 L carbon dioxide gas ($d = 1.9769$ g/L).

Law of Constant Composition

7. In Example 2-1, we established that the mass ratio of magnesium to magnesium oxide is 0.455 g magnesium/ 0.755 g magnesium oxide.
   (a) What is the ratio of oxygen to magnesium oxide, by mass?
   (b) What is the mass ratio of oxygen to magnesium in magnesium oxide?
   (c) What is the percent by mass of magnesium in magnesium oxide?
8. Samples of pure carbon weighing 3.62, 5.91, and 7.07 g were burned in an excess of air. The masses of carbon dioxide obtained (the sole product in each case) were 13.26, 21.66, and 25.91 g, respectively.
   (a) Do these data establish that carbon dioxide has a fixed composition?
   (b) What is the composition of carbon dioxide, expressed in % C and % O, by mass?
9. In one experiment, 2.18 g sodium was allowed to react with 16.12 g chlorine. All the sodium was used up, and 5.54 g sodium chloride (salt) was produced. In a second experiment, 2.10 g chlorine was allowed to react with 10.00 g sodium. All the chlorine was used up, and 3.46 g sodium chloride was produced. Show that these results are consistent with the law of constant composition.
10. When 3.06 g hydrogen was allowed to react with an excess of oxygen, 27.35 g water was obtained. In a second experiment, a sample of water was decomposed by electrolysis, resulting in 1.45 g hydrogen and 11.51 g oxygen. Are these results consistent with the law of constant composition? Demonstrate why or why not.
11. In one experiment, the burning of 0.312 g sulfur produced 0.623 g sulfur dioxide as the sole product of the reaction. In a second experiment, 0.842 g sulfur dioxide was obtained. What mass of sulfur must have been burned in the second experiment?
12. In one experiment, the reaction of 1.00 g mercury and an excess of sulfur yielded 1.16 g of a sulfide of mercury as the sole product. In a second experiment, the same sulfide was produced in the reaction of 1.50 g mercury and 1.00 g sulfur.
   (a) What mass of the sulfide of mercury was produced in the second experiment?
   (b) What mass of which element (mercury or sulfur) remained *unreacted* in the second experiment?

Law of Multiple Proportions

13. Sulfur forms two compounds with oxygen. In the first compound, 1.000 g sulfur is combined with 0.998 g oxygen, and in the second, 1.000 g sulfur is combined with 1.497 g oxygen. Show that these results are consistent with Dalton's law of multiple proportions.
14. Phosphorus forms two compounds with chlorine. In the first compound, 1.000 g of phosphorus is combined with 3.433 g chlorine, and in the second, 2.500 g phosphorus is combined with 14.308 g chlorine. Show that these results are consistent with Dalton's law of multiple proportions.
15. The following data were obtained for compounds of nitrogen and hydrogen:

Compound	Mass of Nitrogen, g	Mass of Hydrogen, g
A	0.500	0.108
B	1.000	0.0720
C	0.750	0.108

   (a) Show that these data are consistent with the law of multiple proportions.
   (b) If the formula of compound B is N_2H_2, what are the formulas of compounds A and C?

16. The following data were obtained for compounds of iodine and fluorine:

Compound	Mass of Iodine, g	Mass of Fluorine, g
A	1.000	0.1497
B	0.500	0.2246
C	0.750	0.5614
D	1.000	1.0480

   (a) Show that these data are consistent with the law of multiple proportions.
   (b) If the formula for compound A is IF, what are the formulas for compounds B, C, and D?
17. There are two oxides of copper. One oxide has 20% oxygen, by mass. The second oxide has a *smaller* percent of oxygen than the first. What is the probable percent of oxygen in the second oxide?
18. The two oxides of carbon described on page 38 were CO and CO_2. Another oxide of carbon has 1.106 g of oxygen in a 2.350 g sample. In what ratio are carbon and oxygen atoms combined in molecules of this third oxide? Explain.

Fundamental Charges and Mass-to-Charge Ratios

19. The following observations were made for a series of five oil drops in an experiment similar to Millikan's (see Figure 2-8). Drop 1 carried a charge of 1.28×10^{-18} C; drops 2 and 3 each carried $\frac{1}{2}$ the charge of drop 1; drop 4 carried $\frac{1}{8}$ the charge of drop 1; drop 5 had a charge four times that of drop 1. Are these data consistent with the value of the electronic charge given in the text? Could Millikan have inferred the charge on the electron from this particular series of data? Explain.
20. In an experiment similar to that described in Exercise 19, drop 1 carried a charge of 6.41×10^{-19} C; drop 2 had $\frac{1}{2}$ the charge of drop 1; drop 3 had twice the charge

of drop 1; drop 4 had a charge of 1.44×10^{-18} C; and drop 5 had $\frac{1}{3}$ the charge of drop 4. Are these data consistent with the value of the electronic charge given in the text? Could Millikan have inferred the charge on the electron from this particular series of data? Explain.

21. Use data from Table 2.1 to verify that
    (a) the mass of electrons is about 1/2000 that of H atoms;

Atomic Number, Mass Number, and Isotopes

23. The following radioactive isotopes have applications in medicine. Write their symbols in the form $_{Z}^{A}E$. (a) cobalt-60; (b) phosphorus-32; (c) iron-59; (d) radium-226.
24. For the isotope ^{202}Hg, express the percentage of the fundamental particles in the nucleus that are neutrons.

(b) the mass-to-charge ratio (m/e) for positive ions is considerably larger than that for electrons.

22. Determine the approximate value of m/e in grams per coulomb for the ions $_{53}^{127}I^{-}$ and $_{16}^{32}S^{2-}$. Why are these values only approximate?

25. Complete the following table. What minimum amount of information is required to completely characterize an atom or ion?
[*Hint:* Not all rows can be completed.]

Name	Symbol	Number Protons	Number Electrons	Number Neutrons	Mass Number
Sodium	$_{11}^{23}Na$	11	11	12	23
Silicon	—	—	—	14	—
—	—	37	—	—	85
—	^{40}K	—	—	—	—
—	—	—	33	42	—
—	$^{20}Ne^{2+}$	—	—	—	—
—	—	—	—	—	80
—	—	—	—	126	—

26. Arrange the following species in order of increasing (a) number of electrons; (b) number of neutrons; (c) mass.

$_{50}^{112}Sn$ $_{18}^{40}Ar$ $_{52}^{122}Te$ $_{29}^{59}Cu$ $_{48}^{120}Cd$ $_{27}^{58}Co$ $_{19}^{39}K$

27. For the atom ^{108}Pd with mass 107.90389 u, determine
    (a) the numbers of protons, neutrons, and electrons in the atom;
    (b) the ratio of the mass of this atom to that of an atom of $_{6}^{12}C$.
28. For the ion $^{228}Ra^{2+}$ with a mass of 228.030 u, determine
    (a) the numbers of protons, neutrons, and electrons in the ion;
    (b) the ratio of the mass of this ion to that of an atom of ^{16}O (refer to page 47).
29. An isotope of silver has a mass that is 6.68374 times that of oxygen-16. What is the mass in u of this isotope? (Refer to page 47.)
30. The ratio of the masses of the two naturally occurring isotopes of indium is 1.0177:1. The heavier of the two isotopes has 7.1838 times the mass of ^{16}O. What are the masses in u of the two isotopes? (Refer to page 47.)
31. The following data on isotopic masses are from a chemical handbook. What is the ratio of each of these masses to that of $_{6}^{12}C$? (a) $_{17}^{35}Cl$, 34.96885 u; (b) $_{12}^{26}Mg$, 25.98259 u; (c) $_{86}^{222}Rn$, 222.0175 u.
32. The following ratios of masses were obtained with a mass spectrometer: $_{9}^{19}F/_{6}^{12}C = 1.5832$; $_{17}^{35}Cl/_{9}^{19}F = 1.8406$; $_{35}^{81}Br/_{17}^{35}Cl = 2.3140$. Determine the mass of a $_{35}^{81}Br$ atom in amu.

33. Which of the following species has
    (a) equal numbers of neutrons and electrons;
    (b) protons, neutrons, and electrons in the ratio 9:11:8;
    (c) a number of neutrons equal to the number of protons plus one-half the number of electrons?

$^{24}Mg^{2+}$, ^{47}Cr, $^{60}Co^{3+}$, $^{35}Cl^{-}$, $^{124}Sn^{2+}$, ^{226}Th, ^{90}Sr

34. Given the same species as listed in Exercise 33, which has
    (a) equal numbers of neutrons and protons;
    (b) protons contributing more than 50% of the mass;
    (c) about 50% more neutrons than protons?
35. An isotope with mass number 44 has four more neutrons than protons. This is an isotope of what element?
36. Identify the isotope X that has one more neutron than protons and a mass number equal to nine times the charge on the ion X^{3+}.
37. Iodine has many radioactive isotopes. Iodine-123 is a radioactive isotope used for obtaining images of the thyroid gland. Iodine-123 is administered to patients in the form of sodium iodide capsules that contain $^{123}I^{-}$ ions. Determine the number of neutrons, protons, and electrons in a single $^{123}I^{-}$ ion.
38. Iodine-131 is a radioactive isotope that has important medical uses. Small doses of iodine-131 are used for treating hyperthyroidism (overactive thyroid) and larger doses are used for treating thyroid cancer. Iodine-131 is administered to patients in the form of sodium iodide capsules that contain $^{131}I^{-}$ ions. Determine the number of neutrons, protons, and electrons in a single $^{131}I^{-}$ ion.

39. Americium-241 is a radioactive isotope that is used in high-precision gas and smoke detectors. How many neutrons, protons, and electrons are there in an atom of americium-241?

40. Some foods are made safer to eat by being exposed to gamma rays from radioactive isotopes, such as cobalt-60. The energy from the gamma rays kills bacteria in the food. How many neutrons, protons, and electrons are there in an atom of cobalt-60?

Atomic Mass Units, Atomic Masses

41. Which statement is probably true concerning the masses of *individual* chlorine atoms: *All have*, *some have*, or *none has* a mass of 35.4527 u? Explain.

42. The mass of a carbon-12 atom is taken to be exactly 12 u. Are there likely to be any other atoms with an *exact* integral (whole number) mass, expressed in u? Explain.

43. There are three naturally occurring isotopes of magnesium. Their masses and percent natural abundances are 23.985042 u, 78.99%; 24.985837 u, 10.00%; and 25.982593 u, 11.01%. Calculate the weighted-average atomic mass of magnesium.

44. There are four naturally occurring isotopes of chromium. Their masses and percent natural abundances are 49.9461 u, 4.35%; 51.9405 u, 83.79%; 52.9407 u, 9.50%; and 53.9389 u, 2.36%. Calculate the weighted-average atomic mass of chromium.

45. The two naturally occurring isotopes of silver have the following abundances: ^{107}Ag, 51.84%; ^{109}Ag,

48.16%. The mass of ^{107}Ag is 106.905092 u. What is the mass of ^{109}Ag?

46. Bromine has two naturally occurring isotopes. One of them, bromine-79, has a mass of 78.918336 u and a natural abundance of 50.69%. What must be the mass and percent natural abundance of the other isotope, bromine-81?

47. The three naturally occurring isotopes of potassium are ^{39}K, 38.963707 u; ^{40}K, 39.963999 u; and ^{41}K. The percent natural abundances of ^{39}K and ^{41}K are 93.2581% and 6.7302%, respectively. Determine the isotopic mass of ^{41}K.

48. What are the percent natural abundances of the two naturally occurring isotopes of boron, ^{10}B and ^{11}B? These isotopes have masses of 10.012937 u and 11.009305 u, respectively.

Mass Spectrometry

49. A mass spectrum of germanium displayed peaks at mass numbers 70, 72, 73, 74, and 76, with relative heights of 20.5, 27.4, 7.8, 36.5, and 7.8, respectively.
    (a) In the manner of Figure 2-14, sketch this mass spectrum.
    (b) Estimate the weighted-average atomic mass of germanium, and state why this result is only approximately correct.

50. Hydrogen and chlorine atoms react to form simple diatomic molecules in a 1:1 ratio, that is, HCl. The

natural abundances of the chlorine isotopes are 75.77% ^{35}Cl and 24.23% ^{37}Cl. The natural abundances of ^{2}H and ^{3}H are 0.015% and less than 0.001%, respectively.
    (a) How many different HCl molecules are possible, and what are their mass numbers (that is, the sum of the mass numbers of the H and Cl atoms)?
    (b) Which is the most abundant of the possible HCl molecules? Which is the second most abundant?

The Periodic Table

51. Refer to the periodic table inside the front cover and identify
    (a) the element that is in group 14 and the fourth period
    (b) one element similar to and one unlike sulfur
    (c) the alkali metal in the fifth period
    (d) the halogen element in the sixth period

52. Refer to the periodic table inside the front cover and identify
    (a) the element that is in group 11 and the sixth period
    (b) an element with atomic number greater than 50 that has properties similar to the element with atomic number 18

    (c) the group number of an element E that forms an ion E^{2-}
    (d) an element M that you would expect to form the ion M^{3+}

53. Assuming that the seventh period of the periodic table has 32 members, what should be the atomic number of **(a)** the noble gas following radon (Rn); **(b)** the alkali metal following francium (Fr)?

54. Find the several pairs of elements that are "out of order" in terms of increasing atomic mass and explain why the reverse order is necessary.

The Avogadro Constant and the Mole

55. What is the total number of atoms in **(a)** 15.8 mol Fe; **(b)** 0.000467 mol Ag; **(c)** 8.5×10^{-11} mol Na?

56. *Without doing detailed calculations*, indicate which of the following quantities contains the greatest number of atoms: 6.022×10^{23} Ni atoms, 25.0 g nitrogen,

52.0 g Cr, 10.0 cm^3 Fe ($d = 7.86$ g/cm^3). Explain your reasoning.

57. Determine
    (a) the number of moles of Zn in a 415.0 g sample of zinc metal

(b) the number of Cr atoms in 147.4 kg chromium

(c) the mass of a one-trillion-atom (1.0×10^{12}) sample of metallic gold

(d) the mass of one fluorine atom

58. Determine

(a) the number of Kr atoms in a 5.25-mg sample of krypton

(b) the molar mass, M, and identity of an element if the mass of a 2.80×10^{22}-atom sample of the element is 2.09 g

(c) the mass of a sample of phosphorus that contains the same number of atoms as 44.75 g of magnesium

59. How many Cu atoms are present in a piece of sterling-silver jewelry weighing 33.24 g? (Sterling silver is a silver–copper alloy containing 92.5% Ag by mass.)

60. How many atoms are present in a 75.0 cm³ sample of plumber's solder, a lead–tin alloy containing 67% Pb by mass and having a density of 9.4 g/cm³?

61. How many ^{204}Pb atoms are present in a piece of lead weighing 215 mg? The percent natural abundance of ^{204}Pb is 1.4%.

62. A particular lead–cadmium alloy is 8.0% cadmium by mass. What mass of this alloy, in grams, must you weigh out to obtain a sample containing 6.50×10^{23} Cd atoms?

63. Medical experts generally believe a level of 30 μg Pb per deciliter of blood poses a significant health risk

(1 dL = 0.1 L). Express this level (a) in the unit mol Pb/L blood; (b) as the number of Pb atoms per milliliter blood.

64. During a severe episode of air pollution, the concentration of lead in the air was observed to be 3.01 μg Pb/m³. How many Pb atoms would be present in a 0.500 L sample of this air (the approximate lung capacity of a human adult)?

65. *Without doing detailed calculations*, determine which of the following samples has the greatest number of atoms:

(a) a cube of iron with a length of 10.0 cm $(d = 7.86 \text{ g/cm}^3)$

(b) 1.00 kg of hydrogen contained in a 10,000 L balloon

(c) a mound of sulfur weighing 20.0 kg

(d) a 76 lb sample of liquid mercury $(d = 13.5 \text{ g/mL})$

66. *Without doing detailed calculations*, determine which of the following samples occupies the largest volume:

(a) 25.5 mol of sodium metal $(d = 0.971 \text{ g/cm}^3)$

(b) 0.725 L of liquid bromine $(d = 3.12 \text{ g/mL})$

(c) 1.25×10^{25} atoms of chromium metal $(d = 9.4 \text{ g/cm}^3)$

(d) 2.15 kg of plumber's solder $(d = 9.4 \text{ g/cm}^3)$, a lead–tin alloy with a 2:1 atom ratio of lead to tin

Integrative and Advanced Exercises

67. A solution was prepared by dissolving 2.50 g potassium chlorate (a substance used in fireworks and flares) in 100.0 mL water at 40 °C. When the solution was cooled to 20 °C, its volume was still found to be 100.0 mL, but some of the potassium chlorate had crystallized (deposited from the solution as a solid). At 40 °C, the density of water is 0.9922 g/mL, and at 20 °C, the potassium chlorate solution had a density of 1.0085 g/mL.

(a) Estimate, to two significant figures, the mass of potassium perchlorate that crystallized.

(b) Why can't the answer in (a) be given more precisely?

68. William Prout (1815) proposed that all other atoms are built up of hydrogen atoms, suggesting that all elements should have integral atomic masses based on an atomic mass of one for hydrogen. This hypothesis appeared discredited by the discovery of atomic masses, such as 24.3 u for magnesium and 35.5 u for chlorine. In terms of modern knowledge, explain why Prout's hypothesis is actually quite reasonable.

69. Fluorine has a single atomic species, ^{19}F. Determine the atomic mass of ^{19}F by summing the masses of its protons, neutrons, and electrons, and compare your results with the value listed on the inside front cover. Explain why the agreement is poor.

70. Use 1×10^{-13} cm as the approximate diameter of the spherical nucleus of the hydrogen-1 atom, together with data from Table 2.1, to estimate the density of matter in a proton.

71. Use fundamental definitions and statements from Chapters 1 and 2 to establish the fact that 6.022×10^{23} u = 1.000 g.

72. In each case, identify the element in question.

(a) The mass number of an atom is 234 and the atom has 60.0% more neutrons than protons.

(b) An ion with a 2+ charge has 10.0% more protons than electrons.

(c) An ion with a mass number of 110 and a 2+ charge has 25.0% more neutrons than electrons.

73. Determine the only possible 2+ ion for which the following two conditions are both satisfied:

• The net ionic charge is *one-tenth* the nuclear charge.

• The number of neutrons is *four* more than the number of electrons.

74. Determine the only possible isotope (E) for which the following conditions are met:

• The mass number of E is 2.50 times its atomic number.

• The atomic number of E is equal to the mass number of another isotope (Y). In turn, isotope Y has a neutron number that is 1.33 times the atomic number of Y and equal to the neutron number of selenium-82.

75. Suppose we redefined the atomic mass scale by arbitrarily assigning to the naturally occurring *mixture* of chlorine isotopes an atomic mass of 35.00000 u.

(a) What would be the atomic masses of helium, sodium, and iodine on this new atomic mass scale?

(b) Why do these three elements have nearly integral (whole-number) atomic masses based on carbon-12, but not based on naturally occurring chlorine?

76. The two naturally occurring isotopes of nitrogen have masses of 14.0031 and 15.0001 u, respectively. Determine the percentage of ^{15}N atoms in naturally occurring nitrogen.

77. The masses of the naturally occurring mercury isotopes are ^{196}Hg, 195.9658 u; ^{198}Hg, 197.9668 u; ^{199}Hg, 198.9683 u; ^{200}Hg, 199.9683 u; ^{201}Hg, 200.9703 u; ^{202}Hg, 201.9706 u; and ^{204}Hg, 203.9735 u. Use these data, together with data from Figure 2-14, to calculate the weighted-average atomic mass of mercury.

78. Germanium has three major naturally occurring isotopes: ^{70}Ge (69.92425 u, 20.85%), ^{72}Ge (71.92208 u, 27.54%), ^{74}Ge (73.92118 u, 36.29%). There are also two minor isotopes: ^{73}Ge (72.92346 u) and ^{76}Ge (75.92140 u). Calculate the percent natural abundances of the two minor isotopes. Comment on the precision of these calculations.

79. From the densities of the lines in the mass spectrum of krypton gas, the following observations were made:
    - Somewhat more than 50% of the atoms were krypton-84.
    - The numbers of krypton-82 and krypton-83 atoms were essentially equal.
    - The number of krypton-86 atoms was 1.50 times as great as the number of krypton-82 atoms.
    - The number of krypton-80 atoms was 19.6% of the number of krypton-82 atoms.
    - The number of krypton-78 atoms was 3.0% of the number of krypton-82 atoms.

    The masses of the isotopes are

    ^{78}Kr, 77.9204 u ^{80}Kr, 79.9164 u ^{82}Kr, 81.9135 u
    ^{83}Kr, 82.9141 u ^{84}Kr, 83.9115 u ^{86}Kr, 85.9106 u

    The weighted-average atomic mass of Kr is 83.80. Use these data to calculate the percent natural abundances of the krypton isotopes.

80. The two naturally occurring isotopes of chlorine are ^{35}Cl (34.9689 u, 75.77%) and ^{37}Cl (36.9658 u, 24.23%). The two naturally occurring isotopes of bromine are ^{79}Br (78.9183 u, 50.69%) and ^{81}Br (80.9163 u, 49.31%). Chlorine and bromine combine to form bromine monochloride, BrCl. Sketch a mass spectrum for BrCl with the relative number of molecules plotted against molecular mass (similar to Figure 2-14).

81. How many atoms are present in a 1.00 m length of 20-gauge copper wire? A 20-gauge wire has a diameter of 0.03196 in., and the density of copper is 8.92 g/cm^3.

82. Monel metal is a corrosion-resistant copper–nickel alloy used in the electronics industry. A particular alloy with a density of 8.80 g/cm^3 and containing 0.022% Si by mass is used to make a rectangular plate 15.0 cm long, 12.5 cm wide, 3.00 mm thick, and has a 2.50 cm diameter hole drilled through its center. How many silicon-30 atoms are found in this plate? The mass of a silicon-30 atom is 29.97376 u, and the percent natural abundance of silicon-30 is 3.10%.

83. Deuterium, ^2H (2.0140 u), is sometimes used to replace the principal hydrogen isotope ^1H in chemical studies. The percent natural abundance of deuterium is 0.015%. If it can be done with 100% efficiency, what mass of naturally occurring hydrogen gas would have to be processed to obtain a sample containing 2.50×10^{21} ^2H atoms?

84. An alloy that melts at about the boiling point of water has Bi, Pb, and Sn atoms in the ratio 10:6:5, respectively. What mass of alloy contains a total of one mole of atoms?

85. A particular silver solder (used in the electronics industry to join electrical components) is to have the *atom* ratio of 5.00 Ag/4.00 Cu/1.00 Zn. What masses of the three metals must be melted together to prepare 1.00 kg of the solder?

86. A low-melting Sn–Pb–Cd alloy called *eutectic alloy* is analyzed. The *mole* ratio of tin to lead is 2.73:1.00, and the *mass* ratio of lead to cadmium is 1.78:1.00. What is the mass percent composition of this alloy?

87. In an experiment, 125 cm^3 of zinc and 125 cm^3 of iodine are mixed together and the iodine is completely converted to 164 cm^3 of zinc iodide. What volume of zinc remains unreacted? The densities of zinc, iodine, and zinc iodide are 7.13 g/cm^3, 4.93 g/cm^3, and 4.74 g/cm^3, respectively.

88. Atoms are spherical and so when silver atoms pack together to form silver metal, they cannot fill all the available space. In a sample of silver metal, approximately 26.0% of the sample is empty space. Given that the density of silver metal is 10.5 g/cm^3, what is the radius of a silver atom? Express your answer in picometers.

Feature Problems

89. The data Lavoisier obtained in the experiment described on page 35 are as follows:

    Before heating: glass vessel + tin + air

    = 13 onces, 2 gros, 2.50 grains

    After heating: glass vessel + tin calx + remaining air

    = 13 onces, 2 gros, 5.62 grains

    How closely did Lavoisier's results conform to the law of conservation of mass? (1 livre = 16 onces; 1 once = 8 gros; 1 gros = 72 grains. In modern terms, 1 livre = 30.59 g.)

90. Some of Millikan's oil-drop data are shown on the next page. The measured quantities were not actual charges on oil drops but were proportional to these charges.

Show that these data are consistent with the idea of a fundamental electronic charge.

Observation	Measured Quantity	Observation	Measured Quantity
1	19.66	8	53.91
2	24.60	9	59.12
3	29.62	10	63.68
4	34.47	11	68.65
5	39.38	12	78.34
6	44.42	13	83.22
7	49.41		

91. Before 1961, the standard for atomic masses was the isotope ^{16}O, to which physicists assigned a value of exactly 16. At the same time, chemists assigned a value of exactly 16 to the naturally occurring mixture of the isotopes ^{16}O, ^{17}O, and ^{18}O. Would you expect atomic masses listed in a 60-year-old text to be the same, generally higher, or generally lower than in this text? Explain.

92. The German chemist Fritz Haber proposed paying off the reparations imposed against Germany after World War I by extracting gold from seawater. Given that (1) the amount of the reparations was $28.8 billion dollars, (2) the value of gold at the time was about $21.25 per troy ounce (1 troy ounce = 31.103 g), and (3) gold occurs in seawater to the extent of 4.67×10^{17} atoms per ton of seawater (1 ton = 2000 lb), how many cubic kilometers of seawater would have had to be processed to obtain the required amount of gold? Assume that the density of seawater is 1.03 g/cm^3.

(Haber's scheme proved to be commercially infeasible, and the reparations were never fully paid.)

93. Mass spectrometry is one of the most versatile and powerful tools in chemical analysis because of its capacity to discriminate between atoms of different masses. When a sample containing a mixture of isotopes is introduced into a mass spectrometer, the ratio of the peaks observed reflects the ratio of the percent natural abundances of the isotopes. This ratio provides an internal standard from which the amount of a certain isotope present in a sample can be determined. This is accomplished by deliberately introducing a known quantity of a particular isotope into the sample to be analyzed. A comparison of the new isotope ratio to the first ratio allows the determination of the amount of the isotope present in the original sample.

An analysis was done on a rock sample to determine its rubidium content. The rubidium content of a portion of rock weighing 0.350 g was extracted, and to the extracted sample was added an additional 29.45 μg of ^{87}Rb. The mass spectrum of this spiked sample showed a ^{87}Rb peak that was 1.12 times as high as the peak for ^{85}Rb. Assuming that the two isotopes react identically, what is the Rb content of the rock (expressed in parts per million by mass)? The natural abundances and isotopic masses are shown in the table.

Isotope	% Natural Abundance	Atomic Mass, u
^{87}Rb	27.83	86.909
^{85}Rb	72.17	84.912

Self-Assessment Exercises

94. In your own words, define or explain these terms or symbols: (a) $^A_Z E$; (b) β particle; (c) isotope; (d) ^{16}O; (e) molar mass.

95. Briefly describe
 (a) the law of conservation of mass
 (b) Rutherford's nuclear atom
 (c) weighted-average atomic mass
 (d) a mass spectrum

96. Explain the important distinctions between each pair of terms:
 (a) cathode rays and X-rays
 (b) protons and neutrons
 (c) nuclear charge and ionic charge
 (d) periods and groups of the periodic table
 (e) metal and nonmetal
 (f) the Avogadro constant and the mole

97. When 10.0 g zinc and 8.0 g sulfur are allowed to react, all the zinc is consumed, 14.9 g zinc sulfide is produced, and the mass of unreacted sulfur remaining is
 (a) 2.0 g
 (b) 3.1 g
 (c) 4.9 g
 (d) impossible to predict from this information alone

98. One oxide of rubidium has 0.187 g O per gram of Rb. A possible O:Rb mass ratio for a second oxide of rubidium is (a) 16:85.5; (b) 8:42.7; (c) 1:2.674; (d) any of these.

99. Cathode rays
 (a) may be positively or negatively charged
 (b) are a form of electromagnetic radiation similar to visible light
 (c) have properties identical to β particles
 (d) have masses that depend on the cathode that emits them

100. The scattering of α particles by thin metal foils established that
 (a) the mass of an atom is concentrated in a positively charged nucleus
 (b) electrons are fundamental particles of all matter
 (c) all electrons carry the same charge
 (d) atoms are electrically neutral

101. Which of the following have the same charge and approximately the same mass?
 (a) an electron and a proton; (b) a proton and a neutron; (c) a hydrogen atom and a proton; (d) a neutron and a hydrogen atom; (e) an electron and an H$^-$ ion.

102. What is the correct symbol for the species that contains 18 neutrons, 17 protons, and 16 electrons?

103. The properties of magnesium will most resemble those of which of the following? **(a)** cesium; **(b)** sodium; **(c)** aluminum; **(d)** calcium; **(e)** manganese.

104. Which group in the main group of elements contains **(a)** no metals or metalloids? **(b)** only one metal or metalloid? **(c)** only one nonmetal? **(d)** only nonmetals?

105. The two species that have the same number of electrons as ^{32}S are **(a)** ^{32}Cl; **(b)** $^{34}S^+$; **(c)** $^{33}P^+$; **(d)** $^{28}Si^{2-}$; **(e)** $^{35}S^{2-}$; **(f)** $^{40}Ar^{2+}$; **(g)** $^{40}Ca^{2+}$.

106. To four significant figures, all of the following masses are possible for an individual titanium atom except one. The exception is **(a)** 45.95 u; **(b)** 46.95 u; **(c)** 47.87 u; **(d)** 47.95 u; **(e)** 48.95 u; **(f)** 49.94 u.

107. The mass of the isotope $^{84}_{36}Xe$ is 83.9115 u. If the atomic mass scale were redefined so that $^{84}_{36}Xe = 84$ u, *exactly*, the mass of the $^{12}_{6}C$ isotope would be **(a)** 11.9115 u; **(b)** 11.9874 u; **(c)** 12 u exactly; **(d)** 12.0127 u; **(e)** 12.0885 u.

108. A 5.585-kg sample of iron (Fe) contains
    (a) 10.0 mol Fe
    (b) twice as many atoms as does 600.6 g C
    (c) 10 times as many atoms as does 52.00 g Cr
    (d) 6.022×10^{24} atoms

109. There are three common iron-oxygen compounds. The one with the greatest proportion of iron has one Fe atom for every O atom and the formula FeO. A second compound has 2.327 g Fe per 1.000 g O, and the third has 2.618 g Fe per 1.000 g O. What are the formulas of these other two iron-oxygen compounds?

110. The four naturally occurring isotopes of strontium have the atomic masses 83.9134 u; 85.9093 u; 86.9089 u; and 87.9056 u. The percent natural abundance of the lightest isotope is 0.56% and of the heaviest, 82.58%. Estimate the percent natural abundances of the other two. Why is this result only a rough approximation?

111. Gold is present in seawater to the extent of 0.15 mg/ton. Assume the density of the seawater is 1.03 g/mL and determine how many Au atoms could conceivably be extracted from 0.250 L of seawater (1 ton = 2.000×10^3 lb; 1 kg = 2.205 lb).

112. Appendix E describes a useful study aid known as concept mapping. Using the method presented in Appendix E, construct a concept map illustrating the different concepts in Sections 2-7 and 2-8.

3

Chemical Compounds

CONTENTS

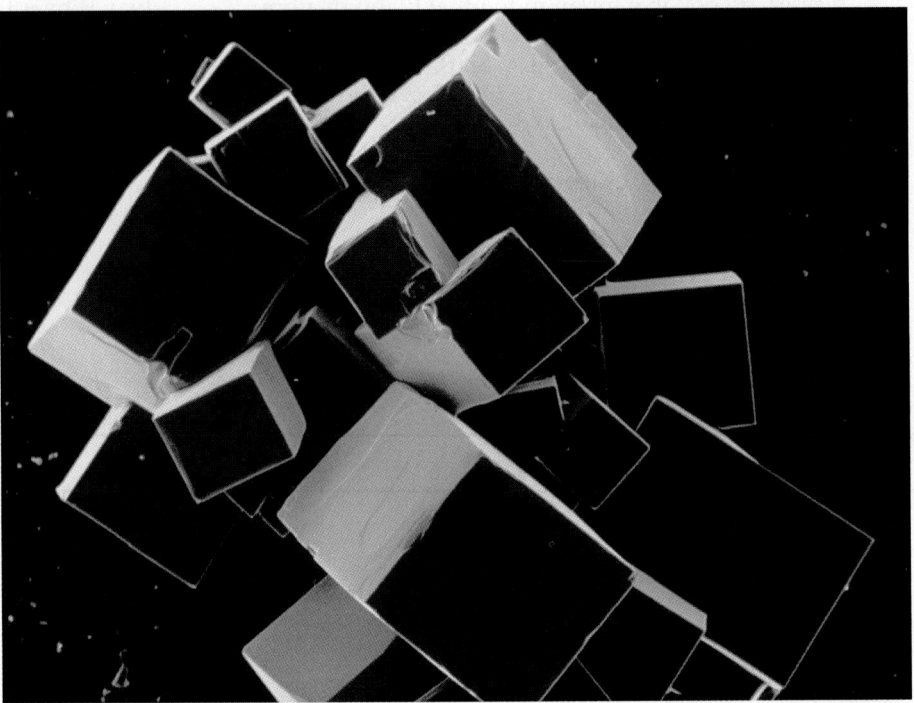

Scanning electron microscope image of sodium chloride crystals. Chemical compounds, their formulas, and their names are topics discussed in this chapter.

Water, ammonia, carbon monoxide, and carbon dioxide—all familiar substances—are rather simple chemical compounds. Only slightly less familiar are sucrose (cane sugar), acetylsalicylic acid (aspirin), and ascorbic acid (vitamin C). They too are chemical compounds. In fact, the study of chemistry is mostly about chemical compounds, and, in this chapter, we will consider a number of ideas about compounds.

The common feature of all compounds is that they are composed of two or more elements. The full range of compounds can be divided into a few broad categories by applying ideas from the periodic table of the elements. Compounds are represented by chemical formulas, which in turn are derived from the symbols of their constituent elements. In this chapter, you will learn how to deduce and write chemical formulas and how to use the information incorporated into chemical formulas. The chapter ends with an overview of the relationship between names and formulas—chemical nomenclature.

3-1   Types of Chemical Compounds and Their Formulas

Generally speaking, two fundamental kinds of chemical bonds hold together the atoms in a compound. *Covalent* bonds, which involve a sharing of electrons between atoms, give rise to molecular compounds. *Ionic* bonds, which involve a transfer of electrons from one atom to another, give rise to ionic compounds. In this section we consider only the basic features of molecular and ionic compounds that we need as background for the early chapters of the text. Our in-depth discussion of chemical bonding will come in Chapters 10 and 11.

◀ In our later study of chemical bonding, we will find that the distinction between covalent and ionic bonding is not as clear-cut as these statements imply, but we will consider this matter in Chapter 10.

Molecular Compounds

A **molecular compound** is made up of discrete units called **molecules**, which typically consist of a small number of *nonmetal* atoms held together by covalent bonds. Molecular compounds are represented by **chemical formulas**, symbolic representations that, at minimum, indicate

- the elements present
- the relative number of atoms of each element

In the formula for water, the constituent elements are denoted by their symbols. The relative numbers of atoms are indicated by *subscripts*. Where no subscript is written, the number 1 is understood.

┌─┬─ The two elements present
↓ ↓
H_2O
↑ └─ Lack of subscript means one atom of O per molecule
Two H atoms per molecule

Another example of a chemical formula is CCl_4, which represents the compound carbon tetrachloride. The formulas H_2O and CCl_4 both represent distinct entities—*molecules*. Thus, we can refer to water and carbon tetrachloride as molecular compounds.

An **empirical formula** is the simplest formula for a compound; it shows the types of atoms present and their relative numbers. The subscripts in an empirical formula are reduced to their simplest whole-number ratio. For example, P_2O_5 is the empirical formula for a compound whose molecules have the formula P_4O_{10}. Generally, the empirical formula does not tell us a great deal about a compound. Acetic acid ($C_2H_4O_2$), formaldehyde (CH_2O, used to make certain plastics and resins), and glucose ($C_6H_{12}O_6$, blood sugar) all have the empirical formula CH_2O.

A **molecular formula** is based on an actual molecule of a compound. In some cases, the empirical and molecular formulas are identical, such as CH_2O for formaldehyde. In other cases, the molecular formula is a multiple of the empirical formula. A molecule of acetic acid, for example, consists of eight atoms—two C atoms, four H atoms, and two O atoms, so the molecular formula of acetic acid is $C_2H_4O_2$. This is twice the number of atoms in the formula unit (CH_2O). Empirical and molecular formulas tell us the combining ratio of the atoms in the compound, but they show nothing about how the atoms are attached to each other. Other types of formulas, however, do convey this information. Figure 3-1 shows several representations of acetic acid, the acid constituent that gives vinegar its sour taste.

A **structural formula** shows the order in which atoms are bonded together in a molecule and by what types of bonds. Thus, the structural formula of acetic acid tells us that three of the four H atoms are bonded to one of the C atoms, and the remaining H atom is bonded to an O atom. Both of the O atoms are bonded to one of the C atoms, and the two C atoms are bonded to each other. The covalent bonds in the structural formula are represented

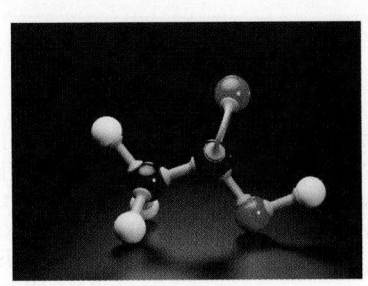

Molecular model
("ball and stick")

Empirical formula: CH_2O

Molecular formula: $C_2H_4O_2$

Structural formula:

$$\begin{array}{ccc} & H & O \\ & | & || \\ H-&C-&C-O-H \\ & | & \\ & H & \end{array}$$

Molecular model
("space filling")

▲ FIGURE 3-1
Several representations of the compound acetic acid
In the molecular model, the black spheres are carbon, the red are oxygen, and the white are hydrogen. To show that one H atom in the molecule is fundamentally different from the other three, the formula of acetic acid is often written as $HC_2H_3O_2$ (see Section 5-3). To show that this H atom is bonded to an O atom, the formulas CH_3COOH and CH_3CO_2H are also used. For a few chemical compounds, you may find different versions of chemical formulas in different sources.

by lines or dashes (—). One of the bonds is represented by a double dash (=) and is called a *double* covalent bond. Differences between single and double bonds are discussed later in the text. For now, just think of a double bond as being a stronger or tighter bond than a single bond.

A **condensed structural formula**, which is written on a single line, is an alternative, less cumbersome way of showing how the atoms of a molecule are connected. Thus, the acetic acid molecule is represented as either CH_3COOH or CH_3CO_2H. With this type of formula, the different ways in which the H atoms are attached are still apparent.

Condensed structural formulas can also be used to show how a group of atoms is attached to another atom. Consider methylpropane, C_4H_{10}, in Figure 3-2(b). The structural formula shows that there is a $—CH_3$ group of atoms attached to the central carbon atom. In the condensed structural formula, this is indicated by enclosing the CH_3 in parentheses to the right of the atom to which it is attached, thus $CH_3CH(CH_3)CH_3$. Alternatively, because the central C atom is bonded to each of the other three C atoms, we can write the condensed structural formula $CH(CH_3)_3$.

Organic compounds are made up principally of carbon and hydrogen, with oxygen and/or nitrogen as important constituents in many of them. Each carbon atom forms *four covalent bonds*. Organic compounds can be very complex, and one way of simplifying their structural formulas is to write structures without showing the C and H atoms explicitly. We do this by using a **line-angle formula** (also referred to as a *line structure*), in which lines represent chemical bonds. A carbon atom exists wherever a line ends or meets another line, and the number of H atoms needed to complete each carbon atom's four bonds are assumed to be present. The symbols of other atoms or groups of atoms and the bond lines joining them to C atoms are written explicitly. The formula of the complex male hormone molecule testosterone, seen in Figure 3-2(c), is a line-angle formula.

Molecules occupy space and have a three-dimensional shape, but empirical and molecular formulas do not convey any information about the spatial arrangements of atoms. Structural formulas can sometimes show this, but usually the only satisfactory way to represent the three-dimensional structure of molecules is with models. In a *ball-and-stick model*, atoms are represented by small balls, and the bonds between atoms by sticks (see Figure 3-1). Such models help us to visualize distances between the nuclei of atoms (bond lengths) and the geometrical shapes of molecules. Ball-and-stick models are easy to draw and interpret, but they can be somewhat misleading. Chemical bonds

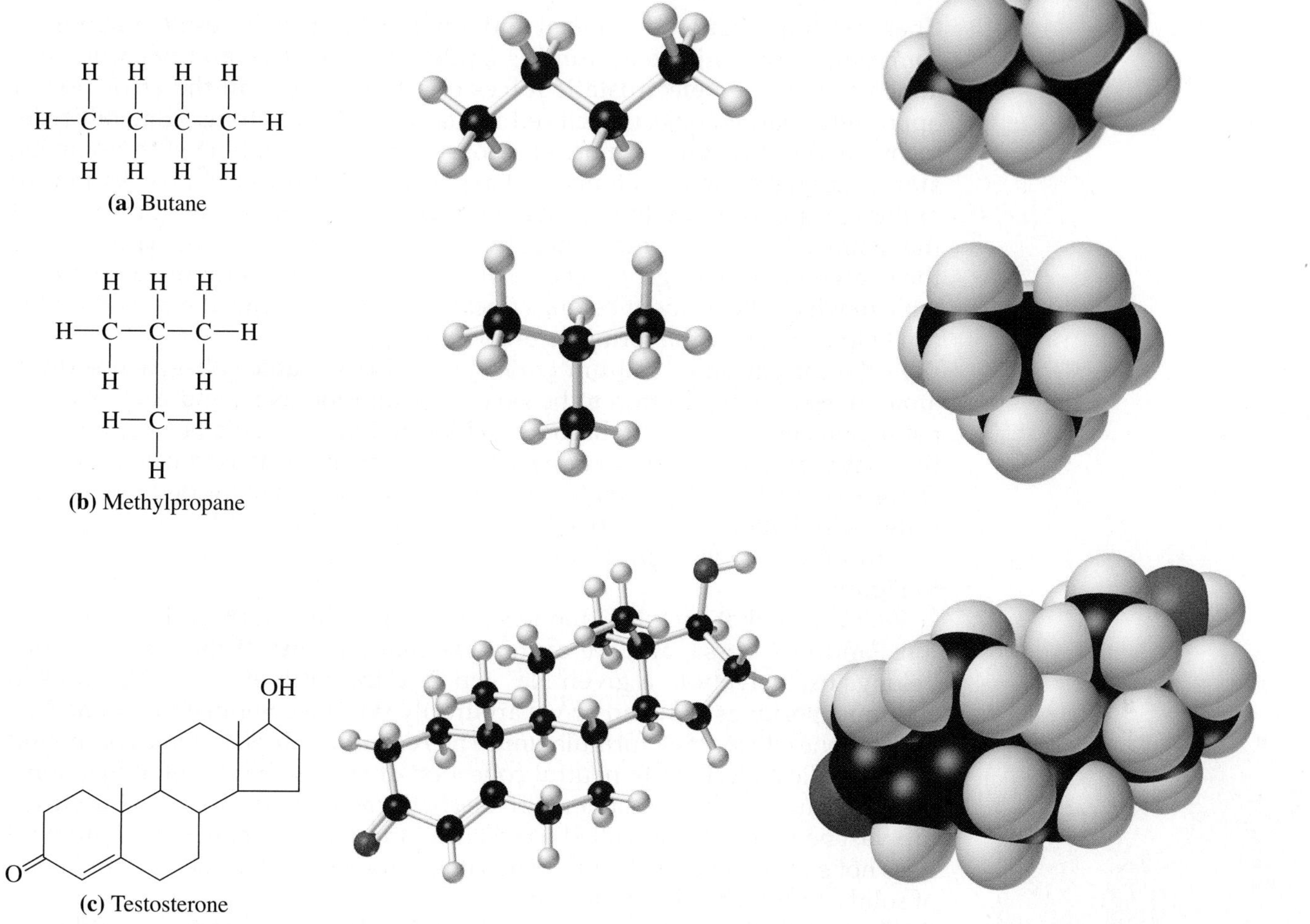

(a) Butane

(b) Methylpropane

OH

O

(c) Testosterone

▲ FIGURE 3-2
Visualizations of (a) butane, (b) methylpropane, and (c) testosterone

are forces that draw atoms in a molecule into direct contact. The atoms are not held apart as implied by a ball-and-stick model.

A *space-filling model* shows that the atoms in a molecule occupy space and that they are in actual contact with one another. Certain computer programs generate images of space-filling models such as those shown in Figures 3-1 and 3-2. A space-filling model is a more accurate representation of the size and shape of a molecule because it is constructed to scale (that is, a nanometer-size molecule is magnified to a millimeter or centimeter scale).

The acetic acid molecule is made up of three types of atoms (C, H, and O) and models of the molecule reflect this fact. Different colors are used to distinguish the various types of atoms in ball-and-stick and space-filling models (see Fig. 3-3). You will notice that the colored spheres are of different sizes, which correspond to the size differences between the various atoms in the periodic table.

The various depictions of molecules just discussed will be used throughout this book. In fact, visualization of the sizes and shapes of molecules and interpretation of the physical and chemical properties in terms of molecular sizes and shapes is one of the most important aspects of modern chemistry.

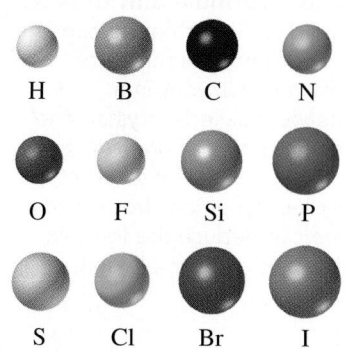

H   B   C   N

O   F   Si   P

S   Cl   Br   I

▲ FIGURE 3-3
Color scheme for use in molecular models
The sizes of atoms, reflected in the various sizes of the colored spheres, are related to the locations of the elements in the periodic table, as discussed in Section 9-3.

🔍 **3-1   CONCEPT ASSESSMENT**

Represent the succinic acid molecule, $HOOCCH_2CH_2COOH$, through empirical, molecular, structural, and line-angle formulas.

Ionic Compounds

Chemical combination of a metal and a nonmetal usually results in an ionic compound. An **ionic compound** is made up of positive and negative ions joined together by electrostatic forces of attraction (recall the attraction of oppositely charged objects pictured in Figure 2-4). The atoms of metallic elements tend to lose one or more electrons when they combine with nonmetal atoms, and the nonmetal atoms tend to gain one or more electrons. As a result of this electron transfer, the metal atom becomes a positive ion, or **cation**, and the nonmetal atom becomes a negative ion, or **anion**. We can usually deduce the charge on a main-group cation or anion from the group of the periodic table to which the element belongs (recall Section 2-6). Thus the periodic table can help us to write the formulas of ionic compounds.

In the formation of sodium chloride—ordinary table salt—each sodium atom gives up one electron to become a sodium ion, Na^+, and each chlorine atom gains one electron to become a chloride ion, Cl^-. This fact conforms to the relationship between locations of the elements in the periodic table and the charges on their simple ions (see page 53). For sodium chloride to be electrically neutral, there must be one Na^+ ion for each Cl^- ion ($+1 - 1 = 0$). Thus, the formula of sodium chloride is NaCl, and its structure is shown in Figure 3-4.

We observe that each Na^+ ion in sodium chloride is surrounded by six Cl^- ions, and vice versa, and we cannot say that any one of these six Cl^- ions belongs exclusively to a given Na^+ ion. Yet, the ratio of Cl^- to Na^+ ions in sodium chloride is 1 : 1, and so we arbitrarily select a combination of one Na^+ ion and one Cl^- ion as a formula unit. The **formula unit** of an ionic compound is the smallest electrically neutral collection of ions. The ratio of atoms (ions) in the formula unit is the same as in the chemical formula. Because it is buried in a vast network of ions, called a crystal, a formula unit of an ionic compound does not exist as a distinct entity. Thus it is inappropriate to call a formula unit of solid sodium chloride a molecule.

The situation with magnesium chloride is similar. In magnesium chloride, found in trace quantities in table salt, magnesium atoms lose two electrons to become magnesium ions, Mg^{2+} (Mg is in group 2). To obtain an electrically neutral formula unit, there must be two Cl^- ions, each with a charge of $1-$, for every Mg^{2+} ion. The formula of magnesium chloride is $MgCl_2$.

The ions Na^+, Mg^{2+}, and Cl^- are *monatomic*, meaning that each consists of a single ionized atom. By contrast, a *polyatomic* ion is made up of two or more atoms. In the nitrate ion, NO_3^-, the subscripts signify that *three* O atoms and *one* N atom are joined by covalent bonds into the single ion NO_3^-. Magnesium nitrate is an ionic compound made up of magnesium and nitrate ions. An electrically neutral formula unit of this compound must consist of one Mg^{2+} ion and two NO_3^- ions. The formula based on this formula unit is denoted by enclosing NO_3 in parentheses, followed by the subscript 2; thus, $Mg(NO_3)_2$. Polyatomic ions are discussed further in Section 3-6.

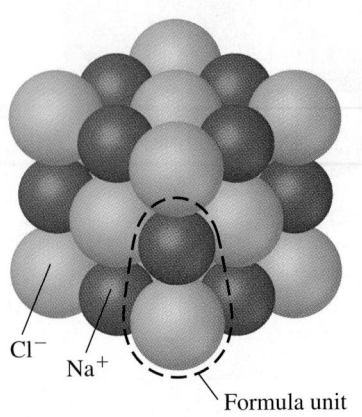

Cl^- Na^+ Formula unit

▲ FIGURE 3-4
Portion of an ionic crystal and a formula unit of NaCl
Solid sodium chloride consists of enormous numbers of Na^+ and Cl^- ions in a network called a crystal. The combination of one Na^+ and one Cl^- ion is the smallest collection of ions from which we can deduce the formula NaCl. It is a formula unit.

3-1 ARE YOU WONDERING...

If a compound can be formed between different metal atoms?

In a metal, electrons of the atoms interact to form *metallic* bonds. The bonded atoms are usually of the same element, but they may also be of different elements, giving rise to *intermetallic* compounds. The metallic bond gives metals and intermetallic compounds their characteristic properties of electrical and heat conductivity. These bonds are described in Chapter 11.

3-2   The Mole Concept and Chemical Compounds

Once we know the chemical formula of a compound, we can determine its formula mass. **Formula mass** is the mass of a *formula unit* in atomic mass units. It is always appropriate to use the term formula mass, but, for a molecular compound, the formula unit is an actual molecule, so we can speak of molecular mass. **Molecular mass** is the mass of a **molecule** in atomic mass units.

Weighted-average formula and molecular masses can be obtained just by adding up weighted-average atomic masses (those on the inside front cover). Thus, for the molecular compound water, H_2O,

$$\text{molecular mass } H_2O = 2(\text{atomic mass H}) + (\text{atomic mass O})$$
$$= 2(1.00794 \text{ u}) + 15.9994 \text{ u}$$
$$= 18.0153 \text{ u}$$

◀ The terms *formula weight* and *molecular weight* are often used in place of formula mass and molecular mass. This is similar to the situation described for atomic mass and atomic weight in the footnote on page 48.

For the ionic compound magnesium chloride, $MgCl_2$,

$$\text{formula mass } MgCl_2 = \text{atomic mass Mg} + 2(\text{atomic mass Cl})$$
$$= 24.3050 \text{ u} + 2(35.453 \text{ u})$$
$$= 95.211 \text{ u}$$

and for the ionic compound magnesium nitrate, $Mg(NO_3)_2$,

$$\text{formula mass } Mg(NO_3)_2 = \text{atomic mass Mg} + 2[\text{atomic mass N} + 3(\text{atomic mass O})]$$
$$= 24.3050 \text{ u} + 2[14.0067 \text{ u} + 3(15.9994 \text{ u})]$$
$$= 148.3148 \text{ u}$$

◀ The terms *formula mass* and *molecular mass* have essentially the same meaning, although when referring to ionic compounds, such as NaCl and $MgCl_2$, formula mass is the proper term.

Mole of a Compound

Recall that in Chapter 2 a mole was defined as an amount of substance having the same number of elementary entities as there are atoms in exactly 12 g of pure carbon-12. This definition carefully avoids saying that the entities to be counted are always atoms. As a result, we can apply the concept of a mole to any quantity that we can represent by a symbol or formula—atoms, ions, formula units, or molecules. Specifically, a *mole of compound* is an amount of compound containing Avogadro's number (6.02214×10^{23}) of formula units or molecules. The **molar mass** is the mass of one mole of compound—one mole of molecules of a molecular compound and one mole of formula units of an ionic compound.

The weighted-average molecular mass of H_2O is 18.0153 u, compared with a mass of exactly 12 u for a carbon-12 atom. If we compare samples of water molecules and carbon atoms by using Avogadro's number of each, we get a mass of 18.0153 g H_2O, compared with exactly 12 g for carbon-12. The molar mass of H_2O is 18.0153 g H_2O/mol H_2O. If we know the formula of a compound, we can equate the following terms, as illustrated for H_2O, $MgCl_2$, and $Mg(NO_3)_2$.

KEEP IN MIND

that although *molecular mass* and *molar mass* sound similar and are related, they are not the same. Molecular mass is the weighted-average mass of one molecule expressed in atomic mass units, u. *Molar mass* is the mass of Avogadro's number of molecules expressed in grams per mole, g/mol. The two terms have the same numerical value but different units.

$$1 \text{ mol } H_2O = 18.0153 \text{ g } H_2O = 6.02214 \times 10^{23} \text{ } H_2O \text{ molecules}$$
$$1 \text{ mol } MgCl_2 = 95.211 \text{ g } MgCl_2 = 6.02214 \times 10^{23} \text{ } MgCl_2 \text{ formula units}$$
$$1 \text{ mol } Mg(NO_3)_2 = 148.3148 \text{ g } Mg(NO_3)_2$$
$$= 6.02214 \times 10^{23} \text{ } Mg(NO_3)_2 \text{ formula units}$$

Such expressions as these provide several different types of conversion factors that can be applied in a variety of problem-solving situations. The strategy that works best for a particular problem will depend, in part, on how the necessary conversions are visualized. As we learned in Section 2-7, the most direct link to an amount in moles is through a mass in grams, so generally the central focus of a problem is the conversion of a mass in grams to an amount in moles, or vice versa. This conversion must often be preceded or followed by other conversions involving volumes, densities, percentages, and so on. As we saw in Chapter 2, one helpful tool in problem solving is to establish a conversion pathway. In Table 3.1, we summarize the roles that density, molar mass, and the Avogadro constant play in a conversion pathway.

TABLE 3.1 Density, Molar Mass, and the Avogadro Constant as Conversion Factors

Density, d	converts from volume to mass
Molar mass, M	converts from mass to amount (mol)
Avogadro constant, N_A	converts from amount (mol) to elementary entities

---

EXAMPLE 3-1   Relating Molar Mass, the Avogadro Constant, and Formula Units of an Ionic Compound

An analytical balance can detect a mass of 0.1 mg. How many ions are present in this minimally detectable quantity of $MgCl_2$?

Analyze

The central focus is the conversion of a measured quantity, 0.1 mg $MgCl_2$, to an amount in moles. After making the mass conversion, mg $\longrightarrow$ g, we can use the molar mass to convert from mass to amount in moles. Then, with the Avogadro constant as a conversion factor, we can convert from amount in moles to number of formula units. The final factor we need is based on the fact that there are *three* ions (*one* Mg^{2+} and *two* Cl^-) per formula unit (fu) of $MgCl_2$. It is often helpful to map out a conversion pathway that starts with the information given and proceeds through a series of conversion factors to the information sought. For this problem, we can begin with milligrams of $MgCl_2$ and make the following conversions:

$$\text{mg} \longrightarrow \text{g} \longrightarrow \text{mol} \longrightarrow \text{fu} \longrightarrow \text{number of ions}$$

Solve

The required conversions can be carried out in a stepwise fashion (as was done in Example 2-9), or they can be combined into a single line calculation. To avoid having to write down intermediate results and to avoid rounding errors, we'll use a single line calculation this time.

$$? \text{ ions} = 0.1 \text{ mg MgCl}_2 \times \frac{1 \text{ g MgCl}_2}{1000 \text{ mg MgCl}_2} \times \frac{1 \text{ mol MgCl}_2}{95 \text{ g MgCl}_2}$$

$$\times \frac{6.0 \times 10^{23} \text{ fu MgCl}_2}{1 \text{ mol MgCl}_2} \times \frac{3 \text{ ions}}{1 \text{ fu MgCl}_2}$$

$$= 2 \times 10^{18} \text{ ions}$$

Assess

The mass of the sample (0.1 mg) is given with one significant figure, and so the final answer is rounded to one significant figure. In the calculation above, the molar mass of $MgCl_2$ and the Avogadro constant are rounded off to two significant figures, that is, with one more significant figure than in the measured quantity.

PRACTICE EXAMPLE A: How many grams of $MgCl_2$ would you need to obtain 5.0×10^{23} Cl^- ions?

PRACTICE EXAMPLE B: How many nitrate ions, NO_3^-, and how many oxygen atoms are present in 1.00 μg of magnesium nitrate, $Mg(NO_3)_2$?

EXAMPLE 3-2   Combining Several Factors in a Calculation Involving Molar Mass

The volatile liquid ethyl mercaptan, C_2H_6S, is one of the most odoriferous substances known. It is sometimes added to natural gas to make gas leaks detectable. How many C_2H_6S molecules are contained in a 1.0 μL sample? The density of liquid ethyl mercaptan is 0.84 g/mL.

Analyze

The central focus is again the conversion of a measured quantity to an amount in moles. Because the density is given in g/mL, it will be helpful to convert the measured volume to milliliters. Then, density can be used as a conversion factor to obtain the mass in grams, and the molar mass can then be used to convert mass to amount in moles. Finally, the Avogadro constant can be used to convert the amount in moles to the number of molecules. In summary, the conversion pathway is $\mu L \rightarrow L \rightarrow g \rightarrow mol \rightarrow$ molecules.

Solve

As always, the required conversions can be combined into a single line calculation. However, it is instructive to break the calculation into three steps: (1) a conversion from volume to mass, (2) a conversion from mass to amount in moles, and (3) a conversion from amount in moles to molecules. These three steps emphasize, respectively, the roles played by density, molar mass, and the Avogadro constant in the conversion pathway. (See Table 3.1.)

Convert from volume to mass.
$$? \text{ g } C_2H_6S = 1.0 \text{ } \mu L \times \frac{1 \times 10^{-6} \text{ L}}{1 \text{ } \mu L} \times \frac{1000 \text{ mL}}{1 \text{ L}} \times \frac{0.84 \text{ g } C_2H_6S}{1 \text{ mL}}$$
$$= 8.4 \times 10^{-4} \text{ g } C_2H_6S$$

Convert from mass to amount in moles.
$$? \text{ mol } C_2H_6S = 8.4 \times 10^{-4} \text{ g } C_2H_6S \times \frac{1 \text{ mol } C_2H_6S}{62.1 \text{ g } C_2H_6S}$$
$$= 1.4 \times 10^{-5} \text{ mol } C_2H_6S$$

Convert from moles to molecules.
$$? \text{ molecules } C_2H_6S = 1.4 \times 10^{-5} \text{ mol } C_2H_6S \times \frac{6.02 \times 10^{23} \text{ molecules } C_2H_6S}{1 \text{ mol } C_2H_6S}$$
$$= 8.1 \times 10^{18} \text{ molecules } C_2H_6S$$

Assess

Remember to store intermediate results in your calculator *without* rounding off. Round off at the end. The answer is rounded to two significant figures because the volume and density are given with two significant figures. Rounding errors are avoided if the required conversions are combined into a single line calculation.

$$? \text{ molecules } C_2H_6S = 1.0 \text{ } \mu L \times \frac{1 \times 10^{-6} \text{ L}}{1 \text{ } \mu L} \times \frac{1000 \text{ mL}}{1 \text{ L}} \times \frac{0.84 \text{ g } C_2H_6S}{1 \text{ mL}}$$
$$\times \frac{1 \text{ mol } C_2H_6S}{62.1 \text{ g } C_2H_6S} \times \frac{6.02 \times 10^{23} \text{ molecules } C_2H_6S}{1 \text{ mol } C_2H_6S}$$
$$= 8.1 \times 10^{18} \text{ molecules } C_2H_6S$$

PRACTICE EXAMPLE A: Gold has a density of 19.32 g/cm^3. A piece of gold foil is 2.50 cm on each side and 0.100 mm thick. How many atoms of gold are in this piece of gold foil?

PRACTICE EXAMPLE B: If the 1.0 μL sample of liquid ethyl mercaptan from Example 3-2 is allowed to evaporate and distribute itself throughout a chemistry lecture room with dimensions 62 ft $\times$ 35 ft $\times$ 14 ft, will the odor of the vapor be detectable in the room? The limit of detectability is 9 $\times$ 10^{-4} μmol/m^3.

Mole of an Element—A Second Look

In Chapter 2, we took one mole of an element to be 6.02214×10^{23} *atoms* of the element. This is the only definition possible for such elements as iron, magnesium, sodium, and copper, in which enormous numbers of individual spherical atoms are clustered together, much like marbles in a can. But the atoms of some elements are joined together to form molecules. Bulk samples of these elements are composed of collections of molecules. The molecules of P_4 and S_8 are represented in Figure 3-5. The molecular formulas of elements that you should become familiar with are

$$H_2 \quad O_2 \quad N_2 \quad F_2 \quad Cl_2 \quad Br_2 \quad I_2 \quad P_4 \quad S_8$$

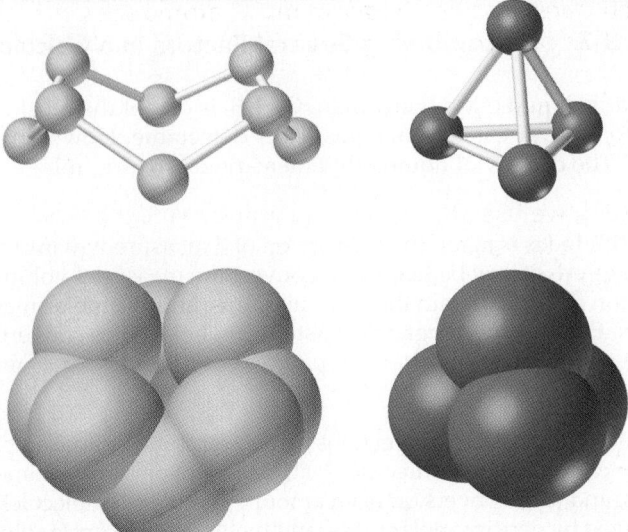

▶ FIGURE 3-5
Molecular forms of elemental sulfur and phosphorus
In a sample of solid sulfur, there are eight sulfur atoms in a sulfur molecule. In solid white phosphorus, there are four phosphorus atoms per molecule.

For these elements, we speak of an *atomic* mass or a *molecular* mass, and molar mass can be expressed in two ways. Hydrogen, for example, has an atomic mass of 1.00794 u and a molecular mass of 2.01588 u; its molar mass can be expressed as 1.00794 g H/mol H or 2.01588 g H_2/mol H_2.

Another phenomenon occasionally encountered is the existence of an element in more than one molecular form, a situation referred to as *allotropy*. Thus, oxygen exists in two allotropic forms, the predominantly abundant diatomic oxygen, O_2, and the much less abundant allotrope *ozone*, O_3. The molar mass of ordinary dioxygen is 31.9988 g O_2/mol O_2, and that of ozone is 47.9982 g O_3/mol O_3.

🔍 **3-2   CONCEPT ASSESSMENT**

Without doing detailed calculations, determine which of the following quantities has the greatest mass and which has the smallest mass: (a) 0.50 mol O_2; (b) 2.0 × 10²³ Cu atoms; (c) 1.0 × 10²⁴ H_2O molecules; (d) a 20.000 g brass weight; (e) 1.0 mol Ne.

3-3   Composition of Chemical Compounds

A chemical formula conveys considerable quantitative information about a compound and its constituent elements. We have already learned how to determine the molar mass of a compound, and, in this section, we consider some other types of calculations based on the chemical formula.

The colorless, volatile liquid halothane has been used as a fire extinguisher and also as an inhalation anesthetic. Both its empirical and molecular formulas are $C_2HBrClF_3$, its molecular mass is 197.382 u, and its molar mass is 197.382 g/mol, as calculated below:

▼ Two representations of halothane.

$$M_{C_2HBrClF_3} = 2M_C + M_H + M_{Br} + M_{Cl} + 3M_F$$
$$= [(2 \times 12.0107) + 1.00794 + 79.904 + 35.453 + (3 \times 18.9984)] \text{ g/mol}$$
$$= 197.382 \text{ g/mol}$$

The molecular formula of $C_2HBrClF_3$ tells us that *per mole* of halothane there are two moles of C atoms, one mole each of H, Br, and Cl atoms, and three moles of F atoms. This factual statement can be turned into conversion factors to answer such questions as, "How many C atoms are present per mole of halothane?" In this case, the factor needed is 2 mol C/mol $C_2HBrClF_3$. That is,

$$? \text{ C atoms} = 1.000 \text{ mol } C_2HBrClF_3 \times \frac{2 \text{ mol C}}{1 \text{ mol } C_2HBrClF_3} \times \frac{6.022 \times 10^{23} \text{ C atoms}}{1 \text{ mol C}}$$

$$= 1.204 \times 10^{24} \text{ C atoms}$$

In Example 3-3, we use another conversion factor derived from the formula for halothane. This factor is shown in blue in the setup, which includes other familiar factors to make the conversion pathway:

$$mL \longrightarrow g \longrightarrow mol \; C_2HBrClF_3 \longrightarrow mol \; F$$

EXAMPLE 3-3   **Using Relationships Derived from a Chemical Formula**

How many moles of F atoms are in a 75.0 mL sample of halothane ($d = 1.871$ g/mL)?

Analyze

The conversion pathway for this problem is given above. First, convert the volume of the sample to mass; this requires density as a conversion factor. Next, convert the mass of halothane to its amount in moles; this requires the inverse of the molar mass as a conversion factor. The final conversion factor is based on the formula of halothane.

Solve

$$? \text{ mol F} = 75.0 \text{ mL } C_2HBrClF_3 \times \frac{1.871 \text{ g } C_2HBrClF_3}{1 \text{ mL } C_2HBrClF_3}$$

$$\times \frac{1 \text{ mol } C_2HBrClF_3}{197.4 \text{ g } C_2HBrClF_3} \times \frac{3 \text{ mol F}}{1 \text{ mol } C_2HBrClF_3}$$

$$= 2.13 \text{ mol F}$$

Assess

If we had been asked for the number of moles of C instead, the final conversion factor in the calculation above would have been (2 mol C/1 mol $C_2HBrClF_3$).

PRACTICE EXAMPLE A: How many grams of Br are contained in 25.00 mL of halothane ($d = 1.871$ g/mL)?

PRACTICE EXAMPLE B: How many milliliters of halothane would contain 1.00×10^{24} Br atoms?

🔍 **3-3   CONCEPT ASSESSMENT**

In hexachlorophene, $C_{13}H_6Cl_6O_2$, a compound used in germicidal soaps, which element contributes the greatest number of atoms and which contributes the greatest mass?

Calculating Percent Composition from a Chemical Formula

When chemists believe that they have synthesized a new compound, a sample is generally sent to an analytical laboratory where its percent composition is determined. This experimentally determined percent composition is then compared with the percent composition calculated from the formula of the expected compound. In this way, chemists can see if the compound obtained could be the one expected.

Equation (3.1) establishes how the mass percent of an element in a compound is calculated. In applying the equation, as in Example 3-4, think in terms of the following steps.

1. **Determine the molar mass of the compound.** This is the *denominator* in equation (3.1).

2. **Determine the contribution of the given element to the molar mass.** This product of the formula subscript and the molar mass of the element appears in the *numerator* of equation (3.1).

3. **Formulate the ratio of the mass of the given element to the mass of the compound as a whole.** This is the ratio of the numerator from step 2 to the denominator from step 1.

4. **Multiply this ratio by 100% to obtain the mass percent of the element.**

$$\text{mass \% element} = \frac{\left(\begin{array}{c}\text{number of}\\\text{atoms of element}\\\text{per formula unit}\end{array}\right) \times \left(\begin{array}{c}\text{molar mass}\\\text{of element}\end{array}\right)}{\text{molar mass of compound}} \times 100\% \qquad (3.1)$$

The mass composition of a compound is the collection of mass percentages of the individual elements in the compound.

EXAMPLE 3-4   Calculating the Mass Percent Composition of a Compound

What is the mass percent composition of halothane, $C_2HBrClF_3$?

Analyze

Apply the four-step method described above. First, determine the molar mass of $C_2HBrClF_3$. Then formulate the mass ratios and convert them to mass percents. If we use molar masses that are rounded to two decimal places, the calculated mass percents will be accurate to two decimal places.

Solve

The molar mass of $C_2HBrClF_3$ is 197.38 g/mol. The mass percents are

$$\% \text{ C} = \frac{\left(2 \text{ mol C} \times \dfrac{12.01 \text{ g C}}{1 \text{ mol C}}\right)}{197.38 \text{ g } C_2HBrClF_3} \times 100\% = 12.17\% \text{ C}$$

$$\% \text{ H} = \frac{\left(1 \text{ mol H} \times \dfrac{1.01 \text{ g H}}{1 \text{ mol H}}\right)}{197.38 \text{ g } C_2HBrClF_3} \times 100\% = 0.51\% \text{ H}$$

$$\% \text{ Br} = \frac{\left(1 \text{ mol Br} \times \dfrac{79.90 \text{ g Br}}{1 \text{ mol Br}}\right)}{197.38 \text{ } C_2HBrClF_3} \times 100\% = 40.48\% \text{ Br}$$

$$\% \text{ Cl} = \frac{\left(1 \text{ mol Cl} \times \dfrac{35.45 \text{ g Cl}}{1 \text{ mol Cl}}\right)}{197.38 \text{ g } C_2HBrClF_3} \times 100\% = 17.96\% \text{ Cl}$$

$$\% \text{ F} = \left(\frac{3 \text{ mol F} \times \dfrac{19.00 \text{ g F}}{1 \text{ mol F}}}{197.38 \text{ g } C_2HBrClF_3}\right) \times 100\% = 28.88\% \text{ F}$$

Thus, the percent composition of halothane is 12.17% C, 0.51% H, 40.48% Br, 17.96% Cl, and 28.88% F.

Assess

The mass ratios appearing above are based on a sample that contains exactly *one mole* of halothane. Another approach is to calculate the mass of each element present in a sample that contains exactly *100 g* of halothane. For example, in a *100 g* sample of halothane,

$$? \text{ g C} = 100 \text{ g } C_2HBrClF_3 \times \frac{1 \text{ mol } C_2HBrClF_3}{197.38 \text{ g } C_2HBrClF_3} \times \frac{2 \text{ mol C}}{1 \text{ mol } C_2HBrClF_3} \times \frac{12.01 \text{ g C}}{1 \text{ mol C}} = 12.17 \text{ g C}$$

and so halothane is 12.17% C. The mass of carbon in a 100 g sample is numerically equal to the mass percent of carbon. If you compare the calculation of g C with that for % C given earlier, you will see that both calculations involve exactly the same factors but in a slightly different order.

PRACTICE EXAMPLE A: Adenosine triphosphate (ATP) is the main energy-storage molecule in cells. Its chemical formula is $C_{10}H_{16}N_5P_3O_{13}$. What is its mass percent composition?

PRACTICE EXAMPLE B: *Without doing detailed calculations*, determine which two compounds from the following list have the same percent oxygen by mass: **(a)** CO; **(b)** CH_3COOH; **(c)** C_2O_3; **(d)** N_2O; **(e)** $C_6H_{12}O_6$; **(f)** $HOCH_2CH_2OH$.

The percentages of the elements in a compound should add up to 100.00%, and we can use this fact in two ways.

1. Check the accuracy of the computations by ensuring that the percentages total 100.00%. As applied to the results of Example 3-4:

$$12.17\% + 0.51\% + 40.48\% + 17.96\% + 28.88\% = 100.00\%$$

2. Determine the percentages of all the elements but one. Obtain that one by difference (subtraction). From Example 3-4:

$$\% \, H = 100.00\% - \% \, C - \% \, Br - \% \, Cl - \% \, F$$
$$= 100.00\% - 12.17\% - 40.48\% - 17.96\% - 28.88\%$$
$$= 0.51\%$$

Establishing Formulas from the Experimentally Determined Percent Composition of Compounds

At times, a chemist isolates a chemical compound—say, from an exotic tropical plant—and has no idea what it is. A report from an analytical laboratory on the percent composition of the compound yields data needed to determine its formula.

Percent composition establishes the relative proportions of the elements in a compound on a *mass* basis. A chemical formula requires these proportions to be on a *mole* basis, that is, in terms of *numbers* of atoms. Consider the following five-step approach to determining a formula from the experimentally determined percent composition of the compound 2-deoxyribose, a sugar that is a basic constituent of DNA (deoxyribonucleic acid). The mass percent composition of 2-deoxyribose is 44.77% C, 7.52% H, and 47.71% O.

1. Although we could choose any sample size, if we take one of *exactly* 100 g, the masses of the elements are numerically equal to their percentages, that is, 44.77 g C, 7.52 g H, and 47.71 g O.

2. Convert the masses of the elements in the 100.00 g sample to amounts in moles.

$$? \, mol \, C = 44.77 \, g \, C \times \frac{1 \, mol \, C}{12.011 \, g \, C} = 3.727 \, mol \, C$$

$$? \, mol \, H = 7.52 \, g \, H \times \frac{1 \, mol \, H}{1.008 \, g \, H} = 7.46 \, mol \, H$$

$$? \, mol \, O = 47.71 \, g \, O \times \frac{1 \, mol \, O}{15.999 \, g \, O} = 2.982 \, mol \, O$$

◀ Use the molar mass with one more significant figure than in the mass of the element.

3. Write a tentative formula based on the numbers of moles just determined.

$$C_{3.727}H_{7.46}O_{2.982}$$

4. Attempt to convert the subscripts in the tentative formula to small whole numbers. This requires dividing each of the subscripts by the smallest one (2.982).

$$C_{\frac{3.727}{2.982}} H_{\frac{7.46}{2.982}} O_{\frac{2.982}{2.982}} = C_{1.25} H_{2.50} O$$

If all subscripts at this point differ only slightly from whole numbers—which is not the case here—round them off to whole numbers, concluding the calculation at this point.

5. If one or more subscripts is still not a whole number—which is the case here—multiply all subscripts by a small whole number that will make them all integral. Thus, multiply by 4 here.

$$C_{(4\times1.25)} H_{(4\times2.50)} O_{(4\times1)} = C_5H_{10}O_4$$

The formula that we get by the method just outlined, $C_5H_{10}O_4$, is the simplest possible formula—the *empirical formula*. The actual *molecular formula* may be equal to, or some multiple of, the empirical formula, such as $C_{10}H_{20}O_8$, $C_{15}H_{30}O_{12}$, $C_{20}H_{40}O_{16}$, and so on. To find the multiplying factor, we must compare the formula mass based on the empirical formula with the true molecular mass of the compound. We can establish the molecular mass from a separate experiment (by methods introduced in Chapters 6 and 13). The experimentally determined molecular mass of 2-deoxyribose is 134 u. The formula mass based on the empirical formula, $C_5H_{10}O_4$, is 134.1 u. The measured molecular mass is the same as the empirical formula mass. The molecular formula is also $C_5H_{10}O_4$.

We outline the five-step approach described above in the flow diagram below and then apply the approach to Example 3-5, where we will find that the empirical formula and the molecular formula are not the same.

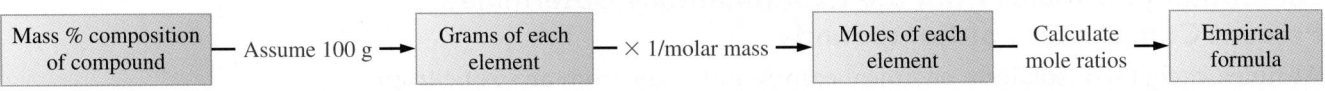

(3.2)

EXAMPLE 3-5   Determining the Empirical and Molecular Formulas of a Compound from Its Mass Percent Composition

Dibutyl succinate is an insect repellent used against household ants and roaches. Its composition is 62.58% C, 9.63% H, and 27.79% O. Its experimentally determined molecular mass is 230 u. What are the empirical and molecular formulas of dibutyl succinate?

Analyze

Use the five-step approach described above.

Solve

1. Determine the mass of each element in a 100.00 g sample.

   62.58 g C, 9.63 g H, 27.79 g O

2. Convert each of these masses to an amount in moles.

   $$? \text{ mol C} = 62.58 \text{ g C} \times \frac{1 \text{ mol C}}{12.011 \text{ g C}} = 5.210 \text{ mol C}$$

   $$? \text{ mol H} = 9.63 \text{ g H} \times \frac{1 \text{ mol H}}{1.008 \text{ g H}} = 9.55 \text{ mol H}$$

   $$? \text{ mol O} = 27.79 \text{ g O} \times \frac{1 \text{ mol O}}{15.999 \text{ g O}} = 1.737 \text{ mol O}$$

3. Write a tentative formula based on these numbers of moles.

   $C_{5.21} H_{9.55} O_{1.74}$

4. Divide each of the subscripts of the tentative formula by the smallest subscript (1.74),

   and round off any subscripts that differ only slightly from whole numbers; that is, round 2.99 to 3.

$$C_{\frac{5.21}{1.74}} \, H_{\frac{9.55}{1.74}} \, O_{\frac{1.74}{1.74}} = C_{2.99} \, H_{5.49} \, O$$

$$C_3 H_{5.49} O$$

5. Multiply all subscripts by a small whole number to make them integral (here by the factor 2), and write the empirical formula.

$$C_{2\times3} \, H_{2\times5.49} \, O_{2\times1} = C_6 \, H_{10.98} \, O_2$$
$$2 \times 5.49 = 10.98 \approx 11$$

Empirical formula: $C_6 H_{11} O_2$

To establish the molecular formula, first determine the empirical formula mass.

$$[(6 \times 12.0) + (11 \times 1.0) + (2 \times 16.0)]\,u = 115 \; u$$

Since the experimentally determined formula mass (230 u) is twice the empirical formula mass, the molecular formula is twice the empirical formula.

Molecular formula: $C_{12} H_{22} O_4$

Assess

Check the result by working the problem in reverse and using numbers that are rounded off slightly. For $C_{12}H_{22}O_4$, % C ≈ (12 × 12 u/230 u) × 100% = 63%; % H ≈ (22 × 1 u/230 u) × 100% = 9.6%; and % O ≈ (4 × 16 u/230 u) × 100% = 28%. The calculated mass percents agree well with those given in the problem, so we can be confident that our answer is correct.

PRACTICE EXAMPLE A: Sorbitol, used as a sweetener in some "sugar-free" foods, has a molecular mass of 182 u and a mass percent composition of 39.56% C, 7.74% H, and 52.70% O. What are the empirical and molecular formulas of sorbitol?

PRACTICE EXAMPLE B: The chlorine-containing narcotic drug pentaerythritol chloral has 21.51% C, 2.22% H, and 17.64% O, by mass, as its other elements. Its molecular mass is 726 u. What are the empirical and molecular formulas of pentaerythritol chloral?

3-4   CONCEPT ASSESSMENT

Explain why, if the percent composition *and* the molar mass of a compound are both known, the molecular formula can be determined much more readily by using the molar mass rather than a 100 g sample in the first step of the five-step procedure in Example 3-5. In this instance, how would you then obtain the empirical formula?

3-2   ARE YOU WONDERING...

How much rounding off to do to get integral subscripts in an empirical formula and what factors to use to convert fractional to whole numbers?

How much rounding off is justified depends on how precisely the elemental analysis is done. As a result, there is no ironclad rule on the matter. For the examples in this text, if you carry all the significant figures allowable in a calculation, you can generally round off a subscript that is within a few hundredths of a whole number (for example, 3.98 rounds off to 4). If the deviation is more than this, you will need to adjust subscripts to integral values by multiplying by the appropriate constant. If the appropriate constant is larger than a simple integer, such as 2, 3, 4, or 5, you may sometimes find it easier to make the adjustment by multiplying twice, for example, by 2 and then by 4 if the necessary constant is 8.

Combustion Analysis

Figure 3-6 illustrates an experimental method for establishing an empirical formula for compounds that are easily burned, such as compounds containing carbon and hydrogen with oxygen, nitrogen, and a few other elements. In *combustion analysis*, a weighed sample of a compound is burned in a stream of oxygen gas. The water vapor and carbon dioxide gas produced in the combustion are absorbed by appropriate substances. The increases in mass of these absorbers correspond to the masses of water and carbon dioxide. We can think of the matter as shown below. (The subscripts x, y, and z are integers whose values we do not know initially.)

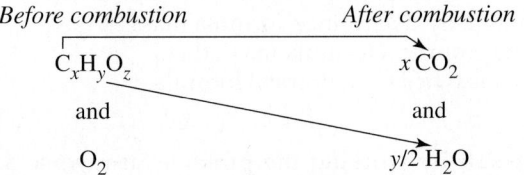

After combustion, all the carbon atoms in the sample are found in the CO_2. All the H atoms are in the H_2O. Moreover, the only source of the carbon and hydrogen atoms was the sample being analyzed. Oxygen atoms in the CO_2 and H_2O could have come partly from the sample and partly from the oxygen gas consumed in the combustion. Thus, the quantity of oxygen in the sample has to be determined indirectly. These ideas are applied in Example 3-6.

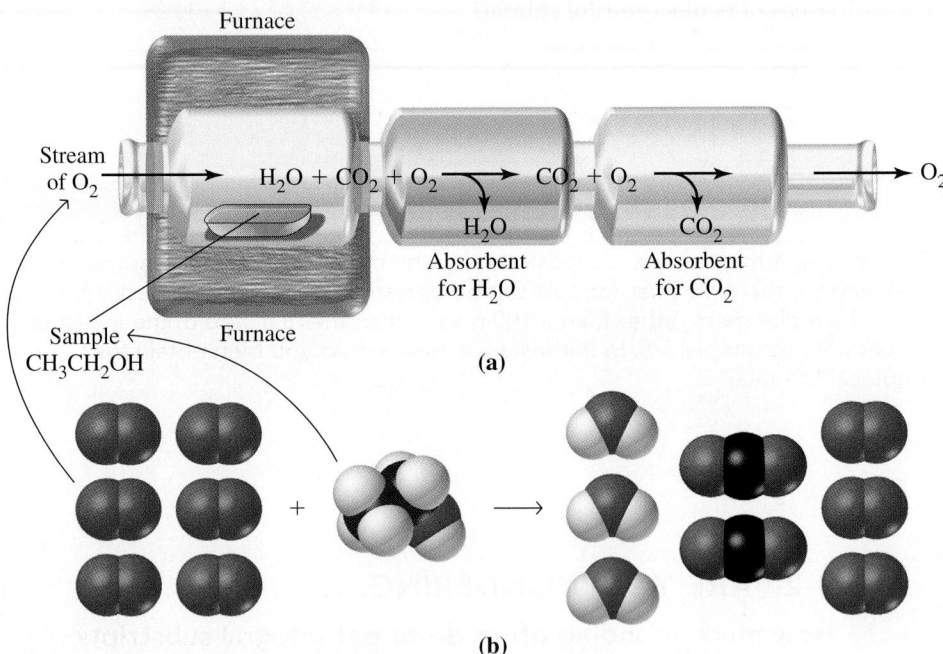

▲ FIGURE 3-6
Apparatus for combustion analysis
(a) Oxygen gas passes through the combustion tube containing the sample being analyzed. This portion of the apparatus is enclosed in a high-temperature furnace. Products of the combustion are absorbed as they leave the furnace—water vapor by magnesium perchlorate, and carbon dioxide gas by sodium hydroxide (producing sodium carbonate). The differences in mass of the absorbers, after and before the combustion, yield the masses of H_2O and CO_2 produced in the combustion reaction.
(b) A molecular picture of the combustion of ethanol. Each molecule of ethanol produces two CO_2 molecules and three H_2O molecules. Combustion takes place in an excess of oxygen, so that oxygen molecules are present at the end of the reaction. Note the conservation of mass.

EXAMPLE 3-6   Determining an Empirical Formula from Combustion Analysis Data

Vitamin C is essential for the prevention of scurvy. Combustion of a 0.2000 g sample of this carbon–hydrogen–oxygen compound yields 0.2998 g CO_2 and 0.0819 g H_2O. What are the percent composition and the empirical formula of vitamin C?

Analyze

After combustion, all the carbon atoms from the vitamin C sample are in CO_2 and all the hydrogen atoms are in H_2O. However, the oxygen atoms in CO_2 and H_2O come partly from the sample and partly from the oxygen gas consumed in the combustion. So, in the determination of the percent composition, we focus first on carbon and hydrogen and then on oxygen. To determine the empirical formula, we must calculate the amounts of C, H, and O in moles, and then calculate the mole ratios.

Solve

Percent Composition
First, determine the mass of carbon in 0.2988 g CO_2, by converting to mol C,

$$? \text{ mol C} = 0.2998 \text{ g CO}_2 \times \frac{1 \text{ mol CO}_2}{44.010 \text{ g CO}_2} \times \frac{1 \text{ mol C}}{1 \text{ mol CO}_2} = 0.006812 \text{ mol C}$$

and then to g C.

$$? \text{ g C} = 0.006812 \text{ mol C} \times \frac{12.011 \text{ g C}}{1 \text{ mol C}} = 0.08182 \text{ g C}$$

Proceed in a similar fashion for 0.0819 g H_2O to obtain

$$? \text{ mol H} = 0.0819 \text{ g H}_2\text{O} \times \frac{1 \text{ mol H}_2\text{O}}{18.02 \text{ g H}_2\text{O}} \times \frac{2 \text{ mol H}}{1 \text{ mol H}_2\text{O}} = 0.00909 \text{ mol H}$$

and

$$? \text{ g H} = 0.00909 \text{ mol H} \times \frac{1.008 \text{ g H}}{1 \text{ mol H}} = 0.00916 \text{ g H}$$

Obtain the mass of O in the 0.2000 g sample as the difference

$$? \text{ g O} = 0.2000 \text{ g sample} - 0.08182 \text{ g C} - 0.00916 \text{ g H} = 0.1090 \text{ g O}$$

Finally, multiply the mass fractions of the three elements by 100% to obtain mass percentages.

$$\% \text{ C} = \frac{0.08182 \text{ g C}}{0.2000 \text{ g sample}} \times 100\% = 40.91\% \text{ C}$$

$$\% \text{ H} = \frac{0.00916 \text{ g H}}{0.2000 \text{ g sample}} \times 100\% = 4.58\% \text{ H}$$

$$\% \text{ O} = \frac{0.1090 \text{ g O}}{0.2000 \text{ g sample}} \times 100\% = 54.50\% \text{ O}$$

Empirical Formula
At this point we can choose either of two alternatives. The first is to obtain the empirical formula from the mass percent composition, in the same manner illustrated in Example 3-5. The second is to note that we have already determined the number of moles of C and H in the 0.2000 g sample. The number of moles of O is

$$? \text{ mol O} = 0.1090 \text{ g O} \times \frac{1 \text{ mol O}}{15.999 \text{ g O}} = 0.006813 \text{ mol O}$$

From the numbers of moles of C, H, and O in the 0.2000 g sample, we obtain the tentative empirical formula

$$C_{0.006812}H_{0.00909}O_{0.006813}$$

Next, divide each subscript by the smallest (0.006812) to obtain

$$CH_{1.33}O$$

(continued)

Finally, multiply all the subscripts by 3 to obtain

Empirical formula of vitamin C: $C_3H_4O_3$

Assess

The determination of the empirical formula does not require determining the mass percent composition as a preliminary calculation. The empirical formula can be based on a sample of any size, as long as the numbers of moles of the different atoms in that sample can be determined.

PRACTICE EXAMPLE A: Isobutyl propionate is the substance that flavors rum extract. Combustion of a 1.152 g sample of this carbon–hydrogen–oxygen compound yields 2.726 g CO_2 and 1.116 g H_2O. What is the empirical formula of isobutyl propionate?

PRACTICE EXAMPLE B: Combustion of a 1.505 g sample of thiophene, a carbon–hydrogen–sulfur compound, yields 3.149 g CO_2, 0.645 g H_2O, and 1.146 g SO_2 as the only products of the combustion. What is the empirical formula of thiophene?

3-5 CONCEPT ASSESSMENT

When combustion is carried out in an excess of oxygen, the quantity producing the greatest mass of both CO_2 and H_2O is **(a)** 0.50 mol $C_{10}H_8$; **(b)** 1.25 mol CH_4; **(c)** 0.500 mol C_2H_5OH; **(d)** 1.00 mol C_6H_5OH.

We have just seen how combustion reactions can be used to analyze chemical substances, but not all samples can be easily burned. Fortunately, several other types of reactions can be used for chemical analyses. Also, modern methods in chemistry rely much more on physical measurements with instruments than on chemical reactions. We will cite some of these methods later in the text.

3-4 Oxidation States: A Useful Tool in Describing Chemical Compounds

▶ The oxidation state (O.S.) can be described as "a sometimes fictional charge." With monatomic ions, O.S. and charge are the same thing. For polyatomic ions and molecules, O.S. is fictional, but equal to what the charge would be if the compounds were entirely ionic.

Most basic concepts in chemistry deal with measurable properties or phenomena. In a few instances, though, a concept has been devised more for convenience than because of any fundamental significance. This is the case with the **oxidation state** (oxidation number),* which is related to the number of electrons that an atom loses, gains, or otherwise appears to use in joining with other atoms in compounds.

Consider NaCl. In this compound an Na atom, a metal, loses one electron to a Cl atom, a nonmetal. The compound consists of the ions Na^+ and Cl^- (see Figure 3-4). Na^+ is in a $+1$ oxidation state and Cl^- is in a -1 state.

In $MgCl_2$, an Mg atom loses two electrons to become Mg^{2+}, and each Cl atom gains one electron to become Cl^-. As in NaCl, the oxidation state of Cl is -1, but that of Mg is $+2$. If we take the *total* of the oxidation states of all the atoms (ions) in a formula unit of $MgCl_2$, we get $+2 - 1 - 1 = 0$.

In the molecule Cl_2, the two Cl atoms are identical and should have the same oxidation state. But if their total is to be zero, each oxidation state must itself be 0. Thus, the oxidation state of an atom can vary, depending on the compounds in which it occurs. In the molecule H_2O, we arbitrarily assign H the oxidation state of $+1$. Then, because the total of the oxidation states of the atoms must be zero, the oxidation state of oxygen must be -2.

---

*Because oxidation state refers to a number, the term oxidation number is often used synonymously. We will use the two terms interchangeably.

From these examples you can see that we need some conventions or rules for assigning oxidation states. The seven rules in Table 3.2 are sufficient to deal with most cases in this text, with this understanding: *The rules must be applied in the numerical order listed, and whenever two rules appear to contradict each other (which they sometimes will), follow the lower numbered rule.* Some examples are given for each rule, and the rules are applied in Example 3-7.

TABLE 3.2   Rules for Assigning Oxidation States

1. *The oxidation state (O.S.) of an individual atom in a free element (uncombined with other elements) is 0.*

   [*Examples:* The O.S. of an isolated Cl atom is 0; the two Cl atoms in the molecule Cl_2 both have an O.S. of 0.]

2. *The total of the O.S. of all the atoms in*

   (a) *neutral species, such as isolated atoms, molecules, and formula units, is 0;*

   [*Examples:* The sum of the O.S. of all the atoms in CH_3OH and of all the ions in $MgCl_2$ is 0.]

   (b) *an ion is equal to the charge on the ion.*

   [*Examples:* The O.S. of Fe in Fe^{3+} is +3. The sum of the O.S. of all atoms in MnO_4^- is −1.]

3. *In their compounds, the group 1 metals have an O.S. of +1 and the group 2 metals have an O.S. of +2.*

   [*Examples:* The O.S. of K is +1 in KCl and K_2CO_3; the O.S. of Mg is +2 in $MgBr_2$ and $Mg(NO_3)_2$.]

4. *In its compounds, the O.S. of fluorine is −1.*

   [*Examples:* The O.S. of F is −1 in HF, ClF_3, and SF_6.]

5. *In its compounds, hydrogen usually has an O.S. of +1.*

   [*Examples:* The O.S. of H is +1 in HI, H_2S, NH_3, and CH_4.]

6. *In its compounds, oxygen usually has an O.S. of −2.*

   [*Examples:* The O.S. of O is −2 in H_2O, CO_2 and $KMnO_4$.]

7. *In binary (two-element) compounds with metals, group 17 elements have an O.S. of −1; group 16 elements, −2; and group 15 elements, −3.*

   [*Examples:* The O.S. of Br is −1 in $MgBr_2$; the O.S. of S is −2 in Li_2S; and the O.S. of N is −3 in Li_3N.]

◀ The principal exceptions to rule 5 occur when H is bonded to metals, as in LiH, NaH, and CaH_2; exceptions to rule 6 occur in compounds with O—F bonds, such as OF_2, and in compounds where O atoms are bonded to one another, as in H_2O_2, and KO_2.

EXAMPLE 3-7   Assigning Oxidation States

What is the oxidation state of the underlined element in **(a)** $\underline{P}_4$; **(b)** $\underline{Al}_2O_3$; **(c)** $\underline{Mn}O_4^-$; **(d)** Na$\underline{H}$; **(e)** $H_2\underline{O}_2$; **(f)** $\underline{Fe}_3O_4$?

Analyze

Apply the rules in Table 3.2.

Solve

(a) P_4: This formula represents a molecule of elemental phosphorus. For an atom of a free element, the O.S. = 0 (rule 1). The O.S. of P in P_4 is 0.

(b) Al_2O_3: The total of the oxidation states of all the atoms in this formula unit is 0 (rule 2). The O.S. of oxygen is −2 (rule 6). The total for three O atoms is −6. The total for two Al atoms is +6. The O.S. of Al is +3.

(c) MnO_4^-: This is the formula for permanganate ion. The total of the oxidation states of all the atoms in the ion is −1 (rule 2). The total for the four O atoms is −8. The O.S. of Mn is +7.

(d) NaH: This is a formula unit of the ionic compound sodium hydride. Rule 3 states that the O.S. of Na is +1. Rule 5 indicates that H should also have an O.S. of +1. If both atoms had an O.S. of +1, the total for the formula unit would be +2. This violates rule 2. *Rules 2 and 3 take precedence over rule 5.* Na has an O.S. of +1; the total for the formula unit is 0; and the O.S. of H must be −1.

(continued)

(e) H_2O_2: This is hydrogen peroxide. Rule 5, stating that H has an O.S. of +1, takes precedence over rule 6 (which says that oxygen has an O.S. of −2). The sum of the oxidation states of the two H atoms is +2 and that of the two O atoms must be −2. The O.S. of O must be −1.

(f) Fe_3O_4: The total of the oxidation states of four O atoms is −8. For three Fe atoms, the total must be +8. The O.S. per Fe atom is $\frac{8}{3}$ or $+2\frac{2}{3}$.

Assess

With practice, you should be able to do the arithmetic associated with assigning oxidation states in your head, that is, without writing down any arithmetic expressions. Also, when you have determined an oxidation state by using arithmetic (as required by rule 2), check your result by making sure the sum of the oxidation states is equal to the charge on the atom, molecule, or ion. For example, in part (c), we determined that the oxidation state of Mn in MnO_4^- is +7. We know this result is correct because the sum of the oxidation states is $7 + 4(-2) = -1$, which is equal to the charge on the MnO_4^- ion.

PRACTICE EXAMPLE A: What is the oxidation state of the underlined element in each of the following: $\underline{S}_8$; $\underline{Cr}_2O_7{}^{2-}$; $\underline{Cl}_2O$; $K\underline{O}_2$?

PRACTICE EXAMPLE B: What is the oxidation state of the underlined element in each of the following: $\underline{S}_2O_3{}^{2-}$; $Hg_2\underline{Cl}_2$; $K\underline{Mn}O_4$; $H_2\underline{C}O$?

In part (f) of Example 3-7, we got the somewhat surprising answer of $+2\frac{2}{3}$ for the oxidation state of the iron atoms in Fe_3O_4. Prior to that, we saw only integral values for oxidation states. How does this fractional value come about? Generally, it comes from the assumption that all the atoms of an element have the same oxidation state in a given compound. Usually they do, but not always. Fe_3O_4, for example, is probably better represented as $FeO \cdot Fe_2O_3$, that is, through a combination of two simpler formula units. In FeO, the O.S. of the Fe atom is +2. In Fe_2O_3, the O.S. of each of *two* Fe atoms is +3. When we *average* the oxidation states over all three Fe atoms, we get a nonintegral value: $(2 + 3 + 3)/3 = \frac{8}{3} = 2\frac{2}{3}$.

Also, we may at times need to "fragment" a formula into its constituent parts before assigning oxidation states. The ionic compound NH_4NO_3, for instance, consists of the ions NH_4^+ and NO_3^-. The oxidation state of N in NH_4^+ is −3, and in NO_3^-, +5, and we do *not* want to average them. It is far more useful to know the oxidation states of the individual N atoms than it is to deal with an average oxidation state of +1 for the two N atoms.

Our first use of oxidation states comes in the naming of chemical compounds in the next section.

🔍 **3-6   CONCEPT ASSESSMENT**

A nitrogen–hydrogen compound with molar mass 32 g/mol has N in a higher oxidation state than in NH_3. What is a plausible formula for that compound?

3-5   Naming Compounds: Organic and Inorganic Compounds

▲ FIGURE 3-7
Two oxides of lead
These two compounds contain the same elements—lead and oxygen—but in different proportions. Their names and formulas must convey this fact: lead(IV) oxide = PbO_2 (red-brown); lead(II) oxide = PbO (yellow).

Throughout this chapter, we have referred to compounds mostly by their formulas, but we do need to give them names. When we know the name of a compound, we can look up its properties in a handbook, locate a chemical on a storeroom shelf, or discuss an experiment with a colleague. Later in the text we will see cases in which different compounds have the same formula. In these instances, we will find it essential to distinguish among compounds by name. We cannot give two substances the same name, yet we do want some similarities in the names of similar substances (Fig. 3-7). If all compounds were referred to by a common or trivial name, such as water (H_2O), ammonia (NH_3), or glucose

($C_6H_{12}O_6$), we would have to learn millions of unrelated names—an impossibility. What we need is a systematic method of assigning names—a system of *nomenclature*. Several systems are used, and we will introduce each at an appropriate point in the text. Compounds formed by carbon and hydrogen or carbon and hydrogen together with oxygen, nitrogen, and a few other elements are **organic compounds**. They are generally considered in a special branch of chemistry—*organic chemistry*—that has its own set of nomenclature rules. Compounds that do not fit this description are **inorganic compounds**. The branch of chemistry that concerns itself with the study of these compounds is called *inorganic chemistry*. In the next section, we will consider the naming of inorganic compounds, and in Section 3-7, we will introduce the naming of organic compounds.

3-6   Names and Formulas of Inorganic Compounds

Binary Compounds of Metals and Nonmetals

Binary compounds are those formed between *two elements*. If one of the elements is a metal and the other a nonmetal, the binary compound is usually made up of ions; that is, it is a binary *ionic* compound. To name a binary compound of a metal and a nonmetal,

- write the *unmodified* name of the metal
- then write the name of the nonmetal, modified to end in *-ide*

The approach is illustrated below for NaCl, MgI_2, and Al_2O_3.

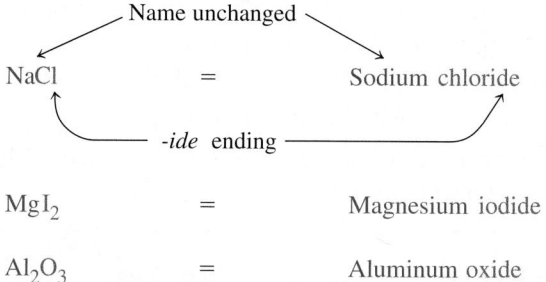

$$MgI_2 \quad = \quad \text{Magnesium iodide}$$

$$Al_2O_3 \quad = \quad \text{Aluminum oxide}$$

Ionic compounds, though made up of positive and negative ions, must be *electrically neutral*. The net, or total, charge of the ions in a formula unit must be zero. This means one Na^+ to one Cl^- in NaCl; one Mg^{2+} to two I^- in MgI_2; two Al^{3+} to three O^{2-} in Al_2O_3; and so on. Table 3.3 lists the names and symbols of simple ions formed by metals and nonmetals. You will find this list useful when writing names and formulas of binary compounds of metals and nonmetals. Because some metals may form several ions, it is important to distinguish between them in naming their compounds. The metal iron, for example, forms *two* common ions, Fe^{2+} and Fe^{3+}. The first is called the *iron(II)* ion, and the second is the *iron(III)* ion. The Roman numeral immediately following the name of the metal indicates its oxidation state or simply the charge on the ion. Thus, $FeCl_2$ is *iron(II)* chloride, while $FeCl_3$ is *iron(III)* chloride.

An earlier system of nomenclature that is still used to some extent applies two different word endings to distinguish between two binary compounds containing the same two elements but in different proportions, such as Cu_2O and CuO. In Cu_2O, the oxidation state of copper is +1, and in CuO it is +2. Cu_2O is assigned the name cup*rous* oxide, and CuO is cup*ric* oxide. Similarly, $FeCl_2$ is fer*rous* chloride, and $FeCl_3$ is fer*ric* chloride. The idea is to use the *-ous* ending for the lower oxidation state of the metal and *-ic* for the higher oxidation state. The *ous/ic* system has several inadequacies though, and we will not

▶ Table 3.3 simplifies considerably once you understand the pattern. All group 1 ions are 1+, group 2 ions are 2+, and Al^{3+}, Zn^{2+}, and Ag^+ have only the one common charge. Most other metal ions have variable charges. Monatomic halogen ions are 1− and oxygen is 2−. Almost everything else can be worked out from these basics and the rule that the sum of all O.S. equals the total charge.

TABLE 3.3   Some Simple Ions

Name	Symbol	Name	Symbol
		Positive ions (cations)	
Lithium ion	Li^+	Chromium(II) ion	Cr^{2+}
Sodium ion	Na^+	Chromium(III) ion	Cr^{3+}
Potassium ion	K^+	Iron(II) ion	Fe^{2+}
Rubidium ion	Rb^+	Iron(III) ion	Fe^{3+}
Cesium ion	Cs^+	Cobalt(II) ion	Co^{2+}
Magnesium ion	Mg^{2+}	Cobalt(III) ion	Co^{3+}
Calcium ion	Ca^{2+}	Copper(I) ion	Cu^+
Strontium ion	Sr^{2+}	Copper(II) ion	Cu^{2+}
Barium ion	Ba^{2+}	Mercury(I) ion	Hg_2^{2+}
Aluminum ion	Al^{3+}	Mercury(II) ion	Hg^{2+}
Zinc ion	Zn^{2+}	Tin(II) ion	Sn^{2+}
Silver ion	Ag^+	Lead(II) ion	Pb^{2+}
		Negative ions (anions)	
Hydride ion	H^-	Iodide ion	I^-
Fluoride ion	F^-	Oxide ion	O^{2-}
Chloride ion	Cl^-	Sulfide ion	S^{2-}
Bromide ion	Br^-	Nitride ion	N^{3-}

use it in this text. For example, the -ous and -ic endings do not help in naming the four oxides of vanadium: VO, V_2O_3, VO_2, and V_2O_5.

Binary Compounds of Two Nonmetals

If the two elements in a binary compound are both nonmetals instead of a metal and a nonmetal, the compound is a molecular compound. The method of naming these compounds is similar to that just discussed. For example,

$$HCl = \text{hydrogen chloride}$$

In both the formula and the name, we write the element with the positive oxidation state first: HCl and *not* ClH.

EXAMPLE 3-8   Writing Formulas When Names of Compounds Are Given

Write formulas for the compounds barium oxide, calcium fluoride, and iron(III) sulfide.

Analyze

In each case, identify the cations and their charges, based on periodic table group numbers or on oxidation states appearing as Roman numerals in names: Ba^{2+}, Ca^{2+}, and Fe^{3+}. Then identify the anions and their charges: O^{2-}, F^-, and S^{2-}. Combine the cations and anions in the relative numbers required to produce electrically *neutral* formula units.

Solve

barium oxide:	*one* Ba^{2+} and *one* O^{2-} = BaO
calcium fluoride:	*one* Ca^{2+} and *two* F^- = CaF_2
iron(III) sulfide:	*two* Fe^{3+} and *three* S^{2-} = Fe_2S_3

Assess

In the first case, an electrically neutral formula unit results from the combination of the charges 2+ and 2−; in the second case, 2+ and 2 × (1−); and in the third case, 2 × (3+) and 3 × (2−).

PRACTICE EXAMPLE A: Write formulas for lithium oxide, tin(II) fluoride, and lithium nitride.

PRACTICE EXAMPLE B: Write formulas for the compounds aluminum sulfide, magnesium nitride, and vanadium(III) oxide.

EXAMPLE 3-9   Naming Compounds When Their Formulas Are Given

Write acceptable names for the compounds Na_2S, AlF_3, Cu_2O.

Analyze

This task is generally easier than that of Example 3-8 because all you need to do is name the ions present. However, you must recognize that copper forms two different ions and that the cation in Cu_2O is Cu^+, copper(I).

Solve

Na_2S: sodium sulfide

AlF_3:   aluminum fluoride

Cu_2O: copper(I) oxide

Assess

Knowing when to use Roman numerals and when not to use them is a tricky aspect of naming ionic compounds. Because the metals of groups 1 and 2 have only one ionic form (one oxidation state), Roman numerals are not used in naming their compounds.

PRACTICE EXAMPLE A: Write acceptable names for CsI, CaF_2, FeO, $CrCl_3$.

PRACTICE EXAMPLE B: Write acceptable names for CaH_2, $CuCl$, Ag_2S, Hg_2Cl_2.

Some pairs of nonmetals form more than one binary molecular compound, and we need to distinguish among them. Generally, we indicate relative numbers of atoms through *prefixes*: *mono* = 1, *di* = 2, *tri* = 3, *tetra* = 4, *penta* = 5, *hexa* = 6, *hepta* = 7, *octa* = 8, *nona* = 9, *deca* = 10, and so on. Thus, for the two principal oxides of sulfur we write

SO_2 = sulfur *di*oxide

SO_3 = sulfur *tri*oxide

and for the following boron–bromine compound

B_2Br_4 = *di*boron *tetra*bromide

Additional examples are given in Table 3.4. Note that in these examples the prefix *mono-* is treated in a special way. We do not use it for the first named

TABLE 3.4   Naming Binary Molecular Compounds

Formula	Name[a]
BCl_3	Boron trichloride
CCl_4	Carbon tetrachloride
CO	Carbon monoxide
CO_2	Carbon dioxide
NO	Nitrogen monoxide
NO_2	Nitrogen dioxide
N_2O	Dinitrogen monoxide
N_2O_3	Dinitrogen trioxide
N_2O_4	Dinitrogen tetroxide
N_2O_5	Dinitrogen pentoxide
PCl_3	Phosphorus trichloride
PCl_5	Phosphorus pentachloride
SF_6	Sulfur hexafluoride

[a]When the prefix ends in *a* or *o* and the element name begins with *a* or *o*, the final vowel of the prefix is dropped for ease of pronunciation. For example, carbon *mono*oxide, not carbon *mono*oxide, and dinitrogen *tetr*oxide, not dinitrogen *tetra*oxide. However, PI_3 is phosphorus *tri*iodide, not phosphorus *tri*odide.

element. Thus, NO is called nitrogen monoxide, *not mono*nitrogen *mono*xide. Finally, several substances have common or trivial names that are so well established that their systematic names are almost never used. For example,

$$H_2O = \text{water} \ (\textit{di}\text{hydrogen} \ \textit{mono}\text{xide})$$

$$NH_3 = \text{ammonia} \ (H_3N = \textit{tri}\text{hydrogen} \ \textit{mono}\text{nitride})$$

Binary Acids

Even though we use names like hydrogen chloride for pure binary molecular compounds, we sometimes want to emphasize that their aqueous solutions are acids. Acids will be discussed in detail later in the text. For now, so that we can recognize these substances and name them when we see their formulas, let us say that an *acid* is a substance that *ionizes* or breaks down in water to produce hydrogen ions (H^+)* and anions. A *binary acid* is a two-element compound of hydrogen and a nonmetal. For example, HCl ionizes into hydrogen ions (H^+) and chloride ions (Cl^-) in water; it is a binary acid. NH_3 in water is *not* an acid. It shows practically no tendency to produce H^+ under any conditions. NH_3 belongs to a complementary category of substances called *bases*. As we will see in Chapter 5, bases yield hydroxide ions (OH^-) in aqueous solutions.

In naming binary acids we use the prefix *hydro-* followed by the name of the other nonmetal modified with an *-ic* ending. The most important binary acids are listed below.

▶ The symbol (aq) signifies a substance in aqueous (water) solution.

$$HF(aq) = \textit{hydro}\text{fluor}\textit{ic} \ \text{acid}$$

$$HBr(aq) = \textit{hydro}\text{brom}\textit{ic} \ \text{acid}$$

$$HCl(aq) = \textit{hydro}\text{chlor}\textit{ic} \ \text{acid}$$

$$HI(aq) = \textit{hydro}\text{iod}\textit{ic} \ \text{acid}$$

$$H_2S(aq) = \textit{hydro}\text{sulfur}\textit{ic} \ \text{acid}$$

Polyatomic Ions

With the exception of Hg_2^{2+}, the ions listed in Table 3.3 are *monatomic*—each consists of a single atom. In **polyatomic ions**, two or more atoms are joined together by covalent bonds. These ions are common, especially among the nonmetals. A number of polyatomic ions and compounds containing them are listed in Table 3.5. From this table, you can see that

1. Polyatomic anions are more common than polyatomic cations. The most familiar polyatomic cation is the ammonium ion, NH_4^+.

2. Very few polyatomic anions carry the *-ide* ending in their names. Of those listed, only OH^- (hydroxide ion) and CN^- (cyanide ion) do. The common endings are *-ite* and *-ate*, and some names carry prefixes, *hypo-* or *per-*.

3. An element common to many polyatomic anions is *oxygen*, usually in combination with another nonmetal. Such anions are called **oxoanions**.

4. Certain nonmetals (such as Cl, N, P, and S) form a series of oxoanions containing different numbers of oxygen atoms. Their names are related to the oxidation state of the nonmetal atom to which the O atoms are bonded, ranging from *hypo-* (lowest) to *per-* (highest) according to the following scheme.

Increasing oxidation state of nonmetal →

hypo___ite	*___ite*	*___ate*	*per___ate*

(3.3)

Increasing number of oxygen atoms →

*The species produced in aqueous solution is actually more complex than the simple ion H^+. The H^+ combines with an H_2O molecule to produce an ion known as the hydronium ion, H_3O^+. Chemists often use H^+ for H_3O^+, and that is what we will do until we discuss this matter more fully in Chapter 5.

TABLE 3.5   Some Common Polyatomic Ions

Name	Formula	Typical Compound
Cation		
Ammonium ion	NH_4^+	NH_4Cl
Anions		
Acetate ion	CH_3COO^-	$NaCH_3COO$
Carbonate ion	CO_3^{2-}	Na_2CO_3
Hydrogen carbonate ion[a] (or bicarbonate ion)	HCO_3^-	$NaHCO_3$
Hypochlorite ion	ClO^-	$NaClO$
Chlorite ion	ClO_2^-	$NaClO_2$
Chlorate ion	ClO_3^-	$NaClO_3$
Perchlorate ion	ClO_4^-	$NaClO_4$
Chromate ion	CrO_4^{2-}	Na_2CrO_4
Dichromate ion	$Cr_2O_7^{2-}$	$Na_2Cr_2O_7$
Cyanide ion	CN^-	$NaCN$
Hydroxide ion	OH^-	$NaOH$
Nitrite ion	NO_2^-	$NaNO_2$
Nitrate ion	NO_3^-	$NaNO_3$
Oxalate ion	$C_2O_4^{2-}$	$Na_2C_2O_4$
Permanganate ion	MnO_4^-	$NaMnO_4$
Phosphate ion	PO_4^{3-}	Na_3PO_4
Hydrogen phosphate ion[a]	HPO_4^{2-}	Na_2HPO_4
Dihydrogen phosphate ion[a]	$H_2PO_4^-$	NaH_2PO_4
Sulfite ion	SO_3^{2-}	Na_2SO_3
Hydrogen sulfite ion[a] (or bisulfite ion)	HSO_3^-	$NaHSO_3$
Sulfate ion	SO_4^{2-}	Na_2SO_4
Hydrogen sulfate ion[a] (or bisulfate ion)	HSO_4^-	$NaHSO_4$
Thiosulfate ion	$S_2O_3^{2-}$	$Na_2S_2O_3$

[a]These anion names are sometimes written as a single word—for example, hydrogencarbonate, hydrogenphosphate, and so forth.

◀ Learn the most common ions first, such as NH_4^+, CO_3^{2-}, OH^-, NO_3^-, PO_4^{3-}, SO_4^{2-}, and ClO_3^-. When you understand the scheme in expression (3.3), the names of several others, such as NO_2^-, SO_3^{2-}, ClO^-, ClO_2^-, and ClO_4^- will become obvious. Over time, the rest will become more familiar to you.

5. All the common oxoanions of Cl, Br, and I carry a charge of $1-$.

6. Some series of oxoanions also contain various numbers of H atoms and are named accordingly. For example, HPO_4^{2-} is the *hydrogen phosphate* ion and $H_2PO_4^-$, the *dihydrogen phosphate* ion.

7. The prefix *thio-* signifies that a sulfur atom has been substituted for an oxygen atom. (The sulfate ion has *one* S and *four* O atoms; thiosulfate ion has *two* S and *three* O atoms.)

Oxoacids

The majority of acids are **ternary compounds**. They contain *three* different elements—hydrogen and two other nonmetals. If one of the nonmetals is oxygen, the acid is called an **oxoacid**. Think of oxoacids as combinations of hydrogen ions (H^+) and oxoanions. The scheme for naming oxoacids is similar to that outlined for oxoanions, except that the ending *-ous* is used instead of *-ite* and *-ic* instead of *-ate*. Several oxoacids are listed in Table 3.6. Also listed are the names and formulas of compounds in which the hydrogen of the oxoacid has been replaced by a metal such as sodium. These compounds are called *salts*; we will say much more about them in later chapters, beginning in Chapter 5. Acids are molecular compounds, and salts are ionic compounds.

TABLE 3.6 Nomenclature of Some Oxoacids and Their Salts

Oxidation State	Formula of Acid[a]	Name of Acid[b]	Formula of Salt[b]	Name of Salt
Cl: +1	$HClO$	*Hypo*chlor*ous* acid	$NaClO$	Sodium *hypo*chlor*ite*
Cl: +3	$HClO_2$	Chlor*ous* acid	$NaClO_2$	Sodium chlor*ite*
Cl: +5	$HClO_3$	Chlor*ic* acid	$NaClO_3$	Sodium chlor*ate*
Cl: +7	$HClO_4$	*Per*chlor*ic* acid	$NaClO_4$	Sodium *per*chlor*ate*
N: +3	HNO_2	Nitr*ous* acid	$NaNO_2$	Sodium nitr*ite*
N: +5	HNO_3	Nitr*ic* acid	$NaNO_3$	Sodium nitr*ate*
S: +4	H_2SO_3	Sulfur*ous* acid	Na_2SO_3	Sodium sulf*ite*
S: +6	H_2SO_4	Sulfur*ic* acid	Na_2SO_4	Sodium sulf*ate*

[a]In all these acids, H atoms are bonded to O atoms, not the central nonmetal atom. Often formulas are written to reflect this fact, for instance, HOCl instead of HClO and HOClO instead of $HClO_2$.

[b]In general, the *-ic* and *-ate* names are assigned to compounds in which the central nonmetal atom has an oxidation state equal to the periodic table group number minus 10. Halogen compounds are exceptional in that the *-ic* and *-ate* names are assigned to compounds in which the halogen has an oxidation state of +5 (even though the group number is 17).

▶ When we refer to perchloric acid, for example, we generally mean an aqueous solution of $HClO_4$.

EXAMPLE 3-10 Applying Various Rules for Naming Compounds

Name the compounds **(a)** $CuCl_2$; **(b)** ClO_2; **(c)** HIO_4; **(d)** $Ca(H_2PO_4)_2$.

Analyze

$CuCl_2$ and $Ca(H_2PO_4)_2$ are ionic compounds. To name these compounds, we must identify and name the ions. ClO_2 and HIO_4 are molecular compounds. ClO_2 is a binary compound of two nonmetals and HIO_4 is an oxoacid.

Solve

(a) In this compound, the oxidation state of Cu is +2. Because Cu can also exist in the oxidation state of +1, we must clearly distinguish between the two possible chlorides. $CuCl_2$ is copper(II) chloride.

(b) Both Cl and O are nonmetals. ClO_2 is a binary molecular compound called chlorine dioxide.

(c) The oxidation state of I is +7. By analogy to the chlorine-containing oxoacids in Table 3.6, we should name this compound periodic acid (pronounced "purr-eye-oh-dic" acid).

(d) The polyatomic anion $H_2PO_4^-$ is dihydrogen phosphate ion. Two of these ions are present for every Ca^{2+} ion in the compound calcium dihydrogen phosphate.

Assess

A fair bit of memorization is associated with naming compounds correctly. Mastery of this subject usually requires a lot of practice.

PRACTICE EXAMPLE A: Name the compounds SF_6, HNO_2, $Ca(HCO_3)_2$, $FeSO_4$.

PRACTICE EXAMPLE B: Name the compounds NH_4NO_3, PCl_3, $HBrO$, $AgClO_4$, $Fe_2(SO_4)_3$.

Some Compounds of Greater Complexity

The copper compound that Joseph Proust used to establish the law of constant composition (page 36) is referred to in different ways. If you look up Proust's compound in a handbook of minerals, you will find it listed as *malachite*, with the formula $Cu_2(OH)_2CO_3$. In a handbook that specializes in pharmaceutical applications, this same compound is listed as *basic cupric carbonate*, with the formula $CH_2Cu_2O_5$. In a chemistry handbook, it is listed as *copper(II) carbonate dihydroxide*, with the formula $CuCO_3 \cdot Cu(OH)_2$. All you need to understand at this point is that regardless of the formula you use, you

▶ In general, centered dots (·) show that a formula is a composite of two or more simpler formulas.

EXAMPLE 3-11   Applying Various Rules in Writing Formulas

Write the formula of the compound (a) tetranitrogen tetrasulfide; (b) ammonium chromate; (c) bromic acid; (d) calcium hypochlorite.

Analyze

We must apply our knowledge of prefixes (such as *tetra-*) and endings (such as *-ic*), as well as the names of common polyatomic ions (such as *ammonium*, *chromate*, and *hypochlorite*).

Solve

(a) Molecules of this compound consist of *four* N atoms and *four* S atoms. The formula is N_4S_4.

(b) Two ammonium ions (NH_4^+) must be present for every chromate ion (CrO_4^{2-}). Place parentheses around NH_4^+, followed by the subscript 2. The formula is $(NH_4)_2CrO_4$. (This formula is read as "N-H-4, taken twice, C-R-O-4.")

(c) The *-ic* acid for the oxoacids of the halogens (group 17) has the halogen in the oxidation state of +5. Bromic acid is $HBrO_3$ (analogous to $HClO_3$ in Table 3.6).

(d) Here there are one Ca^{2+} and two ClO^- ions in a formula unit. This leads to the formula $Ca(ClO)_2$.

Assess

Notice that in writing formulas for compounds containing two or more polyatomic ions of the *same type*, as in (b) and (d), we put parentheses around the formula of the ion (without the charge), followed by a subscript indicating the number of ions of that type. Proper use and placement of parentheses in writing formulas is important.

PRACTICE EXAMPLE A: Write formulas for the compounds (a) boron trifluoride, (b) potassium dichromate, (c) sulfuric acid, (d) calcium chloride.

PRACTICE EXAMPLE B: Write formulas for the compounds (a) aluminum nitrate, (b) tetraphosphorous decoxide, (c) chromium(III) hydroxide, (d) iodic acid.

should obtain the same molar mass (221.116 g/mol), the same mass percent copper (57.48% Cu), the same $H:O$ mole ratio (2 mol H/5 mol O), and so on. In short, you should be able to interpret a formula, no matter how complex its appearance.

Some complex substances you are certain to encounter are known as hydrates. In a **hydrate**, each formula unit of the compound has associated with it a certain number of water molecules. This does not mean that the compounds are "wet," however. The water molecules are incorporated in the solid structure of the compound. The formula shown below signifies six H_2O molecules per formula unit of $CoCl_2$.

$$CoCl_2 \cdot 6\,H_2O$$

The prefix for six is *hexa-* and this compound is cobalt(II) chloride *hexa*hydrate. Its formula mass is that of $CoCl_2$ *plus* that associated with six H_2O: 129.839 u + (6 × 18.0153 u) = 237.931 u. We can speak of the mass percent water in a hydrate; for $CoCl_2 \cdot 6\,H_2O$ this is

$$\% \text{ } H_2O = \frac{(6 \times 18.0153) \text{ g } H_2O}{237.931 \text{ g } CoCl_2 \cdot 6\,H_2O} \times 100\% = 45.43\%$$

The water present in compounds as water of hydration can generally be removed, in part or totally, by heating. When the water is totally removed, the resulting compound is said to be *anhydrous* (without water). Anhydrous compounds can be used as water absorbers, as in the use of anhydrous magnesium perchlorate in combustion analysis (recall Figure 3-6). $CoCl_2$ gains and loses water quite readily and indicates this through a color change. Anhydrous $CoCl_2$ is blue, whereas the hexahydrate is pink. This fact can be used to make a simple moisture detector (Fig. 3-8).

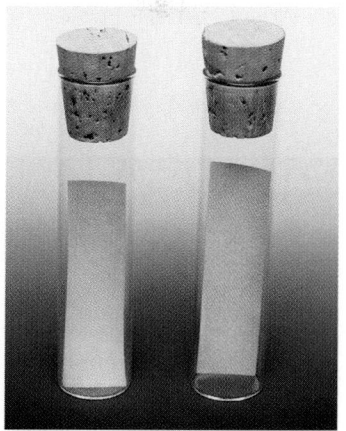

▲ FIGURE 3-8
Effect of moisture on CoCl₂
The piece of filter paper was soaked in a water solution of cobalt(II) chloride and then allowed to dry. When kept in dry air, the paper is blue in color (anhydrous $CoCl_2$). In humid air, the paper changes to pink ($CoCl_2 \cdot 6\,H_2O$).

3-7   Names and Formulas of Organic Compounds

Organic compounds abound in nature. The foods we eat are made up almost exclusively of organic compounds, including not only energy-producing fats and carbohydrates and muscle-building proteins but also trace compounds that impart color, odor, and flavor to these foods. Almost all fuels, whether used to power automobiles, trucks, trains, or airplanes, are mixtures of organic compounds of a type called hydrocarbons. Most of the drugs produced by pharmaceutical companies are complex organic compounds, as are common plastics. The multiplicity of organic compounds is so vast that *organic chemistry* exists as a distinctive field of chemistry.

The great diversity of organic compounds arises from the ability of carbon atoms to combine readily with other carbon atoms and with atoms of a number of other elements. Carbon atoms join together to form a framework of chains or rings to which other atoms are attached. All organic compounds contain carbon atoms; almost all contain hydrogen atoms; and many common ones also have oxygen, nitrogen, or sulfur atoms. These possibilities allow for an almost limitless number of different organic compounds. Organic compounds are mostly molecular; a few are ionic.

There are millions of organic compounds, many comprising highly complex molecules. Their names are equally complicated. A systematic approach to naming these compounds is crucial, and the rules for naming inorganic compounds are of little use here. The usual name, often called the common or *trivial* name, for a familiar sweetener is sucrose (sugar). The systematic name is α-D-glucopyranosyl-β-D-fructofuranoside. At this point, however, we only need to recognize organic compounds and use their common names, together with an occasional systematic name. We will look at the systematic nomenclature of organic compounds in more detail in Chapter 26.

Hydrocarbons

Compounds containing only carbon and hydrogen are called **hydrocarbons**. The simplest hydrocarbon contains one carbon atom and four hydrogen atoms—methane, CH_4 (Fig. 3-9a). As the number of carbon atoms increases, the number of hydrogen atoms also increases in a systematic way, depending on the type of hydrocarbon. The complexity of organic chemistry arises because carbon atoms can form chains and rings, and the nature of the chemical bonds between the carbon atoms can vary. Hydrocarbons containing only single bonds are called **alkanes**. The simplest alkane is methane, followed by ethane, C_2H_6 (Fig. 3-9b), and then propane, C_3H_8 (Fig. 3-9c). The fourth member of the series, butane, C_4H_{10}, was shown in Figure 3-2(a). Notice that each succeeding member of the alkane series is formed by the addition of one C atom and two H atoms to the preceding member.

The names of the alkanes are composed of two parts: a word stem and the ending (suffix) *-ane* indicating that the molecule is an *alkane*. The first four word stems in Table 3.7 reflect common names, while the rest indicate the number of carbon atoms in the alkane. Thus C_5H_{12} is pentane and C_7H_{16} is heptane (hept = 7).

Hydrocarbon molecules with one or more double bonds between carbon atoms are called *alkenes*. The simplest of the alkenes is ethene (Fig. 3-9d); its name consists of the stem *eth-* and the ending *-ene*. Benzene, C_6H_6 (Fig. 3-9e), is a molecule with six carbon atoms arranged in a hexagonal ring. Molecules with structures related to benzene make up a large proportion of known organic compounds.

Refer to Figure 3-2 (page 71) and you will notice that butane and methylpropane have the same molecular formula, C_4H_{10}, but different structural formulas. Butane is based on a four-carbon chain, whereas in methylpropane, a

KEEP IN MIND

that each carbon atom forms four covalent bonds.

TABLE 3.7   Word Stem (or Prefix) Indicating the Number of Carbon Atoms in Simple Organic Molecules

Stem (or prefix)	Number of C Atoms
Meth-	1
Eth-	2
Prop-	3
But-	4
Pent-	5
Hex-	6
Hept-	7
Oct-	8
Non-	9
Dec-	10

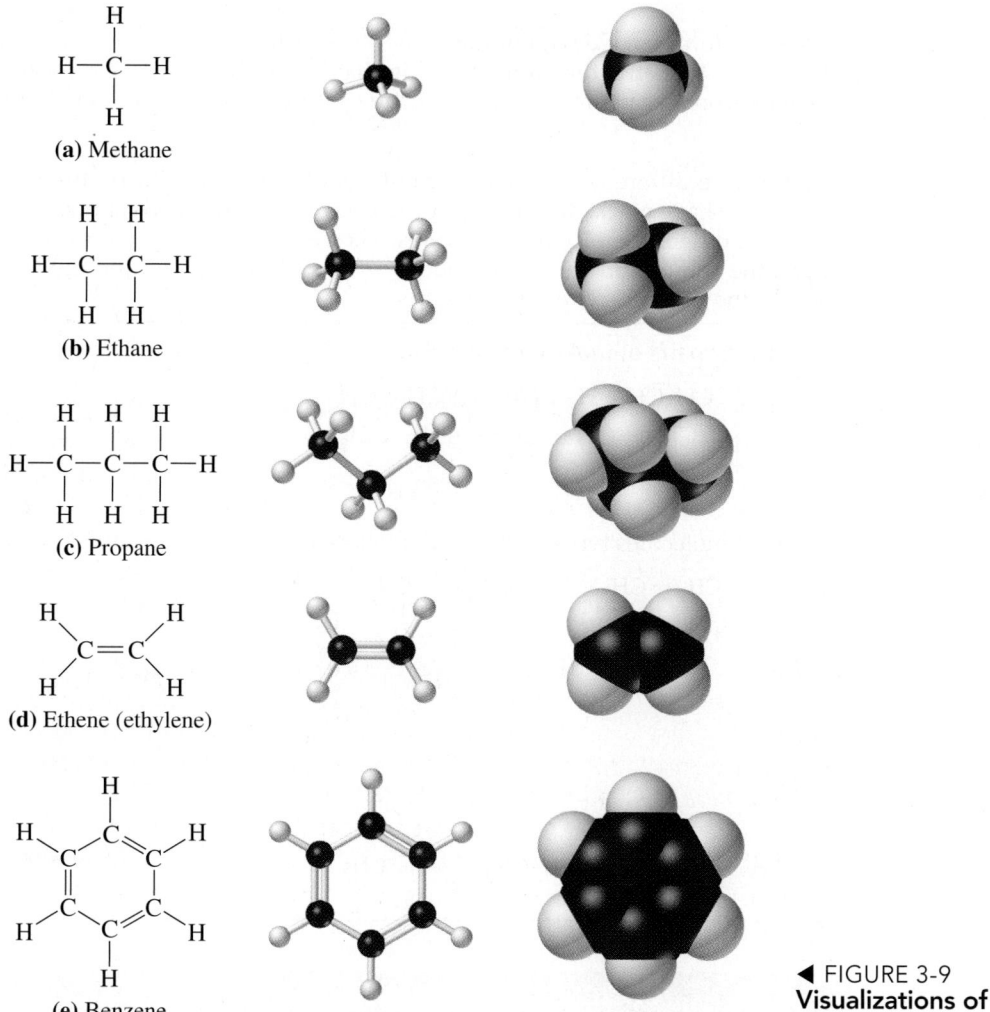

(a) Methane

(b) Ethane

(c) Propane

(d) Ethene (ethylene)

(e) Benzene

◄ FIGURE 3-9
Visualizations of some hydrocarbons

—CH_3 group, called a *methyl* group, is attached to the middle carbon atom of the three-carbon propane chain. Butane and methylpropane are *isomers*. **Isomers** are molecules that have the same molecular formula but different arrangements of atoms in space. As organic molecules become more complex, the possibilities for isomerism increase very rapidly.

EXAMPLE 3-12   Recognizing Isomers

Are the following pairs of molecules isomers or not?

(a) $CH_3CH(CH_3)(CH_2)_3CH_3$ and $CH_3CH_2CH(CH_3)(CH_2)_3CH_3$

(b) CH_3—CH—CH_2—CH_3 and CH_3—CH—CH_2—CH_2—CH_3
          | |
     CH_2—CH_3 CH_3

Analyze

Isomers have the same molecular formula but different structures. We first check to see if the molecular formulas are the same. If the formulas are not the same, they represent different compounds. If the formulas are the same, the compounds *may* be isomers, but only if their structures are different.

Solve

(a) The molecular formula of the first compound is C_7H_{16}, while that of the second compound is C_8H_{18}. The molecules are not isomers.

(continued)

(b) These molecules have the same formula, C_6H_{14}, but they differ in structure. They are isomers. The difference in structure is that in the first structure, a methyl side chain is on the middle carbon atom of a five-carbon chain, and in the second structure it is on the second carbon atom from the end of the chain.

Assess

The compounds shown in part (a) have different molecular formulas and, thus, are clearly different compounds. We expect these compounds to have markedly different properties. The compounds shown in part (b) have the same molecular formula but different molecular structures. They are different compounds and have different properties. For example, the compound shown on the left in part (b) has a slightly higher boiling point than the compound shown on the right (63 °C versus 60 °C).

PRACTICE EXAMPLE A: Are the following pairs of molecules isomers?

(a) $CH_3C(CH_3)_2(CH_2)_3CH_3$ and $CH_3CH(CH_3)CH(CH_3)(CH_2)_3CH_3$

(b) CH_3—CH—CH_2—CH_3 and CH_3—CH—CH_2—CH—CH_3

$\quad\quad\quad\quad\quad\; | \quad\quad\quad\quad\quad\quad\quad\quad\quad\quad | \quad\quad\quad\quad\; |$

$\quad\quad CH_2$—CH_2—$CH_3 \quad\quad\quad\quad\quad CH_3 \quad\quad CH_3$

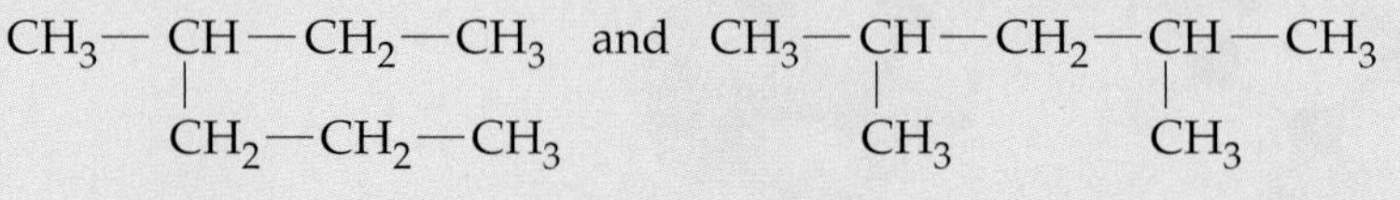

PRACTICE EXAMPLE B: Are the pairs of molecules represented by the following structural formulas isomers?

(a)

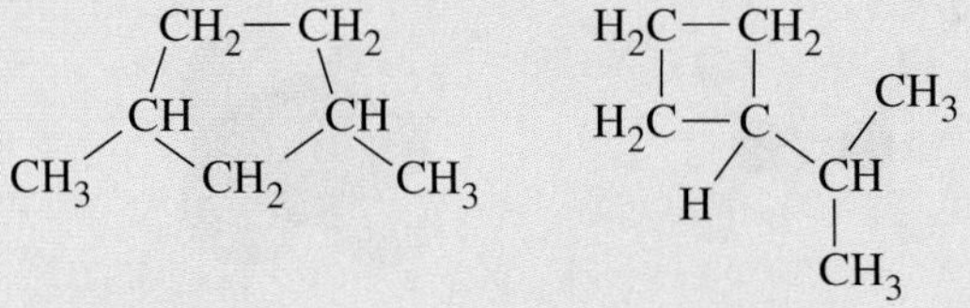

CH_2—CH_2

$CH \quad\quad CH$

$CH_3 \quad CH_2 \quad CH_3$

H_2C—CH_2

H_2C—C — CH_3

$\quad\quad H \quad CH$

$\quad\quad\quad\quad | $

$\quad\quad\quad\quad CH_3$

(b) CH_3 CH_3 CH_3 H

$\quad\quad\quad C=C \quad\quad\quad\quad C=C$

$\quad\quad H \quad\quad H \quad\quad CH_3 \quad\quad CH_3$

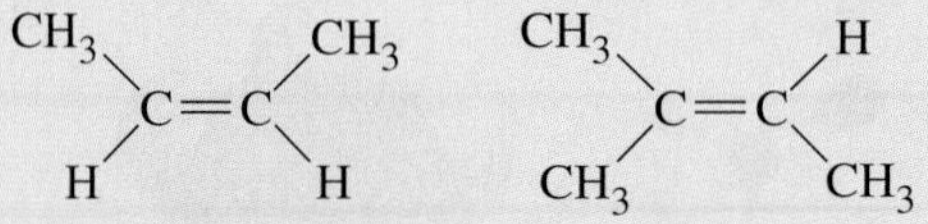

3-7   CONCEPT ASSESSMENT

In the combustion of a hydrocarbon, can the mass of H_2O produced ever exceed that of the CO_2? Explain.

Functional Groups

Carbon chains provide the framework of organic compounds; other atoms or groups of atoms replace one or more of the hydrogen atoms to form different compounds. We can illustrate this with the common alcohol molecule that occurs in beer, wine, and spirits. The molecule is *ethanol*, CH_3CH_2OH, in which one of the H atoms of ethane is replaced by an —OH group (Fig. 3.10a). The systematic name ethanol is derived from the name of the alkane, ethane, with the final *-e* replaced by the suffix *-ol*. The suffix *-ol* designates the presence of the OH group in a class of organic molecules called **alcohols**.

Ethyl alcohol, the common name of ethanol, also indicates attachment of the —OH group to the ethane hydrocarbon chain. To name the alkane chain as a group, replace the final *-e* with *-yl*, so that ethane becomes *ethyl*, thus *ethyl alcohol* for CH_3CH_2OH. It is often the case that the common name of one compound, alcohol in this case, will provide the generic name for a complete class of compounds; that is, all alcohols contain at least one —OH group.

Another common alcohol is *methanol*, or wood alcohol, which has the formula CH_3OH (Fig. 3.10b). The common name for methanol is methyl alcohol. It is interesting to note that wood alcohol is a dangerous poison, whereas the grain alcohol in beer and wine is safe to consume in moderate quantities.

The —OH group in alcohols is one of the many *functional groups* found in organic compounds. **Functional groups** are individual atoms or groupings of

KEEP IN MIND

that alcohols are covalently bonded molecular compounds in which oxygen atoms are covalently bonded to carbon atoms. They are not ionic compounds that contain hydroxide ions.

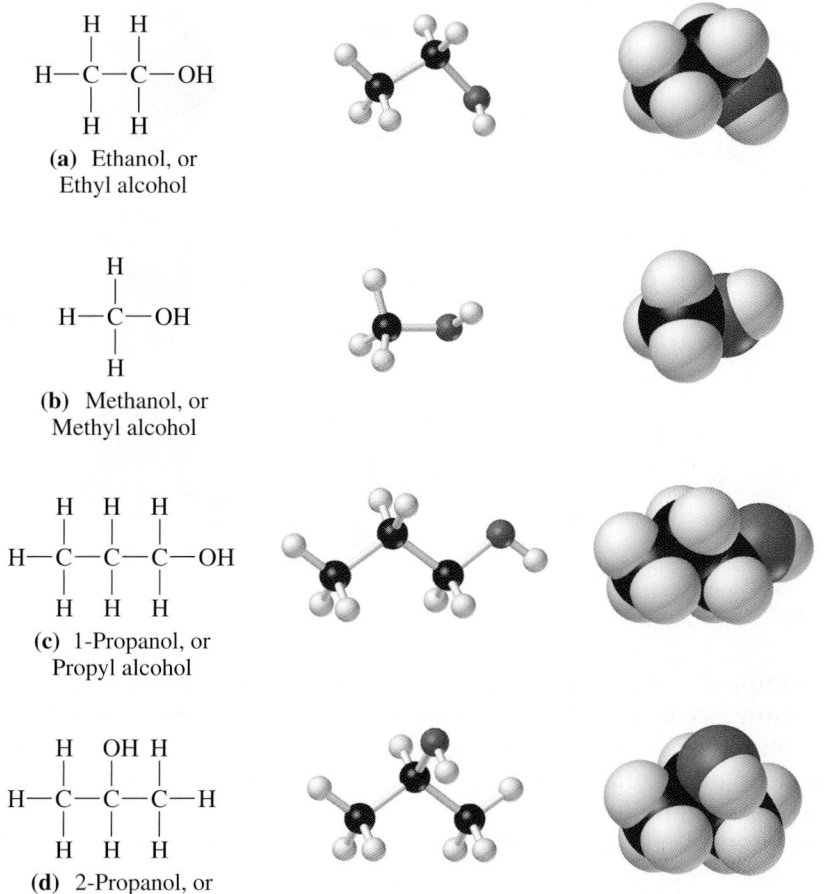

(a) Ethanol, or
Ethyl alcohol

(b) Methanol, or
Methyl alcohol

(c) 1-Propanol, or
Propyl alcohol

(d) 2-Propanol, or
Isopropyl alcohol

◀ FIGURE 3-10
Visualizations of some alcohols

atoms that are attached to the carbon chains or rings of organic molecules and give the molecules their characteristic properties. Compounds with the same functional group generally have similar properties. The —OH group is called the *hydroxyl group*. At this point only a few functional groups will be introduced; functional groups are discussed in more detail in Chapter 26.

The presence of functional groups also increases the possibility of isomers. For example, there is only one propane molecule, C_3H_8. However, if one of the H atoms is replaced by a hydroxyl group, two possibilities exist for the point of attachment: at one of the end C atoms or at the middle C atom (Figs. 3.10c and d). This leads to two isomers. The alcohol with the —OH group attached to the end carbon atom is commonly called propyl alcohol or, systematically, 1-propanol; the prefix 1- indicates that the —OH group is on the first or end C atom. The alcohol with the —OH group attached to the middle carbon atom is commonly called isopropyl alcohol or, systematically, 2-propanol; the prefix 2- indicates that the —OH group is on the second C atom from the end.

Another important functional group is the *carboxyl* group, —COOH, or —CO_2H which confers acidic properties on a molecule. The C atom in the carboxyl group is bound to the two O atoms in two ways. One bond is a single bond to an oxygen atom that is also attached to a H atom, and the other is a double bond to a lone O atom (Fig. 3-11a). The hydrogen attached to one of the O atoms in a carboxyl group is *ionizable* or *acidic*. Compounds containing the carboxyl group are called **carboxylic acids**. The first carboxylic acid based on alkanes is *methanoic* acid, HCOOH (Fig. 3.11b). In the systematic name, the *methan-* indicates one carbon atom and the *-oic* acid indicates a carboxylic acid. The common name for methanoic acid is formic acid, deriving from the Latin word *formica*, meaning "ant." Formic acid is injected by an ant when it bites; this leads to the burning sensation that accompanies the bite.

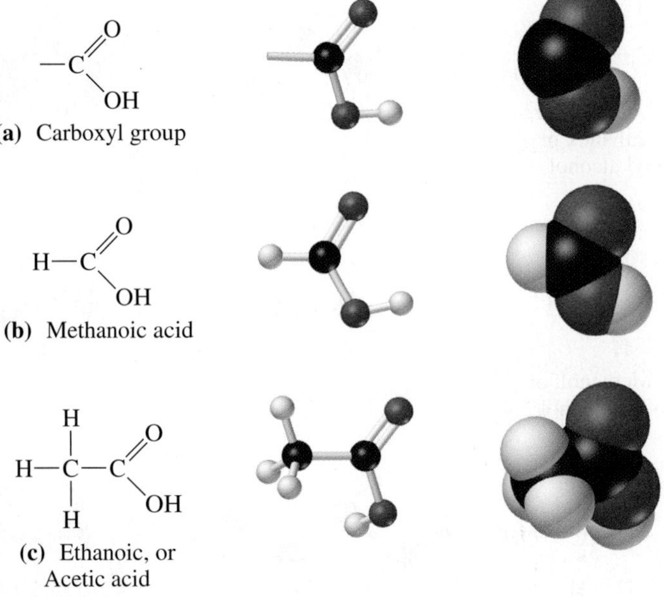

(a) Carboxyl group

(b) Methanoic acid

(c) Ethanoic, or Acetic acid

▶ FIGURE 3-11
The carboxyl group and visualizations of two carboxylic acids

The simplest carboxylic acid containing two carbon atoms is *ethanoic* acid, more commonly known as acetic acid. The molecular formula is CH_3COOH, and the structure is shown in Figure 3-11(c). Vinegar is a solution of acetic acid in water. An additional functional group is introduced in the examples that follow—a halogen atom (F, Cl, Br, I) substituting for one or more H atoms. When present as functional groups, the halogens carry the names, *fluoro-*, *chloro-*, *bromo-*, and *iodo-*.

EXAMPLE 3-13   Recognizing Types of Organic Compounds

What type of compound is each of the following?

(a) $CH_3CH_2CH_2CH_3$ (b) $CH_3CHClCH_2CH_3$

(c) $CH_3CH_2CO_2H$ (d) $CH_3CH_2CH(OH)CH_2CH_3$

Analyze

For each compound, examine the formula to determine which functional group, if any, is present. Also consider whether or not all the carbon–carbon bonds are single bonds.

Solve

(a) The carbon–carbon bonds are all single bonds in this hydrocarbon. This compound is an alkane.

(b) There are only single bonds in its molecules, and one H atom has been replaced by a Cl atom. This compound is a chloroalkane.

(c) The presence of the carboxyl group, $-CO_2H$, in its molecules means that this compound is a carboxylic acid.

(d) The presence of the hydroxyl group, $-OH$, in its molecules means that this compound is an alcohol.

Assess

Carbon–hydrogen and carbon–carbon bonds, especially carbon–carbon single bonds, are relatively unreactive. Because of this, functional groups are most important in determining the characteristic properties of organic compounds. We will encounter other functional groups and examine organic compounds in more detail in Chapters 26 and 27.

PRACTICE EXAMPLE A: What types of compounds correspond to each of the following formulas?

(a) $CH_3CH_2CH_3$ (b) $ClCH_2CH_2CH_3$

(c) $CH_3CH_2CH_2CO_2H$ (d) $CH_3CHCHCH_3$

PRACTICE EXAMPLE B: What types of compounds correspond to these formulas?

(a) $CH_3CH(OH)CH_3$ (b) $CH_3CH(OH)CH_2CO_2H$

(c) $CH_2ClCH_2CO_2H$ (d) $BrCHCHCH_3$

EXAMPLE 3-14   Naming Organic Compounds

Name these compounds.

(a) $CH_3CH_2CH_2CH_2CH_3$ (b) $CH_3CHFCH_2CH_3$

(c) $CH_3CH_2CO_2H$ (d) $CH_3CH_2CH(OH)CH_2CH_3$

Analyze

First, determine the type of compound. Then, count the number of carbon atoms and select the appropriate stem (or prefix) from Table 3.7 to form the name. If the compound is an alcohol, change the final -*e* in the name to -*ol*; if it is a carboxylic acid, change the final -*e* to -*oic acid*.

Solve

(a) The structure is that of an alkane molecule with a five-carbon chain, so the compound is pentane.

(b) The structure is that of a fluoroalkane molecule with the F atom on the *second* C atom of a four-carbon chain. The compound is called 2-fluorobutane.

(c) The carbon chain in this structure is three C atoms long, with the end C atom in a carboxyl group. The compound is propanoic acid.

(d) This structure is that of an alcohol with the hydroxyl group on the *third* C atom of a five-carbon chain. The compound is called 3-pentanol.

Assess

In naming the compound in part (b), we stated that the F atom was bonded to the second C atom in a four-carbon chain. There is some ambiguity in that statement because, as illustrated below, we can number the C atoms in two different ways.

2-fluorobutane
(correct)

3-fluorobutane
(incorrect)

By convention, we always number the carbon atoms so that the position of the functional group is designated by the smallest possible number. Thus, 2-fluorobutane is the correct name.

PRACTICE EXAMPLE A: Name the following compounds: (a) $CH_3CH(OH)CH_3$; (b) $ICH_2CH_2CH_3$; (c) $CH_3CH(CH_3)CH_2CO_2H$; (d) CH_3CHCH_2.

PRACTICE EXAMPLE B: Give plausible names for the molecules that correspond to the following ball-and-stick models.

(a)   (b)   (c)

EXAMPLE 3-15   Writing Structural Formulas from the Names of Organic Compounds

Write the condensed structural formula for the organic compounds: **(a)** butane, **(b)** butanoic acid, **(c)** 1-chloropentane, **(d)** 1-hexanol.

Analyze

First, identify the number of carbon atoms in the chain, and then determine the type and position of the functional group, if any.

Solve

(a) The word stem *but-* indicates a structure with a four-carbon chain, and the suffix *-ane* indicates an alkane. No functional groups are indicated; hence, the condensed structural formula is $CH_3(CH_2)_2CH_3$.

(b) The *-oic* ending indicates that the end carbon atom of the four-carbon chain is part of a carboxylic acid group. The condensed structural formula is $CH_3(CH_2)_2CO_2H$.

(c) The prefix *chloro-* indicates the substitution of a chlorine atom for a H atom, and the 1- designates that it is on the first C atom of the carbon chain. The carbon chain is five C atoms long, as signified by the word stem *pent-*. The condensed structural formula is $CH_3(CH_2)_3CH_2Cl$.

(d) The suffix *-ol* indicates the presence of a hydroxyl group in place of a H atom, and the 1- designates that it is on the first C atom of the carbon chain. The word stem *hex-* signifies that the carbon chain is six C atoms long. The condensed structural formula is $CH_3(CH_2)_4CH_2OH$.

Assess

In summary, to obtain a structural formula from the name, we split the name into its component pieces: stem, prefix, and suffix. All three components provide information about the structure of the molecule.

PRACTICE EXAMPLE A: Write the condensed structural formula for the organic compounds **(a)** pentane, **(b)** ethanoic acid, **(c)** 1-iodooctane (pronounced eye-oh-dough-octane), **(d)** 1-pentanol.

PRACTICE EXAMPLE B: Write the line-angle formula for the organic compounds **(a)** propene, **(b)** 1-heptanol, **(c)** chloroacetic acid, **(d)** hexanoic acid.

Mastering**CHEMISTRY**   www.masteringchemistry.com.

For a discussion of how chemists use mass spectrometry to establish both molecular and structural formulas, go to the Focus On feature for Chapter 3, entitled Mass Spectrometry—Determining Molecular and Structural Formulas, on the MasteringChemistry site.

Summary

3-1 Types of Chemical Compounds and Their Formulas—The two main classes of chemical compounds are **molecular compounds** and **ionic compounds**. The fundamental unit of a molecular compound is a molecule and that of an ionic compound is a **formula unit**. A formula unit is the smallest collection of positively charged ions—called **cations**—and negatively charged ions—called **anions**—that is electrically neutral overall. A **chemical formula** is a symbolic representation of a compound that can be written in several ways (Fig. 3-1). If the formula has the smallest integral subscripts possible, it is an **empirical formula**; if the formula represents an actual molecule, it is a **molecular formula**; and if the formula is written to show how individual atoms are joined together into molecules, it is a **structural formula**. Abbreviated structural formulas, called **condensed structural formulas**, are often used for organic molecules. Also used for organic molecules is the **line-angle formula**, in which all bond lines are shown except those between C and H atoms and in which the symbols C and H are mostly omitted. The relative sizes

and positions of the atoms in molecules can be depicted by ball-and-stick and space-filling molecular models.

3-2 The Mole Concept and Chemical Compounds—In this section, the concept of atomic mass is extended to **molecular mass**, the mass in atomic mass units of a molecule of a molecular compound, and **formula mass**, the mass in atomic mass units of a formula unit of an ionic compound. Likewise, the concept of the Avogadro constant and the mole is now applied to compounds, with emphasis on quantitative applications involving the mass of a mole of compound—the **molar mass** M. For several elements, we can distinguish between a mole of molecules (for example, P_4) and a mole of atoms (that is, P).

3-3 Composition of Chemical Compounds—The mass percent composition of a compound can be established from its formula (equation 3.1). Conversely, a chemical formula can be deduced from the experimentally determined percent composition of the compound. For organic compounds, this often involves combustion analysis (Fig. 3-6). Formulas determined from experimental percent composition data are empirical formulas—the simplest formulas that can be written. Molecular formulas can be related to empirical formulas when experimentally determined molecular masses are available.

3-4 Oxidation States: A Useful Tool in Describing Chemical Compounds

—The **oxidation state** of an atom is roughly related to the number of electrons involved in the formation of a bond between that atom and another atom. Oxidation states are expressed as numbers assigned according to a set of conventions (Table 3.2). The oxidation state concept has several uses, one of which is to aid in naming chemical compounds and in writing chemical formulas.

3-5 Naming Compounds: Organic and Inorganic Compounds

—**Assigning names** to the formulas of chemical compounds—nomenclature—is an important activity in chemistry. The topic is introduced in this section and applied to two broad categories: (1) **organic compounds**, which are compounds formed by the elements carbon and hydrogen, often together with a few other elements, such as oxygen and nitrogen; and (2) **inorganic compounds**, a category that includes the remaining chemical compounds. The subject of nomenclature is extended to additional contexts throughout the text.

3-6 Names and Formulas of Inorganic Compounds

—In this section on the nomenclature of inorganic compounds, the names and formulas of two-element or **binary compounds** are considered first. The names and symbols of some simple ions (Table 3.3) and the names of a few typical binary molecular compounds (Table 3.4) are listed. Also listed are some common ions comprising two or more atoms, **polyatomic ions**, and typical ionic compounds containing them (Table 3.5). The naming of binary acids is related to the names of the corresponding binary hydrogen compounds. The majority of acids, however, are **ternary compounds**; they consist of three elements—hydrogen and two other nonmetals. Typically, one of the elements in ternary acids is oxygen, giving rise to the name **oxoacids**. The polyatomic anions derived from oxoacids are **oxoanions**. A scheme for relating the names and formulas of oxoanions is given (expression 3.3), and the nomenclature of oxoacids and their salts is summarized (Table 3.6). Also described in this section are **hydrates,** ionic compounds having fixed numbers of water molecules associated with their formula units.

3-7 Names and Formulas of Organic Compounds

—Organic compounds are based on the element carbon. **Hydrocarbons** contain only carbon and hydrogen (Fig. 3-9). **Alkane** hydrocarbon molecules contain only single bonds, and alkenes contain at least one double bond. A common phenomenon found in organic compounds is isomerism—the existence of different compounds, called **isomers,** having identical molecular formulas but different structural formulas. **Functional groups** confer distinctive properties on organic molecules when they are substituted for H atoms on carbon chains or rings. The hydroxyl group —OH is present in **alcohols** (Fig. 3-10) and the carboxyl group —COOH, in **carboxylic acids** (Fig. 3-11).

Integrative Example

Molecules of a dicarboxylic acid have *two* carboxyl groups (—COOH). A 2.250 g sample of a dicarboxylic acid was burned in an excess of oxygen and yielded 4.548 g CO_2 and 1.629 g H_2O. In a separate experiment, the molecular mass of the acid was found to be 174 u. From these data, what can we deduce about the structural formula of this acid?

Analyze

Our approach will require several steps: (1) Use the combustion data to determine the percent composition of the compound (similar to Example 3-6). (2) Determine the empirical formula from the percent composition (similar to Example 3-5). (3) Obtain the molecular formula from the empirical formula and the molecular mass. (4) Determine how the C, H, and O atoms represented in the molecular formula might be assembled into a dicarboxylic acid. Use molar masses with (at least) one more significant figure than in the measured masses; store intermediate results in your calculator without rounding off.

Solve

1. *Determine the percent composition.* Calculate the mass of H in 1.629 g H_2O

$$? \text{ g H} = 1.629 \text{ g } H_2O \times \frac{1 \text{ mol } H_2O}{18.015 \text{ g } H_2O}$$

$$\times \frac{2 \text{ mol H}}{1 \text{ mol } H_2O} \times \frac{1.0079 \text{ g H}}{1 \text{ mol H}} = 0.1823 \text{ g H}$$

and then the mass percent H in the 2.250 g sample of the dicarboxylic acid.

$$\% \text{ H} = \frac{0.1823 \text{ g H}}{2.250 \text{ g compd.}} \times 100\% = 8.102\% \text{ H}$$

Also, calculate the mass of C in 4.548 g CO_2,

$$? \text{ g C} = 4.548 \text{ g } CO_2 \times \frac{1 \text{ mol } CO_2}{44.010 \text{ g } CO_2}$$

$$\times \frac{1 \text{ mol C}}{1 \text{ mol } CO_2} \times \frac{12.011 \text{ g C}}{1 \text{ mol C}} = 1.241 \text{ g C}$$

followed by the mass percent C in the 2.250 g sample of dicarboxylic acid.

$$\% \text{ C} = \frac{1.241 \text{ g C}}{2.250 \text{ g compd.}} \times 100\% = 55.16\% \text{ C}$$

The % O in the compound is obtained as a difference, that is,

$$\% \text{ O} = 100.00\% - 55.16\% \text{ C} - 8.102\% \text{ H} = 36.74 \% \text{ O}$$

2. *Obtain the empirical formula from the percent composition.* The masses of the elements in 100.0 g of the compound are

55.16 g C   8.102 g H   36.74 g O

The numbers of moles of the elements in 100.0 g of the compound are

$$55.16 \text{ g C} \times \frac{1 \text{ mol C}}{12.011 \text{ g C}} = 4.592 \text{ mol C}$$

$$8.102 \text{ g H} \times \frac{1 \text{ mol H}}{1.0079 \text{ g H}} = 8.038 \text{ mol H}$$

$$36.74 \text{ g O} \times \frac{1 \text{ mol O}}{15.999 \text{ g O}} = 2.296 \text{ mol O}$$

The tentative formula based on these numbers is

$C_{4.592}H_{8.038}O_{2.296}$

Divide all the subscripts by 2.296 to obtain

$C_2H_{3.50}O$

Multiply all subscripts by two to obtain the empirical formula,

$C_4H_7O_2$

and then determine the empirical formula mass.

$(4 \times 12.0107 \text{ u}) + (7 \times 1.00794 \text{ u}) + (2 \times 15.9994 \text{ u}) = 87.0972 \text{ u}$

3. *Obtain the molecular formula.* The experimentally determined molecular mass of 174 u is twice the empirical formula mass. The molecular formula is

$C_8H_{14}O_4$

4. *Assemble the atoms in $C_8H_{14}O_4$ into a plausible structural formula.* The dicarboxylic acid must contain two —COOH groups. This accounts for the two C atoms, two H atoms, and all four O atoms. The remainder of the structure is based on C_6H_{12}. For example, arrange the six —CH$_2$ segments into a six-carbon chain and attach the —COOH groups at the ends of the chain.

$\text{HOOC}-\text{CH}_2(\text{CH}_2)_4\text{CH}_2-\text{COOH}$

However, there are other possibilities based on shorter chains with branches, for example

$$\text{HOOC}-\text{CH}_2-\overset{\overset{\displaystyle \text{CH}_3}{|}}{\underset{\underset{\displaystyle \text{CH}_3}{|}}{\text{C}}}-\text{CH}_2-\text{CH}_2-\text{COOH}$$

Assess

We have found a plausible structural formula, but there are many other possibilities. For example, the following three isomers have a seven-carbon chain with one methyl group (—CH$_3$) substituted for an H atom on the chain:

HOOCCHCH$_3$(CH$_2$)$_4$COOH;   HOOCCH$_2$CHCH$_3$(CH$_2$)$_3$COOH;   HOOC(CH$_2$)$_2$CHCH$_3$(CH$_2$)$_2$COOH

In conclusion, we cannot identify a specific isomer with only the data given.

PRACTICE EXAMPLE A: A 2.4917 g sample of an unknown solid hydrate was heated to drive off all the water of hydration. The remaining solid, which weighed 1.8558 g, was analyzed and found to be 27.74% Mg, 23.57% P, and 48.69% O, by mass. What is the formula and name of the unknown solid hydrate?

PRACTICE EXAMPLE B: An unknown solid hydrate was analyzed and found to be 17.15% Cu, 19.14% Cl, and 60.45% O, by mass; the remainder was hydrogen. What are the oxidation states of copper and chlorine in this compound? What is the name of this compound?

Exercises

Representing Molecules

1. Refer to the color scheme given in Figure 3-3, and give the molecular formulas for the molecules whose ball-and-stick models are given here.

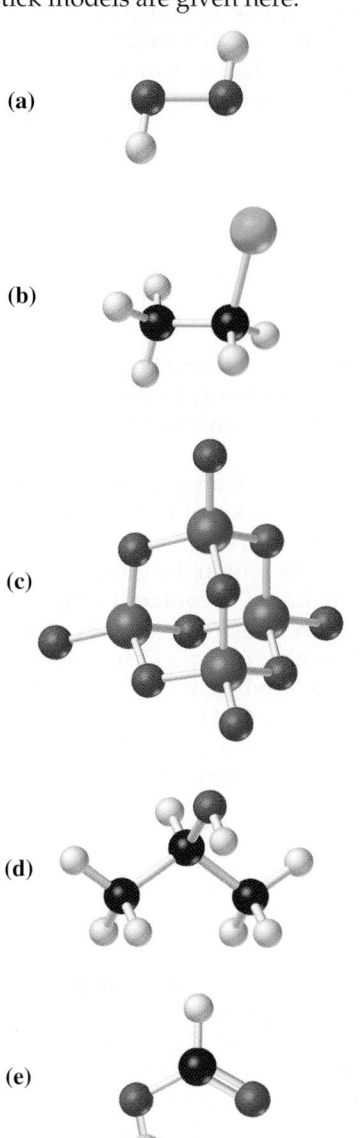

(a)

(b)

(c)

(d)

(e)

2. Give the molecular formulas for the molecules whose ball-and-stick models are given here. Refer to the color scheme in Figure 3-3.

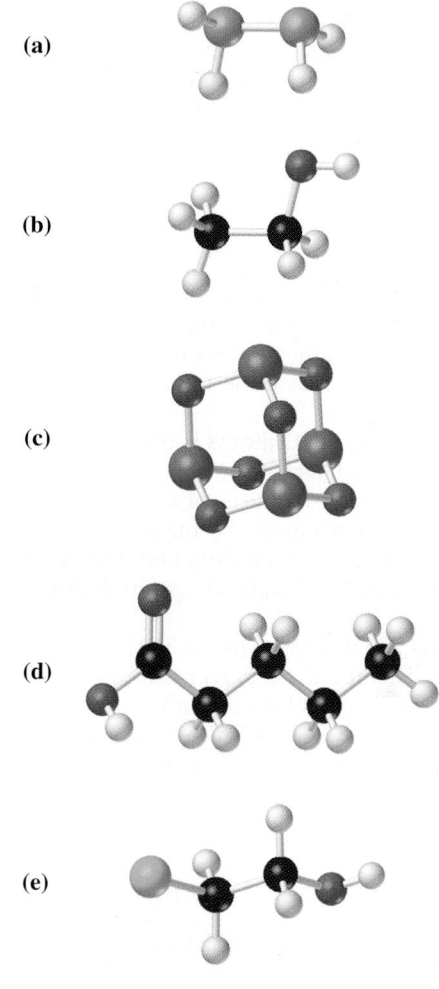

(a)

(b)

(c)

(d)

(e)

3. Give the structural formulas of the molecules shown in Exercise 1 **(b)**, **(d)**, and **(e)**.
4. Give the structural formulas of the molecules shown in Exercise 2 **(b)**, **(d)**, and **(e)**.

The Avogadro Constant and the Mole

5. Calculate the total number of **(a)** atoms in one molecule of trinitrotoluene (TNT), $CH_3C_6H_2(NO_2)_3$; **(b)** atoms in 0.00102 mol $CH_3(CH_2)_4CH_2OH$; **(c)** F atoms in 12.15 mol $C_2HBrClF_3$.
6. Determine the *mass*, in grams, of
   (a) 7.34 mol N_2O_4;
   (b) 3.16×10^{24} O_2 molecules;

   (c) 18.6 mol $CuSO_4 \cdot 5\,H_2O$;
   (d) 4.18×10^{24} molecules of $C_2H_4(OH)_2$.
7. The amino acid methionine, which is essential in human diets, has the molecular formula $C_5H_{11}NO_2S$. Determine **(a)** its molecular mass; **(b)** the number of moles of H atoms per mole of methionine;

(c) the number of grams of C per mole of methionine;
(d) the number of C atoms in 9.07 mol methionine.

8. Determine the number of moles of Br_2 in a sample consisting of **(a)** 8.08×10^{22} Br_2 molecules; **(b)** 2.17×10^{24} Br atoms; **(c)** 11.3 kg bromine; **(d)** 2.65 L liquid bromine ($d = 3.10$ g/mL).

9. *Without doing detailed calculations,* explain which of the following has the greatest number of N atoms **(a)** 50.0 g N_2O; **(b)** 17.0 g NH_3; **(c)** 150 mL of liquid pyridine, C_5H_5N ($d = 0.983$ g/mL); **(d)** 1.0 mol N_2.

10. *Without doing detailed calculations,* determine which of the following has the greatest number of S atoms **(a)** 0.12 mol of solid sulfur, S_8; **(b)** 0.50 mol of gaseous S_2O; **(c)** 65 g of gaseous SO_2; **(d)** 75 mL of liquid thiophene, C_4H_4S ($d = 1.064$ g/mL).

11. Determine the number of *moles* of
    (a) N_2O_4 in a 115 g sample
    (b) N atoms in 43.5 g of $Mg(NO_3)_2$
    (c) N atoms in a sample of $C_7H_5(NO_2)_3$ that has the same number of O atoms as 12.4 g $C_6H_{12}O_6$

12. Determine the *mass,* in grams, of
    (a) 6.25×10^{-2} mol P_4
    (b) 4.03×10^{24} molecules of stearic acid, $C_{18}H_{36}O_2$
    (c) a quantity of the amino acid lysine, $C_6H_{14}N_2O_2$, containing 3.03 mol N atoms

13. The hemoglobin content of blood is about 15.5 g/100 mL blood. The molar mass of hemoglobin is about 64,500 g/mol, and there are four iron (Fe) atoms in a hemoglobin molecule. Approximately how many Fe atoms are present in the 6 L of blood in a typical adult?

14. In white phosphorus, P atoms are joined into P_4 molecules (see Figure 3-5). White phosphorus is commonly supplied in chalk-like cylindrical form. Its density is 1.823 g/cm^3. For a cylinder of white phosphorus 6.50 cm long and 1.22 cm in diameter, determine **(a)** the number of moles of P_4 present; **(b)** the total number of P atoms.

Chemical Formulas

15. Explain which of the following statement(s) is (are) correct concerning glucose (blood sugar), $C_6H_{12}O_6$.
    (a) The percentages, by mass, of C and O are the same as in CO.
    (b) The ratio of C:H:O atoms is the same as in dihydroxyacetone, $(CH_2OH)_2CO$.
    (c) The proportions, by mass, of C and O are equal.
    (d) The highest percentage, by mass, is that of H.

16. Explain which of the following statement(s) is (are) correct for sorbic acid, $C_6H_8O_2$, an inhibitor of mold and yeast.
    (a) It has a C:H:O mass ratio of 3:4:1.
    (b) It has the same mass percent composition as the aquatic herbicide, acrolein, C_3H_4O.
    (c) It has the same empirical formula as aspidinol, $C_{12}H_{16}O_4$, a drug used to kill parasitic worms.

    (d) It has four times as many H atoms as O atoms, but four times as much O as H by mass.

17. For the mineral torbernite, $Cu(UO_2)_2(PO_4)_2 \cdot 8\,H_2O$, determine
    (a) the total number of atoms in one formula unit
    (b) the ratio, by number, of H atoms to O atoms
    (c) the ratio, by mass, of Cu to P
    (d) the element present in the greatest mass percent
    (e) the mass required to contain 1.00 g P

18. For the compound $Ge[S(CH_2)_4CH_3]_4$, determine
    (a) the total number of atoms in one formula unit
    (b) the ratio, by number, of C atoms to H atoms
    (c) the ratio, by mass, of Ge to S
    (d) the number of g S in 1 mol of the compound
    (e) the number of C atoms in 33.10 g of the compound

Percent Composition of Compounds

19. Determine the mass percent H in the hydrocarbon decane, $C_{10}H_{22}$.

20. Determine the mass percent O in the mineral malachite, $Cu_2(OH)_2CO_3$.

21. Determine the mass percent H in the hydrocarbon isooctane, $C(CH_3)_3CH_2CH(CH_3)_2$.

22. Determine the mass percent H_2O in the hydrate $Cr(NO_3)_3 \cdot 9\,H_2O$.

23. Determine the mass percent of each of the elements in the antimalarial drug quinine, $C_{20}H_{24}N_2O_2$.

24. Determine the mass percent of each of the elements in the fungicide copper(II) oleate, $Cu(C_{18}H_{33}O_2)_2$.

25. Determine the percent, by mass, of the indicated element:
    (a) Pb in tetraethyl lead, $Pb(C_2H_5)_4$, once extensively used as an additive to gasoline to prevent engine knocking

    (b) Fe in Prussian blue, $Fe_4[Fe(CN)_6]_3$, a pigment used in paints and printing inks
    (c) Mg in chlorophyll, $C_{55}H_{72}MgN_4O_5$, the green pigment in plant cells

26. All of the following minerals are semiprecious or precious stones. Determine the mass percent of the indicated element.
    (a) Zr in zircon, $ZrSiO_4$
    (b) Be in beryl (emerald), $Be_3Al_2Si_6O_{18}$
    (c) Fe in almandine (garnet), $Fe_3Al_2Si_3O_{12}$
    (d) S in lazurite (lapis lazuli), $Na_4SSi_3Al_3O_{12}$

27. *Without doing detailed calculations,* arrange the following in order of increasing % Cr, by mass, and explain your reasoning: CrO, Cr_2O_3, CrO_2, CrO_3.

28. Without doing detailed calculations, explain which of the following has the greatest mass percent of sulfur: SO_2, S_2Cl_2, Na_2S, $Na_2S_2O_3$, or CH_3CH_2SH.

Chemical Formulas from Percent Composition

29. Two oxides of sulfur have nearly identical molecular masses. One oxide consists of 40.05% S. What are the simplest possible formulas for the two oxides?
30. An oxide of chromium used in chrome plating has a formula mass of 100.0 u and contains *four* atoms per formula unit. Establish the formula of this compound, *with a minimum of calculation.*
31. Diethylene glycol, used to deice aircraft, is a carbon–hydrogen–oxygen compound with 45.27% C and 9.50% H by mass. What is its empirical formula?
32. The food flavor enhancer monosodium glutamate (MSG) has the composition 13.6% Na, 35.5% C, 4.8% H, 8.3% N, 37.8% O, by mass. What is the empirical formula of MSG?
33. Determine the empirical formula of **(a)** the rodenticide (rat killer) warfarin, which consists of 74.01% C, 5.23% H, and 20.76% O, by mass; **(b)** the antibacterial agent sulfamethizole, which consists of 39.98% C, 3.73% H, 20.73% N, 11.84% O, and 23.72% S, by mass.
34. Determine the empirical formula of **(a)** benzo-[*a*]pyrene, a suspected carcinogen found in cigarette smoke, consisting of 95.21% C and 4.79% H, by mass; **(b)** hexachlorophene, used in germicidal soaps, which consists of 38.37% C, 1.49% H, 52.28% Cl, and 7.86% O by mass.
35. A compound of carbon and hydrogen consists of 94.34% C and 5.66% H, by mass. The molecular mass of the compound is found to be 178 u. What is its molecular formula?
36. Selenium, an element used in the manufacture of photoelectric cells and solar energy devices, forms two oxides. One has 28.8% O, by mass, and the other, 37.8% O. What are the formulas of these oxides? Propose acceptable names for them.
37. Indigo, the dye for blue jeans, has a percent composition, by mass, of 73.27% C, 3.84% H, 10.68% N, and the remainder is oxygen. The molecular mass of indigo is 262.3 u. What is the molecular formula of indigo?
38. Adenine, a component of nucleic acids, has the mass percent composition: 44.45% C, 3.73% H, 51.82% N. Its molecular mass is 135.14 u. What is its molecular formula?
39. The element X forms the chloride XCl_4 containing 75.0% Cl, by mass. What is element X?
40. The element X forms the compound $XOCl_2$ containing 59.6% Cl. What is element X?
41. Chlorophyll contains 2.72% Mg by mass. Assuming one Mg atom per chlorophyll molecule, what is the molecular mass of chlorophyll?
42. Two compounds of Cl and X are found to have molecular masses and % Cl, by mass, as follows: 137 u, 77.5% Cl; 208 u, 85.1% Cl. What is element X? What is the formula for each compound?

Combustion Analysis

43. A 0.1888 g sample of a hydrocarbon produces 0.6260 g CO_2 and 0.1602 g H_2O in combustion analysis. Its molecular mass is found to be 106 u. For this hydrocarbon, determine its **(a)** mass percent composition; **(b)** empirical formula; **(c)** molecular formula.
44. *Para*-cresol (*p*-cresol) is used as a disinfectant and in the manufacture of herbicides. A 0.4039 g sample of this carbon–hydrogen–oxygen compound yields 1.1518 g CO_2 and 0.2694 g H_2O in combustion analysis. Its molecular mass is 108.1 u. For *p*-cresol, determine its **(a)** mass percent composition; **(b)** empirical formula; **(c)** molecular formula.
45. Dimethylhydrazine is a carbon–hydrogen–nitrogen compound used in rocket fuels. When burned in an excess of oxygen, a 0.312 g sample yields 0.458 g CO_2 and 0.374 g H_2O. The nitrogen content of a 0.486 g sample is converted to 0.226 g N_2. What is the empirical formula of dimethylhydrazine?
46. The organic solvent thiophene is a carbon–hydrogen–sulfur compound that yields CO_2, H_2O, and SO_2 when burned in an excess of oxygen. When subjected to combustion analysis, a 1.3020 g sample of thiophene produces 2.7224 g CO_2, 0.5575 g H_2O, and 0.9915 g SO_2. What is the empirical formula of thiophene?
47. *Without doing detailed calculations*, explain which of these compounds produces the *greatest* mass of CO_2 when 1.00 mol of the compound is burned in an excess of oxygen: CH_4, C_2H_5OH, $C_{10}H_8$, C_6H_5OH.
48. *Without doing detailed calculations*, explain which of these compounds produces the *greatest* mass of H_2O when 1.00 g of the compound is burned in an excess of oxygen: CH_4, C_2H_5OH, $C_{10}H_8$, C_6H_5OH.
49. A 1.562 g sample of the alcohol $CH_3CHOHCH_2CH_3$ is burned in an excess of oxygen. What masses of CO_2 and H_2O should be obtained?
50. Liquid ethyl mercaptan, C_2H_6S, has a density of 0.84 g/mL. Assuming that the combustion of this compound produces only CO_2, H_2O, and SO_2, what masses of each of these three products would be produced in the combustion of 3.15 mL of ethyl mercaptan?

Oxidation States

51. Indicate the oxidation state of the underlined element in **(a)** $\underline{C}H_4$; **(b)** $\underline{S}F_4$; **(c)** $Na_2\underline{O}_2$; **(d)** $\underline{C}_2H_3O_2^-$; **(e)** $\underline{Fe}O_4^{2-}$.
52. Indicate the oxidation state of S in **(a)** SO_3^{2-}; **(b)** $S_2O_3^{2-}$; **(c)** $S_2O_8^{2-}$; **(d)** HSO_4^-; **(e)** $S_4O_6^{2-}$.
53. Chromium forms three principal oxides. Write appropriate formulas for these compounds in which the oxidation states of Cr are +3, +4, and +6, respectively.
54. Nitrogen forms five oxides in which its oxidation states are +1, +2, +3, +4, and +5, respectively. Write appropriate formulas for these compounds.

55. In many of its compounds, oxygen has an oxidation state of -2. However, there are exceptions. What is the oxidation state of oxygen in each of the following compounds? (a) OF_2; (b) O_2F_2; (c) CsO_2; (d) BaO_2.

56. Hydrogen and oxygen usually have oxidation states of $+1$ and -2, respectively, in their compounds. The following cases serve to remind us that there are exceptions. What are the oxidation states of the atoms in each of the following compounds? (a) MgH_2; (b) CsO_3; (c) HOF; (d) $NaAlH_4$.

Nomenclature

57. Name these compounds: (a) SrO; (b) ZnS; (c) K_2CrO_4; (d) Cs_2SO_4; (e) Cr_2O_3; (f) $Fe_2(SO_4)_3$; (g) $Mg(HCO_3)_2$; (h) $(NH_4)_2HPO_4$; (i) $Ca(HSO_3)_2$; (j) $Cu(OH)_2$; (k) HNO_3; (l) $KClO_4$; (m) $HBrO_3$; (n) H_3PO_3.

58. Name these compounds: (a) $Ba(NO_3)_2$; (b) HNO_2; (c) CrO_2; (d) KIO_3; (e) $LiCN$; (f) KIO; (g) $Fe(OH)_2$; (h) $Ca(H_2PO_4)_2$; (i) H_3PO_4; (j) $NaHSO_4$; (k) $Na_2Cr_2O_7$; (l) $NH_4C_2H_3O_2$; (m) MgC_2O_4; (n) $Na_2C_2O_4$.

59. Assign suitable names to the compounds (a) CS_2; (b) SiF_4; (c) ClF_5; (d) N_2O_5; (e) SF_6; (f) I_2Cl_6.

60. Assign suitable names to the compounds (a) ICl; (b) ClF_3; (c) SF_4; (d) BrF_5; (e) N_2O_4; (f) S_4N_4.

61. Write formulas for the compounds: (a) aluminum sulfate; (b) ammonium dichromate; (c) silicon tetrafluoride; (d) iron(III) oxide; (e) tricarbon disulfide; (f) cobalt(II) nitrate; (g) strontium nitrite; (h) hydrobromic acid; (i) iodic acid; (j) phosphorus dichloride trifluoride.

62. Write formulas for the compounds: (a) magnesium perchlorate; (b) lead(II) acetate; (c) tin(IV) oxide; (d) hydroiodic acid; (e) chlorous acid; (f) sodium hydrogen sulfite; (g) calcium dihydrogen phosphate; (h) aluminum phosphate; (i) dinitrogen tetroxide; (j) disulfur dichloride.

63. Write a formula for (a) the chloride of titanium having Ti in the O.S. $+4$; (b) the sulfate of iron having Fe in the O.S. $+3$; (c) an oxide of chlorine with Cl in the O.S. $+7$; (d) an oxoanion of sulfur in which the apparent O.S. of S is $+7$ and the ionic charge is $2-$.

64. Write a formula for (a) an oxide of nitrogen with N in the O.S. $+5$; (b) an oxoacid of nitrogen with N in the O.S. $+3$; (c) an oxide of carbon in which the apparent O.S. of C is $+4/3$; (d) a sulfur-containing oxoanion in which the apparent O.S. of S is $+2.5$ and the ionic charge is $2-$.

65. Name the acids: (a) $HClO_2$; (b) H_2SO_3; (c) H_2Se; (d) HNO_2.

66. Supply the formula for the acids: (a) hydrofluoric acid; (b) nitric acid; (c) phosphorous acid; (d) sulfuric acid.

67. Name the following compounds and specify which ones are best described as ionic: (a) OF_2; (b) XeF_2; (c) $CuSO_3$; (d) $(NH_4)_2HPO_4$;.

68. Name the following compounds and specify which ones are best described as ionic: (a) KNO_2; (b) BrF_3; (c) S_2Cl_2; (d) $Mg(ClO)_2$; (e) Cl_2O.

Hydrates

69. *Without performing detailed calculations*, indicate which of the following hydrates has the greatest % H_2O by mass: $CuSO_4 \cdot 5\,H_2O$, $Cr_2(SO_4)_3 \cdot 18\,H_2O$, $MgCl_2 \cdot 6\,H_2O$, and $LiC_2H_3O_2 \cdot 2\,H_2O$.

70. *Without performing detailed calculations*, determine the hydrate of Na_2SO_3 that contains almost exactly 50% H_2O, by mass.

71. Anhydrous $CuSO_4$ can be used to dry liquids in which it is insoluble. The $CuSO_4$ is converted to $CuSO_4 \cdot 5\,H_2O$, which can be filtered off from the liquid. What is the minimum mass of anhydrous $CuSO_4$ needed to remove 12.6 g H_2O from a tankful of gasoline?

72. Anhydrous sodium sulfate, Na_2SO_4, absorbs water vapor and is converted to the decahydrate, $Na_2SO_4 \cdot 10\,H_2O$. How much would the mass of 24.05 g of anhydrous Na_2SO_4 increase if converted completely to the decahydrate?

73. A certain hydrate is found to have the composition 20.3% Cu, 8.95% Si, 36.3% F, and 34.5% H_2O by mass. What is the empirical formula of this hydrate?

74. An 8.129 g sample of $MgSO_4 \cdot x\,H_2O$ is heated until all the water of hydration is driven off. The resulting anhydrous compound, $MgSO_4$, weighs 3.967 g. What is the formula of the hydrate?

Organic Compounds and Organic Nomenclature

75. Which of the following names is most appropriate for the molecule with the structure shown below? (a) butyl alcohol; (b) 2-butanol; (c) 1-butanol; (d) isopentyl alcohol.

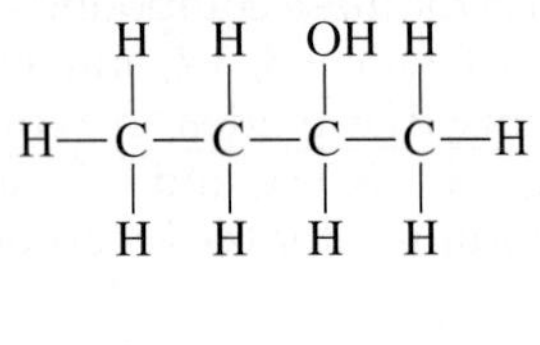

```
      H   H   OH  H
      |   |   |   |
  H—C—C—C—C—H
      |   |   |   |
      H   H   H   H
```

76. Which of the following names is most appropriate for the molecule $CH_3(CH_2)_2COOH$? (a) dimethyleneacetic acid; (b) propanoic acid; (c) butanoic acid; (d) oxobutylalcohol.

77. Which of the following structures are isomers?

(a) CH_3—CH—CH_2—OH
               |
           CH_2—CH_3

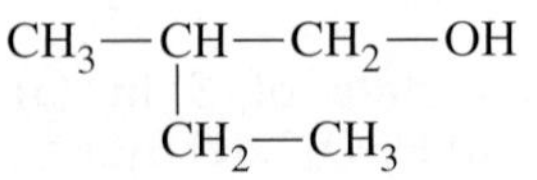

(b) CH$_3$—CH—CH$_2$—CH$_2$—OH
$\qquad\quad$|
$\qquad\quad$CH$_3$

(c) CH$_3$—CH$_2$—CH—CH$_2$—OH
$\qquad\qquad\qquad$|
$\qquad\qquad\quad$CH$_3$

(d) CH$_3$—CH—CH$_2$—O—CH$_3$
$\qquad\quad$|
$\qquad\quad$CH$_3$

(e) CH$_3$—CH—CH$_2$—CH—CH$_3$
$\qquad\quad$|$\qquad\qquad$|
$\qquad\quad$CH$_3$$\qquad\quad$OH

78. Which of the following structures are isomers?

(a) CH$_3$—CH—CH$_2$—Cl
$\qquad\quad$|
$\qquad\quad$CH$_2$—CH$_3$

(b) CH$_3$—CH—CH$_2$—CH$_3$
$\qquad\quad$|
$\qquad\quad$CH$_2$Cl

(c) CH$_3$—CH—CHClCH$_3$
$\qquad\quad$|
$\qquad\quad$CH$_3$

(d) CH$_3$—CH—CH$_2$—CH—CH$_3$
$\qquad\quad$|$\qquad\qquad$|
$\qquad\quad$CH$_3$$\qquad\quad$Cl

79. Write the condensed structural formulas for the organic compounds:
(a) heptane$\qquad\qquad$**(b)** propanoic acid
(c) 2-methyl-1-pentanol$\quad$**(d)** fluoroethane
80. Write the condensed structural formulas for the organic compounds:
(a) octane$\qquad\qquad\quad$**(b)** heptanoic acid
(c) 3-hexanol$\qquad\qquad$**(d)** 2-chlorobutane

81. Give the name, condensed structural formula, and molecular mass of the molecule whose ball-and-stick model is shown. Refer to the color scheme in Figure 3-3.

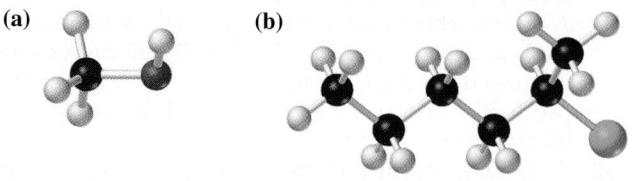

82. Give the name, condensed structural formula, and molecular mass of the molecule whose ball-and-stick model is shown. Refer to the color scheme in Figure 3-3.

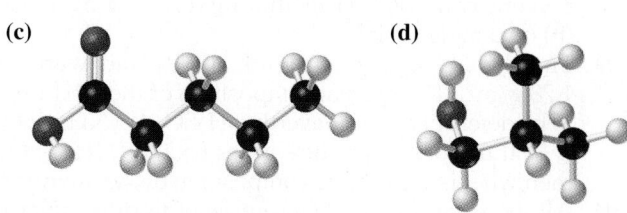

Integrative and Advanced Exercises

83. The mineral spodumene has the empirical formula LiAlSi$_2$O$_6$. Given that the percentage of lithium-6 atoms in naturally occuring lithium is 7.40%, how many lithium-6 atoms are present in a 518 g sample of spodumene?

84. A particular type of brass contains Cu, Sn, Pb, and Zn. A 1.1713 g sample is treated in such a way as to convert the Sn to 0.245 g SnO$_2$, the Pb to 0.115 g PbSO$_4$, and the Zn to 0.246 g Zn$_2$P$_2$O$_7$. What is the mass percent of each element in the sample?

85. A brand of lunchmeat contains 0.10% by mass of sodium benzoate, C$_6$H$_5$COONa. How many mg of Na does a person ingest by eating 2.52 oz of this meat?

86. The important natural sources of boron compounds are the minerals kernite, Na$_2$B$_4$O$_7 \cdot 4$ H$_2$O and borax, Na$_2$B$_4$O$_7 \cdot 10$ H$_2$O. How much *additional* mass of mineral must be processed per kilogram of boron obtained if the mineral is borax rather than kernite?

87. To deposit exactly one mole of Ag from an aqueous solution containing Ag$^+$ requires a quantity of electricity known as one faraday (F). The electrodeposition requires that each Ag$^+$ ion gain one electron to become an Ag atom. Use appropriate physical constants listed on the inside back cover to obtain a precise value of the Avogadro constant, N_A.

88. By analysis, a compound was found to contain 26.58 % K and 35.45 % Cr by mass; the remainder was oxygen. What is the oxidation state of chromium in this compound? What is the name of the compound?

89. Is it possible to have a sample of S$_8$ that weighs 1.00×10^{-23} g? What is the smallest possible mass that a sample of S$_8$ can have? Express your answer to the second question in appropriate SI units so that your answer has a numerical value greater than 1. (See Table 1.2 for a list of SI prefixes.)

90. What is the molecular formula of a hydrocarbon containing n carbon atoms and only one double bond? Can such a hydrocarbon yield a greater mass of H_2O than CO_2 when burned in an excess of oxygen?

91. A hydrocarbon mixture consists of 60.0% by mass of C_3H_8 and 40.0% of C_xH_y. When 10.0 g of this mixture is burned, 29.0 g CO_2 and 18.8 g H_2O are the only products. What is the formula of the unknown hydrocarbon?

92. A 0.732 g mixture of methane, CH_4, and ethane, C_2H_6, is burned, yielding 2.064 g CO_2. What is the percent composition of this mixture (a) by mass; (b) on a mole basis?

93. The density of a mixture of H_2SO_4 and water is 1.78 g/mL. The percent composition of the mixture is to be determined by converting H_2SO_4 to $(NH_4)_2SO_4$. If 32.0 mL of the mixture gives 65.2 g $(NH_4)_2SO_4$, then what is the percent composition of the mixture?

94. All the silver in a 26.39 g sample of impure silver is converted to silver chloride. If 31.56 g of silver chloride are obtained, then what is the mass percent of silver in the sample?

95. In the year 2000, the *Guinness Book of World Records* called ethyl mercaptan, C_2H_6S, the smelliest substance known. The average person can detect its presence in air at levels as low as 9×10^{-4} μmol/m^3. Express the limit of detectability of ethyl mercaptan in parts per billion (ppb). (Note: 1 ppb C_2H_6S means there is 1 g C_2H_6S per billion grams of air.) The density of air is approximately 1.2 g/L at room temperature.

96. Dry air is essentially a mixture of the following entities: N_2, O_2, Ar, and CO_2. The composition of dry air, in *mole percent*, is 78.08% N_2, 20.95% O_2, 0.93% Ar, and 0.04% CO_2. (a) What is the mass, in grams, of a sample of air that contains exactly one mole of the entities? (b) Dry air also contains other entities in much smaller amounts. For example, the mole percent of krypton (Kr) is about 1.14×10^{-4} %. Given that the density of dry air is about 1.2 g/L at room temperature, what mass of krypton could be obtained from exactly one cubic meter of dry air?

97. A public water supply was found to contain 1 part per billion (ppb) by mass of chloroform, $CHCl_3$. (a) How many $CHCl_3$ molecules would be present in a 225 mL glass of this water? (b) If the $CHCl_3$ in part (a) could be isolated, would this quantity be detectable on an ordinary analytical balance that measures mass with a precision of ± 0.0001 g?

98. A sample of the compound MSO_4 weighing 0.1131 g reacts with barium chloride and yields 0.2193 g $BaSO_4$. What must be the atomic mass of the metal M? [*Hint:* All the SO_4^{2-} from the MSO_4 appears in the $BaSO_4$.]

99. The metal M forms the sulfate $M_2(SO_4)_3$. A 0.738 g sample of this sulfate is converted to 1.511 g $BaSO_4$. What is the atomic mass of M?
[*Hint:* Refer to Exercise 98.]

100. A 0.622 g sample of a metal oxide with the formula M_2O_3 is converted to 0.685 g of the sulfide, MS. What is the atomic mass of the metal M?

101. $MgCl_2$ often occurs in table salt (NaCl) and is responsible for caking of the salt. A 0.5200 g sample of table salt is found to contain 61.10% Cl, by mass. What is the % $MgCl_2$ in the sample? Why is the precision of this calculation so poor?

102. When 2.750 g of the oxide of lead Pb_3O_4 is strongly heated, it decomposes and produces 0.0640 g of oxygen gas and 2.686 g of a second oxide of lead. What is the empirical formula of this second oxide?

103. A 1.013 g sample of $ZnSO_4 \cdot x\, H_2O$ is dissolved in water and the sulfate ion precipitated as $BaSO_4$. The mass of pure, dry $BaSO_4$ obtained is 0.8223 g. What is the formula of the zinc sulfate hydrate?

104. The iodide ion in a 1.552 g sample of the ionic compound MI is removed through precipitation. The precipitate is found to contain 1.186 g I. What is the element M?

105. An oxoacid with the formula $H_xE_yO_z$ has a formula mass of 178 u, has 13 atoms in its formula unit, contains 34.80% by mass, and 15.38% by number of atoms, of the element E. What is the element E, and what is the formula of this oxoacid?

106. The insecticide dieldrin contains carbon, hydrogen, oxygen, and chlorine. When burned in an excess of oxygen, a 1.510 g sample yields 2.094 g CO_2 and 0.286 g H_2O. The compound has a molecular mass of 381 u and has half as many chlorine atoms as carbon atoms. What is the molecular formula of dieldrin?

107. A thoroughly dried 1.271 g sample of Na_2SO_4 is exposed to the atmosphere and found to gain 0.387 g in mass. What is the percent, by mass, of $Na_2SO_4 \cdot 10\, H_2O$ in the resulting mixture of anhydrous Na_2SO_4 and the decahydrate?

108. The atomic mass of Bi is to be determined by converting the compound $Bi(C_6H_5)_3$ to Bi_2O_3. If 5.610 g of $Bi(C_6H_5)_3$ yields 2.969 g Bi_2O_3, what is the atomic mass of Bi?

109. A piece of gold (Au) foil measuring 0.25 mm $\times$ 15 mm $\times$ 15 mm is treated with fluorine gas. The treatment converts all the gold in the foil to 1.400 g of a gold fluoride. What is the formula and name of the fluoride? The density of gold is 19.3 g/cm^3.

110. In an experiment, 244 mL of chlorine gas (Cl_2, $d = 2.898$ g/L) combines with iodine to give 1.553 g of a binary compound. In a separate experiment, the molar mass of the compound is found to be about 467 g/mol. What is the molecular formula of this compound?

Feature Problems

111. All-purpose fertilizers contain the essential elements nitrogen, phosphorus, and potassium. A typical fertilizer carries numbers on its label, such as "5-10-5". These numbers represent the % N, % P_2O_5, and % K_2O, respectively. The N is contained in the form of a nitrogen compound, such as $(NH_4)_2SO_4$, NH_4NO_3, or $CO(NH_2)_2$ (urea). The P is generally present as a phosphate, and the K as KCl. The expressions

% P_2O_5 and % K_2O were devised in the nineteenth century, before the nature of chemical compounds was fully understood. To convert from % P_2O_5 to % P and from % K_2O to % K, the factors 2 mol P/mol P_2O_5 and 2 mol K/mol K_2O must be used, together with molar masses.

(a) Assuming three-significant-figure precision, what is the percent composition of the "5-10-5" fertilizer in % N, % P, and % K?
(b) What is the % P_2O_5 in the following compounds (both common fertilizers)? **(i)** $Ca(H_2PO_4)_2$; **(ii)** $(NH_4)_2HPO_4$.
(c) In a similar manner to the "5-10-5" fertilizer described in this exercise, how would you describe a fertilizer in which the mass ratio of $(NH_4)_2HPO_4$ to KCl is 5.00:1.00?
(d) Can a "5-10-5" fertilizer be prepared in which $(NH_4)_2HPO_4$ and KCl are the sole fertilizer components, with or without inert nonfertilizer additives? If so, what should be the proportions of the constituents of the fertilizer mixture? If this "5-10-5" fertilizer cannot be prepared, why not?

112. A hydrate of copper(II) sulfate, when heated, goes through the succession of changes suggested by the photograph. In this photograph, **(a)** is the original fully hydrated copper(II) sulfate; **(b)** is the product obtained by heating the original hydrate to 140 °C; **(c)** is the product obtained by further heating to 400 °C; and **(d)** is the product obtained at 1000 °C.

(a)	**(b)**	**(c)**	**(d)**

A 2.574 g sample of $CuSO_4 \cdot x\,H_2O$ was heated to 140 °C, cooled, and reweighed. The resulting solid was reheated to 400 °C, cooled, and reweighed. Finally, this solid was heated to 1000 °C, cooled, and reweighed for the last time.

Original sample	2.574 g
After heating to 140 °C	1.833 g
After reheating to 400 °C	1.647 g
After reheating to 1000 °C	0.812 g

(a) Assuming that all the water of hydration is driven off at 400 °C, what is the formula of the original hydrate?
(b) What is the formula of the hydrate obtained when the original hydrate is heated to only 140 °C?
(c) The black residue obtained at 1000 °C is an oxide of copper. What is its percent composition and empirical formula?

113. Some substances that are only very slightly soluble in water will spread over the surface of water to produce a film that is called a *monolayer* because it is only one molecule thick. A practical use of this phenomenon is to cover ponds to reduce the loss of water by evaporation. Stearic acid forms a monolayer on water. The molecules are arranged upright and in contact with one another, rather like pencils tightly packed and standing upright in a coffee mug. The model below represents an individual stearic acid molecule in the monolayer.

(a) How many square meters of water surface would be covered by a monolayer made from 10.0 g of stearic acid?
[*Hint:* What is the formula of stearic acid?]
(b) If stearic acid has a density of 0.85 g/cm³, estimate the length (in nanometers) of a stearic acid molecule.
[*Hint:* What is the thickness of the monolayer described in part a?]
(c) A very dilute solution of oleic acid in liquid pentane is prepared in the following way:

1.00 mL oleic acid + 9.00 mL pentane → solution (1);
1.00 mL solution (1) + 9.00 mL pentane → solution (2);
1.00 mL solution (2) + 9.00 mL pentane → solution (3);
1.00 mL solution (3) + 9.00 mL pentane → solution (4).

A 0.10 mL sample of solution (4) is spread in a monolayer on water. The area covered by the monolayer is 85 cm². Assume that oleic acid molecules are arranged in the same way as described for stearic acid, and that the cross-sectional area of the molecule is 4.6×10^{-15} cm². The density of oleic acid is 0.895 g/mL. Use these data to obtain an approximate value of Avogadro's number.

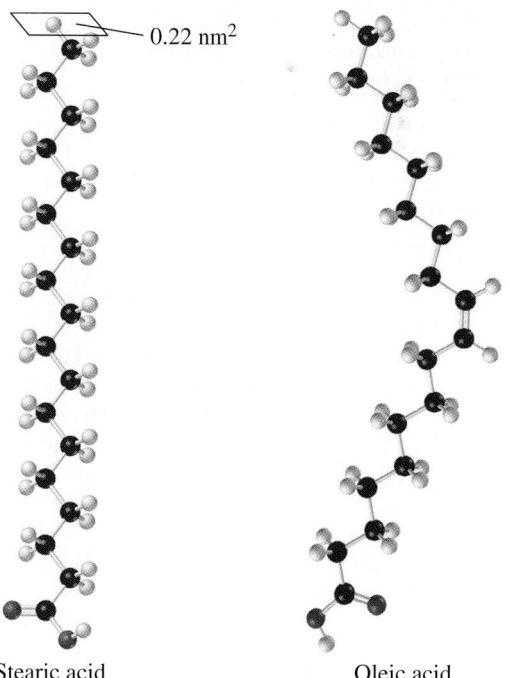

Stearic acid   Oleic acid

Self-Assessment Exercises

114. In your own words, define or explain the following terms or symbols: **(a)** formula unit; **(b)** P_4; **(c)** molecular compound; **(d)** binary compound; **(e)** hydrate.

115. Briefly describe each of the following ideas or methods: **(a)** mole of a compound; **(b)** structural formula; **(c)** oxidation state; **(d)** carbon–hydrogen–oxygen determination by combustion analysis.

116. Explain the important distinctions between each pair of terms: **(a)** molecular mass and molar mass; **(b)** empirical and molecular formulas; **(c)** systematic and trivial, or common, name; **(d)** hydroxyl and carboxyl functional group.

117. Explain each term as it applies to the element nitrogen. **(a)** atomic mass; **(b)** molecular mass; **(c)** molar mass.

118. Which answer is correct? One mole of liquid bromine, Br_2, **(a)** has a mass of 79.9 g; **(b)** contains 6.022×10^{23} Br atoms; **(c)** contains the same number of atoms as in 12.01 g H_2O; **(d)** has twice the mass of 0.500 mole of gaseous Cl_2.

119. Three of the following formulas might be either an empirical or a molecular formula. The formula that must be a molecular formula is **(a)** N_2O; **(b)** N_2H_4; **(c)** NaCl; **(d)** NH_3.

120. The compound $C_7H_7NO_2$ contains **(a)** 17 atoms per mole; **(b)** equal percents by mass of C and H; **(c)** about twice the percent by mass of O as of N; **(d)** about twice the percent by mass of N as of H.

121. The greatest number of N atoms is found in **(a)** 50.0 g N_2O; **(b)** 17.0 g NH_3; **(c)** 150 mL of liquid pyridine, C_5H_5N ($d = 0.983$ g/mL); **(d)** 1.0 mol N_2.

122. XF_3 consists of 65% F by mass. The atomic mass of the element X must be **(a)** 8 u; **(b)** 11u; **(c)** 31 u; **(d)** 35 u.

123. The oxidation state of I in the ion $H_4IO_6^-$ is **(a)** −1; **(b)** +1; **(c)** +7; **(d)** +8.

124. The formula for calcium chlorite is **(a)** $CaClO_2$; **(b)** $Ca(ClO_2)_2$; **(c)** $CaClO_3$; **(d)** $Ca(ClO_4)_2$.

125. A formula unit of the compound $[Cu(NH_3)_4]SO_4$ has nearly equal masses of **(a)** S and O; **(b)** N and O; **(c)** H and N; **(d)** Cu and O.

126. An isomer of the compound $CH_3CH_2CHOHCH_3$ is **(a)** $C_4H_{10}O$; **(b)** $CH_3CHOHCH_2CH_3$; **(c)** $CH_3(CH_2)_2OH$; **(d)** $CH_3CH_2OCH_2CH_3$.

127. A hydrate of Na_2SO_3 contains almost exactly 50% H_2O by mass. What is the formula of this hydrate?

128. Malachite is a common copper-containing mineral with the formula $CuCO_3 \cdot Cu(OH)_2$. **(a)** What is the mass percent copper in malachite? **(b)** When malachite is strongly heated, carbon dioxide and water are driven off, yielding copper(II) oxide as the sole product. What mass of copper(II) oxide is produced per kg of malachite?

129. Acetaminophen, an analgesic and antipyretic drug, has a molecular mass of 151.2 u and a mass percent composition of 63.56% C, 6.00% H, 9.27% N, and 21.17% O. What is the molecular formula of acetaminophen?

130. Ibuprofen is a compound used in painkillers. When a 2.174 g sample is burned in an excess of oxygen, it yields 6.029 g CO_2 and 1.709 g H_2O as the sole products. **(a)** What is the percent composition, by mass, of ibuprofen? **(b)** What is the empirical formula of ibuprofen?

131. Appendix E describes a useful study aid known as concept mapping. Using the method presented in Appendix E, construct a concept map illustrating the different concepts in Sections 3-2 and 3-3.

Chemical
Reactions

The space shuttle *Discovery* lifts off on mission STS-26. Combustion reactions in the solid-fuel rocket engines provide the thrust to lift the shuttle off the launch pad. In this chapter, we learn to write and use balanced chemical equations for a wide variety of chemical reactions, including combustion reactions.

We are all aware that iron rusts and natural gas burns. These processes are chemical reactions. Chemical reactions are the central concern not just of this chapter but of the entire science of chemistry. In this chapter, we will establish quantitative (numerical) relationships among the substances involved in a reaction, a topic known as *reaction stoichiometry*. Because many chemical reactions occur in solution, we will also consider *solution stoichiometry* and introduce a method of describing the composition of a solution called *solution molarity*.

In describing chemical reactions, we often take a microscopic view and focus on the entities—atoms, ions, or molecules—that make up the substances involved. However, when *doing* chemistry, we often think of reactions in more macroscopic terms because, in the laboratory, we handle quantities of substances—grams or liters—that can be easily measured or manipulated. In large part, reaction stoichiometry provides the relationships we need to relate macroscopic amounts of substances to our microscopic view of chemical reactions.

To some, stoichiometry is no more exciting than the law of conservation of mass, but make no mistake—stoichiometry is important. Chemists use stoichiometric principles routinely to plan experiments, analyze their results, and make predictions, all of which contribute to making new discoveries and expanding our knowledge of the microscopic world of atoms, molecules, and ions.

In this chapter, we will first learn to represent chemical reactions by chemical equations, and then we will use chemical equations—and ideas from earlier chapters—to establish the quantitative relationships we seek. Throughout the chapter, we will discuss new aspects of problem solving and more uses for the mole concept.

4-1 Chemical Reactions and Chemical Equations

A **chemical reaction** is a process in which one set of substances, called **reactants**, is converted to a new set of substances, called **products**. In other words, a chemical reaction is the process by which a chemical change occurs. In many cases, though, nothing happens when substances are mixed; each retains its original composition and properties. We need evidence before we can say that a reaction has occurred. Some of the types of physical evidence to look for are shown here:

- a color change (Fig. 4-1)
- formation of a solid (precipitate) within a clear solution (Fig. 4-1)
- evolution of a gas (Fig. 4-2a)
- evolution or absorption of heat (Fig. 4-2b)

Although such observations as these usually signify that a reaction has occurred, conclusive evidence still requires a detailed chemical analysis of the reaction mixture to identify all the substances present. Moreover, a chemical analysis may reveal that a chemical reaction has occurred even in the absence of obvious physical signs.

Just as there are symbols for elements and formulas for compounds, there is a symbolic, or shorthand, way of representing a chemical reaction—the **chemical equation**. In a chemical equation, formulas for the reactants are written on the left side of the equation and formulas for the products are written on the right. The two sides of the equation are joined by an arrow ($\longrightarrow$). We say that the reactants *yield* the products. Consider the reaction of colorless nitrogen

▲ FIGURE 4-1
Precipitation of silver chromate
When aqueous solutions of silver nitrate and potassium chromate are mixed, the disappearance of the distinctive yellow color of chromate ion and the appearance of the red-brown solid, silver chromate, provide physical evidence of a reaction.

▶ FIGURE 4-2
Evidence of a chemical reaction
(a) Evolution of a gas: When a copper penny reacts with nitric acid, the red-brown gas NO_2 is evolved. (b) Evolution of heat: When iron gauze (steel wool) is ignited in an oxygen atmosphere, evolved heat and light provide physical evidence of a reaction.

(a)

(b)

monoxide and oxygen gases to form red-brown nitrogen dioxide gas, a reaction that occurs in the manufacture of nitric acid.

nitrogen monoxide + oxygen $\longrightarrow$ nitrogen dioxide

To complete the shorthand representation of this reaction, we must do two things:

1. Substitute chemical formulas for names, to obtain the following expression.

$NO + O_2 \longrightarrow NO_2$

In this expression, there are three O atoms on the left side (one in the molecule NO and two in the molecule O_2), but only two O atoms (in the molecule NO_2) on the right. Because atoms are neither created nor destroyed in a chemical reaction, this expression needs to be *balanced*.

2. Balance the numbers of atoms of each kind on both sides of the expression to obtain a balanced chemical equation.* In this step, the coefficient two is placed in front of the formulas NO and NO_2. This means that two molecules of NO are consumed and two molecules of NO_2 are produced for every molecule of O_2 consumed. In the balanced equation there are two N atoms and four O atoms on each side. In a **balanced equation**, the total number of atoms of each element present is the same on both sides of the equation. We see this below, both in the symbolic equation and in the molecular representation of the reaction.

$2\,NO + O_2 \longrightarrow 2\,NO_2$

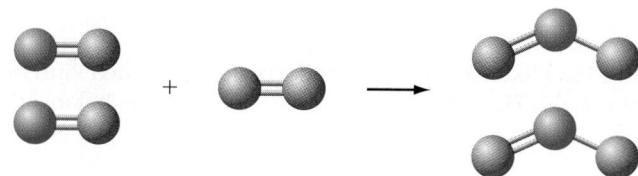

The coefficients required to balance a chemical equation are called **stoichiometric coefficients**. These coefficients are essential in relating the amounts of reactants used and products formed in a chemical reaction, through a variety of calculations. In balancing a chemical equation, keep the following point in mind.

> An equation can be balanced only by adjusting the coefficients of formulas.

The method of equation balancing described above is called *balancing by inspection*. Balancing by inspection means to adjust stoichiometric coefficients by trial and error until a balanced condition is found. Although the elements can generally be balanced in any order, equation balancing need not be a hit-or-miss affair. Here are some useful strategies for balancing equations.

- If an element occurs in only one compound on each side of the equation, try balancing this element first.
- When one of the reactants or products exists as the free element, balance this element last.
- In some reactions, certain groups of atoms (for example, polyatomic ions) remain unchanged. In such cases, balance these groups as a unit.

◀ Sometimes products react, partially or completely, to re-form the original reactants. Such reactions are called *reversible reactions* and are designated by a double arrow ($\rightleftharpoons$). In this chapter, we assume that any reverse reactions are negligible and that reactions go only in the forward direction. We consider reversible reactions in detail in Chapter 15.

◀ The reaction of NO and O_2 is of considerable environmental interest. Automobile engines—and all other combustion devices—make NO. One purpose of a catalytic converter is to reduce the amount of NO emitted from a tailpipe. Once in the atmosphere, NO is converted into brown NO_2 by the reaction illustrated on this page. This is why many urban atmospheres have a brown tinge.

---

*An equation—whether mathematical or chemical—must have the left and right sides equal. We should not call an expression an equation until it is balanced. The term *chemical equation* automatically signifies that this balance exists. Although unnecessary, the term *balanced* is commonly used when referring to a chemical equation.

▶ We will encounter a few situations in Chapters 7 and 19 where some fractional coefficients are actually required.

KEEP IN MIND

that the strategies described here work well for simple reactions. However, some reactions cannot be balanced by inspection and systematic methods must be employed. We will encounter such situations in Chapter 5.

- It is permissible to use fractional as well as integral numbers as coefficients. At times, an equation can be balanced most easily by using one or more fractional coefficients and then, if desired, clearing the fractions by multiplying *all* coefficients by a common multiplier.

In this chapter you should concentrate on learning to write formulas for the reactants and products of a reaction and to balance the equation representing the reaction. A third task, which you will face in later chapters, is to predict the products formed when certain reactants are brought together under the appropriate conditions. Even now, however, based on concepts presented in the previous chapter, you should be able to predict the products of a *combustion reaction*. In a plentiful supply of oxygen gas, the combustion of hydrocarbons and of carbon–hydrogen–oxygen compounds produces carbon dioxide and water as the only products. If the compound contains sulfur as well, sulfur dioxide will also be a product. These ideas and an equation-balancing strategy are illustrated in Examples 4-1 and 4-2.

EXAMPLE 4-1   **Balancing an Equation**

Ammonia and oxygen can react in several different ways, including the following. Balance the equation for this reaction.

$$NH_3 + O_2 \longrightarrow N_2 + H_2O$$

Analyze

We adjust the coefficients of NH_3, O_2, N_2, and H_2O to balance the numbers of atoms of each kind on both sides of the equation. There are three kinds of atoms: N, H, and O. The chemical equation is balanced only if it is balanced with respect to each kind of atom.

Solve

Since there is *one* N atom in NH_3 on the left, and *two* N atoms in N_2 on the right, the coefficient 2 must be added in front of NH_3.

$$2\,NH_3 + O_2 \longrightarrow N_2 + H_2O$$

Since there are *six* H atoms (in $2\,NH_3$) on the left of the equation just written and only *two* on the right (in H_2O), the coefficient 3 must be added in front of H_2O.

$$2\,NH_3 + O_2 \longrightarrow N_2 + 3\,H_2O$$

There are now *two* O atoms on the left (in O_2) and *three* O atoms on the right (in $3\,H_2O$). Three O atoms can be shown on the left by placing the coefficient $\frac{3}{2}$ in front of O_2.

$$2\,NH_3 + \tfrac{3}{2}O_2 \longrightarrow N_2 + 3\,H_2O \quad \text{(balanced)}$$

Multiply all the coefficients by 2 to produce a balanced equation in which all the coefficients are integers.

$$4\,NH_3 + 3\,O_2 \longrightarrow 2\,N_2 + 6\,H_2O \quad \text{(balanced)}$$

Assess

Note that the final equation above is balanced, with four N atoms, twelve H atoms, and six O atoms on each side.

PRACTICE EXAMPLE A: Balance the equations.

(a) $H_3PO_4 + CaO \longrightarrow Ca_3(PO_4)_2 + H_2O$

(b) $C_3H_8 + O_2 \longrightarrow CO_2 + H_2O$

PRACTICE EXAMPLE B: Balance the equations.

(a) $NH_3 + O_2 \longrightarrow NO_2 + H_2O$

(b) $NO_2 + NH_3 \longrightarrow N_2 + H_2O$

EXAMPLE 4-2   Writing and Balancing an Equation: The Combustion of a Carbon–Hydrogen–Oxygen Compound

Liquid triethylene glycol is used as a solvent and plasticizer for vinyl and polyurethane plastics. Write a balanced chemical equation for the combustion of this compound in a plentiful supply of oxygen. A ball-and-stick model of triethylene glycol is shown here.

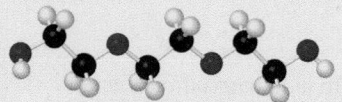

Triethylene glycol

Analyze

We deduce the formula of triethylene glycol from the molecular model. (Refer to the color scheme given on the inside back cover.) We see 6 C atoms (black), 4 O atoms (red), and 14 H atoms. The formula is $C_6H_{14}O_4$. When a carbon–hydrogen–oxygen compound is burned in excess oxygen, O_2, the products are CO_2 and H_2O.

Solve

Having identified the reactants and products, we write down an unbalanced chemical equation for the reaction, showing all reactants and products, and then we balance the equation with respect to each kind of atom.

Starting expression: $C_6H_{14}O_4 + O_2 \longrightarrow CO_2 + H_2O$

Balance C: $C_6H_{14}O_4 + O_2 \longrightarrow 6\,CO_2 + H_2O$

Balance H: $C_6H_{14}O_4 + O_2 \longrightarrow 6\,CO_2 + 7\,H_2O$

At this point, the right side of the expression has 19 O atoms (12 in six CO_2 molecules and 7 in seven H_2O molecules), and the left side, only 4 O atoms (in $C_6H_{14}O_4$). To obtain 15 more O atoms requires a fractional coefficient of $15/2$ for O_2.

Balance O: $C_6H_{14}O_4 + \dfrac{15}{2}O_2 \longrightarrow 6\,CO_2 + 7\,H_2O$ (balanced)

To remove the fractional coefficient, multiply all coefficients by two:

$$2\,C_6H_{14}O_4 + 15\,O_2 \longrightarrow 12\,CO_2 + 14\,H_2O \quad \text{(balanced)}$$

Assess

To check that the equation is balanced, determine the numbers of C, H, and O atoms that appear on the each side of the equation.

Left: $(2 \times 6) = 12$ C; $(2 \times 14) = 28$ H; $[(2 \times 4) + (15 \times 2)] = 38$ O

Right: $(12 \times 1) = 12$ C; $(14 \times 2) = 28$ H; $[(12 \times 2) + (14 \times 1)] = 38$ O

PRACTICE EXAMPLE A: Write a balanced equation to represent the reaction of mercury(II) sulfide and calcium oxide to produce calcium sulfide, calcium sulfate, and mercury metal.

PRACTICE EXAMPLE B: Write a balanced equation for the combustion of thiosalicylic acid, $C_7H_6O_2S$, used in the manufacture of indigo dyes. A ball-and-stick model of $C_7H_6O_2S$ is shown here.

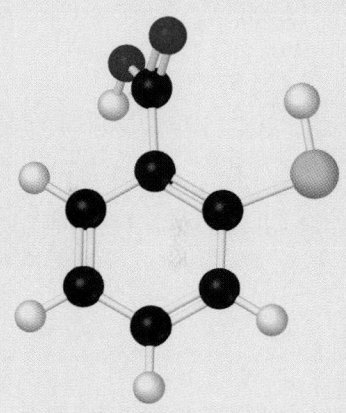

Thiosalicylic acid

States of Matter

Ammonia (Example 4-1) is a gas at 25 °C, but triethylene glycol (Example 4-2) is a liquid. Such facts as these are inconsequential if our interest is only in balancing an equation. Still, we convey a more complete representation of the reaction by including this information, and sometimes it is essential to include such information in a chemical equation. The state of matter or physical form of reactants and products is shown by symbols in parentheses.

(g) gas   (l) liquid   (s) solid

Thus, the equation for combustion of triethylene glycol can be written as

$$2\,C_6H_{14}O_4(l) + 15\,O_2(g) \longrightarrow 12\,CO_2(g) + 14\,H_2O(l)$$

Another commonly used symbol for reactants or products dissolved in water is

(aq) *aqueous* solution

◀ At the high combustion temperature, water is present as $H_2O(g)$. But when the reaction products are returned to the initial temperature, the water condenses to a liquid, $H_2O(l)$.

Reaction Conditions

The equation for a chemical reaction does not provide enough information to enable you to carry out the reaction in a laboratory or chemical plant. An important aspect of modern chemical research involves working out the conditions for a reaction. The reaction conditions are often written above or below the arrow in an equation. For example, the Greek capital letter delta, Δ, means that a high temperature is required—that is, the reaction mixture must be heated, as in the decomposition of silver oxide.

$$2\,Ag_2O(s) \xrightarrow{\Delta} 4\,Ag(s) + O_2(g)$$

▶ In a *decomposition reaction*, a substance is broken down into simpler substances (for instance, into its elements).

▶ In a *synthesis reaction*, a new compound is formed from the reaction of two or more simpler substances, usually called the reactants or starting materials.

An even more explicit statement of reaction conditions is shown below for the BASF (Badische Anilin & Soda-Fabrik) process for the synthesis of methanol from CO and H_2. This reaction occurs at 350 °C, under a total gas pressure that is 340 times as great as the normal pressure of the atmosphere, and on the surface of a mixture of ZnO and Cr_2O_3 acting as a *catalyst*. As we will learn later in the text, a catalyst is a substance that enters into a reaction in such a way that it speeds up the reaction without itself being consumed or changed by the reaction.

$$CO(g) + 2\,H_2(g) \xrightarrow[\substack{340\ \text{atm} \\ ZnO,\ Cr_2O_3}]{350\,°C} CH_3OH(g)$$

It is important to be able to calculate how much of a particular product will be produced when certain quantities of the reactants are consumed. In the next section, we will see how to use chemical equations to set up conversion factors that we can use for these and related calculations.

🔍 4-1 CONCEPT ASSESSMENT

For the elements K, Cl, and O in the following equations, the requirement that the number of atoms be equal on either side of the equation is met. Why is none of them an acceptable balanced equation for the decomposition of solid potassium chlorate yielding solid potassium chloride and oxygen gas?

(a) $KClO_3(s) \longrightarrow KCl(s) + 3\,O(g)$

(b) $KClO_3(s) \longrightarrow KCl(s) + O_2(g) + O(g)$

(c) $KClO_3(s) \longrightarrow KClO(s) + O_2(g)$

4-2 Chemical Equations and Stoichiometry

In Greek, the word *stoicheion* means element. The term **stoichiometry** (stoy-key-om'-eh-tree) means, literally, to measure the elements—but from a practical standpoint, it includes all the quantitative relationships involving atomic and formula masses, chemical formulas, and chemical equations. We considered the quantitative meaning of chemical formulas in Chapter 3, and now we will explore some additional quantitative aspects of chemical equations.

The coefficients in the chemical equation

$$2\,H_2(g) + O_2(g) \longrightarrow 2\,H_2O(l) \tag{4.1}$$

mean that

$$2x \text{ molecules } H_2 + x \text{ molecules } O_2 \longrightarrow 2x \text{ molecules } H_2O$$

Suppose we let $x = 6.02214 \times 10^{23}$ (Avogadro's number). Then x molecules represents *1 mole*. Thus the chemical equation also means that

$$2 \text{ mol } H_2 + 1 \text{ mol } O_2 \longrightarrow 2 \text{ mol } H_2O$$

The coefficients in the chemical equation allow us to make statements such as

- *Two* moles of H_2O are *produced* for every *two* moles of H_2 *consumed*.
- *Two* moles of H_2O are *produced* for every *one* mole of O_2 *consumed*.
- *Two* moles of H_2 are *consumed* for every *one* mole of O_2 *consumed*.

Moreover, we can turn such statements into conversion factors, called stoichiometric factors. A **stoichiometric factor** relates the amounts, on a mole basis, of any two substances involved in a chemical reaction; thus a stoichiometric factor is a mole ratio. In the following examples, stoichiometric factors are printed in blue.

EXAMPLE 4-3   Relating the Numbers of Moles of Reactant and Product

How many moles of CO_2 are produced in the combustion of 2.72 mol of triethylene glycol, $C_6H_{14}O_4$, in an excess of O_2?

Analyze

"An excess of O_2" means that there is more than enough O_2 available to permit the complete conversion of the triethylene glycol to CO_2 and H_2O. The factor for converting from moles of $C_6H_{14}O_4$ to moles of CO_2 is obtained from the balanced equation for the combustion reaction.

Solve

The first step in a stoichiometric calculation is to write a balanced equation for the reaction. The balanced chemical equation for the reaction is given below.

$$2 C_6H_{14}O_4 + 15 O_2 \longrightarrow 12 CO_2 + 14 H_2O$$

Thus, 12 mol CO_2 are produced for every 2 mol $C_6H_{14}O_4$ burned. The production of 12 mol CO_2 is equivalent to the consumption of 2 mol $C_6H_{14}O_4$; thus, the ratio 12 mol CO_2/2 mol $C_6H_{14}O_4$ converts from mol $C_6H_{14}O_4$ to mol CO_2.

$$? \text{ mol } CO_2 = 2.72 \text{ mol } C_6H_{14}O_4 \times \frac{12 \text{ mol } CO_2}{2 \text{ mol } C_6H_{14}O_4} = 16.3 \text{ mol } CO_2$$

Assess

The expression above can be written in terms of two equal ratios:

$$\frac{? \text{ mol } CO_2}{2.72 \text{ mol } C_6H_{14}O_4} = \frac{12 \text{ mol } CO_2}{2 \text{ mol } C_6H_{14}O_4}$$

You may find it easier to set up an expression in terms of ratios and then solve it for the unknown quantity.

PRACTICE EXAMPLE A: How many moles of O_2 are produced from the decomposition of 1.76 moles of potassium chlorate?

$$2 KClO_3(s) \longrightarrow 2 KCl(s) + 3 O_2(g)$$

PRACTICE EXAMPLE B: How many moles of Ag are produced in the decomposition of 1.00 kg of silver(I) oxide?

$$2 Ag_2O(s) \longrightarrow 4 Ag(s) + O_2(g)$$

Reaction stoichiometry problems range from relatively simple to complex, yet all can be solved by the same general strategy. This strategy, outlined in Figure 4-3, yields information about one substance, B, from information given about a second substance, A. Thus, the conversion pathway is from the *known* A to the *unknown* B. The heart of the calculation, shown in blue in Figure 4-3, is conversion from the number of moles of A to the number of moles of B using a

KEEP IN MIND

that it is important to include units and to work from a balanced chemical equation when solving stoichiometry problems.

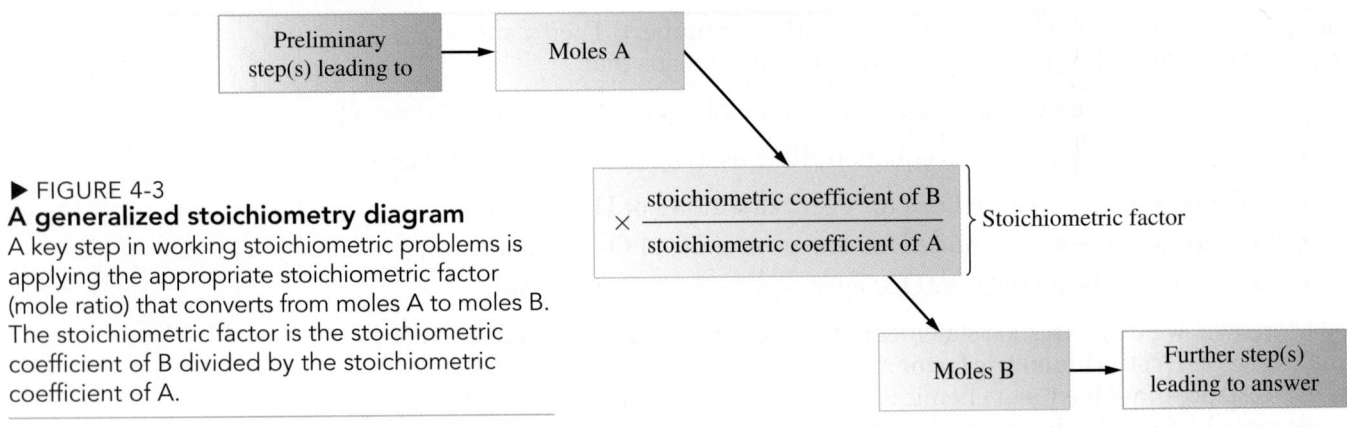

▶ FIGURE 4-3
A generalized stoichiometry diagram
A key step in working stoichiometric problems is applying the appropriate stoichiometric factor (mole ratio) that converts from moles A to moles B. The stoichiometric factor is the stoichiometric coefficient of B divided by the stoichiometric coefficient of A.

factor based on stoichiometric coefficients from the balanced chemical equation—that is, (stoichiometric coefficient of B)/(stoichiometric coefficient of A). Example 4-3 was a single-step calculation requiring only the stoichiometric factor. Examples 4-4 through 4-7 require additional steps, both before and after the stoichiometric factor is employed. In short, this strategy involves the following conversions: "to moles," "between moles," and "from moles." The conversions can be done separately or they can be combined into a single-step calculation, as shown in the following examples.

EXAMPLE 4-4   Relating the Mass of a Reactant and a Product

What mass of CO_2 is formed in the reaction of 4.16 g triethylene glycol, $C_6H_{14}O_4$, with an excess of O_2?

Analyze

The general strategy involves the following conversions: (1) to moles, (2) between moles, and (3) from moles. In this example, the required conversions are g $C_6H_{14}O_4 \xrightarrow{1}$ mol $C_6H_{14}O_4 \xrightarrow{2}$ mol $CO_2 \xrightarrow{3}$ g CO_2. Each numbered arrow refers to a conversion factor that changes the unit on the left to the one on the right.

Solve

The conversions can be carried out by using either a stepwise approach or the conversion pathway approach. Using a *stepwise approach*, we proceed as follows.

Convert from grams of $C_6H_{14}O_4$ to moles of $C_6H_{14}O_4$ by using the molar mass of $C_6H_{14}O_4$ as a conversion factor.

$$? \text{ mol } C_6H_{14}O_4 = 4.16 \text{ g } C_6H_{14}O_4 \times \frac{1 \text{ mol } C_6H_{14}O_4}{150.2 \text{ g } C_6H_{14}O_4}$$
$$= 0.0277 \text{ mol } C_6H_{14}O_4$$

Convert from moles of $C_6H_{14}O_4$ to moles of CO_2 by using the stoichiometric factor.

$$? \text{ mol } CO_2 = 0.0277 \text{ mol } C_6H_{14}O_4 \times \frac{12 \text{ mol } CO_2}{2 \text{ mol } C_6H_{14}O_4}$$
$$= 0.166 \text{ mol } CO_2$$

Convert from moles of CO_2 to grams of CO_2 by using the molar mass of CO_2 as a conversion factor.

$$? \text{ g } CO_2 = 0.166 \text{ mol } CO_2 \times \frac{44.01 \text{ g } CO_2}{1 \text{ mol } CO_2}$$
$$= 7.31 \text{ g } CO_2$$

In the *conversion pathway* approach, the individual steps are combined into a single line calculation, as shown below.

$$? \text{ g } CO_2 = 4.16 \text{ g } C_6H_{14}O_4 \times \underbrace{\frac{1 \text{ mol } C_6H_{14}O_4}{150.2 \text{ g } C_6H_{14}O_4}}_{\substack{\text{converts to moles} \\ \text{of } C_6H_{14}O_4}} \times \underbrace{\frac{12 \text{ mol } CO_2}{2 \text{ mol } C_6H_{14}O_4}}_{\substack{\text{converts to moles} \\ \text{of } CO_2}} \times \underbrace{\frac{44.01 \text{ g } CO_2}{1 \text{ mol } CO_2}}_{\substack{\text{converts to} \\ \text{grams of } CO_2}} = 7.31 \text{ g } CO_2$$

Note the connection between the stepwise approach and the conversion pathway approach for this problem. In the conversion pathway approach, the first conversion factor converts from grams of $C_6H_{14}O_4$ to moles of $C_6H_{14}O_4$. The second conversion factor is the stoichiometric factor, and it converts from moles of $C_6H_{14}O_4$ to moles of CO_2. The third conversion factor converts from moles of CO_2 to grams of CO_2. The conversion pathway approach combines all three calculations into a single line.

Assess

A quick scan of the numbers to the right of 4.16 g $C_6H_{14}O_4$ indicates that the stoichiometric factor has a value of 6; 6×44.01 is between 250 and 300 which, when divided by 150.2, yields a factor somewhat smaller than 2. The mass of CO_2 should be somewhat less than twice that of the $C_6H_{14}O_4$, and it is (compare 7.31 to 2×4.16). Also, note that all units cancel properly in the ultimate conversion from g $C_6H_{14}O_4$ to g CO_2.

PRACTICE EXAMPLE A: How many grams of magnesium nitride, Mg_3N_2, are produced by the reaction of 3.82 g Mg with an excess of N_2?

PRACTICE EXAMPLE B: How many grams of $H_2(g)$ are required to produce 1.00 kg methanol, CH_3OH, by the reaction $CO + 2H_2 \longrightarrow CH_3OH$?

EXAMPLE 4-5   Relating the Masses of Two Reactants to Each Other

What mass of O_2 is consumed in the complete combustion of 6.86 g of triethylene glycol, $C_6H_{14}O_4$?

Analyze

The required conversions are g $C_6H_{14}O_4 \xrightarrow{1}$ mol $C_6H_{14}O_4 \xrightarrow{2}$ mol $O_2 \xrightarrow{3}$ g O_2.

Solve

We will first use a stepwise approach to solve this problem.

Convert from grams of $C_6H_{14}O_4$ to moles of $C_6H_{14}O_4$ by using the molar mass of $C_6H_{14}O_4$ as a conversion factor.

$$? \text{ mol } C_6H_{14}O_4 = 6.86 \text{ g } C_6H_{14}O_4 \times \frac{1 \text{ mol } C_6H_{14}O_4}{150.2 \text{ g } C_6H_{14}O_4}$$

$$= 0.0457 \text{ mol } C_6H_{14}O_4$$

Convert from moles of $C_6H_{14}O_4$ to moles of O_2 by using the stoichiometric factor.

$$? \text{ mol } O_2 = 0.0457 \text{ mol } C_6H_{14}O_4 \times \frac{15 \text{ mol } O_2}{2 \text{ mol } C_6H_{14}O_4}$$

$$= 0.0343 \text{ mol } O_2$$

Convert from moles of O_2 to grams of O_2 by using the molar mass of O_2 as a conversion factor.

$$? \text{ g } O_2 = 0.343 \text{ mol } O_2 \times \frac{32.00 \text{ g } O_2}{1 \text{ mol } O_2}$$

$$= 11.0 \text{ g } O_2$$

As in Example 4-4, the three steps can be combined into a single calculation, as shown below.

$$? \text{ g } O_2 = 6.86 \text{ g } C_6H_{14}O_4 \times \frac{1 \text{ mol } C_6H_{14}O_4}{150.2 \text{ g } C_6H_{14}O_4} \times \frac{15 \text{ mol } O_2}{2 \text{ mol } C_6H_{14}O_6} \times \frac{32.00 \text{ g } O_2}{1 \text{ mol } O_2} = 11.0 \text{ g } O_2$$

Assess

Focus on the single line calculation shown above. A quick scan of the numbers to the right of 6.86 g $C_6H_{14}O_4$ indicates that the stoichiometric factor has a value of 7.5; the product, 7.5×32.00, is about 250, which, when divided by 150.2, yields a factor of about $250/150 = 5/3$. The mass of O_2 should be about 5/3 that of the $C_6H_{14}O_4$, and it is—that is, compare 11.0 with 5/3 of 6.86, which is about 35/3 or somewhat less than 12. As in Example 4-4, note that the proper cancellation of units occurs.

PRACTICE EXAMPLE A: For the reaction in Example 4-1, how many grams of NH_3 are consumed per gram of O_2?

PRACTICE EXAMPLE B: In the combustion of octane, C_8H_{18}, how many grams of O_2 are consumed per gram of octane?

4-2   CONCEPT ASSESSMENT

Which statements are correct for the reaction $2\,H_2S + SO_2 \longrightarrow 3\,S + 2\,H_2O$? Explain your reasoning.

(a) 3 mol S is produced per mole of H_2S.

(b) 3 g S is produced for every gram of SO_2 consumed.

(c) 1 mol H_2O is produced per mole of H_2S consumed.

(d) Two-thirds of the S produced comes from H_2S.

(e) The number of moles of products formed equals the number of moles of reactants consumed.

(f) The number of moles of atoms present after the reaction is the same as the number of moles of atoms before the reaction.

What lends great variety to stoichiometric calculations is that many other conversions may be required before and after the mol A $\longrightarrow$ mol B step at the center of the stoichiometric scheme shown in Figure 4-3. In Examples 4-4 and 4-5, the additional conversions involved molar mass. Other common conversions may require such factors as volume, density, and percent composition. In every case, however, we must always use the appropriate stoichiometric factor from the chemical equation as a key conversion factor.

The reaction between solid aluminum, Al(s), and aqueous hydrochloric acid, HCl(aq), can be used for preparing small volumes of hydrogen gas, $H_2(g)$, in the laboratory. A balanced chemical equation for the reaction is shown below.

$$2\,Al(s) + 6\,HCl(aq) \longrightarrow 2\,AlCl_3(aq) + 3\,H_2(g) \qquad (4.2)$$

A simple laboratory setup for collecting the hydrogen gas is pictured in Figure 4-4. The reaction between Al(s) and HCl(aq) provides a range of possibilities for calculations. Examples 4-6 and 4-7 are based on this reaction.

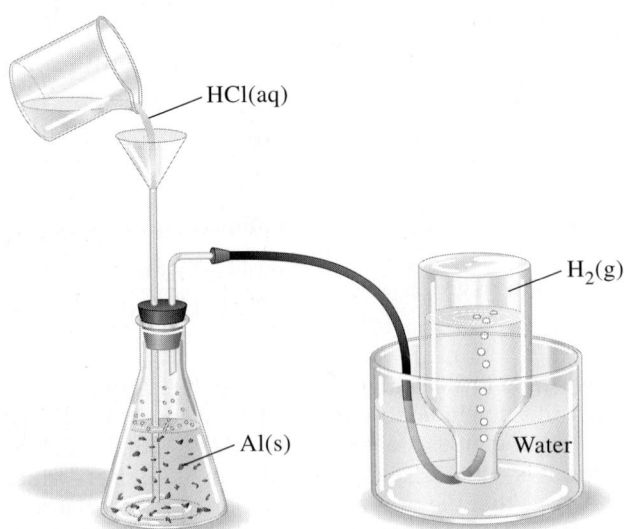

▲ FIGURE 4-4

The reaction $2\,Al(s) + 6\,HCl(aq) \longrightarrow 2\,AlCl_3(aq) + 3\,H_2(g)$

HCl(aq) is added to the flask on the left. The reaction occurs within the flask. The liberated $H_2(g)$ flows into a gas-collection apparatus, where it displaces water. Hydrogen is only very slightly soluble in water.

EXAMPLE 4-6   Additional Conversion Factors in a Stoichiometric Calculation: Volume, Density, and Percent Composition

An alloy used in aircraft structures consists of 93.7% Al and 6.3% Cu by mass. The alloy has a density of 2.85 g/cm^3. A 0.691 cm^3 piece of the alloy reacts with an excess of HCl(aq). If we assume that *all* the Al but *none* of the Cu reacts with HCl(aq), what is the mass of H_2 obtained? Refer to reaction (4.2).

Analyze

A simple approach to this calculation is outlined below. Each numbered arrow refers to a conversion factor that changes the unit on the left to the one on the right.

$$\text{cm}^3 \text{ alloy} \xrightarrow{1} \text{g alloy} \xrightarrow{2} \text{g Al} \xrightarrow{3} \text{mol Al} \xrightarrow{4} \text{mol } H_2 \xrightarrow{5} \text{g } H_2$$

The calculation can be done in five distinct steps, or with a single setup in which the five conversions are performed in sequence.

Solve

Using a stepwise approach, we proceed as follows.

Convert from volume of alloy to grams of alloy by using the density as a conversion factor.

$$? \text{ g alloy} = 0.691 \text{ cm}^3 \text{ alloy} \times \frac{2.85 \text{ g alloy}}{1 \text{ cm}^3 \text{ alloy}}$$
$$= 1.97 \text{ g alloy}$$

Convert from grams of alloy to grams of Al by using the percentage by mass of Al as a conversion factor.

$$? \text{ g Al} = 1.97 \text{ g alloy} \times \frac{93.7 \text{ g Al}}{100 \text{ g alloy}}$$
$$= 1.85 \text{ g Al}$$

Convert from grams of Al to moles of Al by using the molar mass as a conversion factor.

$$? \text{ mol Al} = 1.85 \text{ g Al} \times \frac{1 \text{ mol Al}}{26.98 \text{ g Al}}$$
$$= 0.0684 \text{ mol Al}$$

Convert from moles of Al to moles of H_2 by using the stoichiometric factor.

$$? \text{ mol } H_2 = 0.0684 \text{ mol Al} \times \frac{3 \text{ mol } H_2}{2 \text{ mol Al}}$$
$$= 0.103 \text{ mol } H_2$$

Convert from moles of Al to moles of H_2 by using the stoichiometric factor.

$$? \text{ g } H_2 = 0.103 \text{ mol } H_2 \times \frac{2.016 \text{ g } H_2}{1 \text{ mol } H_2}$$
$$= 0.207 \text{ g } H_2$$

Remember to store intermediate results without rounding off. When all of the steps are combined into a single calculation, we do not have to write down intermediate results and we reduce rounding errors.

$$? \text{ g } H_2 = 0.691 \text{ cm}^3 \text{ alloy} \times \frac{2.85 \text{ g alloy}}{1 \text{ cm}^3 \text{ alloy}} \times \frac{93.7 \text{ g Al}}{100 \text{ g alloy}} \times \frac{1 \text{ mol Al}}{26.98 \text{ g Al}} \times \frac{3 \text{ mol } H_2}{2 \text{ mol Al}} \times \frac{2.016 \text{ g } H_2}{1 \text{ mol } H_2}$$
$$= 0.207 \text{ g } H_2$$

Assess

The units work out properly, but we must evaluate whether the answer is a reasonable one. The molar masses of Al and H_2 are approximately 27 g/mol and 2 g/mol, respectively. Equation (4.2) tells us that 1 mole of Al, which weighs approximately 27 g, produces 1.5 mol of H_2, which weighs $1.5 \times 2 = 3$ g. Thus, 27 g Al produces approximately 3 g H_2 and, thus, 2.7 g Al produces approximately 0.3 g H_2. In this example, we are dealing with less than 2.7 g of Al; therefore, we expect less than 0.3 g of H_2. The answer, 0.207 g H_2, is reasonable.

PRACTICE EXAMPLE A: What volume of the aluminum-copper alloy described in Example 4-6 must be dissolved in an excess of HCl(aq) to produce 1.00 g H_2?

[*Hint:* Think of this as the "inverse" of Example 4-6.]

PRACTICE EXAMPLE B: A fresh sample of the aluminum-copper alloy described in Example 4-6 yielded 1.31 g H_2. How many grams of copper were present in the sample?

EXAMPLE 4-7   Additional Conversion Factors in a Stoichiometric Calculation: Volume, Density, and Percent Composition of a Solution

A hydrochloric acid solution consists of 28.0% HCl by mass and has a density of 1.14 g/mL. What volume of this solution is required to react completely with 1.87 g Al in reaction (4.2)?

Analyze

The first challenge here is to determine where to begin. Although the problem refers to 28.0% HCl and a density of 1.14 g/mL, the appropriate starting point is with the given information—1.87 g Al. The goal of our calculation is a solution volume—mL HCl solution.

$$g\ Al \xrightarrow{1} mol\ Al \xrightarrow{2} mol\ HCl \xrightarrow{3} g\ HCl \xrightarrow{4} g\ HCl\ solution \xrightarrow{5} mL\ HCl\ solution$$

The conversion factors in the calculation involve (1) the molar mass of Al, (2) stoichiometric coefficients from equation (4.2), (3) the molar mass of HCl, (4) the percent composition of the HCl solution, and (5) the density of the HCl solution.

Solve

Using a stepwise approach, we proceed as follows.

Convert from grams of Al to moles of Al by using the molar mass of Al.

$$?\ mol\ Al\ =\ 1.87\ g\ Al\ \times\ \frac{1\ mol\ Al}{26.98\ g\ Al}\ =\ 0.0693\ mol\ Al$$

Convert from moles of Al to moles of HCl by using the stoichiometric factor.

$$?\ mol\ HCl\ =\ 0.0693\ mol\ Al\ \times\ \frac{6\ mol\ HCl}{2\ mol\ Al}\ =\ 0.208\ mol\ HCl$$

Convert from moles of HCl to grams of HCl by using the molar mass of HCl.

$$?\ g\ HCl\ =\ 0.208\ mol\ HCl\ \times\ \frac{36.46\ g\ HCl}{1\ mol\ HCl}\ =\ 7.58\ g\ HCl$$

Convert from grams of HCl to grams of HCl solution by using the percentage by mass.

$$?\ g\ HCl\ soln\ =\ 7.58\ g\ HCl\ \times\ \frac{100\ g\ HCl\ soln}{28.0\ g\ HCl}$$
$$=\ 27.1\ g\ HCl\ soln$$

Convert from grams of HCl solution to milliliters of HCl solution by using the density.

$$?\ mL\ HCl\ soln\ =\ 27.1\ g\ HCl\ soln\ \times\ \frac{1\ mL\ HCl\ soln}{1.14\ g\ HCl\ soln}$$
$$=\ 23.8\ mL\ HCl\ soln$$

In the conversion pathway approach, we combine the individual steps into a single line.

$$(g\ Al \xrightarrow{1} mol\ Al \xrightarrow{2} mol\ HCl \xrightarrow{3} g\ HCl$$

$$?\ mL\ HCl\ soln\ =\ 1.87\ g\ Al\ \times\ \frac{1\ mol\ Al}{26.98\ g\ Al}\ \times\ \frac{6\ mol\ HCl}{2\ mol\ Al}\ \times\ \frac{36.46\ g\ HCl}{1\ mol\ HCl}$$

$$\xrightarrow{4} g\ HCl\ soln \xrightarrow{5} mL\ HCl\ soln)$$

$$\times\ \frac{100.0\ g\ HCl\ soln}{28.0\ g\ HCl}\ \times\ \frac{1\ mL\ HCl\ soln}{1.14\ g\ HCl\ soln}$$

$$=\ 23.8\ mL\ HCl\ soln$$

Assess

Let's attempt to establish whether the answer is reasonable by working the problem in reverse and using numbers that are rounded off slightly. Because the density of the solution is approximately 1 g/mL and the solution is approximately 30% HCl by mass, a 24 mL sample of the solution will contain approximately $24 \times 0.30 = 7.2$ g of HCl. Equation (4.2) tells us that 1 mol Al, or 27 g Al, reacts with 3 mol HCl, or 108 g HCl. Stated another way, 4 g HCl reacts with 1 g Al. In this example, we are using approximately 7.2 g HCl; thus, we consume approximately 7.2 g HCl $\times$ 1 g Al/4 g HCl = 1.8 g Al. This value is close to the actual amount of Al consumed and we conclude that our answer, 23.8 mL of HCl solution, is reasonable.

PRACTICE EXAMPLE A: How many milligrams of H_2 are produced when one drop (0.05 mL) of the hydrochloric acid solution described in Example 4-7 reacts with an excess of aluminum in reaction (4.2)?

PRACTICE EXAMPLE B: A particular vinegar contains 4.0% CH_3COOH by mass. It reacts with sodium carbonate to produce sodium acetate, carbon dioxide, and water. How many grams of carbon dioxide are produced by the reaction of 5.00 mL of this vinegar with an excess of sodium carbonate? The density of the vinegar is 1.01 g/mL.

Without performing detailed calculations, determine which reaction produces the maximum quantity of $O_2(g)$ per gram of reactant.

(a) $2\ NH_4NO_3(s) \xrightarrow{\Delta} 2\ N_2(g) + 4\ H_2O(l) + O_2(g)$

(b) $2\ Ag_2O(s) \xrightarrow{\Delta} 4\ Ag(s) + O_2(g)$

(c) $2\ HgO(s) \xrightarrow{\Delta} 2\ Hg(l) + O_2(g)$

(d) $2\ Pb(NO_3)_2(s) \xrightarrow{\Delta} 2\ PbO(s) + 4\ NO_2(g) + O_2(g)$

4-3   Chemical Reactions in Solution

Most reactions in the general chemistry laboratory are carried out in solutions. This is partly because mixing the reactants in solution helps to achieve the close contact between atoms, ions, or molecules necessary for a reaction to occur. The stoichiometry of reactions in solutions can be described in the same way as the stoichiometry of other reactions, as we saw in Example 4-7. A few new ideas that apply specifically to solution stoichiometry are also helpful.

One component of a solution, called the **solvent**, determines whether the solution exists as a solid, liquid, or gas. In this discussion we will limit ourselves to *aqueous solutions*—solutions in which liquid water is the solvent. The other components of a solution, called **solutes**, are dissolved in the solvent. We use the notation NaCl(aq), for example, to describe a solution in which liquid water is the solvent and NaCl is the solute. The term *aqueous* does not convey any information, however, about the relative proportions of NaCl and H_2O in the solution. For this purpose, the property called *molarity* is commonly used.

Molarity

The composition of a solution may be specified by giving its molarity, which is defined as the amount of solute, in moles, per liter of solution:

$$molarity = \frac{\text{amount of solute (in moles)}}{\text{volume of solution (in liters)}}$$

The expression above can be written more compactly as

$$M = \frac{n}{V} \qquad (4.3)$$

where M is the molarity in moles per liter (mol/L), n is the amount of solute in moles (mol), and V is the volume of the solution in liters (L).

If 0.440 mol urea, $CO(NH_2)_2$, is dissolved in enough water to make 1.000 L of solution, the solution concentration, or molarity, is

$$\frac{0.440\ \text{mol}\ CO(NH_2)_2}{1.000\ \text{L soln}} = 0.440\ M\ CO(NH_2)_2$$

The symbol M stands for the term *molar*, or mol/L. Thus, a solution that has 0.440 mol $CO(NH_2)_2$/L is 0.440 M $CO(NH_2)_2$ or 0.440 molar $CO(NH_2)_2$. Notice that molarity (M) has units mol/L or M. In this text, we use the symbol M (in italics) to represent molarity and the symbol M (no italics) as an abbreviation for the unit mol/L.

Alternatively, if 0.110 mol urea is present in 250.0 mL of solution, the solution is also 0.440 M.

$$\frac{0.110\ \text{mol}\ CO(NH_2)_2}{0.2500\ \text{L soln}} = 0.440\ M\ CO(NH_2)_2$$

◀ The IUPAC-preferred term for molarity is *amount concentration,* and the preferred symbol is C, although many use M. Be sure to maintain a clear distinction between the *property* concentration (molarity) $= C = M = n/V$ and the *unit* in which it is expressed, molarity $=$ M $=$ mol/L.

When calculating concentration, we must determine the amount of solute in moles from other quantities that can be readily measured, such as the mass of solute or the volume of a liquid solute. In Example 4-8, the mass of a liquid solute is related to its volume by using density as a conversion factor. Then, molar mass is used to convert from the mass of solute to an amount of solute in moles. Also, the solution volume is converted from milliliters (mL) to liters (L).

Figure 4-5 illustrates a method commonly used to prepare a solution. A solid sample is weighed out and dissolved in sufficient water to produce a solution of known volume—250.0 mL. To fill a beaker to a 250 mL mark is not nearly precise enough as a volume measurement. As discussed in Section 1-6, there would be a systematic error because the beaker is not calibrated with sufficient precision. (The error in volume could be 10 to 20 mL or more.) To dissolve the solute in water and bring the solution level to the 250 mL mark in a graduated cylinder is also not sufficiently precise. Although the graduated cylinder is calibrated more precisely than the beaker, the error might still be 1 to 2 mL or more. However, when filled to the calibration mark, the volumetric flask pictured in Figure 4-5 contains 250.0 mL with only about 0.1 mL of error.

When the molarity of a solution is known accurately, we can calculate the number of moles of solute in a carefully measured volume of solution by using the equation below.

$$n = M \times V \tag{4.4}$$

The expression above is equivalent to equation (4.3), but equation (4.4) emphasizes that the molarity, M, is a conversion factor for converting from liters of solution to moles of solute. In Example 4-9, we use molarity as a conversion factor in a calculation to determine the mass of solute needed to produce the solution in Figure 4-5.

EXAMPLE 4-8   Calculating Molarity from Measured Quantities

A solution is prepared by dissolving 25.0 mL ethanol, CH_3CH_2OH ($d = 0.789$ g/mL), in enough water to produce 250.0 mL solution. What is the molarity of ethanol in the solution?

Analyze

We must first calculate how many moles of ethanol are in a 25.0 mL sample of pure ethanol. This calculation requires the following conversions: mL ethanol $\xrightarrow{1}$ g ethanol $\xrightarrow{2}$ mol ethanol. The first conversion uses the density as a conversion factor and the second conversion uses molar mass as a conversion factor. The molarity of the solution is then calculated by using equation (4.3).

Solve

The number of moles of ethanol in a 25.0 mL sample of pure ethanol is calculated below in a single line.

$$? \text{ mol } CH_3CH_2OH = 25.0 \text{ mL } CH_3CH_2OH \times \frac{0.789 \text{ g } CH_3CH_2OH}{1 \text{ mL } CH_3CH_2OH} \times \frac{1 \text{ mol } CH_3CH_2OH}{46.07 \text{ g } CH_3CH_2OH}$$

$$= 0.428 \text{ mol } CH_3CH_2OH$$

To apply the definition of molarity given in expression (4.3), note that 250.0 mL = 0.2500 L.

$$\text{molarity} = \frac{0.428 \text{ mol } CH_3CH_2OH}{0.2500 \text{ L soln}} = 1.71 \text{ M } CH_3CH_2OH$$

Assess

It is important to include the units in this calculation to ensure that we obtain the correct units for the final answer. When dealing with liquid solutes, be careful to distinguish between mL solute and mL soln.

PRACTICE EXAMPLE A: A 22.3 g sample of acetone (see the model here) is dissolved in enough water to produce 1.25 L of solution. What is the molarity of acetone in this solution?

PRACTICE EXAMPLE B: If 15.0 mL of acetic acid, CH_3COOH ($d = 1.048$ g/mL), is dissolved in enough water to produce 500.0 mL of solution, then what is the molarity of acetic acid in the solution?

Acetone

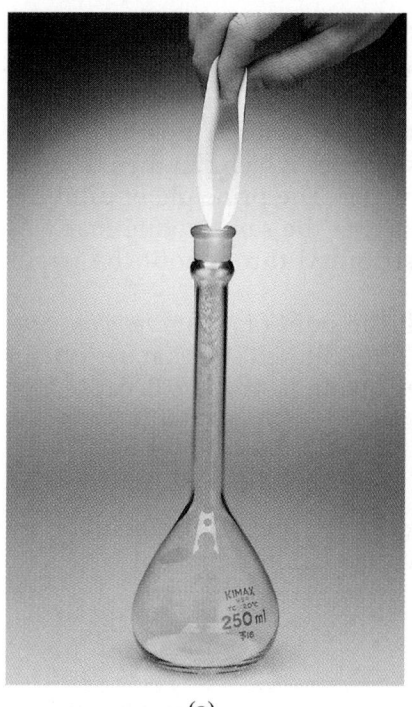
(a)

(b)

(c)

▲ FIGURE 4-5
Preparation of 0.250 M K₂CrO₄—Example 4-9 illustrated
The solution cannot be prepared just by adding 12.1 g K₂CrO₄(s) to 250.0 mL water.
Instead, (a) the weighed quantity of K₂CrO₄(s) is first added to a clean, dry 250 mL
volumetric flask; (b) the K₂CrO₄(s) is dissolved in less than 250 mL of water; and (c) the
flask is filled to the 250.0 mL calibration mark by the careful addition (dropwise) of the
remaining water.

EXAMPLE 4-9   Calculating the Mass of Solute in a Solution of Known Molarity

What mass of K_2CrO_4 is needed to prepare exactly 0.2500 L (250.0 mL) of a 0.250 M K_2CrO_4 solution in water?
(See Figure 4-5.)

Analyze

The conversion pathway is L soln → mol K_2CrO_4 → g K_2CrO_4. The first conversion factor is the molarity
of the solution, shown below in blue, and the second conversion factor is the molar mass of K_2CrO_4.

Solve

$$? \text{ g } K_2CrO_4 = 0.2500 \text{ L soln} \times \frac{0.250 \text{ mol } K_2CrO_4}{1 \text{ L soln}} \times \frac{194.2 \text{ g } K_2CrO_4}{1 \text{ mol } K_2CrO_4}$$

$$= 12.1 \text{ g } K_2CrO_4$$

Assess

The answer has the correct units. We can check whether the answer is reasonable by working the problem in
reverse and using numbers that are rounded off slightly. Because the molar mass of K_2CrO_4 is approximately
200 g/mol and the mass of the sample is approximately 12 g, the number of moles of K_2CrO_4 in the sample is
approximately 12/200 = 6/100 = 0.06 mol. The approximate molarity is 0.06/0.250 = 0.24 mol/L. This estimate is close to the true molarity, and so we are confident that the answer, 12.1 g K_2CrO_4, is correct.

PRACTICE EXAMPLE A: An aqueous solution saturated with $NaNO_3$ at 25 °C is 10.8 M $NaNO_3$. What mass of
$NaNO_3$ is present in 125 mL of this solution at 25 °C?

PRACTICE EXAMPLE B: What mass of $Na_2SO_4 \cdot 10 H_2O$ is needed to prepare 355 mL of 0.445 M Na_2SO_4?

Solution Dilution

A common sight in chemistry storerooms and laboratories is rows of bottles containing solutions for use in chemical reactions. It is not practical, however, to store solutions of every possible concentration. Instead, most labs store fairly concentrated solutions, so-called *stock solutions*, which can then be used to prepare more dilute solutions by adding water. The principle of dilution, which you have probably already inferred, is that the same solute that was present in a sample of stock solution is distributed throughout the larger volume of a diluted solution (see Figure 4-6).

When a volume of a solution is diluted, the amount of solute *remains constant*. If we write equation (4.4) for the initial (i) undiluted solution, we obtain $n_i = M_iV_i$; for the final (f) diluted solution, we obtain $n_f = M_fV_f$. Because n_i is equal to n_f, we obtain the following result:

$$M_iV_i = M_fV_f \qquad (4.5)$$

Figure 4-7 illustrates the laboratory procedure for preparing a solution by dilution. Example 4-10 explains the necessary calculation.

▶ A *concentrated solution* has a relatively large amount of dissolved solute; a *dilute solution* has a relatively small amount.

KEEP IN MIND

that equation (4.5) applies only to dilution problems. An alternative expression uses the subscripts 1 and 2 in place of i and f: $M_1V_1 = M_2V_2$. Some students mistakenly use $M_1V_1 = M_2V_2$ to convert from moles of substance 1 to moles of substance 2 when doing stoichiometry problems. To avoid making this mistake, always use the subscripts "i" and "f" when using equation (4.5), or even better, use the subscripts "dil" (for diluted) and "conc" (for concentrated):
$M_{dil}V_{dil} = M_{conc}V_{conc}.$

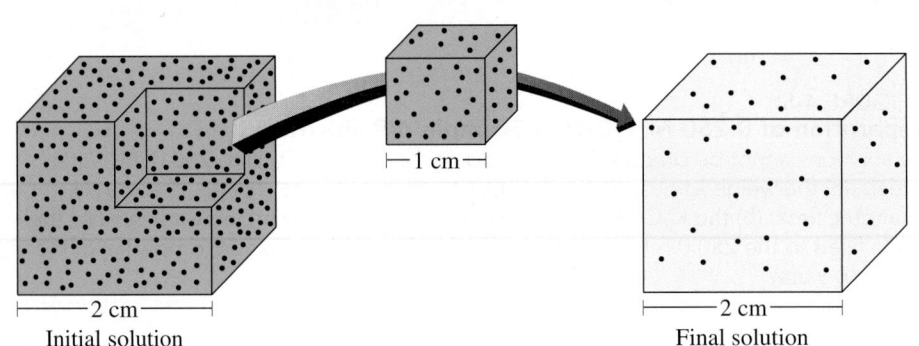

▲ FIGURE 4-6
Visualizing the dilution of a solution
The final solution is prepared by extracting $\frac{1}{8}$ of the initial solution—1 cm³—and diluting it with water to a volume of 8 cm³. The number of dots in the 8 cm³ of final solution, representing the number of solute particles, is the same as in the 1 cm³ of initial solution.

EXAMPLE 4-10   Preparing a Solution by Dilution

A particular analytical chemistry procedure requires 0.0100 M K_2CrO_4. What volume of 0.250 M K_2CrO_4 must be diluted with water to prepare 0.2500 L of 0.0100 M K_2CrO_4?

Analyze

First, we calculate the number of moles K_2CrO_4 that must be present in the final solution. Then, we calculate the volume of 0.250 M K_2CrO_4 that contains this amount of K_2CrO_4.

Solve

First, calculate the amount of solute that must be present in the final solution.

$$? \text{ mol } K_2CrO_4 = 0.2500 \text{ L soln} \times \frac{0.0100 \text{ mol } K_2CrO_4}{1 \text{ L soln}} = 0.00250 \text{ mol } K_2CrO_4$$

Second, calculate the volume of 0.250 M K_2CrO_4 that contains 0.00250 mol K_2CrO_4.

$$? \text{ L soln} = 0.00250 \text{ mol } K_2CrO_4 \times \frac{1 \text{ L soln}}{0.250 \text{ mol } K_2CrO_4} = 0.0100 \text{ L soln}$$

An alternative approach is to use equation (4.5). The known factors include the volume of solution to be prepared, ($V_f = 250.0$ mL) and the concentrations of the final (0.0100 M) and initial (0.250 M) solutions. We must solve for the initial volume, V_i. Note that although in deriving equation (4.5) volumes were expressed in liters, in applying the equation any volume unit can be used as long as we use the same unit for both V_i and V_f (milliliters in the present case). The term needed to convert volumes to liters would appear on both sides of the equation and cancel out.

$$V_i = V_f \times \frac{M_f}{M_i} = 250.0 \text{ mL} \times \frac{0.0100 \text{ M}}{0.250 \text{ M}} = 10.0 \text{ mL}$$

Assess

Let's work the problem in reverse. If we dilute 10.0 mL of 0.250 M K_2CrO_4 to 0.250 L with water, then the concentration of the diluted solution is 0.010 L $\times$ 0.250 mol L^{-1}/0.250 L = 0.010 mol/L. This is the desired molarity; therefore, the answer, 10.0 mL of 0.250 M K_2CrO_4, is correct.

PRACTICE EXAMPLE A: A 15.00 mL sample of 0.450 M K_2CrO_4 is diluted to 100.00 mL. What is the concentration of the new solution?

PRACTICE EXAMPLE B: When left in an open beaker for a period of time, the volume of 275 mL of 0.105 M NaCl is found to decrease to 237 mL because of the evaporation of water. What is the new concentration of the solution?

(a)

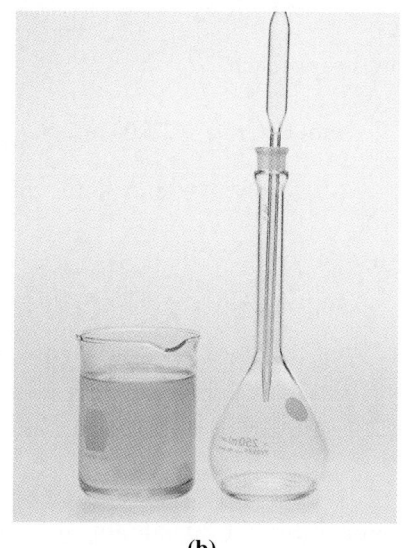
(b)

(c)

▲ FIGURE 4-7
Preparing a solution by dilution—Example 4-10 illustrated
(a) A pipet is used to withdraw a 10.0 mL sample of 0.250 M K_2CrO_4. (b) The pipetful of 0.250 M K_2CrO_4 is discharged into a 250.0 mL volumetric flask. (c) Water is then added to bring the level of the solution to the calibration mark on the neck of the flask. At this point, the solution is 0.0100 M K_2CrO_4.

🔍 4-4 CONCEPT ASSESSMENT

Without doing detailed calculations, and assuming that the volumes of solutions and water are additive, indicate the molarity of the final solution obtained as a result of

(a) adding 200.0 mL of water to 100.0 mL of 0.150 M NaCl

(b) evaporating 50.0 mL water from 250.0 mL of 0.800 M $C_{12}H_{22}O_{11}$

(c) mixing 150.0 mL 0.270 M KCl and 300.0 mL 0.135 M KCl

Stoichiometry of Reactions in Solution

The central conversion factor in Example 4-11 is the same as in previous stoichiometry problems—the appropriate stoichiometric factor. What differs from previous examples is that we use molarity as a conversion factor from solution volume to number of moles of reactant in a preliminary step preceding the stoichiometric factor.

We will consider a number of additional examples of stoichiometric calculations involving solutions in Chapter 5.

EXAMPLE 4-11   **Relating the Mass of a Product to the Volume and Molarity of a Reactant Solution**

A 25.00 mL pipetful of 0.250 M K_2CrO_4 is added to an excess of $AgNO_3(aq)$. What mass of Ag_2CrO_4 will precipitate from the solution?

$$K_2CrO_4(aq) + 2\,AgNO_3(aq) \longrightarrow Ag_2CrO_4(s) + 2\,KNO_3(aq)$$

Analyze

The fact that an excess of $AgNO_3(aq)$ is used tells us that all of the K_2CrO_4 in the 25.00 mL sample of $K_2CrO_4(aq)$ is consumed. The calculation begins with a volume of 25.00 mL and ends with a mass of Ag_2CrO_4 expressed in grams. The conversion pathway is mL soln $\longrightarrow$ L soln $\longrightarrow$ mol K_2CrO_4 $\longrightarrow$ mol Ag_2CrO_4 $\longrightarrow$ g Ag_2CrO_4.

Solve

Let's solve this problem by using a stepwise approach.

Convert the volume of $K_2CrO_4(aq)$ from milliliters to liters, and then use molarity as a conversion factor between volume of solution and moles of solute (as in Example 4-10).

$$? \text{ mol } K_2CrO_4 = 25.00 \text{ mL} \times \frac{1 \text{ L}}{1000 \text{ mL}} \times \frac{0.250 \text{ mol } K_2CrO_4}{1 \text{ L}}$$
$$= 6.25 \times 10^{-3} \text{ mol } K_2CrO_4$$

Use a stoichiometric factor from the equation to convert from moles of K_2CrO_4 to moles of Ag_2CrO_4.

$$? \text{ mol } Ag_2CrO_4 = 6.25 \times 10^{-3} \text{ mol } K_2CrO_4 \times \frac{1 \text{ mol } Ag_2CrO_4}{1 \text{ mol } K_2CrO_4}$$
$$= 6.25 \times 10^{-3} \text{ mol } Ag_2CrO_4$$

Use the molar mass to convert from moles to grams of Ag_2CrO_4.

$$? \text{ g } Ag_2CrO_4 = 6.25 \times 10^{-3} \text{ mol } Ag_2CrO_4 \times \frac{331.7 \text{ g } Ag_2CrO_4}{1 \text{ mol } Ag_2CrO_4}$$
$$= 2.07 \text{ g } Ag_2CrO_4$$

The same final answer can be obtained more directly by combining the steps into a single line calculation.

$$? \text{ g } Ag_2CrO_4 = 25.00 \text{ mL} \times \frac{1 \text{ L}}{1000 \text{ mL}} \times \frac{0.250 \text{ mol } K_2CrO_4}{1 \text{ L}} \times \frac{1 \text{ mol } Ag_2CrO_4}{1 \text{ mol } K_2CrO_4} \times \frac{331.7 \text{ g } Ag_2CrO_4}{1 \text{ mol } Ag_2CrO_4}$$
$$= 2.07 \text{ g } Ag_2CrO_4$$

Assess

The units work out properly, which is always a good sign. As we have done before, let's work the problem in reverse and use numbers that are rounded off slightly. A 2 g sample of Ag_2CrO_4 contains $2/332 = 0.006$ moles of Ag_2CrO_4. The number of moles of K_2CrO_4 in the 25 mL sample of the K_2CrO_4 solution is also approximately 0.006 moles; thus, the molarity of the K_2CrO_4 solution is approximately 0.006 mol/0.025 L = 0.24 M. This result is close to the true molarity (0.250 M), and so we can be confident that our answer for the mass of Ag_2CrO_4 is correct.

PRACTICE EXAMPLE A: How many milliliters of 0.250 M K_2CrO_4 must be added to excess $AgNO_3(aq)$ to produce 1.50 g Ag_2CrO_4?

[*Hint:* Think of this as the inverse of Example 4-11.]

PRACTICE EXAMPLE B: How many milliliters of 0.150 M $AgNO_3$ are required to react completely with 175 mL of 0.0855 M K_2CrO_4? What mass of Ag_2CrO_4 is formed?

4-4   Determining the Limiting Reactant

When all the reactants are completely and simultaneously consumed in a chemical reaction, the reactants are said to be in **stoichiometric proportions**; that is, they are present in the mole ratios dictated by the coefficients in the balanced equation. This condition is sometimes required, for example, in certain chemical analyses. At other times, as in a precipitation reaction, one of the reactants is completely converted into products by using an excess of all the other reactants. The reactant that is completely consumed—the **limiting reactant**—determines the quantities of products formed. In the reaction described in Example 4-11, K_2CrO_4 is the limiting reactant and $AgNO_3$ is present in excess. Up to this point, we have stated which reactant is present in excess and, by implication, which is the limiting reactant. In some cases, however, the limiting reactant will not be indicated explicitly. If the quantities of two or more reactants are given, you must determine which is the limiting reactant, as suggested by the analogy in Figure 4-8.

Sometimes our interest in a limiting reactant problem is in determining how much of an excess reactant remains, as well as how much product is formed. This additional calculation is illustrated in Example 4-13.

◀ By an *excess of a reactant*, we mean that more of the reactant is present than is consumed in the reaction. Some is left over.

◀ Chemicals are often referred to as *reagents*, and the limiting reactant in a reaction is sometimes called the *limiting reagent*.

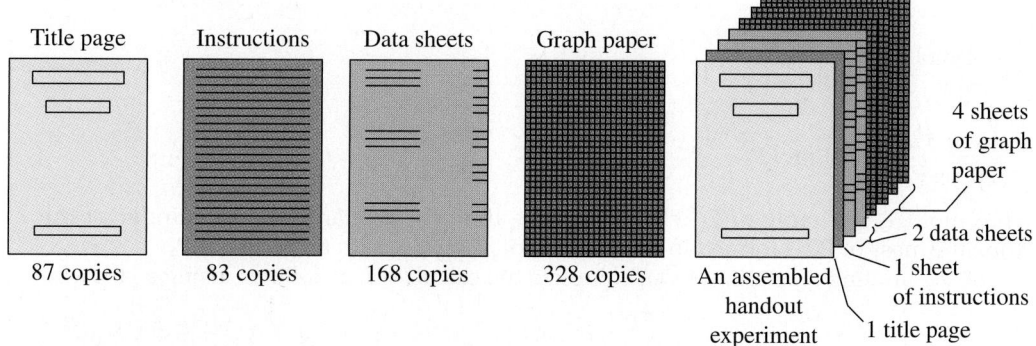

▲ FIGURE 4-8
An analogy to determining the limiting reactant in a chemical reaction—assembling a handout experiment
From the numbers of copies available and the instructions on how to assemble the handout, can you see that only 82 complete handouts are possible and that the graph paper is the limiting reactant?

EXAMPLE 4-12   Determining the Limiting Reactant in a Reaction

Phosphorus trichloride, PCl_3, is a commercially important compound used in the manufacture of pesticides, gasoline additives, and a number of other products. A ball-and-stick model of PCl_3 is shown in the margin. Liquid PCl_3 is made by the direct combination of phosphorus and chlorine.

$$P_4(s) + 6\,Cl_2(g) \longrightarrow 4\,PCl_3(l)$$

What is the maximum mass of PCl_3 that can be obtained from 125 g P_4 and 323 g Cl_2?

Analyze

The following conversions are required: mol limiting reactant → mol PCl_3 → g PCl_3. The key to solving this problem is to correctly identify the limiting reactant. One approach to this problem, outlined in Figure 4-9, is to compare the initial mole ratio of the two reactants with the ratio in which the reactants combine—6 mol Cl_2 to 1 mol P_4. If more than 6 mol Cl_2 is available per mole of P_4, chlorine is in excess and P_4 is the limiting reactant. If fewer than 6 mol Cl_2 is available per mole of P_4, chlorine is the limiting reactant.

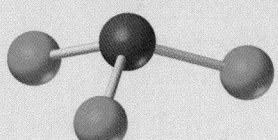

Phosphorus trichloride

(continued)

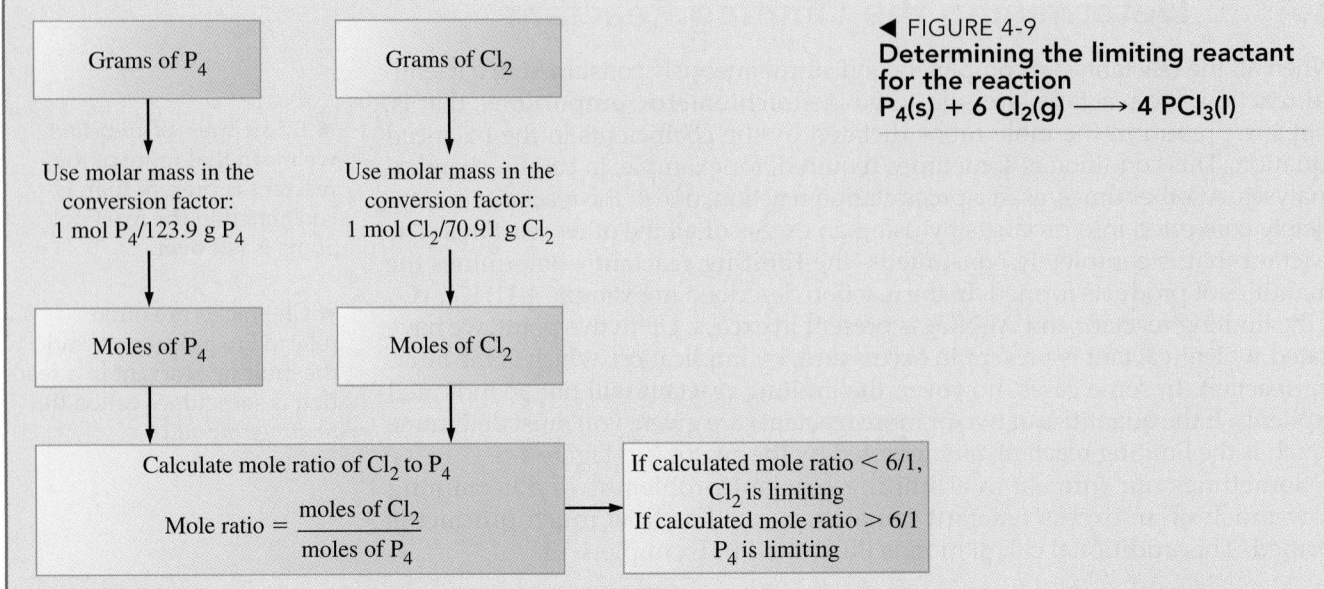

◀ FIGURE 4-9
Determining the limiting reactant for the reaction
$P_4(s) + 6 Cl_2(g) \longrightarrow 4 PCl_3(l)$

Solve

$$? \text{ mol } Cl_2 = 323 \text{ g } Cl_2 \times \frac{1 \text{ mol } Cl_2}{70.91 \text{ g } Cl_2} = 4.56 \text{ mol } Cl_2$$

$$? \text{ mol } P_4 = 125 \text{ g } P_4 \times \frac{1 \text{ mol } P_4}{123.9 \text{ g } P_4} = 1.01 \text{ mol } P_4$$

Since there is less than 6 mol Cl_2 per mole of P_4, chlorine is the limiting reactant. The remainder of the calculation is to determine the mass of PCl_3 formed in the reaction of 323 g Cl_2 with an excess of P_4.

Having identified Cl_2 as the limiting reactant, we can complete the calculation as follows, using a stepwise approach.

Convert from grams of Cl_2 to moles of Cl_2 by using the molar mass of Cl_2.

$$? \text{ mol } Cl_2 = 323 \text{ g } Cl_2 \times \frac{1 \text{ mol } Cl_2}{70.91 \text{ g } Cl_2} = 4.56 \text{ mol } Cl_2$$

Convert from moles of Cl_2 to moles of PCl_3 by using the stoichiometric factor.

$$? \text{ mol } PCl_3 = 4.56 \text{ mol } Cl_2 \times \frac{4 \text{ mol } PCl_3}{6 \text{ mol } Cl_2} = 3.04 \text{ mol } PCl_3$$

Convert from moles of PCl_3 to grams of PCl_3 by using the molar mass of PCl_3.

$$? \text{ g } PCl_3 = 3.04 \text{ mol } PCl_3 \times \frac{137.3 \text{ g } PCl_3}{1 \text{ mol } PCl_3} = 417 \text{ g } PCl_3$$

Using the conversion pathway approach, we combine the steps into a single line calculation.

$$? \text{ g } PCl_3 = 323 \text{ g } Cl_2 \times \frac{1 \text{ mol } Cl_2}{70.91 \text{ g } Cl_2} \times \frac{4 \text{ mol } PCl_3}{6 \text{ mol } Cl_2} \times \frac{137.3 \text{ g } PCl_3}{1 \text{ mol } PCl_3} = 417 \text{ g } PCl_3$$

Assess

A different approach that leads to the same result is to do two separate calculations. First, calculate the mass of PCl_3 produced by the reaction of 323 g Cl_2 with an excess of P_4 (= 417 g PCl_3), and then calculate the mass of PCl_3 produced by the reaction of 125 g P_4 with an excess of Cl_2 (= 554 g PCl_3). Only one answer can be correct, and it must be the *smaller* of the two. An advantage of this approach is that it can be easily generalized to deal with reactions involving many reactants.

PRACTICE EXAMPLE A: If 215 g P_4 is allowed to react with 725 g Cl_2 in the reaction in Example 4-12, how many grams of PCl_3 are formed?

PRACTICE EXAMPLE B: If 1.00 kg each of PCl_3, Cl_2, and P_4O_{10} are allowed to react, how many kilograms of $POCl_3$ will be formed?

$$6 PCl_3(l) + 6 Cl_2(g) + P_4O_{10}(s) \longrightarrow 10 POCl_3(l)$$

EXAMPLE 4-13   Determining the Quantity of Excess Reactant(s) Remaining After a Reaction

What mass of P_4 remains following the reaction in Example 4-12?

Analyze

We established in Example 4-12 that Cl_2 is the limiting reactant and P_4 is the excess reactant. The key to this problem is to calculate the mass of P_4 that is consumed, and we can base this calculation either on the mass of Cl_2 consumed or on the mass of PCl_3 produced. Starting from the mass of Cl_2 consumed, we use the following conversion pathway to calculate the mass of P_4 that is consumed.

$$\text{g } Cl_2 \longrightarrow \text{mol } Cl_2 \longrightarrow \text{mol } P_4 \longrightarrow \text{g } P_4$$

The mass of P_4 that remains is calculated by subtracting the mass of P_4 that is consumed from the total mass of P_4.

Solve

We can calculate the mass of P_4 that is consumed by using the following stepwise approach.

Convert from grams of Cl_2 to moles of Cl_2 by using the molar mass of Cl_2.

$$? \text{ mol } Cl_2 = 323 \text{ g } Cl_2 \times \frac{1 \text{ mol } Cl_2}{70.91 \text{ g } Cl_2} = 4.56 \text{ mol } Cl_2$$

Convert from moles of Cl_2 to moles of P_4 by using the stoichiometric factor.

$$? \text{ mol } P_4 = 4.56 \text{ mol } Cl_2 \times \frac{1 \text{ mol } P_4}{6 \text{ mol } Cl_2} = 0.759 \text{ mol } P_4$$

Convert from moles of P_4 to grams of P_4 by using the molar mass of P_4.

$$? \text{ g } P_4 = 0.759 \text{ mol } P_4 \times \frac{123.9 \text{ g } P_4}{1 \text{ mol } P_4} = 94.1 \text{ g } P_4$$

The single line calculation is as follows.

$$? \text{ g } P_4 = 323 \text{ g } Cl_2 \times \frac{1 \text{ mol } Cl_2}{70.91 \text{ g } Cl_2} \times \frac{1 \text{ mol } P_4}{6 \text{ mol } Cl_2} \times \frac{123.9 \text{ g } P_4}{1 \text{ mol } P_4} = 94.1 \text{ g } P_4$$

The following single line calculation shows that we obtain the same answer starting from the mass of PCl_3 that is produced.

$$? \text{ g } P_4 = 417 \text{ g } PCl_3 \times \frac{1 \text{ mol } PCl_3}{137.3 \text{ g } PCl_3} \times \frac{1 \text{ mol } P_4}{4 \text{ mol } PCl_3} \times \frac{123.9 \text{ g } P_4}{1 \text{ mol } P_4} = 94.1 \text{ g } P_4$$

The mass of P_4 remaining after the reaction is simply the difference between what was originally present and what was consumed; that is,

$$125 \text{ g } P_4 \text{ initially} - 94.1 \text{ g } P_4 \text{ consumed} = 31 \text{ g } P_4 \text{ remaining}$$

Assess

It is possible to use the law of conservation of mass to verify that the calculated result is correct. The total mass of the reactants used in the reaction ($323 \text{ g} + 125 \text{ g} = 448 \text{ g}$) must be equal to the mass of product (417 g) plus the mass of the P_4 that remains. Therefore, the mass of the P_4 that remains is $448 \text{ g} - 417 \text{ g} = 31 \text{ g}$. In performing the subtraction required in this example, the quantity with the fewest digits after the decimal point determines how the answer should be expressed. Because there are no digits after the decimal point in 125, there can be none after the decimal point in 31.

PRACTICE EXAMPLE A: In Practice Example 4-12A, which reactant is in excess and what mass of that reactant remains after the reaction to produce PCl_3?

PRACTICE EXAMPLE B: If 12.2 g H_2 and 154 g O_2 are allowed to react, which gas and what mass of that gas remains after the reaction?

$$2 H_2(g) + O_2(g) \longrightarrow 2 H_2O(l)$$

🔍 4-5   CONCEPT ASSESSMENT

Suppose that the reaction of 1.0 mol $NH_3(g)$ and 1.0 mol $O_2(g)$ is carried to completion, producing $NO(g)$ and $H_2O(l)$ as the only products. Which, if any, of the following statements is correct about this reaction? **(a)** 1.0 mol $NO(g)$ is produced; **(b)** 4.0 mol $NO(g)$ is produced; **(c)** 1.5 mol $H_2O(l)$ is produced; **(d)** all the $O_2(g)$ is consumed; **(e)** $NH_3(g)$ is the limiting reactant.

4-5   Other Practical Matters in Reaction Stoichiometry

In this section we consider a few additional factors in reaction stoichiometry—both in the laboratory and in the manufacturing plant. First, the calculated outcome of a reaction may not be what is actually observed. Specifically, the amount of product may be, unavoidably, less than expected. Second, the route to producing a desired chemical may require several reactions carried out in sequence. And third, in some cases two or more reactions may occur simultaneously.

Theoretical Yield, Actual Yield, and Percent Yield

The **theoretical yield** of a reaction is the calculated quantity of product expected from given quantities of reactants. The quantity of product that is actually produced is called the **actual yield**. The **percent yield** is defined as

$$\text{percent yield} = \frac{\text{actual yield}}{\text{theoretical yield}} \times 100\% \qquad \textbf{(4.6)}$$

In many reactions the actual yield almost exactly equals the theoretical yield, and the reactions are said to be *quantitative*. Such reactions can be used in quantitative chemical analyses. In other reactions the actual yield is less than the theoretical yield, and the percent yield is less than 100%. The reduced yield may occur for a variety of reasons. (1) The product of a reaction rarely appears in a pure form, and some product may be lost during the necessary purification steps, which reduces the yield. (2) In many cases the reactants may participate in reactions other than the one of central interest. These are called **side reactions**, and the unintended products are called **by-products**. To the extent that side reactions occur, the yield of the main product is reduced. (3) If a reverse reaction occurs, some of the expected product may react to re-form the reactants, and again the yield is less than expected.

At times, the apparent yield is greater than 100%. Because we cannot get something from nothing, this situation usually indicates an error in technique. Some products are formed as a precipitate from a solvent. If the product is weighed when it is wet with solvent, the result will be a larger-than-expected mass. More thorough drying of the product would give a more accurate yield determination. Another possibility is that the product is contaminated with an excess reactant or a by-product. This makes the mass of product appear larger than expected. In any case, a product must be purified before its yield is determined.

In Example 4-14, we establish the theoretical, actual, and percent yields of an important industrial process.

If we know that a yield will be less than 100%, we need to adjust the amounts of reactants used to produce the desired amount of product. We cannot simply use the theoretical amounts of reactants; we must use more. This point is illustrated in Example 4-15.

Consecutive, Simultaneous, and Overall Reactions

Both in laboratory work and in manufacturing, the preferred processes are those that yield a product through a single reaction. Often such processes give a higher yield because there is no need to remove products from one reaction mixture for further processing in subsequent reactions. However, in many cases a multistep process is unavoidable. **Consecutive reactions** are reactions carried out one after another in sequence to yield a final product. In **simultaneous reactions**, two or

EXAMPLE 4-14   Determining Theoretical, Actual, and Percent Yields

Billions of kilograms of urea, $CO(NH_2)_2$, are produced annually for use as a fertilizer. A ball-and-stick model of urea is shown here. The reaction used is given below.

$$2\,NH_3(g) + CO_2(g) \longrightarrow CO(NH_2)_2(s) + H_2O(l)$$

The typical starting reaction mixture has a 3:1 mole ratio of NH_3 to CO_2. If 47.7 g urea forms *per mole* of CO_2 that reacts, what is the **(a)** theoretical yield; **(b)** actual yield; and **(c)** percent yield?

Urea

Analyze

The reaction mixture contains fixed amounts of NH_3 and CO_2, and so we must first determine which reactant is the limiting reactant. The stoichiometric proportions are 2 mol NH_3:1 mol CO_2. In the reaction mixture, the mole ratio of NH_3 to CO_2 is 3:1. Therefore, NH_3 is the excess reactant and CO_2 is the limiting reactant. The calculation of the theoretical yield of urea must be based on the amount of CO_2, the limiting reactant. Because the quantity of urea is given per mole of CO_2, we should base the calculation on 1.00 mol CO_2. The following conversions are required: mol $CO_2 \rightarrow$ mol $CO(NH_2)_2 \rightarrow$ g $CO(NH_2)_2$.

Solve

(a) Let's calculate the theoretical yield by using a stepwise approach.

Convert from mol CO_2 to mol $CO(NH_2)_2$ by using the stoichiometric factor.

$$? \text{ mol } CO(NH_2)_2 = 1.00 \text{ mol } CO_2 \times \frac{1 \text{ mol } CO(NH_2)_2}{1 \text{ mol } CO_2}$$
$$= 1.00 \text{ mol } CO(NH_2)_2$$

Convert from mol $CO(NH_2)_2$ to g $CO(NH_2)_2$ by using the molar mass of $CO(NH_2)_2$.

$$? \text{ g } CO(NH_2)_2 = 1.00 \text{ mol } CO(NH_2)_2 \times \frac{60.1 \text{ g } CO(NH_2)_2}{1 \text{ mol } CO(NH_2)_2}$$
$$= 60.1 \text{ g } CO(NH_2)_2$$

Thus, 1.00 mol CO_2 is expected to yield 60.1 g $CO(NH_2)_2$, and so the theoretical yield of $CO(NH_2)_2$ is 60.1 g. As has been the case in all our examples, we could have combined the steps into a single line calculation.

$$\text{theoretical yield} = 1.00 \text{ mol } CO_2 \times \frac{1 \text{ mol } CO(NH_2)_2}{1 \text{ mol } CO_2} \times \frac{60.1 \text{ g } CO(NH_2)_2}{1 \text{ mol } CO(NH_2)_2} = 60.1 \text{ g } CO(NH_2)_2$$

(b) actual yield = 47.7 g $CO(NH_2)_2$

(c) percent yield = $\dfrac{47.7 \text{ g } CO(NH_2)_2}{60.1 \text{ g } CO(NH_2)_2} \times 100\% = 79.4\%$

Assess

A quick way to determine the limiting reactant is to identify the reactant that has the smallest value of (*n*/*coeff.*), where *n* is the number of moles of reactant available and *coeff.* is the coefficient of that reactant in the balanced chemical equation. In this example, the reaction mixture contains 3 mol NH_3 and 1 mol CO_2, and so (*n*/*coeff.*) is $3/2 = 1.5$ for NH_3 and $1/1 = 1$ for CO_2. Because $1 < 1.5$, CO_2 is the limiting reactant. To understand why we can identify the limiting reactant by using this method, it is helpful to consider how we convert from moles of reactant to moles of product: *n* mol reactant $\times \dfrac{\text{coefficient of product}}{\text{coefficient of reactant}}$. The smaller the value of (*n*/*coeff.*), the smaller the amount of product. The limiting reactant is the one that yields the smallest amount of product (as explained in Example 4-12); thus, the limiting reactant is the one with the smallest value of (*n*/*coeff.*)

PRACTICE EXAMPLE A: Formaldehyde, CH_2O, can be made from methanol by the following reaction, using a copper catalyst.

$$CH_3OH(g) \longrightarrow CH_2O(g) + H_2(g)$$

If 25.7 g $CH_2O(g)$ is produced per mole of methanol that reacts, what are **(a)** the theoretical yield, **(b)** the actual yield, and **(c)** the percent yield?

PRACTICE EXAMPLE B: What is the percent yield if the reaction of 25.0 g P_4 and 91.5 g Cl_2 produces 104 g PCl_3? The balanced chemical equation for the reaction is given below.

$$P_4(s) + 6\,Cl_2(g) \longrightarrow 4\,PCl_3(l)$$

EXAMPLE 4-15 **Adjusting the Quantities of Reactants in Accordance with the Percent Yield of a Reaction**

When heated with sulfuric or phosphoric acid, cyclohexanol, $C_6H_{11}OH$, is converted to cyclohexene, C_6H_{10}. Ball-and-stick models of cyclohexanol and cyclohexene are shown here. The balanced chemical equation for the reaction is shown below.

$$C_6H_{11}OH(l) \longrightarrow C_6H_{10}(l) + H_2O(l) \qquad (4.7)$$

If the percent yield is 83%, what mass of cyclohexanol must we use to obtain 25 g of cyclohexene?

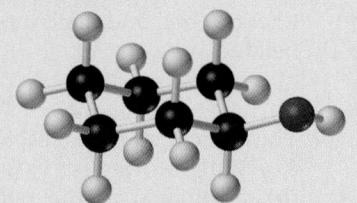

Cyclohexanol

Analyze

We are given the percent yield (83%) and the actual yield of C_6H_{10} (25 g). We can use equation (4.6) to calculate the theoretical yield of C_6H_{10}. Then, we can calculate the quantity of $C_6H_{11}OH$ required to produce the theoretical yield of C_6H_{10} by using the following series of conversions: g C_6H_{10} → mol C_6H_{10} → mol $C_6H_{11}OH$ → g $C_6H_{11}OH$.

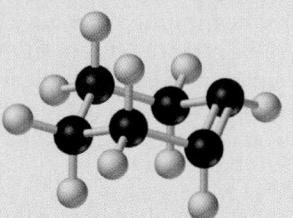

Cyclohexene

Solve

Let's use a stepwise approach.

Rearrange equation (4.6) and calculate the theoretical yield.

$$\text{theoretical yield} = \frac{25 \text{ g} \times 100\%}{83\%} = 3.0 \times 10^1 \text{ g}$$

The theoretical yield is expressed as 3.0×10^1 g rather than as 30 g to emphasize that there are two significant figures in the calculated result. If we expressed the result as 30 g, then the number of significant figures would be ambiguous. (See Section 1-7.) The remaining conversions are as follows.

Convert from g C_6H_{10} to mol C_6H_{10} by using the molar mass of C_6H_{10}.

$$? \text{ mol } C_6H_{10} = 3.0 \times 10^1 \text{ g } C_6H_{10} \times \frac{1 \text{ mol } C_6H_{10}}{82.1 \text{ g } C_6H_{10}} = 0.37 \text{ mol } C_6H_{10}$$

Convert from mol C_6H_{10} to mol $C_6H_{11}OH$ using the stoichiometric factor.

$$? \text{ mol } C_6H_{11}OH = 0.37 \text{ mol } C_6H_{10} \times \frac{1 \text{ mol } C_6H_{11}OH}{1 \text{ mol } C_6H_{10}} = 0.37 \text{ mol } C_6H_{11}OH$$

Convert from mol $C_6H_{11}OH$ to g $C_6H_{11}OH$ by using the molar mass of $C_6H_{11}OH$.

$$? \text{ g } C_6H_{11}OH = 0.37 \text{ mol } C_6H_{11}OH \times \frac{100.2 \text{ g } C_6H_{11}OH}{1 \text{ mol } C_6H_{11}OH} = 37 \text{ g } C_6H_{11}OH$$

We could have combined the last three steps into a single line calculation. The single line calculation is shown below, starting from a theoretical yield of 3.0×10^1 g C_6H_{10}.

$$? \text{ g } C_6H_{11}OH = 30 \text{ g } C_6H_{10} \times \frac{1 \text{ mol } C_6H_{10}}{82.1 \text{ g } C_6H_{10}} \times \frac{1 \text{ mol } C_6H_{11}OH}{1 \text{ mol } C_6H_{10}} \times \frac{100.2 \text{ g } C_6H_{11}OH}{1 \text{ mol } C_6H_{11}OH} = 37 \text{ g } C_6H_{11}OH$$

Assess

Because we have already calculated the molar masses of C_6H_{10} and $C_6H_{11}OH$, it is a simple matter to work the problem in reverse by using numbers that are slightly rounded off. A 37 g sample of $C_6H_{11}OH$ contains approximately $37/100 = 0.37$ moles of $C_6H_{11}OH$, and we expect a maximum of 0.37 moles of C_6H_{10} or $0.37 \times 82 \approx 30$ g C_6H_{10}. We obtain only 25 g C_6H_{10} and so the percent yield for the experiment is approximately $(25/30) \times 100\% = 83\%$. By working the problem in reverse, we verify that the percent yield is 83%; thus, we are confident we have solved the problem correctly.

PRACTICE EXAMPLE A: If the percent yield for the formation of urea in Example 4-14 were 87.5%, what mass of CO_2, together with an excess of NH_3, would have to be used to obtain 50.0 g $CO(NH_2)_2$?

PRACTICE EXAMPLE B: Calculate the mass of cyclohexanol ($C_6H_{11}OH$) needed to produce 45.0 g cyclohexene (C_6H_{10}) by reaction (4.7) if the reaction has a 86.2% yield and the cyclohexanol is 92.3% pure.

more substances react independently of one another in separate reactions occurring at the same time.

Example 4-16 presents an industrial process that is carried out in two consecutive reactions. The key to the calculation is to use a stoichiometric factor for each reaction. Example 4-17 deals with two reactions that occur simultaneously to produce a common product, hydrogen gas.

EXAMPLE 4-16   Calculating the Quantity of a Substance Produced by Reactions Occurring Consecutively

Titanium dioxide, TiO_2, is the most widely used white pigment for paints, having displaced most lead-based pigments, which are environmental hazards. Before it can be used, however, naturally occurring TiO_2 must be freed of colored impurities. One process for doing this converts impure $TiO_2(s)$ to $TiCl_4(g)$, which is then converted back to pure $TiO_2(s)$. The process is based on the following reactions, the first of which generates $TiCl_4$.

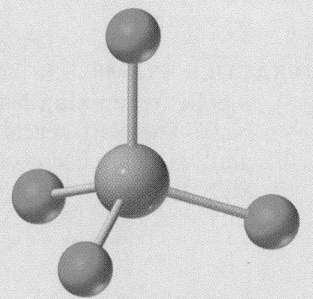

Titanium tetrachloride

$$2\ TiO_2\ (\text{impure}) + 3\ C(s) + 4\ Cl_2(g) \longrightarrow 2\ TiCl_4(g) + CO_2(g) + 2\ CO(g)$$
$$TiCl_4(g) + O_2(g) \longrightarrow TiO_2(s) + 2\ Cl_2(g)$$

What mass of carbon is consumed in producing 1.00 kg of pure $TiO_2(s)$ in this process?

Analyze

In this calculation, we begin with the product, TiO_2, and work backward to one of the reactants, C. The following conversions are required.

$$\text{kg } TiO_2 \longrightarrow \text{g } TiO_2 \longrightarrow \text{mol } TiO_2 \overset{(a)}{\longrightarrow} \text{mol } TiCl_4 \overset{(b)}{\longrightarrow} \text{mol C} \longrightarrow \text{g C}$$

In the conversion from mol TiO_2 to mol $TiCl_4$, labeled (a), we focus on the second reaction. In the conversion from mol $TiCl_4$ to mol C, labeled (b), we focus on the first reaction.

Solve

Using a stepwise approach, we proceed as follows.

Convert from kg TiO_2 to g TiO_2 and then to mol TiO_2 by using the molar mass of TiO_2.

$$? \text{ mol } TiO_2 = 1.00 \text{ kg } TiO_2 \times \frac{1000 \text{ g } TiO_2}{1 \text{ kg } TiO_2} \times \frac{1 \text{ mol } TiO_2}{79.88 \text{ g } TiO_2} = 12.5 \text{ mol } TiO_2$$

Convert from mol TiO_2 to mol $TiCl_4$ by using the stoichiometric factor from the second reaction.

$$? \text{ mol } TiCl_4 = 12.5 \text{ mol } TiO_2 \times \underbrace{\frac{1 \text{ mol } TiCl_4}{1 \text{ mol } TiO_2}}_{(a)} = 12.5 \text{ mol } TiCl_4$$

Convert from mol $TiCl_4$ to mol C by using the stoichiometric factor from the first reaction.

$$? \text{ mol C} = 12.5 \text{ mol } TiCl_4 \times \underbrace{\frac{3 \text{ mol C}}{2 \text{ mol } TiCl_4}}_{(b)} = 18.8 \text{ mol C}$$

Convert from mol C to g C by using the molar mass of C.

$$? \text{ g C} = 18.8 \text{ mol C} \times \frac{12.01 \text{ g C}}{1 \text{ mol C}} = 226 \text{ g C}$$

The conversions given above can be combined into a single line, as shown below.

$$? \text{ g C} = 1.00 \text{ kg } TiO_2 \times \frac{1000 \text{ g } TiO_2}{1 \text{ kg } TiO_2} \times \frac{1 \text{ mol } TiO_2}{79.88 \text{ g } TiO_2} \times \overset{(a)}{\frac{1 \text{ mol } TiCl_4}{1 \text{ mol } TiO_2}} \times \overset{(b)}{\frac{3 \text{ mol C}}{2 \text{ mol } TiCl_4}} \times \frac{12.01 \text{ g C}}{1 \text{ mol C}} = 226 \text{ g C}$$

Assess

To obtain the stoichiometric factor for converting from mol TiO_2 to mol C, we could have reasoned as follows. From the equations for the two reactions, we see that 3 mol C will give 2 mol $TiCl_4$, which then reacts to give an equal amount (2 mol) of TiO_2. That is, 3 mol C will give 2 mol TiO_2, or equivalently, 2 mol TiO_2 requires 3 mol C. Thus, the required stoichiometric factor is (3 mol C)/(2 mol TiO_2), the product of the two factors above labeled (a) and (b).

(continued)

PRACTICE EXAMPLE A: Nitric acid, HNO_3, is produced from ammonia and oxygen by the consecutive reactions

$$4\,NH_3(g) + 5\,O_2(g) \longrightarrow 4\,NO(g) + 6\,H_2O(g)$$
$$2\,NO(g) + O_2(g) \longrightarrow 2\,NO_2(g)$$
$$3\,NO_2(g) + H_2O(l) \longrightarrow 2\,HNO_3(aq) + NO(g)$$

How many grams of nitric acid can be obtained from 1.00 kg $NH_3(g)$, if $NO(g)$ in the third reaction is *not* recycled?

PRACTICE EXAMPLE B: Nitrogen gas is used to inflate automobile airbags. The gas is produced by detonation of a pellet containing NaN_3, KNO_3, and SiO_2 (all solids). When detonated, NaN_3 decomposes into $N_2(g)$ and $Na(l)$. The $Na(l)$ reacts almost instantly with KNO_3 producing more $N_2(g)$ as well as K_2O and Na_2O. The K_2O and Na_2O are each converted independently into harmless salts, Na_2SiO_3 and K_2SiO_3, when they react with SiO_2. Balanced chemical equations for the reactions are given below.

$$2\,NaN_3(s) \longrightarrow 3\,N_2(g) + 2\,Na(l)$$
$$10\,Na(l) + 2\,KNO_3(s) \longrightarrow N_2(s) + K_2O(s) + 5\,Na_2O(s)$$
$$K_2O(s) + SiO_2(s) \longrightarrow K_2SiO_3(s)$$
$$Na_2O(s) + SiO_2(s) \longrightarrow Na_2SiO_3(s)$$

What are the minimum masses of KNO_3 and SiO_2 that must be mixed with 95 g NaN_3 to ensure that no Na, Na_2O, or K_2O remains after the mixture is detonated?

EXAMPLE 4-17   **Calculating the Quantity of a Substance Produced by Reactions Occurring Simultaneously**

Magnesium–aluminum alloys are widely used in aircraft construction. One particular alloy contains 70.0% Al and 30.0% Mg, by mass. How many grams of $H_2(g)$ are produced in the reaction of a 0.710 g sample of this alloy with excess HCl(aq)? Balanced chemical equations are given below for the reactions that occur.

$$2\,Al(s) + 6\,HCl(aq) \longrightarrow 2\,AlCl_3(aq) + 3\,H_2(g)$$
$$Mg(s) + 2\,HCl(aq) \longrightarrow MgCl_2(aq) + H_2(g)$$

Analyze

The two reactions given above are simultaneous reactions. Simultaneous reactions occur independently; thus, we have two conversion pathways to consider:

(1)   g alloy $\longrightarrow$ g Al $\longrightarrow$ mol Al $\xrightarrow{\text{(a)}}$ mol H_2;

(2)   g alloy $\longrightarrow$ g Mg $\longrightarrow$ mol Mg $\xrightarrow{\text{(b)}}$ mol H_2;

Pathways (1) and (2) are based on the first and second reactions, respectively. The total amount of H_2 produced is obtained by adding together the amounts produced by each reaction. The conversion from mol Al to mol H_2 requires a stoichiometric factor, labeled (a). The conversion from mol Mg to mol H_2 requires a different stoichiometric factor, labeled (b).

Solve

Convert from g alloy to mol Al and from g alloy to mol Mg by using the mass percentages of Al and Mg and the molar masses of Al and Mg.

$$? \text{ mol Al} = 0.710 \text{ g alloy} \times \frac{70.0 \text{ g Al}}{100.0 \text{ g alloy}} \times \frac{1 \text{ mol Al}}{26.98 \text{ g Al}}$$
$$= 0.0184 \text{ mol Al}$$

$$? \text{ mol Mg} = 0.710 \text{ g alloy} \times \frac{30.0 \text{ g Mg}}{100.0 \text{ g alloy}} \times \frac{1 \text{ mol Mg}}{24.31 \text{ g Mg}}$$
$$= 8.76 \times 10^{-3} \text{ mol Mg}$$

Convert from mol Al to mol H_2 and from mol Mg to mol H_2 by using stoichiometric factors (a) and (b). The total number of moles of H_2 is obtained by adding together the two independent contributions.

$$? \text{ mol } H_2 = 0.0184 \text{ mol Al} \times \underbrace{\frac{3 \text{ mol } H_2}{2 \text{ mol Al}}}_{(a)} +$$

$$\underbrace{8.76 \times 10^{-3} \text{ mol Mg} \times \frac{1 \text{ mol } H_2}{1 \text{ mol Mg}}}_{(b)} = 0.0364 \text{ mol } H_2$$

Convert from mol H_2 to g H_2 by using the molar mass of H_2 as a conversion factor.

$$? \text{ g } H_2 = 0.0364 \text{ mol } H_2 \times \frac{2.016 \text{ g } H_2}{1 \text{ mol } H_2} = 0.0734 \text{ g } H_2$$

An alternative approach is to combine the steps into a single line calculation, as shown below.

$$? \text{ g } H_2 = \left(0.710 \text{ g alloy} \times \frac{70.0 \text{ g Al}}{100.0 \text{ g alloy}} \times \frac{1 \text{ mol Al}}{26.98 \text{ g Al}} \times \underbrace{\frac{3 \text{ mol } H_2}{2 \text{ mol Al}}}_{(a)} \times \frac{2.016 \text{ g } H_2}{1 \text{ mol } H_2} \right)$$

$$+ \left(0.710 \text{ g alloy} \times \frac{30.0 \text{ g Mg}}{100.0 \text{ g alloy}} \times \frac{1 \text{ mol Mg}}{24.31 \text{ g Mg}} \times \underbrace{\frac{1 \text{ mol } H_2}{1 \text{ mol Mg}}}_{(b)} \times \frac{2.016 \text{ g } H_2}{1 \text{ mol } H_2} \right) = 0.0734 \text{ g } H_2$$

Assess

In this example, the composition of the alloy is given and we solved for the amount of H_2 that is produced. The inverse problem, in which we are given the amount of H_2 produced and are asked to determine the amounts of Al and Mg in the alloy, is a little harder to solve. See Practice Examples A and B below.

PRACTICE EXAMPLE A: A 1.00 g sample of a magnesium-aluminum alloy yields 0.107 g H_2 when treated with an excess of HCl(aq). What is the percentage by mass of Al in the alloy? [*Hint:* This is the inverse of Example 4-17. To solve this problem, let m and $1.00 - m$ be the masses of Al and Mg, respectively, and then use these masses in the setup above to develop an equation that relates m to the total mass of H_2 obtained. Then solve for m.]

PRACTICE EXAMPLE B: A 1.500 g sample of a mixture containing only Cu_2O and CuO was treated with hydrogen to produce copper metal and water. After the water evaporated, 1.2244 g of pure copper metal was recovered. What is the percentage by mass of Cu_2O in the original mixture? Balanced chemical equations are given below for the reactions involved.

$$CuO(s) + H_2(g) \longrightarrow Cu(s) + H_2O(l)$$
$$Cu_2O(s) + H_2(g) \longrightarrow 2\,Cu(s) + H_2O(l)$$

🔍 4-6   CONCEPT ASSESSMENT

Suppose that each of these reactions has a 90% yield.

$$CH_4(g) + Cl_2(g) \longrightarrow CH_3Cl(g) + HCl(g)$$
$$CH_3Cl(g) + Cl_2(g) \longrightarrow CH_2Cl_2(l) + HCl(g)$$

Starting with 50.0 g CH_4 in the first reaction and an excess of $Cl_2(g)$, the number of grams of CH_2Cl_2 formed in the second reaction is

(a) $50.0 \times 0.81 \times (85/16)$ (c) $50.0 \times 0.90 \times 0.90$
(b) 50.0×0.90 (d) $50.0 \times 0.90 \times 0.90 \times (16/50.5)(70.9/85)$

Often, we can combine a series of chemical equations for *consecutive reactions* to obtain a single equation to represent the **overall reaction**. The equation for this overall reaction is the **overall equation**. At times we can use the overall equation for solving problems instead of working with the individual

▶Whereas chemical equations for consecutive reactions can be combined to obtain an equation for the overall reaction, those for a set of simultaneous reactions cannot. If the two equations from Example 4-17 were added, we would obtain the equation $2\,Al + Mg + 8\,HCl \rightarrow 2\,AlCl_3 + MgCl_2 + 4\,H_2$. This equation is clearly incorrect because, for example, it suggests that to make H_2, the reaction mixture must contain both Al and Mg. However, we know that each metal can react independently with HCl to give H_2.

equations. This strategy does not work, however, if the substance of interest is not a starting material or final product but appears only in one of the intermediate reactions. Any substance that is produced in one step and consumed in another step of a multistep process is called an **intermediate**. In Example 4-16, $TiCl_4(g)$ is an intermediate.

To write an overall equation for Example 4-16, multiply the coefficients in the second equation by the factor 2, add the second equation to the first, and cancel any substances that appear on both sides of the overall equation.

$$2\,\cancel{TiO_2}\,(\text{impure}) + 3\,C(s) + 4\,\cancel{Cl_2}(g) \longrightarrow 2\,\cancel{TiCl_4}(g) + CO_2(g) + 2\,CO(g)$$
$$2\,\cancel{TiCl_4}(g) + 2\,O_2(g) \longrightarrow 2\,\cancel{TiO_2}(s) + 4\,\cancel{Cl_2}(g)$$

Overall equation: $\quad 3\,C(s) + 2\,O_2(g) \longrightarrow CO_2(g) + 2\,CO(g)$

The result suggests that (1) we should obtain as much TiO_2 in the second reaction as we started with in the first, (2) the $Cl_2(g)$ produced in the second reaction can be recycled back into the first reaction, and (3) the only substances actually consumed in the overall reaction are $C(s)$ and $O_2(g)$. You can also see that we could not have used this overall equation in the calculation of Example 4-16—$TiO_2(s)$ does not appear in it.

Mastering**CHEMISTRY**   **www.masteringchemistry.com.**

Every day, tons of chemicals are made to be used directly or as reactants in the production of other materials. Because these chemicals are sold for use in everything from food to pharmaceuticals, it is important that they made in high yield and be of high purity. To learn more about the many factors that chemists and engineers must consider when making decisions about building and operating a chemical plant, go to the Focus On feature for Chapter 4, Industrial Chemistry, on the MasteringChemistry site.

Summary

4-1 Chemical Reactions and Chemical Equations—**Chemical reactions** are processes in which one or more starting substances, the **reactants**, form one or more new substances, the **products**. A chemical reaction can be represented symbolically in a **chemical equation**, with formulas of reactants on the left and formulas of products on the right; reactants and products are separated by an arrow. The equation must be balanced. A **balanced equation** reflects the true quantitative relationships between reactants and products. An equation is balanced by placing **stoichiometric coefficients** before formulas to signify that the total number of each kind of atom is the same on each side of the equation. The physical states of reactants and products can be indicated by symbols, such as (s), (l), (g), and (aq), signifying solid, liquid, gas, and aqueous solution, respectively.

4-2 Chemical Equations and Stoichiometry—**Stoichiometry** comprises quantitative relationships based on atomic and molecular masses, chemical formulas, balanced chemical equations, and related matters. The stoichiometry of chemical reactions makes use of conversion factors derived from balanced chemical equations and called **stoichiometric factors** (Fig. 4-3). Stoichiometric calculations usually require molar masses, densities, percent compositions, along with other factors.

4-3 Chemical Reactions in Solution—The **molarity** of a solution is the amount of **solute** (in moles)

per liter of solution (expression 4.3). Molarity can be treated as a conversion factor between solution volume and amount of solute. Molarity as a conversion factor may be applied to individual solutions, to solutions that are mixed or diluted by adding more **solvent** (Fig. 4-6; expression 4.5), and to reactions occurring in solution.

4-4 Determining the Limiting Reactant—In some reactions, one of the reactants is completely consumed, while the other reactants remain in excess. The reactant that is completely consumed determines the amount of product formed and is called the **limiting reactant**. In some reactions, the reactants are consumed simultaneously, with no reactant remaining in excess; such reactants are said to be in **stoichiometric proportions**. In some reactions, the limiting reactant must be determined before a stoichiometric calculation can be completed.

4-5 Other Practical Matters in Reaction Stoichiometry—Stoichiometric calculations sometimes involve additional factors, including the reaction's **actual yield**, the presence of **by-products**, and how the reaction or reactions proceed. For example, some reactions yield exactly the quantity of product calculated—the **theoretical yield**. When the actual yield equals the theoretical yield, the **percent yield** is 100%. In some reactions, the actual yield is less than the theoretical, in which case the percent yield is less than 100%. Lower yields may result from the formation of by-products, substances that replace some of the desired product because of reactions other than the one of interest, called **side reactions**. Some stoichiometric calculations are complicated by the fact that two or more **simultaneous reactions** may occur. In other cases, a final product may be formed through a sequence of **consecutive reactions**. An **intermediate** is any substance that is a product of one reaction and a reactant in a subsequent reaction. The equations for a series of consecutive reactions are often replaced with a single **overall equation** for the **overall reaction**.

Integrative Example

Sodium nitrite is used in the production of dyes for coloring fabrics, as a preservative in meat processing (to prevent botulism), as a bleach for fibers, and in photography. It can be prepared by passing nitrogen monoxide and oxygen gases into an aqueous solution of sodium carbonate. Carbon dioxide gas is another product of the reaction.

In one experimental method, which gives a 95.0% yield, 225 mL of 1.50 M aqueous solution of sodium carbonate, 22.1 g of nitrogen monoxide, and a large excess of oxygen gas are allowed to react. What mass of sodium nitrite is obtained?

Analyze

Five tasks must be performed in this problem: (1) Represent the reaction by a chemical equation in which the names of reactants and products are replaced with formulas. (2) Balance the formula equation by inspection. (3) Determine the limiting reactant. (4) Calculate the theoretical yield of sodium nitrite based on the quantity of limiting reactant. (5) Use expression (4.6) to calculate the actual yield of sodium nitrite.

Solve

Obtain the unbalanced equation: Substitute formulas for names.

$$Na_2CO_3(aq) + NO(g) + O_2(g) \longrightarrow NaNO_2(aq) + CO_2(g)$$

Obtain the balanced equation: Begin by noting that there must be 2 $NaNO_2$ for every Na_2CO_3.

$$Na_2CO_3(aq) + NO(g) + O_2(g) \longrightarrow 2\,NaNO_2(aq) + CO_2(g)$$

To balance the N atoms requires 2 NO on the left side:

$$Na_2CO_3(aq) + 2\,NO(g) + O_2(g) \longrightarrow 2\,NaNO_2(aq) + CO_2(g)$$

To balance the O atoms requires $\frac{1}{2}$ O_2 on the left, resulting in six O atoms on each side:

$$Na_2CO_3(aq) + 2\,NO(g) + \frac{1}{2}O_2(g) \longrightarrow 2\,NaNO_2(aq) + CO_2(g)$$

Multiply coefficients by 2 to make all of them integers in the final balanced equation.

$$2\,Na_2CO_3(aq) + 4\,NO(g) + O_2(g) \longrightarrow 4\,NaNO_2(aq) + 2\,CO_2(g)$$

Determine the limiting reactant: The problem states that oxygen is in excess. To determine the limiting reactant, the number of moles of Na_2CO_3, requiring molarity as a conversion factor, must be compared with the number of moles of NO, requiring a conversion factor based on the molar mass of NO.

$$? \text{ mol } Na_2CO_3 = 0.225 \text{ L soln} \times \frac{1.50 \text{ mol } Na_2CO_3}{1 \text{ L soln}}$$
$$= 0.338 \text{ mol } Na_2CO_3$$
$$? \text{ mol NO} = 22.1 \text{ g NO} \times \frac{1 \text{ mol NO}}{30.01 \text{ g NO}} = 0.736 \text{ mol NO}$$

From the balanced equation, the stoichiometric mole ratio is

$$4 \text{ mol NO} : 2 \text{ mol } Na_2CO_3 = 2 : 1$$

The available mole ratio is

$$0.736 \text{ mol NO} : 0.338 \text{ mol } Na_2CO_3 = 2.18 : 1$$

Because the available mole ratio exceeds 2:1, NO is in excess and Na_2CO_3 is the limiting reactant.

Determine the theoretical yield of the reaction: This calculation is based on the amount of Na_2CO_3, the limiting reactant.

$$? \text{ g } NaNO_2 = 0.338 \text{ mol } Na_2CO_3 \times \frac{4 \text{ mol } NaNO_2}{2 \text{ mol } Na_2CO_3}$$
$$\times \frac{69.00 \text{ g } NaNO_2}{1 \text{ mol } NaNO_2} = 46.6 \text{ g } NaNO_2$$

Determine the actual yield: Use expression (4.6), which relates percent yield (95.0%), theoretical yield (46.6 g $NaNO_2$), and the actual yield.

$$\text{actual yield} = \frac{\text{percent yield} \times \text{theoretical yield}}{100\%}$$
$$= \frac{95.0\% \times 46.6 \text{ g } NaNO_2}{100\%} = 44.3 \text{ g } NaNO_2$$

Assess

After solving a multistep problem, it is important to check over your work. In this example, we should first double-check that the chemical equation is properly balanced. There are 4 Na's on each side, 2 C's, 12 O's, and 4 N's. We can double-check that we correctly identified the limiting reagent by using a different approach. Because 0.338 mol Na_2CO_3 would yield 0.676 mol $NaNO_2$ (in the presence of excess NO and O_2) and 0.736 mol NO would yield 0.736 mol $NaNO_2$ (in the presence of excess Na_2CO_3 and O_2), we conclude that Na_2CO_3 must be the limiting reactant; *the limiting reactant is the one that limits the amount of product obtained*. In checking the calculation of the mass of $NaNO_2$ obtained, we notice that the units work out properly. To check the final step, we can use our final answer to calculate the percent yield for the experiment: $(44.3/46.6) \times 100\% = 95\%$. This is the correct result for the percent yield.

PRACTICE EXAMPLE A: Hexamethylenediamine has the molecular formula $C_6H_{16}N_2$. It is one of the starting materials for the production of nylon. It can be prepared by the following reaction:

$$C_6H_{10}O_4(l) + NH_3(g) + H_2(g) \longrightarrow C_6H_{16}N_2(l) + H_2O(l) \text{ (not balanced)}$$

A large reaction vessel contains 4.15 kg $C_6H_{10}O_4$, 0.547 kg NH_3, and 0.172 kg H_2. If 1.46 kg $C_6H_{16}N_2$ is obtained, then what is the percent yield for the experiment?

PRACTICE EXAMPLE B: Zinc metal and aqueous hydrochloric acid, HCl(aq), react to give hydrogen gas, $H_2(g)$, and aqueous zinc chloride, $ZnCl_2(aq)$. A 0.4000 g sample of *impure* zinc reacts completely when added to 750.0 mL of 0.0179 M HCl. After the reaction, the molarity of the HCl solution is determined to be 0.00403 M. What is the percent by mass of zinc in the sample?

Mastering **CHEMISTRY**

You'll find a link to additional self study questions in the study area on www.masteringchemistry.com

Exercises

Writing and Balancing Chemical Equations

1. Balance the following equations by inspection.
   (a) $SO_3 \longrightarrow SO_2 + O_2$
   (b) $Cl_2O_7 + H_2O \longrightarrow HClO_4$
   (c) $NO_2 + H_2O \longrightarrow HNO_3 + NO$
   (d) $PCl_3 + H_2O \longrightarrow H_3PO_3 + HCl$

2. Balance the following equations by inspection.
   (a) $P_2H_4 \longrightarrow PH_3 + P_4$
   (b) $P_4 + Cl_2 \longrightarrow PCl_3$
   (c) $FeCl_3 + H_2S \longrightarrow Fe_2S_3 + HCl$
   (d) $Mg_3N_2 + H_2O \longrightarrow Mg(OH)_2 + NH_3$

3. Balance the following equations by inspection.
   (a) $PbO + NH_3 \longrightarrow Pb + N_2 + H_2O$
   (b) $FeSO_4 \longrightarrow Fe_2O_3 + SO_2 + O_2$
   (c) $S_2Cl_2 + NH_3 \longrightarrow N_4S_4 + NH_4Cl + S_8$
   (d) $C_3H_7CHOHCH(C_2H_5)CH_2OH + O_2 \longrightarrow CO_2 + H_2O$

4. Balance the following equations by inspection.
   (a) $SO_2Cl_2 + HI \longrightarrow H_2S + H_2O + HCl + I_2$
   (b) $FeTiO_3 + H_2SO_4 + H_2O \longrightarrow FeSO_4 \cdot 7 H_2O + TiOSO_4$
   (c) $Fe_3O_4 + HCl + Cl_2 \longrightarrow FeCl_3 + H_2O + O_2$
   (d) $C_6H_5CH_2SSCH_2C_6H_5 + O_2 \longrightarrow CO_2 + SO_2 + H_2O$

5. Write balanced equations based on the information given.
   (a) solid magnesium + oxygen gas $\longrightarrow$ solid magnesium oxide

   (b) nitrogen monoxide gas + oxygen gas $\longrightarrow$ nitrogen dioxide gas
   (c) gaseous ethane(C_2H_6) + oxygen gas $\longrightarrow$ carbon dioxide gas + liquid water
   (d) aqueous silver sulfate + aqueous barium iodide $\longrightarrow$ solid barium sulfate + solid silver iodide

6. Write balanced equations based on the information given.
   (a) solid magnesium + nitrogen gas $\longrightarrow$ solid magnesium nitride
   (b) solid potassium chlorate $\longrightarrow$ solid potassium chloride + oxygen gas
   (c) solid sodium hydroxide + solid ammonium chloride $\longrightarrow$ solid sodium chloride + gaseous ammonia + water vapor
   (d) solid sodium + liquid water $\longrightarrow$ aqueous sodium hydroxide + hydrogen gas

7. Write balanced equations to represent the complete combustion of each of the following in excess oxygen: **(a)** butane, C_4H_{10}; **(b)** isopropyl alcohol, $CH_3CH(OH)CH_3$; **(c)** lactic acid, $CH_3CH(OH)COOH$.

8. Write balanced equations to represent the complete combustion of each of the following in excess oxygen: **(a)** propylene, C_3H_6; **(b)** thiobenzoic acid, C_6H_5COSH **(c)** glycerol, $CH_2(OH)CH(OH)CH_2OH$.

9. Write balanced equations to represent:
   (a) the decomposition, by heating, of solid ammonium nitrate to produce dinitrogen monoxide gas (laughing gas) and water vapor
   (b) the reaction of aqueous sodium carbonate with hydrochloric acid to produce water, carbon dioxide gas, and aqueous sodium chloride
   (c) the reaction of methane (CH_4), ammonia, and oxygen gases to form gaseous hydrogen cyanide (HCN) and water vapor

10. Write balanced equations to represent:
   (a) the reaction of sulfur dioxide gas with oxygen gas to produce sulfur trioxide gas (one of the reactions involved in the industrial preparation of sulfuric acid)
   (b) the dissolving of limestone (calcium carbonate) in water containing dissolved carbon dioxide to produce calcium hydrogen carbonate (a reaction producing temporary hardness in groundwater)
   (c) the reaction of ammonia and nitrogen monoxide to form nitrogen gas and water vapor

11. Write a balanced chemical equation for the reaction depicted below.

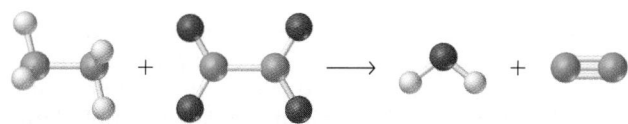

12. Write a balanced chemical equation for the reaction depicted below.

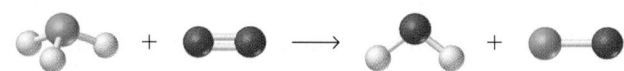

Stoichiometry of Chemical Reactions

13. In an experiment, 0.689 g Cr(s) reacts completely with 0.636 g O_2(g) to form a single solid compound. Write a balanced chemical equation for the reaction.

14. A 3.104 g sample of an oxide of manganese contains 1.142 grams of oxygen. Write a balanced chemical equation for the reaction that produces the compound from Mn(s) and O_2(g).

15. Iron metal reacts with chlorine gas. How many grams of $FeCl_3$ are obtained when 515 g Cl_2 reacts with excess Fe?

$$2\ Fe(s) + 3\ Cl_2(g) \longrightarrow 2\ FeCl_3(s)$$

16. If 46.3 g PCl_3 is produced by the reaction

$$6\ Cl_2(g) + P_4(s) \longrightarrow 4\ PCl_3(l)$$

how many grams each of Cl_2 and P_4 are consumed?

17. A laboratory method of preparing O_2(g) involves the decomposition of $KClO_3$(s).

$$2\ KClO_3(s) \xrightarrow{\Delta} 2\ KCl(s) + 3\ O_2(g)$$

   (a) How many moles of O_2(g) can be produced by the decomposition of 32.8 g $KClO_3$?
   (b) How many grams of $KClO_3$ must decompose to produce 50.0 g O_2?
   (c) How many grams of KCl are formed, together with 28.3 g O_2, in the decomposition of $KClO_3$?

18. A commercial method of manufacturing hydrogen involves the reaction of iron and steam.

$$3\ Fe(s) + 4\ H_2O(g) \xrightarrow{\Delta} Fe_3O_4(s) + 4\ H_2(g)$$

   (a) How many grams of H_2 can be produced from 42.7 g Fe and an excess of H_2O(g) (steam)?
   (b) How many grams of H_2O are consumed in the conversion of 63.5 g Fe to Fe_3O_4?
   (c) If 14.8 g H_2 is produced, how many grams of Fe_3O_4 must also be produced?

19. How many grams of Ag_2CO_3 are decomposed to yield 75.1 g Ag in this reaction?

$$Ag_2CO_3(s) \xrightarrow{\Delta}$$
$$Ag(s) + CO_2(g) + O_2(g)\ \text{(not balanced)}$$

20. How many kilograms of HNO_3 are consumed to produce 125 kg $Ca(H_2PO_4)_2$ in this reaction?

$$Ca_3(PO_4)_2 + HNO_3 \longrightarrow$$
$$Ca(H_2PO_4)_2 + Ca(NO_3)_2\ \text{(not balanced)}$$

21. The reaction of calcium hydride with water can be used to prepare small quantities of hydrogen gas, as is done to fill weather-observation balloons.

$$CaH_2(s) + H_2O(l) \longrightarrow$$
$$Ca(OH)_2(s) + H_2(g)\ \text{(not balanced)}$$

   (a) How many grams of H_2(g) result from the reaction of 127 g CaH_2 with an excess of water?
   (b) How many grams of water are consumed in the reaction of 56.2 g CaH_2?
   (c) What mass of CaH_2(s) must react with an excess of water to produce 8.12×10^{24} molecules of H_2?

22. The reaction of potassium superoxide, KO_2, is used in life-support systems to replace CO_2(g) in expired air with O_2(g). The unbalanced chemical equation for the reaction is given below.

$$KO_2(s) + CO_2(g) \longrightarrow K_2CO_3(s) + O_2(g)$$

   (a) How many moles of O_2(g) are produced by the reaction of 156 g CO_2(g) with excess KO_2(s)?
   (b) How many grams of KO_2(s) are consumed per 100.0 g CO_2(g) removed from expired air?
   (c) How many O_2 molecules are produced per milligram of KO_2 consumed?

23. Iron ore is impure Fe_2O_3. When Fe_2O_3 is heated with an excess of carbon (coke), metallic iron and carbon monoxide gas are produced. From a sample of ore weighing 938 kg, 523 kg of pure iron is obtained. What is the mass percent Fe_2O_3, by mass, in the ore sample, assuming that none of the impurities contain Fe?

24. Solid silver oxide, Ag_2O(s), decomposes at temperatures in excess of 300 °C, yielding metallic silver and oxygen gas. A 3.13 g sample of impure silver oxide yields 0.187 g O_2(g). What is the mass percent Ag_2O in the sample? Assume that Ag_2O(s) is the only source of O_2(g). [*Hint:* Write a balanced equation for the reaction.]

25. Decaborane, $B_{10}H_{14}$, was used as a fuel for rockets in the 1950s. It reacts violently with oxygen, O_2, to produce B_2O_3 and water. Calculate the percentage by mass of $B_{10}H_{14}$ in a fuel mixture designed to ensure that $B_{10}H_{14}$ and O_2 run out at exactly the same time. (Such a mixture minimizes the mass of fuel that a rocket must carry.)

26. The rocket boosters of the space shuttle *Discovery*, launched on July 26, 2005, used a fuel mixture containing primarily solid ammonium perchlorate, $NH_4ClO_4(s)$, and aluminum metal. The unbalanced chemical equation for the reaction is given below.

$$Al(s) + NH_4ClO_4(s) \longrightarrow$$
$$Al_2O_3(s) + AlCl_3(s) + H_2O(l) + N_2(g)$$

What is the minimum mass of NH_4ClO_4 consumed, per kilogram of Al, by the reaction of NH_4ClO_4 and Al? [*Hint*: Balance the elements in the order Cl, H, O, Al, N.]

27. A piece of aluminum foil measuring 10.25 cm $\times$ 5.50 cm $\times$ 0.601 mm is dissolved in excess HCl(aq). What mass of $H_2(g)$ is produced? Use equation (4.2) and $d = 2.70$ g/cm^3 for Al.

28. An excess of aluminum foil is allowed to react with 225 mL of an aqueous solution of HCl ($d = 1.088$ g/mL) that contains 18.0% HCl by mass. What mass of $H_2(g)$ is produced? [*Hint*: Use equation (4.2).]

29. *Without performing detailed calculations*, which of the following metals yields the greatest amount of H_2 per gram of metal reacting with HCl(aq)? **(a)** Na, **(b)** Mg, **(c)** Al, **(d)** Zn. [*Hint*: Write equations similar to (4.2).]

30. *Without performing detailed calculations*, which of the following yields the same mass of $CO_2(g)$ per gram of compound as does ethanol, CH_3CH_2OH, when burned in excess oxygen? **(a)** H_2CO; **(b)** $HOCH_2CH_2OH$; **(c)** $HOCH_2CHOHCH_2OH$; **(d)** CH_3OCH_3; **(e)** C_6H_5OH.

Molarity

31. What are the molarities of the following solutes when dissolved in water?
    (a) 2.92 mol CH_3OH in 7.16 L of solution
    (b) 7.69 mmol CH_3CH_2OH in 50.00 mL of solution
    (c) 25.2 g $CO(NH_2)_2$ in 275 mL of solution

32. What are the molarities of the following solutes when dissolved in water?
    (a) 2.25×10^{-4} mol CH_3CH_2OH in 125 mL of solution
    (b) 57.5 g $(CH_3)_2CO$ in 525 mL of solution
    (c) 18.5 mL of $C_3H_5(OH)_3$ ($d = 1.26$ g/mL) in 375 mL of solution

33. What are the molarities of the following solutes?
    (a) sucrose ($C_{12}H_{22}O_{11}$) if 150.0 g is dissolved per 250.0 mL of water solution
    (b) urea, $CO(NH_2)_2$, if 98.3 mg of the 97.9% pure solid is dissolved in 5.00 mL of aqueous solution
    (c) methanol, CH_3OH, ($d = 0.792$ g/mL) if 125.0 mL is dissolved in enough water to make 15.0 L of solution

34. What are the molarities of the following solutes?
    (a) aspartic acid ($H_2C_4H_5NO_4$) if 0.405 g is dissolved in enough water to make 100.0 mL of solution
    (b) acetone, C_3H_6O, ($d = 0.790$ g/mL) if 35.0 mL is dissolved in enough water to make 425 mL of solution
    (c) diethyl ether, $(C_2H_5)_2O$, if 8.8 mg is dissolved in enough water to make 3.00 L of solution

35. How much
    (a) glucose, $C_6H_{12}O_6$, in grams, must be dissolved in water to produce 75.0 mL of 0.350 M $C_6H_{12}O_6$?
    (b) methanol, CH_3OH ($d = 0.792$ g/mL), in milliliters, must be dissolved in water to produce 2.25 L of 0.485 M CH_3OH?

36. How much
    (a) ethanol, CH_3CH_2OH ($d = 0.789$ g/mL), in liters, must be dissolved in water to produce 200.0 L of 1.65 M CH_3CH_2OH?
    (b) concentrated hydrochloric acid solution (36.0% HCl by mass; $d = 1.18$ g/mL), in milliliters, is required to produce 12.0 L of 0.234 M HCl?

37. In the United States, the concentration of glucose, $C_6H_{12}O_6$, in the blood is reported in units of milligrams per deciliter (mg/dL). In Canada, the United Kingdom, and elsewhere, the blood glucose concentration is reported in millimoles per liter (mmol/L), where 1 mmol = 1×10^{-3} mol. If a person has a blood glucose level of 85 mg/dL, then what is **(a)** the blood glucose level in mmol/L; **(b)** the molarity of glucose in the blood?

38. In many communities, water is fluoridated to prevent tooth decay. In the United States, for example, more than half of the population served by public water systems has access to water that is fluoridated at approximately 1 mg F^- per liter. **(a)** What is the molarity of F^- in water if it contains 1.2 mg F^- per liter? **(b)** How many grams of solid KF should be added to a 1.6×10^8 L water reservoir to give a fluoride concentration of 1.2 mg F^- per liter?

39. Which of the following is a 0.500 M KCl solution? **(a)** 0.500 g KCl/mL solution; **(b)** 36.0 g KCl/L solution; **(c)** 7.46 mg KCl/mL solution; **(d)** 373 g KCl in 10.00 L solution

40. Which two solutions have the same concentration? **(a)** 55.45 g NaCl/L solution; **(b)** 5.545 g NaCl/100 g solution; **(c)** 55.45 g NaCl/kg water; **(d)** 55.45 mg NaCl/1.00 mL solution; **(e)** 5.00 mmol NaCl/5.00 mL solution

41. Which has the higher concentration of sucrose: a 46% sucrose solution by mass ($d = 1.21$ g/mL), or 1.50 M $C_{12}H_{22}O_{11}$? Explain your reasoning.

42. Which has the greater molarity of ethanol: a white wine ($d = 0.95$ g/mL) with 11% CH_3CH_2OH by mass, or the solution described in Example 4-8? Explain your reasoning.

43. A 10.00 mL sample of 2.05 M KNO_3 is diluted to a volume of 250.0 mL. What is the concentration of the diluted solution?

44. What volume of 0.750 M $AgNO_3$ must be diluted with water to prepare 250.0 mL of 0.425 M $AgNO_3$?

45. Water is evaporated from 125 mL of 0.198 M K_2SO_4 solution until the volume becomes 105 mL. What is the molarity of K_2SO_4 in the remaining solution?

46. A 25.0 mL sample of HCl(aq) is diluted to a volume of 500.0 mL. If the concentration of the diluted solution is found to be 0.085 M HCl, what was the concentration of the original solution?

47. Given a 0.250 M K_2CrO_4 stock solution, describe how you would prepare a solution that is 0.0125 M K_2CrO_4. That is, what combination(s) of pipet and volumetric flask would you use? Typical sizes of vol- umetric flasks found in a general chemistry labora- tory are 100.0, 250.0, 500.0, and 1000.0 mL, and typi- cal sizes of volumetric pipets are 1.00, 5.00, 10.00, 25.00, and 50.00 mL.

48. Given two liters of 0.496 M KCl, describe how you would use this solution to prepare 250.0 mL of 0.175 M KCl. Give sufficient details so that another student could follow your instructions.

Chemical Reactions in Solution

49. Consider the reaction below:

$$2\,AgNO_3(aq) + Na_2S(aq) \longrightarrow$$
$$Ag_2S(s) + 2\,NaNO_3(aq)$$

(a) How many grams of $Na_2S(s)$ are required to react completely with 27.8 mL of 0.163 M $AgNO_3$?
(b) How many grams of $Ag_2S(s)$ are obtained from the reaction in part **(a)**?

50. Excess $NaHCO_3$ is added to 525 mL of 0.220 M $Cu(NO_3)_2$. These substances react as follows:

$$Cu(NO_3)_2(aq) + 2\,NaHCO_3(s) \longrightarrow$$
$$CuCO_3(s) + 2\,NaNO_3(aq) + H_2O(l) + CO_2(g)$$

(a) How many grams of the $NaHCO_3(s)$ will be consumed?
(b) How many grams of $CuCO_3(s)$ will be produced?

51. How many milliliters of 0.650 M K_2CrO_4 are needed to precipitate all the silver in 415 mL of 0.186 M $AgNO_3$ as $Ag_2CrO_4(s)$?

$$2\,AgNO_3(aq) + K_2CrO_4(aq) \longrightarrow$$
$$Ag_2CrO_4(s) + 2\,KNO_3(aq)$$

52. Consider the reaction below.

$$Ca(OH)_2(s) + 2\,HCl(aq) \longrightarrow CaCl_2(aq) + 2\,H_2O(l)$$

(a) How many grams of $Ca(OH)_2$ are required to react completely with 415 mL of 0.477 M HCl?
(b) How many kilograms of $Ca(OH)_2$ are required to react with 324 L of a HCl solution that is 24.28% HCl by mass, and has a density of 1.12 g/mL?

53. Exactly 1.00 mL of an aqueous solution of HNO_3 is diluted to 100.0 mL. It takes 29.78 mL of 0.0142 M $Ca(OH)_2$ to convert all of the HNO_3 to $Ca(NO_3)_2$. The other product of the reaction is water. Calculate the molarity of the undiluted HNO_3 solution.

54. A 5.00 mL sample of an aqueous solution of H_3PO_4 requires 49.1 mL of 0.217 M NaOH to convert all of the H_3PO_4 to Na_2HPO_4. The other product of the reaction is water. Calculate the molarity of the H_3PO_4 solution.

55. Refer to Example 4-7 and equation (4.2). For the con- ditions stated in Example 4-7, determine **(a)** the num- ber of moles of $AlCl_3$ and **(b)** the molarity of the $AlCl_3(aq)$ if the solution volume is simply the 23.8 mL calculated in the example.

56. Refer to the Integrative Example on page 139. If 138 g Na_2CO_3 in 1.42 L of aqueous solution is treated with an excess of NO(g) and $O_2(g)$, what is the molarity of the $NaNO_2(aq)$ solution that results? (Assume that the reaction goes to completion.)

57. How many grams of Ag_2CrO_4 will precipitate if excess $K_2CrO_4(aq)$ is added to the 415 mL of 0.186 M $AgNO_3$ in Exercise 51?

58. What volume of 0.0797 M $KMnO_4$ is necessary to con- vert 9.13 g KI to I_2 in the reaction below? Assume that H_2SO_4 is present in excess.

$$2\,KMnO_4 + 10\,KI + 8\,H_2SO_4 \longrightarrow$$
$$6\,K_2SO_4 + 2\,MnSO_4 + 5\,I_2 + 8\,H_2O$$

59. How many grams of sodium must react with 155 mL H_2O to produce a solution that is 0.175 M NaOH? (Assume a final solution volume of 155 mL.)

$$2\,Na(s) + 2\,H_2O(l) \longrightarrow 2\,NaOH(aq) + H_2(g)$$

60. A method of lowering the concentration of HCl(aq) is to allow the solution to react with a small quantity of Mg. How many milligrams of Mg must be added to 250.0 mL of 1.023 M HCl to reduce the solution con- centration to exactly 1.000 M HCl?

$$Mg(s) + 2\,HCl(aq) \longrightarrow MgCl_2(aq) + H_2(g)$$

61. A 0.3126 g sample of oxalic acid, $H_2C_2O_4$, requires 26.21 mL of a particular concentration of NaOH(aq) to complete the following reaction. What is the molarity of the NaOH(aq)?

$$H_2C_2O_4(s) + 2\,NaOH(aq) \longrightarrow$$
$$Na_2C_2O_4(aq) + 2\,H_2O(l)$$

62. A 25.00 mL sample of HCl(aq) was added to a 0.1000 g sample of $CaCO_3$. All the $CaCO_3$ reacted, leaving some excess HCl(aq).

$$CaCO_3(s) + 2\,HCl(aq) \longrightarrow$$
$$CaCl_2(aq) + H_2O(l) + CO_2(g)$$

The excess HCl(aq) required 43.82 mL of 0.01185 M $Ba(OH)_2$ to complete the following reaction. What was the molarity of the original HCl(aq)?

$$2\,HCl(aq) + Ba(OH)_2(aq) \longrightarrow BaCl_2(aq) + 2\,H_2O(l)$$

Determining the Limiting Reactant

63. How many moles of NO(g) can be produced in the reaction of 3.00 mol $NH_3(g)$ and 4.00 mol $O_2(g)$?

$$4\,NH_3(g) + 5\,O_2(g) \xrightarrow{\Delta} 4\,NO(g) + 6\,H_2O(l)$$

64. The reaction of calcium hydride and water produces calcium hydroxide and hydrogen as products. How many moles of $H_2(g)$ will be formed in the reaction between 0.82 mol $CaH_2(s)$ and 1.54 mol $H_2O(l)$?

65. A 0.696 mol sample of Cu is added to 136 mL of 6.0 M HNO_3. Assuming the following reaction is the only one that occurs, will the Cu react completely?

$$3 Cu(s) + 8 HNO_3(aq) \longrightarrow$$
$$3 Cu(NO_3)_2(aq) + 4 H_2O(l) + 2 NO(g)$$

66. How many grams of $H_2(g)$ are produced by the reaction of 1.84 g Al with 75.0 mL of 2.95 M HCl? [*Hint:* Recall equation (4.2).]

67. A side reaction in the manufacture of rayon from wood pulp is

$$3 CS_2 + 6 NaOH \longrightarrow 2 Na_2CS_3 + Na_2CO_3 + 3 H_2O$$

How many grams of Na_2CS_3 are produced in the reaction of 92.5 mL of liquid CS_2 ($d = 1.26$ g/mL) and 2.78 mol NaOH?

68. Lithopone is a brilliant white pigment used in water-based interior paints. It is a mixture of $BaSO_4$ and ZnS produced by the reaction

$$BaS(aq) + ZnSO_4(aq) \longrightarrow \underset{\text{lithopone}}{ZnS(s) + BaSO_4(s)}$$

How many grams of lithopone are produced in the reaction of 315 mL of 0.275 M $ZnSO_4$ and 285 mL of 0.315 M BaS?

69. Ammonia can be generated by heating together the solids NH_4Cl and $Ca(OH)_2$. $CaCl_2$ and H_2O are also formed. **(a)** If a mixture containing 33.0 g each of

NH_4Cl and $Ca(OH)_2$ is heated, how many grams of NH_3 will form? **(b)** Which reactant remains in excess, and in what mass?

70. Chlorine can be generated by heating together calcium hypochlorite and hydrochloric acid. Calcium chloride and water are also formed. **(a)** If 50.0 g $Ca(OCl)_2$ and 275 mL of 6.00 M HCl are allowed to react, how many grams of chlorine gas will form? **(b)** Which reactant, $Ca(OCl)_2$ or HCl, remains in excess, and in what mass?

71. Chromium(II) sulfate, $CrSO_4$, is a reagent that has been used in certain applications to help reduce carbon–carbon double bonds (C=C) in molecules to single bonds (C—C). The reagent can be prepared via the following reaction.

$$4 Zn(s) + K_2Cr_2O_7(aq) + 7 H_2SO_4(aq) \longrightarrow$$
$$4 ZnSO_4(aq) + 2 CrSO_4(aq) + K_2SO_4(aq) + 7 H_2O(l)$$

What is the maximum number of grams of $CrSO_4$ that can be made from a reaction mixture containing 3.2 mol Zn, 1.7 mol $K_2Cr_2O_7$, and 5.0 mol H_2SO_4?

72. Titanium tetrachloride, $TiCl_4$, is prepared by the reaction below.

$$3 TiO_2(s) + 4 C(s) + 6 Cl_2(g) \longrightarrow$$
$$3 TiCl_4(g) + 2 CO_2(g) + 2 CO(g)$$

What is the maximum mass of $TiCl_4$ that can be obtained from 35 g TiO_2, 45 g Cl_2, and 11 g C?

Theoretical, Actual, and Percent Yields

73. In the reaction of 277 g CCl_4 with an excess of HF, 187 g CCl_2F_2 is obtained. What are the **(a)** theoretical, **(b)** actual, and **(c)** percent yields of this reaction?

$$CCl_4 + 2 HF \longrightarrow CCl_2F_2 + 2 HCl$$

74. In the reaction shown, 100.0 g $C_6H_{11}OH$ yielded 64.0 g C_6H_{10}. **(a)** What is the theoretical yield of the reaction? **(b)** What is the percent yield? **(c)** What mass of $C_6H_{11}OH$ would produce 100.0 g C_6H_{10} if the percent yield is that determined in part **(b)**?

$$C_6H_{11}OH \longrightarrow C_6H_{10} + H_2O$$

75. Cryolite, Na_3AlF_6, is an important industrial reagent. It is made by the reaction below.

$$Al_2O_3(s) + 6 NaOH(aq) + 12 HF(g) \longrightarrow$$
$$2 Na_3AlF_6(s) + 9 H_2O(l)$$

In an experiment, 7.81 g Al_2O_3 and excess HF(g) were dissolved in 3.50 L of 0.141 M NaOH. If 28.2 g Na_3AlF_6 was obtained, then what is the percent yield for this experiment?

76. Nitrogen gas, N_2, can be prepared by passing gaseous ammonia over solid copper(II) oxide, CuO, at high temperatures. The other products of the reaction are solid copper, Cu, and water vapor. In a certain experiment, a reaction mixture containing 18.1 g NH_3 and 90.4 g CuO yields 6.63 g N_2. Calculate the percent yield for this experiment.

77. The reaction of 15.0 g C_4H_9OH, 22.4 g NaBr, and 32.7 g H_2SO_4 yields 17.1 g C_4H_9Br in the reaction shown.

What are the **(a)** theoretical yield, **(b)** actual yield, and **(c)** percent yield of this reaction?

$$C_4H_9OH + NaBr + H_2SO_4 \longrightarrow$$
$$C_4H_9Br + NaHSO_4 + H_2O$$

78. Azobenzene, an intermediate in the manufacture of dyes, can be prepared from nitrobenzene by reaction with triethylene glycol in the presence of Zn and KOH. In one reaction, 0.10 L of nitrobenzene ($d = 1.20$ g/mL) and 0.30 L of triethylene glycol ($d = 1.12$ g/mL) yields 55 g azobenzene. What are the **(a)** theoretical yield, **(b)** actual yield, and **(c)** percent yield of this reaction?

$$\underset{\text{nitrobenzene \quad triethylene glycol}}{2 C_6H_5NO_2 + 4 C_6H_{14}O_4} \xrightarrow[\text{KOH}]{\text{Zn}}$$
$$\underset{\text{azobenzene}}{(C_6H_5N)_2 + 4 C_6H_{12}O_4 + 4 H_2O}$$

79. How many grams of commercial acetic acid (97% CH_3COOH by mass) must be allowed to react with an excess of PCl_3 to produce 75 g of acetyl chloride (CH_3COCl), if the reaction has a 78.2% yield?

$$CH_3COOH + PCl_3 \longrightarrow$$
$$CH_3COCl + H_3PO_3 \text{ (not balanced)}$$

80. Suppose that reactions **(a)** and **(b)** each have a 92% yield. Starting with 112 g CH_4 in reaction **(a)** and an excess of $Cl_2(g)$, how many grams of CH_2Cl_2 are formed in reaction **(b)**?
    (a) $CH_4 + Cl_2 \longrightarrow CH_3Cl + HCl$
    (b) $CH_3Cl + Cl_2 \longrightarrow CH_2Cl_2 + HCl$

81. An essentially 100% yield is necessary for a chemical reaction used to *analyze* a compound, but it is almost never expected for a reaction that is used to *synthesize* a compound. Explain this difference.

82. Suppose we carry out the precipitation of $Ag_2CrO_4(s)$ described in Example 4-11. If we obtain 2.058 g of precipitate, we might conclude that it is nearly pure $Ag_2CrO_4(s)$, but if we obtain 2.112 g, we can be quite sure that the precipitate is not pure. Explain this difference.

Consecutive Reactions, Simultaneous Reactions

83. How many grams of HCl are consumed in the reaction of 425 g of a mixture containing 35.2% $MgCO_3$ and 64.8% $Mg(OH)_2$, by mass?

$$Mg(OH)_2 + 2\,HCl \longrightarrow MgCl_2 + 2\,H_2O$$
$$MgCO_3 + 2\,HCl \longrightarrow MgCl_2 + H_2O + CO_2(g)$$

84. How many grams of CO_2 are produced in the complete combustion of 406 g of a bottled gas that consists of 72.7% propane (C_3H_8) and 27.3% butane (C_4H_{10}), by mass?

85. Dichlorodifluoromethane, once widely used as a refrigerant, can be prepared by the reactions shown. How many moles of Cl_2 must be consumed in the first reaction to produce 2.25 kg CCl_2F_2 in the second? Assume that all the CCl_4 produced in the first reaction is consumed in the second.

$$CH_4 + Cl_2 \longrightarrow CCl_4 + HCl \text{ (not balanced)}$$
$$CCl_4 + HF \longrightarrow CCl_2F_2 + HCl \text{ (not balanced)}$$

86. Carbon dioxide gas, $CO_2(g)$, produced in the combustion of a sample of ethane is absorbed in $Ba(OH)_2(aq)$, producing 0.506 g $BaCO_3(s)$. How many grams of ethane (C_2H_6) must have been burned?

$$C_2H_6(g) + O_2(g) \longrightarrow$$
$$CO_2(g) + H_2O(l) \text{ (not balanced)}$$
$$CO_2(g) + Ba(OH)_2(aq) \longrightarrow BaCO_3(s) + H_2O(l)$$

87. The following process has been used to obtain iodine from oil-field brines in California. How many kilograms of silver nitrate are required in the first step for every kilogram of iodine produced in the third step?

sodium iodide + silver nitrate $\longrightarrow$
    silver iodide + sodium nitrate

silver iodide + iron $\longrightarrow$ iron(II) iodide + silver

iron(II) iodide + chlorine gas $\longrightarrow$
    iron(III) chloride + solid iodine

88. Sodium bromide, used to produce silver bromide for use in photography, can be prepared as shown. How many kilograms of iron are consumed to produce 2.50×10^3 kg NaBr?

$$Fe + Br_2 \longrightarrow FeBr_2$$
$$FeBr_2 + Br_2 \longrightarrow Fe_3Br_8 \text{ (not balanced)}$$
$$Fe_3Br_8 + Na_2CO_3 \longrightarrow$$
$$NaBr + CO_2 + Fe_3O_4 \text{ (not balanced)}$$

89. High-purity silicon is obtained using a three-step process. The first step involves heating solid silicon dioxide, SiO_2, with solid carbon to give solid silicon and carbon monoxide gas. In the second step, solid silicon is converted into liquid silicon tetrachloride, $SiCl_4$, by treating it with chlorine gas. In the last step, $SiCl_4$ is treated with hydrogen gas to give ultrapure solid silicon and hydrogen chloride gas.
(a) Write balanced chemical equations for the steps involved in this three-step process.
(b) Calculate the masses of carbon, chlorine, and hydrogen required per kilogram of silicon.

90. The following set of reactions is to be used as the basis of a method for producing nitric acid, HNO_3. Calculate the minimum masses of N_2, H_2, and O_2 required per kilogram of HNO_3.

$$N_2(g) + 3\,H_2(g) \longrightarrow 2\,NH_3(g)$$
$$4\,NH_3(g) + 5\,O_2(g) \longrightarrow 4\,NO(g) + 6\,H_2O(g)$$
$$2\,NO(g) + O_2(g) \longrightarrow 2\,NO_2(g)$$
$$3\,NO_2(g) + H_2O(l) \longrightarrow 2\,HNO_3(aq) + NO(g)$$

91. When a solid mixture of $MgCO_3$ and $CaCO_3$ is heated strongly, carbon dioxide gas is given off and a solid mixture of MgO and CaO is obtained. If a 24.00 g sample of a mixture of $MgCO_3$ and $CaCO_3$ produces 12.00 g CO_2, then what is the percentage by mass of $MgCO_3$ in the original mixture?

92. A mixture of Fe_2O_3 and FeO was analyzed and found to be 72.0% Fe by mass. What is the percentage by mass of Fe_2O_3 in the mixture?

Integrative and Advanced Exercises

93. Write chemical equations to represent the following reactions.
(a) Limestone rock (calcium carbonate) is heated (calcined) and decomposes to calcium oxide and carbon dioxide gas.
(b) Zinc sulfide ore is heated in air (roasted) and is converted to zinc oxide and sulfur dioxide gas. (Note that oxygen gas in the air is also a reactant.)
(c) Propane gas reacts with gaseous water to produce a mixture of carbon monoxide and hydrogen gases. (This mixture, called *synthesis gas*, is used to produce a variety of organic chemicals.)
(d) Sulfur dioxide gas is passed into an aqueous solution containing sodium sulfide and sodium carbonate. The reaction products are carbon dioxide and an aqueous solution of sodium thiosulfate.

94. Write chemical equations to represent the following reactions.

    (a) Calcium phosphate is heated with silicon dioxide and carbon, producing calcium silicate ($CaSiO_3$), phosphorus (P_4), and carbon monoxide. The phosphorus and chlorine react to form phosphorus trichloride, and the phosphorus trichloride and water react to form phosphorous acid.

    (b) Copper metal reacts with gaseous oxygen, carbon dioxide, and water to form green basic copper carbonate, $Cu_2(OH)_2CO_3$ (a reaction responsible for the formation of the green patina, or coating, often seen on outdoor bronze statues).

    (c) White phosphorus and oxygen gas react to form tetraphosphorus decoxide. The tetraphosphorus decoxide reacts with water to form an aqueous solution of phosphoric acid.

    (d) Calcium dihydrogen phosphate reacts with sodium hydrogen carbonate (bicarbonate), producing calcium phosphate, sodium hydrogen phosphate, carbon dioxide, and water (the principal reaction occurring when ordinary baking powder is added to cakes, bread, and biscuits).

95. The three astronauts aboard *Apollo 13*, which was launched in 1970 on April 11 and returned to Earth on April 17, were kept alive during their mission, in part, because of lithium hydroxide (LiOH) canisters that were designed to remove exhaled CO_2 from the air. Solid lithium hydroxide reacts with $CO_2(g)$ to give solid Li_2CO_3 and water. With the assumption that an astronaut exhales approximately 1.00 kg CO_2 per day, what mass of LiOH was required to remove all of the CO_2 exhaled by the three-member crew on their six-day mission?

96. Chalkboard chalk is made from calcium carbonate and calcium sulfate, with minor impurities such as SiO_2. Only the $CaCO_3$ reacts with dilute HCl(aq). What is the mass percent $CaCO_3$ in a piece of chalk if a 3.28-g sample yields 0.981 g $CO_2(g)$?

    $$CaCO_3(s) + 2\,HCl(aq) \longrightarrow$$
    $$CaCl_2(aq) + H_2O(l) + CO_2(g)$$

97. Hydrogen gas, $H_2(g)$, is passed over $Fe_2O_3(s)$ at 400 °C. Water vapor is formed together with a black residue—a compound consisting of 72.3% Fe and 27.7% O. Write a balanced equation for this reaction.

98. A sulfide of iron, containing 36.5% S by mass, is heated in $O_2(g)$, and the products are sulfur dioxide and an oxide of iron containing 27.6% O, by mass. Write a balanced chemical equation for this reaction.

99. Water and ethanol, $CH_3CH_2OH(l)$, are miscible, that is, they can be mixed in all proportions. However, when these liquids are mixed, the total volume of the resulting solution is not equal to the sum of the pure liquid volumes, and we say that the volumes are not additive. For example, when 50.0 mL of water and 50.0 mL of $CH_3CH_2OH(l)$, are mixed at 20 °C, the total volume of the solution is 96.5 mL, not 100.0 mL. (The volumes are not additive because the interactions and packing of water molecules are slightly different from the interactions and packing of CH_3CH_2OH molecules.) Calculate the molarity of CH_3CH_2OH in a solution prepared by mixing 50.0 mL of water and 50.0 mL of $CH_3CH_2OH(l)$ at 20 °C. At this temperature, the densities of water and ethanol are 0.99821 g/mL and 0.7893 g/mL, respectively.

100. When water and methanol, $CH_3OH(l)$, are mixed, the total volume of the resulting solution is not equal to the sum of the pure liquid volumes. (Refer to Exercise 99 for an explanation.) When 72.061 g H_2O and 192.25 g CH_3OH are mixed at 25 °C, the resulting solution has a density of 0.86070 g/mL. At 25 °C, the densities of water and methanol are 0.99705 g/mL and 0.78706 g/mL, respectively.

     (a) Calculate the volumes of the pure liquid samples and the solution, and show that the pure liquid volumes are not additive. [*Hint*: Although the volumes are not additive, the masses are.]

     (b) Calculate the molarity of CH_3OH in this solution.

101. What volume of 0.149 M HCl must be added to 1.00×10^2 mL of 0.285 M HCl so that the resulting solution has a molarity of 0.205 M? Assume that the volumes are additive.

102. What volume of 0.0175 M CH_3OH must be added to 50.0 mL of 0.0248 M CH_3OH so that the resulting solution has a molarity of exactly 0.0200 M? Assume that the volumes are additive.

103. What is the molarity of NaCl(aq) if a solution has 1.52 ppm Na? Assume that NaCl is the only source of Na and that the solution density is 1.00 g/mL. (The unit *ppm* is parts per million; here it can be taken to mean g Na per million grams of solution.)

104. How many milligrams $Ca(NO_3)_2$ must be present in 50.0 L of a solution containing 2.35 ppm Ca? [*Hint*: See also Exercise 103.]

105. A drop (0.05 mL) of 12.0 M HCl is spread over a sheet of thin aluminum foil. Assume that all the acid reacts with, and thus dissolves through, the foil. What will be the area, in cm^2, of the cylindrical hole produced? (Density of Al = 2.70 g/cm^3; foil thickness = 0.10 mm.)

     $$2\,Al(s) + 6\,HCl(aq) \longrightarrow 2\,AlCl_3(aq) + 3\,H_2(g)$$

106. A small piece of zinc is dissolved in 50.00 mL of 1.035 M HCl. At the conclusion of the reaction, the concentration of the 50.00 mL sample is redetermined and found to be 0.812 M HCl. What must have been the mass of the piece of zinc that dissolved?

     $$Zn(s) + 2\,HCl(aq) \longrightarrow ZnCl_2(aq) + H_2(g)$$

107. How many milliliters of 0.715 M NH_4NO_3 solution must be diluted with water to produce 1.00 L of a solution with a concentration of 2.37 mg N/mL?

108. A seawater sample has a density of 1.03 g/mL and 2.8% NaCl by mass. A saturated solution of NaCl in water is 5.45 M NaCl. How many liters of water would have to be evaporated from 1.00×10^6 L of the seawater before NaCl would begin to crystallize? (A saturated solution contains the maximum amount of dissolved solute possible.)

109. A 99.8 mL sample of a solution that is 12.0% KI by mass ($d = 1.093$ g/mL) is added to 96.7 mL of another solution that is 14.0% $Pb(NO_3)_2$ by mass ($d = 1.134$ g/mL). How many grams of PbI_2 should form?

     $$Pb(NO_3)_2(aq) + 2\,KI(aq) \longrightarrow PbI_2(s) + 2\,KNO_3(aq)$$

110. Solid calcium carbonate, $CaCO_3(s)$, reacts with $HCl(aq)$ to form H_2O, $CaCl_2(aq)$, and $CO_2(g)$. If a 45.0 g sample of $CaCO_3(s)$ is added to 1.25 L of $HCl(aq)$ that is 25.7% HCl by mass ($d = 1.13$ g/mL), what will be the molarity of HCl in the solution after the reaction is completed? Assume that the solution volume remains constant.

111. A 2.05 g sample of an iron–aluminum alloy (ferro-aluminum) is dissolved in excess $HCl(aq)$ to produce 0.105 g $H_2(g)$. What is the percent composition, by mass, of the ferroaluminum?

$$Fe(s) + 2\,HCl(aq) \longrightarrow FeCl_2(aq) + H_2(g)$$
$$2\,Al(s) + 6\,HCl(aq) \longrightarrow 2\,AlCl_3(aq) + 3\,H_2(g)$$

112. A 0.155 g sample of an Al–Mg alloy reacts with an excess of $HCl(aq)$ to produce 0.0163 g H_2. What is the percent Mg in the alloy?
[*Hint:* Write equations similar to (4.2).]

113. An organic liquid is either methyl alcohol (CH_3OH), ethyl alcohol (CH_3CH_2OH), or a mixture of the two. A 0.220-g sample of the liquid is burned in an excess of $O_2(g)$ and yields 0.352 g $CO_2(g)$. Is the liquid a pure alcohol or a mixture of the two?

114. The manufacture of ethyl alcohol, CH_3CH_2OH, yields diethyl ether, $(C_2H_5)_2O$ as a by-product. The complete combustion of a 1.005 g sample of the product of this process yields 1.963 g CO_2. What must be the mass percents of (CH_3CH_2OH), and of $(C_2H_5)_2O$ in this sample?

115. A mixture contains only $CuCl_2$ and $FeCl_3$. A 0.7391 g sample of the mixture is completely dissolved in water and then treated with $AgNO_3(aq)$. The following reactions occur.

$$CuCl_2(aq) + 2\,AgNO_3(aq) \longrightarrow$$
$$2\,AgCl(s) + Cu(NO_3)_2(aq)$$
$$FeCl_3(aq) + 3\,AgNO_3(aq) \longrightarrow$$
$$3\,AgCl(s) + Fe(NO_3)_3(aq)$$

If it takes 86.91 mL of 0.1463 M $AgNO_3$ solution to precipitate all the chloride as AgCl, then what is the percentage by mass of copper in the mixture?

116. Under appropriate conditions, copper sulfate, potassium chromate, and water react to form a product containing Cu^{2+}, CrO_4^{2-}, and OH^- ions. Analysis of the compound yields 48.7% Cu^{2+}, 35.6% CrO_4^{2-}, and 15.7% OH^-.
(a) Determine the empirical formula of the compound.
(b) Write a plausible equation for the reaction.

117. Write a chemical equation to represent the complete combustion of malonic acid, a compound with 34.62% C, 3.88% H, and 61.50% O, by mass.

118. Aluminum metal and iron(III) oxide react to give aluminum oxide and iron metal. What is the maximum mass of iron that can be obtained from a reaction mixture containing 2.5 g of aluminum and 9.5 g of iron(III) oxide. What mass of the excess reactant remains?

119. Silver nitrate is a very expensive chemical. For a particular experiment, you need 100.0 mL of 0.0750 M $AgNO_3$, but only 60 mL of 0.0500 M $AgNO_3$ is available. You decide to pipet exactly 50.00 mL of the solution into a 100.0 mL flask, add an appropriate mass of $AgNO_3$, and then dilute the resulting solution to exactly 100.0 mL. What mass of $AgNO_3$ must you use?

120. When sulfur (S_8) and chlorine are mixed in a reaction vessel, disulfur dichloride is the sole product. The starting mixture below is represented by yellow spheres for the S_8 molecules and green spheres for the chlorine molecules.

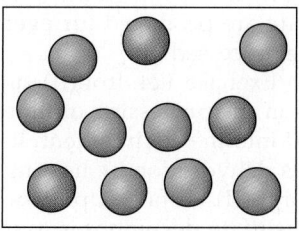

Which of the following is (are) a valid representation(s) of the contents of the reaction vessel after some disulfur dichloride (represented by red spheres) has formed?

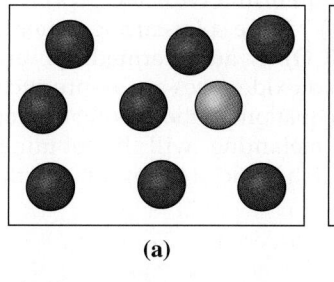

(a)

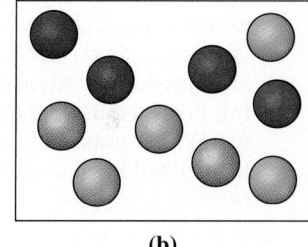

(b)

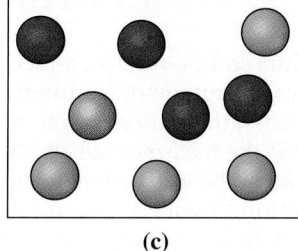

(c)

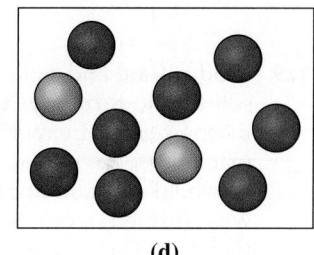

(d)

121. A method for eliminating oxides of nitrogen (e.g., NO_2) from automobile exhaust gases is to pass the exhaust gases over solid cyanuric acid, $C_3N_3(OH)_3$. When the hot exhaust gases come in contact with cyanuric acid, solid $C_3N_3(OH)_3$ decomposes into isocyanic acid vapor, $HNCO(g)$, which then reacts with NO_2 in the exhaust gases to give N_2, CO_2, and H_2O. How many grams of $C_3N_3(OH)_3$ are needed per gram of NO_2 in this method?
[*Hint*: To balance the equation for reaction between HNCO and NO_2, balance with respect to each kind of atom in this order: H, C, O, and N.]

122. For a specific reaction, ammonium dichromate is the only reactant and chromium(III) oxide and water are two of the three products. What is the third product and how many grams of this product are produced per kilogram of ammonium dichromate decomposed?

123. It is desired to produce as large a volume of 1.25 M urea [$CO(NH_2)_2(aq)$] as possible from these three sources: 345 mL of 1.29 M $CO(NH_2)_2$, 485 mL of 0.653 M $CO(NH_2)_2$, and 835 mL of 0.775 M $CO(NH_2)_2$. How

can this be done? What is the maximum volume of this solution obtainable?

124. The mineral ilmenite, $FeTiO_3$, is an important source of titanium dioxide for use as a white pigment. In the first step in its conversion to titanium dioxide, ilmenite is treated with sulfuric acid and water to form $TiOSO_4$ and iron(II) sulfate heptahydrate. Titanium dioxide is obtained in two subsequent steps. How many kilograms of iron(II) sulfate heptahydrate are produced for every 1.00×10^3 kg of ilmenite processed?

125. Refer to Exercise 124. Iron(II) sulfate heptahydrate formed in the processing of ilmenite ore cannot be released into the environment. Its further treatment involves dehydration by heating to produce anhydrous iron(II) sulfate. Upon further heating, the iron(II) sulfate decomposes to iron(III) oxide, and sulfur dioxide and oxygen gases. The iron(III) oxide is used in the production of iron and steel. How many kilograms of iron(III) oxide are obtained for every 1.00×10^3 kg of iron(II) sulfate heptahydrate?

126. Melamine, $C_3N_3(NH_2)_3$, is used in adhesives and resins. It is manufactured in a two-step process in which urea, $CO(NH_2)_2$, is the sole starting material, isocyanic acid (HNCO) is an intermediate, and ammonia and carbon dioxide gases are by-products.
(a) Write a balanced equation for the overall reaction.
(b) What mass of melamine will be obtained from 100.0 kg of urea if the yield of the overall reaction is 84%?

127. Acrylonitrile is used in the production of synthetic fibers, plastics, and rubber goods. It can be prepared from propylene (propene), ammonia, and oxygen in the reaction illustrated below.
(a) Write a balanced chemical equation for this reaction.
(b) The actual yield of the reaction is 0.73 kg acrylonitrile per kilogram of propylene. What is the minimum mass of ammonia required to produce 1.00 metric ton (1000 kg) of acrylonitrile?

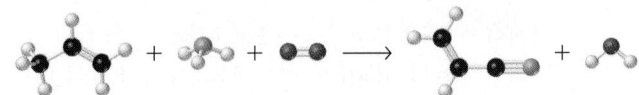

128. It is often difficult to determine the concentration of a species in solution, particularly if it is a biological species that takes part in complex reaction pathways. One way to do this is through a dilution experiment with labeled molecules. Instead of molecules, however, we will use fish.

An angler wants to know the number of fish in a particular pond, and so puts an indelible mark on 100 fish and adds them to the pond's existing population. After waiting for the fish to spread throughout the pond, the angler starts fishing, eventually catching 18 fish. Of these, five are marked. What is the total number of fish in the pond?

Feature Problems

129. Lead nitrate and potassium iodide react in aqueous solution to form a yellow precipitate of lead iodide. In one series of experiments, the masses of the two reactants were varied, but the *total* mass of the two was held constant at 5.000 g. The lead iodide formed was filtered from solution, washed, dried, and weighed. The table gives data for a series of reactions.

Experiment	Mass of Lead Nitrate, g	Mass of Lead Iodide, g
1	0.500	0.692
2	1.000	1.388
3	1.500	2.093
4	3.000	2.778
5	4.000	1.391

(a) Plot the data in a graph of mass of lead iodide versus mass of lead nitrate, and draw the appropriate curve(s) connecting the data points. What is the maximum mass of precipitate that can be obtained?
(b) Explain why the maximum mass of precipitate is obtained when the reactants are in their stoichiometric proportions. What are these stoichiometric proportions expressed as a mass ratio, and as a mole ratio?
(c) Show how the stoichiometric proportions determined in part (b) are related to the balanced equation for the reaction.

130. The emerging field of *green chemistry* comprises all the methods used to minimize the environmental impact of chemical activities. One of the dozen or so factors in green chemistry is *atom economy*—the efficiency in rearranging the atoms in a set of reactants into the desired product. Atom economy (AE) is expressed as a percentage by the equation

$$AE = \frac{\text{formula mass of desired product}}{\substack{\text{sum of formula masses} \\ \text{of all the reactants}}} \times 100\%$$

For example, when $H_2(g)$ burns in $O_2(g)$, the sole product is H_2O. All the atoms of the reactants appear in the product and the atom economy is 100%, as implied by the equation $2\,H_2(g) + O_2(g) \longrightarrow 2\,H_2O(l)$. For the Raschig process described in the Focus On feature for Chapter 4 on www.masteringchemistry.com,
(a) what is the theoretical maximum AE for the synthesis of hydrazine (equation 4B)?
(b) Why is the actual AE less than that calculated in part (a)?
(c) Explain why substituting acetone for water improves the AE.
(d) Propose a reaction for the synthesis of hydrazine that would have a theoretical AE of 100%.

131. Baking soda, $NaHCO_3$, is made from soda ash, a common name for sodium carbonate. The soda ash is obtained in two ways. It can be manufactured in a

process in which carbon dioxide, ammonia, sodium chloride, and water are the starting materials. Alternatively, it is mined as a mineral called *trona* (left photo). Whether the soda ash is mined or manufactured, it is dissolved in water and carbon dioxide is bubbled through the solution. Sodium bicarbonate precipitates from the solution.

As a chemical analyst you are presented with two samples of sodium bicarbonate—one from the manufacturing process and the other derived from trona. You are asked to determine which is purer and are told that the impurity is sodium carbonate. You decide to treat the samples with just sufficient hydrochloric acid to convert all the sodium carbonate and bicarbonate to sodium chloride, carbon dioxide, and water. You then precipitate silver chloride in the reaction of sodium chloride with silver nitrate. A 6.93 g sample of baking soda derived from trona gave 11.89 g of silver chloride. A 6.78 g sample from

manufactured sodium carbonate gave 11.77 g of silver chloride. Which sample is purer, that is, which has the greater mass percent $NaHCO_3$?

Trona
$Na_2CO_3 \cdot NaHCO_3 \cdot 2H_2O$

Baking soda
$NaHCO_3$

Self-Assessment Exercises

132. In your own words, define or explain these terms or symbols.
   (a) $\xrightarrow{\Delta}$
   (b) (aq)
   (c) stoichiometric coefficient (d) overall equation
133. Briefly describe (a) balancing a chemical equation; (b) preparing a solution by dilution; (c) determining the limiting reactant in a reaction.
134. Explain the important distinctions between (a) chemical formula and chemical equation; (b) stoichiometric coefficient and stoichiometric factor; (c) solute and solvent; (d) actual yield and percent yield; (e) consecutive and simultaneous reactions.
135. When the equation below is balanced, the correct set of stoichiometric coefficients is (a) 1, 6 ⟶ 1, 3, 4; (b) 1,4 ⟶ 1, 2, 2; (c) 2, 6 ⟶ 2, 3, 2; (d) 3, 8 ⟶ 3, 4, 2.
   ? $Cu(s)$ + ? $HNO_3(aq)$ ⟶
   ? $Cu(NO_3)_2(aq)$ + ? $H_2O(l)$ + ? $NO(g)$
136. A reaction mixture contains 1.0 mol $CaCN_2$ (calcium cyanamide) and 1.0 mol H_2O. The maximum number of moles of NH_3 produced is (a) 3.0; (b) 2.0; (c) between 1.0 and 2.0; (d) less than 1.0.
   $CaCN_2(s) + 3 H_2O(l) \longrightarrow CaCO_3 + 2 NH_3(g)$
137. Consider the chemical equation below. What is the maximum number of moles of K_2SO_4 that can be obtained from a reaction mixture containing 5.0 moles each of $KMnO_4$, KI, and H_2SO_4? (a) 3.0 mol; (b) 3.8 mol; (c) 5.0 mol; (d) 6.0 mol; (e) 15 mol.
   $2 KMnO_4 + 10 KI + 8 H_2SO_4 \longrightarrow$
   $6 K_2SO_4 + 2 MnSO_4 + 5 I_2 + 8 H_2O$
138. In the decomposition of silver(I) carbonate to form metallic silver, carbon dioxide gas, and oxygen gas, (a) one mol of oxygen gas is formed for every 2 mol of carbon dioxide gas; (b) 2 mol of silver metal is formed for every 1 mol of oxygen gas; (c) equal numbers of

moles of carbon dioxide and oxygen gases are produced; (d) the same number of moles of silver metal are formed as of the silver(I) carbonate decomposed.
139. To obtain a solution that is 1.00 M $NaNO_3$, you should prepare (a) 1.00 L of aqueous solution containing 100 g $NaNO_3$; (b) 1 kg of aqueous solution containing 85.0 g $NaNO_3$; (c) 5.00 L of aqueous solution containing 425 g $NaNO_3$; (d) an aqueous solution containing 8.5 mg $NaNO_3$/mL.
140. To prepare a solution that is 0.50 M KCl starting with 100.0 mL of 0.40 M KCl, you should (a) add 20.0 mL of water; (b) add 0.075 g KCl; (c) add 0.10 mol KCl; (d) evaporate 20.0 mL of water.
141. An aqueous solution that is 5.30% LiBr by mass has a density of 1.040 g/mL. What is the molarity of this solution? (a) 0.563 M; (b) 0.635 M; (c) 0.0635 M; (d) 0.0563 M; (e) 12.0 M.
142. In the reaction of 2.00 mol CCl_4 with an excess of HF, 1.70 mol CCl_2F_2 is obtained.
   $CCl_4 + 2 HF \longrightarrow CCl_2F_2 + 2 HCl$
   (a) The theoretical yield is 1.70 mol CCl_2F_2.
   (b) The theoretical yield is 1.00 mol CCl_2F_2.
   (c) The theoretical yield depends on how large an excess of HF is used.
   (d) The percent yield is 85%.
143. The incomplete combustion of gasoline produces $CO(g)$ as well as $CO_2(g)$. Write an equation for (a) the complete combustion of the gasoline component octane, $C_8H_{18}(l)$, and (b) incomplete combustion of octane with 25% of the carbon appearing as $CO(g)$.
144. The minerals calcite, $CaCO_3$, magnesite, $MgCO_3$, and dolomite, $CaCO_3 \cdot MgCO_3$, decompose when strongly heated to form the corresponding metal oxide(s) and carbon dioxide gas. A 1.000-g sample known to be one of the three minerals was strongly heated and 0.477 g CO_2 was obtained. Which of the three minerals was it?

145. A 1.000 g sample of a mixture of CH_4 and C_2H_6 is analyzed by burning it completely in O_2, yielding 2.776 g CO_2. What is the percentage by mass of CH_4 in the mixture? **(a)** 93%; **(b)** 82%; **(c)** 67% ; **(d)** 36%; **(e)** less than 36%.

146. Nitric acid, HNO_3, can be manufactured from ammonia, NH_3, by using the three reactions shown below.

    Step 1: $4 NH_3(g) + 5 O_2(g) \rightarrow 4 NO(g) + 6 H_2O(l)$

    Step 2: $2 NO(g) + O_2(g) \rightarrow 2 NO_2(g)$

    Step 3: $3 NO_2(g) + H_2O(l) \rightarrow 2 HNO_3(aq) + NO(g)$

    What is the maximum number of moles of HNO_3 that can be obtained from 4.00 moles of NH_3? (Assume that the NO produced in step 3 is not recycled back into step 2.) **(a)** 1.33 mol; **(b)** 2.00 mol; **(c)** 2.67 mol; **(d)** 4.00 mol; **(e)** 6.00 mol.

147. Appendix E describes a useful study aid known as concept mapping. Using the method presented in Appendix E, construct a concept map relating the topics found in Sections 4-3, 4-4, and 4-5.

Introduction to Reactions in Aqueous Solutions

CONTENTS

When clear, colorless aqueous solutions of cobalt(II) chloride and sodium hydroxide are mixed, a blue cloud of solid cobalt(II) hydroxide is formed. Such precipitation reactions are one of the three types of reactions considered in this chapter.

Most reactions in the general chemistry laboratory are carried out in aqueous solutions—solutions for which water is the solvent. Aqueous solutions provide a convenient way of bringing together accurately measured amounts of reactants, and, not surprisingly, aqueous solutions feature prominently in many methods of chemical analysis. In this chapter, we will explore three different classes of reactions that occur in aqueous solutions—precipitation, acid–base, and oxidation–reduction reactions—with the goal of understanding the nature of the substances involved, the changes that occur in these substances, and the way each reaction can be used in the laboratory for analyzing samples.

Precipitation, the formation of a solid when solutions are mixed, is probably the most common evidence of a chemical reaction that general chemistry students see. A practical application of precipitation is in determining the presence of certain ions in solution. If, for example, we are uncertain whether the clear, colorless liquid in an unlabeled bottle is a barium nitrate or a barium chloride solution, we can easily find out by adding a few drops of silver nitrate solution to a small sample of the liquid. If a white solid forms, the sample is a barium chloride solution; if nothing happens, it is barium nitrate. The Ag^+ from the silver nitrate and the Cl^- from the barium chloride combine to produce an insoluble precipitate of AgCl(s). Precipitation reactions are the first reaction type we will study in this chapter. $Mg(OH)_2(s)$ is insoluble in water but soluble in hydrochloric acid, HCl(aq), as a result of an acid–base reaction. This is the reaction by which milk of magnesia neutralizes excess stomach acid. Magnesium hydroxide is a base, and acid–base reactions are the second class of reactions presented in this chapter. The third class of reactions is oxidation–reduction reactions, which are found in all aspects of life, from reactions in organisms to processes for manufacturing chemicals, to such practical matters as bleaching fabrics, purifying water, and destroying toxic chemicals.

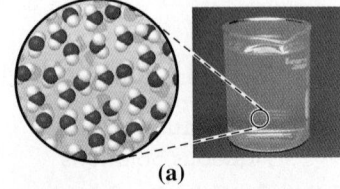

(a)

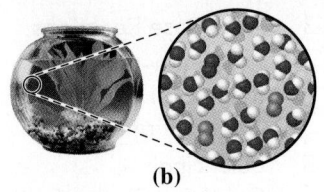

(b)

▲ FIGURE 5-1
Molecular view of water and an aqueous solution of air
(a) Water molecules (red and white) are in close proximity in liquid water. **(b)** Dissolved oxygen (red) and nitrogen (blue) molecules are far apart, separated by water molecules.

▶ If a solute is *completely ionized*, then essentially all the dissolved solute exists as ions. If a solute is *partially ionized*, then only some (not all) of the solute molecules have been converted into ions.

5-1   The Nature of Aqueous Solutions

Let's try to form a mental image of a solution at the molecular level. The solvent molecules, which are rather tightly packed, greatly outnumber all other molecules. Water is the solvent in an aqueous solution, and our mental image of water might look something like Figure 5-1(a). Solute particles—molecules or ions—are present in much smaller number and are randomly distributed among the solvent molecules. Our mental image of an aqueous solution of air might look something like Figure 5-1(b).

Because we will encounter aqueous solutions of ions throughout this chapter, it is useful to examine the nature of such solutions in a bit more detail. An important characteristic of an aqueous solution of ions is that it will conduct electricity, provided the concentration of ions is not too low. An aqueous solution of ions conducts electricity because the ions move essentially independently of each other, each one carrying a certain quantity of charge. (In a metallic conductor, such as copper or tungsten, electrons carry the charge.) The manner in which ions conduct electric current is suggested by Figure 5-2.

Whether or not an aqueous solution is a conductor of electricity depends on the nature of the solute(s). Pure water contains so few ions that it does not conduct electric current. However, some solutes produce ions in solution, thereby making the solution an electrical conductor. Solutes that provide ions when dissolved in water are called **electrolytes**. Solutes that that do not provide ions in water are called **nonelectrolytes**. All electrolytes provide ions in water but not all electrolytes are equal in their tendencies for providing ions. A **strong electrolyte** is a substance that is essentially *completely ionized* in aqueous solution. Stated another way, a strong electrolyte has a strong (or

▶ FIGURE 5-2
Conduction of electricity through a solution
Two graphite rods called electrodes are placed in a solution. The external source of electricity pulls electrons from one rod and forces them onto the other, creating a positive charge on one electrode and a negative charge on the other (right). In the solution, positive ions (cations) are attracted to the negative electrode, the *cathode*; negative ions (anions) are attracted to the positive electrode, the *anode*. Thus, electric charge is carried through the solution by the migration of ions. Other important aspects of electrical conductivity are discussed in Chapter 20.

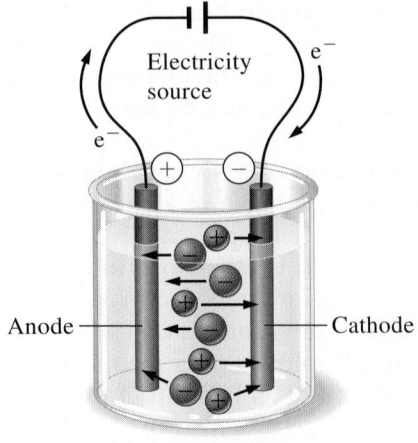

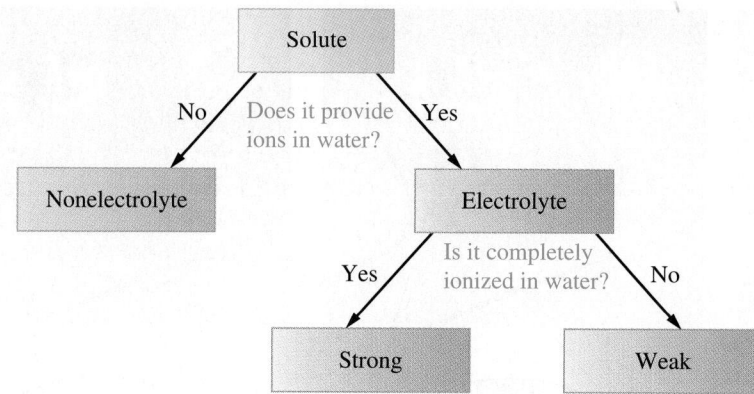

◀ FIGURE 5-3
A classification scheme for solutes

high) tendency for providing ions. A **weak electrolyte** is a substance that is only *partially ionized* in aqueous solution. A weak electrolyte has a weak (or small) tendency for providing ions. One scheme for classifying solutes is summarized in Figure 5-3.

With the apparatus depicted in Figure 5-4, we can detect the presence of ions in an aqueous solution by measuring how well the solution conducts electricity. We can make one of three possible observations:

- *The lamp fails to light up* (Fig. 5-4a). Conclusion: no ions are present (or if some are present, their concentration is extremely low). The solution is either a solution of a nonelectrolyte or a very dilute solution of an electrolyte. Methanol, CH_3OH, is an example of a solute that does not provide ions in water; methanol is a *nonelectrolyte*. The microscopic view in Figure 5-4(a) is for an aqueous solution of methanol, and in this view we see that none of the CH_3OH molecules are ionized in water.

- *The lamp lights up brightly* (Fig. 5-4b). Conclusion: the concentration of ions in solution is high. The solute is a *strong electrolyte*. Magnesium chloride, $MgCl_2$, is an ionic compound that is completely ionized in water. The microscopic view in Figure 5-4(b) shows that an aqueous solution of $MgCl_2$ consists of Mg^{2+} and Cl^- ions in the solvent.

- *The lamp lights up only dimly* (Fig. 5-4c). Conclusion: ions are present in solution but the concentration of ions is low. The solution could be a solution of a weak electrolyte, such as acetic acid (CH_3COOH), or it could be a dilute—but not too dilute—solution of a strong electrolyte. The microscopic view in Figure 5-4(c) is for an aqueous solution of CH_3COOH and it shows that, in water, only some of the CH_3COOH molecules are ionized. An aqueous solution of CH_3COOH is only a weak conductor of electricity.

The following generalization is helpful when deciding whether a particular solute in an aqueous solution is most likely to be a nonelectrolyte, a strong electrolyte, or a weak electrolyte.

KEEP IN MIND

that the electrical conductivity of a solution depends on two factors: (1) the total concentration of the electrolyte, and (2) the extent to which the electrolyte dissociates into ions. For example, a 0.001 M HCl(aq) solution will conduct electricity, but a 1×10^{-6} M HCl(aq) solution will not, even though every HCl molecule dissociates into H^+ and Cl^- ions in both solutions.

- Essentially all soluble ionic compounds and only a relatively few molecular compounds are strong electrolytes.
- Most molecular compounds are either nonelectrolytes or weak electrolytes.

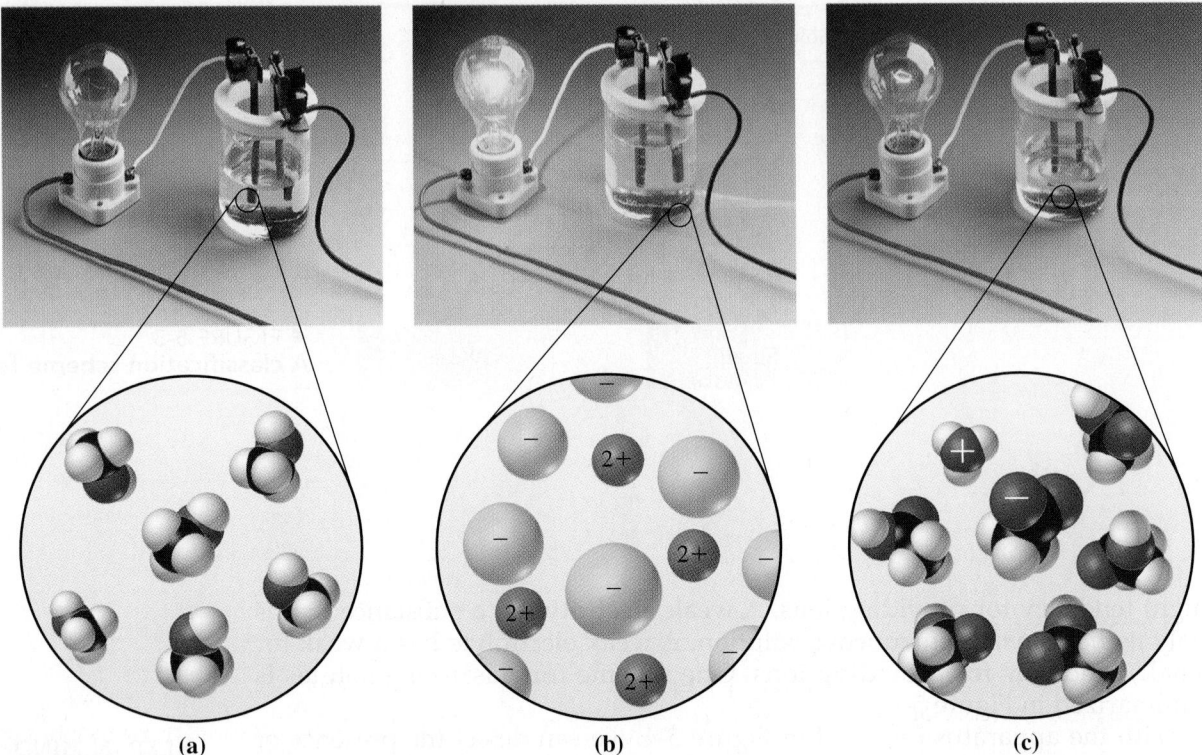

(a)   (b)   (c)

▲ FIGURE 5-4
Three types of electrolytes
In **(a)**, there are no ions present to speak of—only molecules. Methanol (methyl alcohol), CH_3OH, is a *nonelectrolyte* in aqueous solutions. In **(b)**, the solute is present almost entirely as individual ions. $MgCl_2$ is a *strong electrolyte* in aqueous solutions. In **(c)**, although most of the solute is present as molecules, a small fraction of the molecules ionize. CH_3COOH is a *weak electrolyte* in aqueous solution. The CH_3COOH molecules that ionize produce acetate ions CH_3COO^- and H^+ ions, and the H^+ ions attach themselves to water molecules to form hydronium ions, H_3O^+.

KEEP IN MIND

that the solvent molecules are densely packed. In such diagrams as Figure 5-4, we will often show the solvent as a uniformly colored background and depict only the solute particles.

Now let us consider how best to represent these three types of substances in chemical equations. For water-soluble $MgCl_2$, an ionic compound, we can write

$$MgCl_2(aq) \longrightarrow Mg^{2+}(aq) + 2\,Cl^-(aq)$$

This equation means that, in the presence of water, formula units of $MgCl_2$ are dissociated into the separate ions. The best representation of $MgCl_2(aq)$, then, is $Mg^{2+}(aq) + 2\,Cl^-(aq)$.

Hydrogen chloride, HCl, is an example of a molecular compound that is a strong electrolyte. When HCl dissolves in water, the following reaction occurs:

$$HCl(aq) \longrightarrow H^+(aq) + Cl^-(aq)$$

The best representation of hydrochloric acid, HCl(aq), is $H^+(aq) + Cl^-(aq)$.

The hydrogen cation H^+ is an interesting and important species that has been the subject of intensive research. The ion H^+—a bare proton—is a small particle that interacts with the water molecules surrounding it. The simple hydrogen ion, H^+, does not exist in aqueous solutions. Its actual form is as *hydronium ion*, H_3O^+, in which an H^+ ion is attached to an H_2O molecule. The hydronium ion, in turn, interacts with the water molecules surrounding it to

▶ The International Union of Pure and Applied Chemistry (IUPAC) has recommended that the H_3O^+, ion be referred to as the *oxonium* ion. However, this recommendation has not yet been universally adopted by chemists.

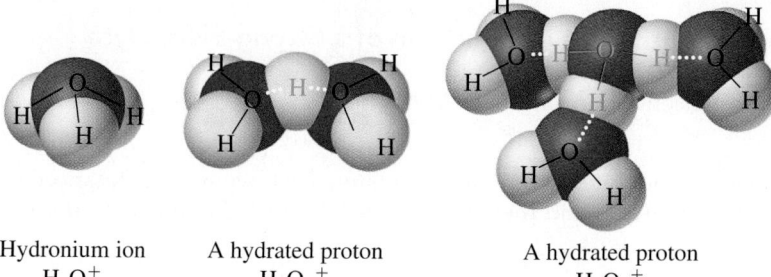

Hydronium ion
H_3O^+

A hydrated proton
$H_5O_2^+$

A hydrated proton
$H_9O_4^+$

◀ FIGURE 5-5
The hydrated proton
The hydronium ion, H_3O^+, interacts with other water molecules through electrostatic attractions.

form additional species, such as $H_5O_2^+$, $H_7O_3^+$, $H_9O_4^+$ (shown in Figure 5-5) and many others. These interactions are called *hydration*, and the hydrated proton is represented as $H^+(aq)$, a shorthand notation. The interaction of the proton with a single water molecule is emphasized by using $H_3O^+(aq)$. The basis for these interactions will be discussed in Chapter 12 and employed extensively in Chapter 16.

For a solution of a weak electrolyte, the reaction that produces ions does not go to completion. In such solutions, only a fraction of the solute molecules in solution are ionized. Equations for solutions of this type are written with double arrows, as shown below.

$$CH_3COOH(aq) \rightleftharpoons H^+(aq) + CH_3COO^-(aq)$$

The double arrows indicate that the process is *reversible*—that is, while some CH_3COOH (acetic acid) molecules ionize, some H^+ and CH_3COO^- ions in solution recombine to form new CH_3COOH molecules. However, the relative proportions of ionized and nonionized (molecular) acid remain fixed. The predominant species is the molecule CH_3COOH, and the solution is best represented by $CH_3COOH(aq)$, *not* $H^+(aq) + CH_3COO^-(aq)$. If we could tag a particular CH_3COO group and watch it over time, we would sometimes see it as the ion CH_3COO^-, but most of the time it would be in a molecule, CH_3COOH.

For a nonelectrolyte, we simply write the molecular formula. Thus, for a solution of methanol in water, we would write $CH_3OH(aq)$.

With this new information about the nature of aqueous solutions, we can introduce a useful notation for solution concentrations. In a solution that is 0.0050 M $MgCl_2$, we assume that the $MgCl_2$ is completely dissociated into ions. Because there are two Cl^- ions for every Mg^{2+} ion, the solution is 0.0050 M Mg^{2+} but 0.0100 M Cl^-. Better still, let us introduce a special symbol for the concentration of a species in solution—the bracket symbol [ ]. The statement $[Mg^{2+}] = 0.0050$ M means that the concentration of the species within the brackets—that is, Mg^{2+}—is 0.0050 mol/L. Thus,

in 0.0050 M MgCl₂: $[Mg^{2+}] = 0.0050$ M; $[Cl^-] = 0.0100$ M; $[MgCl_2] = 0$ M

Although we do not usually write expressions like $[MgCl_2] = 0$, it is done here to emphasize that there is essentially no undissociated $MgCl_2$ in the solution.*

Example 5-1 shows how to calculate the concentrations of ions in a strong electrolyte solution.

> **KEEP IN MIND**
>
> that there are different ways to write the formulas for acetic acid. The molecular formula, $C_2H_4O_2$, makes no distinction between the types of H atoms; $HC_2H_3O_2$ emphasizes that there is only one ionizable H atom; and CH_3COOH and CH_3CO_2H emphasize that the ionizable H atom is part of the carboxyl group. Recall: The carboxyl group and carboxylic acids were introduced in Chapter 3.

*To say that a strong electrolyte is completely dissociated into individual ions in aqueous solution is a good approximation but somewhat of an oversimplification. Some of the cations and anions in solution may become associated into units called *ion pairs*. Generally, though, at the low solution concentrations we will be using, assuming complete dissociation will not seriously affect our results.

EXAMPLE 5-1   Calculating Ion Concentrations in a Solution of a Strong Electrolyte

What are the aluminum and sulfate ion concentrations in 0.0165 M $Al_2(SO_4)_3$?

Analyze

The solute is a strong electrolyte. Thus, it dissociates completely in water. First, we write a balanced chemical equation for the dissociation of $Al_2(SO_4)_3(aq)$, and then set up stoichiometric factors to relate Al^{3+} and SO_4^{2-} to the molarity of $Al_2(SO_4)_3$.

Solve

The dissociation of $Al_2(SO_4)_3$ is represented by the equation below.

$$Al_2(SO_4)_3(s) \xrightarrow{H_2O} 2 Al^{3+}(aq) + 3 SO_4^{2-}(aq)$$

The stoichiometric factors, shown in blue in the following equations, are derived from the fact that 1 mol $Al_2(SO_4)_3$ produces 2 mol Al^{3+} and 3 mol SO_4^{2-}.

$$[Al^{3+}] = \frac{0.0165 \text{ mol } Al_2(SO_4)_3}{1 \text{ L}} \times \frac{2 \text{ mol } Al^{3+}}{1 \text{ mol } Al_2(SO_4)_3} = \frac{0.0330 \text{ mol } Al^{3+}}{1 \text{ L}}$$

$$= 0.0330 \text{ M}$$

$$[SO_4^{2-}] = \frac{0.0165 \text{ mol } Al_2(SO_4)_3}{1 \text{ L}} \times \frac{3 \text{ mol } SO_4^{2-}}{1 \text{ mol } Al_2(SO_4)_3} = \frac{0.0495 \text{ mol } SO_4^{2-}}{1 \text{ L}}$$

$$= 0.0495 \text{ M}$$

Assess

For a strong electrolyte, the concentrations of the ions will always be integer multiples of the electrolyte molarity. For example, in 0.0165 M $MgCl_2$, we have $[Mg^{2+}] = 1 \times 0.0165$ M and $[Cl^-] = 2 \times 0.0165$ M.

PRACTICE EXAMPLE A: The chief ions in seawater are Na^+, Mg^{2+}, and Cl^-. Seawater is approximately 0.438 M NaCl and 0.0512 M $MgCl_2$. What is the molarity of Cl^-—that is, the total $[Cl^-]$—in seawater?

PRACTICE EXAMPLE B: A water treatment plant adds fluoride ion to the water to the extent of 1.5 mg F^-/L.

(a) What is the molarity of fluoride ion in this water?
(b) If the fluoride ion in the water is supplied by calcium fluoride, what mass of calcium fluoride is present in 1.00×10^6 L of this water?

🔍 **5-1   CONCEPT ASSESSMENT**

(1) Which solution is the best electrical conductor? **(a)** 0.50 M CH_3COCH_3; **(b)** 0.50 M CH_3CH_2OH; **(c)** 1.00 M $CH_2OHCHOHCH_2OH$; **(d)** 0.050 M CH_3COOH; **(e)** 0.025 M $RbNO_3$.

(2) Which solution has the highest total molarity of ions? **(a)** 0.008 M $Ba(OH)_2$; **(b)** 0.010 M KI; **(c)** 0.011 M CH_3COOH; **(d)** 0.030 M $HOCH_2CH_2OH$; **(e)** 0.004 M $Al_2(SO_4)_3$?

5-2   Precipitation Reactions

Some metal salts, such as NaCl, are quite soluble in water, while others, such as AgCl, are not very soluble at all. In fact, so little AgCl dissolves in water that this compound is generally considered to be *insoluble*. Precipitation reactions occur when certain cations and anions combine to produce an insoluble ionic solid called a **precipitate**. One laboratory use of precipitation reactions is in identifying the ions present in a solution, as shown in Figure 5-6. In industry, precipitation reactions are used to manufacture numerous chemicals. In the extraction of magnesium metal from seawater, for instance, the first step is to precipitate Mg^{2+} as $Mg(OH)_2(s)$. In this section, the objective is to represent precipitation reactions by chemical equations and to apply some simple rules for predicting precipitation reactions.

Net Ionic Equations

The reaction of silver nitrate and sodium iodide in an aqueous solution yields sodium nitrate in solution and a pale yellow or cream-colored precipitate of silver iodide, as shown in Figure 5-7. Applying the principles of equation writing from Chapter 4, we can write

$$AgNO_3(aq) + NaI(aq) \longrightarrow AgI(s) + NaNO_3(aq) \qquad \textbf{(5.1)}$$

You might note a contradiction, however, between equation (5.1) and something we learned earlier in this chapter. In their aqueous solutions, the soluble ionic compounds $AgNO_3$, NaI, and $NaNO_3$—all *strong* electrolytes—should be represented by their separate ions.

$$Ag^+(aq) + NO_3^-(aq) + Na^+(aq) + I^-(aq) \longrightarrow$$
$$AgI(s) + Na^+(aq) + NO_3^-(aq) \qquad \textbf{(5.2)}$$

We might say that equation (5.1) is the "whole formula" form of the equation, whereas equation (5.2) is the "ionic" form. Notice also that in equation (5.2) $Na^+(aq)$ and $NO_3^-(aq)$ appear on both sides of the equation. These ions are not reactants; they go through the reaction unchanged. We call them **spectator ions**. If we eliminate the spectator ions, all that remains is the net ionic equation:

$$Ag^+(aq) + I^-(aq) \longrightarrow AgI(s) \qquad \textbf{(5.3)}$$

A **net ionic equation** is an equation that includes only the actual participants in a reaction, with each participant denoted by the symbol or formula that best represents it. Symbols are written for individual ions, such as $Ag^+(aq)$, and whole formulas are written for insoluble solids, such as $AgI(s)$. Because net ionic equations include electrically charged species—ions—a net ionic equation must be balanced both for the numbers of atoms of all types and for electric charge. The same net electric charge must appear on both sides of the equation. Throughout the remainder of this chapter, we will represent most chemical reactions in aqueous solution by net ionic equations.

Predicting Precipitation Reactions

Suppose we are asked whether precipitation occurs when the following aqueous solutions are mixed.

$$AgNO_3(aq) + KBr(aq) \longrightarrow ? \qquad \textbf{(5.4)}$$

▲ FIGURE 5-6
Qualitative test for Cl^- in tap water
The test involves the addition of a few drops of $AgNO_3(aq)$ to tap water. The formation of a precipitate of $AgCl(s)$ confirms the presence of Cl^-.

KEEP IN MIND

that although the insoluble solid consists of ions, we don't represent ionic charges in the whole formula. That is, we write $AgI(s)$, not $Ag^+I^-(s)$.

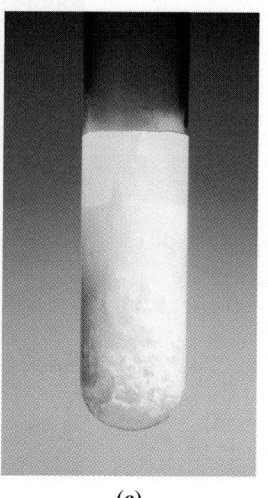

(a)   **(b)**   **(c)**

◀ FIGURE 5-7
A precipitate of silver iodide
When an aqueous solution of $AgNO_3$ **(a)** is added to one of NaI **(b)**, insoluble pale yellow or cream-colored $AgI(s)$ precipitates from solution **(c)**.

A good way to begin is to rewrite expression (5.4) in the ionic form.

$$\text{Ag}^+(\text{aq}) + \text{NO}_3^-(\text{aq}) + \text{K}^+(\text{aq}) + \text{Br}^-(\text{aq}) \longrightarrow ? \qquad \textbf{(5.5)}$$

There are only two possibilities. Either some cation–anion combination leads to an insoluble solid—a precipitate—or no such combination is possible, and there is no reaction at all.

To predict what will happen without doing experiments, we need some information about which sorts of ionic compounds are water soluble and which are water insoluble. We expect the insoluble ones to form when the appropriate ions are mixed in solution. We don't have all-encompassing rules for predicting solubilities, but a few guidelines work for the majority of common ionic solutes. A concise form of these guidelines is presented in Table 5.1.

▶ In principle, all ionic compounds dissolve in water to some extent, though this may be very slight. For practical purposes, we consider a compound to be insoluble if the maximum amount that can dissolve is less than about 0.01 mol/L.

TABLE 5.1 Solubility Guidelines for Common Ionic Solids

Follow the lower-numbered guideline when two guidelines are in conflict. This leads to the correct prediction in most cases.
1. Salts of group 1 cations (with some exceptions for Li^+) and the NH_4^+ cation are soluble.
2. Nitrates, acetates, and perchlorates are soluble.
3. Salts of silver, lead, and mercury(I) are insoluble.
4. Chlorides, bromides, and iodides are soluble.
5. Carbonates, phosphates, sulfides, oxides, and hydroxides are insoluble (sulfides of group 2 cations and hydroxides of $\text{Ca}^{2+}, \text{Sr}^{2+}$, and Ba^{2+} are slightly soluble).
6. Sulfates are soluble except for those of calcium, strontium, and barium.

KEEP IN MIND

that when two ionic compounds form a solid precipitate, they do so by exchanging ions. In the formation of AgBr from KBr and AgNO_3, the following exchange takes place.

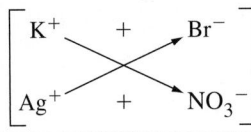

The guidelines from Table 5.1 are applied in the order listed, with the lower-numbered guideline taking precedence in cases of a conflict. According to these guidelines, AgBr(s) is insoluble in water (because rule 3 takes precedence over rule 4) and should precipitate, whereas $\text{KNO}_3(\text{s})$ is soluble (because of rule 1). Written as an ionic equation, expression (5.5) becomes

$$\text{Ag}^+(\text{aq}) + \text{NO}_3^-(\text{aq}) + \text{K}^+(\text{aq}) + \text{Br}^-(\text{aq}) \longrightarrow \text{AgBr(s)} + \text{K}^+(\text{aq}) + \text{NO}_3^-(\text{aq})$$

For the net ionic equation, we have

$$\text{Ag}^+(\text{aq}) + \text{Br}^-(\text{aq}) \longrightarrow \text{AgBr(s)} \qquad \textbf{(5.6)}$$

The three predictions concerning precipitation reactions made in Example 5-2 are verified in Figure 5-8.

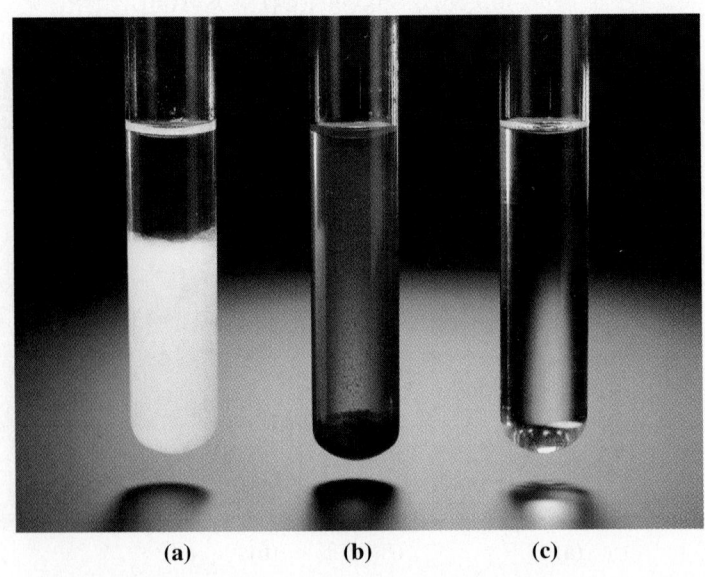

▶ FIGURE 5-8
Verifying the predictions made in Example 5-2
(a) When NaOH(aq) is added to $\text{MgCl}_2(\text{aq})$, a white precipitate of $\text{Mg(OH)}_2(\text{s})$ forms. **(b)** When colorless BaS(aq) is added to blue $\text{CuSO}_4(\text{aq})$, a dark precipitate forms. The precipitate is a mixture of white $\text{BaSO}_4(\text{s})$ and black CuS(s); a slight excess of CuSO_4 remains in solution. **(c)** No reaction occurs when colorless $(\text{NH}_4)_2\text{SO}_4(\text{aq})$ is added to colorless $\text{ZnCl}_2(\text{aq})$.

(a)   (b)   (c)

EXAMPLE 5-2 Using Solubility Guidelines to Predict Precipitation Reactions

Predict whether a reaction will occur in each of the following cases. If so, write a net ionic equation for the reaction.

(a) $NaOH(aq) + MgCl_2(aq) \longrightarrow ?$
(b) $BaS(aq) + CuSO_4(aq) \longrightarrow ?$
(c) $(NH_4)_2SO_4(aq) + ZnCl_2(aq) \longrightarrow ?$

Analyze

All the compounds shown in **(a)**, **(b)**, and **(c)** are soluble and they provide ions in solution. By using the solubility guidelines in Table 5.1, determine whether the positive ions from one compound combine with the negative ions of the other to form soluble or insoluble compounds. If only soluble compounds are formed, then all ions remain in solution (no reaction). If an insoluble compound is formed, then the insoluble compound precipitates from the solution. The net ionic equation for the precipitation reaction is obtained by eliminating the spectator ions from the full ionic equation.

Solve

For each of **(a)**, **(b)** and **(c)**, apply the strategy described above.

(a) In aqueous solution, we get Na^+ and OH^- from NaOH and Mg^{2+} and Cl^- from $MgCl_2$. The combination of Na^+ and Cl^- gives NaCl, a soluble compound; thus, the Na^+ and Cl^- ions remain in solution. However, the Mg^{2+} and OH^- ions combine to produce $Mg(OH)_2$, an insoluble compound. The full, ionic equation is

$$2\,Na^+(aq) + 2\,OH^-(aq) + Mg^{2+}(aq) + 2\,Cl^-(aq) \longrightarrow$$
$$Mg(OH)_2(s) + 2\,Na^+(aq) + 2\,Cl^-(aq)$$

With the elimination of spectator ions, we obtain

$$2\,OH^-(aq) + Mg^{2+}(aq) \longrightarrow Mg(OH)_2(s)$$

(b) In aqueous solution, we get Ba^{2+} and S^{2-} from BaS and Cu^{2+} and SO_4^{2-} from $CuSO_4$. The Ba^{2+} and SO_4^{2-} ions combine to form $BaSO_4$, an insoluble compound, and the Cu^{2+} and S^{2-} ions combine to form CuS, also an insoluble compound. The full ionic equation is

$$Ba^{2+}(aq) + S^{2-}(aq) + Cu^{2+}(aq) + SO_4^{2-}(aq) \longrightarrow BaSO_4(s) + CuS(s)$$

The equation above is also the net ionic equation because there are no spectator ions.

(c) We get $NH_4^+, SO_4^{2-}, Zn^{2+}$, and Cl^- ions in solution. Because all possible combinations of positive and negative ions lead to water soluble compounds, all of the ions remain in solution. No reaction occurs.

Assess

Problems of this type can also be solved by using a diagrammatic approach which is illustrated for part **(a)**.

As you gain experience, you should be able to go directly to a net ionic equation without first having to write an ionic equation that includes spectator ions.

$Mg(OH)_2$ is insoluble.
NaCl is soluble.

PRACTICE EXAMPLE A: Indicate whether a precipitate forms by completing each equation as a net ionic equation. If no reaction occurs, so state.

(a) $AlCl_3(aq) + KOH(aq) \longrightarrow ?$
(b) $K_2SO_4(aq) + FeBr_3(aq) \longrightarrow ?$
(c) $CaI_2(aq) + Pb(NO_3)_2(aq) \longrightarrow ?$

PRACTICE EXAMPLE B: Indicate through a net ionic equation whether a precipitate forms when the following compounds in aqueous solution are mixed. If no reaction occurs, so state.

(a) sodium phosphate + aluminum chloride $\longrightarrow ?$
(b) aluminum sulfate + barium chloride $\longrightarrow ?$
(c) ammonium carbonate + lead nitrate $\longrightarrow ?$

5-1 ARE YOU WONDERING...

Whether an insoluble ionic compound, such as AgCl, is a strong electrolyte or a weak electrolyte?

Silver chloride, AgCl, is an ionic compound. When AgCl dissolves in water, it is 100% dissociated into Ag^+ and Cl^- ions; there are no AgCl ion pairs. If we focus only on the degree of dissociation, then AgCl, like HCl, is a strong electrolyte.

At one time, a strong electrolyte was defined in more practical terms as a substance that when dissolved in water, gives a solution that is a good conductor of electricity. Because AgCl has very low solubility in water, approximately 1×10^{-5} moles per liter, a solution of AgCl is not a good conductor of electricity.

Today, most chemists would argue that AgCl is a strong electrolyte (because it is 100% dissociated in aqueous solution) but some may argue that it is a weak electrolyte (because an aqueous solution of AgCl is not a good conductor of electricity).

Does it matter that chemists may not totally agree on whether AgCl should be called a strong electrolyte or a weak electrolyte? Not at all, because all chemists agree on the following facts: (1) AgCl is essentially 100% dissociated in water; (2) only a small amount of AgCl can be dissolved in water; and (3) an aqueous solution of AgCl is not a good conductor of electricity.

5-3 Acid–Base Reactions

Ideas about acids and bases (or alkalis) date back to ancient times. The word *acid* is derived from the Latin *acidus* (sour). *Alkali* (base) comes from the Arabic *al-qali*, referring to the ashes of certain plants from which alkaline substances can be extracted. The acid–base concept is a major theme in the history of chemistry. In this section, we emphasize the view proposed by Svante Arrhenius in 1884 but also introduce a more modern theory proposed in 1923 by Thomas Lowry and by Johannes Brønsted.

Acids

From a practical standpoint, acids can be identified by their sour taste, their ability to react with a variety of metals and carbonate minerals, and the effect they have on the colors of substances called *acid–base indicators*. Methyl red is an acid–base indicator that appears red in acidic environments and yellow otherwise (see Figure 5-9). From a chemist's point of view, however, an **acid** can be defined as a substance that provides hydrogen ions (H^+) in aqueous solution. This definition was first proposed by Svante Arrhenius in 1884.

Different acids exhibit different tendencies for producing H^+ ions in aqueous solution. **Strong acids** have a strong tendency for producing H^+ ions. Strong acids are molecular compounds that are almost completely ionized into $H^+(aq)$ and accompanying anions when in aqueous solution. Hydrogen chloride, HCl, and nitric acid, HNO_3, are examples of strong acids. When HCl dissolves in water, complete ionization into $H^+(aq)$ and $Cl^-(aq)$ occurs:

$$HCl(aq) \longrightarrow H^+(aq) + Cl^-(aq) \tag{5.7}$$

KEEP IN MIND

that $H^+(aq)$ actually represents a hydrated proton, that is, a proton attached to one H_2O molecule, as in H_3O^+, or to several H_2O molecules, as in $H_9O_4^+$.

◀ FIGURE 5-9
An acid, a base, and an acid–base indicator
The acidic nature of lemon juice is shown by the red color of the acid–base indicator *methyl red*. The basic nature of soap is indicated by the change in color of the indicator from red to yellow.

When HNO_3 dissolves in water, complete ionization into $H^+(aq)$ and $NO_3^-(aq)$ occurs. There are so few common strong acids that they make only a short list. The list of common strong acids is given in Table 5.2. It is imperative that you memorize this list.

Weak acids are molecular compounds that have a weak tendency for producing H^+ ions; weak acids are incompletely ionized in aqueous solution. The vast majority of acids are weak acids. The ionization of a weak acid is best described in terms of a reversible reaction that does not go to completion. As described on page 155, the ionization reaction for acetic acid, CH_3COOH, is

$$CH_3COOH(aq) \rightleftharpoons H^+(aq) + CH_3COO^-(aq) \qquad (5.8)$$

Equation (5.8) has the following interpretation: in aqueous solution, only some of the CH_3COOH molecules are converted into H^+ and CH_3COO^- ions. The fraction of molecules that ionize can be calculated, but the calculation is not simple. We will defer such calculations until Chapter 16.

Equations (5.7) and (5.8) are based on the Arrhenius theory of acids and bases, and these equations might lead you to think that acids simply fall apart into H^+ ions and the accompanying anions when they are dissolved in water. However, plenty of experimental evidence proves that this is not the case. In 1923, Johannes Brønsted in Denmark and Thomas Lowry in Great Britain independently proposed that the key process responsible for the properties of acids (and bases) was the transfer of an H^+ ion (a proton) from one substance to another. For example, when acids dissolve in water, H^+ ions are transferred from acid molecules to water molecules, as shown below for HCl and CH_3COOH.

$$HCl(aq) + H_2O(l) \longrightarrow H_3O^+(aq) + Cl^-(aq) \qquad (5.9)$$
$$CH_3COOH(aq) + H_2O(l) \rightleftharpoons H_3O^+(aq) + CH_3COO^-(aq) \qquad (5.10)$$

In equations (5.9) and (5.10), the acid molecules are acting as proton donors and the water molecules are acting as proton acceptors. According to the Brønsted-Lowry theory, an acid is a **proton donor**.

It is partly a matter of preference whether we include water as a reactant in the equation for the reaction that occurs when an acid is dissolved in water. Some chemists prefer to write the reaction without water as a reactant, as we did in equations (5.7) and (5.8), to eliminate the clutter of "extra" water molecules. If that is also your preference, then you must remember that the H^+ ion is not a free proton in solution but rather is firmly bound to a water molecule and exists as an H_3O^+ ion. The H_3O^+ ion is even further hydrated (see Figure 5-5). Many chemists prefer to include H_2O as a reactant, as we did in equations (5.9) and (5.10), to emphasize that the reactions actually involve the transfer of protons from acid molecules to water molecules.

TABLE 5.2 Common Strong Acids and Strong Bases	
Acids	**Bases**
HCl	LiOH
HBr	NaOH
HI	KOH
$HClO_4$	RbOH
HNO_3	CsOH
H_2SO_4[a]	$Ca(OH)_2$
	$Sr(OH)_2$
	$Ba(OH)_2$

[a]H_2SO_4 ionizes in two distinct steps. It is a strong acid only in its first ionization step (see Section 16-6).

KEEP IN MIND

that a hydrogen atom consists of one proton and one electron. Therefore, a hydrogen ion, H^+, is simply a proton.

Bases

From a practical standpoint, we can identify bases through their bitter taste, slippery feel, and effect on the colors of acid–base indicators (Fig. 5-9). The Arrhenius definition of a **base** is a substance that produces hydroxide ions (OH^-) in aqueous solution. Consider a soluble ionic hydroxide, such as NaOH. In the solid state, this compound consists of Na^+ and OH^- ions. When the solid dissolves in water, the ions dissociate.

$$NaOH(aq) \longrightarrow Na^+(aq) + OH^-(aq)$$

The equation above indicates that NaOH(aq) is best represented as $Na^+(aq)$ plus $OH^-(aq)$.

A base that dissociates completely, or very nearly so, in aqueous solution is a **strong base**. As is true of strong acids, the number of common strong bases is small (see Table 5.2). They are primarily the hydroxides of group 1 and some group 2 metals. Memorize the list.

Certain substances produce OH^- ions by reacting with water, not just by dissolving in it. Such substances, for example, ammonia, are also bases.

$$NH_3(aq) + H_2O(l) \rightleftharpoons NH_4^+(aq) + OH^-(aq) \qquad (5.11)$$

NH_3 is a weak electrolyte; its reaction with water does not go to completion. A base that is incompletely ionized in aqueous solution is a **weak base**. Most basic substances are weak bases.

We can also examine equation (5.11) in terms of the Brønsted-Lowry theory, which focuses on the transfer of protons from one substance to another. According to this theory, a base is a **proton acceptor**. In equation (5.11), NH_3 behaves as a proton acceptor (a Brønsted-Lowry base) and H_2O behaves as a proton donor (a Brønsted-Lowry acid).

KEEP IN MIND

that $NH_4^+(aq)$ is formed by the transfer of a proton from an H_2O to a NH_3 molecule, and NH_4^+ interacts with water in much the same way as the hydronium ion does. A ball-and-stick model of the ammonium ion is shown below.

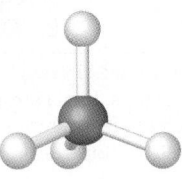

Ammonium ion

Acidic and Basic Solutions

We have seen that when dissolved in water, an acid produces H^+ ions and a base produces OH^- ions. However, experiment shows small numbers of H^+ and OH^- ions are present even in pure water. In pure water, the following reaction occurs to a limited extent, hence the use of a double arrow ($\rightleftharpoons$) rather than a single arrow ($\longrightarrow$).

$$H_2O(l) \rightleftharpoons H^+(aq) + OH^-(aq)$$

Careful measurements show that $[H^+]_{water} = [OH^-]_{water} = 1.0 \times 10^{-7} \, M$ at 25 °C. (The subscripts on the square brackets are there to emphasize that the values given for $[H^+]$ and $[OH^-]$ are for pure water only.) Because an acid produces H^+ ions in solution, we expect that a solution of acid at 25 °C will have $[H^+] > 1.0 \times 10^{-7} \, M$. Such a solution is said to be *acidic*. An acidic solution has a greater concentration of H^+ ions than does pure water. A base produces OH^- ions, and so a solution of base will have $[OH^-] > 1.0 \times 10^{-7} \, M$ at 25 °C. Such a solution is said to be *basic*. These ideas are summarized below.

An *acidic* solution has $[H^+] > [H^+]_{water}$.
A *basic* solution has $[OH^-] > [OH^-]_{water}$.

The statements above can be expressed another way. An acidic solution has an excess of H^+ ions (compared with pure water), and a basic solution has an excess of OH^- ions. We will use these ideas in Section 5-5 and encounter them again in Chapter 16.

Neutralization

Perhaps the most significant property of acids and bases is the ability of each to cancel or neutralize the properties of the other. In a **neutralization reaction**, an acid and a base react to form water and an aqueous solution of an ionic compound called a **salt**. Thus, in molecular form,

$$HCl(aq) + NaOH(aq) \longrightarrow NaCl(aq) + H_2O(l)$$
$$\text{(acid)} + \text{(base)} \longrightarrow \text{(salt)} + \text{(water)}$$

Switching to the ionic form we write,

$$\underbrace{H^+(aq) + Cl^-(aq)}_{\text{(acid)}} + \underbrace{Na^+(aq) + OH^-(aq)}_{\text{(base)}} \longrightarrow \underbrace{Na^+(aq) + Cl^-(aq)}_{\text{salt}} + \underbrace{H_2O(l)}_{\text{water}}$$

When the spectator ions are eliminated, the net ionic equation shows the essential nature of the neutralization of a strong acid by a strong base: H^+ ions from the acid and OH^- ions from the base combine to form water.

$$H^+(aq) + OH^-(aq) \longrightarrow H_2O(l)$$

In a neutralization involving the weak base $NH_3(aq)$, we can think of H^+ from an acid combining directly with NH_3 molecules to form NH_4^+. The neutralization can be represented by an ionic equation, for example,

$$\underbrace{H^+(aq) + Cl^-(aq)}_{\text{(acid)}} + \underbrace{NH_3(aq)}_{\text{(base)}} \longrightarrow \underbrace{NH_4^+(aq) + Cl^-(aq)}_{\text{(salt)}}$$

or by a net ionic equation

$$H^+(aq) + NH_3(aq) \longrightarrow NH_4^+(aq)$$

All the neutralization reactions given above involve a strong acid or a strong base and all of them go essentially to completion, that is, until the limiting reagent is used up. Thus, we use a single arrow ($\longrightarrow$) rather than a double arrow ($\rightleftharpoons$) in the equations for these reactions.

◀ The formula of $NH_3(aq)$ is sometimes written as NH_4OH (ammonium hydroxide) and its ionization represented as

$$NH_4OH(aq) \rightleftharpoons$$
$$NH_4^+(aq) + OH^-(aq)$$

There is no hard evidence for the existence of NH_4OH, however, that is, in the sense of a discrete substance comprising NH_4^+ and OH^- ions. We will use only the formula $NH_3(aq)$.

Recognizing Acids and Bases

Acids contain *ionizable* hydrogen atoms, which are generally identified by the way in which the formula of an acid is written. Ionizable H atoms are separated from other H atoms in the formula either by writing them first in the molecular formula or by indicating where they are found in the molecule. Thus, there are two ways that we can show that one H atom in the acetic acid molecule is ionizable and the other three H atoms are not.

$$\underbrace{HC_2H_3O_2 \text{ or } CH_3COOH}_{\text{acetic acid}}$$

In contrast to acetic acid, methane has four H atoms, but they are not ionizable. CH_4 is neither an acid nor a base.

A substance whose formula indicates a combination of OH^- ions with cations is generally a strong base (for example, NaOH). To identify a weak base, we usually need a chemical equation for the ionization reaction, as in equation (5.11). The main weak base we will work with at present is NH_3. Note that ethanol, CH_3CH_2OH, is not a base. The OH group is not present as OH^-, both in pure ethanol and in its aqueous solutions.

More Acid–Base Reactions

$Mg(OH)_2$ is a base because it contains OH^-, but this compound is quite insoluble in water. Its finely divided solid particles form a suspension in water that is the familiar milk of magnesia, used as an antacid. In this suspension, $Mg(OH)_2(s)$ does dissolve very slightly, producing some OH^- in solution. If an acid is added, H^+ from the acid combines with this OH^- to form water—neutralization occurs. More $Mg(OH)_2(s)$ dissolves to produce more OH^- in solution, which is neutralized by more H^+, and so on. In this way, the neutralization reaction results in the dissolving of otherwise insoluble $Mg(OH)_2(s)$. The net ionic equation for the reaction of $Mg(OH)_2(s)$ with a strong acid is

$$Mg(OH)_2(s) + 2\,H^+(aq) \longrightarrow Mg^{2+}(aq) + 2\,H_2O(l) \qquad \textbf{(5.12)}$$

$Mg(OH)_2(s)$ also reacts with a weak acid, such as acetic acid. In the net ionic equation, acetic acid is written in its molecular form. But remember that some H^+ and CH_3COO^- ions are always present in an acetic acid solution. The H^+ ions react with OH^- ions, as in reaction (5.12), followed by further ionization of CH_3COOH, more neutralization, and so on. If enough acetic acid is present, the $Mg(OH)_2$ will dissolve completely. The equation for the reaction is given below.

$$Mg(OH)_2(s) + 2\,CH_3COOH(aq) \longrightarrow Mg^{2+}(aq) + 2\,CH_3COO^-(aq) + 2\,H_2O(l)$$
$$\textbf{(5.13)}$$

Calcium carbonate, which is present in limestone and marble, is another water-insoluble solid that is soluble in strong and weak acids. Here the solid produces a low concentration of CO_3^{2-} ions, which combine with H^+ to form the weak acid H_2CO_3. This causes more of the solid to dissolve, and so on. Carbonic acid, H_2CO_3, is a very unstable substance that decomposes into H_2O and $CO_2(g)$. The net ionic equation for the reaction of $CaCO_3$ with an acid is given below.

$$CaCO_3(s) + 2\,H^+(aq) \longrightarrow Ca^{2+}(aq) + H_2O(l) + CO_2(g) \qquad \textbf{(5.14)}$$

Thus, a gas is given off when $CaCO_3(s)$ reacts with an acid and dissolves. The reaction represented by equation (5.14) is responsible for the erosion of marble statues by acid rain, such as the one shown in Figure 5-10. Equation (5.14) also shows that $CaCO_3(s)$ has the ability to neutralize acids. Not surprisingly, calcium carbonate, like magnesium hydroxide, is used as an antacid.

The Arrhenius definition recognizes only OH^- as a base, but when we reconsider acids and bases in more detail in Chapter 16, we will see that modern theories identify CO_3^{2-} and many other anions, including OH^-, as bases. Table 5.3 lists several common anions and one cation that produce gases in acid–base reactions.

▲ FIGURE 5-10
Damage caused by acid rain
This marble statue has been eroded by acid rain. Marble consists primarily of $CaCO_3$. Acids react with and dissolve marble through the reaction described in equation (5.14).

TABLE 5.3	Some Common Gas-Forming Reactions
Ion	**Reaction**
HSO_3^-	$HSO_3^- + H^+ \longrightarrow SO_2(g) + H_2O(l)$
SO_3^{2-}	$SO_3^{2-} + 2\,H^+ \longrightarrow SO_2(g) + H_2O(l)$
HCO_3^-	$HCO_3^- + H^+ \longrightarrow CO_2(g) + H_2O(l)$
CO_3^{2-}	$CO_3^{2-} + 2\,H^+ \longrightarrow CO_2(g) + H_2O(l)$
S^{2-}	$S^{2-} + 2\,H^+ \longrightarrow H_2S(g)$
NH_4^+	$NH_4^+ + OH^- \longrightarrow NH_3(g) + H_2O(l)$

EXAMPLE 5-3   Writing Equations for Acid–Base Reactions

Write a net ionic equation to represent the reaction of **(a)** aqueous strontium hydroxide with nitric acid; **(b)** solid aluminum hydroxide with hydrochloric acid.

Analyze

The reactions are neutralization reactions, which means they are of the general form acid + base → salt + water. We can start with the whole formula equation, switch to the ionic equation, and then delete the spectator ions to arrive at the net ionic equation.

Solve

(a) $2 HNO_3(aq) + Sr(OH)_2(aq) \longrightarrow Sr(NO_3)_2(aq) + 2 H_2O(l)$

Ionic form:

$$2 H^+(aq) + 2 NO_3^-(aq) + Sr^{2+}(aq) + 2 OH^-(aq) \longrightarrow Sr^{2+}(aq) + 2 NO_3^-(aq) + 2 H_2O(l)$$

Net ionic equation: Delete the spectator ions (Sr^{2+} and NO_3^-).

$$2 H^+(aq) + 2 OH^-(aq) \longrightarrow 2 H_2O(l)$$

or, more simply,

$$H^+(aq) + OH^-(aq) \longrightarrow H_2O(l)$$

(b) $Al(OH)_3(s) + 3 HCl(aq) \longrightarrow AlCl_3(aq) + 3 H_2O(l)$

Ionic form:

$$Al(OH)_3(s) + 3 H^+(aq) + 3 Cl^-(aq) \longrightarrow Al^{3+}(aq) + 3 Cl^-(aq) + 3 H_2O(l)$$

Net ionic equation: Delete the spectator ion (Cl^-).

$$Al(OH)_3(s) + 3 H^+(aq) \longrightarrow Al^{3+}(aq) + 3 H_2O(l)$$

Assess

In part **(a)**, the net ionic equation is $H^+(aq) + OH^-(aq) \rightarrow H_2O(l)$, as is always the case when the neutralization reaction involves a soluble strong acid and a soluble strong base. In part **(b)**, the base was not soluble; thus, the net ionic equation includes a solid.

PRACTICE EXAMPLE A: Write a net ionic equation to represent the reaction of aqueous ammonia with propionic acid, CH_3CH_2COOH. Assume that the neutralization reaction goes to completion. What is the formula and name of the salt that results from this neutralization?

PRACTICE EXAMPLE B: Calcium carbonate is a major constituent of the hard water deposits found in teakettles and automatic coffeemakers. Vinegar, which is essentially a dilute aqueous solution of acetic acid, is commonly used to remove such deposits. Write a net ionic equation for the reaction that occurs. Assume that the neutralization reaction goes to completion.

5-3   CONCEPT ASSESSMENT

You are given the four solids, K_2CO_3, CaO, $ZnSO_4$, and $BaCO_3$, and three solvents, $H_2O(l)$, HCl(aq), and $H_2SO_4(aq)$. You are asked to prepare four solutions, each containing one of the four cations, that is, one with $K^+(aq)$, one with $Ca^{2+}(aq)$, and so on. Using water as your *first* choice, what solvent would you use to prepare each solution? Explain your choices.

5-4   Oxidation–Reduction Reactions: Some General Principles

Practical applications of oxidation–reduction reactions can be traced back thousands of years to the period in human culture when metal tools were first made. The metal needed to make tools was obtained by heating copper or iron ores, such as cuprite (Cu_2O) or hematite (Fe_2O_3), in the presence of carbon. Since

◀ An ore is a mineral from which a metal can be extracted. Many metal ores are oxides and the metals are obtained from their oxides by the removal of oxygen.

that time, iron has become the most widely used of all metals and it is produced in essentially the same way: by heating Fe_2O_3 in the presence of carbon in a blast furnace. A simplified chemical equation for the reaction is given below.

▶ In a blast furnace, carbon from the coke is converted to CO, which then reacts with Fe_2O_3.

$$Fe_2O_3(s) + 3\,CO(g) \xrightarrow{\Delta} 2\,Fe(l) + 3\,CO_2(g) \qquad (5.15)$$

In this reaction, we can think of the CO(g) as taking O atoms away from Fe_2O_3 to produce $CO_2(g)$ and the free element iron. A commonly used term to describe a reaction in which a substance gains O atoms is *oxidation*, and a reaction in which a substance loses O atoms is *reduction*. In reaction (5.15), CO(g) is oxidized and $Fe_2O_3(s)$ is reduced. Oxidation and reduction must always occur together, and such a reaction is called an **oxidation–reduction**, or **redox**, **reaction**. The oxygen in Fe_2O_3 can also be removed by igniting a finely divided mixture of Fe_2O_3 and Al. The reaction produces a spectacular display, shown in Figure 5-11, and releases a tremendous amount of heat, which causes the iron to melt. Mixtures of finely divided Fe_2O_3 and Al are used by railway workers to produce liquid iron for welding together iron railway tracks.

▶ Because it is easier to say, the term *redox* is often used instead of oxidation–reduction.

Definitions of oxidation and reduction based solely on the transfer of O atoms are too restrictive. By using broader definitions, many reactions in aqueous solution can be described as oxidation–reduction reactions, even when no oxygen is involved.

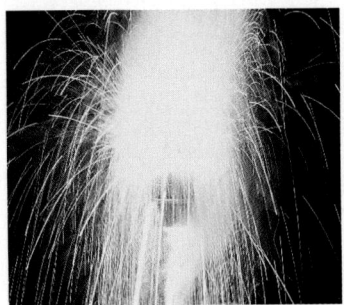

▲ FIGURE 5-11
Thermite reaction
Iron atoms of iron(III) oxide give up O atoms to Al atoms, producing Al_2O_3.

$$Fe_2O_3(s) + 2\,Al(s) \longrightarrow$$
$$Al_2O_3(s) + 2\,Fe(l)$$

Oxidation State Changes

Suppose we rewrite equation (5.15) and indicate the oxidation states (O.S.) of the elements on both sides of the equation by using the rules listed on page 85.

$$\overset{+3\ -2}{Fe_2O_3} + 3\,\overset{+2\ -2}{CO} \longrightarrow 2\,\overset{0}{Fe} + 3\,\overset{+4\ -2}{CO_2}$$

The O.S. of oxygen is −2 everywhere it appears in this equation. That of iron (shown in red) changes. It *decreases* from +3 in Fe_2O_3 to 0 in the free element, Fe. The O.S. of carbon (shown in blue) also changes. It *increases* from +2 in CO to +4 in CO_2. In terms of oxidation state changes, in an oxidation process, the O.S. of some element increases; in a reduction process, the O.S. of some element decreases.

Even though we assess oxidation state changes by element, oxidation and reduction involve the entire species in which the element is found. Thus, for the reaction above, the whole compound Fe_2O_3 is reduced, not just the Fe atoms; and CO is oxidized, not just the C atom.

EXAMPLE 5-4 Identifying Oxidation–Reduction Reactions

Indicate whether each of the following is an oxidation–reduction reaction.

(a) $MnO_2(s) + 4\,H^+(aq) + 2\,Cl^-(aq) \longrightarrow Mn^{2+}(aq) + 2\,H_2O(l) + Cl_2(g)$

(b) $H_2PO_4^-(aq) + OH^-(aq) \longrightarrow HPO_4^{2-}(aq) + H_2O(l)$

Analyze

In each case, indicate the oxidation states of the elements on both sides of the equation, and look for changes.

Solve

(a) The O.S. of Mn decreases from +4 in MnO_2 to +2 in Mn^{2+}. MnO_2 is reduced to Mn^{2+}. The O.S. of O remains at −2 throughout the reaction, and that of H, at +1. The O.S. of Cl increases from −1 in Cl^- to 0 in Cl_2. Cl^- is oxidized to Cl_2. The reaction is an oxidation–reduction reaction.

(b) The O.S. of H is +1 on both sides of the equation. Oxygen remains at O.S. −2 throughout. The O.S. of phosphorus is +5 in $H_2PO_4^-$ and also +5 in HPO_4^{2-}. There are no changes in O.S. This is not an oxidation–reduction reaction. (It is, in fact, an acid–base reaction.)

Assess

Because many redox reactions involve H^+, OH^-, or insoluble ionic compounds, it is easy to confuse a redox reaction with an acid–base or a precipitation reaction. It is important that you remember the defining features of each type of reaction. Precipitation reactions involve the combination of ions in solution to produce an insoluble precipitate, acid–base reactions involve proton (H^+) transfer, and redox reactions involve electron transfer and changes in oxidation states.

PRACTICE EXAMPLE A: Identify whether each of the following is an oxidation–reduction reaction.

(a) $(NH_4)_2SO_4(aq) + Ba(NO_3)_2(aq) \longrightarrow BaSO_4(s) + 2\,NH_4NO_3(aq)$

(b) $2\,Pb(NO_3)_2(s) \longrightarrow 2\,PbO(s) + 4\,NO_2(g) + O_2(g)$

PRACTICE EXAMPLE B: Identify the species that is oxidized and the species that is reduced in the reaction below.

$$5\,VO^{2+}(aq) + MnO_4^-(aq) + H_2O(l) \longrightarrow 5\,VO_2^+(aq) + Mn^{2+}(aq) + 2\,H^+(aq)$$

Oxidation and Reduction Half-Reactions

The reaction illustrated in Figure 5-12 is an oxidation–reduction reaction. The chemical equation for the reaction is given below.

$$Zn(s) + Cu^{2+}(aq) \longrightarrow Zn^{2+}(aq) + Cu(s)$$

We can show that the reaction is an oxidation–reduction reaction by evaluating changes in oxidation state, but there is another especially useful way to establish this. Think of the reaction as involving two **half-reactions** occurring at the same time—an oxidation and a reduction. The overall reaction is the

(a)

(b)

(c)

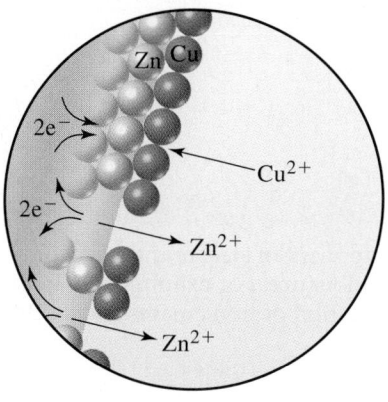

▲ FIGURE 5-12
An oxidation–reduction reaction
(a) A zinc rod above an aqueous solution of copper(II) sulfate. (b) Following immersion of the Zn rod in the $CuSO_4(aq)$ for several hours, the blue color of $Cu^{2+}(aq)$ disappears and a deposit of copper forms on the rod. In the microscopic view on the left, the gray spheres represent Zn atoms and the red spheres represent Cu atoms. In the reaction, Zn atoms lose electrons to the metal surface and enter the solution as Zn^{2+} ions. Cu^{2+} ions from the solution pick up electrons and deposit on the metal surface as atoms of solid copper. (c) The pitted zinc rod (providing evidence that zinc was consumed in a chemical reaction) and the collected copper metal.

sum of the two half-reactions. We can represent the half-reactions by half-equations and the overall reaction by an overall equation.

Oxidation: $$Zn(s) \longrightarrow Zn^{2+}(aq) + 2\,e^-$$ **(5.16)**

Reduction: $$\underline{Cu^{2+}(aq) + 2\,e^- \longrightarrow Cu(s)}$$ **(5.17)**

Overall: $$Zn(s) + Cu^{2+}(aq) \longrightarrow Zn^{2+}(aq) + Cu(s)$$ **(5.18)**

In half-reaction (5.16), Zn is oxidized—its oxidation state increases from 0 to $+2$. This change corresponds to a loss of two electrons by each zinc atom. In half-reaction (5.17), Cu^{2+} is reduced—its oxidation state decreases from $+2$ to 0. This change corresponds to the gain of two electrons by each Cu^{2+} ion. To summarize,

- **Oxidation** is a process in which the O.S. of some element *increases* as electrons are lost. Electrons appear on the *right* side of a half-equation.
- **Reduction** is a process in which the O.S. of some element *decreases* as electrons are gained. Electrons appear on the *left* side of a half-equation.
- Oxidation and reduction half-reactions must always occur together, and the total number of electrons associated with the oxidation must *equal* the total number associated with the reduction.

Redox reactions are similar to acid–base reactions in that both types of reactions involve the transfer of small, fundamental particles; protons are transferred in acid–base reactions and electrons are transferred in redox reactions. However, acid-base reactions are easily identified because the hydrogen atoms and protons (H^+) are shown explicitly in the balanced chemical equation. Redox reactions are much more difficult to identify because the electrons are "hidden." When we write out the half-equations for the oxidation and reduction processes, the electrons are shown explicitly and the key feature of a redox reaction, electron transfer, is emphasized.

Figure 5-13 and Example 5-5 suggest some fundamental questions about oxidation–reduction. For example,

- Why does Fe react with HCl(aq), displacing $H_2(g)$, whereas Cu does not?
- Why does Fe form Fe^{2+} and not Fe^{3+} in this reaction?

EXAMPLE 5-5   **Representing an Oxidation–Reduction Reaction Through Half-Equations and an Overall Equation**

Write equations for the oxidation and reduction processes that occur and the overall equation for the reaction of iron with hydrochloric acid solution to produce $H_2(g)$ and Fe^{2+}. The reaction is shown in Figure 5-13.

Analyze

The reactants are Fe(s) and HCl(aq), and the products are $H_2(g)$ and $FeCl_2(aq)$, a soluble ionic compound. In the reaction, the oxidation state of iron changes from 0 in Fe to $+2$ in $FeCl_2$, and the oxidation state of hydrogen changes from $+1$ in HCl to 0 in H_2. Thus, iron is oxidized and hydrogen is reduced.

Solve

The balanced chemical equations are as follows.

Oxidation: $$Fe(s) \longrightarrow Fe^{2+}(aq) + 2\,e^-$$

Reduction: $$\underline{2\,H^+(aq) + 2\,e^- \longrightarrow H_2(g)}$$

Overall: $$Fe(s) + 2\,H^+(aq) \longrightarrow Fe^{2+}(aq) + H_2(g)$$

Assess

This example illustrates that iron dissolves in acid solution. Iron is a major component of steel and the reaction in this example contributes to the corrosion of steel that is exposed to air and moisture. For example, H^+ ions from acid rain cause Fe atoms in steel to become oxidized to Fe^{2+} ions. The oxidation of iron creates small pits in the steel surface and leads to corrosion.

(a)   (b)   (c)

▲ FIGURE 5-13
Displacement of H⁺ (aq) by iron—Example 5-5 illustrated
(a) An iron nail is wrapped in a piece of copper screen. **(b)** The nail and screen
are placed in HCl(aq). Hydrogen gas is evolved as the nail reacts. **(c)** The nail
reacts completely and produces Fe^{2+}(aq), but the copper does not react.

PRACTICE EXAMPLE A: Represent the reaction of aluminum with hydrochloric acid to produce $AlCl_3$(aq) and
H_2(g) by oxidation and reduction half-equations and an overall equation.

PRACTICE EXAMPLE B: Represent the reaction of chlorine gas with aqueous sodium bromide to produce liquid
bromine and aqueous sodium chloride by oxidation and reduction half-equations and an overall equation.

Even now, you can probably see that the answers to these questions lie in the
relative abilities of Fe and Cu atoms to give up electrons—to become oxidized.
Fe gives up electrons more easily than does Cu; also, Fe is more readily oxidized to Fe^{2+} than it is to Fe^{3+}. We can give more complete answers after we
develop specific criteria describing electron loss and gain in Chapter 20. For
now, the information in Table 5.4 should be helpful. The table lists some common metals that react with acids to displace H_2(g) and a few that do not. As
noted in the table, most of the group 1 and 2 metals are so reactive that they
will react with cold water to produce H_2(g) and a solution of the metal
hydroxide.

TABLE 5.4 Behavior of Some Common Metals with Nonoxidizing Acids[a]	
React to Produce H_2(g)	Do Not React
Alkali metals (group 1)[b]	Cu, Ag, Au, Hg
Alkaline earth metals (group 2)[b]	
Al, Zn, Fe, Sn, Pb	

[a]A nonoxidizing acid (for example, HCl, HBr, HI) is one in which the only possible reduction half-reaction is the reduction of H⁺ to H_2. Additional possibilities
for metal–acid reactions are considered in Chapter 20.
[b]With the exception of Be and Mg, all group 1 and group 2 metals also react
with cold water to produce H_2(g); the metal hydroxide is the other product.

🔍 **5-4 CONCEPT ASSESSMENT**

Disregarding the fact that the equations below are not balanced, is it likely that either represents a reaction that could possibly occur? Explain.

(a) $MnO_4^-(aq) + Cr_2O_7^{2-}(aq) + H^+(aq) \longrightarrow MnO_2(s) + Cr^{3+}(aq) + H_2O(l)$

(b) $Cl_2(g) + OH^-(aq) \longrightarrow Cl^-(aq) + ClO_3^-(aq) + H_2O(l)$

5-5 Balancing Oxidation–Reduction Equations

▶ *Half-reaction* or *half-equation*, which is it? *Reaction* refers to the actual process. An *equation* is a notation we write out to indicate the formulas of the reactants and products, and their mole relationships. Many chemists refer to what we balance here as half-reactions but half-equation is more proper.

In a chemical reaction, atoms are neither created nor lost; they are simply rearranged. We used this idea in Chapter 4 to balance chemical equations by inspection. In a redox reaction, we have additional considerations. Electrons are transferred from one substance to another and so we must keep track of electrons and the charge that these electrons carry. Therefore, in balancing the chemical equation for a redox reaction, we focus equally on three factors: (1) the number of atoms of each type, (2) the number of electrons transferred, and (3) the total charges on reactants and products. We should point out, however, that if we balance the equation with respect to the number of atoms and the number of electrons transferred, then the equation is automatically balanced with respect to the total charges.

Because it is very challenging to deal with all three of these factors simultaneously, only a small proportion of oxidation–reduction equations can be balanced by simple inspection. To make this point clear, consider the following reactions, each of which appears to be balanced. The reactions are balanced with respect to the number of atoms of each type and the total charges on reactants and products, but not with respect to the number of electrons transferred.

▶ For brevity, the physical forms of reactants and products are not included in these equations. For all ions, as well as H_2O_2, the physical form is "(aq)." Of course, the physical forms of oxygen and water are $O_2(g)$ and $H_2O(l)$.

$$2\,MnO_4^- + H_2O_2 + 6\,H^+ \longrightarrow 2\,Mn^{2+} + 3\,O_2 + 4\,H_2O \qquad \textbf{(5.19)}$$

$$2\,MnO_4^- + 3\,H_2O_2 + 6\,H^+ \longrightarrow 2\,Mn^{2+} + 4\,O_2 + 6\,H_2O \qquad \textbf{(5.20)}$$

$$2\,MnO_4^- + 5\,H_2O_2 + 6\,H^+ \longrightarrow 2\,Mn^{2+} + 5\,O_2 + 8\,H_2O \qquad \textbf{(5.21)}$$

$$2\,MnO_4^- + 7\,H_2O_2 + 6\,H^+ \longrightarrow 2\,Mn^{2+} + 6\,O_2 + 10\,H_2O \qquad \textbf{(5.22)}$$

These are just a few of the "balanced" chemical equations we can write for the reaction. However, only one, equation (5.21), is properly balanced because it is the only one that is balanced with respect to the number of electrons transferred.

To balance the chemical equation for a redox reaction, we need to make use of a systematic approach that considers each of the relevant factors in turn. Although several methods are available, we emphasize one that focuses first on balancing separate half-equations and then on combining the two half-equations to obtain the overall chemical equation. The method is summarized below. An alternative method is described in Exercise 98.

The Half-Equation Method

The basic steps in this method of balancing a redox equation are as follows:

- Write and balance separate half-equations for oxidation and reduction.
- Adjust coefficients in the two half-equations so that the same number of electrons appear in each half-equation.
- Add together the two half-equations (canceling out electrons) to obtain the balanced overall equation.

> **TABLE 5.5 Balancing Equations for Redox Reactions in Acidic Aqueous Solutions by the Half-Equation Method: A Summary**
>
> - Write the equations for the oxidation and reduction half-reactions.
> - In each half-equation
>   (1) Balance atoms of all the elements except H and O
>   (2) Balance oxygen by using H_2O
>   (3) Balance hydrogen by using H^+
>   (4) Balance charge by using electrons
> - If necessary, equalize the number of electrons in the oxidation and reduction half-equations by multiplying one or both half-equations by appropriate integers.
> - Add the half-equations, then cancel species common to both sides of the overall equation.
> - Check that both numbers of atoms and charges balance.

This first step in this method actually involves several steps. A detailed description of the method is given in Table 5.5. The method is appropriate for reactions that occur in an acidic solution. Because an acidic solution contains an excess of H^+ ions, the method uses H^+ ions in balancing the half-equations.

EXAMPLE 5-6   Balancing the Equation for a Redox Reaction in an Acidic Solution

The reaction described by expression (5.23) below is used to determine the sulfite ion concentration present in wastewater from a papermaking plant. Use the half-equation method to obtain a balanced equation for this reaction in an acidic solution.

$$SO_3^{2-}(aq) + MnO_4^-(aq) \longrightarrow SO_4^{2-}(aq) + Mn^{2+}(aq) \tag{5.23}$$

Analyze

The reaction occurs in acidic aqueous solution. We can use the method summarized in Table 5.5 to balance it.

Solve

The O.S. of sulfur increases from +4 in SO_3^{2-} to +6 in SO_4^{2-}. The O.S. of Mn decreases from +7 in MnO_4^- to +2 in Mn^{2+}. Thus, SO_3^{2-} is oxidized and MnO_4^- is reduced.

Step 1. *Write skeleton half-equations based on the species undergoing oxidation and reduction.* The half-equations are

$$SO_3^{2-}(aq) \longrightarrow SO_4^{2-}(aq)$$
$$MnO_4^-(aq) \longrightarrow Mn^{2+}(aq)$$

Step 2. *Balance each half-equation for numbers of atoms, in this order:*

- atoms other than H and O
- O atoms, by adding H_2O with the appropriate coefficient
- H atoms, by adding H^+ with the appropriate coefficient

The other atoms (S and Mn) are already balanced in the half-equations. To balance O atoms, we add one H_2O molecule to the left side of the first half-equation and four to the right side of the second.

$$SO_3^{2-}(aq) + H_2O(l) \longrightarrow SO_4^{2-}(aq)$$
$$MnO_4^-(aq) \longrightarrow Mn^{2+}(aq) + 4\,H_2O(l)$$

To balance H atoms, we add two H^+ ions to the right side of the first half-equation and eight to the left side of the second.

$$SO_3^{2-}(aq) + H_2O(l) \longrightarrow SO_4^{2-}(aq) + 2\,H^+(aq)$$
$$MnO_4^-(aq) + 8\,H^+(aq) \longrightarrow Mn^{2+}(aq) + 4\,H_2O(l)$$

(continued)

Step 3. *Balance each half-equation for electric charge.* Add the number of electrons necessary to get the same electric charge on both sides of each half-equation. By doing this, you will see that the half-equation in which electrons appear on the right side is the *oxidation half-equation*. The other half-equation, with electrons on the left side, is the *reduction half-equation*.

Oxidation:
$$SO_3^{2-}(aq) + H_2O(l) \longrightarrow SO_4^{2-}(aq) + 2\,H^+(aq) + 2\,e^-$$
$$\text{(net charge on each side, } -2)$$

Reduction:
$$MnO_4^-(aq) + 8\,H^+(aq) + 5\,e^- \longrightarrow Mn^{2+}(aq) + 4\,H_2O(l)$$
$$\text{(net charge on each side, } +2)$$

Step 4. *Obtain the overall redox equation by combining the half-equations.* Multiply the oxidation half-equation by 5 and the reduction half-equation by 2. This results in $10\,e^-$ on each side of the overall equation. These terms cancel out. *Electrons must not appear in the final equation.*

Overall:
$$5\,SO_3^{2-}(aq) + 5\,H_2O(l) \longrightarrow 5\,SO_4^{2-}(aq) + 10\,H^+(aq) + \cancel{10\,e^-}$$
$$\underline{2\,MnO_4^-(aq) + 16\,H^+(aq) + \cancel{10\,e^-} \longrightarrow 2\,Mn^{2+}(aq) + 8\,H_2O(l)}$$
$$5\,SO_3^{2-}(aq) + 2\,MnO_4^-(aq) + 5\,H_2O(l) + 16\,H^+(aq) \longrightarrow$$
$$5\,SO_4^{2-}(aq) + 2\,Mn^{2+}(aq) + 8\,H_2O(l) + 10\,H^+(aq)$$

Step 5. *Simplify.* The overall equation should not contain the same species on both sides. Subtract $5\,H_2O$ from each side of the equation in step 4. This leaves $3\,H_2O$ on the right. Also subtract $10\,H^+$ from each side, leaving $6\,H^+$ on the left.

$$5\,SO_3^{2-}(aq) + 2\,MnO_4^-(aq) + 6\,H^+(aq) \longrightarrow 5\,SO_4^{2-}(aq) + 2\,Mn^{2+}(aq) + 3\,H_2O(l)$$

Step 6. *Verify.* Check the overall equation to ensure that it is balanced both for numbers of atoms and electric charge. For example, show that in the balanced equation from step 5, the net charge on each side of the equation is -6: $(5 \times 2-) + (2 \times 1-) + (6 \times 1+) = (5 \times 2-) + (2 \times 2+) = -6$.

Assess

The final check completed in step 6 gives us confidence that our result is correct. This is an important step; always take the time to complete it. It is also worth pointing out that, in this example, there was only one atom per formula that was oxidized or reduced. (Refer to the skeleton half-equations given in step 1.) Many students have difficulty balancing half-equations in which more than one atom per formula is oxidized or reduced, as is the case when $Cr_2O_7^{2-}$ is reduced to Cr^{3+}. Had we used $Cr_2O_7^{2-}$ instead of MnO_4^- in equation (5.23), the balanced chemical equation for the reaction would have been $3\,SO_3^{2-} + Cr_2O_7^{2-} + 8\,H^+ \rightarrow 3\,SO_4^{2-} + 2\,Cr^{3+} + 4\,H_2O$.

PRACTICE EXAMPLE A: Balance the equation for this reaction in acidic solution.

$$Fe^{2+}(aq) + MnO_4^-(aq) \longrightarrow Fe^{3+}(aq) + Mn^{2+}(aq)$$

PRACTICE EXAMPLE B: Balance the equation for this reaction in acidic solution.

$$UO^{2+}(aq) + Cr_2O_7^{2-}(aq) \longrightarrow UO_2^{2+}(aq) + Cr^{3+}(aq)$$

Balancing Redox Equations in a Basic Solution

To balance equations for redox reactions in a basic solution, a step or two must be added to the procedure used in Example 5-6. The problem is that in basic solution, OH^-, not H^+, must appear in the final balanced equation. (Recall that, in basic solutions, OH^- ions are present in excess.) Because both OH^- and H_2O contain H and O atoms, at times it is hard to decide on which side of the half-equations to put each one. One simple approach is to treat the reaction as though it were occurring in an acidic solution, and balance it as in Example 5-6. Then, add to each side of the overall redox equation a number of OH^- ions equal to the number of H^+ ions. Where H^+ and OH^- appear on the same side of the equation, combine them to produce H_2O molecules. If H_2O now appears on both sides of the equation, subtract the same number of H_2O molecules from each side, leaving a remainder of H_2O on just one side. This method is illustrated in Example 5-7. For ready reference, the procedure is summarized in Table 5.6.

EXAMPLE 5-7   Balancing the Equation for a Redox Reaction in Basic Solution

Balance the equation for the reaction in which cyanide ion is oxidized to cyanate ion by permanganate ion in a basic solution, and the permanganate is itself reduced to $MnO_2(s)$.

$$MnO_4^-(aq) + CN^-(aq) \longrightarrow MnO_2(s) + OCN^-(aq) \tag{5.24}$$

Analyze

The reaction occurs in basic solution. We can balance it by using the method described in Table 5.6 on the next page. The half-reactions and the overall reaction are initially treated as though they were occurring in an acidic solution and, finally, the overall equation is adjusted to a basic solution.

Solve

Step 1. *Write half-equations for the oxidation and reduction half-reactions, and balance them for Mn, C, and N atoms.*

$$MnO_4^-(aq) \longrightarrow MnO_2(s)$$
$$CN^-(aq) \longrightarrow OCN^-(aq)$$

Step 2. *Balance the half-equations for O and H atoms. Add H_2O and/or H^+ as required.* In the MnO_4^- half-equation, there are four O's on the left and two on the right. Adding $2 H_2O$ balances the O's on the right. Since there are now four H's on the right, it is necessary to add $4 H^+$ on the left side to balance them. In the CN^- half-equation, there is one O on the right but none on the left, so H_2O must be added to the left side and $2 H^+$ to the right.

$$MnO_4^-(aq) + 4 H^+(aq) \longrightarrow MnO_2(s) + 2 H_2O(l)$$
$$CN^-(aq) + H_2O(l) \longrightarrow OCN^-(aq) + 2 H^+(aq)$$

Step 3. *Balance the half-equations for electric charge by adding the appropriate numbers of electrons.*

Reduction: $MnO_4^-(aq) + 4 H^+(aq) + 3 e^- \longrightarrow MnO_2(s) + 2 H_2O(l)$

Oxidation: $CN^-(aq) + H_2O(l) \longrightarrow OCN^-(aq) + 2 H^+(aq) + 2 e^-$

Step 4. *Combine the half-equations to obtain an overall redox equation.* Multiply the reduction half-equation by two and the oxidation half-equation by three to obtain the common multiple $6 e^-$ in each half-equation. Make the appropriate cancellations of H^+ and H_2O.

$$2 MnO_4^-(aq) + 8 H^+(aq) + \cancel{6 e^-} \longrightarrow 2 MnO_2(s) + 4 H_2O(l)$$
$$\underline{3 CN^-(aq) + 3 H_2O(l) \longrightarrow 3 OCN^-(aq) + 6 H^+(aq) + \cancel{6 e^-}}$$

Overall: $2 MnO_4^-(aq) + 3 CN^-(aq) + 2 H^+(aq) \longrightarrow$
$$2 MnO_2(s) + 3 OCN^-(aq) + H_2O(l)$$

Step 5. *Change from an acidic to a basic medium by adding $2 OH^-$ to both sides of the overall equation; combine $2 H^+$ and $2 OH^-$ to form $2 H_2O$, and simplify.*

$$2 MnO_4^-(aq) + 3 CN^-(aq) + 2 H^+(aq) + 2 OH^-(aq) \longrightarrow$$
$$2 MnO_2(s) + 3 OCN^-(aq) + H_2O(l) + 2 OH^-(aq)$$
$$2 MnO_4^-(aq) + 3 CN^-(aq) + 2 H_2O(l) \longrightarrow 2 MnO_2(s) + 3 OCN^-(aq) + H_2O(l) + 2 OH^-(aq)$$

Subtract one H_2O molecule from each side to obtain the overall balanced redox equation for reaction (5.24).

$$2 MnO_4^-(aq) + 3 CN^-(aq) + H_2O(l) \longrightarrow 2 MnO_2(s) + 3 OCN^-(aq) + 2 OH^-(aq)$$

Step 6. *Verify.* Check the final overall equation to ensure that it is balanced both for number of atoms and for electric charge. For example, show that in the balanced equation from step 5, the net charge on each side of the equation is -5.

Assess

We can use the rules for assigning oxidation states (given in Table 3.2) to deduce that manganese is reduced from $+7$ in MnO_4^- to $+4$ in MnO_2. We conclude that the other substance, CN^-, is oxidized. (The rules do not allow us to assign oxidation states to C and N in CN^- or CNO^-.) Even though we cannot identify the oxidation states of C or N, we could still balance the equation for the reaction. That is one advantage of the methods we presented in Tables 5-5 and 5-6. In Chapter 10, we will discuss another method for assigning oxidation states and learn how to determine the oxidation states of C and N in such species as CN^- or CNO^-.

(continued)

PRACTICE EXAMPLE A: Balance the equation for this reaction in basic solution.

$$S(s) + OCl^-(aq) \longrightarrow SO_3^{2-}(aq) + Cl^-(aq)$$

PRACTICE EXAMPLE B: Balance the equation for this reaction in basic solution.

$$MnO_4^-(aq) + SO_3^{2-}(aq) \longrightarrow MnO_2(s) + SO_4^{2-}(aq)$$

▶ As an alternative method for half-equations in basic solutions, add *two* OH^- for every O required to the O-deficient side, and add *one* H_2O to the other side (the net effect is adding one O to the O-deficient side). Then add *one* H_2O for every H required to the H-deficient side, and add *one* OH^- to the other side (the net effect is adding one H to the H-deficient side).

TABLE 5.6 Balancing Equations for Redox Reactions in Basic Aqueous Solutions by the Half-Equation Method: A Summary

- Balance the equation as if the reaction were occurring in acidic medium by using the method for acidic aqueous solutions summarized in Table 5-5.
- Add a number of OH^- ions equal to the number of H^+ ions to both sides of the overall equation.
- On the side of the overall equation containing both H^+ and OH^- ions, combine them to form H_2O molecules. If H_2O molecules now appear on both sides of the overall equation, cancel the same number from each side, leaving a remainder of H_2O on just one side.
- Check that both numbers of atoms and charges balance.

Disproportionation Reactions

In some oxidation–reduction reactions, called **disproportionation reactions**, the same substance is both oxidized and reduced. An example is the decomposition of hydrogen peroxide, H_2O_2, into H_2O and O_2:

$$2\,H_2O_2(aq) \longrightarrow 2\,H_2O(l) + O_2(g) \tag{5.25}$$

In reaction (5.25), the oxidation state of oxygen changes from -1 in H_2O_2 to -2 in H_2O (a reduction) and to 0 in $O_2(g)$ (an oxidation). H_2O_2 is both oxidized and reduced. Reaction (5.25) produces $O_2(g)$, which bubbles out of the solution. (See Figure 5-14.)

Another example is the disproportionation of $S_2O_3^{2-}$ in acid solution:

$$S_2O_3^{2-}(aq) + 2\,H^+(aq) \longrightarrow S(s) + SO_2(g) + H_2O(l) \tag{5.26}$$

The oxidation states of S are $+2$ in $S_2O_3^{2-}$, 0 in S, and $+4$ in SO_2. Thus, $S_2O_3^{2-}$ is simultaneously oxidized and reduced. Solutions of sodium thiosulfate ($Na_2S_2O_3$) are often used in the laboratory in redox reactions, and stock solutions of $Na_2S_2O_3$ sometimes develop small deposits of sulfur, a pale yellow solid, over time.

The same substance appears on the left side in each half-equation for a disproportionation reaction. The balanced half-equations and overall equation for reaction (5.26) are given below.

Oxidation: $S_2O_3^{2-}(aq) + H_2O(l) \longrightarrow 2\,SO_2(g) + 2\,H^+(aq) + 4\,e^-$
Reduction: $S_2O_3^{2-}(aq) + 6\,H^+(aq) + 4\,e^- \longrightarrow 2\,S(s) + 3\,H_2O(l)$
_____
                  $2\,S_2O_3^{2-}(aq) + 4\,H^+(aq) \longrightarrow 2\,S(s) + 2\,SO_2(g) + 2\,H_2O(l)$
Overall: $S_2O_3^{2-}(aq) + 2\,H^+(aq) \longrightarrow S(s) + SO_2(g) + H_2O(l)$

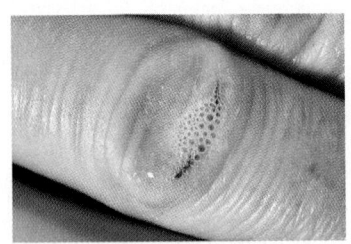

▲ FIGURE 5-14
Antiseptic action of hydrogen peroxide solution
Dilute aqueous solutions of hydrogen peroxide (usually 3% hydrogen peroxide by mass) were once commonly used to clean minor cuts and scrapes. A solution of hydrogen peroxide bubbles when poured over a cut because of the production of gaseous oxygen.

5-5   CONCEPT ASSESSMENT

Is it possible for two different reactants in a redox reaction to yield a single product? Explain.

5-2   ARE YOU WONDERING...

How to balance the equation for a redox reaction that occurs in a medium other than an aqueous solution?

An example of such a reaction is the oxidation of $NH_3(g)$ to $NO(g)$, the first step in the commercial production of nitric acid.

$$NH_3(g) + O_2(g) \longrightarrow NO(g) + H_2O(g)$$

Some people prefer a method called the *oxidation-state change method* (*see below*) for reactions of this type, but the half-equation method works just as well. All that is needed is to treat the reaction as if it were occurring in an aqueous acidic solution; H^+ should appear as both a reactant and a product and cancel out in the overall equation, as seen below.

Oxidation: $4\{NH_3 + H_2O \longrightarrow NO + 5\,H^+ + 5\,e^-\}$

Reduction: $5\{O_2 + 4\,H^+ + 4\,e^- \longrightarrow 2\,H_2O\}$

Overall: $4\,NH_3 + 5\,O_2 \longrightarrow 4\,NO + 6\,H_2O$

Oxidation-state change method

In this method, changes in oxidation states are identified. That of nitrogen increases from -3 in NH_3 to $+2$ in NO, corresponding to a "loss" of five electrons per N atom. That of oxygen decreases from 0 in O_2 to -2 in NO and H_2O, corresponding to a "gain" of two electrons per O atom. The proportion of N to O atoms must be 2 N (loss of $10\,e^-$) to 5 O (gain of $10\,e^-$).

$$2\,NH_3 + \frac{5}{2}\,O_2 \longrightarrow 2\,NO + 3\,H_2O \quad \text{or} \quad 4\,NH_3 + 5\,O_2 \longrightarrow 4\,NO + 6\,H_2O$$

5-6   Oxidizing and Reducing Agents

Chemists frequently use the terms *oxidizing agent* and *reducing agent* to describe certain of the reactants in redox reactions, as in statements like "fluorine gas is a powerful oxidizing agent," or "calcium metal is a good reducing agent." Let us briefly consider the meaning of these terms.

In a redox reaction, the substance that makes it possible for some other substance to be oxidized is called the **oxidizing agent**, or **oxidant**. In doing so, the oxidizing agent is itself reduced. Similarly, the substance that causes some other substance to be reduced is called the **reducing agent**, or **reductant**. In the reaction, the reducing agent is itself oxidized. Or, stated in other ways,

An oxidizing agent (oxidant)

- causes another substance to be oxidized
- contains an element whose oxidation state *decreases* in a redox reaction
- *gains* electrons (electrons are found on the left side of its half-equation)
- is reduced

A reducing agent (reductant)

- causes another substance to be reduced
- contains an element whose oxidation state *increases* in a redox reaction
- *loses* electrons (electrons are found on the right side of its half-equation)
- is oxidized

In general, a substance with an element in one of its highest possible oxidation states is an oxidizing agent. If the element is in one of its lowest possible oxidation states, the substance is a reducing agent.

Species	O.S.
NO_3^-	+5
N_2O_4	+4
NO_2^-	+3
NO	+2
N_2O	+1
N_2	0
NH_2OH	−1
N_2H_4	−2
NH_3	−3

This species cannot be oxidized further.

This species cannot be reduced further.

▶ FIGURE 5-15
Oxidation states of nitrogen: Identifying oxidizing and reducing agents
In NO_3^- and N_2O_4, nitrogen is in one of its highest possible oxidation states (O.S). These species are usually oxidizing agents in redox reactions. In N_2H_4 and NH_3, nitrogen is in one of its lowest oxidation states. These species are usually reducing agents.

Figure 5-15 shows the range of oxidation states of nitrogen and the species to which they correspond. The oxidation state of the nitrogen in dinitrogen tetroxide (N_2O_4) is nearly the maximum value attainable, and hence N_2O_4 is generally an oxidizing agent. Conversely, the nitrogen atom in hydrazine (N_2H_4) is in nearly the lowest oxidation state, and hence hydrazine is generally a reducing agent. When these two liquid compounds are mixed, a vigorous reaction takes place:

$$N_2O_4(l) + 2 N_2H_4(l) \longrightarrow 3 N_2(g) + 4 H_2O(g)$$

In this reaction, N_2O_4 is the oxidizing agent and N_2H_4 is the reducing agent. This reaction releases so much energy that it is used in some rocket propulsion systems.

Certain substances in which the oxidation state of an element is between its highest and lowest possible values may act as oxidizing agents in some instances and reducing agents in others. For example, in the reaction of hydrazine with hydrogen to produce ammonia, hydrazine acts as an oxidizing agent.

$$N_2H_4(l) + H_2(g) \longrightarrow 2 NH_3(g)$$

Permanganate ion, MnO_4^-, is a versatile oxidizing agent that has many uses in the chemical laboratory. In the next section, we describe its use in the quantitative analysis of iron—that is, the determination of the exact (quantitative) amount of iron in an iron-containing material. Ozone, $O_3(g)$, a triatomic form of oxygen, is an oxidizing agent used in water purification, as in the oxidation of the organic compound phenol, C_6H_5OH.

$$C_6H_5OH(aq) + 14 O_3(g) \longrightarrow 6 CO_2(g) + 3 H_2O(l) + 14 O_2(g)$$

▲ FIGURE 5-16
Bleaching action of NaOCl(aq)
A red cloth becomes white when immersed in NaOCl(aq), which oxidizes the red pigment to colorless products.

Aqueous sodium hypochlorite, NaOCl(aq), is a powerful oxidizing agent. It is the active ingredient in many liquid chlorine bleaches. The bleaching action of NaOCl(aq) is associated with the reduction of the OCl^- ion to Cl^-; the electrons required for the reduction come from colored compounds in stains. The bleaching action of NaOCl(aq) is demonstrated in Figure 5-16.

Thiosulfate ion, $S_2O_3^{2-}$, is an important reducing agent. One of its industrial uses is as an antichlor to destroy residual chlorine from the bleaching of fibers.

$$S_2O_3^{2-}(aq) + 4 Cl_2(aq) + 5 H_2O(l) \longrightarrow 2 HSO_4^-(aq) + 8 H^+(aq) + 8 Cl^-(aq)$$

Oxidizing and reducing agents also play important roles in biological systems—in photosynthesis (using solar energy to synthesize glucose), metabolism (oxidizing glucose), and the transport of oxygen.

EXAMPLE 5-8   Identifying Oxidizing and Reducing Agents

Hydrogen peroxide, H_2O_2, is a versatile chemical. Its uses include bleaching wood pulp and fabrics and substituting for chlorine in water purification. One reason for its versatility is that it can be either an oxidizing or a reducing agent. For the following reactions, identify whether hydrogen peroxide is an oxidizing or reducing agent.

(a) $H_2O_2(aq) + 2\,Fe^{2+}(aq) + 2\,H^+(aq) \longrightarrow 2\,H_2O(l) + 2\,Fe^{3+}(aq)$

(b) $5\,H_2O_2(aq) + 2\,MnO_4^-(aq) + 6\,H^+(aq) \longrightarrow 8\,H_2O(l) + 2\,Mn^{2+}(aq) + 5\,O_2(g)$

Analyze

Before we can identify the oxidizing and reducing agents, we must first assign oxidation states, and then identify which substance is being oxidized and which substance is being reduced. The oxidizing agent causes another substance to be oxidized. The reducing agent causes another substance to be reduced.

Solve

(a) Fe^{2+} is oxidized to Fe^{3+} and because H_2O_2 makes this possible, it is an oxidizing agent. Viewed another way, we see that the oxidation state of oxygen in H_2O_2 is -1. In H_2O, it is -2. Hydrogen peroxide is reduced and thereby acts as an oxidizing agent.

(b) MnO_4^- is reduced to Mn^{2+}, and H_2O_2 makes this possible. In this situation, hydrogen peroxide is a reducing agent. Or, the oxidation state of oxygen increases from -1 in H_2O_2 to 0 in O_2. Hydrogen peroxide is oxidized and thereby acts as a reducing agent.

Assess

The versatility of H_2O_2 lies in its ability to act as an oxidizing agent and a reducing agent. When H_2O_2 acts as an oxidizing agent, it is reduced to H_2O, in an acidic solution, as was the case in part (a), or to OH^- in basic solution. When it acts as a reducing agent, it is oxidized to $O_2(g)$, as was the case in part (b).

PRACTICE EXAMPLE A: Is $H_2(g)$ an oxidizing or reducing agent in this reaction? Explain.

$$2\,NO_2(g) + 7\,H_2(g) \longrightarrow 2\,NH_3(g) + 4\,H_2O(g)$$

PRACTICE EXAMPLE B: Identify the oxidizing agent and the reducing agent in the reaction.

$$4\,Au(s) + 8\,CN^-(aq) + O_2(g) + 2\,H_2O(l) \longrightarrow 4[Au(CN)_2]^-(aq) + 4\,OH^-(aq)$$

5-6   CONCEPT ASSESSMENT

A newspaper account of an accidental spill of hydrochloric acid in an area where sodium hydroxide solution was also stored spoke of the potential hazardous release of chlorine gas if the two solutions should come into contact. Was this an accurate accounting of the hazard involved? Explain.

5-7   Stoichiometry of Reactions in Aqueous Solutions: Titrations

If our objective is to obtain the maximum yield of a product at the lowest cost, we would generally choose the most expensive reactant as the limiting reactant and use excess amounts of the other reactants. This is the case in most precipitation reactions. In some instances, as in determining the concentration of a solution, we may not be interested in the products of a reaction but only in the relationship between two reactants. Then we have to carry out the reaction in such a way that neither reactant is in excess. A method that has long been used for doing this is known as *titration*. The glassware typically used in a titration is shown in Figure 5-17.

A solution of one reactant is placed in a small beaker or flask. Another reactant, also in solution and commonly referred to as titrant, is in a *buret*, a long,

(a)   (b)   (c)

▲ FIGURE 5-17
An acid–base titration—Example 5-9 illustrated
(a) A 5.00 mL sample of vinegar, a small quantity of water, and a few drops of
phenolphthalein indicator are added to a flask. **(b)** 0.1000 M NaOH from a previously
filled buret is slowly added. **(c)** As long as the acid is in excess, the solution in the flask
remains colorless. When the acid has been neutralized, an additional drop of NaOH(aq)
causes the solution to become slightly basic. The phenolphthalein indicator turns a
light pink. The first lasting appearance of the pink color is taken to be the equivalence
point of the titration.

▶ The key to a successful
acid–base titration is in selecting the right indicator. We
learn how to do this when we
consider theoretical aspects of
titration in Chapter 17.

graduated tube equipped with a stopcock valve. The second solution is slowly
added to the first by manipulating the stopcock. **Titration** is a reaction carried
out by the carefully controlled addition of one solution to another. The trick is
to stop the titration at the point where both reactants have reacted completely,
a condition called the **equivalence point** of the titration. Key to every titration
is that at the equivalence point, the two reactants have combined in stoichiometric proportions; both have been consumed, and neither remains in excess.

In modern chemical laboratories, appropriate measuring instruments are used
to signal when the equivalence point is reached. Still widely used, though, is a
technique in which a very small quantity of a substance added to the reaction
mixture changes color at or very near the equivalence point. Such substances are
called **indicators**. Figure 5-17 illustrates the neutralization of an acid by a base by
the titration technique. Calculations that use titration data are much the same as
those introduced in Section 4-3, and they are also illustrated in Example 5-9.

Suppose we need a $KMnO_4(aq)$ solution of exactly known molarity, close
to 0.020 M. We cannot prepare this solution by weighing out the required
amount of $KMnO_4(s)$ and dissolving it in water. The solid is not pure, and its
actual purity (that is, the mass percent $KMnO_4$) is *not known*. Conversely, we
can obtain iron wire in essentially pure form and allow the wire to react with
an acid to yield $Fe^{2+}(aq)$. $Fe^{2+}(aq)$ is oxidized to $Fe^{3+}(aq)$ by $KMnO_4(aq)$ in
an acidic solution. By determining the volume of $KMnO_4(aq)$ required to oxidize a known quantity of $Fe^{2+}(aq)$, we can calculate the exact molarity of the
$KMnO_4(aq)$. This procedure, which determines the exact molarity of a solution, is called **standardization of a solution**. It is illustrated in Example 5-10
and Figure 5-18.

▶ *Standardize* means
determine the concentration
of a solution, usually to
three or four significant
figures. It is not so important
that the concentration be a
round number (as 0.1000 vs.
0.1035 M), but rather that the
concentration be accurately
known.

EXAMPLE 5-9   Using Titration Data to Establish the Concentrations of Acids and Bases

Vinegar is a dilute aqueous solution of acetic acid produced by the bacterial fermentation of apple cider, wine, or other carbohydrate material. The legal minimum acetic acid content of vinegar is 4% by mass. A 5.00 mL sample of a particular vinegar is titrated with 38.08 mL of 0.1000 M NaOH. Does this sample exceed the minimum limit? (Vinegar has a density of about 1.01 g/mL.)

Analyze

Acetic acid, CH_3COOH, is a weak acid and NaOH is a strong base. The reaction between CH_3COOH and NaOH is an acid–base neutralization reaction. We start by writing a balanced chemical equation for the reaction. We must convert mL NaOH to CH_3COOH. The necessary conversions are as follows:

$$mL\ NaOH \longrightarrow L\ NaOH \longrightarrow mol\ NaOH \longrightarrow mol\ CH_3COOH \longrightarrow g\ CH_3COOH$$

Solve

The balanced chemical equation for the reaction is given below.

$$CH_3COOH(aq) + NaOH(aq) \longrightarrow NaCH_3COO(aq) + H_2O(l)$$

$$? \text{ g } HC_2H_3O_2 = 38.08 \text{ mL} \times \frac{1\ L}{1000\ mL} \times \frac{0.1000\ mol\ NaOH}{1\ L}$$

$$\times \frac{1\ mol\ CH_3COOH}{1\ mol\ NaOH} \times \frac{60.05\ g\ CH_3COOH}{1\ mol\ CH_3COOH}$$

$$= 0.2287 \text{ g } CH_3COOH$$

This mass of CH_3COOH is found in 5.00 mL of vinegar of density 1.01 g/mL. The percent mass of CH_3COOH is

$$\% \ CH_3COOH = \frac{0.2287\ g\ CH_3COOH}{5.00\ mL\ vinegar} \times \frac{1\ mL\ vinegar}{1.01\ g\ vinegar} \times 100\%$$

$$= 4.53\% \ CH_3COOH$$

The vinegar sample exceeds the legal minimum limit but only slightly. There is also a standard for the maximum amount of acetic acid allowed in vinegar. A vinegar producer might use this titration technique to ensure that the vinegar stays between these limits.

Assess

Such problems as this one involve many steps or conversions. Try to break the problem into simpler ones involving fewer steps or conversions. It may also help to remember that solving a stoichiometry problem involves three steps: (1) converting to moles, (2) converting between moles, and (3) converting from moles. Use molarities and molar masses to carry out volume–mole conversions and gram–mole conversions, respectively, and stoichiometric factors to carry out mole–mole conversions. The stoichiometric factors are constructed from a balanced chemical equation.

PRACTICE EXAMPLE A: A particular solution of NaOH is supposed to be approximately 0.100 M. To determine the exact molarity of the NaOH(aq), a 0.5000 g sample of $KHC_8H_4O_4$ is dissolved in water and titrated with 24.03 mL of the NaOH(aq). What is the actual molarity of the NaOH(aq)?

$$HC_8H_4O_4^-(aq) + OH^-(aq) \longrightarrow C_8H_4O_4^{2-}(aq) + H_2O(l)$$

PRACTICE EXAMPLE B: A 0.235 g sample of a solid that is 92.5% NaOH and 7.5% $Ca(OH)_2$, by mass, requires 45.6 mL of a HCl(aq) solution for its titration. What is the molarity of the HCl(aq)?

🔍 5-7   CONCEPT ASSESSMENT

A 10.00 mL sample of 0.311 M KOH is added to 31.10 mL of 0.100 M HCl.
Is the resulting mixture acidic, basic, or exactly neutral? Explain.

| (a) | (b) | (c) |

▲ FIGURE 5-18
Standardizing a solution of an oxidizing agent through a redox titration—Example 5-10 illustrated
(a) The solution contains a known amount of Fe^{2+}, and the buret is filled with the intensely colored $KMnO_4(aq)$ to be standardized. (b) As it is added to the strongly acidic solution of $Fe^{2+}(aq)$, the $KMnO_4(aq)$ is immediately decolorized as a result of reaction (5.27). (c) When all the Fe^{2+} has been oxidized to Fe^{3+}, additional $KMnO_4(aq)$ has nothing left to oxidize and the solution turns a distinctive pink. Even a fraction of a drop of the $KMnO_4(aq)$ beyond the equivalence point is sufficient to cause this pink coloration.

EXAMPLE 5-10   Standardizing a Solution for Use in Redox Titrations

A piece of iron wire weighing 0.1568 g is converted to $Fe^{2+}(aq)$ and requires 26.24 mL of a $KMnO_4(aq)$ solution for its titration. What is the molarity of the $KMnO_4(aq)$?

$$5\,Fe^{2+}(aq) + MnO_4^-(aq) + 8\,H^+(aq) \longrightarrow 5\,Fe^{3+}(aq) + Mn^{2+}(aq) + 4\,H_2O(l) \qquad \textbf{(5.27)}$$

Analyze

The key to a titration calculation is that the amounts of two reactants consumed in the titration are stoichiometrically equivalent—neither reactant is in excess. We are given an amount of Fe (0.156 g) and must determine the number of moles of $KMnO_4$ in the 26.24 mL sample. The following conversions are required:

$$\text{g Fe} \longrightarrow \text{mol Fe} \longrightarrow \text{mol Fe}^{2+} \longrightarrow \text{mol MnO}_4^- \longrightarrow \text{mol KMnO}_4$$

The third conversion, from mol Fe^{2+} to mol MnO_4^-, requires a stoichiometric factor constructed from the coefficients in equation (5.27).

Solve

First, determine the amount (in moles) of $KMnO_4$ consumed in the titration.

$$? \text{ mol KMnO}_4 = 0.1568 \text{ g Fe} \times \frac{1 \text{ mol Fe}}{55.847 \text{ g Fe}} \times \frac{1 \text{ mol Fe}^{2+}}{1 \text{ mol Fe}}$$

$$\times \frac{1 \text{ mol MnO}_4^-}{5 \text{ mol Fe}^{2+}} \times \frac{1 \text{ mol KMnO}_4}{1 \text{ mol MnO}_4^-}$$

$$= 5.615 \times 10^{-4} \text{ mol KMnO}_4$$

The volume of solution containing the 5.615×10^{-4} mol $KMnO_4$ is 26.24 mL = 0.02624 L, which means that

$$\text{concn } KMnO_4 = \frac{5.615 \times 10^{-4} \text{ mol } KMnO_4}{0.02624 \text{ L}} = 0.02140 \text{ M } KMnO_4$$

Assess

For practical applications, such as for titrations, we use solutions with molarities that are neither very large nor very small. Typically, the molarities lie in the range 0.001 M to 0.1 M. If you calculate a molarity that is significantly larger than 0.1 M, or significantly smaller than 0.001 M, then you must carefully check your calculation for possible errors.

PRACTICE EXAMPLE A: A 0.376 g sample of an iron ore is dissolved in acid, and the iron reduced to $Fe^{2+}(aq)$ and then titrated with 41.25 mL of 0.02140 M $KMnO_4$. Determine the mass percent Fe in the iron ore. [*Hint:* Use equation (5.27).]

PRACTICE EXAMPLE B: Another substance that may be used to standardize $KMnO_4(aq)$ is sodium oxalate. If 0.2482 g $Na_2C_2O_4$ is dissolved in water and titrated with 23.68 mL $KMnO_4$, what is the molarity of the $KMnO_4(aq)$?

$$MnO_4^-(aq) + C_2O_4^{2-}(aq) + H^+(aq) \longrightarrow Mn^{2+}(aq) + H_2O(l) + CO_2(g) \quad \text{(not balanced)}$$

Mastering**CHEMISTRY**   **www.masteringchemistry.com**

Access to a plentiful supply of "pure water" is something most of us take for granted. Still, most of us would agree that purification of water is an important concern. The way in which water is purified depends on how it is to be used or on how it has been used. For a discussion of the removal or destruction of undesirable chemical substances from water, go to the Focus On feature for Chapter 5, entitled Water Treatment, on the MasteringChemistry site.

Summary

5-1 The Nature of Aqueous Solutions—Solutes in aqueous solution are characterized as **nonelectrolytes**, which do not produce ions, or **electrolytes**, which produce ions. **Weak electrolytes** dissociate to a limited extent, and **strong electrolytes** dissociate almost completely into ions. In addition to the molarity based on the solute as a whole, a solution's concentration can be stated in terms of the molarities of the individual solute species present—molecules and ions. Calculating ion concentrations in a strong electrolyte solution is easily done.

5-2 Precipitation Reactions—Some reactions in aqueous solution involve the combination of ions to yield a water-insoluble solid—a **precipitate**. Precipitation reactions are generally represented by **net ionic equations**, a form in which only the reacting ions and solid precipitates are shown, and **spectator ions** are deleted. Precipitation reactions usually can be predicted by using a few simple solubility guidelines (Table 5.1).

5-3 Acid–Base Reactions—According to the Arrhenius theory, a substance that ionizes to produce H^+ ions in aqueous solution is an **acid**. It is a **strong acid** (Table 5.2) if the ionization goes essentially to completion and a **weak acid** if the ionization is limited. Similarly, a **base** produces OH^- ions in aqueous solution and is either a **strong base** (Table 5.2) or a **weak base**, depending on the extent of the ionization. According to the Brønsted-Lowry

theory, in an acid–base reaction, protons are transferred from the acid (the **proton donor**) to the base (the **proton acceptor**). In a typical acid–base, or **neutralization**, **reaction**, H^+ ions from the acid and OH^- ions from the base combine to form HOH (water). The other product of the reaction is an ionic compound, a **salt**. Some reactions in which gases are evolved can also be treated as acid–base reactions (Table 5.3).

5-4 Oxidation–Reduction Reactions: Some General Principles—In an **oxidation–reduction (redox) reaction** certain atoms undergo an increase in oxidation state, a process called **oxidation**. Other atoms undergo a decrease in oxidation state, or **reduction**. Another useful view of redox reactions is as the combination of separate **half-reactions** for the oxidation and the reduction.

5-5 Balancing Oxidation–Reduction Equations—An effective way to balance a redox equation is to break down the reaction into separate half-reactions, write and balance half-equations for these half-reactions, and recombine the balanced half-equations into an overall balanced equation (Table 5.5). A slight variation of this method is used for a reaction that occurs in a basic aqueous solution (Table 5.6). A redox reaction in which the same substance is both oxidized and reduced is called a **disproportionation reaction**.

5-6 Oxidizing and Reducing Agents—The **oxidizing agent (oxidant)** is the key reactant in an oxidation half-reaction and is *reduced* in the redox reaction. The **reducing agent (reductant)** is the key reactant in a reduction half-reaction and is *oxidized* in the redox reaction. Some substances act only as oxidizing agents; others, only as reducing agents. Many can act as either, depending on the reaction (Fig. 5-15).

5-7 Stoichiometry of Reactions in Aqueous Solutions: Titrations—A common laboratory technique applicable to precipitation, acid–base, and redox reactions is **titration**. The key point in a titration is the **equivalence point**, which is assessed with the aid of an **indicator**. Titration data can be used to establish a solution's molarity, called **standardization of a solution**, or to provide other information about the compositions of samples being analyzed.

Integrative Example

Sodium dithionite, $Na_2S_2O_4$, is an important reducing agent. One interesting use is the reduction of chromate ion to insoluble chromium(III) hydroxide by dithionite ion, $S_2O_4^{2-}$, in basic solution. Sulfite ion is another product. The chromate ion may be present in wastewater from a chromium-plating plant, for example. What mass of $Na_2S_2O_4$ is consumed in a reaction with 100.0 L of wastewater having $[CrO_4^{2-}] = 0.0148$ M?

◀ White solid sodium dithionite, $Na_2S_2O_4$, is added to a yellow solution of potassium chromate, $K_2CrO_4(aq)$ (left). A product of the reaction is gray-green chromium(III) hydroxide, $Cr(OH)_3(s)$ (right).

Analyze

The phrase "reduction of chromate" tells us that the reaction between CrO_4^{2-} and $S_2O_4^{2-}$ is a redox reaction. We must obtain a balanced chemical equation for the reaction by using the method summarized in Table 5.6, and then convert 100.0 L of wastewater into grams of $Na_2S_2O_4$. The necessary conversions are as follows:

$$100.0 \text{ L wastewater} \longrightarrow \text{mol } CrO_4^{2-} \longrightarrow \text{mol } S_2O_4^{2-} \longrightarrow \text{mol } Na_2S_2O_4 \longrightarrow \text{g } Na_2S_2O_4$$

Solve

1. *Write an ionic expression representing the reaction.*

$$CrO_4^{2-}(aq) + S_2O_4^{2-}(aq) + OH^-(aq) \longrightarrow$$
$$Cr(OH)_3(s) + SO_3^{2-}(aq)$$

2. *Balance the redox equation.* Begin by writing skeleton half-equations.

$$CrO_4^{2-} \longrightarrow Cr(OH)_3$$
$$S_2O_4^{2-} \longrightarrow SO_3^{2-}$$

Balance the half-equations for Cr, S, O, and H atoms as if the half-reactions occur in acidic solution.

$$CrO_4^{2-} + 5H^+ \longrightarrow Cr(OH)_3 + H_2O$$
$$S_2O_4^{2-} + 2H_2O \longrightarrow 2SO_3^{2-} + 4H^+$$

Balance the half-equations for charge, and label them as oxidation and reduction.

Oxidation: $\quad S_2O_4^{2-} + 2H_2O \longrightarrow 2SO_3^{2-} + 4H^+ + 2e^-$
Reduction: $CrO_4^{2-} + 5H^+ + 3e^- \longrightarrow Cr(OH)_3 + H_2O$

Combine the half-equations into an overall equation.

$$3 \times [S_2O_4^{2-} + 2H_2O \longrightarrow 2SO_3^{2-} + 4H^+ + 2e^-]$$
$$2 \times [CrO_4^{2-} + 5H^+ + 3e^- \longrightarrow Cr(OH)_3 + H_2O]$$
$$\overline{3S_2O_4^{2-} + 2CrO_4^{2-} + 4H_2O \longrightarrow}$$
$$6SO_3^{2-} + 2Cr(OH)_3 + 2H^+$$

3. *Change the conditions to basic solution.* Add $2OH^-$ to each side of the equation for acidic solution, and combine $2H^+$ and $2OH^-$ to form $2H_2O$ on the right.

$$3S_2O_4^{2-} + 2CrO_4^{2-} + 4H_2O + 2OH^- \longrightarrow$$
$$6SO_3^{2-} + 2Cr(OH)_3 + 2H_2O$$

Subtract $2H_2O$ from each side of the equation to obtain the final balanced equation.

$$3S_2O_4^{2-}(aq) + 2CrO_4^{2-}(aq) + 2H_2O(l) + 2OH^-(aq) \longrightarrow$$
$$6SO_3^{2-}(aq) + 2Cr(OH)_3(s)$$

4. *Complete the stoichiometric calculation.* The conversion pathway is

100.0 L waste water $\longrightarrow$ mol $CrO_4{}^{2-}$ $\longrightarrow$

mol $S_2O_4{}^{2-}$ $\longrightarrow$ mol $Na_2S_2O_4$ $\longrightarrow$ g $Na_2S_2O_4$.

$$? \text{ g Na}_2\text{S}_2\text{O}_4 = 100.0 \text{ L} \times \frac{0.0148 \text{ mol CrO}_4{}^{2-}}{1 \text{ L}}$$

$$\times \frac{3 \text{ mol S}_2\text{O}_4{}^{2-}}{2 \text{ mol CrO}_4{}^{2-}} \times \frac{1 \text{ mol Na}_2\text{S}_2\text{O}_4}{1 \text{ mol S}_2\text{O}_4{}^{2-}}$$

$$\times \frac{174.1 \text{ g Na}_2\text{S}_2\text{O}_4}{1 \text{ mol Na}_2\text{S}_2\text{O}_4} = 387 \text{ g Na}_2\text{S}_2\text{O}_4$$

Assess

In solving this problem the major effort was to balance a redox equation for a reaction under basic conditions. This allowed us to find the molar relationship between dithionite and chromate ions. The remainder of the problem was a stoichiometry calculation for a reaction in solution, much like Example 4-11 (page 128). A quick check of the final result involves (1) ensuring that the redox equation is balanced, and (2) noting that the number of moles of $CrO_4{}^{2-}$ is about 1.5 (i.e., 100×0.0148), that the number of moles of $S_2O_4{}^{2-}$ is about 2.25 (i.e., $1.5 \times 3/2$), and that the mass of $Na_2S_2O_4$ is somewhat more than 350 (i.e., 2.25×175).

PRACTICE EXAMPLE A: The amount of potassium chlorate, $KClO_3$, in a 0.1432 g sample was determined as follows. The sample was dissolved in 50.00 mL of 0.09101 M $Fe(NO_3)_2$ and the solution was acidified. The excess Fe^{2+} was back-titrated with 12.59 mL of 0.08362 M $Ce(NO_3)_4$ solution. What is the percentage by mass of $KClO_3$ in the sample? Chemical equations for the reactions involved are as follows:
$ClO_3{}^-(aq) + Fe^{2+}(aq) \rightarrow Cl^-(aq) + Fe^{3+}(aq)$ (not balanced) and $Fe^{2+}(aq) + Ce^{4+}(aq) \rightarrow Fe^{3+}(aq) + Ce^{3+}(aq)$.

PRACTICE EXAMPLE B: The amount of arsenic, As, in a 7.25 g sample was determined by converting all the arsenic in the sample to arsenous acid (H_3AsO_3), and then titrating H_3AsO_3 with 23.77 mL of 0.02144 M $KMnO_4$. What is the percentage by mass of As in the sample? The unbalanced chemical equation for the titration reaction is $H_3AsO_3(aq) + MnO_4{}^-(aq) \rightarrow H_3AsO_4(aq) + Mn^{2+}(aq)$.

Mastering**CHEMISTRY**

You'll find a link to additional self study questions in the study area on www.masteringchemistry.com

Exercises

Strong Electrolytes, Weak Electrolytes, and Nonelectrolytes

1. Using information from this chapter, indicate whether each of the following substances in aqueous solution is a nonelectrolyte, weak electrolyte, or strong electrolyte. **(a)** HC_6H_5O; **(b)** Li_2SO_4; **(c)** MgI_2; **(d)** $(CH_3CH_2)_2O$; **(e)** $Sr(OH)_2$.

2. Select the **(a)** best and **(b)** poorest electrical conductors from the following solutions, and explain the reason for your choices: 0.10 M NH_3; 0.10 M NaCl; 0.10 M CH_3COOH (acetic acid); 0.10 M CH_3CH_2OH (ethanol).

3. What response would you expect in the apparatus of Figure 5-4 if the solution tested were 1.0 M HCl? What response would you expect if the solution were both 1.0 M HCl and 1.0 M CH_3COOH?

4. $NH_3(aq)$ conducts electric current only weakly. The same is true for $CH_3COOH(aq)$. When these solutions are mixed, however, the resulting solution is a good conductor. How do you explain this?

5. Sketches **(a–c)** are molecular views of the solute in an aqueous solution. For each of the sketches, indicate whether the solute is a strong, weak, or nonelectrolyte; and which of these substances it is: sodium chloride, propionic acid, hypochlorous acid, ammonia, barium bromide, ammonium chloride, methanol.

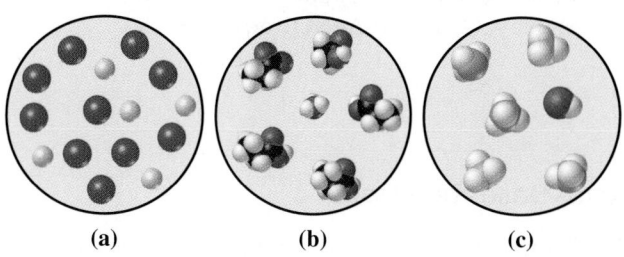

(a)   **(b)**   **(c)**

6. After identifying the three substances represented by the sketches in Exercise 5, sketch molecular views of aqueous solutions of the remaining four substances listed.

Ion Concentrations

7. Determine the concentration of the ion indicated in each solution. **(a)** $[K^+]$ in 0.238 M KNO_3; **(b)** $[NO_3^-]$ in 0.167 M $Ca(NO_3)_2$; **(c)** $[Al^{3+}]$ in 0.083 M $Al_2(SO_4)_3$; **(d)** $[Na^+]$ in 0.209 M Na_3PO_4.

8. Which solution has the greatest $[SO_4^{2-}]$: **(a)** 0.075 M H_2SO_4; **(b)** 0.22 M $MgSO_4$; **(c)** 0.15 M Na_2SO_4; **(d)** 0.080 M $Al_2(SO_4)_3$; **(e)** 0.20 M $CuSO_4$?

9. A solution is prepared by dissolving 0.132 g $Ba(OH)_2 \cdot 8 H_2O$ in 275 mL of water solution. What is $[OH^-]$ in this solution?

10. A solution is 0.126 M KCl and 0.148 M $MgCl_2$. What are $[K^+]$, $[Mg^{2+}]$, and $[Cl^-]$ in this solution?

11. Express the following data for cations in solution as molarities. **(a)** 14.2 mg Ca^{2+}/L; **(b)** 32.8 mg K^+/100 mL; **(c)** $225\mu g$ Zn^{2+}/mL.

12. What molarity of NaF(aq) corresponds to a fluoride ion content of 0.9 mg F^-/L, the federal government's recommended limit for fluoride ion in drinking water?

13. Which of the following aqueous solutions has the highest concentration of K^+? **(a)** 0.0850 M K_2SO_4; **(b)** a solution containing 1.25 g KBr/100 mL; **(c)** a solution having 8.1 mg K^+/mL.

14. Which aqueous solution has the greatest $[H^+]$: **(a)** 0.011 M CH_3COOH; **(b)** 0.010 M HCl; **(c)** 0.010 M H_2SO_4; **(d)** 1.00 M NH_3? Explain your choice.

15. How many milligrams of MgI_2 must be added to 250.0 mL of 0.0876 M KI to produce a solution with $[I^-] = 0.1000$ M?

16. If 25.0 mL H_2O evaporates from 1.00 L of a solution containing 12.0 mg K_2SO_4/mL, what is $[K^+]$ in the solution that remains?

17. Assuming the volumes are additive, what is the $[Cl^-]$ in a solution obtained by mixing 225 mL of 0.625 M KCl and 615 mL of 0.385 M $MgCl_2$?

18. Assuming the volumes are additive, what is the $[NO_3^-]$ in a solution obtained by mixing 275 mL of 0.283 M KNO_3, 328 mL of 0.421 M $Mg(NO_3)_2$, and 784 mL of H_2O?

Predicting Precipitation Reactions

19. Complete each of the following as a net ionic equation, indicating whether a precipitate forms. If no reaction occurs, so state.
    (a) $Na^+ + Br^- + Pb^{2+} + 2 NO_3^- \longrightarrow$
    (b) $Mg^{2+} + 2 Cl^- + Cu^{2+} + SO_4^{2-} \longrightarrow$
    (c) $Fe^{3+} + 3 NO_3^- + Na^+ + OH^- \longrightarrow$

20. Complete each of the following as a net ionic equation. If no reaction occurs, so state.
    (a) $Ca^{2+} + 2 I^- + 2 Na^+ + CO_3^{2-} \longrightarrow$
    (b) $Ba^{2+} + S^{2-} + 2 Na^+ + SO_4^{2-} \longrightarrow$
    (c) $2 K^+ + S^{2-} + Ca^{2+} + 2 Cl^- \longrightarrow$

21. Predict in each case whether a reaction is likely to occur. If so, write a net ionic equation.
    (a) $HI(aq) + Zn(NO_3)_2(aq) \longrightarrow$
    (b) $CuSO_4(aq) + Na_2CO_3(aq) \longrightarrow$
    (c) $Cu(NO_3)_2(aq) + Na_3PO_4(aq) \longrightarrow$

22. Predict in each case whether a reaction is likely to occur. If so, write a net ionic equation.
    (a) $AgNO_3(aq) + CuCl_2(aq) \longrightarrow$
    (b) $Na_2S(aq) + FeCl_2(aq) \longrightarrow$
    (c) $Na_2CO_3(aq) + AgNO_3(aq) \longrightarrow$

23. What reagent solution might you use to separate the cations in the following mixtures, that is, with one ion appearing in solution and the other in a precipitate? [*Hint:* Refer to Table 5.1, and consider water also to be a reagent.]
    (a) $BaCl_2(s)$ and $MgCl_2(s)$
    (b) $MgCO_3(s)$ and $Na_2CO_3(s)$
    (c) $AgNO_3(s)$ and $Cu(NO_3)_2(s)$

24. What reagent solution might you use to separate the cations in each of the following mixtures? [*Hint:* Refer to Exercise 23.]
    (a) $PbSO_4(s)$ and $Cu(NO_3)_2(s)$
    (b) $Mg(OH)_2(s)$ and $BaSO_4(s)$
    (c) $PbCO_3(s)$ and $CaCO_3(s)$

25. You are provided with NaOH(aq), K_2SO_4(aq), $Mg(NO_3)_2$(aq), $BaCl_2$(aq), NaCl(aq), $Sr(NO_3)_2$(aq), $AgNO_3$(aq), and $BaSO_4(s)$. Write net ionic equations to show how you would use one or more of those reagents to obtain **(a)** $SrSO_4(s)$; **(b)** $Mg(OH)_2(s)$; **(c)** KCl(aq).

26. Write net ionic equations to show how you would use one or more of the reagents in Exercise 25 to obtain **(a)** $BaSO_4(s)$; **(b)** AgCl(s); **(c)** KNO_3(aq).

Acid–Base Reactions

27. Complete each of the following as a *net ionic equation*. If no reaction occurs, so state.
    (a) $Ba^{2+} + 2 OH^- + CH_3COOH \longrightarrow$
    (b) $H^+ + Cl^- + CH_3CH_2COOH \longrightarrow$
    (c) $FeS(s) + H^+ + I^- \longrightarrow$
    (d) $K^+ + HCO_3^- + H^+ + NO_3^- \longrightarrow$
    (e) $Mg(s) + H^+ \longrightarrow$

28. Every antacid contains one or more ingredients capable of reacting with excess stomach acid (HCl).

The essential neutralization products are CO_2 and/or H_2O. Write net ionic equations to represent the neutralizing action of the following popular antacids.
    (a) Alka-Seltzer (sodium bicarbonate)
    (b) Tums (calcium carbonate)
    (c) milk of magnesia (magnesium hydroxide)
    (d) Maalox (magnesium hydroxide, aluminum hydroxide)
    (e) Rolaids [$NaAl(OH)_2CO_3$]

29. In this chapter, we described an acid as a substance capable of producing H^+ and a salt as the ionic compound formed by the neutralization of an acid by a base. Write ionic equations to show that sodium hydrogen sulfate has the characteristics of both a salt and an acid (sometimes called an *acid salt*).

30. A neutralization reaction between an acid and a base is a common method of preparing useful salts. Give net ionic equations showing how the following salts could be prepared in this way: **(a)** $(NH_4)_2HPO_4$; **(b)** NH_4NO_3; and **(c)** $(NH_4)_2SO_4$.

31. Which solutions would you use to precipitate Mg^{2+} from an aqueous solution of $MgCl_2$? Explain your choice. **(a)** $KNO_3(aq)$; **(b)** $NH_3(aq)$; **(c)** $H_2SO_4(aq)$; **(d)** $HC_2H_3O_2(aq)$.

32. Determine which of the following react(s) with $HCl(aq)$ to produce a gas, and write a net ionic equation(s) for the reaction(s). **(a)** Na_2SO_4; **(b)** $KHSO_3$; **(c)** $Zn(OH)_2$; **(d)** $CaCl_2$.

Oxidation–Reduction (Redox) Equations

33. Assign oxidation states to the elements involved in the following reactions. Indicate which are redox reactions and which are not.
    (a) $MgCO_3(s) + 2H^+(aq) \longrightarrow$
    $$Mg^{2+}(aq) + H_2O(l) + CO_2(g)$$
    (b) $Cl_2(aq) + 2Br^-(aq) \longrightarrow 2Cl^-(aq) + Br_2(aq)$
    (c) $Ag(s) + 2H^+(aq) + NO_3^-(aq) \longrightarrow$
    $$Ag^+(aq) + H_2O(l) + NO_2(g)$$
    (d) $2Ag^+(aq) + CrO_4^{2-}(aq) \longrightarrow Ag_2CrO_4(s)$

34. Explain why these reactions cannot occur as written.
    (a) $Fe^{3+}(aq) + MnO_4^-(aq) + H^+(aq) \longrightarrow$
    $$Mn^{2+}(aq) + Fe^{2+}(aq) + H_2O(l)$$
    (b) $H_2O_2(aq) + Cl_2(aq) \longrightarrow$
    $$ClO^-(aq) + O_2(g) + H^+(aq)$$

35. Complete and balance these half-equations.
    (a) $SO_3^{2-} \longrightarrow S_2O_3^{2-}$ (acidic solution)
    (b) $HNO_3 \longrightarrow N_2O(g)$ (acidic solution)
    (c) $Al(s) \longrightarrow Al(OH)_4^-$ (basic solution)
    Indicate whether oxidation or reduction is involved.

36. Complete and balance these half-equations.
    (a) $C_2O_4^{2-} \longrightarrow CO_2$ (acidic solution)
    (b) $Cr_2O_7^{2-} \longrightarrow Cr^{3+}$ (acidic solution)
    (c) $MnO_4^- \longrightarrow MnO_2$ (basic solution)
    Indicate whether oxidation or reduction is involved.

37. Balance these equations for redox reactions occurring in acidic solution.
    (a) $MnO_4^- + I^- \longrightarrow Mn^{2+} + I_2(s)$
    (b) $BrO_3^- + N_2H_4 \longrightarrow Br^- + N_2$
    (c) $VO_4^{3-} + Fe^{2+} \longrightarrow VO^{2+} + Fe^{3+}$
    (d) $UO^{2+} + NO_3^- \longrightarrow UO_2^{2+} + NO(g)$

38. Balance these equations for redox reactions occurring in acidic solution.
    (a) $P_4(s) + NO_3^- \longrightarrow H_2PO_4^- + NO(g)$
    (b) $S_2O_3^{2-} + MnO_4^- \longrightarrow SO_4^{2-} + Mn^{2+}$
    (c) $HS^- + HSO_3^- \longrightarrow S_2O_3^{2-}$
    (d) $Fe^{3+} + NH_3OH^+ \longrightarrow Fe^{2+} + N_2O(g)$

39. Balance these equations for redox reactions in basic solution.
    (a) $MnO_2(s) + ClO_3^- \longrightarrow MnO_4^- + Cl^-$
    (b) $Fe(OH)_3(s) + OCl^- \longrightarrow FeO_4^{2-} + Cl^-$
    (c) $ClO_2 \longrightarrow ClO_3^- + Cl^-$
    (d) $Ag(s) + CrO_4^{2-} \longrightarrow Ag^+ + Cr(OH)_3(s)$

40. Balance these equations for redox reactions occurring in basic solution.
    (a) $CrO_4^{2-} + S_2O_4^{2-} \longrightarrow Cr(OH)_3(s) + SO_3^{2-}$
    (b) $[Fe(CN)_6]^{3-} + N_2H_4 \longrightarrow [Fe(CN)_6]^{4-} + N_2(g)$

    (c) $Fe(OH)_2(s) + O_2(g) \longrightarrow Fe(OH)_3(s)$
    (d) $CH_3CH_2OH + MnO_4^- \longrightarrow$
    $$CH_3COO^- + MnO_2(s)$$

41. Balance these equations for disproportionation reactions.
    (a) $Cl_2(g) \longrightarrow Cl^- + ClO_3^-$ (basic solution)
    (b) $S_2O_4^{2-} \longrightarrow S_2O_3^{2-} + HSO_3^-$ (acidic solution)

42. Balance these equations for disproportionation reactions.
    (a) $MnO_4^{2-} \longrightarrow MnO_2(s) + MnO_4^-$ (basic solution)
    (b) $P_4(s) \longrightarrow H_2PO_2^- + PH_3(g)$ (basic solution)
    (c) $S_8(s) \longrightarrow S^{2-} + S_2O_3^{2-}$ (basic solution)
    (d) $As_2S_3 + H_2O_2 \longrightarrow AsO_4^{3-} + SO_4^{2-}$

43. Write a balanced equation for these redox reactions.
    (a) The oxidation of nitrite ion to nitrate ion by permanganate ion, MnO_4^-, in acidic solution (MnO_4^- ion is reduced to Mn^{2+}).
    (b) The reaction of manganese(II) ion and permanganate ion in basic solution to form solid manganese dioxide.
    (c) The oxidation of ethanol by dichromate ion in acidic solution, producing chromium(III) ion, acetaldehyde (CH_3CHO), and water as products.

44. Write a balanced equation for the redox reactions.
    (a) The reaction of aluminum metal with hydroiodic acid.
    (b) The reduction of vanadyl ion (VO^{2+}) to vanadic ion (V^{3+}) in acidic solution with zinc metal as the reducing agent.
    (c) The oxidation of methanol by chlorate ion in acidic solution, producing carbon dioxide gas, water, and chlorine dioxide gas as products.

45. The following reactions do not occur in aqueous solutions. Balance their equations by the half-equation method, as suggested in Are You Wondering 5-2.
    (a) $CH_4(g) + NO(g) \longrightarrow$
    $$CO_2(g) + N_2(g) + H_2O(g)$$
    (b) $H_2S(g) + SO_2(g) \longrightarrow S_8(s) + H_2O(g)$
    (c) $Cl_2O(g) + NH_3(g) \longrightarrow$
    $$N_2(g) + NH_4Cl(s) + H_2O(l)$$

46. The following reactions do not occur in aqueous solutions. Balance their equations by the half-equation method, as suggested in Are You Wondering 5-2.
    (a) $CH_4(g) + NH_3(g) + O_2(g) \longrightarrow$
    $$HCN(g) + H_2O(g)$$
    (b) $NO(g) + H_2(g) \longrightarrow NH_3(g) + H_2O(g)$
    (c) $Fe(s) + H_2O(l) + O_2(g) \longrightarrow Fe(OH)_3(s)$

Oxidizing and Reducing Agents

47. What are the oxidizing and reducing agents in the following redox reactions?
(a) $5\,SO_3^{2-} + 2\,MnO_4^- + 6\,H^+ \longrightarrow$
$$5\,SO_4^{2-} + 2\,Mn^{2+} + 3\,H_2O$$
(b) $2\,NO_2(g) + 7\,H_2(g) \longrightarrow 2\,NH_3(g) + 4\,H_2O(g)$
(c) $2\,[Fe(CN)_6]^{4-} + H_2O_2 + 2\,H^+ \longrightarrow$
$$2\,[Fe(CN)_6]^{3-} + 2\,H_2O$$

48. Thiosulfate ion, $S_2O_3^{2-}$, is a reducing agent that can be oxidized to different products, depending on the strength of the oxidizing agent and other conditions. By adding H^+, H_2O, and/or OH^- as necessary, write redox equations to show the oxidation of $S_2O_3^{2-}$ to
(a) $S_4O_6^{2-}$ by I_2 (iodide ion is another product)
(b) HSO_4^- by Cl_2 (chloride ion is another product)
(c) SO_4^{2-} by OCl^- in basic solution (chloride ion is another product)

Neutralization and Acid–Base Titrations

49. What volume of 0.0962 M NaOH is required to exactly neutralize 10.00 mL of 0.128 M HCl?

50. The exact neutralization of 10.00 mL of 0.1012 M $H_2SO_4(aq)$ requires 23.31 mL of NaOH. What must be the molarity of the NaOH(aq)?

$$H_2SO_4(aq) + 2\,NaOH(aq) \longrightarrow$$
$$Na_2SO_4(aq) + 2\,H_2O(l)$$

51. How many milliliters of 2.155 M KOH are required to titrate 25.00 mL of 0.3057 M CH_3CH_2COOH (propionic acid)?

52. How many milliliters of 0.0844 M $Ba(OH)_2$ are required to titrate 50.00 mL of 0.0526 M HNO_3?

53. An NaOH(aq) solution cannot be made up to an exact concentration simply by weighing out the required mass of NaOH, because the NaOH is not pure. Also, water vapor condenses on the solid as it is being weighed. The solution must be standardized by titration. For this purpose, a 25.00 mL sample of an NaOH(aq) solution requires 28.34 mL of 0.1085 M HCl. What is the molarity of the NaOH(aq)?

$$HCl(aq) + NaOH(aq) \longrightarrow NaCl(aq) + H_2O(l)$$

54. Household ammonia, used as a window cleaner and for other cleaning purposes, is $NH_3(aq)$. The NH_3 present in a 5.00 mL sample is neutralized by 28.72 mL of 1.021 M HCl. The net ionic equation for the neutralization is

$$NH_3(aq) + H^+(aq) \longrightarrow NH_4^+(aq)$$

What is the molarity of NH_3 in the sample?

55. We want to determine the acetylsalicyclic acid content of a series of aspirin tablets by titration with NaOH(aq). Each of the tablets is expected to contain about 0.32 g of $HC_9H_7O_4$. What molarity of NaOH(aq) should we use for titration volumes of about 23 mL? (This procedure ensures good precision and allows the titration of two samples with the contents of a 50 mL buret.)

$$HC_9H_7O_4(aq) + OH^-(aq) \longrightarrow$$
$$C_9H_7O_4^-(aq) + H_2O(l)$$

56. For use in titrations, we want to prepare 20 L of HCl(aq) with a concentration known to four significant figures. This is a two-step procedure beginning with the preparation of a solution of about 0.10 M HCl. A sample of this dilute HCl(aq) is titrated with a NaOH(aq) solution of known concentration.
(a) How many milliliters of concentrated HCl(aq) ($d = 1.19\,g/mL; 38\%$ HCl, by mass) must be diluted with water to 20.0 L to prepare 0.10 M HCl?
(b) A 25.00 mL sample of the approximately 0.10 M HCl prepared in part (a) requires 20.93 mL of 0.1186 M NaOH for its titration. What is the molarity of the HCl(aq)?
(c) Why is a titration necessary? That is, why not prepare a standard solution of 0.1000 M HCl simply by an appropriate dilution of the concentrated HCl(aq)?

57. A 25.00 mL sample of 0.132 M HNO_3 is mixed with 10.00 mL of 0.318 M KOH. Is the resulting solution acidic, basic, or exactly neutralized?

58. A 7.55 g sample of $Na_2CO_3(s)$ is added to 125 mL of a vinegar that is 0.762 M CH_3COOH. Will the resulting solution still be acidic? Explain.

59. Refer to Example 5-9. Suppose the analysis of all vinegar samples uses 5.00 mL of the vinegar and 0.1000 M NaOH for the titration. What volume of the 0.1000 M NaOH would represent the legal minimum 4.0%, by mass, acetic acid content of the vinegar? That is, calculate the volume of 0.1000 M NaOH so that if a titration requires more than this volume, the legal minimum limit is met (less than this volume, and the limit is not met).

60. The electrolyte in a lead storage battery must have a concentration between 4.8 and 5.3 M H_2SO_4 if the battery is to be most effective. A 5.00 mL sample of a battery acid requires 49.74 mL of 0.935 M NaOH for its complete reaction (neutralization). Does the concentration of the battery acid fall within the desired range? [*Hint:* Keep in mind that the H_2SO_4 produces two H^+ ions per formula unit.]

61. Which of the following points in a titration is represented by the molecular view shown in the sketch?
(a) 20% of the necessary titrant added in the titration of $NH_4Cl(aq)$ with HCl(aq)

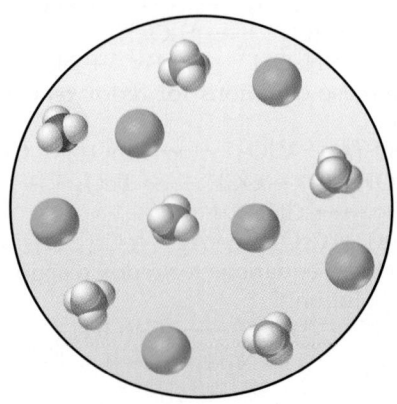

(b) 20% of the necessary titrant added in the titration of $NH_3(aq)$ with $HCl(aq)$
(c) the equivalence point in the titration of $NH_3(aq)$ with $HCl(aq)$
(d) 120% of the necessary titrant added in the titration of $NH_3(aq)$ with $HCl(aq)$

62. Using the sketch in Exercise 61 as a guide, sketch the molecular view of a solution in which
(a) $HCl(aq)$ is titrated to the equivalence point with $KOH(aq)$
(b) $CH_3COOH(aq)$ is titrated halfway to the equivalence point with $NaOH(aq)$.

Stoichiometry of Oxidation–Reduction Reactions

63. A $KMnO_4(aq)$ solution is to be standardized by titration against $As_2O_3(s)$. A 0.1078 g sample of As_2O_3 requires 22.15 mL of the $KMnO_4(aq)$ for its titration. What is the molarity of the $KMnO_4(aq)$?

$$5\,As_2O_3 + 4\,MnO_4^- + 9\,H_2O + 12\,H^+ \longrightarrow$$
$$10\,H_3AsO_4 + 4\,Mn^{2+}$$

64. Refer to Example 5-6. Assume that the only reducing agent present in a particular wastewater is SO_3^{2-}. If a 25.00 mL sample of this wastewater requires 31.46 mL of 0.02237 M $KMnO_4$ for its titration, what is the molarity of SO_3^{2-} in the wastewater?

65. An iron ore sample weighing 0.9132 g is dissolved in $HCl(aq)$, and the iron is obtained as $Fe^{2+}(aq)$. This solution is then titrated with 28.72 mL of 0.05051 M $K_2Cr_2O_7$. What is the mass percent Fe in the ore sample?

$$6\,Fe^{2+} + 14\,H^+ + Cr_2O_7^{2-} \longrightarrow$$
$$6\,Fe^{3+} + 2\,Cr^{3+} + 7\,H_2O$$

66. The concentration of $Mn^{2+}(aq)$ can be determined by titration with $MnO_4^-(aq)$ in basic solution. A 25.00 mL sample of $Mn^{2+}(aq)$ requires 37.21 mL of 0.04162 M $KMnO_4$ for its titration. What is $[Mn^{2+}]$ in the sample?

$$Mn^{2+} + MnO_4^- \longrightarrow MnO_2(s) \quad \text{(not balanced)}$$

67. The titration of 5.00 mL of a saturated solution of sodium oxalate, $Na_2C_2O_4$, at 25 °C requires 25.8 mL of 0.02140 M $KMnO_4$ in acidic solution. What mass of $Na_2C_2O_4$ in grams would be present in 1.00 L of this saturated solution?

$$C_2O_4^{2-} + MnO_4^- \longrightarrow$$
$$Mn^{2+} + CO_2(g) \quad \text{(not balanced)}$$

68. Refer to the Integrative Example. In the treatment of 1.00×10^2 L of a wastewater solution that is 0.0126 M CrO_4^{2-}, how many grams of **(a)** $Cr(OH)_3(s)$ would precipitate; **(b)** $Na_2S_2O_4$ would be consumed?

Integrative and Advanced Exercises

69. Write net ionic equations for the reactions depicted in photo **(a)** sodium metal reacts with water to produce hydrogen; photo **(b)** an excess of aqueous iron(III) chloride is added to the solution in **(a)**; and photo **(c)** the precipitate from **(b)** is collected and treated with an excess of $HCl(aq)$.

(a)   (b)   (c)

70. Following are some laboratory methods occasionally used for the preparation of small quantities of chemicals. Write a balanced equation for each.
(a) preparation of $H_2S(g)$: $HCl(aq)$ is heated with $FeS(s)$
(b) preparation of $Cl_2(g)$: $HCl(aq)$ is heated with $MnO_2(s)$; $MnCl_2(aq)$ and $H_2O(l)$ are other products
(c) preparation of N_2: Br_2 and NH_3 react in aqueous solution; NH_4Br is another product

(d) preparation of chlorous acid: an aqueous suspension of solid barium chlorite is treated with dilute $H_2SO_4(aq)$

71. When concentrated $CaCl_2(aq)$ is added to $Na_2HPO_4(aq)$, a white precipitate forms that is 38.7% Ca by mass. Write a net ionic equation representing the probable reaction that occurs.

72. You have a solution that is 0.0250 M $Ba(OH)_2$ and the following pieces of equipment: 1.00, 5.00, 10.00, 25.00, and 50.00 mL pipets and 100.0, 250.0, 500.0, and 1000.0 mL volumetric flasks. Describe how you would use this equipment to produce a solution in which $[OH^-]$ is 0.0100 M.

73. Sodium hydroxide used to make standard $NaOH(aq)$ solutions for acid–base titrations is invariably contaminated with some sodium carbonate. **(a)** Explain why, except in the most precise work, the presence of this sodium carbonate generally does not seriously affect the results obtained, for example, when $NaOH(aq)$ is used to titrate $HCl(aq)$. **(b)** Conversely, show that if Na_2CO_3 comprises more than 1% to 2% of the solute in $NaOH(aq)$, the titration results are affected.

74. A 110.520 g sample of mineral water is analyzed for its magnesium content. The Mg^{2+} in the sample is first precipitated as $MgNH_4PO_4$, and this precipitate is then converted to $Mg_2P_2O_7$, which is found to weigh 0.0549 g. Express the quantity of magnesium in the sample in parts per million (that is, in grams of Mg per million grams of H_2O).

75. What volume of 0.248 M $CaCl_2$ must be added to 335 mL of 0.186 M KCl to produce a solution with a concentration of 0.250 M Cl^-? Assume that the solution volumes are additive.

76. An unknown white solid consists of two compounds, each containing a different cation. As suggested in the illustration, the unknown is partially soluble in water. The solution is treated with NaOH(aq) and yields a white precipitate. The part of the original solid that is insoluble in water dissolves in HCl(aq) with the evolution of a gas. The resulting solution is then treated with $(NH_4)_2SO_4(aq)$ and yields a white precipitate. **(a)** Is it possible that any of the cations Mg^{2+}, Cu^{2+}, Ba^{2+}, Na^+, or NH_4^+ were present in the original unknown? Explain your reasoning. **(b)** What compounds could be in the unknown mixture (that is, what anions might be present)?

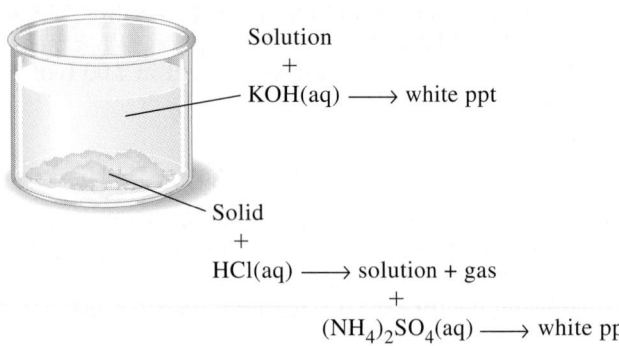

$$Solution$$
$$+$$
$$KOH(aq) \longrightarrow white\ ppt$$

$$Solid$$
$$+$$
$$HCl(aq) \longrightarrow solution + gas$$
$$+$$
$$(NH_4)_2SO_4(aq) \longrightarrow white\ ppt$$

77. Balance these equations for reactions in acidic solution.
   (a) $IBr + BrO_3^- + H^+ \longrightarrow IO_3^- + Br^- + H_2O$
   (b) $C_2H_5NO_3 + Sn \longrightarrow$
   $$NH_2OH + CH_3CH_2OH + Sn^{2+}$$
   (c) $As_2S_3 + NO_3^- \longrightarrow H_3AsO_4 + S + NO$
   (d) $H_5IO_6 + I_2 \longrightarrow IO_3^- + H^+ + H_2O$
   (e) $S_2F_2 + H_2O \longrightarrow S_8 + H_2S_4O_6 + HF$

78. Balance these equations for reactions in basic solution.
   (a) $Fe_2S_3 + H_2O + O_2 \longrightarrow Fe(OH)_3 + S$
   (b) $O_2^- + H_2O \longrightarrow OH^- + O_2$
   (c) $CrI_3 + H_2O_2 \longrightarrow CrO_4^{2-} + IO_4^-$
   (d) $Ag + CN^- + O_2 + OH^- \longrightarrow$
   $$[Ag(CN)_2]^- + H_2O$$
   (e) $B_2Cl_4 + OH^- \longrightarrow BO_2^- + Cl^- + H_2O + H_2$

79. A method of producing phosphine, PH_3, from elemental phosphorus, P_4, involves heating the P_4 with H_2O. An additional product is phosphoric acid, H_3PO_4. Write a balanced equation for this reaction.

80. Iron (Fe) is obtained from rock that is extracted from open pit mines and then crushed. The process used to obtain the pure metal from the crushed rock produces solid waste, called *tailings*, which are stored in disposal areas near the mines. The tailings pose a serious environmental risk because they contain sulfides, such as pyrite (FeS_2), which oxidize in air to produce metal ions and H^+ ions that can enter into surface water or ground water. The oxidation of FeS_2 to Fe^{3+} is described by the unbalanced chemical equation below.

$$FeS_2(s) + O_2(g) + H_2O(l) \longrightarrow$$
$$Fe^{3+}(aq) + SO_4^{2-}(aq) + H^+(aq) \quad (not\ balanced)$$

Thus, the oxidation of pyrite produces Fe^{3+} and H^+ ions that can leach into surface or ground water. The leaching of H^+ ions causes the water to become very acidic. To prevent acidification of nearby ground or surface water, limestone ($CaCO_3$) is added to the tailings to neutralize the H^+ ions:

$$CaCO_3(s) + 2\,H^+(aq) \longrightarrow$$
$$Ca^{2+}(aq) + H_2O(l) + CO_2(g)$$

(a) Balance the equation above for the reaction of FeS_2 and O_2. [*Hint:* Start with the half-equations $FeS_2(s) \rightarrow Fe^{3+}(aq) + SO_4^{2-}(aq)$ and $O_2(g) \rightarrow H_2O(l)$.]
(b) What is the minimum amount of $CaCO_3(s)$ required, per kilogram of tailings, to prevent contamination if the tailings contain 3% S by mass? Assume that all the sulfur in the tailings is in the form FeS_2.

81. A sample of battery acid is to be analyzed for its sulfuric acid content. A 1.00 mL sample weighs 1.239 g. This 1.00 mL sample is diluted to 250.0 mL, and 10.00 mL of this diluted acid requires 32.44 mL of 0.00498 M $Ba(OH)_2$ for its titration. What is the mass percent of H_2SO_4 in the battery acid? (Assume that complete ionization and neutralization of the H_2SO_4 occurs.)

82. A piece of marble (assume it is pure $CaCO_3$) reacts with 2.00 L of 2.52 M HCl. After dissolution of the marble, a 10.00 mL sample of the resulting solution is withdrawn, added to some water, and titrated with 24.87 mL of 0.9987 M NaOH. What must have been the mass of the piece of marble? Comment on the precision of this method; that is, how many significant figures are justified in the result?

83. The reaction below can be used as a laboratory method of preparing small quantities of $Cl_2(g)$. If a 62.6 g sample that is 98.5% $K_2Cr_2O_7$ by mass is allowed to react with 325 mL of HCl(aq) with a density of 1.15 g/mL and 30.1% HCl by mass, how many grams of $Cl_2(g)$ are produced?

$$Cr_2O_7^{2-} + H^+ + Cl^- \longrightarrow$$
$$Cr^{3+} + H_2O + Cl_2(g) \quad (not\ balanced)$$

84. Refer to Example 5-10. Suppose that the $KMnO_4(aq)$ were standardized by reaction with As_2O_3 instead of iron wire. If a 0.1304 g sample that is 99.96% As_2O_3 by mass had been used in the titration, how many milliliters of the $KMnO_4(aq)$ would have been required?

$$As_2O_3 + MnO_4^- + H^+ + H_2O \longrightarrow$$
$$H_3AsO_4 + Mn^{2+} \quad (not\ balanced)$$

85. A new method under development for water treatment uses chlorine dioxide rather than chlorine. One method of producing ClO_2 involves passing $Cl_2(g)$ into a concentrated solution of sodium chlorite. $Cl_2(g)$ and sodium chlorite are the sole reactants, and NaCl(aq) and $ClO_2(g)$ are the sole products. If the reaction has a 97% yield, what mass of ClO_2 is produced per gallon of 2.0 M $NaClO_2(aq)$ treated in this way?

86. The active component in one type of calcium dietary supplement is calcium carbonate. A 1.2450 g tablet of the supplement is added to 50.00 mL of 0.5000 M HCl and allowed to react. After completion of the reaction, the excess HCl(aq) requires 40.20 mL of 0.2184 M NaOH for its titration to the equivalence point. What is the calcium content of the tablet, expressed in milligrams of Ca^{2+}?

87. A 0.4324 g sample of a potassium hydroxide–lithium hydroxide mixture requires 28.28 mL of 0.3520 M HCl for its titration to the equivalence point. What is the mass percent lithium hydroxide in this mixture?

88. Chile saltpeter is a natural source of $NaNO_3$; it also contains $NaIO_3$. The $NaIO_3$ can be used as a source of iodine. Iodine is produced from sodium iodate in a two-step process occurring under acidic conditions:

$$IO_3^-(aq) + HSO_3^-(aq) \longrightarrow$$
$$I^-(aq) + SO_4^{2-}(aq) \quad \text{(not balanced)}$$
$$I^-(aq) + IO_3^-(aq) \longrightarrow$$
$$I_2(s) + H_2O(l) \quad \text{(not balanced)}$$

In the illustration, a 5.00 L sample of a $NaIO_3(aq)$ solution containing 5.80 g $NaIO_3$/L is treated with the stoichiometric quantity of $NaHSO_3$ (no excess of either reactant). Then, a further quantity of the initial $NaIO_3(aq)$ is added to the reaction mixture to bring about the second reaction. **(a)** How many grams of $NaHSO_3$ are required in the first step? **(b)** What additional volume of the starting solution must be added in the second step?

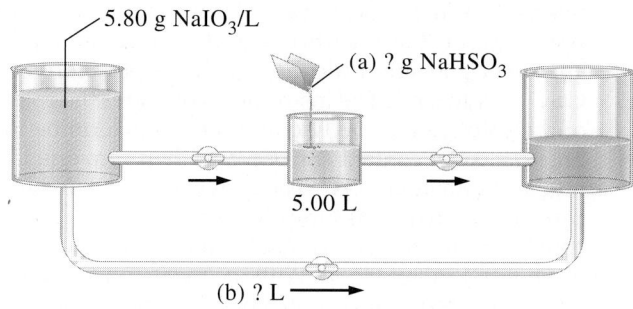

89. The active ingredients in a particular antacid tablet are aluminum hydroxide, $Al(OH)_3$, and magnesium hydroxide, $Mg(OH)_2$. A 5.00×10^2 mg sample of the active ingredients was dissolved in 50.0 mL of 0.500 M HCl. The resulting solution, which was still acidic, required 16.5 mL of 0.377 M NaOH for neutralization. What are the mass percentages of $Al(OH)_3$ and $Mg(OH)_2$ in the sample?

90. A compound contains only Fe and O. A 0.2729 g sample of the compound was dissolved in 50 mL of concentrated acid solution, reducing all the iron to Fe^{2+} ions. The resulting solution was diluted to 100 mL and then titrated with a 0.01621 M $KMnO_4$ solution. The unbalanced chemical equation for reaction between Fe^{2+} and MnO_4^- is given below.

$$MnO_4^-(aq) + Fe^{2+}(aq) \longrightarrow$$
$$Mn^{2+}(aq) + Fe^{3+}(aq) \quad \text{(not balanced)}$$

The titration required 42.17 mL of the $KMnO_4$ solution to reach the pink endpoint. What is the empirical formula of the compound?

91. Warfarin, $C_{19}H_{16}O_4$, is the active ingredient used in some anticoagulant medications. The amount of warfarin in a particular sample was determined as follows. A 13.96 g sample was first treated with an alkaline I_2 solution to convert $C_{19}H_{16}O_4$ to CHI_3. This treatment gives one mole of CHI_3 for every mole of $C_{19}H_{16}O_4$ that was initially present in the sample. The iodine in CHI_3 is then precipitated as AgI(s) by treatment with excess $AgNO_3(aq)$:

$$CHI_3(aq) + 3\,AgNO_3(aq) + H_2O(l) \longrightarrow$$
$$3\,AgI(s) + 3\,HNO_3(aq) + CO(g)$$

If 0.1386 g solid AgI were obtained, then what is the percentage by mass of warfarin in the sample analyzed?

92. Copper refining traditionally involves "roasting" insoluble sulfide ores (CuS) with oxygen. Unfortunately, the process produces large quantities of $SO_2(g)$, which is a major contributor to pollution and acid rain. An alternative process involves treating the sulfide ore with $HNO_3(aq)$, which dissolves the CuS without generating any SO_2. The unbalanced chemical equation for the reaction is given below.

$$CuS(s) + NO_3^-(aq) \longrightarrow$$
$$Cu^{2+}(aq) + NO(g) + HSO_4^-(aq) \quad \text{(not balanced)}$$

What volume of concentrated nitric acid solution is required per kilogram of CuS? Assume that the concentrated nitric acid solution is 70% HNO_3 by mass and has a density of 1.40 g/mL.

93. Phosphorus is essential for plant growth, but an excess of phosphorus can be catastrophic in aqueous ecosystems. Too much phosphorus can cause algae to grow at an explosive rate and this robs the rest of the ecosystem of oxygen. Effluent from sewage treatment plants must be treated before it can be released into lakes or streams because the effluent contains significant amounts of $H_2PO_4^-$ and HPO_4^{2-}. (Detergents are a major contributor to phosphorus levels in domestic sewage because many detergents contain Na_2HPO_4.) A simple way to remove $H_2PO_4^-$ and HPO_4^{2-} from the effluent is to treat it with lime, CaO, which produces Ca^{2+} and OH^- ions in water. The OH^- ions convert $H_2PO_4^-$ and HPO_4^{2-} ions into PO_4^{3-} ions and, finally, Ca^{2+}, OH^-, and PO_4^{3-} ions combine to form a precipitate of $Ca_5(PO_4)_3OH(s)$.
(a) Write balanced chemical equations for the four reactions described above.
[*Hint*: The reactants are CaO and H_2O; $H_2PO_4^-$ and OH^-; HPO_4^{2-} and OH^-; Ca^{2+}, PO_4^{3-}, and OH^-.]
(b) How many kilograms of lime are required to remove the phosphorus from a 1.00×10^4 L holding tank filled with contaminated water, if the water contains 10.0 mg of phosphorus per liter?

Feature Problems

94. Sodium cyclopentadienide, NaC_5H_5, is a common reducing agent in the chemical laboratory, but there is a problem in using it: NaC_5H_5 is contaminated with tetrahydrofuran (THF), C_4H_8O, a solvent used in its preparation. The THF is present as $NaC_5H_5 \cdot (THF)_x$, and it is generally necessary to know exactly how

much of this $NaC_5H_5 \cdot (THF)_x$ is present. This is accomplished by allowing a small amount of the $NaC_5H_5 \cdot (THF)_x$ to react with water,

$$NaC_5H_5 \cdot (C_4H_8O)_x + H_2O \longrightarrow$$
$$NaOH(aq) + C_5H_5 - H + x\,C_4H_8O$$

followed by titration of the $NaOH(aq)$ with a standard acid. From the sample data tabulated below, determine the value of x in the formula $NaC_5H_5 \cdot (THF)_x$.

	Trial 1	Trial 2
Mass of $NaC_5H_5 \cdot (THF)_x$	0.242 g	0.199 g
Volume of 0.1001 M HCl required to titrate NaOH(aq)	14.92 mL	11.99 mL

95. Manganese is derived from pyrolusite ore, an impure manganese dioxide. In the procedure used to analyze a pyrolusite ore for its MnO_2 content, a 0.533 g sample is treated with 1.651 g oxalic acid $(H_2C_2O_4 \cdot 2\,H_2O)$ in an acidic medium. Following this reaction, the excess oxalic acid is titrated with 0.1000 M $KMnO_4$, 30.06 mL being required. What is the mass percent MnO_2 in the ore?

$$H_2C_2O_4 + MnO_2 + H^+ \longrightarrow$$
$$Mn^{2+} + H_2O + CO_2 \quad \text{(not balanced)}$$

$$H_2C_2O_4 + MnO_4^- + H^+ \longrightarrow$$
$$Mn^{2+} + H_2O + CO_2 \quad \text{(not balanced)}$$

96. The Kjeldahl method is used in agricultural chemistry to determine the percent protein in natural products. The method is based on converting all the protein nitrogen to ammonia and then determining the amount of ammonia by titration. The percent nitrogen in the sample under analysis can be calculated from the quantity of ammonia produced. Interestingly, the majority of protein molecules in living matter contain just about 16% nitrogen.

A 1.250 g sample of meat is heated with concentrated sulfuric acid and a catalyst to convert all the nitrogen in the meat to $(NH_4)_2SO_4$. Then excess $NaOH(aq)$ is added to the mixture, which is heated to expel $NH_3(g)$. All the nitrogen from the sample is found in the $NH_3(g)$, which is then absorbed in and neutralized by 50.00 mL of dilute $H_2SO_4(aq)$. The excess $H_2SO_4(aq)$ requires 32.24 mL of 0.4498 M NaOH for its titration. A separate 25.00 mL sample of

the dilute $H_2SO_4(aq)$ requires 22.24 mL of 0.4498 M NaOH for its titration. What is the percent protein in the meat?

97. Blood alcohol content (BAC) is often reported in weight–volume percent (w/v%). For example, a BAC of 0.10% corresponds to 0.10 g CH_3CH_2OH per 100 mL of blood. Estimates of BAC can be obtained from breath samples by using a number of commercially available instruments, including the Breathalyzer for which a patent was issued to R. F. Borkenstein in 1958. The chemistry behind the Breathalyzer is described by the oxidation–reduction reaction below, which occurs in acidic solution:

$$\underset{\text{ethyl alcohol}}{CH_3CH_2OH(g)} + \underset{\text{(yellow-orange)}}{Cr_2O_7^{2-}(aq)} \longrightarrow$$

$$\underset{\text{(green)}}{CH_3COOH(aq) + Cr^{3+}(aq)} \quad \text{(not balanced)}$$

A Breathalyzer instrument contains two ampules, each of which contains 0.75 mg $K_2Cr_2O_7$ dissolved in 3 mL of 9 mol/L $H_2SO_4(aq)$. One of the ampules is used as reference. When a person exhales into the tube of the Breathalyzer, the breath is directed into one of the ampules, and ethyl alcohol in the breath converts $Cr_2O_7^{2-}$ into Cr^{3+}. The instrument compares the colors of the solutions in the two ampules to determine the breath alcohol content (BrAC), and then converts this into an estimate of BAC. The conversion of BrAC into BAC rests on the assumption that 2100 mL of air exhaled from the lungs contains the same amount of alcohol as 1 mL of blood. With the theory and assumptions described in this problem, calculate the molarity of $K_2Cr_2O_7$ in the ampules before and after a breath test in which a person with a BAC of 0.05% exhales 0.500 L of his breath into a Breathalyzer instrument.

98. In this problem, we describe an alternative method for balancing equations for oxidation-reduction reactions. The method is similar to the method given previously in Tables 5.5 and 5.6, but it places more emphasis on the assignment of oxidation states. (The method summarized in Tables 5.5 and 5.6 does not require you to assign oxidation states.) An emphasis on oxidation states is warranted because oxidation states are useful not only for keeping track of electrons but also for predicting chemical properties. The method is summarized in the table below.

A Method for Balancing Equations for Oxidation–Reduction Reactions That Occur in an Acidic or a Basic Aqueous Solution

1. Assign oxidation states to each element in the reaction and identify the species being oxidized and reduced.
2. Write separate, unbalanced equations for the oxidation and reduction half-reactions.
3. Balance the separate half-equations, in this order:
   - first with respect to the element being oxidized or reduced
   - then by adding electrons to one side or the other to account for the number of electrons produced (oxidation) or consumed (reduction)
4. Combine the half-reactions algebraically so that the total number of electrons cancels out.
5. Balance the net charge by either adding OH^- (for basic solutions) or H^+ (for acidic solutions).
6. Balance the O and H atoms by adding H_2O.
7. Check that the final equation is balanced with respect to each type of atom and with respect to charge.

The method offers a couple of advantages. First, the method applies to both acidic and basic environments because we balance charges by using either H^+ (for acidic environments) or OH^- (for basic environments). Second, the method is somewhat more efficient than the method we described previously because, in the method described here, we balance only once for charge and only once for hydrogen and oxygen. In the other method, we focus on the half-equations sepa-rately and must balance twice for charge and twice for hydrogen and oxygen.

Use the alternative method described above to balance the following oxidation-reduction equations.

(a) $Cr_2O_7{}^{2-}(aq) + Cl^-(aq) \longrightarrow$
$$Cr^{3+}(aq) + Cl_2(g) \quad \text{(acidic solution)}$$
(b) $C_2O_4{}^{2-}(aq) + MnO_4{}^-(aq) \longrightarrow$
$$CO_3{}^{2-}(aq) + MnO_2(s) \quad \text{(basic solution)}$$

Self-Assessment Exercises

99. In your own words, define or explain the terms or symbols **(a)** $\rightleftharpoons$ **(b)** [ ]; **(c)** spectator ion; **(d)** weak acid.

100. Briefly describe **(a)** half-equation method of balancing redox equations; **(b)** disproportionation reaction; **(c)** titration; **(d)** standardization of a solution.

101. Explain the important distinctions between **(a)** a strong electrolyte and strong acid; **(b)** an oxidizing agent and reducing agent; **(c)** precipitation reactions and neutralization reactions; **(d)** half-reaction and overall reaction.

102. The number of moles of hydroxide ion in 0.300 L of 0.0050 M $Ba(OH)_2$ is **(a)** 0.0015; **(b)** 0.0030; **(c)** 0.0050; **(d)** 0.010.

103. The highest $[H^+]$ will be found in an aqueous solution that is **(a)** 0.10 M HCl; **(b)** 0.10 M NH_3; **(c)** 0.15 M CH_3COOH; **(d)** 0.10 M H_2SO_4.

104. To precipitate Zn^{2+} from $Zn(NO_3)_2(aq)$, add **(a)** NH_4Cl; **(b)** $MgBr_2$; **(c)** K_2CO_3; **(d)** $(NH_4)_2SO_4$.

105. When treated with dilute HCl(aq), the solid that reacts to produce a gas is **(a)** $BaSO_3$; **(b)** ZnO; **(c)** NaBr; **(d)** Na_2SO_4.

106. What is the net ionic equation for the reaction that occurs when an aqueous solution of KI is added to an aqueous solution of $Pb(NO_3)_2$?

107. When aqueous sodium carbonate, Na_2CO_3, is treated with dilute hydrochloric acid, HCl, the products are sodium chloride, water, and carbon dioxide gas. What is the net ionic equation for this reaction?

108. Describe the synthesis of each of the following ionic compounds, starting from solutions of sodium and nitrate salts. Then write the net ionic equation for each synthesis.
(a) $Zn_3(PO_4)_2$;
(b) $Cu(OH)_2$;
(c) $NiCO_3$.

109. Consider the following redox reaction:
$$4\,NO(g) + 3\,O_2(g) + 2\,H_2O(l) \longrightarrow$$
$$4\,NO_3{}^-(aq) + 4\,H^+(aq)$$
(a) Which species is oxidized?
(b) Which species is reduced?
(c) Which species is the oxidizing agent?
(d) Which species is the reducing agent?
(e) Which species gains electrons?
(f) Which species loses electrons?

110. In the equation
$$?\,Fe^{2+}(aq) + O_2(g) + 4\,H^+(aq) \longrightarrow$$
$$?\,Fe^{3+}(aq) + 2\,H_2O(l)$$
the missing coefficients **(a)** are each 2; **(b)** are each 4; **(c)** can have any values as long as they are the same; **(d)** must be determined by experiment.

111. What is the simplest ratio $a{:}b$ when the equation below is properly balanced?
$$a\,ClO^-(aq) + b\,I_2(aq) \xrightarrow{\text{acidic solution}} c\,Cl^-(aq) + d\,IO_3{}^-(aq)$$
(a) 2:5; **(b)** 5:2; **(c)** 1:5; **(d)** 5:1; **(e)** 2:3.

112. In the half-reaction in which $NpO_2{}^+$ is converted to Np^{4+}, the number of electrons appearing in the half-equation is **(a)** 1; **(b)** 2; **(c)** 3; **(d)** 4.

113. Classify each of the following statements as true or false.
(a) Barium chloride, $BaCl_2$, is a weak electrolyte in aqueous solution.
(b) In the reaction $H^-(aq) + H_2O(l) \rightarrow H_2(g) + OH^-(aq)$, water acts as both an acid and an oxidizing agent.
(c) A precipitate forms when aqueous sodium carbonate, $Na_2CO_3(aq)$, is treated with excess aqueous hydrochloric acid, HCl(aq).
(d) Hydrofluoric acid, HF, is a strong acid in water.
(e) Compared with a 0.010 M solution of $NaNO_3$, a 0.010 M solution of $Mg(NO_3)_2$ is a better conductor of electricity.

114. Which of the following reactions are oxidation-reduction reactions?
(a) $H_2CO_3(aq) \longrightarrow H_2O(l) + CO_2(g)$
(b) $2\,Li(s) + 2\,H_2O(l) \longrightarrow 2\,LiOH(aq) + H_2(g)$
(c) $4\,Ag(s) + PtCl_4(aq) \longrightarrow 4\,AgCl(s) + Pt(s)$
(d) $2\,HClO_4(aq) + Ca(OH)_2(aq) \longrightarrow$
$$2\,H_2O(l) + Ca(ClO_4)_2(aq)$$

115. Similar to Figure 5-4(c), but using the formulas HAc, Ac^-, and H_3O^+, give a more accurate representation of $CH_3COOH(aq)$ in which ionization is 5% complete.

116. Appendix E describes a useful study aid known as concept mapping. Using the method presented in Appendix E, construct a concept map illustrating the different concepts introduced in Sections 5-4, 5-5, and 5-6.

6

Gases

CONTENTS

Hot-air balloons have intrigued people from the time the simple gas laws fundamental to their operation came to be understood more than 200 years ago.

You shouldn't overinflate a bicycle tire, or discard an aerosol can in an incinerator, or search for a gas leak with an open flame. In each case there is a danger of explosion. These and many other observations concerning gases can be explained by concepts considered in this chapter. The behaviors of the bicycle tire and the aerosol can are based on relationships among pressure, temperature, volume, and amount of gas. Other examples of the behavior of gases can be seen in a balloon filled with helium or hot air rising in air and carbon dioxide gas vaporizing from a block of dry ice and sinking to the floor. An understanding of the lifting power of lighter-than-air balloons comes in large part from knowledge of gas densities and their dependence on molar mass, temperature, and pressure. Predicting how far and how fast gas molecules migrate through air requires knowing something about the phenomenon of diffusion.

For a quantitative description of the behavior of gases, we will employ some simple gas laws and a more general expression called the *ideal gas equation*. These laws will be explained by the kinetic-molecular theory of gases. The topics covered in this chapter extend the discussion of reaction stoichiometry from the previous two chapters and lay some groundwork for use

◀ FIGURE 6-1
The gaseous states of three halogens (group 17)
The greenish yellow gas is $Cl_2(g)$; the brownish red gas is $Br_2(g)$ above a small pool of liquid bromine; the violet gas is $I_2(g)$ in contact with grayish-black solid iodine. Most other common gases, such as H_2, O_2, N_2, CO, and CO_2, are colorless.

in the following chapter on thermochemistry. The relationships between gases and the other states of matter—liquids and solids—are discussed in Chapter 12.

6-1 Properties of Gases: Gas Pressure

Some characteristics of gases are familiar to everyone. Gases expand to fill their containers and assume the shapes of their containers. They diffuse into one another and mix in all proportions. We cannot see individual particles of a gas, although we can see the bulk gas if it is colored (Fig. 6-1). Some gases, such as hydrogen and methane, are combustible; whereas others, such as helium and neon, are chemically unreactive.

Four properties determine the physical behavior of a gas: the amount of the gas (in moles) and the volume, temperature, and pressure of the gas. If we know any three of these, we can usually calculate the value of the remaining one by using a mathematical equation called an *equation of state* (such as the ideal gas equation, given on page 204). To some extent we have already discussed the properties of amount, volume, and temperature, but we need to consider the idea of pressure.

The Concept of Pressure

A balloon expands when it is inflated with air, but what keeps the balloon in its distended shape? A plausible hypothesis is that molecules of a gas are in constant motion, frequently colliding with one another and with the walls of their container. In their collisions, the gas molecules exert a force on the container walls. This force keeps the balloon distended. It is not easy, however, to measure the total force exerted by a gas. Instead of focusing on this total force, we consider instead the gas pressure. **Pressure** is defined as a force per unit area, that is, a force divided by the area over which the force is distributed. Figure 6-2 illustrates the idea of pressure exerted by a solid.

In SI, the unit of force is a *newton* (N), which is the force, F, required to produce an acceleration of one meter per second per second $(1\,\mathrm{m\,s^{-2}})$ in a one-kilogram mass (1 kg), that is, $1\,\mathrm{N} = 1\,\mathrm{kg\,m\,s^{-2}}$. The corresponding force per unit area—pressure—is expressed in the unit $\mathrm{N/m^2}$. A pressure of one newton per square meter is defined as one **pascal (Pa)**. Thus, a pressure in pascals is

$$P\,(\mathrm{Pa}) = \frac{F\,(\mathrm{N})}{A\,(\mathrm{m^2})} \qquad \textbf{(6.1)}$$

A pascal is a rather small pressure unit, so the **kilopascal (kPa)** is more commonly used. The pascal honors Blaise Pascal (1623–1662), who studied pressure and its transmission through fluids—the basis of modern hydraulics.

▲ FIGURE 6-2
Illustrating the pressure exerted by a solid
The two cylinders have the same mass and exert the same force on the supporting surface ($F = g \times m$). The tall, thin one has a smaller area of contact, however, and exerts a greater pressure ($P = F/A$).

◀ If you have not taken a physics course, consult Appendix B for a brief discussion of force and work.

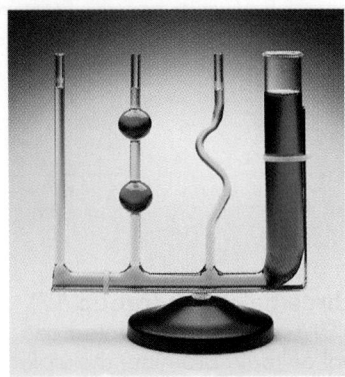

▲ FIGURE 6-3
The concept of liquid pressure
All the interconnected vessels fill to the same height. As a result, the liquid pressures are the same despite the different shapes and volumes of the containers.

Liquid Pressure

Because it is difficult to measure the total force exerted by gas molecules, it is also difficult to apply equation (6.1) to gases. The pressure of a gas is usually measured indirectly, by comparing it with a liquid pressure. Figure 6-3 illustrates the concept of liquid pressure and suggests that the pressure of a liquid depends only on the height of the liquid column and the density of the liquid. To confirm this statement, consider a liquid with density d, contained in a cylinder with cross-sectional area A, filled to a height h.

Now recall that (1) weight is a force, and weight (W) and mass (m) are proportional: $W = g \times m$. (2) The mass of a liquid is the product of its volume and density: $m = V \times d$. (3) The volume of a cylinder is the product of its height and cross-sectional area: $V = h \times A$. We use these facts to derive the equation:

$$P = \frac{F}{A} = \frac{W}{A} = \frac{g \times m}{A} = \frac{g \times V \times d}{A} = \frac{g \times h \times A \times d}{A} = g \times h \times d \quad \textbf{(6.2)}$$

Thus, because g is a constant, *liquid pressure is directly proportional to the liquid density and the height of the liquid column.*

Barometric Pressure

In 1643, Evangelista Torricelli constructed the device pictured in Figure 6-4 to measure the pressure exerted by the atmosphere. This device is called a **barometer**.

If a glass tube that is open at both ends stands upright in a container of mercury (Fig. 6-4a), the mercury levels inside and outside the tube are the same. To create the situation in Figure 6-4(b), we seal one end of a long glass tube, completely fill the tube with Hg(l), cover the open end, and invert the tube into a container of Hg(l). Then we reopen the end that is submerged in the mercury. The mercury level in the tube falls to a certain height and stays there. Something keeps the mercury at a greater height inside the tube than outside. Some tried to ascribe this phenomenon to forces within the tube, but Torricelli understood that the forces involved originated outside the tube.

In the open-end tube (Fig. 6-4a), the atmosphere exerts the same pressure on the surface of the mercury both inside and outside the tube, and the liquid levels are equal. Inside the closed-end tube (Fig. 6-4b), there is no air above the mercury (only a trace of mercury vapor). The atmosphere exerts a force on the surface of the mercury in the outside container. This force is transmitted through the liquid, holding up the mercury column within the tube. The column exerts a downward pressure that depends on its height and the density of Hg(l). When the pressure at the bottom of the mercury column is equal to the pressure of the atmosphere, the column height is maintained.

The height of mercury in a barometer provides a measure of **barometric pressure**. Barometric pressures may be expressed in a unit called **millimeter of mercury (mmHg)**, defined as the pressure exerted by a column of mercury that is exactly 1 mm in height when the density of mercury is equal to 13.5951 g/cm³(0 °C) and the acceleration due to gravity, g, is equal to 9.80655 m/s². Notice that this unit of pressure assumes specific values for the density of mercury and the acceleration due to gravity. This is because the density of Hg(l) depends on temperature and g depends on the specific location on Earth (recall p. 9). Typically, the pressure exerted by the atmosphere can support a column of mercury that is about 760 mm high and thus, atmospheric pressure is typically about 760 mmHg. Let's use equation (6.2) to calculate the pressure exerted by a column of mercury that is exactly 760 mm high when the density of mercury is $d = 13.5951 \text{ g/cm}^3 = 1.35951 \times 10^4 \text{ kg/m}^3$ and $g = 9.80655 \text{ m/s}^2$.

$$P = (9.80665 \text{ m s}^{-2})(0.760000 \text{ m})(1.35951 \, 10^4 \text{ k-g m}^{-3})$$
$$= 1.01325 \, 10^5 \text{ kg m}^{-1} \text{s}^{-2}$$

▶ In this calculation, we have written the units of d and g in the form kg m^{-3} and m s^{-2}, respectively, rather than as kg/m^3 and m/s^2. You must become equally comfortable with using either negative exponents or a slash (/) when working with derived units. For example, the unit $\text{kg m}^{-1} \text{s}^{-2}$ may also be written as $\text{kg/(m s}^2)$.

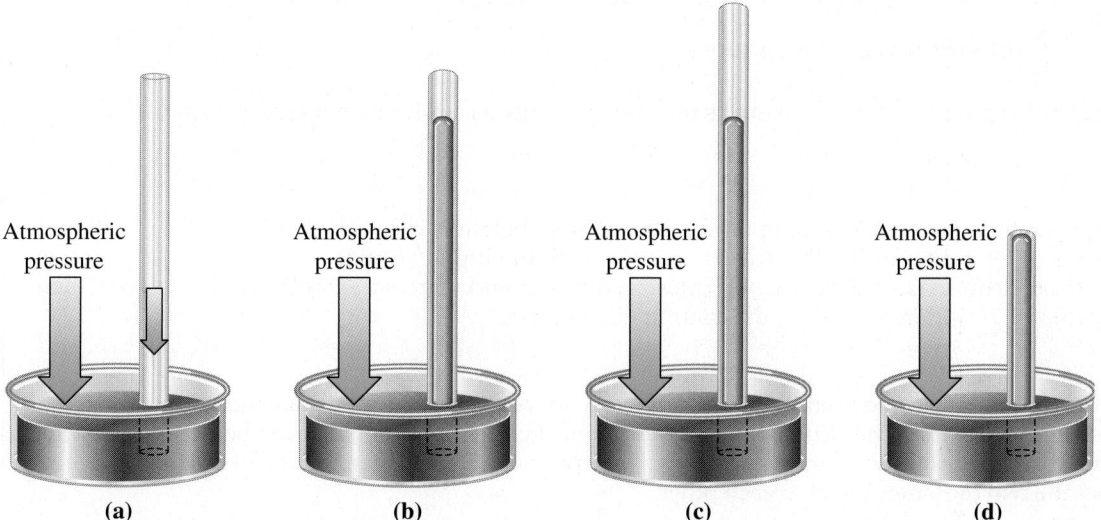

▲ FIGURE 6-4
Measurement of atmospheric pressure with a mercury barometer
Arrows represent the pressure exerted by the atmosphere. **(a)** The liquid mercury levels are equal inside and outside the open-end tube. **(b)** A column of mercury 760 mm high is maintained in the closed-end tube, regardless of the overall height of the tube **(c)** as long as it exceeds 760 mm. **(d)** A column of mercury fills a closed-end tube that is shorter than 760 mm. In the closed-end tubes in (b) and (c), the region above the mercury column is devoid of air and contains only a trace of mercury vapor.

The unit that arises in this calculation, $kg\,m^{-1}\,s^{-2}$, is the SI unit for pressure and, as mentioned earlier, is called the pascal (Pa). A pressure of exactly 101,325 Pa or 101.325 kPa, has special significance because in the SI system of units, **one standard atmosphere (atm)** is defined to be exactly equal to 101,325 Pa, or 101.325 kPa. Another pressure unit that is sometimes encountered is a unit called a **torr** and denoted by the symbol **Torr**. This unit honors Torricelli and is defined as exactly $1/760$ of a standard atmosphere. The following expression shows the relationships among these units.

$$1\,atm = 760\,Torr \approx 760\,mmHg \qquad (6.3)$$

As indicated above, the units torr and millimeters of mercury are not strictly equal. This is because 760 Torr is *exactly* equal to 101,325 Pa but 760 mmHg is only *approximately* equal to 101,325 Pa (that is, to about six or seven significant figures). The difference between a torr and a millimeter of mercury is too small to worry about, except in highly accurate work. Thus, in this text, we will use the pressure units of Torr and mmHg interchangeably.

Mercury is a relatively rare, expensive, and poisonous liquid. Why use it rather than water in a barometer? As we will see in Example 6-1, the extreme height required for a water barometer is a distinct disadvantage. Whereas atmospheric pressure can be measured with a mercury barometer less than 1 m high, a water barometer would have to be as tall as a three-storey building.

When you use a drinking straw, you reduce the air pressure above the liquid inside the straw by inhaling. Atmospheric pressure on the liquid outside the straw then pushes the liquid up the straw and into your mouth. An old-fashioned hand suction pump for pumping water (once common in rural areas) works by the same principle. The result of Example 6-1 indicates, however, that even if all the

EXAMPLE 6-1   Comparing Liquid Pressures

What is the height of a column of water that exerts the same pressure as a column of mercury 76.0 cm (760 mm) high?

Analyze

Equation (6.2) shows that, for a given liquid pressure, the column height is inversely proportional to the liquid density. The lower the liquid density, the greater the height of the liquid column. Mercury is 13.6 times as dense as water ($13.6 \, g/cm^3$ versus $1.00 \, g/cm^3$). If columns of water and mercury exert the same pressure, then the column of water is 13.6 times as high as the column of mercury.

Solve

Although we have already reasoned out the answer, we can arrive at the same conclusion by applying equation (6.2) twice, and then setting the two pressures equal to each other. Equation (6.2) can be used to describe the pressure of the mercury column of known height and the pressure of the water column of unknown height. Then we can set the two pressures equal to each other.

$$\text{pressure of Hg column} = g \times h_{Hg} \times d_{Hg} = g \times 76.0 \, cm \times 13.6 \, g/cm^3$$
$$\text{pressure of H}_2\text{O column} = g \times h_{H_2O} \times d_{H_2O} = g \times h_{H_2O} \times 1.00 \, g/cm^3$$
$$g \times h_{H_2O} \times 1.00 \, g/cm^3 = g \times 76.0 \, cm \times 13.6 \, g/cm^3$$
$$h_{H_2O} = 76.0 \, cm \times \frac{13.6 \, g/cm^3}{1.00 \, g/cm^3} = 1.03 \times 10^3 \, cm = 10.3 \, m$$

Assess

We can think about equation (6.2) in another way. For a column of liquid of fixed height, the greater the density of the liquid, the greater the pressure exerted by the liquid column. A column of mercury that is 760 mm high will exert a pressure 13.6 times as great as a column of water that is 760 mm high.

PRACTICE EXAMPLE A: A barometer is filled with diethylene glycol ($d = 1.118 \, g/cm^3$). The liquid height is found to be 9.25 m. What is the barometric (atmospheric) pressure expressed in millimeters of mercury?

PRACTICE EXAMPLE B: A barometer is filled with triethylene glycol. The liquid height is found to be 9.14 m when the atmospheric pressure is 757 mmHg. What is the density of triethylene glycol?

6-1   CONCEPT ASSESSMENT

Explain how the action of a water siphon is related to that of a suction pump.

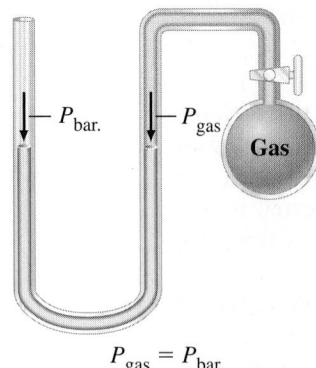

$P_{gas} = P_{bar.}$

(a) The gas pressure is equal to the barometric pressure.

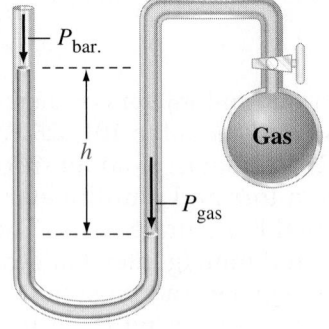

$P_{gas} = P_{bar.} + \Delta P$
$(\Delta P = g \times h \times d > 0)$

(b) The gas pressure is greater than the barometric pressure.

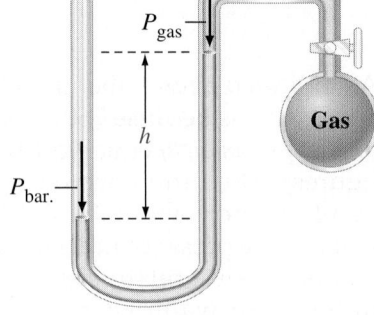

$P_{gas} = P_{bar.} + \Delta P$
$(\Delta P = -g \times h \times d < 0)$

(c) The gas pressure is less than the barometric pressure.

▲ FIGURE 6-5
Measurement of gas pressure with an open-end manometer
The possible relationships between barometric pressure and a gas pressure under measurement are pictured here and described in Example 6–2. If P_{gas} and $P_{bar.}$ are expressed in mmHg, then ΔP is numerically equal to the height h expressed in millimeters.

EXAMPLE 6-2   Using a Manometer to Measure Gas Pressure

When the manometer in Figure 6-5(c) is filled with liquid mercury ($d = 13.6$ g/cm^3), the barometric pressure is 748.2 mmHg, and the difference in mercury levels is 8.6 mmHg. What is the gas pressure P_{gas}?

Analyze

We must first establish which is greater: the barometric pressure or the gas pressure. In Figure 6-5(c), the barometric pressure forces liquid mercury down the tube toward the gas sample. The barometric pressure is greater than the gas pressure. Thus, $\Delta P = P_{gas} - P_{bar.} < 0$.

Solve

The gas pressure is less than the barometric pressure. Therefore, we subtract 8.6 mmHg from the barometric pressure to obtain the gas pressure.

$$P_{gas} = P_{bar.} + \Delta P = 748.2 \, \text{mmHg} - 8.6 \, \text{mmHg} = 739.6 \, \text{mmHg}$$

Assess

Because all pressures are expressed in millimeters of mercury, the pressure difference (ΔP) is numerically equal to the difference in mercury levels. Thus, the density of mercury does not enter into the calculation.

PRACTICE EXAMPLE A: Suppose that the mercury level in Example 6-2 is 7.8 mm higher in the arm open to the atmosphere than in the closed arm. What would be the value of P_{gas}?

PRACTICE EXAMPLE B: Suppose $P_{bar.}$ and P_{gas} are those described in Example 6-2, but the manometer is filled with liquid glycerol ($d = 1.26$ g/cm^3) instead of mercury. What would be the difference in the two levels of the liquid?

---

air could be removed from inside a pipe, atmospheric pressure outside the pipe could not raise water to a height of more than about 10 m. Thus, a suction pump works only for shallow wells. To pump water from a deep well, a mechanical pump is required. The mechanical pump pushes the water upward by using a force that is greater than the force of the atmosphere pushing the water down.

Manometers

Although a mercury barometer is indispensable for measuring the pressure of the atmosphere, it is rarely used alone to measure other gas pressures. The difficulty with a barometer is in placing it inside the container of gas whose pressure is to be measured. However, the pressure of the gas to be measured can be compared with barometric pressure by using a **manometer**. Figure 6-5 illustrates the principle of an open-end manometer. When the gas pressure being measured and the prevailing atmospheric (barometric) pressure are equal, the heights of the mercury columns in the two arms of the manometer are equal. A difference in height of the two arms signifies a difference between the gas pressure and barometric pressure.

Units of Pressure: A Summary

Table 6.1 lists several different units used to express pressure. The units shown in red are used frequently by chemists, even though they are not part of the SI system. The atmosphere is a useful unit because volumes of gases are often measured at the prevailing atmospheric pressure. Typically, the atmospheric pressure is close to 1 atm, or 760 Torr. The units shown in blue are those preferred in the SI system.

The units shown in black in Table 6.1 are based on the unit **bar**. One bar is 100 times as large as a kilopascal. Atmospheric pressure is typically close to 1 bar. The unit *millibar* is commonly used by meteorologists.

Although we can generally choose freely among the pressure units in Table 6.1 when doing calculations involving gases, we will encounter situations that require SI units. This is the case in Example 6-3.

TABLE 6.1 Some Common Pressure Units		
Atmosphere	atm	
Millimeter of mercury	mmHg	1 atm $\simeq$ 760 mmHg
Torr	Torr	= 760 Torr
Pascal	Pa	= 101,325 Pa
Kilopascal	kPa	= 101.325 kPa
Bar	bar	= 1.01325 bar
Millibar	mbar	= 1013.25 mbar

EXAMPLE 6-3   Using SI Units of Pressure

The 1.000 kg red cylinder in Figure 6-2 has a diameter of 4.10 cm. What pressure, expressed in Torr, does this cylinder exert on the surface beneath it?

Analyze

We must apply equation (6.2). It is best to use SI units and obtain a pressure in SI units (Pa), and then convert the pressure to the required units (Torr).

Solve

Expression (6.1) defines pressure as force divided by area.

$$P = \frac{F}{A}$$

The force exerted by the cylinder is its weight.

$$F = W = m \times g$$

The mass is 1.000 kg, and g (the acceleration due to gravity) is $9.81 \, \text{m s}^{-2}$. The product of these two terms is the force in newtons.

$$F = m \times g = 1.000 \, \text{kg} \times 9.81 \, \text{m s}^{-2} = 9.81 \, \text{N}$$

The force is exerted on the area of contact between the cylinder and the underlying surface. This circular area is calculated by using the radius of the cylinder—one-half the 4.10 cm diameter, expressed in meters.

$$A = \pi r^2 = 3.1416 \times \left(2.05 \, \text{cm} \times \frac{1 \, \text{m}}{100 \, \text{cm}} \right)^2 = 1.32 \times 10^{-3} \, \text{m}^2$$

The force divided by the area (in square meters) gives the pressure in pascals.

$$P = \frac{F}{A} = \frac{9.81 \, \text{N}}{1.32 \times 10^{-3} \, \text{m}^2} = 7.43 \times 10^3 \, \text{Pa}$$

The relationship between the units Torr and pascal (Table 6.1) is used for the final conversion.

$$P = 7.43 \times 10^3 \, \text{Pa} \times \frac{760 \, \text{Torr}}{101,325 \, \text{Pa}} = 55.7 \, \text{Torr}$$

Assess

It's difficult to tell at a glance whether this is a reasonable result. To check our result, let us focus instead on a cylindrical column of mercury that is 55.7 mm high and has a diameter of 4.10 cm. This column of mercury also exerts a pressure of 55.7 mmHg = 55.7 Torr. The volume of mercury in this column is $V = \pi r^2 h = \pi \times (2.05 \, \text{cm})^2 \times 5.57 \, \text{cm} = 73.5 \, \text{cm}^3$. The density of mercury is about $13.6 \, \text{g/cm}^3$ (page 194) and thus, the mass of the mercury column is $73.5 \, \text{cm}^3 \times 13.6 \, \text{g/cm}^3 = 1.00 \times 10^3 \, \text{g} = 1.00 \, \text{kg}$. This is exactly the mass of the steel cylinder.

PRACTICE EXAMPLE A: The 1.000 kg green cylinder in Figure 6-2 has a diameter of 2.60 cm. What pressure, expressed in Torr, does this cylinder exert on the surface beneath it?

PRACTICE EXAMPLE B: We want to increase the pressure exerted by the 1.000 kg red cylinder in Example 6-3 to 100.0 mb by placing a weight on top of it. What must be the mass of this weight? Must the added weight have the same cross-sectional area as the cylinder? Explain.

6-2   The Simple Gas Laws

In this section, we consider relationships involving the pressure, volume, temperature, and amount of a gas. Specifically, we will see how one variable depends on another, as the remaining two are held fixed. Collectively, these relationships are referred to as the simple gas laws. You can use these laws in problem solving, but

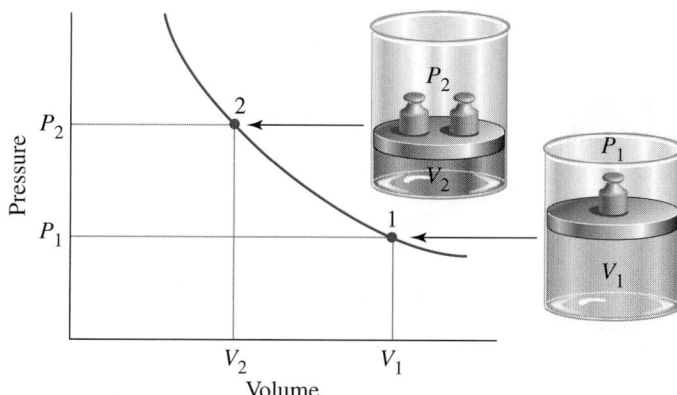

◀ FIGURE 6-6
Relationship between gas volume and pressure—Boyle's law
When the temperature and amount of gas are held constant, gas volume is inversely proportional to the pressure: A doubling of the pressure causes the volume to decrease to one-half its original value.

you will probably prefer the equation developed in the next section—the ideal gas equation. You may find that the greatest use of the simple gas laws is in solidifying your qualitative understanding of the behavior of gases.

Boyle's Law

In 1662, working with air, Robert Boyle discovered the first of the simple gas laws, now known as **Boyle's law**.

> For a fixed amount of gas at a constant temperature, the gas volume is inversely proportional to the gas pressure.   **(6.4)**

Consider the gas in Figure 6-6. It is confined in a cylinder closed off by a freely moving "weightless" piston. The pressure of the gas depends on the total weight placed on top of the piston. This weight (a force) divided by the area of the piston yields the gas pressure. If the weight on the piston is doubled, the pressure doubles and the gas volume decreases to one-half its original value. If the pressure of the gas is tripled, the volume decreases to one-third. Conversely, if the pressure is reduced by one-half, the gas volume doubles, and so on. Mathematically, the inverse relationship between gas pressure and volume is expressed as

$$P \propto \frac{1}{V} \quad \text{or} \quad PV = a \text{ (a constant)} \qquad \textbf{(6.5)}$$

When the proportionality sign ($\propto$) is replaced with an equal sign and a proportionality constant, the product of the pressure and volume of a fixed amount of gas at a given temperature is seen to be a constant (a). The value of a depends on the amount of gas and the temperature. The graph in Figure 6-6 is that of $PV = a$. It is called a hyperbola.

The equation $PV = a$ can be used to derive another equation that is useful for situations in which a gas undergoes a change at constant temperature. If we write equation (6.5) for the initial state (i) and for the final state (f), we get $P_i V_i = a$ and $P_f V_f = a$. Because both PV products are equal to the same value of a, we obtain the result

$$P_i V_i = P_f V_f \qquad (n \text{ constant}, T \text{ constant})$$

The equation above is often used to relate pressure and volume changes.

🔍 **6-2   CONCEPT ASSESSMENT**

A 50.0 L cylinder contains nitrogen gas at a pressure of 21.5 atm. The contents of the cylinder are emptied into an evacuated tank of unknown volume. If the final pressure in the tank is 1.55 atm, then what is the volume of the tank?
(a) $(21.5/1.55) \times 50.0$ L; **(b)** $(1.55/21.5) \times 50.0$ L; **(c)** $21.5/(1.55 \times 50)$ L; **(d)** $1.55/(21.5 \times 50)$ L.

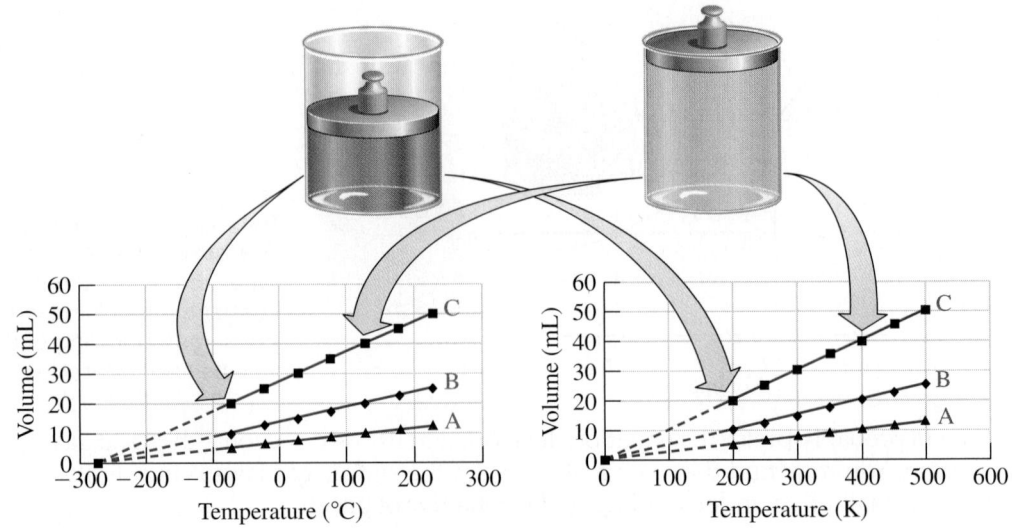

▲ FIGURE 6-7
Gas volume as a function of temperature
Volume is plotted against temperature on two different scales—Celsius and Kelvin. The volumes of three different gases (A, B, and C) are measured at 1 atm and 500 K. As the temperature is lowered, the volume decreases as predicted by Charles's law. Thus, at 250 K (−23 °C), for example, the volume of gas C has become 25 mL, one-half of the original 50 mL. Although the relationship between volume and temperature is linear for both the Celsius and Kelvin temperature scales, the volume is directly proportional only to the absolute temperature. That is, the volume must be zero at a temperature of zero. Only the Kelvin scale meets this requirement.

▶ It is probably fair to say that the absolute zero of temperature was "discovered" by noting that a plot of V (or P) vs. T for any gas extrapolates to −273 °C.

Charles's Law

The relationship between the volume of a gas and temperature was discovered by the French physicists Jacques Charles in 1787 and, independently, by Joseph Louis Gay-Lussac, who published it in 1802.

Figure 6-7 pictures a fixed amount of gas confined in a cylinder. The pressure is held constant at 1 atm while the temperature is varied. The volume of gas increases as the temperature is raised and decreases as the temperature is lowered. The relationship is linear. Figure 6-7 shows the linear dependence of volume on temperature for three gases at three different initial conditions. One point in common to the three lines is their intersection with the temperature axis. Although they differ at every other temperature, the gas volumes all reach a value of zero at the same temperature. The temperature at which the volume of a hypothetical* gas becomes zero is the absolute zero of temperature: −273.15 °C on the Celsius scale or 0 K on the **absolute**, or **Kelvin**, scale. The relationship between the Kelvin temperature, T, and the Celsius temperature, t, is shown below in equation (6.6).

▶ When converting from °C to K, apply the addition and subtraction significant figure rule: the two-significant-figure 25 °C, when added to 273.15, becomes the three-significant-figure 298 K. Similarly, 25.0 °C becomes 298.2 K (that is, 25.0 + 273.15 = 298.15, which rounds to 298.2).

$$T(\text{K}) = t(°\text{C}) + 273.15 \qquad (6.6)$$

The graph on the right hand side of Figure 6-7 shows that, from a volume of zero at 0 K, the gas volume is directly proportional to the Kelvin temperature. The statement of **Charles's law** given below in (6.7) summarizes the relationship between volume and temperature.

▶ Charles's ideas about the effect of temperature on the volume of a gas were probably influenced by his passion for hot-air balloons, a popular craze of the late eighteenth century.

The volume of a fixed amount of gas at constant pressure is directly proportional to the Kelvin (absolute) temperature. (6.7)

*All gases condense to liquids or solids before the temperature approaches absolute zero. Also, when we speak of the volume of a gas, we mean the free volume among the gas molecules, not the volume of the molecules themselves. Thus, the gas we refer to here is *hypothetical*. It is a gas whose molecules have mass but no volume and that does not condense to a liquid or solid.

In mathematical terms, Charles's law is

$$V \propto T \qquad \text{or} \qquad V = bT \text{ (where } b \text{ is a constant)} \qquad \textbf{(6.8)}$$

The value of the constant b depends on the amount of gas and the pressure. It does *not* depend on the identity of the gas.

From either expression (6.7) or (6.8), we see that doubling the Kelvin temperature of a gas causes its volume to double. Reducing the Kelvin temperature by one-half (say, from 300 to 150 K) causes the volume to decrease to one-half, and so on.

Equation (6.8) can be used to derive an equation that is useful for situations in which a gas undergoes a change at constant pressure. If we apply equation (6.8) twice, once for the initial state (i) and once for the final state (f), we get $(V_i/T_i) = b$ and $(V_f/T_f) = b$. Because both (V/T) quotients are equal to the same value of b, we obtain the result

$$\frac{V_i}{T_i} = \frac{V_f}{T_f} \qquad (n \text{ constant, } P \text{ constant})$$

The equation above is often used to relate volume and temperature changes.

▲ Charles experimented with the first hydrogen-filled balloons, much like the one shown here, though smaller. He also invented most of the features of modern ballooning, including the suspended basket and the valve to release gas.

🔍 6-3   CONCEPT ASSESSMENT

A balloon is inflated to a volume of 2.50 L inside a house that is kept at 24 °C. Then it is taken outside on a very cold winter day. If the temperature outside is −25 °C, what will be the volume of the balloon when it is taken outside? Assume that the quantity of air in the balloon and its pressure both remain constant.
(a) $(248/297) \times 2.50$ L; **(b)** $(297/248) \times 2.50$ L; **(c)** $248/(297 \times 2.50)$ L; **(d)** $297/(248 \times 2.50)$ L.

🔍 6-4   CONCEPT ASSESSMENT

Doubling a gas temperature from 100 K to 200 K causes a gas volume to double. Would you expect a similar doubling of the gas volume when a gas is heated from 100 °C to 200 °C? Explain.

Standard Conditions of Temperature and Pressure

Because gas properties depend on temperature and pressure, it is useful to have a set of standard conditions of temperature and pressure that can be used for comparing different gases. The standard temperature for gases is taken to be 0 °C = 273.15 K and standard pressure, 1 bar = 100 kPa = 10^5 Pa. **Standard conditions of temperature and pressure** are usually abbreviated as **STP**. It is important to emphasize that STP was defined differently in the past, and some texts and chemists still use the old definition. The old definition, which was based on a standard pressure of 1 atm, is discouraged. In this text, we use the definition recommended by the International Union of Pure and Applied Chemistry (IUPAC):

◄ Although IUPAC has recommended that one *bar* should replace the *atmosphere* as the standard state condition for gas law and thermodynamic data, use of the atmosphere persists.

Standard Temperature and Pressure (STP):	0 °C and 1 bar = 10^5 Pa

Avogadro's Law

In 1808, Gay-Lussac reported that gases react by volumes in the ratio of small whole numbers. One proposed explanation was that equal volumes of gases at the same temperature and pressure contain equal numbers of atoms. Dalton did not agree with this proposition, however. If Gay-Lussac's proposition

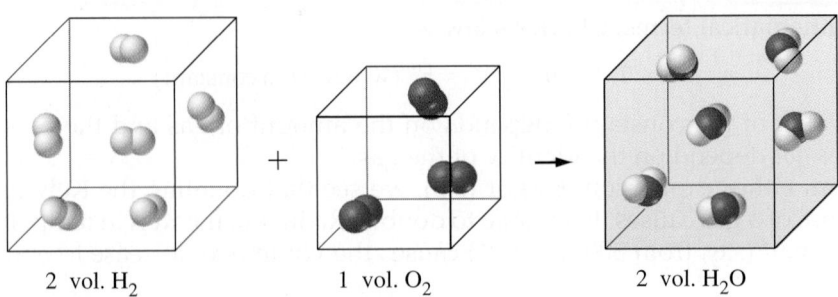

2 vol. H$_2$ 1 vol. O$_2$ 2 vol. H$_2$O

▲ FIGURE 6-8
Formation of water—actual observation and Avogadro's hypothesis
In the reaction 2 H$_2$(g) + 1 O$_2$(g) $\longrightarrow$ 2 H$_2$O(g), only one-half as many O$_2$ molecules are required as are H$_2$ molecules. If equal volumes of gases contain equal numbers of molecules, this means the volume of O$_2$(g) is one-half that of H$_2$(g). The combining ratio by volume is 2 : 1 : 2.

▶ Avogadro's hypothesis and statements derived from it apply only to gases. There is no similar relationship for liquids or solids.

were true, then the reaction of hydrogen and oxygen to form water would be H(g) + O(g) $\longrightarrow$ HO(g), with combining volumes of 1 : 1 : 1, rather than the 2 : 1 : 2 that was observed.

In 1811, Amedeo Avogadro resolved this dilemma by proposing not only the "equal volumes–equal numbers" hypothesis, but also that molecules of a gas may break up into half–molecules when they react. Using modern terminology, we would say that O$_2$ molecules split into atoms, which then combine with molecules of H$_2$ to form H$_2$O molecules. In this way, the volume of oxygen needed is only one-half that of hydrogen. Avogadro's reasoning is outlined in Figure 6-8.

Avogadro's equal volumes–equal numbers hypothesis can be stated in either of two ways.

1. Equal volumes of different gases compared at the same temperature and pressure contain equal numbers of molecules.

2. Equal numbers of molecules of different gases compared at the same temperature and pressure occupy equal volumes.

A relationship that follows from *Avogadro's hypothesis*, often called **Avogadro's law**, is as follows.

> At a fixed temperature and pressure, the volume of a gas is directly proportional to the amount of gas.   **(6.9)**

If the number of moles of gas (n) is doubled, the volume doubles, and so on. A mathematical statement of this fact is

$$V \propto n \quad \text{and} \quad V = c \times n$$

The constant c, which is equal to V/n, is the volume per mole of gas, a quantity we call the molar volume. Molar volumes of gases vary with temperature and pressure but experiment reveals that, for given values of T and P, the molar volumes of all gases are approximately equal. (In Section 6-9, we will discuss why the molar volumes of gases are not exactly equal.) The data in Table 6.2 show that the molar volume of a gas is approximately 22.414 L at 0 °C and 1 atm and 22.711 L at STP. The following statement summarizes these observations.

▶ In general, relating the amount of gas and its volume is best done with the ideal gas equation (Section 6-3).

> 1 mol gas = 22.414 L (at 0 °C, 1 atm) = 22.711 L (at STP)   **(6.10)**

Figure 6-9 should help you to visualize 22.414 L of a gas.

TABLE 6.2   Densities and Molar Volumes of Various Gases

Gas	Molar Mass, $g \, mol^{-1}$	Density (at STP), $g \, L^{-1}$	Molar Volume,[a] $L \, mol^{-1}$ (at STP)	Molar Volume,[a] $L \, mol^{-1}$ (at 0°C, 1 atm)
H_2	2.01588	8.87104×10^{-2}	22.724	22.427
He	4.00260	0.17615	22.722	22.425
Ideal gas	–	–	22.711	22.414
N_2	28.0134	1.23404	22.701	22.404
CO	28.0101	1.23375	22.696	22.399
O_2	31.9988	1.41034	22.689	22.392
CH_4	16.0425	0.70808	22.656	22.360
NF_3	71.0019	3.14234	22.595	22.300
CO_2	44.0095	1.95096	22.558	22.263
N_2O	44.0128	1.95201	22.550	22.255
C_2H_6	30.0690	1.33740	22.483	22.189
NH_3	17.0352	0.76139	22.374	22.081
SF_6	146.0554	6.52800	22.374	22.081
C_3H_8	44.0956	1.98318	22.235	21.944
SO_2	64.064	2.89190	22.153	21.863

Source: The densities are from the National Institute of Standards and Technology (NIST) *Chemistry WebBook*, available online at http://webbook.nist.gov/chemistry/.

[a]The molar volume is equal to the molar mass divided by the density. The molar volume at 0 °C and 1 atm is obtained by dividing the molar volume at STP by 1.01325.

6-5   CONCEPT ASSESSMENT

Without doing an actual calculation, determine which of the following expressions equals the final volume of gas if 20.0 g O_2 is added to 40.0 g O_2 at 0 °C and 1 atm, the temperature is changed to 30 °C, and the pressure to 825 Torr.

(a) $\left(\dfrac{40.0}{20.0} \times \dfrac{1}{32.00} \times 22.4 \times \dfrac{760}{825} \times \dfrac{303}{273} \right) L$

(b) $\left(\dfrac{20.0}{40.0} \times \dfrac{1}{32.00} \times 22.4 \times \dfrac{825}{760} \times \dfrac{273}{303} \right) L$

(c) $\left(60.0 \times \dfrac{1}{32.00} \times 22.4 \times \dfrac{825}{760} \times \dfrac{303}{273} \right) L$

(d) $\left(\dfrac{3}{2} \times 40.0 \times \dfrac{1}{32.00} \times 22.4 \times \dfrac{760}{825} \times \dfrac{303}{273} \right) L$

(e) $\left(\dfrac{2}{3} \times 60.0 \times \dfrac{1}{32.00} \times 22.4 \times \dfrac{760}{825} \times \dfrac{303}{273} \right) L$

◄ FIGURE 6-9
Molar volume of a gas visualized
The wooden cube is 28.2 cm on edge and has approximately the same volume as one mole of gas at 1 atm and 0 °C: 22.414 L. By contrast, the volume of the basketball is 7.5 L; the soccer ball, 6.0 L; and the football, 4.4 L.

▶ The ideal gas equation will probably not be supplied on exams and should be memorized. Values of R given here, however, will be available. There is really only one R but like many properties and constants, its value can be expressed in a variety of units.

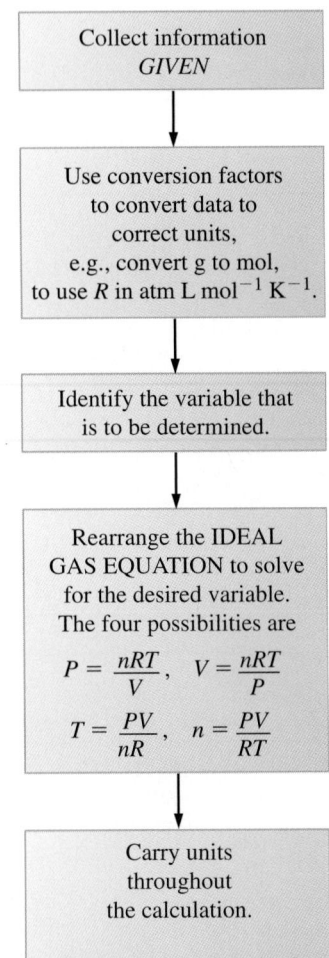

Collect information
GIVEN

↓

Use conversion factors to convert data to correct units, e.g., convert g to mol, to use R in atm L mol^{-1} K^{-1}.

↓

Identify the variable that is to be determined.

↓

Rearrange the IDEAL GAS EQUATION to solve for the desired variable. The four possibilities are

$$P = \frac{nRT}{V}, \quad V = \frac{nRT}{P}$$

$$T = \frac{PV}{nR}, \quad n = \frac{PV}{RT}$$

↓

Carry units throughout the calculation.

This enables you to check your calculation.

▲ **Applying the ideal gas equation**

TABLE 6.3 Five Common Values of R
0.082057 atm L mol^{-1} K^{-1}
0.083145 bar L K^{-1} mol^{-1}
8.3145 kPa L K^{-1} mol^{-1}
8.3145 Pa m^3 mol^{-1} K^{-1}
8.3145 J mol^{-1} K^{-1}

6-3 Combining the Gas Laws: The Ideal Gas Equation and the General Gas Equation

Each of the three simple gas laws describes the effect that changes in one variable have on the gas volume when the other two variables are held constant.

1. Boyle's law describes the effect of pressure, $V \propto 1/P$.
2. Charles's law describes the effect of temperature, $V \propto T$.
3. Avogadro's law describes the effect of the amount of gas, $V \propto n$.

These three laws can be combined into a single equation—the **ideal gas equation**—that includes all four gas variables: volume, pressure, temperature, and amount of gas.

The Ideal Gas Equation

In accord with the three simple gas laws, the volume of a gas is *directly* proportional to the amount of gas, *directly* proportional to the Kelvin temperature, and *inversely* proportional to pressure. That is,

$$V \propto \frac{nT}{P} \quad \text{and} \quad V = \frac{RnT}{P}$$

$$PV = nRT \tag{6.11}$$

A gas whose behavior conforms to the ideal gas equation is called an **ideal**, or **perfect, gas**. Before we can apply equation (6.11), we need a value for the constant R, called the **gas constant**. One way to obtain this is to substitute into equation (6.11) the molar volume of an ideal gas at 0 °C and 1 atm. However, the value of R will then depend on which units are used to express pressure and volume. With a molar volume of 22.4140 L and pressure in atmospheres,

$$R = \frac{PV}{nT} = \frac{1 \text{ atm} \times 22.4140 \text{ L}}{1 \text{ mol} \times 273.15 \text{ K}} = 0.082057 \text{ atm L mol}^{-1}\text{K}^{-1}$$

Using the SI units of m^3 for volume and Pa for pressure gives

$$R = \frac{PV}{nT} = \frac{101{,}325 \text{ Pa} \times 2.24140 \times 10^{-2} \text{ m}^3}{1 \text{ mol} \times 273.15 \text{ K}} = 8.3145 \text{ Pa m}^3 \text{ mol}^{-1}\text{K}^{-1}$$

The units Pa m^3 mol^{-1} K^{-1} also have another significance. The pascal has units kg m^{-1} s^{-2}, so the units m^3 Pa become kg m^2 s^{-2}, which is the SI unit of energy—the joule. Thus R also has the value

$$R = 8.3145 \text{ J mol}^{-1}\text{K}^{-1}$$

We will use this value of R when we consider the energy involved in gas expansion and compression.

Common values of the gas constant are listed in Table 6.3, and you will have a chance to use all of them in the Practice Examples and end-of-chapter exercises in this chapter. A general strategy for applying the ideal gas equation is illustrated in the diagram in the margin.

The General Gas Equation

In Examples 6-4 and 6-5, the ideal gas equation is applied to a single set of conditions ($P, V, n,$ and T). Sometimes a gas is described under two different

EXAMPLE 6-4   Calculating a Gas Volume with the Ideal Gas Equation

What is the volume occupied by 13.7 g $Cl_2(g)$ at 45 °C and 745 mmHg?

Analyze

This is a relatively straightforward application of the ideal gas equation. We are given an amount of gas (in grams), a pressure (in mmHg), and a temperature (in °C). Before using the ideal gas equation, we must express the amount in moles, the pressure in atmospheres, and the temperature in Kelvin. Include units throughout the calculation to ensure that the final result has acceptable units.

Solve

$$P = 745 \text{ mmHg} \times \frac{1 \text{ atm}}{760 \text{ mmHg}} = 0.980 \text{ atm}$$

$$V = ?$$

$$n = 13.7 \text{ g Cl}_2 \times \frac{1 \text{ mol Cl}_2}{70.91 \text{ g Cl}_2} = 0.193 \text{ mol Cl}_2$$

$$R = 0.08206 \text{ atm L mol}^{-1} \text{K}^{-1}$$

$$T = 45 \text{ °C} + 273 = 318 \text{ K}$$

Divide both sides of the ideal gas equation by P to solve for V.

$$\frac{PV}{P} = \frac{nRT}{P} \quad \text{and} \quad V = \frac{nRT}{P}$$

$$V = \frac{nRT}{P} = \frac{0.193 \text{ mol} \times 0.08206 \text{ atm L mol}^{-1} \text{K}^{-1} \times 318 \text{ K}}{0.980 \text{ atm}} = 5.14 \text{ L}$$

Assess

A useful check of the calculated result is to make certain the units cancel properly. In the setup above, all units cancel except for L, a unit of volume. Keep in mind that when canceling units, such a unit as mol^{-1} is the same as $1/\text{mol}$. Thus, $\text{mol} \times \text{mol}^{-1} = 1$ and $\text{K} \times \text{K}^{-1} = 1$.

PRACTICE EXAMPLE A: What is the volume occupied by 20.2 g $NH_3(g)$ at −25°C and 752 mmHg?

PRACTICE EXAMPLE B: At what temperature will a 13.7 g Cl_2 sample exert a pressure of 0.993 bar when confined in a 7.50 L container?

EXAMPLE 6-5   Calculating a Gas Pressure with the Ideal Gas Equation

What is the pressure, in kilopascals, exerted by 1.00×10^{20} molecules of N_2 in a 305 mL flask at 175°C?

Analyze

We are given an amount of gas (in molecules), a volume (in mL), and a temperature (in °C). Before using these quantities in the ideal gas equation, we must express the amount in moles, the volume in liters, and the temperature in Kelvin. Include units throughout the calculation to ensure that the final result has acceptable units.

Solve

Because we seek a pressure in kilopascals, let us use the form of the ideal gas equation having

$$R = 8.3145 \text{ Pa m}^3 \text{ mol}^{-1} \text{K}^{-1}$$

The first step is to convert from molecules to moles of a gas, n.

$$n = 1.00 \times 10^{20} \text{ molecules N}_2 \times \frac{1 \text{ mol N}_2}{6.022 \times 10^{23} \text{ molecules N}_2}$$

$$= 0.000166 \text{ mol N}_2$$

(continued)

Convert from milliliters to liters and then to cubic meters (recall Figure 1-9).

$$V = 305 \text{ mL} \times \frac{1 \text{ L}}{1000 \text{ mL}} \times \frac{1 \text{ m}^3}{1000 \text{ L}} = 3.05 \times 10^{-4} \text{ m}^3$$

Express gas temperature on the Kelvin scale.

$$T = 175°C + 273 = 448 \text{ K}$$

Rearrange the ideal gas equation to the form $P = nRT/V$, and substitute the above data.

$$P = \frac{nRT}{V} = \frac{0.000166 \text{ mol} \times 8.3145 \text{ Pa m}^3 \text{ mol}^{-1} \text{ K}^{-1} \times 448 \text{ K}}{3.05 \times 10^{-4} \text{ m}^3}$$

$$= 2.03 \times 10^3 \text{ Pa}$$

Finally, convert the pressure to the unit kilopascal.

$$P = 2.03 \times 10^3 \text{ Pa} \times \frac{1 \text{ kPa}}{1000 \text{ Pa}} = 2.03 \text{ kPa}$$

Assess

Again, we see from the cancellation of units above that only the desired unit—a pressure unit—remains.

PRACTICE EXAMPLE A: How many moles of He(g) are in a 5.00 L storage tank filled with helium at 10.5 atm pressure at 30.0 °C?

PRACTICE EXAMPLE B: How many molecules of N_2(g) remain in an ultrahigh vacuum chamber of 3.45 m^3 volume when the pressure is reduced to 6.67×10^{-7} Pa at 25 °C?

sets of conditions. Here, the ideal gas equation must be applied twice—to an initial condition and a final condition. That is,

Initial condition (i)	Final condition (f)
$P_i V_i = n_i R T_i$	$P_f V_f = n_f R T_f$
$R = \dfrac{P_i V_i}{n_i T_i}$	$R = \dfrac{P_f V_f}{n_f T_f}$

The above expressions are equal to each other because each is equal to R.

$$\frac{P_i V_i}{n_i T_i} = \frac{P_f V_f}{n_f T_f} \tag{6.12}$$

Expression (6.12) is called the **general gas equation**. It is often applied in cases in which one or two of the gas properties are held constant, and the equation can be simplified by eliminating these constants. For example, if a constant mass of gas is subject to changes in temperature, pressure, and volume, n_i and n_f cancel because they are equal (constant moles); thus we have

$$\frac{P_i V_i}{T_i} = \frac{P_f V_f}{T_f} \quad (n \text{ constant})$$

This equation is sometimes referred to as the *combined gas law*. In Example 6-6, both volume and mass are constant, and this establishes the simple relationship between gas pressure and temperature known as *Amontons's law*: The pressure of a fixed amount of gas confined to a fixed volume is directly proportional to the Kelvin temperature.

Using the Gas Laws

When confronted with a problem involving gases, students sometimes wonder which gas equation to use. Gas law problems can often be thought of in more than one way. When a problem involves a comparison of two gases or two states (initial and final) of a single gas, use the general gas equation (6.12) after eliminating any term (n, P, T, V) that remains constant. Otherwise, use the ideal gas equation (6.11).

EXAMPLE 6-6   Applying the General Gas Equation

The situation pictured in Figure 6-10(a) is changed to that in Figure 6-10(b). What is the gas pressure in Figure 6-10(b)?

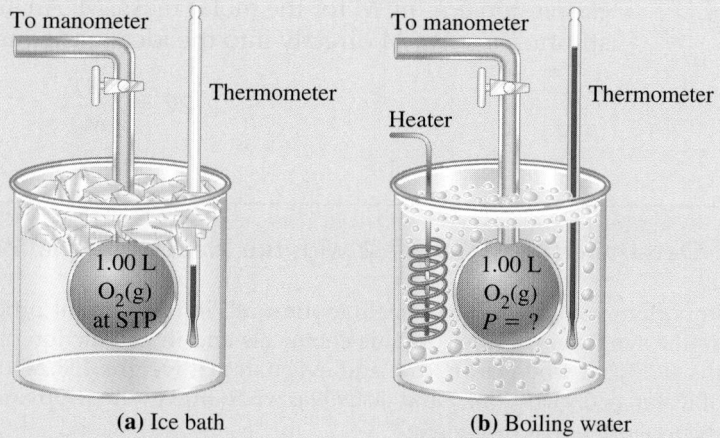

(a) Ice bath   (b) Boiling water

▲ FIGURE 6-10
Pressure of a gas as a function of temperature—Example 6-6 visualized
The amount of gas and volume are held constant. **(a)** 1.00 L $O_2(g)$ at STP; **(b)** 1.00 L $O_2(g)$ at 100 °C.

Analyze

Identify the quantities in the general gas equation that remain constant. Cancel out these quantities and solve the equation that remains.

Solve

In this case, the amount of O_2 is constant ($n_i = n_f$) and the volume is constant ($V_i = V_f$).

$$\frac{P_i V_i}{n_i T_i} = \frac{P_f V_f}{n_f T_f} \quad \text{and} \quad \frac{P_i}{T_i} = \frac{P_f}{T_f} \quad \text{and} \quad P_f = P_i \times \frac{T_f}{T_i}$$

Since $P_i = 1.00$ bar, $T_i = 273$ K, and $T_f = 373$ K, then

$$P_f = 1.00\,\text{bar} \times \frac{373\,\text{K}}{273\,\text{K}} = 1.37\,\text{bar}$$

Assess

We can base our check on a qualitative, intuitive understanding of what happens when a gas is heated in a closed container. Its pressure increases (possibly to the extent that the container bursts). If, by error, we had used the ratio of temperatures 273 K/373 K, the final pressure would have been less than 1.00 bar—an impossible result.

PRACTICE EXAMPLE A: A 1.00 mL sample of $N_2(g)$ at 36.2 °C and 2.14 atm is heated to 37.8 °C, and the pressure changed to 1.02 atm. What volume does the gas occupy at this final temperature and pressure?

PRACTICE EXAMPLE B: Suppose that in Figure 6-10 we want the pressure to remain at 1.00 bar when the $O_2(g)$ is heated to 100 °C. What mass of $O_2(g)$ must we release from the flask?

6-4   Applications of the Ideal Gas Equation

Although the ideal gas equation can always be used as it was presented in equation (6.11), it is useful to recast it into slightly different forms for some applications. We will consider two such applications in this section: determination of molar masses and gas densities.

Molar Mass Determination

If we know the volume of a gas at a fixed temperature and pressure, we can solve the ideal gas equation for the amount of the gas in moles. Because the number of moles of gas (n) is equal to the mass (m) of gas divided by the molar mass (M), if we know the mass and number of moles of gas, we can solve the expression $n = m/M$ for the molar mass, M. An alternative is to make the substitution $n = m/M$ directly into the ideal gas equation.

$$PV = \frac{mRT}{M} \qquad \textbf{(6.13)}$$

EXAMPLE 6-7   Determining a Molar Mass with the Ideal Gas Equation

Propylene is an important commercial chemical (about ninth in the amount produced among manufactured chemicals) used in the synthesis of other organic chemicals and in production of plastics (polypropylene). A glass vessel weighs 40.1305 g when clean, dry, and evacuated; it weighs 138.2410 g when filled with water at 25.0 °C (density of water = 0.9970 g/mL) and 40.2959 g when filled with propylene gas at 740.3 mmHg and 24.0 °C. What is the molar mass of propylene?

Analyze

We are given a pressure (in mmHg), a temperature (in °C), and information that will enable us to determine the amount of gas (in grams) and the volume of the vessel. If we express these quantities in Kelvin, atmospheres, moles, and liters, respectively, then we can use equation (6.13), with $R = 0.08206 \text{ atm L K}^{-1} \text{ mol}^{-1}$, to calculate the molar mass of the gas.

Solve

First determine the mass of water required to fill the vessel.

mass of water to fill vessel = 138.2410 g − 40.1305 g
= 98.1105 g

Use the density of water in a conversion factor to obtain the volume of water (and hence, the volume of the glass vessel).

volume of water (volume of vessel) = $98.1105 \text{ g H}_2\text{O} \times \dfrac{1 \text{ mL H}_2\text{O}}{0.9970 \text{ g H}_2\text{O}}$
= 98.41 mL = 0.09841 L

The mass of the gas is the difference between the weight of the vessel filled with propylene gas and the weight of the empty vessel.

mass of gas = 40.2959 g − 40.1305 g = 0.1654 g

The values of temperature and pressure are given.

$T = 24.0\ °\text{C} + 273.15 = 297.2 \text{ K}$

$P = 740.3 \text{ mmHg} \times \dfrac{1 \text{ atm}}{760 \text{ mmHg}} = 0.9741 \text{ atm}$

Substitute data into the rearranged version of equation (6.13).

$M = \dfrac{mRT}{PV} = \dfrac{0.1654 \text{ g} \times 0.08206 \text{ atm L mol}^{-1} \text{K}^{-1} \times 297.2 \text{ K}}{0.9741 \text{ atm} \times 0.09841 \text{ L}}$

$= 42.08 \text{ g mol}^{-1}$

Assess

Cancellations leave the units g and mol^{-1}. The unit g mol^{-1} or g/mol is that for molar mass, the quantity we are seeking. We can use another approach to solving this problem. We can substitute the pressure (0.9741 atm), temperature (297.2 K), and volume (0.09841 L) into the ideal gas equation to calculate the number of moles in the gas sample (0.003931 mol). Because the sample contains 0.003931 mol and has a mass of 0.165 g, the molar mass is 0.165 g/0.003931 mol = 42.0 g mol^{-1}. The advantage of this alternative approach is that it makes use of only the ideal gas equation; you do not have to memorize or derive equation (6.13) for the cases when you might need it.

PRACTICE EXAMPLE A: The same glass vessel used in Example 6-7 is filled with an unknown gas at 772 mmHg and 22.4 °C. The gas-filled vessel weighs 40.4868 g. What is the molar mass of the gas?

PRACTICE EXAMPLE B: A 1.27 g sample of an oxide of nitrogen, believed to be either NO or N_2O, occupies a volume of 1.07 L at 25 °C and 737 mmHg. Which oxide is it?

Equation (6.13) is used to determine a molar mass in Example 6-7, but note that the equation can also be used when the molar mass of a gas is known and the mass of a particular sample of the gas is sought.

Suppose that we want to determine the formula of an unknown hydrocarbon. With combustion analysis we can establish the mass percent composition, and from this, we can determine the empirical formula. The method of Example 6-7 gives us a molar mass, in g mol^{-1}, which is numerically equal to the molecular mass, in u. This is all the information we need to establish the true molecular formula of the hydrocarbon (see Exercise 96).

Gas Densities

To determine the density of a gas, we can start with the density equation, $d = m/V$. Then we can express the mass of gas as the product of the number of moles of gas and the molar mass: $m = n \times M$. This leads to

$$d = \frac{m}{V} = \frac{n \times M}{V} = \frac{n}{V} \times M$$

Now, with the ideal gas equation, we can replace n/V by its equivalent, P/RT, to obtain

$$d = \frac{m}{V} = \frac{MP}{RT} \tag{6.14}$$

The density of a gas at STP can easily be calculated by dividing its molar mass by the molar volume (22.7 L/mol). For $O_2(g)$ at STP, for example, the density is 32.0 g/22.7 L = 1.41 g/L. Equation (6.14) can be used for other conditions of temperature and pressure.

KEEP IN MIND

that gas densities are typically much smaller than those of liquids and solids. Gas densities are usually expressed in grams per *liter* rather than grams per *milliliter*.

EXAMPLE 6-8   Using the Ideal Gas Equation to Calculate a Gas Density

What is the density of oxygen gas (O_2) at 298 K and 0.987 atm?

Analyze

The gas is identified, and therefore the molar mass can be calculated. We are given a temperature in Kelvin and a pressure in atmospheres, so we can use equation (6.14) directly with $R = 0.08206$ atm L K^{-1} mol^{-1}.

Solve

The molar mass of O_2 is 32.0 g mol^{-1}. Now, use equation (6.14).

$$d = \frac{m}{V} = \frac{MP}{RT} = \frac{32.00 \text{ g mol}^{-1} \times 0.987 \text{ atm}}{0.08206 \text{ atm L mol}^{-1} \text{ K}^{-1} \times 298 \text{ K}} = 1.29 \text{ g/L}$$

Assess

We can solve this problem in another way. To calculate the density of a gas at a certain temperature and pressure, use a 1.00 L sample of the gas. The mass of a 1.00 L sample is equal to the density in grams per liter. To calculate the mass of a 1.00 L sample, first use the ideal gas equation to calculate the number of moles in the sample, and then convert the amount in moles to an amount in grams by using the molar mass as a conversion factor. In the present case, the amount of $O_2(g)$ in a 1.00 L sample at 0.987 atm and 298 K is 0.0404 mol O_2, or 1.29 g O_2. Because a 1.00 L sample of O_2 at this temperature and pressure has a mass of 1.29 g, the density is 1.29 g/L.

PRACTICE EXAMPLE A: What is the density of helium gas at 298 K and 0.987 atm? Based on your answer, explain why we can say that helium is "lighter than air."

PRACTICE EXAMPLE B: The density of a sample of gas is 1.00 g/L at 745 mmHg and 109 °C. What is the molar mass of the gas?

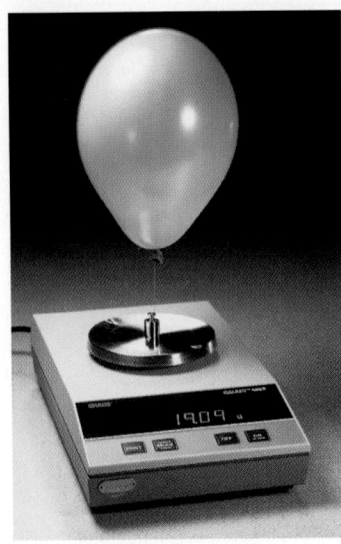

▲ FIGURE 6-11
The helium-filled balloon
exerts a lifting force on the
20.00 g weight, so that the
balloon and the weight
together weigh only 19.09 g.

The density of gases differs from that of solids and liquids in two important ways.

1. Gas densities depend strongly on pressure and temperature, increasing as the gas pressure increases and decreasing as the temperature increases. Densities of liquids and solids also depend somewhat on temperature, but they depend far less on pressure.
2. The density of a gas is directly proportional to its molar mass. No simple relationship exists between density and molar mass for liquids and solids.

An important application of gas densities is in establishing conditions for lighter-than-air balloons. A gas-filled balloon will rise in the atmosphere only if the density of the gas is less than that of the surrounding air. Because gas densities are directly proportional to molar masses, the lower the molar mass of the gas, the greater its lifting power. The lowest molar mass is that of hydrogen, but hydrogen is flammable and forms explosive mixtures with air. The explosion of the dirigible *Hindenburg* in 1937 spelled the end to transoceanic travel by hydrogen-filled airships. Now, airships, such as the Goodyear blimps, use helium, which has a molar mass only twice that of hydrogen (Fig. 6-11) but is inert. Hydrogen is still used for weather and other observational balloons.

Another alternative is to fill a balloon with *hot* air. Equation (6.14) indicates that the density of a gas is *inversely* proportional to temperature. Hot air is less dense than cold air. However, because the density of air decreases rapidly with altitude, there is a limit to how high a hot-air balloon or any gas-filled balloon can rise.

6-5 Gases in Chemical Reactions

Reactions involving gases as reactants or products (or both) are no strangers to us. We now have a new tool to apply to reaction stoichiometry calculations: the ideal gas equation. Specifically, we can now handle information about gases in terms of volumes, temperatures, and pressures, as well as by mass and amount in moles. A practical application is the nitrogen-forming reaction in an automobile air-bag safety system, which utilizes the rapid decomposition of sodium azide.

$$2\,NaN_3(s) \xrightarrow{\Delta} 2\,Na(l) + 3\,N_2(g)$$

The essential components of the system are an ignition device and a pellet containing sodium azide and appropriate additives. When activated, the system inflates an air bag in 20 to 60 ms and converts Na(l) to a harmless solid residue.

For reactions involving gases, we can (1) use stoichiometric factors to relate the amount of a gas to amounts of other reactants or products, and (2) use the ideal gas equation to relate the amount of gas to volume, temperature, and pressure. In Example 6-9, we use this approach to determine the volume of $N_2(g)$ produced in a typical air-bag system.

Law of Combining Volumes

If the reactants and products involved in a stoichiometric calculation are gases, sometimes we can use a particularly simple approach. Consider this reaction.

$$2\,NO(g) + O_2(g) \longrightarrow 2\,NO_2(g)$$
$$2\,mol\,NO(g) + 1\,mol\,O_2(g) \longrightarrow 2\,mol\,NO_2(g)$$

EXAMPLE 6-9   **Using the Ideal Gas Equation in Reaction Stoichiometry Calculations**

What volume of N_2, measured at 735 mmHg and 26 °C, is produced when 75.0 g NaN_3 is decomposed?

$$2\,NaN_3(s) \xrightarrow{\Delta} 2\,Na(l) + 3\,N_2(g)$$

Analyze

The following conversions are required.

$$g\,NaN_3 \longrightarrow mol\,NaN_3 \longrightarrow mol\,N_2 \longrightarrow L\,N_2$$

The molar mass of NaN_3 is used for the first conversion. The second conversion makes use of a stoichiometric factor constructed from the coefficients of the chemical equation. The ideal gas equation is used to complete the final conversion.

Solve

$$?\,mol\,N_2 = 75.0\,g\,NaN_3 \times \frac{1\,mol\,NaN_3}{65.01\,g\,NaN_3} \times \frac{3\,mol\,N_2}{2\,mol\,NaN_3} = 1.73\,mol\,N_2$$

$$P = 735\,mmHg \times \frac{1\,atm}{760\,mmHg} = 0.967\,atm$$

$$V = ?$$
$$n = 1.73\,mol$$
$$R = 0.08206\,atm\,L\,mol^{-1}\,K^{-1}$$
$$T = 26\,°C + 273 = 299\,K$$
$$V = \frac{nRT}{P} = \frac{1.73\,mol \times 0.08206\,atm\,L\,mol^{-1}\,K^{-1} \times 299\,K}{0.967\,atm} = 43.9\,L$$

Assess

75.0 g NaN_3 is slightly more than one mole ($M \approx 65$ g/mol). From this amount of NaN_3 we should expect a little more than 1.5 mol $N_2(g)$. At 0 °C and 1 atm, 1.5 mol $N_2(g)$ would occupy a volume of $1.5 \times 22.4 = 33.6$ L. Because the temperature is higher than 0 °C and the pressure is lower than 1 atm, the sample should have a volume somewhat greater than 33.6 L.

PRACTICE EXAMPLE A: How many grams of NaN_3 are needed to produce 20.0 L of $N_2(g)$ at 30.0 °C and 776 mmHg?

PRACTICE EXAMPLE B: How many grams of $Na(l)$ are produced per liter of $N_2(g)$ formed in the decomposition of sodium azide if the gas is collected at 25 °C and 1.0 bar?

Suppose the gases are compared at the same T and P. Under these conditions, one mole of gas occupies a particular volume, call it V liters; two moles of gas occupy $2V$ liters; and so on.

$$2\,V\,L\,NO(g) + V\,L\,O_2(g) \longrightarrow 2\,V\,L\,NO_2(g)$$

If we divide each coefficient by V, we get the following result:

$$2\,L\,NO(g) + 1\,L\,O_2(g) \longrightarrow 2\,L\,NO_2(g)$$

Thus, the volume ratio of the gases consumed and produced in a chemical is the same as the mole ratio, provided the volumes are all measured at the same temperature and pressure.

What we have just done is to develop, in modern terms, Gay-Lussac's **law of combining volumes**. We previewed this law on page 201 by suggesting that the volumes of gases involved in a reaction are in the ratio of small whole numbers. The small whole numbers are simply the stoichiometric coefficients in the balanced equation. We apply this law in Example 6-10.

EXAMPLE 6-10   Applying the Law of Combining Volumes

Zinc blende, ZnS, is the most important zinc ore. Roasting (strong heating) of ZnS in oxygen is the first step in the commercial production of zinc.

$$2\,ZnS(s) + 3\,O_2(g) \xrightarrow{\Delta} 2\,ZnO(s) + 2\,SO_2(g)$$

What volume of $SO_2(g)$ can be obtained from $1.00\,L\,O_2(g)$ and excess $ZnS(s)$? Both gases are measured at 25 °C and 745 mmHg.

Analyze

The reactant and product being compared are both gases, and both are at the same temperature and pressure. Therefore, we can use the law of combining volumes and treat the coefficients in the balanced chemical equation as if they had units of liters.

Solve

The stoichiometric factor (shown below in blue) converts from $L\,O_2(g)$ to $L\,SO_2(g)$.

$$?\,L\,SO_2(g) = 1.00\,L\,O_2(g) \times \frac{2\,L\,SO_2(g)}{3\,L\,O_2(g)} = 0.667\,L\,SO_2(g)$$

Assess

Some students would solve this problem by using the following sequence of conversions: $L\,O_2 \longrightarrow$ mol $O_2 \longrightarrow$ mol $SO_2 \longrightarrow L\,SO_2$. This approach is acceptable but not as simple as the approach we used.

PRACTICE EXAMPLE A: The first step in making nitric acid is to convert ammonia to nitrogen monoxide. This is done under conditions of high temperature and in the presence of a platinum catalyst. What volume of $O_2(g)$ is consumed per liter of $NO(g)$ formed?

$$4\,NH_3(g) + 5\,O_2(g) \xrightarrow[850\,°C]{Pt} 4\,NO(g) + 6\,H_2O(g)$$

PRACTICE EXAMPLE B: If all gases are measured at the same temperature and pressure, what volume of $NH_3(g)$ is produced when 225 L $H_2(g)$ are consumed in the reaction $N_2(g) + H_2(g) \longrightarrow NH_3(g)$ (not balanced)?

6-6   CONCEPT ASSESSMENT

Would the answer in Example 6-10 be greater than, less than, or equal to 0.677 L if (a) both gases were measured at STP; (b) if the O_2 were measured at STP and the SO_2 were measured at 25 °C and 745 mmHg?

6-6   Mixtures of Gases

The simple gas laws, such as Boyle's and Charles's laws, were based on the behavior of air—a mixture of gases. So, the simple gas laws and the ideal gas equation apply to a *mixture* of nonreactive gases as well as to individual gases. Where possible, the simplest approach to working with gaseous mixtures is just to use for the value of n the *total* number of moles of the gaseous mixture (n_{tot}).

As a specific example, consider a mixture of gases in a vessel of fixed volume V at temperature T. The total pressure of the mixture is determined by the total number of moles:

$$P_{tot} = \frac{n_{tot}\,RT}{V} \qquad (T \text{ constant}, V \text{ constant}) \qquad \textbf{(6.15)}$$

For fixed values of T and P, the total volume of a mixture of gases is also determined by the total number of moles:

$$V_{tot} = \frac{n_{tot}\,RT}{P} \qquad (T \text{ constant}, P \text{ constant})$$

EXAMPLE 6-11   Applying the Ideal Gas Equation to a Mixture of Gases

What is the pressure, in bar, exerted by a mixture of 1.0 g H_2 and 5.00 g He when the mixture is confined to a volume of 5.0 L at 20 °C?

Analyze

For fixed T and V, the total pressure of a mixture of gases is determined by the total number of moles of gas: $P_{tot} = n_{tot}RT/V$.

Solve

$$n_{tot} = \left(1.0 \text{ g } H_2 \times \frac{1 \text{ mol } H_2}{2.02 \text{ g } H_2}\right) + \left(5.00 \text{ g He} \times \frac{1 \text{ mol He}}{4.003 \text{ g He}}\right)$$

$$= 0.50 \text{ mol } H_2 + 1.25 \text{ mol He} = 1.75 \text{ mol gas}$$

$$P = \frac{1.75 \text{ mol} \times 0.0831 \text{ bar L mol}^{-1}\text{K}^{-1} \times 293 \text{ K}}{5.0 \text{ L}} = 8.5 \text{ bar}$$

Assess

It is also possible to solve this problem by starting from equation (6.12). Because 1 mol of ideal gas occupies 22.7 L at 0 °C and 1 bar, the pressure exerted by 1.75 mol of gas in a 5.0 L vessel at 293 K is $(1.75 \text{ mol}/1.00 \text{ mol}) \times (293 \text{ K}/273 \text{ K}) \times (22.7 \text{ L}/5.0 \text{ L}) \times 1.0 \text{ bar} = 8.5 \text{ bar}$.

PRACTICE EXAMPLE A: What will be the total gas pressure if 12.5 g Ne is added to the mixture of gases described in Example 6-11 and the temperature is then raised to 55 °C? [*Hint:* What is the new number of moles of gas? What effect does raising the temperature have on the pressure of a gas at constant volume?]

PRACTICE EXAMPLE B: 2.0 L of $O_2(g)$ and 8.0 L of $N_2(g)$, each at 0.00 °C and 1.00 atm, are mixed together. The nonreactive gaseous mixture is compressed to occupy 2.0 L at 298 K. What is the pressure exerted by this mixture?

John Dalton made an important contribution to the study of gaseous mixtures. He proposed that in a mixture, each gas expands to fill the container and exerts the same pressure (called its **partial pressure**) that it would if it were alone in the container. **Dalton's law of partial pressures** states that the total pressure of a mixture of gases is the sum of the partial pressures of the components of the mixture, as shown in Figure 6-12. For a mixture of gases, A, B, and so on,

$$P_{tot} = P_A + P_B + \cdots \tag{6.16}$$

In a gaseous mixture of n_A moles of A, n_B moles of B, and so on, the volume each gas would individually occupy at a pressure equal to P_{tot} is

$$V_A = n_A RT/P_{tot}; V_B = n_B RT/P_{tot}; \cdots$$

The total volume of the gaseous mixture is

$$V_{tot} = V_A + V_B \cdots$$

and the commonly used expression *percent by volume* is

$$\text{volume \% A} = \frac{V_A}{V_{tot}} \times 100\%; \text{Volume \% B} = \frac{V_B}{V_{tot}} \times 100\%; \cdots$$

We can derive a particularly useful expression from the following ratios,

$$\frac{P_A}{P_{tot}} = \frac{n_A(RT/V_{tot})}{n_{tot}(RT/V_{tot})} = \frac{n_A}{n_{tot}} \quad \text{and} \quad \frac{V_A}{V_{tot}} = \frac{n_A(RT/P_{tot})}{n_{tot}(RT/P_{tot})} = \frac{n_A}{n_{tot}}$$

which means that

$$\frac{n_A}{n_{tot}} = \frac{P_A}{P_{tot}} = \frac{V_A}{V_{tot}} = x_A \tag{6.17}$$

KEEP IN MIND

that when this expression is used, $V_A = V_B = \cdots = V_{tot}$

KEEP IN MIND

that when using this expression, $P_A = P_B = \cdots = P_{tot}$

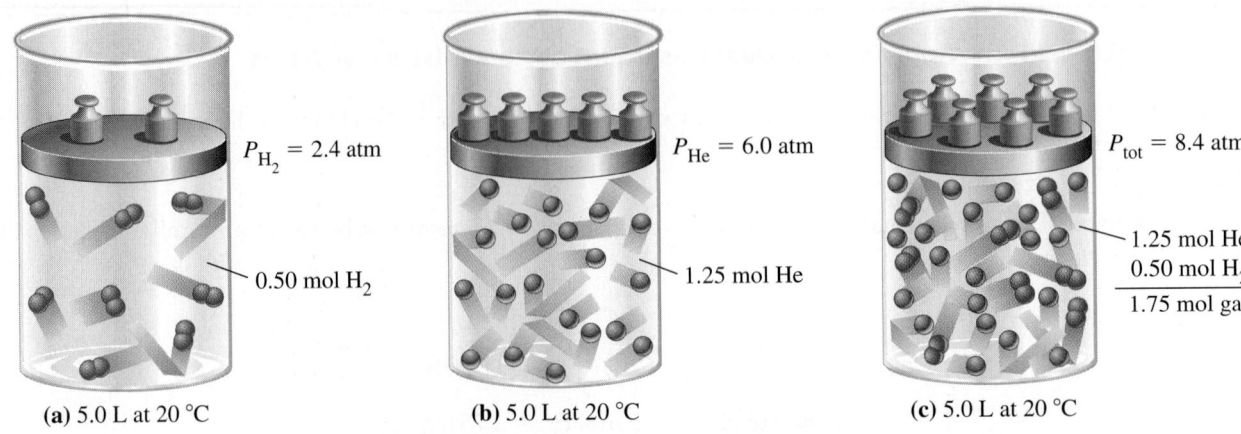

(a) 5.0 L at 20 °C   **(b)** 5.0 L at 20 °C   **(c)** 5.0 L at 20 °C

▲ FIGURE 6-12
Dalton's law of partial pressures illustrated
The pressure of each gas is proportional to the number of moles of gas. The total
pressure is the sum of the partial pressures of the individual gases.

The term n_A/n_{tot} is given a special name, the mole fraction of A, x_A. The **mole
fraction** of a component in a mixture is the fraction of all the molecules in the
mixture contributed by that component. The sum of all the mole fractions in a
mixture is one.

As illustrated in Example 6-12, we can often think about mixtures of gases
in more than one way.

EXAMPLE 6-12   Calculating the Partial Pressures in a Gaseous Mixture

What are the partial pressures of H_2 and He in the gaseous mixture described in Example 6-11?

Analyze

The ideal gas equation can be applied to each gas individually to obtain the partial pressure of each gas.

Solve

One approach involves a direct application of Dalton's law in which we calculate the pressure that each gas
would exert if it were alone in the container.

$$P_{H_2} = \frac{n_{H_2} \times RT}{V} = \frac{0.50 \text{ mol} \times 0.0821 \text{ atm L mol}^{-1} \text{K}^{-1} \times 293 \text{ K}}{5.0 \text{ L}} = 2.4 \text{ atm}$$

$$P_{He} = \frac{n_{He} \times RT}{V} = \frac{1.25 \text{ mol} \times 0.0821 \text{ atm L mol}^{-1} \text{K}^{-1} \times 293 \text{ K}}{5.0 \text{ L}} = 6.0 \text{ atm}$$

Expression (6.17) gives us a simpler way to answer the question because we already know the number of
moles of each gas and the total pressure from Example 6-11 ($P_{tot} = 8.4$ atm).

$$P_{H_2} = \frac{n_{H_2}}{n_{tot}} \times P_{tot} = \frac{0.50}{1.75} \times 8.4 \text{ atm} = 2.4 \text{ atm}$$

$$P_{He} = \frac{n_{He}}{n_{tot}} \times P_{tot} = \frac{1.25}{1.75} \times 8.4 \text{ atm} = 6.0 \text{ atm}$$

Assess

An effective way of checking an answer is to obtain the same answer when the problem is done in different
ways, as was the case here.

PRACTICE EXAMPLE A: A mixture of 0.197 mol $CO_2(g)$ and 0.00278 mol $H_2O(g)$ is held at 30.0 °C and 2.50 atm.
What is the partial pressure of each gas?

PRACTICE EXAMPLE B: The percent composition of air by volume is 78.08% N_2, 20.95% O_2, 0.93% Ar, and 0.036%
CO_2. What are the partial pressures of these four gases in a sample of air at a barometric pressure of 748 mmHg?

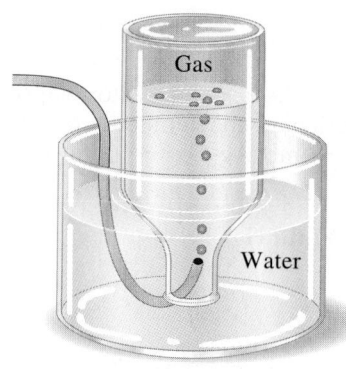 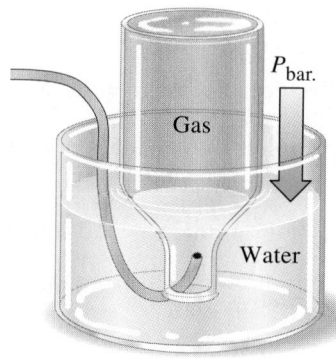

▲ FIGURE 6-13
Collecting a gas over water
The bottle is filled with water and its open end is held below the water level in the container. Gas from a gas-generating apparatus is directed into the bottle. As gas accumulates in the bottle, water is displaced from the bottle into the container. To make the total gas pressure in the bottle equal to barometric pressure, the position of the bottle must be adjusted so that the water levels inside and outside the bottle are the same.

6-7   CONCEPT ASSESSMENT

Without doing a detailed calculation, state the outcome(s) you would expect in Figure 6-12(c) if an additional 0.50 mol H_2 were added to the cylinder in Figure 6-12(a): **(a)** the mole fraction of H_2 would double; **(b)** the partial pressure of He would remain the same; **(c)** the mole fraction of He would remain the same; **(d)** the total gas pressure would increase by 50%; **(e)** the total mass of gas would increase by 1.0 g.

◀ Some early experimenters used mercury in the pneumatic trough so as to be able to collect gases soluble in water.

The device pictured in Figure 6-13, a *pneumatic trough*, played a crucial role in isolating gases in the early days of chemistry. The method works, of course, only for gases that are insoluble in and do not react with the liquid being displaced. Many important gases meet these criteria. For example, H_2, O_2, and N_2 are all essentially insoluble in and unreactive with water.

A gas collected in a pneumatic trough filled with water is said to be *collected over water* and is "wet." It is a mixture of two gases—the desired gas and water vapor. The gas being collected expands to fill the container and exerts its partial pressure, P_{gas}. Water vapor, formed by the evaporation of liquid water, also fills the container and exerts a partial pressure, P_{H_2O}. The pressure of the water vapor depends only on the temperature of the water, as shown in Table 6.4.

According to Dalton's law, the total pressure of the wet gas is the sum of the two partial pressures. The total pressure can be made equal to the prevailing pressure of the atmosphere (barometric pressure) by adjusting the position of the bottle; thus we can write

$$P_{tot} = P_{bar.} = P_{gas} + P_{H_2O} \quad \text{or} \quad P_{gas} = P_{bar.} - P_{H_2O}$$

Once P_{gas} has been established, it can be used in stoichiometric calculations as illustrated in Example 6-13.

TABLE 6.4   Vapor Pressure of Water at Various Temperatures

Temperature, °C	Vapor Pressure, mmHg
15.0	12.79
17.0	14.53
19.0	16.48
21.0	18.65
23.0	21.07
25.0	23.76
30.0	31.82
50.0	92.51

EXAMPLE 6-13   Collecting a Gas over a Liquid (Water)

In the following reaction, 81.2 mL of $O_2(g)$ is collected over water at 23 °C and barometric pressure 751 mmHg. What mass of $Ag_2O(s)$ decomposed? (The vapor pressure of water at 23 °C is 21.1 mmHg.)

$$2\,Ag_2O(s) \longrightarrow 4\,Ag(s) + O_2(g)$$

(continued)

Analyze

The key concept is that the gas collected is wet, that is, a *mixture* of $O_2(g)$ and water vapor. Use $P_{bar.} = P_{O_2} + P_{H_2O}$ to calculate P_{O_2}, and then use the ideal gas equation to calculate the number of moles of O_2. The following conversions are used to complete the calculation: $mol\ O_2 \longrightarrow mol\ Ag_2O \longrightarrow g\ Ag_2O$.

Solve

$$P_{O_2} = P_{bar.} - P_{H_2O} = 751\ mmHg - 21.1\ mmHg = 730\ mmHg$$

$$P_{O_2} = 730\ mmHg \times \frac{1\ atm}{760\ mmHg} = 0.961\ atm$$

$$V = 81.2\ mL = 0.0812\ L$$

$$n = ?$$

$$R = 0.08206\ atm\ L\ mol^{-1}\ K^{-1}$$

$$T = 23\ °C + 273 = 296\ K$$

$$n = \frac{PV}{RT} = \frac{0.961\ atm \times 0.0812\ L}{0.08206\ atm\ L\ mol^{-1}\ K^{-1} \times 296\ K} = 0.00321\ mol$$

From the chemical equation we obtain a factor to convert from moles of O_2 to moles of Ag_2O. The molar mass of Ag_2O provides the final factor.

$$?\ g\ Ag_2O = 0.00321\ mol\ O_2 \times \frac{2\ mol\ Ag_2O}{1\ mol\ O_2} \times \frac{231.7\ g\ Ag_2O}{1\ mol\ Ag_2O} = 1.49\ g\ Ag_2O$$

Assess

The determination of the number of moles of O_2 in the sample is the key calculation. We can quickly estimate the number of moles of O_2 in the sample by using the fact that for typical conditions ($T \approx 298\ K$, $P \approx 760\ mmHg$), the molar volume of an ideal gas is about 24 L. The number of moles of gas (mostly O_2) in the sample is approximately $0.08\ L/24\ L \approx 0.003\ mol$. This estimate is quite close to the value calculated above.

PRACTICE EXAMPLE A: The reaction of aluminum with hydrochloric acid produces hydrogen gas. The balanced chemical equation for the reaction is given below.

$$2\ Al(s) + 6\ HCl(aq) \longrightarrow 2\ AlCl_3(aq) + 3\ H_2(g)$$

If 35.5 mL of $H_2(g)$ is collected over water at 26 °C and a barometric pressure of 755 mmHg, how many moles of HCl must have been consumed? (The vapor pressure of water at 26 °C is 25.2 mmHg.)

PRACTICE EXAMPLE B: An 8.07 g sample of impure Ag_2O decomposes into solid silver and $O_2(g)$. If 395 mL $O_2(g)$ is collected over water at 25 °C and 749.2 mmHg barometric pressure, then what is the percent by mass of Ag_2O in the sample? The vapor pressure of water at 25 °C is 23.8 mmHg.

6-7   Kinetic-Molecular Theory of Gases

Let us apply some terminology that we introduced in Section 1-1 on the scientific method: The simple gas laws and the ideal gas equation are used to predict gas behavior. They are *natural laws*. To explain the gas laws, we need a *theory*. One theory developed during the mid-nineteenth century is called the **kinetic-molecular theory of gases**. It is based on the *model* illustrated in Figure 6-14 and outlined as follows.

- A gas is composed of a very large number of extremely small particles (molecules or, in some cases, atoms) in constant, random, straight-line motion.
- Molecules of a gas are separated by great distances. The gas is mostly empty space. (The molecules are treated as so-called point masses, as though they have mass but no volume.)
- Molecules collide only fleetingly with one another and with the walls of their container, and most of the time molecules are not colliding.
- There are assumed to be no forces between molecules except very briefly during collisions. That is, each molecule acts independently of all the others and is unaffected by their presence, except during collisions.

- Individual molecules may gain or lose energy as a result of collisions. In a collection of molecules at constant temperature, however, *the total energy remains constant.*

The validity of this model can be ascertained only by comparing predictions based on the model with experimental facts. As we will see, the predictions based on this model are consistent with several observed macroscopic properties.

Derivation of Boyle's Law

In this section, we will demonstrate that the kinetic-molecular theory of gases provides a satisfactory explanation of Boyle's law. Boyle's law was stated mathematically in equation (6.5):

$$PV = a$$

The value of a depends on the number, N, of molecules in the sample and the temperature, T. Because pressure is a force per unit area, the key to deriving equation (6.5) is in assessing the forces associated with molecules hitting the walls of the container. Let's focus on a molecule traveling along the x direction toward a wall perpendicular to its path. The speed of the molecule is denoted by u_x. The force exerted on the wall by the molecule depends on the following factors.

1. The *frequency* of molecular collisions—the number of collisions per second. The higher this frequency, the greater the total force on the wall of the container. Collision frequency increases with the number of molecules per unit volume and with molecular speeds.

   collision frequency $\propto$ (molecular speed) $\times$ (molecules per unit volume)

   collision frequency $\propto (u_x) \times (N/V)$

2. The *momentum transfer*, or impulse. When a molecule hits the wall of a vessel, momentum is transferred as the molecule reverses direction. This momentum transfer is called an *impulse*. The magnitude of the impulse is directly proportional to the mass, m, of a molecule and its speed, u:

   impulse (momentum transfer) $\propto$ (mass of particle) $\times$ (molecular speed)

   impulse $\propto (mu_x)$

   The pressure of a gas (P) is the product of impulse and collision frequency. Thus, the complete proportionality expression for factors that affect pressure is

$$P \propto (mu_x) \times (u_x) \times (N/V) \propto (N/V)mu_x^2$$

At any instant, however, the molecules in a gas sample are traveling at different speeds. Therefore, we must replace u_x^2 in the expression above with the average value of u_x^2, which is denoted by $\overline{u_x^2}$. (In this context, the overbar reminds us that we are referring to the average value of u_x^2.) To better understand the concept of the average of the squares of speeds, consider five molecules with speeds 400, 450, 525, 585, and 600 m/s. We find the average of the squares of these speeds by squaring the speeds, adding the squares, and dividing by the number of particles, in this case five.

$$\overline{u^2} = \frac{(400 \text{ m/s})^2 + (450 \text{ m/s})^2 + (525 \text{ m/s})^2 + (585 \text{ m/s})^2 + (600 \text{ m/s})^2}{5}$$

$$= 2.68 \times 10^5 \text{ m}^2/\text{s}^2$$

Thus the proportionality expression for pressure becomes

$$P \propto \frac{N}{V}m\overline{u_x^2}$$

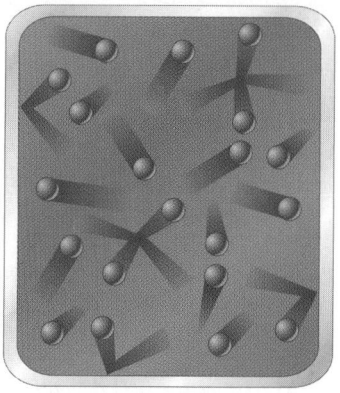

▲ FIGURE 6-14
Visualizing molecular motion
Molecules of a gas are in constant motion and collide with one another and with the container wall.

◀ The bar over a quantity means that the quantity can have a range of values and that the *average* value is intended.

We have one final factor to consider. There is absolutely nothing special about the x direction, so we should expect that $\overline{u_x^2} = \overline{u_y^2} = \overline{u_z^2} = \frac{1}{3}\overline{u^2}$, where the quantity $\overline{u^2} = \overline{u_x^2} + \overline{u_y^2} + \overline{u_z^2}$ is the average value of u^2 taking into account all the molecules, not just those moving in the x direction. $\overline{u^2}$ is called the mean-square speed. When we substitute $\frac{1}{3}\overline{u^2}$ for $\overline{u_x^2}$ in the expression above, we can also convert the expression into an equation because the factor of $\frac{1}{3}$ is, in fact, the proportionality constant. Thus, when all factors are properly considered, the result is

$$P = \frac{1}{3}\frac{N}{V}m\overline{u^2} \qquad (6.18)$$

This is the basic equation of the kinetic-molecular theory of gases. We can rewrite equation (6.18) as $PV = \frac{1}{3}Nm\overline{u^2}$. If it is true that $\overline{u^2}$ depends only on temperature, then equation (6.18) is in fact a mathematical statement of Boyle's law. We will not prove mathematically that $\overline{u^2}$ depends only on temperature. However, the discussion of the next section should help to convince you that this is true. Equation (6.18) leads to some interesting results, as we see next.

Distribution of Molecular Speeds

As pointed out in the derivation of Boyle's law, not all the molecules in a gas travel at the same speed. Because of the large number of molecules, we cannot know the speed of each molecule, but we can make a statistical prediction of how many molecules have a particular speed. The fraction, F, of molecules that have speed u is given by the following equation:

$$F(u) = 4\pi\left(\frac{M}{2\pi RT}\right)^{3/2} u^2 e^{-(Mu^2/2RT)}$$

The equation above is called the Maxwell distribution of speeds, in honor of James Clerk Maxwell who derived it in 1860. We will not attempt to derive the Maxwell distribution, because to do so requires complex mathematics. Instead, we will accept it as valid and use it to help us understand how the distribution of speeds depends on molar mass, M, and temperature, T.

A plot of $F(u)$ versus u is shown in Figure 6-15 for a sample of $H_2(g)$ at 0 °C. The shape of the distribution is easily justified. The distribution depends on the product of two opposing factors: a factor that is proportional to u^2 and an exponential factor, $e^{-Mu^2/2RT}$. As u increases, the u^2 factor increases from a value of zero, while the exponential factor decreases from a value of one. The u^2 factor favors the presence of molecules with high speeds and is responsible for there being few molecules with speeds near zero. The exponential factor favors low speeds and limits the number of molecules that can have high

▶ Maxwell derived his equation in 1860, but it took until 1955 for direct experimental verification of this equation to be made. The experimental determination of the molecular speed distribution is described on page 221.

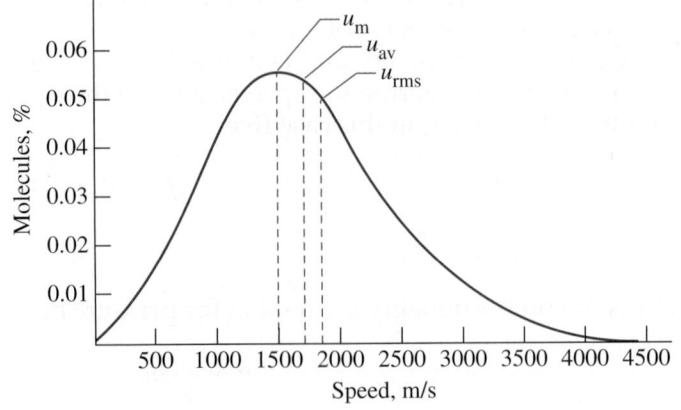

▶ FIGURE 6-15
Distribution of molecular speeds—hydrogen gas at 0 °C
The percentages of molecules with a certain speed are plotted as a function of the speed. Three different speeds are noted on the graph. The most probable speed is approximately 1500 m/s; the average speed is approximately 1700 m/s; and the root-mean-square speed is approximately 1800 m/s. Notice that $u_m < u_{av} < u_{rms}$.

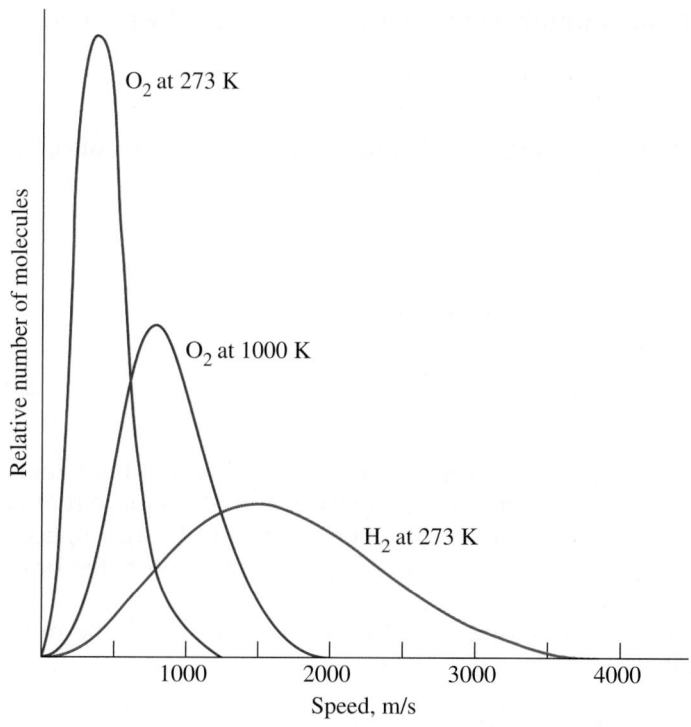

◀ The area under each curve is the same. Each curve is for the same total number of molecules. Raising T (or lowering M) is like stretching a rubber graph. The same number of molecules are spread out over a wider range of speeds.

◀ FIGURE 6-16
Distribution of molecular speeds—the effect of mass and temperature
The relative numbers of molecules with a certain speed are plotted as a function of the speed. Note the effect of temperature on the distribution for oxygen molecules and the effect of mass—oxygen must be heated to a very high temperature to have the same distribution of speeds as does hydrogen at 273 K.

speeds. Because F is the product of these two opposing factors, F increases from a value of zero, reaches a maximum, and then decreases as u increases. Notice that the distribution is not symmetrical about its maximum.

In Figure 6-16, we show how the distribution of molecular speeds depends on temperature and molar mass. When we compare the distributions for $O_2(g)$ at 273 K and 1000 K, we see that the range of speeds broadens as the temperature increases and that the distribution shifts toward higher speeds. The distributions for $O_2(g)$ at 273 K and $H_2(g)$ at 273 K reveal that the lighter the gas, the broader the range of speeds.

Let us focus now on the three characteristic speeds identified in Figure 6-15. The most probable speed, or modal speed, is denoted by u_m; more molecules have this speed than any other speed. The average speed is denoted by u_{av}. The **root-mean-square speed**, u_{rms}, is obtained from the average of u^2: $u_{rms} = \sqrt{\overline{u^2}}$. The root-mean-square speed is of particular interest because we can use the value of u_{rms} to calculate the average kinetic energy, $\overline{e_k}$, of a collection of molecules.

The average kinetic energy of a collection of molecules, each having a mass $m = M/N_A$, is the average of $\frac{1}{2}mu^2$:

$$\overline{e_k} = \tfrac{1}{2}m\overline{u^2}$$

Since $u_{rms} = \sqrt{\overline{u^2}}$, then $u^2_{rms} = \overline{u^2}$ and the equation given above for $\overline{e_k}$ can be rewritten as

$$\overline{e_k} = \tfrac{1}{2}mu^2_{rms}$$

This result shows that we can calculate the average kinetic energy, $\overline{e_k}$, from the values of the molecular mass, m, and the root-mean-square speed, u_{rms}. All we need now is a simple way to calculate u_{rms}.

We can derive an equation for u_{rms} by combining equation (6.18) with the ideal gas equation. Consider 1 mol of an ideal gas. The number of molecules present is $N = N_A$ (Avogadro's number), and the ideal gas equation becomes $PV = RT$ (that is, $n = 1$ in the equation $PV = nRT$). First, replace N by N_A, and multiply both sides of equation (6.18) by V. This leads to

◀ If the postulates of the kinetic-molecular theory hold, a gas is automatically an ideal gas.

$$PV = \frac{1}{3}N_A m\overline{u^2} \qquad\qquad \textbf{(6.19)}$$

Next, replace PV by RT and multiply both sides of the equation by three.

$$3\,RT = N_A m \overline{u^2}$$

Now, note that the product $N_A m$ represents the mass of 1 mol of molecules, the molar mass, M.

$$3\,RT = M\overline{u^2}$$

Finally, solve for $\overline{u^2}$ and then, $\sqrt{\overline{u^2}}$ which is u_{rms}.

$$u_{rms} = \sqrt{\overline{u^2}} = \sqrt{\frac{3\,RT}{M}} \qquad \textbf{(6.20)}$$

Equation (6.20) shows that u_{rms} of a gas is directly proportional to the square root of its Kelvin temperature and inversely proportional to the square root of its molar mass. From this we infer that, on average, (1) molecular speeds increase as the temperature increases, and (2) lighter gas molecules have greater speeds than do heavier ones.

To use equation (6.20) in calculating a root-mean-square speed, we must express the gas constant as

$$R = 8.3145\,\text{J}\,\text{mol}^{-1}\,\text{K}^{-1}$$

The joule has the units $\text{kg}\,\text{m}^2\,\text{s}^{-2}$, and consequently we must express molar mass in *kilograms* per mole, as we show in Example 6-14.

EXAMPLE 6-14   Calculating a Root-Mean-Square Speed

Which is the greater speed, that of a bullet fired from a high-powered M-16 rifle (2180 mi/h) or the root-mean-square speed of H_2 molecules at 25 °C?

Analyze

This is a straightforward application of equation (6.20). We must use SI units: $R = 8.3145\,\text{J}\,\text{K}^{-1}\,\text{mol}^{-1}$ and $M = 2.016 \times 10^{-3}\,\text{kg}\,\text{mol}^{-1}$. Recall that $1\,\text{J} = 1\,\text{kg}\,\text{m}^2\,\text{s}^{-2}$.

Solve

Determine u_{rms} of H_2 with equation (6.20).

$$u_{rms} = \sqrt{\frac{3 \times 8.3145\,\text{kg}\,\text{m}^2\,\text{s}^{-2}\,\text{mol}^{-1}\,\text{K}^{-1} \times 298\,\text{K}}{2.016 \times 10^{-3}\,\text{kg}\,\text{mol}^{-1}}}$$

$$= \sqrt{3.69 \times 10^6\,\text{m}^2/\text{s}^2} = 1.92 \times 10^3\,\text{m/s}$$

The remainder of the problem requires us either to convert $1.92 \times 10^3\,\text{m/s}$ to a speed in miles per hour, or $2180\,\text{mi/h}$ to meters per second. Then we can compare the two speeds. When we do this, we find that $1.92 \times 10^3\,\text{m/s}$ corresponds to $4.29 \times 10^3\,\text{mi/h}$. The root-mean-square speed of H_2 molecules at 25 °C is greater than the speed of the high-powered rifle bullet.

Assess

The cancellation of units yields a result for u_{rms} with the correct units (m/s). Also, Figure 6-15 shows that u_{rms} for H_2 is a bit greater than $1500\,\text{m/s}$ at 273 K. At 298 K, u_{rms} should be slightly greater than it is at 273 K.

PRACTICE EXAMPLE A: Which has the greater root-mean-square speed at 25 °C, $NH_3(g)$ or $HCl(g)$? Calculate u_{rms} for the one with the greater speed.

PRACTICE EXAMPLE B: At what temperature are u_{rms} of H_2 and the speed of the M-16 rifle bullet given in Example 6-14 the same?

6-8 CONCEPT ASSESSMENT

Without performing an actual calculation, indicate which has the greater u_{rms}, He(g) at 1000 K or H_2(g) at 250 K.

6-1 ARE YOU WONDERING...

How the distribution of molecular speeds can be demonstrated experimentally?

This can be done with the apparatus shown in Figure 6-17. An oven and attached evacuated chamber are separated by a wall with a small hole in it. Gas molecules are heated in the oven, emerge through the hole, and pass through a series of slits, called *collimators*, that herd the molecules into a beam. The number of molecules in the beam is kept low so that collisions between them will not disturb the beam.

The molecular beam passes through a series of rotating disks. Each disk has a slit cut in it. The slits on successive disks are offset from each other by a certain angle. A molecule passing through the first rotating disk will pass through the second disk only if its velocity is such that it arrives at the disk at the exact moment that the second slit appears. Thus, for a given rotation speed, only those molecules with the appropriate velocity can pass through the entire series of disks.

The number of molecules that pass through the disks and arrive at the detector is recorded for each chosen speed of rotation. The number of molecules for each speed of rotation is then plotted against the rotation speed. From the dimensions of the apparatus, the rotation speeds of the disks can be converted to molecular speeds, and a plot similar to the one in Figure 6-15 can be obtained.

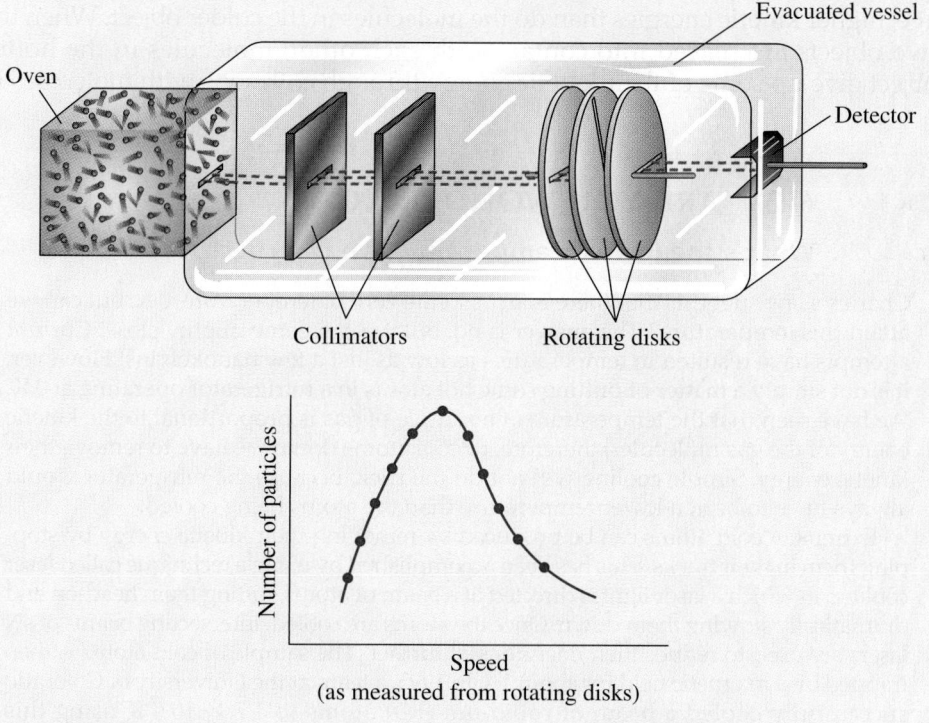

▲ FIGURE 6-17
Distribution of molecular speeds—an experimental determination
Only those molecules with the correct speed to pass through all rotating disks will reach the detector, where they can be counted. By changing the rate of rotation of the disks, the complete distribution of molecular speeds can be determined.

The Meaning of Temperature

We can gain an important insight into the meaning of temperature by starting with equation (6.19), the basic equation of the kinetic-molecular theory written for 1 mol of gas.

We modify it slightly by replacing the fraction $\frac{1}{3}$ with the equivalent product $\frac{2}{3} \times \frac{1}{2}$.

$$PV = \frac{1}{3} N_A \overline{mu^2} = \frac{2}{3} N_A \left(\frac{1}{2} \overline{mu^2} \right)$$

The quantity $\left(\frac{1}{2} \overline{mu^2} \right)$ represents the average translational kinetic energy, $\overline{e_k}$, of a collection of molecules. When we make this substitution and replace PV by RT, we obtain

$$RT = \frac{2}{3} N_A \overline{e_k}$$

Then, we can solve this equation for e_k.

$$\overline{e_k} = \frac{3}{2} \left(\frac{RT}{N_A} \right) \tag{6.21}$$

Because R and N_A are constants, equation (6.21) simply states that $\overline{e_k} = \text{constant} \times T$. This leads to an interesting new idea about temperature.

> The Kelvin temperature (T) of a gas is directly proportional to the average translational kinetic energy ($\overline{e_k}$) of its molecules. **(6.22)**

The idea expressed in equation (6.22) also helps us to understand what is happening at the molecular level when objects with different temperatures are placed into contact with each other. Molecules in the hotter object have, on average, higher kinetic energies than do the molecules in the colder object. When the two objects are placed into contact with each other, molecules in the hotter object give up some of their kinetic energy through collisions with molecules in

6-2 ARE YOU WONDERING...

What's the lowest temperature we can reach?

Charles's law suggests that there is an absolute zero of temperature, 0 K, but can we attain this temperature? The answer is no, but we can come mighty close. Current attempts have resulted in temperatures as low as just a few nanokelvins! However, it is not simply a matter of putting some hot atoms in a refrigerator operating at 0 K. We have seen that the temperature of a sample of gas is proportional to the kinetic energy of the gas molecules; therefore, to cool atoms down we have to remove their kinetic energy. Simple cooling will not do the trick, because the refrigerator would always have to be at a lower temperature than the atoms being cooled.

Extremely cold atoms can be produced by removing their kinetic energy by stopping them in their tracks. This has been accomplished by using a technique called laser cooling, in which a laser light is directed at a beam of atoms, hitting them head-on and dramatically slowing them down. Once the atoms are cooled, intersecting beams of six lasers are used to reduce their energies still further. The sample of cold atoms is then trapped by a magnetic field for about 1 s. In 1995, a team at the University of Colorado successfully cooled a beam of rubidium (Rb) atoms to 1.7×10^{-7} K using this procedure. The coldest human-made temperature, 450 picokelvin, was reported by scientists from the Massachusetts Institute of Technology in 2003.*

*A. E. Leanhardt, T. A. Pasquini, M. Saba, A. Schirotzek, Y. Shin, D. Kielpinski, D. E. Pritchard, and W. Ketterle, Cooling Bose-Einstein Condensates Below 500 Picokelvin, *Science* 2003 vol 301, p. 1513.

the colder object. The transfer of energy continues until the average kinetic energies of the molecules in the two objects become equal, that is until the temperatures become equalized. Finally, the idea expressed in equation (6.22) provides a new way of looking at the absolute zero of temperature: *It is the temperature at which translational molecular motion should cease.*

6-8 Gas Properties Relating to the Kinetic-Molecular Theory

A molecular speed of 1500 m/s corresponds to 5400 km/h or about 3400 mi/h. From this, it might seem that a given gas molecule could travel very long distances over a very short time, but this is not quite the case. Every gas molecule undergoes collisions with other gas molecules and, as a result, keeps changing direction. Gas molecules follow a tortuous path, which slows them down in getting from one point to another. Still, the net rate at which gas molecules move in a particular direction does depend on their average speeds.

Diffusion is the migration of molecules as a result of random molecular motion. Figure 6-18 pictures a common diffusion seen in a chemistry laboratory. The diffusion of two or more gases results in an intermingling of the molecules and, in a closed container, soon produces a homogeneous mixture, as shown in Figure 6-19(a). A related phenomenon, **effusion**, is the escape of gas molecules from their container through a tiny orifice or pinhole. The effusion of a hypothetical mixture of two gases is suggested by Figure 6-19(b).

The rate at which effusion occurs is directly proportional to molecular speeds. That is, molecules with high speeds effuse faster than molecules with low speeds. Let us consider the effusion of two different gases at the same temperature and pressure. We can first compare effusion rates with root-mean-square speeds and then substitute expression (6.20) for these speeds.

$$\frac{\text{rate of effusion of A}}{\text{rate of effusion of B}} = \frac{(u_{\text{rms}})_A}{(u_{\text{rms}})_B} = \sqrt{\frac{3\,RT/M_A}{3\,RT/M_B}} = \sqrt{\frac{M_B}{M_A}} \tag{6.23}$$

Any appropriate units (for example, g/s, mol/min) can be used to express a rate of effusion in equation (6.23) because in the ratio of two rates, the units

▲ FIGURE 6-18
Diffusion of $NH_3(g)$ and $HCl(g)$
$NH_3(g)$ escapes from $NH_3(aq)$ (but labeled ammonium hydroxide in this photograph), and $HCl(g)$ escapes from $HCl(aq)$. The gases diffuse toward each other, and, where they meet, a white cloud of ammonium chloride forms as a result of the following reaction:

$$NH_3(g) + HCl(g) \longrightarrow NH_4Cl(s).$$

Because of their greater average speed, NH_3 molecules diffuse faster than HCl. As a result, the cloud forms close to the mouth of the $HCl(aq)$ container.

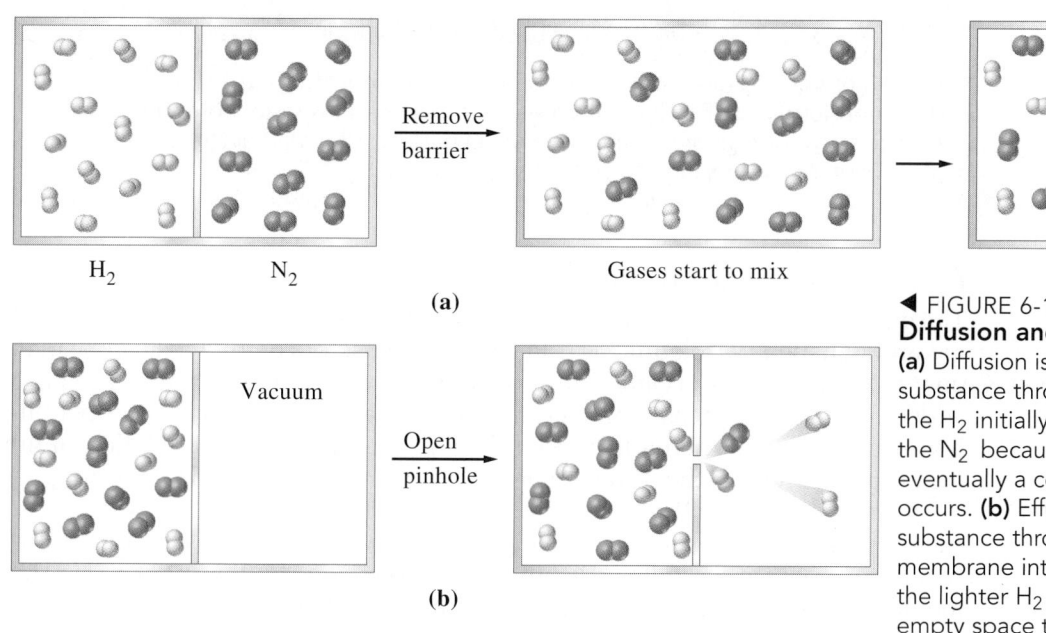

H_2 N_2

(a)

Remove barrier →

Gases start to mix

Gases mixed

Vacuum

Open pinhole →

(b)

◀ FIGURE 6-19
Diffusion and effusion
(a) Diffusion is the passage of one substance through another. In this case, the H_2 initially diffuses farther through the N_2 because it is lighter, although eventually a complete random mixing occurs. **(b)** Effusion is the passage of a substance through a pinhole or porous membrane into a vacuum. In this case, the lighter H_2 effuses faster across the empty space than does the N_2.

cancel. Equation (6.23) is a kinetic-theory statement of a nineteenth-century law called **Graham's law**.

> The rate of effusion of a gas is inversely proportional to the square root of its molar mass. **(6.24)**

Graham's law has serious limitations that you should be aware of. It can be used to describe effusion only for gases at very low pressures, so that molecules escape through an orifice individually, not as a jet of gas. Also, the orifice must be tiny so that no collisions occur as molecules pass through. Graham proposed his law in 1831 to describe the diffusion of gases, but the law actually does not apply to diffusion. Molecules of a diffusing gas undergo collisions with each other and with the gas into which they are diffusing. Some even move in the opposite direction to the net flow. Nevertheless, diffusion does occur, and gases of low molar mass do diffuse faster than those of higher molar mass. We cannot, however, use Graham's law to make quantitative predictions about rates of diffusion.

When compared at the same temperature, two different gases have the same value of $\overline{e_k} = \frac{1}{2}m\overline{u^2}$. This means that molecules with a smaller mass (m) have a higher speed (u_{rms}). When effusion takes place under the restrictions described above, equation (6.23) can be used to determine which of two gases effuses faster, which does so in a shorter time, which travels farther in a given time, and so on. An effective way to do this is to note that in every case, a ratio of effusion rates, times, distances, and so on is equal to the square root of a ratio of molar masses, as indicated in equation (6.25).

$$\text{ratio of} \begin{cases} \text{molecular speeds} \\ \text{effusion rates} \\ \text{effusion times} \\ \text{distances traveled by molecules} \\ \text{amounts of gas effused} \end{cases} = \sqrt{\text{ratio of two molar masses}} \quad \textbf{(6.25)}$$

When using equation (6.25), first reason qualitatively and work out whether the ratio of properties should be greater or less than one. Then set up the ratio of molar masses accordingly. Examples 6-15 and 6-16 illustrate this line of reasoning.

🔍 6-9 CONCEPT ASSESSMENT

Which answer(s) is (are) true when comparing 1.0 mol $H_2(g)$ at STP and 0.50 mol He(g) at STP? The two gases have equal **(a)** average molecular kinetic energies; **(b)** root-mean-square speeds; **(c)** masses; **(d)** volumes; **(e)** densities; **(f)** effusion rates through the same orifice.

Applications of Diffusion

The diffusion of gases into one another has many practical applications. Natural gas and liquefied petroleum gas (LPG) are odorless; for commercial use, a small quantity of a gaseous organic sulfur compound, methyl mercaptan, CH_3SH, is added to them. The mercaptan has an odor that can be detected in parts per billion (ppb) or less. When a leak occurs, which can lead to asphyxiation or an explosion, we rely on the diffusion of this odorous compound for a warning.

During World War II, the Manhattan Project (the secret, U.S. government-run program for developing the atomic bomb) used a method called gaseous diffusion to separate the desired isotope ^{235}U from the predominant ^{238}U. The method is based on the fact that uranium hexafluoride is one of the few compounds of uranium that can be obtained as a gas at moderate temperatures.

EXAMPLE 6-15   Comparing Amounts of Gases Effusing Through an Orifice

If 2.2×10^{-4} mol $N_2(g)$ effuses through a tiny hole in 105 s, then how much $H_2(g)$ would effuse through the same orifice in 105 s?

Analyze

Let us reason qualitatively: H_2 molecules are lighter than N_2 molecules, so $H_2(g)$ should effuse faster than $N_2(g)$ when the gases are compared at the same temperature. Before we set the ratio

$$\frac{\text{mol } H_2 \text{ effused}}{\text{mol } N_2 \text{ effused}}$$

equal to $\sqrt{\text{ratio of molar masses}}$, we must ensure that the ratio of molar masses is *greater than 1*.

Solve

$$\frac{? \text{ mol } H_2}{2.2 \times 10^{-4} \text{ mol } N_2} = \sqrt{\frac{M_{N_2}}{M_{H_2}}} = \sqrt{\frac{28.014}{2.016}} = 3.728$$

$$? \text{ mol } H_2 = 3.728 \times 2.2 \times 10^{-4} = 8.2 \times 10^{-4} \text{ mol } H_2$$

Assess

We could have estimated the result before calculating it. Because the ratio of molar masses is approximately 14, the ratio of effusion rates is approximately $\sqrt{14}$, which is slightly smaller than 4. Therefore, H_2 will effuse almost 4 times as fast as N_2 and almost 4 times as much H_2 will effuse in the same period.

PRACTICE EXAMPLE A: In Example 6-15, how much $O_2(g)$ would effuse through the same orifice in 105 s?

PRACTICE EXAMPLE B: In Example 6-15, how long would it take for 2.2×10^{-4} mol H_2 to effuse through the same orifice as the 2.2×10^{-4} mol N_2?

EXAMPLE 6-16   Relating Effusion Times and Molar Masses

A sample of $Kr(g)$ escapes through a tiny hole in 87.3 s. The same amount of an unknown gas escapes in 42.9 s under identical conditions. What is the molar mass of the unknown gas?

Analyze

Because the unknown gas effuses faster, it must have a smaller molar mass than Kr. Before we set the ratio

$$\frac{\text{effusion time for unknown}}{\text{effusion time for Kr}}$$

equal to $\sqrt{\text{ratio of two molar masses}}$, we must make sure the ratio of molar masses is smaller than one. Thus, the ratio of molar masses must be written with the molar mass of the lighter gas (the unknown gas) in the numerator.

Solve

$$\frac{\text{effusion time for unknown}}{\text{effusion time for Kr}} = \frac{42.9 \text{ s}}{87.3 \text{ s}} = \sqrt{\frac{M_{unk}}{M_{Kr}}} = 0.491$$

$$M_{unk} = (0.491)^2 \times M_{Kr} = (0.491)^2 \times 83.80 = 20.2 \text{ g/mol}$$

Assess

Use the final result and work backward. The molar mass of the unknown is about 4 times as small as that of Kr. Because *effusion rate* $\propto 1/\sqrt{M}$, the unknown will effuse about $\sqrt{4} = 2$ times as fast as Kr. The effusion times show that the unknown does indeed effuse 2 times as fast as Kr.

PRACTICE EXAMPLE A: Under the same conditions as in Example 6-16, another unknown gas requires 131.3 s to escape. What is the molar mass of this unknown gas?

PRACTICE EXAMPLE B: Given all the same conditions as in Example 6-16, how long would it take for a sample of ethane gas, C_2H_6, to effuse?

When high-pressure $UF_6(g)$ is forced through a barrier having millions of submicroscopic holes per square centimeter, molecules containing the isotope ^{235}U pass through the barrier slightly faster than those containing ^{238}U, just as expected from expression (6.25), and therefore $UF_6(g)$ contains a slightly higher ratio of ^{235}U to ^{238}U than it did previously. The gas has become enriched in ^{235}U. Carrying this process through several thousand passes yields a product highly enriched in ^{235}U.

6-9 Nonideal (Real) Gases

The data in Table 6.2 provide us with clear evidence that real gases are not "ideal." We should comment briefly on the conditions under which a real gas is ideal or nearly so and what to do when the conditions lead to nonideal behavior. A useful measure of how much a gas deviates from ideal gas behavior is found in its compressibility factor. The *compressibility factor* of a gas is the ratio PV/nRT. From the ideal gas equation we see that for an ideal gas, $PV/nRT = 1$. For a real gas, the compressibility factor can have values that are significantly different from 1. Values of the compressibility factor are given in Table 6.5 for a variety of gases at 300 K and 10 bar. The data in Table 6.5 show that the deviations from ideal gas behavior can be small or large, depending on the gas. At 300 K and 10 bar, He, H_2, CO, N_2, and O_2 behave almost ideally ($PV/nRT \approx 1$) but NH_3 and SF_6 do not ($PV/nRT \approx 0.88$). In Figure 6-20, the compressibility factor is plotted as a function of pressure for three different gases. The principal conclusion from this plot is that all gases behave ideally at sufficiently low pressures, say, below 1 atm, but that deviations set in at increased pressures. At very high pressures, the compressibility factor is always greater than one.

Nonideal gas behavior can be described as follows: Boyle's law predicts that at very high pressures, a gas volume becomes extremely small and

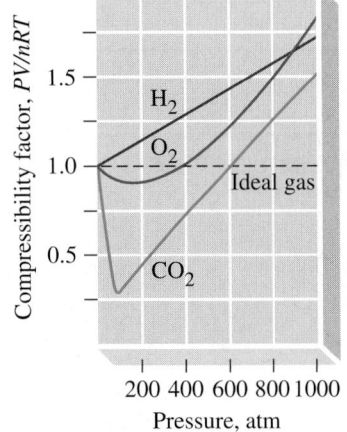

▲ FIGURE 6-20
The behavior of real gases—compressibility factor as a function of pressure at 0 °C
Values of the compressibility factor less than one signify that intermolecular forces of attraction are largely responsible for deviations from ideal gas behavior. Values greater than one are found when the volume of the gas molecules themselves is a significant fraction of the total gas volume.

TABLE 6.5   van der Waals Constants and Compressibility Factors (at 10 bar and 300 K) for Various Gases

Gas	van der Waals Constants		Compressibility Factor
	a, bar L^2 mol^{-2}	b, L mol^{-1}	
H_2	0.2452	0.0265	1.006
He	0.0346	0.0238	1.005
Ideal gas	0	0	1
N_2	1.370	0.0387	0.998
CO	1.472	0.0395	0.997
O_2	1.382	0.0319	0.994
CH_4	2.303	0.0431	0.983
NF_3	3.58	0.0545	0.965
CO_2	3.658	0.0429	0.950
N_2O	3.852	0.0444	0.945
C_2H_6	5.580	0.0651	0.922
NH_3	4.225	0.0371	0.887
SF_6	5.580	0.0651	0.880
C_3H_8	9.39	0.0905	a
SO_2	7.857	0.0879	a

Source: van der Waals constants are from the *CRC Handbook of Chemistry and Physics*, 83rd ed., David R. Lide (ed.), Boca Raton, FL: Taylor & Francis Group, 2002. Compressibility factors are calculated by using data from the National Institute of Standards and Technology (NIST) Chemistry WebBook, available online at http://webbook.nist.gov/chemistry/.
[a]At 10 bar and 300 K, C_3H_8 and SO_2 are liquids.

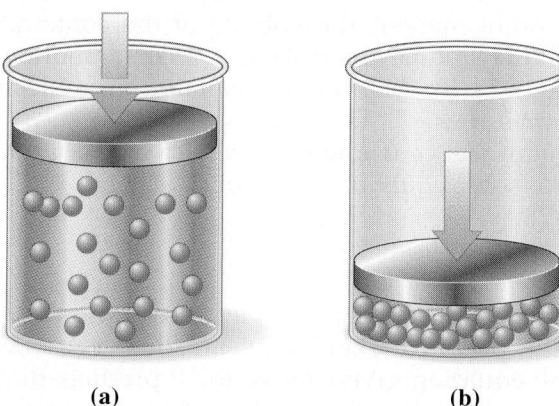

▲ FIGURE 6-21
The effect of finite molecular size
In (a), a significant fraction of the container is empty space and the gas can still be compressed to a smaller volume. In (b), the molecules occupy most of the available space. The volume of the system is only slightly greater than the total volume of the molecules.

approaches zero. This cannot be, however, because the molecules themselves occupy space and are practically incompressible, as suggested in Figure 6-21. Because of the finite size of the molecules, the PV product at high pressures is larger than predicted for an ideal gas, and the compressibility factor is greater than one. Another consideration is that intermolecular forces exist in gases. Figure 6-22 shows that because of attractive forces between the molecules, the force of the collisions of gas molecules with the container walls is less than expected for an ideal gas. Intermolecular forces of attraction account for compressibility factors of less than one. These forces become increasingly important at low temperatures, where translational molecular motion slows down. To summarize:

- Gases tend to behave *ideally* at *high temperatures* and *low pressures*.
- Gases tend to behave *nonideally* at *low temperatures* and *high pressures*.

The van der Waals Equation

A number of equations can be used for real gases, equations that apply over a wider range of temperatures and pressures than the ideal gas equation. Such equations are not as general as the ideal gas equation. They contain terms that have specific, but different, values for different gases. Such equations must correct for the volume associated with the molecules themselves and for intermolecular forces of attraction. Of all the equations that chemists use for modeling the behavior of real gases, the **van der Waals equation**, equation (6.26), is the simplest to use and interpret.

$$\left(P + \frac{an^2}{V^2}\right)(V - nb) = nRT \tag{6.26}$$

The equation incorporates two molecular parameters, a and b, whose values vary from molecule to molecule, as shown in Table 6.5.

The van der Waals equation and the ideal gas equation both have the form *pressure factor* $\times$ *volume factor* $= nRT$. The van der Waals equation uses a modified pressure factor, $P + an^2/V^2$, in place of P and a modified volume factor, $V - nb$, in place of V. In the modified volume factor, the term nb accounts for the volume of the molecules themselves. The parameter b is called the *excluded volume per mole*, and, to a rough approximation, it is the volume that one mole of gas occupies when it condenses to a liquid. Because the

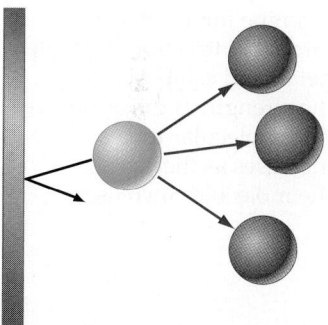

▲ FIGURE 6-22
Intermolecular forces of attraction
Attractive forces of the red molecules for the green molecule cause the green molecule to exert less force when it collides with the wall than if these attractions did not exist.

◀ The van der Waals equation reproduces the observed behavior of gases with moderate accuracy. It is most accurate for gases comprising approximately spherical molecules that have small dipole moments. We will discuss molecular shapes and dipole moments in Chapter 10.

molecules are not point masses, the volume of the container must be no smaller than nb, and the volume available for molecular motion is $V - nb$. As suggested in Figure 6-21(b), the volume available for molecular motion is quite small at high pressures.

To explain the significance of the term an^2/V^2 in the modified pressure factor, it is helpful to solve equation (6.26) for P:

$$P = \frac{nRT}{V - nb} - \frac{an^2}{V^2}$$

Provided V is not too small, the first term in the equation above is approximately equal to the pressure exerted by an ideal gas: $nRT/(V - nb) \approx nRT/V = P_{ideal}$. The equation given above for P predicts that the pressure exerted by a real gas will be less than that of an ideal gas. Figure 6-22 illustrates why. Because of attractive forces, molecules near the container walls are attracted toward the molecules behind them; as a result, the gas exerts less force on the container walls. The term an^2/V^2 takes into account the decrease in pressure caused by intermolecular attractions. In 1873, the Dutch physicist Johannes van der Waals reasoned that the decrease in pressure caused by intermolecular attractions should be proportional to the square of the concentration, and so the decrease in pressure is represented in the form an^2/V^2. The proportionality constant, a, provides a measure of how strongly the molecules attract each other.

▶ In Chapter 12, we will examine intermolecular forces of attraction in greater detail and establish why the strength of the attractive intermolecular forces increases as the sizes of the molecules increase.

A close examination of Table 6.5 shows that the values of both a and b increase as the sizes of the molecules increase. The smaller the values of a and b, the more closely the gas resembles an ideal gas. Deviations from ideality, as measured by the compressibility factor, become more pronounced as the values of a and b increase. In Example 6-17 we calculate the pressure of a real gas by using the van der Waals equation. Solving the equation for either n or V is more difficult, however (see Exercise 121).

EXAMPLE 6-17   **Using the van der Waals Equation to Calculate the Pressure of a Gas**

Use the van der Waals equation to calculate the pressure exerted by 1.00 mol $Cl_2(g)$ confined to a volume of 2.00 L at 273 K. The value of $a = 6.49$ L^2 atm mol^{-2}, and that of $b = 0.0562$ L mol^{-1}.

Analyze

This is a straightforward application of equation (6.26). It is important to include units to make sure the units cancel out properly.

Solve

Solve equation (6.26) for P.

$$P = \frac{nRT}{V - nb} - \frac{n^2a}{V^2}$$

Then substitute the following values into the equation.

$$n = 1.00 \text{ mol}; V = 2.00 \text{ L}; T = 273 \text{ K}; R = 0.08206 \text{ atm L mol}^{-1}\text{K}^{-1}$$

$$n^2a = (1.00)^2 \text{ mol}^2 \times 6.49\frac{L^2 \text{ atm}}{\text{mol}^2} = 6.49 \text{ L}^2 \text{ atm}$$

$$nb = 1.00 \text{ mol} \times 0.0562 \text{ L mol}^{-1} = 0.0562 \text{ L}$$

$$P = \frac{1.00 \text{ mol} \times 0.08206 \text{ atm L mol}^{-1}\text{K}^{-1} \times 273 \text{ K}}{(2.00 - 0.0562)\text{L}} - \frac{6.49 \text{ L}^2 \text{ atm}}{(2.00)^2 \text{ L}^2}$$

$$P = 11.5 \text{ atm} - 1.62 \text{ atm} = 9.9 \text{ atm}$$

Assess

The pressure calculated with the ideal gas equation is 11.2 atm. By including only the b term in the van der Waals equation, we get a value of 11.5 atm. Including the a term reduces the calculated pressure by 1.62 atm. Under the conditions of this problem, intermolecular forces of attraction are the main cause of the departure from ideal behavior. Although the deviation from ideality here is rather large, in problem-solving situations, you can generally assume that the ideal gas equation will give satisfactory results.

PRACTICE EXAMPLE A: Substitute $CO_2(g)$ for $Cl_2(g)$ in Example 6-17, given the values $a = 3.66 \, L^2 \, bar \, mol^{-2}$ and $b = 0.0427 \, L \, mol^{-1}$. Which gas, CO_2 or Cl_2, shows the greater departure from ideal gas behavior? [*Hint:* For which gas do you find the greater difference in calculated pressures, first using the ideal gas equation and then the van der Waals equation?]

PRACTICE EXAMPLE B: Substitute $CO(g)$ for $Cl_2(g)$ in Example 6-17, given the values $a = 1.47 \, L^2 \, bar \, mol^{-2}$ and $b = 0.0395 \, L \, mol^{-1}$. Including CO_2 from Practice Example 6-17A, which of the three gases—Cl_2, CO_2, or CO—shows the greatest departure from ideal gas behavior?

🔍 6-10 CONCEPT ASSESSMENT

Following are the measured densities at 20.0 °C and 1 atm pressure of three gases: O_2, 1.331 g/L; OF_2, 2.26 g/L; NO, 1.249 g/L. Arrange them in the order of increasing adherence to the ideal gas equation. [*Hint:* What property can you calculate and compare with a known value?]

Mastering**CHEMISTRY** **www.masteringchemistry.com**

The blanket of gases surrounding Earth forms our atmosphere. It not only protects us from harmful radiation but also plays an essential role in moving water, essential to life, from the oceans to the land. For a discussion of the regions and composition of Earth's atmosphere, go to the Focus On feature for Chapter 6 on the MasteringChemistry site.

Summary

6-1 Properties of Gases: Gas Pressure—A gas is described in terms of its **pressure**, temperature, volume, and amount. Gas pressure is most readily measured by comparing it with the pressure exerted by a liquid column, usually mercury (equation 6.2). The pressure exerted by a column of mercury in a **barometer** and called the **barometric pressure** is equal to the prevailing pressure of the atmosphere (Fig. 6-4). Other gas pressures can be measured with a **manometer** (Fig. 6-5). Pressure can be expressed in a variety of units (Table 6.1), including the SI units **pascal (Pa)** and **kilopascal (kPa)**. Also commonly used are **bar; millimeter of mercury (mmHg)**; **torr (Torr)**, where 1 Torr = 1 mmHg; and **atmosphere (atm)**, where 1 atm = 760 mmHg = 760 Torr.

6-2 The Simple Gas Laws—The most common simple gas laws are **Boyle's law** relating gas pressure and volume (equation 6.5, Fig. 6-6); **Charles's law** relating gas volume and temperature (equation 6.8, Fig. 6-7); and **Avogadro's law**, relating volume and amount of gas. Some important ideas that originate from the simple gas laws are the **Kelvin** (**absolute**) scale of temperature (equation 6.6), the **standard conditions of temperature and pressure (STP)**, and the molar volume of a gas at STP— 22.7 L/mol (expression 6.10).

6-3 Combining the Gas Laws: The Ideal Gas Equation and the General Gas Equation—The simple gas laws can be combined into the **ideal gas equation**, $PV = nRT$ (equation 6.11), where R is called the **gas constant**. A gas whose behavior can be predicted with this equation is known as an **ideal**, or **perfect, gas**. The ideal gas equation can be solved for any one of the variables when all the others are known. The **general gas equation** (equation 6.12) is a useful variant of the ideal gas equation for describing the behavior of a gas when certain variables are held constant and others are allowed to change.

6-4 Applications of the Ideal Gas Equation—An important application of the ideal gas equation is its use in determining molecular masses (equation 6.13) and gas densities (equation 6.14).

6-5 Gases in Chemical Reactions—Because it relates the volume of a gas at a given temperature and pressure to the amount of gas, the ideal gas equation often enters into stoichiometric calculations for reactions involving gases. In calculations based on the volumes of two gaseous reactants and/or products measured at the same temperature and pressure, the **law of combining volumes** is generally applicable.

6-6 Mixtures of Gases—The ideal gas equation applies to mixtures of ideal gases as well as to pure gases. The enabling principle, known as **Dalton's law of partial pressures**, is that each gas expands to fill the container, exerting the same pressure as if it were alone in the container (Fig. 6-12). The total pressure is the sum of these **partial pressures** (equation 6.16). A useful concept in dealing with mixtures of gases is that of **mole fraction**, the fraction of the molecules in a mixture contributed by each component (equation 6.17). In the common procedure of collecting a gas over water (Fig. 6-13), the particular gas being isolated is mixed with water vapor.

6-7 Kinetic-Molecular Theory of Gases—The **kinetic-molecular theory of gases** yields a basic expression (equation 6.18) from which other relationships can be established between the **root-mean-square speed** (u_{rms}) of molecules, temperature, and molar mass of a gas (equation 6.20), and the average molecular translational kinetic energy and Kelvin temperature (equation 6.21). An important aspect of the kinetic-molecular theory is the concept of a distribution of molecular speeds (Figs. 6-15 and 6-16).

6-8 Gas Properties Relating to the Kinetic-Molecular Theory—The **diffusion** and **effusion** of gases (Fig. 6-19) can be described by the kinetic-molecular theory. Using an approximation known as **Graham's law**, molar masses can be determined by measuring rates of effusion (equation 6.23).

6-9 Nonideal (Real) Gases—Because of finite molecular size and intermolecular forces of attraction (Figs. 6-21 and 6-22), real gases generally behave ideally only at high temperatures and low pressures. Other equations of state, such as the **van der Waals equation** (equation 6.26), take into account the factors causing nonideal behavior and often work when the ideal gas equation fails.

Integrative Example

Combustion of 1.110 g of a gaseous hydrocarbon yields 3.613 g CO_2 and 1.109 g H_2O, and no other products. A 0.288 g sample of the hydrocarbon occupies a volume of 131 mL at 24.8 °C and 753 mmHg. Write a plausible structural formula for a hydrocarbon corresponding to these data.

Analyze
Use the combustion data for the 1.110 g sample of hydrocarbon and the method of Example 3-6 on page 83 to determine the empirical formula. Use the $P - V - T$ data in equation (6.13) for the 0.288 g sample to determine the molar mass and molecular mass of the hydrocarbon. By comparing the empirical formula mass and the molecular mass, establish the molecular formula. Now write a structural formula consistent with the molecular formula.

Solve

Calculate the number of moles of C and H in the 1.110 g sample of hydrocarbon based on the masses of CO_2 and H_2O obtained in its combustion.

$$? \text{ mol C} = 3.613 \text{ g CO}_2 \times \frac{1 \text{ mol CO}_2}{44.01 \text{ g CO}_2} \times \frac{1 \text{ mol C}}{1 \text{ mol CO}_2} = 0.08209 \text{ mol C}$$

$$? \text{ mol H} = 1.109 \text{ g H}_2\text{O} \times \frac{1 \text{ mol H}_2\text{O}}{18.02 \text{ g H}_2\text{O}} \times \frac{2 \text{ mol H}}{1 \text{ mol H}_2\text{O}} = 0.1231 \text{ mol H}$$

Use these numbers of moles as the provisional subscripts in the formula.

$$C_{0.08209}H_{0.1231}$$

Divide each provisional subscript by the smaller of the two to obtain the empirical formula.

$$C_{\frac{0.08209}{0.08209}} H_{\frac{0.1231}{0.08209}}$$
$$CH_{1.500} = C_2H_3$$

To determine the molar mass, use a modified form of equation (6.13).

$$M = \frac{mRT}{PV} = \frac{0.288 \text{ g} \times 0.08206 \text{ atm L K}^{-1}\text{mol}^{-1} \times (24.8 + 273.2) \text{ K}}{(753 \text{ mmHg} \times 1 \text{ atm}/760 \text{ mmHg}) \times 0.131 \text{ L}}$$
$$= 54.3 \text{ g mol}^{-1}$$

The empirical formula mass is

$$\left(2 \text{ C atoms} \times \frac{12.0 \text{ u}}{1 \text{ C atom}}\right) + \left(3 \text{ H atoms} \times \frac{1.01 \text{ u}}{1 \text{ H atom}}\right) = 27.0 \text{ u}$$

The molar mass based on the empirical formula, 27.0 g mol^{-1}, is almost exactly one-half the observed molar mass of 54.3 g mol^{-1}. The molecular formula of the hydrocarbon is

$$C_{2\times2}H_{2\times3} = C_4H_6$$

The four-carbon alkane is butane, C_4H_{10}. Removal of 4 H atoms to obtain the formula C_4H_6 is achieved by inserting two C-to-C double bonds.

$$H_2C{=}CH{-}CH{=}CH_2 \quad \text{or} \quad H_2C{=}C{=}CH{-}CH_3$$

Two other possibilities involve the presence of a C-to-C triple bond.

$H_3C\!-\!C\!\equiv\!C\!-\!CH_3$ or $H_3C\!-\!CH_2\!-\!C\!\equiv\!CH$

Assess

The combination of combustion data and gas-law data yields a molecular formula with certainty. However, because of isomerism the exact structural formula cannot be pinpointed. All that we can say is that the hydrocarbon might have any one of the four structures shown, but it might be still another structure, for example, based on a ring of C atoms rather than a straight chain.

PRACTICE EXAMPLE A: When a 0.5120 g sample of a gaseous hydrocarbon was burned in excess oxygen, 1.687 g CO_2 and 0.4605 g H_2O were obtained. The density of the compound, in its vapor form, is 1.637 g/L at 25 °C and 101.3 kPa. Determine the molecular formula of the hydrocarbon, and draw a plausible structural formula for the molecule.

PRACTICE EXAMPLE B: An organic compound contains only C, H, N, and O. When the compound is burned in oxygen, with appropriate catalysts, nitrogen gas (N_2), carbon dioxide (CO_2), and water vapor (H_2O) are produced. A 0.1023 g sample of the compound yielded 151.2 mg CO_2, 69.62 mg H_2O, and 9.62 mL of $N_2(g)$ at 0.00 °C and 1.00 atm. The density of the compound, in its vapor form, was found to be 3.57 g L^{-1} at 127 °C and 748 mmHg. What is the molecular formula of the compound?

Exercises

Pressure and Its Measurement

1. Convert each pressure to an equivalent pressure in atmospheres. **(a)** 736 mmHg; **(b)** 0.776 bar; **(c)** 892 Torr; **(d)** 225 kPa.

2. Calculate the height of a mercury column required to produce a pressure **(a)** of 0.984 atm; **(b)** of 928 Torr; **(c)** equal to that of a column of water 142 ft high.

3. Calculate the height of a column of liquid benzene ($d = 0.879$ g/cm^3), in meters, required to exert a pressure of 0.970 atm.

4. Calculate the height of a column of liquid glycerol ($d = 1.26$ g/cm^3), in meters, required to exert the same pressure as 3.02 m of $CCl_4(l)$ ($d = 1.59$ g/cm^3).

5. What is the pressure (in mmHg) of the gas inside the apparatus below if $P_{bar.} = 740$ mmHg, $h_1 = 30$ mm and $h_2 = 50$ mm?

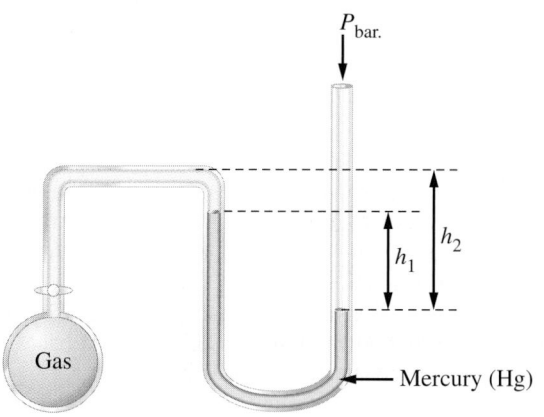

6. What is the pressure (in mmHg) of the gas inside the apparatus below if $P_{bar.} = 740$ mmHg, $h_1 = 30$ mm and $h_2 = 40$ mm?

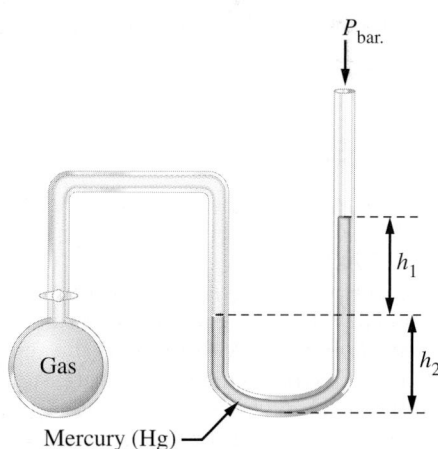

7. At times, a pressure is stated in units of *mass* per unit area rather than *force* per unit area. Express $P = 1$ atm in the unit kg/cm^2.
   [*Hint:* How is a mass in kilograms related to a force?]

8. Express $P = 1$ atm in pounds per square inch (psi).
   [*Hint:* Refer to Exercise 7.]

The Simple Gas Laws

9. A sample of $O_2(g)$ has a volume of 26.7 L at 762 Torr. What is the new volume if, with the temperature and amount of gas held constant, the pressure is **(a)** lowered to 385 Torr; **(b)** increased to 3.68 atm?

10. An 886 mL sample of Ne(g) is at 752 mmHg and 26 °C. What will be the new volume if, with the pressure and amount of gas held constant, the temperature is **(a)** increased to 98 °C; **(b)** lowered to −20 °C?

11. If 3.0 L of oxygen gas at 177 °C is cooled at constant pressure until the volume becomes 1.50 L, then what is the final temperature?

12. We want to change the volume of a fixed amount of gas from 725 mL to 2.25 L while holding the temperature constant. To what value must we change the pressure if the initial pressure is 105 kPa?

13. A 35.8 L cylinder of Ar(g) is connected to an evacuated 1875 L tank. If the temperature is held constant and the final pressure is 721 mmHg, what must have been the original gas pressure in the cylinder, in atmospheres?

14. A sample of $N_2(g)$ occupies a volume of 42.0 mL under the existing barometric pressure. Increasing the pressure by 85 mmHg reduces the volume to 37.7 mL. What is the prevailing barometric pressure, in millimeters of mercury?

15. A weather balloon filled with He gas has a volume of $2.00 \times 10^3 \ m^3$ at ground level, where the atmospheric pressure is 1.000 atm and the temperature 27 °C. After the balloon rises high above Earth to a point where the atmospheric pressure is 0.340 atm, its volume increases to $5.00 \times 10^3 \ m^3$. What is the temperature of the atmosphere at this altitude?

16. The photographs show the contraction of an argon-filled balloon when it is cooled by liquid nitrogen. To what approximate fraction of its original volume will the balloon shrink when it is cooled from a room temperature of 22 °C to a final temperature of about −22 °C?

17. What is the mass of argon gas in a 75.0 mL volume at STP?

18. What volume of gaseous chlorine at STP would you need to obtain a 250.0 g sample of gas?

19. A 27.6 mL sample of $PH_3(g)$ (used in the manufacture of flame-retardant chemicals) is obtained at STP.
    (a) What is the mass of this gas, in milligrams?
    (b) How many molecules of PH_3 are present?

20. A 5.0×10^{17} atom sample of radon gas is obtained.
    (a) What is the mass of this sample, in micrograms?
    (b) What is the volume of this sample at STP, in microliters?

21. You purchase a bag of potato chips at an ocean beach to take on a picnic in the mountains. At the picnic, you notice that the bag has become inflated, almost to the point of bursting. Use your knowledge of gas behavior to explain this phenomenon.

22. Scuba divers know that they must not ascend quickly from deep underwater because of a condition known as the *bends*, discussed in Chapter 13. Another concern is that they must constantly exhale during their ascent to prevent damage to the lungs and blood vessels. Describe what would happen to the lungs of a diver who inhaled compressed air at a depth of 30 m and held her breath while rising to the surface.

General Gas Equation

23. A sample of gas has a volume of 4.25 L at 25.6 °C and 748 mmHg. What will be the volume of this gas at 26.8 °C and 742 mmHg?

24. A 10.0 g sample of a gas has a volume of 5.25 L at 25 °C and 762 mmHg. If 2.5 g of the same gas is added to this *constant* 5.25 L volume and the temperature raised to 62 °C, what is the new gas pressure?

25. A constant-volume vessel contains 12.5 g of a gas at 21 °C. If the pressure of the gas is to remain constant as the temperature is raised to 210 °C, how many grams of gas must be released?

26. A 34.0 L cylinder contains 305 g $O_2(g)$ at 22 °C. How many grams of $O_2(g)$ must be released to reduce the pressure in the cylinder to 1.15 atm if the temperature remains constant?

Ideal Gas Equation

27. What is the volume, in milliliters, occupied by 89.2 g $CO_2(g)$ at 37 °C and 737 mmHg?

28. A 12.8 L cylinder contains 35.8 g O_2 at 46 °C. What is the pressure of this gas, in atmospheres?

29. Kr(g) in a 18.5 L cylinder exerts a pressure of 11.2 atm at 28.2 °C. How many grams of gas are present?

30. A 72.8 L constant-volume cylinder containing 7.41 g He is heated until the pressure reaches 3.50 atm. What is the final temperature in degrees Celsius?

31. A laboratory high vacuum system is capable of evacuating a vessel to the point that the amount of gas remaining is 5.0×10^9 molecules per cubic meter. What is the residual pressure in pascals?

32. What is the pressure, in pascals, exerted by 1242 g CO(g) when confined at −25 °C to a cylindrical tank 25.0 cm in diameter and 1.75 m high?

33. What is the molar volume of an ideal gas at **(a)** 25 °C and 1.00 atm; **(b)** 100 °C and 748 Torr?

34. At what temperature is the molar volume of an ideal gas equal to 22.4 L, if the pressure of the gas is 2.5 atm?

Determining Molar Mass

35. A 0.418 g sample of gas has a volume of 115 mL at 66.3 °C and 743 mmHg. What is the molar mass of this gas?

36. What is the molar mass of a gas found to have a density of 0.841 g/L at 415 K and 725 Torr?

37. What is the molecular formula of a gaseous fluoride of sulfur containing 70.4% F and having a density of approximately 4.5 g/L at 20 °C and 1 atm?

38. A 2.650 g sample of a gaseous compound occupies 428 mL at 24.3 °C and 742 mmHg. The compound consists of 15.5% C, 23.0% Cl, and 61.5% F, by mass. What is its molecular formula?

39. A gaseous hydrocarbon weighing 0.231 g occupies a volume of 102 mL at 23 °C and 749 mmHg. What is the molar mass of this compound? What conclusion can you draw about its molecular formula?

40. A 132.10 mL glass vessel weighs 56.1035 g when evacuated and 56.2445 g when filled with the gaseous hydrocarbon acetylene at 749.3 mmHg and 20.02 °C. What is the molar mass of acetylene? What conclusion can you draw about its molecular formula?

Gas Densities

41. A particular application calls for $N_2(g)$ with a density of 1.80 g/L at 32 °C. What must be the pressure of the $N_2(g)$ in millimeters of mercury? What is the molar volume under these conditions?

42. Monochloroethylene is used to make polyvinylchloride (PVC). It has a density of 2.56 g/L at 22.8 °C and 756 mmHg. What is the molar mass of monochloroethylene? What is the molar volume under these conditions?

43. In order for a gas-filled balloon to rise in air, the density of the gas in the balloon must be less than that of air.
    (a) Consider air to have a molar mass of 28.96 g/mol; determine the density of air at 25 °C and 1 atm, in g/L.

    (b) Show by calculation that a balloon filled with carbon dioxide at 25 °C and 1 atm could not be expected to rise in air at 25 °C.

44. Refer to Exercise 43, and determine the minimum temperature to which the balloon described in part **(b)** would have to be heated before it could begin to rise in air. (Ignore the mass of the balloon itself.)

45. The density of phosphorus vapor is 2.64 g/L at 310 °C and 775 mmHg. What is the molecular formula of the phosphorus under these conditions?

46. A particular gaseous hydrocarbon that is 82.7% C and 17.3% H by mass has a density of 2.33 g/L at 23 °C and 746 mmHg. What is the molecular formula of this hydrocarbon?

Gases in Chemical Reactions

47. What volume of $O_2(g)$ is consumed in the combustion of 75.6 L $C_3H_8(g)$ if both gases are measured at STP?

48. How many liters of $H_2(g)$ at STP are produced per gram of Al(s) consumed in the following reaction?

    $$2\,Al(s) + 6\,HCl(aq) \longrightarrow 2\,AlCl_3(aq) + 3\,H_2(g)$$

49. A particular coal sample contains 3.28% S by mass. When the coal is burned, the sulfur is converted to $SO_2(g)$. What volume of $SO_2(g)$, measured at 23 °C and 738 mmHg, is produced by burning 1.2×10^6 kg of this coal?

50. One method of removing $CO_2(g)$ from a spacecraft is to allow the CO_2 to react with LiOH. How many liters of $CO_2(g)$ at 25.9 °C and 751 Torr can be removed per kilogram of LiOH consumed?

    $$2\,LiOH(s) + CO_2(g) \longrightarrow Li_2CO_3(s) + H_2O(l)$$

51. A 3.57 g sample of a KCl–KClO$_3$ mixture is decomposed by heating and produces 119 mL $O_2(g)$, measured at 22.4 °C and 738 mmHg. What is the mass percent of KClO$_3$ in the mixture?

    $$2\,KClO_3(s) \longrightarrow 2\,KCl(s) + 3\,O_2(g)$$

52. Hydrogen peroxide, H_2O_2, is used to disinfect contact lenses. How many milliliters of $O_2(g)$ at 22 °C and 752 mmHg can be liberated from 10.0 mL of an aqueous solution containing 3.00% H_2O_2 by mass? The density of the aqueous solution of H_2O_2 is 1.01 g/mL.

    $$2\,H_2O_2(aq) \longrightarrow 2\,H_2O(l) + O_2(g)$$

53. Calculate the volume of $H_2(g)$, measured at 26 °C and 751 Torr, required to react with 28.5 L CO(g), measured at 0 °C and 760 Torr, in this reaction.

    $$3\,CO(g) + 7\,H_2(g) \longrightarrow C_3H_8(g) + 3\,H_2O(l)$$

54. The Haber process is the principal method for fixing nitrogen (converting N_2 to nitrogen compounds).

    $$N_2(g) + 3\,H_2(g) \longrightarrow 2\,NH_3(g)$$

    Assume that the reactant gases are completely converted to $NH_3(g)$ and that the gases behave ideally.
    (a) What volume of $NH_3(g)$ can be produced from 152 L $N_2(g)$ and 313 L of $H_2(g)$ if the gases are measured at 315 °C and 5.25 atm?
    (b) What volume of $NH_3(g)$, measured at 25 °C and 727 mmHg, can be produced from 152 L $N_2(g)$ and 313 L $H_2(g)$, measured at 315°C and 5.25 atm?

Mixtures of Gases

55. What is the volume, in liters, occupied by a mixture of 15.2 g Ne(g) and 34.8 g Ar(g) at 7.15 atm pressure and 26.7 °C?

56. A balloon filled with $H_2(g)$ at 0.0 °C and 1.00 atm has a volume of 2.24 L. What is the final gas volume if 0.10 mol He(g) is added to the balloon and the temperature is then raised to 100 °C while the pressure and amount of gas are held constant?

57. A gas cylinder of 53.7 L volume contains $N_2(g)$ at a pressure of 28.2 atm and 26 °C. How many grams of Ne(g) must we add to this same cylinder to raise the total pressure to 75.0 atm?

58. A 2.35 L container of $H_2(g)$ at 762 mmHg and 24 °C is connected to a 3.17 L container of He(g) at 728 mmHg and 24 °C. After mixing, what is the total gas pressure, in millimeters of mercury, with the temperature remaining at 24 °C?

59. Which actions would you take to establish a pressure of 2.00 atm in a 2.24 L cylinder containing 1.60 g $O_2(g)$ at 0 °C? **(a)** add 1.60 g O_2; **(b)** release 0.80 g O_2; **(c)** add 2.00 g He; **(d)** add 0.60 g He.

60. A mixture of 4.0 g $H_2(g)$ and 10.0 g He(g) in a 4.3 L flask is maintained at 0 °C.
(a) What is the total pressure in the container?
(b) What is the partial pressure of each gas?

61. A 2.00 L container is filled with Ar(g) at 752 mmHg and 35 °C. A 0.728 g sample of C_6H_6 vapor is then added.
(a) What is the total pressure in the container?
(b) What is the partial pressure of Ar and of C_6H_6?

62. The chemical composition of air that is exhaled (expired) is different from ordinary air. A typical analysis of expired air at 37 °C and 1.00 atm, expressed as percent by volume, is 74.2% N_2, 15.2% O_2, 3.8% CO_2, 5.9% H_2O, and 0.9% Ar. The composition of ordinary air is given in Practice Example 6-12B.
(a) What is the ratio of the partial pressure of $CO_2(g)$ in expired air to that in ordinary air?
(b) Would you expect the density of expired air to be greater or less than that of ordinary air at the same temperature and pressure? Explain.
(c) Confirm your expectation by calculating the densities of ordinary air and expired air at 37 °C and 1.00 atm.

63. In the drawing below, 1.00 g $H_2(g)$ is maintained at 1 atm pressure in a cylinder closed off by a freely moving piston. Which sketch, (a), (b), or (c), best represents the mixture obtained when 1.00 g He(g) is added? Explain.

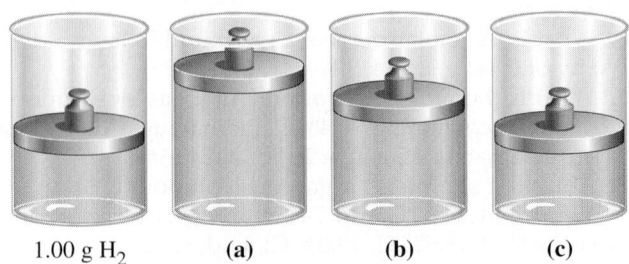

1.00 g H_2 **(a)**   **(b)**   **(c)**

64. In the drawing above, 1.00 g $H_2(g)$ at 300 K is maintained at 1 atm pressure in a cylinder closed off by a freely moving piston. Which sketch, **(a)**, **(b)**, or **(c)**, best represents the mixture obtained when 0.50 g $H_2(g)$ is added and the temperature is reduced to 275 K? Explain your answer.

65. A 4.0 L sample of O_2 gas has a pressure of 1.0 atm. A 2.0 L sample of N_2 gas has a pressure of 2.0 atm. If these two samples are mixed and then compressed in a 2.0 L vessel, what is the final pressure of the mixture? Assume that the temperature remains unchanged.

66. The following figure shows the contents and pressures of three vessels of gas that are joined by a connecting tube.

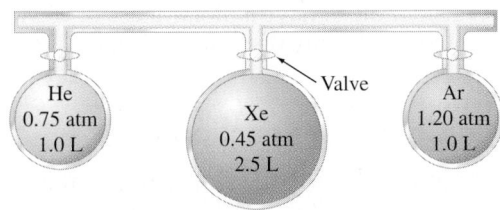

He
0.75 atm
1.0 L

Xe
0.45 atm
2.5 L

Valve

Ar
1.20 atm
1.0 L

After the valves on the vessels are opened, the final pressure is measured and found to be 0.675 atm. What is the total volume of the connecting tube? Assume that the temperature remains constant.

Collecting Gases over Liquids

67. A 1.65 g sample of Al reacts with excess HCl, and the liberated H_2 is collected over water at 25 °C at a barometric pressure of 744 mmHg. What volume of gaseous mixture, in liters, is collected?

$$2 \, Al(s) + 6 \, HCl(aq) \longrightarrow 2 \, AlCl_3(aq) + 3 \, H_2(g)$$

68. An 89.3 mL sample of wet $O_2(g)$ is collected over water at 21.3 °C at a barometric pressure of 756 mmHg (vapor pressure of water at 21.3 °C = 19 mmHg).
(a) What is the partial pressure of $O_2(g)$ in the sample collected, in millimeters of mercury? **(b)** What is the volume percent O_2 in the gas collected? **(c)** How many grams of O_2 are present in the sample?

69. A sample of $O_2(g)$ is collected over water at 24 °C. The volume of gas is 1.16 L. In a subsequent experiment, it is determined that the mass of O_2 present is 1.46 g. What must have been the barometric pressure at the time the gas was collected? (Vapor pressure of water = 22.4 Torr.)

70. A 1.072 g sample of He(g) is found to occupy a volume of 8.446 L when collected over hexane at 25.0 °C and 738.6 mmHg barometric pressure. Use these data to determine the vapor pressure of hexane at 25 °C.

71. At elevated temperatures, solid sodium chlorate ($NaClO_3$) decomposes to produce sodium chloride, NaCl, and O_2 gas. A 0.8765 g sample of impure sodium chlorate was heated until the production of oxygen ceased. The oxygen gas was collected over water and occupied a volume of 57.2 mL at 23.0 °C and 734 Torr. Calculate the mass percentage of $NaClO_3$ in the original sample. Assume that none of the impurities produce oxygen on heating. The vapor pressure of water is 21.07 Torr at 23 °C.

72. When solid $KClO_3$ is heated strongly, it decomposes to form solid potassium chloride, KCl, and O_2 gas. A 0.415 g sample of impure $KClO_3$ is heated strongly and the O_2 gas produced by the decomposition is collected over water. When the wet O_2 gas is cooled back to 26 °C, the total volume is 229 mL and the total pressure is 323 Torr. What is the mass percentage of $KClO_3$ in the original sample? Assume that none of the impurities produce oxygen on heating. The vapor pressure of water is 25.22 Torr at 26 °C.

Kinetic-Molecular Theory

73. Calculate u_{rms}, in meters per second, for $Cl_2(g)$ molecules at 30 °C.
74. The u_{rms} of H_2 molecules at 273 K is 1.84×10^3 m/s. At what temperature is u_{rms} for H_2 twice this value?
75. Refer to Example 6-14. What must be the molecular mass of a gas if its molecules are to have a root-mean-square speed at 25 °C equal to the speed of the M-16 rifle bullet?
76. Refer to Example 6-14. Noble gases (group 18) exist as atoms, not molecules (they are monatomic). Cite one noble gas whose u_{rms} at 25 °C is higher than the speed of the rifle bullet and one whose u_{rms} is lower.

77. At what temperature will u_{rms} for Ne(g) be the same as u_{rms} for He at 300 K?
78. Determine u_m, $\bar{u}$, and u_{rms} for a group of ten automobiles clocked by radar at speeds of 38, 44, 45, 48, 50, 55, 55, 57, 58, and 60 mi/h, respectively.
79. Calculate the average kinetic energy, $\bar{e_k}$, for $O_2(g)$ at 298 K and 1.00 atm.
80. Calculate the total kinetic energy, in joules, of 155 g $N_2(g)$ at 25 °C and 1.00 atm. [*Hint:* First calculate the average kinetic energy, $\bar{e_k}$.]

Diffusion and Effusion of Gases

81. If 0.00484 mol $N_2O(g)$ effuses through an orifice in a certain period of time, how much $NO_2(g)$ would effuse in the same time under the same conditions?
82. A sample of $N_2(g)$ effuses through a tiny hole in 38 s. What must be the molar mass of a gas that requires 64 s to effuse under identical conditions?
83. What are the ratios of the diffusion rates for the pairs of gases (a) N_2 and O_2; (b) H_2O and D_2O (D = deuterium, i.e., $_1^2H$); (c) $^{14}CO_2$ and $^{12}CO_2$; (d) $^{235}UF_6$ and $^{238}UF_6$?
84. Which of the following visualizations best represents the distribution of O_2 and SO_2 molecules near an orifice some time after effusion occurs in the direction indicated by the arrows? The initial condition was one of equal numbers of O_2 molecules (●) and SO_2 molecules (●) on the left side of the orifice. Explain.

(a)   (b)   (c)   (d)

85. It takes 22 hours for a neon-filled balloon to shrink to half its original volume at STP. If the same balloon had been filled with helium, then how long would it have taken for the balloon to shrink to half its original volume at STP?
86. The molar mass of radon gas was first estimated by comparing its diffusion rate with that of mercury vapor, Hg(g). What is the molar mass of radon if mercury vapor diffuses 1.082 times as fast as radon gas? Assume that Graham's law holds for diffusion.

Nonideal Gases

87. Refer to Example 6-17. Recalculate the pressure of $Cl_2(g)$ by using both the ideal gas equation and the van der Waals equation at the temperatures (a) 100 °C; (b) 200 °C; (c) 400 °C. From the results, confirm the statement that a gas tends to be more ideal at high temperatures than at low temperatures.
88. Use both the ideal gas equation and the van der Waals equation to calculate the pressure exerted by 1.50 mol of $SO_2(g)$ when it is confined at 298 K to a volume of (a) 100.0 L, (b) 50.0 L, (c) 20.0 L, (d) 10.0 L. Under which of these conditions is the pressure calculated with the ideal gas equation within a few percent of that calculated with the van der Waals equation? Use values of a and b from Table 6.5.

89. Use the value of the van der Waals constant b for He(g), given in Table 6.5, to estimate the radius, r, of a single helium atom. Give your answer in picometers. [*Hint:* The volume of a sphere of radius r is $4\pi r^3/3$.]
90. (a) Use the value of the van der Waals constant b for $CH_4(g)$, given in Table 6.5, to estimate the radius of the CH_4 molecule. (See Exercise 89.) How does your estimate of the radius compare with the value $r = 228$ pm, obtained experimentally from an analysis of the structure of solid methane? (b) The density of $CH_4(g)$ is 66.02 g mL^{-1} at 100 bar and 325 K. What is the value of compressibility factor at this temperature and pressure?

Integrative and Advanced Exercises

91. Explain why it is necessary to include the density of Hg(l) and the value of the acceleration due to gravity, g, in a precise definition of a millimeter of mercury (page 194).

92. Assume the following initial conditions for the graphs labeled A, B, and C in Figure 6-7. (A) 10.0 mL at 400 K; (B) 20.0 mL at 400 K; (C) 40.0 mL at 400 K. Use Charles's law to calculate the volume of each gas at 0, −100, −200, −250, and −270 °C. Show that the volume of each gas becomes zero at −273.15 °C.

93. Consider the diagram below. The "initial" sketch illustrates, both at the macroscopic and molecular levels, an initial condition: 1 mol of a gas at 273 K and 1.00 atm. With as much detail as possible, illustrate the final condition after each of the following changes.
(a) The pressure is changed to 250 mmHg while standard temperature is maintained.
(b) The temperature is changed to 140 K while standard pressure is maintained.
(c) The pressure is changed to 0.5 atm while the temperature is changed to 550 K.
(d) An additional 0.5 mol of gas is introduced into the cylinder, the temperature is changed to 135 °C, and the pressure is changed to 2.25 atm.

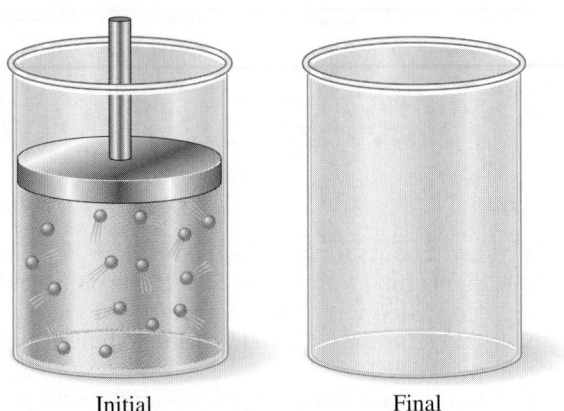

Initial   Final

94. Two evacuated bulbs of equal volume are connected by a tube of negligible volume. One of the bulbs is placed in a constant-temperature bath at 225 K and the other bulb is placed in a constant-temperature bath at 350 K. Exactly 1 mol of an ideal gas is injected into the system. Calculate the final number of moles of gas in each bulb.

95. A compound is 85.6% carbon by mass. The rest is hydrogen. When 10.0 g of the compound is evaporated at 50.0 °C, the vapor occupies 6.30 L at 1.00 atm pressure. What is the molecular formula of the compound?

96. A 0.7178 g sample of a hydrocarbon occupies a volume of 390.7 mL at 65.0 °C and 99.2 kPa. When the sample is burned in excess oxygen, 2.4267 g CO_2 and 0.4967 g H_2O are obtained. What is the molecular formula of the hydrocarbon? Write a plausible structural formula for the molecule.

97. A 3.05 g sample of $NH_4NO_3(s)$ is introduced into an evacuated 2.18 L flask and then heated to 250 °C.

What is the total gas pressure, in atmospheres, in the flask at 250 °C when the NH_4NO_3 has completely decomposed?

$$NH_4NO_3(s) \longrightarrow N_2O(g) + 2\,H_2O(g)$$

98. Ammonium nitrite, NH_4NO_2, decomposes according to the chemical equation below.

$$NH_4NO_2(s) \longrightarrow N_2(g) + 2\,H_2O(g)$$

What is the total volume of products obtained when 128 g NH_4NO_2 decomposes at 819 °C and 101 kPa?

99. A mixture of 1.00 g H_2 and 8.60 g O_2 is introduced into a 1.500 L flask at 25 °C. When the mixture is ignited, an explosive reaction occurs in which water is the only product. What is the total gas pressure when the flask is returned to 25 °C? (The vapor pressure of water at 25 °C is 23.8 mmHg.)

100. In the reaction of $CO_2(g)$ and solid sodium peroxide (Na_2O_2), solid sodium carbonate (Na_2CO_3) and oxygen gas are formed. This reaction is used in submarines and space vehicles to remove expired $CO_2(g)$ and to generate some of the $O_2(g)$ required for breathing. Assume that the volume of gases exchanged in the lungs equals 4.0 L/min, the CO_2 content of expired air is 3.8% CO_2 by volume, and the gases are at 25 °C and 735 mmHg. If the $CO_2(g)$ and $O_2(g)$ in the above reaction are measured at the same temperature and pressure, **(a)** how many milliliters of $O_2(g)$ are produced per minute and **(b)** at what rate is the $Na_2O_2(s)$ consumed, in grams per hour?

101. What is the partial pressure of $Cl_2(g)$, in millimeters of mercury, at 0.00 °C and 1.00 atm in a gaseous mixture that consists of 46.5% N_2, 12.7% Ne, and 40.8% Cl_2, by mass?

102. A gaseous mixture of He and O_2 has a density of 0.518 g/L at 25 °C and 721 mmHg. What is the mass percent He in the mixture?

103. When working with a mixture of gases, it is sometimes convenient to use an *apparent molar mass* (a weighted-average molar mass). Think in terms of replacing the mixture with a hypothetical single gas. What is the apparent molar mass of air, given that air is 78.08% N_2, 20.95% O_2, 0.93% Ar, and 0.036% CO_2, by volume?

104. A mixture of $N_2O(g)$ and $O_2(g)$ can be used as an anesthetic. In a particular mixture, the partial pressures of N_2O and O_2 are 612 Torr and 154 Torr, respectively. Calculate **(a)** the mass percentage of N_2O in this mixture, and **(b)** the apparent molar mass of this anesthetic. [*Hint:* For part (b), refer to Exercise 103.]

105. Gas cylinder A has a volume of 48.2 L and contains $N_2(g)$ at 8.35 atm at 25 °C. Gas cylinder B, of unknown volume, contains He(g) at 9.50 atm and 25 °C. When the two cylinders are connected and the gases mixed, the pressure in each cylinder becomes 8.71 atm. What is the volume of cylinder B?

106. The accompanying sketch is that of a closed-end manometer. Describe how the gas pressure is measured. Why is a measurement of $P_{bar.}$ not necessary when using this manometer? Explain why the closed-end manometer is more suitable for measuring low pressures and the open-end manometer more suitable for measuring pressures nearer atmospheric pressure.

Gas

107. Producer gas is a type of fuel gas made by passing air or steam through a bed of hot coal or coke. A typical producer gas has the following composition in percent by volume: 8.0% CO_2, 23.2% CO, 17.7% H_2, 1.1% CH_4, and 50.0% N_2.
(a) What is the density of this gas at 23 °C and 763 mmHg, in grams per liter?
(b) What is the partial pressure of CO in this mixture at 0.00 °C and 1 atm?
(c) What volume of air, measured at 23 °C and 741 Torr, is required for the complete combustion of 1.00×10^3 L of this producer gas, also measured at 23 °C and 741 Torr?
[*Hint:* Which three of the constituent gases are combustible?]

108. The heat required to sustain animals while they hibernate comes from the biochemical combustion of fatty acids, such as arachidonic acid, $C_{20}H_{32}O_2$. What volume of air, measured at 298 K and 1.00 atm, is required to burn 2.00 kg $C_{20}H_{32}O_2$? Air is approximately 78.1% N_2 and 20.9% O_2, by volume. Other gases make up the remaining 1.0%.

109. A mixture of $H_2(g)$ and $O_2(g)$ is prepared by electrolyzing 1.32 g water, and the mixture of gases is collected over water at 30 °C and 748 mmHg. The volume of "wet" gas obtained is 2.90 L. What must be the vapor pressure of water at 30 °C?

$$2\,H_2O(l) \xrightarrow{\text{electrolysis}} 2\,H_2(g) + O_2(g)$$

110. Aluminum (Al) and iron (Fe) each react with hydrochloric acid solution (HCl) to produce a chloride salt and hydrogen gas, $H_2(g)$. A 0.1924 g sample of a mixture of Al and Fe is treated with excess HCl solution. A volume of 159 mL of H_2 gas is collected over water at 19.0 °C and 841 Torr. What is the percent (by mass) of Fe in the mixture? The vapor pressure of water at 19.0 °C is 16.5 Torr.

111. A 0.168 L sample of $O_2(g)$ is collected over water at 26 °C and a barometric pressure of 737 mmHg. In the gas that is collected, what is the percent water vapor (a) by volume; (b) by number of molecules; (c) by mass? (Vapor pressure of water at 26 °C = 25.2 mmHg.)

112. A breathing mixture is prepared in which He is substituted for N_2. The gas is 79% He and 21% O_2, by volume. (a) What is the density of this mixture in grams per liter at 25 °C and 1.00 atm? (b) At what pressure would the He–O_2 mixture have the same

density as that of air at 25 °C and 1.00 atm? See Exercise 103 for the composition of air.

113. Chlorine dioxide, ClO_2, is sometimes used as a chlorinating agent for water treatment. It can be prepared from the reaction below:

$$Cl_2(g) + 4\,NaClO(aq) \longrightarrow 4\,NaCl(aq) + 2\,ClO_2(g)$$

In an experiment, 1.0 L $Cl_2(g)$, measured at 10.0 °C and 4.66 atm, is dissolved in 0.750 L of 2.00 M NaClO(aq). If 25.9 g of pure ClO_2 is obtained, then what is the percent yield for this experiment?

114. The amount of ozone, O_3, in a mixture of gases can be determined by passing the mixture through a solution of excess potassium iodide, KI. Ozone reacts with the iodide ion as follows:

$$O_3(g) + 3I^-(aq) + H_2O(l) \longrightarrow$$
$$O_2(g) + I_3^-(aq) + 2OH^-(aq)$$

The amount of I_3^- produced is determined by titrating with thiosulfate ion, $S_2O_3^{2-}$:

$$I_3^-(aq) + 2\,S_2O_3^{2-}(aq) \longrightarrow 3I^-(aq) + S_4O_6^{2-}(aq)$$

A mixture of gases occupies a volume of 53.2 L at 18 °C and 0.993 atm. The mixture is passed slowly through a solution containing an excess of KI to ensure that all the ozone reacts. The resulting solution requires 26.2 mL of 0.1359 M $Na_2S_2O_3$ to titrate to the end point. Calculate the mole fraction of ozone in the original mixture.

115. A 0.1052 g sample of $H_2O(l)$ in an 8.050 L sample of dry air at 30.1 °C evaporates completely. To what temperature must the air be cooled to give a relative humidity of 80.0%? Vapor pressures of water: 20 °C, 17.54 mmHg; 19 °C, 16.48 mmHg; 18 °C, 15.48 mmHg; 17 °C, 14.53 mmHg; 16 °C, 13.63 mmHg; 15 °C, 12.79 mmHg. [*Hint:* Go to Focus On feature for Chapter 6 on the MasteringChemistry site, www.masteringchemistry.com, for a discussion of relative humidity.]

116. An alternative to Figure 6-6 is to plot P against $1/V$. The resulting graph is a straight line passing through the origin. Use Boyle's data from Feature Problem 125 to draw such a straight-line graph. What factors would affect the *slope* of this straight line? Explain.

117. We have noted that atmospheric pressure depends on altitude. Atmospheric pressure as a function of altitude can be calculated with an equation known as the barometric formula:

$$P = P_0 \times 10^{-Mgh/2.303RT}$$

In this equation, P and P_0 can be in any pressure units, for example, Torr. P_0 is the pressure at sea level, generally taken to be 1.00 atm or its equivalent. The units in the exponential term must be SI units, however. Use the barometric formula to
(a) estimate the barometric pressure at the top of Mt. Whitney in California (altitude: 14,494 ft; assume a temperature of 10 °C)
(b) show that barometric pressure decreases by one-thirtieth in value for every 900-ft increase in altitude

118. Consider a sample of $O_2(g)$ at 298 K and 1.0 atm. Calculate (a) u_{rms} and (b) the fraction of molecules that have speed equal to u_{rms}.

119. A nitrogen molecule (N_2) having the average kinetic energy at 300 K is released from Earth's surface to travel upward. If the molecule could move upward without colliding with other molecules, then how high would it go before coming to rest? Give your answer in kilometers. [*Hint*: When the molecule comes to rest, the potential energy of the molecule will be mgh, where m is the molecular mass in kilograms, $g = 9.81$ m s^{-2} is the acceleration due to gravity, and h is the height, in meters, above Earth's surface.]

120. For $H_2(g)$ at 0 °C and 1 atm, calculate the percentage of molecules that have speed **(a)** 0 m s^{-1}; **(b)** 500 m s^{-1}; **(c)** 1000 m s^{-1}; **(d)** 1500 m s^{-1}; **(e)** 2000 m s^{-1}; **(f)** 2500 m s^{-1}; **(g)** 3500 m s^{-1}. Graph your results to obtain your own version of Figure 6-15.

121. If the van der Waals equation is solved for volume, a cubic equation is obtained.

(a) Derive the equation below by rearranging equation (6.26).

$$V^3 - n\left(\frac{RT + bP}{P}\right)V^2 + \left(\frac{n^2a}{P}\right)V - \frac{n^3ab}{P} = 0$$

(b) What is the volume, in liters, occupied by 185 g $CO_2(g)$ at a pressure of 12.5 atm and 286 K? For $CO_2(g)$, $a = 3.61$ L^2 atm mol^{-2} and $b = 0.0429$ L mol^{-1}.
[*Hint*: Use the ideal gas equation to obtain an estimate of the volume. Then refine your estimate, either by trial and error, or using the method of successive approximations. See Appendix A, pages A5–A6, for a description of the method of successive approximations.]

122. According to the *CRC Handbook of Chemistry and Physics* (83rd ed.), the molar volume of $O_2(g)$ is 0.2168 L mol^{-1} at 280 K and 10 MPa. (Note: 1 MPa = 1×10^6 Pa.)

(a) Use the van der Waals equation to calculate the pressure of one mole of $O_2(g)$ at 280 K if the volume is 0.2168 L. What is the % error in the calculated pressure? The van der Waals constants are $a = 1.382$ L^2 bar mol^{-2} and $b = 0.0319$ L mol^{-1}.

(b) Use the ideal gas equation to calculate the volume of one mole of $O_2(g)$ at 280 K and 10 MPa. What is the % error in the calculated volume?

123. A particular equation of state for $O_2(g)$ has the form

$$P\overline{V} = RT\left(1 + \frac{B}{\overline{V}} + \frac{C}{\overline{V}^2}\right)$$

where $\overline{V}$ is the molar volume, $B = -21.89$ cm^3/mol and $C = 1230$ cm^6/mol.

(a) Use the equation to calculate the pressure exerted by 1 mol $O_2(g)$ confined to a volume of 500 cm^3 at 273 K.

(b) Is the result calculated in part **(a)** consistent with that suggested for $O_2(g)$ by Figure 6-20? Explain.

124. A 0.156 g sample of a magnesium–aluminum alloy dissolves completely in an excess of HCl(aq). The liberated $H_2(g)$ is collected over water at 5 °C when the barometric pressure is 752 Torr. After the gas is collected, the water and gas gradually warm to the prevailing room temperature of 23 °C. The pressure of the collected gas is again equalized against the barometric pressure of 752 Torr, and its volume is found to be 202 mL. What is the percent composition of the magnesium–aluminum alloy? (Vapor pressure of water: 6.54 mmHg at 5 °C and 21.07 mmHg at 23 °C).

Feature Problems

125. Shown to the right is a diagram of Boyle's original apparatus. At the start of the experiment, the length of the air column (A) on the left was 30.5 cm and the heights of mercury in the arms of the tube were equal. When mercury was added to the right arm of the tube, a difference in mercury levels (B) was produced, and the entrapped air on the left was compressed into a shorter length of the tube (smaller volume) as shown in the illustration for A = 27.9 cm and B = 7.1 cm. Boyle's values of A and B, in centimeters, are listed as follows:

A:	30.5	27.9	25.4	22.9	20.3
B:	0.0	7.1	15.7	25.7	38.3
A:	17.8	15.2	12.7	10.2	7.6
B:	53.8	75.4	105.6	147.6	224.6

Barometric pressure at the time of the experiment was 739.8 mmHg. Assuming that the length of the air column (A) is proportional to the volume of air, show that these data conform reasonably well to Boyle's law.

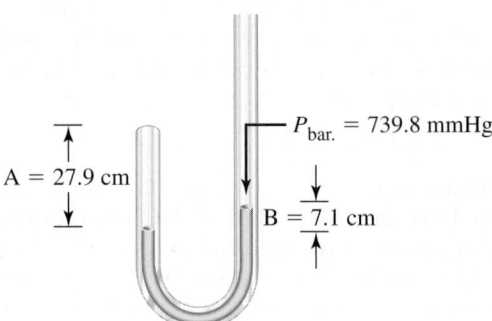

$P_{\text{bar.}} = 739.8$ mmHg
A = 27.9 cm
B = 7.1 cm

126. In 1860, Stanislao Cannizzaro showed how Avogadro's hypothesis could be used to establish the atomic masses of elements in gaseous compounds. Cannizzaro took the atomic mass of hydrogen to be exactly one and assumed that hydrogen exists as H_2 molecules (molecular mass = 2). Next, he determined the volume of $H_2(g)$ at 0.00 °C and 1.00 atm that has a mass of exactly 2 g. This volume is 22.4 L. Then he assumed that 22.4 L of any other gas would have the same number of molecules as in 22.4 L of $H_2(g)$. (Here is where Avogadro's hypothesis entered in.) Finally, he reasoned that the ratio of the mass of 22.4 L of any other gas to the mass of 22.4 L of $H_2(g)$ should be the same as the ratio of their molecular masses. The sketch below illustrates Cannizzaro's reasoning in establishing the atomic weight of oxygen as 16. The gases in the table all contain the element X. Their molecular masses were determined by Cannizzaro's method. Use the percent composition data to deduce the atomic mass of X, the number of atoms of X in each of the gas molecules, and the identity of X.

Compound	Molecular Mass, u	Mass Percent X, %
Nitryl fluoride	65.01	49.4
Nitrosyl fluoride	49.01	32.7
Thionyl fluoride	86.07	18.6
Sulfuryl fluoride	102.07	31.4

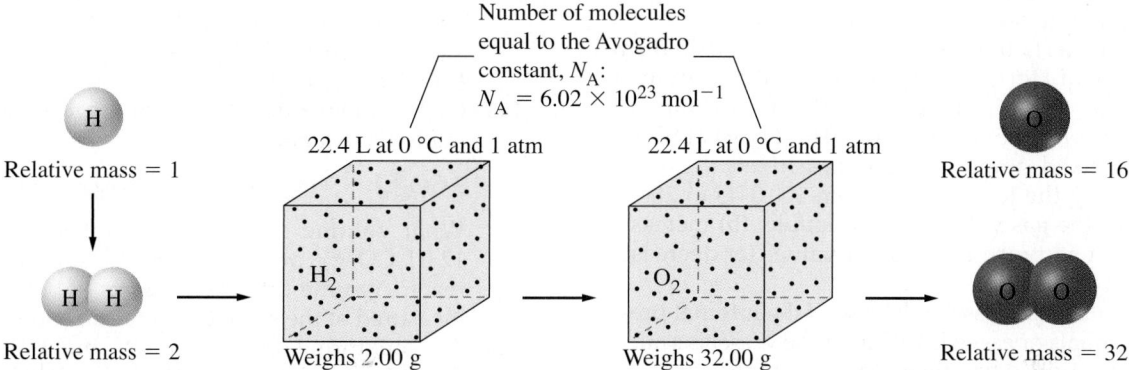

127. In research that required the careful measurement of gas densities, John Rayleigh, a physicist, found that the density of $O_2(g)$ had the same value whether the gas was obtained from air or derived from one of its compounds. The situation with $N_2(g)$ was different, however. The density of $N_2(g)$ had the same value when the $N_2(g)$ was derived from any of various compounds, but a *different* value if the $N_2(g)$ was extracted from air. In 1894, Rayleigh enlisted the aid of William Ramsay, a chemist, to solve this apparent mystery; in the course of their work they discovered the noble gases.
(a) Why do you suppose that the $N_2(g)$ extracted from liquid air did *not* have the same density as $N_2(g)$ obtained from its compounds?
(b) Which gas do you suppose had the greater density: $N_2(g)$ extracted from air or $N_2(g)$ prepared from nitrogen compounds? Explain.
(c) The way in which Ramsay proved that nitrogen gas extracted from air was itself a mixture of gases involved allowing this nitrogen to react with magnesium metal to form magnesium nitride. Explain the significance of this experiment.
(d) Calculate the *percent difference* in the densities at 0.00 °C and 1.00 atm of Rayleigh's $N_2(g)$ extracted from air and $N_2(g)$ derived from nitrogen compounds. [The volume percentages of the major components of air are 78.084% N_2, 20.946% O_2, 0.934% Ar, and 0.0379% CO_2.]
128. The equation $d/P = M/RT$, which can be derived from equation (6.14), suggests that the ratio of the density (d) to pressure (P) of a gas at constant temperature should be a constant. The gas density data at the end of this question were obtained for $O_2(g)$ at various pressures at 273.15 K.

(a) Calculate values of d/P, and with a graph or by other means determine the ideal value of the term d/P for $O_2(g)$ at 273.15 K.
[*Hint:* The ideal value is that associated with a perfect (ideal) gas.]
(b) Use the value of d/P from part (a) to calculate a precise value for the atomic mass of oxygen, and compare this value with that listed on the inside front cover.

P, mmHg:	760.00	570.00	380.00	190.00
d, g/L:	1.428962	1.071485	0.714154	0.356985

129. A sounding balloon is a rubber bag filled with $H_2(g)$ and carrying a set of instruments (the payload). Because this combination of bag, gas, and payload has a smaller mass than a corresponding volume of air, the balloon rises. As the balloon rises, it expands. From the table below, estimate the maximum height to which a spherical balloon can rise given the mass of balloon, 1200 g; payload, 1700 g: quantity of $H_2(g)$ in balloon, 120 ft^3 at 0.00 °C and 1.00 atm; diameter of balloon at maximum height, 25 ft. Air pressure and temperature as functions of altitude are:

Altitude, km	Pressure, mb	Temperature, K
0	1.0×10^3	288
5	5.4×10^2	256
10	2.7×10^2	223
20	5.5×10^1	217
30	1.2×10^1	230
40	2.9×10^0	250
50	8.1×10^{-1}	250
60	2.3×10^{-1}	256

Self-Assessment Exercises

130. In your own words, define or explain each term or symbol. **(a)** atm; **(b)** STP; **(c)** R; **(d)** partial pressure; **(e)** u_{rms}.

131. Briefly describe each concept or process: **(a)** absolute zero of temperature; **(b)** collection of a gas over water; **(c)** effusion of a gas; **(d)** law of combining volumes.

132. Explain the important distinctions between **(a)** barometer and manometer; **(b)** Celsius and Kelvin temperature; **(c)** ideal gas equation and general gas equation; **(d)** ideal gas and real gas.

133. Which exerts the greatest pressure, **(a)** a 75.0 cm column of Hg(l) ($d = 13.6$ g/mL); **(b)** a column of air 10 mi high; **(c)** a 5.0 m column of CCl_4(l) ($d = 1.59$ g/mL); **(d)** 10.0 g H_2(g) at STP?

134. For a fixed amount of gas at a fixed pressure, changing the temperature from 100.0 °C to 200 K causes the gas volume to **(a)** double; **(b)** increase, but not to twice its original value; **(c)** decrease; **(d)** stay the same.

135. A fragile glass vessel will break if the internal pressure equals or exceeds 2.0 bar. If the vessel is sealed at 0 °C and 1.0 bar, then at what temperature will the vessel break? Assume that the vessel does not expand when heated.

136. Which of the following choices represents the molar volume of an ideal gas at 25 °C and 1.5 atm? **(a)** $(298 \times 1.5/273) \times 22.4$ L; **(b)** 22.4 L; **(c)** $(273 \times 1.5/298) \times 22.4$ L; **(d)** $[298/(273 \times 1.5)] \times 22.4$ L; **(e)** $[273/(298 \times 1.5)] \times 22.4$ L.

137. The gas with the greatest density at STP is **(a)** N_2O; **(b)** Kr; **(c)** SO_3; **(d)** Cl_2.

138. If the Kelvin temperature of a sample of ideal gas doubles (e.g., from 200 K to 400 K), what happens to the root-mean-square speed, u_{rms}? **(a)** u_{rms} increases by a factor of $\sqrt{2}$; **(b)** u_{rms} increases by a factor of 2; **(c)** u_{rms} decreases by a factor of 2; **(d)** u_{rms} increases by a factor of 4; **(e)** u_{rms} decreases by a factor of 4.

139. Consider the statements (a) to (e) below. Assume that H_2(g) and O_2(g) behave ideally. State whether each of the following statements is true or false. For each false statement, explain how you would change it to make it a true statement.
(a) Under the same conditions of temperature and pressure, the average kinetic energy of O_2 molecules is less than that of H_2 molecules.
(b) Under the same conditions of temperature and pressure, H_2 molecules move faster, on average, than O_2 molecules.
(c) The volume of 1.00 mol of H_2(g) at 25.0 °C, 1.00 atm is 22.4 L.
(d) The volume of 2.0 g H_2(g) is equal to the volume of 32.0 g O_2(g), at the same temperature and pressure.
(e) In a mixture of H_2 and O_2 gases, with partial pressures P_{H_2} and P_{O_2}, respectively, the total pressure is the larger of P_{H_2} and P_{O_2}.

140. A sample of O_2(g) is collected over water at 23 °C and a barometric pressure of 751 Torr. The vapor pressure of water at 23 °C is 21 mmHg. The partial pressure of O_2(g) in the sample collected is **(a)** 21 mmHg; **(b)** 751 Torr; **(c)** 0.96 atm; **(d)** 1.02 atm.

141. At 0 °C and 0.500 atm, 4.48 L of gaseous NH_3 **(a)** contains 6.02×10^{22} molecules; **(b)** has a mass of 17.0 g; **(c)** contains 0.200 mol NH_3; **(d)** has a mass of 3.40 g.

142. To establish a pressure of 2.00 atm in a 2.24 L cylinder containing 1.60 g O_2(g) at 0 °C, **(a)** add 1.60 g O_2; **(b)** add 0.60 g He(g); **(c)** add 2.00 g He(g); **(d)** release 0.80 g O_2(g).

143. Carbon monoxide, CO, and hydrogen react according to the equation below.

$$3\,CO(g) + 7\,H_2(g) \longrightarrow C_3H_8(g) + 3\,H_2O(g)$$

What volume of which reactant gas remains if 12.0 L CO(g) and 25.0 L H_2(g) are allowed to react? Assume that the volumes of both gases are measured at the same temperature and pressure.

144. A mixture of 5.0×10^{-5} mol H_2(g) and 5.0×10^{-5} mol SO_2(g) is placed in a 10.0 L container at 25 °C. The container has a pinhole leak. After a period of time, the partial pressure of H_2(g) in the container **(a)** is less than that of the SO_2(g); **(b)** is equal to that of the SO_2(g); **(c)** exceeds that of the SO_2(g); **(d)** is the same as in the original mixture.

145. Under which conditions is Cl_2 most likely to behave like an ideal gas? Explain. **(a)** 100 °C and 10.0 atm; **(b)** 0 °C and 0.50 atm; **(c)** 200 °C and 0.50 atm; **(d)** 400 °C and 10.0 atm.

146. Without referring to Table 6.5, state which species in each of the following pairs has the greater value for the van der Waals constant a, and which one has the greater value for the van der Waals constant b. **(a)** He or Ne; **(b)** CH_4 or C_3H_8; **(c)** H_2 or Cl_2.

147. Explain why the height of the mercury column in a barometer is independent of the diameter of the barometer tube.

148. A gaseous hydrocarbon that is 82.7% C and 17.3% H by mass has a density of 2.35 g/L at 25 °C and 752 Torr. What is the molecular formula of this hydrocarbon?

149. Draw a box to represent a sample of air containing N_2 molecules (represented as squares) and O_2 molecules (represented as circles) in their correct proportions. How many squares and circles would you need to draw to also represent the CO_2(g) in air through a single mark? What else should you add to the box for this more complete representation of air? [*Hint:* See Exercise 103.]

150. Appendix E describes a useful study aid known as concept mapping. Using the method presented in Appendix E, construct a concept map illustrating the different concepts to show the relationships among all the gas laws described in this chapter.

Thermochemistry

Potassium reacts with water, liberating sufficient heat to ignite the hydrogen evolved. The transfer of heat between substances in chemical reactions is an important aspect of thermochemistry.

Natural gas consists mostly of methane, CH_4. As we learned in Chapter 4, the combustion of a hydrocarbon, such as methane, yields carbon dioxide and water as products. More important, however, is another "product" of this reaction, which we have not previously mentioned: heat. This heat can be used to produce hot water in a water heater, to heat a house, or to cook food.

Thermochemistry is the branch of chemistry concerned with the heat effects that accompany chemical reactions. To understand the relationship between heat and chemical and physical changes, we must start with some basic definitions. We will then explore the concept of heat and the methods used to measure the transfer of energy across boundaries. Another form of energy transfer is work, and, in combination with heat, we will define the first law of thermodynamics. At this point, we will establish the relationship between heats of reaction and changes in internal energy and enthalpy. We will see that the tabulation of the change in internal energy and change in enthalpy can be used to calculate, directly or indirectly, energy changes during chemical and physical changes. Finally, concepts introduced in this chapter will answer a host of practical questions, such as

◀ Thermochemistry is a subfield of a larger discipline called *thermodynamics*. The broader aspects of thermodynamics are considered in Chapters 19 and 20.

241

why natural gas is a better fuel than coal and why the energy value of fats is greater than that of carbohydrates and proteins.

7-1   Getting Started: Some Terminology

In this section, we introduce and define some very basic terms. Most are discussed in greater detail in later sections, and your understanding of these terms should grow as you proceed through the chapter.

Let us think of the universe as being comprised of a system and its surroundings. A **system** is the part of the universe chosen for study, and it can be as large as all the oceans on Earth or as small as the contents of a beaker. Most of the systems we will examine will be small and we will look, particularly, at the transfer of *energy* (as heat and work) and *matter* between the system and its surroundings. The **surroundings** are that part of the universe outside the system with which the system interacts. Figure 7-1 pictures three common systems: first, as we see them and, then, in an abstract form that chemists commonly use. An **open system** freely exchanges energy and matter with its surroundings (Fig. 7-1a). A **closed system** can exchange energy, but not matter, with its surroundings (Fig. 7-1b). An **isolated system** does not interact with its surroundings (approximated in Figure 7-1c).

The remainder of this section says more, in a general way, about energy and its relationship to work. Like many other scientific terms, *energy* is derived from Greek. It means "work within." **Energy** is the capacity to do work. **Work** is done when a force acts through a distance. Moving objects do work when they slow down or are stopped. Thus, when one billiard ball strikes another and sets it in motion, work is done. The energy of a moving object is called **kinetic energy** (the word *kinetic* means "motion" in Greek). We can see the relationship between work and energy by comparing the units for these two quantities. The kinetic energy (e_k) of an object is based on its mass (m) and

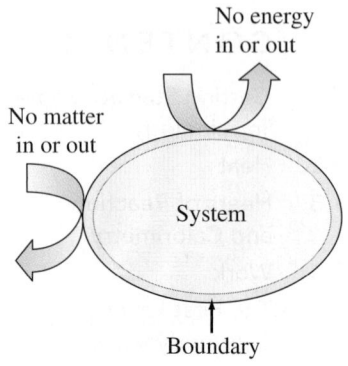

▲ **Isolated system**
Neither energy nor matter is transferred between the system and its surroundings.

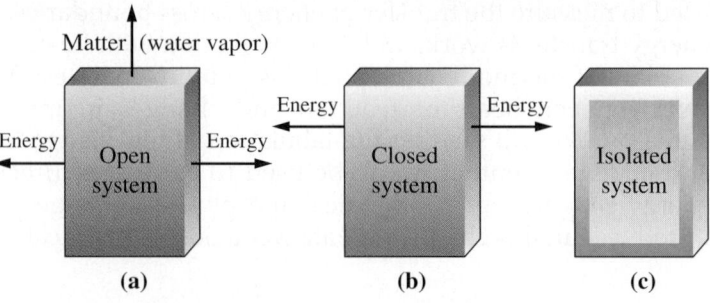

▶ FIGURE 7-1
Systems and their surroundings
(a) *Open system*. The beaker of hot coffee transfers energy to the surroundings—it loses heat as it cools. Matter is also transferred in the form of water vapor. **(b)** *Closed system*. The flask of hot coffee transfers energy (heat) to the surroundings as it cools. Because the flask is stoppered, no water vapor escapes and no matter is transferred. **(c)** *Isolated system*. Hot coffee in an insulated container approximates an isolated system. No water vapor escapes, and, for a time at least, little heat is transferred to the surroundings. (Eventually, though, the coffee in the container cools to room temperature.)

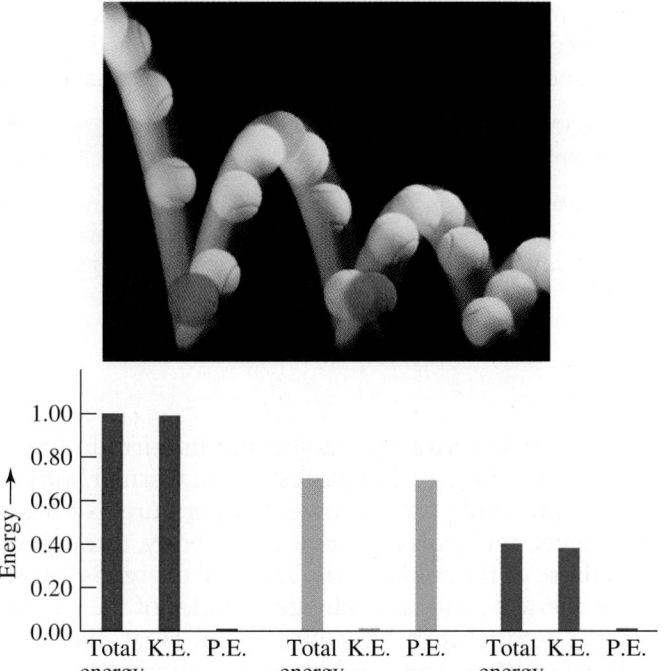

◀ FIGURE 7-2
Potential energy (P.E.) and kinetic energy (K.E.)
The energy of the bouncing tennis ball changes continuously from potential to kinetic energy and back again. The maximum potential energy is at the top of each bounce, and the maximum kinetic energy occurs at the moment of impact. The sum of P.E. and K.E. decreases with each bounce as the thermal energies of the ball and the surroundings increase. The ball soon comes to rest. The bar graph below the bouncing balls illustrates the relative contributions that the kinetic and potential energy make to the total energy for each ball position. The red bars correspond to the red ball, green bars correspond to the green ball and the blue bars correspond to the blue ball.

velocity (u) through the first equation below; work (w) is related to force [mass (m) $\times$ acceleration (a)] and distance (d) by the second equation.

$$e_k = \tfrac{1}{2}mu^2$$
$$w = m \times a \times d \tag{7.1}$$

When mass, speed, acceleration, and distance are expressed in SI units, the units of both kinetic energy and work will be kg m^2 s^{-2}, which is the SI unit of energy—the *joule* (J). That is, 1 J = 1 kg m^2 s^{-2}.

The bouncing ball in Figure 7-2 suggests something about the nature of energy and work. First, to lift the ball to the starting position, we have to apply a force through a distance (to overcome the force of gravity). The work we do is "stored" in the ball as energy. This stored energy has the potential to do work when released and is therefore called potential energy. **Potential energy** is energy resulting from condition, position, or composition; it is an energy associated with forces of attraction or repulsion between objects.

When we release the ball, it is pulled toward Earth's center by the force of gravity—it falls. Potential energy is converted to kinetic energy during this fall. The kinetic energy reaches its maximum just as the ball strikes the surface. On its rebound, the kinetic energy of the ball decreases (the ball slows down), and its potential energy increases (the ball rises). If the collision of the ball with the surface were perfectly *elastic*, like collisions between molecules in the kinetic-molecular theory, the sum of the potential and kinetic energies of the ball would remain constant. The ball would reach the same maximum height on each rebound, and it would bounce forever. But we know this doesn't happen—the bouncing ball soon comes to rest. All the energy originally invested in the ball as potential energy (by raising it to its initial position) eventually appears as additional kinetic energy of the atoms and molecules that make up the ball, the surface, and the surrounding air. This kinetic energy associated with random molecular motion is called **thermal energy**.

In general, thermal energy is proportional to the temperature of a system, as suggested by the kinetic theory of gases. The more vigorous the motion of the molecules in the system, the hotter the sample and the greater is its thermal energy. However, the thermal energy of a system also depends on the number of particles present, so that a small sample at a high temperature (for example, a cup of coffee at 75 °C) may have less thermal energy than a larger sample at a lower temperature (for example, a swimming pool at 30 °C). Thus, temperature

◀ As discussed in Appendix B-1, the SI unit for acceleration is m s^{-2}. We encountered this unit previously (page 194)—the acceleration due to gravity was given as $g = 9.80665$ m s^{-2}.

◀ A unit of work, heat, and energy is the joule, but work and heat are not forms of energy but *processes* by which the energy of a system is changed.

and thermal energy must be carefully distinguished. Equally important, we need to distinguish between energy changes produced by the action of forces through distances—*work*—and those involving the transfer of thermal energy—*heat*.

🔍 **7-1   CONCEPT ASSESSMENT**

Consider the following situations: a stick of dynamite exploding deep within a mountain cavern, the titration of an acid with base in a laboratory, and a cylinder of a steam engine with all of its valves closed. To what type of thermodynamic systems do these situations correspond?

7-2   Heat

▲ **James Joule (1818–1889)—an amateur scientist**
Joule's primary occupation was running a brewery, but he also conducted scientific research in a home laboratory. His precise measurements of quantities of heat formed the basis of the law of conservation of energy.

Heat is energy transferred between a system and its surroundings as a result of a temperature difference. Energy that passes from a warmer body (with a higher temperature) to a colder body (with a lower temperature) is transferred as heat. At the molecular level, molecules of the warmer body, through collisions, lose kinetic energy to those of the colder body. Thermal energy is transferred—"heat flows"—until the average molecular kinetic energies of the two bodies become the same, until the temperatures become equal. Heat, like work, describes energy in transit between a system and its surroundings.

Not only can heat transfer cause a change in temperature but, in some instances, it can also change a state of matter. For example, when a solid is heated, the molecules, atoms, or ions of the solid move with greater vigor and eventually break free from their neighbors by overcoming the attractive forces between them. Energy is required to overcome these attractive forces. During the process of melting, the temperature remains constant as a thermal energy transfer (heat) is used to overcome the forces holding the solid together. A process occurring at a constant temperature is said to be *isothermal*. Once a solid has melted completely, any further heat flow will raise the temperature of the resulting liquid.

Although we commonly use expressions like "heat is lost," "heat is gained," "heat flows," and "the system loses heat to the surroundings," you should not take these statements to mean that a system contains heat. It does not. The energy content of a system, as we shall see in Section 7-5, is a quantity called the *internal energy*. Heat is simply a form in which a quantity of energy may be *transferred* across a boundary between a system and its surroundings.

It is reasonable to expect that the quantity of heat, q, required to change the temperature of a substance depends on

- how much the temperature is to be changed
- the quantity of substance
- the nature of the substance (type of atoms or molecules)

Historically, the quantity of heat required to change the temperature of one gram of water by one degree Celsius has been called the **calorie (cal)**. The calorie is a small unit of energy, and the unit *kilocalorie* (kcal) has also been widely used. The SI unit for heat is simply the basic SI energy unit, the joule (J).

$$1 \, \text{cal} = 4.184 \, \text{J} \qquad\qquad (7.2)$$

Although the joule is used almost exclusively in this text, the calorie is widely encountered in older scientific literature. In the United States, the kilocalorie is commonly used for measuring the energy content of foods (see Focus On feature for Chapter 7 on www.masteringchemistry.com).

The quantity of heat required to change the temperature of a system by one degree is called the **heat capacity** of the system. If the system is a mole of substance, the term *molar heat capacity* is applicable. If the system is one gram of substance, the applicable term is *specific heat capacity*, or more commonly, **specific heat** (sp ht).* The specific heats of substances are somewhat temperature dependent. At 25 °C, the specific heat of water is

$$\frac{4.18\,\text{J}}{\text{g\,°C}} = 4.18\,\text{J\,g}^{-1}\,°\text{C}^{-1} \tag{7.3}$$

In Example 7-1, the objective is to calculate a quantity of heat based on the amount of a substance, the specific heat of that substance, and its temperature change.

EXAMPLE 7-1   **Calculating a Quantity of Heat**

How much heat is required to raise the temperature of 7.35 g of water from 21.0 to 98.0 °C? (Assume the specific heat of water is $4.18\,\text{J\,g}^{-1}\,°\text{C}^{-1}$ throughout this temperature range.)

Analyze

To answer this question, we begin by multiplying the specific heat capacity by the mass of water to obtain the heat capacity of the system. To find the amount of heat required to produce the desired temperature change we multiply the heat capacity by the temperature difference.

Solve

The specific heat is the heat capacity of 1.00 g water:

$$\frac{4.18\,\text{J}}{\text{g water\,°C}}$$

The heat capacity of the system (7.35 g water) is

$$7.35\,\text{g water} \times \frac{4.18\,\text{J}}{\text{g water\,°C}} = 30.7\,\frac{\text{J}}{°\text{C}}$$

The required temperature change in the system is

$$(98.0 - 21.0)\,°\text{C} = 77.0\,°\text{C}$$

The heat required to produce this temperature change is

$$30.7\,\frac{\text{J}}{°\text{C}} \times 77.0\,°\text{C} = 2.36 \times 10^3\,\text{J}$$

Assess

Remember that specific heat is a quantity that depends on the amount of material. Also note that the change in temperature is determined by subtracting the initial temperature from the final temperature. This will be important in determining the sign on the value you determine for heat, as will become apparent in the next section.

PRACTICE EXAMPLE A: How much heat, in kilojoules (kJ), is required to raise the temperature of 237 g of cold water from 4.0 to 37.0 °C (body temperature)?

PRACTICE EXAMPLE B: How much heat, in kilojoules (kJ), is required to raise the temperature of 2.50 kg Hg(l) from −20.0 to −6.0 °C? Assume a density of 13.6 g/mL and a molar heat capacity of $28.0\,\text{J\,mol}^{-1}\,°\text{C}^{-1}$ for Hg(l).

The line of reasoning used in Example 7-1 can be summarized in equation (7.5), which relates a quantity of heat to the mass of a substance, its specific heat, and the temperature change.

$$\text{quantity of heat} = \underbrace{\text{mass of substance} \times \text{specific heat}}_{\text{heat capacity} = C} \times \text{temperature change} \tag{7.4}$$

◀ The Greek letter delta, Δ, indicates a *change* in some quantity.

$$q = m \times \text{specific heat} \times \Delta T = C \times \Delta T \tag{7.5}$$

*The original meaning of specific heat was that of a *ratio*: the quantity of heat required to change the temperature of a mass of substance divided by the quantity of heat required to produce the same temperature change in the same mass of water—this definition would make specific heat dimensionless. The meaning given here is more commonly used.

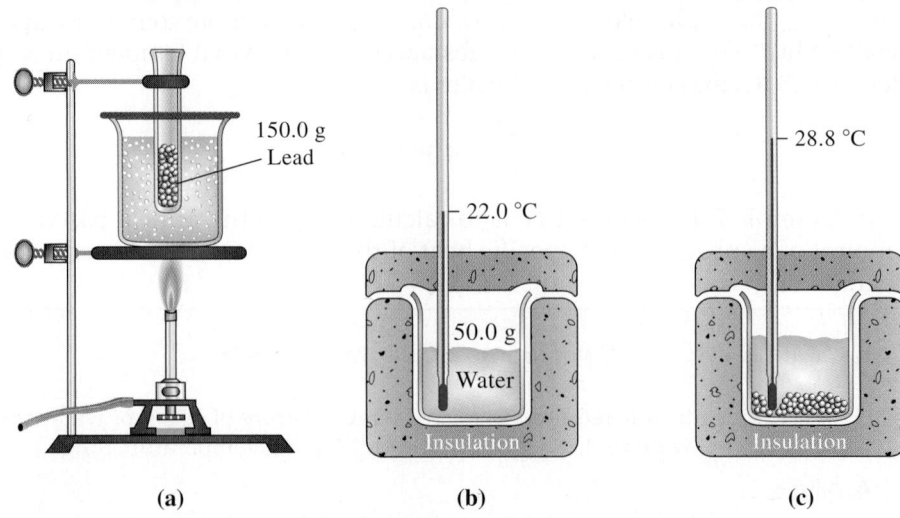

▶ FIGURE 7-3
Determining the specific heat of lead—Example 7-2 illustrated
(a) A 150.0 g sample of lead is heated to the temperature of boiling water (100.0 °C). **(b)** A 50.0 g sample of water is added to a thermally insulated beaker, and its temperature is found to be 22.0 °C. **(c)** The hot lead is dumped into the cold water, and the temperature of the final lead–water mixture is 28.8 °C.

▶ The symbol > means "greater than," and < means "less than."

In equation (7.5), the temperature change is expressed as $\Delta T = T_f - T_i$, where T_f is the final temperature and T_i is the initial temperature. When the temperature of a system increases ($T_f > T_i$), ΔT is *positive*. A positive q signifies that heat is absorbed or *gained* by the system. When the temperature of a system decreases ($T_f < T_i$), ΔT is *negative*. A negative q signifies that heat is evolved or *lost* by the system.

Another idea that enters into calculations of quantities of heat is the **law of conservation of energy**: In interactions between a system and its surroundings, the total energy remains *constant*—energy is neither created nor destroyed. Applied to the exchange of heat, this means that

$$q_{system} + q_{surroundings} = 0 \qquad (7.6)$$

Thus, heat *gained* by a system is *lost* by its surroundings, and vice versa.

$$q_{system} = -q_{surroundings} \qquad (7.7)$$

Experimental Determination of Specific Heats

Let us consider how the law of conservation of energy is used in the experiment outlined in Figure 7-3. The object is to determine the specific heat of lead. The transfer of energy, as heat, from the lead to the cooler water causes the temperature of the lead to decrease and that of the water to increase, until the lead and water are at the same temperature. Either the lead or the water can be considered the system. If we consider lead to be the system, we can write $q_{lead} = q_{system}$. Furthermore, if the lead and water are maintained in a thermally insulated enclosure, we can assume that $q_{water} = q_{surroundings}$. Then, applying equation (7.7), we have

$$q_{lead} = -q_{water} \qquad (7.8)$$

We complete the calculation in Example 7-2.

EXAMPLE 7-2   Determining a Specific Heat from Experimental Data

Use data presented in Figure 7-3 to calculate the specific heat of lead.

Analyze

Keep in mind that if we know any four of the five quantities—q, m, specific heat, T_f, T_i—we can solve equation (7.5) for the remaining one. We know from Figure 7-3 that a known quantity of lead is heated and then dumped into a known amount of water at a known temperature, which is the initial temperature. Once the system comes to equilibrium, the water temperature is the final temperature. In this type of question, we will use equation (7.5).

Solve

First, use equation (7.5) to calculate q_{water}.

$$q_{water} = 50.0 \text{ g water} \times \frac{4.18 \text{ J}}{\text{g water } °C} \times (28.8 - 22.0) °C = 1.4 \times 10^3 \text{ J}$$

From equation (7.8) we can write

$$q_{lead} = -q_{water} = -1.4 \times 10^3 \text{ J}$$

Now, from equation (7.5) again, we obtain

$$q_{lead} = 150.0 \text{ g lead} \times \text{specific heat of lead} \times (28.8 - 100.0) °C = -1.4 \times 10^3 \text{ J}$$

$$\text{specific heat of lead} = \frac{-1.4 \times 10^3 \text{ J}}{150.0 \text{ g lead} \times (28.8 - 100.0) °C} = \frac{-1.4 \times 10^3 \text{ J}}{150.0 \text{ g lead} \times -71.2 °C} = 0.13 \text{ J g}^{-1}°C^{-1}$$

Assess

The key concept to recognize is that energy, in the form of heat, flowed from the lead, which is our system, to the water, which is part of the surroundings. A quick way to make sure that we have done the problem correctly is to check the sign on the final answer. For specific heat, the sign should always be positive and have the units of $J \text{ g}^{-1}°C^{-1}$.

PRACTICE EXAMPLE A: When 1.00 kg lead (specific heat $= 0.13 \text{ J g}^{-1}°C^{-1}$) at 100.0 °C is added to a quantity of water at 28.5 °C, the final temperature of the lead–water mixture is 35.2 °C. What is the mass of water present?

PRACTICE EXAMPLE B: A 100.0 g copper sample (specific heat $= 0.385 \text{ J g}^{-1}°C^{-1}$) at 100.0 °C is added to 50.0 g water at 26.5 °C. What is the final temperature of the copper–water mixture?

🔍 7-2   CONCEPT ASSESSMENT

With a minimum of calculation, estimate the final temperature reached when 100.0 mL of water at 10.00 °C is added to 200.0 mL of water at 70.00 °C. What basic principle did you use and what assumptions did you make in arriving at this estimate?

Specific Heats of Some Substances

Table 7.1 lists specific heats of some substances. For many substances, the specific heat is less than $1 \text{ J g}^{-1} °C^{-1}$. A few substances, $H_2O(l)$ in particular, have specific heats that are substantially larger. Can we explain why liquid water has a high specific heat? The answer is most certainly yes, but the explanation relies on concepts we have not yet discussed. The fact that water molecules form hydrogen bonds (which we discuss in Chapter 12) is an important part of the reason why water has a large specific heat value.

Because of their greater complexity at the molecular level, compounds generally have more ways of storing internal energy than do the elements; they tend to have higher specific heats. Water, for example, has a specific heat that is more than 30 times as great as that of lead. We need a much larger quantity of heat to change the temperature of a sample of water than of an equal mass of a metal.

An environmental consequence of the high specific heat of water is found in the effect of large lakes on local climates. Because a lake takes much longer to heat up in summer and cool down in winter than other types of terrain, lakeside communities tend to be cooler in summer and warmer in winter than communities more distant from the lake.

🔍 7-3   CONCEPT ASSESSMENT

Two objects of the same mass absorb the same amount of heat when heated in a flame, but the temperature of one object increases more than the temperature of the other. Which object has the greater specific heat?

TABLE 7.1   Some Specific Heat Values, $J \text{ g}^{-1}°C^{-1}$

Solids	
$Pb(s)$	0.130
$Cu(s)$	0.385
$Fe(s)$	0.449
$S_8(s)$	0.708
$P_4(s)$	0.769
$Al(s)$	0.897
$Mg(s)$	1.023
$H_2O(s)$	2.11
Liquids	
$Hg(l)$	0.140
$Br_2(l)$	0.474
$CCl_4(l)$	0.850
$CH_3COOH(l)$	2.15
$CH_3CH_2OH(l)$	2.44
$H_2O(l)$	4.18
Gases	
$CO_2(g)$	0.843
$N_2(g)$	1.040
$C_3H_8(g)$	1.67
$NH_3(g)$	2.06
$H_2O(g)$	2.08

Source: CRC Handbook of Chemistry and Physics, 90th ed., David R. Lide (ed.), Boca Raton, FL: Taylor & Francis Group, 2010.

7-3   Heats of Reaction and Calorimetry

In Section 7-1, we introduced the notion of *thermal energy*—kinetic energy associated with random molecular motion. Another type of energy that contributes to the internal energy of a system is **chemical energy**. This is energy associated with chemical bonds and intermolecular attractions. If we think of a chemical reaction as a process in which some chemical bonds are broken and others are formed, then, in general, we expect the chemical energy of a system to change as a result of a reaction. Furthermore, we might expect some of this energy change to appear as heat. A **heat of reaction**, q_{rxn}, is the quantity of heat exchanged between a system and its surroundings when a chemical reaction occurs within the system at *constant temperature*. One of the most common reactions studied is the combustion reaction. This is such a common reaction that we often refer to the *heat of combustion* when describing the heat released by a combustion reaction.

If a reaction occurs in an *isolated* system, that is, one that exchanges no matter or energy with its surroundings, the reaction produces a change in the thermal energy of the system—the temperature either increases or decreases. Imagine that the previously isolated system is allowed to interact with its surroundings. The heat of reaction is the quantity of heat exchanged between the system and its surroundings as the system is restored to its initial temperature (Fig. 7-4). In actual practice, we do not physically restore the system to its initial temperature. Instead, we calculate the quantity of heat that *would be* exchanged in this restoration. To do this, a probe (thermometer) is placed within the system to record the temperature change produced by the reaction. Then, we use the temperature change and other system data to calculate the heat of reaction that would have occurred at constant temperature.

Two widely used terms related to heats of reaction are exothermic and endothermic reactions. An **exothermic reaction** is one that produces a temperature increase in an isolated system or, in a nonisolated system, gives off heat to the surroundings. For an exothermic reaction, the heat of reaction is a negative quantity ($q_{rxn} < 0$). In an **endothermic reaction**, the corresponding situation is a temperature decrease in an isolated system or a gain of heat from the surroundings by a nonisolated system. In this case, the heat of reaction is a positive quantity ($q_{rxn} > 0$). Heats of reaction are experimentally determined in a **calorimeter**, a device for measuring quantities of heat. We will consider two types of calorimeters in this section, and we will treat both of them as *isolated* systems.

▶ FIGURE 7-4
Conceptualizing a heat of reaction at constant temperature
The solid lines indicate the initial temperature and the **(a)** maximum and **(b)** minimum temperature reached in an isolated system, in an exothermic and an endothermic reaction, respectively. The broken lines represent pathways to restoring the system to the initial temperature. The heat of reaction is the heat lost or gained by the system in this restoration.

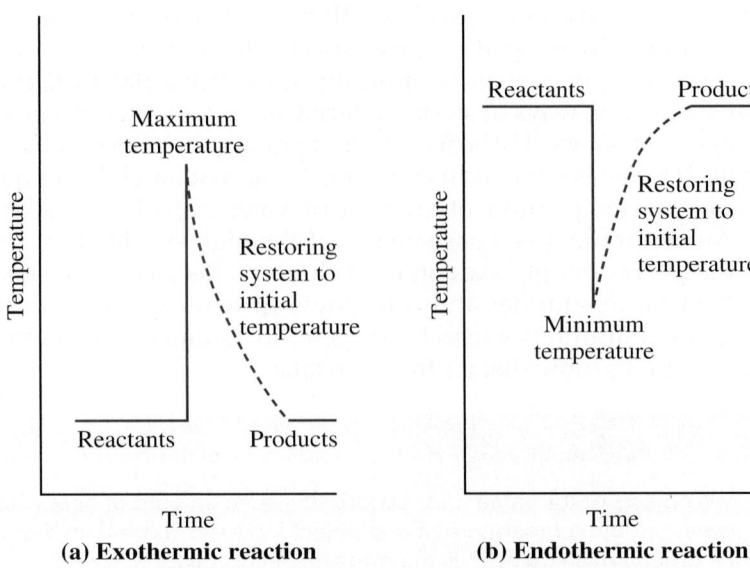

(a) Exothermic reaction

(b) Endothermic reaction

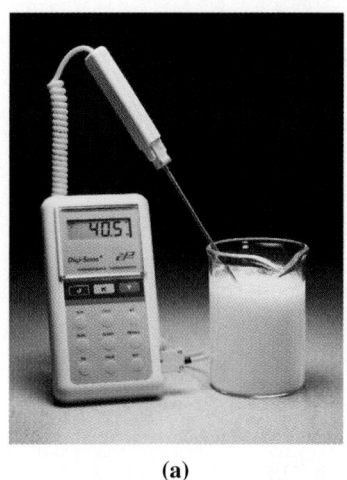
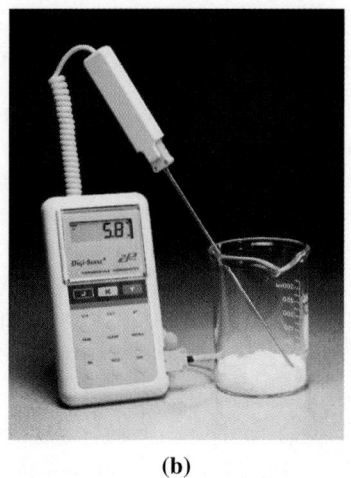

(a)   **(b)**

◀ **Exothermic and endothermic reactions**
(a) An exothermic reaction. Slaked lime, $Ca(OH)_2$, is produced by the action of water on quicklime, (CaO). The reactants are mixed at room temperature, but the temperature of the mixture rises to 40.5 °C.

$$CaO(s) + H_2O(l) \longrightarrow Ca(OH)_2 (s)$$

(b) An endothermic reaction. $Ba(OH)_2 \cdot 8 H_2O(s)$ and $NH_4Cl(s)$ are mixed at room temperature, and the temperature falls to 5.8 °C in the reaction.

$$Ba(OH)_2 \cdot 8 H_2O(s) + 2 NH_4Cl(s) \longrightarrow$$
$$BaCl_2 \cdot 2H_2O(s) + 2 NH_3(aq) + 8 H_2O(l)$$

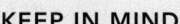

Bomb Calorimetry

Figure 7-5 shows a **bomb calorimeter**, which is ideally suited for measuring the heat evolved in a combustion reaction. The system is everything within the double-walled outer jacket of the calorimeter. This includes the bomb and its contents, the water in which the bomb is immersed, the thermometer, the stirrer, and so on. The system is *isolated* from its surroundings. When the combustion reaction occurs, chemical energy is converted to thermal energy, and the temperature of the system rises. The heat of reaction, as described earlier, is the quantity of heat that the system would have to *lose* to its surroundings to be restored to its initial temperature. This quantity of heat, in turn, is just the *negative* of the thermal energy gained by the calorimeter and its contents ($q_{calorim}$).

$$q_{rxn} = -q_{calorim} \text{ (where } q_{calorim} = q_{bomb} + q_{water} \cdots \text{)} \qquad \textbf{(7.9)}$$

If the calorimeter is assembled in exactly the same way each time we use it—that is, use the same bomb, the same quantity of water, and so on—we can define a *heat capacity of the calorimeter*. This is the quantity of heat required to raise the temperature of the calorimeter assembly by one degree Celsius. When this heat capacity is multiplied by the observed temperature change, we get $q_{calorim}$.

$$q_{calorim} = \text{heat capacity of calorim} \times \Delta T \qquad \textbf{(7.10)}$$

And from $q_{calorim}$, we then establish q_{rxn}, as in Example 7-3, where we determine the heat of combustion of sucrose (table sugar).

KEEP IN MIND

that the temperature of a reaction mixture usually changes during a reaction, so the mixture must be returned to the initial temperature (actually or hypothetically) before we assess how much heat is exchanged with the surroundings.

◀ The heat capacity of a bomb calorimeter must be determined by experiment.

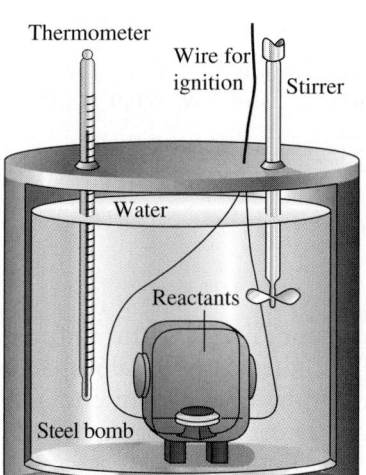

◀ FIGURE 7-5
A bomb calorimeter assembly
An iron wire is embedded in the sample in the lower half of the bomb. The bomb is assembled and filled with $O_2(g)$ at high pressure. The assembled bomb is immersed in water in the calorimeter, and the initial temperature is measured. A short pulse of electric current heats the sample, causing it to ignite. The final temperature of the calorimeter assembly is determined after the combustion. Because the bomb confines the reaction mixture to a fixed volume, the reaction is said to occur at *constant volume*. The significance of this fact is discussed on page 259.

EXAMPLE 7-3   Using Bomb Calorimetry Data to Determine a Heat of Reaction

The combustion of 1.010 g sucrose, $C_{12}H_{22}O_{11}$, in a bomb calorimeter causes the temperature to rise from 24.92 to 28.33 °C. The heat capacity of the calorimeter assembly is 4.90 kJ/°C. **(a)** What is the heat of combustion of sucrose expressed in kilojoules per mole of $C_{12}H_{22}O_{11}$? **(b)** Verify the claim of sugar producers that one teaspoon of sugar (about 4.8 g) contains only 19 Calories.

Analyze

We are given a specific heat and two temperatures, the initial and the final, which indicate that we are to use equation (7.5). In these kinds of experiments one obtains the amount of heat generated by the reaction by measuring the temperature change in the surroundings. This means that $q_{rxn} = -q_{calorim}$.

Solve

(a) Calculate $q_{calorim}$ with equation (7.10).

$$q_{calorim} = 4.90 \text{ kJ/°C} \times (28.33 - 24.92) \text{ °C} = (4.90 \times 3.41) \text{ kJ} = 16.7 \text{ kJ}$$

Now, using equation (7.9), we get

$$q_{rxn} = -q_{calorim} = -16.7 \text{ kJ}$$

This is the heat of combustion of the 1.010 g sample.
Per gram $C_{12}H_{22}O_{11}$:

$$q_{rxn} = \frac{-16.7 \text{ kJ}}{1.010 \text{ g } C_{12}H_{22}O_{11}} = -16.5 \text{ kJ/g } C_{12}H_{22}O_{11}$$

Per mole $C_{12}H_{22}O_{11}$:

$$q_{rxn} = \frac{-16.5 \text{ kJ}}{\text{g } C_{12}H_{22}O_{11}} \times \frac{342.3 \text{ g } C_{12}H_{22}O_{11}}{1 \text{ mol } C_{12}H_{22}O_{11}} = -5.65 \times 10^3 \text{ kJ/mol } C_{12}H_{22}O_{11}$$

(b) To determine the caloric content of sucrose, we can use the heat of combustion per gram of sucrose determined in part **(a)**, together with a factor to convert from kilojoules to kilocalories. (Because 1 cal = 4.184 J, 1 kcal = 4.184 kJ.)

$$? \text{ kcal} = \frac{4.8 \text{ g } C_{12}H_{22}O_{11}}{\text{tsp}} \times \frac{-16.5 \text{ kJ}}{\text{g } C_{12}H_{22}O_{11}} \times \frac{1 \text{ kcal}}{4.184 \text{ kJ}} = \frac{-19 \text{ kcal}}{\text{tsp}}$$

1 food Calorie (1 Calorie with a capital C) is actually 1000 cal, or 1 kcal. Therefore, 19 kcal = 19 Calories. The claim is justified.

Assess

A combustion reaction is an exothermic reaction, which means that energy flows, in the form of heat, from the reaction system to the surroundings. Therefore, the q for a combustion reaction is negative.

PRACTICE EXAMPLE A: Vanillin is a natural constituent of vanilla. It is also manufactured for use in artificial vanilla flavoring. The combustion of 1.013 g of vanillin, $C_8H_8O_3$, in the same bomb calorimeter as in Example 7-3 causes the temperature to rise from 24.89 to 30.09 °C. What is the heat of combustion of vanillin, expressed in kilojoules per mole?

PRACTICE EXAMPLE B: The heat of combustion of benzoic acid is −26.42 kJ/g. The combustion of a 1.176 g sample of benzoic acid causes a temperature *increase* of 4.96°C in a bomb calorimeter assembly. What is the heat capacity of the assembly?

The "Coffee-Cup" Calorimeter

In the general chemistry laboratory you are much more likely to run into the simple calorimeter pictured in Figure 7-6 (on page 252) than a bomb calorimeter. We mix the reactants (generally in aqueous solution) in a Styrofoam cup and measure the temperature change. Styrofoam is a good heat insulator, so there is very little heat transfer between the cup and the surrounding air. We treat the system—the cup and its contents—as an *isolated* system.

As with the bomb calorimeter, the heat of reaction is defined as the quantity of heat that would be exchanged with the surroundings in restoring the calorimeter to its initial temperature. But, again, the calorimeter is not physically restored to its initial conditions. We simply take the heat of reaction to be the *negative* of the quantity of heat producing the temperature change in the calorimeter. That is, we use equation (7.9): $q_{rxn} = -q_{calorim}$.

In Example 7-4, we make certain assumptions to simplify the calculation, but for more precise measurements, these assumptions would not be made (see Exercise 25).

EXAMPLE 7-4   Determining a Heat of Reaction from Calorimetric Data

In the neutralization of a strong acid with a strong base, the essential reaction is the combination of $H^+(aq)$ and $OH^-(aq)$ to form water (recall page 165).

$$H^+(aq) + OH^-(aq) \longrightarrow H_2O(l)$$

Two solutions, 25.00 mL of 2.50 M HCl(aq) and 25.00 mL of 2.50 M NaOH(aq), both initially at 21.1 °C, are added to a Styrofoam-cup calorimeter and allowed to react. The temperature rises to 37.8 °C. Determine the heat of the neutralization reaction, expressed per mole of H_2O formed. Is the reaction endothermic or exothermic?

Analyze

In addition to assuming that the calorimeter is an isolated system, assume that all there is in the system to absorb heat is 50.00 mL of water. This assumption ignores the fact that 0.0625 mol each of NaCl and H_2O are formed in the reaction, that the density of the resulting NaCl(aq) is not exactly 1.00 g/mL, and that its specific heat is not exactly $4.18 \, J \, g^{-1} \, °C^{-1}$. Also, ignore the small heat capacity of the Styrofoam cup itself.

Because the reaction is a neutralization reaction, let us call the heat of reaction q_{neutr}. Now, according to equation (7.9), $q_{neutr} = -q_{calorim}$, and if we make the assumptions described above, we can solve the problem.

Solve

We begin with

$$q_{calorim} = 50.00 \, \text{mL} \times \frac{1.00 \, \text{g}}{\text{mL}} \times \frac{4.18 \, \text{J}}{\text{g} \, °\text{C}} \times (37.8 - 21.1) \, °\text{C} = 3.5 \times 10^3 \, \text{J}$$

$$q_{neutr} = -q_{calorim} = -3.5 \times 10^3 \, \text{J} = -3.5 \, \text{kJ}$$

In 25.00 mL of 2.50 M HCl, the amount of H^+ is

$$? \, \text{mol} \, H^+ = 25.00 \, \text{mL} \times \frac{1 \, \text{L}}{1000 \, \text{mL}} \times \frac{2.50 \, \text{mol}}{1 \, \text{L}} \times \frac{1 \, \text{mol} \, H^+}{1 \, \text{mol HCl}} = 0.0625 \, \text{mol} \, H^+$$

Similarly, in 25.00 mL of 2.50 M NaOH there is 0.0625 mol OH^-. Thus, the H^+ and the OH^- combine to form 0.0625 mol H_2O. (The two reactants are in *stoichiometric* proportions; neither is in excess.)

The amount of heat produced per mole of H_2O is

$$q_{neutr} = \frac{-3.5 \, \text{kJ}}{0.0625 \, \text{mol} \, H_2O} = -56 \, \text{kJ/mol} \, H_2O$$

Assess

Because q_{neutr} is a *negative* quantity, the neutralization reaction is *exothermic*. Even though, in this example, we considered a specific reaction, the result $q_{neutr} = -56$ kJ/mol is more general. We will obtain the same value of q_{neutr} by considering any strong acid-strong base reaction because the net ionic equation is the same for all strong acid-strong base reactions.

PRACTICE EXAMPLE A: Two solutions, 100.0 mL of 1.00 M $AgNO_3$(aq) and 100.0 mL of 1.00 M NaCl(aq), both initially at 22.4 °C, are added to a Styrofoam-cup calorimeter and allowed to react. The temperature rises to 30.2 °C. Determine q_{rxn} per mole of AgCl(s) in the reaction.

$$Ag^+(aq) + Cl^-(aq) \longrightarrow AgCl(s)$$

PRACTICE EXAMPLE B: Two solutions, 100.0 mL of 1.020 M HCl and 50.0 mL of 1.988 M NaOH, both initially at 24.52 °C, are mixed in a Styrofoam-cup calorimeter. What will be the final temperature of the mixture? Make the same assumptions, and use the heat of neutralization established in Example 7-4. [*Hint:* Which is the limiting reactant?]

▲ FIGURE 7-6
A Styrofoam "coffee-cup" calorimeter
The reaction mixture is in the inner cup. The outer cup provides additional thermal insulation from the surrounding air. The cup is closed off with a cork stopper through which a thermometer and a stirrer are inserted and immersed into the reaction mixture. The reaction in the calorimeter occurs under the *constant pressure* of the atmosphere. We consider the difference between constant-volume and constant-pressure reactions in Section 7-6.

🔍 **7-4   CONCEPT ASSESSMENT**

How do we determine the specific heat of the bomb calorimeter or the solution calorimeter (coffee-cup calorimeter)?

7-4   Work

We have just learned that heat effects generally accompany chemical reactions. In some reactions, work is also involved—that is, the system may do work on its surroundings or vice versa. Consider the decomposition of potassium chlorate to potassium chloride and oxygen. Suppose that this decomposition is carried out in the strange vessel pictured in Figure 7-7. The walls of the container resist moving under the pressure of the expanding $O_2(g)$ except for the piston that closes off the cylindrical top of the vessel. The pressure of the $O_2(g)$ exceeds the atmospheric pressure and the piston is lifted—the system does work on the surroundings. Can you see that even if the piston were removed, work still would be done as the expanding $O_2(g)$ pushed aside other atmospheric gases? Work involved in the expansion or compression of gases is called **pressure–volume work**. Pressure–volume, or *P–V*, work is the type of work performed by explosives and by the gases formed in the combustion of gasoline in an automobile engine.

Now let us switch to a somewhat simpler situation to see how to calculate a quantity of *P–V* work.

In the hypothetical apparatus pictured in Figure 7-8(a), a weightless piston is attached to a weightless wire support, to which is attached a weightless pan. On the pan are two identical weights just sufficient to stop the gas from expanding. The gas is confined by the cylinder walls and piston, and the space above the piston is a vacuum. The cylinder is contained in a constant-temperature water bath, which keeps the temperature of the gas constant. Now imagine that one of the two weights is removed, leaving half the original mass on the pan. Let us call this remaining mass M. The gas will expand and the remaining weight will move against gravity, the situation represented by Figure 7-8(b). After the expansion, we find that the piston has risen through a vertical distance, Δh; that the volume of gas has doubled; and that the pressure of the gas has decreased.

Now let us see how pressure and volume enter into calculating how much *pressure–volume* work the expanding gas does. First we can calculate the work done by the gas in moving the weight of mass M through a displacement Δh. Recall from equation (7.1) that the work can be calculated by

$$\text{work } (w) = \text{force } (M \times g) \times \text{distance } (\Delta h) = -M \times g \times \Delta h$$

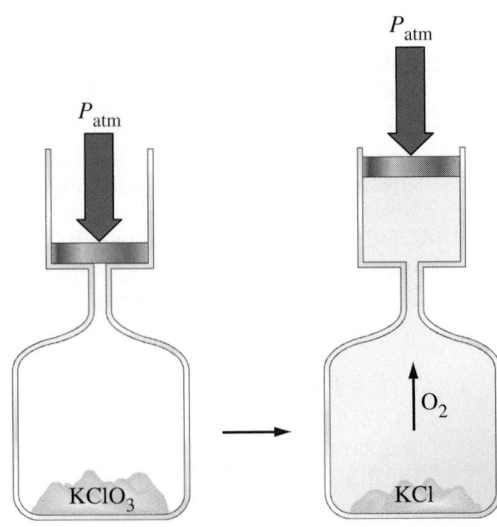

▶ FIGURE 7-7
Illustrating work (expansion) during the chemical reaction
2 KClO$_3$(s) $\longrightarrow$ 2 KCl(s) + 3 O$_2$(g)
The oxygen gas that is formed pushes back the weight and, in doing so, does work on the surroundings.

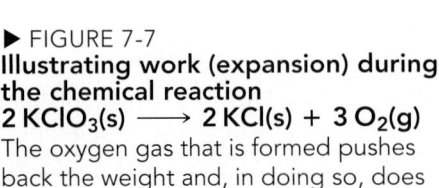

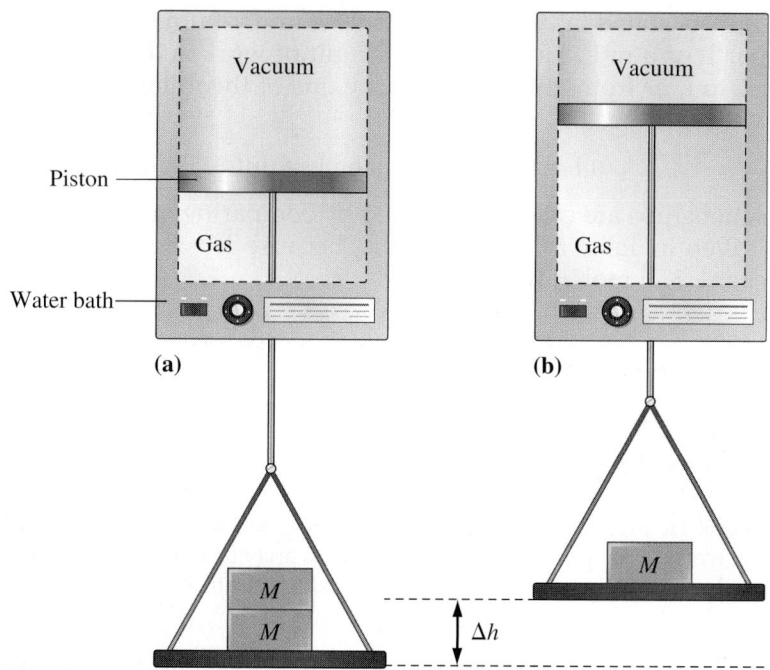

▲ FIGURE 7-8
Pressure–volume work
(a) In this hypothetical apparatus, a gas is confined by a massless piston of area A. A massless wire is attached to the piston and the gas is held back by two weights with a combined mass of 2M resting on the massless pan. The cylinder is immersed in a large water bath in order to keep the gas temperature constant. The initial state of the gas is $P_i = 2Mg/A$ with a volume V_i at temperature, T. **(b)** When the external pressure on the confined gas is suddenly lowered by removing one of the weights the gas expands, pushing the piston up by the distance, Δh. The increase in volume of the gas (ΔV) is the product of the cross-sectional area of the cylinder (A) and the distance (Δh). The final state of the gas is $P_f = Mg/A$, V_f, and T.

The magnitude of the force exerted by the weight is $M \times g$, where g is the acceleration due to gravity. The negative sign appears because the force is acting in a direction opposite to the piston's direction of motion.

Now recall equation (6.1)—pressure = force $(M \times g)$/area (A)—so that if the expression for work is multiplied by A/A we get

$$w = -\frac{M \times g}{A} \times \Delta h \times A = -P_{ext}\Delta V \qquad (7.11)$$

The "pressure" part of the pressure–volume work is seen to be the external pressure (P_{ext}) on the gas, which in our thought experiment is equal to the weight pulling down on the piston and is given by Mg/A. Note that the product of the area (A) and height (Δh) is equal to a volume—the volume change, ΔV, produced by the expansion.

Two significant features to note in equation (7.11) are the *negative* sign and the factor P_{ext}. The negative sign is necessary to conform to sign conventions that we will introduce in the next section. When a gas expands, ΔV is positive and w is negative, signifying that energy leaves the system as work. When a gas is compressed, ΔV is negative and w is positive, signifying that energy (as work) enters the system. P_{ext} is the *external* pressure—the pressure against which a system expands or the applied pressure that compresses a system. In many instances the internal pressure in a system will be essentially equal to the external pressure, in which case the pressure in equation (7.11) is expressed simply as P.

◀ Work is negative when energy is transferred out of the system and is positive when energy is transferred into the system. This is consistent with the signs associated with the heat of a reaction (q) during endothermic and exothermic processes.

▶ The unit atm L, often written as L atm, is the liter-atmosphere. The use of this unit still persists.

If pressure is stated in bars or atmospheres and volume in liters, the unit of work is bar L or atm L. However, the SI unit of work is the joule. To convert from bar L to J, or from atm L to J, we use one of the following relationships, both of which are *exact*.

$$1 \text{ bar L} = 100 \text{ J} \quad 1 \text{ atm L} = 101.325 \text{ J}$$

These relationships are easily established by comparing values of the gas constant, R, given in Table 6.3. For example, because $R = 8.3145 \text{ J K}^{-1} \text{ mol}^{-1} = 0.083145 \text{ bar L K}^{-1} \text{ mol}^{-1}$, we have

$$\frac{8.3145 \text{ J K}^{-1} \text{ mol}^{-1}}{0.083145 \text{ bar L K}^{-1} \text{ mol}^{-1}} = 100 \frac{\text{J}}{\text{bar L}}$$

EXAMPLE 7-5   **Calculating Pressure–Volume Work**

Suppose the gas in Figure 7-8 is 0.100 mol He at 298 K, the two weights correspond to an external pressure of 2.40 atm in Figure 7-8(a), and the single weight in Figure 7-8(b) corresponds to an external pressure of 1.20 atm. How much work, in joules, is associated with the gas expansion at constant temperature?

Analyze

We are given enough data to calculate the initial and final gas volumes (note that the identity of the gas does not enter into the calculations because we are assuming ideal gas behavior). With these volumes, we can obtain ΔV. The external pressure in the pressure–volume work is the *final* pressure: 1.20 atm. The product $-P_{ext} \times \Delta V$ must be multiplied by a factor to convert work in liter-atmospheres to work in joules.

Solve

First calculate the initial and final volumes.

$$V_{initial} = \frac{nRT}{P_i} = \frac{0.100 \text{ mol} \times 0.0821 \text{ L atm mol}^{-1} \text{K}^{-1} \times 298 \text{ K}}{2.40 \text{ atm}} = 1.02 \text{ L}$$

$$V_{final} = \frac{nRT}{P_f} = \frac{0.100 \text{ mol} \times 0.0821 \text{ L atm mol}^{-1} \text{K}^{-1} \times 298 \text{ K}}{1.20 \text{ atm}} = 2.04 \text{ L}$$

$$\Delta V = V_f - V_i = 2.04 \text{ L} - 1.02 \text{ L} = 1.02 \text{ L}$$

$$w = -P_{ext} \times \Delta V = -1.20 \text{ atm} \times 1.02 \text{ L} \times \frac{101 \text{ J}}{1 \text{ L atm}} = -1.24 \times 10^2 \text{ J}$$

Assess

The negative value signifies that the expanding gas (i.e., the system) does work on its surroundings. Keep in mind that the ideal gas equation embodies Boyle's law: The volume of a fixed amount of gas at a fixed temperature is inversely proportional to the pressure. Thus, in Example 7-5 we could simply write that

$$V_f = 1.02 \text{ L} \times \frac{2.40 \text{ atm}}{1.20 \text{ atm}}$$

$$V_f = 2.04 \text{ L}$$

PRACTICE EXAMPLE A: How much work, in joules, is involved when 0.225 mol N_2 at a constant temperature of 23 °C is allowed to expand by 1.50 L in volume against an external pressure of 0.750 atm? [*Hint:* How much of this information is required?]

PRACTICE EXAMPLE B: How much work is done, in joules, when an external pressure of 2.50 atm is applied, at a constant temperature of 20.0 °C, to 50.0 g $N_2(g)$ in a 75.0 L cylinder? The cylinder is like that shown in Figure 7-8.

🔍 **7-5   CONCEPT ASSESSMENT**

A gas in a 1.0 L closed cylinder has an initial pressure of 10.0 bar. It has a final pressure of 5.0 bar. The volume of the cylinder remained constant during this time. What form of energy was transferred across the boundary to cause this change? In which direction did the energy flow?

This result confirms that 1 bar L = 100 J. How do we establish that 1 atm L is exactly 101.325 J? Recall that 1 atm is exactly 1.01325 bar (see Table 6.1). Thus, 1 atm L = 1.01325 bar L = 1.01325 × 100 J = 101.325 J.

7-5   The First Law of Thermodynamics

The absorption or evolution of heat and the performance of work require changes in the energy of a system and its surroundings. When considering the energy of a system, we use the concept of internal energy and how heat and work are related to it.

Internal energy, U, is the total energy (both kinetic and potential) in a system, including *translational kinetic energy* of molecules, the energy associated with molecular rotations and vibrations, the energy stored in chemical bonds and intermolecular attractions, and the energy associated with electrons in atoms. Some of these forms of internal energy are illustrated in Figure 7-9. Internal energy also includes energy associated with the interactions of protons and neutrons in atomic nuclei, although this component is unchanged in chemical reactions. A system contains *only* internal energy. A system does not contain energy in the form of heat or work. Heat and work are the means by which a system exchanges energy with its surroundings. *Heat and work exist only during a change in the system.* The relationship between heat (q), work (w), and changes in internal energy (ΔU) is dictated by the law of conservation of energy, expressed in the form known as the **first law of thermodynamics**.

$$\Delta U = q + w \qquad (7.12)$$

An isolated system is unable to exchange either heat or work with its surroundings, so that $\Delta U_{\text{isolated system}} = 0$, and we can say

The energy of an isolated system is constant.

In using equation (7.12) we must keep these important points in mind.

- Any energy *entering* the system carries a *positive* sign. Thus, if heat is *absorbed* by the system, $q > 0$. If work is done *on* the system, $w > 0$.

- Any energy *leaving* the system carries a *negative* sign. Thus, if heat is *given off* by the system, $q < 0$. If work is done *by* the system, $w < 0$.

- In general, the internal energy of a system changes as a result of energy entering or leaving the system as heat and/or work. If, on balance, more energy enters the system than leaves, ΔU is *positive*. If more energy leaves than enters, ΔU is *negative*.

- A consequence of $\Delta U_{\text{isolated system}} = 0$ is that $\Delta U_{\text{system}} = -\Delta U_{\text{surroundings}}$; that is, energy is conserved.

These ideas are summarized in Figure 7-10 and illustrated in Example 7-6.

Translational

 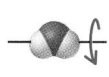

Rotational

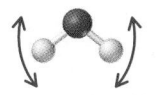 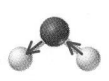 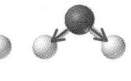

Vibrational

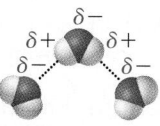

Electrostatic
(Intermolecular attractions)

▲ FIGURE 7-9
Some contributions to the internal energy of a system
The models represent water molecules, and the arrows represent the types of motion they can undergo. In the intermolecular attractions between water molecules, the symbols $\delta+$ and $\delta-$ signify a separation of charge, producing centers of positive and negative charge that are smaller than ionic charges. These intermolecular attractions are discussed in Chapter 12.

KEEP IN MIND

that heat is the disordered flow of energy and work is the ordered flow of energy.

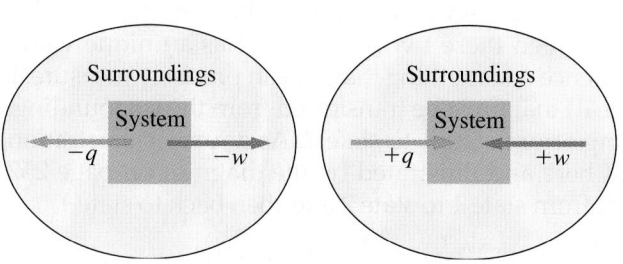

◄ FIGURE 7-10
Illustration of sign conventions used in thermodynamics
Arrows represent the direction of heat flow (⟶) and work (⟶). In the left diagram, the minus (−) signs signify energy leaving the system and entering the surroundings. In the right diagram the plus (+) signs refer to energy entering the system from the surroundings. These sign conventions are consistent with the expression $\Delta U = q + w$.

EXAMPLE 7-6   Relating ΔU, q, and w Through the First Law of Thermodynamics

A gas, while expanding (recall Figure 7-8), absorbs 25 J of heat and does 243 J of work. What is ΔU for the gas?

Analyze

The key to problems of this type lies in assigning the correct signs to the quantities of heat and work. Because heat is absorbed by (enters) the system, q is *positive*. Because work done *by* the system represents energy *leaving* the system, w is *negative*. You may find it useful to represent the values of q and w, with their correct signs, within parentheses. Then complete the algebra.

Solve

$$\Delta U = q + w = (+25\,\text{J}) + (-243\,\text{J}) = 25\,\text{J} - 243\,\text{J} = -218\,\text{J}$$

Assess

The negative sign for the change in internal energy, ΔU, signifies that the system, in this case the gas, has lost energy.

PRACTICE EXAMPLE A: In compressing a gas, 355 J of work is done on the system. At the same time, 185 J of heat escapes from the system. What is ΔU for the system?

PRACTICE EXAMPLE B: If the internal energy of a system *decreases* by 125 J at the same time that the system *absorbs* 54 J of heat, does the system do work or have work done on it? How much?

Q   7-6   CONCEPT ASSESSMENT

When water is injected into a balloon filled with ammonia gas, the balloon shrinks and feels warm. What are the sources of heat and work, and what are the signs of q and w in this process?

Functions of State

To describe a system completely, we must indicate its temperature, its pressure, and the kinds and amounts of substances present. When we have done this, we have specified the *state* of the system. Any property that has a unique value for a specified state of a system is said to be a **function of state**, or a **state function**. For example, a sample of pure water at 20 °C (293.15 K) and under a pressure of 100 kPa is in a specified state. The density of water in this state is 0.99820 g/mL. We can establish that this density is a unique value—a function of state—in the following way: Obtain three different samples of water—one purified by extensive distillation of groundwater; one synthesized by burning pure $H_2(g)$ in pure $O_2(g)$; and one prepared by driving off the water of hydration from $CuSO_4 \cdot 5\,H_2O$ and condensing the gaseous water to a liquid. The densities of the three different samples for the state that we specified will all be the same: 0.99820 g/mL. Thus, the value of a function of state depends on the state of the system, and not on how that state was established.

The internal energy of a system is a function of state, although there is no simple measurement or calculation that we can use to establish its value. That is, we cannot write down a value of U for a system in the same way that we can write $d = 0.99820$ g/mL for the density of water at 20 °C. Fortunately, we don't need to know actual values of U. Consider, for example, heating 10.0 g of ice at 0 °C to a final temperature of 50 °C. The internal energy of the ice at 0 °C has one unique value, U_1, while that of the liquid water at 50 °C has another, U_2. The *difference* in internal energy between these two states also has a unique value, $\Delta U = U_2 - U_1$, and this difference *is* something that we can precisely measure. It is the quantity of energy (as heat) that must be transferred from the surroundings to the system during the change from state 1 to state 2. As a further illustration, consider the scheme outlined here and illustrated by the diagram on page 257. Imagine that a system changes from state 1 to state 2 and then back to state 1.

$$\text{State 1 }(U_1) \xrightarrow{\ \Delta U\ } \text{State 2 }(U_2) \xrightarrow{\ -\Delta U\ } \text{State 1 }(U_1)$$

$$U_2 \text{ — State 2}$$

$$\Delta U = (U_2 - U_1) \qquad -\Delta U = (U_1 - U_2)$$

$$U_1 \text{ — State 1}$$

$$\Delta U_{\text{overall}} = U_2 - U_1 + U_1 - U_2 = 0$$

Because U has a unique value in each state, ΔU also has a unique value; it is $U_2 - U_1$. The change in internal energy when the system is returned from state 2 to state 1 is $-\Delta U = U_1 - U_2$. Thus, the *overall* change in internal energy is

$$\Delta U + (-\Delta U) = (U_2 - U_1) + (U_1 - U_2) = 0$$

This means that the internal energy returns to its initial value of U_1, which it must do, since it is a function of state. It is important to note here that when we reverse the direction of change, we change the sign of ΔU.

Path-Dependent Functions

Unlike internal energy and changes in internal energy, heat (q) and work (w) are *not* functions of state. Their values depend on the path followed when a system undergoes a change. We can see why this is so by considering again the process described by Figure 7-8 and Example 7-5. Think of the 0.100 mol of He at 298 K and under a pressure of 2.40 atm as *state 1*, and under a pressure of 1.20 atm as *state 2*. The change from state 1 to state 2 occurred in a single step. Suppose that in another instance, we allowed the expansion to occur through an intermediate stage pictured in Figure 7-11. That is, suppose the external pressure on the gas was first reduced from 2.40 atm to 1.80 atm (at which point, the gas volume would be 1.36 L). Then, in a second stage, reduced from 1.80 atm to 1.20 atm, thereby arriving at state 2.

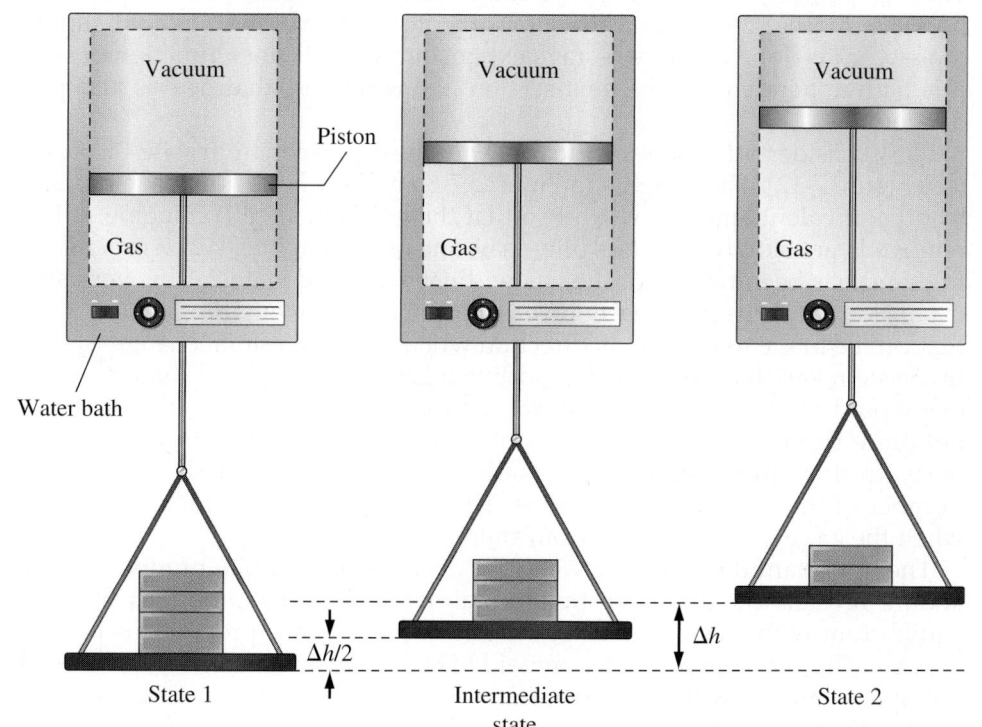

◀ FIGURE 7-11
A two-step expansion for the gas shown in Figure 7-8
In the initial state there are four weights of mass $M/2$ holding the gas back. In the intermediate state one of these weights has been removed and in the final state a second weight of mass $M/2$ has been removed. The initial and final states in this figure are the same as in Figure 7-8. This two-step expansion helps us to establish that the work of expansion depends on the path taken.

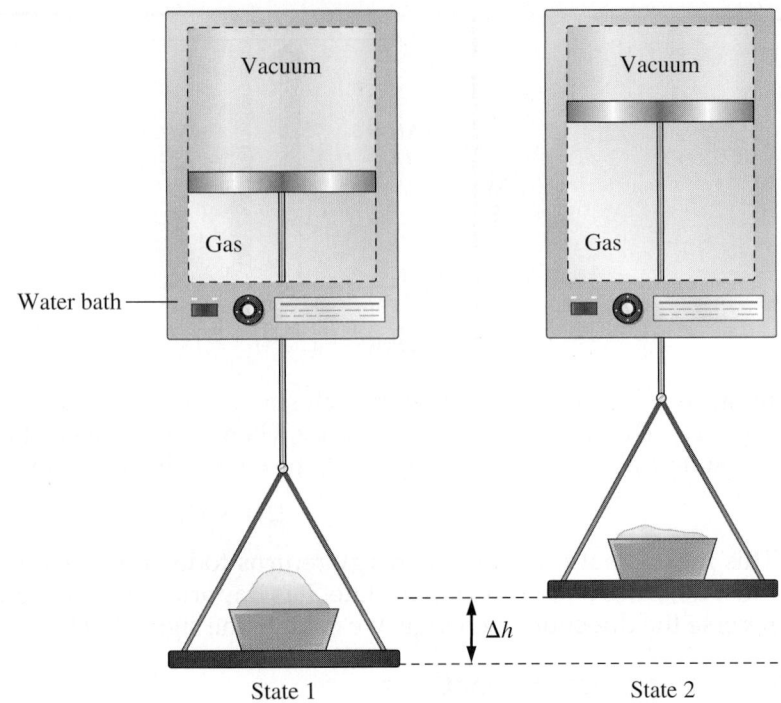

State 1   State 2

▶ FIGURE 7-12
A different method of achieving the expansion of a gas
In this expansion process, the weights in Figures 7-8 and 7-11 have been replaced by a pan containing sand, which has a mass of 2*M*, equivalent to that of the weights in the initial state. In the final state the mass of the sand has been reduced to *M*.

We calculated the amount of work done by the gas in a single-stage expansion in Example 7-5; it was $w = -1.24 \times 10^2$ J. The amount of work done in the two-stage process is the sum of two terms: the pressure–volume work for each stage of the expansion.

$$w = -1.80 \, \text{atm} \times (1.36 \, \text{L} - 1.02 \, \text{L}) - 1.20 \, \text{atm} \times (2.04 \, \text{L} - 1.36 \, \text{L})$$

$$= -0.61 \, \text{L atm} - 0.82 \, \text{L atm}$$

$$= -1.43 \, \text{L atm} \times \frac{101 \, \text{J}}{1 \, \text{L atm}} = -1.44 \times 10^2 \, \text{J}$$

KEEP IN MIND

that if *w* differs in the two expansion processes, *q* must also differ, and in such a way that $q + w = \Delta U$ has a unique value, as required by the first law of thermodynamics.

The value of ΔU is the same for the single- and two-stage expansion processes because internal energy is a function of state. However, we see that slightly more work is done in the two-stage expansion. Work is not a function of state; it is path dependent. In the next section, we will stress that heat is also path dependent.

Now consider a different way to carry out the expansion from state 1 to state 2 (see Figure 7-12). The weights in Figures 7-8 and 7-11 have now been replaced by an equivalent amount of sand so that the gas is in state 1. Imagine sand is removed very slowly from this pile—say, one grain at a time. When exactly half the sand has been removed, the gas will have reached state 2. This very slow expansion proceeds in a nearly reversible fashion. A **reversible process** is one that can be made to reverse its direction when an infinitesimal change is made in a system variable. For example, adding a grain of sand rather than removing one would reverse the expansion we are describing. However, the process is not quite reversible because grains of sand have more than an infinitesimal mass. In this approximately reversible process we have made a very large number of intermediate expansions. This process provides more work than when the gas expands directly from state 1 to state 2.

The important difference between the expansion in a finite number of steps and the reversible expansion is that the gas in the reversible process is always in equilibrium with its surroundings whereas in a stepwise process this is never the case. The stepwise processes are said to be **irreversible** because the system is not in equilibrium with the surroundings, and the process cannot be reversed by an infinitesimal change in a system variable.

In comparing the quantity of work done in the two different expansions (Figs. 7-8 and 7-11), we found them to be different, thereby proving that work is not a state function. Additionally, the quantity of work performed is greater in the two-step expansion (Fig. 7-11) than in the single-step expansion (Fig. 7-8). We leave it to the interested student to demonstrate, through Feature Problem 125, that the maximum possible work is that done in a reversible expansion (Fig. 7-12).

🔍 **7-7   CONCEPT ASSESSMENT**

A sample can be heated very slowly or very rapidly. The darker shading in the illustration indicates a higher temperature. Which of the two sets of diagrams do you think corresponds to reversible heating and which to spontaneous, or irreversible, heating?

◀ Although no perfectly reversible process exists, the melting and freezing of a substance at its transition temperature is an example of a process that is nearly reversible: pump in heat (melts), take out heat (freezes).

7-6   Heats of Reaction: ΔU and ΔH

Think of the reactants in a chemical reaction as the initial state of a system and the products as the final state.

$$\text{reactants} \longrightarrow \text{products}$$
$$\text{(initial state)} \qquad \text{(final state)}$$
$$U_i \qquad\qquad U_f$$
$$\Delta U = U_f - U_i$$

According to the first law of thermodynamics, we can also say that $\Delta U = q + w$. We have previously identified a heat of reaction as q_{rxn}, and so we can write

$$\Delta U = q_{rxn} + w$$

Now consider again a combustion reaction carried out in a bomb calorimeter (see Figure 7-5). The original reactants and products are confined within the bomb, and we say that the reaction occurs at *constant volume*. Because the volume is constant, $\Delta V = 0$, and no work is done. That is, $w = -P\Delta V = 0$. Denoting the heat of reaction for a constant-volume reaction as q_V, we see that $\Delta U = q_V$.

$$\Delta U = q_{rxn} + w = q_{rxn} + 0 = q_{rxn} = q_V \qquad (7.13)$$

The heat of reaction measured in a bomb calorimeter is equal to ΔU.

Chemical reactions are not ordinarily carried out in bomb calorimeters. The metabolism of sucrose occurs under the conditions present in the human body. The combustion of methane (natural gas) in a water heater occurs in an open flame. This question then arises: How does the heat of a reaction measured in a

▶ FIGURE 7-13
Two different paths leading to the same internal energy change in a system
In path **(a)**, the volume of the system remains constant and no internal energy is converted into work—think of burning gasoline in a bomb calorimeter. In path **(b)**, the system does work, so some of the internal energy change is used to do work—think of burning gasoline in an automobile engine to produce heat and work.

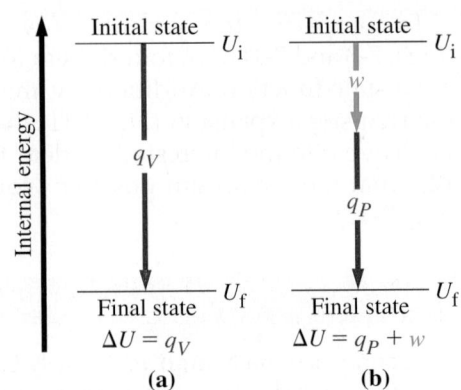

bomb calorimeter compare with the heat of reaction if the reaction is carried out in some other way? The usual other way is in beakers, flasks, and other containers open to the atmosphere and under the *constant pressure* of the atmosphere. We live in a world of constant pressure! The neutralization reaction of Example 7-4 is typical of this more common method of conducting chemical reactions.

In many reactions carried out at constant pressure, a small amount of pressure–volume work is done as the system expands or contracts (recall Figure 7-7). In these cases, the heat of reaction, q_P, is different from q_V. We know that the change in internal energy (ΔU) for a reaction carried out between a given initial and a given final state has a unique value. Furthermore, for a reaction at constant volume, $\Delta U = q_V$. From Figure 7-13 and the first law of thermodynamics, we see that for the same reaction at constant pressure $\Delta U = q_P + w$, which means $\Delta U = q_V = q_P + w$. Thus, unless $w = 0$, q_V and q_P must be different. The fact that q_V and q_P for a reaction may differ, even though ΔU has the same value, underscores that U is a function of state and q and w are not.

The relationship between q_V and q_P can be used to devise another state function that represents the heat flow for a process at constant pressure. To do this, we begin by writing

$$q_V = q_P + w$$

Now, using $\Delta U = q_V$, $w = -P\Delta V$ and rearranging terms, we obtain

$$\Delta U = q_P - P\Delta V$$
$$q_P = \Delta U + P\Delta V$$

The quantities $U, P,$ and V are all state functions, so it should be possible to derive the expression $\Delta U + P\Delta V$ from yet another state function. This state function, called **enthalpy, H,** is the sum of the internal energy and the pressure–volume product of a system: $H = U + PV$. The **enthalpy change, ΔH,** for a process between initial and final states is

$$\Delta H = H_f - H_i = (U_f + P_f V_f) - (U_i + P_i V_i)$$
$$\Delta H = (U_f - U_i) + (P_f V_f - P_i V_i)$$
$$\Delta H = \Delta U + \Delta PV$$

If the process is carried out at a constant temperature and pressure ($P_i = P_f$) and with work limited to pressure–volume work, the enthalpy change is

$$\Delta H = \Delta U + P\Delta V$$

and the heat flow for the process under these conditions is

$$\Delta H = q_P \tag{7.14}$$

🔍 7-8   CONCEPT ASSESSMENT

Suppose a system is subjected to the following changes: a 40 kJ quantity of heat is added and the system does 15 kJ of work; then the system is returned to its original state by cooling and compression. What is the value of ΔH?

Enthalpy (ΔH) and Internal Energy (ΔU) Changes in a Chemical Reaction

We have noted that the heat of reaction at constant pressure, ΔH, and the heat of reaction at constant volume, ΔU, are related by the expression

$$\Delta U = \Delta H - P\Delta V \qquad (7.15)$$

The last term in this expression is the energy associated with the change in volume of the system under a constant external pressure. To assess just how significant pressure–volume work is, consider the following reaction, which is also illustrated in Figure 7-14.

$$2\,CO(g) + O_2(g) \longrightarrow 2\,CO_2(g)$$

If the heat of this reaction is measured under constant-pressure conditions at a constant temperature of 298 K, we get -566.0 kJ, indicating that 566.0 kJ of energy has left the system as heat: $\Delta H = -566.0$ kJ. To evaluate the pressure–volume work, we begin by writing

$$P\Delta V = P(V_f - V_i)$$

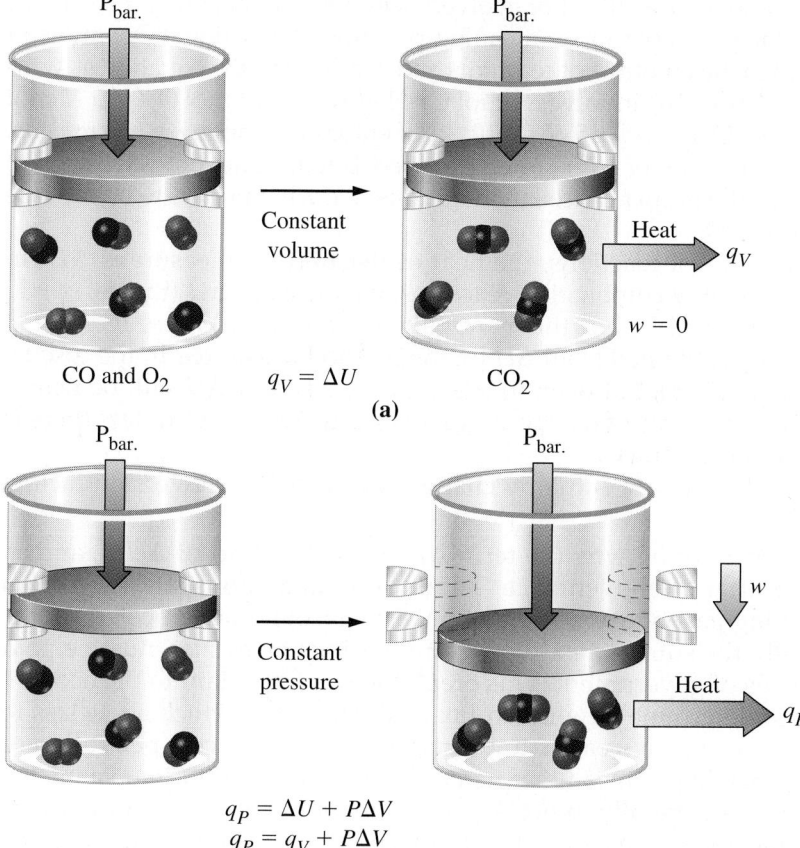

CO and O$_2$ $q_V = \Delta U$ CO$_2$

(a)

$q_P = \Delta U + P\Delta V$
$q_P = q_V + P\Delta V$

(b)

◀ FIGURE 7-14
Comparing heats of reaction at constant volume and constant pressure for the reaction 2 CO(g) + O$_2$(g) $\longrightarrow$ 2 CO$_2$(g)
(a) No work is performed at constant volume because the piston cannot move because of the stops placed through the cylinder walls; $q_V = \Delta U = -563.5$ kJ. (b) When the reaction is carried out at constant pressure, the stops are removed. This allows the piston to move and the surroundings do work on the system, causing it to shrink into a smaller volume. More heat is evolved than in the constant-volume reaction; $q_P = \Delta H = -566.0$ kJ.

Then we can use the ideal gas equation to write this alternative expression.

$$P\Delta V = RT(n_f - n_i)$$

Here, n_f is the number of moles of gas in the products (2 mol CO_2) and n_i is the number of moles of gas in the reactants (2 mol CO + 1 mol O_2). Thus,

$$P\Delta V = 0.0083145 \text{ kJ mol}^{-1}\text{K}^{-1} \times 298 \text{ K} \times [2 - (2+1)] \text{ mol} = -2.5 \text{ kJ}$$

The change in internal energy is

$$\begin{aligned}\Delta U &= \Delta H - P\Delta V \\ &= -566.0 \text{ kJ} - (-2.5 \text{ kJ}) \\ &= -563.5 \text{ kJ}\end{aligned}$$

This calculation shows that the $P\Delta V$ term is quite small compared to ΔH and that ΔU and ΔH are almost the same. An additional interesting fact here is that the volume of the system decreases as a consequence of the work done on the system by the surroundings.

In the combustion of sucrose at a fixed temperature, the heat of combustion turns out to be the same, whether at constant volume (q_V) or constant pressure (q_P). Only heat is transferred between the reaction mixture and the surroundings; no pressure–volume work is done. This is because the volume of a system is almost entirely determined by the volume of gas and because 12 mol $CO_2(g)$ occupies the same volume as 12 mol $O_2(g)$. There is no change in volume in the combustion of sucrose: $q_P = q_V$. Thus, the result of Example 7-3 can be represented as

$$C_{12}H_{22}O_{11}(s) + 12 O_2(g) \longrightarrow 12 CO_2(g) + 11 H_2O(l) \quad \Delta H = -5.65 \times 10^3 \text{ kJ} \textbf{ (7.16)}$$

That is, 1 mol $C_{12}H_{22}O_{11}(s)$ reacts with 12 mol $O_2(g)$ to produce 12 mol $CO_2(g)$, 11 mol $H_2O(l)$, and 5.65×10^3 kJ of evolved heat. Strictly speaking, the unit for ΔH should be kilojoules per mole, meaning per mole of reaction. "One mole of reaction" relates to the amounts of reactants and products in the equation as written. Thus, reaction (7.16) involves 1 mol $C_{12}H_{22}O_{11}(s)$, 12 mol $O_2(g)$, 12 mol $CO_2(g)$, 11 mol $H_2O(l)$, and -5.65×10^3 kJ of enthalpy change *per mol reaction*. The mol^{-1} part of the unit of ΔH is often dropped, but there are times we need to carry it to achieve the proper cancellation of units. We will find this to be the case in Chapters 19 and 20.

In summary, in most reactions, the heat of reaction we measure is ΔH. In some reactions, notably combustion reactions, we measure ΔU (that is, q_V). In reaction (7.16), $\Delta U = \Delta H$, but this is not always the case. Where it is not, a value of ΔH can be obtained from ΔU by the method illustrated in the discussion of expression (7.15), but even in those cases, ΔH and ΔU will be nearly equal. In this text, all heats of reactions are treated as ΔH values unless there is an indication to the contrary.

Example 7-7 shows how enthalpy changes can provide conversion factors for problem solving.

You may be wondering why the term ΔH is used instead of $\Delta U, q$, and w. It's mainly a matter of convenience. Think of an analogous situation from daily life—buying gasoline at a gas station. The gasoline price posted on the pump is actually the sum of a base price and various taxes that must be paid to different levels of government. This breakdown is important to the accountants who must determine how much tax is to be paid to which agencies. To the consumer, however, it's easier to be given just the total cost per gallon or liter. After all, this determines what he or she must pay. In thermochemistry, our chief interest is generally in heats of reaction, not pressure–volume work. And because most reactions are carried out under atmospheric pressure, it's

---

EXAMPLE 7-7   Stoichiometric Calculations Involving Quantities of Heat

How much heat is associated with the complete combustion of 1.00 kg of sucrose, $C_{12}H_{22}O_{11}$?

Analyze

Equation (7.16) represents the combustion of 1 mol of sucrose. In that reaction the amount of heat generated is given as $\Delta H = -5.65 \times 10^3$ kJ/mol. The first step is to determine the number of moles in 1.00 kg of sucrose, and then use that value with the change in enthalpy for the reaction.

Solve

Express the quantity of sucrose in moles.

$$? \text{ mol} = 1.00 \text{ kg } C_{12}H_{22}O_{11} \times \frac{1000 \text{ g } C_{12}H_{22}O_{11}}{1 \text{ kg } C_{12}H_{22}O_{11}} \times \frac{1 \text{ mol } C_{12}H_{22}O_{11}}{342.3 \text{ g } C_{12}H_{22}O_{11}} = 2.92 \text{ mol } C_{12}H_{22}O_{11}$$

Formulate a conversion factor (shown in blue) based on the information in equation (7.16)—that is, -5.65×10^3 kJ of heat is associated with the combustion of 1 mol $C_{12}H_{22}O_{11}$.

$$? \text{ kJ} = 2.92 \text{ mol } C_{12}H_{22}O_{11} \times \frac{-5.65 \times 10^3 \text{ kJ}}{1 \text{ mol } C_{12}H_{22}O_{11}} = -1.65 \times 10^4 \text{ kJ}$$

The negative sign denotes that heat is given off in the combustion.

Assess

As discussed on page 249, the heat produced by a combustion reaction is not immediately transferred to the surroundings. Use data from Table 7.1 to show that the heat released by this reaction is *more* than that required to raise the temperature of the products to 100 °C.

PRACTICE EXAMPLE A: What mass of sucrose must be burned to produce 1.00×10^3 kJ of heat?

PRACTICE EXAMPLE B: A 25.0 mL sample of 0.1045 M HCl(aq) was neutralized by NaOH(aq). Use the result of Example 7-4 to determine the heat evolved in this neutralization.

---

helpful to have a function of state, enthalpy, H, whose change is exactly equal to something we can measure: q_P.

Enthalpy Change (ΔH) Accompanying a Change in State of Matter

When a liquid is in contact with the atmosphere, energetic molecules at the surface of the liquid can overcome forces of attraction to their neighbors and pass into the gaseous, or vapor, state. We say that the liquid *vaporizes*. If the temperature of the liquid is to remain constant, the liquid must absorb heat from its surroundings to replace the energy carried off by the vaporizing molecules. The heat required to vaporize a fixed quantity of liquid is called the enthalpy (or heat) of vaporization. Usually the fixed quantity of liquid chosen is one mole, and we can call this quantity the *molar enthalpy of vaporization*. For example,

$$H_2O(l) \longrightarrow H_2O(g) \qquad \Delta H = 44.0 \text{ kJ at 298 K}$$

We described the melting of a solid in a similar fashion (page 244). The energy requirement in this case is called the enthalpy (or heat) of fusion. For the melting of one mole of ice, we can write

$$H_2O(s) \longrightarrow H_2O(l) \qquad \Delta H = 6.01 \text{ kJ at 273.15 K}$$

We can use the data represented in these equations, together with other appropriate data, to answer questions like those posed in Example 7-8 and its accompanying Practice Examples.

EXAMPLE 7-8   Enthalpy Changes Accompanying Changes in States of Matter

Calculate ΔH for the process in which 50.0 g of water is converted from liquid at 10.0 °C to vapor at 25.0 °C.

Analyze

The key to this calculation is to view the process as proceeding in two steps: first raising the temperature of liquid water from 10.0 to 25.0 °C, and then completely vaporizing the liquid at 25.0 °C. The total enthalpy change is the sum of the changes in the two steps. For a process at constant pressure, $\Delta H = q_P$, so we need to calculate the heat absorbed in each step.

Solve

HEATING WATER FROM 10.0 TO 25.0°C

This heat requirement can be determined by the method shown in Example 7-1; that is, we apply equation (7.5).

$$? \, kJ = 50.0 \, g \, H_2O \times \frac{4.18 \, J}{g \, H_2O \, °C} \times (25.0 - 10.0) \, °C \times \frac{1 \, kJ}{1000 \, J} = 3.14 \, kJ$$

VAPORIZING WATER AT 25.0°C

For this part of the calculation, the quantity of water must be expressed in moles so that we can then use the molar enthalpy of vaporization at 25 °C: 44.0 kJ/mol.

$$? \, kJ = 50.0 \, g \, H_2O \times \frac{1 \, mol \, H_2O}{18.02 \, g \, H_2O} \times \frac{44.0 \, kJ}{1 \, mol \, H_2O} = 122 \, kJ$$

TOTAL ENTHALPY CHANGE

$$\Delta H = 3.14 \, kJ + 122 \, kJ = 125 \, kJ$$

Assess

Note that the enthalpy change is positive, which reflects that the system (i.e., the water) gains energy. The reverse would be true for condensation of water at 25.0 °C and cooling it to 10.0 °C.

PRACTICE EXAMPLE A: What is the enthalpy change when a cube of ice 2.00 cm on edge is brought from −10.0 °C to a final temperature of 23.2 °C? For ice, use a density of 0.917 g/cm³, a specific heat of 2.01 J g⁻¹ °C⁻¹, and an enthalpy of fusion of 6.01 kJ/mol.

PRACTICE EXAMPLE B: What is the maximum mass of ice at −15.0 °C that can be completely converted to water vapor at 25.0 °C if the available heat for this transition is 5.00×10^3 kJ?

Standard States and Standard Enthalpy Changes

The measured enthalpy change for a reaction has a unique value *only* if the initial state (reactants) and final state (products) are precisely described. If we define a particular state as *standard* for the reactants and products, we can then say that the standard enthalpy change is the enthalpy change in a reaction in which the reactants and products are in their standard states. This so-called **standard enthalpy of reaction** is denoted with a degree symbol, $\Delta H°$.

The **standard state** of a solid or liquid substance is the pure element or compound at a pressure of *1 bar* $(10^5 \, Pa)$* and at the temperature of interest. For a gas, the standard state is the pure gas behaving as an (hypothetical) ideal gas at a pressure of 1 bar and the temperature of interest. Although temperature is not part of the definition of a standard state, it still must be specified in tabulated values of $\Delta H°$, because $\Delta H°$ depends on temperature. The values given in this text are all for 298.15 K (25 °C) unless otherwise stated.

*The International Union of Pure and Applied Chemistry (IUPAC) recommended that the standard-state pressure be changed from 1 atm to 1 bar about 25 years ago, but some data tables are still based on the 1 atm standard. Fortunately, the differences in values resulting from this change in standard-state pressure are very small—almost always small enough to be ignored.

In the rest of this chapter, we will mostly use standard enthalpy changes. We will explore the details of nonstandard conditions in Chapter 19.

Enthalpy Diagrams

The negative sign of ΔH in equation (7.16) means that the enthalpy of the products is lower than that of the reactants. This *decrease* in enthalpy appears as heat evolved to the surroundings. The combustion of sucrose is an exothermic reaction. In the reaction

$$N_2(g) + O_2(g) \longrightarrow 2\,NO(g) \qquad \Delta H° = 180.50\,kJ \qquad \textbf{(7.17)}$$

the products have a *higher* enthalpy than the reactants; ΔH is positive. To produce this increase in enthalpy, heat is absorbed from the surroundings. The reaction is endothermic. An **enthalpy diagram** is a diagrammatic representation of enthalpy changes in a process. Figure 7-15 shows how exothermic and endothermic reactions can be represented through such diagrams.

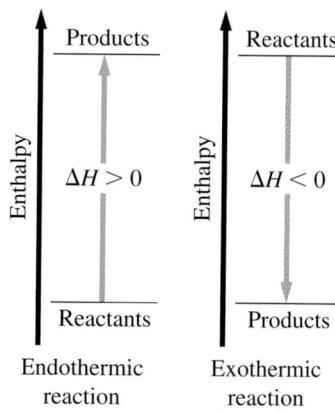

▲ FIGURE 7-15
Enthalpy diagrams
Horizontal lines represent absolute values of enthalpy. The higher a horizontal line, the greater the value of H that it represents. Vertical lines or arrows represent changes in enthalpy (ΔH). Arrows pointing up signify increases in enthalpy—endothermic reactions. Arrows pointing down signify decreases in enthalpy—exothermic reactions.

7-1   ARE YOU WONDERING...

Why ΔH depends on temperature?

The difference in ΔH for a reaction at two different temperatures is determined by the amount of heat involved in changing the reactants and products from one temperature to the other under constant pressure. These quantities of heat can be calculated with the help of equation (7.5): q_P = heat capacity × temperature change = $C_P \times \Delta T$. We write an expression of this type for each reactant and product and combine these expressions with the measured ΔH value at one temperature to obtain the value of ΔH at another. This method is illustrated in Figure 7-16 and applied in Exercise 117.

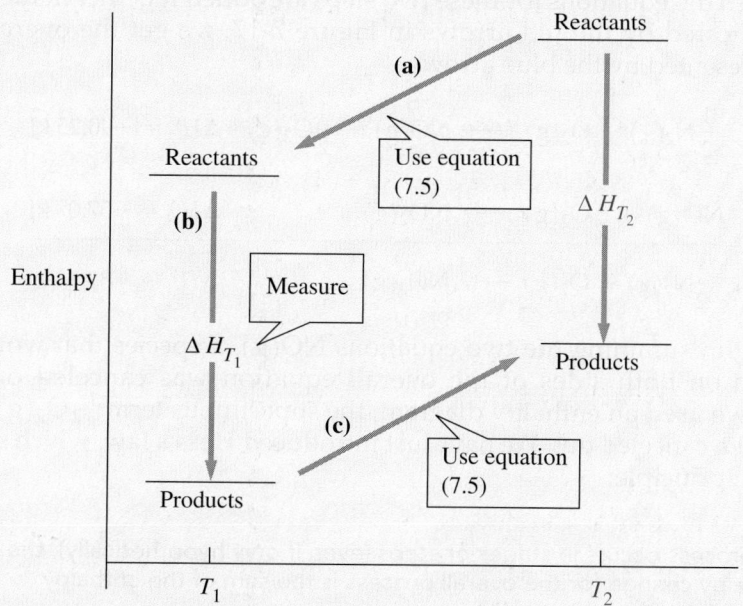

▲ FIGURE 7-16
Conceptualizing ΔH as a function of temperature
In the three-step process outlined here, **(a)** the reactants are cooled from the temperature T_2 to T_1. **(b)** The reaction is carried out at T_1, and **(c)** the products are warmed from T_1 to T_2. When the quantities of heat associated with each step are combined, the result is the same as if the reaction had been carried out at T_2, that is, ΔH_{T_2}.

7-7   Indirect Determination of ΔH: Hess's Law

One of the reasons that the enthalpy concept is so useful is that a large number of heats of reaction can be calculated from a small number of measurements. The following features of enthalpy change (ΔH) make this possible.

- **ΔH Is an Extensive Property.** Consider the standard enthalpy change in the formation of NO(g) from its elements at 25 °C.

$$N_2(g) + O_2(g) \longrightarrow 2\,NO(g) \qquad \Delta H° = 180.50\ kJ$$

To express the enthalpy change in terms of *one mole* of NO(g), we divide all coefficients *and the ΔH value* by *two*.

$$\frac{1}{2}N_2(g) + \frac{1}{2}O_2(g) \longrightarrow NO(g) \qquad \Delta H° = \frac{1}{2} \times 180.50 = 90.25\ kJ$$

Enthalpy change is directly proportional to the amounts of substances in a system.

- **ΔH Changes Sign When a Process Is Reversed.** As we learned on page 257, if a process is reversed, the change in a function of state reverses sign. Thus, ΔH for the *decomposition* of one mole of NO(g) is $-\Delta H$ for the *formation* of one mole of NO(g).

$$NO(g) \longrightarrow \frac{1}{2}N_2(g) + \frac{1}{2}O_2(g) \quad \Delta H° = -90.25\ kJ$$

- **Hess's Law of Constant Heat Summation.** To describe the standard enthalpy change for the formation of $NO_2(g)$ from $N_2(g)$ and $O_2(g)$,

$$\frac{1}{2}N_2(g) + O_2(g) \longrightarrow NO_2(g) \quad \Delta H° = ?$$

we can think of the reaction as proceeding in two steps: First we form NO(g) from $N_2(g)$ and $O_2(g)$, and then $NO_2(g)$ from NO(g) and $O_2(g)$. When the equations for these two steps are added together in the manner suggested by the red arrows in Figure 7-17, we get the overall result represented by the blue arrow.

$$\frac{1}{2}N_2(g) + O_2(g) \longrightarrow NO(g) + \frac{1}{2}O_2(g) \quad \Delta H° = +90.25\ kJ$$

$$NO(g) + \frac{1}{2}O_2(g) \longrightarrow NO_2(g) \qquad\qquad \Delta H° = -57.07\ kJ$$

$$\overline{\frac{1}{2}N_2(g) + O_2(g) \longrightarrow NO_2(g) \qquad\qquad \Delta H° = +33.18\ kJ}$$

Note that in summing the two equations NO(g), a species that would have appeared on both sides of the overall equation was canceled out. Also, because we used an enthalpy diagram, the superfluous term $\frac{1}{2}O_2(g)$ entered in and then canceled out. We have just introduced **Hess's law**, which states the following principle:

> If a process occurs in stages or steps (even if only hypothetically), the enthalpy change for the overall process is the sum of the enthalpy changes for the individual steps.

Hess's law is simply a consequence of the state function property of enthalpy. Regardless of the path taken in going from the initial state to the final state, ΔH (or $\Delta H°$ if the process is carried out under standard conditions) has the same value.

Suppose we want the standard enthalpy change for the reaction

$$3\,C(\text{graphite}) + 4\,H_2(g) \longrightarrow C_3H_8(g) \qquad \Delta H° = ? \qquad \textbf{(7.18)}$$

▶ Although we have avoided fractional coefficients previously, we need them here. The coefficient of NO(g) must be one.

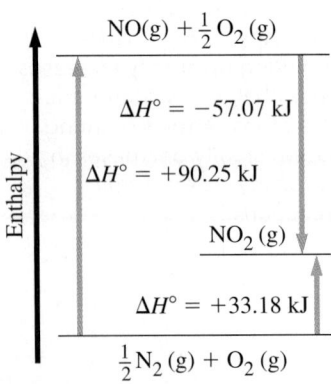

▲ FIGURE 7-17
An enthalpy diagram illustrating Hess's law
Whether the reaction occurs through a single step (blue arrow) or in two steps (red arrows), the enthalpy change is $\Delta H° = 33.18\ kJ$ for the overall reaction $\frac{1}{2}N_2(g) + O_2(g) \longrightarrow NO_2(g)$.

KEEP IN MIND

that $\Delta H°$ is an extensive property. In a chemical equation, the stoichiometric coefficients specify the amounts involved, and the unit kJ suffices for $\Delta H°$. When $\Delta H°$ is not accompanied by an equation, the amount involved must somehow be specified, such as per mole of $C_3H_8(g)$ in the expression $\Delta H°_{comb} = -2219.9\ kJ/mol\ C_3H_8(g)$.

How should we proceed? If we try to get graphite and hydrogen to react, a slight reaction will occur, but it will not go to completion. Furthermore, the product will not be limited to propane (C_3H_8); several other hydrocarbons will form as well. The fact is that we cannot directly measure $\Delta H°$ for reaction (7.18). Instead, we must resort to an *indirect calculation* from $\Delta H°$ values that can be established by experiment. Here is where Hess's law is of greatest value. It permits us to calculate ΔH values that we cannot measure directly. In Example 7-9,

---

EXAMPLE 7-9   Applying Hess's Law

Use the heat of combustion data from page 268 to determine $\Delta H°$ for reaction (7.18)
$$3\,C(\text{graphite}) + 4\,H_2(g) \longrightarrow C_3H_8(g) \qquad \Delta H° = ?$$

Analyze

To determine an enthalpy change with Hess's law, we need to combine the appropriate chemical equations. A good starting point is to write chemical equations for the given combustion reactions based on *one mole* of the indicated reactant. Recall (see page 114) that the products of the combustion of carbon–hydrogen–oxygen compounds are $CO_2(g)$ and $H_2O(l)$.

Solve

Begin by writing the following equations

(a) $C_3H_8(g) + 5\,O_2(g) \longrightarrow 3\,CO_2(g) + 4\,H_2O(l)$ $\Delta H° = -2219.9\,\text{kJ}$
(b) $C(\text{graphite}) + O_2(g) \longrightarrow CO_2(g)$ $\Delta H° = -393.5\,\text{kJ}$

(c) $H_2(g) + \dfrac{1}{2}O_2(g) \longrightarrow H_2O(l)$ $\Delta H° = -285.8\,\text{kJ}$

Because our objective in reaction (7.18) is to *produce* $C_3H_8(g)$, the next step is to find a reaction in which $C_3H_8(g)$ is formed—the *reverse* of reaction **(a)**.

$-$**(a)**: $3\,CO_2(g) + 4\,H_2O(l) \longrightarrow C_3H_8(g) + 5\,O_2(g)$ $\Delta H° = -(-2219.9)\,\text{kJ} = +2219.9\,\text{kJ}$

Now, we turn our attention to the reactants, C(graphite) and $H_2(g)$. To get the proper number of moles of each, we must multiply equation **(b)** by three and equation **(c)** by four.

$3 \times$ **(b)**: $3\,C(\text{graphite}) + 3\,O_2(g) \longrightarrow 3\,CO_2(g)$ $\Delta H° = 3(-393.5\,\text{kJ}) = -1181\,\text{kJ}$
$4 \times$ **(c)**: $4\,H_2(g) + 2\,O_2(g) \longrightarrow 4\,H_2O(l)$ $\Delta H° = 4(-285.8\,\text{kJ}) = -1143\,\text{kJ}$

Here is the overall change we have described: 3 mol C(graphite) and 4 mol $H_2(g)$ have been consumed, and 1 mol $C_3H_8(g)$ has been produced. This is exactly what is required in equation (7.18). We can now combine the three modified equations.

$-$(a): $3\,CO_2(g) + 4\,H_2O(l) \longrightarrow C_3H_8(g) + 5\,O_2(g)$ $\Delta H° = +2219.9\,\text{kJ}$
$3 \times$ (b): $3\,C(\text{graphite}) + 3\,O_2(g) \longrightarrow 3\,CO_2(g)$ $\Delta H° = -1181\,\text{kJ}$
$4 \times$ (c): $4\,H_2(g) + 2\,O_2(g) \longrightarrow 4\,H_2O(l)$ $\Delta H° = -1143\,\text{kJ}$
_____
        $3\,C(\text{graphite}) + 4\,H_2(g) \longrightarrow C_3H_8(g)$ $\Delta H° = -104\,\text{kJ}$

Assess

Hess's law is a powerful technique to determine the enthalpy of reaction by using a series of unrelated reactions, along with their enthalpies of reaction. In this example, we took three unrelated combustion reactions and were able to determine the enthalpy of reaction of another reaction.

PRACTICE EXAMPLE A: The standard heat of combustion of propene, $C_3H_6(g)$, is $-2058\,\text{kJ/mol}\,C_3H_6(g)$. Use this value and other data from this example to determine $\Delta H°$ for the hydrogenation of propene to propane.

$$CH_3CH{=}CH_2(g) + H_2(g) \longrightarrow CH_3CH_2CH_3(g) \qquad \Delta H° = ?$$

PRACTICE EXAMPLE B: From the data in Practice Example 7-9A and the following equation, determine the standard enthalpy of combustion of one mole of 2-propanol, $CH_3CH(OH)CH_3(l)$.

$$CH_3CH{=}CH_2(g) + H_2O(l) \longrightarrow CH_3CH(OH)CH_3(l) \qquad \Delta H° = -52.3\,\text{kJ}$$

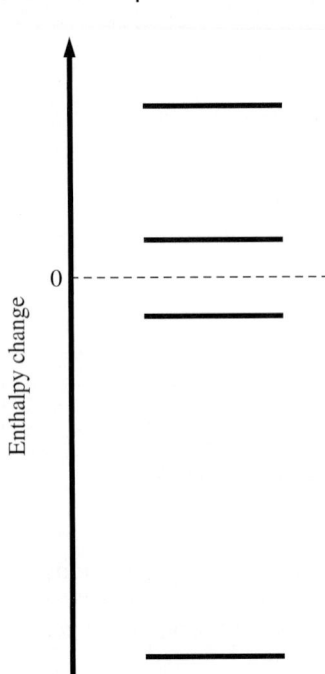

Enthalpy change

0

we use the following standard heats of combustion ΔH°_{comb} to calculate ΔH° for reaction (7.18).

$$\Delta H^\circ_{comb} \quad C_3H_8(g) = -2219.9 \text{ kJ/mol } C_3H_8(g)$$
$$C(\text{graphite}) = -393.5 \text{ kJ/mol } C(\text{graphite})$$
$$H_2(g) = -285.8 \text{ kJ/mol } H_2(g)$$

🔍 7-9 CONCEPT ASSESSMENT

The heat of reaction between carbon (graphite) and the corresponding stoichiometric amounts of hydrogen gas to form $C_2H_2(g)$, $C_2H_4(g)$, and $C_2H_6(g)$ are 226.7, 52.3 and -84.7 kJ mol^{-1}, respectively. Relate these values to the enthalpy diagram shown in the margin. Indicate on the diagram the standard enthalpy change for the reaction $C_2H_2(g) + 2 H_2(g) \longrightarrow C_2H_6(g)$.

7-8 Standard Enthalpies of Formation

In the enthalpy diagrams we have drawn, we have not written any numerical values on the enthalpy axis. This is because we cannot determine *absolute* values of enthalpy, *H*. However, enthalpy *is* a function of state, so *changes* in enthalpy, ΔH, have unique values. We can deal just with these changes. Nevertheless, as with many other properties, it is still useful to have a starting point, a zero value.

Consider a map-making analogy: What do we list as the height of a mountain? Do we mean by this the vertical distance between the mountaintop and the center of Earth? Between the mountaintop and the deepest trench in the ocean? No. By agreement, we mean the vertical distance between the mountaintop and mean sea level. We *arbitrarily* assign to mean sea level an elevation of zero, and all other points on Earth are relative to this zero elevation. The elevation of Mt. Everest is $+8848$ m; that of Badwater, Death Valley, California, is -86 m. We do something similar with enthalpies. We relate our zero to the enthalpies of certain forms of the elements and determine the enthalpies of other substances relative to this zero.

The **standard enthalpy of formation** (ΔH°_f) of a substance is the enthalpy *change* that occurs in the formation of one mole of the substance in the standard state from the *reference* forms of the elements in their standard states. The reference forms of the elements in all but a few cases are the most stable forms of the elements at one bar and the given temperature. The degree symbol denotes that the enthalpy change is a standard enthalpy change, and the subscript "f" signifies that the reaction is one in which a substance is formed from its elements. Because the formation of the most stable form of an element from itself is no change at all,

> **KEEP IN MIND**
>
> that we use the expression "standard enthalpy of formation" even though what we are describing is actually a standard enthalpy *change*.

the standard enthalpy of formation of a pure element in its reference form is 0.

Listed here are the most stable forms of several elements at 298.15 K, the temperature at which thermochemical data are commonly tabulated.

$$\text{Na(s)} \quad H_2(g) \quad N_2(g) \quad O_2(g) \quad C(\text{graphite}) \quad Br_2(l)$$

The situation with carbon is an interesting one. In addition to graphite, carbon also exists naturally in the form of diamond. However, because there is a measurable enthalpy difference between them, they cannot both be assigned $\Delta H^\circ_f = 0$.

$$C(\text{graphite}) \longrightarrow C(\text{diamond}) \quad \Delta H^\circ = 1.9 \text{ kJ}$$

We choose as the reference form the more stable form, the one with the lower enthalpy. Thus, we assign $\Delta H^\circ_f(\text{graphite}) = 0$, and $\Delta H^\circ_f(\text{diamond}) =$

▲ Diamond and graphite.

1.9 kJ/mol. Although we can obtain bromine in either the gaseous or liquid state at 298.15 K, $Br_2(l)$ is the most stable form. $Br_2(g)$, if obtained at 298.15 K and 1 bar pressure, immediately condenses to $Br_2(l)$.

$$Br_2(l) \longrightarrow Br_2(g) \qquad \Delta H_f^\circ = 30.91 \text{ kJ}$$

The enthalpies of formation are $\Delta H_f^\circ[Br_2(l)] = 0$ and $\Delta H_f^\circ[Br_2(g)] = 30.91$ kJ/mol.

A rare case in which the reference form is not the most stable form is the element phosphorus. Although over time it converts to solid red phosphorus, solid white phosphorus has been chosen as the reference form.

$$P(s, \text{white}) \longrightarrow P(s, \text{red}) \qquad \Delta H_f^\circ = -17.6 \text{ kJ}$$

The standard enthalpies of formation are $\Delta H_f^\circ[P(s, \text{white})] = 0$ and $\Delta H_f^\circ[P(s, \text{red})] = -17.6$ kJ/mol.

Standard enthalpies of formation of some common substances are presented in Table 7.2. Figure 7-18 emphasizes that both positive and negative standard enthalpies of formation are possible. It also suggests that standard enthalpies of formation are related to molecular structure.

We will use standard enthalpies of formation in a variety of calculations. Often, the first thing we must do is write the chemical equation to which a ΔH_f° value applies, as in Example 7-10.

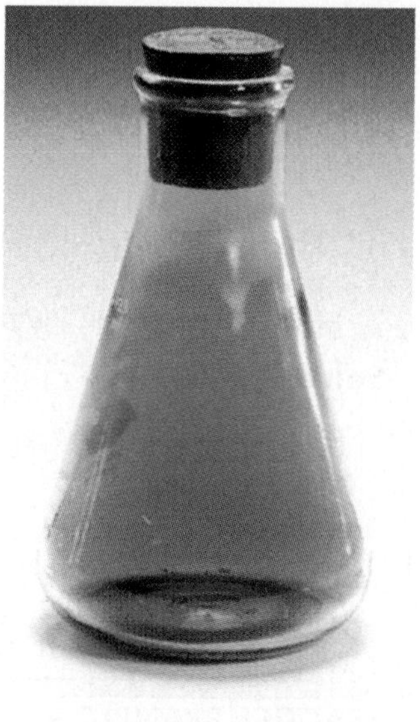

▲ Liquid bromine vaporizing.

TABLE 7.2 Some Standard Molar Enthalpies of Formation, ΔH_f° at 298.15 K

Substance	kJ/mol[a]	Substance	kJ/mol[a]
$CO(g)$	−110.5	$HBr(g)$	−36.40
$CO_2(g)$	−393.5	$HI(g)$	26.48
$CH_4(g)$	−74.81	$H_2O(g)$	−241.8
$C_2H_2(g)$	226.7	$H_2O(l)$	−285.8
$C_2H_4(g)$	52.26	$H_2S(g)$	−20.63
$C_2H_6(g)$	−84.68	$NH_3(g)$	−46.11
$C_3H_8(g)$	−103.8	$NO(g)$	90.25
$C_4H_{10}(g)$	−125.6	$N_2O(g)$	82.05
$CH_3OH(l)$	−238.7	$NO_2(g)$	33.18
$C_2H_5OH(l)$	−277.7	$N_2O_4(g)$	9.16
$HF(g)$	−271.1	$SO_2(g)$	−296.8
$HCl(g)$	−92.31	$SO_3(g)$	−395.7

[a]Values are for reactions in which one mole of substance is formed. Most of the data have been rounded off to four significant figures.

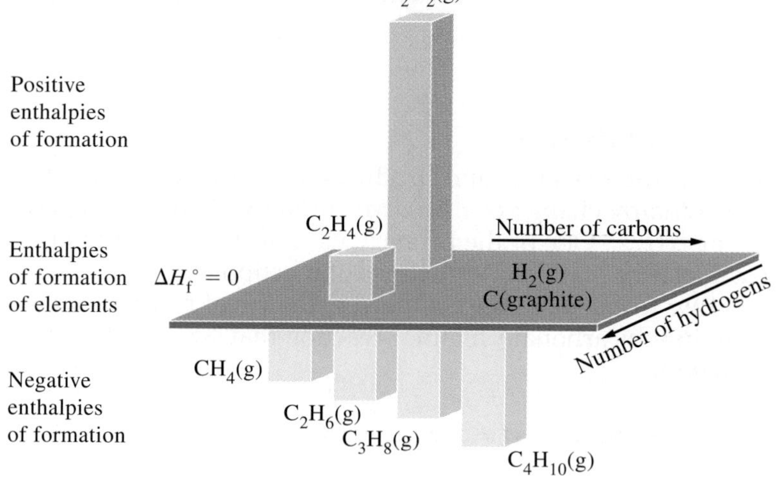

◀ FIGURE 7-18
Some standard enthalpies of formation at 298.15 K
Standard enthalpies of formation of elements are shown in the central plane, with $\Delta H_f^\circ = 0$. Substances with positive enthalpies of formation are above the plane, while those with negative enthalpies of formation are below the plane.

EXAMPLE 7-10   Relating a Standard Enthalpy of Formation to a Chemical Equation

The enthalpy of formation of formaldehyde is $\Delta H_f^\circ = -108.6$ kJ/mol HCHO(g) at 298 K. Write the chemical equation to which this value applies.

Analyze

The equation must be written for the formation of one mole of gaseous HCHO. The most stable forms of the elements at 298.15 K and 1 bar are gaseous H_2 and O_2 and solid carbon in the form of graphite (Fig. 7-19). Note that we need one fractional coefficient in this equation.

Solve

$$H_2(g) + \frac{1}{2}O_2(g) + C(\text{graphite}) \longrightarrow HCHO(g) \qquad \Delta H_f^\circ = -108.6 \text{ kJ}$$

Assess

When answering these types of problems, we must remember to use the elements in their most stable form under the given conditions. In this example, the stated conditions were 298 K and 1 bar.

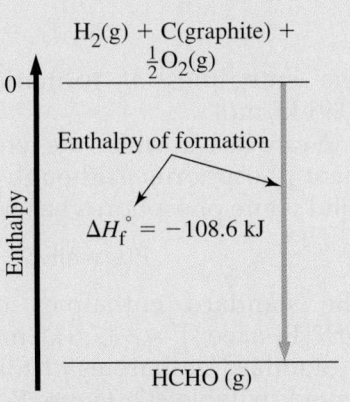

▲ FIGURE 7-19
Standard enthalpy of formation of formaldehyde, HCHO(g)
The formation of HCHO(g) from its elements in their standard states is an exothermic reaction. The heat evolved per mole of HCHO(g) formed is the standard enthalpy (heat) of formation.

PRACTICE EXAMPLE A: The standard enthalpy of formation for the amino acid leucine is -637.3 kJ/mol $C_6H_{13}O_2N(s)$. Write the chemical equation to which this value applies.

PRACTICE EXAMPLE B: How is ΔH° for the following reaction related to the standard enthalpy of formation of $NH_3(g)$ listed in Table 7.2? What is the value of $\Delta H^\circ = ?$

$$2\,NH_3(g) \longrightarrow N_2(g) + 3\,H_2(g) \qquad \Delta H^\circ = ?$$

7-2   ARE YOU WONDERING...

What is the significance of the sign of a ΔH_f° value?

A compound having a positive value of ΔH_f° is formed from its elements by an endothermic reaction. If the reaction is reversed, the compound decomposes into its elements in an exothermic reaction. We say that the compound is unstable with respect to its elements. This does not mean that the compound cannot be made, but it does suggest a tendency for the compound to enter into chemical reactions yielding products with lower enthalpies of formation.

When no other criteria are available, chemists sometimes use enthalpy change as a rough indicator of the likelihood of a chemical reaction occurring—exothermic reactions generally being more likely to occur unassisted than endothermic ones. We'll present much better criteria later in the text.

Standard Enthalpies of Reaction

We have learned that if the reactants and products of a reaction are in their standard states, the enthalpy change is a *standard* enthalpy change, which we can denote as ΔH° or ΔH_{rxn}°. One of the primary uses of standard enthalpies of formation is in calculating standard enthalpies of reaction.

Let us use Hess's law to calculate the standard enthalpy of reaction for the decomposition of sodium bicarbonate, a minor reaction that occurs when baking soda is used in baking.

$$2\,NaHCO_3(s) \longrightarrow Na_2CO_3(s) + H_2O(l) + CO_2(g) \qquad \Delta H^\circ = ? \qquad \textbf{(7.19)}$$

From Hess's law, we see that the following four equations yield equation (7.19) when added together.

(a) $2\,NaHCO_3(s) \longrightarrow 2\,Na(s) + H_2(g) + 2\,C\,(graphite) + 3\,O_2(g)$

$$\Delta H^\circ = -2 \times \Delta H_f^\circ[NaHCO_3(s)]$$

(b) $2\,Na(s) + C(graphite) + \dfrac{3}{2}O_2(g) \longrightarrow Na_2CO_3(s)$

$$\Delta H^\circ = \Delta H_f^\circ[Na_2CO_3(s)]$$

(c) $H_2(g) + \dfrac{1}{2}O_2(g) \longrightarrow H_2O(l)$ $\qquad\qquad \Delta H^\circ = \Delta H_f^\circ[H_2O(l)]$

(d) $C(graphite) + O_2(g) \longrightarrow CO_2(g)$ $\qquad\qquad \Delta H^\circ = \Delta H_f^\circ[CO_2(g)]$

$2\,NaHCO_3(s) \longrightarrow Na_2CO_3(s) + H_2O(l) + CO_2(g)$ $\qquad \Delta H^\circ = ?$

Equation (a) is the *reverse* of the equation representing the formation of two moles of $[NaHCO_3(s)]$ from its elements. This means that ΔH° for reaction (a) is the *negative* of twice $\Delta H_f^\circ[NaHCO_3(s)]$. Equations (b), (c) and (d) represent the formation of *one* mole each of $Na_2CO_3(s), CO_2(g)$ and $H_2O(l)$. Thus, we can express the value of ΔH° for the decomposition reaction as

$$\Delta H^\circ = \Delta H_f^\circ[Na_2CO_3(s)] + \Delta H_f^\circ[H_2O(l)] +$$
$$\Delta H_f^\circ[CO_2(g)] - 2 \times \Delta H_f^\circ[NaHCO_3(s)] \quad \textbf{(7.20)}$$

We can use the enthalpy diagram in Figure 7-20 to visualize the Hess's law procedure and to show how the state function property of enthalpy enables us to arrive at equation (7.20). Imagine the decomposition of sodium bicarbonate taking place in two steps. In the first step, suppose a vessel contains 2 mol $NaHCO_3$, which is allowed to decompose into 2 mol $Na(s)$, 2 mol $C(graphite)$, 1 mol $H_2(g)$, and 3 mol $O_2(g)$, as in equation (a) above. In the second step, recombine the 2 mol $Na(s)$, 2 mol $C(graphite)$, 1 mol $H_2(g)$, and 3 mol $O_2(g)$ to form the products according to equations (b), (c), and (d) above.

The pathway shown in Figure 7-20 *is not* how the reaction actually occurs. This does not matter, though, because enthalpy is a state function and the change of any state function is independent of the path chosen. The enthalpy change for the overall reaction is the sum of the standard enthalpy changes of the individual steps.

$$\Delta H^\circ = \Delta H_{decomposition}^\circ + \Delta H_{recombination}^\circ$$
$$\Delta H_{decomposition}^\circ = -2 \times \Delta H_f^\circ[NaHCO_3(s)]$$
$$\Delta H_{recombination}^\circ = \Delta H_f^\circ[Na_2CO_3(s)] + \Delta H_f^\circ[H_2O(l)] + \Delta H_f^\circ[CO_2(g)]$$

so that

$$\Delta H^\circ = \Delta H_f^\circ[Na_2CO_3(s)] + \Delta H_f^\circ[H_2O(l)] + \Delta H_f^\circ[CO_2(g)] - 2 \times \Delta H_f^\circ[NaHCO_3(s)]$$

Equation (7.20) is a specific application of the following more general relationship for a standard enthalpy of reaction.

$$\Delta H^\circ = \sum v_p \Delta H_f^\circ(\text{products}) - \sum v_r \Delta H_f^\circ(\text{reactants}) \quad \textbf{(7.21)}$$

The symbol Σ (Greek, sigma) means "the sum of." The terms that are added together are the products of the standard enthalpies of formation (ΔH_f°) and their stoichiometric coefficients, v. One sum is required for the reaction products (subscript p), and another for the initial reactants (subscript r). The enthalpy change of the reaction is the sum of terms for the products *minus* the sum of terms for the reactants. Equation (7.21) avoids the manipulation of a number of chemical equations. The state function basis for equation (7.21) is shown in Figure 7-21 and is applied in Example 7-11.

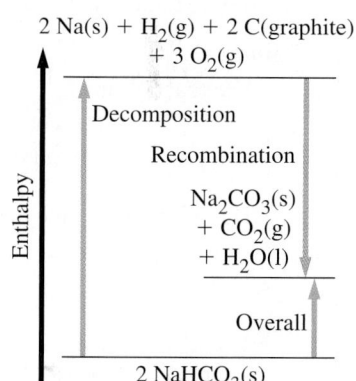

$2\,Na(s) + H_2(g) + 2\,C(graphite) + 3\,O_2(g)$

Decomposition

Recombination

$Na_2CO_3(s) + CO_2(g) + H_2O(l)$

Overall

$2\,NaHCO_3(s)$

Enthalpy

▲ FIGURE 7-20
Computing heats of reaction from standard enthalpies of formation
Enthalpy is a state function, hence ΔH° for the overall reaction $2\,NaHCO_3(s) \longrightarrow Na_2CO_3(s) + CO_2(g) + H_2O(l)$ is the sum of the enthalpy changes for the two steps shown.

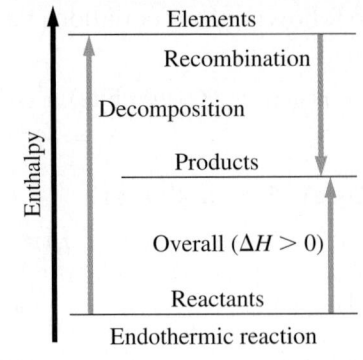

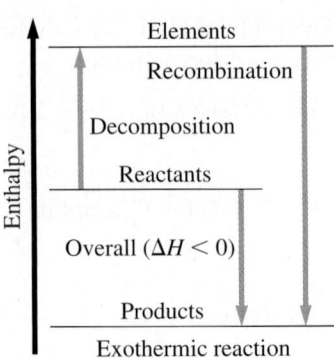

▶ FIGURE 7-21
Diagrammatic representation of equation (7.21)

EXAMPLE 7-11   Calculating $\Delta H°$ from Tabulated Values of ΔH_f°

Let us apply equation (7.21) to calculate the standard enthalpy of combustion of ethane, $C_2H_6(g)$, a component of natural gas.

Analyze

This type of problem is a straightforward application of equation (7.21). Appendix D has a table of thermodynamic data which includes the standard enthalpy of formation for a number of compounds.

Solve

The reaction is

$$C_2H_6(g) + \frac{7}{2}O_2(g) \longrightarrow 2\,CO_2(g) + 3\,H_2O(l)$$

The relationship we need is equation (7.21). The data we substitute into the relationship are from Table 7.2.

$$\Delta H° = \{2\,mol\,CO_2 \times \Delta H_f^\circ[CO_2(g)] + 3\,mol\,H_2O \times \Delta H_f^\circ[H_2O(l)]\}$$

$$- \{1\,mol\,C_2H_6 \times \Delta H_f^\circ[C_2H_6(g)] + \frac{7}{2}mol\,O_2 \times \Delta H_f^\circ[O_2(g)]\}$$

$$= 2\,mol\,CO_2 \times (-393.5\,kJ/mol\,CO_2) + 3\,mol\,H_2O \times (-285.8\,kJ/mol\,H_2O)$$

$$- 1\,mol\,C_2H_6 \times (-84.7\,kJ/mol\,C_2H_6) - \frac{7}{2}mol\,O_2 \times 0\,kJ/mol\,O_2$$

$$= -787.0\,kJ - 857.4\,kJ + 84.7\,kJ = -1559.7\,kJ$$

Assess

In these types of problems, we must make sure to subtract the sum of the products' standard enthalpies of formation from the sum of the reactants' standard enthalpies of formation. We must also keep in mind that the standard enthalpy of formation of an element in its reference form is zero. Thus, we can drop the term involving $\Delta H_f^\circ[O_2(g)]$ at any time in the calculation.

PRACTICE EXAMPLE A: Use data from Table 7.2 to calculate the standard enthalpy of combustion of ethanol, $CH_3CH_2OH(l)$, at 298.15 K.

PRACTICE EXAMPLE B: Calculate the standard enthalpy of combustion at 298.15 K *per mole* of a gaseous fuel that contains C_3H_8 and C_4H_{10} in the mole fractions 0.62 and 0.38, respectively.

A type of calculation as important as the one illustrated in Example 7-11 is the determination of an unknown ΔH_f° value from a set of known ΔH_f° values and a known standard enthalpy of reaction, $\Delta H°$. As shown in Example 7-12, the essential step is to rearrange expression (7.21) to isolate the unknown ΔH_f° on one side of the equation. Also shown is a way of organizing the data that you may find helpful.

EXAMPLE 7-12   Calculating an Unknown ΔH_f° Value

Use the data here and in Table 7.2 to calculate ΔH_f° of benzene, $C_6H_6(l)$.

$$2\,C_6H_6(l) + 15\,O_2(g) \longrightarrow 12\,CO_2(g) + 6\,H_2O(l) \qquad \Delta H^\circ = -6535\,\text{kJ}$$

Analyze

We have a chemical equation and know the standard enthalpy of reaction. We are asked to determine a standard enthalpy of formation. Equation (7.21) relates a standard enthalpy of reaction to standard enthalpy of formations for reactants and products. To begin, we organize the data needed in the calculation by writing the chemical equation for the reaction with ΔH_f° data listed under the chemical formulas.

Solve

$$2\,C_6H_6(l) + 15\,O_2(g) \longrightarrow 12\,CO_2(g) + 6\,H_2O(l) \qquad \Delta H^\circ = -6535\,\text{kJ}$$

ΔH_f°, kJ/mol   ?   0 -393.5 -285.8

Now, we can substitute known data into expression (7.21) and rearrange the equation to obtain a lone term on the left: $\Delta H_f^\circ[C_6H_6(l)]$. The remainder of the problem simply involves numerical calculations.

$$\Delta H^\circ = \{12\,\text{mol}\,CO_2 \times -(393.5\,\text{kJ/mol}\,CO_2) + 6\,\text{mol}\,H_2O \times (-285.8\,\text{kJ/mol}\,H_2O)\}$$
$$- 2\,\text{mol}\,C_2H_6 \times \Delta H_f^\circ[C_6H_6(l)] = -6535\,\text{kJ}$$

$$\Delta H_f^\circ[C_6H_6(l)] = \frac{\{-4722\,\text{kJ} - 1715\,\text{kJ}\} + 6535\,\text{kJ}}{2\,\text{mol}\,C_6H_6} = 49\,\text{kJ/mol}\,C_6H_6(l)$$

Assess

By organizing the data as shown, we were able to identify what is unknown and see how to use equation (7.21). To obtain the correct answer, we also needed to use the correct states for the compounds. In combustion reactions, the water in the product is always liquid. If we had used the standard enthalpy of formation for gaseous water, we would have obtained the wrong answer.

PRACTICE EXAMPLE A: The overall reaction that occurs in photosynthesis in plants is

$$6\,CO_2(g) + 6\,H_2O(l) \longrightarrow C_6H_{12}O_6(s) + 6\,O_2(g) \qquad \Delta H^\circ = 2803\,\text{kJ}$$

Determine the standard enthalpy of formation of glucose, $C_6H_{12}O_6(s)$, at 298 K.

PRACTICE EXAMPLE B: A handbook lists the standard enthalpy of combustion of gaseous dimethyl ether at 298 K as $-31.70\,\text{kJ/g}(CH_3)_2O(g)$. What is the standard molar enthalpy of formation of dimethyl ether at 298 K?

Ionic Reactions in Solutions

Many chemical reactions in aqueous solution are best thought of as reactions between ions and best represented by net ionic equations. Consider the neutralization of a strong acid by a strong base. Using a somewhat more accurate enthalpy of neutralization than we obtained in Example 7-4, we can write

$$H^+(aq) + OH^-(aq) \longrightarrow H_2O(l) \qquad \Delta H^\circ = -55.8\,\text{kJ} \qquad \textbf{(7.22)}$$

We should also be able to calculate this enthalpy of neutralization by using enthalpy of formation data in expression (7.21), but this requires us to have enthalpy of formation data for individual ions. And there is a slight problem in getting these. We cannot create ions of a single type in a chemical reaction. We always produce cations and anions simultaneously, as in the reaction of sodium and chlorine to produce Na^+ and Cl^- in NaCl. We must choose a particular ion to which we assign an enthalpy of formation of *zero* in its aqueous solutions. We then compare the enthalpies of formation of other ions to this reference ion. The ion we arbitrarily choose for our zero is $H^+(aq)$. Now let us see how we can use expression (7.21) and data from equation (7.22) to determine the enthalpy of formation of $OH^-(aq)$.

$$\Delta H° = 1 \, mol \, H_2O \times \Delta H_f°[H_2O(l)] - \{1 \, mol \, H^+ \times \Delta H_f°[H^+(aq)]$$
$$+ \, 1 \, mol \, OH^- \times \Delta H_f°[OH^-(aq)]\} = -55.8 \, kJ$$

$$\Delta H_f°[OH^-(aq)] =$$
$$\frac{55.8 \, kJ + (1 \, mol \, H_2O \times \Delta H_f°[H_2O(l)]) - (1 \, mol \, H^+ \times \Delta H_f°[H^+(aq)])}{1 \, mol \, OH^-}$$

$$\Delta H_f°[OH^-(aq)] = \frac{55.8 \, kJ - 285.8 \, kJ - 0 \, kJ}{1 \, mol \, OH^-} = -230.0 \, kJ/mol \, OH^-$$

Table 7.3 lists data for several common ions in aqueous solution. Enthalpies of formation in solution depend on the solute concentration. These data are representative for *dilute* aqueous solutions (about 1 M), the type of solution that we normally deal with. Some of these data are used in Example 7-13.

TABLE 7.3 Some Standard Molar Enthalpies of Formation, $\Delta H_f°$ of Ions in Aqueous Solution at 298.15 K

Ion	kJ/mol	Ion	kJ/mol
H^+	0	OH^-	−230.0
Li^+	−278.5	Cl^-	−167.2
Na^+	−240.1	Br^-	−121.6
K^+	−252.4	I^-	−55.19
NH_4^+	−132.5	NO_3^-	−205.0
Ag^+	105.6	CO_3^{2-}	−677.1
Mg^{2+}	−466.9	S^{2-}	33.05
Ca^{2+}	−542.8	SO_4^{2-}	−909.3
Ba^{2+}	−537.6	$S_2O_3^{2-}$	−648.5
Cu^{2+}	64.77	PO_4^{3-}	−1277
Al^{3+}	−531		

EXAMPLE 7-13   Calculating the Enthalpy Change in an Ionic Reaction

Given that $\Delta H_f°[BaSO_4(s)] = -1473 \, kJ/mol$, what is the standard enthalpy change for the precipitation of barium sulfate?

Analyze

First, write the net ionic equation for the reaction and introduce the relevant data. Then make use of equation (7.21).

Solve

Start by organizing the data in a table.

$$Ba^{2+}(aq) + SO_4^{2-}(aq) \longrightarrow BaSO_4(s) \qquad \Delta H° = ?$$

| $\Delta H_f°$, kJ/mol | −537.6 | −909.3 | −1473 |

Then substitute data into equation (7.21).

$$\Delta H° = 1 \, mol \, BaSO_4 \times \Delta H_f°[BaSO_4(s)] - 1 \, mol \, Ba^{2+} \times \Delta H_f°[Ba^{2+}(aq)] - 1 \, mol \, SO_4^{2-} \times \Delta H_f°[SO_4^{2-}(aq)]$$
$$= 1 \, mol \, BaSO_4 \times (-1473 \, kJ/mol \, BaSO_4) - 1 \, mol \, Ba^{2+} \times (-537.6 \, kJ/mol \, Ba^{2+})$$
$$- \, 1 \, mol \, SO_4^{2-} \times (-909.3 \, kJ/mol \, SO_4^{2-})$$
$$= -1473 \, kJ + 537.6 \, kJ + 909.3 \, kJ = -26 \, kJ$$

Assess

The standard enthalpy of reaction determined here is the heat given off by the system (i.e., the ionic reaction).

PRACTICE EXAMPLE A: Given that $\Delta H_f°[AgI(s)] = -61.84 \, kJ/mol$, what is the standard enthalpy change for the precipitation of silver iodide?

PRACTICE EXAMPLE B: The standard enthalpy change for the precipitation of $Ag_2CO_3(s)$ is −39.9 kJ per mole of $Ag_2CO_3(s)$ formed. What is $\Delta H_f°[Ag_2CO_3(s)]$?

7-10   CONCEPT ASSESSMENT

Is it possible to calculate a heat of reaction at 373.15 K by using standard enthalpies of formation at 298.15 K? If so, explain how you would do this, and indicate any additional data you might need.

7-9   Fuels as Sources of Energy

One of the most important uses of thermochemical measurements and calculations is in assessing materials as energy sources. For the most part, these materials, called fuels, liberate heat through the process of combustion. We will briefly survey some common fuels, emphasizing matters that a thermochemical background helps us to understand.

Fossil Fuels

The bulk of current energy needs are met by petroleum, natural gas, and coal—so-called fossil fuels. These fuels are derived from plant and animal life of millions of years ago. The original source of the energy locked into these fuels is solar energy. In the process of *photosynthesis*, CO_2 and H_2O, in the presence of enzymes, the pigment chlorophyll, and sunlight, are converted to *carbohydrates*. These are compounds with formulas $C_m(H_2O)_n$, where m and n are integers. For example, in the sugar glucose $m = n = 6$, that is, $C_6(H_2O)_6 = C_6H_{12}O_6$. Its formation through photosynthesis is an *endothermic* process, represented as

$$6\,CO_2(g) + 6\,H_2O(l) \xrightarrow[\text{sunlight}]{\text{chlorophyll}} C_6H_{12}O_6(s) + 6\,O_2(g) \quad \Delta H^\circ = +2.8 \times 10^3\ kJ$$

$$(7.23)$$

◀ Although the formula $C_m(H_2O)_n$ suggests a "hydrate" of carbon, in carbohydrates, there are no H_2O units, as there are in hydrates, such as $CuSO_4 \cdot 5\,H_2O$. H and O atoms are simply found in the same numerical ratio as in H_2O.

When reaction (7.23) is reversed, as in the combustion of glucose, heat is evolved. The combustion reaction is *exothermic*.

The complex carbohydrate cellulose, with molecular masses ranging up to 500,000 u, is the principal structural material of plants. When plant life decomposes in the presence of bacteria and out of contact with air, O and H atoms are removed and the approximate carbon content of the residue increases in the progression

Peat $\longrightarrow$ lignite (32% C) $\longrightarrow$ sub-bituminous coal (40% C) $\longrightarrow$

bituminous coal (60% C) $\longrightarrow$ anthracite coal (80% C)

For this process to progress all the way to anthracite coal may take about 300 million years. Coal, then, is a combustible organic rock consisting of carbon, hydrogen, and oxygen, together with small quantities of nitrogen, sulfur, and mineral matter (ash). (One proposed formula for a "molecule" of bituminous coal is $C_{153}H_{115}N_3O_{13}S_2$.)

Petroleum and natural gas formed in a different way. The remains of plants and animals living in ancient seas fell to the ocean floor, where they were decomposed by bacteria and covered with sand and mud. Over time, the sand and mud were converted to sandstone by the weight of overlying layers of sand and mud. The high pressures and temperatures resulting from this overlying sandstone rock formation transformed the original organic matter into petroleum and natural gas. The ages of these deposits range from about 250 million to 500 million years.

A typical natural gas consists of about 85% methane (CH_4), 10% ethane (C_2H_6), 3% propane (C_3H_8), and small quantities of other combustible and noncombustible gases. A typical petroleum consists of several hundred different hydrocarbons that range in complexity from C_1 molecules (CH_4) to C_{40} or higher (such as $C_{40}H_{82}$).

TABLE 7.4 Approximate Heats of Combustion of Some Fuels

Heat of Combustion

Fuel	kJ/g
Municipal waste	−12.7
Cellulose	−17.5
Pinewood	−21.2
Methanol	−22.7
Peat	−20.8
Bituminous coal	−28.3
Isooctane (a component of gasoline)	−47.8
Natural gas	−49.5

▶ Environmental issues associated with oxides of sulfur and nitrogen are discussed more fully in later chapters.

One way to compare different fuels is through their heats of combustion: In general, *the higher the heat of combustion, the better the fuel*. Table 7.4 lists approximate heats of combustion for the fossil fuels. These data show that *biomass* (living matter or materials derived from it—wood, alcohols, municipal waste) is a viable fuel, but that fossil fuels yield more energy per unit mass.

Problems Posed by Fossil Fuel Use There are two fundamental problems with the use of fossil fuels. First, fossil fuels are essentially *nonrenewable* energy sources. The world consumption of fossil fuels is expected to increase for the foreseeable future (Fig. 7-22), but when will Earth's supply of these fuels run out? There is currently a debate about whether oil production has peaked now and is about to decline, or whether it will peak more toward the middle of the this century. The second problem with fossil fuels is their environmental effect. Sulfur impurities in fuels produce oxides of sulfur. The high temperatures associated with combustion cause the reaction of N_2 and O_2 in air to form oxides of nitrogen. Oxides of sulfur and nitrogen are implicated in air pollution and are important contributors to the environmental problem known as acid rain. Another inevitable product of the combustion of fossil fuels is carbon dioxide, one of the "greenhouse" gases leading to *global warming* and potential changes in Earth's climate.

Global Warming—An Environmental Issue Involving Carbon Dioxide We do not normally think of CO_2 as an air pollutant because it is essentially non-toxic and is a natural and necessary component of air. Its ultimate effect on the environment, however, could be very significant. A buildup of $CO_2(g)$ in the atmosphere may disturb the energy balance on Earth.

Earth's atmosphere, discussed in Focus On 6 on the Mastering Chemistry website, is largely transparent to visible and UV radiation from the sun.

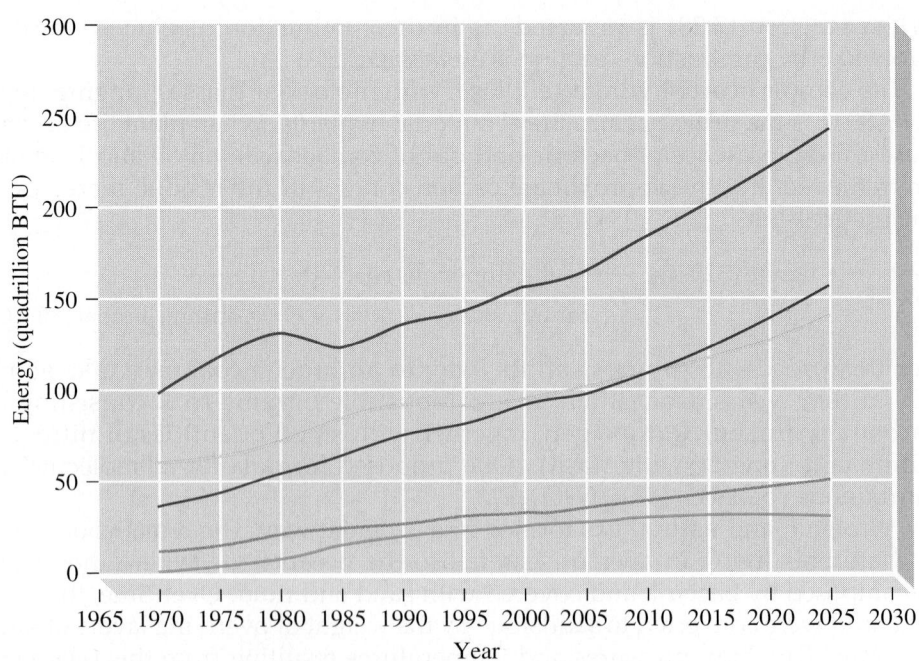

▲ FIGURE 7-22
World primary energy consumption by energy source
These graphs show the history of energy consumption since 1970, with predictions to 2025. Petroleum (dark blue line) is seen to be the major source of energy for the foreseeable future, followed by coal (yellow) and natural gas (pink), which are about the same. Other sources of energy included are wind power (purple) and nuclear power (light blue). The unit BTU is a measure of energy and stands for British thermal unit (see Exercise 95). [*Source: www.eia.doe.gov/oiaf/ieo/pdf/ieoreftab_2.pdf*]

This radiation is absorbed at Earth's surface, which is warmed by it. Some of this absorbed energy is reradiated as infrared radiation. Certain atmospheric gases, primarily CO_2, methane, and water vapor, absorb some of this infrared radiation, and the energy thus retained in the atmosphere produces a warming effect. This process, outlined in Figure 7-23, is often compared to the retention of thermal energy in a greenhouse and is called the "greenhouse effect."* The natural greenhouse effect is essential in maintaining the proper temperature for life on Earth. Without it, Earth would be permanently covered with ice.

Over the past 400,000 years, the atmospheric carbon dioxide concentration has varied from 180 to 300 parts per million with the preindustrial-age concentration at about 285 ppm. By 2005, the level had increased to about 376 ppm and is still rising (Fig. 7-24). Increasing atmospheric carbon dioxide concentrations result from the burning of carbon-containing fuels such as wood, coal, natural gas, and gasoline (Fig. 7-24) and from the deforestation of tropical regions (plants, through photosynthesis, consume CO_2 from the atmosphere). The expected effect of a CO_2 buildup is an increase in Earth's average temperature, a **global warming**. Some estimates are that a doubling of the CO_2 content over that of preindustrial times could occur before the end of the present century and that this doubling could produce an average global temperature increase of 1.5 to 4.5 °C.

Predicting the probable effects of a CO_2 buildup in the atmosphere is done largely through computer models, and it is very difficult to know all the factors that should be included in these models and the relative importance of these factors. For example, global warming could lead to the increased evaporation of water and increased cloud formation. In turn, an increased cloud cover could reduce the amount of solar radiation reaching Earth's surface and, to some extent, offset global warming.

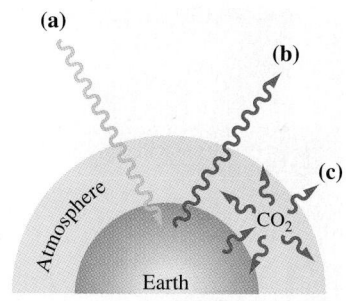

▲ FIGURE 7-23
The "greenhouse" effect
(a) Some incoming radiation from sunlight is reflected back into space by the atmosphere, and some, such as certain UV light, is absorbed by stratospheric ozone. Much of the radiation from sunlight, however, reaches Earth's surface. **(b)** Earth's surface re-emits some of this energy as infrared radiation. **(c)** Some of the infrared radiation leaving the Earth's surface is absorbed by CO_2 and other greenhouse gases and is redirected back towards the Earth's surface. The redirected infrared radiation warms the atmosphere.

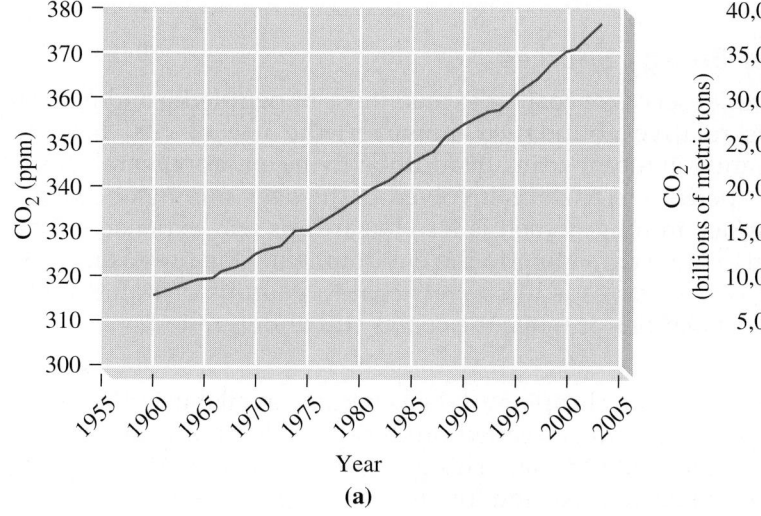

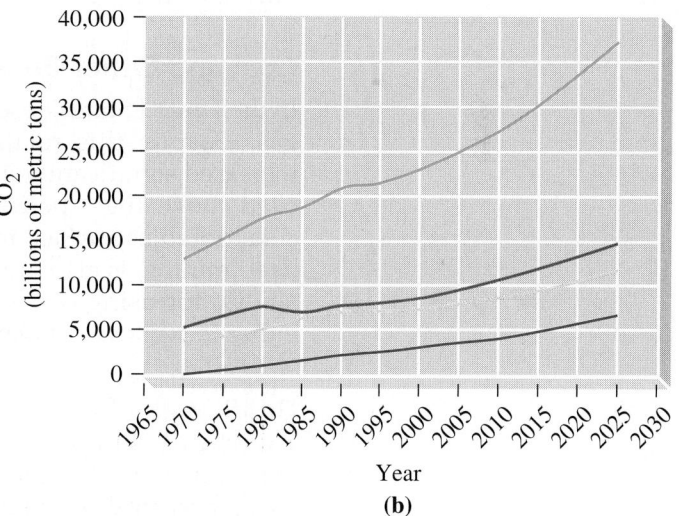

▲ FIGURE 7-24
Increasing carbon dioxide content of the atmosphere
(a) The global average atmospheric carbon dioxide level over a 50-year span, expressed in parts per million by volume, as measured by a worldwide cooperative sampling network. **(b)** The actual and predicted CO_2 emissions for a 55-year span due to the combustion of natural gas (pink line), coal (yellow), and petroleum (dark blue), together with the total of all CO_2 emissions (light blue). The CO_2 content of the atmosphere continues to increase, from approximately 375 ppm in 2003 to 385 ppm in 2008.

*Glass, like CO_2, is transparent to visible and some UV light but absorbs infrared radiation. The glass in a greenhouse, though, acts primarily to prevent the bulk flow of warm air out of the greenhouse.

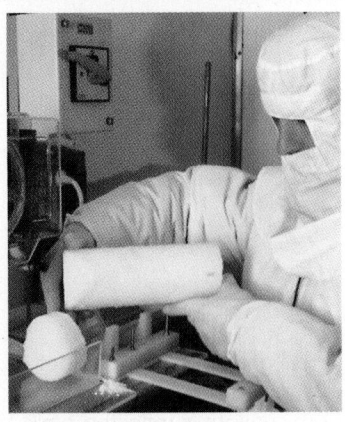

▲ An ice core from the ice sheet in Antarctica is cut into sections in a refrigerated clean room. The ice core is then analyzed to determine the amount and type of trapped gases and trace elements it contains. These data provide information regarding past changes in climate and current trends in the pollution of the atmosphere.

Some of the significant possible effects of global warming are

- local temperature changes. The average annual temperature for Alaska and Northern Canada has increased by 1.9 °C over the past 50 years. Alaskan winter temperatures have increased by an average of 3.5 °C over this same time period.
- a rise in sea level caused by the thermal expansion of seawater and increased melting of continental ice caps. A potential increase in sea level of up to 1 m by 2100 would displace tens of millions of inhabitants in Bangladesh alone.
- the migration of plant and animal species. Vegetation now characteristic of certain areas of the globe could migrate into regions several hundred kilometers closer to the poles. The areas in which diseases, such as malaria, are endemic could also expand.

Although some of the current thinking involves speculation, a growing body of evidence supports the likelihood of global warming, also called climate change. For example, analyses of tiny air bubbles trapped in the Antarctic ice cap show a strong correlation between the atmospheric CO_2 content and temperature for the past 160,000 years—low temperatures during periods of low CO_2 levels and higher temperatures with higher levels of CO_2.

CO_2 is not the only greenhouse gas. Several gases are even stronger infrared absorbers—specifically, methane (CH_4), ozone (O_3), nitrous oxide (N_2O), and chlorofluorocarbons (CFCs). Furthermore, atmospheric concentrations of some of these gases have been growing at a faster rate than that of CO_2. No strategies beyond curtailing the use of chlorofluorocarbons and fossil fuels have emerged for countering a possible global warming. Like several other major environmental issues, some aspects of climate change are not well understood, and research, debate, and action are all likely to occur simultaneously for a long time to come.

Coal and Other Energy Sources

In the United States, reserves of coal far exceed those of petroleum and natural gas. Despite this relative abundance, however, the use of coal has not increased significantly in recent years. In addition to the environmental effects cited above, the expense and hazards involved in the deep mining of coal are considerable. Surface mining, which is less hazardous and expensive than deep mining, is also more damaging to the environment. One promising possibility for using coal reserves is to convert coal to gaseous or liquid fuels, either in surface installations or while the coal is still underground.

Gasification of Coal Before cheap natural gas became available in the 1940s, gas produced from coal (variously called producer gas, town gas, or city gas) was widely used in the United States. This gas was manufactured by passing steam and air through heated coal and involved such reactions as

$$C(\text{graphite}) + H_2O(g) \longrightarrow CO(g) + H_2(g) \qquad \Delta H° = +131.3 \text{ kJ} \qquad \textbf{(7.24)}$$

$$CO(g) + H_2O(g) \longrightarrow CO_2(g) + H_2(g) \qquad \Delta H° = -41.2 \text{ kJ} \qquad \textbf{(7.25)}$$

$$2C(\text{graphite}) + O_2(g) \longrightarrow 2CO(g) \qquad \Delta H° = -221.0 \text{ kJ} \qquad \textbf{(7.26)}$$

$$C(\text{graphite}) + 2H_2(g) \longrightarrow CH_4(g) \qquad \Delta H° = -74.8 \text{ kJ} \qquad \textbf{(7.27)}$$

The principal gasification reaction (7.24) is highly endothermic. The heat requirements for this reaction are met by the carefully controlled partial burning of coal (reaction 7.26).

A typical producer gas consists of about 23% CO, 18% H_2, 8% CO_2, and 1% CH_4 by volume. It also contains about 50% N_2 because air is used in its production. Because the N_2 and CO_2 are noncombustible, producer gas has only about 10% to 15% of the heat value of natural gas. Modern gasification processes include several features:

1. They use $O_2(g)$ instead of air, thereby eliminating $N_2(g)$ in the product.
2. They provide for the removal of noncombustible $CO_2(g)$ and sulfur impurities. For example,

$$CaO(s) + CO_2(g) \longrightarrow CaCO_3(s)$$
$$2\,H_2S(g) + SO_2(g) \longrightarrow 3\,S(s) + 2\,H_2O(g)$$

3. They include a step (called *methanation*) to convert CO and H_2, in the presence of a catalyst, to CH_4.

$$CO(g) + 3\,H_2(g) \xrightarrow{\text{catalyst}} CH_4(g) + H_2O(l)$$

The product is called *substitute natural gas* (SNG), a gaseous mixture with composition and heat value similar to that of natural gas.

Liquefaction of Coal The first step in obtaining liquid fuels from coal generally involves gasification of coal, as in reaction (7.24). This step is followed by catalytic reactions in which liquid hydrocarbons are formed.

$$n\,CO + (2n + 1)H_2 \longrightarrow C_nH_{2n+2} + n\,H_2O$$

In still another process, liquid methanol is formed.

$$CO(g) + 2\,H_2(g) \longrightarrow CH_3OH(l) \qquad \textbf{(7.28)}$$

In 1942, some 32 million gallons of aviation fuel were made from coal in Germany. In South Africa, the Sasol process for coal liquefaction has been a major source of gasoline and a variety of other petroleum products and chemicals for more than 50 years.

Methanol

Methanol, CH_3OH, can be obtained from coal by reaction (7.28). It can also be produced by thermal decomposition (pyrolysis) of wood, manure, sewage, or municipal waste. The heat of combustion of methanol is only about one-half that of a typical gasoline on a mass basis, but methanol has a high octane number—106—compared with 100 for the gasoline hydrocarbon isooctane and about 92 for premium gasoline. Methanol has been tested and used as a fuel in internal combustion engines and is cleaner burning than gasoline. Methanol can also be used for space heating, electric power generation, fuel cells, and as a reactant to make a variety of other organic compounds.

Ethanol

Ethanol, C_2H_5OH, is produced mostly from ethylene, C_2H_4, which in turn is derived from petroleum. Current interest centers on the production of ethanol by the fermentation of organic matter, a process known throughout recorded history. Ethanol production by fermentation is probably most advanced in Brazil, where sugarcane and cassava (manioc) are the plant matter (biomass) used. In the United States, corn-based ethanol is used chiefly as an additive to gasoline to improve its octane rating and reduce air pollution. Also, a 90% gasoline–10% ethanol mixture is used as an automotive fuel under the name *gasohol*.

Biofuels

Biofuels are renewable energy sources that are similar to fossil fuels. Biofuels are fuels derived from dead biological material, most commonly plants. Fossil fuels are derived from biological material that has been dead for a very long time. The use of biofuels is not new; several car inventors had envisioned their vehicles running on such fuels as peanut oil, hemp-derived fuel, and ethanol. Reacting vegetable oil with a base–alcohol mixture produces a compound commonly called a biodiesel. A typical petro–diesel compound is the hydrocarbon cetane ($C_{16}H_{34}$), and the typical biodiesel compound contains oxygen atoms, as illustrated in the figure below. The standard enthalpies of combustion of the petro–diesel and the biodiesel are very similar.

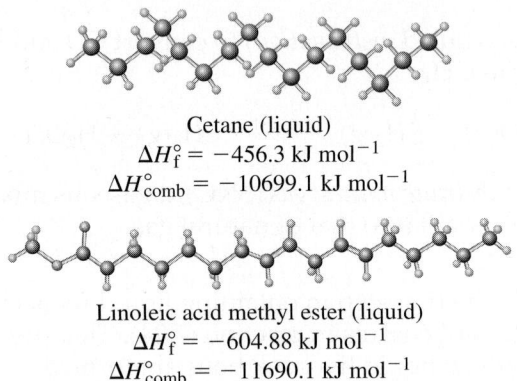

Cetane (liquid)
$\Delta H_f^\circ = -456.3$ kJ mol^{-1}
$\Delta H_{comb}^\circ = -10699.1$ kJ mol^{-1}

Linoleic acid methyl ester (liquid)
$\Delta H_f^\circ = -604.88$ kJ mol^{-1}
$\Delta H_{comb}^\circ = -11690.1$ kJ mol^{-1}

Although biofuels are appealing replacements for fossil fuels, their widespread adoption has several potential drawbacks. One major concern is the food-versus-fuel issue. Typical plants used for food (e.g., sugar cane) are sources of biofuels, which drives up the cost of food. A positive aspect of biofuels is that they are carbon neutral; that is, the $CO_2(g)$ produced by the burning of a biofuel is then used by plants for new growth, resulting in no net gain of carbon in the atmosphere. Biofuels and their use have many other advantages and disadvantages. Importantly, chemical knowledge of these compounds is needed to address these issues.

Hydrogen

Another fuel with great potential is hydrogen. Its most attractive features are that

- on a per gram basis, its heat of combustion is more than twice that of methane and about three times that of gasoline;
- the product of its combustion is H_2O, not CO and CO_2 as with gasoline.

Currently, the bulk of hydrogen used commercially is made from petroleum and natural gas, but for hydrogen to be a significant fuel of the future, efficient methods must be perfected for obtaining hydrogen from other sources, especially water. Alternative methods of producing hydrogen and the prospects of developing an economy based on hydrogen are discussed later in the text.

Alternative Energy Sources

Combustion reactions are only one means of extracting useful energy from materials. An alternative, for example, is to carry out reactions that yield the same products as combustion reactions in electrochemical cells called *fuel cells*. The energy is released as electricity rather than as heat (see Section 20-5). Solar energy can be used directly, without recourse to photosynthesis. Nuclear processes can be used in place of chemical reactions (Chapter 25). Other alternative sources in various stages of development and use include hydroelectric energy, geothermal energy, and tidal and wind power.

Summary

7-1 Getting Started: Some Terminology—The subject of a thermochemical study is broken down into the **system** of interest and the portions of the universe with which the system may interact, the **surroundings**. An **open system** can exchange both energy and matter with its surroundings. A **closed system** can exchange only energy and not matter. An **isolated system** can exchange neither energy nor matter with its surroundings (Fig. 7-1). **Energy** is the capacity to do work, and **work** is performed when a force acts through a distance. Energy can be further characterized (Fig. 7-2) as **kinetic energy** (energy associated with matter in motion) or **potential energy** (energy resulting from the position or composition of matter). Kinetic energy associated with random molecular motion is sometimes called **thermal energy**.

7-2 Heat—**Heat** is energy transferred between a system and its surroundings as a result of a temperature difference between the two. In some cases, heat can be transferred at constant temperature, as in a change in state of matter in the system. A quantity of heat is the product of the heat capacity of the system and the temperature change (equation 7.5). In turn, **heat capacity** is the product of mass and **specific heat**, the amount of heat required to change the temperature of one gram of substance by one degree Celsius. Historically, the unit for measuring heat has been the **calorie (cal)**, but the SI unit of heat is the joule, the same as for other forms of energy (equation 7.2). Energy transfers between a system and its surroundings must conform to the **law of conservation of energy**, meaning that all heat lost by a system is gained by its surroundings (equation 7.6).

7-3 Heats of Reaction and Calorimetry—In a chemical reaction, a change in the **chemical energy** associated with the reactants and products may appear as heat. The **heat of reaction** is the quantity of heat exchanged between a system and its surroundings when the reaction occurs at a constant temperature. In an **exothermic reaction**, heat is given off by the system; in an **endothermic reaction** the system absorbs heat. Heats of reaction are determined in a **calorimeter**, a device for measuring quantities of heat (equation 7.10). Exothermic combustion reactions are usually studied in a **bomb calorimeter** (Fig. 7-5). A common type of calorimeter used in the general chemistry laboratory is constructed from ordinary Styrofoam cups (Fig. 7-6).

7-4 Work—In some reactions an energy transfer between a system and its surroundings occurs as work. This is commonly the work involved in the expansion or compression of gases (Fig. 7-8) and is called **pressure–volume work** (equation 7.11).

7-5 The First Law of Thermodynamics—**Internal energy** (U) is the total energy (both kinetic and potential) in a system. The **first law of thermodynamics** relates changes in the internal energy of a system (ΔU) to the quantities of heat (q) and work (w) exchanged between the system and its surroundings. The relationship is $\Delta U = q + w$ (equation 7.12) and requires that a set of sign conventions be consistently followed. A **function of state** (**state function**) has a value that depends only on the exact condition or state in which a system is found and not on how that state was reached. Internal energy is a state function. A path-dependent function, such as heat or work, depends on how a change in a system is achieved. A change that is accomplished through an infinite number of infinitesimal steps is a **reversible process** (Fig. 7-12), whereas a change accomplished in one step or a series of finite steps is **irreversible**.

7-6 Heats of Reaction: ΔU and ΔH—In a chemical reaction with work limited to pressure–volume work and conducted at constant volume, the heat of reaction is equal to the change in internal energy (equation 7.13). For reactions at constant pressure a more useful function is **enthalpy** (H), defined as the internal energy (U) of a system plus the pressure–volume product (PV). The **enthalpy change** (ΔH) in a reaction proves to be the heat of reaction at constant pressure (equation 7.14). Most heats of reaction are reported as ΔH values. A substance under a pressure of 1 bar (10^5 Pa) and at the temperature of interest is said to be in its **standard state**. If the reactants and products of a reaction are in their standard states, the enthalpy change in a reaction is called the **standard enthalpy of reaction** and designated as $\Delta H°$. Enthalpy changes can be represented schematically through **enthalpy diagrams** (Fig. 7-15).

7-7 Indirect Determination of ΔH: Hess's Law—Often an unknown ΔH value can be established indirectly through **Hess's law**, which states that an overall enthalpy change is the sum of the enthalpy changes of the individual steps leading to the overall process (Fig. 7-17).

7-8 Standard Enthalpies of Formation—By arbitrarily assigning an enthalpy of zero to the reference forms of the elements in their standard states, the enthalpy change in the formation of a compound from its elements becomes a **standard enthalpy of formation** ($\Delta H_f°$). Using tabulated standard enthalpies of formation (Table 7.2), it is possible to calculate standard enthalpies of reactions without having to perform additional experiments (equation 7.21).

7-9 Fuels as Sources of Energy—One of the chief applications of thermochemistry is in the study of the combustion of fuels as energy sources. Currently, the principal fuels are the fossil fuels, but potential alternative fuels are also mentioned in this chapter and discussed in more depth later in the text. One of the problems with the use of fossil fuels is the potential for **global warming**.

Integrative Example

When charcoal is burned in a limited supply of oxygen in the presence of H_2O, a mixture of CO, H_2, and other noncombustible gases (mostly CO_2) is obtained. Such a mixture is called *synthesis gas*. This gas can be used to synthesize organic compounds, or it can be burned as a fuel. A typical synthesis gas consists of 55.0% $CO(g)$, 33.0% $H_2(g)$, and 12.0% noncombustible gases (mostly CO_2), *by volume*. To what temperature can 25.0 kg water at 25.0 °C be heated with the heat liberated by the combustion of 0.205 m^3 of this typical synthesis gas, measured at 25.0 °C and 102.6 kPa pressure?

Analyze

First, use the ideal gas equation to calculate the total number of moles of gas, and then use equation (6.17) to establish the number of moles of each combustible gas. Next, write an equation for the combustion of each gas. Use these equations and enthalpy of formation data to calculate the total amount of heat released by the combustion. Finally, use equation (7.5) to calculate the temperature increase when this quantity of heat is absorbed by the 25.0 kg of water. The final water temperature is then easily established.

Solve

Substitute the applicable data into the ideal gas equation using SI units, with $R = 8.3145 \, m^3 \, Pa \, mol^{-1} K^{-1}$. Solve for n.

$$n = \frac{PV}{RT} = \frac{102.6 \, kPa \times 1000 \, Pa/1 \, kPa \times 0.205 \, m^3}{8.3145 \, m^3 \, Pa \, mol^{-1} K^{-1} \times 298.2 \, K}$$
$$= 8.48 \, mol \, gas$$

Now, apportion the 8.48 moles among the three gases present, converting the volume percents to mole fractions and using equation (6.17).

$$n_{CO} = n_{tot} \times x_{CO} = 8.48 \, mol \times 0.550 = 4.66 \, mol \, CO$$
$$n_{H_2} = n_{tot} \times x_{H_2} = 8.48 \, mol \times 0.330 = 2.80 \, mol \, H_2$$

(remaining gas noncombustible)

Write an equation for the combustion of $CO(g)$, list ΔH_f° data beneath the equation, and determine ΔH_{comb}° per mole of $CO(g)$.

$$CO(g) + \frac{1}{2}O_2(g) \longrightarrow CO_2(g)$$

ΔH_f°: $-110.5 \, kJ/mol$ $-393.5 \, kJ/mol$

$\Delta H_{comb}^\circ = 1 \, mol \, CO_2 \times (-393.5 \, kJ/mol \, CO_2) - 1 \, mol \, CO \times (-110.5 \, kJ/mol \, CO)$
$= -283.0 \, kJ$

Write another equation for the combustion of $H_2(g)$, again listing ΔH_f° data beneath the equation, and determining ΔH_{comb}° per mole of $H_2(g)$.

$$H_2(g) + \frac{1}{2}O_2(g) \longrightarrow H_2O(l)$$

ΔH_f°: $0 \, kJ/mol$ $-285.8 \, kJ/mol$

$\Delta H_{comb}^\circ = 1 \, mol \, H_2O \times (-285.8 \, kJ/mol \, H_2O) = -285.8 \, kJ$

Determine the total heat released in the combustion of the amounts of CO and H_2 in the 0.205 m^3 of gas.

$4.66 \, mol \, CO \times (-283.0 \, kJ/mol \, CO) + 2.80 \, mol \, H_2 \times (-285.8 \, kJ/mol \, H_2)$
$= -2.12 \times 10^3 \, kJ$

The quantity of heat absorbed by the 25.0 kg of water is

$q_{water} = -q_{comb}$

$$= -\left(-2.12 \times 10^3 \, kJ \times \frac{1000 \, J}{1 \, kJ}\right) = 2.12 \times 10^6 \, J$$

Rearrange equation (7.5) to solve for the temperature change in the 2.50×10^4 g (25.0 kg) of water.

$$\Delta T = \frac{q_{water}}{mass \, water \times sp \, ht \, water}$$

$$\Delta T = \frac{2.12 \times 10^6 \, J}{\left(2.50 \times 10^4 \, g \, H_2O \times \frac{4.18 \, J}{g \, H_2O°C}\right)} = 20.3 °C$$

From the initial temperature and the temperature change, determine the final temperature.

$T_f = T_i + \Delta T = 25.0 °C + 20.3 °C = 45.3 °C$

Assess

The assumption that the gas sample obeys the ideal gas law is probably valid since the temperature of the gas (25.0 °C) is not particularly low and the gas pressure, about 1 atm, is not particularly high. However, the implicit assumption that all the heat of combustion could be transferred to the water was probably not valid. If the transfer were to occur in an ordinary gas-fired water heater, some of the heat would undoubtedly be lost through the exhaust vent. Thus, our calculation was of the highest temperature that could possibly be attained. Note that in using the ideal gas equation the simplest approach was to work with SI units because those were the units of the data that were given.

PRACTICE EXAMPLE A: The enthalpy of combustion for 1-hexadecene, $C_{16}H_{32}$, is $-10539.0\ kJ\ mol^{-1}$, and that of hexadecane, $C_{16}H_{34}$, is $-10699.1\ kJ\ mol^{-1}$. What is the enthalpy of hydrogenation of 1-hexadecene to hexadecane?

PRACTICE EXAMPLE B: A chemist mixes 56 grams of CaO, powdered lime, with 100 mL of water at 20 °C. After the completion of the reaction, $CaO(s) + H_2O(l) \rightarrow Ca(OH)_2(s)$, what are the contents of the reaction vessel? [*Hint:* Assume that the heat released by the reaction is absorbed by the water.]

Mastering **CHEMISTRY**

You'll find a link to additional self study questions in the study area on www.masteringchemistry.com

Exercises

Heat Capacity (Specific Heat)

1. Calculate the quantity of heat, in kilojoules, **(a)** required to raise the temperature of 9.25 L of water from 22.0 to 29.4 °C; **(b)** associated with a 33.5 °C decrease in temperature in a 5.85 kg aluminum bar (specific heat of aluminum = $0.903\ J\ g^{-1}\ {}^{\circ}C^{-1}$).

2. Calculate the final temperature that results when **(a)** a 12.6 g sample of water at 22.9 °C absorbs 875 J of heat; **(b)** a 1.59 kg sample of platinum at 78.2 °C gives off 1.05 kcal of heat (sp ht of Pt = $0.032\ cal\ g^{-1}\ {}^{\circ}C^{-1}$).

3. Refer to Example 7-2. The experiment is repeated with several different metals substituting for the lead. The masses of metal and water and the initial temperatures of the metal and water are the same as in Figure 7-3. The final temperatures are **(a)** Zn, 38.9 °C; **(b)** Pt, 28.8 °C; **(c)** Al, 52.7 °C. What is the specific heat of each metal, expressed in $J\ g^{-1}\ {}^{\circ}C^{-1}$?

4. A 75.0 g piece of Ag metal is heated to 80.0 °C and dropped into 50.0 g of water at 23.2 °C. The final temperature of the Ag–H$_2$O mixture is 27.6 °C. What is the specific heat of silver?

5. A 465 g chunk of iron is removed from an oven and plunged into 375 g water in an insulated container. The temperature of the water increases from 26 to 87 °C. If the specific heat of iron is $0.449\ J\ g^{-1}\ {}^{\circ}C^{-1}$, what must have been the original temperature of the iron?

6. A piece of stainless steel (sp ht = $0.50\ J\ g^{-1}\ {}^{\circ}C^{-1}$) is transferred from an oven at 183 °C into 125 mL of water at 23.2 °C. The water temperature rises to 51.5 °C. What is the mass of the steel? How precise is this method of mass determination? Explain.

7. A 1.00 kg sample of magnesium at 40.0 °C is added to 1.00 L of water maintained at 20.0 °C in an insulated container. What will be the final tempera-ture of the Mg–H$_2$O mixture (specific heat of Mg = $1.024\ J\ g^{-1}\ {}^{\circ}C^{-1}$)?

8. Brass has a density of $8.40\ g/cm^3$ and a specific heat of $0.385\ J\ g^{-1}\ {}^{\circ}C^{-1}$. A 15.2 cm^3 piece of brass at an initial temperature of 163 °C is dropped into an insulated container with 150.0 g water initially at 22.4 °C. What will be the final temperature of the brass–water mixture?

9. A 74.8 g sample of copper at 143.2 °C is added to an insulated vessel containing 165 mL of glycerol, $C_3H_8O_3(l)$ ($d = 1.26\ g/mL$), at 24.8 °C. The final temperature is 31.1°C. The specific heat of copper is $0.385\ J\ g^{-1}\ {}^{\circ}C^{-1}$. What is the heat capacity of glycerol in $J\ mol^{-1}\ {}^{\circ}C^{-1}$?

10. What volume of 18.5 °C water must be added, together with a 1.23 kg piece of iron at 68.5 °C, so that the temperature of the water in the insulated container shown in the figure remains constant at 25.6 °C?

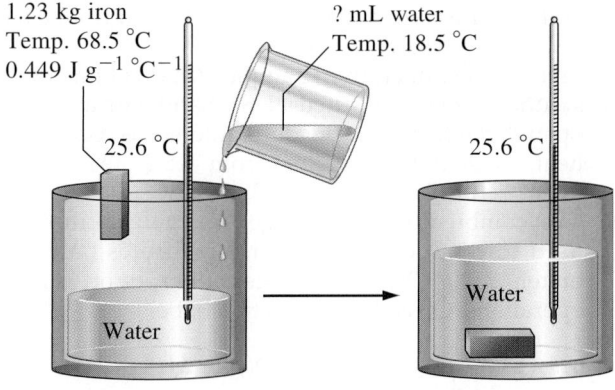

1.23 kg iron
Temp. 68.5 °C
$0.449\ J\ g^{-1}\ {}^{\circ}C^{-1}$

? mL water
Temp. 18.5 °C

25.6 °C

25.6 °C

Water

Water

11. In the form of heat, 6.052 J of energy is transferred to a 1.0 L sample of air ($d = 1.204 \, mg/cm^3$) at 20.0 °C. The final temperature of the air is 25.0 °C. What is the heat capacity of air in J/K?

Heats of Reaction

13. How much heat, in kilojoules, is associated with the production of 283 kg of slaked lime, $Ca(OH)_2$?

$$CaO(s) + H_2O(l) \longrightarrow$$
$$Ca(OH)_2(s) \quad \Delta H° = -65.2 \, kJ$$

14. The standard enthalpy change in the combustion of the hydrocarbon octane is $\Delta H° = -5.48 \times 10^3 \, kJ/mol$ $C_8H_{18}(l)$. How much heat, in kilojoules, is liberated *per gallon* of octane burned? (Density of octane = 0.703 g/mL; 1 gal = 3.785 L.)

15. How much heat, in kilojoules, is evolved in the complete combustion of **(a)** 1.325 g $C_4H_{10}(g)$ at 25 °C and 1 atm; **(b)** 28.4 L $C_4H_{10}(g)$ at STP; **(c)** 12.6 L $C_4H_{10}(g)$ at 23.6 °C and 738 mmHg? Assume that the enthalpy change for the reaction does not change significantly with temperature or pressure. The complete combustion of butane, $C_4H_{10}(g)$, is represented by the equation

$$C_4H_{10}(g) + \frac{13}{2} O_2(g) \longrightarrow 4 CO_2(g) + 5 H_2O(l)$$
$$\Delta H° = -2877 \, kJ$$

16. Upon complete combustion, the indicated substances evolve the given quantities of heat. Write a balanced equation for the combustion of 1.00 mol of each substance, including the enthalpy change, ΔH, for the reaction.
    (a) 0.584 g of propane, $C_3H_8(g)$, yields 29.4 kJ
    (b) 0.136 g of camphor, $C_{10}H_{16}O(s)$, yields 5.27 kJ
    (c) 2.35 mL of acetone, $(CH_3)_2CO(l)$ ($d = 0.791$ g/mL), yields 58.3 kJ

17. The combustion of methane gas, the principal constituent of natural gas, is represented by the equation

$$CH_4(g) + 2 O_2(g) \longrightarrow CO_2(g) + 2 H_2O(l)$$
$$\Delta H° = -890.3 \, kJ$$

    (a) What mass of methane, in kilograms, must be burned to liberate $2.80 \times 10^7 \, kJ$ of heat?
    (b) What quantity of heat, in kilojoules, is liberated in the complete combustion of 1.65×10^4 L of $CH_4(g)$, measured at 18.6 °C and 768 mmHg?
    (c) If the quantity of heat calculated in part **(b)** could be transferred with 100% efficiency to water, what volume of water, in liters, could be heated from 8.8 to 60.0 °C as a result?

18. Refer to the Integrative Example. What volume of the synthesis gas, measured at STP and burned in an open flame (constant-pressure process), is required to heat 40.0 gal of water from 15.2 to 65.0 °C? (1 gal = 3.785 L.)

19. The combustion of hydrogen–oxygen mixtures is used to produce very high temperatures (approximately 2500 °C) needed for certain types of welding operations. Consider the reaction to be

$$H_2(g) + \frac{1}{2} O_2(g) \longrightarrow H_2O(g) \quad \Delta H° = -241.8 \, kJ$$

What is the quantity of heat evolved, in kilojoules, when a 180 g mixture containing equal parts of H_2 and O_2 by mass is burned?

20. Thermite mixtures are used for certain types of welding, and the thermite reaction is highly exothermic.

$$Fe_2O_3(s) + 2 Al(s) \longrightarrow Al_2O_3(s) + 2 Fe(s)$$
$$\Delta H° = -852 \, kJ$$

1.00 mol of granular Fe_2O_3 and 2.00 mol of granular Al are mixed at room temperature (25 °C), and a reaction is initiated. The liberated heat is retained within the products, whose combined specific heat over a broad temperature range is about $0.8 \, J \, g^{-1} °C^{-1}$. (The melting point of iron is 1530 °C.) Show that the quantity of heat liberated is more than sufficient to raise the temperature of the products to the melting point of iron.

21. A 0.205 g pellet of potassium hydroxide, KOH, is added to 55.9 g water in a Styrofoam coffee cup. The water temperature rises from 23.5 to 24.4 °C. [Assume that the specific heat of dilute KOH(aq) is the same as that of water.]
    (a) What is the approximate heat of solution of KOH, expressed as kilojoules per mole of KOH?
    (b) How could the precision of this measurement be improved *without* modifying the apparatus?

22. The heat of solution of KI(s) in water is +20.3 kJ/mol KI. If a quantity of KI is added to sufficient water at 23.5 °C in a Styrofoam cup to produce 150.0 mL of 2.50 M KI, what will be the final temperature? (Assume a density of 1.30 g/mL and a specific heat of $2.7 \, J \, g^{-1} °C^{-1}$ for 2.50 M KI.)

23. You are planning a lecture demonstration to illustrate an endothermic process. You want to lower the temperature of 1400 mL water in an insulated container from 25 to 10 °C. Approximately what mass of $NH_4Cl(s)$ should you dissolve in the water to achieve this result? The heat of solution of NH_4Cl is +14.7 kJ/mol NH_4Cl.

24. Care must be taken in preparing solutions of solutes that liberate heat on dissolving. The heat of solution of NaOH is −44.5 kJ/mol NaOH. To what maximum temperature may a sample of water, originally at 21 °C, be raised in the preparation of 500 mL of 7.0 M NaOH? Assume the solution has a density of 1.08 g/mL and specific heat of $4.00 \, J \, g^{-1} °C^{-1}$.

25. Refer to Example 7-4. The product of the neutralization is 0.500 M NaCl. For this solution, assume a density of 1.02 g/mL and a specific heat of $4.02 \, J \, g^{-1} °C^{-1}$. Also, assume a heat capacity for the Styrofoam cup of 10 J/°C, and recalculate the heat of neutralization.

26. The heat of neutralization of HCl(aq) by NaOH(aq) is −55.84 kJ/mol H_2O produced. If 50.00 mL of 1.05 M NaOH is added to 25.00 mL of 1.86 M HCl, with both solutions originally at 24.72 °C, what will be the final solution temperature? (Assume that no heat is lost to the surrounding air and that the solution produced in the neutralization reaction has a density of 1.02 g/mL and a specific heat of $3.98 \, J \, g^{-1} °C^{-1}$.)

12. What is the final temperature (in °C) of 1.24 g of water with an initial temperature of 20.0 °C after 6.052 J of heat is added to it?

27. Acetylene (C_2H_2) torches are used in welding. How much heat (in kJ) evolves when 5.0 L of C_2H_2 ($d = 1.0967 \, kg/m^3$) is mixed with a stoichiometric amount of oxygen gas? The combustion reaction is

$$C_2H_2(g) + \frac{5}{2}O_2(g) \longrightarrow 2 CO_2(g) + H_2O(l)$$
$$\Delta H° = -1299.5 \, kJ$$

28. Propane (C_3H_8) gas ($d = 1.83 \, kg/m^3$) is used in most gas grills. What volume (in liters) of propane is needed to generate 273.8 kJ of heat?

$$C_3H_8(g) + 5 O_2(g) \longrightarrow 3 CO_2(g) + 4 H_2O(l)$$
$$\Delta H° = -2219.9 \, kJ$$

Enthalpy Changes and States of Matter

29. What mass of ice can be melted with the same quantity of heat as required to raise the temperature of 3.50 mol $H_2O(l)$ by 50.0 °C? [$\Delta H°_{fusion} = 6.01 \, kJ/mol$ $H_2O(s)$]

30. What will be the final temperature of the water in an insulated container as the result of passing 5.00 g of steam, $H_2O(g)$, at 100.0 °C into 100.0 g of water at 25.0 °C? ($\Delta H°_{vap} = 40.6 \, kJ/mol \, H_2O$)

31. A 125 g stainless steel ball bearing (sp ht = 0.50 $J \, g^{-1} \, °C^{-1}$) at 525 °C is dropped into 75.0 mL of water at 28.5 °C in an open Styrofoam cup. As a result, the water is brought to a boil when the temperature reaches 100.0 °C. What mass of water vaporizes while the boiling continues? ($\Delta H°_{vap} = 40.6 \, kJ/mol \, H_2O$)

32. If the ball bearing described in Exercise 31 is dropped onto a large block of ice at 0 °C, what mass of liquid water will form? ($\Delta H°_{fusion} = 6.01 \, kJ/mol \, H_2O(s)$)

33. The enthalpy of sublimation (solid $\rightarrow$ gas) for dry ice (i.e., CO_2) is $\Delta H°_{sub} = 571 \, kJ/kg$ at -78.5 °C. If 125.0 J of heat is transferred to a block of dry ice that is -78.5°C, what volume of CO_2 gas ($d = 1.98 \, g/L$) will be generated?

34. The enthalpy of vaporization for $N_2(l)$ is 5.56 kJ/mol. How much heat (in J) is required to produce 1.0 L of $N_2(g)$ at 77.36 K and 1.0 atm?

Calorimetry

35. A sample gives off 5228 cal when burned in a bomb calorimeter. The temperature of the calorimeter assembly increases by 4.39 °C. Calculate the heat capacity of the calorimeter, in kilojoules per degree Celsius.

36. The following substances undergo complete combustion in a bomb calorimeter. The calorimeter assembly has a heat capacity of 5.136 kJ/°C. In each case, what is the final temperature if the initial water temperature is 22.43 °C?
(a) 0.3268 g caffeine, $C_8H_{10}O_2N_4$ (heat of combustion = $-1014.2 \, kcal/mol$ caffeine);
(b) 1.35 mL of methyl ethyl ketone, $C_4H_8O(l)$, $d = 0.805 \, g/mL$ (heat of combustion = $-2444 \, kJ/mol$ methyl ethyl ketone).

37. A bomb calorimetry experiment is performed with xylose, $C_5H_{10}O_5(s)$, as the combustible substance. The data obtained are

mass of xylose burned:	1.183 g
heat capacity of calorimeter:	4.728 kJ/°C
initial calorimeter temperature:	23.29 °C
final calorimeter temperature:	27.19 °C

(a) What is the heat of combustion of xylose, in kilojoules per mole? **(b)** Write the chemical equation for the complete combustion of xylose, and represent the value of ΔH in this equation. (Assume for this reaction that $\Delta U \approx \Delta H$.)

38. A coffee-cup calorimeter contains 100.0 mL of 0.300 M HCl at 20.3 °C. When 1.82 g Zn(s) is added, the temperature rises to 30.5 °C. What is the heat of reaction per mol Zn? Make the same assumptions as in Example 7-4, and also assume that there is no heat lost with the $H_2(g)$ that escapes.

$$Zn(s) + 2 H^+(aq) \longrightarrow Zn^{2+}(aq) + H_2(g)$$

39. A 0.75 g sample of KCl is added to 35.0 g H_2O in a Styrofoam cup and stirred until it dissolves. The temperature of the solution drops from 24.8 to 23.6 °C.
(a) Is the process endothermic or exothermic?
(b) What is the heat of solution of KCl expressed in kilojoules per mole of KCl?

40. The heat of solution of potassium acetate in water is $-15.3 \, kJ/mol \, KC_2H_3O_2$. What will be the final temperature when 0.136 mol $KC_2H_3O_2$ is dissolved in 525 mL water that is initially at 25.1 °C?

41. A 1.620 g sample of naphthalene, $C_{10}H_8(s)$, is completely burned in a bomb calorimeter assembly and a temperature increase of 8.44 °C is noted. If the heat of combustion of naphthalene is $-5156 \, kJ/mol \, C_{10}H_8$, what is the heat capacity of the bomb calorimeter?

42. Salicylic acid, $C_7H_6O_3$, has been suggested as a calorimetric standard. Its heat of combustion is $-3.023 \times 10^3 \, kJ/mol \, C_7H_6O_3$. From the following data determine the heat capacity of a bomb calorimeter assembly (that is, the bomb, water, stirrer, thermometer, wires, and so forth).

mass of salicylic acid burned:	1.201 g
initial calorimeter temperature:	23.68 °C
final calorimeter temperature:	29.82 °C

43. Refer to Example 7-3. Based on the heat of combustion of sucrose established in the example, what should be the temperature change (ΔT) produced by the combustion of 1.227 g $C_{12}H_{22}O_{11}$ in a bomb calorimeter assembly with a heat capacity of 3.87 kJ/°C?

44. A 1.397 g sample of thymol, $C_{10}H_{14}O(s)$ (a preservative and a mold and mildew preventative), is burned in a bomb calorimeter assembly. The temperature increase is 11.23 °C, and the heat capacity of the bomb calorimeter is 4.68 kJ/°C. What is the heat of combustion of thymol, expressed in kilojoules per mole of $C_{10}H_{14}O$?

45. A 5.0 g sample of NaCl is added to a Styrofoam cup of water, and the change in water temperature is 5.0 °C. The heat of solution of NaCl is 3.76 kJ/mol. What is the mass (in g) of water in the Styrofoam cup?

46. We can determine the purity of solid materials by using calorimetry. A gold ring (for pure gold, specific heat = $0.1291 \, J \, g^{-1} \, K^{-1}$) with mass of 10.5 g is heated to 78.3 °C and immersed in 50.0 g of 23.7 °C water in a constant-pressure calorimeter. The final temperature of the water is 31.0 °C. Is this a pure sample of gold?

Pressure–Volume Work

47. Calculate the quantity of work associated with a 3.5 L expansion of a gas (ΔV) against a pressure of 748 mmHg in the units (a) atm L; (b) joules (J); (c) calories (cal).

48. Calculate the quantity of work, in joules, associated with the compression of a gas from 5.62 L to 3.37 L by a constant pressure of 1.23 atm.

49. A 1.00 g sample of Ne(g) at 1 atm pressure and 27 °C is allowed to expand into an *evacuated* vessel of 2.50 L volume. Does the gas do work? Explain.

50. Compressed air in aerosol cans is used to free electronic equipment of dust. Does the air do any work as it escapes from the can?

51. In each of the following processes, is any work done when the reaction is carried out at constant pressure in a vessel open to the atmosphere? If so, is work done by the reacting system or on it? (a) Neutralization of $Ba(OH)_2(aq)$ by HCl(aq); (b) conversion of gaseous nitrogen dioxide to gaseous dinitrogen tetroxide; (c) decomposition of calcium carbonate to calcium oxide and carbon dioxide gas.

52. In each of the following processes, is any work done when the reaction is carried out at constant pressure in a vessel open to the atmosphere? If so, is work done by the reacting system or on it? (a) Reaction of nitrogen monoxide and oxygen gases to form gaseous nitrogen dioxide; (b) precipitation of magnesium hydroxide by the reaction of aqueous solutions of NaOH and $MgCl_2$; (c) reaction of copper(II) sulfate and water vapor to form copper(II) sulfate pentahydrate.

53. If 325 J of work is done by a system at a pressure of 1.0 atm and 298 K, what is the change in the volume of the system?

54. A movable piston in a cylinder holding 5.0 L $N_2(g)$ is used to lift a 1.23 kg object to a height of 4.5 meters. How much work (in J) was done by the gas?

First Law of Thermodynamics

55. What is the change in internal energy of a system if the system (a) absorbs 58 J of heat and does 58 J of work; (b) absorbs 125 J of heat and does 687 J of work; (c) evolves 280 cal of heat and has 1.25 kJ of work done on it?

56. What is the change in internal energy of a system if the *surroundings* (a) transfer 235 J of heat and 128 J of work to the system; (b) absorb 145 J of heat from the system while doing 98 J of work on the system; (c) exchange no heat, but receive 1.07 kJ of work from the system?

57. The internal energy of a fixed quantity of an ideal gas depends only on its temperature. A sample of an ideal gas is allowed to expand at a constant temperature (isothermal expansion). (a) Does the gas do work? (b) Does the gas exchange heat with its surroundings? (c) What happens to the temperature of the gas? (d) What is ΔU for the gas?

58. In an *adiabatic* process, a system is thermally insulated— there is no exchange of heat between system and surroundings. For the adiabatic expansion of an ideal gas (a) does the gas do work? (b) Does the internal energy of the gas increase, decrease, or remain constant? (c) What happens to the temperature of the gas? [*Hint*: Refer to Exercise 57.]

59. Do you think the following observation is in any way possible? An ideal gas is expanded isothermally and is observed to do twice as much work as the heat absorbed from its surroundings. Explain your answer. [*Hint*: Refer to Exercises 57 and 58.]

60. Do you think the following observation is any way possible? A gas absorbs heat from its surroundings while being compressed. Explain your answer. [*Hint*: Refer to Exercises 55 and 56.]

61. There are other forms of work besides P–V work. For example, electrical work is defined as the potential × change in charge, $w = \phi \, dq$. If a charge in a system is changed from 10 C to 5 C in a potential of 100 V and 45 J of heat is liberated, what is the change in the internal energy? (Note: 1 V = 1 J/C)

62. Another form of work is extension, defined as the tension × change in length, $w = f \, \Delta l$. A piece of DNA has an approximate tension of $f = 10 \, pN$. What is the change in the internal energy of the adiabatic stretching of DNA by 10 pm?

Relating ΔH and ΔU

63. Only one of the following expressions holds true for the heat of a chemical reaction, *regardless of how the reaction is carried out*. Which is the correct expression and why? (a) q_V; (b) q_P; (c) $\Delta U - w$; (d) ΔU; (e) ΔH.

64. Determine whether ΔH is equal to, greater than, or less than ΔU for the following reactions. Keep in mind that "greater than" means more positive or less negative, and "less than" means less positive or more negative. Assume that the only significant change in volume during a reaction at constant pressure is that associated with changes in the amounts of gases.

(a) The complete combustion of one mole of 1-butanol(l).

(b) The complete combustion of one mole of glucose, $C_6H_{12}O_6(s)$.

(c) The decomposition of solid ammonium nitrate to produce liquid water and gaseous dinitrogen monoxide.

65. The heat of combustion of 2-propanol at 298.15 K, determined in a bomb calorimeter, is -33.41 kJ/g. For the combustion of one mole of 2-propanol, determine **(a)** ΔU, and **(b)** ΔH.

66. Write an equation to represent the combustion of thymol referred to in Exercise 44. Include in this equation the values for ΔU and ΔH.

Hess's Law

67. The standard enthalpy of formation of $NH_3(g)$ is -46.11 kJ/mol NH_3. What is $\Delta H°$ for the following reaction?

$$\frac{2}{3}NH_3(g) \longrightarrow \frac{1}{3}N_2(g) + H_2(g) \qquad \Delta H° =$$

68. Use Hess's law to determine $\Delta H°$ for the reaction
$$CO(g) + \frac{1}{2}O_2(g) \longrightarrow CO_2(g), \text{ given that}$$

$$C(graphite) + \frac{1}{2}O_2(g) \longrightarrow CO(g)$$
$$\Delta H° = -110.54 \text{ kJ}$$
$$C(graphite) + O_2(g) \longrightarrow CO_2(g)$$
$$\Delta H° = -393.51 \text{ kJ}$$

69. Use Hess's law to determine $\Delta H°$ for the reaction $C_3H_4(g) + 2H_2(g) \longrightarrow C_3H_8(g)$, given that

$$H_2(g) + \frac{1}{2}O_2(g) \longrightarrow H_2O(l) \quad \Delta H° = -285.8 \text{ kJ}$$
$$C_3H_4(g) + 4O_2(g) \longrightarrow 3CO_2(g) + 2H_2O(l)$$
$$\Delta H° = -1937 \text{ kJ}$$
$$C_3H_8(g) + 5O_2(g) \longrightarrow 3CO_2(g) + 4H_2O(l)$$
$$\Delta H° = -2219.1 \text{ kJ}$$

70. Given the following information:
$$\frac{1}{2}N_2(g) + \frac{3}{2}H_2(g) \longrightarrow NH_3(g) \qquad \Delta H_1°$$

$$NH_3(g) + \frac{5}{4}O_2(g) \longrightarrow NO(g) + \frac{3}{2}H_2O(l) \quad \Delta H_2°$$

$$H_2(g) + \frac{1}{2}O_2(g) \longrightarrow H_2O(l) \qquad \Delta H_3°$$

Determine $\Delta H°$ for the following reaction, expressed in terms of $\Delta H_1°$, $\Delta H_2°$, and $\Delta H_3°$.
$$N_2(g) + O_2(g) \longrightarrow 2NO(g) \qquad \Delta H° = ?$$

71. For the reaction $C_2H_4(g) + Cl_2(g) \longrightarrow C_2H_4Cl_2(l)$, determine $\Delta H°$, given that
$$4HCl(g) + O_2(g) \longrightarrow 2Cl_2(g) + 2H_2O(l)$$
$$\Delta H° = -202.4 \text{ kJ}$$
$$2HCl(g) + C_2H_4(g) + \frac{1}{2}O_2(g) \longrightarrow$$
$$C_2H_4Cl_2(l) + H_2O(l) \quad \Delta H° = -318.7 \text{ kJ}$$

72. Determine $\Delta H°$ for this reaction from the data below.
$$N_2H_4(l) + 2H_2O_2(l) \longrightarrow N_2(g) + 4H_2O(l)$$
$$N_2H_4(l) + O_2(g) \longrightarrow N_2(g) + 2H_2O(l)$$
$$\Delta H° = -622.2 \text{ kJ}$$
$$H_2(g) + \frac{1}{2}O_2(g) \longrightarrow H_2O(l) \qquad \Delta H° = -285.8 \text{ kJ}$$
$$H_2(g) + O_2(g) \longrightarrow H_2O_2(l) \qquad \Delta H° = -187.8 \text{ kJ}$$

73. Substitute natural gas (SNG) is a gaseous mixture containing $CH_4(g)$ that can be used as a fuel. One reaction for the production of SNG is

$$4CO(g) + 8H_2(g) \longrightarrow$$
$$3CH_4(g) + CO_2(g) + 2H_2O(l) \quad \Delta H° = ?$$

Use appropriate data from the following list to determine $\Delta H°$ for this SNG reaction.

$$C(graphite) + \frac{1}{2}O_2(g) \longrightarrow CO(g)$$
$$\Delta H° = -110.5 \text{ kJ}$$
$$CO(g) + \frac{1}{2}O_2(g) \longrightarrow CO_2(g) \quad \Delta H° = -283.0 \text{ kJ}$$

$$H_2(g) + \frac{1}{2}O_2(g) \longrightarrow H_2O(l) \quad \Delta H° = -285.8 \text{ kJ}$$

$$C(graphite) + 2H_2(g) \longrightarrow CH_4(g)$$
$$\Delta H° = -74.81 \text{ kJ}$$
$$CH_4(g) + 2O_2(g) \longrightarrow CO_2(g) + 2H_2O(l)$$
$$\Delta H° = -890.3 \text{ kJ}$$

74. CCl_4, an important commercial solvent, is prepared by the reaction of $Cl_2(g)$ with a carbon compound. Determine $\Delta H°$ for the reaction

$$CS_2(l) + 3Cl_2(g) \longrightarrow CCl_4(l) + S_2Cl_2(l).$$

Use appropriate data from the following listing.
$$CS_2(l) + 3O_2(g) \longrightarrow CO_2(g) + 2SO_2(g)$$
$$\Delta H° = -1077 \text{ kJ}$$
$$2S(s) + Cl_2(g) \longrightarrow S_2Cl_2(l) \quad \Delta H° = -58.2 \text{ kJ}$$
$$C(s) + 2Cl_2(g) \longrightarrow CCl_4(l) \quad \Delta H° = -135.4 \text{ kJ}$$
$$S(s) + O_2(g) \longrightarrow SO_2(g) \quad \Delta H° = -296.8 \text{ kJ}$$
$$SO_2(g) + Cl_2(g) \longrightarrow SO_2Cl_2(l) \quad \Delta H° = +97.3 \text{ kJ}$$
$$C(s) + O_2(g) \longrightarrow CO_2(g) \quad \Delta H° = -393.5 \text{ kJ}$$
$$CCl_4(l) + O_2(g) \longrightarrow COCl_2(g) + Cl_2O(g)$$
$$\Delta H° = -5.2 \text{ kJ}$$

75. Use Hess's law and the following data
$$CH_4(g) + 2O_2(g) \longrightarrow CO_2(g) + 2H_2O(g)$$
$$\Delta H° = -802 \text{ kJ}$$
$$CH_4(g) + CO_2(g) \longrightarrow 2CO(g) + 2H_2(g)$$
$$\Delta H° = +247 \text{ kJ}$$
$$CH_4(g) + H_2O(g) \longrightarrow CO(g) + 3H_2(g)$$
$$\Delta H° = +206 \text{ kJ}$$

to determine $\Delta H°$ for the following reaction, an important source of hydrogen gas

$$CH_4(g) + \frac{1}{2}O_2(g) \longrightarrow CO(g) + 2H_2(g)$$

76. The standard heats of combustion ($\Delta H°$) per mole of 1,3-butadiene, $C_4H_6(g)$; butane, $C_4H_{10}(g)$; and $H_2(g)$ are -2540.2, -2877.6, and -285.8 kJ, respectively. Use these data to calculate the heat of hydrogenation of 1,3-butadiene to butane.

$$C_4H_6(g) + 2\,H_2(g) \longrightarrow C_4H_{10}(g) \qquad \Delta H° = ?$$

[*Hint*: Write equations for the combustion reactions. In each combustion, the products are $CO_2(g)$ and $H_2O(l)$.]

77. One glucose molecule, $C_6H_{12}O_6(s)$, is converted to two lactic acid molecules, $CH_3CH(OH)COOH(s)$, during glycolysis. Given the combustion reactions of glucose and lactic acid, determine the standard enthalpy for glycolysis.

$$C_6H_{12}O_6(s) + 6\,O_2(g) \longrightarrow 6\,CO_2(g) + 6\,H_2O(l)$$
$$\Delta H° = -2808\text{ kJ}$$

$$CH_3CH(OH)COOH(s) + 3\,O_2(g) \longrightarrow$$
$$3\,CO_2(g) + 3\,H_2O(l) \quad \Delta H° = -1344\text{ kJ}$$

78. The standard enthalpy of fermentation of glucose to ethanol is

$$C_6H_{12}O_6(s) \rightarrow 2\,CH_3CH_2OH(l) + 2\,CO_2(g)$$
$$\Delta H° = -72\text{ kJ}$$

Use the standard enthalpy of combustion for glucose to calculate the enthalpy of combustion for ethanol.

Standard Enthalpies of Formation

79. Use standard enthalpies of formation from Table 7.2 and equation (7.21) to determine the standard enthalpy changes in the following reactions.
  (a) $C_3H_8(g) + H_2(g) \longrightarrow C_2H_6(g) + CH_4(g)$;
  (b) $2\,H_2S(g) + 3\,O_2(g) \longrightarrow 2\,SO_2(g) + 2\,H_2O(l)$.

80. Use standard enthalpies of formation from Tables 7.2 and 7.3 and equation (7.21) to determine the standard enthalpy change in the following reaction.

$$NH_4^+(aq) + OH^-(aq) \longrightarrow H_2O(l) + NH_3(g).$$

81. Use the information given here, data from Appendix D, and equation (7.21) to calculate the standard enthalpy of formation per mole of $ZnS(s)$.

$$2\,ZnS(s) + 3\,O_2(g) \longrightarrow 2\,ZnO(s) + 2\,SO_2(g)$$
$$\Delta H° = -878.2\text{ kJ}$$

82. Use the data in Figure 7-18 and information from Section 3-7 to establish possible relationships between the molecular structure of the hydrocarbons and their standard enthalpies of formation.

83. Use standard enthalpies of formation from Table 7.2 to determine the enthalpy change at 25 °C for the following reaction.

$$2\,Cl_2(g) + 2\,H_2O(l) \longrightarrow 4\,HCl(g) + O_2(g)$$
$$\Delta H° = ?$$

84. Use data from Appendix D to calculate the standard enthalpy change for the following reaction at 25 °C.

$$Fe_2O_3(s) + 3\,CO(g) \longrightarrow 2\,Fe(s) + 3\,CO_2(g)$$
$$\Delta H° = ?$$

85. Use data from Table 7.2 to determine the standard heat of combustion of $C_2H_5OH(l)$, if reactants and products are maintained at 25 °C and 1 bar.

86. Use data from Table 7.2, together with the fact that $\Delta H° = -3509$ kJ for the complete combustion of one mole of pentane, $C_5H_{12}(l)$, to calculate $\Delta H°$ for the synthesis of 1 mol $C_5H_{12}(l)$ from $CO(g)$ and $H_2(g)$.

$$5\,CO(g) + 11\,H_2(g) \longrightarrow C_5H_{12}(l) + 5\,H_2O(l)$$
$$\Delta H° = ?$$

87. Use data from Table 7.2 and $\Delta H°$ for the following reaction to determine the standard enthalpy of formation of $CCl_4(g)$ at 25 °C and 1 bar.

$$CH_4(g) + 4\,Cl_2(g) \longrightarrow CCl_4(g) + 4\,HCl(g)$$
$$\Delta H° = -397.3\text{ kJ}$$

88. Use data from Table 7.2 and $\Delta H°$ for the following reaction to determine the standard enthalpy of formation of hexane, $C_6H_{14}(l)$, at 25 °C and 1 bar.

$$2\,C_6H_{14}(l) + 19\,O_2(g) \longrightarrow 12\,CO_2(g) + 14\,H_2O(l)$$
$$\Delta H° = -8326\text{ kJ}$$

89. Use data from Table 7.3 and Appendix D to determine the standard enthalpy change in the following reaction.

$$Al^{3+}(aq) + 3\,OH^-(aq) \longrightarrow Al(OH)_3(s) \quad \Delta H° = ?$$

90. Use data from Table 7.3 and Appendix D to determine the standard enthalpy change in the following reaction.

$$Mg(OH)_2(s) + 2\,NH_4^+(aq) \longrightarrow$$
$$Mg^{2+}(aq) + 2\,H_2O(l) + 2\,NH_3(g) \quad \Delta H° = ?$$

91. The decomposition of limestone, $CaCO_3(s)$, into quicklime, $CaO(s)$, and $CO_2(g)$ is carried out in a gas-fired kiln. Use data from Appendix D to determine how much heat is required to decompose 1.35×10^3 kg $CaCO_3(s)$. (Assume that heats of reaction are the same as at 25 °C and 1 bar.)

92. Use data from Table 7.2 to calculate the volume of butane, $C_4H_{10}(g)$, measured at 24.6 °C and 756 mmHg, that must be burned to liberate 5.00×10^4 kJ of heat.

93. Ants release formic acid (HCOOH) when they bite. Use the data in Table 7.2 and the standard enthalpy of combustion for formic acid ($\Delta H° = -255$ kJ/mol) to calculate the standard enthalpy of formation for formic acid.

94. Calculate the enthalpy of combustion for lactic acid by using the data in Table 7.2 and the standard enthalpy of formation for lactic acid [$CH_3CH(OH)COOH(s)$]: $\Delta H_f° = -694.0$ kJ/mol.

Integrative and Advanced Exercises

95. A British thermal unit (Btu) is defined as the quantity of heat required to change the temperature of 1 lb of water by 1 °F. Assume the specific heat of water to be independent of temperature. How much heat is required to raise the temperature of the water in a 40 gal water heater from 48 to 145 °F in **(a)** Btu; **(b)** kcal; **(c)** kJ?

96. A 7.26 kg shot (as used in the sporting event, the shot put) is dropped from the top of a building 168 m high. What is the maximum temperature increase that could occur in the shot? Assume a specific heat of $0.47 \, J \, g^{-1} \, {}^{\circ}C^{-1}$ for the shot. Why would the actual measured temperature increase likely be less than the calculated value?

97. An alternative approach to bomb calorimetry is to establish the heat capacity of the calorimeter, *exclusive* of the water it contains. The heat absorbed by the water and by the rest of the calorimeter must be calculated separately and then added together. A bomb calorimeter assembly containing 983.5 g water is calibrated by the combustion of 1.354 g anthracene. The temperature of the calorimeter rises from 24.87 to 35.63 °C. When 1.053 g citric acid is burned in the same assembly, but with 968.6 g water, the temperature increases from 25.01 to 27.19 °C. The heat of combustion of anthracene, $C_{14}H_{10}(s)$, is $-7067 \, kJ/mol$ $C_{14}H_{10}$. What is the heat of combustion of citric acid, $C_6H_8O_7$, expressed in kJ/mol?

98. The method of Exercise 97 is used in some bomb calorimetry experiments. A 1.148 g sample of benzoic acid is burned in excess $O_2(g)$ in a bomb immersed in 1181 g of water. The temperature of the water rises from 24.96 to 30.25 °C. The heat of combustion of benzoic acid is $-26.42 \, kJ/g$. In a second experiment, a 0.895 g powdered coal sample is burned in the same calorimeter assembly. The temperature of 1162 g of water rises from 24.98 to 29.81 °C. How many metric tons (1 metric ton = 1000 kg) of this coal would have to be burned to release $2.15 \times 10^9 \, kJ$ of heat?

99. A handbook lists two different values for the heat of combustion of hydrogen: $33.88 \, kcal/g \, H_2$ if $H_2O(l)$ is formed, and $28.67 \, kcal/g \, H_2$ if $H_2O(g)$ is formed. Explain why these two values are different, and indicate what property this difference represents. Devise a means of verifying your conclusions.

100. Determine the missing values of $\Delta H°$ in the diagram shown below.

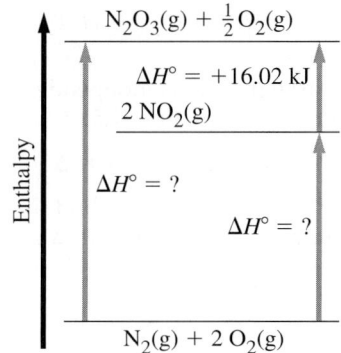

101. A particular natural gas consists, in mole percents, of 83.0% CH_4, 11.2% C_2H_6, and 5.8% C_3H_8. A 385 L sample of this gas, measured at 22.6 °C and 739 mmHg, is burned at constant pressure in an excess of oxygen gas. How much heat, in kilojoules, is evolved in the combustion reaction?

102. An overall reaction for a coal gasification process is

$$2 \, C(\text{graphite}) + 2 \, H_2O(g) \longrightarrow CH_4(g) + CO_2(g)$$

Show that this overall equation can be established by an appropriate combination of equations from Section 7-9.

103. Which of the following gases has the greater fuel value on a per liter (STP) basis? That is, which has the greater heat of combustion? [*Hint:* The only combustible gases are CH_4, C_3H_8, CO, and H_2.]
(a) coal gas: 49.7% H_2, 29.9% CH_4, 8.2% N_2, 6.9% CO, 3.1% C_3H_8, 1.7% CO_2, and 0.5% O_2, by volume.
(b) sewage gas, 66.0% CH_4, 30.0% CO_2, and 4.0% N_2, by volume.

104. A calorimeter that measures an exothermic heat of reaction by the quantity of ice that can be melted is called an ice calorimeter. Now consider that 0.100 L of methane gas, $CH_4(g)$, at 25.0 °C and 744 mmHg is burned at constant pressure in air. The heat liberated is captured and used to melt 9.53 g ice at 0 °C (ΔH_{fusion} of ice = 6.01 kJ/mol).
(a) Write an equation for the complete combustion of CH_4, and show that combustion is incomplete in this case.
(b) Assume that CO(g) is produced in the incomplete combustion of CH_4, and represent the combustion as best you can through a single equation with small whole numbers as coefficients. ($H_2O(l)$ is another product of the combustion.)

105. For the reaction

$$C_2H_4(g) + 3 \, O_2(g) \longrightarrow 2 \, CO_2(g) + 2 \, H_2O(l)$$
$$\Delta H° = -1410.9 \, kJ$$

if the H_2O were obtained as a gas rather than a liquid, **(a)** would the heat of reaction be greater (more negative) or smaller (less negative) than that indicated in the equation? **(b)** Explain your answer. **(c)** Calculate the value of $\Delta H°$ in this case.

106. Some of the butane, $C_4H_{10}(g)$, in a 200.0 L cylinder at 26.0 °C is withdrawn and burned at a constant pressure in an excess of air. As a result, the pressure of the gas in the cylinder falls from 2.35 atm to 1.10 atm. The liberated heat is used to raise the temperature of 132.5 L of water in a heater from 26.0 to 62.2 °C. Assume that the combustion products are $CO_2(g)$ and $H_2O(l)$ exclusively, and determine the efficiency of the water heater. (That is, what percent of the heat of combustion was absorbed by the water?)

107. The metabolism of glucose, $C_6H_{12}O_6$, yields $CO_2(g)$ and $H_2O(l)$ as products. Heat released in the process is converted to useful work with about 70% efficiency. Calculate the mass of glucose metabolized by a 58.0 kg person in climbing a mountain with an elevation gain of 1450 m. Assume that the work performed in the climb is about four times that required to simply lift 58.0 kg by 1450 m. (ΔH_f° of $C_6H_{12}O_6(s)$ is -1273.3 kJ/mol.)

108. An alkane hydrocarbon has the formula C_nH_{2n+2}. The enthalpies of formation of the alkanes decrease (become more negative) as the number of C atoms increases. Starting with butane, $C_4H_{10}(g)$, for each additional CH_2 group in the formula, the enthalpy of formation, ΔH_f°, changes by about -21 kJ/mol. Use this fact and data from Table 7.2 to estimate the heat of combustion of heptane, $C_7H_{16}(l)$.

109. Upon complete combustion, a 1.00 L sample (at STP) of a natural gas gives off 43.6 kJ of heat. If the gas is a mixture of $CH_4(g)$ and $C_2H_6(g)$, what is its percent composition, *by volume*?

110. Under the entry H_2SO_4, a reference source lists many values for the standard enthalpy of formation. For example, for pure $H_2SO_4(l)$, $\Delta H_f^\circ = -814.0$ kJ/mol; for a solution with 1 mol H_2O per mole of H_2SO_4, -841.8; with 10 mol H_2O, -880.5; with 50 mol H_2O; -886.8; with 100 mol H_2O, -887.7; with 500 mol H_2O, -890.5; with 1000 mol H_2O, -892.3; with 10,000 mol H_2O, -900.8; and with 100,000 mol H_2O, -907.3.
   (a) Explain why these values are not all the same.
   (b) The value of $\Delta H_f^\circ[H_2SO_4(aq)]$ in an infinitely dilute solution is -909.3 kJ/mol. What data from this chapter can you cite to confirm this value? Explain.
   (c) If 500.0 mL of 1.00 M $H_2SO_4(aq)$ is prepared from pure $H_2SO_4(l)$, what is the approximate change in temperature that should be observed? Assume that the $H_2SO_4(l)$ and $H_2O(l)$ are at the same temperature initially and that the specific heat of the $H_2SO_4(aq)$ is about $4.2\,J\,g^{-1}\,{}^\circ C^{-1}$.

111. Refer to the discussion of the gasification of coal (page 278), and show that some of the heat required in the gasification reaction (equation 7.24) can be supplied by the *methanation* reaction. This fact contributes to the success of modern processes that produce *synthetic natural gas* (SNG).

112. A 1.103 g sample of a gaseous carbon–hydrogen–oxygen compound that occupies a volume of 582 mL at 765.5 Torr and 25.00 °C is burned in an excess of $O_2(g)$ in a bomb calorimeter. The products of the combustion are 2.108 g $CO_2(g)$, 1.294 g $H_2O(l)$, and enough heat to raise the temperature of the calorimeter assembly from 25.00 to 31.94 °C. The heat capacity of the calorimeter is 5.015 kJ/°C. Write an equation for the combustion reaction, and indicate ΔH° for this reaction at 25.00 °C.

113. Several factors are involved in determining the cooking times required for foods in a microwave oven. One of these factors is specific heat. Determine the approximate time required to warm 250 mL of chicken broth from 4 °C (a typical refrigerator temperature) to 50 °C in a 700 W microwave oven. Assume that the density of chicken broth is about 1 g/mL and that its specific heat is approximately $4.2\,J\,g^{-1}\,{}^\circ C^{-1}$.

114. Suppose you have a setup similar to the one depicted in Figure 7-8 except that there are two different weights rather than two equal weights. One weight is a steel cylinder 10.00 cm in diameter and 25 cm long, the other weight produces a pressure of 745 Torr. The temperature of the gas in the cylinder in which the expansion takes place is 25.0 °C. The piston restraining the gas has a diameter of 12.00 cm, and the height of the piston above the base of the gas expansion cylinder is 8.10 cm. The density of the steel is 7.75 g/cm³. How much work is done when the steel cylinder is suddenly removed from the piston?

115. When one mole of sodium carbonate decahydrate (washing soda) is gently warmed, 155.3 kJ of heat is absorbed, water vapor is formed, and sodium carbonate heptahydrate remains. On more vigorous heating, the heptahydrate absorbs 320.1 kJ of heat and loses more water vapor to give the monohydrate. Continued heating gives the anhydrous salt (soda ash) while 57.3 kJ of heat is absorbed. Calculate ΔH for the conversion of one mole of washing soda into soda ash. Estimate ΔU for this process. Why is the value of ΔU only an estimate?

116. The oxidation of $NH_3(g)$ to $NO(g)$ in the Ostwald process must be very carefully controlled in terms of temperature, pressure, and contact time with the catalyst. This is because the oxidation of $NH_3(g)$ can yield any one of the products $N_2(g)$, $N_2O(g)$, $NO(g)$, and $NO_2(g)$, depending on conditions. Show that oxidation of $NH_3(g)$ to $N_2(g)$ is the most exothermic of the four possible reactions.

117. In the Are You Wondering 7-1 box, the temperature variation of enthalpy is discussed, and the equation
q_P = heat capacity × temperature change = $C_P \times \Delta T$
was introduced to show how enthalpy changes with temperature for a constant-pressure process. Strictly speaking, the heat capacity of a substance at constant pressure is the slope of the line representing the variation of enthalpy (H) with temperature, that is

$$C_P = \frac{dH}{dT} \quad \text{(at constant pressure)}$$

where C_P is the heat capacity of the substance in question. Heat capacity is an extensive quantity and heat capacities are usually quoted as molar heat capacities $C_{P,m}$, the heat capacity of one mole of substance; an intensive property. The heat capacity at constant pressure is used to estimate the change in enthalpy due to a change in temperature. For infinitesimal changes in temperature,

$$dH = C_P dT \quad \text{(at constant pressure)}$$

To evaluate the change in enthalpy for a particular temperature change, from T_1 to T_2, we write

$$\int_{H(T_1)}^{H(T_2)} dH = H(T_2) - H(T_1) = \int_{T_1}^{T_2} C_P dT$$

If we assume that C_P is independent of temperature, then we recover equation (7.5)

$$\Delta H = C_P \times \Delta T$$

On the other hand, we often find that the heat capacity is a function of temperature; a convenient empirical expression is

$$C_{P,m} = a + bT + \frac{c}{T^2}$$

What is the change in molar enthalpy of N_2 when it is heated from 25.0 °C to 100.0 °C? The molar heat capacity of nitrogen is given by

$$C_{P,m} = 28.58 + 3.77 \times 10^{-3}\,T - \frac{0.5 \times 10^5}{T^2}\,\text{J K}^{-1}\,\text{mol}^{-1}$$

118. How much heat is required to convert 10.0 g of ice at −5.0 °C to steam at 100.0 °C? The temperature-dependent constant-pressure specific heat of ice is $C_p(T)/(\text{kJ kg}^{-1}\,\text{K}^{-1}) = 1.0187T - 1.49 \times 10^{-2}$. The temperature-dependent constant-pressure specific heat for water is $C_p(T)/(\text{kJ kg}^{-1}\,\text{K}^{-1}) = -1.0 \times 10^{-7}T^3 + 1.0 \times 10^{-4}T^2 - 3.92 \times 10^{-2}T + 8.7854$.

119. The standard enthalpy of formation of gaseous H_2O at 298.15 K is −241.82 kJ mol^{-1}. Using the ideas contained in Figure 7-16, estimate its value at 100.0 °C given the following values of the molar heat capacities at constant pressure: $H_2O(g)$: 33.58 J K^{-1} mol^{-1}; $H_2(g)$: 28.84 J K^{-1} mol^{-1}; $O_2(g)$: 29.37 J K^{-1} mol^{-1}. Assume the heat capacities are independent of temperature.

Feature Problems

120. James Joule published his definitive work related to the first law of thermodynamics in 1850. He stated that "the quantity of heat capable of increasing the temperature of one pound of water by 1 °F requires for its evolution the expenditure of a mechanical force represented by the fall of 772 lb through the space of one foot." Validate this statement by relating it to information given in this text.

121. Based on specific heat measurements, Pierre Dulong and Alexis Petit proposed in 1818 that the specific heat of an element is inversely related to its atomic weight (atomic mass). Thus, by measuring the specific heat of a new element, its atomic weight could be readily established.
    (a) Use data from Table 7.1 and inside the front cover to plot a *straight-line* graph relating atomic mass and specific heat. Write the equation for this straight line.
    (b) Use the measured specific heat of 0.23 J g^{-1} °C^{-1} and the equation derived in part (a) to obtain an approximate value of the atomic mass of cadmium, an element discovered in 1817.
    (c) To raise the temperature of 75.0 g of a particular metal by 15 °C requires 450 J of heat. What might this metal be?

122. We can use the heat liberated by a neutralization reaction as a means of establishing the stoichiometry of the reaction. The data in the table are for the reaction of 1.00 M NaOH with 1.00 M citric acid, $C_6H_8O_7$, in a total solution volume of 60.0 mL.

mL 1.00 M NaOH Used	mL 1.00 M Citric Acid Used	ΔT, °C
20.0	40.0	4.7
30.0	30.0	6.3
40.0	20.0	8.2
50.0	10.0	6.7
55.0	5.0	2.7

    (a) Plot ΔT versus mL 1.00 M NaOH, and identify the exact stoichiometric proportions of NaOH and citric acid at the equivalence point of the neutralization reaction.
    (b) Why is the temperature change in the neutralization greatest when the reactants are in their exact stoichiometric proportions? That is, why not use an excess of one of the reactants to ensure that the neutralization has gone to completion to achieve the maximum temperature increase?
    (c) Rewrite the formula of citric acid to reflect more precisely its acidic properties. Then write a balanced net ionic equation for the neutralization reaction.

123. In a student experiment to confirm Hess's law, the reaction

$$NH_3(\text{concd aq}) + HCl(aq) \longrightarrow NH_4Cl(aq)$$

was carried out in two different ways. First, 8.00 mL of concentrated $NH_3(aq)$ was added to 100.0 mL of 1.00 M HCl in a calorimeter. [The $NH_3(aq)$ was slightly in excess.] The reactants were initially at 23.8 °C, and the final temperature after neutralization was 35.8 °C. In the second experiment, air was bubbled through 100.0 mL of concentrated $NH_3(aq)$, sweeping out $NH_3(g)$ (see sketch). The $NH_3(g)$ was neutralized in 100.0 mL of 1.00 M HCl. The temperature of the concentrated $NH_3(aq)$ fell from 19.3 to 13.2 °C. At the same time, the temperature of the 1.00 M HCl rose from 23.8 to 42.9 °C as it was neutralized by $NH_3(g)$. Assume that all solutions have densities of 1.00 g/mL and specific heats of 4.18 J g^{-1} °C^{-1}.
    (a) Write the two equations and ΔH values for the processes occurring in the second experiment. Show that the sum of these two equations is the same as the equation for the reaction in the first experiment.
    (b) Show that, within the limits of experimental error, ΔH for the overall reaction is the same in the two experiments, thereby confirming Hess's law.

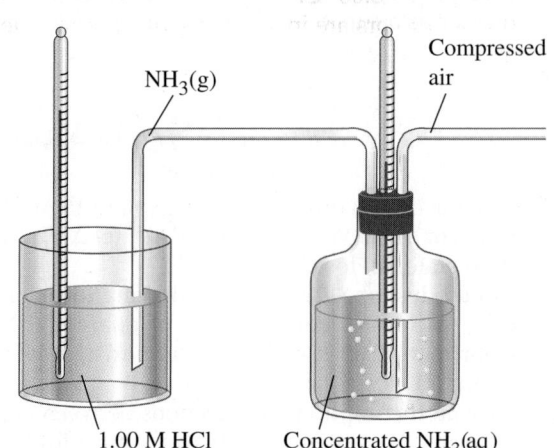

$NH_3(g)$ Compressed air

1.00 M HCl   Concentrated $NH_3(aq)$

124. When an ideal gas is heated, the change in internal energy is limited to increasing the average translational kinetic energy of the gas molecules. Thus, there is a simple relationship between ΔU of the gas and the change in temperature that occurs. Derive this relationship with the help of ideas about the kinetic-molecular theory of gases developed in Chapter 6. After doing so, obtain numerical values (in $J\,mol^{-1}\,K^{-1}$) for the following molar heat capacities.
    (a) The heat capacity, C_V, for one mole of gas under constant-volume conditions
    (b) The heat capacity, C_P, for one mole of gas under constant-pressure conditions

125. Refer to Example 7-5 dealing with the work done by 0.100 mol He at 298 K in expanding in a single step from 2.40 to 1.20 atm. Review also the two-step expansion (2.40 atm $\longrightarrow$ 1.80 atm $\longrightarrow$ 1.20 atm) described on page 257 (see Figure 7-11).
    (a) Determine the total work that would be done if the He expanded in a series of steps, at 0.10 atm intervals, from 2.40 to 1.20 atm.

(b) Represent this total work on the graph below, in which the quantity of work done in the two-step expansion is represented by the sum of the colored rectangles.
(c) Show that the maximum amount of work would occur if the expansion occurred in an infinite number of steps. To do this, express each infinitesimal quantity of work as $dw = P\,dV$ and use the methods of integral calculus (integration) to sum these quantities. Assume ideal behavior for the gas.
(d) Imagine reversing the process, that is, compressing the He from 1.20 to 2.40 atm. What are the maximum and minimum amounts of work required to produce this compression? Explain.
(e) In the isothermal compression described in part (d), what is the change in internal energy assuming ideal gas behavior? What is the value of q?
(f) Using the formula for the work derived in part (c), obtain an expression for q/T. Is this new function a state function? Explain.

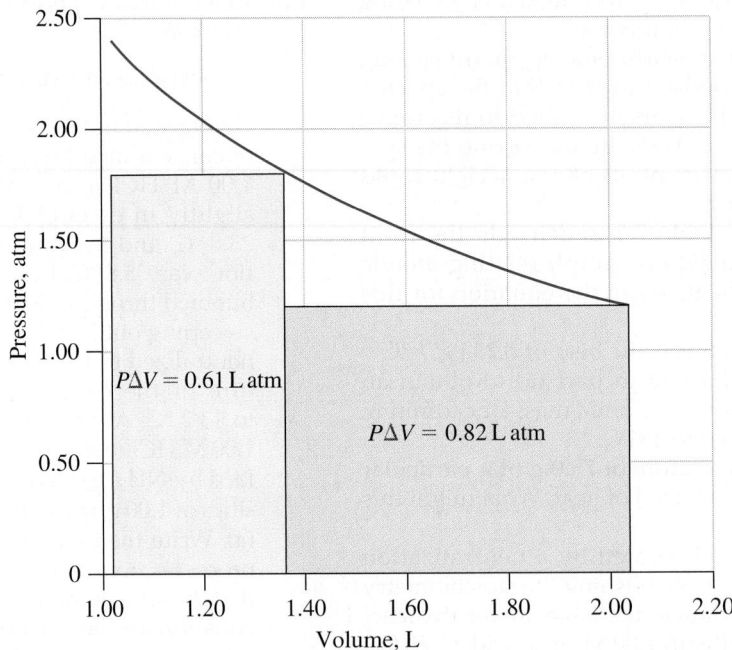

126. Look up the specific heat of several elements, and plot the products of the specific heats and atomic masses as a function of the atomic masses. Based on the plot, develop a hypothesis to explain the data. How could you test your hypothesis?

Self-Assessment Exercises

127. In your own words, define or explain the following terms or symbols: (a) ΔH; (b) $P\Delta V$; (c) ΔH_f°; (d) standard state; (e) fossil fuel.

128. Briefly describe each of the following ideas or methods: (a) law of conservation of energy; (b) bomb calorimetry; (c) function of state; (d) enthalpy diagram; (e) Hess's law.

129. Explain the important distinctions between each pair of terms: (a) system and surroundings; (b) heat and work; (c) specific heat and heat capacity; (d) endothermic and exothermic; (e) constant-volume process and constant-pressure process.

130. The temperature increase of 225 mL of water at 25 °C contained in a Styrofoam cup is noted when a 125 g sample of a metal at 75 °C is added. With reference to Table 7.1, the greatest temperature increase will be noted if the metal is (a) lead; (b) aluminum; (c) iron; (d) copper.

131. A plausible final temperature when 75.0 mL of water at 80.0 °C is added to 100.0 mL of water at 20 °C is **(a)** 28 °C; **(b)** 40 °C; **(c)** 46 °C; **(d)** 50 °C.

132. $\Delta U = 100$ J for a system that gives off 100 J of heat and **(a)** does no work; **(b)** does 200 J of work; **(c)** has 100 J of work done on it; **(d)** has 200 J of work done on it.

133. The heat of solution of NaOH(s) in water is -41.6 kJ/mol NaOH. When NaOH(s) is dissolved in water the solution temperature **(a)** increases; **(b)** decreases; **(c)** remains constant; **(d)** either increases or decreases, depending on how much NaOH is dissolved.

134. The standard molar enthalpy of formation of $CO_2(g)$ is equal to **(a)** 0; **(b)** the standard molar heat of combustion of graphite; **(c)** the sum of the standard molar enthalpies of formation of $CO(g)$ and $O_2(g)$; **(d)** the standard molar heat of combustion of $CO(g)$.

135. Which two of the following statements are false? **(a)** $q_V = q_P$ for the reaction $N_2(g) + O_2(g) \longrightarrow 2\,NO(g)$; **(b)** $\Delta H > 0$ for an endothermic reaction; **(c)** By convention, the most stable form of an element must always be chosen as the reference form and assigned the value $\Delta H_f^\circ = 0$; **(d)** ΔU and ΔH for a reaction can never have the same value; **(e)** $\Delta H < 0$ for the neutralization of a strong acid by a strong base.

136. A 1.22 kg piece of iron at 126.5 °C is dropped into 981 g water at 22.1 °C. The temperature rises to 34.4 °C. What will be the final temperature if this same piece of iron at 99.8 °C is dropped into 325 mL of glycerol, $HOCH_2CH(OH)CH_2OH(l)$ at 26.2 °C? For glycerol, $d = 1.26$ g/mL; $C_p = 219$ J K^{-1} mol^{-1}.

137. Write the balanced chemical equations for reactions that have the following as their standard enthalpy changes.
(a) $\Delta H_f^\circ = +82.05$ kJ/mol $N_2O(g)$
(b) $\Delta H_f^\circ = -394.1$ kJ/mol $SO_2Cl_2(l)$
(c) $\Delta H_{comb}^\circ = -1527$ kJ/mol $CH_3CH_2COOH(l)$

138. The standard molar heats of combustion of C(graphite) and CO(g) are -393.5 and -283 kJ/mol, respectively. Use those data and that for the following reaction

$$CO(g) + Cl_2(g) \longrightarrow COCl_2(g) \qquad \Delta H^\circ = -108 \text{ kJ}$$

to calculate the standard molar enthalpy of formation of $COCl_2(g)$.

139. Can a chemical compound have a standard enthalpy of formation of zero? If so, how likely is this to occur? Explain.

140. Is it possible for a chemical reaction to have $\Delta U < 0$ and $\Delta H > 0$? Explain.

141. Use principles from this chapter to explain the observation that professional chefs prefer to cook with a gas stove rather than an electric stove.

142. Hot water and a piece of cold metal come into contact in an isolated container. When the final temperature of the metal and water are identical, is the total energy change in this process **(a)** zero; **(b)** negative; **(c)** positive; **(d)** not enough information.

143. A clay pot containing water at 25 °C is placed in the shade on a day in which the temperature is 30 °C. The outside of the clay pot is kept moist. Will the temperature of the water inside the clay pot **(a)** increase; **(b)** decrease; **(c)** remain the same?

144. Construct a concept map encompassing the ideas behind the first law of thermodynamics.

145. Construct a concept map to show the use of enthalpy for chemical reactions.

146. Construct a concept map to show the interrelationships between path-dependent and path-independent quantities in thermodynamics.

8

Electrons in Atoms

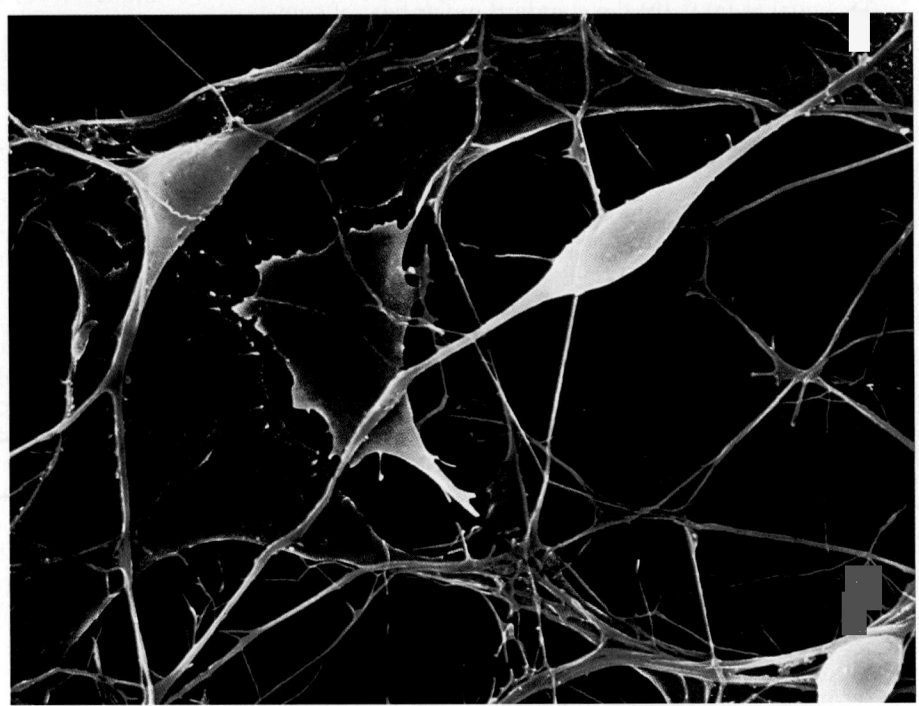

This image of two neurons (gray objects) is produced by an electron microscope that relies on the wave properties of electrons discussed in this chapter.

At the end of the nineteenth century, some observers of the scientific scene believed that it was nearly time to close the books on the field of physics. They thought that with the accumulated knowledge of the previous two or three centuries, the main work left to be done was to apply this body of physics—classical physics—to such fields as chemistry and biology.

Only a few fundamental problems remained, including an explanation of certain details of light emission and a phenomenon known as the photo-electric effect. But the solution to these problems, rather than marking an end in the study of physics, spelled the beginning of a new golden age of physics. These problems were solved through a bold new proposal—the quantum theory—a scientific breakthrough of epic proportions. In this chapter, we will see that to explain phenomena at the atomic and molecular level, classical physics is inadequate—only the quantum theory will do.

The aspect of quantum mechanics emphasized in this chapter is how electrons are described through features known as quantum numbers and electron orbitals. The model of atomic structure developed here will explain many of the topics discussed in the next several chapters: periodic trends in the physical and chemical properties of the elements, chemical bonding, and intermolecular forces.

8-1   Electromagnetic Radiation

Our understanding of the electronic structures of atoms will be gained by studying the interactions of electromagnetic radiation and matter. The chapter begins with background information about electromagnetic radiation, and then turns to connections between electromagnetic radiation and atomic structure. The best approach to learning material in this chapter is to concentrate on the basic ideas relating to atomic structure, many of which are illustrated through the in-text examples. At the same time, pursue further details of interest in some of the Are You Wondering features and portions of Sections 8-6, 8-8, and 8-10.

Electromagnetic radiation is a form of energy transmission in which electric and magnetic fields are propagated as waves through empty space (a vacuum) or through a medium, such as glass. A **wave** is a disturbance that transmits energy through space or a material medium. Anyone who has sat in a small boat on a large body of water has experienced wave motion. The wave moves across the surface of the water, and the disturbance alternately lifts the boat and allows it to drop. Although water waves may be more familiar, let us use a simpler example to illustrate some important ideas and terminology about waves—a traveling wave in a rope.

Imagine tying one end of a long rope to a post and holding the other end in your hand (Fig. 8-1). Imagine also that you have marked one small segment of the rope with red ink. As you move your hand up and down, you set up a wave motion in the rope. The wave travels along the rope toward the distant post, but the colored segment simply moves up and down. In relation to the center line (the broken line in Figure 8-1), the wave consists of *crests*, or high points, where the rope is at its greatest height above the center line, and *troughs*, or low points, where the rope is at its greatest depth below the center line. The maximum height of the wave above the center line or the maximum depth below is called the **amplitude**. The distance between the tops of two successive crests (or the bottoms of two troughs) is called the **wavelength**, designated by the Greek letter lambda, λ.

Wavelength is one important characteristic of a wave. Another feature, **frequency**, designated by the Greek letter nu, ν, is the number of crests or troughs that pass through a given point per unit of time. Frequency has the unit, time^{-1}, usually s^{-1} (per second), meaning the number of events or cycles per second. The product of the length of a wave (λ) and the frequency (ν) shows how far the wave front travels in a unit of time. This is the speed of the wave. Thus, if the wavelength in Figure 8-1 were 0.5 m and the frequency, $3\,\text{s}^{-1}$ (meaning three complete up-and-down hand motions per second), the speed of the wave would be $0.5\,\text{m} \times 3\,\text{s}^{-1} = 1.5\,\text{m/s}$.

We cannot actually see an electromagnetic wave as we do the traveling wave in a rope, but we can try to represent it as in Figure 8-2. As the figure shows, the magnetic field component lies in a plane perpendicular to the electric field component. An electric field is the region around an electrically charged particle. The presence of an electric field can be detected by measuring the force on an electrically charged object when it is brought into the field. A magnetic field is found in the region surrounding a magnet. According to a theory proposed by James Clerk Maxwell (1831–1879) in 1865, electromagnetic radiation—a propagation of electric and magnetic fields—is produced by an accelerating electrically charged particle (a charged particle whose velocity changes). Radio waves, for example, are a form of electromagnetic radiation produced by causing oscillations (fluctuations) of the electric current in a specially designed electrical circuit. With visible light, another

◀ Water waves, sound waves, and seismic waves (which produce earthquakes) are unlike electromagnetic radiation. They require a material medium for their transmission.

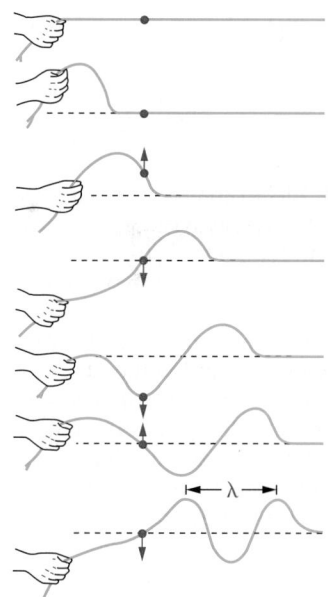

▲ FIGURE 8-1
The simplest wave motion—traveling wave in a rope
As a result of the up-and-down hand motion (top to bottom), waves pass along the long rope from left to right. This one-dimensional moving wave is called a traveling wave. The wavelength of the wave, λ— the distance between two successive crests—is identified.

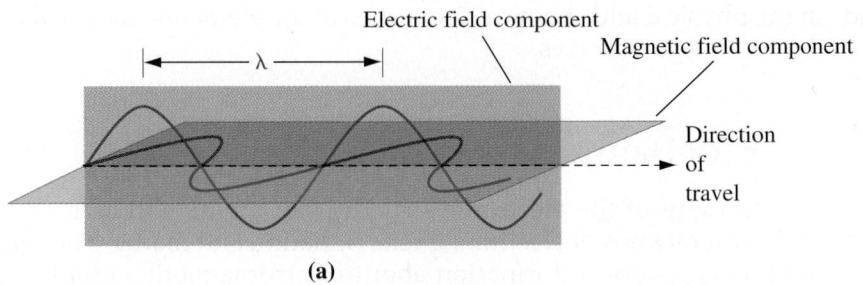

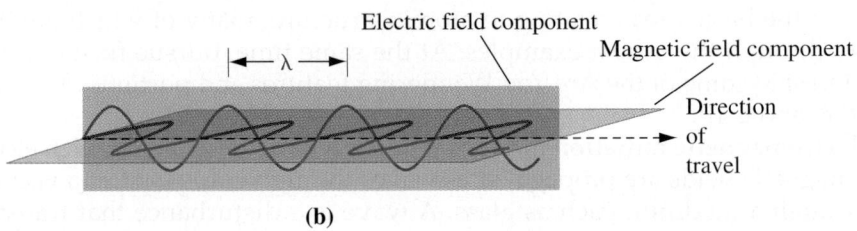

▲ FIGURE 8-2
Electromagnetic waves
This sketch of two different electromagnetic waves shows the propagation of mutually perpendicular oscillating electric and magnetic fields. For a given wave, the wavelengths, frequencies, and amplitudes of the electric and magnetic field components are identical. If these views are of the same instant of time, we would say that **(a)** has the longer wavelength and lower frequency, and **(b)** has the shorter wavelength and higher frequency.

form of electromagnetic radiation, the accelerating charged particles are the electrons in atoms or molecules.

Frequency, Wavelength, and Speed of Electromagnetic Radiation

▶ Electromagnetic waves are *transverse* waves—the electric and magnetic fields are *perpendicular* to the perceived direction of motion. So, to a first approximation, are water waves. Sound waves, by contrast, are *longitudinal*. This effect is the result of small pulses of pressure that move in the *same* direction as the sound travels.

The SI unit for frequency, s^{-1}, is the **hertz (Hz)**, and the basic SI wavelength unit is the meter (m). Because many types of electromagnetic radiation have very short wavelengths, however, smaller units, including those listed below, are also used. The angstrom, named for the Swedish physicist Anders Ångström (1814–1874), is not an SI unit.

$$1 \text{ centimeter (cm)} = 1 \times 10^{-2} \text{ m}$$
$$1 \text{ millimeter (mm)} = 1 \times 10^{-3} \text{ m}$$
$$1 \text{ micrometer } (\mu\text{m}) = 1 \times 10^{-6} \text{ m}$$
$$1 \text{ nanometer (nm)} = 1 \times 10^{-9} \text{ m} = 1 \times 10^{-7} \text{ cm} = 10 \text{ Å}$$
$$1 \text{ angstrom (Å)} = 1 \times 10^{-10} \text{ m} = 1 \times 10^{-8} \text{ cm} = 100 \text{ pm}$$
$$1 \text{ picometer (pm)} = 1 \times 10^{-12} \text{ m} = 1 \times 10^{-10} \text{ cm} = 10^{-2} \text{ Å}$$

▶ The speed of light is commonly rounded off to $3.00 \times 10^{8} \text{ m s}^{-1}$.

A distinctive feature of electromagnetic radiation is its *constant* speed of $2.99792458 \times 10^{8} \text{ m s}^{-1}$ in a vacuum, often referred to as the **speed of light**. The speed of light is represented by the symbol c, and the relationship between this speed and the frequency and wavelength of electromagnetic radiation is

$$c = \nu \times \lambda \qquad (8.1)$$

Figure 8-3 indicates the wide range of possible wavelengths and frequencies for some common types of electromagnetic radiation and illustrates this important fact: The wavelength of electromagnetic radiation is shorter for high frequencies and longer for low frequencies. Example 8-1 illustrates the use of equation (8.1).

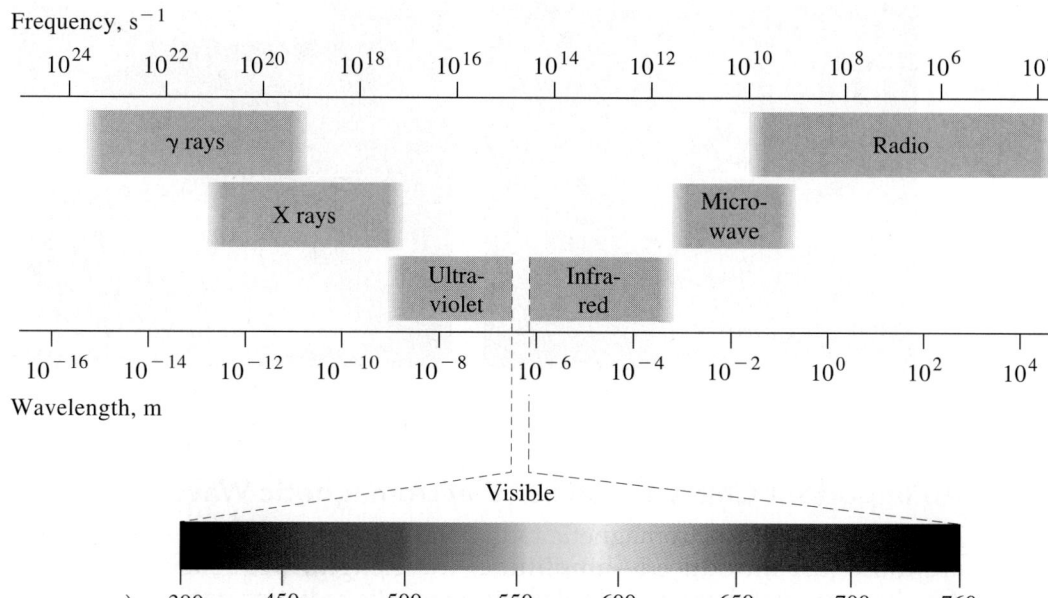

▲ FIGURE 8-3
The electromagnetic spectrum
The visible region, which extends from violet at the shortest wavelength to red at the
longest wavelength, is only a small portion of the entire spectrum. The approximate
wavelength and frequency ranges of some other forms of electromagnetic radiation
are also indicated.

EXAMPLE 8-1   Relating Frequency and Wavelength of Electromagnetic Radiation

Most of the light from a sodium vapor lamp has a wavelength of 589 nm. What is the frequency of this radiation?

Analyze

To use equation (8.1), we first convert the wavelength of the light from nanometers to meters, since the speed of light is in $m\ s^{-1}$. Then, we rearrange it to the form $\nu = c/\lambda$ and solve for ν.

Solve

Change the units of λ from nanometers to meters.

$$\lambda = 589\ \text{nm} \times \frac{1 \times 10^{-9}\ \text{m}}{1\ \text{nm}} = 5.89 \times 10^{-7}\ \text{m}$$

$$c = 2.998 \times 10^{8}\ \text{m}\ \text{s}^{-1}$$

$$\nu = ?$$

Rearrange equation (8.1) to the form $\nu = c/\lambda$, and solve for ν.

$$\nu = \frac{c}{\lambda} = \frac{2.998 \times 10^{8}\ \text{m}\ \text{s}^{-1}}{5.89 \times 10^{-7}\ \text{m}} = 5.09 \times 10^{14}\ \text{s}^{-1} = 5.09 \times 10^{14}\ \text{Hz}$$

Assess

The essential element here is to recognize the need to change the units of λ. This change is often needed when converting wavelength to frequency and vice versa.

PRACTICE EXAMPLE A: The light from red LEDs (light-emitting diodes) is commonly seen in many electronic devices. A typical LED produces 690 nm light. What is the frequency of this light?

PRACTICE EXAMPLE B: An FM radio station broadcasts on a frequency of 91.5 megahertz (MHz). What is the wavelength of these radio waves in meters?

▶ FIGURE 8-4
Examples of interference
(a) Stones and ripples.
(b) CD reflection.

(a)

(b)

An Important Characteristic of Electromagnetic Waves

The properties of electromagnetic radiation that we will use most extensively are those just introduced—amplitude, wavelength, frequency, and speed. Another essential characteristic of electromagnetic radiation, which will underpin our discussion of atomic structure later in the chapter, is described next.

▶ The *wave nature of light* is demonstrated by its ability to be dispersed by diffraction and refraction.

If two pebbles are dropped close together into a pond, ripples (waves) emerge from the points of impact of the two stones. The two sets of waves intersect, and there are places where the waves disappear and places where the waves persist, creating a crisscross pattern (Fig. 8-4a). Where the waves are "in step" upon meeting, their crests coincide, as do their troughs. The waves combine to produce the highest crests and deepest troughs in the water. The waves are said to be *in phase*, and the addition of the waves is called *constructive interference* (Fig. 8-5a). Where the waves meet in such a way that the peak of one wave occurs at the trough of another, the waves cancel and the water is flat (Fig. 8-5b). These out-of-step waves are said to be *out of phase*, and the cancellation of the waves is called *destructive interference*.

KEEP IN MIND

that destructive interference occurs when waves are out of phase by one-half wavelength. If waves are out of phase by more or less than this, but also not completely in phase, then only partial destructive interference occurs.

An everyday illustration of interference involving electromagnetic waves is seen in the rainbow of colors that shine from the surface of a compact disc (Fig. 8-4b). White light, such as sunlight, contains all the colors of the rainbow. The colors differ in wavelength (and frequency), and when these different wavelength components are reflected off the tightly spaced grooves of the CD,

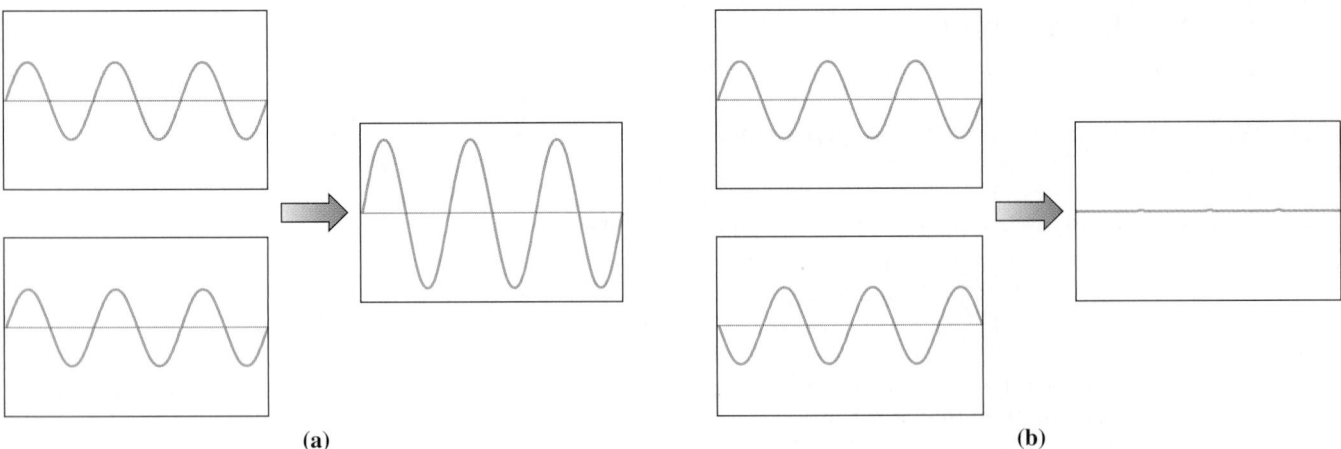

(a)

(b)

▲ FIGURE 8-5
Interference in two overlapping light waves
(a) In constructive interference, the troughs and crests are in step (in phase), leading to addition of the two waves. (b) In destructive interference, the troughs and crests are out of step (out of phase), leading to cancellation of the two waves.

8-1   ARE YOU WONDERING...

What happens to the energy of an electromagnetic wave when interference takes place?

As noted in Figure 8-2, an electromagnetic wave is made up of oscillating electric (E) and magnetic (B) fields. The magnitudes of E and B continuously oscillate between positive and negative. These oscillating fields create an oscillating electromagnetic force. A charged particle will interact with the electromagnetic force and will oscillate back and forth with a constantly changing velocity. The changing velocity gives the particle a changing kinetic energy that is proportional to the square of the velocity (that is, $e_k = \frac{1}{2}mu^2$). Consequently the energy of a wave depends not on the values of E and B alone, but on the sum of their squares, that is, on $E^2 + B^2$. The energy is also related to the *intensity* (I) of a wave, a quantity which, in turn, is related to the *square* of the wave amplitude.

Suppose we let the amplitude of the waves = 1. Then for each wave, whether a pair of waves is in phase or out of phase, the energy is proportional to $1^2 + 1^2 = 2$. The average energy of the pair of waves is also proportional to two [that is, $(2 + 2)/2$]. In constructive interference, the amplitudes become two, so that the energy is proportional to four. In destructive interference, the amplitude is zero and the energy is zero. Note, however, that the *average* between the two situations is still two [that is, $(4 + 0)/2$] so that energy is conserved, as it must be.

they travel slightly different distances. This creates phase differences that depend on the angle at which we hold the CD to the light source. The light waves in the beam interfere with each other, and, for a given angle between the incoming and reflected light, all colors cancel except one. Light waves of that color interfere constructively and reinforce one another. Thus, as we change the angle of the CD to the light source, we see different colors. The dispersion of different wavelength components of a light beam through the interference produced by reflection from a grooved surface is called **diffraction**.

Diffraction is a phenomenon that can be explained only as a property of waves. Both the physical picture and mathematics of interference and diffraction are the same for water waves and electromagnetic waves.

The Visible Spectrum

The speed of light is lower in any medium than it is in a vacuum. Also, the speed is different in different media. As a consequence, light is refracted, or bent, when it passes from one medium to another (Fig. 8-6). Moreover, although electromagnetic waves all have the same speed in a vacuum, waves of different wavelengths have slightly different speeds in air and other media. Thus, when a beam of white light is passed through a transparent medium, the wavelength components are refracted differently. The light is dispersed into a band of colors, a *spectrum*. In Figure 8-7(a), a beam of white light (for example, sunlight) is dispersed by a glass prism into a continuous band of colors corresponding to all the wavelength components from red to violet. This is the visible spectrum shown in Figure 8-3 and also seen in a rainbow, where the medium that disperses the sunlight is droplets of water (Fig. 8-7b).

▲ FIGURE 8-6
Refraction of light
Light is refracted (bent) as it passes from air into the glass prism, and again as it emerges from the prism into air. This photograph shows that red light is refracted the least and blue light the most. The blue light strikes the prism at such an angle that the beam undergoes an internal reflection before it emerges from the prism.

 8-1   CONCEPT ASSESSMENT

Red laser light is passed through a device called a frequency doubler. What is the approximate color of the light that exits the frequency doubler? How are the wavelengths of the original light and the frequency-doubled light related?

▶ The *importance of light* to chemistry is that light is a form of energy and that by studying light–matter interactions we can detect energy changes in atoms and molecules. Another means of monitoring the energy of a system is through observations of heat transfer. Light can be more closely controlled and thus gives us more detailed information than can be obtained with heat measurements.

(a)   (b)

▲ FIGURE 8-7
The spectrum of "white" light
(a) Dispersion of light through a prism. Red light is refracted the least and violet light the most when "white" light is passed through a glass prism. The other colors of the visible spectrum are found between red and violet. **(b)** Rainbow near a waterfall. Here, water droplets are the dispersion medium.

8-2   Atomic Spectra

The visible spectrum in Figure 8-7 is said to be a *continuous spectrum* because the light being diffracted consists of many wavelength components. If the source of a spectrum produces light having only a relatively small number of wavelength components, then a *discontinuous spectrum* is observed. For example, if the light source is an electric discharge passing through a gas, only certain colors are seen in the spectrum (Fig. 8-8a and b). Or, if the light source is a gas flame into which an ionic compound has been introduced, the flame may acquire a distinctive color indicative of the metal ion present (Fig. 8-8c–e). In each of these cases, the emitted light produces a spectrum consisting of only a limited number of discrete wavelength components, observed as colored lines with dark spaces between them. These discontinuous spectra are called **atomic**, or **line**, **spectra**.

The production of the line spectrum of helium is illustrated in Figure 8-9. The light source is a lamp containing helium gas at a low pressure. When an electric discharge is passed through the lamp, helium atoms absorb energy, which they then emit as light. The light is passed through a narrow slit and then dispersed by a prism. The colored components of the light are detected and recorded on photographic film. Each wavelength component appears as

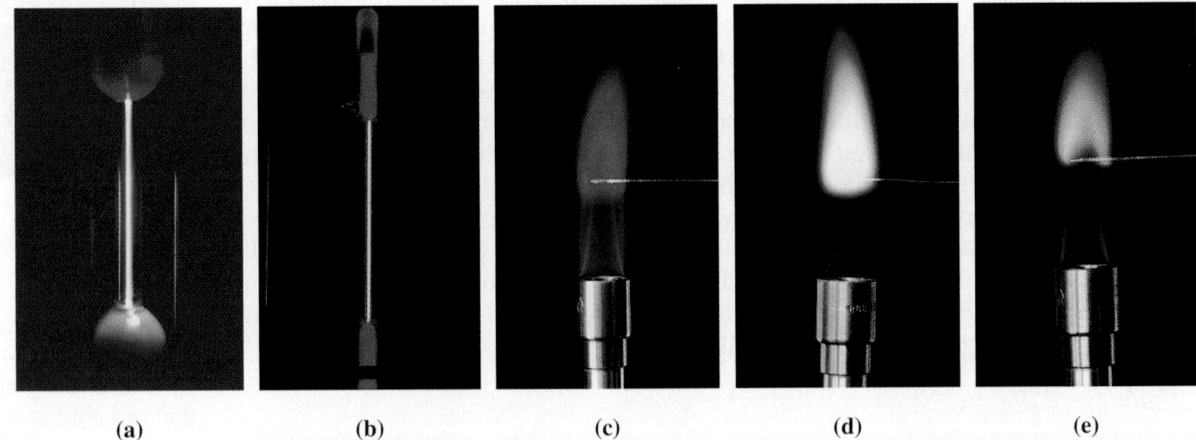

(a)   (b)   (c)   (d)   (e)

▲ FIGURE 8-8
Sources for light emission
Light emitted by an electric discharge through **(a)** hydrogen gas and **(b)** neon gas. Light emitted when compounds of the alkali metals are excited in the gas flames: **(c)** lithium, **(d)** sodium, and **(e)** potassium.

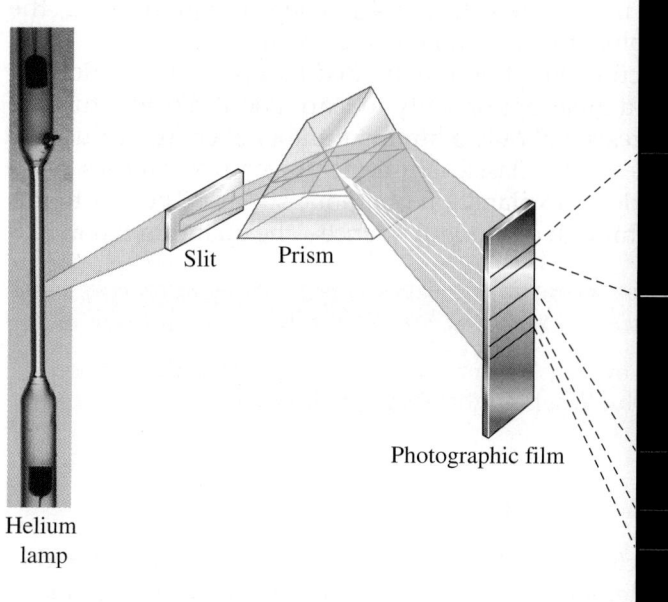

◀ FIGURE 8-9
The atomic, or line, spectrum of helium
The apparatus pictured here, in which the spectral lines are photographed, is called a *spectrograph*. If the observations are made by visual sighting alone, the device is called a *spectroscope*. If the positions and brightness of the lines are measured and recorded by other than visual or photographic means, the term generally used is *spectrometer*.

an image of the slit—a thin line. In all, there are five lines in the spectrum of helium that can be seen with the unaided eye.

Each element has its own distinctive line spectrum—a kind of atomic fingerprint. Robert Bunsen (1811–1899) and Gustav Kirchhoff (1824–1887) developed the first spectroscope and used it to identify elements. In 1860, they discovered a new element and named it cesium (Latin, *caesius*, sky blue) because of the distinctive blue lines in its spectrum. They discovered rubidium in 1861 in a similar way (Latin, *rubidius*, deepest red). Still another element characterized by its unique spectrum is helium (Greek, *helios*, the sun). Its spectrum was observed during the solar eclipse of 1868, but helium was not isolated on Earth for another 27 years.

Among the most extensively studied atomic spectra has been the hydrogen spectrum. Light from a hydrogen lamp appears to the eye as a reddish purple color (Fig. 8-8a). The principal wavelength component of this light is red light of wavelength 656.3 nm. Three other lines appear in the visible spectrum of atomic hydrogen, however: a greenish blue line at 486.1 nm, a violet line at 434.0 nm, and another violet line at 410.1 nm. The visible atomic spectrum of hydrogen is shown in Figure 8-10. In 1885, Johann Balmer, apparently through trial and error, deduced a formula for the wavelengths of these spectral lines. Balmer's equation, rearranged to a form based on frequency, is

$$\nu = 3.2881 \times 10^{15}\,\text{s}^{-1}\left(\frac{1}{2^2} - \frac{1}{n^2}\right) \tag{8.2}$$

In this equation, ν is the frequency of the spectral line, and n must be an *integer* (whole number) *greater than two*. If $n = 3$ is substituted into the equation, the

◀ Bunsen designed a special gas burner for his spectroscopic studies. This burner, the common laboratory Bunsen burner, produces very little background radiation to interfere with spectral observations.

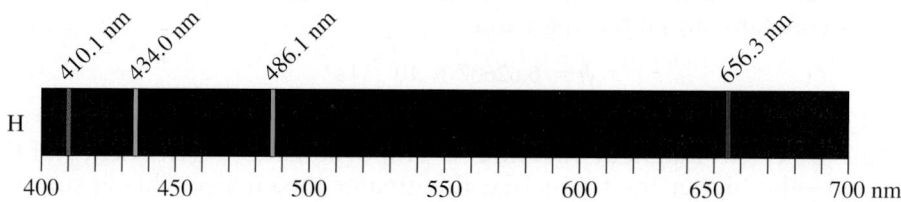

▲ FIGURE 8-10
The Balmer series for hydrogen atoms—a line spectrum
The four lines shown are the only ones visible to the unaided eye. Additional, closely spaced lines lie in the ultraviolet (UV) region.

▲ Light emission by molten iron.

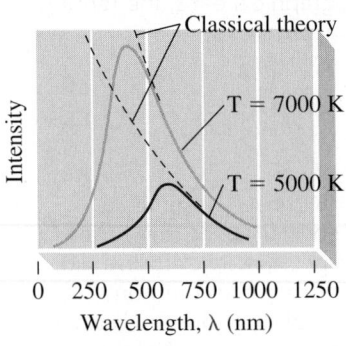

▲ FIGURE 8-11
Spectrum of radiation given off by a heated body
A red-hot object has a spectrum that peaks around 675 nm, whereas a white-hot object has a spectrum that has comparable intensities for all wavelengths in the visible region. The sun has a blackbody temperature of about 5750 K. Objects emit radiation at *all* temperatures, not just at high temperatures. For example, night-vision goggles makes infrared radiation emitted by objects visible in the dark.

▶ Planck's equation can be used to develop relationships among frequency, wavelength, and energy. By using this information, the relative energies of radiation on the electromagnetic spectrum can be compared.

frequency of the red line is obtained. If $n = 4$ is used in equation 8.2, the frequency of the greenish blue line is obtained, and so on.

The fact that atomic spectra consist of only limited numbers of well-defined wavelength lines provides a great opportunity to learn about the structures of atoms. For example, it suggests that only a limited number of energy values are available to excited gaseous atoms. Classical (nineteenth-century) physics, however, was not able to provide an explanation of atomic spectra. The key to this puzzle lay in a great breakthrough of modern science—the quantum theory.

🔍 **8-2   CONCEPT ASSESSMENT**

When comet Schumacher-Levy crashed into Jupiter's surface, scientists viewed the event with spectrographs. What did they hope to discover?

8-3   Quantum Theory

We are aware that hot objects emit light of different colors, from the dull red of an electric-stove heating element to the bright white of a light bulb filament. Light emitted by a hot radiating object can be dispersed by a prism to produce a continuous color spectrum. As seen in Figure 8-11, the light intensity varies smoothly with wavelength, peaking at a wavelength fixed by the source temperature. As with atomic spectra, classical physics could not provide a complete explanation of light emission by heated solids, a process known as *blackbody radiation*. Classical theory predicts that the intensity of the radiation emitted would increase indefinitely, as indicated by the dashed lines in Figure 8-11. In 1900, to explain the fact that the intensity does not increase indefinitely, Max Planck (1858–1947) made a revolutionary proposal: *Energy, like matter, is discontinuous.* Here, then, is the essential difference between the classical physics of Planck's time and the new quantum theory that he proposed: Classical physics places no limitations on the amount of energy a system may possess, whereas quantum theory limits this energy to a discrete set of specific values. The *difference* between any two allowed energies of a system also has a specific value, called a **quantum** of energy. This means that when the energy increases from one allowed value to another, it increases by a tiny jump, or quantum. Here is a way of thinking about a quantum of energy: It bears a similar relationship to the total energy of a system as a single atom does to an entire sample of matter.

The model Planck used for the emission of electromagnetic radiation was that of a group of atoms on the surface of the heated object oscillating together with the same frequency. Planck's assumption was that the group of atoms, the oscillator, must have an energy corresponding to the equation

$$\epsilon = nh\nu$$

where ϵ is the energy, n is a positive integer, ν is the oscillator frequency, and h is a constant that had to be determined by experiment. By using his theory and experimental data for the distribution of frequencies with temperature, Planck established the following value for the constant h. We now call it **Planck's constant**, and it has the value

$$h = 6.62607 \times 10^{-34} \, \text{J s}$$

Planck's postulate can be rephrased in this more general way: The energy of a quantum of electromagnetic radiation is proportional to the frequency of the radiation—the higher the frequency, the greater the energy. This is summarized by what we now call Planck's equation.

$$E = h\nu \tag{8.3}$$

8-2 ARE YOU WONDERING...

How Planck's ideas account for the fact that the intensity of blackbody radiation drops off at higher frequencies?

Planck was aware of the work of Ludwig Boltzmann, who, with James Maxwell, had derived an equation to account for the distribution of molecular speeds. Boltzmann had shown that the relative chance of finding a molecule with a particular speed was related to its energy by the following expression.

$$\text{relative chance} \propto e^{\left(-\frac{\text{kinetic energy}}{k_{\mathrm{B}}T}\right)}$$

where k_{B} is the Boltzmann constant, and T is the Kelvin temperature. You will also notice that the curve of intensity versus wavelength in Figure 8-11 bears a strong resemblance to the distribution of molecular speeds in Figure 6-15. Planck assumed that the energies of the substance oscillating to emit blackbody radiation were distributed according to the Boltzmann distribution law. That is, the relative chance of an oscillator having the energy $nh\nu$ is proportional to $e^{-nh\nu/k_{\mathrm{B}}T}$, where n is an integer, 1, 2, 3, and so on. So this expression shows that the chance of an oscillator having a high frequency is lower than for oscillators having lower frequencies because as n increases, $e^{-nh\nu/k_{\mathrm{B}}T}$, decreases. The assumption that the energy of the oscillators in the light-emitting source cannot have continuous values leads to excellent agreement between theory and experiment.

▲ Max Planck (1858–1947)

At the time Planck made his quantum hypothesis, scientists had had no previous experience with macroscopic physical systems that required the existence of separate energy levels and that energy may only be emitted or absorbed in specific quanta. Their experience was that there were no theoretical limits on the energy of a system and that the transfer of energy was continuous. Thus it is not surprising that scientists, including Planck himself, were initially skeptical of the quantum hypothesis. It had been designed to explain radiation from heated bodies and certainly could not be accepted as a general principle until it had been tested on other applications.

Only after the quantum hypothesis was successfully applied to phenomena other than blackbody radiation did it acquire status as a great new scientific theory. The first of these successes came in 1905 with Albert Einstein's quantum explanation of the photoelectric effect.

The Photoelectric Effect

In 1888, Heinrich Hertz discovered that when light strikes the surface of certain metals, electrons are ejected. This phenomenon is called the **photoelectric effect** and its salient feature is that electron emission only occurs when the frequency of the incident light exceeds a particular threshold value (ν_0). When this condition is met,

- the number of electrons emitted depends on the intensity of the incident light, but
- the kinetic energies of the emitted electrons depend on the frequency of the light.

These observations, especially the dependency on frequency, could not be explained by classical wave theory. However, Albert Einstein showed that they are exactly what would be expected with a particle interpretation of radiation. In 1905, Einstein proposed that electromagnetic radiation has particle-like qualities and that "particles" of light, subsequently called **photons** by G. N. Lewis, have a characteristic energy given by Planck's equation, $E = h\nu$.

In the particle model, a photon of energy $h\nu$ strikes a bound electron, which absorbs the photon energy. If the photon energy, $h\nu$, is greater than the energy

◄ Light–matter interactions usually involve *one photon per atom or electron.* Thus, to escape from a photoelectric surface, an electron must do so with the energy from a single photon collision. The electron cannot accumulate the energy from several hits by photons.

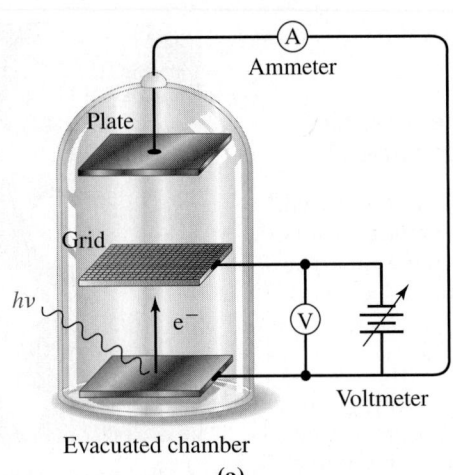

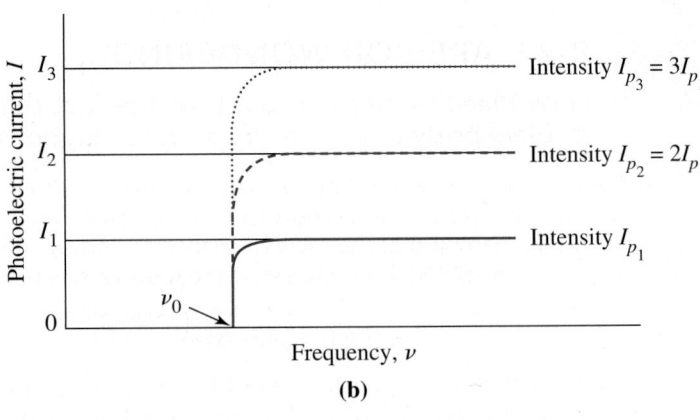

(a) Evacuated chamber

(b)

▲ FIGURE 8-12
The photoelectric effect
(a) Schematic diagram of the apparatus for photoelectric effect measurements.
(b) The photoelectric current, I_p, appears only if the frequency (ν) is greater
than the threshold value (ν_0). For $\nu > \nu_0$, the current (I_p) increases as the
intensity of the light is increased. (c) Stopping voltage of photoelectrons as a
function of frequency of incident radiation. The stopping voltage (V_s) is plotted
against the frequency of the incident radiation. The threshold frequency (ν_0) of
the metal is found by extrapolation.

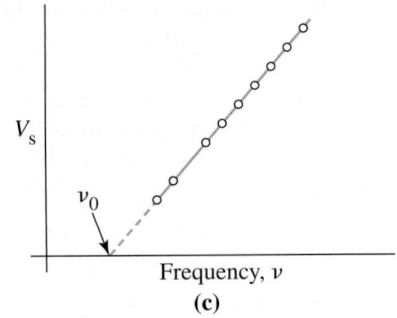

(c)

▶ With the advent of lasers
we have observed the
simultaneous absorption of
two photons by one electron.
Instances of two adjacent
molecules cooperatively
absorbing one photon are
also known. Such occurrences
are exceptions to the more
normal one photon/one
electron phenomena.

binding the electron to the surface (a quantity known as the *work function*), a
photoelectron is liberated. Thus, the lowest frequency light producing the
photoelectric effect is the threshold frequency, and any energy in excess of the
work function appears as kinetic energy in the emitted photoelectrons.

In the discussion that follows, based on the experimental setup shown in
Figure 8-12, we will see how the threshold frequency and work function are
evaluated. Also, we will see that the photoelectric effect provides an independent evaluation of Planck's constant, h.

In Figure 8-12, light (designated $h\nu$) is allowed to shine on a piece of metal in
an evacuated chamber. The electrons emitted by the metal (photoelectrons)
travel to the upper plate and complete an electric circuit set up to measure the
photoelectric current through an ammeter. Figure 8-12(b) illustrates the variation
of the photoelectric current, I_p, detected by the ammeter as the frequency (ν) and
intensity of the incident light is increased. We see that no matter how intense the
light, no current flows if the frequency is below the threshold frequency, ν_0, and
no photoelectric current is produced. In addition no matter how weak the light,
there is a photoelectric current if $\nu > \nu_0$. The magnitude of the photoelectric current is, as shown in Figure 8-12(b), directly proportional to the intensity of the
light, so that the number of photoelectrons increases with the intensity of the incident light. Therefore, we can associate light intensity with the number of photons
arriving at a point per unit time.

A second circuit is set up to measure the velocity of the photoelectrons, and
hence their kinetic energy. In this circuit, a potential difference (voltage) is maintained between the photoelectric metal and an open-grid electrode placed below
the upper plate. For electric current to flow, electrons must pass through the
openings in the grid and onto the upper plate. The negative potential on the grid
acts to slow down the approaching electrons. As the potential difference between
the grid and the emitting metal is increased, a point is reached at which the photoelectrons are stopped at the grid and the current ceases to flow through the
ammeter. The potential difference at this point is called the *stopping voltage, V_s*. At
the stopping voltage, the kinetic energy of the photoelectrons has been converted

to potential energy, expressed through the following equation (in which m, u, and e are the mass, speed, and charge of an electron, respectively).

$$\frac{1}{2} mu^2 = eV_s$$

As a result of experiments of the type just described, we find that V_s is proportional to the frequency of the incident light but independent of the light intensity. Also, as shown in Figure 8-12, if the frequency, ν, is below the *threshold frequency*, ν_0, no photoelectric current is produced. At frequencies greater than ν_0, the empirical equation for the stopping voltage is

$$V_s = k(\nu - \nu_0)$$

The constant k is independent of the metal used, but ν_0 varies from one metal to another. Although there is no relation between V_s and the light intensity, the

◀ Albert Einstein received a Nobel Prize for his work on the photoelectric effect. He is better known for his development of the theory of relativity, and $E = mc^2$.

▲ **Albert Einstein (1879–1955)**

8-3 ARE YOU WONDERING...

How the energy of a photon is manifested?

We begin with an important relationship between the mass and velocity of a particle given by Einstein. Let the mass of a particle when the particle and the measuring device are at rest be denoted as m_0. If we remeasure the mass when the particle moves with a velocity u, we find that its mass increases according to the equation

$$m = \frac{m_0}{\sqrt{1 - u^2/c^2}}$$

where m is the particle mass, referred to as the *relativistic mass*, and c is the speed of light. For particles moving at speeds less than 90% of the speed of light, the relativistic mass (m) is essentially the same as the rest mass (m_0).

We have seen that the kinetic energy of a particle is given by

$$E_K = \frac{1}{2} mu^2$$

However, because photons travel at the speed of light they must have zero rest mass (otherwise their relativistic mass m would become infinite). So where is their energy?

Although photons have zero rest mass, they do possess momentum, which is defined as the relativistic mass times the velocity of the particle, because they are in motion. Einstein's theory of special relativity states that a particle's energy and momentum ($p = mu$, recall page 217) are related by the expression

$$E^2 = (pc)^2 + (m_0c^2)^2$$

where m_0 is the rest mass of the particle. For photons traveling at the speed of light c, the rest mass is zero. Hence

$$E = pc = h\nu$$

$$p = \frac{h\nu}{c} = \frac{h}{\lambda}$$

Photons possess momentum, and it is this momentum that is transferred to an electron in a collision. In all collisions between photons and electrons, momentum is conserved. Thus, we see that the wave and the photon models are intimately connected. The energy of a photon is related to the frequency of the wave by Planck's equation, and the momentum of the photon is related to the wavelength of the wave by the equation just derived! When a photon collides with an electron, it transfers momentum to the electron, which accelerates to a new velocity. The energy of the photon decreases, and, as a consequence, its wavelength increases. This phenomenon, called the Compton effect, was discovered in 1923 and confirmed the particulate nature of light.

photoelectric current, I_p, is proportional to the intensity of the light as illustrated in Figure 8-12b.

The work function is a quantity of work and, hence, of energy. One way to express this quantity is as the product of Planck's constant and the threshold frequency: $E = h\nu_0$. Another way is as the product of the charge on the electron, e, and the potential, V_0, that has to be overcome in the metal: $E = eV_0$. Thus, the threshold frequency for the photoelectric effect is given by the expression

$$\nu_0 = \frac{eV_0}{h}$$

Since the work function (eV_0) is a characteristic of the metal used in the experiment, then ν_0 is also a characteristic of the metal, as confirmed by experiment.

When a photon of energy $h\nu$ strikes an electron, the electron overcomes the work function eV_0 and is liberated with kinetic energy $\left(\dfrac{1}{2}\right) mu^2$. Thus, by the law of conservation of energy, we have

$$\frac{1}{2} mu^2 + eV_0 = h\nu$$

which gives

$$eV_s = \frac{1}{2} mu^2 = h\nu - eV_0$$

which is identical to the empirically determined equation for V_s with $k = h/e$ when $h\nu_0 = eV_0$. Careful experiments showed that the constant h had the same value as determined by Planck for blackbody radiation. The additional fact that the number of photoelectrons increases with the intensity of light indicates that we should associate light intensity with the number of photons arriving at a point per unit time.

8-3 CONCEPT ASSESSMENT

The wavelength of light needed to eject electrons from hydrogen atoms is 91.2 nm. When light of 80.0 nm is shone on a sample of hydrogen atoms, electrons are emitted from the hydrogen gas. If, in a different experiment, the wavelength of the light is changed to 70.0 nm, what is the effect compared to the use of 80.0 nm light? Are more electrons emitted? If not, what happens?

Photons of Light and Chemical Reactions

Chemical reactions that are induced by light are called *photochemical reactions*. Because they are essential to these reactions, we can think of photons as "reactants" and we can designate them in chemical equations by the symbol $h\nu$. The reactions by which ozone molecules, O_3, are produced from oxygen molecules, O_2, are represented below.

$$O_2 + h\nu \longrightarrow O + O$$
$$O_2 + O + M \longrightarrow O_3 + M^*$$

The radiation required in the first reaction is UV radiation with wavelength less than 242.4 nm. O atoms from the first reaction then combine with O_2 to form O_3. In the second reaction, a "third body," M, such as $N_2(g)$, is needed to carry away excess energy to prevent immediate dissociation of O_3 molecules.

Photochemical reactions involving ozone are the subject of Example 8-2. There we see that the product of Planck's constant h and frequency (ν) yields the energy of a single photon of electromagnetic radiation in the unit joule. Invariably, this energy is only a tiny fraction of a joule. Often it is useful to deal with the much larger energy of a mole of photons (6.02214×10^{23} photons).

▶ The reactions shown here describing the formation ozone from oxygen gas are the reactions that occur in the atmosphere to produce ozone.

EXAMPLE 8-2   Using Planck's Equation to Calculate the Energy of Photons of Light

For radiation of wavelength 242.4 nm, the longest wavelength that will bring about the photodissociation of O_2, what is the energy of **(a)** one photon, and **(b)** a mole of photons of this light?

Analyze

To use Planck's equation, we need the frequency of the radiation. We can get this from equation (8.1) after first expressing the wavelength in meters. Planck's equation is written for one photon of light. We emphasize this by including the unit in the value of h. Once we have the energy per photon, we can multiply it by the Avogadro constant to convert to a per-mole basis.

Solve

(a) First, calculate the frequency of the radiation.

$$\nu = \frac{c}{\lambda} = \frac{2.998 \times 10^8 \, \text{m s}^{-1}}{242.4 \times 10^{-9} \, \text{m}} = 1.237 \times 10^{15} \, \text{s}^{-1}$$

Then, calculate the energy of a single photon.

$$E = h\nu = 6.626 \times 10^{-34} \times \frac{\text{J s}}{\text{photon}} \times 1.237 \times 10^{15} \, \text{s}^{-1}$$
$$= 8.196 \times 10^{-19} \, \text{J/photon}$$

(b) Calculate the energy of a mole of photons.

$$E = 8.196 \times 10^{-19} \, \text{J/photon} \times 6.022 \times 10^{23} \, \text{photons/mol}$$
$$= 4.936 \times 10^5 \, \text{J/mol}$$

Assess

We can see from this example that when the energy of a single photon is expressed in SI units, the energy is rather small and perhaps difficult to interpret. However, the amount of energy carried by a *mole* of photons is something we can easily relate to. As shown above, light with a wavelength of 242.4 nm has an energy content of 493.6 kJ/mol, which is similar in magnitude to the internal energy and enthalpy changes of chemical reactions (see Chapter 7).

PRACTICE EXAMPLE A: The protective action of ozone in the atmosphere comes through ozone's absorption of UV radiation in the 230 to 290 nm wavelength range. What is the energy, in kilojoules per mole, associated with radiation in this wavelength range?

PRACTICE EXAMPLE B: Chlorophyll absorbs light at energies of 3.056×10^{-19} J/photon and 4.414×10^{-19} J/photon To what color and frequency do these absorptions correspond?

8-4   The Bohr Atom

The Rutherford model of a nuclear atom (Section 2-3) does not indicate how electrons are arranged outside the nucleus of an atom. According to classical physics, stationary, negatively charged electrons would be pulled into the positively charged nucleus. This suggests that the electrons in an atom must be in motion, like the planets orbiting the sun. However, again according to classical physics, orbiting electrons should be constantly accelerating and should radiate energy. By losing energy, the electrons would be drawn ever closer to the nucleus and soon spiral into it. In 1913, Niels Bohr (1885–1962) resolved this problem by using Planck's quantum hypothesis. In an interesting blend of classical and quantum theory, Bohr postulated that for a hydrogen atom:

1. The electron moves in circular orbits about the nucleus with the motion described by classical physics.

2. The electron has only a fixed set of allowed orbits, called *stationary states*. The allowed orbits are those in which certain properties of the electron have unique values. Even though classical theory would predict otherwise, *as long as an electron remains in a given orbit, its energy is constant and no energy is emitted*. The particular property of the electron having only certain allowed values, leading to only a discrete set of allowed orbits, is called the *angular momentum*. Its possible

▲ **Niels Bohr (1885–1962)**
In addition to his work on the hydrogen atom, Bohr headed the Institute of Theoretical Physics in Copenhagen, which became a mecca for theoretical physicists in the 1920s and 1930s.

▶ FIGURE 8-13
Bohr model of the hydrogen atom
A portion of the hydrogen atom is pictured. The nucleus is at the center, and the electron is found in one of the discrete orbits, $n = 1, 2$, and so on. Excitation of the atom raises the electron to higher-numbered orbits, as shown with black arrows. Light is emitted when the electron falls to a lower-numbered orbit. Two transitions that produce lines in the Balmer series of the hydrogen spectrum are shown in the approximate colors of the spectral lines.

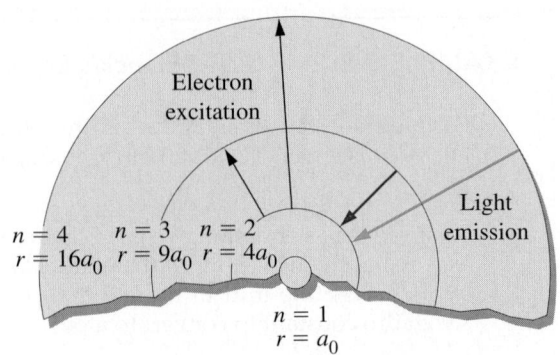

KEEP IN MIND

that momentum (p) is the product of the mass and the velocity of a particle. If the particle undergoes a circular motion, then the particle possesses angular momentum.

values are $nh/2\pi$, where n must be an integer. Thus the quantum numbers progress: $n = 1$ for the first orbit; $n = 2$ for the second orbit; and so on.

3. An electron can pass only from one allowed orbit to another. In such transitions, fixed discrete quantities of energy (quanta) are involved—either absorbed or emitted.

The atomic model of hydrogen based on these ideas is pictured in Figure 8-13. The allowed states for the electron are numbered, $n = 1, n = 2, n = 3$, and so on. These *integral* numbers, which arise from Bohr's assumption that only certain values are allowed for the angular momentum of the electron, are called **quantum numbers**.

The Bohr theory predicts the radii of the allowed orbits in a hydrogen atom.

$$r_n = n^2 a_0, \text{ where } n = 1, 2, 3, \dots \quad \text{and} \quad a_0 = 53 \text{ pm } (0.53 \text{ Å}) \qquad \textbf{(8.4)}$$

▶ According to the Bohr model, lower energy orbits are closer to the nucleus and electrons associated with low energy orbits must absorb more energy to be removed from the atom.

The theory also allows us to calculate the electron velocities in these orbits and, most important, the energy. When the electron is free of the nucleus, by convention, it is said to be at a *zero* of energy. When a free electron is attracted to the nucleus and confined to the orbit n, the electron energy becomes negative, with its value lowered to

▶ Think of a person on a stairway going up steps (excitation) or down steps (emission). The person must stop on a step—in-between levels are not available.

$$E_n = \frac{-R_H}{n^2} \qquad \textbf{(8.5)}$$

R_H is a numerical constant with a value of 2.179×10^{-18} J.

With expression (8.5), we can calculate the energies of the allowed energy states, or *energy levels*, of the hydrogen atom. These levels can be represented schematically as in Figure 8-14. This representation is called an **energy-level diagram**. Example 8-3 shows how Bohr's model can be used to predict whether certain energy levels are possible (allowed) or impossible (not allowed).

Normally, the electron in a hydrogen atom is found in the orbit closest to the nucleus ($n = 1$). This is the lowest allowed energy, or the **ground state**. When the electron gains a quantum of energy, it moves to a higher level ($n = 2, 3$, and so on) and the atom is in an **excited state**. When the electron drops from a higher to a lower numbered orbit, a unique quantity of energy is emitted—the difference in energy between the two levels. Equation (8.5) can be used to derive an expression for the difference in energy between two levels, where n_f is the final level and n_i is the initial one.

▶ Notice the resemblance of this equation to the Balmer equation (8.2). In addition to developing a theory of atomic structure to account for Rutherford's atomic model, Bohr sought a theoretical explanation of the Balmer equation.

$$\Delta E = E_f - E_i = \frac{-R_H}{n_f^2} - \frac{-R_H}{n_i^2} = R_H\left(\frac{1}{n_i^2} - \frac{1}{n_f^2}\right) = 2.179 \times 10^{-18} \text{J}\left(\frac{1}{n_i^2} - \frac{1}{n_f^2}\right) \qquad \textbf{(8.6)}$$

The energy of the photon, E_{photon}, either absorbed or emitted, is equal to the magnitude of this energy difference. Because $E_{photon} = h\nu$ and $E_{photon} = |\Delta E|$, we can write

$$|\Delta E| = E_{photon} = h\nu$$

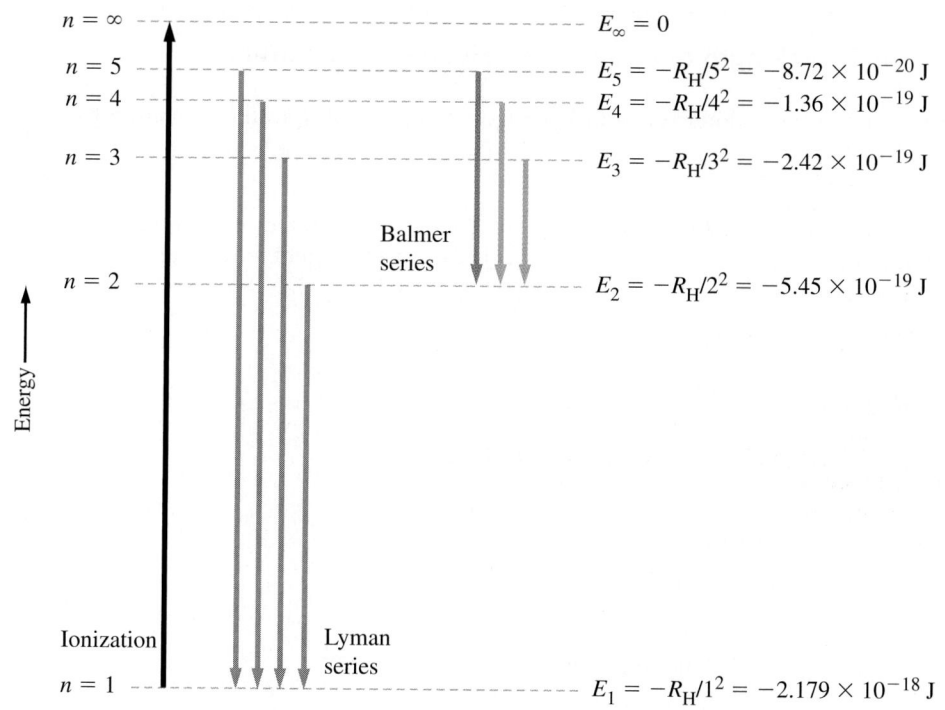

◀ **FIGURE 8-14**
Energy-level diagram for the hydrogen atom
If the electron acquires 2.179×10^{-18} J of energy, it moves to the orbit $n = \infty$; ionization of the H atom occurs (black arrow). Energy emitted when the electron falls from higher-numbered orbits to the orbit $n = 1$ is in the form of ultraviolet light, which produces a spectral series called the Lyman series (gray lines). Electron transitions to the orbit $n = 2$ yield lines in the Balmer series (recall Figure 8-10); three of the lines are shown here (in color). Transitions to $n = 3$ yield spectral lines in the infrared.

EXAMPLE 8-3 Understanding the Meaning of Quantization of Energy

Is it likely that there is an energy level for the hydrogen atom, $E_n = -1.00 \times 10^{-20}$ J?

Analyze

Rearrange equation (8.5) for n^2 and solve for n. If the value of n is an integer, then the given energy corresponds to an energy level for the hydrogen atom.

Solve

Let us rearrange equation (8.5), solve for n^2, and then for n.

$$n^2 = \frac{-R_H}{E_n}$$

$$= \frac{-2.179 \times 10^{-18}\,\text{J}}{-1.00 \times 10^{-20}\,\text{J}} = 2.179 \times 10^2 = 217.9$$

$$n = \sqrt{217.9} = 14.76$$

Because the value of n is not an integer, this is not an allowed energy level for the hydrogen atom.

Assess

Equation (8.5) places a severe restriction on the energies allowed for a hydrogen atom.

PRACTICE EXAMPLE A: Is there an energy level for the hydrogen atom, $E_n = -2.69 \times 10^{-20}$ J?

PRACTICE EXAMPLE B: The energy of an electron in a hydrogen atom is -4.45×10^{-20} J. What level does it occupy?

---

where $|\Delta E|$ represents the *magnitude* of the energy difference between the energy levels involved in the electronic transition.

Example 8-4 uses equation (8.6) as a basis for calculating the lines in the hydrogen emission spectrum. Because the differences between energy levels are limited in number, so too are the energies of the emitted photons. Therefore, only certain wavelengths (or frequencies) are observed for the spectral lines.

EXAMPLE 8-4   **Calculating the Wavelength of a Line in the Hydrogen Spectrum**

Determine the wavelength of the line in the Balmer series of hydrogen corresponding to the transition from $n = 5$ to $n = 2$.

Analyze

This problem is an application of equation (8.6). After the energy difference is calculated, we can obtain the photon frequency by rearranging $|\Delta E| = E_{photon} = h\nu$. Equation (8.1) is then used to get the wavelength.

Solve

The specific data for equation (8.6) are $n_i = 5$ and $n_f = 2$.

$$\Delta E = 2.179 \times 10^{-18}\,\text{J}\left(\frac{1}{5^2} - \frac{1}{2^2}\right)$$

$$= 2.179 \times 10^{-18} \times (0.04000 - 0.25000)$$

$$= -4.576 \times 10^{-19}\,\text{J}$$

Rearranging $E_{photon} = \Delta E = h\nu$ gives the frequency

$$\nu = \frac{E_{photon}}{h} = \frac{4.576 \times 10^{-19}\,\text{J photon}^{-1}}{6.626 \times 10^{-34}\,\text{J s photon}^{-1}} = 6.906 \times 10^{14}\,\text{s}^{-1}$$

Rearranging $c = \lambda\nu$ for the wavelength gives the following result:

$$\lambda = \frac{c}{\nu} = \frac{2.998 \times 10^8\,\text{m s}^{-1}}{6.906 \times 10^{14}\,\text{s}^{-1}} = 4.341 \times 10^{-7}\,\text{m} = 434.1\,\text{nm}$$

Assess

Note the good agreement between this result and the data in Figure 8-10. The color of the spectral line is determined by the energy difference, ΔE, while the intensity is determined by the number of hydrogen atoms undergoing this transition. The greater the number of atoms undergoing the same transition, the greater the number of emitted photons, resulting in greater intensity.

PRACTICE EXAMPLE A: Determine the wavelength of light absorbed in an electron transition from $n = 2$ to $n = 4$ in a hydrogen atom.

PRACTICE EXAMPLE B: Refer to Figure 8-14 and determine which transition produces the longest wavelength line in the Lyman series of the hydrogen spectrum. What is the wavelength of this line in nanometers and in angstroms?

The Bohr Theory and Spectroscopy

As shown in Example 8-4, the Bohr theory provides a model for understanding the emission spectra of atoms. Emission spectra are obtained when the individual atoms in a collection of atoms (roughly 10^{20} of them) are excited to the various possible excited states of the atom. The atoms then relax to states of lower energy by emitting photons of frequency given by

$$\nu_{photon} = \frac{E_i - E_f}{h} \tag{8.7}$$

Thus, the quantization of the energy states of atoms leads to line spectra.

Earlier in this chapter, we learned how the emission spectrum of a sample can be measured by dispersing emitted light through a prism and determining the wavelengths of the individual components of that light. We can conceive of an alternate technique in which we pass electromagnetic radiation, such as white light, through sample of atoms in their ground states and then pass the emerging light through a prism. Now we observe which frequencies of light the atoms *absorb*. This form of spectroscopy is called *absorption spectroscopy*. The two types of spectroscopy—emission and absorption—are illustrated in Figure 8-15.

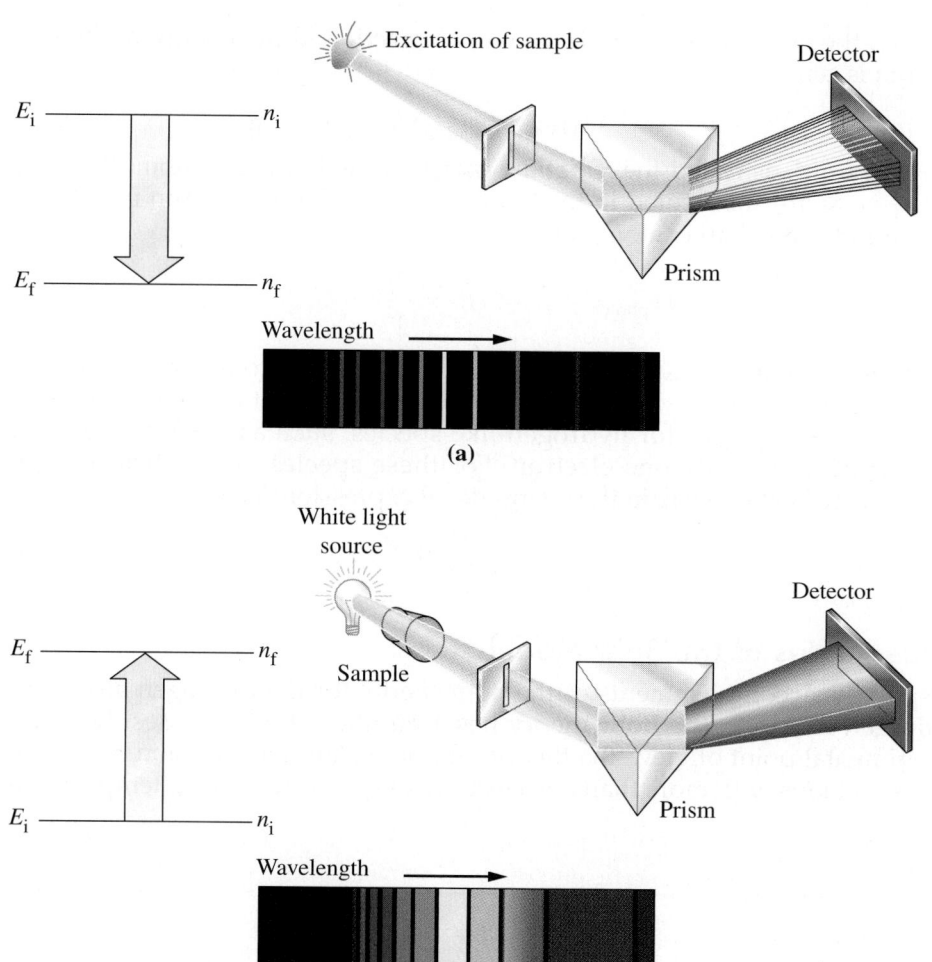

◀ FIGURE 8-15
Emission and absorption spectroscopy
(a) Emission spectroscopy. Bright lines are observed on a dark background of the photographic plate. (b) Absorption spectroscopy. Dark lines are observed on a bright background on the photographic plate.

For absorption of a photon to take place, the energy of the photon must exactly match the energy difference between the final and initial states, that is,

$$\nu = \frac{E_f - E_i}{h} \tag{8.8}$$

Note that the farther apart the energy levels, the shorter the wavelength of the photon needed to induce a transition.

You may have also noticed that in equation (8.7) the energy difference is expressed as $E_i - E_f$, whereas in equation (8.8) it is $E_f - E_i$. This is done to signify that energy is conserved during photon absorption and emission. That is, during emission $E_f = E_i - h\nu$, so that $\nu = (E_i - E_f)/h$. During photon absorption, $E_f = E_i + h\nu$, so that $\nu = (E_f - E_i)/h$.

Spectroscopic techniques have been used extensively in the study of molecular structures. Other forms of spectroscopy available to chemists are described elsewhere in the text.

Emission spectra are generally more complicated than absorption spectra. An excited sample will contain atoms in a variety of states, each being able to drop down to any of several lower states. An absorbing sample generally is cool and transitions are possible only from the ground state. The hydrogen Balmer lines are not seen, for example, in absorption from cold hydrogen atoms.

The Bohr Theory and the Ionization Energy of Hydrogen

The Bohr model of the atom helps to clarify the mechanism of formation of cations. In the special case where the energy of a photon interacting with a hydrogen atom is just enough to remove an electron from the ground state

($n = 1$), the electron is freed, the atom is ionized, and the energy of the free electron is zero.

$$h\nu_{photon} = E_i = -E_1$$

The quantity E_i is called the *ionization energy* of the hydrogen atom. If $n_i = 1$ and $n_f = \infty$ are substituted in the Bohr expression for an electron initially in the ground state of an H atom, then

$$h\nu_{photon} = E_i = -E_1 = \frac{R_H}{1^2} = R_H$$

The ideas just developed about the ionization of atoms are applied in Example 8-5, where they are coupled with another aspect of the Bohr model: The model also works for hydrogen-like species, such as the ions He$^+$ and Li^{2+}, which have only one electron. For these species, the nuclear charge (atomic number) appears in the energy-level expression. That is,

$$E_n = \frac{-Z^2 R_H}{n^2} \qquad \text{(8.9)}$$

Inadequacies of the Bohr Model

Despite the accomplishments of the Bohr model for the hydrogen atom and hydrogen-like ions, the Bohr theory has a number of weaknesses. From an experimental point of view, the theory cannot explain the emission spectra of atoms and ions with more than one electron, despite numerous attempts to do

---

EXAMPLE 8-5   Using the Bohr Model

Determine the kinetic energy of the electron ionized from a Li^{2+} ion in its ground state, using a photon of frequency $5.000 \times 10^{16}\,\text{s}^{-1}$.

Analyze

When a photon of a given energy ionizes a species, any excess energy is transferred as kinetic energy to the electron; that is, $E_{photon} = IE + KE_{electron}$. The energy of the electron in the Li^{2+} ion is calculated by using equation (8.9), and the energy of the photon is calculated by using Planck's relationship. The difference is the kinetic energy of the electron.

Solve

$$E_1 = \frac{-3^2 \times 2.179 \times 10^{-18}\,\text{J}}{1^2} = -1.961 \times 10^{-17}\,\text{J}$$

The energy of a photon of frequency $5.000 \times 10^{16}\,\text{s}^{-1}$ is

$$E = h\nu = 6.626 \times 10^{-34} \times \frac{\text{J s}}{\text{photon}} \times 5.000 \times 10^{16}\,\text{s}^{-1} = 3.313 \times 10^{-17}\,\text{J photon}^{-1}$$

The kinetic energy of the electron is given by $KE_{electron} = E_{photon} - IE$; that is,

$$\text{kinetic energy} = 3.313 \times 10^{-17}\,\text{J} - 1.961 \times 10^{-17}\,\text{J} = 1.352 \times 10^{-17}\,\text{J}$$

Assess

Notice the similarity between the energy conservation expression used in solving this problem ($E_{photon} = IE + KE_{electron}$) and the one used in explaining the photoelectric effect ($E_{photon} = eV_0 + KE_{electron}$).

PRACTICE EXAMPLE A: Determine the wavelength of light emitted in an electron transition from $n = 5$ to $n = 3$ in a Be^{3+} ion.

PRACTICE EXAMPLE B: The frequency of the $n = 3$ to $n = 2$ transition for an unknown hydrogen-like ion occurs at a frequency 16 times that of the hydrogen atom. What is the identity of the ion?

so. In addition, the theory cannot explain the effect of magnetic fields on emission spectra. From a fundamental standpoint, the Bohr theory is an uneasy mixture of classical and nonclassical physics. Bohr understood at the time that there is no fundamental basis for the postulate of quantized angular momentum forcing an electron into a circular orbit. He made the postulate only so that his theory would agree with experiment.

Modern quantum mechanics replaced the Bohr theory in 1926. The quantization of energy and angular momentum arose out of the postulates of this new quantum theory and required no extra assumptions. Moreover, the circular orbits of the Bohr theory do not occur in quantum mechanics. In summary, the Bohr theory gave the paradigm shift—the quantum leap—from classical physics to the new quantum physics, and we must not underestimate its importance as a scientific development.

8-4   CONCEPT ASSESSMENT

Which of the following electronic transitions in a hydrogen atom will lead to the emission of a photon with the shortest wavelength, $n = 1$ to $n = 4$, $n = 4$ to $n = 2$, $n = 3$ to $n = 2$?

8-5   Two Ideas Leading to a New Quantum Mechanics

In the previous section, we examined some successes of the Bohr theory and pointed out its inability to deal with multielectron atoms. A decade or so after Bohr's work on hydrogen, two landmark ideas stimulated a new approach to quantum mechanics. Those ideas are considered in this section and the new quantum mechanics—wave mechanics—in the next.

Wave–Particle Duality

To explain the photoelectric effect, Einstein suggested that light has particle-like properties, which are displayed through photons. Other phenomena, however, such as the dispersion of light into a spectrum by a prism, are best understood in terms of the wave theory of light. Light, then, appears to have a *dual* nature.

In 1924, Louis de Broglie, considering the nature of light and matter, offered a startling proposition: *Small particles of matter may at times display wave-like properties.* How did de Broglie come up with such a suggestion? He was aware of Einstein's famous equation

$$E = mc^2$$

where m is the relativistic mass of the photon and c is the speed of light. He combined this equation with the Planck relationship for the energy of a photon $E = h\nu$ as follows

$$h\nu = mc^2$$

$$\frac{h\nu}{c} = mc = p$$

where p is the momentum of the photon. Using $\nu\lambda = c$, we have

$$p = \frac{h}{\lambda}$$

In order to use this equation for a material particle, such as an electron, de Broglie substituted for the momentum, p, its equivalent—the product of the

▲ **Louis de Broglie (1892–1987)**
De Broglie conceived of the wave–particle duality of small particles while working on his doctorate degree. He was awarded the Nobel Prize in physics 1929 for this work.

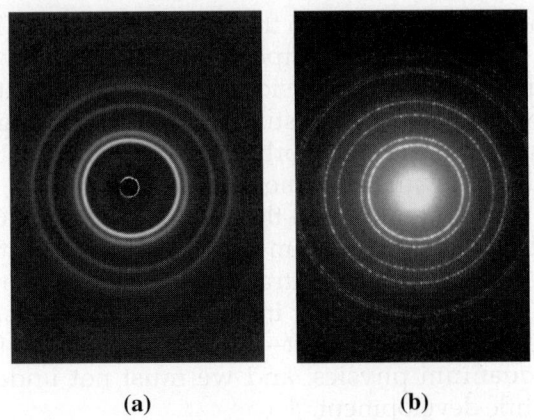

▶ FIGURE 8-16
Wave properties of electrons demonstrated.
(a) Diffraction of X-rays by metal foil.
(b) Diffraction of electrons by metal foil, confirming the wave-like properties of electrons.

KEEP IN MIND

that in equation (8.10), wavelength is in meters, mass is in kilograms, and velocity is in meters per second. Planck's constant must also be expressed in units of mass, length, and time. This requires replacing the joule by the equivalent units $kg\,m^2\,s^{-2}$.

mass of the particle, m, and its velocity, u. When this is done, we arrive at de Broglie's famous relationship.

$$\lambda = \frac{h}{p} = \frac{h}{mu} \qquad\qquad (8.10)$$

De Broglie called the waves associated with material particles "matter waves." If matter waves exist for small particles, then beams of particles, such as electrons, should exhibit the characteristic properties of waves, namely diffraction (recall page 298). If the distance between the objects that the waves scatter from is about the same as the wavelength of the radiation, diffraction occurs and an interference pattern is observed. For example, X-rays are highly energetic photons with an associated wavelength of about 1 Å (100 pm). X-rays are scattered by the regular array of atoms in the metal aluminum, where the atoms are about 2 Å (200 pm) apart, producing the diffraction pattern shown in Figure 8-16.

In 1927, C. J. Davisson and L. H. Germer of the United States showed that a beam of slow electrons is diffracted by a crystal of nickel. In a similar experiment in that same year, G. P. Thomson of Scotland directed a beam of electrons at a thin metal foil. He obtained the same pattern for the diffraction of electrons by aluminum foil as with X-rays of the same wavelength (Fig. 8-16).

Thomson and Davisson shared the 1937 Nobel Prize in physics for their electron diffraction experiments. George P. Thomson was the son of J.J. Thomson, who had won the Nobel Prize in physics in 1906 for his discovery of the electron. It is interesting to note that Thomson the father showed that the electron is a particle, and Thomson the son showed that the electron is a wave. Father and son together demonstrated the **wave–particle duality** of electrons.

The wavelength calculated in Example 8-6, 24.2 pm, is about one-half the radius of the first Bohr orbit of a hydrogen atom. It is only when wavelengths are comparable to atomic or nuclear dimensions that wave–particle duality is important. The concept has little meaning when applied to large (macroscopic) objects, such as baseballs and automobiles, because their wavelengths are too small to measure. For these macroscopic objects, the laws of classical physics are quite adequate.

The Uncertainty Principle

The laws of classical physics permit us to make precise predictions. For example, we can calculate the exact point at which a rocket will land after it is fired. The more precisely we measure the variables that affect the rocket's trajectory (path), the more accurate our calculation (prediction) will be. In effect, there is no limit to the accuracy we can achieve. In classical physics, nothing is left to chance—physical behavior can be predicted with certainty.

EXAMPLE 8-6   Calculating the Wavelength Associated with a Beam of Particles

What is the wavelength associated with electrons traveling at one-tenth the speed of light?

Analyze

To calculate the wavelength, we use equation (8.10). To use it, we have to collect the electron mass, the electron velocity, and Planck's constant, and then adjust the units so that they are expressed in terms of kg, m, and s.

Solve

The electron mass, expressed in kilograms, is 9.109×10^{-31} kg (recall Table 2.1).
The electron velocity is $u = 0.100 \times c = 0.100 \times 3.00 \times 10^8 \, \text{m s}^{-1} = 3.00 \times 10^7 \, \text{m s}^{-1}$.
Planck's constant $h = 6.626 \times 10^{-34} \, \text{J s} = 6.626 \times 10^{-34} \, \text{kg m}^2 \, \text{s}^{-2} \, \text{s} = 6.626 \times 10^{-34} \, \text{kg m}^2 \, \text{s}^{-1}$.
Substituting these data into equation (8.10), we obtain

$$\lambda = \frac{6.626 \times 10^{-34} \, \text{kg m}^2 \, \text{s}^{-1}}{(9.109 \times 10^{-31} \, \text{kg})(3.00 \times 10^7 \, \text{m s}^{-1})}$$
$$= 2.42 \times 10^{-11} \, \text{m} = 24.2 \, \text{pm}$$

Assess

By converting the unit J to $\text{kg m}^2 \, \text{s}^{-2}$, we are able to obtain the wavelength in meters.

PRACTICE EXAMPLE A: Assuming Superman has a mass of 91 kg, what is the wavelength associated with him if he is traveling at one-fifth the speed of light?

PRACTICE EXAMPLE B: To what velocity (speed) must a beam of protons be accelerated to display a de Broglie wavelength of 10.0 pm? Obtain the proton mass from Table 2.1.

During the 1920s, Niels Bohr and Werner Heisenberg considered hypothetical experiments to establish just how precisely the behavior of subatomic particles can be determined. The two variables that must be measured are the position of the particle (x) and its momentum ($p = mu$). The conclusion they reached is that there must *always* be uncertainties in measurement such that the product of the uncertainty in position, Δx, and the uncertainty in momentum, Δp, is

$$\Delta x \Delta p \geq \frac{h}{4\pi} \qquad (8.11)$$

The significance of this expression, called the **Heisenberg uncertainty principle**, is that we cannot measure position and momentum with great precision simultaneously. An experiment designed to locate the position of a particle with great precision cannot also measure the momentum of the particle precisely, and vice versa. In simpler terms, if we know precisely where a particle is, we cannot also know precisely where it has come from or where it is going. If we know precisely how a particle is moving, we cannot also know precisely where it is. In the subatomic world, things must always be "fuzzy." Why should this be so?

The de Broglie relationship (equation 8.10) implies that for a wavelength λ, the momentum of the associated particle is precisely known. However, since the wave itself is spread out over all space, we do not know exactly where the particle is! To get around this inability to locate the particle, we can combine several waves of different wavelengths into a "wave packet" to produce an interference pattern that tends to localize the wave, as suggested in Figure 8-17. However, because each wavelength in the wave packet corresponds to a specific, but different, momentum, the momentum of the particle corresponding to the collection of waves has become uncertain. So as we combine more and more waves to localize the particle, the momentum becomes more and more uncertain. And, conversely, the more precisely we want to know the momentum of a particle the fewer and fewer wavelengths we should combine, meaning that the position

▲ **Werner Heisenberg (1901–1976)**
In addition to his enunciation of the uncertainty principle, for which he won the Nobel Prize in physics in 1932, Heisenberg also developed a mathematical description of the hydrogen atom that gave the same results as Schrödinger's equation (page 323). Heisenberg (left) is shown here dining with Niels Bohr.

▶ The uncertainty principle is not easy for most people to accept. Einstein spent a good deal of time from the middle 1920s until his death in 1955 attempting, unsuccessfully, to disprove it.

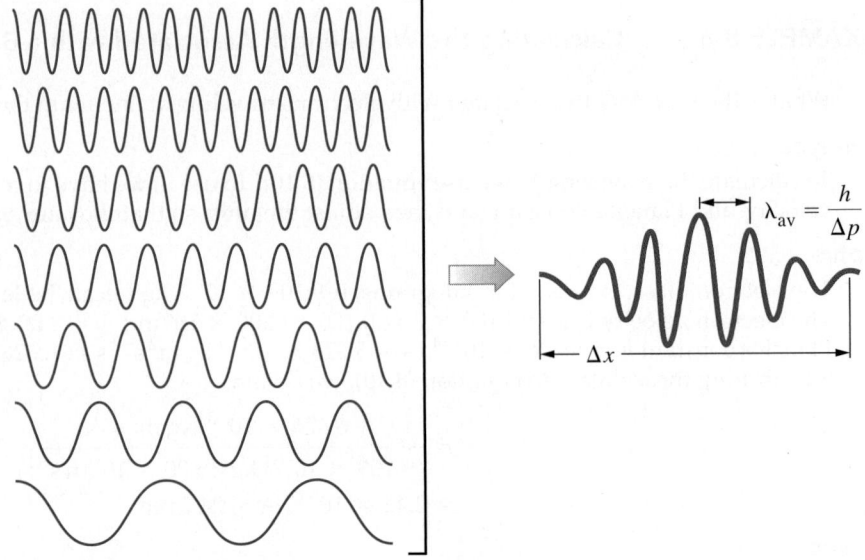

▲ FIGURE 8-17
The uncertainty principle interpreted graphically
A collection of waves with varying wavelengths (left) can combine into a "wave packet" (right). The superposition of the different wavelengths yields an average wavelength (λ_{av}) and causes the wave packet to be more localized (Δx) than the individual waves. The greater the number of wavelengths that combine, the more precisely an associated particle can be located, that is, the smaller Δx. However, because each of the wavelengths corresponds to a different value of momentum according to the de Broglie relationship, the greater is the uncertainty in the resultant momentum.

becomes more spread out. Thus as we proceed downward in size to atomic dimensions, it is no longer valid to consider a particle to be like a hard sphere. Instead, it becomes more and more wavelike, and it is no longer possible to determine with precision both its position and its momentum.

Once we understand that the consequence of wave–particle duality is the uncertainty principle, we realize that a fundamental error of the Bohr model was to constrain an electron to a one-dimensional orbit (1-D in the sense that the electron cannot move off a circular path of a fixed radius). We are now ready to turn our attention to a modern description of electrons in atoms.

The unit electron-volt (eV). One electron-volt is the energy acquired by an electron as it falls through an electric potential difference of 1 volt. ▼

EXAMPLE 8-7   Calculating the Uncertainty of the Position of an Electron

A 12 eV electron can be shown to have a speed of 2.05×10^6 m/s. Assuming that the precision (uncertainty) of this value is 1.5%, with what precision can we simultaneously measure the position of the electron?

Analyze

When given an uncertainty as a percentage, we have to convert it to a fraction by dividing by 100%. The uncertainty of the velocity is then obtained by multiplying this number by the actual velocity.

Solve

The uncertainty in the electron speed is

$$\Delta u = 0.015 \times 2.05 \times 10^6 \, \text{m s}^{-1} = 3.1 \times 10^4 \, \text{m s}^{-1}$$

The electron mass, 9.109×10^{-31} kg (recall Table 2.1), is known much more precisely than the electron speed, which means that

$$\Delta p = m\Delta u = 9.109 \times 10^{-31} \, \text{kg} \times 3.1 \times 10^4 \, \text{m s}^{-1}$$
$$= 2.8 \times 10^{-26} \, \text{kg m s}^{-1}$$

From equation (8.11), the uncertainty in the electron's position is

$$\Delta x = \frac{h}{4\pi \Delta p} = \frac{6.63 \times 10^{-34}\,\text{kg m}^2\,\text{s}^{-1}}{4 \times 3.14 \times 2.8 \times 10^{-26}\,\text{kg m s}^{-1}} = 1.9 \times 10^{-9}\,\text{m} = 1.9 \times 10^{3}\,\text{pm}$$

Assess

The uncertainty of the electron's position is about 10 atomic diameters. Given the uncertainty in its speed, there is no way to pin down the electron's position with any greater precision.

PRACTICE EXAMPLE A: Superman has a mass of 91 kg and is traveling at one-fifth the speed of light. If the speed at which Superman travels is known with a precision of 1.5%, what is the uncertainty in his position?

PRACTICE EXAMPLE B: What is the uncertainty in the speed of a beam of protons whose position is known with the uncertainty of 24 nm?

🔍 8-5   CONCEPT ASSESSMENT

An electron has a mass approximately 1/2000th of the mass of a proton. Assuming that a proton and an electron have similar wavelengths, how would their speeds compare?

8-6   Wave Mechanics

De Broglie's relationship suggests that electrons are matter waves and thus should display wavelike properties. A consequence of this wave–particle duality is the limited precision in determining an electron's position and momentum imposed by the Heisenberg uncertainty principle. How then are we to view electrons in atoms? To answer this question, we must begin by identifying two types of waves.

Standing Waves

On an ocean, the wind produces waves on the surface whose crests and troughs travel great distances. These are called *traveling waves*. In the traveling wave shown in Figure 8-1, every portion of a very long rope goes through an identical up-and-down motion. The wave transmits energy along the entire length of the rope. An alternative form of a wave is seen in the vibrations in a plucked guitar string, suggested by Figure 8-18.

Segments of the string experience up-and-down displacements with time, and they oscillate or vibrate between the limits set by the blue curves. The important aspect of these waves is that the crests and troughs of the wave occur at fixed positions and the amplitude of the wave at the fixed ends is zero. Of special interest is the fact that the magnitudes of the oscillations differ from point to point along the wave, including certain points, called *nodes*, that undergo no displacement at all. A wave with these characteristics is called a **standing wave**.

We might say that the permitted wavelengths of standing waves are quantized. They are equal to twice the path length (L) divided by a whole number (n), that is,

$$\lambda = \frac{2L}{n} \quad \text{where } n = 1, 2, 3, \ldots \text{ and the total number of nodes} = n + 1 \quad \textbf{(8.12)}$$

The plucked guitar string represents a one-dimensional standing wave, and so does an electron in a Bohr orbit. Bohr surmised that for an electron to be stable in a circular orbit, it has to be represented by a standing wave and that an integral number of wavelengths have to fit the circumference of the orbit (Fig. 8-19). Also, the fact that Bohr orbits are one dimensional (they are defined by a single

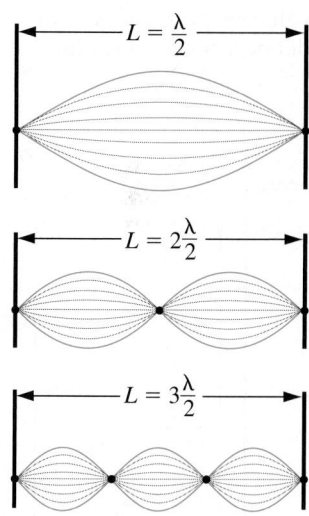

▲ FIGURE 8-18
Standing waves in a string
The string can be set into motion by plucking it. The blue boundaries outline the range of displacements at each point for each standing wave. The relationships between the wavelength, string length, and the number of nodes—points that are not displaced—are given by equation (8.12). The nodes are marked by bold dots.

▶ Beating a drum produces a two-dimensional standing wave, and ringing a spherical bell produces a three-dimensional standing wave.

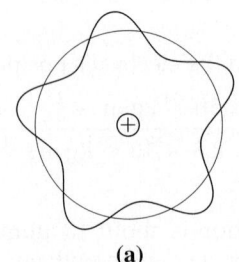 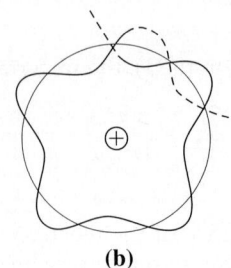

(a)   (b)

▲ FIGURE 8-19
The electron as a matter wave
These patterns are two-dimensional cross-sections of a much more complicated three-dimensional wave. The wave pattern in **(a)**, a *standing wave*, is an acceptable representation. It has an integral number of wavelengths (five) about the nucleus; successive waves reinforce one another. The pattern in **(b)** is unacceptable. The number of wavelengths is nonintegral, and successive waves tend to cancel each other; that is, the crest in one part of the wave overlaps a trough in another part of the wave, and there is no resultant wave at all.

dimension, the radius) also points up a serious deficiency in the Bohr model: The matter waves of electrons in the hydrogen atom must be three dimensional.

Particle in a Box: Standing Waves, Quantum Particles, and Wave Functions

In 1927, Erwin Schrödinger, an expert on the theory of vibrations and standing waves, suggested that an electron (or any other particle) exhibiting wavelike properties should be describable by a mathematical equation called a **wave function**. The wave function, denoted by the Greek letter psi, ψ, should correspond to a standing wave within the boundary of the system being described. The simplest system for which we can write a wave function is another one-dimensional system, that of a quantum particle confined to move in a one-dimensional box, a line. The wave function for this so-called "particle in a box" looks like those of a guitar string (Fig. 8-18), but now it represents the matter waves of a particle. Since the particle is constrained to be in the box, the waves also must be in the box, as illustrated in Figure 8-20.

If the length of the box is L and the particle moves along the x direction, then the equation for the standing wave is

$$\psi_n(x) = \sqrt{\frac{2}{L}}\sin\left(\frac{n\pi x}{L}\right) \quad n = 1, 2, 3, \ldots \qquad \textbf{(8.13)}$$

where the quantum number, n, labels the wave function.

This wave function is a sine function. To illustrate, consider the case where $n = 2$.

When

$x = 0$,	$\sin 2\pi x/L = \sin 0 = 0$,	and $\psi_n(x) = 0$
$x = L/4$,	$\sin 2\pi(L/4)/L = \sin \pi/2 = 1$,	and $\psi_n(x) = (2/L)^{1/2}$
$x = L/2$,	$\sin 2\pi(L/2)/L = \sin \pi = 0$,	and $\psi_n(x) = 0$
$x = 3L/4$	$\sin 2\pi(3L/4)/L = \sin 3\pi/2 = -1$	and $\psi_n(x) = -(2/L)^{1/2}$
$x = L$	$\sin 2\pi(L)/L = \sin 2\pi = 0$,	and $\psi_n(x) = 0$

At one end of the box ($x = 0$), both the sine function and the wave function are zero. At one-fourth the length of the box ($x = L/4$), the sine function and the wave function both reach their maximum values. At the midpoint of the box, both are again zero; the wave function has a node. At three-fourths the box length, both functions reach their minimum values (negative quantities), and at the farther end of the box, both functions are again zero.

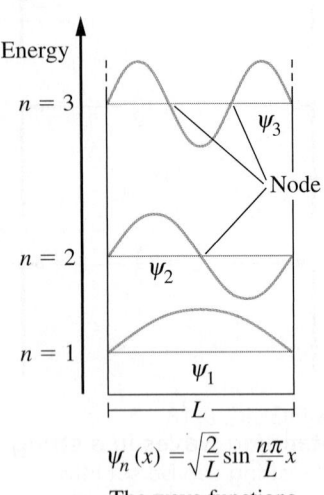

$$\psi_n(x) = \sqrt{\frac{2}{L}}\sin\frac{n\pi}{L}x$$
The wave functions

▲ FIGURE 8-20
The standing waves of a particle in a one-dimensional box
The first three wave functions and their energies are shown in relation to the position of the particle within the box. The wave function changes sign at the nodes.

8-4   ARE YOU WONDERING...

How did we arrive at Equation (8.13)?

The answer to how we arrived at equation (8.13) lies in the equation that gives the form of the wave function and the boundaries to which the quantum mechanical particle is confined. The particle-in-a-box model assumes that the electron is free in the box but is unable to get out of the box. This means that the wave function will be a standing wave inside the box. If you are familiar with differential calculus, you will recognize the equation below as a differential equation. Specifically, it describes a one-dimensional standing wave for the simple system of a particle in a box. The solution to this equation is the wave function for the system.

$$\frac{d^2\psi}{dx^2} = -\left(\frac{2\pi}{\lambda}\right)^2 \psi$$

Notice the form of the wave equation: By differentiating the wave function twice, we obtain the wave function times a constant. Many functions satisfy this requirement. For example, two trigonometric functions that have this property are the sine and cosine functions. First, let us consider the function $\psi = A\cos(ax)$ and differentiate twice with respect to x:

$$\frac{d\psi}{dx} = -aA\sin(ax)$$

$$\frac{d^2\psi}{dx^2} = -a^2A\cos(ax) = -a^2\psi$$

where we can identify $a = (2\pi/\lambda)$ and A is an arbitrary factor to be determined. Second, for the sine function,

$$\frac{d\psi}{dx} = aA\cos(ax)$$

$$\frac{d^2\psi}{dx^2} = -a^2A\sin(ax) = -a^2\psi$$

Therefore, both functions are acceptable from this point of view. However, the function must form a standing wave in the box, with a value of zero at the edges of the box. When $x = 0$, $\cos 0 = 1$ and $\sin 0 = 0$, and so the sine function is the appropriate function. The determination of A is not quite as straightforward, and we need to know how to interpret the wave function to determine the value of A. In carrying out this procedure, we have used the boundary conditions of the system to help decide on the correct form of the wave function, which is a common procedure when solving quantum mechanical problems.

Finally, if we identify $a = (2\pi/\lambda)$ and use the standing wave requirement in equation (8.12), then

$$a = \frac{2\pi}{\lambda} = \frac{2\pi}{2L/n} = \frac{n\pi}{L}$$

and the function is

$$\psi_n = A\sin\left(\frac{n\pi x}{L}\right)$$

where n is identified as a quantum number, $n = 1, 2, 3, 4, \ldots$

(continued)

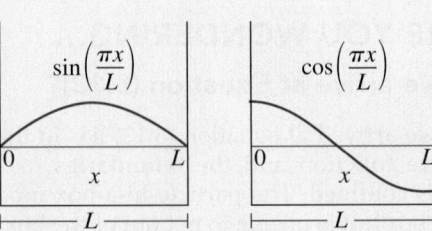

▲ Illustration of why a cosine function is an unacceptable solution for the particle in a box. The sine function correctly goes to zero at the edge of the box, but the cosine function does not.

What sense can we make of the wave function and the quantum number? First, consider the quantum number, n. What can we relate it to? The particle that we are considering is freely moving (not acted upon by any outside forces) with a kinetic energy given by the expression

$$E_k = \frac{1}{2}mu^2 = \frac{m^2u^2}{2m} = \frac{p^2}{2m} \tag{8.14}$$

Now, to associate this kinetic energy with a wave, we can use de Broglie's relationship ($\lambda = h/p$) to get

$$E_k = \frac{p^2}{2m} = \frac{h^2}{2m\lambda^2}$$

The wavelengths of the matter wave have to fit the standing wave conditions described earlier for the standing waves of a guitar string (equation 8.12). Substituting the wavelength of the matter wave from equation (8.12) into the equation for the energy of the wave yields

$$E_k = \frac{h^2}{2m\lambda^2} = \frac{h^2}{2m(2L/n)^2} = \frac{n^2h^2}{8mL^2}$$

So we see that the standing wave condition naturally gives rise to quantization of the wave's energy, with the allowable values determined by the value of n. Note also that as we decrease the size of the box, the kinetic energy of the particle increases, and according to the uncertainty principle, our knowledge of the momentum must decrease. A final noteworthy point is that the energy of the particle *cannot be zero*. The lowest possible energy, corresponding to $n = 1$, is called the **zero-point energy**. Because the zero-point energy is not zero, the particle cannot be at rest. This observation is consistent with the uncertainty principle because the position and momentum both must be uncertain, and there is nothing uncertain about a particle at rest.

The particle-in-a-box model helps us see the origin of the quantization of energy, but how are we to interpret the wave function, ψ? What does it mean that the value of the wave function can be positive or negative? Actually, unlike the trajectory of a classical particle, the wave function of a particle has no physical significance. We need to take a different approach, one suggested by the German physicist Max Born in 1926. From the electron-as-particle standpoint, we have a special interest in the *probability* that the electron is at some particular point; from the electron-as-wave standpoint, our interest is in *electron charge density*. In a classical wave (such as visible light), the amplitude of the wave corresponds to ψ, and the intensity of the wave to ψ^2. The intensity relates to the photon density—the number of photons present in a region. For an electron wave, then, ψ^2 relates to electron charge density. Electron probability is proportional to electron charge density, and both these quantities are associated with ψ^2. Thus, in Born's interpretation of the wave function, the total probability of finding an electron in a small volume of space is the product of the square of the wave function, ψ^2, and the volume of interest. The factor ψ^2 is called the electron probability density.

Now let us return to a particle constrained to a one-dimensional path in a box and look at the probabilities for the wave functions. These are shown in Figure 8-21. First, notice that even where the wave function is negative, the probability density is positive, as it should be in all cases. Next, look at the probability density for the wave function corresponding to $n = 1$. The highest value of ψ^2 is at the center of the box; that is, the particle is most likely to be found there. The probability density for the state with $n = 2$ indicates that the particle is most likely to be found between the center of the box and the walls.

A final consideration of the particle-in-a-box model concerns its extension to a three-dimensional box. In this case, the particle can move in all three directions—x, y, and z—and the quantization of energy is described by the following expression.

$$E_{n_x n_y n_z} = \frac{h^2}{8m}\left[\frac{n_x^2}{L_x^2} + \frac{n_y^2}{L_y^2} + \frac{n_z^2}{L_z^2}\right]$$

where there is one quantum number for each dimension. Thus, a three-dimensional system needs three quantum numbers. With these particle-in-a-box ideas, we can now discuss solving the quantum mechanical problem of the hydrogen atom.

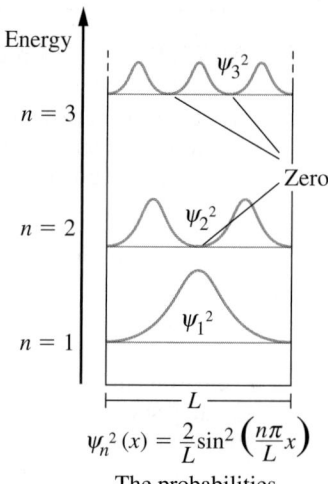

$$\psi_n^2(x) = \frac{2}{L}\sin^2\left(\frac{n\pi}{L}x\right)$$

The probabilities

▲ FIGURE 8-21
The probabilities of a particle in a one-dimensional box
The squares of the first three wave functions and their energies are shown in relation to the position of the particle within the box. There is no chance of finding the particle at the points where $\psi^2 = 0$.

EXAMPLE 8-8   Using the Wave Functions of a Particle in a One-Dimensional Box

What is the fraction, as a percentage, of the total probability of finding, between points at 0 pm and 30 pm, an electron in the $n = 5$ level of a one-dimensional box 150 pm long?

Analyze

If an electron is in the $n = 5$ level, then we have a 100% chance of finding it in that level. The $n = 5$ wave function has 4 nodes 30 pm apart, and there are five maxima in ψ^2 at 15 pm, 45 pm, 75 pm, 105 pm, and 135 pm for a one-dimensional box 150 pm long.

Solve

The position at 30 pm corresponds to a node in the wave function, and there are four of these (not counting the nodes at each end of the box). The total area between 0 and 30 pm of ψ^2 represents 25% of the total probability, and so between 0 pm and 30 pm, we expect to find 25% of the probability.

Assess

We must remember that the particle we are considering exhibits wave–particle duality, making it inappropriate to ask a question about how it gets from one side of the node to the other (but that is an appropriate question for a *classical* particle). All we know is that in the $n = 5$ state, for example, the particle is in the box somewhere. When we make a measurement, we'll find the particle on one side of a node or the other. Between 0 and 30 pm, we have a 25% chance of finding the particle, and the maximum chance occurs at 15 pm.

PRACTICE EXAMPLE A: What is the fraction, as a percentage of the total probability of finding, between points at 50 pm and 75 pm, an electron in the $n = 6$ level of a one-dimensional box 150 pm long?

PRACTICE EXAMPLE B: A particle is confined to a one-dimensional box 300 pm long. For the state having $n = 3$, at what points (not counting the ends of the box) does the particle have zero probability of being found?

EXAMPLE 8-9   Calculating Transition Energy and Photon Wavelengths for the Particle in a Box

What is the energy difference between the ground state and the first excited state of an electron contained in a one-dimensional box 1.00×10^2 pm long? Calculate the wavelength of the photon that could excite the electron from the ground state to the first excited state.

Analyze

The energy of an electron (E_n) in level n is

$$E_n = \frac{n^2 h^2}{8mL^2}$$

We can write expressions for E_n and E_{n+1}, subtract them, and then substitute the values for h, m, and L. The ground state corresponds to $n = 1$, and the first excited state corresponds to $n = 2$. Finally, we can calculate the wavelength of the photon from the Planck relationship and $c = \lambda \nu$.

Solve

The energies for the states $n = 1$ and $n = 2$ are

$$E_{\text{ground state}} = E_1 = \frac{h^2}{8mL^2}(1^2)$$

$$E_{\text{first excited state}} = E_2 = \frac{h^2}{8mL^2}(2^2)$$

The energy difference is

$$\Delta E = E_2 - E_1 = \frac{3h^2}{8mL^2}$$

The electron mass is 9.109×10^{-31} kg, Planck's constant $h = 6.626 \times 10^{-34}$ J s, and the length of the box is 1.00×10^{-10} m. (Recall: 1 pm = 10^{-12} m.) Substituting these data into the equation, we obtain

$$\Delta E = \frac{3(6.626 \times 10^{-34} \text{ J s})^2}{8(9.109 \times 10^{-31} \text{ kg})(1.00 \times 10^{-10} \text{ m})^2}$$
$$= 1.81 \times 10^{-17} \text{ J}$$

By using Planck's constant and this value as the energy of a photon, we can calculate the frequency of the photon and then the wavelength. Combining these steps,

$$\lambda = \frac{hc}{E_{\text{photon}}} = \frac{hc}{\Delta E} = \frac{6.626 \times 10^{-34} \text{ J s} \times 3.00 \times 10^8 \text{ m s}^{-1}}{1.81 \times 10^{-17} \text{ J}} = 11.0 \times 10^{-9} \text{ m} = 11.0 \text{ nm}$$

Assess

If we needed the energy of the photon in kJ mol^{-1}, we would have had to multiply 1.8×10^{-17} J by 10^{-3} kJ/J and $N_A = 6.022 \times 10^{23}$ mol^{-1}.

PRACTICE EXAMPLE A: Calculate the wavelength of the photon emitted when an electron in a box 5.0×10^1 pm long falls from the $n = 5$ level to the $n = 3$ level.

PRACTICE EXAMPLE B: A photon of wavelength 24.9 nm excites an electron in a one-dimensional box from the ground state to the first excited state. Estimate the length of the box.

🔍 **8-6   CONCEPT ASSESSMENT**

For a particle in a one-dimensional box, in which state (value of n) is the greatest probability of finding the particle at one-quarter the length of the box from either end?

Wave Functions of the Hydrogen Atom

In 1927, Schrödinger showed that the wave functions of a quantum mechanical system can be obtained by solving a wave equation that has since become known as the **Schrödinger equation**. We will not go into the details of its

8-5   ARE YOU WONDERING...

What is the Schrödinger equation for the hydrogen atom?

To obtain the Schrödinger equation, we start with the equation for a standing wave in one dimension:

$$\frac{d^2\psi}{dx^2} = -\left(\frac{2\pi}{\lambda}\right)^2 \psi$$

The next step is to substitute de Broglie's relationship for the wavelength of a matter wave.

$$\frac{d^2\psi}{dx^2} = -\left(\frac{2\pi}{h}p\right)^2 \psi$$

Finally, we use the relationship between momentum and kinetic energy, equation 8.14, to obtain

$$-\frac{h^2}{8\pi^2 m}\frac{d^2\psi}{dx^2} = E_k\psi$$

This is the Schrödinger equation of a free particle moving in one dimension. It is customary to replace E_k by $E - V(x)$, where E is the total energy (a constant) and $V(x)$ is the potential energy, which in general depends upon the particle's position. We obtain

$$-\frac{h^2}{8\pi^2 m}\frac{d^2\psi}{dx^2} + V(x)\psi = E\psi$$

Extending this treatment to three dimensions, we obtain the Schrödinger equation for the hydrogen atom or hydrogen-like ion, where we understand $V(r)$ to be $(-e)(Ze)/4\pi\epsilon_0 r)$, the potential energy associated with the interaction of the electron (charge $= -e$), and the nucleus of the one electron atom or ion (charge $= Ze$). (See Appendix B, page B-4.)

$$-\frac{h^2}{8\pi^2 m_e}\left(\frac{\partial^2\psi}{\partial x^2} + \frac{\partial^2\psi}{\partial y^2} + \frac{\partial^2\psi}{\partial z^2}\right) - \frac{Ze^2}{r}\psi = E\psi$$

This is the equation that Schrödinger obtained. In the equation above, $\partial^2\psi/\partial x^2$ means that we differentiate ψ twice with respect to x, treating the other variables (y and z) as constants. The notation $\partial^2\psi/\partial x^2$ is used instead of $d^2\psi/dx^2$ because ψ depends on more than one variable.

Following a suggestion by Eugene Wigner, Schrödinger used spherical polar coordinates to solve it rather than the Cartesian coordinates shown here. That is, he substituted the values of x, y, and z in terms of spherical polar coordinates given in the caption for Figure 8-22 and performed the necessary lengthy algebra to collect the variables r, θ, and ϕ. The equation he obtained is

$$-\frac{h^2}{8\pi^2\mu r^2}\left[\frac{\partial}{\partial r}\left(r^2\frac{\partial\psi}{\partial r}\right) + \frac{1}{\sin\theta}\frac{\partial}{\partial\theta}\left(\sin\theta\frac{\partial\psi}{\partial\theta}\right) + \frac{1}{\sin^2\theta}\frac{\partial^2\psi}{\partial\phi^2}\right] - \frac{Ze^2}{r}\psi = E\psi$$

where the mass of the electron has been replaced by the more correct reduced mass of the atom, μ, given by

$$\frac{1}{\mu} = \frac{1}{m_e} + \frac{1}{m_{\text{nucleus}}}$$

This is the Schrödinger equation in spherical polar coordinates for a hydrogen-like ion of atomic number Z or the hydrogen atom if $Z = 1$. The solutions are shown in Table 8.1 on page 327.

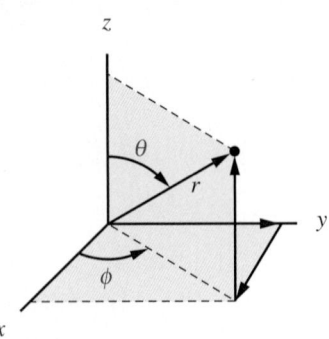

Spherical polar coordinates

$x^2 + y^2 + z^2 = r^2$

$x = r \sin \theta \cos \phi$

$y = r \sin \theta \sin \phi$

$z = r \cos \theta$

▲ FIGURE 8-22
The relationship between spherical polar coordinates and Cartesian coordinates
The coordinates x, y, and z are expressed in terms of the distance r and the angles θ and ϕ.

solution but just describe and interpret the solution using ideas introduced in the previous discussion.

Solutions of the Schrödinger equation for the hydrogen atom give the wave functions for the electron in the hydrogen atom. These wave functions are called **orbitals** to distinguish them from the orbits of the Bohr theory. The mathematical form of these orbitals is more complex than for the particle in a box, but nonetheless they can be interpreted in a straightforward way.

Wave functions are most easily analyzed in terms of the three variables required to define a point with respect to the nucleus. In the usual Cartesian coordinate system, these three variables are the x, y, and z dimensions. In the spherical polar coordinate system, they are r, the distance of the point from the nucleus, and the angles θ (theta) and ϕ (phi), which describe the orientation of the distance line, r, with respect to the x, y, and z axes (Fig. 8-22). Either coordinate system could be used in solving the Schrödinger equation. However, whereas in the Cartesian coordinate system the orbitals would involve all three variables, x, y, and z, in the spherical polar system the orbitals can be expressed in terms of one function R that depends only on r, and a second function Y that depends on θ and ϕ. That is,

$$\psi(r, \theta, \phi) = R(r)Y(\theta, \phi)$$

The function $R(r)$ is called the **radial wave function**, and the function $Y(\theta, \phi)$ is called the **angular wave function**. Each orbital has three quantum numbers to define it since the hydrogen atom is a three-dimensional system. The particular set of quantum numbers confers a particular functional form to $R(r)$ and $Y(\theta, \phi)$.

Probability densities and the spatial distribution of these densities can be derived from these functional forms. We will first discuss quantum numbers and the orbitals they define, and then the distribution of probability densities associated with the orbitals.

8-7 Quantum Numbers and Electron Orbitals

In the preceding section we stated that by specifying three quantum numbers in a wave function ψ, we obtain an orbital. Here, we explore the combinations of quantum numbers that produce different orbitals. First, though, we need to learn more about the nature of these three quantum numbers.

Assigning Quantum Numbers

The following relationships involving the three quantum numbers arise from the solution of the Schrödinger wave equation for the hydrogen atom. In this solution the values of the quantum numbers are fixed *in the order listed*.

The first number to be fixed is the *principal quantum number, n*, which may have only a positive, nonzero integral value.

$$n = 1, 2, 3, 4, \ldots \tag{8.15}$$

Second is the *orbital angular momentum quantum number, ℓ*, which may be zero or a positive integer, but not larger than $n - 1$ (where n is the principal quantum number).

$$\ell = 0, 1, 2, 3, \ldots, n - 1 \tag{8.16}$$

Third is the *magnetic quantum number, m_ℓ*, which may be a negative or positive integer, including zero, and ranging from $-\ell$ to $+\ell$ (where ℓ is the orbital angular momentum quantum number).

$$m_\ell = -\ell, (-\ell + 1), \ldots, -2, -1, 0, 1, 2 \ldots, (\ell - 1), +\ell \tag{8.17}$$

EXAMPLE 8-10   Applying Relationships Among Quantum Numbers

Can an orbital have the quantum numbers $n = 2$, $\ell = 2$, and $m_\ell = 2$?

Analyze

We must determine whether the given set of quantum numbers is allowed by the rules expressed in equations (8.15), (8.16), and (8.17).

Solve

No. The ℓ quantum number cannot be greater than $n - 1$. Thus, if $n = 2$, ℓ can be only 0 or 1. And if ℓ can be only 0 or 1, m_ℓ cannot be 2; m_ℓ must be 0 if $\ell = 0$ and may be $-1, 0$, or $+1$ if $\ell = 1$.

Assess

It is important that we remember the physical significance of the various quantum numbers, as well as the rules interrelating their values. Quantum number n determines the radial distribution and the average *distance* of the electron and, thus, is most important in determining the *energy* of an electron. Quantum number ℓ determines the angular distribution or *shape* of an orbital. As we will soon see, the relationships among the quantum numbers impart a logical organization of orbitals into shells and subshells.

PRACTICE EXAMPLE A: Can an orbital have the quantum numbers $n = 3$, $\ell = 0$, and $m_\ell = 0$?

PRACTICE EXAMPLE B: For an orbital with $n = 3$ and $m_\ell = 1$, what is (are) the possible value(s) of ℓ?

Principal Shells and Subshells

All orbitals with the same value of n are in the same **principal electronic shell** or **principal level**, and all orbitals with the same n and ℓ values are in the same **subshell**, or **sublevel**.

Principal electronic shells are numbered according to the value of n. The first principal shell consists of orbitals with $n = 1$; the second principal shell of orbitals with $n = 2$; and so on. The value of n relates to the energy and most probable distance of an electron from the nucleus. The higher the value of n, the greater the electron energy and the farther, on average, the electron is from the nucleus. The principal quantum number, therefore, has a physical significance, as do the other quantum numbers. The quantum number ℓ determines the angular distribution, or *shape*, of an orbital and m_ℓ determines the *orientation* of the orbital.

The number of subshells in a principal electronic shell is the same as the number of allowed values of the orbital angular momentum quantum number, ℓ. In the first principal shell, with $n = 1$, the only allowed value of ℓ is 0, and there is a single subshell. The second principal shell ($n = 2$), with the allowed ℓ values of 0 and 1, consists of two subshells; the third principal shell ($n = 3$) has three subshells ($\ell = 0, 1$, and 2); and so on. Or, to put the matter in another way, because there are n possible values of the ℓ quantum number, that is, $0, 1, 2, \ldots (n - 1)$, the number of subshells in a principal shell is equal to the principal quantum number. As a result, there is one subshell in the principal shell with $n = 1$, two subshells in the principal shell with $n = 2$, and so on. The name given to a subshell, regardless of the principal shell in which it is found, depends on the value of the ℓ quantum number. The first four subshells are

s subshell	*p* subshell	*d* subshell	*f* subshell
$\ell = 0$	$\ell = 1$	$\ell = 2$	$\ell = 3$

The number of orbitals in a subshell is the same as the number of allowed values of m_ℓ for the particular value of ℓ. Recall that the allowed values of m_ℓ are $0, \pm 1, \pm 2, \ldots \pm \ell$, and thus the total number of orbitals in a subshell is $2\ell + 1$. The names of the orbitals are the same as the names of the subshells in which they appear.

s orbitals	p orbitals	d orbitals	f orbitals
$\ell = 0$	$\ell = 1$	$\ell = 2$	$\ell = 3$
$m_\ell = 0$	$m_\ell = 0, \pm 1$	$m_\ell = 0, \pm 1, \pm 2$	$m_\ell = 0, \pm 1, \pm 2, \pm 3$
one s orbital in an s subshell	three p orbitals in a p subshell	five d orbitals in a d subshell	seven f orbitals in an f subshell

To designate the particular principal shell in which a given subshell or orbital is found, we use a combination of a number and a letter. For example, the symbol 2p is used to designate both the p subshell of the second principal shell and any of the three p orbitals in that subshell.

The energies of the orbitals for a hydrogen atom, in joules, are given by an equation with a familiar appearance.

$$E_n = -2.178 \times 10^{-18}\left(\frac{1}{n^2}\right) \text{J}$$

▶ In Section 8-10 and in Chapter 24, we will see that orbital energies of multielectron atoms also depend on the quantum numbers ℓ and m_ℓ.

It is the same as equation (8.5), the formula derived by Bohr. Orbital energies for a hydrogen atom depend only on the principal quantum number n. This means that all the subshells within a principal electronic shell have the same energy, as do all the orbitals within a subshell. Orbitals at the same energy level are said to be **degenerate**. Figure 8-23 shows an energy-level diagram and the arrangement of shells and subshells for a hydrogen atom.

Some of the points discussed in the preceding paragraphs are illustrated in Example 8-11.

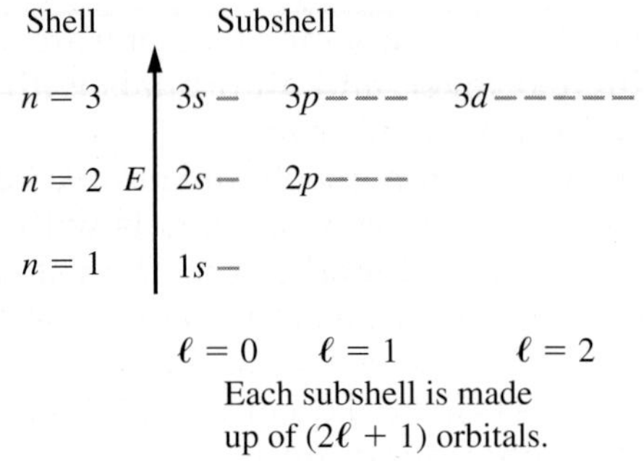

▶ FIGURE 8-23
Shells and subshells of a hydrogen atom
The hydrogen atom orbitals are organized into shells and subshells.

Shell	Subshell
$n = 3$	$3s -$ $\quad$ $3p ---$ $\quad$ $3d -----$
$n = 2$ E	$2s -$ $\quad$ $2p ---$
$n = 1$	$1s -$

$\ell = 0$ $\quad$ $\ell = 1$ $\quad$ $\ell = 2$
Each subshell is made up of $(2\ell + 1)$ orbitals.

EXAMPLE 8-11   Relating Orbital Designations and Quantum Numbers

Write an orbital designation corresponding to the quantum numbers $n = 4, \ell = 2, m_\ell = 0$.

Analyze

To write orbital designations you need to recall the conventions associated with the quantum numbers n and ℓ. For the quantum number n we use only the number while for the quantum number ℓ we use the following letters $\ell = 0, s; \ell = 1, p; \ell = 2, d;$ and so on.

Solve

The magnetic quantum number, m_ℓ, is not reflected in the orbital designation. The type of orbital is determined by the ℓ quantum number. Because $\ell = 2$, the orbital is of the d type. Because $n = 4$, the orbital designation is 4d.

Assess

This is another type of problem in which we need to have memorized the quantum number rules and their designations. This information will be important in the later chapters.

PRACTICE EXAMPLE A: Write an orbital designation corresponding to the quantum numbers $n = 3, \ell = 1$, and $m_\ell = 1$.

PRACTICE EXAMPLE B: Write all the combinations of quantum numbers that define hydrogen-atom orbitals with the same energy as the 3s orbital.

Interpreting and Representing the Orbitals of the Hydrogen Atom

Our major undertaking in this section will be to describe the three-dimensional probability density distributions obtained for the various orbitals in the hydrogen atom. Through the Born interpretation of wave functions (page 320), we will represent the probability densities of the orbitals of the hydrogen atom as surfaces that encompass most of the electron probability. We will see that the probability density for each type of orbital has its own distinctive shape, and like all waves, probability densities exhibit nodes and differing phase behavior. In studying this section, it is important for you to remember that, even though we will offer some additional quantitative information about orbitals, your primary concern should be to acquire a broad qualitative understanding. It is this qualitative understanding that you can apply in our later discussion of how orbitals enter into a description of chemical bonding.

Throughout this discussion, recall that orbitals are wave functions, mathematical solutions of the Schrödinger wave equation. The wave function itself has no physical significance. However, the square of the wave function, ψ^2, is a quantity that is related to probabilities. Probability density distributions based on ψ^2 are three-dimensional, and it is these three-dimensional regions that we mean when we refer to the shape of an orbital.

The forms of the radial wave function $R(r)$ and the angular wave function $Y(\theta, \phi)$ for a one-electron, hydrogen-like atom are shown in Table 8.1. The first thing to note is that the angular part of the wave function for an s orbital,

◄ In Chapter 11, we will discover important uses of the wave function, ψ, itself as a basis for discussing bonding between atoms.

TABLE 8.1   The Angular and Radial Wave Functions of a Hydrogen-Like Atom

Angular Part $Y(\theta, \phi)$	Radial Part $R_{n,\ell}(r)$
$Y(s) = \left(\dfrac{1}{4\pi}\right)^{1/2}$	$R(1s) = 2\left(\dfrac{Z}{a_0}\right)^{3/2} e^{-\sigma/2}$
	$R(2s) = \dfrac{1}{2\sqrt{2}}\left(\dfrac{Z}{a_0}\right)^{3/2}(2-\sigma)e^{-\sigma/2}$
	$R(3s) = \dfrac{1}{9\sqrt{3}}\left(\dfrac{Z}{a_0}\right)^{3/2}(6-6\sigma+\sigma^2)e^{-\sigma/2}$
$Y(p_x) = \left(\dfrac{3}{4\pi}\right)^{1/2}\sin\theta\cos\phi$	$R(2p) = \dfrac{1}{2\sqrt{6}}\left(\dfrac{Z}{a_0}\right)^{3/2}\sigma e^{-\sigma/2}$
$Y(p_y) = \left(\dfrac{3}{4\pi}\right)^{1/2}\sin\theta\sin\phi$	$R(3p) = \dfrac{1}{9\sqrt{6}}\left(\dfrac{Z}{a_0}\right)^{3/2}(4-\sigma)\sigma e^{-\sigma/2}$
$Y(p_z) = \left(\dfrac{3}{4\pi}\right)^{1/2}\cos\theta$	
$Y(d_{z^2}) = \left(\dfrac{5}{16\pi}\right)^{1/2}(3\cos^2\theta - 1)$	$R(3d) = \dfrac{1}{9\sqrt{30}}\left(\dfrac{Z}{a_0}\right)^{3/2}\sigma^2 e^{-\sigma/2}$
$Y(d_{x^2-y^2}) = \left(\dfrac{15}{16\pi}\right)^{1/2}\sin^2\theta\cos 2\phi$	
$Y(d_{xy}) = \left(\dfrac{15}{16\pi}\right)^{1/2}\sin^2\theta\sin 2\phi$	$\sigma = \dfrac{2Zr}{na_0}$
$Y(d_{xz}) = \left(\dfrac{15}{4\pi}\right)^{1/2}\sin\theta\cos\theta\cos\phi$	
$Y(d_{yz}) = \left(\dfrac{15}{4\pi}\right)^{1/2}\sin\theta\cos\theta\sin\phi$	

$\left(\dfrac{1}{4\pi}\right)^{1/2}$, is always the same, regardless of the principal quantum number. Next, note that the angular parts of the p and d orbitals are also independent of the quantum number n. Therefore all orbitals of a given type (s, p, d, f) have the same angular behavior. Also note that the equations in Table 8.1 are in a general form where the atomic number Z is included. This means that the equations apply to any one-electron atom, that is, to a hydrogen atom or a hydrogen-like ion. Finally, note that the term σ appearing throughout the table is equal to $2Zr/na_0$.

To obtain the wave function for a particular state, we simply multiply the radial part by the angular part. We will now illustrate this by looking at the three major types of orbitals.

s Orbitals

To obtain a complete wave function for the hydrogen $1s$ orbital, we use $Z = 1$ and $n = 1$, and combine the angular and radial wave functions where the red (radial) and blue (angular) colors indicate the origin of the two parts of the wave function.

$$\psi(1s) = R(r) \times Y(\theta, \phi) = \frac{2e^{-r/a_0}}{a_0^{3/2}} \times \frac{1}{\sqrt{4\pi}} = \frac{e^{-r/a_0}}{\sqrt{(\pi a_0^3)}}$$

The term a_0 has the same significance as in the Bohr theory; it is the first Bohr radius—53 pm. By squaring $\psi(1s)$ we obtain an expression for the probability density of finding a $1s$ electron at a distance r from the nucleus in a hydrogen atom.

$$\psi^2(1s) = \frac{1}{\pi}\left(\frac{1}{a_0}\right)^3 e^{-2r/a_0} \tag{8.18}$$

▶ $\psi^2(1s)$ is the probability density for a $1s$ electron at *one* point a distance r from the nucleus. Equally important is the probability density distribution, which gives the total probability for *all* points at a distance r from the nucleus. In Section 8-10 we will see that this distribution is given by $4\pi r^2\psi^2$.

How can we represent ψ^2 in expression (8.18)? One way is to pass a plane through the nucleus (for example, the xy plane) and plot a graph of electron probability densities (ψ^2) as perpendicular heights above the many points in the plane at which the electron might be found. The resultant graph, seen in Figure 8-24(a), looks like a symmetrical, cone-shaped "hill" (think of a volcano)

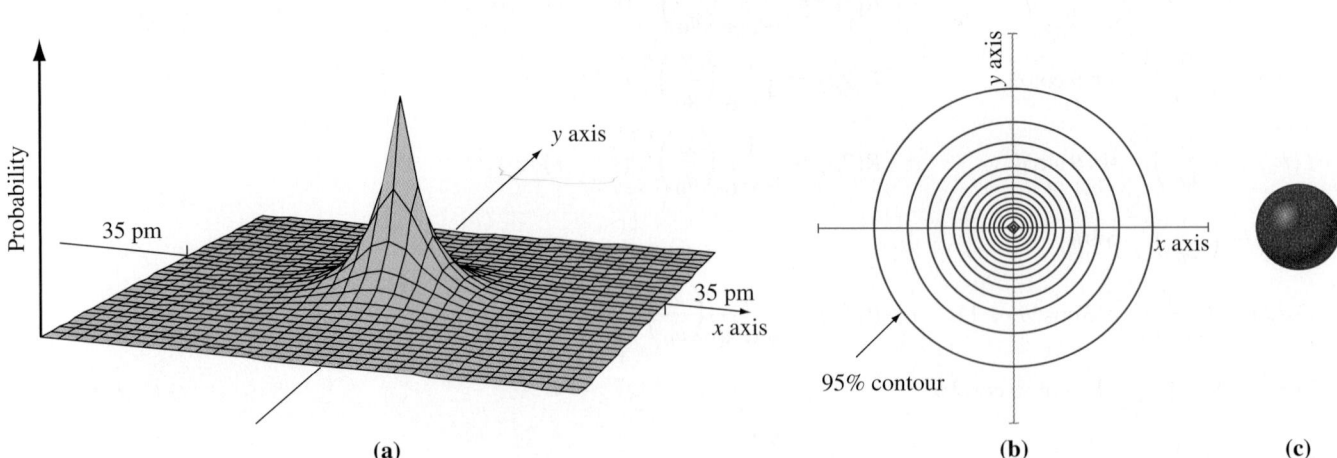

(a)   (b)   (c)

▲ FIGURE 8-24
Three representations of the electron probability density for the 1s orbital
(a) In this diagram the probability density is represented by the height above the xy plane (the xy plane is an arbitrary choice, any plane could have been chosen). **(b)** A contour map of the $1s$ orbital probability density in the xy plane, pointing out the 95% contour. **(c)** A reduced scale 3D representation of the 95% contour of a $1s$ orbital.

of electron probability densities with its peak directly above the nucleus. As in topographical maps of Earth's surface, we can project the three-dimensional surface onto a two-dimensional contour map. The contour map of the "hill" of electron probability densities is shown in Figure 8-24(b). The circular contour lines join points of equal electron probability density. The contours close to the nucleus join points of high probability of finding the electron, those farther away correspond to a lower probability. A simpler way to display the electron probability is to select just one large contour that together with all the contours within it, encompasses an area of high probability of finding the electron. The contour usually chosen is the one outlining an area in which the chance of finding the electron is 95%. The 95% contour just described is for a plane passing through the nucleus, but an electron in a 1s orbital moves in three-dimensional space. The complete 95% probability surface is a sphere, as seen in Figure 8-24(c).

Now let's look at the wave function of the 2s orbital.

$$\psi(2s) = R(r) \times Y(\theta,\phi) = \frac{1}{2\sqrt{2}} \frac{1}{a_0^{3/2}} \left(2 - \frac{r}{a_0}\right)e^{-r/2a_0} \times \frac{1}{\sqrt{4\pi}} = \frac{1}{4}\left(\frac{1}{2\pi a_0^3}\right)^{1/2}\left(2 - \frac{r}{a_0}\right)e^{-r/2a_0}$$

The electron probability density for the 2s orbital is given by

$$\psi^2(2s) = \frac{1}{8\pi}\left(\frac{1}{a_0}\right)^3\left(2 - \frac{r}{a_0}\right)^2 e^{-r/a_0} \tag{8.19}$$

which, when compared to expression (8.18) for the 1s orbital, shows that the 2s electron tends to stay farther from the nucleus than the 1s electron, because the exponential has changed from $-2r/a_0$ for the 1s (equation 8.18) to $-r/a_0$ for the 2s orbital (equation 8.19). The exponential of the 2s orbital decays more slowly than that of the 1s.

The factor $\left(2 - \frac{r}{a_0}\right)$ in the 2s wave function controls the sign of the function. For small values of r, r/a_0 is smaller than two and the wave function is positive, but for large values of r, r/a_0 is larger than two and the wave function is negative. At $r = 2a_0$, the pre-exponential factor is zero and the wave function is said to have a *radial node*. The wave function changes phase (sign) at this radial node.

The fact that the electron probability density of the 2s orbital extends farther from the nucleus than that of the 1s orbital, together with the presence of the node, means that the 95% electron probability density sphere of a 2s orbital is bigger than that of a 1s orbital and contains a sphere of zero probability due to the radial node. These features are illustrated in Figure 8-25, which compares the 1s, 2s, and 3s orbitals. Note that the 3s orbital exhibits two radial nodes and is larger than both the 1s and 2s orbitals. The number of nodes increases as the energy is increased—a characteristic of high-energy standing waves. To highlight the change in phase of an orbital in progressing outward from the nucleus, we have adopted the modern usage of different colors to represent a change in phase. Thus in Figure 8-25 the 1s orbital is a single red color throughout, whereas the 2s orbital starts out red and then switches to blue; and finally, the 3s starts out red, changes to blue and then back to red, reflecting the presence of two radial nodes. We now turn our attention to the p orbitals.

p Orbitals

The radial part of $\psi(2p)$ for a hydrogen atom is

$$R(2p) = \frac{1}{2\sqrt{6}}\left(\frac{1}{a_0}\right)^{3/2}\frac{r}{a_0}e^{-r/2a_0}$$

3s

2s

1s

▲ FIGURE 8-25
Three-dimensional representations of the 95% electron probability density for the 1s, 2s, and 3s orbitals
The first three s orbitals of the hydrogen atom. Note the increasing size of the 95% probability density contour in proceeding from 1s to 2s and on to 3s.

Thus, the 2*p* orbital has no radial nodes at finite values of *r*. In contrast to the *s* orbitals, which are nonzero at $r = 0$, the *p* orbitals vanish at $r = 0$. This difference will have an important consequence when we consider multi-electron atoms.

In contrast to the angular function of the 2*s* orbital, the angular part of the 2*p* orbital is not a constant, but a function of θ and ϕ. This means that the electron probability density distribution of a *p* orbital is not spherically symmetric; that is, it does not have a spherical shape. We see this most easily in Table 8.1 in the functional form of the angular part of the $2p_z$ wave function; it is proportional to $\cos \theta$. Thus the $2p_z$ wave function has an angular maximum along the positive *z* axis, for there $\theta = 0$ and $\cos 0 = +1$. Along the negative *z* axis, the p_z wave function has its most negative value, for there $\theta = \pi$ and $\cos \pi = -1$. That the angular part has its maximum magnitude along the *z* axis is the reason for the designation p_z. Everywhere in the *xy* plane $\theta = \pi/2$ and $\cos \theta = 0$, so the *xy* plane is a node. Because this node arises in the angular function, it is called an *angular node*. A similar analysis of the p_x and p_y orbitals shows that they are similar to the p_z orbital, but with angular nodes in the *yz* and *xz* planes, respectively.

Figure 8-26 shows the two ways of representing the angular part of the p_z wave function. In Figure 8-26(a), the function $\cos \theta$ is plotted as a function of θ and results in two tangential circles. In Figure 8-26(b), the function $\cos^2 \theta$, which is related to the angular electron probability density, is plotted as a function of θ, resulting in a double teardrop shape. Both these representations are used. What is important to note in part (a) of the figure is the phase of the plot of $\cos \theta$ and in (b), the lack of phase of $\cos^2 \theta$, which is always positive. We shall see later in the text that the phase of the orbital is important in understanding chemical bonding.

The simultaneous display of both the radial and angular parts of $\psi^2(2p)$ is more difficult to achieve, but Figure 8-27 attempts to do this for the 95% probability surface of the p_z orbital. All three of the *p* orbitals are shown in Figure 8-28 and are seen to be directed along the three perpendicular axes of the Cartesian system. Again we have used different colors to represent the phase alternation in these orbitals. However, we must remember that these refer only to the phases of the original wave function, *not* to ψ^2.

▶ The points at which a wave function changes sign are nodes. However, even though the 2*p* wave function becomes zero at $r = 0$ and $r = \infty$, these points are not true nodes because the function does not change sign at these points. These points are sometimes called *trivial nodes*.

d Orbitals

The *d* orbitals occur for the first time when $n = 3$. The angular function in these cases possesses two angular (or planar) nodes. Let's illustrate this with the orbital that has an angular function proportional to

$$\sin^2 \theta \cos 2\phi$$

▶ FIGURE 8-26
Two representations of the p orbital angular function
(a) A plot of $\cos \theta$ in the *zx* plane, representing the angular part of the $2p_z$ wave function. Note the difference in the color of the function in the two lobes representing the phase of the angular wave function. (b) A plot of $\cos^2 \theta$ in the *zx* plane, representing the square of the wave function and proportional to the angular probability density of finding the electron.

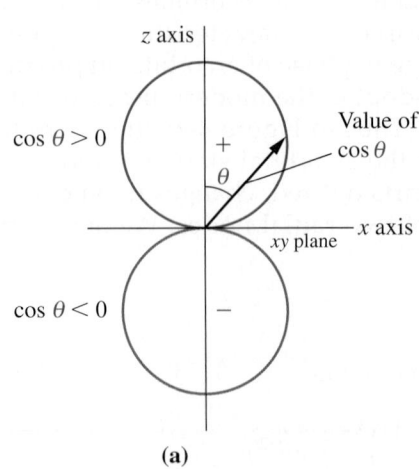

(a)

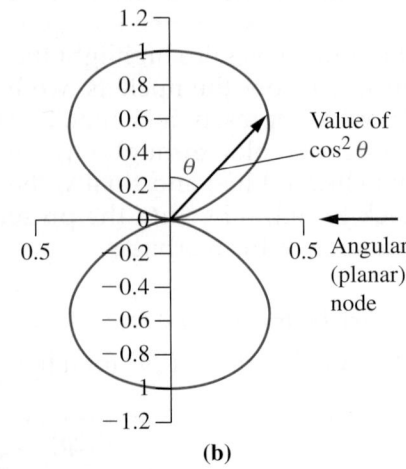

(b)

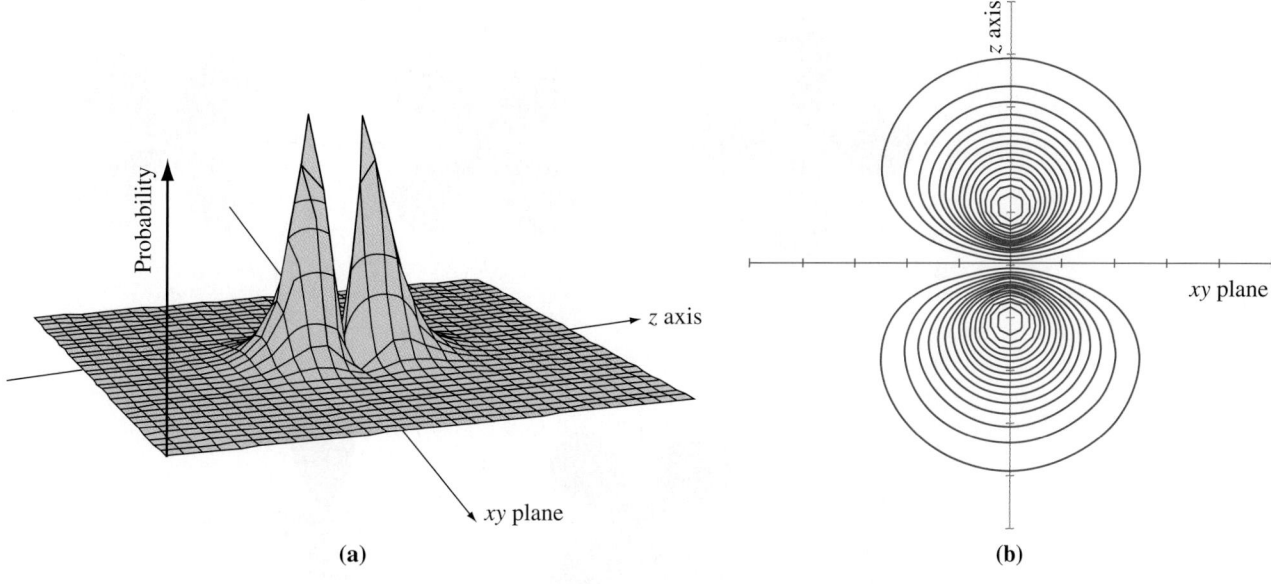

(a)   **(b)**

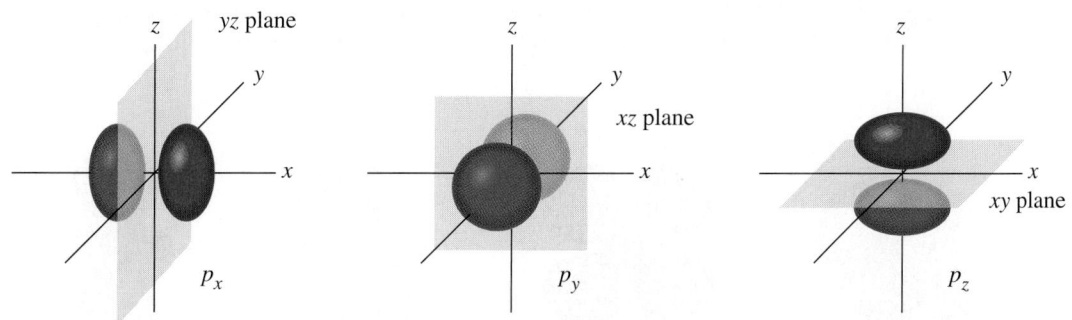

(c)

▲ FIGURE 8-27
Three representations of electron probability for a 2p orbital
(a) The value of ψ^2 is plotted as a height above a plane passing through the nucleus, such as the xy plane. The value of ψ^2 is zero at the nucleus, rises to a maximum on either side, and then falls off with distance (r) along a line through the nucleus (that is, along the x, y, or z axis). **(b)** A contour plot of the electron probability in a plane passing through the nucleus, for example, the xz plane. **(c)** Electron probabilities and charge densities represented in three dimensions. The greatest probability of finding an electron is within the two lobes of the dumbbell-shaped region. Note that this region is *not* spherically symmetric. Note also that the probability drops to zero in the shaded plane—the nodal plane (the xy plane). As with the 2s and 3s orbitals, we have indicated changes of phase through different colors.

▲ FIGURE 8-28
The three 2p orbitals
The p orbitals are usually represented as directed along the perpendicular x, y, and z axes, and the symbols p_x, p_y and p_z are often used. The p_z orbital has $m_\ell = 0$. The situation with p_x and p_y is more complex, however. Each of these orbitals has contributions from both $m_\ell = 1$ and $m_\ell = -1$. Our main concern is just to recognize that p orbitals occur in sets of three and can be represented in the orientation shown here. In higher-numbered shells, p orbitals have a somewhat different appearance, but we will use these general shapes for all p orbitals. The colors of the lobes signify the different phases of the original wave function.

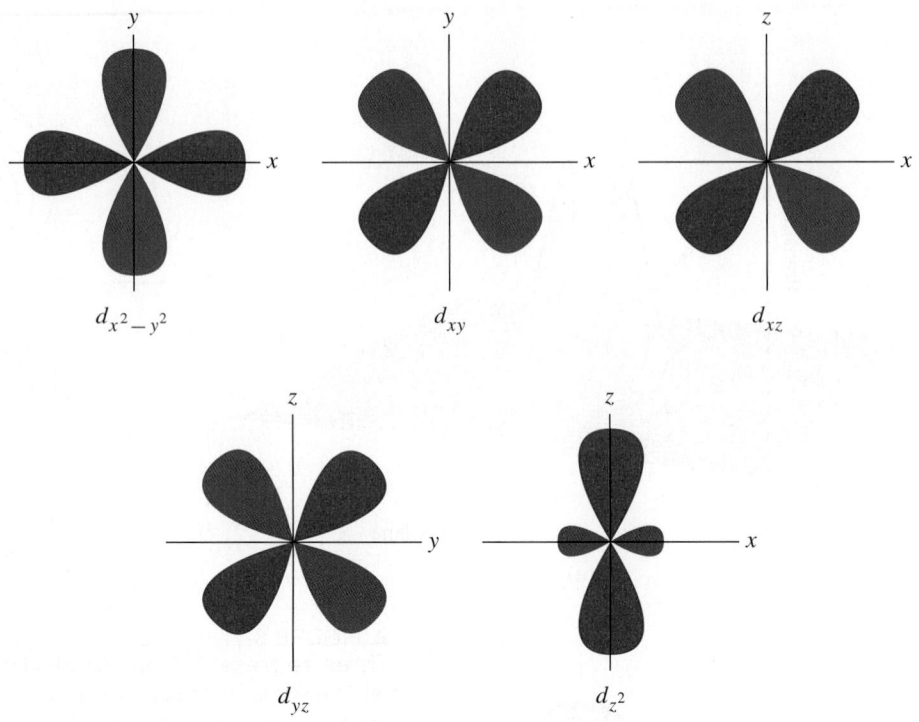

▲ FIGURE 8-29
Cross sections of the five *d* orbitals
The two-dimensional cross sections of the angular functions of the five *d* orbitals in the planes indicated.

How should we visualize this function? We can proceed by setting $\theta = \pi/2$ and plotting the function $\cos 2\phi$. Study Figure 8-22 (page 324) and you will see that the angle $\theta = \pi/2$ corresponds to the *xy* plane, yielding the cross section in the upper left of Figure 8-29. The wave function exhibits positive and negative lobes, indicated by the red and blue lobes, respectively, along the *x* and *y* axes. This orbital, in common with all the other *d* orbitals, is a function of two of the three variables (*x*, *y*, and *z*). It is designated $d_{x^2-y^2}$. The other *d* orbitals, d_{xy}, d_{xz}, d_{yz}, and d_{z^2}, are also displayed in Figure 8-29. We observe that four of them have the same basic shape except for orientation with respect to the axes and that d_{z^2} has quite a different shape.

The 95% probability surfaces of the five *d* orbitals are shown in Figure 8-30. Two of the *d* orbitals ($d_{x^2-y^2}$ and d_{z^2}) are seen to be directed along the three

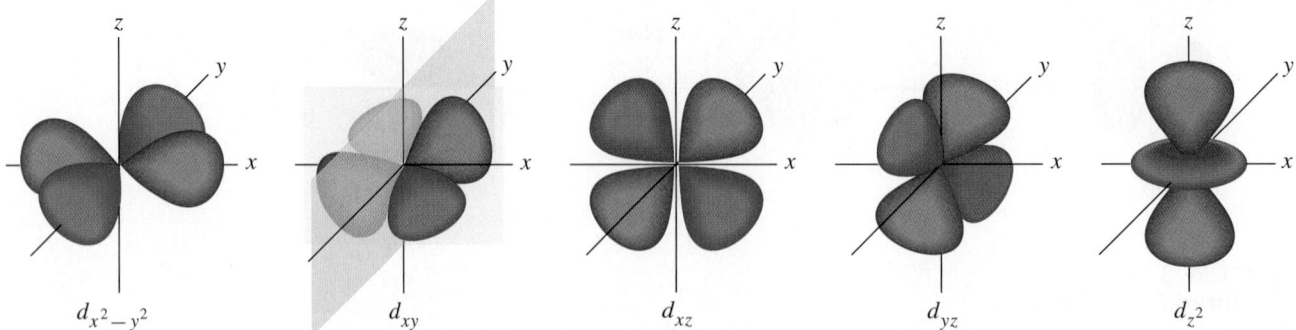

▲ FIGURE 8-30
Representations of the five *d* orbitals
The designations *xy*, *xz*, *yz*, and so on, are related to the values of the quantum number m_ℓ, but this is a detail that we will not pursue in the text. The number of nodal surfaces for an orbital is equal to the ℓ quantum number. For *d* orbitals, there are two such surfaces. The nodal planes for the d_{xy} orbital are shown here. (The nodal surfaces for the d_{z^2} orbital are actually cone-shaped.)

perpendicular axes of the Cartesian system, and the remaining three (d_{xy}, d_{xz}, d_{yz}) are seen to point between these Cartesian axes. Again, the relative phases of the lobes of the original wave function are indicated by the different colors. The d orbitals are important in understanding the chemistry of the transition elements, as we will see in Chapter 23.

8-6   ARE YOU WONDERING...

What does a $3p_z$ and a $4d_{xy}$ orbital look like?

When considering the shapes of the atomic orbitals with higher principal quantum number, we can draw on what we have discussed already and include the extra radial nodes that occur. For example, the $3p_z$ orbital has a total of two nodes (number of nodes $= n - 1$); one of these nodes is taken up with the angular node of the angular part of the $3p_z$ wave function, so that the radial part of the wave function will also have a node. Figure 8-31 shows a contour plot of the value of the $3p_z$ wave function in the xz plane in the manner of Figure 8-27(b). We notice that the $3p_z$ orbital has the same general shape as a $2p_z$ orbital due to the angular node, but the radial node has appeared as a circle (dashed in Figure 8-31). The appearance of the $3p_z$ orbital is that of a smaller p orbital inside a larger one. Similarly the $4d_{xy}$ orbital appears as a smaller d_{xy} inside a larger one. In Figure 8-31 the radial node is indicated by the dashed circle and the presence of the node is indicated by the alternation in color. This idea can be extended to enable us to sketch orbitals of increasing principal quantum number.

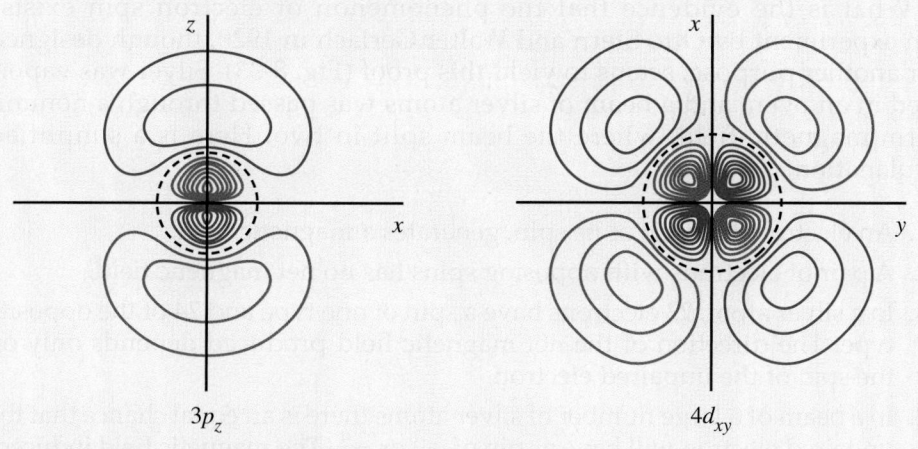

$3p_z$ $4d_{xy}$

▲ FIGURE 8-31
Contour plots for $3p_z$ and the $4d_{xy}$ orbital
The relative phases in these orbitals are shown by the colors red and blue.
The radial nodes are represented by the dashed circles.

 8-7   CONCEPT ASSESSMENT

What type of orbital has three angular nodes and one radial node?

8-9   Electron Spin: A Fourth Quantum Number

Wave mechanics provides three quantum numbers with which we can develop a description of electron orbitals. However, in 1925, George Uhlenbeck and Samuel Goudsmit proposed that some unexplained features of

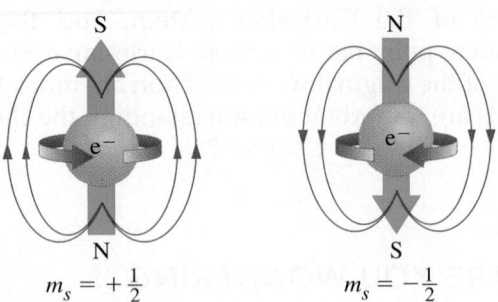

▲ FIGURE 8-32
Electron spin visualized
Two possibilities for electron spin are shown with their associated magnetic fields. Two electrons with opposing spins have opposing magnetic fields that cancel, leaving no net magnetic field for the pair.

the hydrogen spectrum could be understood by assuming that an electron acts as if it spins, much as Earth spins on its axis. As suggested by Figure 8-32, there are two possibilities for **electron spin**. Thus, these two possibilities require a fourth quantum number, the electron spin quantum number m_s. The electron spin quantum number may have a value of $+\frac{1}{2}$ (also denoted by the arrow ↑) or $-\frac{1}{2}$ (denoted by the arrow ↓); the value of m_s does not depend on any of the other three quantum numbers.

What is the evidence that the phenomenon of electron spin exists? An experiment by Otto Stern and Walter Gerlach in 1920, though designed for another purpose, seems to yield this proof (Fig. 8-33). Silver was vaporized in an oven, and a beam of silver atoms was passed through a nonuniform magnetic field, where the beam split in two. Here is a simplified explanation.

1. An electron, because of its spin, generates a magnetic field.
2. A pair of electrons with opposing spins has no net magnetic field.
3. In a silver atom, 23 electrons have a spin of one type and 24 of the opposite type. The direction of the net magnetic field produced depends only on the spin of the unpaired electron.
4. In a beam of a large number of silver atoms there is an equal chance that the unpaired electron will have a spin of $+\frac{1}{2}$ or $-\frac{1}{2}$. The magnetic field induced by the silver atoms interacts with the nonuniform field, and the beam of silver atoms splits into two beams.

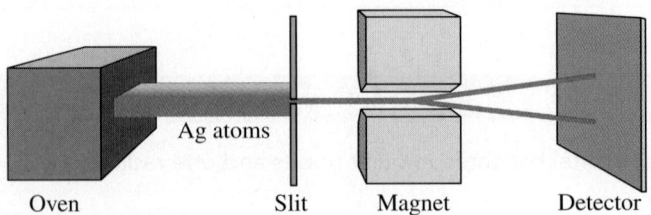

▲ FIGURE 8-33
The Stern–Gerlach experiment
Ag atoms vaporized in the oven are collimated into a beam by the slit, and the beam is passed through a nonuniform magnetic field. The beam splits in two. The beam of atoms would not experience a force if the magnetic field were uniform. The field strength must be stronger in certain directions than in others.

Electronic Structure of the H Atom: Representing the Four Quantum Numbers

Now that we have described the four quantum numbers, we are in a position to bring them together into a description of the electronic structure of the hydrogen atom. The electron in a ground-state hydrogen atom is found at the lowest energy level. This corresponds to the principal quantum number $n = 1$, and because the first principal shell consists only of an s orbital, the orbital quantum number $\ell = 0$. The only possible value of the magnetic quantum number is $m_\ell = 0$. Either spin state is possible for the electron, and we do not know which it is unless we do an experiment like that of Uhlenbeck and Goudsmit's. Thus,

$$n = 1 \qquad \ell = 0 \qquad m_\ell = 0 \qquad m_s = +\frac{1}{2} \text{ or } -\frac{1}{2}$$

Chemists often say that the electron in the ground-state hydrogen atom is in the 1s orbital, or that it is a 1s electron, and they represent this by the notation

$$1s^1$$

where the superscript 1 indicates one electron in the 1s orbital. Either spin state is allowed, but we do not designate the spin state in this notation.

In the excited states of the hydrogen atom, the electron occupies orbitals with higher values of n. Thus, when excited to the level with $n = 2$, the electron can occupy either the 2s or one of the 2p orbitals; all have the same energy. Because the probability density extends farther from the nucleus in the 2s and 2p orbitals than in the 1s orbital, the excited-state atom is larger than is the ground-state atom. The excited states just described can be represented as

$$2s^1 \text{ or } 2p^1$$

In the remaining sections of the chapter this discussion will be extended to the electronic structures of atoms having more than one electron—*multielectron* atoms.

KEEP IN MIND

that orbitals are mathematical functions and not themselves physical regions in space. However, it is customary to refer to an electron that is described by a particular orbital as being "in the orbital."

EXAMPLE 8-12   **Choosing an Appropriate Combination of the Four Quantum Numbers: n, ℓ, m_ℓ, and m_s**

From the following sets of quantum numbers (n, ℓ, m_ℓ, m_s), identify the set that is correct, and state the orbital designation for those quantum numbers:

$$(2,1,0,0) \quad \left(2,0,1,\frac{1}{2}\right) \quad \left(2,2,0,\frac{1}{2}\right) \quad \left(2,-1,0,\frac{1}{2}\right) \quad \left(2,1,0,-\frac{1}{2}\right)$$

Analyze

We know that if $n = 2$, ℓ has two possible values: 0 or 1. The range of values for m_ℓ is given by equation (8.16), and $m_s = \pm\frac{1}{2}$. By using this information, we can judge which combination is correct.

Solve

(n, ℓ, m_ℓ, m_s)	Comment
$(2,1,0,0)$	The value of m_s is incorrect.
$\left(2,0,1,\dfrac{1}{2}\right)$	The value of m_ℓ is incorrect.
$\left(2,2,0,\dfrac{1}{2}\right)$	The value of ℓ is incorrect.
$\left(2,-1,0,\dfrac{1}{2}\right)$	The value of ℓ is incorrect.
$\left(2,1,0,-\dfrac{1}{2}\right)$	All the quantum numbers are correct.

(continued)

The correct combination of quantum numbers has $n = 2$, $\ell = 1$, $m_\ell = 0$, and $m_s = -\frac{1}{2}$, which corresponds to a 2p orbital.

Assess

The combination of quantum numbers identified above for an electron in a 2p orbital is one of six possible combinations. The other five combinations for an electron in a 2p orbital are $\left(2, 1, 0, \frac{1}{2}\right)$, $\left(2, 1, -1, -\frac{1}{2}\right)$, $\left(2, 1, -1, \frac{1}{2}\right)$, $\left(2, 1, 1, -\frac{1}{2}\right)$, and $\left(2, 1, 1, \frac{1}{2}\right)$.

PRACTICE EXAMPLE A: Determine which set of the following quantum numbers (n, ℓ, m_ℓ, m_s) is wrong and indicate why:

$$(3, 2, -2, 1) \quad \left(3, 1, -2, \frac{1}{2}\right) \quad \left(3, 0, 0, \frac{1}{2}\right) \quad \left(2, 3, 0, \frac{1}{2}\right) \quad \left(1, 0, 0, -\frac{1}{2}\right) \quad \left(2, -1, -1, \frac{1}{2}\right)$$

PRACTICE EXAMPLE B: Identify the error in each set of quantum numbers below:

$$(2, 1, 1, 0) \quad \left(1, 1, 0, \frac{1}{2}\right) \quad \left(3, -1, 1, -\frac{1}{2}\right) \quad \left(0, 0, 0, -\frac{1}{2}\right) \quad \left(2, 1, 2, \frac{1}{2}\right)$$

8-10 Multielectron Atoms

Schrödinger developed his wave equation for the hydrogen atom—an atom containing just one electron. For multielectron atoms, a new factor arises: mutual repulsion between electrons. The repulsion between the electrons means that the electrons in a multielectron atom tend to stay away from one another, and their motions become inextricably entangled. The approximate approach taken to solve this many-particle problem is to consider the electrons, one by one, in the environment established by the nucleus and the other electrons. When this is done, the electron orbitals obtained are of the same types as those obtained for the hydrogen atom; they are called *hydrogen-like* orbitals. Compared with the hydrogen atom, the angular parts of the orbitals of a multielectron atom are unchanged, but the radial parts are different.

We have seen that the solution of the Schrödinger equation for a hydrogen atom gives the energies of the orbitals and that all orbitals with the same principal quantum number n are degenerate—they have the same energy. In a hydrogen atom, the orbitals 2s and 2p are degenerate, as are 3s, 3p, and 3d.

In multielectron atoms, the attractive force of the nucleus for a given electron increases as the nuclear charge increases. As a result, we find that orbital energies become lower (more negative) with increasing atomic number of the atom. Also, orbital energies in multielectron atoms depend on the type of orbital; the orbitals with different values of ℓ within a principal shell are not degenerate.

KEEP IN MIND

that orbital-wave functions extend farther out from the nucleus as n increases. Thus, an electron in a 3s or 3p orbital has a higher probability of being farther from the nucleus than does an electron in a 1s orbital.

Penetration and Shielding

Think about the attractive force of the atomic nucleus for one particular electron some distance from the nucleus. Electrons in orbitals closer to the nucleus *screen* or *shield* the nucleus from electrons farther away. In effect, the screening electrons reduce the effectiveness of the nucleus in attracting the particular more-distant electron. They effectively reduce the nuclear charge.

The magnitude of the reduction of the nuclear charge depends on the types of orbitals the inner electrons are in and the type of orbital that the screened electron is in. We have seen that electrons in s orbitals have a high probability density at the nucleus, whereas p and d orbitals have zero probability densities at the nucleus. Thus, electrons in *s* orbitals are more effective at screening the nucleus from outer electrons than are electrons in *p* or *d* orbitals. This ability of electrons in *s* orbitals that allows them to get close to the nucleus is called *penetration*. An electron in an orbital with good penetration is better at screening than one with low penetration.

We must consider a different kind of probability distribution to describe the penetration to the nucleus by orbital electrons. Rather than considering the probability at a point, which we did to ascribe three-dimensional shapes to orbitals, we need to consider the probability of finding the electron anywhere in a spherical shell of radius *r* and an infinitesimal thickness. This type of probability is called a *radial probability distribution* and is found by multiplying the radial probability density, $R^2(r)$, by the factor $4\pi r^2$, the area of a sphere of radius *r*. Figure 8-34 offers a dartboard analogy that might help clarify the distinction between probability at a point and probability in a region of space.

The quantity $4\pi r^2 \times R^2(r)$ provides a different insight into the behavior of the electron. The radial probability distributions for some hydrogenic (hydrogen-like) orbitals are plotted in Figure 8-35. The radial probability density, $R^2(r)$, for a 1s orbital predicts that the maximum probability for a 1s electron is *at* the nucleus. However, because the volume of this region is vanishingly small ($r = 0$), the radial probability distribution $[4\pi r^2 \times R^2(r)]$ is zero at the nucleus. The electron in a hydrogen atom is most likely to be found 53 pm from the nucleus; this is where the radial probability distribution reaches a maximum. This is the same radius as the first Bohr orbit. The boundary surface within which there is a 95% probability of finding an electron (see Figure 8-25, p. 329) is a much larger sphere, one with a radius of about 141 pm.

In comparing the radial probability curves for the 1s, 2s, and 3s orbitals, we find that a 1s electron has a greater probability of being close to the nucleus than a 2s electron does, which in turn has a greater probability than does a 3s electron. In comparing 2s and 2p orbitals, a 2s electron has a greater chance of being close to the nucleus than a 2p electron does. The 2s electron exhibits

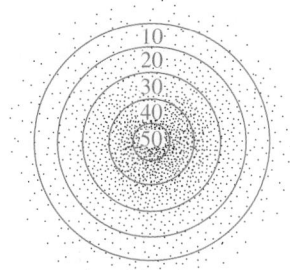

▲ FIGURE 8-34
Dartboard analogy to a 1s orbital
Imagine that a single dart (electron) is thrown at a dartboard 1500 times. The board contains 95% of all the holes; it is analogous to the 1s orbital. Where is a thrown dart most likely to hit? The number of holes per unit area is greatest in the "50" region—that is, the 50 region has the greatest probability density. The most likely score is "30," however, because the most probable area hit is in the 30 ring and not the 50 ring, which is smaller than the 30 ring. The 30 *ring* on the dartboard is analogous to a spherical *shell* of 53 pm radius within the larger sphere representing the 1s orbital.

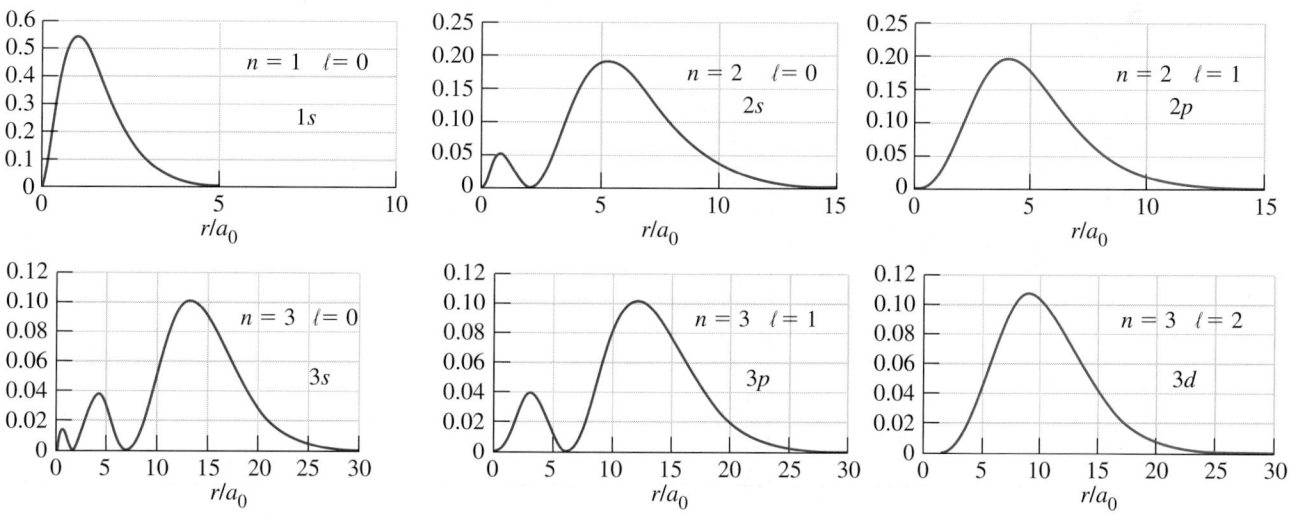

▲ FIGURE 8-35
Radial probability distributions
Graphs of the value of $4\pi r^2 R^2(r)$ as a function of *r* for the orbitals in the first three principal shells. Note that the smaller the orbital angular momentum quantum number, the more closely an electron approaches the nucleus. Thus, *s* orbital electrons penetrate more, and are less shielded from the nucleus, than electrons in other orbitals with the same value of *n*.

greater penetration than the 2*p* electron. Electrons having a high degree of penetration effectively "block the view" of an electron in an outer orbital "looking" for the nucleus.

The nuclear charge that an electron would experience if there were no intervening electrons is Z, the atomic number. The nuclear charge that an electron actually experiences is reduced by intervening electrons to a value of Z_{eff}, called the **effective nuclear charge**. The less of the nuclear charge that an outer electron "sees" (that is, the smaller the value of Z_{eff}), the smaller is the attraction of the electron to the nucleus, and hence the *higher* is the energy of the orbital in which the electron is found.

To summarize, compared with a *p* electron in the same principal shell, an *s* electron is more penetrating and not as well screened. The *s* electron experiences a higher Z_{eff}, is held more tightly, and is at a lower energy than a *p* electron. Similarly, the *p* electron is at a lower energy than a *d* electron in the same principal shell. Thus, the energy level of a principal shell is split into separate levels for its subshells. There is no further splitting of energies within a subshell, however, because all the orbitals in the subshell have the same radial characteristics and thereby experience the same effective nuclear charge, Z_{eff}. As a result, all three *p* orbitals of a principal shell have the same energy; all five *d* orbitals have the same energy; and so on.

In a few instances, the combined effect of the decreased spacing between successive energy levels at higher quantum numbers (because of the energy dependence on $1/n^2$) and the splitting of subshell energy levels (because of shielding and penetration) causes some energy levels to overlap. For example, because of the extra penetration of a 4*s* electron over that of a 3*d* electron, the 4*s* energy level is below the 3*d* level despite its higher principal quantum number *n* (Fig. 8-36). We will see some of the consequences of this energy-level splitting in the next two sections, where we consider the relationship between the electronic structures of atoms and their positions in the periodic table.

KEEP IN MIND

that, similar to the situation in equation (8.9), the energy of an orbital (E_n) is given by the proportionality

$$E_n \propto -\frac{Z_{eff}^2}{n^2}.$$

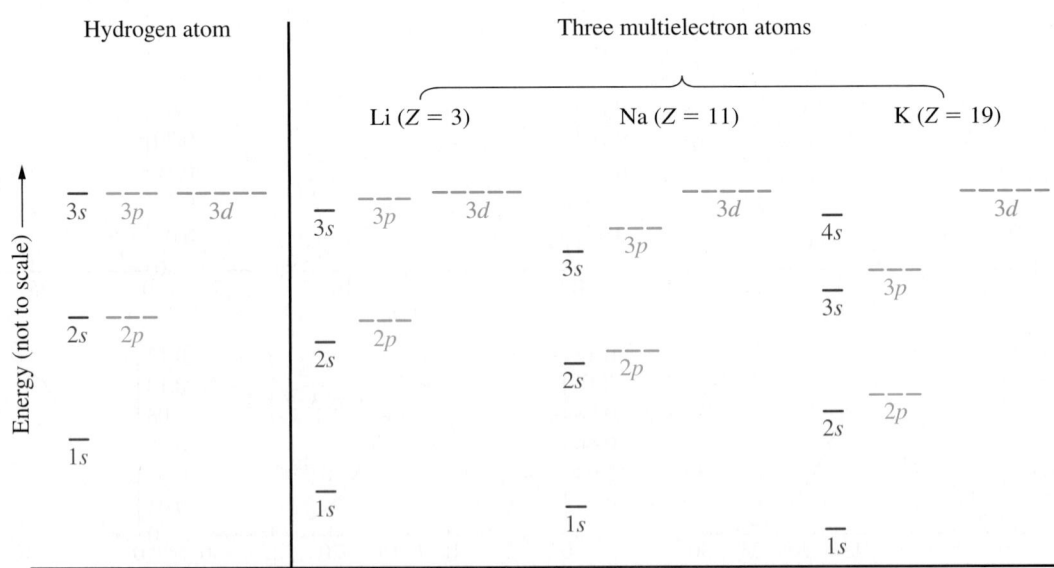

▲ FIGURE 8-36
Orbital energy-level diagram for the first three electronic shells
Energy levels are shown for a hydrogen atom (left) and three typical multielectron atoms (right). Each multielectron atom has its own energy-level diagram. Note that for the hydrogen atom, orbital energies within a principal shell—for example, 3*s*, 3*p*, 3*d*—are alike (degenerate), but in a multielectron atom they become rather widely separated. Another feature of the diagram is the steady decrease in all orbital energies with increasing atomic number. Finally, note that the 4*s* orbital is at a lower energy than 3*d*.

8-11   Electron Configurations

The **electron configuration** of an atom is a designation of how electrons are distributed among various orbitals in principal shells and subshells. In later chapters, we will find that many of the physical and chemical properties of elements can be correlated with electron configurations. In this section, we will see how the results of wave mechanics, expressed as a set of rules, can help us to write probable electron configurations for the elements.

Rules for Assigning Electrons to Orbitals

1. **Electrons occupy orbitals in a way that minimizes the energy of the atom.** Figure 8-36 suggests the order in which electrons occupy the subshells in the principal electronic shells; first the $1s$, then $2s$, $2p$, and so on. The exact order of filling of orbitals has been established by experiment, principally through spectroscopy and magnetic studies, and it is this order based on experiment that we must follow in assigning electron configurations to the elements. With only a few exceptions, the order in which orbitals fill is

$$1s, 2s, 2p, 3s, 3p, 4s, 3d, 4p, 5s, 4d, 5p, 6s, 4f, 5d, 6p, 7s, 5f, 6d, 7p \qquad \textbf{(8.20)}$$

   Some students find the diagram pictured in Figure 8-37 a useful way to remember this order, but the best method of establishing the order of filling of orbitals is based on the periodic table, as we will see in Section 8-12.

2. **No two electrons in an atom can have all four quantum numbers alike—the Pauli exclusion principle.** In 1926, Wolfgang Pauli explained complex features of emission spectra associated with atoms in magnetic fields by proposing that no two electrons in an atom can have all four quantum numbers alike. The first three quantum numbers, n, ℓ, and m_ℓ, determine a specific orbital. Two electrons may have these three quantum numbers alike; but if they do, they must have different values of m_s, the spin quantum number. Another way to state this result is that *only two electrons may occupy the same orbital, and these electrons must have opposing spins.*

   Because of this limit of two electrons per orbital, the capacity of a subshell for electrons can be obtained by doubling the number of orbitals in the subshell. Thus, the s subshell consists of *one* orbital with a capacity of *two* electrons; the p subshell consists of *three* orbitals with a total capacity of *six* electrons; and so on.

3. **When orbitals of identical energy (degenerate orbitals) are available, electrons initially occupy these orbitals singly.** In line with this rule, known as **Hund's rule**, an atom tends to have as many unpaired electrons as possible. This behavior can be rationalized by saying that electrons, because they all carry the same electric charge, try to get as far apart as possible. They do this by seeking out empty orbitals of similar energy in preference to pairing up with an electron in a half-filled orbital.

Representing Electron Configurations

Before we assign electron configurations to atoms of the different elements, we need to introduce methods of representing these configurations. The electron configuration of a carbon atom is shown in three different ways:

spdf notation (condensed):   C $1s^2 2s^2 2p^2$

spdf notation (expanded):   C $1s^2 2s^2 2p_x^1 2p_y^1$

orbital diagram: C

↑↓	↑↓	↑	↑	

    $1s$ $2s$ $2p$

In each of these methods we assign six electrons because the atomic number of carbon is 6. Two of these electrons are in the $1s$ subshell, two in the $2s$, and

◀ This order of filling corresponds roughly to the order of increasing orbital energy, but the overriding principle governing the order of filling of orbitals is that the energy of the atom as a whole be kept at a minimum.

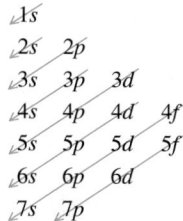

▲ FIGURE 8-37
The order of filling of electronic subshells
Beginning with the top line, follow the arrows, and the order obtained is the same as in expression (8.20).

▶ When listed in tables, as in Appendix D, electron configurations are usually written in the condensed *spdf* notation.

two in the 2*p*. The condensed *spdf* **notation** denotes only the total number of electrons in each subshell; it does not show how electrons are distributed among orbitals of equal energy. In the expanded *spdf* **notation,** Hund's rule is reflected in the assignment of electrons to the 2*p* subshell—two 2*p* orbitals are singly occupied and one remains empty. The **orbital diagram** breaks down each subshell into individual orbitals (drawn as boxes). This notation is similar to an energy-level diagram except that the direction of increasing energy is from left to right instead of vertically.

Electrons in orbitals are shown as arrows. An arrow pointing up corresponds to one type of spin $(+\frac{1}{2})$, and an arrow pointing down to the other $(-\frac{1}{2})$. Electrons in the same orbital with opposing (opposite) spins are said to be *paired* $(\uparrow\downarrow)$. The electrons in the 1*s* and 2*s* orbitals of the carbon atom are paired. Electrons in different, singly occupied orbitals of the same subshell have the same, or *parallel*, spins (arrows pointing in the same direction). This is conveyed in the orbital diagram for carbon, where we write $[\uparrow][\uparrow][\ ]$ rather than $[\uparrow][\downarrow][\ ]$ for the 2*p* subshell. Both experiment and theory confirm that an electron configuration in which electrons in singly occupied orbitals have parallel spins is a better representation of the lowest energy state of an atom than any other electron configuration that we can write. The configuration represented by the orbital diagram $[\uparrow][\downarrow][\ ]$ is, in fact, an excited state of carbon; any orbital diagram with unpaired spins that are not parallel constitutes an excited state.

The most stable or the most energetically favorable configurations for isolated atoms, those discussed here, are called *ground-state electron configurations*. Later in the text we will briefly mention some electron configurations that are not the most stable. Atoms with such configurations are said to be in an *excited state*.

The Aufbau Process

To write electron configurations we will use the **aufbau process.** *Aufbau* is a German word that means "building up," and what we do is assign electron configurations to the elements in order of increasing atomic number. To proceed from one atom to the next, we add a proton and some neutrons to the nucleus and then describe the orbital into which the added electron goes.

Z = 1, H. The lowest energy state for the electron is the 1*s* orbital. The electron configuration is 1*s*^{1}.

Z = 2, He. A second electron goes into the 1*s* orbital, and the two electrons have opposing spins, 1*s*^{2}.

Z = 3, Li. The third electron cannot be accommodated in the 1*s* orbital (Pauli exclusion principle). It goes into the lowest energy orbital available, 2*s*. The electron configuration is 1*s*^{2}2*s*^{1}.

Z = 4, Be. The configuration is 1*s*^{2}2*s*^{2}.

Z = 5, B. Now the 2*p* subshell begins to fill: 1*s*^{2}2*s*^{2}2*p*^{1}.

Z = 6, C. A second electron also goes into the 2*p* subshell, but into one of the remaining empty *p* orbitals (Hund's rule) with a spin parallel to the first 2*p* electron. (See figure to the left.)

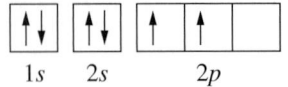

1*s*   2*s*   2*p*

Z = 7–10, N through Ne In this series of four elements, the filling of the subshell is completed. The number of unpaired electrons reaches a maximum (three) with nitrogen and then decreases to zero with neon.

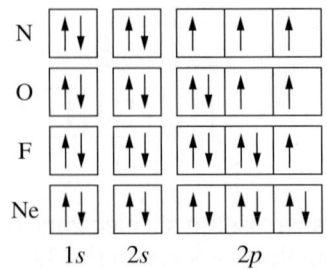

1*s*   2*s*   2*p*

Z = 11–18, Na through Ar. The filling of orbitals for this series of eight elements closely parallels the eight elements from Li through Ne, except that electrons go into 3*s* and 3*p* subshells. Each element has the 1*s*, 2*s*, and 2*p* subshells filled. Because the configuration $1s^22s^22p^6$ is that of neon, we will call this the neon core, represent it as [Ne], and concentrate on the electrons beyond the core. Electrons that are added to the electronic shell of highest principal quantum number (the outermost, or *valence shell*) are called **valence electrons**. The electron configuration of Na is written below in a form called a *noble-gas-core-abbreviated electron configuration*, consisting of [Ne] as the noble gas core and $3s^1$ as the configuration of the valence electron. For the other third-period elements, only the valence-shell electron configurations are shown.

Na	Mg	Al	Si	P	S	Cl	Ar
$[Ne]3s^1$	$3s^2$	$3s^23p^1$	$3s^23p^2$	$3s^23p^3$	$3s^23p^4$	$3s^23p^5$	$3s^23p^6$

Z = 19 and 20, K and Ca. After argon, instead of 3*d*, the next subshell to fill is 4*s*. Using the symbol [Ar] to represent the noble gas core, $1s^22s^22p^63s^23p^6$, we get the electron configurations shown below for K and Ca.

$$\text{K: [Ar]}4s^1 \quad \text{and} \quad \text{Ca: [Ar]}4s^2$$

Z = 21–30, Sc through Zn. In this next series of elements, electrons fill the *d* orbitals of the third shell. The *d* subshell has a total capacity of ten electrons—ten elements are involved. There are two possible ways to write the electron configuration of scandium.

$$\text{(a) Sc: [Ar]}3d^14s^2 \quad \text{or} \quad \text{(b) Sc: [Ar]}4s^23d^1$$

◀ Although method (b) conforms better to the order in which orbitals fill, method (a) better represents the order in which electrons are lost on ionization, as we will see in the next chapter.

Both methods are commonly used. Method (a) groups together all the subshells of a principal shell and places subshells of the highest principal quantum level last. Method (b) lists orbitals in the apparent order in which they fill. In this text, we will use method (a).

The electron configurations of this series of ten elements are listed below in both the orbital diagram and the *spdf* notation.

		3*d*	3*d*	3*d*	3*d*	3*d*	4*s*	
Sc:	[Ar]	↑					↑↓	$[Ar]3d^14s^2$
Ti:	[Ar]	↑	↑				↑↓	$[Ar]3d^24s^2$
V:	[Ar]	↑	↑	↑			↑↓	$[Ar]3d^34s^2$
Cr:	[Ar]	↑	↑	↑	↑	↑	↑	$[Ar]3d^54s^1$
Mn:	[Ar]	↑	↑	↑	↑	↑	↑↓	$[Ar]3d^54s^2$
Fe:	[Ar]	↑↓	↑	↑	↑	↑	↑↓	$[Ar]3d^64s^2$
Co:	[Ar]	↑↓	↑↓	↑	↑	↑	↑↓	$[Ar]3d^74s^2$
Ni:	[Ar]	↑↓	↑↓	↑↓	↑	↑	↑↓	$[Ar]3d^84s^2$
Cu:	[Ar]	↑↓	↑↓	↑↓	↑↓	↑↓	↑	$[Ar]3d^{10}4s^1$
Zn:	[Ar]	↑↓	↑↓	↑↓	↑↓	↑↓	↑↓	$[Ar]3d^{10}4s^2$

The *d* orbitals fill in a fairly regular fashion in this series, but there are two exceptions: chromium (Cr) and copper (Cu). These exceptions are usually explained in terms of a special stability for configurations in which a $3d$ subshell is half-filled with electrons, as with Cr ($3d^5$), or completely filled, as with Cu ($3d^{10}$).

Z = 31–36, Ga through Kr. In this series of six elements, the $4p$ subshell is filled, ending with krypton.

$$\text{Kr: } [\text{Ar}]3d^{10}4s^24p^6$$

Z = 37–54, Rb to Xe. In this series of 18 elements, the subshells fill in the order $5s, 4d$, and $5p$, ending with the configuration of xenon.

$$\text{Xe: } [\text{Kr}]4d^{10}5s^25p^6$$

Z = 55–86, Cs to Rn. In this series of 32 elements, with a few exceptions, the subshells fill in the order $6s, 4f, 5d, 6p$. The configuration of radon is

$$\text{Rn: } [\text{Xe}]4f^{14}5d^{10}6s^26p^6$$

Z = 87–?, Fr to ? Francium starts a series of elements in which the subshells that fill are $7s, 5f, 6d$, and presumably $7p$, although atoms in which filling of the $7p$ subshell is expected have only recently been discovered and are not yet characterized.

Appendix D gives a complete listing of probable electron configurations.

8-7 ARE YOU WONDERING...

Why chromium and copper have "anomalous" electron configurations?

First we need to recognize that the balance between electron–electron repulsions, electron–nuclear attractions, and other electron–electron interactions, due to the fact that electron motions in a multielectron atom are correlated, determines the most stable electron configuration. Next, recall Hund's rule that the subshell electron configuration with the greatest number of unpaired electrons is the most stable. On page 340, this led us to the conclusion that the ground-state electron configuration of carbon is $1s^22s^22p^2$, with two unpaired electrons in the $2p$ subshell. Calculations on simple atoms and ions have shown that, although there is a larger electron–electron repulsion in the state with the unpaired electrons, this repulsion is more than offset by a larger electron–nucleus attraction because the electrons are closer to the nucleus in the configuration with unpaired electrons. Thus, configurations with unpaired electrons are favored.

In determining the ground-state electron configuration of Cr, we have to choose between $[\text{Ar}]3d^44s^2$ and $[\text{Ar}]3d^54s^1$, because the energies of the $3d$ and $4s$ orbitals are very similar. In this case, we can use Hund's rule to decide that the most stable electron configuration is that with the most unpaired electrons, that is, $[\text{Ar}]3d^54s^1$.

For Cu we have to choose between $[\text{Ar}]3d^94s^2$ and $[\text{Ar}]3d^{10}4s^1$, and there is no difference between the number of unpaired electrons—Hund's rule is no help here. So where does the extra stability for the configuration $[\text{Ar}]3d^{10}4s^1$ come from? A filled (or half-filled) subshell has a spherically symmetrical charge density that leads to a more stable electron configuration; and the greater the number of electrons in that subshell, the greater the stabilization. The $[\text{Ar}]3d^{10}4s^1$ electron configuration is more stable than $[\text{Ar}]3d^94s^2$ because of its filled $3d$ subshell. Thus nearly all the anomalous configurations contain either filled or half-filled subshells.

EXAMPLE 8-13   Recognizing Correct and Incorrect Ground State and Excited State Atomic Orbital Diagrams

Which of the following orbital diagrams is incorrect? Explain. Which of the correct diagrams corresponds to an excited state and which to the ground state of the neutral atom?

Analyze

When faced with a set of orbital diagrams, the best strategy is to investigate each one and apply Hund's rule and the Pauli exclusion principle, the former to decide on ground or excited states, and the latter for the correctness of the diagram.

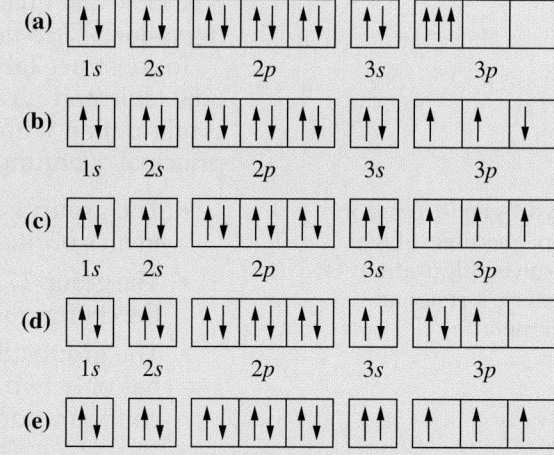

Solve

(a) By scanning diagram (a), we see that all the orbitals $1s, 2s, 2p$, and $3s$ are filled with two electrons of opposite spin, conforming to the Pauli exclusion principle. However, the $3p$ orbital contains three electrons, which violates this principle.

(b) In diagram (b), the orbitals $1s, 2s, 2p$, and $3s$ are filled with two electrons of opposite spin, which is correct. The $3p$ level contains three electrons in separate orbitals, conforming to Hund's rule, but two of them have opposite spin to the other; consequently, this is an excited state of the element.

(c) When we compare diagram (c) with diagram (b), we see that all the three electrons in the $3p$ subshell have the same spin, and so this is the ground state.

(d) When we compare diagram (d) with diagram (b), we see that of the three electrons in the $3p$ subshell, two are paired and one is not. Again, this is an excited state.

(e) By scanning diagram (e), we see that all the orbitals $1s, 2s$, and $2p$ are filled with two electrons of opposite spin. However, the $3p$ orbital contains two electrons with the same spin, which violates the Pauli principle. This diagram is incorrect.

Assess

Orbital diagrams are a useful way to display electronic configurations, but we must take care to obey Hund's rule and the Pauli exclusion principle.

PRACTICE EXAMPLE A: Which two of the following orbital diagrams are equivalent?

(a) [↑↓] [↑↓] [↓] [ ] [↓]
     $1s$ $2s$ $2p$

(b) [↑↓] [↑↓] [↑↓] [ ] [ ]
     $1s$ $2s$ $2p$

(c) [↑↓] [↑↓] [↑] [↑] [ ]
     $1s$ $2s$ $2p$

(d) [↑↓] [↑↓] [↑] [↓] [ ]
     $1s$ $2s$ $2p$

PRACTICE EXAMPLE B: Does the following orbital diagram for a neutral species correspond to the ground state or an excited state?

[Ar] [↑↓]   [↑↓ ↑↓ ↑↓ ↑↓ ↑↓]   [↑↓ ↑]
      $4s$ $3d$ $4p$

8-12 Electron Configurations and the Periodic Table

We have just described the aufbau process of making probable assignments of electrons to the orbitals in atoms. Although electron configurations may seem rather abstract, they actually lead us to a better understanding of the periodic table. Around 1920, Niels Bohr began to promote the connection between the periodic table and quantum theory. The chief link, he pointed out, is in electron configurations. *Elements in the same group of the table have similar electron configurations.*

To construct Table 8.2, we have taken three groups of elements from the periodic table and written their electron configurations. The similarity in electron configuration within each group is readily apparent. If the shell of the highest principal quantum number—the outermost, or valence, shell—is labeled n, then

▶ Hydrogen is found in group 1 because of its electron configuration, $1s^1$. However, it is not an alkali metal.

- The group 1 atoms (alkali metals) have *one* outer-shell (valence) electron in an s orbital, that is, ns^1.
- The group 17 atoms (halogens) have *seven* outer-shell (valence) electrons, in the configuration ns^2np^5.
- The group 18 atoms (noble gases)—with the exception of helium, which has only two electrons—have outermost shells with *eight* electrons, in the configuration ns^2np^6.

Although it is not correct in all details, Figure 8-38 relates the aufbau process to the periodic table by dividing the table into the following four blocks of elements according to the subshells being filled:

- *s block.* The s orbital of highest principal quantum number (n) fills. The s block consists of groups 1 and 2 (plus He in group 18).
- *p block.* The p orbitals of highest quantum number (n) fill. The p block consists of groups 13, 14, 15, 16, 17, and 18 (except He).
- *d block.* The d orbitals of the electronic shell $n - 1$ (the next to outermost) fill. The d block includes groups 3, 4, 5, 6, 7, 8, 9, 10, 11, and 12.
- *f block.* The f orbitals of the electronic shell $n - 2$ fill. The f-block elements are the lanthanides and the actinides.

TABLE 8.2 Electron Configurations of Some Groups of Elements

Group	Element	Configuration
1	H	$1s^1$
	Li	$[\text{He}]2s^1$
	Na	$[\text{Ne}]3s^1$
	K	$[\text{Ar}]4s^1$
	Rb	$[\text{Kr}]5s^1$
	Cs	$[\text{Xe}]6s^1$
	Fr	$[\text{Rn}]7s^1$
17	F	$[\text{He}]2s^22p^5$
	Cl	$[\text{Ne}]3s^23p^5$
	Br	$[\text{Ar}]3d^{10}4s^24p^5$
	I	$[\text{Kr}]4d^{10}5s^25p^5$
	At	$[\text{Xe}]4f^{14}5d^{10}6s^26p^5$
18	He	$1s^2$
	Ne	$[\text{He}]2s^22p^6$
	Ar	$[\text{Ne}]3s^23p^6$
	Kr	$[\text{Ar}]3d^{10}4s^24p^6$
	Xe	$[\text{Kr}]4d^{10}5s^25p^6$
	Rn	$[\text{Xe}]4f^{14}5d^{10}6s^26p^6$

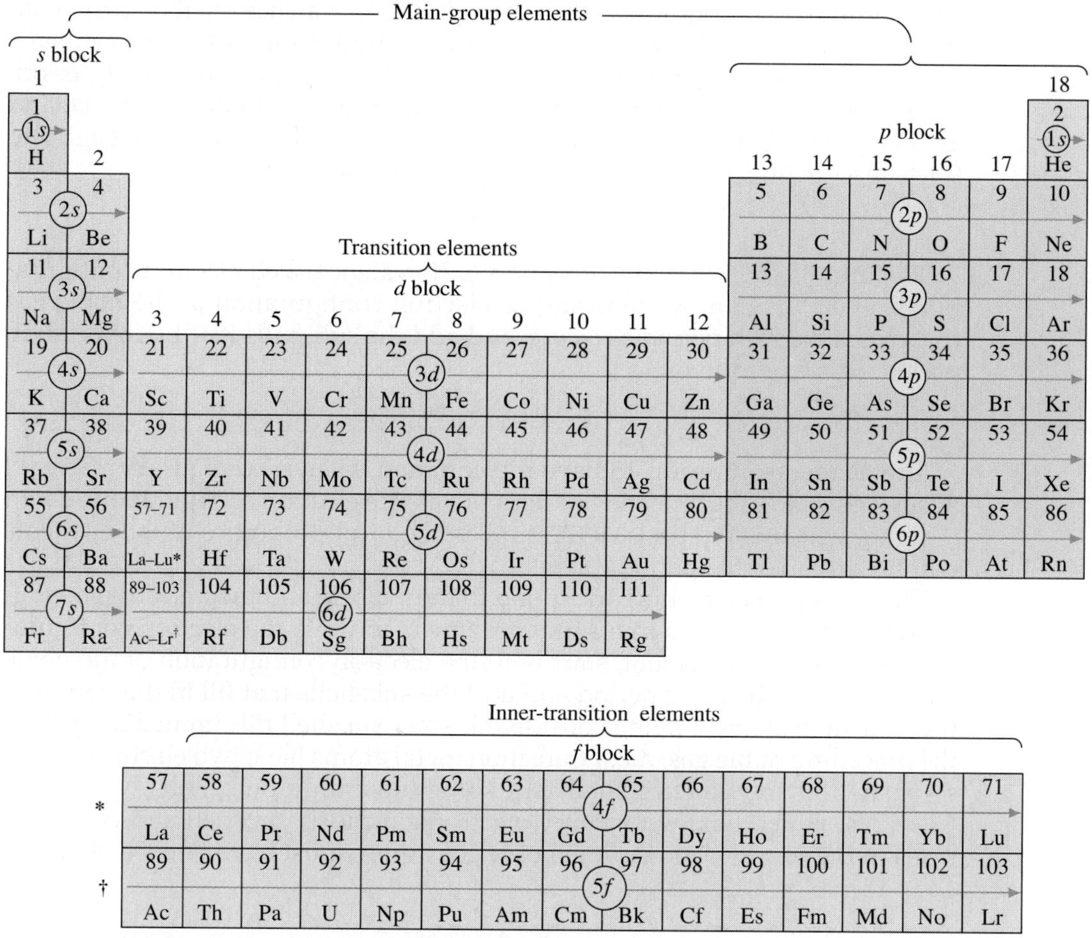

▲ FIGURE 8-38
Electron configurations and the periodic table
To use this figure as a guide to the aufbau process, locate the position of an element in the table. Subshells listed ahead of this position are filled. For example, germanium (Z = 32) is located in group 14 of the blue $4p$ row. The filled subshells are $1s^2$, $2s^2$, $2p^6$, $3s^2$, $3p^6$, $4s^2$, and $3d^{10}$. At (Z = 32), a second electron has entered the $4p$ subshell. The electron configuration of Ge is $[Ar]3d^{10}4s^24p^2$. Exceptions to the orderly filling of subshells suggested here are found among a few of the d-block and some of the f-block elements.

Another point to notice from Table 8.2 is that the electron configuration consists of a noble-gas core corresponding to the noble gas from the previous period plus the additional electrons required to satisfy the atomic number. Recognizing this and dividing the periodic table into blocks can simplify the task of assigning electron configurations. For example, strontium is in group 2, the second s-block group, so that its valence-shell configuration is $5s^2$ since it is in the fifth period. The remaining electrons are in the krypton core configuration (the noble gas in the previous period); thus the electron configuration of Sr is

$$\text{Sr:} [Kr]5s^2$$

For the p-block elements in groups 13 to 18, the number of valence electrons is from 1 to 6. For example, aluminum is in period 3 and group 13, its valence-shell electron configuration is $3s^23p^1$. We use $n = 3$ since Al is in the third period and we have to accommodate three electrons after the neon core, which contains 10 electrons. Thus the electron configuration of Al is

$$\text{Al:} [Ne]3s^23p^1$$

Gallium is also in group 13, but in period 4. Its valence-shell electron configuration is $4s^2 4p^1$. To write the electron configuration of Ga, we can start with the electron configuration of the noble gas that closes the third period, argon, and we add to it the subshells that fill in the fourth period: $4s$, $3d$, and $4p$. The $3d$ subshell must fill with 10 electrons before the $4p$ subshell begins to fill. Consequently, the electron configuration of gallium must be

$$\text{Ga: } [\text{Ar}]3d^{10}4s^2 4p^1$$

Thallium is in group 13 and period 6. Its valence-shell electron configuration is $6s^2 6p^1$. Again, we indicate the electron configuration of the noble gas that closes the fifth period as a core, and add the subshells that fill in the sixth period: $6s$, $4f$, $5d$, and $6p$.

$$\text{Tl: } [\text{Xe}]4f^{14}5d^{10}6s^2 6p^1$$

The elements in group 13 have the common valence configuration $ns^2 np^1$, again illustrating the repeating pattern of valence electron configurations down a group, which is the basis of the similar chemical properties of the elements within a group of the periodic table.

The transition elements correspond to the d block, and their electron configurations are established in a similar manner. To write the electron configuration of a transition element, start with the electron configuration of the noble gas that closes the prior period and add the subshells that fill in the period of the transition element being considered. The s subshell fills immediately after the preceding noble gas; most transition metal atoms have two electrons in the s subshell of the valence shell, but some have only one. Thus, vanadium ($Z = 23$), which has two valence electrons in the $4s$ subshell and core electrons in the configuration of the noble gas argon, must have *three* $3d$ electrons ($2 + 18 + 3 = 23$).

$$\text{V: } [\text{Ar}]3d^3 4s^2$$

Chromium ($Z = 24$), as we have seen before, has only one valence electron in the $4s$ subshell and core electrons in the argon configuration. Consequently it must have *five* $3d$ electrons ($1 + 18 + 5 = 24$).

$$\text{Cr: } [\text{Ar}]3d^5 4s^1$$

Copper ($Z = 29$) also has only one valence electron in the $4s$ subshell in addition to its argon core, so the copper atom must have *ten* $3d$ electrons ($1 + 18 + 10 = 29$).

$$\text{Cu: } [\text{Ar}]3d^{10}4s^1$$

Chromium and copper are two exceptions to the straightforward filling of atomic subshells in the first d-block row. An examination of the electron configurations of the heavier elements (Appendix D) will reveal that there are other special cases that are not easily explained—for example, gadolinium has the configuration $[\text{Xe}]4f^7 6d^1 6s^2$. Examples 8-14 through 8-16 provide several more illustrations of the assignment of electron configurations using the ideas presented here.

▶ The electron configurations for the *lower d- and f-block* elements contain many exceptions that need not be memorized. Few people know all of them. Anyone needing any of these configurations can look them up when needed in tables, such as in Appendix D.

🔍 **8-8   CONCEPT ASSESSMENT**

The following orbital diagram represents an excited state of an atom. Identify the atom and give the orbital diagram corresponding to its ground state orbital diagram.

[Ar]   4s   3d   4p

EXAMPLE 8-14   Using *spdf* Notation for an Electron Configuration

(a) Identify the element having the electron configuration

$$1s^2 2s^2 2p^6 3s^2 3p^5$$

(b) Write the electron configuration of arsenic.

Analyze

The total number of electrons in a neutral atomic species is equal to the atomic number of the element. All electrons must be accounted for in an electron configuration.

Solve

(a) Add the superscript numerals $(2 + 2 + 6 + 2 + 5)$ to obtain the atomic number 17. The element with this atomic number is chlorine.

(b) Arsenic $(Z = 33)$ is in period 4 and group 15. Its valence-shell electron configuration is $4s^2 4p^3$. The noble gas that closes the third period is Ar $(Z = 18)$, and the subshells that fill in the fourth period are $4s, 3d,$ and $4p,$ in that order. Note that we account for 33 electrons in the configuration

$$\text{As: } [\text{Ar}]3d^{10}4s^2 4p^3$$

Assess

As long as we count the number of electrons accurately and know the order of the orbitals, we should be able to interpret or write the correct electronic configuration.

PRACTICE EXAMPLE A: Identify the element having the electron configuration $1s^2 2s^2 2p^6 3s^2 3p^6 3d^2 4s^2$.

PRACTICE EXAMPLE B: Use *spdf* notation to show the electron configuration of iodine. How many electrons does the I atom have in its $3d$ subshell? How many unpaired electrons are there in an I atom?

EXAMPLE 8-15   Representing Electron Configurations

Write (a) the electron configuration of mercury, and (b) an orbital diagram for the electron configuration of tin.

Analyze

To write the electronic configuration, we locate the element on the periodic table and then ascertain which subshells are filled. We must be careful, with high-atomic-number elements, to take into account the lanthanide and actinide elements.

Solve

(a) Mercury, in period 6 and group 12, is the transition element at the end of the third transition series, in which the $5d$ subshell fills $(5d^{10})$. The noble gas that closes period 5 is xenon, and the lanthanide series intervenes between xenon and mercury, in which the $4f$ subshell fills $(4f^{14})$. When we put all these facts together, we conclude that the electron configuration of mercury is

$$[\text{Xe}]4f^{14}5d^{10}6s^2$$

(b) Tin is in period 5 and group 14. Its valence-shell electron configuration is $5s^2 5p^2$. The noble gas that closes the fourth period is Kr $(Z = 36)$, and the subshells that fill in the fifth period are $5s, 4d,$ and $5p$. Note that all subshells are filled in the orbital diagram except for $5p$. Two of the $5p$ orbitals are occupied by single electrons with parallel spins; one $5p$ orbital remains empty.

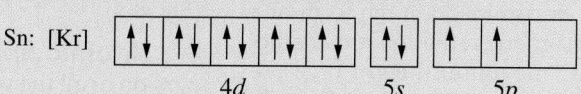

Sn: [Kr]   4d   5s   5p

Assess

Using the periodic table helps when writing electronic configurations.

PRACTICE EXAMPLE A: Represent the electron configuration of iron with an orbital diagram.

PRACTICE EXAMPLE B: Represent the electron configuration of bismuth with an orbital diagram.

EXAMPLE 8-16 Relating Electron Configurations to the Periodic Table

Indicate the number of **(a)** valence electrons in an atom of bromine; **(b)** $5p$ electrons in an atom of tellurium; **(c)** unpaired electrons in an atom of indium; **(d)** $3d$ and $4d$ electrons in a silver atom.

Analyze

Determine the atomic number and the periodic table location of each element. Then, explain the significance of its location.

Solve

(a) Bromine ($Z = 35$) is in group 17. There are seven outer-shell, or valence, electrons in all atoms in this group.

(b) Tellurium ($Z = 52$) is in period 5 and group 16. There are six outer-shell electrons, two of them are s, and the other four are p. The valence-shell electron configuration of tellurium is $5s^25p^4$; the tellurium atom has four $5p$ electrons.

(c) Indium ($Z = 49$) is in period 5 and group 13. The electron configuration of its inner shells is $[Kr]4d^{10}$. All the electrons in this inner-shell configuration are paired. The valence-shell electron configuration is $5s^25p^1$. The two $5s$ electrons are paired, and the $5p$ electron is unpaired. The In atom has one unpaired electron.

(d) Ag ($Z = 47$) is in period 5 and group 11. The noble gas that closes period 4 is krypton, and the $4d$ subshell fills $4(d^{10})$. There is one electron in the $5s$ orbital; thus the electron configuration of silver is

$$\text{Ag: } [Kr]4d^{10}5s^1$$

There are ten $3d$ electrons and ten $4d$ electrons in a silver atom.

Assess

Again, the relationship between the periodic table and electronic configurations is evident.

PRACTICE EXAMPLE A: For an atom of Sn, indicate the number of **(a)** electronic shells that are either filled or partially filled; **(b)** $3p$ electrons; **(c)** $5d$ electrons; and **(d)** unpaired electrons.

PRACTICE EXAMPLE B: Indicate the number of **(a)** $3d$ electrons in Y atoms; **(b)** $4p$ electrons in Ge atoms; and **(c)** unpaired electrons in Au atoms.

Mastering**CHEMISTRY** **www.masteringchemistry.com**

Laser devices are in use everywhere—in compact disc players, bar-code scanners, laboratory instruments, and in cosmetic, dental, and surgical procedures. Lasers produce light with highly desirable properties by a process called stimulated emission. For a discussion of how lasers work, go to the Focus On feature for Chapter 8, Helium-Neon Lasers, on the MasteringChemistry site.

Summary

8-1 Electromagnetic Radiation—Electromagnetic radiation is a type of energy transmission in the form of a **wave**. The waves of electromagnetic radiation are characterized by an **amplitude**, the maximum height of wave crests and maximum depth of wave troughs, a **wavelength**, λ, the distance between wave crests and **frequency**, ν, which signifies how often the fluctuations occur. Frequency is measured in **hertz**, **Hz** (cycles per second). Wavelength and frequency are related by the equation (8.1): $c = \lambda\nu$, where c is the **speed of light**. The wave character of electromagnetic radiation means that the waves can be dispersed into individual components of different wavelengths, a

diffraction pattern, by striking a closely grooved surface (Fig. 8-4).

8-2 Atomic Spectra—A rainbow results from the dispersion of all the wavelength components of visible light by a prism or raindrops; it is an example of a continuous spectrum (Fig. 8-7). The spectra produced by light emitted from excited atoms and ions are called **atomic spectra** or **line spectra**, because only certain frequencies are observed.

8-3 Quantum Theory—The study of electromagnetic radiation emitted from hot objects led to Planck's theory, which postulates that quantities of energy can have only certain discrete values, with the smallest unit of energy being that of a **quantum**. The energy of a quantum is given by equation (8.3): $E = h\nu$, where h is **Planck's constant**. Einstein's interpretation of the **photoelectric effect**—the ability of light to eject electrons when striking certain surfaces (Fig. 8-12)—led to a new interpretation of electromagnetic radiation: Light has a particle-like nature in addition to its wave-like properties. Light particles are called **photons**. The energy of a photon is related to the frequency of the radiation by $E_{photon} = h\nu$.

8-4 The Bohr Atom

The first attempt to explain atomic (line) spectra was made by Niels Bohr who postulated that an electron in a hydrogen atom exists in a circular orbit designated by a **quantum number**, n, that describes the energy of the electron in the orbit. The state of the electron with the lowest quantum number, $n = 1$, is called the **ground state**. An **excited state** of a hydrogen atom corresponds to those states with $n > 1$. The Bohr theory also provided a means for constructing an **energy-level diagram** (Fig. 8-14) so that emission spectra could be understood.

8-5 Two Ideas Leading to a New Quantum Mechanics

Louis de Broglie postulated a **wave–particle duality** in which particles of matter such as protons and electrons would at times display wave-like properties (equation 8.10). Because of an inherent uncertainty of the position and momentum of a wave-like particle, Heisenberg postulated that we cannot simultaneously know a subatomic particle's precise momentum and its position, a proposition referred to as the **Heisenberg uncertainty principle** (equation 8.11).

8-6 Wave Mechanics

The application of the concept of wave–particle duality requires that we view the electron in a system through a **wave function** that corresponds to a **standing wave** within the boundary of the system (Figs. 8-18 and 8-19). Application of these ideas to a particle in a one-dimensional box shows that at the lowest energy level the energy of the particle is nonzero that is, the system has a **zero-point energy**. The solution of the **Schrödinger equation** for the hydrogen atom provides wave functions called **orbitals**, which are the product of an **angular wave function**, $Y(\theta, \phi)$, and a **radial wave function**, $R(r)$.

8-7 Quantum Numbers and Electron Orbitals

The three quantum numbers arising from the Schrödinger wave equation are the *principal quantum number*, n, the *orbital angular momentum quantum number*, ℓ, and the *magnetic quantum number*, m_ℓ. All orbitals with the same value of n are in the same **principal electronic shell** (**principal level**), and all orbitals with the same values of n and ℓ are in the same **subshell (sublevel)**. The orbitals with different values of ℓ (0, 1, 2, 3, and so on) are designated s, p, d, f (Fig. 8-23). Orbitals in the same subshell of a hydrogen-like species have the same energy and are said to be **degenerate**.

8-8 Interpreting and Representing the Orbitals of the Hydrogen Atom

Interpreting the solutions to the Schrödinger equation for the hydrogen atom leads to a description of the shapes of the electron probability distributions for electrons in the s, p, and d orbitals. The number of nodes $(n - 1)$ in an orbital increases as n increases. Nodes are where the wave function changes sign.

8-9 Electron Spin: A Fourth Quantum Number

Stern and Gerlach demonstrated that electrons possess a quality called **electron spin** (Figs. 8-32 and 8-33). The electron spin quantum number, m_s, takes the value $+\frac{1}{2}$ or $-\frac{1}{2}$.

8-10 Multielectron Atoms

In multielectron atoms, orbitals with different values of ℓ are not degenerate. The loss of degeneracy within a principal shell is a result of the different **effective nuclear charge**, Z_{eff}, experienced by electrons in different subshells.

8-11 Electron Configurations

Electron configuration describes how the electrons are distributed among the various orbitals in principal shells and subshells of an atom. Electrons fill orbitals from the lowest energy to the highest (Fig. 8-37), ensuring that the energy of the atom is at a minimum. The **Pauli exclusion principle** states that a maximum of two electrons may occupy an orbital. **Hund's rule** says that when degenerate orbitals are available, electrons initially occupy these orbitals singly. Electron configurations are represented by either expanded or condensed *spdf* notation or an **orbital diagram** (page 339). The **aufbau process** is used to assign electron configurations to the elements of the periodic table. Electrons added to the shell of highest quantum number in the aufbau process are called **valence electrons**.

8-12 Electron Configurations and the Periodic Table

Elements in the same group of the periodic table have similar electron configurations. Groups 1 and 2 correspond to the *s* **block** with filled or partially filled valence-shell s orbitals. Groups 13 through 18 correspond to the *p* **block** with filled or partially filled valence-shell p orbitals. The *d* **block** corresponds to groups 3 through 12 as the $n - 1$ energy level is being filled—that is, having filled or partially filled d orbitals. In the *f*-**block** elements, also called the lanthanides and actinides, the $n - 2$ shell fills with electrons; that is, they have filled or partially filled f orbitals.

Integrative Example

Microwave ovens have become increasingly popular in kitchens around the world. They are also useful in the chemical laboratory, particularly in drying samples for chemical analysis. A typical microwave oven uses microwave radiation with a wavelength of 12.2 cm.

Are there any electronic transitions in the hydrogen atom that could *conceivably* produce microwave radiation of wavelength 12.2 cm? Estimate the principal quantum levels between which the transition occurs.

Analyze

Use the wavelength of microwaves to calculate the frequency of the radiation. Calculate the energy of the photon that has this frequency. Estimate where in the emission spectrum for a hydrogen atom such a photon emission might be found, by using equation (8.6).

Solve

1. Calculate the frequency of the microwave radiation. Microwaves are a form of electromagnetic radiation and thus travel at the speed of light, $2.998 \times 10^8 \, \text{m s}^{-1}$. Convert the wavelength to meters, and then use the equation

$$\nu = c/\lambda$$

$$\nu = \frac{2.998 \times 10^8 \, \text{m s}^{-1}}{12.2 \, \text{cm} \times 1 \, \text{m}/100 \, \text{cm}} = 2.46 \times 10^9 \, \text{Hz}$$

2. Calculate the energy associated with one photon of the microwave radiation. This is a direct application of Planck's equation.

$$E = h\nu = 6.626 \times 10^{-34} \, \text{J s} \times 2.46 \times 10^9 \, \text{s}^{-1}$$
$$= 1.63 \times 10^{-24} \, \text{J}$$

3. Determine whether there are any electronic transitions in the hydrogen atom with an energy per photon of $1.63 \times 10^{-24} \, \text{J}$. Let the principal quantum number of the final state (n_f) be n, and that of the initial state (n_i) is then $n + 1$. Substitute these two values into equation (8.9).

$$E_{\text{photon}} = \Delta E = 2.179 \times 10^{-18} \, \text{J} \left(\frac{1}{(n+1)^2} - \frac{1}{n^2} \right)$$

Solving for n

$$E_{\text{photon}} = \Delta E = 2.179 \times 10^{-18} \, \text{J} \left(\frac{n^2 - (n+1)^2}{n^2(n+1)^2} \right)$$

$$= -2.179 \times 10^{-18} \, \text{J} \left(\frac{2n+1}{n^2(n+1)^2} \right)$$

4. The negative sign indicates that a photon is emitted and now can be ignored. Substitute the value for the energy of the photon and rearrange the equation

$$E_{\text{photon}} = 1.63 \times 10^{-24} \, \text{J} = 2.179 \times 10^{-18} \, \text{J} \left(\frac{2n+1}{n^2(n+1)^2} \right)$$

$$\frac{1.63 \times 10^{-24} \, \text{J}}{2.179 \times 10^{-18} \, \text{J}} = 7.48 \times 10^{-7} = \left(\frac{2n+1}{n^2(n+1)^2} \right)$$

5. Look at Figure 8-14, the *energy*-level diagram for the Bohr hydrogen atom. Energy differences between the low-lying levels are of the order 10^{-19} to $10^{-20} \, \text{J}$. These are orders of magnitude (10^4 to 10^5 times) greater than the energy per photon of $1.63 \times 10^{-24} \, \text{J}$ from part 2. Note, however, that the energy differences become progressively smaller for high-numbered orbits. As n approaches ∞, the energy differences approach zero, and some transitions between high-numbered orbits should correspond to microwave radiation. Thus we expect n to be large, so that to a good approximation we can neglect one with respect to n and write

$$7.48 \times 10^{-7} = \left(\frac{2n+1}{n^2(n+1)^2} \right) \simeq \left(\frac{2n}{n^2 n^2} \right) \simeq \left(\frac{2}{n^3} \right)$$

Solving for n

$$n \simeq \left(\frac{2}{7.48 \times 10^{-7}} \right)^{1/3} \simeq 138.8$$

6. We can check this result by substituting this value of $n = 139$ into the exact expression

$$7.48 \times 10^{-7} = \frac{2n+1}{n^2(n+1)^2} = \frac{2(139)+1}{139^2(139+1)^2} = 7.37 \times 10^{-7}$$

The agreement is not very good, so let's try $n = 138$

$$7.48 \times 10^{-7} = \frac{2n+1}{n^2(n+1)^2} = \frac{2(138)+1}{138^2(138+1)^2} = 7.53 \times 10^{-7}$$

This provides closer agreement. The value of the principal quantum number is $n = 138$.

Assess

Using equation (8.6) we have shown that the emission of a photon for the deexcitation of an electron from $n = 139$ to $n = 138$ produces a wavelength for that photon in the microwave region. However, we might question whether the $n = 139$ state is still a bound state or whether the energy required to create this state causes ionization (see Exercise 106).

PRACTICE EXAMPLE A: Calculate the de Broglie wavelength of a helium atom at 25 °C and moving at the root-mean-square velocity. At what temperature would the average helium atom be moving fast enough to have a de Broglie wavelength comparable to that of the size of a typical atom, about 300 pm?

PRACTICE EXAMPLE B: By using a two-photon process (that is, two sequential excitations), a chemist is able to excite the electron in a hydrogen atom to the 5d level. Not all excitations are possible; they are governed by a selection rule. The selection rule states that the allowed excitations must have $\Delta\ell = +1$ or -1 and Δn can have any value. Use this selection rule to identify the possible intermediate levels (more than one are possible) involved, and calculate the frequencies of the two photons involved in each process. Identify the transitions allowed when a sample of hydrogen atoms excited to the 5d level exhibits an emission spectrum. When a sample of gaseous sodium atoms is similarly excited to the 5d level, what would be the difference in the emission spectrum observed?

Mastering**CHEMISTRY**

You'll find a link to additional self study questions in the study area on:
www.masteringchemistry.com

Exercises

Electromagnetic Radiation

1. A hypothetical electromagnetic wave is pictured here. What is the wavelength of this radiation?

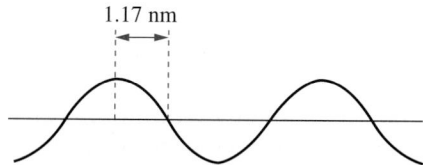

1.17 nm

2. For the electromagnetic wave described in Exercise 1, what are **(a)** the frequency, in hertz, and **(b)** the energy, in joules per photon?

3. The magnesium spectrum has a line at 266.8 nm. Which of these statements about this radiation is (are) correct? Explain.
(a) It has a higher frequency than radiation with wavelength 402 nm.
(b) It is visible to the eye.
(c) It has a greater speed in a vacuum than does red light of wavelength 652 nm.
(d) Its wavelength is longer than that of X-rays.

4. The most intense line in the cerium spectrum is at 418.7 nm.
(a) Determine the frequency of the radiation producing this line.
(b) In what part of the electromagnetic spectrum does this line occur?
(c) Is it visible to the eye? If so, what color is it? If not, is this line at higher or lower energy than visible light?

5. *Without doing detailed calculations,* determine which of the following wavelengths represents light of the highest frequency: **(a)** 6.7×10^{-4} cm; **(b)** 1.23 mm; **(c)** 80 nm; **(d)** 6.72 μm.

6. *Without doing detailed calculations,* arrange the following electromagnetic radiation sources in order of increasing frequency: **(a)** a red traffic light, **(b)** a 91.9 MHz radio transmitter, **(c)** light with a frequency of 3.0×10^{14} s^{-1}, **(d)** light with a wavelength of 49 nm.

7. How long does it take light from the sun, 93 million miles away, to reach Earth?

8. In astronomy, distances are measured in *light-years,* the distance that light travels in one year. What is the distance of one light-year expressed in kilometers?

Atomic Spectra

9. Use the Balmer equation (8.2) to determine
(a) the frequency, in s^{-1}, of the radiation corresponding to $n = 5$;
(b) the wavelength, in nanometers, of the line in the Balmer series corresponding to $n = 7$;
(c) the value of n corresponding to the Balmer series line at 380 nm.

10. How would the Balmer equation (8.2) have to be modified to predict lines in the infrared spectrum of hydrogen? [*Hint:* Compare equations (8.2) and (8.6).]

11. Use Planck's equation (8.3) to determine
(a) the energy, in joules per photon, of radiation of frequency 7.39×10^{15} s^{-1};
(b) the energy, in kilojoules per mole, of radiation of frequency 1.97×10^{14} s^{-1}.

12. Use Planck's equation (8.3) to determine
(a) the frequency, in hertz, of radiation having an energy of 8.62×10^{-21} J/photon;
(b) the wavelength, in nanometers, of radiation with 360 kJ/mol of energy.

13. What is ΔE for the transition of an electron from $n = 6$ to $n = 3$ in a Bohr hydrogen atom? What is the frequency of the spectral line produced?

14. What is ΔE for the transition of an electron from $n = 5$ to $n = 2$ in a Bohr hydrogen atom? What is the frequency of the spectral line produced?

15. To what value of n in equation (8.2) does the line in the Balmer series at 389 nm correspond?

16. The Lyman series of the hydrogen spectrum can be represented by the equation

$$\nu = 3.2881 \times 10^{15}\,s^{-1}\left(\frac{1}{1^2} - \frac{1}{n^2}\right) (\text{where } n = 2, 3, \ldots)$$

(a) Calculate the maximum and minimum wavelength lines, in nanometers, in this series.

(b) What value of n corresponds to a spectral line at 95.0 nm?

(c) Is there a line at 108.5 nm? Explain.

17. Calculate the wavelengths, in nanometers, of the first four lines of the Balmer series of the hydrogen spectrum, starting with the *longest* wavelength component.

18. A line is detected in the hydrogen spectrum at 1880 nm. Is this line in the Balmer series? Explain.

Quantum Theory

19. A certain radiation has a wavelength of 574 nm. What is the energy, in joules, of (a) one photon; (b) a mole of photons of this radiation?

20. What is the wavelength, in nanometers, of light with an energy content of 1979 kJ/mol? In what portion of the electromagnetic spectrum is this light?

21. *Without doing detailed calculations*, indicate which of the following electromagnetic radiations has the greatest energy per photon and which has the least: (a) 662 nm; (b) 2.1×10^{-5} cm; (c) 3.58 μm; (d) 4.1×10^{-6} m.

22. *Without doing detailed calculations*, arrange the following forms of electromagnetic radiation in *increasing*

order of energy per mole of photons: (a) radiation with $\nu = 3.0 \times 10^{15}\,s^{-1}$, (b) an infrared heat lamp, (c) radiation having $\lambda = 7000$ Å, (d) dental X-rays.

23. In what region of the electromagnetic spectrum would you expect to find radiation having an energy per photon 100 times that associated with 988 nm radiation?

24. High-pressure sodium vapor lamps are used in street lighting. The two brightest lines in the sodium spectrum are at 589.00 and 589.59 nm. What is the difference in energy per photon of the radiations corresponding to these two lines?

The Photoelectric Effect

25. The lowest-frequency light that will produce the photoelectric effect is called the *threshold frequency*.

(a) The threshold frequency for indium is $9.96 \times 10^{14}\,s^{-1}$. What is the energy, in joules, of a photon of this radiation?

(b) Will indium display the photoelectric effect with UV light? With infrared light? Explain.

26. The minimum energy required to cause the photoelectric effect in potassium metal is 3.69×10^{-19} J. Will photoelectrons be produced when visible light shines on the surface of potassium? If 400 nm radiation is shone on potassium, what is the velocity of the ejected electrons?

The Bohr Atom

27. Use the description of the Bohr atom given in the text to determine (a) the radius, in nanometers, of the sixth Bohr orbit for hydrogen; (b) the energy, in joules, of the electron when it is in this orbit.

28. Calculate the increase in (a) distance from the nucleus and (b) energy when an electron is excited from the first to the third Bohr orbit.

29. What are the (a) frequency, in s^{-1}, and (b) wavelength, in nanometers, of the light emitted when the electron in a hydrogen atom drops from the energy level $n = 7$ to $n = 4$? (c) In what portion of the electromagnetic spectrum is this light?

30. *Without doing detailed calculations*, indicate which of the following electron transitions requires the greatest amount of energy to be *absorbed* by a hydrogen atom: from (a) $n = 1$ to $n = 2$; (b) $n = 2$ to $n = 4$; (c) $n = 3$ to $n = 9$; (d) $n = 10$ to $n = 1$.

31. For the Bohr hydrogen atom determine
(a) the radius of the orbit $n = 4$
(b) whether there is an orbit having a radius of 4.00 Å
(c) the energy level corresponding to $n = 8$
(d) whether there is an energy level at -2.5×10^{-17} J

32. *Without doing detailed calculations*, indicate which of the following electron transitions in the hydrogen atom results in the emission of light of the longest wavelength. (a) $n = 4$ to $n = 3$; (b) $n = 1$ to $n = 2$; (c) $n = 1$ to $n = 6$; (d) $n = 3$ to $n = 2$.

33. What electron transition in a hydrogen atom, starting from the orbit $n = 7$, will produce light of wavelength 410 nm?

34. What electron transition in a hydrogen atom, ending in the orbit $n = 3$, will produce light of wavelength 1090 nm?

35. The emission spectrum below for a one-electron (hydrogen-like) species in the gas phase shows all the lines, before they merge together, resulting from transitions to the ground state from higher energy states. Line A has a wavelength of 103 nm.

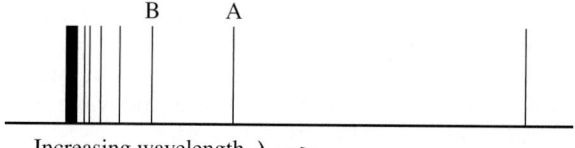

Increasing wavelength, λ ⟶

(a) What are the upper and lower principal quantum numbers corresponding to the lines labeled A and B?
(b) Identify the one-electron species that exhibits the spectrum.

36. The emission spectrum below for a one-electron (hydrogen-like) species in the gas phase shows all the lines, before they merge together, resulting from transitions to the first excited state from higher energy states. Line A has a wavelength of 434 nm.

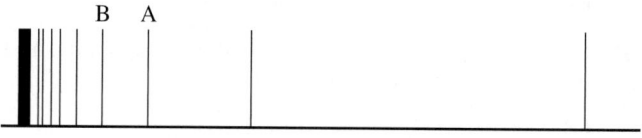

Increasing wavelength, λ ⟶

(a) What are the upper and lower principal quantum numbers corresponding to the lines labeled A and B?
(b) Identify the one-electron species that exhibits the spectrum.

37. The emission spectrum below for a one-electron (hydrogen-like) species in the gas phase shows all the lines, before they merge together, resulting from transitions to the first excited state from higher energy states. Line A has a wavelength of 27.1 nm.

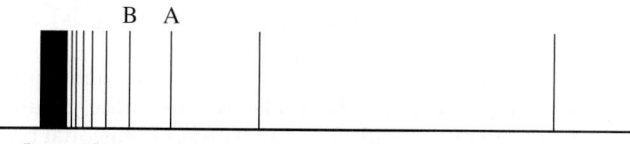

Increasing wavelength, λ ⟶

(a) What are the upper and lower principal quantum numbers corresponding to the lines labeled A and B?
(b) Identify the one-electron species that exhibits the spectrum.

38. The emission spectrum below for a one-electron (hydrogen-like) species in the gas phase shows all the lines, before they merge together, resulting from transitions to the ground state from higher energy states. Line A has a wavelength of 10.8 nm.

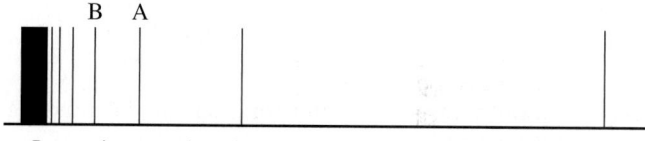

Increasing wavelength, λ ⟶

(a) What are the upper and lower principal quantum numbers corresponding to the lines labeled A and B?
(b) Identify the one-electron species that exhibits the spectrum.

Wave–Particle Duality

39. Which must possess a greater velocity to produce matter waves of the same wavelength (such as 1 nm), protons or electrons? Explain your reasoning.
40. What must be the velocity, in meters per second, of a beam of electrons if they are to display a de Broglie wavelength of 1 μm?
41. Calculate the de Broglie wavelength, in nanometers, associated with a 145 g baseball traveling at a speed of 168 km/h. How does this wavelength compare with typical nuclear or atomic dimensions?
42. What is the wavelength, in nanometers, associated with a 1000 kg automobile traveling at a speed of 25 m s^{-1}, that is, considering the automobile to be a matter wave? Comment on the feasibility of an experimental measurement of this wavelength.

The Heisenberg Uncertainty Principle

43. Describe how the Bohr model of the hydrogen atom appears to violate the Heisenberg uncertainty principle.
44. Although Einstein made some early contributions to quantum theory, he was never able to accept the Heisenberg uncertainty principle. He stated, "God does not play dice with the Universe." What do you suppose Einstein meant by this remark? In reply to Einstein's remark, Niels Bohr is supposed to have said, "Albert, stop telling God what to do." What do you suppose Bohr meant by this remark?
45. A proton is accelerated to one-tenth the velocity of light, and this velocity can be measured with a precision of 1%. What is the uncertainty in the position of this proton?
46. Show that the uncertainty principle is not significant when applied to large objects such as automobiles. Assume that m is precisely known; assign a reasonable value to either the uncertainty in position or the uncertainty in velocity, and estimate a value of the other.
47. What must be the velocity of electrons if their associated wavelength is to equal the radius of the first Bohr orbit of the hydrogen atom?
48. What must be the velocity of electrons if their associated wavelength is to equal the *longest* wavelength line in the Lyman series? [*Hint:* Refer to Figure 8-14.]

Wave Mechanics

49. A standing wave in a string 42 cm long has a total of six nodes (including those at the ends). What is the wavelength, in centimeters, of this standing wave?

50. What is the length of a string that has a standing wave with four nodes (including those at the ends) and $\lambda = 17$ cm?

51. Calculate the wavelength of the electromagnetic radiation required to excite an electron from the ground state to the level with $n = 4$ in a one-dimensional box 50. pm long.

52. An electron in a one-dimensional box requires a wavelength of 618 nm to excite an electron from the $n = 2$ level to the $n = 4$ level. Calculate the length of the box.

53. An electron in a 20.0 nm box is excited from the ground state into a higher energy state by absorbing a photon of wavelength 8.60×10^{-5} m. Determine the final energy state.

54. Calculate the wavelength of the electromagnetic radiation required to excite a proton from the ground state to the level with $n = 4$ in a one-dimensional box 50. pm long.

55. Describe some of the differences between the orbits of the Bohr atom and the orbitals of the wave mechanical atom. Are there any similarities?

56. The greatest probability of finding the electron in a small-volume element of the $1s$ orbital of the hydrogen atom is at the nucleus. Yet the most probable distance of the electron from the nucleus is 53 pm. How can you reconcile these two statements?

Quantum Numbers and Electron Orbitals

57. Select the correct answer and explain your reasoning. An electron having $n = 3$ and $m_\ell = 0$ **(a)** must have $m_s = +\frac{1}{2}$; **(b)** must have $\ell = 1$; **(c)** may have $\ell = 0, 1,$ or 2; **(d)** must have $\ell = 2$.

58. Write an acceptable value for each of the missing quantum numbers.
(a) $n = 3, \ell = ?, m_\ell = 2, m_s = +\frac{1}{2}$
(b) $n = ?, \ell = 2, m_\ell = 1, m_s = -\frac{1}{2}$
(c) $n = 4, \ell = 2, m_\ell = 0, m_s = ?$
(d) $n = ?, \ell = 0, m_\ell = ?, m_s = ?$

59. What type of orbital (i.e., $3s, 4p, \ldots$) is designated by these quantum numbers?
(a) $n = 5, \ell = 1, m_\ell = 0$
(b) $n = 4, \ell = 2, m_\ell = -2$
(c) $n = 2, \ell = 0, m_\ell = 0$

60. Which of the following statements is (are) correct for an electron with $n = 4$ and $m_\ell = 2$? Explain.
(a) The electron is in the fourth principal shell.
(b) The electron may be in a d orbital.

(c) The electron may be in a p orbital.
(d) The electron must have $m_s = +\frac{1}{2}$.

61. Concerning the electrons in the shells, subshells, and orbitals of an atom, how many can have
(a) $n = 4, \ell = 2, m_\ell = 1,$ and $m_s = +\frac{1}{2}$?
(b) $n = 4, \ell = 2,$ and $m_\ell = 1$?
(c) $n = 4$ and $\ell = 2$?
(d) $n = 4$?
(e) $n = 4, \ell = 2,$ and $m_s = +\frac{1}{2}$?

62. Concerning the concept of subshells and orbitals,
(a) How many subshells are found in the $n = 3$ level?
(b) What are the names of the subshells in the $n = 3$ level?
(c) How many orbitals have the values $n = 4$ and $\ell = 3$?
(d) How many orbitals have the values $n = 3, \ell = 2,$ and $m_\ell = -2$?
(e) What is the total number of orbitals in the $n = 4$ level?

The Shapes of Orbitals and Radial Probabilities

63. Calculate the finite value of r, in terms of a_0, at which the node occurs in the wave function of the $2s$ orbital of a hydrogen atom.

64. Calculate the finite value of r, in terms of a_0, at which the node occurs in the wave function of the $2s$ orbital of a Li^{2+} ion.

65. Show that the probability of finding a $2p_y$ electron in the xz plane is zero.

66. Show that the probability of finding a $3d_{xz}$ electron in the xy plane is zero.

67. Prepare a two-dimensional plot of $Y(\theta, \phi)$ for the p_y orbital in the xy plane.

68. Prepare a two-dimensional plot of $Y^2(\theta, \phi)$ for the p_y orbital in the xy plane.

69. Using a graphical method, show that in a hydrogen atom the radius at which there is a maximum probability of finding an electron is a_0 (53 pm).

70. Use a graphical method or some other means to show that in a Li^{2+} ion, the radius at which there is a maximum probability of finding an electron is $\frac{a_0}{3}$ (18 pm).

71. Identify the orbital that has **(a)** one radial node and one angular node; **(b)** no radial nodes and two angular nodes; **(c)** two radial nodes and three angular nodes.

72. Identify the orbital that has **(a)** two radial nodes and one angular node; **(b)** five radial nodes and zero angular nodes; **(c)** one radial node and four angular nodes.

73. A contour map for an atomic orbital of hydrogen is shown at the top of page 355 for the xy and xz planes. Identify the orbital.

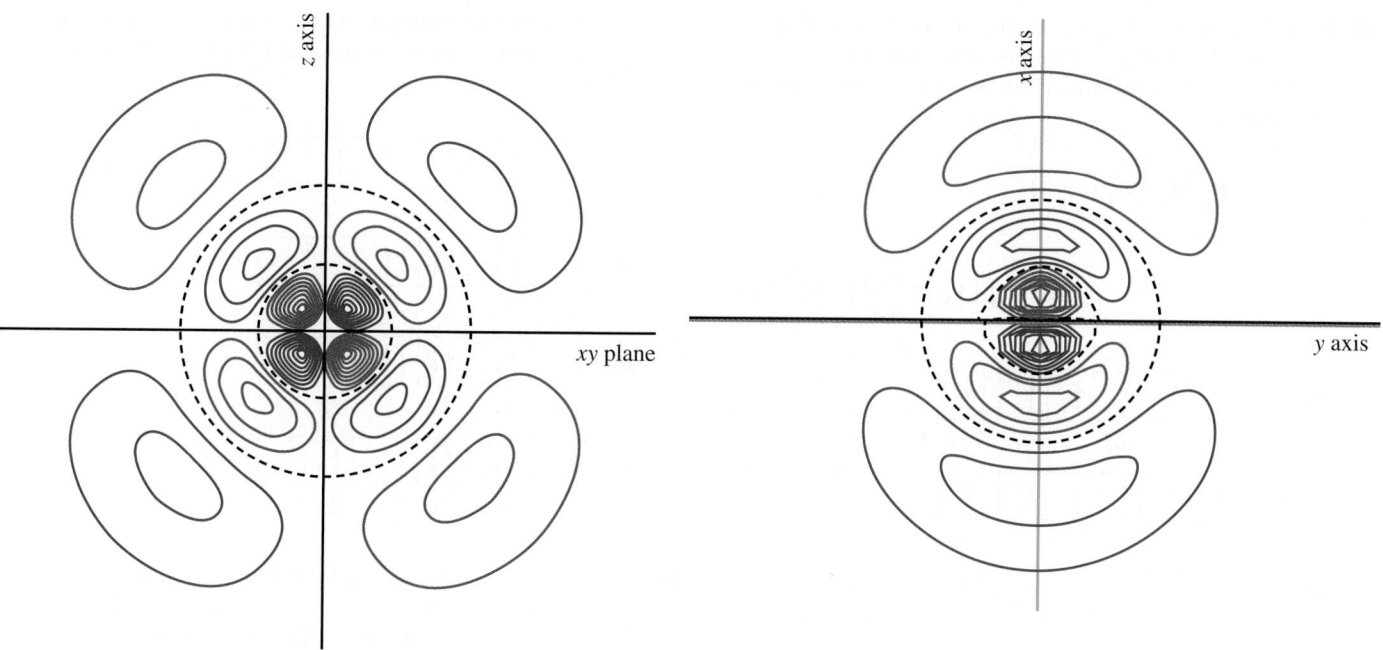

74. A contour map for an atomic orbital of hydrogen is shown below for the xy and xz planes. Identify the type $(s, p, d, f, g \ldots)$ of orbital.

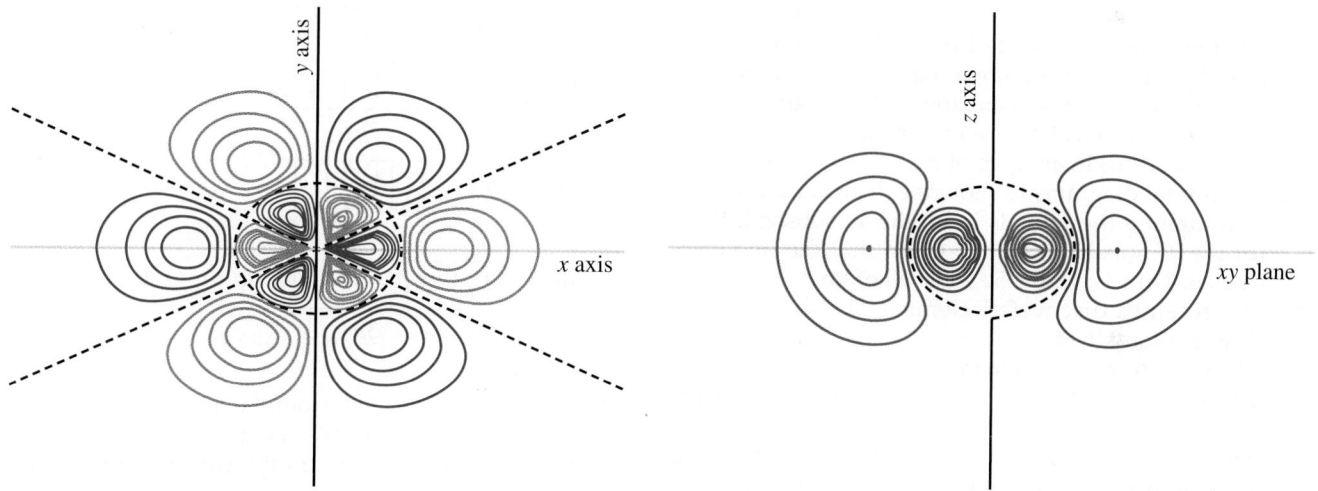

Electron Configurations

75. On the basis of the periodic table and rules for electron configurations, indicate the number of **(a)** $2p$ electrons in N; **(b)** $4s$ electrons in Rb; **(c)** $4d$ electrons in As; **(d)** $4f$ electrons in Au; **(e)** unpaired electrons in Pb; **(f)** elements in group 14 of the periodic table; **(g)** elements in the sixth period of the periodic table.

76. Based on the relationship between electron configurations and the periodic table, give the number of **(a)** outer-shell electrons in an atom of Sb; **(b)** electrons in the fourth principal electronic shell of Pt; **(c)** elements whose atoms have six outer-shell electrons; **(d)** unpaired electrons in an atom of Te; **(e)** transition elements in the sixth period.

77. Which of the following is the correct orbital diagram for the ground-state electron configuration of phosphorus? Explain what is wrong with each of the others.

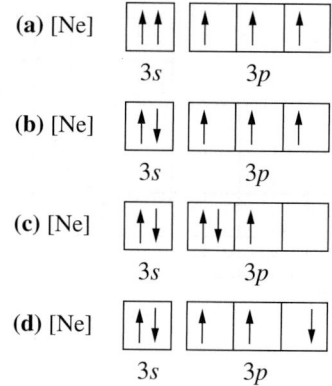

78. Which of the following is the correct orbital diagram for the ground-state electron configuration of molybdenum? Explain what is wrong with each of the others.

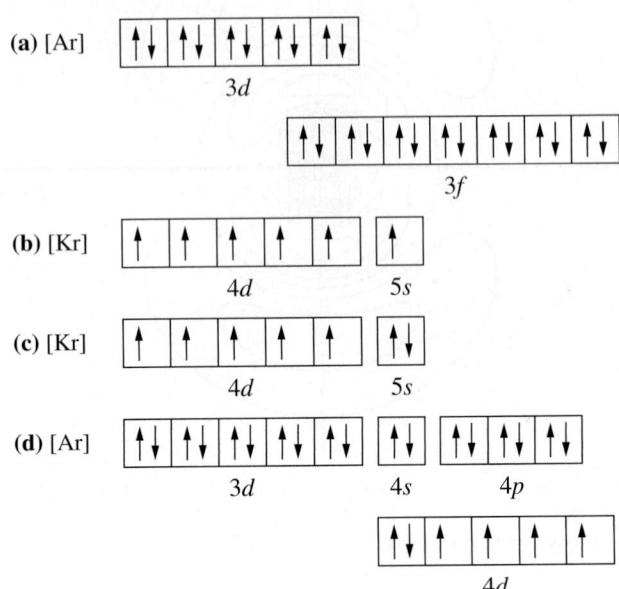

79. Use the basic rules for electron configurations to indicate the number of **(a)** unpaired electrons in an atom of P; **(b)** $3d$ electrons in an atom of Br; **(c)** $4p$ electrons in an atom of Ge; **(d)** $6s$ electrons in an atom of Ba; **(e)** $4f$ electrons in an atom of Au.

80. Use orbital diagrams to show the distribution of electrons among the orbitals in **(a)** the $4p$ subshell of Br; **(b)** the $3d$ subshell of Co^{2+}, given that the two electrons lost are $4s$; **(c)** the $5d$ subshell of Pb.

81. The recently discovered element 114 should most closely resemble Pb.
(a) Write the electron configuration of Pb.
(b) Propose a plausible electron configuration for element 114.

82. Without referring to any tables or listings in the text, mark an appropriate location in the blank periodic table provided for each of the following: **(a)** the fifth-period noble gas; **(b)** a sixth-period element whose atoms have three unpaired p electrons; **(c)** a d-block element having one $4s$ electron; **(d)** a p-block element that is a metal.

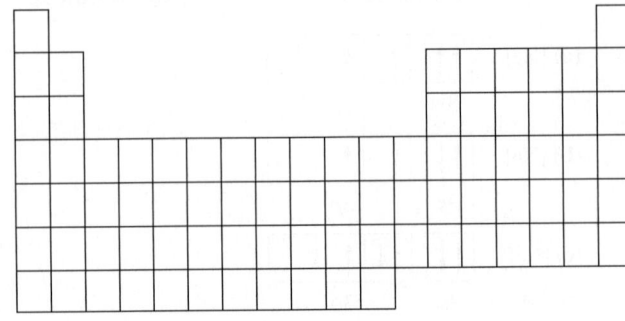

83. Which of the following electron configurations corresponds to the ground state and which to an excited state?

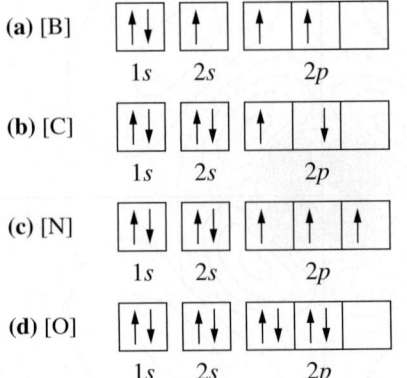

84. To what neutral atom do the following valence-shell configurations correspond? Indicate whether the configuration corresponds to the ground state or an excited state.

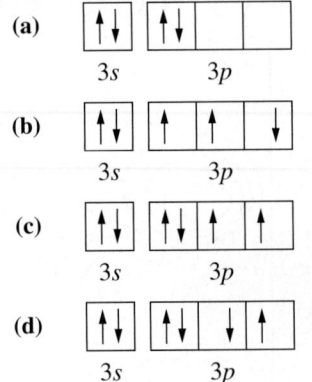

85. What is the expected ground-state electron configuration for each of the following elements? **(a)** mercury; **(b)** calcium; **(c)** polonium; **(d)** tin; **(e)** tantalum; **(f)** iodine.

86. What is the expected ground-state electron configuration for each of the following elements? **(a)** tellurium; **(b)** cesium; **(c)** selenium; **(d)** platinum; **(e)** osmium; **(f)** chromium.

87. The following electron configurations correspond to the ground states of certain elements. Name each element. **(a)** $[Rn]7s^2 6d^2$; **(b)** $[He]2s^2 2p^2$; **(c)** $[Ar]3d^3 4s^2$; **(d)** $[Kr]4d^{10}5s^2 5p^4$; **(e)** $[Xe]4f^2 6s^2 6p^1$.

88. The following electron configurations correspond to the ground states of certain elements. Name each element. **(a)** $[Ar]3d^{10}4s^2 4p^3$; **(b)** $[Ne]3s^2 3p^4$; **(c)** $[Ar]3d^1 4s^2$; **(d)** $[Kr]4d^6 5s^2$; **(e)** $[Xe]4f^{12}6s^2$.

Integrative and Advanced Exercises

89. Derive the Balmer equation from equation (8.6).
90. Electromagnetic radiation can be transmitted through a vacuum or empty space. Can heat be similarly transferred? Explain.
91. The *work function* is the energy that must be supplied to cause the release of an electron from a photoelectric material. The corresponding photon frequency is the threshold frequency. The higher the energy of the incident light, the more kinetic energy the electrons have in moving away from the surface. The work function for mercury is equivalent to 435 kJ/mol photons.
    (a) Can the photoelectric effect be obtained with mercury by using visible light? Explain.
    (b) What is the kinetic energy, in joules, of the ejected electrons when light of 215 nm strikes a mercury surface?
    (c) What is the velocity, in meters per second, of the ejected electrons in part (b)?
92. Infrared lamps are used in cafeterias to keep food warm. How many photons per second are produced by an infrared lamp that consumes energy at the rate of 95 W and is 14% efficient in converting this energy to infrared radiation? Assume that the radiation has a wavelength of 1525 nm.
93. In 5.0 s, a 75 watt light source emits 9.91×10^{20} photons of a monochromatic (single wavelength) radiation. What is the color of the emitted light?
94. In everyday usage, the term "quantum jump" describes a change of a very significant magnitude compared to more gradual, incremental changes; it is similar in meaning to the term "a sea change." Does quantum jump have the same meaning when applied to events at the atomic or molecular level? Explain.
95. The Pfund series of the hydrogen spectrum has as its *longest* wavelength component a line at 7400 nm. Describe the electron transitions that produce this series. That is, give a Bohr quantum number that is common to this series.
96. Between which two orbits of the Bohr hydrogen atom must an electron fall to produce light of wavelength 1876 nm?
97. Use appropriate relationships from the chapter to determine the wavelength of the line in the emission spectrum of He^+ produced by an electron transition from $n = 5$ to $n = 2$.
98. Draw an energy-level diagram that represents all the possible lines in the emission spectrum of hydrogen atoms produced by electron transitions, in one or more steps, from $n = 5$ to $n = 1$.
99. An atom in which just one of the outer-shell electrons is excited to a very high quantum level n is called a "high Rydberg" atom. In some ways, all these atoms resemble a Bohr hydrogen atom with its electron in a high-numbered orbit. Explain why you might expect this to be the case.
100. If all other rules governing electron configurations were valid, what would be the electron configuration of cesium if (a) there were *three* possibilities for electron spin; (b) the quantum number ℓ could have the value n?
101. Ozone, O_3, absorbs ultraviolet radiation and dissociates into O_2 molecules and O atoms: $O_3 + h\nu \longrightarrow O_2 + O$. A 1.00 L sample of air at 22 °C and 748 mmHg contains 0.25 ppm of O_3. How much energy, in joules, must be absorbed if all the O_3 molecules in the sample of air are to dissociate? Assume that each photon absorbed causes one O_3 molecule to dissociate, and that the wavelength of the radiation is 254 nm.
102. Radio signals from *Voyager 1* in the 1970s were broadcast at a frequency of 8.4 GHz. On Earth, this radiation was received by an antenna able to detect signals as weak as 4×10^{-21} W. How many photons per second does this detection limit represent?
103. Certain metal compounds impart colors to flames—sodium compounds, yellow; lithium, red; barium, green—and flame tests can be used to detect these elements. (a) At a flame temperature of 800 °C, can collisions between gaseous atoms with average kinetic energies supply the energies required for the emission of visible light? (b) If not, how do you account for the excitation energy?
104. The angular momentum of an electron in the Bohr hydrogen atom is *mur*, where *m* is the mass of the electron, *u*, its velocity, and *r*, the radius of the Bohr orbit. The angular momentum can have only the values $nh/2\pi$, where *n* is an integer (the number of the Bohr orbit). Show that the *circumferences* of the various Bohr orbits are integral multiples of the de Broglie wavelengths of the electron treated as a matter wave.
105. A molecule of chlorine can be dissociated into atoms by absorbing a photon of sufficiently high energy. Any excess energy is translated into kinetic energy as the atoms recoil from one another. If a molecule of chlorine at rest absorbs a photon of 300 nm wavelength, what will be the velocity of the two recoiling atoms? Assume that the excess energy is equally divided between the two atoms. The bond energy of Cl_2 is 242.6 kJ mol^{-1}.
106. Refer to the Integrative Example. Determine whether or not $n = 138$ is a bound state. If it is, what sort of state is it? What is the radius of the orbit and how many revolutions per second does the electron make about the nucleus?
107. Using the relationships given in Table 8.1, find the finite values of *r*, in terms of a_0, of the nodes for a 3*s* orbital.
108. Use a graphical method or some other means to determine the radius at which the probability of finding a 2*s* orbital is maximum.
109. Using the relationships in Table 8.1, prepare a sketch of the 95% probability surface of a $4p_x$ orbital.
110. Show that the volume of a spherical shell of radius *r* and thickness *dr* is $4\pi r^2 dr$. [*Hint:* This exercise requires calculus.]
111. In the ground state of a hydrogen atom, what is the probability of finding an electron anywhere in a sphere of radius (a) a_0, or (b) $2a_0$? [*Hint:* This exercise requires calculus.]
112. When atoms in excited states collide with unexcited atoms they can transfer their excitation energy to those atoms. The most efficient energy transfer occurs when the excitation energy matches the energy of an excited

state in the unexcited atom. Assuming that we have a collection of excited hydrogen atoms in the $2s^1$ excited state, are there any transitions of He^+ that could be most efficiently excited by the hydrogen atoms?

Feature Problems

113. We have noted that an emission spectrum is a kind of "atomic fingerprint." The various steels are alloys of iron and carbon, usually containing one or more other metals. Based on the principal lines of their atomic spectra, which of the metals in the table below are likely to be present in a steel sample whose hypothetical emission spectrum is pictured? Is it likely that still other metals are present in the sample? Explain.

Principal Spectral Lines of Some Period 4 Transition Metals (nm)							
V	306.64	309.31	318.40	318.54	327.11	437.92	438.47	439.00
Cr	357.87	359.35	360.53	361.56	425.44	427.48	428.97	520.45
Mn	257.61	259.37	279.48	279.83	403.08	403.31	403.45	
Fe	344.06	358.12	372.00	373.49	385.99			
Ni	341.48	344.63	345.85	346.17	349.30	351.51	352.45	361.94

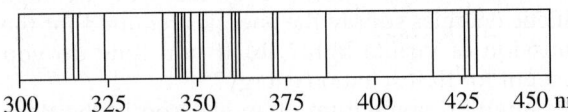

300   325   350   375   400   425   450 nm

▲ **Hypothetical emission spectrum**
In a real spectrum, the photographic images of the spectral lines would differ in depth and thickness depending on the strengths of the emissions producing them. Some of the spectral lines would not be seen because of their faintness.

114. Balmer seems to have deduced his formula for the visible spectrum of hydrogen just by manipulating numbers. A more common scientific procedure is to graph experimental data and then find a mathematical equation to describe the graph. Show that equation (8.2) describes a straight line. Indicate which variables must be plotted, and determine the numerical values of the slope and intercept of this line. Use data from Figure 8-10 to confirm that the four lines in the visible spectrum of hydrogen fall on the straight-line graph.

115. Follow up on the dartboard analogy of Figure 8-34 by plotting a graph of the scoring summary tabulated below. That is, plot the number of hits as a function of the scoring ring: 50, 40, ... What illustration in the text does this plot most resemble? Explain similarities and differences between the two.

Summary of Scoring (1500 darts)	
Darts	Score
200	"50"
300	"40"
400	"30"
250	"20"
200	"10"
150	off the board

116. Emission and absorption spectra of the hydrogen atom exhibit line spectra characteristic of quantized systems. In an absorption experiment, a sample of hydrogen atoms is irradiated with light with wavelengths ranging from 100 to 1000 nm. In an emission spectrum experiment, the hydrogen atoms are excited through an energy source that provides a range of energies from 1230 to 1240 kJ mol^{-1} to the atoms. Assume that the absorption spectrum is obtained at room temperature, when all atoms are in the ground state.
(a) Calculate the position of the lines in the absorption spectrum.
(b) Calculate the position of the lines in the emission spectrum.
(c) Compare the line spectra observed in the two experiments. In particular, will the number of lines observed be the same?

117. Diffraction of radiation takes place when the distance between the scattering centers is comparable to the wavelength of the radiation.
(a) What velocity must helium atoms possess to be diffracted by a film of silver atoms in which the spacing is 100 pm?
(b) Electrons accelerated through a certain potential are diffracted by a thin film of gold. Would you expect a beam of protons accelerated through the same potential to be diffracted when it strikes the film of gold? If not, what would you expect to see instead?

118. The emission spectrum below is for hydrogen atoms in the gas phase. The spectrum is of the first few emission lines from principal quantum number 6 down to all possible lower levels.

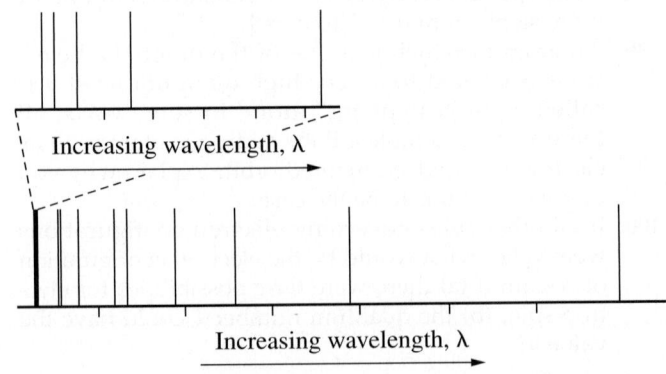

Increasing wavelength, λ

Increasing wavelength, λ

Not all possible de-excitations are possible; the transitions are governed by what is known as a *selection rule*. The selection rule states that the allowed transitions correspond to Δn being arbitrary while $\Delta \ell = \pm 1$. This selection rule can be understood in terms of conservation of momentum, since a photon has an angular momentum of unity. Using this selection rule, identify the transitions, in terms of the types of orbital (s, p, d, f), involved, that are observed in the spectrum shown on page 358.

In the presence of a magnetic field, the lines split into more lines according to the magnetic quantum number. The selection rule for transitions between different magnetic quantum numbers is $\Delta m_\ell = 0$ and ± 1. Identify the line(s) in the spectrum that splits into the greatest number of lines.

Self-Assessment Exercises

119. In your own words, define the following terms or symbols: **(a)** λ; **(b)** ν; **(c)** h; **(d)** ψ; **(e)** principal quantum number, n.

120. Briefly describe each of the following ideas or phenomena: **(a)** atomic (line) spectrum; **(b)** photoelectric effect; **(c)** matter wave; **(d)** Heisenberg uncertainty principle; **(e)** electron spin; **(f)** Pauli exclusion principle; **(g)** Hund's rule; **(h)** orbital diagram; **(i)** electron charge density; **(j)** radial electron density.

121. Explain the important distinctions between each pair of terms: **(a)** frequency and wavelength; **(b)** ultraviolet and infrared light; **(c)** continuous and discontinuous spectra; **(d)** traveling and standing waves; **(e)** quantum number and orbital; **(f)** *spdf* notation and orbital diagram; **(g)** *s* block and *p* block; **(h)** main group and transition element; **(i)** the ground state and excited state of a hydrogen atom.

122. Describe two ways in which the orbitals of multielectron atoms resemble hydrogen orbitals and two ways in which they differ from hydrogen orbitals.

123. Explain the phrase *effective nuclear charge*. How is this related to the shielding effect?

124. With the help of sketches, explain the difference between a p_x, p_y, and p_z orbital.

125. With the help of sketches, explain the difference between a $2p_z$ and $3p_z$ orbital.

126. If traveling at equal speeds, which of the following matter waves has the longest wavelength? Explain. **(a)** electron; **(b)** proton; **(c)** neutron; **(d)** α particle (He^{2+}).

127. For electromagnetic radiation transmitted through a vacuum, state whether each of the following properties is directly proportional to, inversely proportional to, or independent of the frequency: **(a)** velocity; **(b)** wavelength; **(c)** energy per mole. Explain.

128. Sir James Jeans described the photoelectric effect in this way: "It not only prohibits killing two birds with one stone but also the killing of one bird with two stones." By referring to the margin note on page 303, comment on the appropriateness of this analogy.

129. Construct a concept map representing the ideas of modern quantum mechanics.

130. Construct a concept map representing the atomic orbitals of hydrogen and their properties.

131. Construct a concept map for the configurations of multielectron atoms.

9

The Periodic Table and Some Atomic Properties

CONTENTS

A scanning tunneling microscope image of 48 iron atoms adsorbed onto a surface of copper atoms. The iron atoms were moved into position with the tip of the scanning tunneling microscope in order to create a barrier that forced some electrons of the copper atoms into a quantum state seen here as circular rings of electron density. The colors are from the computer rendering of the image. In this chapter we discuss the periodic table and the properties of atoms and ions.

Chemists value the periodic table as a means of organizing their field, and they would have continued to use it even if they had never figured out why it works. The underlying rationale of the periodic table was discovered about 50 years after the table was proposed.

The basis of the periodic table is the electron configurations of the elements, a topic we studied in Chapter 8. In this chapter, we will use the table as a backdrop for a discussion of some properties of elements, including atomic radii, ionization energies, and electron affinities. These atomic properties also arise in the discussion of chemical bonding in the following two chapters, and the periodic table itself will be our indispensable guide throughout much of the remainder of the text.

9-1   Classifying the Elements: The Periodic Law and the Periodic Table

In 1869, Dmitri Mendeleev and Lothar Meyer independently proposed the **periodic law**:

> When the elements are arranged in order of increasing atomic mass, certain sets of properties recur periodically.

Meyer based his periodic law on the property called atomic volume—the atomic mass of an element divided by the density of its solid form. We now just call this property *molar volume*.

$$\text{atomic (molar) volume (cm}^3\text{/mol)} = \text{molar mass (g/mol)} \times 1/d \text{ (cm}^3\text{/g)} \quad \textbf{(9.1)}$$

Meyer presented his results as a graph of atomic volume against atomic mass. Now it is customary to plot his results as molar volume against atomic number, as seen in Figure 9-1. Notice how high atomic volumes recur periodically for the alkali metals Li, Na, K, Rb, and Cs. Later, Meyer examined other physical properties of the elements and their compounds, such as hardness, compressibility, and boiling points, and found that these also vary periodically.

Mendeleev's Periodic Table

As previously described, the periodic table is a tabular arrangement of the elements that groups similar elements together. Mendeleev's work attracted more attention than Meyer's for two reasons: He left blank spaces in his table for undiscovered elements, and he corrected some atomic mass values. The blanks in his table came at atomic masses 44, 68, 72, and 100 for the elements we now know as scandium, gallium, germanium, and technetium. Two of the atomic mass values he corrected were those of indium and uranium.

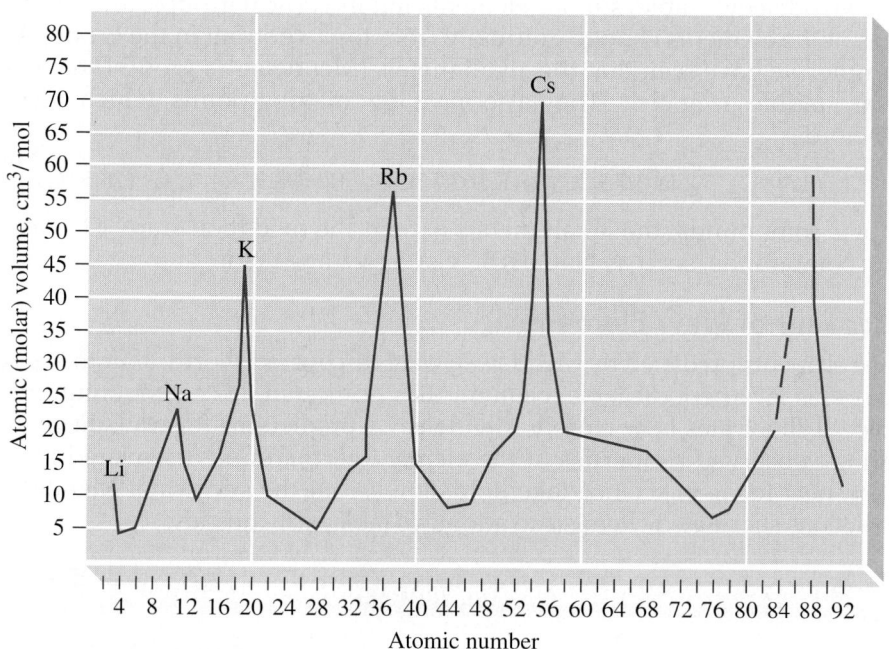

▲ FIGURE 9-1
An illustration of the periodic law—variation of atomic volume with atomic number
This adaptation of Meyer's 1870 graph plots atomic volumes against atomic numbers. Of course, a number of elements, such as the noble gases, were undiscovered in Meyer's time. The graph shows peaks at the alkali metals (Li, Na, K, and so on). Nonmetals fall on the ascending portions of the curve and metals at the peaks, on the descending portions, and in the valleys.

Reihen	Gruppe I. — R^2O	Gruppe II. — RO	Gruppe III. — R^2O^3	Gruppe IV. RH^4 RO^2	Gruppe V. RH^3 R^2O^5	Gruppe VI. RH^2 RO^3	Gruppe VII. RH R^2O^7	Gruppe VIII. — RO^4
1	H = 1							
2	Li = 7	Be = 9,4	B = 11	C = 12	N = 14	O = 16	F = 19	
3	Na = 23	Mg = 24	Al = 27,3	Si = 28	P = 31	S = 32	Cl = 35,5	
4	K = 39	Ca = 40	– = 44	Ti = 48	V = 51	Cr = 52	Mn = 55	Fe = 56, Co = 59, Ni = 59, Cu = 63.
5	(Cu = 63)	Zn = 65	– = 68	– = 72	As = 75	Se = 78	Br = 80	
6	Rb = 85	Sr = 87	?Yt = 88	Zr = 90	Nb = 94	Mo = 96	– = 100	Ru = 104, Rh = 104, Pd = 106, Ag = 108
7	(Ag = 108)	Cd = 112	In = 113	Sn = 118	Sb = 122	Te = 125	J = 127	
8	Cs = 133	Ba = 137	?Di = 138	?Ce = 140	–	–	–	– – – –
9	(–)	–	–	–	–	–	–	
10	–	–	?Er = 178	?La = 180	Ta = 182	W = 184	–	Os = 195, Ir = 197, Pt = 198, Au = 199
11	(Au = 199)	Hg = 200	Tl = 204	Pb = 207	Bi = 208			
12	–	–	–	Th = 231	–	U = 240		

▲ **Dmitri Mendeleev (1834–1907)**

Mendeleev's discovery of the periodic table came from his attempts to systematize properties of the elements for presentation in a chemistry textbook. His highly influential book went through eight editions in his lifetime and five more after his death.

In his periodic table Mendeleev arranged the elements into eight groups (Gruppe) and twelve rows (Reihen). The formulas are written as Mendeleev wrote them. R^2O, RO, and so on, are formulas of the element oxides (such as Li_2O, MgO, . . .); RH^4, RH^3, and so forth, are formulas of the element hydrides (such as CH_4, NH_3, . . .).

▶ Other properties of the alkali metals are discussed in Section 9-7.

In Mendeleev's table, similar elements fall in vertical groups, and the properties of the elements change gradually from top to bottom in the group. As an example, we have seen that the alkali metals (Mendeleev's group I) have high molar volumes (Fig. 9-1). They also have low melting points, which decrease in the order

$$\text{Li} \ (174 \,°\text{C}) > \text{Na} \ (97.8 \,°\text{C}) > \text{K} \ (63.7 \,°\text{C}) > \text{Rb} \ (38.9 \,°\text{C}) > \text{Cs} \ (28.5 \,°\text{C})$$

In their compounds, the alkali metals exhibit the oxidation state +1, forming ionic compounds, such as NaCl, KBr, CsI, Li_2O, and so on.

Discovery of New Elements

Three elements predicted by Mendeleev were discovered shortly after the appearance of his 1871 periodic table (gallium, 1875; scandium, 1879; germanium, 1886). Table 9.1 illustrates how closely Mendeleev's predictions for eka-silicon agree with the observed properties of the element germanium, discovered in 1886. Often, new ideas in science take hold slowly, but the success of Mendeleev's predictions stimulated chemists to adopt his table fairly quickly.

▶ The term *eka* is derived from Sanskrit and means "first." That is, eka-silicon means, literally, "first comes silicon" (and then comes the unknown element).

TABLE 9.1 Properties of Germanium: Predicted and Observed

Property	Predicted Eka-silicon (1871)	Observed Germanium (1886)
Atomic mass	72	72.6
Density, g/cm³	5.5	5.47
Color	dirty gray	grayish white
Density of oxide, g/cm³	EsO_2: 4.7	GeO_2: 4.703
Boiling point of chloride	$EsCl_4$: below 100 °C	$GeCl_4$: 86 °C
Density of chloride, g/cm³	$EsCl_4$: 1.9	$GeCl_4$: 1.887

One group of elements that Mendeleev did not anticipate was the noble gases. He left no blanks for them. William Ramsay, their discoverer, proposed placing them in a separate group of the table. Because argon, the first noble gas discovered (1894), had an atomic mass greater than that of chlorine and comparable to that of potassium, Ramsay placed the new group, which he called group 0, between the halogen elements (group VII) and the alkali metals (group I).

Atomic Number as the Basis for the Periodic Law

Mendeleev placed certain elements out of the order of increasing atomic mass to get them into the proper groups of his periodic table. He assumed this was because of errors in atomic masses. With improved methods of determining atomic masses and with the discovery of argon (group 0, atomic mass 39.9), which was placed ahead of potassium (group I, atomic mass 39.1), it became clear that a few elements might always remain "out of order." At the time, these out-of-order placements were justified by chemical evidence. Elements were placed in the groups that their chemical behavior dictated. There was no theoretical explanation for this reordering. Matters changed in 1913 as a result of some research by Henry G. J. Moseley on the X-ray spectra of the elements.

As we learned in Chapter 2, X-rays are a high-frequency form of electromagnetic radiation produced when a cathode-ray (electron) beam strikes the anode of a cathode-ray tube (see Figure 9-2a). The anode is called the *target*. Moseley was familiar with Bohr's atomic model, and explained X-ray emission in the following way. If the bombarding electrons have sufficient energy, they can eject electrons from the inner orbitals of target metal atoms. Electrons from higher orbitals then drop down to fill the vacancies, emitting X-ray photons with energies corresponding to the difference in energy between the originating level and the vacancy level (see Figure 8-13). Moseley reasoned that because the energies of electron orbitals depend on the nuclear charge, the frequencies of emitted X-rays should depend on the nuclear charges of atoms in the target. Using techniques newly developed by the father–son team of W. Henry Bragg and W. Lawrence Bragg, Moseley obtained photographic images of X-ray spectra and assigned frequencies to the spectral lines. His spectra for the elements from Ca to Zn are reproduced in Figure 9-2(b).

Moseley was able to correlate X-ray frequencies to numbers equal to the nuclear charges and corresponding to the positions of elements in Mendeleev's periodic table. For example, aluminum, the thirteenth element in the table, was assigned an *atomic number* of 13. Moseley's equation is $\nu = A(Z - b)^2$, where ν is the X-ray frequency, Z is the atomic number, and A and b are constants. Moseley used this relationship to predict three new elements ($Z = 43, 61$, and 75), which were discovered in 1937, 1945, and 1925, respectively. Also, he proved that in the portion of the periodic table with which he worked (from $Z = 13$ to $Z = 79$), there could be no additional new elements beyond those three. All available atomic numbers had been assigned. From the standpoint of Moseley's work, then, we should restate the periodic law.

▲ **Henry G. J. Moseley (1887–1915)**
Moseley was one of a group of brilliant scientists whose careers were launched under Ernest Rutherford. He was tragically killed at Gallipoli, in Turkey, during World War I.

> Similar properties recur periodically when elements are arranged according to increasing atomic number.

Description of a Modern Periodic Table: The Long Form

Mendeleev's periodic table consisted of 8 groups, but most modern periodic tables are arranged in 18 groups of elements. Let us briefly review the description of the periodic table given in Section 2-6.

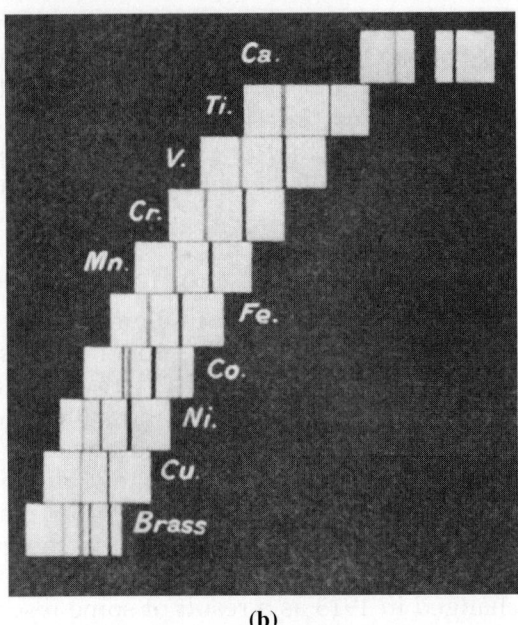

Evacuated tube

Target

X-rays

Electrons

Cathode

Current

(a)

(b)

▲ FIGURE 9-2

Schematic of an X-ray tube and Moseley's X-ray spectra of several elements

(a) A heated filament emits electrons by a process called thermionic emission. The electrons are accelerated by a high voltage, and collide with the metal target. The highly energetic electrons ionize electrons from the inner shells of the metal atoms of the target. Subsequently, electrons from higher orbitals drop down to occupy the vacancies and in doing so, emit X-ray photons that correspond to the energy difference between the two orbitals. (b) In this photograph from Moseley's 1913 paper, you can see two lines for each element, beginning with Ca at the top. With each successive element, the lines are displaced to the left, the direction of increasing X-ray frequency in these experiments. Where more than two lines appear, the sample contained one or more other elements, or impurities. Notice, for example, that one line in the Co spectrum matches a line in the Fe spectrum, and another matches a line in the Ni spectrum. Brass, which is an alloy of copper and zinc, shows two lines for Cu and two for Zn.

In the periodic table (see inside front cover), the vertical groups bring together elements with similar properties. The horizontal periods of the table are arranged in order of increasing atomic number from left to right. The groups are numbered at the top, and the periods at the extreme left. The first two groups—the s block—and the last six groups—the p block—together constitute the *main-group elements*. Because they come between the s block and the p block, the d block elements are known as the *transition elements*. The f block elements, sometimes called the *inner transition elements*, would extend the table to a width of 32 members if incorporated in the main body of the table. The table would generally be too wide to fit on a printed page, and so the f block elements are extracted from the table and placed at the bottom. The 15 elements following barium ($Z = 56$) are called the *lanthanides*, and the 15 following radon ($Z = 88$) are called the *actinides*.

9-2 Metals and Nonmetals and Their Ions

In Section 2-6, we established two categories of elements, *metals* and *nonmetals*, and described them in terms of physical properties. Most metals are good conductors of heat and electricity, are malleable and ductile, and have moderate to high melting points. In general, nonmetals are nonconductors of heat and electricity and are nonmalleable (brittle) solids, though a number of non-metals are gases at room temperature.

Through the color scheme of the periodic table on the inside front cover, we see that the majority of the elements are metals (orange) and that non-metals (blue) are confined to the right side of the table. The noble gases (purple) are treated as a special group of nonmetals. Metals and nonmetals are often separated by a stairstep diagonal line, and several elements near this line are often called metalloids (green). **Metalloids** are elements that look like metals and in some ways behave like metals but also have some nonmetallic properties.

◄ Metalloids (such as silicon) are semiconductors and materials composed of metalloids play an important role in microcomputer technology.

In the original periodic table, the positions of the elements were based on readily observable physical and chemical properties. In Chapter 8, we learned of the close correlation between electron configurations and the positions of the elements in the table. Thus, it appears that the physical and chemical properties of an element are determined largely by its electron configuration, particularly that of the *valence* (outermost) electronic shell. Adjacent members of a series of main-group elements in the same period (such as P, S, and Cl) have significantly different properties because they differ in their valence-electron configurations. Within a transition series, differences in electron configurations are mostly in inner shells, and so a transition element has some similarities to neighboring transition elements in the same period. In particular we find many similar properties for adjacent members of the same period within the *f* block. In fact, the strong similarities among the lanthanide elements presented a particular challenge to the nineteenth-century chemists who tried to separate and identify them.

Let's now briefly explore a few of the links between electron configurations and some other observations about the elements, starting with the noble gases.

Noble Gases

Atoms of the noble gases have the maximum number of electrons permitted in the valence shell of an atom, two in helium ($1s^2$) and eight in the other noble gas atoms (ns^2np^6). These electron configurations are very difficult to alter and seem to confer a high degree of chemical inertness to the noble gases. It is interesting to note, then, that the s-block metals, together with Al in group 13, tend to lose enough electrons to acquire the electron configurations of the noble gases. Conversely, nonmetals tend to gain enough electrons to achieve the same configurations.

◄ Compounds containing radon, xenon and krypton have been prepared. Recently, compounds containing argon have also been prepared.

Main-Group Metal Ions

The atoms of elements of groups 1 and 2—the most active metals—have electron configurations that differ from those of the noble gas of the preceding period by only one and two electrons in the s orbital of a new electron shell. If a K atom is stripped of its outer-shell electron, it becomes the *positive ion* K^+ with the electron configuration [Ar]. A Ca atom acquires the [Ar] configuration following the removal of two electrons.

$$K\,([Ar]4s^1) \longrightarrow K^+\,([Ar]) + e^-$$
$$Ca\,([Ar]4s^2) \longrightarrow Ca^{2+}([Ar]) + 2\,e^-$$

Although metal atoms do not lose electrons spontaneously, the energy required to bring about ionization is often provided by other processes occurring at the same time (such as an attraction between positive and negative ions). Aluminum is the only *p*-block metal that forms an ion with a noble gas electron configuration—Al^{3+}. This is because all other *p*-block elements would have to remove 10 *d* electrons to attain the electron configuration of the previous noble gas. The electron configurations of the other *p*-block metal ions are summarized in Table 9.2.

▲ Metals tend to lose electrons to attain noble gas electron configurations.

TABLE 9.2 Electron Configurations of Some Metal Ions[a]					
"Noble Gas"		"Pseudo-Noble Gas"[b]		"18 + 2"[c]	Other
Li^+	Be^{2+}	Ga^{3+}		In^+	Cr^{2+}, Cr^{3+}
Na^+	Mg^{2+}	Tl^{3+}		Tl^+	Mn^{2+}, Fe^{2+}
K^+	Ca^{2+}	Cu^+		Sn^{2+}	Fe^{3+}, Co^{2+}
Rb^+	Sr^{2+}	Ag^+, Au^+		Pb^{2+}	Ni^{2+}, Cu^{2+}
Cs^+	Ba^{2+}	Zn^{2+}		Sb^{3+}	
Fr^+	Ra^{2+}			Bi^{3+}	
Al^{3+}					

[a]Main-group metal ions are printed in black and transition metal ions in blue.
[b]In the configuration labeled "pseudo-noble gas," all electrons of the outermost shell have been lost. The next-to-outermost electron shell of the atom becomes the outermost shell of the ion and contains 18 electrons, for example, Ga^{3+}: $[Ne]3s^23p^63d^{10}$.
[c]In the configuration labeled "18 + 2" all outer-shell electrons except the two s electrons are lost, producing an ion with 18 electrons in the next-to-outermost shell and 2 electrons in the outermost, for example, Sn^{2+}: $[Ar]3d^{10}4s^24p^64d^{10}5s^2$.

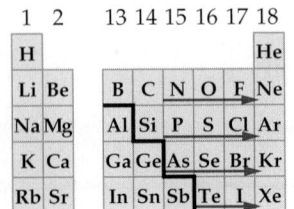

▲ Nonmetals tend to gain electrons to attain noble-gas electron configurations.

▶ A useful mnemonic is that the electron configuration of a cation can be obtained from the electron configuration of the parent atom by removing those electrons in orbitals with the *highest* quantum number first.

Main-Group Nonmetal Ions

The atoms of groups 17 and 16—the most active nonmetals—have one and two electrons fewer than the noble gas at the end of the period. Groups 17 and 16 atoms can acquire the electron configurations of noble gas atoms by *gaining* the appropriate numbers of electrons.

$$Cl\ ([Ne]3s^23p^5) + e^- \longrightarrow Cl^-\ ([Ar])$$
$$S\ ([Ne]3s^23p^4) + 2\,e^- \longrightarrow S^{2-}\ ([Ar])$$

In most cases, a nonmetal atom will gain a single electron spontaneously, but energy is required to force it to accept more than one. The necessary energy is often supplied by other processes that occur simultaneously (such as an attraction between positive and negative ions). Nonmetal ions with a charge of 3− are rare. However, some metal nitrides containing the nitride ion, N^{3-}, and some metal phosphides containing the phosphide ion, P^{3-}, are known.

Transition Metal Ions

In the aufbau process, the *ns* subshell fills before electrons enter the $(n - 1)d$ subshell (page 340), but the energy levels of these two subshells are nearly the same (Fig. 8-36). Thus, it is not surprising that when transition metal atoms ionize, the *ns* subshell is emptied. For example, the electron configuration of Ti is $[Ar]3d^24s^2$, and that of Ti^{2+} is $[Ar]3d^2$. Moreover, in some cases one or more $(n - 1)d$ electrons might be lost together with the *ns* electrons. This happens in the formation of Ti^{4+}, which has the electron configuration [Ar].

A few transition metal atoms acquire noble-gas electron configurations when forming cations, as do Sc in Sc^{3+} and Ti in Ti^{4+}, but most transition metal atoms do not (see Table 9.2). An iron atom does not acquire a noble-gas electron configuration when it loses its $4s^2$ electrons to form the ion Fe^{2+},

$$Fe\ ([Ar]3d^64s^2) \longrightarrow Fe^{2+}\ ([Ar]3d^6) + 2\,e^-$$

nor does it with the loss of an additional 3d electron to form the ion Fe^{3+}.

$$Fe\ ([Ar]3d^64s^2) \longrightarrow Fe^{3+}\ ([Ar]3d^5) + 3\,e^-$$

The 3d subshell in Fe^{3+} is half-filled, a fact that helps to account for the observed ease of oxidation of iron(II) to iron(III) compounds. Electron configurations with half-filled or filled *d* or *f* subshells have a special stability, and a number of transition metal ions have such configurations.

Hydrogen

Although all the other elements have a definite place in the periodic table, hydrogen does not. Its uniqueness stems from the fact that its atoms have only one electron, in the configuration $1s^1$. This single electron is the reason that we put hydrogen in group 1, even though we classify it as a nonmetal. Hydrogen does appear to become metallic when subjected to pressures of about 2 million bar, but these are hardly ordinary laboratory conditions. Because hydrogen, like the halogens, is one electron short of having a noble-gas electron configuration, it is sometimes placed in group 17; however, hydrogen does not resemble the halogens very much. For example, F_2 and Cl_2 are excellent oxidizing agents, but H_2 is a reducing agent. Still another alternative places hydrogen by itself at the top of the periodic table and near the center.

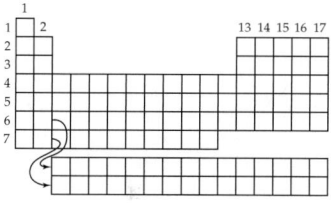

🔍 **9-1   CONCEPT ASSESSMENT**

On the blank periodic table in the margin, indicate where elements that satisfy the following descriptions occur.

(a) A noble gas having the same electron configuration as the Na^+ cation

(b) An anion of a nonmetal with a charge of 3− that has a noble-gas electron configuration

(c) A transition metal ion with a charge of 2+ and no valence shell s and p electrons

---

9-3   Sizes of Atoms and Ions

In earlier chapters we discovered the importance of atomic masses in matters relating to stoichiometry. To understand certain physical and chemical properties, we need to know something about atomic sizes. In this section we describe atomic radius, the first of a group of atomic properties that we will examine in this chapter.

Atomic Radius

Unfortunately, atomic radius is hard to define. The probability of finding an electron decreases with increasing distance from the nucleus, but nowhere does the probability fall to zero, so there is no precise outer boundary to an atom. We might describe an *effective* atomic radius as, say, the distance from the nucleus within which 95% of all the electron charge density is found, but in fact, all that we can measure is the distance between the nuclei of adjacent atoms (internuclear distance). Even though it varies, depending on whether atoms are chemically bonded or merely in contact without forming a bond, we define atomic radius in terms of internuclear distance.

Because we are primarily interested in bonded atoms, we will emphasize an atomic radius based on the distance between the nuclei of two atoms joined by a chemical bond. The **covalent radius** is one-half the distance between the nuclei of two identical atoms joined by a single covalent bond. The **ionic radius** is based on the distance between the nuclei of ions joined by an ionic bond. Because the ions are not identical in size, this distance must be properly apportioned between the cation and anion. One way to apportion the electron density between the ions is to define the radius of one ion and then infer the radius of the other ion. The convention we have chosen to use is to assign O^{2-} an ionic radius of 140 pm. An alternative apportioning scheme is to use F^- as the reference ionic radius. When using ionic radii data, one should carefully note which convention is used and not mix radii from the different conventions. Starting with a radius of 140 pm for O^{2-}, the radius of Mg^{2+} can be obtained

▶ Despite the SI convention, the angstrom unit, Å, is still widely used by X-ray crystallographers and others who work with atomic and molecular dimensions.

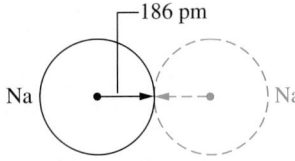

Covalent radius:

Metallic radius:

Ionic radius:

▲ FIGURE 9-3
Covalent, metallic, and ionic radii compared
Atomic radii are represented by the solid arrows. The covalent radius is based on the diatomic molecule $Na_2(g)$, found only in gaseous sodium. The metallic radius is based on adjacent atoms in solid sodium, Na(s). The value of the ionic radius of Na^+ is obtained by the comparative method described in the text.

from the internuclear distance in MgO, the radius of Cl^- from the internuclear distance in $MgCl_2$, and the radius of Na^+ from the internuclear distance in NaCl. For metals, we define a **metallic radius** as one-half the distance between the nuclei of two atoms in contact in the crystalline solid metal. Similarly in a solid sample of a noble gas the distance between the centers of neighboring atoms is called the **van der Waals radius**. There is much debate about the values of the atomic radii of noble gases because the experimental determination of the van der Waals radii is difficult; consequently, the atomic radii of noble gases are left out of the discussion of trends in atomic radii.

The angstrom unit, Å, has long been used for atomic dimensions ($1\ Å = 10^{-10}$ m). The angstrom, however, is not a recognized SI unit. The SI units are the nanometer (nm) and picometer (pm).

$$1\ nm = 1 \times 10^{-9}\ m;\ 1\ pm = 1 \times 10^{-12}\ m;\ 1\ nm = 1000\ pm \qquad (9.2)$$

Figure 9-3 illustrates the definitions of covalent, ionic, and metallic radii by comparing these three radii for sodium. Figure 9-4 is a plot of atomic radius against atomic number for a large number of elements. In this plot metallic radii are used for metals and covalent radii for nonmetals. Figure 9-4 suggests certain trends in atomic radii, for example, large radii for group 1, decreasing across the periods to smaller radii for group 17. To interpret these trends, let us first return to a topic introduced in Chapter 8.

Screening and Penetration

In Section 8-10, penetration was described as a gauge of how close an electron gets to the nucleus. When interpreting the radial probability distributions, we saw that s electrons, by virtue of their extra humps of probability close to the nucleus (Fig. 8-35), penetrate better than p electrons, which in turn penetrate better than d electrons. Screening, or shielding, reflects how an outer electron is blocked from the nuclear charge by inner electrons.

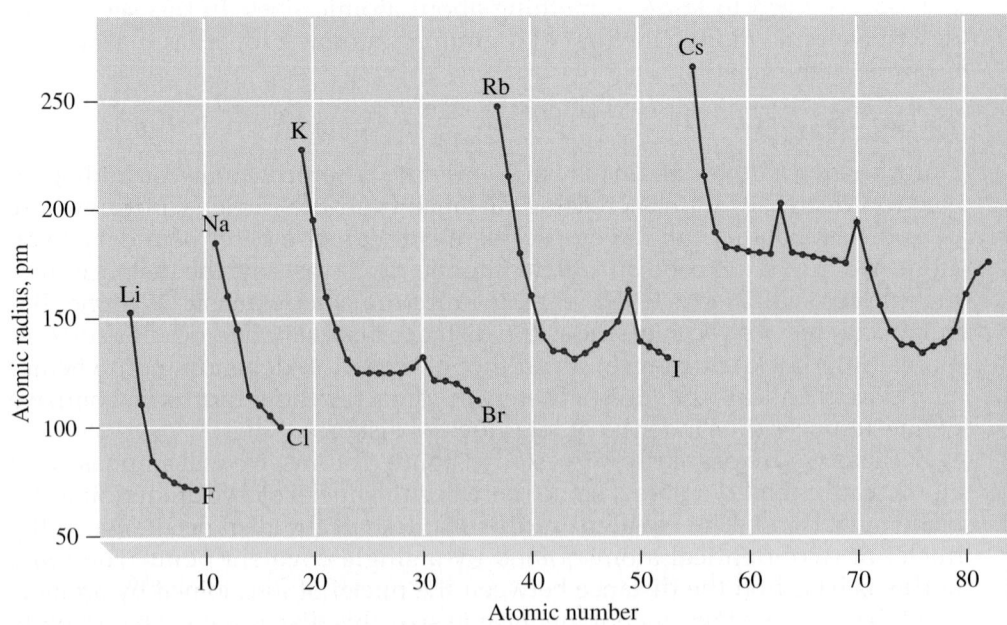

▲ FIGURE 9-4
Atomic radii
The values plotted are metallic radii for metals and covalent radii for nonmetals. Data for the noble gases are not included because of the difficulty of measuring covalent radii for these elements (only Kr and Xe compounds are known). The explanations usually given for the several small peaks in the middle of some periods are beyond the scope of this discussion.

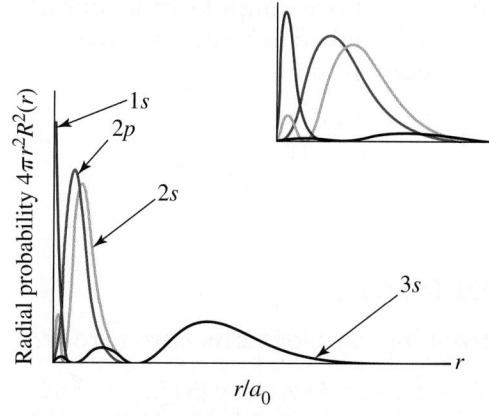

◀ FIGURE 9-5
Radial probability distributions for magnesium
Graphs of $4\pi r^2 R^2(r)$ as a function of r for the 1s, 2s, and 3s orbitals of magnesium. The graphs were obtained by using the radial functions from Table 8.1 with $Z_{eff} = 11.61$ (for 1s); $Z_{eff} = 7.39$ (for 2s); $Z_{eff} = 7.83$ (for 2p); and $Z_{eff} = 3.31$ (for 3s). The inset plot has a smaller range of r values in order to highlight the behavior of the probability functions closer to $r = 0$.

Consider the hypothetical process of building up each atom in the third period from the atom preceding it, beginning with sodium. In this process, the number of inner-shell, or core, electrons is fixed at ten in the configuration $1s^2 2s^2 2p^6$. As a first approximation, let us *assume* that the core electrons completely cancel an equivalent charge on the nucleus. In this way, the core electrons shield, or screen, the outer-shell electrons from the full attractive force of the nucleus. Let us also assume that the outer-shell electrons do not screen one another. Finally, let us redefine an **effective nuclear charge, Z_{eff}**, first introduced in Section 8-10, as the true nuclear charge minus the charge that is screened out by electrons.

$$Z_{eff} = Z - S \qquad (9.3)$$

Think of S as representing the number of inner electrons that appear to screen or shield an outer electron. Based on the two assumptions just stated, in sodium ($Z = 11$) the ten core electrons would screen out 10 units of nuclear charge (that is, $S = 10$), leaving an effective nuclear charge of $11 - 10 = +1$. In magnesium ($Z = 12$), Z_{eff} would be $+2$. In aluminum, Z_{eff} would be $+3$, and so on across the period.

Actually, neither assumption we made above—full screening by inner-shell electrons and no screening by outer-shell electrons—is correct. These assumptions ignore the fact that the electrons, both inner and outer, occupy orbitals with different radial probability distributions and, consequently, different degrees of penetration (Fig. 9-5). Thus, an s electron, with its greater penetration, will be screened by inner electrons less than will a p electron. Similarly a p electron is shielded less than a d electron, which has a much lower penetration. In sodium, the 10 core electrons cancel only about 8.5 units of charge; thus, Z_{eff} is about 2.5+, not 1+. Also, outer-shell electrons do screen one another somewhat because of penetration effects. Each outer-shell electron is about one-third effective in screening the other outer-shell electrons. Thus, the Z_{eff} experienced by each of the two outer-shell electrons in magnesium is about $12 - 8.5 - \frac{1}{3} = 3.2+$, not 2+ (Fig. 9-6). As we will see, trends in properties of atoms and ions in the periodic table are largely governed by the effective nuclear charge, Z_{eff}.

The Effects of Penetration and Screening

The wave function of a multielectron atom provides a qualitative understanding of the effects of penetration and screening. In this simplified picture, the nuclear charge is replaced by Z_{eff}, so that the orbital energy is approximated by

$$E_n = -R_H \frac{Z_{eff}^2}{n^2} \qquad (9.4)$$

where Z_{eff} is the effective nuclear charge in the shell corresponding to the value of n. All other symbols have their usual meaning (see Chapter 8). Equation (9.4) has the same form as the energy of the hydrogen atom obtained as a solution to the Schrödinger equation. The multielectron atom has been

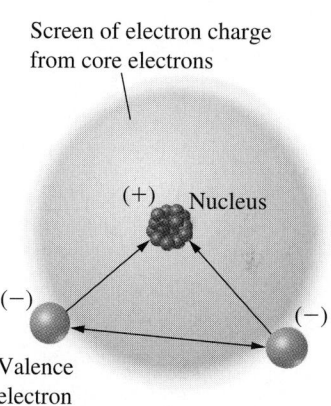

Screen of electron charge
from core electrons

(+) Nucleus

(−)   (−)

Valence
electron

▲ FIGURE 9-6
The shielding effect and effective nuclear charge, Z_{eff}
Two valence electrons (blue) are attracted to the nucleus of a Mg atom. The atom's 12+ nuclear charge is screened by the 10 core electrons (gray), but not perfectly. The valence electrons also screen each other somewhat. The result is an *effective* nuclear charge, Z_{eff}, closer to 3+ than to 2+.

▶ The leading term of equation (9.5) is the Bohr radius of the nth Bohr orbit. The corresponding formula for the orbitals of the H atom can be obtained by setting $Z_{\text{eff}} = 1$.

reduced to a one-electron approximation, a great oversimplification, but useful nevertheless. The average size of an orbital is taken to be the average value of the distance, $\bar{r}_{nl}$, of the electron in that orbital from the nucleus.

$$\bar{r}_{nl} = \frac{n^2 a_0}{Z_{\text{eff}}}\left\{1 + \frac{1}{2}\left[1 - \frac{\ell(\ell + 1)}{n^2}\right]\right\} \tag{9.5}$$

9-1 ARE YOU WONDERING...

Where estimates of the screening by electrons come from?

These estimates come from an analysis of the wave functions of multielectron atoms. An exact solution of the Schrödinger equation can be obtained for the H atom, but for multielectron atoms, only approximate solutions are possible. The principle of the calculation is to assume each electron in the atom occupies an orbital much like those of the hydrogen atom. However, the functional form of the orbital is based on another assumption: that the electron moves in an effective or average field dictated by all the other electrons. With this assumption, the complicated multielectron Schrödinger equation is converted into a set of simultaneous equations—one for each electron. Each equation contains the unknown effective field and the unknown functional form of the orbital for the electron. The approach to solving such a set of equations is to guess at the functional forms of the orbitals, calculate an average potential for each electron to move in, and then solve for a new set of orbitals—one for each electron. The expectation is that the new orbitals are better than the initial guess. The new orbitals are then used to calculate a new effective field for the electrons, and the whole process is repeated until the calculated orbitals do not change much.

This iterative procedure, called the *self-consistent field (SCF) method*, was devised by Douglas Hartree in 1936, before the advent of computers. Currently, the wave functions of atoms and molecules are obtained by implementing the SCF procedures on computers. The use of computers to calculate molecular properties from a wave function by the SCF procedure has lead to the term molecular modeling. Molecular modeling has become a tool in modern chemical research.

The atomic orbitals obtained from SCF calculations closely resemble the atomic orbitals of the hydrogen atom in many ways. The angular dependence of the orbitals is identical, so that we can identify s, p, d, f orbitals by their characteristic shapes. The radial functions of the orbitals are different because the effective field is different from the one in the hydrogen atom, but the principal quantum number can still be defined. Thus, each electron in a multielectron atom has associated with it the four quantum numbers n, ℓ, m_ℓ, and m_s. Estimates of screening constants are based on an analysis of the radial functions obtained from SCF calculations.

In equation (9.5), all the symbols have their usual meaning. Again, equation (9.5) is the equation for the hydrogen atom or hydrogen-like ions with the nuclear charge replaced by Z_{eff} to approximate multielectron effects. Equations (9.4) and (9.5) are approximate but provide for a very useful semiquantitative interpretation of atomic properties. We consider next the three most important trends among atomic radii in relation to the periodic table.

KEEP IN MIND

that s- and p-valence electrons have some probability of being near the nucleus (Fig. 8-35). These electrons penetrate the inner core of electrons and experience a greater attraction to the nucleus than otherwise expected.

1. **Variation of Atomic Radii Within a Group of the Periodic Table.** Radial probability densities extend farther out from the nucleus as n increases, a fact seen both in Figure 8-35 and in equation (9.5). Thus, we should expect that the more electron shells occupied by electrons, the larger the atom. This idea works for the group members of lower atomic numbers, where the increase in radius from one period to the next is large (as from Li to Na to K in group 1). At higher atomic numbers, the increase in radius is smaller (as from K to Rb to Cs in group 1). In these elements of higher atomic number, outer-shell electrons are held somewhat more tightly than

expected because inner-shell electrons in d and f subshells are less effective than s and p electrons in screening outer-shell electrons from the nucleus; that is, Z_{eff} is larger than expected. Nevertheless, in general, the following is true.

> The more electronic shells in an atom, the larger is the atom. Atomic radius increases from top to bottom through a group of elements.

2. **Variation of Atomic Radii Within a Period of the Periodic Table.** From Figure 9-4, we see that, in general, atomic radius decreases from left to right across a period. A careful look at the figure suggests that this trend does not apply to the transition elements. Let us look first at the general trend of decreasing radii and then at what is special about the transition elements.

   Across a period, the atomic number increases by one for each succeeding element. For the main-group elements, each increase in atomic number is accompanied by the addition of one electron to the valence shell. The valence-shell electrons, being in the same shell, shield each other poorly from the increasing nuclear charge. The Z_{eff} for the $2s$ electron of Li is 1.3, and that for Be is 1.9; thus, Z_{eff} increases as Z increases across the main-group portions of a period. Across a period, the principal quantum number stays constant, so that whether we use Z_{eff} or just the nuclear charge Z in equation (9.5), the result is as follows.

◀ Values of Z_{eff} can be estimated using the rules set out in Feature Problem 73.

> The atomic radius decreases from left to right through a period of elements.

3. **Variation in Atomic Radius Within a Transition Series.** With the transition elements, the situation is a little different from that described above. In Figure 9-4, it is apparent that the atomic radii of transition elements tend to be about the same across a period but with a few unusual peaks. It is beyond the scope of this text to explain the exceptions; however, the general trend is not difficult to understand. In a series of transition elements, additional electrons go into an *inner* electron shell, where they participate in shielding outer-shell electrons from the nucleus. At the same time, the number of electrons in the *outer* shell tends to remain constant. Thus, the outer-shell electrons experience a roughly comparable force of attraction to the nucleus throughout a transition series. Consider Fe, Co, and Ni. Fe has 26 protons in the nucleus and 24 inner-shell electrons. In Co ($Z = 27$), there are 25 inner-shell electrons, and in Ni ($Z = 28$), there are 26. In each case, the two outer-shell electrons are under the influence of about the same net charge (about +2). That is, Z_{eff} for the $4s$ electrons of the first transition series is approximately constant. Thus, atomic radii do not change very much for this series of three elements, namely, 124 pm for Fe and 125 pm for Co and Ni.

The graph in the margin represents the variation of Z_{eff} and atomic radius with atomic number. Which axis and correspondingly colored line corresponds to Z_{eff} and which to atomic radius?

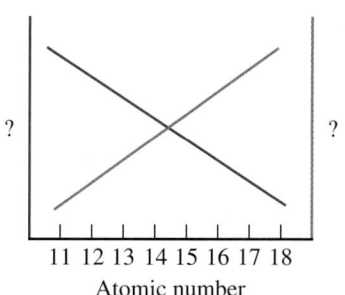

11 12 13 14 15 16 17 18
Atomic number

EXAMPLE 9-1   Relating Atomic Size to Position in the Periodic Table

Refer only to the periodic table on the inside front cover, and determine which is the largest atom: Sc, Ba, or Se.

Analyze

We first place the element in the periodic table and decide whether or not the elements are in the same period and whether they are on the right or left of the periodic table. We can then use the rules noted above to decide on the relative sizes of atoms (or ions).

Solve

Sc and Se are both in the fourth period, and we would expect Sc to be larger than Se because atomic sizes decrease from left to right in a period. Ba is in the sixth period and so has more electronic shells than either Sc or Se. Furthermore, it lies even closer to the left side of the table (group 2) than does Sc (group 3). We can say with confidence that the Ba atom should be the largest of the three.

Assess

By using the procedure outlined above, we have been able to show that $r_{Ba} > r_{Sc} > r_{Se}$. The actual atomic radii are Se, 117 pm; Sc, 161 pm; and Ba, 217 pm.

PRACTICE EXAMPLE A: Use the periodic table on the inside front cover to predict which is the smallest atom: As, I, or S.

PRACTICE EXAMPLE B: Which of the following atoms do you think is closest in size to the Na atom: Br, Ca, K, or Al? Explain your reasoning, and do not use any tabulated data from the chapter in reaching your conclusion.

Ionic Radius

When a metal atom loses one or more electrons to form a positive ion, the positive nuclear charge exceeds the negative charge of the electrons in the resulting cation. The nucleus draws the electrons in closer, and, as a consequence, the following holds true.

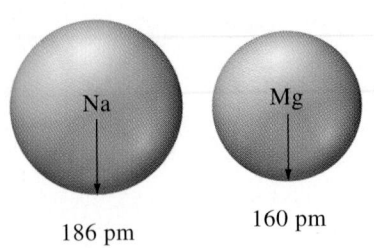

186 pm   160 pm

> **Cations are smaller than the atoms from which they are formed.**

Figure 9-7 compares four species: the atoms Na and Mg and the ions Na^+ and Mg^{2+}. As expected, the Mg atom is smaller than the Na atom, and the cations are smaller than the corresponding atoms. Na^+ and Mg^{2+} are **isoelectronic**—they have equal numbers of electrons (10) in identical configurations, $1s^2 2s^2 2p^6$. Mg^{2+} is smaller than Na^+ because its nuclear charge is larger (+12, compared with +11 for Na).

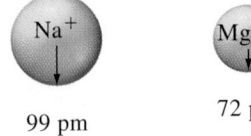

99 pm   72 pm

▲ FIGURE 9-7
A comparison of atomic and ionic sizes
Metallic radii are shown for Na and Mg and ionic radii for Na^+ and Mg^{2+}.

> **For isoelectronic cations, the more positive the ionic charge, the smaller the ionic radius.**

When a nonmetal atom gains one or more electrons to form a negative ion (anion), the nuclear charge remains constant, but Z_{eff} is reduced because of the additional electron(s). The electrons are not held as tightly. Repulsions among the electrons increase. The electrons spread out more, and the size of the atom increases, as suggested in Figure 9-8.

> **Anions are larger than the atoms from which they are formed. For isoelectronic anions, the more negative the charge, the larger the ionic radius.**

Knowledge of atomic and ionic radii can be used to vary certain physical properties. One example concerns strengthening glass. Normal window glass contains Na^+ and Ca^{2+} ions. The glass is brittle and shatters easily when struck a hard blow. One way to strengthen the glass is to replace the Na^+ ions at the surface with K^+ ions. The K^+ ions are larger and fill up the surface sites, leaving less opportunity for cracking than with the smaller Na^+ ions. The result is a shatter-resistant glass.

Another example is the striking result when Cr^{3+} ions replace about 1% of the Al^{3+} ions in aluminum oxide, Al_2O_3. This substitution is possible because Cr^{3+} ions are only slightly larger (by 9 pm) than Al^{3+} ions. Pure aluminum oxide is colorless, but with this small amount of chromium(III) ion, it is a beautiful red color. This impure Al_2O_3 is the gem known as a ruby. Rubies and other gemstones can be made artificially and are used as jewelry and in devices such as lasers. The color of the ruby is further discussed in Chapter 24.

Figure 9-9, arranged in the format of the periodic table, shows relative sizes of typical atoms and ions, and summarizes the generalizations described in this section.

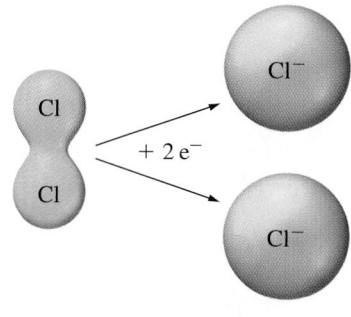

Covalent radius	Ionic radius
99 pm	181 pm

▲ FIGURE 9-8
Covalent and anionic radii compared
The two Cl atoms in a Cl_2 molecule gain one electron each to form two Cl^- ions.

EXAMPLE 9-2   Comparing the Sizes of Cations and Anions

Refer only to the periodic table on the inside front cover, and arrange the following species in order of increasing size: K^+, Cl^-, S^{2-}, and Ca^{2+}.

Analyze

The key lies in recognizing that the four species are *isoelectronic*, having the electron configuration of argon: $1s^2 2s^2 2p^6 3s^2 3p^6$. When considering isoelectronic cations, the higher the charge on the ion, the smaller the ion.

Solve

The larger charge on the calcium ion means that Ca^{2+} is smaller than K^+. Because K^+ has a higher nuclear charge than Cl^- ($Z = 19$, compared with $Z = 17$), it is smaller than Cl^-. For isoelectronic anions, the higher the charge, the larger the ion. S^{2-} is larger than Cl^-. The order of increasing size is

$$Ca^{2+} < K^+ < Cl^- < S^{2-}$$

Assess

We can summarize the generalizations about isoelectronic atoms and ions into a single statement: Among isoelectronic species, the greater the atomic number, the smaller the size.

PRACTICE EXAMPLE A: Refer only to the periodic table on the inside front cover, and arrange the following species in order of increasing size: $Ti^{2+}, V^{3+}, Ca^{2+}, Br^-$, and Sr^{2+}.

PRACTICE EXAMPLE B: Refer only to the periodic table on the inside front cover, and determine which species is in the *middle* position when the following five are ranked according to size: the atoms N, Cs, and As and the ions Mg^{2+} and Br^-.

9-3   CONCEPT ASSESSMENT

On the blank periodic table in the margin, locate the following:

(a) The smallest group 13 atom

(b) The smallest period 3 atom

(c) The largest anion of a nonmetal in period 3

(d) The largest group 13 cation

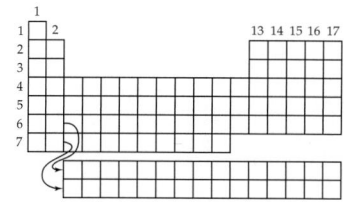

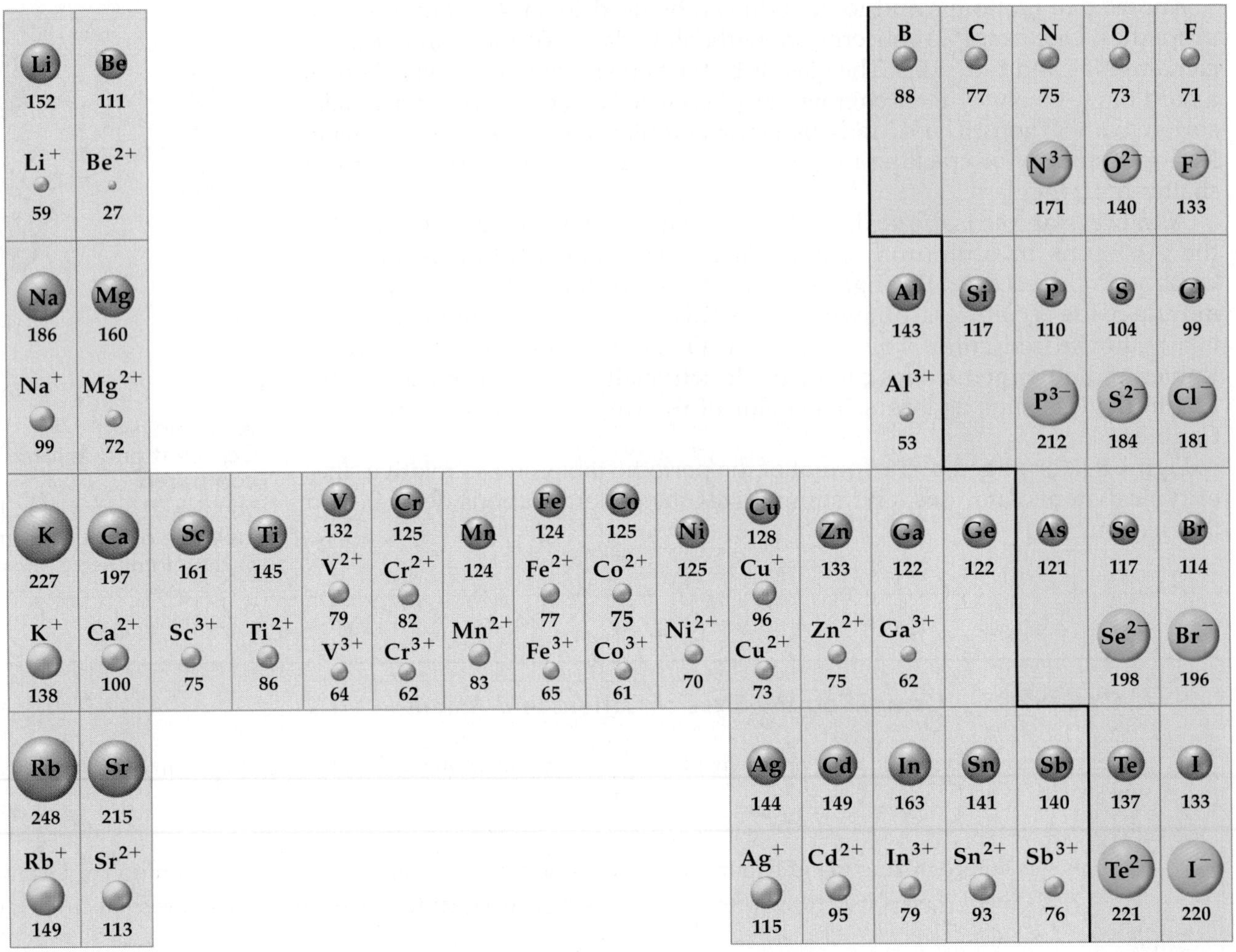

▲ FIGURE 9-9
A comparison of some atomic and ionic radii
The values given, in picometers (pm), are metallic radii for metals, single covalent radii for nonmetals, and ionic radii for the ions indicated.

9-4   Ionization Energy

In discussing metals, we talked about metal atoms losing electrons and thereby altering their electron configurations. But atoms do not eject electrons spontaneously. Electrons are attracted to the positive charge on the nucleus of an atom, and energy is needed to overcome that attraction. The more easily its electrons are lost, the more metallic an atom is considered to be. The **ionization energy**, *I*, is the quantity of energy a *gaseous* atom must absorb to be able to expel an electron. The electron that is lost is the one that is most loosely held.

Ionization energies are usually measured through experiments based on the photoelectric effect in which gaseous atoms at low pressures are bombarded with photons of sufficient energy to eject an electron from the atom. Here are two typical values.

$$Mg(g) \longrightarrow Mg^+(g) + e^- \qquad I_1 = 738 \text{ kJ/mol}$$
$$Mg^+(g) \longrightarrow Mg^{2+}(g) + e^- \qquad I_2 = 1451 \text{ kJ/mol}$$

The symbol I_1 stands for the *first* ionization energy—the energy required to strip one electron from a neutral gaseous atom.* I_2 is the *second* ionization energy—the energy to strip an electron from a gaseous ion with a charge of 1+. Further ionization energies are I_3, I_4, and so on. Each succeeding ionization energy is invariably larger than the preceding one. In the case of magnesium, for example, in the second ionization, the electron, once freed, has to move away from an ion with a charge of 2+ (Mg^{2+}). More energy must be invested than for a freed electron to move away from an ion with a charge of 1+ (Mg^+). This is a direct consequence of Coulomb's law, which states, in part, that the force of attraction between oppositely charged particles is directly proportional to the magnitudes of the charges.

First ionization energies (I_1) for many of the elements are plotted in Figure 9-10. In general, the farther an electron is from the nucleus, the more easily it can be extracted.

◀ A distinction between valence electrons and core electrons can be made based on the ionization energies for removing electrons one by one. The ionization energies of valence electrons are much smaller and show a big jump when the first core electron is removed.

Ionization energies decrease as atomic radii increase.

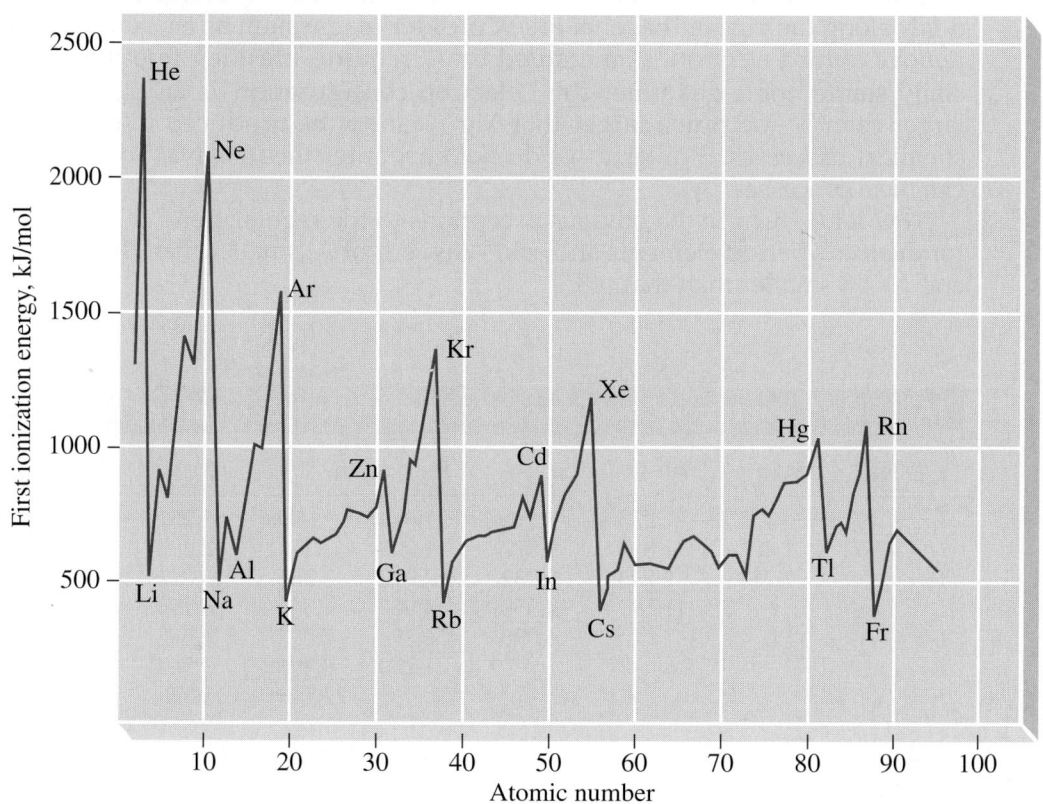

▲ FIGURE 9-10
First ionization energies as a function of atomic number
Because their electron configurations are so stable, more energy is required to ionize noble gas atoms than to ionize atoms of the elements immediately preceding or following them. The maxima on the graph come at the atomic numbers of the noble gases. The alkali metals are the most easily ionized of all groups. The minima in the graph come at their atomic numbers.

---

*Ionization energies are sometimes expressed in the unit electron-volt (eV). One electron-volt is the energy acquired by an electron as it falls through an electric potential difference of 1 volt. It is a very small energy unit, especially suited to describing processes involving individual atoms. When ionization is based on a *mole* of atoms, kJ/mol is the preferred unit (1 eV/atom = 96.49 kJ/mol). Sometimes the term *ionization potential* is used instead of ionization energy. Further, the quantities I_1, I_2, . . . , may be replaced by enthalpy changes, ΔH_1, ΔH_2, and so on.

▶ This generalization works well for main-group elements but less so for transition elements, where there are several exceptions.

This observation that ionization energies decrease as atomic radii increase reflects the effect of n and Z_{eff} on the ionization energy (I). Equation (9.4) suggests that the ionization energy is given by

$$I = R_H \times \frac{Z_{eff}^2}{n^2} \qquad (9.6)$$

so that across a period, as Z_{eff} increases and the valence-shell principal quantum number n remains constant, the ionization energy should increase. And down a group, as n increases and Z_{eff} increases only slightly, the ionization energy should decrease. Thus, atoms lose electrons more easily (become more metallic) as we move from top to bottom in a group of the periodic table. The decreases in ionization energy and the parallel increases in atomic radii are outlined in Table 9.3 for group 1.

Table 9.4 lists ionization energies for the third-period elements. With minor exceptions, the trend in moving across a period (follow the colored stripe) is that atomic radii decrease, ionization energies increase, and the elements become less metallic, or more nonmetallic, in character. Table 9.4 lists stepwise ionization energies (I_1, I_2, and so forth.). Note particularly the large breaks that occur along the zigzag diagonal line. Consider magnesium as an example. To remove a third electron, as measured by I_3, requires breaking into the especially stable noble-gas inner-shell electron configuration $2s^2 2p^6$. I_3 is *much* larger than I_2—so much larger that Mg^{3+} cannot be produced in ordinary chemical processes. Similarly, we do not encounter the ions Na^{2+} or Al^{4+} in chemical processes.

Now let us turn to the obvious exceptions to the regular trend in I_1 values for the third-period elements and ask, Why is I_1 of Al smaller than that of Mg and I_1 of S smaller than that of P?

TABLE 9.3 Atomic Radii and First Ionization Energies of the Alkali Metal (Group 1) Elements

	Atomic Radius, pm	Ionization Energy (I_1), kJ/mol
Li	152	520.2
Na	186	495.8
K	227	418.8
Rb	248	403.0
Cs	265	375.7

TABLE 9.4 Ionization Energies of the Third-Period Elements (in kJ/mol)

	Na	Mg	Al	Si	P	S	Cl	Ar
I_1	495.8	737.7	577.6	786.5	1,012	999.6	1,251.1	1,520.5
I_2	4,562	1,451	1,817	1,577	1,903	2,251	2,297	2,666
I_3		7,733	2,745	3,232	2,912	3,361	3,822	3,931
I_4			11,580	4,356	4,957	4,564	5,158	5,771
I_5				16,090	6,274	7,013	6,542	7,238
I_6					21,270	8,496	9,362	8,781
I_7						27,110	11,020	12,000

9-2 ARE YOU WONDERING...

If ionization energies can be used to estimate the effective nuclear charge?

One of the earliest estimates of effective nuclear charge was obtained by analyzing ionization energies in terms of equation (9.6). Thus, for example, the ionization energy of Li in its ground state is 519 kJ mol^{-1} and from equation (9.6), we have

$$I_1 = 1312.1 \frac{Z_{eff}^2}{n^2} \text{ kJ mol}^{-1}$$

so that

$$519 \text{ kJ mol}^{-1} = 1312.1 \frac{Z_{\text{eff}}^2}{2^2} \text{ kJ mol}^{-1}$$

and we get

$$Z_{\text{eff}} = 1.26$$

The Z_{eff} obtained from the ionization energy of the first excited state of Li ($1s^2 2p$), 339 kJ mol^{-1}, is 1.02. The value of Z_{eff} is very close to one because the inner $1s^2$ core almost perfectly screens the $2p$ electron. By contrast, the penetration of the $2s$ electron leads to the somewhat larger $Z_{\text{eff}} = 1.26$ for the ground state.

Consider the orbital diagrams for Mg, Al, P, and S shown in the margin. We expect I_1 of Al to be larger than for Mg. The reversal occurs because of the particular electrons lost. Mg loses a $3s$ electron, while Al loses a $3p$ electron. We expect that *more* energy is required to strip an electron from the lower energy $3s$ orbital in Mg ([Ne]$3s^2$) than from a half-filled $3p$ orbital in Al ([Ne]$3s^2 3p^1$). I_1 for S is slightly lower than for P for a different reason. Although the orbitals in the $3p$ subshell are degenerate, we can think of repulsion between electrons in the filled $3p$ orbital of a S atom ([Ne]$3s^2 3p^4$) as making it easier to remove one of those electrons than an electron from the half-filled $3p$ subshell of a P atom ([Ne]$3s^2 3p^3$).

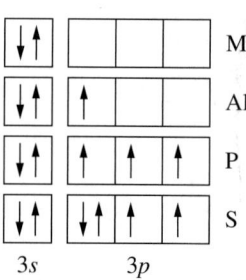

▲ Orbital diagram showing the valence electron configuration of magnesium, aluminum, phosphorus, and sulfur.

EXAMPLE 9-3   Relating Ionization Energies

Refer to the periodic table on the inside front cover, and arrange the following in the expected order of increasing first ionization energy, I_1: As, Sn, Br, Sr.

Analyze

Ionization energies decrease as atomic radii increase. Thus, if we arrange these four atoms according to decreasing radius, we will likely have arranged them according to increasing ionization energy. The largest atoms are to the left and the bottom of the periodic table. The smallest atoms are to the right and toward the top of the periodic table.

Solve

Of the four atoms, the one that best fits the large-atom category is Sr. Although none of the four atoms is particularly close to the top of the table, Br is the farthest to the right. This fixes the two extremes: Sr with the lowest ionization energy and Br with the highest. A tin atom should be larger than an arsenic atom, and thus Sn should have a lower ionization energy than As. The expected order of *increasing* ionization energies is Sr < Sn < As < Br.

Assess

The generalization that ionization energies decrease as atomic radii increase ignores the exceptions that occur when making comparisons between atoms of groups 2 and 13, as well as atoms of groups 15 and 16.

PRACTICE EXAMPLE A: Refer to the periodic table on the inside front cover, and arrange the following in the expected order of increasing first ionization energy, I_1: Cl, K, Mg, S.

PRACTICE EXAMPLE B: Refer to the periodic table on the inside front cover, and determine which element is most likely in the *middle* position when the following five elements are arranged according to first ionization energy, I_1. Rb, As, Sb, Br, Sr.

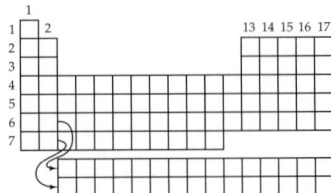

On the blank periodic table in the margin, locate the following:

(a) The group 14 element with the highest first ionization energy

(b) The element with the greatest first ionization energy in period 4

(c) A *p*-block element in period 4 that has a lower first ionization energy than the element immediately preceding it and the element directly following it

9-5 Electron Affinity

Ionization energy is the energy change for the removal of an electron. Let's consider the energy change associated with the addition of an electron. The thermochemical equation for the addition of an electron to a fluorine atom is

$$F(g) + e^- \rightarrow F^-(g) \qquad \Delta H_{ea} = -328 \text{ kJ mol}^{-1}$$

Notice that the process above is *exothermic*, meaning that energy is given off when an F atom gains an electron. **Electron affinity**, *EA*, can be defined as the enthalpy change, ΔH_{ea}, that occurs when an atom in the gas phase gains an electron. According to this definition, the electron affinity of fluorine is a negative quantity.

We have defined electron affinity to reflect the tendency for a neutral atom to gain an electron. An alternative definition refers to the energy change in the process: $X^-(g) \longrightarrow X(g) + e^-$; that is, reflecting the tendency of an anion to lose an electron. This alternative definition leads to the opposite signs for *EA* values from those written in this text. You should be prepared to see electron affinities expressed in both ways in the chemical literature.

Some representative electron affinities are listed in Figure 9-11. It is more difficult to make generalizations about electron affinities than about ionization energies. The smaller atoms to the right of the periodic table (for example, group 17) tend to have large, negative electron affinities.* Electron affinities

1							18
H −72.8	2	13	14	15	16	17	**He** >0
Li −59.6	**Be** >0	**B** −26.7	**C** −121.8	**N** +7	**O** −141.0	**F** −328.0	**Ne** >0
Na −52.9	**Mg** >0	**Al** −42.5	**Si** −133.6	**P** −72	**S** −200.4	**Cl** −349.0	**Ar** >0
K −48.4	**Ca** −2.37	**Ga** −28.9	**Ge** −119.0	**As** −78	**Se** −195.0	**Br** −324.6	**Kr** >0
Rb −46.9	**Sr** −5.03	**In** −28.9	**Sn** −107.3	**Sb** −103.2	**Te** −190.2	**I** −295.2	**Xe** >0
Cs −45.5	**Ba** −13.95	**Tl** −19.2	**Pb** −35.1	**Bi** −91.2	**Po** −186	**At** −270	**Rn** >0

▶ FIGURE 9-11
Electron affinities of main-group elements
Values are in kilojoules per mole for the process $X(g) + e^- \longrightarrow X^-(g)$.

*It is somewhat awkward to speak of larger and smaller with the term *electron affinity*. A strong tendency to gain an electron, which implies a high "affinity" for an electron, as with F and Cl, is reflected through a *low* value of *EA*—a large *negative* value.

tend to become less negative in progressing toward the bottom of a group, with the notable exception of the second-period members of groups 15, 16, and 17 (namely, N, O, and F). It is likely that for these small atoms, an incoming electron encounters strong repulsive forces from other electrons in the atom and is thereby not as tightly bound as we might otherwise expect.

Some atoms have no tendency to gain an electron. This is the case with the noble gases, where an added electron would have to enter the empty s orbital of the next electronic shell. Other cases include the groups 2 and 12 elements, where the electron would have to enter the p subshell of the valence shell and a few other elements, such as Mn, where the electron would have to enter either the p subshell of the valence shell or a half-filled $3d$ subshell.

In considering the gain of a second electron by a nonmetal atom, we encounter positive electron affinities. Here the electron to be added is approaching not a neutral atom, but a negative ion. There is a strong repulsive force between the electron and the ion, and the energy of the system increases. Thus, for an element like oxygen, the first electron affinity is negative and the second is positive.

$$O(g) + e^- \longrightarrow O^-(g) \qquad EA_1 = -141.0 \text{ kJ/mol}$$
$$O^-(g) + e^- \longrightarrow O^{2-}(g) \qquad EA_2 = +744 \text{ kJ/mol}$$

The high positive value of EA_2 makes the formation of *gaseous* O^{2-} seem very unlikely. The ion O^{2-} can exist, however, in ionic compounds, such as MgO(s), where formation of the ion is accompanied by other energetically favorable processes.

🔍 9-5   CONCEPT ASSESSMENT

On the blank periodic table in the margin locate the group expected to have:

(a) the most negative electron affinities in each period

(b) the least negative electron affinities in each period

(c) all positive electron affinities in each period

9-6   Magnetic Properties

An important property related to the electron configurations of atoms and ions is their behavior in a magnetic field. A spinning electron is an electric charge in motion. It induces a magnetic field (recall the discussion on page 334). In a **diamagnetic** atom or ion, all electrons are paired and the individual magnetic effects cancel out. A diamagnetic species is weakly repelled by a magnetic field. A **paramagnetic** atom or ion has unpaired electrons, and the individual magnetic effects do not cancel out. The unpaired electrons possess a magnetic moment that causes the atom or ion to be attracted to an external magnetic field. The more unpaired electrons present, the stronger is this attraction.

Manganese has a paramagnetism corresponding to five unpaired electrons, which is consistent with the electron configuration

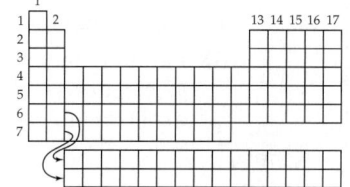

KEEP IN MIND

that it was the effect of the magnetic moments associated with the two different possibilities for an electron spin quantum number (equal in magnitude and opposite in sign) that allowed Stern and Gerlach to detect the presence of electron spin by using a magnetic field.

When a manganese atom loses two electrons, it becomes the ion Mn^{2+}, which is paramagnetic, and the strength of its paramagnetism corresponds to five unpaired electrons.

$$Mn^{2+}: \text{[Ar]} \quad \boxed{\uparrow \,|\, \uparrow \,|\, \uparrow \,|\, \uparrow \,|\, \uparrow} \quad \boxed{}$$

$$3d \qquad\qquad 4s$$

When a third electron is lost to produce Mn^{3+}, the ion has a paramagnetism corresponding to four unpaired electrons. The third electron lost is one of the unpaired $3d$ electrons.

$$Mn^{3+}: \text{[Ar]} \quad \boxed{\uparrow \,|\, \uparrow \,|\, \uparrow \,|\, \uparrow \,|\, } \quad \boxed{}$$

$$3d \qquad\qquad 4s$$

EXAMPLE 9-4   **Determining the Magnetic Properties of an Atom or Ion**

Which of the following would you expect to be diamagnetic and which paramagnetic?

(a) Na atom   **(b)** Mg atom   **(c)** Cl^- ion   **(d)** Ag atom

Analyze

To determine whether or not an atom or ion is paramagnetic, we need to determine the electronic configuration of the species.

Solve

(a) Paramagnetic. The Na atom has a single $3s$ electron outside the Ne core. This electron is unpaired.

(b) Diamagnetic. The Mg atom has *two* $3s$ electrons outside the Ne core. They must be paired, as are all the other electrons in the Ne core.

(c) Diamagnetic. Cl^- is isoelectronic with Ar, and Ar has all electrons paired ($1s^2 2s^2 2p^6 3s^2 3p^6$).

(d) Paramagnetic. We do not need to work out the exact electron configuration of Ag. Because the atom has 47 electrons—an odd number—at least one of the electrons must be unpaired (recall the Stern–Gerlach experiment, page 334).

Assess

We see that a quick method of determining the magnetic properties of an atom or an ion is to use that atomic number and add or subtract for anions and cations. If the resultant number is odd, then the species is paramagnetic. However, if the number is even, the species may or may not be diamagnetic, depending on the electronic configuration—for example, consider Ti.

PRACTICE EXAMPLE A: Which of the following are paramagnetic and which are diamagnetic: Zn, Cl, K^+, O^{2-}, and Al?

PRACTICE EXAMPLE B: Which has the greater number of unpaired electrons, Cr^{2+} or Cr^{3+}? Explain.

Q 9-6   CONCEPT ASSESSMENT

On the blank periodic table in the margin locate the following:

(a) The period 4 transition element having a cation in the +3 oxidation state that is diamagnetic

(b) The period 5 element existing in the −2 oxidation state as an anion that is diamagnetic

(c) The period 4 transition element having a 2+ cation that is paramagnetic and has a half-filled d subshell

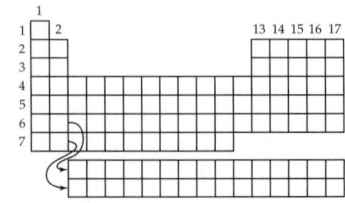

9-7   Periodic Properties of the Elements

As we noted at the beginning of the chapter, we can use the periodic law and the periodic table to predict some of the atomic, physical, and chemical properties of elements and compounds.

Atomic Properties

In this chapter we have learned how some atomic properties—atomic radius and ionization energy—vary within groups and periods of elements. We summarize these trends in relation to the periodic table in Figure 9-12. Trends are generally easy to apply within a group: The atomic radius of Sr is greater than that of Mg; both elements are in group 2. Usually, there is no difficulty in applying trends within a period either: The first ionization energy of P is greater than that of Mg; both elements are in the third period. However, comparing elements that are not within the same group or period can be difficult. The atomic radius of Sr is greater than that of P. Sr is farther down in its group of the periodic table and much farther to the left in its period than is P. Each of these directions is that of increasing atomic radius. We cannot, however, easily predict whether Mg or I has the larger atomic radius. The position of Mg to the left in its period suggests that Mg should have the larger radius, but the position of I toward the bottom of its group argues for I. Despite this limitation, you should find Figure 9-12 helpful in most cases.

Variation of Physical Properties Within a Group   Table 9.5 lists some properties of three of the halogens (group 17). The table has two blank spaces for bromine. We fill in these blanks in Example 9-5 by making an assumption that works often enough to make it useful:

> The value of a property often changes uniformly from the top to the bottom of a group of elements in the periodic table.

TABLE 9.5   Some Properties of Three Halogen (Group 17) Elements

	Atomic Number	Atomic Mass, u	Molecular Form	Melting Point, K	Boiling Point, K
Cl	17	35.45	Cl_2	172	239
Br	35	79.90	Br_2	?	?
I	53	126.90	I_2	387	458

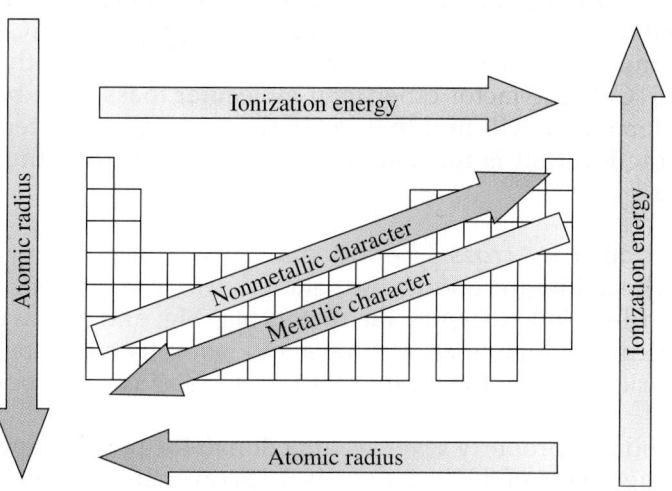

◀ FIGURE 9-12
Atomic properties and the periodic table—a summary
Atomic radius refers to metallic radius for metals and covalent radius for nonmetals. Ionization energies refer to first ionization energy. Metallic character relates generally to the ability to lose electrons, and nonmetallic character to the ability to gain electrons.

EXAMPLE 9-5   Using the Periodic Table to Estimate Physical Properties

Use data from Table 9.5 to estimate the boiling point of bromine.

Analyze

Remember that boiling points increase going down a group. As an initial guess we can consider the average between two elements.

Solve

The atomic number of bromine (35) is between the atomic numbers of chlorine (17) and iodine (53). Its atomic mass (79.90 u) is also intermediate to those of chlorine and iodine. (The average of the atomic masses of Cl and I is 81.18 u.) It is reasonable to expect that the boiling point of liquid bromine might also be intermediate to the boiling points of chlorine and iodine.

$$\text{bp Br}_2 \approx \frac{239 \text{ K} + 458 \text{ K}}{2} = 349 \text{ K}$$

Assess

The observed boiling point is 332 K, which is close to the calculated result. A better estimate of the boiling point could possibly be made by plotting the known boiling points as a function of atomic mass, and then fitting the data.

PRACTICE EXAMPLE A: Estimate the melting point of bromine.

PRACTICE EXAMPLE B: Estimate the boiling point of astatine, At.

▲ FIGURE 9-13
Three halogen elements
Chlorine is a yellow-green gas. Bromine is a dark red liquid. Iodine is a grayish black solid.

TABLE 9.6   Melting Points of Two Series of Compounds

	Molecular Mass, u	Melting Point, °C
CF$_4$	88.0	−183.7
CCl$_4$	153.8	−22.9
CBr$_4$	331.6	90.1
CI$_4$	519.6	171
HF	20.0	−83.6
HCl	36.5	−114.2
HBr	80.9	−86.8
HI	127.9	−50.8

First, let us make some predictions about fluorine, the halogen not listed in Table 9.5. Its closest neighbor in group 17 is chlorine, which has a boiling point of 239 K (−34 °C); chlorine is a *gas* at room temperature (about 298 K). The other halogens are *liquid* bromine and *solid* iodine (Fig. 9-13). We would expect fluorine to have a lower melting point (mp) and lower boiling point (bp) than chlorine and also to be a gas at room temperature. (Observed values for F$_2$: mp − 53 K; bp = 85 K.)

The generalization that a property varies uniformly within a group of the periodic table can work for compounds as well as for elements. Table 9.6 lists the melting points of two sets of compounds, binary carbon–halogen compounds and the *hydrogen halides*, HX (where X = F, Cl, Br, or I). We see that the melting points increase fairly uniformly with increasing molecular mass for the carbon–halogen compounds. This trend between melting point (and boiling point) and molecular mass can be explained in terms of intermolecular forces, as we will see in Chapter 12. Based on the melting points of HCl, HBr, and HI, the melting point of HF should be about −145 °C, but the observed value is −83.6 °C. Some factor other than molecular mass must be involved here. In Chapter 12, we will find that in HF there is a special intermolecular force of attraction that is missing or unimportant in the other compounds in Table 9.6.

Variation of Physical Properties Across a Period   A few properties vary regularly across a period. The ability to conduct heat and electricity are two that do. Thus, among the third-period elements, the metals Na, Mg, and Al have good thermal and electrical conductivities. The metalloid Si is only a fair conductor of heat and electricity, while the nonmetals P, S, Cl, and Ar are poor conductors.

In some cases, the trend in a property reverses direction in the period (similar to the trend in melting points of the hydrogen halides reversing direction

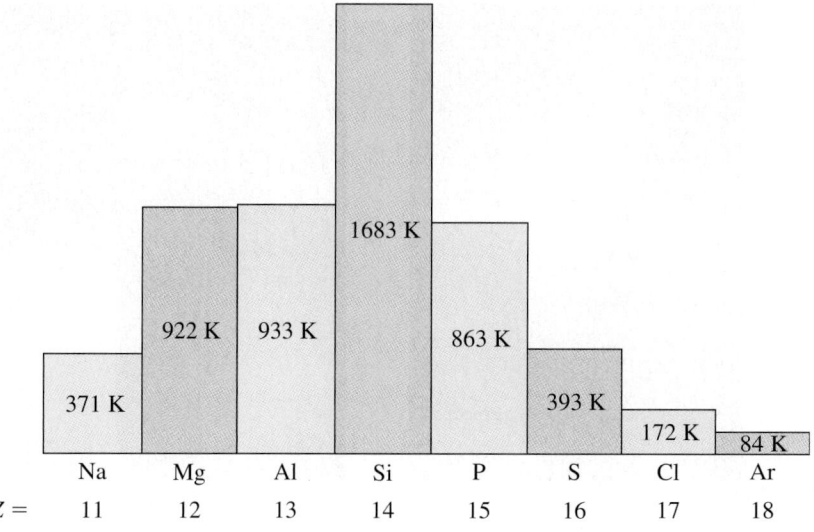

◀ FIGURE 9-14
Melting points of the third-period elements
Sometimes the trend in a property reverses direction within a period, as illustrated by this bar graph.

within a group). Consider, for example, the melting points of the third-period elements shown in a bar graph in Figure 9-14. Melting involves destruction of the orderly arrangement of the atoms or molecules in a crystalline solid. The amount of thermal energy needed for melting to occur, and hence the melting-point temperature, depends on the strength of the attractive forces between the atoms or molecules in the solid. For the metals Na, Mg, and Al, these forces are *metallic bonds*, which, roughly speaking, become stronger as the number of electrons available to participate in the bonding increases. Sodium, therefore, has the lowest melting point (371 K) of the third-period metals. With silicon, the forces between atoms are strong *covalent bonds* extending throughout the crystalline solid. Silicon has the highest melting point (1683 K) of the third-period elements. Phosphorus, sulfur, and chlorine exist as discrete molecules (P_4, S_8, and Cl_2). The bonds between atoms within molecules are strong, but *intermolecular forces*, the attractive forces between molecules, become progressively weaker across the period, and the melting points decrease. Argon atoms do not form molecules, and the forces between Ar atoms in solid argon are especially weak. Argon's melting point is the lowest for the entire period (84 K). The property of hardness also depends on forces between atoms and molecules in a solid. So the hardness of the solid third-period elements varies in much the same way as their melting points. Thus, on a 10-point scale in which solids are rated according to their abilities to scratch or abrade one another, sodium has a hardness of about 0.5; magnesium, 2; aluminum, 3; silicon, 7; and phosphorus and sulfur 1 to 2. Silicon has the greatest hardness.

◀ When evaluating trends, it is often useful to sketch a graph showing the variation of the property.

◀ Metallic bonds are described in Section 11-7, covalent bonding in substances like silicon is discussed in Section 12-7, and the topic of intermolecular forces is examined throughout Chapter 12.

Reducing Abilities of Group 1 and 2 Metals We learned in Chapter 5 that a reducing agent makes possible a reduction half-reaction. The reducing agent itself, by losing electrons, is oxidized. In the following reactions, M, a group 1 or 2 metal, is the reducing agent and H_2O is the substance that is reduced.

$$2\,M(s) + 2\,H_2O(l) \longrightarrow 2\,M^+(aq) + 2\,OH^-(aq) + H_2(g) \qquad (M = \text{group 1 metal})$$
$$M(s) + 2\,H_2O(l) \longrightarrow M^{2+}(aq) + 2\,OH^-(aq) + H_2(g) \qquad (M = \text{Ca, Sr, Ba, or Ra})$$

At first guess, we might think that the lower the energy requirement for extracting electrons—the lower the ionization energy—the better the metal is as a reducing agent and the more vigorous its reaction with water. Potassium, for instance, has a lower ionization energy ($I_1 = 419$ kJ/mol) than does the next member of the fourth period, calcium ($I_1 = 590$; $I_2 = 1145$ kJ/mol). Our expectation is that potassium should react more vigorously with water than

(a)   (b)

▶ FIGURE 9-15
A comparison of the reactions of potassium and calcium with water
(a) Potassium, a group 1 metal, reacts so rapidly that the hydrogen evolved bursts into flame. Notice that the metal is less dense than water.
(b) Calcium, a group 2 metal, reacts more slowly than does potassium. Also, calcium is denser than water. The pink color of the acid–base indicator phenolphthalein signals the buildup of OH^- ions.

calcium does. This, indeed, is the case (Fig. 9-15). Mg and Be do not react with cold water as do the other alkaline earth metals. This might be explained in terms of the higher ionization energies for those two metals (Mg: $I_1 = 738$, $I_2 = 1451$ kJ/mol; Be: $I_1 = 900$, $I_2 = 1757$ kJ/mol).

Attributing the reactivity of these group 1 and 2 metals just to their ionization energies is an oversimplification, however. As long as the differences in ionization energies are very large, it is possible to make comparisons by considering only this factor. Where differences in ionization energies are smaller, though, other factors must also be considered, as we will see elsewhere in the text.

Oxidizing Abilities of the Halogen Elements (Group 17) An oxidizing agent gains the electrons that are lost in an oxidation half-reaction. The oxidizing agent, by gaining electrons, is itself reduced. Electron affinity is the atomic property introduced in this chapter that is related to the gain of electrons. We might expect an atom with a strong tendency to gain electrons (a large *negative* electron affinity) to take electrons away from atoms with low ionization energies—metals. In these terms, it is understandable that active metals form ionic compounds with active nonmetals. If M is a group 1 metal and X a group 17 nonmetal (halogen), this exchange of an electron leads to the formation of M^+ and X^- ions. In some cases, the reaction is especially vigorous (Fig. 9-16).

$$2\,M + X_2 \longrightarrow 2\,MX \qquad [\text{e.g., } 2\,Na(s) + Cl_2\,(g) \longrightarrow 2\,NaCl(s)]$$

Another interesting oxidation–reduction reaction involving the halogens is a *displacement reaction*. Two halogens, one in molecular form and the other in ionic form, exchange places, as in this reaction (Fig. 9-17).

$$Cl_2(g) + 2\,I^-(aq) \longrightarrow I_2(aq) + 2\,Cl^-(aq)$$

We might think of this reaction as involving a competition between Cl and I atoms for an extra electron that only the I atoms (as I^-) have initially. The Cl atoms win out because they have a more negative electron affinity. (This is an oversimplified explanation, however, because strictly speaking electron affinities apply only to the behavior of isolated gaseous atoms and not to atoms in molecules or ions in solution.) By similar reasoning, can you see why no reaction occurs for this combination?

$$Br_2(l) + Cl^-(aq) \longrightarrow \text{no reaction}$$

▲ FIGURE 9-16
Reaction of sodium metal and chlorine gas
The contents of the flask glow in this exothermic reaction between Na(s) and $Cl_2(g)$. The product is the ionic solid NaCl(s).

Such predictions as those in the preceding paragraph work well for the halogens Cl_2, Br_2, and I_2, but not for F_2. We cannot account for the observed fact that F_2 is the strongest oxidizing agent among all chemical substances, just by considering electron affinities.

Acid–Base Nature of Element Oxides. Some metal oxides, such as Li_2O, react with water to produce the metal hydroxide.

$$Li_2O(s) \ + H_2O(l) \longrightarrow 2\,Li^+(aq) + 2\,OH^-(aq)$$

A basic oxide   Lithium hydroxide

These metal oxides are called *basic oxides* or *base anhydrides*. The term **anhydride** means "without water." A "base without water" becomes a base when the water is added. Thus, the base anhydride Li_2O becomes the base LiOH, and BaO becomes $Ba(OH)_2$ after reaction with water. Moving from top to bottom down a group, the elements become more metallic, and their oxides become more basic.

Some nonmetal oxides react with water to produce an acidic solution. These are *acidic oxides* or *acid anhydrides*. $SO_2(g)$ reacts with water to produce H_2SO_3, a weak acid.

$$SO_2(g) + H_2O(l) \longrightarrow H_2SO_3(aq)$$

An acidic oxide   Sulfurous acid

Now let us examine the acid–base properties of the oxides of the third-period elements. We expect the metal oxides at the left of the period to be basic and the nonmetal oxides at the right to be acidic, but where and how does the changeover occur? Na_2O and MgO yield basic solutions in water. Cl_2O, SO_2, and P_4O_{10} produce acidic solutions. SiO_2 (quartz) does not dissolve in water. However, it does dissolve slightly in strongly basic solutions to produce silicates (similar to the carbonates formed by CO_2 in basic solutions). For this reason, we consider SiO_2 to be an acidic oxide.

Aluminum, a good conductor of heat and electricity, is clearly metallic in its physical properties. Al_2O_3, however, can act as either an acidic or a basic oxide. Oxides with this ability are called **amphoteric** (from the Greek word *amphos*, meaning "both"). Al_2O_3 is insoluble in water but exhibits its amphoterism by reacting with both acidic and basic solutions.

$$Al_2O_3(s) \ + 6\,HCl(aq) \longrightarrow 2\,AlCl_3(aq) \ + 3\,H_2O(l)$$

Base   Acid

$$Al_2O_3(s) + 2\,NaOH(aq) + 3\,H_2O(l) \longrightarrow 2\,Na[Al(OH)_4](aq)$$

Acid   Base   Sodium aluminate

The amphoterism of Al_2O_3 signifies the point at which a changeover from basic to acidic oxides occurs in the third period of elements. Figure 9-18 summarizes the acid–base properties of the oxides of the main-group elements.

▲ FIGURE 9-17
Displacement of I^-(aq) by Cl_2(g)
(a) Cl_2(g) is bubbled through colorless, dilute I^-(aq). (b) The I_2 produced is extracted into CCl_4(l), in which it is much more soluble (purple layer).

1	2	13	14	15	16	17
Li	Be	B	C	N	O	F
Na	Mg	Al	Si	P	S	Cl
K	Ca	Ga	Ge	As	Se	Br
Rb	Sr	In	Sn	Sb	Te	I
Cs	Ba	Tl	Pb	Bi	Po	At

▲ FIGURE 9-18
Acidic, basic, and amphoteric oxides of the s- and p-block elements
The acidic oxides are pink, the basic oxides are blue, and the amphoteric oxides are tan.

9-7 CONCEPT ASSESSMENT

On the blank periodic table in the margin locate the following:

(a) The group 13 element that is expected to form the most basic oxide

(b) The group 15 element that is expected to form the most acidic oxide

(c) The period 5 element that is expected to form the most basic oxide

(d) The period 5 element that is expected to form the most acidic oxide

(e) The period 3 element that exhibits amphoteric behavior

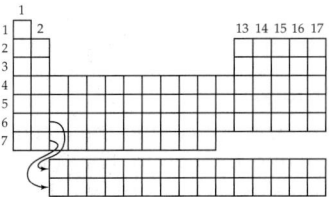

Summary

9-1 Classifying the Elements: The Periodic Law and the Periodic Table—The experimental basis of the periodic table of the elements is the **periodic law**: Certain properties recur periodically when the elements are arranged by increasing atomic number. The theoretical basis is that the properties of an element are related to the electron configuration of its atoms, and elements in the same group of the periodic table have similar electron configurations.

9-2 Metals and Nonmetals and Their Ions— The three classes of elements of the periodic table are the *nonmetals*, *metals*, and *metalloids*. **Metalloids** have some properties characteristic of metals and some characteristics of nonmetals. The nonmetals are further divided into the noble gases and the remainder of the main-group nonmetals, while the metals include the main-group metals and transition elements.

9-3 Sizes of Atoms and Ions—Types of atomic radii include **covalent radii**, **metallic radii**, and **van der Waals radii** (Fig. 9-3). In general, atomic radii decrease across a period and increase down a group of the periodic table (Figs. 9-4 and 9-9), mirroring the variation in **effective nuclear charge**, Z_{eff}, (equation 9.3) across a period and down a group. The **ionic radii** of positive ions are smaller than the neutral atom, whereas negative ions are larger than the parent atom (Figs. 9-7 and 9-8). Ionic radii exhibit adherence to the periodic law similar to that of atomic radii. When atoms or ions have the same number of electrons, they are said to be **isoelectronic**. When the radii of isoelectronic species are compared, the more negative the charge, the larger the radius of the ion or atom.

9-4 Ionization Energy—A study of **ionization energies**, I, shows that the periodic relationship observed is governed by the variation of Z_{eff}—that is, the ionization energy decreases down a group and increases across a period (Fig. 9-10, Tables 9.3 and 9.4).

9-5 Electron Affinity—**Electron affinity**, EA, is the energy change when an electron is added to a gaseous atom. Electron affinity does not exhibit clear-cut trends (Fig. 9-11).

9-6 Magnetic Properties—The magnetic properties of an atom or ion stem from the presence or absence of unpaired electrons. **Paramagnetic** atoms and ions have one or more unpaired electrons. In **diamagnetic** atoms and ions, all electrons are paired.

9-7 Periodic Properties of the Elements—The metallic, nonmetallic, and metalloid characteristics of atoms can be related to a set of atomic properties. In general, large atomic radii and low ionization energies are associated with metals; small atomic radii, high ionization energies, and large negative electron affinities are associated with nonmetals. Metalloids occur first in the third row (at Si) where the break between acid–base properties becomes less defined. Metals usually form oxides that give basic solutions, whereas nonmetal oxides form acidic solutions. The nonmetal oxide is called an **anhydride** since the addition of water gives an acid. At the break between clearly basic and acidic properties, some metals and metalloids exhibit **amphoteric** behavior because they react with both acids and bases (Fig. 9-18).

Integrative Example

When the ionization energies of a series of isoelectronic atoms and ions are compared, an interesting relationship is observed for some of them. In particular, if the square root of the ionization energy (in kJ mol^{-1}) for the series Li, Be$^+$, B^{2+}, C^{3+}, N^{4+}, O^{5+}, and F^{6+} is plotted against the atomic number (Z) of the species, a linear relationship is obtained. The corresponding graph for the series Na, Mg$^+$, Al^{2+}, Si^{3+}, P^{4+}, S^{5+}, and Cl^{6+}, is also linear. The graph is shown on page 387.

The equations for the two lines joining the points are

$$\text{Second-row elements:} \qquad \sqrt{I} = 18.4Z - 32.0 \qquad \textbf{(9.7)}$$
$$\text{Third-row elements:} \qquad \sqrt{I} = 13.5Z - 124 \qquad \textbf{(9.8)}$$

Explain the origin of these relationships and the differences in the numerical coefficients.

Analyze

We first notice that the electron configuration of the second-row atoms and ions is $1s^2 2s^1$, that is, a single electron ($2s^1$) beyond the helium core ($1s^2$). Similarly, for the third-row atoms and ions the electron configuration is $1s^2 2s^2 2p^6 3s^1$, that is, a single electron ($3s^1$) beyond the neon core ($1s^2 2s^2 2p^6$).

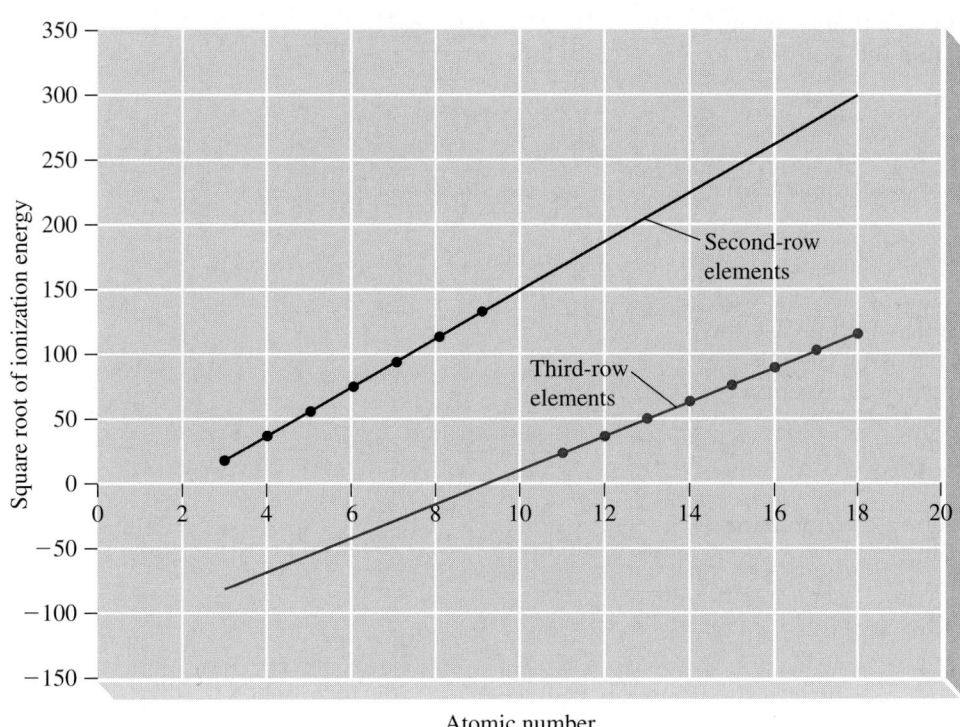

◀ Variation of $\sqrt{I}$ with atomic number for the second-row elements (black line) and the third-row elements (red line).

In both series of atoms and ions, the inner-core electrons screen the single valence-shell electron from the nucleus. These species are all reminiscent of the Bohr atom so that as an approximation we can use the expression for the energy levels for hydrogen-like atoms or hydrogen-like ions (equation 8.9). Specifically, we should be able to use equation (8.9) to derive equations for the energy required to remove the electron from the valence shells of a hydrogen-like species, that is, the ionization energy (I). Once we have these equations, we can compare them with the equations for the two straight-line graphs.

Solve

Equation (8.9), with the substitution $Z = Z_{eff}$, produces equation (9.4); and with the further substitution $R_H = 2.178 \times 10^{-18}$ J, we have

$$E_n = -2.178 \times 10^{-18}\left(\frac{Z_{eff}^2}{n^2}\right) \text{J}$$

E_n in equations (8.9) and (9.4) has the unit J atom^{-1}, which must be converted to kJ mol^{-1}—the unit of I in the two straight-line equations given to us.

$$E_n = -2.178 \times 10^{-18}\frac{\text{J}}{\text{atom}} \times 6.022 \times 10^{23}\frac{\text{atom}}{\text{mol}} \times \frac{Z_{eff}^2}{n^2}$$

$$= -1.3116 \times 10^6 \frac{Z_{eff}^2}{n^2} \text{J mol}^{-1} \times \frac{1 \text{ kJ}}{1000 \text{ J}}$$

The energy required to remove an electron from an orbital with principal quantum number n in a hydrogen-like species—the ionization energy—is the *negative* of that shown above, that is,

$$I = -E_n = 1311.6 \times \frac{Z_{eff}^2}{n^2} \text{ kJ mol}^{-1} \qquad (9.9)$$

Equation (9.9) shows that the ionization energy (I) is a linear function of $(Z_{eff})^2$ and the straight-line graphs (equations 9.7 and 9.8) show that $\sqrt{I}$ is a linear function of Z. We are on the right track. However, we must now take into account that we are considering one-electron systems with a nucleus shielded by a closed shell.

First consider the second-row series (Li, Be$^+$, B^{2+}, C^{3+}, N^{4+}, O^{5+}, and F^{6+}), all members of which have the electron configuration $1s^2 2s^1$. If we assume that the closed shell $1s^2$ perfectly screens the outer electron $2s^1$, the value of Z_{eff} for this series will be $Z - 2$. Thus we should substitute $Z_{eff} = Z - 2$ into equation (9.9), and also $n = 2$ since ionization occurs from the $2s$ orbital, to obtain

$$I = 1311.6 \times \frac{(Z-2)^2}{n^2} = 1311.6 \times \frac{(Z-2)^2}{2^2}$$

Taking the square root of both sides and clearing the fraction gives

$$\sqrt{I} = 36.22 \times \left(\frac{Z-2}{2}\right) = 18.11Z - 36.22 \qquad \text{(9.10)}$$

Let us now look at the third-row series. In this case the configuration of the isoelectronic series is $1s^2 2s^2 2p^6 3s^1$ so that if we assume perfect screening of the $3s^1$ electron by the ten inner-core electrons we have

$$Z_{eff} = Z - 10$$

Proceeding as before and remembering that ionization occurs from the $n = 3$ level, we obtain

$$I = 1311.6 \times \frac{(Z-10)^2}{n^2} = 1311.6 \times \frac{(Z-10)^2}{3^2}$$

and

$$\sqrt{I} = 36.22 \times \left(\frac{Z-10}{3}\right) = 12.07Z - 120.7 \qquad \text{(9.11)}$$

Assess

In comparing equations (9.10) and (9.11), we find that the difference in the slope (coefficient for Z) is due to the difference in the principal quantum number of the orbital from which the ionization occurs. The difference in the intercepts is due both to the principal quantum number from which the ionization occurs and to the number of electrons screening the valence-shell electron.

Equation (9.10) for the second-row elements is in remarkable agreement with the empirically observed equation (9.7) at the beginning of this example, especially considering our use of the modified Bohr model.

Although the general form of equation (9.11) is correct for the third-row series, the agreement between the numerical constants is not as good. This is to be expected because we have assumed perfect screening by the core electrons, which completely ignores the different characteristics of the electrons composing the inner core. The intricate, correlated motions of the electrons in the core leads to a complicated combination of screening and penetration that cannot be accounted for with our simple model.

PRACTICE EXAMPLE A: Francium ($Z = 87$) is an extremely rare radioactive element formed when actinium ($Z = 89$) undergoes alphaparticle emission. Francium occurs in natural uranium minerals, but estimates are that little more than 15 g of francium exists in the top 1 km of Earth's crust. Few of francium's properties have been measured, but some can be inferred from its position in the periodic table. Estimate the melting point, density, and atomic (metallic) radius of francium.

PRACTICE EXAMPLE B: Discuss the likelihood that element 168, should it ever be synthesized in sufficient quantity, would be a "noble liquid" at 298 K and 1 bar. Some data that might be useful appear in the table below. Could element 168 be a "noble solid" at 298 K and 1 bar? Use *spdf* notation to show the electron configuration you would expect for element 168.

Element	Atomic Mass, u	mp, K	bp, K
Argon	39.948	83.95	87.45
Helium	4.0026		4.25
Krypton	83.80	116.5	120.9
Neon	20.179	24.48	27.3
Radon	222	202	211.4
Xenon	131.29	161.3	166.1

Exercises

The Periodic Law

1. Use data from Figure 9-1 and equation (9.1) to estimate the density of the recently discovered element 114. Assume a mass number of 298.
2. Suppose that lanthanum ($Z = 57$) were a newly discovered element having a density of $6.145 \, g/cm^3$. Estimate its molar mass.
3. The following densities, in grams per cubic centimeter, are for the listed elements in their standard states at 298 K. Show that density is a periodic property of these elements: Al, 2.699; Ar, 0.0018; As, 5.778; Br, 3.100; Ca, 1.550; Cl, 0.0032; Ga, 5.904; Ge, 5.323; Kr, 0.0037; Mg, 1.738; P, 1.823; K, 0.856; Se, 4.285; Si, 2.336; Na, 0.968; S, 2.069.
4. The following melting points are in degrees Celsius. Show that melting point is a periodic property of these elements: Al, 660; Ar, -189; Be, 1278; B, 2300; C, 3350; Cl, -101; F, -220; Li, 179; Mg, 651; Ne, -249; N, -210; O, -218; P, 590; Si, 1410; Na, 98; S, 119.

The Periodic Table

5. Mendeleev's periodic table did not preclude the possibility of a new group of elements that would fit within the existing table, as was the case with the noble gases. Moseley's work did preclude this possibility. Explain this difference.
6. Explain why the several periods in the periodic table do not all have the same number of members.
7. Assuming that the seventh period is 32 members long, what should be the atomic number of the noble gas following radon (Rn)? Of the alkali metal following francium (Fr)? What would you expect their approximate atomic masses to be?
8. Concerning the incomplete seventh period of the periodic table, what should be the atomic number of the element (a) for which the filling of the $6d$ subshell is completed; (b) that should most closely resemble bismuth; (c) that should be a noble gas?

Atomic Radii and Ionic Radii

9. For each of the following pairs, indicate the atom that has the *larger* size: (a) Te or Br; (b) K or Ca; (c) Ca or Cs; (d) N or O; (e) O or P; (f) Al or Au.
10. Indicate the *smallest* and the *largest* species (atom or ion) in the following group: Al atom, F atom, As atom, Cs^+ ion, I^- ion, N atom.
11. Explain why the radii of atoms do not simply increase uniformly with increasing atomic number.
12. The masses of individual atoms can be determined with great precision, yet there is considerable uncertainty about the exact size of an atom. Explain why this is the case.
13. Which is (a) the smallest atom in group 13; (b) the smallest of the following atoms: Te, In, Sr, Po, Sb? Why?
14. How would you expect the sizes of the hydrogen ion, H^+, and the hydride ion, H^-, to compare with that of the H atom and the He atom? Explain.
15. Arrange the following in expected order of increasing radius: Br, Li^+, Se, I^-. Explain your answer.
16. Explain why the generalizations presented in Figure 9-12 cannot be used to answer the question, Which is larger, an Al atom or an I atom?
17. Among the following ions, several pairs are *isoelectronic*. Identify these pairs. Fe^{2+}, Sc^{3+}, Ca^{2+}, F^-, Co^{2+}, Co^{3+}, Sr^{2+}, Cu^+, Zn^{2+}, Al^{3+}.
18. The following species are isoelectronic with the noble gas krypton. Arrange them in order of increasing radius and comment on the principles involved in doing so: Rb^+, Y^{3+}, Br^-, Sr^{2+}, Se^{2-}.
19. All the isoelectronic species illustrated in the text had the electron configurations of noble gases. Can two ions be isoelectronic *without* having noble-gas electron configurations? Explain.
20. Is it possible for two different atoms to be isoelectronic? two different cations? two different anions? a cation and an anion? Explain.

Ionization Energies; Electron Affinities

21. Use principles established in this chapter to arrange the following atoms in order of *increasing* value of the first ionization energy: Sr, Cs, S, F, As.

22. Are there any atoms for which the second ionization energy (I_2) is smaller than the first (I_1)? Explain.

23. Some electron affinities are negative quantities, and some are zero or positive. Why is this not also the case with ionization energies?

24. How much energy, in joules, must be absorbed to convert to Na^+ all the atoms present in 1.00 mg of *gaseous* Na? The first ionization energy of Na is 495.8 kJ/mol.

25. How much energy, in kilojoules, is required to remove all the third-shell electrons in a mole of gaseous silicon atoms?

26. What is the maximum number of Cs^+ ions that can be produced per joule of energy absorbed by a sample of gaseous Cs atoms?

27. The production of gaseous bromide ions from bromine molecules can be considered a two-step process in which the first step is

$$Br_2(g) \longrightarrow 2\,Br(g) \qquad \Delta H = +193\,kJ$$

Is the formation of $Br^-(g)$ from $Br_2(g)$ an endothermic or exothermic process?

28. Use ionization energies and electron affinities listed in the text to determine whether the following reaction is endothermic or exothermic.

$$Mg(g) + 2\,F(g) \longrightarrow Mg^{2+}(g) + 2\,F^-(g)$$

29. The Na^+ ion and the Ne atom are isoelectronic. The ease of loss of an electron by a gaseous Ne atom, I_1, has a value of 2081 kJ/mol. The ease of loss of an electron from a gaseous Na^+ ion, I_2, has a value of 4562 kJ/mol. Why are these values not the same?

30. From the data in Figure 9-11, the formation of a gaseous anion Li^- appears energetically favorable. That is, energy is given off when gaseous Li atoms accept electrons. Comment on the likelihood of forming a stable compound containing the Li^- ion, such as $Li^+ Li^-$ or $Na^+ Li^-$.

31. Compare the elements Al, Si, S, and Cl.
    (a) Place the elements in order of increasing ionization energy.
    (b) Place the elements in order of increasing electron affinity.

32. Compare the elements Na, Mg, O, and P.
    (a) Place the elements in order of increasing ionization energy.
    (b) Place the elements in order of increasing electron affinity.

Magnetic Properties

33. Unpaired electrons are found in only one of the following species. Indicate which one, and explain why: $F^-, Ca^{2+}, Fe^{2+}, S^{2-}$.

34. Which of the following species has the greatest number of unpaired electrons (a) Ge; (b) Cl; (c) Cr^{3+}; (d) Br^-?

35. Which of the following species would you expect to be diamagnetic and which paramagnetic? (a) K^+; (b) Cr^{3+}; (c) Zn^{2+}; (d) Cd; (e) Co^{3+}; (f) Sn^{2+}; (g) Br.

36. Write electron configurations consistent with the following data on numbers of unpaired electrons: Ni^{2+}, 2; Cu^{2+}, 1; Cr^{3+}, 3.

37. Must all atoms with an odd atomic number be paramagnetic? Must all atoms with an even atomic number be diamagnetic? Explain.

38. Neither Co^{2+} nor Co^{3+} has 4s electrons in its electron configuration. How many unpaired electrons would you expect to find in each of these ions? Explain.

Predictions Based on the Periodic Table

39. Use ideas presented in this chapter to indicate (a) three metals that you would expect to exhibit the photoelectric effect with visible light and three that you would not; (b) the noble gas element that should have the highest density in the liquid state; (c) the approximate I_1 of fermium ($Z = 100$); (d) the approximate density of solid radium ($Z = 88$).

40. The heat of atomization for an element is the amount of energy required to convert an appropriate amount of an element in its standard state to a mole of atoms in the gaseous state. The heats of atomization, in kilojoules per mole, for three of the group 14 elements are carbon, 717; silicon, 452; and tin, 302 kJ/mol. Predict the heat of atomization of germanium.

41. Gallium is a commercially important element (used to make gallium arsenide for the semiconductor industry). Although gallium was unknown in Mendeleev's time, he predicted properties of this element. Predict the following for gallium: (a) density; (b) the formula and percent composition of its oxide. [*Hint*: Use Figure 9-1, equation (9.1), and Mendeleev's periodic table (page 362).]

42. For the following groups of elements, select the one that has the property noted:
    (a) the largest atom: Mg, Mn, Mo, Ba, Bi, Br
    (b) the lowest first ionization energy: B, Sr, Al, Br, Mg, Pb

(c) the most negative electron affinity: As, B, Cl, K, Mg, S
(d) the largest number of unpaired electrons: F, N, S^{2-}, Mg^{2+}, Sc^{3+}, Ti^{3+}

43. The boiling points of the noble gases increase down the group as follows: He, 4.2; Ne, 27.1; Ar, 87.3; Kr, 119.7; and Xe, 165 K. What would you predict to be the boiling point of radon?
44. Estimate the missing boiling point in the following series of compounds.
   (a) CH_4, $-164\,°C$; SiH_4, $-112\,°C$; GeH_4, $-90\,°C$; SnH_4, ? $°C$
   (b) H_2O, ? $°C$; H_2S, $-61\,°C$; H_2Se, $-41\,°C$; H_2Te, $-2\,°C$
   Does your estimate in part (b) agree with the known value?
45. Match each of the lettered items on the left with an appropriate numbered item on the right. All the numbered items should be used, and some more than once.
   (a) $Z = 32$
   (b) $Z = 8$
   (c) $Z = 53$
   (d) $Z = 38$
   (e) $Z = 48$
   (f) $Z = 20$

   1. two unpaired p electrons
   2. diamagnetic
   3. more negative electron affinity than elements on either side of it in the same period
   4. first ionization energy lower than that of Ca but greater than that of Cs

46. Match each of the lettered items in the column on the left with the most appropriate numbered item(s) in the column on the right. Some of the numbered items may be used more than once and some not at all.
   (a) Tl
   (b) $Z = 70$
   (c) Ni
   (d) $[Ar]4s^2$
   (e) a metalloid
   (f) a nonmetal

   1. an alkaline earth metal
   2. element in period 5 and group 15
   3. largest atomic radius of all the elements
   4. an element in period 4 and group 16
   5. $3d^8$
   6. one p electron in the shell of highest n
   7. lowest ionization energy of all the elements
   8. an f-block element

47. Which of the following ions are unlikely to be found in chemical compounds: K^+, Ga^{4+}, Fe^{6+}, S^{2-}, Ge^{5+}, or Br^-? Explain briefly.
48. Which of the following ions are likely to be found in chemical compounds: Na^{2+}, Li^+, Al^{4+}, F^{2-}, or Te^{2-}? Explain briefly.

Integrative and Advanced Exercises

49. Complete and balance the following equations. If no reaction occurs, so state.
   (a) $Rb(s) + H_2O(l) \longrightarrow$
   (b) $I_2(s) + Na^+(aq) + Br^-(aq) \longrightarrow$
   (c) $SrO(s) + H_2O(l) \longrightarrow$
   (d) $SO_3(g) + H_2O(l) \longrightarrow$
50. Write balanced equations to represent
   (a) the displacement of a halide anion from aqueous solution by liquid bromine
   (b) the reaction with water of an alkali metal with $Z > 50$
   (c) the reaction of tetraphosphorus decoxide with water
   (d) the reaction of aluminum oxide with aqueous sulfuric acid
51. Four atoms and/or ions are sketched below in accordance with their relative atomic and/or ionic radii.

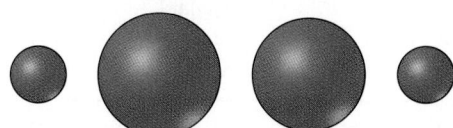

   Which of the following sets of species are compatible with the sketch? Explain. **(a)** C, Ca^{2+}, Cl^-, Br^-; **(b)** Sr, Cl, Br^-, Na^+; **(c)** Y, K, Ca, Na^+; **(d)** Al, Ra^{2+}, Zr^{2+}, Mg^{2+}; **(e)** Fe, Rb, Co, Cs.
52. Sketch a periodic table that would include *all* the elements in the main body of the table. How many "numbers" wide would the table be?
53. In Mendeleev's time, indium oxide, which is 82.5% In by mass, was thought to be InO. If this were the case, in which group of Mendeleev's table (page 362) should indium be placed?

54. Instead of accepting the atomic mass of indium implied by the data in Exercise 53, Mendeleev proposed that the formula of indium oxide is In_2O_3. Show that this assumption places indium in the proper group of Mendeleev's periodic table on page 362.
55. Listed below are two atomic properties of the element germanium. Refer only to the periodic table on the inside front cover and indicate probable values for each of the following elements, expressed as greater than, about equal to, or less than the value for Ge.

Element	Atomic Radius	First Ionization Energy
Ge	122 pm	762 kJ/mol
Al	?	?
In	?	?
Se	?	?

56. In estimating the boiling point and melting point of bromine in Example 9-5, could we have used Celsius or Fahrenheit instead of Kelvin temperature? Explain.
57. In the formula X_2, if the two X atoms are the same halogen, the substance is a halogen element (for example, Cl_2, Br_2). If the two X atoms are different halogens (such as Cl and Br), the substance is an *interhalogen compound*. Use data from Table 9.5 to estimate the melting points and boiling points of the interhalogen compounds BrCl and ICl.
58. Refer to Figure 9-9 and explain why the difference between the ionic radii of the 1− and 2− anions does not remain constant from top to bottom of the periodic table.

59. Explain why the third ionization energy of Li(g) is an easier quantity to calculate than either the first or second ionization energies. Calculate I_3 for Li, and express the result in kJ/mol.

60. Two elements, A and B, have the electron configurations shown.

$$A = [Ar]4s^1 \quad B = [Ar]3d^{10}4s^24p^3$$

(a) Which element is a metal?
(b) Which element has the greater ionization energy?
(c) Which element has the larger atomic radius?
(d) Which element has the greater electron affinity?

61. Two elements, A and B, have the electron configurations shown.

$$A = [Kr]4s^2 \quad B = [Ar]3d^{10}4s^24p^5$$

(a) Which element is a metal?
(b) Which element has the greater ionization energy?
(c) Which element has the larger atomic radius?
(d) Which element has the greater electron affinity?

62. Studies done in 1880 showed that a chloride of uranium had 37.34% Cl by mass and an approximate formula mass of 382 u. Other data indicated the specific heat of uranium to be 0.0276 cal $g^{-1}\,^\circ C^{-1}$. Are these data in agreement with the atomic mass of uranium assigned by Mendeleev, 240 u? [*Hint:* Refer to Feature Problem 121 of Chapter 7.]

63. Assume that atoms are hard spheres, and use the metallic radius of 186 pm for Na to estimate the volumes of one Na atom and of one mole of Na atoms. How does your result compare with the atomic volume found in Figure 9-1? Why is there so much disagreement between the two values?

64. When sodium chloride is strongly heated in a flame, the flame takes on the yellow color associated with the emission spectrum of sodium atoms. The reaction that occurs in the *gaseous* state is

$$Na^+(g) + Cl^-(g) \longrightarrow Na(g) + Cl(g).$$

Calculate ΔH for this reaction.

65. Use information from Chapters 8 and 9 to calculate the *second* ionization energy for the He atom. Compare your result with the tabulated value of 5251 kJ/mol.

66. Refer only to the periodic table on the inside front cover, and arrange the following ionization energies in order of increasing value: I_1 for F; I_2 for Ba; I_3 for Sc; I_2 for Na; I_3 for Mg. Explain the basis of any uncertainties.

67. Refer to the footnote on page 375. Then use values of basic physical constants and other data from the appendices to show that 1 eV/atom = 96.49 kJ/mol.

68. Plot a graph of the square roots of the ionization energies versus the nuclear charge for the two series Li, Be^+, B^{2+}, C^{3+}, and Na, Mg^+, Al^{2+}, Si^{3+}. Explain the observed relationship with the aid of Bohr's expression for the binding energy of an electron in a one-electron atom.

Feature Problems

69. The work functions for a number of metals are given in the following table. How do the work functions vary
(a) down a group?
(b) across a period?
(c) Estimate the work function for potassium and compare it with a published value.
(d) What periodic property is the work function most like?

Metal	Work Function, J × 10^{19}
Al	6.86
Cs	3.45
Li	4.6
Mg	5.86
Na	4.40
Rb	3.46

70. The following are a few elements and their characteristic X-ray wavelengths:

Element	X-ray Wavelength, pm
Mg	987
S	536
Ca	333
Cr	229
Zn	143
Rb	93

Use these data to determine the constants A and b in Moseley's relationship (page 363). Compare your value of A with the value obtained from Bohr's theory for the frequencies emitted by one-electron atoms. Suggest a reasonable interpretation of the quantity b.

71. Gaseous sodium atoms absorb quanta with the energies shown in the table below.

Energy of Quanta, kJ mol^{-1}	Electron Configuration
0	$[Ne]3s^1$
203	$[Ne]3p^1$
308	$[Ne]4s^1$
349	$[Ne]3d^1$
362	$[Ne]4p^1$

(a) The ionization energy of the ground state is 496 kJ mol^{-1}. Calculate the ionization energies for each of the states given in the table.
(b) Calculate Z_{eff} for each state.
(c) Calculate $\bar{r}_{nl}$ for each state.
(d) Interpret the results obtained from parts (b) and (c) in terms of penetration and screening.

72. A method for estimating electron affinities is to extrapolate Z_{eff} values for atoms and ions that contain the same number of electrons as the negative ion of interest. Use the data in the table on the next page to answer the questions that follow.

Atom or Ion: $I(kJ\,mol^{-1})$	Atom or Ion: $I(kJ\,mol^{-1})$	Atom or Ion: $I(kJ\,mol^{-1})$
Ne: 2080	F: 1681	O: 1314
Na^+: 4565	Ne^+: 3963	F^+: 3375
Mg^{2+}: 7732	Na^{2+}: 6912	Ne^{2+}: 6276
Al^{3+}: 11,577	Mg^{3+}: 10,548	Na^{3+}: 9540

(a) Estimate the electron affinity of F, and compare it with the experimental value.
(b) Estimate the electron affinities of O and N.
(c) Examine your results in terms of penetration and screening.

73. We have seen that the wave functions of hydrogen-like atoms contain the nuclear charge Z for hydrogen-like atoms and ions, but modified through equation (9.3) to account for the phenomenon of shielding or screening. In 1930, John C. Slater devised the following set of empirical rules to calculate a shielding constant for a designated electron in the orbital ns or np:
   (i) Write the electron configuration of the element, and group the subshells as follows: $(1s)$, $(2s, 2p)$, $(3s, 3p)$, $(3d)$, $(4s, 4p)$, $(4d)$, $(4f)$, $(5s, 5p)$, etc.
   (ii) Electrons in groups to the right of the (ns, np) group contribute nothing to the shielding constant for the designated electron.
   (iii) All the other electrons in the (ns, np) group shield the designated electron to the extent of 0.35 each.
   (iv) All electrons in the $n - 1$ shell shield to the extent of 0.85 each.

   (v) All electrons in the $n - 2$ shell, or lower, shield completely—their contributions to the shielding constant are 1.00 each.
When the designated electron being shielded is in an nd or nf group, rules (ii) and (iii) remain the same but rules (iv) and (v) are replaced by
   (vi) Each electron in a group lying to the left of the nd or nf group contributes 1.00 to the shielding constant.
These rules are a simplified generalization based on the average behavior of different types of electrons. Use these rules to do the following:
(a) Calculate Z_{eff} for a valence electron of oxygen.
(b) Calculate Z_{eff} for the $4s$ electron in Cu.
(c) Calculate Z_{eff} for a $3d$ electron in Cu.
(d) Evaluate the Z_{eff} for the valence electrons in the group 1 elements (including H), and show that the ionization energies observed for this group are accounted for by using the Slater rules. [*Hint:* Do not overlook the effect of n on the orbital energy.]
(e) Evaluate Z_{eff} for a valence electron in the elements Li through Ne, and use the results to explain the observed trend in first ionization energies for these elements.
(f) Using the radial functions given in Table 8.1 and Z_{eff} estimated with the Slater rules, compare plots of the radial probability for the $3s$, $3p$, and $3d$ orbitals for the H atom and the Na atom. What do you observe from these plots regarding the effect of shielding on radial probability distributions?

Self-Assessment Exercises

74. In your own words, define the following terms: **(a)** isoelectronic; **(b)** valence-shell electrons; **(c)** metal; **(d)** nonmetal; **(e)** metalloid.
75. Briefly describe each of the following ideas or phenomena: **(a)** the periodic law; **(b)** ionization energy; **(c)** electron affinity; **(d)** paramagnetism.
76. Explain the important distinctions between each pair of terms: **(a)** actinide and lanthanide element; **(b)** covalent and metallic radius; **(c)** atomic number and effective nuclear charge; **(d)** ionization energy and electron affinity; **(e)** paramagnetic and diamagnetic.
77. The element whose atoms have the electron configuration $[Kr]4d^{10}5s^25p^3$ **(a)** is in group 13 of the periodic table; **(b)** bears a similarity to the element Bi; **(c)** is similar to the element Te; **(d)** is a transition element.
78. The fourth-period element with the largest atom is **(a)** K; **(b)** Br; **(c)** Pb; **(d)** Kr.
79. The largest of the following is **(a)** an Ar atom; **(b)** a K^+ ion; **(c)** a Ca^{2+} ion; **(d)** a Cl^- ion.
80. The highest first ionization energy of the following is that of **(a)** Cs; **(b)** Cl; **(c)** I; **(d)** Li.
81. The most negative electron affinity of the following elements is that of **(a)** Br; **(b)** Sn; **(c)** Ba; **(d)** Li.
82. An ion that is isoelectronic with Se^{2-} is **(a)** S^{2-}; **(b)** I^-; **(c)** Xe; **(d)** Sr^{2+}.

83. Write electron configurations to show the first two ionizations for Cs. Explain why the second ionization energy is much greater than the first.
84. Explain why the first ionization energy of Mg is greater that of Na, whereas the second ionization of Na is greater than that of Mg.
85. Answer each of the following questions:
   (a) Which of the elements P, As, and S has the largest atomic radius?
   (b) Which of the following has the smallest radius: Xe, O^{2-}, N^{3-}, or F^-?
   (c) Which should have the largest difference between the first and second ionization energy: Al, Si, P, or Cl?
   (d) Which has the largest ionization energy: C, Si, or Sn?
   (e) Which has the largest electron affinity: Na, B, Al, or C?
86. The first ionization energies of Si, P, S, and Cl are given in Table 9.4. Briefly provide an explanation for this trend.
87. Find three pairs of elements that are out of order in the periodic table in terms of their atomic masses. Why is it necessary to invert their order in the table?
88. For the atom $^{119}_{50}Sn$, indicate the number of **(a)** protons in the nucleus; **(b)** neutrons in the nucleus; **(c)** $4d$ electrons; **(d)** $3s$ electrons; **(e)** $5p$ electrons; **(f)** electrons in the valence shell.

89. Refer to the periodic table on the inside front cover and indicate **(a)** the most nonmetallic element; **(b)** the transition metal with lowest atomic number; **(c)** a metalloid whose atomic number is exactly midway between those of two noble gas elements.

90. Give the symbol of the element **(a)** in group 14 that has the smallest atoms; **(b)** in period 5 that has the largest atoms; **(c)** in group 17 that has the lowest first ionization energy.

91. Refer only to the periodic table on the inside front cover and indicate which of the atoms, Bi, S, Ba, As, and Ca, **(a)** is most metallic; **(b)** is most nonmetallic; **(c)** has the intermediate value when the five are arranged in order of increasing first ionization energy.

92. Arrange the following elements in order of decreasing metallic character: Sc, Fe, Rb, Br, O, Ca, F, Te.

93. In multielectron atoms many of the periodic trends can be explained in terms of Z_{eff}. Consider the following statements and discuss whether or not the statement is true or false.
(a) Electrons in a p orbital are more effective than electrons in the s orbitals in shielding other electrons from the nuclear charge.
(b) Z_{eff} for an electron in an s orbital is lower than that for an electron in a p orbital in the same shell.

(c) Z_{eff} is usually less than Z.
(d) Electrons in orbitals having $\ell = 1$ penetrate better than those with $\ell = 2$.
(e) Z_{eff} for the orbitals of the elements Na($3s$), Mg($3s$), Al($3p$), P($3p$), and S($3p$) are in the order Z_{eff}(Na) $<$ Z_{eff}(Mg) $>$ Z_{eff}(Al) $<$ Z_{eff}(P) $>$ Z_{eff}(S).

94. Consider a nitrogen atom in the ground state and comment on whether the following statements are true or false.
(a) Z_{eff} for an electron in a $2s$ orbital is greater than that for the $1s$ orbital.
(b) The Z_{eff} for the $2p$ and $2s$ orbitals is the same.
(c) More energy is required to remove an electron from a $2s$ orbital than from the $2p$ orbital.
(d) The $2s$ electron is less shielded than the $2p$ electron.

95. Describe how the ionization energies of the ions He^-, Li^-, Be^-, B^-, C^-, N^-, O^-, and F^- vary with atomic number.

96. Describe how the ionization energies of the ions Be^+, B^+, C^+, N^+, O^+, F^+, Ne^+, and Na^+ vary with atomic number.

97. Construct a concept map (see Appendix E) connecting the ideas that govern the periodic law and the periodic variation of atomic properties.

Chemical Bonding I: Basic Concepts

CONTENTS

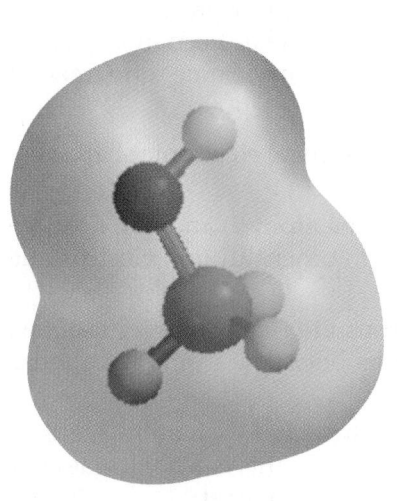
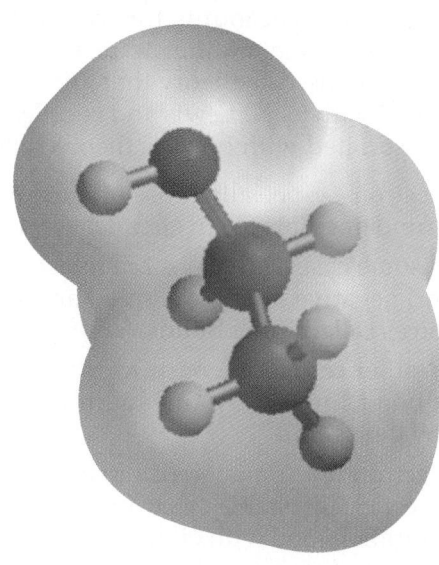

Computer-generated electrostatic potential maps of methanol (CH_3OH) and ethanol (CH_3CH_2OH). The surface encompassing each molecule shows the extent of electron charge density while the colors show the distribution of charge in the molecule. In this chapter, we study ideas that enable us to predict the geometric shapes and polarity of molecules.

Consider all that we already know about chemical compounds. We can determine their compositions and write their formulas. We can represent the reactions of compounds by chemical equations and perform stoichiometric and thermochemical calculations based on these equations. And we can do all this without really having to consider the ultimate structure of matter—the structure of atoms and molecules. Yet the shape of a molecule—that is, the arrangement of its atoms in space—often defines its chemistry. If water had a different shape, its properties would be significantly different, and life as we know it would not be possible.

In this chapter, we will describe the interactions between atoms called *chemical bonds*. Most of the discussion centers on the Lewis theory, which provides one of the simplest methods of representing chemical bonding. We will also explore another relatively simple theory, one for predicting probable molecular shapes. Throughout the chapter, we will try to relate these theories to what is known about molecular structures from experimental measurements. In Chapter 11 we will examine the subject of chemical bonding in greater depth, and in Chapter 12 we will describe intermolecular forces—forces between molecules—and explore further the relationship between molecular shape and the properties of substances.

10-1   Lewis Theory: An Overview

▶ Since 1962, a number of compounds of Xe and Kr have been synthesized. As we will see in this chapter, a focus on noble-gas electron configurations can still be useful, even if the idea that they confer complete inertness is invalid.

In the period from 1916 to 1919, two Americans, G. N. Lewis and Irving Langmuir, and a German, Walther Kossel, advanced an important proposal about chemical bonding: Something unique in the electron configurations of noble gas atoms accounts for their inertness, and atoms of other elements combine with one another to acquire electron configurations like those of noble gas atoms. The theory that grew out of this model has been most closely associated with G. N. Lewis and is called the **Lewis theory**. Some fundamental ideas associated with Lewis's theory follow:

1. Electrons, especially those of the outermost (valence) electronic shell, play a fundamental role in chemical bonding.

2. In some cases, electrons are *transferred* from one atom to another. Positive and negative ions are formed and attract each other through electrostatic forces called **ionic bonds**.

▶ The term *covalent* was introduced by Irving Langmuir.

3. In other cases, one or more pairs of electrons are *shared* between atoms. A bond formed by the sharing of electrons between atoms is called a **covalent bond**.

4. Electrons are transferred or shared in such a way that each atom acquires an especially stable electron configuration. Usually this is a noble gas configuration, one with eight outer-shell electrons, or an **octet**.

Lewis Symbols and Lewis Structures

Lewis developed a special set of symbols for his theory. A **Lewis symbol** consists of a chemical symbol to represent the nucleus and *core* (inner-shell) *electrons* of an atom, together with dots placed around the symbol to represent the *valence* (outer-shell) *electrons*. Thus, the Lewis symbol for silicon, which has the electron configuration $[Ne]3s^2 3p^2$, is

$$\cdot \overset{\cdot\cdot}{\underset{\cdot}{Si}} \cdot$$

▲ **Gilbert Newton Lewis (1875–1946)**
Lewis's contribution to the study of chemical bonding is evident throughout this text. Equally important, however, was his pioneering introduction of thermodynamics into chemistry.

Electron spin had not yet been proposed when Lewis framed his theory, and so he did not show that two of the valence electrons $(3s^2)$ are paired and two $(3p^2)$ are unpaired. We will write Lewis symbols in the way Lewis did. We will place single dots on the sides of the symbol, up to a maximum of four. Then we will pair up dots until we reach an octet. Lewis symbols are commonly written for main-group elements but much less often for transition elements. Lewis symbols for several main-group elements are written in Example 10-1.

A **Lewis structure** is a combination of Lewis symbols that represents either the transfer or the sharing of electrons in a chemical bond.

Ionic bonding (transfer of electrons):	$Na\times + \cdot\overset{\cdot\cdot}{\underset{\cdot\cdot}{Cl}}: \longrightarrow [Na]^+ [\times\overset{\cdot\cdot}{\underset{\cdot\cdot}{Cl}}:]^-$	**(10.1)**
	Lewis symbols Lewis structure	

Covalent bonding (sharing of electrons):	$H\times + \cdot\overset{\cdot\cdot}{\underset{\cdot\cdot}{Cl}}: \longrightarrow H\overset{\cdot\cdot}{\underset{\cdot\cdot}{\times Cl}}:$	**(10.2)**
	Lewis symbols Lewis structure	

EXAMPLE 10-1   Writing Lewis Symbols

Write Lewis symbols for the following elements: **(a)** N, P, As, Sb, Bi; **(b)** Al, I, Se, Ar.

Analyze

The position of the element in the periodic table determines the number of valence electrons in the Lewis symbol. For main-group elements, the number of valence electrons, and hence the number of dots appearing in a Lewis symbol, is equal to the group number for the s-block elements and to the group number minus 10 for the p-block elements.

Solve

(a) These are group 15 elements, and their atoms all have five valence electrons (ns^2np^3). The Lewis symbols all have five dots.

$$\cdot\ddot{\text{N}}\cdot \qquad \cdot\ddot{\text{P}}\cdot \qquad \cdot\ddot{\text{As}}\cdot \qquad \cdot\ddot{\text{Sb}}\cdot \qquad \cdot\ddot{\text{Bi}}\cdot$$

(b) Al is in group 13; I, in group 17; Se, in group 16; Ar, in group 18.

$$\cdot\dot{\text{Al}}\cdot \qquad :\ddot{\text{I}}\cdot \qquad :\ddot{\text{Se}}\cdot \qquad :\ddot{\text{Ar}}:$$

Assess

This example, although very straightforward, is very important. The accurate counting of valence electrons is essential for many aspects of chemical bonding.

PRACTICE EXAMPLE A: Write Lewis symbols for Mg, Ge, K, and Ne.

PRACTICE EXAMPLE B: Write the Lewis symbols expected for Sn, Br^-, Tl^+, and S^{2-}.

In these two examples, we designated the electrons involved in bond formation differently—(×) from one atom and (•) from the other. This helps to emphasize that an electron is transferred in ionic bonding and that a pair of electrons is shared in covalent bonding. Of course, it is impossible to distinguish between electrons, and henceforth we will use only dots (•) to represent electrons in Lewis structures. In Lewis theory, we use square brackets to identify ions, as we did in equation (10.1). The charge on the ion is given as a superscript.

Lewis's work dealt mostly with covalent bonding, which we will emphasize throughout this chapter. However, Lewis's ideas also apply to ionic bonding, and we briefly describe this application next.

Lewis Structures for Ionic Compounds

In Section 3-2, we learned that the formula unit of an ionic compound is the simplest electrically neutral collection of cations and anions from which the chemical formula of the compound can be established. The Lewis structure of sodium chloride (structure 10.1) represents its formula unit. For an ionic compound of a main-group element, (1) the Lewis symbol of the metal ion has no dots if all the valence electrons are lost, and (2) the ionic charges of both cations and anions are shown. These ideas are further illustrated through Example 10-2.

◀ No bond is 100% ionic. All ionic bonds have some covalent character.

EXAMPLE 10-2   Writing Lewis Structures of Ionic Compounds

Write Lewis structures for the following compounds: **(a)** BaO; **(b)** $MgCl_2$; **(c)** aluminum oxide.

Analyze

Our approach here is to write the Lewis symbol and determine how many electrons each atom must gain or lose to acquire a noble-gas-electron configuration.

(continued)

Solve

(a) Ba loses two electrons, and O gains two. In the equation below, we use curved red arrows, each with half an arrowhead, to show the movement of single electrons.

$$Ba\cdot + \cdot \ddot{O}\!: \longrightarrow [Ba]^{2+}[:\ddot{\underset{..}{O}}\!:]^{2-}$$

Lewis structure

(b) A Cl atom can accept only one electron because it already has seven valence electrons. One more will give it a complete octet. Conversely, a Mg atom must lose two electrons to have the electron configuration of the noble gas neon. So two Cl atoms are required for each Mg atom.

$$Mg\cdot \; + \; \begin{matrix} \cdot\ddot{\underset{..}{Cl}}\!: \\[6pt] \cdot\ddot{\underset{..}{Cl}}\!: \end{matrix} \longrightarrow [Mg]^{2+}2[:\ddot{\underset{..}{Cl}}\!:]^{-}$$

Lewis structure

(c) The formula of aluminum oxide follows directly from the Lewis structure. The combination of one Al atom, which loses three electrons, and one O atom, which gains two, leaves an excess of one lost electron. To match the numbers of electrons lost and gained, the formula unit must be based on *two* Al atoms and *three* O atoms.

$$\begin{matrix} Al\cdot & \cdot\ddot{O}\!: \\[6pt] + & \cdot\ddot{O}\!: \\[6pt] Al\cdot & \cdot\ddot{O}\!: \end{matrix} \longrightarrow 2[Al]^{3+}3[:\ddot{\underset{..}{O}}\!:]^{2-}$$

Lewis structure

Assess

We almost never write Lewis structures for ionic compounds, except when we want to emphasize the ratio in which the ions combine. The structures of ionic compounds are much more complicated than is suggested by the Lewis structure. See, for example, the structure of NaCl shown in Figure 10-1.

PRACTICE EXAMPLE A: Write plausible Lewis structures for **(a)** Na_2S and **(b)** Mg_3N_2.

PRACTICE EXAMPLE B: Write plausible Lewis structures for **(a)** calcium iodide; **(b)** barium sulfide; **(c)** lithium oxide.

The compounds described in Example 10-2 are *binary ionic compounds* consisting of monatomic cations and monatomic anions. Commonly encountered *ternary ionic compounds* consist of monatomic and polyatomic ions. Bonding between atoms within the polyatomic ions is covalent. Some ternary ionic compounds are considered later in the chapter.

With the exception of ion pairs, such as (Na^+Cl^-), that may be found in the *gaseous* state, formula units of solid ionic compounds do not exist as separate entities. Instead, each cation is surrounded by anions and each anion by cations. These very large numbers of ions are arranged in an orderly network called an *ionic crystal* (Fig. 10-1). Ionic crystal structures and the energy changes accompanying the formation of ionic crystals are described in Chapter 12.

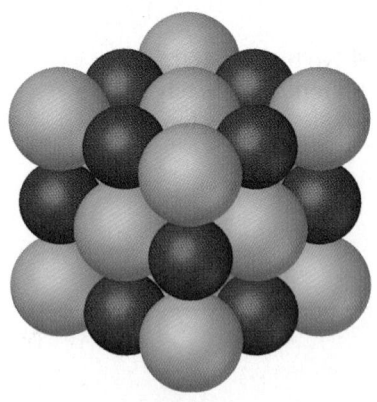

▲ FIGURE 10-1
Portion of an ionic crystal
This structure of alternating Na^+ and Cl^- ions extends in all directions and involves countless numbers of ions.

10-1 CONCEPT ASSESSMENT

How many valence electrons do the Lewis symbols for the elements in group 16 have? Which of the following are correct Lewis symbols for sulfur?

$$:\ddot{\underset{..}{S}}\!: \qquad :\dot{\underset{..}{S}}\!: \qquad \cdot\dot{\underset{.}{S}}\!\cdot \qquad \cdot\ddot{\underset{..}{S}}\!:$$

10-2   Covalent Bonding: An Introduction

A chlorine atom shows a tendency to gain an electron, as indicated by its electron affinity (-349 kJ/mol). From which atom, sodium or hydrogen, can the electron most readily be extracted? Neither atom gives up an electron freely, but the energy required to extract an electron from Na ($I_1 = 496$ kJ/mol) is much smaller than that for H ($I_1 = 1312$ kJ/mol). In Chapter 9 we learned that the lower its ionization energy, the more metallic an element is; sodium is much more metallic than hydrogen (recall Figure 9-12). In fact, hydrogen is considered to be a nonmetal. A hydrogen atom in the gaseous state does not give up an electron to another nonmetal atom. Bonding between a hydrogen atom and a chlorine atom involves the sharing of electrons, which leads to a *covalent bond*.

To emphasize the sharing of electrons, let us think of the Lewis structure of HCl in this manner.

The broken circles represent the outermost electron shells of the bonded atoms. The number of dots lying on or within each circle represents the effective number of electrons in each valence shell. The H atom has two dots, as in the electron configuration of He. The Cl atom has eight dots, corresponding to the outer-shell configuration of Ar. Note that we counted the two electrons between H and Cl (⁚) *twice*. These two electrons are shared by the H and Cl atoms. This shared pair of electrons constitutes the covalent bond. Written below are two additional Lewis structures of simple molecules.

$$\text{H·} + \text{·}\overset{..}{\underset{..}{O}}\text{·} + \text{·H} \longrightarrow \text{H}\overset{..}{\underset{..}{O}}\text{H} \quad \text{and} \quad \overset{..}{:}\overset{}{\underset{..}{Cl}}\text{·} + \text{·}\overset{..}{\underset{..}{O}}\text{·} + \text{·}\overset{..}{\underset{..}{Cl}}: \longrightarrow \overset{..}{:}\underset{..}{Cl}\overset{..}{\underset{..}{O}}\underset{..}{Cl}:$$

Water   Dichlorine monoxide

As was the case for Cl in HCl, the O atom in the Lewis structure of H_2O and in Cl_2O is surrounded by eight electrons (when the bond-pair electrons are double counted). In attaining these eight electrons, the O atom conforms to the **octet rule**—a requirement of eight valence-shell electrons for the atoms in a Lewis structure. Note, however, that the H atom is an exception to this rule. The H atom can accommodate only two valence-shell electrons.

Lewis theory helps us to understand why elemental hydrogen and chlorine exist as diatomic molecules, H_2 and Cl_2. In each case, a pair of electrons is shared between the two atoms. The sharing of a single pair of electrons between bonded atoms produces a **single covalent bond**. To underscore the importance of electron pairs in the Lewis theory the term **bond pair** applies to a pair of electrons in a covalent bond, while **lone pair** applies to electron pairs that are not involved in bonding. Also, in writing Lewis structures it is customary to replace bond pairs with lines (—). These features are shown in the following Lewis structures.

$$\text{H·} + \text{·H} \longrightarrow \text{H}\overset{.}{\underset{.}{\,}}\text{H} \quad \text{or} \quad \text{H}-\text{H} \qquad \qquad \textbf{(10.3)}$$

Bond pair

$$\overset{..}{:}\overset{}{\underset{..}{Cl}}\text{·} + \text{·}\overset{..}{\underset{..}{Cl}}: \longrightarrow \overset{..}{:}\underset{..}{Cl}\overset{.}{\underset{.}{\,}}\underset{..}{Cl}: \quad \text{or} \quad \overset{..}{:}\underset{..}{Cl}-\underset{..}{Cl}: \Longleftarrow \text{Lone pairs} \qquad \textbf{(10.4)}$$

Bond pair

◀ The Lewis structures for H_2O and Cl_2O suggest that these molecules have a linear shape. They do not. Lewis theory by itself does not address the question of molecular shape (see Section 10-7).

Coordinate Covalent Bonds

The Lewis theory of bonding describes a covalent bond as the sharing of a pair of electrons, but this does not necessarily mean that each atom contributes an electron to the bond. A covalent bond in which a single atom contributes both of the electrons to a shared pair is called a **coordinate covalent bond**.

EXAMPLE 10-3   Writing Simple Lewis Structures

Write a Lewis structure for the ammonia molecule, NH₃.

Analyze

To write a Lewis structure we must know the number of valence electrons associated with each atom.

Solve

The valence electrons can then be represented in the Lewis symbols, as shown here.

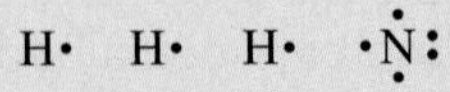

$$H\cdot \quad H\cdot \quad H\cdot \quad \cdot \overset{\cdot}{\underset{\cdot}{N}}\colon$$

Now we can assemble one N and three H atoms into a structure that gives the N atom a valence-shell octet and each of the H atoms two valence electrons (producing the electron configuration of He).

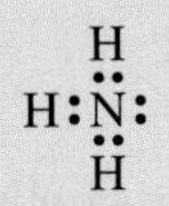

$$\begin{array}{c} H \\ H\colon \overset{\cdot\cdot}{N}\colon \\ H \end{array}$$

Assess

The application of the octet rule has led us to the correct Lewis structure for ammonia, but, as we will see later in this text, many molecules do not obey the octet rule.

PRACTICE EXAMPLE A: Write Lewis structures for Br₂, CH₄, and HOCl.

PRACTICE EXAMPLE B: Write Lewis structures for NI₃, N₂H₄, and C₂H₆.

If we attempt to attach a fourth H atom to the Lewis structure of NH₃ shown in Example 10-3, we encounter a difficulty. The electron brought by the fourth H atom would raise the total number of valence electrons around the N atom to *nine*, so there would no longer be an octet. The *molecule* NH₄ does not form, but the *ammonium ion*, NH₄⁺, does, as suggested in Figure 10-2. That is, the lone pair of electrons on a NH₃ molecule extracts an H atom from a HCl molecule, and the electrons in the H—Cl bond remain on the Cl atom. The result is equivalent to a H⁺ ion joining with the NH₃ molecule to form the NH₄⁺ ion,

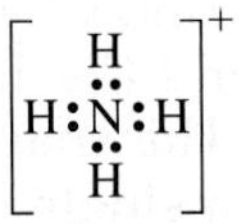

$$\left[\begin{array}{c} H \\ H\colon \overset{\cdot\cdot}{N}\colon H \\ H \end{array}\right]^{+} \tag{10.5}$$

As shown in Figure 10-2, the electron pair from the H—Cl bond remains on the Cl atom, converting it to a Cl⁻ ion.

The bond formed between the N atom of NH₃ and the H⁺ ion in structure (10.5) is a *coordinate covalent bond*. It is important to note, however, that once the bond has formed, it is impossible to say which of the four N—H bonds is the coordinate covalent bond. Thus, a coordinate covalent bond is indistinguishable from a regular covalent bond.

Another example of coordinate covalent bonding is found in the familiar hydronium ion.

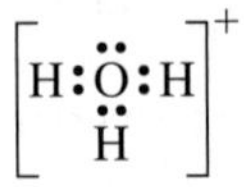

$$\left[H\colon \overset{\cdot\cdot}{O}\colon H \right]^{+} \tag{10.6}$$

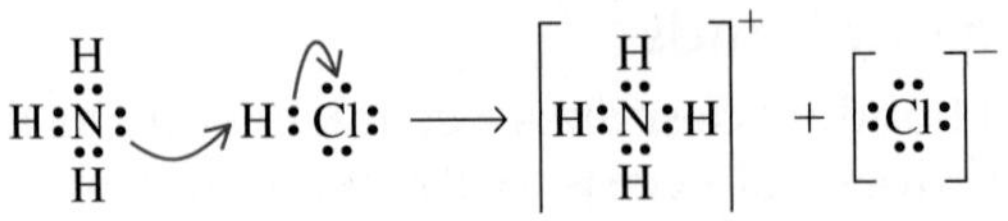

▶ FIGURE 10-2
Formation of the ammonium ion, NH₄⁺
The H atom of HCl leaves its electron with the Cl atom and, as H⁺, attaches itself to the NH₃ molecule through the lone-pair electrons on the N atom. The ions NH₄⁺ and Cl⁻ are formed.

Multiple Covalent Bonds

In the preceding description of the Lewis model for covalent chemical bonding, we have used a single pair of electrons between two atoms to describe a single covalent bond. Often, however, more than one pair of electrons must be shared if an atom is to attain an octet (noble gas electron configuration). CO_2 and N_2 are two molecules in which atoms share more than one pair of electrons.

First, let's apply the ideas about Lewis structures to CO_2. From the Lewis symbols, we see that the C atom can share a valence electron with each O atom, thus forming two carbon-to-oxygen single bonds.

$$:\!\ddot{\text{O}}\cdot \quad \cdot \dot{\text{C}}\cdot \quad \cdot \ddot{\text{O}}\!: \longrightarrow :\!\ddot{\text{O}}\!:\!\dot{\text{C}}\!:\!\ddot{\text{O}}\!:$$

But this leaves the C atom and both O atoms still shy of an octet. The problem is solved by shifting the unpaired electrons into the region of the bond, as indicated by the red arrows.

$$:\!\ddot{\text{O}}\!:\!\text{C}\!:\!\ddot{\text{O}}\!: \longrightarrow :\!\ddot{\text{O}}\!:\!:\!\text{C}\!:\!:\!\text{O}\!: \longrightarrow :\!\ddot{\text{O}}\!=\!\text{C}\!=\!\ddot{\text{O}}\!: \qquad \textbf{(10.7)}$$

In Lewis structure (10.7), the bonded atoms are seen to share *two* pairs of electrons (a total of four electrons) between them—a **double covalent bond** ($=$).

Now let's try our hand at writing a Lewis structure for the N_2 molecule. Our first attempt might again involve a single covalent bond and the incorrect structure shown below.

$$:\!\dot{\text{N}}\cdot + \cdot \dot{\text{N}}\!: \longrightarrow :\!\dot{\text{N}}\!:\!\dot{\text{N}}\!: \quad (\textit{Incorrect})$$

Each N atom appears to have only six outer-shell electrons, not the expected eight. The situation can be corrected by bringing the four unpaired electrons into the region between the N atoms and using them for additional bond pairs. In all, we now show the sharing of *three* pairs of electrons between the N atoms. The bond between the N atoms in N_2 is a **triple covalent bond** ($\equiv$). Double and triple covalent bonds are known as **multiple covalent bonds**.

$$:\!\text{N}\!:\!\text{N}\!: \longrightarrow :\!\text{N}\!\equiv\!\text{N}\!: \qquad \textbf{(10.8)}$$

The triple covalent bond in N_2 is a very strong bond that is difficult to break in a chemical reaction. The unusual strength of this bond makes $N_2(g)$ quite inert. As a result, $N_2(g)$ coexists with $O_2(g)$ in the atmosphere and forms oxides of nitrogen only in trace amounts at high temperatures. The lack of reactivity of N_2 with O_2 is an essential condition for life on Earth. The inertness of $N_2(g)$ also makes it difficult to synthesize nitrogen compounds.

Another molecule whose Lewis structure features a multiple bond is O_2, which has a double bond.

$$:\!\ddot{\text{O}}\cdot + \cdot \ddot{\text{O}}\!: \longrightarrow :\!\ddot{\text{O}}\!:\!\ddot{\text{O}}\!: \longrightarrow :\!\ddot{\text{O}}\!=\!\ddot{\text{O}}\!: \ ? \qquad \textbf{(10.9)}$$

The blue question mark suggests that there is some doubt about the validity of structure (10.9), and the source of the doubt is illustrated in Figure 10-3. The structure fails to account for the *paramagnetism* of oxygen—the O_2 molecule must have *unpaired* electrons. Unfortunately, no completely satisfactory Lewis structure is possible for O_2, but in Chapter 11, bonding in the O_2 molecule is described in a way that accounts for both the double bond and the observed paramagnetism.

We could continue applying ideas introduced in this section, but our ability to write plausible Lewis structures will be greatly aided by a couple of new ideas that we introduce in Section 10-3.

◀ Throughout this chapter, we use curved red arrows to help us visualize the movement of electrons. IUPAC recommends we use an arrow with a half arrowhead ⤻ when a single electron is moved and an arrow with a full arrowhead ⤺ when a pair of electrons is moved.

▲ FIGURE 10-3
Paramagnetism of oxygen
Liquid oxygen is attracted into the magnetic field of a large magnet.

KEEP IN MIND

that merely being able to write a plausible Lewis structure does not prove that it is the correct electronic structure. Proof can come only through confirming experimental evidence.

10-3 Polar Covalent Bonds and Electrostatic Potential Maps

We have introduced ionic and covalent bonds as though they are of two distinctly different types: ionic bonds involving a *complete transfer* of electrons and covalent bonds involving an *equal sharing* of electron pairs. Such is not the case, however, and most chemical bonds fall between the two extremes of 100% ionic and 100% covalent. A covalent bond in which electrons are not shared equally between two atoms is called a **polar covalent bond**. In such a bond, electrons are displaced toward the more nonmetallic element. The unequal sharing of the electrons leads to a partial negative charge on the more nonmetallic element, signified by $\delta-$, and a corresponding partial positive charge on the more metallic element, designated by $\delta+$. Thus we can represent the polar bond in HCl by a Lewis structure in which the partial charges $\delta+$ and $\delta-$ indicate that the bond pair of electrons lies closer to the Cl than to the H.

$$^{\delta+}\text{H} \;\; :\underset{\cdot\cdot}{\overset{\cdot\cdot}{\text{Cl}}}:^{\delta-}$$

The advent of inexpensive, fast computers has allowed chemists to develop methods for displaying the electron distribution within molecules. This distribution is obtained, in principle, by solving the Schrödinger equation for a molecule. Although the solution can be obtained only by using approximate methods, these methods provide an **electrostatic potential map**, a way to visualize the charge distribution within a molecule.

Before discussing these maps let us first review the notion of electron density, or charge density, introduced in Chapter 8. There we saw that the behavior of electrons in atoms can be described by mathematical functions called orbitals. The probability of finding an electron at some point in the three-dimensional region associated with an orbital is related to the square of an atomic orbital function. Typically, we refer to the region encompassing 95% of the probability of finding the electron as the shape of the orbital. In a similar way we can map the total electron density throughout a molecule, that is, not just the density of a single orbital. The electron density surface that encompasses 95% of the charge density in ammonia is depicted in Figure 10-4.

The *electrostatic potential* is the work done in moving a unit of positive charge at a constant speed from one region of a molecule to another. The electrostatic potential map is obtained by hypothetically probing an electron density surface with a positive point charge. The positive point charge will be attracted to an electron-rich region—a region of excess negative charge when all the charges of the nuclei and electrons have been taken into account—and the electrostatic potential will be *negative*. Conversely, if the point charge is placed in an electron-poor region, a region of excess positive charge, the positive point charge will be repelled, and the electrostatic potential will be *positive*. The procedure for making an electrostatic potential map is illustrated in Figure 10-4, which shows the distribution of electron density in ammonia.

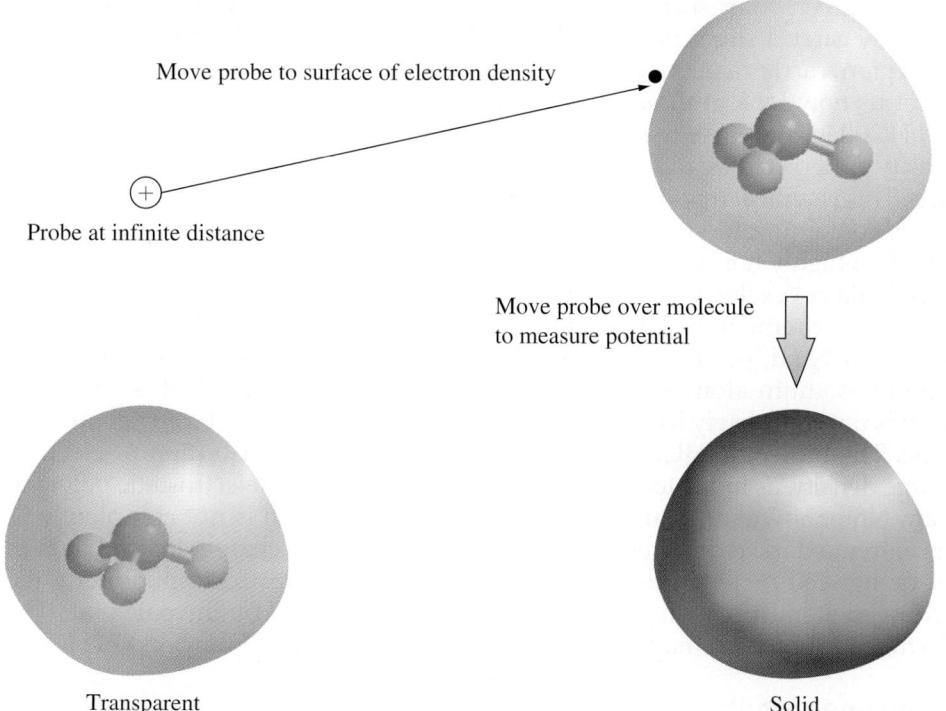

Move probe to surface of electron density

Probe at infinite distance

Move probe over molecule
to measure potential

Transparent

Solid

◀ FIGURE 10-4
Determination of the electrostatic potential map for ammonia
The electrostatic potential at any point on the charge density surface of a molecule is defined as the change in energy that occurs when a unit positive charge is brought to this point, starting from another point that is infinitely far removed from the molecule. The surface encompassing the ammonia molecule is analogous to the 95% surface of electron charge density for atomic orbitals discussed in Chapter 8. The electrostatic potential map gives information about the distribution of electron charge within this surface.

An electrostatic potential map gives information about the distribution of electron charge in a molecule. For example, in a neutral molecule, if the potential at a point is positive, it is likely that an atom at this point carries a net positive charge. An arbitrary "rainbow" color scheme is adopted in the display of an electrostatic potential map. Red, the low-energy end of the spectrum, is used for regions of the most negative electrostatic potential, and blue is used to color regions of the most positive electrostatic potential. Intermediate colors represent intermediate values of the electrostatic potential. Thus, the potential increases from red through yellow to blue, as seen in the scale in Figure 10-5. For example,

◀ Electrostatic potential is the work done in moving a unit of positive charge at a constant speed from one region of a molecule to another.

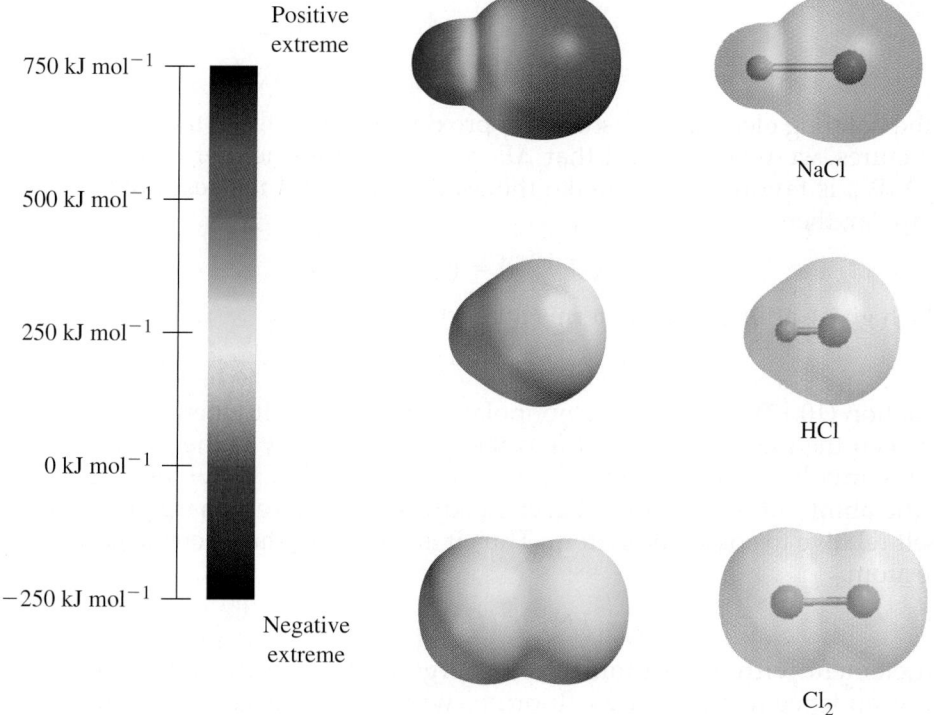

Positive
extreme

750 kJ mol^{-1}

500 kJ mol^{-1}

250 kJ mol^{-1}

0 kJ mol^{-1}

−250 kJ mol^{-1}

Negative
extreme

NaCl

HCl

Cl$_2$

◀ FIGURE 10-5
The electrostatic potential maps for sodium chloride, hydrogen chloride, and chlorine
The dark red and dark blue on the electrostatic potential map correspond to the extremes of the electrostatic potential, negative to positive, for the particular molecule for which the map is calculated. To get a reliable comparison of different molecules, the values of the extremes in electrostatic potential (in kJ mol^{-1}) must be the same for all of the molecules compared. In the maps shown here the range is −250 to 750 kJ mol^{-1}.

the blue-green color surrounding the hydrogen atoms in Figure 10-4 suggests that they carry a slight positive charge. The nitrogen atom, being closest to the red region, carries a net negative charge.

Let us now look at the computed electrostatic potential maps for NaCl, Cl_2, and HCl (Fig. 10-5). We see that Cl_2 has a uniform distribution of electron charge density as depicted by the uniform color distribution in the electrostatic potential map. This is typical for a nonpolar covalent bond and occurs in all diatomic molecules containing identical atoms. The sodium chloride molecule, conversely, exhibits a highly nonuniform distribution of electron charge density. The sodium atom is almost exclusively in the blue extreme of positive charge and the chlorine in the red extreme of negative charge. This electrostatic potential map is typical of an ionic bond, yet it is clear from the map that the transfer of electron density from the sodium atom to the chlorine atom is not complete. That is, the NaCl bond is not completely ionic. Experiments show that the bond is only about 80% ionic. The molecule HCl also has an unsymmetrical distribution of electron charge density, as indicated by the gradation of color in the electrostatic potential map. Note, however, that in this case the chlorine atom is not completely in the extreme dark red corresponding to a large negative charge. Instead, it is in the orange-red region indicating a partial negative charge. Correspondingly, the hydrogen atom has a partial positive charge, as indicated by the pale blue. The electrostatic potential map clearly depicts the polar nature of the bond in HCl.

Electronegativity

We expect the H—Cl bond to be polar because the Cl atom has a greater affinity for electrons than does the H atom. Electron affinity is an atomic property, however, and more meaningful predictions about bond polarities are those based on a molecular property, one that relates to the ability of atoms to lose or gain electrons when they are part of a molecule rather than isolated from other atoms.

Electronegativity (EN) describes an atom's ability to compete for electrons with other atoms to which it is bonded. As such, electronegativity is related to ionization energy (I) and electron affinity (EA). To see how they are related consider the reaction between two hypothetical elements, A and B, which could give the products A^+B^- or A^-B^+. We represent these two reactions by the expressions

$$A + B \longrightarrow A^+B^- \qquad \Delta E_1 = (I_A + EA_B) \qquad \text{(10.10)}$$

$$A + B \longrightarrow A^-B^+ \qquad \Delta E_2 = (I_B + EA_A) \qquad \text{(10.11)}$$

If the bonding electrons are shared approximately equally in these hypothetic structures, we would expect that $\Delta E_1 = \Delta E_2$ because neither extreme (A^+B^- or A^-B^+) is favored. If we make the assumption that the resultant bond is nonpolar, then

$$(I_A + EA_B) = (I_B + EA_A)$$

which gives, after collecting terms for each atom,

$$(I_A - EA_A) = (I_B - EA_B) \qquad \text{(10.12)}$$

Equation (10.12) tells us that a nonpolar bond will result when the difference between the ionization energy and the electron affinity is the same for both atoms involved in the bond. The quantity $(I - EA)$ provides a measure of the ability of an atom to attract electrons (or electron charge density) to itself relative to some other atom. Thus it is related to the electronegativity of the atom.

$$EN_A \propto (I_A - EA_A)$$

An element with a high ionization energy and an electron affinity that is large and negative, such as fluorine, will have a large electronegativity

1	2	3	4	5	6	7	8	9	10	11	12	13	14	15	16	17
H 2.1																
Li 1.0	Be 1.5											B 2.0	C 2.5	N 3.0	O 3.5	F 4.0
Na 0.9	Mg 1.2											Al 1.5	Si 1.8	P 2.1	S 2.5	Cl 3.0
K 0.8	Ca 1.0	Sc 1.3	Ti 1.5	V 1.6	Cr 1.6	Mn 1.5	Fe 1.8	Co 1.8	Ni 1.8	Cu 1.9	Zn 1.6	Ga 1.6	Ge 1.8	As 2.0	Se 2.4	Br 2.8
Rb 0.8	Sr 1.0	Y 1.2	Zr 1.4	Nb 1.6	Mo 1.8	Tc 1.9	Ru 2.2	Rh 2.2	Pd 2.2	Ag 1.9	Cd 1.7	In 1.7	Sn 1.8	Sb 1.9	Te 2.1	I 2.5
Cs 0.8	Ba 0.9	La–Lu*	Hf 1.3	Ta 1.5	W 2.4	Re 1.9	Os 2.2	Ir 2.2	Pt 2.2	Au 2.4	Hg 1.9	Tl 1.8	Pb 1.8	Bi 1.9	Po 2.0	At 2.2
Fr 0.7	Ra 0.9	Ac–Lr†														

Legend:
- below 1.0
- 1.0–1.4
- 1.5–1.9
- 2.0–2.4
- 2.5–2.9
- 3.0–4.0

*Lanthanides: 1.1–1.3
†Actinides: 1.3–1.5

▲ FIGURE 10-6
Electronegativities of the elements
As a general rule, electronegativities *decrease* from *top to bottom* in a group and *increase* from *left to right* in a period of elements. The values are from L. Pauling, *The Nature of the Chemical Bond*, 3rd ed., Cornell University, Ithaca, NY, 1960, page 93. Values may be somewhat different when based on other electronegativity scales.

relative to an atom with a low ionization energy and a small electron affinity, such as sodium.

There are several methods for converting qualitative comparisons to actual numerical values of the electronegativities of the elements. One widely used electronegativity scale, with values given in Figure 10-6, is that devised by Linus Pauling (1901–1994). Pauling's EN values range from about 0.7 to 4.0. In general, the lower its EN, the more metallic the element is, and the higher the EN, the more nonmetallic it is. From Figure 10-6 we also see that electronegativity decreases from top to bottom in a group and increases from left to right in a period of the periodic table. These are the expected trends when we interpret electronegativity in terms of the quantity $(I - EA)$. That is, as the ionization energy (I) increases across the period we expect the electronegativity to increase. The distinction between electron affinity and electronegativity is clearly seen when we consider the electron affinities of F ($-328\ \text{kJ mol}^{-1}$) and chlorine ($-349\ \text{kJ mol}^{-1}$): Although the electron affinity of Cl ($-349\ \text{kJ/mol}$) is somewhat more negative than that of F ($-328\ \text{kJ/mol}$), the EN of Cl (3.0) is significantly lower than that of F (4.0) because of the decreased ionization energy of chlorine (1251 kJ/mol) relative to fluorine (1681 kJ/mol).

🔍 10-4 CONCEPT ASSESSMENT

With the aid of only a periodic table, decide which is the most electronegative atom of each of the following sets of elements: **(a)** As, Se, Br, I; **(b)** Li, Be, Rb, Sr; **(c)** Ge, As, P, Sn.

Electronegativity values allow an insight into the amount of polar character in a covalent bond based on **electronegativity difference, ΔEN**—the absolute value of the difference in EN values of the bonded atoms. If ΔEN for two

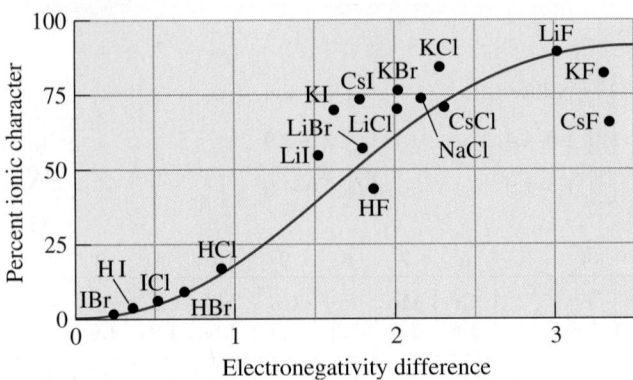

► FIGURE 10-7
Percent ionic character of a chemical bond as a function of electronegativity difference

atoms is very small, the bond between them is essentially covalent. If ΔEN is large, the bond is essentially ionic. For intermediate values of ΔEN, the bond is described as polar covalent. A useful rough relationship between ΔEN and percent ionic character of a bond is presented in Figure 10-7.

Large EN differences are found between the more metallic and the more nonmetallic elements. Combinations of these elements are expected to produce bonds that are essentially ionic. Small EN differences are expected for two nonmetal atoms, and the bond between them should be essentially covalent. Thus, even without a compilation of EN values at hand, you should be able to predict the essential character of a bond between two atoms. Simply assess the metallic/nonmetallic characters of the bonded elements from the periodic table (recall Figure 9-12).

► Although Figure 10-7 suggests that the bond between two identical metal atoms should be covalent [as it is in $Li_2(g)$, for example], in *solid* metals, where bonding extends throughout a network of many, many atoms, the bonding is of a type called *metallic* (explored in the next chapter).

EXAMPLE 10-4 Assessing Electronegativity Differences and the Polarity of Bonds

(a) Which bond is more polar, H—Cl or H—O?
(b) What is the percent ionic character of each of these bonds?

Analyze

To decide which bond is more polar, look up EN values for H, Cl, and O in Figure 10-6, and then compute electronegativity differences, ΔEN, for H—Cl and H—O bonds. The greater the electronegativity difference, the more polar the bond. To determine the percentage ionic character, we use the curve in Figure 10-7.

Solve

(a) $EN_H = 2.1$; $EN_{Cl} = 3.0$; $EN_O = 3.5$. For the H—Cl bond, $\Delta EN = 3.0 - 2.1 = 0.9$. For the H—O bond, $\Delta EN = 3.5 - 2.1 = 1.4$. Because its ΔEN is somewhat greater, we expect the H—O bond to be the more polar bond.

(b) Determine the percent ionic character from Figure 10-7.

$$H—Cl \text{ bond: } \Delta EN = 0.9 \quad \approx 20\% \text{ ionic}$$
$$H—O \text{ bond: } \Delta EN = 1.4 \quad \approx 35\% \text{ ionic}$$

Assess

In this example, we used EN values to decide which of two bonds is more polar. EN values can also be used to decide which end of a given bond will be slightly negative. For example, because EN_{Cl} is greater than EN_H, we conclude that the Cl end of the H—Cl bond will be slightly negative; thus, the H end will be slightly positive.

PRACTICE EXAMPLE A: Which of the following bonds are the most polar, that is, have the greatest ionic character: H—Br, N—H, N—O, P—Cl?

PRACTICE EXAMPLE B: Which is the most polar bond: C—S, C—P, P—O, or O—F?

To illustrate the variation of bond polarity with electronegativity using electrostatic potential maps, consider the electrostatic potential maps for HCl, HBr and HI displayed below.

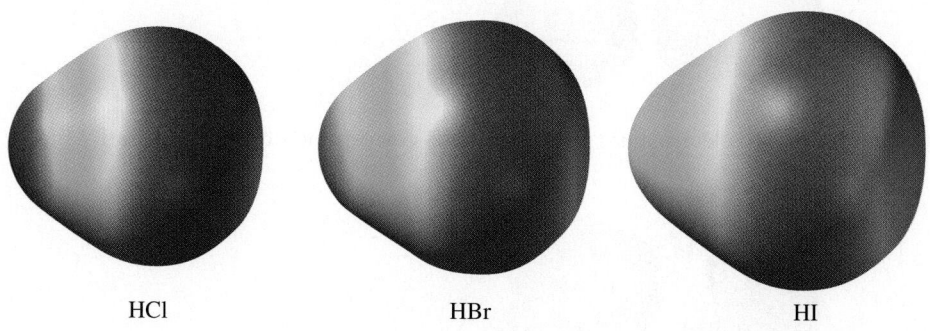

HCl HBr HI

◀ You may have noticed that the electrostatic potential map shown here for HCl has a different range of colors than the one shown in Figure 10-5. In Fig. 10-5, the electrostatic potential map for NaCl shows the greatest range of colors because, of the molecules shown there, NaCl has the greatest ionic character. Here, the electrostatic potential map for HCl shows the greatest range of colors because, compared to HBr and HI, HCl has the greatest ionic character. (See Fig. 10-7.)

The gradation of the charge on the H atom ranges from quite positive in HCl to less positive in HBr and HI, as is seen in the gradation of color on the H atom from dark blue to pale blue. This trend of decreasing charge separation in these three molecules corresponds to the decrease in electronegativity of the halogen atoms from Cl to Br to I. Correspondingly, the halogen atom becomes less red—signifying a decreasing negative charge in going from chlorine to iodine. Electrostatic potential maps are a powerful way of displaying the variation of polarity within a group of related molecules. We will use computed electrostatic potential maps later in this chapter and in subsequent chapters, whenever charge separation within a molecule contributes significantly to understanding the topic at hand.

EXAMPLE 10-5   Identifying a Molecular Structure Using Electronegativity and Electrostatic Potential Maps

Two electrostatic potential maps are shown below. One corresponds to NaF and the other to NaH. Which map corresponds to which molecule?

Analyze

Look up EN values for H, F, and Na in Figure 10-6, and then compute electronegativity differences for NaF and NaH bonds. The bond with the greatest electronegativity difference will be more polar, and its electrostatic potential map will show a greater range of colors.

Solve

$EN_H = 2.1$; $EN_{Na} = 0.9$; $EN_F = 4.0$. For the H—Na bond, $\Delta EN = 2.1 - 0.9 = 1.2$. For the F—Na bond, $\Delta EN = 4.0 - 0.9 = 3.1$. Because ΔEN for NaF is greater, we expect the F—Na bond to be the more polar bond. We conclude that the electrostatic potential map on the left represents NaF.

Assess

It may seem surprising that, in the electrostatic potential maps shown for NaF and NaH, the H "atom" in NaH appears larger than the F "atom" in NaF. Bear in mind that the bonds in both molecules have significant ionic character, and so, when comparing the electrostatic potentials maps for these two molecules, it is more appropriate to think in terms of F^- and H^- ions. Various studies suggest that the NaH bond is probably between 50% and 80% ionic and that the NaF bond is about 90% ionic. Studies on solid NaH suggest that the radius of a H^- ion is somewhere between that of F^- (133 pm) and that of Cl^- (181 pm).

(continued)

PRACTICE EXAMPLE A: Which of the following electrostatic potential maps corresponds to IF, and which to IBr?

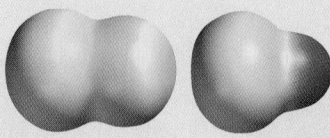

PRACTICE EXAMPLE B: Which of the following electrostatic potential maps corresponds to CH₃OH, and which to CH₃SH?

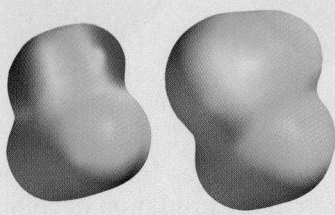

10-4   Writing Lewis Structures

In this section we combine the ideas introduced in the preceding three sections with a few new concepts to write a variety of Lewis structures. Let us begin with a reminder of some of the essential features of Lewis structures that we have already encountered.

- *All* the valence electrons of the atoms in a Lewis structure must appear in the structure.
- *Usually*, all the electrons in a Lewis structure are paired.
- *Usually*, each atom acquires an outer-shell octet of electrons. Hydrogen, however, is limited to two outer-shell electrons.
- *Sometimes*, multiple covalent bonds (double or triple bonds) are needed. Multiple covalent bonds are formed most readily by C, N, O, P, and S atoms.

Skeletal Structures

The usual starting point in writing a Lewis structure is to designate the **skeletal structure**—all the atoms in the structure arranged in the order in which they are bonded to one another. In a skeletal structure with more than two atoms, we generally need to distinguish between central and terminal atoms. A **central atom** is bonded to two or more atoms, and a **terminal atom** is bonded to just one other atom. As an example, consider ethanol, CH_3CH_2OH. Its skeletal structure is the same as the following structural formula. In this structure, the *central atoms*—both C atoms and the O atom—are printed in red. The *terminal atoms*—all six H atoms—are printed in blue.

$$
\begin{array}{ccc}
 & H & H \\
 & | & | \\
H- & C-C & -O-H \\
 & | & | \\
 & H & H
\end{array}
$$

(10.13)

Here are a few additional facts about central atoms, terminal atoms, and skeletal structures.

- *Hydrogen atoms are always terminal atoms.* This is because an H atom can accommodate only two electrons in its valence shell, so it can form only one bond to another atom. (An interesting and rare exception occurs in some boron–hydrogen compounds.)

- *Central atoms are generally those with the lowest electronegativity.* In the skeletal structure (10.13), the atoms of lowest electronegativity (EN = 2.1) happen to be H atoms, but as noted above, H atoms can be only terminal atoms. Next lowest in electronegativity (EN = 2.5) are the C atoms, and these are central atoms. The O atom has the highest electronegativity (3.5) but nevertheless is also a central atom. For O to be a terminal atom in structure (10.13) would require it to exchange places with an H atom, but this would make the H atom a central atom and that is not possible. The chief cases where O atoms are central atoms are in structures with a *peroxo* linkage (—O—O—) or a *hydroxy* group (—O—H). Otherwise, expect an O atom to be a *terminal* atom.

- *Carbon atoms are always central atoms.* This is a useful fact to keep in mind when writing Lewis structures of organic molecules.

- Except for the very large number of chain-like organic molecules, *molecules and polyatomic ions generally have compact, symmetrical structures.* Thus, of the two skeletal structures below, the more compact structure on the right is the one actually observed for phosphoric acid, H_3PO_4.

$$
\begin{array}{cc}
\begin{array}{c}
\text{H} \\
| \\
\text{H—O—O—P—O—O—H}
\end{array}
&
\begin{array}{c}
\text{O—H} \\
| \\
\text{H—O—P—O—H} \\
| \\
\text{O}
\end{array}
\\
(\textit{Incorrect}) & (\textit{Correct})
\end{array}
$$

A Strategy for Writing Lewis Structures

At this point, let us incorporate a number of the ideas that we have considered so far into a specific approach to writing Lewis structures. This strategy is designed to give you a place to begin, as well as consecutive steps to follow to achieve a plausible Lewis structure.

1. Determine the total number of valence electrons that must appear in the structure.

   Examples: In the *molecule* CH_3CH_2OH, there are *4* valence electrons for each C atom, or *8* for the two C atoms; *1* for each H atom, or *6* for the six H atoms; and *6* for the lone O atom. The total number of valence electrons in the Lewis structure of CH_3CH_2OH is

   $$8 + 6 + 6 = 20$$

   In the *polyatomic ion* PO_4^{3-}, there are *5* valence electrons for the P atom and *6* for each O atom, or *24* for all four O atoms. To produce the charge of 3−, an additional *3* valence electrons must be brought into the structure. The total number of valence electrons in the Lewis structure of PO_4^{3-} is

   $$5 + 24 + 3 = 32$$

   In the *polyatomic ion* NH_4^+, there are *5* valence electrons for the N atom and *1* for each H atom, or *4* for all four H atoms. To account for the charge of 1+, one of the electrons must be *lost*. The total number of valence electrons in NH_4^+ is

   $$5 + 4 - 1 = 8$$

2. Identify the central atoms(s) and terminal atoms.

3. Write a plausible skeletal structure. Join the atoms in the skeletal structure by *single* covalent bonds (single dashes, representing two electrons each).

4. For each bond in the skeletal structure, subtract *two* from the total number of valence electrons.

5. With the valence electrons remaining, *first* complete the octets of the terminal atoms. *Then*, to the extent possible, complete the octets of the central

atom(s). If there are just enough valence electrons to complete octets for all the atoms, the structure at this point is a satisfactory Lewis structure.

6. If one or more central atoms are left with an incomplete octet after step 5, move lone-pair electrons from one or more terminal atoms to form *multiple* covalent bonds to central atoms. Do this to the extent necessary to give all atoms complete octets, thereby producing a satisfactory Lewis structure.

Figure 10-8 summarizes this procedure for writing Lewis structures.

▶ It requires a lot of practice to become proficient at writing Lewis structures. Begin by writing structures of molecules that have only one central atom before trying to write the structures of more complicated molecules.

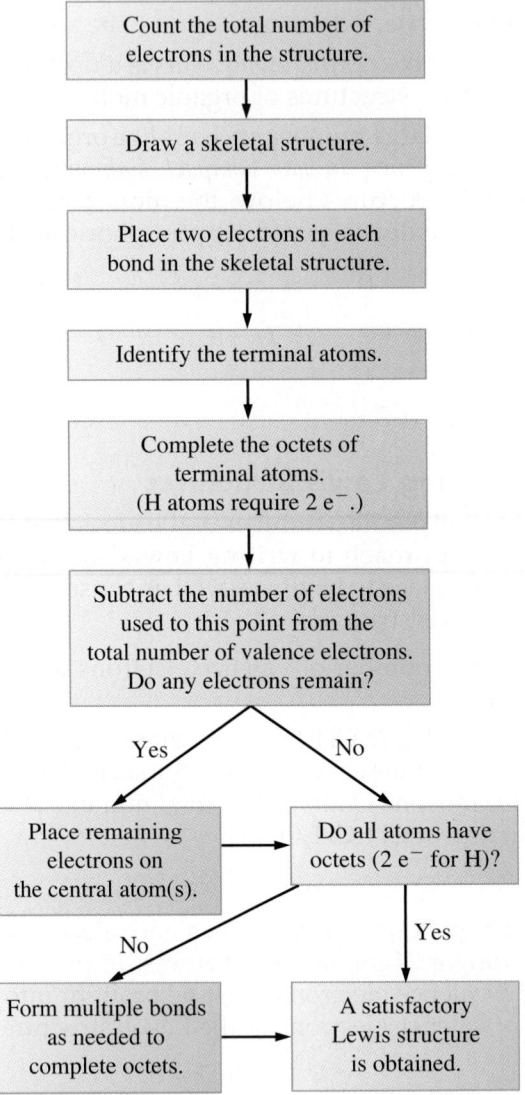

▶ FIGURE 10-8
Summary scheme for drawing Lewis structures

EXAMPLE 10-6   Applying the General Strategy for Writing Lewis Structures

Write a plausible Lewis structure for *cyanogen*, C_2N_2, a poisonous gas used as a fumigant and rocket propellant.

Analyze

Here, we apply the scheme for constructing Lewis structures (Fig. 10-8).

Solve

Step 1. Determine the total number of valence electrons. Each of the two C atoms (group 14) has *four* valence electrons, and each of the two N atoms (group 15) has *five*. The total number of valence electrons is *4 + 4 + 5 + 5 = 18*.

Step 2.   Identify the central atom(s) and terminal atoms. Because the C atoms have a lower electronegativity (2.5) than do the N atoms (3.0), C atoms are central atoms, and N atoms are terminal atoms.

Step 3.   Write a plausible skeletal structure by joining atoms through *single* covalent bonds.

$$N-C-C-N$$

Step 4.   Subtract *two* electrons for each bond in the skeletal structure. The three bonds in this structure account for *6* of the 18 valence electrons. This leaves *12* valence electrons to be assigned.

Step 5.   Complete octets for the terminal N atoms, and to the extent possible, the central C atoms. The remaining *12* valence electrons are sufficient only to complete the octets of the N atoms.

$$:\ddot{N}-C-C-\ddot{N}:$$

Step 6.   Move lone pairs of electrons from the terminal N atoms to form multiple bonds to the central C atoms. Each C atom has only four electrons in its valence shell and needs four more to complete an octet. Thus, each C atom requires two additional pairs of electrons, which it acquires if we move two lone pairs from each N atom into its bond with a C atom, as shown below.

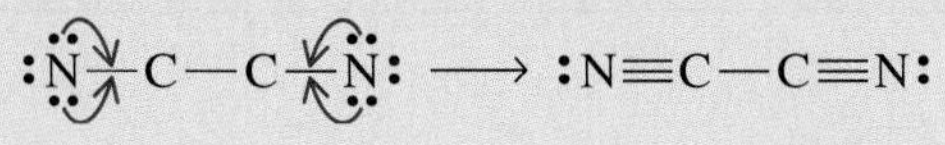

$$:\ddot{N}-C-C-\ddot{N}: \longrightarrow :N{\equiv}C-C{\equiv}N:$$

Assess

The construction of correct Lewis structures is an important skill that all chemists have to master. It is imperative to be able to apply the scheme without referring to the steps on page 409 or in Figure 10-8.

PRACTICE EXAMPLE A: Write plausible Lewis structures for **(a)** CS_2, **(b)** HCN, and **(c)** $COCl_2$.

PRACTICE EXAMPLE B: Write plausible Lewis structures for **(a)** formic acid, HCOOH, and **(b)** acetaldehyde, CH_3CHO.

EXAMPLE 10-7   Writing a Lewis Structure for a Polyatomic Ion

Write the Lewis structure for the *nitronium* ion, NO_2^+.

Analyze

Again, we use the strategy illustrated in Figure 10-8.

Solve

Step 1.   Determine the total number of valence electrons. The N atom (group 15) has *five* valence electrons, and each of the two O atoms (group 16) has *six*. However, *one* valence electron must be removed to produce the charge of 1+.

The total number of valence electrons is $5 + 6 + 6 - 1 = 16$

Step 2.   Identify the central atom(s) and terminal atoms. The N atom has a lower electronegativity (3.0) than the O atoms (3.5). N is the central atom, and the O atoms are the terminal atoms.

Step 3.   Write a plausible skeletal structure by joining atoms through *single* covalent bonds.

$$O-N-O$$

Step 4.   Subtract *two* electrons for each bond in the skeletal structure. The two bonds in this structure account for *4* of the 16 valence electrons. This leaves *12* valence electrons to be assigned.

Step 5.   Complete octets for the terminal O atoms, and to the extent possible, the central N atom. The remaining *12* valence electrons are sufficient only to complete the octets of the O atoms.

$$\left[:\ddot{O}-N-\ddot{O}:\right]^+$$

(continued)

Step 6. Move lone pairs of electrons from the terminal O atoms to form multiple bonds to the central N atom. The N atom has only four electrons in its valence shell and needs four more to complete an octet. Thus, the N atom requires two additional pairs of electrons, which it acquires if we move one lone pair from each O atom into its bond with the N atom, as shown below.

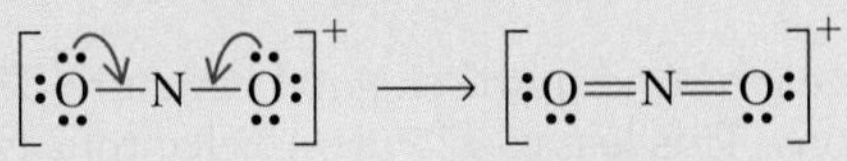

$$\left[:\ddot{\text{O}}\curvearrowright\!\text{N}\!\curvearrowleft\ddot{\text{O}}:\right]^{+} \longrightarrow \left[:\text{O}\!=\!\text{N}\!=\!\ddot{\text{O}}:\right]^{+} \qquad\qquad (10.14)$$

Assess

After drawing a Lewis structure, and before moving on to the next step of a problem or to the next exercise, check the structure. Each atom is surrounded by 8 electrons (each atom has an octet), and the structure has a total valence of 16 (we have not inadvertently added or dropped electrons). In assessing the structure, we must remember that each line represents *two* electrons (a bonding pair).

PRACTICE EXAMPLE A: Write plausible Lewis structures for the following ions: **(a)** NO^{+}; **(b)** $N_2H_5^{+}$; **(c)** O^{2-}.

PRACTICE EXAMPLE B: Write plausible Lewis structures for the following ions: **(a)** BF_4^{-}; **(b)** NH_3OH^{+}; **(c)** NCO^{-}.

Formal Charge

Instead of writing Lewis structure (10.14) for the nitronium ion in Example 10-6, we might have written the following structure.

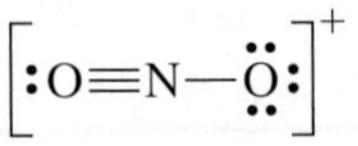

$$\left[:\text{O}\!\equiv\!\text{N}\!-\!\ddot{\text{O}}:\right]^{+} \qquad\qquad (\textit{Improbable}) \quad (10.15)$$

Despite the fact that this structure satisfies the usual requirements—the correct number of valence electrons and an octet for each atom—we have marked it improbable because it fails in one additional requirement. Have you noticed that in our strategy for writing Lewis structures, once the total number of valence electrons has been determined, there is no need to keep track of which electrons came from which atoms? Nevertheless, after we have a plausible Lewis structure, we can go back and assess where each electron apparently came from, and in this way we can evaluate formal charges. **Formal charges (FC)** are apparent charges on certain atoms in a Lewis structure that arise when atoms have not contributed equal numbers of electrons to the covalent bonds joining them. In cases where more than one Lewis structure seems possible, formal charges are used to ascertain which sequence of atoms and arrangement of bonds is most satisfactory.

The formal charge on an atom in a Lewis structure is the number of valence electrons in the free (uncombined) atom minus the number of electrons assigned to that atom in the Lewis structure, with the electrons assigned in the following way.

- Count *lone-pair electrons* as belonging entirely to the atom on which they are found.
- Divide *bond-pair electrons* equally between the bonded atoms.

Assigning electrons (e^{-}) in this way is equivalent to writing that

e^{-} assigned to a bonded atom in a Lewis structure

$$= \text{number lone-pair } e^{-} + \frac{1}{2}\text{number bond-pair } e^{-}$$

Because formal charge is the difference between the assignment of valence electrons to a free (uncombined) atom and to the atom in a Lewis structure, it can be expressed as

$$FC = \tag{10.16}$$

$$\text{number valence } e^- \text{ in free atom } - \text{ number lone-pair } e^- - \frac{1}{2} \text{ number bond-pair } e^-$$

Now, let us assign formal charges to the atoms in structure (10.15), proceeding from left to right.

$:O\!\equiv\ \ \ $ FC = *6* valence e^- in O − *2* lone-pair e^- − $\frac{1}{2}$ (*6* bond-pair e^-) = 6 − 2 − 3 = +1

$\equiv\!N\!-\ \ \ $ FC = *5* valence e^- in N − *0* lone-pair e^- − $\frac{1}{2}$ (*8* bond-pair e^-) = 5 − 0 − 4 = +1

$-\ddot{\underset{\cdot\cdot}{O}}\!:\ \ \ $ FC = *6* valence e^- in O − *6* lone-pair e^- − $\frac{1}{2}$ (*2* bond-pair e^-) = 6 − 6 − 1 = −1

Formal charges in a Lewis structure can be shown by using small, encircled numbers.

$$\left[\overset{\oplus 1}{:O} \equiv \overset{\oplus 1}{N} - \overset{\ominus 1}{\ddot{\underset{\cdot\cdot}{O}}:} \right]^{+} \tag{10.17}$$

The following are general rules that can help to determine the plausibility of a Lewis structure based on its formal charges.

- The sum of the formal charges in a Lewis structure must equal *zero* for a neutral molecule and must equal the magnitude of the charge for a polyatomic ion. [Thus for structure (10.17), this sum is +1 + 1 − 1 = +1.]
- Where formal charges are required, they should be as small as possible.
- Negative formal charges usually appear on the most electronegative atoms; positive formal charges, on the least electronegative atoms.
- Structures having formal charges of the same sign on adjacent atoms are unlikely.

◄ We will see some exceptions to the idea that formal charges should be kept to a minimum in Section 10-6.

Lewis structure (10.17) conforms to the first two rules, but is not in good accordance with the third rule. Despite the fact that O is the most electronegative element in the structure, one of the O atoms has a positive formal charge. The greatest failing, though, is in the fourth rule. Both the O atom on the left and the N atom adjacent to it have positive formal charges. Structure (10.17) is not the most satisfactory Lewis structure. By contrast, the Lewis structure of NO_2^+ derived in Example 10-7 has only one formal charge, +1, on the central N atom. It conforms to the rules completely and is the most satisfactory Lewis structure.

EXAMPLE 10-8   Using Formal Charges in Writing Lewis Structures

Write the most plausible Lewis structure of nitrosyl chloride, NOCl, one of the oxidizing agents present in *aqua regia*, a mixture of concentrated nitric and hydrochloric acids capable of dissolving gold.

Analyze

Although the formula is written as NOCl, we can reject the skeletal structure N—O—Cl because it places the most electronegative atom as the central atom. (We are asked to consider N—O—Cl in Practice Example A.) Having ruled out N—O—Cl as a possible skeletal structure, we are left with the following as possibilities:

$$O\!-\!Cl\!-\!N \text{ and } O\!-\!N\!-\!Cl$$

(continued)

To determine the best structure, we must first complete the skeletal structures and then assign formal charges. The best structure will have the fewest and smallest formal charges.

Solve

Regardless of the skeletal structure chosen, the number of valence electrons (dots) that must appear in the final Lewis structure is

$$5 \text{ from N} + 6 \text{ from O} + 7 \text{ from Cl} = 18$$

When we apply the four steps listed below to the two possible skeletal structures, we obtain a total of four Lewis structures—two for each skeletal structure. This doubling occurs because in step 4, there are two ways to complete the octets of the central atoms. The final Lewis structures obtained are labeled (a_1), (a_2), (b_1), and (b_2).

(a)

O—Cl—N   1. Assign four electrons.

$:\ddot{\text{O}}—\text{Cl}—\ddot{\text{N}}:$ 2. Assign twelve more electrons.

$:\ddot{\text{O}}—\ddot{\text{Cl}}—\ddot{\text{N}}:$ 3. Assign the last two electrons.

(b)

O—N—Cl

$:\ddot{\text{O}}—\text{N}—\ddot{\text{Cl}}:$

$:\ddot{\text{O}}—\ddot{\text{N}}—\ddot{\text{Cl}}:$

4. Complete the octet on the central atom.

(a_1) (a_2) (b_1) (b_2)

$:\ddot{\text{O}}=\ddot{\text{Cl}}—\ddot{\text{N}}:$ $:\ddot{\text{O}}—\ddot{\text{Cl}}=\text{N}:$ $:\ddot{\text{O}}=\ddot{\text{N}}—\ddot{\text{Cl}}:$ $:\ddot{\text{O}}—\ddot{\text{N}}=\ddot{\text{Cl}}:$

Evaluate formal charges by using equation (10.16). In structure (a_1),

for the N atom,

$$FC = 5 - 6 - \frac{1}{2}(2) = -2$$

for the O atom,

$$FC = 6 - 4 - \frac{1}{2}(4) = 0$$

for the Cl atom,

$$FC = 7 - 2 - \frac{1}{2}(6) = +2$$

Proceed in a similar manner for the other three structures. Summarize the formal charges for the four structures.

	(a_1)	(a_2)	(b_1)	(b_2)
N:	−2	−1	0	0
O:	0	−1	0	−1
Cl:	+2	+2	0	+1

Select the best Lewis structure in terms of the formal-charge rules. First, note that all four structures obey the requirement that formal charges of a neutral molecule add up to zero. In structure (a_1), the formal charges are large (+2 on Cl and −2 on N) and the negative formal charge is not on the most electronegative atom. Structure (a_2) has formal charges on all atoms, one of them large (+2 on Cl). Structure (b_1) is the ideal we seek—no formal charges. In structure (b_2), we again have formal charges. The best Lewis structure of nitrosyl chloride is

$$:\ddot{\text{O}}=\ddot{\text{N}}—\ddot{\text{Cl}}:$$

Assess

Based on structure (b_1), ONCl is a better way to write the formula of nitrosyl chloride.

PRACTICE EXAMPLE A: Write a Lewis structure for nitrosyl chloride based on the skeletal structure N—O—Cl, and show that this structure is not as plausible as the one obtained in Example 10-8.

PRACTICE EXAMPLE B: Write two Lewis structures for cyanamide, NH_2CN, an important chemical of the fertilizer and plastics industries. Use the formal charge concept to choose the more plausible structure.

 10-5 CONCEPT ASSESSMENT

For molecules, the most satisfactory Lewis structure may have no formal charges (FC = 0) in some cases and formal charges in others. For polyatomic ions, minimally the most satisfactory Lewis structure has a formal charge on at least one atom. Explain the basis of these observations.

10-1 ARE YOU WONDERING...

Do formal charges represent actual charges on the atoms?

Formal charges are not actual charges, which can be seen from a comparison of the electrostatic potential map of the HCN molecule and the formal charges derived from the Lewis structure.

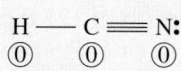

Although the formal charges are all zero, the electrostatic potential map shows that the H atom in the HCN molecule is slightly positive (blue) and that the nitrogen atom is slightly negative (red). In molecules, the true charges on atoms are usually, but not always, between +1 and −1. For example, in the HCl molecule, the charge on H is about +0.17 and that on Cl is about −0.17. (See page 431.)

The method used for assigning formal charges is really just a form of "electron bookkeeping." In this text, we have now discussed two different concepts—oxidation states and formal charges—that are used for electron bookkeeping. Oxidation states and formal charges are both very useful. They are compared in the table below.

	Interpretation	Comments
Oxidation state	The charge an atom would have if the bonding electrons in each bond were *transferred* to the more electronegative atom.	• The oxidation state concept tends to exaggerate the ionic character of the bonding between atoms. • Oxidation states are used to predict and rationalize chemical properties of compounds.
Formal charge	The charge an atom would have if the bonding electrons in each bond were *divided equally* between the two atoms involved.	• The formal charge concept tends to exaggerate the covalent character of the bonding between atoms. • Formal charges are used to assess which Lewis structure is the most satisfactory representation of the true structure.

For many molecules, the bonding is closer to being "pure covalent" than it is to being "pure ionic" and so the formal charge on an atom is often—but not always—numerically closer to the true charge. That's why we focus on formal charges when assessing the relative importance of different Lewis structures. That being said, it is important to emphasize that chemists still question and debate whether it is true that the best structure is the one having the fewest and smallest formal charges.

10-5   Resonance

The ideas presented in the previous section allow us to write many Lewis structures, but some structures still present problems. We describe these problems in the next two sections.

Although we usually think of the formula of oxygen as O_2, there are actually two different oxygen molecules. Familiar oxygen is *dioxygen*, O_2; the other molecule is *trioxygen*—ozone, O_3. The term used to describe the existence of two or more forms of an element that differ in their bonding and molecular structure is *allotropy*—O_2 and O_3 are allotropes of oxygen. Ozone is found naturally in the stratosphere and is also produced in the lower atmosphere as a constituent of smog.

When we apply the usual rules for Lewis structures for ozone, we come up with these *two* possibilities.

$$:\ddot{O}=\ddot{O}-\ddot{\underset{..}{O}}: \qquad :\ddot{\underset{..}{O}}-\ddot{O}=\ddot{O}:$$

▶ Bond lengths are discussed more fully in Section 10-8.

Each structure suggests that one oxygen-to-oxygen bond is single and the other is double. Yet experimental evidence indicates that the two oxygen-to-oxygen bonds are the same; each has a length of 127.8 pm. This bond length is shorter than the O—O single-bond length of 147.5 pm in hydrogen peroxide, $H-\ddot{\underset{..}{O}}-\ddot{\underset{..}{O}}-H$, but it is longer than the double-bond length of 120.74 pm in diatomic oxygen, $:\ddot{O}=\ddot{O}:$. The bonds in ozone are intermediate between a single and a double bond. The difficulty is resolved if we say that the true Lewis structure of O_3 is *neither* of the previously proposed structures but a composite, or *hybrid*, of the two, a fact that we can represent as

Lone pair becomes a bond pair.   Bond pair becomes a lone pair.

$$:\ddot{\underset{..}{O}}\overset{\frown}{}\ddot{O}=\ddot{\underset{..}{O}}: \quad \longleftrightarrow \quad :\ddot{\underset{..}{O}}=\ddot{O}\overset{\frown}{}\ddot{\underset{..}{O}}: \qquad \textbf{(10.18)}$$

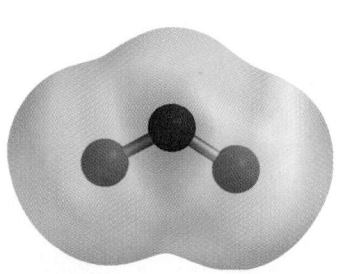

▲ Electrostatic potential map of ozone
All the atoms in an ozone molecule have the same electronegativity and yet the distribution of electron density is nonuniform. The reason will become apparent when we describe in a more sophisticated way the bonding in this molecule.

The situation in which two or more plausible Lewis structures contribute to the "correct" structure is called **resonance**. The true structure is a *resonance hybrid* of plausible contributing structures. Acceptable contributing structures to a resonance hybrid must all have the same skeletal structure (the atomic positions cannot change); they can differ only in how electrons are distributed within the structure. In expression (10.18), the two contributing structures are joined by a double-headed arrow. The arrow does *not* mean that the molecule has one structure part of the time and the other structure the rest of the time. *It has the same structure all the time.* By averaging the single bond in one structure with the double bond in the other, we might say that the oxygen-to-oxygen bonds in ozone are halfway between a single and double bond, that is, 1.5 bonds. The fact that the electrons in ozone are distributed over the whole molecule so as to produce two equivalent bonds is readily seen in the electrostatic potential map of ozone, shown in the margin.

The two resonance structures in expression (10.18) are equivalent; that is, they contribute equally to the structure of the resonance hybrid. In many cases, several contributing resonance structures do not contribute equally. For example, consider the azide anion, N_3^-, for which three resonance structures are given below.

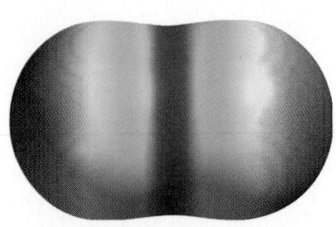

▲ Electrostatic potential map of the azide anion

$$\left[:\ddot{\underset{..}{N}}-N\equiv N:\right]^- \quad \longleftrightarrow \quad \left[\ddot{\underset{..}{N}}=N=\ddot{\underset{..}{N}}\right]^- \quad \longleftrightarrow \quad \left[:N\equiv N-\ddot{\underset{..}{N}}:\right]^-$$
$$\underset{-2}{}\underset{+1}{}\underset{0}{} \qquad\qquad \underset{-1}{}\underset{+1}{}\underset{-1}{} \qquad\qquad \underset{0}{}\underset{+1}{}\underset{-2}{}$$

We can decide which resonance structure likely contributes most to the hybrid by applying the general rules for formal charges (page 413). The central resonance structure avoids the unlikely large formal charge of -2 found on an N atom in the other two structures. Consequently, we expect that structure to contribute most to the resonance hybrid of the azide anion.

EXAMPLE 10-9   Representing the Lewis Structure of a Resonance Hybrid

Write the Lewis structure of the acetate ion, CH_3COO^-.

Analyze

A key concept is that resonance structures differ only in how electrons are distributed within the structure. We cannot change the positions of the atoms. First, we draw a skeletal structure (see the electrostatic potential map shown in the margin), and then we complete it by using the strategy we've used previously. Finally, we generate additional structures (resonance structures) by moving electron pairs.

Solve

The skeletal structure has the three H atoms as terminal atoms bonded to a C atom as a central atom. The second C atom is also a central atom bonded to the first. The two O atoms are terminal atoms bonded to the second C atom.

$$
\begin{array}{ccc}
\text{H} & & \text{O} \\
| & & | \\
\text{H}-\text{C}- & \text{C} & -\text{O} \\
| & & \\
\text{H} & &
\end{array}
$$

The number of valence electrons (dots) that must appear in the Lewis structure is

$$(3 \times 1) + (2 \times 4) + (2 \times 6) + 1 = 3 + 8 + 12 + 1 = 24$$

From H   From C   From O   To establish charge of $1-$

Twelve of the valence electrons are used in the bonds in the skeletal structure, and the remaining twelve are distributed as lone-pair electrons on the two O atoms.

$$
\left[
\begin{array}{ccc}
\text{H} & & \overset{\ominus}{:\ddot{\text{O}}:} \\
| & & || \\
\text{H}-\text{C}- & \text{C} & -\underset{\ominus}{\ddot{\text{O}}:} \\
| & \oplus & \\
\text{H} & &
\end{array}
\right]^{-}
$$

In completing the octet of the C atom on the right, we discover that we can write two completely equivalent Lewis structures, depending on which of the two O atoms furnishes the lone pair of electrons to form a carbon-to-oxygen double bond. The true Lewis structure is a resonance hybrid of the following two contributing structures.

$$
\left[
\begin{array}{ccc}
\text{H} & & :\ddot{\text{O}}: \\
| & & \| \\
\text{H}-\text{C}- & \text{C} & \overset{\ominus}{-\ddot{\text{O}}:} \\
| & & \\
\text{H} & &
\end{array}
\right]^{-}
\longleftrightarrow
\left[
\begin{array}{ccc}
\text{H} & & \overset{\ominus}{:\ddot{\text{O}}:} \\
| & & | \\
\text{H}-\text{C}- & \text{C} & =\ddot{\text{O}} \\
| & & \\
\text{H} & &
\end{array}
\right]^{-}
\qquad \textbf{(10.19)}
$$

Assess

Even though the formal process of converting one resonance structure to another moves electrons, resonance is not meant to indicate the motion of electrons. The acetate anion has a structure that is a composite of the two resonance forms that we have constructed.

▲ Acetate anion

PRACTICE EXAMPLE A: Draw Lewis structures to represent the resonance hybrid for the SO_2 molecule.

PRACTICE EXAMPLE B: Draw Lewis structures to represent the resonance hybrid for the nitrate ion.

Is resonance possible in the acetic acid (CH_3CO_2H) molecule? Explain.

10-6 Exceptions to the Octet Rule

The octet rule has been our mainstay in writing Lewis structures, and it will continue to be one. Yet at times, we must depart from the octet rule, as we will see in this section.

Odd-Electron Species

The molecule NO has 11 valence electrons, an odd number. If the number of valence electrons in a Lewis structure is odd, there must be an unpaired electron somewhere in the structure. Lewis theory deals with electron pairs and does not tell us where to put the unpaired electron; it could be on either the N or the O atom. To obtain a structure free of formal charges, however, we will put the unpaired electron on the N atom.

$$\cdot \ddot{N} = \ddot{O} :$$

The presence of unpaired electrons causes odd-electron species to be paramagnetic. NO is paramagnetic. Molecules with an even number of electrons are expected to have all electrons paired and to be diamagnetic. An important exception is seen in the case of O_2, which is paramagnetic despite having 12 valence electrons. Lewis theory does not provide a good electronic structure for O_2, but the molecular orbital theory that we will consider in the next chapter is much more successful.

The number of stable odd-electron molecules is quite limited. More common are **free radicals**, or simply *radicals*, highly reactive molecular fragments with one or more unpaired electrons. The formulas of free radicals are usually written with a dot to emphasize the presence of an unpaired electron, such as in the *methyl* radical, $\cdot CH_3$, and the *hydroxyl* radical, $\cdot OH$. The Lewis structures of these two free radicals are

$$\begin{array}{c} \text{H} \\ | \\ \text{H} - \underset{\bullet}{\text{C}} - \text{H} \end{array} \qquad \cdot \ddot{\text{O}} - \text{H}$$

Both of these free radicals are commonly encountered as transitory species in flames. In addition, $\cdot OH$ is formed in the atmosphere in trace amounts as a result of photochemical reactions.

$$\cdot OH + CO \longrightarrow CO_2 + \cdot H$$

Many important atmospheric reactions involve free radicals as reactants, such as in the above oxidation of CO to CO_2. Free radicals, because of their unpaired electron, are highly reactive species. The hydroxyl radical, for example, is implicated in DNA damage that can lead to cancer.

Incomplete Octets

Our initial attempt to write the Lewis structure of boron trifluoride leads to a structure in which the B atom has only *six* electrons in its valence shell—an *incomplete octet*.

$$: \ddot{\text{F}} - \text{B} - \ddot{\text{F}} : \qquad (10.20)$$
$$\overset{|}{\underset{:\ddot{\text{F}}:}{}}$$

▶ Experimental evidence for the paramagnetism of O_2 is shown in Figure 10-3.

We have learned to complete the octets of central atoms by shifting lone-pair electrons from terminal atoms to form multiple bonds. One of three equivalent structures with a boron-to-fluorine double bond is shown below.

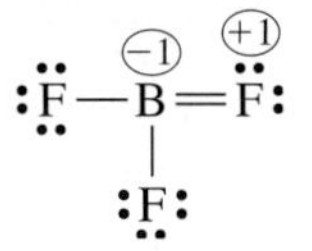

$$:\ddot{\underset{\cdot\cdot}{F}}\overset{\ominus}{-}\underset{\underset{\underset{\cdot\cdot}{\ddot{F}}:}{|}}{B}=\overset{\oplus}{\ddot{F}}: \qquad (10.21)$$

An observation in support of structure (10.21) is that the B—F bond length in BF_3 (130 pm) is less than expected for a single bond. A shorter bond suggests that more than two electrons are present, that is, that there is multiple-bond character in the bond. However, the placement of formal charges in structure (10.18) breaks an important rule—negative formal charge should be found on the more electronegative atom in the bond. In this structure, the positive formal charge is on the most electronegative of all atoms—F.

The high electronegativity of fluorine (4.0) and the much lower one of boron (2.0) suggest an appreciable ionic character to the boron-to-fluorine bond (see Figure 10-7). This suggests the possibility of such ionic structures as the following.

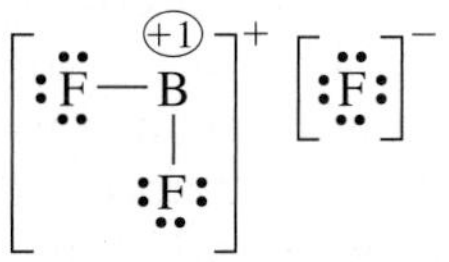

$$\left[:\ddot{\underset{\cdot\cdot}{F}}-\underset{\underset{\underset{\cdot\cdot}{\ddot{F}}:}{|}}{\overset{\oplus}{B}}\right]^{+}\left[:\ddot{\underset{\cdot\cdot}{F}}:\right]^{-} \qquad (10.22)$$

In view of its molecular properties and chemical behavior, the best representation of BF_3 appears to be a resonance hybrid of structures (10.20, 10.21, and 10.22), with perhaps the most important contribution made by the structure with an incomplete octet (10.20). Whichever BF_3 structure we choose to emphasize, an important characteristic of BF_3 is its strong tendency to form a coordinate covalent bond with a species capable of donating an electron pair to the B atom. This can be seen in the formation of the BF_4^- ion.

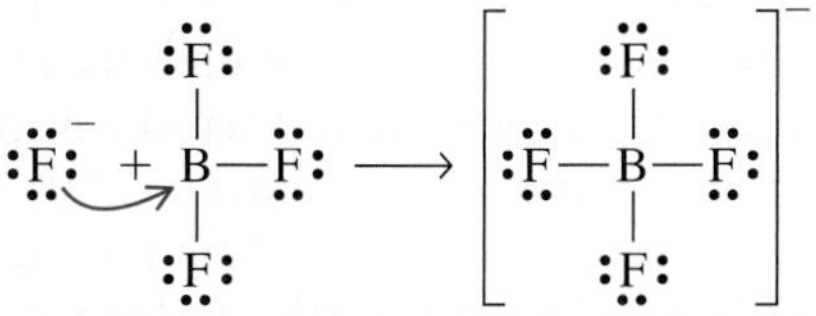

$$:\ddot{\underset{\cdot\cdot}{F}}:^{-} + \underset{\underset{\underset{\cdot\cdot}{\ddot{F}}:}{|}}{\overset{\overset{:\ddot{F}:}{|}}{B}}-\ddot{\underset{\cdot\cdot}{F}}: \longrightarrow \left[:\ddot{\underset{\cdot\cdot}{F}}-\underset{\underset{\underset{\cdot\cdot}{\ddot{F}}:}{|}}{\overset{\overset{:\ddot{F}:}{|}}{B}}-\ddot{\underset{\cdot\cdot}{F}}:\right]^{-}$$

In BF_4^-, the bonds are single bonds and the bond length is 145 pm.

The number of species with incomplete octets is limited to some beryllium, boron, and aluminum compounds. Perhaps the best examples are the boron hydrides. Bonding in the boron hydrides will be discussed in Chapter 22.

Expanded Valence Shells

We have consistently tried to write Lewis structures in which all atoms except H have a complete octet, that is, in which each atom has eight valence electrons. There are a few Lewis structures that break this rule by having 10 or even 12 valence electrons around the central atom, creating what is called an **expanded valence shell**. Describing bonding in these structures is an area of active interest among chemists.

Molecules with expanded valence shells typically involve nonmetal atoms of the third period and beyond that are bonded to highly electronegative atoms. For example, phosphorus forms two chlorides, PCl_3 and PCl_5. We can write a Lewis structure for PCl_3 with the octet rule. In PCl_5, with five Cl atoms bonded directly to the central P atom, the outer shell of the P atom appears to have *ten* electrons. We might say that the valence shell has expanded to ten electrons. In the SF_6 molecule, the valence shell appears to expand to 12.

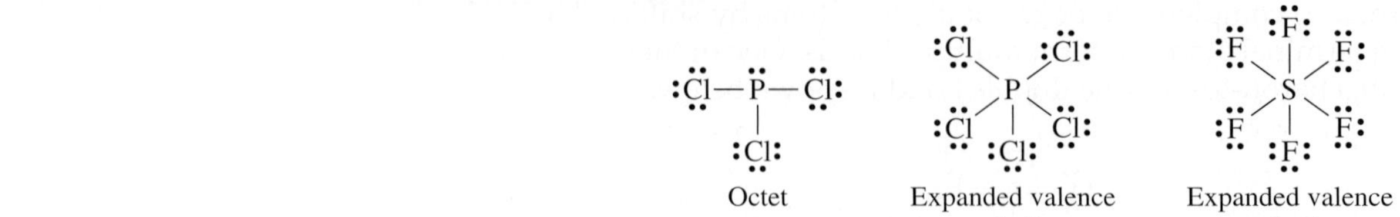

Octet | Expanded valence shell | Expanded valence shell

Expanded valence shells have also been used in cases where they appear to give a better Lewis structure than strict adherence to the octet rule, as suggested by the two Lewis structures for the sulfate ion that follow.

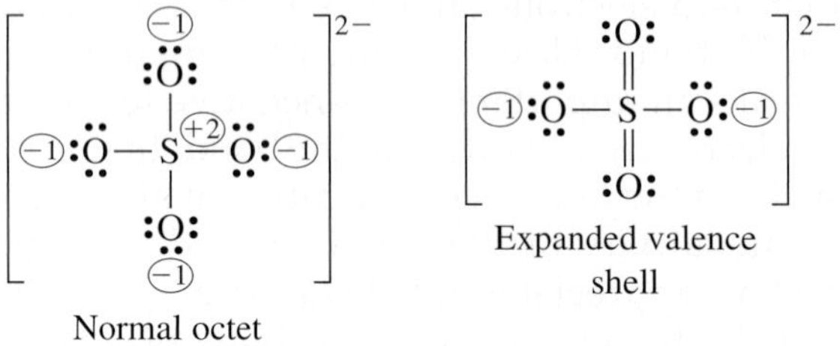

Normal octet | Expanded valence shell

The argument for including the expanded valence-shell structure is that it reduces formal charges. Also, the experimentally determined sulfur-to-oxygen bond lengths in SO_4^{2-} and H_2SO_4 are in agreement with this idea. The experimental results for H_2SO_4, summarized in structure (10.23), indicate that the S—O bond with O as a central atom and with an attached H atom is longer than the S—O bond with O as a terminal atom.

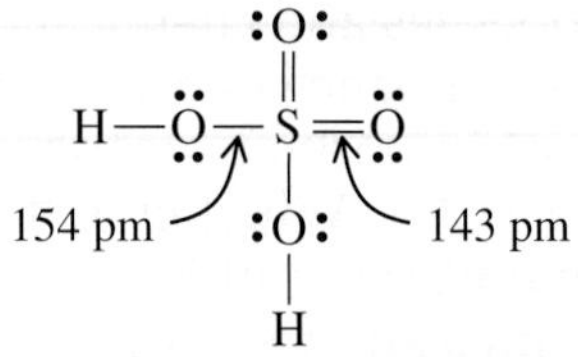

$$H—\ddot{O}—\overset{\displaystyle :O:}{\underset{\displaystyle :O:}{\overset{\|}{S}}}=\ddot{O} \qquad (10.23)$$

154 pm   :O: 143 pm
          |
          H

Experimental evidence appears to support using an expanded valence shell in the Lewis structure of sulfuric acid. The experimentally determined S—O bond length in the sulfate anion—149 pm—lies between the two S—O bond lengths found in sulfuric acid, suggesting a partial double-bond character. The expanded valence-shell structure is suggestive of this partial double-bond character, whereas the octet structure is not. For the sulfate anion, best agreement with the observed S—O bond lengths is found in a resonance hybrid having strong contributions from a series of resonance structures (10.24) based on expanded valence shells.

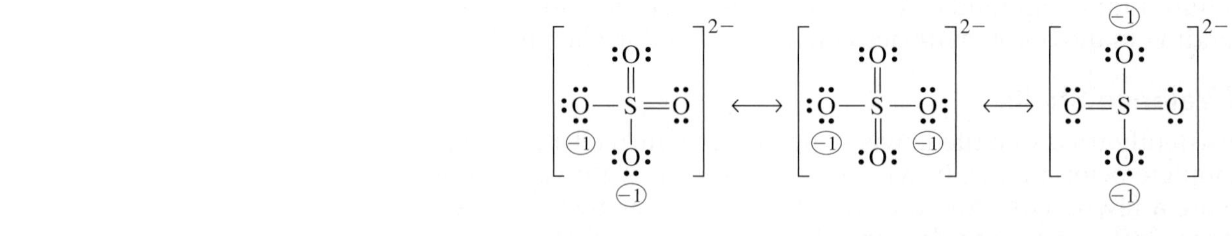

$$(10.24)$$

The problem with expanded valence-shell structures is, of course, to explain where the "extra" electrons go. This expansion has been rationalized by assuming that after the $3s$ and $3p$ subshells of the central atom fill to capacity (eight electrons), extra electrons go into the empty $3d$ subshell. If we assume that the energy difference between the $3p$ and $3d$ levels is not very large, the valence-shell expansion scheme seems reasonable. But is this a valid assumption? The use of the $3d$ orbitals for valence-shell expansion is a matter of scientific dispute.* Although unresolved questions about the expanded

*L. Suidan et al., *J. Chem. Educ.*, **72**, 583 (1995); G. H. Purser, *J. Chem. Educ.*, **78**, 981 (2001).

valence-shell concept may be unsettling, the point to keep in mind is that the unmodified octet rule works perfectly well for most uses of Lewis structures. We will return to this topic, together with several other unsettled issues, in the concluding section of Chapter 11.

🔍 10-7   CONCEPT ASSESSMENT

On page 416, we reasoned that because of resonance, oxygen–oxygen bonds in O_3 were halfway between single and double bonds, that is, 1.5 bonds. Do you expect the sulfur–oxygen bonds in SO_2 to be single, double, or 1.5 bonds? Explain your answer, bearing in mind the ability of sulfur to expand its valence shell.

10-7   Shapes of Molecules

The Lewis structure for water gives the impression that the constituent atoms are arranged in a straight line.

$$H-\ddot{\underset{\cdot\cdot}{O}}-H$$

However, the experimentally determined shape of the molecule is not linear. The molecule is *bent*, as shown in Figure 10-9. Does it really matter that the H_2O molecule is bent rather than linear? The answer is, decidedly, yes. As we will learn in Chapter 12, the bent shape of water molecules helps to account for the fact that water is a liquid rather than a gas at room temperature. In Chapter 13, we will find that it also accounts for the ability of liquid water to dissolve so many different substances.

What we seek in this section is a simple model for predicting the approximate shape of a molecule. Unfortunately, Lewis theory tells us nothing about the shapes of molecules, but it is an excellent place to begin. The next step is to use an idea based on repulsions between valence-shell electron pairs. We will discuss this idea after defining a few terms.

By molecular shape, we mean the geometric figure we get when joining the nuclei of bonded atoms by straight lines. Figure 10-9 depicts the *triatomic* (three-atom) water molecule using a ball-and-stick model. The balls represent the three atoms in the molecule, and the straight lines (sticks), the bonds between atoms. In reality, the atoms in the molecule are in close contact, but for clarity we show only the centers of the atoms. To have a complete description of the shape of a molecule, we need to know not only the bond lengths, the distances between the nuclei of bonded atoms, but also the **bond angles**, the angles between adjacent lines representing bonds. We will concentrate on bond angles in this section and bond lengths in Section 10-8.

A diatomic molecule has only one bond and no bond angle. Because the geometric shape determined by two points is a straight line, *all diatomic molecules are linear*. A triatomic molecule has two bonds and one bond angle. If the bond angle is 180°, the three atoms lie on a straight line, and the molecule is *linear*. For any other bond angle, a triatomic molecule is said to be *angular, bent, or V-shaped*. Some polyatomic molecules with more than three atoms have planar or even linear shapes. More commonly, however, the centers of the atoms in these molecules define a three-dimensional geometric figure.

Valence-Shell Electron-Pair Repulsion (VSEPR) Theory

The shape of a molecule is established by experiment or by a quantum mechanical calculation confirmed by experiment. The results of these experiments and calculations are generally in good agreement with the **valence-shell electron-pair repulsion theory (VSEPR)**. In VSEPR theory, we focus on *pairs* of electrons in the *valence* electron shell of a central atom in a structure.

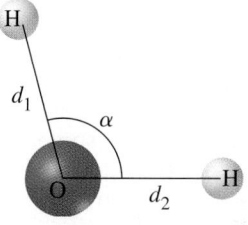

▲ FIGURE 10-9
Geometric shape of a molecule
To establish the shape of the triatomic H_2O molecule shown here, we need to determine the distances between the nuclei of the bonded atoms and the angle between adjacent bonds. In H_2O, the bond lengths $d_1 = d_2 = 95.8\,pm$ and the bond angle $\alpha = 104.45°$.

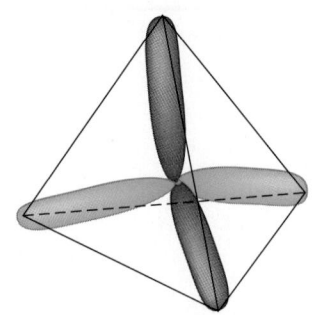

▲ FIGURE 10-10
Balloon analogy to valence-shell electron-pair repulsion
When two elongated balloons are twisted together, they separate into four lobes. To minimize interferences, the lobes spread out into a tetrahedral pattern. (A regular tetrahedron has four faces, each an equilateral triangle.) The lobes are analogous to valence-shell electron pairs.

Electron pairs repel each other, whether they are in chemical bonds (bond pairs) or unshared (lone pairs). Electron pairs assume orientations about an atom to minimize repulsions.

This, in turn, results in particular geometric shapes for molecules.

Another aspect of VSEPR theory is a focus not just on electron pairs but on electron *groups*. A group of electrons can be a pair, either a lone pair or a bond pair, or it can be a single unpaired electron on an atom with an incomplete octet, as in NO. A group can also be a double or triple bond between two atoms. Thus in the $\ddot{O}=C=\ddot{O}$ molecule, the central C atom has *two* electron groups in its valence shell. Each of the double bonds with its two electron pairs is treated as *one* electron group.

Consider the methane molecule, CH_4, in which the central C atom has acquired the electron configuration of Ne by forming covalent bonds with four H atoms.

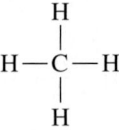

What orientation will the four electron groups (bond pairs) assume? The balloon analogy of Figure 10-10 suggests that electron-group repulsions will force the groups as far apart as possible—to the corners of a tetrahedron having the C atom at its center. The VSEPR method predicts, correctly, that CH_4 is a *tetrahedral* molecule.

Having established that the molecular shape of methane is tetrahedral, the following question arises: How can we represent the three-dimensional shape of molecule on a sheet of paper? In the diagram in the margin, we have enclosed a methane molecule in a tetrahedron (red lines).

We see that the C—H bonds point to the vertices of the tetrahedron. Any two of the C—H bonds define a plane, and in the figures below we choose two C—H bonds (shown in blue) to lie in the plane of the page.

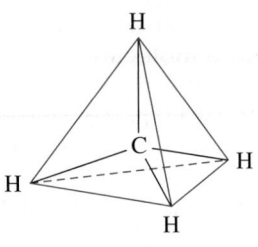

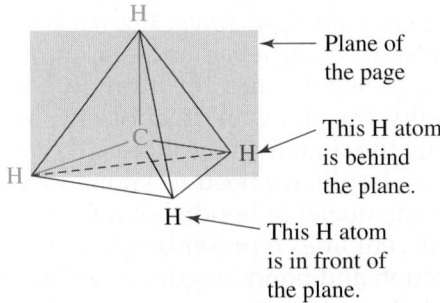

When examining the figure above, we can see that the other two C—H bonds do not lie in the plane of the page. One of the C—H bonds points out of the plane of the page (toward us) and the other points behind (away from us). In the figure in the lower margin, we use a solid wedge to represent the bond that points toward us and a dashed wedge to represent the bond that points away.

The *dash and wedge* symbolism described above is routinely used by chemists to represent three-dimensional structures of molecules. The symbolism is based on the following conventions:

- Ordinary lines are used to represent bonds that lie in the plane of the paper.
- Solid wedges are used to represent bonds that point toward the viewer, that is, in front of the plane of the paper.
- Dashed wedges are used to represent bonds that point away from the viewer, that is, behind the plane of the paper.

The dashed wedge represents a bond that points behind the plane of the page.

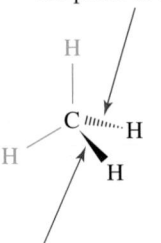

The solid wedge represents a bond that points out of the plane of the page.

In NH_3 and H_2O, the central atom is also surrounded by four groups of electrons, but these molecules *do not* have a tetrahedral shape.

$$H-\underset{\overset{\cdot\cdot}{|}}{N}-H \quad \text{and} \quad \underset{\cdot\cdot}{:}\overset{\overset{H}{|}}{O}-H$$

Here is the situation: VSEPR theory predicts the distribution of electron groups, and in these molecules, electron groups are arranged tetrahedrally about the central atom. The shape of a molecule, however, is determined by the location of the atomic nuclei. To avoid confusion, we will call the geometric distribution of electron groups the **electron-group geometry** and the geometric arrangement of the atomic nuclei—the actual determinant of the molecular shape—the **molecular geometry**.

In the NH_3 molecule, only three of the electron groups are bond pairs; the fourth is a lone pair. Joining the N nucleus to the H nuclei by straight lines outlines a pyramid with the N atom at the apex and the three H atoms at the base; it is called a *trigonal pyramid*. We say that the electron-group geometry is tetrahedral and the molecular geometry is trigonal-pyramidal.

In the H_2O molecule, two of the four electron groups are bond pairs and two are lone pairs. The molecular shape is obtained by joining the two H nuclei to the O nucleus with straight lines. For H_2O, the electron-group geometry is tetrahedral and the molecular geometry is V-shaped, or bent. In the diagram below, the Lewis structure for water is drawn in two ways.

In the first diagram, which is how we usually draw the structure, all the atoms lie on the plane of the paper. In the second structure, we use dash and wedge symbols to indicate that one of the bonds points toward us and the other points away.

The geometric shapes of CH_4, NH_3, and H_2O are summarized in Figure 10-11. In the VSEPR notation used in Figure 10-11, A is the central atom, X

◀ According to IUPAC recommendations, the hashed wedge bond should have the narrow end at the central atom, with the interpretation that the bond projects below the plane of the paper. However, IUPAC indicates it is acceptable to reverse the directionality of the hashed wedge to emphasize the perspective that the atom bonded to the central atom points behind the plane of the page.

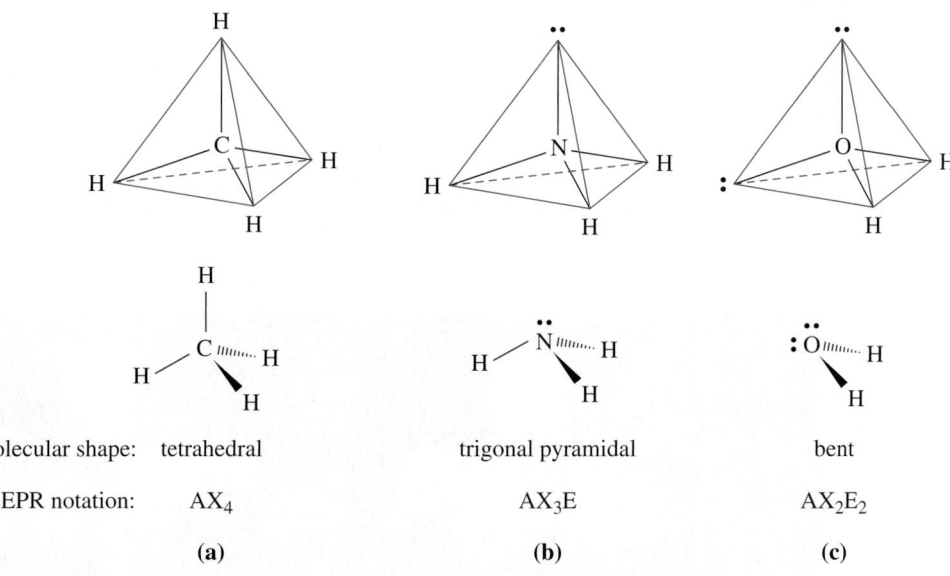

Molecular shape:	tetrahedral	trigonal pyramidal	bent
VSEPR notation:	AX_4	AX_3E	AX_2E_2
	(a)	**(b)**	**(c)**

▲ FIGURE 10-11
Molecular shapes based on tetrahedral electron-group geometry of CH_4, NH_3, and H_2O
All three molecules have a tetrahedral arrangement of electron groups around the central atom. However, molecular shapes (or molecular geometries) are established by focusing only on the positions of the atoms. Pretend that lone pairs are invisible when visualizing the molecular shapes. In (a), there are no lone pairs and the C atom sits at the center of a tetrahedron; the molecular shape is tetrahedral. Pretending the lone pair on the N atom in (b) is invisible, the N atom is the vertex in a pyramid having a triangular base; the molecular shape is called trigonal pyramidal. In (c), the O and H atoms form a V-shape and the molecular shape is V-shaped or bent.

is a terminal atom or group of atoms bonded to the central atom, and E is a lone pair of electrons. Thus, the symbol AX_2E_2 signifies that *two* atoms or groups (X) are bonded to the central atom (A). The central atom also has *two* lone pairs of electrons (E). H_2O is an example of a molecule of the AX_2E_2 type.

For tetrahedral electron-group geometry, we expect bond angles of $109.5°$, known as the *tetrahedral* bond angle. In the CH_4 molecule, the measured bond angles are, in fact, $109.5°$. The bond angles in NH_3 and H_2O are slightly smaller: $107°$ for the H—N—H bond angle and $104.5°$ for the H—O—H bond angle. We can explain these less-than-tetrahedral bond angles by the fact that the charge cloud of the lone-pair electrons spreads out. This forces the bond-pair electrons closer together and reduces the bond angles.

VSEPR theory works best for second-period elements. The predicted bond angle of $109.5°$ for H_2O is close to the measured angle of $104.5°$. For H_2S, however, the predicted value of $109.5°$ is not in good agreement with the observed value, $92°$. Even though VSEPR theory does not give an accurate prediction for the angle in H_2S, it does provide an indication that the molecule is bent.

Possibilities for Electron-Group Distributions

The most common situations are those in which central atoms have two, three, four, five, or six electron groups distributed around them.

Electron-group geometries

- two electron groups: linear
- three electron groups: trigonal planar
- four electron groups: tetrahedral
- five electron groups: trigonal bipyramidal
- six electron groups: octahedral

Figure 10-12 extends the balloon analogy to these cases. The cases for five- and six-electron groups are typified by PCl_5 and SF_6, molecules with expanded valence shells.

The molecular geometry is the same as the electron-group geometry *only* when all electron groups are bond pairs. These are for the VSEPR notation AX_n (that is, AX_2, AX_3, AX_4, and so on). In Table 10.1, the AX_n cases are illustrated by

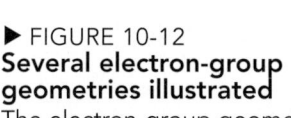

▶ FIGURE 10-12
Several electron-group geometries illustrated
The electron-group geometries pictured are trigonal-planar (orange), tetrahedral (gray), trigonal-bipyramidal (pink), and octahedral (yellow). The atoms at the ends of the balloons are not shown and are not important in this model.

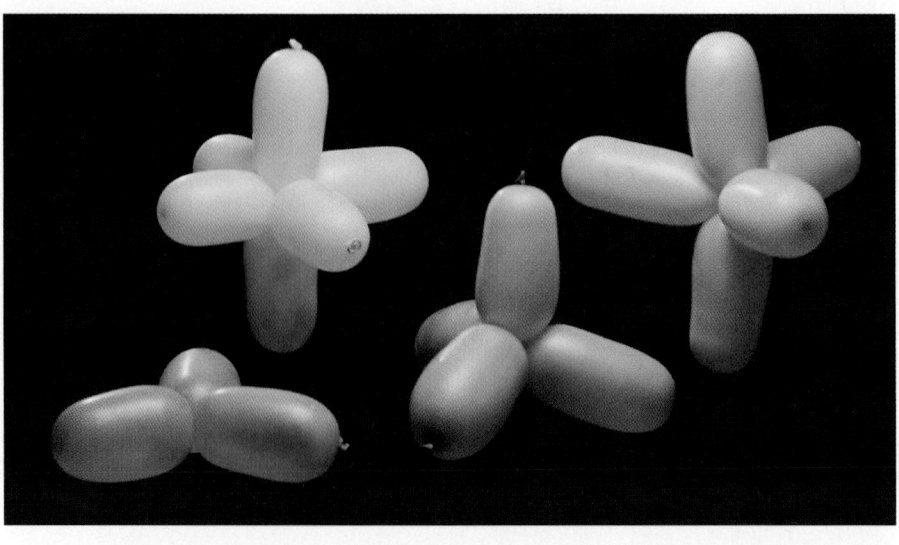

| TABLE 10.1 | Molecular Geometry as a Function of Electron-Group Geometry | | | | | |

Number of Electron Groups	Electron-Group Geometry	Number of Lone Pairs	VSEPR Notation	Molecular Geometry	Ideal Bond Angles	Example
2	linear	0	AX_2	(linear)	180°	$BeCl_2$
3	trigonal planar	0	AX_3	(trigonal planar)	120°	BF_3
	trigonal planar	1	AX_2E	(bent)	120°	SO_2^a
4	tetrahedral	0	AX_4	(tetrahedral)	109.5°	CH_4
	tetrahedral	1	AX_3E	(trigonal pyramidal)	109.5°	NH_3
	tetrahedral	2	AX_2E_2	(bent)	109.5°	OH_2
5	trigonal bipyramidal	0	AX_5	(trigonal bipyramidal)	90°, 120°	PCl_5

(continued)

TABLE 10.1   Molecular Geometry as a Function of Electron-Group Geometry   (Continued)

Number of Electron Groups	Electron-Group Geometry	Number of Lone Pairs	VSEPR Notation	Molecular Geometry	Ideal Bond Angles	Example
	trigonal bipyramidal	1	AX_4E^b	(seesaw)	90°, 120°	SF_4
	trigonal bipyramidal	2	AX_3E_2	(T-shaped)	90°	ClF_3
	trigonal bipyramidal	3	AX_2E_3	(linear)	180°	XeF_2
6	octahedral	0	AX_6	(octahedral)	90°	SF_6
	octahedral	1	AX_5E	(square pyramidal)	90°	BrF_5
	octahedral	2	AX_4E_2	(square planar)	90°	XeF_4

[a]For a discussion of the structure of SO_2, see page 428.
[b]For a discussion of the placement of the lone-pair electrons in this structure, see page 427.

photographs of ball-and-stick models. If one or more electron groups are lone pairs, the molecular geometry is different from the electron-group geometry, although still derived from it. The relationship between electron-group geometry and molecular geometry is summarized in Table 10.1. To understand all the cases in Table 10.1, we need two more ideas.

- *The closer together two groups of electrons are forced, the stronger the repulsion between them.* The repulsion between two electron groups is much stronger at an angle of 90° than at 120° or 180°.
- *Lone-pair electrons spread out more than do bond-pair electrons.* As a result, the repulsion of one lone pair of electrons for another lone pair is greater than, say, between two bond pairs. The order of repulsive forces, from strongest to weakest, is

lone pair–lone pair > lone pair–bond pair > bond pair–bond pair

Consider SF_4, with the VSEPR notation AX_4E. Two possibilities for its structure are presented in the margin, but only one is correct. The correct structure (top)

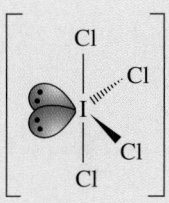

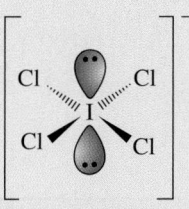

EXAMPLE 10-10   Using VSEPR Theory to Predict a Geometric Shape

Predict the molecular geometry of the polyatomic anion ICl_4^-.

Analyze

To solve this problem, apply the four steps outlined on page 428.

Solve

Step 1. Write the Lewis structure. The number of valence electrons is

From I	From Cl	To establish ionic charge of −1	
(1×7)	$(4 \times 7) +$	1	$= 36$

To join 4 Cl atoms to the central I atom and to provide octets for all the atoms, we need 32 electrons. In order to account for all 36 valence electrons, we need to place an *additional* four electrons around the I atom as lone pairs. That is, we are forced to expand the valence shell of the I atom to accommodate all the electrons required in the Lewis structure.

Step 2. There are six electron groups around the I atom, four *bond pairs* and two *lone pairs*.

Step 3. The electron-group geometry (the orientation of six electron groups) is *octahedral*.

Step 4. The ICl_4^- anion is of the type AX_4E_2, which according to Table 10.1, leads to a molecular geometry that is square planar.

Assess

Figure 10-13 suggests two possibilities for distributing bond pairs and lone pairs in ICl_4^-. The square planar structure is correct because the lone pair–lone pair interaction is kept at 180°. In the incorrect structure, this interaction is at 90°, which results in a strong repulsion.

▲ FIGURE 10-13
Example 10-10 illustrated
The observed structure of ICl_4^- is square planar. The iodine atom has a formal charge of −1 (not shown in the structures above).

PRACTICE EXAMPLE A: Predict the molecular geometry of nitrogen trichloride.

PRACTICE EXAMPLE B: Predict the molecular geometry of phosphoryl chloride, $POCl_3$, an important chemical in the manufacture of gasoline additives, hydraulic fluids, and fire retardants.

places a lone pair of electrons in the central plane of the bipyramid. As a result, *two* lone pair–bond pair interactions are 90°. In the incorrect structure (bottom), the lone pair of electrons is at the bottom of the bipyramid and results in *three* lone pair–bond pair interactions of 90°. This is a less favorable arrangement.

Applying VSEPR Theory

The following four-step strategy can be used for predicting the shapes of molecules.

1. Draw a plausible Lewis structure of the species (molecule or polyatomic ion).
2. Determine the number of electron groups around the central atom, and identify them as being either *bond-pair* electron groups or *lone pairs* of electrons.
3. Establish the electron-group geometry around the central atom—linear, trigonal planar, tetrahedral, trigonal bipyramidal, or octahedral.
4. Determine the molecular geometry from the positions of the atoms bonded directly to the central atom. (Refer to Table 10.1.)

10-8   CONCEPT ASSESSMENT

The ions ICl_2^- and ICl_2^+ differ by only two electrons. Would you expect them to have the same geometric shape? Explain.

Structures with Multiple Covalent Bonds

In a multiple covalent bond, all electrons in the bond are confined to the region between the bonded atoms, and together constitute one group of electrons. Let us test this idea by predicting the molecular geometry of sulfur dioxide. S is the central atom, and the total number of valence electrons is $3 \times 6 = 18$. The Lewis structure is the resonance hybrid of the three contributing structures shown below.

Because we count the electrons in the double covalent bond as one group, the electron-group geometry around the central S atom is that of *three* electron groups—*trigonal-planar*. Of the three electron groups, two are bonding groups and one is a lone pair. This is the case of AX_2E (see Table 10.1). The molecular shape is *angular,* or *bent,* with an expected bond angle of 120°. (The measured bond angle in SO_2 is 119°.)

EXAMPLE 10-11   Using VSEPR Theory to Predict the Shape of a Molecule with a Multiple Covalent Bond

Predict the molecular geometry of formaldehyde, H_2CO, used to make a number of polymers, such as melamine resins. The Lewis structure of the H_2CO molecule is shown here.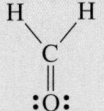

Analyze

We see from the Lewis structure that the carbon–oxygen bond is a double bond. When considering molecules with double or triple bonds, we treat a double or triple bond as a single electron group.

Solve

There are three electron groups around the C atom, two groups in the carbon-to-hydrogen single bonds and the third group in the carbon-to-oxygen double bond. The electron-group geometry for three electron groups is *trigonal-planar*. Because all the electron groups are involved in bonding, the VSEPR notation for this molecule is AX_3. The molecular geometry is also trigonal-planar.

Assess

Because the geometry is trigonal planar, we expect the angle between the two H—C bonds to be close to 120°.

PRACTICE EXAMPLE A: Predict the shape of the COS molecule.

PRACTICE EXAMPLE B: Nitrous oxide, N_2O, is the familiar laughing gas used as an anesthetic in dentistry. Predict the shape of the N_2O molecule.

Molecules with More Than One Central Atom

Although many of the structures of interest to us have only one central atom, VSEPR theory can also be applied to molecules or polyatomic anions with more than one central atom. In such cases, the geometric distribution of terminal atoms around *each* central atom must be determined and the results then combined into a single description of the molecular shape. We use this idea in Example 10-12.

EXAMPLE 10-12   Applying VSEPR Theory to a Molecule with More Than One Central Atom

Methyl isocyanate, CH_3NCO, is used in the manufacture of insecticides, such as carbaryl (Sevin). In the CH_3NCO molecule, the three H atoms and the O atom are terminal atoms and the two C and one N atom are central atoms. Draw the structure of this molecule, using dash and wedge symbols, and indicate the various bond angles.

Analyze

We must first draw a plausible Lewis structure. Then, we determine the electron group geometry around each atom and estimate the angles between pairs of bonds.

Solve

The number of valence electrons in the structure is

From C	From N	From O	From H
(2×4)	(1×5)	(1×6)	$(3 \times 1) = 22$

In drawing the skeletal structure and assigning valence electrons, we first obtain a structure with incomplete octets. By shifting the indicated electrons, we can give each atom an octet.

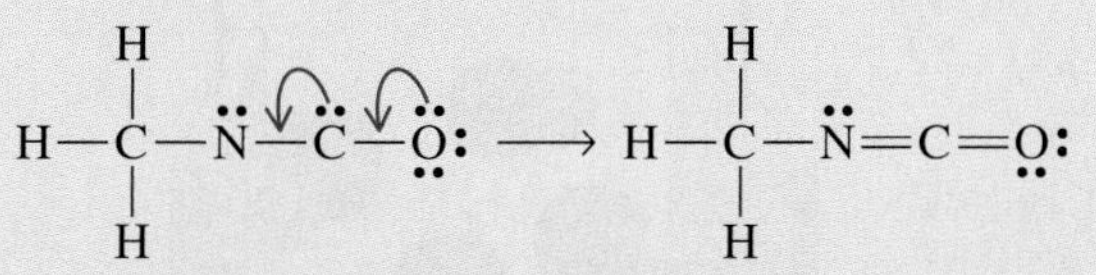

The C atom on the left has four electron groups around it—all bond pairs. The shape of this end of the molecule is *tetrahedral*. The C atom to the right, by forming two double bonds, is treated as having two groups of electrons around it. This distribution is *linear*. For the N atom, three groups of electrons are distributed in a *trigonal planar* manner. The C—N—C bond angle should be about 120°.

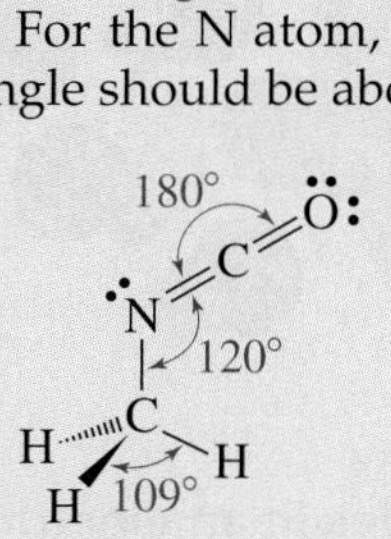

Assess

The strategy outlined above can be applied to molecules of varying complexity.

(continued)

PRACTICE EXAMPLE A: Sketch, by using dash and wedge symbols, the methanol molecule, CH_3OH. Indicate the bond angles in this molecule.

PRACTICE EXAMPLE B: Glycine, an amino acid, has the formula H_2NCH_2COOH. Sketch, by using dash and wedge symbols, the glycine molecule, and indicate the various bond angles.

🔍 10-9 CONCEPT ASSESSMENT

Methyl isocynate, CH_3NCO, can be represented as a hybrid of three Lewis structures. The most satisfactory structure is the one given above in Example 10-12. Draw the other two structures. On the basis of formal charges, which of the structures is least satisfactory?

Molecular Shapes and Dipole Moments

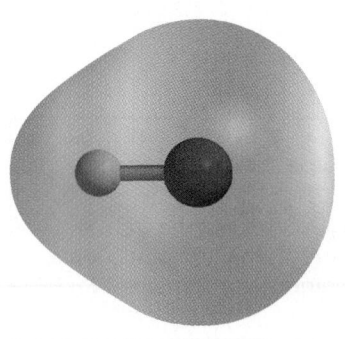

Let us recall some facts that we learned about polar covalent bonds in Section 10-3. In the HCl molecule, the Cl atom is more electronegative than the H atom. Electrons are displaced toward the Cl atom, as shown in the electrostatic potential map in the margin. The HCl molecule is a **polar molecule**. In the representation below, we use a cross-base arrow ($\longmapsto$) that points to the atom that attracts electrons more strongly.

$$^{\delta+}H \longmapsto Cl^{\delta-}$$

The extent of the charge displacement in a polar covalent bond is given by the **dipole moment**, μ. The dipole moment is the product of a partial charge (δ) and distance (d).

$$\mu = \delta \times d \qquad (10.25)$$

If the product, $\delta \times d$, has a value of 3.34×10^{-30} coulomb $\cdot$ meter (C $\cdot$ m), the dipole moment, μ, has a value called 1 *debye*, D (pronounced duh-bye). One experimental method of determining dipole moments is based on the behavior of polar molecules in an electric field, suggested in Figure 10-14.

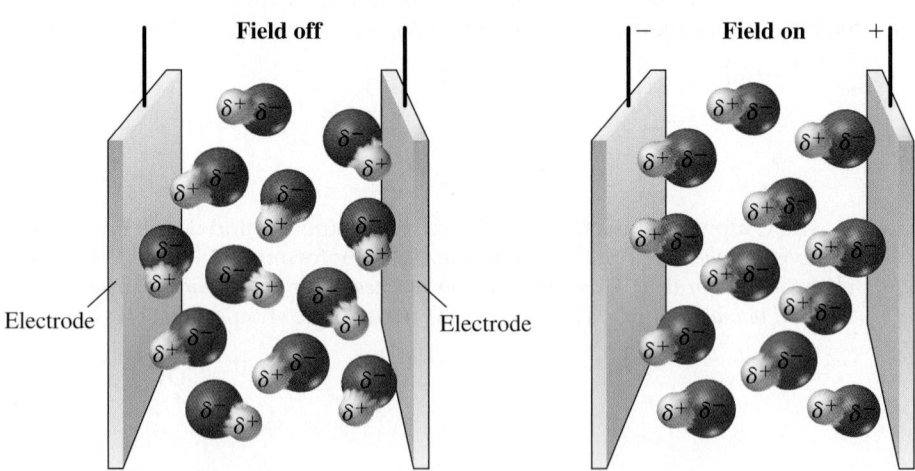

▲ FIGURE 10-14
Polar molecules in an electric field
The device pictured is called an electrical condenser (or capacitor). It consists of a pair of electrodes separated by a medium that does not conduct electricity but consists of polar molecules. (a) When the field is off, the molecules orient randomly. (b) When the electric field is turned on, the polar molecules orient in the field between the charged plates so that the negative ends of the molecules are toward the positive plate and vice versa.

The polarity of the H—Cl bond, as demonstrated on page 402, involves a shift of the electron charge density toward the Cl atom, and this produces a separation of the centers of positive and negative charge. Suppose, instead of a shift in electron charge density, we think of an equivalent situation—the transfer of a *fraction* of the charge of an electron from the H atom to the Cl atom through the entire internuclear distance. Let us determine the magnitude of this partial charge, δ. To do this, we need the measured dipole moment, 1.03 D; the H—Cl bond length, 127.4 pm; and equation (10.25) rearranged to

$$\delta = \frac{\mu}{d} = \frac{1.03 \text{ D} \times 3.34 \times 10^{-30} \text{ C} \cdot \text{m/D}}{127.4 \times 10^{-12} \text{ m}} = 2.70 \times 10^{-20} \text{ C}$$

This charge is about 17% of the charge on an electron (1.602×10^{-19} C) and suggests that HCl is about 17% ionic. This assessment of the percent ionic character of the H—Cl bond agrees well with the 20% we made based on electronegativity differences (recall Example 10-4).

CO_2. Carbon dioxide molecules are *nonpolar*. To understand this observation, we need to distinguish between the displacement of electron charge density in a particular bond and in the molecule as a whole. The electronegativity difference between C and O causes a displacement of electron charge density toward the O atom in each carbon-to-oxygen bond and gives rise to a *bond dipole*. However, because the two bond dipoles are equal in magnitude and point in opposite directions, they cancel each other and lead to a *resultant* dipole moment of zero for the molecule. The symmetrical nature of the electron charge density is clear in the electrostatic potential map for CO_2 shown in the margin.

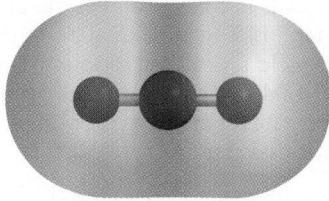

▲ Electrostatic potential map of carbon dioxide

$$\overset{\longleftarrow}{\text{O}}=\overset{\longrightarrow}{\text{C}}=\text{O} \qquad \mu = 0$$

The fact that CO_2 is nonpolar is experimental proof that the three atoms in the molecule lie along a straight line in the order O—C—O. Of course, we can also predict that CO_2 is a linear molecule with the VSEPR theory, based on the Lewis structure

$$:\ddot{\text{O}}=\text{C}=\ddot{\text{O}}:$$

H_2O. Water molecules are *polar*. They have bond dipoles because of the electronegativity difference between H and O, and the bond dipoles combine to produce a resultant dipole moment of 1.84 D. The electrostatic potential map for water provides visual evidence of a net dipole moment on the water molecule. The molecule cannot be linear, for this would lead to a cancellation of bond dipoles, just as with CO_2. We have predicted with the VSEPR theory that the H_2O molecule is bent, and the observation that it is a polar molecule simply confirms the prediction.

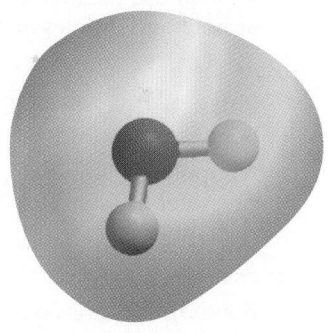

▲ Electrostatic potential map of water

$$\overset{\text{O}\overset{\longleftarrow}{\text{H}}}{\underset{\text{H}}{\diagup}}\ 104°$$

CCl_4. Carbon tetrachloride molecules are *nonpolar*. Based on the electronegativity difference between Cl and C, we expect a bond dipole for the C—Cl bond. The fact that the resultant dipole moment is *zero* means that the bond dipoles must be oriented in such a way that they cancel. The tetrahedral molecular geometry of CCl_4 provides the symmetrical distribution of bond dipoles that leads to this cancellation, as shown in Figure 10-15(a). Can you see that the molecule will be polar if one of the Cl atoms is replaced by an atom with a different electronegativity, say H? In the molecule, $CHCl_3$, there is a resultant dipole moment (Fig. 10-15b).

KEEP IN MIND

that the lack of a molecular dipole moment cannot distinguish between the two possible molecular geometries: tetrahedral and square planar. To do this, other experimental evidence, such as X-ray diffraction, is required.

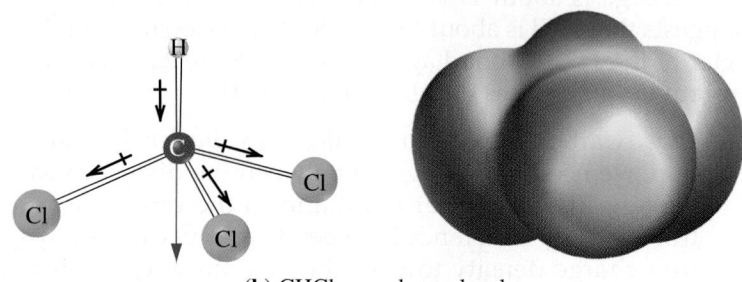

(a) CCl_4: a nonpolar molecule

▶ FIGURE 10-15
Molecular shapes and dipole moments
(a) The resultant of two of the C—Cl bond dipoles is shown as a red arrow, and that of the other two, as a blue arrow. The red and blue arrows point in opposite directions and cancel. The CCl_4 molecule is nonpolar. The balance of the charge distribution in CCl_4 is clearly seen in the electrostatic potential map. (b) The individual bond dipoles do combine to yield a resultant dipole moment (red arrow) of 1.04 D. The electrostatic potential map indicates that the hydrogen atom has a partial positive charge.

(b) $CHCl_3$: a polar molecule

EXAMPLE 10-13 **Determining the Relationship Between Geometric Shapes and the Resultant Dipole Moments of Molecules**

Which of these molecules would you expect to be polar: Cl_2, ICl, BF_3, NO, SO_2?

Analyze

We will use the methods described above to determine the shape of the molecule, and then ascertain whether or not bond dipoles, if present, produce a net permanent dipole moment.

Solve

Polar: ICl, NO, SO_2. ICl and NO are diatomic molecules with an electronegativity difference between the bonded atoms. SO_2 is a bent molecule with an electronegativity difference between the S and O atoms.

Nonpolar: Cl_2 and BF_3. Cl_2 is a diatomic molecule of identical atoms; hence no electronegativity difference. For BF_3, refer to Table 10.1. BF_3 is a symmetrical planar molecule (120° bond angles). The B—F bond dipoles cancel each other.

Assess

Bond dipoles are vector quantities. When adding them together, we must add them as vectors, that is, "head-to-tail," as illustrated below.

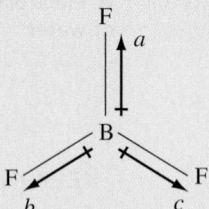

The bond dipoles are labeled a, b, and c.

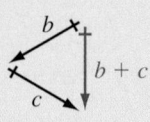

Add b and c "head-to-tail" to form the vector sum $b + c$.

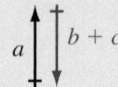

The vector sum of a and $b + c$ is a vector of zero length (no dipole).

PRACTICE EXAMPLE A: Only one of the following molecules is polar. Which is it, and why? SF_6, H_2O_2, C_2H_4.

PRACTICE EXAMPLE B: Only one of the following molecules is *nonpolar*. Which is it, and why? Cl_3CCH_3, PCl_5, CH_2Cl_2, NH_3.

The molecule NH_3 has the dipole moment $\mu = 1.47$ D, whereas for the similar molecule NF_3, $\mu = 0.24$ D. Why do you suppose there is such a large difference in these two values? [*Hint:* What is the effect of the lone-pair electrons on the N atom?]

10-8   Bond Order and Bond Lengths

The term **bond order** describes whether a covalent bond is *single* (bond order = 1), *double* (bond order = 2), or *triple* (bond order = 3). Think of electrons as the "glue" that binds atoms together in covalent bonds. The higher the bond order—that is, the more electrons present—the more glue and the more tightly the atoms are held together.

Bond length is the distance between the centers of two atoms joined by a covalent bond. A double bond between atoms is shorter than a single bond, and a triple bond is shorter still. You can see this relationship clearly in Table 10.2 by comparing the three different bond lengths for the nitrogen-to-nitrogen bond. For example, the measured length of the nitrogen-to-nitrogen triple bond in N_2 is 109.8 pm, whereas the nitrogen-to-nitrogen single bond in hydrazine, H_2N-NH_2, is 147 pm.

Perhaps you can now also better understand the meaning of covalent radius that we introduced in Section 9-3. The single covalent radius is one-half the distance between the centers of identical atoms joined by a *single* covalent bond. Thus, the single covalent radius of chlorine in Figure 9-9 (99 pm) is one-half the bond length given in Table 10.2, that is, $\frac{1}{2} \times 199$ pm. Furthermore, as a rough generalization,

the length of the covalent bond between two atoms can be approximated as the sum of the covalent radii of the two atoms.

Some of these ideas about bond length are applied in Example 10-14.

TABLE 10.2   Some Average Bond Lengths[a]

Bond	Bond Length, pm	Bond	Bond Length, pm	Bond	Bond Length, pm
H—H	74.14	C—C	154	N—N	145
H—C	110	C=C	134	N=N	123
H—N	100	C≡C	120	N≡N	109.8
H—O	97	C—N	147	N—O	136
H—S	132	C=N	128	N=O	120
H—F	91.7	C≡N	116	O—O	145
H—Cl	127.4	C—O	143	O=O	121
H—Br	141.4	C=O	120	F—F	143
H—I	160.9	C—Cl	178	Cl—Cl	199
				Br—Br	228
				I—I	266

[a]Most values (C—H, N—H, C—H, and so on) are averaged over a number of species containing the indicated bond and may vary by a few picometers. Where a diatomic molecule exists, the value given is the actual bond length in that molecule (H_2, N_2, HF, and so on) and is known more precisely.

EXAMPLE 10-14   Estimating Bond Lengths

Provide the best estimate you can of these bond lengths for the **(a)** the nitrogen-to-hydrogen bonds in NH_3: **(b)** the bromine-to-chlorine bond in BrCl.

Analyze

If no bond length is listed for a particular bond, A—B, then look up the bond lengths for A—A and B—B. The A—B bond length can then be estimated as one-half the A—A bond length plus one-half the B—B bond length.

Solve

(a) The Lewis structure of ammonia (page 400) shows the nitrogen-to-hydrogen bonds as single bonds. The value listed in Table 10.2 for the N—H bond is 100 pm, so this is the value we would predict. (The measured N—H bond length in NH_3 is 101.7 pm.)

(b) There is no bromine-to-chlorine bond length in Table 10.2, so we need to calculate an approximate bond length using the relationship between bond length and covalent radii. BrCl contains a Br—Cl *single* bond [imagine substituting one Br atom for one Cl atom in structure (10.4)]. The length of the Br—Cl bond is one-half the Cl—Cl bond length *plus* one-half the Br—Br bond length: $\left(\frac{1}{2} \times 199 \text{ pm}\right) + \left(\frac{1}{2} \times 228 \text{ pm}\right) = 214 \text{ pm}$. (The measured bond length is 213.8 pm.)

Assess

The data in Table 10.2 can be used to make estimates of bond lengths in a variety of molecules.

PRACTICE EXAMPLE A: Estimate the bond lengths of the carbon-to-hydrogen bonds and the carbon-to-bromine bond in CH_3Br.

PRACTICE EXAMPLE B: In the thiocyanate ion, SCN^-, the length of the carbon-to-nitrogen bond is 115 pm. Write a plausible Lewis structure for this ion and describe its geometric shape.

An interesting situation arises for molecules in which resonance is present. In such molecules, fractional bond orders are possible. Consider, for example, the carbonate anion, CO_3^{2-}, shown below.

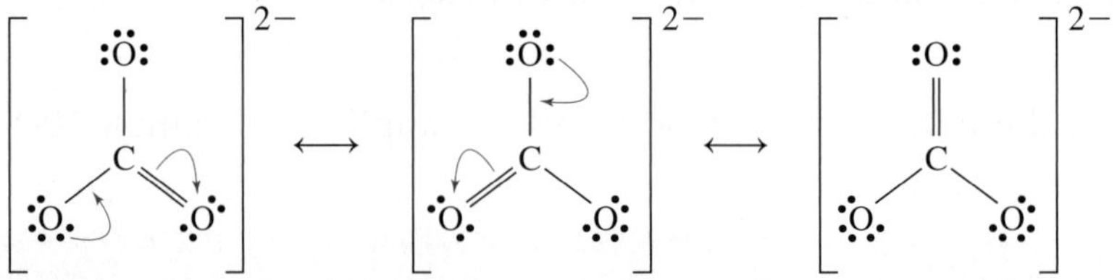

Each resonance form has one double bond and two single bonds. Because the actual structure of the carbonate anion is an average of these three structures, the average bond order is $\frac{1}{3}(1 + 1 + 2) = 1\frac{1}{3}$. The CO bond distance in the carbonate anion is 129 pm, which is intermediate between a C—O single bond (143 pm) and a C=O double bond (120 pm), as we might expect for a fractional bond order.

🔍 **10-11 CONCEPT ASSESSMENT**

NO_2^- and NO_2^+ are made up of the same atoms. How would you expect the nitrogen-to-oxygen bond lengths in these two ions to compare?

10-9 Bond Energies

Together with bond lengths, bond energies can be used to assess the suitability of a proposed Lewis structure. Bond energy, bond length, and bond order are interrelated properties in this sense: the higher the bond order, the shorter the bond between two atoms and the greater the bond energy.

Energy is *released* when isolated atoms join to form a covalent bond, and energy must be *absorbed* to break apart covalently bonded atoms. **Bond-dissociation energy**, D, is the quantity of energy required to break one mole of covalent bonds in a *gaseous* species. The SI units are kilojoules per mole of bonds (kJ/mol).

We can think of bond-dissociation energy as an enthalpy change or a heat of reaction, as discussed in Chapter 7. For example,

Bond breakage: $H_2(g) \longrightarrow 2\,H(g)$ $\Delta H = D(H—H) = +435.93\ \text{kJ/mol}$

Bond formation: $2\,H(g) \longrightarrow H_2(g)$ $\Delta H = -D(H—H) = -435.93\ \text{kJ/mol}$

It is not hard to picture the meaning of bond energy for a diatomic molecule, because there is only one bond in the molecule. It is also not difficult to see that the bond-dissociation energy of a diatomic molecule can be expressed rather precisely, as is that of $H_2(g)$. With a polyatomic molecule, such as H_2O, the situation is different (Fig. 10-16). The energy needed to dissociate one mole of H atoms by breaking one O—H bond per H_2O molecule,

$$H—OH(g) \longrightarrow H(g) + OH(g) \quad \Delta H = D(H—OH) = +498.7\ \text{kJ/mol}$$

is different from the energy required to dissociate one mole of H atoms by breaking the bonds in OH(g):

$$O—H(g) \longrightarrow H(g) + O(g) \quad \Delta H = D(O—H) = +428.0\ \text{kJ/mol}$$

The two O—H bonds in H_2O are identical; therefore, they should have identical energies. This energy, which we can call the O—H bond energy in H_2O, is the *average* of the two values listed above: 463.4 kJ/mol. The O—H bond energy in other molecules containing the OH group will be somewhat different from that in H—O—H. For example, in methanol, CH_3OH, the O—H bond-dissociation energy, which we can represent as $D(H—OCH_3)$, is 436.8 kJ/mol. The usual method of tabulating bond energies (Table 10.3) is as *averages*. An **average bond energy** is the average of bond-dissociation energies for a number of different species containing the particular bond. Understandably, average bond energies cannot be stated as precisely as specific bond-dissociation energies.

As you can see from Table 10.3, double bonds have higher bond energies than do single bonds between the same atoms, but they are *not* twice as large. Triple bonds are stronger still, but their bond energies are *not* three times as

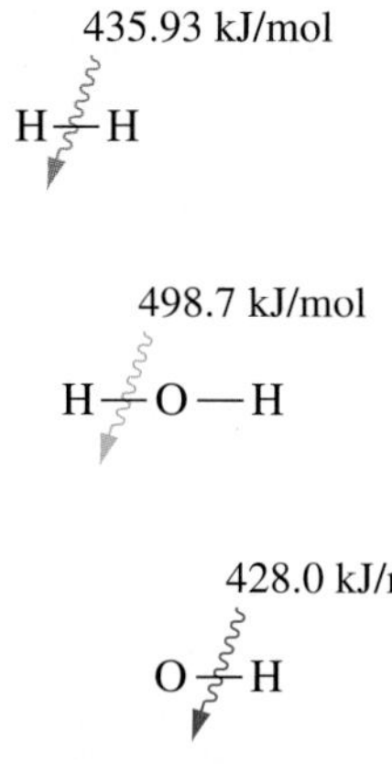

435.93 kJ/mol

H—H

498.7 kJ/mol

H—O—H

428.0 kJ/mol

O—H

▲ FIGURE 10-16
Some bond energies compared
The same quantity of energy, 435.93 kJ/mol, is required to break all H—H bonds. In H_2O, more energy is required to break the first bond (498.7 kJ/mol) than to break the second (428.0 kJ/mol). The second bond broken is that in the OH radical. The O—H bond energy in H_2O is the average of the two values: 463.4 kJ/mol.

KEEP IN MIND

that tabulated bond energies are for isolated molecules in the gaseous state. They do not apply to molecules in close contact in liquids and solids.

TABLE 10.3 Some Average Bond Energies[a]

Bond	Bond Energy, kJ/mol	Bond	Bond Energy, kJ/mol	Bond	Bond Energy, kJ/mol
H—H	436	C—C	347	N—N	163
H—C	414	C=C	611	N=N	418
H—N	389	C≡C	837	N≡N	946
H—O	464	C—N	305	N—O	222
H—S	368	C=N	615	N=O	590
H—F	565	C≡N	891	O—O	142
H—Cl	431	C—O	360	O=O	498
H—Br	364	C=O	736[b]	F—F	159
H—I	297	C—Cl	339	Cl—Cl	243
				Br—Br	193
				I—I	151

[a]Although all data are listed with about the same precision (three significant figures), some values are actually known more precisely. Specifically, the values for the diatomic molecules H_2, HF, HCl, HBr, HI, N_2 (N≡N), O_2 (O=O), F_2, Cl_2, Br_2, and I_2 are actually bond-dissociation energies, rather than average bond energies.
[b]The value for the C=O bonds in CO_2 is 799 kJ/mol.

large as single bonds between the same atoms. This observation about bond order and bond energy will seem quite reasonable after multiple bonds are more fully described in the next chapter.

Bond energies also have some interesting uses in thermochemistry. For a reaction involving *gases*, visualize the process

$$\text{gaseous reactants} \longrightarrow \text{gaseous atoms} \longrightarrow \text{gaseous products}$$

In this hypothetical process, we first break all the bonds in reactant molecules and form gaseous atoms. For this step, the enthalpy change is ΔH (bond breakage) $= \Sigma BE$ (reactants), where BE stands for bond energy. Next, we allow the gaseous atoms to recombine into product molecules. In this step, bonds are formed and ΔH (bond formation) $= -\Sigma BE$ (products). The enthalpy change of the reaction, then, is

KEEP IN MIND

that the difference in equation (10.26) is calculated as reactants minus products and not products minus reactants.

$$\Delta H = \Delta H \text{ (bond breakage)} + \Delta H \text{ (bond formation)}$$
$$\approx \Sigma BE \text{ (reactants)} - \Sigma BE \text{ (products)}$$

(10.26)

The approximately equal sign ($\approx$) in expression (10.26) signifies that some of the bond energies used are likely to be *average* bond energies rather than true bond-dissociation energies. Also, a number of terms often cancel out because some of the same types of bonds appear in the products as in the reactants. We can base the calculation of ΔH just on the *net* number and types of bonds broken and formed, as illustrated in Example 10-15.

EXAMPLE 10-15   Calculating an Enthalpy of Reaction from Bond Energies

The reaction of methane (CH_4) and chlorine produces a mixture of products called chloromethanes. One of these is monochloromethane, CH_3Cl, used in the preparation of silicones. Calculate ΔH for the reaction

$$CH_4(g) + Cl_2(g) \longrightarrow CH_3Cl(g) + HCl(g)$$

Analyze

To identify which bonds are broken and formed, it helps to draw structural formulas (or Lewis structures), as in Figure 10-17. To apply expression (10.26) literally, we would break *four* C—H bonds and *one* Cl—Cl bond and form *three* C—H bonds, *one* C—Cl bond, and *one* H—Cl bond. The *net* change, however, is the breaking of *one* C—H bond and *one* Cl—Cl bond, followed by the formation of *one* C—Cl bond and *one* H—Cl bond.

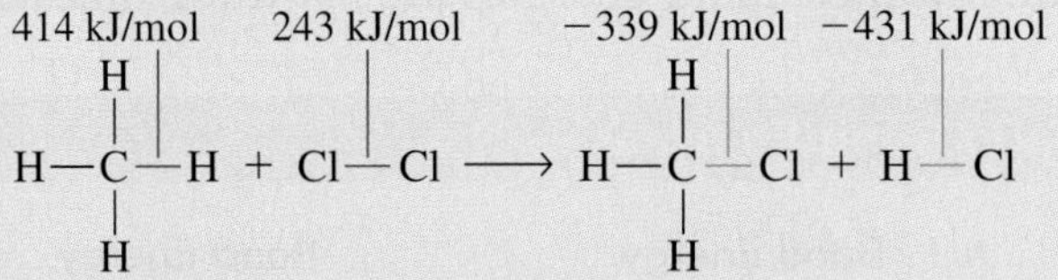

▲ FIGURE 10-17
Net bond breakage and formation in a chemical reaction—Example 10-15 illustrated
Bonds that are broken are shown in red and bonds that are formed, in blue. Bonds that remain unchanged are black. The net change is that *one* C—H and *one* Cl—Cl bond break and *one* C—Cl and *one* H—Cl bond form.

Solve

ΔH for net bond breakage:	1 mol C—H bonds	+414 kJ
	1 mol Cl—Cl bonds	+243 kJ
	sum:	+657 kJ
ΔH for net bond formation:	1 mol C—Cl bonds	−339 kJ
	1 mol H—Cl bonds	−431 kJ
	sum:	−770 kJ
Enthalpy of reaction:	$\Delta H = 657 - 770 = -113$ kJ	

Assess

A number of terms cancel out because some of the same types of bonds appear in both reactants and products. Such a situation is not uncommon.

PRACTICE EXAMPLE A: Use bond energies to estimate the enthalpy change for the reaction

$$2\,H_2(g) + O_2(g) \longrightarrow 2\,H_2O(g)$$

PRACTICE EXAMPLE B: Use bond energies to estimate the enthalpy of formation of $NH_3(g)$.

In Chapter 7, we learned how to calculate ΔH for any reaction using ΔH_f° values—see equation (7.21). Equation (10.26) gives us another way to calculate ΔH for a gas-phase reaction. There is no advantage to using bond energies over enthalpy-of-formation data. Enthalpies of formation are generally known rather precisely, whereas bond energies are only average values. But when enthalpy-of-formation data are lacking, bond energies can prove particularly useful.

Another way to use bond energies is in predicting whether a reaction will be *endothermic* or *exothermic*. In general, if

$$\underset{\text{(reactants)}}{\text{weak bonds}} \longrightarrow \underset{\text{(products)}}{\text{strong bonds}} \qquad \Delta H < 0 \qquad \text{(exothermic)}$$

and

$$\underset{\text{(reactants)}}{\text{strong bonds}} \longrightarrow \underset{\text{(products)}}{\text{weak bonds}} \qquad \Delta H > 0 \qquad \text{(endothermic)}$$

Example 10-16 applies this idea to a reaction involving highly reactive, unstable species for which enthalpies of formation are not normally listed.

EXAMPLE 10-16   Using Bond Energies to Predict Exothermic and Endothermic Reactions

One of the steps in the formation of monochloromethane (Example 10-15) is the reaction of a gaseous chlorine *atom* (a chlorine radical) with a molecule of methane. The products are an unstable methyl radical and $HCl(g)$. Is this reaction endothermic or exothermic?

$$CH_4 + \cdot Cl(g) \longrightarrow \cdot CH_3(g) + HCl(g)$$

Analyze

In the reaction, one $C-H$ bond is broken for every $H-Cl$ bond formed. Thus, we must compare the bond energies for the $C-H$ and $H-Cl$ bonds to decide whether the reaction is endothermic or exothermic.

Solve

For every molecule of CH_4 that reacts, *one* $C-H$ bond *breaks*, requiring 414 kJ per mole of bonds; and *one* $H-Cl$ bond *forms*, releasing 431 kJ per mole of bonds. Because more energy is released in forming new bonds than is absorbed in breaking old ones, we predict that the reaction is exothermic.

Assess

In the example above, we had to break only $C-H$ bonds and form only $H-Cl$ bonds. Most reactions involve breaking and forming several types of bonds, and so it is usually not obvious whether the reaction will be exothermic or endothermic. In such cases, we must calculate ΔH, by using equation (10.26), to see whether $\Delta H > 0$ or $\Delta H < 0$.

PRACTICE EXAMPLE A: Is the following reaction endothermic or exothermic?

$$CH_3COCH_3(g) + H_2(g) \longrightarrow (CH_3)_2CH(OH)(g)$$

PRACTICE EXAMPLE B: Predict whether the following reaction should be exothermic or endothermic:

$$H_2O(g) + Cl_2(g) \longrightarrow \frac{1}{2}O_2(g) + 2\,HCl(g).$$

Summary

10-1 Lewis Theory: An Overview

—A **Lewis symbol** represents the valence electrons of an atom by using dots placed around the chemical symbol. A **Lewis structure** is a combination of Lewis symbols used to represent chemical bonding. Normally, all the electrons in a Lewis structure are paired, and each atom in the structure acquires an **octet**—that is, there are eight electrons in the valence shell. In **Lewis theory**, chemical bonds are classified as **ionic bonds**, which are formed by electron transfer between atoms, or **covalent bonds**, which are formed by electrons shared between atoms. Most bonds, however, have partial ionic and partial covalent characteristics.

10-2 Covalent Bonding: An Introduction

—Atoms in molecules are often surrounded by eight valence-shell electrons (an octet) and thus conform to the **octet rule**. In covalent bonds, pairs of electrons shared between two atoms to form the bonds are called **bond pairs**, while pairs of electrons not shared in a chemical bond are called **lone pairs**. A single pair of electrons shared between two atoms constitutes a **single covalent bond**. When both electrons in a covalent bond between two atoms are provided by only one of the atoms, the bond is called a **coordinate covalent bond**. To construct a Lewis structure for a molecule in which all atoms obey the octet rule, it is often necessary for atoms to share more than one pair of electrons, thus forming **multiple covalent bonds**. Two shared electron pairs constitute a **double covalent bond** and three shared pairs a **triple covalent bond**.

10-3 Polar Covalent Bonds and Electrostatic Potential Maps

—A covalent bond in which the electron pair is not shared equally by the bonded atoms is called a **polar covalent bond**. Whether a bond is polar or not can be predicted by comparing the **electronegativity (EN)** of the atoms involved (Fig. 10-6). The greater the **electronegativity difference** (Δ**EN**) between two atoms in a chemical bond, the more polar the bond and the more ionic its character. The electron charge distribution in a molecule can be visualized by computing an **electrostatic potential map** (Figs. 10-4 and 10-5). The variation of charge in the molecule is represented by a color spectrum in which red is the most negative and blue the most positive. Electrostatic potential maps are a powerful way of representing electron charge distribution in both polar and nonpolar molecules.

10-4 Writing Lewis Structures

—To draw the Lewis structure of a covalent molecule, one needs to know the **skeletal structure**—that is, which is the **central atom** and what atoms are bonded to it. Atoms that are bonded to just one other atom are called **terminal atoms** (structure 10.13). Typically, the atom with the lowest electronegativity is a central atom. At times, the concept of **formal charge** (expression 10.16) is useful in selecting a skeletal structure and assessing the plausibility of a Lewis structure.

10-5 Resonance

—Often, more than one plausible Lewis structure can be written for a species; this situation is called **resonance**. In these cases the true structure is a resonance hybrid of two or more contributing structures.

10-6 Exceptions to the Octet Rule

—There are often exceptions to the octet rule. (1) Odd-electron species, such as NO, have an unpaired electron and are paramagnetic. Many of these species are reactive molecular fragments, such as OH, called **free radicals**. (2) A few molecules have incomplete octets in their Lewis structures, that is, not enough electrons to provide an octet for every atom. (3) **Expanded valence shells** occur in some compounds of nonmetals of the third period and beyond. In these, the valence shell of the central atom must be expanded to 10 or 12 electrons in order to write a Lewis structure.

10-7 Shapes of Molecules

—A powerful method for predicting the **molecular geometry**, or molecular shape, of a species is the **valence-shell electron-pair repulsion theory (VSEPR)**. The shape of a molecule or polyatomic ion depends on the geometric distribution of valence-shell electron groups—the **electron-group geometry**—and whether these groups contain bonding electrons or lone pairs (Fig. 10-11, Table 10.1). The angles between the electron groups provide a method for predicting **bond angles** in a molecule. An important use of information about the shapes of molecules is in establishing whether bonds in a molecule combine to produce a resultant **dipole moment** (Fig. 10-15). Molecules with a resultant dipole moment are **polar molecules**; those with no resultant dipole moment are nonpolar.

10-8 Bond Order and Bond Lengths

—Single, double, and triple covalent bonds are said to have a **bond order** of 1, 2, and 3, respectively. **Bond length** is the distance between the centers of two atoms joined by a covalent bond. The greater the bond order, the shorter the bond length (Table 10.2).

10-9 Bond Energies

—**Bond-dissociation energy**, D, is the quantity of energy required to break one mole of covalent bonds in a gaseous molecule. **Average bond energies** (Table 10.3) can be used to estimate enthalpy changes for reactions involving gases.

Integrative Example

Nitryl fluoride is a reactive gas useful in rocket propellants. Its mass percent composition is 21.55% N, 49.23% O, and 29.23% F. Its density is 2.7 g/L at 20 °C and 1.00 atm pressure. Describe the nitryl fluoride molecule as completely as possible—that is, its formula, Lewis structure, molecular shape, and polarity.

Analyze

First, determine the empirical formula of nitryl fluoride from the composition data and the molar mass based on that formula. Next, determine the true molar mass from the vapor density data. Now the two results can be compared to establish the molecular formula. Then, write a plausible Lewis structure based on the molecular formula, and apply VSEPR theory to the Lewis structure to predict the molecular shape. Finally, assess the polarity of the molecule from the molecular shape and electronegativity values.

Solve

To determine the empirical formula, use the method of Example 3-5 (page 80). In 100.0 g of the compound,

$$mol\ N = 21.55\ g\ N \times \frac{1\ mol\ N}{14.007\ g\ N} = 1.539\ mol\ N$$

$$mol\ O = 49.23\ g\ O \times \frac{1\ mol\ O}{15.999\ g\ O} = 3.077\ mol\ O$$

$$mol\ F = 29.23\ g\ F \times \frac{1\ mol\ F}{18.998\ g\ F} = 1.539\ mol\ F$$

The empirical formula is

$$N_{1.539}O_{3.077}F_{1.539} = NO_2F$$

The molar mass based on this formula is

$$14 + 32 + 19 = 65\ g/mol$$

Use the method of Example 6-7 (page 208) to determine the molar mass of the gas.

$$molar\ mass = \frac{mRT}{PV} = \frac{dRT}{P}$$

$$= \frac{2.7\ g/L \times 0.0821\ L\ atm\ mol^{-1}K^{-1} \times 293\ K}{1.00\ atm}$$

$$= 65\ g/mol$$

Because the two molar mass results are the same, the molecular and empirical formulas are the same: NO_2F.

Because N has the lowest electronegativity, it should be the lone central atom in the Lewis structure. There are two equivalent contributing structures to a resonance hybrid.

Three electron groups around the N atom produce a trigonal planar electron-group geometry. All the electron groups participate in bonding, so the molecular geometry is also trigonal planar. The predicted bond angles are 120°. (The experimentally determined F—N—O bond angle is 118°.)

The molecule has a symmetrical shape, but because the electronegativity of F is different from that of O, we should expect the electron charge distribution in the molecule to be nonsymmetrical, leading to a small resultant dipole moment. NO_2F is a polar molecule.

Assess

In the contributing structures to the resonance hybrid Lewis structure, the F atom has a formal charge of zero and the two O atoms have an average formal charge of $-\frac{1}{2}$. As a result, we might expect the oxygen-atom region of the molecule to be the most negative in electron charge density and the fluorine region to be more neutral. This conclusion is confirmed by the electrostatic potential map for NO_2F. Notice also the positive charge

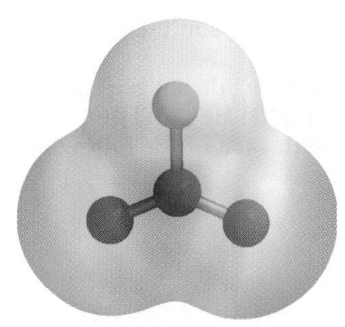

density around the nitrogen nucleus, which is in accord with the fact that the nitrogen is the least electronegative of the three elements.

PRACTICE EXAMPLE A: Phosphorus pentachloride, $PCl_5(g)$, can be made from the reaction of $PCl_3(g)$ and $Cl_2(g)$. By using only data from Appendix D, estimate the average bond energy for the P—Cl bond. Assume that the P—Cl bond energies are the same in PCl_3 and PCl_5. With reference to the geometries of PCl_3 and PCl_5, explain why the P—Cl bond energies are probably not the same in these two molecules.

PRACTICE EXAMPLE B: The condensed structural formulas of formamide and formaldoxime are H_2NCHO and H_2CNOH, respectively. One of these molecules is much more stable than the other. **(a)** Sketch the structures of these molecules, using dash and wedge symbols, indicate the various bond angles, and use bond energies to determine which molecule is more stable. **(b)** Experiment shows that the H—N—H angle in formamide is 119° and the N—C—O angle is 124°. Draw a Lewis structure for formamide that is consistent with this structural information.

Mastering**CHEMISTRY**

You'll find a link to additional self study questions in the study area on:
www.masteringchemistry.com

Exercises

Lewis Theory

1. Write Lewis symbols for the following atoms. **(a)** Kr; **(b)** Ge; **(c)** N; **(d)** Ga; **(e)** As; **(f)** Rb.
2. Write Lewis symbols for the following ions. **(a)** H^-; **(b)** Sn^{2+}; **(c)** K^+; **(d)** Br^-; **(e)** Se^{2-}; **(f)** Sc^{3+}.
3. Write plausible Lewis structures for the following molecules that contain only single covalent bonds. **(a)** FCl; **(b)** I_2; **(c)** SF_2; **(d)** NF_3; **(e)** H_2Te.
4. Each of the following molecules contains at least one multiple (double or triple) covalent bond. Give a plausible Lewis structure for **(a)** OCS; **(b)** CH_3CHO; **(c)** F_2CO; **(d)** Cl_2SO; **(e)** C_2H_2.
5. By means of Lewis structures, represent bonding between the following pairs of elements: **(a)** Cs and Br; **(b)** H and Sb; **(c)** B and Cl; **(d)** Cs and Cl; **(e)** Li and O; **(f)** Cl and I. Your structures should show whether the bonding is essentially ionic or covalent.
6. Which of the following have Lewis structures that *do not* obey the octet rule: NF_3, $AlCl_3$, SiF_6^{2-}, SO_3, PH_4^+, PO_4^{3-}, ClO_2?
7. Give several examples for which the following statement proves to be incorrect. "All atoms in a Lewis structure have an octet of electrons in their valence shells."
8. Suggest reasons why the following do not exist as stable molecules: **(a)** H_3; **(b)** HHe; **(c)** He_2; **(d)** H_3O.
9. Describe what is wrong with each of the following Lewis structures.

   (a) H—H—N̈—Ö—H

   (b) Ca—Ö̈:

10. Describe what is wrong with each of the following Lewis structures.

    (a) :Ö—C̈l—Ö:

    (b) [·C̈=N̈:]⁻

11. Only one of the following Lewis structures is correct. Select that one and indicate the errors in the others.

    (a) cyanate ion   [:Ö—C=N̈:]⁻

    (b) carbide ion   [C≡C:]²⁻

    (c) hypochlorite ion   [:C̈l—Ö:]⁻

    (d) nitrogen(II) oxide :N̈=Ö:

12. Indicate what is wrong with each of the following Lewis structures. Replace each one with a more acceptable structure.

    (a) Mg :Ö:

    (b) [:Ö—N=Ö:]⁺

    (c) [:C̈l]⁺[:Ö:]²⁻[C̈l:]⁺

    (d) [:S̈—C=N̈:]⁻

Ionic Bonding

13. Write Lewis structures for the following ionic compounds: **(a)** calcium chloride; **(b)** barium sulfide; **(c)** lithium oxide; **(d)** sodium fluoride.

14. Under appropriate conditions, both hydrogen and nitrogen can form monatomic anions. What are the Lewis symbols for these ions? What are the Lewis structures of the compounds **(a)** lithium hydride; **(b)** calcium hydride; **(c)** magnesium nitride?

15. Derive the correct formulas for the following ionic compounds by writing Lewis structures. **(a)** lithium sulfide; **(b)** sodium fluoride; **(c)** calcium iodide; **(d)** scandium chloride.

16. Each of the following ionic compounds consists of a combination of monatomic and polyatomic ions. Represent these compounds with Lewis structures. **(a)** $Al(OH)_3$; **(b)** $Ca(CN)_2$; **(c)** NH_4F; **(d)** $KClO_3$; **(e)** $Ba_3(PO_4)_2$.

Formal Charge

17. Assign formal charges to each of the atoms in the following structures.

    (a) $[H\!-\!C\!\equiv\!C\!:]^-$

    (b)

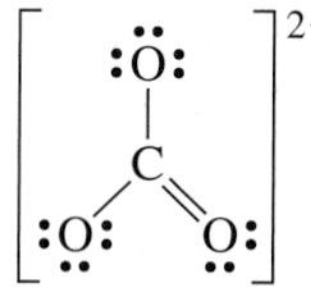

    (c) $[CH_3\!-\!CH\!-\!CH_3]^+$

18. Assign formal charges to each of the atoms in the following structures.

    (a)

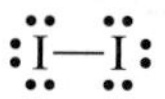

    (b)

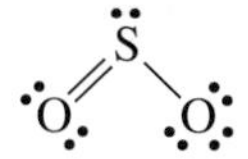

    (c)

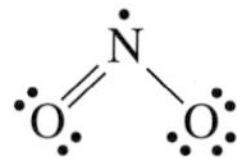

19. Both oxidation state and formal charge involve conventions for assigning valence electrons to bonded atoms in compounds, but clearly they are not the same. Describe several ways in which these concepts differ.

20. Although the notion that a Lewis structure in which formal charges are zero or held to a minimum seems to apply in most instances, describe several significant situations in which this appears not to be the case.

21. What is the formal charge of the indicated atom in each of the following structures?
    (a) the central O atom in O_3
    (b) Al in AlH_4^-
    (c) Cl in ClO_3^-
    (d) Si in SiF_6^{2-}
    (e) Cl in ClF_3

22. Assign formal charges to the atoms in the following species, and then select the more likely skeletal structure.
    (a) H_2NOH or H_2ONH
    (b) SCS or CSS
    (c) NFO or FNO
    (d) $SOCl_2$ or $OSCl_2$ or OCl_2S
    (e) F_3SN and F_3NS

23. The concept of formal charge helped us to choose the more plausible of the Lewis structures for NO_2^+ given in expressions (10.14) and (10.15). Can it similarly help us to choose a single Lewis structure as most plausible for CO_2H^+? Explain.

24. Show that the idea of minimizing the formal charges in a structure is at times in conflict with the observation that compact, symmetrical structures are more commonly observed than elongated ones with many central atoms. Use ClO_4^- as an illustrative example.

Lewis Structures

25. Write acceptable Lewis structures for the following molecules: **(a)** H_2NNH_2; **(b)** $HOClO$; **(c)** $(HO)_2SO$; **(d)** HOOH **(e)** SO_4^{2-}.

26. Two molecules that have the same formulas but different structures are said to be isomers. (In isomers, the same atoms are present but linked together in different ways.) Draw acceptable Lewis structures for *two* isomers of S_2F_2.

27. The following polyatomic anions involve covalent bonds between O atoms and the central nonmetal atom. Propose an acceptable Lewis structure for each.
    (a) SO_3^{2-}; **(b)** NO_2^-; **(c)** CO_3^{2-}; **(d)** HO_2^-.

28. Represent the following ionic compounds by Lewis structures: **(a)** barium hydroxide; **(b)** sodium nitrite; **(c)** magnesium iodate; **(d)** aluminum sulfate.

29. Write a plausible Lewis structure for crotonaldehyde, $CH_3CHCHCHO$, a substance used in tear gas and insecticides.

30. Write a plausible Lewis structure for C_3O_2, a substance known as carbon suboxide.

31. Write Lewis structures for the molecules represented by the following molecular models.

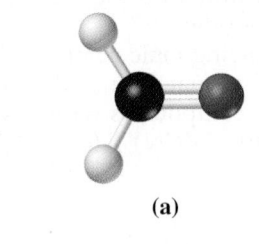

(a)

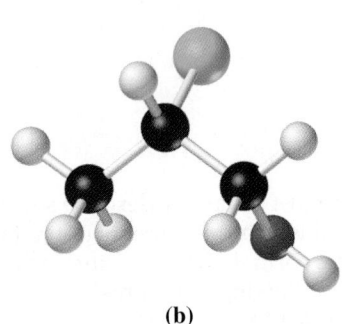

(b)

32. Write Lewis structures for the molecules represented by the following molecular models.

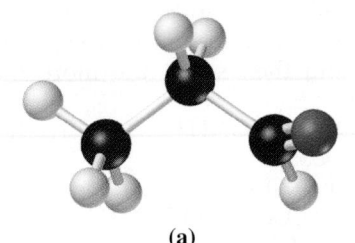

(a)

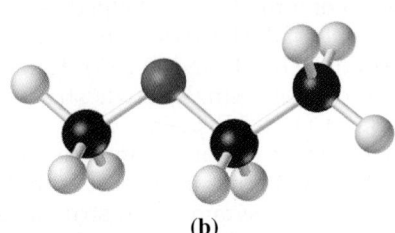

(b)

33. Write Lewis structures for the molecules represented by the following line-angle formulas. [*Hint:* Recall page 70 and Figure 3-2.]

(a)

Cl   O-H

(b) HO   O   OH

34. Write Lewis structures for the molecules represented by the following line-angle formulas. [*Hint:* Recall page 70 and Figure 3-2.]

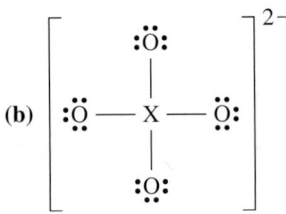

(a)

O

(b) Cl   -NH₂

35. Identify the main group that the element X belongs to in each of the following Lewis structures. For the types of molecule shown, give an example that exists.

(a) $\left[\ddot{\ddot{X}}-\ddot{\ddot{X}}\right]^{2-}$

(b) $\left[\begin{array}{c}:\ddot{O}: \\ | \\ :\ddot{O}-X-\ddot{O}: \\ | \\ :\ddot{O}: \end{array}\right]^{2-}$

(c) $\left[:\ddot{O}-\overset{:\ddot{O}:}{\underset{}{X}}-\ddot{O}:\right]^{-}$

(d) $\left[\begin{array}{c}H \\ | \\ H-X-H \\ | \\ H \end{array}\right]^{-}$

36. Identify the main group that the element X belongs to in each of the following Lewis structures. For the types of molecule shown, give an example that exists.

(a) $\ddot{O}=X=\ddot{O}$

(b) $\left[\ddot{O}=X-\ddot{O}:\right]^{-}$

(c) 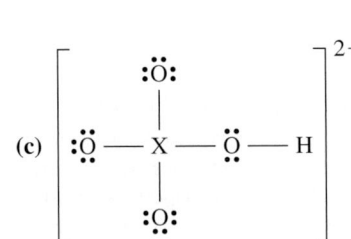 $\left[:\ddot{O}-\overset{:\ddot{O}:}{\underset{:\ddot{O}:}{X}}-\ddot{O}-H\right]^{2-}$

(d) 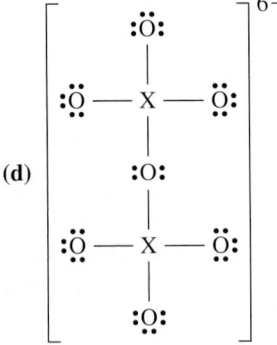 $\left[\begin{array}{c}:\ddot{O}: \\ | \\ :\ddot{O}-X-\ddot{O}: \\ | \\ :O: \\ | \\ :\ddot{O}-X-\ddot{O}: \\ | \\ :O: \end{array}\right]^{6-}$

Polar Covalent Bonds and Electrostatic Potential Maps

37. Use your knowledge of electronegativities, but *do not* refer to tables or figures in the text, to arrange the following bonds in terms of *increasing* ionic character: C—H, F—H, Na—Cl, Br—H, K—F.

38. Which of the following molecules would you expect to have a resultant dipole moment (μ); **(a)** F₂, **(b)** NO₂, **(c)** BF₃, **(d)** HBr, **(e)** H₂CCl₂, **(f)** SiF₄, **(g)** OCS? Explain.

39. What is the percent ionic character of each of the following bonds: **(a)** S—H; **(b)** O—Cl; **(c)** Al—O; **(d)** As—O?

40. Plot the data of Figure 10-6 as a function of atomic number. Does the property of electronegativity conform to the periodic law? Do you think it should?

41. Use a cross-base arrow ($\longleftrightarrow$) to represent the polarity of the bond in each of the following diatomic molecules. Then use the data below to calculate, in the manner described on page 431, the partial charges (δ) on the atoms in each molecule. Express the partial charges as a decimal fraction of the elementary charge, e = 1.602×10^{-19} C, for example $\delta = +0.17e$ or $\delta = -0.17e$.

	Bond Length, pm	Dipole Moment, D
ClF	162.8	0.8881
RbF	227.0	8.547
SnO	183.3	4.3210
BaO	194.0	7.954

42. Use a cross-base arrow ($\longleftrightarrow$) to represent the polarity of the bond in each of the following diatomic molecules. Then use the data below to calculate the partial charges (δ) on the atoms in each molecule. Express the partial charges in the manner described in Exercise 41.

	Bond Length, pm	Dipole Moment, D
OH	98.0	1.66
CH	131.1	1.46
CN	117.5	1.45
CS	194.4	1.96

43. Which electrostatic potential map corresponds to $F_2C=O$, and which to $H_2C=O$?

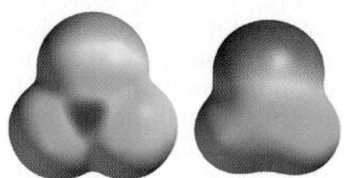

44. Match the correct electrostatic potential map corresponding to HOCl, FOCl, and HOF.

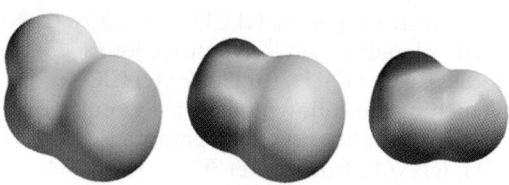

45. Two electrostatic potential maps are shown, one corresponding to a molecule containing only S and F, the other Si and F. Match them. What are the molecular formulas of the compounds?

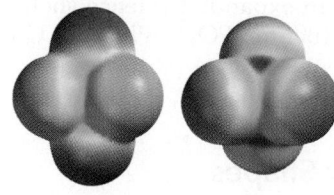

46. Two electrostatic potential maps are shown, one corresponding to a molecule containing only Cl and F, the other P and F. Match them. What are the molecular formulas of the compounds?

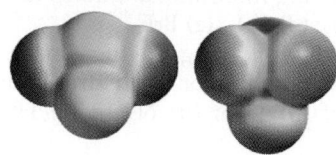

Resonance

47. Through appropriate Lewis structures, show that the phenomenon of resonance is involved in the nitrite ion.
48. Which of the following species requires a resonance hybrid for its Lewis structure: (a) CO_2, (b) OCl^-, (c) CO_3^{2-}, or (d) OH^-? Explain.
49. Dinitrogen oxide (nitrous oxide, or "laughing gas") is sometimes used as an anesthetic. Here are some data about the N_2O molecule: N—N bond length = 113 pm; N—O bond length = 119 pm. Use these data and other information from the chapter to comment on the plausibility of each of the following Lewis structures shown. Are they all valid? Which ones do you think contribute most to the resonance hybrid?

$$:N\equiv N-\ddot{\underset{..}{O}}: \qquad :\ddot{N}=N=\ddot{\underset{..}{O}}:$$
$$\textbf{(a)} \qquad\qquad\qquad \textbf{(b)}$$

$$:\ddot{\underset{..}{N}}-N\equiv O: \qquad :\ddot{N}=O=\ddot{N}:$$
$$\textbf{(c)} \qquad\qquad\qquad \textbf{(d)}$$

50. The Lewis structure of nitric acid, $HONO_2$, is a resonance hybrid. How important do you think the

contribution of the following structure is to the resonance hybrid? Explain.

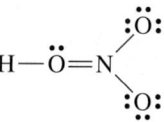

$$H-\ddot{\underset{..}{O}}=N\begin{matrix}:\ddot{O}:\\ \diagup\\ \diagdown\\ :\ddot{\underset{..}{O}}:\end{matrix}$$

51. Draw Lewis structures for the following species, indicating formal charges and resonance where applicable:
    (a) HCO_2^-
    (b) HCO_3^-
    (c) FSO_3^-
    (d) $N_2O_3^{2-}$ (the nitrogen atoms are joined centrally with one oxygen atom on one N and two on the other)
52. Draw Lewis structures for the following species, indicating formal charges and resonance where applicable:
    (a) $HOSO_3^-$
    (b) H_2NCN
    (c) FCO_2^-
    (d) S_2N_2 (a cyclic structure with S and N alternating)

Odd-Electron Species

53. Write plausible Lewis structures for the following odd-electron species: **(a)** CH_3; **(b)** ClO_2; **(c)** NO_3.
54. Write plausible Lewis structures for the following free radicals: **(a)** $\cdot C_2H_5$; **(b)** $HO_2 \cdot$; **(c)** $ClO \cdot$.
55. Which of the following species would you expect to be diamagnetic and which paramagnetic: **(a)** OH^-; **(b)** OH; **(c)** NO_3; **(d)** SO_3; **(e)** SO_3^{2-}; **(f)** HO_2?

56. Write a plausible Lewis structure for NO_2, and indicate whether the molecule is diamagnetic or paramagnetic. Two NO_2 molecules can join together (*dimerize*) to form N_2O_4. Write a plausible Lewis structure for N_2O_4, and comment on the magnetic properties of the molecule.

Expanded Valence Shells

57. In which of the following species is it *necessary* to employ an expanded valence shell to represent the Lewis structure: PO_4^{3-}, PI_3, ICl_3, $OSCl_2$, SF_4, ClO_4^-? Explain your choices.

58. Describe the carbon-to-sulfur bond in H_2CSF_4. That is, is it most likely a single, double, or triple bond?

Molecular Shapes

59. Use VSEPR theory to predict the geometric shapes of the following molecules and ions: **(a)** N_2; **(b)** HCN; **(c)** NH_4^+; **(d)** NO_3^-; **(e)** NSF.
60. Use VSEPR theory to predict the geometric shapes of the following molecules and ions: **(a)** PCl_3; **(b)** SO_4^{2-}; **(c)** $SOCl_2$; **(d)** SO_3; **(e)** BrF_4^+.
61. Each of the following is either linear, angular (bent), planar, tetrahedral, or octahedral. Indicate the correct shape of **(a)** H_2S; **(b)** N_2O_4; **(c)** HCN; **(d)** $SbCl_6^-$; **(e)** BF_4^-.
62. Predict the geometric shapes of **(a)** CO; **(b)** $SiCl_4$; **(c)** PH_3; **(d)** ICl_3; **(e)** $SbCl_5$; **(f)** SO_2; **(g)** AlF_6^{3-}.
63. One of the following ions has a *trigonal-planar* shape: SO_3^{2-}; PO_4^{3-}; PF_6^-; CO_3^{2-}. Which ion is it? Explain.
64. Two of the following have the same shape. Which two, and what is their shape? What are the shapes of the other two? NI_3, HCN, SO_3^{2-}, NO_3^-.
65. Each of the following molecules contains one or more multiple covalent bonds. Draw plausible Lewis structures to represent this fact, and predict the shape of each molecule. **(a)** CO_2; **(b)** Cl_2CO; **(c)** $ClNO_2$.
66. Sketch the probable geometric shape of a molecule of **(a)** N_2O_4 (O_2NNO_2); **(b)** C_2N_2 ($NCCN$); **(c)** C_2H_6 (H_3CCH_3); **(d)** C_2H_6O (H_3COCH_3).
67. Use the VSEPR theory to predict the shapes of the anions **(a)** ClO_4^-; **(b)** $S_2O_3^{2-}$ (that is, SSO_3^{2-}); **(c)** PF_6^-; **(d)** I_3^-.

68. Use the VSEPR theory to predict the shape of **(a)** the molecule OSF_2; **(b)** the molecule O_2SF_2; **(c)** the ion SF_5^-; **(d)** the ion ClO_4^-; **(e)** the ion ClO_3^-.
69. The molecular shape of BF_3 is planar (see Table 10.1). If a fluoride ion is attached to the B atom of BF_3 through a coordinate covalent bond, the ion BF_4^- results. What is the shape of this ion?
70. Explain why it is not necessary to find the Lewis structure with the smallest formal charges to make a successful prediction of molecular geometry in the VSEPR theory. For example, write Lewis structures for SO_2 having different formal charges, and predict the molecular geometry based on these structures.
71. Comment on the similarities and differences in the molecular structure of the following triatomic species: CO_2, NO_2^-, O_3, and ClO_2^-.
72. Comment on the similarities and differences in the molecular structure of the following four-atom species: NO_3^-, CO_3^{2-}, SO_3^{2-}, and ClO_3^-.
73. Draw a plausible Lewis structure for the following series of molecules and ions: **(a)** ClF_2^-; **(b)** ClF_3; **(c)** ClF_4^-; **(d)** ClF_5. Describe the electron group geometry and molecular structure of these species.
74. Draw a plausible Lewis structure for the following series of molecules and ions: **(a)** SiF_6^{2-}; **(b)** PF_5; **(c)** SF_4; **(d)** XeF_4. Describe the electron group geometry and molecular structure of these species.

Shapes of Molecules with More Than One Central Atom

75. Sketch the propyne molecule, $CH_3C{\equiv}CH$. Indicate the bond angles in this molecule. What is the maximum number of atoms that can be in the same plane?
76. Sketch the propene molecule, $CH_3CH{=}CH_2$. Indicate the bond angles in this molecule. What is the maximum number of atoms that can be in the same plane?
77. Lactic acid has the formula $CH_3CH(OH)COOH$. Sketch the lactic acid molecule, and indicate the various bond angles.

78. Levulinic acid has the formula $CH_3(CO)CH_2CH_2COOH$. Sketch the levulinic acid molecule, and indicate the various bond angles.
79. Sketch, by using the dash and wedge symbolism, the H_2NCH_2CHO molecule, and indicate the various bond angles.
80. One of the isomers of chloromethanol has the formula $ClCH_2OH$. Sketch, by using the dash and wedge symbolism, this isomer of chloromethanol, and indicate the various bond angles.

Polar Molecules

81. Predict the shapes of the following molecules, and then predict which would have resultant dipole moments: (a) SO_2; (b) NH_3; (c) H_2S; (d) C_2H_4; (e) SF_6; (f) CH_2Cl_2.

82. Which of the following molecules would you expect to be polar: (a) HCN; (b) SO_3; (c) CS_2; (d) OCS; (e) $SOCl_2$; (f) SiF_4; (g) POF_3? Give reasons for your conclusions.

83. The molecule H_2O_2 has a resultant dipole moment of 2.2 D. Can this molecule be linear? If not, describe a shape that might account for this dipole moment.

84. Refer to the Integrative Example. A compound related to nitryl fluoride is nitrosyl fluoride, FNO. For this molecule, indicate (a) a plausible Lewis structure and (b) the geometric shape. (c) Explain why the measured resultant dipole moment for FNO is larger than the value for FNO_2.

Bond Lengths

85. *Without* referring to tables in the text, indicate which of the following bonds you would expect to have the greatest bond length, and give your reasons. (a) O_2; (b) N_2; (c) Br_2; (d) BrCl.

86. Estimate the lengths of the following bonds and indicate whether your estimate is likely to be too high or too low: (a) I—Cl; (b) C—F.

87. A relationship between bond lengths and single covalent radii of atoms is given on page 443. Use this relationship together with appropriate data from Table 10.2 to estimate these single-bond lengths. (a) I—Cl; (b) O—Cl; (c) C—F; (d) C—Br.

88. In which of the following molecules would you expect the oxygen-to-oxygen bond to be the *shortest*: (a) H_2O_2, (b) O_2, (c) O_3? Explain.

89. Refer to the Integrative Example. Use data from the chapter to estimate the length of the N—F bond in FNO_2.

90. Write a Lewis structure of the hydroxylamine molecule, H_2NOH. Then, with data from Table 10.2, determine all the bond lengths.

Bond Energies

91. A reaction involved in the formation of ozone in the upper atmosphere is $O_2 \longrightarrow 2\,O$. *Without* referring to Table 10.3, indicate whether this reaction is endothermic or exothermic. Explain.

92. Use data from Table 10.3, but *without performing detailed calculations*, determine whether each of the following reactions is exothermic or endothermic.
    (a) $CH_4(g) + I(g) \longrightarrow \cdot CH_3(g) + HI(g)$
    (b) $H_2(g) + I_2(g) \longrightarrow 2\,HI(g)$

93. Use data from Table 10.3 to estimate the enthalpy change (ΔH) for the following reaction.

$$C_2H_6(g) + Cl_2(g) \longrightarrow C_2H_5Cl(g) + HCl(g)$$
$$\Delta H = ?$$

94. One of the chemical reactions that occurs in the formation of photochemical smog is $O_3 + NO \longrightarrow NO_2 + O_2$. Estimate the enthalpy change of this reaction by using appropriate Lewis structures and data from Table 10.3.

95. Estimate the standard enthalpies of formation at 25 °C and 1 bar of (a) OH(g); (b) $N_2H_4(g)$. Write Lewis structures and use data from Table 10.3, as necessary.

96. Use ΔH for the reaction in Example 10-15 and other data from Appendix D to estimate $\Delta H_f^\circ[CH_3Cl(g)]$.

97. Use bond energies from Table 10.3 to estimate the enthalpy change (ΔH) for the following reaction.

$$C_2H_2(g) + H_2(g) \longrightarrow C_2H_4(g) \qquad \Delta H = ?$$

98. Equations (1) and (2) can be combined to yield the equation for the formation of $CH_4(g)$ from its elements.

(1)	$C(s)$	$\longrightarrow C(g)$	$\Delta H = 717\,kJ$
(2)	$C(g) + 2\,H_2(g)$	$\longrightarrow CH_4(g)$	$\Delta H = ?$
Overall:	$C(s) + 2\,H_2(g)$	$\longrightarrow CH_4(g)$	$\Delta H_f^\circ = -75$ kJ/mol

Use the preceding data and a bond energy of 436 kJ/mol for H_2 to estimate the C—H bond energy. Compare your result with the value listed in Table 10.3.

99. One reaction involved in the sequence of reactions leading to the destruction of ozone is

$$NO_2(g) + O(g) \longrightarrow NO(g) + O_2(g)$$

Calculate ΔH° for this reaction by using the thermodynamic data in Appendix D. Use your ΔH° value, plus data from Table 10.3, to estimate the nitrogen–oxygen bond energy in NO_2. [*Hint:* The structure of nitrogen dioxide, NO_2, is best represented as a resonance hybrid of two equivalent Lewis structures.]

100. A reaction involved in the sequence of reactions leading to the destruction of ozone is

$$O_3(g) + O(g) \longrightarrow 2\,O_2(g) \qquad \Delta H^\circ = -394\,kJ$$

Estimate the oxygen-oxygen bond energy in ozone by using the oxygen–oxygen bond energy in dioxygen from Table 10.3. Compare this value with the O—O and O=O bond energies in Table 10.3. How could you explain any differences?

Integrative and Advanced Exercises

101. Given the bond-dissociation energies: nitrogen-to-oxygen bond in NO, 631 kJ/mol; H—H in H_2, 436 kJ/mol; N—H in NH_3, 389 kJ/mol; O—H in H_2O, 463 kJ/mol; calculate ΔH for the reaction.

$$2\,NO(g) + 5\,H_2(g) \longrightarrow 2\,NH_3(g) + 2\,H_2O(g)$$

102. The following statements are not made as carefully as they might be. Criticize each one.
    (a) Lewis structures with formal charges are incorrect.
    (b) Triatomic molecules have a planar shape.
    (c) Molecules in which there is an electronegativity difference between the bonded atoms are polar.

103. A compound consists of 47.5% S and 52.5% Cl, by mass. Write a Lewis structure based on the empirical formula of this compound, and comment on its deficiencies. Write a more plausible structure with the *same* ratio of S to Cl.

104. A 0.325 g sample of a gaseous hydrocarbon occupies a volume of 193 mL at 749 mmHg and 26.1 °C. Determine the molecular mass, and write a plausible condensed structural formula for this hydrocarbon.

105. A 1.24 g sample of a hydrocarbon, when completely burned in an excess of $O_2(g)$, yields 4.04 g CO_2 and 1.24 g H_2O. Draw a plausible structural formula for the hydrocarbon molecule. [*Hint:* There is more than one possible arrangement of the C and H atoms.]

106. Draw Lewis structures for two different molecules with the formula C_3H_4. Is either of these molecules linear? Explain.

107. Sodium azide, NaN_3, is the nitrogen gas-forming substance used in automobile air-bag systems. It is an ionic compound containing the azide ion, N_3^-. In this ion, the two nitrogen-to-nitrogen bond lengths are 116 pm. Describe the resonance hybrid Lewis structure of this ion.

108. Use the bond-dissociation energies of $N_2(g)$ and $O_2(g)$ in Table 10.3, together with data from Appendix D, to estimate the bond-dissociation energy of NO(g).

109. Hydrogen azide, HN_3, is a liquid that explodes violently when subjected to physical shock. In the HN_3 molecule, one nitrogen-to-nitrogen bond length is 113 pm, and the other is 124 pm. The H—N—N bond angle is 112°. Draw Lewis structures and a sketch of the molecule consistent with these facts.

110. A few years ago the synthesis of a salt containing the N_5^+ ion was reported. What is the likely shape of this ion—linear, bent, zigzag, tetrahedral, seesaw, or square-planar? Explain your choice.

111. Carbon suboxide has the formula C_3O_2. The carbon-to-carbon bond lengths are 130 pm and carbon-to-oxygen, 120 pm. Propose a plausible Lewis structure to account for these bond lengths, and predict the shape of the molecule.

112. In certain polar solvents, PCl_5 undergoes an ionization reaction in which a Cl^- ion leaves one PCl_5 molecule and attaches itself to another. The products of the ionization are PCl_4^+ and PCl_6^-. Draw a sketch showing the changes in geometric shapes that occur in this ionization (that is, give the shapes of PCl_5, PCl_4^+, and PCl_6^-).

$$2\,PCl_5 \rightleftharpoons PCl_4^+ + PCl_6^-$$

113. Estimate the enthalpy of formation of HCN using bond energies from Table 10.3, data from elsewhere in the text, and the reaction scheme outlined as follows.

(1) $C(s)$ → $C(g)$ $\Delta H° = ?$
(2) $C(g) + \frac{1}{2}N_2(g) + \frac{1}{2}H_2(g)$ → $HCN(g)$ $\Delta H° = ?$

Overall: $C(g) + \frac{1}{2}N_2(g) + \frac{1}{2}H_2(g)$ → $HCN(g)$ $\Delta H_f° = ?$

114. The enthalpy of formation of $H_2O_2(g)$ is -136 kJ/mol. Use this value, with other appropriate data from the text, to estimate the oxygen-to-oxygen single-bond energy. Compare your result with the value listed in Table 10.3.

115. Use the VSEPR theory to predict a probable shape of the molecule F_4SCH_2, and explain the source of any ambiguities in your prediction.

116. The enthalpy of formation of methanethiol, $CH_3SH(g)$, is -22.9 kJ/mol. Methanethiol can be synthesized by the reaction of gaseous methanol and $H_2S(g)$. Water vapor is another product. Use this information and data from elsewhere in the text to estimate the carbon-to-sulfur bond energy in methanethiol.

117. For LiBr, the dipole moment (measured in the gas phase) and the bond length (measured in the solid state) are 7.268 D and 217 pm, respectively. For NaCl, the corresponding values are 9.001 D and 236.1 pm. (a) Calculate the percent ionic character for each bond. (b) Compare these values with the expected ionic character based on differences in electronegativity (see Figure 10-7). (c) Account for any differences in the values obtained in these two different ways.

118. One possibility for the electron-group geometry for *seven* electron groups is pentagonal-bipyramidal, as found in the IF_7 molecule. Write the VSEPR notation for this molecule. Sketch the structure of the molecule, labeling all the bond angles.

119. The extent to which an acid (HA) dissociates in water depends upon the stability of the anion (A^-); the more stable the anion, the more extensive is the dissociation of the acid. The anion is most stable when the negative charge is distributed over the whole anion rather than localized at one particular atom. Consider the following acids: acetic acid, fluoroacetic acid, cyanoacetic acid, and nitroacetic acid. Draw Lewis structures for their anions, including contributing resonance structures. Rank the acids in order of increasing extent of dissociation. Electrostatic potential maps for the four anions are provided on the next page. Identify which map corresponds to which anion, and discuss whether the maps confirm conclusions based on Lewis structures.

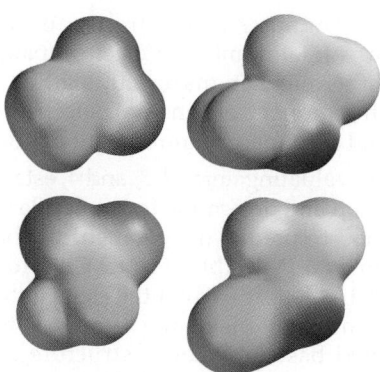

(d) $:\!\overset{..}{\underset{..}{Cl}}\!-\!\overset{..}{S}\!\equiv\!S\!-\!\overset{..}{\underset{..}{Cl}}\!:$

(e) $:\!\overset{..}{\underset{..}{Cl}}\!-\!\overset{..}{\underset{..}{S}}\!-\!\overset{..}{\underset{..}{S}}\!-\!\overset{..}{\underset{..}{Cl}}\!:$

120. R. S. Mulliken proposed that the electronegativity (χ) of an atom is given by

$$\chi = k \times (I - EA)$$

where I and EA are the ionization energy and electron affinity of the atom, respectively. Using the electron affinities and ionization energy values for the halogen atoms up to iodine, estimate the value of k by employing the electronegativity values in Figure 10-6. Estimate the electron affinity of At.

121. When molten sulfur reacts with chlorine gas, a vile-smelling orange liquid forms. When analyzed, the liquid compound has the empirical formula SCl. Several possible Lewis structures are shown below. Criticize these structures and choose the best one.

(a) $:\!\overset{..}{\underset{..}{Cl}}\!-\!\overset{..}{S}\!=\!\overset{..}{S}\!-\!\overset{..}{\underset{..}{Cl}}\!:$

(b) $:\!\overset{..}{\underset{..}{Cl}}\!=\!S\!-\!\overset{..}{S}\!=\!\overset{..}{\underset{..}{Cl}}\!:$

(c) $:\!\overset{..}{\underset{..}{S}}\!=\!\overset{..}{\underset{..}{Cl}}\!-\!\overset{..}{\underset{..}{Cl}}\!=\!\overset{..}{\underset{..}{S}}\!:$

122. Hydrogen azide, HN_3, can exist in two forms. One form has the three nitrogen atoms connected in a line; and the nitrogen atoms form a triangle in the other. Construct Lewis structures for these isomers and describe their shapes. Other interesting derivatives are nitrosyl azide (N_4O) and trifluoromethyl azide (CF_3N_3). Describe the shapes of these molecules based on a line of nitrogen atoms.

123. A pair of isoelectronic species for C and N exist with the formula X_2O_4 in which there is an X—X bond. A corresponding fluoride of boron also exists. Draw Lewis structures for these species and describe their shapes.

124. Acetone $(CH_3)C\!=\!O$, a ketone, will react with a strong base (A^-) to produce the enolate anion, $CH_3(C\!=\!O)CH_2^-$. Draw the Lewis structure of the enolate anion, and describe the relative contributions of any resonance structures.

125. The species PBr_4^- has been synthesized and has been described as a tetrahedral anion. Comment on this description.

126. One of the allotropes of sulfur is a ring of eight sulfur atoms. Draw the Lewis structure for the S_8 ring. Is the ring likely to be planar? The S_8 ring can be oxidized to produce S_8O. In S_8O, the oxygen atom is bonded to one of the S atoms and the S_8 ring is still intact. Draw the Lewis structure for S_8O.

127. One of the allotropes of phosphorus consists of four phosphorus atoms at the corners of a tetrahedron. Draw a Lewis structure for this allotrope that satisfies the octet rule. The P_4 molecule can be oxidized to P_4O_6, where the oxygen atoms insert between the phosphorus atoms. Draw the Lewis structure of this oxide. Are the P—O—P bonds linear?

Feature Problems

128. Pauling's reasoning in establishing his original electronegativity scale went something like this: If we assume that the bond A—B is nonpolar, its bond energy is the average of the bond energies of A—A and B—B. The *difference* between the calculated and measured bond energies of the bond A—B is attributable to the partial ionic character of the bond and is called the *ionic resonance energy* (IRE). If the IRE is expressed in kilojoules per mole, the relationship between IRE and the electronegativity difference is $(\Delta EN)^2 = IRE/96$. To test this basis for an electronegativity scale,
    (a) Use data from Table 10.3 to determine IRE for the H—Cl bond.
    (b) Determine ΔEN for the H—Cl bond.

    (c) Establish the approximate percent ionic character in the H—Cl bond by using the result of part (b) and Figure 10-7. Compare this result with that obtained in Example 10-4.

129. On page 431, the bond angle in the H_2O molecule is given as 104° and the resultant dipole moment as $\mu = 1.84$ D.
    (a) By an appropriate geometric calculation, determine the value of the H—O bond dipole in H_2O.
    (b) Use the same method as in part (a) to estimate the bond angle in H_2S, given that the H—S bond dipole is 0.67 D and that the resultant dipole moment is $\mu = 0.93$ D.
    (c) Refer to Figure 10-15. Given the bond dipoles 1.87 D for the C—Cl bond and 0.30 D for the

C—H bond, together with $\mu = 1.04$ D, estimate the H—C—Cl bond angle in $CHCl_3$.

130. Alternative strategies to the one used in this chapter have been proposed for applying the VSEPR theory to molecules or ions with a single central atom. In general, these strategies do not require writing Lewis structures. In one strategy, we write

(1) the total number of electron pairs = [(number of valence electrons) ± (electrons required for ionic charge)]/2
(2) the number of bonding electron pairs = (number of atoms) − 1

(3) the number of electron pairs around central atom = total number of electron pairs − 3 × [number of terminal atoms (excluding H)]
(4) the number of lone-pair electrons = number of central atom pairs − number of bonding pairs

After evaluating items 2, 3, and 4, establish the VSEPR notation and determine the molecular shape. Use this method to predict the geometrical shapes of the following: **(a)** PCl_5; **(b)** NH_3; **(c)** ClF_3; **(d)** SO_2; **(e)** ClF_4^-; **(f)** PCl_4^+. Justify each of the steps in the strategy, and explain why it yields the same results as the VSEPR method based on Lewis structures. How does the strategy deal with multiple bonds?

Self-Assessment Exercises

131. In your own words, define the following terms: **(a)** valence electrons; **(b)** electronegativity; **(c)** bond-dissociation energy; **(d)** double covalent bond; **(e)** coordinate covalent bond.
132. Briefly describe each of the following ideas: **(a)** formal charge; **(b)** resonance; **(c)** expanded valence shell; **(d)** bond energy.
133. Explain the important distinctions between **(a)** ionic and covalent bonds; **(b)** lone-pair and bond-pair electrons; **(c)** molecular geometry and electron-group geometry; **(d)** bond dipole and resultant dipole moment; **(e)** polar molecule and nonpolar molecule.
134. Of the following species, the one with a triple covalent bond is **(a)** NO_3^-; **(b)** CN^-; **(c)** CO_2; **(d)** $AlCl_3$.
135. The formal charges on the O atoms in the ion $[ONO]^+$ is **(a)** −2; **(b)** −1; **(c)** 0; **(d)** +1.
136. Which molecule is nonlinear? **(a)** SO_2; **(b)** CO_2; **(c)** HCN; **(d)** NO.
137. Which molecule is nonpolar? **(a)** SO_3; **(b)** CH_2Cl_2; **(c)** NH_3; **(d)** FNO.
138. The highest bond-dissociation energy is found in **(a)** O_2; **(b)** N_2; **(c)** Cl_2; **(d)** I_2.
139. The greatest bond length is found in **(a)** O_2; **(b)** N_2; **(c)** Br_2; **(d)** BrCl.
140. Draw plausible Lewis structures for the following species; use expanded valence shells where necessary. **(a)** Cl_2O; **(b)** PF_3; **(c)** CO_3^{2-}; **(d)** BrF_5.
141. Predict the shapes of the following sulfur-containing species. **(a)** SO_2; **(b)** SO_3^{2-}; **(c)** SO_4^{2-}.

142. Without referring to tables or figures in the text other than the periodic table, indicate which of the following atoms, Bi, S, Ba, As, or Mg, has the intermediate value when they are arranged in order of increasing electronegativity.
143. Use data from Tables 10.2 and 10.3 to determine for each bond in this following structure **(a)** the bond length and **(b)** the bond energy.

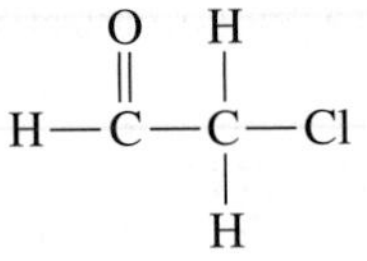

$$
\begin{array}{ccc}
\text{O} & \text{H} & \\
\parallel & \mid & \\
\text{H}-\text{C}-\text{C}-\text{Cl} \\
\mid & \\
\text{H} &
\end{array}
$$

144. What is the VSEPR theory? On what physical basis is the VSEPR theory founded?
145. Use the NH_3 molecule as an example to explain the difference between molecular geometry and electron-group geometry.
146. If you have four electron pairs around a central atom, under what circumstances can you have a pyramidal molecule? Similarly, how can you have a bent molecule? What are the expected bond angles in each case?
147. Draw three resonance structures for the sulfine molecule, H_2CSO. Do not consider ring structures.
148. Construct a concept map illustrating the connections between Lewis dot structures, the shapes of molecules, and polarity.

Chemical Bonding II: Additional Aspects

11

CONTENTS

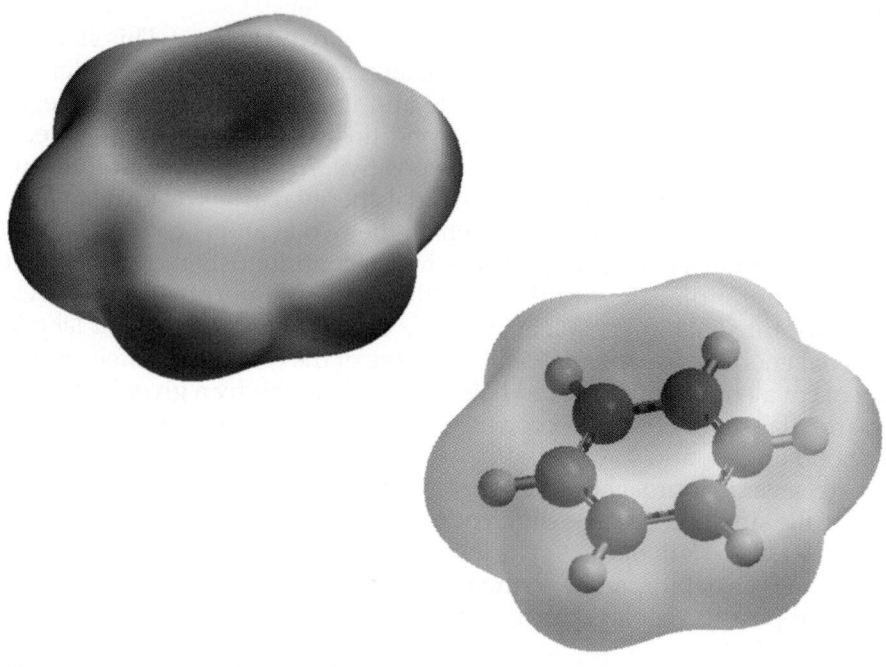

Electrostatic potential maps of benzene (one solid and one transparent) showing the negative charge density caused by the π molecular orbitals of benzene.

Although the Lewis theory has been useful in our discussion of chemical bonding, it does have shortcomings. For example, it does not help explain why metals conduct electricity or how a semiconductor works. Although we will continue to use the Lewis theory for most purposes, some cases require more sophisticated approaches. One such approach involves the familiar *s*, *p*, and *d* atomic orbitals, or mixed-orbital types called *hybrid orbitals*. A second approach involves the creation of a set of orbitals that belongs to a molecule as a whole. Electrons are then assigned to these *molecular orbitals*.

Our purpose in this chapter is not to try to master theories of covalent bonding in all their details. We want simply to discover how these theories provide models that yield deeper insights into the nature of chemical bonding than do Lewis structures alone.

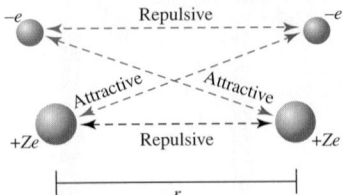

▲ FIGURE 11-1
Type of interactions between two hydrogen atoms
The types of interactions that occur as two hydrogen atoms, infinitely separated, approach each other. The dashed lines represent the types of interactions (red for attractive interactions and blue and black for repulsive interactions). The solid black line represents the internuclear distance, r, between the two hydrogen atoms.

11-1 What a Bonding Theory Should Do

The hydrogen molecule is a simple model for discussing bonding theories. Imagine bringing together two H atoms that are initially very far apart. When the H atoms are infinitely far apart, the two H atoms do not interact with each other, and by convention the net energy of interaction between the H atoms is zero. As the two H atoms approach each other, three types of interactions occur: (1) each electron is attracted to the other nucleus (illustrated by a red dashed line in Fig. 11-1); (2) the electrons repel each other (illustrated by a blue dashed line in Fig. 11-1); and (3) the two nuclei repel each other (illustrated by a black dashed line in Fig. 11-1).

We can plot potential energy—the net energy of interaction of the atoms—as a function of the distance between the atomic nuclei.

Figure 11-2 shows the net energy of interaction of two H atoms. This starts at zero when the atoms are very far apart (condition a). At intermediate distances (condition b), attractive forces predominate and the potential energy is negative. At very small internuclear distances (condition d), repulsive forces exceed attractive forces and the potential energy is positive. At one particular internuclear distance (74 pm, condition c) the potential energy reaches its lowest value (-436 kJ/mol). This is the condition in which the two H atoms combine into a H_2 molecule through a covalent bond. The nuclei continuously move back and forth; that is, the molecule vibrates, but the average internuclear distance remains constant. This internuclear distance corresponds to the *bond length*. The potential energy corresponds to the negative of the *bond-dissociation energy*. A theory of covalent bonding should help us understand why a given molecule has its particular set of observed properties—bond-dissociation energies, bond lengths, bond angles, and so on.

There are several approaches to understanding bonding. The approach used depends on the situation because different methods have different strengths and weaknesses. The strength of the Lewis theory is in the ease with which it can be applied; a Lewis structure can be written rather quickly. VSEPR theory makes it possible to propose molecular shapes that are generally in good agreement with

▶ FIGURE 11-2
Energy of interaction of two hydrogen atoms plotted for internuclear separations from zero to infinity
(a) The potential energy is defined as zero when the two H atoms are infinitely separated. **(b)** Where the curve slopes downward from right to left, the net interaction is attractive. When the interaction is attractive, the potential energy decreases as the internuclear distance, r, decreases. **(c)** Where the curve is flat, at the bottom of the potential energy well, there is no tendency either to attract or to repel. The molecule is most stable at this internuclear distance (74 pm). **(d)** Where the curve slopes upward from right to left, the interaction is repulsive. When the interaction is repulsive, the potential energy increases as the internuclear distance, r, decreases.

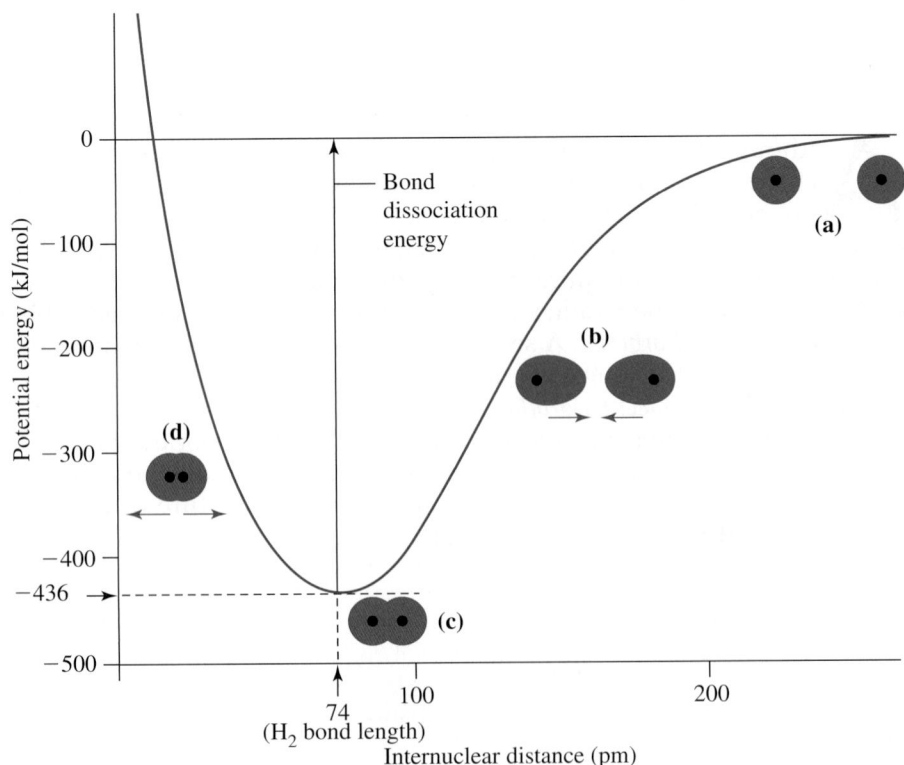

experimental results. However, neither method yields quantitative information about bond energies and bond lengths, and the Lewis theory has problems with odd-electron species and situations in which it is not possible to represent a molecule through a single structure (resonance).

11-2 Introduction to the Valence-Bond Method

Recall the region of high electron probability in a H atom that we described in Chapter 8 through the mathematical function called a $1s$ orbital (page 328). As the two H atoms pictured in Figure 11-2 approach each other, these regions begin to interpenetrate. We say that the two orbitals overlap. Furthermore, we can say that a bond is produced between the two atoms because of the high electron density probability found in the region between the atomic nuclei where the $1s$ orbitals overlap. The increased electron density, with its negative charge, attracts the two positively charged nuclei. In this way, a covalent bond is formed between the two H atoms in the H_2 molecule.

A description of covalent bond formation in terms of atomic orbital overlap is called the **valence-bond method**. The creation of a covalent bond in the valence-bond method is normally based on the overlap of half-filled orbitals, but sometimes such an overlap involves a filled orbital on one atom and an empty orbital on another. The valence-bond method gives a *localized* electron model of bonding: Core electrons and lone-pair valence electrons retain the same orbital locations as in the separated atoms, and the charge density of the bonding electrons is concentrated in the region of orbital overlap.

Figure 11-3 shows the imagined overlap of atomic orbitals in the formation of hydrogen-to-sulfur bonds in hydrogen sulfide. Note especially that maximum overlap between the $1s$ orbital of a H atom and a $3p$ orbital of a S atom occurs along a line joining the centers of the H and S atoms. The two half-filled sulfur

◀ What we are calling "overlap" is actually an interpenetration of two orbitals.

KEEP IN MIND

the actual orbital shapes are pointed out on page 331. The orbitals here, as well as those throughout the rest of the book, are schematic representations.

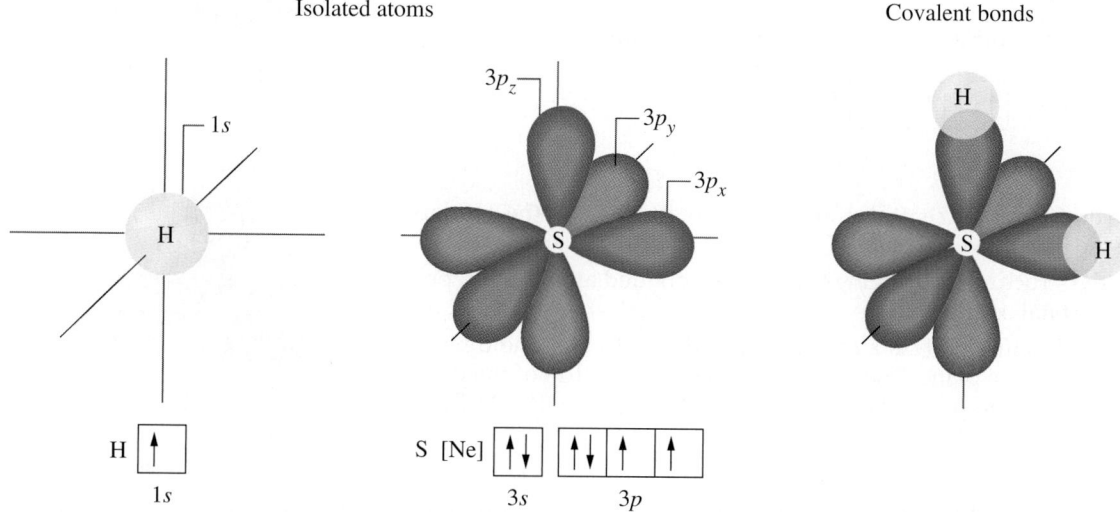

▲ FIGURE 11-3
Bonding in H₂S represented by atomic orbital overlap
For S, only $3p$ orbitals are shown. The phases of the lobes of the sulfur $3p$ orbitals are shown in red and blue for positive and negative. However, we do not know which lobe is positive or negative; all we know is that they are opposite. Although the angular part of the p orbital is circular, when multiplied by the radial part of the wave function, each lobe is more pear shaped, similar to the probability distribution shown in Figure 8-27(c). Bond formation occurs between orbitals that are in phase (same color), although the hydrogen $1s$ orbital is colored yellow for clarity.

3p orbitals that overlap in H₂S are perpendicular to each other, and the valence-bond method suggests a H—S—H bond angle of 90°. This is in good agreement with the observed angle of 92°.

11-1 ARE YOU WONDERING...

Why the overlap of orbitals leads to a chemical bond?

The origin of this extra stability comes from the overlap of the two orbitals in which the two atomic wave functions are *in phase*, leading to constructive interference of the wave functions between the two nuclei and hence increased electron density between the two nuclei. The increased electron density, with its negative charge, attracts the two positively charged nuclei, leading to an energy that is lower than that of the two separated atoms. Thus, the increased electron charge density between the nuclei produces the chemical bond. We will say more about the interaction between orbitals later in this chapter.

EXAMPLE 11-1   Using the Valence-Bond Method to Describe a Molecular Structure

Describe the phosphine molecule, PH₃. by the valence-bond method.

Analyze

We use four steps when applying the valence-bond method. First, we identify the valence orbitals of the central atom. Second, we sketch the valence orbitals. Third, we bring in the atoms to be bonded to the central atom and sketch the orbital overlap. Finally, we describe the resulting structure. These steps are illustrated in Figure 11-4.

Bonding orbitals of P atom

Solve

Step 1. Draw valence-shell orbital diagrams for the separate atoms.

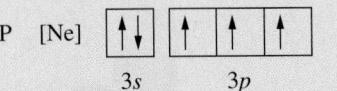

Step 2. Sketch the orbitals of the central atom (P) that are involved in the overlap. These are the half-filled 3p orbitals (Fig. 11-4).

Step 3. Complete the structure by bringing together the bonded atoms and representing the orbital overlap.

Step 4. Describe the structure. PH₃ is a *trigonal-pyramidal* molecule. The three H atoms lie in the same plane. The P atom is situated at the top of the pyramid above the plane of the H atoms, and the three H—P—H bond angles are 90°.

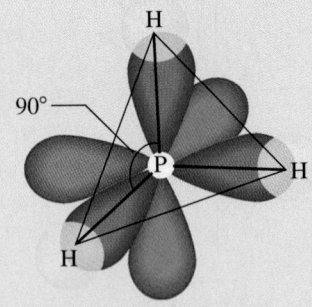

Covalent bonds formed

Assess

The predicted H—P—H bond angle is 90°, and the experimentally measured bond angles are 93° to 94°. These are in good agreement.

PRACTICE EXAMPLE A: Use the valence-bond method to describe bonding and the expected molecular geometry in nitrogen triiodide, NI₃.

PRACTICE EXAMPLE B: Describe the molecular geometry of NH₃, first using the VSEPR method and then using the valence-bond method described above. How do your answers differ? Which method seems to be more appropriate in this case? Explain.

▲ FIGURE 11-4
Bonding and structure of the PH₃ molecule— Example 11-1 illustrated
Only bonding orbitals are shown. The 1s orbitals (yellow) of three H atoms overlap with the three 3p orbitals of the P atom.

11-3   Hybridization of Atomic Orbitals

If we try to extend the unmodified valence-bond method of Section 11-2 to a greater number of molecules, we are quickly disappointed. In most cases, our descriptions of molecular geometry based on the simple overlap of unmodified atomic orbitals do not conform to observed measurements. For example, based on the *ground-state* electron configuration of the valence shell of carbon

Ground state   C   [↑↓] [↑][↑][ ]
                  2s   2p

and employing only half-filled orbitals, we expect the existence of a molecule with the formula CH_2 and a bond angle of 90°. The CH_2 molecule is a highly reactive molecule observed only under specially designed circumstances.

The simplest hydrocarbon observed under normal laboratory conditions is methane, CH_4. This is a stable, unreactive molecule with a molecular formula consistent with the octet rule of the Lewis theory. To obtain this molecular formula by the valence-bond method, we need an orbital diagram for carbon in which there are four unpaired electrons so that orbital overlap leads to four C—H bonds. To get such a diagram, imagine that one of the 2s electrons in a ground-state C atom absorbs energy and is promoted to the empty 2p orbital. The resulting electron configuration is that of an *excited state*.

Excited state   C   [↑] [↑][↑][↑]
                  2s   2p

The electron configuration of this excited state suggests a molecule with three mutually perpendicular C—H bonds based on the 2p orbitals of the C atom (90° bond angles). The fourth bond would be directed to whatever position in the molecule could accommodate the fourth H atom. This description, however, does not agree with the experimentally determined H—C—H bond angles, all four of which are found to be 109.5°, the same as predicted by VSEPR theory (Fig. 11-5). A bonding scheme based on the excited-state electron configuration does a poor job of explaining the bond angles in CH_4.

The problem is not with the theory but with the way the situation has been defined. We have been describing *bonded* atoms as though they have the same kinds of orbitals (that is, *s, p,* and so on) as isolated, *nonbonded* atoms. This assumption worked rather well for H_2S and PH_3, but we have no reason to expect these unmodified pure atomic orbitals to work equally well in all cases.

One way to deal with this problem is to modify the atomic orbitals of the bonded atoms. Recall that atomic orbitals are mathematical expressions of the electron waves in an atom. An algebraic combination of the wave equations of the 2s and three 2p orbitals of the carbon atom produces a new set of four identical orbitals. These new orbitals, which are directed in a tetrahedral fashion, have energies that are intermediate between those of the 2s and 2p orbitals. This mathematical process of replacing pure atomic orbitals with reformulated atomic orbitals for bonded atoms is called **hybridization**, and the new orbitals are called **hybrid orbitals**. Figure 11-6 pictures the hybridization of one s and three p orbitals into a new set of four sp^3 **hybrid orbitals**. Each sp^3 hybrid orbital has 25% s character and 75% p character.

In a hybridization scheme, *the number of hybrid orbitals equals the total number of atomic orbitals that are combined*. The symbols identify the numbers and kinds of orbitals involved. Thus, sp^3 signifies that *one s* and *three p* orbitals are

▲ FIGURE 11-5
Ball-and-stick model of methane, CH_4
The molecule has a tetrahedral structure, and the H—C—H bond angles are 109.5°.

◀ The algebraic combination of wave functions is, in fact, a linear combination of atomic orbitals; that is, they are simply added or subtracted. The resultant linear combinations are solutions to the Schrödinger equation of the molecule.

◀ **Hybridization** occurs only when the bonds are being formed.

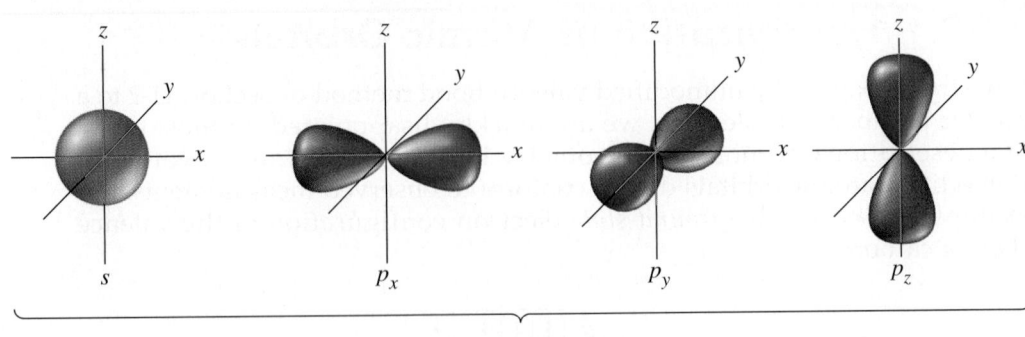

Combine to generate
four sp^3 orbitals

Which are represented
as the set

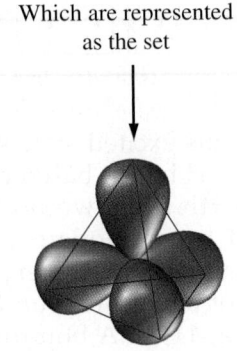

▲ FIGURE 11-6
The sp^3 hybridization scheme

combined. A useful representation of sp^3 hybridization of the valence-shell orbitals of carbon is

Note that the three p orbitals move down by $\frac{1}{4}$ of the energy difference between the s orbitals and the p orbitals and that the s orbital moves up by $\frac{3}{4}$ of that energy difference; that is, energy is conserved. Figure 11-7 shows sp^3 hybrid orbitals and bond formation in methane.

The objective of a hybridization scheme is an after-the-fact rationalization of the experimentally observed shape of a molecule. Hybridization is not an actual physical phenomenon. We cannot observe electron charge distributions changing from those of pure orbitals to those of hybrid orbitals. Moreover, for some covalent bonds no single hybridization scheme works well. Nevertheless,

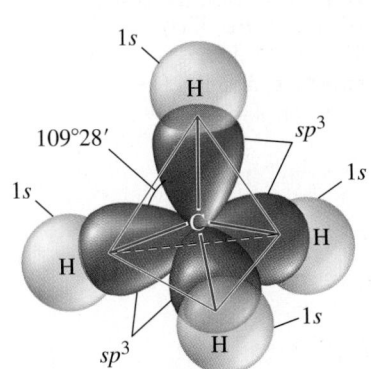

▲ FIGURE 11-7
Bonding and structure of CH$_4$
The four carbon orbitals are sp^3 hybrid orbitals (blue). Those of the hydrogen atoms (yellow) are 1s. The structure is tetrahedral, with H—C—H bond angles of 109.5° (more precisely, 109.471). Remember that the hydrogen orbitals and the carbon hybrid orbitals have the same phase, but we have colored the hydrogen orbitals yellow for clarity.

the concept of hybridization works very well for carbon-containing molecules and is therefore used a great deal in organic chemistry.

Bonding in H₂O and NH₃

Applied to H_2O and NH_3, VSEPR theory describes a tetrahedral electron-group geometry for *four* electron groups. This, in turn, requires an sp^3 hybridization scheme for the central atoms in H_2O and NH_3. This scheme suggests angles of 109.5° for the H—O—H bond in water and the H—N—H bonds in NH_3. These angles are in reasonably good agreement with the experimentally observed bond angles of 104.5° in water and 107° in NH_3. Bonding in NH_3, for example, can be described in terms of the following valence-shell orbital diagram for nitrogen.

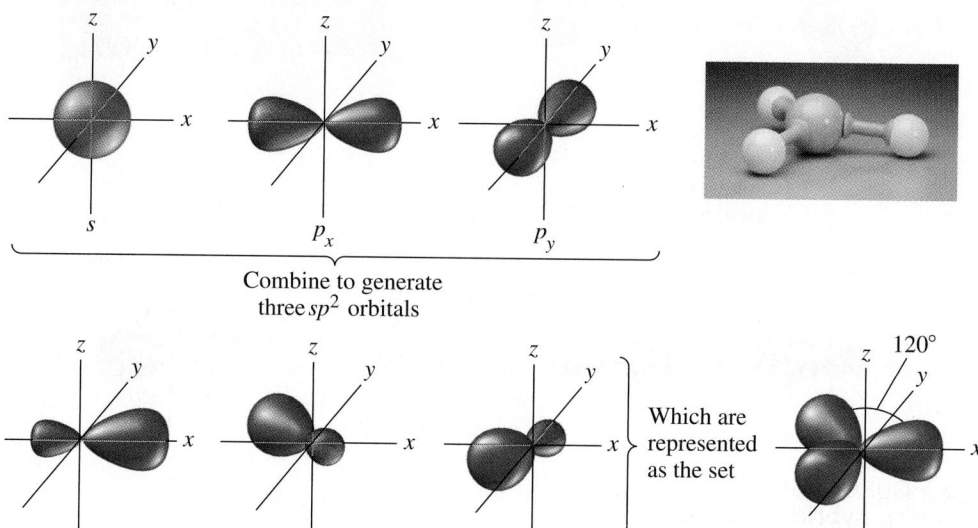

Because one of the sp^3 orbitals is occupied by a lone pair of electrons, only the three half-filled sp^3 orbitals are involved in bond formation. This suggests the trigonal-pyramidal molecular geometry depicted in Figure 11-8, just as does VSEPR theory.

Even though the sp^3 hybridization scheme seems to work quite well for H_2O and NH_3, both theoretical and experimental (spectroscopic) evidence favors a description based on *unhybridized p* orbitals of the central atoms. The H—O—H and H—N—H bond angle expected for $1s$ and $2p$ atomic orbital overlaps is 90°, which does not conform to the observed bond angles. One possible explanation is that because O—H and N—H bonds have considerable ionic character, repulsions between the positive partial charges associated with the H atoms force the H—O—H and H—N—H bonds to "open up" to values greater than 90°. The issue of how best to describe the bonding orbitals in H_2O and NH_3 is still unsettled and underscores the occasional difficulty of finding a single theory that is consistent with all the available evidence.

sp² Hybrid Orbitals

Carbon's group 13 neighbor, boron, has *four* orbitals but only *three* electrons in its valence shell. For most boron compounds, the appropriate hybridization scheme combines the $2s$ and two $2p$ orbitals into *three* **sp^2 hybrid orbitals** and leaves one p orbital unhybridized. Valence-shell orbital diagrams for this hybridization scheme for boron are shown here, and the scheme is further outlined in Figure 11-9.

◀ Notice that hybrid orbitals can accommodate lone-pair electrons as well as bonding electrons.

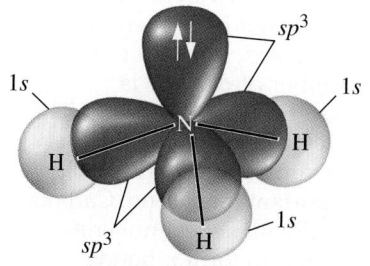

▲ FIGURE 11-8
sp^3 hybrid orbitals and bonding in NH₃

An sp^3 hybridization scheme conforms to a molecular geometry in close agreement with experimental observations. Excluding the orbital occupied by a lone pair of electrons, the centers of the atoms form a trigonal pyramid. The hydrogen orbitals are colored yellow for clarity, but they have the same phase as the nitrogen hybrid orbitals.

Combine to generate three sp^2 orbitals

Which are represented as the set

120°

◀ FIGURE 11-9
The sp^2 hybridization scheme

The sp^2 hybridization scheme corresponds to trigonal-planar electron-group geometry and 120° bond angles, as in BF_3. Note again that in the hybridization schemes of valence-bond theory, the number of orbitals is conserved; that is, in an sp^2 hybridized atom there are still four orbitals: three sp^2 hybrids and an unhybridized p orbital.

sp Hybrid Orbitals

Boron's group 2 neighbor, beryllium, has *four* orbitals and only *two* electrons in its valence shell. In the hybridization scheme that best describes certain *gaseous* beryllium compounds, the 2s and one 2p orbital of Be are hybridized into *two* **sp hybrid orbitals**, and the remaining two 2p orbitals are left unhybridized. Valence-shell orbital diagrams of beryllium in this hybridization scheme are shown here, and the scheme is further outlined in Figure 11-10.

► In hybridization, molecular orbital, and valence-bond theory, not only is energy conserved, but also the number of orbitals is conserved. sp^2 hybrid orbitals still have one p orbital left over and this is particularly important for carbon. Carbon readily uses the leftover p orbitals to form π bonds (see page 461). In contrast, silicon, the element one below carbon, does not use the p orbitals as readily because silicon is larger and the p orbitals do not project out far enough to form double and triple bonds.

The sp hybridization scheme corresponds to a linear electron-group geometry and a 180° bond angle, as in $BeCl_2(g)$.

11-1   CONCEPT ASSESSMENT

Criticize the following statement: The hybridization of the C atom in CH_3^+ and CH_3^- are both expected to be the same as in CH_4.

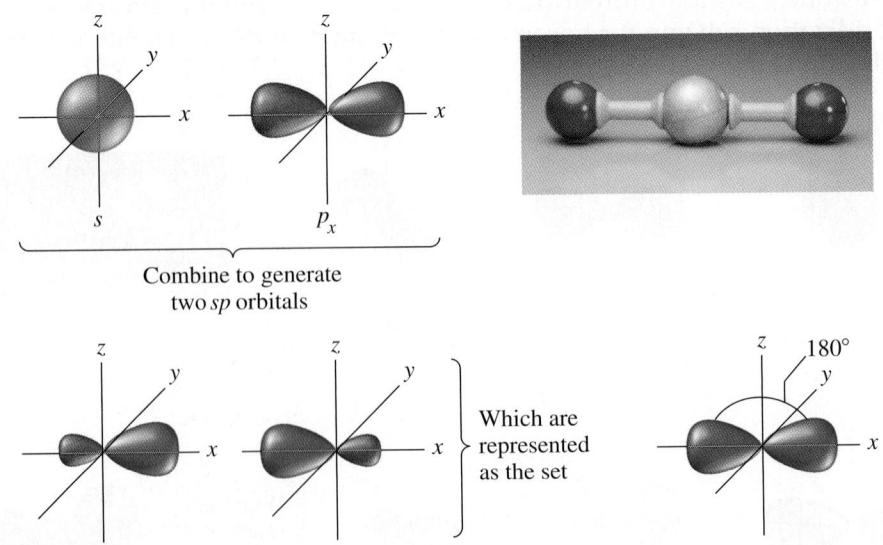

▲ FIGURE 11-10
The sp hybridization scheme

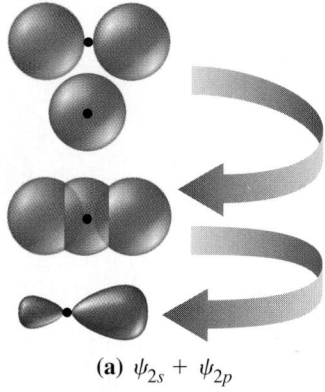

(a) $\psi_{2s} + \psi_{2p}$

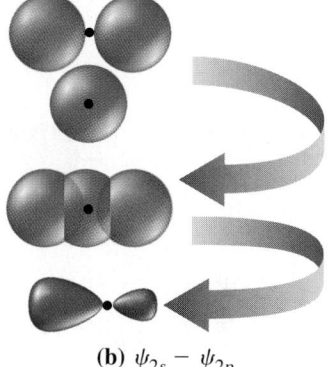

(b) $\psi_{2s} - \psi_{2p}$

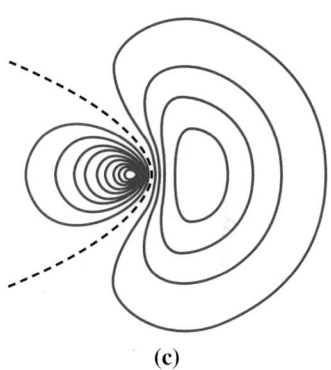

(c)

▲ FIGURE 11-11
The formation of an *sp* hybrid orbital
In **(a)** and **(b)**, the *p*, *s*, and *sp* hybrid orbitals themselves are depicted rather than their probability-density plots. **(c)** Contour map for one of the *sp* hybrids. The contour maps for the *sp*² and *sp*³ hybrids are similar.

11-2 ARE YOU WONDERING...

How the atomic orbitals mix to form a hybrid orbital?

A hybrid atomic orbital is the result of a mathematical combination (algebraic addition and subtraction) of the wave functions describing two or more atomic orbitals. When the algebraic functions that represent *s* and *p* orbitals are added, a new function is produced; this is an *sp* hybrid. When the same algebraic functions are subtracted, another new function is produced; this is a second *sp* hybrid. The hybridization process is shown in Figure 11-11, where we see the consequence of the phase of the *p* orbital when we add the *s* and *p* orbitals: The negative phase of the *p* orbital cancels part of the positive *s* orbital. This leads to the teardrop-shaped orbital pointing in the direction of the positive lobe of the *p* orbital. As shown in Figure 11-11, subtraction of the two orbitals reverses this situation. The two ways of combining an *s* and a *p* orbital generate the two equivalent *sp* hybrid orbitals, each having its greatest amplitude (or electron density if we square the amplitude) in a direction 180° from the other. A similar procedure is used to construct the three *sp*² and the four *sp*³ hybrid orbitals, although the combinations of orbitals are slightly more complicated.

sp³d and *sp³d²* Hybrid Orbitals

To describe hybridization schemes that correspond to the 5- and 6-electron-group geometries of VSEPR theory, we need to go beyond the *s* and *p* subshells of the valence shell, and traditionally this has meant including *d*-orbital contributions. We can achieve the *five* half-filled orbitals of phosphorus to account for the five P—Cl bonds in PCl_5 and its trigonal-bipyramidal molecular geometry through the hybridization of the *s*, three *p*, and one *d* orbital of the valence shell into *five **sp³d** hybrid orbitals*.

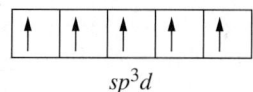

sp^3d

We can achieve the *six* half-filled orbitals of sulfur to account for the six S—F bonds in SF_6 and its octahedral molecular geometry through the hybridization of the *s*, three *p*, and two *d* orbitals of the valence shell into *six sp³d² hybrid orbitals*.

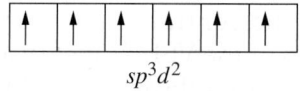

sp^3d^2

The *sp³d* and *sp³d²* hybrid orbitals and two of the molecular geometries in which they can occur are featured in Figure 11-12.

We have previously stated that hybridization is not a real phenomenon, but an after-the-fact rationalization of an experimentally determined result. Perhaps there is no better illustration of this point than the issue of the *sp³d* and *sp³d²* hybrid orbitals. In discussing the concept of the expanded valence shell in Chapter 10, we noted that valence-shell expansion would seem to require *d* electrons in bonding schemes, but recent theoretical considerations cast serious doubt on *d*-electron participation. The same doubt, of course, extends to the use of *d* orbitals in hybridization schemes.

Despite the difficulty posed by hybridization schemes involving *d* orbitals, the *sp*, *sp*², and *sp*³ hybridization schemes are well established and very commonly encountered, particularly among the second-period elements.

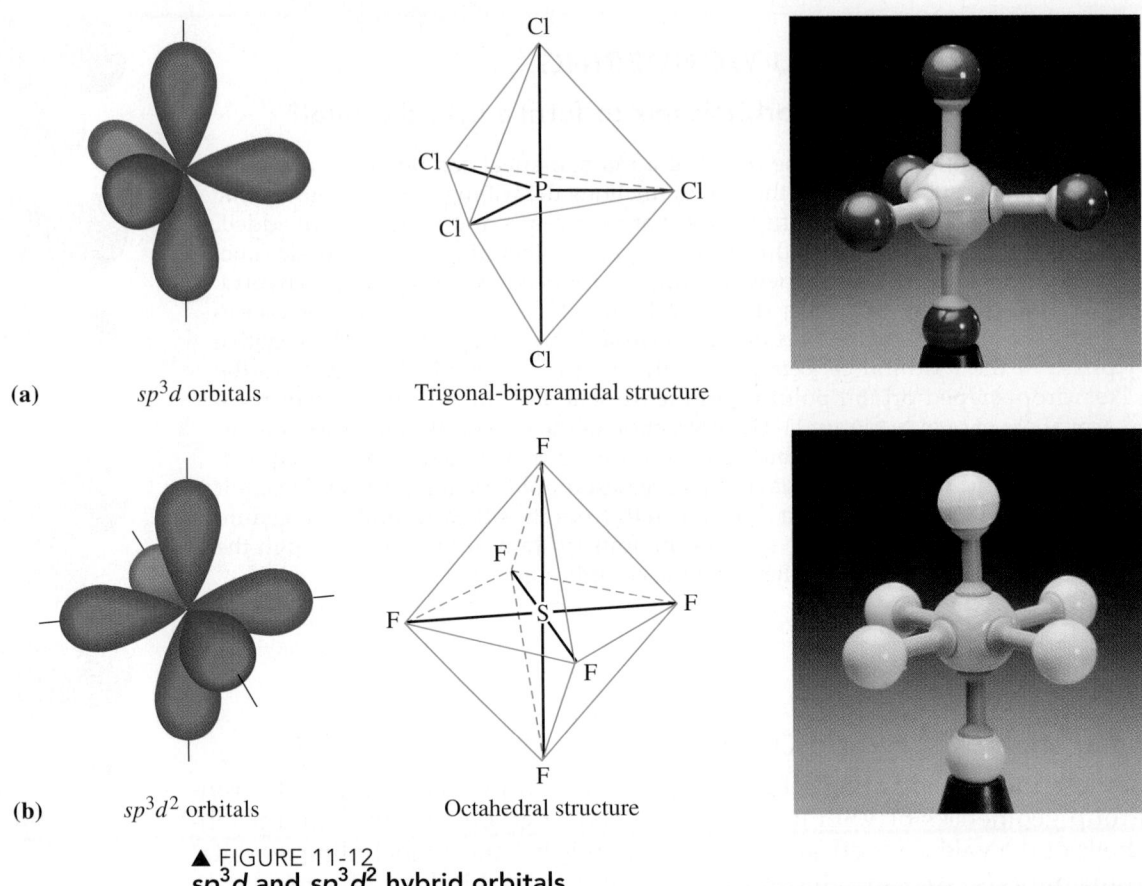

(a) sp^3d orbitals   Trigonal-bipyramidal structure

(b) sp^3d^2 orbitals   Octahedral structure

▲ FIGURE 11-12
sp^3d and sp^3d^2 hybrid orbitals

🔍 **11-2   CONCEPT ASSESSMENT**

Give the formula of a compound or ion composed of arsenic and fluorine in which the arsenic atom has a sp^3d^2 hybridization state.

Hybrid Orbitals and the Valence-Shell Electron-Pair Repulsion (VSEPR) Theory

In the previous section, we used either the experimental geometry or the geometry predicted by VSEPR theory to help us decide on the appropriate hybridization scheme for the central atom. The concept of hybridization arose before the formulation of the VSEPR theory as we used it in Chapter 10. In 1931, Linus Pauling introduced the concept of hybridization of orbitals to account for the known geometries of CH_4, H_2O, and NH_3. It was first suggested by Nevil Vincent Sidgwick and Herbert Marcus Powell in 1940 that molecular geometry was determined by the arrangement of electron pairs in the valence shell, and this suggestion was subsequently developed into the set of rules known as VSEPR by Ronald Gillespie and Ronald Nyholm in 1957. The advantage of VSEPR is that it has a predictive capability based on Lewis structures, whereas hybridization schemes, as described here, require a prior knowledge of the molecular geometry. So how should we proceed to describe the bonding in molecules? We can choose the likely hybridization scheme for a central atom in a structure in the valence-bond method by

- writing a plausible Lewis structure for the species of interest
- using VSEPR theory to predict the probable electron-group geometry of the central atom

- selecting the hybridization scheme corresponding to the electron-group geometry

The procedure outlined above is illustrated in Figure 11-13 using the molecule SF_4 as an example.

As suggested by Table 11.1, the hybridization scheme adopted for a central atom should be the one producing the same number of hybrid orbitals as there are valence-shell electron groups, and in the same geometric orientation. Thus, an sp^3 hybridization scheme for the central atom predicts that four hybrid orbitals are distributed in a tetrahedral fashion. This results in molecular structures that are tetrahedral, trigonal-pyramidal, or angular, depending on how many hybrid orbitals are involved in orbital overlap and how many contain lone-pair electrons, corresponding to the VSEPR notations AX_4, AX_3E, and AX_2E_2 respectively.

The s and p orbital hybridization schemes are especially important in organic compounds, whose principal elements are C, O, and N, in addition to H. We will consider some important applications to organic chemistry in the next section.

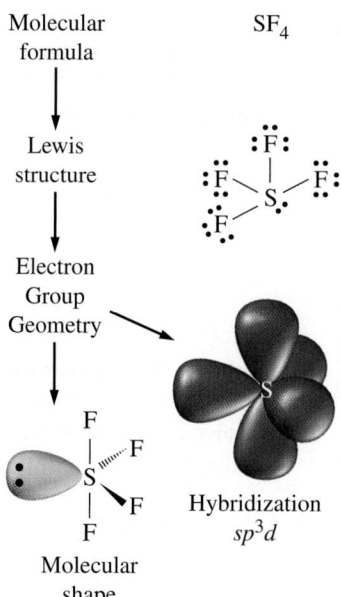

▲ FIGURE 11-13
Using electron-group geometry to determine hybrid orbitals

TABLE 11.1 Some Hybrid Orbitals and Their Geometric Orientations

Hybrid Orbitals	Geometric Orientation	Example
sp	Linear	$BeCl_2$
sp^2	Trigonal-planar	BF_3
sp^3	Tetrahedral	CH_4
sp^3d	Trigonal-bipyramidal	PCl_5
sp^3d^2	Octahedral	SF_6

EXAMPLE 11-2   Proposing a Hybridization Scheme to Account for the Shape of a Molecule

Predict the shape of the XeF_4 molecule and a hybridization scheme consistent with this prediction.

Analyze
Follow the procedure outlined in Figure 11-13.

Solve

1. **Write a plausible Lewis structure.** The Lewis structure we write must account for 36 valence electrons—*eight* from the Xe atom and *seven* each from the *four* F atoms. To place this many electrons in the Lewis structure, we must expand the valence shell of the Xe atom to accommodate 12 electrons. The Lewis structure is

$$:\ddot{F}—\overset{\displaystyle :\ddot{F}:}{\underset{\displaystyle :\ddot{F}:}{Xe}}—\ddot{F}:$$

2. **Use the VSEPR theory to establish the electron-group geometry of the central atom.** From the Lewis structure, we see that there are *six* electron groups around the Xe atom. Four electron groups are bond pairs and two are lone pairs. The electron-group geometry for *six* electron groups is *octahedral*.

(continued)

3. **Describe the molecular geometry.** The VSEPR notation for XeF_4 is AX_4E_2, and the molecular geometry is *square-planar* (see Table 10.1). The four pairs of bond electrons are directed to the corners of a square, and the lone pairs of electrons are found above and below the plane of the Xe and F atoms, as shown here.

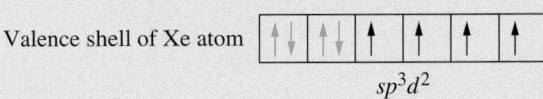

4. **Select a hybridization scheme that corresponds to the VSEPR prediction.** The only hybridization scheme consistent with an octahedral distribution of *six* electron groups is sp^3d^2. The orbital diagram for this scheme shows clearly that *four* of the eight valence electrons of the central Xe atom singly occupy four of the sp^3d^2 orbitals. The remaining *four* valence electrons of the atom occupy the remaining two sp^3d^2 orbitals as lone pairs. These are the lone pairs of electrons situated above and below the plane of the Xe and F atoms in the above sketch.

Valence shell of Xe atom

sp^3d^2

Assess

Note that the number of electron pairs in the electron-group geometry dictates how many orbitals are used in the hybridization scheme. We see that a combination of VSEPR and hybridization theory is an appealing way to describe the shape of and bonding in a molecule.

PRACTICE EXAMPLE A: Describe the molecular geometry and propose a plausible hybridization scheme for the central atom in the ion Cl_2F^+.

PRACTICE EXAMPLE B: Describe the molecular geometry and propose a plausible hybridization scheme for the central atom in the ion BrF_4^+.

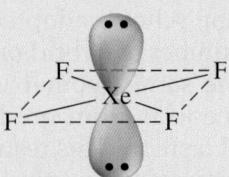

11-3 ARE YOU WONDERING...

Whether to use the VSEPR method (Section 10-7) or the valence-bond method in rationalizing the geometric shape of a molecule?

There is no "correct" method for describing molecular structures. The only correct information is the experimental evidence from which the structure is established. Once this experimental evidence is in hand, you may find it easier to rationalize this evidence by one method or another. For H_2S, the valence-bond method, which suggests a bond angle of 90°, seems to do a better job of explaining the observed 92° bond angle than does the VSEPR theory. For the Lewis structure, $H-\overset{..}{\underset{..}{S}}-H$, VSEPR theory predicts a tetrahedral electron-group geometry, which in turn suggests a tetrahedral bond angle—that is, 109.5°. However, by modifying this initial VSEPR prediction to accommodate lone-pair–lone-pair and lone-pair–bond-pair repulsions (see page 427), the predicted bond angle is less than 109.5°.

VSEPR theory gives reasonably good results in the majority of cases. Unless you have specific information to suggest otherwise, describing a molecular shape with the VSEPR theory is a good bet. It is important to remember that both the VSEPR and valence-bond methods are simply models we use to rationalize the shapes and bonding of polyatomic molecules and as such, should be viewed with a critical eye, always keeping experimental results in sight.

KEEP IN MIND

that the VSEPR method uses empirical data to give an approximate molecular geometry, whereas the valence-bond method relates to the orbitals used in bonding based on a given geometry.

11-4   Multiple Covalent Bonds

Two different types of orbital overlap occur when multiple bonds are described by the valence-bond method. In our discussion we will use as specific examples the carbon-to-carbon double bond in ethylene, C_2H_4, and the carbon-to-carbon triple bond in acetylene, C_2H_2.

Bonding in C_2H_4

Ethylene has a carbon-to-carbon double bond in its Lewis structure.

$$\begin{array}{c}
\text{H} \quad \text{H} \\
| \quad\quad | \\
\text{C}=\text{C} \\
| \quad\quad | \\
\text{H} \quad \text{H}
\end{array}$$

Ethylene is a planar molecule with 120° H—C—H and H—C—C bond angles. VSEPR theory treats each C atom as being surrounded by *three* electron groups in a trigonal-planar arrangement. VSEPR theory does *not* dictate that the two —CH₂ groups be coplanar, but as we will see, valence-bond theory does.

The hybridization scheme that produces a set of hybrid orbitals with a trigonal-planar orientation is sp^2. The valence-shell orbital diagrams of carbon for this scheme are

The $sp^2 + p$ orbital set is pictured in Figure 11-14. One of the bonds between the carbon atoms results from the overlap of sp^2 hybrid orbitals from each atom. This overlap occurs along the line joining the nuclei of the

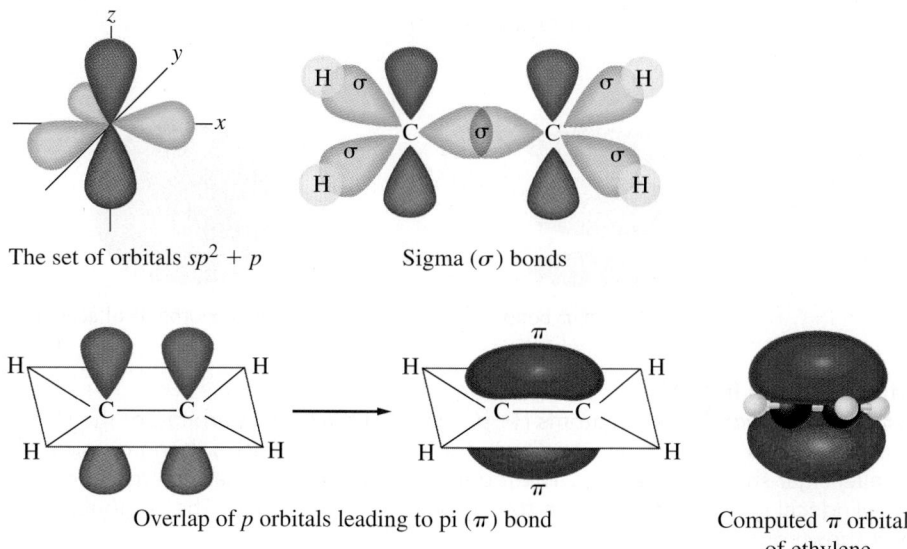

The set of orbitals $sp^2 + p$

Sigma (σ) bonds

Overlap of p orbitals leading to pi (π) bond

Computed π orbitals of ethylene

◀ FIGURE 11-14
Sigma (σ) and pi (π) bonding in C_2H_4
The purple orbitals are sp^2 hybrid orbitals; the red and blue orbitals are $2p$, with the colors indicating their phase. The sp^2 hybrid orbitals overlap along the line joining the bonded atoms—a σ bond. The $2p$ orbitals overlap in a side-to-side fashion and form a π bond. Notice that the phase of the p orbitals is retained.

▲ FIGURE 11-15
Ball-and-stick model of ethylene, C_2H_4
The H—C—H and C—C—H bond angles are 120°. The model also distinguishes between the σ bond between the C atoms (the straight plastic tube) and the π bond extending above and below the plane of the molecule. The picture of the π bond suggested by the white plastic "arches" is somewhat distorted, but the model does convey the idea that the π bond places a high electron charge density above and below the plane of the molecule.

KEEP IN MIND

that only one of the bonds in a multiple bond is a σ bond; the others are π bonds—one π bond in a double bond and two in a triple bond.

two atoms. Orbitals that overlap in this end-to-end fashion produce a **sigma bond**, designated σ **bond**. Figure 11-14 shows that a second bond between the C atoms results from the overlap of the unhybridized p orbitals. In this bond, there is a region of high electron charge density above and below the plane of the carbon and hydrogen atoms. The bond produced by this side-to-side overlap of two parallel orbitals is called a **pi bond**, designated π **bond**.

The ball-and-stick model in Figure 11-15 illustrates bonding in ethylene. It helps to show that

- the shape of a molecule is determined only by the orbitals forming σ bonds (the σ-bond framework).
- rotation about the double bond is severely restricted. In the ball-and-stick model, we could easily twist or rotate the terminal H atoms about the s bonds that join them to a C atom. To twist one —CH_2 group out of the plane of the other, however, would reduce the amount of overlap of the p orbitals and weaken the π bond. The double bond is rigid, and the C_2H_4 molecule is planar.

Additionally, in carbon-to-carbon multiple bonds, the σ bond involves more extensive overlap than does the π bond. As a result, a carbon-to-carbon double bond ($\sigma + \pi$) is stronger than a single bond (σ), but not twice as strong (from Table 10.3, C—C, 347 kJ/mol; C=C, 611 kJ/mol; C≡C, 837 kJ/mol).

Bonding in C_2H_2

Bonding in acetylene, C_2H_2, is similar to that in C_2H_4, but with these differences: The Lewis structure of C_2H_2 features a triple covalent bond, H—C≡C—H. The molecule is *linear*, as found by experiment and as expected from VSEPR theory. A hybridization scheme to produce hybrid orbitals in a linear orientation is *sp*. The valence-shell orbital diagrams representing *sp* hybridization are

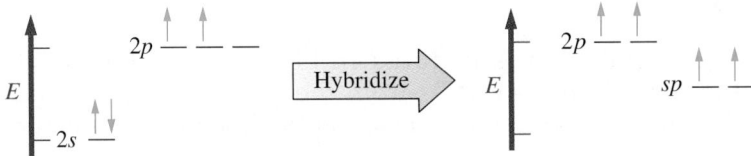

In the triple bond in C_2H_2, one of the carbon-to-carbon bonds is a σ bond and *two* are π bonds, as suggested in Figure 11-16.

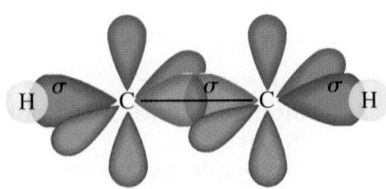

Formation of σ bonds

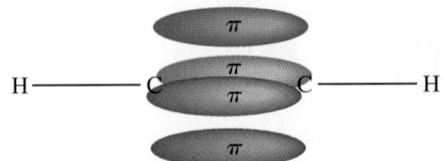

Formation of π bonds

Computed π orbitals of acetylene

▲ FIGURE 11-16
σ and π bonding in C_2H_2
The σ-bond framework joins the atoms H—C—C—H through 1s orbitals of the H atoms and sp orbitals of the C atoms. There are two π bonds. Each π bond consists of two parallel cigar-shaped segments. The four segments shown actually merge into a hollow cylindrical shell with the carbon-to-carbon σ bond as its axis. The computed model is shown on the right.

EXAMPLE 11-3   Proposing Hybridization Schemes Involving σ and π Bonds

Formaldehyde gas, H_2CO, is used in the manufacture of plastics; in aqueous solution, it is the familiar biological preservative called formalin. Describe the molecular geometry and a bonding scheme for the H_2CO molecule.

Analyze

The number of electron groups around central atoms dictates the number of atomic orbitals that undergo hybridization. The carbon atom in formaldehyde contains three electron groups (remember, a double bond is counted as one group), which means that three atomic orbitals are hybridized to form three sp^2 orbitals. We will follow the procedure shown in Figure 11-13.

Solve

1. **Write the Lewis structure.** C is the central atom, and H and O are terminal atoms. The total number of valence electrons is 12. Note that this structure requires a carbon-to-oxygen double bond.

H
|
H—C═Ö:

2. **Determine the electron-group geometry of the central C atom.** The σ-bond framework is based on three electron groups around the central C atom. VSEPR theory, based on the distribution of three electron groups, suggests a trigonal-planar molecule with 120° bond angles.

3. **Identify the hybridization scheme that conforms to the electron-group geometry.** A trigonal-planar orientation of orbitals is associated with sp^2 hybrid orbitals.

4. **Identify the orbitals of the central atom that are involved in orbital overlap.** The C atom is hybridized to produce the orbital set $sp^2 + p$, as in C_2H_4. Two of the sp^2 hybrid orbitals are used to form σ bonds with the H atoms. The remaining sp^2 hybrid orbital is used to form a σ bond with oxygen. The unhybridized p orbital of the C atom is used to form a π bond with O.

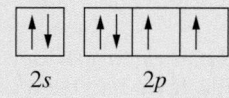

Valence shell of the C atom: sp^2 ↑ ↑ ↑ $2p$ ↑

5. **Sketch the bonding orbitals of the central and terminal atoms.** The bonding orbitals of the central C atom described above are pictured in Figure 11-17(a). The sp^2 hybrid orbitals are shown in lavender, and the pure p orbitals, in blue and red. The H atoms have only $1s$ orbitals available for bonding. For oxygen, the half-filled $2p$ orbital can be used for end-to-end overlap in the σ bond to carbon, and a half-filled $2p$ orbital can participate in the side-to-side overlap leading to a π bond. Thus, the valence-shell orbital diagram we can use for oxygen is

↑↓ | ↑↓ ↑ ↑

$2s$ $2p$

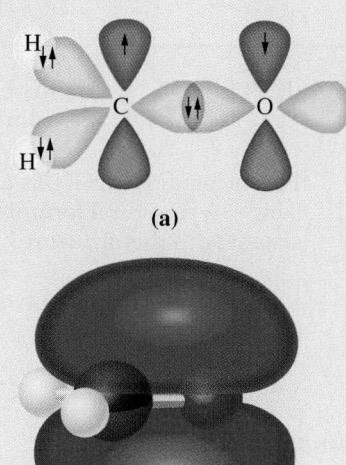

(a)

(b)

▲ FIGURE 11-17
Bonding and structure of the H_2CO molecule—Example 11-3 illustrated
(a) The orbital set $sp^2 + p$ is used for the C atom, $1s$ orbitals for H, and two half-filled $2p$ orbitals for O. For simplicity, only bonding orbitals of the valence shells are shown. **(b)** Computed π orbital for the CO group.

Assess

The bonding and structure of the formaldehyde molecule are suggested by the three-dimensional sketch in Figure 11-17(b).

One of the main purposes of hybridizing orbitals is to describe molecular geometry (for example, bond angles), and so we generally apply hybridization schemes only to central atoms, although hybridization of terminal atoms (except H) may also be invoked.

The H_2CO molecule has four bonds around carbon (three σ bonds and one π bond). These four bonds are used to bond three groups: two H atoms and one O atom. In determining the electron-group geometry and bond angles around a particular atom, we focus on the number of σ bonds and lone pairs, if any.

PRACTICE EXAMPLE A: Describe a plausible bonding scheme and the molecular geometry of dimethyl ether, CH_3OCH_3.

PRACTICE EXAMPLE B: Acetic acid, the acidic component of vinegar, has the formula CH_3COOH. Describe the molecular geometry and a bonding scheme for this molecule.

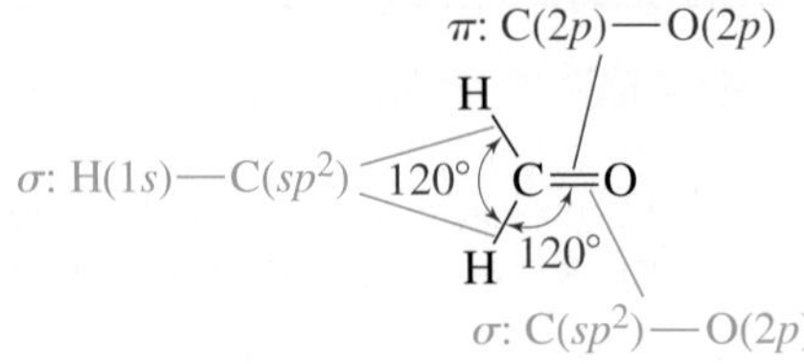

π: C(2p)—O(2p)

σ: H(1s)—C(sp^2) 120° C=O

σ: C(sp^2)—O(2p)

▶ FIGURE 11-18
Bonding in H₂CO—a schematic representation

Drawing three-dimensional sketches to show orbital overlaps, as in Figure 11-17(b), is not easy. A simpler, two-dimensional representation of the bonding scheme for formaldehyde is shown in Figure 11-18. Bonds between atoms are drawn as straight lines. They are labeled σ or π, and the orbitals that overlap are indicated.

We have stressed a Lewis structure as the first step in describing a bonding scheme. Sometimes the starting point is a description of the species obtained by experiment. Example 11-4 illustrates such a case.

EXAMPLE 11-4   Using Experimental Data to Assist in Selecting a Hybridization and Bonding Scheme

Formic acid, HCOOH, is an irritating substance released by ants when they bite (*formica* is Latin, meaning "ant"). A structural formula with bond angles is given here. Propose a hybridization and bonding scheme consistent with this structure.

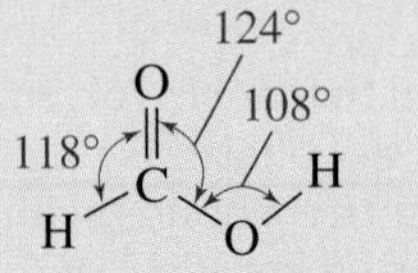

124°
O 108°
118°
H—C—H
H   O

Analyze

When we are given the bond angles for a molecule, we know the hybridization scheme is dictated by the bond angles of that atom; for example, if the bond angle is close to 109.5°, we expect sp^3 hybridization.

Solve

The 118° H—C—O bond angle on the left is very nearly the 120° angle for a trigonal-planar distribution of three groups of electrons. This requires an sp^2 hybridization scheme for the C atom. The 124° O—C—O bond angle is also close to the 120° expected for sp^2 hybridization. The C—O—H bond angle of 108° is close to the tetrahedral angle—109.5°. The O atom on the right employs an sp^3 hybridization scheme. The four σ and one π bonds and the orbital overlaps producing them are indicated in Figure 11-19.

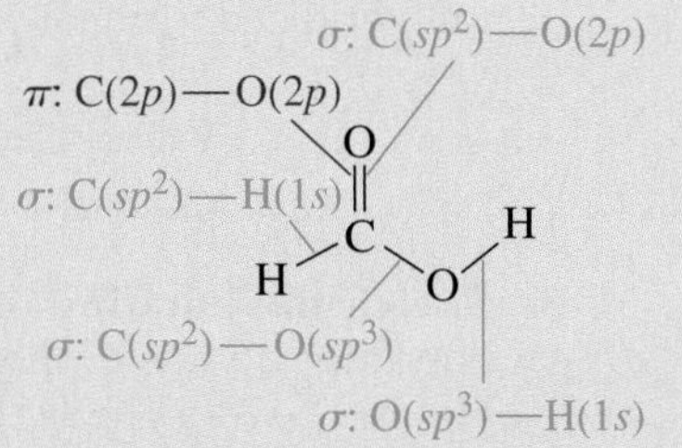

σ: C(sp^2)—O(2p)
π: C(2p)—O(2p)
σ: C(sp^2)—H(1s)
O
C
H
H   O   H
σ: C(sp^2)—O(sp^3)
σ: O(sp^3)—H(1s)

▲ FIGURE 11-19
Bonding and structure of HCOOH—Example 11-4 illustrated

Assess

The structure shown in Figure 11-19 does not show the lone pair electrons; they are there implicitly. We can infer the presence of the lone pairs from the observed bond angles. If we had been given just the skeletal structure of this molecule, we would have deduced the sp^3 hybridization at the oxygen atom by using a Lewis structure and VSEPR theory.

PRACTICE EXAMPLE A: Acetonitrile is an industrial solvent. Propose a hybridization and bonding scheme consistent with its structure.

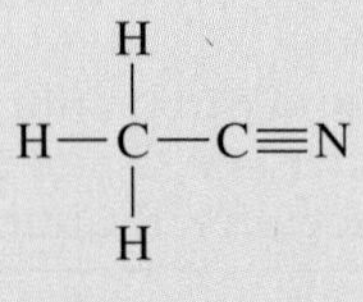

H
|
H—C—C≡N
|
H

PRACTICE EXAMPLE B: A reference source on molecular structures lists the following data for dinitrogen monoxide (nitrous oxide), N₂O: Bond lengths: N—N = 113 pm; N—O = 119 pm; bond angle = 180°. Show that the Lewis structure of N₂O is a resonance hybrid of two contributing structures, and describe a plausible hybridization and bonding scheme for each.

The molecule diazine has the molecular formula N_2H_2. What is the hybridization of the nitrogen and does the molecule contain a double or triple bond?

11-5   Molecular Orbital Theory

Lewis structures, VSEPR theory, and the valence-bond method make a potent combination for describing covalent bonding and molecular structures. They are satisfactory for most of our purposes. Sometimes, however, chemists need a greater understanding of molecular structures and properties than these methods provide. None of these methods, for instance, provides an explanation of the electronic spectra of molecules, why oxygen is paramagnetic, or why H_2^+ is a stable species. To address these questions, we need a different method of describing chemical bonding.

This method, called **molecular orbital theory**, starts with a simple picture of molecules, but it quickly becomes complex in its details. We can provide only an overview here. The theory assigns the electrons in a molecule to a series of orbitals that belong to the molecule as a whole. These are called molecular orbitals. Like atomic orbitals, molecular orbitals are mathematical functions, but we can relate them to the probability of finding electrons in certain regions of a molecule. Also like an atomic orbital, a molecular orbital can accommodate just two electrons, and the electrons must have opposing spins.

Earlier in this chapter, we described the approach of two H atoms toward each other to form a chemical bond (see Figure 11-2). What happens to the atomic orbitals as the two H atoms merge to form a chemical bond? As the atoms approach, the two $1s$ wave functions combine; they do this by interfering constructively or destructively. Constructive interference corresponds to adding the two mathematical functions (the positive sign puts the waves in phase), while destructive interference corresponds to subtracting the two mathematical functions (the minus sign puts the waves out of phase). These two types of combination are illustrated in Figure 11-20.

How do we interpret these two different combinations of wave functions? The constructive interference (addition) of the two wave functions leads to a greater probability of finding the electron between the nuclei. The increased electron charge density between the nuclei causes them to draw closer together, forming a chemical bond. The electron probability or electron charge density in the σ_{1s} orbital is $(1s_A + 1s_B)^2$, the square of the new function $(1s_A + 1s_B)$, where $1s_A$ and $1s_B$ are the two $1s$ orbitals on the two H atoms. The square is $1s_A^2 + 1s_B^2$ *plus* the extra term $2 \times 1s_A1s_B$, which is the extra charge density between the nuclei.

The result of this constructive interference is a **bonding molecular orbital** because it places a high electron charge density between the two nuclei. A high electron charge density between atomic nuclei reduces repulsions between the positively charged nuclei and promotes a strong bond. This bonding molecular orbital, designated σ_{1s}, is at a *lower* energy than the $1s$ atomic orbitals.

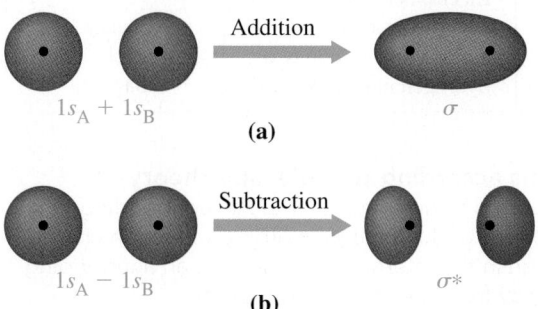

Addition

$1s_A + 1s_B$ σ
(a)

Subtraction

$1s_A - 1s_B$ σ^*
(b)

◀ FIGURE 11-20
Formation of bonding and antibonding orbitals
(a) The addition of two $1s$ orbitals in phase to form a σ_{1s} molecular orbital. This orbital produces electron density between the nuclei, leading to a chemical bond. **(b)** The addition of two $1s$ orbitals out of phase to produce a σ_{1s}^* antibonding orbital. This orbital has a nodal plane perpendicular to the internuclear axis, as do all antibonding orbitals.

The molecular orbital formed by the destructive interference (subtraction) of the two 1s orbitals leads to reduced electron probability between the nuclei. This produces an **antibonding molecular orbital**, designated by a superscript asterisk (*), because it places a very low electron charge density between the two nuclei. The electron probability or electron charge density in the σ_{1s}^* orbital is $(1s_A - 1s_B)^2$, the square of the new function $(1s_A - 1s_B)$, where $1s_A$ and $1s_B$ are the 1s orbitals on the two H atoms. The square is $1s_A^2 + 1s_B^2$ *minus* the extra term $2 \times 1s_A 1s_B$ which is the loss of charge density between the nuclei. Notice that $-2 \times 1s_A 1s_B$ here exactly balances the extra density $(+2 \times 1s_A 1s_B)$ in the molecular orbital formed by addition of the atomic functions.

With a low electron charge density between atomic nuclei, the nuclei are not screened from each other, strong repulsions occur, and the bond is weakened (hence the term "antibonding"). This antibonding molecular orbital, designated σ_{1s}^*, is at a *higher* energy than the 1s atomic orbitals.

The combination of two 1s orbitals of H atoms into two molecular orbitals in a H$_2$ molecule is summarized in Figure 11-21.

Basic Ideas Concerning Molecular Orbitals

Here are some useful ideas about molecular orbitals and how electrons are assigned to them.

> ▶ Molecular orbital theory states that the number of molecular orbitals formed is equal to the number of atomic orbitals combined.

1. The number of molecular orbitals (MOs) formed is equal to the number of atomic orbitals combined.

2. Of the two MOs formed when two atomic orbitals are combined, one is a *bonding* MO at a *lower* energy than the original atomic orbitals. The other is an *antibonding* MO at a *higher* energy.

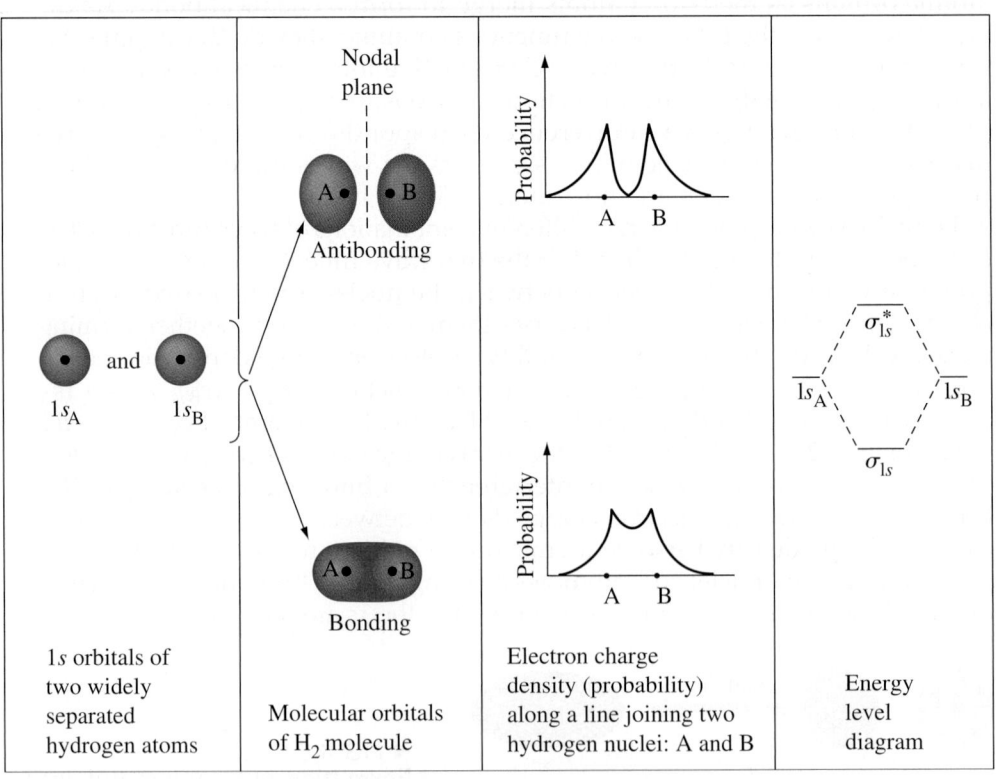

▲ FIGURE 11-21
The interaction of two hydrogen atoms according to molecular theory
The energy of the bonding σ_{1s} molecular orbital is lower and that of the antibonding σ_{1s}^* molecular orbital is higher than the energies of the 1s atomic orbitals. Electron charge density in a bonding molecular orbital is high in the internuclear region. In an antibonding orbital, it is high in parts of the molecule away from the internuclear region.

3. In ground-state configurations, electrons enter the lowest energy MOs available.

4. The maximum number of electrons in a given MO is two (Pauli exclusion principle, page 339).

5. In ground-state configurations, electrons enter MOs of identical energies *singly* before they pair up (Hund's rule, page 339).

A stable molecular species has more electrons in bonding orbitals than in antibonding orbitals. For example, if the excess of bonding over antibonding electrons is *two*, this corresponds to a *single* covalent bond in Lewis theory. In molecular orbital theory, we say that the bond order is 1. **Bond order** is one-half the difference between the number (no.) of bonding and antibonding electrons (e^-), that is,

$$\text{bond order} = \frac{\text{no. of } e^- \text{ in bonding MOs} - \text{no. of } e^- \text{ in antibonding MOs}}{2} \quad \textbf{(11.1)}$$

KEEP IN MIND

that equation (11.1) is only valid for diatomic molecules

Diatomic Molecules of the First-Period Elements

Let's use the ideas just outlined to describe some molecular species of the first-period elements, H and He (Fig. 11-22).

H_2^+ This species has a single electron. It enters the σ_{1s} orbital, a bonding molecular orbital. Using equation (11.1), we see that the bond order is $(1 - 0)/2 = \frac{1}{2}$. This is equivalent to a one-electron, or *half*, bond, a bond type that is not easily described by the Lewis theory.

H_2 This molecule has two electrons, both in the σ_{1s} orbital. The bond order is $(2 - 0)/2 = 1$. With Lewis theory and the valence-bond method, we describe the bond in H_2 as single covalent.

He_2^+ This ion has three electrons. Two electrons are in the σ_{1s} orbital, and one is in the σ_{1s}^* orbital. This species exists as a stable ion with a bond order of $(2 - 1)/2 = \frac{1}{2}$.

He_2 Two electrons are in the σ_{1s} orbital, and two are in the σ_{1s}^*. The bond order is $(2 - 2)/2 = 0$. No bond is produced—He_2 is not a stable species.

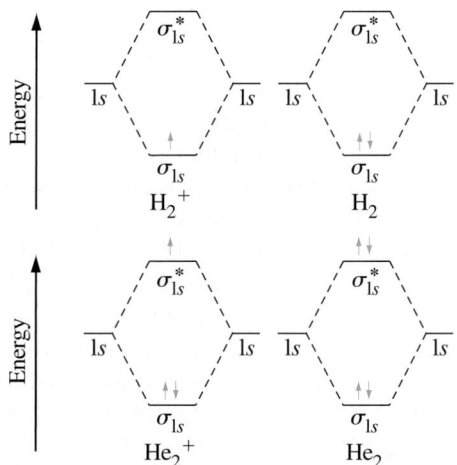

▲ FIGURE 11-22
Molecular orbital diagrams for the diatomic molecules and ions of the first-period elements
The 1s energy levels of the isolated atoms are shown to the left and right of each diagram. The line segments in the middle represent the molecular orbital energy levels—lower than the 1s levels for σ_{1s} and higher than the 1s levels for σ_{1s}^*.

EXAMPLE 11-5 Relating Bond Energy and Bond Order

The bond energy of H_2 is 436 kJ/mol. Estimate the bond energies of H_2^+ and He_2^+.

Analyze

The strength of a bond is directly proportional to its bond order. If we double the bond order, we double the strength (approximately).

Solve

The bond order in H_2 is 1, equivalent to a single bond. In both H_2^+ and He_2^+, the bond order is $\frac{1}{2}$. We should expect the bonds in these two species to be only about half as strong as in H_2—about 220 kJ/mol.

Assess

The actual bond energies are 255 kJ/mol and 251 kJ/mol for H_2^+ and He_2^+, respectively, showing that our approximation is reasonable.

PRACTICE EXAMPLE A: The bond energy of Li_2 is 106 kJ/mol. Estimate the bond energy of Li_2^+.

PRACTICE EXAMPLE B: Do you think the ion H_2^- is stable? Explain.

🔍 **11-5 CONCEPT ASSESSMENT**

A ground state H_2 molecule can absorb electromagnetic radiation to form the ion H_2^+ or an excited state with an electron promoted to the σ_{1s}^* orbital. Which process requires the greater amount of energy? Which species is most stable?

Molecular Orbitals of the Second-Period Elements

For diatomic molecules and ions of H and He, we had to combine only $1s$ orbitals. In the second period, the situation is more interesting because we must work with both $2s$ and $2p$ orbitals. This results in *eight* molecular orbitals. Let's see how this comes about.

The molecular orbitals formed by combining $2s$ atomic orbitals are similar to those from $1s$ atomic orbitals, except they are at a higher energy. The situation for combining $2p$ atomic orbitals, however, is different. Two possible ways for $2p$ atomic orbitals to combine into molecular orbitals are shown in Figure 11-23: end-to-end and side-to-side. The best overlap for p orbitals is along a straight line (that is, end-to-end). This combination produces σ-type molecular orbitals: σ_{2p} and σ_{2p}^*. In forming the bonding and antibonding combinations along the internuclear axis, we must take into account the phase of the $2p$ orbitals. We set up the atomic orbitals as shown in Figure 11-23(a), with the positive (blue) lobe of each function pointing to the internuclear region. Then, since the wave functions are in phase, the addition of the two wave functions leads to an increase of electron density in the internuclear region and produces a σ_{2p} orbital. When the two atomic orbitals are set up as shown in Figure 11-23(b), with lobes of opposite phase pointing into the internuclear region, a nodal plane midway between the nuclei is formed, leading to an antibonding σ_{2p}^* orbital.

Only one pair of p orbitals can combine in an end-to-end fashion. The other two pairs must combine in a parallel or side-to-side fashion to produce π-type molecular orbitals: π_{2p} and π_{2p}^*. The two possible ways for the side-to-side combination of a pair of $2p$ orbitals are shown in Figure 11-23(c and d). The π_{2p} bonding orbital (Fig. 11-23c) is formed by adding the p orbital on one nucleus to a p orbital on the other nucleus, in such a way that the positive and negative lobes of one orbital are in phase with the positive and negative lobes of the other p orbital on the other nucleus. This produces additional electron density between the nuclei, but in a much less direct way than in the σ orbital because the additional electron density is not found along the internuclear axis. Typically, the π bond is weaker than the σ bond. The π_{2p}^* antibonding orbital is

► In Figure 11-23, the simplified p orbitals are used rather than their probability-density plots because our emphasis is on the phases of the orbitals that are being combined.

KEEP IN MIND

that the different colors of the orbitals depicted in these various figures represent the phases of the orbitals.

KEEP IN MIND

that subtracting two wave functions that are in phase is equivalent to *adding* the same functions when they are out of phase.

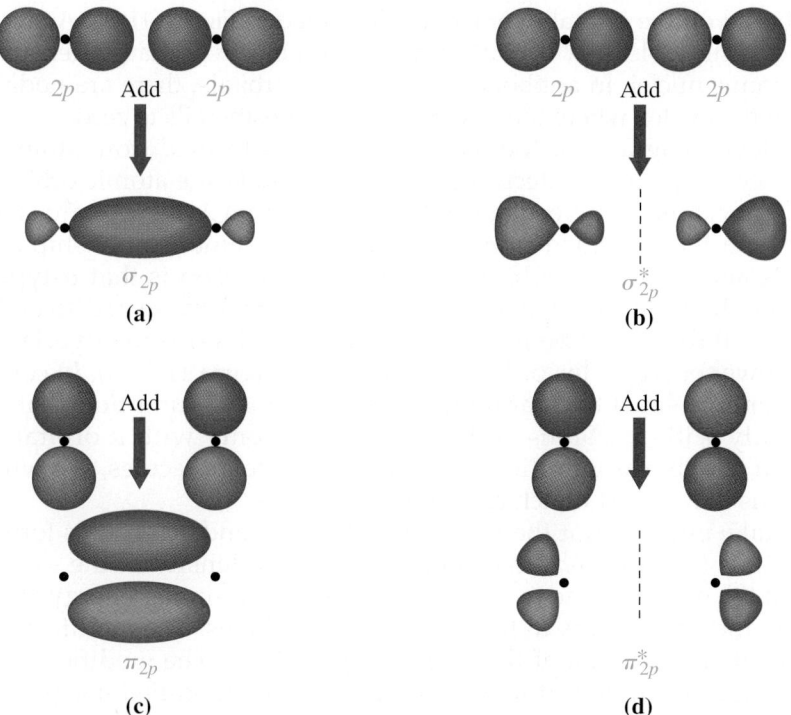

◄ **FIGURE 11-23**
Formation of bonding and antibonding orbitals from 2p orbitals
(a) The addition of two 2p orbitals in phase along the internuclear axis to form a σ_{2p} molecular orbital. This orbital produces electron density between the nuclei, leading to a chemical bond. **(b)** The addition of two 2p orbitals out of phase to produce an antibonding σ_{2p}^* orbital. This orbital has a nodal plane perpendicular to the internuclear axis, as do all antibonding orbitals. **(c)** The addition of two 2p orbitals in phase perpendicular to the internuclear axis to form a π_{2p} molecular orbital. This orbital produces electron density between the nuclei, contributing to a multiple chemical bond. **(d)** The addition of two 2p orbitals out of phase to produce a π_{2p}^* antibonding orbital with a nodal plane.

formed by subtracting the two *p* orbitals perpendicular to the internuclear axes, as shown in Figure 11-23(d). Now, in addition to the nodal plane that contains the nuclei, a node is formed between the nuclei, and this is a characteristic of antibonding character. There are actually *four* π-type molecular orbitals (two bonding and two antibonding) because there are *two* pairs of 2p atomic orbitals arranged in a parallel fashion. Figure 11-24 depicts the approximate probability

◄ Note that assigning blue to the positive lobe is quite arbitrary. The important aspect is that the orbitals are in phase to create a bonding orbital.

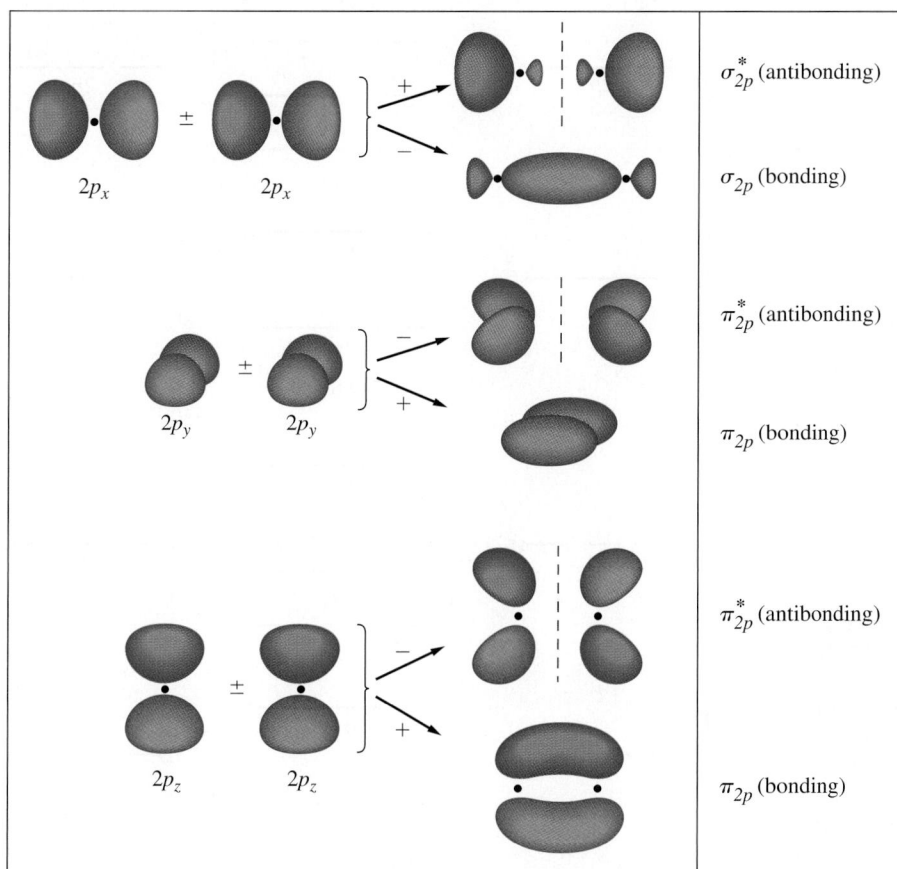

◄ **FIGURE 11-24**
Combining 2p atomic orbitals
These diagrams suggest the electron charge distributions for several orbitals. They are not exact in all details. A dashed line is used to represent a nodal plane that is perpendicular to the internuclear axis.

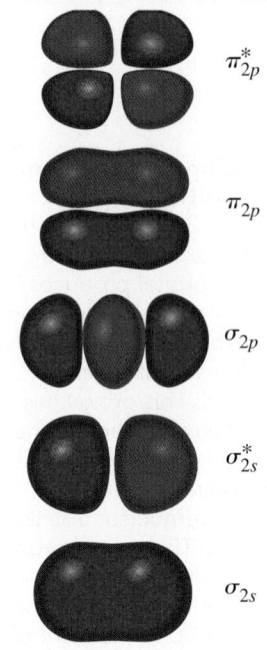

Some computed molecular orbitals of F_2

distributions based on the orbital interactions described in Figure 11-23. Again, we see that bonding molecular orbitals place a high electron charge density between the atomic nuclei. In antibonding molecular orbitals, there are nodal planes between the nuclei, where the electron charge density falls to zero.

The energy-level diagram for the molecular orbitals formed from atomic orbitals of the second principal electronic shell is related to the atomic orbital energy levels. For example, molecular orbitals formed from $2s$ orbitals are at a lower energy than those formed from $2p$ orbitals—the same relationship as between the $2s$ and $2p$ atomic orbitals. Another expectation is that σ-type bonding orbitals should have lower energies than π-type because end-to-end overlap of $2p$ orbitals should be more extensive than side-to-side overlap, resulting in a lower energy. This ordering is shown in Figure 11-25(a). In constructing this energy-level diagram, we have made the assumption that s orbitals mix only with s orbitals and p orbitals mix only with p orbitals. However, if we use this assumption for some diatomic molecules, we will make predictions that do not match experimental results.

We need to take into account the fact that both the $2s$ and $2p$ orbitals form molecular orbitals (σ_{2s} and σ_{2p}) that produce electron density in the same region between the nuclei. These two σ orbitals are of such similar energy and shape that they themselves mix to form modified σ orbitals. The modified σ orbitals each contain a fraction of the original σ_{2s} and σ_{2p}. The modified σ_{2s} (with some σ_{2p} mixed in) goes down in energy, and the modified σ_{2p} (with some σ_{2s} mixed in) goes up in energy, producing a different ordering of energy levels. The important aspect of this mixing is that the modified σ_{2p} is pushed up in energy above the π_{2p} orbitals (Fig. 11-25b).

For the molecular orbitals in O_2 and F_2, the situation is as expected because the energy difference between the $2s$ and $2p$ orbitals is large, and little s and p mixing takes place; that is, the σ_{2s} and σ_{2p} orbitals are *not* modified as

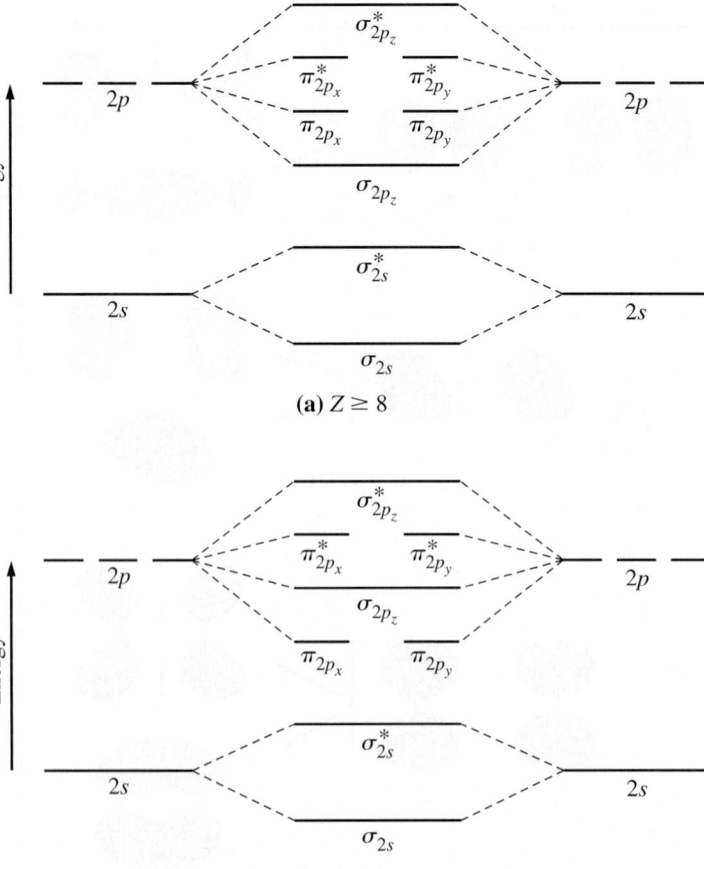

▶ FIGURE 11-25
The two possible molecular orbital energy-level schemes for diatomic molecules of the second-period elements
(a) The expected ordering when σ_{2p} lies below the π_{2p}. This is the ordering for elements with $Z \geq 8$.
(b) The modified ordering due to s and p orbital mixing when σ_{2p} lies above the π_{2p}. This is the ordering for elements with $Z \leq 7$. In this figure we have assumed that the z-axis is the internuclear axis.

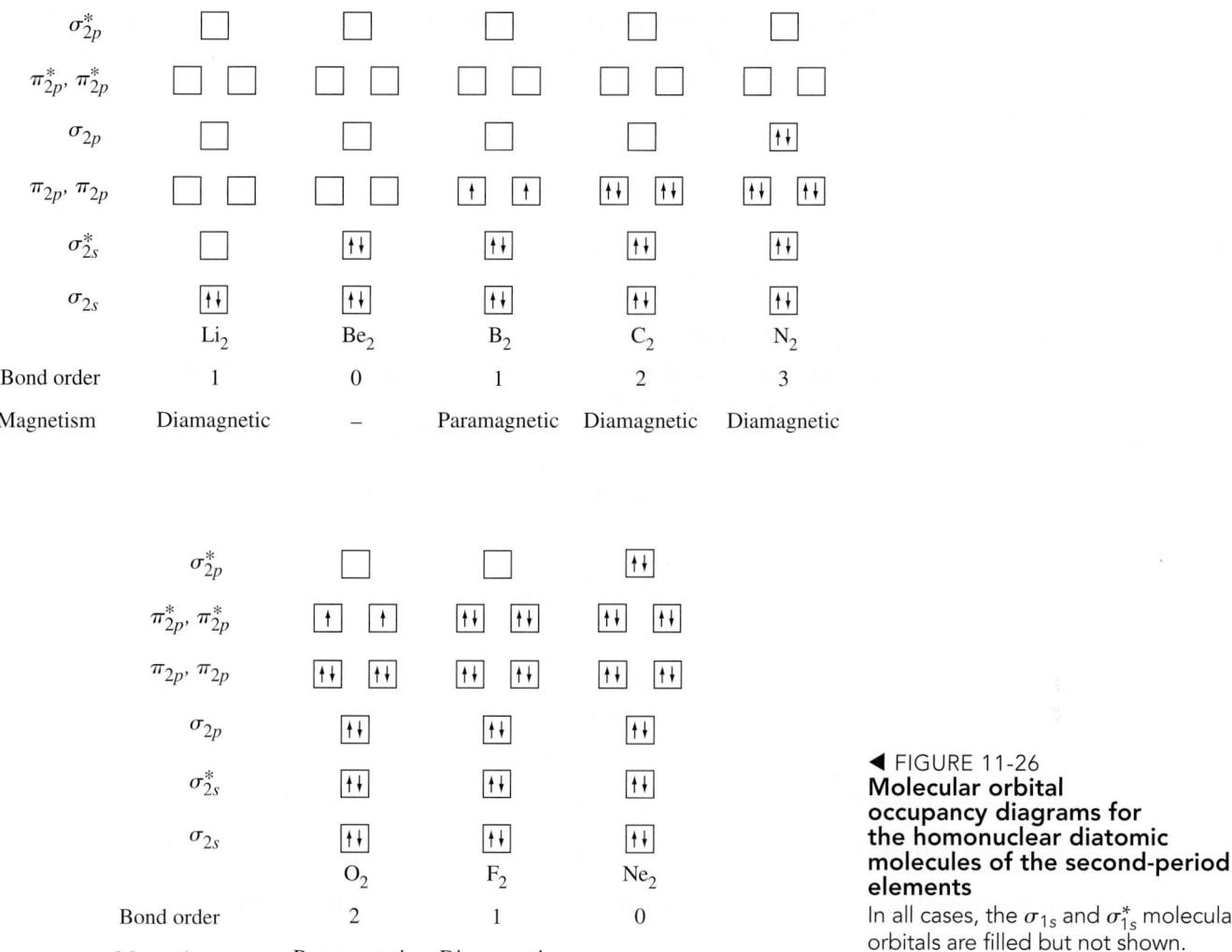

◀ FIGURE 11-26
Molecular orbital occupancy diagrams for the homonuclear diatomic molecules of the second-period elements
In all cases, the σ_{1s} and σ_{1s}^* molecular orbitals are filled but not shown.

described above. For other diatomic molecules of the second-period elements (for example, C_2 and N_2), the π_{2p} orbitals are at a lower energy than σ_{2p} because the energy difference between the $2s$ and $2p$ orbitals is smaller, and $2s–2p$ orbital interactions affect the way in which atomic orbitals combine. This leads to the modified σ_{2s} and σ_{2p} orbitals described above.

Here is how we assign electrons to the molecular orbitals of the diatomic molecules of the second-period elements: We start with the σ_{1s} and σ_{1s}^* orbitals filled. Then we add electrons, in order of increasing energy, to the available molecular orbitals of the second principal shell. Figure 11-26 shows the electron assignments for the homonuclear diatomic molecules of the second-period elements. Some molecular properties are also listed in the figure.

Just as we might arrange the valence-shell atomic orbitals of an atom, we can arrange the second-shell molecular orbitals of a diatomic molecule in the order of increasing energy. Then we can assign electrons to these orbitals, thereby obtaining a *molecular orbital diagram*. If we assign the eight valence electrons of the molecule C_2 to the diagram in Figure 11-25(a), we obtain

$$\boxed{\uparrow\downarrow}\ \boxed{\uparrow\downarrow}\ \boxed{\uparrow\downarrow}\quad \boxed{\uparrow}\ \boxed{\uparrow}\quad \boxed{\ }\ \boxed{\ }\quad \boxed{\ }$$
$$\quad \sigma_{2s}\ \ \sigma_{2s}^*\ \ \sigma_{2p}\qquad\ \pi_{2p}\qquad\quad \pi_{2p}^*\qquad\ \sigma_{2p}^*$$

Experiment shows that the C_2 molecule is diamagnetic, not paramagnetic, and the configuration just described is *incorrect*. So here we see the importance

◀ A *homonuclear* diatomic molecule (X_2) is one in which both atoms are of the same kind. A *heteronuclear* diatomic molecule (XY) is one in which the two atoms are different.

KEEP IN MIND

that Hund's rule applies to molecules as well as atoms.

▶ Just as we wrote an orbital diagram for an atom and a diatomic molecule, we can also write electron configurations for an atom and a diatomic molecule. Thus, for the ground state of C_2, we have (omitting the σ orbitals formed by the 1s orbitals) $\sigma_{2s}^2 \sigma_{2s}^{*2} \pi_{2p}^4$.

▶ Instead of writing the molecular orbital diagram, we can write the ground state configuration of O_2 as $\sigma_{2s}^2 \sigma_{2s}^{*2} \sigma_{2p}^2 \pi_{2p}^4 \pi_{2p}^{*2}$. When writing the electron configuration for a molecule, we follow the convention we used when writing the electron configuration for an atom. That is, we write the orbital designation by placing the number of electrons as a superscript.

of the modified energy-level diagram in Figure 11-25(b). Assignment of eight electrons to the following molecular orbital diagram is consistent with the observation that C_2 is diamagnetic.

$$\boxed{\uparrow\downarrow}\quad\boxed{\uparrow\downarrow}\quad\boxed{\uparrow\downarrow\,|\,\uparrow\downarrow}\quad\boxed{}\quad\boxed{\,|\,}\quad\boxed{}$$
$$\quad\;\sigma_{2s}\quad\;\;\sigma_{2s}^*\quad\;\;\;\pi_{2p}\qquad\;\sigma_{2p}\qquad\;\pi_{2p}^*\qquad\;\;\sigma_{2p}^*$$

This modified energy-level diagram is used for homonuclear diatomic molecules involving elements with atomic numbers from three through seven.

A Special Look at O_2

Molecular orbital theory helps us understand some of the previously unexplained features of the O_2 molecule. Each O atom brings six valence electrons to the diatomic molecule, O_2. In the molecular orbital diagram below, we see that when 12 valence electrons are assigned to molecular orbitals, the molecule has *two unpaired electrons*. This explains the paramagnetism of O_2 (see page 401).

$$O_2\quad\boxed{\uparrow\downarrow}\quad\boxed{\uparrow\downarrow}\quad\boxed{\uparrow\downarrow}\quad\boxed{\uparrow\downarrow\,|\,\uparrow\downarrow}\quad\boxed{\uparrow\,|\,\uparrow}\quad\boxed{}$$
$$\qquad\quad\sigma_{2s}\qquad\sigma_{2s}^*\qquad\sigma_{2p}\qquad\pi_{2p}\qquad\;\pi_{2p}^*\qquad\sigma_{2p}^*$$

(11.2)

There are eight valence electrons in bonding orbitals and four in antibonding orbitals so the bond order is two. A bond order of two corresponds to a covalent double bond.

EXAMPLE 11-6   Writing a Molecular Orbital Diagram and Determining Bond Order

Represent bonding in O_2^+ with a molecular orbital diagram, and determine the bond order in this ion.

Analyze

The O_2^+ ion has 11 valence electrons. We can assign these to the available molecular orbitals in accordance with the ideas stated on page 466. Alternatively, we can remove one electron from an orbital in the molecular orbital diagram given in equation (11.2).

Solve

In the following diagram, there is an excess of *five* bonding electrons over antibonding ones. The bond order is 2.5.

$$O_2^+\quad\boxed{\uparrow\downarrow}\quad\boxed{\uparrow\downarrow}\quad\boxed{\uparrow\downarrow}\quad\boxed{\uparrow\downarrow\,|\,\uparrow\downarrow}\quad\boxed{\uparrow\,|\,}\quad\boxed{}$$
$$\qquad\quad\sigma_{2s}\qquad\sigma_{2s}^*\qquad\sigma_{2p}\qquad\pi_{2p}\qquad\;\pi_{2p}^*\qquad\sigma_{2p}^*$$

Assess

The electronic configuration of O_2^+ is $\sigma_{2s}^2 \sigma_{2s}^{*2} \sigma_{2p}^2 \pi_{2p}^4 \pi_{2p}^{*1}$. Note that in this diatomic molecule, the molecular orbital diagram without 2s–2p mixing is used, because the 2s–2p separation in oxygen is large.

PRACTICE EXAMPLE A: Refer to Figure 11-26. Write a molecular orbital diagram, determine the bond order, and write the electronic configurations of **(a)** N_2^+; **(b)** Ne_2^+; **(c)** C_2^{2-}.

PRACTICE EXAMPLE B: The bond lengths for O_2^+, O_2, O_2^-, and O_2^{2-} are 112, 121, 128, and 149 pm, respectively. Are these bond lengths consistent with the bond order determined from the molecular orbital diagram? Explain.

🔍 **11-6**   **CONCEPT ASSESSMENT**

In valence bond theory, π bonds are always accompanied by a σ bond. Can the same also be said of the molecular orbital theory for diatomic molecules?

A Look at Heteronuclear Diatomic Molecules

The ideas that we have developed for homonuclear diatomic species can be extended, with some care, to give us an idea of the bonding in heteronuclear diatomic species. To illustrate how to construct a molecular orbital diagram for a heteronuclear diatomic molecule, let us consider the molecule carbon monoxide, CO. The first task is to decide whether or not there is $2s–2p$ mixing, based on our knowledge of how Z_{eff} varies across the period. The Z_{eff} of carbon is much lower than that of oxygen, and so the $2s$ and $2p$ orbitals of carbon are higher in energy than those of oxygen. This means that the contributions by each atom to the molecular orbitals will be unequal (we will discuss this in more detail later). In addition, the $2s–2p$ separation in carbon is much lower than in oxygen, and so we expect $2s–2p$ mixing to occur in the σ orbitals.

The molecular orbitals for CO are shown in Figure 11-27. As a result of $2s$ and $2p$ orbital mixing, the third σ orbital, 3σ, has an orbital energy greater than that of the π_{2p} orbitals, and therefore the *highest occupied molecular orbital* (HOMO) is 3σ. Figure 11-27 does not show the individual contributions made by the carbon and oxygen atomic orbitals to the 3σ or other σ orbitals. To visualize the σ molecular orbitals, we must use the results of quantum mechanical calculations. The results of such calculations are shown in Figure 11-27. Two interesting features are evident. First, the HOMO (3σ) is dominated by the oxygen $2s$ orbital and is essentially nonbonding. Second, the 1σ orbital is dominated by the oxygen $2s$ orbital and is also essentially nonbonding. A *nonbonding molecular orbital* has the same energy as the atomic orbitals from which it is formed, and it neither adds to nor detracts from bond formation. The bonding in CO comes from two electrons in the 2σ orbital and two electron pairs in the degenerate 1π orbitals, producing an effective bond order of 3. The configuration of CO (ignoring the $1s$ orbitals) can therefore be written as follows:

$$1\sigma^2 2\sigma^2 1\pi^4 3\sigma^2$$

where, as indicated above, we have lost the designations, such as σ_{2s}, through $2s$ and $2p$ mixing. As an approximation, we can pretend we still have these designations and write the configuration as follows:

$$\sigma_{2s}^2 \sigma_{2s}^{*2} 1\pi^4 \sigma_{2p}^2$$

This still gives a bond order of 3, but the source of chemical bonding is misassigned. Bearing this in mind, we can still use this approach to estimate bond orders. When we consider the free radical NO, the extra electron is in the 2π orbital; thus, the configuration is

$$1\sigma^2 2\sigma^2 1\pi^4 3\sigma^2 2\pi^1$$

◄ For a discussion of the orbital structure of CO, see the Focus On feature for Chapter 11, Photoelectron Spectroscopy on the MasteringChemistry website, www.masteringchemistry.com.

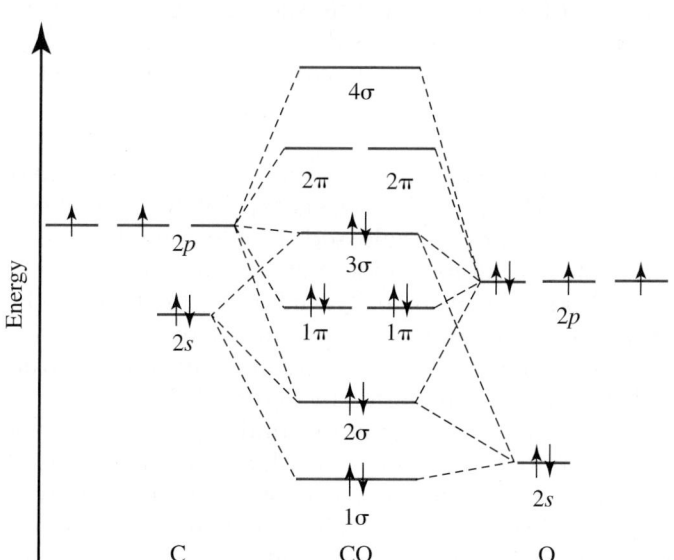

◄ FIGURE 11-27
The molecular orbital diagram of CO
Notice that the CO molecular orbitals are ordered in the same manner as the molecular orbitals in Figure 11-25(b). The σ orbitals for CO are labeled 1σ, 2σ, 3σ, and 4σ, because these orbitals are mixtures of $2s$ and $2p$ orbitals from both C and O.

The bond order becomes 2.5. We expect the bond energy in NO to be less than in CO, as is observed experimentally.

EXAMPLE 11-7   **Writing Electron Configurations for Heteronuclear Diatomic Species**

Write the molecular orbital for the cyanide ion, CN^-, and determine the bond order for this ion.

Analyze

The ion CN^- is isoelectronic with CO. Both elements have a low $2s$–$2p$ gap so that we expect $2s$–$2p$ mixing. The 3σ orbital is higher in energy than the 1π orbitals.

Solve

The number of valence electrons to be assigned to the molecular orbitals is $10(4 + 5 + 1)$. Because of the $2s$–$2p$ mixing, we use the modified order of molecular orbitals and write the configuration of CN^- as

$$1\sigma^2 2\sigma^2 1\pi^4 3\sigma^2$$

Or, as an approximation,

$$\sigma_{2s}^2 \sigma_{2s}^{*2} 1\pi^4 \sigma_{2p}^2$$

which gives a bond order of $(2 + 4 + 2 - 2)/2 = 3$.

Assess

As expected the bond order in CN^- is 3, as it is in the isoelectronic molecule CO. In addition the Lewis structure also gives a triple bond.

PRACTICE EXAMPLE A: Write the electron configuration for CN^+, and determine the bond order.

PRACTICE EXAMPLE B: Write the electron configuration for BN, and determine the bond order.

🔍 **11-7   CONCEPT ASSESSMENT**

Would you expect NeO to be a stable molecule?

▶ The term *aromatic* relates to the fragrant aromas associated with some (but by no means all) of these compounds.

▲ Dame Kathleen Lonsdale first determined the X-ray crystal structure of benzene. Her experiment demonstrated that the benzene molecule is flat, as predicted by theorists.

11-6   Delocalized Electrons: Bonding in the Benzene Molecule

In Section 11-4 we discussed localized π bonds, such as those in ethylene, C_2H_4. Some molecules, such as benzene (C_6H_6) and substances related to it—aromatic compounds—have a network of π bonds. In this section we will use the bonding theories we have studied thus far to consider bonding in benzene; the conclusions we reach will help us understand other cases of bonding as well.

Bonding in Benzene

In 1865, Friedrich Kekulé advanced the first good proposal for the structure of benzene. He suggested that the C_6H_6 molecule consists of a flat, hexagonal ring of six carbon atoms joined by alternating single and double covalent bonds. Each C atom is joined to two other C atoms and to one H atom. To explain the fact that the carbon-to-carbon bonds are all alike, Kekulé suggested that the single and double bonds continually oscillate from one position to the other. Today, we say that the two possible Kekulé structures are actually contributing structures to a resonance hybrid. This view is suggested by Figure 11-28.

We can gain a more thorough understanding of bonding in the benzene molecule by combining the valence-bond and molecular orbital methods. A σ-bond framework for the observed planar structure can be constructed with 120° bond angles by using sp^2 hybridization at each carbon atom. End-to-end overlap of the sp^2 orbitals produces σ bonds. The six remaining $2p$ orbitals are

(a)   (b)

(c)

◄ FIGURE 11-28
Resonance in the benzene molecule and the Kekulé structures
(a) Lewis structure for C_6H_6 showing alternate carbon-to-carbon single and double bonds. (b) Two equivalent Kekulé structures for benzene. A carbon atom is at each corner of the hexagonal structure, and a hydrogen atom is bonded to each carbon atom. (The symbols for carbon and hydrogen, as well as the C—H bonds, are customarily omitted in these structures.) (c) A space-filling model.

used to construct the delocalized π bonds. Figure 11-29 gives a valence-bond theory representation of bonding in C_6H_6.

We do not need to think in terms of an oscillation between two structures (Kekulé) or of a resonance hybrid for benzene. The π bonds are not localized between specific carbon atoms but are spread out around the six-membered ring. To represent this *delocalized* π bonding, the symbol for benzene is often written as a hexagon with an inscribed circle (Fig. 11-29c).

We can best understand delocalized π bonds through molecular orbital theory. Six $2p$ atomic orbitals (see Figure 11-29b) of the C atoms combine to form six molecular orbitals of the π type. Three of these π-type molecular orbitals are bonding, and three are antibonding. What do these π molecular orbitals look like? Recall that as the energy of orbitals increases, the number of nodes increases, so we expect that the lowest π molecular orbital will not possess a node, and this is the case (Fig. 11-30a). All six $2p$ orbitals are in phase as indicated by the fact that all the blue lobes are on one side of the σ framework. The next two π-bonding molecular orbitals each have one node (dashed line in Figure 11-30a) and consequently have the same energy; that is, they are degenerate (Fig. 11-30b). The next pair of orbitals, which are antibonding π orbitals, have two nodes, and the final orbital has three nodes (Fig. 11-30a). The computed π molecular orbitals of benzene are also included in Figure 11-30(c).

The three bonding orbitals fill with six electrons (one $2p$ electron from each C atom), and the three antibonding orbitals remain empty. The bond order associated with the six electrons in π-bonding molecular orbitals is $(6 - 0)/2 = 3$. The three bonds are distributed among the six C atoms, which amounts to $3/6$, or a half-bond, between each pair of C atoms. Add to this the σ bonds in the σ-bond framework, and we have a bond order of 1.5 for each carbon-to-carbon bond. This is exactly what we also get by averaging the two Kekulé structures of Figure 11-28.

▼ FIGURE 11-29
Bonding in benzene, C_6H_6, by the valence-bond method
(a) Carbon atoms use sp^2 and p orbitals. Each carbon atom forms three σ bonds, two with neighboring C atoms in the hexagonal ring and a third with a H atom. (b) The unhybridized $2p$ orbitals on the carbon atoms of benzene, which produce the delocalized π bonds in benzene. (c) Because the π bonding is delocalized around the benzene ring, the molecule is often represented by a hexagon with an inscribed circle.

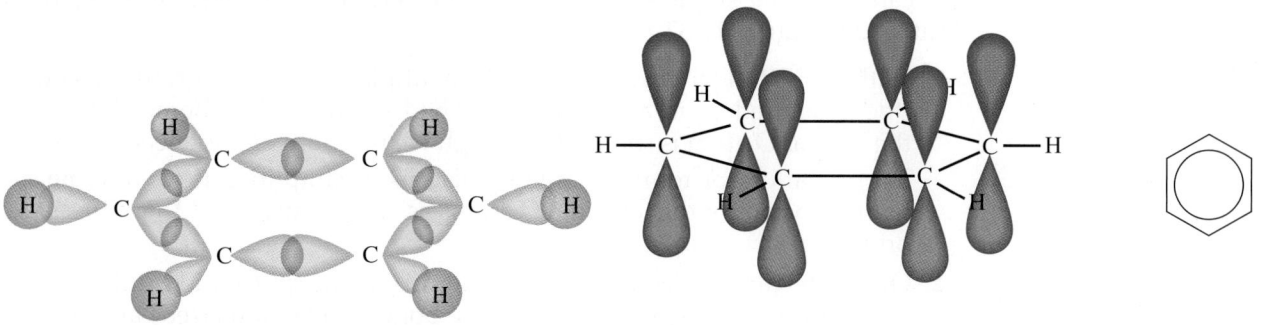

(a) σ bond framework   (b) Carbon $2p$ orbitals to be used in π bonding   (c) Symbolic representation

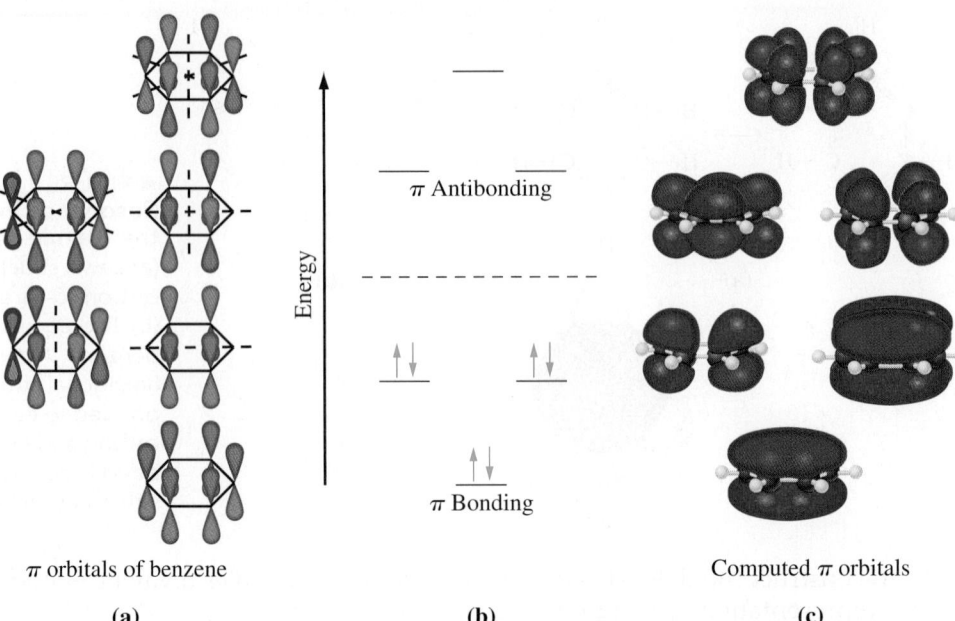

π orbitals of benzene

(a)

π Antibonding

- - - - - - - - - - - -

π Bonding

(b)

Computed π orbitals

(c)

▲ FIGURE 11-30
π molecular orbital diagram for C_6H_6
Of the six π molecular orbitals, three are bonding orbitals and each of these is filled with an electron pair. The three antibonding molecular orbitals at higher energy remain empty.

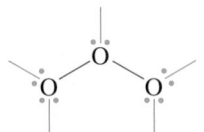

▲ FIGURE 11-31
Molecular orbital representation of π bonding in benzene
The computed π molecular orbitals of benzene.

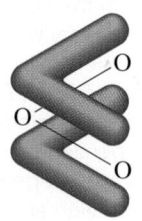

(a) σ bond framework

(b) Delocalized π molecular orbital

▲ FIGURE 11-32
Structure of the ozone molecule, O_3
(a) The σ-bond framework and the assignment of bond-pair (l) and lone-pair (:) electrons to sp^2 hybrid orbitals are discussed in points 1 and 2.
(b) The π molecular orbitals and assignments of electrons to them are discussed in points 3 and 4.

The three bonding π molecular orbitals in C_6H_6 describe the distribution of π electron charge in the molecule. We can think of this in terms of two doughnut-shaped regions: one above and one below the plane of the C and H atoms. Because they are spread out among all six C atoms instead of being concentrated between pairs of C atoms, these molecular orbitals are called **delocalized molecular orbitals**. Figure 11-31 pictures these delocalized molecular orbitals in the benzene molecule. The electrostatic potential maps of benzene displayed at the opening of this chapter also show the buildup of negative charge density above and below the plane of the benzene molecule.

Other Structures with Delocalized Molecular Orbitals

By using delocalized bonding schemes, we can avoid writing two or more contributing structures to a resonance hybrid, as is so often required in the Lewis theory. Consider the ozone molecule, O_3, that we used to introduce the concept of resonance in Section 10-5. In place of the resonance hybrid based on these contributing structures,

we can write the single structure shown in Figure 11-32. Here are the ideas that lead to Figure 11-32.

1. With VSEPR theory, we predict a trigonal-planar electron-group geometry (the measured bond angle is 117°). The hybridization scheme chosen for the central O atom is sp^2, and although we normally do not need to invoke hybridization for terminal atoms, this case is simplified if we assume sp^2 hybridization for the terminal O atoms as well. Thus, each O atom uses the orbital set $sp^2 + p$.

2. Of the 18 valence electrons in O_3, assign 14 to the sp^2 hybrid orbitals of the σ-bond framework. Four of these are bonding electrons (red) and ten are lone-pair electrons (blue).

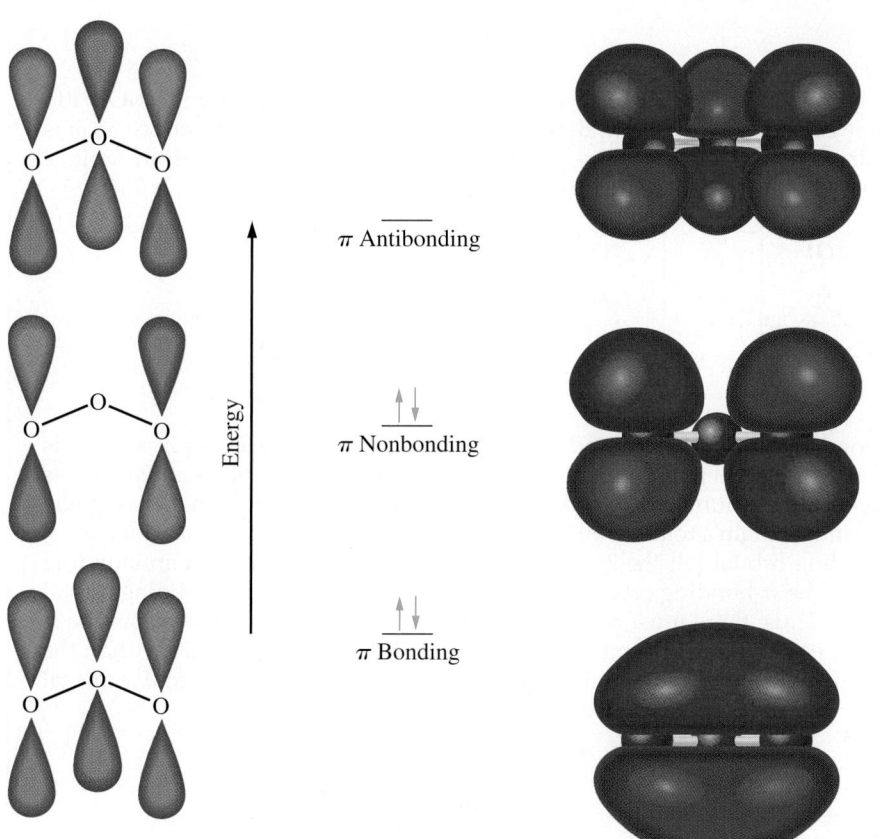

◀ FIGURE 11-33
π bonding orbitals of the ozone molecule O_3

The π-bonding molecular orbital has all the $2p$ orbitals in phase. The π-antibonding molecular orbital has all the three $2p$ orbitals out of phase. The π-nonbonding molecular orbital has a single node and makes zero contribution to the wave function from the central atom. This orbital is called *nonbonding* because there is no region of increased electron density between the central atom and its neighbors. The nonbonding orbital is at the same energy as the original $2p$ orbitals on the oxygen atoms, whereas the π-bonding molecular orbital is stabilized with respect to the original orbitals and the π-antibonding molecular orbital is destabilized by an equal amount with respect to the original $2p$ orbitals. The energy level diagram is shown in the middle of this figure, with the computed π molecular orbitals shown on the right.

3. The three unhybridized $2p$ orbitals combine to form *three* molecular orbitals of the π type (Fig. 11-33). One of these orbitals is a bonding molecular orbital, and the second is antibonding. The third is a type we described on page 473—a *nonbonding molecular orbital*. A nonbonding molecular orbital has the same energy as the atomic orbitals from which it is formed, and it neither adds to nor detracts from bond formation.

4. The remaining four valence electrons are assigned to the π molecular orbitals. Two go into the bonding orbital and two into the nonbonding orbital. The antibonding orbital remains empty.

5. The bond order associated with the π molecular orbitals is $(2 - 0)/2 = 1$. This π bond is distributed between the two O—O bonds and amounts to one-half of a π bond for each.

The points listed here lead to a total bond order of 1.5 for the O—O bonds in O_3. This is equivalent to averaging the two Lewis structures. The O—O bond length suggested by this method was described in Section 10-5.

EXAMPLE 11-8   Representing Delocalized Molecular Orbitals with Atomic Orbital Diagrams

Represent π molecular orbital formation, bonding, and orbital occupancy in the nitrate ion, NO_3^-.

Analyze

We proceed in the manner described on page 476. First, we focus on electrons associated with the σ-bond framework and determine hybridization schemes for the atoms. Then, we determine the number of electrons in the π system, and the number of $2p$ orbitals involved forming the π molecular orbitals. The key to solving this problem is reasoning out the number of each type of molecular orbital (bonding, antibonding, nonbonding) in the π system. The final step is to assign electrons in the π system to the appropriate molecular orbitals.

(continued)

Solve

The total number of electrons in the NO_3^- ion is $(3 \times 6 + 5 + 1) = 24$. Resonance structures for the NO_3^- ion and the corresponding σ bond framework are shown below. Because the N atom is bonded to three atoms, it is sp^2-hybridized. Each O atom may also be considered to be sp^2-hybridized, with one of its sp^2 hybrid orbitals used to form a σ bond with N and other two accommodating the lone pairs.

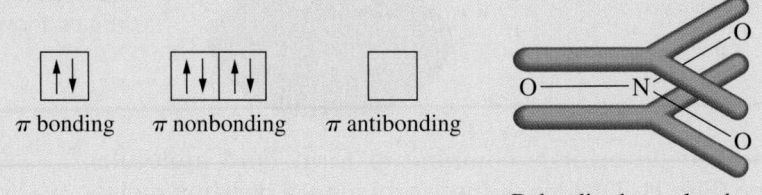

sp^2 σ framework

The number of electrons in the σ bond framework is 18, and so the number of electrons in the π system is $24 - 18 = 6$. These 6 electrons must be assigned to π molecular orbitals obtained from combining four $2p$ orbitals (one on each atom). We must now reason out the number and types of molecular orbitals there are in the π system.

Starting from four $2p$ orbitals, we must obtain a total of four π molecular orbitals. As was the case for ozone, one of the molecular orbitals is a bonding orbital (all the $2p$ orbitals in phase) and another is an antibonding orbital (all the $2p$ orbitals out of phase). The π-bonding orbital is lower in energy and the π-antibonding orbital is higher in energy than the $2p$ orbitals. Thus, there must be $4 - 2 = 2$ *nonbonding* molecular orbitals in the π system. These nonbonding orbitals are equal in energy to the $2p$ orbitals, and thus higher in energy than the bonding molecular orbital but lower in energy than the antibonding molecular orbital. The molecular orbital diagram below illustrates how the six π electrons are assigned to the π molecular orbitals.

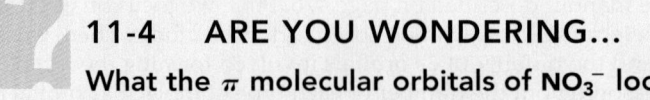

π bonding π nonbonding π antibonding

Delocalized π molecular orbitals of NO_3^-

Thus, the overall π bond order in the NO_3^- ion is one.

Assess

The overall π bond order for this ion is one, and as suggested by the diagram above, the π bond is spread equally over three nitrogen–oxygen bonds. Are You Wondering 11-4 provides schematic representations of the π molecular orbitals of NO_3^-.

PRACTICE EXAMPLE A: Represent chemical bonding in the molecule SO_3 by using a combination of localized and delocalized orbitals.

PRACTICE EXAMPLE B: Represent chemical bonding in the ion NO_2^- by using a combination of localized and delocalized orbitals.

🔍 **11-8   CONCEPT ASSESSMENT**

Would you expect the delocalized π-bonding framework in HCO_2^- to be similar to that in ozone or to that in the nitrate anion?

11-4   ARE YOU WONDERING...

What the π molecular orbitals of NO_3^- look like?

The construction of these orbitals is shown in Figure 11-34. The π-bonding molecular orbital has the four $2p$ orbitals (one on the N atom and one each on the three oxygen atoms) all in phase. The corresponding antibonding π orbital has two nodes and is at the highest energy. Since we are combining four $2p$ orbitals we expect to get four molecular orbitals. The remaining two molecular orbitals will each have one node. The only way to create a molecular orbital with one node in NO_3^- is for the molecular orbital to have a node at the nitrogen atom. Such a molecular orbital must be nonbonding with respect to nitrogen. That the nitrogen does not contribute to the degenerate nonbonding orbitals can be clearly seen in the computed molecular orbitals depicted in Figure 11-34.

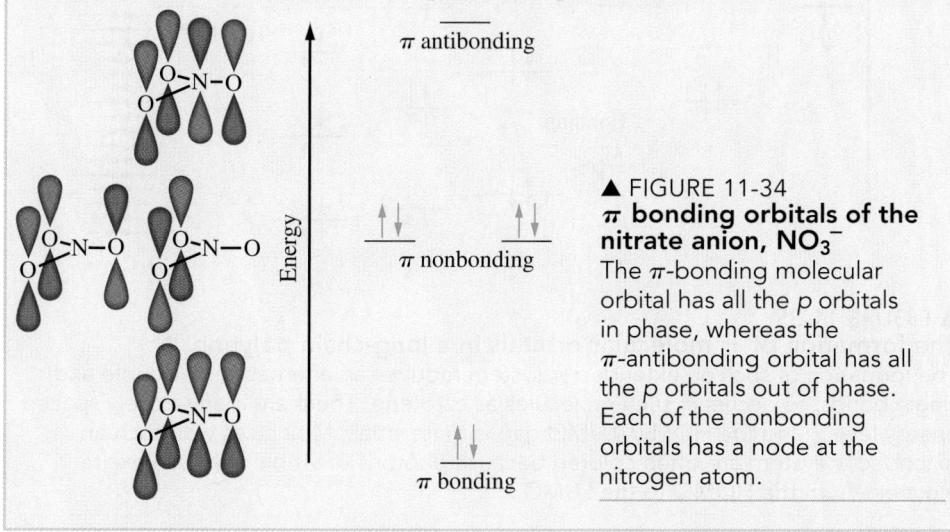

▲ FIGURE 11-34
π bonding orbitals of the nitrate anion, NO_3^-
The π-bonding molecular orbital has all the p orbitals in phase, whereas the π-antibonding orbital has all the p orbitals out of phase. Each of the nonbonding orbitals has a node at the nitrogen atom.

To illustrate the importance of molecular orbital theory, we will see how it is used to explain the colors of plants. Two pigment molecules typically isolated from vegetables are β-carotene

β-Carotene, $C_{40}H_{56}$

found in carrots and leaves, and lycopene

Lycopene, $C_{40}H_{58}$

which is present in tomatoes. The common feature of these molecules is the extended π system. The many p orbitals on the trigonal-planar carbon atoms contribute to many π molecular orbitals. As a consequence, the molecules have many π molecular energy levels that become very closely spaced (see Fig. 11-35).

In molecules with long extended π systems, the *highest occupied molecular orbital* (HOMO) is very close in energy to the *lowest unoccupied molecular orbital*

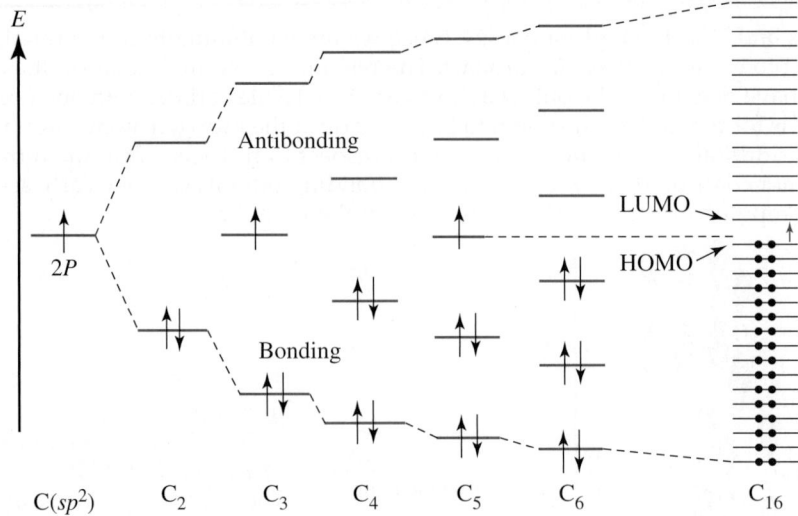

▲ FIGURE 11-35
The formation of π molecular orbitals in a long-chain polyene
The formation of such an extended π system requires an alternation of double and single bonds, as occurs in such molecules as carotene. There are many closely spaced energy levels, and the HOMO–LUMO gap is quite small. Molecules with such an extended π system are often colored because photons of visible light can excite electrons from the HOMO to the LUMO.

(LUMO). As a result it takes very little energy to excite an electron from the HOMO to the LUMO (Fig. 11-35). Photons of visible light have enough energy to excite the electrons across the energy gap between the HOMO and the LUMO, and the absorption of these photons is responsible for the colors that we see.

11-7 Bonding in Metals

In nonmetal atoms, the valence shells generally have more electrons than they do orbitals. To illustrate, a F atom has *four* valence-shell orbitals ($2s$, $2p_x$, $2p_y$, $2p_z$) and *seven* valence-shell electrons. Whether fluorine exists as a solid, liquid, or gas, F atoms join in pairs to form F_2 molecules. One pair of electrons is shared in the F—F bond, and the other electron pairs are lone pairs, as seen in the Lewis structure $:\ddot{F}—\ddot{F}:$. By contrast, the metal atom Li has the same four valence-shell orbitals as F but only *one* valence-shell electron ($2s^1$). This may account for the formation of the *gaseous* molecule Li:Li, but in the solid metal, each Li atom is somehow bonded to *eight* neighbors. The challenge to a bonding theory for metals is to explain how so much bonding can occur with so few electrons. Also, the theory should account for certain properties that metals display to a far greater extent than nonmetals—such as a lustrous appearance, an ability to conduct electricity, and ease of deformation (metals are easily flattened into sheets and drawn into wires).

The Electron Sea Model

An oversimplified theory, which can explain some of the properties of metals just cited, pictures a solid metal as a network of positive ions immersed in a "sea of electrons." In lithium, for instance, the ions are Li^+ and one electron per atom is contributed to the sea of electrons. Electrons in the sea are *free* (not attached to any particular ion), and they are mobile. Thus, if electrons from an external source enter a metal wire at one end, free electrons pass through the wire and leave the other end at the same rate. In this way, electrical conductivity is explained.

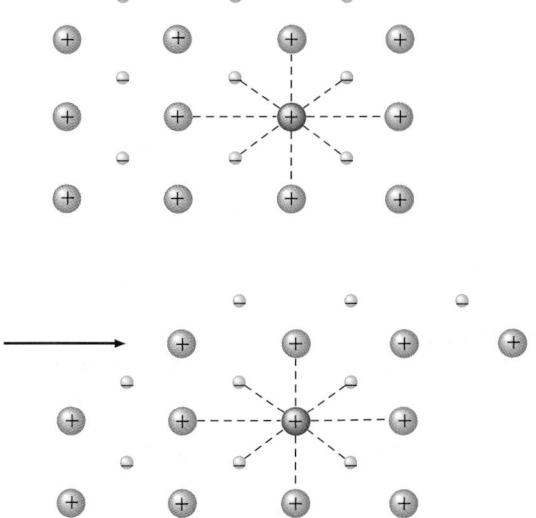

◀ FIGURE 11-36
The electron sea model of metals
A network of positive ions is immersed in a "sea of electrons," derived from the valence shells of the metal atoms and belonging to the crystal as a whole. One particular ion (red), its nearest neighboring ions (brown), and nearby electrons in the electron sea (blue) are emphasized. At the bottom of the figure, a force is applied (from left to right). The highlighted cation is unaffected; its immediate environment is unchanged. The electron sea model explains the ease of deformation of metals.

Free electrons (those in the electron sea) are not limited in their ability to absorb photons of visible light as are electrons bound to an atom. Thus, metals absorb visible light; they are opaque. Electrons at the surface of a metal are able to reradiate—at the same frequency—light that strikes the surface, which explains the lustrous appearance of metals. The ease of deformation of metals can be explained as follows: If one layer of metal ions is forced across another, perhaps by hammering, no bonds are broken, the internal structure of the metal remains essentially unchanged, and the sea of electrons rapidly adjusts to the new situation (Fig. 11-36).

◀ A dull metal surface usually signifies that the surface is coated with a compound of the metal (for example, an oxide, a sulfide, or a carbonate).

Band Theory

The electron-sea model is a simple qualitative description of the metallic state, but for most purposes, the theory of metallic bonding used is a form of molecular orbital theory called **band theory**.

Recall the formation of molecular orbitals and the bonding between two Li atoms (see Figure 11-26). Each Li atom contributes one 2s orbital to the production of two molecular orbitals: σ_{2s} and σ_{2s}^*. The electrons originally described as the $2s^1$ electrons of the Li atoms enter and half-fill these molecular orbitals. That is, they fill the σ_{2s} orbital and leave the σ_{2s}^* empty. If we extend this combination of Li atoms to a third Li atom, three molecular orbitals are formed, containing a total of three electrons. Again, the set of molecular orbitals is half-filled. We can extend this process to an enormously large number (N) of atoms—the total number of atoms in a crystal of Li. Here is the result we get: a set of N molecular orbitals with an extremely small energy separation between each pair of successive levels. This collection of very closely spaced molecular orbital energy levels is called an *energy band* (Fig. 11-37).

In the band just described, there are N electrons (a 2s electron from each Li atom) occupying, in pairs, N/2 molecular orbitals of lowest energy. These are the electrons responsible for bonding the Li atoms together. They are valence electrons, and the band in which they are found is called a *valence band*. Because the energy differences between the occupied and unoccupied levels in the valence band are so small, however, electrons can be easily excited from the highest filled levels to the unfilled levels that lie immediately above them in energy. This excitation, which has the effect of producing mobile electrons, can be accomplished by applying a small electric potential difference across the crystal. This is how the band theory explains the ability of metals to conduct electricity. The essential feature for electrical conductivity, then, is *an energy band that is only partly filled with electrons*. Such an energy band is called a *conduction band*. In lithium, the 2s band is both a valence band and a conduction band (Fig. 11-38a).

◀ Even though bands are formed in metals, the electrons still have to fill according to Pauli's exclusion principle.

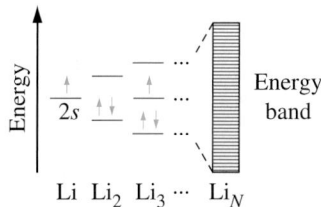

▲ FIGURE 11-37
Formation of an energy band in lithium metal
As more and more Li atoms are added to the growing "molecule," Li₂, Li₃, and so on, additional energy levels are added and the spacing between levels becomes increasingly smaller. In an entire crystal of N atoms, the energy levels merge into a band of N closely spaced levels. The lowest N/2 levels are filled with electrons, and the upper N/2 levels are empty.

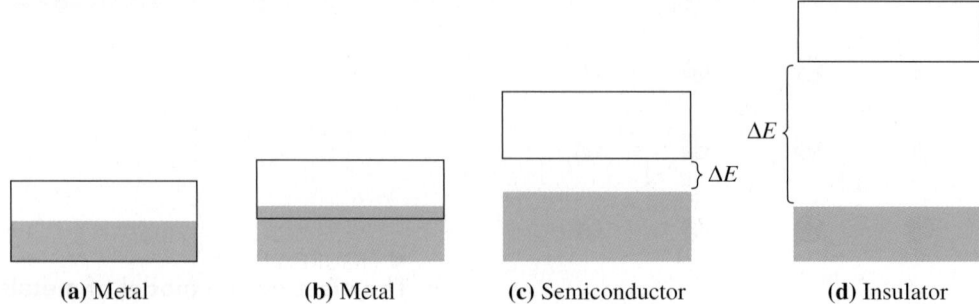

(a) Metal   (b) Metal   (c) Semiconductor   (d) Insulator

▲ FIGURE 11-38
Metals, semiconductors, and insulators as viewed by band theory
(a) In some metals, the valence band (blue) is only partially filled (for example, the half-filled 3s band in Na). The valence band also serves as a conduction band (outlined in black). **(b)** In other metals, the valence band is full, but a conduction band overlaps it (for example, the empty 2p band of Be overlaps the full 2s valence band). **(c)** In a semiconductor, the valence band is full and the conduction band is empty. The energy gap (ΔE) between the two is small enough, however, that some electrons make the transition between them just by acquiring thermal energy. **(d)** In an insulator, the valence band is filled with electrons and a large energy gap (ΔE) separates the valence band from the conduction band. Few electrons can make the transition between bands, and the insulator does not conduct electricity.

Let's extend our discussion to N atoms of beryllium, which has the electron configuration $1s^2 2s^2$. We expect the band formed from 2s atomic orbitals to be filled—N molecular orbitals and $2N$ electrons. But how can we reconcile this with the fact that Be is a good electric conductor? At the same time that 2s orbitals are being combined into a 2s band, 2p orbitals combine to form an *empty* 2p band. The lowest levels of the 2p band are at a *lower* energy than the highest levels of the 2s band. The bands overlap (Fig. 11-38b). As a consequence, empty molecular orbitals are available to the valence electrons in beryllium.

In an *electrical insulator*, like diamond or silica (SiO_2), not only is the valence band filled but there is a large energy gap between the valence band and the conduction band (Fig. 11-38d). Very few electrons are able to make the transition between the two.

Semiconductors

Much of modern electronics depends on the use of semiconductor materials. Light-emitting diodes (LEDs), transistors, and solar cells are among the familiar electronic components that use semiconductors. Such semiconductors as cadmium yellow (CdS) and vermilion (HgS) are brilliantly colored, and artists use them in paints.

What determines the electronic properties of a semiconductor is the energy gap (band gap) between the valence band and the conduction band (Fig. 11-38c). In some materials, such as CdS, this gap is of a fixed size. These materials are called *intrinsic semiconductors*. When white light interacts with the semiconductor, electrons are excited (promoted) to the conduction band. CdS absorbs violet light and some blue light, but other frequencies contain less energy than is needed to excite an electron above the energy gap. The frequencies that are not absorbed are reflected, and the color we see is yellow. Some semiconductors, such as GaAs and PbS, have a sufficiently small band gap that all frequencies of visible light are absorbed. There is no reflected visible light, and the materials have a black color.

In a semiconductor, such as silicon or germanium, the filled valence band and empty conduction band are separated by only a small energy gap. Electrons in the valence band may acquire enough thermal energy to jump to a level in the conduction band. The greater the thermal energy, the more electrons can make

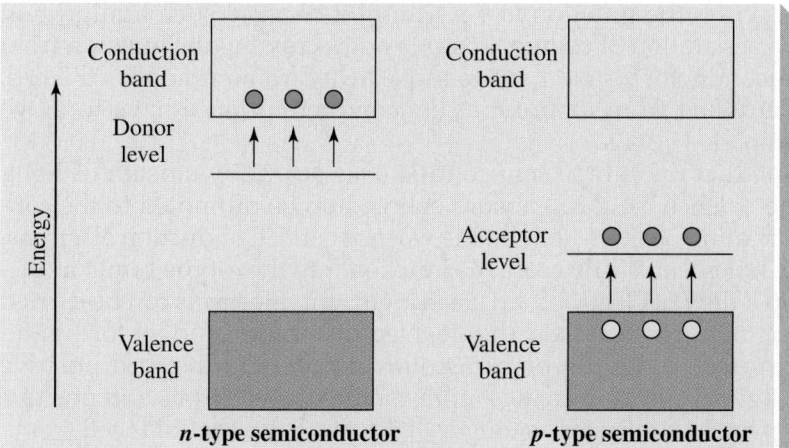

◀ FIGURE 11-39
p- and n-type semiconductors
In a semiconductor with donor atoms (for example, P in Si), the donor level lies just beneath the conduction band. Electrons (●) are easily promoted into the conduction band. The semiconductor is of the *n*-type. In a semiconductor with acceptor atoms (for example, Al in Si), the acceptor level lies just above the valence band. Electrons (●) are easily promoted to the acceptor level, leaving positive holes (○) in the valence band. The semiconductor is of the *p*-type.

the transition. In this way, band theory explains the observation that the electrical conductivity of semiconductors increases with temperature.

In many semiconductors, called *extrinsic semiconductors*, the size of the band gap is controlled by carefully adding impurities—a process called *doping*. Let's consider what doping does to one of the most common semiconductors, silicon.

When silicon is doped with phosphorus, the energy level of the P atoms lies just below the conduction band of the silicon, as shown in Figure 11-39. Each P atom uses *four* of its five valence electrons to form bonds to four neighboring Si atoms, and thermal energy alone is enough to cause the "extra" valence electron to be promoted to the conduction band, leaving behind an immobile positive P^+ ion. The P atoms are called *donor* atoms, and electrical conductivity in this type of semiconductor involves primarily the movement of electrons from donor atoms through the conduction band. This type of semiconductor is called an *n-type*, where *n* refers to negative—the type of electric charge carried by electrons.

When silicon is doped with aluminum, the energy level of the Al atoms, called *acceptor* atoms, lies just above the valence band of the silicon (Fig. 11-39). Because an Al atom has only *three* valence electrons, it forms regular electron-pair bonds with three neighboring Si atoms but only a one-electron bond with a fourth Si atom. An electron is easily promoted from the valence band to an Al atom in the acceptor level, however, forming an immobile negative Al^- ion. When this occurs, a *positive hole* is created in the valence band. Because electrical conductivity in this type of semiconductor consists primarily of the migration of positive holes, it is called a *p-type* semiconductor.

Figure 11-40 suggests how semiconductors are used in photovoltaic (solar) cells. A thin layer of a *p*-type semiconductor is in contact with an *n*-type semiconductor in a region called the *junction*. Normally, migration of electrons and

◀ A natural semiconductor is called an intrinsic semiconductor. Semimetals that become semiconductors because of doping are called extrinsic semiconductors.

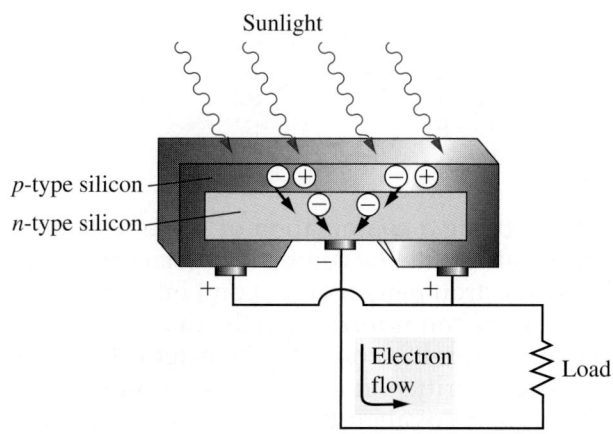

◀ FIGURE 11-40
A photovoltaic (solar) cell using silicon-based semiconductors

positive holes across the junction is very limited because such a migration would lead to a separation of charge: Positive holes crossing the junction from the p-type semiconductor would have to move away from immobile Al^- ions, and electrons crossing from the n-type semiconductor would have to move away from immobile P^+ ions.

Now imagine that the p-type semiconductor is struck by a beam of light. Electrons in the valence band can absorb energy and be promoted to the conduction band, creating positive holes in the valence band. Conduction electrons, unlike positive holes, can easily cross the junction into the n-type semiconductor. This sets up a flow of electrons, an electric current. Electrons can be carried by wires through an external load (lights, electric motors, and so forth) and eventually returned to the p-type semiconductor, where they fill positive holes. Further light absorption creates more conduction electrons and positive holes, and the process continues as long as light shines on the solar cell.

11-9 CONCEPT ASSESSMENT

Do you think GaN is a semiconductor?

11-8 Some Unresolved Issues: Can Electron Charge-Density Plots Help?

In Chapters 10 and 11 we have presented a wide range of views of chemical bonding, from simple Lewis theory to the more advanced valence-bond and molecular orbital approaches. We must emphasize, however, that each of these models has its deficiencies, and their uncritical use can lead to incorrect conclusions.

Here are some of the unresolved issues we have encountered: Employing expanded valence shells in Lewis structures created the quandary of where to accommodate the extra valence electrons in such molecules as SF_6 and SF_4 (page 419), specifically, are d-orbitals used in the bonding description? A related issue is whether to use expanded valence shells to minimize formal charges in anions such as SO_4^{2-} (page 420). Still another issue is whether VSEPR theory or hybridization schemes of the valence-bond method gives the more fundamental view of molecular shapes. Finally, we might wonder how the valence-bond and molecular orbital theories are related. In this section, we will attempt to provide answers to these questions—stressing the significance of electron charge density calculations.

Bonding in the Molecule SF_6

First let's see if it is possible to describe the bonding in SF_6 while maintaining the octet rule. One proposal has been to introduce resonance structures of the form

with the actual electronic structure being a combination of these resonance structures. The resonance structures illustrate the concept of *hyperconjugation*, a situation in which the number of electron pairs used to bond other atoms to a central atom is less than the number of bonds formed. In the case of SF_6, four electron pairs bond six F atoms to a central S atom and the octet rule is preserved. When resonance structures are written in this way, the assumption is that the bond lines represent nonpolar covalent bonds, whereas the other bonds are fully ionic. The molecule SF_6 has equivalent bonds and consequently would require a total of 15 structures of the type shown. This description implies a charge of 2+ on the sulfur and a charge of 1/3− on each F atom, corresponding to a collective charge of 2− on the six F atoms. This description of bonding, then, has the appeal of describing the polarity of the bonds while getting around the problem that arises when the Lewis structure is written as shown below, namely, where do the "extra" electrons go on the S atom?

$$
\begin{array}{ccc}
 & F & \\
 & | & \\
F \cdots\cdots & S & \cdots\cdots F \\
 & | & \\
F & & F \\
 & | & \\
 & F &
\end{array}
$$

Should we use hyperconjugation to describe the bonding in SF_6? One answer is to compare the suggested charges on the S and F atoms with those obtained from a quantum-mechanical calculation. The calculation gives a charge of 3.17+ on sulfur and 0.53− on each fluorine. To describe bonding through hyperconjugation that is in better agreement with the quantum-mechanical calculation, we would have to use additional resonance structures with higher charges and fewer covalent bonds. Such an approach is clearly cumbersome, and adoption of this large number of structures is not justified just to satisfy the octet rule.

The problem in describing molecules with expanded valence shells, so-called *hypervalent* molecules, is that there is no generally accepted way of denoting polar bonds in a structure. Furthermore, we must remember that Lewis devised his "rule of eight" in an era when only a few molecules, such as, PCl_5 and SF_6, were known, and he did not consider these exceptions to be of any great significance. Why was that? Lewis viewed the "rule of two" to be of greater fundamental importance than the rule of eight; that is, the electron charge density caused by the electron pair is paramount in understanding bonding. Bonding in molecules with expanded valence shells is not a consequence of a special type of bonding. Bonds in these molecules are similar to those in other molecules and can vary from predominantly covalent to predominantly ionic.

If we do not use hyperconjugation, how are we to describe where the "extra" electrons go? This question arises because we seem implicitly to think about bonding in terms of hybridization of atomic orbitals. Thus, in order to describe bonding in the methane molecule consistent with its tetrahedral geometry we introduced the concept of sp^3 hybrid orbitals. That is, the hybridization was introduced *because* of the geometry. The geometry of a molecule can be determined only experimentally or estimated by using VSEPR theory.

We extended the concept of hybridization of orbitals to molecules with expanded valence shells to include *d* orbitals, that is, sp^3d and sp^3d^2 hybrid orbitals to accommodate five- and six-electron pairs, respectively. Although this is an appealing idea, it has come into question because of the relatively high energy of the *d* orbitals, and quantum-mechanical calculations have shown that the wave functions contain very little contribution from *d* orbitals. Thus, it appears that to describe bonding in SF_6 we should avoid using *d* orbitals in hybridization schemes.

How are we to proceed? Where are the electrons in the SF_6 molecule? Recall that we have already employed the results of quantum-mechanical calculations in constructing electrostatic potential maps (page 402). Let's turn again to the results of such calculations to improve our understanding of the bonding in SF_6 and similar molecules.

Bonding in the Molecule SCl_2

To continue our discussion consider first a simpler molecule—SCl_2. Figure 11-41 shows how the electron charge density (ρ) varies in the plane that contains the sulfur atom and the two chlorine atoms. The most striking feature of this

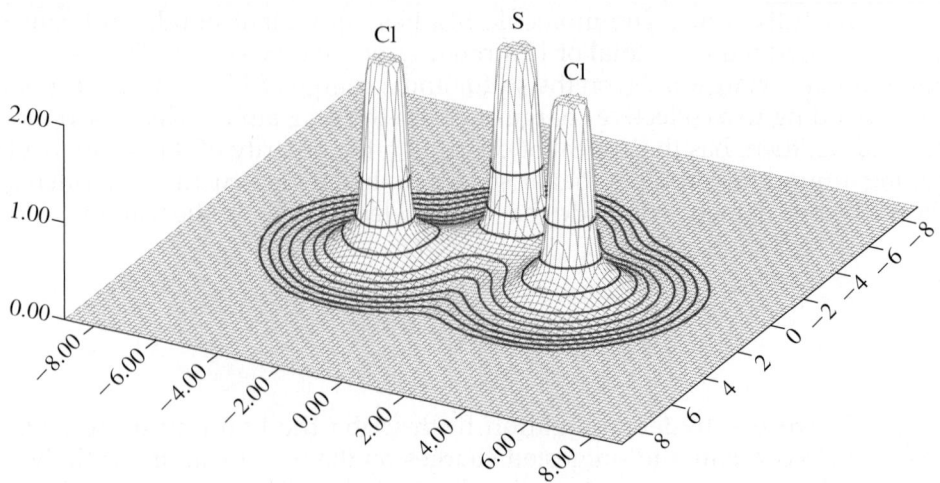

▲ FIGURE 11-41
Isodensity contour map of the electron density of SCl_2 in the plane of the nuclei
The isodensity contour lines, in atomic units (au), are shown in color, in the order 0.001, 0.002, 0.004, 0.008 (four outermost contours in blue); 0.2, 0.4, 0.8 (next three). Densities are truncated at 2.00 au (innermost red contour). The atomic unit of electron charge density $= e/a_0^3 = 1.081 \times 10^{12}$ C m^3, where a_0 is the Bohr radius (adapted from Matta and Gillespie, *J. Chem. Ed.*, 79, 1141, 2002).

diagram is that the electron density is very high at each nucleus; in fact, we have truncated the very large maxima in order to show other features in the diagram. An especially significant feature is the small ridge of increased electron density between the sulfur and each of the chlorine atoms. Despite its modest height, this ridge of electron density is responsible for an attractive force between the nuclei.

An alternative representation of the electron density distribution is a contour map. Such a contour map for SCl_2 is shown in Figure 11-42. The lines drawn between the S atom and each Cl atom in Figure 11-42 represent the lines along the top of the small ridge of electron charge density between these atoms that we saw in Figure 11-41. The lines between the chlorine atoms and the sulfur atom are called the *bond paths* and represent the chemical bond between the S and one Cl atom as we would normally draw it in a Lewis structure. In Figure 11-42, notice the vertical line above the S atom that splits in two just before the sulfur atom is reached, with each segment intersecting the bond paths for the two sulfur-chlorine bonds between the S and a Cl atom. This

KEEP IN MIND

that a contour map represents changes in topology of a surface; similar maps are used by mountaineers to plan their ascent of a mountain.

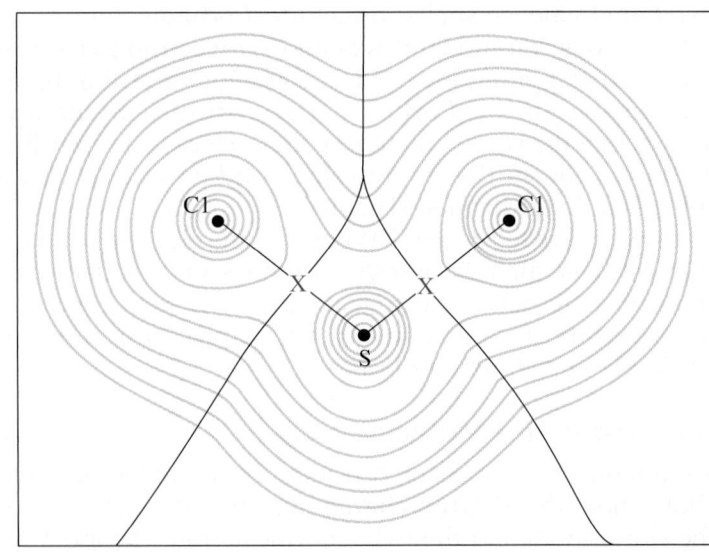

▶ FIGURE 11-42
Contour map of the electron density in SCl_2
The electron charge density increases from the outermost isodensity contour at 0.001 au in incremental steps of 2×10^{-3} au, 4×10^{-2} au, 8×10^{-1} au, 16×10^0 au, and so on. The lines connecting the nuclei are the bond paths. The bond critical points are depicted by red Xs (adapted from Matta and Gillespie, *J. Chem. Ed.*, 79, 1141, 2002).

bifurcated line represents a path tracing the minimum in the electron density, analogous to a path along the valley floor between the "mountains" of electron density. The point where this line intersects each bond line is called the *bond critical point*. The electron density at the bond critical point can be used to describe the type of bond connecting a pair of atoms in a molecule. The greater the electron density, the higher the bond order is.

Bonding in the Molecule H_2SO_4 and the Anion SO_4^{2-}

To decide whether or not to use expanded valence shells to minimize formal charges, we will consider the sulfuric acid molecule and the sulfate anion. Figure 11-43 is a three-dimensional representation of the electron charge distribution surfaces at a value of 0.001 atomic units (au); they correspond to the surface encompassing about 98% of the electron charge density in H_2SO_4 and SO_4^{2-}. If we choose a surface with a higher charge density value, then we include less of the electron charge distribution. Below the 98% surfaces are two electron charge density plots at increasing densities for the surfaces being calculated, also shown in Figure 11-43. What do these tell us? When we reach a density for the calculated surface just greater than the density at the bond critical point, the electron density in that bond disappears and we have established the amount of electron charge density in that bond. If there are bonds in the molecule that have more electron density at their bond critical points, then electron density will still appear in the three-dimensional representation. To illustrate this point, observe that at 0.22 au of electron charge density, the electron density between the sulfur atom and the two oxygen atoms attached to hydrogen atoms has disappeared (the stick from the ball-and-stick model is

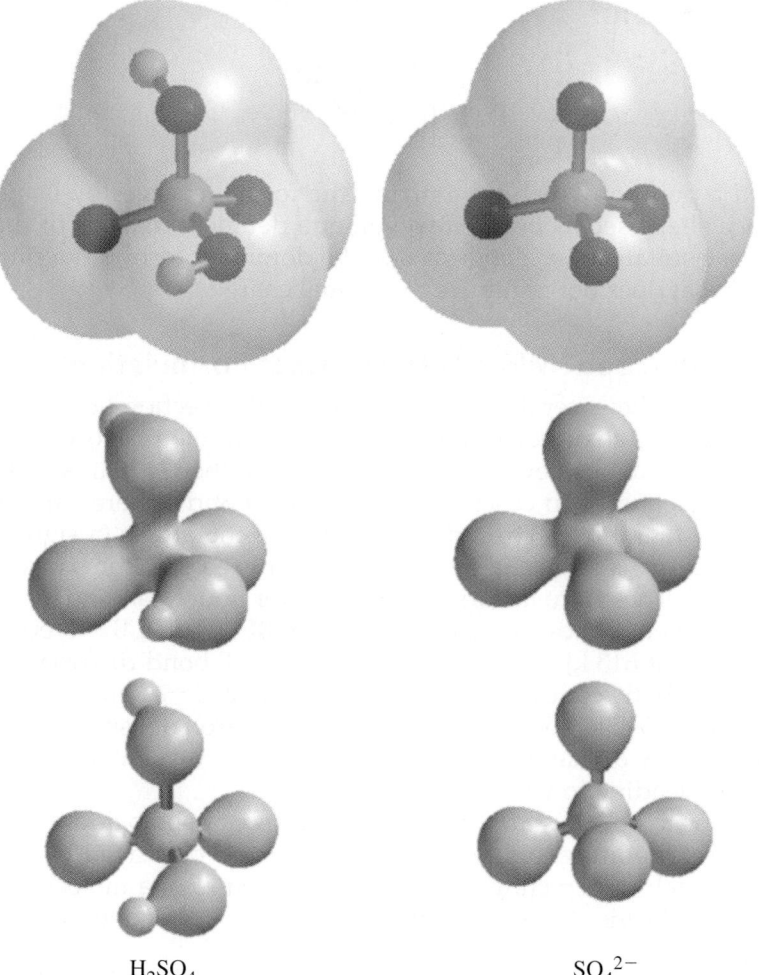

H_2SO_4 SO_4^{2-}

◀ FIGURE 11-43
Three-dimensional plots of the electron density in H_2SO_4 and SO_4^{2-}
The values for the outer isodensity envelope are set at 0.002 au, 0.22 au, and 0.28 au in the three figures for both species.

apparent), but between the sulfur and the two oxygen atoms that do not have a hydrogen atom, it is still visible. The electron density between the sulfur atom and the nonprotonated oxygen (the oxygen atom that does not have a hydrogen attached) atoms does not disappear until the density of the calculated surface is increased to 0.33 au. We conclude that there is more electron density in the bonds between the sulfur atom and the nonprotonated oxygen atoms than in the bonds between sulfur and oxygen attached to a proton. This extra electron density can be represented in a Lewis structure that places a double bond between the central S atom and the two terminal O atoms, thereby reducing the formal charges seen in the octet Lewis structure of H_2SO_4.

In considering the electron density surface for the sulfate anion, we see that the bond critical density is 0.28 au for the four bonds between the sulfur atom and the oxygen atoms. This is similar to that for the bond between sulfur and the nonprotonated oxygen atoms and greater than that for the sulfur–oxygen bonds attached to protonated oxygen atoms. The presence of this higher bond critical point in the sulfate anion suggests that maybe the Lewis structure that minimizes formal charges is the best. All four sulfur–oxygen bonds have the same bond critical point, which corresponds to the possible resonance structures that can be written.

We have reached a point at which minimizing the formal charges by using expanded valence shells seems best. However, one detailed analysis of the wavefunction of the sulfate anion suggests that the dominant form is the simple octet structure that does not minimize the formal charge.*

How Should We Proceed? What Is the Correct Formulation?

Perhaps the answer to the several questions posed in this section lies in the work of R. J. Gillespie, a coauthor of the VSEPR method. He proposes that Lewis structures be written as Lewis would have written them—that is, with no expanded valence shells. In cases of highly polar bonds, there can be, simultaneously, considerable electron density between the atoms (a significantly covalent bond) *and* a large charge separation between the atoms (a significantly ionic bond). These two factors provide a better understanding of the strength of polar covalent bonds in molecules such as BF_3 (with a B—F bond dissociation enthalpy of 613 kJ/mol) or SiF_4 (with a Si—F bond dissociation enthalpy of 567 kJ/mol). By contrast, the nonpolar covalent C—C bond dissociation enthalpy is only 345 kJ/mol. An analysis of the electron densities in BF_3 and SiF_4 shows that they do have a combination of highly covalent and ionic bond characteristics, leading to very strong bonds.

The controversy as to how best to write Lewis structures will no doubt continue in the chemical literature, but you should not be too dismayed by this situation. Our approach to depicting the electronic structure of a molecule is based on the simplest Lewis structure and its concomitant use in determining

*L. Suidan, J. K. Badenhoop, E. D. Glendenning, and F. Weinhold, *J. Chem. Educ.*, **72**, 583 (1995).

the shape of a molecule through VSEPR theory. In order to probe more deeply into the nature of a chemical bond—for example, to understand experimental results, such as bond enthalpy values—we must analyze a computed electron density map for that molecule rather than rely just on the Lewis structure.

Mastering**CHEMISTRY**   **www.masteringchemistry.com**

The orbital structures of molecules are studied using photoelectron spectroscopy. The method involves passing high-energy photons through a gaseous sample of molecules, and measuring the kinetic energies of the ejected electrons. For a discussion of Photoelectron Spectroscopy, go to the Focus On feature for Chapter 11 on the MasteringChemistry site.

Summary

11-1 What a Bonding Theory Should Do—A basic requirement of a bonding theory is that it provide a better description of the electronic structure of molecules than the simple ideas of the Lewis model.

11-2 Introduction to the Valence-Bond Method—**Valence-bond method** considers a covalent bond in terms of the overlap of atomic orbitals of the bonded atoms.

11-3 Hybridization of Atomic Orbitals—Some molecules can be described in terms of the overlap of simple orbitals, but often orbitals that are a composite of simple orbitals—**hybrid orbitals**—are needed. The **hybridization** scheme chosen is the one that produces an orientation of hybrid orbitals to match the electron-group geometry predicted by the VSEPR theory (Fig. 11-13). *sp* **hybrid orbitals** (Fig. 11-10) are associated with linear electron-group geometries; *sp*^{2} **hybrid orbitals** (Fig. 11-9) with trigonal planar geometries; *sp*^{3} **hybrid orbitals** (Fig. 11-6) with tetrahedral geometries; *sp*^{3}*d* **hybrid orbitals** with trigonal bipyramidal geometries; and *sp*^{3}*d*^{2} **hybrid orbitals** (Fig. 11-12) with octahedral geometries.

11-4 Multiple Covalent Bonds—End-to-end overlap of orbitals produces *σ* **(sigma) bonds**. Side-to-side overlap of two *p* orbitals produces a *π* **(pi) bond**. Single covalent bonds are *σ* bonds. A double bond consists of one *σ* bond and one *π* bond (Fig. 11-14). A triple bond consists of one *σ* bond and two *π* bonds (Fig. 11-16). The geometric shape of a species determines the *σ*-bond framework, and *π* bonds are added as required to complete the bonding description.

11-5 Molecular Orbital Theory—In **molecular orbital theory**, electrons are assigned to molecular orbitals. The numbers and kinds of molecular orbitals are related to the atomic orbitals used to generate them. Electron charge density between atoms is high in **bonding molecular orbitals**

and very low in **antibonding orbitals** (Fig. 11-21). **Bond order** is one-half the difference between the numbers of electrons in bonding molecular orbitals and in antibonding molecular orbitals (equation 11.1). Molecular orbital energy-level diagrams and an aufbau process can be used to describe the electronic structure of a molecule; this is similar to what was done for atomic electron configurations in Chapter 8.

11-6 Delocalized Electrons: Bonding in the Benzene Molecule—Bonding in the benzene molecule, C_6H_6, is partly based on the concept of **delocalized molecular orbitals**. These are regions of high electron charge density that extend over several atoms in a molecule (Fig. 11-31). Delocalized molecular orbitals also provide an alternative to the concept of resonance in other molecules and ions.

11-7 Bonding in Metals—Molecular orbital theory, in the form called **band theory**, can be applied to metals, semiconductors, and insulators (Fig. 11-38). Band theory explains thermal and electric conductivity, the ease of deformation, and the characteristic luster of metals. It also explains the colors of semiconductors and the fact that their electric conductivities increase with temperature.

11-8 Some Unresolved Issues—Can Electron Charge-Density Plots Help?—Charge-density plots can be used as a guide in understanding the bonding in molecules that don't necessarily have simple Lewis structures. In molecules such as SF6, we employed the concept of hyperconjugation. For SCl$_2$ the electron density was analyzed in terms of bond paths and bond critical points. For H_2SO_4 and SO_4^{2-}, bond critical points derived from charge density point out that drawing Lewis structures with the lowest formal charge is one of the best ways to represent the bonding in these compounds. In general one should write Lewis structures without expanded valence shells.

Integrative Example

Hydrogen azide, HN_3, and its salts (metal azides) are unstable substances used in detonators for high explosives. Sodium azide, NaN_3, is used in air-bag safety systems in automobiles (see page 210). A reference source lists the following data for HN_3. (The subscripts a, b, and c distinguish the three N atoms from one another.) Bond lengths: $N_a—N_b = 124$ pm; $N_b—N_c = 113$ pm. Bond angles: $H—N_a—N_b = 112.7°$; $N_a—N_b—N_c = 180°$.

Write two contributing structures to the resonance hybrid for HN_3, and describe a plausible hybridization and bonding scheme for each structure.

Analyze

We can use data from Table 10.2 to estimate the bond order for the two nitrogen-to-nitrogen bonds. From this information we can write plausible Lewis structures, and by applying VSEPR theory to the Lewis structures, we can predict a likely geometric shape of the molecule. Finally, with this information we can propose hybridization schemes for the central atoms and an overall bonding scheme for the molecule.

Solve

From Table 10.2, the average bond lengths for N-to-N bonds are 145 pm for a single bond, 123 pm for a double bond, and 110 pm for a triple bond. Thus it is likely that the N_a—N_b bond (124 pm) has a considerable double-bond character, and the N_b—N_c bond (113 pm) has a considerable triple-bond character.

The HN_3 molecule has a total of 16 valence electrons in 8 electron pairs. The plausible Lewis structures have N_a and N_b atoms as central atoms, the N_c and H atoms as terminal atoms, and bonds reflecting the observed bond lengths.

(I) $H-\ddot{N}_a{=}N_b{=}\ddot{N}_c$ (II) $H-\ddot{N}_a-N_b{\equiv}\ddot{N}_c$

According to VSEPR theory, in both structures (I) and (II) the electron-group geometry around N_b is linear. This corresponds to sp hybridization. In structure (I), the electron-group geometry around N_a is trigonal-planar, corresponding to sp^2 hybridization; in structure (II), the electron-group geometry around N_a is tetrahedral, corresponding to sp^3 hybridization. These hybridization schemes, the orbital overlaps, and the geometric structures of the two resonance structures are indicated on the right.

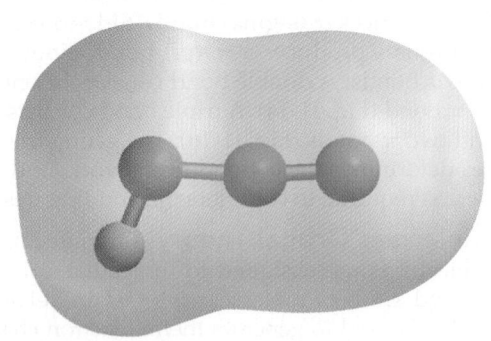

σ: H$(1s)$—N$_a(sp^2)$ π: N$_a(2p)$—N$_b(2p)$

π: N$_b(2p)$—N$_c(2p)$

σ: N$_a(sp^2)$—N$_b(sp)$ σ: N$_b(sp)$—N$_c(2p)$

Structure (I)

σ: H$(1s)$—N$_a(sp^3)$ σ: N$_a(sp^3)$—N$_b(sp)$

π: N$_b(2p)$—N$_c(2p)$

π: N$_b(2p)$—N$_c(2p)$

σ: N$_b(sp)$—N$_c(2p)$

Structure (II)

Assess

Of these two resonance structures, which is the most favorable? We could try to use formal charges, but we find that in structure (I), the formal charges on N_a, N_b, and N_c, are 0, +1, and −1, respectively; correspondingly in structure (II), the formal charges are −1, +1, and 0, respectively. Because the formal charges are so similar, we cannot make any definitive conclusion as to which structure is favored. The surest way to decide is to compare the observed molecular structure with that suggested by the two hybridization-schemes given above. In structure (I), N_a is sp^2 hybridized so that a $H-N_a-N_b$ angle is expected to be close to 120°; whereas in structure (II), the hybridization on N_a is sp^3 so that the $H-N_a-N_b$ angle is expected to be close to 109°. Experimentally it is found that the $H-N_a-N_b$ angle is 109°, so that the hybridization scheme in structure (II) is to be preferred. The electrostatic potential map for HN_3 is shown on the right, and we observe that N_a is relatively negatively charged as compared to N_b and N_c—in accord with our preferred structure (II).

PRACTICE EXAMPLE A: Melamine is a carbon–hydrogen–nitrogen compound used in the manufacture of adhesives, protective coatings, and textile finishing (such as in wrinkle-free, wash-and-wear fabrics). Its mass percent composition is 28.57% C, 4.80% H, and 66.64% N. The melamine molecule features a six-member ring with alternating carbon and nitrogen atoms. Half the nitrogen atoms and all the H atoms are outside the ring. For melamine, **(a)** write a plausible Lewis structure, **(b)** describe bonding in the molecule by the valence-bond method, and **(c)** describe bonding in the ring system through molecular orbital theory.

PRACTICE EXAMPLE B: Dimethylglyoxime (DMG) is a carbon–hydrogen–nitrogen–oxygen compound with a molecular mass of 116.12 u. In a combustion analysis, a 2.464 g sample of DMG yields 3.735 g CO_2 and 1.530 g H_2O. In a separate experiment, the nitrogen in a 1.868 g sample of DMG is converted to $NH_3(g)$ and the NH_3 is neutralized by passing it into 50.00 mL of 0.3600 M $H_2SO_4(aq)$. After neutralization of the NH_3 the excess $H_2SO_4(aq)$ requires 18.63 mL of 0.2050 M NaOH(aq) for its neutralization. Using these data, determine for dimethylglyoxime **(a)** the most plausible Lewis structure, and **(b)** in the manner of Figure 11-18, a plausible bonding scheme.

Mastering**CHEMISTRY**

You'll find a link to additional self study questions in the study area on www.masteringchemistry.com

Exercises

Valence-Bond Method

1. Indicate several ways in which the valence-bond method is superior to Lewis structures in describing covalent bonds.

2. Explain why it is necessary to hybridize atomic orbitals when applying the valence-bond method— that is, why are there so few molecules that can be described by the overlap of pure atomic orbitals only?

3. Describe the molecular geometry of H_2O suggested by each of the following methods: **(a)** Lewis theory; **(b)** valence-bond method using simple atomic orbitals; **(c)** VSEPR theory; **(d)** valence-bond method using hybridized atomic orbitals.

4. Describe the molecular geometry of CCl_4 suggested by each of the following methods: **(a)** Lewis theory; **(b)** valence-bond method using simple atomic orbitals; **(c)** VSEPR theory; **(d)** valence-bond method using hybridized atomic orbitals.

5. In which of the following, CO_3^{2-}, SO_2, CCl_4, CO, NO_2^-, would you expect to find sp^2 hybridization of the central atom? Explain.

6. In the manner of Example 11-1, describe the probable structure and bonding in **(a)** HI; **(b)** BrCl; **(c)** H_2Se; **(d)** OCl_2.

7. For each of the following species, identify the central atom(s) and propose a hybridization scheme for those atom(s): **(a)** CO_2; **(b)** $HONO_2$; **(c)** ClO_3^-; **(d)** BF_4^-.

8. Propose a plausible Lewis structure, geometric structure, and hybridization scheme for the NSF molecule.

9. Describe a hybridization scheme for the central Cl atom in the molecule ClF_3 that is consistent with the geometric shape pictured in Table 10.1. Which orbitals of the Cl atom are involved in overlaps, and which are occupied by lone-pair electrons?

10. Describe a hybridization scheme for the central S atom in the molecule SF_4 that is consistent with the geometric shape pictured in Table 10.1. Which orbitals of the S atom are involved in overlaps, and which are occupied by lone-pair electrons?

11. Match each of the following species with one of these hybridization schemes: sp, sp^2, sp^3, sp^3d, sp^3d^2. **(a)** PF_6^-; **(b)** COS; **(c)** $SiCl_4$; **(d)** NO_3^-; **(e)** AsF_5.

12. Propose a hybridization scheme to account for bonds formed by the central carbon atom in each of the following molecules: **(a)** hydrogen cyanide, HCN;

(b) methyl alcohol, CH_3OH; **(c)** acetone, $(CH_3)_2CO$; **(d)** carbamic acid,

13. Indicate which of the following molecules and ions are linear, which are planar, and which are neither. Then propose hybridization schemes for the central atoms. **(a)** $Cl_2C{=}CCl_2$; **(b)** $N{\equiv}C{-}C{\equiv}N$; **(c)** $F_3C{-}C{\equiv}N$; **(d)** $[S{-}C{\equiv}N]^-$.

14. In the manner of Figure 11-17, indicate the structures of the following molecules in terms of the overlap of simple atomic orbitals and hybrid orbitals: **(a)** CH_2Cl_2; **(b)** OCN^-; **(c)** BF_3.

15. Write Lewis structures for the following molecules, and then label each σ and π bond. **(a)** HCN; **(b)** C_2N_2; **(c)** $CH_3CHCHCCl_3$; **(d)** HONO.

16. Represent bonding in the carbon dioxide molecule, CO_2, by **(a)** a Lewis structure and **(b)** the valence-bond method. Identify σ and π bonds, the necessary hybridization scheme, and orbital overlap.

17. Use the method of Figure 11-18 to represent bonding in each of the following molecules: **(a)** CCl_4; **(b)** ONCl; **(c)** HONO; **(d)** $COCl_2$.

18. Use the method of Figure 11-18 to represent bonding in each of the following ions: **(a)** NO_2^-; **(b)** I_3^-; **(c)** $C_2O_4^{2-}$; **(d)** HCO_3^-.

19. The molecular model below represents citric acid, an acidic component of citrus juices. Represent bonding in the citric acid molecule using the method of Figure 11-18 to indicate hybridization schemes and orbital overlaps.

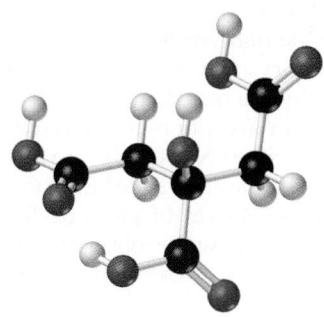

20. Malic acid is a common organic acid found in unripe apples and other fruit. With the help of the molecular model shown below, represent bonding in the malic acid molecule, using the method of Figure 11-18 to indicate hybridization schemes and orbital overlaps.

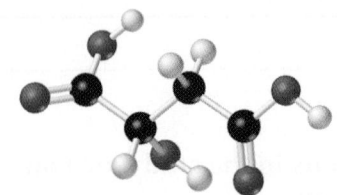

21. Shown below are ball-and-stick models. Describe hybridization and orbital-overlap schemes consistent with these structures.

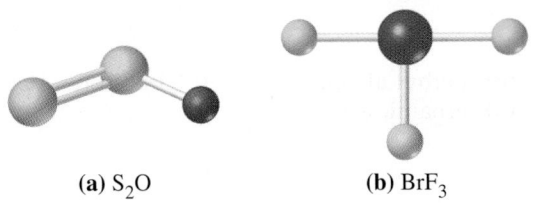

(a) S₂O   **(b)** BrF₃

22. Shown below are ball-and-stick models. Describe hybridization and orbital-overlap schemes consistent with these structures.

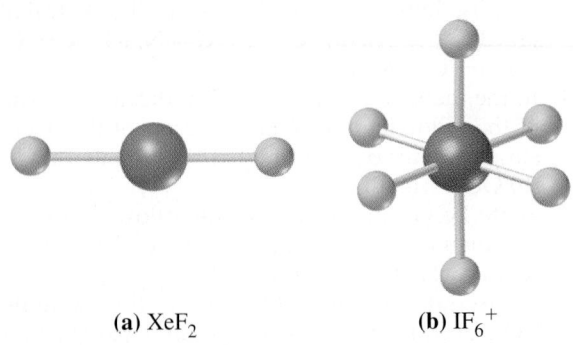

(a) XeF₂   **(b)** IF₆⁺

23. Propose a bonding scheme that is consistent with the structure for propynal. [*Hint:* Consult Table 10.2 to assess the multiple-bond character in some of the bonds.]

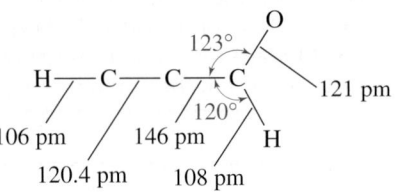

24. The structure of the molecule allene, CH_2CCH_2, is shown here. Propose hybridization schemes for the C atoms in this molecule.

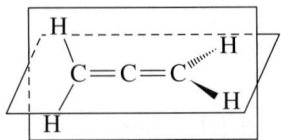

25. Angelic acid, shown below, occurs in sumbol root, a herb used as a stimulant.

$$\begin{array}{ccc} H & & CH_3 \\ & \diagdown \quad \diagup & \\ & C=C & \\ H_3C & & COOH \end{array}$$

Represent the bonding in the angelic acid molecule by using the method in Figure 11-18 to indicate hybridization schemes and orbital overlaps. What is the maximum number of atoms that can lie in the same plane?

26. Dimethylolpropionic acid, shown below, is used in the preparation of resins.

$$\begin{array}{ccc} O & CH_3 & \\ \| & | & \\ HO-C-C & -CH_2OH \\ & | & \\ & CH_2OH & \end{array}$$

Represent the bonding in the dimethylolpropionic acid molecule by using the method in Figure 11-18 to indicate hybridization schemes and orbital overlaps. What is the maximum number of atoms that can lie in the same plane?

Molecular Orbital Theory

27. Explain the essential difference in how the valence-bond method and molecular orbital theory describe a covalent bond.

28. Describe the bond order of diatomic carbon, C₂, with Lewis theory and molecular orbital theory, and explain why the results are different.

29. N₂(g) has an exceptionally high bond energy. Would you expect either N_2^- or N_2^{2-} to be a stable diatomic species in the gaseous state? Explain.

30. The paramagnetism of gaseous B₂ has been established. Explain how this observation confirms that the π_{2p} orbitals are at a lower energy than the σ_{2p} orbital for B₂.

31. In our discussion of bonding, we have not encountered a bond order higher than triple. Use the energy-level diagrams of Figure 11-26 to show why this is to be expected.

32. Is it correct to say that when a diatomic molecule loses an electron, the bond energy always decreases (that is, that the bond is always weakened)? Explain.

33. For the following pairs of molecular orbitals, indicate the one you expect to have the lower energy, and state the reason for your choice. **(a)** σ_{1s} or σ_{1s}^*; **(b)** σ_{2s} or σ_{2p}; **(c)** σ_{1s}^* or σ_{2s}; **(d)** σ_{2p} or σ_{2p}^*.

34. For each of the species C_2^+, O_2^-, F_2^+, and NO^+,
   (a) Write the molecular orbital diagram (as in Example 11-6).
   (b) Determine the bond order, and state whether you expect the species to be stable or unstable.
   (c) Determine if the species is diamagnetic or paramagnetic; and if paramagnetic, indicate the number of unpaired electrons.

35. Write plausible molecular orbital diagrams for the following *heteronuclear* diatomic species: **(a)** NO; **(b)** NO^+; **(c)** CO; **(d)** CN; **(e)** CN^-; **(f)** CN^+; **(g)** BN.

36. We have used the term "isoelectronic" to refer to atoms with identical electron configurations. In molecular orbital theory, this term can be applied to molecular species as well. Which of the species in Exercise 35 are isoelectronic?

37. Consider the molecules NO^+ and N_2^+ and use molecular orbital theory to answer the following:
   (a) Write the molecular orbital configuration of each ion (ignore the 1s electrons).
   (b) Predict the bond order of each ion.

(c) Which of these ions is paramagnetic? Which is diamagnetic?
(d) Which of these ions do you think has the greater bond length? Explain.

38. Consider the molecules CO^+ and CN^- and use molecular orbital theory to answer the following:
   (a) Write the molecular orbital configuration of each ion (ignore the 1s electrons).
   (b) Predict the bond order of each ion.
   (c) Which of these ions is paramagnetic? Which is diamagnetic?
   (d) Which of these ions do you think has the greater bond length? Explain.

39. Construct the molecular orbital diagram for CF. Would you expect the bond length of CF^+ to be longer or shorter than that of CF?

40. Construct the molecular orbital diagram for CaF. Would you expect the bond length of CaF^+ to be longer or shorter than that of CaF?

Delocalized Molecular Orbitals

41. Explain why the concept of delocalized molecular orbitals is essential to an understanding of bonding in the benzene molecule, C_6H_6.

42. Explain how it is possible to avoid the concept of resonance by using molecular orbital theory.

43. In which of the following molecules would you expect to find delocalized molecular orbitals: **(a)** C_2H_4; **(b)** SO_2; **(c)** H_2CO? Explain.

44. In which of the following ions would you expect to find delocalized molecular orbitals: **(a)** HCO_2^-; **(b)** CO_3^{2-}; **(c)** CH_3^+? Explain.

Metallic Bonding

45. Which of the following factors are especially important in determining whether a substance has metallic properties: **(a)** atomic number; **(b)** atomic mass; **(c)** number of valence electrons; **(d)** number of vacant atomic orbitals; **(e)** total number of electronic shells in the atom? Explain.

46. Based on the ground-state electron configurations of the atoms, how would you expect the melting points and hardnesses of sodium, iron, and zinc to compare? Explain.

47. How many energy levels are present in the 3s conduction band of a single crystal of sodium weighing 26.8 mg? How many electrons are present in this band?

48. Magnesium is an excellent electrical conductor even though it has a full 3s subshell with the electron configuration: $[Ne]3s^2$. Use band theory to explain why magnesium conducts electricity.

Semiconductors

49. From this list of terms—electrical conductor, insulator, semiconductor—choose the one that best characterizes each of the following materials: **(a)** stainless steel; **(b)** solid sodium chloride; **(c)** sulfur; **(d)** germanium; **(e)** seawater; **(f)** solid iodine.

50. In what type of material is the energy gap between the valence band and the conduction band greatest: metal, semiconductor, or insulator? Explain.

51. Which of the following substances, when added in trace amounts to silicon, would produce a *p*-type semiconductor: **(a)** sulfur, **(b)** arsenic, **(c)** lead, **(d)** boron, **(e)** gallium arsenide, **(f)** gallium? Explain.

52. Which of the following substances, when added in trace amounts to germanium, would produce an *n*-type semiconductor: **(a)** sulfur, **(b)** aluminum, **(c)** tin, **(d)** cadmium sulfide, **(e)** arsenic, **(f)** gallium arsenide? Explain.

53. The effect of temperature change on the electrical conductivity of ultrapure silicon is quite different from that on silicon containing a minute trace of arsenic. Why is this so?

54. Explain why the electrical conductivity of a semiconductor is significantly increased if trace amounts of either donor or acceptor atoms are present, but is unchanged if both are present in equal number.

55. The energy gap, ΔE, for silicon is 110 kJ/mol. What is the minimum wavelength of light that can promote an electron from the valence band to the conduction band in silicon? In what region of the electromagnetic spectrum is this light?

56. Explain why the solar cell in Figure 11-40 operates over a broad range of wavelengths rather than at a single wavelength (often the case when quantum effects are involved)?

Integrative and Advanced Exercises

57. The Lewis structure of N_2 indicates that the nitrogen-to-nitrogen bond is a triple covalent bond. Other evidence suggests that the σ bond in this molecule involves the overlap of sp hybrid orbitals.
    (a) Draw orbital diagrams for the N atoms to describe bonding in N_2.
    (b) Can this bonding be described by either sp^2 or sp^3 hybridization of the N atoms? Can bonding in N_2 be described in terms of unhybridized orbitals? Explain.

58. Show that both the valence-bond method and molecular orbital theory provide an explanation for the existence of the covalent molecule Na_2 in the gaseous state. Would you predict Na_2 by the Lewis theory?

59. A group of spectroscopists believe that they have detected one of the following species: NeF, NeF^+, or NeF^-. Assume that the energy-level diagrams of Figure 11-26 apply, and describe bonding in these species. Which of these species would you expect the spectroscopists to have observed?

60. Lewis theory is satisfactory to explain bonding in the ionic compound K_2O, but it does not readily explain formation of the ionic compounds potassium superoxide, KO_2, and potassium peroxide, K_2O_2.
    (a) Show that molecular orbital theory can provide this explanation.
    (b) Write Lewis structures consistent with the molecular orbital explanation.

61. The compound potassium sesquoxide has the empirical formula K_2O_3. Show that this compound can be described by an appropriate combination of potassium, peroxide, and superoxide ions. Write a Lewis structure for a formula unit of the compound.

62. Draw a Lewis structure for the urea molecule, $CO(NH_2)_2$, and predict its geometric shape with the VSEPR theory. Then revise your assessment of this molecule, given the fact that all the atoms lie in the same plane, and all the bond angles are 120°. Propose a hybridization and bonding scheme consistent with these experimental observations.

63. Methyl nitrate, CH_3NO_3, is used as a rocket propellant. The skeletal structure of the molecule is CH_3ONO_2. The N and three O atoms all lie in the same plane, but the CH_3 group is not in the same plane as the NO_3 group. The bond angle C—O—N is 105°, and the bond angle O—N—O is 125°. One nitrogen-to-oxygen bond length is 136 pm, and the other two are 126 pm.
    (a) Draw a sketch of the molecule showing its geometric shape.
    (b) Label all the bonds in the molecule as σ or π, and indicate the probable orbital overlaps involved.
    (c) Explain why all three nitrogen-to-oxygen bond lengths are not the same.

64. Fluorine nitrate, $FONO_2$, is an oxidizing agent used as a rocket propellant. A reference source lists the following data for FO_aNO_2. (The subscript "a" shows that this O atom is different from the other two.)

    Bond lengths: N—O = 129 pm;
    N—O_a = 139 pm; O_a—F = 142 pm

Bond angles: O—N—O = 125°;
F—O_a—N = 105°
NO_aF plane is perpendicular to the O_2NO_a plane

Use these data to construct a Lewis structure(s), a three-dimensional sketch of the molecule, and a plausible bonding scheme showing hybridization and orbital overlaps.

65. Draw a Lewis structure(s) for the nitrite ion, NO_2^-. Then propose a bonding scheme to describe the σ and π bonding in this ion. What conclusion can you reach about the number and types of π molecular orbitals in this ion? Explain.

66. Think of the reaction shown here as involving the transfer of a fluoride ion from ClF_3 to AsF_5 to form the ions ClF_2^+ and AsF_6^-. As a result, the hybridization scheme of each central atom must change. For each reactant molecule and product ion, indicate (a) its geometric structure and (b) the hybridization scheme for its central atom.

$$ClF_3 + AsF_5 \longrightarrow (ClF_2^+)(AsF_6^-)$$

67. In the gaseous state, HNO_3 molecules have two nitrogen-to-oxygen bond distances of 121 pm and one of 140 pm. Draw a plausible Lewis structure(s) to represent this fact, and propose a bonding scheme in the manner of Figure 11-18.

68. He_2 does not exist as a stable molecule, but there is evidence that such a molecule can be formed between electronically excited He atoms. Write a molecular orbital diagram to account for this.

69. The molecule formamide, $HCONH_2$, has the approximate bond angles H—C—O, 123°; H—C—N, 113°; N—C—O, 124°; C—N—H, 119°; H—N—H, 119°. The C—N bond length is 138 pm. Two Lewis structures can be written for this molecule, with the true structure being a resonance hybrid of the two. Propose a hybridization and bonding scheme for each structure.

70. Pyridine, C_5H_5N, is used in the synthesis of vitamins and drugs. The molecule can be thought of in terms of replacing one CH unit in benzene with a N atom. Draw orbital diagrams to show the orbitals of the C and N atoms involved in the σ and π bonding in pyridine. How many bonding and antibonding π-type molecular orbitals are present? How many delocalized electrons are present?

71. One of the characteristics of antibonding molecular orbitals is the presence of a nodal plane. Which of the bonding molecular orbitals considered in this chapter have nodal planes? Explain how a molecular orbital can have a nodal plane and still be a bonding molecular orbital.

72. The ion F_2Cl^- is linear, but the ion F_2Cl^+ is bent. Describe hybridization schemes for the central Cl atom consistent with this difference in structure.

73. Ethyl cyanoacetate, a chemical used in the synthesis of dyes and pharmaceuticals, has the mass percent composition: 53.09% C, 6.24% H, 12.39% N, and 28.29% O. In the manner of Figure 11-18, show a bonding scheme for this substance. The scheme

should designate orbital overlaps, σ and π bonds, and expected bond angles.

74. A certain monomer used in the production of polymers has one nitrogen atom and the mass composition 67.90% C, 5.70% H, and 26.40% N. Sketch the probable geometric structure of this molecule, labeling all the expected bond lengths and bond angles.

75. A solar cell that is 15% efficient in converting solar to electric energy produces an energy flow of 1.00 kW/m² when exposed to full sunlight.
(a) If the cell has an area of 40.0 cm², what is the power output of the cell, in watts?
(b) If the power calculated in part (a) is produced at 0.45 V, how much current does the cell deliver?

76. Toluene-2,4-diisocyanate is used in the manufacture of polyurethane foam. Its structural formula is shown

here. Describe the hybridization scheme for the atoms marked with an asterisk, and indicate the values of the bond angles marked α and β.

77. The anion $I_4{}^{2-}$ is linear, and the anion $I_5{}^-$ is V-shaped, with a 95° angle between the two arms of the V. For the central atoms in these ions, propose hybridization schemes that are consistent with these observations.

78. Pentadiene, C_5H_8, has three isomers, depending on the position of the two double bonds. Determine the shape of these isomers by using VSEPR theory. Describe the bonding in these molecules by using the valence-bond method. Do the shapes agree in the two theories? Use molecular orbital theory to decide which of these molecules has a delocalized π system. Sketch the molecular orbital and an energy-level diagram.

79. A conjugated hydrocarbon has an alternation of double and single bonds. Draw the molecular orbitals of the π system of 1,3,5-hexatriene. If the energy required to excite an electron from the HOMO to the LUMO corresponds to a wavelength of 256 nm, do you expect the wavelength for the corresponding excitation in 1,3,5,7-octatetraene to be a longer or shorter wavelength?

Feature Problems

80. Resonance energy is the difference in energy between a real molecule—a resonance hybrid—and its most important contributing structure. To determine the resonance energy for benzene, we can determine an energy change for benzene and the corresponding change for one of the Kekulé structures. The resonance energy is the difference between these two quantities.
(a) Use data from Appendix D to determine the enthalpy of hydrogenation of liquid benzene to liquid cyclohexane.
(b) Use data from Appendix D to determine the enthalpy of hydrogenation of liquid cyclohexene to liquid cyclohexane.

For the enthalpy of formation of liquid cyclohexene, use $\Delta H_f^\circ = -38.5$ kJ/mol.
(c) Assume that the enthalpy of hydrogenation of 1,3,5-cyclohexatriene is three times as great as that of cyclohexene, and calculate the resonance energy of benzene.
(d) Another way to assess resonance energy is through bond energies. Use bond energies from Table 10.3 (page 435) to determine the total enthalpy change required to break all the bonds in a Kekulé structure of benzene. Next, determine the enthalpy change for the dissociation of $C_6H_6(g)$ into its gaseous atoms by using data from Table 10.3 and Appendix D. Then calculate the resonance energy of benzene.

81. The 60-cycle alternating electric current (AC) commonly used in households changes direction 120 times per second. That is, in a one-second time period

a terminal at an electric outlet is positive 60 times and negative 60 times. In direct electric current (DC), the flow between terminals is in one direction only. A *rectifier* is a device that converts alternating to direct current. One type of rectifier is the *p–n* junction rectifier. It is commonly incorporated in adapters required to operate electronic devices from ordinary house current. In the operation of this rectifier, a *p*-type semiconductor and an *n*-type semiconductor are in contact along a boundary, or junction. Each semiconductor is connected to one of the terminals in an AC electrical outlet. Describe how this rectifier works. That is, show that when the semiconductors are connected to the terminals in an AC outlet, half the time a large flow of charge occurs and half the time essentially no charge flows across the *p–n* junction.

82. Furan, C_4H_4O, is a substance derivable from oat hulls, corn cobs, and other cellulosic waste. It is a starting material for the synthesis of other chemicals used as pharmaceuticals and herbicides. The furan molecule is planar and the C and O atoms are bonded into a five-membered pentagonal ring. The H atoms are attached to the C atoms. The chemical behavior of the molecule suggests that it is a resonance hybrid of several contributing structures. These structures show that the double bond character is associated with the entire ring in the form of a π electron cloud.
(a) Draw Lewis structures for the several contributing structures to the resonance hybrid mentioned above.
(b) Draw orbital diagrams to show the orbitals that are involved in the σ and π bonding in furan. [*Hint:* You need use only one of the contributing structures, such as the one with no formal charges.]
(c) How many π electrons are there in the furan molecule? Show that this number of π electrons is the

same, regardless of the contributing structure you use for this assessment.

83. As discussed in the Are You Wondering feature on page 457, the sp hybrid orbitals are algebraic combinations of the s and p orbitals. The required combinations of $2s$ and $2p$ orbitals are

$$\psi_1(sp) = \frac{1}{\sqrt{2}}[\psi(2s) + \psi(2p_z)]$$

$$\psi_2(sp) = \frac{1}{\sqrt{2}}[\psi(2s) - \psi(2p_z)]$$

(a) By combining the appropriate functions given in Table 8.1, construct a polar plot in the manner of Figure 8-26 for each of the above functions in the xz plane. In a polar plot, the value of r/a_0 is set at a fixed value (for example, 1). Describe the shapes and phases of the different portions of the hybrid orbitals, and compare them with those shown in Figure 11-11.
(b) Convince yourself that the combinations employing the $2p_x$ or $2p_y$ orbital also give similar hybrid orbitals but pointing in different directions.
(c) The combinations for the sp^2 hybrids in the xy plane are

$$\psi_1(sp^2) = \frac{1}{\sqrt{3}}\psi(2s) + \frac{\sqrt{2}}{\sqrt{3}}\psi(2p_x)$$

$$\psi_2(sp^2) = \frac{1}{\sqrt{3}}\psi(2s) - \frac{1}{\sqrt{6}}\psi(2p_x) + \frac{1}{\sqrt{2}}\psi(2p_y)$$

$$\psi_3(sp^2) = \frac{1}{\sqrt{3}}\psi(2s) - \frac{1}{\sqrt{6}}\psi(2p_x) - \frac{1}{\sqrt{2}}\psi(2p_y)$$

By constructing polar plots (in the xy plane), show that these functions correspond to the sp^2 hybrids depicted in Figure 11-9.

84. In Chapter 10, we saw that electronegativity differences determine whether bond dipoles exist in a molecule and that molecular shape determines whether bond dipoles cancel (nonpolar molecules) or combine to produce a resultant dipole moment (polar molecules). Thus, the ozone molecule, O_3, has no bond dipoles because all the atoms are alike. Yet, O_3 *does* have a resultant dipole moment: $\mu = 0.534$ D. The electrostatic potential map for ozone is shown below. Use the electrostatic potential map to decide the direction of the dipole. Using the ideas of delocalized bonding in molecules, can you rationalize this electrostatic potential map?

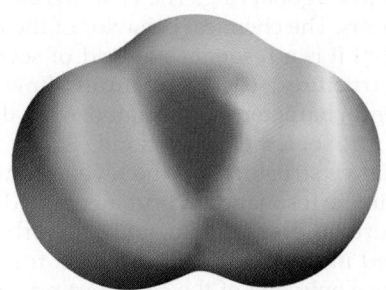

85. Borazine, $B_3N_3H_6$ is often referred to as inorganic benzene because of its similar structure. Like benzene,

borazine has a delocalized π system. Describe the molecular orbitals of the π system. Identify the highest occupied molecular orbital (HOMO) and the lowest unoccupied molecular orbital (LUMO). How many nodes does the LUMO possess?

86. Which of the following combinations of orbitals give rise to bonding molecular orbitals? For those combinations that do, label the resulting bonding molecular orbital as s or p.

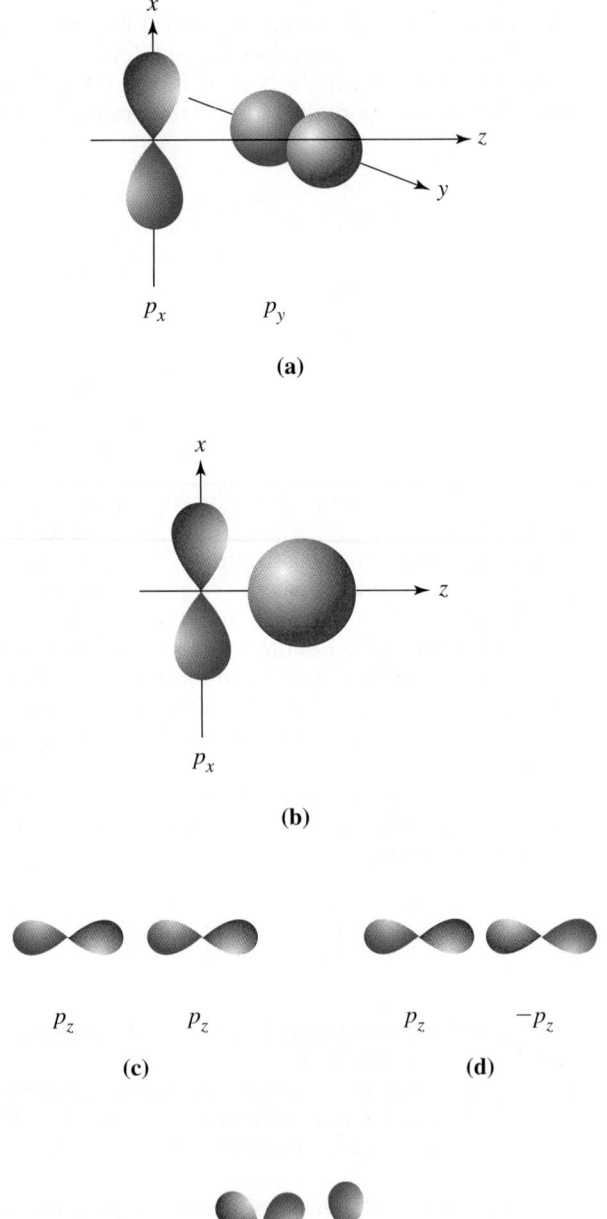

87. Construct a molecular orbital diagram for HF, and label the molecular orbitals as bonding, antibonding, or nonbonding.

Self-Assessment Exercises

88. In your own words, define the following terms or symbols: **(a)** sp^2; **(b)** σ_{2p}^*; **(c)** bond order; **(d)** π bond.

89. Briefly describe each of the following ideas: **(a)** hybridization of atomic orbitals; **(b)** σ-bond framework; **(c)** Kekulé structures of benzene, C_6H_6; **(d)** band theory of metallic bonding.

90. Explain the important distinctions between the terms in each of the following pairs: **(a)** σ and π bonds; **(b)** localized and delocalized electrons; **(c)** bonding and antibonding molecular orbitals; **(d)** metal and semiconductor.

91. A molecule in which sp^2 hybrid orbitals are used by the central atom in forming covalent bonds is **(a)** PCl_5; **(b)** N_2; **(c)** SO_2; **(d)** He_2.

92. The bond angle in H_2Se is best described as **(a)** between 109° and 120°; **(b)** less than in H_2S; **(c)** less than in H_2S, but not less than 90°; **(d)** less than 90°.

93. The hybridization scheme for the central atom includes a d orbital contribution in **(a)** I_3^-; **(b)** PCl_3; **(c)** NO_3^-; **(d)** H_2Se.

94. Of the following, the species with a bond order of 1 is **(a)** H_2^+; **(b)** Li_2; **(c)** He_2; **(d)** H_2^-.

95. The hybridization scheme for Xe in XeF_2 is **(a)** sp; **(b)** sp^3; **(c)** sp^3d; **(d)** sp^3d^2.

96. Delocalized molecular orbitals are found in **(a)** H_2; **(b)** HS^-; **(c)** CH_4; **(d)** CO_3^{2-}.

97. The best electrical conductor of the following materials is **(a)** $Li(s)$; **(b)** $Br_2(l)$; **(c)** $Ge(s)$; **(d)** $Si(s)$.

98. A substance in which the valence and conduction bands overlap is **(a)** a semiconductor; **(b)** a metalloid; **(c)** a metal; **(d)** an insulator.

99. Explain why the molecular structure of BF_3 cannot be adequately described through overlaps involving pure s and p orbitals.

100. Why does the hybridization sp^3d not account for bonding in the molecule BrF_5? What hybridization scheme does work? Explain.

101. What is the total number of **(a)** σ bonds and **(b)** π bonds in the molecule CH_3NCO?

102. Which of the following species are paramagnetic? **(a)** B_2; **(b)** B_2^-; **(c)** B_2^+. Which species has the strongest bond?

103. Use the valence molecular orbital configuration to determine which of the following species is expected to have the lowest ionization energy: **(a)** C_2^+; **(b)** C_2; **(c)** C_2^-.

104. Use the valence molecular orbital configuration to determine which of the following species is expected to have the greatest electron affinity: **(a)** C_2^+; **(b)** Be_2; **(c)** F_2; **(d)** B_2^+.

105. Which of these diatomic molecules do you think has the greater bond energy, Li_2 or C_2? Explain.

106. Construct a concept map that embodies the ideas of valence bond theory.

107. Construct a concept map that connects the ideas of molecular orbital theory.

108. Construct a concept map that describes the interconnection between valence-bond theory and molecular orbital theory in the description of resonance structures.

12

Intermolecular Forces: Liquids and Solids

CONTENTS

▶ Two of the many natural phenomena described in this chapter include the more ordered structure of the solid compared with the liquid state and the variation of density with the state of matter.

In this scene from Antarctica, water exists in all three states of matter—solid in the ice, liquid in the sea, and gas in the atmosphere. Solids, liquids, and gases were compared at the macroscopic and microscopic levels in Chapter 1 (Fig. 1-7).

When we make ice cubes by placing water in a tray in a freezer, energy is removed from the water molecules, which gradually slow down. Attractive (intermolecular) forces between the molecules take over, and the water solidifies into ice. When an ice cube melts, energy from the surroundings is absorbed by the water molecules, which overcome the intermolecular forces within the ice cube and enter the liquid state. In our study of gases, we intentionally sought conditions in which the intermolecular forces were negligible. This approach allowed us to describe gases with the ideal gas equation and to explain their behavior with the kinetic-molecular theory of gases. To describe the other states of matter—liquids and solids—we must first be able to identify the various intermolecular forces and then find situations in which the intermolecular forces are significant. We then consider some interesting properties of liquids and solids related to the strengths of these forces.

12-1   Intermolecular Forces

In our study of gases, we noted that at high pressures and low temperatures intermolecular forces cause gas behavior to depart from ideality. When these forces are sufficiently strong, a gas condenses to a liquid. That is, the intermolecular forces keep the molecules in such close proximity that they are confined to a definite volume, as expected for the liquid state.

Intermolecular forces are important in establishing the form and behavior of matter. The origin of intermolecular forces, those interactions between molecules, arises from the permanent and momentary unequal distribution of electron density within molecules.

Van der Waals Forces

Because helium forms no stable chemical bonds, we might expect it to remain a gas right down to 0 K. Although helium remains gaseous to very low temperatures, it does condense to a liquid at 4 K and freeze to a solid (at 25 atm pressure) at 1 K. These data suggest that intermolecular forces, even though very weak, must exist among He atoms. If the temperature is sufficiently low, these forces overcome thermal agitation and cause helium to condense. In this section, we will examine the types of intermolecular forces known collectively as **van der Waals forces**. The intermolecular forces contributing to the term $a(n/V)^2$ in the van der Waals equation for nonideal gases (equation 6.26) are of this type.

Instantaneous and Induced Dipoles

In describing electronic structures, we speak of electron charge density, or the probability that an electron is in a certain region at a given time. One probability is that at some particular instant—purely by chance—electrons are concentrated in one region of an atom or a molecule. This displacement of electrons causes a normally nonpolar species to become momentarily polar. An *instantaneous dipole* is formed. That is, the molecule has an instantaneous dipole moment. After this, electrons in a neighboring atom or molecule may be displaced to also produce a dipole. This is a process of induction (Fig. 12-1), and the newly formed dipole is called an *induced dipole*.

Taken together, these two events lead to an intermolecular force of attraction (Fig. 12-2). We can call this an instantaneous dipole–induced dipole attraction, but the names more commonly used are **dispersion force** and **London force**. (In 1928, Fritz London offered a theoretical explanation of these forces.)

Polarizability is the term used to describe the relative tendency for a charge distribution to distort from its normal shape in an atom or a molecule. The greater this tendency, the more polarizable an atom or a molecule is said to be. Polarizability increases with atomic or molecular size (see the table on the next page), which is defined by the volume of the electron cloud around a substance. Also, in large molecules, some electrons, being farther from atomic nuclei, are less firmly held. These electrons are more easily displaced, and the polarizability

▲ FIGURE 12-1
The phenomenon of induction
The attraction of a balloon to a surface is a commonplace example of induction. The balloon is charged by rubbing, and the charged balloon induces an opposite charge on the surface. (See also Appendix B.)

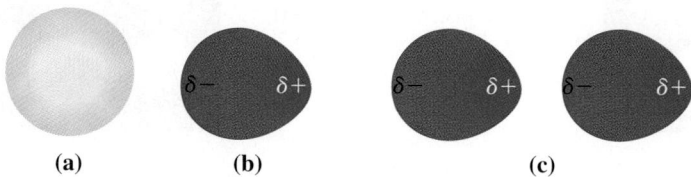

| (a) | (b) | (c) |

▲ FIGURE 12-2
Instantaneous and induced dipoles
(a) In the *normal condition*, a nonpolar molecule has a symmetrical charge distribution. **(b)** In the *instantaneous condition*, a displacement of the electronic charge produces an instantaneous dipole with a charge separation represented as $\delta+$ and $\delta-$. **(c)** In an *induced dipole*, the instantaneous dipole on the left induces a charge separation in the molecule on the right. The result is an instantaneous dipole–induced dipole attraction.

KEEP IN MIND

that a bond dipole results from the separation of centers of positive and negative charge in a covalent bond, that a resultant dipole moment, μ, is a summation of bond dipoles taking into account their magnitudes and directions, and that a *polar* molecule (Section 10-7) is one that has a permanent resultant dipole moment.

Compound	Polarizability,* 10^{-25} cm^3	Molar Mass, amu	Boiling Point, K
H$_2$	7.90	2.0158	20.35
O$_2$	16.0	31.9988	90.19
N$_2$	17.6	28.0134	77.35
CH$_4$	26.0	16.04	109.15
C$_2$H$_6$	44.7	30.07	184.55
Cl$_2$	46.1	70.906	238.25
C$_3$H$_8$	62.9	44.11	231.05
CCl$_4$	105	153.81	349.95

*Sometimes polarizability is referred to as *polarizability volume*. Note that the units of polarizability given above have the units of volume. That provides a measure of the atomic or molecular volume.

▶ Recall from Chapter 3, a molecular substance is made up of molecules. The molecules interact with each other through relatively weak intermolecular forces. The atoms of a given molecule are held together by relatively strong covalent bonds.

of the molecule increases. Because dispersion forces become stronger (more attractive) as polarizability increases, and because polarizability generally increases with molecular mass, the melting points and boiling points of molecular substances generally increase with increasing molecular mass. For instance, helium (atomic mass, 4 u) has a boiling point of 4 K, whereas radon (atomic mass, 222 u) has a boiling point of 211 K. The melting points and boiling points of the halogens increase in a similar way in the series F$_2$, Cl$_2$, Br$_2$, I$_2$ (recall Table 9.5).

The strength of dispersion forces also depends on *molecular shape*. Electrons in elongated molecules are more easily displaced than are those in small, compact, symmetrical molecules; the elongated molecules are more polarizable. Two substances with identical numbers and kinds of atoms but different molecular shapes (*isomers*) may have different properties. This idea is illustrated in Figure 12-3.

Dipole–Dipole Interactions

In a *polar* substance, the molecules have permanent dipole moments, so the molecules tend to line up with the positive end of one dipole directed toward the negative ends of neighboring dipoles (Fig. 12-4). This additional partial ordering of molecules can cause a substance to persist as a solid or liquid at temperatures higher than otherwise expected. Consider N$_2$, O$_2$, and NO. There are no electronegativity differences in N$_2$ and O$_2$, and both substances are nonpolar. In NO, conversely, there is an electronegativity difference, and the molecule has a slight dipole moment. Considering only dispersion forces, we would expect the boiling point of NO(l) to be intermediate to those of N$_2$(l) and O$_2$(l), but in the comparison on the next page, we see that it is not. NO(l) has the highest boiling point of the three because of its additional permanent dipole.

(a) Neopentane
bp = 9.5 °C

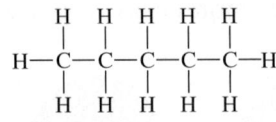

(b) Pentane
bp = 36.1 °C

▲ FIGURE 12-3
Molecular shapes and polarizability
The elongated pentane molecule is more polarized than is the compact neopentane molecule. Intermolecular forces are stronger in pentane than they are in neopentane. As a result, pentane boils at a higher temperature than neopentane.

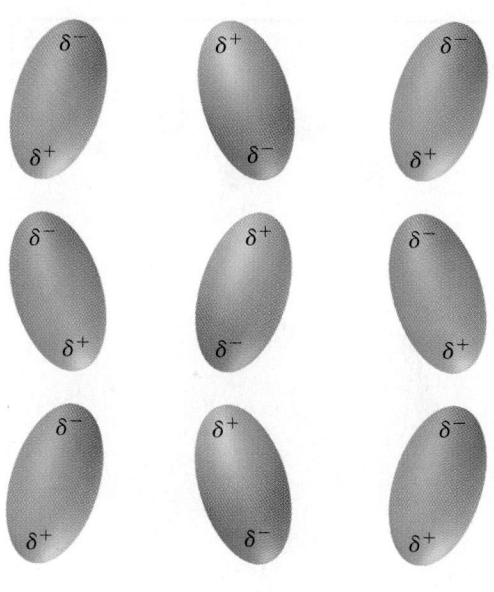

◀ FIGURE 12-4
Dipole–dipole interactions
Dipoles tend to arrange themselves with the positive end of one dipole pointed toward the negative end of a neighboring dipole. Ordinarily, thermal motion upsets this orderly array. Nevertheless, this tendency for dipoles to align themselves can affect physical properties, such as the melting points of solids and the boiling points of liquids.

N_2	NO	O_2
$\mu = 0$ (nonpolar)	$\mu = 0.153$ D (polar)	$\mu = 0$ (nonpolar)
mol. mass = 28 u	mol. mass = 30 u	mol. mass = 32 u
bp = 77.34 K	bp = 121.39 K	bp = 90.19 K

◀ The SI unit for dipole moment, C m, is inconvenient for expressing molecular dipole moments. The non-SI unit, debye (D), is often used. One debye is approximately 3.34×10^{-30} C m.

Hydrogen Bonding

Figure 12-5, in which the boiling points of a series of similar compounds are plotted as a function of molecular mass, demonstrates some features that we cannot explain by the types of intermolecular forces considered to this point. The hydrogen compounds (hydrides) of the group 14 elements display normal behavior; that is, the boiling points increase regularly as the molecular mass increases. But there are three striking exceptions in groups 15, 16, and 17. The boiling points of NH_3, H_2O, and HF are as high or higher than those of any other hydride in their group—not lowest, as we might expect. A special type of intermolecular force

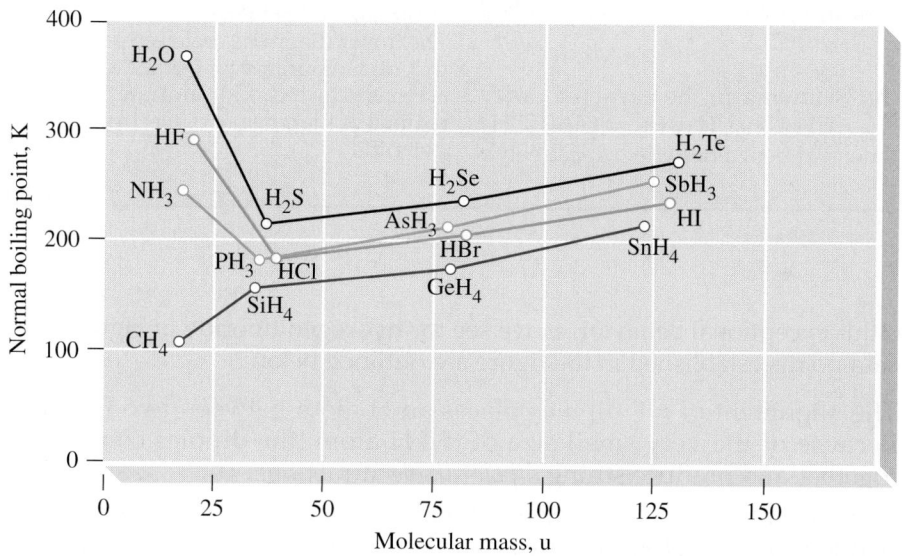

▲ FIGURE 12-5
Comparison of boiling points of some hydrides of the elements of groups 14, 15, 16, and 17
The values for NH_3, H_2O, and HF are unusually high compared with those of other members of their groups.

EXAMPLE 12-1   Comparing Physical Properties of Polar and Nonpolar Substances

Which would you expect to have the higher boiling point, the hydrocarbon fuel butane, C_4H_{10}, or the organic solvent acetone, $(CH_3)_2CO$?

Analyze

Ordinarily, the first clue comes in a comparison of molecular masses. However, because the two substances have the same molecular mass (58 u), we have to look elsewhere for a factor on which to base our prediction.

    The next consideration is the polarity of the molecules. The electronegativity difference between C and H is so small that we generally expect hydrocarbons, such as butane, to be nonpolar. However, we notice that one of the molecules contains a carbon-oxygen bond, and thus, a strong carbon-to-oxygen dipole. At times, it is helpful to sketch the structure of a molecule to see whether symmetrical features cause bond dipoles to cancel. It is not necessary to sketch the structure of the acetone molecule to deduce that it is a polar molecule. The $C{=}O$ bond dipole in acetone cannot be offset by other bond dipoles. Thus, acetone is polar.

Solve

Given two substances with the same molecular mass, one polar and one nonpolar, we expect the polar substance—acetone—to have the higher boiling point. (The measured boiling points are butane, $-0.5\,°C$; acetone, $56.2\,°C$.)

Assess

In general, when comparing the properties of different substances, we must consider the various types of intermolecular forces and the factors that affect the strength of each type of force. Although it wasn't important here, the three-dimensional shape of a molecule is usually a very important consideration and it is usually necessary to sketch the molecular structure to see how molecular shape plays a role.

PRACTICE EXAMPLE A: Which of the following substances would you expect to have the highest boiling point: C_3H_8, CO_2, CH_3CN? Explain.

PRACTICE EXAMPLE B: Arrange the following in the expected order of increasing boiling point: C_8H_{18}, $CH_3CH_2CH_2CH_3$, $(CH_3)_3CH$, C_6H_5CHO (octane, butane, isobutane, and benzaldehyde respectively).

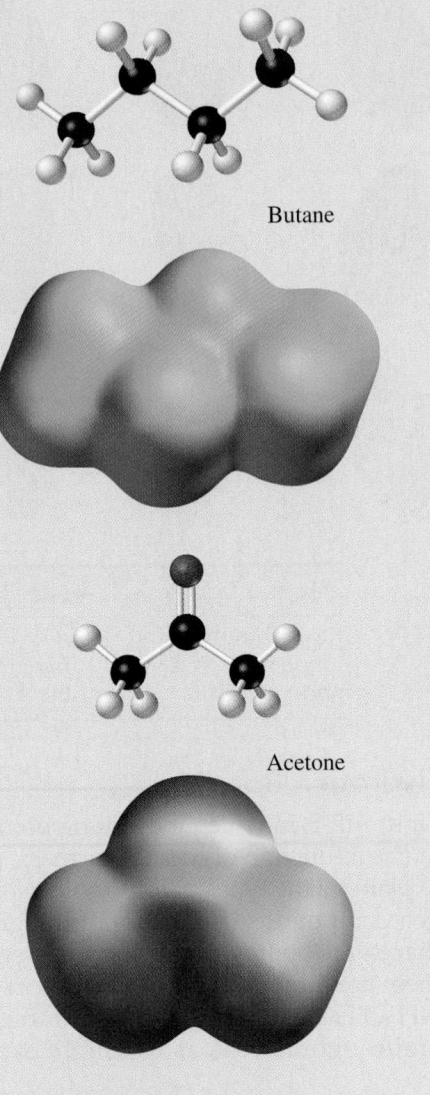

Butane

Acetone

▲ **Butane and acetone**
The lower diagrams are electrostatic potential diagrams for butane and acetone. The red color indicates regions of high negative electrostatic potential.

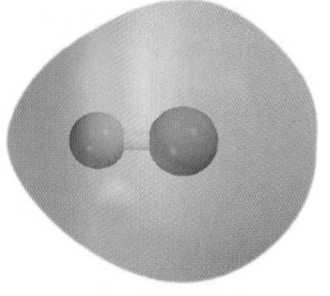

▲ **Electrostatic potential map of HF**

causes this exceptional behavior, as we see for hydrogen fluoride in Figure 12-6. The main points established in the figure are outlined below.

- The alignment of HF dipoles places an H atom between two F atoms. Because of the very small size of the H atom, the dipoles come close together and produce strong *dipole–dipole* attractions.

- Although an H atom is covalently bonded to one F atom, it is also weakly bonded to the F atom of a nearby HF molecule. This occurs through a lone pair of electrons on the F atom. Each H atom acts as a bridge between two F atoms.

- The bond angle between two F atoms bridged by an H atom (that is, the angle $F{-}HF\cdots F$) is about $180°$.

The type of intermolecular force just described is called a hydrogen bond, although it is simply an electrostatic attraction and not an actual chemical bond like a covalent bond. In a **hydrogen bond** an H atom is covalently bonded to a highly electronegative atom, which attracts electron density away from the H nucleus. This in turn allows the H nucleus, a proton, to be simultaneously attracted to a lone pair of electrons on a highly electronegative atom in a neighboring molecule.

Hydrogen bonds are possible only with certain hydrogen-containing compounds because all atoms other than H have inner-shell electrons to shield their nuclei from attraction by lone-pair electrons of nearby atoms. Only F, O, and N easily meet the requirements for hydrogen-bond formation. Weak hydrogen bonding is occasionally encountered between an H atom of one molecule and a Cl or S atom in a neighboring molecule. Compared with other intermolecular forces, hydrogen bonds are relatively strong, having energies of the order of 15 to 40 kJ mol^{-1}. By contrast, single covalent bonds (also known as intramolecular bonds) are much stronger still—greater than 150 kJ mol^{-1}. (See Table 10.3 for further comparisons.)

Hydrogen Bonding in Water

Ordinary water is certainly the most common substance in which hydrogen bonding occurs. Figure 12-7 shows how one water molecule is held to four neighbors in a tetrahedral arrangement by hydrogen bonds. In ice, hydrogen bonds hold the water molecules in a rigid but rather open structure. As ice melts, only a fraction of the hydrogen bonds are broken. One indication of this is the relatively low heat of fusion of ice (6.01 kJ mol^{-1}). It is much less than we would expect if all the hydrogen bonds were to break during melting.

The open structure of ice shown in Figure 12-7(b) gives ice a low density. When ice melts, some of the hydrogen bonds are broken. This allows the water

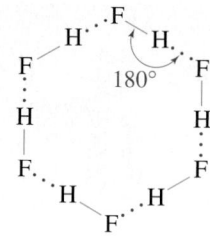

▲ FIGURE 12-6
Hydrogen bonding in gaseous hydrogen fluoride
In gaseous hydrogen fluoride, many of the HF molecules are associated into cyclic (HF)$_6$ structures of the type pictured here. Each H atom is bonded to one F atom by a single covalent bond (—) and to another F atom through a hydrogen bond ($\cdots$).

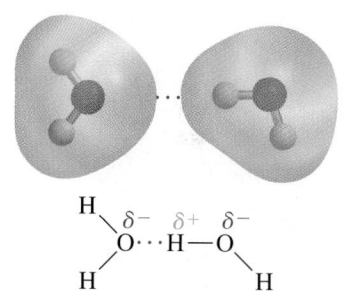

▲ Hydrogen bonding between H$_2$O molecules

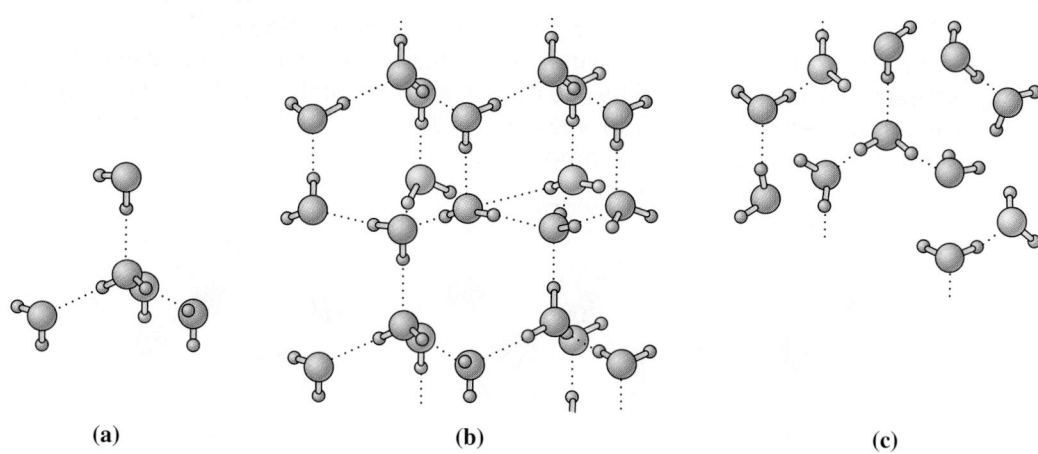

(a)	(b)	(c)

▲ FIGURE 12-7
Hydrogen bonding in water
(a) Each water molecule is linked to four others through hydrogen bonds. The arrangement is tetrahedral. Each H atom is situated along a line joining two O atoms, but closer to one O atom (100 pm) than to the other (180 pm). (b) For the crystal structure of ice, H atoms lie between pairs of O atoms, again closer to one O atom than to the other. (Molecules behind the plane of the page are light blue.) O atoms are arranged in bent hexagonal rings arranged in layers. This characteristic pattern is similar to the hexagonal shapes of snowflakes. (c) In the liquid, water molecules have hydrogen bonds to only some of their neighbours. This allows the water molecules to pack more densely in the liquid than in the solid.

▲ FIGURE 12-8
Solid and liquid densities compared
The sight of ice cubes floating on liquid water (left) is a familiar one; ice is less dense than liquid water. The more common situation, however, is that of paraffin wax (right). Solid paraffin is denser than the liquid and sinks to the bottom of the beaker.

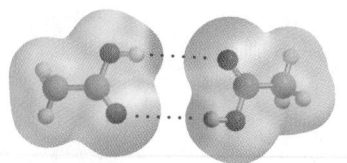

▲ FIGURE 12-9
An acetic acid dimer
Electrostatic potential maps showing hydrogen bonding.

molecules to be more compactly arranged, accounting for the increase in density when ice melts. That is, the number of H_2O molecules per unit volume is greater in the liquid than in the solid.

As liquid water is heated above the melting point, hydrogen bonds continue to break. The molecules become even more closely packed, and the density of the liquid water continues to increase. Liquid water attains its maximum density at $3.98\,°C$. Above this temperature, the water behaves in a "normal" fashion: Its density decreases as temperature increases. The unusual freezing-point behavior of water explains why a freshwater lake freezes from the top down. When the water temperature falls below $4\,°C$, the denser water sinks to the bottom of the lake and the colder surface water freezes. The ice over the top of the lake then tends to insulate the water below from further heat loss. This allows fish to survive the winter in a lake that has been frozen over. Without hydrogen bonding, all lakes would freeze from the bottom up; and fish, small bottom-feeding animals, and aquatic plants would not survive the winter. The density relationship between liquid water and ice is compared in Figure 12-8 with the more common liquid–solid density relationship.

Other Properties Affected by Hydrogen Bonding

Water is one example of a substance whose properties are affected by hydrogen bonding. There are numerous others. In acetic acid, CH_3COOH, pairs of molecules tend to join together into *dimers* (double molecules), both in the liquid and in the vapor states (Fig. 12-9). Not all the hydrogen bonds are disrupted when liquid acetic acid vaporizes, and, as a result, the heat of vaporization is abnormally low.

Certain trends in viscosity can also be explained by hydrogen bonding. In alcohols, the H atom in a —OH group in one molecule can form a hydrogen bond to the O atom in a neighboring alcohol molecule. An alcohol molecule with two —OH groups (a *diol*) has more possibilities for hydrogen-bond formation than a comparable alcohol with a single —OH group. Having stronger intermolecular forces, we expect the diol to flow more slowly, that is, to have a greater viscosity, than the simple alcohol. When still more —OH groups are present (*polyols*), we expect a further increase in viscosity. These comparisons are illustrated by the three common alcohols below. (The unit cP is a centipoise. The SI unit of viscosity is $1\,N\,s\,m^{-2} = 10\,P$. The Greek letter eta, η, is typically used as a symbol for viscosity.)

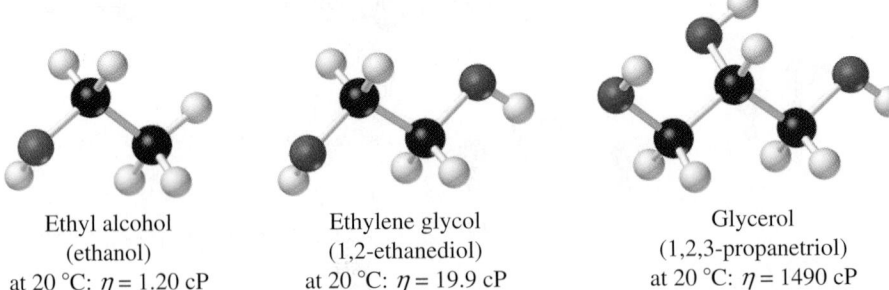

| Ethyl alcohol (ethanol) at 20 °C: $\eta = 1.20$ cP | Ethylene glycol (1,2-ethanediol) at 20 °C: $\eta = 19.9$ cP | Glycerol (1,2,3-propanetriol) at 20 °C: $\eta = 1490$ cP |

Intermolecular and *Intramolecular* Hydrogen Bonding

All the examples of hydrogen bonding presented to this point have involved an intermolecular force *between two molecules*, and this is called an *intermolecular hydrogen bond*. Another possibility occurs in molecules with an H atom covalently bonded to one highly electronegative atom (for example, O or N) and with another highly electronegative atom nearby in the same molecule. This type of hydrogen bonding *within a molecule* is called *intramolecular hydrogen bonding*. As shown in the molecular model of salicylic acid on the facing page, an intramolecular hydrogen bond (represented by a dotted line) joins the —OH group to the doubly bonded oxygen atom of the —COOH group on the same molecule. To underscore the importance of molecular geometry in establishing the conditions necessary for intramolecular hydrogen bonding, we need only turn to an isomer of salicylic acid called *para*-hydroxybenzoic acid. In this molecule, the H atom of the —COOH group is too close to the doubly bonded O atom of the same group to form a hydrogen bond, and the H atom of the —OH group on the opposite side of the molecule is too far away. Intramolecular hydrogen bonding does not occur in this situation.

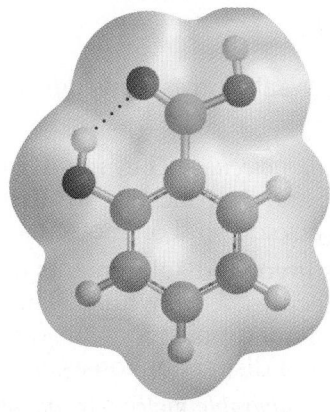

▲ Electrostatic potential map of salicylic acid showing intramolecular hydrogen bonding. The double bond character of certain bonds are not shown in the ball-and-stick model.

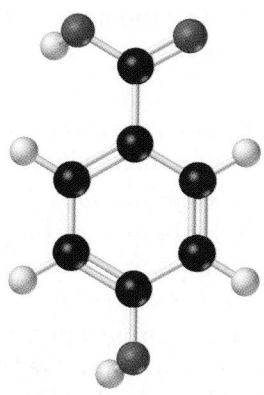

▲ There is no intramolecular hydrogen bonding in *para*-hydroxybenzoic acid.

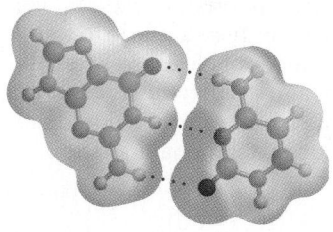

▲ Hydrogen bonding between guanine (left) and cytosine(right)in DNA

Hydrogen Bonding In Living Matter

Some chemical reactions in living matter involve complex structures, such as proteins and DNA, and in these reactions certain bonds must be easily broken and re-formed. Hydrogen bonding is the only type of bonding with energies of just the right magnitude to allow this, as we will discover in Chapter 27. We will also find that both intra- and intermolecular hydrogen bonding is involved in these complex structures.

Hydrogen bonding seems to provide an answer to the puzzle of how some trees are able to grow to great heights. In Chapter 6 (page 181), we learned that atmospheric pressure is capable of pushing a column of water to a maximum height of only about 10 m—other factors must be involved in transporting water to the tops of redwood trees up to 100 m tall. Hydrogen bonding seems to be a factor in transporting water in trees. Thin columns of water (in xylem, a plant tissue) extend from the roots to the leaves in the very tops of trees. In these columns, the water molecules are hydrogen-bonded to one another, with each water molecule acting like a link in a cohesive chain. When one water molecule evaporates from a leaf, another molecule in the chain moves to take its place and all the other molecules are pulled up the chain. Ultimately, a new water molecule joins the chain in the root system. In the next chapter, we will learn about another factor in transporting water in trees: osmotic pressure and its ability to force water through a membrane.

▲ **Sequoia trees**
The mystery of how these trees can bring water to leaves that are hundreds of feet up may be explained by hydrogen bonding.

🔍 12-1 CONCEPT ASSESSMENT

What are the types of intermolecular interactions in $CH_3CH_2NH_2(l)$, and which is the strongest?

TABLE 12.1	Intermolecular Forces and Properties of Selected Substances					
	Molecular Mass, u	Dipole Moment, D	Van der Waals Forces		ΔH_{vap}, kJ mol^{-1}	Boiling Point, K
			% Dispersion	% Dipole		
F_2	38.00	0	100	0	6.86	85.01
HCl	36.46	1.08	81.4	18.6	16.15	188.11
HBr	80.92	0.82	94.5	5.5	17.61	206.43
HI	127.91	0.44	99.5	0.5	19.77	237.80

Summary of van der Waals Forces

When assessing the importance of van der Waals forces, consider the following statements.

- *Dispersion (London) forces exist between all molecules.* They involve displacements of all the electrons in molecules, and they increase in strength with increasing molecular mass. The forces also depend on molecular shapes.
- *Forces associated with permanent dipoles* involve displacements of electron pairs in bonds rather than in molecules as a whole. These forces are found only in substances with resultant dipole moments (polar molecules). Their existence *adds* to the effect of dispersion forces also present.
- *When comparing substances of roughly comparable molecular masses,* dipole forces can produce significant differences in properties such as melting point, boiling point, and enthalpy of vaporization.
- *When comparing substances of widely different molecular masses,* dispersion forces are usually more significant than dipole forces.

▶ In our discussion of London forces, we use molar mass only as a guide to the number of electrons present in a molecule; the forces holding a molecule in a liquid are not gravitational in nature.

Let's see how these statements relate to the data in Table 12.1, which includes a rough breakdown of van der Waals forces into dispersion forces and forces caused by dipoles. HCl and F_2 have comparable molecular masses, but because HCl is polar, it has a significantly larger ΔH_{vap} and a higher boiling point. Within the series HCl, HBr, and HI, molecular mass increases sharply and ΔH_{vap} and boiling points increase in the order HCl < HBr < HI. The more polar nature of HCl and HBr relative to HI is not sufficient to reverse the trends produced by the increasing molecular masses—dispersion forces are the predominant intermolecular forces.

12-1 ARE YOU WONDERING...

Are there other types of intermolecular forces?

There are other types of intermolecular forces in addition to London dispersion, dipole–dipole, and hydrogen bonding. Table 12.2 summarizes some of the more commonly encountered intermolecular forces, along with their typical strengths.

In Table 12.2 one of the strongest forces between ions or molecules arises from the electrostatic attraction of opposite charge (ion–ion). This type of intermolecular force is what gives rise to the high melting points of ionic solids, along with their brittle nature. In such molecules as biomolecules, interactions between different charged groups, also known as salt bridges, increase their stability.

As we discovered in Chapter 11 molecules have dipole moments because of electronegativity differences between atoms. The dipole moment in a molecule will interact with a charged ion to form an ion–dipole interaction. Ion–dipole interactions are important in understanding the dissolution of salts.

In the absence of charges and dipole moments, other higher-order moments (e.g., quadrupole moments) become dominant. An example of quadrupole–quadrupole interactions would be between two CO_2 molecules.

KEEP IN MIND

that the term *intermolecular forces* is used to classify a specific set of noncovalent interactions.

TABLE 12.2 Summary of Noncovalent Interactions

Force	Energy,[a] kJ/mol	Example	Model
Intermolecular			
London dispersion	0.05–40	$CH_4 \cdots CH_4$	
Dipole–induced dipole	2–10	$CH_3(CO)CH_3 \ldots CH_5H_{12}$	
Ion–induced dipole	3–15	$Li^+ \ldots C_5H_{12}$	
Dipole–dipole	5–25	$H_2O \ldots CO$	
Hydrogen bond	10–40	$CH_3OH \ldots H_2O$	
Ion–dipole	40–600	$K^+ \ldots H_2O$	
Ion–ion	400–4000	$Lys^+ \ldots Glu^-$	
Interatomic			
London dispersion	0.05–40	$Ar \ldots Ar$	
Ion–ion	400–4000	$Na^+ \ldots Cl^-$	
Metallic	100–1000	$Ag \ldots Ag$	

[a]These are gas phase values.

12-2   Some Properties of Liquids

Surface Tension

The observation of a needle floating on water, as pictured in Figure 12-10, is puzzling. Steel is much denser than water and should not float. Something must overcome the force of gravity on the needle, allowing it to remain suspended on the surface of the water. What is this special quality associated with the surface of liquid water?

Figure 12-11 suggests an important difference in the forces experienced by molecules within the bulk of a liquid and by those at the surface. Interior molecules have more neighbors and experience more attractive intermolecular interactions than surface molecules. The increased number of attractions by neighboring molecules places an interior molecule in a more stable environment (lower energy state) than a surface molecule. Consequently, as many molecules as possible tend to enter the bulk of a liquid, while as few as possible remain at the surface. Thus, liquids tend to maintain a minimum surface area. To increase the surface area of a liquid requires that molecules be moved from the interior to the surface of a liquid, and this requires that work be done. The steel needle of Figure 12-10 remains suspended on the surface of the water because energy is required to spread the surface of the water over the top of the needle.

Surface tension is the energy, or work, required to increase the surface area of a liquid. Surface tension is often represented by the Greek letter gamma (γ) and has the units of energy per unit area, typically joules per square meter ($\mathrm{J\ m^{-2}}$). As the temperature—and hence the intensity of molecular motion—increases, intermolecular forces become less effective. Less work is required to extend the surface of a liquid, meaning that surface tension *decreases* with *increased* temperature.

When a drop of liquid spreads into a film across a surface, we say that the liquid *wets* the surface. Whether a drop of liquid wets a surface or retains its spherical shape and stands on the surface depends on the strengths of two types of intermolecular forces. The forces exerted between molecules holding them together in the drop are **cohesive forces**, and the forces between liquid molecules and the surface are **adhesive forces**. If cohesive forces are strong compared with adhesive forces, a drop maintains its shape. If adhesive forces are strong enough, the energy requirement for spreading the drop into a film is met through the work done by the collapsing drop.

Water wets many surfaces, such as glass and certain fabrics. This characteristic is essential to its use as a cleaning agent. If glass is coated with a film of oil

▲ FIGURE 12-10
An effect of surface tension illustrated
Despite being denser than water, the needle is supported on the surface of the water. The property of surface tension accounts for this unexpected behaviour.

▶ The surface tension of water at 20 °C, for example, is $7.28 \times 10^{-2}\,\mathrm{J\ m^{-2}}$, and that of mercury is more than six times as large, at $47.2 \times 10^{-2}\,\mathrm{J\ m^{-2}}$.

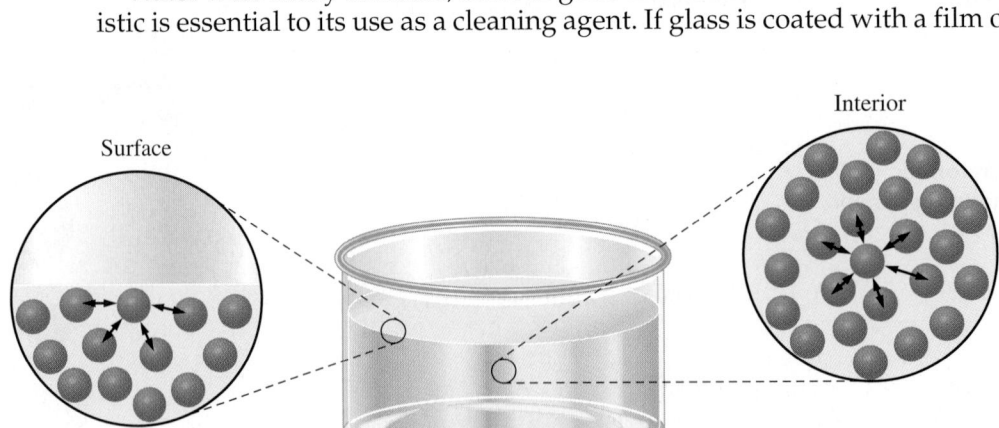

▲ FIGURE 12-11
Intermolecular forces in a liquid
Molecules at the surface are attracted only by other surface molecules and by molecules below the surface. Molecules in the interior experience forces from neighbouring molecules in all directions.

▲ FIGURE 12-12
Wetting of a surface
Water spreads into a thin film on a clean glass surface (left). If the glass is coated with oil or grease, the adhesive forces between the water and oil are not strong enough to spread the water, and droplets stand on the surface (right).

▲ FIGURE 12-13
Meniscus formation
Water wets glass (left). The meniscus is concave—the bottom of the meniscus is below the level of the water–glass contact line. Mercury does not wet glass. The meniscus is convex—the top of the meniscus is above the mercury–glass contact line.

or grease, water no longer wets the surface and water droplets stand on the glass, as shown in Figure 12-12. When we clean glassware in the laboratory, we have done a good job if water forms a uniform thin film on the glass. When we wax a car, we have done a good job if water uniformly beads up all along the surface.

Adding a detergent to water has two effects: The detergent solution dissolves grease to expose a clean surface, and the detergent lowers the surface tension of water. Lowering the surface tension means lowering the energy required to spread drops into a film. Substances that reduce the surface tension of water and allow it to spread more easily are known as *wetting agents*. They are used in applications ranging from dish washing to industrial processes.

Figure 12-13 illustrates another familiar observation. If the liquid in the glass tube is water, the water is drawn slightly up the walls of the tube by adhesive forces between water and glass. The interface between the water and the air above it, called a *meniscus*, is concave, or curved in. With liquid mercury, the meniscus is convex, or curved out. Cohesive forces in mercury, consisting of metallic bonds between Hg atoms, are strong; mercury does not wet glass. The effect of meniscus formation is greatly magnified in tubes of small diameter, called *capillary tubes*. In the *capillary action* shown in Figure 12-14, the water level inside the capillary tube is noticeably higher than outside. The soaking action of a sponge depends on the rise of water into capillaries of a fibrous material, such as cellulose. The penetration of water into soils also depends in part on capillary action. Conversely, mercury—with its strong cohesive forces and weaker adhesive forces—does not show a capillary rise. Rather, mercury in a glass capillary tube will have a lower level than the mercury outside the capillary.

▲ FIGURE 12-14
Capillary action
A thin film of water spreads up the inside walls of the capillary because of strong adhesive forces between water and glass (water wets glass). The pressure below the meniscus falls slightly. Atmospheric pressure then pushes a column of water up the tube to eliminate the pressure difference. The *smaller* the diameter of the capillary, the *higher* the liquid rises. Because its magnitude is also directly proportional to surface tension, capillary rise provides a simple experimental method of determining surface tension, described in Exercise 119.

Viscosity

Another property at least partly related to intermolecular forces is **viscosity**—a liquid's resistance to flow. The stronger the intermolecular forces of attraction, the greater the viscosity. When a liquid flows, one portion of the liquid moves with respect to neighboring portions. Cohesive forces within the liquid create an internal friction, which reduces the rate of flow. In liquids of low viscosity, such as ethyl alcohol and water, the effect is weak, and they flow easily. Liquids such as honey and heavy motor oil flow much more sluggishly. We say that they are *viscous*. One method of measuring viscosity is to time the fall of a steel ball through a certain depth of liquid (Fig. 12-15). The greater the viscosity of the liquid, the longer it takes for the ball to fall. Because intermolecular forces of attraction can be offset by higher molecular kinetic energies, viscosity generally *decreases* with *increased* temperature for liquids.

◀ For liquids, viscosity decreases with increasing temperature, but for gases, the viscosity increases with increasing temperature.

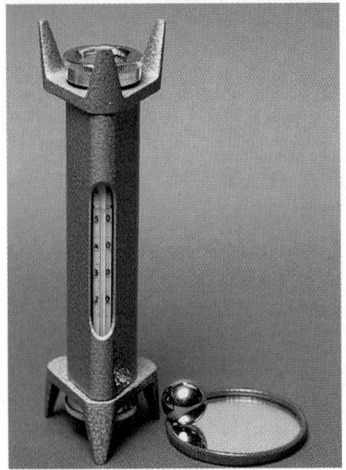

▲ FIGURE 12-15
Measuring viscocity
By measuring the velocity of
a ball dropping through a
liquid, a measure of the liquid
viscosity can be obtained.

12-2 CONCEPT ASSESSMENT

The viscosity of automotive motor oil is designated by its SAE number, such as 40 W. When compared in a ball viscometer (Fig. 12-15), the ball drops much faster through 10 W oil than through 40 W oil. Which of these two oils provides better winter service in the Arctic region of Canada? Which is best suited for summer use in the American Southwest? Which oil has the stronger intermolecular forces of attraction?

Enthalpy of Vaporization

In our study of the kinetic-molecular theory (Section 6-7), we saw that the speeds and kinetic energies of molecules vary over a wide range at any given temperature (Fig. 6-16). Then, in Chapter 7, we learned that molecules having kinetic energies sufficiently above the average value are able to overcome intermolecular forces of attraction and escape from the surface of the liquid into the gaseous state. This passage of molecules from the surface of a liquid into the gaseous, or vapor, state is called **vaporization** or **evaporation**. Vaporization occurs more readily with

- *increased temperature*—more molecules have sufficient kinetic energy to overcome intermolecular forces of attraction in the liquid.
- *increased surface area* of the liquid—a greater proportion of the liquid molecules are at the surface.
- *decreased strength of intermolecular forces*—the kinetic energy needed to overcome intermolecular forces of attraction is less, and more molecules have enough energy to escape.

Because the molecules lost through evaporation are much more energetic than average, the average kinetic energy of the remaining molecules decreases. The temperature of the evaporating liquid falls. This accounts for the cooling sensation you feel when a volatile liquid, such as ethyl alcohol, evaporates on your skin.

To vaporize a liquid at constant temperature, we must replace the excess kinetic energy carried away by the vaporizing molecules by adding heat to the liquid. The *enthalpy of vaporization* is the quantity of heat that must be absorbed if a certain quantity of liquid is vaporized at a constant temperature. Stated in another way,

$$\Delta H_{\text{vaporization}} = H_{\text{vapor}} - H_{\text{liquid}}$$

Because vaporization is an *endothermic* process, $\Delta H_{\text{vaporization}}$ (or ΔH_{vap} as it is more commonly denoted) is always positive. We will generally express enthalpies of vaporizations in terms of one mole of liquid vaporized, as seen in Table 12.3. The differences in enthalpies of vaporization in Table 12.3 are the result of intermolecular forces. For example, the intermolecular forces of diethyl ether are dispersion, while those of methyl alcohol are a combination of dipole and hydrogen bonding, creating the difference in their enthalpies of vaporization. Likewise, the higher dispersion in ethyl alcohol than in methyl alcohol yields a stronger interaction between ethyl alcohol's molecules and therefore

TABLE 12.3 Some Enthalpies of Vaporization at 298 K[a]	
Liquid	ΔH_{vap}, kJ mol^{-1}
Diethyl ether, $(C_2H_5)_2O$	29.1
Methyl alcohol, CH_3OH	38.0
Ethyl alcohol, CH_3CH_2OH	42.6
Water, H_2O	44.0

[a]ΔH_{vap} values are somewhat temperature-dependent (see Exercise 93).

an increase in its enthalpy of vaporization. The reason that water has a much higher enthalpy of vaporization than the alcohols is that water forms four hydrogen bonds, whereas the alcohols can form only three hydrogen bonds.

The conversion of a gas or vapor to a liquid is called **condensation**. From a thermochemical standpoint, condensation is the reverse of vaporization.

$$\Delta H_{condensation} = H_{liquid} - H_{vapor} = -\Delta H_{vap}$$

Because it is opposite in sign but equal in magnitude to ΔH_{vap}, $\Delta H_{condensation}$ (ΔH_{cond}) is always negative. Condensation is an *exothermic* process. This explains why burns produced by a given mass of steam (vaporized water) are much more severe than burns produced by the same mass of hot water. Hot water burns only by releasing heat as it cools. Steam releases a large quantity of heat when it condenses to liquid water, followed by the further release of heat as the hot water cools.

KEEP IN MIND

that absolute enthalpies, such as H_{vapor} and H_{liquid}, cannot be measured (page 268). However, because enthalpy is a function of state, the difference between the absolute enthalpies has a unique value, and it *can* be measured.

EXAMPLE 12-2   Estimating the Heat Evolved in the Condensation of Steam

A 0.750 L sample of steam obtained at the normal boiling point of water was allowed to condense on a slightly cooler surface. Estimate the quantity of heat evolved. Why is the result only an estimate?

Analyze

First, let us describe the steam sample a bit more precisely. Steam is water vapor, $H_2O(g)$; and when in equilibrium with liquid water at its normal boiling point, the steam is at 1.000 atm pressure.

Solve

We are given a volume of gas (0.750 L) at a fixed temperature (100.00 °C) and pressure (1.000 atm), and so we use the ideal gas equation to calculate the number of moles of $H_2O(g)$. That is,

$$n_{H_2O} = \frac{PV}{RT} = \frac{1.000 \text{ atm} \times 0.750 \text{ L}}{0.08206 \text{ L atm mol}^{-1}\text{K}^{-1} \times (273.15 + 100.00) \text{ K}}$$

$$= 0.0245 \text{ mol}$$

On a molar basis, estimate the enthalpy of condensation of the $H_2O(g)$ to be the *negative* of the value of ΔH_{vap} of H_2O given in Table 12.3, that is, -44.0 kJ mol^{-1}. For the 0.0245 mol sample of steam,

$$\Delta H_{cond} = 0.0245 \text{ mol} \times (-44.0 \text{ kJ mol}^{-1}) = -1.08\text{kJ}$$

Assess

This result is only an estimate for two reasons: (1) ΔH_{cond}, like ΔH_{vap}, is temperature dependent. The value used was for 298 K, whereas it should have been for 373 K, a value that was not given. (2) The condensed liquid water was at a temperature lower than 373 K ("slightly cooler surface"). An additional small quantity of heat was liberated as the condensed steam cooled to that lower temperature.

PRACTICE EXAMPLE A: How much heat is required to vaporize a 2.35 g sample of diethyl ether at 298 K?

PRACTICE EXAMPLE B: Calculate a more accurate answer to Example 12-2 by using $\Delta H_{vap} = 40.7$ kJ mol^{-1} for water at 100 °C, 85.0 °C as the temperature of the surface on which the steam condenses, and 4.21 J g^{-1} °C^{-1} as the average specific heat of $H_2O(l)$ in the temperature range 85 to 100 °C.

Vapor Pressure

Water left in an open beaker evaporates completely. A different condition results if the beaker with the water is placed in a closed container. As shown in Figure 12-16, in a container with both liquid and vapor present, vaporization and condensation occur simultaneously. If sufficient liquid is present, eventually a condition is reached in which the amount of vapor remains constant. This condition is one of *dynamic equilibrium.* Dynamic equilibrium always implies that two opposing processes are occurring simultaneously and at equal rates.

▶ FIGURE 12-16
Establishing liquid–vapor equilibrium
(a) A liquid is allowed to evaporate into a closed container. Initially, only vaporization occurs.
(b) Condensation begins. The rate at which molecules evaporate is greater than the rate at which they condense, and the number of molecules in the vapor state continues to increase.
(c) The rate of condensation is equal to the rate of vaporization. The number of vapor molecules remains constant over time, as does the pressure exerted by this vapor.

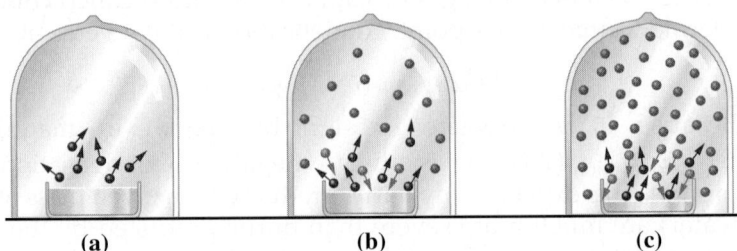

● Molecules in vapor state
●► Molecules undergoing vaporization
●► Molecules undergoing condensation

(a)   **(b)**   **(c)**

As a result, there is no net change with time once equilibrium has been established. A symbolic representation of the liquid–vapor equilibrium is shown below.

$$\text{liquid} \underset{\text{condensation}}{\overset{\text{vaporization}}{\rightleftharpoons}} \text{vapor}$$

▶ Gasoline is a mixture of volatile hydrocarbons and is an important precursor of smog, whether it is vaporized from oil refineries, filling-station operations, automobile gas tanks, or power lawn mowers.

The pressure exerted by a vapor in dynamic equilibrium with its liquid is called the **vapor pressure**. Liquids with high vapor pressures at room temperature are said to be *volatile,* and those with very low vapor pressures are *nonvolatile.* Whether a liquid is volatile or not is determined primarily by the strengths of its intermolecular forces—the weaker these forces, the more volatile the liquid (the higher its vapor pressure). Diethyl ether and acetone are volatile liquids; at 25 °C their vapor pressures are 534 and 231 mmHg, respectively. Water at ordinary temperatures is a moderately volatile liquid; at 25 °C, its vapor pressure is 23.8 mmHg. Mercury is essentially a nonvolatile liquid; at 25 °C, its vapor pressure is 0.0018 mmHg.

As an excellent first approximation, the vapor pressure of a liquid depends only on the particular liquid and its temperature. Vapor pressure depends on neither the amount of liquid nor the amount of vapor, as long as some of each is present at equilibrium. These statements are illustrated in Figure 12-17. A graph of vapor pressure as a function of temperature is known as a **vapor pressure curve**. Vapor pressure curves always have the appearance of those in Figure 12-18: *Vapor pressure increases with temperature.* Vapor pressures of water at different temperatures are presented in Table 12.4.

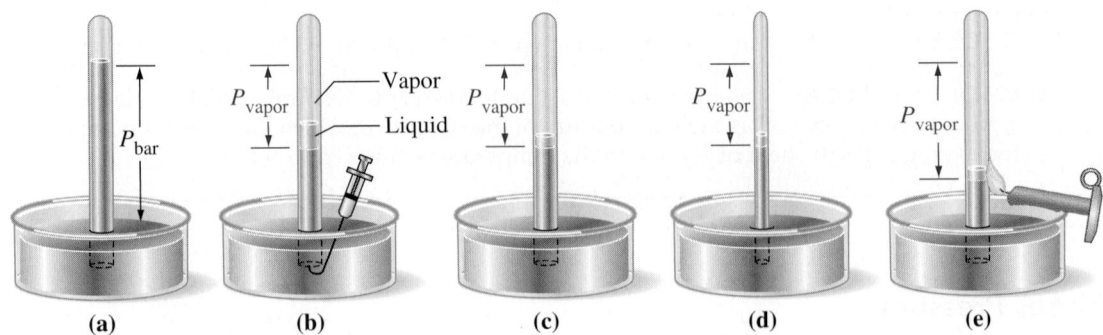

(a)   **(b)**   **(c)**   **(d)**   **(e)**

▲ FIGURE 12-17
Vapour pressure illustrated
(a) A mercury barometer. **(b)** The pressure exerted by the vapor in equilibrium with a liquid injected to the top of the mercury column depresses the mercury level. **(c)** Compared with (b), the vapor pressure is independent of the volume of liquid injected. **(d)** Compared with (c), the vapor pressure is independent of the volume of vapor present. **(e)** Vapor pressure increases with an increase in temperature.

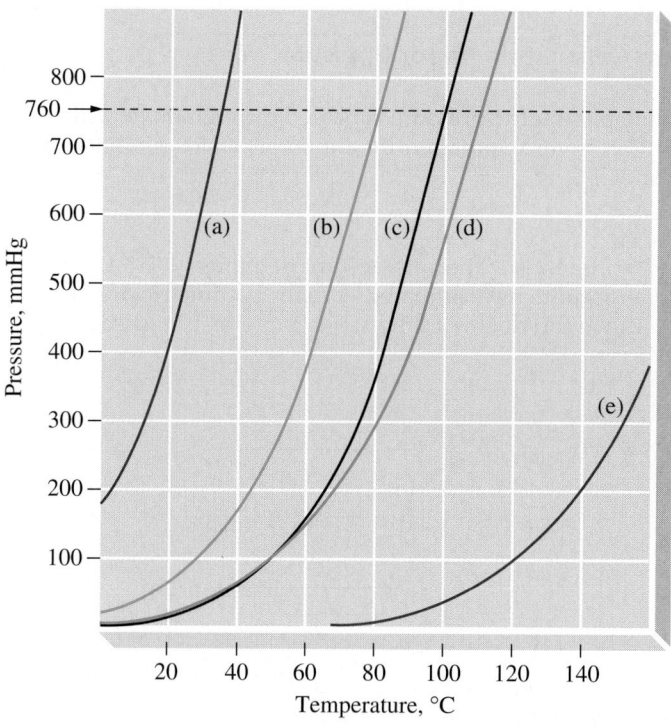

◀ FIGURE 12-18
Vapor pressure curves of several liquids
(a) Diethyl ether, $C_4H_{10}O$; **(b)** benzene, C_6H_6; **(c)** water, H_2O; **(d)** toluene, C_7H_8; **(e)** aniline, C_6H_7N. The normal boiling points are the temperatures at the intersection of the dashed line at $P = 760$ mmHg with the vapor pressure curves.

TABLE 12.4   Vapor Pressure of Water at Various Temperatures

Temperature, °C	Pressure, mmHg	Temperature, °C	Pressure, mmHg	Temperature, °C	Pressure, mmHg
0.0	4.6	29.0	30.0	93.0	588.6
10.0	9.2	30.0	31.8	94.0	610.9
20.0	17.5	40.0	55.3	95.0	633.9
21.0	18.7	50.0	92.5	96.0	657.6
22.0	19.8	60.0	149.4	97.0	682.1
23.0	21.1	70.0	233.7	98.0	707.3
24.0	22.4	80.0	355.1	99.0	733.2
25.0	23.8	90.0	525.8	100.0	760.0
26.0	25.2	91.0	546.0	110.0	1074.6
27.0	26.7	92.0	567.1	120.0	1489.1
28.0	28.3				

Measuring Vapor Pressure

Figure 12-17 suggests one method of determining vapor pressure—inject a small sample of the target liquid at the top of a mercury barometer, and measure the depression of the mercury level. The method does not give very precise results, however, and it is not useful for measuring vapor pressures that are either very low or quite high. Better results are obtained with methods in which the pressure above a liquid is continuously varied and measured, and the liquid–vapor equilibrium temperature is recorded. In short, the boiling point of the liquid changes in accordance with the change in the pressure above the liquid, and the vapor pressure curve of the liquid can be traced. The pressure measurements are made with either a closed-end or open-end manometer (page 195). A method that is useful for determining very low vapor pressures is based on the rate of effusion of a gas through a tiny orifice. In this method, equations from the kinetic-molecular theory (Section 6-7) are applied. Example 12-3 illustrates a method (called the transpiration method) in which an inert gas is saturated with the vapor under study. Then the ideal gas equation is used to calculate the vapor pressure.

EXAMPLE 12-3   Using the Ideal Gas Equation to Calculate a Vapor Pressure

A sample of 113 L of helium gas at 1360 °C and prevailing barometric pressure is passed through molten silver at the same temperature. The gas becomes saturated with silver vapor, and the liquid silver loses 0.120 g in mass. What is the vapor pressure of liquid silver at 1360 °C?

Analyze

Let's assume that after the gas has become saturated with silver vapor, its volume remains at 113 L. This assumption will be valid if the vapor pressure of the silver is quite low compared with the barometric pressure. According to Dalton's law of partial pressures (page 213) we can deal with the silver vapor as if it were a single gas occupying a volume of 113 L.

Solve

The data required in the ideal gas equation are listed below.

$P = ?$ $V = 113$ L

$R = 0.08206 \text{ L atm mol}^{-1}\text{K}^{-1}$ $T = 1360 + 273.15 = 1633$ K

$$n = 0.120 \text{ g Ag} \times \frac{1 \text{ mol Ag}}{107.9 \text{ g Ag}} = 0.00111 \text{ mol Ag}$$

$$P = \frac{nRT}{V}$$

$$P = \frac{0.00111 \text{ mol} \times 0.08206 \text{ L atm mol}^{-1}\text{K}^{-1} \times 1633 \text{ K}}{113 \text{ L}}$$

$$= 1.32 \times 10^{-3} \text{ atm } (1.00 \text{ Torr})$$

Assess

The assumption we made appears to be valid, because the experimental vapor pressure of liquid silver at 1360 °C is 1 mmHg or 1.32×10^{-3} atm.

PRACTICE EXAMPLE A: Equilibrium is established between liquid hexane, C_6H_{14}, and its vapor at 25.0 °C. A sample of the vapor is found to have a density of 0.701 g/L. Calculate the vapor pressure of hexane at 25.0 °C, expressed in Torr.

PRACTICE EXAMPLE B: With the help of Figure 12-18, estimate the density of the vapor in equilibrium with liquid diethyl ether at 20.0 °C.

Using Vapor Pressure Data

One use of vapor pressure data is in calculations dealing with the collection of gases over liquids, particularly water (Section 6-6). Another use, illustrated in Example 12-4, is in predicting whether a substance exists solely as a gas (vapor) or as a liquid and vapor in equilibrium.

EXAMPLE 12-4   Making Predictions with Vapor Pressure Data

As a result of a chemical reaction, 0.132 g H_2O is produced and maintained at a temperature of 50.0 °C in a closed flask of 525 mL volume. Will the water be present as liquid only, vapor only, or liquid and vapor in equilibrium (Fig. 12-19)?

Analyze

Let's consider each of the three possibilities in the order that they are given.

Solve

LIQUID ONLY

With a density of about 1 g/mL, a 0.132 g sample of H_2O has a volume of only about 0.13 mL. There is no way that the sample could completely fill a 525 mL flask. The condition of liquid only is *impossible*.

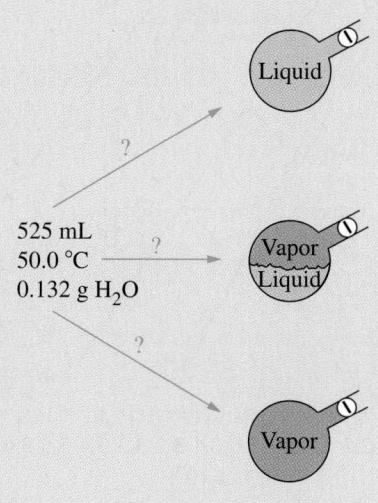

525 mL
50.0 °C
0.132 g H_2O

◀ FIGURE 12-19
Predicting states of matter—Example 12.4 illustrated
For the conditions given on the left, which of the final conditions pictured on the right will result?

VAPOR ONLY

The portion of the flask that is not occupied by liquid water must be filled with something (it cannot remain a vacuum). That something is water vapor. The question is, will the sample vaporize completely, leaving no liquid? Let's use the ideal gas equation to calculate the pressure that would be exerted if the entire 0.132 g H_2O were present in the gaseous state.

$$P = \frac{nRT}{V}$$

$$= \frac{0.132 \text{ g } H_2O \times \dfrac{1 \text{ mol } H_2O}{18.02 \text{ g } H_2O} \times 0.08206 \text{ L atm mol}^{-1} \text{K}^{-1} \times 323.2 \text{ K}}{0.525 \text{ L}}$$

$$= 0.370 \text{ atm} \times \frac{760 \text{ mmHg}}{1 \text{ atm}} = 281 \text{ mmHg}$$

Now compare this calculated pressure with the vapor pressure of water at 50.0 °C (Table 12.4). The calculated pressure—281 mmHg—greatly exceeds the vapor pressure—92.5 mmHg. Water formed in the reaction as $H_2O(g)$ condenses to $H_2O(l)$ when the gas pressure reaches 92.5 mmHg, for this is the pressure at which the liquid and vapor are in equilibrium at 50.0 °C. The condition of vapor only is *impossible*.

LIQUID AND VAPOR

This is the only possibility for the final condition in the flask. Liquid water and water vapor coexist in equilibrium at 50.0 °C and 92.5 mmHg.

Assess

We found the solution to this problem through the application of the ideal gas equation and our understanding of vapor pressure. Note that in the first two steps, we considered the two extremes, with the first being just liquid water and the second all vapor.

PRACTICE EXAMPLE A: If the reaction described in this example resulted in H_2O produced and maintained at 80.0 °C, would the water be present as vapor only or as liquid and vapor in equilibrium? Explain.

PRACTICE EXAMPLE B: For the situation described in Example 12.4, what mass of water is present as liquid and what mass as vapor?

An Equation for Expressing Vapor Pressure Data

If you look for vapor pressure data on a liquid in a handbook or in data tables, you are unlikely to find graphs like Figure 12-18. Also, with the exception of a few liquids, such as water and mercury, you are unlikely to find data tables like Table 12.4. What you will find, instead, are mathematical equations relating vapor pressures and temperatures. Such equations can summarize in one line data that might otherwise take a full page. Equation (12.1) is a particularly common form of vapor pressure equation. It expresses the natural logarithm

▶ FIGURE 12-20
Vapor pressure data plotted as ln *P* versus 1/*T*
Pressures are in millimeters of mercury, and temperatures are in Kelvin. Data from Figure 12-18 have been recalculated and replotted as in the following example: For benzene at 60 °C, the vapor pressure is 400 mmHg;
ln P = ln 400 = 5.99.
T = 60 + 273 = 333 K;
$1/T$ = 1/333 = 0.00300 = 3.00 × 10^{-3}; $1/T \times 10^3$ = 3.00 × $10^{-3} \times 10^3$ = 3.00. The point corresponding to (3.00, 5.99) is marked by the black arrow.

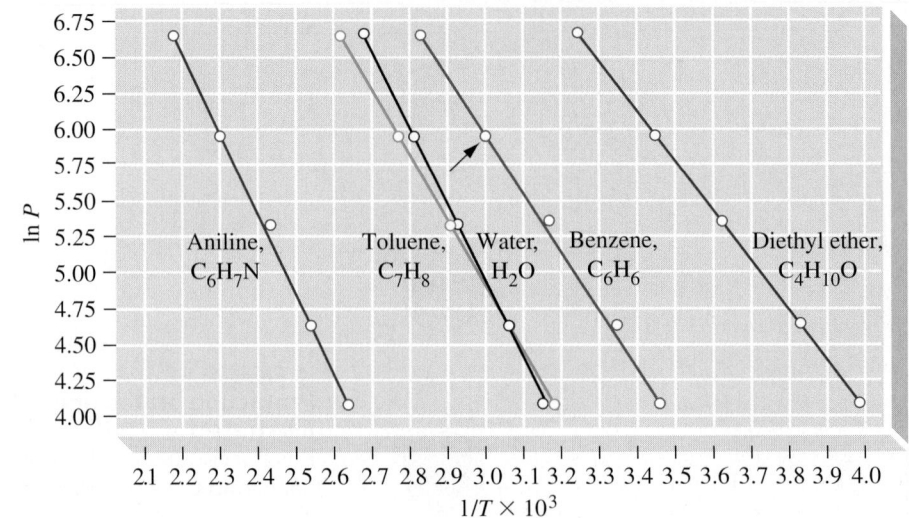

(ln) of vapor pressure as a function of the reciprocal of the Kelvin temperature (1/*T*). The relationship is that of a straight line, and the straight-line plots for the liquids featured in Figure 12-18 are drawn in Figure 12-20.

▶ Refer to Appendix A to see how the constant *B* is eliminated to give equation (12.2).

Equation of straight line:

$$\ln P = \underbrace{-A\left(\frac{1}{T}\right)}_{} + \underbrace{B}_{}$$
$$y = m \times x + b$$

(12.1)

To use equation (12.1), we need values for the two constants, *A* and *B*. The constant *A* is related to the enthalpy of vaporization of the liquid: $A = \Delta H_{vap}/R$, where ΔH_{vap} is expressed in the unit $J\,mol^{-1}$ and the value used for *R* is $8.3145\,J\,mol^{-1}K^{-1}$. It is customary to eliminate *B* by rewriting equation (12.1) for two different temperatures, in a form called the Clausius–Clapeyron equation.

KEEP IN MIND

that the heat of vaporization in this equation cannot be ΔH°_{vap}, since in general the pressure is not 1 bar.

$$\ln\left(\frac{P_2}{P_1}\right) = -\frac{\Delta H_{vap}}{R}\left(\frac{1}{T_2} - \frac{1}{T_1}\right)$$

(12.2)

We apply equation (12.2) in Example 12-5.

EXAMPLE 12-5 Applying the Clausius–Clapeyron Equation

Calculate the vapor pressure of water at 35.0 °C using data from Tables 12.3 and 12.4.

Analyze

Starting with the Clausius-Clapeyron equation, we recognize that we need four pieces of data to solve for the fifth. Since we are asked to calculate a vapor pressure, we will need two temperatures, a pressure, and the enthalpy of vaporization.

Solve

Designate the unknown vapor pressure as P_1 at the temperature T_1. That is,

P_1 = ? $\qquad\qquad T_1$ = (35.0 + 273.15) K = 308.2 K

For P_2 and T_2 choose known data for a temperature close to 35.0 °C, for example, 40.0 °C.

P_2 = 55.3 mmHg $\qquad T_2$ = (40.0 + 273.15) K = 313.2 K

For ΔH_{vap}, let's assume that the value given in Table 12.3 applies throughout the temperature range from 30.0 °C to 40.0 °C.

$$\Delta H_{vap} = 44.0\,kJ\,mol^{-1} \times \frac{1000\,J}{1\,kJ} = 44.0 \times 10^3\,J\,mol^{-1}$$

Now substitute these values into equation (12.2) to obtain

$$\ln\left(\frac{55.3 \text{ mmHg}}{P_1}\right) = -\frac{44.0 \times 10^3 \text{ J mol}^{-1}}{8.3145 \text{ J mol}^{-1} \text{ K}^{-1}}\left(\frac{1}{313.2} - \frac{1}{308.2}\right)\text{K}^{-1}$$

$$= -5.29 \times 10^3 (0.003193 - 0.003245) = 0.28$$

Next, determine that $e^{0.28} = 1.32$ (see Appendix A). Thus,

$$\frac{55.3 \text{ mmHg}}{P_1} = e^{0.28} = 1.32$$

$$P_1 = 55.3 \text{ mmHg}/1.32 = 41.9 \text{ mmHg}$$

Assess

Here, P_1 must be *smaller than* P_2 because $T_1 < T_2$. Thus, regardless of how we write equation (12.2)—different formulations are possible—or choose (T_1, P_1) and (T_2, P_2), we are guided by the fact that vapor pressure always increases with temperature. One way to check our answer is to repeat the calculation by using this pressure as the known pressure to see whether we obtain the pressure given in the problem. We can also check this against the experimentally determined vapor pressure of water at 35.0 °C, which is 42.175 mmHg.

PRACTICE EXAMPLE A: A handbook lists the vapor pressure of methyl alcohol as 100 mmHg at 21.2 °C. What is its vapor pressure at 25.0 °C?

PRACTICE EXAMPLE B: A handbook lists the normal boiling point of isooctane, a gasoline component, as 99.2 °C and its enthalpy of vaporization (ΔH_{vap}) as 35.76 kJ mol^{-1} C$_8$H$_{18}$. Calculate the vapor pressure of isooctane at 25 °C.

Boiling and the Boiling Point

When a liquid is heated in a container *open to the atmosphere*, there is a particular temperature at which vaporization occurs throughout the liquid rather than simply at the surface. Vapor bubbles form within the bulk of the liquid, rise to the surface, and escape. The pressure exerted by escaping molecules equals that exerted by molecules of the atmosphere, and **boiling** is said to occur. During boiling, energy absorbed as heat is used only to convert molecules of liquid to vapor. The temperature remains constant until all the liquid has boiled away, as is dramatically illustrated in Figure 12-21. The temperature at which the vapor pressure of a liquid is equal to standard atmospheric pressure (1 atm = 760 mmHg) is the **normal boiling point**. In other words, the normal boiling point is the boiling point of a liquid at 1 atm pressure. The normal boiling points of several liquids can be determined from the intersection of the dashed line in Figure 12-18 with the vapor pressure curves for the liquids.

◀ When a pan of water is put on the stove to boil, small bubbles are usually observed as the water begins to warm. These are bubbles of dissolved air being expelled. Once the water boils, however, all the dissolved air is expelled and the bubbles consist only of water vapor.

◀ **FIGURE 12-21**
Boiling water in a paper cup
An empty paper cup heated over a Bunsen burner quickly bursts into flame. If a paper cup is filled with water, it can be heated for an extended time as the water boils. This is possible for three reasons: (1) Because of the high heat capacity of water, heat from the burner goes primarily into heating the water, not the cup. (2) As the water boils, large quantities of heat (ΔH_{vap}) are required to convert the liquid to its vapor. (3) The temperature of the cup does not rise above the boiling point of water as long as liquid water remains. The boiling point of 99.9 °C instead of 100.0 °C suggests that the prevailing barometric pressure was slightly below 1 atm.

▶ A more extreme case is that on the summit of Mt. Everest, where a climber would barely be able to heat a cup of tea to 70 °C.

▲ **A liquid boils at low pressure**
Water boils when its vapor pressure equals the pressure on its surface. Bubbles form throughout the liquid.

▶ Although the term *gas* can be used exclusively, sometimes the term *vapor* is used for the gaseous state at temperatures *below* T_c and gas at temperatures *above* T_c.

Figure 12-18 also helps us see that the boiling point of a liquid varies significantly with barometric pressure. Shift the dashed line shown at $P = 760$ mmHg to higher or lower pressures, and the new points of intersection with the vapor pressure curves come at different temperatures. Barometric pressures below 1 atm are commonly encountered at high altitudes. At an altitude of 1609 m (that of Denver, Colorado), barometric pressure is about 630 mmHg. The boiling point of water at this pressure is 95 °C (203 °F). It takes longer to cook foods under conditions of lower boiling-point temperatures. A three–minute boiled egg takes longer than three minutes to cook. We can counteract the effect of high altitudes by using a pressure cooker. In a pressure cooker, the cooking water is maintained under higher-than-atmospheric pressure and its boiling temperature increases, for example, to about 120 °C at 2 atm pressure.

🔍 12-3   CONCEPT ASSESSMENT

Why does a three-minute boiled egg take longer than three minutes to cook in Switzerland and not on Manhattan Island in New York City?

The Critical Point

In describing boiling, we made an important qualification: Boiling occurs "in a container open to the atmosphere." If a liquid is heated in a *sealed* container, boiling does not occur. Instead, the temperature and vapor pressure rise continuously. Pressures many times atmospheric pressure may be attained. If just the right quantity of liquid is sealed in a glass tube and the tube is heated, as in Figure 12-22, the following phenomena can be observed:

- The density of the liquid decreases, that of the vapor increases, and eventually the two densities become equal.
- The surface tension of the liquid approaches zero. The interface between the liquid and vapor becomes less distinct and eventually disappears.

The **critical point** is the point at which these conditions are reached and the liquid and vapor become indistinguishable. The temperature at the critical point is the critical temperature, T_c, and the pressure is the critical pressure, P_c. The critical point is the highest point on a vapor pressure curve and represents the highest temperature at which the liquid can exist. Several critical temperatures and pressures are listed in Table 12.5.

A gas can be liquefied only at temperatures *below* its critical temperature, T_c. If room temperature is *below* T_c, this liquefaction can be accomplished just by applying sufficient pressure. If room temperature is *above* T_c, however, added pressure *and* a lowering of temperature to a value below T_c are required. We will comment further on the liquefaction of gases on page 523.

▶ FIGURE 12-22
Attainment of the critical point for benzene
In a sealed container, the meniscus separating a liquid from its vapor is just barely visible at the instant the critical point is reached. At the critical point—the liquid and vapor become indistinguishable.

About 10 °C
below T_c

About 1 °C
below T_c

Critical
temp. T_c

TABLE 12.5   Some Critical Temperatures, T_c, and Critical Pressures, P_c

Substance	T_c, K	P_c, atm
"Permanent" gases[a]		
H_2	33.3	12.8
N_2	126.2	33.5
O_2	154.8	50.1
CH_4	191.1	45.8
"Nonpermanent" gases[b]		
CO_2	304.2	72.9
HCl	324.6	82.1
NH_3	405.7	112.5
SO_2	431.0	77.7
H_2O	647.3	218.3

[a]Permanent gases cannot be liquefied at 25 °C (298 K).
[b]Nonpermanent gases can be liquefied at 25 °C.

12-4   CONCEPT ASSESSMENT

Compare the critical temperatures of NH_3 and N_2 (Table 12.5). Which gas has the stronger intermolecular forces?

EXAMPLE 12-6   Relating Intermolecular Forces and Physical Properties

Arrange the following substances in the order in which you would expect their boiling points to increase: CCl_4, Cl_2, ClNO, N_2.

Analyze

Recall that boiling point trends are related to intermolecular forces. We should begin by identifying the types and strengths of intermolecular forces at work.

Solve

Three of the substances are nonpolar. For these, the strengths of dispersion forces, and hence the boiling points, should increase with increasing molecular mass, that is, $N_2 < Cl_2 < CCl_4$. ClNO has a molecular mass (65.5 u) comparable to that of Cl_2 (70.9 u), but the ClNO molecule is polar (bond angle $\approx 120°$). This suggests stronger intermolecular forces and a higher boiling point for ClNO than for Cl_2. We should not expect the boiling point of ClNO to be higher than that of CCl_4, however, because of the large difference in their molecular masses (65.5 u compared with 154 u). The expected order is $N_2 < Cl_2 < $ ClNO $ < CCl_4$. (The observed boiling points are 77.3, 239.1, 266.7, and 349.9 K, respectively.)

Assess

Even though one molecule (ClNO) is polar, it does not have the highest boiling point, indicating that dispersion forces can be stronger than dipole–dipole forces.

PRACTICE EXAMPLE A: Arrange the following in the expected order of increasing boiling point: Ne, He, Cl_2, $(CH_3)_2CO$, O_2, O_3.

PRACTICE EXAMPLE B: Following are some values of ΔH_{vap} for several liquids at their normal boiling points: H_2, 0.92 kJ mol^{-1}; CH_4, 8.16 kJ mol^{-1}; C_6H_6, 31.0 kJ mol^{-1}; CH_3NO_2, 34.0 kJ mol^{-1}. Explain the differences among these values.

Q 12-5 CONCEPT ASSESSMENT

Explain why CCl_4 has a higher boiling point than CH_3Cl, despite the polarity of CH_3Cl.

12-3 Some Properties of Solids

We mentioned some properties of solids (for example, malleability, ductility) at the beginning of this text, and we will continue to consider additional properties. For now, we will comment on some properties that allow us to think of solids in relation to the other states of matter—liquids and gases.

Melting, Melting Point, and Heat of Fusion

As a crystalline solid is heated, its atoms, ions, or molecules vibrate more vigorously. Eventually a temperature is reached at which these vibrations disrupt the ordered crystalline structure. The atoms, ions, or molecules can slip past one another, and the solid loses its definite shape and is converted to a liquid. This process is called **melting**, or fusion, and the temperature at which it occurs is the **melting point**. The reverse process, the conversion of a liquid to a solid, is called **freezing**, or solidification, and the temperature at which it occurs is the **freezing point**. The melting point of a solid and the freezing point of its liquid are identical. At this temperature, solid and liquid coexist in equilibrium.

If we add heat uniformly to a solid–liquid mixture at equilibrium, the temperature remains constant while the solid melts. Only when all the solid has melted does the temperature begin to rise. Conversely, if we remove heat uniformly from a solid–liquid mixture at equilibrium, the liquid freezes at a constant temperature. The quantity of heat required to melt a solid is the *enthalpy of fusion*, ΔH_{fus}. Some typical enthalpies of fusion, expressed in kilojoules per mole, are listed in Table 12.6. Perhaps the most familiar example of a melting (and freezing) point is that of water, $0\,°C$. This is the temperature at which liquid and solid water, in contact with air and under standard atmospheric pressure, are in equilibrium. The enthalpy of fusion of water is $6.01\ kJ\ mol^{-1}$, which we can express as

$$H_2O(s) \longrightarrow H_2O(l) \qquad \Delta H_{fus} = +6.01\ kJ\ mol^{-1} \qquad \textbf{(12.3)}$$

Here is an easy way to determine the freezing point of a liquid. Allow the liquid to cool, and measure the liquid temperature as it falls with time. When freezing begins, the temperature *remains constant* until all the liquid has frozen. Then the temperature is again free to fall as the solid cools. If we plot temperatures against time, we get a graph known as a *cooling curve*. Figure 12-23 is a

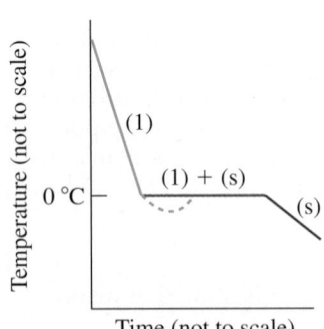

▲ FIGURE 12-23
Cooling curve for water
The broken-line portion represents the condition of supercooling that occasionally occurs.
(l) = liquid; (s) = solid.

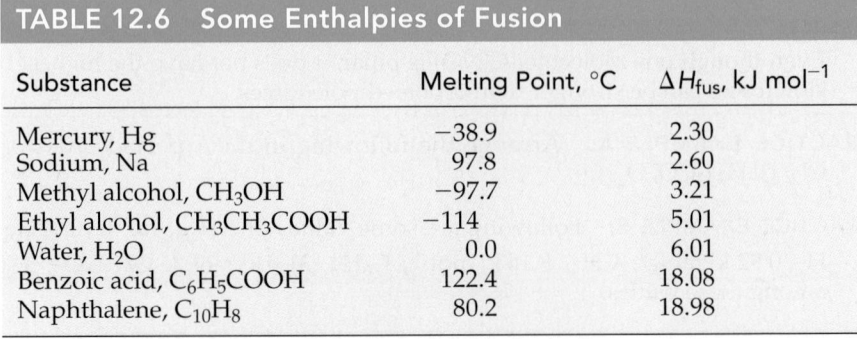

TABLE 12.6 Some Enthalpies of Fusion		
Substance	Melting Point, °C	ΔH_{fus}, kJ mol^{-1}
Mercury, Hg	−38.9	2.30
Sodium, Na	97.8	2.60
Methyl alcohol, CH_3OH	−97.7	3.21
Ethyl alcohol, CH_3CH_3COOH	−114	5.01
Water, H_2O	0.0	6.01
Benzoic acid, C_6H_5COOH	122.4	18.08
Naphthalene, $C_{10}H_8$	80.2	18.98

cooling curve for water. We can also run this process backward, that is, by starting with the solid and adding heat. Now the temperature remains constant while melting occurs. This temperature–time plot is called a *heating curve*. Generally speaking, the appearance of the heating curve is that of a cooling curve that has been flipped from left to right. A heating curve for water is sketched in Figure 12-24.

Often, an experimentally determined cooling curve does not look quite like the solid–line plot in Figure 12-23. The temperature may drop below the freezing point without any solid appearing. This condition is known as *supercooling*. For a crystalline solid to start forming from a liquid at the freezing point, the liquid must contain some small particles (for example, suspended dust particles) on which crystals can form. If a liquid contains a very limited number of particles on which crystals can grow, it may supercool for a time before freezing. When a supercooled liquid does begin to freeze, however, the temperature rises back to the normal freezing point while freezing is completed. We can always recognize supercooling through a slight dip in a cooling curve just before the horizontal portion.

Sublimation

Like liquids, solids can also give off vapors, although because of the stronger intermolecular forces present, solids are generally not as volatile as liquids at a given temperature. The direct passage of molecules from the solid to the vapor state is called **sublimation**. The reverse process, the passage of molecules from the vapor to the solid state, is called **deposition**. When sublimation and deposition occur at equal rates, a dynamic equilibrium exists between a solid and its vapor. The vapor exerts a characteristic pressure called the *sublimation pressure*. A plot of sublimation pressure as a function of temperature is called a *sublimation curve*. The *enthalpy of sublimation* (ΔH_{sub}) is the quantity of heat needed to convert a solid to vapor. At the sublimation point, sublimation (solid $\longrightarrow$ vapor) is equivalent to melting (solid $\longrightarrow$ liquid) followed by vaporization (liquid $\longrightarrow$ vapor). This suggests the following relationship among ΔH_{fus}, ΔH_{vap}, and ΔH_{sub} at the melting point.

$$\Delta H_{sub} = \Delta H_{fus} + \Delta H_{vap} \qquad \text{(12.4)}$$

The value of ΔH_{sub} obtained with equation (12.4) can replace the enthalpy of vaporization in the Clausius–Clapeyron equation (12.2), so that sublimation pressures can be calculated as a function of temperature.

Two familiar solids with significant sublimation pressures are ice and dry ice (solid carbon dioxide). If you live in a cold climate, you are aware that snow may disappear from the ground even though the temperature may fail to rise above $0\,°C$. Under these conditions, the snow does not melt; it sublimes. The sublimation pressure of ice at $0\,°C$ is 4.58 mmHg. That is, the solid ice has a vapor pressure of 4.58 mmHg at $0\,°C$. If the air is not already saturated with water vapor, the ice will sublime. The sublimation and deposition of iodine are pictured in Figure 12-25.

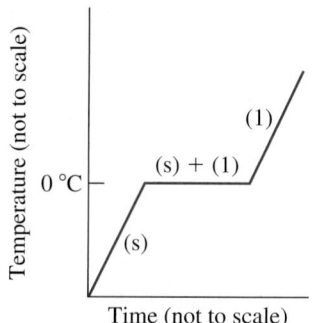

▲ FIGURE 12-24
Heating curve for water
This curve traces the changes that occur as ice is heated from below the melting point to produce liquid water somewhat above the melting point.

◀ Examples of supercooled substances are water droplets in the sky. They remain liquid at temperatures well below the freezing point. When they find a bit of dust on which they can nucleate, the droplets spontaneously turn to ice.

▲ FIGURE 12-25
Sublimation of iodine
Even at temperatures well below its melting point of $114\,°C$, solid iodine exhibits an appreciable sublimation pressure. Here, purple iodine vapor is produced at about $70\,°C$. Deposition of the vapor to solid iodine occurs on the colder walls of the flask.

🔍 **12-6   CONCEPT ASSESSMENT**

Recall the discussion of dew and frost formation (see the *Focus On* feature for Chapter 6, *Earth's Atmosphere*, at www.masteringchemistry.com). Do the surroundings absorb or lose heat when water vapor condenses to dew or frost? Is the quantity of heat per gram of $H_2O(g)$ condensed the same whether the condensate is dew or frost? Explain.

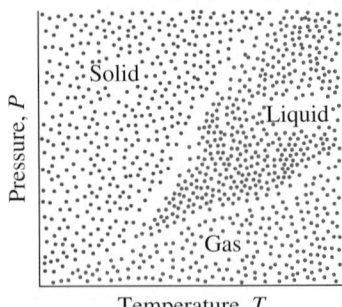

▲ FIGURE 12-26
Temperature, pressure, and states of matter
The outline of a phase diagram is suggested by the distribution of points. The red points identify the temperatures and pressures at which solid is the stable phase; the blue points identify the temperatures and pressures at which liquid is the stable phase; and the brown points represent the temperatures and pressures at which gas is the stable phase. (See also Figures 12-27 and 12-28.)

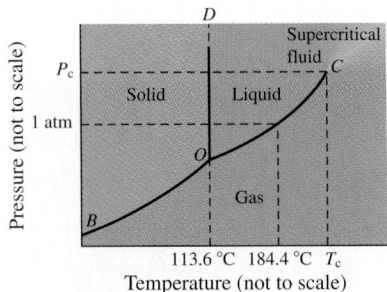

▲ FIGURE 12-27
Phase diagram for iodine
Note that the melting point and triple point temperatures for iodine are essentially the same. Generally, large pressure increases are required to produce even small changes in solid–liquid equilibrium temperatures. The pressure and temperature axes on a phase diagram are generally not drawn to scale so that the significant features of the diagram can be more readily emphasized.

12-4 Phase Diagrams

Imagine constructing a pressure–temperature graph in which each point on the graph represents a condition under which a substance might be found. At low temperatures and high pressures, such as the red points in Figure 12-26, we expect the atoms, ions, or molecules of a substance to be in a close orderly arrangement—a solid. At high temperatures and low pressures—the brown points in Figure 12-26—we expect the gaseous state; and at intermediate temperatures and pressures, we expect a liquid (blue points in Figure 12-26).

Figure 12-26 is a **phase diagram**, a graphical representation of the conditions of temperature and pressure at which solids, liquids, and gases (vapors) exist, either as single phases, or states, of matter or as two or more phases in equilibrium with one another. The different regions of the diagram correspond to single phases, or states, of matter. Straight or curved lines where single-phase regions adjoin represent two phases in equilibrium.

Iodine

One of the simplest phase diagrams is that of iodine shown in Figure 12-27. The curve OC is the vapor pressure curve of liquid iodine, and C is the critical point. OB is the sublimation curve of solid iodine. The nearly vertical line OD represents the effect of pressure on the melting point of iodine; it is called the *fusion curve*. The point O has a special significance. It defines the *unique* temperature and pressure at which the *three* states of matter, solid, liquid, and gas, coexist in equilibrium. It is called a **triple point**. For iodine, the triple point is at 113.6 °C and 91.6 mmHg. The normal melting point (113.6 °C) and the boiling point (184.4 °C) are the temperatures at which a line at $P = 1$ atm intersects the fusion and vapor pressure curves, respectively. Melting is essentially unaffected by pressure in the limited range from 91.6 mmHg to 1 atm, and the normal melting point and the triple point are at almost the same temperature.

The sublimation curve for iodine in Figure 12-27 appears to be a continuation of the vapor pressure curve, but if the data are plotted to scale, a discontinuity is seen at the triple point O. Moreover, this must *always* be the case. If these two curves were continuous, then the lines representing the variation of ln P with $1/T$ (Fig. 12-20) would have the same slope—but this is not possible. The value of ΔH_{vap} determines the slope of the vapor pressure line (recall equation 12.1), whereas ΔH_{sub} determines the slope of the sublimation line. However, these two enthalpy changes can never be the same, because $\Delta H_{sub} = \Delta H_{vap} + \Delta H_{fus}$.

The extreme range of temperatures and pressures required for the entire phase diagram precludes plotting it to scale. This is why the axes are labeled "not to scale."

Carbon Dioxide

The case of carbon dioxide, shown in Figure 12-28, differs from that of iodine in one important respect—the pressure at the triple point O is greater than 1 atm. A line at $P = 1$ atm intersects the *sublimation curve*, not the vapor pressure curve. If solid CO_2 is heated in an open container, it sublimes away at a constant temperature of -78.5 °C. It *does not melt* at atmospheric pressure (and so is called "dry ice"). Because it maintains a low temperature and does not produce a liquid by melting, dry ice is widely used in freezing and preserving foods.

Liquid CO_2 can be obtained at pressures above 5.1 atm and it is most frequently encountered in CO_2 fire extinguishers. All three states of matter are involved in the action of these fire extinguishers. When the liquid CO_2 is released, most of it quickly vaporizes. The heat required for this vaporization is extracted from the remaining $CO_2(l)$, which has its temperature lowered to the point that it freezes and falls as a $CO_2(s)$ "snow." In turn, the $CO_2(s)$ quickly sublimes to $CO_2(g)$. All of this helps to quench a fire by displacing the air around the fire with a "blanket" of $CO_2(g)$ and by cooling the area somewhat.

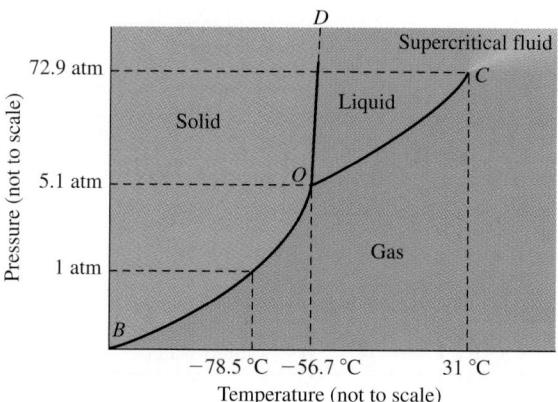

▲ FIGURE 12-28
Phase diagram for carbon dioxide
Several aspects of this diagram are described in the text. An additional feature not shown here is the curvature of the fusion curve OD to the right at very high pressures, ultimately reaching temperatures above the critical temperature.

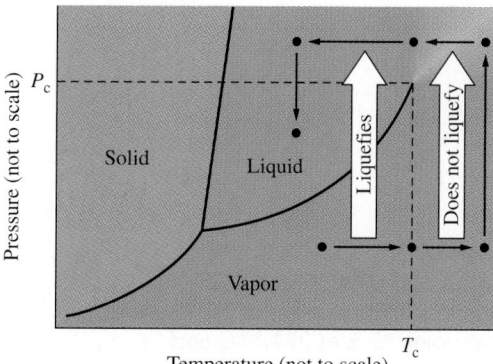

▲ FIGURE 12-29
Critical point and critical isotherm
Applying pressure to a gas at temperatures below the critical isotherm, T_c, causes a liquid to form with the appearance of a meniscus, a discontinuous phase change. Applying pressure above the critical isotherm simply increases the density of the supercritical fluid. In a path traced by the small arrows, gas changes to liquid without exhibiting a discontinuous phase transition.

Supercritical Fluids

Because the liquid and gaseous states become identical and indistinguishable at the critical point, it is difficult to know what to call the state of matter at temperatures and pressures above the critical point. For example, this state of matter has the high density of a liquid and the low viscosity of a gas. The term that is now commonly used is *supercritical fluid* (SCF). Above the critical temperature, no amount of pressure can liquefy a supercritical fluid. Consider the generic phase diagram in Figure 12-29. The path of dots starting with a vapor below the critical isotherm takes us to a low-density gas above the isotherm. When the pressure is greatly increased, we produce a supercritical fluid of much greater density. If, while the pressure exceeds the critical pressure, P_c, the temperature is reduced below the critical temperature, we obtain a liquid. Even with further reduction of pressure, the sample remains in the liquid phase. In following the path described, we have gone from a gas to a liquid without observing a liquid–gas interface. The only way to observe the liquid–vapor interface is to cross the phase boundary below the critical temperature. Note that in the present case we could observe the liquid–vapor interface by lowering the pressure on the liquid to a point on the vapor pressure curve.

Although we do not ordinarily think of liquids or solids as being soluble in gases, *volatile* liquids and solids are. The mole fraction solubility is simply the ratio of the vapor pressure (or the sublimation pressure) to the total gas pressure. And liquids and solids become much more soluble in a gas that is above its critical pressure and temperature, mostly because the density of the SCF is high and approaches that of a liquid. Molecules in supercritical fluids, being in much closer proximity than in ordinary gases, can exert strong attractive forces on the molecules of a liquid or solid solute. SCFs display solvent properties similar to ordinary liquid solvents. To vary the pressure of an SCF means to vary its density and also its solvent properties. Thus, a given SCF, such as carbon dioxide, can be made to behave like many different solvents.

Until recently, the principal method of decaffeinating coffee was to extract the caffeine with a solvent, such as methylene chloride (CH_2Cl_2). This solvent is objectionable because it is hazardous in the workplace and difficult to completely remove from the coffee. Now, supercritical fluid CO_2 is used. In one process, green coffee beans are brought into contact with CO_2 at about 90 °C and 160 to 220 atm. The caffeine content of the coffee is reduced from its

▲ Decaffeinated coffee
"Naturally" decaffeinated coffee is made through a process that uses supercritical fluid CO_2 as a solvent to dissolve the caffeine in green coffee beans. Afterward, the beans are roasted and sold to consumers.

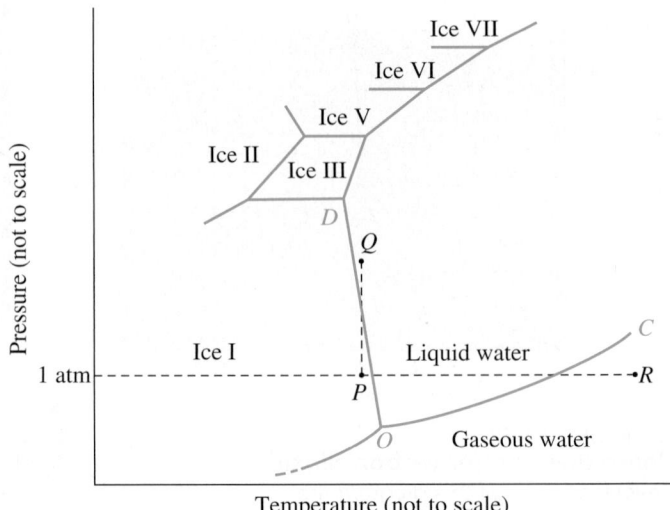

▶ FIGURE 12-30
Phase diagram for water
Point *O*, the triple point, is at 0.0098 °C and 4.58 mmHg. (The normal melting point is at exactly 0 °C and 760 mmHg.) The critical point, *C*, is at 374.1 °C and 218.2 atm. At point *D* the temperature is −22.0 °C and the pressure is 2045 atm. The negative slope of the fusion curve, *OD* (greatly exaggerated here), and the significance of the broken straight lines are discussed in the text.

normal 1% to 3% to about 0.02%. When the temperature and pressure of the CO_2 are reduced, the caffeine precipitates and the CO_2 is recycled.

Water

▶ Since the L-S line in the phase diagram of water is negative, that means that ice floats. If ice did not float, then the polar seas would fill with ice that would never melt. Most solids are more dense than their liquid state. Water is, for us, a happy exception.

The phase diagram of water (Fig. 12-30) presents several new features. One is that the fusion curve *OD* has a *negative* slope; that is, it slopes toward the pressure axis. The melting point of ice *decreases* with an increase in pressure, and this is rather unusual behavior for a solid (bismuth and antimony also behave in this way). However, because large changes in pressure are required to produce even small decreases in the melting point, we do not commonly observe this melting behavior of ice. One example that has been given comes from ice-skating. Presumably, the pressure of the skate blades melts the ice, and the skater skims along on a thin lubricating film of liquid water. This explanation is unlikely, however, because the pressure of the blades doesn't produce a significant lowering of the melting point and certainly cannot explain the ability to skate on ice at temperatures much below the freezing point. (Recent experimental evidence suggests that molecules in a very thin surface layer on ice are mobile in the same way as in liquid water, and this mobility persists even at very low temperatures.)

Another feature illustrated in the phase diagram of water is **polymorphism**, the existence of a solid substance in more than one form. Ordinary ice, called ice I, exists under ordinary pressures. The other forms exist only at high pressures. Polymorphism is more the rule than the exception among solids. Where it occurs, a phase diagram has triple points in addition to the usual solid-liquid-vapor triple point. For example at point *D* in Figure 12-30, ice I, ice III, and liquid H_2O are in equilibrium at −22.0 °C and 2045 atm. Note that the fusion curves for the forms of ice other than ice I have *positive* slopes. Thus, the triple point with ice VI, ice VII, and liquid water is at 81.6 °C and 21,700 atm.

▶ An increase in pressure to 125 atm lowers the freezing point of water by only about 1 °C.

Phases and Phase Transitions

What's the difference between a phase and a state of matter? These terms tend to be used synonymously, but there is a small distinction between them. As we have already noted, there are just *three* states of matter: solid, liquid, and gas. A *phase* is any sample of matter with definite composition and uniform properties that is distinguishable from other phases with which it is in contact. Thus, we can describe liquid water in equilibrium with its vapor as a two-phase mixture. The liquid is one phase and the gas, or vapor, is the other. In this case, the phases (liquid and gas) are the same as the states of matter present (liquid and gas).

We can describe the equilibrium mixture at the triple point *D* in Figure 12-30 as a *three-phase* mixture, even though only *two* states of matter are present (solid and liquid). Two of the phases are in the solid state—the polymorphic forms ice I and ice III. For mixtures of two or more components, different phases may exist in the liquid state as well as in the solid state. For example, most mixtures of triethylamine, $N(CH_2CH_3)_3$, and water at 25 °C separate into two physically distinct liquid phases. One is a saturated solution of triethylamine in water and the other, a saturated solution of water in triethylamine. Because the pressure–temperature diagrams we have been describing can accommodate all the phases in a system, we call them *phase* diagrams. We call the crossing of a two-phase curve in a phase diagram a *phase transition*.

Listed below are six common names assigned to phase transitions.

melting(s $\longrightarrow$ l)   freezing (l $\longrightarrow$ s)

vaporization (l $\longrightarrow$ g)   condensation (g $\longrightarrow$ l)

sublimation (s $\longrightarrow$ g)   deposition (g $\longrightarrow$ s)

Following are two useful generalizations about the changes that occur when crossing a two-phase equilibrium curve in a phase diagram.

- From lower to higher temperatures along a *constant-pressure* line (an isobar), enthalpy *increases*. (Heat is absorbed.)
- From lower to higher pressures along a *constant-temperature* line (an isotherm), volume *decreases*. (The phase at the higher pressure has the higher density.)

The second generalization helps us to understand why a fusion curve usually has a positive slope. Typical behavior is for a solid to have a greater density than the corresponding liquid. Example 12-7 illustrates how we can use a phase diagram to describe the phase transitions that a substance can undergo.

◀ Because ice I is less dense than $H_2O(l)$, the fusion curve *OD* in Figure 12-30 has a negative slope.

EXAMPLE 12-7   Interpreting a Phase Diagram

A sample of ice is maintained at 1 atm and at a temperature represented by point *P* in Figure 12-30. Describe what happens when (a) the temperature is raised, at constant pressure, to point *R*, and (b) the pressure is raised, at constant temperature, to point *Q*. The sketches Figure 12-31 suggest the conditions at points *P*, *Q*, and *R*.

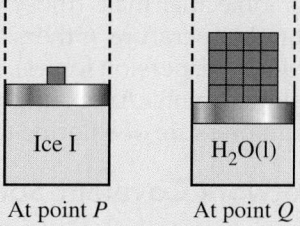

At point *P*   At point *Q*

Analyze

Recall that the lines separating the different phases represent coexistence lines. At these coexistence lines, the system is a mixture of both phases. On either side of those lines, the system is in that particular phase. Also recall that as the system moves from one phase to another at the coexistence lines, the temperature remains constant until all of one phase is converted to another.

Solve

(a) When the temperature reaches a point on the fusion curve *OD* (0 °C), ice begins to melt. The temperature remains constant as ice is converted to liquid. When melting is complete, the temperature

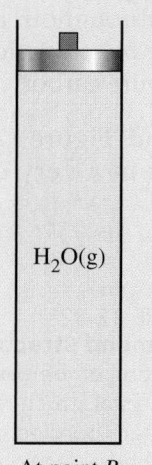

At point *R*

◀ FIGURE 12-31
Example 12-7 illustrated
A sample of pure water is confined in a cylinder by a freely moving piston surmounted by weights to establish the confining pressure. Sketched here are conditions at the points labeled *P*, *Q*, and *R* in Figure 12-30. The transition from point *P* to *Q* is accomplished by changing the pressure at constant temperature (isothermal). The transition from point *P* to *R* is accomplished by changing the temperature at constant pressure (isobaric).

(continued)

again increases. No vapor appears in the cylinder until the temperature reaches 100 °C, at which point the vapor pressure is 1 atm. When all the liquid has vaporized, the temperature is again free to rise to a final value of R.

(b) Because solids are not very compressible, very little change occurs until the pressure reaches the point of intersection of the constant-temperature line PQ with the fusion curve OD. Here melting begins. A significant *decrease* in volume occurs (about 10%) as ice is converted to liquid water. After melting, additional pressure produces very little change because liquids are not very compressible.

Assess

Phase diagrams are very useful for understanding the conditions needed to observe the different phases of matter. We should now be able to use the phase diagram in Figure 12-30 to determine the pressure required to observe sublimation instead of melting.

PRACTICE EXAMPLE A: With as much detail as possible, describe the phase changes that would occur if a sample of water represented by point R in Figure 12-30 were brought first to point P and then to point Q.

PRACTICE EXAMPLE B: Draw a sketch showing the condition prevailing along the line PR when 1.00 mol of water has been brought to the point where exactly one-half of it has vaporized. Compare this to the condition at point R in Figure 12-31, assuming that this is also based on 1.00 mol of water. For example, is the volume of the system the same as that in Figure 12-31? If not, is it larger or smaller, and by how much? Assume that the temperature at point R is the same as the critical temperature of water and that water vapor behaves as an ideal gas.

🔍 12-7 CONCEPT ASSESSMENT

One method of restoring water-damaged books after a fire is extinguished in a library is by "freeze drying" them in evacuated chambers. Describe how this method might work.

12-5 Network Covalent Solids and Ionic Solids

In most covalent substances, intermolecular forces are quite weak compared with the bonds between atoms within molecules. This is why covalent substances of low molecular mass (those with weak dispersion forces) are generally gaseous at room temperature. Others, usually of somewhat higher molecular masses (with stronger dispersion forces), are liquids. Still others are solids with moderately low melting points. An example of such a trend are the room-temperature states at 1 atm pressure of chlorine (a gas), bromine (a liquid), and iodine (a solid).

Network Covalent Solids

In a few substances, known as **network covalent solids**, covalent bonds extend throughout a crystalline solid. In these cases, the entire crystal is held together by strong forces. Consider, for example, two of the allotropic forms in which pure carbon occurs—diamond and graphite.

▶ Another silicon-containing network covalent solid is ordinary silica—silicon dioxide, SiO_2.

Diamond Figure 12-32 shows one way that carbon atoms can bond one to another in a very extensive array or crystal. The two-dimensional Lewis

KEEP IN MIND

that four bonds directed from a central atom to the corners of a tetrahedron correspond to the sp^3 hybridization scheme.

▶ FIGURE 12-32
The diamond structure
(a) A portion of the Lewis structure, (b) The crystal structure shows each carbon atom bonded to four others in a tetrahedral fashion. The segment of the entire crystal shown here is called a unit cell.

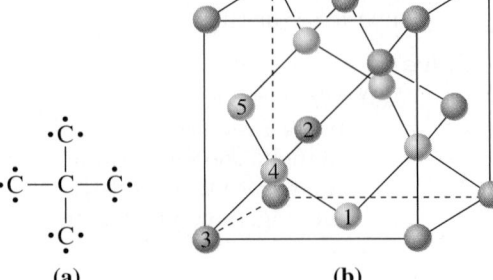

(a)

(b)

structure (Fig. 12-32a) is useful only in suggesting that the bonding scheme involves ever-increasing numbers of C atoms leading to a giant molecule. It does not give any insight into the three-dimensional structure of the molecule. For this, we need the portion of the crystal shown in Figure 12-32(b). Each atom is bonded to four others. Atoms 1, 2, and 3 lie in a plane, with atom 4 above the plane. Atoms 1, 2, 3, and 5 define a tetrahedron with atom 4 at its center. When viewed from a particular direction, a nonplanar hexagonal arrangement of carbon atoms (gray) is also seen.

If silicon atoms are substituted for half the carbon atoms in this structure, the resulting structure is that of silicon carbide (carborundum). Both diamond and silicon carbide are extremely hard, and this accounts for their extensive use as abrasives. In fact, diamond is the hardest substance known. To scratch or break diamond or silicon carbide crystals, covalent bonds must be broken. These two materials are also nonconductors of electricity and do not melt or sublime except at very high temperatures. SiC sublimes at 2700 °C, and diamond melts above 3500 °C.

Graphite Carbon atoms can bond together to produce a solid with properties very different from those of diamond. In graphite, bonding involves the orbital set $sp^2 + p$. The three sp^2 orbitals are directed in a plane at angles of 120 °C, and the p orbitals overlap in the manner described for carbon atoms in benzene, C_6H_6 (see Figure 11-29). Thus, the p electrons are delocalized—that is, not restricted to the region between two C atoms but shared among many C atoms within a plane of C atoms. This type of bonding produces the crystal structure shown in Figure 12-33. Each carbon atom forms strong covalent bonds with three neighboring carbon atoms in the same plane, giving rise to layers of carbon atoms in a hexagonal arrangement. Bonding within layers is strong, but the intermolecular forces between layers are the much weaker van der Waals forces. We can see this through bond distances. The C — C bond distance within a layer is 142 pm (compared with 139 pm in benzene); between layers, it is 335 pm.

Its unique crystal structure gives graphite some distinctive properties. Because bonding between layers is weak, the layers can glide over one another rather easily. As a result, graphite is a good lubricant, either in dry form or in an oil suspension.* If a mild pressure is applied to a piece of graphite, layers of the graphite flake off; this is what happens when we use a graphite pencil. Also, because the p electrons are delocalized, they migrate through the planes of carbon atoms when an electric field is applied; graphite conducts electricity. An important use of graphite is as electrodes in batteries and in industrial electrolysis. Diamond is not an electrical conductor because all its valence electrons are localized or permanently fixed into single covalent bonds.

Other Allotropes of Carbon

In 1985, the first of what is now known to be an extensive series of allotropes of carbon was discovered. In experiments designed to mimic conditions found near red-giant stars, a number of carbon-containing molecules were discovered and characterized through mass spectroscopy. The strongest peak in the mass spectrum came at 720 u, corresponding to the molecule C_{60}. For a time, proposing a plausible structure for this molecule proved to be a challenge. Neither diamond- nor graphite-type structures could account for a molecule with 60 carbon atoms because "dangling" bonds would remain at the edges of the structures. The structure that was finally proposed, and confirmed by X-ray crystallography, is that of a *truncated icosahedron*—a three-dimensional figure composed of 12 pentagonal and 20 hexagonal faces, with a carbon atom at each of its 60 vertices (Fig. 12-34). This figure resembles a soccer ball and

▲ **Graphite conducts electricity**
Delocalized electons in graphite allow the conduction of electricity. In the photo, pencil "lead,"a mixture of graphite and clay, is used as electodes to complete the circuit.The beaker contains a solution of ions that carry the current between the pencil electrodes.

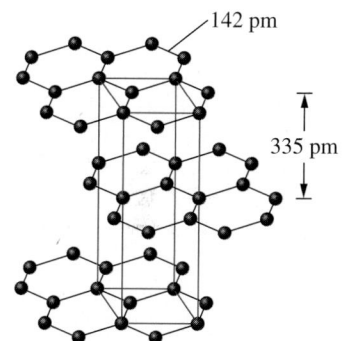

142 pm

335 pm

▲ FIGURE 12-33
The graphite structure

*The lubricating properties of graphite also appear to depend on the presence of molecules between the layers of carbon atoms. When graphite is strongly heated in a vacuum it becomes a much poorer lubricant.

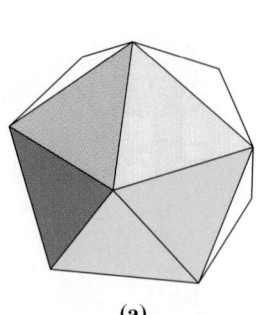
(a)

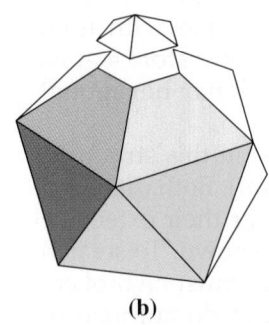
(b)

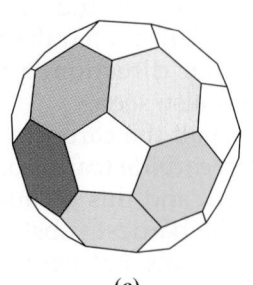
(c)

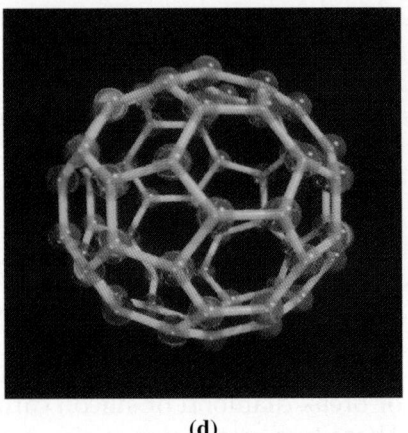
(d)

▲ FIGURE 12-34
Fullerenes
(a) An icosahedron, a shape formed by 20 equilateral triangles. Five triangles meet at each of the 12 vertices. (b) Truncating or cutting off a vertex reveals a new pentagonal face. (c) The truncated icosahedron. Twelve pentagons have replaced the original 12 vertices, and the 20 equilateral triangles have been replaced by 20 hexagons. (d) The C_{60} molecule.

also certain geodesic domes. In fact, the resemblance to the geodesic dome led first to the proposed name "buckminsterfullerenes," then simply, *fullerenes*, and finally the colloquial expression "buckyballs." (The geodesic dome is an architectural form pioneered by R. Buckminster Fuller.) Since 1985, many other fullerenes have been discovered, including C_{70}, C_{74}, and C_{82}. Fullerenes can also form compounds, some by attaching atoms or groups of atoms to their surfaces, others by encasing an atom inside the fullerene structure. To date, several thousand fullerene compounds have been prepared.

Research on fullerenes has led to the discovery of a related type of carbon allotrope—*nanotubes*. A nanotube can be thought of as a two-dimensional array of hexagonal rings of carbon atoms, which is called a graphene sheet. An analogous macroscopic structure is a sheet of chicken wire. Now imagine rolling the graphene sheet into a cylinder (something that a section cut from a roll of chicken wire seems to do so naturally). Finally, cap off each end of the cylindrical graphene sheet with half of a fullerene (Fig. 12-35). The diameters of these tubes are of the order of a few nanometers (hence the name *nano*tube). Their lengths can vary from several nanometers to a micrometer or more. Nanotubes possess

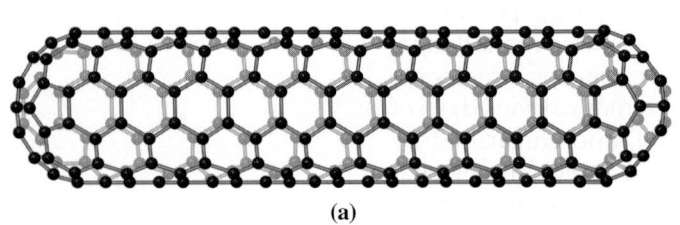
(a)

▲ FIGURE 12-35
Nanotubes
(a) Ball-and-stick model of a small nanotube. (b) A bundle of single-wall nanotubes.

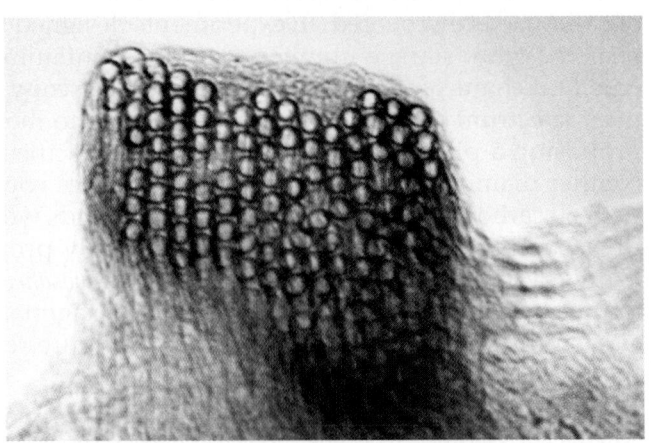
(b)

unusual electronic and mechanical properties that offer the promise of some applications in the macroscopic world and probably many more in the submicroscopic world of *nanotechnology*. For example, nanotubes might one day be used to form the molecular wires of nanoscale electronic devices.

Ionic Solids

When predicting properties of an ionic solid, we often face this question: How difficult is it to break up an ionic crystal and separate its ions? This question is addressed by the *lattice energy* of a crystal. Let us define **lattice energy** as the energy given off when separated *gaseous* ions—positive and negative—come together to form *one mole* of a *solid* ionic compound. Lattice energies can be useful in making predictions about the melting points and water solubilities of ionic compounds. We will examine how to calculate lattice energies in Section 12-9. At times, however, we need to make only *qualitative* comparisons of *interionic forces*, and the following generalization works quite well.

> The attractive force between a pair of oppositely charged ions increases with increased charge on the ions and with decreased ionic sizes.

This idea is based on Coulomb's law (Appendix B) and illustrated in Figure 12-36.

For most ionic compounds, lattice energies are great enough that ions do not readily detach themselves from the crystal and pass into the gaseous state. Ionic solids do not sublime at ordinary temperatures. We can melt ionic solids by supplying enough thermal energy to disrupt the crystalline lattice. In general, the higher the lattice energy of an ionic compound, the higher is its melting point.

The energy required to break up an ionic crystal when it dissolves results from the interaction of ions in the crystal with molecules of the solvent. The extent to which an ionic solid dissolves in a solvent, however, depends only in part on the lattice energy of the ionic solid. As a rough rule, though, the lower the lattice energy, the greater the quantity of an ionic solid that can be dissolved in a given quantity of solvent.

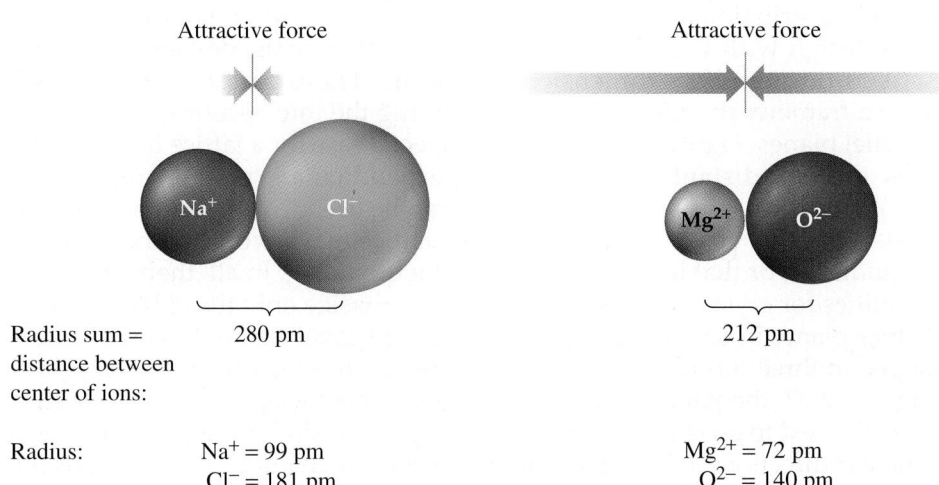

| | Attractive force | | Attractive force |

Radius sum = distance between center of ions: 280 pm   212 pm

Radius: $Na^+ = 99$ pm $Mg^{2+} = 72$ pm
                   $Cl^- = 181$ pm $O^{2-} = 140$ pm

▲ FIGURE 12-36
Interionic forces of attraction
Because of the higher charges on the ions and the closer proximity of their centers, the interionic attractive force between Mg^{2+} and O^{2-} is about seven times as great as between Na^+ and Cl^-.

EXAMPLE 12-8   **Predicting Physical Properties of Ionic Compounds**

Which has the higher melting point, KI or CaO?

Analyze

Trends in melting points, just as we observed for boiling points, depend on intermolecular forces. For ionic compounds the melting points are dependent on the interionic forces, which are related to Coulomb's law.

Solve

Ca^{2+} and O^{2-} are more highly charged than K^+ and I^-. Also, Ca^{2+} is smaller than K^+, and O^{2-} is smaller than I^-. We would expect the interionic forces in crystalline CaO to be much larger than in KI. CaO should have the higher melting point. (The observed melting points are 677 °C for KI and 2590 °C for CaO.)

Assess

We see that two factors contribute to the interionic forces in this problem. The first is the charge and the second is the radius of each ion.

PRACTICE EXAMPLE A: Cite one ionic compound that you would expect to have a lower melting point than KI and one with a higher melting point than CaO.

PRACTICE EXAMPLE B: Which would you expect to have the greater solubility in water, NaI or $MgCl_2$? Explain.

12-6   Crystal Structures

Crystals, solid structures having plane surfaces, sharp edges, and regular geometric shapes, have aroused interest from earliest times, whether as ice, rock salt, quartz, or gemstones. Only in relatively recent times have we come to a fundamental understanding of the crystalline state. This understanding started with the invention of the optical microscope and was greatly expanded following the discovery of X-rays. The key idea, now supported by countless experiments, is that the regularity observed in crystals at the macroscopic level is due to an underlying regular pattern in the arrangement of atoms, ions, or molecules.

Crystal Lattices

You can probably think of a number of situations in which you have had to deal with repeating patterns in one or two dimensions. These might include projects like stringing beads to make a necklace, wallpapering a room, or creating a design with floor tiles. The structures of crystals, however, must be described through *three*-dimensional patterns. These patterns are outlined against a framework called a *lattice*, comprising the intersections of three sets of parallel planes. Figure 12-37 shows the special case for a lattice in which the planes are equidistant and mutually perpendicular (intersecting at 90° angles). This is called a *cubic* lattice, and it can be used to describe a number of crystals. For other crystals, the appropriate lattice may involve planes that are not equidistant or that intersect at angles other than 90°. In all, there are seven possibilities for crystal lattices, but we will emphasize only the cubic lattice.

Lattice planes intersect to produce three-dimensional figures having six faces arranged in three sets of parallel planes. These figures are called *parallelepipeds*. In Figure 12-37, the parallelepipeds are cubes. A parallelepiped called the **unit cell** can be used to generate the entire lattice by replicating it along the three perpendicular directions. (Consider a two-dimensional analogy: A single floor tile is like a unit cell and, wherever the first tile is placed, the entire floor can be covered by adding identical tiles in the two perpendicular directions from the initial one.) Where possible, we arrange the three-dimensional space lattice so that the centers of the structural particles of the crystal (atoms, ions, or molecules) are situated at lattice points. If a unit cell has structural particles only at its corners, it is called a *primitive* or simple unit cell, because it is the simplest unit cell we

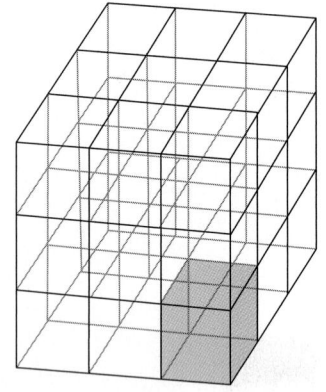

▲ FIGURE 12-37
The cubic lattice
One parallelepiped formed by the intersection of mutually perpendicular planes is shaded in green—it is a cube. An endless lattice can be generated by simple displacements of the green cube in the three perpendicular directions (that is, left and right, up and down, and forward and backward).

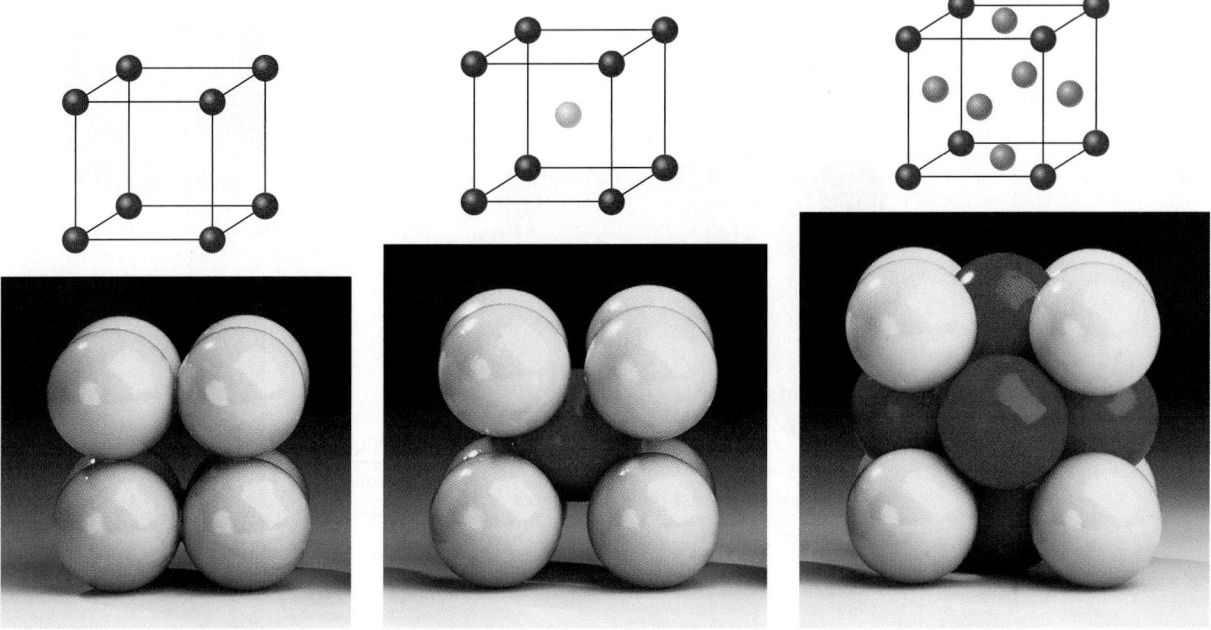

Simple cubic   Body-centered cubic   Face-centered cubic

▲ FIGURE 12-38
Unit cells in the cubic crystal system
In the line-and-ball drawings in the top row, only the centers of spheres (atoms) are shown at their respective positions in the unit cells. The space-filling models in the bottom row show contacts between spheres (atoms). In the simple cubic cell, spheres come into contact along each edge. In the body-centered cubic (bcc) cell, contact of the spheres is along the cube diagonal. In the face-centered cubic (fcc) cell, contact is along the diagonal of each face. The spheres shown here are identical atoms; color is used only for emphasis.

can consider. But some unit cells have more structural particles than those found at its corners. In the **body-centered cubic (bcc)** structure, there is a structural particle at the center of the cube as well as at each corner of the unit cell. In a **face-centered cubic (fcc)** structure, there is a structural particle at the center of each face as well as at each corner. These unit cells are shown in Figure 12-38.

Closest Packed Structures

Unlike boxes, which can be stacked to fill all space, when spheres are stacked together, there must always be some unfilled space. In some arrangements, however, the spheres come into as close contact as possible, and the volume of the holes, or voids, is at a minimum. These arrangements are known as closest packed structures and are the basis of a number of crystal structures.

To analyze the closest packed structures in Figure 12-39(a), imagine one layer of spheres, layer A (red), in which each sphere is in contact with six others arranged in a hexagonal fashion around it. Among the spheres, there are holes. The hole between three spheres resembling a triangle is called a *trigonal hole* (Fig. 12-39b). Once the first sphere is placed in the next layer, layer B (yellow), the entire pattern for that layer is fixed. Again, there are holes in layer B, but the holes are of two different types. *Tetrahedral holes* fall directly over *spheres* in layer A, and *octahedral holes* fall directly over *holes* in layer A (Fig. 12-39b).

There are two possibilities for C, the third layer (Fig. 12-39a). In one arrangement, called **hexagonal closest packed (hcp)**, all the tetrahedral holes are covered. Layer C is identical to layer A, and the structure begins to repeat itself. In the other arrangement, called **cubic closest packed**, all the octahedral holes are covered. The spheres in layer C (blue) are out of line with those in layer A. Only when the fourth layer is added does the structure begin to repeat itself.

▲ A closest packed pyramid of cannonballs. Oranges at a fruit stand are often packed in cubic closest packed pyramids so that they will not slip.

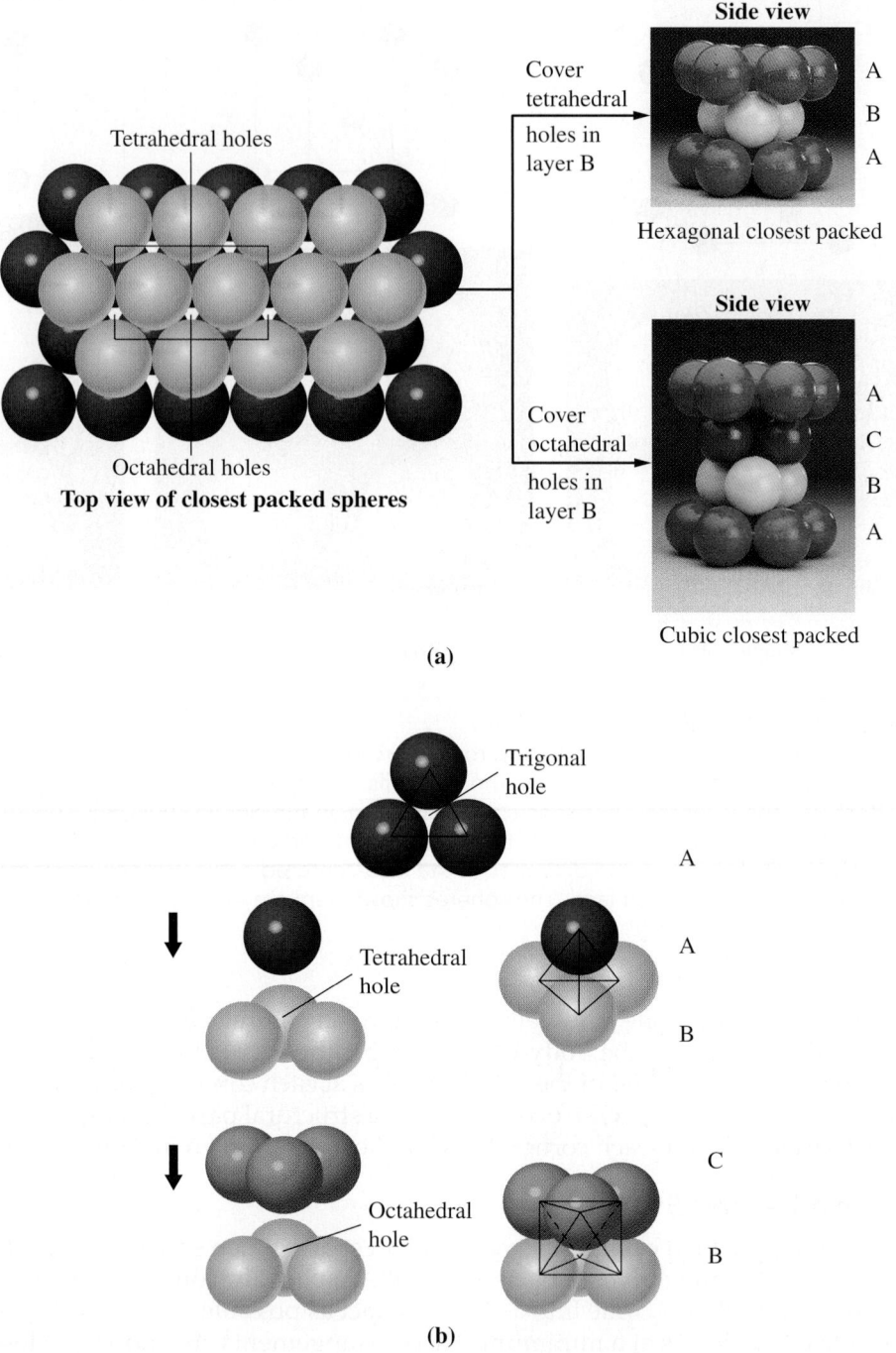

▲ FIGURE 12-39
Closest packed structures
(a) Spheres in layer A are red. Those in layer B are yellow, and in layer C, blue. (b) The holes in closest packed structures. The trigonal hole is formed by three spheres in one of the layers. The tetrahedral hole is formed when a sphere in the upper layer sits in the dimple of the lower layer. The octahedral hole is formed between two groups of three spheres in two layers.

Study Figure 12-40 and you will see that the cubic closest packed structure has a face-centered cubic unit cell. The unit cell of the hexagonal closest packed structure is shown in Figure 12-41. In both the hcp and fcc structures, holes account for only 25.96% of the total volume. Another possibility in which spheres are closely packed but not in a closest packed arrangement has a body-centered cubic unit cell. In this structure, holes account for 31.98% of

  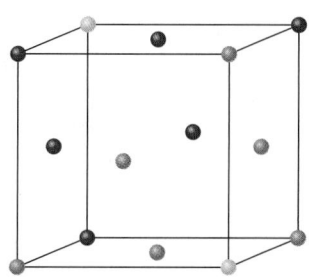

▲ FIGURE 12-40
A face-centered cubic unit cell for the cubic closest packing of spheres
The 14 spheres on the left are extracted from a larger array of spheres in a cubic closest packed structure. The two middle layers each have six atoms; the top and bottom layers, one. Rotation of the group of 14 spheres reveals the fcc unit cell (right).

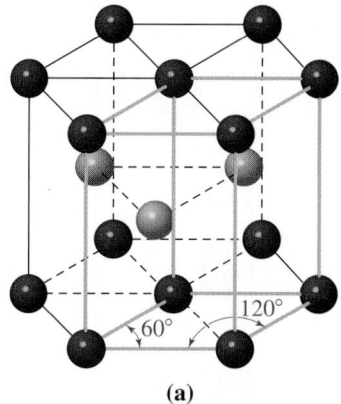

TABLE 12.7   Four Ways of Packing Identical Spheres in Cubic System

Unit Cell	Coordination Number	Atoms per Unit Cell	Volume Occupied, %	Examples
Simple cubic	6	1	52	Po
Body-centered cubic (bcc)	8	2	68	Fe, Na, K, W
Hexagonal closest packed (hcp)	12	2	74	Cd, Mg, Ti, Zn
Face-centered cubic (fcc)	12	4	74	Ag, Cu, Pb

Note: Face-centered cubic (fcc) is equivalent to cubic closest packed (ccp).

the total volume. The best examples of crystal structures based on the packing of spheres are found among the metals. Some examples are listed in Table 12.7.

Coordination Number and Number of Atoms per Unit Cell

In crystals with atoms as their structural units, each atom is in contact with several others. For example, can you see in Figure 12-38 that the center atom in the bcc unit cell is in contact with each corner atom? The number of atoms with which a given atom is in contact is called its *coordination number*. For the bcc structure, this is 8; for the fcc and hcp structures, the coordination number is 12. The easiest way to visualize the coordination number 12 is from the layering of spheres described in Figure 12-39. Each sphere is in contact with *six* others in the same layer, *three* in the layer above, and *three* in the layer below.

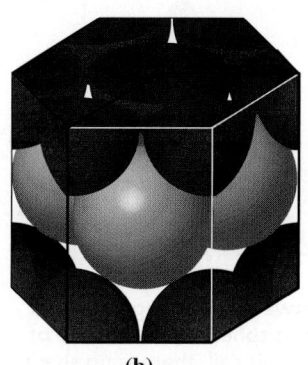

▲ FIGURE 12-41
The hexagonal closest packed (hcp) crystal structure
(a) A unit cell is highlighted in heavy green lines. The atoms that are part of that cell are joined in solid lines. Note that the unit cell is a parallelepiped but not a cube. Three adjoining unit cells are depicted. The highlighted unit cell and broken-line regions together show the layering (ABA) described in Figure 12-39. **(b)** The hexagonal prism showing parts of the shared spheres at the corners and the single sphere at the center of the unit cell.

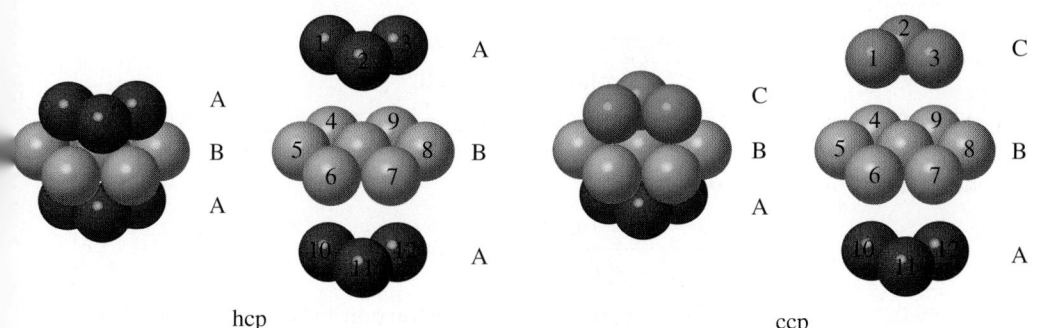

▲ Illustrating the coordination number for the hcp and ccp structures

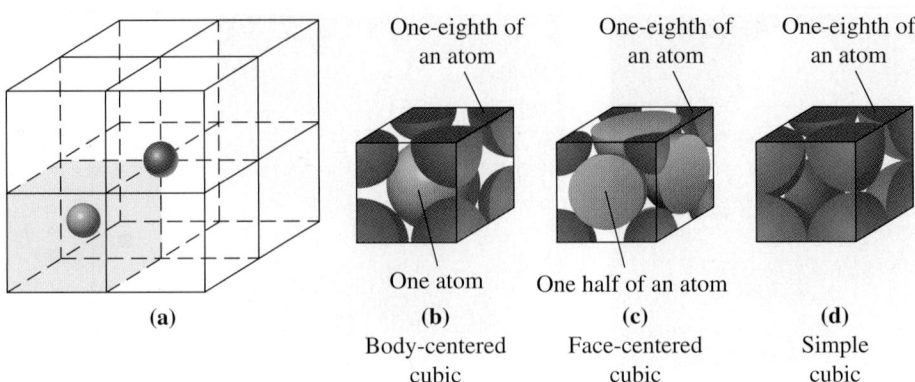

One-eighth of an atom

One-eighth of an atom

One-eighth of an atom

One atom

One half of an atom

(a)

(b)
Body-centered cubic

(c)
Face-centered cubic

(d)
Simple cubic

▲ FIGURE 12-42
Apportioning atoms among cubic unit cells
(a) Eight unit cells are outlined. Attention is directed to the blue-shaded unit cell. For clarity, only the centers of two atoms are pictured. The atom in the center of the blue cell belongs entirely to that cell. The corner atom is seen to be shared by all eight unit cells. **(b)** The shared spheres of a body-centered unit cell. **(c)** The shared spheres of a face-centered unit cell. **(d)** The shared spheres of a simple cubic unit cell. The spheres shown here are identical atoms; color is used only for emphasis.

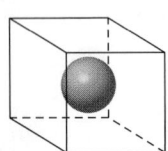
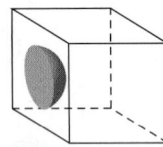
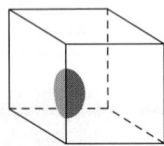
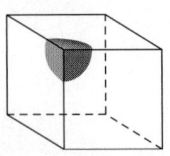

▲ How spheres are shared between or among unit cells. For a sphere in the middle of the unit cell, there is no sharing; on a face 1/2 of the sphere is in the unit cell; at an edge only 1/4 of the sphere is in the unit cell; and in a corner only 1/8 is contained within the unit cell.

Although it takes nine atoms to draw the bcc unit cell, it is wrong to conclude that the unit cell consists of nine atoms. As shown in Figure 12-42(a), only the center atom belongs *entirely* to the bcc unit cell. The other atoms are shared with other unit cells. The corner atoms are shared among eight adjoining unit cells. Only one-eighth of each corner atom should be thought of as belonging entirely to a given unit cell (Fig. 12-42b). Thus, the eight corner atoms collectively contribute the equivalent of *one* atom to the unit cell. The total number of atoms in a bcc unit cell, then, is *two* [that is, $1 + (8 \times \frac{1}{8})$]. For the hcp unit cell of Figure 12-41, we also get *two* atoms per unit cell if we use the correct counting procedure. The corner atoms account for $\frac{1}{8} \times 8 = 1$ atom, and the central atom belongs entirely to the unit cell. In the fcc unit cell, the corner atoms account for $\frac{1}{8} \times 8 = 1$ atom, and those in the center of the faces for $\frac{1}{2} \times 6 = 3$ atoms. The fcc unit cell contains *four* atoms (Fig. 12-42c). The simple cubic unit cell contains only *one* atom per unit cell (Fig. 12-42d).

12-2 ARE YOU WONDERING...

How to calculate the volume of the holes in a structure?

To illustrate this, consider the bcc structure. The ratio of the occupied volume to the unit cell volume is

$$f_V = \frac{\text{volume of spheres in unit cell}}{\text{volume of unit cell}}$$

If the radius of the atom is r, the volume of a sphere is $(4/3)\pi r^3$, and as shown in Figure 12-45 on page 536 the cube edge l is $r(4/\sqrt{3})$. Based on two complete spheres in the unit cell, we have

$$f_V = \frac{2 \times (4/3)\pi r^3}{[r(4/\sqrt{3})]^3} = 0.6802$$

Thus, 68.02% of the unit cell is occupied and 31.98% of the unit cell is empty. Note also that this percentage of empty space is the same regardless of the radius of the sphere.

X-Ray Diffraction

We can see macroscopic objects by using visible light and our eyes. To "see" how atoms, ions, or molecules are arranged in a crystal requires light of much shorter wavelength. When a beam of X-rays encounters atoms, the X-rays interact with electrons in the atoms and the original beam is scattered in all directions. The pattern of this scattered radiation is related to the distribution of electronic charge in the atoms and/or molecules. The scattered X-rays can produce a visible pattern, as on a photographic film, and it is then possible to infer the microscopic structure of the substance from this visible pattern. How successful we are in making inferences depends on the amount of the scattered radiation recovered, that is, on how much "information" is gathered. The power of the X-ray diffraction method has been greatly increased by the use of high-speed computers to process vast amounts of X-ray data.

Figure 12-43 suggests a method of scattering X-rays from a crystal. X-ray data can be explained by a geometric analysis proposed by W. H. Bragg and W. L. Bragg in 1912 and illustrated in Figure 12-44. The figure shows two rays in a monochromatic (single-wavelength) X-ray beam, labeled *a* and *b*. Wave *a* is diffracted, or scattered, by one plane of atoms or ions in a crystal and wave *b* from the next plane below. Wave *b* travels a greater distance than wave *a*. The additional distance is $2d \sin \theta$. The intensity of the scattered radiation will be greatest if waves *a* and *b* reinforce each other, that is, if their crests and troughs line up. To satisfy this requirement, the additional distance traveled by wave *b* must be an integral multiple of the wavelength of the X-rays.

$$n\lambda = 2d \sin \theta \qquad (12.5)$$

From the measured angle θ yielding the maxmum intensity for the scattered X-rays, and the X-ray wavelength (λ), we can calulate the spacing (*d*) between

◀ The X-ray diffraction method was originated by Max von Laue (Nobel Prize, 1914) but carried further by the Braggs. William Lawrence Bragg was only 25 years old when he and his father, William Henry Bragg, won the Nobel Prize in 1915.

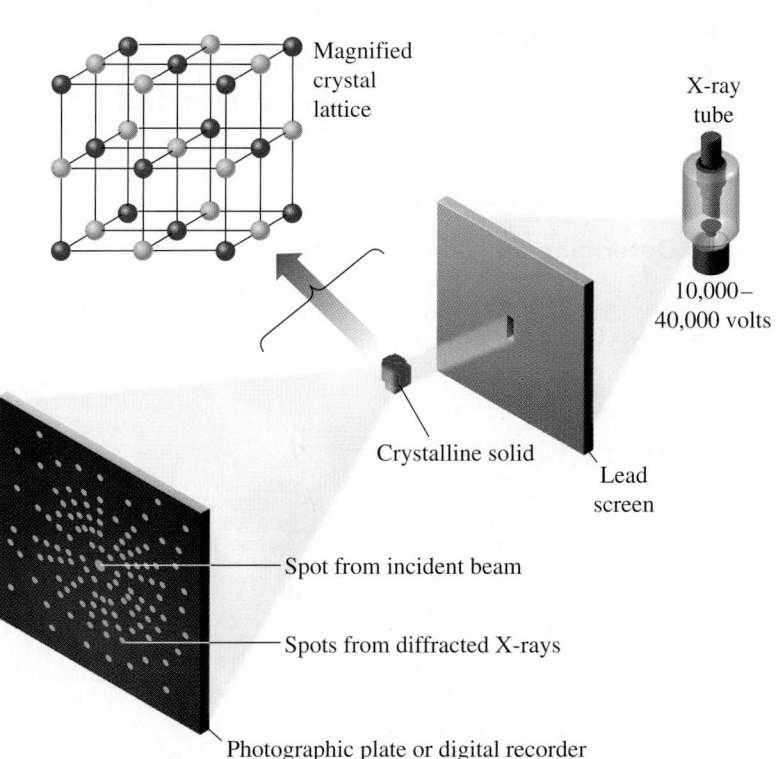

Magnified crystal lattice

X-ray tube

10,000– 40,000 volts

Crystalline solid

Lead screen

Spot from incident beam

Spots from diffracted X-rays

Photographic plate or digital recorder

▲ FIGURE 12-43
Diffraction of X-rays by a crystal
In X-ray diffraction, the scattering is usually from no more than 20 planes deep in a crystal. The size of the single crystal is to have enough of the surface available for diffraction, yet the diffraction is dominated by a few of the surface planes.

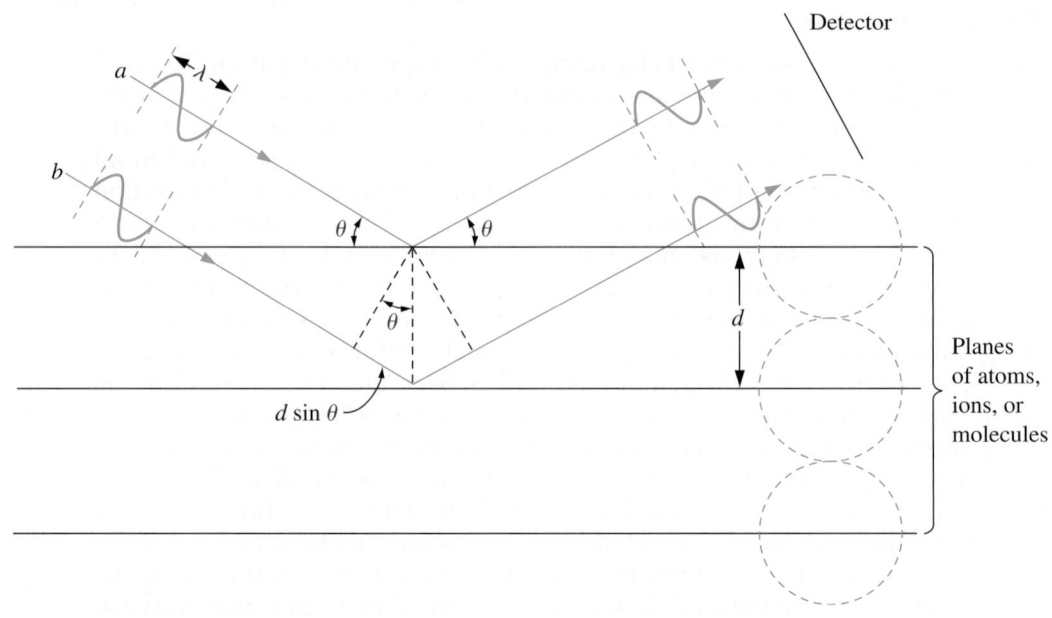

▲ FIGURE 12-44
Determination of crystal structure by X-ray diffraction
The two triangles outlined by dashed lines are identical. The hypotenuse of each triangle is equal to the interatomic distance, d. The side opposite the angle θ thus has a length of $d \sin \theta$. Wave b travels farther than wave a by the distance $2d \sin \theta$.

atomic planes. With different orientations of the crystal, we can determine atomic spacings and electron densities for different directions through the crystal, in short, the crystal structure.

Once a crystal structure is known, certain other properties can be determined by calculation. Example 12-9 shows the determination of a metallic

---

EXAMPLE 12-9   Using X-Ray Data to Determine an Atomic Radius

At room remperature, iron crystallizes in a bcc structure. By X-ray diffraction, the edge of the cubic cell corresponding to Figure 12-45 is found to be 287 pm. What is the radius of an iron atom?

Analyze

Nine atoms are associated with a bcc unit cell. One atom is located at each of the eight corners of the cube and one at the center. The three atoms along a diagonal through the cube are in contact. The length of the cube diagonal (the distance from the farthest upper-right corner to the nearest lower-left corner) is four times the atomic radius. Also shown in Figure 12-45 is the fact the diagonal of a cube is equal to $\sqrt{3} \times l$. The length of an edge, l, is what is given.

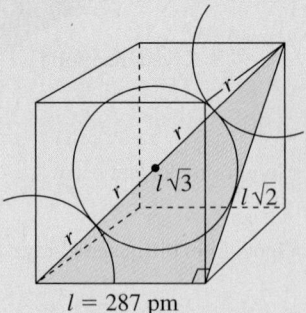
$l = 287$ pm

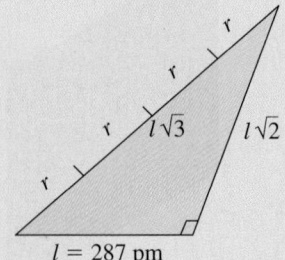
$l = 287$ pm

▲ FIGURE 12-45
Determination of the atomic radius of iron—Example 12-9 illustrated
The right triangle must conform to the Pythagorean formula $a^2 + b^2 = c^2$. That is, with l as an edge of the cube, $(l)^2 + (l\sqrt{2})^2 = (l\sqrt{3})^2$, or $(l)^2 + 2(l)^2 = 3(l)^2$.

Solve

Setting the length of the cube diagonal, in terms of atomic radii, equal to the expression relating the diagonal to the cube, we have

$$4r = l\sqrt{3}$$

which is used to solve for the atomic radius of an iron atom:

$$r = \frac{\sqrt{3} \times 287\,\text{pm}}{4} = \frac{1.732 \times 287\,\text{pm}}{4} = 124\,\text{pm}$$

Assess

We see that it is important to know the atomic arrangement for each basic unit cell and to know that atoms at the corners are shared between unit cells.

PRACTICE EXAMPLE A: Potassium crystallizes in the bcc structure. What is the length of the unit cell in this structure? Use the metallic radius of potassium given in Figure 9-9.

PRACTICE EXAMPLE B: Aluminum crystallizes in an fcc structure. Given that the atomic radius of Al is 143.1 pm, what is the volume of a unit cell?

radius, and Example 12-10 estimates the density of a crystalline solid. For both of these calculations, we need to sketch, or in some way visualize, a unit cell of the crystal. In particular, we need to see which atoms are in direct contact.

EXAMPLE 12-10   Relating Density to Crystal Structure Data

Use data from Example 12-9, together with the molar mass of Fe and the Avogadro constant, to calculate the density of iron.

Analyze

To calculate the density, we need the mass of the unit cell (in grams) and its volume (in cm^3). From Table 12.7, we find that there are *two* Fe atoms per bcc unit cell, which we can use to calculate the mass of the unit cell. In Example 12-9, we saw that the length of a unit cell is $l = 287\,\text{pm} = 287 \times 10^{-12}\,\text{m} = 2.87 \times 10^{-8}\,\text{cm}$, which we can use to find the density of iron.

Solve

We need the mass of these two atoms, and the key to getting this is a conversion factor based on the fact that $1\,\text{mol Fe} = 6.022 \times 10^{23}\,\text{Fe atoms} = 55.85\,\text{g Fe}$.

$$m = 2\,\text{Fe atoms} \times \frac{55.85\,\text{g Fe}}{6.022 \times 10^{23}\,\text{Fe atoms}} = 1.855 \times 10^{-22}\,\text{g Fe}$$

The volume of the unit cell is $V = l^3 = (2.87 \times 10^{-8})^3\,\text{cm}^3$. Density is the ratio of mass to volume.

$$\text{density of Fe} = \frac{m}{V} = \frac{1.855 \times 10^{-22}\,\text{g Fe}}{2.36 \times 10^{-23}\,\text{cm}^3} = 7.86\,\text{g Fe cm}^{-3}$$

Assess

The use of crystal structure data is another way we can determine the density of materials. This method is especially useful when we are studying new compounds or materials of which we have only a small quantity. The type of unit cell adopted by a metal can be deduced by using the experimental density and experimentally determined cell edge length.

PRACTICE EXAMPLE A: Use the result of Practice Example 12-9A, the molar mass of K and the Avogadro constant to calculate the density of potassium.

PRACTICE EXAMPLE B: Use the result of Practice Example 12-9B, the molar mass of Al and its density ($2.6984\,\text{g cm}^{-3}$) to evaluate the Avogadro constant, N_A.

[*Hint:* From the volume of a unit cell and the density of Al, you can determine the mass of a unit cell. Knowing the number of Al atoms in the fcc unit cell, you can determine the mass per Al atom.]

🔍 **12-8 CONCEPT ASSESSMENT**

Suppose a first-order diffraction is observed at an angle θ for a particular plane of atoms by using X-rays of wavelength λ. In order to observe a second-order diffraction from the same plane of atoms at the same angle θ, the wavelength of the X-rays must be a multiple of λ. What is that multiple?

KEEP IN MIND

that cations almost always have smaller radii than anions. This was discussed in Chapter 9.

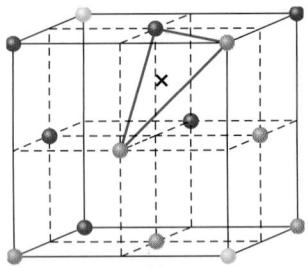

(a) Trigonal hole

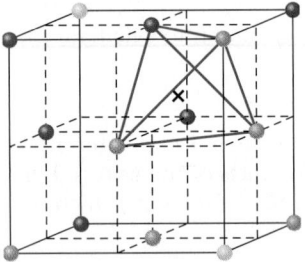

(b) Tetrahedral hole

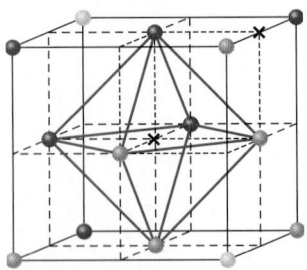

(c) Octahedral hole

▲ FIGURE 12-46
Holes in a face-centered cubic unit cell
(a) The trigonal hole is formed by two face-centered spheres and one corner sphere. **(b)** The tetrahedral hole is formed by three face-centered spheres and one corner sphere. **(c)** The octahedral hole is formed by all six face-centered spheres in the cube.

Ionic Crystal Structures

If we try to apply the packing-of-spheres model to an ionic crystal, we run into two complications: (1) Some of the ions are positively charged and some are negatively charged, and (2) the cations and anions are of different sizes. What we can expect, however, is that oppositely charged ions will come into close proximity. Generally we think of them as being in contact. Like-charged ions, because of mutual repulsions, are not in direct contact. We can think of some ionic crystals as a fairly closely packed arrangement of ions of one type with holes filled by ions of the opposite charge. The relative sizes of cations and anions are important in establishing a particular packing arrangement.

A common arrangement in binary ionic solids is the face-centered cubic arrangement. Very often one of the ions, usually the anion, can be viewed as adopting the face-centered cubic structure while the cation occupies one of the holes between the closest-packed spheres. The three types of holes of the cubic closest packed structure—trigonal, tetrahedral, and octahedral—are shown in Figure 12-46. The size of the holes is related to the radius, R, of the anions used to form the structure. Figure 12-47 shows a cross section through an octahedral hole. The radius of the cation, r, that can just fit into the hole can be found by using the Pythagorean formula, as follows:

$$(2R)^2 + (2R)^2 = (2R + 2r)^2$$
$$2\sqrt{2}R = 2R + 2r$$
$$(2\sqrt{2} - 2)R = 2r$$
$$(\sqrt{2} - 1)R = r$$
$$r = 0.414\,R$$

Similar calculations can be used for tetrahedral and trigonal holes, for which $r = 0.225\,R$ and $r = 0.155\,R$, respectively. These calculations show that in the cubic closest packed structure, the octahedral hole is bigger than the tetrahedral hole.

Another arrangement adopted by binary ionic solids is the simple cubic arrangement. The simple cubic arrangement is not a closest packed structure and has larger holes than the cubic closest packed arrangement. The simple cubic structure has a cubic hole at the center of the unit cell. The size of the cubic hole is $r = 0.732\,R$; of the cubic unit cells considered here, this is the largest hole.

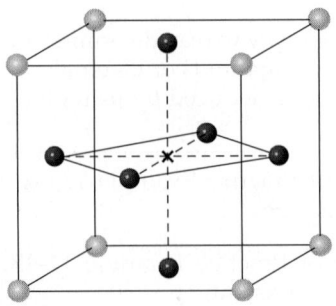
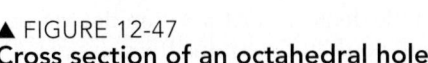

▲ FIGURE 12-47
Cross section of an octahedral hole

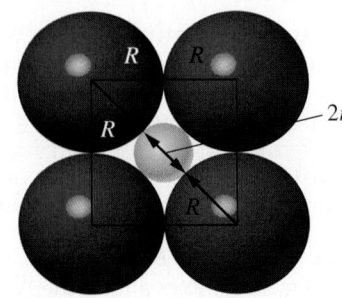

Which hole does the cation occupy in a closest-packed array of anions? The cation occupies a hole that maximizes the attractions between the cation and anion and minimizes the repulsions between the anions. This can be accomplished by accommodating cations into holes that are slightly smaller than the actual size of the ion. This pushes the anions of the closest packed array slightly apart, reducing repulsions, while the anion and cation are in contact, maximizing attractions. Therefore, if a cation is to occupy a tetrahedral hole, the ion should be bigger than the tetrahedral hole but smaller than the octahedral hole; that is,

$$0.225\, R_{anion} < r_{cation} < 0.414\, R_{anion}$$

or in terms of the *radius ratio* of the cation (r) to the anion (R)

$$0.225 < (r_{cation}/R_{anion}) < 0.414$$

Similarly, if a cation is to occupy an octahedral hole, the radii will be governed by the radius ratio inequality

$$0.414 < (r_{cation}/R_{anion}) < 0.732$$

where the upper limit of the inequality corresponds to the hole in a simple cubic lattice. When the cation is too large, that is, bigger than $0.732\, R$, the anions adopt a simple cubic structure, which allows the cation to be accommodated in the cubic hole of the lattice.

To summarize, if

$0.225 < (r_{cation}/R_{anion}) < 0.414$	tetrahedral hole of fcc array of anions occupied by the cation
$0.414 < (r_{cation}/R_{anion}) < 0.732$	octahedral hole of fcc array of anions occupied by the cation
$0.732 < (r_{cation}/R_{anion})$	cubic hole of simple cubic array of anions occupied by the cation

The criteria given here provide a useful way of rationalizing the structures of binary ionic solids. However, as with all simplified models, we must be aware of the limitations of the model. In developing the criteria given, we have assumed that there are no interactions other than coulombic attractions between the ions. The criteria will fail if this is not the case. Nonetheless, we will find the criteria to be very useful.

In defining a unit cell of an ionic crystal we must choose a unit cell that

- by translation in three dimensions generates the entire crystal
- is consistent with the formula of the compound
- indicates the coordination numbers of the ions

Unit cells of crystalline NaCl and CsCl are pictured in Figures 12-48 and 12-49, respectively. We can investigate these structures for their consistency with the formula of the compound and the type of hole the cation occupies.

The radius ratio for NaCl is

$$\frac{r_{Na^+}}{R_{Cl^-}} = \frac{99\ pm}{181\ pm} = 0.55$$

We expect Na^+ ions to occupy the octahedral holes of the cubic closest packed arrays of Cl^- ions. The sodium chloride unit cell is shown in Figure 12-48. To establish the formula of the compound, we must apportion the 27 ions in Figure 12-48 among the unit cell and its neighboring unit cells. Recall from Figure 12-42 on how this apportioning is done. Each Cl^- ion in a corner position is shared by *eight* unit cells, and each Cl^- in the center of a face is shared by *two* unit cells. This leads to a total number of Cl^- ions in the unit cell of $(8 \times \frac{1}{8}) + (6 \times \frac{1}{2}) = 1 + 3 = 4$. There are $2\, Na^+$ ions along the edges of the unit cell, and each edge is shared by *four* unit cells. The Na^+ ion in the very center of the

KEEP IN MIND

that a cubic closest packed array of spheres produces a face-centered cubic unit cell.

▶ FIGURE 12-48
The sodium chloride unit cell
For clarity, only the centers of the ions are shown. Oppositely charged ions are actually in contact. We can think of this structure as an fcc lattice of Cl^- ions, with Na^+ ions filling the octahedral holes.

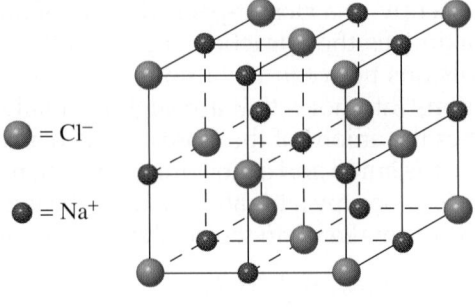

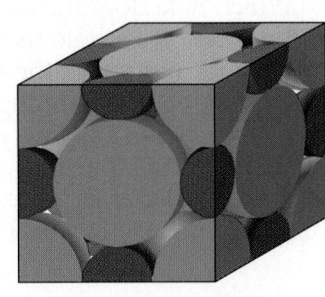

⬤ $= Cl^-$

⬤ $= Na^+$

▶ FIGURE 12-49
The cesium chloride unit cell
The Cs^+ ion is in the center of the cube, with Cl^- ions at the corners. In reality, each Cl^- is in contact with the Cs^+ ion. An alternative unit cell has Cl^- at the center and Cs^+ at the corners.

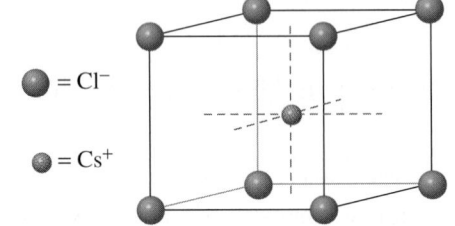

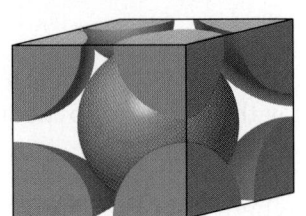

⬤ $= Cl^-$

⬤ $= Cs^+$

unit cell belongs entirely to that cell. Thus, the total number of Na^+ ions in a unit cell is $(12 \times \frac{1}{4}) + (1 \times 1) = 3 + 1 = 4$. The unit cell has the equivalent of $4 Na^+$ and $4 Cl^-$ ions. The ratio of Na^+ to Cl^- is $4:4 = 1:1$, corresponding to the formula NaCl.

To establish the coordination number in an ionic crystal, count the number of nearest neighbor ions of opposite charge to any given ion in the crystal. In NaCl, each Na^+ is surrounded by *six* Cl^- ions. The coordination numbers of both Na^+ and Cl^- are *six*. By contrast, the coordination numbers of Cs^+ and Cl^- in Figure 12-49 are *eight*. The difference in the structure of CsCl from that of NaCl can be accounted for in terms of the radius ratio for this compound.

$$\frac{r_{Cs^+}}{R_{Cl^-}} = \frac{169 \text{ pm}}{181 \text{ pm}} = 0.934$$

EXAMPLE 12-11   Relating Ionic Radii and the Dimensions of a Unit Cell of an Ionic Crystal

The ionic radii of Na^+ and Cl^- in NaCl are 99 and 181 pm, respectively. What is the length of the unit cell of NaCl?

Analyze

Again, the key to solving this problem lies in understanding geometric relationships in the unit cell. Along each edge of the unit cell (see Figure 12-48), two Cl^- ions are in contact with one Na^+. The edge length is equal to the radius of one Cl^- plus the diameter of Na^+, plus the radius of another Cl^-.

Solve

The solution is,

$$\begin{aligned} \text{Length} &= (r_{Cl^-}) + (r_{Na^+}) + (r_{Na^+}) + (r_{Cl^-}) \\ &= 2(r_{Na^+}) + 2(r_{Cl^-}) \\ &= (2 \times 99) + (2 \times 181) = 560 \text{ pm} \end{aligned}$$

Assess

As we saw earlier, we need to remember the atomic arrangement of the unit cell to determine its the geometric relationships.

PRACTICE EXAMPLE A: The ionic radius of Cs^+ is 167 pm. Use Figure 12-49 and information in Examples 12-9 and 12-11 to determine the length of the unit cell of CsCl.

PRACTICE EXAMPLE B: Use the length of the unit cell of NaCl obtained in Example 12-11, together with the molar mass of NaCl and the Avogadro constant, to estimate the density of NaCl.

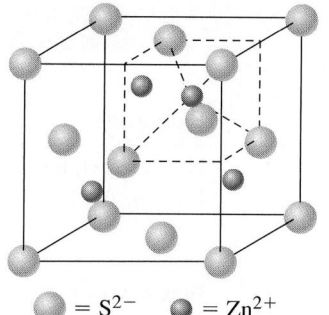

= S^{2-} = Zn^{2+}

(a) Unit cell of ZnS,
the zinc blend structure

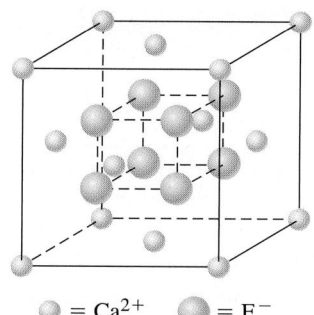

= Ca^{2+} = F^-

(b) Unit cell of CaF_2,
the fluorite structure

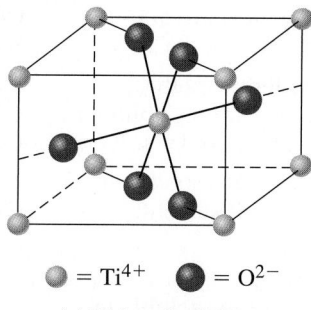

= Ti^{4+} = O^{2-}

(c) Unit cell of TiO_2,
the rutile structure

◀ FIGURE 12-50
Some unit cells of greater complexity

We expect from the radius ratio inequalities above that the Cs^+ ion will occupy a cubic hole in a simple cubic lattice of Cl^- ions. This unit cell is in accord with the one-to-one ratio of Cs^+ to Cl^- ions since there is one Cs^+ in the center of the unit cell and $8 \times (\frac{1}{8})$ Cl^- ions at the corners.

Ionic compounds of the type $M^{2+}X^{2-}$ (for example, MgO, BaS, CaO) may form crystals of the NaCl type. However, if the cation is small enough, as in the case of Zn^{2+}, it can occupy the tetrahedral holes. The radius ratio for ZnS is 0.35, so to satisfy the stoichiometry only half of the tetrahedral holes (there are eight of them) are occupied, to correspond to the four S^{2-} forming the face-centered cubic array (Fig. 12-50a). For substances with the formulas MX_2 or M_2X, the crystal structures are more complex. Because the cations and anions occur in unequal numbers, the crystals have *two* coordination numbers, one for the cation and another for the anion.

CaF_2 (the fluorite structure) has twice as many fluoride ions as calcium ions (Fig. 12-50b). The coordination number of Ca^{2+} is *eight,* and that of F^- is *four.* This is easiest to see by looking at the Ca^{2+} ion in the middle of a face. There are four F^- ions within the unit cell that are nearest neighbors. In addition, the four F^- ions in the next unit cell (the one that shares the face-centered Ca^{2+} ion) are also nearest neighbors. This gives a coordination number of eight for the Ca^{2+}. The F^- ions each have one corner Ca^{2+} ion and three face-centered Ca^{2+} ions as nearest neighbors, giving a coordination number of four. In TiO_2 (the rutile structure, Figure 12-50c), Ti^{4+} has a coordination number of *six* and O^{2-}, *three.* In this structure, two of the O^{2-} ions are within the interior of the cell, two are in the top face, and two in the bottom face of the cell. Ti^{4+} ions are at the corners and the center of the cell.

🔍 **12-9   CONCEPT ASSESSMENT**

Buckminsterfullerene C_{60} crystallizes in a face-centered cubic array. If potassium atoms fill all the tetrahedral and octahedral holes, what is the formula of the resulting compound?

TABLE 12.8	Characteristics of Crystalline Solids			
Types	Structural Particles	Strongest Contributing Forces[a]	Typical Properties	Examples
Metallic	Cations and delocalized electrons	Metallic bonds	Hardness varies from soft to very hard; melting point varies from low to very high; lustrous; ductile; malleable; very good conductors of heat and electricity	Na, Mg, Al, Fe, Sn, Cu, Ag, W
Ionic	Cations and anions	Electrostatic attractions	Hard; moderate to very high melting points; nonconductors as solids, but good electric conductors as liquids; many are soluble in polar solvents such as water	NaCl, MgO, NaNO$_3$
Network covalent	Atoms	Covalent bonds	Most are very hard and either sublime or melt at very high temperatures; most are non-conductors of electricity	C(diamond), C(graphite), SIC, AlN, SiO$_2$
Molecular *Nonpolar*	Atoms or nonpolar molecules	Dispersion forces	Soft; extremely low to moderate melting points (depending on molar mass); sublime in some cases; soluble in some nonpolar solvents	He, Ar, H$_2$, CO$_2$, CCl$_4$, CH$_4$, I$_2$
Polar	Polar molecules	Dispersion forces and dipole–dipole attractions	Low to moderate melting points; soluble in some polar and some nonpolar solvents	(CH$_3$)$_2$O, CHCl$_3$, HCl
Hydrogen-Bonded	Molecules with H bonded to N, O, or F	Hydrogen bonds	Low to moderate melting points; soluble in some hydrogen-bonded solvents and some polar solvents	H$_2$O, NH$_3$

[a] Generally speaking, more than one type of force contributes. For each case above, we have listed only the strongest of all the contributing forces.

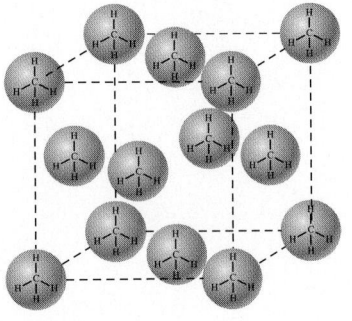

▲ **A sketch of the unit cell of methane**
The unit cell of methane is a face-centered cubic array of CH$_4$ molecules. In the diagram, the spheres containing the methane molecules are meant to emphasize the fcc array of molecules.

Types of Crystalline Solids: A Summary

In this section, we have emphasized crystal structures composed of metal atoms and those composed of ions. But the structural particles of crystalline solids can be atoms, ions, or molecules. To picture a crystal structure having molecules as its structural units, consider solid methane, CH$_4$. The crystal structure is fcc, which means that the unit cell has a CH$_4$ molecule at each corner and at the center of each face. Each CH$_4$ molecule occupies a volume equivalent to a sphere with a radius of 228 pm. Substances that adopt fcc structures are relatively uncommon. Most complex molecules adopt less symmetric unit cells, but a further discussion of this matter is beyond the scope of this text.

The intermolecular forces operating among the structural units of a crystal may be metallic bonds, interionic attractions, van der Waals forces, hydrogen bonds, or covalent bonds. Table 12.8 lists the basic types of crystalline solids, the intermolecular forces within them, some of their characteristic properties, and examples of each.

12-7 Energy Changes in the Formation of Ionic Crystals

The concept of lattice energy, which we introduced qualitatively in Section 12-5, is most useful when stated in quantitative terms. It is difficult, however, to calculate a lattice energy directly. The problem is that oppositely charged ions attract one another and like-charged ions repel one another, and these interactions must be considered at the same time. More commonly, lattice energy is determined *indirectly* through an application of Hess's law known as the Born–Fajans–Haber cycle, named after its originators Max Born, Kasimir Fajans, and Fritz Haber. The crux of the method is to design a sequence of steps in

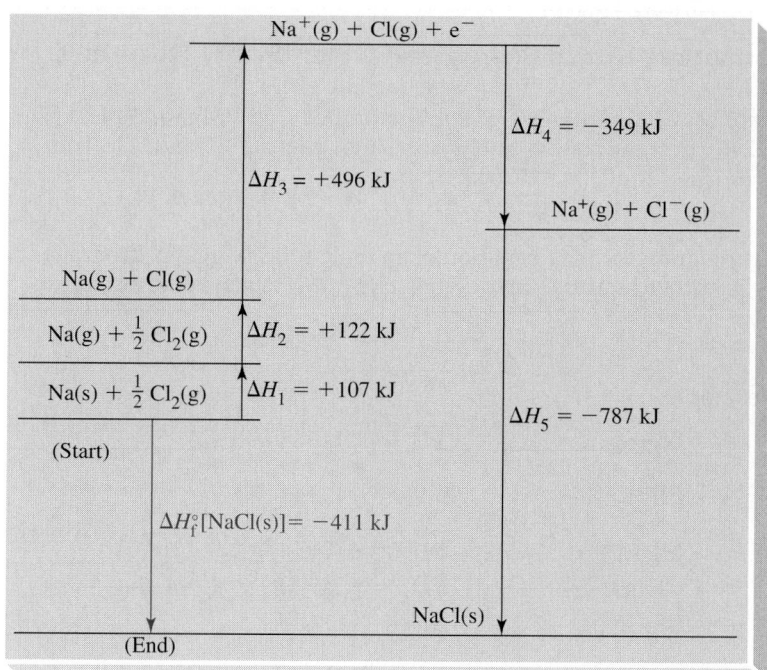

◀ FIGURE 12-51
Enthalpy diagram for the formation of an ionic crystal
Shown here is a five-step sequence for the formation of one mole of NaCl(s) from its elements in their standard states. The sum of the five enthalpy changes gives $\Delta H_f^\circ[\text{NaCl(s)}]$. The equivalent one-step reaction for the formation of NaCl(s) directly from Na(s) and $Cl_2(g)$ is shown in color. (The vertical arrows representing ΔH values are not to scale.)

which enthalpy changes are known for all the steps but one—the step in which a crystal lattice is formed from gaseous ions.

Figure 12-51 illustrates a five-step method for finding the lattice energy of NaCl.

Step 1. Sublime one mole of solid Na.

Step 2. Dissociate 0.5 mole of $Cl_2(g)$ into one mole of Cl(g).

Step 3. Ionize one mole of Na(g) to $Na^+(g)$.

Step 4. Convert one mole of Cl(g) to $Cl^-(g)$.

Step 5. Allow the $Na^+(g)$ and $Cl^-(g)$ to form one mole of NaCl(s).

◀ NaCl has a lattice energy of about $\frac{1}{4}$ that of MgO. This is because the coulombic attraction between the ions is proportional to $(-1)(+1)$ in the former and $(-2)(+2)$ in the latter.

The overall change in these five steps is the same as the reaction in which NaCl(s) is formed from its elements in their standard states—that is, $\Delta H_{\text{overall}} = \Delta H_f^\circ[\text{NaCl(s)}]$. From Appendix D, we see that $\Delta H_f^\circ[\text{NaCl(s)}] = -411 \text{ kJ mol}^{-1}$, so in the following setup, the lattice energy of NaCl is the only unknown.

1. $\text{Na(s)} \longrightarrow \text{Na(g)}$ $\Delta H_1 = \Delta H_{\text{sublimation}} = +107 \text{ kJ}$

2. $\dfrac{1}{2}Cl_2(g) \longrightarrow \text{Cl(g)}$ $\Delta H_2 = \dfrac{1}{2}\text{Cl—Cl bond energy} = +122 \text{ kJ}$

3. $\text{Na(g)} \longrightarrow Na^+(g) + e^-$ $\Delta H_3 = \text{1st ioniz. energy} = +496 \text{ kJ}$

4. $\text{Cl(g)} + e^- \longrightarrow Cl^-(g)$ $\Delta H_4 = \text{electron affinity of Cl} = -349 \text{ kJ}$

5. $Na^+(g) + Cl^-(g) \longrightarrow \text{NaCl(s)}$ $\Delta H_5 = \text{lattice energy of NaCl} = ?$

overall: $\text{Na(s)} + \dfrac{1}{2}Cl_2(g) \longrightarrow \text{NaCl(s)}$

$$\Delta H_{\text{overall}} = -411 \text{ kJ} = \Delta H_1 + \Delta H_2 + \Delta H_3 + \Delta H_4 + \Delta H_5$$
$$-411 \text{ kJ} = 107 \text{ kJ} + 122 \text{ kJ} + 496 \text{ kJ} - 349 \text{ kJ} + \Delta H_5$$
$$\Delta H_5 = \text{lattice energy} = (-411 - 107 - 122 - 496 + 349) \text{ kJ} = -787 \text{ kJ}$$

◀ A commonly used convention defines lattice energy in terms of the breakup of a crystal rather than its formation. By that convention, all lattice energies are positive quantities and $\Delta H_5 = -(\text{lattice energy})$.

One way to use the concept of lattice energy is in making predictions about the possibility of synthesizing ionic compounds. In Example 12-12, we predict the likelihood of obtaining the compound MgCl(s) by evaluating its enthalpy of formation.

EXAMPLE 12-12   Relating Enthalpy of Formation, Lattice Energy, and Other Energy Quantities

With the following data, calculate ΔH_f° per mol MgCl(s): Enthalpy of sublimation of 1 mol Mg(s): +146 kJ; enthalpy of dissociation of $\frac{1}{2}$ mol $Cl_2(g)$: +122 kJ; first ionization energy of 1 mol Mg(g): +738 kJ; electron affinity of 1 mol Cl(g): −349 kJ; lattice energy of 1 mol MgCl(s): −676 kJ.

Analyze

We begin this problem by drawing an enthalpy diagram for the formation of an ionic solid, MgCl(s). From the diagram we find that the lattice energy (ΔH_5) is known, and the unknown is $\Delta H_{overall}$, which is the enthalpy of formation of MgCl(s).

Solve

The various reactions are given as part of the example:

$$Mg(s) \longrightarrow Mg(g) \qquad\qquad \Delta H_1 = +146 \text{ kJ}$$

$$\frac{1}{2}Cl_2(g) \longrightarrow Cl(g) \qquad\qquad \Delta H_2 = +122 \text{ kJ}$$

$$Mg(g) \longrightarrow Mg^+(g) + e^- \qquad \Delta H_3 = +738 \text{ kJ}$$

$$Cl(g) + e^- \longrightarrow Cl^-(g) \qquad \Delta H_4 = -349 \text{ kJ}$$

$$Mg^+(g) + Cl^-(g) \longrightarrow MgCl(s) \qquad \Delta H_5 = -676 \text{ kJ}$$

$$\textit{overall:} \quad Mg(s) + \frac{1}{2}Cl_2(g) \longrightarrow MgCl(s)$$

$$\Delta H_{overall} = \Delta H_f^\circ[MgCl(s)] = \Delta H_1 + \Delta H_2 + \Delta H_3 + \Delta H_4 + \Delta H_5$$
$$= 146 \text{ kJ} + 122 \text{ kJ} + 738 \text{ kJ} - 349 \text{ kJ} - 676 \text{ kJ} = -19 \text{ kJ}$$

Assess

These types of problems are another form of Hess's law. The enthalpy diagram is just a way for us to keep the equations straight.

PRACTICE EXAMPLE A: The enthalpy of sublimation of cesium is 78.2 kJ mol^{-1}, and $\Delta H_f^\circ[CsCl(s)] =$ −442.8 kJ mol^{-1}. Use these values, together with other data from the text, to calculate the lattice energy of CsCl(s).

PRACTICE EXAMPLE B: Given the following data, together with data included in Example 12-12, calculate ΔH_f° per mol $CaCl_2(s)$: Enthalpy of sublimation of Ca(s), +178.2 kJ mol^{-1}; first ionization energy of Ca(g), +590 kJ mol^{-1}; second ionization energy of Ca(g), +1145 kJ mol^{-1}; lattice energy of $CaCl_2(s)$, −2223 kJ mol^{-1}.

Example 12-12 suggests that we can obtain MgCl(s) as a stable compound—it has a slightly negative enthalpy of formation. Why have we been writing $MgCl_2$ all this time instead of MgCl? You might think that because MgCl has a Mg-to-Cl ratio of 1 : 1 and $MgCl_2$ has a ratio of 1 : 2, MgCl should form if Mg(s) reacts with a limited amount of $Cl_2(g)$. But this is not the case. No matter how limited the amount of $Cl_2(g)$ available, the only compound that forms is $MgCl_2$. To understand this, repeat the calculation of Example 12-12 for the formation of $MgCl_2(s)$, and you will obtain an enthalpy of formation that is very much more negative than that for MgCl(s) (see Exercise 87). Even though the energy requirement to produce Mg^{2+} is larger than to produce Mg^+, the lattice energy is very much greater for $MgCl_2(s)$ than for MgCl(s). This is because the *doubly* charged Mg^{2+} ions exert a much stronger force on Cl^- ions than do *singly* charged Mg^+ ions. The reaction between Mg and Cl atoms does not stop at MgCl but continues on to the more stable $MgCl_2$.

Is $NaCl_2$ a stable compound? Here, the answer is no. The additional lattice energy associated with $NaCl_2$ over NaCl is not nearly enough to compensate for the very high second ionization energy of sodium (see Exercise 114).

Summary

12-1 Intermolecular Forces—The intermolecular forces that occur between molecules are collectively known as **van der Waals forces**. The most common intermolecular forces of attraction are those between instantaneous and induced dipoles (**dispersion forces**, or **London forces**). The magnitudes of dispersion forces depend on how easily electron displacements within molecules cause a temporary imbalance of electron charge distribution, that is, on the **polarizability** of the molecule. In polar substances, there are also dipole–dipole forces. Some hydrogen-containing substances exhibit significant intermolecular attractions called **hydrogen bonds,** in which H atoms bonded to highly electronegative atoms—N, O, or F—in a molecule are simultaneously attracted to other highly electronegative atoms in the same molecule or in different molecules. Hydrogen bonding has a profound effect on physical properties, such as boiling points (Fig. 12-5) and is a vital intermolecular force in living systems.

12-2 Some Properties of Liquids—**Surface tension**, the energy required to extend the surface of a liquid, and **viscosity,** a liquid's resistance to flow, are properties related to intermolecular forces. Familiar phenomena such as drop shape, meniscus formation, and capillary action depend on surface tension. Specifically, these phenomena are influenced by the balance between **cohesive forces**, intermolecular forces between molecules in a liquid, and **adhesive forces**, intermolecular forces between liquid molecules and a surface. **Vapor pressure,** the pressure exerted by a vapor in equilibrium with a liquid, is a measure of the volatility of a liquid and is related to the strength of intermolecular forces. The conversion of a liquid to a vapor is called **vaporization** or **evaporation**; the reverse process is called **condensation**. The dependence of vapor pressure on temperature is represented by a **vapor pressure curve** (Fig. 12-18) and can be expressed in the Clausius–Clapeyron equation (Equation 12.2). When the pressure exerted by the escaping molecules from the surface of the liquid equals the pressure exerted by the molecules in the atmosphere, **boiling** is said to occur. The **normal boiling point** is the temperature at which the vapor pressure of the liquid equals 1 atm. The **critical point** is the condition of temperature and pressure at which a liquid and its vapor become indistinguishable (Fig. 12-22).

12-3 Some Properties of Solids—When crystalline solids are heated, a temperature is reached where the solid state is converted to a liquid—**melting** occurs. The temperature at which this occurs is the **melting point**. When liquids are cooled, the crystalline material will form during the process of **freezing**, and the temperature at which this occurs is the **freezing point**. Under certain conditions solids can directly convert into vapor by the process of **sublimation**. The reverse process is called **deposition**. Among the properties of a solid affected by intermolecular forces are its sublimation (vapor) pressure and its melting point.

12-4 Phase Diagrams—A **phase diagram** (Figs. 12-24 to 12-28) is a graphical plot of conditions under which solids, liquids, and gases (vapors) exist, as single phases or as two or more phases in equilibrium with one another. Significant points on a phase diagram are the **triple point** (where all three phases coexist), melting point, boiling point, and critical point, beyond which a supercritical fluid is possible. Some substances can exist in different forms in the solid state; such behavior is called **polymorphism** (Fig. 12-28).

12-5 Network Covalent Solids and Ionic Solids—In **network covalent solids**, chemical bonds extend throughout a crystalline structure. For these substances, the chemical bonds are themselves intermolecular forces. **Lattice energy** is the energy released when separated gaseous ions come together to form one mole of an ionic solid.

12-6 Crystal Structures—Some crystal structures can be described in terms of the packing of spheres. Depending on the way in which the spheres are packed, different **unit cells** are obtained (Fig. 12-37). The hexagonal unit cell is obtained with **hexagonal closest packed (hcp)** spheres; a **face-centered cubic (fcc)** unit cell is obtained with **cubic closest packed** spheres (Figs. 12-38 and 12-39). A **body-centered cubic (bcc)** (Fig. 12-38) unit cell is found in some cases where spheres are not packed as closely as in the hcp and fcc structures. The dimensions of the unit cell can be determined by X-ray diffraction, and these dimensions can be used in calculating atomic radii and densities. An important consideration with ionic crystals is that the ions are not all of the same size or charge. Ionic crystals can often be viewed as an array of anions with the cations fitting in the holes within the anion array.

12-7 Energy Changes in the Formation of Ionic Crystals—Lattice energies of ionic crystals can be related to certain atomic and thermodynamic properties by means of the Born–Fajans–Haber cycle (Fig. 12-51).

Integrative Example

Use data from the table of physical properties of hydrazine, N_2H_4, to calculate the partial pressure of $N_2H_4(g)$ when a container filled with an equilibrium mixture of $N_2H_4(g)$ and $N_2H_4(l)$ at 25.0 °C is cooled to the temperature of an ice–water bath.

Property	Value
Freezing point	2.0 °C
Boiling point	113.5 °C
Critical temperature	380 °C
Critical pressure	145.4 atm
Enthalpy of fusion	12.66 kJ mol^{-1}
Heat capacity of liquid	98.84 J mol^{-1} °C^{-1}
Density of liquid at 25.0 °C	1.0036 g mL^{-1}
Vapor pressure at 25.0 °C	14.4 Torr

Analyze

At a temperature below its freezing point of 2.0 °C, the hydrazine will be present as a solid in equilibrium with its vapor. We are seeking the sublimation pressure of $N_2H_4(s)$ at the melting point of ice, 0 °C. At its freezing point of 2.0 °C, the hydrazine coexists in three phases—liquid, solid, and vapor. We must first determine the vapor pressure of hydrazine at 2.0 °C. Then we can then use the Clausius–Clapeyron equation (12.2) to calculate the vapor (sublimation) pressure at 0 °C. Our principal task will be to identify the data needed to apply the Clausius–Clapeyron equation, three times in all, as detailed in the stepwise solution to the problem.

Solve

To determine a value of ΔH_{vap}, choose the vapor pressure data at 25.0 °C for T_1 and P_1 and at the normal boiling point for T_2 and P_2 for substitution into equation (12.2).

$$\ln\frac{P_2}{P_1} = -\frac{\Delta H_{vap}}{R}\left(\frac{1}{T_2} - \frac{1}{T_1}\right)$$

$$\ln\frac{760 \text{ Torr}}{14.4 \text{ Torr}} = -\frac{\Delta H_{vap}}{8.3145 \text{ J mol}^{-1}\text{K}^{-1}} \times \left(\frac{1}{386.7 \text{ K}} - \frac{1}{298.2 \text{ K}}\right) = 3.967$$

$$\Delta H_{vap} = -\frac{8.3145 \text{ J mol}^{-1}\text{K}^{-1} \times 3.967}{(0.002586 - 0.003353) \text{ K}^{-1}} = 4.30 \times 10^4 \text{ J mol}^{-1}$$

Return to the Clausius–Clapeyron equation (12.2) by using the value of ΔH_{vap} just obtained. Also use the same data as in the first calculation for T_1 and P_1, but now with $T_2 = 2.0 + 273.15 = 275.2 \ K$ and P_2 as an unknown. Solve the equation for P_2.

$$\ln\frac{P_2}{14.4} = -\frac{4.30 \times 10^4 \text{ J mol}^{-1}}{8.3145 \text{ J mol}^{-1}\text{K}^{-1}} \times \left(\frac{1}{275.2 \text{ K}} - \frac{1}{298.2 \text{ K}}\right)$$

$$= -5.17 \times 10^3 \times (2.80 \times 10^{-4}) = -1.45$$

$$P_2/14.4 = e^{-1.45} = 0.235$$

$$P_2 = 0.235 \times 14.4 = 3.38 \text{ Torr}$$

We now have the triple point data for hydrazine. The triple point temperature is 2.0 °C (275.2 K) and the triple point pressure is 3.38 Torr. In our final application of equation (12.2), we use those data as T_2 and P_2. The temperature T_1 is 0 °C (273.2 K) and the unknown sublimation pressure is P_1. The enthalpy change needed in this final calculation must be the enthalpy of sublimation, which is

$$\Delta H_{sub} = \Delta H_{fus} + \Delta H_{vap}$$

$$= \left(12.66\frac{\text{kJ}}{\text{mol}} \times \frac{1000 \text{ J}}{1 \text{ kJ}}\right) + 4.30 \times 10^4 \text{ J mol}^{-1} = 5.57 \times 10^4 \text{ J mol}^{-1}$$

Finally, we substitute these data into equation (12.2) and solve for P_1, the sublimation pressure of hydrazine at $0\,°C$.

$$\ln\frac{3.38\text{ Torr}}{P_1} = -\frac{5.57\times10^4\text{J mol}^{-1}}{8.3145\text{J mol}^{-1}\text{K}^{-1}}\times\left(\frac{1}{275.2}-\frac{1}{273.2}\right)\text{K}^{-1}$$

$$\ln\frac{3.38\text{ Torr}}{P_1} = -6.70\times10^3\times(-2.66\times10^{-5}) = 0.178$$

$$\frac{3.38\text{ Torr}}{P_1} = e^{0.178} = 1.19$$

$$P_1 = \frac{3.38\text{ Torr}}{1.19} = 2.84\text{ Torr}$$

Assess

Observe that, compared with the vapor pressure at $25\,°C$ (14.4 Torr), the calculated triple point pressure (3.38 Torr) is smaller; and the sublimation pressure at $0\,°C$ (2.84 Torr) is smaller still. This is certainly the trend expected for the three values. In the three situations in which equation (12.2) is used, the first one is the most subject to error because the difference between T_2 and T_1 is $89\,°C$, while in the other two it is $23\,°C$ and $2\,°C$, respectively. Both ΔH_{vap} and ΔH_{sub} are undoubtedly temperature-dependent.

PRACTICE EXAMPLE A: The normal boiling point of isooctane (a gasoline component with a high octane rating) is $99.2\,°C$, and its ΔH_{vap} is 35.76 kJ mol^{-1}. Because isooctane and water have nearly identical boiling points, will they have nearly equal vapor pressures at room temperature? If not, which would you expect to be more volatile? Explain.

PRACTICE EXAMPLE B: We cannot measure the second electron affinity of oxygen directly:

$$O^-(g) + e^- \longrightarrow O_2{}^-(g)\quad EA_2 = ?$$

The $O_2{}^-$ ion can exist in the solid state, however, where the high energy requirement for its formation is offset by the large lattice energies of ionic oxides.

(a) Show that EA_2 can be calculated from the enthalpy of formation and lattice energy of $MgO(s)$, the enthalpy of sublimation of $Mg(s)$, the ionization energies of Mg, the bond energy of O_2, and the EA_1 for $O(g)$.

(b) The lattice energy of MgO is -3925 kJ mol^{-1}. Combine this with other values in the text to estimate EA_2 for oxygen.

Mastering**CHEMISTRY**

You'll find a link to additional self study questions in the study area on www.masteringchemistry.com

Exercises

Intermolecular Forces

1. For each of the following substances describe the importance of dispersion (London) forces, dipole–dipole interactions, and hydrogen bonding: **(a)** HCl; **(b)** Br_2; **(c)** ICl; **(d)** HF; **(e)** CH_4.
2. When another atom or group of atoms is substituted for one of the hydrogen atoms in benzene, C_6H_6, the boiling point changes. Explain the order of the following boiling points: C_6H_6, $80\,°C$; C_6H_5Cl, $132\,°C$; C_6H_5Br, $156\,°C$; C_6H_5OH, $182\,°C$.
3. Arrange the liquids represented by the following molecular models in the expected order of increasing viscosity at $25\,°C$.

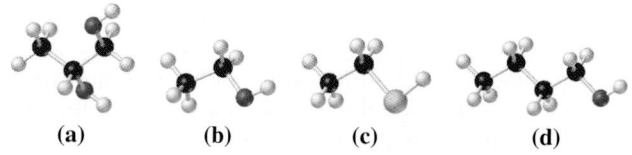

(a) (b) (c) (d)

4. Arrange the liquids represented by the following molecular models in the expected order of increasing normal boiling point.

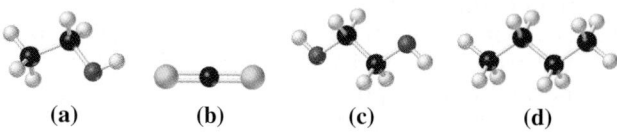

(a) (b) (c) (d)

5. One of the following substances is a liquid at room temperature and the others are gaseous: CH_3OH; C_3H_8; N_2; N_2O. Which do you think is the liquid? Explain.

6. In which of the following compounds do you think that intramolecular hydrogen bonding is an important factor: (a) $CH_3CH_2CH_2CH_3$; (b) $HOOCCH_2CH_2CH_2CH_2COOH$; (c) CH_3COOH; (d) *ortho*-phthalic acid? Explain.

ortho-Phthalic acid

7. How many water molecules can hydrogen bond to methanol?

8. What is the maximum number of hydrogen bonds that can form between two acetic acid molecules?

9. In DNA the nucleic acid bases form hydrogen bonds between them, which are responsible for the formation of the double-stranded helix. Arrange the bases guanine and cytosine to give the maximum number of hydrogen bonds.

Guanine   Cytosine

10. Water molecules will form small, stable clusters. Draw one possible water cluster by using six water molecules and maximizing the number of hydrogen bonds for each water molecule.

Surface Tension and Viscosity

11. Silicone oils, such as $H_3C[SiO(CH_3)_2]_n Si(CH_3)$, are used in water repellents for treating tents, hiking boots, and similar items. Explain how silicone oils function.

12. Surface tension, viscosity, and vapor pressure are all related to intermolecular forces. Why do surface tension and viscosity decrease with temperature, whereas vapor pressure increases with temperature?

13. Is there any scientific basis for the colloquial expression "slower than molasses in January"? Explain.

14. A television commercial claims that a product makes water "wetter." Can there be any basis to this claim? Explain.

15. Rank the following in order of increasing surface tension (at room temperature): (a) CH_3OH; (b) CCl_4; (c) $CH_3CH_2OCH_2CH_3$.

16. Would you predict the surface tension of t-butyl alcohol, $(CH_3)_3COH$, to be greater than or less than that of n-butyl alcohol, $CH_3CH_2CH_2CH_2OH$? Explain.

17. Butanol and pentane have approximately the same mass, however, the viscosity (at 20 °C) of butanol is $\eta = 2.948$ cP, and the viscosity of pentane is $\eta = 0.240$ cP. Explain this difference.

18. Carbon tetrachloride (CCl_4) and mercury have similar viscosities at 20 °C. Explain.

Vaporization

19. As a liquid evaporated from an open container, its temperature was observed to remain roughly constant. When the same liquid evaporated from a thermally insulated container (a vacuum bottle or Dewar flask), its temperature was observed to drop. How would you account for this difference?

20. Explain why vaporization occurs only at the surface of a liquid until the boiling point temperature is reached. That is, why does vapor not form throughout the liquid at all temperatures?

21. The enthalpy of vaporization of benzene, $C_6H_6(l)$, is 33.9 kJ mol^{-1} at 298 K. How many liters of $C_6H_6(g)$, measured at 298 K and 95.1 mmHg, are formed when 1.54 kJ of heat is absorbed by $C_6H_6(l)$ at a constant temperature of 298 K?

22. A vapor volume of 1.17 L forms when a sample of liquid acetonitrile, CH_3CN, absorbs 1.00 kJ of heat at its normal boiling point (81.6 °C and 1 atm). What is ΔH_{vap} in kilojoules per mole of CH_3CN?

23. Use data from the Integrative Example (page 546) to determine how much heat is required to convert 25.00 mL of liquid hydrazine at 25.0 °C to hydrazine vapor at its normal boiling point.

24. How much heat is required to raise the temperature of 215 g $CH_3OH(l)$ from 20.0 to 30.0 °C and then vaporize it at 30.0 °C? Use data from Table 12.3 and a molar heat capacity of $CH_3OH(l)$ of 81.1 J mol^{-1}K^{-1}.

25. How many liters of $CH_4(g)$, measured at 23.4 °C and 768 mmHg, must be burned to provide the heat needed to vaporize 3.78 L of water at 100 °C? $\Delta H_{combustion} = -8.90 \times 10^2$ kJ mol^{-1} CH_4. For $H_2O(l)$ at 100 °C, $d = 0.958$ g cm^{-3}, and $\Delta H_{vap} = 40.7$ kJ mol^{-1}.

26. A 50.0 g piece of iron at 152 °C is dropped into 20.0 g $H_2O(l)$ at 89 °C in an open, thermally insulated container. How much water would you expect to vaporize, assuming no water splashes out? The specific heats of iron and water are 0.45 and 4.21 J g^{-1}°C^{-1}, respectively, and $\Delta H_{vap} = 40.7$ kJ mol^{-1} H_2O.

Vapor Pressure and Boiling Point

27. From Figure 12-18, estimate (a) the vapor pressure of C_6H_7N at 100 °C; (b) the normal boiling point of C_7H_8.

28. Use data in Figure 12-20 to estimate (a) the normal boiling point of aniline; (b) the vapor pressure of diethyl ether at 25 °C.

29. Equilibrium is established between $Br_2(l)$ and $Br_2(g)$ at 25.0 °C. A 250.0 mL sample of the vapor weighs 0.486 g. What is the vapor pressure of bromine at 25.0 °C, in millimeters of mercury?

30. The density of acetone vapor in equilibrium with liquid acetone, $(CH_3)_2CO$, at 32 °C is 0.876 g L^{-1}. What is the vapor pressure of acetone at 32 °C, expressed in kilopascals?

31. A double boiler is used when a careful control of temperature is required in cooking. Water is boiled in an outside container to produce steam, and the steam condenses on the outside walls of an inner container in which cooking occurs. (A related laboratory device is called a steam bath.) (a) How is heat energy conveyed to the food to be cooked in a double boiler? (b) What is the maximum temperature that can be reached in the inside container?

32. One popular demonstration in chemistry labs is performed by boiling a small quantity of water in a metal can (such as a used soda can), picking up the can with tongs and quickly submerging it upside down in cold water. The can collapses with a loud and satisfying pop. Give an explanation of this crushing of the can. (Note: If you try this demonstration, do not heat the can over an open flame.)

33. Pressure cookers achieve a high cooking temperature to speed the cooking process by heating a small amount of water under a constant pressure. If the pressure is set at 2 atm, what is the boiling point of the water? Use information from Table 12.4.

34. Use data from Table 12.4 to estimate (a) the boiling point of water in Santa Fe, New Mexico, if the prevailing atmospheric pressure is 640 mmHg; (b) the prevailing atmospheric pressure at Lake Arrowhead, California, if the observed boiling point of water is 94 °C.

35. A 25.0 L volume of He(g) at 30.0 °C is passed through 6.220 g of liquid aniline ($C_6H_5NH_2$) at 30.0 °C. The liquid remaining after the experiment weighs 6.108 g. Assume that the He(g) becomes saturated with aniline vapor and that the total gas volume and temperature remain constant. What is the vapor pressure of aniline at 30.0 °C?

36. A 7.53 L sample of $N_2(g)$ at 742 mmHg and 45.0 °C is bubbled through $CCl_4(l)$ at 45.0 °C. Assuming the gas becomes saturated with $CCl_4(g)$, what is the volume of the resulting gaseous mixture, if the total pressure remains at 742 mmHg and the temperature remains at 45 °C? The vapor pressure of CCl_4 at 45 °C is 261 mmHg.

37. Some vapor pressure data for Freon-12, CCl_2F_2, once a common refrigerant, are −12.2 °C, 2.0 atm; 16.1 °C, 5.0 atm; 42.4 °C, 10.0 atm; 74.0 °C, 20.0 atm. Also, bp = −29.8 °C, T_c = 111.5 °C, P_c = 39.6 atm. Use these data to plot the vapor pressure curve of Freon-12. What approximate pressure would be required in the compressor of a refrigeration system to convert Freon-12 vapor to liquid at 25.0 °C?

38. A 10.0 g sample of liquid water is sealed in a 1515 mL flask and allowed to come to equilibrium with its vapor at 27 °C. What is the mass of $H_2O(g)$ present when equilibrium is established? Use vapor pressure data from Table 12.4.

The Clausius–Clapeyron Equation

39. Cyclohexanol has a vapor pressure of 10.0 mmHg at 56.0 °C and 100.0 mmHg at 103.7 °C. Calculate its enthalpy of vaporization, ΔH_{vap}.

40. The vapor pressure of methyl alcohol is 40.0 mmHg at 5.0 °C. Use this value and other information from the text to estimate the normal boiling point of methyl alcohol.

41. The normal boiling point of acetone, an important laboratory and industrial solvent, is 56.2 °C and its ΔH_{vap} is 25.5 kJ mol^{-1}. At what temperature does acetone have a vapor pressure of 375 mmHg?

42. The vapor pressure of trichloromethane (chloroform) is 40.0 Torr at −7.1 °C. Its enthalpy of vaporization is 29.2 kJ mol^{-1}. Calculate its normal boiling point.

43. Benzaldehyde, C_6H_5CHO, has a normal boiling point of 179.0 °C and a critical point at 422 °C and 45.9 atm. Estimate its vapor pressure at 100.0 °C.

44. With reference to Figure 12-20, which is the more volatile liquid, benzene or toluene? At approximately what temperature does the less volatile liquid have the same vapor pressure as the more volatile one at 65 °C?

Critical Point

45. Which substances listed in Table 12.5 can exist as liquids at room temperature (about 20.0 °C)? Explain.

46. Can SO_2 be maintained as a liquid under a pressure of 100 atm at 0 °C? Can liquid methane be obtained under the same conditions?

Melting and Freezing

47. The normal melting point of copper is 1357 K, and ΔH_{fus} of Cu is 13.05 kJ mol^{-1}. (a) How much heat, in kilojoules, is evolved when a 3.78 kg sample of molten Cu freezes?

(b) How much heat, in kilojoules, must be absorbed at 1357 K to melt a bar of copper that is 75 cm × 15 cm × 12 cm? (Assume d = 8.92 g/cm^3 for Cu.)

48. An ice calorimeter measures quantities of heat by the quantity of ice melted. How many grams of ice would be melted by the heat released in the complete com-bustion of 1.35 L of propane gas, $C_3H_8(g)$, measured at 25.0 °C and 748 mmHg? [*Hint:* What is the standard molar enthalpy of combustion of $C_3H_8(g)$?]

States of Matter and Phase Diagrams

49. An 80.0 g piece of dry ice, $CO_2(s)$, is placed in a 0.500 L container, and the container is sealed. If this container is held at 25 °C, what state(s) of matter must be present? [*Hint:* Refer to Table 12.5 and Figure 12-28.]

50. Sketch a plausible phase diagram for hydrazine (N_2H_4) from the following data: triple point (2.0 °C and 3.4 mmHg), the normal melting point (2 °C), the normal boiling point (113.5 °C), and the critical point (380 °C and 145 atm). The density of the liquid is less than that of the solid. Label significant data points on this diagram. Are there any features of the diagram that remain uncertain? Explain.

51. Shown here is a portion of the phase diagram for phosphorus.
(a) Indicate the phases present in the regions labeled with a question mark.
(b) A sample of solid red phosphorus cannot be melted by heating in a container open to the atmosphere. Explain why this is so.
(c) Trace the phase changes that occur when the pressure on a sample is reduced from point A to B, at constant temperature.

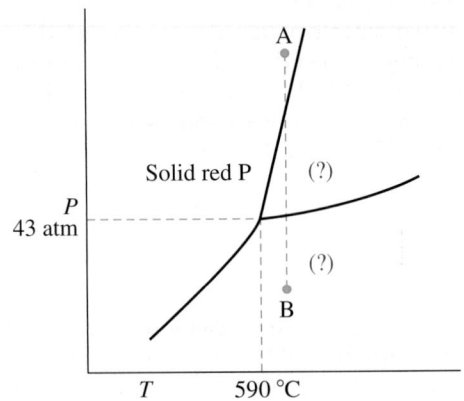

52. Describe what happens to the following samples in situations like those pictured in Figure 12-31. Be as specific as you can about the temperatures and pressures at which changes occur.
(a) A sample of water is heated from −20 to 200 °C at a constant pressure of 600 Torr.
(b) The pressure on a sample of iodine is increased from 90 mmHg to 100 atm at a constant temperature of 114 °C.

(c) A sample of carbon dioxide at 35 °C is cooled to −100 °C at a constant pressure of 50 atm. [*Hint:* Refer also to Table 12.5.]

53. A 0.240 g sample of $H_2O(l)$ is sealed into an evacuated 3.20 L flask. What is the pressure of the vapor in the flask if the temperature is **(a)** 30.0 °C; **(b)** 50.0 °C; **(c)** 70.0 °C?

54. A 2.50 g sample of $H_2O(l)$ is sealed in a 5.00 L flask at 120.0 °C.
(a) Show that the sample exists completely as vapor.
(b) Estimate the temperature to which the flask must be cooled before liquid water condenses.

55. Use appropriate phase diagrams and data from Table 12.5 to determine whether any of the following is likely to occur naturally at or near Earth's surface anywhere on Earth: **(a)** $CO_2(s)$; **(b)** $CH_4(l)$; **(c)** $SO_2(g)$; **(d)** $I_2(l)$; **(e)** $O_2(l)$. Explain.

56. Trace the phase changes that occur as a sample of $H_2O(g)$, originally at 1.00 mmHg and −0.10 °C, is compressed at constant temperature until the pressure reaches 100 atm.

57. To an insulated container with 100.0 g $H_2O(l)$ at 20.0 °C, 175 g steam at 100.0 °C and 1.65 kg of ice at 0.0 °C are added.
(a) What mass of ice remains unmelted after equilibrium is established?
(b) What *additional* mass of steam should be introduced into the insulated container to just melt all of the ice?

58. A 54 cm^3 ice cube at −25.0 °C is added to a thermally insulated container with 400.0 mL $H_2O(l)$ at 32.0 °C. What will be the final temperature in the container and what state(s) of matter will be present? (Specific heats: $H_2O(s)$, 2.01 J g^{-1} °C^{-1}; $H_2O(l)$, 4.18 J g^{-1} °C^{-1}. Densities: $H_2O(s)$, 0.917 g/cm^3; $H_2O(l)$, 0.998 g/cm^3. Also, ΔH_{fus} of ice = 6.01 kJ mol^{-1}.)

59. You decide to cool a can of soda pop quickly in the freezer compartment of a refrigerator. When you take out the can, the soda pop is still liquid; but when you open the can, the soda pop immediately freezes. Explain why this happens.

60. Why is the triple point of water (ice–liquid–vapor) a better fixed point for establishing a thermometric scale than either the melting point of ice or the boiling point of water?

Network Covalent Solids

61. Based on data presented in the text, would you expect diamond or graphite to have the greater density? Explain.

62. Diamond is often used as a cutting medium in glass cutters. What property of diamond makes this possible? Could graphite function as well?

63. Silicon carbide, SiC, crystallizes in a form similar to diamond, whereas boron nitride, BN, crystallizes in a form similar to graphite.
(a) Sketch the SiC structure as in Figure 12-32(b).
(b) Propose a bonding scheme for BN.

64. Are the fullerenes network covalent solids? What makes them different from diamond and graphite? It has been shown that carbon can form chains in which every other carbon atom is bonded to the next carbon atom by a triple bond. Is this allotrope of carbon a network covalent solid? Explain.

Ionic Bonding and Properties

65. The melting points of NaF, NaCl, NaBr, and NaI are 988, 801, 755, and 651 °C, respectively. Are these data consistent with ideas developed in Section 12-5? Explain.

66. Use Coulomb's law (see Appendix B) to verify the conclusion concerning the relative strengths of the attractive forces in the ion pairs Na^+Cl^- and $Mg^{2+}O^{2-}$ presented in Figure 12-36.

67. The hardness of crystals is rated based on Mohs hardness values. The higher the Mohs value, the harder the material is to scratch. Which crystal will have the highest Mohs value: NaF, NaCl, or KCl?

68. Will the mineral villaumite (NaF) or periclase (MgO) have a higher Mohs hardness value (see Exercise 67)?

Crystal Structures

69. Explain why there are *two* arrangements for the closest packing of spheres rather than a single one.

70. Argon, copper, sodium chloride, and carbon dioxide all crystallize in the fcc structure. How can this be when their physical properties are so different?

71. Consider the two-dimensional lattice shown here.

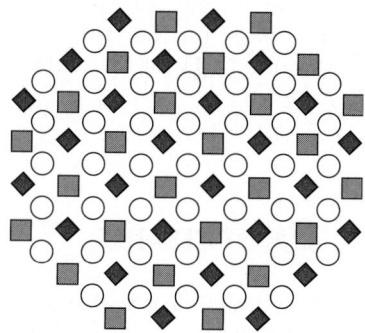

(a) Identify a unit cell.
(b) How many of each of the following elements are in the unit cell: ◆, ▇, and ○ ?
(c) Indicate some simpler units than the unit cell, and explain why they cannot function as a unit cell.

72. As we saw in Section 12-6, stacking spheres always leaves open space. Consider the corresponding situation in two dimensions: Squares can be arranged to cover all the area, but circles cannot. For the arrangement of circles pictured here, what percentage of the area remains uncovered?

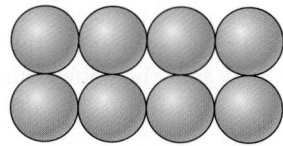

73. Tungsten has a body-centered cubic crystal structure. Using a metallic radius of 139 pm for the W atom, calculate the density of tungsten.

74. Magnesium crystallizes in the hcp arrangement shown in Figure 12-41. The dimensions of the unit cell are height, 520 pm; length on an edge, 320 pm. Calculate the density of Mg(s), and compare with the measured value of 1.738 g/cm^3.

75. Polonium (Po) is the largest member of group 16 and is the only element known to take on the simple cubic crystal system. The distance between nearest neighbor Po atoms in this structure is 335 pm.
(a) What is the diameter of a Po atom?
(b) What is the density of Po metal?
(c) At what angle (in degrees) to the parallel faces of the Po unit cells would first-order diffraction be observed when using X-rays of wavelength 1.785×10^{-10} m?

76. Germanium has a cubic unit cell with a side edge of 565 pm. The density of germanium is 5.36 g/cm^3. What is the crystal system adopted by germanium?

77. Silicon tetrafluoride molecules are arranged in a body-centered cubic unit cell. How many silicon atoms are in the unit cell?

78. Two views, a top and side view, for the unit cell for rutile (TiO_2) are shown here. **(a)** How many titanium atoms (blue) are in this unit cell? **(b)** How many oxygen atoms (red) are in this unit cell?

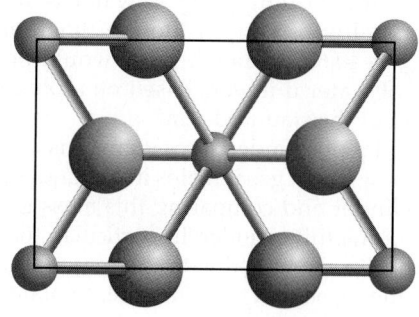

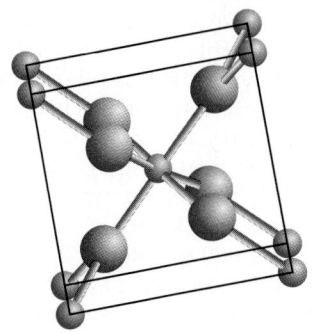

Ionic Crystal Structures

79. Show that the unit cells for CaF_2 and TiO_2 in Figure 12-50 are consistent with their formulas.
80. Using methods similar to Examples 12-10 and 12-11, calculate the density of CsCl. Use 169 pm as the radius of Cs^+.
81. The crystal structure of magnesium oxide, MgO, is of the NaCl type (Fig. 12-48). Use this fact, together with ionic radii from Figure 9-8, to establish the following.
    (a) the coordination numbers of Mg^{2+} and O^{2-};
    (b) the number of formula units in the unit cell;
    (c) the length and volume of a unit cell;
    (d) the density of MgO.

82. Potassium chloride has the same crystal structure as NaCl. Careful measurement of the internuclear distance between K^+ and Cl^- ions gave a value of 314.54 pm. The density of KCl is $1.9893 \ g/cm^3$. Use these data to evaluate the Avogadro constant, N_A.
83. Use data from Figure 9-9 to predict the type of cubic unit cell adopted by **(a)** CaO; **(b)** CuCl; **(c)** LiO_2 (the radius of the O_2^- ion is 128 pm).
84. Use data from Figure 9-9 to predict the type of cubic unit cell adopted by **(a)** BaO; **(b)** CuI; **(c)** LiS_2. (The radii of Ba^{2+} and S_2^- ions are 135 and 198 pm, respectively.)

Lattice Energy

85. *Without doing calculations*, indicate how you would expect the lattice energies of LiCl(s), KCl(s), RbCl(s), and CsCl(s) to compare with the value of $-787 \ kJ \ mol^{-1}$ determined for NaCl(s) on page 543. [*Hint:* Assume that the enthalpies of sublimation of the alkali metals are comparable in value. What atomic properties from Chapter 9 should you compare?]
86. Determine the lattice energy of KF(s) from the following data: $\Delta H_f^\circ[KF(s)] = -567.3 \ kJ \ mol^{-1}$; enthalpy of sublimation of K(s), $89.24 \ kJ \ mol^{-1}$; enthalpy of dissociation of $F_2(g)$, $159 \ kJ \ mol^{-1} \ F_2$; I_1 for K(g), $418.9 \ kJ \ mol^{-1}$; EA for F(g), $-328 kJ \ mol^{-1}$.
87. Refer to Example 12-12. Together with data given there, use the data here to calculate ΔH_f° for 1 mol

$MgCl_2(s)$. Explain why you would expect $MgCl_2$ to be a much more stable compound than MgCl. (Second ionization energy of Mg, $I_2 = 1451 \ kJ \ mol^{-1}$; lattice energy of $MgCl_2(s) = -2526 \ kJ \ mol^{-1} \ MgCl_2$.)
88. In ionic compounds with certain metals, hydrogen exists as the hydride ion, H^-. Determine the electron affinity of hydrogen; that is, ΔH for the process $H(g) + e^- \rightarrow H^-(g)$. To do so, use data from Section 12-7; the bond energy of $H_2(g)$ from Table 10.3; $-812 \ kJ \ mol^{-1}$ NaH for the lattice energy of NaH(s); and $-57 \ kJ \ mol^{-1}$ NaH for the enthalpy of formation of NaH(s).

Integrative and Advanced Exercises

89. When a wax candle is burned, the fuel consists of *gaseous* hydrocarbons appearing at the end of the candle wick. Describe the phase changes and processes by which the solid wax is ultimately consumed.
90. The normal boiling point of water is 100.00 °C and the enthalpy of vaporization at this temperature is $\Delta H_{vap} = 40.657 \ kJ \ mol^{-1}$. What would be the boiling point of water if it were based on a pressure of 1 bar instead of the standard atm?
91. A supplier of cylinder gases warns customers to determine how much gas remains in a cylinder by weighing the cylinder and comparing this mass to the original mass of the full cylinder. In particular, the customer is told not to try to estimate the mass of gas available from the measured gas pressure. Explain the basis of this warning. Are there cases where a measurement of the gas pressure *can* be used as a measure of the remaining available gas? If so, what are they?
92. Use the following data and data from Appendix D to determine the quantity of heat needed to convert 15.0 g of solid mercury at −50.0 °C to mercury vapor at 25 °C. Specific heats: Hg(s), $24.3 J \ mol^{-1} \ K^{-1}$; Hg(l), $28.0 J \ mol^{-1} \ K^{-1}$. Melting point of Hg(s), −38.87 °C. Heat of fusion, $2.33 \ kJ \ mol^{-1}$.
93. To vaporize 1.000 g water at 20 °C requires 2447 J of heat. At 100 °C, 10.00 kJ of heat will convert 4.430 g $H_2O(l)$ to $H_2O(g)$. Do these observations conform to your expectations? Explain.

94. Estimate how much heat is absorbed when 1.00 g of Instant Car Kooler vaporizes. Comment on the effectiveness of this spray in cooling the interior of a car. Assume the spray is 10% $C_2H_5OH(aq)$ by mass, the temperature is 55 °C, the heat capacity of air is $29 J \ mol^{-1} \ K^{-1}$, and use ΔH_{vap} data from Table 12.3.
95. Because solid *p*-dichlorobenzene, $C_6H_4Cl_2$, sublimes rather easily, it has been used as a moth repellent. From the data given, estimate the sublimation pressure of $C_6H_4Cl_2(s)$ at 25 °C. For $C_6H_4Cl_2$; mp = 53.1 °C; vapor pressure of $C_6H_4Cl_2(l)$ at 54.8 °C is 10.0 mmHg; $\Delta H_{fus} = 17.88 \ kJ \ mol^{-1}$; $\Delta H_{vap} = 72.22 \ kJ \ mol^{-1}$.
96. A 1.05 mol sample of $H_2O(g)$ is compressed into a 2.61 L flask at 30.0 °C. Describe the point(s) in Figure 12-30 representing the final condition.
97. One handbook lists the sublimation pressure of *solid* benzene as a function of *Kelvin* temperature, T, as $\log P \ (mmHg) = 9.846 - 2309/T$. Another handbook lists the vapor pressure of *liquid* benzene as a function of *Celsius* temperature, t, as $\log P \ (mmHg) = 6.90565 - 1211.033/(220.790 + t)$. Use these equations to estimate the normal melting point of benzene, and compare your result with the listed value of 5.5 °C.
98. By the method used to graph Figure 12-20, plot $\ln P$ versus $1/T$ for liquid white phosphorus, and estimate

(a) its normal boiling point and **(b)** its enthalpy of vaporization, ΔH_{vap}, in kJ mol^{-1}. Vapor pressure data: 76.6 °C, 1 mmHg; 128.0 °C, 10 mmHg; 166.7 °C, 40 mmHg; 197.3 °C, 100 mmHg; 251.0 °C, 400 mmHg.

99. Assume that a skater has a mass of 80 kg and that his skates make contact with 2.5 cm^2 of ice. **(a)** Calculate the pressure in atm exerted by the skates on the ice. **(b)** If the melting point of ice decreases by 1.0 °C for every 125 atm of pressure, what would be the melting point of the ice under the skates?

100. Estimate the boiling point of water in Leadville, Colorado, elevation 3170 m. To do this, use the barometric formula relating pressure and altitude: $P = P_0 \times 10^{-Mgh/2.303\,RT}$ (where P = pressure in atm; $P_0 = 1$ atm; g = acceleration due to gravity; molar mass of air, $M = 0.02896$ kg mol^{-1}; $R = 8.3145$ J mol^{-1} K^{-1}; and T is the Kelvin temperature). Assume the air temperature is 10.0 °C and that $\Delta H_{vap} = 41$ kJ mol^{-1} H$_2$O.

101. Inspection of the straight-line graphs in Figure 12-20 suggests that the graphs for benzene and water intersect at a point that falls off the page. At this point, the two liquids have the same vapor pressure. Estimate the temperature and the vapor pressure at this point by a calculation based on data obtainable from the graphs.

102. A cylinder containing 151 lb Cl$_2$ has an inside diameter of 10 in. and a height of 45 in. The gas pressure is 100 psi (1 atm = 14.7 psi) at 20 °C. Cl$_2$ melts at -103 °C, boils at -35 °C, and has its critical point at 144 °C and 76 atm. In what state(s) of matter does the Cl$_2$ exist in the cylinder?

103. In acetic acid vapor, some molecules exist as monomers and some as dimers (see Figure 12-9). If the density of the vapor at 350 K and 1 atm is 3.23 g/L, what percentage of the molecules must exist as dimers? Would you expect this percent to increase or decrease with temperature?

104. A 685 mL sample of Hg(l) at 20 °C is added to a large quantity of liquid N$_2$ kept at its boiling point in a thermally insulated container. What mass of N$_2$(l) is vaporized as the Hg is brought to the temperature of the liquid N$_2$? For the specific heat of Hg(l) from 20 to -39 °C use 0.138 J g^{-1} °C^{-1}, and for Hg(s) from -39 to -196 °C, 0.126 J g^{-1} °C^{-1}. The density of Hg(l) is 13.6 g/mL, its melting point is -39 °C, and its enthalpy of fusion is 2.30 kJ mol^{-1}. The boiling point of N$_2$(l) is -196 °C, and its ΔH_{vap} is 5.58 kJ mol^{-1}.

105. Sketched here are two hypothetical phase diagrams for a substance, but neither of these diagrams is possible. Indicate what is wrong with each of them.

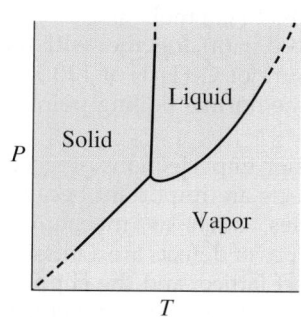
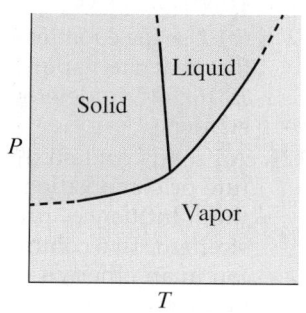

106. A chemistry handbook lists the following equation for the vapor pressure of NH$_3$(l) as a function of temperature. What is the normal boiling point of NH$_3$(l)?

$$\log_{10} P(\text{mmHg}) = 9.95028 - 0.003863T - \frac{1473.17}{T}$$

107. The triple point temperature of bismuth is 544.5 K and the normal boiling point is 1832 K. Imagine that a 1.00 mol sample of bismuth is heated at a constant rate of 1.00 kJ min^{-1} in an apparatus in which the sample is maintained under a constant pressure of 1 atm. In the manner shown in Figure 12-24 and as much to scale as possible, that is in terms of times and temperatures, sketch the heating curve that would be obtained in heating the sample from 300 K to 2000 K. Use the following data. $\Delta H_{fus} = 10.9$ kJ mol^{-1} Bi(s); $\Delta H_{vap} = 151.5$ kJ mol^{-1} Bi(l); average molar heat capacities, in J mol^{-1} K^{-1}, 28 for Bi(s), 31 for Bi(l), and 21 for Bi(g). [*Hint:* Under the conditions described, no vapor appears until the normal boiling point is reached.]

108. The crystal structure of lithium sulfide (Li$_2$S), is pictured here. The length of the unit cell is 5.88×10^2 pm. For this structure, determine

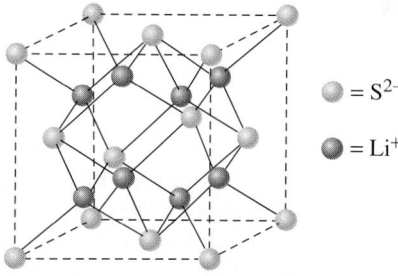

(a) the coordination numbers of Li$^+$ and S^{2-}
(b) the number of formula units in the unit cell
(c) the density of Li$_2$S

109. Refer to Figure 12-44 and Figure 12-48. Suppose that the two planes of ions pictured in Figure 12-44 correspond to the top and middle planes of ions in the NaCl unit cell in Figure 12-48. If the X-rays used have a wavelength of 154.1 pm, at what angle θ would the diffracted beam have its greatest intensity? [*Hint:* Use $n=1$ in equation (12.5).]

110. Use the analyses of a bcc structure on page 534 and the fcc structure in Exercise 139 to determine the percent voids in the packing-of-spheres arrangement found in the fcc crystal structure.

111. One way to describe ionic crystal structures is in terms of cations filling voids among closely packed anions. Show that in order for cations to fill the tetrahedral voids in a close packed arrangement of anions, the radius ratio of cation, r_c, to anion, r_a, must fall between the following limits $0.225 < r_c:r_a < 0.414$.

112. Use the unit cell of diamond in Figure 12-32(b) and a carbon-to-carbon bond length of 154.45 pm, together with other relevant data from the text, to calculate the density of diamond.

113. The enthalpy of formation of NaI(s) is -288 kJ mol^{-1}. Use this value, together with other data in the text, to

calculate the lattice energy of NaI(s). [*Hint:* Use data from Appendix D also.]

114. Show that the formation of $NaCl_2$(s) is very unfavorable; that is, $\Delta H_f^\circ[NaCl_2(s)]$ is a large *positive* quantity. To do this, use data from Section 12-7 and assume that the lattice energy for $NaCl_2$ would be about the same as that of $MgCl_2$, -2.5×10^3 kJ mol^{-1}.

115. A crystalline solid contains three types of ions, Na^+, O^{2-}, and Cl^-. The solid is made up of cubic unit cells that have O^{2-} ions at each corner, Na^+ ions at the center of each face, and Cl^- ions at the center of the cells. What is the chemical formula of the compound? What are the coordination numbers for the O^{2-} and Cl^- ions? If the length of one edge of the unit cell is a, what is the shortest distance from the center of a Na^+ ion to the center of an O^{2-} ion? Similarly, what is the shortest distance from the center of a Cl^- ion to the center of an O^{2-} ion?

116. A certain mineral has a cubic unit cell with calcium at each corner, oxygen at the center of each face, and titanium at its body center. What is the formula of the mineral? An alternate way of drawing the unit cell has calcium at the center of each cubic unit cell. What are the positions of titanium and oxygen in such a representation of the unit cell? How many

oxygen atoms surround a particular titanium atom in either representation?

117. Calculate the radius ratio (r_+/r_-) for CaF_2. Suggest an alternative structure to that shown in Figure 12-50(b) that better conforms to the radius ratio you compute.

118. In some barbecue grills the electric lighter consists of a small hammer-like device striking a small crystal, which generates voltage and causes a spark between wires that are attached to opposite surfaces of the crystal. The phenomenon of causing an electric potential through mechanical stress is known as the piezoelectric effect. One type of crystal that exhibits the piezoelectric effect is lead zirconate titanate. In this perovskite crystal structure, a titanium(IV) ion sits in the middle of a tetragonal unit cell with dimensions of 0.403 nm × 0.398 nm × 0.398 nm. At each corner is a lead(II) ion, and at the center of each face is an oxygen anion. Some of the Ti(IV) are replaced by Zr(IV). This substitution, along with Pb(II), results in the piezoelectic behavior.
    (a) How many oxygen ions are in the unit cell?
    (b) How many lead(II) ions are in the unit cell?
    (c) How many titanium(IV) ions are in the unit cell?
    (d) What is the density of the unit cell?

Feature Problems

119. In a capillary rise experiment, the height (h) to which a liquid rises depends on the density (d) and surface tension (γ) of the liquid and the radius of the capillary (r). The equation relating these quantities and the acceleration due to gravity (g) is $h = 2\gamma/dgr$. The sketch provides data obtained with ethanol. What is the surface tension of ethanol?

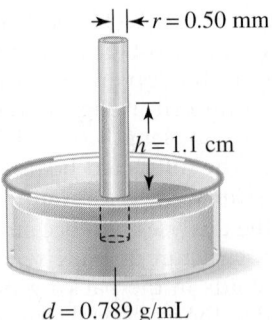

→| |← $r = 0.50$ mm

$h = 1.1$ cm

$d = 0.789$ g/mL

120. We have learned that the enthalpy of vaporization of a liquid is generally a function of temperature. If we wish to take this temperature variation into account, we cannot use the Clausius–Clapeyron equation in the form given in the text (that is, equation 12.2). Instead, we must go back to the differential equation upon which the Clausius–Clapeyron equation is based and reintegrate it into a new expression. Our starting point is the following equation describing the rate of change of vapor pressure with temperature in terms of the enthalpy of vaporization, the difference in

molar volumes of the vapor (V_g), and liquid (V_l), and the temperature.

$$\frac{dP}{dT} = \frac{\Delta H_{vap}}{T(V_g - V_l)}$$

Because in most cases the volume of one mole of vapor greatly exceeds the molar volume of liquid, we can treat the V_l term as if it were zero. Also, unless the vapor pressure is unusually high, we can treat the vapor as if it were an ideal gas; that is, for one mole of vapor, $PV = RT$. Make appropriate substitutions into the above expression, and separate the P and dP terms from the T and dT terms. The appropriate substitution for ΔH_{vap} means expressing it as a function of temperature. Finally, integrate the two sides of the equation between the limits P_1 and P_2 on one side and T_1 and T_2 on the other.
    (a) Derive an equation for the vapor pressure of C_2H_4(l) as a function of temperature, if $\Delta H_{vap} = 15{,}971 + 14.55\,T - 0.160\,T^2$ (in J mol^{-1}).
    (b) Use the equation derived in (a), together with the fact that the vapor pressure of C_2H_4(l) at 120 K is 10.16 Torr, to determine the normal boiling point of ethylene.

121. All solids contain defects or imperfections of structure or composition. Defects are important because they influence properties, such as mechanical strength. Two common types of defects are a missing ion in an otherwise perfect lattice, and the slipping of an ion from its normal site to a hole in the lattice.

The holes discussed in this chapter are often called *interstitial sites*, since the holes are in fact interstices in the array of spheres. The two types of defects described here are called *point defects* because they occur within specific sites. In the 1930s, two solid-state physicists, W. Schottky and J. Fraenkel, studied the two types of point defects: A Schottky defect corresponds to a missing ion in a lattice, while a Fraenkel defect corresponds to an ion that is displaced into an interstitial site.

(a) An example of a Schottky defect is the absence of a Na^+ ion in the NaCl structure. The absence of a Na^+ ion means that a Cl^- ion must also be absent to preserve electrical neutrality. If one NaCl unit is missing per unit cell, does the overall stoichiometry change, and what is the change in density?

(b) An example of a Fraenkel defect is the movement of a Ag^+ ion to a tetrahedral interstitial site from its normal octahedral site in AgCl, which has a structure like NaCl. Does the overall stoichiometry of the compound change, and do you expect the density to change?

(c) Titanium monoxide (TiO) has a sodium chloride-like structure. X-ray diffraction data show that the edge length of the unit cell is 418 pm. The density of the crystal is 4.92 g/cm³ Do the data indicate the presence of vacancies? If so, what type of vacancies?

122. In an ionic crystal lattice each cation will be attracted by anions next to it and repulsed by cations near it. Consequently the coulomb potential leading to the lattice energy depends on the type of crystal. To get the total lattice energy you must sum all of the electrostatic interactions on a given ion. The general form of the electrostatic potential is

$$V = \frac{Q_1 Q_2 e^2}{d_{12}}$$

where Q_1 and Q_2 are the charges on ions 1 and 2, d_{12} is the distance between them in the crystal lattice. and e is the charge on the electron.

(a) Consider the linear "crystal" shown below.

The distance between the centers of adjacent spheres is R. Assume that the blue sphere and the green spheres are cations and that the red spheres are anions. Show that the total electrostatic energy is

$$V = -\frac{Q^2 e^2}{d} \times \ln 2$$

(b) In general, the electrostatic potential in a crystal can be written as

$$V = -k_M \frac{Q^2 e^2}{R}$$

where k_M is a geometric constant, called the Madelung constant, for a particular crystal system under consideration. Now consider the NaCl crystal structure and let R be the distance between the centers of sodium and chloride ions. Show that by

considering three layers of nearest neighbors to a central chloride ion, k_M is given by

$$k_M = \left(6 - \frac{12}{\sqrt{2}} + \frac{8}{\sqrt{3}} - \frac{6}{\sqrt{4}} \cdots \right)$$

(c) Carry out the same calculation for the CsCl structure. Are the Madelung constants the same?

123. Plot the following data first as boiling point versus polarizability, and then as boiling point versus molecular mass. What conclusions can you draw from these plots?

Compound	Polarizability, 10^{-25} cm³	Mass, amu	Boiling Point, K
H_2	7.90	2.0158	20.35
N_2	17.6	28.0134	77.35
O_2	16.0	31.9988	90.188
Cl_2	46.1	70.906	238.25
HF	24.6	20.01	292.69
HCl	26.3	36.46	188.25
HBr	36.1	80.92	206.15
HI	54.4	127.91	237.77
N_2O	30.0	44.01	184.65
CO	19.5	28.01	81.65
SO_2	37.2	64.06	263.15
H_2S	37.8	34.08	212.45
CS_2	87.4	76.14	319.45
NH_3	22.6	17.03	239.8
HCN	25.9	27.03	299.15
CH_4	26.0	16.04	109.15
C_2H_6	44.7	30.07	184.55
$CH_2{=}CH_2$	42.6	28.05	169.45
$CH{=}CH$	33.3	26.01	189.15
C_3H_8	62.9	44.11	231.05
C_6H_6	103	78.12	353.25
CH_3Cl	45.6	50.94	248.95
CH_2Cl_2	64.8	84.93	313.15
$CHCl_3$	82.3	119.38	334.85
CCl_4	105	153.81	349.95
CH_3OH	32.3	32.04	338.15

124. The Born–Fajans–Haber cycle uses thermodynamic cycles to determine lattice energy. An alternative to the Born–Fajans–Haber method is one based on fundamental principles. Because the dominant interactions in an ionic crystal are Coulomb interactions, we can use the theory of electrostatics to calculate the lattice energy. Kapustinskii used these ideas and proposed the following equation:

$$U = \frac{120{,}250 \, v \, Z^+ Z^-}{r_0} \left(1 - \frac{34.5}{r_0} \right) (kJ \, mol^{-1})$$

where the number of ions per formula unit is given by v and r_0 is equal to the sum of the ionic radii, $r_+ + r_-$(pm). Use the equation to complete the following table:

Compound	Lattice Energy, KJ mol⁻¹	r_-, pm	r_+, pm
NaCl		181	99
LaF₃		133	117
Na₂SO₄	−3389		99

Self-Assessment Exercises

125. In your own words, define or explain the following terms or symbols: **(a)** ΔH_{vap}; **(b)** T_c; **(c)** instantaneous dipole; **(d)** coordination number; **(e)** unit cell.

126. Briefly describe each of the following phenomena or methods: **(a)** capillary action; **(b)** polymorphism; **(c)** sublimation; **(d)** supercooling; **(e)** determining the freezing point of a liquid from a cooling curve.

127. Explain the important distinctions between each pair of terms: **(a)** adhesive and cohesive forces; **(b)** vaporization and condensation; **(c)** triple point and critical point; **(d)** face-centered and body-centered cubic unit cell; **(e)** tetrahedral and octahedral hole.

128. The magnitude of one of the following properties must always increase with temperature; that one is **(a)** surface tension; **(b)** density; **(c)** vapor pressure; **(d)** ΔH_{vap}.

129. Of the compounds HF, CH_4, CH_3OH, N_2H_4, and $CHCl_3$, hydrogen bonding is an important intermolecular force in **(a)** none of these; **(b)** two of these; **(c)** three of these; **(d)** all but one of these; **(e)** all of these.

130. A metal that crystallizes in the body-centered cubic (bcc) structure has a crystal coordination number of **(a)** 6; **(b)** 8; **(c)** 12; **(d)** any even number between 4 and 12.

131. A unit cell of an ionic crystal **(a)** shares some ions with other unit cells; **(b)** is the same as the formula unit; **(c)** is any portion of the crystal that has a cubic shape; **(d)** must contain the same number of cations and anions.

132. If the triple point pressure of a substance is greater than 1 atm, which two of the following conclusions are valid?
    (a) The solid and liquid states of the substance cannot coexist at equilibrium.
    (b) The melting point and boiling point of the substance are identical.
    (c) The liquid state of the substance cannot exist.
    (d) The liquid state cannot be maintained in a beaker open to air at 1 atm pressure.
    (e) The melting point of the solid must be greater than 0 °C.
    (f) The gaseous state at 1 atm pressure cannot be condensed to the solid at the triple point temperature.

133. In each of the following pairs, which would you expect to have the higher boiling point? **(a)** C_7H_{16} or $C_{10}H_{22}$; **(b)** C_3H_8 or $(CH_3)_2O$; **(c)** CH_3CH_2SH or CH_3CH_2OH.

134. One of the substances is out of order in the following list based on increasing boiling point. Identify it, and put it in its proper place: N_2, O_3, F_2, Ar, Cl_2. Explain your reasoning.

135. Arrange the following substances in the expected order of increasing melting point: KI, Ne, K_2SO_4, C_3H_8, CH_3CH_2OH, MgO, $CH_2OHCHOHCH_2OH$.

136. Is it possible to obtain a sample of ice from liquid water without ever putting the water in a freezer or other enclosure at a temperature below 0 °C? If so, how might this be done?

137. The phenomena in Figure 12-22 will be seen at the critical temperature only if the proper amount of liquid is placed in the sealed tube initially. Why should this be the case? What would you expect to see if too little liquid was present initially? If too much liquid was present?

138. The following data are given for CCl_4. Normal melting point, −23 °C; normal boiling point, 77 °C; density of liquid 1.59 g/mL; $\Delta H_{fus}=3.28$ kJ mol^{-1}; vapor pressure at 25 °C, 110 Torr.
    (a) What phases—solid, liquid, and/or gas—are present if 3.50 g CCl_4 is placed in a closed 8.21 L container at 25 °C?
    (b) How much heat is required to vaporize 2.00 L of $CCl_4(l)$ at its normal boiling point?

139. The fcc unit cell is a cube with atoms at each of the corners and in the center of each face, as shown here. Copper has the fcc crystal structure. Assume an atomic radius of 128 pm for a Cu atom.

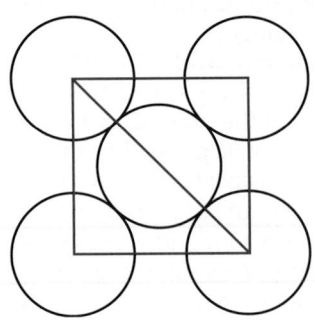

    (a) What is the length of the unit cell of Cu?
    (b) What is the volume of the unit cell?
    (c) How many atoms belong to the unit cell?
    (d) What percentage of the volume of the unit cell is occupied?
    (e) What is the mass of a unit cell of copper?
    (f) Calculate the density of copper.

140. Of the following liquids at 20 °C, which has the smallest surface tension? **(a)** CH_3OH; **(b)** CH_3CH_2OH; **(c)** $CH_3CH_2CH_2OH$; **(d)** $CH_3CH_2CH_2CH_2OH$.

141. Of the following liquids at 20 °C, which has the smallest viscosity? **(a)** Dodecane, $C_{12}H_{26}$; **(b)** n-nonane, C_9H_{20}; **(c)** n-heptane C_7H_{16}; **(d)** n-pentane C_5H_{12}.

142. Would you expect an ionic solid or a network covalent solid to have the higher melting point?

143. In the lithium iodide crystal, the Li–I distance is 3.02 Å. Calculate the iodide radius, assuming that the iodide ions are in contact.

144. Which of the following phase transitions is most likely to occur when the pressure on a metallic solid increases? **(a)** bcc to sc; **(b)** fcc to sc; **(c)** bcc to fcc; **(d)** fcc to sc.

145. Construct a concept map representing the different types of intermolecular forces and their origin.

146. Construct a concept map using the ideas of packing of spheres and the structure of metal and ionic crystals.

147. Construct a concept map showing the ideas contained in a phase diagram.

Solutions and Their Physical Properties

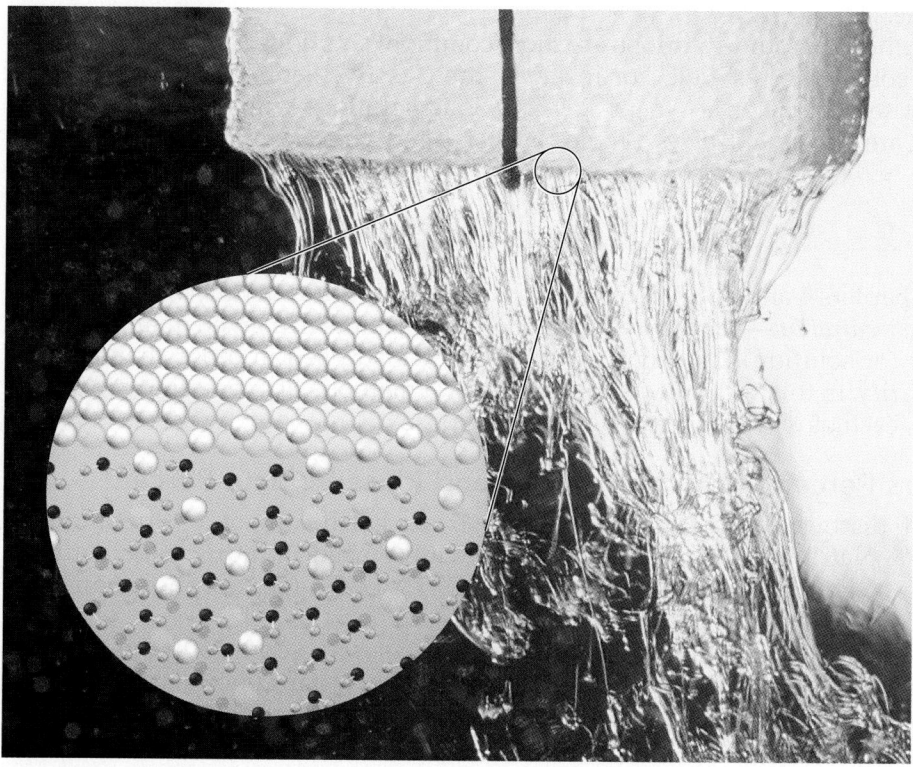

The dissolving of a cube of sugar (sucrose) is seen here as swirls of a higher-density sucrose solution falling through the lower-density water.

CONTENTS

R esidents of cold climate regions know they must add antifreeze to the water in the cooling system of an automobile in the winter. The antifreeze–water mixture has a much lower freezing point than does pure water. In this chapter we will learn why.

To restore body fluids to a dehydrated individual by intravenous injection, pure water cannot be used. A solution with just the right value of a physical property known as osmotic pressure is necessary, and this requires a solution of a particular concentration of solutes. Again, in this chapter we will learn why.

Altogether, we will explore several solution properties whose values depend on solution concentration. Our emphasis will be on describing solution phenomena and their applications and explaining these phenomena at the molecular level.

13-1   Types of Solutions: Some Terminology

In Chapters 1 and 4 we learned that a solution is a *homogeneous mixture*. It is *homogeneous* because its composition and properties are uniform, and it is a *mixture* because it contains two or more substances in proportions that can be varied. The **solvent** is the component that is present in the greatest quantity or that determines the state of matter in which a solution exists. Other solution components, called **solutes**, are said to be dissolved in the solvent. A *concentrated* solution has a relatively large quantity of dissolved solute(s), and a *dilute* solution has only a small quantity. Consider solutions containing sucrose (cane sugar) as one of the solutes in the solvent water: Pancake syrup is a concentrated solution, whereas a sweetened cup of coffee is much more dilute.

Although liquid solutions are most common, solutions can exist in gaseous and solid states as well. For instance, the U.S. five-cent coin, the nickel, is a solid solution of 75% Cu and 25% Ni. Solid solutions with a metal as the solvent are also called *alloys*.* Table 13.1 lists a few common solutions.

13-2   Solution Concentration

In Chapters 4 and 5 we learned that to describe a solution fully we must know its *concentration*—a measure of the quantity of solute in a given quantity of solvent (or solution). The concentration unit we stressed in those chapters was molarity. In this section, we describe several methods of expressing concentration, each of which serves a different purpose.

Mass Percent, Volume Percent, and Mass/Volume Percent

If we dissolve 5.00 g NaCl in 95.0 g H_2O, we get 100.0 g of a solution that is 5.00% NaCl, by *mass*. Mass percent is widely used in industrial chemistry. Thus, we might read that the action of 78% $H_2SO_4(aq)$ on phosphate rock $[3\ Ca_3(PO_4)_2 \cdot CaF_2]$ produces 46% $H_3PO_4(aq)$.

Because liquid volumes are so easily measured, some solutions are prepared on a *volume* percent basis. For example, a handbook lists a freezing point of $-15.6\ °C$ for a methyl alcohol–water antifreeze solution that is 25.0%

TABLE 13.1 Some Common Solutions	
Solution	Components
Gaseous solutions	
Air	N_2, O_2, and several others
Natural gas	CH_4, C_2H_6, and several others
Liquid solutions	
Seawater	H_2O, NaCl, and many others
Vinegar	H_2O, CH_3COOH (acetic acid)
Soda pop	H_2O, CO_2, $C_{12}H_{22}O_{11}$ (sucrose), and several others
Solid solutions	
Yellow brass	Cu, Zn
Palladium–Hydrogen	Pd, H_2

*The term *alloy* can also apply to certain heterogeneous mixtures, such as the common two-phase solid mixture of lead and tin known as *solder*, or to intermetallic compounds, such as the silver–tin compound Ag_3Sn that, in the past, was mixed with mercury in *dental amalgam*.

CH_3OH, by volume. Such a solution could be prepared by dissolving 25.0 mL $CH_3OH(l)$ with water until the total solution volume is 100.0 mL.

Another possibility is to express the *mass* of solute and *volume* of solution. An aqueous solution with 0.9 g NaCl in 100.0 mL of solution is said to be 0.9% NaCl (*mass/volume*). Mass/volume percent is extensively used in the medical and pharmaceutical fields.

Parts per Million, Parts per Billion, and Parts per Trillion

In solutions where the mass or volume percent of a component is very low, we often switch to other units to describe solution concentration. For example, 1 mg solute/L solution amounts to only 0.001 g/L. A solution that is this dilute will have the same density as water, approximately 1 g/mL; therefore, the solution concentration is 0.001 g solute/1000 g solution, which is the same as 1 g solute/1,000,000 g solution. We can describe the solute concentration more succinctly as 1 *part per million* (ppm). For a solution with only 1 μg solute/L solution, the situation is 1×10^{-6} g solute/1000 g solution, or 1.0 g solute/ 1×10^9 g solution. Here, the solute concentration is 1 *part per billion* (ppb). If the solute concentration is only 1 μg solute/L solution, the concentration is 1 *part per trillion* (ppt).

Because these terms are widely used in environmental reporting, they may be more familiar than other units that chemists use. For example, a consumer in California might read in an annual water quality report from the municipal water department that the maximum contaminant level allowed for nitrate ion is 45 ppm and for carbon tetrachloride, 0.5 ppb.

◀ A part per billion: The average head of hair has 10,000 hairs. Therefore, 1 part per billion is one hair out of 100,000 people.

KEEP IN MIND

that 1 ppm = 1 mg/L, 1 ppb = 1 μg/L, and 1 ppt = 1 ng/L.

Mole Fraction and Mole Percent

To relate certain physical properties (such as vapor pressure) to solution concentration, we need a unit in which all solution components are expressed on a mole basis. We can do this with the mole fraction. The *mole fraction* of component i, designated x_i, is the fraction of all the molecules in a solution that are of type i. The mole fraction of component j is x_j, and so on. The mole fraction of a solution component is defined as

$$x_i = \frac{\text{amount of component } i \text{ (in moles)}}{\text{total amount of all solution components (in moles)}}$$

The sum of the mole fractions of all the solution components is 1.

$$x_i + x_j + x_k + \cdots = 1$$

The *mole percent* of a solution component is the percent of all the molecules in solution that are of a given type. Mole percents are mole fractions multiplied by 100%.

🔍 13-1   CONCEPT ASSESSMENT

In one mole of a solution with a mole fraction of 0.5 water, how many water molecules would there be?

Molarity

In Chapters 4 and 5 we introduced molarity to provide a conversion factor relating the amount of solute and the volume of solution. We used it in various stoichiometric calculations. As we learned at that time,

$$\text{molarity } (M) = \frac{\text{amount of solute (in moles)}}{\text{volume of solution (in liters)}}$$

Molality

Suppose we prepare a solution at 20 °C by using a volumetric flask calibrated at 20 °C. Then suppose we warm this solution to 25 °C. As the temperature increases from 20 to 25 °C, the amount of solute remains constant, but the solution volume increases slightly (by about 0.1%). The number of moles of solute per liter—the molarity—*decreases* slightly (by about 0.1%). This temperature dependence of molarity can be a problem in experiments demanding a high precision. That is, the solution might be used at a temperature different from the one at which it was prepared, and so its molarity is not exactly the one written on the label. A concentration unit that is *independent* of temperature, and also proportional to mole fraction in dilute solutions, is **molality** (*m*)—the number of moles of solute per kilogram of *solvent* (not of solution). A solution in which 1.00 mol of urea, $CO(NH_2)_2$, is dissolved in 1.00 kg of water is described as a 1.00 molal solution and designated as 1.00 *m* $CO(NH_2)_2$. Molality is defined as

▶ Molal units, being the number of moles per kg of solvent, are independent of temperature in contrast to molar units, being the number of moles per liter.

$$\text{molality } (m) = \frac{\text{amount of solute (in moles)}}{\text{mass of solvent (in kilograms)}}$$

The concentration of a solution is expressed in several different ways in Example 13-1. The calculation in Example 13-2 is perhaps more typical: A concentration is converted from one unit (molarity) to another (mole fraction).

EXAMPLE 13-1   Expressing a Solution Concentration in Various Units

An ethanol–water solution is prepared by dissolving 10.00 mL of ethanol, CH_3CH_2OH ($d = 0.789$ g/mL), in a sufficient volume of water to produce 100.0 mL of a solution with a density of 0.982 g/mL (Fig. 13-1). What is the concentration of ethanol in this solution expressed as **(a)** volume percent; **(b)** mass percent; **(c)** mass/volume percent; **(d)** mole fraction; **(e)** mole percent; **(f)** molarity; **(g)** molality?

Analyze

Each part of this problem uses an equation presented in the text. Expressing concentrations in these different units will illustrate the similarities and differences among volume percent, mass percent, mass/volume percent, mole fraction, mole percent, molarity, and molality.

Solve

(a) Volume percent ethanol

$$\text{volume percent ethanol} = \frac{10.00 \text{ mL ethanol}}{100.0 \text{ mL solution}} \times 100\% = 10.00\%$$

(b) Mass percent ethanol

$$\text{mass ethanol} = 10.00 \text{ mL ethanol} \times \frac{0.789 \text{ g ethanol}}{1.00 \text{ mL ethanol}}$$
$$= 7.89 \text{ g ethanol}$$

$$\text{mass soln} = 100.0 \text{ mL soln} \times \frac{0.982 \text{ g soln}}{1.0 \text{ mL solution}} = 98.2 \text{ g soln}$$

$$\text{mass percent ethanol} = \frac{7.89 \text{ g ethanol}}{98.2 \text{ g solution}} \times 100\% = 8.03\%$$

(c) Mass/volume percent ethanol

$$\text{mass/volume percent ethanol} = \frac{7.89 \text{ g ethanol}}{100.0 \text{ mL solution}} \times 100\% = 7.89\%$$

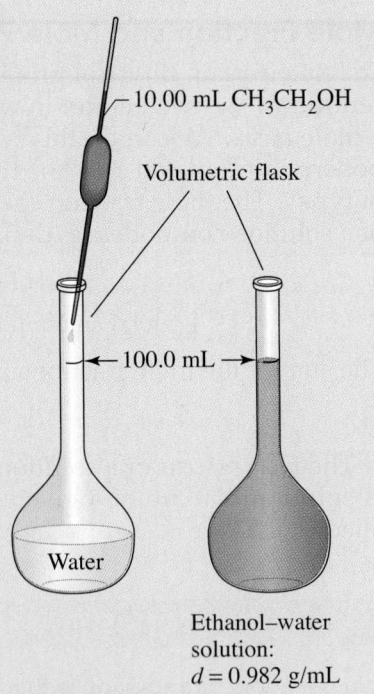

▲ FIGURE 13-1
Preparation of an ethanol–water solution— Example 13-1 illustrated
A 10.00 mL sample of CH_3CH_2OH is added to some water in the volumetric flask. The solution is mixed, and more water is added to bring the total volume to 100.0 mL.

(d) Mole fraction of ethanol
Convert the mass of ethanol from part (b) to an amount in moles.

$$? \text{ mol CH}_3\text{CH}_2\text{OH} = 7.89 \text{ g CH}_3\text{CH}_2\text{OH} \times \frac{1 \text{ mol CH}_3\text{CH}_2\text{OH}}{46.07 \text{ g CH}_3\text{CH}_2\text{OH}}$$

$$= 0.171 \text{ mol CH}_3\text{CH}_2\text{OH}$$

Determine the mass of water present in 100.0 mL of solution.

$$98.2 \text{ g soln} - 7.89 \text{ g ethanol} = 90.3 \text{ g water}$$

Convert the mass of water to the number of moles present.

$$? \text{ mol H}_2\text{O} = 90.3 \text{ g H}_2\text{O} \times \frac{1 \text{ mol H}_2\text{O}}{18.02 \text{ g H}_2\text{O}} = 5.01 \text{ mol H}_2\text{O}$$

$$x_{\text{CH}_3\text{CH}_2\text{OH}} = \frac{0.171 \text{ mol CH}_3\text{CH}_2\text{OH}}{0.171 \text{ mol CH}_3\text{CH}_2\text{OH} + 5.01 \text{ mol H}_2\text{O}} = \frac{0.171}{5.18} = 0.0330$$

(e) Mole percent ethanol

$$\text{mole percent CH}_3\text{CH}_2\text{OH} = x_{\text{CH}_3\text{CH}_2\text{OH}} \times 100\% = 0.0330 \times 100\% = 3.30\%$$

(f) Molarity of ethanol
Divide the number of moles of ethanol from part (d) by the solution volume, 100.0 mL = 0.1000 L.

$$\text{molarity} = \frac{0.171 \text{ mol CH}_3\text{CH}_2\text{OH}}{0.1000 \text{ L soln}} = 1.71 \text{ M CH}_3\text{CH}_2\text{OH}$$

(g) Molality of ethanol
First, convert the mass of water present in 100.0 mL of solution [from part (d)] to the unit kg.

$$? \text{ kg H}_2\text{O} = 90.3 \text{ g H}_2\text{O} \times \frac{1 \text{ kg H}_2\text{O}}{1000 \text{ g H}_2\text{O}} = 0.0903 \text{ kg H}_2\text{O}$$

Use this result and the number of moles of CH_3CH_2OH from part (d) to establish the molality.

$$\text{molality} = \frac{0.171 \text{ mol CH}_3\text{CH}_2\text{OH}}{0.0903 \text{ kg H}_2\text{O}} = 1.89 \text{ } m \text{ CH}_3\text{CH}_2\text{OH}$$

Assess

For the same solution, the volume percent, mass percent, and mass/volume percent are not necessarily the same. Molarity and molality are also not the same values, because molarity is based on the volume of solution and molality is based on the mass of the solvent.

PRACTICE EXAMPLE A: A solution that is 20.0% ethanol, by volume, is found to have a density of 0.977 g/mL. Use this fact, together with data from Example 13-1, to determine the mass percent ethanol in the solution.

PRACTICE EXAMPLE B: A 11.3 mL sample of CH_3OH ($d = 0.793$ g/mL) is dissolved in enough water to produce 75.0 mL of a solution with a density of 0.980 g/mL. What is the solution concentration expressed as **(a)** mole fraction H_2O; **(b)** molarity of CH_3OH; **(c)** molality of CH_3OH?

EXAMPLE 13-2   Converting Molarity to Mole Fraction

Laboratory ammonia is 14.8 M $NH_3(aq)$ with a density of 0.8980 g/mL. What is x_{NH_3} in this solution?

Analyze

In this problem we note that no volume of solution is stated, suggesting that our calculation can be based on any fixed volume of our choice. A convenient volume to work with is one liter. We need to determine the number of moles of NH_3 and of H_2O in one liter of the solution.

(continued)

Solve

Find the number of moles of NH_3 by using the definition of molarity.

$$\text{moles of } NH_3 = 1.00 \text{ L} \times \frac{14.8 \text{ mol } NH_3}{1 \text{ L}} = 14.8 \text{ mol } NH_3$$

For moles of H_2O, first find mass of the solution by using solution density.

$$\text{mass of soln} = 1000.0 \text{ mL soln} \times \frac{0.8980 \text{ g soln}}{1.0 \text{ mL solution}} = 898.0 \text{ g soln}$$

Then use moles of NH_3 and molar mass to find the mass of NH_3.

$$\text{mass of } NH_3 = 14.8 \text{ mol } NH_3 \times \frac{17.03 \text{ g } NH_3}{1 \text{ mol } NH_3} = 252 \text{ g } NH_3$$

Find the mass of H_2O by subtracting the mass of NH_3 from the solution mass.

$$\text{mass of } H_2O = 898.0 \text{ g soln} - 252 \text{ g } NH_3 = 646 \text{ g } H_2O$$

Find moles of H_2O by multiplying by the inverse of the molar mass for H_2O.

$$\text{moles of } H_2O = 646 \text{ g } H_2O \times \frac{1 \text{ mol } H_2O}{18.02 \text{ g } H_2O} = 35.8 \text{ mol } H_2O$$

Find the mole fraction of ammonia x_{NH_3} by dividing moles NH_3 by the total number of moles of NH_3 and H_2O in the solution.

$$x_{NH_3} = \frac{14.8 \text{ mol } NH_3}{14.8 \text{ mol } NH_3 + 35.8 \text{ mol } H_2O} = 0.292$$

Assess

By using the solution concentration definitions, we were able to convert from one concentration unit to another. This skill is used frequently by chemists.

PRACTICE EXAMPLE A: A 16.00% aqueous solution of glycerol, $HOCH_2CH(OH)CH_2OH$, by mass, has a density of 1.037 g/mL. What is the mole fraction of glycerol in this solution?

PRACTICE EXAMPLE B: A 10.00% aqueous solution of sucrose, $C_{12}H_{22}O_{11}$, by mass, has a density of 1.040 g/mL. What is **(a)** the molarity; **(b)** the molality; and **(c)** the mole fraction of $C_{12}H_{22}O_{11}$, in this solution?

🔍 13-2 CONCEPT ASSESSMENT

Which of the several concentration units described in Section 13-2 are temperature-dependent and which are not? Explain.

13-3 Intermolecular Forces and the Solution Process

If there is even a little water in the fuel tank of an automobile, the engine will misfire. This problem would not occur if water were soluble in gasoline. Why does water not form solutions with gasoline? We can often understand a process if we analyze its energy requirements; this approach can help us to explain why some substances mix to form solutions and others do not. In this section, we focus on the behavior of molecules in solution, specifically on intermolecular forces and their contribution to the energy required for the dissolution process.

Enthalpy of Solution

In the formation of some solutions, heat is given off to the surroundings; in other cases, heat is absorbed. An enthalpy of solution, ΔH_{soln}, can be rather easily measured—for example, in the coffee-cup calorimeter of Figure 7-6—but why should some solution processes be exothermic, whereas others are endothermic?

Let's think in terms of a three-step approach to ΔH_{soln}. First, solvent molecules must be separated from one another to make room for the solute molecules. Some energy is required to overcome the forces of attraction between solvent molecules. As a result, this step should be an endothermic one: $\Delta H > 0$. Second, the solute molecules must be separated from one another.

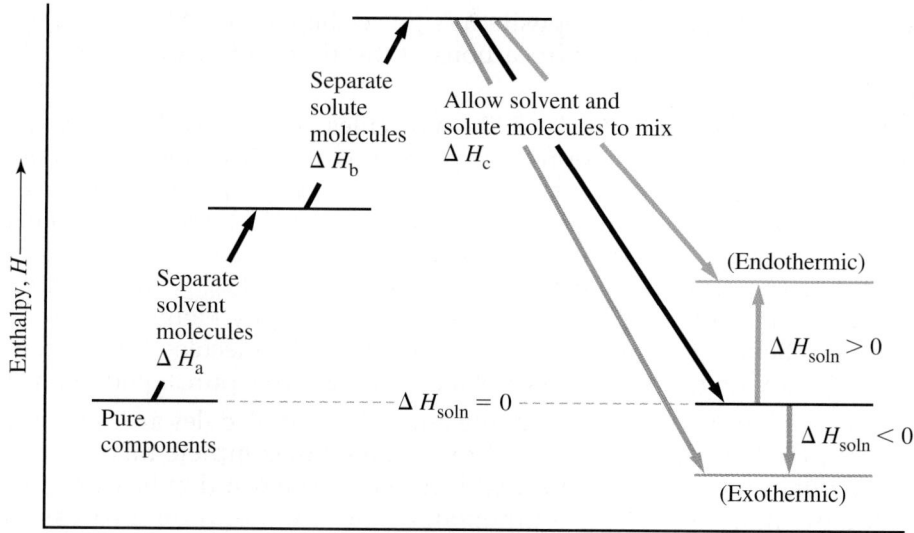

▲ FIGURE 13-2
Enthalpy diagram for solution formation
The solution process can be endothermic (blue arrow), exothermic (red arrow), or has $\Delta H_{soln} = 0$ (black arrow), depending on the magnitude of the enthalpy change in the mixing step.

This step, too, will take energy and should be endothermic. Finally, we can imagine that we allow the separated solvent and solute molecules to be attracted to one another. These attractions will bring the molecules closer together and energy should be released. This is an exothermic step: $\Delta H < 0$. The enthalpy of solution is the sum of the three enthalpy changes just described, and depending on their relative values, ΔH_{soln} is either positive (endothermic) or negative (exothermic). This three-step process is summarized by equation (13.1) and Figure 13-2.

(a)   pure solvent $\longrightarrow$ separated solvent molecules $\Delta H_a > 0$

(b)   pure solute $\longrightarrow$ separated solute molecules $\Delta H_b > 0$

(c)   separated solvent
and solute molecules $\longrightarrow$ solution $\Delta H_c < 0$

Overall: pure solvent + pure solute $\longrightarrow$ solution

$$\Delta H_{soln} = \Delta H_a + \Delta H_b + \Delta H_c \qquad \textbf{(13.1)}$$

Intermolecular Forces in Mixtures

We see from equation (13.1) that the magnitude and sign of ΔH_{soln} depends on the values of the three terms ΔH_a, ΔH_b, and ΔH_c. These, in turn, depend on the strengths of the *three* kinds of intermolecular forces of attraction represented in Figure 13-3. Four possibilities for the relative strengths of these intermolecular forces are described in the discussion that follows.

1. If the intermolecular forces of attraction shown in Figure 13-3 are of the same type and of equal strength, the solute and solvent molecules mix randomly. A homogeneous mixture or solution results. Because properties of solutions of this type can generally be predicted from the properties of the pure components, they are called **ideal solutions**. There is no overall enthalpy change in the formation of an ideal solution from its components, and $\Delta H_{soln} = 0$. This means that ΔH_c in equation (13.1) is

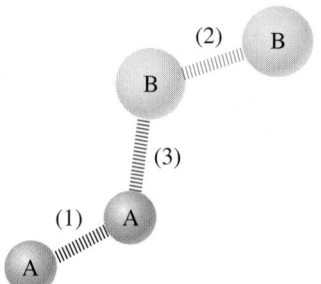

▲ FIGURE 13-3
Intermolecular forces in a solution
The intermolecular forces of attraction, represented here by dashed lines, are between: (1) solvent molecules, A–A; (2) solute molecules, B–B; and (3) solvent and solute molecules, A–B.

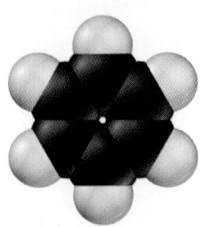

(a)

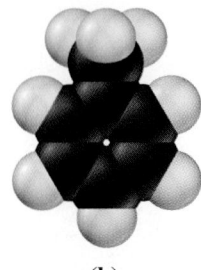

(b)

▲ FIGURE 13-4
Two components of a nearly ideal solution
Think of the —CH₃ group in toluene **(b)** as a small "bump" on the planar benzene ring **(a)**. Substances with similar molecular structures have similar intermolecular forces of attraction.

equal in magnitude and opposite in sign to the sum of ΔH_a and ΔH_b. Many mixtures of liquid hydrocarbons fit this description, or very nearly so (Fig. 13-4).

2. If forces of attraction between unlike molecules *exceed* those between like molecules, a solution also forms. The properties of such solutions generally cannot be predicted, however, and they are called *nonideal solutions*. Interactions between solute and solvent molecules (ΔH_c) release more heat than the heat absorbed to separate the solvent and solute molecules ($\Delta H_a + \Delta H_b$). The solution process is exothermic ($\Delta H_{soln} < 0$). Solutions of acetone and chloroform fit this type. As suggested by Figure 13-5, weak hydrogen bonding occurs between the two kinds of molecules, but the conditions for hydrogen bonding are not met in either of the pure liquids alone.*

3. If forces of attraction between solute and solvent molecules are *somewhat weaker* than between molecules of the same kind, complete mixing may still occur, but the solution formed is *nonideal*. The solution has a higher enthalpy than the pure components, and the solution process is endothermic. This type of behavior is observed in mixtures of carbon disulfide (CS_2), a nonpolar liquid, and acetone, a polar liquid. In these mixtures, the acetone molecules are attracted to other acetone molecules by dipole–dipole interactions and hence show a preference for other acetone molecules as neighbors. A possible explanation of how a solution process can be endothermic and still occur is found on page 567.

4. Finally, if forces of attraction between unlike molecules are *much weaker* than those between like molecules, the components remain segregated in a *heterogeneous mixture*. Dissolution does not occur to any significant extent. In a mixture of water and octane (a constituent of gasoline), strong hydrogen bonds hold water molecules together in clusters. The nonpolar octane molecules cannot exert a strong attractive force on the water molecules, and the two liquids do not mix. Thus, we now have an answer to the question posed at the beginning of this section of why water does not dissolve in gasoline or vice versa.

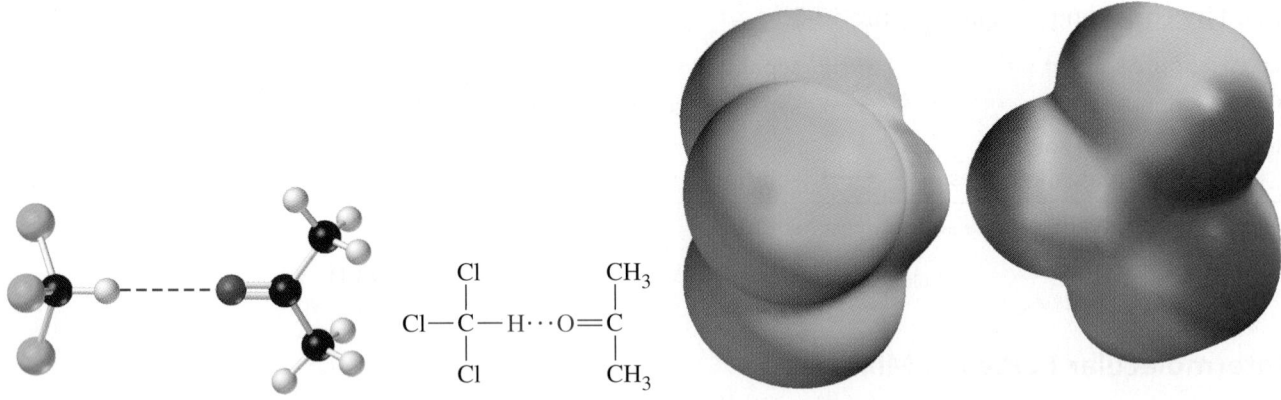

▲ FIGURE 13-5
Intermolecular force between unlike molecules leading to a nonideal solution
The interaction between these molecules is illustrated by using three different representations: ball and stick, line representation, and electrostatic potential maps. Hydrogen bonding between CHCl₃ (chloroform) and (CH₃)₂CO (acetone) molecules produces forces of attraction between unlike molecules that exceed those between like molecules.

*In most cases, H atoms bonded to C atoms cannot participate in hydrogen bonding. In a molecule like CHCl₃, however, the three Cl atoms have a strong electron-withdrawing effect on electrons in the C—H bond ($\mu = 1.01$ D). The H atom is then attracted to a lone pair of electrons on the O atom of (CH₃)₂CO (but not to Cl atoms in other CHCl₃ molecules).

As an oversimplified summary of the four cases described in the preceding paragraphs, we can say that "like dissolves like." That is, substances with similar molecular structures are likely to exhibit similar intermolecular forces of attraction and to be soluble in one another. Substances with dissimilar structures are likely not to form solutions. Of course, in many cases, parts of the structures may be similar and parts may be dissimilar. Then it is a matter of trying to establish which are the more important parts, a matter we explore in Example 13-3.

◀ Remember that "like dissolves like."

EXAMPLE 13-3   Using Intermolecular Forces to Predict Solution Formation

Predict whether or not a solution will form in each of the following mixtures and whether the solution is likely to be ideal: **(a)** ethyl alcohol, CH_3CH_2OH, and water; **(b)** the hydrocarbons hexane, $CH_3(CH_2)_4CH_3$, and octane, $CH_3(CH_2)_6CH_3$; **(c)** octanol, $CH_3(CH_2)_6CH_2OH$, and water.

Analyze

Keep in mind that ideal or nearly ideal solutions are not too common. They require the solvent and solute(s) to be quite similar in structure.

(a) If we think of water as H—OH, ethyl alcohol is similar to water. (Just substitute the group CH_3CH_2— for one of the H atoms in water.) Both molecules meet the requirements of hydrogen bonding as an important intermolecular force. The strengths of the hydrogen bonds between like molecules and between unlike molecules are likely to differ, however.

(b) In hexane, the carbon chain is six atoms long, and in octane it is eight. Both substances are virtually nonpolar, and intermolecular attractive forces (of the dispersion type) should be quite similar both in the pure liquids and in the solution.

(c) At first sight, this case may seem similar to (a), with the substitution of a hydrocarbon group for a H atom in H—OH. Here, however, the carbon chain is *eight* members long. This long carbon chain is much more important than the terminal —OH group in establishing the physical properties of octanol. Viewed from this perspective, octanol and water are quite *dissimilar*.

Solve

(a) We expect ethyl alcohol and water to form *nonideal* solutions.

(b) We expect a solution to form, and it should be nearly *ideal*.

(c) We do not expect a solution to form.

Assess

In these types of problems a strong understanding of both molecular structure and intermolecular forces is required. Keep in mind the statement "like dissolves like."

In our answer to part (c), we observed that octanol does not form a solution; however, alcohols, such as butyl alcohol, $CH_3CH_2CH_2CH_2OH$, have a limited solubility in water (9 grams per 100 grams of water). The aqueous solubilities of alcohols fall off fairly rapidly as the hydrocarbon chain length increases beyond four.

PRACTICE EXAMPLE A: Which of the following organic compounds do you think is most readily soluble in water? Explain.

(a) Toluene   **(b)** Oxalic acid   **(c)** Benzaldehyde

PRACTICE EXAMPLE B: In which solvent is solid iodine likely to be more soluble, water or carbon tetrachloride? Explain.

13-1 ARE YOU WONDERING...

What is the nature of the intermolecular forces in a mixture of carbon disulfide and acetone?

Carbon disulfide is a nonpolar molecule, and so in the pure substance the only intermolecular forces are weak London dispersion forces; carbon disulfide is a volatile liquid. Acetone is a polar molecule, and in the pure substance dipole–dipole forces are strong. Acetone is somewhat less volatile than carbon disulfide. In a solution of acetone in carbon disulfide (case 3 on page 564), the dipoles of acetone molecules polarize carbon disulfide molecules, giving rise to *dipole–induced dipole* interactions.

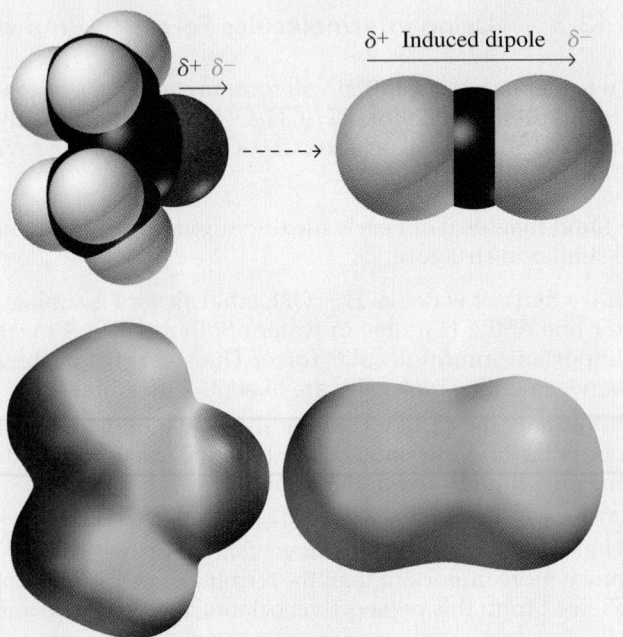

The dipole–induced dipole forces between acetone and carbon disulfide molecules are weaker than the dipole–dipole interactions among acetone molecules, causing the acetone molecules to be relatively less stable in their solutions with carbon disulfide than they are in pure acetone. As a result, acetone–carbon disulfide mixtures are nonideal solutions.

Formation of Ionic Solutions

To assess the energy requirements for the formation of aqueous solutions of ionic compounds, we turn to the process pictured in Figure 13-6. Water dipoles are shown clustered around ions at the surface of a crystal. The negative ends of water dipoles are pointed toward the positive ions, and the positive ends of water dipoles toward negative ions. The interaction between an ion and a dipole is an intermolecular force known as an ion–dipole force. If these ion–dipole forces of attraction are strong enough to overcome the interionic forces of attraction in the crystal, dissolving will occur. Moreover, these ion–dipole forces also persist in the solution. An ion surrounded by a cluster of water molecules is said to be *hydrated*. Energy is *released* when ions become hydrated. The greater the hydration energy compared with the energy needed to separate ions from the ionic crystal, the more likely that the ionic solid will dissolve in water.

We can again use a *hypothetical* three-step process to describe the dissolution of an ionic solid. The energy requirement to dissociate a mole of an ionic solid into separated gaseous ions, an endothermic process, is the *negative of* the

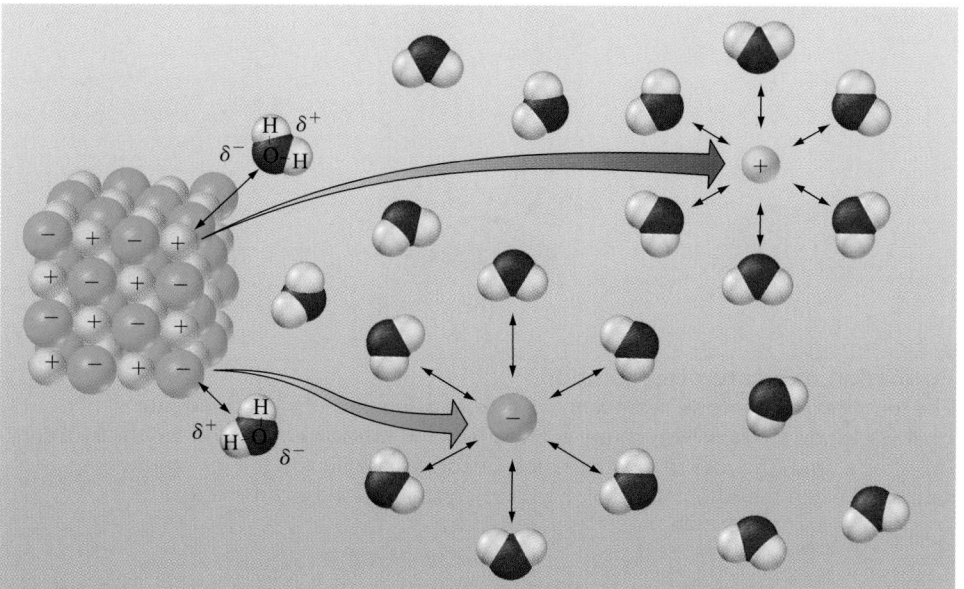

▲ FIGURE 13-6
An ionic crystal dissolving in water
Clustering of water dipoles around the surface of the ionic crystal and the formation of hydrated ions in solution are the key factors in the dissolution process.

lattice energy. Energy is released in the next two steps—hydration of the gaseous cations and anions. The enthalpy of solution is the sum of these three ΔH values, described below for NaCl.

◀ We discussed lattice energy in Section 12-7.

$$\text{NaCl(s)} \longrightarrow \text{Na}^+(g) + \text{Cl}^-(g) \qquad \Delta H_1 = (-\text{lattice energy of NaCl}) > 0$$

$$\text{Na}^+(g) \xrightarrow{\text{H}_2\text{O}} \text{Na}^+(aq) \qquad \Delta H_2 = (\text{hydration energy of Na}^+) < 0$$

$$\text{Cl}^-(g) \xrightarrow{\text{H}_2\text{O}} \text{Cl}^-(aq) \qquad \Delta H_3 = (\text{hydration energy of Cl}^-) < 0$$

$$\text{NaCl(s)} \xrightarrow{\text{H}_2\text{O}} \text{Na}^+(aq) + \text{Cl}^-(aq) \qquad \Delta H_{\text{soln}} = \Delta H_1 + \Delta H_2 + \Delta H_3 \approx +5 \text{ kJ/mol}$$

The dissolution of sodium chloride in water is *endothermic*, and this is also the case for the vast majority (about 95%) of soluble ionic compounds. Why does NaCl dissolve in water if the process is endothermic? It might appear that an endothermic process would not occur because of the increase in enthalpy. Because NaCl does actually dissolve in water, there must be another factor involved. In fact, *two* factors must be considered in determining whether a process will occur spontaneously. Enthalpy change is only one of them. The other factor, called *entropy* (see page 821), concerns the natural tendency for microscopic particles—atoms, ions, or molecules—to spread themselves out in the space available to them. The dispersed condition of the microscopic particles in NaCl(aq) compared with pure NaCl(s) and $H_2O(l)$ offsets the +5 kJ/mol increase in enthalpy in the solution process. In summary, if the hypothetical three-step process for solution formation is *exothermic*, we expect dissolution to occur; but we also expect a solution to form for an endothermic solution process, as long as ΔH_{soln} is not too large.

◀ A common misconception is that an endothermic process cannot be spontaneous.

13-4 Solution Formation and Equilibrium

In the previous section, we described what happens at the molecular (microscopic) level when solutions form. In this section, we will describe solution formation in terms of phenomena that we can actually observe, that is, a macroscopic view.

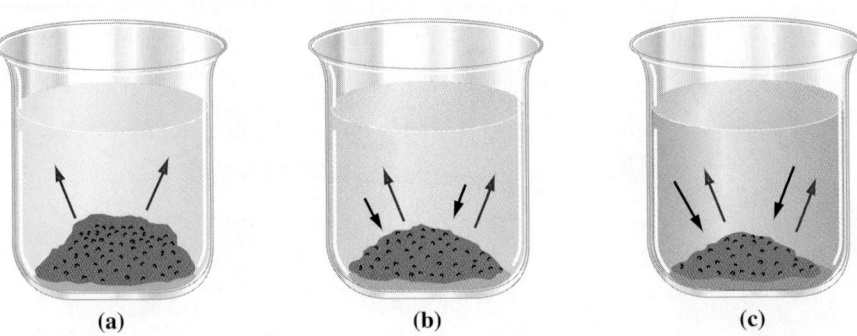

▲ FIGURE 13-7
Formation of a saturated solution
The lengths of the arrows represent the rate of dissolution (↑) and the rate of crystallization (↓). **(a)** When solute is first placed in the solvent, only dissolution occurs. **(b)** After a time, the rate of crystallization becomes significant. **(c)** The solution is saturated when the rates of dissolution and crystallization become equal.

Figure 13-7 suggests what happens when a solid solute and liquid solvent are mixed. At first, only dissolution occurs, but soon the reverse process of crystallization becomes increasingly important; and some dissolved atoms, ions, or molecules return to the undissolved state. When dissolution and crystallization occur at the same rate, the solution is in a state of dynamic equilibrium. The quantity of dissolved solute remains constant with time, and the solution is said to be a **saturated solution**. The concentration of the saturated solution is called the **solubility** of the solute in the given solvent. Solubility varies with temperature, and a solubility–temperature graph is called a *solubility curve*. Some typical solubility curves are shown in Figure 13-8.

If, in preparing a solution, we start with less solute than would be present in the saturated solution, the solute completely dissolves, and the solution is

▶ FIGURE 13-8
Aqueous solubility of several salts as a function of temperature
Solubilities can be expressed in many ways: molarities, mass percent, or, as in this figure, grams of solute per 100 g H_2O. For each solubility curve (as shown here for $KClO_4$), points on the curve (S) represent saturated solutions. Regions above the curve (1) correspond to supersaturated solutions and below the curve (2), to unsaturated solutions.

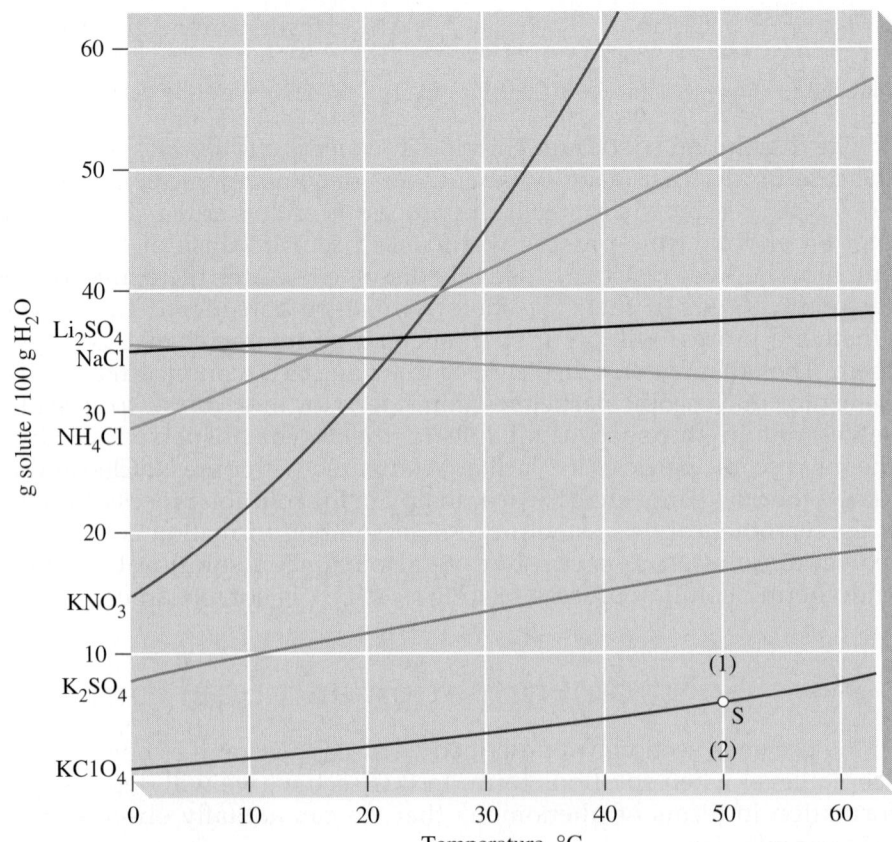

an **unsaturated solution**. But suppose we prepare a saturated solution at one temperature and then change the temperature to a value at which the solubility is lower (this generally means a lower temperature). Usually, the excess solute crystallizes from solution, but occasionally all the solute may remain in solution. In these cases, because the quantity of solute is greater than in a saturated solution, the solution is said to be a **supersaturated solution**. A supersaturated solution is unstable, and if a few crystals of solute are added to serve as particles on which crystallization can occur, the excess solute crystallizes. Figure 13-8 shows how unsaturated and supersaturated solutions can be represented with a solubility curve.

Solubility as a Function of Temperature

As a general observation, the solubilities of ionic substances (about 95% of them) *increase* with increasing temperature. Exceptions to this generalization tend to be found among compounds containing the anions SO_3^{2-}, SO_4^{2-}, AsO_4^{3-}, and PO_4^{3-}.

In Chapter 15 we will learn to predict how an equilibrium condition changes with such variables as temperature and pressure by using an idea known as *Le Châtelier's principle*. One statement of the principle is that heat added to a system at equilibrium stimulates the heat-absorbing, or endothermic, reaction. This suggests that when $\Delta H_{soln} > 0$, raising the temperature stimulates dissolving and *increases* the solubility of the solute. Conversely, if $\Delta H_{soln} < 0$ (exothermic), the solubility *decreases* with increasing temperature. In this case, crystallization—being endothermic—is favored over dissolving.

We must be careful in applying the relationship we just described. The particular value of ΔH_{soln} that establishes whether solubility increases or decreases with increased temperature is that associated with dissolving a small quantity of solute in a solution that is already very nearly saturated. In some cases, this heat effect is altogether different from what is observed when a solute is added to the pure solvent. For example, when NaOH is dissolved in water, there is a sharp increase in temperature—*an exothermic* process. This fact suggests that the solubility of NaOH in water should decrease as the temperature is raised. What is observed, though, is that the solubility of NaOH in water *increases* with increased temperature. This is because when a small quantity of NaOH is added to a solution that is already nearly saturated, heat is *absorbed*, not evolved.*

Fractional Crystallization

Compounds synthesized in chemical reactions are generally impure, but the fact that the solubilities of most solids increase with increased temperature provides the basis for one simple method of purification. Usually, the impure solid consists of a high proportion of the desired compound and lesser proportions of the impurities. Suppose that both the compound and its impurities are soluble in a particular solvent and that we prepare a concentrated solution at a high temperature. Then we let the concentrated solution cool. At lower temperatures, the solution becomes saturated in the desired compound. The excess compound crystallizes from solution. The impurities remain in solution because the temperature is still too high for these to crystallize.† This method of purifying a solid, called **fractional crystallization**, or **recrystallization**, is pictured in Figure 13-9. Example 13-4 illustrates how solubility curves can be used to predict the outcome of a fractional crystallization.

▲ FIGURE 13-9
Recrystallization of KNO₃
Colorless crystals of KNO$_3$ separated from an aqueous solution of KNO$_3$ and CuSO$_4$ (an impurity). The pale blue color of the solution is produced by Cu^{2+}, which remains in solution.

---

*The solid in equilibrium with saturated NaOH(aq) over a range of temperatures around 25 °C is NaOH · H$_2$O(s). It is actually the temperature dependence of the solubility of this hydrate that we have been discussing.

†This is the usual behavior, but at times, one or more impurities may form a solid solution with the compound being recrystallized. In these cases simple recrystallization does not work as a method of purification.

EXAMPLE 13-4 **Applying Solubility Data in Fractional Crystallization**

A solution is prepared by dissolving 95 g NH_4Cl in 200.0 g H_2O at 60 °C. **(a)** What mass of NH_4Cl will recrystallize when the solution is cooled to 20 °C? **(b)** How might we improve the yield of NH_4Cl?

Analyze

We need to know the solubility of NH_4Cl at 20°C and at 95 °C. We obtain the required data from Figure 13-8, which shows the solubility of several salts as a function of temperature.

Solve

(a) Using Figure 13-8, we estimate that the solubility of NH_4Cl at 20 °C is 37 g NH_4Cl/100 g H_2O. The quantity of NH_4Cl in the saturated solution at 20 °C is

$$200.0 \text{ g } H_2O \times \frac{37 \text{ g } NH_4Cl}{100 \text{ g } H_2O} = 74 \text{ g } NH_4Cl$$

The mass of NH_4Cl recrystallized is $95 - 74 = 21 \text{ g } NH_4Cl$.

(b) The yield of NH_4Cl in (a) is rather poor—21 g out of 95 g, or 22%. We can do better: (1) The solution at 60 °C, although concentrated, is not saturated. Using Figure 13-8, we estimate that a saturated solution at 60 °C has 55 g NH_4Cl/100 g H_2O. Thus, the 95 g NH_4Cl requires less than 200.0 g H_2O to make a saturated solution. At 20 °C, a smaller quantity of saturated solution would contain less NH_4Cl than in (a), and the yield of recrystallized NH_4Cl would be greater. (2) Instead of cooling the solution to 20 °C, we might cool it to 0 °C. Here the solubility of NH_4Cl is less than at 20 °C, and more solid would recrystallize. (3) Still another possibility is to start with a solution at a temperature higher than 60 °C, say closer to 100 °C. The mass of water needed for the saturated solution would be less than at 60 °C. Note that options (1) and (3) both require changing the conditions by using a different amount of water from that originally specified.

Assess

The amount of dissolved salt can be increased by increasing the volume of solvent or by increasing the temperature. Keep in mind that fractional crystallization works best when the quantities of impurities are small and the solubility curve of the desired solute rises steeply with temperature.

PRACTICE EXAMPLE A: Calculate the quantity of NH_4Cl that would be obtained if suggestions (1) and (2) in Example 13-4 (b) were followed. [*Hint:* Use data from Figure 13-8. What mass of water is needed to produce a saturated solution containing 95 g NH_4Cl at 60 °C?]

PRACTICE EXAMPLE B: Use Figure 13-8 to examine the solubility curves for the three potassium salts: $KClO_4$, K_2SO_4, and KNO_3. If saturated solutions of these salts at 40 °C are cooled to 20 °C, rank the salts in order of highest percent yield for the recrystallization.

13-5 Solubilities of Gases

Why does a freshly opened can of soda pop fizz, and why does the soda go flat after a time? To answer questions like these requires an understanding of the solubilities of gases. As discussed in this section, the effect of temperature on the solubility of gases is generally different from that on solid solutes. Additionally, the pressure of a gas strongly affects its solubility.

Effect of Temperature

We cannot make an all-inclusive generalization about the effect of temperature on the solubilities of gases in solvents. It is certainly true, though, that the solubilities of most gases in water *decrease* with an increase in temperature. This is true of $N_2(g)$ and $O_2(g)$—the major components of air—and of air itself (Fig. 13-10). This fact helps to explain why many types of fish can survive only in cold water. There is not enough dissolved air (oxygen) in warm water to sustain them.

For solutions of gases in organic solvents, the situation is often the reverse of that just described; that is, gases may become more soluble at higher temperatures. The solubility behavior of the noble gases in water is more complex. The solubility of each gas decreases with an increase in temperature, reaching a minimum at a certain temperature; then the solubility trend reverses direction, with the gas becoming more soluble with an increase in temperature. For example, for helium at 1 atm pressure, this minimum solubility in water comes at 35 °C.

Effect of Pressure

Pressure affects the solubility of a gas in a liquid much more than does temperature. The English chemist William Henry (1775–1836) found that *the solubility of a gas increases with increasing pressure.* A mathematical statement of **Henry's law** is

$$C = k \times P_{gas} \tag{13.2}$$

In this equation, C represents the solubility of a gas in a particular solvent at a fixed temperature, P_{gas} is the partial pressure of the gas above the solution, and k is a proportionality constant. To evaluate the proportionality constant k, we need to have one measurement of the solubility of the gas at a known pressure and temperature. For example, the aqueous solubility of $N_2(g)$ at 0 °C and 1.00 atm is 23.54 mL N_2 per liter. The Henry's law constant, k, is

$$k = \frac{C}{P_{gas}} = \frac{23.54 \text{ mL } N_2/\text{L}}{1.00 \text{ atm}}$$

Suppose we want to increase the solubility of the $N_2(g)$ to a value of 100.0 mL N_2 per liter. Equation (13.2) suggests that to do so, we must increase the pressure of $N_2(g)$ above the solution. That is,

$$P_{N_2} = \frac{C}{k} = \frac{100.0 \text{ mL } N_2/\text{L}}{(23.54 \text{ mL } N_2/\text{L})/1.00 \text{ atm}} = 4.25 \text{ atm}$$

At times, we are required to change the units used to express a gas solubility at the same time that the pressure is changed. This variation is illustrated in Example 13-5.

We can rationalize Henry's law as follows: In a saturated solution, the rate of evaporation of gas molecules from solution and the rate of condensation of gas molecules into the solution are equal. Both of these rates depend on the number of molecules per unit volume. With increasing pressure on the system, the number of molecules per unit volume in the gaseous state increases (through an increase in the gas pressure), and the number of molecules per unit volume must also increase in the solution (through an increase in concentration). Figure 13-11 illustrates this rationalization.

We see a practical application of Henry's law in carbonated beverages. The dissolved gas is carbon dioxide, and the higher the gas pressure maintained above the soda pop, the more CO_2 that dissolves. When a bottle of soda is opened, some gas is released. As the gas pressure above the solution drops, dissolved CO_2 is expelled, usually fast enough to cause fizzing. In sparkling wines, the dissolved CO_2 is also under pressure, but rather than being added artificially as in soda pop, the CO_2 is produced by a fermentation process within the bottle.

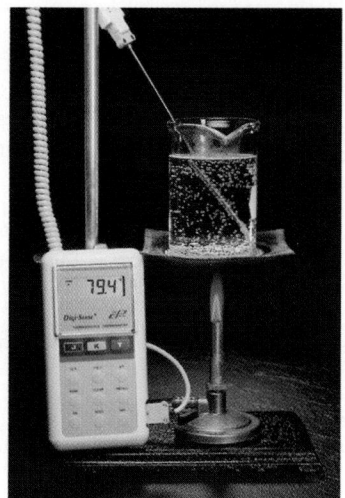

▲ FIGURE 13-10
Effect of temperature on the solubilities of gases
Dissolved air is released as water is heated, even at temperatures well below the boiling point.

▲ The unopened bottle of soda water is under a high pressure of $CO_2(g)$. When a similar bottle is opened, the pressure quickly drops and some of the $CO_2(g)$ is released from solution (bubbles).

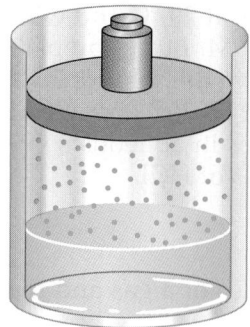 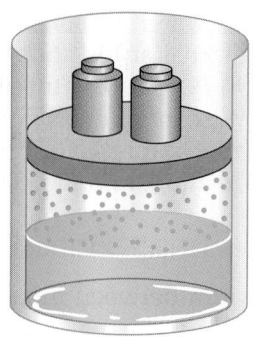 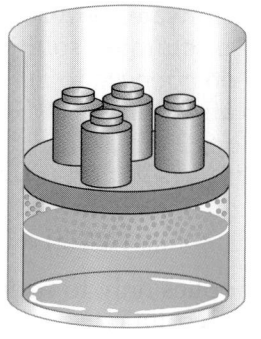

◀ FIGURE 13-11
Effect of pressure on the solubility of a gas
The concentration of dissolved gas (suggested by the depth of color) is proportional to the pressure of the gas above the solution (suggested by the density of the dots).

EXAMPLE 13-5 Using Henry's Law

At 0 °C and an O_2 pressure of 1.00 atm, the aqueous solubility of $O_2(g)$ is 48.9 mL O_2 per liter. What is the molarity of O_2 in a saturated water solution when the O_2 is under its normal partial pressure in air, 0.2095 atm?

Analyze

Think of this as a two-part problem. (1) Determine the molarity of the saturated O_2 solution at 0 °C and 1 atm. (2) Use Henry's law in the manner just outlined.

Solve

Determine the molarity of O_2 at 0 °C when $P_{O_2} = 1$ atm.

$$\text{molarity} = \frac{0.0489 \text{ L } O_2 \times \dfrac{1 \text{ mol } O_2}{22.4 \text{ L } O_2 \text{ (STP)}}}{1 \text{ L soln}} = 2.18 \times 10^{-3} \text{ M } O_2$$

Evaluate the Henry's law constant.

$$k = \frac{C}{P_{gas}} = \frac{2.18 \times 10^{-3} \text{ M } O_2}{1.00 \text{ atm}}$$

Apply Henry's law.

$$C = k \times P_{gas} = \frac{2.18 \times 10^{-3} \text{ M } O_2}{1.00 \text{ atm}} \times 0.2095 \text{ atm} = 4.57 \times 10^{-4} \text{ M } O_2$$

Assess

When working problems involving gaseous solutes in a solution in which the solute is at very low concentration, use Henry's law.

PRACTICE EXAMPLE A: Use data from Example 13-5 to determine the partial pressure of O_2 above an aqueous solution at 0 °C known to contain 5.00 mg O_2 per 100.0 mL of solution.

PRACTICE EXAMPLE B: A handbook lists the solubility of carbon monoxide in water at 0 °C and 1 atm pressure as 0.0354 mL CO per milliliter of H_2O. What pressure of $CO(g)$ must be maintained above the solution to obtain 0.0100 M CO?

▲ To avoid the painful and dangerous condition of the bends, divers must not surface too quickly from great depths.

Deep-sea diving provides us with still another example of Henry's law. Divers must carry a supply of air to breathe while underwater. If they are to stay submerged for any period of time, they must breathe compressed air. High-pressure air, however, is much more soluble in the blood and other body fluids than is air at normal pressures. When a diver returns to the surface, excess dissolved $N_2(g)$ is released as tiny bubbles from body fluids. When the ascent to the surface is made too quickly, N_2 diffuses out of the blood too quickly, causing severe pain in the limbs and joints, probably by interfering with the nervous system. This dangerous condition, known as "the bends," can be avoided if the diver ascends very slowly or spends time in a decompression chamber. Another effective method is to substitute a helium–oxygen mixture for compressed air. Helium is less soluble in blood than is nitrogen.

Henry's law (equation 13.2) fails for gases at high pressures; it also fails if the gas ionizes in water or reacts with water. For example, at 20 °C and with $P_{HCl} = 1$ atm, a saturated solution of HCl(aq) is about 20 M. But to prepare 10 M HCl, we do not need to maintain $P_{HCl} = 0.5$ atm above the solution, nor is $P_{HCl} = 0.05$ atm above 1 M HCl. We cannot even detect HCl(g) above 1 M HCl by its odor. The reason we cannot is that HCl ionizes in aqueous solutions, and in dilute solutions there are almost no molecules of HCl.

$$HCl \text{ (g)} \xrightarrow{\text{H}_2\text{O}} H^+(aq) + Cl^-(aq)$$

Henry's law applies only to equilibrium between molecules of a gas and the same *molecules* in solution.

13-6   Vapor Pressures of Solutions

Separating compounds from one another is a task that chemists commonly face. If the compounds are volatile liquids, this separation often can be achieved by *distillation.* To understand how distillation works, we need to know something about the vapor pressures of solutions. This knowledge will also enable us to deal with other important solution properties, such as boiling points, freezing points, and osmotic pressures.

To simplify the following discussion we will consider only solutions with two components, solvent A and solute B. In the 1880s, the French chemist F. M. Raoult found that a dissolved solute *lowers* the vapor pressure of the solvent. **Raoult's law** states that the partial pressure exerted by solvent vapor above an ideal solution, P_A, is the product of the mole fraction of solvent in the solution, x_A, and the vapor pressure of the pure solvent at the given temperature, P_A°.

$$P_A = x_A P_A^\circ \qquad (13.3)$$

Equation (13.3) relates to Raoult's observation that a dissolved solute lowers the vapor pressure of the solvent because if $x_A + x_B = 1.00$, x_A must be less than 1.00, and P_A must be smaller than P_A°. Strictly speaking, Raoult's law applies only to ideal solutions and to all volatile components of the solutions. However, even in nonideal solutions, the law often works reasonably well for the *solvent* in *dilute* solutions, for example, solutions in which $x_{solv} > 0.98$. A more detailed discussion of Raoult's law requires the notion of entropy, which was briefly mentioned on page 567. Rather than attempt the explanation now, however, we will wait until Section 19-3, after we have said more about entropy.

Liquid–Vapor Equilibrium: Ideal Solutions

The results of Examples 13-6 and 13-7, together with similar data for other benzene–toluene solutions, are plotted in Figure 13-12. This figure consists of four lines—three straight and one curved—spanning the entire concentration range.

The red line shows how the vapor pressure of benzene varies with the solution composition. Because benzene in benzene–toluene solutions obeys Raoult's law, the red line has the equation $P_{benz} = x_{benz}P_{benz}^\circ$. The blue line shows how the vapor pressure of toluene varies with solution composition and indicates that toluene also obeys Raoult's law. The dashed black line shows how the *total* vapor pressure varies with the solution composition. Can you see that each pressure on this black line is the sum of the pressures on the two straight lines that lie below it? Thus, point 3 represents the total vapor pressure (point 1 + point 2) of a benzene–toluene solution in which $x_{benz} = 0.500$ (see Example 13-6).

EXAMPLE 13-6   **Predicting Vapor Pressures of Ideal Solutions**

The vapor pressures of pure benzene and pure toluene at 25 °C are 95.1 and 28.4 mmHg, respectively. A solution is prepared in which the mole fractions of benzene and toluene are both 0.500. What are the partial pressures of the benzene and toluene above this solution? What is the total vapor pressure?

Analyze

We saw in Figure 13-4 that benzene–toluene solutions should be ideal. We expect Raoult's law to apply to both solution components.

Solve

$$P_{benz} = x_{benz}P°_{benz} = 0.500 \times 95.1 \text{ mmHg} = 47.6 \text{ mmHg}$$
$$P_{tol} = x_{tol}P°_{tol} = 0.500 \times 28.4 \text{ mmHg} = 14.2 \text{ mmHg}$$
$$P_{total} = P_{benz} + P_{tol} = 47.6 \text{ mmHg} + 14.2 \text{ mmHg} = 61.8 \text{ mmHg}$$

Assess

In this example we assumed these to be ideal solutions, which allowed us to use Raoult's law. We observe that the vapor pressure of each component is lowered because of the presence of the other component.

PRACTICE EXAMPLE A: The vapor pressure of pure hexane and pentane at 25 °C are 149.1 mmHg and 508.5 mmHg, respectively. If a hexane–pentane solution has a mole fraction of hexane of 0.750, what are the vapor pressures of hexane and pentane above the solution? What is the total vapor pressure?

PRACTICE EXAMPLE B: Calculate the vapor pressures of benzene, C_6H_6, and toluene, C_7H_8, and the total pressure at 25 °C above a solution with equal *masses* of the two liquids. Use the vapor pressure data given in Example 13-6.

EXAMPLE 13-7   **Calculating the Composition of Vapor in Equilibrium with a Liquid Solution**

What is the composition of the vapor in equilibrium with the benzene–toluene solution of Example 13-6?

Analyze

We are being asked to find the mole fraction of benzene and of toluene in the vapor. From Example 13-6 we know the vapor pressure of pure benzene and pure toluene. We have already calculated the partial vapor pressures; now we need to apply the definition of mole fraction.

Solve

The ratio of each partial pressure to the total pressure is the mole fraction of that component in the vapor. (This is another application of equation 6.17.) The mole-fraction composition of the vapor is

$$x_{benz} = \frac{P_{benz}}{P_{total}} = \frac{47.6 \text{ mmHg}}{61.8 \text{ mmHg}} = 0.770$$

$$x_{tol} = \frac{P_{tol}}{P_{total}} = \frac{14.2 \text{ mmHg}}{61.8 \text{ mmHg}} = 0.230$$

Assess

The mole fraction of benzene in the vapor is 0.770, whereas in the liquid the mole fraction of benzene is 0.5. For toluene the mole fraction in the vapor is 0.230, whereas in the liquid the mole fraction of toluene is 0.5. This difference in mole-fraction vapor composition caused by the difference in vapor pressures of the two components is the central concept of fractional distillation, which is discussed next.

PRACTICE EXAMPLE A: What is the composition of the vapor in equilibrium with the hexane–pentane solution described in Practice Example 13-6A?

PRACTICE EXAMPLE B: What is the composition of the vapor in equilibrium with the benzene–toluene solution described in Practice Example 13-6B?

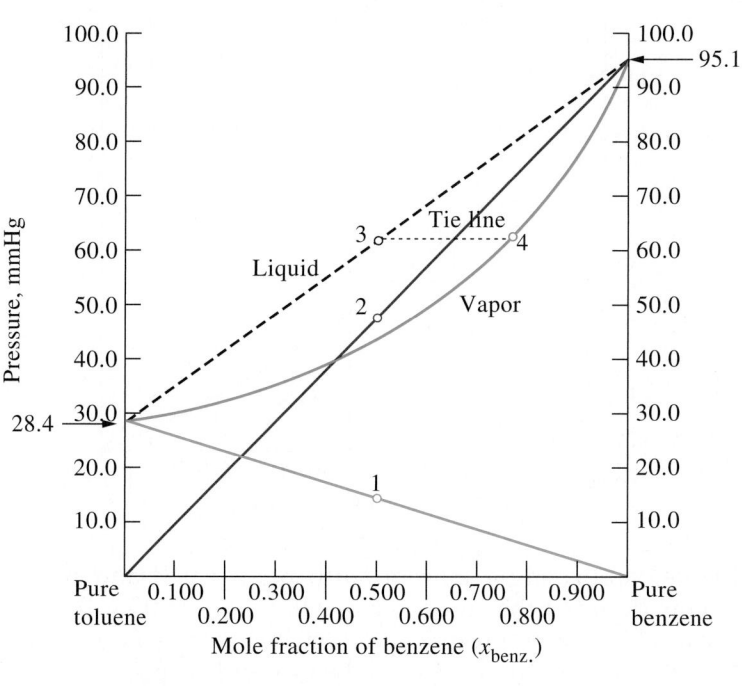

◀ FIGURE 13-12
Liquid–vapor equilibrium for benzene–toluene mixtures at 25 °C
In this diagram, partial pressures and the total pressure of the vapor are plotted as a function of the solution and vapor compositions.

As we calculated in Example 13-7, the vapor in equilibrium with a solution in which $x_{benz} = 0.500$ is richer still in benzene. The vapor has $x_{benz} = 0.770$ (point 4). The line joining points 3 and 4 is called a *tie line*. Imagine establishing a series of tie lines throughout the composition range. The vapor ends of these tie lines can be joined to form the green curve in Figure 13-12. From the relative placement of the liquid and vapor curves, we see that for ideal solutions of two components, *the vapor phase is richer in the more volatile component than is the liquid phase.*

🔍 **13-5   CONCEPT ASSESSMENT**

Describe a case in which the liquid and vapor curves in a diagram such as Figure 13-12 would converge into a single curve. Is such a case likely to exist?

Fractional Distillation

Let's look at liquid–vapor equilibrium in benzene–toluene mixtures in a somewhat different way. Instead of plotting vapor pressures as a function of the solution and vapor compositions, let's plot normal boiling temperature—the temperature at which the *total* vapor pressure of the solution is 1 atm. The resulting graph is shown in Figure 13-13. This graph is useful in explaining **fractional distillation**, a procedure for separating volatile liquids from one another.

Notice that the graph starts at a high temperature—110.6 °C, the boiling point of toluene—and ends at a lower temperature—80.0 °C, the boiling point of benzene. This is the reverse of the situation in Figure 13-12. Also, the vapor curve lies above the liquid curve in Figure 13-13, not below, as is the case in Figure 13-12.

Figure 13-13 indicates that a benzene–toluene solution with $x_{benz} = 0.30$ boils at a temperature of 98.6 °C and is in equilibrium with a vapor in which $x_{benz} = 0.51$. Imagine extracting some of this vapor and cooling it to the point where it condenses to a liquid. The new liquid will have $x_{benz} = 0.51$ and represents the conclusion of stage 1 in Figure 13-13. Now imagine repeating the

KEEP IN MIND

that the placement of the two curves in liquid–vapor equilibrium diagrams is such that the vapor is richer in the more volatile component than is the liquid. The more volatile component is the one with the higher vapor pressure or lower boiling point.

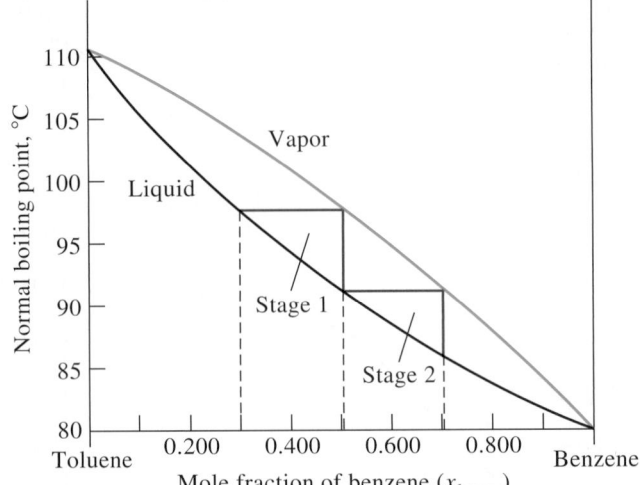

▶ FIGURE 13-13
Liquid–vapor equilibrium for benzene–toluene mixtures at 1 atm
In this diagram, the normal boiling points of solutions are plotted as a function of solution and vapor compositions.

▲ Fractional distillation is used in many industrial processes.

process, that is, vaporizing a solution with $x_{benz} = 0.51$ and condensing the vapor. The new liquid at the end of stage 2 has $x_{benz} = 0.71$. By repeating the cycle, the vapor becomes progressively richer in benzene. As pictured in Figure 13-14, boiling solutions in equilibrium with vapor can be spread out over a long column, called a *fractionating column*, in which the equilibrium temperatures range from lowest at the top of the column to highest at the bottom. The most volatile component in the solution emerges from the top of the column as a vapor that is condensed to a liquid and removed. The least volatile component concentrates in the pot at the bottom of the column. Fractional distillation of a solution of many volatile components, such as petroleum, can be carried out in such a way that the components are withdrawn from the top of the column and condensed, one by one.

Liquid–Vapor Equilibrium: Nonideal Solutions

We cannot construct a liquid–vapor equilibrium diagram for nonideal solutions in the simple manner illustrated in Figure 13-12. For example, vapor pressures in acetone–chloroform solutions are *lower* than we would predict for ideal solutions and boiling temperatures are correspondingly *higher*. In acetone–carbon disulfide solutions, conversely, vapor pressures are higher

▶ FIGURE 13-14
Fractional distillation
The fractionating column is packed with glass beads or stainless steel turnings. Initially, as vapor rises from the pot and encounters these cooler objects, it condenses to a liquid. As the beads or turnings heat up, the liquid–vapor equilibrium front moves progressively up the column. Soon, liquid–vapor equilibrium occurs throughout the column, but with the equilibrium temperature changing continuously from the hottest regions at the bottom of the column to the coolest at the top. The vapor emerging from the top of the column is condensed to a liquid in the water-cooled condenser. The first fraction collected contains the most volatile component (lowest boiling point). Later fractions are less volatile liquids. The least volatile (highest boiling point) components remain as a residue in the distillation pot.

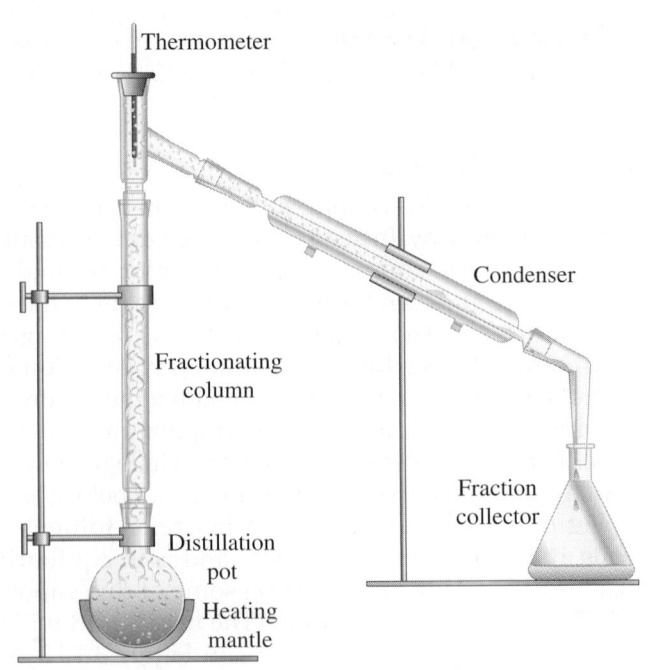

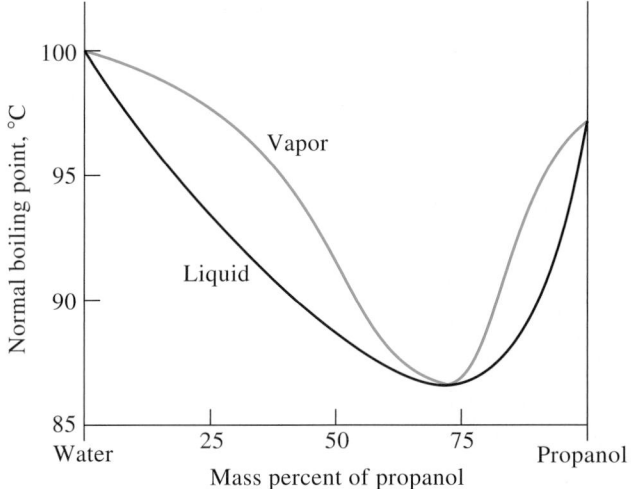

A minimum boiling-point azeotrope
A solution of propanol in water having 71.69% $CH_3CH_2CH_2OH$ by mass—an azeotrope—has a lower boiling point than any other solution of these two components. In fractional distillation, solutions having less than 71.69% of the alcohol yield the azeotrope and water as ultimate products. Solutions with more than 71.69% of the alcohol yield the azeotrope and propanol. In each case, the azeotrope is drawn off through the condenser (Fig. 13-14), and the other component remains in the pot.

than predicted and boiling temperatures are correspondingly lower. In Figure 13-5, we saw that the forces of attraction between unlike molecules are greater than those between like molecules in acetone–chloroform mixtures. It is not unreasonable to expect the components in such solutions to show a reduced tendency to vaporize and to have lower-than-predicted vapor pressures. With acetone–carbon disulfide solutions, the situation is the reverse: Forces of attraction between unlike molecules are weaker than between like molecules. This leads to greater tendencies for vaporization and higher vapor pressures than predicted by Raoult's law.

If the departures from ideal solution behavior are sufficiently great, some solutions may have vapor pressures that pass through either a maximum or a minimum in vapor-pressure-composition graphs. Correspondingly, their boiling points pass through either a minimum or maximum in boiling-point-composition graphs. The solutions corresponding to these maxima or minima boil at a constant temperature and produce a vapor having the *same* composition as the liquid. These solutions are called **azeotropes**. The boiling-point diagram of a minimum boiling-point azeotrope is illustrated in Figure 13-15.

One of the most familiar azeotropes consists of 96.0% ethanol (C_2H_5OH) and 4.0% water, by mass, and has a boiling point of 78.174 °C. Pure ethanol has a boiling point of 78.3 °C. Dilute ethanol–water solutions can be distilled to produce the azeotrope, but the remaining water cannot be removed by ordinary distillation. As a result, most ethanol used in the laboratory or in industry is only 96.0% C_2H_5OH. To obtain absolute, or 100%, C_2H_5OH requires special measures.

13-7 Osmotic Pressure

In the previous section our primary emphasis was on solutions containing a volatile solvent and volatile solute. Another common type of solution is one with a volatile solvent, such as water, but one or more *nonvolatile* solutes, such as glucose, sucrose, or urea. Raoult's law still applies to the solvent in such solutions—the vapor pressure of the solvent is lowered.

Figure 13-16(a) pictures two aqueous solutions of a nonvolatile solute within the same enclosure. They are labeled A and B. The curved arrow indicates that water vaporizes from A and condenses into B. What is the driving force behind this? It must be that the vapor pressure of H_2O above A is greater than that above B. Solution A is more dilute; it has a higher mole fraction of H_2O. How long will this transfer of water continue? Solution A becomes more concentrated as it loses water, and solution B becomes more dilute as it gains water. When the mole fraction of H_2O is the same in both solutions, the *net* transfer of H_2O stops.

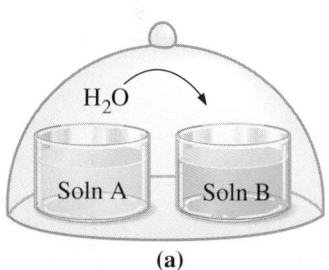

(a)

(b)

▲ FIGURE 13-16
Observing the direction of flow of water vapor
(a) Water passes, as vapor, from the more dilute solution (higher mole fraction of H_2O) to the more concentrated solution. (b) Water vapor in air condenses onto solid calcium chloride hexahydrate, $CaCl_2 \cdot 6\,H_2O$. The liquid water dissolves some of the solid. The eventual result could be an unsaturated solution.

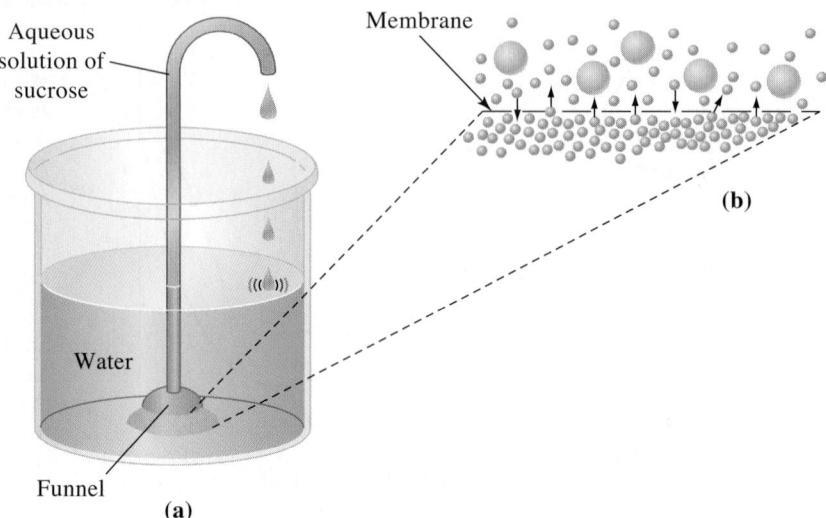

Aqueous
solution of
sucrose

Membrane

(b)

Water

Funnel

(a)

▲ FIGURE 13-17
Osmosis
(a) Water molecules pass through pores in the membrane and create a pressure within the funnel that causes the sucrose solution to rise, overflow, and fall into the pure water. After a time, the solution inside the funnel becomes more dilute and the pure water in the beaker becomes a sucrose solution. Liquid flow stops when the compositions of the solutions separated by the membrane have become nearly equal. **(b)** Enlarged cross-section of the membrane demonstrating its semi-permeable properties: water molecules (represented as small blue spheres) freely cross the membrane, while the sucrose molecule (represented as large gray spheres) cannot cross the membrane.

A related phenomenon occurs when $CaCl_2 \cdot 6\,H_2O(s)$ is exposed to air (Fig. 13-16b). Water vapor from the air condenses on the solid, and the solid begins to dissolve, a phenomenon known as *deliquescence*. For a solid to deliquesce, the partial pressure of water vapor in the air must be greater than the vapor pressure of water above a saturated aqueous solution of the solid. This requirement is often met for certain solids under conditions of appropriate relative humidity. The deliquescence of $CaCl_2 \cdot 6\,H_2O$ occurs when the relative humidity exceeds 32%. (Relative humidity is described in Focus On 6: Earth's Atmosphere.)

Like the case just described, Figure 13-17 also pictures the flow of solvent molecules. Here, however, the flow is not through the vapor phase. An aqueous sucrose (sugar) solution in a long glass tube is separated from pure water by a semipermeable membrane (permeable to water only). Water molecules can pass through the membrane in either direction, and they do. But because the concentration of water molecules is greater in the pure water than in the solution, there is a net flow from the pure water into the solution. This net flow, called **osmosis**, causes the solution to rise in the tube. The more concentrated the sucrose solution, the higher the solution level rises.

▶ Semipermeable membranes are materials containing submicroscopic holes, such as a pig's bladder, parchment, or cellophane. The holes permit the passage of solvent molecules but not those of the solute.

🔍 **13-6   CONCEPT ASSESSMENT**

Describe the similarities and differences between the phenomena depicted in Figures 13-16(a) and 13-17.

Applying pressure to the sucrose solution slows down the net flow of water across the membrane into the solution. With a sufficiently high pressure, the net influx of water can be stopped altogether. The necessary pressure to stop osmotic flow is called the **osmotic pressure** of the solution. For a 20% sucrose solution, this pressure is about 15 atm. The magnitude of osmotic pressure depends only on the number of solute particles per unit volume of solution. It does not depend on the identity of the solute. Properties of this sort, whose values depend only on the concentration of solute particles in solution and *not* on what the solute is, are called **colligative properties**. The following equation works quite well for calculating osmotic pressures of *dilute* solutions of non-electrolytes. The osmotic pressure is represented by the symbol π, R is the gas constant $(0.08206 \text{ L atm mol}^{-1} \text{ K}^{-1})$; and T is the Kelvin temperature. The term n represents the amount of solute (in moles), and V is the volume (in liters) of solution. Notice that this equation is similar to the equation for the ideal gas law. In this case, however, it is convenient to rearrange terms to yield equation (13.4). The ratio, n/V, then, is the *molarity* of the solution, represented by the symbol M.

◀ Vapor-pressure lowering, as expressed through Raoult's law for ideal solutions, is also a colligative property.

$$\pi V = nRT$$

$$\pi = \frac{n}{V}RT = M \times RT \qquad \textbf{(13.4)}$$

◀ The adjustment required to apply equation (13.4) to electrolyte solutions is discussed in Section 13-9.

The pressure difference of 18 mmHg that we calculate in Example 13-8 is easy to measure. (It corresponds to a solution height of about 25 cm.) This means that we can easily use the measurement of osmotic pressure for determining molar masses when we are dealing with very dilute solutions or solutes with high molar masses (or both). Example 13-9 shows how osmotic pressure measurements can be used to determine molar mass.

EXAMPLE 13-8   Calculating Osmotic Pressure

What is the osmotic pressure at 25 °C of an aqueous solution that is 0.0010 M $C_{12}H_{22}O_{11}$ (sucrose)?

Analyze

We just need to substitute the data into equation (13.4).

Solve

$$\pi = \frac{0.0010 \text{ mol} \times 0.08206 \text{ atm L mol}^{-1} \text{ K}^{-1} \times 298 \text{ K}}{1 \text{ L}}$$

$$\pi = 0.024 \text{ atm (18 mmHg)}$$

Assess

At a very low concentration, there is an appreciable amount of osmotic pressure. This fact is used when measuring the molar mass of polymers and biopolymers.

PRACTICE EXAMPLE A: What is the osmotic pressure at 25 °C of an aqueous solution that contains 1.50 g $C_{12}H_{22}O_{11}$ in 125 mL of solution?

PRACTICE EXAMPLE B: What mass of urea $[CO(NH_2)_2]$ would you dissolve in 225 mL of solution to obtain an osmotic pressure of 0.015 atm at 25 °C?

EXAMPLE 13-9   Establishing a Molar Mass from a Measurement of Osmotic Pressure

A 50.00 mL sample of an aqueous solution contains 1.08 g of human serum albumin, a blood-plasma protein. The solution has an osmotic pressure of 5.85 mmHg at 298 K. What is the molar mass of the albumin?

Analyze

We need to use osmotic pressure to determine the molar mass of a protein, human serum albumin, in a solution.

(continued)

Solve

Express osmotic pressure in atmospheres.

$$\pi = 5.85 \text{ mmHg} \times \frac{1 \text{ atm}}{760 \text{ mmHg}} = 7.70 \times 10^{-3} \text{ atm}$$

Modify the basic equation for osmotic pressure, by showing moles of solute (n) as the mass of solute (m) divided by its molar mass (M), and solve for M.

$$\pi V = nRT \qquad \pi V = \frac{m}{M} RT \qquad M = \frac{mRT}{\pi V}$$

Obtain the value of M by substituting the given data into the last equation above, ensuring that units cancel to yield g/mol as the unit for M.

$$M = \frac{1.08 \text{ g} \times 0.08206 \text{ atm L mol}^{-1} \text{K}^{-1} \times 298 \text{ K}}{7.70 \times 10^{-3} \text{ atm} \times 0.0500 \text{ L}} = 6.86 \times 10^4 \text{ g/mol}$$

Assess

Even though the solution is relatively dilute, knowing the osmotic pressure helped us determine the molar mass of human serum albumin.

PRACTICE EXAMPLE A: Creatinine is a by-product of nitrogen metabolism and can be used to provide an indication of renal function. A 4.04 g sample of creatinine is dissolved in enough water to make 100.0 mL of solution. The osmotic pressure of the solution is 8.73 mmHg at 298 K. What is the molar mass of creatinine?

PRACTICE EXAMPLE B: What would be the osmotic pressure of a solution containing 2.12 g of human serum albumin in 75.00 mL of water at 37.0 °C? Use the molar mass determined in Example 13-9.

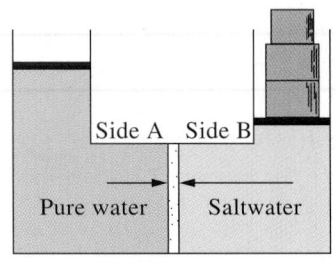

▲ FIGURE 13-18
Desalination of saltwater by reverse osmosis
The membrane is permeable to water but not to ions. The normal flow of water is from side A to side B. If a pressure is exerted on side B that exceeds the osmotic pressure of the saltwater, a net flow of water occurs in the *reverse* direction—from the saltwater to the pure water. The lengths of the arrows suggest the magnitudes of the flow of water molecules in each direction.

Practical Applications

Some of the best examples of osmosis are those associated with living organisms. For instance, if red blood cells are placed in pure water, the cells expand and eventually burst as a result of water that enters through osmosis. The osmotic pressure associated with the fluid inside the cell is equivalent to that of 0.92% (mass/volume) NaCl(aq). Thus, if cells are placed in a sodium chloride (saline) solution of this concentration, there is no net flow of water through the cell membrane, and the cell remains stable. Such a solution is said to be *isotonic*. If cells are placed in a solution with a concentration greater than 0.92% NaCl, water flows out of the cells, and the cells shrink. Such a solution is said to be *hypertonic*. If the NaCl concentration is less than 0.92%, the solution is *hypotonic*, and water flows into the cells. Fluids that are intravenously injected into patients to combat dehydration or to supply nutrients must be adjusted so that they are isotonic with blood. The osmotic pressure of the fluids must be the same as that of 0.92% (mass/volume) NaCl.

One recent application of osmosis goes to the very definition of osmotic pressure. Suppose in the device shown in Figure 13-18, we apply a pressure to the right side (side B) that is less than the osmotic pressure of the saltwater. The net flow of water molecules through the membrane will be from side A to side B, and ordinary osmosis occurs. If we apply a pressure greater than the

▶ A red blood cell in a hypertonic solution (left), an isotonic solution (center), and a hypotonic solution (right).

osmotic pressure to side B, we can cause a net flow of water in the *reverse* direction, from the saltwater into the pure water. This is the condition known as **reverse osmosis**. Reverse osmosis can be used in the *desalination* of seawater to supply drinking water for emergency situations or as an actual source of municipal water. Another application of reverse osmosis is the removal of dissolved materials from industrial or municipal wastewater before it is discharged into the environment.

▲ A small reverse-osmosis unit used to desalinize seawater.

13-8 Freezing-Point Depression and Boiling-Point Elevation of Nonelectrolyte Solutions

In Section 13-6 we examined the lowering of the vapor pressure of a solvent produced by a dissolved solute. Vapor pressure lowering is not measured as frequently as certain properties directly related to it. In Figure 13-19 the blue curves represent the vapor pressure, fusion, and sublimation curves in the phase diagram for a pure solvent. The red curves represent the vapor pressure and fusion curves of the solvent in a solution. The sublimation curve for the solid solvent that freezes from the solution is shown in purple. Two assumptions are implicit in Figure 13-19. One is that the solute is nonvolatile, the other is that the solid that freezes from a solution is pure solvent. For many mixtures, these requirements are easily met.*

The vapor pressure curve of the solution (red) intersects the sublimation curve at a lower temperature than is the case for the pure solvent. The solid-liquid fusion curve, because it originates at the intersection of the sublimation and vapor pressure curves, is also displaced to lower temperatures. Now recall how we establish normal melting points and boiling points in a phase diagram. They are the temperatures at which a line at $P = 1$ atm intersects the fusion and vapor pressure curves, respectively. Four points of intersection are highlighted in Figure 13-19—the freezing points and the boiling points of the pure solvent and of the solvent in a solution. The freezing point of the solvent in solution is *depressed*, and the boiling point is *elevated*.

The extent to which the freezing point is lowered or the boiling point raised is proportional to the mole fraction of solute (just as is vapor pressure lowering).

◀ In Figure 13-19, the solid and vapor free energies do not change with the addition of a solute to the liquid phase. However, adding solute to the liquid does lower the free energy of the solution. Hence, to re-establish equilibrium, the S–L and the L–V curves shift to respectively lower and higher T. This can be thought of as an example of Le Châtelier's principle.

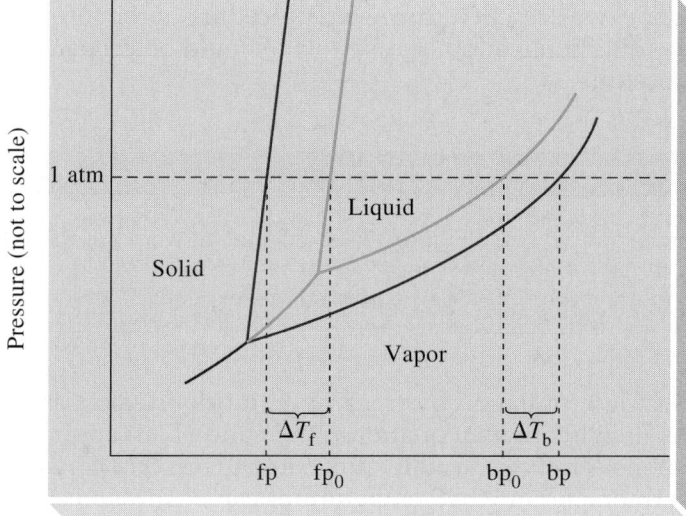

Temperature (not to scale)

◀ FIGURE 13-19
Vapor-pressure lowering by a nonvolatlle solute
The normal freezing point and normal boiling point of the pure solvent are fp_0 and bp_0, respectively. The corresponding points for the solution are fp and bp. The freezing-point depression, ΔT_f, and the boiling-point elevation, ΔT_b, are indicated. Because the solute is assumed to be insoluble in the solid solvent, the sublimation curve of the solvent is unaffected by the presence of solute in the liquid solution phase. That is, the sublimation curve is the same for the two phase diagrams.

*Actually, the equation for freezing-point depression (13.5) applies even if the solute is volatile.

TABLE 13.2 Freezing-Point Depression and Boiling-Point Elevation Constants[a]

Solvent	Normal Freezing Point,°C	K_f, °C m^{-1}	Normal Boiling Point, °C	K_b, °C m^{-1}
Acetic acid	16.6	3.90	118	3.07
Benzene	5.53	5.12	80.10	2.53
Nitrobenzene	5.7	8.1	210.8	5.24
Phenol	41	7.27	182	3.56
Water	0.00	1.86	100.0	0.512

[a]Values correspond to freezing-point depressions and boiling-point elevations, in degrees Celsius, caused by 1 mol of solute particles dissolved in 1 kg of solvent in an ideal solution.

In *dilute* solutions, the solute mole fraction is proportional to its molality, and so we can write

$$\Delta T_f = -K_f \times m \qquad (13.5)$$

$$\Delta T_b = K_b \times m \qquad (13.6)$$

KEEP IN MIND

that freezing-point depression (ΔT_f) is defined as $T - T_f$, where T is the freezing point of the solution and T_f is the freezing point of the pure solvent, and similarly the boiling-point elevation (ΔT_b) is defined as $T - T_b$ where T_b is the boiling point of the pure solvent. So the need for the negative sign in equation (13.5) is evident.

In these equations, ΔT_f and ΔT_b are the freezing-point depression and boiling-point elevation, respectively; m is the solute molality; and K_f and K_b are proportionality constants. The value of K_f depends on the melting point, enthalpy of fusion, and molar mass of the solvent. The value of K_b depends on the boiling point, enthalpy of vaporization, and molar mass of the solvent. The units of K_f and K_b are °C m^{-1}, and you can think of their values as representing the freezing-point depression and boiling-point elevation for a 1 m solution. In practice, though, equations (13.5) and (13.6) often fail for solutions as concentrated as 1 m. Table 13.2 lists some typical values of K_f and K_b.

Historically, chemists have used the group of colligative properties—vapor pressure lowering, freezing-point depression, boiling-point elevation, and osmotic pressure—for molecular mass determinations. In Example 13-9, we showed how this could be accomplished with osmotic pressure. Example 13-10 shows how freezing-point depression can be used to determine a molar mass and, with other information, a molecular formula. To help you understand how this is done, we present a three-step procedure in the form of answers to three separate questions. In other cases, you should be prepared to work out your own procedure.

🔍 13-7 CONCEPT ASSESSMENT

In what important way would Figure 13-19 change if it were based on the phase diagram of water rather than for the general case shown? Would a boiling-point elevation and a freezing-point depression still be expected?

▶ The adjustment required to apply these equations to electrolyte solutions is discussed in Section 13-9.

Molar mass determination by freezing-point depression or boiling-point elevation has its limitations. Equations (13.5) and (13.6) apply only to dilute solutions of nonelectrolytes, usually much less than 1 m. This requires the use of special thermometers so that temperatures can be measured very precisely, say to ±0.001 °C. Because boiling points depend on barometric pressure, precise measurements require that pressure be held constant. As a consequence, boiling-point elevation is not much used. The precision of the freezing-point depression method can be improved by using a solvent with a larger K_f value

EXAMPLE 13-10 Establishing a Molecular Formula with Freezing-Point Data

Nicotine, extracted from tobacco leaves, is a liquid completely miscible with water at temperatures below 60 °C. **(a)** What is the *molality* of nicotine in an aqueous solution that starts to freeze at −0.450 °C? **(b)** If this solution is obtained by dissolving 1.921 g of nicotine in 48.92 g H_2O, what must be the molar mass of nicotine? **(c)** Combustion analysis shows nicotine to consist of 74.03% C, 8.70% H, and 17.27% N, by mass. What is the molecular formula of nicotine?

Analyze

(a) We can establish the molality of the nicotine by using equation (13.5) with the value of K_f for water listed in Table 13.2. **(b)** Once we know the molality, we can use the definition of molality, but with a known molality (from part a) and an unknown molar mass of solute (M). **(c)** To establish the empirical formula of nicotine, we need to use the method of Example 3-5.

Solve

(a) Note that $T_f = -0.450$ °C, and that $\Delta T_f = -0.450$ °C $- 0.000$ °C $= -0.450$ °C.

$$\text{molality} = \frac{\Delta T_f}{-K_f} = \frac{-0.450 \text{ °C}}{-1.86 \text{ °C } m^{-1}} = 0.242 \ m$$

(b) Let's represent the molar mass as x g/mol. The amount, in moles, contained in a 1.921g sample is 1.921g $\times$ (1 mol/x g) $= (1.921/x)$ mol. Thus,

$$\text{molality} = \frac{1.921 \ x^{-1}}{0.04892 \text{ kg water}} = 0.242 \text{ mol (kg water)}^{-1}$$

$$x = 162$$

The molar mass is 162 g/mol.

(c) This calculation is left as an exercise for you to do. The result you should obtain is C_5H_7N. The formula mass based on this empirical formula is 81 u. The molecular mass obtained from the molar mass in part (b) is exactly twice this value—162 u. The molecular formula is twice C_5H_7N, or $C_{10}H_{14}N_2$.

Assess

Using the freezing-point data is another experimental technique that can be used to obtain the chemical properties of a substance. In this example we were able to determine the molecular formula from the freezing-point depression and a known amount of substance dissolved in a solvent. Note that water was used as the solvent here; however, other solvents can be used in freezing-point experiments.

PRACTICE EXAMPLE A: Vitamin B_2, riboflavin, is soluble in water. If 0.833 g of riboflavin is dissolved in 18.1 g H_2O, the resulting solution has a freezing point of −0.227 °C. **(a)** What is the molality of the solution? **(b)** What is the molar mass of riboflavin? **(c)** What is the molecular formula of riboflavin if combustion analysis shows it to consist of 54.25% C, 5.36% H, 25.51% O, and 14.89% N?

PRACTICE EXAMPLE B: An aqueous solution that is 0.205 m urea $[CO(NH_2)_2]$ is found to boil at 100.025 °C. Is the prevailing barometric pressure above or below 760.0 mmHg? [*Hint:* At what temperature should the solution begin to boil if atmospheric pressure is 760.0 mmHg?]

than that of water. For example, for cyclohexane $K_f = 20.0$ °C m^{-1} and for camphor $K_f = 37.7$ °C m^{-1}.

Practical Applications

The typical automobile antifreeze is ethylene glycol, $HOCH_2CH_2OH$. It is a good idea to leave the ethylene glycol–water mixture in the cooling system at all times to provide all-weather protection. In summer, the ethylene glycol helps by raising the boiling point of water and preventing the cooling system from boiling over.

Citrus growers faced with an impending freeze know they must take preventive measures only if the temperature drops below 0 °C by several degrees. The juice in the fruit has enough dissolved solutes to lower the freezing point by a degree or two. The growers also know they must protect lemons sooner

▲ Water sprayed on citrus fruit releases its heat of fusion as it freezes into a layer of ice that acts as a thermal insulator. For a time, the temperature remains at 0 °C. The juice of the fruit, having a freezing point below 0 °C, is protected from freezing.

▲ A typical aircraft deicer is propylene glycol, $CH_3CH(OH)CH_2OH$, diluted with water and applied as a hot, high-pressure spray.

than oranges because lemons have a lower concentration of dissolved solutes (sugars) than do oranges.

Salts, such as NaCl, can be used to prepare a *slush bath*, a mixture used to cool or freeze something. One example is the mixture of ice and NaCl(s) used to freeze ice cream in a home ice-cream maker. Because the slush bath is at a temperature well below 0 °C, it is easy to freeze the sugar-and-milk mixture that makes up the ice cream. NaCl is also useful for deicing roads. It is effective in melting ice at temperatures as low as −21 °C (−6 °F). This is the lowest freezing point of a NaCl(aq) solution.

🔍 **13-8   CONCEPT ASSESSMENT**

Why do you suppose that the freezing point of NaCl(aq) is depressed no further than −21 °C, regardless of how much more NaCl(s) is added to water?

▲ Lowering the freezing point of water on roads.

13-9   Solutions of Electrolytes

The discussion of the electrical conductivities of solutions in Section 5-1 retraced some of the work done by the Swedish chemist Svante Arrhenius for his doctoral dissertation (1883). Prevailing opinion at the time was that ions form only with the passage of electric current. Arrhenius, however, reached the conclusion that ions exist in a solid substance and become dissociated from each other when the solid dissolves in water. Such is the case with NaCl, for example. In other cases, as with HCl, ions do not exist in the substance but are formed when it dissolves in water. In any case, electricity is not required to produce ions.

Although Arrhenius developed his theory of electrolytic dissociation to explain the electrical conductivities of solutions, he was able to apply it more widely. One of his first successes came in explaining certain anomalous values of colligative properties described by the Dutch chemist Jacobus van't Hoff (1852–1911).

Anomalous Colligative Properties

Certain solutes produce a greater effect on colligative properties than expected. For example, consider a 0.0100 m aqueous solution. The predicted freezing-point depression of this solution is

$$\Delta T_f = -K_f \times m = -1.86 \,°C \, m^{-1} \times 0.0100 \, m = -0.0186 \,°C$$

We expect the solution to have a freezing point of $-0.0186\ °C$. If the $0.0100\ m$ solution is $0.0100\ m$ urea, the measured freezing point is just about $-0.0186\ °C$. If the solution is $0.0100\ m$ NaCl, however, the measured freezing point is about $-0.0361\ °C$.

Van't Hoff defined the factor i as the ratio of the measured value of a colligative property to the expected value if the solute is an electrolyte. For $0.0100\ m$ NaCl,

$$i = \frac{\text{measured } \Delta T_f}{\text{expected } \Delta T_f} = \frac{-0.0361\ °C}{-1.86\ °C\ m^{-1} \times 0.0100\ m} = 1.94$$

◀ Later in this section, we explain why the experimentally determined i for $0.0100\ m$ NaCl is 1.94 instead of 2.

13-2 ARE YOU WONDERING...

If there is a molecular interpretation of the freezing-point depression of a solvent by a solute?

When the solid and liquid phases of a pure substance coexist at a particular temperature, two processes occur. First, liquid molecules that collide with the solid sometimes are captured and added to the solid phase. At the same time, molecules on the surface of the solid sometimes become detached and enter the liquid phase. There is a state of dynamic equilibrium in which, at any given time, the number of molecules leaving the surface of the solid matches the number of molecules that enter the solid from the liquid phase. There is no net change even though individual molecules continue to move back and forth between the phases.

Now, imagine adding a solute to the liquid coexisting with its solid phase. In the solution formed above the solid phase, solute molecules replace some of the solvent molecules, and as a consequence, a given volume of solution contains a smaller number of solvent molecules than does the same volume of pure solvent. The dynamic equilibrium between liquid and solid solvent that existed in the pure solvent is disrupted since fewer solvent molecules in the liquid can get to the surface of the solid in a given time. The rate at which molecules leave the liquid phase is reduced and is no longer equal to the rate at which molecules leave the pure solid phase. However, cooling the solution restores the dynamic equilibrium because it simultaneously reduces the number of molecules that have sufficient energy to break away from the surface of the solid and increases the number of molecules in the liquid phase with sufficiently low kinetic energy to be captured by the solid. The lowered equilibrium temperature corresponds to the freezing-point depression.

Although this kinetic-molecular interpretation of freezing-point depression is an appealing one, the phenomenon is better described by the thermodynamic concept of entropy, which we will introduce in Chapter 19.

▲ **Svante Arrhenius (1859–1927)**
At the time Arrhenius was awarded the Nobel Prize in chemistry (1903), his results were described thus: "Chemists would not recognize them as chemistry; nor physicists as physics. They have in fact built a bridge between the two." The field of physical chemistry had its origins in Arrhenius's work.

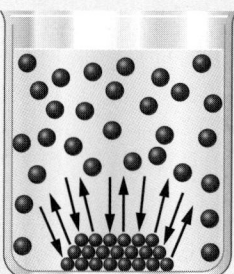

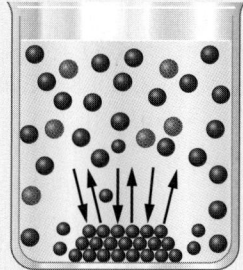

Equilibrium at freezing point of solvent   Equilibrium disrupted by solute added to solvent   Equilibrium restored at a lower temperature

▲ Molecular view of freezing-point depression. The addition of solute to the liquid solvent does not change the rate of escape of molecules from the solid phase, but it does decrease the rate at which solvent molecules can enter the solid phase. The dynamic equilibrium between solid and liquid solvent is disrupted, and can be re-established only at a lower temperature.

Arrhenius's theory of electrolytic dissociation allows us to explain different values of the van't Hoff factor i for different solutes. For such solutes as urea, glycerol, and sucrose (all nonelectrolytes), $i = 1$. For a strong electrolyte such as NaCl, which produces *two* moles of ions in solution per mole of solute dissolved, we would expect the effect on freezing-point depression to be twice as great as for a nonelectrolyte. We would expect that $i = 2$. Similarly, for $MgCl_2$, our expectation would be that $i = 3$. For the weak acid CH_3COOH (acetic acid), which is only slightly ionized in aqueous solution, we expect i to be slightly larger than one but not nearly equal to two.

This discussion suggests that equations (13.4), (13.5), and (13.6) should all be rewritten in the form

$$\pi = i \times M \times RT$$
$$\Delta T_f = -i \times K_f \times m$$
$$\Delta T_b = i \times K_b \times m$$

If these equations are used for nonelectrolytes, simply substitute $i = 1$. For strong electrolytes, predict a value of i as suggested in Example 13-11.

Interionic Attractions

Despite its initial successes, there were apparent deficiencies in Arrhenius's theory. The electrical conductivities of concentrated solutions of strong electrolytes are not as great as expected, and values of the van't Hoff factor i depend on the solution concentrations, as shown in Table 13.3. For strong

▶ Pure water does not conduct electricity. So why should we be careful with electricity when near water? It is not the water but the electrolytes that are dissolved in water.

EXAMPLE 13-11 Predicting Colligative Properties for Electrolyte Solutions

Predict the freezing point of aqueous 0.00145 m $MgCl_2$.

Analyze

We will use a modified freezing-point depression equation in which the van't Hoff factor i is included. We first note that $MgCl_2$ is a salt that completely dissociates when it is dissolved in water. So we determine the value of i for $MgCl_2$. We can do this by writing an equation to represent the dissociation of $MgCl_2(s)$. Then we use the appropriate freezing-point depression expression.

Solve

$$MgCl_2(s) \xrightarrow{H_2O} Mg^{2+}(aq) + 2\,Cl^-(aq)$$

Because *three* moles of ions are obtained per mole of formula units dissolved, we expect the value $i = 3$.
Now use the expression

$$\Delta T_f = -i \times K_f \times m$$
$$= -3 \times 1.86\,°C\,m^{-1} \times 0.00145\,m$$
$$= -0.0081\,°C$$

The predicted freezing point is $-0.0081\,°C$.

Assess

Because the value of i for $MgCl_2$ is not exactly 3, we are not justified in carrying more than one or two significant figures in our answer. If we had ignored the fact that $MgCl_2$ is a strong electrolyte, then our calculated freezing-point depression would have been three times as small as the experimental value. Always remember to include the van't Hoff factor when ionic compounds are given as part of the problem.

PRACTICE EXAMPLE A: What is the expected osmotic pressure of a 0.0530 M $MgCl_2$ solution at 25 °C?

PRACTICE EXAMPLE B: You want to prepare an aqueous solution that has a freezing point of -0.100 °C. How many milliliters of 12.0 M HCl would you use to prepare 250.0 mL of such a solution? [*Hint:* Note that in a dilute aqueous solution, molality and molarity are essentially numerically equal.]

TABLE 13.3 Variation of the van't Hoff Factor, i, with Solution Molality						
	Molality, m					
Solute	1.0	0.10	0.010	0.0010	$\cdots$	Inf. dil.[a]
NaCl	1.81	1.87	1.94	1.97	$\cdots$	2
MgSO$_4$	1.09	1.21	1.53	1.82	$\cdots$	2
Pb(NO$_3$)$_2$	1.31	2.13	2.63	2.89	$\cdots$	3

[a]The limiting values: i = 2, 2, and 3 are reached when the solution is infinitely dilute. Note that a solute whose ions are singly charged (for example, NaCl) approaches its limiting value more quickly than does a solute whose ions carry higher charges. Interionic attractions are greater in solutes with more highly charged ions.

electrolytes that exist completely in ionic form in aqueous solutions, we would expect i = 2 for NaCl, i = 3 for MgCl$_2$, and so on, regardless of the solution concentration.

These difficulties can be resolved with a theory of electrolyte solutions proposed by Peter Debye and Erich Hückel in 1923. This theory continues to view strong electrolytes as existing only in ionic form in aqueous solution, but the ions in solution *do not behave independently of one another*. Instead, each cation is surrounded by a cluster of ions in which anions predominate, and each anion is surrounded by a cluster in which cations predominate. In short, each ion is enveloped by an *ionic atmosphere* with a net charge opposite its own (Fig. 13-20).

In an electric field, the mobility of each ion is reduced because of the attraction or drag exerted by its ionic atmosphere. Similarly, the magnitudes of colligative properties are reduced. This explains why, for example, the value of i for 0.010 m NaCl is 1.94 rather than 2.00. What we can say is that each type of ion in an aqueous solution has two "concentrations." One is called the *stoichiometric concentration* and is based on the amount of solute dissolved. The other is the "effective" concentration, called the *activity*, which takes into account interionic attractions. Stoichiometric calculations of the type presented in Chapters 4 and 5 can be made with great accuracy using stoichiometric concentrations. However, no calculations involving solution properties are 100% accurate if stoichiometric concentrations are used. Activities are needed instead. The activity of a solution is related to its stoichiometric concentration through a factor called an *activity coefficient*. How much do we need to know about activities? First, it is worth noting that in sufficiently dilute solutions, activity coefficients approximate one: Activities and stoichiometric concentrations are the same. In doing calculations, we will work only with stoichiometric concentrations because a quantitative treatment of activity coefficients is beyond the scope of this text. Mainly, you should be aware of the distinction between activity and stoichiometric concentration because it helps to explain a number of chemical phenomena. We will discuss the importance of activities in more detail in Chapters 15 and 19.

(a)

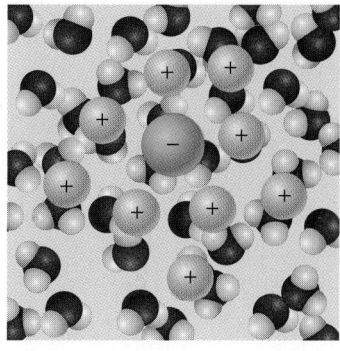

(b)

▲ FIGURE 13-20
Interionic attractions in aqueous solution
(a) A positive ion in aqueous solution is surrounded by a shell of negative ions.
(b) A negative ion attracts positive ions to its immediate surroundings.

13-10 Colloidal Mixtures

In a mixture of sand and water, the sand (silica, SiO$_2$) quickly settles to the bottom of the container. Yet mixtures can be prepared containing up to 40% by mass of SiO$_2$, and the silica may remain dispersed in the aqueous medium for many years. In these mixtures, the silica is *not* present as ions or molecules. Rather, much larger particles of silica are present, though they are still submicroscopic in size. The mixture is said to be a **colloid**. To be classified *colloidal*, a material must have one or more of its dimensions (length, width, or thickness)

◀ Colloids are extremely important to the food industry. Gelatins are colloids, and many foods are colloidal mixtures for which the food industry exerts considerable research effort to prevent their separation.

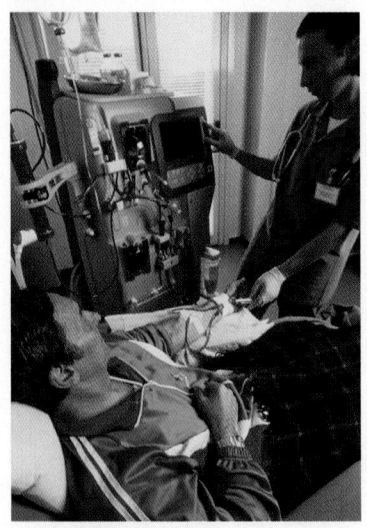

▲ **Hemodialysis**
An artificial kidney machine cleaning the blood of patients with impaired kidney function.

► FIGURE 13-21
The Tyndall effect
The flashlight beam is not visible as it passes through a true solution (left), but it is readily seen as it passes through the colloidal dispersion of Fe_2O_3 (right).

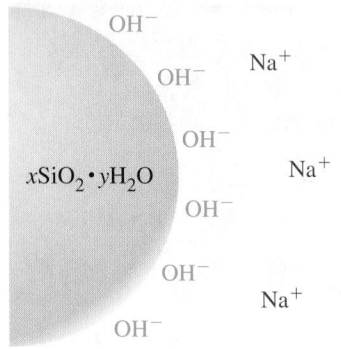

▲ FIGURE 13-22
Surface of SiO_2 particle in colloidal silica
The points made in this simplified drawing are: (1) The SiO_2 particles are hydrated; (2) OH^- ions are preferentially adsorbed on the surface; (3) In the immediate vicinity of the particle, negative ions outnumber positive ions and the particle carries a net negative charge. Not illustrated here are the facts that some of the negative charge comes from silicate anions (for example, SiO_3^{2-}), and as a whole, the solution in which these particles are found is electrically neutral.

in the approximate range of 1–1000 nm. If all the dimensions are smaller than 1 nm, the particles are of molecular size. If all the dimensions exceed 1000 nm, the particles are of ordinary, or macroscopic, size (even if they are visible only under a microscope). One method of determining whether a mixture is a true solution or a colloid is illustrated in Figure 13-21. When light passes through a true solution, an observer viewing from a direction perpendicular to the light beam sees no light. In a colloidal dispersion, light is scattered in many directions and is readily seen. This effect, first studied by John Tyndall in 1869, is known as the *Tyndall effect*. A common example is the scattering of light by dust particles in a flashlight beam.

The particles in colloidal silica have a spherical shape. Some colloidal particles are rod-shaped, and some, like gamma globulin in human blood plasma, have a disc-like shape. Thin films, like an oil slick on water, are colloidal. And some colloids, such as cellulose fibers, are randomly coiled filaments.

What keeps the SiO_2 particles suspended in colloidal silica? The most important factor is that the surfaces of the particles *adsorb*, or attach to themselves, ions from the solution, and they preferentially adsorb one type of ion over others. In the case of SiO_2, the preferred ions that are adsorbed are OH^- (see Figure 13-22). As a result, the particles acquire a net negative charge. Having like charges, the particles repel one another. These mutual repulsions overcome the force of gravity, and the particles remain suspended indefinitely.

Although electric charge can be important in stabilizing a colloid, a high concentration of ions can also bring about the *coagulation*, or precipitation, of a colloid (Fig. 13-23). The ions responsible for the coagulation are those carrying a charge opposite to that on the colloidal particles. *Dialysis*, a process similar to osmosis, can be used to remove excess ions from a colloidal mixture. As suggested by Figure 13-24, molecules of solvent and molecules or ions of solute pass through a semipermeable membrane, but the much larger colloidal particles do not. In some cases, the process is more effective when carried out in an electric field. In *electrodialysis*, ions are attracted out of a colloidal mixture by an electrode carrying the opposite charge. A human kidney dialyzes blood, a colloidal mixture, to remove excess electrolytes produced by metabolic processes. Certain diseases cause the kidneys to lose this ability, but a dialysis machine, external to the body, can function for the kidneys.

Table 13.4 lists some common colloids, and as so aptly put by Wilder Bancroft, an American pioneer in the field of colloid chemistry, ". . . colloid chemistry is essential to anyone who really wishes to understand . . . oils, greases, soaps, . . . glue, starch, adhesives, . . . paints, varnishes, lacquers, . . . cream, butter, cheese, . . . cooking, washing, dyeing, . . . colloid chemistry is the chemistry of life."

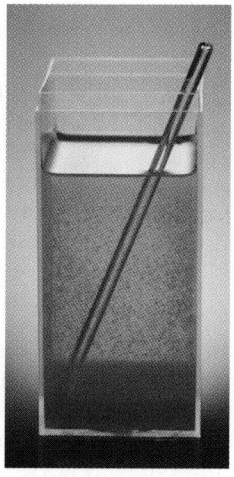

▲ FIGURE 13-23
Coagulation of colloidal iron oxide
On the left is red colloidal hydrous Fe_2O_3, obtained by adding $FeCl_3(aq)$ to boiling water. When a few drops of $Al_2(SO_4)_3(aq)$ are added, the suspended particles rapidly coagulate into a precipitate of $Fe_2O_3(s)$, as shown on the right.

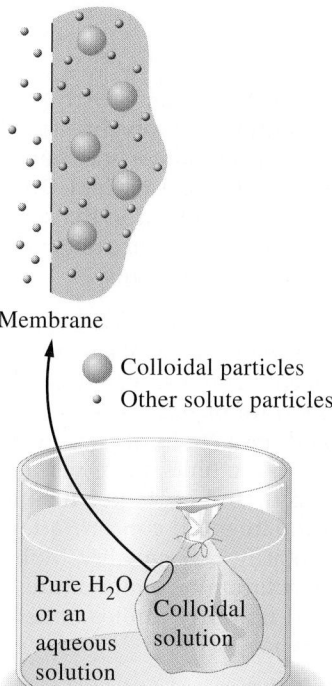

▲ FIGURE 13-24
The principles of dialysis
Water molecules, other solute molecules, and dissolved ions are all free to pass through the pores of the membrane (for example, a film of cellophane) in either direction. The direction of net flow of these species depends on their relative concentrations on either side of the membrane. Colloidal particles, however, cannot pass through the pores of the membrane.

TABLE 13.4 Some Common Types of Colloids

Dispersed Phase	Dispersion Medium	Type	Examples
Solid	Liquid	Sol	Clay sols,[a] colloidal silica, colloidal gold
Liquid	Liquid	Emulsion	Oil in water, milk, mayonnaise
Gas	Liquid	Foam	Soap and detergent suds, whipped cream, meringues
Solid	Gas	Aerosol[b]	Smoke, dust-laden air[c]
Liquid	Gas	Aerosol[b]	Fog, mist (as in aerosol products)
Solid	Solid	Solid sol	Ruby glass, certain natural and synthetic gems, blue rock salt, black diamond
Liquid	Solid	Solid emulsion	Opal, pearl
Gas	Solid	Solid foam	Pumice, lava, volcanic ash

[a]In water purification, it is sometimes necessary to precipitate clay particles or other suspended colloidal materials. This is often done by treating the water with an appropriate electrolyte. Clay sols are suspected of adsorbing organic substances, such as pesticides, and distributing them in the environment.
[b]Smogs are complex materials that are at least partly colloidal. The suspended particles are both solid and liquid. Other constituents of smog are molecular, for example, sulfur oxides, nitrogen oxides, and ozone.
[c]The bluish haze of tobacco smoke and the brilliant sunsets in desert regions are both attributable to the scattering of light by colloidal particles suspended in air.

Mastering**CHEMISTRY**   **www.masteringchemistry.com**

The separation of complex mixtures, such as gasoline, into their individual components can be accomplished using a technique known as chromatography. There are many variations of the technique, but they all take advantage of differences in the way molecules of different compounds interact with a common material. To find out more, go to the Focus On feature for Chapter 13, Chromatography, on the MasteringChemistry site.

Summary

13-1 Types of Solutions: Some Terminology—In a solution, the **solvent**—usually the component present in greatest amount—determines the state of matter in which the solution exists (Table 13.1). **A solute** is a solution component dissolved in the solvent. Dilute solutions contain relatively small amounts of solute and concentrated solutions, large amounts.

13-2 Solution Concentration—Any description of the composition of a solution must indicate the quantities of solute and solvent (or solution) present. Solution concentrations expressed as mass percent, volume percent, and mass/volume percent all have practical importance, as do the units, parts per million **(ppm)**, parts per billion **(ppb)**, and parts per trillion **(ppt)**. However, the more fundamental concentration units are mole fraction, molarity, and molality. Molarity (moles of solute per liter of solution) is temperature dependent, but mole fraction and **molality** (moles of solute per kilogram of solvent) are not.

13-3 Intermolecular Forces and the Solution Process—Predictions about whether two substances will mix to form a solution involve knowledge of intermolecular forces between like and unlike molecules (Figs. 13-2 and 13-3). This approach makes it possible to identify an **ideal solution**, one whose properties can be predicted from properties of the individual solution components. Most solutions are nonideal.

13-4 Solution Formation and Equilibrium—Generally, a solvent has a limited ability to dissolve a solute. A solution containing the maximum amount of solute possible is a **saturated solution**. A solution with less than this maximum amount is an **unsaturated solution**. Under certain conditions a solution can be prepared that contains more solute than a normal saturated solution; such solutions are called **supersaturated**. **Solubility** refers to the concentration of solute in a saturated solution and depends on temperature. Graphs of solute solubility versus temperature (Fig. 13-8) can be used to devise conditions for recovering a pure solute from a solution of several solutes through **fractional crystallization** **(recrystallization)** (Fig. 13-9).

13-5 Solubilities of Gases—The solubilities of gases depend on pressure as well as temperature, and many familiar phenomena are related to gas solubilities. **Henry's law** (equation 13.2) relates the concentration of a gas in solution to its pressure above the solution.

13-6 Vapor Pressures of Solutions—The vapor pressure of a solution depends on the vapor pressures of its pure components. If the solution is ideal, **Raoult's law** (equation 13.3) can be used to calculate the solution vapor pressure. Liquid–vapor equilibrium curves showing either solution vapor pressures (Fig. 13-12) or solution boiling points (Fig. 13-13) as a function of solution composition help us to visualize **fractional distillation**, a common method of separating the volatile components of a solution. Such curves also illustrate the formation of azeotropes in some nonideal solutions. **Azeotropes** are solutions that boil at a constant temperature and produce vapor of the same composition as the liquid; they have boiling points that in some cases are greater than the boiling points of the pure components and in some cases, less (Fig. 13-15).

13-7 Osmotic Pressure—**Osmosis** is the spontaneous flow of solvent through a semipermeable membrane separating two solutions of different concentration. The net flow is from the less to the more concentrated solution (Fig. 13-17). Osmotic flow can be stopped by applying a pressure, called the **osmotic pressure**, to the more concentrated solution. In **reverse osmosis**, the direction of flow is reversed by applying a pressure that exceeds the osmotic pressure to the more concentrated solution. Both osmosis and reverse osmosis have important practical applications. Osmotic pressure can be calculated with a simple relationship (equation 13.4) resembling the ideal gas equation. **Colligative properties** are certain properties that depend only on the concentration of solute particles in a solution, and not on the identity of the solute. Vapor-pressure lowering (Section 13-6) is one such property, and osmotic pressure is another.

13-8 Freezing-Point Depression and Boiling-Point Elevation of Nonelectrolyte Solutions—Freezing-point depression and boiling-point elevation (Fig. 13-19) are colligative properties having many familiar practical applications. For reasonably dilute solutions, their values are proportional to the molality of the solution (equations 13.5 and 13.6). The proportionality constants are K_f and K_b, respectively (Table 13.2). Historically, freezing-point depression was a common method for determining molar masses.

13-9 Solutions of Electrolytes—Calculating colligative properties of electrolyte solutions is more difficult than for solutions of nonelectrolytes. The solute particles in electrolyte solutions are ions or ions and molecules. Calculations using equations (13.5) and (13.6) must be based on the total number of particles present, and the van't Hoff factor is introduced into these equations to reflect this number. In all but the most dilute solutions, composition must be in terms of activities—effective concentrations that take into account interionic forces.

13-10 Colloidal Mixtures—**Colloids** are an important intermediate state between a true solution and a heterogeneous mixture. Colloidal mixtures are responsible for some unusual phenomena (Fig. 13.21) and are encountered in a broad range of contexts, from fluids in living organisms to pollutants in large air masses (Table 13-4).

Integrative Example

A 50.00 g sample of a solution of naphthalene [$C_{10}H_8(s)$] in benzene [$C_6H_6(l)$] has a freezing point of 4.45 °C. Calculate the mass percent $C_{10}H_8$ and the boiling point of this solution.

Analyze

Equation (13.5) relates freezing-point depression to the molality of a solution, but we will have to devise an algebraic method that yields the mass of each solution component and consequently the mass percent composition. The boiling point of the solution can be determined with a minimum of calculation through equation (13.6).

Solve

Substitute data for benzene from Table 13.2 ($K_f = 5.12$ °C m^{-1}, fp = 5.53 °C) and the measured freezing point of the solution into equation (13.5), to obtain

$$\Delta T_f = -K_f \times m$$
$$(4.45 - 5.53) \text{ °C} = -5.12 \text{ kg mol}^{-1} \text{ °C} \times m$$
$$1.08 = 5.12 \text{ kg mol}^{-1} \times m$$

Express the molality of the solution in terms of the masses of naphthalene, x, and benzene, y, and the molar mass, 128.2 g $C_{10}H_8$/mol.

$$m = \frac{\dfrac{x \text{ g } C_{10}H_8}{128.2 \text{ g } C_{10}H_8/\text{mol}}}{y \text{ g } C_6H_6 \times 1 \text{ kg}/1000 \text{ g } C_6H_6}$$

Because the total mass is 50.00 g, $x + y = 50.00$ and the expression above reduces to

$$m = \frac{\text{moles } C_{10}H_8}{\text{kg } C_6H_6} = \frac{(x/128.2)}{(50.00 - x)/1000} \text{ mol kg}^{-1}$$

Now, substituting this expression of m into the expression derived from equation (13.5), we obtain

$$1.08 = 5.12 \times \frac{(x/128.2)}{(50.00 - x)/1000}$$

which we solve for x, the mass of naphthalene in grams.

$$\frac{1.08 \times 128.2}{5.12 \times 1000} = \frac{x}{50.00 - x}$$

$$0.0270 = \frac{x}{50.00 - x}$$

$$1.0270x = 0.0270 \times 50.00$$

$$x = 1.31$$

The mass percent naphthalene in the solution is

$$\%C_{10}H_8 = \frac{1.31 \text{ g } C_{10}H_8}{50.00 \text{ g soln}} \times 100\% = 2.62\%$$

A simple way to find the boiling point of the solution is to first solve equation (13.5) for the molality of the solution.

$$\text{molality} = \frac{\Delta T_f}{-K_f} = \frac{-1.08 \text{ °C}}{-5.12 \text{ °C } m^{-1}} = 0.211 \text{ } m$$

Because the molality at the boiling point is the same as at the freezing point, substitute 0.211 m into equation (13.6) and solve for ΔT_b.

$$\Delta T_b = K_b \times m = 2.53 \text{ °C } m^{-1} \times 0.211 \text{ } m = 0.534 \text{ °C}$$

The boiling point of the solution is 0.534 °C higher than the normal boiling point of benzene (80.10 °C), that is,

$$80.10 \text{ °C} + 0.53 \text{ °C} = 80.63 \text{ °C}$$

Assess

The mass of solution can easily be determined to the nearest 0.01 g and expressed with four significant figures. The freezing point of the solution can also be established to the nearest 0.01 °C with good precision. However, the freezing point depression (−1.08 °C), the difference between two numbers of comparable magnitude, is valid only to about one part per hundred. Although significant-figure rules permit three significant figures in the remainder of the calculation, the actual precision of the calculated quantities is still only about one part per hundred. The final estimate of the boiling point (80.63 °C) seems reasonably good since it required expressing ΔT_b only to two significant figures. We assume that the 0.211 m solution is dilute enough and close enough to ideal in behavior to make equations (13.5) and (13.6) applicable.

PRACTICE EXAMPLE A: Water and phenol are completely miscible at temperatures above 66.8 °C but only partially miscible at temperatures below 66.8 °C. In a mixture prepared at 29.6 °C from 50.0 g water and 50.0 g phenol, 32.8 g of a phase consisting of 92.50% water and 7.50% phenol by mass is obtained. This is a saturated solution of phenol in water. What is the mass percent of water in the second phase—a saturated solution of water in phenol? What is the mole fraction of phenol in the mixture at temperatures above 66.8 °C?

PRACTICE EXAMPLE B: At a constant temperature of 25.00 °C, a current of dry air was passed through pure water and then through a drying tube, D_1, followed by passage through 1.00 m sucrose and another drying tube, D_2. After the experiment, D_1 had gained 11.7458 g in mass and D_2 had gained 11.5057 g. Given that the vapor pressure of water is 23.76 mmHg at 25.00 °C, **(a)** what was the vapor pressure lowering in the 1.00 m sucrose, and **(b)** what was the expected lowering?

Mastering**CHEMISTRY**

You'll find a link to additional self study questions in the study area on www.masteringchemistry.com

Exercises

Homogeneous and Heterogeneous Mixtures

1. Which of the following do you expect to be most water soluble, and why? $C_{10}H_8(s)$, $NH_2OH(s)$, $C_6H_6(l)$, $CaCO_3(s)$.

2. Which of the following is moderately soluble both in water and in benzene [$C_6H_6(l)$], and why? **(a)** 1-butanol, $CH_3(CH_2)_2CH_2OH$; **(b)** naphthalene, $C_{10}H_8$; **(c)** hexane, C_6H_{14}; **(d)** NaCl(s).

3. Substances that dissolve in water generally do not dissolve in benzene. Some substances are moderately soluble in both solvents, however. One of the following is such a substance. Which do you think it is and why?

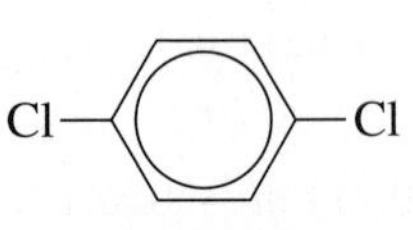

(a) *para*-Dichlorobenzene
(a moth repellent)

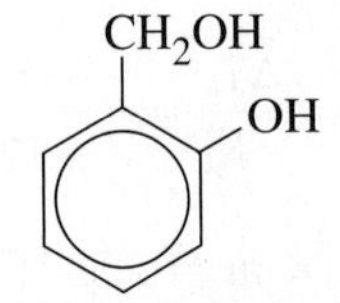

(b) Salicyl alcohol
(a local anesthetic)

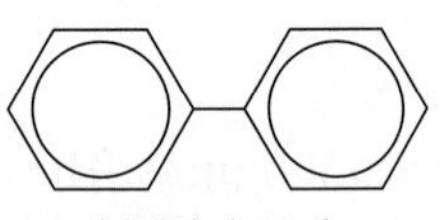

(c) Diphenyl
(a heat transfer agent)

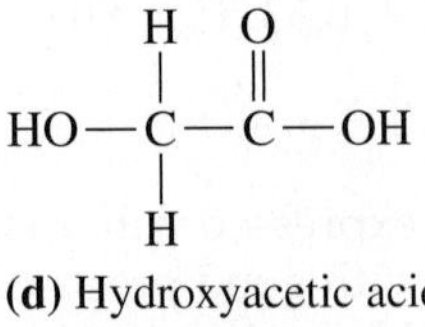

(d) Hydroxyacetic acid
(used in textile dyeing)

4. Some vitamins are water soluble and some are fat soluble. (Fats are substances whose molecules have long hydrocarbon chains.) The structural formulas of two vitamins are shown here—one is water soluble and one is fat soluble. Identify which is which, and explain your reasoning.

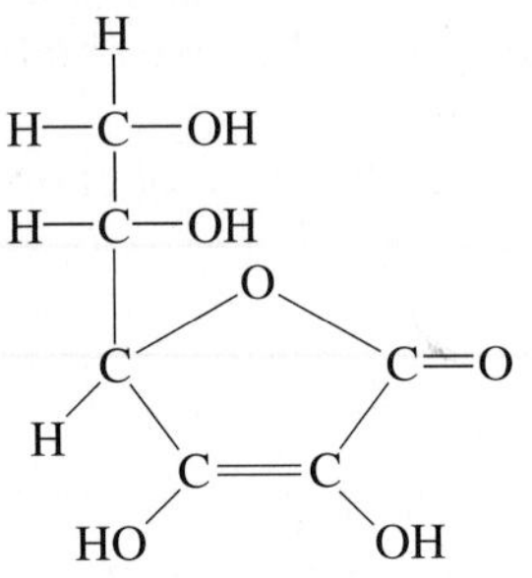

Vitamin C

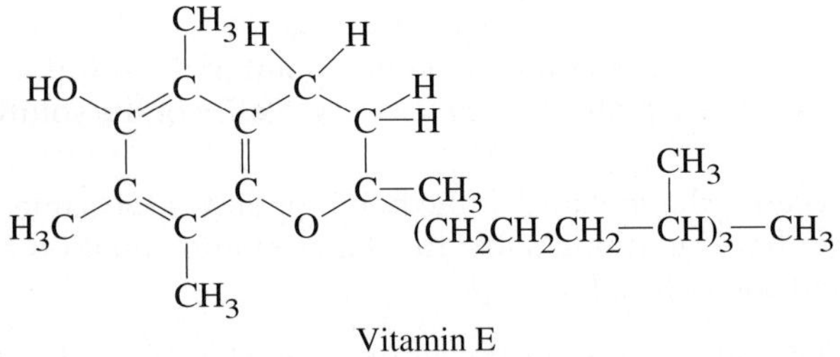

Vitamin E

5. Two of the substances listed here are highly soluble in water, two are only slightly soluble in water, and two are insoluble in water. Indicate the situation you expect for each one.
(a) iodoform, CHI_3

(b) benzoic acid,

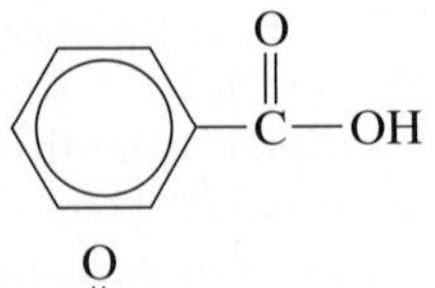

(c) formic acid, $H-\overset{O}{\overset{\|}{C}}-OH$
(d) 1-butanol, $CH_3CH_2CH_2CH_2OH$

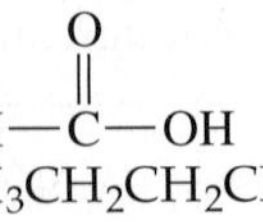

(e) chlorobenzene,

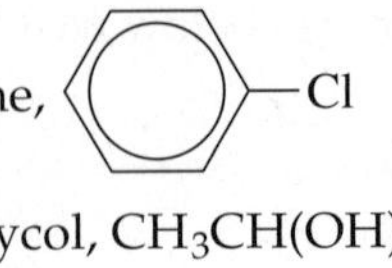

(f) propylene glycol, $CH_3CH(OH)CH_2OH$

6. Benzoic acid, C_6H_5COOH, is much more soluble in $NaOH(aq)$ than it is in pure water. Can you suggest a reason for this? The structural formula for benzoic acid is given in Exercise 5(b).

7. In light of the factors outlined on pages 566–567, which of the following ionic fluorides would you expect to be most water soluble on a moles-per-liter basis: MgF_2, NaF, KF, CaF_2? Explain your reasoning.

8. Explain the observation that all metal nitrates are water soluble but many metal sulfides are not. Among metal sulfides, which would you expect to be most soluble?

Percent Concentration

9. A saturated aqueous solution of NaBr at 20 °C contains 116 g $NaBr/100$ g H_2O. Express this composition in the more conventional percent by mass, that is, as grams of NaBr per 100 grams of solution.

10. An aqueous solution with density 0.988 g/mL at 20 °C is prepared by dissolving 12.8 mL $CH_3CH_2CH_2OH$ ($d = 0.803$ g/mL) in enough water to produce 75.0 mL of solution. What is the percent $CH_3CH_2CH_2OH$ expressed as (a) percent by volume; (b) percent by mass; (c) percent (mass/volume)?

11. A certain brine has 3.87% NaCl by mass. A 75.0 mL sample weighs 76.9 g. How many liters of this solution should be evaporated to dryness to obtain 725 kg NaCl?

12. You are asked to prepare 125.0 mL of 0.0321 M $AgNO_3$. How many grams would you need of a sample known to be 99.81% $AgNO_3$ by mass?

13. According to Example 13-1, the mass percent ethanol in a particular aqueous solution is less than the volume percent in the same solution. Explain why this is

also true for *all* aqueous solutions of ethanol. Would it be true of all ethanol solutions, regardless of the other component? Explain.

14. Blood cholesterol levels are generally expressed as milligrams of cholesterol per deciliter of blood. What is the approximate mass percent cholesterol in a blood sample having a cholesterol level of 176? Why can you not give a more precise answer?

15. A certain vinegar is 6.02% acetic acid (CH_3COOH) by mass. How many grams of CH_3COOH are contained in a 355 mL bottle of vinegar? Assume a density of 1.01 g/mL.

16. 6.00 M sulfuric acid, $H_2SO_4(aq)$, has a density of 1.338 g/mL. What is the percent by mass of sulfuric acid in this solution?

17. The sulfate ion level in a municipal water supply is given as 46.1 ppm. What is $[SO_4^{2-}]$ in this water?

18. A water sample is found to have 9.4 ppb of chloroform, $CHCl_3$. How many grams of $CHCl_3$ would be found in a glassful (250 mL) of this water?

Molarity

19. An aqueous solution is 6.00% methanol (CH_3OH) by mass, with $d = 0.988$ g/mL. What is the molarity of CH_3OH in this solution?

20. A typical commercial grade aqueous phosphoric acid is 75% H_3PO_4 by mass and has a density of 1.57 g/mL. What is the molarity of H_3PO_4 in this solution?

21. How many milliliters of the ethanol–water solution described in Example 13-1 should be diluted with water to produce 825 mL of 0.235 M CH_3CH_2OH?

22. A 30.00%-by-mass solution of nitric acid, HNO_3, in water has a density of 1.18 g/cm^3 at 20 °C. What is the molarity of HNO_3 in this solution?

23. What is the molarity of CO_2 in a liter of ocean water at 25 °C that contains approximately 280 ppm of CO_2? The density of ocean water is 1027 kg/m^3.

24. At 25 °C and 0% salinity the amount of oxygen in the ocean is 5.77 mL/L. What is the molarity of oxygen in the ocean at these conditions?

Molality

25. What is the molality of *para*-dichlorobenzene in a solution prepared by dissolving 2.65 g $C_6H_4Cl_2$ in 50.0 mL benzene ($d = 0.879$ g/mL)?

26. What is the molality of the sulfuric acid solution described in Exercise 16?

27. How many grams of iodine, I_2, must be dissolved in 725 mL of carbon disulfide, CS_2 ($d = 1.261$ g/mL), to produce a 0.236 *m* solution?

28. How many grams of water would you add to 1.00 kg of 1.38 *m* $CH_3OH(aq)$ to reduce the molality to 1.00 *m* CH_3OH?

29. An aqueous solution is 34.0% H_3PO_4 by mass and has a density of 1.209 g/mL. What are the molarity and molality of this solution?

30. A 10.00%-by-mass solution of ethanol, CH_3CH_2OH, in water has a density of 0.9831 g/mL at 15 °C and 0.9804 g/mL at 25 °C. Calculate the molality of the ethanol–water solution at these two temperatures. Does the molality differ at the two temperatures (that is, at 15 and 25 °C)? Would you expect the molarities to differ? Explain.

Mole Fraction, Mole Percent

31. A solution is prepared by mixing 1.28 mol C_7H_{16}, 2.92 mol C_8H_{18}, and 2.64 mol C_9H_{20}. What is the (a) mole fraction and (b) mole percent of each component of the solution?

32. Calculate the mole fraction of solute in the following aqueous solutions: (a) 21.7% CH_3CH_2OH, by mass; (b) 0.684 *m* $CO(NH_2)_2$ (urea).

33. Calculate the mole fraction of the solute in the following aqueous solutions: **(a)** 0.112 M $C_6H_{12}O_6$ (d = 1.006 g/mL); **(b)** 3.20% ethanol, by volume (d = 0.993 g/mL; pure CH_3CH_2OH, d = 0.789 g/mL).

34. Refer to Example 13-1. How many grams of CH_3CH_2OH must be added to 100.0 mL of the solution described in part (d) to increase the mole fraction of CH_3CH_2OH to 0.0525?

35. What volume of glycerol, $CH_3CH(OH)CH_2OH$ (d = 1.26 g/mL), must be added per kilogram of water to produce a solution with 4.85 mol % glycerol?

36. Two aqueous solutions of sucrose, $C_{12}H_{22}O_{11}$, are mixed. One solution is 0.1487 M $C_{12}H_{22}O_{11}$ and has

d = 1.018 g/mL; the other is 10.00% $C_{12}H_{22}O_{11}$ by mass and has d = 1.038 g/mL. Calculate the mole percent $C_{12}H_{22}O_{11}$ in the mixed solution.

37. The Environmental Protection Agency has a limit of 15 ppm for the amount of lead in drinking water. If a 1.000 mL sample of water at 20 °C contains 15 ppm of lead, how many lead ions are there in this sample of water? What is the mole fraction of lead ion in solution?

38. The amount of CO_2 in the ocean is approximately 280 ppm. What is the mole fraction of CO_2 in a liter of ocean water?

Solubility Equilibrium

39. Refer to Figure 13-8 and determine the molality of NH_4Cl in a saturated aqueous solution at 40 °C.

40. Refer to Figure 13-8 and estimate the temperature at which a saturated aqueous solution of $KClO_4$ is 0.200 m.

41. A solution of 20.0 g $KClO_4$ in 500.0 g of water is brought to a temperature of 40 °C.
    (a) Refer to Figure 13-8 and determine whether the solution is unsaturated or supersaturated at 40 °C.
    (b) Approximately what mass of $KClO_4$, in grams, must be added to saturate the solution (if originally

unsaturated), or what mass of $KClO_4$ can be crystallized (if originally supersaturated)?

42. One way to recrystallize a solute from a solution is to change the temperature. Another way is to evaporate solvent from the solution. A 335 g sample of a saturated solution of $KNO_3(s)$ in water is prepared at 25.0 °C. If 55 g H_2O is evaporated from the solution at the same time as the temperature is reduced from 25.0 to 0.0 °C, what mass of $KNO_3(s)$ will recrystallize? (Refer to Figure 13-8.)

Solubility of Gases

43. Under an $O_2(g)$ pressure of 1.00 atm, 28.31 mL of $O_2(g)$ dissolves in 1.00 L H_2O at 25 °C. What will be the molarity of O_2 in the saturated solution at 25 °C when the O_2 pressure is 3.86 atm? (Assume that the solution volume remains at 1.00 L.)

44. Using data from Exercise 43, determine the molarity of O_2 in an aqueous solution at equilibrium with air at normal atmospheric pressure. The volume percent of O_2 in air is 20.95%. [*Hint:* Recall equation (6.17).]

45. Natural gas consists of about 90% methane, CH_4. Assume that the solubility of natural gas at 20 °C and 1 atm gas pressure is about the same as that of CH_4, 0.02 g/kg water. If a sample of natural gas under a pressure of 20 atm is kept in contact with 1.00×10^3 kg of water, what mass of natural gas will dissolve?

46. At 1.00 atm, the solubility of O_2 in water is 2.18×10^{-3} M at 0 °C and 1.26×10^{-3} M at 25 °C. What volume of $O_2(g)$, measured at 25 °C and 1.00 atm, is expelled when 515 mL of water saturated with O_2 is heated from 0 to 25 °C?

47. The aqueous solubility at 20 °C of Ar at 1.00 atm is equivalent to 33.7 mL Ar(g), measured at STP, per liter

of water. What is the molarity of Ar in water that is saturated with air at 1.00 atm and 20 °C? Air contains 0.934% Ar by volume. Assume that the volume of water does not change when it becomes saturated with air.

48. The aqueous solubility of CO_2 at 20 °C and 1.00 atm is equivalent to 87.8 mL $CO_2(g)$, measured at STP, per 100 mL of water. What is the molarity of CO_2 in water that is at 20 °C and saturated with air at 1.00 atm? The volume percent of CO_2 in air is 0.0360%. Assume that the volume of the water does not change when it becomes saturated with air.

49. Henry's law can be stated this way: The mass of a gas dissolved by a given quantity of solvent at a fixed temperature is directly proportional to the pressure of the gas. Show how this statement is related to equation (13.2).

50. Another statement of Henry's law is: At a fixed temperature, a given quantity of liquid dissolves the same volume of gas at all pressures. What is the connection between this statement and the one given in Exercise 49? Under what conditions is this second statement not valid?

Raoult's Law and Liquid–Vapor Equilibrium

51. What are the partial and total vapor pressures of a solution obtained by mixing 35.8 g benzene, C_6H_6, and 56.7 g toluene, $C_6H_5CH_3$, at 25 °C? At 25 °C the vapor pressure of C_6H_6 = 95.1 mmHg; the vapor pressure of $C_6H_5CH_3$ = 28.4 mmHg.

52. Determine the composition of the vapor above the benzene-toluene solution described in Exercise 51.

53. Calculate the vapor pressure at 25 °C of a solution containing 165 g of the *nonvolatile* solute, glucose, $C_6H_{12}O_6$, in 685 g H_2O. The vapor pressure of water at 25 °C is 23.8 mmHg.

54. Calculate the vapor pressure at 20 °C of a saturated solution of the *nonvolatile* solute, urea, $CO(NH_2)_2$, in methanol, CH_3OH. The solubility is 17 g urea/100 mL methanol. The density of methanol is 0.792 g/mL, and its vapor pressure at 20 °C is 95.7 mmHg.

55. Styrene, used in the manufacture of polystyrene plastics, is made by the extraction of hydrogen atoms from ethylbenzene. The product obtained contains about 38% styrene ($C_6H_5CH=CH_2$) and 62% ethylbenzene ($C_6H_5CH_2CH_3$), by mass. The mixture is separated by fractional distillation at 90 °C. Determine the composition of the vapor in equilibrium with this 38%–62% mixture at 90 °C. The vapor pressure of ethylbenzene is 182 mmHg and that of styrene is 134 mmHg.

56. Calculate $x_{C_6H_6}$ in a benzene–toluene liquid solution that is in equilibrium at 25 °C with a vapor phase that contains 62.0 mol % C_6H_6. (Use data from Exercise 51.)

57. A benzene–toluene solution with $x_{benz} = 0.300$ has a normal boiling point of 98.6 °C. The vapor pressure of pure toluene at 98.6 °C is 533 mmHg. What must be the vapor pressure of pure benzene at 98.6 °C? (Assume ideal solution behavior.)

58. The two NaCl(aq) solutions pictured are at the same temperature.

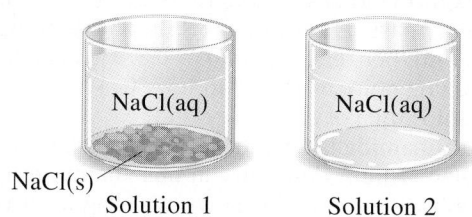

NaCl(s)   Solution 1   Solution 2

(a) Above which solution is the vapor pressure of water, P_{H_2O}, greater? Explain.
(b) Above one of these solutions, the vapor pressure of water, P_{H_2O}, remains *constant*, even as water evaporates from solution. Which solution is this? Explain.
(c) Which of these solutions has the higher boiling point? Explain.

Osmotic Pressure

59. A 0.72 g sample of polyvinyl chloride (PVC) is dissolved in 250.0 mL of a suitable solvent at 25 °C. The solution has an osmotic pressure of 1.67 mmHg. What is the molar mass of the PVC?

60. Verify that a 20% aqueous solution by mass of sucrose ($C_{12}H_{22}O_{11}$) would rise to a height of about 150 m in an apparatus of the type pictured in Figure 13-17.

61. When the stems of cut flowers are held in concentrated NaCl(aq), the flowers wilt. In a similar solution a fresh cucumber shrivels up (becomes pickled). Explain the basis of these phenomena.

62. Some fish live in saltwater environments and some in freshwater, but in either environment they need water to survive. Saltwater fish drink water, but freshwater fish do not. Explain this difference between the two types of fish.

63. In what volume of water must 1 mol of a nonelectrolyte be dissolved if the solution is to have an osmotic pressure of 1 atm at 273 K? Which of the gas laws does this result resemble?

64. The molecular mass of hemoglobin is 6.86×10^4 u. What mass of hemoglobin must be present per 100.0 mL of a solution to exert an osmotic pressure of 7.25 mmHg at 25 °C?

65. At 25 °C a 0.50 g sample of polyisobutylene (a polymer used in synthetic rubber) in 100.0 mL of benzene solution has an osmotic pressure that supports a 5.1 mm column of solution ($d = 0.88$ g/mL). What is the molar mass of the polyisobutylene? (For Hg, $d = 13.6$ g/mL.)

66. Use the concentration of an isotonic saline solution, 0.92% NaCl (mass/volume), to determine the osmotic pressure of blood at body temperature, 37.0 °C. [*Hint:* Assume that NaCl is completely dissociated in aqueous solutions.]

67. What approximate pressure is required in the reverse osmosis depicted in Figure 13-18 if the saltwater contains 2.5% NaCl, by mass? [*Hint:* Assume that NaCl is completely dissociated in aqueous solutions. Also, assume a temperature of 25 °C.]

68. The two solutions pictured here are separated by a semipermeable membrane that permits only the passage of water molecules. In what direction will a net flow of water occur, that is, from left to right or right to left? Glycerol is $HOCH_2CH(OH)CH_2OH$; sucrose is $C_{12}H_{22}O_{11}$.

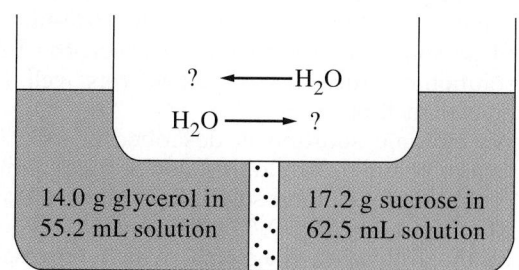

Freezing-Point Depression and Boiling-Point Elevation

69. 1.10 g of an unknown compound reduces the freezing point of 75.22 g benzene from 5.53 to 4.92 °C. What is the molar mass of the compound?

70. The freezing point of a 0.010 *m* aqueous solution of a nonvolatile solute is −0.072 °C. What would you expect the normal boiling point of this same solution to be?

71. Adding 1.00 g of benzene, C_6H_6, to 80.00 g cyclohexane, C_6H_{12}, lowers the freezing point of the cyclohexane from 6.5 to 3.3 °C.
(a) What is the value of K_f for cyclohexane?
(b) Which is the better solvent for molar mass determinations by freezing-point depression, benzene or cyclohexane? Explain.

72. The boiling point of water at 749.2 mmHg is 99.60 °C. What mass percent sucrose ($C_{12}H_{22}O_{11}$) should be present in an aqueous sucrose solution to raise the boiling point to 100.00 °C at this pressure?

73. A compound is 42.9% C, 2.4% H, 16.7% N, and 38.1% O, by mass. Addition of 6.45 g of this compound to 50.0 mL benzene, C_6H_6 (d = 0.879 g/mL), lowers the freezing point from 5.53 to 1.37 °C. What is the molecular formula of this compound?

74. Nicotinamide is a water-soluble vitamin important in metabolism. A deficiency in this vitamin results in the debilitating condition known as pellagra. Nicotinamide is 59.0% C, 5.0% H, 22.9% N, 13.1% O, by mass. Addition of 3.88 g of nicotinamide to 30.0 mL nitrobenzene, $C_6H_5NO_2$ (d = 1.204 g/mL), lowers the freezing point from 5.7 to −1.4 °C. What is the molecular formula of this compound?

75. Thiophene (fp = −38.3; bp = 84.4 °C) is a sulfur-containing hydrocarbon sometimes used as a solvent in place of benzene. Combustion of a 2.348 g sample of thiophene produces 4.913 g CO_2, 1.005 g H_2O, and 1.788 g SO_2. When a 0.867 g sample of thiophene is dissolved in 44.56 g of benzene (C_6H_6), the freezing point is lowered by 1.183 °C. What is the molecular formula of thiophene?

76. Coniferin is a glycoside (a derivative of a sugar) found in conifers, such as fir trees. When a 1.205 g sample of coniferin is subjected to combustion analysis (recall Section 3-3), the products are 0.698 g H_2O and 2.479 g CO_2. A 2.216 g sample is dissolved in 48.68 g H_2O, and the normal boiling point of this solution is found to be 100.068 °C. What is the molecular formula of coniferin?

77. Cooks often add some salt to water before boiling it. Some people say this helps the cooking process by raising the boiling point of the water. Others say not enough salt is usually added to make any noticeable difference. Approximately how many grams of NaCl must be added to a liter of water at 1 atm pressure to raise the boiling point by 2 °C? Is this a typical amount of salt that you might add to cooking water?

78. An important test for the purity of an organic compound is to measure its melting point. Usually, if the compound is not pure, it begins to melt at a *lower* temperature than the pure compound.
    (a) Why is this the case, rather than the melting point being higher in some cases and lower in others?
    (b) Are there any conditions under which the melting point of the impure compound is *higher* than that of the pure compound? Explain.

79. The freezing point of Arctic Ocean waters is about −1.94 °C. What is the molality of ions for a liter of ocean water?

80. If ocean water consisted of 3.5% salt, what would be the freezing point of an ocean?

Strong Electrolytes, Weak Electrolytes, and Nonelectrolytes

81. Predict the approximate freezing points of 0.10 m solutions of the following solutes dissolved in water:
    (a) $CO(NH_2)_2$ (urea); (b) NH_4NO_3; (c) HCl; (d) $CaCl_2$; (e) $MgSO_4$; (f) C_2H_5OH (ethanol); (g) $HC_2H_3O_2$ (acetic acid).

82. Calculate the van't Hoff factors of the following weak electrolyte solutions:
    (a) 0.050 m $HCHO_2$, which begins to freeze at −0.0986 °C;
    (b) 0.100 M HNO_2, which has a hydrogen ion (and nitrite ion) concentration of 6.91×10^{-3} M.

83. NH_3(aq) conducts electric current only weakly. The same is true for acetic acid, $HC_2H_3O_2$(aq). When these solutions are mixed, however, the resulting solution conducts electric current very well. Propose an explanation.

84. An isotonic solution is described as 0.92% NaCl (mass/volume). Would this also be the required concentration for isotonic solutions of other salts, such as KCl, $MgCl_2$, or $MgSO_4$? Explain.

85. In the following diagrams, which representation demonstrates a weak electrolyte?

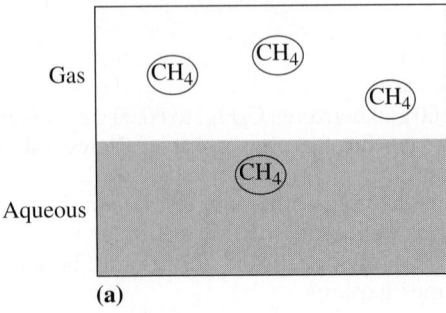

(a)

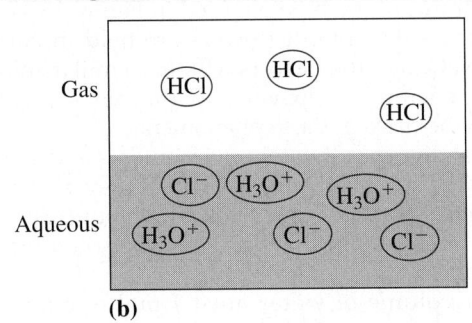

(b)

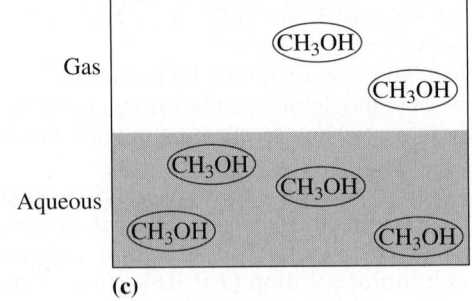

(c)

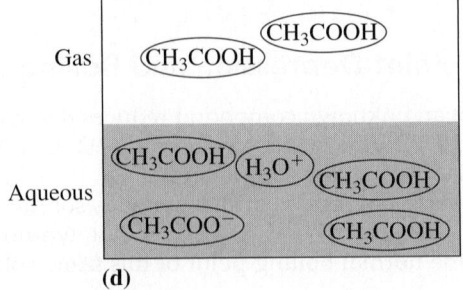

(d)

86. In the following diagrams, which representation demonstrates a strong electrolyte?

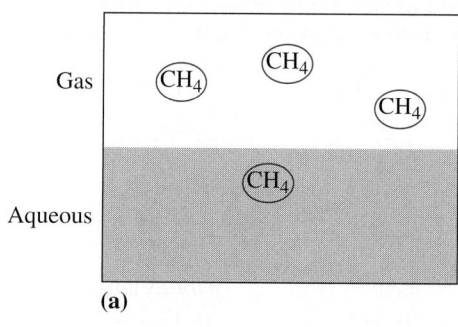

(a)

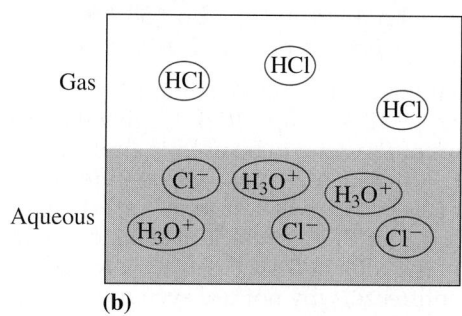

(b)

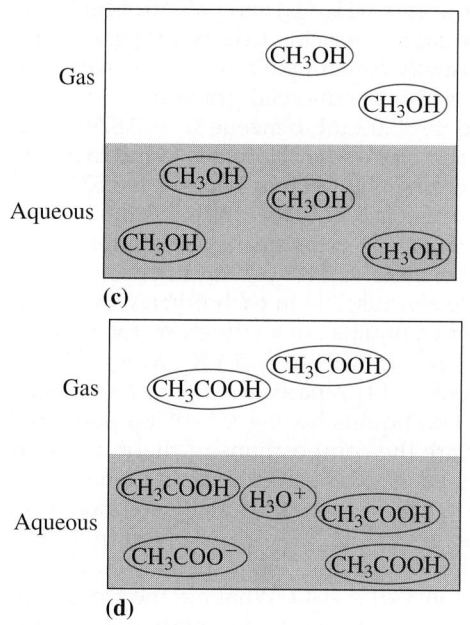

(c)

(d)

Integrative and Advanced Exercises

87. A typical root beer contains 0.13% of a 75% H_3PO_4 solution by mass. How many milligrams of phosphorus are contained in a 12 oz can of this root beer? Assume a solution density of 1.00 g/mL; also, 1 oz = 29.6 mL.

88. An aqueous solution has 109.2 g KOH/L solution. The solution density is 1.09 g/mL. Your task is to use 100.0 mL of this solution to prepare 0.250 m KOH. What mass of which component, KOH or H_2O, would you add to the 100.0 mL of solution?

89. The term "proof," still used to describe the ethanol content of alcoholic beverages, originated in seventeenth-century England. A sample of whiskey was poured on gunpowder and set afire. If the gunpowder ignited after the whiskey had burned off, this "proved" that the whiskey had not been watered down. The minimum ethanol content for a positive test was about 50%, by volume. The 50% ethanol solution became known as "100 proof." Thus, an 80-proof whiskey would be 40% CH_3CH_2OH by volume. Listed in the table below are some data for several aqueous solutions of ethanol. *With a minimum amount of calculation*, determine which of the solutions are more than 100 proof. Assume that the density of pure ethanol is 0.79 g/mL.

Molarity of Ethanol, M	Density of Solution, g/mL
4.00	0.970
5.00	0.963
6.00	0.955
7.00	0.947
8.00	0.936
9.00	0.926
10.00	0.913

90. Four aqueous solutions of acetone, CH_3COCH_3, are prepared at different concentrations: **(a)** 0.100% CH_3COCH_3, by mass; **(b)** 0.100 M CH_3COCH_3; **(c)** 0.100 m CH_3COCH_3; and **(d)** $\chi_{acetone} = 0.100$. Estimate the highest partial pressure of water at 25 °C to be found in the equilibrium vapor above these solutions. Also, estimate the lowest freezing point to be found among these solutions.

91. A solid mixture consists of 85.0% KNO_3 and 15.0% K_2SO_4, by mass. A 60.0 g sample of this solid is added to 130.0 g of water at 60 °C. Refer to Figure 13-8.
(a) Will all the solid dissolve at 60 °C?
(b) If the resulting solution is cooled to 0 °C, what mass of KNO_3 should crystallize?
(c) Will K_2SO_4 also crystallize at 0 °C?

92. Suppose you have available 2.50 L of a solution (d = 0.9767 g/mL) that is 13.8% ethanol (CH_3CH_2OH), by mass. From this solution you would like to make the *maximum* quantity of ethanol-water antifreeze solution that will offer protection to −2.0 °C. Would you add more ethanol or more water to the solution? What mass of liquid would you add?

93. Hydrogen chloride is a colorless gas, yet when a bottle of concentrated hydrochloric acid [HCl(conc aq)] is opened, mist-like fumes are often seen to escape from the bottle. How do you account for this?

94. Use the following information to confirm that the triple point temperature of water is about 0.0098 °C.
(a) The slope of the fusion curve of water in the region of the normal melting point of ice (Fig. 12-30) is −0.00750 °C/atm.
(b) The solubility of air in water at STP is 0.02918 mL of air per mL of water.

95. Stearic acid ($C_{18}H_{36}O_2$) and palmitic acid ($C_{16}H_{32}O_2$) are common fatty acids. Commercial grades of stearic acid usually contain palmitic acid as well. A 1.115 g sample of a commercial-grade stearic acid is dissolved in 50.00 mL benzene ($d = 0.879$ g/mL). The freezing point of the solution is found to be 5.072 °C. The freezing point of pure benzene is 5.533 °C, and K_f for benzene is 5.12 °C m^{-1}. What is the mass percent of palmitic acid in the stearic acid sample?

96. Nitrobenzene, $C_6H_5NO_2$, and benzene, C_6H_6, are completely miscible in each other. Other properties of the two liquids are *nitrobenzene:* fp = 5.7 °C, K_f = 8.1 °C m^{-1}; *benzene:* fp = 5.5 °C, K_f = 5.12 °C m^{-1}. It is possible to prepare *two different* solutions with these two liquids having a freezing point of 0.0 °C. What are the compositions of these two solutions, expressed as mass percent nitrobenzene?

97. Refer to Figure 13-16(a). Initially, solution A contains 0.515 g urea, $CO(NH_2)_2$, dissolved in 92.5 g H_2O; solution B contains 2.50 g sucrose, $C_{12}H_{22}O_{11}$, dissolved in 85.0 g H_2O. What are the compositions of the two solutions when equilibrium is reached, that is, when the two have the same vapor pressure?

98. In Figure 13-17, why does the net transfer of water stop when the two solutions are of *nearly* equal concentrations rather than of *exactly* equal concentrations?

99. Shown below is a typical cooling curve for an aqueous solution. Why is there no straight-line portion comparable to that seen in the cooling curve for pure water in Figure 12-23?

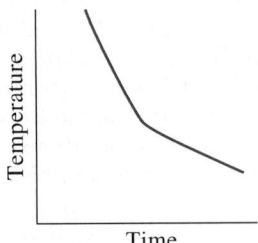

100. Suppose that 1.00 mg of gold is obtained in a colloidal dispersion in which the gold particles are spherical, with a radius of 1.00×10^2 nm. (The density of gold is 19.3 g/cm³.)
(a) What is the total surface area of the particles?
(b) What is the surface area of a single cube of gold of mass 1.00 mg?

101. At 20 °C, liquid benzene has a density of 0.879 g/cm³; liquid toluene, 0.867 g/cm³. Assume ideal solutions.
(a) Calculate the densities of solutions containing 20, 40, 60, and 80 volume percent benzene.
(b) Plot a graph of density versus volume percent composition.
(c) Write an equation that relates the density (d) to the volume percent benzene (V) in benzene-toluene solutions at 20 °C.

102. The two compounds whose structures are depicted here are isomers. When derived from petroleum, they always occur mixed together. *meta*-Xylene is used in aviation fuels and in the manufacture of dyes and insecticides. The principal use of *para*-xylene is in the manufacture of polyester resins and fibers (for example, Dacron). Comment on the effectiveness of fractional distillation as a method of separating these two xylenes. What other method(s) might be used to separate them?

meta-xylene
fp, 47.9 °C
bp, 139.1 °C
d, 0.864 g/mL

para-xylene
fp, 13.3 °C
bp, 138.4 °C
d, 0.861 g/mL

103. Instructions on a container of antifreeze (ethylene glycol; fp, −12.6 °C, bp, 197.3 °C) give the following volumes of Prestone to be used in protecting a 12 qt cooling system against freeze-up at different temperatures (the remaining liquid is water): 10 °F, 3 qt; 0 °F, 4 qt; −15 °F, 5 qt; −34 °F, 6 qt. Since the freezing point of the coolant is successively lowered by using more antifreeze, why not use even more than 6 qt of antifreeze (and proportionately less water) to ensure the maximum protection against freezing?

104. Demonstrate that
(a) for a *dilute aqueous* solution, the numerical value of the molality is essentially equal to that of the molarity.
(b) in a *dilute* solution, the solute mole fraction is proportional to the molality.
(c) in a *dilute aqueous* solution, the solute mole fraction is proportional to the molarity.

105. At 25 °C and under an $O_2(g)$ pressure of 1 atm, the solubility of $O_2(g)$ in water is 28.31 mL/1.00 L H_2O. At 25 °C and under an $N_2(g)$ pressure of 1 atm, the solubility of $N_2(g)$ in water is 14.34 mL/1.00 L H_2O. The composition of the atmosphere is 78.08% N_2 and 20.95% O_2, by volume. What is the composition of air dissolved in water expressed as volume percents of N_2 and O_2?

106. We noted in Figure 13-13 that the liquid and vapor curves taken together outline a lens-shaped region when the normal boiling points of benzene-toluene solutions are plotted as a function of mole fraction of benzene. That is, unlike Figure 13-12, the liquid curve is not a straight line. Use data from Figure 13-13 and show by calculation that this should be the case.

107. A saturated solution prepared at 70 °C contains 32.0 g $CuSO_4$ per 100.0 g solution. A 335 g sample of this solution is then cooled to 0 °C and $CuSO_4 \cdot 5 H_2O$ crystallizes out. If the concentration of a saturated solution at 0 °C is 12.5 g $CuSO_4$/100 g soln, what mass of $CuSO_4 \cdot 5 H_2O$ would be obtained? [*Hint:* Note that the solution composition is stated in terms of $CuSO_4$ and that the solid that crystallizes is the hydrate $CuSO_4 \cdot 5 H_2O$.]

108. The concentration of Ar in the ocean at 25 °C is 11.5 μM. The Henry's law constant for Ar is 1.5×10^{-3} mol L⁻¹ atm⁻¹. Calculate the mass of Ar in a liter of ocean water. Calculate the partial pressure of Ar in the atmosphere.

109. The concentration of N_2 in the ocean at 25 °C is 445 μM. The Henry's law constant for N_2 is 0.61×10^{-3} mol L^{-1} atm^{-1}. Calculate the mass of N_2 in a liter of ocean water. Calculate the partial pressure of N_2 in the atmosphere.

110. A solution contains 750 g of ethanol and 85.0 g of sucrose (180 g mol^{-1}). The volume of the solution is 810.0 mL. Determine
(a) the density of the solution
(b) the percent of sucrose in the solution
(c) the mole fraction of sucrose
(d) the molality of the solution
(e) the molarity of the solution

111. What volume of ethylene glycol (HOCH$_2$CH$_2$OH, density = 1.12 g mL^{-1}) must be added to 20.0 L of water (K_f = 1.86 °C/m) to produce a solution that freezes at -10 °C?

112. In the figure below, the open squares represent solvent molecules and the filled squares represent solute molecules.

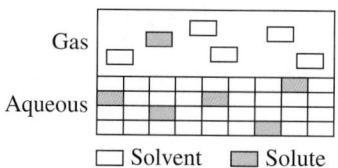

(a) What is the mole fraction of solute in the liquid phase?
(b) Which component has the higher vapor pressure?
(c) What is the percent solute in the vapor?

Feature Problems

113. Cinnamaldehyde is the chief constituent of cinnamon oil, which is obtained from the twigs and leaves of cinnamon trees grown in tropical regions. Cinnamon oil is used in the manufacture of food flavorings, perfumes, and cosmetics. The normal boiling point of cinnamaldehyde, C$_6$H$_5$CH=CHCHO, is 246.0 °C, but at this temperature it begins to decompose. As a result, cinnamaldehyde cannot be easily purified by ordinary distillation. A method that can be used instead is *steam distillation*. A heterogeneous mixture of cinnamaldehyde and water is heated until the sum of the vapor pressures of the two liquids is equal to barometric pressure. At this point, the temperature remains constant as the liquids vaporize. The mixed vapor condenses to produce two immiscible liquids; one liquid is essentially pure water and the other, pure cinnamaldehyde. The following vapor pressures of cinnamaldehyde are given: 1 mmHg at 76.1 °C; 5 mmHg at 105.8 °C; and 10 mmHg at 120.0 °C. Vapor pressures of water are given in Table 13.2.
(a) What is the approximate temperature at which the steam distillation occurs?
(b) The proportions of the two liquids condensed from the vapor is *independent* of the composition of the boiling mixture, as long as both liquids are present in the boiling mixture. Explain why this is so.
(c) Which of the two liquids, water or cinnamaldehyde, condenses in the greater quantity, by mass? Explain.

114. The phase diagram shown is for mixtures of HCl and H$_2$O at a pressure of 1 atm. The red curve represents the normal boiling points of solutions of HCl(aq) of various mole fractions. The blue curve represents the compositions of the vapors in equilibrium with boiling solutions.
(a) As a solution containing x_{HCl} = 0.50 begins to boil, will the vapor have a mole fraction of HCl equal to, less than, or greater than 0.50? Explain.
(b) In the boiling of a pure liquid, there is no change in composition. However, as a solution of HCl(aq)

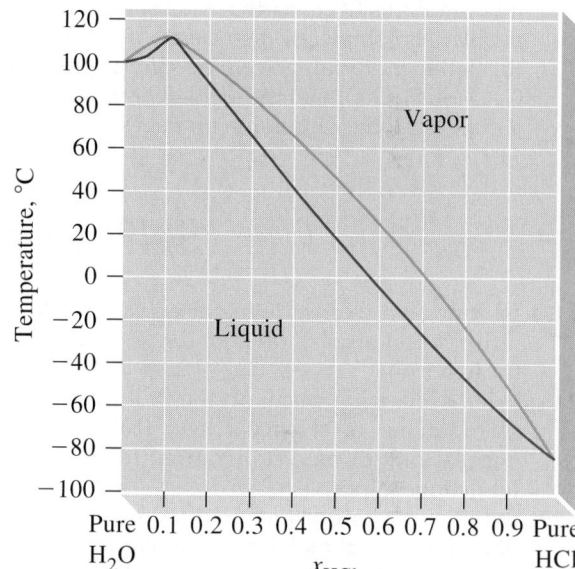

boils in an open container, the composition changes. Explain why this is so.
(c) One particular solution (the azeotrope) is an exception to the observation stated in part (b); that is, its composition remains *unchanged* during boiling. What are the approximate composition and boiling point of this solution?
(d) A 5.00 mL sample of the azeotrope (d = 1.099 g/mL) requires 30.32 mL of 1.006 M NaOH for its titration in an acid-base reaction. Use these data to determine a more precise value of the composition of the azeotrope, expressed as the mole fraction of HCl.

115. The laboratory device pictured on the following page is called a *desiccator*. It can be used to maintain a constant relative humidity within an enclosure. The material(s) used to control the relative humidity are placed in the bottom compartment, and the substance being subjected to a controlled relative humidity is placed on the platform in the container.

(a) If the material in the bottom compartment is a saturated solution of NaCl(aq) in contact with NaCl(s), what will be the approximate relative humidity in the container at a temperature of 20 °C? Obtain the solubility of NaCl from Figure 13-8 and vapor pressure data for water from Table 12.4; use the definition of Raoult's law from page 573 and that of relative humidity from Focus On 6: Earth's Atmosphere; assume that the NaCl is completely dissociated into its ions.
(b) If the material placed on the platform is dry $CaCl_2 \cdot 6\,H_2O(s)$, will the solid deliquesce? Explain. [*Hint:* Recall the discussion on pages 577–578.]

(c) To maintain $CaCl_2 \cdot 6\,H_2O(s)$ in the dry state in a desiccator, should the substance in the saturated solution in the bottom compartment be one with a high or a low water solubility? Explain.

116. Every year, oral rehydration therapy (ORT)—the feeding of an electrolyte solution—saves the lives of countless children worldwide who become severely dehydrated as a result of diarrhea. One requirement of the solution used is that it be *isotonic* with human blood.
(a) One definition of an isotonic solution given in the text is that it have the same osmotic pressure as 0.92% NaCl(aq) (mass/volume). Another definition is that the solution have a freezing point of -0.52 °C. Show that these two definitions are in reasonably close agreement given that we are using solution concentrations rather than activities.
(b) Use the freezing-point definition from part (a) to show that an ORT solution containing 3.5 g NaCl, 1.5 g KCl, 2.9 g $Na_3C_6H_5O_7$ (sodium citrate), and 20.0 g $C_6H_{12}O_6$ (glucose) per liter meets the requirement of being isotonic. [*Hint:* Which of the solutes are nonelectrolytes, and which are strong electrolytes?]

Self-Assessment Exercises

117. In your own words, define or explain the following terms or symbols: **(a)** x_B; **(b)** P_A°; **(c)** K_f; **(d)** i; **(e)** activity.
118. Briefly describe each of the following ideas or phenomena: **(a)** Henry's law; **(b)** freezing-point depression; **(c)** recrystallization; **(d)** hydrated ion; **(e)** deliquescence.
119. Explain the important distinctions between each pair of terms: **(a)** molality and molarity; **(b)** ideal and nonideal solution; **(c)** unsaturated and supersaturated solution; **(d)** fractional crystallization and fractional distillation; **(e)** osmosis and reverse osmosis.
120. An aqueous solution is 0.010 M CH_3OH. The concentration of this solution is also very nearly **(a)** 0.010% CH_3OH (mass/volume); **(b)** 0.010 *m* CH_3OH; **(c)** $x_{CH_3OH} = 0.010$; **(d)** 0.990 M H_2O.
121. The most likely of the following mixtures to be an ideal solution is **(a)** $NaCl$–H_2O; **(b)** CH_3CH_2OH–C_6H_6; **(c)** C_7H_{16}–H_2O; **(d)** C_7H_{16}–C_8H_{18}.
122. The solubility of a nonreactive gas in water increases with **(a)** an increase in gas pressure; **(b)** an increase in temperature; **(c)** increases in both temperature and pressure; **(d)** an increase in the volume of gas in equilibrium with the available water.
123. Of the following aqueous solutions, the one with the lowest freezing point is **(a)** 0.010 *m* $MgSO_4$; **(b)** 0.011 *m* NaCl; **(c)** 0.018 *m* CH_3CH_2OH; **(d)** 0.0080 *m* $MgCl_2$.
124. An ideal liquid solution has two volatile components. In the vapor in equilibrium with the solution, the mole fractions of the components are **(a)** both 0.50; **(b)** equal, but not necessarily 0.50; **(c)** not very likely to be equal; **(d)** 1.00 for the solvent and 0.00 for the solute.
125. A solution prepared by dissolving 1.12 mol NH_4Cl in 150.0 g H_2O is brought to a temperature of 30 °C. Use Figure 13-8 to determine whether the solution is unsaturated or whether excess solute will crystallize.
126. NaCl(aq) isotonic with blood is 0.92% NaCl (mass/volume). For this solution, what is **(a)** $[Na^+]$; **(b)** the total molarity of ions; **(c)** the osmotic pressure at 37 °C; **(d)** the approximate freezing point? (Assume that the solution has a density of 1.005 g/mL.)
127. A solution ($d = 1.159$ g/mL) is 62.0% glycerol, $HOCH_2CH(OH)CH_2OH$, and 38.0% H_2O, by mass. Determine **(a)** the molarity of glycerol with H_2O as the solvent; **(b)** the molarity of H_2O with glycerol as the solvent; **(c)** the molality of H_2O in glycerol; **(d)** the mole fraction of glycerol; **(e)** the mole percent of H_2O.
128. Which aqueous solution from the column on the right has the property listed on the left? Explain your choices.

Property	Solution
1. lowest electrical conductivity	**a.** 0.10 *m* KCl(aq)
2. lowest boiling point	**b.** 0.15 *m* $C_{12}H_{22}O_{11}$(aq)
3. highest vapor pressure of water at 25 °C	**c.** 0.10 *m* CH_3COOH(aq)
4. lowest freezing point	**d.** 0.05 *m* NaCl

129. Which of the following represents MgCl$_2$ in solution?

(a)

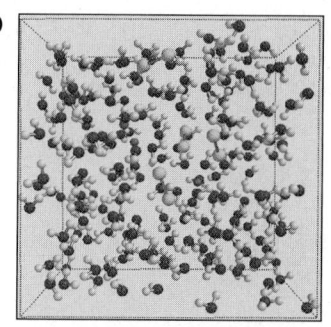

(b)

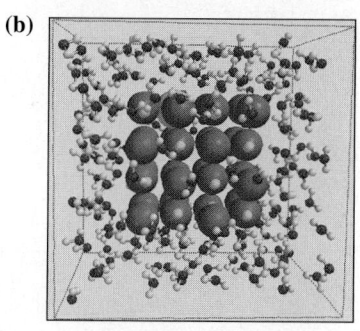

(c)

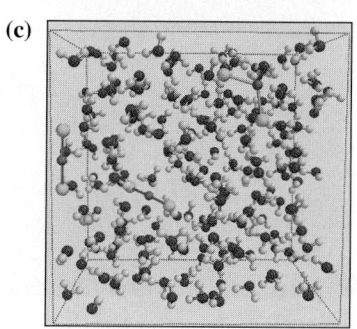

(d)

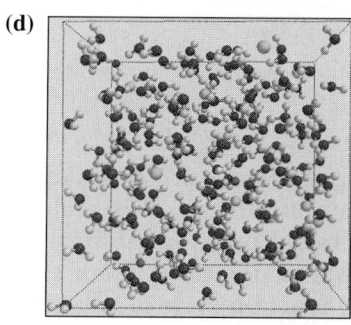

Legend: O ● H ● Cl ● Mg ●

130. Which of the following ions has the greater charge density? **(a)** Na$^+$; **(b)** F$^-$; **(c)** K$^+$; **(d)** Cl$^-$.

131. When NH$_4$Cl dissolves in a test tube of water, the test tube becomes colder. Is the magnitude of $\Delta H_{lattice}$ for NH$_4$Cl larger or smaller than the sum of $\Delta H_{hydration}$ of the ions?

132. In a saturated solution at 25 °C and 1 bar, for the following solutes, which condition will increase solubility? **(a)** Ar(g), decrease temperature; **(b)** NaCl(s), increase pressure; **(c)** N$_2$, decrease pressure; **(d)** CO$_2$, increase volume.

133. Which of the following represents a nonvolatile solute?

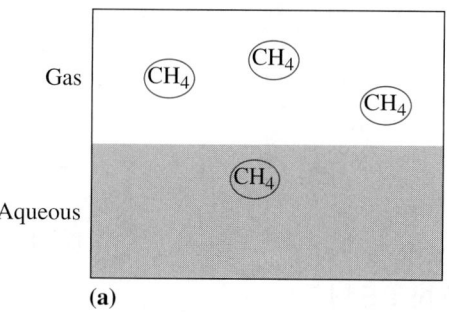

(a)

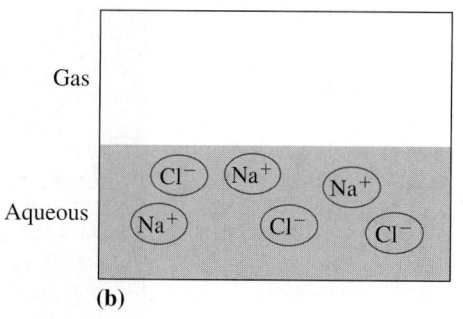

(b)

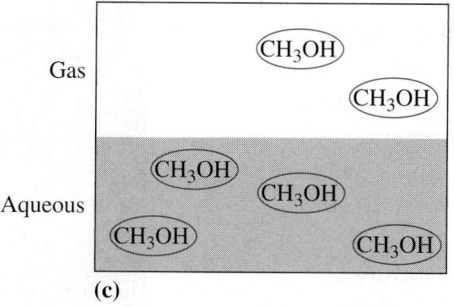

(c)

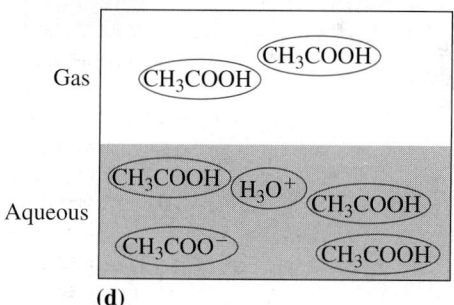

(d)

134. What is the mole fraction of a nonvolatile solute in a hexane solution that has a vapor pressure of 600 mmHg at 68.7 °C, hexane's normal boiling point? **(a)** 0.21; **(b)** 0.11; **(c)** 0.27; **(d)** 0.79.

135. 15.2 L of a 0.312 M starch solution displays which of the following osmotic pressures in bars at 75 °C? **(a)** 13721.1; **(b)** 194.5; **(c)** 355.7; **(d)** 0.0016; **(e)** 9.03.

136. What is the weight percent of 23.4 g of CaF$_2$ if dissolved in 10.5 mol of water? **(a)** 0.028; **(b)** 1.59; **(c)** 11.0; **(d)** 12.4; **(e)** none of these.

137. Using the method of concept mapping presented in Appendix E, construct a concept map showing the relationships among the various concentration units presented in Section 13-2.

138. Construct a concept map for the relationships that exist among the colligative properties of Sections 13-6, 13-7, 13-8, and 13-9.

14

Chemical Kinetics

Although stable at room temperature, ammonium dichormate decomposes very rapidly once ignited:

$$(NH_4)_2Cr_2O_7(s) \xrightarrow{\Delta} N_2(g) + 4H_2O(g) + Cr_2O_3(s)$$

The rates of chemical reactions and the effect of temperature on those rates are among several key concepts explored in this chapter.

Rocket fuel is designed to give a rapid release of gaseous products and energy to provide a rocket maximum thrust. Milk is stored in a refrigerator to slow down the chemical reactions that cause it to spoil. Current strategies to reduce the rate of deterioration of the ozone layer try to deprive the ozone-consuming reaction cycle of key intermediates that come from chlorofluorocarbons (CFCs). Catalysts are used to reduce the harmful emissions from internal combustion engines that contribute to smog. These examples illustrate the importance of the rates of chemical reactions. Moreover, how fast a reaction occurs depends on the reaction mechanism—the step-by-step molecular pathway leading from reactants to products. Thus, *chemical kinetics* concerns how rates of chemical reactions are measured, how they can be predicted, and how reaction-rate data are used to deduce probable reaction mechanisms.

The chapter begins with a discussion of the meaning of a rate of reaction and some ideas about measuring rates of reaction. This is followed by the introduction of mathematical equations, called rate laws, that relate the rates of reactions to the concentrations of the reactants. Finally, with this information as background, rate laws are related to plausible reaction mechanisms, which are a central focus of this chapter.

14-1 The Rate of a Chemical Reaction

Rate, or speed, refers to something that happens in a unit of time. A car traveling at 60 mph, for example, covers a distance of 60 miles in one hour. For chemical reactions, the rate of reaction describes how fast the concentration of a reactant or product changes with time.

To illustrate, let's consider the reaction that begins immediately after the ions Fe^{3+} and Sn^{2+} are simultaneously introduced into water.

$$2\,Fe^{3+}(aq) + Sn^{2+}(aq) \longrightarrow 2\,Fe^{2+}(aq) + Sn^{4+}(aq) \qquad (14.1)$$

Suppose that 38.5 s after the reaction starts, $[Fe^{2+}]$ is found to be 0.0010 M. During the period of time, $\Delta t = 38.5\,s$, the *change* in concentration of Fe^{2+}, which we can designate as $\Delta[Fe^{2+}]$, is $\Delta[Fe^{2+}] = 0.0010\,M - 0 = 0.0010\,M$. The *average* rate at which Fe^{2+} is formed in this interval is the change in concentration of Fe^{2+} divided by the change in time.

◀ Recall that the symbol [ ] means "concentration." Also, Δ means "the change in," that is, the final value minus the initial value.

$$\text{rate of formation of } Fe^{2+} = \frac{\Delta[Fe^{2+}]}{\Delta t} = \frac{0.0010\,M}{38.5\,s} = 2.6 \times 10^{-5}\,M\,s^{-1}$$

How has the concentration of Sn^{4+} changed during the 38.5 s we were monitoring the Fe^{2+}? Can you see that in 38.5 s, $\Delta[Sn^{4+}]$ will be 0.00050 M − 0 = 0.00050 M? Because only *one* Sn^{4+} ion is produced for every *two* Fe^{2+} ions, the buildup of $[Sn^{4+}]$ will be only one-half that of $[Fe^{2+}]$. Consequently the rate of formation of Sn^{4+} is

$$\text{rate of formation of } Sn^{4+} = \frac{0.00050\,M}{38.5\,s} = 1.3 \times 10^{-5}\,M\,s^{-1}$$

We can also follow the course of the reaction by monitoring the concentrations of the starting reactants. Thus, the amount of Fe^{3+} consumed is the same as the amount of Fe^{2+} produced. The *change* in concentration of Fe^{3+} is $\Delta[Fe^{3+}] = -0.0010\,M$. Thus,

$$\frac{\Delta[Fe^{3+}]}{\Delta t} = \frac{-0.0010\,M}{38.5\,s} = -2.6 \times 10^{-5}\,M\,s^{-1}$$

The quantity above is the average rate of change of change of $[Fe^{3+}]$ in this interval. It is a negative quantity because $[Fe^{3+}]$ decreases with time. The average rate of disappearance of Fe^{3+} is defined as follows.

$$\text{rate of disappearance of } Fe^{3+} = -\frac{\Delta[Fe^{3+}]}{\Delta t} = 2.6 \times 10^{-5}\,M\,s^{-1}$$

Why is a negative sign incorporated into the definition of rate in this case? It is because the term "rate of disappearance" implies that $[Fe^{3+}]$ decreases with time. When told the rate of disappearance of Fe^{3+} is $2.6 \times 10^{-5}\,M\,s^{-1}$, we know the rate of change of *concentration* must be $-2.6 \times 10^{-5}\,M\,s^{-1}$.

In the same way that we related the rate of formation of Sn^{4+} to that of Fe^{2+}, we can relate the rate of disappearance of Sn^{2+} to that of Fe^{3+}. That is, the rate of disappearance of Sn^{2+} is half that of Fe^{3+}, giving

$$\text{rate of disappearance of } Sn^{2+} = 1.3 \times 10^{-5}\,M\,s^{-1}$$

When we refer to the rate of reaction (14.1), which of the four quantities described here should we use? To avoid confusion in this matter, the International Union of Pure and Applied Chemistry (IUPAC) recommends that we use a *general* rate of reaction, which, for the hypothetical reaction represented by the balanced equation,

$$aA + bB \longrightarrow gG + hH$$

is

$$\text{rate of reaction} = -\frac{1}{a}\frac{\Delta[A]}{\Delta t} = -\frac{1}{b}\frac{\Delta[B]}{\Delta t} = \frac{1}{g}\frac{\Delta[G]}{\Delta t} = \frac{1}{h}\frac{\Delta[H]}{\Delta t} \quad \textbf{(14.2)}$$

In this expression, we take the negative value of $\Delta[X]/\Delta t$, when X refers to a reactant to ensure that the **rate of reaction** is a positive quantity. To obtain a single, positive quantity it is necessary to divide all rates by the appropriate stoichiometric coefficients. If we apply this expression to reaction (14.1), we obtain

$$\text{rate of reaction} = -\frac{1}{2}\frac{\Delta[Fe^{3+}]}{\Delta t} = -\frac{\Delta[Sn^{2+}]}{\Delta t}$$

$$= \frac{1}{2}\frac{\Delta[Fe^{2+}]}{\Delta t} = \frac{\Delta[Sn^{4+}]}{\Delta t} = 1.3 \times 10^{-5}\,M\,s^{-1}$$

EXAMPLE 14-1 **Expressing the Rate of a Reaction**

Suppose that at some point in the reaction

$$A + 3B \longrightarrow 2C + 2D$$

$[B] = 0.9986\,M$, and that 13.20 min later $[B] = 0.9746\,M$. What is the average rate of reaction during this time period, expressed in $M\,s^{-1}$?

Analyze

This is a straightforward application of the definition of reaction rate, expression (14.2). To formulate the rate, we use $\Delta[B] = 0.9746\,M - 0.9986\,M = -0.0240\,M$ and $\Delta t = 13.20$ min.

Solve

The solution is

$$\text{average rate of reaction} = -\frac{1}{3}\frac{\Delta[B]}{\Delta t} = -\frac{1}{3} \times \frac{-0.0240\,M}{13.20\,\text{min}} = 6.06 \times 10^{-4}\,M\,\text{min}$$

To express the rate of reaction in moles per liter per second, we must convert from min^{-1} to s^{-1}. We can do this with the conversion factor 1 min/60 s.

$$\text{rate of reaction} = 6.06 \times 10^{-4}\,M\,\text{min}^{-1} \times \frac{1\,\text{min}}{60\,s} = 1.01 \times 10^{-5}\,M\,s^{-1}$$

Alternatively, we could have converted 13.20 min to 792 s and used $\Delta t = 792$ s in evaluating the rate of reaction.

Assess

By monitoring changes in concentration over a time period, we can obtain the average rate of reaction. Remember that the rate of reaction can be defined in terms of any reactant or product.

PRACTICE EXAMPLE A: At some point in the reaction $2A + B \longrightarrow C + D$, $[A] = 0.3629\,M$. At a time 8.25 min later $[A] = 0.3187\,M$. What is the average rate of reaction during this time interval, expressed in $M\,s^{-1}$?

PRACTICE EXAMPLE B: In the reaction $2A \longrightarrow 3B$, $[A]$ drops from 0.5684 M to 0.5522 M in 2.50 min. What is the average rate of formation of B during this time interval, expressed in $M\,s^{-1}$?

🔍 14-1 CONCEPT ASSESSMENT

In the reaction of gaseous nitrogen and hydrogen to form gaseous ammonia, what are the relative rates of disappearance of the two reactants? How is the rate of formation of the product related to the rates of disappearance of the reactants?

14-2   Measuring Reaction Rates

To determine a rate of reaction, we need to measure changes in concentration over time. A change in time can be measured with a stopwatch or other timing device, but how do we measure concentration changes during a chemical reaction? Also, why is the *term average* used in referring to a rate of reaction? These are two of the questions that are answered in this section.

Following a Chemical Reaction

A 3% aqueous solution of hydrogen peroxide is a common antiseptic. Its antiseptic action results from the release of $O_2(g)$ as the H_2O_2 decomposes. $O_2(g)$ escapes from the $H_2O_2(aq)$, and ultimately the reaction goes to completion.*

$$H_2O_2(aq) \longrightarrow H_2O(l) + \frac{1}{2}O_2(g) \qquad \textbf{(14.3)}$$

We can follow the progress of the reaction by focusing either on the formation of $O_2(g)$ or on the disappearance of H_2O_2. For example, we can

- Measure the volumes of $O_2(g)$ produced at different times and relate these volumes to decreases in concentration of H_2O_2 (Fig. 14-1).
- Remove small samples of the reaction mixture from time to time, and analyze these samples for their H_2O_2 content. One way to do this is by titration with $KMnO_4$ in acidic solution. The net ionic equation for this oxidation–reduction reaction is

$$2\,MnO_4^-(aq) + 5\,H_2O_2(aq) + 6\,H^+(aq) \longrightarrow 2\,Mn^{2+}(aq) + 8\,H_2O(l) + 5\,O_2(g)$$
$$\textbf{(14.4)}$$

Table 14.1 lists typical data for the decomposition of H_2O_2, and Figure 14-2 displays comparable data graphically.

Rate of Reaction Expressed as Concentration Change over Time

Some data extracted from Figure 14-2 are listed in Table 14.2 (in blue). Column III lists the molarities of H_2O_2 at the times shown in column I. Column II lists the *arbitrary* time interval between data points—400 s. Column IV reports the concentration changes that occur for each 400 s interval. The rates of reaction, expressed as the rate of disappearance of H_2O_2, are shown in column V. The data show that the reaction rate is not constant—the lower the remaining concentration of H_2O_2, the more slowly the reaction proceeds.

Rate of Reaction Expressed as the Slope of a Tangent Line

When the rate of reaction is expressed as $-\Delta[H_2O_2]/\Delta t$, the result is an *average* value for the time interval Δt. For example, in the interval from 1200 to 1600 s, the rate averages $6.3 \times 10^{-4}\,M\,s^{-1}$ (fourth entry in column V of Table 14.2). This can be thought of as the reaction rate at about the middle of the interval—1400 s. We could just as well have chosen $\Delta t = 200$ s in Table 14.2. In this case, to obtain the rate of reaction at $t = 1400$ s, we would use concentration data at $t = 1300$ s and $t = 1500$ s. The rate of reaction would be slightly less than $6.3 \times 10^{-4}\,M\,s^{-1}$. The smaller the time interval used, the closer the result is to the actual rate at $t = 1400$ s. As Δt approaches 0 s, the rate of reaction approaches the *negative*

---

*Although reaction (14.3) goes to completion, it does so very slowly. Generally, a catalyst is used to speed up the reaction. We describe the catalysis in this reaction in Section 14-11. When $H_2O_2(aq)$ is applied to an open wound, the enzyme catalase in blood catalyzes its decomposition.

KEEP IN MIND

that according to expression (14.2), the rate of this reaction can be expressed either as rate of disappearance of H_2O_2 or $2 \times$ rate of formation of $O_2(g)$.

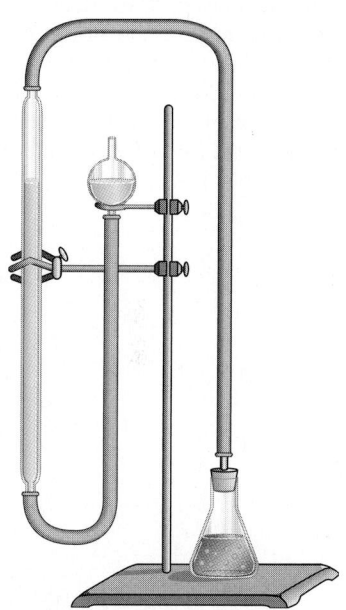

▲ FIGURE 14-1
Experimental setup for determining the rate of decomposition of H_2O_2
Oxygen gas given off by the reaction mixture is trapped, and its volume is measured in the gas buret. The amount of H_2O_2 consumed and the remaining concentration of H_2O_2 can be calculated from the measured volume of $O_2(g)$.

TABLE 14.1 Decomposition of H_2O_2	
Time, s	$[H_2O_2]$, M
0	2.32
200	2.01
400	1.72
600	1.49
1200	0.98
1800	0.62
3000	0.25

► FIGURE 14-2
Graphical representation of kinetic data for the reaction:

$$H_2O_2(aq) \longrightarrow H_2O(l) + \frac{1}{2}O_2(g)$$

This is the usual form in which concentration–time data are plotted. Reaction rates are determined from the slopes of the tangent lines. The blue line has a slope of $-1.70\,M/2800\,s = -6.1 \times 10^{-4}\,M\,s^{-1}$. The slope of the black line and its relation to the initial rate of reaction are described in Example 14-2.

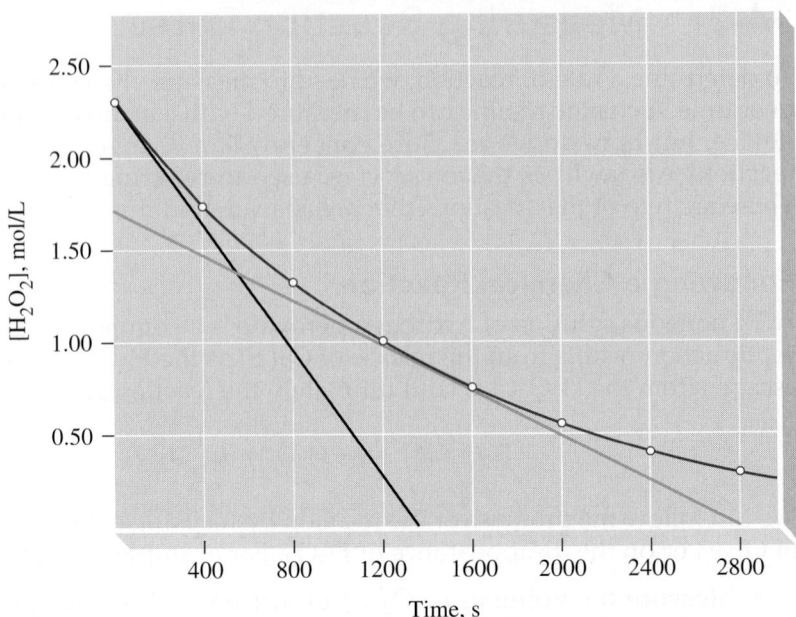

TABLE 14.2   Decomposition of H_2O_2—Derived Rate Data

I	II	III	IV	V
				Reaction Rate $-\Delta[H_2O_2]/\Delta t$,
Time, s	Δt, s	$[H_2O_2]$, M	$\Delta[H_2O_2]$, M	$M\,s^{-1}$
0		2.32		
	400		−0.60	15.0×10^{-4}
400		1.72		
	400		−0.42	10.5×10^{-4}
800		1.30		
	400		−0.32	8.0×10^{-4}
1200		0.98		
	400		−0.25	6.3×10^{-4}
1600		0.73		
	400		−0.19	4.8×10^{-4}
2000		0.54		
	400		−0.15	3.8×10^{-4}
2400		0.39		
	400		−0.11	2.8×10^{-4}
2800		0.28		

of the slope of the tangent line to the curve of Figure 14-2. The rate of reaction determined from the slope of a tangent line to a concentration–time curve is the **instantaneous rate of reaction** at the point where the tangent line touches the curve. The best estimate of the rate of reaction of H_2O_2 at 1400 s, then, is obtained as follows from the slope of the blue tangent line in Figure 14-2: $-(-6.1 \times 10^{-4}\,M\,s^{-1}) = 6.1 \times 10^{-4}\,M\,s^{-1}$.

To better understand the difference between average and instantaneous reaction rates, think of taking a 108 mi highway trip in 2.00 h. The *average* speed is 54.0 mph. The *instantaneous* speed is the speedometer reading at any instant.

► Your instantaneous speed can result in a speeding ticket even if the average speed never exceeds the speed limit.

14-1 ARE YOU WONDERING...

How to differentiate between the average rate of reaction and the instantaneous rate?

If you are familiar with differential calculus, you probably know the answer. If the rate of reaction (14.3) is written as

$$\lim_{\Delta t \longrightarrow 0} \frac{-\Delta[H_2O_2]}{\Delta t}$$

the delta quantities can be replaced by the differentials $-d[H_2O_2]$ and dt, leading to the expression

$$\frac{-d[H_2O_2]}{dt}$$

Thus, the delta notation (not taken to the limit of $\Delta t \longrightarrow 0$) signifies an *average* rate and the differential notation, an *instantaneous* rate.

Initial Rate of Reaction

Sometimes we simply want to find the rate of reaction when the reactants are first brought together—the **initial rate of reaction**. This rate can be obtained from the tangent line to the concentration-time curve at $t = 0$. An alternative way is to measure the concentration of the chosen reactant as soon as possible after mixing, in this way obtaining $\Delta[\text{reactant}]$ for a very short time interval (Δt) at essentially $t = 0$. These two approaches give the same result if the time interval used is limited to that in which the tangent line and the concentration-time curve practically coincide. In Figure 14-2 this condition occurs for about the first 200 s.

EXAMPLE 14-2   Determining and Using an Initial Rate of Reaction

From the data in Table 14.1 and Figure 14-2 for the decomposition of H_2O_2, determine **(a)** the initial rate of reaction, and **(b)** $[H_2O_2]_t$ at $t = 100$ s, assuming that the initial rate is constant for at least 100 s.

Analyze

To determine the initial rate of reaction from the slope of the tangent line in part **(a)**, we use the intersection of the black tangent line with the axis. In part **(b)** we will assume that the rate determined in **(a)** remains essentially constant for at least 100 s.

Solve

(a) $t = 0$, $[H_2O_2] = 2.32$ M; $t = 1360$ s, $[H_2O_2] = 0$.

$$\text{initial rate of reaction} = -(\text{slope of tangent line}) = \frac{-(0 - 2.32)\,\text{M}}{(1360 - 0)\,\text{s}}$$

$$= 1.71 \times 10^{-3}\,\text{M s}^{-1}$$

An alternative method is to use data from Table 14.1: $[H_2O_2] = 2.32$ M at $t = 0$ and $[H_2O_2] = 2.01$ M at $t = 200$ s.

$$\text{initial rate} = \frac{-\Delta[H_2O_2]}{\Delta t} = \frac{-(2.01 - 2.32)\,\text{M}}{200\,\text{s}}$$

$$= 1.6 \times 10^{-3}\,\text{M s}^{-1}$$

(continued)

(b) Since

$$\text{rate of reaction} = \frac{-\Delta[H_2O_2]}{\Delta t}$$

then

$$1.71 \times 10^{-3}\,M\,s^{-1} = \frac{-\Delta[H_2O_2]}{100\,s}$$

$$-(1.71 \times 10^{-3}\,M\,s^{-1})(100\,s) = \Delta[H_2O_2] = [H_2O_2]_t - [H_2O_2]_0$$

$$-1.71 \times 10^{-1}\,M = [H_2O_2]_t - 2.32\,M$$

$$[H_2O_2]_t = 2.32\,M - 0.17\,M = 2.15\,M$$

Assess

In part **(a)** the agreement between the two methods is fairly good, although it might be better if the time interval were shorter than 200 s. Of the two results, the one based on the tangent line is presumably more precise because it is expressed with more significant figures. However, the reliability of the tangent line depends on how carefully the tangent line is constructed. Another reason for favoring a graphical method is that it tends to minimize the effect of errors that may be found in individual data points. In part **(b)** if we had used initial rate $= 1.6 \times 10^{-3}\,M\,s^{-1}$, we would have calculated $[H_2O_2]_t = 2.16\,M$.

PRACTICE EXAMPLE A: For reaction (14.3), determine **(a)** the *instantaneous* rate of reaction at 2400 s and **(b)** $[H_2O_2]$ at 2450 s. [*Hint:* Assume that the instantaneous rate of reaction at 2400 s holds constant for the next 50 s.]

PRACTICE EXAMPLE B: Use data *only* from Table 14.2 to determine $[H_2O_2]$ at $t = 100$ s. Compare this value with the one calculated in Example 14-2(b). Explain the reason for the difference.

🔍 14-2   CONCEPT ASSESSMENT

As shown later in the chapter, for certain reactions the initial and instantaneous rates of reaction are equal throughout the course of the reaction. What must be the shape of the concentration–time graph for such a reaction?

14-3   Effect of Concentration on Reaction Rates: The Rate Law

▶ Most rate laws are obtained experimentally and are empirical.

One of the goals in a chemical kinetics study is to derive an equation that can be used to predict the relationship between the rate of reaction and the concentrations of reactants. Such an experimentally determined equation is called a **rate law**, or **rate equation**.

Consider the hypothetical reaction

$$a A + b B \cdots \longrightarrow g G + h H \cdots \qquad \textbf{(14.5)}$$

where $a, b, \ldots$ stand for coefficients in the balanced equation. We can often express the rate of such a reaction as*

$$\text{rate of reaction} = k[A]^m[B]^n \cdots \qquad \textbf{(14.6)}$$

▶ The exponents in the rate law are not necessarily equal to the stoichiometric coefficients in the chemical equation. In addition, the value of k varies with temperature even though it is called a rate *constant*.

The terms $[A], [B], \ldots$ represent reactant molarities. The required exponents, $m, n, \ldots$ are generally small, positive whole numbers, although in some cases

*We assume that reaction (14.5) goes to completion. If it is reversible, the rate equation is more complex than (14.6). Even for reversible reactions, though, equation (14.6) applies to the *initial* rate of reaction because in the early stages of the reaction there are not enough products for a significant reverse reaction to occur.

they may be zero, fractional, or negative. They must be determined by experiment and are generally *not* related to stoichiometric coefficients $a, b, \ldots$. That is, often $m \neq a$, $n \neq b$, and so on.

The term *order* is related to the exponents in the rate law and is used in two ways: (1) If $m = 1$, we say that the reaction is *first order in A*. If $n = 2$, the reaction is *second order in B*, and so on. (2) The *overall* **order of reaction** is the sum of all the exponents: $m + n + \cdots$. The proportionality constant k relates the rate of reaction to reactant concentrations and is called the **rate constant** of the reaction. Its value depends on the specific reaction, the presence of a catalyst (if any), and the temperature. *The larger the value of k, the faster a reaction goes.* The order of the reaction establishes the general form of the rate law and the appropriate units of k (that is, depending on the values of the exponents).

With the rate law for a reaction, we can

• calculate rates of reaction for known concentrations of reactants
• derive an equation that expresses a reactant concentration as a function of time

But how do we establish the rate law? We need to use *experimental* data of the type described in Section 14-2. The method we describe next works especially well.

◀ A common misconception is that rate and k are the same. As equation (14.6) clearly shows, rate $\neq k$ in general. The rate k is one of the factors affecting the rate of reaction but not the only one.

Method of Initial Rates

As its name implies, this method requires us to work with *initial* rates of reaction. As an example, let's look at a specific reaction: that between mercury(II) chloride and oxalate ion.

$$2\,HgCl_2(aq) + C_2O_4{}^{2-}(aq) \longrightarrow 2\,Cl^-(aq) + 2\,CO_2(g) + Hg_2Cl_2(s) \quad \textbf{(14.7)}$$

The tentative rate law that we can write for this reaction is

$$\text{rate of reaction} = k[HgCl_2]^m[C_2O_4{}^{2-}]^n \quad \textbf{(14.8)}$$

We can follow the reaction by measuring the quantity of $Hg_2Cl_2(s)$ formed as a function of time. Some representative data are given in Table 14.3, which we can assume are based on either the rate of formation of Hg_2Cl_2 or the rate of disappearance of $C_2O_4{}^{2-}$. In Example 14-3, we will use some of these data to illustrate the method of initial rates.

TABLE 14.3   Kinetic Data for the Reaction: $2\,HgCl_2 + C_2O_4{}^{2-} \longrightarrow 2\,Cl^- + 2\,CO_2 + Hg_2Cl_2$

Experiment	$[HgCl_2]$, M	$[C_2O_4{}^{2-}]$, M	Initial Rate, M min^{-1}
1	$[HgCl_2]_1 = 0.105$	$[C_2O_4{}^{2-}]_1 = 0.15$	$R_1 = 1.8 \times 10^{-5}$
2	$[HgCl_2]_2 = 0.105$	$[C_2O_4{}^{2-}]_2 = 0.30$	$R_2 = 7.1 \times 10^{-5}$
3	$[HgCl_2]_3 = 0.052$	$[C_2O_4{}^{2-}]_3 = 0.30$	$R_3 = 3.5 \times 10^{-5}$

EXAMPLE 14-3   Establishing the Order of a Reaction by the Method of Initial Rates

Use data from Table 14.3 to establish the order of reaction (14.7) with respect to $HgCl_2$ and $C_2O_4{}^{2-}$ and also the overall order of the reaction.

Analyze

We need to determine the values of m and n in equation (14.8). In comparing Experiment 2 with Experiment 3, note that $[HgCl_2]$ is essentially doubled ($0.105\,M \approx 2 \times 0.052\,M$) while $[C_2O_4{}^{2-}]$ is held constant (at 0.30 M). Note also that $R_2 = 2 \times R_3$ ($7.1 \times 10^{-5} \approx 2 \times 3.5 \times 10^{-5}$). Rather than use the actual concentrations and rates in the following rate equation, let's work initially with their symbolic equivalents.

(continued)

Solve

We begin by writing

$$R_2 = k \times [HgCl_2]_2^m \times [C_2O_4^{2-}]_2^n = k \times (2 \times [HgCl_2]_3)^m \times [C_2O_4^{2-}]_3^n$$

$$R_3 = k \times [HgCl_2]_3^m \times [C_2O_4^{2-}]_3^n$$

$$\frac{R_2}{R_3} = \frac{2 \times R_3}{R_3} = 2 = \frac{k \times 2^m \times [HgCl_2]_3^m \times [C_2O_4^{2-}]_3^n}{k \times [HgCl_2]_3^m \times [C_2O_4^{2-}]_3^n} = 2^m$$

In order that $2^m = 2$, $m = 1$.

To determine the value of n, we can form the ratio R_2/R_1. Now, $[C_2O_4^{2-}]$ is doubled and $[HgCl_2]$ is held constant. This time, let's use actual concentrations instead of symbolic equivalents. Also, we now have the value $m = 1$.

$$R_2 = k \times [HgCl_2]_2^1 \times [C_2O_4^{2-}]_2^n = k \times (0.105)^1 \times (2 \times 0.15)^n$$

$$R_1 = k \times [HgCl_2]_1^1 \times [C_2O_4^{2-}]_1^n = k \times (0.105)^1 \times (0.15)^n$$

$$\frac{R_2}{R_1} = \frac{7.1 \times 10^{-5}}{1.8 \times 10^{-5}} \approx 4 = \frac{k \times (0.105)^1 \times 2^n \times (0.15)^n}{k \times (0.105)^1 \times (0.15)^n} = 2^n$$

In order that $2^n = 4$, $n = 2$.

Assess

In summary, the reaction is *first* order in $HgCl_2$ ($m = 1$), *second* order in $C_2O_4^{2-}$ ($n = 2$), and *third* order overall ($m + n = 1 + 2 = 3$). In this example we found n by solving the equation $2^n = 4$. In some cases we may need to solve equations containing a noninteger number, such as in $2^n = 1.4143$. To solve this kind of equation, we take the logarithm of both sides, for example, $\log(2^n) = \log(1.4143)$, and rearrange to get $n = \log(1.4143)/\log(2)$. The answer is $n = 0.5$. Consult Appendix A-2 if you are unfamiliar with logarithms.

PRACTICE EXAMPLE A: The decomposition of N_2O_5 is given by the following equation:

$$2\,N_2O_5 \longrightarrow 4\,NO_2 + O_2$$

At an initial $[N_2O_5] = 3.15\,M$, the initial rate of reaction $= 5.45 \times 10^{-5}\,M\,s^{-1}$, and when $[N_2O_5] = 0.78\,M$, the initial rate of reaction $= 1.35 \times 10^{-5}\,M\,s^{-1}$. Determine the order of this decomposition reaction.

PRACTICE EXAMPLE B: Consider a hypothetical Experiment 4 in Table 14.3, in which the initial conditions are $[HgCl_2]_4 = 0.025\,M$ and $[C_2O_4^{2-}]_4 = 0.045\,M$. Predict the initial rate of reaction.

We made an important observation in Example 14-3: If a reaction is *first order* in one of the reactants, doubling the initial concentration of that reactant causes the initial rate of reaction to double. Following is the general effect of *doubling* the initial concentration of a particular reactant (with other reactant concentrations held constant).

▶ Note that rates that are too fast for traditional methods of analysis are followed by spectroscopic methods.

- *Zero order* in the reactant—there is *no effect* on the initial rate of reaction.
- *First order* in the reactant—the initial rate of reaction *doubles*.
- *Second order* in the reactant—the initial rate of reaction *quadruples*.
- *Third order* in the reactant—the initial rate of reaction *increases eightfold*.

As previously mentioned, the order of a reaction, as indicated through the rate law, establishes the units of the rate constant, k. That is, if on the left side of the rate law the rate of reaction has the units $M\,(time)^{-1}$, on the right side, the units of k must provide for the cancellations that also lead to $M\,(time)^{-1}$. Thus, for the rate law established in Example 14-3,

rate law: rate of reaction $= k \times [HgCl_2] \times [C_2O_4^{2-}]^2$

units: $M\,min^{-1}$ $M^{-2}\,min^{-1}$ M M^2

Once we have the exponents in a rate equation, we can determine the value of the rate constant, k. To do this, all we need is the rate of reaction corresponding to known initial concentrations of reactants, as illustrated in Example 14-4.

EXAMPLE 14-4   Using the Rate Law

Use the results of Example 14-3 and data from Table 14.3 to establish the value of k in the rate law (14.8).

Analyze

We can use data from any one of the three experiments of Table 14.3, together with the values $m = 1$ and $n = 2$.

Solve

First, we solve equation (14.8) for k.

$$k = \frac{R_1}{[HgCl_2][C_2O_4^{2-}]^2} = \frac{1.8 \times 10^{-5}\,M\,min^{-1}}{0.105\,M \times (0.15)^2\,M^2}$$

$$= 7.6 \times 10^{-3}\,M^{-2}\,min^{-1}$$

Assess

If the rate data in Table 14.3 were based on the disappearance of $HgCl_2$ instead of $C_2O_4^{2-}$, R_1 in this setup would be twice as great and k for the general rate of reaction would be based on $-\frac{1}{2} \times \Delta[HgCl_2]/\Delta t$. Note the units on the rate constant are $M^{-2}\,min^{-1}$, which are appropriate units for a third-order rate constant. Checking the units in an answer is one way to ensure that we have not made any mistakes.

PRACTICE EXAMPLE A: A reaction has the rate law: rate $= k[A]^2[B]$. When $[A] = 1.12\,M$ and $[B] = 0.87\,M$, the rate of reaction $= 4.78 \times 10^{-2}\,M\,s^{-1}$. What is the value of the rate constant, k?

PRACTICE EXAMPLE B: What is the rate of reaction (14.7) at the point where $[HgCl_2] = 0.050\,M$ and $[C_2O_4^{2-}] = 0.025\,M$?

14-3   CONCEPT ASSESSMENT

The rate of decomposition of gaseous acetaldehyde, CH_3CHO, to gaseous methane and carbon monoxide is found to increase by a factor of 2.83 when the initial concentration of acetaldehyde is doubled. What is the order of this reaction?

14-4   Zero-Order Reactions

An overall **zero-order reaction** has a rate law in which the sum of the exponents, $m + n \cdots$, is equal to 0. As an example, let's take a reaction in which a single reactant A decomposes to products.

$$A \longrightarrow products$$

If the reaction is zero order, the rate law is

$$\text{rate of reaction} = k[A]^0 = k = \text{constant} \qquad (14.9)$$

Other features of this zero-order reaction are

- The concentration–time graph is a straight line with a *negative* slope (Fig. 14-3).
- The rate of reaction, which is equal to k and remains constant throughout the reaction, is the *negative* of the slope of this line.
- The units of k are the same as the units of the rate of a reaction: mol L^{-1} (time)$^{-1}$, for example, mol $L^{-1}\,s^{-1}$ or $M\,s^{-1}$.

Equation (14.9) is the rate law for a zero-order reaction. Another useful equation, called an **integrated rate law**, expresses the concentration of a reactant as

▶ The dependence of k on temperature is discussed in Section 14-9.

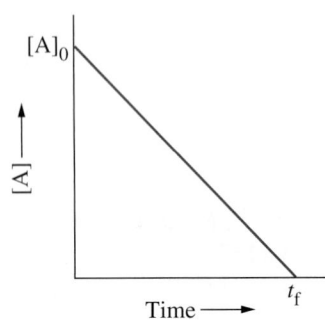

▲ FIGURE 14-3
A zero-order reaction:
A ⟶ products
The initial concentration of the reactant A is $[A]_0$, that is, $[A] = [A]_0$ at $t = 0$. $[A]$ decreases in concentration until the reaction stops. This occurs at the time, t_f, where $[A] = 0$. The slope of the line is

$$\frac{(0 - [A]_0)}{(t_f - 0)} = -\frac{[A]_0}{t_f}.$$

The rate constant is the *negative* of the slope:

$$k = -\text{slope} = \frac{[A]_0}{t_f}.$$

▶ Perhaps the most important examples of zero-order reactions are found in the action of enzymes. Enzyme-catalyzed reactions are discussed in Section 14-11.

14-2 ARE YOU WONDERING...

About the difference between the rate of a reaction and the rate *constant* of a reaction?

Many students have difficulty with this distinction. Remember, a *rate of reaction* tells how the amount of a reactant or product changes with time and is usually expressed as moles per liter per time. A rate of reaction can be established through an expression of the type $-\Delta[A]/\Delta t$, from a tangent to a concentration–time curve, and by calculation from a rate law. In most cases, the rate of a reaction strongly depends on reactant concentrations and therefore changes continuously during a reaction.

The *rate constant* of a reaction (k) relates the rate of a reaction to reactant concentrations. Generally, it is not itself a rate of reaction, but it can be used it to calculate rates of reaction. Once the value of k at a given temperature has been established, *this value stays fixed* for the given reaction, regardless of the reactant concentrations. Whereas the units of rates of reaction do not depend on the order of a reaction, those of k do, as we shall see in this and the following two sections.

a function of time. This equation can be established rather easily from the graph in Figure 14-3. Let's start with the general equation for a straight line

$$y = mx + b$$

and substitute $y = [A]_t$ (the concentration of A at some time t); $x = t$ (time); $b = [A]_0$ (the initial concentration of A at time $t = 0$); and $m = -k$ (m, the slope of the straight line, is obtained as indicated in the caption to Figure 14-3).

$$[A]_t = -kt + [A]_0 \tag{14.10}$$

14-3 ARE YOU WONDERING...

If the term "integrated" rate law has anything to do with integral calculus?

Not surprisingly, it does. The rate of reaction in a rate law, such as (14.9), is an *instantaneous* rate, which we learned in Are You Wondering 14-1 can be represented through differentials. When $-d[A]/dt$ is substituted for the rate of reaction in the rate law for a zero-order reaction, we get this equation, $-d[A]/dt = k$. We can separate the differentials to obtain $d[A] = -kdt$. At this point we can apply the calculus procedure of integration to obtain, successively, the following expressions. The final one is the integrated rate law for a zero-order reaction (14.10).

$$\int_{[A]_0}^{[A]_t} d[A] = -k \int_0^t dt, \qquad [A]_t - [A]_0 = -kt, \qquad [A]_t = -kt + [A]_0 \tag{14.10}$$

14-5 First-Order Reactions

An overall **first-order reaction** has a rate law in which the sum of the exponents, $m + n + \cdots$, is equal to 1. A particularly common type of first-order reaction, and the only type we will consider, is one in which a single reactant decomposes into products. Reaction (14.3), the decomposition of H_2O_2 that we described in Section 14-2, is a first-order reaction.

$$H_2O_2(aq) \longrightarrow H_2O(l) + \frac{1}{2}O_2(g)$$

The rate of reaction depends on the concentration of H_2O_2 raised to the *first* power, that is,

$$\text{rate of reaction} = k[H_2O_2] \qquad \textbf{(14.11)}$$

It is easy to establish that reaction (14.3) is first order by the method of initial rates, but there are also other ways of recognizing a first-order reaction.

An Integrated Rate Law for a First-Order Reaction

Let us begin our discussion of first-order reactions as we did zero-order reactions, by examining a hypothetical reaction

$$A \longrightarrow \text{products}$$

for which the rate law is

$$\text{rate of reaction} = -\frac{\Delta[A]}{\Delta t} = k[A] \qquad \textbf{(14.12)}$$

We can obtain the *integrated* rate law for this first-order reaction by applying the calculus technique of *integration* to equation (14.12). The result of this derivation (shown in Are You Wondering 14-4) is

$$\ln\frac{[A]_t}{[A]_0} = -kt \quad \text{or} \quad \ln[A]_t = -kt + \ln[A]_0 \qquad \textbf{(14.13)}$$

$[A]_t$ is the concentration of A at time t, $[A]_0$ is its concentration at $t = 0$, and k is the rate constant. Because the logarithms of numbers are dimensionless (have no units), the product $-k \times t$ must also be dimensionless. This means that the unit of k in a first-order reaction is $(\text{time})^{-1}$, such as s^{-1} or min^{-1}. Equation (14.13) is that of a straight line.

$$\underbrace{\ln[A]_t}_{y} = \underbrace{(-k)t}_{m \cdot x} + \underbrace{\ln[A]_0}_{b}$$

Equation of straight line

◀ Most processes in nature follow first-order chemical kinetics. If you had a bank that compounded interest continuously, then your funds would grow exponentially.

KEEP IN MIND

that scatter of experimental data on a straight-line graph often make a value of k calculated from equation (14.13) less reliable than a value obtained from the slope of the straight line.

14-4   ARE YOU WONDERING...

How to obtain the integrated rate law for a first-order reaction?

In differential form, the rate law for the reaction $A \longrightarrow$ products is $d[A]/dt = -k[A]$.

Separation of the differentials leads to the expression $d[A]/[A] = -kdt$.

Integration of this expression between the limits $[A]_0$ at time $t = 0$ and $[A]_t$ at time t is indicated through the expression

$$\int_{[A]_0}^{[A]_t} \frac{d[A]}{[A]} = -k\int_0^t dt, \qquad \text{yielding } \ln\frac{[A]_t}{[A]_0} = -kt$$

The result of the integration is the integrated rate law.

$$\ln\frac{[A]_t}{[A]_0} = -kt \qquad \textbf{(14.13)}$$

EXAMPLE 14-5   Using the Integrated Rate Law for a First-Order Reaction

$H_2O_2(aq)$, initially at a concentration of 2.32 M, is allowed to decompose. What will $[H_2O_2]$ be at $t = 1200$ s? Use $k = 7.30 \times 10^{-4}\,s^{-1}$ for this first-order decomposition.

Analyze

We have values for three of the four quantities in equation (14.13):

$k = 7.30 \times 10^{-4}\,s^{-1}$ $t = 1200$ s

$[H_2O_2]_0 = 2.32$ M $[H_2O_2]_t = ?$

We need to solve for the fourth quantity.

Solve

We substitute into the expression

$$\ln[H_2O_2]_t = -kt + \ln[H_2O_2]_0$$
$$= -(7.30 \times 10^{-4}\,s^{-1} \times 1200\,s) + \ln 2.32$$
$$= \quad\quad -0.876 \quad\quad + 0.842 = -0.034$$
$$[H_2O_2]_t = e^{-0.034} = 0.967\,M$$

Assess

This calculated value agrees well with the experimentally determined value of 0.98 M, shown below.

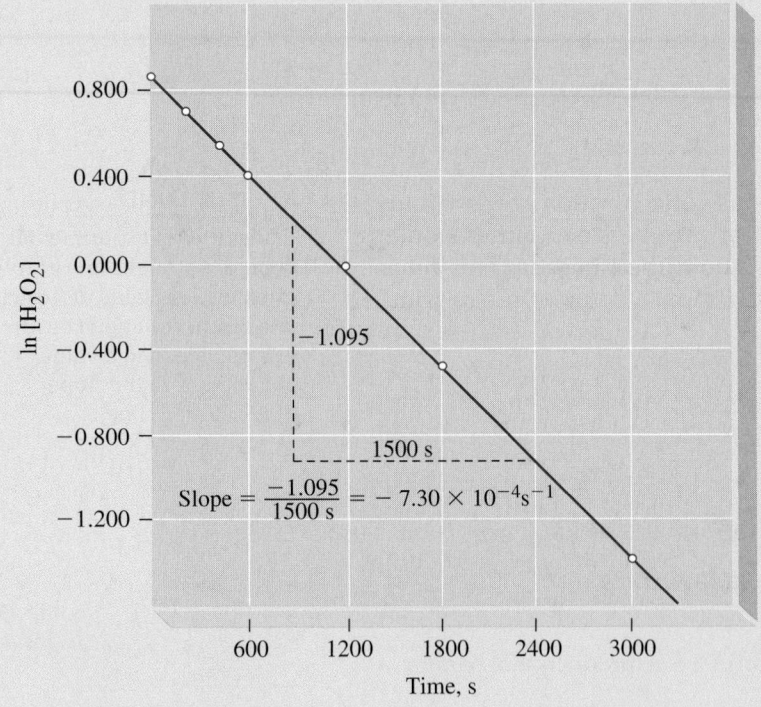

◀ FIGURE 14-4
Test for a first-order reaction: Decomposition of $H_2O_2(aq)$
Plot of ln H_2O_2 versus t. The data are based on Table 14.1 and are listed below. The slope of the line is used in the text.

t, s	[H$_2$O$_2$], M	ln[H$_2$O$_2$]
0	2.32	0.842
200	2.01	0.698
400	1.72	0.542
600	1.49	0.399
1200	0.98	−0.020
1800	0.62	−0.48
3000	0.25	−1.39

PRACTICE EXAMPLE A: The reaction $A \longrightarrow 2B + C$ is first order. If the initial $[A] = 2.80$ M and $k = 3.02 \times 10^{-3}\,s^{-1}$, what is the value of $[A]$ after 325 s?

PRACTICE EXAMPLE B: Use data tabulated in Figure 14-4, together with equation (14.13), to show that the decomposition of H_2O_2 is a first-order reaction. [*Hint:* Use a pair of data points for $[H_2O_2]_0$ and $[H_2O_2]_t$ and their corresponding times to solve for k. Repeat this calculation using other sets of data. How should the results compare?]

An easy test for a first-order reaction is to plot the natural logarithm of a reactant concentration versus time and see if the graph is linear. The data from Table 14.1 are plotted in Figure 14-4, and the rate constant k is derived from the slope of the line: $k = -\text{slope} = -(-7.30 \times 10^{-4}\,\text{s}^{-1}) = 7.30 \times 10^{-4}\,\text{s}^{-1}$. An alternative, nongraphical approach, illustrated in Practice Example 14-5B, is to substitute data points into equation (14.13) and solve for k.

Although until now we have used only molar concentrations in kinetics equations, we can sometimes work directly with the masses of reactants. Another possibility is to work with a fraction of reactant consumed, as is done in the concept of half-life.

The **half-life** of a reaction is the time required for one-half of a reactant to be consumed. It is the time during which the amount of reactant or its concentration decreases to one-half of its initial value. That is, at $t = t_{1/2}, [A]_t = \frac{1}{2}[A]_0$. At this time, equation (14.13) takes the form

$$\ln\frac{[A]_t}{[A]_0} = \ln\frac{\frac{1}{2}[A]_0}{[A]_0} = \ln\frac{1}{2} = -\ln 2 = -k \times t_{1/2}$$

$$t_{1/2} = \frac{\ln 2}{k} = \frac{0.693}{k} \qquad\qquad \textbf{(14.14)}$$

Equation (14.14) is valid only for first-order reactions. We will derive half-life expressions for other types of reactions as we encounter them.

For the decomposition of $H_2O_2(aq)$, the reaction described in Section 14-2, we conclude that the half-life is

$$t_{1/2} = \frac{0.693}{7.30 \times 10^{-4}\,\text{s}^{-1}} = 9.49 \times 10^2\,\text{s} = 949\,\text{s}$$

Equation (14.14) indicates that *the half-life is constant for a first-order reaction.* Thus, regardless of the value of $[A]_0$ at the time we begin to follow a reaction, at $t = t_{1/2}, [A] = \frac{1}{2}[A]_0$. After *two* half-lives, that is, at $t = 2 \times t_{1/2}. [A] = \frac{1}{2} \times \frac{1}{2}[A]_0 = \frac{1}{4}[A]_0$. At $t = 3 \times t_{1/2}, [A] = \frac{1}{8}[A]_0$, and so on.

The constancy of the half-life and its independence of the initial concentration can be used as a test for a first-order reaction. Try it with the simple concentration-time graph of Figure 14-2. That is, starting with $[H_2O_2] = 2.32\,\text{M}$ at $t = 0$, at what time is $[H_2O_2]$ equal to 1.16 M? 0.58 M? 0.29 M? Starting with $[H_2O_2] = 1.50\,\text{M}$ at $t = 600\,\text{s}$, at what time is $[H_2O_2] = 0.75\,\text{M}$?

In the discussion above we made the assumption that the stoichiometric coefficient $a = 1$. What would the rate equations look like if $a \neq 1$? In this case we would write

$$aA \longrightarrow \text{products}$$

for which the rate law is

$$\text{rate of reaction} = -\frac{1}{a}\frac{\Delta[A]}{\Delta t} = k[A]$$

which can be rearranged to give

$$-\frac{\Delta[A]}{\Delta t} = ak[A]$$

The expression above is very similar to equation (14.12) except that ak has taken the place of k in that equation. As a result, the integrated rate law and the half-life for the reaction $aA \longrightarrow$ products can be obtained from equations (14.13) and (14.14) by replacing k with ak. We obtain

$$\ln[A]_t = -akt + \ln[A]_0 \quad \text{and} \quad t_{1/2} = \frac{0.693}{ak}$$

As illustrated in Example 14-6, a first-order reaction can also be described in terms of the percent of a reactant consumed or remaining.

◀ Half-life is constant for first-order reactions but not constant for zero-order or second-order reactions. For a zero-order reaction half-life decreases with decreasing reactant concentration. For a second-order reaction, half-life increases with decreasing reactant concentration.

EXAMPLE 14-6 Expressing Fraction (or Percent) of Reactant Consumed in a First-Order Reaction

Use a value of $k = 7.30 \times 10^{-4} \, \text{s}^{-1}$ for the first-order decomposition of $H_2O_2(aq)$ to determine the percent H_2O_2 that has decomposed in the first 500.0 s after the reaction begins.

Analyze

The ratio $[H_2O_2]_t/[H_2O_2]_0$ represents the fractional part of the initial amount of H_2O_2 that remains *unreacted* at time t. Our problem is to evaluate this ratio at $t = 500.0$ s. This is done by making use of equation (14.13), which relates the concentration at $t = 0$ to the concentration at some other time.

Solve

Substituting into equation (14.13),

$$\ln \frac{[H_2O_2]_t}{[H_2O_2]_0} = -kt = -7.30 \times 10^{-4} \, \text{s}^{-1} \times 500.0 \, \text{s} = -0.365$$

$$\frac{[H_2O_2]_t}{[H_2O_2]_0} = e^{-0.365} = 0.694 \quad \text{and} \quad [H_2O_2]_t = 0.694[H_2O_2]_0$$

The fractional part of the H_2O_2 remaining is 0.694, or 69.4%. The percent of H_2O_2 that has decomposed is $100.0\% - 69.4\% = 30.6\%$.

Assess

The integrated form of the rate law is used for determining concentrations as a function of time. The integrated form of the rate law for first-order reactions is useful for many types of reactions and events. We will see this equation later when we talk about radioactivity.

PRACTICE EXAMPLE A: Consider the first-order reaction A $\longrightarrow$ products, with $k = 2.95 \times 10^{-3} \, \text{s}^{-1}$. What percent of A remains after 150 s?

PRACTICE EXAMPLE B: At what time after the start of the reaction is a sample of $H_2O_2(aq)$ two-thirds decomposed?

Reactions Involving Gases

For gaseous reactions, rates are often measured in terms of gas pressures. For the hypothetical reaction, A(g) $\longrightarrow$ products, the initial partial pressure, $(P_A)_0$, and the partial pressure at some time t, $(P_A)_t$, are related through the expression

$$\ln \frac{(P_A)_t}{(P_A)_0} = -kt \qquad \textbf{(14.15)}$$

To see how this equation is derived, start with the ideal gas equation written for reactant A: $P_A V = n_A RT$. Note that the ratio n_A/V is the same as [A]. So, $[A]_0 = (P_A)_0/RT$ and $[A]_t = (P_A)_t/RT$. Substitute these terms into equation (14.13), and note that the RT terms cancel in the numerator and denominator and leave the simple ratio $(P_A)_t/(P_A)_0$.

Di-t-butyl peroxide (DTBP) is used as a catalyst in the manufacture of polymers. In the gaseous state, DTBP decomposes into acetone and ethane by a first-order reaction.

$$C_8H_{18}O_2(g) \longrightarrow 2 \, CH_3COCH_3(g) + CH_3CH_3(g) \qquad \textbf{(14.16)}$$

$$\text{DTBP} \qquad\qquad \text{acetone} \qquad\qquad \text{ethane}$$

Partial pressures of DTBP are plotted as a function of time in Figure 14-5, and the half-life of the reaction is indicated.

As shown in Are You Wondering 14-5, the partial pressure of the reactant DTBP can be obtained from two experimentally determined quantities: the initial pressure P_0, and the total pressure P_{total}.

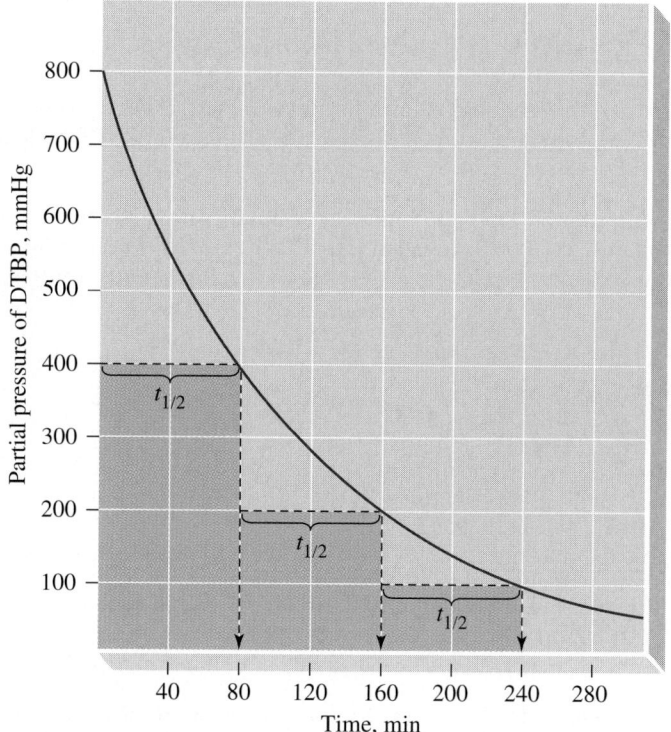

▲ FIGURE 14-5
Decomposition of di-*t*-butyl peroxide (DTBP) at 147 °C
The decomposition reaction is described through equation (14.16). In this graph of the partial pressure of DTBP as a function of time, three successive half-life periods of 80 min each are indicated. This constancy of the half-life is proof that the reaction is first order.

14-5   ARE YOU WONDERING...

How the partial pressure can be obtained experimentally in a reaction involving gases?

In the study of a reaction like the decomposition of DTBP, the *total* pressure is typically measured as a function of time. At any time t, the total pressure is

$$(P_{total})_t = (P_{DTBP})_t + (P_{acetone})_t + (P_{ethane})_t$$

If the initial presure of DTBP is P_0, then using the stoichiometry of the balanced chemical equation, the partial pressure of DTBP is $P_{DTBP} = (P_0 - P_{ethane})$ since for every mole of DTBP that decomposes, a mole of ethane is produced. The partial pressure of acetone is $P_{acetone} = 2 P_{ethane}$, again using the stoichiometry of the reaction. The total pressure is then given by

$$P_{total} = (P_0 - P_{ethane}) + 2 P_{ethane} + P_{ethane} = P_0 + 2 P_{ethane}$$

and

$$P_{ethane} = \frac{P_{total} - P_0}{2}$$

So that

$$P_{DTBP} = (P_0 - P_{ethane}) = \frac{2 P_0 - (P_{total} - P_0)}{2} = \frac{3 P_0 - P_{total}}{2}$$

◀ Note that the total pressure in reaction (14.16) increases from P_0 at the start of the reaction to $3 P_0$ when P_{DTBP} has fallen to zero.

EXAMPLE 14-7 Applying First-Order Kinetics to a Reaction Involving Gases

Reaction (14.16) is started with pure DTBP at 147 °C and 800.0 mmHg pressure in a flask of constant volume. **(a)** What is the value of the rate constant k? **(b)** At what time will the partial pressure of DTBP be 50.0 mmHg?

Analyze

In part **(a)** we observe from Figure 14-5 that $t_{1/2} = 8.0 \times 10^1$ min. Recall that for a first-order reaction the relationship between $t_{1/2}$ and the rate constant is $t_{1/2} = 0.693/k$. In part **(b)** the final DTBP partial pressure of 50.0 mmHg is $\frac{1}{16}$ of the starting pressure of 800.0 mmHg; that is, $P_{DTBP} = (\frac{1}{2})^4 \times 800.0 = 50.0$ mmHg. The reaction must go through *four* half-lives.

Solve

(a) Substituting in the appropriate values we have

$$k = 0.693/t_{1/2} = 0.693/8.0 \times 10^1 \text{ min} = 8.7 \times 10^{-3} \text{ min}^{-1}$$

(b) Therefore, $t = 4 \times t_{1/2} = 4 \times 8.0 \times 10^1 \text{ min} = 3.2 \times 10^2 \text{ min}$.

Assess

This type of analysis is useful only if we have the half-life data. If we do not, we will need to use equation (14.13).

PRACTICE EXAMPLE A: Start with DTBP at a pressure of 800.0 mmHg at 147 °C. What will be the pressure of DTBP at $t = 125$ min, if $t_{1/2} = 8.0 \times 10^1$ min? [*Hint:* Because 125 min is not an exact multiple of the half-life, you must use equation (14.15). Can you see that the answer is between 200 and 400 mmHg?]

PRACTICE EXAMPLE B: Use data from Table 14.4 to determine **(a)** the partial pressure of ethylene oxide, and **(b)** the total gas pressure after 30.0 h in a reaction vessel at 415 °C if the initial partial pressure of $(CH_2)_2O(g)$ is 782 mmHg.

TABLE 14.4 Some Typical First-Order Processes

Process	Half-Life, $t_{1/2}$	Rate Constant k, s^{-1}
Radioactive decay of $^{238}_{92}U$	4.51×10^9 yr	4.87×10^{-18}
Radioactive decay of $^{14}_{6}C$	5.73×10^3 yr	3.83×10^{-12}
Radioactive decay of $^{32}_{15}P$	14.3 d	5.61×10^{-7}
Radioactive decay of $^{131}_{53}I$	8.04 d	9.98×10^{-7}
$C_{12}H_{22}O_{11}(aq) + H_2O(l) \xrightarrow{15\,°C} C_6H_{12}O_6(aq) + C_6H_{12}O_6(aq)$ sucrose $\qquad$ glucose $\qquad$ fructose	8.4 h	2.3×10^{-5}
$(CH_2)_2O(g) \xrightarrow{415\,°C} CH_4(g) + CO(g)$ ethylene oxide	56.3 min	2.05×10^{-4}
$N_2O_5(g) \xrightarrow[45\,°C]{\text{In } CCl_4} N_2O_4(g) + \frac{1}{2}O_2(g)$	18.6 min	6.21×10^{-4}
$CH_3COOH(aq) \longrightarrow H^+(aq) + CH_3COO^-(aq)$	8.9×10^{-7} s	7.8×10^5

Examples of First-Order Reactions

▶ We will explore radioactive decay in some detail in Chapter 25.

One of the most familiar examples of a first-order process is radioactive decay. For example, the radioactive isotope iodine-131, used in treating thyroid disorders, has a half-life of 8.04 days. Whatever number of iodine-131 atoms are in a sample at a given moment, there will be half that number in 8.04 days; one-quarter of that number in 8.04 + 8.04 = 16.08 days; and so on. The rate constant for the decay is $k = 0.693/t_{1/2}$, and in equation (14.13) we can use numbers of atoms, that is, N_t for $[A]_t$ and N_0 for $[A]_0$. Table 14.4 lists several examples of first-order processes. Note the great range of values of $t_{1/2}$ and k. The processes range from very slow to ultrafast.

14-6   Second-Order Reactions

An overall **second-order reaction** has a rate law with the sum of the exponents, $m + n \cdots$, equal to 2. As with zero- and first-order reactions, our discussion will be limited to reactions involving the decomposition of a single reactant

$$A \longrightarrow \text{products}$$

that follow the rate law

$$\text{rate of reaction} = k[A]^2 \qquad \textbf{(14.17)}$$

Again, our primary interest will be in the integrated rate law that is derived from the rate law. This proves to be the equation of a straight-line graph.

$$\frac{1}{[A]_t} = kt + \frac{1}{[A]_0} \qquad \textbf{(14.18)}$$

Figure 14-6 is a plot of $1/[A]_t$ against time. The slope of the line is k, and the intercept is $1/[A]_0$. From the graph, we can see that the units of k must be the *reciprocal* of concentration divided by time: $M^{-1}/(\text{time})$ or $M^{-1}(\text{time})^{-1}$—for example, $M^{-1} s^{-1}$ or $M^{-1} \min^{-1}$. We can reach this same conclusion by determining the units of k that produce the required units for the rate of a reaction, that is, moles per liter per time. From equation (14.17):

rate law: rate of reaction $= k \times [A]^2$

units: $M \, \text{time}^{-1}$ $M^{-1} \text{time}^{-1} M^2$

For the half-life of the second-order reaction $A \longrightarrow$ products, we can substitute $t = t_{1/2}$ and $[A] = \frac{1}{2}[A]_0$ into equation (14.18).

$$\frac{1}{[A]_0/2} = kt_{1/2} + \frac{1}{[A]_0}; \qquad \frac{2}{[A]_0} = kt_{1/2} + \frac{1}{[A]_0} \qquad \textbf{(14.19a)}$$

and

$$t_{1/2} = \frac{1}{k[A]_0} \qquad \textbf{(14.19b)}$$

From equation (14.19a and 14.19b), we see that the half-life depends on both the rate constant *and* the initial concentration $[A]_0$. The *half-life is not constant.* Its value depends on the concentration of reactant at the start of each half-life interval. Because the starting concentration is always one-half that of the previous half-life, each successive half-life is twice as long as the one before it.

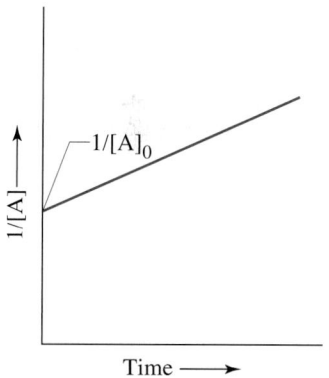

▲ FIGURE 14-6
A straight-line plot for the second order reaction $A \longrightarrow$ products
The reciprocal of the concentration, $1/[A]$, is plotted against time. As the reaction proceeds, $[A]$ decreases and $1/[A]$ increases in a linear fashion. The slope of the line is the rate constant k.

Pseudo-First-Order Reactions

At times, it is possible to simplify the kinetic study of complex reactions by getting them to behave like reactions of a lower order. Then their rate laws become easier to work with. Consider the hydrolysis of ethyl acetate, which is second order overall.

$$CH_3COOCH_2CH_3 + H_2O \longrightarrow CH_3COOH + CH_3CH_2OH$$
Ethyl acetate   Acetic acid   Ethanol

Suppose we follow the hydrolysis of 1 L of aqueous 0.01 M ethyl acetate to completion. $[CH_3COOCH_2CH_3]$ decreases from 0.01 M to essentially zero. This means that 0.01 mol $CH_3COOCH_2CH_3$ is consumed, and along with it, 0.01 mol H_2O. Now consider what happens to the molarity of the H_2O. Initially, the solution contains about 1000 g H_2O, or about 55.5 mol H_2O. When the reaction is completed, there is still 55.5 mol H_2O (that is, $55.5 - 0.01 \approx 55.5$). The molarity of the water remains essentially constant throughout the reaction—55.5 M. The rate of reaction does not appear to depend on $[H_2O]$. So, the reaction appears to be *zero* order in H_2O, *first* order in $CH_3COOCH_2CH_3$, and *first* order overall. A second-order reaction that is made to behave like a first-order reaction by holding one reactant concentration constant is called a *pseudo*-first-order reaction. We can treat the reaction with the methods of first-order reaction kinetics. Other reactions of higher order can be made to behave like reactions of lower order under certain conditions. Thus, a third-order reaction might be converted to pseudo-second order or even to pseudo-first order.

▶ The rate constant obtained from a study that used an excess of one reactant, sometimes called Ostwald's isolation method after the kineticist who invented it, is a pseudo-rate constant and is concentration-dependent.

14-6 ARE YOU WONDERING...

How we obtain the integrated rate law for the second-order reaction A $\longrightarrow$ products?

In differential form, the rate law for the reaction A $\longrightarrow$ products, is $d[A]/dt = -k[A]^2$.

Separation of the differentials leads to the expression $d[A]/[A]^2 = -kdt$.

Integration of this expression between the limits $[A]_0$ at time $t = 0$ and $[A]_t$ at time t is indicated through the expression

$$\int_{[A]_0}^{[A]_t} \frac{d[A]}{[A]^2} = -\int_0^t kdt$$

The result of the integration is the integrated rate law.

$$-\frac{1}{[A]_t} + \frac{1}{[A]_0} = -kt \quad \text{or} \quad \frac{1}{[A]_t} = kt + \frac{1}{[A]_0} \qquad (14.18)$$

14-7 Reaction Kinetics: A Summary

Let's pause briefly to review what we have learned about rates of reaction, rate constants, and reaction orders. Although a problem often can be solved in several different ways, these approaches are generally most direct.

1. To calculate a rate of reaction when the rate law is known, use this expression: rate of reaction $= k[A]^m[B]^n \cdots$

2. To determine a rate of reaction when the rate law is not given, use
   - the slope of an appropriate tangent line to the graph of $[A]$ versus t
   - the expression $-\Delta[A]/\Delta t$, with a short time interval Δt

TABLE 14.5   Reaction Kinetics: A Summary for the Hypothetical Reaction $aA \longrightarrow$ Products

Order	Rate Law[a]	Integrated Rate Equation	Straight Line	$k =$	Units of k	Half-Life
0	rate $= k$	$[A]_t = -akt + [A]_0$	$[A]$ v. time	$-$slope	mol L^{-1} s^{-1}	$\dfrac{[A]_0}{2\,ak}$
1	rate $= k[A]$	$\ln[A]_t = -akt + \ln[A]_0$	$\ln[A]$ v. time	$-$slope	s^{-1}	$\dfrac{0.693}{ak}$
2	rate $= k[A]^2$	$\dfrac{1}{[A]_t} = akt + \dfrac{1}{[A]_0}$	$\dfrac{1}{[A]}$ v. time	slope	L mol^{-1} s^{-1}	$\dfrac{1}{ak[A]_0}$

[a]rate $= -\left(\dfrac{1}{a}\right)\dfrac{\Delta[A]}{\Delta t}$

3. To determine the order of a reaction, use one of the following methods.
   - Use the method of initial rates if the experimental data are given in the form of reaction rates at different initial concentrations.
   - Find the graph of rate data that yields a straight line (Table 14.5).
   - Test for the constancy of the half-life (good only for first-order).
   - Substitute rate data into integrated rate laws to find the one that gives a constant value of k.

4. To find the rate constant k for a reaction, use one of the following methods.
   - Obtain k from the slope of a straight-line graph.
   - Substitute concentration–time data into the appropriate integrated rate law.
   - Obtain k from the half-life of the reaction (good only for a first-order reaction).

5. To relate reactant concentrations and times, use the appropriate integrated rate law after first determining k.

EXAMPLE 14-8   Graphing Data to Determine the Order of a Reaction

The data listed in Table 14.6 were obtained for the decomposition reaction A $\longrightarrow$ products. **(a)** Establish the order of the reaction. **(b)** What is the rate constant, k? **(c)** What is the half-life, $t_{1/2}$, if $[A]_0 = 1.00$ M?

TABLE 14.6   Kinetic Data for Example 14-8

Time, min	[A], M	ln[A]	1/[A]
0	1.00	0.00	1.00
5	0.63	−0.46	1.6
10	0.46	−0.78	2.2
15	0.36	−1.02	2.8
25	0.25	−1.39	4.0

Analyze

To determine the order of the reaction, we need to plot the data according to the integrated rate law for the different reaction orders. The plot yielding a straight line shows the overall reaction order. Recall that the slope of the straight line is related to the rate constant. The half-life of the reaction is also dependent on the overall reaction order.

Solve

(a) Plot the following three graphs.

   1. $[A]$ versus time. (If a straight line, reaction is zero order.)

(continued)

2. $\ln[A]$ versus time. (If a straight line, reaction is first order.)

3. $1/[A]$ versus time. (If a straight line, reaction is second order.)

These graphs are plotted in Figure 14-7. The reaction is second order.

(b) The slope of graph 3 in Figure 14-7 is

$$k = \frac{(4.00 - 1.00)\text{L/mol}}{25 \text{ min}} = 0.12 \text{ M}^{-1}\text{min}^{-1}$$

(c) According to equation (14.19),

$$t_{1/2} = \frac{1}{k[A]_0} = \frac{1}{0.12 \text{ M}^{-1}\text{min}^{-1} \times 1.00 \text{ M}} = 8.3 \text{ min}$$

Assess

We can check the result in this example by examining the half-life results on the plot of $[A]$ versus time. The plot shows that $[A]$ decreases to 0.5 M between $t = 5$ min and $t = 10$ min, and so $t_{1/2} = 8.3$ min is a reasonable answer.

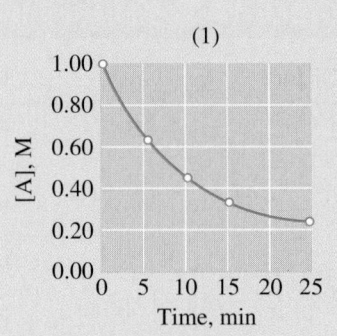

(1)

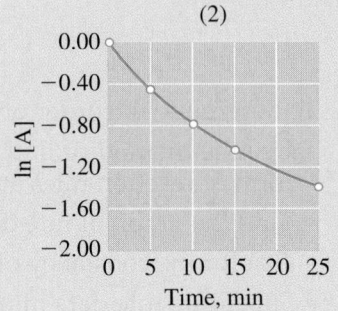

(2)

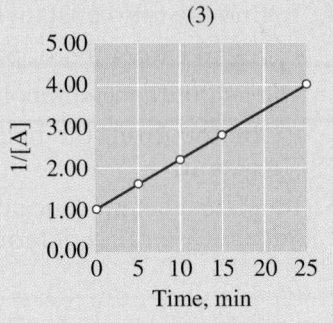

(3)

PRACTICE EXAMPLE A: In the decomposition reaction $B \longrightarrow$ products, the following data are obtained: $t = 0$ s, $[B] = 0.88$ M; $t = 25$ s, 0.74 M; $t = 50$ s, 0.62 M; $t = 75$ s, 0.52 M; $t = 100$ s, 0.44 M; $t = 150$ s, 0.31 M; $t = 200$ s, 0.22 M; $t = 250$ s, 0.16 M. What are the order of this reaction and its rate constant k?

PRACTICE EXAMPLE B: The following data are obtained for the reaction $A \longrightarrow$ products: $t = 0$ min, $[A] = 0.250$ M; $t = 4.22$ min, $[A] = 0.210$ M; $t = 6.60$ min, $[A] = 0.188$ M; $t = 10.61$ min, $[A] = 0.150$ M; $t = 14.48$ min, $[A] = 0.114$ M; $t = 18.00$ min, $[A] = 0.083$ M. What are the order of this reaction and its rate constant, k?

▶ FIGURE 14-7
Testing for the order of a reaction—Example 14-8 illustrated
The straight-line plot is obtained for $1/[A]$ versus t, graph (3). The reaction is second order.

🔍 **14-5** CONCEPT ASSESSMENT

How could you ascertain, by examining only a plot of $[A]$ versus time, whether a reaction is **(a)** zero order; **(b)** first order; **(c)** second order?

14-8 Theoretical Models for Chemical Kinetics

Practical aspects of reaction kinetics—rate laws, rate constants—can be described without considering the behavior of individual molecules. However, insight into the processes involved requires examination at the molecular level. For example, experiments show that the decomposition of H_2O_2 is first order, but *why* is this so? The remainder of the chapter considers the theoretical aspects of chemical kinetics that help answer such questions.

Collision Theory

In our discussion of kinetic-molecular theory in Chapter 6 our emphasis was on molecular speeds. A further aspect of the theory with relevance to chemical kinetics is collision density, the number of collisions per unit volume per unit time.

▶ Not all collisions are effective in causing a reaction. The average time between collisions in a gas at STP is about 10^{-10} s and the time that the collision is actually taking place is about 10^{13} s.

In a typical gas-phase reaction, the calculated collision density is of the order of 10^{32} collisions per liter per second. If each collision yielded product molecules, the rate of reaction would be about $10^6\,M\,s^{-1}$, an extremely rapid rate. The typical gas-phase reaction would go essentially to completion in a fraction of a second. Gas-phase reactions generally proceed at a much slower rate, perhaps on the order of $10^{-4}\,M\,s^{-1}$. This must mean that, generally, *only a fraction of the collisions among gaseous molecules lead to chemical reaction*. This is a reasonable conclusion; we should not expect every collision to result in a reaction.

For a reaction to occur following a collision between molecules, there must be a redistribution of energy that puts enough energy into certain key bonds to break them. We would not expect two slow-moving molecules to bring enough kinetic energy into their collision to permit bond breakage. We would expect two fast-moving molecules to do so, however, or perhaps one extremely fast molecule colliding with a slow-moving one. The **activation energy** of a reaction is the minimum kinetic energy that molecules must bring to their collisions for a chemical reaction to occur.

The kinetic-molecular theory can be used to establish the fraction of all the molecules in a mixture that possess certain kinetic energies. The results of this calculation are depicted in Figure 14-8. On this graph, a hypothetical energy is noted and the fraction of all molecules having energies in excess of this value is identified. Let's assume that these are the molecules whose molecular collisions are most likely to lead to chemical reaction. The rate of a reaction, then, depends on the product of the collision frequency *and* the fraction of these "activated" molecules—in other words, on how often molecules with sufficient kinetic energy to react are likely to collide with each other. Because the fraction of high-energy molecules is generally so small, the rate of reaction is usually much smaller than the collision frequency. Moreover, the *higher* the activation energy of a reaction, the *smaller* is the fraction of energetic collisions and the slower the reaction.

Another factor that can strongly affect the rate of a reaction is the *orientation* of molecules at the time of their collision. In a reaction in which two hydrogen atoms combine to form a hydrogen molecule (see margin) no bonds are broken and a H—H bond forms

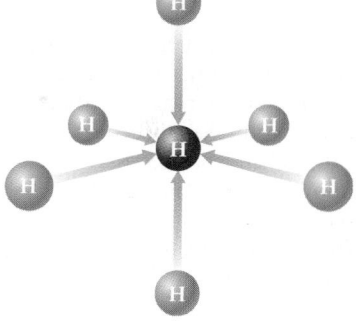

$$H\cdot \;+\; \cdot H \longrightarrow H_2$$

The H atoms are spherically symmetrical, and all approaches of one H atom to another prior to collision are equivalent. Orientation is *not* a factor, and the reaction occurs about as rapidly as the atoms collide. Orientation of the colliding molecules, however, is a crucial matter in the reaction of N_2O and NO, represented here in an equation highlighting chemical bonds.

$$N\equiv N-O \;+\; N=O \longrightarrow N\equiv N \;+\; O-N\overset{\displaystyle O}{\diagup\!\!\!\diagup} \qquad \textbf{(14.20)}$$

The fundamental changes that occur during a successful collision are that the N—O bond in N_2O breaks and a new O—N bond is established to the NO

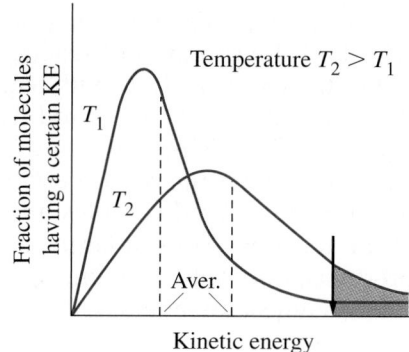

◀ FIGURE 14-8
Distribution of molecular kinetic energies
At both temperatures, the fraction of all molecules having kinetic energies in excess of the value marked by the heavy black arrow is small. (Note the shaded areas on the right.) At the higher temperature T_2 (red), however, this fraction is considerably larger than at the lower temperature T_1 (blue).

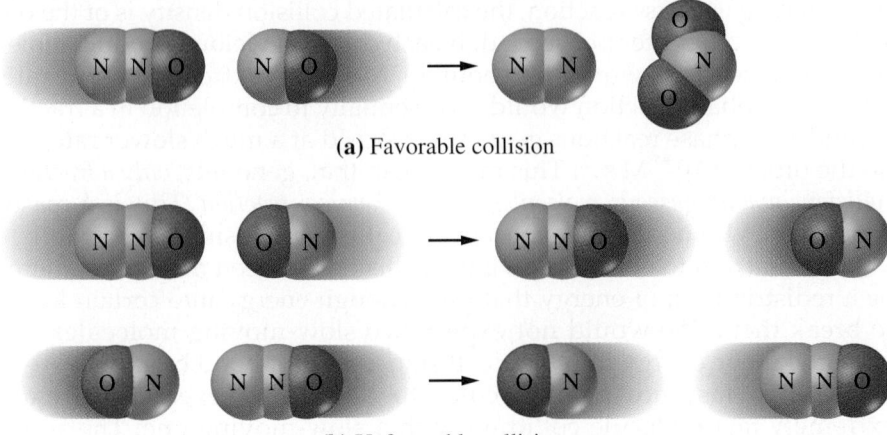

(a) Favorable collision

(b) Unfavorable collisions

▶ FIGURE 14-9
Molecular collisions and chemical reactions
(a) A favorable collision between N_2O and NO molecules, resulting in the products N_2 and NO_2. **(b)** Two unfavorable collisions between N_2O and NO molecules; no reaction follows the collisions.

molecule. As a result of the collision, the molecules N_2 and NO_2 are formed. As suggested by Figure 14-9, a favorable collision requires the N atom of the NO molecule to strike the O atom of N_2O during a collision. Other orientations, such as the N atom of NO striking the terminal N atom of N_2O, do not produce a reaction. The number of unfavorable collisions in the reaction mixture exceeds the number of favorable ones.

Transition State Theory

In a theory proposed by Henry Eyring (1901–1981) and others, special emphasis is placed on a hypothetical species believed to exist in a transitory state that lies between the reactants and the products. We call this state the **transition state**, and the hypothetical species, the **activated complex**. The activated complex, formed through collisions, either dissociates back into the original reactants or forms product molecules. We can represent an activated complex for reaction (14.20) in this way.

▶ Modern Raman laser spectroscopic methods can be used to study reactions that occur in the femtosecond (10^{-15}) time scale. This allows all motions, except electronic transitions, to be frozen. Such techniques are used to study activated complexes.

$$N\equiv N{-}O \ + \ N{=}O \ \rightleftharpoons \ N\equiv N\cdots O\cdots N{\overset{\displaystyle O}{\diagup}} \ \longrightarrow \ N\equiv N \ + \ O{-}N{\overset{\displaystyle O}{\diagup}}$$

Reactants   Activated complex   Products

In the reactants, there is no bond between the O atom of N_2O and the N atom of NO. In the activated complex, the O atom has been partially removed from the N_2O molecule and partially joined to the NO molecule, as indicated by the *partial bonds* ($\cdots$). The formation of the activated complex is a reversible process. Once formed, some molecules of the activated complex may dissociate back into the reactants, but others may dissociate into the product molecules, where the partial bond of the O atom in N_2O has been completely severed and the partial bond between the O atom and NO has become a complete bond.

Figure 14-10, called a reaction profile, is a graphical way of looking at activation energy. In a **reaction profile**, energies are plotted on the vertical axis against a quantity called "progress of reaction," or simply reaction progress, on the horizontal axis. Think of the progress of reaction as representing the extent of the reaction. That is, the reaction starts with reactants on the left, progresses through a transition state, and ends with products on the right.

The difference in energies between the reactants and products is ΔH for the reaction. Reaction (14.20) is an exothermic reaction with $\Delta H = -139$ kJ. The difference in energy between the activated complex and the reactants, 209 kJ, is the *activation energy* of the reaction. Thus, a large energy barrier separates the reactants from the products, and only very energetic molecules can pass over this barrier. Figure 14-11 suggests an analogy to activation energy and the reaction profile.

KEEP IN MIND

that the difference in potential energy between reactants and product is ΔU of a reaction. For this reaction $\Delta U = \Delta H$ because the number of product gas molecules is equal to the number of reactant gas molecules. As we learned in Chapter 7, even when this is not the case, the differences between ΔU and ΔH are usually quite small.

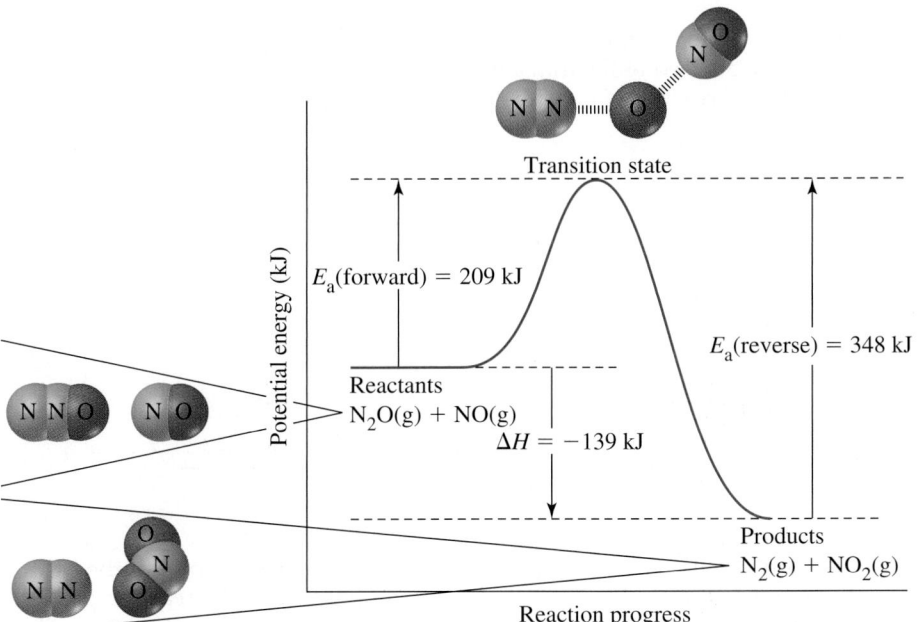

▲ FIGURE 14-10
A reaction profile for the reaction

$$N_2O(g) + NO(g) \longrightarrow N_2(g) + NO_2(g)$$

The simplified reaction profile traces energy changes during the course of reaction (14.20). The reactant and product molecules are depicted by the molecular models, as is the activated complex.

▲ FIGURE 14-11
An analogy for a reaction profile and activation energy
A hike (red path) is taken from the valley on the left (reactants) over the ridge to the valley on the right (products). The ridge above the starting point corresponds to the transition state. It is probably the height of this ridge (activation energy) more than anything else that determines how many people are willing to take the hike, regardless of the fact that it is all downhill on the other side.

Figure 14-10 describes both the forward reaction and its reverse—the reaction of N_2 and NO_2 to form N_2O and NO. The activation energy for the reverse reaction is 348 kJ; this reverse reaction is highly endothermic. Figure 14-10 also illustrates two useful ideas. (1) The enthalpy change of a reaction is equal to the difference in activation energies of the forward and reverse reactions. (2) For an

14-7 ARE YOU WONDERING...

Is there a molecular interpretation of reaction progress?

Simply stated, the reaction progress refers to the minimum energy path (MEP) that leads from reactants to products. To understand MEP we must know how the potential energy (PE) of the system depends on the relative positions of all the particles in the reaction. Let us consider a very simple exothermic gas phase reaction: $A_2(g) + B(g) \longrightarrow AB(g) + A(g)$. The PE as a function of two degrees of freedom for this reaction can be visualized as shown on the right. Such a representation of the PE is commonly referred to as a potential energy surface (PES). In this case the reaction progress is highlighted by the blue line on the PES, and this represents the MEP.

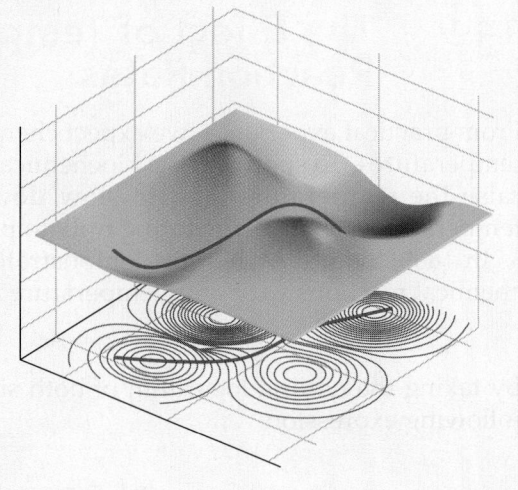

(continued)

The blue line on the PES is the reaction progress and represents the PE of the reactants as they proceed to the transition state and end at the products. The reaction profile for this reaction is given below.

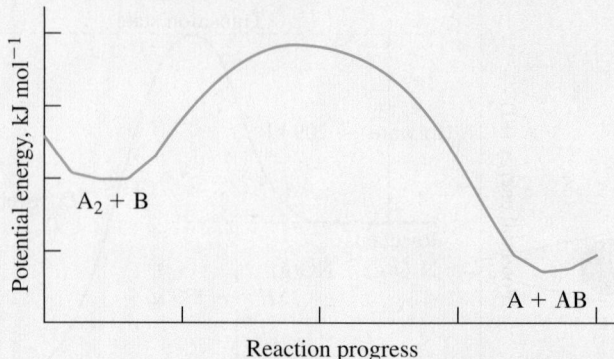

The reaction profile represents a simplified, one-dimensional plot of the MEP on the three-dimensional PES. Note that the highest point on the reaction profile (i.e., along the reaction path) symbolizes the transition state and is not necessarily the highest point on the PES for this reaction.

endothermic reaction, the activation energy must be equal to or greater than the enthalpy of reaction (and usually it is greater).

Attempts at purely theoretical predictions of rate constants have not been very successful. The principal value of reaction-rate theories is to help us explain experimentally observed reaction-rate data. For example, in the next section we will see how the concept of activation energy enters into a discussion of the effect of temperature on reaction rates.

14-6 CONCEPT ASSESSMENT

Indicate whether the following conditions can exist for a chemical reaction: **(a)** $\Delta H < 0 < E_a$; **(b)** $0 < \Delta H < E_a$; **(c)** $0 < E_a < \Delta H$; **(d)** $0 = \Delta H < E_a$; **(e)** $E_a < \Delta H < 0$. Where the conditions can exist, describe the reaction profile.

14-9 The Effect of Temperature on Reaction Rates

From practical experience, we expect chemical reactions to go faster at higher temperatures. To speed up the biochemical reactions involved in cooking, we raise the temperature, and to slow down other reactions, we lower the temperature—as in refrigerating milk to prevent it from souring.

In 1889, Svante Arrhenius demonstrated that the rate constants of many chemical reactions vary with temperature in accordance with the expression

$$k = Ae^{-E_a/RT} \tag{14.21}$$

By taking the natural logarithm of both sides of this equation, we obtain the following expression.

$$\ln k = -\frac{E_a}{RT} + \ln A$$

▶ This equation indicates that a rate constant *increases* as the temperature *increases* and as the activation energy *decreases*.

A graph of ln k versus $1/T$ is a straight line, thus giving us a graphical method for determining the activation energy of a reaction, as shown in

Figure 14-12. We can also derive a useful variation of the equation by writing it twice—each time for a different value of k and the corresponding temperature—and then eliminating the constant $\ln A$. The result, also called the Arrhenius equation, is

◀ This is the same technique used for the Clausius–Clapeyron equation on page 516 and illustrated in Appendix A-4.

$$\ln \frac{k_2}{k_1} = \frac{E_a}{R}\left(\frac{1}{T_1} - \frac{1}{T_2}\right) \qquad \textbf{(14.22)}$$

In equation (14.22)* T_2 and T_1 are two Kelvin temperatures; k_2 and k_1 are the rate constants at these temperatures; and E_a is the activation energy in joules per mole. R is the gas constant expressed as $8.3145\ \mathrm{J\ mol^{-1}\ K^{-1}}$.

Equation (14.22) can be rewritten in several ways.

$$\ln \frac{k_2}{k_1} = -\frac{E_a}{R}\left(\frac{1}{T_2} - \frac{1}{T_1}\right) = -\frac{E_a}{R}\left(\frac{T_1 - T_2}{T_2 T_1}\right) = \frac{E_a}{R}\left(\frac{T_2 - T_1}{T_2 T_1}\right)$$

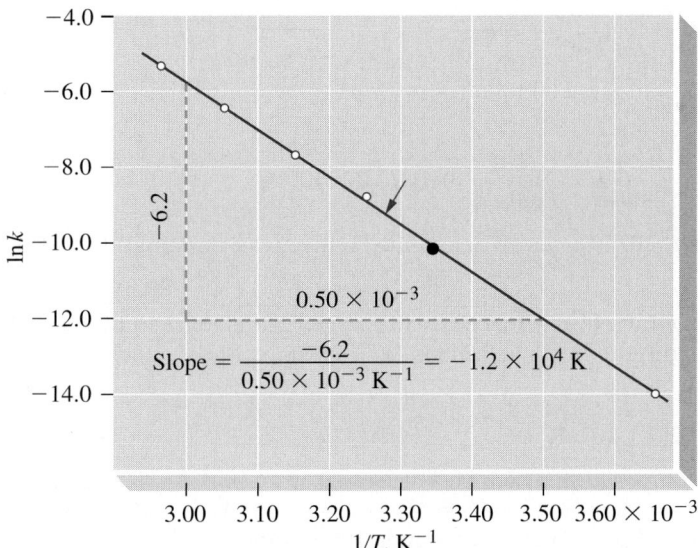

▲ FIGURE 14-12
Temperature dependence of the rate constant k for the reaction

$$N_2O_5\ (\text{in } CCl_4) \longrightarrow N_2O_4\ (\text{in } CCl_4) + \frac{1}{2}\,O_2(g)$$

Data are plotted as follows, for the representative point in black.

$$t = 25\,°C = 298\ K$$
$$1/T = 1/298 = 0.00336 = 3.36 \times 10^{-3}\ K^{-1}$$
$$k = 3.46 \times 10^{-5}\ s^{-1}; \ln k = \ln 3.46 \times 10^{-5} = -10.272$$

To evaluate E_a,

$$\text{slope of line} = -E_a/R = -1.2 \times 10^4\ K$$
$$E_a = 8.3145\ J\ mol^{-1}\ K^{-1} \times 1.2 \times 10^4\ K$$
$$= 1.0 \times 10^5\ J/mol = 1.0 \times 10^2\ kJ/mol$$

(A more precise plot yields a value of $E_a = 106\ kJ/mol$. The arrow points to data referred to in Example 14-9.)

---

*To see that this expression is dimensionally correct, note that on the right side the units in the numerator are $J\ mol^{-1}$ (for E_a) and K^{-1} (for the quantity in parentheses) and that in the denominator the units are $J\ mol^{-1}\ K^{-1}$ (for R). Cancellation yields a dimensionless quantity on the right, to match the dimensionless logarithmic term on the left.

The first form brings out the minus sign and reverses the T_1 and T_2, and the next two combine the fractions within the parentheses. Also, the equation can be written in exponential form.

$$\frac{k_2}{k_1} = e^{-\frac{E_a}{R}\left(\frac{1}{T_2} - \frac{1}{T_1}\right)}$$

Which equation is used is a matter of calculational convenience.

Arrhenius established equation (14.22) by fitting experimental data into his equation. This was before the collision theory of chemical reactions had been developed, but his equation is consistent with the collision theory. In the preceding section, we discussed the importance of (1) the frequency of molecular

---

EXAMPLE 14-9   Applying the Arrhenius Equation

Use data from Figure 14-12 to determine the temperature at which $t_{1/2}$ for the first-order decomposition of N_2O_5 in CCl_4 is 2.00 h.

Analyze

First, find the rate constant k corresponding to a 2.00 h half-life. This can be done by using the half-life for a *first-order reaction*,

$$k = \frac{\ln 2}{t_{1/2}} = \frac{0.693}{2.00\,\text{h}} = \frac{0.693}{7200\,\text{s}} = 9.63 \times 10^{-5}\,\text{s}^{-1}$$

Now, the temperature at which $k = 9.63 \times 10^{-5}\,\text{s}^{-1}$ can be determined in two ways: graphically and analytically.

Solve

Graphical Method. The temperature at which $\ln k = \ln 9.63 \times 10^{-5} = -9.248$ is marked by the red arrow in Figure 14-12. The value of $1/T$ corresponding to $\ln k = -9.248$ is $1/T = 3.28 \times 10^{-3}\,\text{K}^{-1}$, which means that

$$T = \left(\frac{1}{3.28 \times 10^{-3}\,\text{K}^{-1}}\right) = 305\,\text{K} = 32\,°\text{C}$$

Analytical Method. Take T_2 to be the temperature at which $k = k_2 = 9.63 \times 10^{-5}\,\text{s}^{-1}$. T_1 is some other temperature at which a value of k is known. Suppose we take $T_1 = 298\,\text{K}$ and $k_1 = 3.46 \times 10^{-5}\,\text{s}^{-1}$, a point referred to in the caption of Figure 14-12. The activation energy is $106\,\text{kJ/mol} = 1.06 \times 10^5\,\text{J/mol}$ (the more precise value given in Figure 14-12). Now we can solve equation (14.22) for T_2. (For simplicity, we have omitted units below, but the temperature is obtained in kelvins.)

$$\ln \frac{k_2}{k_1} = \frac{E_a}{R}\left(\frac{1}{T_1} - \frac{1}{T_2}\right)$$

$$\ln \frac{9.63 \times 10^{-5}}{3.46 \times 10^{-5}} = \frac{1.06 \times 10^5}{8.3145}\left(\frac{1}{298} - \frac{1}{T_2}\right)$$

$$1.024 = 1.27 \times 10^4 \left(0.00336 - \frac{1}{T_2}\right) = 42.7 - \frac{1.27 \times 10^4}{T_2}$$

$$\frac{1.27 \times 10^4}{T_2} = 42.7 - 1.024 = 41.7$$

$$T_2 = \frac{(1.27 \times 10^4)}{41.7} = 305\,\text{K}$$

Assess

Both methods agree extremely well in this case. Depending on the circumstance one method may be preferred over the other.

PRACTICE EXAMPLE A: What is the half-life of the first-order decomposition of N_2O_5 at 75.0 °C? Use data from Example 14-9.

PRACTICE EXAMPLE B: At what temperature will it take 1.50 h for two-thirds of a sample of N_2O_5 in CCl_4 to decompose in Example 14-9?

collisions, (2) the fraction of collisions energetic enough to produce a reaction, and (3) the need for favorable orientations during collisions. Let's represent the collision frequency by the symbol Z_0. From kinetic-molecular theory, the fraction of sufficiently energetic collisions proves to be $e^{-E_a/RT}$. The probability of favorable orientations of colliding molecules is p and commonly referred to as the steric factor. Typical steric factors are given in the table below. Notice that the steric factor decreases as the complexity of the reactant molecules increases. This reflects the fact that fewer collisions will occur with the proper orientation to produce chemical reactions.

In collision theory, the rate constant of a reaction can be expressed as the product of Z_0, $e^{-E_a/RT}$, and p. If the product $Z_0 \times p$ is replaced by A, collision theory yields a result identical to Arrhenius's experimentally determined equation.

$$k = Z_0 \cdot p \cdot e^{-E_a/RT} = Ae^{-E_a/RT}$$

Steric Factors (p) for Selected Reactions

Reaction	Steric Factor, p
$H + H \longrightarrow H_2(g)$	1
$O(g) + N_2(g) \longrightarrow N_2O(g)$	0.8
$2\,CH_3(g) \longrightarrow C_2H_6(g)$	0.073
$SO(g) + O_2(g) \longrightarrow SO_2(g) + O(g)$	2.4×10^{-3}
$CH_3(g) + C_2H_6(g) \longrightarrow CH_4(g) + C_2H_5(g)$	7.3×10^{-4}
$H_2(g) + C_2H_4(g) \longrightarrow C_2H_6(g)$	2.5×10^{-6}

▲ Both the rate of chirping of tree crickets and the flashing of fireflies roughly double for a 10 °C temperature rise. This corresponds to an activation energy of about 50 kJ/mol and suggests that the physiological processes governing these phenomena involve chemical reactions.

🔍 14-7   CONCEPT ASSESSMENT

Without doing calculations, use a 1n k versus $1/T$ graph to explain whether the rate of change of the rate constant with temperature is affected more strongly by a low or high energy of activation.

14-10   Reaction Mechanisms

$NO_2(g)$ is known to play a key role in the formation of photochemical smog (see page 635), but it is unlikely that very much of this gas is formed in the atmosphere by the direct reaction

$$2\,NO(g) + O_2(g) \longrightarrow 2\,NO_2(g) \qquad \textbf{(14.23)}$$

For this reaction to occur in a single step in the manner suggested by equation (14.23), *three* molecules would have to collide simultaneously, or very nearly so. A three-molecule collision is an unlikely event. The reaction appears to follow a different mechanism or pathway. One of the main purposes in determining rate laws of chemical reactions is to relate them to probable reaction mechanisms.

A **reaction mechanism** is a step-by-step detailed description of a chemical reaction. Each step in a mechanism is called an **elementary process**, which describes any molecular event that significantly alters a molecule's energy or geometry or produces a new molecule. Two requirements of a plausible reaction mechanism are that it must

◀ Although much progress has been made in the theoretical understanding of reaction mechanisms, by far most of the data and rate constants are obtained experimentally.

- be consistent with the stoichiometry of the overall reaction
- account for the *experimentally determined* rate law

▲ FIGURE 14-13
The San Ysidro/Tijuana border station: An analogy for a rate-determining step
Note the tie-up of traffic on the Mexican side of the border (top) and the relatively few cars on the United States side (bottom). This station is a bottleneck and hence the rate-determining part of the trip by car from Tijuana, Mexico, to San Diego, California, two cities located only about a dozen miles apart.

In this section, we will first explore the nature of elementary processes and then apply these processes to two simple types of reaction mechanisms.

Elementary Processes

The characteristics of elementary processes are as follows:

1. Elementary processes are either **unimolecular**—a process in which a single molecule dissociates—or **bimolecular**—a process involving the collision of two molecules. A *termolecular* process, which would involve the simultaneous collision of three molecules, is relatively rare as an elementary process.

2. The exponents of the concentration terms in the rate law for an elementary process are the *same* as the stoichiometric coefficients in the balanced equation for the process. (Note that this is unlike the case of the overall rate law, for which the exponents are *not* necessarily related to the stoichiometric coefficients in the overall equation.)

3. Elementary processes are reversible, and some may reach a condition of equilibrium in which the rates of the forward and reverse processes are equal.

4. Certain species are produced in one elementary process and consumed in another. In a proposed reaction mechanism, such intermediates must not appear in either the overall chemical equation or the overall rate law.

5. One elementary process may occur much more slowly than all the others, and in some cases may determine the rate of the overall reaction. Such a process is called the **rate-determining step** (Fig. 14-13).

Keep these characteristics in mind as we will apply them in our analysis of different mechanisms below.

A Mechanism with a Slow Step Followed by a Fast Step

The reaction between gaseous iodine monochloride and gaseous hydrogen produces iodine and hydrogen chloride as gaseous products.

$$H_2(g) + 2\,ICl(g) \longrightarrow I_2(g) + 2\,HCl(g)$$

The experimentally determined rate law for this reaction is

$$\text{rate of reaction} = k[H_2][ICl]$$

Let's begin with a mechanism that seems plausible, such as the following two-step mechanism.

(1) Slow:	$H_2 + ICl \longrightarrow HI + HCl$	
(2) Fast:	$HI + ICl \longrightarrow I_2 + HCl$	
Overall:	$H_2 + 2\,ICl \longrightarrow I_2 + 2\,HCl$	

This scheme seems plausible for two reasons: (1) The sum of the two steps yields the experimentally observed *overall* reaction. (2) As we have noted, unimolecular and bimolecular elementary processes are most plausible, and each step in the above mechanism is bimolecular. Because each step is an elementary process, we can write

$$\text{rate}\,(1) = k_1[H_2][ICl] \quad \text{and} \quad \text{rate}\,(2) = k_2[HI][ICl]$$

Now, note that our mechanism proposes that step (1) occurs slowly but step (2) occurs rapidly. This suggests that HI is consumed in the second step just as fast as it is formed in the first. The first step is the rate-determining step, and

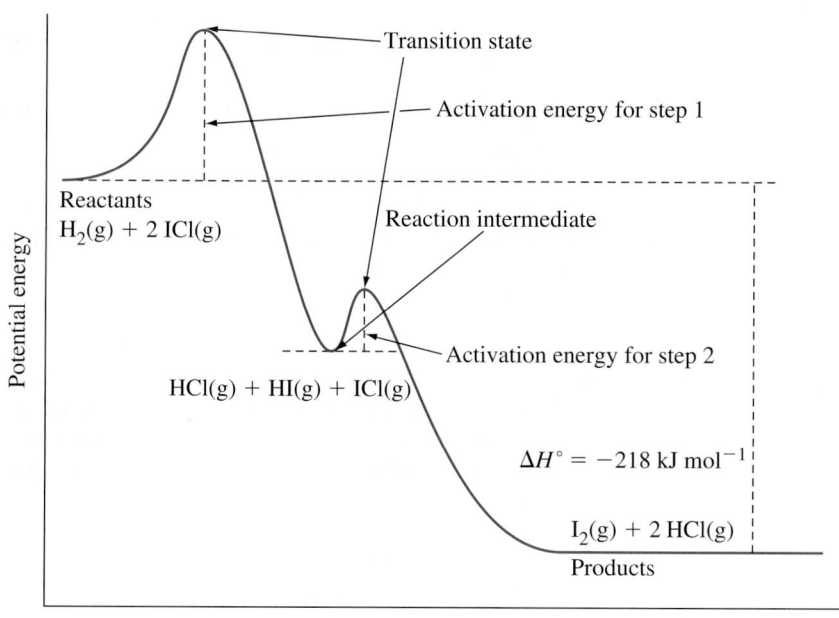

Potential energy

Transition state

Activation energy for step 1

Reactants
$H_2(g) + 2\,ICl(g)$

Reaction intermediate

$HCl(g) + HI(g) + ICl(g)$

Activation energy for step 2

$\Delta H° = -218\ \text{kJ mol}^{-1}$

$I_2(g) + 2\,HCl(g)$

Products

Reaction progress

◀ FIGURE 14-14
A reaction profile for a two-step mechanism

the rate of the overall reaction is governed just by the rate at which HI is formed in this first step, that is, by rate (1). This explains why the observed rate law for the net reaction is rate of reaction $= k[H_2][ICl]$. The proposed mechanism gives a rate law that is in agreement with experiment, as it should if we have made a reasonable proposal.

The species HI is called a **reaction intermediate**, and it does not appear in the experimental rate law. In this case, the intermediate species is a well-known stable molecule. Often, when postulating mechanisms, we have to invoke less well-known and less stable species; and in these instances, we have to rely on the chemical reasonableness of the basic assumptions. The presence of a reaction intermediate leads to a slightly more complicated reaction profile. The reaction profile for the two steps in the proposed mechanism is shown in Figure 14-14. We see that there are two transition states and one reaction intermediate. The activation energy for the first step is greater than that for the second step, which we would expect if the first step in the reaction mechanism is the slowest. It is important to understand the difference between a transition state (activated complex) and a reaction intermediate. The transition state represents the highest energy structure involved in a reaction (or step in a mechanism). While transition states exist only momentarily and can never be isolated, reaction intermediates can sometimes be isolated. Transition states have *partially formed bonds*, whereas reaction intermediates have *fully formed bonds*.

◀ A transition state is not a "real" species. It is only hypothetical.

A Mechanism with a Fast Reversible First Step Followed by a Slow Step

The rate law for the reaction of $NO(g)$ and $O_2(g)$

$$2\,NO(g) + O_2(g) \longrightarrow 2\,NO_2(g) \qquad \textbf{(14.23)}$$

is found to be

$$\text{rate of reaction} = k[NO]^2[O_2] \qquad \textbf{(14.24)}$$

Even though it is consistent with this rate law, we have already noted that the one-step *termolecular* mechanism suggested by equation (14.23) is highly improbable. Let's explore instead the following mechanism.

▶ The IUPAC convention for labeling rate constants is as follows: The rate constants for the steps in a mechanism are numbered sequentially, k_1, k_2, k_3, and so on. If step n is a reversible step, then the rate constant for the reverse reaction is denoted k_{-n}. This is just a labeling convention and no mathematical relationship is implied.

$$\text{Fast:} \qquad 2\,NO(g) \underset{k_{-1}}{\overset{k_1}{\rightleftharpoons}} N_2O_2(g) \qquad (14.25)$$

$$\text{Slow:} \qquad N_2O_2(g) + O_2(g) \overset{k_2}{\longrightarrow} 2\,NO_2(g) \qquad (14.26)$$

$$\text{Overall:} \qquad 2\,NO(g) + O_2(g) \longrightarrow 2\,NO_2(g) \qquad (14.23)$$

In this mechanism, there is a rapid equilibrium as the first step, but some of the N_2O_2 is slowly drawn off and consumed in the second, slow step. The rate law for the slow, or *rate-determining*, step (14.26) is

$$\text{rate of reaction} = k_2[N_2O_2][O_2] \qquad (14.27)$$

Because N_2O_2 is an *intermediate*, however, we must eliminate it from the rate law. We are told that the first step of the mechanism consists of a fast reversible reaction, so we can assume that this step progresses rapidly to equilibrium. If this is the case, the forward and reverse rates of reaction in the first step become equal and we write

$$\text{rate of forward reaction} = \text{rate of reverse reaction}$$

$$k_1[NO]^2 = k_{-1}[N_2O_2]$$

Now, let us arrange this equation into an expression having a ratio of rate constants on one side and a ratio of concentration terms on the other. Also, we can replace the ratio of rate constants by a single constant, which we will represent as K_1.

$$K_1 = \frac{k_1}{k_{-1}} = \frac{[N_2O_2]}{[NO]^2}$$

▶ Equilibrium constant expressions are of fundamental importance throughout chemistry. Here, we see their significance in chemical kinetics. In Chapter 15, we describe the experimental basis of equilibrium constant expressions and their application to the stoichiometry of reversible reactions. In Chapter 19, we explore their thermodynamic basis.

The above expression is known as an equilibrium constant expression; the numerical constant, K_1, is an equilibrium constant. Next, we rearrange the expression to solve for the term $[N_2O_2]$.

$$[N_2O_2] = K_1[NO]^2$$

Then, substituting this into equation (14.27) we obtain the experimentally observed rate law.

$$\text{rate of reaction} = k_2[N_2O_2][O_2] = k_2K_1[NO]^2[O_2]$$

The experimentally observed rate constant, k, is related to the other constants in the proposed mechanism, as follows:

$$k = k_2K_1 = k_2 \times \frac{k_1}{k_{-1}}$$

The type of mechanism described here with a *rapid pre-equilibrium* is a very common mechanism and is to be expected when the overall stoichiometry suggests an unlikely termolecular collision.

We have just shown that the proposed mechanism is consistent with (a) the reaction stoichiometry and (b) the experimentally determined rate law. Whether this mechanism is the actual reaction path, we cannot say, however. All that we can say is that it is *plausible*; it has not been ruled out by kinetics.

The Steady-State Approximation

The reaction mechanisms considered so far have had one particular rate-determining step, and the rate law of the reaction could be deduced from the rate of this step after the relationships for the concentrations of any intermediates were established. In complex multistep reaction mechanisms, however, more than one step may control the rate of a reaction.

EXAMPLE 14-10   Testing a Reaction Mechanism

An alternative mechanism of the reaction $2\,NO(g) + O_2(g) \longrightarrow 2\,NO_2(g)$ follows. Show that this mechanism is consistent with the rate law (14.24).

$$\textit{Fast:} \qquad NO(g) + O_2(g) \underset{k_{-1}}{\overset{k_1}{\rightleftharpoons}} NO_3(g)$$

$$\textit{Slow:} \qquad NO_3(g) + NO(g) \xrightarrow{k_2} 2\,NO_2(g)$$

$$\textit{Overall:} \qquad 2\,NO(g) + O_2(g) \longrightarrow 2\,NO_2(g)$$

Analyze

In this type of problem we begin by identifying the slow step (which is typically given) and using it to write the rate of reaction. Because the fast step shown above is given as an equilibrium, we can assume that the equilibrium is rapidly established. The common species between the two reactions, NO_3, also called an intermediate, is not found in the rate law, and so it can be eliminated by using the reaction equilibrium constant expression for the fast step.

Solve

The rate equation for the rate-determining step is

$$\text{rate of reaction} = k_2[NO_3][NO]$$

Eliminate $[NO_3]$, by assuming that the pre-equilibrium is rapidly established.

$$\text{rate of forward reaction} = \text{rate of reverse reaction}$$
$$k_1[NO][O_2] = k_{-1}[NO_3]$$

Rearrange the previous equation to develop an expression for an equilibrium constant (K) in terms of the rate constants k_1 and k_{-1}.

$$K = \frac{k_1}{k_{-1}} = \frac{[NO_3]}{[NO][O_2]}$$

Rearrange this expression for K to solve for $[NO_3]$.

$$[NO_3] = K[NO][O_2]$$

Finally, substitute the value of $[NO_3]$ into the rate equation to obtain the observed rate law (equation 14.24).

$$\text{rate} = k_2[NO_3][NO]$$
$$\text{rate of reaction} = k_2K[NO]^2[O_2] = k_2\frac{k_1}{k_{-1}}[NO]^2[O_2] = k[NO]^2[O_2]$$

Assess

The final rate law obtained, rate $= k[NO]^2[O_2]$, is consistent with the experimental rate law. This alternative mechanism is plausible based on this analysis. However, it does not mean that this is the reaction mechanism.

PRACTICE EXAMPLE A: In a proposed two-step mechanism for the reaction $CO(g) + NO_2(g) \longrightarrow CO_2(g) + NO(g)$, the second, fast step is $NO_3(g) + CO(g) \longrightarrow NO_2(g) + CO_2(g)$. What must be the *slow* step? What would you expect the rate law of the reaction to be? Explain.

PRACTICE EXAMPLE B: Show that the proposed mechanism for the reaction $2\,NO_2(g) + F_2(g) \longrightarrow 2\,NO_2F(g)$ is plausible. The rate law is rate $= k[NO_2][F_2]$.

$$\textit{Fast:} \qquad NO_2(g) + F_2(g) \rightleftharpoons NO_2F_2(g)$$

$$\textit{Slow:} \qquad NO_2F_2(g) \longrightarrow NO_2F(g) + F(g)$$

$$\textit{Fast:} \qquad F(g) + NO_2(g) \longrightarrow NO_2F(g)$$

To illustrate, let's reconsider the first mechanism presented for the reaction of nitric oxide with oxygen, but this time we will make no assumptions about the relative rates of the steps in the mechanism. The proposed mechanism is

$$NO + NO \xrightarrow{k_1} N_2O_2$$

$$N_2O_2 \xrightarrow{k_{-1}} NO + NO$$

$$N_2O_2 + O_2 \xrightarrow{k_2} 2\,NO_2$$

For clarity, the first, reversible reaction is written as two separate steps.

We choose one of the steps of the mechanism that provides a convenient relationship to the observed rate of reaction. In this case, we will use the last step, since it involves the disappearance of O_2. Therefore, the rate of the reaction from this mechanism is

KEEP IN MIND

that if the rate of change of the concentration of a substance is zero, then the concentration of that substance is constant.

$$\text{rate of reaction} = k_2[N_2O_2][O_2] \qquad \textbf{(14.27)}$$

As before, the intermediate N_2O_2 must be eliminated from this rate law. We can do this by assuming that the $[N_2O_2]$ reaches a *steady-state condition* in which N_2O_2 is produced and consumed at equal rates. That is, $[N_2O_2]$ remains constant throughout most of the reaction. We can use the steady-state assumption to express $[N_2O_2]$ in terms of $[NO]$.

$$\Delta[N_2O_2]/\Delta t = \text{rate of formation of } N_2O_2 - \text{rate of disappearance of } N_2O_2 = 0$$

$$\text{rate of formation of } N_2O_2 = \text{rate of disappearance of } N_2O_2$$

The rate of disappearance of N_2O_2 is made up of two parts—the reverse step of equation (14.25) and the forward step of (14.26)—so we write

$$\text{rate of disappearance of } N_2O_2 = k_{-1}[N_2O_2] + k_2[N_2O_2][O_2]$$

The two rates for the steps depleting the concentration of N_2O_2 have been added. Now, as dictated by the steady-state assumption, the rate of disappearance of N_2O_2 is equated with the rate of appearance of N_2O_2, which is $k_1[NO]^2$.

$$k_1[NO]^2 = k_{-1}[N_2O_2] + k_2[N_2O_2][O_2] = [N_2O_2](k_{-1} + k_2[O_2])$$

Rearranging to solve for $[N_2O_2]$, we have

$$[N_2O_2] = \frac{k_1[NO]^2}{k_{-1} + k_2[O_2]}$$

We now substitute this into equation (14.27) to obtain

$$\text{rate} = k_2[O_2][N_2O_2] = k_2[O_2]\left(\frac{k_1[NO]^2}{k_{-1} + k_2[O_2]}\right)$$

or

▶ Note that this rate does not follow the usual nth order reaction rate law as seen in Table 14.5.

$$\text{rate} = \frac{k_1k_2[O_2][NO]^2}{k_{-1} + k_2[O_2]}$$

This is the rate law for the proposed mechanism based on our steady-state analysis. This rate law is more complicated than the observed rate law. What happened? In carrying out the steady-state calculation, we did not make any assumptions about the relative rates of the three steps in the mechanism. If we now make the assumption that the rate of disappearance of N_2O_2 in the second step of the proposed mechanism is greater than the rate of disappearance of N_2O_2 in the third step of the proposed mechanism, then

$$k_{-1}[N_2O_2] > k_2[N_2O_2][O_2]$$

which means

$$k_{-1} > k_2[O_2]$$

and

$$k_{-1} + k_2[O_2] \approx k_{-1}$$

so that

$$[N_2O_2] = \frac{k_1[NO]^2}{k_{-1}}$$

If we substitute this value of $[N_2O_2]$ into the rate equation (14.27) and replace k_1k_2/k_{-1} by k, we obtain for the overall reaction

$$\text{rate of reaction} = \frac{k_1k_2}{k_{-1}}[\text{NO}]^2[\text{O}_2] = k[\text{NO}]^2[\text{O}_2] \qquad \textbf{(14.24)}$$

The result of a steady-state analysis of any mechanism in which no rate-determining step can be identified will often be a complicated rate law. The use of this type of rate law is illustrated in the section on enzyme catalysis later in this chapter.

Smog—An Environmental Problem with Roots in Chemical Kinetics

About 100 years ago, a new word entered the English language—**smog**. It referred to a condition, common in London, in which a combination of *smoke* and *fog* obscured visibility and produced health hazards (including death). These conditions are often associated with heavy industry, and this type of smog is now called *industrial* smog.

A more familiar form of air pollution commonly thought of as smog results from the action of sunlight on the products of combustion. Chemical reactions brought about by light are called *photochemical* reactions, and the smog formed primarily as a result of photochemical reactions is photochemical smog. This type of smog is associated with high-temperature combustion processes, such as those that occur in internal combustion engines. Because the combustion of motor fuels takes place in air rather than in pure oxygen, oxides of nitrogen, principally $NO(g)$, are inevitably found in the exhaust from motor vehicles. Other products found in the exhaust are hydrocarbons (unburned gasoline) and partially oxidized hydrocarbons. These, then, are the starting materials—the precursors—of photochemical smog.

Many substances have been identified in smoggy air, including NO, NO_2, O_3 (ozone, an allotrope of oxygen introduced on page 416 and discussed further in Section 22-4), and a variety of organic compounds derived from gasoline hydrocarbons. Ozone is very reactive and is largely responsible for the breathing difficulties that some people experience during smog episodes. Another noxious substance found in smog is an organic compound known as peroxyacetyl nitrate (PAN). PAN is a powerful lacrimator—that is, it causes tear formation in the eyes. Photochemical smog components cause heavy crop damages (to oranges, for example) and the deterioration of rubber goods. And, of course, the best-known symbol of photochemical smog is the hazy brown air that results in reduced visibility (Fig. 14-15).

Chemists who have been studying photochemical smog formation over the past several decades have determined that certain precursors are converted to the observable smog components through the action of sunlight. Because the chemical reactions involved are very complex and still not totally understood, we will give only a very brief, simplified reaction mechanism showing how photochemical smog is formed.

Smog formation begins with $NO(g)$, produced by reaction (14.28).

$$N_2(g) + O_2(g) \xrightarrow{\Delta} 2\,NO(g) \qquad \textbf{(14.28)}$$

$NO(g)$ is subsequently converted to $NO_2(g)$, which then absorbs ultraviolet radiation from sunlight and decomposes.

$$NO_2 + \text{sunlight} \longrightarrow NO + O \qquad \textbf{(14.29)}$$

This is followed by a reaction forming ozone, O_3.

$$O + O_2 \longrightarrow O_3 \qquad \textbf{(14.30)}$$

◀ Air quality in London has been greatly improved by control measures, such as elimination of coal as a household fuel, introduced after a severe smog episode in 1952.

▲ FIGURE 14-15
Smog in Mexico City
At times, the topographical features, climatic conditions, traffic congestion, and heavy industrial pollution combine to create severe smog conditions in Mexico City.

Thus, a large buildup of ozone in photochemical smog requires a plentiful source of NO_2. At one time, this source was believed to be reaction (14.31).

$$2 NO + O_2 \longrightarrow 2 NO_2 \qquad \textbf{(14.31)}$$

It is now well established, however, that reaction (14.31) occurs much too slowly to yield the required levels of NO_2 in photochemical smog. Instead, NO is rapidly converted to NO_2 when it reacts with O_3,

$$O_3 + NO \longrightarrow O_2 + NO_2 \qquad \textbf{(14.32)}$$

Even though reaction (14.32) accounts for the formation of NO_2, it leads to the destruction of ozone. Thus, photochemical smog formation cannot occur just through the reaction sequence: (14.28), (14.29), (14.30), and (14.31). The ozone would be consumed as quickly as it was formed, and there would be no ozone buildup at all.

It is now known that organic compounds, particularly unburned hydrocarbons in automotive exhaust, provide a pathway for the conversion of NO to NO_2. The reaction sequence that follows involves some highly reactive molecular fragments. Recall (page 418) that these fragments are known as *free radicals* and are represented by formulas written with a bold dot. RH represents a hydrocarbon molecule, and $R\cdot$ is a free-radical fragment of a hydrocarbon molecule. Oxygen atoms, fragments of the O_2 molecule, are also represented as free radicals, as are hydroxyl groups, fragments of the H_2O molecule.

$$RH + O\cdot \longrightarrow R\cdot + \cdot OH$$
$$RH + \cdot OH \longrightarrow R\cdot + H_2O$$
$$R\cdot + O_2 \longrightarrow RO_2\cdot$$
$$RO_2\cdot + NO \longrightarrow RO\cdot + NO_2$$

The final step in this reaction mechanism accounts for the rapid conversion of NO to NO_2 that seems essential to smog formation.

The role of NO_2 in the formation of the smog component PAN is suggested by the equation

$$\underset{}{CH_3\overset{\overset{\displaystyle O}{\|}}{C}-O-O\cdot} + NO_2 \longrightarrow \underset{PAN}{CH_3\overset{\overset{\displaystyle O}{\|}}{C}-O-ONO_2}$$

The details of smog formation have been worked out in part through the use of smog chambers. By varying experimental conditions in these chambers, scientists have been able to create polluted atmospheres very similar to smog. For example, they have found that if hydrocarbons are omitted from the starting materials in the smog chamber, no ozone is formed. The reaction scheme just proposed is consistent with this observation. Figure 14-16 gives a typical result from a smog chamber.

To control smog, automobiles are now provided with *catalytic converters*. CO and hydrocarbons are oxidized to CO_2 and H_2O in the presence of an oxidation catalyst such as platinum or palladium metal. NO must be reduced to N_2, and this requires a reduction catalyst. A dual-catalyst system uses both types of catalysts. Alternatively, the air-fuel ratio of the engine is set to produce some CO and unburned hydrocarbons; these then act as reducing agents to reduce NO to N_2.

$$2 CO(g) + 2 NO(g) \longrightarrow 2 CO_2(g) + N_2(g)$$

Next, the exhaust gases are passed through an oxidation catalyst to oxidize the remaining hydrocarbons and CO to CO_2 and H_2O. Future smog-control measures may include the use of alternative fuels, such as methanol or hydrogen, and the development of electric-powered automobiles.

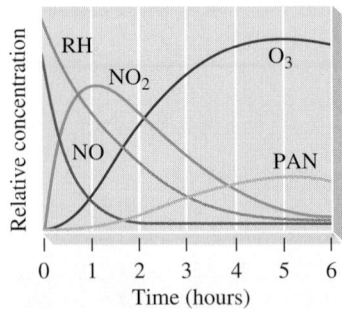

▲ FIGURE 14-16
Smog component profile
Data from a smog chamber show how the concentrations of smog components change with time. For example, the concentrations of hydrocarbon (RH) and nitrogen monoxide (NO) fall continuously, whereas that of nitrogen dioxide (NO_2) rises to a maximum and then drops off. The concentrations of ozone (O_3) and peroxyacetyl nitrate (PAN) build up more slowly. Any reaction scheme proposed to explain smog formation must be consistent with observations such as these. Under actual smog conditions, the pattern of concentration changes shown here repeats itself on a daily basis.

14-11 Catalysis

A reaction can generally be made to go faster by increasing the temperature. Another way to speed up a reaction is to use a catalyst. A **catalyst** provides an alternative reaction pathway of lower activation energy. The catalyst participates in a chemical reaction but does not itself undergo a permanent change. As a result, the formula of a catalyst does not appear in the overall chemical equation (its formula is generally placed over the reaction arrow).

The success of a chemical process often hinges on finding the right catalyst, as in the manufacture of nitric acid. By conducting the oxidation of $NH_3(g)$ very quickly (less than 1 ms) in the presence of a Pt–Rh catalyst, $NO(g)$ can be obtained as a product instead of $N_2(g)$. The formation of $HNO_3(aq)$ from $NO(g)$ then follows easily.

In this section, the two basic types of catalysis—homogeneous and heterogeneous—are described first. This is followed by discussions of the catalyzed decomposition of $H_2O_2(aq)$ and the biological catalysts called enzymes.

Homogeneous Catalysis

Figure 14-17 shows reaction profiles for the decomposition of formic acid (HCOOH). In the uncatalyzed reaction, a H atom must be transferred from one part of the formic acid molecule to another, shown by the arrow. Then a C—O bond breaks. Because the energy requirement for this atom transfer is high, the activation energy is high and the reaction is slow. In the acid-catalyzed decomposition of formic acid, a hydrogen ion from solution attaches itself to the O atom that is singly bonded to the C atom to form $[HCOOH_2]^+$. The C—O bond breaks, and a H atom attached to a carbon atom in the intermediate species $[HCO]^+$ is released to the solution as H^+.

$$\underset{\substack{\| \\ H-C-O-H}}{O} + H^+ \longrightarrow \left[\underset{\substack{\| \quad | \\ H-C-O-H}}{O \quad H} \right]^+ \longrightarrow \left[\underset{\substack{\| \\ H-C}}{O} \right]^+ + H_2O$$
$$H^+ + C\equiv O$$

This catalyzed reaction pathway does not require a H atom to be transferred within the formic acid molecule. It has a lower activation energy than does the uncatalyzed reaction and proceeds at a faster rate. Because the reactants and products of this reaction are all present throughout the solution, or homogeneous mixture, this type of catalysis is called **homogeneous catalysis**.

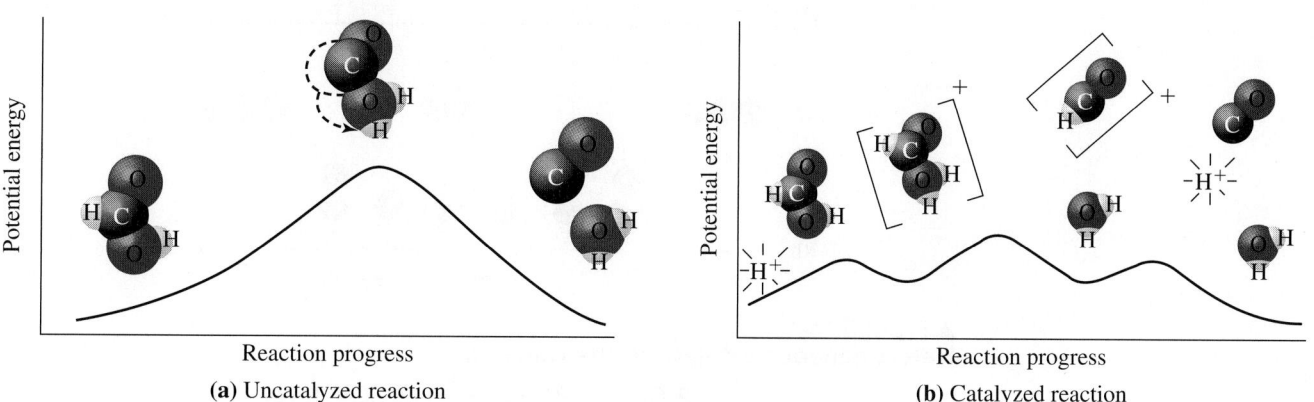

Potential energy — Reaction progress
(a) Uncatalyzed reaction

Potential energy — Reaction progress
(b) Catalyzed reaction

▲ FIGURE 14-17
An example of homogeneous catalysis
The activation energy is lowered in the presence of H^+, a catalyst for the decomposition of HCOOH.

Heterogeneous Catalysis

Many reactions can be catalyzed by allowing them to occur on an appropriate solid surface. Essential reaction intermediates are found on the surface. This type of catalysis is called **heterogeneous catalysis** because the catalyst is present in a different phase of matter than are the reactants and products. Catalytic activity is associated with many transition elements and their compounds. The precise mechanism of heterogeneous catalysis is not totally understood, but in many cases the availability of electrons in *d* orbitals in surface atoms may play a role.

A key feature of heterogeneous catalysis is that reactants from a gaseous or solution phase are adsorbed, or attached, to the surface of the catalyst. Not all surface atoms are equally effective for catalysis; those that are effective are called **active sites**. Basically, heterogeneous catalysis involves (1) *adsorption* of reactants; (2) *diffusion* of reactants along the surface; (3) *reaction* at an active site to form adsorbed product; and (4) *desorption* of the product.

An interesting reaction is the oxidation of CO to CO_2 and the reduction of NO to N_2 in automotive exhaust gases as a smog-control measure. Figure 14-18 shows how this reaction is thought to occur on the surface of rhodium metal in a catalytic converter. In general, the reaction profile for a surface–catalyzed reaction resembles that shown in Figure 14-19.

The Catalyzed Decomposition of Hydrogen Peroxide

As previously noted (see the footnote on page 605), the decomposition of $H_2O_2(aq)$ is a slow reaction and generally must be catalyzed. Iodide ion is a good catalyst that seems to function by means of the following two-step mechanism.

$$\textit{Slow:} \qquad H_2O_2 + I^- \longrightarrow OI^- + H_2O$$

$$\textit{Fast:} \qquad H_2O_2 + OI^- \longrightarrow H_2O + I^- + O_2(g)$$

$$\overline{\textit{Overall:} \qquad 2\,H_2O_2 \longrightarrow 2\,H_2O + O_2(g)}$$

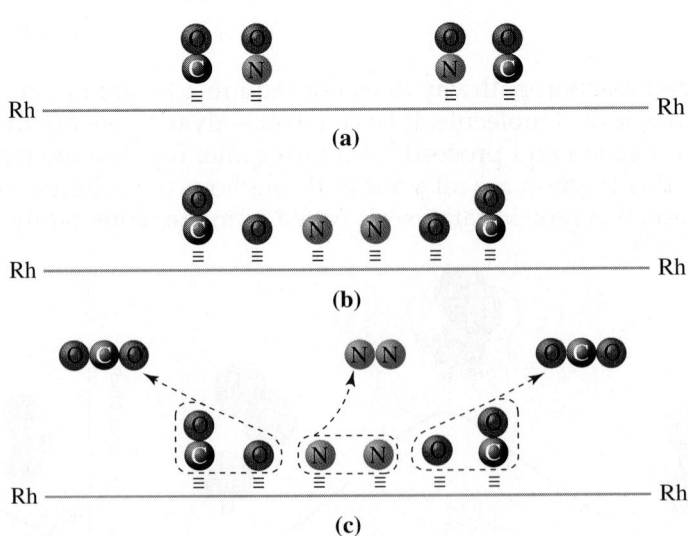

▲ FIGURE 14-18
Heterogeneous catalysis in the reaction

$$2\,CO\ +\ 2\,NO\ \xrightarrow{\ Rh\ }\ 2\,CO_2\ +\ N_2$$

(a) Molecules of CO and NO are adsorbed on the rhodium surface. **(b)** The adsorbed NO molecules dissociate into adsorbed N and O atoms. **(c)** Adsorbed CO molecules and O atoms combine to a form CO_2 molecules, which desorb into the gaseous state. Two N atoms combine and are desorbed as a N_2 molecule.

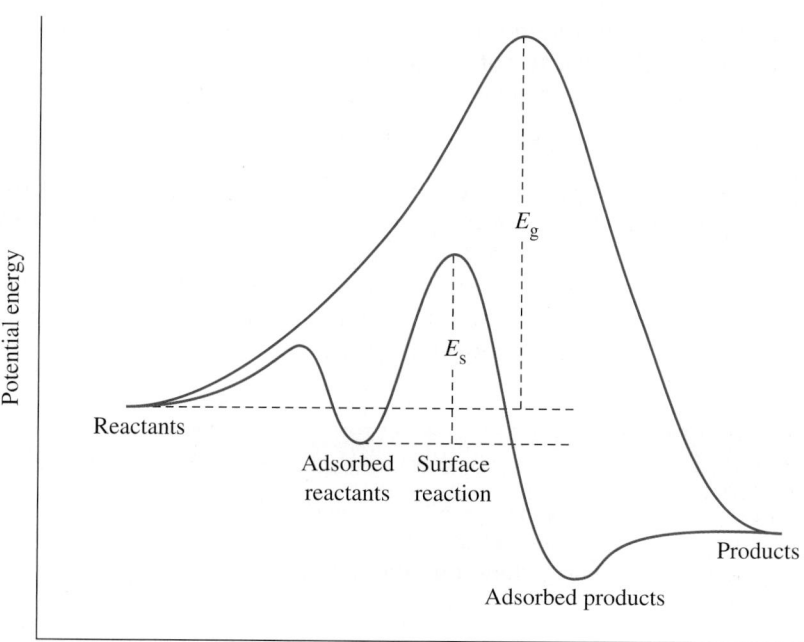

▲ FIGURE 14-19
Reaction profile for a surface-catalyzed reaction
In the reaction profile (blue) for the surface-catalyzed reaction, the activation energy
for the reaction step, E_s, is considerably less than in the reaction profile (red) for the
uncatalyzed gas-phase reaction, E_g.

As required for a catalyzed reaction, the formula of the catalyst does not
appear in the overall equation. Neither does the intermediate species OI^-. The
rate of reaction of H_2O_2 is determined by the rate of the slow first step.

$$\text{rate of reaction } H_2O_2 = k[H_2O_2][I^-] \qquad \textbf{(14.33)}$$

◀ Although catalysts are
not in the overall chemical
equation, they do participate
in the chemical reaction and
do appear in the mechanism.

◀ The decomposition of hydrogen
peroxide, H_2O_2, to H_2O and O_2 is a
highly exothermic reaction that is
catalyzed by platinum metal.

Because I^- is constantly regenerated, its concentration is constant throughout a given reaction. If the product of the constant terms $k[I^-]$ is replaced by a new constant, k', the rate law can be rewritten as

$$\text{rate of reaction of } H_2O_2 = k'[H_2O_2] \qquad \textbf{(14.34)}$$

Equation (14.33) indicates that the rate of decomposition of $H_2O_2(aq)$ is affected by the initial concentration of I^-. For each initial concentration of I^-, we obtain a different rate constant, k', in equation (14.34).

We have just described the homogeneous catalysis of the decomposition of hydrogen peroxide. The decomposition can also be catalyzed by heterogeneous catalysis, as seen at the bottom of page 639.

🔍 14-8 CONCEPT ASSESSMENT

In the production of ammonia from nitrogen and hydrogen, the rate of reaction can be increased by using a catalyst or by increasing the temperature. Is the means by which these two different methods increase the rate of reaction the same? Explain.

▶ Many people lose the ability to produce lactase when they become adults. In such cases, lactose passes through the small intestine into the colon, where it ferments and may cause severe gastric disturbances.

Enzymes as Catalysts

Unlike platinum, which catalyzes a wide variety of reactions, the catalytic action of high-molar-mass proteins known as **enzymes** is very specific. For example, in the digestion of milk, lactose, a more complex sugar, breaks down into two simpler ones, glucose and galactose. This occurs in the presence of the enzyme *lactase*.

$$\text{lactose} \xrightarrow{\text{lactase}} \text{glucose} + \text{galactose}$$

"Milk sugar"

Biochemists describe enzyme activity with the "lock-and-key" model (Fig. 14-20). The reacting substance, the **substrate** (S), attaches itself to the enzyme (E) at a particular point called an *active site* to form the enzyme–substrate complex (ES). The complex decomposes to form products (P) and regenerate the enzyme.

$$E + S \underset{k_{-1}}{\overset{k_1}{\rightleftharpoons}} ES$$

$$ES \xrightarrow{k_2} E + P$$

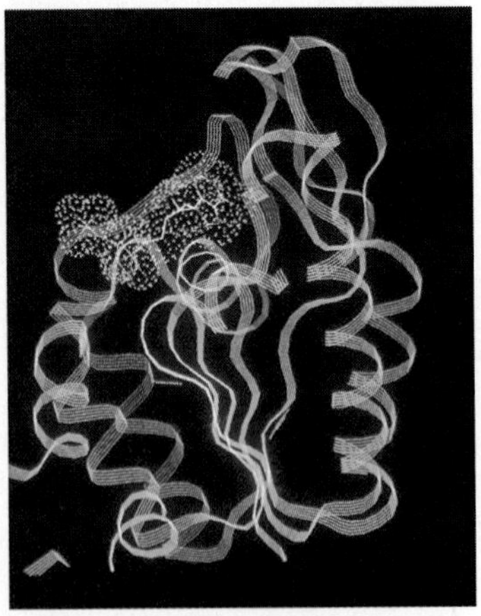

▲ A computer graphics representation of the enzyme phosphoglycerate kinase (carbon backbone shown as blue ribbon). A molecule of ATP, the substrate, is shown in green.

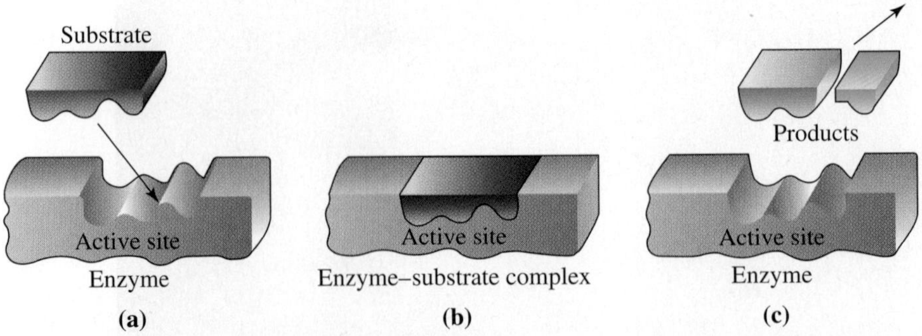

▲ FIGURE 14-20
Lock-and-key model of enzyme action
(a) The substrate attaches itself to an active site on an enzyme molecules. (b) Reaction occurs. (c) Product molecules detach themselves from the site, freeing the enzyme molecule to attach another molecule of substrate. The substrate and enzyme must have complementary structures to produce a complex, hence the term *lock and key*.

Most human enzyme-catalyzed reactions proceed fastest at about 37 °C (body temperature). If the temperature is raised much higher than that, the structure of the enzyme changes, the active sites become distorted, and the catalytic activity is lost.

Determining the rates of enzyme-catalyzed reactions is an important part of enzyme studies. Figure 14-21, a plot of reaction rate against substrate concentration, illustrates what we generally observe. Along the rising portion of the graph, the rate of reaction is proportional to the substrate concentration, [S]. The reaction is first order: rate of reaction = k[S]. At high substrate concentrations, the rate is independent of [S]. The reaction follows a zero-order rate equation: rate of reaction = k.

This behavior can be understood in terms of the three-step mechanism given above. The rate at which the product appears, which is often called the velocity (V) of the reaction by biochemists, is given by

$$\text{rate of production of P} = V = k_2[\text{ES}]$$

To proceed further, we need an expression for the enzyme–substrate complex, and we can get this by applying the steady-state approximation to the concentration of the enzyme–substrate complex.

$$\text{rate of formation of ES} = \text{rate of destruction of ES}$$

$$k_1[\text{E}][\text{S}] = (k_{-1} + k_2)[\text{ES}] \qquad \textbf{(14.35)}$$

We can solve this equation for [ES], but the solution contains the concentration of free enzyme E, which is unknown. However, we do know the *total* concentration of enzyme in an experiment, $[\text{E}]_0$. Then, by the condition known as *material balance*, we have

$$[\text{E}]_0 = [\text{E}] + [\text{ES}]$$

Solving this equation for [E] and substituting the result into equation (14.35), we get

$$k_1[\text{S}([\text{E}]_0 - [\text{ES}]) = (k_{-1} + k_2)[\text{ES}]$$

$$[\text{ES}] = \frac{k_1[\text{E}]_0[\text{S}]}{(k_{-1} + k_2) + k_1[\text{S}]}$$

for the concentration of the enzyme–substrate complex. Substituting this value into the rate of reaction, we get

$$V = \frac{k_2 k_1[\text{E}]_0[\text{S}]}{(k_{-1} + k_2) + k_1[\text{S}]}$$

This equation can be put into a more convenient form by dividing the numerator and denominator by k_1 and replacing a ratio of rate constants by the single constant K_M.

$$K_\text{M} = \frac{k_{-1} + k_2}{k_1}$$

Our final result is

$$V = \frac{k_2[\text{E}]_0[\text{S}]}{K_\text{M} + [\text{S}]} \qquad \textbf{(14.36)}$$

Now, let's test whether the reaction velocity V, given by equation (14.36), depends on the substrate concentration in the manner suggested by Figure 14-21. At sufficiently low concentrations of S, we have the inequality

$$K_\text{M} \gg [\text{S}]$$

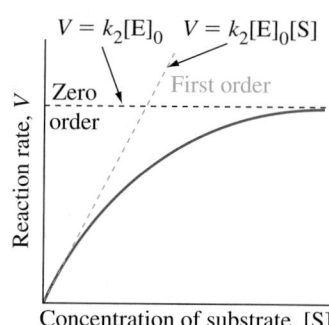

▲ FIGURE 14-21
Effect of substrate concentration on the rate of an enzyme reaction

◀ The reaction mechanism outlined here was proposed by Michaelis and Menten in 1913, accounting for the subscript M in the constant K_M.

We can ignore [S] with respect to K_M in the denominator and obtain the following for the reaction velocity.

$$V = \frac{k_2}{K_M}[E]_0[S]$$

Because the total enzyme concentration is constant, the rate law is first order with respect to substrate, as observed experimentally.

The other limiting case is that in which the velocity of the reaction becomes independent of substrate concentration. At sufficiently high concentrations of substrate,

$$[S] \gg K_M$$

and

$$V = k_2[E]_0$$

Here, for a chosen concentration of enzyme, the reaction velocity is constant and is the maximum reaction velocity attainable for the particular enzyme. This velocity corresponds to the experimentally observed plateau at high substrate concentrations in a plot such as Figure 14-21. Thus, there is a satisfactory agreement between the predictions of the postulated mechanism and the experimental results. As is typical of the scientific method, postulated mechanisms are continually tested by subsequent experiment and modified when necessary.

Mastering**CHEMISTRY**   www.masteringchemistry.com

A typical explosion is a combustion reaction that proceeds at an ever-increasing rate. For a discussion of how combustion reactions can become explosive and ways to prevent such occurrences, go to the Focus On feature for Chapter 14, Combustion and Explosions, on the MasteringChemistry site.

Summary

14-1 Rate of a Chemical Reaction—The **rate of reaction** reflects the rate of change in the concentrations of the reactants and products of a reaction. A *general rate of reaction* (equation 14.2) is defined so that the same value is obtained no matter which reactant or product serves as the basis of kinetic measurements.

14-2 Measuring Reaction Rates—An **initial rate of reaction** is the rate measured over a short time interval at the start of a reaction. An **instantaneous rate of reaction** is assessed over an infinitesimal time interval at any point in the reaction, often through the slope of a tangent line to a graph of concentration versus time (Fig. 14-2). Reaction rates measured over longer time intervals are simply *average* rates, since in almost all cases reaction rates continuously decrease as a reaction proceeds.

14-3 Effect of Concentration on Reaction Rates: The Rate Law—The relationship between the rate of a reaction and the concentrations of the reactants is called the **rate law (rate equation)** (equation 14.6); it has the form: rate of reaction = $k[A]^m[B]^n \cdots$. The **order of a reaction** refers to the exponents $m, n, \ldots$ in the rate law. If $m = 2$, the reaction is second order in A; if $n = 1$, the reaction is first order in B; and so on. The *overall order* of a reaction is given by the sum $m + n + \cdots$. One method of

establishing the rate law of a reaction is by the *method of initial* rates. The **rate constant** relates the rate of reaction to reactant concentrations.

14-4 Zero-Order Reactions—A reaction having $m + n + \cdots = 0$ is a **zero-order reaction** (equation 14.9). A useful equation expressing concentration of a reactant as a function of time is called an **integrated rate law** (equation 14.10). A plot of concentration as a function of time for the zero-order reaction A $\longrightarrow$ products is a straight line with a slope of $-k$ (Fig. 14-3).

14-5 First-Order Reactions—A reaction having $m + n + \cdots = 1$ is a **first-order reaction** (equations 14.12 and 14.13). A graph of $\ln[A]$ versus time for the first-order reaction A $\longrightarrow$ products is a straight line with a slope of $-k$ (Fig. 14-4). The **half-life** of a reaction is the time required for the amount of a reactant to be reduced to one-half its initial value. For a first-order reaction, the half-life is a constant (equation 14.14). Many important reactions are first-order processes, including the decay of radioactive nuclides (Table 14.4).

14-6 Second-Order Reactions—A reaction having $m + n + \cdots = 2$ is a **second-order reaction** (equations 14.17 and 14.18). A graph of $1/[A]$ versus time for the second-order reaction, A $\longrightarrow$ products, is a straight line with a slope of k (Fig. 14-6). For a second-order reaction, the half-life is *not* constant; each successive half-life period is twice as long as the one preceding it (equation 14.19a and 14.19b).

Some second-order reactions, called pseudo-first-order reactions, can be treated as first-order if one of the reactants is present at such a high concentration that its concentration remains essentially constant during the reaction.

14-7 Reaction Kinetics: A Summary—A helpful summary of some basic ideas of reaction kinetics can be found on pages 620 to 622 and in Table 14.5.

14-8 Theoretical Models for Chemical Kinetics—
The rate of reaction depends on the number of molecular collisions per unit volume per unit time, the proportion of molecules having energies in excess of the **activation energy** (Fig. 14-8) and proper orientation of molecules for effective collisions (Fig. 14-9). **A reaction profile** (Fig. 14-10) traces the progress of a reaction, highlighting the energy states of the reactants, the products, and the **activated complex**, a hypothetical transitory species that exists in a high-energy **transition state** between reactants and products.

14-9 The Effect of Temperature on Reaction Rates—The principal basis for describing the effect of temperature on the rate of a chemical reaction is the Arrhenius equation (14.21) or a variant of it (14.22). A graph of $\ln k$ versus $1/T$ is linear, with the slope of the line equal to $-E_a/R$ (Fig. 14-12).

14-10 Reaction Mechanisms—A **reaction mechanism** is a step-by-step description of a chemical reaction consisting of a series of **elementary processes**. Rate laws are written for the elementary processes and combined into a rate law for the overall reaction. To be plausible, the reaction mechanism must be consistent with the stoichiometry of the overall reaction and its experimentally determined rate law.

The most common elementary processes are **unimolecular** (one molecule dissociates) and **bimolecular** (two molecules collide). Some of the species in elementary processes may be **reaction intermediates**, species produced in one elementary process and consumed in another. One of the elementary processes may be the **rate-determining step**. When a single rate-determining step cannot be identified, the mechanism can often be established by the *steady-state approximation*. Reaction mechanisms can also be depicted through reaction profiles (Fig. 14-14).

Photochemical **smog** forms through the action of sunlight on the combustion products of internal combustion engines. The mechanism of its formation has been extensively studied by the methods of chemical kinetics.

14-11 Catalysis—A **catalyst** changes a reaction mechanism to one with a lower activation energy, thereby speeding up the overall reaction, but the catalyst is not itself changed by the reaction. In **homogeneous catalysis**, the catalytic reaction occurs within a single phase (Fig. 14-17). In **heterogeneous catalysis**, the catalytic action occurs on a surface separating two phases (Figs. 14-18 and 14-19). In biochemical reactions, the catalysts are high-molecular-mass proteins called **enzymes**. The reactant, called the **substrate**, attaches to the **active site** on the enzyme, where reaction occurs (Fig. 14-20). The rate of an enzyme-catalyzed reaction can be calculated with an equation (14.36) based on a generally accepted mechanism of enzyme catalysis.

Integrative Example

Peroxyacetyl nitrate (PAN) is an air pollutant produced in photochemical smog by the reaction of hydrocarbons, oxides of nitrogen, and sunlight. PAN is unstable and dissociates into peroxyacetyl radicals and $NO_2(g)$. Its presence in polluted air is like a reservoir for NO_2 storage.

$$
\begin{array}{ccc}
\overset{\displaystyle O}{\overset{\displaystyle \|}{CH_3COONO_2}} & \longrightarrow & \overset{\displaystyle O}{\overset{\displaystyle \|}{CH_3COO\cdot}} + NO_2 \\
\text{PAN} & & \text{Peroxyacetyl} \\
& & \text{radical}
\end{array}
$$

The first-order decomposition of PAN has a half-life of 35 h at 0 °C and 30.0 min at 25 °C. At what temperature will a sample of air containing 5.0×10^{14} PAN molecules per liter decompose at the rate of 1.0×10^{12} PAN molecules per liter per minute?

Analyze

This problem, which requires four principal tasks, is centered on the relationship between rate constants and temperature (equation 14.22) and between a rate constant and the rate of a reaction (equation 14.6). Specifically, we will need to (1) convert the two half-lives to values of k; (2) use those values of k and their related temperatures to determine the activation energy of the reaction; (3) find the value of k corresponding to the decomposition rate specified; and (4) calculate the temperature at which k has the value determined in (3).

Solve

Determine the value of k at 0 °C for the first-order reaction.

$$k = 0.693/t_{1/2}$$

$$k = \frac{0.693}{35 \text{ h}} \times \frac{1 \text{ h}}{60 \text{ min}} = 3.3 \times 10^{-4} \text{ min}^{-1}$$

At 25 °C,

$$k = \frac{0.693}{30.0 \text{ min}} = 2.31 \times 10^{-2} \text{ min}^{-1}$$

To determine the activation energy of the reaction, substitute these data

$$k_2 = 2.31 \times 10^{-2} \text{ min}^{-1}; T_2 = 25\,°C = 298 \text{ K}$$
$$k_1 = 3.3 \times 10^{-4} \text{ min}^{-1}; T_1 = 0\,°C = 273 \text{ K}$$

into equation (14.22)

$$\ln\frac{k_2}{k_1} = \frac{E_a}{R}\left(\frac{1}{T_1} - \frac{1}{T_2}\right)$$

$$\ln\frac{2.31 \times 10^{-2} \text{ min}^{-1}}{3.3 \times 10^{-4} \text{ min}^{-1}} = \frac{E_a}{8.3145 \text{ J mol}^{-1} \text{ K}^{-1}} \times \left(\frac{1}{273 \text{ K}} - \frac{1}{298 \text{ K}}\right)$$

$$= \frac{E_a}{8.3145 \text{ J mol}^{-1}}(0.00366 - 0.00336) = 4.25$$

$$E_a = \frac{8.3145 \text{ J mol}^{-1} \times 4.25}{0.00030} = 1.2 \times 10^5 \text{ J mol}^{-1}$$

Because the reaction is first order, the rate law is rate $= k[\text{PAN}]$, which can be rearranged to give

$$k = \text{rate}/[\text{PAN}]$$

Because mol/L and molecules/L are related through the Avogadro constant, $[\text{PAN}]$ can be expressed as molecules/L.

$$k = \frac{\text{rate of reaction}}{[\text{PAN}]} = \frac{1.0 \times 10^{12} \text{ molecules L}^{-1} \text{ min}^{-1}}{5.0 \times 10^{14} \text{ molecules L}^{-1}}$$

$$= 2.0 \times 10^{-3} \text{ min}^{-1}$$

To determine the unknown temperature, choose one of the known combinations of k and T as k_2 and T_2. For example,

$$k_2 = 2.31 \times 10^{-2} \text{ min}^{-1}; T_2 = 298 \text{ K}$$

The unknown combination is

$$k_1 = 2.0 \times 10^{-3} \text{ min}^{-1}; T_1 = ?$$

Substitute these values into equation (14.22), together with the known value, $E_a = 1.2 \times 10^5$ J mol^{-1} to obtain

$$\ln\frac{2.31 \times 10^{-2} \text{ min}^{-1}}{2.0 \times 10^{-3} \text{ min}^{-1}} = \frac{1.2 \times 10^5 \text{ J mol}^{-1}}{8.3145 \text{ J mol}^{-1} \text{ K}^{-1}} \times \left(\frac{1}{T_1} - \frac{1}{298}\right) \text{K}^{-1}$$

$$\ln 12 = 1.4 \times 10^4\left(\frac{1}{T_1} - 3.36 \times 10^{-3}\right)$$

$$2.5 = \frac{1.4 \times 10^4}{T_1} - 47$$

$$50T_1 = 1.4 \times 10^4$$

$$T_1 = 2.8 \times 10^2 \text{ K}$$

Assess
The best check on the reasonableness of the final answer is to note that based on the values $k = 3.3 \times 10^{-4}$ min^{-1} at 273 K and $k = 2.31 \times 10^{-2}$ min^{-1} at 298 K, the temperature at which $k = 2.0 \times 10^{-3}$ min^{-1} should be somewhere between 273 K and 298 K. Our result is just that: 2.8×10^2 K.

PRACTICE EXAMPLE A: At room temperature (20 °C), milk turns sour in about 64 hours. In a refrigerator at 3 °C, milk can be stored three times as long before it sours. **(a)** Estimate the activation energy of the reaction that causes the souring of milk. **(b)** How long should it take milk to sour at 40 °C?

PRACTICE EXAMPLE B: The following mechanism can be used to account for the change in apparent order of unimolecular reactions, such as the conversion of cyclopropane (A) into propene (P), where A* is an energetic form of cyclopropane that can either react or return to unreacted cyclopropane.

$$A + A \underset{k_{-1}}{\overset{k_1}{\rightleftharpoons}} A^* + A$$

$$A^* \xrightarrow{k_2} P$$

Show that at low pressures of cyclopropane, the rate law is second order in A and at high pressures, it is first order in A.

Mastering**CHEMISTRY**

You'll find a link to additional self study questions in the study area on www.masteringchemistry.com

Exercises

Rates of Reactions

1. In the reaction $2\,A + B \longrightarrow C + 3\,D$, reactant A is found to disappear at the rate of $6.2 \times 10^{-4}\,M\,s^{-1}$. **(a)** What is the rate of reaction at this point? **(b)** What is the rate of disappearance of B? **(c)** What is the rate of formation of D?

2. From Figure 14-2 estimate the rate of reaction at **(a)** $t = 800\,s$; **(b)** the time at which $[H_2O_2] = 0.50\,M$.

3. In the reaction $A \longrightarrow$ products, $[A]$ is found to be 0.485 M at $t = 71.5\,s$ and 0.474 M at $t = 82.4\,s$. What is the average rate of the reaction during this time interval?

4. In the reaction $A \longrightarrow$ products, at $t = 0$, $[A] = 0.1565\,M$. After 1.00 min, $[A] = 0.1498\,M$, and after 2.00 min, $[A] = 0.1433\,M$.
   (a) Calculate the average rate of the reaction during the first minute and during the second minute.
   (b) Why are these two rates not equal?

5. In the reaction $A \longrightarrow$ products, 4.40 min after the reaction is started, $[A] = 0.588\,M$. The rate of reaction at this point is rate $= -\Delta[A]/\Delta t = 2.2 \times 10^{-2}\,M\,min^{-1}$. Assume that this rate remains constant for a short period of time.
   (a) What is $[A]$ 5.00 min after the reaction is started?
   (b) At what time after the reaction is started will $[A] = 0.565\,M$?

6. Refer to Experiment 2 of Table 14.3 and to reaction (14.7) and rate law (14.8). Exactly 1.00 h after the reaction is started, what are **(a)** $[HgCl_2]$ and **(b)** $[C_2O_4{}^{2-}]$ in the mixture?

7. For the reaction $A + 2\,B \longrightarrow 2\,C$, the rate of reaction is $1.76 \times 10^{-5}\,M\,s^{-1}$ at the time when $[A] = 0.3580\,M$.

(a) What is the rate of formation of C?
(b) What will $[A]$ be 1.00 min later?
(c) Assume the rate remains at $1.76 \times 10^{-5}\,M\,s^{-1}$. How long would it take for $[A]$ to change from 0.3580 to 0.3500 M?

8. If the rate of reaction (14.3) is $5.7 \times 10^{-4}\,M\,s^{-1}$, what is the rate of production of $O_2(g)$ from 1.00 L of the $H_2O_2(aq)$, expressed as **(a)** mol $O_2\,s^{-1}$; **(b)** mol $O_2\,min^{-1}$; **(c)** mL O_2(STP) min^{-1}?

9. In the reaction $A(g) \longrightarrow 2\,B(g) + C(g)$, the *total* pressure increases while the *partial* pressure of A(g) decreases. If the initial pressure of A(g) in a vessel of constant volume is 1.000×10^3 mmHg,
   (a) What will be the total pressure when the reaction has gone to completion?
   (b) What will be the total gas pressure when the partial pressure of A(g) has fallen to 8.00×10^2 mmHg?

10. At 65 °C, the half-life for the first-order decomposition of $N_2O_5(g)$ is 2.38 min.

$$N_2O_5(g) \longrightarrow 2\,NO_2(g) + \frac{1}{2}O_2(g)$$

If 1.00 g of N_2O_5 is introduced into an evacuated 15 L flask at 65 °C,
(a) What is the initial partial pressure, in mmHg, of $N_2O_5(g)$?
(b) What is the partial pressure, in mmHg, of $N_2O_5(g)$ after 2.38 min?
(c) What is the total gas pressure, in mmHg, after 2.38 min?

Method of Initial Rates

11. The initial rate of the reaction $A + B \longrightarrow C + D$ is determined for different initial conditions, with the results listed in the table.
(a) What is the order of reaction with respect to A and to B?
(b) What is the overall reaction order?
(c) What is the value of the rate constant, k?

Expt	[A], M	[B], M	Initial Rate, $M\,s^{-1}$
1	0.185	0.133	3.35×10^{-4}
2	0.185	0.266	1.35×10^{-3}
3	0.370	0.133	6.75×10^{-4}
4	0.370	0.266	2.70×10^{-3}

12. For the reaction $A + B \longrightarrow C + D$, the following initial rates of reaction were found. What is the rate law for this reaction?

Expt	[A], M	[B], M	Initial Rate, $M\,min^{-1}$
1	0.50	1.50	4.2×10^{-3}
2	1.50	1.50	1.3×10^{-2}
3	3.00	3.00	5.2×10^{-2}

13. The following rates of reaction were obtained in three experiments with the reaction $2\,NO(g) + Cl_2(g) \longrightarrow 2\,NOCl(g)$.

Expt	Initial [NO], M	Initial $[Cl_2]$, M	Initial Rate of Reaction, $M\,s^{-1}$
1	0.0125	0.0255	2.27×10^{-5}
2	0.0125	0.0510	4.55×10^{-5}
3	0.0250	0.0255	9.08×10^{-5}

What is the rate law for this reaction?

14. The following data are obtained for the initial rates of reaction in the reaction $A + 2\,B + C \longrightarrow 2\,D + E$.

Expt	Initial [A], M	Initial [B], M	[C], M	Initial Rate
1	1.40	1.40	1.00	R_1
2	0.70	1.40	1.00	$R_2 = \frac{1}{2} \times R_1$
3	0.70	0.70	1.00	$R_3 = \frac{1}{4} \times R_2$
4	1.40	1.40	0.50	$R_4 = 16 \times R_3$
5	0.70	0.70	0.50	$R_5 = ?$

(a) What are the reaction orders with respect to A, B, and C?
(b) What is the value of R_5 in terms of R_1?

First-Order Reactions

15. One of the following statements is true and the other is false regarding the first-order reaction $2A \longrightarrow B + C$. Identify the true statement and the false one, and explain your reasoning.
(a) The rate of the reaction decreases as more and more of B and C form.
(b) The time required for one-half of substance A to react is directly proportional to the quantity of A present initially.

16. One of the following statements is true and the other is false regarding the first-order reaction $2A \longrightarrow B + C$. Identify the true statement and the false one, and explain your reasoning.
(a) A graph of [A] versus time is a straight line.
(b) The rate of the reaction is one-half the rate of disappearance of A.

17. The first-order reaction $A \longrightarrow$ products has $t_{1/2} = 180\,s$.
(a) What percent of a sample of A remains *unreacted* 900 s after a reaction has been started?
(b) What is the rate of reaction when $[A] = 0.50\,M$?

18. The reaction $A \longrightarrow$ products is first order in A. Initially, $[A] = 0.800\,M$; and after 54 min, $[A] = 0.100\,M$.
(a) At what time is $[A] = 0.025\,M$?
(b) What is the rate of reaction when $[A] = 0.025\,M$?

19. The reaction $A \longrightarrow$ products is first order in A.
(a) If 1.60 g A is allowed to decompose for 38 min, the mass of A remaining undecomposed is found to be 0.40 g. What is the half-life, $t_{1/2}$, of this reaction?
(b) Starting with 1.60 g A, what is the mass of A remaining undecomposed after 1.00 h?

20. In the first-order reaction $A \longrightarrow$ products, $[A] = 0.816\,M$ initially and 0.632 M after 16.0 min.
(a) What is the value of the rate constant, k?
(b) What is the half-life of this reaction?
(c) At what time will $[A] = 0.235\,M$?
(d) What will [A] be after 2.5 h?

21. In the first-order reaction $A \longrightarrow$ products, it is found that 99% of the original amount of reactant A decomposes in 137 min. What is the half-life, $t_{1/2}$, of this decomposition reaction?

22. The half-life of the radioactive isotope phosphorus-32 is 14.3 days. How long does it take for a sample of phosphorus-32 to lose 99% of its radioactivity?

23. Acetoacetic acid, CH_3COCH_2COOH, a reagent used in organic synthesis, decomposes in acidic solution, producing acetone and $CO_2(g)$.

$$CH_3COCH_2COOH(aq) \longrightarrow CH_3COCH_3(aq) + CO_2(g)$$

This first-order decomposition has a half-life of 144 min.

(a) How long will it take for a sample of acetoacetic acid to be 65% decomposed?

(b) How many liters of $CO_2(g)$, measured at 24.5 °C and 748 Torr, are produced as a 10.0 g sample of CH_3COCH_2COOH decomposes for 575 min? [Ignore the aqueous solubility of $CO_2(g)$.]

24. The following first-order reaction occurs in $CCl_4(l)$ at 45 °C: $N_2O_5 \longrightarrow N_2O_4 + \frac{1}{2}O_2(g)$. The rate constant is $k = 6.2 \times 10^{-4}\,s^{-1}$. An 80.0 g sample of N_2O_5 in $CCl_4(l)$ is allowed to decompose at 45 °C.

(a) How long does it take for the quantity of N_2O_5 to be reduced to 2.5 g?

(b) How many liters of O_2, measured at 745 mmHg and 45 °C, are produced up to this point?

25. For the reaction A $\longrightarrow$ products, the following data give [A] as a function of time: $t = 0\,s, [A] = 0.600\,M$; 100 s, 0.497 M; 200 s, 0.413 M; 300 s, 0.344 M; 400 s, 0.285 M; 600 s, 0.198 M; 1000 s, 0.094 M.

(a) Show that the reaction is first order.

(b) What is the value of the rate constant, k?

(c) What is [A] at $t = 750\,s$?

26. The decomposition of dimethyl ether at 504 °C is

$$(CH_3)_2O(g) \longrightarrow CH_4(g) + H_2(g) + CO(g)$$

The following data are partial pressures of dimethyl ether (DME) as a function of time: $t = 0\,s, P_{DME} = 312\,mmHg$; 390 s, 264 mmHg; 777 s, 224 mmHg; 1195 s, 187 mmHg; 3155 s, 78.5 mmHg.

(a) Show that the reaction is first order.

(b) What is the value of the rate constant, k?

(c) What is the total gas pressure at 390 s?

(d) What is the total gas pressure when the reaction has gone to completion?

(e) What is the total gas pressure at $t = 1000\,s$?

Reactions of Various Orders

Three different sets of data of [A] versus time are given in the following table for the reaction A $\longrightarrow$ products. [*Hint:* There are several ways of arriving at answers for each of the following six questions.]

Data for Exercises 27–32					
I		II		III	
Time, s	[A], M	Time, s	[A], M	Time, s	[A], M
0	1.00	0	1.00	0	1.00
25	0.78	25	0.75	25	0.80
50	0.61	50	0.50	50	0.67
75	0.47	75	0.25	75	0.57
100	0.37	100	0.00	100	0.50
150	0.22			150	0.40
200	0.14			200	0.33
250	0.08			250	0.29

27. Which of these sets of data corresponds to a (a) zero-order, (b) first-order, (c) second-order reaction?

28. What is the value of the rate constant k of the zero-order reaction?

29. What is the approximate half-life of the first-order reaction?

30. What is the approximate initial rate of the second-order reaction?

31. What is the approximate rate of reaction at $t = 75\,s$ for the (a) zero-order, (b) first-order, (c) second-order reaction?

32. What is the approximate concentration of A remaining after 110 s in the (a) zero-order, (b) first-order, (c) second-order reaction?

33. The reaction A + B $\longrightarrow$ C + D is second order in A and zero order in B. The value of k is $0.0103\,M^{-1}\,min^{-1}$. What is the rate of this reaction when $[A] = 0.116\,M$ and $[B] = 3.83\,M$?

34. A reaction is 50% complete in 30.0 min. How long after its start will the reaction be 75% complete if it is (a) first order; (b) zero order?

35. The decomposition of HI(g) at 700 K is followed for 400 s, yielding the following data: at $t = 0, [HI] = 1.00\,M$; at $t = 100\,s, [HI] = 0.90\,M$; at $t = 200\,s, [HI] = 0.81\,M; t = 300\,s, [HI] = 0.74\,M;$ at $t = 400\,s, [HI] = 0.68\,M$. What are the reaction order and the rate constant for the reaction:

$$HI(g) \longrightarrow \frac{1}{2}H_2(g) + \frac{1}{2}I_2(g)?$$

Write the rate law for the reaction at 700 K.

36. For the disproportionation of *p*-toluenesulfinic acid,

$$3\,ArSO_2H \longrightarrow ArSO_2SAr + ArSO_3H + H_2O$$

(where Ar = p-$CH_3C_6H_4$—), the following data were obtained: $t = 0\,min, [ArSO_2H] = 0.100\,M$; 15 min, 0.0863 M; 30 min, 0.0752 M; 45 min, 0.0640 M; 60 min, 0.0568 M; 120 min, 0.0387 M; 180 min, 0.0297 M; 300 min, 0.0196 M.

(a) Show that this reaction is second order.

(b) What is the value of the rate constant, k?

(c) At what time would $[ArSO_2H] = 0.0500\,M$?

(d) At what time would $[ArSO_2H] = 0.0250\,M$?

(e) At what time would $[ArSO_2H] = 0.0350\,M$?

37. For the reaction A $\longrightarrow$ products, the following data were obtained: $t = 0\,s, [A] = 0.715\,M$; 22 s, 0.605 M; 74 s, 0.345 M; 132 s, 0.055 M. (a) What is the order of this reaction? (b) What is the half-life of the reaction?

38. The following data were obtained for the dimerization of 1,3-butadiene, $2\,C_4H_6(g) \longrightarrow C_8H_{12}(g)$, at 600 K: $t = 0\,min, [C_4H_6] = 0.0169\,M$; 12.18 min, 0.0144 M; 24.55 min, 0.0124 M; 42.50 min, 0.0103 M; 68.05 min, 0.00845 M.

(a) What is the order of this reaction?

(b) What is the value of the rate constant, k?

(c) At what time would $[C_4H_6] = 0.00423\,M$?

(d) At what time would $[C_4H_6] = 0.0050\,M$?

39. For the reaction A $\longrightarrow$ products, the data tabulated below are obtained.

(a) Determine the initial rate of reaction (that is, $-\Delta[A]/\Delta t$) in each of the two experiments.

(b) Determine the order of the reaction.

First Experiment	
$[A]$ = 1.512 M	t = 0 min
$[A]$ = 1.490 M	t = 1.0 min
$[A]$ = 1.469 M	t = 2.0 min

Second Experiment	
$[A]$ = 3.024 M	t = 0 min
$[A]$ = 2.935 M	t = 1.0 min
$[A]$ = 2.852 M	t = 2.0 min

40. For the reaction A $\longrightarrow$ 2 B + C, the following data are obtained for $[A]$ as a function of time: t = 0 min, $[A]$ = 0.80 M; 8 min, 0.60 M; 24 min, 0.35 M; 40 min, 0.20 M.

(a) By suitable means, establish the order of the reaction.

(b) What is the value of the rate constant, k?

(c) Calculate the rate of formation of B at t = 30 min.

41. In three different experiments, the following results were obtained for the reaction A $\longrightarrow$ products: $[A]_0$ = 1.00 M, $t_{1/2}$ = 50 min; $[A]_0$ = 2.00 M, $t_{1/2}$ = 25 min; $[A]_0$ = 0.50 M, $t_{1/2}$ = 100 min. Write the rate equation for this reaction, and indicate the value of k.

42. Ammonia decomposes on the surface of a hot tungsten wire. Following are the half-lives that were obtained at 1100 °C for different initial concentrations of NH_3: $[NH_3]_0$ = 0.0031 M, $t_{1/2}$ = 7.6 min; 0.0015 M, 3.7 min; 0.00068 M, 1.7 min. For this decomposition reaction, what is (a) the order of the reaction; (b) the rate constant, k?

43. The half-lives of both zero-order and second-order reactions depend on the initial concentration, as well as on the rate constant. In one case, the half-life gets longer as the initial concentration increases, and in the other it gets shorter. Which is which, and why isn't the situation the same for both?

44. Consider three hypothetical reactions, A $\longrightarrow$ products, all having the same numerical value of the rate constant k. One of the reactions is zero order, one is first order, and one is second order. What must be the initial concentration $[A]_0$ if (a) the zero- and first-order; (b) the zero- and second-order; (c) the first- and second-order reactions are to have the same half-life?

Collision Theory; Activation Energy

45. Explain why

(a) A reaction rate cannot be calculated from the collision frequency alone.

(b) The rate of a chemical reaction may increase dramatically with temperature, whereas the collision frequency increases much more slowly.

(c) The addition of a catalyst to a reaction mixture can have such a pronounced effect on the rate of a reaction, even if the temperature is held constant.

46. If even a tiny spark is introduced into a mixture of $H_2(g)$ and $O_2(g)$, a highly exothermic explosive reaction occurs. Without the spark, the mixture remains unreacted indefinitely.

(a) Explain this difference in behavior.

(b) Why is the nature of the reaction independent of the size of the spark?

47. For the reversible reaction A + B $\rightleftharpoons$ C + D, the enthalpy change of the forward reaction is +21 kJ/mol. The activation energy of the forward reaction is 84 kJ/mol.

(a) What is the activation energy of the reverse reaction?

(b) In the manner of Figure 14-10, sketch the reaction profile of this reaction.

48. By an appropriate sketch, indicate why there is some relationship between the enthalpy change and the activation energy for an endothermic reaction but not for an exothermic reaction.

49. By inspection of the reaction profile for the reaction A to D given below, answer the following questions.

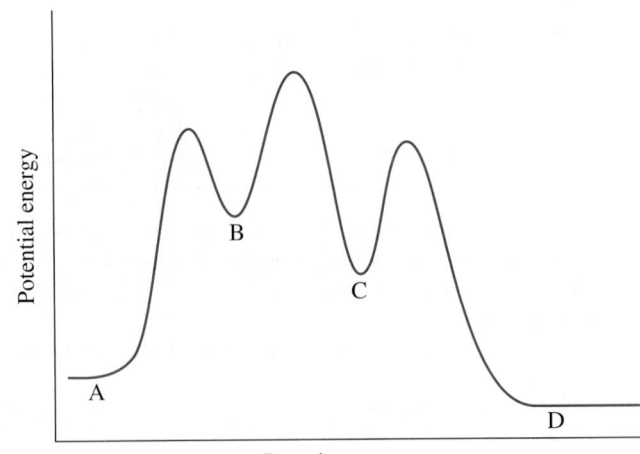

(a) How many intermediates are there in the reaction?

(b) How many transition states are there?

(c) Which is the fastest step in the reaction?

(d) Which step has the smallest rate constant?

(e) Is the first step of the reaction exothermic or endothermic?

(f) Is the overall reaction exothermic or endothermic?

50. By inspection of the reaction profile for the reaction A to D given, answer the following questions.
    (a) How many intermediates are there in the reaction?
    (b) How many transition states are there?
    (c) Which is the fastest step in the reaction?
    (d) Which step has the smallest rate constant?
    (e) Is the first step of the reaction exothermic or endothermic?
    (f) Is the overall reaction exothermic or endothermic?

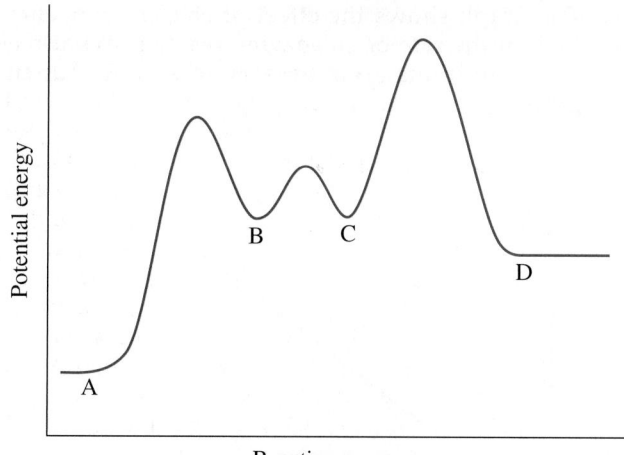

Reaction progress

Effect of Temperature on Rates of Reaction

51. The rate constant for the reaction $H_2(g) + I_2(g) \longrightarrow 2\,HI(g)$ has been determined at the following temperatures: 599 K, $k = 5.4 \times 10^{-4}\,M^{-1}\,s^{-1}$; 683 K, $k = 2.8 \times 10^{-2}\,M^{-1}\,s^{-1}$. Calculate the activation energy for the reaction.

52. At what temperature will the rate constant for the reaction in Exercise 51 have the value $k = 5.0 \times 10^{-3}\,M^{-1}\,s^{-1}$?

53. A treatise on atmospheric chemistry lists the following rate constants for the decomposition of PAN described in the Integrative Example on page 643: 0 °C, $k = 5.6 \times 10^{-6}\,s^{-1}$; 10 °C, $k = 3.2 \times 10^{-5}\,s^{-1}$; 20 °C, $k = 1.6 \times 10^{-4}\,s^{-1}$; 30 °C, $k = 7.6 \times 10^{-4}\,s^{-1}$.
    (a) Construct a graph of $\ln k$ versus $1/T$.
    (b) What is the activation energy, E_a, of the reaction?
    (c) Calculate the half–life of the decomposition reaction at 40 °C.

54. The reaction $C_2H_5I + OH^- \longrightarrow C_2H_5OH + I^-$ was studied in an ethanol (C_2H_5OH) solution, and the following rate constants were obtained: 15.83 °C, $k = 5.03 \times 10^{-5}$; 32.02 °C, 3.68×10^{-4}; 59.75 °C, 6.71×10^{-3}; 90.61 °C, $0.119\,M^{-1}\,s^{-1}$.
    (a) Determine E_a for this reaction by a graphical method.
    (b) Determine E_a by the use of equation (14.22).
    (c) Calculate the value of the rate constant k at 100.0 °C.

55. The first-order reaction A $\longrightarrow$ products has a half-life, $t_{1/2}$, of 46.2 min at 25 °C and 2.6 min at 102 °C.

    (a) Calculate the activation energy of this reaction.
    (b) At what temperature would the half-life be 10.0 min?

56. For the first-order reaction

$$N_2O_5(g) \longrightarrow 2\,NO_2(g) + \frac{1}{2}O_2(g)$$

$t_{1/2} = 22.5\,h$ at 20 °C and 1.5 h at 40 °C.
    (a) Calculate the activation energy of this reaction.
    (b) If the Arrhenius constant $A = 2.05 \times 10^{13}\,s^{-1}$, determine the value of k at 30 °C.

57. A commonly stated rule of thumb is that reaction rates double for a temperature increase of about 10 °C. (This rule is very often wrong.)
    (a) What must be the approximate activation energy for this statement to be true for reactions at about room temperature?
    (b) Would you expect this rule of thumb to apply at room temperature for the reaction profiled in Figure 14-10? Explain.

58. Concerning the rule of thumb stated in Exercise 57, estimate how much faster cooking occurs in a pressure cooker with the vapor pressure of water at 2.00 atm instead of in water under normal boiling conditions. [*Hint:* Refer to Table 12.2.]

Catalysis

59. The following statements about catalysis are not stated as carefully as they might be. What slight modifications would you make in them?
    (a) A catalyst is a substance that speeds up a chemical reaction but does not take part in the reaction.
    (b) The function of a catalyst is to lower the activation energy for a chemical reaction.

60. The following substrate concentration [S] versus time data were obtained during an enzyme-catalyzed

reaction: $t = 0\,min$, [S] $= 1.00\,M$; 20 min, 0.90 M; 60 min, 0.70 M; 100 min, 0.50 M; 160 min, 0.20 M. What is the order of this reaction with respect to S in the concentration range studied?

61. What are the similarities and differences between the catalytic activity of platinum metal and of an enzyme?

62. Certain gas-phase reactions on a heterogeneous catalyst are first order at low gas pressures and zero order at high pressures. Can you suggest a reason for this?

63. The graph shows the effect of enzyme concentration on the rate of an enzyme reaction. What reaction conditions are necessary to account for this graph?

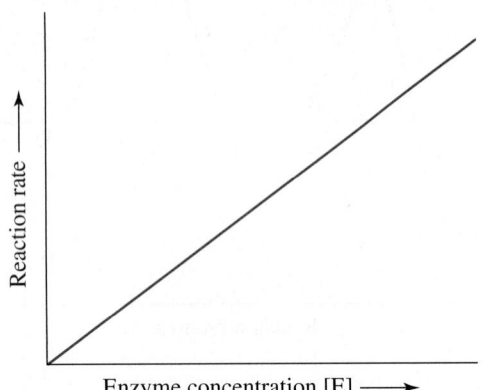

64. The graph shows the effect of temperature on enzyme activity. Explain why the graph has the general shape shown. For human enzymes, at what temperature would you expect the maximum in the curve to appear?

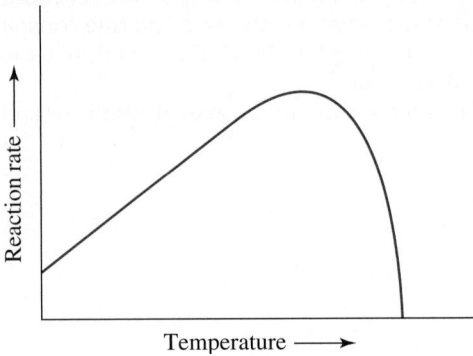

Reaction Mechanisms

65. We have used the terms *order of a reaction* and *molecularity of an elementary process* (that is, unimolecular, bimolecular). What is the relationship, if any, between these two terms?

66. According to collision theory, chemical reactions occur through molecular collisions. A unimolecular elementary process in a reaction mechanism involves dissociation of a *single* molecule. How can these two ideas be compatible? Explain.

67. The reaction $2 NO + 2 H_2 \longrightarrow N_2 + 2 H_2O$ is second order in $[NO]$ and first order in $[H_2]$. A *three-step* mechanism has been proposed. The first, fast step is the elementary process given as equation (14.25) on page 632. The third step, also fast, is $N_2O + H_2 \longrightarrow N_2 + H_2O$. Propose an entire three-step mechanism, and show that it conforms to the experimentally determined reaction order.

68. The mechanism proposed for the reaction of $H_2(g)$ and $I_2(g)$ to form $HI(g)$ consists of a fast reversible first step involving $I_2(g)$ and $I(g)$, followed by a slow step. Propose a two-step mechanism for the reaction $H_2(g) + I_2(g) \longrightarrow 2 HI(g)$, which is known to be first order in H_2 and first order in I_2.

69. The reaction $2 NO + Cl_2 \longrightarrow 2 NOCl$ has the rate law: rate of reaction $= k[NO]^2[Cl_2]$. Propose a two-step mechanism for this reaction consisting of a fast reversible first step, followed by a slow step.

70. A simplified rate law for the reaction $2 O_3(g) \longrightarrow 3 O_2(g)$ is

$$\text{rate} = k \, \frac{[O_3]^2}{[O_2]}$$

For this reaction, propose a two-step mechanism that consists of a fast, reversible first step, followed by a slow second step.

71. One proposed mechanism for the formation of a double helix in DNA is given by

$$(S_1 + S_2) = (S_1{:}S_2)^* \qquad \text{(fast)}$$
$$(S_1{:}S_2)^* \longrightarrow S_1{:}S_2 \qquad \text{(slow)}$$

where S_1 and S_2 represent strand 1 and 2, and $(S_1{:}S_2)^*$ represents an unstable helix. Write the rate of reaction expression for the formation of the double helix.

72. One proposed mechanism for the condensation of propanone, $(CH_3)_2CO$, is as follows:

$(CH_3)_2CO(aq) + OH^-(aq) = CH_3C(O)CH_2^-(aq) + H_2O(l)$
$CH_3C(O)CH_2^-(aq) + (CH_3)_2CO(aq) \longrightarrow$ product

Use the steady-state approximation to determine the rate of formation for the product.

Integrative and Advanced Exercises

73. Suppose that the reaction in Example 14-8 is first order with a rate constant of $0.12 \, \text{min}^{-1}$. Starting with $[A]_0 = 1.00 \, M$, will the curve for $[A]$ versus t for the first-order reaction cross the curve for the second-order reaction at some time after $t = 0$? Will the two curves cross if $[A]_0 = 2.00 \, M$? In each case, if the curves are found to cross, at what time will this happen?

74. $[A]_t$ as a function of time for the reaction $A \longrightarrow$ products is plotted in the following graph. Use data from this graph to determine **(a)** the order of the reaction; **(b)** the rate constant, k; **(c)** the rate of the reaction at $t = 3.5 \, \text{min}$, using the results of parts (a) and (b); **(d)** the rate of the reaction at $t = 5.0 \, \text{min}$, from the slope of the tangent line; **(e)** the initial rate of the reaction.

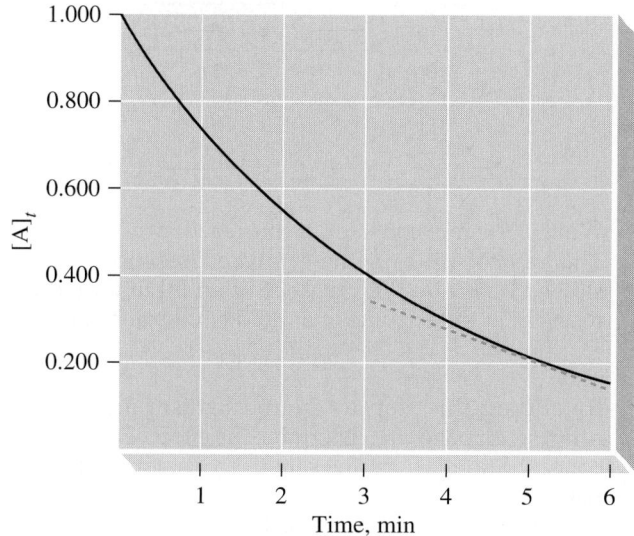

75. Exactly 300 s after decomposition of $H_2O_2(aq)$ begins (reaction 14.3), a 5.00 mL sample is removed and immediately titrated with 37.1 mL of 0.1000 M $KMnO_4$. What is $[H_2O_2]$ at this 300 s point in the reaction?

76. Use the method of Exercise 75 to determine the volume of 0.1000 M $KMnO_4$ required to titrate 5.00 mL samples of $H_2O_2(aq)$ for each of the entries in Table 14.1. Plot these volumes of $KMnO_4(aq)$ as a function of time, and show that from this graph you can get the same rate of reaction at 1400 s as that obtained in Figure 14-2.

77. The initial rate of reaction (14.3) is found to be $1.7 \times 10^{-3}\,M\,s^{-1}$. Assume that this rate holds for 2 minutes. Start with 175 mL of 1.55 M $H_2O_2(aq)$ at $t = 0$. How many milliliters of $O_2(g)$, measured at 24 °C and 757 mmHg, are released from solution in the first minute of the reaction?

78. We have seen that the unit of k depends on the overall order of a reaction. Derive a general expression for the units of k for a reaction of any overall order, based on the order of the reaction (o) and the units of concentration (M) and time (s).

79. Hydroxide ion is involved in the mechanism of the following reaction but is not consumed in the overall reaction.

$$OCl^- + I^- \xrightarrow{OH^-} OI^- + Cl^-$$

(a) From the data given, determine the order of the reaction with respect to OCl^-, I^-, and OH^-.
(b) What is the overall reaction order?
(c) Write the rate equation, and determine the value of the rate constant, k.

$[OCl^-]$, M	$[I^-]$, M	$[OH^-]$, M	Rate Formation OI^-, $M\,s^{-1}$
0.0040	0.0020	1.00	4.8×10^{-4}
0.0020	0.0040	1.00	5.0×10^{-4}
0.0020	0.0020	1.00	2.4×10^{-4}
0.0020	0.0020	0.50	4.6×10^{-4}
0.0020	0.0020	0.25	9.4×10^{-4}

80. The half-life for the first-order decomposition of nitramide, $NH_2NO_2(aq) \longrightarrow N_2O(g) + H_2O(l)$, is 123 min at 15 °C. If 165 mL of a 0.105 M NH_2NO_2 solution is allowed to decompose, how long must the reaction proceed to yield 50.0 mL of $N_2O(g)$ collected over water at 15 °C and a barometric pressure of 756 mmHg? (The vapor pressure of water at 15 °C is 12.8 mmHg.)

81. The decomposition of ethylene oxide at 690 K is monitored by measuring the *total* gas pressure as a function of time. The data obtained are $t = 10$ min, $P_{tot} = 139.14$ mmHg; 20 min, 151.67 mmHg; 40 min, 172.65 mmHg; 60 min, 189.15 mmHg; 100 min, 212.34 mmHg; 200 min, 238.66 mmHg; ∞, 249.88 mmHg. What is the order of the reaction $(CH_2)_2O(g) \longrightarrow CH_4(g) + CO(g)$?

82. Refer to Example 14-7. For the decomposition of di-*t*-butyl peroxide (DTBP), determine the time at which the *total* gas pressure is 2100 mmHg.

83. The following data are for the reaction $2\,A + B \longrightarrow$ products. Establish the order of this reaction with respect to A and to B.

Expt 1, [B] = 1.00 M		Expt 2, [B] = 0.50 M	
Time, min	[A], M	Time, Min	[A], M
0	1.000×10^{-3}	0	1.000×10^{-3}
1	0.951×10^{-3}	1	0.975×10^{-3}
5	0.779×10^{-3}	5	0.883×10^{-3}
10	0.607×10^{-3}	10	0.779×10^{-3}
20	0.368×10^{-3}	20	0.607×10^{-3}

84. Show that the following mechanism is consistent with the rate law established for the iodide–hypochlorite reaction in Exercise 79.

$$\textit{Fast:} \quad OCl^- + H_2O \underset{k_{-1}}{\overset{k_1}{\rightleftharpoons}} HOCl + OH^-$$

$$\textit{Slow:} \quad I^- + HOCl \xrightarrow{k_2} HOI + Cl^-$$

$$\textit{Fast:} \quad HOI + OH^- \underset{k_{-3}}{\overset{k_3}{\rightleftharpoons}} H_2O + OI^-$$

85. In the hydrogenation of a compound containing a carbon-to-carbon triple bond, two products are possible, as in the reaction

$$CH_3-C\equiv C-CH_3 + H_2 \longrightarrow$$

$$\underset{(I)}{\overset{CH_3}{\underset{H}{>}}C=C\overset{CH_3}{\underset{H}{<}}} \quad \text{or} \quad \underset{(II)}{\overset{CH_3}{\underset{H}{>}}C=C\overset{H}{\underset{CH_3}{<}}}$$

The amount of each product can be controlled by using an appropriate catalyst. The Lindlar catalyst is a *heterogeneous* catalyst that produces only one of these products. Which product, I or II, do you think is produced and why? Draw a sketch of how the reaction might occur.

86. Derive a plausible mechanism for the following reaction in aqueous solution, $Hg_2^{2+} + Tl^{3+} \longrightarrow 2 Hg^{2+} + Tl^+$, for which the observed rate law is: rate = $k[Hg_2^{2+}][Tl^{3+}]/[Hg^{2+}]$.

87. The following three-step mechanism has been proposed for the reaction of chlorine and chloroform.

(1) $Cl_2(g) \underset{k_{-1}}{\overset{k_1}{\rightleftharpoons}} 2 Cl(g)$

(2) $Cl(g) + CHCl_3(g) \xrightarrow{k_2} HCl(g) + CCl_3(g)$

(3) $CCl_3(g) + Cl(g) \xrightarrow{k_3} CCl_4(g)$

The numerical values of the rate constants for these steps are $k_1 = 4.8 \times 10^3$; $k_{-1} = 3.6 \times 10^3$; $k_2 = 1.3 \times 10^{-2}$; $k_3 = 2.7 \times 10^2$. Derive the rate law and the magnitude of k for the overall reaction.

88. For the reaction A $\longrightarrow$ products, derive the integrated rate law and an expression for the half-life if the reaction is third order.

89. The reaction A + B $\longrightarrow$ products is first order in A, first order in B, and second order overall. Consider that the starting concentrations of the reactants are $[A]_0$ and $[B]_0$, and that x represents the decrease in these concentrations at the time t. That is, $[A]_t = [A]_0 - x$ and $[B]_t = [B]_0 - x$. Show that the integrated rate law for this reaction can be expressed as shown below.

$$\ln \frac{[A]_0 \times [B]_t}{[B]_0 \times [A]_t} = ([B]_0 - [A]_0) \times kt$$

90. The rate of the reaction

$$2 CO(g) \longrightarrow CO_2(g) + C(s)$$

was studied by injecting $CO(g)$ into a reaction vessel and measuring the total pressure at constant volume.

P_{total} (Torr)	Time (s)
250	0
238	398
224	1002
210	1801

What is the rate constant of this reaction?

91. The kinetics of the decomposition of phosphine at 950 K

$$4 PH_3(g) \longrightarrow P_4(g) + 6 H_2(g)$$

was studied by injecting $PH_3(g)$ into a reaction vessel and measuring the total pressure at constant volume.

P_{total} (Torr)	Time (s)
100	0
150	40
167	80
172	120

What is the rate constant of this reaction?

92. The rate of an enzyme-catalyzed reaction can be slowed down by the presence of an inhibitor (I) that reacts with the enzyme in a rapid equilibrium process.

$$E + I \rightleftharpoons EI$$

By adding this step to the mechanism for enzyme catalysis on page 641, determine the effect of adding the concentration $[I_0]$ on the rate of an enzyme-catalyzed reaction.

93. By taking the reciprocal of both sides of equation 14.36, obtain an expression for $1/V$. Using the resulting equation, suggest a strategy for determining the Michaelis–Menten constant, K_M, and the value of k_2.

94. You want to test the following proposed mechanism for the oxidation of HBr.

$$HBr + O_2 \xrightarrow{k_1} HOOBr$$

$$HOOBr + HBr \xrightarrow{k_{-1}} 2 HOBr$$

$$HOBr + HBr \xrightarrow{k_2} H_2O + Br_2$$

You find that the rate is first order with respect to HBr and to O_2. You cannot detect HOBr among the products.
(a) If the proposed mechanism is correct, which must be the rate-determining step?
(b) Can you prove the mechanism from these observations?
(c) Can you disprove the mechanism from these observations?

95. The decomposition of nitric oxide occurs through two parallel reactions:

$$NO(g) \longrightarrow \frac{1}{2} N_2(g) + \frac{1}{2} O_2(g) \qquad k_1 = 25.7 \, s^{-1}$$

$$NO(g) \longrightarrow \frac{1}{2} N_2O(g) + \frac{1}{4} O_2(g) \qquad k_2 = 18.2 \, s^{-1}$$

(a) What is the reaction order for these reactions?
(b) Which reaction is the slow reaction?
(c) If the initial concentration of $NO(g)$ is 2.0 M, what is the concentration of $N_2(g)$ after 0.1 seconds?
(d) If the initial concentration of $NO(g)$ is 4.0 M, what is the concentration of $N_2O(g)$ after 0.025 seconds?

Feature Problems

96. Benzenediazonium chloride decomposes by a first-order reaction in water, yielding $N_2(g)$ as one product.

$$C_6H_5N_2Cl \longrightarrow C_6H_5Cl + N_2(g)$$

The reaction can be followed by measuring the volume of $N_2(g)$ as a function of time. The data in the table were obtained for the decomposition of a 0.071 M solution at 50 °C, where $t = \infty$ corresponds to the completed reaction.

Time, min	$N_2(g)$, mL	Time, min	$N_2(g)$, mL
0	0	18	41.3
3	10.8	21	44.3
6	19.3	24	46.5
9	26.3	27	48.4
12	32.4	30	50.4
15	37.3	∞	58.3

(a) Convert the information given here into a table with one column for time and the other for $[C_6H_5N_2Cl]$.
(b) Construct a table similar to Table 14.2, in which the time interval is $\Delta t = 3$ min.
(c) Plot graphs similar to Figure 14-2, showing both the formation of $N_2(g)$ and the disappearance of $C_6H_5N_2Cl$ as a function of time.
(d) From the graph of part (c), determine the rate of reaction at $t = 21$ min, and compare your result with the reported value of 1.1×10^{-3} M min^{-1}.
(e) Determine the initial rate of reaction.
(f) Write the rate law for the first-order decomposition of $C_6H_5N_2Cl$, and estimate a value of k based on the rate determined in parts (d) and (e).
(g) Determine $t_{1/2}$ for the reaction by estimation from the graph of the rate data and by calculation.
(h) At what time would the decomposition of the sample be three-fourths complete?
(i) Plot $\ln[C_6H_5N_2Cl]$ versus time, and show that the reaction is indeed first order.
(j) Determine k from the slope of the graph of part (i).

97. The object is to study the kinetics of the reaction between peroxodisulfate and iodide ions.

(a) $S_2O_8^{2-}(aq) + 3\,I^-(aq) \longrightarrow$
$$2\,SO_4^{2-}(aq) + I_3^-(aq)$$

The I_3^- formed in reaction (a) is actually a complex of iodine, I_2, and iodide ion, I^-. Thiosulfate ion, $S_2O_3^{2-}$, also present in the reaction mixture, reacts with I_3^- just as fast as it is formed.

(b) $2\,S_2O_3^{2-}(aq) + I_3^-(aq) \longrightarrow S_4O_6^{2-} + 3\,I^-(aq)$

When all of the thiosulfate ion present initially has been consumed by reaction (b), a third reaction occurs between $I_3^-(aq)$ and starch, which is also present in the reaction mixture.

(c) $I_3^-(aq) + \text{starch} \longrightarrow \text{blue complex}$

The rate of reaction (a) is inversely related to the time required for the blue color of the starch–iodine complex to appear. That is, the faster reaction (a) proceeds, the

more quickly the thiosulfate ion is consumed in reaction (b), and the sooner the blue color appears in reaction (c). One of the photographs shows the initial colorless solution and an electronic timer set at $t = 0$; the other photograph shows the very first appearance of the blue complex (after 49.89 s). Tables I and II list some actual student data obtained in this study.

TABLE I

Reaction conditions at 24 °C: 25.0 mL of the $(NH_4)_2S_2O_8(aq)$ listed, 25.0 mL of the KI(aq) listed, 10.0 mL of 0.010 M $Na_2S_2O_3(aq)$, and 5.0 mL starch solution are mixed. The time is that of the first appearance of the starch–iodine complex.

Experiment	Initial Concentrations, M		
	$(NH_4)_2S_2O_8$	KI	Time, s
1	0.20	0.20	21
2	0.10	0.20	42
3	0.050	0.20	81
4	0.20	0.10	42
5	0.20	0.050	79

TABLE II

Reaction conditions: those listed in Table I for Experiment 4, but at the temperatures listed.

Experiment	Temperature, °C	Time, s
6	3	189
7	13	88
8	24	42
9	33	21

(a) Use the data in Table I to establish the order of reaction (a) with respect to $S_2O_8^{2-}$ and to I^-. What is the overall reaction order? [*Hint:* How are the times required for the blue complex to appear related to the actual rates of reaction?]
(b) Calculate the initial rate of reaction in Experiment 1, expressed in M s^{-1}. [*Hint:* You must take into account the dilution that occurs when the various solutions are mixed, as well as the reaction stoichiometry indicated by equations (a), (b), and (c).]
(c) Calculate the value of the rate constant, k, based on experiments 1 and 2.
(d) Calculate the rate constant, k, for the four different temperatures in Table II.
(e) Determine the activation energy, E_a, of the peroxodisulfate–iodide ion reaction.
(f) The following mechanism has been proposed for reaction (a). The first step is slow, and the others are fast.

$$I^- + S_2O_8^{2-} \longrightarrow IS_2O_8^{3-}$$
$$IS_2O_8^{3-} \longrightarrow 2\,SO_4^{2-} + I^+$$
$$I^+ + I^- \longrightarrow I_2$$
$$I_2 + I^- \longrightarrow I_3^-$$

Show that this mechanism is consistent with both the stoichiometry and the rate law of reaction (a). Explain why it is reasonable to expect the first step in the mechanism to be slower than the others.

Self-Assessment Exercises

98. In your own words, define or explain the following terms or symbols: (a) $[A]_0$; (b) k; (c) $t_{1/2}$; (d) zero-order reaction; (e) catalyst.

99. Briefly describe each of the following ideas, phenomena, or methods: (a) the method of initial rates; (b) activated complex; (c) reaction mechanism; (d) heterogeneous catalysis; (e) rate-determining step.

100. Explain the important distinctions between each pair of terms: (a) first-order and second-order reactions; (b) rate law and integrated rate law; (c) activation energy and enthalpy of reaction; (d) elementary process and overall reaction; (e) enzyme and substrate.

101. The rate equation for the reaction $2 A + B \longrightarrow C$ is found to be rate $= k[A][B]$. For this reaction, we can conclude that (a) the unit of $k = s^{-1}$; (b) $t_{1/2}$ is constant; (c) the value of k is independent of the values of $[A]$ and $[B]$; (d) the rate of formation of C is twice the rate of disappearance of A.

102. A first-order reaction, $A \longrightarrow$ products, has a half-life of 75 s, from which we can draw two conclusions. Which of the following are those two (a) the reaction goes to completion in 150 s; (b) the quantity of A remaining after 150 s is half of what remains after 75 s; (c) the same quantity of A is consumed for every 75 s of the reaction; (d) one-quarter of the original quantity of A is consumed in the first 37.5 s of the reaction; (e) twice as much A is consumed in 75 s when the initial amount of A is doubled; (f) the amount of A consumed in 150 s is twice as much as is consumed in 75 s.

103. A first order reaction $A \longrightarrow$ products has a half-life of 13.9 min. The rate at which this reaction proceeds when $[A] = 0.40 M$ is (a) $0.020 \, mol \, L^{-1} \, min^{-1}$; (b) $5.0 \times 10^{-2} \, mol \, L^{-1} \, min^{-1}$; (c) $8.0 \, mol \, L^{-1} \, min^{-1}$; (d) $0.125 \, mol \, L^{-1} \, min^{-1}$.

104. The reaction $A \longrightarrow$ products is second order. The initial rate of decomposition of A when $[A]_0 = 0.50 M$ is (a) the same as the initial rate for any other value of $[A]_0$; (b) half as great as when $[A]_0 = 1.00 M$; (c) five times as great as when $[A]_0 = 0.10 M$; (d) four times as great as when $[A]_0 = 0.25 M$.

105. The rate of a chemical reaction generally increases rapidly, even for small increases in temperature, because of a rapid increase in (a) collision frequency; (b) fraction of reactant molecules with very high kinetic energies; (c) activation energy; (d) average kinetic energy of the reactant molecules.

106. For the reaction $A + B \longrightarrow 2 C$, which proceeds by a single-step bimolecular elementary process, (a) $t_{1/2} = 0.693/k$; (b) rate of appearance of C = $-$rate of disappearance of A; (c) rate of reaction = $k[A][B]$; (d) $\ln[A]_t = -kt + \ln[A]_0$.

107. In the first-order decomposition of substance A the following concentrations are found at the indicated times: $t = 0 \, s$, $[A] = 0.88 \, M$; $t = 50 \, s$, $[A] = 0.62 \, M$; $t = 100 \, s$, $[A] = 0.44 \, M$; $t = 150 \, s$, $[A] = 0.31 \, M$. Calculate the instantaneous rate of decomposition at $t = 100 \, s$.

108. A reaction is 50% complete in 30.0 min. How long after its start will the reaction be 75% complete if it is (a) first order; (b) zero order?

109. A kinetic study of the reaction $A \longrightarrow$ products yields the data: $t = 0 \, s$, $[A] = 2.00 \, M$; 500 s, 1.00 M; 1500 s, 0.50 M; 3500 s, 0.25 M. *Without performing detailed calculations*, determine the order of this reaction and indicate your method of reasoning.

110. For the reaction $A \longrightarrow$ products the following data are obtained.

Experiment 1		Experiment 2	
$[A] = 1.204 \, M$	$t = 0 \, min$	$[A] = 2.408 \, M$	$t = 0 \, min$
$[A] = 1.180 \, M$	$t = 1.0 \, min$	$[A] = ?$	$t = 1.0 \, min$
$[A] = 0.602 \, M$	$t = 35 \, min$	$[A] = ?$	$t = 30 \, min$

(a) Determine the initial rate of reaction in Experiment 1.
(b) If the reaction is second order, what will be $[A]$ at $t = 1.0 \, min$ in Experiment 2?
(c) If the reaction is first order, what will be $[A]$ at 30 min in Experiment 2?

111. For the reaction $A + 2 B \longrightarrow C + D$, the rate law is rate of reaction $= k[A][B]$.
(a) Show that the following mechanism is consistent with the stoichiometry of the overall reaction and with the rate law.

$$A + B \longrightarrow I \quad (slow)$$
$$I + B \longrightarrow C + D \quad (fast)$$

(b) Show that the following mechanism is consistent with the stoichiometry of the overall reaction, but *not* with the rate law.

$$2 B \underset{k_{-1}}{\overset{k_1}{\rightleftharpoons}} B_2 \quad (fast)$$
$$A + B_2 \overset{k_2}{\longrightarrow} C + D \quad (slow)$$

112. If the plot of the reactant concentration versus time is nonlinear, but the concentration drops by 50% every 10 seconds, then the order of the reaction is (a) zero order; (b) first order; (c) second order; (d) third order.

113. If the plot of the reactant concentration versus time is linear, then the order of the reaction is (a) zero order; (b) first order; (c) second order; (d) third order.

114. For a given reaction, doubling the temperature will increase the reaction rate by (a) 2; (b) 0.61; (c) 2.7; (d) 1.6.

115. One example of a zero-order reaction is the decomposition of ammonia on a hot platinum wire, $2 NH_3(g) \longrightarrow N_2(g) + 3 H_2(g)$. If the concentration of ammonia is doubled, the rate of the reaction will (a) be zero; (b) double; (c) remain the same; (d) exponentially increase.

116. Using the method presented in Appendix E, construct a concept map illustrating the concepts in Sections 14-8 and 14-9.

Principles of Chemical Equilibrium

A vital natural reaction is in progress in the lightning bolt seen here: $N_2(g) + O_2(g) \rightleftharpoons 2NO(g)$. Usually this reversible reaction does not occur to any significant extent in the forward direction, but in the high-temperature lightning bolt it does. At equilibrium at high temperatures, measurable conversion of $N_2(g)$ and $O_2(g)$ to $NO(g)$ occurs. In this chapter we study the equilibrium condition in a reversible reaction and the factors affecting it.

Until now, we have stressed reactions that go to completion and the concepts of stoichiometry that allow us to calculate the outcomes of such reactions. We have made occasional references to situations involving both a forward and a reverse reaction—*reversible reactions*—but in this chapter, we will look at them in a detailed and systematic way.

Our emphasis will be on the equilibrium condition reached when forward and reverse reactions proceed at the same rate. Our main tool in dealing with equilibrium will be the equilibrium constant. We will begin with some key relationships involving equilibrium constants; then we will make qualitative predictions about the condition of equilibrium; and finally we will do various equilibrium calculations. As we will discover throughout the remainder of the text, the equilibrium condition plays a role in numerous natural phenomena and affects the methods used to produce many important industrial chemicals.

15-1 Dynamic Equilibrium

Let's begin by describing three simple physical and one chemical phenomena that will help us to establish the core attribute of a system at **equilibrium**—two opposing processes take place at equal rates.

1. When a liquid vaporizes within a closed container, after a time, vapor molecules condense to the liquid state at the same rate at which liquid molecules vaporize. Even though molecules continue to pass back and forth between liquid and vapor (a *dynamic* process), the pressure exerted by the vapor remains constant with time. *The vapor pressure of a liquid is a property resulting from an equilibrium condition.*

2. When a solute is added to a solvent, the system may reach a point at which the rate of dissolution is just matched by the rate at which dissolved solute crystallizes—that is, the solution is saturated. Even though solute particles continue to pass back and forth between the saturated solution and the undissolved solute, the concentration of dissolved solute remains constant. *The solubility of a solute is a property resulting from an equilibrium condition.*

3. When an aqueous solution of iodine, I_2, is shaken with pure carbon tetrachloride, $CCl_4(l)$, the I_2 molecules move into the CCl_4 layer. As the concentration of I_2 builds up in the CCl_4, the rate of return of I_2 to the water layer becomes significant. When I_2 molecules pass between the two liquids at equal rates—a condition of dynamic equilibrium—the concentration of I_2 in each layer remains constant. At this point, the concentration of I_2 in the CCl_4 is about 85 times as great as in the H_2O (Fig. 15-1). The ratio of concentrations of a solute in two immiscible solvents is called the distribution coefficient. *The distribution coefficient, which represents the partitioning of a solute between two immiscible solvents, is a property resulting from an equilibrium condition.*

4. When gaseous phosphorus pentachloride is heated, it decomposes to phosphorus trichloride and chlorine gases: $PCl_5(g) \longrightarrow PCl_3(g) + Cl_2(g)$. Consider a sample of $PCl_5(g)$ initially exerting a pressure of 1.0 atm in a closed container at 250 °C. The gas pressure in the container first rises rapidly and then ever more slowly, reaching a maximum, unchanging pressure of 1.7 atm. Because two moles of gas are produced for each mole of $PCl_5(g)$ that decomposes, if the reaction went to completion the final pressure would have been 2.0 atm. We conclude that *the decomposition of PCl_5 is a reversible reaction that reaches an equilibrium condition.*

The properties in the first three situations just described—vapor pressure, solubility, and distribution coefficient—are examples of physical equilibria. The fourth situation is an example of chemical equilibrium. All four are described through a general quantity known as an *equilibrium constant*, the subject of the next section.

(a) **(b)**

▲ FIGURE 15-1
Dynamic equilibrium in a physical process
(a) A yellow-brown saturated solution of I_2 in water (top layer) is brought into contact with colorless $CCl_4(l)$ (bottom layer), **(b)** I_2 molecules distribute themselves between the H_2O and CCl_4. When equilibrium is reached, $[I_2]$ in the CCl_4 (violet, bottom layer) is about 85 times as great as in the water (colorless, top layer).

🔍 **15-1 CONCEPT ASSESSMENT**

For each chemical equation state whether it represents physical or chemical equilibrium: **(a)** $CaCO_3(s) \rightleftharpoons Ca^{2+}(aq) + CO_3^{2-}(aq)$; **(b)** $I_2(s) \rightleftharpoons I_2(g)$; **(c)** $Fe(s) + 4H_2O(l) \rightleftharpoons Fe_3O_4(s) + 4H_2(g)$.

15-2 The Equilibrium Constant Expression

The oxidation–reduction reaction of copper(II) and tin(II) in aqueous solution is a reversible reaction, which means that at the same time as copper(I) and tin(IV) are being formed,

$$2\,Cu^{2+}(aq) + Sn^{2+}(aq) \longrightarrow 2\,Cu^+(aq) + Sn^{4+}(aq) \qquad \textbf{(15.1)}$$

copper(I) and tin(IV) ions elsewhere in the solution are being consumed to form copper(II) and tin(II):

$$2\,Cu^+(aq) + Sn^{4+}(aq) \longrightarrow 2\,Cu^{2+}(aq) + Sn^{2+}(aq) \qquad \textbf{(15.2)}$$

15-1 ARE YOU WONDERING...

How we know that an equilibrium is dynamic—that forward and reverse reactions continue even after equilibrium is reached?

Suppose we have an equilibrium mixture of AgI(s) and its saturated aqueous solution.

$$AgI(s) \rightleftharpoons Ag^+(aq) + I^-(aq)$$

Now let's add to this mixture some saturated solution of AgI made from AgI containing radioactive iodine-131 as iodide ion, as illustrated in Figure 15-2. If both the forward and reverse processes stopped at equilibrium, radioactivity would be confined to the solution. What we find, though, is that radioactivity shows up in the solid in contact with the saturated solution. Over time, the radioactive "hot" spots distribute themselves throughout the solution and undissolved solid. The only way this can happen is if the dissolving of the solid solute and its crystallization from the saturated solution continue indefinitely. The equilibrium condition is *dynamic*.

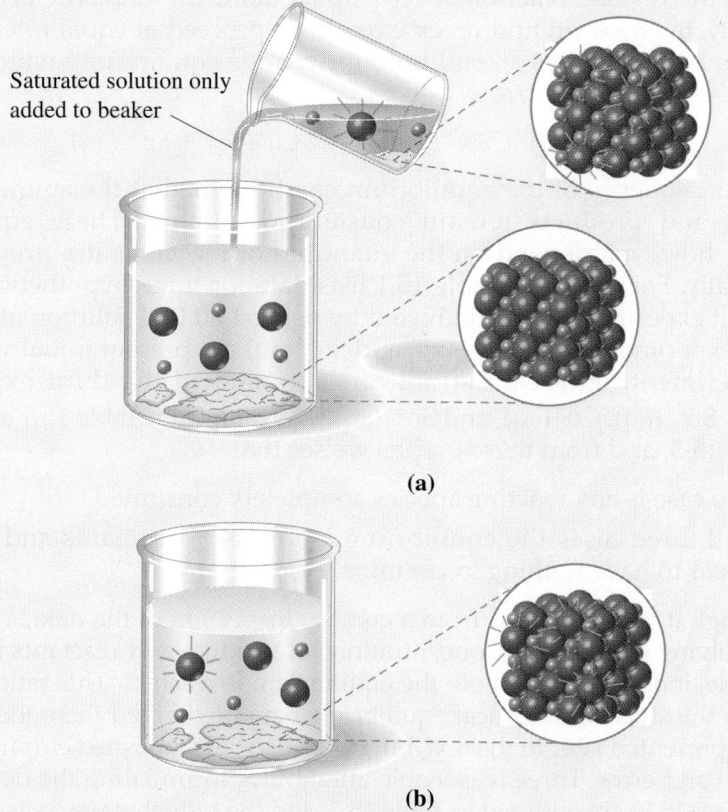

Saturated solution only added to beaker

(a)

(b)

▲ FIGURE 15-2
Dynamic equilibrium illustrated
(a) A saturated solution of radioactive AgI is added to a saturated solution of AgI. **(b)** The radioactive iodide ions distribute themselves throughout the solution and the solid AgI, showing that the equilibrium is dynamic.

▶ The equilibrium state can be obtained by starting with any combination of reactants and products.

TABLE 15.1 Three Approaches to Equilibrium in the Reaction[a] $2\,Cu^{2+}(aq) + Sn^{2+}(aq) \rightleftharpoons 2\,Cu^+(aq) + Sn^{4+}(aq)$

	$Cu^{2+}(aq)$	$Sn^{2+}(aq)$	$Cu^+(aq)$	$Sn^{4+}(aq)$
Experiment 1				
Initial amounts, mol	0.100	0.100	0.000	0.000
Equilibrium amounts, mol	0.0360	0.0680	0.0640	0.0320
Equilibrium concentrations, mol/L	0.0360	0.0680	0.0640	0.0320
Experiment 2				
Initial amounts, mol	0.000	0.000	0.100	0.100
Equilibrium amounts, mol	0.0567	0.0283	0.0433	0.0717
Equilibrium concentrations, mol/L	0.0567	0.0283	0.0433	0.0717
Experiment 3				
Initial amounts, mol	0.100	0.100	0.100	0.100
Equilibrium amounts, mol	0.0922	0.0961	0.1078	0.1039
Equilibrium concentrations, mol/L	0.0922	0.0961	0.1078	0.1039

The concentrations printed in blue are used in the calculations in Table 15.2.
[a] Reaction carried out in 1.00 L of solution at 298 K.

▶ The rates of reactions (forward and reverse) are affected by the concentration of reactants.

Initially, only the forward reaction (15.1) occurs, but as soon as some Cu^+ and Sn^{4+} form, the reverse reaction (15.2) begins. With passing time, the forward reaction slows because of the decreasing concentrations of Cu^{2+} and Sn^{2+} and the reverse reaction speeds up as more Cu^+ and Sn^{4+} accumulate. Eventually, the forward and reverse reactions proceed at equal rates, and the reaction mixture reaches a condition of dynamic equilibrium, which we can represent with a double arrow $\rightleftharpoons$.

$$2\,Cu^{2+}(aq) + Sn^{2+}(aq) \rightleftharpoons 2\,Cu^+(aq) + Sn^{4+}(aq) \qquad (15.3)$$

One consequence of the equilibrium condition is that the amounts of the reactants and products remain constant with time. These equilibrium amounts, however, depend on the quantities of reactants and products present initially. For example, Table 15.1 lists data for three hypothetical experiments. All experiments are conducted by using 1.00 L of solution at 298 K. In the first experiment, only $Cu^{2+}(aq)$ and $Sn^{2+}(aq)$ are present initially; in a second experiment, only $Cu^+(aq)$ and $Sn^{4+}(aq)$; and in the third experiment, $Cu^{2+}(aq)$, $Sn^{2+}(aq)$, $Cu^+(aq)$, and $Sn^{4+}(aq)$. The data from Table 15.1 are plotted in Figure 15-3, and from these graphs we see that

- in no case is any reacting species completely consumed.
- in all three cases the equilibrium amounts of reactants and products appear to have nothing in common.

Although it is not obvious from a cursory inspection of the data, a particular ratio involving equilibrium concentrations of product and reactants has a constant value, independent of how the equilibrium is reached. This ratio, which is central to the study of chemical equilibrium, can be derived theoretically using concepts presented later in the text, but it can also be established empirically, that is, by trial and error. Three reasonable attempts at formulating the desired ratio for reaction (15.3) are outlined in Table 15.2, and the ratio that works is identified.

For the copper–tin oxidation–reduction reaction, the ratio of equilibrium concentrations in the following equation has a constant value of 1.48 at 300 K.

▶ The representation of the equilibrium expression in terms of concentrations is valid only at low concentrations, usually less than a few moles per liter.

$$K = \frac{\left[Cu^+\right]^2_{eq}\left[Sn^{4+}\right]_{eq}}{\left[Cu^{2+}\right]^2_{eq}\left[Sn^{2+}\right]_{eq}} = 1.48 \qquad (15.4)$$

This ratio is called the **equilibrium constant expression** and its numerical value is the **equilibrium constant**.

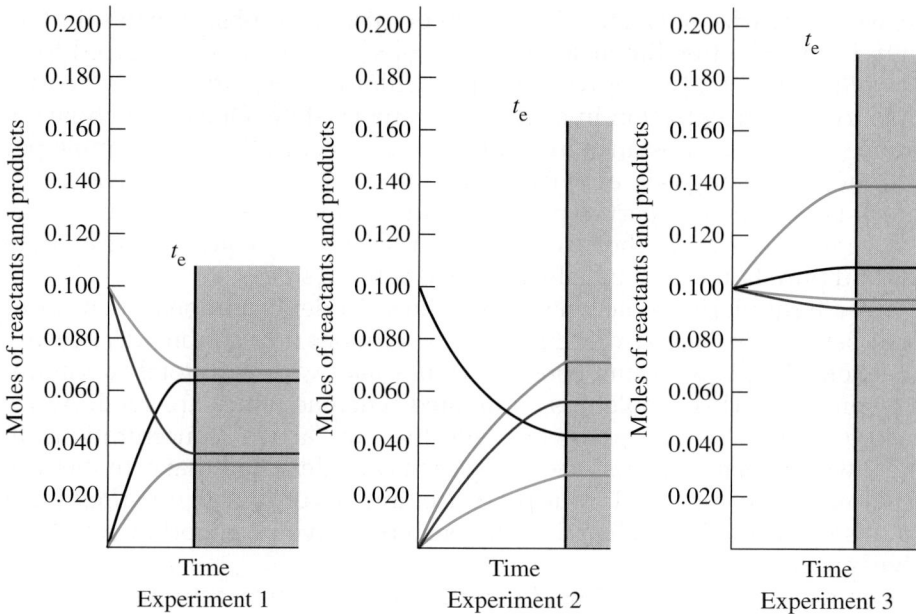

t_e = time for equilibrium to be reached
—— mol Cu^{2+}
—— mol Sn^{2+}
—— mol Cu^+
—— mol Sn^{4+}

▲ FIGURE 15-3
Three approaches to equilibrium in the reaction

$$2\ Cu^{2+}(aq) + Sn^{2+}(aq) \rightleftharpoons 2\ Cu^+(aq) + Sn^{4+}(aq)$$

The initial and equilibrium amounts for each of these three cases are listed in Table 15.1. t_e = time for equilibrium to be reached.

The Equilibrium Constant and Activities

We have discovered that the equilibrium constant expression is written as a ratio of the product concentrations divided by the reactant concentrations. G. N. Lewis* proposed a more appropriate way to represent the equilibrium constant expression that uses the activities of the reactants and products. **Activity** is a thermodynamic concept that we will see again in Chapter 19.

◀ "Equilibrium" means equal rates of forward and reverse reactions, *not* equal concentrations of reactants and products.

TABLE 15.2			
Expt	Trial 1: $\dfrac{[Cu^+]\,[Sn^{4+}]}{[Cu^{2+}]\,[Sn^{2+}]}$	Trial 2: $\dfrac{(2 \times [Cu^+])\,[Sn^{4+}]}{(2 \times [Cu^{2+}])\,[Sn^{2+}]}$	Trial 3: $\dfrac{[Cu^+]^2\,[Sn^{4+}]}{[Cu^{2+}]^2\,[Sn^{2+}]}$
1	$\dfrac{0.0640 \times 0.0320}{0.0360 \times 0.0680} = 0.837$	$\dfrac{(2 \times 0.0640) \times 0.0320}{(2 \times 0.0360) \times 0.0680} = 0.837$	$\dfrac{0.0640^2 \times 0.0320}{0.0360^2 \times 0.0680} = 1.49$
2	$\dfrac{0.0433 \times 0.0717}{0.0567 \times 0.0283} = 1.93$	$\dfrac{(2 \times 0.0433) \times 0.0717}{(2 \times 0.0567) \times 0.0283} = 1.93$	$\dfrac{0.0433^2 \times 0.0717}{0.0567^2 \times 0.0283} = 1.48$
3	$\dfrac{0.1078 \times 0.1039}{0.0922 \times 0.0961} = 1.26$	$\dfrac{(2 \times 0.1078) \times 0.1039}{(2 \times 0.0922) \times 0.0961} = 1.26$	$\dfrac{0.1078^2 \times 0.1039}{0.0922^2 \times 0.0961} = 1.48$

Equilibrium concentration data are from Table 15.1. In Trial 1, the equilibrium concentration of Cu^+ and Sn^{2+} are placed in the numerator and the equilibrium concentration of Cu^{2+} and Sn^{4+}, in the denominator. In Trial 2, each concentration is multiplied by its stoichiometric coefficient. In Trial 3, each concentration is raised to a power equal to its stoichiometric coefficient. Trial 3 has essentially the same value for each experiment. This value is the equilibrium constant K.

*G. N. Lewis and M. Randall, *Thermodynamics*, McGraw Hill, New York, 1923.

▶Because of interactions between particles, the "effective" or "active" concentration (activity) of a substance is usually different from the stoichiometric concentration (see page 587). The activity coefficient, γ, takes into account the effect of particle interactions.

Here, we summarize only a few important points. For a substance in solution, activity is equal to the dimensionless ratio $\gamma[X]/c°$, where γ is referred to as the *activity coefficient*, $[X]$ represents a particular concentration and $c°$ corresponds to the concentration in a chosen reference state. Our usual choice of reference state for a substance in solution is a concentration of one mole per liter (1 mol L^{-1}). For gases, activity is equal to the dimensionless ratio $\gamma P/P°$, where P is a particular partial pressure and $P°$ is the partial pressure in the reference state; our usual choice of $P°$ is 1 bar (essentially equal to 1 atm). Pure solids and pure liquids are usually assigned activities of 1.

Why base equilibrium constants on activities? Under conditions where gases do not obey the ideal gas law (Section 6-9) or solutions depart from ideal behavior (Section 13-3), equilibrium constant values may vary with total concentration or pressure. This problem is eliminated when activities are used. As we learned in Section 13-9, activities are "effective" or "active" concentrations. In this text we will generally assume that systems are ideal, and that activities can be replaced by the numerical values of concentrations or partial pressures. Later on in this text we will establish the relationship between equilibrium and thermodynamic quantities.

Let us reconsider expression (15.4), this time by using activities. We will replace equation (15.4) with the following expression:

$$K = \frac{a_{Cu^+}^2 a_{Sn^{4+}}}{a_{Cu^{2+}}^2 a_{Sn^{2+}}} = 1.48 \qquad \textbf{(15.5)}$$

To establish the relationship between an equilibrium constant expressed in activities and the corresponding one expressed in concentration, we begin by writing the activity of each species, using the [ ] symbol for the equilibrium concentration and $c°$ for the concentration in the reference state.

$$a_{Cu^+} = \frac{\gamma_{Cu^+}[Cu^+]}{c°}; \qquad a_{Sn^{4+}} = \frac{\gamma_{Sn^{4+}}[Sn^{4+}]}{c°}$$

$$a_{Cu^{2+}} = \frac{\gamma_{Cu^{2+}}[Cu^{2+}]}{c°}; \qquad a_{Sn^{2+}} = \frac{\gamma_{Sn^{2+}}[Sn^{2+}]}{c°}$$

▶ The activity coefficients for ions will not be equal to 1 because the interactions between the ions are significant, even at low concentrations.

Then we choose the value $c° = 1 \text{ mol L}^{-1}$, we substitute these relationships into the equilibrium constant expression (15.5), and we have

$$K = \frac{\left(\dfrac{\gamma_{Cu^+}[Cu^+]_{eq}}{c°}\right)^2 \dfrac{\gamma_{Sn^{4+}}[Sn^{4+}]_{eq}}{c°}}{\left(\dfrac{\gamma_{Cu^{2+}}[Cu^{2+}]_{eq}}{c°}\right)^2 \dfrac{\gamma_{Sn^{2+}}[Sn^{2+}]_{eq}}{c°}} \approx \frac{[Cu^+]_{eq}^2 [Sn^{4+}]_{eq}}{[Cu^{2+}]_{eq}^2 [Sn^{2+}]_{eq}} = 1.48 \qquad \textbf{(15.6)}$$

The approximation made in writing expression (15.6) is that the values of all the γ are equal to one. This is equivalent to assuming that the ions behave ideally in solution (see page 587). Notice that we have arrived at exactly the expression in equation (15.4).

15-2 CONCEPT ASSESSMENT

How would you write the equilibrium constant expression for $Cu(s) + 2 H^+(aq) \longrightarrow Cu^{+2}(aq) + H_2(g)$? Write this first in terms of activities and then convert to pressures and concentrations.

A General Expression for *K*

Before proceeding to other matters, let us emphasize that the equilibrium constant expression for the oxidation–reduction reaction of copper(II) and tin(II)

EXAMPLE 15-1   Relating Equilibrium Concentrations of Reactants and Products

These equilibrium concentrations are measured in reaction (15.3) at 300 K: $[Cu^+]_{eq} = 0.148$ M, $[Sn^{2+}]_{eq} = 0.124$ M, and $[Sn^{4+}]_{eq} = 0.176$ M. What is the equilibrium concentration of $Cu^{2+}(aq)$?

Analyze

First, we write the equilibrium constant expression for reaction (15.3) in terms of activities, along with the value of the equilibrium constant. Then, we convert the activities to concentrations and replace the concentrations with the given data.

Solve

Write the equilibrium constant expression.

$$K = \frac{a_{Cu^+}^2 a_{Sn^{4+}}}{a_{Cu^{2+}}^2 a_{Sn^{2+}}} = 1.48$$

Assume that the reaction conditions are such that the activities can be replaced by their concentration values, allowing concentration units to be canceled, as in expression (15.6).

$$K = \frac{[Cu^+]_{eq}^2 [Sn^{4+}]_{eq}}{[Cu^{2+}]_{eq}^2 [Sn^{2+}]_{eq}} = 1.48$$

Substitute the known equilibrium concentrations into the equilibrium constant expression.

$$K = \frac{[Cu^+]_{eq}^2 [Sn^{4+}]_{eq}}{[Cu^{2+}]_{eq}^2 [Sn^{2+}]_{eq}} = \frac{(0.148)^2 (0.176)}{[Cu^{2+}]_{eq}^2 (0.124)} = 1.48$$

Solve for the unknown concentration, $[Cu^{2+}]$. (An implicit calculation to restore the concentration unit is $[Cu^{2+}] = a_{Cu^{2+}} \times c° = 0.145 \times 1.000$ M $= 0.145$ M.)

$$[Cu^{2+}]^2 = \frac{0.148^2 \times 0.176}{0.124 \times 1.48} = 0.0210$$

$$[Cu^{2+}] = \sqrt{0.0210} = 0.145 \text{ M}$$

Assess

When solving equilibrium problems, we should examine the results to ensure they make sense. We can do this easily by placing the solution back into the equilibrium constant expression and calculating the equilibrium constant to see whether it agrees with the value stated in the problem, as shown below.

$$K = \frac{0.148^2 \times 0.176}{0.145^2 \times 0.124} = 1.48$$

PRACTICE EXAMPLE A: In another experiment, equal concentrations of $[Cu^+]$, $[Sn^{4+}]$, and $[Sn^{2+}]$ are found to be in equilibrium in reaction (15.3). What must be the equilibrium concentration of $[Cu^{2+}]$?

PRACTICE EXAMPLE B: At 25 °C, $K = 9.14 \times 10^{-6}$ for the reaction $2 Fe^{3+}(aq) + Hg_2^{2+}(aq) \rightleftharpoons 2 Fe^{2+}(aq) + 2 Hg^{2+}(aq)$. If the equilibrium concentrations of Fe^{3+}, and Hg^{2+} are 0.015 M, 0.0025 M and 0.0018 M, Fe^{2+} respectively, what is the equilibrium concentration of Hg_2^{2+}?

---

summarized through expression (15.6) is just a specific example of a more general case. For the hypothetical, generalized reaction

$$a A(aq) + b B(aq) \cdots \rightleftharpoons g G(aq) + h H(aq) \cdots$$

The equilibrium constant expression has the form

$$K = \frac{(a_G)^g (a_H)^h \cdots}{(a_A)^a (a_B)^b \cdots} \approx \left(\frac{1}{c°}\right)^{\Delta n} \frac{[G]^g [H]^h \cdots}{[A]^a [B]^b \cdots} = \left(\frac{1}{c°}\right)^{\Delta n} K_c \quad \textbf{(15.7)}$$

where $\Delta n = (g + h + \cdots) - (a + b + \cdots)$ and K_c is an equilibrium constant expression written in terms of concentrations. The factor $(1/c°)^{\Delta n}$ ensures that K is a dimensionless quantity.

The *numerator* of an equilibrium constant expression is the product of the activities of the species on the *right* side of the equation ($a_G, a_H, \ldots$), with each activity raised to a power given by the stoichiometric coefficient ($g, h, \ldots$). The *denominator* is the product of the activities of the species on the *left* side of the

◀ In principle, when using equation (15.7), we can express concentrations in any unit. However, the use of a unit other than mol/L will require an "extra" calculation, that is, the calculation of the value of $(1/c°)^{\Delta n}$. When we express concentrations, in mol/L, the factor $(1/c°)^{\Delta n}$ has a numerical value of 1. For this reason, when using equation (15.7), we will express concentrations in mol/L and substitute their values (without units) into the expression to obtain the correct value of K.

Consider a hypothetical reaction in which one molecule, A, is converted to its isomer, B, that is, the reversible reaction A $\rightleftharpoons$ B. Start with a flask containing 54 molecules of A, represented by open circles. Convert the appropriate number of open circles to filled circles to represent the isomer B and portray the equilibrium condition if $K = 0.02$. Repeat the process for $K = 0.5$ and then for $K = 1$.

equation $(a_A, a_B, \ldots)$, and again, with each activity raised to a power given by the stoichiometric coefficient $(a, b, \ldots)$. As previously noted, where equilibrium systems are sufficiently close to ideal in their behavior, equilibrium concentrations are acceptable approximations to true activities.

The numerical value of an equilibrium constant, K, depends on the particular reaction and on the temperature. We will explore the significance of these numerical values in Section 15-4.

Relationship Between the Equilibrium Constant and Rate Constants

Given the requirement that the rates of the forward and reverse reactions become equal at equilibrium, it seems that a relationship should exist between the equilibrium constant and the rate constants for the forward and reverse reactions. That such a relationship does exist can be demonstrated easily for elementary reactions. Consider again the hypothetical generalized reaction

$$aA + bB + \cdots \underset{k_{-1}}{\overset{k_1}{\rightleftharpoons}} gG + hH + \cdots$$

k_1 and k_{-1} are the rate constants for the forward and reverse reactions. *With the assumption that both the forward and reverse reactions are elementary reactions*, we can write

$$\text{rate of forward reaction} = k_1[A]^a[B]^b \cdots$$
$$\text{rate of reverse reaction} = k_{-1}[G]^g[H]^h \ldots$$

At equilibrium, these two rates become equal; thus,

$$k_1[A]^a[B]^b \cdots = k_{-1}[G]^g[H]^h \cdots$$

which can be rearranged into an expression having rate constants on one side and concentrations on the other:

$$\frac{k_1}{k_{-1}} = \frac{[G]^g[H]^h \cdots}{[A]^a[B]^b \cdots}$$

Because the right side of the equation above is the equilibrium constant expression for the reaction, we arrive at the following result:

$$\frac{k_1}{k_{-1}} = K$$

Keep in mind that this result is based on the assumption that the forward and reverse reactions are elementary reactions. For reactions that involve a multistep mechanism, the relationship between K and the rate constants is more complicated. For a mechanism involving n steps, it can be demonstrated (in the manner described in Exercise 95) that the relationship between K and the rate constants is

$$\frac{k_1}{k_{-1}} \times \frac{k_2}{k_{-2}} \times \cdots \times \frac{k_n}{k_{-n}} = K$$

Although the expression above shows that there is indeed a relationship between the equilibrium constant and rate constants, it is generally easier to obtain K directly from measurements on equilibrium conditions than to attempt a calculation based on rate constants. In Chapters 19 and 20, we will learn about much more direct measurements and calculations leading to values of equilibrium constants.

15-3 Relationships Involving Equilibrium Constants

Before assessing an equilibrium situation, it may be necessary to make some preliminary calculations or decisions to get the appropriate equilibrium constant expression. This section presents some useful ideas in working with equilibrium constants.

Relationship of K to the Balanced Chemical Equation

We must always make certain that the expression for K matches the corresponding balanced equation. In doing so, the following hold true.

- When we *reverse* an equation, we *invert* the value of K.
- When we *multiply* the coefficients in a balanced equation by a common factor $(2, 3, \dots)$, we raise the equilibrium constant to the *corresponding power* $(2, 3, \dots)$.
- When we *divide* the coefficients in a balanced equation by a common factor $(2, 3, \dots)$, we take the *corresponding root* of the equilibrium constant (square root, cube root, ...).

To illustrate these points, let us consider the synthesis of methanol (methyl alcohol) from a carbon monoxide–hydrogen mixture called synthesis gas. This reaction is likely to become increasingly important as methanol and its mixtures with gasoline find greater use as motor fuels. The balanced reaction is

$$CO(g) + 2\,H_2(g) \rightleftharpoons CH_3OH(g) \quad K = 9.23 \times 10^{-3}$$

Suppose that in discussing the synthesis of CH_3OH from CO and H_2, we had written the reverse reaction, that is,

$$CH_3OH(g) \rightleftharpoons CO(g) + 2\,H_2(g) \quad K' = ?$$

Now, according to the generalized equilibrium constant expression (15.7), we should write

$$K' = \frac{a_{CO}\,a_{H_2}^2}{a_{CH_3OH}} = \frac{1}{\dfrac{a_{CH_3OH}}{a_{CO}\,a_{H_2}^2}} = \frac{1}{K} = \frac{1}{9.23 \times 10^{-3}} = 1.08 \times 10^2$$

▲ Methanol is actively being considered as an alternative fuel to gasoline.

In the preceding expression, the equilibrium constant expression and K value for the reaction, as originally written, are printed in blue. We see that $K' = 1/K$.

Suppose that for a certain application we want an equation based on synthesizing *two* moles of $CH_3OH(g)$.

$$2\,CO(g) + 4\,H_2(g) \rightleftharpoons 2\,CH_3OH(g) \quad K'' = ?$$

Here, $K'' = K^2$. That is,

$$K'' = \frac{a_{CH_3OH}^2}{a_{CO}^2\,a_{H_2}^4} = \left(\frac{a_{CH_3OH}}{a_{CO}\,a_{H_2}^2}\right)^2 = (K)^2 = (9.23 \times 10^{-3})^2 = 8.52 \times 10^{-5}$$

🔍 15-4 CONCEPT ASSESSMENT

Can you conclude whether the numerical value of K for the reaction $2\,ICl(g) \rightleftharpoons I_2(g) + Cl_2(g)$ is greater or less than the numerical value of K for the reaction $ICl(g) \rightleftharpoons \frac{1}{2}I_2(g) + \frac{1}{2}Cl_2(g)$? Explain.

EXAMPLE 15-2 Relating K to the Balanced Chemical Equation

The following K value is given at 298 K for the synthesis of $NH_3(g)$ from its elements.

$$N_2(g) + 3H_2(g) \rightleftharpoons 2NH_3(g) \qquad K = 5.8 \times 10^5$$

What is the value of K at 298 K for the following reaction?

$$NH_3(g) \rightleftharpoons \frac{1}{2}N_2(g) + \frac{3}{2}H_2(g) \qquad K = ?$$

Analyze

The solution to this problem lies in recognizing that the reaction is the reverse and one-half of the given reaction. In this example we apply two of the rules given above that relate K to balanced chemical reactions.

Solve

First, reverse the given equation. This puts $NH_3(g)$ on the left side of the equation, where we need it.

$$2NH_3(g) \rightleftharpoons N_2(g) + 3H_2(g)$$

The equilibrium constant K' becomes

$$K' = 1/(5.8 \times 10^5) = 1.7 \times 10^{-6}$$

Then, to base the equation on 1 mol $NH_3(g)$, divide all coefficients by 2.

$$NH_3(g) \rightleftharpoons \frac{1}{2}N_2(g) + \frac{3}{2}H_2(g)$$

This requires the square root of K'.

$$K = \sqrt{1.7 \times 10^{-6}} = 1.3 \times 10^{-3}$$

Assess

Because the rules given on page 663 are used extensively throughout this book, memorizing them will be helpful.

PRACTICE EXAMPLE A: Use data from Example 15-2 to determine the value of K at 298 K for the reaction

$$\frac{1}{3}N_2(g) + H_2(g) \rightleftharpoons \frac{2}{3}NH_3(g)$$

PRACTICE EXAMPLE B: For the reaction $NO(g) + \frac{1}{2}O_2(g) \rightleftharpoons NO_2(g)$ at 184 °C, $K = 1.2 \times 10^2$. What is the value of K at 184 °C for the reaction $2NO_2(g) \rightleftharpoons 2NO(g) + O_2(g)$?

Combining Equilibrium Constant Expressions

In Section 7-7, through Hess's law, we showed how to combine a series of equations into a single overall equation. The enthalpy change of the overall reaction was obtained by adding together the enthalpy changes of the individual reactions. A similar procedure can be used with equilibrium constants, but with this important difference:

> When individual equations are combined (that is, added), their equilibrium constants are *multiplied* to obtain the equilibrium constant for the overall reaction.

Suppose we need the equilibrium constant for the reaction

$$N_2O(g) + \frac{1}{2}O_2(g) \rightleftharpoons 2NO(g) \qquad K = ? \tag{15.8}$$

and know the K values of these two equilibria.

$$N_2(g) + \frac{1}{2}O_2(g) \rightleftharpoons N_2O(g) \qquad K = 5.4 \times 10^{-19} \tag{15.9}$$

$$N_2(g) + O_2(g) \rightleftharpoons 2NO(g) \qquad K = 4.6 \times 10^{-31} \tag{15.10}$$

Equation (15.8) is obtained by reversing equation (15.9) and adding it to (15.10). This requires that we also take the *reciprocal* of the K value of equation (15.9).

(a)
$$N_2O(g) \rightleftharpoons N_2(g) + \frac{1}{2}O_2(g) \qquad K(a) = 1/(5.4 \times 10^{-19})$$
$$= 1.9 \times 10^{18}$$

(b)
$$N_2(g) + O_2(g) \rightleftharpoons 2\,NO(g) \qquad K(b) = 4.6 \times 10^{-31}$$

Overall:
$$N_2O(g) + \frac{1}{2}O_2(g) \rightleftharpoons 2\,NO(g) \qquad K(\text{overall}) = ?$$

The overall equation is expression (15.8), and according to the general expression (15.7),

$$K(\text{overall}) = \frac{a_{NO}^2}{a_{N_2O}\,a_{O_2}^{1/2}} = \underbrace{\frac{a_{N_2}a_{O_2}^{1/2}}{a_{N_2O}}}_{K(a)} \times \underbrace{\frac{a_{NO}^2}{a_{N_2}\,a_{O_2}}}_{K(b)} = K(a) \times K(b)$$

$$= 1.9 \times 10^{18} \times 4.6 \times 10^{-31} = 8.5 \times 10^{-13}$$

🔍 15-5   CONCEPT ASSESSMENT

You want to calculate K for the reaction
$$CH_4(g) + 2\,H_2O(g) \rightleftharpoons CO_2(g) + 4\,H_2(g)$$
and you have available a K value for the reaction
$$CO_2(g) + H_2(g) \rightleftharpoons CO(g) + H_2O(g)$$
What additional K value do you need, assuming that all K values are at the same temperature?

Equilibria Involving Gases

Mixtures of gases are as much solutions as are mixtures in a liquid solvent. Thus, concentrations in a gaseous mixture can be expressed on a mole-per-liter basis. However, as mentioned on page 660, the activity of a gas is not defined in terms of its concentration, but rather, in terms of its partial pressure relative to a reference pressure of $P° = 1$ bar. Let us investigate equilibria involving gases by considering a specific gas-phase reaction.

A key step in the manufacture of sulfuric acid is the following reversible reaction.

$$2\,SO_2(g) + O_2(g) \rightleftharpoons 2\,SO_3(g) \qquad \textbf{(15.11)}$$

◀ The reaction of sulfur dioxide with oxygen gas to form sulfur trioxide is a reaction that takes place in the atmosphere. This results in the formation of sulfuric acid, which contributes to acid rain.

Since all the reaction species are in the gas phase it seems reasonable to use a partial-pressure reference state. We begin by writing the equilibrium constant expression in terms of activities

$$K = \left(\frac{(a_{SO_3})^2}{(a_{SO_2})^2 a_{O_2}} \right)_{eq} \qquad \textbf{(15.12)}$$

where the activities are

$$a_{SO_2} = \frac{\gamma_{SO_2} P_{SO_2}}{P°_{SO_2}}; \qquad a_{O_2} = \frac{\gamma_{O_2} P_{O_2}}{P°_{O_2}}; \qquad a_{SO_3} = \frac{\gamma_{SO_3} P_{SO_3}}{P°_{SO_3}}$$

The reference-state partial pressure is $P° = 1$ bar, which we will take to be essentially the same as 1 atm. Substituting these relationships into equation (15.12) and setting all the γ values equal to 1, we obtain

$$K = \left(\frac{\left(\frac{P_{SO_3}}{P^\circ} \right)^2}{\left(\frac{P_{SO_2}}{P^\circ} \right)^2 \frac{P_{O_2}}{P^\circ}} \right)_{eq} = P^\circ \left(\frac{(P_{SO_3})^2}{(P_{SO_2})^2 P_{O_2}} \right)_{eq} = P^\circ \times K_p \qquad (15.13)$$

As always, the equilibrium constant, K, is dimensionless. The quantity multiplying P° on the right hand side of equation (15.13) is an equilibrium constant expressions written in terms of partial pressures; it is given the symbol K_p:

$$K_p = \left(\frac{(P_{SO_3})^2}{(P_{SO_2})^2 P_{O_2}} \right)_{eq}$$

To establish the equilibrium constant based on concentrations, we first use the ideal gas law, $PV = nRT$, to relate gas concentrations and partial pressures,

$$[SO_2] = \frac{n}{V} = \frac{P_{SO_2}}{RT}; \qquad [O_2] = \frac{n}{V} = \frac{P_{O_2}}{RT}; \qquad [SO_3] = \frac{n}{V} = \frac{P_{SO_3}}{RT}$$

and then substitute into the expression for K_p. We obtain

$$K_p = \left(\frac{(P_{SO_3})^2}{(P_{SO_2})^2 P_{O_2}} \right)_{eq} = \left(\frac{([SO_3]RT)^2}{([SO_2]RT)^2 [O_2]RT} \right)_{eq} = \frac{1}{RT} \left(\frac{[SO_3]^2}{[SO_2]^2 [O_2]} \right)_{eq} \qquad (15.14)$$

The final parenthetical factor in equation (15.14) is given the symbol K_c, where the subscript "c" emphasizes that this equilibrium constant expression is written in terms of concentrations. Equation (15.14) makes it clear that, in general, K_p and K_c have different values. This result is not wrong. It stems from the fact that for gases, we can choose to express the equilibrium constants in terms of either pressures or concentrations.

Thus, the relationship between K_p and K_c for reaction (15.11) is

$$K_p = K_c(RT)^{-1}$$

If we carried out a similar derivation for the general reaction,

$$aA(g) + bB(g) + \cdots \rightleftharpoons gG(g) + hH(g) + \cdots$$

the results would be

$$K = \frac{(a_G)^g (a_H)^h \cdots}{(a_A)^a (a_B)^b \cdots} \approx \left(\frac{1}{P^\circ} \right)^{\Delta n} \frac{P_G^g P_H^h \cdots}{P_A^a P_B^b \cdots} = \left(\frac{1}{P^\circ} \right)^{\Delta n} K_p \qquad (15.15)$$

and

$$K_p = K_c(RT)^{\Delta n_{gas}} \qquad (15.16)$$

where Δn_{gas} is the difference in the stoichiometric coefficients of *gaseous* products and reactants; that is, $\Delta n_{gas} = (g + h + \cdots) - (a + b + \cdots)$. In reaction (15.11), $\Delta n_{gas} = 2 - (2 + 1) = -1$, and thus, $K_p = K_c(RT)^{-1}$, as we established previously. It is important to note that, when applying equation (15.16), we must use $R = 0.08314472$ bar L K^{-1} mol^{-1}. This value of R is required because, as discussed previously, partial pressures are to be expressed in bar and concentrations in mol/L.

Let us address another very important fact concerning the equilibrium constants we have encountered. The equilibrium constant K, expressed in terms of activities, is a dimensionless quantity. Unlike K, the equilibrium constants K_p and K_c have units of $(bar)^{\Delta n}$ and $(mol/L)^{\Delta n}$, respectively. The factors $(1/c^\circ)^{\Delta n}$ and $(1/P^\circ)^{\Delta n}$ appearing in equations (15.7) and (15.15) ensure that K is a dimensionless quantity. Throughout this text, we will use K_p and K_c expressions in solving problems. To avoid the clutter of units in these calculations, and to simplify the

---

EXAMPLE 15-3   Illustrating the Dependence of K on the Reference State

Complete the calculation of K_p for reaction (15.11) knowing that $K_c = 2.8 \times 10^2$ (at 1000 K).

Analyze

We use equation (15.16), with $R = 0.08314$ bar L K^{-1} mol^{-1}. For reasons outlined on page 666, units are omitted from our calculations.

Solve

Write the equation relating the two equilibrium constants with different reference states.

$$K_c = RT \times K_p$$

Rearrange the expression to obtain the quantity desired, K_p.

$$K_p = \frac{K_c}{RT}$$

Substitute the given data and solve.

$$K_p = \frac{2.8 \times 10^2}{0.08314 \times 1000} = 3.4$$

Assess

In this example the equilibrium constant depends on the reference state used. When working this type of problem, keep in mind that Δn is equal to the sum of coefficients of products in the gas phase minus the sum of coefficients of reactants in the gas phase. That is, for the purposes of converting a K_p value to K_c value (or vice versa), Δn is obtained from the coefficients in the chemical equation, not from the actual amounts of gases present in an equilibrium mixture.

PRACTICE EXAMPLE A: For the reaction $2\,NH_3(g) \rightleftharpoons N_2(g) + 3\,H_2(g)$ at 298 K, $K_c = 2.8 \times 10^{-9}$. What is the value of K_p for this reaction?

PRACTICE EXAMPLE B: At 1065 °C, for the reaction $2\,H_2S(g) \rightleftharpoons 2\,H_2(g) + S_2(g)$, $K_p = 1.2 \times 10^{-2}$. What is the value of K_c for the reaction $H_2(g) + \frac{1}{2}S_2(g) \rightleftharpoons H_2S(g)$ at 1065 °C?

---

conversion of a K_c or K_p value to a K value, we will always use the numerical values of partial pressures or concentrations in the expressions without including the units bar or mol/L explicitly.

Equilibria Involving Pure Liquids and Solids

Up to this point in the chapter, all our examples have involved gas phase reactions or reactions occurring in aqueous solutions. Gas-phase reactions and reactions in aqueous solution are *homogeneous* reactions: They occur within a single phase. Let's extend our coverage now to include reactions involving one or more condensed phases—solids and liquids—in contact with a gas or solution phase. These are called *heterogeneous* reactions. One of the most important ideas about heterogeneous reactions is that

> Equilibrium constant expressions *do not* contain concentration terms for solid or liquid phases of a single component (that is, for pure solids and liquids).

We can think about this statement in either of two ways: (1) An equilibrium constant expression includes terms only for reactants and products whose concentrations and/or partial pressures *can change* during a chemical reaction. The concentration of the single component within a pure solid or liquid phase *cannot change*. (2) Alternatively, recall that the activities of pure liquids and solids are set equal to 1; thus the effect on the numerical value of the equilibrium constant is the same as not including terms for pure solids and liquids at all.

The water–gas reaction, used to make combustible gases from coal, has reacting species in both gaseous and solid phases.

$$C(s) + H_2O(g) \rightleftharpoons CO(g) + H_2(g)$$

◄ Still another way to think about solids and liquids is through their densities. Density, the mass per unit volume of a substance, can be expressed in moles per liter by converting the unit volume from milliliter to liter and dividing the mass in grams by the molar mass. The resultant molar density (mol/L) is a concentration term and, at a fixed temperature, is a constant that would be incorporated in the K_c value.

▶ FIGURE 15-4
Equilibrium in the reaction

$$CaCO_3(s) \rightleftharpoons CaO(s) + CO_2(g)$$

(a) Decomposition of $CaCO_3(s)$ upon heating in a closed vessel yields a few granules $CaO(s)$, together with $CO_2(g)$, which soon exerts its equilibrium partial pressure. **(b)** Introduction of additional $CaCO_3(s)$ and/or more $CaO(s)$ has no effect on the partial pressure of the $CO_2(g)$, which remains the same as in (a).

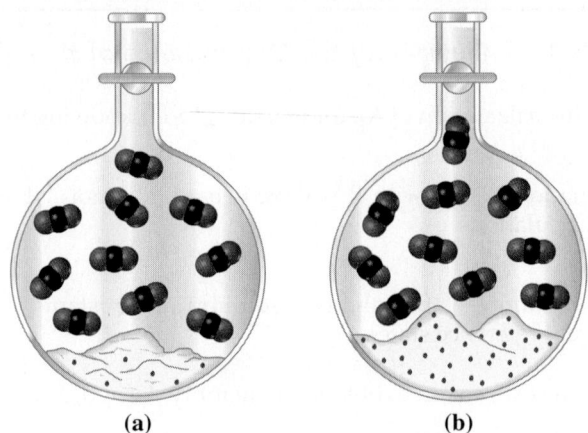

(a)　　　(b)

Although solid carbon must be present for the reaction to occur, the equilibrium constant expression contains terms only for the species in the homogeneous gas phase: H_2O, CO, and H_2.

$$K = \frac{a_{CO}\, a_{H_2}}{a_{C(s)}\, a_{H_2O}} \approx \frac{P_{CO}\, P_{H_2}}{P_{H_2O}} = K_p$$

The activity of solid carbon is $a_{C(s)} = 1$, and we have implicitly divided through each of the remaining pressures by the reference-state pressure, $P° = 1$ bar, to obtain a dimensionless K.

The decomposition of calcium carbonate (limestone) is also a heterogeneous reaction. The equilibrium constant expression, K_c, contains just a single term.

$$CaCO_3(s) \rightleftharpoons CaO(s) + CO_2(g) \qquad K_c = [CO_2] \qquad \textbf{(15.17)}$$

We can write K_p for reaction (15.17) by using equation (15.16), with $\Delta n_{gas} = 1$.

$$K_p = P_{CO_2} \quad \text{and} \quad K_p = K_c(RT) \qquad \textbf{(15.18)}$$

Equation (15.18) indicates that the equilibrium pressure of $CO_2(g)$ in contact with $CaCO_3(s)$ and $CaO(s)$ is a constant equal to K_p. Its value is *independent* of the quantities of $CaCO_3$ and CaO (as long as both solids are present). Figure 15-4 offers a conceptualization of this decomposition reaction.

One of our examples in Section 15-1 was liquid–vapor equilibrium. This is a physical equilibrium because no chemical reactions are involved. Consider the liquid–vapor equilibrium for water.

$$H_2O(l) \rightleftharpoons H_2O(g)$$

$$K_c = [H_2O(g)] \qquad K_p = P_{H_2O} \qquad K_p = K_c RT$$

So, equilibrium vapor pressures such as P_{H_2O} are just values of K_p. As we have seen before, these values do *not* depend on the quantities of liquid or vapor at equilibrium, as long as some of each is present.

EXAMPLE 15-4　　**Writing Equilibrium Constant Expressions for Reactions Involving Pure Solids or Liquids**

At equilibrium in the following reaction at 60 °C, the partial pressures of the gases are found to be $P_{HI} = 3.70 \times 10^{-3}$ bar and $P_{H_2S} = 1.01$ bar. What is the value of K for the reaction?

$$H_2S(g) + I_2(s) \rightleftharpoons 2\,HI(g) + S(s) \qquad K = ?$$

Analyze

We need to first write the equilibrium constant expression in terms of activities, and then eliminate the activities of pure solids and pure liquids by setting their activities to 1.

Solve

Write the equilibrium constant expression in terms of activities. Note that activities for the iodine and sulfur are not included, since the activity of a pure solid is 1.

$$K = \frac{(a_{HI})^2}{(a_{H_2S})}$$

The partial pressures are given in bar. The activity of each gas is equal to the numerical value of its partial pressure.

$a_{HI} = 3.70 \times 10^{-3}$ and $a_{H_2S} = 1.01$

Substitute the given equilibrium data into the equilibrium constant expression.

$$K = \frac{(3.70 \times 10^{-3})^2}{1.01} = 1.36 \times 10^{-5}$$

Assess

Note that the equilibrium constant, K, has no units. You must remember that activities are dimensionless quantities, and that when the partial pressure of a gas is expressed in bar, the activity is equal to the numerical value of its pressure.

PRACTICE EXAMPLE A: Teeth are made principally from the mineral hydroxyapatite, $Ca_5(PO_4)_3OH$, which can be dissolved in acidic solution such as that produced by bacteria in the mouth. The reaction that occurs is $Ca_5(PO_4)_3OH(s) + 4\,H^+(aq) \rightleftharpoons 5\,Ca^{2+}(aq) + 3\,HPO_4^{2-}(aq) + H_2O(l)$. Write the equilibrium constant expression K_c for this reaction.

PRACTICE EXAMPLE B: The steam–iron process is used to generate $H_2(g)$, mostly for use in hydrogenating oils. Iron metal and steam $[H_2O(g)]$ react to produce $Fe_3O_4(s)$ and $H_2(g)$. Write expressions for K_c and K_p for this reversible reaction. How are the values of K_c and K_p related to each other? Explain.

15-4   The Magnitude of an Equilibrium Constant

In principle, every chemical reaction has an equilibrium constant, but often the constants are not used. Why is this so? Table 15.3 lists equilibrium constants for several reactions mentioned in this chapter or previously in the text. The first of these reactions is the synthesis of H_2O from its elements. We have always assumed that this reaction goes to completion, that is, that the reverse reaction is negligible and the overall reaction proceeds only in the forward direction. If a reaction goes to completion, one (or more) of the reactants is used up. A term in the *denominator* of the equilibrium constant expression approaches *zero* and makes the value of the equilibrium constant very large. A very large numerical value of K signifies that the forward reaction, as written, *goes to completion* or very nearly so. Because the value of K_p for the water synthesis reaction is 1.4×10^{83}, we are entirely justified in saying that the reaction goes to completion at 298 K.

If the equilibrium constant is so large, why is a mixture of hydrogen and oxygen gases stable at room temperature? The value of the equilibrium constant

TABLE 15.3   Equilibrium Constants of Some Common Reactions

Reaction	Equilibrium Constant, K_p
$2\,H_2(g) + O_2(g) \rightleftharpoons 2\,H_2O\,(l)$	1.4×10^{83} at 298 K
$CaCO_3(s) \rightleftharpoons CaO(s) + CO_2(g)$	1.9×10^{-23} at 298 K
	1.0 at about 1200 K
$2\,SO_2(g) + O_2(g) \rightleftharpoons 2\,SO_3(g)$	3.4 at 1000 K
$C(s) + H_2O(g) \rightleftharpoons CO(g) + H_2(g)$	1.6×10^{-21} at 298 K
	10.0 at about 1100 K

▶ The products formed in a reaction under *kinetic control* are determined by reaction rates. The products formed in a reaction under *thermodynamic control* will depend on the stability of the products.

relates to thermodynamic stability: $H_2O(l)$ is much more thermodynamically stable than a mixture of $H_2(g)$ and $O_2(g)$ because it lies at a lower energy state. As noted in Chapter 14, however, the rate of a chemical reaction is strongly governed by the activation energy, E_a. Because E_a is very high for the synthesis of $H_2O(l)$ from $H_2(g)$ and $O_2(g)$, the rate of reaction is inconsequential at 298 K. To get the reaction to occur at a measurable rate, we must either raise the temperature or use a catalyst. A chemist would say that the synthesis of $H_2O(l)$ at 298 K is a *kinetically* controlled reaction (as opposed to *thermodynamically* controlled).

From Table 15.3, we see that K_p for the decomposition of $CaCO_3(s)$ (limestone) is very small at 298 K (only 1.9×10^{-23}). To account for a very small numerical value of an equilibrium constant, the *numerator* must be very small (approaching zero). A very small numerical value of K signifies that the forward reaction, as written, *does not occur to any significant extent*. Although limestone does not decompose at ordinary temperatures, the partial pressure of $CO_2(g)$ in equilibrium with $CaCO_3(s)$ and $CaO(s)$ increases with temperature. It becomes 1 atm at about 1200 K. An important application of this decomposition reaction is in the commercial production of quicklime (CaO).

The conversion of $SO_2(g)$ and $O_2(g)$ to $SO_3(g)$ at 1000 K has an equilibrium constant such that we expect significant amounts of both reactants and products to be present at equilibrium (see Table 15.3). Both the forward and reverse reactions are important. A similar situation exists for the reaction of $C(s)$ and $H_2O(g)$ at 1100 K, but not at 298 K where the forward reaction does not occur to any significant extent ($K_p = 1.6 \times 10^{-21}$).

In light of the several cases from Table 15.3, we can conclude the following:

> We consider a reaction going to completion if $K > 10^{10}$ or a reaction not occurring in the forward direction if $K < 10^{-10}$.

Thus, we see that equilibrium calculations are not required for all reactions. At times, we can use simple stoichiometric calculations to determine the outcome of a reaction, and in some cases there may be no reaction at all.

🔍 15-6 CONCEPT ASSESSMENT

Why is having a balanced equation a necessary condition for predicting the outcome of a chemical reaction, but often not a sufficient condition?

15-5 The Reaction Quotient, Q: Predicting the Direction of Net Change

Let's return briefly to the set of three experiments that we discussed in Section 15-2, involving the reaction

$$2\,Cu^{2+}(aq) + Sn^{2+}(aq) \longrightarrow 2\,Cu^{+}(aq) + Sn^{4+}(aq) \qquad K_c = 1.48 \qquad \textbf{(15.19)}$$

Experiment 1 starts with just the reactants, Cu^{2+} and Sn^{2+}. An overall, or net, change has to occur in which some Cu^{+} and Sn^{4+} forms. Only in this way can an equilibrium condition be reached in which all reacting species are present. We say that a net change occurs in the *forward* direction (*to the right*).

Experiment 2 starts with just the product, Cu^{+} and Sn^{4+}. Here, some of the Cu^{+} and Sn^{4+} must decompose back to Cu^{2+} and Sn^{2+} before equilibrium can be established. We say that a net change occurs in the *reverse* direction (*to the left*).

Experiment 3 starts with all the reacting species present: Cu^{2+}, Sn^{2+}, Cu^+, and Sn^{4+}. In this system, it is not obvious in what direction a net change occurs to establish equilibrium.

The ability to predict the direction of net change in establishing equilibrium is important for two reasons.

- At times we do not need detailed equilibrium calculations. We may need only a qualitative description of the changes that occur in establishing equilibrium from a given set of initial conditions.

- In some equilibrium calculations, it is helpful to determine the direction of net change as a first step.

For any set of *initial* activities in a reaction mixture, we can set up a ratio of activities having the same form as the equilibrium constant expression. This ratio is called the **reaction quotient** and is designated Q. For a hypothetical generalized reaction, the reaction quotient, first written in terms of activities, and then as concentrations assuming a concentration reference state, is

$$Q_c = \frac{(a_{init})^g (a_{init})^h}{(a_{init})^a (a_{init})^b} \qquad Q_c = \frac{[G]_{init}^g [H]_{init}^h}{[A]_{init}^a [B]_{init}^b} \qquad \textbf{(15.20)}$$

If $Q = K$, a reaction is at equilibrium, but our primary interest in the relationship between Q and K is for a reaction mixture that is not at equilibrium. To see what this relationship is, let's turn again to the experiments in Table 15.1.

In Experiment 1, the initial concentrations of Cu^{2+} and Sn^{2+} are 0.100 mol/ 1.00 L = 0.100 M. Initially there is no Cu^+ and Sn^{4+}. The expression for Q_c is

$$Q_c = \frac{[Cu^+]_{init}^2 [Sn^{4+}]_{init}}{[Cu^{2+}]_{init}^2 [Sn^{2+}]_{init}} = \frac{0 \times 0}{(0.100)^2 (0.100)} = 0 \qquad \textbf{(15.21)}$$

We know that a net reaction occurs *to the right*, producing some Cu^+ and Sn^{4+}. As it does, the numerator in expression (15.21) increases, the denominator decreases, and the value of Q_c increases; eventually $Q_c = K_c$.

If $Q_c < K_c$, then a net change occurs from left to right (the direction of the forward reaction).

In Experiment 2, the initial concentrations of Cu^+ and Sn^{4+} are 0.100 mol/ 1.00 L = 0.100 M. Initially, there is no Cu^{2+} and Sn^{2+}. The expression for Q_c is

$$Q_c = \frac{[Cu^+]_{init}^2 [Sn^{4+}]_{init}}{[Cu^{2+}]_{init}^2 [Sn^{4+}]_{init}} = \frac{(0.100)^2 (0.100)}{0 \times 0} = \infty \qquad \textbf{(15.22)}$$

We know that a net reaction occurs *to the left*, producing some Cu^{2+} and Sn^{2+}. As it does, the numerator in expression (15.22) decreases, the denominator increases, and the value of Q_c decreases; eventually $Q_c = K_c$.

If $Q_c > K_c$, a net change occurs from right to left (the direction of the reverse reaction).

◀ Strictly speaking, we cannot evaluate Q_c in this case. Any value divided by zero is undefined. By writing $Q_c = \infty$, we mean the following: As the concentrations of reactants approach a value of zero, Q_c approaches ∞.

Now let us turn to a case where direction of net change is not immediately obvious. In Experiment 3, the initial concentrations of all four species are 0.100 mol/1.00 L = 0.100 M. The value of Q_c is

$$Q_c = \frac{[Cu^+]_{init}^2 [Sn^{4+}]_{init}}{[Cu^{2+}]_{init}^2 [Sn^{2+}]_{init}} = \frac{(0.100)^2 (0.100)}{(0.100)^2 (0.100)} = 1.00$$

Because $Q_c < K_c$ (1.00 compared with 1.48), a net change occurs in the *forward direction*. Note that you can verify this conclusion from Figure 15-3. The

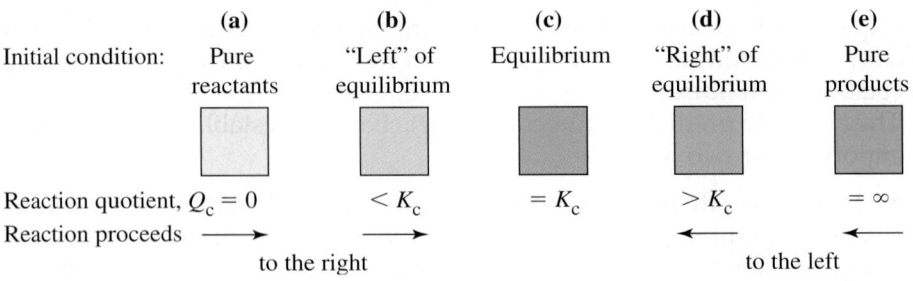

▲ FIGURE 15-5
Predicting the direction of net change in a reversible reaction
Five possibilities for the relationship of initial and equilibrium conditions are shown.
From Table 15.1 and Figure 15-3, Experiment 1 corresponds to initial condition (a);
Experiment 2 to condition (e); and Experiment 3 to (d). The situation in Example 15-5
also corresponds to condition (d).

amounts of Cu^{2+} and Sn^{2+} at equilibrium are less than they were initially, and
the amounts of Cu^+ and Sn^{4+} are greater.

The criteria for predicting the direction of a net chemical change in a
reversible reaction are summarized in Figure 15-5 and applied in Example 15-5.

EXAMPLE 15-5   **Predicting the Direction of a Net Chemical Change in Establishing Equilibrium**

To increase the yield of $H_2(g)$ in the water–gas reaction—the reaction of $C(g)$ and $H_2O(g)$ to form $CO(g)$ and
$H_2(g)$—a follow-up reaction called the "water–gas shift reaction" is generally used. In this reaction, some of
the $CO(g)$ of the water gas is replaced by $H_2(g)$.

$$CO(g) + H_2O(g) \rightleftharpoons CO_2(g) + H_2(g)$$

$K_c = 1.00$ at about 1100 K. The following amounts of substances are brought together and allowed to react at
this temperature: 1.00 mol CO, 1.00 mol H_2O, 2.00 mol CO_2, and 2.00 mol H_2. Compared with their initial
amounts, which of the substances will be present in a greater amount and which in a lesser amount when equilibrium is established?

Analyze

Our task is to determine the direction of net change by evaluating Q_c and comparing it to K_c.

Solve

Write down the expression for Q_c.

$$Q_c = \frac{[CO_2][H_2]}{[CO][H_2O]}$$

Substitute concentrations into the expression for
Q_c, by assuming an arbitrary volume V (which
cancels out in the calculation).

$$Q_c = \frac{(2.00/V)(2.00/V)}{(1.00/V)(1.00/V)} = 4.00$$

Compare Q_c to K_c.

$$4.00 > 1.00$$

Because $Q_c > K_c$ (that is, $4.00 > 1.00$), a net change occurs to the *left*. When equilibrium is established, the
amounts of CO and H_2O will be greater than the initial quantities and the amounts of CO_2 and H_2 will be less.

Assess

It is important to be able to determine the direction of reaction. As we will see in Section 15-7, this step must
be completed before we attempt to determine what the equilibrium amounts will be.

PRACTICE EXAMPLE A: In Example 15-5, equal masses of CO, H_2O, CO_2, and H_2 are mixed at a temperature of
about 1100 K. When equilibrium is established, which substance(s) will show an increase in quantity and which
will show a decrease compared with the initial quantities?

PRACTICE EXAMPLE B: For the reaction $PCl_5(g) \rightleftharpoons PCl_3(g) + Cl_2(g)$, $K_c = 0.0454$ at 261 °C. If a vessel is filled
with these gases such that the initial partial pressures are $P_{PCl_3} = 2.19$ atm, $P_{Cl_2} = 0.88$ atm, $P_{PCl_5} = 19.7$ atm,
in which direction will a net change occur?

🔍 **15-7   CONCEPT ASSESSMENT**

A mixture of 1.00 mol each of $CO(g)$, $H_2O(g)$, and $CO_2(g)$ is placed in a 10.0 L flask at a temperature at which $K_p = 10.0$ in the reaction

$$CO(g) + H_2O(g) \rightleftharpoons CO_2(g) + H_2(g)$$

When equilibrium is established, **(a)** the amount of $H_2(g)$ will be 1.00 mol; **(b)** the amounts of all reactants and products will be greater than 1.00 mol; **(c)** the amounts of all reactants and products will be less than 1.00 mol; **(d)** the amount of $CO_2(g)$ will be greater than 1.00 mol and the amounts of $CO(g)$, $H_2O(g)$, and $H_2(g)$ will be less than 1.00 mol; **(e)** the amounts of reactants and products cannot be predicted and can only be determined by analyzing the equilibrium mixture.

15-6   Altering Equilibrium Conditions: Le Châtelier's Principle

At times, we want only to make qualitative statements about a reversible reaction: the direction of a net change, whether the amount of a substance will have increased or decreased when equilibrium is reached, and so on. Also, we may not have the data needed for a quantitative calculation. In these cases, we can use a statement attributed to the French chemist Henri Le Châtelier (1884). **Le Châtelier's principle** is hard to state unambiguously, but its essential meaning is stated here.

> When an equilibrium system is subjected to a change in temperature, pressure, or concentration of a reacting species, the system responds by attaining a new equilibrium that *partially* offsets the impact of the change.

As we will see in the examples that follow, it is generally not difficult to predict the outcome of changing one or more variables in a system at equilibrium.

Effect of Changing the Amounts of Reacting Species on Equilibrium

Let's return to reaction (15.11)

$$2\,SO_2(g) + O_2(g) \rightleftharpoons 2\,SO_3(g) \qquad K_c = 2.8 \times 10^2 \text{ at } 1000 \text{ K}$$

Suppose we start with certain equilibrium amounts of SO_2, O_2, and SO_3, as suggested by Figure 15-6(a). Now let's create a disturbance in the equilibrium mixture by forcing an additional 1.00 mol SO_3 into the 10.0 L flask (Fig. 15-6b). How will the amounts of the reacting species change to re-establish equilibrium?

According to Le Châtelier's principle, if the system is to partially offset an action that increases the equilibrium concentration of one of the reacting species, it must do so by favoring the reaction in which that species is consumed. In this case, this is the *reverse* reaction—conversion of some of the added SO_3 to SO_2 and O_2. In the new equilibrium, there are greater amounts of all the substances than in the original equilibrium, but the additional amount of SO_3 is less than the 1.00 mol that was added.

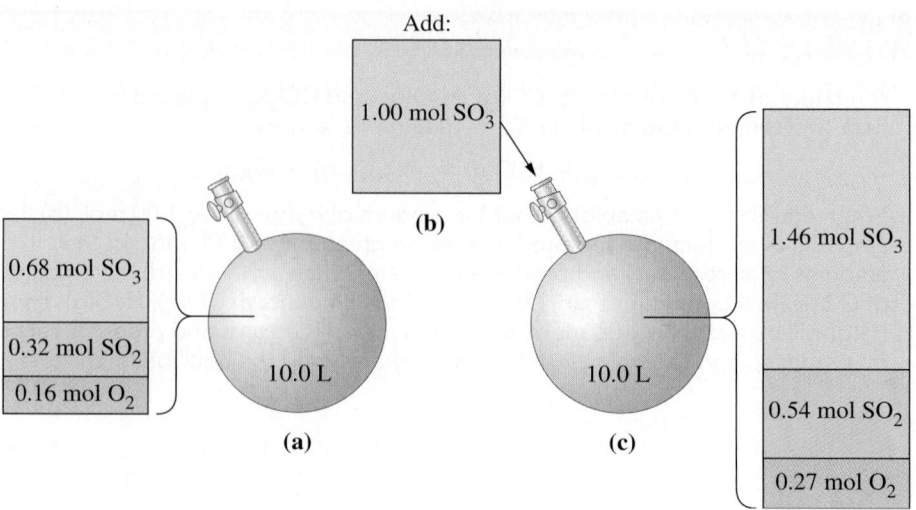

▲ FIGURE 15-6
Changing equilibrium conditions by changing the amount of a reactant

$$2\,SO_2(g) + O_2(g) \rightleftharpoons 2\,SO_3(g),\ K_c = 2.8 \times 10^2 \text{ at 1000 K}$$

(a) The original equilibrium condition. **(b)** Disturbance caused by adding 1.00 mol SO_3. **(c)** The new equilibrium condition. The amount of SO_3 in the new equilibrium mixture, 1.46 mol, is greater than the original 0.68 mol but it is not as great as immediately after the addition of 1.00 mol SO_3. The effect of adding SO_3 to an equilibrium mixture is *partially* offset when equilibrium is restored.

Another way to look at the matter is to evaluate the reaction quotient immediately after adding the SO_3.

Original equilibrium   *Immediately following disturbance*

$$Q_c = \frac{[SO_3]}{[SO_2]^2[O_2]} = K_c \qquad Q_c = \frac{[SO_3]}{[SO_2]^2[O_2]} > K_c$$

EXAMPLE 15-6   Applying Le Châtelier's Principle: Effect of Adding More of a Reactant to an Equilibrium Mixture

Predict the effect of adding more $H_2(g)$ to a constant-volume equilibrium mixture of N_2, H_2, and NH_3.

$$N_2(g) + 3\,H_2(g) \rightleftharpoons 2\,NH_3(g)$$

Analyze

When a system at equilibrium is disturbed by adding more of one reactant, the system responds by using up (consuming) some of the added reactant.

Solve

Increasing $[H_2]$ stimulates the forward reaction and a shift in the equilibrium condition to the right. However, only a portion of the added H_2 is consumed in this reaction.

Assess

When equilibrium is re-established, there will be more H_2 than was present originally, and also more NH_3, but the amount of N_2 will be *smaller*. Some of the original N_2 must be consumed in converting some of the added H_2 to NH_3.

PRACTICE EXAMPLE A: Given the reaction $2\,CO(g) + O_2(g) \rightleftharpoons 2\,CO_2(g)$, what is the effect of adding $O_2(g)$ to a constant-volume equilibrium mixture?

PRACTICE EXAMPLE B: Calcination of limestone (decomposition by heating), $CaCO_3(s) \rightleftharpoons CaO(s) + CO_2(g)$, is the commercial source of quicklime, $CaO(s)$. After this equilibrium has been established in a constant-temperature, constant-volume container, what is the effect on the equilibrium amounts of materials caused by *adding* some **(a)** $CaO(s)$; **(b)** $CO_2(g)$; **(c)** $CaCO_3(s)$?

Adding any quantity of SO_3 to a constant-volume equilibrium mixture makes Q_c larger than K_c. A net change occurs in the direction that reduces $[SO_3]$, that is, to the left, or in the reverse direction. Notice that reaction in the reverse direction increases $[SO_2]$ and $[O_2]$, further decreasing the value of Q_c.

Effect of Changes in Pressure or Volume on Equilibrium

There are three ways to change the pressure of a constant-temperature equilibrium mixture.

1. **Add or remove a gaseous reactant or product.** The effect of these actions on the equilibrium condition is simply that caused by adding or removing a reaction component, as described previously.

2. **Add an inert gas to the constant-volume reaction mixture.** This has the effect of increasing the *total* pressure, but the partial pressures of the reacting species are all unchanged. An inert gas added to a constant-volume equilibrium mixture has no effect on the equilibrium condition.

3. **Change the pressure by changing the volume of the system.** Decreasing the volume of the system increases the pressure, and increasing the system volume decreases the pressure. Thus, the effect of this type of pressure change is simply that of a volume change.

Let's explore the third situation first. Consider, again, the formation of $SO_3(g)$ from $SO_2(g)$ and $O_2(g)$.

$$2\,SO_2(g) + O_2(g) \rightleftharpoons 2\,SO_3(g) \qquad K_c = 2.8 \times 10^2 \text{ at } 1000 \text{ K}$$

The equilibrium mixture in Figure 15-7(a) has its volume reduced to one-tenth of its original value by increasing the external pressure. To see how the equilibrium amounts of the gases change, let's first rearrange the equilibrium constant expression to the form

$$K_c = \frac{[SO_3]^2}{[SO_2]^2[O_2]} = \frac{(n_{SO_3}/V)^2}{(n_{SO_2}/V)^2(n_{O_2}/V)} = \frac{(n_{SO_3})^2}{(n_{SO_2})^2(n_{O_2})} \times V = 2.8 \times 10^2 \quad \textbf{(15.23)}$$

From equation (15.23), we see that if V is *reduced* by a factor of 10, the ratio

$$\frac{(n_{SO_3})^2}{(n_{SO_2})^2(n_{O_2})}$$

must *increase* by a factor of 10. In this way, the value of K_c is restored, as it must be to restore equilibrium. There is only one way in which the ratio of moles of gases will increase in value: The number of moles of SO_3 must increase, and

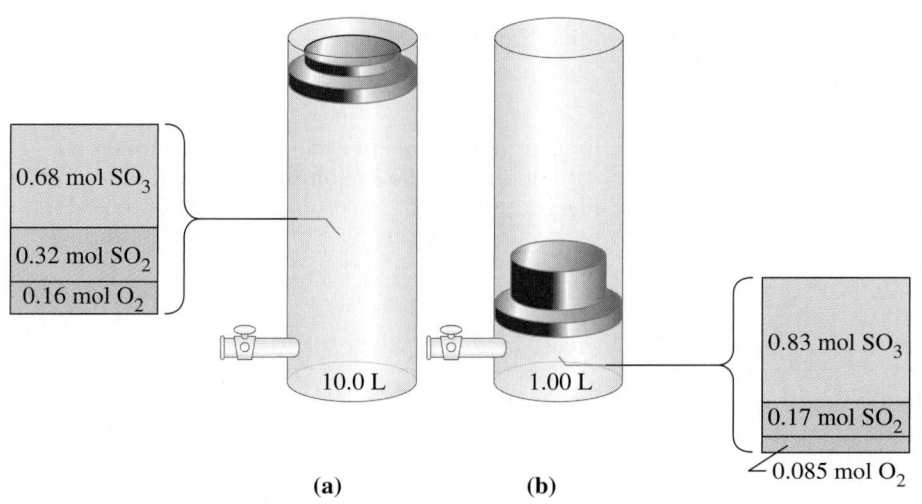

◀ FIGURE 15-7
Effect of pressure change on equilibrium in the reaction

$$2SO_2(g) + O_2(g) \rightleftharpoons 2\,SO_3(g)$$

An increase in external pressure causes a decrease in the reaction volume and a shift in equilibrium "to the right." (See Exercise 77 for a calculation of the new equilibrium amounts.)

the numbers of moles of SO_2 and O_2 must decrease. The equilibrium shifts in the direction producing more SO_3—to the right.

Notice that three moles of *gas* on the left produce two moles of *gas* on the right in reaction (15.11). When compared at the same temperature and pressure, two moles of $SO_3(g)$ occupy a smaller volume than does a mixture of two moles of $SO_2(g)$ and one mole of $O_2(g)$. Given this fact and the observation from equation (15.23) that a decrease in volume favors the production of additional SO_3, we can formulate a statement that is especially easy to apply.

> When the volume of an equilibrium mixture of gases is *reduced*, a net change occurs in the direction that produces *fewer moles of gas*. When the volume is *increased*, a net change occurs in the direction that produces *more moles of gas*.

KEEP IN MIND

that an inert gas has no effect on an equilibrium condition if the gas is added to a system maintained at constant volume, but it can have an effect if added at constant pressure.

Figure 15-7 suggests a way of decreasing the volume of gaseous mixture at equilibrium—by increasing the external pressure. One way to increase the volume is to lower the external pressure. Another way is to transfer the equilibrium mixture from its original container to one of larger volume. A third method is to add an inert gas at *constant pressure*; the volume of the mixture must increase to make room for the added gas. The effect on the equilibrium, however, is the same for all three methods: Equilibrium shifts in the direction of the reaction producing the greater number of moles of gas.

Equilibria between condensed phases are not affected much by changes in external pressure because solids and liquids are not easily compressible. Also, we cannot assess whether the forward or reverse reaction is favored by these changes by examining only the chemical equation.

15-8   CONCEPT ASSESSMENT

To the hypothetical reaction $A(g) + B(g) \rightleftharpoons C(g)$ 0.100 mol of the inert gas argon is added. In addition, the volume of the container is decreased. According to Le Châtelier's principle, will the reaction shift to the right or left? Explain.

EXAMPLE 15-7   Applying Le Châtelier's Principle: The Effect of Changing Volume

An equilibrium mixture of $N_2(g)$, $H_2(g)$, and $NH_3(g)$ is transferred from a 1.50 L flask to a 5.00 L flask. In which direction does a net change occur to restore equilibrium?

$$N_2(g) + 3\,H_2(g) \rightleftharpoons 2\,NH_3(g)$$

Analyze

Because the volume has increased, the reaction will move in the direction that increases the number of moles of gas.

Solve

When the gaseous mixture is transferred to the larger flask, the partial pressure of each gas and the total pressure drop. Whether we think in terms of a decrease in pressure or an increase in volume, we reach the same conclusion. Equilibrium shifts in such a way as to produce a larger number of moles of gas. Some of the NH_3 originally present decomposes back to N_2 and H_2. A net change occurs in the direction of the reverse reaction—to the left—in restoring equilibrium.

Assess

Whether we think in terms of a decrease in pressure or an increase in volume, the conclusion is the same.

PRACTICE EXAMPLE A: The reaction $N_2O_4(g) \rightleftharpoons 2\,NO_2(g)$ is at equilibrium in a 3.00 L cylinder. What would be the effect on the concentrations of $N_2O_4(g)$ and $NO_2(g)$ if the pressure were doubled (that is, cylinder volume decreased to 1.50 L)?

PRACTICE EXAMPLE B: How is the equilibrium amount of $H_2(g)$ produced in the water–gas shift reaction affected by changing the total gas pressure or the system volume? Explain.

$$CO(g) + H_2O(g) \rightleftharpoons CO_2(g) + H_2(g)$$

15-9 CONCEPT ASSESSMENT

The following reaction is brought to equilibrium at 700 °C.

$$2 H_2S(g) + CH_4(g) \rightleftharpoons CS_2(g) + 4 H_2(g)$$

Indicate whether each of the following statements is true, false, or not possible to evaluate from the information given.

(a) If the equilibrium mixture is allowed to expand into an evacuated larger container, the mole fraction of H_2 will increase.

(b) If several moles of $Ar(g)$ are forced into the reaction container, the amounts of H_2S and CH_4 will increase.

(c) If the equilibrium mixture is cooled to 100 °C, the mole fractions of the four gases will likely change.

(d) If the equilibrium mixture is forced into a slightly smaller container, the partial pressures of the four gases will all increase.

Effect of Temperature on Equilibrium

We can think of changing the temperature of an equilibrium mixture in terms of adding heat (raising the temperature) or removing heat (lowering the temperature). According to Le Châtelier's principle, adding heat favors the reaction in which heat is absorbed (*endothermic* reaction). Removing heat favors the reaction in which heat is evolved (*exothermic* reaction). Stated in terms of changing temperature,

> *Raising the temperature* of an equilibrium mixture shifts the equilibrium condition in the direction of the *endothermic* reaction. *Lowering the temperature* causes a shift in the direction of the *exothermic* reaction.

The principal effect of temperature on equilibrium is in changing the value of the equilibrium constant. In Chapter 19, we will learn how to calculate equilibrium constants as a function of temperature. For now, we will limit ourselves to making qualitative predictions.

EXAMPLE 15-8 Applying Le Châtelier's Principle: Effect of Temperature on Equilibrium

Consider the reaction

$$2 SO_2(g) + O_2(g) \rightleftharpoons 2 SO_3(g) \qquad \Delta H° = -197.8 \text{ kJ}$$

Will the amount of $SO_3(g)$ formed from given amounts of $SO_2(g)$ and $O_2(g)$ be greater at high or low temperatures?

Analyze

We must think of the impact made by changing the temperature. In general, an increase in temperature causes a shift in the direction of the endothermic reaction.

Solve

The sign of $\Delta H°$ tells us that the forward reaction is exothermic. Thus, the reverse reaction is endothermic. In this case, increasing the temperature will favor the reverse reaction and lowering the temperature will favor the forward reaction. The conversion of SO_2 to SO_3 is favored at *low* temperatures.

(continued)

Assess

Be sure not to confuse shifts in equilibrium with changes in reaction rates that result from temperature changes. That is, equilibria of exothermic and endothermic reactions will shift *differently* when temperatures are increased, but the rates of exothermic and endothermic reactions *both increase* with increasing temperature. Changing the temperature is somewhat different than other changes we have discussed in this section. Changing the temperature causes a shift in the equilibrium position and changes the value of the equilibrium constant. We will discuss the temperature dependence of K in Chapter 19.

PRACTICE EXAMPLE A: The reaction $N_2O_4(g) \rightleftharpoons 2\,NO_2(g)$ has $\Delta H° = +57.2$ kJ. Will the amount of $NO_2(g)$ formed from $N_2O_4(g)$ be greater at high or low temperatures?

PRACTICE EXAMPLE B: The enthalpy of formation of NH_3 is $\Delta H_f°[NH_3(g)] = -46.11$ kJ/mol NH_3. Will the concentration of NH_3 in an equilibrium mixture with its elements be greater at 100 or at 300 °C? Explain.

▲ Sulfuric acid is produced from SO_3

$$SO_3(g) + H_2O(l) \rightleftharpoons H_2SO_4(aq)$$

The catalyst used to speed up the conversion of SO_2 to SO_3 in the commercial production of sulfuric acid is $V_2O_5(s)$. What appears to be smoke coming from the cooling tower (in the rear) is in fact just water vapor.

Effect of a Catalyst on Equilibrium

Adding a catalyst to a reaction mixture speeds up both the forward and reverse reactions. Equilibrium is achieved more rapidly, but the equilibrium amounts are unchanged by the catalyst. Consider again reaction (15.11)

$$2\,SO_2(g) + O_2(g) \rightleftharpoons 2\,SO_3(g) \qquad K_c = 2.8 \times 10^2 \text{ at } 1000 \text{ K}$$

For a given set of reaction conditions, the equilibrium amounts of $SO_2, O_2,$ and SO_3 have fixed values. This is true whether the reaction is carried out by a slow homogeneous reaction, catalyzed in the gas phase, or conducted as a heterogeneous reaction on the surface of a catalyst. Stated another way, the presence of a catalyst does not change the numerical value of the equilibrium constant.

We now have two thoughts about a catalyst to reconcile: one from the preceding chapter and one from this discussion.

- A catalyst changes the mechanism of a reaction to one with a lower activation energy.
- A catalyst has no effect on the condition of equilibrium in a reversible reaction.

Taken together, these two statements must mean that an equilibrium condition is *independent* of the reaction mechanism. Thus, even though equilibrium has been described in terms of opposing reactions occurring at equal rates, we do not have to be concerned with the kinetics of chemical reactions to work with the equilibrium concept. This observation is still another indication that the equilibrium constant is a thermodynamic quantity, as we shall describe more fully in Chapter 19.

🔍 15-10   CONCEPT ASSESSMENT

Two students are performing the same experiment in which an endothermic reaction rapidly attains a condition of equilibrium. Student A does the reaction in a beaker resting on the surface of the lab bench while student B holds the beaker in which the reaction occurs. Assuming that all other environmental variables are the same, which student should end up with more product? Explain.

15-7   Equilibrium Calculations: Some Illustrative Examples

We are now ready to tackle the problem of describing, in quantitative terms, the condition of equilibrium in a reversible reaction. Part of the approach we use may seem unfamiliar at first—it has an algebraic look to it. But as you adjust to this "new look," do not lose sight of the fact that we continue to use some familiar and important ideas—molar masses, molarities, and stoichiometric factors from the balanced equation, for example.

The five numerical examples that follow apply the general equilibrium principles described earlier in the chapter. The first four involve gases, while the fifth deals with equilibrium in an aqueous solution. (The study of equilibria in aqueous solutions is the principal topic of the next three chapters.) Each example includes an assessment that summarizes the essential features of equilibrium calculations exemplified by that type of problem. You may find it helpful to return to these assessments from time to time as you encounter new equilibrium situations in later chapters.

Example 15-9 is relatively straightforward. It demonstrates how to determine the equilibrium constant of a reaction when the equilibrium concentrations of the reactants and products are known.

Example 15-10 is somewhat more involved than Example 15-9. We are still interested in determining the equilibrium constant for a reaction, but we do not have the same sort of information as in Example 15-9. We are given the initial concentrations of all the reactants and products, but the equilibrium concentration of only one substance. This case requires a little algebra and some careful bookkeeping. We will introduce a tabular system, sometimes called an **ICE table**, for keeping track of changing concentrations of reactants and products. The table contains the *initial, change in,* and *equilibrium concentration* of each species. It is a helpful device that we will use throughout the next three chapters.

EXAMPLE 15-9   **Determining a Value of K_c from the Equilibrium Quantities of Substances**

Dinitrogen tetroxide, $N_2O_4(l)$, is an important component of rocket fuels—for example, as an oxidizer of liquid hydrazine in the Titan rocket. At 25 °C, N_2O_4 is a colorless gas that partially dissociates into NO_2, a red-brown gas. The color of an equilibrium mixture of these two gases depends on their relative proportions, which in turn depends on the temperature (Fig. 15-8).

(continued)

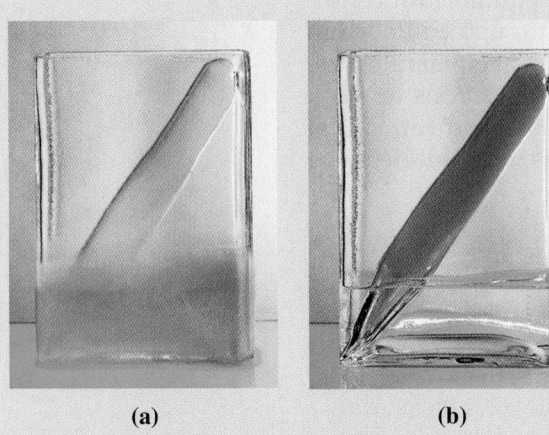

(a)   (b)

▲ FIGURE 15-8
The equilibrium $N_2O_4(g) \rightleftharpoons 2\ NO_2(g)$
(a) At dry ice temperatures, N_2O_4 exists as a solid. The gas in equilibrium with the solid is mostly colorless N_2O_4, with only a trace of brown NO_2.
(b) When warmed to room temperature and above, the N_2O_4 melts and vaporizes. The proportion of $NO_2(g)$ at equilibrium increases over that at low temperatures, and the equilibrium mixture of $N_2O_4(g)$ and $NO_2(g)$ has a red-brown color.

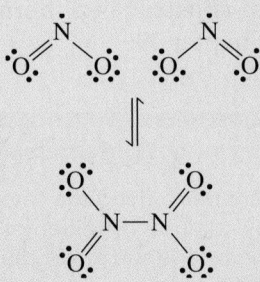

▲ **The Lewis structures of N_2O_4 and $NO_2(g)$**
Nitrogen dioxide is a free radical that combines exothermically to form dinitrogen tetroxide.

Equilibrium is established in the reaction $N_2O_4(g) \rightleftharpoons 2\ NO_2(g)$ at 25 °C. The quantities of the two gases present in a 3.00 L vessel are 7.64 g N_2O_4 and 1.56 g NO_2. What is the value of K_c for this reaction?

Analyze

We are given the equilibrium amounts (in terms of mass) of the reactants and products, along with the volume of the reaction vessel. We use these values to determine the equilibrium concentrations and plug them into the equilibrium constant expression.

Solve

Convert the mass of N_2O_4 to moles.
$$\text{mol } N_2O_4 = 7.64 \text{ g } N_2O_4 \times \frac{1 \text{ mol } N_2O_4}{92.01 \text{ g } N_2O_4} = 8.303 \times 10^{-2} \text{ mol}$$

Convert moles of N_2O_4 to mol/L.
$$[N_2O_4] = \frac{8.303 \times 10^2 \text{ mol}}{3.00 \text{ L}} = 0.0277 \text{ M}$$

Convert the mass of NO_2 to moles.
$$\text{mol } NO_2 = 1.56 \text{ g } NO_2 \times \frac{1 \text{ mol } NO_2}{46.01 \text{ g } NO_2} = 3.391 \times 10^{-2} \text{ mol}$$

Convert moles of NO_2 to mol/L.
$$[NO_2] = \frac{3.391 \times 10^{-2}}{3.00 \text{ L}} = 0.0113 \text{ M}$$

Write the equilibrium constant expression, substitute the equilibrium concentrations, and solve for K_c.
$$K_c = \frac{[NO_2]^2}{[N_2O_4]} = \frac{(0.0113)^2}{(0.0277)} = 4.61 \times 10^{-3}$$

Assess

The quantities required in an equilibrium constant expression, K_c, are equilibrium *concentrations in moles per liter*, not simply equilibrium amounts in moles or masses in grams. It is helpful to organize all the equilibrium data and carefully label each item.

PRACTICE EXAMPLE A: Equilibrium is established in a 3.00 L flask at 1405 K for the reaction $2\ H_2S(g) \rightleftharpoons 2\ H_2(g) + S_2(g)$. At equilibrium, there is 0.11 mol $S_2(g)$, 0.22 mol $H_2(g)$, and 2.78 mol $H_2S(g)$. What is the value of K_c for this reaction?

PRACTICE EXAMPLE B: Equilibrium is established at 25 °C in the reaction $N_2O_4(g) \rightleftharpoons 2\ NO_2(g)$, $K_c = 4.61 \times 10^{-3}$. If $[NO_2] = 0.0236$ M in a 2.26 L flask, how many grams of N_2O_4 are also present?

EXAMPLE 15-10   **Determining a Value of K_p from Initial and Equilibrium Amounts of Substances: Relating K_c and K_p**

The equilibrium condition for $SO_2(g)$, $O_2(g)$, and $SO_3(g)$ is important in sulfuric acid production. When a 0.0200 mol sample of SO_3 is introduced into an evacuated 1.52 L vessel at 900 K, 0.0142 mol SO_3 is present at equilibrium. What is the value of K_p for the dissociation of $SO_3(g)$ at 900 K?

$$2\,SO_3(g) \rightleftharpoons 2\,SO_2(g) + O_2(g) \qquad K_p = ?$$

Analyze

Let's first determine K_c and then convert to K_p by using equation (15.16). In the ICE table below, the key term leading to the other data is the change in amount of SO_3: In progressing from 0.0200 mol SO_3 to 0.0142 mol SO_3, 0.0058 mol SO_3 is dissociated. The *negative sign* (-0.0058 mol) indicates that this amount of SO_3 is consumed in establishing equilibrium. In the row labeled "changes," the changes in amounts of SO_2 and O_2 must be related to the change in amount of SO_3. For this, we use the stoichiometric coefficients from the balanced equation: 2, 2, and 1. That is, *two* moles of SO_2 and *one* mole of O_2 are produced for every *two* moles of SO_3 that dissociate.

Solve

The reaction:	$2\,SO_3(g)$	$\rightleftharpoons$	$2\,SO_2(g)$	$+$	$O_2(g)$
initial amounts:	0.0200 mol		0.00 mol		0.00 mol
changes:	−0.0058 mol		+0.0058 mol		+0.0029 mol
equil amounts:	0.0142 mol		0.0058 mol		0.0029 mol

$$[SO_3] = \frac{0.0142\ \text{mol}}{1.52\ \text{L}}; \qquad [SO_2] = \frac{0.0058\ \text{mol}}{1.52\ \text{L}}; \qquad [O_2] = \frac{0.0029\ \text{mol}}{1.52\ \text{L}}$$

$$[SO_3] = 9.34 \times 10^{-3}\ \text{M}; \qquad [SO_2] = 3.8 \times 10^{-3}\ \text{M}; \qquad [O_2] = 1.9 \times 10^{-3}\ \text{M}$$

$$K_c = \frac{[SO_2]^2[O_2]}{[SO_3]^2} = \frac{(3.8 \times 10^{-3})^2(1.9 \times 10^{-3})}{(9.34 \times 10^{-3})^2} = 3.1 \times 10^{-4}$$

$$K_p = K_c(RT)^{\Delta n_{\text{gas}}} = 3.1 \times 10^{-4}\,(0.0821 \times 900)^{(2+1)-2}$$

$$= 3.1 \times 10^{-4}\,(0.0821 \times 900)^1 = 2.3 \times 10^{-2}$$

Assess

The chemical equation for a reversible reaction serves both to establish the form of the equilibrium constant expression and to provide the conversion factors (stoichiometric factors) to relate the equilibrium quantity of one species to equilibrium quantities of the others.

For equilibria involving gases, we can use either K_c or K_p. In general, if the data given involve amounts of substances and volumes, it is easier to work with K_c. If data are given as partial pressures, then work with K_p. Whether working with K_c or K_p or the relationship between them, we must always base these expressions on the given chemical equation, not on equations we may have used in other situations.

PRACTICE EXAMPLE A: A 5.00 L evacuated flask is filled with 1.86 mol NOBr. At equilibrium at 25 °C, there is 0.082 mol of Br_2 present. Determine K_c and K_p for the reaction $2\,NOBr(g) \rightleftharpoons 2\,NO(g) + Br_2(g)$.

PRACTICE EXAMPLE B: 0.100 mol SO_2 and 0.100 mol O_2 are introduced into an evacuated 1.52 L flask at 900 K. When equilibrium is reached, the amount of SO_3 found is 0.0916 mol. Use these data to determine K_p for the reaction $2\,SO_3(g) \rightleftharpoons 2\,SO_2(g) + O_2(g)$.

The methods used in Examples 15-9 and 15-10 are summarized in Figure 15-9. Example 15-11 demonstrates that we can often determine several pieces of useful information about an equilibrium system from just the equilibrium constant and the reaction equation.

Example 15-12 brings back the ICE format, but with a twist. This time the known values include the equilibrium constant and an initial amount of the reactant, but no information is given about the equilibrium amount of the reactant or the product. That means that we do not know how much the initial value will change. We show this by using an "x" in that part of the table. The setup will be quite algebraic; in fact, we must use the quadratic formula to obtain a solution.

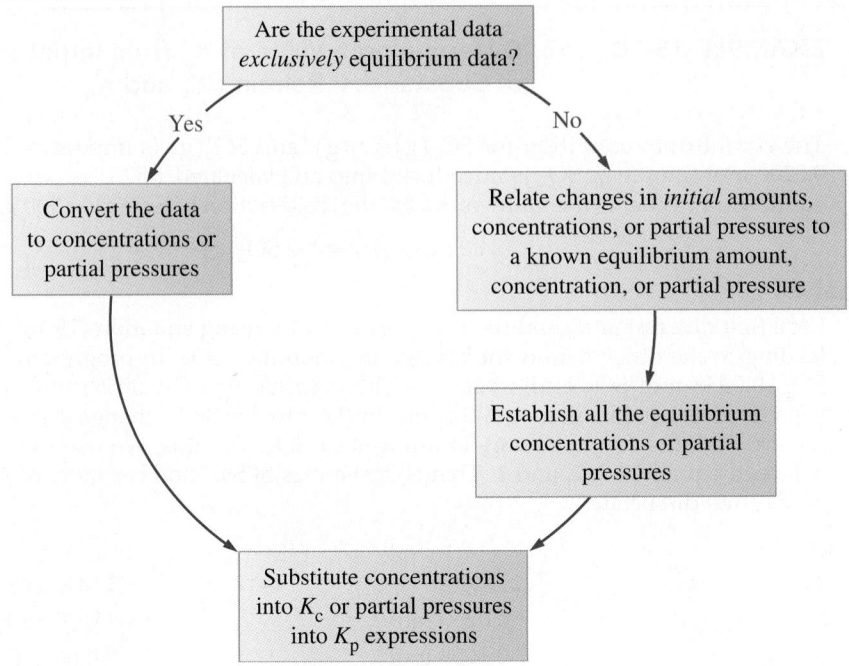

▲ FIGURE 15-9
Determining K_c or K_p from experimental data

EXAMPLE 15-11   Determining Equilibrium Partial and Total Pressures from a Value of K_p

Ammonium hydrogen sulfide, $NH_4HS(s)$, used as a photographic developer, is unstable and dissociates at room temperature.

$$NH_4HS(s) \rightleftharpoons NH_3(g) + H_2S(g) \qquad K_p = 0.108 \text{ at } 25\ °C$$

A sample of $NH_4HS(s)$ is introduced into an evacuated flask at 25 °C. What is the total gas pressure at equilibrium?

Analyze

We begin by writing the equilibrium constant expression in terms of pressure. The key step is to recognize that the pressure of ammonia is equal to the pressure of hydrogen sulfide. This will then allow us to determine the pressure of ammonia and hydrogen sulfide.

Solve

K_p for this reaction is just the product of the equilibrium partial pressures of $NH_3(g)$ and $H_2S(g)$, each stated in atmospheres. (There is no term for NH_4HS because it is a solid.) Because these gases are produced in equimolar amounts, $P_{NH_3} = P_{H_2S}$.

$$K_p = (P_{NH_3})(P_{H_2S}) = 0.108$$
$$K_p = (P_{NH_3})(P_{H_2S}) = (P_{NH_3})(P_{NH_3}) = (P_{NH_3})^2 = 0.108$$

Find P_{NH_3}. (Note that the unit atm appears because in the equilibrium expression the reference pressure $P°$ was implicitly included.)

$$P_{NH_3} = \sqrt{0.108} = 0.329 \text{ atm} \qquad P_{H_2S} = P_{NH_3} = 0.329 \text{ atm}$$

The total pressure is

$$P_{tot} = P_{NH_3} + P_{H_2S} = 0.329 \text{ atm} + 0.329 \text{ atm} = 0.658 \text{ atm}$$

Assess

When using K_p expressions, look for relationships among partial pressures of the reactants. If we need to relate the total pressure to the partial pressures of the reactants, we should be able to do this with some equations presented in Chapter 6 (for example, equations 6.15, 6.16, and 6.17).

PRACTICE EXAMPLE A: Sodium hydrogen carbonate (baking soda) decomposes at elevated temperatures and is one of the sources of $CO_2(g)$ when this compound is used in baking.

$$2\,NaHCO_3(s) \rightleftharpoons Na_2CO_3(s) + H_2O(g) + CO_2(g) \qquad K_p = 0.231 \text{ at } 100\ °C$$

What is the partial pressure of $CO_2(g)$ when this equilibrium is established starting with $NaHCO_3(s)$?

PRACTICE EXAMPLE B: If enough additional $NH_3(g)$ is added to the flask in Example 15-11 to raise its partial pressure to 0.500 atm at equilibrium, what will be the *total* gas pressure when equilibrium is re-established?

EXAMPLE 15-12   Calculating Equilibrium Concentrations from Initial Conditions

A 0.0240 mol sample of $N_2O_4(g)$ is allowed to come to equilibrium with $NO_2(g)$ in a 0.372 L flask at 25 °C. Calculate the amount of N_2O_4 present at equilibrium (Fig. 15-10).

$$N_2O_4(g) \rightleftharpoons 2\,NO_2(g) \qquad K_c = 4.61 \times 10^{-3} \text{ at } 25 °C$$

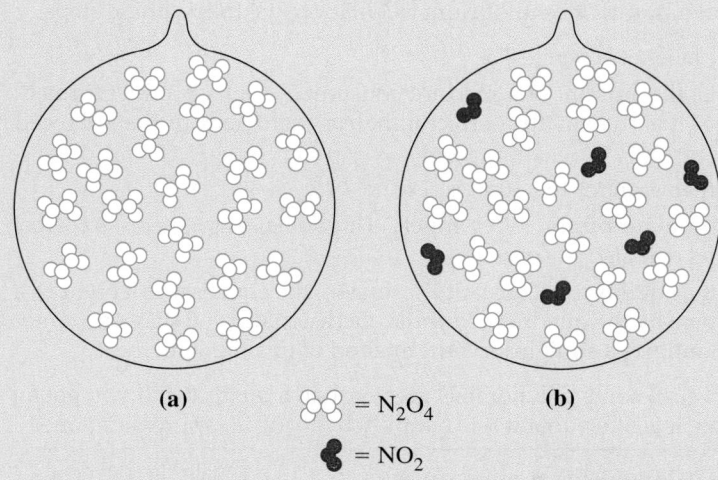

◀ FIGURE 15-10
Equilibrium in the reaction

$$N_2O_4(g) \rightleftharpoons 2\,NO_2(g)$$

at 25 °C—Example 15-12 illustrated
Each "molecule" illustrated represents 0.001 mol. **(a)** Initially, the bulb contains 0.024 mol N_2O_4, represented by 24 "molecules." **(b)** At equilibrium, some "molecules" of N_2O_4 have dissociated to NO_2. The 21 "molecules" of N_2O_4 and 6 of NO_2 correspond to 0.021 mol N_2O_4 and 0.006 mol NO_2 at equilibrium.

(a)   ⬡⬡ $= N_2O_4$ **(b)**

⬤⬤ $= NO_2$

Analyze

We need to determine the amount of N_2O_4 that dissociates to establish equilibrium. For the first time, we introduce an algebraic unknown, x. Suppose we let $x =$ the number of moles of N_2O_4 that dissociate. In the following ICE table, we enter the value $-x$ into the row labeled "changes." The amount of NO_2 produced is $+2x$ because the stoichiometric coefficient of NO_2 is 2 and that of N_2O_4 is 1.

Solve

The reaction:	$N_2O_4(g)$	$\rightleftharpoons$	$2\,NO_2(g)$
initial amounts:	0.0240 mol		0.00 mol
changes:	$-x$ mol		$+2x$ mol
equil amounts:	$(0.0240 - x)$ mol		$2x$ mol
equil concns:	$[N_2O_4] = (0.0240 - x) \text{ mol}/0.372 \text{ L}$		$[NO_2] = 2x \text{ mol}/0.372 \text{ L}$

$$K_c = \frac{[NO_2]^2}{[N_2O_4]} = \frac{\left(\dfrac{2x}{0.372}\right)^2}{\left(\dfrac{0.0240 - x}{0.372}\right)} = \frac{4x^2}{0.372(0.0240 - x)} = 4.61 \times 10^{-3}$$

$$4x^2 = 4.12 \times 10^{-5} - (1.71 \times 10^{-3})x$$

$$x^2 + (4.28 \times 10^{-4})x - 1.03 \times 10^{-5} = 0$$

$$x = \frac{-4.28 \times 10^{-4} \pm \sqrt{(4.28 \times 10^{-4})^2 + 4 \times 1.03 \times 10^{-5}}}{2}$$

$$= \frac{-4.28 \times 10^{-4} \pm \sqrt{(1.83 \times 10^{-7}) + 4.12 \times 10^{-5}}}{2}$$

$$x = \frac{-4.28 \times 10^{-4} \pm \sqrt{4.14 \times 10^{-5}}}{2}$$

$$= \frac{-4.28 \times 10^{-4} \pm 6.43 \times 10^{-3}}{2}$$

$$= \frac{-4.28 \times 10^{-4} + 6.43 \times 10^{-3}}{2} = \frac{6.00 \times 10^{-3}}{2}$$

$$= 3.00 \times 10^{-3} \text{ mol } N_2O_4$$

The amount of N_2O_4 at equilibrium is $(0.0240 - x) = (0.0240 - 0.0030) = 0.0210 \text{ mol } N_2O_4$.

Assess

When we need to introduce an algebraic unknown, x, into an equilibrium calculation, we follow these steps.

- Introduce x into the ICE setup in the row labeled "changes."
- Decide which change to label as x, that is, the amount of a reactant consumed or of a product formed. Usually, we base this on the species that has the smallest stoichiometric coefficient in the balanced chemical equation.
- Use stoichiometric factors to relate the other changes to x (that is, $2x, 3x, \dots$).
- Consider that equilibrium amounts = initial amounts + "changes." (If you have assigned the correct signs to the changes, equilibrium amounts will also be correct.)
- After substitutions have been made into the equilibrium constant expression, the equation will often be a quadratic equation in x, which you can solve by the quadratic formula. Occasionally you may encounter a higher-degree equation. Appendix A-3 outlines a straightforward method of dealing with these.

Most of us can solve quadratic equations, but few can solve polynomials greater than a quadratic. If you get an equation that is a cubic or higher degree equation, it is likely that it will simplify by using an approximation.

PRACTICE EXAMPLE A: If 0.150 mol $H_2(g)$ and 0.200 mol $I_2(g)$ are introduced into a 15.0 L flask at 445 °C and allowed to come to equilibrium, how many moles of $HI(g)$ will be present?

$$H_2(g) + I_2(g) \rightleftharpoons 2\,HI(g) \qquad K_c = 50.2 \text{ at } 445\,°C$$

PRACTICE EXAMPLE B: Suppose the equilibrium mixture of Example 15-12 is transferred to a 10.0 L flask. **(a)** Will the equilibrium amount of N_2O_4 increase or decrease? Explain. **(b)** Calculate the number of moles of N_2O_4 in the new equilibrium condition.

Our final example is similar to the previous one, but with this slight complication: Initially, we don't know whether a net change occurs to the right or to the left to establish equilibrium. We can find out, though, by using the reaction quotient, Q_c, and proceeding in the manner suggested in Figure 15-11. Also, because the reactants and products are in solution, we can work exclusively with concentrations in formulating the K_c expression.

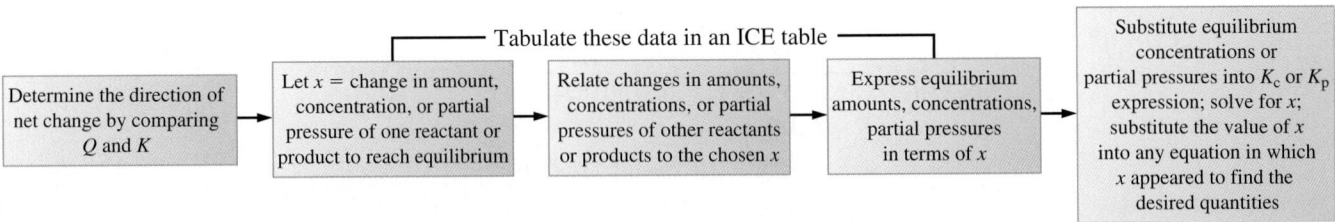

▲ FIGURE 15-11
Determining equilibrium concentrations and partial pressures

EXAMPLE 15-13 Using the Reaction Quotient, Q_c, in an Equilibrium Calculation

Solid silver is added to a solution with these initial concentrations: $[Ag^+] = 0.200$ M, $[Fe^{2+}] = 0.100$ M, and $[Fe^{3+}] = 0.300$ M. The following reversible reaction occurs.

$$Ag^+(aq) + Fe^{2+}(aq) \rightleftharpoons Ag(s) + Fe^{3+}(aq) \qquad K_c = 2.98$$

What are the ion concentrations when equilibrium is established?

(continued)

Analyze

Because all reactants and products are present initially, we need to use the reaction quotient Q_c to determine the direction in which a net change occurs.

$$Q_c = \frac{[Fe^{3+}]}{[Ag^+][Fe^{2+}]} = \frac{0.300}{(0.200)(0.100)} = 15.0$$

Because Q_c (15.0) is larger than K_c (2.98), a net change must occur in the direction of the reverse reaction, *to the left*. Let's define x as the change in molarity of Fe^{3+}. Because the net change occurs *to the left*, we designate the changes for the species on the left side of the equation as positive and those on the right side as negative.

Solve

The reaction:	**$Ag^+(aq)$**	+	**$Fe^{2+}(aq)$**	$\rightleftharpoons$	**$Ag(s) + Fe^{3+}(aq)$**
initial concns:	0.200 M		0.100 M		0.300 M
changes:	$+x$ M		$+x$ M		$-x$ M
equil concns:	$(0.200 + x)$ M		$(0.100 + x)$ M		$(0.300 - x)$ M

$$K_c = \frac{[Fe^{3+}]}{[Ag^+][Fe^{2+}]} = \frac{(0.300 - x)}{(0.200 + x)(0.100 + x)} = 2.98$$

This equation, which is solved in Appendix A-3, is a quadratic equation for which the acceptable root is $x = 0.11$. To obtain the equilibrium concentrations, we substitute this value of x into the terms shown in the table of data.

$$[Ag^+]_{equil} = 0.200 + 0.11 = 0.31 \text{ M}$$
$$[Fe^{2+}]_{equil} = 0.100 + 0.11 = 0.21 \text{ M}$$
$$[Fe^{3+}]_{equil} = 0.300 - 0.11 = 0.19 \text{ M}$$

Assess

If we have done the calculation correctly, we should obtain a value very close to that given for K_c when we substitute the *calculated* equilibrium concentrations into the reaction quotient, Q_c. We do.

$$Q_c = \frac{[Fe^{3+}]}{[Ag^+][Fe^{2+}]} = \frac{(0.19)}{(0.31)(0.21)} = 2.9 \qquad (K_c = 2.98)$$

PRACTICE EXAMPLE A: Excess $Ag(s)$ is added to 1.20 M $Fe^{3+}(aq)$. Given that

$$Ag^+(aq) + Fe^{2+}(aq) \rightleftharpoons Ag(s) + Fe^{3+}(aq) \qquad K_c = 2.98$$

what are the equilibrium concentrations of the species in solution?

PRACTICE EXAMPLE B: A solution is prepared with $[V^{3+}] = [Cr^{2+}] = 0.0100$ M and $[V^{2+}] = [Cr^{3+}] = 0.150$ M. The following reaction occurs.

$$V^{3+}(aq) + Cr^{2+}(aq) \rightleftharpoons V^{2+}(aq) + Cr^{3+}(aq) \qquad K_c = 7.2 \times 10^2$$

What are the ion concentrations when equilibrium is established? [*Hint:* The algebra can be greatly simplified by extracting the square root of both sides of an equation at the appropriate point.]

Mastering**CHEMISTRY**   www.masteringchemistry.com

Reversible reactions play an important role in the conversion of elemental nitrogen, $N_2(g)$, into nitrogen compounds, both in Nature and in the chemical industry. For a discussion of both natural and industrial processes that convert elemental nitrogen to nitrogen compounds, go to the Focus On feature for Chapter 15, The Nitrogen Cycle and the Synthesis of Nitrogen Compounds, on the MasteringChemistry site.

Summary

15-1 Dynamic Equilibrium—**Equilibrium** is the condition in which the forward and reverse reaction rates of reversible processes are equal. Chemical and physical processes in equilibrium are dynamic by nature.

15-2 The Equilibrium Constant Expression—This condition of dynamic equilibrium is described through an **equilibrium constant expression**. The form of the equilibrium constant expression is established from the balanced chemical equation using activities to express the "effective" concentrations (equation 15.7). The numerical value obtained from the equilibrium constant expression is referred to as the **equilibrium constant**, K. Equilibrium constants are unitless.

15-3 Relationships Involving Equilibrium Constants—When the equation for a reversible reaction is written in the reverse order, the equilibrium constant expression and the value of K are both inverted from their original form. When two or more reactions are coupled together, the equilibrium constant for the overall reaction is the product of the K values of the individual reactions. The equilibrium constant of a reaction can have different values depending on the reference state used. For K_c a concentration reference state is used, while for K_p, a pressure reference state is used. The relationship between K_c and K_p is given by equation (15.16).

15-4 The Magnitude of an Equilibrium Constant—The magnitude of the equilibrium constant can be used to determine the outcome of a reaction. For large values of K the reaction goes to completion, with all reactants converted to products. A very small equilibrium constant, for example, a large negative power of ten, indicates that practically none of the reactants have been converted to products. Finally, equilibrium constants of an intermediate value, for example, between 10^{-10} and 10^{10},

indicate that some of the reactants have been converted to products.

15-5 The Reaction Quotient, Q: Predicting the Direction of Net Change—The **reaction quotient**, Q (equation 15.20), has the same form as the equilibrium constant expression; however, its numerical value is determined using the initial reaction activities. A comparison of the reaction quotient with the equilibrium constant makes it possible to predict the direction of net change leading to equilibrium (Fig. 15-5). If $Q < K$, the forward reaction is favored, meaning that when equilibrium is established the amounts of products will have increased and the amounts of reactants will have decreased. If $Q > K$, the reverse reaction is favored until equilibrium is established. If $Q = K$, neither the forward nor reverse reaction is favored. The initial conditions are in fact equilibrium conditions.

15-6 Altering Equilibrium Conditions: Le Châtelier's Principle—**Le Châtelier's principle** is used to make qualitative predictions of the effects of different variables on an equilibrium condition. This principle describes how an equilibrium condition is modified, or "shifts," in response to the addition or removal of reactants or changes in reaction volume, external pressure, or temperature. Catalysts, by speeding up the forward and reverse reactions equally, have no effect on an equilibrium condition.

15-7 Equilibrium Calculations Some Illustrative Examples—For quantitative equilibrium calculations, a few basic principles and algebraic techniques are required. A useful method employs a tabular system, called an **ICE table**, for keeping track of the *initial* concentrations of the reactants and products, *changes* in these concentrations, and the *equilibrium* concentrations.

Integrative Example

In the manufacture of ammonia, the chief source of hydrogen gas is the following reaction for the reforming of methane at high temperatures.

$$CH_4(g) + 2\,H_2O(g) \rightleftharpoons CO_2(g) + 4\,H_2(g) \tag{15.24}$$

The following data are also given.

(a) $CO(g) + H_2O(g) \rightleftharpoons CO_2(g) + H_2(g)$ $\Delta H^\circ = -40$ kJ; $K_c = 1.4$ at 1000 K

(b) $CO(g) + 3\,H_2(g) \rightleftharpoons H_2O(g) + CH_4(g)$ $\Delta H^\circ = -230$ kJ; $K_c = 190$ at 1000 K

At 1000 K, 1.00 mol each of CH_4 and H_2O are allowed to come to equilibrium in a 10.0 L vessel. Calculate the number of moles of H_2 present at equilibrium. Would the yield of H_2 increase if the temperature were raised above 1000 K?

Analyze

First, we should assemble the data needed to solve this problem. The amounts of substances and a reaction volume are given, so we should be able to work with a K_c expression. However, because the K_c value for the reaction of interest is not given, we will have to derive this value by combining the two equations for which data are given. This will yield values of both K_c and ΔH for the reaction of interest.

To calculate the number of moles of H_2 at equilibrium we can use the ICE method, and to assess the effect of temperature on the equilibrium yield of H_2 we can apply Le Châtelier's principle.

Solve

We combine equations (a) and (b) to obtain the data needed in this problem.

(a) $\qquad CO(g) + H_2O(g) \rightleftharpoons CO_2(g) + H_2(g) \qquad \Delta H = -40 \text{ kJ} \quad K_c = 1.4$

(b) $\qquad \underline{CH_4(g) + H_2O(g) \rightleftharpoons CO(g) + 3\,H_2(g)} \qquad \Delta H = 230 \text{ kJ} \quad K_c = 1/190$

Overall: $\quad CH_4(g) + 2\,H_2O(g) \rightleftharpoons CO_2(g) + 4\,H_2(g) \qquad \Delta H = 190 \text{ kJ} \quad K_c = 1.4/190 = 7.4 \times 10^{-3}$

Next we set up an ICE table in which x represents the number of moles of CH_4 consumed in reaching equilibrium.

The reaction:	$CH_4(g)$	+	$2\,H_2O(g)$	$\rightleftharpoons$	$CO_2(g)$	+	$4\,H_2(g)$
initial amounts:	1.00 mol		1.00 mol		0.00 mol		0.00 mol
changes:	$-x$ mol		$-2x$ mol		x mol		$4x$ mol
equil amounts:	$(1.00 - x)$ mol		$(1.00 - 2x)$ mol		x mol		$4x$ mol
equil concns, M:	$(1.00 - x)/10.0$		$(1.00 - 2x)/10.0$		$x/10.0$		$4x/10.0$

Now we set up K_c and make substitutions into the expression.

$$K_c = \frac{[CO_2][H_2]^4}{[CH_4][H_2O]^2}$$

$$= \frac{(x/10.0)(4x/10.0)^4}{[(1.00-x)/10.0][(1.00-2x)/10.0]^2}$$

$$= \frac{x(4x)^4}{100(1.00-x)(1.00-2x)^2} = 7.4 \times 10^{-3}$$

The above equation reduces to $\qquad 256x^5 = 0.74[(1.00-x)(1.00-2x)^2]$

and then to $\qquad 256x^5 - 0.74[(1.00-x)(1.00-2x)^2] = 0 \qquad$ **(15.25)**

The solution to this equation is $x = 0.23$ mol. The number of moles of H_2 at equilibrium is $4x = 0.92$ mol.

Because the reaction is endothermic ($\Delta H = 190$ kJ), the forward reaction is favored at higher temperatures. The equilibrium yield of H_2 will increase if the temperature is raised above 1000 K.

Assess

Equation (15.25) looks impossibly difficult to solve, but it is not. It can be solved for x rather simply by the method of successive approximations. This is done in Appendix A-3, equation (A.2). An important clue as to the possible range of values for x can be found in the ICE table. Note that the equilibrium amount of $H_2O(g)$ is $1.00 - 2x$, meaning that $x < 0.50$, or else all of the $H_2O(g)$ would be consumed. This marks a good place to start the approximations.

PRACTICE EXAMPLE A: Glycolysis involves ten biochemical reactions. The first two reactions of the glycolysis cycle are

$$C_6H_{12}O_6(aq) + ATP(aq) \rightleftharpoons G6P(aq) + ADP(aq) \qquad \Delta H° = -19.74 \text{ kJ mol}^{-1}$$
$$G6P(aq) \rightleftharpoons F6P(aq) \qquad \Delta H° = 2.84 \text{ kJ mol}^{-1}$$

Calculate the equilibrium concentration of F6P(aq) generated in the glycolysis cycle at normal body temperature, 37 °C, starting with $[C_6H_{12}O_6(aq)] = 1.20 \times 10^{-6}$ M; $[ATP(aq)] = 10^{-4}$ M; and $[ADP(aq)] = 10^{-2}$ M. The equilibrium constant for the first reaction is 4.630×10^3; for the second reaction it is 2.76×10^{-1}. During a fever body temperature increases. Will [G6P] increase or decrease with an increase with temperature?

PRACTICE EXAMPLE B: A procedure calls for adding 0.100 mol of $Br_2(g)$ at 25 °C to a reaction. The only source of bromine in the laboratory is a bottle of liquid bromine. **(a)** Given the following data, what size container (in liters) must be used to extract enough $Br_2(g)$ for this reaction? The equilibrium constant at 298 K for the reaction $Br_2(g) \rightleftharpoons 2Br(g)$ is $K = 3.30 \times 10^{-29}$. The vapor pressure of liquid bromine is 0.289 atm. **(b)** At 1000 K the equilibrium constant for the reaction $Br_2(g) \rightleftharpoons 2Br(g)$ is 3.4×10^{-5}. What size vessel (in liters) will be needed if the temperature of the vapor is raised to 1000 K?

Exercises

Writing Equilibrium Constants Expressions

1. Based on these descriptions, write a balanced equation and the corresponding K_c expression for each reversible reaction.
   (a) Carbonyl fluoride, $COF_2(g)$, decomposes into gaseous carbon dioxide and gaseous carbon tetrafluoride.
   (b) Copper metal displaces silver(I) ion from aqueous solution, producing silver metal and an aqueous solution of copper(II) ion.
   (c) Peroxodisulfate ion, $S_2O_8^{2-}$, oxidizes iron(II) ion to iron(III) ion in aqueous solution and is itself reduced to sulfate ion.

2. Based on these descriptions, write a balanced equation and the corresponding K_p expression for each reversible reaction.
   (a) Oxygen gas oxidizes gaseous ammonia to gaseous nitrogen and water vapor.
   (b) Hydrogen gas reduces gaseous nitrogen dioxide to gaseous ammonia and water vapor.
   (c) Nitrogen gas reacts with the solid sodium carbonate and carbon to produce solid sodium cyanide and carbon monoxide gas.

3. Write equilibrium constant expressions, K_c, for the reactions
   (a) $2 NO(g) + O_2(g) \rightleftharpoons 2 NO_2(g)$
   (b) $Zn(s) + 2 Ag^+(aq) \rightleftharpoons Zn^{2+}(aq) + 2 Ag(s)$
   (c) $Mg(OH)_2(s) + CO_3^{2-}(aq) \rightleftharpoons$
   $$MgCO_3(s) + 2 OH^-(aq)$$

4. Write equilibrium constant expressions, K_p, for the reactions
   (a) $CS_2(g) + 4 H_2(g) \rightleftharpoons CH_4(g) + 2 H_2S(g)$
   (b) $Ag_2O(s) \rightleftharpoons 2 Ag(s) + \frac{1}{2} O_2(g)$
   (c) $2 NaHCO_3(s) \rightleftharpoons$
   $$Na_2CO_3(s) + CO_2(g) + H_2O(g)$$

5. Write an equilibrium constant, K_c, for the formation from its gaseous elements of (a) 1 mol $HF(g)$; (b) 2 mol $NH_3(g)$; (c) 2 mol $N_2O(g)$; (d) 1 mol $ClF_3(l)$.

6. Write an equilibrium constant, K_p, for the formation from its gaseous elements of (a) 1 mol $NOCl(g)$; (b) 2 mol $ClNO_2(g)$; (c) 1 mol $N_2H_4(g)$; (d) 1 mol $NH_4Cl(s)$.

7. Determine values of K_c from the K_p values given.
   (a) $SO_2Cl_2(g) \rightleftharpoons SO_2(g) + Cl_2(g)$
   $$K_p = 2.9 \times 10^{-2} \text{ at } 303 \text{ K}$$
   (b) $2 NO(g) + O_2(g) \rightleftharpoons 2 NO_2(g)$
   $$K_p = 1.48 \times 10^4 \text{ at } 184 \text{ °C}$$
   (c) $Sb_2S_3(s) + 3 H_2(g) \rightleftharpoons 2 Sb(s) + 3 H_2S(g)$
   $$K_p = 0.429 \text{ at } 713 \text{ K}$$

8. Determine the values of K_p from the K_c values given.
   (a) $N_2O_4(g) \rightleftharpoons 2 NO_2(g)$
   $$K_c = 4.61 \times 10^{-3} \text{ at } 25 \text{ °C}$$
   (b) $2 CH_4(g) \rightleftharpoons C_2H_2(g) + 3 H_2(g)$
   $$K_c = 0.154 \text{ at } 2000 \text{ K}$$
   (c) $2 H_2S(g) + CH_4(g) \rightleftharpoons 4 H_2(g) + CS_2(g)$
   $$K_c = 5.27 \times 10^{-8} \text{ at } 973 \text{ K}$$

9. The vapor pressure of water at 25 °C is 23.8 mmHg. Write K_p for the vaporization of water, with pressures in atmospheres. What is the value of K_c for the vaporization process?

10. If $K_c = 5.12 \times 10^{-3}$ for the equilibrium established between liquid benzene and its vapor at 25 °C, what is the vapor pressure of C_6H_6 at 25 °C, expressed in millimeters of mercury?

11. Determine K_c for the reaction
   $$\frac{1}{2} N_2(g) + \frac{1}{2} O_2(g) + \frac{1}{2} Br_2(g) \rightleftharpoons NOBr(g)$$
   from the following information (at 298 K).
   $$2 NO(g) \rightleftharpoons N_2(g) + O_2(g) \quad K_c = 2.1 \times 10^{30}$$
   $$NO(g) + \frac{1}{2} Br_2(g) \rightleftharpoons NOBr(g) \qquad K_c = 1.4$$

12. Given the equilibrium constant values
   $$N_2(g) + \frac{1}{2} O_2(g) \rightleftharpoons N_2O(g) \quad K_c = 2.7 \times 10^{-18}$$
   $$N_2O_4(g) \rightleftharpoons 2 NO_2(g) \quad K_c = 4.6 \times 10^{-3}$$
   $$\frac{1}{2} N_2(g) + O_2(g) \rightleftharpoons NO_2(g) \quad K_c = 4.1 \times 10^{-9}$$
   Determine a value of K_c for the reaction
   $$2 N_2O(g) + 3 O_2(g) \rightleftharpoons 2 N_2O_4(g)$$

13. Use the following data to estimate a value of K_p at 1200 K for the reaction $2 H_2(g) + O_2(g) \rightleftharpoons 2 H_2O(g)$
   $$C(graphite) + CO_2(g) \rightleftharpoons 2 CO(g) \qquad K_c = 0.64$$
   $$CO_2(g) + H_2(g) \rightleftharpoons CO(g) + H_2O(g) \quad K_c = 1.4$$
   $$C(graphite) + \frac{1}{2} O_2(g) \rightleftharpoons CO(g) \qquad K_c = 1 \times 10^8$$

14. Determine K_c for the reaction $N_2(g) + O_2(g) + Cl_2(g) \rightleftharpoons 2 NOCl(g)$, given the following data at 298 K.
   $$\frac{1}{2} N_2(g) + O_2(g) \rightleftharpoons NO_2(g) \qquad K_p = 1.0 \times 10^{-9}$$

$$NOCl(g) + \frac{1}{2}O_2(g) \rightleftharpoons NO_2Cl(g) \qquad K_p = 1.1 \times 10^2$$

$$NO_2(g) + \frac{1}{2}Cl_2(g) \rightleftharpoons NO_2Cl(g) \qquad K_p = 0.3$$

15. An important environmental and physiological reaction is the formation of carbonic acid, $H_2CO_3(aq)$, from carbon dioxide and water. Write the equilibrium constant expression for this reaction in terms of activities. Convert that expression into an equilibrium constant expression containing concentrations and pressures.

16. Rust, $Fe_2O_3(s)$, is caused by the oxidation of iron by oxygen. Write the equilibrium constant expression first in terms of activities, and then in terms of concentration and pressure.

Experimental Determination of Equilibrium Constants

17. 1.00×10^{-3} mol PCl_5 is introduced into a 250.0 mL flask, and equilibrium is established at 284 °C: $PCl_5(g) \rightleftharpoons PCl_3(g) + Cl_2(g)$. The quantity of $Cl_2(g)$ present at equilibrium is found to be 9.65×10^{-4} mol. What is the value of K_c for the dissociation reaction at 284 °C?

18. A mixture of 1.00 g H_2 and 1.06 g H_2S in a 0.500 L flask comes to equilibrium at 1670 K: $2 H_2(g) + S_2(g) \rightleftharpoons 2 H_2S(g)$. The equilibrium amount of $S_2(g)$ found is 8.00×10^{-6} mol. Determine the value of K_p at 1670 K.

19. The two common chlorides of phosphorus, PCl_3 and PCl_5, both important in the production of other phosphorus compounds, coexist in equilibrium through the reaction

$$PCl_3(g) + Cl_2(g) \rightleftharpoons PCl_5(g)$$

At 250 °C, an equilibrium mixture in a 2.50 L flask contains 0.105 g PCl_5, 0.220 g PCl_3, and 2.12 g Cl_2. What are the values of **(a)** K_c and **(b)** K_p for this reaction at 250 °C?

20. A 0.682 g sample of $ICl(g)$ is placed in a 625 mL reaction vessel at 682 K. When equilibrium is reached between the $ICl(g)$ and $I_2(g)$ and $Cl_2(g)$ formed by its dissociation, 0.0383 g I_2 is present. What is K_c for this reaction?

21. Write the equilibrium constant expression for the following reaction,

$$Fe(OH)_3 + 3H^+(aq) \rightleftharpoons Fe^{3+}(aq) + 3H_2O(l)$$

$$K = 9.1 \times 10^3$$

and compute the equilibrium concentration for $[Fe^{3+}]$ at pH = 7 (i.e., $[H^+] = 1.0 \times 10^{-7}$).

22. Write the equilibrium constant expression for the dissolution of ammonia in water:

$$NH_3(g) \rightleftharpoons NH_3(aq) \qquad K = 57.5$$

Use this equilibrium constant expression to estimate the partial pressure of $NH_3(g)$ over a solution containing 5×10^{-9} M $NH_3(aq)$. These are conditions similar to that found for acid rains with a high ammonium ion concentration.

Equilibrium Relationships

23. Equilibrium is established at 1000 K, where $K_c = 281$ for the reaction $2 SO_2(g) + O_2(g) \rightleftharpoons 2 SO_3(g)$. The equilibrium amount of $O_2(g)$ in a 0.185 L flask is 0.00247 mol. What is the ratio of $[SO_2]$ to $[SO_3]$ in this equilibrium mixture?

24. For the dissociation of $I_2(g)$ at about 1200 °C, $I_2(g) \rightleftharpoons 2 I(g)$, $K_c = 1.1 \times 10^{-2}$. What volume flask should we use if we want 0.37 mol I to be present for every 1.00 mol I_2 at equilibrium?

25. In the Ostwald process for oxidizing ammonia, a variety of products is possible—N_2, N_2O, NO, and NO_2—depending on the conditions. One possibility is

$$NH_3(g) + \frac{5}{4}O_2(g) \rightleftharpoons NO(g) + \frac{3}{2}H_2O(g)$$

$$K_p = 2.11 \times 10^{19} \text{ at 700 K}$$

For the decomposition of NO_2 at 700 K,

$$NO_2(g) \rightleftharpoons NO(g) + \frac{1}{2}O_2(g) \qquad K_p = 0.524$$

(a) Write a chemical equation for the oxidation of $NH_3(g)$ to $NO_2(g)$.
(b) Determine K_p for the chemical equation you have written.

26. At 2000 K, $K_c = 0.154$ for the reaction $2 CH_4(g) \rightleftharpoons C_2H_2(g) + 3 H_2(g)$. If a 1.00 L equilibrium mixture at 2000 K contains 0.10 mol each of $CH_4(g)$ and $H_2(g)$,
(a) what is the mole fraction of $C_2H_2(g)$ present?
(b) Is the conversion of $CH_4(g)$ to $C_2H_2(g)$ favored at high or low pressures?
(c) If the equilibrium mixture at 2000 K is transferred from a 1.00 L flask to a 2.00 L flask, will the number of moles of $C_2H_2(g)$ increase, decrease, or remain unchanged?

27. An equilibrium mixture at 1000 K contains 0.276 mol H_2, 0.276 mol CO_2, 0.224 mol CO, and 0.224 mol H_2O.

$$CO_2(g) + H_2(g) \rightleftharpoons CO(g) + H_2O(g)$$

(a) Show that for this reaction, K_c is independent of the reaction volume, V.
(b) Determine the value of K_c and K_p.

28. For the reaction $CO(g) + H_2O(g) \rightleftharpoons CO_2(g) + H_2(g)$, $K_p = 23.2$ at 600 K. Explain which of the following situations might be found at equilibrium:
(a) $P_{CO} = P_{H_2O} = P_{CO_2} = P_{H_2}$; **(b)** $P_{H_2}/P_{H_2O} = P_{CO_2}/P_{CO}$; **(c)** $(P_{CO_2})(P_{H_2}) = (P_{CO})(P_{H_2O})$; **(d)** $P_{CO_2}/P_{H_2O} = P_{H_2}/P_{CO}$.

Direction and Extent of Chemical Change

29. Can a mixture of 2.2 mol O_2, 3.6 mol SO_2, and 1.8 mol SO_3 be maintained indefinitely in a 7.2 L flask at a temperature at which $K_c = 100$ in this reaction? Explain.

$$2 SO_2(g) + O_2(g) \rightleftharpoons 2 SO_3(g)$$

30. Is a mixture of 0.0205 mol $NO_2(g)$ and 0.750 mol $N_2O_4(g)$ in a 5.25 L flask at 25 °C, at equilibrium? If not, in which direction will the reaction proceed—toward products or reactants?

$$N_2O_4(g) \rightleftharpoons 2 NO_2(g) \qquad K_c = 4.61 \times 10^{-3} \text{ at } 25 \text{ °C}$$

31. In the reaction $2 SO_2(g) + O_2(g) \rightleftharpoons 2 SO_3(g)$, 0.455 mol SO_2, 0.183 mol O_2, and 0.568 mol SO_3 are introduced simultaneously into a 1.90 L vessel at 1000 K.
  (a) If $K_c = 2.8 \times 10^2$, is this mixture at equilibrium?
  (b) If not, in which direction will a net change occur?

32. In the reaction $CO(g) + H_2O(g) \rightleftharpoons CO_2(g) + H_2(g)$, $K = 31.4$ at 588 K. Equal masses of each reactant and product are brought together in a reaction vessel at 588 K.
  (a) Can this mixture be at equilibrium?
  (b) If not, in which direction will a net change occur?

33. A mixture consisting of 0.150 mol H_2 and 0.150 mol I_2 is brought to equilibrium at 445 °C, in a 3.25 L flask. What are the equilibrium amounts of H_2, I_2, and HI?

$$H_2(g) + I_2(g) \rightleftharpoons 2 HI (g) \qquad K_c = 50.2 \text{ at } 445 \text{ °C}$$

34. Starting with 0.280 mol $SbCl_3$ and 0.160 mol Cl_2, how many moles of $SbCl_5$, $SbCl_3$, and Cl_2 are present when equilibrium is established at 248 °C in a 2.50 L flask?

$$SbCl_5(g) \rightleftharpoons SbCl_3(g) + Cl_2(g)$$
$$K_c = 2.5 \times 10^{-2} \text{ at } 248 \text{ °C}$$

35. Starting with 0.3500 mol $CO(g)$ and 0.05500 mol $COCl_2(g)$ in a 3.050 L flask at 668 K, how many moles of $Cl_2(g)$ will be present at equilibrium?

$$CO(g) + Cl_2(g) \rightleftharpoons COCl_2(g)$$
$$K_c = 1.2 \times 10^3 \text{ at } 668 \text{ K}$$

36. 1.00 g *each* of CO, H_2O, and H_2 are sealed in a 1.41 L vessel and brought to equilibrium at 600 K. How many grams of CO_2 will be present in the equilibrium mixture?

$$CO(g) + H_2O(g) \rightleftharpoons CO_2(g) + H_2(g) \qquad K_c = 23.2$$

37. Equilibrium is established in a 2.50 L flask at 250 °C for the reaction

$$PCl_5(g) \rightleftharpoons PCl_3(g) + Cl_2(g) \qquad K_c = 3.8 \times 10^{-2}$$

How many moles of PCl_5, PCl_3, and Cl_2 are present at equilibrium, if
  (a) 0.550 mol each of PCl_5 and PCl_3 are initially introduced into the flask?
  (b) 0.610 mol PCl_5 alone is introduced into the flask?

38. For the following reaction, $K_c = 2.00$ at 1000 °C.

$$2 COF_2(g) \rightleftharpoons CO_2(g) + CF_4(g)$$

If a 5.00 L mixture contains 0.145 mol COF_2, 0.262 mol CO_2, and 0.074 mol CF_4 at a temperature of 1000 °C,
  (a) Will the mixture be at equilibrium?
  (b) If the gases are not at equilibrium, in what direction will a net change occur?
  (c) How many moles of each gas will be present at equilibrium?

39. In the following reaction, $K_c = 4.0$.

$$C_2H_5OH + CH_3COOH \rightleftharpoons CH_3COOC_2H_5 + H_2O$$

A reaction is allowed to occur in a mixture of 17.2 g C_2H_5OH, 23.8 g CH_3COOH, 48.6 g $CH_3COOC_2H_5$, and 71.2 g H_2O.
  (a) In what direction will a net change occur?
  (b) How many grams of each substance will be present at equilibrium?

40. The N_2O_4–NO_2 equilibrium mixture in the flask on the left in the figure is allowed to expand into the evacuated flask on the right. What is the composition of the gaseous mixture when equilibrium is re-established in the system consisting of the two flasks?

$$N_2O_4(g) \rightleftharpoons 2 NO_2(g) \qquad K_c = 4.61 \times 10^{-3} \text{ at } 25 \text{ °C}$$

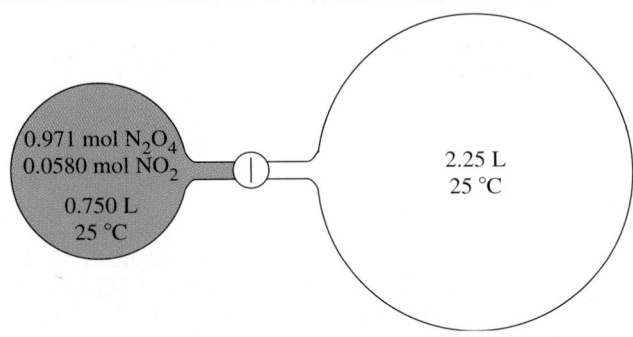

0.971 mol N_2O_4
0.0580 mol NO_2
0.750 L
25 °C

2.25 L
25 °C

41. Formamide, used in the manufacture of pharmaceuticals, dyes, and agricultural chemicals, decomposes at high temperatures.

$$HCONH_2(g) \rightleftharpoons NH_3(g) + CO(g)$$
$$K_c = 4.84 \text{ at } 400 \text{ K}$$

If 0.186 mol $HCONH_2(g)$ dissociates in a 2.16 L flask at 400 K, what will be the *total* pressure at equilibrium?

42. A mixture of 1.00 mol $NaHCO_3(s)$ and 1.00 mol $Na_2CO_3(s)$ is introduced into a 2.50 L flask in which the partial pressure of CO_2 is 2.10 atm and that of $H_2O(g)$ is 715 mmHg. When equilibrium is established at 100 °C, will the partial pressures of $CO_2(g)$ and $H_2O(g)$ be greater or less than their initial partial pressures? Explain.

$$2 NaHCO_3(s) \rightleftharpoons Na_2CO_3(s) + CO_2(g) + H_2O(g)$$
$$K_p = 0.23 \text{ at } 100 \text{ °C}$$

43. Cadmium metal is added to 0.350 L of an aqueous solution in which $[Cr^{3+}] = 1.00$ M. What are the concentrations of the different ionic species at equilibrium? What is the minimum mass of cadmium metal required to establish this equilibrium?

$$2\,Cr^{3+}(aq) + Cd(s) \rightleftharpoons 2\,Cr^{2+}(aq) + Cd^{2+}(aq)$$
$$K_c = 0.288$$

44. Lead metal is added to 0.100 M $Cr^{3+}(aq)$. What are $[Pb^{2+}]$, $[Cr^{2+}]$, and $[Cr^{3+}]$ when equilibrium is established in the reaction?

$$Pb(s) + 2\,Cr^{3+}(aq) \rightleftharpoons Pb^{2+}(aq) + 2\,Cr^{2+}(aq)$$
$$K_c = 3.2 \times 10^{-10}$$

45. One sketch below represents an initial nonequilibrium mixture in the reversible reaction

$$SO_2(g) + Cl_2(g) \rightleftharpoons SO_2Cl_2(g) \qquad K_c = 4.0$$

Which of the other three sketches best represents an equilibrium mixture? Explain.

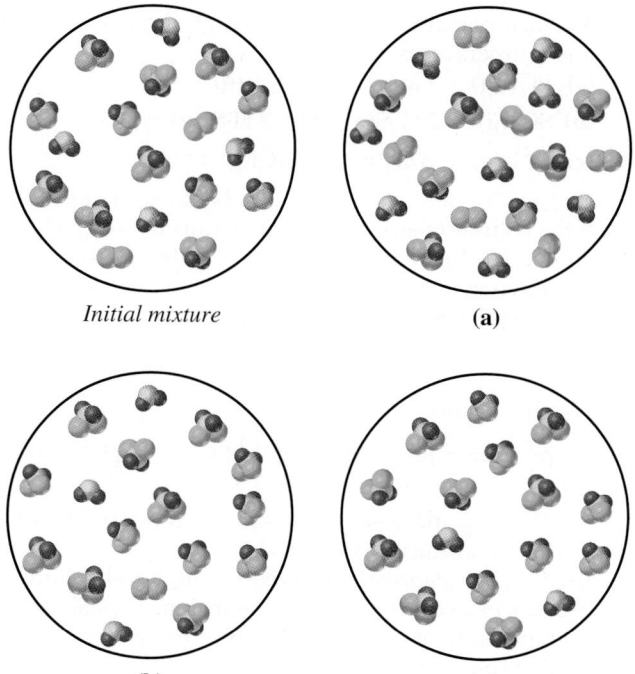

Initial mixture   **(a)**

(b)   **(c)**

46. One sketch below represents an initial nonequilibrium mixture in the reversible reaction

$$2\,NO(g) + Br_2(g) \rightleftharpoons 2\,NOBr(g) \qquad K_c = 3.0$$

Which of the other three sketches best represents an equilibrium mixture? Explain.

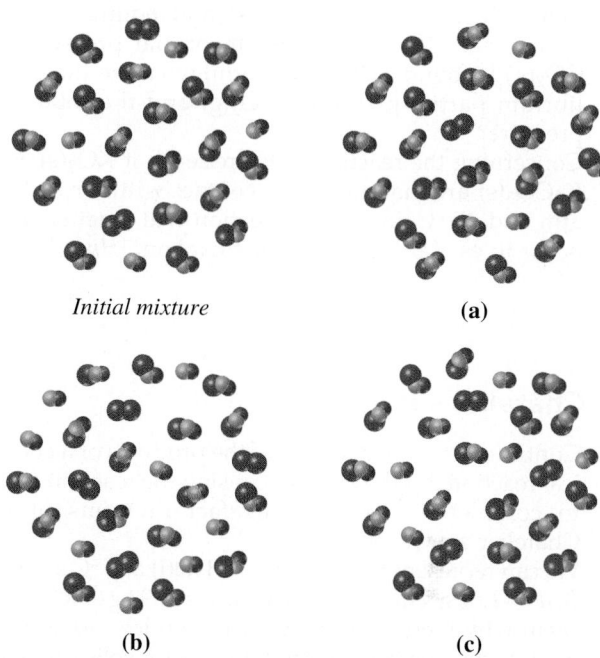

Initial mixture   **(a)**

(b)   **(c)**

47. One important reaction in the citric acid cycle is

$$\text{citrate(aq)} \rightleftharpoons \text{aconitate(aq)} + H_2O(l) \qquad K = 0.031$$

Write the equilibrium constant expression for the above reaction. Given that the concentrations of [citrate(aq)] = 0.00128 M, [aconitate(aq)] = 4.0×10^{-5} M, and $[H_2O] = 55.5$ M, calculate the reaction quotient. Is this reaction at equilibrium? If not, in which direction will it proceed?

48. The following reaction is an important reaction in the citric acid cycle:

$$\text{citrate(aq)} + \text{NAD}_{ox}(aq) + H_2O(l) \rightleftharpoons$$
$$CO_2(aq) + \text{NAD}_{red} + \text{oxoglutarate(aq)} \qquad K = 0.387$$

Write the equilibrium constant expression for the above reaction. Given the following data for this reaction, [citrate] = 0.00128 M, $[\text{NAD}_{ox}] = 0.00868$, $[H_2O] = 55.5$ M, $[CO_2] = 0.00868$ M, $[\text{NAD}_{red}] = 0.00132$ M, and [oxoglutarate] = 0.00868 M, calculate the reaction quotient. Is this reaction at equilibrium? If not, in which direction will it proceed?

Partial Pressure Equilibrium Constant, K_p

49. Refer to Example 15-4. $H_2S(g)$ at 747.6 mmHg pressure and a 1.85 g sample of $I_2(s)$ are introduced into a 725 mL flask at 60 °C. What will be the total pressure in the flask at equilibrium?

$$H_2S(g) + I_2(s) \rightleftharpoons 2\,HI(g) + S(s)$$
$$K_p = 1.34 \times 10^{-5} \text{ at } 60\,°C$$

50. A sample of $NH_4HS(s)$ is placed in a 2.58 L flask containing 0.100 mol $NH_3(g)$. What will be the total gas pressure when equilibrium is established at 25 °C?

$$NH_4HS(s) \rightleftharpoons NH_3(g) + H_2S(g)$$
$$K_p = 0.108 \text{ at } 25\,°C$$

51. The following reaction is used in some self-contained breathing devices as a source of $O_2(g)$.

$$4 KO_2(s) + 2 CO_2(g) \rightleftharpoons 2 K_2CO_3(s) + 3 O_2(g)$$
$$K_p = 28.5 \text{ at } 25 °C$$

Suppose that a sample of $CO_2(g)$ is added to an evacuated flask containing $KO_2(s)$ and equilibrium is established. If the equilibrium partial pressure of $CO_2(g)$ is found to be 0.0721 atm, what are the equilibrium partial pressure of $O_2(g)$ and the total gas pressure?

52. Concerning the reaction in Exercise 51, if $KO_2(s)$ and $K_2CO_3(s)$ are maintained in contact with air at 1.00 atm and 25 °C, in which direction will a net change occur to establish equilibrium? Explain. [*Hint:* Recall equation (6.17). Air is 20.946% O_2 and 0.0379% CO_2 by volume.]

53. 1.00 mol *each* of CO and Cl_2 are introduced into an evacuated 1.75 L flask, and the following equilibrium is established at 668 K.

$$CO(g) + Cl_2(g) \longrightarrow COCl_2(g) \qquad K_p = 22.5$$

For this equilibrium, calculate **(a)** the partial pressure of $COCl_2(g)$; **(b)** the total gas pressure.

54. For the reaction $2 NO_2(g) \rightleftharpoons 2 NO(g) + O_2(g)$, $K_c = 1.8 \times 10^{-6}$ at 184 °C. What is the value of K_p for this reaction at 184 °C?

$$NO(g) + \frac{1}{2}O_2(g) \rightleftharpoons NO_2(g)$$

Le Châtelier's Principle

55. Continuous removal of one of the products of a chemical reaction has the effect of causing the reaction to go to completion. Explain this fact in terms of Le Châtelier's principle.

56. We can represent the freezing of $H_2O(l)$ at 0 °C as H_2O (l, $d = 1.00 \text{ g/cm}^3$) $\rightleftharpoons$ $H_2O(s, d = 0.92 \text{ g/cm}^3)$. Explain why increasing the pressure on ice causes it to melt. Is this the behavior you expect for solids in general? Explain.

57. Explain how each of the following affects the amount of H_2 present in an equilibrium mixture in the reaction

$$3 Fe(s) + 4 H_2O(g) \rightleftharpoons Fe_3O_4(s) + 4 H_2(g)$$
$$\Delta H° = -150 \text{ kJ}$$

(a) Raising the temperature of the mixture; **(b)** introducing more $H_2O(g)$; **(c)** doubling the volume of the container holding the mixture; **(d)** adding an appropriate catalyst.

58. In the gas phase, iodine reacts with cyclopentene (C_5H_8) by a free radical mechanism to produce cyclopentadiene (C_5H_6) and hydrogen iodide. Explain how each of the following affects the amount of HI(g) present in the equilibrium mixture in the reaction

$$I_2(g) + C_5H_8(g) \rightleftharpoons C_5H_6(g) + 2 HI(g)$$
$$\Delta H° = 92.5 \text{ kJ}$$

(a) Raising the temperature of the mixture; **(b)** introducing more $C_5H_6(g)$; **(c)** doubling the volume of the container holding the mixture; **(d)** adding an appropriate catalyst; **(e)** adding an inert gas such as He to a constant-volume reaction mixture.

59. The reaction $N_2(g) + O_2(g) \rightleftharpoons 2 NO(g)$, $\Delta H° = +181$ kJ, occurs in high-temperature combustion processes carried out in air. Oxides of nitrogen produced from the nitrogen and oxygen in air are intimately involved in the production of photochemical smog. What effect does increasing the temperature have on **(a)** the equilibrium production of NO(g); **(b)** the rate of this reaction?

60. Use data from Appendix D to determine whether the forward reaction is favored by high temperatures or low temperatures.
   (a) $PCl_3(g) + Cl_2(g) \rightleftharpoons PCl_5(g)$
   (b) $SO_2(g) + 2 H_2S(g) \rightleftharpoons 2 H_2O(g) + 3 S(s)$
   (c) $2 N_2(g) + 3 O_2(g) + 4 HCl(g) \rightleftharpoons$
   $$4 NOCl(g) + 2 H_2O(g)$$

61. If the volume of an equilibrium mixture of $N_2(g)$, $H_2(g)$, and $NH_3(g)$ is reduced by doubling the pressure, will P_{N_2} have increased, decreased, or remained the same when equilibrium is re established? Explain.

$$N_2(g) + 3 H_2(g) \rightleftharpoons 2 NH_3(g)$$

62. For the reaction

$$A(s) \rightleftharpoons B(s) + 2 C(g) + \frac{1}{2}D(g) \qquad \Delta H° = 0$$

(a) Will K_p increase, decrease, or remain constant with temperature? Explain.
(b) If a *constant-volume* mixture at equilibrium at 298 K is heated to 400 K and equilibrium re-established, will the number of moles of D(g) increase, decrease, or remain constant? Explain.

63. What effect does increasing the volume of the system have on the equilibrium condition in each of the following reactions?
   (a) $C(s) + H_2O(g) \rightleftharpoons CO(g) + H_2(g)$
   (b) $Ca(OH)_2(s) + CO_2(g) \rightleftharpoons CaCO_3(s) + H_2O(g)$
   (c) $4 NH_3(g) + 5 O_2(g) \rightleftharpoons 4 NO(g) + 6 H_2O(g)$

64. For which of the following reactions would you expect the extent of the forward reaction to increase with increasing temperatures? Explain.

(a) $NO(g) \rightleftharpoons \frac{1}{2}N_2(g) + \frac{1}{2}O_2(g)$ $\Delta H° = -90.2$ kJ

(b) $SO_3(g) \rightleftharpoons SO_2(g) + \frac{1}{2}O_2(g)$ $\Delta H° = +98.9$ kJ

(c) $N_2H_4(g) \rightleftharpoons N_2(g) + 2 H_2(g)$ $\Delta H° = -95.4$ kJ

(d) $COCl_2(g) \rightleftharpoons CO(g) + Cl_2(g)$ $\Delta H° = +108.3$ kJ

65. The following reaction represents the binding of oxygen by the protein hemoglobin (Hb):

$$Hb(aq) + O_2(aq) \rightleftharpoons Hb:O_2(aq) \qquad \Delta H < 0$$

Explain how each of the following affects the amount of $Hb:O_2$: **(a)** increasing the temperature; **(b)** decreasing the pressure of O_2; **(c)** increasing the amount of hemoglobin.

66. In the human body, the enzyme carbonic anahydrase catalyzes the interconversion of CO_2 and HCO_3^- by either adding or removing the hydroxide anion. The overall reaction is endothermic. Explain how the following affect the amount of carbon dioxide: **(a)** increasing the amount of bicarbonate anion; **(b)** increasing the pressure of carbon dioxide; **(c)** increasing the amount of carbonic anhydrase; **(d)** decreasing the temperature.

67. A crystal of dinitrogen tetroxide (melting point, $-9.3\ ^\circ C$; boiling point, $21.3\ ^\circ C$) is added to an equilibrium mixture of dintrogen tetroxide and nitrogen dioxide that is at $20.0\ ^\circ C$. Will the pressure of nitrogen dioxide increase, decrease, or remain the same? Explain.

68. When hydrogen iodide is heated, the degree of dissociation increases. Is the dissociation reaction exothermic or endothermic? Explain.

69. The standard enthalpy of reaction for the decomposition of calcium carbonate is $\Delta H^\circ = 813.5\ \text{kJ mol}^{-1}$. As temperature increases, does the concentration of calcium carbonate increase, decrease, or remain the same? Explain.

70. Would you expect that the amount of N_2 to increase, decrease, or remain the same in a scuba diver's body as he or she descends below the water surface?

Integrative and Advanced Exercises

71. Explain why the percent of molecules that dissociate into atoms in reactions of the type $I_2(g) \rightleftharpoons 2\,I(g)$ *always* increases with an increase in temperature.

72. A 1.100 L flask at $25\ ^\circ C$ and 1.00 atm pressure contains $CO_2(g)$ in contact with 100.0 mL of a saturated aqueous solution in which $[CO_2(aq)] = 3.29 \times 10^{-2}$ M.
    (a) What is the value of K_c at $25\ ^\circ C$ for the equilibrium $CO_2(g) \rightleftharpoons CO_2(aq)$?
    (b) If 0.01000 mol of radioactive $^{14}CO_2$ is added to the flask, how many moles of the $^{14}CO_2$ will be found in the gas phase and in the aqueous solution when equilibrium is re-established? [*Hint:* The radioactive $^{14}CO_2$ distributes itself between the two phases in exactly the same manner as the nonradioactive $^{12}CO_2$.]

73. Refer to Example 15-13. Suppose that 0.100 L of the equilibrium mixture is diluted to 0.250 L with water. What will be the new concentrations when equilibrium is re-established?

74. In the equilibrium described in Example 15-12, the percent dissociation of N_2O_4 can be expressed as

$$\frac{3.00 \times 10^{-3}\ \text{mol}\ N_2O_4}{0.0240\ \text{mol}\ N_2O_4\ \text{initially}} \times 100\% = 12.5\%$$

What must be the total pressure of the gaseous mixture if $N_2O_4(g)$ is to be 10.0% dissociated at 298 K?

$$N_2O_4 \rightleftharpoons 2\,NO_2(g) \qquad K_p = 0.113 \text{ at } 298\ K$$

75. Starting with $SO_3(g)$ at 1.00 atm, what will be the total pressure when equilibrium is reached in the following reaction at 700 K?

$$2\,SO_3(g) \rightleftharpoons 2\,SO_2(g) + O_2(g) \qquad K_p = 1.6 \times 10^{-5}$$

76. A sample of air with a mole ratio of N_2 to O_2 of $79:21$ is heated to 2500 K. When equilibrium is established in a closed container with air initially at 1.00 atm, the mole percent of NO is found to be 1.8%. Calculate K_p for the reaction.

$$N_2(g) + O_2(g) \rightleftharpoons 2\,NO(g)$$

77. Derive, by calculation, the equilibrium amounts of SO_2, O_2, and SO_3 listed in **(a)** Figure 15-6(c); **(b)** Figure 15-7(b).

78. The decomposition of salicylic acid to phenol and carbon dioxide was carried out at $200.0\ ^\circ C$, a temperature at which the reactant and products are all gaseous. A 0.300 g sample of salicylic acid was introduced into a 50.0 mL reaction vessel, and equilibrium was established. The equilibrium mixture was rapidly cooled to condense salicylic acid and phenol as solids; the $CO_2(g)$ was collected over mercury and its volume was measured at $20.0\ ^\circ C$ and 730 mmHg. In two identical experiments, the volumes of $CO_2(g)$ obtained were 48.2 and 48.5 mL, respectively. Calculate K_p for this reaction.

79. One of the key reactions in the gasification of coal is the methanation reaction, in which methane is produced from synthesis gas—a mixture of CO and H_2.

$$CO(g) + 3\,H_2(g) \rightleftharpoons CH_4(g) + H_2O(g)$$
$$\Delta H = -230\ \text{kJ};\ K_c = 190 \text{ at } 1000\ K$$

    (a) Is the equilibrium conversion of synthesis gas to methane favored at higher or lower temperatures? Higher or lower pressures?
    (b) Assume you have 4.00 mol of synthesis gas with a 3:1 mol ratio of $H_2(g)$ to $CO(g)$ in a 15.0 L flask. What will be the mole fraction of $CH_4(g)$ at equilibrium at 1000 K?

80. A sample of pure $PCl_5(g)$ is introduced into an evacuated flask and allowed to dissociate.

$$PCl_5(g) \rightleftharpoons PCl_3(g) + Cl_2(g)$$

If the fraction of PCl_5 molecules that dissociate is denoted by α, and if the total gas pressure is P, show that

$$K_p = \frac{\alpha^2 P}{1 - \alpha^2}$$

81. Nitrogen dioxide obtained as a cylinder gas is always a mixture of $NO_2(g)$ and $N_2O_4(g)$. A 5.00 g sample obtained from such a cylinder is sealed in a 0.500 L flask at 298 K. What is the mole fraction of NO_2 in this mixture?

$$N_2O_4(g) \rightleftharpoons 2\,NO_2(g) \qquad K_c = 4.61 \times 10^{-3}$$

82. What is the apparent molar mass of the gaseous mixture that results when $COCl_2(g)$ is allowed to dissociate at 395 °C and a total pressure of 3.00 atm?

$COCl_2(g) \rightleftharpoons CO(g) + Cl_2(g)$

$$K_p = 4.44 \times 10^{-2} \text{ at } 395 \text{ °C}$$

Think of the apparent molar mass as the molar mass of a hypothetical single gas that is equivalent to the gaseous mixture.

83. Show that in terms of mole fractions of gases and *total* gas pressure the equilibrium constant expression for

$$N_2(g) + 3\,H_2(g) \rightleftharpoons 2\,NH_3(g)$$

is

$$K_p = \frac{(x_{NH_3})^2}{(x_{N_2})(x_{H_2})^2} \times \frac{1}{(P_{tot})^2}$$

84. For the synthesis of ammonia at 500 K, $N_2(g) + 3\,H_2(g) \rightleftharpoons 2\,NH_3(g)$, $K_p = 9.06 \times 10^{-2}$. Assume that N_2 and H_2 are mixed in the mole ratio 1:3 and that the total pressure is maintained at 1.00 atm. What is the mole percent NH_3 at equilibrium? [*Hint*: Use the equation from Exercise 83.]

85. A mixture of $H_2S(g)$ and $CH_4(g)$ in the mole ratio 2:1 was brought to equilibrium at 700 °C and a total pressure of 1 atm. On analysis, the equilibrium mixture was found to contain 9.54×10^{-3} mol H_2S. The CS_2 present at equilibrium was converted successively to H_2SO_4 and then to $BaSO_4$; 1.42×10^{-3} mol $BaSO_4$ was obtained. Use these data to determine K_p at 700 °C for the reaction

$2\,H_2S(g) + CH_4(g) \rightleftharpoons CS_2(g) + 4\,H_2(g)$

$$K_p \text{ at } 700 \text{ °C} = ?$$

86. A solution is prepared having these initial concentrations: $[Fe^{3+}] = [Hg_2^{2+}] = 0.5000$ M; $[Fe^{2+}] = [Hg^{2+}] = 0.03000$ M. The following reaction occurs among the ions at 25 °C.

$2\,Fe^{3+}(aq) + Hg_2^{2+}(aq) \rightleftharpoons 2\,Fe^{2+}(aq) + 2\,Hg^{2+}(aq)$

$$K_c = 9.14 \times 10^{-6}$$

What will be the ion concentrations at equilibrium?

87. Refer to the Integrative Example. A gaseous mixture is prepared containing 0.100 mol each of $CH_4(g)$, $H_2O(g)$, $CO_2(g)$, and $H_2(g)$ in a 5.00 L flask. Then the mixture is allowed to come to equilibrium at 1000 K in reaction (15.24). What will be the equilibrium amount, in moles, of each gas?

88. Concerning the reaction in Exercise 26 and the situation described in part (c) of that exercise, will the mole fraction of $C_2H_2(g)$ increase, decrease, or remain unchanged when equilibrium is re-established? Explain.

89. The formation of nitrosyl chloride is given by the following equation: $2\,NO(g) + Cl_2(g) \rightleftharpoons 2\,NOCl(g)$; $K_c = 4.6 \times 10^4$ at 298 K. In a 1.50 L flask, there are 4.125 mol of NOCl and 0.1125 mol of Cl_2 present at equilibrium (298 K).
   (a) Determine the partial pressure of NO at equilibrium.
   (b) What is the total pressure of the system at equilibrium?

90. At 500 K, a 10.0 L equilibrium mixture contains 0.424 mol N_2, 1.272 mol H_2, and 1.152 mol NH_3. The mixture is quickly chilled to a temperature at which the NH_3 liquefies, and the $NH_3(l)$ is completely removed. The 10.0 L gaseous mixture is then returned to 500 K, and equilibrium is re-established. How many moles of $NH_3(g)$ will be present in the new equilibrium mixture?

$N_2(g) + 3\,H_2(g) \rightleftharpoons 2\,NH_3 \qquad K_c = 152 \text{ at } 500 \text{ K}$

91. Recall the formation of methanol from synthesis gas, the reversible reaction at the heart of a process with great potential for the future production of automotive fuels (page 663).

$CO(g) + 2\,H_2(g) \rightleftharpoons CH_3OH(g)$

$$K_c = 14.5 \text{ at } 483 \text{ K}$$

A particular synthesis gas consisting of 35.0 mole percent $CO(g)$ and 65.0 mole percent $H_2(g)$ at a total pressure of 100.0 atm at 483 K is allowed to come to equilibrium. Determine the partial pressure of $CH_3OH(g)$ in the equilibrium mixture.

Feature Problems

92. A classic experiment in equilibrium studies dating from 1862 involved the reaction in solution of ethanol (C_2H_5OH) and acetic acid (CH_3COOH) to produce ethyl acetate and water.

$C_2H_5OH + CH_3COOH \rightleftharpoons CH_3COOC_2H_5 + H_2O$

The reaction can be followed by analyzing the equilibrium mixture for its acetic acid content.

$2\,CH_3COOH(aq) + Ba(OH)_2(aq) \rightleftharpoons$
$$Ba(CH_3COO)_2(aq) + 2\,H_2O(l)$$

In one experiment, a mixture of 1.000 mol acetic acid and 0.5000 mol ethanol is brought to equilibrium. A sample containing exactly one-hundredth of the equilibrium mixture requires 28.85 mL 0.1000 M $Ba(OH)_2$ for its titration. Calculate the equilibrium constant, K_c, for the ethanol-acetic acid reaction based on this experiment.

93. The decomposition of HI(g) is represented by the equation

$$2 \, HI(g) \rightleftharpoons H_2(g) + I_2(g)$$

HI(g) is introduced into five identical 400 cm³ glass bulbs, and the five bulbs are maintained at 623 K. Each bulb is opened after a period of time and analyzed for I_2 by titration with 0.0150 M $Na_2S_2O_3(aq)$.

$$I_2(aq) + 2 \, Na_2S_2O_3(aq) \longrightarrow$$
$$Na_2S_4O_6(aq) + 2 \, NaI(aq)$$

Data for this experiment are provided in the table below. What is the value of K_c at 623 K?

Bulb Number	Initial Mass of HI(g), g	Time Bulb Opened, h	Volume 0.0150 M Na₂S₂O₃ Required for Titration, in mL
1	0.300	2	20.96
2	0.320	4	27.90
3	0.315	12	32.31
4	0.406	20	41.50
5	0.280	40	28.68

94. In one of Fritz Haber's experiments to establish the conditions required for the ammonia synthesis reaction, pure $NH_3(g)$ was passed over an iron catalyst at 901 °C and 30.0 atm. The gas leaving the reactor was bubbled through 20.00 mL of a HCl(aq) solution. In this way, the $NH_3(g)$ present was removed by reaction with HCl. The remaining gas occupied a volume of 1.82 L at STP. The 20.00 mL of HCl(aq) through which the gas had been bubbled required 15.42 mL of 0.0523 M KOH for its titration. Another 20.00 mL sample of the same HCl(aq) through which no gas had been bubbled required 18.72 mL of 0.0523 M KOH for its titration. Use these data to obtain a value of K_p at 901 °C for the reaction $N_2(g) + 3 \, H_2(g) \rightleftharpoons 2 \, NH_3(g)$.

95. The following is an approach to establishing a relationship between the equilibrium constant and rate constants mentioned in the section on page 660.
    • Work with the detailed mechanism for the reaction.
    • Use the principle of microscopic reversibility, the idea that every step in a reaction mechanism is reversible. (In the presentation of elementary reactions in Chapter 14, we treated some reaction

steps as reversible and others as going to completion. However, as noted in Table 15.3, every reaction has an equilibrium constant even though a reaction is generally considered to go to completion if its equilibrium constant is very large.)
    • Use the idea that when equilibrium is attained in an overall reaction, it is also attained in each step of its mechanism. Moreover, we can write an equilibrium constant expression for each step in the mechanism, similar to what we did with the steady-state assumption in describing reaction mechanisms.
    • Combine the K_c expressions for the elementary steps into a K_c expression for the overall reaction. The numerical value of the overall K_c can thereby be expressed as a ratio of rate constants, k.

Use this approach to establish the equilibrium constant expression for the overall reaction,

$$H_2(g) + I_2(g) \rightleftharpoons 2 \, HI(g)$$

The mechanism of the reaction appears to be the following:

Fast: $I_2(g) \rightleftharpoons 2 \, I(g)$
Slow: $2 \, I(g) + H_2(g) \rightleftharpoons 2 \, HI(g)$

96. The following two equilibrium reactions can be written for aqueous carbonic acid, $H_2CO_3(aq)$:

$$H_2CO_3(aq) \rightleftharpoons H^+(aq) + HCO_3^-(aq) \quad K_1$$
$$HCO_3^-(aq) \rightleftharpoons H^+(aq) + CO_3^{2-}(aq) \quad K_2$$

For each reaction write the equilibrium constant expression. By using Le Châtelier's principle we may naively predict that by adding H_2CO_3 to the system, the concentration of CO_3^{2-} would increase. What we observe is that after adding H_2CO_3 to the equilibrium mixture, an increase in the concentration of CO_3^{2-} occurs when $[CO_3^{2-}] \ll K_2$ however, the concentration of CO_3^{2-} will decrease when $[CO_3^{2-}] \gg K_2$. Show that this is true by considering the ratio of $[H^+]/[HCO_3^-]$ before and after adding a small amount of H_2CO_3 to the solution, and by using that ratio to calculate the $[CO_3^{2-}]$.

97. In organic synthesis many reactions produce very little yield, that is $K \ll 1$. Consider the following hypothetical reaction: A(aq) + B(aq) $\longrightarrow$ C(aq), $K = 1 \times 10^{-2}$. We can extract product, C, from the aqueous layer by adding an organic layer in which C(aq) $\longrightarrow$ C(or), $K = 15$. Given initial concentrations of [A] = 0.1 M, [B] = 0.1, and [C] = 0.1, calculate how much C will be found in the organic layer. If the organic layer was not present, how much C would be produced?

Self-Assessment Exercises

98. In your own words, define or explain the following terms or symbols: (a) K_p; (b) Q_c; (c) Δn_{gas}.
99. Briefly describe each of the following ideas or phenomena: (a) dynamic equilibrium; (b) direction of a net chemical change; (c) Le Châtelier's principle; (d) effect of a catalyst on equilibrium.

100. Explain the important distinctions between each pair of terms: (a) reaction that goes to completion and reversible reaction; (b) K_c and K_p; (c) reaction quotient (Q) and equilibrium constant expression (K); (d) homogeneous and heterogeneous reaction.

101. In the reversible reaction $H_2(g) + I_2(g) \rightleftharpoons 2\,HI(g)$, an initial mixture contains 2 mol H_2 and 1 mol I_2. The amount of HI expected at equilibrium is (a) 1 mol; (b) 2 mol; (c) less than 2 mol; (d) more than 2 mol but less than 4 mol.

102. Equilibrium is established in the reaction $2\,SO_2(g) + O_2(g) \rightleftharpoons 2\,SO_3(g)$ at a temperature where $K_c = 100$. If the number of moles of $SO_3(g)$ in the equilibrium mixture is the same as the number of moles of $SO_2(g)$, (a) the number of moles of $O_2(g)$ is also equal to the number of moles of $SO_2(g)$; (b) the number of moles of $O_2(g)$ is half the number of moles of SO_2; (c) $[O_2]$ may have any of several values; (d) $[O_2] = 0.010$ M.

103. The volume of the reaction vessel containing an equilibrium mixture in the reaction $SO_2Cl_2(g) \rightleftharpoons SO_2(g) + Cl_2(g)$ is increased. When equilibrium is re-established, (a) the amount of Cl_2 will have increased; (b) the amount of SO_2 will have decreased; (c) the amounts of SO_2 and Cl_2 will have remained the same; (d) the amount of SO_2Cl_2 will have increased.

104. For the reaction $2\,NO_2(g) \rightleftharpoons 2\,NO(g) + O_2(g)$, $K_c = 1.8 \times 10^{-6}$ at 184 °C. At 184 °C, the value of K_c for the reaction $NO(g) + \frac{1}{2}O_2(g) \rightleftharpoons NO_2(g)$ is (a) 0.9×10^6; (b) 7.5×10^2; (c) 5.6×10^5; (d) 2.8×10^5.

105. For the dissociation reaction $2\,H_2S(g) \rightleftharpoons 2\,H_2(g) + S_2(g)$, $K_p = 1.2 \times 10^{-2}$ at 1065 °C. For this same reaction at 1000 K, (a) K_c is less than K_p; (b) K_c is greater than K_p; (c) $K_c = K_p$; (d) whether K_c is less than, equal to, or greater than K_p depends on the total gas pressure.

106. The following data are given at 1000 K: $CO(g) + H_2O(g) \rightleftharpoons CO_2(g) + H_2(g)$; $\Delta H^\circ = -42$ kJ; $K_c = 0.66$. After an initial equilibrium is established in a 1.00 L container, the equilibrium amount of H_2 can be increased by (a) adding a catalyst; (b) increasing the temperature; (c) transferring the mixture to a 10.0 L container; (d) in some way other than (a), (b), or (c).

107. Equilibrium is established in the reversible reaction $2\,A + B \rightleftharpoons 2\,C$. The equilibrium concentrations are $[A] = 0.55$ M, $[B] = 0.33$ M, $[C] = 0.43$ M. What is the value of K_c for this reaction?

108. The Deacon process for producing chlorine gas from hydrogen chloride is used in situations where HCl is available as a by-product from other chemical processes.

$$4\,HCl(g) + O_2(g) \rightleftharpoons 2\,H_2O(g) + 2\,Cl_2(g)$$
$$\Delta H^\circ = -114 \text{ kJ}$$

A mixture of HCl, O_2, H_2O, and Cl_2 is brought to equilibrium at 400 °C. What is the effect on the equilibrium amount of $Cl_2(g)$ if
(a) additional $O_2(g)$ is added to the mixture at constant volume?
(b) HCl(g) is removed from the mixture at constant volume?
(c) the mixture is transferred to a vessel of twice the volume?
(d) a catalyst is added to the reaction mixture?
(e) the temperature is raised to 500 °C?

109. For the reaction $SO_2(g) \rightleftharpoons SO_2(aq)$, $K = 1.25$ at 25 °C. Will the amount of $SO_2(g)$ be greater than or less than the amount of $SO_2(aq)$?

110. In the reaction $H_2O_2(g) \rightleftharpoons H_2O_2(aq)$, $K = 1.0 \times 10^5$ at 25 °C. Would you expect a greater amount of product or reactant?

111. An equilibrium mixture of SO_2, SO_3, and O_2 gases is maintained in a 2.05 L flask at a temperature at which $K_c = 35.5$ for the reaction

$$2\,SO_2(g) + O_2(g) \rightleftharpoons 2\,SO_3(g)$$

(a) If the numbers of moles of SO_2 and SO_3 in the flask are equal, how many moles of O_2 are present?
(b) If the number of moles of SO_3 in the flask is twice the number of moles of SO_2, how many moles of O_2 are present?

112. Using the method in Appendix E, construct a concept map of Section 15-6, illustrating the shift in equilibrium caused by the various types of disturbances discussed in that section.

Acids and Bases

Citrus fruit derives its acidic qualities from citric acid, $H_3C_6H_5O_7$, a type of acid (polyprotic) discussed in Section 16-6. Another important constituent of citrus fruit is ascorbic acid, or vitamin C (page 592), a dietary requirement to prevent scurvy.

The general public is familiar with the concepts of acids and bases. The environmental problem of acid rain is a popular topic in newspapers and magazines, and television commercials mention pH in relation to such products as deodorants, shampoos, and antacids.

Chemists have been classifying substances as acids and bases for a long time. Antoine Lavoisier thought that the common element in all acids was oxygen, a fact conveyed by its name. (*Oxygen* means "acid former" in Greek). In 1810, Humphry Davy showed that hydrogen instead is the element that acids have in common. In 1884, Svante Arrhenius developed the theory of acids and bases that we introduced in Chapter 5. There we emphasized the stoichiometry of acid–base reactions.

Some of the topics we will study in this chapter are modern acid–base theories, factors affecting the strengths of acids and bases, the pH scale, and the calculation of ion concentrations in solutions of weak acids and bases. We bring some of these ideas together in a discussion of acid rain in the Focus On discussion for this chapter (see the MasteringChemistry site).

▶ Descriptions of acids and bases become more general, less restrictive, as we move from the Arrhenius theory, to the Brønsted–Lowry theory, to the Lewis theory.

16-1 Arrhenius Theory: A Brief Review

Some aspects of the behavior of acids and bases can be explained adequately with the theory developed by Arrhenius as part of his studies of electrolytic dissociation (Section 13-9). Arrhenius proposed that in aqueous solutions, strong electrolytes exist only in the form of ions, whereas weak electrolytes exist partly as ions and partly as molecules. When the acid HCl dissolves in water, the HCl molecules ionize completely, yielding hydrogen ions, H^+, as one of the products.

$$HCl \xrightarrow{H_2O} HCl(aq) \rightarrow H^+(aq) + Cl^-(aq)$$

When the base NaOH dissolves in water, the Na^+ and OH^- ions in the solid become dissociated from one another through the action of H_2O molecules (see Figure 13-6).

$$NaOH(s) \xrightarrow{H_2O} Na^+(aq) + OH^-(aq)$$

The neutralization reaction of HCl and NaOH can be represented with the ionic equation

$$\underset{\text{An acid}}{H^+(aq) + Cl^-(aq)} + \underset{\text{A base}}{Na^+(aq) + OH^-(aq)} \longrightarrow \underset{\text{A salt}}{Na^+(aq) + Cl^-(aq)} + \underset{\text{Water}}{H_2O(l)}$$

or, perhaps better still, with the net ionic equation

$$\underset{\text{An acid}}{H^+(aq)} + \underset{\text{A base}}{OH^-(aq)} \longrightarrow \underset{\text{Water}}{H_2O(l)} \qquad \textbf{(16.1)}$$

Equation (16.1) illustrates an essential idea of the Arrhenius theory: *A neutralization reaction involves the combination of hydrogen ions and hydroxide ions to form water.*

Despite its early successes and continued usefulness, the Arrhenius theory does have limitations. One of the most glaring is in its treatment of the weak base ammonia, NH_3. The Arrhenius theory suggests that all bases contain OH^-. Where is the OH^- in NH_3? To get around this difficulty, chemists began to think of aqueous solutions of NH_3 as containing the compound ammonium hydroxide, NH_4OH, which as a weak base is partially ionized into NH_4^+ and OH^- ions:

$$NH_3(g) + H_2O(l) \longrightarrow NH_4OH(aq)$$
$$NH_4OH(aq) \rightleftharpoons NH_4^+(aq) + OH^-(aq)$$

The problem with this formulation is that there is no compelling evidence that NH_4OH exists in aqueous solutions. We should always question a hypothesis or theory that postulates the existence of hypothetical substances. As we will see in Section 16-2, the essential failure of the Arrhenius theory is in not recognizing the key role of the *solvent* in the ionization of a solute.

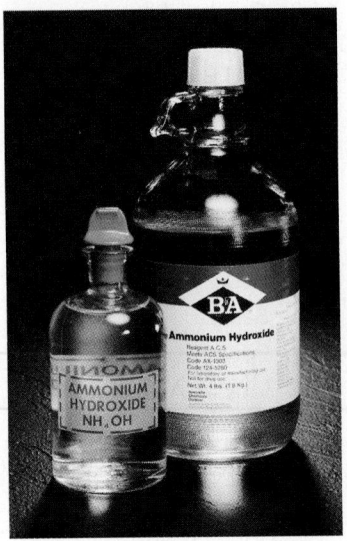

▲ A holdover of the Arrhenius theory. Although there is no compelling evidence that NH_4OH molecules exist in $NH_3(aq)$, solutions are commonly labeled as if they do.

KEEP IN MIND

that a "proton donor" is a donor of H^+ ions. That is, a hydrogen atom consists of one proton and one electron, and the hydrogen ion, H^+, is simply a proton.

16-2 Brønsted–Lowry Theory of Acids and Bases

In 1923, J. N. Brønsted in Denmark and T. M. Lowry in Great Britain independently proposed a new acid–base theory. According to their theory, an acid is a **proton donor** and a base is a **proton acceptor**. To describe the behavior of ammonia as a base, which we found difficult to do with the Arrhenius theory, we can write

$$\underset{\text{Base}}{NH_3} + \underset{\text{Acid}}{H_2O} \longrightarrow NH_4^+ + OH^- \qquad \textbf{(16.2)}$$

In reaction (16.2), H_2O acts as an *acid*. It gives up a proton, H^+, which is taken up by NH_3, a *base*. As a result of this transfer, the polyatomic ions NH_4^+ and OH^- are formed—the same ions produced by the ionization of the hypothetical

NH_4OH of the Arrhenius theory. Because NH_3 is a weak base, we need also to consider the reverse of reaction (16.2). In the reverse reaction, NH_4^+ is an acid and OH^- is a base.

$$NH_4^+ + OH^- \longrightarrow NH_3 + H_2O \qquad \text{(16.3)}$$
$$\text{Acid} \qquad \text{Base}$$

The conventional way to represent a reversible reaction is to use the double arrow notation. In identifying the species in this reversible ionization reaction, we have used the number "1" for the related pair NH_3 and NH_4^+ and the number "2" for the related pair H_2O and OH^-.

$$NH_3 + H_2O \rightleftharpoons NH_4^+ + OH^- \qquad \text{(16.4)}$$
$$\text{Base(1)} \quad \text{Acid(2)} \qquad \text{Acid(1)} \quad \text{Base(2)}$$

An acid and a base that are related to each other as the pair NH_3/NH_4^+ or the pair H_2O/OH^- in reaction (16.4) are referred to as a *conjugate pair*. Thus, when considering an NH_3 molecule as a base, an NH_4^+ ion is the **conjugate acid** of NH_3. Similarly, in reaction (16.4) H_2O is an acid and OH^- is its **conjugate base**. Figure 16-1 illustrates the proton transfer involved in the forward and reverse reactions of (16.4).

On the basis of what we learned in Chapter 15, we might write as the equilibrium constant expression for reaction (16.4)

$$K = \frac{a_{NH_4^+} a_{OH^-}}{a_{NH_3} a_{H_2O}}$$

In an aqueous solution the activities of NH_4^+, OH^-, and NH_3 are approximately equal to $[NH_4^+], [OH^-]$, and $[NH_3]$, respectively. The expression above can be written as

$$K = \frac{[NH_4^+][OH^-]}{[NH_3]} = 1.8 \times 10^{-5} = K_b$$

The equilibrium constant K_b is called the **base ionization constant**.

The ionization of acetic acid can be expressed as

$$CH_3COOH + H_2O \rightleftharpoons CH_3COO^- + H_3O^+$$
$$\text{Acid(1)} \qquad \text{Base(2)} \qquad \text{Base(1)} \qquad \text{Acid(2)}$$

Here, acetate ion, CH_3COO^-, is the conjugate base of the acid CH_3COOH. This time, H_2O acts as a base. Its conjugate acid is the **hydronium ion, H_3O^+**. We first

◀ Brønsted–Lowry theory is not restricted to the dissociation of acids and bases in water, but is valid for any solvent.

KEEP IN MIND

that in designating conjugate pairs, it does not matter which conjugate pair we call (1) and which we call (2). Nor does it matter in what order the acid and base are written on each side of the equation.

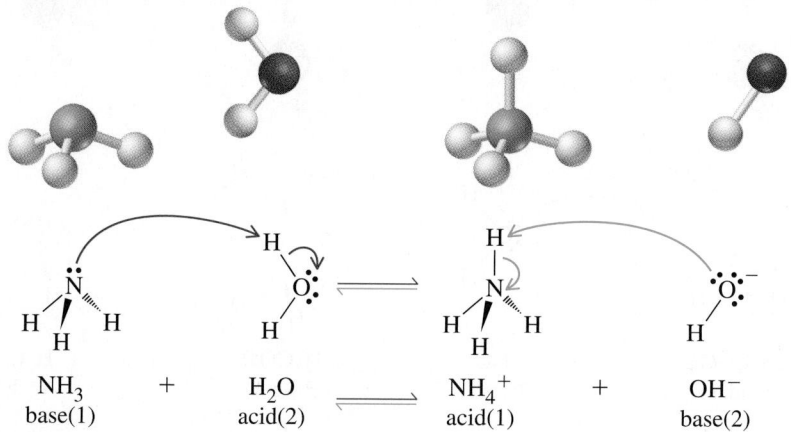

$$\quad \text{NH}_3 \qquad + \qquad \text{H}_2\text{O} \qquad \rightleftharpoons \qquad \text{NH}_4^+ \qquad + \qquad \text{OH}^-$$
$$\text{base(1)} \qquad\qquad \text{acid(2)} \qquad\qquad \text{acid(1)} \qquad\qquad \text{base(2)}$$

▲ FIGURE 16-1

Brønsted–Lowry acid–base reaction: weak base

The curved arrows summarize our visualization of how electrons flow to form and break bonds. The red arrows represent the forward reaction; the blue arrows, the reverse reaction. Because NH_4^+ is a stronger acid than H_2O and OH^- is a stronger base than NH_3, the reverse reaction proceeds to a greater extent than does the forward reaction. Hence, NH_3 is only slightly ionized.

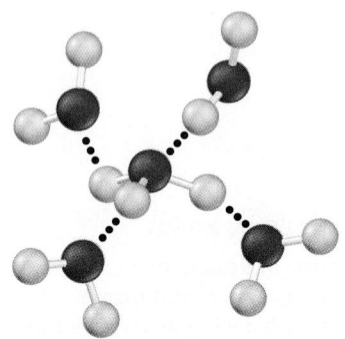

A hydrated hydronium ion
This species, $H_{11}O_5^+$, consists of a central H_3O^+ ion hydrogen-bonded to four H_2O molecules.

discussed the formation of the hydronium ion in Chapter 5 (page 154). Because the simple H^+ ion is tiny, the positive charge of this ion is concentrated in a very small region; the ion has a high positive charge density. We should expect H^+ ions (protons) to seek out centers of negative charge with which to bond. When a H^+ ion attaches to a lone pair of electrons in an O atom in H_2O, the resulting hydronium ion forms hydrogen bonds with several water molecules (Fig. 16-2). Figure 16-3 illustrates the proton transfer involved in the forward and reverse reactions of the ionization of acetic acid.

Using the same approach as for $NH_3(aq)$, the ionization of acetic acid can be described in the following way.

$$K_a = \frac{[CH_3COO^-][H_3O^+]}{[CH_3COOH]} = 1.8 \times 10^{-5}$$

The equilibrium constant K_a is called the **acid ionization constant**. (The fact that K_a of acetic acid and K_b of ammonia have the same value is just a coincidence.)

We can represent the ionization of HCl in the same way that we did for acetic acid. In this case, however, because K_a is so large (about 10^6), we can treat the ionization of HCl as a reaction that goes to completion. We denote this by writing the ionization equation with a single arrow.

$$HCl + H_2O \longrightarrow Cl^- + H_3O^+ \qquad \text{(16.5)}$$

Figure 16-4 illustrates the proton transfer involved in the complete ionization of hydrochloric acid.

In Example 16-1, we identify acids and bases in some typical acid–base reactions. In working through this example, notice the following additional features: (1) Any species that is an acid according to the Arrhenius theory is also an acid according to the Brønsted–Lowry theory; the same is true of bases. (2) Certain species, even though they do not contain the OH group, produce OH^- in aqueous solution—for example, OCl^-. As such, they are Brønsted–Lowry bases. (3) The Brønsted–Lowry theory accounts for substances that can act either as an acid or a base; such substances are said to be **amphiprotic**. The Arrhenius theory does not account for the behavior of amphiprotic substances.

▶ The term *amphiprotic* is similar to the term *amphoteric*, which indicates the ability of a substance to behave as both an acid and a base. Amphiprotic conveys the notion of proton transfer embodied in the Brønsted–Lowry theory of acids and bases.

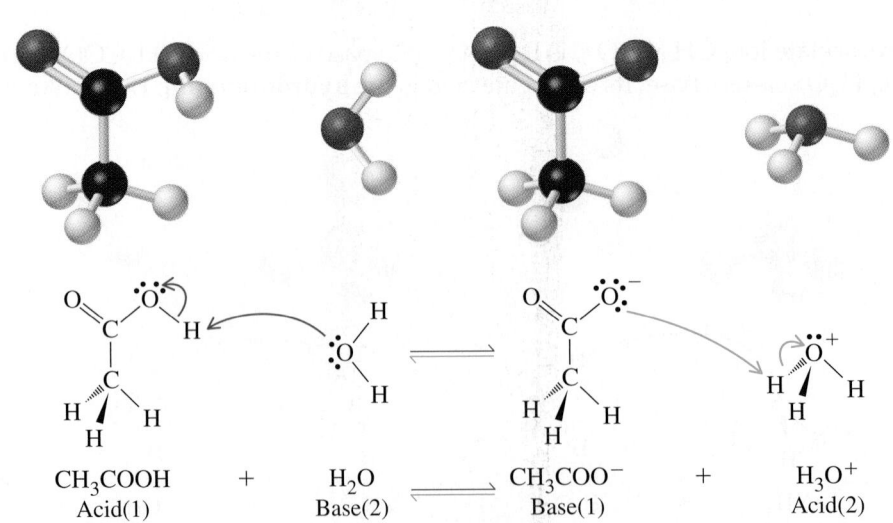

CH₃COOH	+	H₂O		CH₃COO⁻	+	H₃O⁺
Acid(1)		Base(2)		Base(1)		Acid(2)

▲ FIGURE 16-3
Brønsted–Lowry acid–base reaction: weak acid
The curved arrows summarize our visualization of how electrons flow to form and break bonds in the ionization of acetic acid. The red arrows represent the forward reaction; the blue arrows, the reverse reaction. Because H_3O^+ is a stronger acid than CH_3COOH and CH_3COO^- is a stronger base than H_2O, the reverse reaction proceeds to a greater extent than does the forward reaction. Hence, CH_3COOH is only slightly ionized.

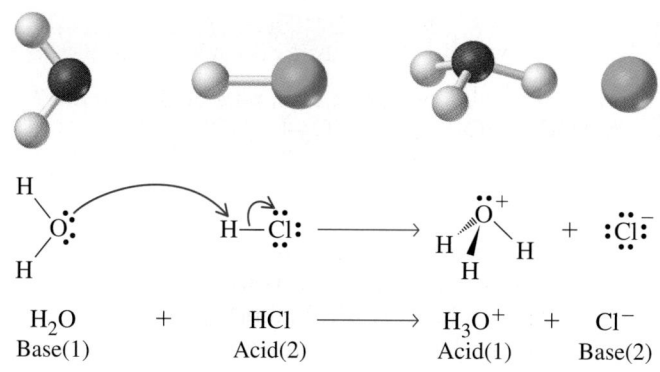

◀ FIGURE 16-4
Brønsted–Lowry acid–base reaction: strong acid
The red arrows summarize our visualization of how electrons flow to form and break bonds in the ionization of hydrochloric acid. Because H_3O^+ is a weaker acid than HCl and Cl^- is a much weaker base than H_2O, the forward reaction proceeds almost to completion. Hence, HCl is essentially completely ionized.

EXAMPLE 16-1   Identifying Brønsted–Lowry Acids and Bases and Their Conjugates

For each of the following, identify the acids and bases in both the forward and reverse reactions in the manner shown in equation (16.4).

(a) $HClO_2 + H_2O \rightleftharpoons ClO_2^- + H_3O^+$

(b) $OCl^- + H_2O \rightleftharpoons HOCl + OH^-$

(c) $NH_3 + H_2PO_4^- \rightleftharpoons NH_4^+ + HPO_4^{2-}$

(d) $HCl + H_2PO_4^- \rightleftharpoons Cl^- + H_3PO_4$

Analyze

Recall that a Brønsted–Lowry acid is one that gives up a proton and a Brønsted–Lowry base is one that takes a proton. Consider $HClO_2$ in reaction (a). It gives up a proton, H^+, to become ClO_2^-. Therefore, $HClO_2$ is an acid, and ClO_2^- is its conjugate base. Now consider H_2O. It takes the proton from $HClO_2$ and becomes H_3O^+. Thus, H_2O is a base, and H_3O^+ is its conjugate acid. In reaction (b), OCl^- is a base and gains a proton from water. OH^- produced in this reaction is the conjugate base of H_2O.

Solve

(a) $HClO_2 + H_2O \rightleftharpoons ClO_2^- + H_3O^+$
    Acid(1)   Base(2)   Base(1)   Acid(2)

(b) $OCl^- + H_2O \rightleftharpoons HOCl + OH^-$
    Base(1) Acid(2)   Acid(1)   Base(2)

(c) $NH_3 + H_2PO_4^- \rightleftharpoons NH_4^+ + HPO_4^{2-}$
    Base(1) Acid(2)   Acid(1)   Base(2)

(d) $HCl + H_2PO_4^- \rightleftharpoons Cl^- + H_3PO_4$
    Acid(1) Base(2)   Base(1)   Acid(2)

Assess

Notice that in (c), $H_2PO_4^-$ is acting as an acid but in (d), it is acting as a base. The conjugate base of $H_2PO_4^-$ is HPO_4^{2-} (the deprotonated form of $H_2PO_4^-$), and the conjugate acid of $H_2PO_4^-$ is H_3PO_4 (the protonated form of $H_2PO_4^-$). This is an example of the general rule that in a conjugate pair, the acid is the protonated form and the base is the deprotonated form.

PRACTICE EXAMPLE A: For each of the following reactions, identify the acids and bases in both the forward and reverse directions.

(a) $HF + H_2O \rightleftharpoons F^- + H_3O^+$

(b) $HSO_4^- + NH_3 \rightleftharpoons SO_4^{2-} + NH_4^+$

(c) $CH_3COO^- + HCl \rightleftharpoons CH_3COOH + Cl^-$

PRACTICE EXAMPLE B: Of the following species, one is acidic, one is basic, and one is amphiprotic in their reactions with water: HNO_2, PO_4^{3-}, HCO_3^-. Write the *four* equations needed to represent these facts.

Is it appropriate to describe each of the following as a conjugate acid–base pair? Explain. **(a)** $HCO_3^- - CO_3^{2-}$; **(b)** $HSO_3^- - SO_4^{2-}$; **(c)** $H_2CO_3 - H_2C_2O_4$; **(d)** $HClO - ClO^-$; **(e)** $H_2S - S^{2-}$.

The ionization of HCl in aqueous solution (reaction 16.5) goes to completion because HCl is a strong acid; it readily gives up protons to H_2O. At the same time, Cl^- ion, the conjugate base of HCl, has very little tendency to take a proton from H_3O^+; Cl^- is a very weak base. This observation suggests the generalization that follows.

> In an acid–base reaction, the favored direction of the reaction is from the stronger to the weaker member of a conjugate acid–base pair.

With this generalization, we can predict that the neutralization of HCl by OH^- should go to completion.

$$HCl + OH^- \longrightarrow Cl^- + H_2O$$

Acid(1)	Base(2)	Base(1)	Acid(2)
strong	strong	weak	weak

And we would predict that the following reaction should occur almost exclusively in the *reverse* direction.

$$H_2O + I^- \longleftarrow OH^- + HI$$

Acid(1)	Base(2)	Base(1)	Acid(2)
weak	weak	strong	strong

▶ The acid–base strengths listed in Table 16.1 are the result of experiments carried out by many chemists.

To be able to apply the generalization more broadly, though, we need a tabulation of acid and base strengths, such as that in Table 16.1. The strongest acids are at the top of the column on the left, and the strongest bases are at the

TABLE 16.1   Relative Strengths of Some Common Brønsted–Lowry Acids and Bases

Acid			Conjugate Base		
↑ Increasing acid strength	Perchloric acid	$HClO_4$	Perchlorate ion	ClO_4^-	Increasing base strength ↓
	Hydroiodic acid	HI	Iodide ion	I^-	
	Hydrobromic acid	HBr	Bromide ion	Br^-	
	Hydrochloric acid	HCl	Chloride ion	Cl^-	
	Sulfuric acid	H_2SO_4	Hydrogen sulfate ion	HSO_4^-	
	Nitric acid	HNO_3	Nitrate ion	NO_3^-	
	Hydronium ion[a]	H_3O^+	Water[a]	H_2O	
	Hydrogen sulfate ion	HSO_4^-	Sulfate ion	SO_4^{2-}	
	Nitrous acid	HNO_2	Nitrite ion	NO_2^-	
	Acetic acid	CH_3COOH	Acetate ion	CH_3COO^-	
	Carbonic acid	H_2CO_3	Hydrogen carbonate ion	HCO_3^-	
	Ammonium ion	NH_4^+	Ammonia	NH_3	
	Hydrogen carbonate ion	HCO_3^-	Carbonate ion	CO_3^{2-}	
	Water	H_2O	Hydroxide ion	OH^-	
	Methanol	CH_3OH	Methoxide ion	CH_3O^-	
	Ammonia	NH_3	Amide ion	NH_2^-	

[a]The hydronium ion–water combination refers to the ease with which a proton is passed from one water molecule to another; that is, $H_3O^+ + H_2O \rightleftharpoons H_2O + H_3O^+$.

bottom of the column on the right. It is important to note that *the stronger an acid, the weaker its conjugate base.*

Both HCl and $HClO_4$ are strong acids because H_2O is a sufficiently strong base to take protons from either acid in a reaction that goes to completion. Because both HCl and $HClO_4$ react to completion with water, yielding H_3O^+ (the strongest acid possible in water), the solvent water is said to have a *leveling effect* on these two acids. How can we ascertain that $HClO_4$ is a stronger acid than HCl as we indicate in Table 16.1?

To determine whether $HClO_4$ or HCl is the stronger acid, we need to use a solvent that is a weaker base than water—a solvent that will take protons from the stronger of the two acids more readily than from the weaker one. In the solvent $(C_2H_5)_2O$, diethyl ether, $HClO_4$ is completely ionized, but HCl is only partially ionized. Thus, $HClO_4$ is a stronger acid than is HCl, and ClO_4^- is a weaker base than Cl^-.

$$HClO_4 + C_2H_5-\overset{..}{\underset{..}{O}}-C_2H_5 \longrightarrow ClO_4^- + C_2H_5-\overset{\overset{H}{|}}{\underset{..}{O^+}}-C_2H_5$$

$$HCl + C_2H_5-\overset{..}{\underset{..}{O}}-C_2H_5 \rightleftharpoons Cl^- + C_2H_5-\overset{\overset{H}{|}}{\underset{..}{O^+}}-C_2H_5$$

16-3   Self-Ionization of Water and the pH Scale

Even when it is pure, water contains a very low concentration of ions that can be detected in precise electrical conductivity measurements. The ions form as a result of the amphiprotic nature of water; some water molecules donate protons and others accept protons. In the **self-ionization** (or *autoionization*) of water, for each H_2O molecule that acts as an acid, another H_2O molecule acts as a base, and hydronium (H_3O^+) and hydroxide (OH^-) ions are formed. The reaction is reversible, and in the reverse reaction, H_3O^+ releases a proton to OH^-. The reverse reaction is far more significant than the forward reaction. *Equilibrium is displaced far to the left.* In reaction (16.6), acid(1) and base(2) are *much* stronger than are acid(2) and base(1).

Base(1)   Acid(2)   Acid(1)   Base(2)   **(16.6)**

Again, we follow the approach we used in writing equilibrium constants for the ionization of NH_3 and CH_3COOH, namely we assume an activity of one for H_2O molecules and replace activities of other species by their molarities. For the self-ionization of water

$$H_2O + H_2O \rightleftharpoons H_3O^+ + OH^-$$

we can write

$$K = [H_3O^+][OH^-]$$

Equation (16.6) indicates that $[H_3O^+]$ and $[OH^-]$ are equal in pure water. There are several experimental methods of determining these concentrations. All lead to this result.

At 25 °C in pure water: $[H_3O^+] = [OH^-] = 1.0 \times 10^{-7}\,M$

◀ A common misconception is that K_w is K_c for the self-ionization reaction. Strictly speaking, it is the ionic product. The equilibrium constant $K_c = [H_3O^+][OH^-]/[H_2O]$ has the value of $10^{-14}/55.55 = 1.8 \times 10^{-16}$ at 25 °C. In Chapter 19, we see that

$$K = (a_{H_3O^+})(a_{OH^-})/(a_{H_2O})^2$$
$$= 1.0 \times 10^{-14}$$

The equilibrium condition for the self-ionization of water is called the **ion product of water**. It is symbolized as K_w. At 25 °C,

$$K_w = [H_3O^+][OH^-] = 1.0 \times 10^{-14} \qquad (16.7)$$

Since K_w is an equilibrium constant, the product of the concentrations of the hydronium and hydroxide ions must always equal 10^{-14}. If the concentration of H_3O^+ is increased by the addition of an acid, then the concentration of OH^- must decrease to maintain the value of K_w. If the concentration of OH^- is increased by the addition of a base, then the concentration of H_3O^+ must decrease. Equation (16.7) connects the concentrations of H_3O^+ and OH^- and applies to *all* aqueous solutions, not just to pure water, as we shall see shortly.

pH and pOH

Because their product in an aqueous solution is only 1.0×10^{-14}, we expect $[H_3O^+]$ and $[OH^-]$ also to be small. Typically, they are less than 1 M—often very much less. Exponential notation is useful in these situations; for example, $[H_3O^+] = 2.2 \times 10^{-13}$ M. But we now want to consider an even more convenient way to describe hydronium and hydroxide ion concentrations.

In 1909, the Danish biochemist Søren Sørensen proposed the term **pH** to refer to the "potential of hydrogen ion." He defined pH as the *negative of the logarithm of* $[H^+]$. Restated in terms of $[H_3O^+]$,*

$$pH = -\log[H_3O^+] \qquad (16.8)$$

> **KEEP IN MIND**
>
> that this definition of pH is one of the few scientific expressions that uses logarithms to the base 10 (log) rather than natural logarithms (ln).

Thus, in a solution that is 0.0025 M HCl,

$$[H_3O^+] = 2.5 \times 10^{-3} \text{ M} \quad \text{and} \quad pH = -\log(2.5 \times 10^{-3}) = 2.60$$

To determine the $[H_3O^+]$ that corresponds to a particular pH value, we do an inverse calculation. In a solution with pH = 4.50,

$$\log[H_3O^+] = -4.50 \quad \text{and} \quad [H_3O^+] = 10^{-4.50} = 3.2 \times 10^{-5} \text{ M}$$

> ▶ The determination of logarithms and inverse logarithms (antilogarithms) is discussed in Appendix A. Significant figure rules for logarithms are also presented there.

The quantity **pOH** can be defined as

$$pOH = -\log[OH^-] \qquad (16.9)$$

Still another useful expression can be derived by taking the *negative logarithm* of the K_w expression (written for 25 °C) and introducing the symbol pK_w.

$$K_w = [H_3O^+][OH^-] = 1.0 \times 10^{-14}$$
$$-\log K_w = -(\log[H_3O^+][OH^-]) = -\log(1.0 \times 10^{-14})$$
$$pK_w = -(\log[H_3O^+] + \log[OH^-]) = -(-14.00)$$
$$= -\log[H_3O^+] - \log[OH^-] = 14.00$$

$$pK_w = pH + pOH = 14.00 \qquad (16.10)$$

*Strictly speaking, we should use the *activity* of H_3O^+, $a_{H_3O^+}$, a dimensionless quantity. But we will not use activities here, just as we did not use them in Chapter 15. We will substitute the numerical value of the molarity of H_3O^+ for its activity and recognize that some pH calculations may be only approximations.

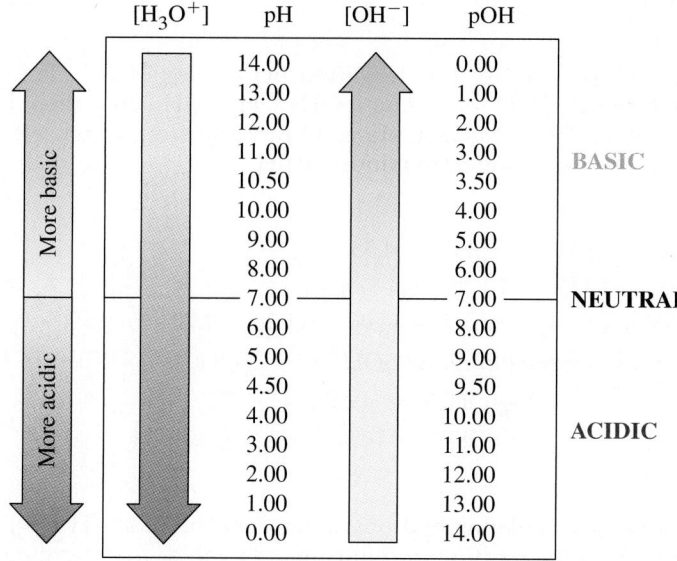

▲ FIGURE 16-5
Relating [H$_3$O$^+$], pH, [OH$^-$], and pOH
In aqueous solutions, the sum of the pH and pOH values always gives pK_w = 14 because of the self-ionization equilibrium of water.

An aqueous solution with $[H_3O^+]$ = $[OH^-]$ is said to be *neutral*. In pure water at 25 °C, $[H_3O^+]$ = $[OH^-]$ = 1.0×10^{-7} M and pH = 7.00. Thus at 25 °C, all aqueous solutions with pH = 7.00 are neutral. If the pH is less than 7.00, the solution is *acidic*; if the pH is greater than 7.00, the solution is *basic*, or alkaline. Equation 16.10 is a restatement of the interrelationship between $[H_3O^+]$, $[OH^-]$, and K_w in terms of pH, pOH, and pK_w. If we know the value of either $[H_3O^+]$ or $[OH^-]$, we can calculate the value of the other (Fig. 16-5).

16-2 CONCEPT ASSESSMENT

Figure 16-5 indicates that pH and pOH *decrease* as [H$^+$] and [OH$^-$] increase. If pH and pOH were defined as pH = log[H$_3$O$^+$] and pOH = log[OH$^-$], pH and pOH would *increase* as [H$^+$] and [OH$^-$] increase. Why do you suppose that this alternate definition was not adopted? Logarithmic functions appear frequently in chemistry (for example, in chemical kinetics and thermodynamics), but these are based on natural logarithms, ln. Why do you suppose that the definitions pH = −ln[H$_3$O$^+$] and pOH = −ln[OH$^-$] were not adopted?

The pH values of a number of materials are depicted in Figure 16-6. These values and the many examples in this chapter and the next should familiarize you with the pH concept. Later, we will consider two methods for measuring pH: by means of acid–base indicators (Section 17-3) and electrical measurements (Section 20-4).

EXAMPLE 16-2 Relating [H$_3$O$^+$], [OH$^-$], pH, and pOH

In a laboratory experiment, students measured the pH of samples of rainwater and household ammonia. Determine **(a)** $[H_3O^+]$ in the rainwater, with pH measured at 4.35; **(b)** $[OH^-]$ in the ammonia, with pH measured at 11.28.

(continued)

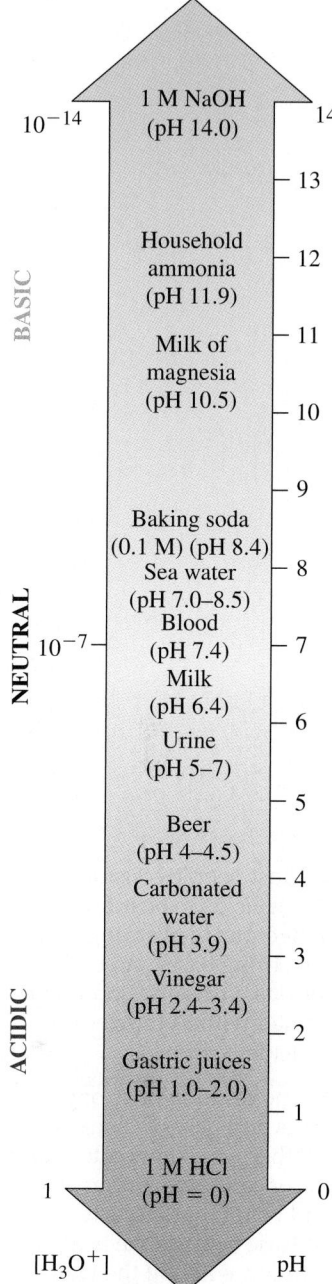

▲ FIGURE 16-6
The pH scale and pH values of some common materials
The scale shown here ranges from pH 0 to pH 14. Slightly negative pH values, perhaps to about −1 (corresponding to [H$_3$O$^+$] ≈ 10 M), are possible. Also possible are pH values up to about 15 (corresponding to [OH$^-$] ≈ 10 M). For practical purposes, however, the pH scale is useful only in the range 2 < pH < 12, because the molarities of H$_3$O$^+$ and OH$^-$ in concentrated acids and bases may differ significantly from their true activities.

Analyze

In this example we use the definition pH = $-\log[H_3O^+]$. For pOH we first determine pOH by using pOH = 14 − pH, and then by using pOH = $-\log[OH^-]$. To calculate concentration from a pH, we take the antilogarithm by raising 10 to minus the pH.

Solve

(a) $\log[H_3O^+] = -\text{pH} = -4.35$

$[H_3O^+] = 10^{-4.35} = 4.5 \times 10^{-5}\,\text{M}$

(b) pOH = $14.00 - \text{pH} = 14.00 - 11.28 = 2.72$

Now, use the definition pOH = $-\log[OH^-]$.

$\log[OH^-] = -\text{pOH} = -2.72$

$[OH^-] = 10^{-2.72} = 1.9 \times 10^{-3}\,\text{M}$

Assess

The process of calculating hydronium ion concentration, $[H_3O^+]$, from pH is simply the antilogarithm of minus the pH value. To determine the concentration of $[OH^-]$ from a pH, we first calculated pOH, which is 14 − pH, and then calculated $[OH^-]$. Be careful when working these types of problems, and keep straight what you have to determine.

PRACTICE EXAMPLE A: Students found that a yogurt sample had a pH of 2.85. What are the $[H^+]$ and $[OH^-]$ of the yogurt?

PRACTICE EXAMPLE B: The pH of a solution of HCl in water is found to be 2.50. What volume of water would you add to 1.00 L of this solution to raise the pH to 3.10?

▶ Most people consider strong acids to be more dangerous than strong bases, but strong bases can cause serious burns and should be treated with as much care as acids.

▶ The dilution of strong acids and bases is usually exothermic. Never add water to concentrated strong acids and bases (particularly sulfuric acid) as the heat of dilution will boil the water and spatter concentrated acid.

16-4 Strong Acids and Strong Bases

As equation (16.5) indicated, the ionization of HCl in dilute aqueous solutions

$$HCl + H_2O \longrightarrow Cl^- + H_3O^+$$

goes essentially to completion.* In contrast, equation (16.6) suggested that the self-ionization of water occurs only to a very slight extent. As a result, we conclude that in calculating $[H_3O^+]$ in an aqueous solution of a strong acid, the strong acid is the only significant source of H_3O^+. The contribution due to the self-ionization of water can generally be ignored *unless the solution is extremely dilute.*

EXAMPLE 16-3 Calculating Ion Concentrations in an Aqueous Solution of a Strong Acid

Calculate $[H_3O^+]$, $[Cl^-]$, and $[OH^-]$ in 0.015 M HCl(aq).

Analyze

Because HCl is a strong acid, all the HCl dissociates. The hydronium ion concentration is equal to the molarity of the solution. The hydroxide concentration is determined by using the water equilibrium because the product of the hydronium ion concentration and hydroxide concentration must equal $K_w = 1.0 \times 10^{-14}$.

*In very concentrated aqueous solutions, HCl does not exist exclusively as the separated ions H_3O^+ and Cl^-. One indication of this is that we can smell HCl in the vapor above such solutions.

Solve

Therefore,

$$[H_3O^+] = 0.015\,M$$

Because one Cl^- ion is produced for every H_3O^+ ion,

$$[Cl^-] = [H_3O^+] = 0.015\,M$$

To calculate $[OH^-]$, we must use the following facts.

1. All the OH^- is derived from the self-ionization of water, by reaction (16.6).
2. $[OH^-]$ and $[H_3O^+]$ must have values consistent with K_w for water.

$$K_w = [H_3O^+][OH^-] = 1.0 \times 10^{-14}$$

So we have

$$[OH^-] = \frac{1.0 \times 10^{-14}}{1.5 \times 10^{-2}} = 6.7 \times 10^{-13}\,M$$

Assess

The self-ionization of water contributes equal amounts of OH^- and H_3O^+ to the solution. The results of this example show that the self-ionization of water contributes only a small amount ($6.7 \times 10^{-13}\,M$) of OH^- and H_3O^+. The self-ionization of water usually, but not always, plays a very minor role in determining the pH of a solution.

PRACTICE EXAMPLE A: A 0.0025 M solution of HI(aq) has $[H_3O^+] = 0.0025\,M$. Calculate $[I^-]$, $[OH^-]$, and the pH of the solution.

PRACTICE EXAMPLE B: If 535 mL of *gaseous* HCl, at 26.5 °C and 747 mmHg, is dissolved in enough water to prepare 625 mL of solution, what is the pH of this solution?

The common strong bases are ionic hydroxides. When these bases dissolve in water, H_2O molecules completely dissociate the cations and anions (OH^-) of the base from each other. The self-ionization of water, because it occurs to so very limited an extent, is an inconsequential source of OH^-. This means that in calculating $[OH^-]$ in an aqueous solution of a strong base, the strong base is the only significant source of OH^- *unless the solution is extremely dilute.*

As noted in Chapter 5, the number of common strong acids and strong bases is quite small. Memorize the listing in Table 16.2.

EXAMPLE 16-4   Calculating the pH of an Aqueous Solution of a Strong Base

Calcium hydroxide (slaked lime), $Ca(OH)_2$, is the cheapest strong base available. It is generally used for industrial operations in which a high concentration of OH^- is not required. $Ca(OH)_2(s)$ is soluble in water only to the extent of 0.16 g $Ca(OH)_2$/100.0 mL solution at 25 °C. What is the pH of saturated $Ca(OH)_2(aq)$ at 25 °C?

Analyze

Because the volume of solution is not specified, let's assume it is 100.0 mL = 0.1000 L. The resulting solution will be basic, so we should focus on the hydroxide ion. To solve this problem, we first calculate the molarity of the solution, and then determine the concentration of hydroxide ion in this solution. Finally, we calculate pOH and then pH.

Solve

Express the solubility of $Ca(OH)_2$ on a molar basis.

$$\text{molarity} = \frac{0.16\,\text{g}\,Ca(OH)_2 \times \dfrac{1\,\text{mol}\,Ca(OH)_2}{74.1\,\text{g}\,Ca(OH)_2}}{0.1000\,L} = 0.022\,M\,Ca(OH)_2$$

(continued)

Relate the molarity of OH^- to the molarity of $Ca(OH)_2$.

$$[OH^-] = \frac{0.022 \text{ mol } Ca(OH)_2}{1 \text{ L}} \times \frac{2 \text{ mol } OH^-}{1 \text{ mol } Ca(OH)_2} = 0.044 \text{ M } OH^-$$

Calculate the pOH and, from it, the pH.

$$pOH = -\log[OH^-] = -\log 0.044 = 1.36$$
$$pH = 14.00 - pOH = 14.00 - 1.36 = 12.64$$

Assess

A common error is to neglect the factor $2 \text{ mol } OH^-/1 \text{ mol } Ca(OH)_2$ in determining $[OH^-]$. When solving problems involving basic solutions, we often solve first for pOH. We must remember to finish the problem and convert from pOH to pH. Finally, although $Ca(OH)_2$ is a slightly soluble hydroxide salt, we observe that the pH of the solution is quite high.

PRACTICE EXAMPLE A: Milk of magnesia is a saturated solution of $Mg(OH)_2$. Its solubility is 9.63 mg $Mg(OH)_2/100.0$ mL solution at 20 °C. What is the pH of saturated $Mg(OH)_2$ at 20 °C?

PRACTICE EXAMPLE B: Calculate the pH of an aqueous solution that is 3.00% KOH, by mass, and has a density of 1.0242 g/mL.

TABLE 16.2
The Common Strong Acids and Strong Bases

Acids	Bases
HCl	LiOH
HBr	NaOH
HI	KOH
$HClO_4$	RbOH
HNO_3	CsOH
H_2SO_4[a]	$Mg(OH)_2$
	$Ca(OH)_2$
	$Sr(OH)_2$
	$Ba(OH)_2$

[a]H_2SO_4 ionizes in two distinct steps. It is a strong acid only in its first ionization (see page 720).

16-1 ARE YOU WONDERING...

How to calculate $[H_3O^+]$ in an extremely dilute solution of a strong acid?

The method of Example 16-3 won't work for calculating the pH of a solution as dilute as 1.0×10^{-8} M HCl. We would write $[H_3O^+] = 1.0 \times 10^{-8}$ M, and pH = 8.00. But how can a solution of a strong acid, no matter how dilute, have a pH greater than 7? The difficulty is that at this extreme dilution, we must consider two sources of H_3O^+. The sources of H_3O^+ and the ion concentrations from both sources are indicated as follows:

$$H_2O + H_2O \rightleftharpoons H_3O^+ + OH^-$$
Molarity: $\qquad\qquad\qquad\qquad x \qquad\quad x$

$$HCl + H_2O \longrightarrow H_3O^+ + Cl^-$$
Molarity: $\qquad\qquad\quad 1.0 \times 10^{-8} \quad 1.0 \times 10^{-8}$

To satisfy the K_w expression for water in this solution, we use equation (16.7) to get

$$[H_3O^+][OH^-] = (x + 1.0 \times 10^{-8})x = 1.0 \times 10^{-14}$$

This expression rearranges to the quadratic form

$$x^2 + (1.0 \times 10^{-8}x) - (1.0 \times 10^{-14}) = 0$$

The solution to this equation is $x = 9.5 \times 10^{-8}$ M. Therefore, we combine $[H_3O^+]$ from both sources to get $[H_3O^+] = (9.5 \times 10^{-8}) + (1.0 \times 10^{-8}) = 1.05 \times 10^{-7}$ M, and pH = 6.98.

From this result, we conclude that the pH is slightly less than 7, as expected for a very dilute acid, and that the self-ionization of water contributes nearly ten times as much hydronium ion to the solution as does the strong acid.

16-5 Weak Acids and Weak Bases

Figure 16-7 illustrates two ways of showing that ionization has occurred in an aqueous solution of an acid: One is by the color of an acid–base indicator; the other, the response of a pH meter. The pink color of the solution in Figure 16-7 tells us the pH of 0.1 M HCl is *less than 1.2*. The pH meter registers a value of 1.20, about what we expect for a strong acid solution with $[H_3O^+]$ approximately equal to 0.1 M. The yellow color of the solution in Figure 16-7 indicates that the pH of 0.1 M CH_3COOH (acetic acid) is *2.8 or greater*. The pH meter registers 2.80.

So we see that two acids can have the same molarity but different pH values. The acid's molarity simply indicates what was put into the solution, but $[H_3O^+]$ and pH depend on what *happens* in the solution. In both solutions, some self-ionization of water occurs, but this reaction is negligible. Ionization of HCl, a strong acid, can be assumed to go to completion, as indicated in equation (16.5). As we previously noted, ionization of CH_3COOH, a weak acid, is a reversible reaction that reaches a condition of equilibrium.*

$$CH_3COOH + H_2O \rightleftharpoons H_3O^+ + CH_3COO^- \qquad \textbf{(16.11)}$$

The equilibrium constant expression for reaction (16.11) is

$$K_a = \frac{[H_3O^+][CH_3COO^-]}{[CH_3COOH]} = 1.8 \times 10^{-5} \qquad \textbf{(16.12)}$$

Just as pH is a convenient shorthand designation related to $[H_3O^+]$, pK is related to an equilibrium constant. That is, $\textbf{p}K = -\log K$. Thus, for acetic acid,

$$pK_a = -\log K_a = -\log(1.8 \times 10^{-5}) = -(-4.74) = 4.74$$

As with other equilibrium constants, the larger the value of K_a (or K_b for a base), the farther the equilibrium condition lies in the direction of the forward reaction. And the more extensive the ionization, the greater are the concentrations of the ions produced. Ionization constants must be determined *by experiment*.

◀ Equilibrium constants (and therefore pK_a and pK_b values) vary with temperature.

◀ The $pK = -\log K$ is introduced so that very large and very small numbers that arise for K can be more easily handled.

◀ Most laboratory pH meters can be read to the nearest 0.01 unit. Some pH meters for research work can be read to 0.001 unit, but unless unusual precautions are taken, the reading of the meter may not correspond to the true pH.

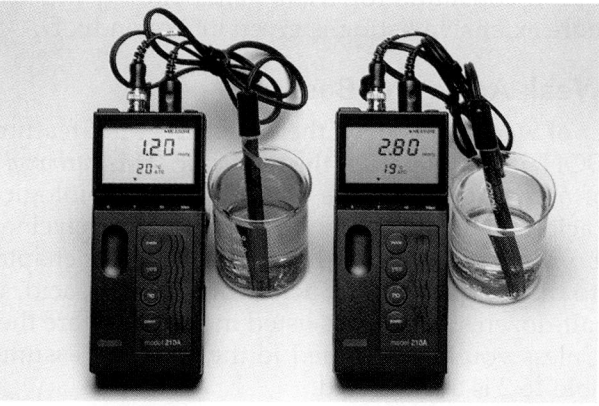

▲ FIGURE 16-7
Strong and weak acids compared
The color of thymol blue indicator, which is present in both solutions, depends on the pH of the solution.

pH < 1.2 < pH < 2.8 < pH
Red   Orange   Yellow

The principle of the pH meter is discussed in Section 20-4. (Left) 0.1 M HCl has pH ≈ 1. The pH meter shows a value of 1.20 rather than 1.00 because the molarity of the HCl solution is, in fact, slightly less than 0.1 M. (Right) 0.1 M CH_3COOH has pH ≈ 2.8.

---

*We have been writing ionization equations in the form: acid(1) + base(2) $\rightleftharpoons$ base(1) + acid(2). Here we have written acid(1) + base(2) $\rightleftharpoons$ acid(2) + base(1) to highlight H_3O^+, the species that is usually the subject of a calculation.

TABLE 16.3 Ionization Constants of Some Weak Acids and Weak Bases in Water at 25 °C

	Ionization Equilibrium	Ionization Constant K	pK
Acid		$K_a =$	p$K_a =$
Iodic acid	$HIO_3 + H_2O \rightleftharpoons H_3O^+ + IO_3^-$	1.6×10^{-1}	0.80
Chlorous acid	$HClO_2 + H_2O \rightleftharpoons H_3O^+ + ClO_2^-$	1.1×10^{-2}	1.96
Chloroacetic acid	$ClCH_2COOH + H_2O \rightleftharpoons H_3O^+ + ClCH_2COO^-$	1.4×10^{-3}	2.85
Nitrous acid	$HNO_2 + H_2O \rightleftharpoons H_3O^+ + NO_2^-$	7.2×10^{-4}	3.14
Hydrofluoric acid	$HF + H_2O \rightleftharpoons H_3O^+ + F^-$	6.6×10^{-4}	3.18
Formic acid	$HCOOH + H_2O \rightleftharpoons H_3O^+ + HCOO^-$	1.8×10^{-4}	3.74
Benzoic acid	$C_6H_5COOH + H_2O \rightleftharpoons H_3O^+ + C_6H_5COO^-$	6.3×10^{-5}	4.20
Hydrazoic acid	$HN_3 + H_2O \rightleftharpoons H_3O^+ + N_3^-$	1.9×10^{-5}	4.72
Acetic acid	$CH_3COOH + H_2O \rightleftharpoons H_3O^+ + CH_3COO^-$	1.8×10^{-5}	4.74
Hypochlorous acid	$HOCl + H_2O \rightleftharpoons H_3O^+ + OCl^-$	2.9×10^{-8}	7.54
Hydrocyanic acid	$HCN + H_2O \rightleftharpoons H_3O^+ + CN^-$	6.2×10^{-10}	9.21
Phenol	$HOC_6H_5 + H_2O \rightleftharpoons H_3O^+ + C_6H_5O^-$	1.0×10^{-10}	10.00
Hydrogen peroxide	$H_2O_2 + H_2O \rightleftharpoons H_3O^+ + HO_2^-$	1.8×10^{-12}	11.74
Base		$K_b =$	p$K_b =$
Diethylamine	$(C_2H_5)_2NH + H_2O \rightleftharpoons (C_2H_5)_2NH_2^+ + OH^-$	6.9×10^{-4}	3.16
Ethylamine	$C_2H_5NH_2 + H_2O \rightleftharpoons C_2H_5NH_3^+ + OH^-$	4.3×10^{-4}	3.37
Ammonia	$NH_3 + H_2O \rightleftharpoons NH_4^+ + OH^-$	1.8×10^{-5}	4.74
Hydroxylamine	$HONH_2 + H_2O \rightleftharpoons HONH_3^+ + OH^-$	9.1×10^{-9}	8.04
Pyridine	$C_5H_5N + H_2O \rightleftharpoons C_5H_5NH^+ + OH^-$	1.5×10^{-9}	8.82
Aniline	$C_6H_5NH_2 + H_2O \rightleftharpoons C_6H_5NH_3^+ + OH^-$	7.4×10^{-10}	9.13

Acid strength

Base strength

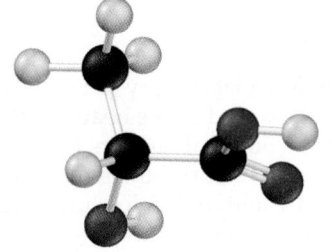

▲ Lactic acid, $CH_3CH(OH)COOH$.

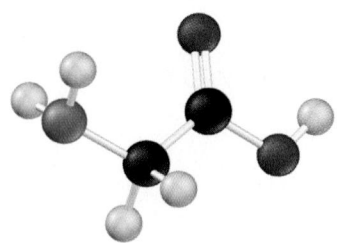

▲ Glycine, NH_2CH_2COOH.

A few ionization constants for weak acids and weak bases are listed in Table 16.3. A more extensive listing is given in Appendix D.

Identifying Weak Acids and Bases

A large number of weak acids have the same structural feature as acetic acid: that is, a —COOH group as part of the molecule. This *carboxyl group* is a common feature of many organic acids, including such biologically important acids as lactic acid and all the amino acids, including glycine. We will use a number of carboxylic acids as examples in this and later chapters.

In general, to distinguish a weak acid from a strong acid, you need recall only that the half-dozen strong acids listed in Table 16.2 are the most common strong acids. Unless you are informed to the contrary, assume that any acid not listed in Table 16.2 is a weak acid.

At first glance, weak bases seem more difficult to identify than weak acids: There is no distinctive element, such as H written first in the formula. Yet, if you study the bases in Table 16.3, you will see that all but one of them (pyridine) can be viewed as an ammonia molecule in which some other group (—C_6H_5, —C_2H_5, —OH, or —CH_3) has been substituted for one of the H atoms. The substitution of a methyl group, —CH_3, for a H atom is suggested in the following structural formulas.

$$H-\overset{\displaystyle |}{\underset{\displaystyle |}{N}}-H \qquad H-\overset{\displaystyle \overset{H}{|}}{\underset{\displaystyle \overset{|}{H}}{C}}-\overset{\displaystyle |}{\underset{\displaystyle |}{N}}-H$$

Ammonia   Methylamine

We can represent the ionization of methylamine in this way.

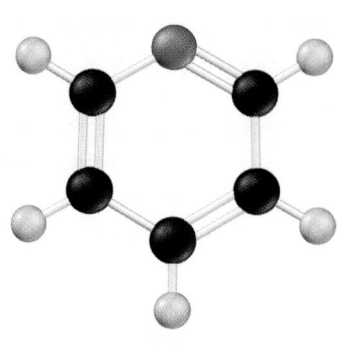

▲ Pyridine, C_5H_5N.

The ionization constant expression is

$$K_b = \frac{[CH_3NH_3^+][OH^-]}{[CH_3NH_2]} = 4.2 \times 10^{-4} \qquad \textbf{(16.13)}$$

Not all weak bases contain N. Yet, so many of them do that the similarity to NH_3 outlined here is well worth remembering. These weak bases derived from ammonia are known as *amines*.

🔍 16-3   CONCEPT ASSESSMENT

Is it possible for a weak acid solution to have a lower pH than a strong acid solution? If not, why not? If it is possible, under what conditions?

Illustrative Examples

For some students, solution equilibrium calculations are among the most challenging in general chemistry. At times, the difficulty is in sorting out what is relevant to a given problem. The number of types of calculations seems very large, although in fact it is quite limited. The key to solving solution equilibrium problems is to be able to imagine what is going on. Here are some questions to ask yourself.

- Which are the principal species in solution?
- What are the chemical reactions that produce them?
- Can some reactions (for example, the self-ionization of water) be ignored?
- Can you make any assumptions that allow you to simplify the equilibrium calculations?
- What is a reasonable answer to the problem? For instance, should the final solution be acidic (pH $<$ 7) or basic (pH $>$ 7)?

In short, first think through a problem *qualitatively*. At times, you may not even have to do a calculation. Next, organize the relevant data in a clear, logical manner. In this way, many problems that at first appear new to you will take on a familiar pattern. Look for other helpful hints as you proceed through this chapter and the following two chapters.

Example 16-6 presents a common problem involving weak acids and weak bases: calculating the pH of a solution of known molarity. The calculation invariably involves a quadratic equation, but very often we can make a simplifying assumption that leads to a shortcut that saves both time and effort.

EXAMPLE 16-5 Determining a Value of K_a from the pH of a Solution of a Weak Acid

Butyric acid, $CH_3(CH_2)_2COOH$, is used to make compounds employed in artificial flavorings and syrups. A 0.250 M aqueous solution of butyric acid is found to have a pH of 2.72. Determine K_a for butyric acid.

$$CH_3(CH_2)_2COOH + H_2O \rightleftharpoons H_3O^+ + CH_3(CH_2)_2COO^- \qquad K_a = ?$$

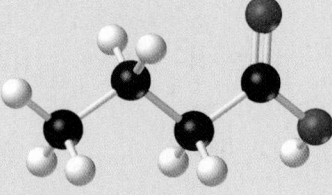

▲ Butyric acid, $CH_3CH_2CH_2COOH$.

Analyze

For $CH_3(CH_2)_2COOH$, K_a is likely to be much larger than K_w. Therefore, we can assume that self-ionization of water is unimportant and that ionization of the butyric acid is the only source of H_3O^+. Let's treat the situation as if $CH_3(CH_2)_2COOH$ first dissolves in molecular form, and then the molecules ionize until equilibrium is reached. That is, we write the balanced chemical equation and use it as the basis for an ICE table, as discussed in Chapter 15 (page 681). We will represent the concentrations of H_3O^+ and $CH_3(CH_2)_2COO^-$ at equilibrium as x M.

Solve

$$CH_3(CH_2)_2COOH + H_2O \rightleftharpoons H_3O^+ + CH_3(CH_2)_2COO^-$$

initial concns:	0.250 M	—	—
changes:	$-x$ M	$+x$ M	$+x$ M
equil concns:	$(0.250 - x)$ M	x M	x M

But x is known. It is the $[H_3O^+]$ in solution, which we can determine from the pH.

$$\log[H_3O^+] = -pH = -2.72$$
$$[H_3O^+] = 10^{-2.72} = 1.9 \times 10^{-3} = x$$

Now we can solve the following expression for K_a, substituting in the known value for x.

$$K_a = \frac{[H_3O^+][CH_3(CH_2)_2COO^-]}{[CH_3(CH_2)_2COOH]} = \frac{x \cdot x}{0.250 - x}$$
$$= \frac{(1.9 \times 10^{-3})(1.9 \times 10^{-3})}{0.250 - (1.9 \times 10^{-3})} = 1.5 \times 10^{-5}$$

Assess

Notice that our original assumption was correct: K_a is much larger than K_w.

PRACTICE EXAMPLE A: Hypochlorous acid, HOCl, is used in water treatment and as a disinfectant in swimming pools. A 0.150 M solution of HOCl has a pH of 4.18. Determine K_a for hypochlorous acid.

PRACTICE EXAMPLE B: The much-abused drug cocaine is an alkaloid. Alkaloids are noted for their bitter taste, an indication of their basic properties. Cocaine, $C_{17}H_{21}O_4N$, is soluble in water to the extent of 0.17 g/100 mL solution, and a saturated solution has a pH = 10.08. What is the value of K_b for cocaine?

$$C_{17}H_{21}O_4N + H_2O \rightleftharpoons C_{17}H_{21}O_4NH^+ + OH^- \qquad K_b = ?$$

EXAMPLE 16-6 Calculating the pH of a Weak Acid Solution

Show by calculation that the pH of 0.100 M CH_3COOH should be about the value shown on the pH meter in Figure 16-7; that is, pH ≈ 2.8.

Analyze

Here, we know that K_a is much larger than K_w. Let's again treat the situation as if CH_3COOH first dissolves in molecular form and then ionizes until equilibrium is reached. In this case, the quantity x is an unknown that must be obtained by an algebraic solution.

Solve

$$CH_3COOH + H_2O \rightleftharpoons H_3O^+ + CH_3COO^-$$

initial concns:	0.100 M	—	—
changes:	$-x$ M	$+x$ M	$+x$ M
equil concns:	$(0.100 - x)$ M	x M	x M

$$K_a = \frac{[H_3O^+][CH_3COO^-]}{[CH_3COOH]} = \frac{x \cdot x}{0.100 - x} = 1.8 \times 10^{-5}$$

We could solve this equation by using the quadratic formula, but let's instead make a simplifying assumption that is often valid. Assume that x is very small compared with 0.100. That is, assume that $(0.100 - x) \approx 0.100$.

$$x^2 = 0.100 \times 1.8 \times 10^{-5} = 1.8 \times 10^{-6}$$

$$x = [H_3O^+] = \sqrt{1.8 \times 10^{-6}} = 1.3 \times 10^{-3} \, M$$

Now, we must check our assumption: $0.100 - 0.0013 = 0.099 \approx 0.100$. Our assumption is good to about 1 part per 100 (1%) and is valid for a calculation involving two or three significant figures.
Finally,

$$pH = -\log[H_3O^+] = -\log(1.3 \times 10^{-3}) = -(-2.89) = 2.89$$

Assess

We observe that our answer is very close to the number on the pH meter. Therefore, our assumption to simplify the calculation was reasonable. This type of assumption may not always work and so we need to check the final answer until we are comfortable with knowing when and when not to apply the assumption to simplify the calculations.

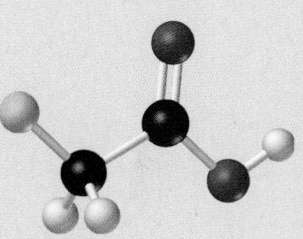

▲ Fluoroacetic acid, CH_2FCOOH.

PRACTICE EXAMPLE A: Substituting halogen atoms for hydrogen atoms bound to carbon increases the strength of carboxylic acids. Show that the pH of 0.100 M CH_2FCOOH, fluoroacetic acid, is lower than that calculated in Example 16-6 for 0.100 M CH_3COOH.

$$CH_2FCOOH + H_2O \rightleftharpoons H_3O^+ + CH_2FCOO^- \qquad K_a = 2.6 \times 10^{-3}$$

PRACTICE EXAMPLE B: Acetylsalicylic acid, $HC_9H_7O_4$, is the active component in aspirin. This acid is the cause of the stomach upset some people get when taking aspirin. Two extra-strength aspirin tablets, each containing 500 mg of acetylsalicylic acid, are dissolved in 325 mL of water. What is the pH of this solution?

$$HC_9H_7O_4 + H_2O \rightleftharpoons H_3O^+ + C_9H_7O_4^- \qquad K_a = 3.3 \times 10^{-4}$$

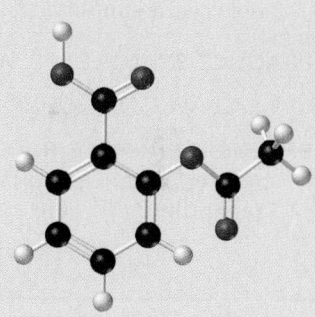

▲ Acetylsalicylic acid, $C_6H_4(OOCCH_3)COOH$.

EXAMPLE 16-7   **Dealing with the Failure of a Simplifying Assumption**

What is the pH of a solution that is 0.00250 M $CH_3NH_2(aq)$? For methylamine, $K_b = 4.2 \times 10^{-4}$.

Analyze

In this example we will apply the same techniques as we did in Example 16-6. We will work the problem twice to see that the simplifying assumption breaks down for weak acids and weak bases at very low concentrations.

(continued)

Solve

$$CH_3NH_2 + H_2O \rightleftharpoons CH_3NH_3^+ + OH^-$$

initial concns:	0.00250 M	—	—
changes:	$-x$ M	$+x$ M	$+x$ M
equil concns:	$(0.00250 - x)$ M	x M	x M

$$K_b = \frac{[CH_3NH_3^+][OH^-]}{[CH_3NH_2]} = \frac{x \cdot x}{0.00250 - x} = 4.2 \times 10^{-4}$$

Now let's assume that x is very much less than 0.00250 and that $0.00250 - x \approx 0.00250$.

$$\frac{x^2}{0.00250} = 4.2 \times 10^{-4} \qquad x^2 = 1.1 \times 10^{-6} \qquad [OH^-] = x = 1.0 \times 10^{-3}\,M$$

The value of x is nearly half as large as 0.00250—too large to ignore. This means using the *quadratic* formula.

$$\frac{x^2}{0.00250 - x} = 4.2 \times 10^{-4}$$

$$x^2 + (4.2 \times 10^{-4}x) - (1.1 \times 10^{-6}) = 0$$

$$x = \frac{(-4.2 \times 10^{-4}) \pm \sqrt{(4.2 \times 10^{-4})^2 + 4 \times 1.1 \times 10^{-6}}}{2}$$

$$x = [OH^-] = \frac{(-4.2 \times 10^{-4}) \pm (2.1 \times 10^{-3})}{2} = 8.4 \times 10^{-4}\,M$$

$$pOH = -\log[OH^-] = -\log(8.4 \times 10^{-4}) = 3.08$$

$$pH = 14.00 - pOH = 14.00 - 3.08 = 10.92$$

Assess

After applying the simplifying assumption, if the value of x is a significant percentage of the initial concentration (for example, greater than 5%), then we should not use the simplifying assumption to obtain the hydronium concentration.

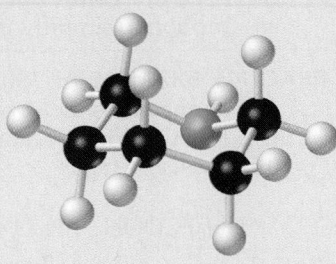

▲ Piperidine, $C_5H_{10}NH$.

PRACTICE EXAMPLE A: What is the pH of 0.015 M $CH_2FCOOH(aq)$?

$$CH_2FCOOH + H_2O \rightleftharpoons H_3O^+ + CH_2FCOO^- \qquad K_a = 2.6 \times 10^{-3}$$

PRACTICE EXAMPLE B: Piperidine is a base found in small amounts in black pepper. What is the pH of 315 mL of an aqueous solution containing 114 mg piperidine?

$$C_5H_{11}N + H_2O \rightleftharpoons C_5H_{11}NH^+ + OH^- \qquad K_b = 1.6 \times 10^{-3}$$

16-2 ARE YOU WONDERING...

How to calculate the pH of a very dilute solution of a weak acid?

Think of this as a companion question to the one posed in Are You Wondering 16-1, except here the acid in question is weak (represented as HA) rather than strong. The initial approach is similar: Write two equations representing the sources of H_3O^+, and indicate the concentrations of various species in the solution.

$$H_2O + H_2O \rightleftharpoons H_3O^+ + OH^-$$

Molarity: x x

$$HA + H_2O \rightleftharpoons H_3O^+ + A^-$$

Molarity: $M - y$ y y

Our ultimate objective is to determine $[H_3O^+]$ in the solution, and this is $x + y$. From $[H_3O^+]$, we can easily get the pH.

Our principal task is to solve a pair of equations simultaneously for x and y. The two equations are

$$K_w = [H_3O^+][OH^-] = (x + y) \times x = 1.0 \times 10^{-14}$$

$$K_a = \frac{[H_3O^+][A^-]}{[HA]} = \frac{(x + y) \times y}{(M - y)}$$

Note the following three points in these equations. (1) There can be only one value of $[H_3O^+]$ in the solution, and it is this value $(x + y)$ that appears in each equation. (2) The stoichiometric concentration of the acid is M (its molarity), and its *equilibrium* concentration is $[HA] = M - y$. The numerical value of M depends on the particular case considered. (3) Similarly, the numerical value of K_a depends on the particular case. When we solve the K_a expression for x, we obtain

$$x = \frac{K_a(M - y)}{y} - y \quad \text{and} \quad x + y = \frac{K_a(M - y)}{y}$$

And when we substitute these values of x and $x + y$ into the K_w equation, we get

$$K_w = \frac{K_a(M - y)}{y} \times \left(\frac{K_a(M - y)}{y} - y \right) = 1.0 \times 10^{-14}$$

The value of y satisfying this equation is not difficult to obtain by the method of successive approximations, illustrated in Appendix A.

The solution of this problem is left for you to do (see Exercise 89), but you will find that for 1.0×10^{-5} M HCN ($K_a = 6.2 \times 10^{-10}$), $y \cong 4.8 \times 10^{-8}$, $(x + y) \cong 1.3 \times 10^{-7}$, and pH $\cong 6.90$. This certainly seems like a reasonable pH for a very dilute solution of a weak acid in water—just below the neutral pH of 7.00.

Finally, we can use the results of the previous discussion to establish a criterion for ignoring the self-ionization of water in calculations. When we do this, we are assuming that $y \gg x$, so that $y \approx x + y$, and $yx \approx (x + y)x = [H_3O^+][OH^-] = K_w$. Also, if $y \gg x$, then $y^2 \gg K_w$. For what values of y can we say that $y \gg x$? Let's take the maximum value of x consistent with ignoring the self-ionization of water to be $1/100$ of y which, as shown below, means that $[H_3O^+]$ must be greater than 10^{-6} M.

$$y^2 > y \times x = y \times \frac{y}{100} = K_w = 1 \times 10^{-14}$$

$$y^2 > 1 \times 10^{-12} \quad \text{and} \quad y > 1 \times 10^{-6}$$

In acid–base calculations, we should first use the simplifying assumptions (ignore K_w and assume that [HA] can be replaced by the molarity of the acid). Then check the answer to see whether $[H_3O^+]$ is greater than 1×10^{-6} M and the 5% criterion presented below is also met.

More on Simplifying Assumptions

The usual simplifying assumption is that of treating a weak acid or weak base as though it remains essentially nonionized (so that $M - x \approx M$). In general, this assumption will work if the molarity of the weak acid, M_A, or that of the weak base, M_B, exceeds the value of K_a or K_b by a factor of at least 100. That is,

$$\frac{M_A \text{ (or } M_B)}{K_a \text{ (or } K_b)} > 100$$

◀ Be sure to maintain a clear distinction between the *property* (molarity, M) and the *unit* in which it is expressed, M = mol/L.

In any case, it is important to test the validity of any assumption that you make. If the assumption is good to within a few percent (say, less than 5%), then it is generally valid. In Example 16-6, the simplifying assumption was good to about 1%, but in Example 16-7 it was off by 40%.

Percent Ionization

The extent of ionization of a weak acid or weak base can be described in terms of the degree of ionization or percent ionization. It is convenient to introduce the generic symbol HA for any weak acid, and A⁻ as the conjugate base of the acid HA.

For the ionization $HA + H_2O \rightleftharpoons H_3O^+ + A^-$, the degree of ionization is the fraction of the acid molecules that ionize. Thus, if in 1.00 M HA, ionization produces $[H_3O^+] = [A^-] = 0.05$ M, the degree of ionization = 0.05 M/1.00 M = 0.05. **Percent ionization** gives the proportion of ionized molecules on a percentage basis.

$$\text{percent ionization} = \frac{\text{molarity of } H_3O^+ \text{ derived from HA}}{\text{initial molarity of HA}} \times 100\% \quad \textbf{(16.14)}$$

If, for example, the degree of ionization is 0.05, then the percent of ionization is 5%.

Figure 16-8 compares percent ionization and solution molarity for a weak acid and a strong acid. Example 16-8 shows by calculation that the percent ionization of a weak acid or a weak base *increases* as the solution becomes *more dilute*, a fact that we can also demonstrate by a simple analysis of the ionization reaction

$$HA + H_2O \rightleftharpoons H_3O^+ + A^-$$

At equilibrium, n_{HA} moles of the acid HA, $n_{H_3O^+}$ moles of H_3O^+, and n_{A^-} moles of A⁻ are present in a volume of V liters. The K_a expression is

$$K_a = \frac{[H_3O^+][A^-]}{[HA]} = \frac{(n_{H_3O^+}/V)(n_{A^-}/V)}{n_{HA}/V} = \frac{(n_{H_3O^+})(n_{A^-})}{n_{HA}} \times \frac{1}{V}$$

When we dilute the solution, V increases, 1/V decreases, and the ratio $(n_{H_3O^+})(n_{A^-})/n_{HA}$ must increase to maintain the constant value of K_a. In turn, $n_{H_3O^+}$ and n_{A^-} must increase and n_{HA} must decrease, signifying an increase in the percent ionization.

▶ FIGURE 16-8
Percent ionization of an acid as a function of concentration
Over the concentration range shown, HCl, a strong acid, is essentially 100% ionized. The percent ionization of CH_3COOH, a weak acid, increases from about 4% in 0.010 M to 20% in 3.6×10^{-5} M. For solutions of acetic acid more dilute than 3.6×10^{-5} M, the percent ionization rises sharply with increasing dilution. The pH of 1.0×10^{-7} M CH_3COOH is about 6.79, the same as for 1.0×10^{-7} M HCl.

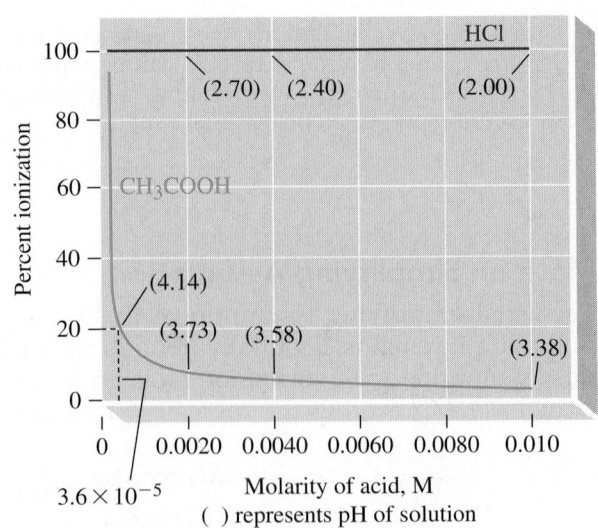

EXAMPLE 16-8   Determining Percent Ionization as a Function of Weak Acid Concentration

What is the percent ionization of acetic acid in 1.0 M, 0.10 M, and 0.010 M CH_3COOH?

Analyze

The percent ionization is determined by dividing the amount of ionized acid by the initial acid concentration and multiplying by 100%.

Solve

Use the ICE format to describe 1.0 M CH_3COOH:

$$CH_3COOH + H_2O \rightleftharpoons H_3O^+ + CH_3COO^-$$

initial concns:	1.0 M	—	—
changes:	$-x$ M	$+x$ M	$+x$ M
equil concns:	$(1.0 - x)$ M	x M	x M

We need to calculate $x = [H_3O^+] = [CH_3COO^-]$. In doing so, let's make the usual assumption: $1.0 - x \approx 1.0$.

$$K_a = \frac{[H_3O^+][CH_3COO^-]}{[CH_3COOH]} = \frac{x \cdot x}{1.0 - x} = \frac{x^2}{1.0} = 1.8 \times 10^{-5}$$

$$x = [H_3O^+] = [C_2H_3O_2^-] = \sqrt{1.8 \times 10^{-5}} = 4.2 \times 10^{-3} \, M$$

The percent ionization of 1.0 M $HC_2H_3O_2$ is

$$\% \text{ ionization} = \frac{[H_3O^+]}{[CH_3COOH]} \times 100\% = \frac{4.2 \times 10^{-3} \, M}{1.0 \, M} \times 100\% = 0.42\%$$

The assumption that x is small compared to 1.0 is clearly valid: x is only 0.42% of 1.0 M. The calculations for 0.10 M CH_3COOH and 0.010 M CH_3COOH are very similar. In 0.10 M CH_3COOH, 1.3% of the acetic acid molecules are ionized and in 0.010 M CH_3COOH, 4.2% are ionized.

Assess

The purpose of calculating the percent ionization for three acetic acid solutions was to confirm the very important point made on page 716: For a weak acid, percent ionization increases with increasing dilution (Fig. 16-8). For very dilute solutions, the calculation of percent ionization is more complicated. (See Are You Wondering 16-2 on page 714.)

PRACTICE EXAMPLE A: What is the percent ionization of hydrofluoric acid in 0.20 M HF and in 0.020 M HF?

PRACTICE EXAMPLE B: An 0.0284 M aqueous solution of lactic acid, a carboxylic acid that accumulates in the blood and muscles during physical activity, is found to be 6.7% ionized. Determine K_a for lactic acid.

$$CH_3CH(OH)COOH + H_2O \rightleftharpoons H_3O^+ + CH_3CH(OH)COO^- \qquad K_a = ?$$

🔍 16-4   CONCEPT ASSESSMENT

You are given two bottles, each of which contains a 0.1 M solution of an unidentified acid. One bottle is labeled $K_a = 7.2 \times 10^{-4}$, and the other is labeled $K_a = 1.9 \times 10^{-5}$. Which bottle contains the more acidic solution? Which bottle has the acid with the larger pK_a?

▲ Phosphoric acid, H_3PO_4.

16-6   Polyprotic Acids

All the acids listed in Table 16.3 are weak *monoprotic acids*, meaning that their molecules have only one ionizable H atom, even though several of these acids contain more than one H atom. But some acids have more than one ionizable H atom per molecule. These are **polyprotic acids**. Table 16.4 lists ionization constants for several polyprotic acids. Additional listings can be found in Appendix D. We will focus on phosphoric acid, H_3PO_4.

◀ Phosphoric acid ranks second only to sulfuric acid among the important commercial acids. It is used in the manufacture of phosphate fertilizers. Various sodium, potassium, and calcium phosphates are used in the food industry.

TABLE 16.4 Ionization Constants of Some Polyprotic Acids

Acid	Ionization Equilibria	Ionization Constants, K	pK
Hydrosulfuric[a]	$H_2S + H_2O \rightleftharpoons H_3O^+ + HS^-$	$K_{a_1} = 1.0 \times 10^{-7}$	p$K_{a_1} = 7.00$
	$HS^- + H_2O \rightleftharpoons H_3O^+ + S^{2-}$	$K_{a_2} = 1 \times 10^{-19}$	p$K_{a_2} = 19.0$
Carbonic[b]	$H_2CO_3 + H_2O \rightleftharpoons H_3O^+ + HCO_3^-$	$K_{a_1} = 4.4 \times 10^{-7}$	p$K_{a_1} = 6.36$
	$HCO_3^- + H_2O \rightleftharpoons H_3O^+ + CO_3^{2-}$	$K_{a_2} = 4.7 \times 10^{-11}$	p$K_{a_2} = 10.33$
Citric	$H_3C_6H_5O_7 + H_2O \rightleftharpoons H_3O^+ + H_2C_6H_5O_7^-$	$K_{a_1} = 7.5 \times 10^{-4}$	p$K_{a_1} = 3.12$
	$H_2C_6H_5O_7^- + H_2O \rightleftharpoons H_3O^+ + HC_6H_5O_7^{2-}$	$K_{a_2} = 1.7 \times 10^{-5}$	p$K_{a_2} = 4.77$
	$HC_6H_5O_7^{2-} + H_2O \rightleftharpoons H_3O^+ + C_6H_5O_7^{3-}$	$K_{a_3} = 4.0 \times 10^{-7}$	p$K_{a_3} = 6.40$
Phosphoric	$H_3PO_4 + H_2O \rightleftharpoons H_3O^+ + H_2PO_4^-$	$K_{a_1} = 7.1 \times 10^{-3}$	p$K_{a_1} = 2.15$
	$H_2PO_4^- + H_2O \rightleftharpoons H_3O^+ + HPO_4^{2-}$	$K_{a_2} = 6.3 \times 10^{-8}$	p$K_{a_2} = 7.20$
	$HPO_4^{2-} + H_2O \rightleftharpoons H_3O^+ + PO_4^{3-}$	$K_{a_3} = 4.2 \times 10^{-13}$	p$K_{a_3} = 12.38$
Oxalic	$H_2C_2O_4 + H_2O \rightleftharpoons H_3O^+ + HC_2O_4^-$	$K_{a_1} = 5.6 \times 10^{-2}$	p$K_{a_1} = 1.25$
	$HC_2O_4^- + H_2O \rightleftharpoons H_3O^+ + C_2O_4^{2-}$	$K_{a_2} = 5.4 \times 10^{-5}$	p$K_{a_2} = 4.27$
Sulfurous[c]	$H_2SO_3 + H_2O \rightleftharpoons H_3O^+ + HSO_3^-$	$K_{a_1} = 1.3 \times 10^{-2}$	p$K_{a_1} = 1.89$
	$HSO_3^- + H_2O \rightleftharpoons H_3O^+ + SO_3^{2-}$	$K_{a_2} = 6.2 \times 10^{-8}$	p$K_{a_2} = 7.21$
Sulfuric[d]	$H_2SO_4 + H_2O \rightleftharpoons H_3O^+ + HSO_4^-$	K_{a_1} = very large	p$K_{a_1} < 0$
	$HSO_4^- + H_2O \rightleftharpoons H_3O^+ + SO_4^{2-}$	$K_{a_2} = 1.1 \times 10^{-2}$	p$K_{a_2} = 1.96$

Acid strength →

[a]The value for K_{a_2} of H_2S most commonly found in older literature is about 1×10^{-14}, but current evidence suggests that the value is considerably smaller.

[b]H_2CO_3 cannot be isolated. It is in equilibrium with H_2O and dissolved CO_2. The value given for K_{a_1} is actually for the reaction

$$CO_2(aq) + 2\,H_2O \rightleftharpoons H_3O^+ + HCO_3^-$$

Generally, aqueous solutions of CO_2 are treated *as if* the $CO_2(aq)$ were first converted to H_2CO_3, followed by ionization of the H_2CO_3.

[c]H_2SO_3 is a hypothetical, nonisolatable species. The value listed for K_{a_1} is actually for the reaction

$$SO_2(aq) + 2\,H_2O \rightleftharpoons H_3O^+ + HSO_3^-$$

[d]H_2SO_4 is completely ionized in the first step.

The H_3PO_4 molecule has *three* ionizable H atoms; it is a *triprotic acid*. It ionizes in three steps. For each step, we can write an ionization equation and an acid ionization constant expression with a distinctive value of K_a.

▶ A common error is to assume that the $[H_3O^+]$ is the same for all three ionizations. In fact, the pH is dominated by only the first ionization as seen from the different K_a values here. This is only true, of course, for weak acids. The examples here illustrate the points well.

(1) $H_3PO_4 + H_2O \rightleftharpoons H_3O^+ + H_2PO_4^-$ $K_{a_1} = \dfrac{[H_3O^+][H_2PO_4^-]}{[H_3PO_4]} = 7.1 \times 10^{-3}$

(2) $H_2PO_4^- + H_2O \rightleftharpoons H_3O^+ + HPO_4^{2-}$ $K_{a_2} = \dfrac{[H_3O^+][HPO_4^{2-}]}{[H_2PO_4^-]} = 6.3 \times 10^{-8}$

(3) $HPO_4^{2-} + H_2O \rightleftharpoons H_3O^+ + PO_4^{3-}$ $K_{a_3} = \dfrac{[H_3O^+][PO_4^{3-}]}{[HPO_4^{2-}]} = 4.2 \times 10^{-13}$

There is a ready explanation for the relative magnitudes of the ionization constants—that is, for the fact that $K_{a_1} > K_{a_2} > K_{a_3}$. When ionization occurs in step (1), a proton (H^+) moves away from an ion with a 1− charge ($H_2PO_4^-$). In step (2), the proton moves away from an ion with a 2− charge (HPO_4^{2-}), a more difficult separation. As a result, the ionization constant in the second step is smaller than that in the first. Ionization is more difficult still in step (3).

We can make three key statements about the ionization of phosphoric acid, as illustrated in Example 16-9.

Phosphoric acid

1. K_{a_1} is so much larger than K_{a_2} and K_{a_3} that essentially all the H_3O^+ is produced in the first ionization step.

2. So little of the $H_2PO_4^-$ forming in the first ionization step ionizes any further that we can assume $[H_2PO_4^-] = [H_3O^+]$ in the solution.

3. $[HPO_4^{2-}] \approx K_{a_2}$, regardless of the molarity of the acid.*

   Although statement (1) seems essential if statements (2) and (3) are to be valid, it is not as crucial as might first appear. Even for polyprotic acids whose K_a values do not differ greatly between successive ionizations, H_3O^+ is often still determined almost exclusively by the K_{a_1} expression, and statements (2) and (3) remain valid. As long as the polyprotic acid is weak in its first ionization step, the concentration of the anion produced in this step will be so much less than the molarity of the acid that additional $[H_3O^+]$ produced in the second ionization remains negligible.

EXAMPLE 16-9   Calculating Ion Concentrations in a Polyprotic Acid Solution

For a 3.0 M H_3PO_4 solution, calculate **(a)** $[H_3O^+]$; **(b)** $[H_2PO_4^-]$; **(c)** $[HPO_4^{2-}]$; and **(d)** $[PO_4^{3-}]$.

Analyze

For a solution of a *weak* polyprotic acid, the first ionization step produces essentially all the H_3O^+ in solution, so we begin as we would for a weak monoprotic acid solution. The concentrations of the other species are obtained by using the expressions for K_{a2} and K_{a3}.

Solve

(a) For the reasons discussed above, let's assume that all the H_3O^+ forms in the *first* ionization step. This is equivalent to thinking of H_3PO_4 as though it were a monoprotic acid, ionizing only in the first step.

$$H_3PO_4 + H_2O \rightleftharpoons H_3O^+ + H_2PO_4^-$$

initial concns:	3.0 M	—	—
changes:	$-x$ M	$+x$ M	$+x$ M
after first ionization:	$(3.0 - x)$ M	x M	x M

Following the usual assumption that x is much smaller than 3.0 and that $3.0 - x \approx 3.0$, we obtain

$$K_{a_1} = \frac{[H_3O^+][H_2PO_4^-]}{[H_3PO_4]} = \frac{x \cdot x}{(3.0 - x)} = \frac{x^2}{3.0} = 7.1 \times 10^{-3}$$

$$x^2 = 0.021 \qquad x = [H_3O^+] = 0.14 \text{ M}$$

In the assumption $3.0 - x \approx 3.0$, $x = 0.14$, which is 4.7% of 3.0. This is about the maximum error that can be tolerated for an acceptable assumption.

(b) From part (a), $x = [H_2PO_4^-] = [H_3O^+] = 0.14$ M.

(c) To determine $[H_3O^+]$ and $[H_2PO_4^-]$, we assumed that the second ionization is insignificant. Here we must consider the second ionization, no matter how slight; otherwise we would have no source of the ion HPO_4^{2-}. We can represent the second ionization, as shown in the following table. *Note especially how the results of the first ionization enter in.* We start with a solution in which $[H_2PO_4^-] = [H_3O^+] = 0.14$ M.

$$H_2PO_4^- + H_2O \rightleftharpoons H_3O^+ + HPO_4^{2-}$$

from first ionization:	0.14 M	0.14 M	—
changes:	$-y$ M	$+y$ M	$+y$ M
after second ionization:	$(0.14 - y)$ M	$(0.14 + y)$ M	y M

If we assume that y is much smaller than 0.14, then $(0.14 + y) \approx (0.14 - y) \approx 0.14$.

(continued)

---

*If we assume that $[H_2PO_4^-] = [H_3O^+]$, the second ionization expression reduces to

$$\frac{[H_3O^+][HPO_4^{2-}]}{[H_2PO_4^-]} = K_{a_2}$$

We then get

$$K_{a_2} = \frac{[H_3O^+][HPO_4^{2-}]}{[H_2PO_4^-]} = \frac{(0.14 + y)(y)}{(0.14 - y)} = \frac{(0.14)(y)}{(0.14)} = 6.3 \times 10^{-8}$$

$$y = [HPO_4^{2-}] = 6.3 \times 10^{-8} \, M$$

Note that the assumption is valid.

(d) The PO_4^{3-} ion forms only in the third ionization step. When we write this acid ionization constant expression, we see that we have already calculated the ion concentrations other than $[PO_4^{3-}]$. We can simply solve the K_{a_3} expression for $[PO_4^{3-}]$.

$$K_{a_3} = \frac{[H_3O^+][PO_4^{3-}]}{[HPO_4^{2-}]} = \frac{0.14 \times [PO_4^{3-}]}{6.3 \times 10^{-8}} = 4.2 \times 10^{-13}$$

$$[PO_4^{3-}] = \frac{4.2 \times 10^{-13} \times 6.3 \times 10^{-8}}{0.14} = 1.9 \times 10^{-19} \, M$$

Assess

In this example the major source of hydronium ions is from the first ionization step. In the second step the amount of hydronium ions is around 10^{-8} M, which is negligible compared with 0.14 M.

PRACTICE EXAMPLE A: Malonic acid, $HOOCCH_2COOH$, is a diprotic acid used in the manufacture of barbiturates.

$$HOOCCH_2COOH + H_2O \rightleftharpoons H_3O^+ + HOOCCH_2COO^- \qquad K_{a_1} = 1.4 \times 10^{-3}$$
$$HOOCCH_2COO^- + H_2O \rightleftharpoons H_3O^+ + {}^-OOCCH_2COO^- \qquad K_{a_2} = 2.0 \times 10^{-6}$$

Calculate $[H_3O^+]$, $[HOOCCH_2COO^-]$, and $[{}^-OOCCH_2COO^-]$ in a 1.00 M solution of malonic acid.

PRACTICE EXAMPLE B: Oxalic acid, found in the leaves of rhubarb and other plants, is a diprotic acid.

$$H_2C_2O_4 + H_2O \rightleftharpoons H_3O^+ + HC_2O_4^- \qquad K_{a_1} = ?$$
$$HC_2O_4^- + H_2O \rightleftharpoons H_3O^+ + C_2O_4^{2-} \qquad K_{a_2} = ?$$

An aqueous solution that is 1.05 M $H_2C_2O_4$ has pH = 0.67. The free oxalate ion concentration in this solution is $[C_2O_4^{2-}] = 5.3 \times 10^{-5}$ M. Determine K_{a_1} and K_{a_2} for oxalic acid.

▲ Sulfuric acid, H_2SO_4.

A Somewhat Different Case: H_2SO_4

Sulfuric acid differs from most polyprotic acids in this important respect: It is a *strong* acid in its first ionization and a *weak* acid in its second. Ionization is complete in the first step, which means that in most $H_2SO_4(aq)$ solutions, $[H_2SO_4] \approx 0$ M. Thus, if a solution is 0.50 M H_2SO_4, we can treat it as though it were 0.50 M H_3O^+ and 0.50 M HSO_4^- initially. Then we can determine the extent to which ionization of HSO_4^- produces additional H_3O^+ and SO_4^{2-}, as illustrated in Example 16-10.

EXAMPLE 16-10 **Calculating Ion Concentrations in Sulfuric Acid Solutions: Strong Acid Ionization Followed by Weak Acid Ionization**

Calculate $[H_3O^+]$, $[HSO_4^-]$, and $[SO_4^{2-}]$ in 0.50 M H_2SO_4.

Analyze

We will modify the approach we used in Example 16-9 to incorporate the fact that for H_2SO_4 the first ionization step goes to completion.

Solve

$$H_2SO_4 + H_2O \longrightarrow H_3O^+ + HSO_4^-$$

initial concn:	0.50 M	—	—
changes:	−0.50 M	+0.50 M	+0.50 M
after first ionization:	≈ 0	0.50 M	0.50 M

$$HSO_4^- + H_2O \rightleftharpoons H_3O^+ + SO_4^{2-}$$

from first ionization:	0.50 M	0.50 M	—
changes:	−x M	+x M	+x M
after second ionization:	(0.50 − x) M	(0.50 + x) M	x M

We need to deal only with the ionization constant expression for K_{a_2}. If we assume that x is much smaller than 0.50, then $(0.50 + x) \approx (0.50 - x) \approx 0.50$ and

$$K_{a_2} = \frac{[H_3O^+][SO_4^{2-}]}{[HSO_4^-]} = \frac{(0.50 + x) \cdot x}{(0.50 - x)} = \frac{0.50 \cdot x}{0.50} = 1.1 \times 10^{-2}$$

Our results, then, are

$$[H_3O^+] = 0.50 + x = 0.51 \text{ M}; \qquad [HSO_4^-] = 0.50 - x = 0.49 \text{ M}$$

$$[SO_4^{2-}] = x = K_{a_2} = 0.011 \text{ M}$$

Assess

In obtaining these results, we assumed that x was much smaller than 0.50. This assumption is appropriate because $x = 0.011$ is only 2.2% of 0.50. Had x been greater than 5% of 0.50, then the assumption would not have been appropriate. Such a situation arises when dealing with more dilute solutions of H_2SO_4.

PRACTICE EXAMPLE A: Calculate $[H_3O^+]$, $[HSO_4^-]$, and $[SO_4^{2-}]$ in 0.20 M H_2SO_4.

PRACTICE EXAMPLE B: Calculate $[H_3O^+]$, $[HSO_4^-]$, and $[SO_4^{2-}]$ in 0.020 M H_2SO_4.

[*Hint:* Is the assumption that $[HSO_4^-] = [H_3O^+]$ valid?]

A General Approach to Solution Equilibrium Calculations

Suppose we were required to determine the stoichiometric molarity of the $H_2SO_4(aq)$ that is required to produce a solution with pH = 2.15. We could start by determining the hydronium ion concentration in the solution: $[H_3O^+] = 10^{-pH} = 10^{-2.15} = 7.1 \times 10^{-3}$ M. What next? We cannot follow the approach used in Example 16-6 because H_2SO_4 is not a monoprotic acid. Instead, we have to use a method like that outlined in Are You Wondering 16-1 and 16-2. An alternative approach worth considering has the appeal of getting us on the right track for all kinds of solution equilibrium calculations. That method has the following format.

1. Identify the species present in any significant amount in the solution (excluding H_2O molecules). Consider the concentrations of these species as unknowns.

2. Write equations that include these species. The number of equations involving these species should match the number of unknowns. The equations are of three types.

   (a) equilibrium constant expressions

   (b) material balance equations

   (c) an electroneutrality condition

3. Solve the system of equations for the unknowns.

Let's apply this approach to the $H_2SO_4(aq)$ solution mentioned in the first sentence of this section.

Possible Species:

$$H_2SO_4, H_3O^+, HSO_4^-, SO_4^{2-}, OH^-$$

We can eliminate H_2SO_4 because its ionization goes to completion in the first step. We can also eliminate OH^- because $[OH^-]$ is exceedingly small in an *acidic* solution that has a pH = 2.15.

Unknowns:

$$[H_3O^+], [HSO_4^-], [SO_4^{2-}], \text{ and } M \text{ [the molarity of the } H_2SO_4(aq)]$$

▶ Calculations like this are not as complicated when performed for other polyprotic acids (phosphoric acid, carbonic acid, etc.) because essentially all the hydronium ions are produced in the first ionization step.

We can eliminate $[H_3O^+]$ because we essentially know its value from the outset. A pH = 2.15 corresponds to $[H_3O^+]$ = 0.0071 M. Thus, we are left with three unknowns, and we need three equations.

Equations:

(a) The K_a expression for the ionization $HSO_4^- + H_2O \rightleftharpoons H_3O^+ + SO_4^{2-}$ is

$$K_{a_2} = \frac{[H_3O^+][SO_4^{2-}]}{[HSO_4^-]} = 1.1 \times 10^{-2}$$

(b) The following material balance equation accounts for the fact that the sum of the concentrations of the sulfur-containing species must equal the stoichiometric molarity of the $H_2SO_4(aq)$.

$$[HSO_4^-] + [SO_4^{2-}] = M$$

(c) The electroneutrality condition simply verifies that the solution carries no net charge. The sum of the positive charges must equal the sum of the negative charges. We can sum these charges on a mol/liter basis. For example, because there is one positive charge for every H_3O^+ ion, the number of moles per liter of positive charge is the same as the number of moles per liter of H_3O^+, 0.0071 M. We multiply $[SO_4^{2-}]$ by *two* because each SO_4^{2-} ion carries two units of negative charge.

$$[H_3O^+] = [HSO_4^-] + (2 \times [SO_4^{2-}]) = 0.0071$$

Solving the Equations:

Solve equation (c) for $[HSO_4^-]$: $[HSO_4^-] = 0.0071 - 2[SO_4^{2-}]$. Substitute this result, together with $[H_3O^+]$ = 0.0071, into equation (a) and obtain the expression

$$\frac{0.0071 \times [SO_4^{2-}]}{0.0071 - 2[SO_4^{2-}]} = 1.1 \times 10^{-2}$$

KEEP IN MIND

that although this method gives you a quick way to set up a solution equilibrium calculation, more information may be needed for you to arrive at an answer without undue effort. For example, answers to the general questions posed on page 711 may point the way to simplifying the algebraic solution.

Solve this equation to find that $[SO_4^{2-}]$ = 0.0027 M. Then substitute this result into equation (c) to obtain $[HSO_4^-]$ = 0.0017 M. Finally, according to equation (b), $[HSO_4^-] + [SO_4^{2-}]$ = 0.0044 M. The required molarity is 0.0044 M H_2SO_4.

Check:

There are usually ways to check the reasonableness of an answer obtained by this method. In this case, we can easily determine the possible pH range for 0.0044 M H_2SO_4. If the acid ionized only in the first step, $[H_3O^+]$ = 0.0044 M (pH = 2.36); if the second ionization step also went to completion, $[H_3O^+]$ = 0.0088 M (pH = 2.06). The observed pH, 2.15, falls squarely in this range.

The alternative method outlined here is ideal for computerized calculation. Moreover, because the additional manipulations required to convert stoichiometric concentrations to activities can be incorporated into the calculations, the solutions obtained are generally both more accurate and more readily obtained than are those derived by traditional methods.

16-7   Ions as Acids and Bases

In our discussion to this point, we have emphasized the behavior of electrically neutral molecules as acids (for example, HCl, CH_3COOH, H_3PO_4) or as bases (for example, NH_3, CH_3NH_2). We have also seen, however, that ions can act as acids or bases. For instance, in the second ionization step of H_3PO_4 (part c of Example 16-9), the $H_2PO_4^-$ ion acts as an acid.

Let's think about how each of the following can be described as an acid–base reaction.

$$\underset{\text{Acid(1)}}{NH_4^+} + \underset{\text{Base(2)}}{H_2O} \rightleftharpoons \underset{\text{Base(1)}}{NH_3} + \underset{\text{Acid(2)}}{H_3O^+} \qquad \textbf{(16.15)}$$

$$\underset{\text{Base(1)}}{CH_3COO^-} + \underset{\text{Acid(2)}}{H_2O} \rightleftharpoons \underset{\text{Acid(1)}}{CH_3COOH} + \underset{\text{Base(2)}}{OH^-} \qquad \textbf{(16.16)}$$

In reaction (16.15), NH_4^+ is an *acid*, giving up a proton to water, a *base*. Equilibrium in this reaction is described by means of the *acid ionization constant* of the ammonium ion, NH_4^+.

$$K_a = \frac{[NH_3][H_3O^+]}{[NH_4^+]} = ? \qquad \textbf{(16.17)}$$

Two of the concentrations in equation (16.17)—$[NH_3]$ and $[NH_4^+]$—are the same as in the K_b expression for NH_3, the conjugate base of NH_4^+. It seems that K_a for NH_4^+ and K_b for NH_3 should bear some relationship to each other, and they do. The easiest way to see this is to multiply both the numerator and the denominator of (16.17) by $[OH^-]$. The product $[H_3O^+] \times [OH^-]$ is the ion product of water, K_w, shown in red. The other concentrations, shown in blue, represent the *inverse* of K_b for NH_3. The value obtained, 5.6×10^{-10}, is the missing value of K_a in expression (16.17).

$$K_a = \frac{[NH_3][H_3O^+][OH^-]}{[NH_4^+][OH^-]} = \frac{K_w}{K_b} = \frac{1.0 \times 10^{-14}}{1.8 \times 10^{-5}} = 5.6 \times 10^{-10}$$

This result is an important consequence of the Brønsted–Lowry theory.

> The product of the ionization constants of an acid and its conjugate base equals the ion product of water.
>
> $$K_a \text{ (acid)} \times K_b \text{ (its conjugate base)} = K_w$$
> $$K_b \text{ (base)} \times K_a \text{ (its conjugate acid)} = K_w \qquad \textbf{(16.18)}$$

◀ In many tabulations of ionization constants, only K_a values are listed, whether for neutral molecules or for ions. Equation (16.18) can be used to obtain the values of their conjugates.

In reaction (16.16), CH_3COO^- acts as a *base* by taking a proton from water, an *acid*. Here, equilibrium is described by means of the *base ionization constant* of the acetate ion, CH_3COO^-. With expression (16.18) we can evaluate K_b.

$$K_b = \frac{[CH_3COOH][OH^-]}{[CH_3COO^-]} = \frac{K_w}{K_a(CH_3COOH)} = \frac{1.0 \times 10^{-14}}{1.8 \times 10^{-5}} = 5.6 \times 10^{-10}$$

From equation (16.18), we deduce that (1) the stronger the acid, the weaker its conjugate base; and (2) the weaker the acid, the stronger its conjugate base. It is easy to misinterpret the second statement. It does *not* mean that the conjugate base of a weak acid is a strong base. When we compare the values of the ionization constants for CH_3COOH and CH_3COO^-, it is clear that *the conjugate base of a weak acid is a weak base*. It is also true that the conjugate acid of a weak base is a weak acid. The following statement summarizes these relationships.

> The conjugate of *weak* is *weak*.

A handbook that lists only pK_a values for weak electrolytes has the following entry for 1,2-ethanediamine, $NH_2CH_2CH_2NH_2$: $pK_1 = 6.85(+2)$; $pK_2 = 9.92(+1)$, and 2-aminopropanoic acid, $NH_2CH(CH_3)COOH$: $pK_1 = 2.34(+1)$; $pK_2 = 9.87(0)$. Interpret these handbook entries by writing equations for the ionization reactions to which these pK values apply. What are the corresponding values of the base ionization constants K_{b_1} and K_{b_2}?

Hydrolysis

In pure water at 25 °C, $[H_3O^+] = [OH^-] = 1.0 \times 10^{-7}$ M and pH = 7.00. *Pure water is pH neutral.* When NaCl dissolves in water at 25 °C, complete dissociation into Na^+ and Cl^- ions occurs, and the pH of the solution remains 7.00. We can represent this fact with the equation

$$Na^+ + Cl^- + H_2O \longrightarrow \text{no reaction}$$

As shown in Figure 16-9, when NH_4Cl is added to water, the pH falls below 7. This means that $[H_3O^+] > [OH^-]$ in the solution. A reaction producing H_3O^+ must occur.

$$Cl^- + H_2O \longrightarrow \text{no reaction}$$
$$NH_4^+ + H_2O \rightleftharpoons NH_3 + H_3O^+$$

KEEP IN MIND

that many students find hydrolysis problems challenging. The equilibrium calculations are actually quite straightforward. The challenging aspect of these problems is recognizing *when* a hydrolysis reaction is the one on which to base the calculations.

The reaction between NH_4^+ and H_2O is fundamentally no different from other acid–base reactions. A reaction between an ion and water, however, is often called a **hydrolysis** reaction. We say that ammonium ion *hydrolyzes* (and chloride ion does not).

When sodium acetate is dissolved in water, the pH rises above 7 (see Figure 16-9). This means that $[OH^-] > [H_3O^+]$ in the solution. Here, acetate ion hydrolyzes.

$$Na^+ + H_2O \longrightarrow \text{no reaction}$$
$$CH_3COO^- + H_2O \rightleftharpoons CH_3COOH + OH^-$$

▲ FIGURE 16-9
Ions as acids and bases
Each of these 1 M solutions contains bromthymol blue indicator, which has the following colors:

pH < 7	pH = 7	pH > 7
Yellow	Green	Blue

(Left) $NH_4Cl(aq)$ is acidic. (Center) NaCl(aq) is neutral. (Right) $NaCH_3COO(aq)$ is basic.

The pH of Salt Solutions

We are now in a position to make both qualitative predictions and quantitative calculations concerning the pH values of aqueous solutions of salts. Whichever of these tasks is called for, note that hydrolysis takes place only if there is a chemical reaction producing a weak acid or weak base. The following generalizations are useful.

- Salts of strong bases and strong acids (for example, NaCl) *do not hydrolyze*: for the solution, pH = 7.
- Salts of strong bases and *weak* acids (for example, $NaCH_3COO$) *hydrolyze*: pH > 7. (The *anion* acts as a *base*.)
- Salts of *weak* bases and strong acids (for example, NH_4Cl) *hydrolyze*: pH < 7. (The *cation* acts as an *acid*.)
- Salts of *weak* bases and *weak* acids (for example, NH_4CH_3COO) *hydrolyze*. (The cations are acids, and the anions are bases. Whether the solution is acidic or basic, however, depends on the relative values of K_a and K_b for the ions.)

EXAMPLE 16-11   Making Qualitative Predictions About Hydrolysis Reactions

Predict whether each of the following solutions is acidic, basic, or pH neutral: **(a)** NaOCl(aq); **(b)** KCl(aq); **(c)** NH_4NO_3(aq).

Analyze

We need to recognize that all three salts are strong electrolytes and completely dissociate in water. Then, we can consider the ions separately and ask which will react (either as an acid or as a base) with water. Recall that the anions from strong acids (e.g., Cl^-) and the cations from strong bases (e.g., Na^+) do not participate in hydrolysis.

Solve

(a) The ions present are Na^+, which does not hydrolyze, and OCl^-, which does. OCl^- is the conjugate base of HOCl and forms a basic solution.

$$OCl^- + H_2O \rightleftharpoons HOCl + OH^-$$

(b) Neither K^+ nor Cl^- hydrolyzes. KCl(aq) is neutral—that is, pH = 7.

(c) NH_4^+ hydrolyzes, but NO_3^- does not (HNO_3 is a strong acid).

$$NH_4^+ + H_2O \rightleftharpoons NH_3 + H_3O^+$$

In this case, $[H_3O^+] > [OH^-]$, and the solution is acidic.

Assess

Recognizing that certain ions in solution can undergo hydrolysis in water will be an important concept in the next chapter. It is important to learn and understand here how this concept works.

PRACTICE EXAMPLE A: Predict whether each of the following 1.0 M solutions is acidic, basic, or pH neutral: **(a)** $CH_3NH_3^+NO_3^-$(aq); **(b)** NaI(aq); **(c)** $NaNO_2$(aq).

PRACTICE EXAMPLE B: An aqueous solution containing $H_2PO_4^-$ has a pH of about 4.7. Write equations for *two* reactions of $H_2PO_4^-$ with water, and explain which reaction occurs to the greater extent.

16-6   CONCEPT ASSESSMENT

Write a chemical equation showing how an HPO_4^{2-} ion can act as both an acid and a base in aqueous solution. Without doing any pH calculations, determine whether 0.10 M Na_2HPO_4 is acidic, basic, or pH neutral. What about 0.10 M NaH_2PO_4?

EXAMPLE 16-12 Evaluating Ionization Constants for Hydrolysis Reactions

Both sodium nitrite, $NaNO_2$, and sodium benzoate, NaC_6H_5COO, are used as food preservatives. If separate solutions of these two salts have the same molarity, which solution will have the *higher* pH?

Analyze

Each of these substances is the salt of a strong base (NaOH) and a *weak acid*. The anions should ionize as *bases*, making their solutions somewhat basic. The hydrolysis will lead to a weak acid species, for which we know the K_a value; however, the expression we write will be for K_b.

$$NO_2^- + H_2O \rightleftharpoons HNO_2 + OH^- \qquad\qquad K_b(NO_2^-) = ?$$
$$C_6H_5COO^- + H_2O \rightleftharpoons C_6H_5COOH + OH^- \qquad K_b(C_6H_5COO^-) = ?$$

Therefore, we will need to recall the relationship between K_a and K_b.

Solve

Our task is to determine the K_b values, neither of which is listed in a table in this chapter. Table 16.3 does list K_a for the conjugate acids, however. Equation (16.18) can be used to write

$$K_b \text{ of } NO_2^- = \frac{K_w}{K_a(HNO_2)} = \frac{1.0 \times 10^{-14}}{7.2 \times 10^{-4}} = 1.4 \times 10^{-11}$$

$$K_b \text{ of } C_6H_5COO^- = \frac{K_w}{K_a(C_6H_5COOH)} = \frac{1.0 \times 10^{-14}}{6.3 \times 10^{-5}} = 1.6 \times 10^{-10}$$

Because the K_b of $C_6H_5COO^-$ is larger than that of NO_2^-, the benzoate ion will hydrolyze to a greater extent than the nitrite ion and will give a solution with a higher $[OH^-]$. A sodium benzoate solution is more basic and has a higher pH than a sodium nitrite solution of the same concentration.

Assess

We could have reasoned out the answer without performing any calculations by focusing instead on the conjugate acids. Because HNO_2 is a stronger acid than C_6H_5COOH, the NO_2^- ion must be a weaker base than the $C_6H_5COO^-$ ion. This is all the information we need to decide which of the two solutions is more basic.

PRACTICE EXAMPLE A: The organic bases cocaine ($pK_b = 8.41$) and codeine ($pK_b = 7.95$) react with hydrochloric acid to form salts (similar to the formation of NH_4Cl by the reaction of NH_3 and HCl). If solutions of the following salts have the same molarity, which solution would have the higher pH: cocaine hydrochloride, $C_{17}H_{21}O_4NH^+Cl^-$, or codeine hydrochloride, $C_{18}H_{21}ClO_3NH^+Cl^-$?

PRACTICE EXAMPLE B: Predict whether the solution $NH_4CN(aq)$ is acidic, basic, or neutral; and explain the basis of your prediction.

EXAMPLE 16-13 Calculating the pH of a Solution in Which Hydrolysis Occurs

Sodium cyanide, NaCN, is extremely poisonous, but it has very useful applications in gold and silver metallurgy and in the electroplating of metals. Aqueous solutions of cyanides are especially hazardous if they become acidified, because toxic hydrogen cyanide gas, HCN(g), is released. Are NaCN(aq) solutions normally acidic, basic, or pH neutral? What is the pH of 0.50 M NaCN(aq)? Note that solutions containing cyanide ion must be handled with extreme caution. They should be handled only in a fume hood by an operator wearing protective clothing.

Analyze

Na^+ does not hydrolyze, but as represented below, CN^- does hydrolyze, producing a basic solution. The question now becomes a hydrolysis equilibrium problem.

Solve

In the tabulation of the concentrations of the species involved in the hydrolysis reaction, let $[OH^-] = x$.

	$CN^- + H_2O \rightleftharpoons HCN + OH^-$		
initial concns:	0.50 M	—	—
changes:	$-x$ M	$+x$ M	$+x$ M
equil concns:	$(0.50 - x)$ M	x M	x M

Use equation (16.18) to obtain a value of K_b.

$$K_b = \frac{K_w}{K_a(HCN)} = \frac{1.0 \times 10^{-14}}{6.2 \times 10^{-10}} = 1.6 \times 10^{-5}$$

Now return to the tabulated data.

$$K_b = \frac{[HCN][OH^-]}{[CN^-]} = \frac{x \cdot x}{0.50 - x} = \frac{x^2}{0.50 - x} = 1.6 \times 10^{-5}$$

Assume: $x \ll 0.50$ and $0.50 - x \approx 0.50$.

$$x^2 = 0.50 \times 1.6 \times 10^{-5} = 0.80 \times 10^{-5} = 8.0 \times 10^{-6}$$

$$x = [OH^-] = (8.0 \times 10^{-6})^{1/2} = 2.8 \times 10^{-3}$$

$$pOH = -\log[OH^-] = -\log(2.8 \times 10^{-3}) = 2.55$$

$$pH = 14.00 - pOH = 14.00 - 2.55 = 11.45$$

Assess

We see that in this example, the simplifying assumption works. We also note that the solution is fairly basic for a relatively dilute solution of a salt of a weak acid and a strong base.

PRACTICE EXAMPLE A: Sodium fluoride, NaF, is found in some toothpaste formulations as an anticavity agent. What is the pH of 0.10 M NaF(aq)?

PRACTICE EXAMPLE B: The pH of an aqueous solution of NaCN is 10.38. What is $[CN^-]$ in this solution?

16-8   Molecular Structure and Acid–Base Behavior

We have now dealt with a number of aspects of acid–base chemistry, both qualitatively and quantitatively. Yet some very fundamental questions still remain to be answered, such as these: Why is HCl a strong acid, whereas HF is a weak acid? Why is acetic acid (CH_3COOH) a stronger acid than ethanol (CH_3CH_2OH) but a weaker acid than chloroacetic acid ($ClCH_2COOH$)?

These questions involve relative acid strengths. In this section, we will examine the relationship between molecular structure and the strengths of acids and bases.

Strengths of Binary Acids

Because the behavior of acids requires the loss of a proton through bond breakage, acid strength and bond strength appear to be related. In general, the stronger the H—X bond, the *weaker* the acid is. Stronger bonds are characterized by short bond lengths and high bond dissociation energies. The appropriate bond dissociation energy to use is the ionization of the H—X bond in the gas phase in equation (16.19):

$$HX(g) \longrightarrow H^+(g) + X^-(g) \qquad \textbf{(16.19)}$$

The bond dissociation energy for the gas phase ionization reaction (equation 16.19) can be obtained by using the following thermodynamic cycle:

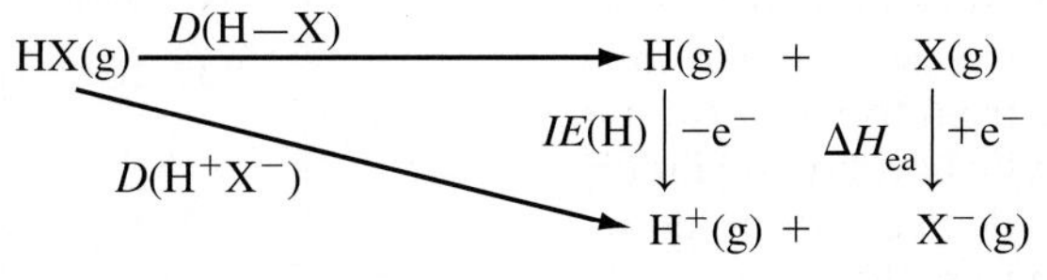

We can write $D(H^+X^-) = D(H—X) + IE(H) + \Delta H_{ea}$, where $D(H—X)$ is the bond dissociation energy for $HX(g) \rightarrow H(g) + X(g)$, $IE(H)$ is the ionization energy of the hydrogen atom, and ΔH_{ea} is the electron affinity of X, as defined on page 378. $D(H^+X^-)$ is called the *heterolytic bond dissociation energy*.

▶ The dissociation of a gas-phase molecule, AB, into A^+ and B^- is called *heterolysis* and the energy change for this process is called the *heterolytic bond dissociation energy*. The dissociation of a gas-phase molecule, AB, into A and B is called *homolysis*. Thus, the bond dissociation energy (*D*), introduced in Chapter 10, is more precisely called the *homolytic bond dissociation energy*.

KEEP IN MIND

that electronegativity increases as we move from left to right across a period and decreases from top to bottom in a given group. Thus, the polarity of the H—X bond increases from left to right across a row in Figure 16-10 and decreases as we move from top to bottom down a column. Atomic radii show the opposite trend (decrease from left to right and increase from top to bottom) and so H—X bond lengths decrease from left to right and increase from top to bottom in Figure 16-10.

Figure 16-10 shows $D(H^+X^-)$ values of binary acids formed by several elements. For binary acids, acid strength increases as the heterolytic bond dissociation energy decreases. Intuitively, this makes a lot of sense. The lower the energy requirement for converting an H—X molecule into H^+ and X^- ions, the greater the acid strength. Can we explain the trend in acid strengths in terms of (homolytic) bond dissociation energies, $D(H—X)$? Not really. For example, $D(H—X)$ values tend to increase from left to right in Figure 16-10, suggesting that the acid strength should decrease across the row. But they don't. The energy requirements for converting an H—X molecule into H and X atoms are not reliable for predicting trends in acid strengths.

Trends in the strengths of binary acids are often explained by considering variations in bond length and bond polarity. Such rationalizations are possible but a little tricky. Intuitively, we expect the acid strength of H—X to increase as the length and polarity of the bond increase. Longer bonds are weaker and easier to break. Polar H—X bonds more readily produce H^+ and X^- ions because such bonds already have partial ionic charges on the H and X atoms. As we move from left to right across a row in Figure 16-10, the H—X bond length decreases whereas the polarity of the bond increases. Because the acid strength (K_a value) increases across the row, we arrive at the following conclusion.

> When comparing binary acids of elements *in the same row* of the periodic table, acid strength increases as the polarity of the bond increases.

We arrive at a different conclusion if we compare binary acids from the same column in Figure 16-10. As we move from top to bottom in a column, both the bond length and acid strength of H—X increase whereas the polarity of the H—X bond decreases. The following statement summarizes the situation.

> When comparing binary acids of elements *in the same group* of the periodic table, acid strength increases as the length of the bond increases.

That HF is a weaker acid than the other hydrogen halides is expected, but that it should be so much weaker has always seemed an anomaly. Explanations

increasing acid strength →

	H—CH$_3$	**H—NH$_2$**	**H—OH**	**H—F**
K_a	1×10^{-60}	1×10^{-34}	1.8×10^{-16}	6.6×10^{-4}
$D(H—X)$	414	389	464	565
$D(H^+X^-)$	**1717**	**1630**	**1598**	**1549**
			H—SH	**H—Cl**
			1.0×10^{-7}	1×10^6
			368	431
			1458	**1394**
			H—SeH	**H—Br**
			1.3×10^{-4}	1×10^8
			335	364
			1434	**1351**
			H—TeH	**H—I**
			2.3×10^{-3}	1×10^9
			277	297
			1386	**1314**

▶ FIGURE 16-10
Bond dissociation energies (kJ mol^{-1}) and K_a values for some binary acids
Homolytic bond dissociation energies, $D(X—H)$, tend to increase from left to right and decrease from top to bottom in this table. Heterolytic bond dissociation energies, $D(H^+X^-)$, decrease from left to right and from top to bottom in this table. The arrows indicate that acid strengths (K_a values) increase from left to right and from top to bottom. The K_a values for NH_3 and CH_4 are very small. These molecules do not behave as acids in water.

of this behavior center on the tendency for hydrogen bonding in HF (recall Figure 12-6). For example, in HF(aq), ion pairs are held together by strong hydrogen bonds, which keeps the concentration of free H_3O^+ from being as large as otherwise expected.

$$HF + H_2O \longrightarrow (^-F\cdots H_3O^+) \rightleftharpoons H_3O^+ + F^-$$

Ion pair

CH_4 and NH_3 do not have acidic properties in water, but HF is an acid of moderate strength ($K_a = 6.6 \times 10^{-4}$).

Strengths of Oxoacids

To describe the relative strengths of oxoacids, we must focus on the attraction of electrons from the O—H bond toward the central atom. The following factors promote this electron withdrawal from O—H bonds: (1) a high electronegativity (EN) of the central atom and (2) a large number of terminal O atoms in the acid molecule.

◀ The term *oxoacid* was defined in Chapter 3.

Neither HOCl nor HOBr has a terminal O atom. The major difference between the two acids is one of electronegativity—Cl is slightly more electronegative than Br. As expected, HOCl is more acidic than HOBr.

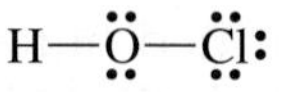

H—Ö—C̈l: H—Ö—B̈r:

$EN_{Cl} = 3.0$ $EN_{Br} = 2.8$

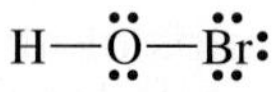

$K_a = 2.9 \times 10^{-8}$ $K_a = 2.1 \times 10^{-9}$

To compare the acid strengths of H_2SO_4 and H_2SO_3, we must look beyond the central atom, which is S in each acid.

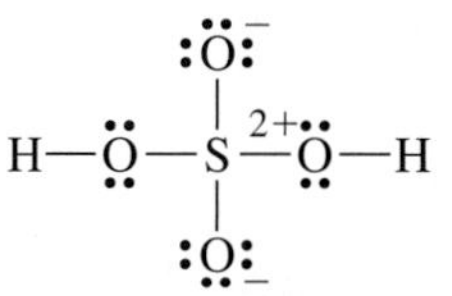

$K_{a_1} \approx 10^3$ 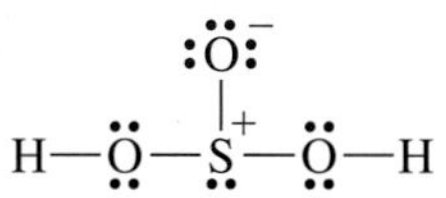 $K_{a_1} = 1.3 \times 10^{-2}$

A highly electronegative terminal O atom tends to withdraw electrons from the O—H bonds, weakening the bonds and increasing the acidity of the molecule. Because H_2SO_4 has *two* terminal O atoms to only one in H_2SO_3, the electron-withdrawing effect is greater in H_2SO_4. As a result, H_2SO_4 is a stronger acid than H_2SO_3.

Strengths of Organic Acids

This discussion of the relationship between molecular structure and acid strength concludes with a brief consideration of some organic compounds. Consider first the case of acetic acid and ethanol. Both have an O—H group bonded to a carbon atom, but acetic acid is a much stronger acid than ethanol.

Acetic acid   Ethanol

$K_a = 1.8 \times 10^{-5}$ $K_a = 1.3 \times 10^{-16}$

◀ Review the concept of resonance. Compounds or ions with more resonance structures are more stable.

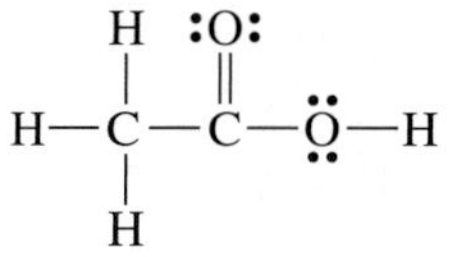

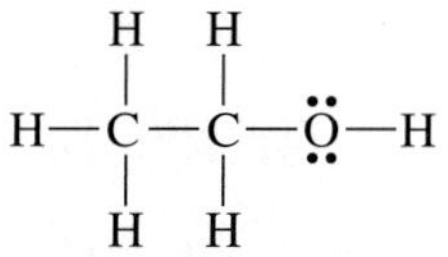

One possible explanation for the large difference in acidity of these two compounds is that the highly electronegative terminal O atom in acetic acid withdraws electrons from the O—H bond. The bond is weakened, and a proton (H^+) is more readily taken from a molecule of the acid by a base. A more satisfactory explanation focuses on the anions formed in the ionization.

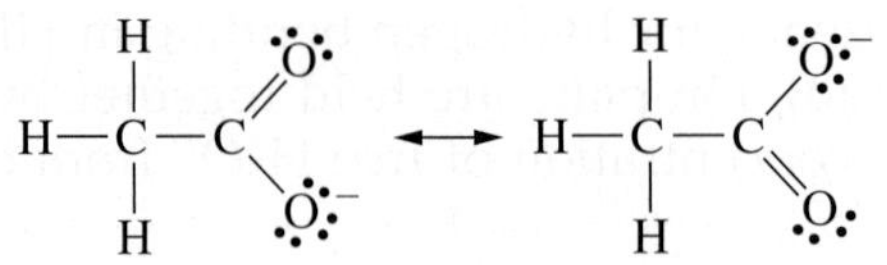

Acetate ion 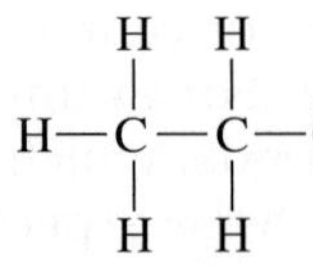   Ethoxide ion

There are two plausible structures for the acetate ion. These structures suggest that each carbon-to-oxygen bond is a "$\frac{3}{2}$" bond and that each O atom carries "$\frac{1}{2}$" unit of negative charge. In short, the excess unit of negative charge in CH_3COO^- is spread out. This arrangement reduces the ability of either O atom to attach a proton and makes acetate ion only a weak Brønsted–Lowry base. In ethoxide ion, conversely, the unit of negative charge is localized on the single O atom. Ethoxide ion is a much stronger base than is acetate ion. The stronger the conjugate base, the weaker the corresponding acid.

The length of the carbon chain in a carboxylic acid has little effect on the acid strength, as in a comparison of acetic acid and octanoic acid.

CH_3COOH $CH_3(CH_2)_6COOH$

Acetic acid   Octanoic acid

$K_a = 1.8 \times 10^{-5}$ $K_a = 1.3 \times 10^{-5}$

Yet, other atoms or groups of atoms substituted onto the carbon chain may strongly affect acid strength. If a Cl atom is substituted for one of the H atoms that is bonded to carbon in acetic acid, the result is chloroacetic acid.

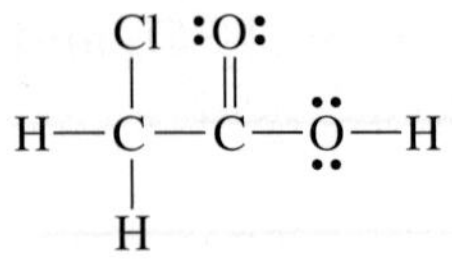

Chloroacetic acid
$K_a = 1.4 \times 10^{-3}$

The highly electronegative Cl atom helps draw electrons away from the O—H bond. The O—H bond is weakened, the proton is lost more readily, and the acid is a stronger acid than acetic acid. This effect falls off rapidly as the distance increases between the substituted atom or group and the O—H bond in an organic acid.

Example 16-14 illustrates some of the factors affecting acid strength that are discussed in this section.

EXAMPLE 16-14   Identifying Factors That Affect the Strengths of Acids

Explain which member of each of the following pairs is the stronger acid.

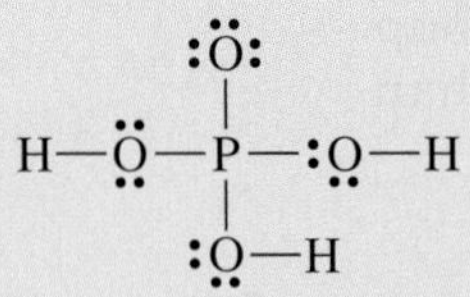

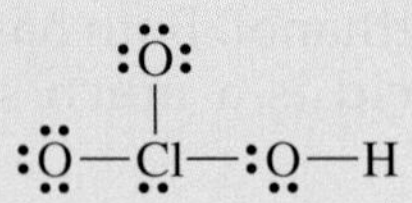

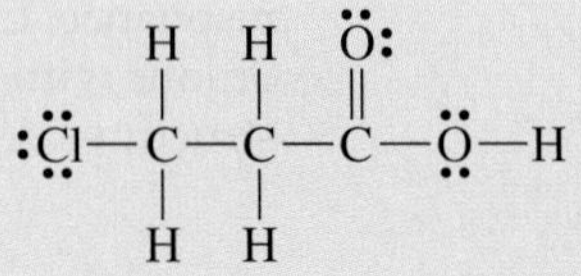

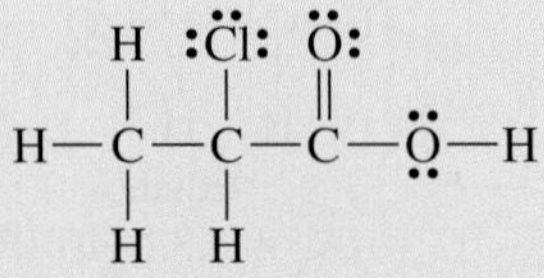

Analyze

In these types of questions we first identify the acidic proton(s), and then look for electronegative atoms or groups that pull electron density away from the acidic proton(s). The more electron density that is pulled away from the proton, the more acidic it is.

(a) Phosphoric acid, H_3PO_4, has four O atoms to the three in $HClO_3$, but it is the number of *terminal* O atoms that we must consider, not just the total number of O atoms in the molecule. $HClO_3$ has *two* terminal O atoms and H_3PO_4 has *one*. Also, the Cl atom (EN = 3.0) is considerably more electronegative than the P atom (EN = 2.1). These facts point to chloric acid (II) as being the stronger of the two acids. ($K_a \approx 5 \times 10^2$ for $HClO_3$ and $K_{a_1} = 7.1 \times 10^{-3}$ for H_3PO_4.)

(b) The Cl atom withdraws electrons more strongly when it is directly adjacent to the carboxyl group. Compound (II), 2-chloropropanoic acid ($K_a = 1.4 \times 10^{-3}$), is a stronger acid than compound (I), 3-chloropropanoic acid ($K_a = 1.0 \times 10^{-4}$).

Assess

This type of analysis is important in organic chemistry. To successfully solve these types of problems, we must know how to draw Lewis structures and we must understand the concept of electronegativity.

PRACTICE EXAMPLE A: Explain which is the stronger acid, HNO_3 or $HClO_4$; CH_2FCOOH or $CH_2BrCOOH$. [*Hint:* Draw plausible Lewis structures.]

PRACTICE EXAMPLE B: Explain which is the stronger acid, H_3PO_4 or H_2SO_3; CCl_3CH_2COOH or CCl_2FCH_2COOH. [*Hint:* Draw plausible Lewis structures.]

Strengths of Amines as Bases

The fundamental factor affecting the strength of an amine as a base concerns the ability of the lone pair of electrons on the N atom to bind a proton taken from an acid. When an atom or group of atoms more electronegative than H replaces one of the H atoms of NH_3, the electronegative group withdraws electron density from the N atom. The lone-pair electrons cannot bind a proton as strongly, and the base is weaker. Thus, bromamine, in which the electronegative Br atom is attached to the amine group (NH_2), is a *weaker* base than ammonia.

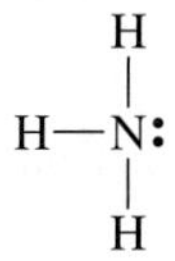

Ammonia
$pK_b = 4.74$

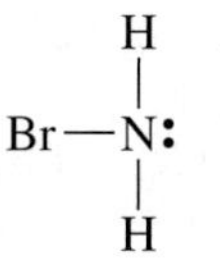

Bromamine
NH_2Br, $pK_b = 7.61$

Hydrocarbon chains have no electron-withdrawing ability. When they are attached to the amine group, the pK_b values are lower than for ammonia due to the electron-donating ability of CH_3 and CH_2CH_3.

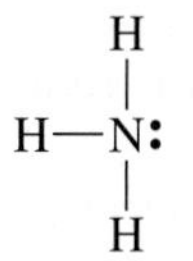

Ammonia
$pK_b = 4.74$

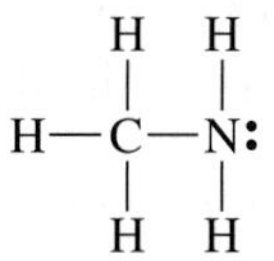

Methylamine
CH_3NH_2, $pK_b = 3.38$

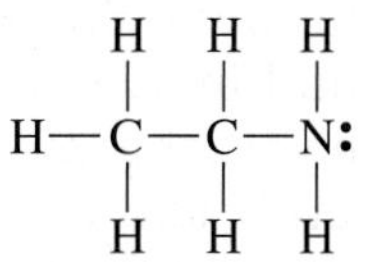

Ethylamine
$CH_3CH_2NH_2$, $pK_b = 3.37$

An additional electron-withdrawing effect is seen in amines that are based on the benzene ring or related structures. Such amines are called *aromatic* amines. Aniline, $C_6H_5NH_2$, is based on benzene, C_6H_6, which, as we learned in Section 11-7 and depicted in several different ways in Figures 11-28, 11-29 and 11-30, is a six-carbon ring molecule with unsaturation in the carbon-to-carbon bonds. The electrons associated with this unsaturation are said to be *delocalized*. As suggested by the following structures, to some extent even the lone-pair electrons of the NH_2 group participate in the "spreading out" of delocalized electrons. (The curved arrows suggest the progressive movement of electrons around the ring.)

◀ Note that these are actually resonance Lewis structures.

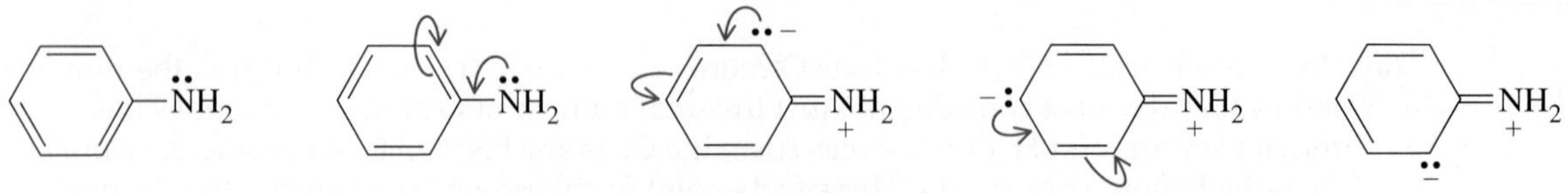

The withdrawal of electron charge density from the NH$_2$ group causes aniline to be a much weaker base than is cyclohexylamine. (H atoms bonded to ring carbon atoms are not shown in the following structures.)

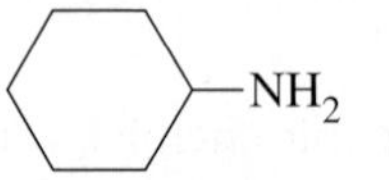

Cyclohexylamine, pK_b = 3.36 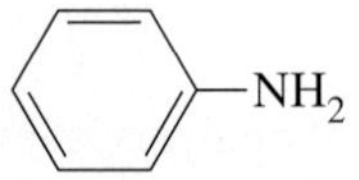   Aniline, pK_b = 9.13

▶ The pK_b for *meta*-chloroaniline is 10.66.

Replacement of a ring-bound H atom in aniline with an atom or group that has a high electronegativity causes even more electron density to be drawn away from the NH$_2$ group, further reducing the base strength. Also, the closer this ring substituent is to the NH$_2$ group, the greater is the effect.

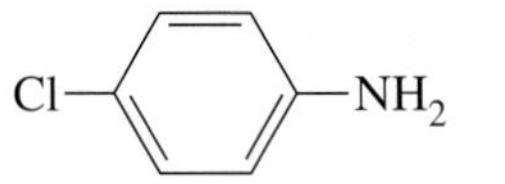

para-Chloroaniline, pK_b = 10.01 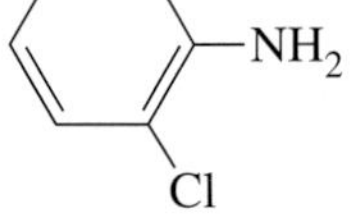   *ortho*-Chloroaniline, pK_b = 11.36

🔍 **16-7   CONCEPT ASSESSMENT**

Would you expect pK_a of *ortho*-chlorophenol to be greater than, less than, or nearly the same as that of phenol? Explain.

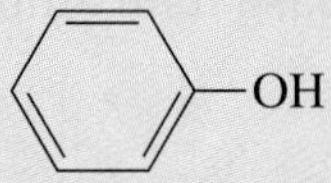

Phenol, pK_a = 10.00 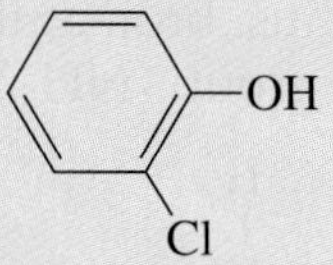   *ortho*-Chlorophenol, pK_a = ?

16-9   Lewis Acids and Bases

In the previous section, we presented ideas about the molecular structures of acids and bases. In 1923, G. N. Lewis proposed an acid–base theory closely related to bonding and structure. The Lewis acid–base theory is not limited to reactions involving H$^+$ and OH$^-$: It extends acid–base concepts to reactions in gases and in solids. It is especially important in describing certain reactions between organic molecules.

A **Lewis acid** is a species (an atom, ion, or molecule) that is an *electron-pair acceptor*, and a **Lewis base** is a species that is an *electron-pair donor*. A reaction between a Lewis acid (A) and a Lewis base (B:) results in the formation of a covalent bond between them. The product of a Lewis acid–base reaction is called an **adduct** (or *addition compound*). The reaction can be represented as

$$B{:} + A \longrightarrow B - A$$

where B:A is the adduct. The formation of a covalent chemical bond by one species donating a pair of electrons to another is called *coordination*, and the bond joining the Lewis acid and Lewis base is called a *coordinate covalent bond* (see page 399). *Lewis acids* are species with vacant orbitals that can accommodate electron pairs; *Lewis bases* are species that have lone-pair electrons available for sharing.

By these definitions, OH$^-$, a Brønsted–Lowry base, is also a Lewis base because lone-pair electrons are present on the O atom. So too is NH$_3$ a Lewis base. HCl, conversely, is not a Lewis acid: It is not an electron-pair acceptor. We can think of HCl as producing H^+, however, and H^+ is a Lewis acid. H^+ forms a coordinate covalent bond with an available electron pair.

Species with an incomplete valence shell are Lewis acids. When the Lewis acid forms a coordinate covalent bond with a Lewis base, the octet is completed. A good example of octet completion is the reaction of BF_3 and NH_3.

$$H-\overset{\overset{\displaystyle H}{|}}{\underset{\underset{\displaystyle H}{|}}{N}}{:} \quad \overset{:\ddot{F}:}{\underset{:\ddot{F}:}{B}}-\ddot{F}: \longrightarrow H-\overset{\overset{\displaystyle H}{|}}{\underset{\underset{\displaystyle H}{|}}{N}}-\overset{:\ddot{F}:}{\underset{:\ddot{F}:}{B}}-\ddot{F}:$$

The reaction of lime (CaO) with sulfur dioxide is an important reaction for reducing SO_2 emissions from coal-fired power plants. This reaction between a solid and a gas underscores that Lewis acid–base reactions can occur in all states of matter. The smaller curved red arrow in reaction (16.20) suggests that an electron pair in the Lewis structure is rearranged.

$$Ca^{2+}:\ddot{\underset{..}{O}}:^{2-} + \overset{\overset{\displaystyle \ddot{O}:}{\|}}{\underset{\underset{\displaystyle :\ddot{O}:}{}}{S}} \longrightarrow Ca^{2+}\left[:\ddot{\underset{..}{O}}-\overset{\overset{\displaystyle :\ddot{O}:}{}}{\underset{\underset{\displaystyle :\ddot{O}:}{}}{S}}:\right]^{2-} \quad \textbf{(16.20)}$$

An important application of the Lewis acid–base theory involves the formation of *complex ions*. Complex ions are polyatomic ions that contain a central metal ion to which other ions or small molecules are attached. *Hydrated metal ions* form in aqueous solution because the water acts as a Lewis base and the metal ion as a Lewis acid. The water molecules attach themselves to the metal ion by means of coordinate covalent bonds. Thus, for example, when anhydrous $AlCl_3$ is added to water, the resultant solution becomes hot because of the heat evolved in the formation of the hydrated metal ion $[Al(H_2O)_6]^{3+}(aq)$ (Fig. 16-11).

The interaction between the metal ion and the water molecules is so strong that when the salt is crystallized from the solution, the water molecules crystallize along with the metal ion, forming the hydrated metal salt $AlCl_3 \cdot 6\ H_2O$. In aqueous solution, the hydrated metal ions can act as Brønsted acids. For instance, the hydrolysis of hydrated Al^{3+} is given by

$$[Al(H_2O)_6]^{3+} + H_2O \rightleftharpoons [Al(OH)(H_2O)_5]^{2+} + H_3O^+$$

In the hydrated metal ion, the OH bond in a water molecule becomes weakened. This happens because, in forming the coordinate covalent bond with the O atom of the water, the metal ion causes electron density to be drawn toward it; hence, electron density is drawn away from the OH bond. As a consequence, the coordinated H_2O molecule can donate a H^+ to a solvent H_2O molecule (Fig. 16-12). The H_2O molecule that has ionized is converted to OH^-, which remains attached to the Al^{3+}; the charge on the complex ion is reduced from 3+ to 2+. The extent of ionization of $[Al(H_2O)_6]^{3+}$, measured by its K_a value and as pictured in Figure 16-13, is essentially the same as that of acetic acid ($K_a = 1.8 \times 10^{-5}$). Many other metal ions hydrolyze, especially the transition metal ions. These and other hydrated metal ions acting as acids are discussed in later chapters.

◀ Bonding in the H_3N-BF_3 adduct can be described by the overlap of sp^3 orbitals on the N and B atoms, with the two electrons donated by the N atom.

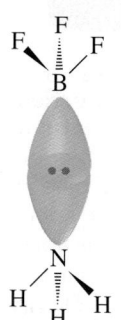

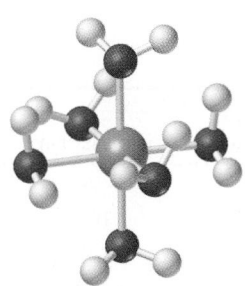

▲ FIGURE 16-11
The Lewis structure of $[Al(H_2O)_6]^{3+}$ and a ball-and-stick representation

◀ FIGURE 16-12
Hydrolysis of $[Al(H_2O)_6]^{3+}$ to produce H_3O^+
An uncoordinated water molecule removes a proton from a coordinated water molecule.

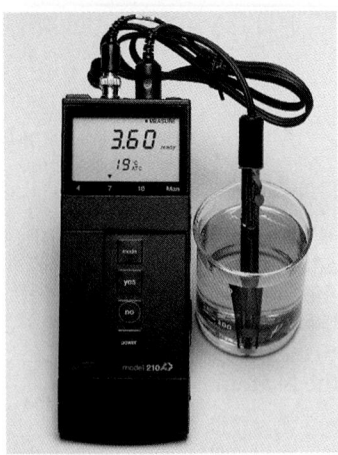

▲ FIGURE 16-13
Acidic properties of hydrated metal ions
The yellow color of bromthymol blue indicator in $Al_2(SO_4)_3(aq)$ denotes that the solution is acidic. The pH meter gives a more precise indication of the pH.

Complex ions can also form between transition metal ions and other Lewis bases, such as NH_3. For instance, Zn^{2+} combines with NH_3 to form the complex ion $[Zn(NH_3)_4]^{2+}$. The central Zn^{2+} ion accepts electrons from the Lewis base NH_3. to form coordinate covalent bonds; it is a Lewis acid. We will discuss the application of Lewis acid–base theory to complex ions in Chapter 24.

16-3 ARE YOU WONDERING...

Why $Na^+(aq)$ does not act as an acid in aqueous solution?

Whether an aqueous solution of a metal ion is acidic depends on two principal factors. The first is the amount of charge on the cation; the second is the size of the ion. The greater the charge on the cation, the greater is the ability of the metal ion to draw electron density away from the O—H bond in a H_2O molecule in its hydration sphere, favoring the release of a H^+ ion. The smaller the cation, the more highly concentrated is the positive charge. Hence, for a given positive charge, the smaller the cation, the more acidic the solution.

The ratio of the charge on the cation to the volume of the cation is called the *charge density*.

$$\rho = \text{charge density} = \frac{\text{ionic charge}}{\text{ionic volume}}$$

The greater its charge density, the more effective a metal ion is at pulling electron density from the O—H bond and the more acidic is the hydrated cation (see the table and plot below). A highly concentrated positive charge on a small cation is better able to pull electron density from the O—H bond than is a less concentrated positive charge on a larger cation.

Thus the small, highly charged Al^{3+} ion produces acidic solutions, but the larger Na^+ cation, with a charge of just 1+, does not increase the concentration of H_3O^+. In fact, none of the group 1 cations produces appreciably acidic solutions, and only Be^{2+} of the group 2 elements is small enough to do so ($pK_a = 5.4$).

Metal Cation	Ionic Radius, pm	$\rho \times 10^7$, Charge pm^{-3}	pK_a
Li^+	76	3.27	13.6
Na^+	102	1.53	14.2
K^+	138	0.680	14.5
Be^{2+}	45	23.2	5.4
Cu^{2+}	66	9.33	8.0
Ni^{2+}	69	8.35	9.9
Mg^{2+}	72	7.51	11.4
Zn^{2+}	74	7.00	9.0
Co^{2+}	74	7.00	9.7
Mn^{2+}	83	5.23	10.6
Ca^{2+}	100	3.22	12.8
Al^{3+}	53	23.8	5.0
Cr^{3+}	61	17.0	4.0
Ti^{3+}	67	13.5	2.2
Fe^{3+}	78	9.19	2.2

The pK_a of H_3O^+ is -1.7, and the pK_a of water is 15.7.

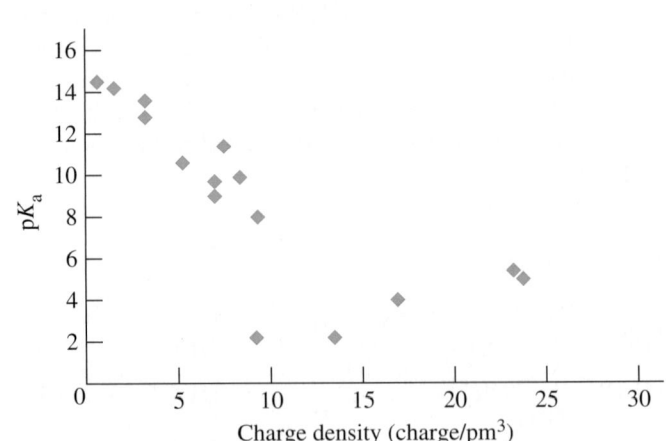

EXAMPLE 16-15   Identifying Lewis Acids and Bases

According to the Lewis theory, each of the following is an acid–base reaction. Which species is the acid and which is the base?

(a) $BF_3 + F^- \longrightarrow BF_4^-$

(b) $OH^-(aq) + CO_2(aq) \longrightarrow HCO_3^-(aq)$

Analyze

Recall that in Lewis theory an acid–base reaction involves the movement of electrons. The Lewis acid accepts electrons and the Lewis base donates electrons. In this example, we need to identify the species that is accepting the electrons and the one that is donating electrons.

Solve

(a) In BF_3, the B atom has a vacant orbital and an incomplete octet. The fluoride ion has an outer-shell octet of electrons. BF_3 is the electron-pair acceptor—the acid. F^- is the electron-pair donor—the base.

(b) We have already identified OH^- as a Lewis base, so we might suspect that it is the base and that $CO_2(aq)$ is the Lewis acid. The following Lewis structures show this to be so. As in reaction (16.20), a rearrangement of an electron pair at one of the double bonds is also required, as indicated by the smaller red arrow.

Assess

Typically, those species that have filled orbitals are Lewis bases, and those with vacant orbitals are Lewis acids. The transfer of electron density from a Lewis base to a vacant orbital on a Lewis acid is a recurring concept in chemistry. We will make use of this concept in the later chapters, as well as in organic chemistry. To describe the reaction in (b) in this way requires us to consider the electronic structure of CO_2 in terms of molecular orbital theory. Some of the $2p$ orbitals on the carbon and oxygen atoms in CO_2 combine to give bonding and antibonding π-type molecular orbitals. We describe a similar situation in Chapter 11 (see Figure 11-33). The vacant orbital in CO_2 that accepts the lone pair from OH^- is an antibonding π-type orbital.

PRACTICE EXAMPLE A: Identify the Lewis acids and bases in these reactions.

(a) $BF_3 + NH_3 \longrightarrow F_3BNH_3$

(b) $Cr^{3+} + 6\,H_2O \longrightarrow [Cr(H_2O)_6]^{3+}$

PRACTICE EXAMPLE B: Identify the Lewis acids and bases in these reactions.

(a) $Al(OH)_3 + OH^- \longrightarrow [Al(OH)_4]^-$

(b) $SnCl_4 + 2\,Cl^- \longrightarrow [SnCl_6]^{2-}$

16-8   CONCEPT ASSESSMENT

Liquid bromine in the presence of iron(III) tribromide forms a bromonium:iron(III) tribromide adduct. Propose a plausible mechanism for adduct formation and identify the Lewis acid and Lewis base. [*Hint*: What is the iron(III) electron configuration?]

Mastering**CHEMISTRY**   www.masteringchemistry.com

Pure water has a pH of 7, but not rainwater. What causes rainwater to be acidic? Rainwater is acidic, in part, because carbon dioxide from the atmosphere reacts with water to form carbonic acid, H_2CO_3, a diprotic acid. For a discussion of the natural sources of acidity in rainwater, and how human activities also contribute, go to the Focus On feature for Chapter 16, Acid Rain, on the MasteringChemistry site.

Summary

16-1 Arrhenius Theory—The central concept in Arrhenius theory describes a neutralization reaction as the combination of hydrogen ions and hydroxide ions to form water (expression 16.1). A major failing of this theory is that it does not account for the key role of the solvent in the ionization of a solute.

16-2 Brønsted–Lowry Theory of Acids and Bases—The Brønsted–Lowry theory describes an acid as a **proton donor** and a base as a **proton acceptor**. In an acid–base reaction, a base takes a proton (H^+) from an acid. In general, acid–base reactions are reversible, but equilibrium is displaced in the direction from the *stronger* acids and bases to their *weaker* conjugates. The **conjugate base** (A^-) is derived from the acid HA while the **conjugate acid** (HB^+) is derived from the base (B). The combinations of HA/A^- and B/HB^+ are known as conjugate acid–base pairs. The conjugate acid of the base H_2O is the **hydronium ion**, H_3O^+. The equilibrium constants representing the ionization of an acid or a base in water are commonly referred to as the **acid ionization constant** and **base ionization constant**, respectively. Certain substances, for example water, are said to be **amphiprotic**. They can act as an acid or base.

16-3 Self-Ionization of Water and the pH Scale—In pure water and in aqueous solutions, **self-ionization** of the water occurs to a very slight extent, producing H_3O^+ and OH^-, as described by the equilibrium constant K_w, known as the **ion product of water** (expression 16.7). The designations **pH** (expression 16.8) and **pOH** (expression 16.9) are often used to describe the concentrations of H_3O^+ and OH in aqueous solutions.

16-4 Strong Acids and Strong Bases—In aqueous solutions, strong acids ionize completely to produce H_3O^+, and strong bases dissociate completely to produce OH^-. Common strong acids and bases are given in Table 16.2 and can be easily memorized.

16-5 Weak Acids and Weak Bases—The ionizations of weak acids and weak bases are reversible reactions, and the extent of their ionization can be related to the

ionization constants K_a and K_b or their logarithmic equivalents $pK_a = -\log K_a$ and $pK_b = -\log K_b$ (Table 16.3). Another method used to indicate the degree of ionization is **percent ionization** (expression 16.14). The acidity of many weak acids is associated with the carboxylic acid group, —COOH (page 97). Weak bases typically contain one or more nitrogen atoms. Calculations involving ionization equilibria are in many ways similar to those introduced in Chapter 15, although some additional considerations are necessary for polyprotic acids.

16-6 Polyprotic Acids—**Polyprotic acids** are acids with more than one ionizable H atom that undergo a stepwise ionization and have a different ionization constant, $K_{a_1}, K_{a_2}, \ldots$, for each ionization step.

16-7 Ions as Acids and Bases—In reactions between ions and water—**hydrolysis** reactions—the ions react as weak acids or weak bases. The pH of salt solutions depends on the anions and/or cations present. Anions from weak acids lead to solutions with pH > 7 while cations from weak bases lead to pH < 7.

16-8 Molecular Structure and Acid–Base Behavior—Molecular composition and structure are the keys to determining whether a substance is acidic, basic, or amphiprotic. In addition, molecular structure affects whether an acid or base is strong or weak. In assessing acid strength, for example, factors that affect the strength of the bond that must be broken to release H^+ must be considered. In assessing base strength, factors that affect the ability of lone-pair electrons to bind a proton are of primary concern.

16-9 Lewis Acids and Bases—The Lewis acid–base theory views an electron-pair acceptor as a **Lewis acid** and an electron-pair donor as a **Lewis base**. The addition compound of a Lewis acid–base reaction is referred to as an **adduct**. The theory is most useful in situations that cannot be described by means of proton transfers, for example, in reactions involving gases and solids and in reactions between organic compounds (considered in Chapter 26).

Integrative Example

Bromoacetic acid, $BrCH_2COOH$, has $pK_a = 2.902$. Calculate the expected values of **(a)** the freezing point of 0.0500 M $BrCH_2COOH(aq)$ and **(b)** the osmotic pressure at 25 °C of 0.00500 M $BrCH_2COOH(aq)$.

Analyze

Freezing point and osmotic pressure are both colligative properties. As we saw in Chapter 13, the values of these properties depend on the total concentrations of particles (molecules and ions) in a solution, but not on the identity of those particles. We can use the ICE method for equilibrium calculations (Chapter 15) to determine the total concentrations of particles (molecules and ions) in a weak electrolyte solution, as we learned to do in this chapter. Once we have those results we can turn to equations (13.5) and (13.4) to do the calculations required in parts (a) and (b).

Solve

A good place to begin is to convert the pK_a for bromoacetic acid to K_a.

$$pK_a = 2.902 = -\log K_a$$

$$K_a = 10^{-2.902} = 1.25 \times 10^{-3}$$

Next, write the equation for the reversible ionization reaction and the equilibrium constant expression.

$$BrCH_2COOH + H_2O \rightleftharpoons H_3O^+ + BrCH_2COO^-$$

$$K_a = \frac{[H_3O^+][BrCH_2COO^-]}{[BrCH_2COOH]} = 1.25 \times 10^{-3}$$

(a) Enter the relevant data into the ICE format under the equation for the ionization reaction.

$$BrCH_2COOH + H_2O \rightleftharpoons H_3O^+ + BrCH_2COO^-$$

	$BrCH_2COOH$		H_3O^+	$BrCH_2COO^-$
initial concns:	0.0500 M		—	—
changes:	$-x$ M		$+x$ M	$+x$ M
equil concns:	$(0.0500 - x)$ M		x M	x M

The equilibrium constant expression based on these data is

$$K_a = \frac{[H_3O^+][BrCH_2COO^-]}{[BrCH_2COOH]} = \frac{x \cdot x}{0.0500 - x} = 1.25 \times 10^{-3}$$

Solve for x.

$$x^2 + 1.25 \times 10^{-3}x - 6.25 \times 10^{-5} = 0$$

$$x = \frac{-1.25 \times 10^{-3} \pm \sqrt{(1.25 \times 10^{-3})^2 + 4 \times 6.25 \times 10^{-5}}}{2}$$

$$x = \frac{-1.25 \times 10^{-3} \pm 1.59 \times 10^{-2}}{2} = 7.3 \times 10^{-3} \, M$$

The total concentration of molecules and ions at equilibrium is

$$(0.0500 - x) \, M + x \, M + x \, M = (0.0500 + x) \, M = 0.0573 \, M$$

Assuming that $0.0573 \, M = 0.0573 \, m$, the freezing point depression of water caused by 0.0573 mol/L of particles is

$$\Delta T_f = -K_f \times m = -1.86 \, °C \, m^{-1} \times 0.0573 \, m = -0.107 \, °C$$

The freezing point of $0.0500 \, M \, BrCH_2COOH(aq)$ is $0.107 \, °C$ below the freezing point of water $(0.000 \, °C)$, that is, $-0.107 \, °C$.

(b) Enter the relevant data into the ICE format under the equation for the ionization reaction.

$$BrCH_2COOH + H_2O \rightleftharpoons H_3O^+ + BrCH_2COO^-$$

	$BrCH_2COOH$		H_3O^+	$BrCH_2COO^-$
initial concns:	0.00500 M		—	—
changes:	$-x$ M		$+x$ M	$+x$ M
equil concns:	$(0.00500 - x)$ M		x M	x M

The equilibrium constant expression based on these data is

$$K_a = \frac{[H_3O^+][BrCH_2COO^-]}{[BrCH_2COOH]} = \frac{x \cdot x}{0.00500 - x} = 1.25 \times 10^{-3}$$

Solve for x.

$$x^2 + 1.25 \times 10^{-3}x - 6.25 \times 10^{-6} = 0$$

$$x = \frac{-1.25 \times 10^{-3} \pm \sqrt{(1.25 \times 10^{-3})^2 + 4 \times 6.25 \times 10^{-6}}}{2}$$

$$x = \frac{-1.25 \times 10^{-3} \pm 5.15 \times 10^{-3}}{2} = 1.95 \times 10^{-3} \, M$$

The total concentration of molecules and ions at equilibrium is

$$(0.00500 - x) \, M + x \, M + x \, M = (0.00500 + x) \, M = 0.00695 \, M$$

At 25.00 °C, the osmotic pressure of an aqueous solution with 0.00695 mol/L of particles (molecules and ions) is

$$\pi = M \times RT = 0.00695 \, mol \, L^{-1} \times 0.08206 \, L \, atm \, mol^{-1} K^{-1} \times$$
$$298.15 \, K = 0.170 \, atm$$

Assess

The pK_a is stated more precisely than in most previous equilibrium calculations, and this permitted us to carry three significant figures rather than the usual two in most of the calculations. The assumption in part (a) that $0.0573 \text{ M} = 0.0573 \, m$ is reasonable for a dilute aqueous solution with a density of essentially 1.00 g/mL. The mass of solvent (water) in one liter of solution is very close to one kilogram, so that molarity (mol solute/L solution) and molality (mol solute/kg solvent) are essentially the same. The calculation in part (b) could have been done more easily by assuming that the concentration of solute particles would be just 10% of that in part (a), that is, 0.00573 M compared to 0.0573 M. However, this would have been a false assumption. Because the percent ionization of the acid is a function of its concentration, the total particle concentration in part (b) was about 12% of that found in part (a), not 10%.

PRACTICE EXAMPLE A: The solubility of $CO_2(g)$ in H_2O at 25 °C and under a $CO_2(g)$ pressure of 1 atm is 1.45 g CO_2/L. Air contains 0.037% CO_2 by volume. Use this information, together with data from Table 16.4, to show that rainwater saturated with CO_2 has a pH $\approx$ 5.6 (the normal pH for rainwater). [*Hint:* Recall Henry's law. What is the partial pressure of $CO_2(g)$ in air?]

PRACTICE EXAMPLE B: Often the following generalization applies to oxoacids with the formula $EO_m(OH)_n$ (where E is the central atom): If $m = 0$, $K_a \approx 10^{-7}$; if $m = 1$, $K_a \approx 10^{-2}$; if $m = 2$, K_a is large; and if $m = 3$, K_a is very large.

(a) Show that this generalization works well for the oxoacids of chlorine: HOCl, $pK_a = 7.52$; HOClO, $pK_a = 1.92$; HOClO$_2$, $pK_a = -3$; HOClO$_3$, $pK_a = -8$.

(b) Estimate the value of K_{a_1} for H_3AsO_4.

(c) Write a Lewis structure for hypophosphorous acid, H_3PO_2, for which $pK_a = 1.1$.

Mastering **CHEMISTRY**

You'll find a link to additional self study questions in the study area on www.masteringchemistry.com

Exercises

Brønsted–Lowry Theory of Acids and Bases

1. According to the Brønsted–Lowry theory, label each of the following as an acid or a base. (a) HNO_2; (b) OCl^-; (c) NH_2^-; (d) NH_4^+; (e) $CH_3NH_3^+$

2. Write the formula of the conjugate base in the reaction of each acid with water. (a) HIO_3; (b) C_6H_5COOH; (c) HPO_4^{2-}; (d) $C_2H_5NH_3^+$

3. For each of the following, identify the acids and bases involved in both the forward and reverse directions.
   (a) $HOBr + H_2O \rightleftharpoons OBr^- + H_3O^+$
   (b) $HSO_4^- + H_2O \rightleftharpoons SO_4^{2-} + H_3O^+$
   (c) $HS^- + H_2O \rightleftharpoons H_2S + OH^-$
   (d) $C_6H_5NH_3^+ + OH^- \rightleftharpoons C_6H_5NH_2 + H_2O$

4. Which of the following species are *amphiprotic* in aqueous solution? For such a species, write one equation showing it acting as an acid, and another equation showing it acting as a base. $OH^-, NH_4^+, H_2O, HS^-, NO_2^-, HCO_3^-, HBr$.

5. With which of the following bases will the ionization of acetic acid, CH_3COOH, proceed furthest toward completion (to the right): (a) H_2O; (b) NH_3; (c) Cl^-; (d) NO_3^-? Explain your answer.

6. In a manner similar to equation (16.6), represent the self-ionization of the following liquid solvents: (a) NH_3; (b) HF; (c) CH_3OH; (d) CH_3COOH; (e) H_2SO_4.

7. With the aid of Table 16.1, predict the direction (forward or reverse) favored in each of the following acid–base reactions.
   (a) $NH_4^+ + OH^- \rightleftharpoons H_2O + NH_3$
   (b) $HSO_4^- + NO_3^- \rightleftharpoons HNO_3 + SO_4^{2-}$
   (c) $CH_3OH + CH_3COO^- \rightleftharpoons CH_3COOH + CH_3O^-$

8. With the aid of Table 16.1, predict the direction (forward or reverse) favored in each of the following acid–base reactions.
   (a) $CH_3COOH + CO_3^{2-} \rightleftharpoons HCO_3^- + CH_3COO^-$
   (b) $HNO_2 + ClO_4^- \rightleftharpoons HClO_4 + NO_2^-$
   (c) $H_2CO_3 + CO_3^{2-} \rightleftharpoons HCO_3^- + HCO_3^-$

Strong Acids, Strong Bases, and pH

9. Calculate $[H_3O^+]$ and $[OH^-]$ for each solution: (a) 0.00165 M HNO_3; (b) 0.0087 M KOH; (c) 0.00213 M $Sr(OH)_2$; (d) 5.8×10^{-4} M HI.

10. What is the pH of each of the following solutions? (a) 0.0045 M HCl; (b) 6.14×10^{-4} M HNO_3; (c) 0.00683 M NaOH; (d) 4.8×10^{-3} M $Ba(OH)_2$.

11. Calculate $[H_3O^+]$ and pH in saturated $Ba(OH)_2(aq)$, which contains 3.9 g $Ba(OH)_2 \cdot 8\,H_2O$ per 100 mL of solution.

12. A saturated aqueous solution of $Ca(OH)_2$ has a pH of 12.35. What is the solubility of $Ca(OH)_2$, expressed in milligrams per 100 mL of solution?

13. What is $[H_3O^+]$ in a solution obtained by dissolving 205 mL HCl(g), measured at 23 °C and 751 mmHg, in 4.25 L of aqueous solution?

14. What is the pH of the solution obtained when 125 mL of 0.606 M NaOH is diluted to 15.0 L with water?

15. How many milliliters of concentrated HCl(aq) (36.0% HCl by mass, $d = 1.18$ g/mL) are required to produce 12.5 L of a solution with pH = 2.10?

16. How many milliliters of a 15.0%, by mass solution of KOH(aq) $(d = 1.14$ g/mL) are required to produce 25.0 L of a solution with pH = 11.55?

17. What volume of 6.15 M HCl(aq) is required to exactly neutralize 1.25 L of 0.265 M $NH_3(aq)$?

$$NH_3(aq) + H_3O^+(aq) \longrightarrow NH_4^+(aq) + H_2O$$

18. A 28.2 L volume of HCl(g), measured at 742 mmHg and 25.0 °C, is dissolved in water. What volume of $NH_3(g)$, measured at 762 mmHg and 21.0 °C, must be absorbed by the same solution to neutralize the HCl?

19. 50.00 mL of 0.0155 M HI(aq) is mixed with 75.00 mL of 0.0106 M KOH(aq). What is the pH of the final solution?

20. 25.00 mL of a $HNO_3(aq)$ solution with a pH of 2.12 is mixed with 25.00 mL of a KOH(aq) solution with a pH of 12.65. What is the pH of the final solution?

Weak Acids, Weak Bases, and pH

(Use data from Table 16.3 as necessary.)

21. What are the $[H_3O^+]$ and pH of 0.143 M HNO_2?
22. What are the $[H_3O^+]$ and pH of 0.085 M $C_2H_5NH_2$?
23. For the ionization of phenylacetic acid,

$$C_6H_5CH_2CO_2H + H_2O \rightleftharpoons H_3O^+ + C_6H_5CH_2CO_2^-$$
$$K_a = 4.9 \times 10^{-5}$$

   (a) What is $[C_6H_5CH_2CO_2^-]$ in 0.186 M $C_6H_5CH_2CO_2H$?
   (b) What is the pH of 0.121 M $C_6H_5CH_2CO_2H$?

24. A 625 mL sample of an aqueous solution containing 0.275 mol propionic acid, $CH_3CH_2CO_2H$, has $[H_3O^+] = 0.00239$ M. What is the value of K_a for propionic acid?

$$CH_3CH_2CO_2H + H_2O \rightleftharpoons H_3O^+ + CH_3CH_2CO_2^-$$
$$K_a = ?$$

25. Fluoroacetic acid occurs in gifblaar, one of the most poisonous of all plants. A 0.318 M solution of the acid is found to have a pH = 1.56. Calculate K_a of fluoroacetic acid.

$$CH_2FCOOH(aq) + H_2O \rightleftharpoons$$
$$H_3O^+(aq) + CH_2FCOO^-(aq) \quad K_a = ?$$

26. Caproic acid, $HC_6H_{11}O_2$, found in small amounts in coconut and palm oils, is used in making artificial flavors. A saturated aqueous solution of the acid contains 11 g/L and has pH = 2.94. Calculate K_a for the acid.

$$HC_6H_{11}O_2 + H_2O \rightleftharpoons H_3O^+ + C_6H_{11}O_2^- \quad K_a = ?$$

27. What mass of benzoic acid, C_6H_5COOH, would you dissolve in 350.0 mL of water to produce a solution with a pH = 2.85?

$$C_6H_5COOH + H_2O \rightleftharpoons H_3O^+ + C_6H_5COO^-$$
$$K_a = 6.3 \times 10^{-5}$$

28. What must be the molarity of an aqueous solution of trimethylamine, $(CH_3)_3N$, if it has a pH = 11.12?

$$(CH_3)_3N + H_2O \rightleftharpoons (CH_3)_3NH^+ + OH^-$$
$$K_b = 6.3 \times 10^{-5}$$

29. What are $[H_3O^+]$, $[OH^-]$, pH, and pOH of 0.55 M M $HClO_2$?

30. What are $[H_3O^+]$, $[OH^-]$, pH, and pOH of 0.386 M CH_3NH_2?

31. The solubility of 1-naphthylamine, $C_{10}H_7NH_2$, a substance used in the manufacture of dyes, is given in a handbook as 1 g per 590 g H_2O. What is the approximate pH of a saturated aqueous solution of 1-naphthylamine?

$$C_{10}H_7NH_2 + H_2O \rightleftharpoons C_{10}H_7NH_3^+ + OH^-$$
$$pK_b = 3.92$$

32. A saturated aqueous solution of o-nitrophenol, $HOC_6H_4NO_2$, has pH = 4.53. What is the solubility of o-nitrophenol in water, in grams per liter?

$$HOC_6H_4NO_2 + H_2O \rightleftharpoons H_3O^+ + {}^-OC_6H_4NO_2$$
$$pK_a = 7.23$$

33. A particular vinegar is found to contain 5.7% acetic acid, CH_3COOH, by mass. What mass of this vinegar should be diluted with water to produce 0.750 L of a solution with pH = 4.52?

34. A particular household ammonia solution $(d = 0.97$ g/mL) is 6.8% NH_3 by mass. How many milliliters of this solution should be diluted with water to produce 625 mL of a solution with pH = 11.55?

35. A 275 mL sample of vapor in equilibrium with 1-propylamine at 25.0 °C is removed and dissolved in 0.500 L H_2O. For 1-propylamine, $pK_b = 3.43$ and v.p. = 316 Torr.
   (a) What should be the pH of the aqueous solution?
   (b) How many mg of NaOH dissolved in 0.500 L of water give the same pH?

36. One handbook lists a value of 9.5 for pK_b of quinoline, C_9H_7N, a weak base used as a preservative for anatomical specimens and to make dyes. Another handbook lists the solubility of quinoline in water at 25 °C as 0.6 g/100 mL. Use this information to calculate the pH of a saturated solution of quinoline in water.

37. The sketch on the far left represents the $[H_3O^+]$ present in an acetic acid solution of molarity M. If the molarity of the solution is doubled, which of the sketches below best represents the resulting solution?

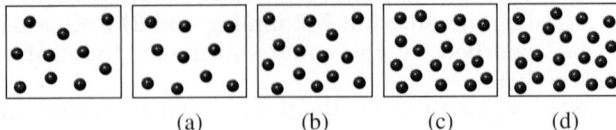

(a)   (b)   (c)   (d)

38. The sketch on the far left represents the $[OH^-]$ present in an ammonia solution of molarity M. If the solution is diluted to half its original molarity, which of the sketches below best represents the resulting solution?

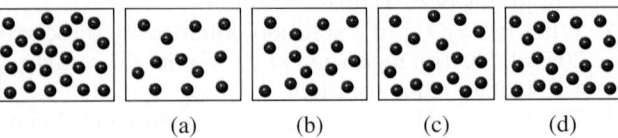

(a)   (b)   (c)   (d)

Percent Ionization

39. What is the (a) degree of ionization and (b) percent ionization of propionic acid in a solution that is 0.45 M $CH_3CH_2CO_2H$?

$$CH_3CH_2CO_2H + H_2O \rightleftharpoons H_3O^+ + CH_3CH_2CO_2^-$$
$$pK_a = 4.89$$

40. What is the (a) degree of ionization and (b) percent ionization of ethylamine, $C_2H_5NH_2$, in an 0.85 M aqueous solution?

41. What must be the molarity of an aqueous solution of NH_3 if it is 4.2% ionized?

42. What must be the molarity of an acetic acid solution if it has the same percent ionization as 0.100 M $CH_3CH_2CO_2H$ (propionic acid, $K_a = 1.3 \times 10^{-5}$)?

43. Continuing the dilutions described in Example 16-8, should we expect the percent ionization to be 13% in 0.0010 M CH_3COOH and 42% in 0.00010 M CH_3COOH? Explain.

44. What is the (a) degree of ionization and (b) percent ionization of trichloroacetic acid in a 0.035 M CCl_3COOH solution?

$$CCl_3COOH + H_2O \rightleftharpoons H_3O^+ + CCl_3COO^-$$
$$pK_a = 0.52$$

Polyprotic Acids

(Use data from Table 16.4 as necessary.)

45. Explain why $[PO_4^{3-}]$ in 1.00 M H_3PO_4 is *not* simply $\frac{1}{3}[H_3O^+]$, but much, much less than $\frac{1}{3}[H_3O^+]$.

46. Cola drinks have a phosphoric acid content that is described as "from 0.057 to 0.084% of 75% phosphoric acid, by mass." Estimate the pH range of cola drinks corresponding to this range of H_3PO_4 content.

47. Determine $[H_3O^+]$, $[HS^-]$, and $[S^{2-}]$ for the following $H_2S(aq)$ solutions: (a) 0.075 M H_2S; (b) 0.0050 M H_2S; (c) 1.0×10^{-5} M H_2S.

48. For 0.045 M H_2CO_3, a weak diprotic acid, calculate (a) $[H_3O^+]$, (b) $[HCO_3^-]$, and (c) $[CO_3^{2-}]$. Use data from Table 16.4 as necessary.

49. Calculate $[H_3O^+]$, $[HSO_4^-]$, and $[SO_4^{2-}]$ in (a) 0.75 M H_2SO_4; (b) 0.075 M H_2SO_4; (c) 7.5×10^{-4} M H_2SO_4. [*Hint:* Check any assumptions that you make.]

50. Adipic acid, $HOOC(CH_2)_4COOH$, is among the top 50 manufactured chemicals in the United States (nearly 1 million metric tons annually). Its chief use is in the manufacture of nylon. It is a *diprotic* acid having $K_{a_1} = 3.9 \times 10^{-5}$ and $K_{a_2} = 3.9 \times 10^{-6}$. A saturated solution of adipic acid is about 0.10 M $HOOC(CH_2)_4COOH$. Calculate the concentration of each ionic species in this solution.

51. The antimalarial drug quinine, $C_{20}H_{24}O_2N_2$, is a *diprotic base* with a water solubility of 1.00 g/1900 mL of solution.
    (a) Write equations for the ionization equilibria corresponding to $pK_{b_1} = 6.0$ and $pK_{b_2} = 9.8$.
    (b) What is the pH of saturated aqueous quinine?

52. For hydrazine, N_2H_4, $pK_{b_1} = 6.07$ and $pK_{b_2} = 15.05$. Draw a structural formula for hydrazine, and write equations to show the ionization of hydrazine in two distinctive steps. Calculate the pH of 0.245 M $N_2H_4(aq)$.

Ions as Acids and Bases (Hydrolysis)

53. Codeine, $C_{18}H_{21}O_3N$, is an opiate, has analgesic and antidiarrheal properties, and is widely used. In water, codeine is a weak base. A handbook gives $pK_a = 6.05$ for protonated codeine, $C_{18}H_{21}O_3NH^+$. Write the reaction for $C_{18}H_{21}O_3NH^+$ and calculate pK_b for codeine.

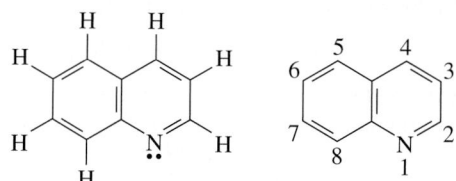

Codeine

54. Approximately 4 metric tons of quinoline, C_9H_7N, is produced annually. The principal source of quinoline is coal tar. Quinoline is a weak base in water. A handbook gives $K_a = 6.3 \times 10^{-10}$ for protonated quinoline, $C_9H_7NH^+$. Write the ionization reaction for $C_9H_7NH^+$ and calculate pK_b for quinoline.

Quinoline

55. Complete the following equations in those instances in which a reaction (hydrolysis) will occur. If no reaction occurs, so state.
(a) $NH_4^+(aq) + NO_3^-(aq) + H_2O \longrightarrow$
(b) $Na^+(aq) + NO_2^-(aq) + H_2O \longrightarrow$
(c) $K^+(aq) + C_6H_5COO^-(aq) + H_2O \longrightarrow$

Molecular Structure and Acid–Base Behavior

65. Predict which is the stronger acid: **(a)** $HClO_2$ or $HClO_3$; **(b)** H_2CO_3 or HNO_2; **(c)** H_2SiO_3 or H_3PO_4. Explain.
66. Explain why trichloroacetic acid, CCl_3COOH, is a stronger acid than acetic acid, CH_3COOH.
67. Which is the stronger acid of each of the following pairs of acids? Explain your reasoning. **(a)** HBr or HI; **(b)** HOClO or HOBr; **(c)** $I_3CCH_2CH_2COOH$ or $CH_3CH_2CCl_2COOH$.
68. Indicate which of the following is the *weakest* acid, and give reasons for your choice: HBr; $CH_2ClCOOH$; CH_3CH_2COOH; CH_2FCH_2COOH; Cl_3COOH.

(d) $K^+(aq) + Cl^-(aq) + Na^+(aq) + I^-(aq) +$ $H_2O \longrightarrow$
(e) $C_6H_5NH_3^+(aq) + Cl^-(aq) + H_2O \longrightarrow$
56. From data in Table 16.3, determine **(a)** K_a for $C_5H_5NH^+$; **(b)** K_b for $HCOO^-$; **(c)** K_b for $C_6H_5O^-$.
57. Predict whether a solution of each of the following salts is acidic, basic, or pH neutral: **(a)** KCl; **(b)** KF; **(c)** $NaNO_3$; **(d)** $Ca(OCl)_2$; **(e)** NH_4NO_2.
58. Arrange the following 0.010 M solutions in order of *increasing* pH: $NH_3(aq)$, $HNO_3(aq)$, $NaNO_2(aq)$, $CH_3COOH(aq)$, $NaOH(aq)$, $NH_4CH_3COO(aq)$, $NH_4ClO_4(aq)$.
59. What is the pH of an aqueous solution that is 0.089 M NaOCl?
60. What is the pH of an aqueous solution that is 0.123 M NH_4Cl?
61. Sorbic acid, $CH_2CH=CH=CHCH_2CO_2H$ ($pK_a = 4.77$), is widely used in the food industry as a preservative. For example, its potassium salt (potassium sorbate) is added to cheese to inhibit the formation of mold. What is the pH of 0.37 M potassium sorbate solution?
62. Pyridine, C_5H_5N ($pK_b = 8.82$), forms a salt, pyridinium chloride, as a result of a reaction with HCl. Write an ionic equation to represent the hydrolysis of the pyridinium ion, and calculate the pH of 0.0482 M $C_5H_5NH^+Cl^-(aq)$.
63. For each of the following ions, write two equations— one showing its ionization as an acid and the other as a base: **(a)** HSO_3^-; **(b)** HS^-; **(c)** HPO_4^-. Then use data from Table 16.4 to predict whether each ion makes the solution acidic or basic.
64. Suppose you wanted to produce an aqueous solution of pH = 8.65 by dissolving one of the following salts in water. Which salt would you use, and at what molarity? **(a)** NH_4Cl; **(b)** $KHSO_4$; **(c)** KNO_2; **(d)** $NaNO_3$.

69. From the following bases, select the one with the *smallest* K_b and the one with the *largest* K_b, and give reasons for your choices.

(a) ⬡—NH_2 (with Cl) **(b)** H_3C—⬡—NH_2

(c) $CH_3CH_2CH_2NH_2$ **(d)** $N\equiv CCH_2NH_2$
70. For the molecular models shown, write the formula of the species that is the most acidic and the one that is most basic, and give reasons for your choices.

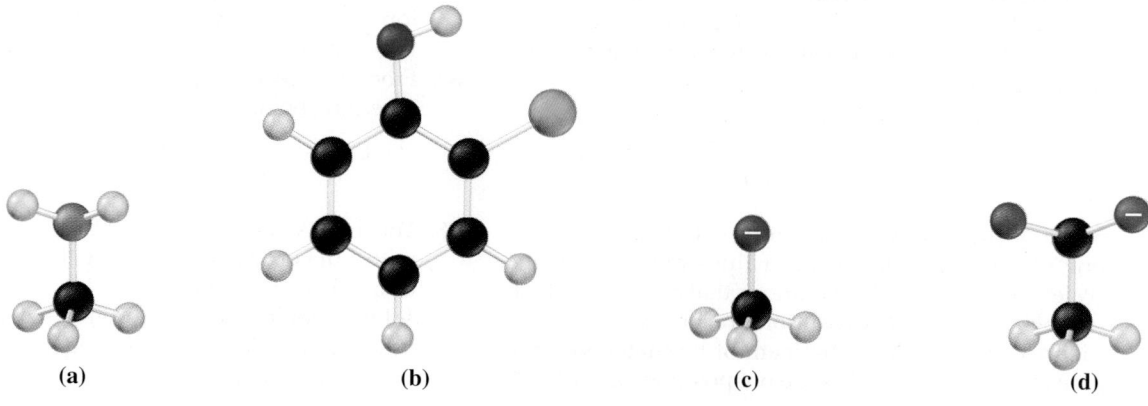

(a)   (b)   (c)   (d)

Lewis Theory of Acids and Bases

71. For each reaction draw a Lewis structure for each species and indicate which is the acid and which is the base:
(a) $CO_2 + H_2O \longrightarrow H_2CO_3$
(b) $H_2O + BF_3 \longrightarrow H_2OBF_3$
(c) $O^{2-} + H_2O \longrightarrow 2\,OH^-$
(d) $S^{2-} + SO_3 \longrightarrow S_2O_3^{2-}$

72. In the following reactions indicate which is the Lewis acid and which is the Lewis base:
(a) $SOI_2 + BaSO_3 \longrightarrow Ba^{2+} + 2\,I^- + 2\,SO_2$
(b) $HgCl_3^- + Cl^- \longrightarrow HgCl_4^{2-}$

73. Indicate whether each of the following is a Lewis acid or base. (a) OH^-; (b) $(C_2H_5)_3B$; (c) CH_3NH_2.

74. Each of the following is a Lewis acid–base reaction. Which reactant is the acid, and which is the base? Explain.
(a) $SO_3 + H_2O \longrightarrow H_2SO_4$
(b) $Zn(OH)_2(s) + 2\,OH^-(aq) \longrightarrow [Zn(OH)_4]^{2-}(aq)$

75. The three following reactions are acid–base reactions according to the Lewis theory. Draw Lewis structures, and identify the Lewis acid and Lewis base in each reaction.
(a) $B(OH)_3 + OH^- \longrightarrow [B(OH)_4]^-$
(b) $N_2H_4 + H_3O^+ \longrightarrow N_2H_5^+ + H_2O$
(c) $(C_2H_5)_2O + BF_3 \longrightarrow (C_2H_5)_2OBF_3$

76. $CO_2(g)$ can be removed from confined quarters (such as a spacecraft) by allowing it to react with an alkali metal hydroxide. Show that this is a Lewis acid–base reaction. For example,

$$CO_2(g) + LiOH(s) \longrightarrow LiHCO_3(s)$$

77. The molecular solid $I_2(s)$ is only slightly soluble in water but will dissolve to a much greater extent in an aqueous solution of KI, because the I_3^- anion forms. Write an equation for the formation of the I_3^- anion, and indicate the Lewis acid and Lewis base.

78. The following very strong acids are formed by the reactions indicated:

$$HF + SbF_5 \longrightarrow HSbF_6$$
(called "super acid," hexafluoroantimonic acid)

$$HF + BF_3 \longrightarrow HBF_4$$
(tetrafluoroboric acid)

(a) Identify the Lewis acids and bases.
(b) To which atom is the H atom bonded in each acid?

79. Use Lewis structures to diagram the following reaction in the manner of reaction (16.20).

$$H_2O + SO_2 \longrightarrow H_2SO_3$$

Identify the Lewis acid and Lewis base.

80. Use Lewis structures to diagram the following reaction in the manner of reaction (16.19).

$$2\,NH_3 + Ag^+ \longrightarrow [Ag(NH_3)_2]^+$$

Identify the Lewis acid and Lewis base.

Integrative and Advanced Exercises

81. The Brønsted–Lowry theory can be applied to acid–base reactions in nonaqueous solvents, where the relative strengths of acids and bases can differ from what they are in aqueous solutions. Indicate whether each of the following would be an acid, a base, or amphiprotic in pure liquid acetic acid, CH_3COOH, as a solvent. (a) CH_3COO^-; (b) H_2O; (c) CH_3COOH; (d) $HClO_4$. [*Hint:* Refer to Table 16.1.]

82. The pH of saturated $Sr(OH)_2(aq)$ is found to be 13.12. A 10.0 mL sample of saturated $Sr(OH)_2(aq)$ is diluted to 250.0 mL in a volumetric flask. A 10.0 mL sample of the diluted $Sr(OH)_2(aq)$ is transferred to a beaker, and some water is added. The resulting solution requires 25.1 mL of a HCl solution for its titration. What is the molarity of this HCl solution?

83. Several approximate pH values are marked on the following pH scale.

Some of the following solutions can be matched to one of the approximate pH values marked on the scale; others cannot. For solutions that can be matched to a pH value, identify each solution and its pH value. Identify the solutions that cannot be matched, and give reasons why matches are not possible. (a) 0.010 M H_2SO_4; (b) 1.0 M NH_4Cl; (c) 0.050 M KI; (d) 0.0020 M CH_3NH_2; (e) 1.0 M NaOCl; (f) 0.10 M C_6H_5OH; (g) 0.10 M HOCl; (h) 0.050 M $ClCH_2COOH$; (i) 0.050 M HCOOH.

84. Show that when $[H_3O^+]$ is reduced to half its original value, the pH of a solution increases by 0.30 unit, *regardless of the initial pH.* Is it also true that when any solution is diluted to half its original concentration, the pH increases by 0.30 unit? Explain.

85. Explain why $[H_3O^+]$ in a strong acid solution *doubles* as the total acid concentration doubles, whereas in a weak acid solution, $[H_3O^+]$ increases only by about a factor of $\sqrt{2}$.

86. Use data from Appendix D to determine whether the ion product of water, K_w, increases, decreases, or remains unchanged with increasing temperature.

87. From the observation that 0.0500 M vinylacetic acid has a freezing point of $-0.096\,°C$, determine K_a for this acid.

$$CH_2{=}CHCH_2CO_2H + H_2O \rightleftharpoons H_3O^+ + CH_2{=}CHCH_2CO_2^-$$

88. You are asked to prepare a 100.0 mL sample of a solution with a pH of 5.50 by dissolving the appropriate amount of a solute in water with pH = 7.00. Which of these solutes would you use, and in what quantity? Explain your choice. (a) 15 M $NH_3(aq)$; (b) 12 M $HCl(aq)$; (c) $NH_4Cl(s)$; (d) glacial (pure) acetic acid, CH_3COOH.

89. Determine the pH of (a) 1.0×10^{-5} M HCN and (b) 1.0×10^{-5} M $C_6H_5NH_2$ (aniline).

90. It is possible to write simple equations to relate pH, pK, and molarities (M) of various solutions. Three such equations are shown here.

Weak acid: $pH = \dfrac{1}{2}pK_a - \dfrac{1}{2}\log M$

Weak base: $pH = 14.00 - \dfrac{1}{2}pK_b + \dfrac{1}{2}\log M$

Salt of weak
acid (pK_a)
and strong
base: $pH = 14.00 - \dfrac{1}{2}pK_w + \dfrac{1}{2}pK_a + \dfrac{1}{2}\log M$

(a) Derive these three equations, and point out the assumptions involved in the derivations.
(b) Use these equations to determine the pH of 0.10 M $CH_3COOH(aq)$, 0.10 M $NH_3(aq)$, and 0.10 M $NaCH_3COO$. Verify that the equations give correct results by determining these pH values in the usual way.

91. A handbook lists the following formula for the percent ionization of a weak acid.

$$\% \text{ ionized} = \dfrac{100}{1 + 10^{(pK - pH)}}$$

(a) Derive this equation. What assumptions must you make in this derivation?
(b) Use the equation to determine the percent ionization of a formic acid solution, HCOOH(aq), with a pH of 2.50.
(c) A 0.150 M solution of propionic acid, CH_3CH_2COOH, has a pH of 2.85. What is K_a for propionic acid?

$$CH_3CH_2CO_2H + H_2O \rightleftharpoons H_3O^+ + CH_3CH_2CO_2^-$$

92. Oxalic acid, HOOCCOOH, a weak diprotic acid, has $pK_{a_1} = 1.25$ and $pK_{a_2} = 3.81$. A related diprotic acid, suberic acid, $HOOC(CH_2)_8COOH$ has $pK_{a_1} = 4.21$ and $pK_{a_2} = 5.40$. Offer a plausible reason as to why the *difference* between pK_{a_1} and pK_{a_2} is so much greater for oxalic acid than for suberic acid.

93. Here is a way to test the validity of the statement made on page 719 in conjunction with the three key

ideas governing the ionization of polyprotic acids. Determine the pH of 0.100 M succinic acid in two ways: first by assuming that H_3O^+ is produced only in the first ionization step, and then by allowing for the possibility that some H_3O^+ is also produced in the second ionization step. Compare the results, and discuss the significance of your finding.

$$H_2C_4H_4O_4 + H_2O \rightleftharpoons H_3O^+ + HC_4H_4O_4^-$$
$$K_{a_1} = 6.2 \times 10^{-5}$$

$$HC_4H_4O_4^- + H_2O \rightleftharpoons H_3O^+ + C_4H_4O_4^{2-}$$
$$K_{a_2} = 2.3 \times 10^{-6}$$

94. What mass of acetic acid, CH_3COOH, must be dissolved per liter of aqueous solution if the solution is to have the same freezing point as 0.150 M $ClCH_2COOH$ (chloroacetic acid)?

95. What is the pH of a solution that is 0.68 M H_2SO_4 and 1.5 M HCOOH (formic acid)?

96. An aqueous solution of two weak acids has a stoichiometric molarity, M, in each acid. If one acid has a K_a value twice as large as the other, show that the pH of the solution is given by the equation $pH = -\frac{1}{2}\log 3\,M\,K_a$. Assume that the criteria for the simplifying assumption on page 715 are met.

97. Use the concept of hybrid orbitals to describe the bonding in the strong acids given in Exercise 78.

98. Phosphorous acid is listed in Appendix D as a *diprotic* acid. Propose a Lewis structure for phosphorous acid that is consistent with this fact.

99. The following four equilibria lie to the right: $N_2H_5^+ + CH_3NH_2 \longrightarrow N_2H_4 + CH_3NH_3^+$; $H_2SO_3 + F^- \longrightarrow HSO_3^- + HF$; $CH_3NH_3^+ + OH^- \longrightarrow CH_3NH_2 + H_2O$; and $HF + N_2H_4 \longrightarrow F^- + N_2H_5^+$.
(a) Rank all the acids involved in order of decreasing acid strength.
(b) Rank all the bases involved in order of decreasing base strength.
(c) State whether each of the following two equilibria lies primarily to the right or to the left: (i) $HF + OH^- \longrightarrow F^- + H_2O$; (ii) $CH_3NH_3^+ + HSO_3^- \longrightarrow CH_3NH_2 + H_2SO_3$.

Feature Problems

100. Maleic acid is a carbon–hydrogen–oxygen compound used in dyeing and finishing fabrics and as a preservative of oils and fats. In a combustion analysis, a 1.054 g sample of maleic acid yields 1.599 g CO_2 and 0.327 g H_2O. In a freezing-point depression experiment, a 0.615 g sample of maleic acid dissolved in 25.10 g of glacial acetic acid, $CH_3COOH(l)$ (which has the freezing-point depression constant $K_f = 3.90\ °C\ m^{-1}$ and in which maleic acid does not ionize), lowers the freezing point by 0.82°C. In a titration experiment, a 0.4250 g sample of maleic acid is dissolved in water and requires 34.03 mL of 0.2152 M KOH for its complete neutralization. The pH of a 0.215 g sample of maleic

acid dissolved in 50.00 mL of aqueous solution is found to be 1.80.
(a) Determine the empirical and molecular formulas of maleic acid. [*Hint:* Which experiment(s) provide the necessary data?]
(b) Use the results of part (a) and the titration data to rewrite the molecular formula to reflect the number of ionizable H atoms in the molecule.
(c) Given that the ionizable H atom(s) is(are) associated with the carboxyl group(s), write the plausible condensed structural formula of maleic acid.
(d) Determine the ionization constant(s) of maleic acid. If the data supplied are insufficient, indicate what additional data would be needed.

(e) Calculate the expected pH of a 0.0500 M aqueous solution of maleic acid. Indicate any assumptions required in this calculation.

101. In Example 16-7, rather than use the quadratic formula to solve the quadratic equation, we could have proceeded in the following way. Substitute the value yielded by our failed assumption—$x = 0.0010$—into the *denominator* of the quadratic equation; that is, use $(0.00250 - 0.0010)$ as the value of $[CH_3NH_2]$ and solve for a new value of x. Use this second value of x to re-evaluate $[CH_3NH_2]$: $[CH_3NH_2] = (0.00250 - \text{second value of } x)$. Solve the simple quadratic equation for a third value of x, and so on. After three or four trials, you will find that the value of x no longer changes. This is the answer you are seeking. **(a)** Complete the calculation of the pH of 0.00250 M CH_3NH_2 by this method, and show that the result is the same as that obtained by using the quadratic formula. **(b)** Use this method to determine the pH of 0.500 M $HClO_2$.

102. Apply the general method for solution equilibrium calculations outlined on page 720 to determine the pH values of the following solutions. In applying the method, look for valid assumptions that may simplify the numerical calculations.
   (a) a solution that is 0.315 M CH_3COOH and 0.250 M $HCOOH$
   (b) a solution that contains 1.55 g CH_3NH_2 and 12.5 g NH_3 in 375 mL
   (c) 1.0 M $NH_4CN(aq)$

Self-Assessment Exercises

103. In your own words, define or explain the following terms or symbols: **(a)** K_w; **(b)** pH; **(c)** pK_a; **(d)** hydrolysis; **(e)** Lewis acid.

104. Briefly describe each of the following ideas or phenomena: **(a)** conjugate base; **(b)** percent ionization of an acid or base; **(c)** self-ionization; **(d)** amphiprotic behavior.

105. Explain the important distinctions between each pair of terms: **(a)** Brønsted–Lowry acid and base; **(b)** $[H_3O^+]$ and pH; **(c)** K_a for NH_4^+ and K_b for NH_3; **(d)** leveling effect and electron-withdrawing effect.

106. Of the following, the amphiprotic ion is **(a)** HCO_3^-; **(b)** CO_3^{2-}; **(c)** NH_4^+; **(d)** $CH_3NH_3^+$; **(e)** ClO_4^-.

107. The pH in 0.10 M $CH_3CH_2COOH(aq)$ must be **(a)** equal to $[H_3O^+]$ in 0.10 M $HNO_2(aq)$; **(b)** less than the pH in 0.10 M $HI(aq)$; **(c)** greater than the pH in 0.10 M $HBr(aq)$; **(d)** equal to 1.0.

108. In 0.10 M $CH_3NH_2(aq)$, **(a)** $[H_3O^+] = 0.10$ M; **(b)** $[OH^-] = 0.10$ M; **(c)** pH < 7; **(d)** pH < 13.

109. The reaction of $CH_3COOH(aq)$ proceeds furthest toward completion with a base when that base is **(a)** H_2O; **(b)** $CH_3NH_3^+$; **(c)** NH_4^+; **(d)** Cl^-; **(e)** CO_3^{2-}.

110. In 0.10 M $H_2SO_4(aq)$, $[H_3O^+]$ is equal to **(a)** 0.050 M; **(b)** 0.10 M; **(c)** 0.11 M; **(d)** 0.20 M.

111. For $H_2SO_3(aq)$, $K_{a_1} = 1.3 \times 10^{-2}$ and $K_{a_2} = 6.3 \times 10^{-8}$. In 0.10 M $H_2SO_3(aq)$, **(a)** $[HSO_3^-] = 0.013$ M; **(b)** $[SO_3^{2-}] = 6.3 \times 10^{-8}$ M; **(c)** $[H_3O^+] = 0.10$ M; **(d)** $[H_3O^+] = 0.013$ M; **(e)** $[SO_3^{2-}] = 0.036$ M.

112. What is the pH of the solution obtained by mixing 24.80 mL of 0.248 M HNO_3 and 15.40 mL of 0.394 M KOH?

113. How many milliliters of a concentrated acetic acid solution (35.0% CH_3COOH by mass; $d = 1.044$ g/mL) must be diluted with water to produce 12.5 L of solution with pH 3.25?

114. Determine the pH of 2.05 M $NaCH_2ClCOO$. (Use data from Table 16.3, as necessary.)

115. Several aqueous solutions are prepared. *Without consulting any tables in the text*, arrange these ten solutions in order of *increasing* pH: 1.0 M NaBr, 0.05 M CH_3COOH, 0.05 M NH_3, 0.02 M KCH_3COO, 0.05 M $Ba(OH)_2$, 0.05 M H_2SO_4, 0.10 M HI, 0.06 M NaOH, 0.05 M NH_4Cl, and 0.05 M $CH_2ClCOOH$.

116. A solution is found to have pH = 5 × pOH. Is this solution acidic or basic? What is $[H_3O^+]$ in the solution? Which of the following could be the solute in this solution: NH_3, CH_3COOH, or NH_4CH_3COO, and what would be its molarity?

117. Propionic acid, CH_3CH_2COOH, is 0.42% ionized in 0.80 M solution. The K_a for this acid is **(a)** 1.42×10^{-5}; **(b)** 1.42×10^{-7}; **(c)** 1.77×10^{-5}; **(d)** 6.15×10^4; **(e)** none of these.

118. The conjugate acid of HPO_4^{2-} is **(a)** PO_4^{3-}; **(b)** $H_2PO_4^-$; **(c)** H_3PO_4; **(d)** H_3O^+; **(e)** none of these.

119. The equilibria $OH^- + HClO \longrightarrow H_2O + ClO^-$ and $ClO^- + HNO_2 \longrightarrow HClO + NO_2^-$ both lie to the right. Which of the following is a list of acids ranked in order of decreasing strength?
   (a) $HClO > HNO_2 > H_2O$
   (b) $ClO^- > NO_2^- > OH^-$
   (c) $NO_2^- > ClO^- > OH^-$
   (d) $HNO_2 > HClO > H_2O$
   (e) none of these

120. 3.00 mol of calcium chlorite is dissolved in enough water to produce 2.50 L of solution. $K_a = 2.9 \times 10^{-8}$ for HClO, and $K_a = 1.1 \times 10^{-2}$ for $HClO_2$. Compute the pH of the solution.

121. Appendix E describes a useful study aid known as concept mapping. Using the method presented in Appendix E, construct a concept map that summarizes the material discussed in Section 16-8.

Additional Aspects of Acid–Base Equilibria

17

CONTENTS

NaOH(aq) is slowly added to an aqueous solution containing HCl(aq) and the indicator phenolphthalein. The indicator color changes from colorless to red as the pH changes from 8.0 to 10.0. The equivalence point of the neutralization is reached when the solution turns a lasting pink (the pink seen here disappears when the flask is swirled to mix the reactants). The selection of indicators for acid–base titrations is one of the topics considered in this chapter.

I n our study of acid rain (see Focus On 16-1 on the MasteringChemistry website), we learned that a very small amount of atmospheric $CO_2(g)$ dissolves in rainwater. Yet this amount is sufficient to lower the pH of rainwater by nearly 2 units. And when acid-forming air pollutants, such as SO_2, SO_3, and NO_2, also dissolve in rainwater, it becomes even more acidic. A chemist would say that water has no "buffer capacity"—that is, its pH changes sharply when even small quantities of acids or bases are dissolved in it.

One of the main topics of this chapter is buffer solutions—solutions that can resist a change in pH when acids or bases are added to them. We will consider how such solutions are prepared, how they maintain a nearly constant pH, and how they are used. At the end of the chapter, we will consider perhaps the most important buffer system to humans: the buffer system that maintains the constant pH of blood.

▲ FIGURE 17-1
A weak acid–strong acid mixture
The solution pictured is 0.100 M CH_3COOH and 0.100 M HCl. The reading on the pH meter (1.0) indicates that essentially all the H_3O^+ comes from the strong acid HCl. The red color of the solution is that of thymol blue indicator. Compare this photo with Figure 16-7, in which the separate acids are shown.

▶ The common-ion effect is not restricted to weak acids and weak bases. Buffers are the most important examples of the common-ion effect in weak acids and weak bases.

A second topic that we will explore is acid–base titrations. Here, our aim will be to calculate how pH changes during a titration. We can use this information to select an appropriate indicator for a titration and to determine, in general, which acid–base titrations work well and which do not. For the most part, we will find the calculations in this chapter to be extensions of those in Chapter 16.

17-1   Common-Ion Effect in Acid–Base Equilibria

The questions answered in Chapter 16 were mostly of the type, "What is the pH of 0.10 M $HC_2H_3O_2$, of 0.10 M NH_3, of 0.10 M H_3PO_4, of 0.10 M NH_4Cl?" In each of these cases, we think of dissolving a *single* substance in aqueous solution and determining the concentrations of the species present at equilibrium. In most situations in this chapter, a solution of a weak acid or weak base initially contains a second source of one of the ions produced in the ionization of the acid or base. The added ions are said to be *common* to the weak acid or weak base. The presence of a common ion can have some important consequences.

Solutions of Weak Acids and Strong Acids

Consider a solution that is at the same time 0.100 M CH_3COOH and 0.100 M HCl. We can write separate equations for the ionizations of the acids, one weak and the other strong, and indicate the expected concentrations of molecules and ions in the solution.

$$CH_3COOH + H_2O \rightleftharpoons H_3O^+ + CH_3COO^- \qquad K_a = 1.8 \times 10^{-5}$$
$$(0.100 - x)\,M \qquad\qquad x\,M \qquad x\,M$$
$$HCl + H_2O \longrightarrow H_3O^+ + Cl^-$$
$$0.100\,M \qquad 0.100\,M$$

Of course, there can be only a single concentration of H_3O^+ in the solution, and this must be $[H_3O^+] = (0.100 + x)$ M. Because H_3O^+ is formed in both ionization processes, we say that it is a *common ion*. The weak acid–strong acid mixture described here is pictured in Figure 17-1. Although it might seem that the pH would be lower than 1.0, the figure indicates that this is not the case.

Concentrations of the species present in this mixture of a weak acid and a strong acid are calculated in Example 17-1, followed by comments on the significance of the result.

EXAMPLE 17-1   **Demonstrating the Common-Ion Effect: A Solution of a Weak Acid and a Strong Acid**

(a) Determine $[H_3O^+]$ and $[CH_3COO^-]$ in 0.100 M CH_3COOH. (b) Then determine these same quantities in a solution that is 0.100 M in both CH_3COOH and HCl.

Analyze

In part (a) we must determine the species in a weak acid solution, and in part (b) we investigate the effects of the addition of a strong acid. The two acids have the common ion H_3O^+.

Solve

(a) This calculation was done in Example 16-6 (page 712). We found that in 0.100 M CH_3COOH, $[H_3O^+] = [CH_3COO^-] = 1.3 \times 10^{-3}$ M.

(b) Instead of writing two separate ionization equations, as we did just prior to this example, let's write only the ionization equation for CH_3COOH and enter information about the common ion, H_3O^+, in the following format.

	CH_3COOH	$+$	H_2O	$\rightleftharpoons$	H_3O^+	$+$	CH_3COO^-
initial concns:							
weak acid:	0.100 M				—		—
strong acid:	—				0.100 M		—
changes:	$-x$ M				$+x$ M		$+x$ M
equil concns:	$(0.100 - x)$ M				$(0.100 + x)$ M		x M

As is customary, we begin with the assumption that x is very small compared with 0.100. Thus, $0.100 - x \approx 0.100 + x \approx 0.100$.

$$K_a = \frac{[H_3O^+][CH_3COO^-]}{[CH_3COOH]} = \frac{(0.100 + x) \cdot x}{0.100 - x} = \frac{0.100 \cdot x}{0.100} = 1.8 \times 10^{-5}$$

$$x = [CH_3COO^-] = 1.8 \times 10^{-5} \text{ M} \quad 0.100 + x = [H_3O^+] = 0.100 \text{ M}$$

Notice that x is only 0.018% of 0.100, and so the assumption that x is small compared with 0.100 is valid.

Assess

In this example the acetate ion concentration goes down after the addition of hydronium ion. In other words, ionization of acetic acid is suppressed by the addition of strong acid. The common-ion effect is not restricted to weak acids and weak bases. Buffers are the most important examples of the common-ion effect in weak acids and weak bases. This shift in equilibrium concentration because of the addition of a common ion is another illustration of Le Châtelier's principle.

PRACTICE EXAMPLE A: Determine $[H_3O^+]$ and $[HF]$ in 0.500 M HF. Then determine these concentrations in a solution that is 0.100 M HCl and 0.500 M HF.

PRACTICE EXAMPLE B: How many drops of 12 M HCl would you add to 1.00 L of 0.100 M CH_3COOH to make $[CH_3COO^-] = 1.0 \times 10^{-4}$ M? Assume that 1 drop = 0.050 mL and that the volume of solution remains 1.00 L after the 12 M HCl is diluted. [*Hint*: What must be the $[H_3O^+]$ in the solution?]

Now we see the consequence of adding a strong acid (HCl) to a weak acid (CH_3COOH): The concentration of the anion $[CH_3COO^-]$ is greatly reduced. Between parts (a) and (b) of Example 17-1, $[CH_3COO^-]$ is lowered from 10^{-3} M to 1.8×10^{-5} M—almost a 100-fold decrease. Another way to state this result is through Le Châtelier's principle (see Chapter 15, page 673). Increasing the concentration of one of the *products* of a reaction—the common ion—shifts the equilibrium condition in the *reverse* direction. The **common-ion effect** is the suppression of the ionization of a weak electrolyte caused by adding more of an ion that is a product of this ionization. The common-ion effect of H_3O^+ on the ionization of acetic acid is suggested as follows.

When a strong acid supplies the common ion H_3O^+, the equilibrium shifts to form more CH_3COOH.

Added H_3O^+

$$CH_3COOH + H_2O \rightleftharpoons H_3O^+ + CH_3COO^- \quad K_a = 1.8 \times 10^{-5}$$

Equilibrium shifts to form more CH_3COOH

(a) **(b)**

▲ FIGURE 17-2
A mixture of a weak acid and its salt
Bromphenol blue indicator is present in both solutions. Its color dependence on pH is

pH < 3.0 < pH < 4.6 < pH

| Yellow | Green | Blue-Violet |

(a) 0.100 M CH_3COOH has a calculated pH of 2.89, but **(b)** if the solution is also 0.100 M in $NaCH_3COO$, the calculated pH is 4.74. (The readability of the pH meters used here is 0.1 unit, and their accuracy is probably somewhat less than that. The discrepancy between 4.74 and the 4.9 value shown here is a result of their limited accuracy.)

The ionization of a weak base, such as NH_3, is suppressed when a strong base, such as NaOH, is added. Here, OH^- is the common ion, and its increased concentration shifts the equilibrium to the left.

When a strong base supplies the common ion OH^-,
the equilibrium shifts to form more NH_3.

Added OH^-

$$NH_3 + H_2O \rightleftharpoons NH_4^+ + OH^- \qquad K_b = 1.8 \times 10^{-5}$$

Equilibrium shifts to
form more NH_3

Solutions of Weak Acids and Their Salts

The salt of a weak acid is a strong electrolyte—its ions become completely dissociated from one another in aqueous solution. One of the ions, the *anion*, is an ion common to the ionization equilibrium of the weak acid. The presence of this common ion suppresses the ionization of the weak acid. For example, we can represent the effect of acetate salts on the acetic acid equilibrium as

When a salt supplies the common anion CH_3COO^-,
the equilibrium shifts to form more CH_3COOH.

$$NaCH_3COO(aq) \longrightarrow Na^+ + CH_3COO^-$$

Added CH_3COO^-

$$CH_3COOH + H_2O \rightleftharpoons H_3O^+ + CH_3COO^- \qquad K_a = 1.8 \times 10^{-5}$$

Equilibrium shifts to
form more CH_3COOH

▶ When concentrations of a weak acid and its conjugate base are the same, $K_a = [H_3O^+]$ and pH = pK_a.

The common-ion effect of acetate ion on the ionization of acetic acid is depicted in Figure 17-2 and demonstrated in Example 17-2. In solving common-ion problems, such as Example 17-2, assume that ionization of the weak acid (or base) does not begin until both the weak acid (or base) and its salt have been placed in solution. Then consider that ionization occurs until equilibrium is reached.

EXAMPLE 17-2 **Demonstrating the Common-Ion Effect: A Solution of a Weak Acid and a Salt of That Weak Acid**

Calculate $[H_3O^+]$ and $[CH_3COO^-]$ in a solution that is 0.100 M in both CH_3COOH and $NaCH_3COO$.

Analyze

This example is very similar to Example 17-1; however, in this case we will be adding a salt of a weak acid and observing the shift in equilibrium. The setup shown here is very similar to that in Example 17-1(b), except that $NaCH_3COO$ is the source of the common ion. This is another illustration of Le Châtelier's principle.

Solve

Begin by setting up the ICE table.

	CH_3COOH	+	H_2O	$\rightleftharpoons$	H_3O^+	+	CH_3COO^-
initial concns:							
weak acid:	0.100 M				—		—
salt:	—				—		0.100 M
changes:	$-x$ M				$+x$ M		$+x$ M
equil concns:	$(0.100 - x)$ M				x M		$(0.100 + x)$ M

Because the salt suppresses the ionization of $HC_2H_3O_2$, we expect $[H_3O^+] = x$ to be very small and $0.100 - x \approx 0.100 + x \approx 0.100$. This proves to be a valid assumption.

$$K_a = \frac{[H_3O^+][CH_3COO^-]}{[CH_3COOH]} = \frac{x \cdot (0.100 + x)}{0.100 - x} = \frac{x \cdot 0.100}{0.100} = 1.8 \times 10^{-5}$$

$$x = [H_3O^+] = 1.8 \times 10^{-5}\,M \qquad 0.100 + x = [CH_3COO^-] = 0.100\,M$$

Assess

The ionization of CH_3COOH is reduced about 100-fold because of the salt that was added. The calculations we performed in this example are very similar to those we did in Example 17-1(b). An important difference, however, is that here we solved for $x = [H_3O^+]$. In Example 17-1(b), we solved for $x = [CH_3COO^-]$. Note that when the concentrations of a weak acid and its conjugate base are the same, $K_a = [H_3O^+]$ and $pH = pK_a$.

PRACTICE EXAMPLE A: Calculate $[H_3O^+]$ and $[HCOO^-]$ in a solution that is 0.100 M HCOOH and 0.150 M NaHCOO.

PRACTICE EXAMPLE B: What mass of $NaCH_3COO$ should be added to 1.00 L of 0.100 M CH_3COOH to produce a solution with pH = 5.00? Assume that the volume remains 1.00 L.

Solutions of Weak Bases and Their Salts

The common-ion effect of a salt of a weak base is similar to the weak acid–anion situation just described. The suppression of the ionization of NH_3 by the common *cation*, NH_4^+, is pictured in Figure 17-3 and represented as follows.

When a salt supplies the common cation NH_4^+, the equilibrium shifts to form more NH_3.

$$NH_4Cl(aq) \longrightarrow NH_4^+ + Cl^-$$

Added NH_4^+

$$NH_3 + H_2O \rightleftharpoons NH_4^+ + OH^- \quad K_b = 1.8 \times 10^{-5}$$

Equilibrium shifts to form more NH_3

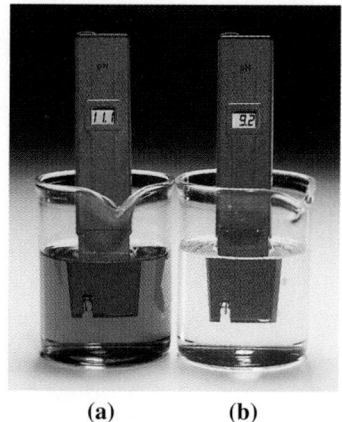

▲ FIGURE 17-3
A mixture of a weak base and its salt
Thymolphthalein indicator is blue if pH > 10 and colorless if pH < 10. **(a)** The pH of 0.100 M NH_3 is above 10 (calculated value: 11.11). **(b)** If the solution is also 0.100 M NH_4Cl, the pH drops below 10 (calculated value: 9.26). The ionization of NH_3 is suppressed in the presence of added NH_4^+. $[OH^-]$ decreases, $[H_3O^+]$ increases, and the pH is lowered.

🔍 17-1   CONCEPT ASSESSMENT

Without doing detailed calculations, determine which of the following will *raise* the pH when added to 1.00 L of 0.100 M NH_3(aq): **(a)** 0.010 mol NH_4Cl(s); **(b)** 0.010 mol $(CH_3CH_2)_2NH$(l); **(c)** 0.010 mol HCl(g); **(d)** 1.00 L of 0.050 M NH_3(aq); **(e)** 1.00 g $Ca(OH)_2$(s). For NH_3, $pK_b = 4.74$; for $(CH_3CH_2)_2NH$, $pK_b = 3.16$.

17-2 Buffer Solutions

Figure 17-4 illustrates a statement made in the chapter introduction: Pure water has no buffer capacity. There are some aqueous solutions, however, called **buffer** (or buffered) **solutions**, whose pH values change only very slightly on the addition of small amounts of either an acid or a base.

What buffer solutions require are two components; one component is able to neutralize acids, and the other is able to neutralize bases. But the two components must not neutralize each other. This requirement rules out mixtures of a strong acid and a strong base. Instead, common buffer solutions are described as combinations of

▶ Recall the Brønsted–Lowry theory and the conjugate acid–base definitions.

- a weak acid and its conjugate base, or
- a weak base and its conjugate acid

To show that such mixtures function as buffer solutions, let's consider a solution that has the equilibrium concentrations $[CH_3COOH] = [CH_3COO^-]$. As summarized in expression (17.1), in this solution $[H_3O^+] = K_a = 1.8 \times 10^{-5}$ M.

$$K_a = \frac{[H_3O^+][CH_3COO^-]}{[CH_3COOH]} = 1.8 \times 10^{-5}$$

$$[H_3O^+] = K_a \times \frac{[\cancel{CH_3COOH}]}{[\cancel{CH_3COO^-}]} = 1.8 \times 10^{-5} \text{ M} \qquad \textbf{(17.1)}$$

As a result, pH $= -\log[H_3O^+] = -\log K_a = -\log 1.8 \times 10^{-5} = 4.74$.

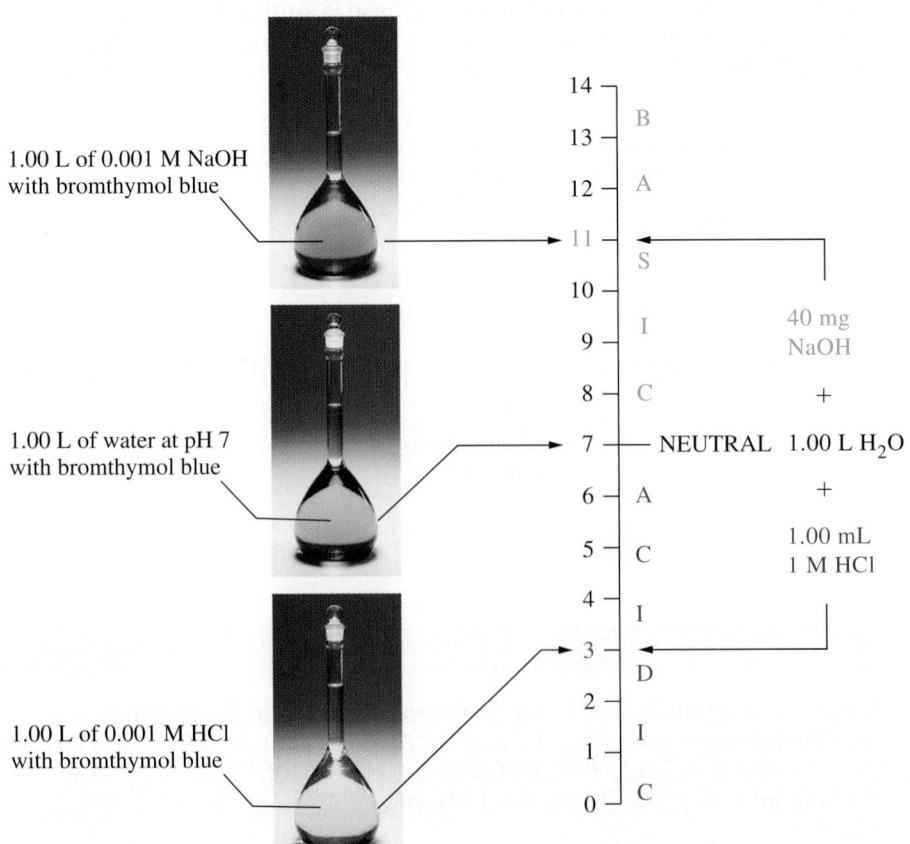

1.00 L of 0.001 M NaOH with bromthymol blue

1.00 L of water at pH 7 with bromthymol blue

1.00 L of 0.001 M HCl with bromthymol blue

14
13 — B
12 — A
11 —
  — S
10 —
9 — I
8 — C
7 — NEUTRAL
6 — A
5 — C
4 — I
3 — D
2 —
1 — I
0 — C

40 mg
NaOH

+

1.00 L H₂O

+

1.00 mL
1 M HCl

▶ FIGURE 17-4
Pure water has no buffering ability
Bromthymol blue indicator is blue at pH > 7, green at pH = 7, and yellow at pH < 7. Pure water has pH = 7.0. The addition of 0.001 mol H_3O^+ (1.00 mL of 1 M HCl) to 1.00 L water produces $[H_3O^+] =$ 0.001 M and pH = 3.0. The addition of 0.001 mol OH^- (40 mg NaOH) to 1.00 L of water produces $[OH^-] = 0.001$ M and pH = 11.0.

Now, imagine adding a *small* amount of a strong acid to this buffer solution. A reaction occurs in which a *small* amount of the base CH_3COO^- is converted to its conjugate acid CH_3COOH.

$$CH_3COO^- + H_3O^+ \longrightarrow CH_3COOH + H_2O$$

After the neutralization of the added H_3O^+, we find that in expression (17.1) $[CH_3COOH]$ has increased *slightly* and $[CH_3COO^-]$ has decreased *slightly*. The ratio $[CH_3COOH]/[CH_3COO^-]$ is only *slightly* greater than 1, and $[H_3O^+]$ has barely changed. The buffer solution has resisted a change in pH following the addition of a small amount of acid; the pH remains close to the original 4.74.

Next, imagine adding a *small* amount of a strong base to the original buffer solution with $[CH_3COOH] = [CH_3COO^-]$. A reaction occurs in which a *small* amount of the weak acid CH_3COOH is converted to its conjugate base CH_3COO^-.

$$CH_3COOH + OH^- \longrightarrow CH_3COO^- + H_2O$$

Here we find that $[CH_3COO^-]$ has increased *slightly*, and $[CH_3COOH]$ has decreased *slightly*. The ratio $[CH_3COOH]/[CH_3COO^-]$ is only *slightly* smaller than 1, and again $[H_3O^+]$ has barely changed. The buffer solution has resisted a change in pH following the addition of a small amount of base; again, the pH remains close to the original 4.74. The variation of the concentration of the weak acid and its conjugate base are illustrated in Figure 17-5.

Later in this section, we will be more specific about what constitutes *small* additions of an acid or a base and *slight* changes in the concentrations of the buffer components and pH. Also, we will discover that an acetic acid–sodium acetate buffer is good only for maintaining a nearly constant pH in a range of about 2 pH units centered on a pH $= pK_a = 4.74$. A buffer solution that maintains a nearly constant pH outside this range requires different buffer components, as suggested in Example 17-3.

We commonly need to calculate the pH of a buffer solution. At a minimum, such a calculation requires use of the ionization constant expression for a weak acid or weak base. Aspects of solution stoichiometry may also be required.

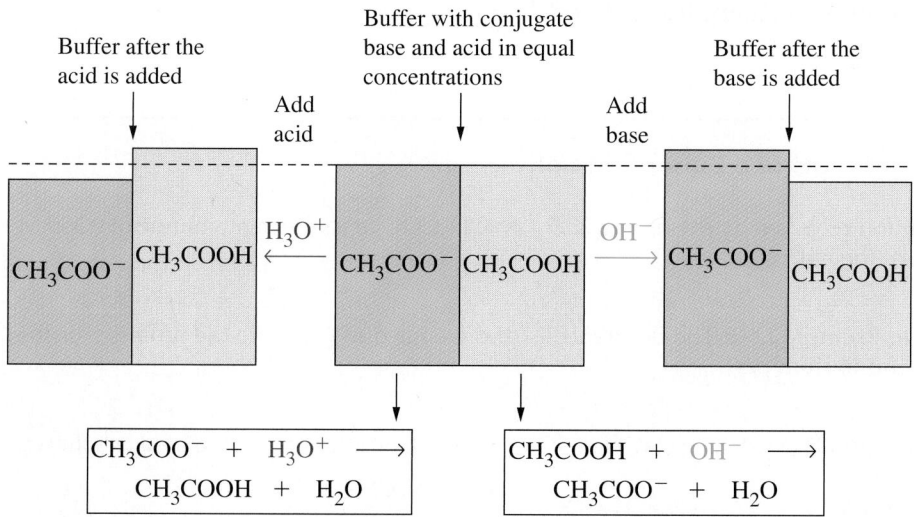

▲ FIGURE 17-5
How a buffer works
Acetate ion, the conjugate base of acetic acid, acts as a proton sink when strong acid is added. In this way, the ratio [conjugate base]/[acid] is kept approximately constant, so there is a minimal change in pH. Similarly, acetic acid acts as a proton donor when strong base is added, keeping the ratio [conjugate base]/[acid] approximately constant and minimizing the change in pH.

EXAMPLE 17-3 **Predicting Whether a Solution Is a Buffer Solution**

Show that an NH_3–NH_4Cl solution is a buffer solution. Over what pH range would you expect it to function?

Analyze

To show that a solution has buffer properties, first identify a component in the solution that neutralizes acids and a component that neutralizes bases.

Solve

In this example, these components are NH_3 and NH_4^+, respectively.

$$NH_3 + H_3O^+ \longrightarrow NH_4^+ + H_2O$$
$$NH_4^+ + OH^- \longrightarrow NH_3 + H_2O$$

In *all* aqueous solutions containing NH_3 and NH_4^+, we know that

$$NH_3 + H_2O \rightleftharpoons NH_4^+ + OH^- \quad \text{and}$$
$$K_b = \frac{[NH_4^+][OH^-]}{[NH_3]} = 1.8 \times 10^{-5}$$

If a solution has approximately equal concentrations of NH_4^+ and NH_3, then $[OH^-] \approx 1.8 \times 10^{-5}$ M; $pOH \approx 4.74$; and $pH \approx 9.26$. Ammonia–ammonium chloride solutions are *basic* buffer solutions that function over the approximate pH range of 8 to 10.

Assess

As we will soon see, not all NH_3–NH_4Cl buffer solutions will be effective buffers. The best buffers have large values for $[NH_3]$ and $[NH_4^+]$, with $[NH_3] \approx [NH_4^+]$.

PRACTICE EXAMPLE A: Describe how a mixture of a strong acid (such as HCl) and the salt of a weak acid (such as $NaCH_3COO$) can be a buffer solution. [*Hint:* What is the reaction that produces CH_3COOH?]

PRACTICE EXAMPLE B: Describe how a mixture of NH_3 and HCl can result in a buffer solution.

In Example 17-4, we first determine the stoichiometric concentrations of the buffer components. Then we perform an equilibrium calculation in the same fashion as in Examples 17-1 and 17-2.

EXAMPLE 17-4 **Calculating the pH of a Buffer Solution**

What is the pH of a buffer solution prepared by dissolving 25.5 g $NaCH_3COO$ in a sufficient volume of 0.550 M CH_3COOH to make 500.0 mL of the buffer?

Analyze

This example is very similar to Example 17-2. The difference is that we need to calculate the molarity of the acetate ion before solving the equilibrium part.

Solve

The molarity of CH_3COO^- corresponding to 25.5 g $NaCH_3COO$ in 500.0 mL of solution is calculated as follows.

$$\text{amount of } CH_3COO^- = 25.5 \text{ g } NaCH_3COO \times \frac{1 \text{ mol } NaCH_3COO}{82.04 \text{ g } NaCH_3COO}$$
$$\times \frac{1 \text{ mol } CH_3COO^-}{1 \text{ mol } NaCH_3COO}$$
$$= 0.311 \text{ mol } CH_3COO^-$$
$$[CH_3COO^-] = \frac{0.311 \text{ mol } CH_3COO^-}{0.500 \text{ L}} = 0.622 \text{ M } CH_3COO^-$$

Equilibrium Calculation:

	CH_3COOH	$+$	H_2O	$\rightleftharpoons$	H_3O^+	$+$	CH_3COO^-
initial concns:							
weak acid:	0.550 M				—		—
salt:	—				—		0.622 M
changes:	$-x$ M				$+x$ M		$+x$ M
equil concns:	$(0.550 - x)$ M				$+x$ M		$(0.622 + x)$ M

Let's assume that x is very small, so $0.550 - x \approx 0.550$ and $0.622 + x \approx 0.622$. We will find this assumption to be valid.

$$K_a = \frac{[H_3O^+][CH_3COO^-]}{[CH_3COOH]} = \frac{(x)(0.622)}{0.550} = 1.8 \times 10^{-5}$$

$$x = [H_3O^+] = \frac{0.550}{0.622} \times 1.8 \times 10^{-5} = 1.6 \times 10^{-5}$$

$$pH = -\log[H_3O^+] = -\log(1.6 \times 10^{-5}) = 4.80$$

Notice that $x = 1.6 \times 10^{-5}$ is only 0.003% of 0.550, and so the assumption that x is small was justified.

Assess

We have seen that $pH = pK_a = 4.74$ when acetic acid and acetate ion are present in equal concentrations. Here the concentration of the conjugate *base* (acetate ion) is greater than that of the acetic acid. The solution should be somewhat more basic (less acidic) than $pH = 4.74$. A pH of 4.80 is a reasonable answer.

PRACTICE EXAMPLE A: What is the pH of a buffer solution prepared by dissolving 23.1 g NaHCOO in a sufficient volume of 0.432 M HCOOH to make 500.0 mL of the buffer?

PRACTICE EXAMPLE B: A handbook states that to prepare 100.0 mL of a particular buffer solution, mix 63.0 mL of 0.200 M CH_3COOH with 37.0 mL of 0.200 M $NaCH_3COO$. What is the pH of this buffer?

An important point worth noting in Example 17-4 is that if a solution is to be an effective buffer, the assumptions $(c - x) \approx c$ and $(c + x) \approx c$ will always be valid (c represents the numerical part of an expression of molarity). That is, the *equilibrium* concentrations of the buffer components will be very nearly the same as their *stoichiometric* concentrations. As a result, in Example 17-4 we could have gone directly from the stoichiometric concentrations of the buffer components to the expression

KEEP IN MIND

that stoichiometric concentration is based on the amount of solute dissolved.

$$K_a = \frac{[H_3O^+][CH_3COO^-]}{[CH_3COOH]} = \frac{[H_3O^+](0.622)}{0.550} = 1.8 \times 10^{-5}$$

without setting up the ICE table. This procedure can be formalized through the special equation that is introduced next.

An Equation for Buffer Solutions: The Henderson–Hasselbalch Equation

Although we can continue to use the format demonstrated in Example 17-4 for buffer calculations, it is often useful to describe a buffer solution by means of an equation known as the **Henderson–Hasselbalch equation**. Biochemists and molecular biologists commonly use this equation. To derive this variation of the ionization constant expression, let's consider a mixture of a hypothetical weak acid, HA (such as CH_3COOH), and its salt, NaA (such as $NaCH_3COO$). We start with the familiar expressions

$$HA + H_2O \rightleftharpoons H_3O^+ + A^-$$

$$K_a = \frac{[H_3O^+][A^-]}{[HA]}$$

and rearrange the right side of the K_a expression to obtain

$$K_a = [H_3O^+] \times \frac{[A^-]}{[HA]}$$

Next, we take the *negative logarithm* of each side of this equation.

$$-\log K_a = -\log[H_3O^+] - \log\frac{[A^-]}{[HA]}$$

Now, recall that $pH = -\log[H_3O^+]$ and that $pK_a = -\log K_a$, which gives

$$pK_a = pH - \log\frac{[A^-]}{[HA]}$$

Solve for pH by rearranging the equation.

$$pH = pK_a + \log\frac{[A^-]}{[HA]}$$

A^- is the conjugate base of the weak acid HA, so we can write the more general equation (17.2), the Henderson–Hasselbalch equation.

$$pH = pK_a + \log\frac{[\text{conjugate base}]}{[\text{acid}]} \qquad (17.2)$$

▶ When the [conjugate base] = [acid], then $pH = pK_a$. When we get to titrations, an indicator is chosen to change color at $pH = pK_a$ of the indicator.

To apply this equation to an acetic acid–sodium acetate buffer, we use pK_a for CH_3COOH and these concentrations: $[CH_3COOH]$ for [acid] and $[CH_3COO^-]$ for [conjugate base]. To apply it to an ammonia–ammonium chloride buffer, we use pK_a for NH_4^+ and these concentrations: $[NH_4^+]$ for [acid] and $[NH_3]$ for [conjugate base].

Equation (17.2) is useful only when we can substitute *stoichiometric* or initial concentrations for equilibrium concentrations to give

$$pH = pK_a + \log\frac{[\text{conjugate base}]_{\text{initial}}}{[\text{acid}]_{\text{initial}}}$$

KEEP IN MIND

that the Henderson–Hasselbalch equation is very useful but should probably not be committed to memory; it is easy to get the conjugate base and acid terms inverted. It is most important to understand the principles that lead to this equation, thereby avoiding the pitfalls of using the equation incorrectly or when it is not valid.

thus avoiding the need to set up an ICE table. This constraint places important limitations on the equation's validity, however. Later, we will see that there are also conditions that must be met if a mixture is to be an effective buffer solution. Although the following rules may be overly restrictive in some cases, a reasonable approach to the twin concerns of effective buffer action and the validity of equation (17.2) is to ensure that

1. the ratio [conjugate base]/[acid] is within the limits

$$0.10 < \frac{[\text{conjugate base}]}{[\text{acid}]} < 10 \qquad (17.3)$$

2. the molarity of each buffer component exceeds the value of K_a by a factor of at least 100

Viewed another way, equation (17.2) works only for cases in which the assumption $c - x \approx c$ is valid. If a quadratic equation is required to solve the equilibrium constant expression, equation (17.2) will likely fail.

Preparing Buffer Solutions

Suppose we need a buffer solution with $pH = 5.09$. Equation (17.2) suggests two alternatives. One is to find a weak acid, HA, that has $pK_a = 5.09$ and prepare a solution with equal molarities of the acid and its salt.

$$pH = pK_a + \log\frac{\cancel{[A^-]}}{\cancel{[HA]}} = 5.09 + \log 1 = 5.09$$

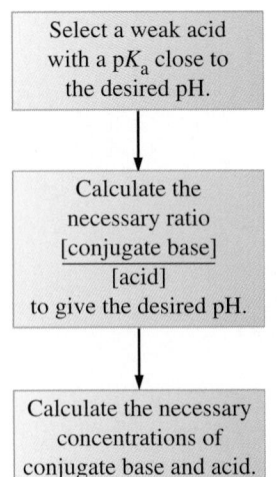

Select a weak acid with a pK_a close to the desired pH.

↓

Calculate the necessary ratio $\dfrac{[\text{conjugate base}]}{[\text{acid}]}$ to give the desired pH.

↓

Calculate the necessary concentrations of conjugate base and acid.

▲ One procedure to follow in making a buffer solution with a desired pH.

Although this alternative is simple in concept, generally, it is not practical. We are not likely to find a readily available, water-soluble weak acid with exactly $pK_a = 5.09$. The second alternative, summarized in the margin, is to use a cheap, common weak acid such as acetic acid, CH_3COOH ($pK_a = 4.74$), and establish an appropriate ratio of $[CH_3COO^-]/[CH_3COOH]$ to obtain a pH of 5.09. Example 17-5 demonstrates this second alternative.

EXAMPLE 17-5   Preparing a Buffer Solution of a Desired pH

What mass of $NaCH_3COO$ must be dissolved in 0.300 L of 0.25 M CH_3COOH to produce a solution with pH = 5.09? (Assume that the solution volume remains constant at 0.300 L.)

Analyze

Equilibrium among the buffer components is expressed by the equation

$$CH_3COOH + H_2O \rightleftharpoons H_3O^+ + CH_3COO^- \qquad K_a = 1.8 \times 10^{-5}$$

and by the ionization constant expression for acetic acid.

$$K_a = \frac{[H_3O^+][CH_3COO^-]}{[CH_3COOH]} = 1.8 \times 10^{-5}$$

Each of the three concentration terms appearing in a K_a expression should be an equilibrium concentration. The $[H_3O^+]$ corresponding to a pH of 5.09 is the equilibrium concentration. For $[CH_3COOH]$, we will assume that the equilibrium concentration is equal to the stoichiometric or initial concentration. The value of $[CH_3COO^-]$ that we calculate with the K_a expression is the equilibrium concentration, and we will assume that it is also the same as the stoichiometric concentration. Thus, we assume that neither the ionization of CH_3COOH to form CH_3COO^- nor the hydrolysis of CH_3COO^- to form CH_3COOH produces much of a difference between the stoichiometric (initial) and equilibrium concentrations of the buffer components. These assumptions work well if the conditions stated in expression (17.3) are met.

Solve

The relevant concentration terms, then, are

$$[H_3O^+] = 10^{-pH} = 10^{-5.09} = 8.1 \times 10^{-6}\,M$$
$$[CH_3COOH] = 0.25\,M$$
$$[CH_3COO^-] = ?$$

The required acetate ion concentration in the buffer solution is

$$[CH_3COO^-] = K_a \times \frac{[CH_3COOH]}{[H_3O^+]} = 1.8 \times 10^{-5} \times \frac{0.25}{8.1 \times 10^{-6}} = 0.56\,M$$

We complete the calculation of the mass of sodium acetate with some familiar ideas of solution stoichiometry.

$$mass = 0.300\,L \times \frac{0.56\,mol\,CH_3COO^-}{1\,L} \times \frac{1\,mol\,NaCH_3COO}{1\,mol\,CH_3COO^-}$$
$$\times \frac{82.0\,g\,NaCH_3COO}{1\,mol\,NaCH_3COO} = 14\,g\,NaCH_3COO$$

Assess

We check the answer by inserting the acetate ion and acetic acid concentrations, along with the pK_a of acetic acid, into equation (17.2) to obtain pH = 5.09. The method described in Example 17-5 is one way to obtain a buffer solution. Another approach involves adding an appropriate amount of strong base (e.g., 0.052 mol NaOH) to 0.300 L of 0.025 M $CH_3COOH(aq)$.

PRACTICE EXAMPLE A: How many grams of $(NH_4)_2SO_4$ must be dissolved in 0.500 L of 0.35 M NH_3 to produce a solution with pH = 9.00? (Assume that the solution volume remains at 0.500 L.)

PRACTICE EXAMPLE B: In Practice Example 17-3A, we established that an appropriate mixture of a strong acid and the salt of a weak acid is a buffer solution. Show that a solution made by adding 33.05 g $NaCH_3COO \cdot 3\,H_2O(s)$ to 300 mL of 0.250 M HCl should have pH $\approx$ 5.1.

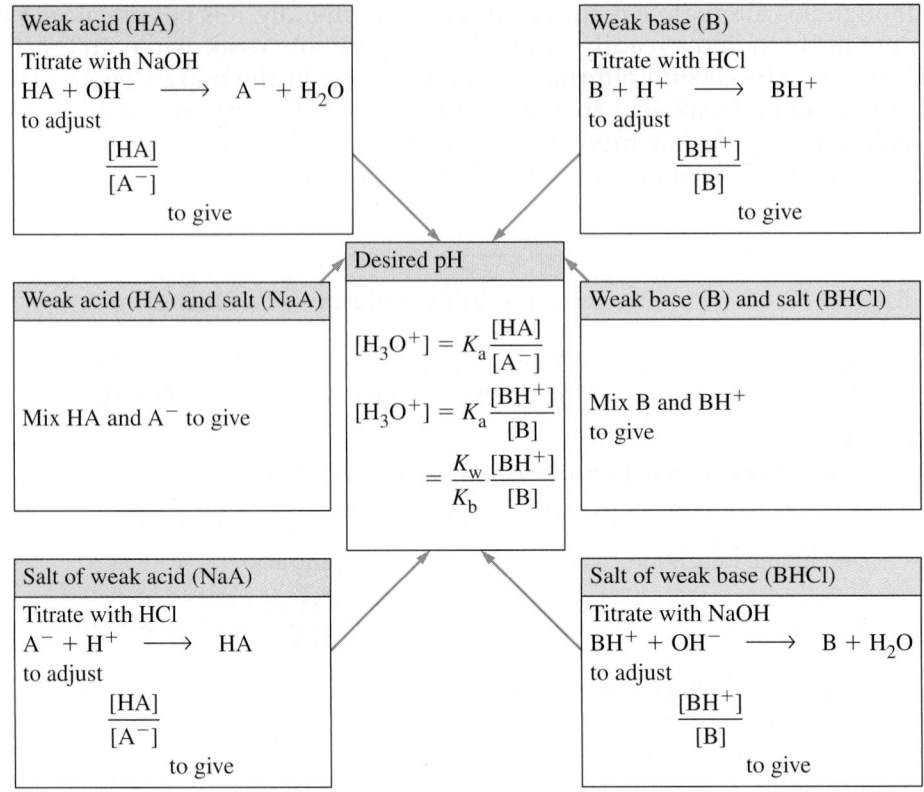

▲ FIGURE 17-6
Six methods for preparing buffer solutions
Depending on the pH range required and the type of experiment the buffer is to be used for, either a weak acid or a weak base can be used to prepare a buffer solution.

In Example 17-5, we achieved the desired ratio of $[CH_3COO^-]/[CH_3COOH]$ by adding 14 g of sodium acetate to the previously prepared 0.25 M CH_3COOH solution. This is a common method of obtaining a buffer solution. Other methods are sometimes useful as well. Sufficient NaOH(aq) could be added to CH_3COOH(aq) to neutralize the acid partially, *producing* CH_3COO^- as a product. Or enough $NaCH_3COO$(s) could be added to HCl(aq) to neutralize all the HCl, producing some CH_3COOH and leaving some CH_3COO^- in excess. As we saw in Chapter 16, amines are weak bases, so an aqueous mixture of an amine and its conjugate acid is a buffer solution. Buffer solutions based on amines can be prepared in ways analogous to those based on weak acids. The methods available for making buffer solutions are summarized in Figure 17-6.

Calculating pH Changes in Buffer Solutions

To calculate how the pH of a buffer solution changes when small amounts of a strong acid or base are added, we must first use *stoichiometric* principles to establish how much of one buffer component is consumed and how much of the other component is produced. Then the new concentrations of weak acid (or weak base) and its salt can be used to calculate the pH of the buffer solution. Essentially, this problem is solved in two steps. First, we assume that the neutralization reaction proceeds to *completion* and determine new stoichiometric concentrations. Then these new stoichiometric concentrations are substituted into the equilibrium constant expression and the expression is solved for $[H_3O^+]$, which is converted to pH. This method is applied in Example 17-6 and illustrated in Figure 17-7.

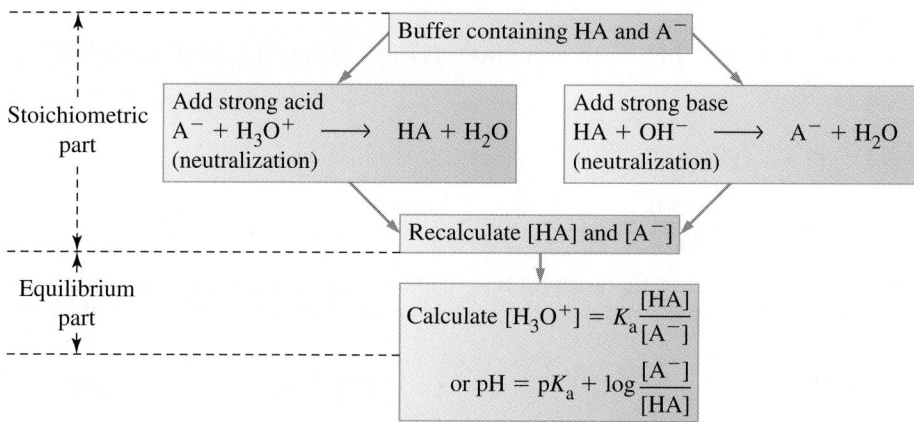

▲ FIGURE 17-7
Calculation of the new pH of a buffer after strong acid or base is added
The stoichiometric and equilibrium parts of the calculation are indicated. This scheme
can also be applied to the conjugate acid–base pair BH⁺/B, where B is a base.

EXAMPLE 17-6   Calculating pH Changes in a Buffer Solution

What are the effects on the pH of adding **(a)** 0.0060 mol HCl and **(b)** 0.0060 mol NaOH to 0.300 L of a buffer solution that is 0.250 M CH_3COOH and 0.560 M $NaCH_3COO$?

Analyze

In parts (a) and (b) we complete essentially the same calculations. We should recognize that we are adding a strong acid or a strong base to the buffer solution. To investigate this effect we make a stoichiometric calculation, followed by an equilibrium calculation. The stoichiometric calculation is necessary to account for the neutralization of the base or acid components of the buffer.

Solve

To judge the effect of adding either (a) acid or (b) base on the pH of the buffer, the value we must keep in mind is the pH of the original buffer. Because the initial (or stoichiometric) concentrations of CH_3COOH and CH_3COO^- are large and not too different, the initial pH of the buffer can be obtained by substituting the initial concentrations into equation (17.2). (See the discussion following equation (17.2) on page 754.)

$$pH = pK_a + \log\frac{[CH_3COO^-]}{[CH_3COOH]}$$

$$= 4.74 + \log\frac{0.560}{0.250} = 4.74 + 0.35 = 5.09$$

(a) Stoichiometric Calculation: Let's calculate amounts in moles, and assume that the neutralization goes to completion. Essentially, this is a limiting reactant calculation, but perhaps simpler than many of those in Chapter 4. In neutralizing the added H_3O^+, 0.0060 mol CH_3COO^- is converted to 0.0060 mol CH_3COOH.

	CH_3COO^-	+	H_3O^+	⟶	CH_3COOH	+	H_2O
original buffer:	$\underbrace{0.300\ L \times 0.560\ M}_{0.168\ mol}$				$\underbrace{0.300\ L \times 0.250\ M}_{0.0750\ mol}$		
add:			0.0060 mol				
changes:	−0.0060 mol		−0.0060 mol		+0.0060 mol		
final buffer:							
amounts:	0.162 mol		≈ 0		0.0810 mol		
concns:	$\underbrace{0.162\ mol/0.300\ L}_{0.540\ M}$		≈ 0		$\underbrace{0.0810\ mol/0.300\ L}_{0.270\ M}$		

(continued)

Equilibrium Calculation: We can calculate the pH with equation (17.2), using the new equilibrium concentrations.

$$pH = pK_a + \log\frac{[CH_3COO^-]}{[CH_3COOH]}$$

$$= 4.74 + \log\frac{0.540}{0.270} = 4.74 + 0.30 = 5.04$$

(b) Stoichiometric Calculation: In neutralizing the added OH^-, 0.0060 mol CH_3COOH is converted to 0.0060 mol CH_3COO^-. The calculation of the new stoichiometric concentrations is shown on the last line of the following table.

	CH_3COOH	+	OH^-	$\longrightarrow$	CH_3COO^-	+	H_2O
original buffer:	$\underbrace{0.300\,L \times 0.250\,M}_{0.0750\,mol}$				$\underbrace{0.300\,L \times 0.560\,M}_{0.168\,mol}$		
add:			0.0060 mol				
changes:	−0.0060 mol		−0.0060 mol		+0.0060 mol		
final buffer:							
amounts:	0.0690 mol		≈0		0.174 mol		
concns:	$\underbrace{0.0690\,mol/0.300\,L}_{0.230\,M}$		≈0		$\underbrace{0.174\,mol/0.300\,L}_{0.580\,M}$		

Equilibrium Calculation: This is the same type of calculation as in part (a), but with slightly different concentrations.

$$pH = 4.74 + \log\frac{0.580}{0.230} = 4.74 + 0.40 = 5.14$$

Assess

The addition of 0.0060 mol HCl *lowers* the pH from 5.09 to 5.04, which is only a small change in pH. The addition of 0.0060 mol OH^- *raises* the pH from 5.09 to 5.14—another small change. Had we instead added 0.0060 mol HCl or 0.0060 mol NaOH to 0.300 L of water, the pH would have changed by more than 5 pH units. The most important factors to confirm in a calculation of this type are that the magnitude of the pH change is small and that the change occurs in the correct direction: *lowering* of the pH by an acid and *raising* of the pH by a base. The results are indeed reasonable.

PRACTICE EXAMPLE A: A 1.00 L volume of buffer is made with concentrations of 0.350 M NaHCOO (sodium formate) and 0.550 M HCOOH (formic acid). **(a)** What is the initial pH? **(b)** What is the pH after the addition of 0.0050 mol HCl(aq)? (Assume that the volume remains 1.00 L.) **(c)** What would be the pH after the addition of 0.0050 mol NaOH to the original buffer?

PRACTICE EXAMPLE B: How many milliliters of 6.0 M HNO_3 would you add to 300.0 mL of the buffer solution of Example 17-6 to change the pH from 5.09 to 5.03?

▶ Dilute and concentrated buffers will have the same pH, but as mentioned in the next section, a given volume of a dilute buffer will have a lower buffer capacity than the same volume of a more concentrated buffer.

Perhaps you have already noticed a way to simplify the calculation in Example 17-6. Because the buffer components are always present in the same solution of volume V, the numbers of moles can be substituted directly into equation (17.2) without regard for the particular value of V. Thus, in Example 17-6(b),

$$pH = 4.74 + \log\frac{[CH_3COO^-]}{[CH_3COOH]} = 4.74 + \log\frac{0.174\ \cancel{mol}/\cancel{V}}{0.0690\ \cancel{mol}/\cancel{V}} = 4.74 + 0.40 = 5.14$$

This expression is also consistent with the observation that, on dilution, buffer solutions resist pH changes. Diluting a buffer solution means increasing its volume V by adding water. This action produces the same change in the numerator and the denominator of the ratio [conjugate base]/[acid]. The ratio itself remains unchanged, as does the pH.

Buffer Capacity and Buffer Range

It is not difficult to see that if we add more than 0.0750 mol OH^- to the buffer solution described in Example 17-6, the 0.0750 mol CH_3COOH will be completely converted to 0.0750 mol CH_3COO^-. An excess of OH^- will remain, and the solution will become rather strongly basic.

Buffer capacity refers to the amount of acid or base that a buffer can neutralize before its pH changes appreciably. In general, the maximum buffer capacity exists when the concentrations of a weak acid and its conjugate base are kept *large* and *approximately equal to each other*. The **buffer range** is the pH range over which a buffer effectively neutralizes added acids and bases and maintains a fairly constant pH. As equation (17.2) suggests,

$$pH = pK_a + \log\frac{[\text{conjugate base}]}{[\text{acid}]}$$

when the ratio $[\text{conjugate base}]/[\text{acid}] = 1$, $pH = pK_a$. When the ratio falls to 0.10, the pH *decreases* by 1 pH unit from pK_a because $\log 0.10 = -1$. If the ratio increases to a value of 10, the pH *increases* by 1 unit because $\log 10 = 1$. For practical purposes, this range of 2 pH units is the maximum range to which a buffer solution should be exposed. For acetic acid–sodium acetate buffers, the effective range is about pH 3.7–5.7; for ammonia–ammonium chloride buffers, it is about pH 8.3–10.3.

◀ Our eyes can see an indicator color change over a range of about 2 pH units.

KEEP IN MIND

that as a rule of thumb, the amounts of the buffer components should be at least ten times as great as the amount of acid or base to be neutralized.

◀ The buffer range for the ammonia–ammonium chloride solution is based on the pK_a of NH_4^+, 9.26.

Applications of Buffer Solutions

An important example of a buffered system is that found in blood, which must be maintained at a pH of 7.4 in humans. We will consider the buffering of blood in the Focus On feature for this chapter on the MasteringChemistry website. But buffers have other important applications, too.

Protein studies often must be performed in buffered media because the structures of protein molecules, including the magnitude and kind of electric charges they carry, depend on the pH (see Section 28-4). The typical enzyme is a protein capable of catalyzing a biochemical reaction, so enzyme activity is closely linked to protein structure and hence to pH. Most enzymes in the body have their maximum activity between pH 6 and pH 8. Studying enzyme activity in the laboratory usually means working with media buffered in this pH range.

The control of pH is often important in industrial processes. For example, in the mashing of barley malt, the first step of making beer, the pH of the solution must be maintained at 5.0 to 5.2, so that the protease and peptidase enzymes can hydrolyze the proteins from the barley. The inventor of the pH scale, Søren Sørensen, was a research scientist in a brewery.

We will consider the importance of buffer solutions in solubility/precipitation processes in Chapter 18.

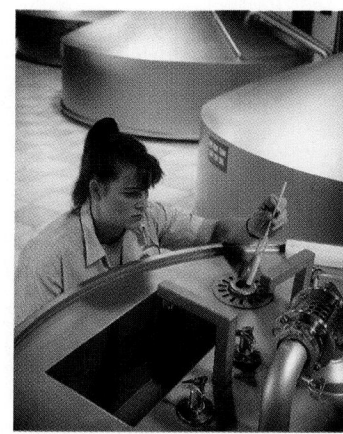

▲ A master brewer inspecting wort temperature and pH in the making of beer.

🔍   **17-2   CONCEPT ASSESSMENT**

You are asked to make a buffer with a pH value close to 4 that would best resist an increase in pH. You can select one of the following acid–conjugate base pairs: acetic acid–acetate, $K_a = 1.8 \times 10^{-5}$; ammonium ion–ammonia, $K_a = 5.6 \times 10^{-10}$; or benzoic acid–benzoate, $K_a = 6.3 \times 10^{-5}$; and you can mix them in the following acid–conjugate base ratios: 1:1, 2:1, or 1:2. What combination would make the best buffer?

17-3   Acid–Base Indicators

An **acid–base indicator** is a substance whose color depends on the pH of the solution to which it is added. Several of the photographs in this and the preceding chapter have shown acid–base indicators in use. The indicator chosen

◀ Two of the most important biological buffers are the phosphate and bicarbonate buffer systems. Proteins and nucleotides also function as buffers on the cellular level.

depended on just how acidic or basic the solution was. In this section, we will consider how an acid–base indicator works and how an appropriate indicator is selected for a pH measurement.

Acid–base indicators exist in two forms: (1) a weak acid, represented symbolically as HIn and having one color, and (2) its conjugate base, represented as In^- and having a different color. When just a small amount of indicator is added to a solution, the indicator does not affect the pH of the solution. Instead, the ionization equilibrium of the indicator is itself affected by the prevailing $[H_3O^+]$ in solution.

$$HIn \; + \; H_2O \; \rightleftharpoons \; H_3O^+ \; + \; In^-$$

Acid color   Base color

From Le Châtelier's principle, we see that *increasing* $[H_3O^+]$ in a solution displaces the equilibrium to the left, increasing the proportion of HIn and hence the acid color. *Decreasing* $[H_3O^+]$ in a solution displaces the equilibrium to the right, increasing the proportion of In^- and hence the base color. The color of the solution depends on the relative proportions of the acid and base. The pH of the solution can be related to these relative proportions and to the pK_a of the indicator by means of an equation similar to equation (17.2).

$$pH = pK_{HIn} + \log\frac{[In^-]}{[HIn]} \tag{17.4}$$

▶ The acid "color" of a few indicators is colorless.

In general, if 90% or more of an indicator is in the form HIn, the solution will take on the acid color. If 90% or more is in the form In^-, the solution takes on the base (or anion) color. If the concentrations of HIn and In^- are about equal, the indicator is in the process of changing from one form to the other and has an intermediate color. The complete change in color occurs over a range of about *2 pH units*, with pH = pK_{HIn} at about the middle of the range. The colors and pH ranges of several acid–base indicators are shown in Figure 17-8. A summary of these ideas is presented in Table 17.1, and an example of their use is given below.

Bromthymol blue, $pK_{HIn} = 7.1$

pH < 6.1 (yellow)   pH ≈ 7.1 (green)   pH > 8.1 (blue)

An acid–base indicator is usually prepared as a solution (in water, ethanol, or some other solvent). In acid–base titrations, a few drops of the indicator solution are added to the solution being titrated. In other applications, porous paper is impregnated with an indicator solution and dried. When this paper is moistened with the solution being tested, it acquires a color determined by the pH of the solution. This paper is usually called *pH test paper*.

TABLE 17.1 pH and the Colors of Acid–Base Indicators

Acid Color	Intermediate Color	Base Color
$[In^-]/[HIn] < 0.10$	$[In^-]/[HIn] \approx 1$	$[In^-]/[HIn] > 10$
$pH < pK_{HIn} + \log 0.10$	$pH \approx pK_{HIn} + \log 1$	$pH > pK_{HIn} + \log 10$
$pH < pK_{HIn} - 1$	$pH \approx pK_{HIn}$	$pH > pK_{HIn} + 1$

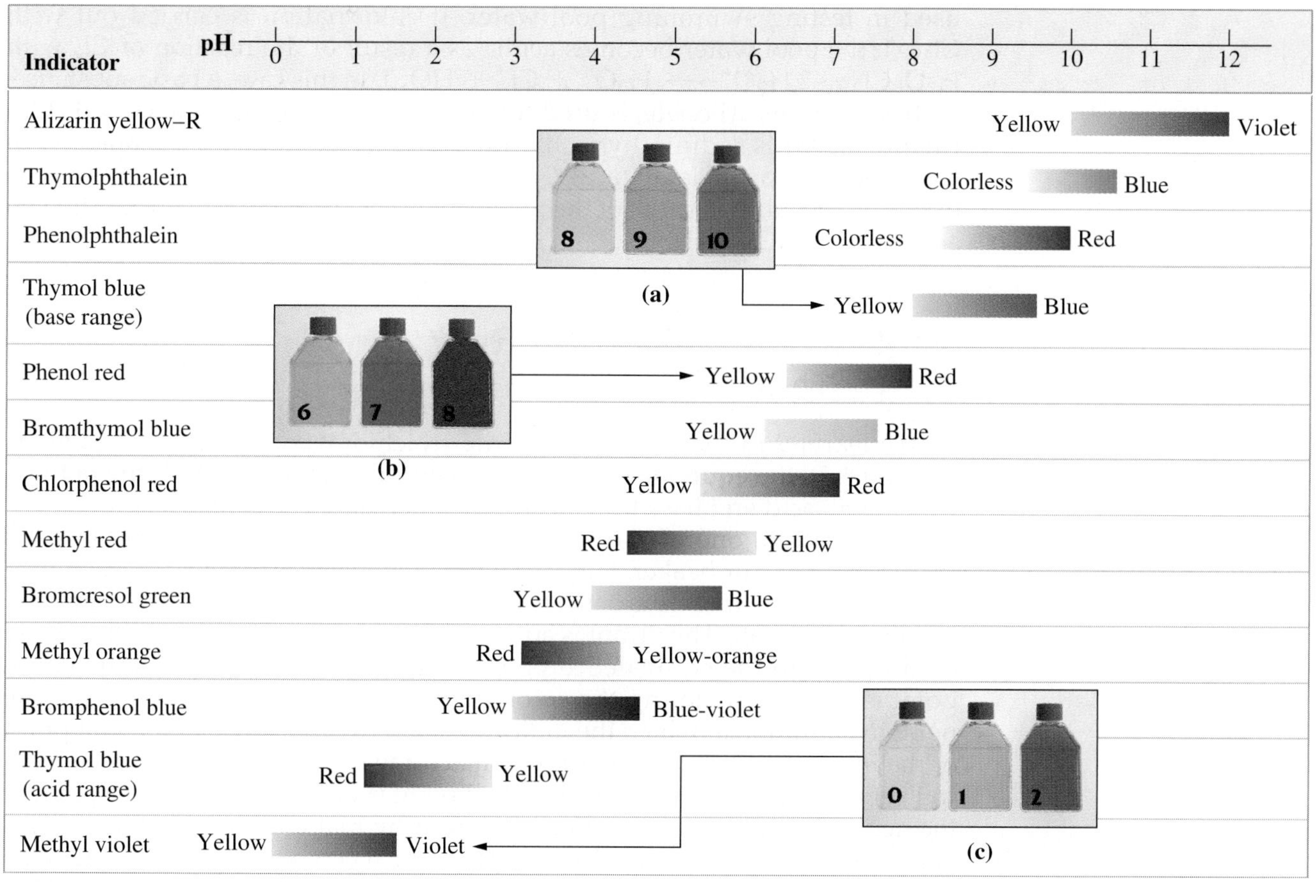

Indicator	pH												
Alizarin yellow–R										Yellow ▬ Violet			
Thymolphthalein										Colorless ▬ Blue			
Phenolphthalein									Colorless ▬ Red				
Thymol blue (base range)								Yellow ▬ Blue					
Phenol red						Yellow ▬ Red							
Bromthymol blue						Yellow ▬ Blue							
Chlorphenol red					Yellow ▬ Red								
Methyl red				Red ▬ Yellow									
Bromcresol green			Yellow ▬ Blue										
Methyl orange		Red ▬ Yellow-orange											
Bromphenol blue		Yellow ▬ Blue-violet											
Thymol blue (acid range)	Red ▬ Yellow												
Methyl violet	Yellow ▬ Violet												

(a) bottles labeled 8, 9, 10

(b) bottles labeled 6, 7, 8

(c) bottles labeled 0, 1, 2

▲ FIGURE 17-8
pH and color changes for some common acid–base indicators
The indicators pictured and the pH values at which they change color are **(a)** thymol blue (pH 8–10); **(b)** phenol red (pH 6–8); and **(c)** methyl violet (pH 0–2).

🔍   17-3   CONCEPT ASSESSMENT

(1) Given that an indicator is itself a weak acid or base, why does adding it to a solution not change the nature of the equilibrium?

(2) Starting with about 10 mL of dilute NaCl(aq) containing a couple of drops of phenol red indicator, what color would the solution be for each of the following actions: **(a)** first 10 drops of 1.0 M HCl(aq) are added to the solution; **(b)** next 15 drops of 1.0 M NaCH$_3$COO(aq) are added to solution **(a)**; **(c)** then one drop of 1.0 M KOH(aq) is added to solution **(b)**; and **(d)** 10 more drops of 1.0 M KOH(aq) are added to solution **(c)**.

Applications

Acid–base indicators are most useful when only an approximate pH determination is needed. For example, they are used in soil-testing kits to establish the approximate pH of soils. Soils are usually acidic in regions of high rainfall and heavy vegetation, and they are alkaline in more arid regions. The pH can vary considerably with local conditions, however. If a soil is found to be too acidic for a certain crop, its pH can be raised by adding slaked lime [Ca(OH)$_2$]. To reduce the pH of a soil, organic matter might be added.

In swimming pools, chlorinating agents are most effective at a pH of about 7.4. At this pH, the growth of algae is avoided, and the corrosion of pool plumbing is minimized. Phenol red (see Figure 17-8) is a common indicator

▲ Testing swimming pool water for its chlorine content and pH.

used in testing swimming pool water. If chlorination is carried out with $Cl_2(g)$, the pool water becomes acidic as a result of the reaction of Cl_2 with H_2O: $Cl_2 + 2 H_2O \longrightarrow H_3O^+ + Cl^- + HOCl$. In this case, a basic substance, such as sodium carbonate, is used to raise the pH. Another widely used chlorinating agent is sodium hypochlorite, $NaOCl(aq)$, made by the reaction of $Cl_2(g)$ with excess $NaOH(aq)$: $Cl_2 + 2 OH^- \longrightarrow Cl^- + OCl^- + H_2O$. The excess NaOH raises the pH of the pool water. The pH is adjusted by adding an acid, such as HCl or H_2SO_4.

17-4 Neutralization Reactions and Titration Curves

As we learned in the discussion of the stoichiometry of titration reactions (Section 5-7), the **equivalence point** of a neutralization reaction is the point at which both acid and base have been consumed and *neither* is in excess.

In a titration, one of the solutions to be neutralized—say, the acid—is placed in a flask or beaker, together with a few drops of an acid–base indicator. The other solution (the base) used in a titration is added from a buret and is called the **titrant**. The titrant is added to the acid, first rapidly and then drop by drop, up to the equivalence point (recall Figure 5-17). The equivalence point is located by noting the color change of the acid–base indicator. The point in a titration at which the indicator changes color is called the **end point** of the indicator. The end point must match the equivalence point of the neutralization. That is, if the indicator's end point is near the equivalence point of the neutralization, the color change marked by that end point will signal the attainment of the equivalence point. This match can be achieved by use of an indicator whose color change occurs over a pH range that includes the pH of the equivalence point.

A graph of pH versus volume of titrant (the solution in the buret) is called a **titration curve**. Titration curves are most easily constructed by measuring the pH during a titration with a pH meter and plotting the data with a recorder. In this section we will emphasize calculating the pH at various points in a titration. These calculations will serve as a review of aspects of acid–base equilibria considered earlier in this chapter and in the preceding chapter.

The Millimole

In a typical titration, the volume of solution delivered from a buret is less than 50 mL (usually about 20–25 mL). The molarity of the solution used for the titration is generally less than 1 M. The typical amount of OH^- (or H_3O^+) delivered from the buret during a titration is only a few thousandths of a mole—for example, 5.00×10^{-3} mol. In calculations it is often easier to work with millimoles instead of moles. The symbol **mmol** stands for a **millimole**, which is one thousandth of a mole, or 10^{-3} mol.

Recall from Chapter 4 that molarity is defined as the number of moles per liter. We can use an alternative definition of molarity by converting from moles to millimoles and from liters to milliliters.

$$M = \frac{mol}{L} = \frac{mol/1000}{L/1000} = \frac{mmol}{mL}$$

Thus, the expression from Chapter 4 that the amount of solute is the product of molarity and solution volume (page 126) can be based either on mol/L × L = mol or on mmol/mL × mL = mmol.

Titration of a Strong Acid with a Strong Base

Suppose that 25.00 mL of 0.100 M HCl (a strong acid) is placed in a small flask or beaker and that 0.100 M NaOH (a strong base) is added to it from a buret.

The pH of the accumulated solution can be calculated at different points in the titration, and these pH values can be plotted against the volume of NaOH added. From this titration curve we can establish the pH at the equivalence point and identify an appropriate indicator for the titration. Some typical calculations are outlined in Example 17-7.

EXAMPLE 17-7   Calculating Points on a Titration Curve: Strong Acid Titrated with a Strong Base

What is the pH at each of the following points in the titration of 25.00 mL of 0.100 M HCl with 0.100 M NaOH?

(a) before the addition of any NaOH (*initial pH*)

(b) after the addition of 24.00 mL 0.100 M NaOH (*before the equivalence point*)

(c) after the addition of 25.00 mL 0.100 M NaOH (*the equivalence point*)

(d) after the addition of 26.00 mL 0.100 M NaOH (*beyond the equivalence point*)

Analyze

Parts (a) to (d) correspond to four different stages of the titration. In part (a) we calculate the initial pH of the HCl solution before the titration begins. In part (b) most but not all the acid has been neutralized. In part (c) all the acid is neutralized, which corresponds to the equivalence point. In part (d) we are past the equivalence point and are dealing with a solution containing an unreacted strong base.

Solve

First, let's write the titration equation in the ionic and net ionic form.

Ionic form: $H_3O^+(aq) + Cl^-(aq) + Na^+(aq) + OH^-(aq) \longrightarrow Na^+(aq) + Cl^-(aq) + 2\,H_2O(l)$

Net ionic form: $H_3O^+(aq) + OH^-(aq) \longrightarrow 2\,H_2O(l)$

(a) Before any NaOH is added, we are dealing with 0.100 M HCl. This solution has $[H_3O^+] = 0.100$ M and pH = 1.00.

(b) The number of millimoles of H_3O^+ to be titrated is

$$25.00\ \text{mL} \times \frac{0.100\ \text{mmol } H_3O^+}{1\ \text{mL}} = 2.50\ \text{mmol } H_3O^+$$

The number of millimoles of OH^- present in 24.00 mL of 0.100 M NaOH is

$$24.00\ \text{mL} \times \frac{0.100\ \text{mmol}}{1\ \text{mL}} = 2.40\ \text{mmol } OH^-$$

Now we can represent the net ionic equation of the neutralization reaction in a familiar format.

	H_3O^+	+	OH^-	$\longrightarrow$	$2\,H_2O$
initially present:	2.50 mmol		—		
add:			2.40 mmol		
changes:	−2.40 mmol		−2.40 mmol		
after reaction:	0.10 mmol		≈0		

The remaining 0.10 mmol of H_3O^+ is present in 49.00 mL of solution (25.00 mL original + 24.00 mL added base).

$$[H_3O^+] = \frac{0.10\ \text{mmol } H_3O^+}{49.00\ \text{mL}} = 2.0 \times 10^{-3}\ \text{M}$$

$$pH = -\log[H_3O^+] = -\log(2.0 \times 10^{-3}) = 2.70$$

(c) The equivalence point is the point at which the HCl is completely neutralized and no excess NaOH is present. As seen in the ionic form of the equation for the neutralization reaction, the solution at the equivalence point is simply NaCl(aq). And, as we learned in Section 16-7, because neither Na^+ nor Cl^- hydrolyzes in water, pH = 7.00.

(continued)

(d) To determine the pH of the solution beyond the equivalence point, we can return to the format in (b), except that now OH^- is in excess. The amount of OH^- added is $26.00 \text{ mL} \times 0.100 \text{ mmol/L} = 2.60 \text{ mmol}$.

$$H_3O^+ \quad + \quad OH^- \quad \longrightarrow \quad 2\,H_2O$$

initially present:	2.50 mmol	—	
add:		2.60 mmol	
changes:	−2.50 mmol	−2.50 mmol	
after reaction:	≈0	0.10 mmol	

The excess 0.10 mmol of NaOH is present in 51.00 mL of solution (25.00 mL original acid + 26.00 mL added base). The concentration of OH^- in this solution is

$$[OH^-] = \frac{0.10 \text{ mmol } OH^-}{51.00 \text{ mL}} = 2.0 \times 10^{-3} \text{ M}$$

$$pOH = -\log(2.0 \times 10^{-3}) = 2.70 \qquad pH = 14.00 - 2.70 = 11.30$$

Assess

In strong acid–strong base titrations, changes in pH are abrupt and occur mostly right before and right after the equivalence point. For a strong acid–strong base titration, the pH at the equivalence point is equal to 7.

PRACTICE EXAMPLE A: For the titration of 25.00 mL of 0.150 M HCl with 0.250 M NaOH, calculate **(a)** the initial pH; **(b)** the pH when neutralization is 50.0% complete; **(c)** the pH when neutralization is 100.0% complete; and **(d)** the pH when 1.00 mL of NaOH is added beyond the equivalence point.

PRACTICE EXAMPLE B: For the titration of 50.00 mL of 0.00812 M Ba(OH) with 0.0250 M HCl, calculate **(a)** the initial pH; **(b)** the pH when neutralization is 50.0% complete; **(c)** the pH when neutralization is 100.0% complete.

Figure 17-9 presents pH versus volume data and the titration curve for the HCl—NaOH titration. From this figure, we can establish these principal features of the titration curve for the titration of a *strong acid with a strong base*.

- The pH has a low value at the beginning of the titration.
- The pH changes slowly until just before the equivalence point.

Titration Data

mL NaOH(aq)	pH
0.00	1.00
10.00	1.37
20.00	1.95
22.00	2.19
24.00	2.70
25.00	7.00
26.00	11.30
28.00	11.75
30.00	11.96
40.00	12.36
50.00	12.52

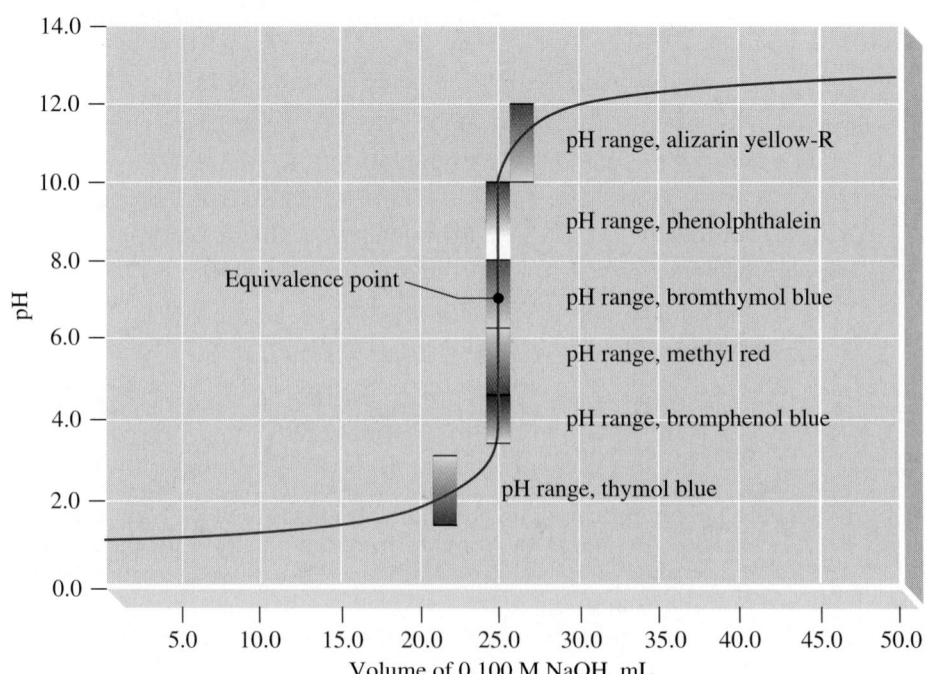

▶ FIGURE 17-9
Titration curve for the titration of a strong acid with a strong base—25.00 mL of 0.100 M HCl with 0.100 M NaOH
All indicators whose color ranges fall along the steep portion of the titration curve are suitable for this titration. Thymol blue changes color too soon; alizarin yellow-R, too late.

- At the equivalence point, the pH rises very sharply, perhaps by 6 units for an addition of only 0.10 mL (2 drops) of base.
- Beyond the equivalence point, the pH again rises only slowly.
- Any acid–base indicator whose color changes in the pH range from about 4 to 10 is suitable for this titration.

In the titration of a strong base with a strong acid, we can obtain a titration curve essentially identical to Figure 17-9 by plotting pOH against volume of titrant (the strong acid). Also, we can make a set of statements similar to those listed above, except that pOH would be substituted for pH. Alternatively, if pH is plotted against volume of titrant (the strong acid), the titration curve looks like Figure 17-9 flipped over from top to bottom, as shown in Figure 17-10.

Titration of a Weak Acid with a Strong Base

Several important differences exist between the titration of a weak acid with a strong base and a strong acid with a strong base, but one feature is *unchanged* when we compare the two titrations.

> For equal volumes of acid solutions of the same molarity, the volume of base required to titrate to the equivalence point is independent of the strength of the acid.

We can think of the neutralization of a weak acid, such as CH_3COOH as involving the direct transfer of protons from CH_3COOH molecules to OH^- ions. In the neutralization of a strong acid, the protons are transferred from H_3O^+ ions. In either case, the acid and base react in a $1:1$ mole ratio.

$$CH_3COOH + OH^- \longrightarrow H_2O + CH_3COO^-$$
$$H_3O^+ + OH^- \longrightarrow H_2O + H_2O$$

Calculations for the titration of a weak acid with a strong base can be divided into a stoichiometric part and an equilibrium part to take into account the partial ionization of the weak acid. The calculation strategy is analogous to that adopted when we considered the addition of a strong base to a buffer solution. In Example 17-8 and in Figure 17-11, we consider the titration of 25.00 mL of 0.100 M CH_3COOH with 0.100 M NaOH.

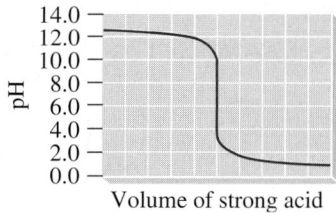

▲ FIGURE 17-10
Titration curve for the titration of a strong base with a strong acid

Titration Data	
mL NaOH(aq)	pH
0.00	2.89
5.00	4.14
10.00	4.57
12.50	4.74
15.00	4.92
20.00	5.35
24.00	6.12
25.00	8.72
26.00	11.30
30.00	11.96
40.00	12.36
50.00	12.52

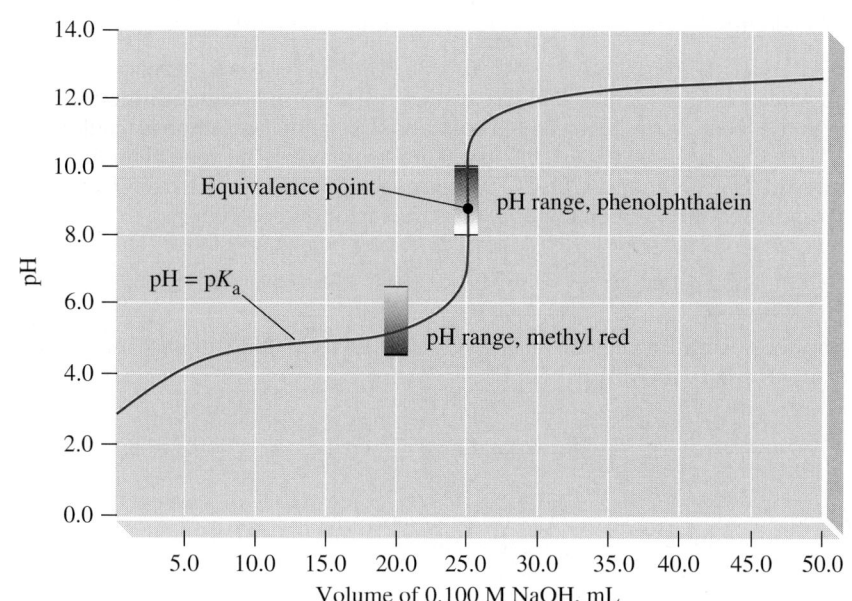

◀ FIGURE 17-11
Titration curve for the titration of a weak acid with a strong base— 25.00 mL of 0.100 M CH_3COOH with 0.100 M NaOH
Phenolphthalein is a suitable indicator for this titration, but methyl red is not. When exactly half of the acid is neutralized, $[CH_3COOH] = [CH_3COO^-]$ and pH = pK_a 4.74.

EXAMPLE 17-8 Calculating Points on a Titration Curve: Weak Acid Titrated with a Strong Base

What is the pH at each of the following points in the titration of 25.00 mL of 0.100 M CH_3COOH with 0.100 M NaOH?

(a) before the addition of any NaOH (*initial pH*)

(b) after the addition of 10.00 mL 0.100 M NaOH (*before equivalence point*)

(c) after the addition of 12.50 mL 0.100 M NaOH (*half-neutralization*)

(d) after the addition of 25.00 mL 0.100 M NaOH (*equivalence point*)

(e) after the addition of 26.00 mL 0.100 M NaOH (*beyond equivalence point*)

Analyze

Titrations between weak acids and strong bases or strong acids and weak bases have four regions of interest. The first is the initial pH, which we calculate in the same way we would calculate the pH for a solution of a weak acid or weak base. The second is the buffer region; the third is the hydrolysis region; and the fourth is beyond the equivalence point.

Solve

(a) The initial $[H_3O^+]$ is obtained by the calculation in Example 16-6 (page 712). pH = $-\log(1.3 \times 10^{-3})$ = 2.89.

(b) The number of millimoles of CH_3COOH to be neutralized is

$$25.00 \text{ mL} \times \frac{0.100 \text{ mmol } CH_3COOH}{1 \text{ mL}} = 2.50 \text{ mmol } CH_3COOH$$

At this point in the titration, the number of millimoles of OH^- added is

$$10.00 \text{ mL} \times \frac{0.100 \text{ mmol } OH^-}{1 \text{ mL}} = 1.00 \text{ mmol } OH^-$$

The total solution volume = 25.00 mL original acid + 10.00 mL added base = 35.00 mL. We enter this information at appropriate points into the following setup.

Stoichiometric Calculation:

	CH_3COOH	+	OH^-	$\longrightarrow$	CH_3COO^-	+	H_2O
initially present:	2.50 mmol		—		—		
add:			1.00 mmol				
changes:	−1.00 mmol		−1.00 mmol		+1.00 mmol		
after reaction:							
mmol:	1.50 mmol				1.00 mmol		
concns:	1.50 mmol/35.00 mL		≈ 0		1.00 mmol/35.00 mL		
	0.0429 M				0.0286 M		

Equilibrium Calculation: The most direct approach is to recognize that the acetic acid–sodium acetate solution is a buffer solution whose pH can be calculated with the Henderson–Hasselbalch equation. Use of that equation is justified for two reasons: (1) The ratio: $[CH_3COO^-]/[CH_3COOH]$ = 0.0286/0.0429 = 0.667 (satisfying the requirement on page 754 that it be between 0.10 and 10), and (2) CH_3COO^- and CH_3COOH exceed $K_a(1.8 \times 10^{-5})$ by the factors 1.6×10^3 and 2.4×10^3, respectively (satisfying the requirement on page 715 that the factor exceed 100). So

$$pH = pK_a + \log\frac{[A^-]}{[HA]} = 4.74 + \log\frac{0.0286}{0.0429} = 4.74 - 0.18 = 4.56$$

Simpler still would be to substitute the numbers of millimoles of CH_3COO^- and CH_3COOH directly into the Henderson–Hasselbalch equation, without converting to molarities. That is,

$$pH = pK_a + \log\frac{[A^-]}{[HA]}$$

$$= 4.74 + \log\frac{1.00 \text{ mmol}/V}{1.50 \text{ mmol}/V} = 4.74 - 0.18 = 4.56$$

(c) When we have added 12.50 mL of 0.100 M NaOH, we have added $12.50 \times 0.100 = 1.25$ mmol OH^-. As the following setup shows, this is enough base to neutralize exactly *half* of the acid.

	CH_3COOH	$+$	OH^-	$\longrightarrow$	CH_3COO^-	$+$	H_2O
initially present:	2.50 mmol		—		—		
add:			1.25 mmol				
changes:	−1.25 mmol		−1.25 mmol		+1.25 mmol		
after reaction:	1.25 mmol				1.25 mmol		

Again, applying the Henderson–Hasselbalch equation, we get

$$pH = pK_a + \log\frac{[CH_3COO^-]}{[CH_3COOH]}$$
$$= 4.74 + \log\frac{1.25 \; \text{mmol}/V}{1.25 \; \text{mmol}/V} = 4.74 + \log 1 = 4.74$$

(d) At the equivalence point, neutralization is complete and 2.50 mmol $NaCH_3COO$ has been produced in 50.00 mL of solution, leading to 0.0500 M $NaCH_3COO$ The question becomes, "What is the pH of 0.0500 M $NaCH_3COO$?" To answer this question, we must recognize that CH_3COO^- hydrolyzes (and Na^+ does not). The hydrolysis reaction and value of K_b are

$$CH_3COO^- + H_2O \rightleftharpoons CH_3COOH + OH^-$$
$$K_b = \frac{K_w}{K_a} = \frac{1.0 \times 10^{-14}}{1.8 \times 10^{-5}} = 5.6 \times 10^{-10}$$

With a format similar to that used in the hydrolysis calculation of Example 16-13 (page 726), we obtain the following expression, where $x = [OH^-]$ and $x \ll 0.0500$.

$$K_b = \frac{[CH_3COOH]}{[CH_3COO^-]} = \frac{x \cdot x}{0.0500 - x} = 5.6 \times 10^{-10}$$
$$x^2 = 2.8 \times 10^{-11} \qquad x = [OH^-] = 5.3 \times 10^{-6} \text{ M}$$
$$pOH = -\log(5.3 \times 10^{-6}) = 5.28$$
$$pH = 14.00 - pOH = 14.00 - 5.28 = 8.72$$

(e) The amount of OH^- added is $26.00 \text{ mL} \times 0.100 \text{ mmol/mL} = 2.60$ mmol. The volume of solution is 25.00 mL acid + 26.00 mL base = 51.00 mL. The 2.60 mmol OH^- neutralizes the 2.50 mmol of available acid, and 0.10 mmol OH^- remains in *excess*. Beyond the equivalence point, the pH of the solution is determined by the excess strong base.

$$[OH^-] = \frac{0.10 \text{ mmol } OH^-}{51.00 \text{ mL}} = 2.0 \times 10^{-3} \text{ M}$$
$$pOH = -\log(2.0 \times 10^{-3}) = 2.70 \qquad pH = 14.00 - 2.70 = 11.30$$

Assess

In this example the pH at the equivalence point is *not* 7 as it was in the strong acid–strong base problem. We could have predicted this result before performing a single calculation. At the equivalence point, the principal species in solution is CH_3COO^-, the conjugate base of a weak acid (CH_3COOH). Recalling that the conjugate of weak is weak, we conclude that CH_3COO^- is a weak base and thus, the pH at the equivalence point must be greater than 7.

PRACTICE EXAMPLE A: A 20.00 mL sample of 0.150 M HF solution is titrated with 0.250 M NaOH. Calculate **(a)** the initial pH and the pH when neutralization is **(b)** 25.0%, **(c)** 50.0%, **(d)** 100.0% complete. [*Hint*: What is the initial amount of HF, and what amount remains unneutralized at the points in question?]

PRACTICE EXAMPLE B: For the titration of 50.00 mL of 0.106 M NH_3 with 0.225 M HCl, calculate **(a)** the initial pH and the pH when neutralization is **(b)** 25.0% complete; **(c)** 50.0% complete; **(d)** 100.0% complete.

▶ FIGURE 17-12
Constructing the titration curve for a weak acid with a strong base
The calculations needed to plot this graph, illustrated in Example 17-8, can be divided into *four* types.

1. pH of a pure weak acid (initial pH)
2. pH of a buffer solution of a weak acid and its salt (over a broad range before the equivalence point)
3. pH of a salt solution undergoing hydrolysis (equivalence point)
4. pH of a solution of a strong base (over a broad range beyond the equivalence point)

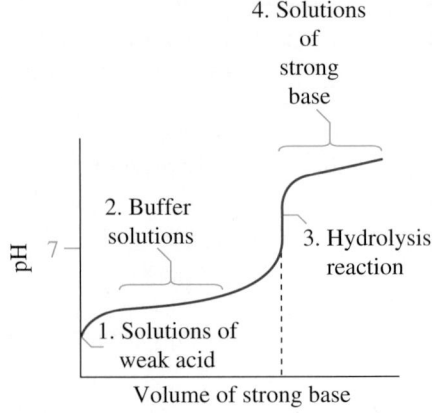

Here are the principal features of the titration curve for a weak acid titrated with a strong base (Fig. 17-11).

- The initial pH is higher (less acidic) than in the titration of a strong acid. (The weak acid is only partially ionized.)
- There is an initial rather sharp increase in pH at the start of the titration. (The anion produced by the neutralization of the weak acid is a common ion that reduces the extent of ionization of the acid.)
- Over a long section of the curve preceding the equivalence point, the pH changes only gradually. (Solutions corresponding to this portion of the curve are buffer solutions.)
- Because $[HA] = [A^-]$ at the point of half-neutralization, $pH = pK_a$.
- At the equivalence point, $pH > 7$. (The conjugate base of a weak acid hydrolyzes, producing OH^-.)
- Beyond the equivalence point, the titration curve is identical to that of a strong acid with a strong base. (In this portion of the titration, the pH is established entirely by the concentration of unreacted OH^-.)
- The steep portion of the titration curve at the equivalence point occurs over a relatively short pH range (from about pH 7 to pH 10).
- The selection of indicators available for the titration is more limited than in a strong acid–strong base titration. (Indicators that change color below pH 7 cannot be used.)

As illustrated in Example 17-8 and suggested by Figure 17-12, the necessary calculations for a weak acid–strong base titration curve are of four distinct types, depending on the portion of the titration curve being described. One type of titration that generally cannot be performed successfully is that of a weak acid with a weak base (or vice versa). The equivalence point cannot be located precisely because the change in pH with volume of titrant is too gradual.

🔍 **17-4   CONCEPT ASSESSMENT**

To raise the pH of 1.00 L of 0.50 M HCl(aq) *significantly*, which of the following would you add to the solution and why? **(a)** 0.50 mol CH_3COOH; **(b)** 1.00 mol NaCl; **(c)** 0.60 mol $NaCH_3COO$; **(d)** 0.40 mol NaOH.

▶ This stepwise neutralization is observed only if successive ionization constants (K_{a_1}, K_{a_2}, ...) differ significantly in magnitude (for example, by a factor of 10^3 or more). Otherwise, the second neutralization step begins before the first step is completed, and so on.

Titration of a Weak Polyprotic Acid

The most striking evidence that a polyprotic acid ionizes in distinct steps comes by way of its titration curve. For a polyprotic acid, we expect to see a separate equivalence point for each acidic hydrogen. Thus, we expect to see three equivalence points when H_3PO_4 is titrated with NaOH(aq). In the neutralization of phosphoric acid by sodium hydroxide, essentially all the H_3PO_4 molecules are

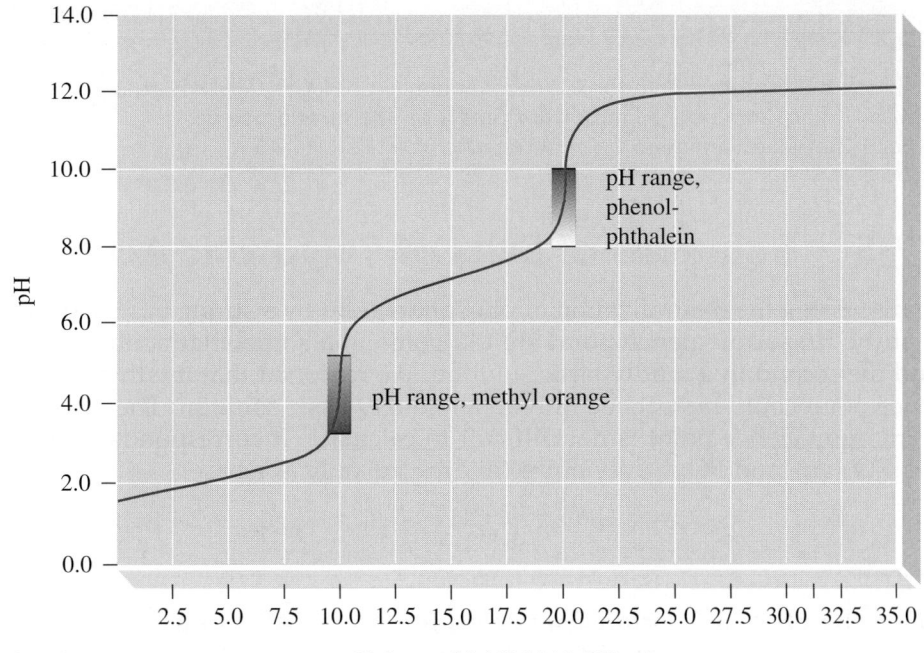

▲ FIGURE 17-13
Titration of a weak polyprotic acid—10.0 mL of 0.100 M H_3PO_4 with 0.100 M NaOH
A 10.0 mL volume of 0.100 M NaOH is required to reach the first equivalence point. The additional volume of 0.100 M NaOH required to reach the second equivalence point is also 10.0 mL. The pH does not increase sharply in the vicinity of the third equivalence point (30 mL).

first converted to the salt, NaH_2PO_4. Then all the NaH_2PO_4 is converted to Na_2HPO_4; and finally the Na_2HPO_4 is converted to Na_3PO_4.

The titration of 10.0 mL of 0.100 M H_3PO_4 with 0.100 M NaOH is pictured in Figure 17-13. Notice that the first two equivalence points come at equal intervals on the volume axis, at 10.0 mL and at 20.0 mL. Although we expect a third equivalence point at 30.0 mL, it is not realized in this titration. The pH of the strongly hydrolyzed Na_3PO_4 solution at the third equivalence point—approaching pH 13—is higher than can be reached by adding 0.100 M NaOH to water. $Na_3PO_4(aq)$ is nearly as basic as the NaOH(aq) used in the titration (as we will see in Section 17-5).

Let's focus on a few details of this titration. For each mole of H_3PO_4, 1 mol NaOH is required to reach the first equivalence point. At this first equivalence point, the solution is essentially $NaH_2PO_4(aq)$. This is an *acidic* solution because $K_{a_2} > K_b$ for $H_2PO_4^-$: the reaction that produces H_3O^+ predominates over the one that produces OH^-.

$$H_2PO_4^- + H_2O \rightleftharpoons H_3O^+ + HPO_4^{2-} \qquad K_{a_2} = 6.3 \times 10^{-8}$$
$$H_2PO_4^- + H_2O \rightleftharpoons H_3PO_4 + OH^- \qquad K_b = 1.4 \times 10^{-12}$$

The pH at the equivalence point falls within the pH range over which the color of methyl orange indicator changes from red to orange.

An additional mole of NaOH is required to convert 1 mol $H_2PO_4^-$ to 1 mol HPO_4^{2-}. At this second equivalence point in the titration of H_3PO_4, the solution is basic because $K_b > K_{a_3}$ for HPO_4^{2-}.

$$HPO_4^{2-} + H_2O \rightleftharpoons H_2PO_4^- + OH^- \qquad K_b = 1.6 \times 10^{-7}$$
$$HPO_4^{2-} + H_2O \rightleftharpoons H_3O^+ + PO_4^{3-} \qquad K_{a_3} = 4.2 \times 10^{-13}$$

Phenolphthalein is an appropriate indicator for this equivalence point; the color of this indicator changes from colorless to light pink.

Sketch the titration curve for 1,2-ethanediamine, $NH_2CH_2CH_2NH_2$(aq), with HCl(aq) and label all important points on the titration curve. For 1,2-ethanediamine, $pK_{b_1} = 4.08$; $pK_{b_2} = 7.15$.

17-5 Solutions of Salts of Polyprotic Acids

In discussing the neutralization of phosphoric acid by a strong base, we found that the first equivalence point should come in a somewhat acidic solution and the second in a mildly basic solution. We reasoned that the third equivalence point could be reached only in a strongly basic solution. The pH at this third equivalence point is not difficult to calculate. It corresponds to that of Na_3PO_4(aq), and PO_4^{3-} can ionize (hydrolyze) only as a base.

$$PO_4^{3-} + H_2O \rightleftharpoons HPO_4^{2-} + OH^- \qquad K_b = K_w/K_{a_3} = 2.4 \times 10^{-2}$$

EXAMPLE 17-9 **Determining the pH of a Solution Containing the Anion (A^{n-}) of a Polyprotic Acid**

Sodium phosphate, Na_3PO_4, is an ingredient of some preparations used to clean painted walls before they are repainted. What is the pH of 0.025 M Na_3PO_4(aq)?

Analyze

PO_4^{3-} is a weak base that will react with water to form HPO_4^{2-} and OH^-, thereby making the solution basic. We don't have to consider additional reactions, such as the reaction of HPO_4^{2-} and H_2O to give $H_2PO_4^-$ and OH^-, because as discussed on page 719, most of the OH^- in solution will come from the reaction of PO_4^{3-} and H_2O. The value of K_b for PO_4^{3-} is large enough, however, that we will *not* be able to make the usual simplifying approximation that $0.025 - x \approx 0.025$.

Solve

In the usual fashion, we can write

$$PO_4^{3-} + H_2O \rightleftharpoons HPO_4^{2-} + OH^- \qquad K_b = 2.4 \times 10^{-2}$$

initial concns:	0.025 M	—	—
changes:	$-x$ M	$+x$ M	$+x$ M
equil concns:	$(0.025 - x)$ M	x M	x M

$$K_b = \frac{[HPO_4^{2-}][OH^-]}{[PO_4^{3-}]} = \frac{x \cdot x}{0.025 - x} = 2.4 \times 10^{-2}$$

Solving the quadratic equation $x^2 + 0.024x - (0.025)(0.024) = 0$ yields $x = [OH^-] = 0.015$ M. Thus,

$$pOH = -\log[OH^-] = -\log(0.015) = 1.82$$
$$pH = 14.00 - 1.82 = 12.18$$

Assess

Notice that more than half (about 61%) of the PO_4^{3-} reacts. In solving this problem, we assumed that most of the OH^- in solution comes from the reaction of PO_4^{3-} and H_2O. That is, the subsequent reaction of HPO_4^{2-} and H_2O does not contribute a significant amount of OH^-. Can we test the validity of this approximation? Of course we can. On page 769, we saw that

$$HPO_4^{2-} + H_2O \rightleftharpoons H_2PO_4^- + OH^- \qquad K_b = [H_2PO_4^-][OH^-]/[HPO_4^{2-}] = 1.6 \times 10^{-7}$$

Using $[HPO_4^{2-}] \approx [OH^-] \approx 0.015$ M (from above), we obtain $[H_2PO_4^-] \approx 1.6 \times 10^{-7}$. The amount of $H_2PO_4^-$ (and OH^-) generated from the reaction of HPO_4^{2-} and H_2O is, as predicted, very small. The approximation is valid.

PRACTICE EXAMPLE A: Using data from Table 16.4, calculate the pH of 1.0 M Na_2CO_3.

PRACTICE EXAMPLE B: Using data from Table 16.4, calculate the pH of 0.500 M Na_2SO_3.

It is more difficult to calculate the pH values of $NaH_2PO_4(aq)$ and $Na_2HPO_4(aq)$ than of $Na_3PO_4(aq)$. This is because with both $H_2PO_4^-$ and HPO_4^{2-}, two equilibria must be considered *simultaneously*: ionization as an acid and ionization as a base (hydrolysis). For solutions that are reasonably concentrated (say, 0.10 M or greater), the pH values prove to be *independent* of the solution concentration. Shown here (with pK_a values from Table 16.4) are general expressions, printed in blue, and their application to $H_2PO_4^-(aq)$ and $HPO_4^{2-}(aq)$:

$$for\ H_2PO_4^-: \qquad pH = \frac{1}{2}(pK_{a_1} + pK_{a_2}) = \frac{1}{2}(2.15 + 7.20) = 4.68 \qquad \textbf{(17.5)}$$

$$for\ HPO_4^{2-}: \qquad pH = \frac{1}{2}(pK_{a_2} + pK_{a_3}) = \frac{1}{2}(7.20 + 12.38) = 9.79 \qquad \textbf{(17.6)}$$

17-1   ARE YOU WONDERING...

How to derive equations (17.5) and (17.6)?

Here is good place to use the general problem-solving method introduced in Section 16-6 (page 721). Consider a solution of NaH_2PO_4, of molarity M. The principal concentrations that we must account for are $[Na^+]$, $[H_3O^+]$, $[H_3PO_4]$, $[H_2PO_4^-]$, $[HPO_4^{2-}]$, and $[OH^-]$. Of these, two have very simple values: $[Na^+] = M$, and $[OH^-] = K_w/[H_3O^+]$. Additionally, we can write the following equations.

1. *Acid Ionization*: $H_2PO_4^- + H_2O \rightleftharpoons H_3O^+ + HPO_4^{2-}$

$$K_{a_2} = \frac{[H_3O^+][HPO_4^{2-}]}{[H_2PO_4^-]}$$

2. *Hydrolysis*: $H_2PO_4^- + H_2O \rightleftharpoons H_3PO_4 + OH^-$

$$K_b = K_w/K_{a_1} = \frac{[H_3PO_4][OH^-]}{[H_2PO_4^-]}$$

3. *Material Balance*: The total concentration of the phosphorus-containing species is the stoichiometric molarity, M. Also, $[Na^+] = M$. We can write
$[H_3PO_4] + [H_2PO_4^-] + [HPO_4^{2-}] + [PO_4^{3-}] = M = [Na^+]$

4. *Electroneutrality Condition*: $[H_3O^+] + [Na^+] = [H_2PO_4^-] + 2 \times [HPO_4^{2-}] + 3 \times [PO_4^{3-}] + [OH^-]$.

The material balance equation and the electroneutrality condition can both be simplified by neglecting the terms involving $[PO_4^{3-}]$ and $[OH^-]$.

Solving the set of equations: Begin by substituting equation (3) into equation (4).

$[H_3O^+] = [H_2PO_4^-] + 2 \times [HPO_4^{2-}] - [H_3PO_4] - [H_2PO_4^-] - [HPO_4^{2-}]$
$\qquad\ = [HPO_4^{2-}] - [H_3PO_4]$

To continue from here, rearrange equation (1) to obtain $[HPO_4^{2-}]$ in terms of $[H_3O^+]$, $[H_2PO_4^-]$, and K_{a_2}; then rearrange equation (2) to obtain $[H_3PO_4]$ in terms of $[H_3O^+]$, $[H_2PO_4^-]$, and K_{a_1}. Next, substitute the results into the expression, $[H_3O^+] = [HPO_4^{2-}] - [H_3PO_4]$. At this point, you will have an equation in terms of $[H_3O^+]$, $[H_2PO_4^-]$, K_{a_1}, and K_{a_2} and both $[PO_4^{3-}]$ and $[OH^-]$. Assume that $K_{a_1} + [H_2PO_4^-] \approx [H_2PO_4^-]$ and you will get an equation from which you can derive equation (17.5). The remainder of this derivation and the derivation of equation (17.6) are left for you to do (see Exercise 79).

◀ The K_a and K_b values for $H_2PO_4^-$ (see page 769) tell us that there is a greater tendency for $H_2PO_4^-$ to act as an acid. The solution will be acidic and $[OH^-]$ will be less than 10^{-7} M. Also, we expect $[PO_4^{3-}]$ to be very small. To obtain PO_4^{3-}, $H_2PO_4^-$ must ionize twice. However, because $H_2PO_4^-$ is a weak acid, only a small fraction will ionize to HPO_4^{2-} and an even smaller fraction will ionize further to PO_4^{3-}.

17-6   Acid–Base Equilibrium Calculations: A Summary

In this and the preceding chapter, we have considered a variety of acid–base equilibrium calculations. When you are faced with a new problem-solving situation, you might find it helpful to relate the new problem to a type that you have encountered before. It is best not to rely exclusively on "labeling" a problem, however. Some problems might not fit a recognizable category. Instead, keep in mind some principles that apply regardless of the particular problem, as suggested by these questions.

1. **Which species are potentially present in solution, and how large are their concentrations likely to be?**

    In a solution containing similar amounts of HCl and CH_3COOH, the only significant *ionic* species are H_3O^+ and Cl^-. HCl is a completely ionized strong acid, and in the presence of a strong acid, the weak acid $HC_2H_3O_2$ is only very slightly ionized because of the common-ion effect. In a mixture containing similar amounts of two *weak* acids of similar strengths, such as CH_3COOH and HNO_2, each acid partially ionizes. All of these concentrations would be significant: $[CH_3COOH]$, $[CH_3COO^-]$, $[HNO_2]$, $[NO_2^-]$, and $[H_3O^+]$. In a solution containing phosphoric acid or a phosphate salt (or both), H_3PO_4, $H_2PO_4^-$, HPO_4^{2-}, PO_4^{3-}, OH^-, H_3O^+, and possibly other cations might be present. If the solution is simply $H_3PO_4(aq)$, however, the only species present in significant concentrations are those associated with the first ionization: H_3PO_4, H_3O^+, and $H_2PO_4^-$. If the solution is instead described as $Na_3PO_4(aq)$, the significant species are Na^+, PO_4^{3-}, and the ions associated with the hydrolysis of PO_4^{3-}, that is, HPO_4^{2-} and OH^-.

2. **Are reactions possible among any of the solution components; if so, what is their stoichiometry?**

    Suppose that you are asked to calculate $[OH^-]$ in a solution that is prepared to be 0.10 M NaOH and 0.20 M NH_4Cl. Before you answer that $[OH^-] = 0.10$ M, consider whether a solution can be *simultaneously* 0.10 M in OH^- and 0.20 M in NH_4^+. It cannot; any solution containing both NH_4^+ and OH^- must also contain NH_3. The OH^- and NH_4^+ react in a 1:1 mole ratio until OH^- is almost totally consumed:

    $$NH_4^+ + OH^- \longrightarrow NH_3 + H_2O$$

    You are now dealing with the buffer solution 0.10 M NH_3–0.10 M NH_4^+.

3. **Which equilibrium equations apply to the particular situation? Which are the most significant?**

    One equation that applies to all acids and bases in aqueous solutions is $K_w = [H_3O^+][OH^-] = 1.0 \times 10^{-14}$. In many calculations, however, this equation is not significant compared with others. One situation in which it is significant is in calculating $[OH^-]$ in an *acidic* solution or $[H_3O^+]$ in a *basic* solution. After all, an acid does not produce OH^-, and a base does not produce H_3O^+. Another situation in which K_w is likely to be significant is in a solution with a pH near 7.

▶ Body temperature is 37 °C and at that temperature K_w does not equal 1.0×10^{-14}. Refer to Exercise 83 on page 781.

Often you will find that the ionization equilibrium with the largest K value is the most significant, but this will not always be the case. The amounts of the various species in solution are also an important consideration. When one drop of 1.00 M H_3PO_4 ($K_{a_1} = 7.1 \times 10^{-3}$) is added to 1.00 L of 0.100 M CH_3COOH ($K_a = 1.8 \times 10^{-5}$), the acetic acid ionization is most important in establishing the pH of the solution. The solution contains far more acetic acid than it does phosphoric acid.

🔍 **17-6   CONCEPT ASSESSMENT**

A solution is formed by mixing 200.0 mL of 0.100 M KOH with 100.0 mL of a solution that is both 0.200 M in CH_3COOH and 0.050 M in HI. *Without doing detailed calculations,* identify in the final solution **(a)** all the solute species present, **(b)** the two most abundant solute species, and **(c)** the two least abundant species.

Summary

17-1 Common-Ion Effect in Acid–Base Equilibria—
The ionization of a weak electrolyte is suppressed by the addition of an ion that is the product of the ionization and is known as the **common-ion effect**. This effect is a manifestation of Le Châtelier's principle, introduced in Chapter 15.

17-2 Buffer Solutions—
Solutions that resist changes in pH upon the addition of small amounts of an acid or base are referred to as **buffer solutions**. The finite amount of acid or base that a buffer solution can neutralize is known as the **buffer capacity**, while the pH range over which the buffer solution neutralizes the added acid or base is referred to as **buffer range**. Pure water has no buffer capacity (Fig. 17-4). Key to the functioning of a buffer solution is the presence of either a weak acid and its conjugate base or a weak base and its conjugate acid (Fig. 17-5). Calculating the pH of a buffer solution can be accomplished by using the ICE method developed in Chapter 15 or by application of the **Henderson–Hasselbalch equation** (expression 17.2). Determination of the pH of a buffer solution after the addition of a strong acid or base requires a stoichiometric calculation followed by an equilibrium calculation (Fig. 17-7).

17-3 Acid-Base Indicators—
Substances whose colors depend on the pH of a solution are known as **acid–base indicators**. Acid–base indicators exist in solution as a weak acid (HIn) and its conjugate base (In^-). Each form has a different color and the proportions of the two forms determine the color of the solution, which in turn depends on the pH of the solution. The pH range over which an acid–base indicator changes color is determined by the K_a of the specific indicator (Fig. 17-8).

17-4 Neutralization Reactions and Titration Curves—
As described in Chapter 5, the concentration of an acidic or basic solution of unknown concentration can be determined by titration with a base or acid of precisely known concentration. In this process a precisely measured volume of the solution of known concentration, the **titrant**, is added through a buret into a precisely measured quantity of the "unknown" contained in a beaker or flask. Typically, the amounts of reactants in a titration are of the order of 10^{-3} mol, that is, **millimoles (mmol)**. A **titration curve** is a graph of a measured property of the reaction mixture as a function of the volume of titrant added—pH for an acid–base titration (Fig. 17-9). The point at which neither reactant is in excess in a titration is known as the **equivalence point**. The **end point** of an acid–base titration can be located through the change in color of an indicator. The indicator must be chosen such that its color change occurs as close to the equivalence point as possible. Strong acid-strong base titration curves (Figs. 17-9 and 17-10) are different from weak acid-strong base titration curves (Fig. 17-11). The two main differences are seen in the latter type, a buffer region and a hydrolysis reaction at the equivalence point (Fig. 17-12).

17-5 Solutions of Salts of Polyprotic Acids—
Calculating the pH of solutions containing the salts of polyprotic acids is made difficult by the fact that two or more equilibria occur simultaneously. Yet for certain solutions, the calculations can be reduced to a simple form (expressions 17.5 and 17.6).

17-6 Acid–Base Equilibrium Calculations: A Summary—
As a general summary of acid–base equilibrium calculations, the essential factors are identifying all the species in solution, their concentrations, the possible reactions between them, and the stoichiometry and equilibrium constants of those reactions.

Integrative Example

The structural formula shown is *para*-hydroxybenzoic acid, a weak diprotic acid used as a food preservative. Titration of 25.00 mL of a dilute aqueous solution of this acid requires 16.24 mL of 0.0200 M NaOH to reach the first equivalence point. The measured pH after the addition of 8.12 mL of the base is 4.57; after 16.24 mL, the pH is 7.02. Determine the values of pK_{a_1} and pK_{a_2} of *para*-hydroxybenzoic acid and the pH values for the two equivalence points in the titration.

Analyze

This is a titration of a polyprotic weak acid with a strong base, and the titration curve for this problem should look very similar to Figure 17-13. In the titration of a weak acid by a strong base we know that at the point of half-neutralization, $pH = pK_a$ and therefore pK_{a_1} should be the pH at 8.12 mL. For pK_{a_2}, we will use expression 17.5 since at this point in the titration we will have an aqueous solution of HOC_6H_4COONa, which is a salt of a polyprotic acid. The pH of the first equivalence point is given and to find the pH of the second equivalence point we must perform an ICE calculation similar to the one in Example 16-13.

Solve

The volume of base needed to reach the first equivalence point is 16.24 mL; at this point, the pH = 7.02. A volume of 8.12 mL is needed to half-neutralize the acid in its first ionization step; at this point the pH = pK_{a_1}, that is, $pK_{a_1} = 4.57$.

At the first equivalence point, the solution is $HOC_6H_4COONa(aq)$ with pH = 7.02. Recognizing this as the salt produced in neutralizing a polyprotic acid in its first ionization, we use equation (17.5) to solve for pK_{a_2}. That is, the pH of an aqueous solution of the ion $HOC_6H_6COO^-$ is given by the expression

$$pH = \tfrac{1}{2}(pK_{a_1} + pK_{a_2}) = \tfrac{1}{2}(4.57 + pK_{a_2}) = 7.02$$
$$pK_{a_2} = (2 \times 7.02) - 4.57 = 9.47$$

Determining the pH at the second equivalence point involves additional calculations. We begin by noting that at the second equivalence point the solution is one of $NaOC_6H_4COONa$. The pH of the solution is established by the hydrolysis of $^-OC_6H_4COO^-$.

$$^-OC_6H_4COO^- + H_2O \rightleftharpoons HOC_6H_4COO^- + OH^-$$
$$K_b = K_w/K_{a_2}$$

To evaluate K_b, let's first obtain K_{a_2} from pK_{a_2}.

$$pK_{a_2} = -\log K_{a_2} = 9.47 \text{ and } K_{a_2} = 10^{-9.47} = 3.4 \times 10^{-10}$$
$$K_b = K_w/K_{a_2} = 1.0 \times 10^{-14}/3.4 \times 10^{-10} = 2.9 \times 10^{-5}$$

We can get the pH of this solution by first calculating $[OH^-]$ and pOH. However, to do this, we still need one more piece of data—the molarity of the $NaOC_6H_4COONa(aq)$. We can get this from data for titration to the first equivalence point.

$$? \text{ mmol } OH^- = 16.24 \text{ mL} \times 0.0200 \text{ mmol } OH^-/\text{mL}$$
$$= 0.325 \text{ mmol } OH^-$$
$$? \text{ mmol } HOC_6H_4COOH = 0.325 \text{ mmol } OH^- \times 1 \text{ mmol}$$
$$HOC_6H_4COOH/\text{mmol } OH^- = 0.325 \text{ mmol } HOC_6H_4COOH$$

The amount of $^-OC_6H_4COO^-$ at the second equivalence point is the same as the amount of acid at the start of the titration.

$$0.325 \text{ mmol } ^-OC_6H_4COO^-$$

The volume of solution at the second equivalence point is

$$25.00 \text{ mL} + 16.24 \text{ mL} + 16.24 \text{ mL} = 57.48 \text{ mL.}$$

Thus,

$$[^-OC_6H_4COO^-] = 0.325 \text{ mmol } ^-OC_6H_4COO^-/57.48 \text{ mL}$$
$$= 5.65 \times 10^{-3} \text{ M.}$$

Now we can return to the hydrolysis equation and the expression for K_b, using the method of Example 16-13.

$$^-OC_6H_4COO^- + H_2O \rightleftharpoons HOC_6H_4COO^- + OH^-$$

initial concns:	5.65×10^{-3} M	—	—
changes:	$-x$ M	$+x$ M	$+x$ M
equil concns:	$(5.65 \times 10^{-3} - x)$ M	x M	x M

$$K_b = \frac{x \cdot x}{(5.65 \times 10^{-3} - x)} = 2.9 \times 10^{-5}$$

The solution to this quadratic equation is $x = [OH^-] = 3.9 \times 10^{-4}$, corresponding to pOH = 3.41 and pH = 10.59.

Assess

The polyprotic acid, *para*-hydroxybenzoic acid, has two functional groups, a carboxylic acid and a phenolic group. Each group has an ionizable proton. Given just two basic pieces of titration data, we used concepts from this and the preceding chapter to determine the pK_a values for both ionizable groups as well as the pH at the two equivalence points. As a check, note that the pK_a values of these groups are comparable to the values for their parent compounds, acetic acid and phenol (see Table 16.3).

PRACTICE EXAMPLE A: 7.500 g of a weak acid HA is added to sufficient distilled water to produce 500.0 mL of solution with pH = 2.716. This solution is titrated with NaOH(aq). Halfway to the equivalence point, pH = 4.602. What is the freezing point of the solution?

PRACTICE EXAMPLE B: The following titration curve was obtained as part of a general chemistry laboratory experiment for an unknown that weighed 0.8 g. The titrant was either a 0.2 M strong base or 0.2 M strong acid. Estimate the molar mass of the unknown and its ionization constant.

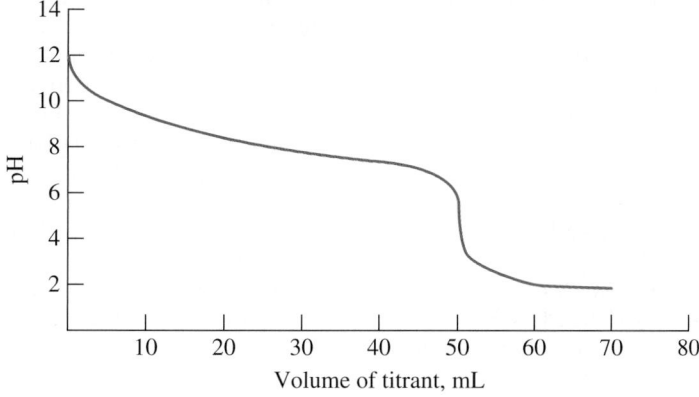

Mastering**CHEMISTRY**

You'll find a link to additional self study questions in the study area on www.masteringchemistry.com

Exercises

The Common-Ion Effect

(*Use data from Table 16.3 as necessary.*)

1. For a solution that is 0.275 M CH_3CH_2COOH (propionic acid, $K_a = 1.3 \times 10^{-5}$) and 0.0892 M HI, calculate **(a)** $[H_3O^+]$; **(b)** $[OH^-]$; **(c)** CH_3CH_2COO; **(d)** $[I^-]$.
2. For a solution that is 0.164 M NH_3 and 0.102 M NH_4Cl, calculate **(a)** $[OH^-]$; **(b)** $[NH_4^+]$; **(c)** $[Cl^-]$; **(d)** $[H_3O^+]$.
3. Calculate the *change* in pH that results from adding **(a)** 0.100 mol $NaNO_2$ to 1.00 L of 0.100 M $HNO_2(aq)$;
(b) 0.100 mol $NaNO_3$ to 1.00 L of 0.100 M $HNO_3(aq)$. Why are the changes not the same? Explain.
4. In Example 16-8, we calculated the percent ionization of CH_3COOH in **(a)** 1.0 M; **(b)** 0.10 M; and **(c)** 0.010 M CH_3COOH solutions. Recalculate those percent ionizations if each solution also contains 0.10 M $NaCH_3COO$. Explain why the results are different from those of Example 16-8.

5. Calculate $[H_3O^+]$ in a solution that is **(a)** 0.035 M HCl and 0.075 M HOCl; **(b)** 0.100 M $NaNO_2$ and 0.0550 M HNO_2; **(c)** 0.0525 M HCl and 0.0768 M $NaCH_3COO$.

6. Calculate $[OH^-]$ in a solution that is **(a)** 0.0062 M $Ba(OH)_2$ and 0.0105 M $BaCl_2$; **(b)** 0.315 M $(NH_4)_2SO_4$ and 0.486 M NH_3; **(c)** 0.196 M NaOH and 0.264 M NH_4Cl.

Buffer Solutions

(*Use data from Tables 16.3 and 16.4 as necessary.*)

7. What concentration of formate ion, $[HCOO^-]$, should be present in 0.366 M HCOOH to produce a buffer solution with pH = 4.06?

$$HCOOH + H_2O \rightleftharpoons H_3O^+ + HCOO^-$$
$$K_a = 1.8 \times 10^{-4}$$

8. What concentration of ammonia, $[NH_3]$, should be present in a solution with $[NH_4^+]$ = 0.732 M to produce a buffer solution with pH = 9.12? For NH_3, $K_b = 1.8 \times 10^{-5}$.

9. Calculate the pH of a buffer that is
   (a) 0.012 M $C_6H_5COOH(K_a = 6.3 \times 10^{-5})$ and 0.033 M NaC_6H_5COO; .
   (b) 0.408 M NH_3 and 0.153 M NH_4Cl.

10. Lactic acid, CH_3CH_2COOH, is found in sour milk. A solution containing 1.00 g $NaCH_3CH_2COO$ in 100.0 mL of 0.0500 M $HC_3H_5O_3$ has a pH = 4.11. What is K_a of lactic acid?

11. Indicate which of the following aqueous solutions are buffer solutions, and explain your reasoning. [*Hint:* Consider any reactions that might occur between solution components.]
    (a) 0.100 M NaCl
    (b) 0.100 M NaCl–0.100 M NH_4Cl
    (c) 0.100 M CH_3NH_2–0.150 M $CH_3NH_3^+Cl^-$
    (d) 0.100 M HCl–0.050 M $NaNO_2$
    (e) 0.100 M HCl–0.200 M $NaCH_3COO$
    (f) 0.100 M CH_3COOH–0.125 M $NaCH_3CH_2COO$

12. The $H_2PO_4^-$–HPO_4^{2-} combination plays a role in maintaining the pH of blood.
    (a) Write equations to show how a solution containing these ions functions as a buffer.
    (b) Verify that this buffer is most effective at pH 7.2.
    (c) Calculate the pH of a buffer solution in which $[H_2PO_4^-]$ = 0.050 M and $[HPO_4^{2-}]$ = 0.150 M. [*Hint:* Focus on the second step of the phosphoric acid ionization.]

13. What is the pH of a solution obtained by adding 1.15 mg of aniline hydrochloride ($C_6H_5NH_3^+Cl^-$) to 3.18 L of 0.105 M aniline ($C_6H_5NH_2$)? [*Hint:* Check any assumptions that you make.]

14. What is the pH of a solution prepared by dissolving 8.50 g of aniline hydrochloride ($C_6H_5NH_3^+Cl^-$) in 750 mL of 0.215 M aniline, ($C_6H_5NH_2$)? Would this solution be an effective buffer? Explain.

15. You wish to prepare a buffer solution with pH = 9.45.
    (a) How many grams of $(NH_4)_2SO_4$ would you add to 425 mL of 0.258 M NH_3 to do this? Assume that the solution's volume remains constant.
    (b) Which buffer component, and how much (in grams), would you add to 0.100 L of the buffer in part (a) to change its pH to 9.30? Assume that the solution's volume remains constant.

16. You prepare a buffer solution by dissolving 2.00 g each of benzoic acid, C_6H_5COOH, and sodium benzoate, NaC_6H_5COO, in 750.0 mL of water.
    (a) What is the pH of this buffer? Assume that the solution's volume is 750.0 mL.
    (b) Which buffer component, and how much (in grams), would you add to the 750.0 mL of buffer solution to change its pH to 4.00?

17. If 0.55 mL of 12 M HCl is added to 0.100 L of the buffer solution in Exercise 15(a), what will be the pH of the resulting solution?

18. If 0.35 mL of 15 M NH_3 is added to 0.750 L of the buffer solution in Exercise 16(a), what will be the pH of the resulting solution?

19. You are asked to prepare a buffer solution with a pH of 3.50. The following solutions, all 0.100 M, are available to you: HCOOH, CH_3COOH, H_3PO_4, NaCHOO, $NaCH_3COO$, and NaH_2PO_4. Describe how you would prepare this buffer solution. [*Hint:* What volumes of which solutions would you use?]

20. You are asked to reduce the pH of the 0.300 L of buffer solution in Example 17-5 from 5.09 to 5.00. How many milliliters of which of these solutions would you use: 0.100 M NaCl, 0.150 M HCl, 0.100 M $NaCH_3COO$, 0.125 M NaOH? Explain your reasoning.

21. Given 1.00 L of a solution that is 0.100 M CH_3CH_2COOH and 0.100 M CH_3CH_2COO,
    (a) Over what pH range will this solution be an effective buffer?
    (b) What is the buffer capacity of the solution? That is, how many millimoles of strong acid or strong base can be added to the solution before any significant change in pH occurs?

22. Given 125 mL of a solution that is 0.0500 M CH_3NH_2 and 0.0500 M $CH_3NH_3^+Cl^-$,
    (a) Over what pH range will this solution be an effective buffer?
    (b) What is the buffer capacity of the solution? That is, how many millimoles of strong acid or strong base can be added to the solution before any significant change in pH occurs?

23. A solution of volume 75.0 mL contains 15.5 mmol $HCHO_2$ and 8.50 mmol NaHCOO.
    (a) What is the pH of this solution?
    (b) If 0.25 mmol $Ba(OH)_2$ is added to the solution, what will be the pH?
    (c) If 1.05 mL of 12 M HCl is added to the original solution, what will be the pH?

24. A solution of volume 0.500 L contains 1.68 g NH_3 and 4.05 g $(NH_4)_2SO_4$.
    (a) What is the pH of this solution?
    (b) If 0.88 g NaOH is added to the solution, what will be the pH?
    (c) How many milliliters of 12 M HCl must be added to 0.500 L of the original solution to change its pH to 9.00?

25. A handbook lists various procedures for preparing buffer solutions. To obtain a pH = 9.00, the handbook says to mix 36.00 mL of 0.200 M NH_3 with 64.00 mL of 0.200 M NH_4Cl.
    (a) Show by calculation that the pH of this solution is 9.00.
    (b) Would you expect the pH of this solution to remain at pH = 9.00 if the 100.00 mL of buffer solution were diluted to 1.00 L? To 1000 L? Explain.
    (c) What will be the pH of the original 100.00 mL of buffer solution if 0.20 mL of 1.00 M HCl is added to it?
    (d) What is the maximum volume of 1.00 M HCl that can be added to 100.00 mL of the original buffer solution so that the pH does not drop below 8.90?

26. An acetic acid–sodium acetate buffer can be prepared by the reaction

    $$CH_3COO^- + H_3O^+ \longrightarrow CH_3COOH + H_2O$$
    (From $NaCH_3COO$)(From HCl)

    (a) If 12.0 g $NaCH_3COO$ is added to 0.300 L of 0.200 M HCl, what is the pH of the resulting solution?
    (b) If 1.00 g $Ba(OH)_2$ is added to the solution in part (a), what is the new pH?
    (c) What is the maximum mass of $Ba(OH)_2$ that can be neutralized by the buffer solution of part (a)?
    (d) What is the pH of the solution in part (a) following the addition of 5.50 g $Ba(OH)_2$?

Acid–Base Indicators

(*Use data from Tables 16.3 and 16.4 as necessary.*)

27. A handbook lists the following data:

Indicator	K_{HIn}	Color Change Acid → Anion
Bromphenol blue	1.4×10^{-4}	yellow → blue
Bromcresol green	2.1×10^{-5}	yellow → blue
Bromthymol blue	7.9×10^{-8}	yellow → blue
2,4-Dinitrophenol	1.3×10^{-4}	colorless → yellow
Chlorphenol red	1.0×10^{-6}	yellow → red
Thymolphthalein	1.0×10^{-10}	colorless → blue

   (a) Which of these indicators change color in acidic solution, which in basic solution, and which near the neutral point?
   (b) What is the approximate pH of a solution if bromcresol green indicator turns green? if chlorphenol red turns orange?

28. With reference to the indicators listed in Exercise 27, what would be the color of each combination?
    (a) 2,4-dinitrophenol in 0.100 M HCl(aq)
    (b) chlorphenol red in 1.00 M NaCl(aq)
    (c) thymolphthalein in 1.00 M NH_3(aq)
    (d) bromcresol green in seawater (recall Figure 17-8)

29. In the use of acid–base indicators,
    (a) Why is it generally sufficient to use a *single* indicator in an acid–base titration, but often necessary to use *several* indicators to establish the approximate pH of a solution?
    (b) Why must the quantity of indicator used in a titration be kept as small as possible?

30. The indicator methyl red has a pK_{HIn} = 4.95. It changes from red to yellow over the pH range from 4.4 to 6.2.

   (a) If the indicator is placed in a buffer solution of pH = 4.55, what percent of the indicator will be present in the acid form, HIn, and what percent will be present in the base or anion form, In^-?
   (b) Which form of the indicator has the "stronger" (that is, more visible) color—the acid (red) form or base (yellow) form? Explain.

31. Phenol red indicator changes from yellow to red in the pH range from 6.6 to 8.0. *Without making detailed calculations*, state what color the indicator will assume in each of the following solutions: **(a)** 0.10 M KOH; **(b)** 0.10 M CH_3COOH; **(c)** 0.10 M NH_4NO_3; **(d)** 0.10 M HBr; **(e)** 0.10 M NaCN; **(f)** 0.10 M CH_3COOH–0.10 M $NaCH_3COO$.

32. Thymol blue indicator has *two* pH ranges. It changes color from red to yellow in the pH range from 1.2 to 2.8, and from yellow to blue in the pH range from 8.0 to 9.6. What is the color of the indicator in each of the following situations?
    (a) The indicator is placed in 350.0 mL of 0.205 M HCl.
    (b) To the solution in part (a) is added 250.0 mL of 0.500 M $NaNO_2$.
    (c) To the solution in part (b) is added 150.0 mL of 0.100 M NaOH.
    (d) To the solution in part (c) is added 5.00 g $Ba(OH)_2$.

33. In the titration of 10.00 mL of 0.04050 M HCl with 0.01120 M $Ba(OH)_2$ in the presence of the indicator 2,4-dinitrophenol, the solution changes from colorless to yellow when 17.90 mL of the base has been added. What is the approximate value of pK_{HIn} for 2,4-dinitrophenol? Is this a good indicator for the titration?

34. Solution **(a)** is 100.0 mL of 0.100 M HCl and solution **(b)** is 150.0 mL of 0.100 M $NaCH_3COO$. A few drops of thymol blue indicator are added to each solution. What is the color of each solution? What is the color of the solution obtained when these two solutions are mixed?

Neutralization Reactions

35. A 25.00 mL sample of H_3PO_4(aq) requires 31.15 mL of 0.2420 M KOH for titration to the second equivalence point. What is the molarity of the H_3PO_4(aq)?

36. A 20.00 mL sample of H_3PO_4(aq) requires 18.67 mL of 0.1885 M NaOH for titration from the first to the second equivalence point. What is the molarity of the H_3PO_4(aq)?

37. Two aqueous solutions are mixed: 50.0 mL of 0.0150 M H_2SO_4 and 50.0 mL of 0.0385 M NaOH. What is the pH of the resulting solution?

38. Two solutions are mixed: 100.0 mL of HCl(aq) with pH 2.50 and 100.0 mL of NaOH(aq) with pH 11.00. What is the pH of the resulting solution?

Titration Curves

39. Calculate the pH at the points in the titration of 25.00 mL of 0.160 M HCl when (a) 10.00 mL and (b) 15.00 mL of 0.242 M KOH have been added.
40. Calculate the pH at the points in the titration of 20.00 mL of 0.275 M KOH when (a) 15.00 mL and (b) 20.00 mL of 0.350 M HCl have been added.
41. Calculate the pH at the points in the titration of 25.00 mL of 0.132 M HNO_2 when (a) 10.00 mL and (b) 20.00 mL of 0.116 M NaOH have been added. For HNO_2, $K_a = 7.2 \times 10^{-4}$.

$$HNO_2 + OH^- \longrightarrow H_2O + NO_2^-$$

42. Calculate the pH at the points in the titration of 20.00 mL of 0.318 M NH_3 when (a) 10.00 mL and (b) 15.00 mL of 0.475 M HCl have been added. For NH_3, $K_b = 1.8 \times 10^{-5}$.

$$NH_3(aq) + HCl(aq) \longrightarrow NH_4^+(aq) + Cl^-(aq)$$

43. Explain why the volume of 0.100 M NaOH required to reach the equivalence point in the titration of 25.00 mL of 0.100 M HA is the same regardless of whether HA is a strong or a weak acid, yet the pH at the equivalence point is not the same.
44. Explain whether the equivalence point of each of the following titrations should be below, above, or at pH 7: (a) $NaHCO_3$(aq) titrated with NaOH(aq); (b) HCl(aq) titrated with NH_3(aq); (c) KOH(aq) titrated with HI(aq).
45. Sketch the titration curves of the following mixtures. Indicate the initial pH and the pH corresponding to the equivalence point. Indicate the volume of titrant required to reach the equivalence point, and select a suitable indicator from Figure 17-8.
(a) 25.0 mL of 0.100 M KOH with 0.200 M HI
(b) 10.0 mL of 1.00 M NH_3 with 0.250 M HCl
46. Determine the following characteristics of the titration curve for 20.0 mL of 0.275 M NH_3(aq) titrated with 0.325 M HI(aq).
(a) the initial pH
(b) the volume of 0.325 M HI(aq) at the equivalence point
(c) the pH at the half-neutralization point
(d) the pH at the equivalence point
47. In the titration of 20.00 mL of 0.175 M NaOH, calculate the number of milliliters of 0.200 M HCl that must

be added to reach a pH of (a) 12.55, (b) 10.80, (c) 4.25. [*Hint:* Solve an algebraic equation in which the number of milliliters is x. Which reactant is in excess at each pH?]

48. In the titration of 25.00 mL of 0.100 M CH_3COOH, calculate the number of milliliters of 0.200 M NaOH that must be added to reach a pH of (a) 3.85, (b) 5.25, (c) 11.10. [*Hint:* Solve an algebraic equation in which the number of milliliters is x. Which reactant is in excess at each pH?]

49. Sketch a titration curve (pH versus mL of titrant) for each of the following three hypothetical weak acids when titrated with 0.100 M NaOH. Select suitable indicators for the titrations from Figure 17-8. [*Hint:* Select a few key points at which to estimate the pH of the solution.]
(a) 10.00 mL of 0.100 M HX; $K_a = 7.0 \times 10^{-3}$
(b) 10.00 mL of 0.100 M HY; $K_a = 3.0 \times 10^{-4}$
(c) 10.00 mL of 0.100 M HZ; $K_a = 2.0 \times 10^{-8}$

50. Sketch a titration curve (pH versus mL of titrant) for each of the following hypothetical weak bases when titrated with 0.100 M HCl. (Think of these bases as involving the substitution of organic groups, R, for one of the H atoms of NH_3.) Select suitable indicators for the titrations from Figure 17-8. [*Hint:* Select a few key points at which to estimate the pH of the solution.]
(a) 10.00 mL of 0.100 M RNH_2; $K_b = 1 \times 10^{-3}$
(b) 10.00 mL of 0.100 M $R'NH_2$; $K_b = 3 \times 10^{-6}$
(c) 10.00 mL of 0.100 M $R''NH_2$; $K_b = 7 \times 10^{-8}$

51. For the titration of 25.00 mL of 0.100 M NaOH with 0.100 M HCl, calculate the pOH at a few representative points in the titration, sketch the titration curve of pOH versus volume of titrant, and show that it has exactly the same form as Figure 17-9. Then, using this curve and the simplest method possible, sketch the titration curve of pH versus volume of titrant.

52. For the titration of 25.00 mL 0.100 M NH_3 with 0.100 M HCl, calculate the pOH at a few representative points in the titration, sketch the titration curve of pOH versus volume of titrant, and show that it has exactly the same form as Figure 17-11. Then, using this curve and the simplest method possible, sketch the titration curve of pH versus volume of titrant.

Salts of Polyprotic Acids

(*Use data from Table 16.4 or Appendix D as necessary.*)

53. Is a solution that is 0.10 M Na_2S(aq) likely to be acidic, basic, or pH neutral? Explain.
54. Is a solution of sodium dihydrogen citrate, NaH_2Cit, likely to be acidic, basic, or neutral? Explain. Citric acid, H_3Cit, is $H_3C_6H_5O_7$.
55. Sodium phosphate, Na_3PO_4, is made commercially by first neutralizing phosphoric acid with sodium carbonate to obtain Na_2HPO_4. The Na_2HPO_4 is further neutralized to Na_3PO_4 with NaOH.

(a) Write net ionic equations for these reactions.
(b) Na_2CO_3 is a much cheaper base than is NaOH. Why do you suppose that NaOH must be used as well as Na_2CO_3 to produce Na_3PO_4?
56. Both sodium hydrogen carbonate (sodium bicarbonate) and sodium hydroxide can be used to neutralize acid spills. What is the pH of 1.00 M $NaHCO_3$(aq) and of 1.00 M NaOH(aq)? On a per-liter basis, do these two solutions have an equal capacity to neutralize acids? Explain. On a per-gram basis, do the two

solids, $NaHCO_3(s)$ and $NaOH(s)$, have an equal capacity to neutralize acids? Explain. Why do you suppose that $NaHCO_3$ is often preferred to NaOH in neutralizing acid spills?

57. The pH of a solution of 19.5 g of malonic acid in 0.250 L is 1.47. The pH of a 0.300 M solution of sodium hydrogen malonate is 4.26. What are the values of K_{a_1} and K_{a_2} for malonic acid?

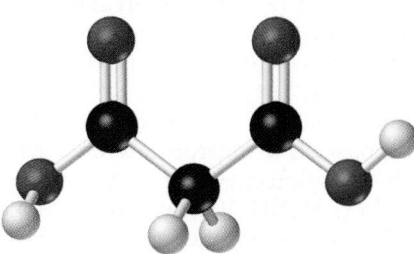

Malonic acid

58. The ionization constants of *ortho*-phthalic acid are $K_{a_1} = 1.1 \times 10^{-3}$ and $K_{a_2} = 3.9 \times 10^{-6}$.

1. $C_6H_4(COOH)_2 + H_2O \rightleftharpoons H_3O^+ + HC_8H_4O_4^-$
2. $HC_8H_4O_4^- + H_2O \rightleftharpoons H_3O^+ + C_6H_4(COO^-)_2$

What are the pH values of the following aqueous solutions: **(a)** 0.350 M potassium hydrogen *ortho*-phthalate; **(b)** a solution containing 36.35 g potassium *ortho*-phthalate per liter?

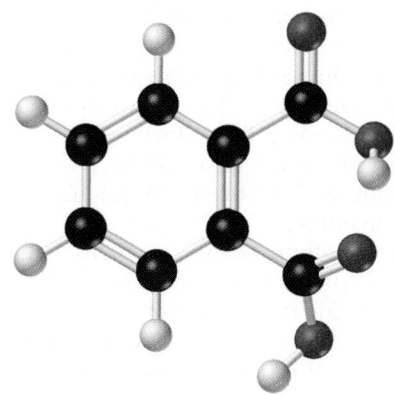

ortho-Phthalic acid

General Acid–Base Equilibria

59. What stoichiometric concentration of the indicated substance is required to obtain an aqueous solution with the pH value shown: **(a)** $Ba(OH)_2$ for pH = 11.88; **(b)** CH_3COOH in 0.294 M $NaCH_3COO$ for pH = 4.52?

60. What stoichiometric concentration of the indicated substance is required to obtain an aqueous solution with the pH value shown: **(a)** aniline, $C_6H_5NH_2$, for pH = 8.95; **(b)** NH_4Cl for pH = 5.12?

61. Using appropriate equilibrium constants but *without doing detailed calculations*, determine whether a solution can be simultaneously:
    (a) 0.10 M NH_3 and 0.10 M NH_4Cl, with pH = 6.07
    (b) 0.10 M $NaC_2H_3O_2$ and 0.058 M HI
    (c) 0.10 M KNO_2 and 0.25 M KNO_3
    (d) 0.050 M $Ba(OH)_2$ and 0.65 M NH_4Cl
    (e) 0.018 M C_6H_5COOH and 0.018 M NaC_6H_5COO, with pH = 4.20

(f) 0.68 M KCl, 0.42 M KNO_3, 1.2 M NaCl, and 0.55 M $NaCH_3COO$, with pH = 6.4

62. This single equilibrium equation applies to different phenomena described in this or the preceding chapter.

$$CH_3COOH + H_2O \rightleftharpoons H_3O^+ + CH_3COO^-$$

Of these four phenomena, ionization of pure acid, common-ion effect, buffer solution, and hydrolysis, indicate which occurs if
(a) $[H_3O^+]$ and $[CH_3COOH]$ are high, but $[CH_3COO^-]$ is very low.
(b) $[CH_3COO^-]$ is high, but $[CH_3COOH]$ and $[H_3O^+]$ are very low.
(c) $[CH_3COOH]$ is high, but $[H_3O^+]$ and $[CH_3COO^-]$ are low.
(d) $[CH_3COOH]$ and $[CH_3COO^-]$ are high, but $[H_3O^+]$ is low.

Integrative and Advanced Exercises

63. Sodium hydrogen sulfate, $NaHSO_4$, is an acidic salt with a number of uses, such as metal pickling (removal of surface deposits). $NaHSO_4$ is made by the reaction of H_2SO_4 with NaCl. To determine the percent NaCl impurity in $NaHSO_4$, a 1.016 g sample is titrated with NaOH(aq); 36.56 mL of 0.225 M NaOH is required.
    (a) Write the net ionic equation for the neutralization reaction.
    (b) Determine the percent NaCl in the sample titrated.
    (c) Select a suitable indicator(s) from Figure 17-8.

64. You are given 250.0 mL of 0.100 M CH_3CH_2COOH (propionic acid, $K_a = 1.35 \times 10^{-5}$). You want to adjust

its pH by adding an appropriate solution. What volume would you add of **(a)** 1.00 M HCl to lower the pH to 1.00; **(b)** 1.00 M $NaCH_3CH_2COO$ to raise the pH to 4.00; **(c)** water to raise the pH by 0.15 unit?

65. Even though the carbonic acid–hydrogen carbonate buffer system is crucial to the maintenance of the pH of blood, it has no practical use as a laboratory buffer solution. Can you think of a reason(s) for this? [*Hint:* Refer to data in Practice Example A of the Integrative Example in Chapter 16.]

66. Thymol blue in its acid range is not a suitable indicator for the titration of HCl by NaOH. Suppose that a

student uses thymol blue by mistake in the titration of Figure 17-9 and that the indicator end point is taken to be pH = 2.0.
(a) Would there be a sharp color change, produced by the addition of a single drop of NaOH(aq)?
(b) Approximately what percent of the HCl remains unneutralized at pH = 2.0?

67. Rather than calculate the pH for different volumes of titrant, a titration curve can be established by calculating the volume of titrant required to reach certain pH values. Determine the volumes of 0.100 M NaOH required to reach the following pH values in the titration of 20.00 mL of 0.150 M HCl: pH = **(a)** 2.00; **(b)** 3.50; **(c)** 5.00; **(d)** 10.50; **(e)** 12.00. Then plot the titration curve.

68. Use the method of Exercise 67 to determine the volume of titrant required to reach the indicated pH values in the following titrations.
(a) 25.00 mL of 0.250 M NaOH titrated with 0.300 M HCl; pH = 13.00, 12.00, 10.00, 4.00, 3.00
(b) 50.00 mL of 0.0100 M benzoic acid (C_6H_5COOH) titrated with 0.0500 M KOH: pH = 4.50, 5.50, 11.50 ($K_a = 6.3 \times 10^{-5}$)

69. A buffer solution can be prepared by starting with a weak acid, HA, and converting some of the weak acid to its salt (for example, NaA) by titration with a strong base. The *fraction* of the original acid that is converted to the salt is designated f.
(a) Derive an equation similar to equation (17.2) but expressed in terms of f rather than concentrations.
(b) What is the pH at the point in the titration of phenol, C_6H_5OH, at which $f = 0.27$ (pK_a of phenol = 10.00)?

70. You are asked to prepare a KH_2PO_4–Na_2HPO_4 solution that has the same pH as human blood, 7.40.
(a) What should be the ratio of concentrations $[HPO_4{}^{2-}]/[H_2PO_4{}^-]$ in this solution?
(b) Suppose you have to prepare 1.00 L of the solution described in part (a) and that this solution must be isotonic with blood (have the same osmotic pressure as blood). What masses of KH_2PO_4 and of $Na_2HPO_4 \cdot 12H_2O$ would you use? [*Hint:* Refer to the definition of isotonic on page 580. Recall that a solution of NaCl with 9.2 g NaCl/L solution is isotonic with blood, and assume that NaCl is completely ionized in aqueous solution.]

71. You are asked to bring the pH of 0.500 L of 0.500 M NH_4Cl(aq) to 7.00. How many drops (1 drop = 0.05 mL) of which of the following solutions would you use: 10.0 M HCl or 10.0 M NH_3?

72. Because an acid–base indicator is a weak acid, it can be titrated with a strong base. Suppose you titrate 25.00 mL of a 0.0100 M solution of the indicator *p*-nitrophenol, $HOC_6H_4NO_2$, with 0.0200 M NaOH. The pK_a of *p*-nitrophenol is 7.15, and it changes from colorless to yellow in the pH range from 5.6 to 7.6.
(a) Sketch the titration curve for this titration.
(b) Show the pH range over which *p*-nitrophenol changes color.
(c) Explain why *p*-nitrophenol cannot serve as its own indicator in this titration.

73. The neutralization of NaOH by HCl is represented in equation (1), and the neutralization of NH_3 by HCl in equation (2).

1. $OH^- + H_3O^+ \rightleftharpoons 2\,H_2O$ $K = ?$
2. $NH_3 + H_3O^+ \rightleftharpoons NH_4{}^+ + H_2O$ $K = ?$
(a) Determine the equilibrium constant K for each reaction.
(b) Explain why each neutralization reaction can be considered to go to completion.

74. The titration of a weak acid by a weak base is not a satisfactory procedure because the pH does not increase sharply at the equivalence point. Demonstrate this fact by sketching a titration curve for the neutralization of 10.00 mL of 0.100 M CH_3COOH with 0.100 M NH_3.

75. At times, a salt of a weak base can be titrated by a strong base. Use appropriate data from the text to sketch a titration curve for the titration of 10.00 mL of 0.0500 M $C_6H_5NH_3{}^+Cl^-$ with 0.100 M NaOH.

76. Sulfuric acid is a diprotic acid, strong in the first ionization step and weak in the second ($K_{a_2} = 1.1 \times 10^{-2}$). By using appropriate calculations, determine whether it is feasible to titrate 10.00 mL of 0.100 M H_2SO_4 to two distinct equivalence points with 0.100 M NaOH.

77. Carbonic acid is a weak diprotic acid (H_2CO_3) with $K_{a_1} = 4.43 \times 10^{-7}$ and $K_{a_2} = 4.73 \times 10^{-11}$. The equivalence points for the titration come at approximately pH 4 and 9. Suitable indicators for use in titrating carbonic acid or carbonate solutions are methyl orange and phenolphthalein.
(a) Sketch the titration curve that would be obtained in titrating a sample of $NaHCO_3$(aq) with 1.00 M HCl.
(b) Sketch the titration curve for Na_2CO_3(aq) with 1.00 M HCl.
(c) What volume of 0.100 M HCl is required for the complete neutralization of 1.00 g $NaHCO_3$(s)?
(d) What volume of 0.100 M HCl is required for the complete neutralization of 1.00 g Na_2CO_3(s)?
(e) A sample of NaOH contains a small amount of Na_2CO_3. For titration to the phenolphthalein end point, 0.1000 g of this sample requires 23.98 mL of 0.1000 M HCl. An additional 0.78 mL is required to reach the methyl orange end point. What is the percent Na_2CO_3, by mass, in the sample?

78. Piperazine is a diprotic weak base used as a corrosion inhibitor and an insecticide. Its ionization is described by the following equations.

$$HN(C_4H_8)\,NH + H_2O \rightleftharpoons$$
$$[HN(C_4H_8)\,NH_2]^+ + OH^- \quad pK_{b_1} = 4.22$$

$$[HN(C_4H_8)\,NH_2]^+ + H_2O \rightleftharpoons$$
$$[H_2N(C_4H_8)\,NH_2]^{2+} + OH^- \quad pK_{b_2} = 8.67$$

The piperazine used commercially is a hexahydrate, $C_4H_{10}N_2 \cdot 6H_2O$. A 1.00-g sample of this hexahydrate is dissolved in 100.0 mL of water and titrated with 0.500 M HCl. Sketch a titration curve for this titration, indicating **(a)** the initial pH; **(b)** the pH at the half-neutralization point of the first neutralization; **(c)** the volume of HCl(aq) required to reach the first equivalence point; **(d)** the pH at the first equivalence point; **(e)** the pH at the point at which the second step of the neutralization is half-completed; **(f)** the volume of 0.500 M HCl(aq) required to reach the second equivalence point of the titration; **(g)** the pH at the second equivalence point.

79. Complete the derivation of equation (17.5) outlined in Are You Wondering 17-1. Then derive equation (17.6).

80. Explain why equation (17.5) fails when applied to dilute solutions—for example, when you calculate the pH of 0.010 M NaH_2PO_4. [*Hint:* Refer also to Exercise 79.]

81. A solution is prepared that is 0.150 M CH_3COOH and 0.250 M NaHCOO.
    (a) Show that this is a buffer solution.
    (b) Calculate the pH of this buffer solution.
    (c) What is the final pH if 1.00 L of 0.100 M HCl is added to 1.00 L of this buffer solution?

82. A series of titrations of lactic acid, $CH_3CH(OH)COOH$ ($pK_a = 3.86$) is planned. About 1.00 mmol of the acid will be titrated with NaOH(aq) to a final volume of about 100 mL at the equivalence point. (a) Which acid–base indicator from Figure 17-8 would you select for the titration? To assist in locating the equivalence point in the titration, a buffer solution is to be prepared having the same pH as that at the equivalence point. A few drops of the indicator in this buffer will produce the color to be matched in the titrations. (b) Which of the following combinations would be suitable for the buffer solutions: CH_3COOH–CH_3COO^-, $H_2PO_4^-$–HPO_4^{2-}, or NH_4^+–NH_3? (c) What ratio of conjugate base to acid is required in the buffer?

83. Hydrogen peroxide, H_2O_2, is a somewhat stronger acid than water. Its ionization is represented by the equation

    $$H_2O_2 + H_2O \rightleftharpoons H_3O^+ + HO_2^-$$

    In 1912, the following experiments were performed to obtain an approximate value of pK_a for this ionization at 0 °C. A sample of H_2O_2 was shaken together with a mixture of water and 1-pentanol. The mixture settled into two layers. At equilibrium, the hydrogen peroxide had distributed itself between the two layers such that the water layer contained 6.78 times as much H_2O_2 as the 1-pentanol layer. In a second experiment, a sample of H_2O_2 was shaken together with 0.250 M NaOH(aq) and 1-pentanol. At equilibrium, the concentration of H_2O_2 was 0.00357 M in the 1-pentanol layer and 0.259 M in the aqueous layer. In a third experiment, a sample of H_2O_2 was brought to equilibrium with a mixture of 1-pentanol and 0.125 M NaOH(aq); the concentrations of the hydrogen peroxide were 0.00198 M in the 1-pentanol and 0.123 M in the aqueous layer. For water at 0 °C, $pK_w = 14.94$. Find an approximate value of pK_a for H_2O_2 at 0 °C. [*Hint:* The hydrogen peroxide concentration in the aqueous layers is the total concentration of H_2O_2 and HO_2^-. Assume that the 1-pentanol solutions contain no ionic species.]

84. Sodium ammonium hydrogen phosphate, $NaNH_4HPO_4$, is a salt in which one of the ionizable H atoms of H_3PO_4 is replaced by Na^+, another is replaced by NH_4^+, and the third remains in the anion HPO_4^{2-}. Calculate the pH of 0.100 M $NaNH_4HPO_4$(aq).
    [*Hint:* You can use the general method introduced on page 720. First, identify all the species that could be present and the equilibria involving these species. Then identify the two equilibrium expressions that will predominate and eliminate all the species whose concentrations are likely to be negligible. At that point, only a few algebraic manipulations are required.]

85. Consider a solution containing two weak monoprotic acids with dissociation constants K_{HA} and K_{HB}. Find the charge balance equation for this system, and use it to derive an expression that gives the concentration of H_3O^+ as a function of the concentrations of HA and HB and the various constants.

86. Calculate the pH of a solution that is 0.050 M acetic acid and 0.010 M phenylacetic acid.

87. A very common buffer agent used in the study of biochemical processes is the weak base TRIS, $(HOCH_2)_3 CNH_2$, which has a pK_b of 5.91 at 25 °C. A student is given a sample of the hydrochloride of TRIS together with standard solutions of 10 M NaOH and HCl.
    (a) Using TRIS, how might the student prepare 1 L of a buffer of pH = 7.79?
    (b) In one experiment, 30 mmol of protons are released into 500 mL of the buffer prepared in part (a). Is the capacity of the buffer sufficient? What is the resulting pH?
    (c) Another student accidentally adds 20 mL of 10 M HCl to 500 mL of the buffer solution prepared in part (a). Is the buffer ruined? If so, how could the buffer be regenerated?

88. The Henderson–Hasselbalch equation can be written as

    $$pH = pK_a - \log\left(\frac{1}{\alpha} - 1\right) \text{ where } \alpha = \frac{[A^-]}{[A^-] + [HA]}.$$

    Thus, the degree of ionization (α) of an acid can be determined if both the pH of the solution and the pK_a of the acid are known.
    (a) Use this equation to plot the pH versus the degree of ionization for the second ionization constant of phosphoric acid ($K_a = 6.3 \times 10^{-8}$).
    (b) If pH = pK_a what is the degree of ionization?
    (c) If the solution had a pH of 6.0 what would the value of α be?

89. The pH of ocean water depends on the amount of atmospheric carbon dioxide. The dissolution of carbon dioxide in ocean water can be approximated by the following chemical reactions (Henry's Law constant for CO_2 is $K_H = [CO_2(aq)]/[CO_2(g)] = 0.8317$.):

    $$CO_2(g) \rightleftharpoons CO_2(aq) \tag{1}$$
    $$CaCO_3(s) \rightleftharpoons Ca^{2+}(aq) + CO_3^-(aq) \tag{2}$$
    $$H_3O^+(aq) + CO_3^-(aq) \rightleftharpoons HCO_3^-(aq) + H_2O(l) \tag{3}$$
    $$H_3O^+(aq) + HCO_3^-(aq) \rightleftharpoons CO_2(aq) + 2 H_2O(l) \tag{4}$$

    (a) Use the equations above to determine the hydronium ion concentration as a function of $[CO_2(g)]$ and $[Ca^{2+}]$.
    (b) During preindustrial conditions, we will assume that the equilibrium concentration of $[CO_2(g)] = 280$ ppm and $[Ca^{2+}] = 10.24$ mM. Calculate the pH of a sample of ocean water.

90. A sample of water contains 23.0 g L^{-1} of Na^+(aq), 10.0 g L^{-1} of Ca^{2+}(aq), 40.2 g L^{-1} CO_3^{2-}(aq), and 9.6 g L^{-1} SO_4^{2-}(aq). What is the pH of the solution if the only other ions present are H_3O^+ and OH^-?

91. In 1922 Donald D. van Slyke (*J. Biol. Chem.*, **52**, 525) defined a quantity known as the buffer index: $\beta = dC_b/d(pH)$, where dC_b represents the increment of moles of strong base to one liter of the buffer. For the addition of a strong acid, he wrote $\beta = -dC_a/d(pH)$.

By applying this idea to a monoprotic acid and its conjugate base, we can derive the following expression:

$$\beta = 2.303\left(\frac{K_w}{[H_3O^+]} + [H_3O^+] + \frac{CK_a[H_3O^+]}{(K_a + [H_3O^+])^2}\right)$$

where C is the total concentration of monoprotic acid and conjugate base.

(a) Use the above expression to calculate the buffer index for the acetic acid buffer with a total acetic acid and acetate ion concentration of 2.0×10^{-2} and a pH = 5.0.

(b) Use the buffer index from part (a) and calculate the pH of the buffer after the addition of of a strong acid. (*Hint:* Let $dC_a/d(pH) \approx \Delta C_a/\Delta pH$.)
(c) Make a plot of β versus pH for a 0.1 M acetic acid buffer system. Locate the maximum buffer index as well as the minimum buffer indices.

Feature Problems

92. The graph below, which is related to a titration curve, shows the fraction (f) of the stoichiometric amount of acetic acid present as non-ionized CH₃COOH and as acetate ion, CH₃COO⁻, as a function of the pH of the solution containing these species.
(a) Explain the significance of the point at which the two curves cross. What are the fractions and the pH at that point?
(b) Sketch a comparable set of curves for carbonic acid, H₂CO₃. [*Hint:* How many carbonate-containing species should appear in the graph? How many points of intersection should there be? at what pH values?]
(c) Sketch a comparable set of curves for phosphoric acid, H₃PO₄. [*Hint:* How many phosphate-containing species should appear in the graph? How many points of intersection should there be? at what pH values?]

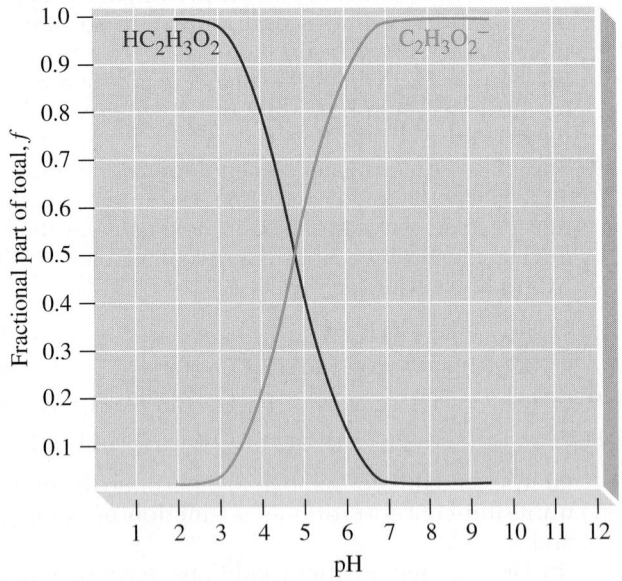

93. In some cases, the titration curve for a mixture of *two* acids has the same appearance as that for a single acid; in other cases it does not.
(a) Sketch the titration curve (pH versus volume of titrant) for the titration with 0.200 M NaOH of 25.00 mL of a solution that is 0.100 M in HCl and 0.100 M in HNO₃. Does this curve differ in any way from what would be obtained in the titration of 25.00 mL of 0.200 M HCl with 0.200 M NaOH? Explain.
(b) The titration curve shown was obtained when 10.00 mL of a solution containing both HCl and

H₃PO₄ was titrated with 0.216 M NaOH. From this curve, determine the stoichiometric molarities of both the HCl and the H₃PO₄.
(c) A 10.00 mL solution that is 0.0400 M H₃PO₄ and 0.0150 M NaH₂PO₄ is titrated with 0.0200 M NaOH. Sketch the titration curve.

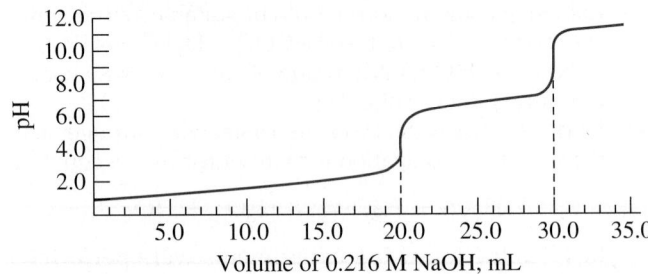

94. Amino acids contain both an acidic carboxylic acid group (—COOH) and a basic amino group (—NH₂). The amino group can be *protonated* (that is, it has an extra proton attached) in a strongly acidic solution. This produces a diprotic acid of the form H₂A⁺, as exemplified by the protonated amino acid alanine.

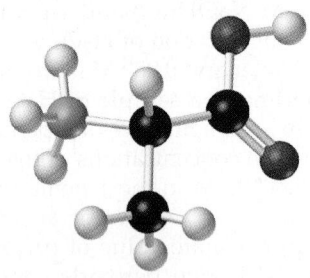

Protonated alanine

The protonated amino acid has two ionizable protons that can be titrated with OH⁻.

$$\underset{\substack{\text{Protonated form}\\\text{of alanine (H}_2\text{A}^+)}}{\overset{O}{\underset{CH_3}{\overset{\|}{^+H_3NCHCOH}}}} \xrightarrow{OH^-} \underset{\substack{\text{Neutral form}\\\text{of alanine (HA)}}}{\overset{O}{\underset{CH_3}{\overset{\|}{^+H_3NCHCO^-}}}} \xrightarrow{OH^-} \underset{\substack{\text{Anionic form}\\\text{of alanine (A}^-)}}{\overset{O}{\underset{CH_3}{\overset{\|}{H_2NCHCO^-}}}}$$

For the —COOH group, $pK_{a_1} = 2.34$; for the —NH₃⁺ group, $pK_{a_2} = 9.69$. Consider the titration of a 0.500 M solution of alanine hydrochloride with

0.500 M NaOH solution. What is the pH of **(a)** the 0.500 M alanine hydrochloride; **(b)** the solution at the first half-neutralization point; **(c)** the solution at the first equivalence point?

The dominant form of alanine present at the first equivalence point is electrically neutral despite the positive charge and negative charge it possesses. The point at which the neutral form is produced is called the *isoelectric point*. Confirm that the pH at the isoelectric point is

$$pH = \frac{1}{2}(pK_{a_1} + pK_{a_2})$$

What is the pH of the solution **(d)** halfway between the first and second equivalence points? **(e)** at the second equivalence point?

(f) Calculate the pH values of the solutions when the following volumes of the 0.500 M NaOH have been added to 50 mL of the 0.500 M alanine hydrochloride solution: 10.0 mL, 20.0 mL, 30.0 mL, 40.0 mL, 50.0 mL, 60.0 mL, 70.0 mL, 80.0 mL, 90.0 mL, 100.0 mL, and 110.0 mL.

(g) Sketch the titration curve for the 0.500 M solution of alanine hydrochloride, and label significant points on the curve.

Self-Assessment Exercises

95. In your own words, define or explain the following terms or symbols: **(a)** mmol; **(b)** HIn; **(c)** equivalence point of a titration; **(d)** titration curve.

96. Briefly describe each of the following ideas, phenomena, or methods: **(a)** the common-ion effect; **(b)** the use of a buffer solution to maintain a constant pH; **(c)** the determination of pK_a of a weak acid from a titration curve; **(d)** the measurement of pH with an acid–base indicator.

97. Explain the important distinctions between each pair of terms: **(a)** buffer capacity and buffer range; **(b)** hydrolysis and neutralization; **(c)** first and second equivalence points in the titration of a weak diprotic acid; **(d)** equivalence point of a titration and end point of an indicator.

98. Write equations to show how each of the following buffer solutions reacts with a small added amount of a strong acid or a strong base: **(a)** $HCOOH$–$KHCOO$; **(b)** $C_6H_5NH_2$–$C_6H_5NH_3^+Cl^-$; **(c)** KH_2PO_4– Na_2HPO_4.

99. Sketch the titration curves that you would expect to obtain in the following titrations. Select a suitable indicator for each titration from Figure 17-8. **(a)** $NaOH(aq)$ titrated with $HNO_3(aq)$ **(b)** $NH_3(aq)$ titrated with $HCl(aq)$ **(c)** $CH_3COOH(aq)$ titrated with $KOH(aq)$ **(d)** $NaH_2PO_4(aq)$ titrated with $KOH(aq)$

100. A 25.00-mL sample of 0.0100 M C_6H_5COOH ($K_a = 6.3 \times 10^{-5}$) is titrated with 0.0100 M $Ba(OH)_2$. Calculate the pH **(a)** of the initial acid solution; **(b)** after the addition of 6.25 mL of 0.0100 M $Ba(OH)_2$; **(c)** at the equivalence point; **(d)** after the addition of a total of 15.00 mL of 0.0100 M $Ba(OH)_2$.

101. To *repress* the ionization of formic acid, $HCOOH(aq)$, which of the following should be added to the solution? **(a)** NaCl; **(b)** NaOH; **(c)** NaHCOO; **(d)** $NaNO_3$

102. To *increase* the ionization of formic acid, $HCOOH(aq)$, which of the following should be added to the solution? **(a)** NaCl; **(b)** NaHCOO; **(c)** H_2SO_4; **(d)** $NaHCO_3$

103. To convert $NH_4^+(aq)$ to $NH_3(aq)$, **(a)** add H_3O^+; **(b)** raise the pH; **(c)** add $KNO_3(aq)$; **(d)** add NaCl.

104. During the titration of equal concentrations of a weak base and a strong acid, at what point would the pH = pK_a? **(a)** the initial pH; **(b)** halfway to the equivalence point; **(c)** at the equivalence point; **(d)** past the equivalence point.

105. Calculate the pH of the buffer formed by mixing equal volumes $[C_2H_5NH_2] = 1.49$ M with $[HClO_4] = 1.001$ M. $K_b = 4.3 \times 10^{-4}$.

106. Calculate the pH of a 0.5 M solution of $Ca(HSe)_2$, given that H_2Se has $K_{a_1} = 1.3 \times 10^{-4}$ and $K_{a_2} = 1 \times 10^{-11}$.

107. The effect of adding 0.001 mol KOH to 1.00 L of a solution that is 0.10 M NH_3–0.10 M NH_4Cl is to **(a)** raise the pH very slightly; **(b)** lower the pH very slightly; **(c)** raise the pH by several units; **(d)** lower the pH by several units.

108. The most acidic of the following 0.10 M salt solutions is **(a)** Na_2S; **(b)** $NaHSO_4$; **(c)** $NaHCO_3$; **(d)** Na_2HPO_4.

109. If an indicator is to be used in an acid–base titration having an equivalence point in the pH range 8 to 10, the indicator must **(a)** be a weak base; **(b)** have $K_a = 1 \times 10^{-9}$; **(c)** ionize in two steps; **(d)** be added to the solution only after the solution has become alkaline.

110. Indicate whether you would expect the equivalence point of each of the following titrations to be below, above, or at pH **7**. Explain your reasoning. **(a)** $NaHCO_3(aq)$ is titrated with $NaOH(aq)$; **(b)** $HCl(aq)$ is titrated with $NH_3(aq)$; **(c)** $KOH(aq)$ is titrated with $HI(aq)$.

111. Using the method presented in Appendix E, construct a concept map relating the concepts in Sections 17-2, 17-3, and 17-4.

18

Solubility and Complex-Ion Equilibria

CONTENTS

Stalactite and stalagmite formations in a cavern in the Guangxi Province, People's Republic of China. Stalactites and stalagmites are formed from calcium salts deposited as underground water seeps into the cavern and evaporates.

The dissolution and precipitation of limestone ($CaCO_3$) underlie a variety of natural phenomena, such as the formation of limestone caverns. Whether a solution containing Ca^{2+} and CO_3^{2-} ions undergoes precipitation depends on the concentrations of these ions. In turn, the CO_3^{2-} ion concentration depends on the pH of the solution. To develop a better understanding of the conditions under which $CaCO_3$ dissolves or precipitates, we must consider equilibrium relationships between Ca^{2+} and CO_3^{2-}, and between CO_3^{2-}, H_3O^+, and HCO_3^-. This requirement suggests a need to combine ideas about acid–base equilibria from Chapters 16 and 17 with ideas about the new types of equilibria to be introduced in this chapter.

Silver chloride is a familiar precipitate in the general chemistry laboratory, yet it does *not* precipitate from a solution that has a moderate to high concentration of $NH_3(aq)$. The silver ion and ammonia combine instead to form a species, called a *complex ion*, that remains in solution. Complex-ion formation and equilibria involving complex ions are additional topics discussed in this chapter.

18-1   Solubility Product Constant, K_{sp}

Gypsum ($CaSO_4 \cdot 2\,H_2O$) is a calcium mineral that is slightly soluble in water. Groundwater that comes into contact with gypsum often contains some dissolved calcium sulfate. This water cannot be used for certain applications, such as in evaporative cooling systems in power plants because the calcium sulfate might precipitate from the water and block pipes. The equilibrium between $Ca^{2+}(aq)$ and $SO_4^{2-}(aq)$ and undissolved $CaSO_4(s)$ can be represented as

$$CaSO_4(s) \rightleftharpoons Ca^{2+}(aq) + SO_4^{2-}(aq)$$

We can write the equilibrium constant expression for this equilibrium as we learned to do in Chapter 15 (page 659)—that is,

$$K = \frac{a_{Ca^{2+}} a_{SO_4^{2-}}}{a_{CaSO_4}}$$

The activity of a pure solid is 1. Because the concentrations of the ions are very small, we can set their activities equal to their molar concentrations. With these substitutions, we obtain expression (18.1).

$$K_{sp} = [Ca^{2+}][SO_4^{2-}] = 9.1 \times 10^{-6}\ (\text{at } 25\,°C) \qquad \textbf{(18.1)}$$

The **solubility product constant**, K_{sp}, is the constant for the equilibrium established between a solid solute and its ions in a saturated solution. Table 18.1 lists several K_{sp} values and the solubility equilibria to which they apply.

◀ Solubility products vary with temperature.

TABLE 18.1   Several Solubility Product Constants at 25 °C[a]

Solute	Solubility Equilibrium	K_{sp}
Aluminum hydroxide	$Al(OH)_3(s) \rightleftharpoons Al^{3+}(aq) + 3\,OH^-(aq)$	1.3×10^{-33}
Barium carbonate	$BaCO_3(s) \rightleftharpoons Ba^{2+}(aq) + CO_3^{2-}(aq)$	5.1×10^{-9}
Barium sulfate	$BaSO_4(s) \rightleftharpoons Ba^{2+}(aq) + SO_4^{2-}(aq)$	1.1×10^{-10}
Calcium carbonate	$CaCO_3(s) \rightleftharpoons Ca^{2+}(aq) + CO_3^{2-}(aq)$	2.8×10^{-9}
Calcium fluoride	$CaF_2(s) \rightleftharpoons Ca^{2+}(aq) + 2\,F^-(aq)$	5.3×10^{-9}
Calcium sulfate	$CaSO_4(s) \rightleftharpoons Ca^{2+}(aq) + SO_4^{2-}(aq)$	9.1×10^{-6}
Chromium(III) hydroxide	$Cr(OH)_3(s) \rightleftharpoons Cr^{3+}(aq) + 3\,OH^-(aq)$	6.3×10^{-31}
Iron(III) hydroxide	$Fe(OH)_3(s) \rightleftharpoons Fe^{3+}(aq) + 3\,OH^-(aq)$	4×10^{-38}
Lead(II) chloride	$PbCl_2(s) \rightleftharpoons Pb^{2+}(aq) + 2\,Cl^-(aq)$	1.6×10^{-5}
Lead(II) chromate	$PbCrO_4(s) \rightleftharpoons Pb^{2+}(aq) + CrO_4^{2-}(aq)$	2.8×10^{-13}
Lead(II) iodide	$PbI_2(s) \rightleftharpoons Pb^{2+}(aq) + 2\,I^-(aq)$	7.1×10^{-9}
Magnesium carbonate	$MgCO_3(s) \rightleftharpoons Mg^{2+}(aq) + CO_3^{2-}(aq)$	3.5×10^{-8}
Magnesium fluoride	$MgF_2(s) \rightleftharpoons Mg^{2+}(aq) + 2\,F^-(aq)$	3.7×10^{-8}
Magnesium hydroxide	$Mg(OH)_2(s) \rightleftharpoons Mg^{2+}(aq) + 2\,OH^-(aq)$	1.8×10^{-11}
Magnesium phosphate	$Mg_3(PO_4)_2(s) \rightleftharpoons 3\,Mg^{2+}(aq) + 2\,PO_4^{3-}(aq)$	1×10^{-25}
Mercury(I) chloride	$Hg_2Cl_2(s) \rightleftharpoons Hg_2^{2+}(aq) + 2\,Cl^-(aq)$	1.3×10^{-18}
Silver bromide	$AgBr(s) \rightleftharpoons Ag^+(aq) + Br^-(aq)$	5.0×10^{-13}
Silver carbonate	$Ag_2CO_3(s) \rightleftharpoons 2\,Ag^+(aq) + CO_3^{2-}(aq)$	8.5×10^{-12}
Silver chloride	$AgCl(s) \rightleftharpoons Ag^+(aq) + Cl^-(aq)$	1.8×10^{-10}
Silver chromate	$Ag_2CrO_4(s) \rightleftharpoons 2\,Ag^+(aq) + CrO_4^{2-}(aq)$	1.1×10^{-12}
Silver iodide	$AgI(s) \rightleftharpoons Ag^+(aq) + I^-(aq)$	8.5×10^{-17}
Strontium carbonate	$SrCO_3(s) \rightleftharpoons Sr^{2+}(aq) + CO_3^{2-}(aq)$	1.1×10^{-10}
Strontium sulfate	$SrSO_4(s) \rightleftharpoons Sr^{2+}(aq) + SO_4^{2-}(aq)$	3.2×10^{-7}

[a]A more extensive listing of K_{sp} values is given in Appendix D.

EXAMPLE 18-1   Writing Solubility Product Constant Expressions for Slightly Soluble Solutes

Write the solubility product constant expression for the solubility equilibrium of

(a) Calcium fluoride, CaF_2 (one of the substances used when a fluoride treatment is applied to teeth)

(b) Copper arsenate, $Cu_3(AsO_4)_2$ (used as an insecticide and fungicide).

Analyze

The equation for the solubility equilibrium is written for one mole of the slightly soluble solute. That is, the coefficient "1" is understood for the slightly soluble solute. The coefficients for the ions in solution are whatever is needed to balance the equation. The coefficients then establish the powers to which the ion concentrations are raised in the K_{sp} expression.

Solve

(a) $CaF_2(s) \rightleftharpoons Ca^{2+}(aq) + 2\,F^-(aq)$ $K_{sp} = [Ca^{2+}][F^-]^2$

(b) $Cu_3(AsO_4)_2(s) \rightleftharpoons 3\,Cu^{2+}(aq) + 2\,AsO_4^{3-}(aq)$ $K_{sp} = [Cu^{2+}]^3[AsO_4^{3-}]^2$

Assess

Note that the concentration of the reactant is not included in the K_{sp} expression. Recall that for the activity of a solid or a liquid we do not include its concentration in the equilibrium constant expression.

PRACTICE EXAMPLE A: Write the solubility product constant expression for **(a)** $MgCO_3$ (one of the components of dolomite, a form of limestone) and **(b)** Ag_3PO_4 (used in photographic emulsions).

PRACTICE EXAMPLE B: A handbook lists $K_{sp} = 1 \times 10^{-7}$ for calcium hydrogen phosphate, a substance used in dentifrices and as an animal feed supplement. Write **(a)** the equation for the solubility equilibrium and **(b)** the solubility product constant expression for this slightly soluble solute.

▲ Some calcium salts, such as calcium fluoride and calcium hydrogen phosphate, have beneficial uses, but another calcium salt, calcium oxalate (CaC_2O_4), can be harmful. The photo is a scanning electron microscope image of calcium oxalate crystals, a common type of kidney stone that can form in the human kidney.

🔍 18-1   CONCEPT ASSESSMENT

A large excess of $MgF_2(s)$ is maintained in contact with 1.00 L of pure water to produce a saturated solution of MgF_2. When an additional 1.00 L of pure water is added to the mixture and equilibrium re-established, compared with the original saturated solution, will $[Mg^{2+}]$ be (a) the same; (b) twice as large; (c) half as large; (d) some unknown fraction of the original $[Mg^{2+}]$? Explain.

18-2   Relationship Between Solubility and K_{sp}

Is there a relationship between the solubility product constant, K_{sp}, of a solute and the solute's *molar solubility*—its molarity in a saturated aqueous solution? As shown in Examples 18-2 and 18-3, there is a definite relationship between them. As discussed in Section 18-4, calculations involving K_{sp} are generally more subject to error than are those involving other equilibrium constants, but the results are suitable for many purposes. In Example 18-2, we start with an experimentally determined solubility and obtain a value of K_{sp}.

The "inverse" of Example 18-2 is the calculation of the solubility of a solute from its K_{sp} value. When this is done, as in Example 18-3, the result is always a molar solubility—a molarity. Additional conversions are required to obtain solubility in units other than moles per liter, as in Practice Example 18-3B.

EXAMPLE 18-2   Calculating K_{sp} of a Slightly Soluble Solute from Its Solubility

A handbook lists the aqueous solubility of $CaSO_4$ at 25 °C as 0.20 g $CaSO_4$/100 mL. What is the K_{sp} of $CaSO_4$ at 25 °C?

$$CaSO_4(s) \rightleftharpoons Ca^{2+}(aq) + SO_4^{2-}(aq) \qquad K_{sp} = ?$$

Analyze

We need to construct a conversion pathway that begins with finding $[Ca^{2+}]$ and $[SO_4^{2-}]$, which we can then substitute into the K_{sp} expression.

$$g\ CaSO_4/100\ mL \longrightarrow mol\ CaSO_4/L \longrightarrow [Ca^{2+}]\ and\ [SO_4^{2-}] \longrightarrow K_{sp}$$

Solve

The first step is to convert the mass of $CaSO_4$ in a 100 mL volume to molar solubility. This is accomplished by using the inverse of the molar mass of $CaSO_4$ and replacing 100 mL with 0.100 L.

$$\frac{mol\ CaSO_4}{L\ satd\ soln} = \frac{0.20\ g\ CaSO_4}{0.100\ L\ soln} \times \frac{1\ mol\ CaSO_4}{136\ g\ CaSO_4}$$

$$= 0.015\ M\ CaSO_4$$

Using stoichiometric ratios (shown in blue), determine $[Ca^{2+}]$ and $[SO_4^{2-}]$.

$$[Ca^{2+}] = \frac{0.015\ mol\ CaSO_4}{1\ L} \times \frac{1\ mol\ Ca^{2+}}{1\ mol\ CaSO_4} = 0.015\ M$$

$$[SO_4^{2-}] = \frac{0.015\ mol\ CaSO_4}{1\ L} \times \frac{1\ mol\ SO_4^{2-}}{1\ mol\ CaSO_4} = 0.015\ M$$

Substitute these ion concentrations into the solubility product expression.

$$K_{sp} = [Ca^{2+}][SO_4^{2-}] = (0.015)(0.015) = 2.3 \times 10^{-4}$$

Assess

The K_{sp} result determined here is significantly different from the value in equation (18.1). The reason for this discrepancy is that ion activities need to be taken into account.

PRACTICE EXAMPLE A: A handbook lists the aqueous solubility of AgOCN as 7 mg/100 mL at 20 °C. What is the K_{sp} of AgOCN at 20 °C?

PRACTICE EXAMPLE B: A handbook lists the aqueous solubility of lithium phosphate at 18 °C as 0.034 g Li_3PO_4/100 mL soln. What is the K_{sp} of Li_3PO_4 at 18 °C?

EXAMPLE 18-3   Calculating the Solubility of a Slightly Soluble Solute from Its K_{sp} Value

Lead(II) iodide, PbI_2, is a dense, golden yellow, "insoluble" solid used in bronzing and in ornamental work requiring a golden color (such as mosaic gold). Calculate the molar solubility of lead(II) iodide in water at 25 °C, given that $K_{sp} = 7.1 \times 10^{-9}$.

Analyze

The solubility equilibrium equation

$$PbI_2(s) \rightleftharpoons Pb^{2+}(aq) + 2\,I^-(aq)$$

shows that for each mole of PbI_2 that dissolves, *one* mole of Pb^{2+} and *two* moles of I^- appear in solution. If we let s represent the number of moles of PbI_2 dissolved per liter of saturated solution, we have

$$[Pb^{2+}] = s \quad and \quad [I^-] = 2s$$

(continued)

Solve

These concentrations must also satisfy the K_{sp} expression.

$$K_{sp} = [Pb^{2+}][I^-]^2 = (s)(2s)^2 = 7.1 \times 10^{-9}$$
$$4s^3 = 7.1 \times 10^{-9}$$
$$s^3 = 1.8 \times 10^{-9}$$
$$s = (1.8 \times 10^{-9})^{1/3} = 1.2 \times 10^{-3}$$
$$= \text{molar solubility of } PbI_2 = 1.2 \times 10^{-3} \, M$$

Assess

One key part of this problem is making sure that we account for the correct number of moles of each species. In this case we had 2 moles of iodide ion for every 1 mole of lead.

PRACTICE EXAMPLE A: The K_{sp} of $Cu_3(AsO_4)_2$ at 25 °C is 7.6×10^{-36}. What is the molar solubility of $Cu_3(AsO_4)_2$ in H_2O at 25 °C?

PRACTICE EXAMPLE B: How many milligrams of $BaSO_4$ are dissolved in a 225 mL sample of saturated $BaSO_4(aq)$? $K_{sp} = 1.1 \times 10^{-10}$.

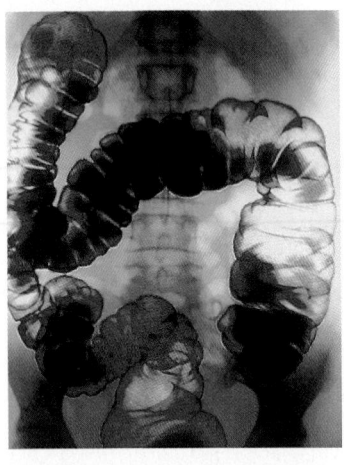

▲ Barium sulfate, $BaSO_4$, is a good absorber of X-rays. As a component of a "barium milk shake," $BaSO_4$ coats the intestinal tract so that this soft tissue will show up when X-rayed. Even though $Ba^{2+}(aq)$ is poisonous, $BaSO_4(s)$ is harmless because its aqueous solubility is very low.

18-1 ARE YOU WONDERING...

When comparing molar solubilities, is a solute with a larger value of K_{sp} always more soluble than one with a smaller value?

If the solutes being compared are of the same type (for example, all are of the type MX or all are of the type, MX_2), their molar solubilities will be related in the same way to their K_{sp} values. Then, the solute with the largest K_{sp} value will have the greatest molar solubility. Thus, AgCl ($K_{sp} = 1.8 \times 10^{-10}$) is more soluble than AgBr ($K_{sp} = 5.0 \times 10^{-13}$). For these particular solutes, the molar solubility is $s = \sqrt{K_{sp}}$.

If the solutes are *not* of the same type, you'll have to calculate, or at least estimate, each molar solubility and compare the results. Thus, even though its solubility product constant is smaller, Ag_2CrO_4 ($K_{sp} = 1.1 \times 10^{-12}$) is more soluble than AgCl ($K_{sp} = 1.8 \times 10^{-10}$). For Ag_2CrO_4, the molar solubility is $s = (K_{sp}/4)^{1/3} = 6.5 \times 10^{-5} \, M$, whereas for AgCl, it is $s = \sqrt{K_{sp}} = 1.3 \times 10^{-5} \, M$.

 18-2 CONCEPT ASSESSMENT

Of the compounds CaF_2, $CaCl_2$, AgF, and AgCl, which would be considered insoluble? Explain.

18-3 Common-Ion Effect in Solubility Equilibria

In Examples 18-2 and 18-3, the ions in the saturated solutions came from a single source, the pure solid solute. Suppose that to the saturated solution of PbI_2 in Example 18-3, we add some I^-—a *common ion*—from a source such as KI(aq). The situation is similar to our first encounter with the common-ion effect in Chapter 17.

According to Le Châtelier's principle, an equilibrium mixture responds to a forced increase in the concentration of one of its reactants by shifting in the direction in which that reactant is consumed. In the lead(II) iodide solubility equilibrium, addition of the common ion, I⁻, causes the reverse reaction to be favored, leading to a new equilibrium.

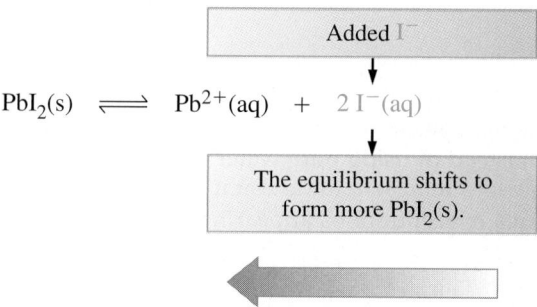

The addition of the common ion shifts the equilibrium of a slightly soluble ionic compound toward the undissolved compound, causing more to precipitate. Thus, the solubility of the compound is reduced.

> The solubility of a slightly soluble ionic compound is lowered in the presence of a second solute that furnishes a common ion.

The common-ion effect is illustrated in Figure 18-1, and it is applied quantitatively in Example 18-4.

The solubility of PbI_2 in the presence of 0.10 M I⁻, as calculated in Example 18-4, is about 2000 times less than its value in pure water (Example 18-3). If you work out Practice Example 18-4A, you will see that the effect of added Pb^{2+} in reducing the solubility of PbI_2 is not as striking as that of I⁻, but it is significant nevertheless.

◀ Here K⁺ does not take part in the process and is generally not included in the solubility equations. Such ions must always be present to ensure electrical neutrality of the solution. These ions are sometimes called "spectator ions."

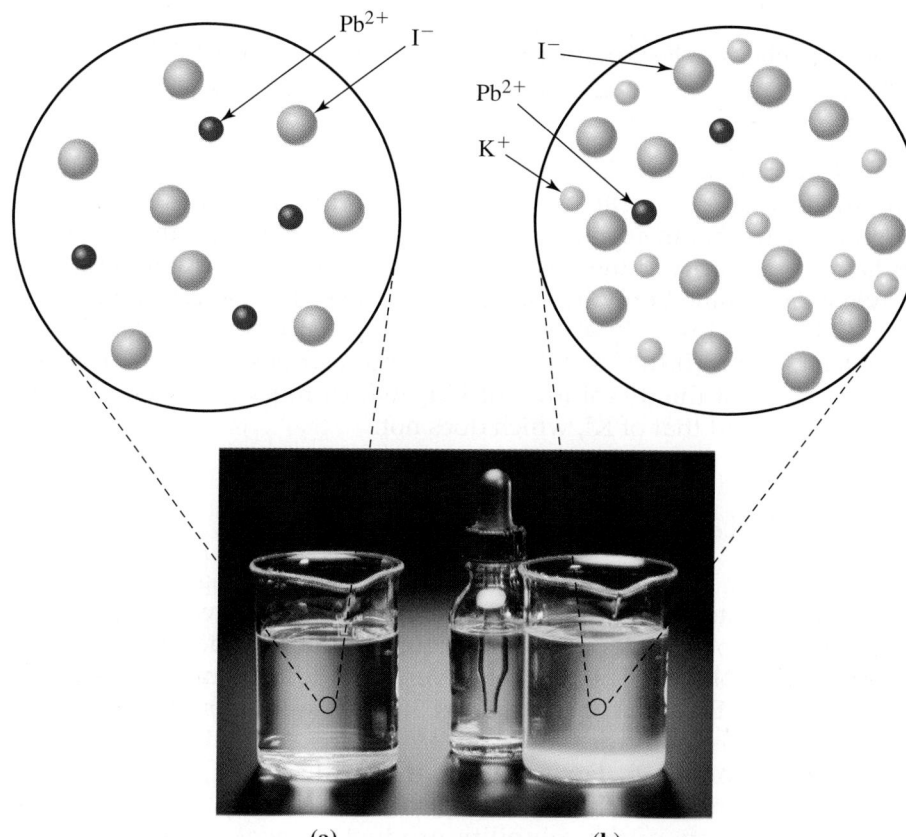

◀ FIGURE 18-1
The common-ion effect in solubility equilibrium
(a) A clear saturated solution of lead(II) iodide from which excess undissolved solute has been filtered off. (b) When a small volume of a concentrated solution of KI (containing the common ion, I⁻) is added, a small quantity of $PbI_2(s)$ precipitates. A common ion reduces the solubility of a sparingly soluble solute.

EXAMPLE 18-4 Calculating the Solubility of a Slightly Soluble Solute in the Presence of a Common Ion

What is the molar solubility of PbI_2 in 0.10 M KI(aq)?

Analyze

To solve this problem, let's set up an ICE table with s instead of x to represent changes in concentrations. Think of producing a saturated solution of PbI_2, but instead of using pure water as the solvent, we will use 0.10 M KI(aq). Thus, we begin with $[I^-] = 0.10$ M. Now let s represent the amount of PbI_2, in moles, that dissolves to produce 1 L of saturated solution. The additional concentrations appearing in this solution are s mol Pb^{2+}/L and $2s$ mol I^-/L.

Solve

The ICE table is

$$PbI_2(s) \rightleftharpoons Pb^{2+}(aq) + 2I^-(aq)$$

initial concns, M:			0.10
from PbI_2, M:		s	$2s$
equil concns, M:		s	$(0.10 + 2s)$

The usual K_{sp} relationship must be satisfied.

$$K_{sp} = [Pb^{2+}][I^-]^2 = (s)(0.10 + 2s)^2 = 7.1 \times 10^{-9}$$

To simplify the solution to this equation, let's assume that s is much smaller than 0.10 M, so that $0.10 + 2s \approx 0.10$.

$$s(0.10)^2 = 7.1 \times 10^{-9}$$

$$s = \frac{7.1 \times 10^{-9}}{(0.10)^2} = 7.1 \times 10^{-7} \text{ M} = \text{molar solubility of } PbI_2$$

Assess

Our assumption is well justified: 7.1×10^{-7} is much smaller than 0.10.

PRACTICE EXAMPLE A: What is the molar solubility of PbI_2 in 0.10 M $Pb(NO_3)_2$(aq)? [*Hint:* To which ion concentration should the solubility be related?]

PRACTICE EXAMPLE B: What is the molar solubility of $Fe(OH)_3$ in a buffered solution with pH = 8.20?

A typical error in such problems as Example 18-4 is to double the common-ion concentration—that is, to write $[I^-] = (2 \times 0.10)$ M instead of $[I^-] = 0.10$ M. Although it is true that in *any* aqueous solution of PbI_2, the $[I^-]$ derived from PbI_2 is *twice* the molarity of the PbI_2, the $[I^-]$ that comes from a soluble strong electrolyte is determined only by the molarity of the strong electrolyte. Thus, $[I^-]$ in 0.10 M KI(aq) is 0.10 M. In 0.10 M KI(aq) that is also saturated with PbI_2, the *total* $[I^-] = (0.10 + 2s)$ M. In short, no relationship exists between the stoichiometry of the dissolution of PbI_2, which requires a factor of 2 in establishing $[I^-]$, and that of KI, which does not.

▶ The difference between stoichiometric concentration and effective concentration increases as more solute is dissolved in the solvent. There are no interactions between solute particles in an ideal solution where stoichiometric and effective concentrations are equal. These interactions increase as more solute is dissolved.

18-4 Limitations of the K_{sp} Concept

We have repeatedly used the term *slightly soluble* in describing the solutes for which we have written K_{sp} expressions. You might wonder if we can write K_{sp} expressions for moderately or highly soluble ionic compounds, such as NaCl, KNO_3, and NaOH. The answer is yes, but the K_{sp} must be based on ion *activities* rather than on concentrations. In ionic solutions of moderate to high concentrations, activities and concentrations are generally far from equal (recall Section 13-9). For example, in 0.1 M KCl(aq) the activity is roughly 24% of the molarity. If we cannot use molarities in place of activities, much of the simplicity of the solubility product concept is lost. Thus, K_{sp}

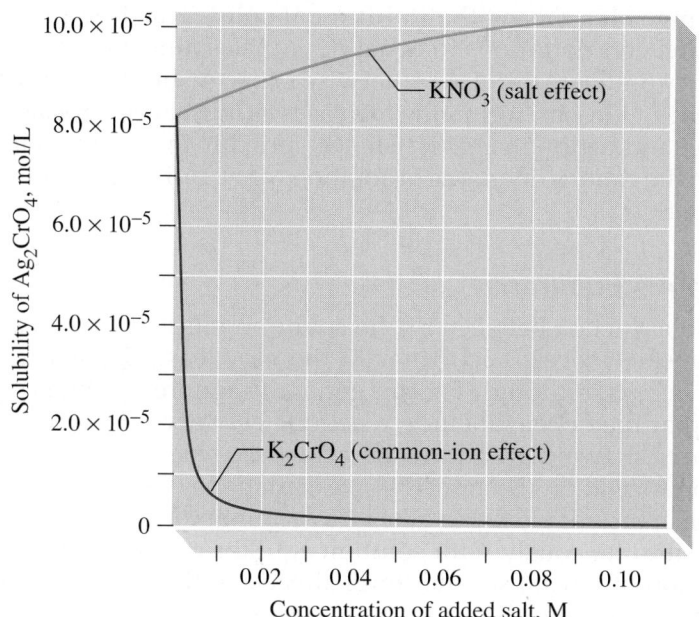

Comparison of the common-ion effect and the salt effect on the molar solubility of Ag_2CrO_4
The presence of CrO_4^{2-} ions, derived from $K_2CrO_4(aq)$, reduces the solubility of Ag_2CrO_4 by a factor of about 35 over the concentration range shown (from 0 to 0.10 M added salt). Over the same concentration range, the solubility of Ag_2CrO_4 is increased by the presence of the diverse ions from KNO_3, but only by about 25%

values are usually limited to slightly soluble (essentially insoluble) solutes, and ion molarities are used in place of activities. In addition, the K_{sp} concept has several other limitations, which are discussed in the following sections.

The Diverse Noncommon Ion Effect: The Salt Effect

We have explored the effect of common ions on a solubility equilibrium, but what effect do ions different from those involved in the equilibrium have on solute solubilities? The effect of "noncommon" or *diverse* ions is not as striking as the common-ion effect. Moreover, diverse ions tend to increase rather than decrease solubility. As the total ionic concentration of a solution increases, interionic attractions become more important. *Activities* (effective concentrations) *become smaller* than the stoichiometric concentration. For the ions involved in the solution process, this means that higher concentrations must appear in solution before equilibrium is established—the *solubility increases*. Figure 18-2 compares the effects of common ions and diverse ions.

The diverse ion effect is more commonly called the **salt effect**. As a result of the salt effect, the numerical value of a K_{sp} based on molarities will vary depending on the ionic atmosphere. Most tabulated values of K_{sp} are based on activities rather than on molarities, thus avoiding the problem of the salt effect.

Incomplete Dissociation of Solute into Ions

In performing calculations involving K_{sp} values and solubilities, we have assumed that all the dissolved solute appears in solution as separated cations and anions. However, this assumption is often not valid. The solute might not be 100% ionic, and some of the solute might enter solution in molecular form. Alternatively, some ions in solution might join together into **ion pairs**. An ion pair is two oppositely charged ions that are held together by the electrostatic attraction between the ions. For example, in a saturated solution of magnesium fluoride, although most of the solute exists as Mg^{2+} and F^- ions, some exists as the ion pair MgF^+.

To the extent that a solution contains cations and anions of a solute as ion pairs, the concentrations of the dissociated ions are reduced from the stoichiometric expectations. Thus, although the measured solubility of MgF_2 is about 4×10^{-3} M, we cannot safely assume that $[Mg^{2+}] = 4 \times 10^{-3}$ M, and that $[F^-] = 8 \times 10^{-3}$ M, because some of the Mg^{2+} and F^- ions are involved

Ion pair

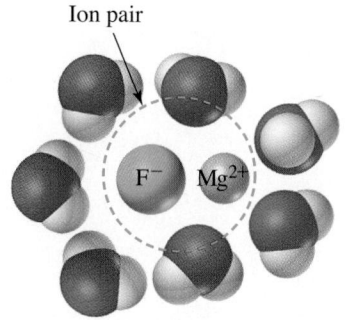

▲ An ion pair, MgF^+, in a magnesium fluoride aqueous solution. The water molecules surrounding the ion pair form what is referred to as a solvent cage.

in ion pairing. This means that additional solute must be present for the product of ion concentrations to equal K_{sp} of the solute, making the true solubility of the solute greater than otherwise expected on the basis of K_{sp}.

The degree of ion-pair formation increases as mutual electrostatic attraction of the anion and cation increases. Hence, ion-pair formation is increasingly likely when the cations or anions in solution carry multiple charges (for example, Mg^{2+} or SO_4^{2-}).

Simultaneous Equilibria

The reversible reaction between a solid solute and its ions in aqueous solution is never the sole process occurring. At the very least, the self-ionization of water also occurs, although we can generally ignore it. Other equilibrium processes that may occur include reactions between solute ions and other solution species. Two possibilities are acid–base reactions (Section 18-7) and complex-ion formation (Section 18-8). Calculations based on the K_{sp} expression may be in error if we fail to take into account other equilibrium processes that occur simultaneously with solution equilibrium. We have encountered the dissolution of $PbI_2(s)$ several times already in this chapter. However, the dissolution process is, in fact, a lot more complex than we've shown. As suggested below, there are many competing processes:

$$PbI_4^{2-}(aq) \underset{+I^-}{\overset{-I^-}{\rightleftharpoons}} PbI_3^-(aq) \underset{+I^-}{\overset{-I^-}{\rightleftharpoons}} PbI_2(s) \rightleftharpoons Pb^{2+}(aq) + 2\,I^-(aq)$$

$$PbI^+(aq) + I^-(aq)$$

$$PbI_2(aq)$$

Assessing the Limitations of K_{sp}

Let's assess the importance of the effects discussed in this section, some of which apply to $CaSO_4$. Recall that in Example 18-2 we calculated K_{sp} for $CaSO_4$ on the basis of its measured solubility. Our result was $K_{sp} = 2.3 \times 10^{-4}$. This value is about 25 times larger than the value listed in Table 18.1, which is $K_{sp} = 9.1 \times 10^{-6}$.

These conflicting results for $CaSO_4$ are understandable. The K_{sp} value listed in Table 18.1 is based on ion activities, whereas the K_{sp} value calculated from the experimentally determined solubility is based on ion concentrations, assuming complete dissociation of the solute into ions and no ion-pair formation. We will continue to substitute molarities for activities of ions, and the case of $CaSO_4$ simply suggests that some of our results, although of the appropriate general magnitude (that is, within a factor of 10 or 100), may not be highly accurate. These order-of-magnitude results, however, still allow us to make some correct predictions and to apply the K_{sp} concept in useful ways.

18-5 Criteria for Precipitation and Its Completeness

Silver iodide is a light-sensitive compound used in photographic film and also in cloud seeding to produce rain. Its solubility equilibrium and K_{sp} are represented as

$$AgI(s) \rightleftharpoons Ag^+(aq) + I^-(aq)$$

$$K_{sp} = [Ag^+][I^-] = 8.5 \times 10^{-17}$$

Suppose we mix solutions of $AgNO_3(aq)$ and $KI(aq)$ to obtain a mixed solution that has $[Ag^+] = 0.010\,M$ and $[I^-] = 0.015\,M$. Is this solution unsaturated, saturated, or supersaturated?

Recall the reaction quotient, Q, that was introduced in Chapter 15. It has the same form as an equilibrium constant expression but uses initial concentrations rather than equilibrium concentrations. Initially,

$$Q_{sp} = [Ag^+]_{init} \times [I^-]_{init} = (0.010)(0.015) = 1.5 \times 10^{-4} > K_{sp}$$

The fact that $Q_{sp} > K_{sp}$ indicates that the concentrations of Ag^+ and I^- are already higher than they would be in a saturated solution and that a net change should occur to the left. The solution is *supersaturated*. As is generally the case with supersaturated solutions, excess AgI should precipitate from solution. If we had found $Q_{sp} < K_{sp}$, the solution would have been *unsaturated*. No precipitate would form from such a solution.

When applied to solubility equilibria, Q_{sp} is generally called the **ion product** because its form is that of the product of ion concentrations raised to appropriate powers. The criteria for determining whether ions in a solution will combine to form a precipitate require us to compare the ion product with K_{sp}.

- Precipitation *should occur* if $Q_{sp} > K_{sp}$.
- Precipitation *cannot occur* if $Q_{sp} < K_{sp}$.
- A solution is just saturated if $Q_{sp} = K_{sp}$.

These criteria are illustrated in Figure 18-3 and Example 18-5. The example emphasizes the important point that *any possible dilutions must be considered before the criteria for precipitation are applied.*

Precipitation of a solute is considered to be complete only if the amount of solute remaining in solution is very small. An arbitrary rule of thumb is that

(a)   (b)

▲ FIGURE 18-3
Applying the criteria for precipitation from solution—Example 18-5 illustrated
(a) When three drops of 0.20 M KI are first added to 100.0 mL of 0.010 M $Pb(NO_3)_2$, a precipitate forms because K_{sp} is exceeded in the immediate vicinity of the drops.
(b) When the KI becomes uniformly mixed in the $Pb(NO_3)_2(aq)$, K_{sp} is no longer exceeded and the precipitate redissolves. The criteria for precipitation must be applied *after* dilution has occurred.

precipitation is complete if 99.9% or more of a particular ion has precipitated, leaving less than 0.1% of the ion in solution. In Example 18-6, we will calculate the concentration of Mg^{2+} that remains in a solution from which $Mg(OH)_2(s)$ has precipitated. We will compare this remaining $[Mg^{2+}]$ to the initial $[Mg^{2+}]$ to determine the completeness of the precipitation.

EXAMPLE 18-5 Applying the Criteria for Precipitation of a Slightly Soluble Solute

Three drops of 0.20 M KI are added to 100.0 mL of 0.010 M $Pb(NO_3)_2$. Will a precipitate of lead(II) iodide form? (Assume 1 drop = 0.05 mL.)

$$PbI_2(s) \rightleftharpoons Pb^{2+}(aq) + 2\,I^-(aq) \qquad K_{sp} = 7.1 \times 10^{-9}$$

Analyze

We need to compare the product $[Pb^{2+}][I^-]^2$ formulated for the initial concentrations with the K_{sp} for PbI_2. For $[Pb^{2+}]$, the dilution caused by adding 3 drops (0.15 mL = 0.00015 L) to 0.100 L of solution is negligible and so we can simply use 0.010 M. For $[I^-]$, however, we must consider the great reduction in concentration that occurs when the three drops of 0.20 M KI are diluted to 100.0 mL.

Solve

Dilution Calculation:

$$\text{amount } I^- = 3 \text{ drops} \times \frac{0.05 \text{ mL}}{1 \text{ drop}} \times \frac{1 \text{ L}}{1000 \text{ mL}} \times \frac{0.20 \text{ mol KI}}{L} \times \frac{1 \text{ mol } I^-}{1 \text{ mol KI}} = 3 \times 10^{-5} \text{ mol } I^-$$

$$\text{Total volume} = 0.1000 \text{ L} + \left(3 \text{ drops} \times \frac{0.05 \text{ mL}}{\text{drop}} \times \frac{1 \text{ L}}{1000 \text{ mL}} \right) = 0.1002 \text{ L}$$

$$[I^-] = \frac{3 \times 10^{-5} \text{ mol } I^-}{0.1002 \text{ L}} = 3 \times 10^{-4} \text{ M}$$

Applying Precipitation Criteria:

$$Q_{sp} = [Pb^{2+}][I^-]^2 = (0.010)(3 \times 10^{-4})^2 = 9 \times 10^{-10}$$

Because a Q_{sp} of 9×10^{-10} is *smaller than* a K_{sp} of 7.1×10^{-9}, we conclude that $PbI_2(s)$ should *not* precipitate.

Assess

In this calculation, as well as problems similar to this one, any possible dilutions must be considered *before* the criteria for precipitation are applied. Note that the addition of the three drops of KI(aq) does not change the value of $[Pb^{2+}]$: 0.010 M $\times$ (0.1000 L/0.1002 L) = 0.010 M (two significant figures).

PRACTICE EXAMPLE A: Three drops of 0.20 M KI are added to 100.0 mL of a 0.010 M solution of $AgNO_3$. Will a precipitate of silver iodide form?

PRACTICE EXAMPLE B: We saw in Example 18-5 that a 3-drop volume of 0.20 M KI is insufficient to cause precipitation in 100.0 mL of 0.010 M $Pb(NO_3)_2$. What minimum number of drops would be required to produce the first precipitate?

EXAMPLE 18-6 Assessing the Completeness of a Precipitation Reaction

The first step in a commercial process in which magnesium is obtained from seawater involves precipitating Mg^{2+} as $Mg(OH)_2(s)$. The magnesium ion concentration in seawater is about 0.059 M. If a seawater sample is treated so that its $[OH^-]$ is maintained at 2.0×10^{-3} M, **(a)** what will be $[Mg^{2+}]$ remaining in solution when precipitation stops ($K_{sp} = 1.8 \times 10^{-11}$)? **(b)** Is the precipitation of $Mg(OH)_2(s)$ complete under these conditions?

Analyze

In part (a) of this example we compare the Q_{sp} of the solution with the known K_{sp} for $Mg(OH)_2$ to determine whether precipitation will occur. If it will, then precipitation of $Mg(OH)_2(s)$ will continue as long as the ion product exceeds K_{sp} and will stop when that product is equal to K_{sp}. At the point at which the ion product equals K_{sp}, whatever Mg^{2+} is in solution remains in solution. In part (b) we need to compare the amount of Mg^{2+} remaining after precipitation with the original amount.

Solve

(a) There is no question that precipitation will occur, because the ion product $Q_{sp} = [Mg^{2+}][OH^-]^2 = (0.059)(2.0 \times 10^{-3})^2 = 2.4 \times 10^{-7}$ exceeds K_{sp}.

$$[Mg^{2+}][OH^-]^2 = [Mg^{2+}](2.0 \times 10^{-3})^2 = 1.8 \times 10^{-11} = K_{sp}$$

$$[Mg^{2+}]_{remaining} = \frac{1.8 \times 10^{-11}}{(2.0 \times 10^{-3})^2} = 4.5 \times 10^{-6}\,M$$

(b) $[Mg^{2+}]$ in seawater is reduced from 0.059 M to 4.5×10^{-6} M as a result of the precipitation reaction. Expressed as a percentage,

$$\%[Mg^{2+}]_{remaining} = \frac{4.5 \times 10^{-6}\,M}{0.059\,M} \times 100\% = 0.0076\%$$

Because less than 0.1% of the Mg^{2+} remains, we conclude that precipitation is essentially complete.

Assess

Because $[OH^-]$ is maintained at a constant value, the calculation of $[Mg^{2+}]_{remaining}$ is straightforward. (A method of maintaining a constant $[OH^-]$ during a precipitation is to carry out the precipitation from a buffer solution.) If $[OH^-]$ was initially 2.0×10^{-3} M but with no source of OH^- to replenish it, $[Mg^{2+}]_{remaining}$ would be 0.058 M. Verify this result yourself.

PRACTICE EXAMPLE A: A typical Ca^{2+} concentration in seawater is 0.010 M. Will the precipitation of $Ca(OH)_2$ be complete from a seawater sample in which $[OH^-]$ is maintained at 0.040 M?

PRACTICE EXAMPLE B: What $[OH^-]$ should be maintained in a solution if, after precipitation of Mg^{2+} as $Mg(OH)_2(s)$, the remaining Mg^{2+} is to be at a level of 1 $\mu g\,Mg^{2+}/L$?

18-2   ARE YOU WONDERING...

What conditions favor completeness of precipitation?

The key factors in determining whether the target ion is essentially completely removed from solution in a precipitation are (1) the value of K_{sp}, (2) the initial concentration of the target ion, and (3) the concentration of the common ion. In general, completeness of precipitation is favored by

- *a very small value of K_{sp}.* (The concentration of the target ion remaining in solution will be very small.)

- *a high initial concentration of the target ion.* (The concentration of the target ion remaining in solution will be only a very small fraction of the initial value.)

- *a concentration of common ion much larger than that of the target ion.* (The common-ion concentration will remain nearly constant during the precipitation.)

18-6   Fractional Precipitation

If a large excess of $AgNO_3(s)$ is added to a solution containing the ions CrO_4^{2-} and Br^-, a mixed precipitate of $Ag_2CrO_4(s)$ and $AgBr(s)$ is obtained. There is a method of adding $AgNO_3$, however, that will cause $AgBr(s)$ to precipitate but leave CrO_4^{2-} in solution. That method is fractional precipitation.

Fractional precipitation is a technique in which two or more ions in solution, each capable of being precipitated by the same reagent, are separated by the proper use of that reagent: *One ion is precipitated, while the other(s) remains in solution.* The primary condition for a successful fractional precipitation is

◀ Fractional precipitation is also called *selective precipitation*.

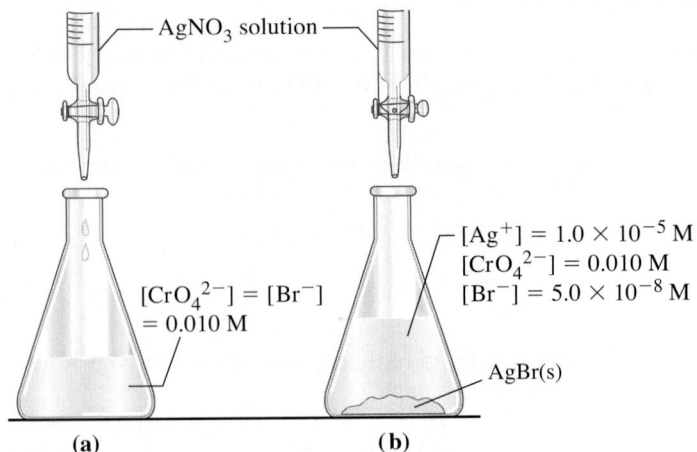

▶ FIGURE 18-4
Fractional precipitation—Example 18-7 illustrated
(a) $AgNO_3(aq)$ is slowly added to a solution that is 0.010 M in Br^- and 0.010 M in CrO_4^{2-}. (b) Essentially all the Br^- has precipitated as pale yellow AgBr(s), with $[Br^-]$ in solution $= 5.0 \times 10^{-8}$ M. Red-brown $Ag_2CrO_4(s)$ is just about to precipitate.

In figure: $AgNO_3$ solution

$[CrO_4^{2-}] = [Br^-]$
$= 0.010$ M

$[Ag^+] = 1.0 \times 10^{-5}$ M
$[CrO_4^{2-}] = 0.010$ M
$[Br^-] = 5.0 \times 10^{-8}$ M

AgBr(s)

(a)   (b)

that there be a significant difference in the solubilities of the substances being separated. (Usually this means a significant difference in their K_{sp} values.) The key to the technique is the slow addition of a concentrated solution of the precipitating reagent to the solution from which precipitation is to occur, as from a buret (Fig. 18-4).

Example 18-7 considers the separation of $CrO_4^{2-}(aq)$ and $Br^-(aq)$ through the use of $Ag^+(aq)$.

EXAMPLE 18-7   **Separating Ions by Fractional Precipitation**

$AgNO_3(aq)$ is slowly added to a solution that has $[CrO_4^{2-}] = 0.010$ M and $[Br^-] = 0.010$ M.

   (a) Show that AgBr(s) should precipitate before $Ag_2CrO_4(s)$ does.
   (b) When $Ag_2CrO_4(s)$ begins to precipitate, what is $[Br^-]$ remaining in solution?
   (c) Is complete separation of $Br^-(aq)$ and $CrO_4^{2-}(aq)$ by fractional precipitation feasible?

Analyze

First, write equations for the solubility equilibria and look up the corresponding K_{sp} values.

$$Ag_2CrO_4(s) \rightleftharpoons 2\,Ag^+(aq) + CrO_4^{2-}(aq) \qquad K_{sp} = 1.1 \times 10^{-12}$$
$$AgBr(s) \rightleftharpoons Ag^+(aq) + Br^-(aq) \qquad K_{sp} = 5.0 \times 10^{-13}$$

Next we determine the concentration of silver ion needed to precipitate either AgBr(s) or $Ag_2CrO_4(s)$. The one that requires the least amount of silver ion will precipitate first. The formation of the first precipitate helps keep $[Ag^+]$ below that required to form the second precipitate; $[Ag^+]$ will, however, slowly increase and, eventually, the second precipitate will form. We can use the value of $[Ag^+]$ at the point at which the second precipitate starts to form to calculate $[Br^-]$ and $[CrO_4^{2-}]$ and determine whether complete separation is possible (i.e., when 99.9% of the bromide ion has precipitated).

Solve

   (a) The required values of $[Ag^+]$ for precipitation to start are

   AgBr ppt: $Q_{sp} = [Ag^+][Br^-] = [Ag^+](0.010) = 5.0 \times 10^{-13} = K_{sp}$
   $[Ag^+] = 5.0 \times 10^{-11}$ M

   Ag_2CrO_4 ppt: $Q_{sp} = [Ag^+]^2[CrO_4^{2-}] = [Ag^+]^2(0.010)$
   $= 1.1 \times 10^{-12} = K_{sp}$
   $[Ag^+]^2 = 1.1 \times 10^{-10}$ and $[Ag^+] = 1.0 \times 10^{-5}$ M

Because the $[Ag^+]$ required to start the precipitation of AgBr(s) is much less than that for $Ag_2CrO_4(s)$, AgBr(s) precipitates first. As long as AgBr(s) is forming, the silver ion concentration can only slowly approach the value required for the precipitation of $Ag_2CrO_4(s)$.

(b) As AgBr(s) precipitates, $[Br^-]$ gradually decreases, and this permits $[Ag^+]$ to increase. When $[Ag^+]$ reaches 1.0×10^{-5} M, precipitation of $Ag_2CrO_4(s)$ begins. To determine $[Br^-]$ at the point at which $[Ag^+] = 1.0 \times 10^{-5}$ M, we use K_{sp} for AgBr and solve for $[Br^-]$.

$$K_{sp} = [Ag^+][Br^-] = (1.0 \times 10^{-5})[Br^-] = 5.0 \times 10^{-13}$$

$$[Br^-] = \frac{5.0 \times 10^{-13}}{1.0 \times 10^{-5}} = 5.0 \times 10^{-8} \text{ M}$$

(c) Before $Ag_2CrO_4(s)$ begins to precipitate, $[Br^-]$ will have been reduced from 1.0×10^{-2} M to 5.0×10^{-8} M. Essentially, all the Br^- will have precipitated from solution as AgBr(s), whereas the CrO_4^{2-} remains in solution. Fractional precipitation is feasible for separating mixtures of Br^- and CrO_4^{2-}.

Assess

Even though the solubility products of these two compounds have similar values, we were able to separate the two ions. The concept of fractional precipitation is used to separate and identify unknown ions in solution.

PRACTICE EXAMPLE A: $AgNO_3(aq)$ is slowly added to a solution with $[Cl^-] = 0.115$ M and $[Br^-] = 0.264$ M. What percent of the Br^- remains unprecipitated at the point at which AgCl(s) begins to precipitate?

$$\text{AgCl:} \quad K_{sp} = 1.8 \times 10^{-10} \qquad \text{AgBr:} \quad K_{sp} = 5.0 \times 10^{-13}$$

PRACTICE EXAMPLE B: A solution has $[Ba^{2+}] = [Sr^{2+}] = 0.10$ M. Use data from Appendix D to choose the best precipitating agent to separate these two ions. What is the concentration of the first ion to precipitate when the second ion begins to precipitate?

18-3 CONCEPT ASSESSMENT

In the fractional precipitation pictured in Figure 18-4 and described in Example 18-7, no mention was made of the concentration of the $AgNO_3(aq)$ used. Is this concentration immaterial? Explain.

18-7 Solubility and pH

The pH of a solution can affect the solubility of a salt to a large degree. That is especially true when the anion of the salt is the conjugate base of a weak acid or the base OH^- itself. An interesting example is the highly insoluble $Mg(OH)_2(s)$, which, when suspended in water, is the popular antacid known as milk of magnesia. A **suspension** is a heterogeneous fluid containing solid particles that are sufficiently large for sedimentation and, unlike colloids, will settle. Hydroxide ions from the dissolved magnesium hydroxide react with hydronium ions (in stomach acid) to form water.

$$Mg(OH)_2(s) \rightleftharpoons Mg^{2+}(aq) + 2\,OH^-(aq) \qquad \textbf{(18.2)}$$

$$OH^-(aq) + H_3O^+(aq) \longrightarrow 2\,H_2O(l) \qquad \textbf{(18.3)}$$

According to Le Châtelier's principle, we expect reaction (18.2) to be displaced to the right—that is, additional $Mg(OH)_2$ would dissolve to replace OH^- ions drawn off by the neutralization reaction (18.3). The overall net ionic equation can be obtained by doubling equation (18.3) and adding it to equation (18.2). At the same time, we can apply the method of combining equilibrium constants (page 664). The result is

$$Mg(OH)_2(s) \rightleftharpoons Mg^{2+}(aq) + 2\,\cancel{OH^-}(aq) \quad K_{sp} = 1.8 \times 10^{-11}$$

$$\underline{2\,\cancel{OH^-}(aq) + 2\,H_3O^+(aq) \rightleftharpoons 4\,H_2O(l) \qquad\qquad K' = 1/K_w^{\,2} = 1.0 \times 10^{28}}$$

$$Mg(OH)_2(s) + 2\,H_3O^+ \rightleftharpoons Mg^{2+}(aq) + 4\,H_2O(l) \qquad \textbf{(18.4)}$$

$$K = K_{sp}/K_w^{\,2} = 1.8 \times 10^{-11} \times 1.0 \times 10^{28} = 1.8 \times 10^{17}$$

▲ Milk of magnesia, an aqueous suspension of $Mg(OH)_2(s)$.

▶ In general, a water-insoluble salt of a weak acid can be dissolved by acid, but not the water-insoluble salt of a strong acid.

▶ Equilibrium constant expressions can be combined. See Section 15-3.

The large value of K for reaction (18.4) indicates that a greater amount of $Mg(OH)_2$ will react (dissolve) in acidic solution than in pure water.

Other slightly soluble solutes having basic anions (such as $ZnCO_3$, MgF_2, and CaC_2O_4) also dissolve to a greater extent in acidic solutions. For these solutes, we can write overall net ionic equations for solubility equilibria and corresponding K values based on K_{sp} for the solutes and K_a for the conjugate acids of the anions.

Although $Mg(OH)_2$ dissolves in acidic solution, in moderately or strongly basic solutions it does not. In Example 18-8, we calculate $[OH^-]$ in a solution of the weak base NH_3 and then use the criteria for precipitation to see if $Mg(OH)_2(s)$ will precipitate. In Example 18-9, we determine how to adjust $[OH^-]$ to prevent precipitation of $Mg(OH)_2(s)$. This adjustment is made by adding NH_4^+ to the $NH_3(aq)$, thereby converting it to a buffer solution.

🔍 18-4 CONCEPT ASSESSMENT

Which will be affected more by the addition of a strong acid or a strong base: the solubility of CaF_2 or the solubility of $CaCl_2$? Explain.

EXAMPLE 18-8 **Determining Whether a Precipitate Will Form in a Solution in Which There Is Also an Ionization Equilibrium**

Should $Mg(OH)_2(s)$ precipitate from a solution that is 0.010 M $MgCl_2$ and also 0.10 M NH_3?

Analyze

The key here is in understanding that $[OH^-]$ is established by the ionization of $NH_3(aq)$.

$$NH_3(aq) + H_2O(l) \rightleftharpoons NH_4^+(aq) + OH^-(aq) \qquad K_b = 1.8 \times 10^{-5}$$

Solve

If we set this up in the usual way, the equilibrium values are $x = [NH_4^+] = [OH^-]$ and $[NH_3] = (0.10 - x) \approx 0.10$. Then we obtain

$$K_b = \frac{[NH_4^+][OH^-]}{[NH_3]} = \frac{x \cdot x}{0.10} = 1.8 \times 10^{-5}$$

$$x^2 = 1.8 \times 10^{-6} \qquad x = [OH^-] = 1.3 \times 10^{-3}\,M$$

Now we can rephrase the original question: Should $Mg(OH)_2(s)$ precipitate from a solution in which $[Mg^{2+}] = 1.0 \times 10^{-2}\,M$ and $[OH^-] = 1.3 \times 10^{-3}\,M$? We must compare the ion product, Q_{sp}, with K_{sp}.

$$Q_{sp} = [Mg^{2+}][OH^-]^2 = (1.0 \times 10^{-2})(1.3 \times 10^{-3})^2$$
$$= 1.7 \times 10^{-8} > K_{sp} = 1.8 \times 10^{-11}$$

Precipitation should occur.

Assess

In this example the ionization equilibrium was easy to identify. Always look for anions and cations of salts that can establish an ionization equilibrium (e.g., through hydrolysis).

PRACTICE EXAMPLE A: Should $Mg(OH)_2(s)$ precipitate from a solution that is 0.010 M $MgCl_2(aq)$ and also 0.10 M $NaC_2H_3O_2$? $K_{sp}[Mg(OH)_2] = 1.8 \times 10^{-11}$; $K_a(HC_2H_3O_2) = 1.8 \times 10^{-5}$. [*Hint:* What equilibrium expression establishes $[OH^-]$ in the solution?]

PRACTICE EXAMPLE B: Will a precipitate of $Fe(OH)_3$ form from a solution that is 0.013 M Fe^{3+} in a buffer solution that is 0.150 M $HC_2H_3O_2$ and 0.250 M $NaC_2H_3O_2$?

EXAMPLE 18-9   Controlling an Ion Concentration, Either to Cause Precipitation or to Prevent It

What $[NH_4^+]$ must be maintained to prevent precipitation of $Mg(OH)_2(s)$ from a solution that is 0.010 M $MgCl_2$ and 0.10 M NH_3?

Analyze

The maximum value of the ion product, Q_{sp}, before precipitation occurs is 1.8×10^{-11}, the value of K_{sp} for $Mg(OH)_2$. This fact allows us to determine the maximum concentration of OH^- that can be tolerated.

Solve

$$[Mg^{2+}][OH^-]^2 = (1.0 \times 10^{-2})[OH^-]^2 = 1.8 \times 10^{-11}$$
$$[OH^-]^2 = 1.8 \times 10^{-9}$$
$$[OH^-] = 4.2 \times 10^{-5} M$$

Next let's determine what $[NH_4^+]$ must be present in 0.10 M NH_3 to maintain $[OH^-] = 4.2 \times 10^{-5} M$.

$$NH_3(aq) + H_2O(l) \rightleftharpoons NH_4^+(aq) + OH^-(aq) \qquad K_b = 1.8 \times 10^{-5}$$

$$K_b = \frac{[NH_4^+][OH^-]}{[NH_3]} = \frac{[NH_4^+](4.2 \times 10^{-5})}{0.10} = 1.8 \times 10^{-5}$$

$$[NH_4^+] = \frac{0.10 \times 1.8 \times 10^{-5}}{4.2 \times 10^{-5}} = 0.043 \ M$$

To keep the $[OH^-]$ at 4.2×10^{-5} M *or less*, and thus prevent the precipitation of $Mg(OH)_2(s)$, $[NH_4^+]$ should be maintained at 0.043 M or *greater*.

Assess

The reason for keeping $[NH_4^+]$ *greater* than 0.043 M is that at higher concentrations of NH_4^+, the equilibrium for the ionization of NH_3 shifts to the left, reducing the amount of OH^-. Practically, one would choose 0.10 M since the ratio $[NH_3]/[NH_4^+]$ would be 1.0 and the buffer capacity of the solution is at a maximum.

PRACTICE EXAMPLE A: What minimum $[NH_4^+]$ must be present to prevent precipitation of $Mn(OH)_2(s)$ from a solution that is 0.0050 M $MnCl_2$ and 0.025 M NH_3? For $Mn(OH)_2$, $K_{sp} = 1.9 \times 10^{-13}$.

PRACTICE EXAMPLE B: What is the molar solubility of $Mg(OH)_2(s)$ in a solution that is 0.250 M NH_3 and 0.100 M NH_4Cl? [*Hint:* Use equation (18.4).]

🔍 18-5   CONCEPT ASSESSMENT

Determine $[Mg^{2+}]$ in a saturated solution of $Mg(OH)_2$ at **(a)** pH = 10.00 and **(b)** pH = 5.00. Is each of these a plausible quantity? Explain.

18-8   Equilibria Involving Complex Ions

As shown in Figure 18-5, when moderately concentrated $NH_3(aq)$ is added to a saturated solution of silver chloride in contact with undissolved $AgCl(s)$, the solid dissolves. The key to this dissolving action is that Ag^+ ions from AgCl combine with NH_3 molecules to form the ions $[Ag(NH_3)_2]^+$, which, together with Cl^- ions, remain in solution as the soluble compound $Ag(NH_3)_2Cl$.

$$AgCl(s) + 2 NH_3(aq) \longrightarrow [Ag(NH_3)_2]^+(aq) + Cl^-(aq) \qquad (18.5)$$

Here $[Ag(NH_3)_2]^+$ is called a complex ion, and $Ag(NH_3)_2Cl$ is called a coordination compound. A **complex ion** is a polyatomic cation or anion composed of a central metal ion to which other groups (molecules or ions) called *ligands* are

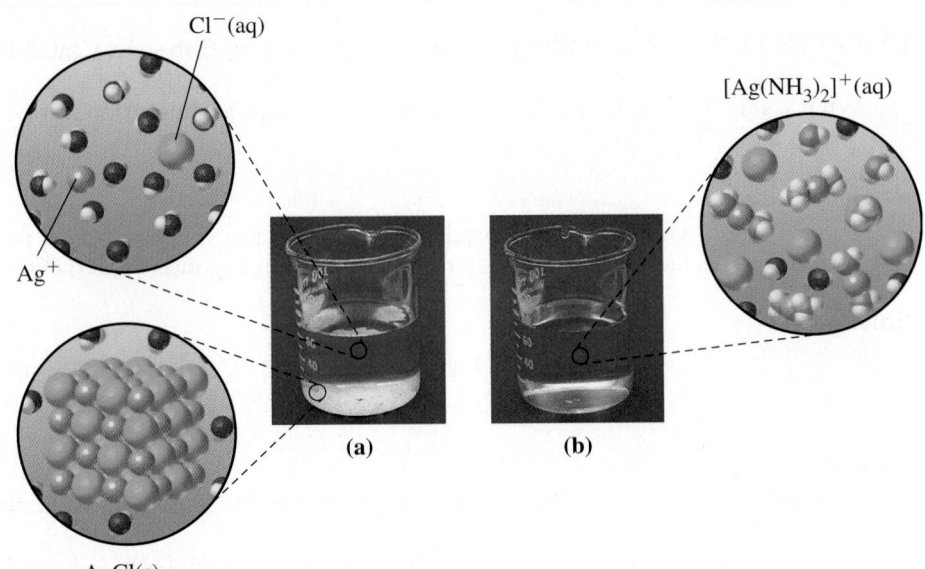

▲ FIGURE 18-5
Complex-ion formation: dissolution of AgCl(s) in NH₃(aq)
(a) A saturated solution of silver chloride in contact with excess AgCl(s). (b) When NH₃(aq) is added, the excess AgCl(s) dissolves through the formation of the complex ion $[Ag(NH_3)_2]^+$.

bonded. **Coordination compounds** are substances containing complex ions. It helps to think of reaction (18.5) as involving two simultaneous equilibria.

$$AgCl(s) \rightleftharpoons Ag^+(aq) + Cl^-(aq) \qquad \textbf{(18.6)}$$

$$Ag^+(aq) + 2\,NH_3(aq) \rightleftharpoons [Ag(NH_3)_2]^+(aq) \qquad \textbf{(18.7)}$$

The equilibrium in reaction (18.7) is shifted far to the right—$[Ag(NH_3)_2]^+$ is a stable complex ion. The equilibrium concentration of $Ag^+(aq)$ in (18.7) is kept so low that the ion product $[Ag^+][Cl^-]$ fails to exceed K_{sp} and AgCl remains in solution. Let's first apply additional qualitative reasoning of this sort in Example 18-10. Then we can turn to some of the quantitative calculations that are also possible.

To describe the ionization of a weak acid, we use the ionization constant K_a. For a solubility equilibrium, we use the solubility product constant K_{sp}. The equilibrium constant that is used to deal with a complex-ion equilibrium is called the formation constant. The **formation constant, K_f,** of a complex ion is the equilibrium constant describing the formation of a complex ion from a central ion and its attached groups. For reaction (18.7) this equilibrium constant expression is

$$K_f = \frac{[[Ag(NH_3)_2]^+]}{[Ag^+][NH_3]^2} = 1.6 \times 10^7$$

Table 18.2 lists some representative formation constants, K_f.

One feature that distinguishes K_f from most other equilibrium constants previously considered is that K_f values are usually *large* numbers. This fact can affect the way we approach certain calculations. With large K values, it is sometimes convenient to solve a problem in two steps. First, assume that the forward reaction goes to completion; second, assume that a small change occurs in the reverse direction to establish the equilibrium. This approach is demonstrated in Example 18-11.

A coordination compound,
$[Co(NH_3)_6]Cl_3$

Complex cation   Anions

$[Co(NH_3)_6]^{3+}$ $3\,Cl^-$

Central
ion   Ligands

EXAMPLE 18-10 Predicting Reactions Involving Complex Ions

Predict what will happen if nitric acid is added to a solution of $[Ag(NH_3)_2]Cl$ in $NH_3(aq)$.

Analyze

We consider the equilibria represented by equations (18.6) and (18.7) and use Le Chatelier's principle to assess the effect of adding HNO_3 to the solution. An important consideration is that nitric acid will protonate the base, NH_3.

Solve

Since $HNO_3(aq)$ is a strong acid, we will represent the acid as $H_3O^+(aq)$ and write the protonation reaction as

$$H_3O^+(aq) + NH_3(aq) \longrightarrow NH_4^+(aq) + H_2O(l)$$

The formation reaction is

$$Ag^+(aq) + 2\,NH_3(aq) \rightleftharpoons [Ag(NH_3)_2]^+(aq)$$

To replace free NH_3 lost in this neutralization, equilibrium in the formation reaction shifts to the *left*. As a result, $[Ag^+]$ increases. When $[Ag^+]$ increases to the point at which the ion product $[Ag^+][Cl^-]$ exceeds K_{sp}, $AgCl(s)$ precipitates (Fig. 18-6).

Assess

The addition of any acid to a solution of $[Ag(NH_3)_2]Cl$ will result in precipitation.

PRACTICE EXAMPLE A: Copper(II) ion forms both an insoluble hydroxide and the complex ion $[Cu(NH_3)_4]^{2+}$. Write equations to represent the expected reaction when **(a)** $CuSO_4(aq)$ and $NaOH(aq)$ are mixed; **(b)** an excess of $NH_3(aq)$ is added to the product of part (a); and **(c)** an excess of $HNO_3(aq)$ is added to the product of part (b).

PRACTICE EXAMPLE B: Zinc(II) ion forms both an insoluble hydroxide and the complex ions $[Zn(OH)_4]^{2-}$ and $[Zn(NH_3)_4]^{2+}$. Write four equations to represent the reactions of **(a)** $NH_3(aq)$ with $ZnSO_4(aq)$, followed by **(b)** enough $HNO_3(aq)$ to make the product of part (a) acidic; **(c)** enough $NaOH(aq)$ to make the product of part (b) slightly basic; and **(d)** enough $NaOH(aq)$ to make the product of part (c) strongly basic.

▲ FIGURE 18-6
Reprecipitating AgCl(s)
The reagent being added to the solution containing $[Ag(NH_3)_2]^+$ and Cl^- is $HNO_3(aq)$. H_3O^+ from the acid reacts with $NH_3(aq)$ to form $NH_4^+(aq)$. As a result, equilibrium between $[Ag(NH_3)_2]^+$, Ag^+, and NH_3 is upset. The complex ion is destroyed, $[Ag^+]$ quickly rises to the point at which K_{sp} for AgCl is exceeded, and a precipitate forms.

TABLE 18.2 Formation Constants for Some Complex Ions[a]

Complex Ion	Equilibrium Reaction[b]	K_f
$[Co(NH_3)_6]^{3+}$	$Co^{3+} + 6\,NH_3 \rightleftharpoons [Co(NH_3)_6]^{3+}$	4.5×10^{33}
$[Cu(NH_3)_4]^{2+}$	$Cu^{2+} + 4\,NH_3 \rightleftharpoons [Cu(NH_3)_4]^{2+}$	1.1×10^{13}
$[Fe(CN)_6]^{4-}$	$Fe^{2+} + 6\,CN^- \rightleftharpoons [Fe(CN)_6]^{4-}$	1×10^{37}
$[Fe(CN)_6]^{3-}$	$Fe^{3+} + 6\,CN^- \rightleftharpoons [Fe(CN)_6]^{3-}$	1×10^{42}
$[Pb(OH)_3]^-$	$Pb^{2+} + 3\,OH^- \rightleftharpoons [Pb(OH)_3]^-$	3.8×10^{14}
$[PbCl_3]^-$	$Pb^{2+} + 3\,Cl^- \rightleftharpoons [PbCl_3]^-$	2.4×10^1
$[Ag(NH_3)_2]^+$	$Ag^+ + 2\,NH_3 \rightleftharpoons [Ag(NH_3)_2]^+$	1.6×10^7
$[Ag(CN)_2]^-$	$Ag^+ + 2\,CN^- \rightleftharpoons [Ag(CN)_2]^-$	5.6×10^{18}
$[Ag(S_2O_3)_2]^{3-}$	$Ag^+ + 2\,S_2O_3^{2-} \rightleftharpoons [Ag(S_2O_3)_2]^{3-}$	1.7×10^{13}
$[Zn(NH_3)_4]^{2+}$	$Zn^{2+} + 4\,NH_3 \rightleftharpoons [Zn(NH_3)_4]^{2+}$	4.1×10^8
$[Zn(CN)_4]^{2-}$	$Zn^{2+} + 4\,CN^- \rightleftharpoons [Zn(CN)_4]^{2-}$	1×10^{18}
$[Zn(OH)_4]^{2-}$	$Zn^{2+} + 4\,OH^- \rightleftharpoons [Zn(OH)_4]^{2-}$	4.6×10^{17}

[a]A more extensive tabulation is given in Appendix D.
[b]Tabulated here are *overall* formation reactions and the corresponding *overall* formation constants. In Section 24-8, we describe the formation of complex ions in a *stepwise* fashion and introduce formation constants for individual steps.

EXAMPLE 18-11 Determining Whether a Precipitate Will Form in a Solution Containing Complex Ions

A 0.10 mol sample of $AgNO_3$ is dissolved in 1.00 L of 1.00 M NH_3. If 0.010 mol NaCl is added to this solution, will AgCl(s) precipitate?

Analyze

We begin by first assuming that because the value of K_f for $[Ag(NH_3)_2]^+$ is very large, the formation reaction goes to completion. Then we use those results to perform an equilibrium calculation. Finally, from the equilibrium calculation we can determine a Q_{sp} for the reaction and determine whether precipitation will occur.

Solve

Assuming the formation reaction goes to completion, we obtain

$$Ag^+(aq) + 2\,NH_3(aq) \longrightarrow [Ag(NH_3)_2]^+(aq)$$

initial concns, M:	0.10	1.00	
changes, M:	−0.10	−0.20	+0.10
after reaction, M:	≈0	0.80	0.10

However, the concentration of uncomplexed silver ion, though very small, is not zero. To determine the value of $[Ag^+]$, let's start with $[[Ag(NH_3)_2]^+]$ and $[NH_3]$ in solution and establish $[Ag^+]$ at equilibrium.

$$Ag^+ + 2\,NH_3 \rightleftharpoons [Ag(NH_3)_2]^+$$

initial concns, M:	0	0.80	0.10
changes, M:	+x	+2x	−x
equil concns, M:	x	0.80 + 2x	0.10 − x

When substituting into the following expression, we make the assumption that $x \ll 0.10$, which we will find to be the case.

$$\frac{[[Ag(NH_3)_2]^+]}{[Ag^+][NH_3]^2} = \frac{0.10 - x}{x(0.80 + 2x)^2} \approx \frac{0.10}{x(0.80)^2} = 1.6 \times 10^7$$

$$x = [Ag^+] = \frac{0.10}{(1.6 \times 10^7)(0.80)^2} = 9.8 \times 10^{-9}\,M$$

Finally, we must compare $Q_{sp} = [Ag^+][Cl^-]$ with K_{sp} for AgCl (that is, 1.8×10^{-10}). $[Ag^+]$ is the value of x that we just calculated. Because the solution contains 0.010 mol NaCl/L, $[Cl^-] = 0.010\,M = 1.0 \times 10^{-2}\,M$, and

$$Q_{sp} = (9.8 \times 10^{-9})(1.0 \times 10^{-2}) = 9.8 \times 10^{-11} < 1.8 \times 10^{-10}$$

AgCl will not precipitate.

Assess

The formation reaction goes almost to completion, with only a small amount of silver ion remaining in solution. Our initial assumption that the formation reaction goes to completion was valid. The amount of silver ion remaining is not enough to cause precipitation.

PRACTICE EXAMPLE A: Will AgCl(s) precipitate from 1.50 L of a solution that is 0.100 M $AgNO_3$ and 0.225 M NH_3 if 1.00 mL of 3.50 M NaCl is added? [*Hint:* What are $[Ag^+]$ and $[Cl^-]$ immediately after the addition of the 1.00 mL of 3.50 M NaCl? Take into account the dilution of the NaCl(aq), but assume the total volume remains at 1.50 L.]

PRACTICE EXAMPLE B: A solution is prepared that is 0.100 M in $Pb(NO_3)_2$ and 0.250 M in the ethylenediaminetetraacetate anion, $EDTA^{4-}$. Together, Pb^{2+} and $EDTA^{4-}$ form the complex ion $[PbEDTA]^{2-}$. If the solution is also made 0.10 M in I^-, will PbI_2(s) precipitate? For PbI_2, $K_{sp} = 7.1 \times 10^{-9}$; for $[PbEDTA]^{2-}$, $K_f = 2 \times 10^{18}$.

Just as some precipitation reactions can be controlled by using a buffer solution (Example 18-9), precipitation from a solution of complex ions can be controlled by fixing the concentration of the complexing agent. Such a case is demonstrated for AgCl in Example 18-12.

EXAMPLE 18-12   Controlling a Concentration to Cause or Prevent Precipitation from a Solution of Complex Ions

What is the *minimum* concentration of NH_3 needed to prevent $AgCl(s)$ from precipitating from 1.00 L of a solution containing 0.10 mol $AgNO_3$ and 0.010 mol NaCl?

Analyze

To prevent $AgCl(s)$ from precipitating, we must ensure that the solubility product for AgCl is not exceeded. We are given a fixed amount of chloride ion in solution, and so that means we need to determine the maximum concentration of silver ion that can exist without precipitation occurring. Finally, we can solve for the amount of NH_3 necessary to complex all the silver ion.

Solve

Solve for the amount of silver ion that can be in solution without precipitation in a solution containing 1.0×10^{-2} M Cl^-. That is, $[Ag^+][Cl^-] \leq K_{sp}$.

$$[Ag^+](1.0 \times 10^{-2}) \leq K_{sp} = 1.8 \times 10^{-10} \qquad [Ag^+] \leq 1.8 \times 10^{-8}\,M$$

Thus, the maximum concentration of *uncomplexed* Ag^+ in solution is 1.8×10^{-8} M. This means that essentially all the Ag^+ (0.10 mol/L) must be tied up (complexed) in the complex ion $[Ag(NH_3)_2]^+$. We need to solve the expression at the right for $[NH_3]$.

$$K_f = \frac{[[Ag(NH_3)_2]^+]}{[Ag^+][NH_3]^2} = \frac{1.0 \times 10^{-1}}{1.8 \times 10^{-8}[NH_3]^2} = 1.6 \times 10^7$$

$$[NH_3]^2 = \frac{1.0 \times 10^{-1}}{1.8 \times 10^{-8} \times 1.6 \times 10^7} = 0.35 \qquad [NH_3] = 0.59\,M$$

The concentration just calculated is that of *free, uncomplexed* NH_3. Considering as well the 0.20 mol NH_3/L complexed in the 0.10 M $[Ag(NH_3)_2]^+$, we see that the total concentration of $NH_3(aq)$ required is

$$[NH_3]_{tot} = 0.59\,M + 0.20\,M = 0.79\,M$$

Assess

In this example the precipitation from a solution is controlled by using a sufficiently large concentration of a complexing agent (i.e., NH_3).

PRACTICE EXAMPLE A: What $[NH_3]_{tot}$ is necessary to keep AgCl from precipitating from a solution that is 0.13 M $AgNO_3$ and 0.0075 M NaCl?

PRACTICE EXAMPLE B: What minimum concentration of thiosulfate ion, $S_2O_3^{2-}$, should be present in 0.10 M $AgNO_3(aq)$ so that $AgCl(s)$ does not precipitate when the solution is also made 0.010 M in Cl^-? For AgCl, $K_{sp} = 1.8 \times 10^{-10}$; for $[Ag(S_2O_3)_2]^{3-}$, $K_f = 1.7 \times 10^{13}$.

On page 799, there is a qualitative description of how the solubility of AgCl increases in the presence of $NH_3(aq)$. Example 18-13 shows how to calculate the actual solubility of AgCl in $NH_3(aq)$.

🔍 18-6   CONCEPT ASSESSMENT

(1) The measured molar solubility of AgCl in water is 1.3×10^{-5} M. In the presence of $Cl^-(aq)$ as a common ion at different concentrations, the measured solubilities of AgCl are as listed below.

$[Cl^-]$, M:	0.0039	0.036	0.35	1.4	2.9	3.8
Solubility of AgCl, M:	7.2×10^{-7}	1.9×10^{-6}	1.7×10^{-5}	1.8×10^{-4}	1.0×10^{-2}	2.5×10^{-2}

Give a plausible explanation for the trends in the molar solubilities of the AgCl.

(2) When we add sodium hydroxide to a lead(II) chloride solution, what species are formed?

EXAMPLE 18-13   Determining the Solubility of a Solute When Complex Ions Form

What is the molar solubility of AgCl in 0.100 M NH_3(aq)?

Analyze

We need to have the equilibrium constant for the reaction that forms $[Ag(NH_3)_2]^+$ from AgCl(s) and NH_3(aq). We can determine the equilibrium constant from the product of K_{sp} for AgCl(s) and K_f for the formation of $[Ag(NH_3)_2]^+$. By using that equilibrium constant, we can then determine the molar solubility.

Solve

As we have already seen, equation (18.5) describes the solubility equilibrium.

$$AgCl(s) + 2\,NH_3(aq) \rightleftharpoons [Ag(NH_3)_2]^+(aq) + Cl^-(aq) \tag{18.5}$$

Let's base our calculation on the equilibrium constant K for reaction (18.5). There are two ways to obtain this value. By one method, equation (18.5) is written as the sum of equations (18.6) and (18.7) on page 800. Then its K value is obtained as the product of a K_{sp} and a K_f.

$$AgCl(s) \rightleftharpoons Ag^+(aq) + Cl^-(aq) \qquad K_{sp} = 1.8 \times 10^{-10}$$
$$\underline{Ag^+(aq) + 2\,NH_3(aq) \rightleftharpoons [Ag(NH_3)_2]^+(aq) \qquad K_f = 1.6 \times 10^7}$$
$$AgCl(s) + 2\,NH_3(aq) \rightleftharpoons [Ag(NH_3)_2]^+(aq) + Cl^-(aq)$$

$$K = K_{sp} \times K_f = 1.8 \times 10^{-10} \times 1.6 \times 10^7 = 2.9 \times 10^{-3}$$

By a second method, the equilibrium constant expression for reaction (18.5) is written first, and then the numerator and denominator are multiplied by $[Ag^+]$.

$$K = \frac{[Ag(NH_3)_2^+][Cl^-]}{[NH_3]^2} = \frac{[Ag(NH_3)_2^+][Cl^-][Ag^+]}{[NH_3]^2[Ag^+]} = K_f \times K_{sp} = 2.9 \times 10^{-3}$$

The expression printed in red is K_f for $[Ag(NH_3)_2]^+$; the one in blue is K_{sp} for AgCl. The K value for reaction (18.5) is the product of the two.

According to equation (18.5), if s mol AgCl/L dissolves (the molar solubility), the expected concentrations of $[Ag(NH_3)_2]^+$ and Cl^- are also equal to s.

$$K = \frac{[[Ag(NH_3)_2]^+][Cl^-]}{[NH_3]^2} = \frac{s \cdot s}{(0.100 - 2s)^2} = \left(\frac{s}{0.100 - 2s}\right)^2 = 2.9 \times 10^{-3}$$

We can solve this equation by taking the square root of both sides.

$$\frac{s}{0.100 - 2s} = \sqrt{2.9 \times 10^{-3}} = 5.4 \times 10^{-2}$$

The molar solubility of AgCl(s) in 0.100 M NH_3(aq) is 4.9×10^{-3} M.

$$s = 5.4 \times 10^{-3} - 0.11s$$
$$1.11s = 5.4 \times 10^{-3}$$
$$s = 4.9 \times 10^{-3}$$

Assess

The usual simplifying assumption, that is, $(0.100 - 2s) \approx 0.100$, would not have worked well in this calculation. If the simplifying assumption had been made, the value of s obtained would have been 5.4×10^{-3} M and $2s$ would have been 10.8% of 0.100. That is, $0.100 - (2 \times 0.0054) \neq 0.100$. Also, the molar solubility is actually the *total* concentration of silver in solution: $[Ag^+] + [[Ag(NH_3)_2]^+]$. Only when K_f is large and the concentration of complexing agent sufficiently high can we ignore the concentration of uncomplexed metal ion, as was the case here.

PRACTICE EXAMPLE A: What is the molar solubility of $Fe(OH)_3$ in a solution containing 0.100 M $C_2O_4^{2-}$? For $[Fe(C_2O_4)_3]^{3-}$, $K_f = 2 \times 10^{20}$.

PRACTICE EXAMPLE B: *Without doing detailed calculations*, show that the order of *decreasing* solubility in 0.100 M NH_3(aq) should be AgCl > AgBr > AgI.

18-9   Qualitative Cation Analysis

In *qualitative analysis*, we determine what substances are present in a mixture but *not* their quantities. An analysis that aims at identifying the cations present in a mixture is called **qualitative cation analysis**. Qualitative cation analysis provides us with many examples of precipitation (and dissolution) equilibria, acid–base equilibria, and oxidation–reduction reactions. Also, in the general chemistry laboratory it offers the challenge of unraveling a mystery—solving a qualitative analysis "unknown."

In the scheme in Figure 18-7, about 25 common cations are divided into five groups, depending on differing solubilities of their compounds. The first cations separated, Pb^{2+}, Hg_2^{2+}, and Ag^+, are those with insoluble *chlorides*. The reagent used is HCl(aq). All other cations remain in solution because their chlorides are soluble. After the chloride group precipitate is removed, the remaining solution is treated with H_2S in an acidic medium. Under these conditions, a group of sulfides precipitates. They are known as the hydrogen sulfide group. Next, the solution containing the remaining cations is treated with H_2S in a buffer mixture of ammonia and ammonium ion, yielding a mixture of insoluble hydroxides and sulfides.

◀ In the Hg_2^{2+} ion, a covalent bond links a Hg atom to a Hg^{2+} ion.

This group is called the ammonium sulfide group. The sulfide of aluminum(III) and chromium(III) are unstable, reacting with water to form the hydroxides.

Treatment of the ammonium sulfide group filtrate with CO_3^{2-} yields the fourth group precipitate, which consists of the carbonates of Mg^{2+}, Ca^{2+}, Sr^{2+}, and Ba^{2+}. It is called the carbonate group because the aqueous carbonate anion is the key reagent. At the end of this series of precipitations, the resulting

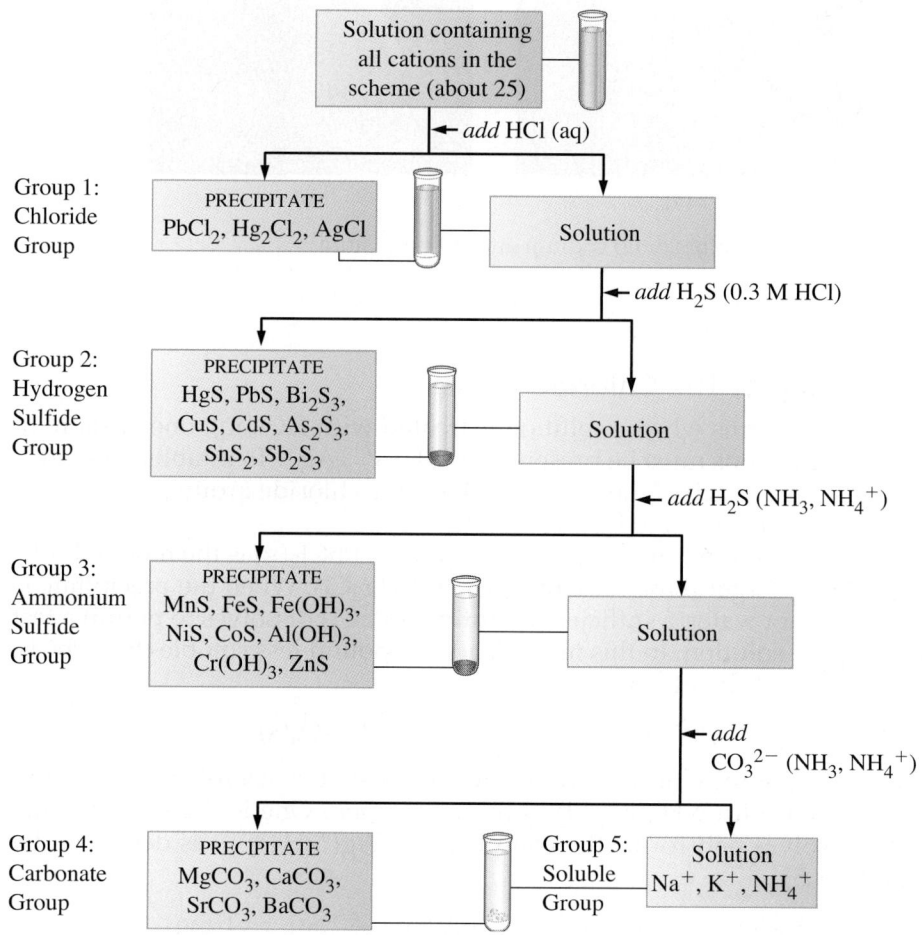

◀ FIGURE 18-7
Outline of a qualitative cation analysis
Various aspects of this scheme are described in the text. A sample containing all 25 cations can be separated into five groups by the indicated reagents.

solution contains only Na^+, K^+, and NH_4^+, all of whose common salts are water soluble.

In this section, we will discuss the chemistry of the chloride and sulfide groups. The chemistry of metal carbonates is discussed in Chapter 21.

18-3 ARE YOU WONDERING...

How to test for the presence of Na^+, K^+, and NH_4^+ cations?

Tests by precipitation are difficult because the salts of these cations exhibit near-universal solubility. Both Na^+ and K^+ ions are most easily detected with a flame test. When a solution containing sodium ions is brought into contact with a flame, the characteristic orange-yellow color of the emission spectrum of sodium atoms is observed. For potassium atoms, a pale violet color results. To detect the presence of NH_4^+, we use the fact that the ammonium ion is the conjugate acid of the weak volatile base ammonia. Heating some of the original solution (not the final solution, which contains NH_4^+ ions added in the fractional precipitation scheme) with excess strong base will liberate ammonia.

$$NH_4^+(aq) + OH^- \longrightarrow NH_3(g) + H_2O(l)$$

The NH_3 is detected by its characteristic odor and its effect on the color of an acid–base indicator such as litmus.

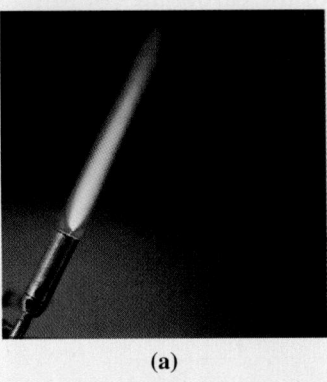 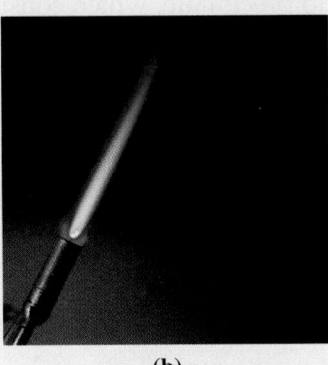

    (a)   (b)

▲ Flame colors of (a) sodium and (b) potassium.

▶ The mechanism of light emission in flame tests is presented on page 921.

  (a)   (b)   (c)

▲ FIGURE 18-8
Chloride group precipitates
(a) Group 1 precipitate: a mixture of $PbCl_2$ (white), Hg_2Cl_2 (white), and AgCl (white). (b) Test for Hg_2^{2+}: a mixture of Hg (black) and $HgNH_2Cl$ (white). (c) Test for Pb^{2+}: a yellow precipitate of $PbCrO_4$(s).

Cation Group 1: The Chloride Group

If a precipitate forms when a solution is treated with HCl(aq), one or more of the following cations must be present: Pb^{2+}, Hg_2^{2+}, Ag^+. To establish the presence or absence of each of these three cations, the chloride group precipitate is filtered off and subjected to further testing.

Of the three chlorides in the group precipitate, $PbCl_2$(s) is the most soluble; its K_{sp} is much larger than those of AgCl and Hg_2Cl_2. When the precipitate is washed with hot water, a sufficient quantity of $PbCl_2$ dissolves to permit a test for Pb^{2+} in the solution. In this test, a lead compound less soluble than $PbCl_2$, such as lead(II) chromate, precipitates (Fig. 18-8).

$$Pb^{2+}(aq) + CrO_4^{2-}(aq) \longrightarrow PbCrO_4(s)$$

The portion of the chloride group precipitate that is insoluble in hot water is then treated with NH_3(aq). Two things happen. One is that any AgCl(s) present dissolves and forms the complex ion $[Ag(NH_3)_2^+]$, as described by equation (18.5).

$$AgCl(s) + 2 NH_3(aq) \longrightarrow [Ag(NH_3)_2]^+(aq) + Cl^-(aq)$$

At the same time, any $Hg_2Cl_2(s)$ present undergoes an oxidation–reduction reaction. One of the products of the reaction is finely divided black mercury. A black color is common for finely divided metals.

$$Hg_2Cl_2(s) + 2\,NH_3(aq) \longrightarrow \underbrace{Hg(l) + HgNH_2Cl(s)}_{\text{Dark gray}} + NH_4{}^+(aq) + Cl^-(aq)$$

The appearance of a dark gray mixture of black mercury and white $HgNH_2Cl$ [mercury(II) amidochloride] is the qualitative analysis confirmation of mercury(I) ion (see Figure 18-8).

When the solution from reaction (18.5) is acidified with $HNO_3(aq)$, any silver ion present precipitates as $AgCl(s)$. This is the reaction predicted in Example 18-10 and pictured in Figure 18-6.

18-7 CONCEPT ASSESSMENT

A qualitative analysis chloride group unknown forms a white precipitate with HCl(aq). The precipitate is treated with hot water, yielding a solution that gives a yellow precipitate with K_2CrO_4(aq). There is no change in color when the undissolved portion of the precipitate is treated with NH_3(aq). Indicate for each chloride group cation whether it is present or absent, or about which there is some doubt.

Cation Groups 2 and 3: Equilibria Involving Hydrogen Sulfide

Figure 18-7 suggests that aqueous hydrogen sulfide (hydrosulfuric acid) is the key reagent in the analysis of cation groups 2 and 3. In aqueous solution, H_2S is a *weak diprotic acid*.

$$H_2S(aq) + H_2O(l) \rightleftharpoons H_3O^+(aq) + HS^-(aq) \qquad K_{a_1} = 1.0 \times 10^{-7}$$
$$HS^-(aq) + H_2O(l) \rightleftharpoons H_3O^+(aq) + S^{2-}(aq) \qquad K_{a_2} = 1 \times 10^{-19}$$

The extremely small value of K_{a_2} suggests that sulfide ion is a very strong base, as seen in the following hydrolysis reaction and K_b value.

$$S^{2-} + H_2O \rightleftharpoons HS^- + OH^- \qquad K_b = K_w/K_{a_2}$$
$$= (1.0 \times 10^{-14})/(1 \times 10^{-19}) = 1 \times 10^5$$

The hydrolysis of S^{2-} should go nearly to completion, which means that very little S^{2-} can exist in an aqueous solution and that sulfide ion is probably not the precipitating agent for sulfides.

One way to handle the precipitation and dissolution of sulfide precipitates is to restrict the discussion to acidic solutions. Then we can write an equilibrium constant expression in which the concentration terms for HS^- and S^{2-} are eliminated. This approach is reasonable because most sulfide separations are carried out in acidic solution.

Consider (1) the solubility equilibrium equation for PbS written to reflect hydrolysis of S^{2-}, (2) an equation written for the reverse of the first ionization of H_2S, and (3) an equation that is the reverse of the self-ionization of water. These three equations can be combined into an overall equation that shows the dissolving of PbS(s) in an acidic solution. The equilibrium constant for this overall equation is generally referred to as K_{spa}.*

> ◀ $H_2S(g)$ has a familiar rotten egg odor, especially noticeable in volcanic areas and near sulfur hot springs. Because of the smell, any gas containing a significant amount of H_2S (4 ppm or more) is called "sour gas."

(1) $\quad$ $PbS(s) + H_2O(l) \rightleftharpoons Pb^{2+}(aq) + HS^-(aq) + OH^-(aq)$ $\qquad K_{sp} = 3 \times 10^{-28}$

(2) $\quad H_3O^+(aq) + HS^-(aq) \rightleftharpoons H_2S(aq) + H_2O(l)$ $\qquad 1/K_{a_1} = 1/1.0 \times 10^{-7}$

(3) $\quad H_3O^+(aq) + OH^-(aq) \rightleftharpoons H_2O(l) + H_2O(l)$ $\qquad 1/K_w = 1/1.0 \times 10^{-14}$

Overall: $PbS(s) + 2\,H_3O^+(aq) \rightleftharpoons Pb^{2+}(aq) + H_2S(aq) + 2\,H_2O(l)$ $\qquad K_{spa} = ?$

*See Myers, R. J. *J. Chem. Educ.* **63**, 687 (1986).

▶ $PbCl_2$ is sufficiently soluble that enough Pb^{2+} ions from cation group 1 remain in solution to precipitate again as $PbS(s)$ in cation group 2.

$$K_{spa} = \frac{K_{sp}}{K_{a_1} \times K_w} = \frac{3 \times 10^{-28}}{(1.0 \times 10^{-7})(1.0 \times 10^{-14})} = 3 \times 10^{-7}$$

Example 18-14 illustrates the use of K_{spa} in the type of calculation needed to sort the sulfides into two qualitative analysis groups. Pb^{2+} is in qualitative analysis cation group 2, and Fe^{2+} is in group 3. The conditions cited in the example are those generally used.

🔍 18-8 CONCEPT ASSESSMENT

The dissolution of $PbS(s)$ in water is given by

$$PbS(s) + H_2O(l) = Pb^{2+}(aq) + HS^- = OH^-(aq)$$

In an apparent violation of Le Châtelier's principle, the addition of NaOH to the solution leads more PbS to dissolve. Explain.

EXAMPLE 18-14 Separating Metal Ions by Selective Precipitation of Metal Sulfides

Show that $PbS(s)$ will precipitate but $FeS(s)$ will not precipitate from a solution that is 0.010 M in Pb^{2+}, 0.010 M in Fe^{2+}, saturated in H_2S (0.10 M H_2S), and maintained with $[H_3O^+] = 0.30$ M. For PbS, $K_{spa} = 3 \times 10^{-7}$; for FeS, $K_{spa} = 6 \times 10^2$.

Analyze

We must determine whether, for the stated conditions, equilibrium is displaced in the forward or reverse direction in reactions of the type

$$MS(s) + 2 H_3O^+(aq) \rightleftharpoons M^{2+}(aq) + H_2S(aq) + 2 H_2O(l) \qquad \textbf{(18.8)}$$

where M represents either Pb or Fe. In each case, we can compare the Q_{spa} expression to the appropriate K_{spa} value for reaction (18.8). If $Q_{spa} > K_{spa}$, a net change will occur to the *left* and $MS(s)$ will precipitate. If $Q_{spa} < K_{spa}$, a net change will occur to the *right*.

Solve

Calculating Q_{spa} we have

$$Q_{spa} = \frac{[M^{2+}][H_2S]}{[H_3O^+]^2} = \frac{0.010 \times 0.10}{(0.30)^2} = 1.1 \times 10^{-2}$$

For PbS: Q_{spa} of $1.1 \times 10^{-2} > K_{spa}$ of 3×10^{-7}. $PbS(s)$ will precipitate.

For FeS: Q_{spa} of $1.1 \times 10^{-2} < K_{spa}$ of 6×10^2. $FeS(s)$ will not precipitate.

Assess

The results here show that using hydrogen sulfide in an acidic medium is a good way to separate cations of groups 2 and 3. You may be wondering how the value for the K_{spa} for FeS was determined. The value of K_{spa} of FeS can be derived from K_{sp} of FeS (6×10^{-19}) by the method outlined for PbS.

PRACTICE EXAMPLE A: Show that $Ag_2S(s)$ ($K_{spa} = 6 \times 10^{-30}$) should precipitate and that $FeS(s)$ ($K_{spa} = 6 \times 10^2$) should not precipitate from a solution that is 0.010 M Ag^+ and 0.020 M Fe^{2+}, but otherwise under the same conditions as in Example 18-14.

PRACTICE EXAMPLE B: What is the minimum pH of a solution that is 0.015 M Fe^{2+} and saturated in H_2S (0.10 M) from which $FeS(s)$ ($K_{spa} = 6 \times 10^2$) can be precipitated?

🔍 18-9 CONCEPT ASSESSMENT

Both Cu^{2+} and Ag^+ are present in the same aqueous solution. Explain which of the following reagents would work best in separating these ions, precipitating one and leaving the other in solution: $(NH_4)_2CO_3(aq)$, $HNO_3(aq)$, $H_2S(aq)$, $HCl(aq)$, $NH_3(aq)$, or $NaOH(aq)$.

Dissolving Metal Sulfides

In the qualitative cation analysis, it is necessary to both precipitate and redissolve sulfides. Here, we look at several methods of dissolving metal sulfides. One way to increase the solubility of any sulfide is to allow it to react with an acid, as equation (18.8) suggests. According to Le Châtelier's principle, the solubility increases as the solution is made more acidic—equilibrium is shifted to the right. As a result, some water-insoluble sulfides, such as FeS, are readily soluble in strongly acidic solutions. Others, such as PbS and HgS, cannot be dissolved in acidic solutions because their K_{sp} values are too low. In these cases, $[H_3O^+]$ cannot be made high enough to force reaction (18.8) very far to the right.

Another way to promote the dissolving of metal sulfides is to use an *oxidizing* acid such as $HNO_3(aq)$. In this case, sulfide ion is oxidized to elemental sulfur and the free metal ion appears in solution, as in the dissolving of $CuS(s)$.

$$3\,CuS(s) + 8\,H^+(aq) + 2\,NO_3^-(aq) \longrightarrow$$
$$3\,Cu^{2+}(aq) + 3\,S(s) + 2\,NO(g) + 4\,H_2O(l) \quad \textbf{(18.9)}$$

To render the $Cu^{2+}(aq)$ more visible, it is converted to the deeply colored complex ion $[Cu(NH_3)_4]^{2+}(aq)$ in a reaction in which NH_3 molecules replace H_2O molecules in a complex ion (Fig. 18-9).

$$[(Cu(H_2O)_4]^{2+}(aq) + 4\,NH_3(aq) \longrightarrow [Cu(NH_3)_4]^{2+}(aq) + 4\,H_2O(l)$$

Pale blue   Deep violet

A few metal sulfides dissolve in a basic solution with a high concentration of HS^-, just as acidic oxides dissolve in solutions with a high concentration of OH^-. This property is used to advantage in separating the eight sulfides of cation group 2, the hydrogen sulfide group, into two subgroups. The subgroup consisting of HgS, PbS, CuS, CdS, and Bi_2S_3 remains unchanged after treatment with an alkaline solution with an excess of HS^-, but As_2S_3, SnS_2, and Sb_2S_3 dissolve.

▲ FIGURE 18-9
Complex-ion formation: A test for $Cu^{2+}(aq)$
Dilute $CuSO_4(aq)$ (left) derives its pale blue color from the complex ions $[Cu(H_2O)_4]^{2+}$. When $NH_3(aq)$ is added (here labeled "conc ammonium hydroxide"), the color changes to a deep violet, signaling the presence of $[Cu(NH_3)_4]^{2+}$ (right). The deep violet color is detectable at much lower concentrations than the pale blue; the formation of $[Cu(NH_3)_4]^{2+}$ is a sensitive test for the presence of Cu^{2+}.

KEEP IN MIND

that $Cu^{2+}(aq)$ means that the copper (II) ion will be coordinated to several water molecules (see page 733).

The formation of some of the Earth's minerals has been by chemical precipitation. Biological precipitation is responsible for the formation of certain marine shells. The Focus On feature for Chapter 18, Shells, Teeth, and Fossils, on the MasteringChemistry site describes the relationship between shells, teeth, and fossils.

Summary

18-1 Solubility Product Constant, K_{sp}—The equilibrium constant for the equilibrium between a solid ionic solute and its ions in a saturated aqueous solution is expressed through the **solubility product constant, K_{sp}** (expression 18.1, Table 18.1).

18-2 Relationship Between Solubility and K_{sp}—A solute's molarity in a saturated aqueous solution is known as its molar solubility. Molar solubility and K_{sp} are related to each other in a way that makes it possible to calculate one when the other is known.

18-3 Common-Ion Effect in Solubility Equilibria—The solubility of a slightly soluble ionic solute is greatly reduced in an aqueous solution containing an ion in common with the solubility equilibrium—a common ion.

18-4 Limitations of the K_{sp} Concept—The solubility product concept is most useful for slightly soluble ionic solutes, and especially in cases where activities can be replaced by molar concentrations. Moderately or highly soluble ionic compounds require the use of activities. The presence of common ions reduces the solubility of a slightly soluble ionic solute, generally to a significant degree.

Other factors also affect solute solubility. The presence of noncommon or diverse ions generally increases solute solubility, an effect called the **salt effect**. Solute activities, and hence solute solubilities, are strongly influenced by interionic attractions, which become especially significant at higher concentrations and for highly charged ions. Dissociation of ionic solutes may not be 100% complete, leading to the formation of **ion pairs** which act as single units, such as MgF^+ in $MgF_2(aq)$.

18-5 Criteria for Precipitation and Its Completeness—To determine whether a slightly soluble solute will precipitate from a solution, the **ion product, Q_{sp}**, is compared with the solubility product constant, K_{sp}. Q_{sp} is based on the initial ion concentrations in a solution. K_{sp}, on the other hand, is based on the equilibrium ion concentrations in a saturated solution. If $Q_{sp} > K_{sp}$, precipitation will

occur; if $Q_{sp} < K_{sp}$, the solution will remain unsaturated. When $Q_{sp} = K_{sp}$, the solution is just saturated.

18-6 Fractional Precipitation—A comparison of K_{sp} values is a factor in determining the feasibility of **fractional precipitation**, a process in which one ionic species is removed by precipitation while others remain in solution.

18-7 Solubility and pH—The solubility of a slightly soluble solute is affected by pH if the anion is OH^- or derived from a weak acid. The solubility increases as the pH is lowered or decreases as the pH is raised. This can be illustrated through Le Châtelier's principle. Also, an equilibrium constant for the dissolution reaction can be obtained by combining the solubility equilibrium equation and the ionization equilibrium equation of the weak electrolyte. Some slightly soluble solutions form **suspensions**, which are heterogeneous fluids containing solid particles that will eventually settle.

18-8 Equilibria Involving Complex Ions—A **complex ion** is a polyatomic ion composed of a central metal ion bonded to two or more molecules or ions called ligands. The formation of a complex ion is an equilibrium process with an equilibrium constant called the **formation constant, K_f**. In general, if the formation constant is large, the concentration of uncomplexed metal ion in equilibrium with the complex ion is very small. Complex-ion formation can render certain insoluble materials quite soluble in appropriate aqueous solutions, such as $AgCl$ in $NH_3(aq)$. A complex ion is either a cation or an anion depending upon the particular central metal ion and the particular ligands. When a complex ion and an oppositely charged ion combine they form a **coordination compound**.

18-9 Qualitative Cation Analysis—Precipitation, acid–base, oxidation–reduction, and complex-ion formation reactions are all used extensively in **qualitative cation analysis**. Such an analysis can provide a rapid means of determining the presence or absence of certain cations in an unknown material.

Integrative Example

Lime (quicklime), CaO, is obtained from the high-temperature decomposition of limestone ($CaCO_3$). Quicklime is the cheapest source of basic substances, but it is water insoluble. It does react with water, however, producing $Ca(OH)_2$ (slaked lime). Unfortunately, $Ca(OH)_2(s)$ has limited solubility, so it cannot be used to prepare aqueous solutions of high pH.

$$Ca(OH)_2(s) \rightleftharpoons Ca^{2+}(aq) + 2\,OH^-(aq) \qquad K_{sp} = 5.5 \times 10^{-6}$$

When $Ca(OH)_2(s)$ reacts with a *soluble* carbonate, such as $Na_2CO_3(aq)$, however, the solution produced has a much higher pH. Equilibrium is displaced to the right in reaction (18.10) because $CaCO_3$ is much less soluble than $Ca(OH)_2$.

$$Ca(OH)_2(s) + CO_3^{2-}(aq) \rightleftharpoons CaCO_3(s) + 2\,OH^-(aq) \qquad \textbf{(18.10)}$$

Assume an initial $[CO_3^{2-}] = 1.0\,M$ in reaction (18.10), and show that the equilibrium pH should indeed be higher than that in saturated $Ca(OH)_2(aq)$.

Analyze

(1) Determine the pH of the saturated $Ca(OH)_2(aq)$. (2) Find the equilibrium constant, K, for reaction (18.10). (3) Calculate the equilibrium $[OH^-]$ in reaction (18.10). (4) Convert $[OH^-]$ to pOH and then to pH. Compare the pH in steps (1) and (4).

Solve

Write the K_{sp} expression for $Ca(OH)_2$. Let s be the molar solubility, so the $[OH^-] = 2s$.

$$K_{sp} = [Ca^{2+}][OH^-]^2 = (s)(2s)^2 = 4s^3 = 5.5 \times 10^{-6}$$
$$s = (5.5 \times 10^{-6}/4)^{1/3} = 0.011 \text{ M}$$
$$[OH^-] = 2s = 0.022 \text{ M}$$

Calculate the pOH.

$$pOH = -\log[OH^-] = -\log 0.022$$

Determine the pH from the pOH by subtracting pOH from 14.00.

$$pOH = 1.66 \quad pH = 14.00 - pOH = 14.00 - 1.66 = 12.34$$

Combine solubility equilibrium equations for $Ca(OH)_2$ and $CaCO_3$ to obtain the net ionic equation (18.10).

$$Ca(OH)_2(s) \rightleftharpoons Ca^{2+}(aq) + 2OH^-(aq) \quad K_{sp} = 5.5 \times 10^{-6}$$
$$Ca^{2+}(aq) + CO_3^{2-}(aq) \rightleftharpoons CaCO_3(s) \quad 1/K_{sp} = 1/(2.8 \times 10^{-9})$$
$$Ca(OH)_2(s) + CO_3^{2-}(aq) \rightleftharpoons CaCO_3(s) + 2OH^-(aq) \quad \textbf{(18.10)}$$

Use the K_{sp} values of $Ca(OH)_2$ and $CaCO_3$ to obtain the overall K.

$$K = \frac{K_{sp}[Ca(OH)_2]}{K_{sp}[CaCO_3]}$$
$$K = 5.5 \times 10^{-6}/2.8 \times 10^{-9} = 2.0 \times 10^3$$

Next, calculate the equilibrium $[OH^-]$ in reaction (18.10), starting with $[CO_3^{2-}] = 1.0$ M, and proceeding in the familiar fashion to obtain a quadratic equation. The solution, to two significant figures, is

	$Ca(OH)_2(s) + CO_3^{2-}(aq) \rightleftharpoons$	$CaCO_3(s) + 2OH^-(aq)$
initial concns, M:	1.0	≈ 0
changes, M:	$-x$	$+2x$
equil concns, M:	$1.0 - x$	$2x$

$$K = \frac{[OH^-]^2}{[CO_3^{2-}]} = \frac{(2x)^2}{(1.0 - x)} = 2.0 \times 10^3$$
$$4x^2 + (2.0 \times 10^3)x - 2.0 \times 10^3 = 0$$
$$x = 1.0 \text{ M}$$

From the value of x, obtain $[OH^-]$, pOH, and pH.

$$[OH^-] = 2x = 2.0 \text{ M}$$
$$pOH = -\log[OH^-] = -\log 2.0 = -0.30$$
$$pH = 14.00 - pOH = 14.00 + 0.30 = 14.30$$

Assess

We have succeeded in showing that the solution produced by reaction (18.10) has a higher pH than that found in saturated $Ca(OH)_2(aq)$, that is, a pH of 14.30 compared with 12.34. An interesting aspect of this calculation concerns solution of the quadratic equation $4x^2 + (2.0 \times 10^3)x - 2.0 \times 10^3 = 0$. By using the quadratic formula with no rounding of intermediate results, the value of $x = 0.998$, which rounds off to 1.0. Alternatively, note that by assuming that $x = 1.0$, the second and third terms of the equation cancel, yielding the result $4(1.0^2) + (2.0 \times 10^3)(1.0) - 2.0 \times 10^3 = 4.0$. We conclude that x must be very slightly less than 1.0. A value of $x = 0.998$ on the left side of the equation yields the result -0.016, very close to the 0 required for an exact solution.

PRACTICE EXAMPLE A: Heavy fertilizer use can lead to phosphate pollution in lakes, causing an explosion of plant growth, particularly algae. Excess algae deplete the lake of the oxygen necessary for other plant growth and animal life. A lake in the middle of a large farm was found to contain the phosphate ion in the concentration of 5.13×10^{-4} M. The lake measured 300 m $\times$ 150 m $\times$ 5 m. One method to neutralize PO_4^{3-} is to add a calcium salt. What mass of $Ca(NO_3)_2$ must be added to lower the phosphate ion concentration to 1.00×10^{-12} M? $K_{sp} = 1.30 \times 10^{-32}$ for $Ca_3(PO_4)_2$.

PRACTICE EXAMPLE B: In a laboratory procedure you are instructed to mix 350.0 mL of 0.200 M $AgNO_3(aq)$ and 250.0 mL of 0.240 M $Na_2SO_4(aq)$. What is the precipitate that you observe? To this mixture you add 400.0 mL of 0.500 M $Na_2S_2O_3$, causing the precipitate to dissolve. What is the mass of the remaining precipitate?

Exercises

(Use data from Chapters 16 and 18 and Appendix D, as needed.)

K_{sp} and Solubility

1. Write K_{sp} expressions for the following equilibria. For example, for the reaction $AgCl(s) \rightleftharpoons Ag^+(aq) + Cl^-(aq)$, $K_{sp} = [Ag^+][Cl^-]$.
   (a) $Ag_2SO_4(s) \rightleftharpoons 2 Ag^+(aq) + SO_4^{2-}(aq)$
   (b) $Ra(IO_3)_2(s) \rightleftharpoons Ra^{2+}(aq) + 2 IO_3^-(aq)$
   (c) $Ni_3(PO_4)_2(s) \rightleftharpoons 3 Ni^{2+}(aq) + 2 PO_4^{3-}(aq)$
   (d) $PuO_2CO_3(s) \rightleftharpoons PuO_2^{2+}(aq) + CO_3^{2-}(aq)$

2. Write solubility equilibrium equations that are described by the following K_{sp} expressions. For example, $K_{sp} = [Ag^+][Cl^-]$ represents $AgCl(s) \rightleftharpoons Ag^+(aq) + Cl^-(aq)$.
   (a) $K_{sp} = [Fe^{3+}][OH^-]^3$
   (b) $K_{sp} = [BiO^+][OH^-]$
   (c) $K_{sp} = [Hg_2^{2+}][I^-]^2$
   (d) $K_{sp} = [Pb^{2+}]^3[AsO_4^{3-}]^2$

3. The following K_{sp} values are found in a handbook. Write the solubility product expression to which each one applies. For example, $K_{sp}(AgCl) = [Ag^+][Cl^-] = 1.8 \times 10^{-10}$.
   (a) $K_{sp}(CrF_3) = 6.6 \times 10^{-11}$
   (b) $K_{sp}[Au_2(C_2O_4)_3] = 1 \times 10^{-10}$
   (c) $K_{sp}[Cd_3(PO_4)_2] = 2.1 \times 10^{-33}$
   (d) $K_{sp}(SrF_2) = 2.5 \times 10^{-9}$

4. Calculate the aqueous solubility, in moles per liter, of each of the following.
   (a) $BaCrO_4$, $K_{sp} = 1.2 \times 10^{-10}$
   (b) $PbBr_2$, $K_{sp} = 4.0 \times 10^{-5}$
   (c) CeF_3, $K_{sp} = 8 \times 10^{-16}$
   (d) $Mg_3(AsO_4)_2$, $K_{sp} = 2.1 \times 10^{-20}$

5. Arrange the following solutes in order of increasing molar solubility in water: $AgCN$, $AgIO_3$, AgI, $AgNO_2$, Ag_2SO_4. Explain your reasoning.

6. Which of the following saturated aqueous solutions would have the highest $[Mg^{2+}]$: (a) $MgCO_3$; (b) MgF_2; (c) $Mg_3(PO_4)_2$? Explain.

7. Fluoridated drinking water contains about 1 part per million (ppm) of F^-. Is CaF_2 sufficiently soluble in water to be used as the source of fluoride ion for the fluoridation of drinking water? Explain. [*Hint:* Think of 1 ppm as signifying 1 g F^- per 10^6 g solution.]

8. In the qualitative cation analysis procedure, Bi^{3+} is detected by the appearance of a white precipitate of bismuthyl hydroxide, $BiOOH(s)$:

   $BiOOH(s) \rightleftharpoons BiO^+(aq) + OH^-(aq)$
   $$K_{sp} = 4 \times 10^{-10}$$

   Calculate the pH of a saturated aqueous solution of $BiOOH$.

9. A solution is saturated with magnesium palmitate $[Mg(C_{16}H_{31}O_2)_2$, a component of bathtub ring] at 50 °C. How many milligrams of magnesium palmitate will precipitate from 965 mL of this solution when it is cooled to 25 °C? For $Mg(C_{16}H_{31}O_2)_2$, $K_{sp} = 4.8 \times 10^{-12}$ at 50 °C and 3.3×10^{-12} at 25 °C.

10. A 725 mL sample of a saturated aqueous solution of calcium oxalate, CaC_2O_4, at 95 °C is cooled to 13 °C. How many milligrams of calcium oxalate will precipitate? For CaC_2O_4, $K_{sp} = 1.2 \times 10^{-8}$ at 95 °C and 2.7×10^{-9} at 13 °C.

11. A 25.00 mL sample of a clear *saturated* solution of PbI_2 requires 13.3 mL of a certain $AgNO_3(aq)$ for its titration. What is the molarity of this $AgNO_3(aq)$?

    $$I^-(satd\ PbI_2) + Ag^+(from\ AgNO_3) \longrightarrow AgI(s)$$

12. A 250 mL sample of saturated $CaC_2O_4(aq)$ requires 4.8 mL of 0.00134 M $KMnO_4(aq)$ for its titration in an acidic solution. What is the value of K_{sp} for CaC_2O_4 obtained with these data? In the titration reaction, $C_2O_4^{2-}$ is oxidized to CO_2 and MnO_4^- is reduced to Mn^{2+}.

13. To precipitate as $Ag_2S(s)$, all the Ag^+ present in 338 mL of a saturated solution of $AgBrO_3$ requires 30.4 mL of $H_2S(g)$ measured at 23 °C and 748 mmHg. What is K_{sp} for $AgBrO_3$?

14. Excess $Ca(OH)_2(s)$ is shaken with water to produce a saturated solution. A 50.00 mL sample of the clear saturated solution is withdrawn and requires 10.7 mL of 0.1032 M HCl for its titration. What is K_{sp} for $Ca(OH)_2$?

The Common-Ion Effect

15. Calculate the molar solubility of $Mg(OH)_2$ ($K_{sp} = 1.8 \times 10^{-11}$) in (a) pure water; (b) 0.0862 M $MgCl_2$; (c) 0.0355 M KOH(aq).

16. How would you expect the presence of each of the following solutes to affect the molar solubility of $CaCO_3$ in water: (a) Na_2CO_3; (b) HCl; (c) $NaHSO_4$? Explain.

17. Describe the effects of the salts KI and $AgNO_3$ on the solubility of AgI in water.

18. Describe the effect of the salt KNO_3 on the solubility of AgI in water, and explain why it is different from the effects noted in Exercise 17.

19. A 0.150 M Na_2SO_4 solution that is saturated with Ag_2SO_4 has $[Ag^+] = 9.7 \times 10^{-3}$ M. What is the value of K_{sp} for Ag_2SO_4 obtained with these data?

20. If 100.0 mL of 0.0025 M $Na_2SO_4(aq)$ is saturated with $CaSO_4$, how many grams of $CaSO_4$ would be present in the solution? [*Hint: Does the usual simplifying assumption hold?*]

21. What $[Pb^{2+}]$ should be maintained in $Pb(NO_3)_2(aq)$ to produce a solubility of 1.5×10^{-4} mol PbI_2/L when $PbI_2(s)$ is added?

22. What $[I^-]$ should be maintained in KI(aq) to produce a solubility of 1.5×10^{-5} mol PbI_2/L when $PbI_2(s)$ is added?

23. Can the solubility of Ag_2CrO_4 be lowered to 5.0×10^{-8} mol Ag_2CrO_4/L by using $CrO_4{}^{2-}$ as the common ion? by using Ag^+? Explain.

24. A handbook lists the K_{sp} values 1.1×10^{-10} for $BaSO_4$ and 5.1×10^{-9} for $BaCO_3$. When saturated $BaSO_4(aq)$ is also made with 0.50 M $Na_2CO_3(aq)$, a precipitate of $BaCO_3(s)$ forms. How do you account for this fact, given that $BaCO_3$ has a larger K_{sp} than does $BaSO_4$?

25. A particular water sample that is saturated in CaF_2 has a Ca^{2+} content of 115 ppm (that is, 115 g Ca^{2+} per 10^6 g of water sample). What is the F^- ion content of the water in ppm?

26. Assume that, to be visible to the unaided eye, a precipitate must weigh more than 1 mg. If you add 1.0 mL of 1.0 M NaCl(aq) to 100.0 mL of a clear saturated aqueous AgCl solution, will you be able to see AgCl(s) precipitated as a result of the common-ion effect? Explain.

Criteria for Precipitation from Solution

27. Will precipitation of $MgF_2(s)$ occur if a 22.5 mg sample of $MgCl_2 \cdot 6\,H_2O$ is added to 325 mL of 0.035 M KF?

28. Will $PbCl_2(s)$ precipitate when 155 mL of 0.016 M KCl(aq) are added to 245 mL of 0.175 M $Pb(NO_3)_2(aq)$?

29. What is the minimum pH at which $Cd(OH)_2(s)$ will precipitate from a solution that is 0.0055 M in $Cd^{2+}(aq)$?

30. What is the minimum pH at which $Cr(OH)_3(s)$ will precipitate from a solution that is 0.086 M in $Cr^{3+}(aq)$?

31. Will precipitation occur in the following cases?
    (a) 0.10 mg NaCl is added to 1.0 L of 0.10 M $AgNO_3(aq)$.
    (b) One drop (0.05 mL) of 0.10 M KBr is added to 250 mL of a saturated solution of AgCl.
    (c) One drop (0.05 mL) of 0.0150 M NaOH(aq) is added to 3.0 L of a solution with 2.0 mg Mg^{2+} per liter.

32. The electrolysis of $MgCl_2(aq)$ can be represented as

    $$Mg^{2+}(aq) + 2\,Cl^-(aq) + 2\,H_2O(l) \longrightarrow$$
    $$Mg^{2+}(aq) + 2\,OH^-(aq) + H_2(g) + Cl_2(g)$$

    The electrolysis of a 315 mL sample of 0.185 M $MgCl_2$ is continued until 0.652 L $H_2(g)$ at 22 °C and 752 mmHg has been collected. Will $Mg(OH)_2(s)$ precipitate when electrolysis is carried to this point? [*Hint: Notice that $[Mg^{2+}]$ remains constant throughout the electrolysis, but $[OH^-]$ increases.*]

33. Determine whether 1.50 g $H_2C_2O_4$ (oxalic acid: $K_{a_1} = 5.2 \times 10^{-2}$, $K_{a_2} = 5.4 \times 10^{-5}$) can be dissolved in 0.200 L of 0.150 M $CaCl_2$ without the formation of $CaC_2O_4(s)$ ($K_{sp} = 1.3 \times 10^{-9}$).

34. If 100.0 mL of a clear saturated solution of Ag_2SO_4 is added to 250.0 mL of a clear saturated solution of $PbCrO_4$, will any precipitate form? [*Hint: Take into account the dilutions that occur. What are the possible precipitates?*]

Completeness of Precipitation

35. When 200.0 mL of 0.350 M $K_2CrO_4(aq)$ are added to 200.0 mL of 0.0100 M $AgNO_3(aq)$, what percentage of the Ag^+ is left *unprecipitated*?

36. What percentage of the original Ag^+ remains in solution when 175 mL 0.0208 M $AgNO_3$ is added to 250 mL 0.0380 M K_2CrO_4?

37. If a constant $[Cl^-] = 0.100$ M is maintained in a solution in which the initial $[Pb^{2+}] = 0.065$ M, what

percentage of the Pb^{2+} will remain in solution after $PbCl_2(s)$ precipitates? What $[Cl^-]$ should be maintained to ensure that only 1.0% of the Pb^{2+} remains unprecipitated?

38. The ancient Romans added calcium sulfate to wine to clarify it and to remove dissolved lead. What is the maximum $[Pb^{2+}]$ that might be present in wine to which calcium sulfate has been added?

Fractional Precipitation

39. Assume that the seawater sample described in Example 18-6 contains approximately 440 g Ca^{2+} per metric ton (1 metric ton = 10^3 kg; density of seawater = 1.03 g/mL).

(a) Should $Ca(OH)_2(s)$ precipitate from seawater under the stated conditions, that is, with $[OH^-] = 2.0 \times 10^{-3}$ M?
(b) Is the separation of Ca^{2+} from Mg^{2+} in seawater feasible?

40. Which one of the following solutions can be used to separate the cations in an aqueous solution in which $[Ba^{2+}] = [Ca^{2+}] = 0.050\,M$: $0.10\,M\ NaCl(aq)$, $0.05\,M\ Na_2SO_4(aq)$, $0.001\,M\ NaOH(aq)$, or $0.50\,M\ Na_2CO_3(aq)$? Explain why.

41. $KI(aq)$ is slowly added to a solution with $[Pb^{2+}] = [Ag^+] = 0.10\,M$. For PbI_2, $K_{sp} = 7.1 \times 10^{-9}$; for AgI, $K_{sp} = 8.5 \times 10^{-17}$.
   (a) Which precipitate should form first, PbI_2 or AgI?
   (b) What $[I^-]$ is required for the *second* cation to begin to precipitate?
   (c) What concentration of the first cation to precipitate remains in solution at the point at which the second cation begins to precipitate?
   (d) Can $Pb^{2+}(aq)$ and $Ag^+(aq)$ be effectively separated by fractional precipitation of their iodides?

42. A solution is $0.010\,M$ in both CrO_4^{2-} and SO_4^{2-}. To this solution, $0.50\,M\ Pb(NO_3)_2(aq)$ is slowly added.
   (a) Which anion will precipitate first from solution?
   (b) What is $[Pb^{2+}]$ at the point at which the second anion begins to precipitate?
   (c) Are the two anions effectively separated by this fractional precipitation?

43. An aqueous solution that $2.00\,M$ in $AgNO_3$ is slowly added from a buret to an aqueous solution that is $0.0100\,M$ in Cl^- and $0.250\,M$ in I^-.
   (a) Which ion, Cl^- or I^-, is the first to precipitate?
   (b) When the second ion begins to precipitate, what is the remaining concentration of the first ion?
   (c) Is the separation of Cl^- and I^- feasible by fractional precipitation in this solution?

44. $AgNO_3(aq)$ is slowly added to a solution that is $0.250\,M\ NaCl$ and also $0.0022\,M\ KBr$.
   (a) Which anion will precipitate first, Cl^- or Br^-?
   (b) What is $[Ag^+]$ at the point at which the second anion begins to precipitate?
   (c) Can the Cl^- and Br^- be separated effectively by this fractional precipitation?

Solubility and pH

45. Which of the following solids is (are) more soluble in an acidic solution than in pure water: KCl, $MgCO_3$, FeS, $Ca(OH)_2$, or C_6H_5COOH? Explain.

46. Which of the following solids is (are) more soluble in a basic solution than in pure water: $BaSO_4$, $H_2C_2O_4$, $Fe(OH)_3$, $NaNO_3$, or MnS? Explain.

47. The solubility of $Mg(OH)_2$ in a particular buffer solution is $0.65\,g/L$. What must be the pH of the buffer solution?

48. To $0.350\,L$ of $0.150\,M\ NH_3$ is added $0.150\,L$ of $0.100\,M\ MgCl_2$. How many grams of $(NH_4)_2SO_4$ should be present to prevent precipitation of $Mg(OH)_2(s)$?

49. For the equilibrium

$$Al(OH)_3(s) \rightleftharpoons Al^{3+}(aq) + 3\,OH^-(aq)$$
$$K_{sp} = 1.3 \times 10^{-33}$$

   (a) What is the *minimum pH* at which $Al(OH)_3(s)$ will precipitate from a solution that is $0.075\,M$ in Al^{3+}?
   (b) A solution has $[Al^{3+}] = 0.075\,M$ and $[CH_3COOH] = 1.00\,M$. What is the maximum quantity of $NaCH_3COO$ that can be added to $250.0\,mL$ of this solution before precipitation of $Al(OH)_3(s)$ begins?

50. Will the following precipitates form under the given conditions?
   (a) $PbI_2(s)$, from a solution that is $1.05 \times 10^{-3}\,M\ HI$, $1.05 \times 10^{-3}\,M\ NaI$, and $1.1 \times 10^{-3}\,M\ Pb(NO_3)_2$.
   (b) $Mg(OH)_2(s)$, from $2.50\,L$ of $0.0150\,M\ Mg(NO_3)_2$ to which is added 1 drop ($0.05\,mL$) of $6.00\,M\ NH_3$.
   (c) $Al(OH)_3(s)$ from a solution that is $0.010\,M$ in Al^{3+}, $0.010\,M\ CH_3COOH$, and $0.010\,M\ NaCH_3COO$.

Complex-Ion Equilibria

51. $PbCl_2(s)$ is considerably more soluble in $HCl(aq)$ than in pure water, but its solubility in $HNO_3(aq)$ is not much different from what it is in water. Explain this difference in behavior.

52. Which of the following would be most effective, and which would be least effective, in reducing the concentration of the complex ion $[Zn(NH_3)_4]^{2+}$ in a solution: HCl, NH_3, or NH_4Cl? Explain your choices.

53. In a solution that is $0.0500\,M$ in $[Cu(CN)_4]^{3-}$ and $0.80\,M$ in free CN^-, the concentration of Cu^+ is $6.1 \times 10^{-32}\,M$. Calculate K_f of $[Cu(CN)_4]^{3-}$.

$$Cu^+(aq) + 4\,CN^-(aq) \rightleftharpoons [Cu(CN)_4]^{3-}(aq) \quad K_f = ?$$

54. Calculate $[Cu^{2+}]$ in a $0.10\,M\ CuSO_4(aq)$ solution that is also $6.0\,M$ in free NH_3.

$$Cu^{2+}(aq) + 4\,NH_3(aq) \rightleftharpoons [Cu(NH_3)_4]^{2+}(aq)$$
$$K_f = 1.1 \times 10^{13}$$

55. Can the following ion concentrations be maintained in the same solution without a precipitate forming: $[[Ag(S_2O_3)_2]^{3-}] = 0.048\,M$, $[S_2O_3^{2-}] = 0.76\,M$, and $[I^-] = 2.0\,M$?

56. A solution is $0.10\,M$ in *free* NH_3, $0.10\,M$ in NH_4Cl, and $0.015\,M$ in $[Cu(NH_3)_4]^{2+}$. Will $Cu(OH)_2(s)$ precipitate from this solution? K_{sp} of $Cu(OH)_2$ is 2.2×10^{-20}.

57. A $0.10\,mol$ sample of $AgNO_3(s)$ is dissolved in $1.00\,L$ of $1.00\,M\ NH_3$. How many grams of KI can be dissolved in this solution without a precipitate of $AgI(s)$ forming?

58. A solution is prepared that has $[NH_3] = 1.00\,M$ and $[Cl^-] = 0.100\,M$. How many grams of $AgNO_3$ can be dissolved in $1.00\,L$ of this solution without a precipitate of $AgCl(s)$ forming?

Precipitation and Solubilities of Metal Sulfides

59. Can Fe^{2+} and Mn^{2+} be separated by precipitating FeS(s) and not MnS(s)? Assume $[Fe^{2+}] = [Mn^{2+}] = [H_2S] = 0.10$ M. Choose a $[H_3O^+]$ that ensures maximum precipitation of FeS(s) but not MnS(s). Will the separation be complete? For FeS, $K_{spa} = 6 \times 10^2$; for MnS, $K_{spa} = 3 \times 10^7$.

60. A solution is 0.05 M in Cu^{2+}, in Hg^{2+}, and in Mn^{2+}. Which sulfides will precipitate if the solution is made to be 0.10 M H_2S(aq) and 0.010 M HCl(aq)? For CuS, $K_{spa} = 6 \times 10^{-16}$; for HgS, $K_{spa} = 2 \times 10^{-32}$; for MnS, $K_{spa} = 3 \times 10^7$.

61. A buffer solution is 0.25 M $CH_3COOH - 0.15$ M $NaCH_3COO$, saturated in H_2S (0.10 M), and with $[Mn^{2+}] = 0.15$ M.
(a) Show that MnS will *not* precipitate from this solution (for MnS, $K_{spa} = 3 \times 10^7$).

Qualitative Cation Analysis

63. Suppose you did a group 1 qualitative cation analysis and treated the chloride precipitate with NH_3(aq) without first treating it with hot water. What might you observe, and what valid conclusions could you reach about cations present, cations absent, and cations in doubt?

64. Show that in qualitative cation analysis group 1, if you obtain 1.00 mL of saturated $PbCl_2$(aq) at 25 °C, sufficient Pb^{2+} should be present to produce a precipitate of $PbCrO_4$(s). Assume that you use 1 drop (0.05 mL) of 1.0 M K_2CrO_4 for the test.

65. The addition of HCl(aq) to a solution containing several different cations produces a white precipitate. The filtrate is removed and treated with H_2S(aq) in 0.3 M HCl. No precipitate forms. Which of the following conclusions is (are) valid? Explain.

(b) Which buffer component would you increase in concentration, and to what minimum value, to ensure that precipitation of MnS(s) begins? Assume that the concentration of the other buffer component is held constant. [*Hint*: Recall equation (18.8).]

62. The following expressions pertain to the precipitation or dissolving of metal sulfides. Use information about the qualitative cation analysis scheme to predict whether a reaction proceeds to a significant extent in the forward direction and what the products are in each case.
(a) Cu^{2+}(aq) + H_2S (satd aq) $\longrightarrow$
(b) Mg^{2+}(aq) + H_2S (satd aq) $\xrightarrow{0.3\,M\,HCl}$
(c) PbS(s) + HCl (0.3 M) $\longrightarrow$
(d) ZnS(s) + HNO_3(aq) $\longrightarrow$

(a) Ag^+ or Hg_2^{2+} (or both) is probably present.
(b) Mg^{2+} is probably not present.
(c) Pb^{2+} is probably not present.
(d) Fe^{2+} is probably not present.

66. Write net ionic equations for the following qualitative cation analysis procedures.
(a) precipitation of $PbCl_2$(s) from a solution containing Pb^{2+}
(b) dissolution of $Zn(OH)_2$(s) in a solution of NaOH(aq)
(c) dissolution of $Fe(OH)_3$(s) in HCl(aq)
(d) precipitation of CuS(s) from an acidic solution of Cu^{2+} and H_2S

Integrative and Advanced Exercises

(Use data from Chapters 16 and 18 and Appendix D as needed.)

67. A particular water sample has 131 ppm of $CaSO_4$ (131 g $CaSO_4$ per 10^6 g water). If this water is boiled in a teakettle, approximately what fraction of the water must be evaporated before $CaSO_4$(s) begins to precipitate? Assume that the solubility of $CaSO_4$(s) does not change much in the temperature range 0 to 100 °C.

68. A handbook lists the solubility of $CaHPO_4$ as 0.32 g $CaHPO_4 \cdot 2\,H_2O/L$ and lists K_{sp} as 1×10^{-7}.

$$CaHPO_4(s) \rightleftharpoons Ca^{2+}(aq) + HPO_4^{2-}(aq)$$

(a) Are these data consistent? (That is, are the molar solubilities the same when derived in two different ways?)
(b) If there is a discrepancy, how do you account for it?

69. A 50.0 mL sample of 0.0152 M Na_2SO_4(aq) is added to 50.0 mL of 0.0125 M $Ca(NO_3)_2$(aq). What percentage of the Ca^{2+} remains unprecipitated?

70. What percentage of the Ba^{2+} in solution is precipitated as $BaCO_3$(s) if equal volumes of 0.0020 M Na_2CO_3(aq) and 0.0010 M $BaCl_2$(aq) are mixed?

71. Determine the molar solubility of lead(II) azide, $Pb(N_3)_2$, in a buffer solution with pH = 3.00, given that

$$Pb(N_3)_2(s) \rightleftharpoons Pb^{2+}(aq) + 2\,N_3^-(aq)$$
$$K_{sp} = 2.5 \times 10^{-9}$$

$$HN_3(aq) + H_2O(l) \rightleftharpoons H_3O^+(aq) + N_3^-(aq)$$
$$K_a = 1.9 \times 10^{-5}$$

72. Calculate the molar solubility of $Mg(OH)_2$ in 1.00 M NH_4Cl(aq).

73. The chief compound in marble is $CaCO_3$. Marble has been widely used for statues and ornamental work on buildings. However, marble is readily attacked by acids. Determine the solubility of marble (that is, $[Ca^{2+}]$ in a saturated solution) in **(a)** normal rainwater of pH = 5.6; **(b)** acid rainwater of pH = 4.20. Assume that the overall reaction that occurs is

$$CaCO_3(s) + H_3O^+(aq) \rightleftharpoons$$
$$Ca^{2+}(aq) + HCO_3^-(aq) + H_2O(l)$$

74. What is the solubility of MnS, in grams per liter, in a buffer solution that is 0.100 M CH_3COOH − 0.500 M $NaCH_3COO$? For MnS, $K_{spa} = 3 \times 10^7$.

75. Write net ionic equations for each of the following observations.
   (a) When concentrated $CaCl_2(aq)$ is added to $Na_2HPO_4(aq)$, a white precipitate forms that is 38.7% Ca by mass.
   (b) When a piece of dry ice, $CO_2(s)$, is placed in a clear dilute solution of limewater $[Ca(OH)_2(aq)]$, bubbles of gas evolve. At first, a white precipitate forms, but then it redissolves.

76. Concerning the reactions described in Exercise 75(b),
   (a) Will the same observations be made if $Ca(OH)_2(aq)$ is replaced by $CaCl_2(aq)$? Explain.
   (b) Show that the white precipitate will redissolve if the $Ca(OH)_2(aq)$ is about 0.005 M, but not if the solution is saturated.

77. Reaction (18.10), described in the Integrative Example, is called a *carbonate transposition*. In such a reaction, anions of a slightly soluble compound (for example, hydroxides and sulfates) are obtained in a sufficient concentration in aqueous solution that they can be identified by qualitative analysis tests. Suppose that 3 M Na_2CO_3 is used and that an anion concentration of 0.050 M is sufficient for its detection. Predict whether carbonate transposition will be effective for detecting **(a)** SO_4^{2-} from $BaSO_4(s)$; **(b)** Cl^- from $AgCl(s)$; **(c)** F^- from $MgF_2(s)$.

78. For the titration in Example 18-7, verify the assertion that $[Ag^+]$ increases very rapidly between the point at which AgBr has finished precipitating and Ag_2CrO_4 is about to begin.

79. Aluminum compounds are soluble in acidic solution, where aluminum(III) exists as the complex ion $[Al(H_2O)_6]^{3+}$, which is generally represented simply as $Al^{3+}(aq)$. They are also soluble in basic solutions, where the aluminum(III) is present as the complex ion $[Al(OH)_4]^-$. At certain intermediate pH values, the concentration of aluminum(III) that can exist in solution is at a minimum. Thus, a plot of the total concentration of aluminum(III) in solution as a function of pH yields a U-shaped curve. Demonstrate that this is the case with a few calculations.

80. The solubility of AgCN(s) in 0.200 M $NH_3(aq)$ is 8.8×10^{-6} mol/L. Calculate K_{sp} for AgCN.

81. The solubility of $CdCO_3(s)$ in 1.00 M KI(aq) is 1.2×10^{-3} mol/L. Given that K_{sp} of $CdCO_3$ is 5.2×10^{-12}, what is K_f for $[CdI_4]^{2-}$?

82. Use K_{sp} for $PbCl_2$ and K_f for $[PbCl_3]^-$ to determine the molar solubility of $PbCl_2$ in 0.10 M HCl(aq). [*Hint:* What is the total concentration of lead species in solution?]

83. A mixture of $PbSO_4(s)$ and $PbS_2O_3(s)$ is shaken with pure water until a saturated solution is formed. Both solids remain in excess. What is $[Pb^{2+}]$ in the saturated solution? For $PbSO_4$, $K_{sp} = 1.6 \times 10^{-8}$; for PbS_2O_3, $K_{sp} = 4.0 \times 10^{-7}$.

84. Use the method of Exercise 83 to determine $[Pb^{2+}]$ in a saturated solution in contact with a mixture of $PbCl_2(s)$ and $PbBr_2(s)$.

85. A 2.50 g sample of $Ag_2SO_4(s)$ is added to a beaker containing 0.150 L of 0.025 M $BaCl_2$.
   (a) Write an equation for any reaction that occurs.
   (b) Describe the final contents of the beaker—that is, the masses of any precipitates present and the concentrations of the ions in solution.

86. How many moles of solid sodium fluoride should be added to 1.0 L of a saturated solution of barium fluoride, BaF_2, at 25 °C to raise the fluoride concentration to 0.030 mol/L? What mass of BaF_2 precipitates? You may ignore the hydrolysis of fluoride ion. [*Hint:* This first part of this problem is most easily solved by first writing down an electroneutrality condition. See pages 721–722.]

Feature Problems

87. In an experiment to measure K_{sp} of $CaSO_4$ [D. Masterman, *J. Chem. Educ.*, **64**, 409 (1987)], a saturated solution of $CaSO_4(aq)$ is poured into the ion-exchange column pictured (and described in Chapter 21). As the solution passes through the column, Ca^{2+} is retained by the ion-exchange medium and H_3O^+ is released; two H_3O^+ ions appear in the effluent solution for every Ca^{2+} ion. As the drawing suggests, a 25.00 mL sample is added to the column, and the effluent is collected and diluted to 100.0 mL in a volumetric flask. A 10.00 mL portion of the diluted solution requires 8.25 mL of 0.0105 M NaOH for its titration. Use these data to obtain a value of K_{sp} for $CaSO_4$.

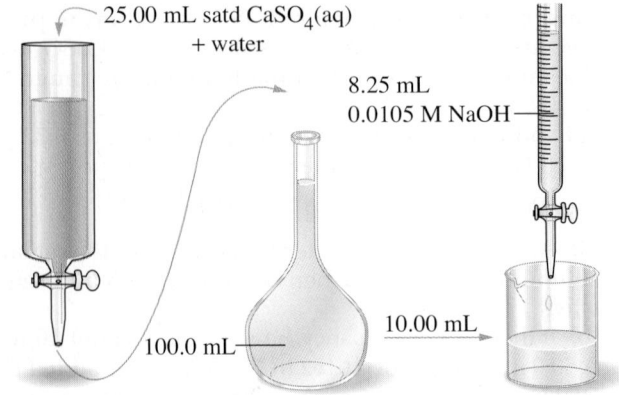

88. In the *Mohr titration*, $Cl^-(aq)$ is titrated with $AgNO_3(aq)$ in solutions that are at about pH = 7. Thus, it is suitable for determining the chloride ion content of drinking water. The indicator used in the titration is $K_2CrO_4(aq)$. A red-brown precipitate of $Ag_2CrO_4(s)$ forms after all the Cl^- has precipitated. The titration reaction is $Ag^+(aq) + Cl^-(aq) \longrightarrow AgCl(s)$. At the equivalence point of the titration, the titration mixture consists of AgCl(s) and a solution having neither Ag^+ nor Cl^- in excess. Also, no $Ag_2CrO_4(s)$ is present, but it forms immediately after the equivalence point.
(a) How many milliliters of 0.01000 M $AgNO_3(aq)$ are required to titrate 100.0 mL of a municipal water sample having 29.5 mg Cl^-/L?
(b) What is $[Ag^+]$ at the equivalence point of the Mohr titration?
(c) What is $[CrO_4^{2-}]$ in the titration mixture to meet the requirement of no precipitation of $Ag_2CrO_4(s)$ until immediately after the equivalence point?
(d) Describe the effect on the results of the titration if $[CrO_4^{2-}]$ were (1) greater than that calculated in part (c) or (2) less than that calculated?
(e) Do you think the Mohr titration would work if the reactants were exchanged—that is, with $Cl^-(aq)$ as the titrant and $Ag^+(aq)$ in the sample being analyzed? Explain.

89. The accompanying drawing suggests a series of manipulations starting with saturated $Mg(OH)_2(aq)$. Calculate $[Mg^{2+}(aq)]$ at each of the lettered stages.

(a) 0.500 L of saturated $Mg(OH)_2(aq)$ is in contact with $Mg(OH)_2(s)$.
(b) 0.500 L of H_2O is added to the 0.500 L of solution in part (a), and the solution is vigorously stirred. Undissolved $Mg(OH)_2(s)$ remains.
(c) 100.0 mL of the clear solution in part (b) is removed and added to 0.500 L of 0.100 M HCl(aq).
(d) 25.00 mL of the clear solution in part (b) is removed and added to 250.0 mL of 0.065 M $MgCl_2(aq)$.
(e) 50.00 mL of the clear solution in part (b) is removed and added to 150.0 mL of 0.150 M KOH(aq).

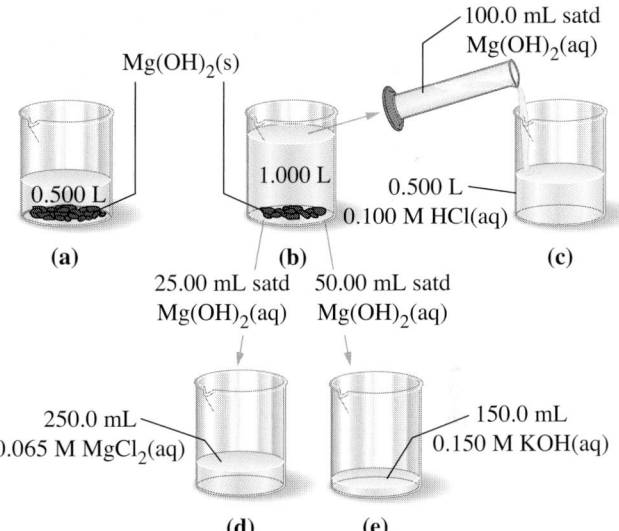

Self-Assessment Exercises

90. In your own words, define the following terms or symbols: **(a)** K_{sp}; **(b)** K_f; **(c)** Q_{sp}; **(d)** complex ion.
91. Briefly describe each of the following ideas, methods, or phenomena: **(a)** common-ion effect in solubility equilibrium; **(b)** fractional precipitation; **(c)** ion-pair formation; **(d)** qualitative cation analysis.
92. Explain the important distinction between each pair of terms: **(a)** solubility and solubility product constant; **(b)** common-ion effect and salt effect; **(c)** ion pair and ion product.
93. Pure water is saturated with slightly soluble PbI_2. Which of the following is a correct statement concerning the lead ion concentration in the solution, and what is wrong with the others? **(a)** $[Pb^{2+}] = [I^-]$; **(b)** $[Pb^{2+}] = K_{sp}$ of PbI_2; **(c)** $[Pb^{2+}] = \sqrt{K_{sp}}$ of PbI_2; **(d)** $[Pb^{2+}] = 0.5[I^-]$.
94. Adding 1.85 g Na_2SO_4 to 500.0 mL of saturated aqueous $BaSO_4$: **(a)** reduces $[Ba^{2+}]$; **(b)** reduces $[SO_4^{2-}]$; **(c)** increases the solubility of $BaSO_4$; **(d)** has no effect.
95. The slightly soluble solute Ag_2CrO_4 is *most* soluble in **(a)** pure water; **(b)** 0.10 M K_2CrO_4; **(c)** 0.25 M KNO_3; **(d)** 0.40 M $AgNO_3$.
96. Cu^{2+} and Pb^{2+} are both present in an aqueous solution. To precipitate one of the ions and leave the other in solution, add **(a)** $H_2S(aq)$; **(b)** $H_2SO_4(aq)$; **(c)** $HNO_3(aq)$; **(d)** $NH_4NO_3(aq)$.

97. All but two of the following solutions yield a precipitate when the solution is also made 2.00 M in NH_3. Those two are **(a)** $MgCl_2(aq)$; **(b)** $FeCl_3(aq)$; **(c)** $(NH_4)_2SO_4(aq)$; **(d)** $Cu(NO_3)_2(aq)$; **(e)** $Al_2(SO_4)_3(aq)$.
98. To increase the molar solubility of $CaCO_3(s)$ in a saturated aqueous solution, add **(a)** ammonium chloride; **(b)** sodium carbonate; **(c)** ammonia; **(d)** more water.
99. The best way to ensure complete precipitation from saturated $H_2S(aq)$ of a metal ion, M^{2+}, as its sulfide, MS(s), is to **(a)** add an acid; **(b)** increase $[H_2S]$ in the solution; **(c)** raise the pH; **(d)** heat the solution.
100. Which of the following solids are likely to be more soluble in acidic solution and which in basic solution? Which are likely to have a solubility that is independent of pH? Explain. **(a)** $H_2C_2O_4$; **(b)** $MgCO_3$; **(c)** CdS; **(d)** KCl; **(e)** $NaNO_3$; **(f)** $Ca(OH)_2$.
101. Both Mg^{2+} and Cu^{2+} are present in the same aqueous solution. Which of the following reagents would work best in separating these ions, precipitating one and leaving the other in solution: NaOH(aq), HCl(aq), $NH_4Cl(aq)$, or $NH_3(aq)$? Explain your choice.
102. Will $Al(OH)_3(s)$ precipitate from a buffer solution that is 0.45 M CH_3COOH and 0.35 M $NaCH_3COO$ and also 0.275 M in $Al^{3+}(aq)$? For $Al(OH)_3$, $K_{sp} = 1.3 \times 10^{-33}$; for CH_3COOH, $K_a = 1.8 \times 10^{-5}$.

103. Saturated solutions of sodium phosphate, copper(II) chloride, and ammonium acetate are mixed together. The precipitate is **(a)** copper(II) acetate; **(b)** copper(II) phosphate; **(c)** sodium chloride; **(d)** ammonium phosphate; **(e)** nothing precipitates.

104. Which of the following has the highest molar solubility? **(a)** MgF_2, $K_{sp} = 3.7 \times 10^{-8}$; **(b)** $MgCO_3$, $K_{sp} = 3.5 \times 10^{-8}$; **(c)** $Mg_3(PO_4)_2$, $K_{sp} = 1 \times 10^{-25}$; **(d)** Li_3PO_4, $K_{sp} = 3.2 \times 10^{-9}$.

105. Lead(II) chloride is most soluble in **(a)** 0.100 M NaCl; **(b)** 0.100 $Na_2S_2O_3$; **(c)** 0.100 M $Pb(NO_3)_2$; **(d)** 0.100 M $NaNO_3$; **(e)** 0.100 $MnSO_4$.

106. Given the following ions in solution, Hg^{2+}, I^-, Ag^+, and NO_3^-, does the formation of a complex ion increase or decrease the amount of precipitate?

107. Will AgI(s) precipitate from a solution with $[[Ag(CN)_2]^-] = 0.012$ M, $[CN^-] = 1.05$ M, and $[I^-] = 2.0$ M? For AgI, $K_{sp} = 8.5 \times 10^{-17}$; for $[Ag(CN)_2]^-$, $K_f = 5.6 \times 10^{18}$.

108. Without performing detailed calculations, indicate whether either of the following compounds is appreciably soluble in $NH_3(aq)$: **(a)** CuS, $K_{sp} = 6.3 \times 10^{-36}$; **(b)** $CuCO_3$, $K_{sp} = 1.4 \times 10^{-10}$. Also use the fact that K_f for $[Cu(NH_3)_4]^{2+}$ is 1.1×10^{13}.

109. Appendix E describes a useful study aid known as concept mapping. Using the methods presented in Appendix E, construct a concept map that links the various factors affecting the solubility of slightly soluble solutes.

Spontaneous Change: Entropy and Gibbs Energy

19

Thermodynamics originated in the early nineteenth century with attempts to improve the efficiency of steam engines. However, the laws of thermodynamics are widely useful throughout the field of chemistry and in biology as well, as we discover in this chapter.

◀ Throughout the text we have adopted the IUPAC recommended name of "Gibbs energy" to replace the terms *free energy* or *Gibbs free energy*.

In Chapter 15, we noted that the reaction of nitrogen and oxygen gases, which does not occur appreciably in the forward direction at room temperature, produces significant equilibrium amounts of NO(g) at *high* temperatures.

$$N_2(g) + O_2(g) \rightleftharpoons 2\,NO(g)$$

Another reaction involving oxides of nitrogen is the conversion of NO(g) to $NO_2(g)$:

$$2\,NO(g) + O_2(g) \rightleftharpoons 2\,NO_2(g)$$

This reaction, unlike the first, yields its greatest equilibrium amounts of $NO_2(g)$ at *low* temperatures.

What is there about these two reactions that causes the forward reaction of one to be favored at high temperatures but the forward reaction of the other to be favored at low temperatures? Our primary objective in this chapter is to develop concepts to help us answer questions like this. This chapter, taken together with ideas from Chapter 7, shows the great power of thermodynamics to provide explanations of many chemical phenomena.

▲ When the spring powering this mechanical toy unwinds, the toy stops moving. The spring cannot spontaneously rewind.

▶ Spontaneous: "proceeding from natural feeling or native tendency without external constraint . . . ; developing without apparent external influence, force, cause, or treatment." *Merriam-Webster's Collegiate Dictionary*, online, 2000.

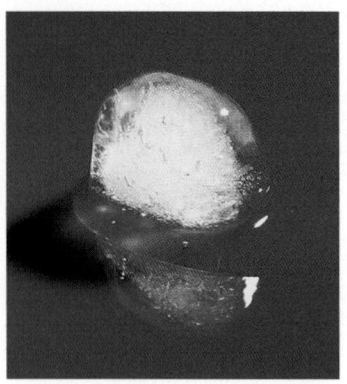

▲ The melting of an ice cube occurs spontaneously at temperatures above 0 °C.

19-1 Spontaneity: The Meaning of Spontaneous Change

Most of us have played with spring-wound toys, whether a toy automobile, top, or music box. In every case, once the wound-up toy is released, it keeps running until the stored energy in the spring has been released; then the toy stops. The toy never rewinds itself. Human intervention is necessary (winding by hand). The running down of a wound-up spring is an example of a *spontaneous* process. The rewinding of the spring is a *nonspontaneous* process. Let's explore the scientific meaning of these two terms.

A **spontaneous process** is a process that occurs in a system left to itself; once started, no action from outside the system (external action) is necessary to make the process continue. Conversely, a **nonspontaneous process** will not occur *unless* some external action is continuously applied. Consider the rusting of an iron pipe exposed to the atmosphere. Although the process occurs slowly, it does so continuously. As a result, the amount of iron decreases and the amount of rust increases until a final state of equilibrium is reached in which essentially all the iron has been converted to iron(III) oxide. We say that the reaction

$$4 \, Fe(s) + 3 \, O_2(g) \longrightarrow 2 \, Fe_2O_3(s)$$

is *spontaneous*. Now consider the reverse situation: the extraction of pure iron from iron(III) oxide. We should not say that the process is impossible, but it is certainly *nonspontaneous*. In fact, this nonspontaneous reverse process is involved in the manufacture of iron from iron ore.

We will consider specific quantitative criteria for spontaneous change later in the chapter, but even now we can identify some spontaneous processes intuitively. For example, in the neutralization of NaOH(aq) with HCl(aq), the net change that occurs is

$$H_3O^+(aq) + OH^-(aq) \longrightarrow 2 \, H_2O(l)$$

There is very little tendency for the reverse reaction (self-ionization) to occur, so the neutralization reaction is a spontaneous reaction. The melting of ice, however, is spontaneous at temperatures above 0 °C but nonspontaneous below 0 °C.

From our discussion of spontaneity to this point, we can reach these conclusions at ambient pressures.

- If a process is spontaneous, the reverse process is nonspontaneous.
- Both spontaneous and nonspontaneous processes are possible, but only spontaneous processes will occur *without intervention*. Nonspontaneous processes require the system to be acted on by an external agent.
- Some spontaneous processes occur very slowly and others occur rather rapidly. For example, the melting of an ice cube that has been dropped into cold water at 1 °C is a spontaneous process that occurs slowly. The melting of an ice cube that has been dropped into hot water at 99 °C is a spontaneous process that occurs rapidly. The main point is that spontaneous does not mean fast.

We would like to do more, however. We want to be able to predict whether the forward or the reverse direction is the direction of spontaneous change in a process, so we need a criterion for spontaneous change. To begin, let's look to mechanical systems for a clue. A ball rolls downhill, and water flows to a lower level. A common feature of these processes is that *potential energy decreases*.

For chemical systems, the property analogous to the potential energy of a mechanical system is the internal energy (U) or the closely related property enthalpy (H). In the 1870s, P. Berthelot and J. Thomsen proposed that the direction of spontaneous change is the direction in which the enthalpy of a system decreases. In a system in which enthalpy decreases, heat is given off by the system to the surroundings. Bertholet and Thomsen concluded that exothermic reactions should be spontaneous. In fact, many exothermic processes are spontaneous, but some are not. Also, some endothermic reactions *are* spontaneous. Thus, we cannot predict whether a process is spontaneous from its enthalpy change alone. Here are three examples of spontaneous, *endothermic* processes:

- the melting of ice at room temperature
- the evaporation of liquid diethyl ether from an open beaker
- the dissolving of ammonium nitrate in water

We will have to look to thermodynamic functions other than enthalpy change (ΔH) as criteria for spontaneous change.

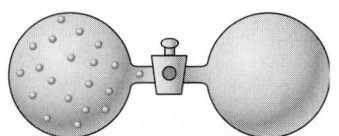

(a) Initial condition

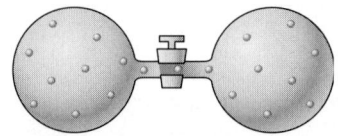

(b) After expansion into vacuum

▲ FIGURE 19-1
Expansion of an ideal gas into a vacuum
(a) Initially, an ideal gas is confined to the bulb on the left at 1.00 atm pressure.
(b) When the stopcock is opened, the gas expands into the identical bulb on the right. The final condition is one in which the gas is equally distributed between the two bulbs at a pressure of 0.50 atm.

 19-1   CONCEPT ASSESSMENT

Is it correct to say that a spontaneous process is a fast process and a nonspontaneous process is a very slow one? Explain.

19-2   The Concept of Entropy

To continue our search for criteria for spontaneous change, consider Figure 19-1, which depicts two identical glass bulbs joined by a stopcock. Initially, the bulb on the left contains an ideal gas at 1.00 atm pressure, and the bulb on the right is evacuated. When the valve is opened, the gas immediately expands into the evacuated bulb. After this expansion, the molecules are dispersed throughout the apparatus, with essentially equal numbers of molecules in both bulbs and a pressure of 0.50 atm. What causes this spontaneous expansion of the gas at a constant temperature?

One of the characteristics of an ideal gas is that its internal energy (U) does not depend on the gas pressure, but only on the temperature. Therefore, in this expansion $\Delta U = 0$. Also, the enthalpy change is zero: $\Delta H = 0$. This means that the expansion is not caused by the system dropping to a lower energy state. A convenient mental image to "explain" the expansion is that the gas molecules tend to spread out into the larger volume available to them at the reduced pressure. A more fundamental description of the underlying cause is that, for the same total energy, in the expanded volume there are more available translational energy levels among which the gas molecules can be distributed. The tendency is for the energy of the system to spread out over a larger number of energy levels.

A similar situation—the mixing of ideal gases—is depicted in Figure 19-2. In this case, the two bulbs initially are filled with different ideal gases at 1.00 atm. When the stopcock is opened, the gases mix. The resulting change is essentially that of the expansion of the ideal gas pictured in Figure 19-1, but twice over. That is, each gas expands into the new volume available to it, without regard for the other gas (recall Dalton's law of partial pressures, page 213). Again, each expanded gas has more translational energy levels available to its molecules—the energy of the system has spread out. And again, the internal energy and enthalpy of the system are not changed by the expansion.

KEEP IN MIND

that U and H are related as follows: $H = U + PV$, and $\Delta H = \Delta U + \Delta(PV)$. For a fixed amount of an ideal gas at a constant temperature, $PV = $ constant, $\Delta(PV) = 0$, and $\Delta H = \Delta U$.

Entropy

The thermodynamic property related to the way in which the energy of a system is distributed among the available microscopic energy levels is called **entropy**.

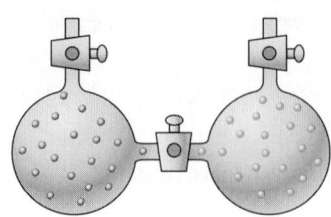

(a) Before mixing

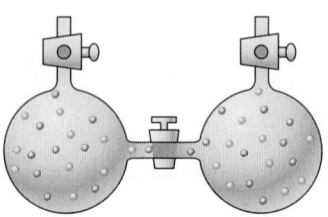

(b) After mixing

· Gas A   · Gas B

▲ FIGURE 19-2
The mixing of ideal gasses
The total volume of the system and the total gas pressure remain fixed. The net change is that **(a)** before mixing, each gas is confined to half the total volume (a single bulb) at a pressure of 1.00 atm, and **(b)** after mixing, each gas has expanded into the total volume (both bulbs) and exerts a partial pressure of 0.50 atm.

▲ **A bust marking Ludwig Boltzmann's tomb in Vienna**
Boltzmann's famous equation is inscribed on the tomb. At the time of Boltzmann's death, the term *log* was used for both natural logarithms and logarithms to the base ten; the symbol "ln" had not yet been adopted.

> The greater the number of arrangements (i.e., microstates) of the microscopic particles (atoms, ions, molecules) among the energy levels in a particular state of a system, the greater the entropy of the system.

Entropy is denoted by the symbol S. Like internal energy and enthalpy, entropy is a function of state (see page 256). It has a unique value for a system whose temperature, pressure, and composition are specified. The **entropy change**, ΔS, is the difference in entropy between two states of a system, and it also has a unique value.

In the gas expansion of Figure 19-1, the entropy of the gas *increases* and $\Delta S > 0$. In the mixing of gases as carried out in Figure 19-2, entropy also increases, a fact that we can represent symbolically.

$$A(g) + B(g) \longrightarrow \text{mixture of A(g) and B(g)}$$
$$\Delta S = S_{\text{mix of gases}} - [S_{A(g)} + S_{B(g)}] > 0$$

Because both of these expansions occur spontaneously and neither is accompanied by a change in internal energy or enthalpy, it seems possible that *increases in entropy underlie spontaneous processes*. This is a proposition that we will have to examine more closely later, but let's accept it for now.

The Boltzmann Equation for Entropy

The connection between macroscopic changes, such as the mixing of gases, and the microscopic nature of matter was enunciated by Ludwig Boltzmann. The conceptual breakthrough that Boltzmann made was to associate the number of energy levels in the system with the number of ways of arranging the particles (atoms, ions, or molecules) in these energy levels. The microscopic energy levels are also called *states*, and the particular way a number of particles is distributed among these states is called a *microstate*. The more states a given number of particles can occupy, the more microstates the system has. The more microstates that exist, the greater the entropy. Boltzmann derived the relationship

$$S = k \ln W \qquad (19.1)$$

where S is the entropy, k is the Boltzmann constant, and W is the number of microstates. We can think of the Boltzmann constant as the gas constant per molecule; that is, $k = R/N_A$. [Although we didn't specifically introduce k in the discussion of kinetic-molecular theory, R/N_A appears in equation (6.21).] The number of microstates, W, is the number of ways that the atoms or molecules can be positioned in the states available and still give rise to the same total energy. Each permissible arrangement of the particles constitutes one of the microstates, so W is the total number of microstates that correspond to the same energy.

Let us explore the connection between the number of microstates (W) and entropy (S) a little further by using a small one-dimensional crystal of four nitrous oxide molecules, NO. At $T = 0$ K, the four NO molecules are aligned as $\text{NO} \cdots \text{NO} \cdots \text{NO} \cdots \text{NO}$, which is the only microstate possible. Therefore, $W = 1$ and $S = 0$. Now consider what happens when the temperature is raised just enough to allow a single NO to rotate. Such an increase in temperature corresponds to an increase in the internal energy of the system. Now four microstates are possible, $\text{NO} \cdots \text{NO} \cdots \text{NO} \cdots \text{ON}$; $\text{NO} \cdots \text{NO} \cdots \text{ON} \cdots \text{NO}$; $\text{NO} \cdots \text{ON} \cdots \text{NO} \cdots \text{NO}$; and $\text{ON} \cdots \text{NO} \cdots \text{NO} \cdots \text{NO}$; W equals 4. By using Boltzmann's equation, we find that $S = k \ln 4 = (1.3807 \times 10^{-23} \text{ J K}^{-1}) \ln 4 = 1.9141 \times 10^{-23} \text{ J K}^{-1}$. As the energy of the system increases, the number of microstates increases; therefore, entropy of the system will increase.

How can we use Boltzmann's equation to think about the distribution of microscopic particles among the energy levels of a system? Let's consider again the particle-in-a-box model for a matter wave (page 318). Specifically, we can use the equation $E = n^2h^2/8mL^2$ to calculate a few energy levels for a matter wave in a one-dimensional box. Representative energy levels are shown in the energy-level diagrams in Figure 19-3(a) for a particle in boxes of lengths L, $2L$, and $3L$. We see the following relationship between the length of the box and the number of levels: L, *three* levels; $2L$, *six* levels; and $3L$, *nine* levels. As the boundaries of the box are expanded, the separation between the energy levels decreases resulting in an increase of accessible energy levels. Extending this model to three-dimensional space and large numbers of gas molecules, we find less crowding of molecules into a limited number of energy levels when the pressure of the gas drops and the molecules expand into a larger volume. Thus, there is a greater spreading of the energy, and the entropy increases.

We can use the same particle-in-a-box model to understand the effect of raising the temperature of a substance on the entropy of the system. We will consider a gas, but our conclusions are equally valid for a liquid or solid. At low temperatures, the molecules have a low energy, and the gas molecules can occupy only a few of the energy levels; the value of W is small, and the entropy is low. As the temperature is raised, the energy of the molecules increases and the molecules have access to a larger number of energy levels. Thus the number of accessible microstates (W) increases and the entropy rises (Fig. 19-3(b)).

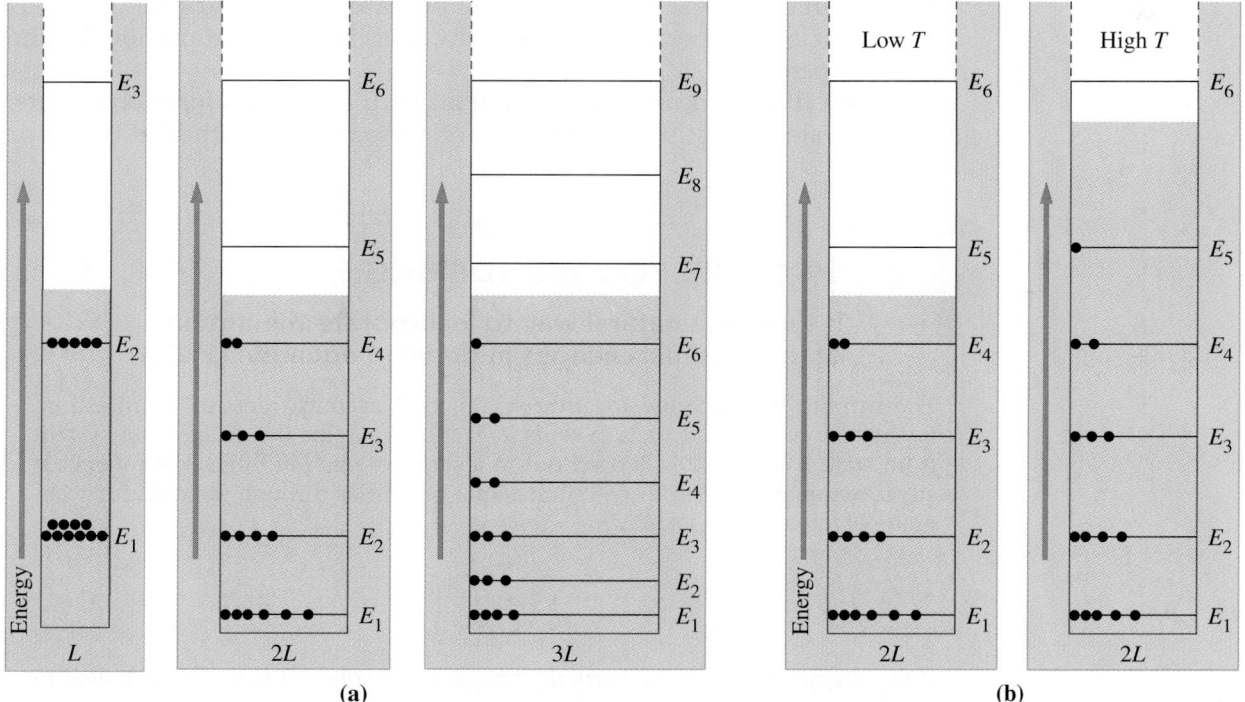

▲ FIGURE 19-3
Energy levels for a particle in a one-dimensional box
(a) The energy levels of a particle in a box become more numerous and closer together as the length of the box increases. The range of thermally accessible levels is indicated by the tinted band. The solid circles signify a system consisting of 15 particles. Each drawing represents a single microstate of the system. Can you see that as the box length increases there are many more microstates available to the particles for a given amount of thermal energy? As the number of possible microstates for a given total energy increases, so does the entropy. **(b)** More energy levels become accessible in a box of fixed length as the temperature is raised. Because the average energy of the particles also increases, the internal energy and entropy both increase as the temperature is raised.

In summary, the state of a thermodynamic system can be described in two ways: the macroscopic description, in terms of state functions P, V, and T; and the microscopic description, requiring a knowledge of the position and velocity of every particle (atom or molecule) in the system. Boltzmann's equation provides the connection between the two.

Entropy Change

▶ The number of accessible states, W, is determined by the properties of a system. When a process takes place, the coupling must make the system larger; hence W must increase as the system spontaneously moves to a different equilibrium.

An entropy change is based on two measurable quantities: heat (q) and temperature (T). Both of these factors affect the availability of energy levels to the microscopic particles of a system. The following equation relates these factors to an entropy change,

$$\Delta S = \frac{q_{\text{rev}}}{T} \qquad \text{(19.2)}$$

where T is the Kelvin temperature. Notice that ΔS is directly proportional to the quantity of heat because the more energy added to a system (as heat), the greater the number of energy levels available to the microscopic particles. Raising the temperature also increases the availability of energy levels, but for a given quantity of heat the proportional increase in number of energy levels is greatest at low temperatures. That is why ΔS is inversely proportional to the Kelvin temperature.

Equation (19.2) appears simple, but it is not. If S is to be a function of state, ΔS for a system must be independent of the path by which heat is lost or

19-1 ARE YOU WONDERING...

If there is a natural way to incorporate the notion of infinitesimal changes in deriving equation (19.2)?

The infinitesimal change in entropy, dS, that accompanies an infinitesimal reversible heat flow, dq_{rev}, is $dS = dq_{\text{rev}}/T$. Now imagine the change in a system from state 1 to state 2 is carried out in a series of such infinitesimal reversible steps. Summation of all these infinitesimal quantities through the calculus technique of integration yields ΔS.

$$\Delta S = \int \frac{dq_{\text{rev}}}{T}$$

If the change of state is isothermal (carried out at constant temperature), we can write

$$\Delta S = \int \frac{dq_{\text{rev}}}{T} = \frac{1}{T} \int dq_{\text{rev}} = \frac{q_{\text{rev}}}{T}$$

and we have recovered the definition of entropy change in (19.2.)

Starting with equivalent expressions for dq_{rev}, ΔS can be related to other system properties. For the isothermal, reversible expansion of an ideal gas, $dq_{\text{rev}} = -dw_{\text{rev}}$, leading to equation (19.10) on page 839, which describes ΔS in terms of gas volumes. For a reversible change in temperature, $dq_{\text{rev}} = C_p \, dT$, which leads to the entropy change for a change in temperature (see Exercise 86).

gained. Conversely, because the value of q ordinarily depends on the path chosen (recall page 260), equation (19.2) holds only for a carefully defined path. The path must be *reversible*, that is, $q = q_{rev}$. As we learned in Chapter 7, a reversible process can be made to reverse its direction when just an infinitesimal change in the opposite direction is made in a system property (review Figure 7-12). Because q has the unit J and $1/T$ has the unit K^{-1}, the unit of entropy change, ΔS, is J/K, or J K^{-1}.

In some instances, it is difficult to construct mental pictures to assess how the entropy of a system changes during a process. However, in many cases, an increase or a decrease in the accessibility of energy levels for the microscopic particles of a system parallels an increase or decrease in the *number* of microscopic particles and the *space* available to them. As a consequence, we can often make qualitative predictions about entropy change by focusing on those two factors. Let's test this idea by considering again the three spontaneous, endothermic processes listed at the conclusion of Section 19-1 and illustrated in Figure 19-4.

In the melting of ice, a crystalline solid is replaced by a less structured liquid. Molecules that were relatively fixed in position in the solid, being limited to vibrational motion, are now free to move about a bit. The molecules have gained some translational and rotational motion. The number of accessible microscopic energy levels has increased, and so has the entropy.

In the vaporization process, a liquid is replaced by the even less structured gas. Molecules in the gaseous state, because they can move within a large free volume, have many more accessible energy levels than do those in the liquid state. In the gas, energy can be spread over a much greater number of microscopic energy levels than in the liquid. The entropy of the gaseous state is much higher than that of the liquid state.

In the dissolving of ammonium nitrate in water, for example, a crystalline solid and a pure liquid are replaced by a mixture of ions and water molecules in the liquid (solution) state. This situation is somewhat more involved than the first two because some decrease in entropy is associated with the clustering of water molecules around the ions because of ion–dipole forces. The increase in entropy that accompanies the destruction of the solid's crystalline lattice predominates, however, and for the overall dissolution process, $\Delta S > 0$.

In each of the three spontaneous, endothermic processes discussed here, the increase in entropy ($\Delta S > 0$) outweighs the fact that heat must be absorbed ($\Delta H > 0$), and each process is spontaneous.

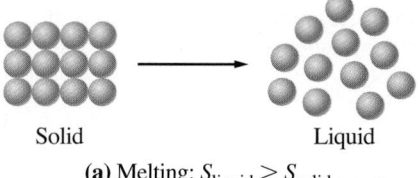

(a) Melting: $S_{liquid} > S_{solid}$

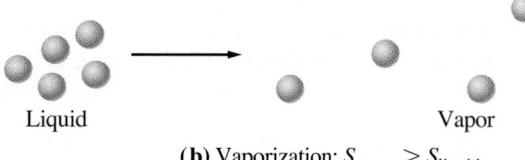

(b) Vaporization: $S_{vapor} > S_{liquid}$

▲ FIGURE 19-4
Three processes in which entropy increases
Each of the processes pictured—**(a)** the melting of a solid, **(b)** the evaporation of a liquid, and **(c)** the dissolving of a solute—results in an increase in entropy. For part (c), the generalization works best for nonelectrolyte solutions, in which ion–dipole forces do not exist.

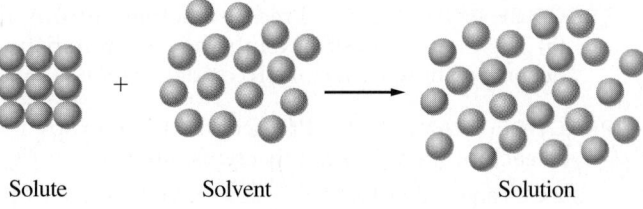

Solute   Solvent   Solution

(c) Dissolving: $S_{soln} > (S_{solvent} + S_{solute})$

In summary, four situations generally produce an *increase* in entropy:

- Pure liquids or liquid solutions are formed from solids.
- Gases are formed from either solids or liquids.
- The number of molecules of gas increases as a result of a chemical reaction.
- The temperature of a substance increases. (Increased temperature means an increased number of accessible energy levels for the increased molecular motion, whether it be vibrational motion of atoms or ions in a solid, or translational and rotational motion of molecules in a liquid or gas.)

We apply these generalizations in Example 19-1.

EXAMPLE 19-1   Making Qualitative Predictions of Entropy Changes in Physical and Chemical Processes

Predict whether each of the following processes involves an increase or a decrease in entropy or whether the outcome is uncertain.

(a) The decomposition of ammonium nitrate (a fertilizer and a highly explosive compound): $2 NH_4NO_3(s) \longrightarrow 2 N_2(g) + 4 H_2O(g) + O_2(g)$.

(b) The conversion of SO_2 to SO_3 (a key step in the manufacture of sulfuric acid): $2 SO_2(g) + O_2(g) \longrightarrow 2 SO_3(g)$.

(c) The extraction of sucrose from cane sugar juice: $C_{12}H_{22}O_{11}(aq) \longrightarrow C_{12}H_{22}O_{11}(s)$.

(d) The "water gas shift" reaction (involved in the gasification of coal): $CO(g) + H_2O(g) \longrightarrow CO_2(g) + H_2(g)$.

Analyze

Apply the generalizations summarized above. Three of the processes are chemical reactions, and for those processes, we should first consider whether the number of molecules of gas increases or decreases.

Solve

(a) Here, a solid yields a large quantity of gas. Entropy increases.

(b) Three moles of gaseous reactants produce two moles of gaseous products. The loss of one mole of gas indicates a loss of volume available to a smaller number of gas molecules. This loss reduces the number of possible configurations for the molecules in the system and the number of accessible microscopic energy levels. Entropy decreases.

(c) The sucrose molecules are reduced in mobility and in the number of forms in which their energy can be stored when they leave the solution and arrange themselves into a crystalline state. Entropy decreases.

(d) The entropies of the four gases are likely to be different because their molecular structures are different. The number of moles of gases is the same on both sides of the equation, however, so the entropy change is likely to be small if the temperature is constant. On the basis of just the generalizations listed above, we cannot determine whether entropy increases or decreases.

Assess

As we shall soon see, the ability to predict increases or decreases in entropy will help us to understand when a process will proceed spontaneously in the forward direction.

PRACTICE EXAMPLE A: Predict whether entropy increases or decreases in each of the following reactions. **(a)** The Claus process for removing H_2S from natural gas: $2 H_2S(g) + SO_2(g) \longrightarrow 3 S(s) + 2 H_2O(g)$; **(b)** the decomposition of mercury(II) oxide: $2 HgO(s) \longrightarrow 2 Hg(l) + O_2(g)$.

PRACTICE EXAMPLE B: Predict whether entropy increases or decreases or whether the outcome is uncertain in each of the following reactions. **(a)** $Zn(s) + Ag_2O(s) \longrightarrow ZnO(s) + 2 Ag(s)$; **(b)** the chlor-alkali process, $2 Cl^-(aq) + 2 H_2O(l) \xrightarrow{\text{electrolysis}} 2 OH^-(aq) + H_2(g) + Cl_2(g)$.

19-3 Evaluating Entropy and Entropy Changes

The difficulty of calculating an entropy change with equation (19.2) was mentioned on page 824. The emphasis then shifted to making qualitative predictions about entropy changes. In this section we will see that in a few instances a simple direct calculation of ΔS is possible. Also, we will find that, unlike the case with internal energy and enthalpy, it is possible to determine *absolute* entropy values.

Phase Transitions

In the equilibrium between two phases, the exchange of heat can be carried out reversibly, and the quantity of heat proves to be equal to the enthalpy change for the transition, ΔH_{tr}. In this case, equation (19.2) can be written as

$$\Delta S_{tr} = \frac{\Delta H_{tr}}{T_{tr}} \qquad (19.3)$$

Rather than use the general symbol "tr" to represent a transition, we can be more specific about just which phases are involved, such as "fus" for the melting of a solid and "vap" for the vaporization of a liquid. If the transitions involve standard state conditions (1 bar $\approx$ 1 atm pressure), we also use the degree sign (°). Thus for the melting (fusion) of ice at its normal melting point,

$$H_2O(s, 1 \text{ atm}) \rightleftharpoons H_2O(l, 1 \text{ atm}) \qquad \Delta H^\circ_{fus} = 6.02 \text{ kJ mol}^{-1} \text{ at } 273.15 \text{ K}$$

the standard entropy change is

$$\Delta S^\circ_{fus} = \frac{\Delta H^\circ_{fus}}{T_{mp}} = \frac{6.02 \text{ kJ mol}^{-1}}{273.15 \text{ K}} = 2.20 \times 10^{-2} \text{ kJ mol}^{-1} \text{ K}^{-1}$$
$$= 22.0 \text{ J mol}^{-1} \text{ K}^{-1}$$

Entropy changes depend on the quantities of substances involved and are usually expressed on a per-mole basis.

A useful generalization known as **Trouton's rule** states that for many liquids at their normal boiling points, the standard molar *entropy of vaporization* has a value of about 87 J mol^{-1} K^{-1}.

$$\Delta S^\circ_{vap} = \frac{\Delta H^\circ_{vap}}{T_{bp}} \approx 87 \text{ J mol}^{-1} \text{ K}^{-1} \qquad (19.4)$$

For instance, the values of ΔS°_{vap} for benzene (C_6H_6) and octane (C_8H_{18}) are 87.1 and 86.2 J mol^{-1} K^{-1}, respectively. If the increased accessibility of microscopic energy levels produced in transferring one mole of molecules from liquid

KEEP IN MIND

that the normal melting point and normal boiling point are determined at 1 atm pressure. The difference between 1 atm and the standard state pressure of 1 bar is so small that we can usually ignore it.

19-2 ARE YOU WONDERING...

If the thermodynamic entropy (equation 19.2), $\Delta S = q_{rev}/T$, can be derived from the statistical entropy (equation 19.1), $S = k \ln W$?

It can, but a rigorous derivation is very complex. However, by using a system of evenly spaced energy levels for the surroundings, a simplified derivation can be shown. Consider just two energy levels from the numerous energy levels representing the surroundings. Now, as suggested in Figure 19-5, construct two states, **A** and **B**, which are the most probable initial state (state **A**) and final state (state **B**) after the addition of an infinitesimal amount of heat (q_{rev}).

The temperature of **B** is slightly higher than the temperature of **A**. The amount of heat (q) used is just the energy difference ($\Delta \varepsilon$) between the two levels and therefore a single particle jumps from level i to level j. The number of **A** microstates, W_A, is almost exactly the same as the number of **B** microstates, W_B. The entropy change for the surroundings is

$$\Delta S_{sur} = k \ln \left(\frac{W_B}{W_A} \right)$$

However, the ratio of probabilities is related to the population of each state as shown in the following relationship:

$$\frac{W_B}{W_A} = \frac{n_i}{n_j + 1} \approx \frac{n_i}{n_j}$$

Boltzmann derived the following distribution law

$$\frac{n_j}{n_i} = e^{-\Delta \varepsilon / kT}$$

which relates the population of level i to that of level j. We use this to determine the entropy change of the surroundings along with the substitution of q_{rev} for $\Delta \varepsilon$.

$$\Delta S_{sur} = k \ln \left(\frac{W_B}{W_A} \right) = k \ln \left(e^{q_{rev}/kT} \right) = k \frac{q_{rev}}{kT} = \frac{q_{rev}}{T}$$

We have shown the relationship of the statistical entropy (equation 19.1) to the thermodynamic entropy (equation 19.2) to be true for the entropy change of the surroundings; however, this can be shown to hold for the system as well.

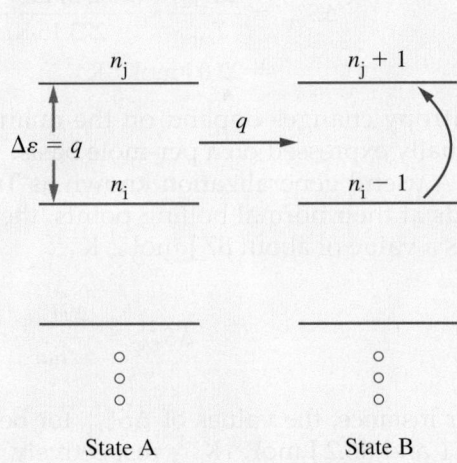

▶ FIGURE 19-5
Energy levels and populations representing the initial state of the surroundings (State A) and the state of the surroundings (State B) upon the addition of a quantity of heat (q). The spacing between the energy levels is $\Delta \varepsilon$. The populations of the energy levels are indicated by n_i and n_j.

EXAMPLE 19-2   Determining the Entropy Change for a Phase Transition

What is the standard molar entropy for the vaporization of water at 373 K given that the standard molar enthalpy of vaporization is 40.7 kJ mol^{-1}?

Analyze

This is an example of a phase transition, which means we can make use of $\Delta S^\circ = \dfrac{\Delta H^\circ_{transition}}{T_{transition}}$.

Solve

Although a chemical equation is not necessary, writing one can help us see the process we use to find the value of ΔS°_{vap}.

$$H_2O(l, 1 \text{ atm}) \rightleftharpoons H_2O(g, 1 \text{ atm}) \qquad\qquad \Delta H^\circ_{vap} = 40.7 \text{ kJ/mol } H_2O$$
$$\Delta S^\circ_{vap} = ?$$

$$\Delta S^\circ_{vap} = \frac{\Delta H^\circ_{vap}}{T_{bp}} = \frac{40.7 \text{ kJ mol}^{-1}}{373 \text{ K}} = 0.109 \text{ kJ mol}^{-1} \text{K}^{-1}$$
$$= 109 \text{ J mol}^{-1} \text{ K}$$

Assess

When solving this type of problem, we should check the sign of ΔS. Here, we expect an increase in entropy (ΔS is positive) because, as discussed on page 825, the entropy of a gas is much higher than that of a liquid.

PRACTICE EXAMPLE A: What is the standard molar entropy of vaporization, ΔS°_{vap}, for CCl_2F_2, a chlorofluorocarbon that once was heavily used in refrigeration systems? Its normal boiling point is -29.79 °C, and $\Delta H^\circ_{vap} = 20.2$ kJ mol^{-1}.

PRACTICE EXAMPLE B: The entropy change for the transition from solid rhombic sulfur to solid monoclinic sulfur at 95.5 °C is $\Delta S^\circ_{tr} = 1.09$ J mol^{-1} K^{-1}. What is the standard molar enthalpy change, ΔH°_{tr}, for this transition?

---

to vapor at 1 atm pressure is roughly comparable for different liquids, then we should expect similar values of ΔS°_{vap}.

Instances in which Trouton's rule fails are also understandable. In water and in ethanol, for example, hydrogen bonding among molecules produces a lower entropy than would otherwise be expected in the liquid state. Consequently, the entropy increase in the vaporization process is greater than normal, and so $\Delta S^\circ_{vap} > 87$ J mol^{-1} K^{-1}.

The entropy concept helps explain Raoult's law (Section 13-6). Recall that for an ideal solution, $\Delta H_{soln} = 0$ and intermolecular forces of attraction are the same as in the pure liquid solvent (page 564). Thus, we expect the molar ΔH_{vap} to be the same whether vaporization of solvent occurs from an ideal solution or from the pure solvent at the same temperature. So, too, should ΔS_{vap} be the same because $\Delta S_{vap} = \Delta H_{vap}/T$. When one mole of solvent is transferred from liquid to vapor at the equilibrium vapor pressure P°, entropy increases by the amount ΔS_{vap}. As shown in Figure 19-6, because the entropy of the ideal solution is greater than that of the pure solvent, the entropy of the vapor produced by the vaporization of solvent from the ideal solution is also greater than the entropy of the vapor obtained from the pure solvent. For the vapor above the solution to have the higher entropy, its molecules must have a greater number of accessible microscopic energy levels. In turn, the vapor must be present in a larger volume and, hence, must be at a lower pressure than the vapor coming from the pure solvent. This relationship corresponds to Raoult's law: $P_A = x_A P^\circ_A$.

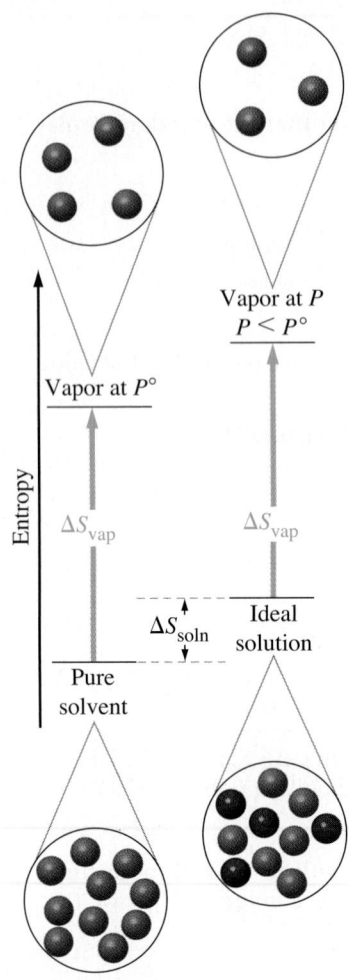

▲ FIGURE 19-6
An entropy-based rationale of Raoult's law
If ΔS_{vap} has the same value for vaporization from the pure solvent and from an ideal solution, the equilibrium vapor pressure is lower above the solution: $P < P°$.

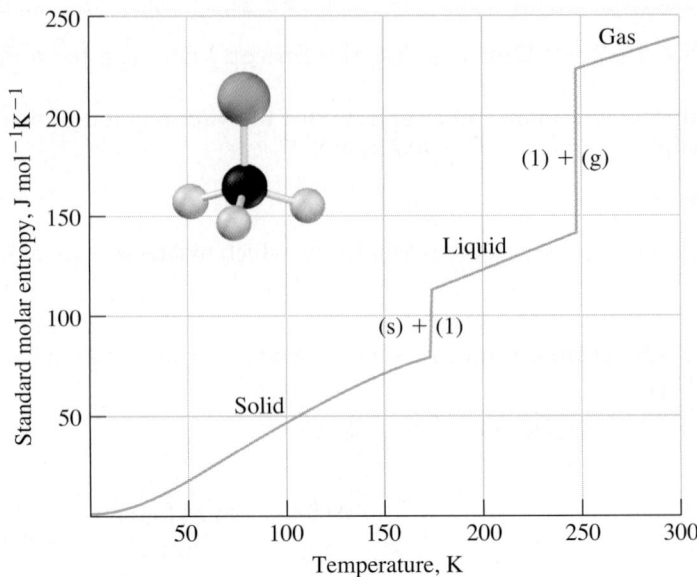

▲ FIGURE 19-7
Molar entropy as a function of temperature
The standard molar entropy of methyl chloride, CH_3Cl, is plotted at various temperatures from 0 to 300 K, with the phases noted. The vertical segment between the solid and liquid phases corresponds to ΔS_{fus}; the other vertical segment, to ΔS_{vap}. By the third law of thermodynamics, an entropy of *zero* is expected at 0 K. Experimental methods cannot be carried to that temperature, however, so an extrapolation is required.

Absolute Entropies

To establish an *absolute* value of the entropy of a substance, we look for a condition in which the substance is in its lowest possible energy state, called the *zero-point energy*. The entropy of this state is taken to be zero. Then we evaluate entropy changes as the substance is brought to other conditions of temperature and pressure. We add together these entropy changes and obtain a numerical value of the absolute entropy. The principle that permits this procedure is the **third law of thermodynamics,** which can be stated as follows:

The entropy of a pure perfect crystal at 0 K is zero.

Figure 19-7 illustrates the method outlined in the preceding paragraph for determining absolute entropy as a function of temperature. Where phase transitions occur, equation (19.3) is used to evaluate the corresponding entropy changes. Over temperature ranges in which there are no transitions, $\Delta S°$ values are obtained from measurements of specific heats as a function of temperature.

The absolute entropy of one mole of a substance in its standard state is called the **standard molar entropy, $S°$.** Standard molar entropies of a number of substances at 25 °C are tabulated in Appendix D. To use these values to calculate the entropy change of a reaction, we use an equation with a familiar form (recall equation 7.21).

$$\Delta S° = \left[\sum \nu_p S°(\text{products}) - \sum \nu_r S°(\text{reactants}) \right]$$ **(19.5)**

KEEP IN MIND

that there are no units of the stoichiometric coefficient, ν.

The symbol $\sum$ means "the sum of," and the terms added are the products of the standard molar entropies and the corresponding stoichiometric coefficients, ν. Example 19-3 shows how to use this equation.

EXAMPLE 19-3   Calculating Entropy Changes from Standard Molar Entropies

Use data from Appendix D to calculate the standard molar entropy change for the conversion of nitrogen monoxide to nitrogen dioxide (a step in the manufacture of nitric acid).

$$2\,NO(g) + O_2(g) \longrightarrow 2\,NO_2(g) \qquad \Delta S^\circ_{298\,K} = ?$$

Analyze

In this example we are applying equation (19.5).

Solve

Equation (19.5) takes the form

$$\Delta S^\circ = 2S^\circ_{NO_2(g)} - 2S^\circ_{NO(g)} - S^\circ_{O_2(g)}$$
$$= (2 \times 240.1) - (2 \times 210.8) - 205.1 = -146.5\,J\,K^{-1}$$

Assess

Some qualitative reasoning can be applied as a useful check on this calculation. Because three moles of gaseous reactants produce only two moles of gaseous products, the entropy should decrease; that is, ΔS° should be negative.

PRACTICE EXAMPLE A: Use data from Appendix D to calculate the standard molar entropy change for the synthesis of ammonia from its elements.

$$N_2(g) + 3\,H_2(g) \longrightarrow 2\,NH_3(g) \qquad \Delta S^\circ_{298\,K} = ?$$

PRACTICE EXAMPLE B: N_2O_3 is an unstable oxide that readily decomposes. The decomposition of 1.00 mol of N_2O_3 to nitrogen monoxide and nitrogen dioxide at 25 °C is accompanied by the entropy change $\Delta S^\circ = 138.5\,J\,K^{-1}$. What is the standard molar entropy of $N_2O_3(g)$ at 25 °C?

Example 19-3 used the standard molar entropies of $NO_2(g)$ and $NO(g)$. Why is the value for $NO_2(g)$, 240.1 J mol^{-1} K^{-1}, greater than that of $NO(g)$, 210.8 J mol^{-1} K^{-1}? Entropy increases when a substance absorbs heat (recall that $\Delta S = q_{rev}/T$), and some of this heat goes into raising the average translational kinetic energies of molecules. But there are other ways for heat energy to be used. One possibility, pictured in Figure 19-8, is that the vibrational energies of molecules can be increased. In the *diatomic* molecule $NO(g)$, only one type of vibration is possible; in the *triatomic* molecule $NO_2(g)$, *three* types are possible. Because there are more possible ways of distributing energy among NO_2 molecules than among NO molecules, $NO_2(g)$ has a higher molar entropy than does $NO(g)$ at the same temperature. Thus the following statement should be added to the generalizations about entropy on page 826.

In general, as seen on page 832, more complex molecules (e.g., with more atoms per molecule) produce greater molar entropies.

◄ In general, at low temperatures, because the quanta of energy are so small, translational energies are most important in establishing the entropy of gaseous molecules. As the temperature increases and the quanta of energy become larger, first rotational energies become important, and finally, at still higher temperatures, vibrational modes of motion start to contribute to the entropy.

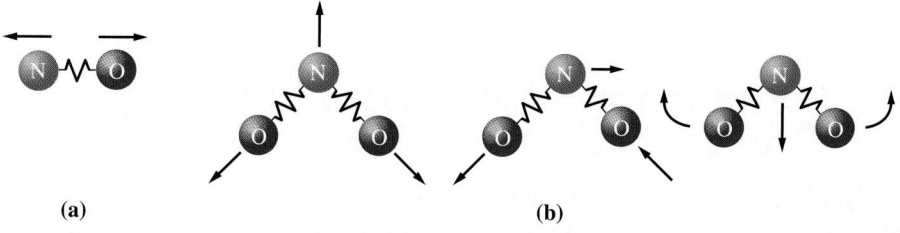

(a)   (b)

▲ FIGURE 19-8
Vibrational energy and entropy
The movement of atoms is suggested by the arrows. **(a)** The NO molecule has only one type of vibrational motion, whereas **(b)** the NO_2 molecule has three. This difference helps account for the fact that the molar entropy of $NO_2(g)$ is greater than that of $NO(g)$.

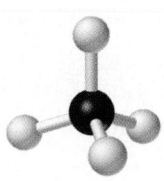

Methane, CH_4
$S° = 186.3 \text{ J mol}^{-1} \text{ K}^{-1}$

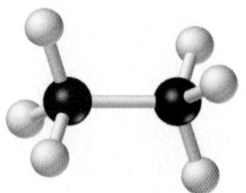

Ethane, C_2H_6
$S° = 229.6 \text{ J mol}^{-1} \text{ K}^{-1}$

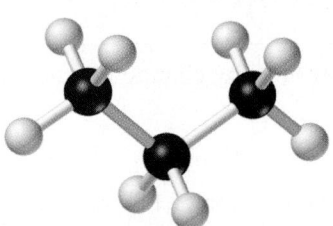

Propane, C_3H_8
$S° = 270.3 \text{ J mol}^{-1} \text{ K}^{-1}$

19-3 ARE YOU WONDERING...

Whether there is a correlation between standard entropy and enthalpy?

The answer to this question is yes, at least for solids heated to 298.15 K at 1 bar. To see why this is so we begin with the temperature variation of entropy and enthalpy written as

$$S° = \int_0^{298.15} \frac{C_p(T)}{T} dT$$

$$\Delta H° = \int_0^{298.15} C_p(T)dT$$

If we assume that the constant pressure heat capacity, C_p, of a solid is a linear function of temperature, $C_p = AT$, then the ratio of the entropy to enthalpy is

$$\frac{S°}{\Delta H°} = \frac{\int_{0\text{ K}}^{298.15\text{ K}} \frac{C_p(T)}{T}dT}{\int_{0\text{ K}}^{298.15\text{ K}} T\frac{C_p(T)}{T}dT} \approx \frac{\int_{0\text{ K}}^{298.15\text{ K}} AdT}{\int_{0\text{ K}}^{298.15\text{ K}} TAdT} = \frac{1}{149.075\text{ K}} = 0.0067\text{ K}^{-1}$$

This result suggests that a simple (linear) correlation exists between entropy and enthalpy. Do experimental facts support this conclusion? The following figure is a plot of standard entropy versus standard enthalpy change at 298.15 K and 1 bar for several monatomic solids.

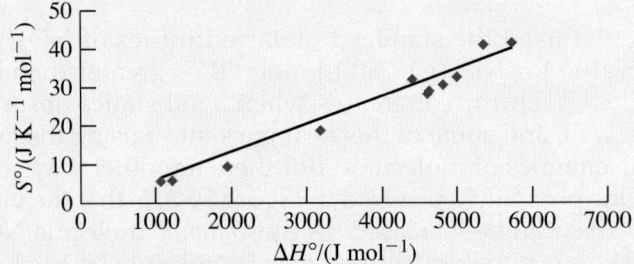

Standard Entropy Versus Enthalpy

The slope of the line is equal to the predicted ratio (0.0067 K^{-1}). Notice that the points lie close to this line. So what does this mean? As we heat the solid from nearly absolute zero to 298.15 K, the solid gains energy, represented here as a standard enthalpy change. This energy is dispersed throughout the various energy levels of the solid. The standard entropy value, $S°$, of the solid at 298.15 K and 1 bar is proportional to the quantity of energy stored in the solid at that temperature with the proportionality constant of 0.0067 K^{-1}.

19-4 Criteria for Spontaneous Change: The Second Law of Thermodynamics

In Section 19-2, we came to the tentative conclusion that processes in which the entropy of a system increases should be spontaneous and that processes in which entropy decreases should be nonspontaneous. But this statement can present difficulties, for example, of how to explain the spontaneous freezing of water at −10 °C. Because crystalline ice has a lower molar entropy than does liquid water, the freezing of water is a process for which entropy *decreases*. The way out of this dilemma is to recognize that *two* entropy changes must always be considered simultaneously—the entropy change of the system itself and the entropy change of the surroundings. The criteria for spontaneous change must be based on the sum of the two, called the entropy change of the "universe":

$$\Delta S_{total} = \Delta S_{universe} = \Delta S_{system} + \Delta S_{surroundings} \tag{19.6}$$

Although it is beyond the scope of this discussion to verify the following expression, the expression provides the basic criterion for spontaneous change. In a spontaneous change,

$$\Delta S_{univ} = \Delta S_{sys} + \Delta S_{surr} > 0 \tag{19.7}$$

Equation (19.7) is one way of stating the **second law of thermodynamics**. Another way is through the following statement.

> All spontaneous processes produce an increase in the entropy of the universe.

According to expression (19.7), if a process produces positive entropy changes in both the system and its surroundings, the process is surely spontaneous. And if both these entropy changes are negative, the process is just as surely nonspontaneous. If one of the entropy changes is positive, and the other negative, whether the sum of the two is positive or negative depends on the relative magnitudes of the two changes. The freezing of water produces a negative entropy change in the system, but in the surroundings, which absorb heat, the entropy change is positive. As long as the temperature is below 0 °C, the entropy of the surroundings increases more than the entropy of the system decreases. Because the *total* entropy change is positive, the freezing of water below 0 °C is indeed spontaneous.

Gibbs Energy and Gibbs Energy Change

We could use expression (19.7) as the basic criterion for spontaneity (spontaneous change), but it would be very difficult to apply. To evaluate a total entropy change (ΔS_{univ}), we always have to evaluate ΔS for the surroundings. At best, this process is tedious, and in many cases it is not even possible, because we cannot figure out all the interactions between a system and its surroundings. Surely it is preferable to have a criterion that can be applied to *the system itself*, without having to worry about changes in the surroundings.

To develop this new criterion, let us explore a hypothetical process conducted at constant temperature and pressure and with work limited to pressure–volume work. This process is accompanied by a heat effect, q_p, which is equal to ΔH for the system (ΔH_{sys}) as seen in Section 7-6. The heat effect experienced by the surroundings is the *negative* of that for the system: $q_{surr} = -q_p = -\Delta H_{sys}$. Furthermore, if the hypothetical surroundings are large enough, the path by which heat enters or leaves the surroundings can be made *reversible*. That is, the quantity of heat can be made to produce only an infinitesimal change in the temperature of the surroundings. In this case, according to equation (19.2), the entropy change in the surroundings is $\Delta S_{surr} = -\Delta H_{sys}/T$.* Now substitute this value of ΔS_{surr} into equation (19.6), then multiply by T to obtain

$$T \Delta S_{univ} = T \Delta S_{sys} - \Delta H_{sys} = -(\Delta H_{sys} - T \Delta S_{sys})$$

---

*We cannot similarly substitute $\Delta H_{sys}/T$ for ΔS_{sys}. A process that occurs spontaneously is generally far removed from an equilibrium condition and is therefore *irreversible*. We cannot substitute q for an irreversible process into equation (19.2).

▲ J. Willard Gibbs (1839–1903)—a great "unknown" scientist
Gibbs, a Yale University professor of mathematical physics, spent most of his career without recognition, partly because his work was abstract and partly because his important publications were in little-read journals. Yet today, Gibbs's ideas serve as the basis of most of chemical thermodynamics.

▶ Sometimes the term "$T \Delta S$" is referred to as "organizational energy," because ΔS is related to the way the energy of a system is distributed among the available energy levels.

Finally, multiply by -1 (change signs).

$$-T \Delta S_{univ} = \Delta H_{sys} - T \Delta S_{sys} \qquad (19.8)$$

This is the significance of equation (19.8). The right side of this equation has terms involving *only the system*. On the left side appears the term ΔS_{univ}, which embodies the criterion for spontaneous change, that for a spontaneous process, $\Delta S_{univ} > 0$.

Equation (19.8) is generally cast in a somewhat different form that requires a new thermodynamic function, called the **Gibbs energy**, G. The Gibbs energy for a system is defined by the equation

$$G = H - TS$$

The **Gibbs energy change, ΔG,** for a process at constant T is

$$\Delta G = \Delta H - T \Delta S \qquad (19.9)$$

In equation (19.9), all the terms apply to measurements *on the system*. All reference to the surroundings has been eliminated. Also, when we compare equations (19.8) and (19.9), we get

$$\Delta G = -T \Delta S_{univ}$$

Now, by noting that ΔG is *negative* when ΔS_{univ} is *positive*, we have our final criterion for spontaneous change based on properties of only the system itself.

For a process occurring at constant T and P, these statements hold true.

- If $\Delta G < 0$ (*negative*), the process is *spontaneous*.
- If $\Delta G > 0$ (*positive*), the process is *nonspontaneous*.
- If $\Delta G = 0$ (*zero*), the process is *at equilibrium*.

Evaluation of the units in equation (19.9) shows that Gibbs energy is indeed an energy term. ΔH has the unit joules per mole ($J\ mol^{-1}$), and the product $T \Delta S$ has the units $K \times J\ mol^{-1}\ K^{-1} = J\ mol^{-1}$. ΔG is the difference in two quantities with units of energy.

Applying the Gibbs Energy Criteria for Spontaneous Change

Later, we will look at quantitative applications of equation (19.9), but for now we can use the equation to make some qualitative predictions. Altogether there are four possibilities for ΔG on the basis of the signs of ΔH and ΔS. These possibilities are outlined in Table 19.1 and demonstrated in Example 19-4.

If ΔH is *negative* and ΔS is *positive*, the expression $\Delta G = \Delta H - T \Delta S$ is negative at all temperatures. The process is spontaneous at all temperatures. This corresponds to the situation noted previously in which both ΔS_{sys} and ΔS_{surr} are positive and ΔS_{univ} is also positive.

Unquestionably, if a process is accompanied by an *increase* in enthalpy (heat is absorbed) and a *decrease* in entropy, ΔG is positive at all temperatures

TABLE 19.1 Criteria for Spontaneous Change: $\Delta G - \Delta H - T\Delta S$

Case	ΔH	ΔS	ΔG	Result	Example
1.	−	+	−	spontaneous at all temp.	$2\ N_2O(g) \longrightarrow 2\ N_2(g) + O_2(g)$
2.	−	−	− +	spontaneous at low temp. nonspontaneous at high temp.	$H_2O(l) \longrightarrow H_2O(s)$
3.	+	+	+ −	nonspontaneous at low temp. spontaneous at high temp.	$2\ NH_3(g) \longrightarrow N_2(g) + 3\ H_2(g)$
4.	+	−	+	nonspontaneous at all temp.	$3\ O_2(g) \longrightarrow 2\ O_3(g)$

and the process is nonspontaneous. This corresponds to a situation in which both ΔS_{sys} and ΔS_{surr} are negative and ΔS_{univ} is also negative.

The questionable cases are those in which the entropy and enthalpy factors work in opposition—that is, with ΔH and ΔS both negative or *both* positive. In these cases, whether a reaction is spontaneous or not (that is, whether ΔG is negative or positive) depends on temperature. In general, if a reaction has negative values for both ΔH and ΔS, it is spontaneous at *lower* temperatures, whereas if ΔH and ΔS are both positive, the reaction is spontaneous at *higher* temperatures.

◀ For cases 2 and 3, there is a particular temperature at which a process switches from being spontaneous to being nonspontaneous. Section 19-6 explains how to determine such a temperature.

EXAMPLE 19-4   **Using Enthalpy and Entropy Changes to Predict the Direction of Spontaneous Change**

Under what temperature conditions would the following reactions occur spontaneously?

(a) $2\,NH_4NO_3(s) \longrightarrow 2\,N_2(g) + 4\,H_2O(g) + O_2(g)$ $\Delta H° = -236.0\ kJ\ mol^{-1}$

(b) $I_2(g) \longrightarrow 2\,I(g)$

Analyze

(a) The reaction is exothermic, and in Example 19-1(a) we concluded that $\Delta S > 0$ because large quantities of gases are produced.

(b) Because one mole of gaseous reactant produces two moles of gaseous product, we expect entropy to increase. But what is the sign of ΔH? We could calculate ΔH from enthalpy of formation data, but there is no need to. In the reaction, covalent bonds in $I_2(g)$ are broken and no new bonds are formed. Because energy is absorbed to break bonds, ΔH must be positive. With $\Delta H > 0$ and $\Delta S > 0$, case 3 in Table 19.1 applies.

Solve

(a) With $\Delta H < 0$ and $\Delta S > 0$, this reaction should be spontaneous at all temperatures (case 1 in Table 19.1). $NH_4NO_3(s)$ exists only because the decomposition has very high activation energy.

(b) ΔH is larger than $T\,\Delta S$ at low temperatures, and the reaction is nonspontaneous. At high temperatures, the $T\,\Delta S$ term becomes larger than ΔH, ΔG becomes negative, and the reaction is spontaneous.

Assess

We observe that reaction spontaneity depends on a balance of enthalpy, entropy, and temperature. Table 19.1 is a good summary of the conditions in which reactions will be spontaneous or nonspontaneous.

PRACTICE EXAMPLE A: Which of the four cases in Table 19.1 would apply to each of the following reactions:

(a) $N_2(g) + 3\,H_2(g) \longrightarrow 2\,NH_3(g),$ $\Delta H° = -92.22\ kJ;$

(b) $2\,C(graphite) + 2\,H_2(g) \longrightarrow C_2H_4(g),$ $\Delta H° = 52.26\ kJ?$

PRACTICE EXAMPLE B: Under what temperature conditions would the following reactions occur spontaneously? **(a)** The decomposition of calcium carbonate into calcium oxide and carbon dioxide. **(b)** The "roasting" of zinc sulfide in oxygen to form zinc oxide and sulfur dioxide. This exothermic reaction releases 439.1 kJ for every mole of zinc sulfide that reacts.

Example 19-4(b) illustrates why there is an upper temperature limit for the stabilities of chemical compounds. No matter how positive the value of ΔH for dissociation of a molecule into its atoms, the term $T\,\Delta S$ will eventually exceed ΔH in magnitude as the temperature increases. Known temperatures range from near absolute zero to the interior temperatures of stars (about 3×10^7 K). Molecules exist only at limited temperatures (up to about 1×10^4 K or about 0.03% of this total temperature range).

▲ Only because of the high activation energy for the $NH_4NO_3(s)$ decomposition reaction does $NH_4NO_3(s)$ exist at all.

◀ A related observation is that only a small fraction of the mass of the universe is in molecular form.

The method of Example 19-4 is adequate for making predictions about the sign of ΔG, but we will also want to use equation (19.9) to calculate numerical values. We will do that in Section 19-5.

Gibbs Energy Change and Work

We might think that the quantity of energy available to do work in the surroundings as a result of a chemical process is $-\Delta H$. This would be the same as the quantity of heat that an exothermic reaction releases to the surroundings. (In thinking along those lines, we would say that an endothermic reaction is incapable of doing work.) However, that quantity of heat must be adjusted for the heat requirement in producing the entropy change in the system ($q_{rev} = T\,\Delta S$). If an exothermic reaction is accompanied by an *increase* in entropy, the amount of energy available to do work in the surroundings is *greater* than $-\Delta H$. If entropy *decreases* in the exothermic reaction, the amount of energy available to do work is *less* than $-\Delta H$. But notice that in either case, this amount of energy is equal to $-\Delta G$. Thus, the amount of work that can be extracted from a chemical process is $-\Delta G$. Because $-\Delta G$ represents the energy freely available for doing work, G was once called *Gibbs free energy*, or *simply free energy*, by most chemists. Notice also that this interpretation of Gibbs energy allows for the possibility of work being done in an endothermic process if $T\,\Delta S$ exceeds ΔH. In Chapter 20, we will see how the Gibbs energy change of a reaction can be converted to electrical work. In any case, do not think of Gibbs energy as being free. Costs are always involved in tapping an energy source.

19-5   Standard Gibbs Energy Change, $\Delta G°$

Because Gibbs energy is related to enthalpy ($G = H - TS$), we cannot establish absolute values of G any more than we can for H. We must work with Gibbs energy changes, ΔG. We will find a special use for the Gibbs energy change corresponding to reactants and products in their standard states—the **standard Gibbs energy change, $\Delta G°$**. The standard state conventions were introduced and applied to enthalpy change in Chapter 7.

The **standard Gibbs energy of formation**, $\Delta G_f°$, is the Gibbs energy change for a reaction in which a substance in its standard state is formed from its elements in their reference forms in their standard states. And, as was the case when we established enthalpies of formation in Section 7-8, this definition leads to values of *zero* for the Gibbs energies of formation of the elements in their reference forms at a pressure of 1 bar. Other Gibbs energies of formation are related to this condition of zero and are generally tabulated per mole of substance (see Appendix D).

Some additional relationships involving Gibbs energy changes are similar to those presented for enthalpy in Section 7-7: (1) ΔG changes sign when a process is reversed; and (2) ΔG for an overall process can be obtained by summing the ΔG values for the individual steps. The two expressions that follow are useful in calculating $\Delta G°$ values, depending on the data available. We can use the first expression at any temperature for which $\Delta H°$ and $\Delta S°$ values are known. We can use the second expression only at temperatures at which $\Delta G_f°$ values are known. The only temperature at which tabulated data are commonly given is 298.15 K. The first expression is applied in Example 19-5 and Practice Example 19-5A, and the second expression in Practice Example 19-5B.

$$\Delta G° = \Delta H° - T\Delta S°$$

$$\Delta G° = \left[\sum \nu_p\,\Delta G_f°(\text{products}) - \sum \nu_r\Delta G_f°(\text{reactants})\right]$$

KEEP IN MIND

that a pressure of 1 bar is very nearly the same as 1 atm. The differences in these two pressures on the values of properties is generally so small that we can use the two pressure units almost interchangeably.

▶ Many substances do not exist at the standard conditions that define their Gibbs energy. This does not matter since we can calculate the standard Gibbs energy from nonstandard conditions. Dealing with nonstandard conditions is discussed on page 839. Gibbs energies are listed under standard conditions for ease and conciseness.

EXAMPLE 19-5   Calculating $\Delta G°$ for a Reaction

Determine $\Delta G°$ at 298.15 K for the reaction

$$2\,NO(g) + O_2(g) \longrightarrow 2\,NO_2(g) \quad (\text{at } 298.15\ K) \qquad \Delta H° = -114.1\ kJ$$
$$\Delta S° = -146.5\ J\,K^{-1}$$

Analyze

Because we have values of $\Delta H°$ and $\Delta S°$, the most direct method of calculating $\Delta G°$ is to use the expression $\Delta G° = \Delta H° - T\,\Delta S°$.

Solve

First we must convert all the data to a common energy unit (for instance, kJ).

$$\Delta G° = -114.1\ kJ - (298.15\ K \times -0.1465\ kJ\,K^{-1})$$
$$= -114.1\ kJ + 43.68\ kJ$$
$$= -70.4\ kJ$$

Assess

In this type of problem, one of the most common mistakes is not keeping the units straight. Note that the unit for the standard molar enthalpy is $kJ\,mol^{-1}$, and for the standard molar entropy it is $J\,mol^{-1}\,K^{-1}$. This example says that all the reactants and products are maintained at 25 °C and 1 bar pressure. Under these conditions, the Gibbs energy change is -70.4 kJ for oxidizing two moles of NO to two moles of NO_2. To do this, it is necessary to replenish the NO so as to maintain the standard conditions.

PRACTICE EXAMPLE A: Determine $\Delta G°$ at 298.15 K for the reaction $4\,Fe(s) + 3O_2(g) \longrightarrow 2\,Fe_2O_3(s)$. $\Delta H° = -1648$ kJ and $\Delta S° = -549.3\ J\,K^{-1}$.

PRACTICE EXAMPLE B: Determine $\Delta G°$ for the reaction in Example 19-5 by using data from Appendix D. Compare the two results.

19-6   Gibbs Energy Change and Equilibrium

We have seen that $\Delta G < 0$ for spontaneous processes and that $\Delta G > 0$ for nonspontaneous processes. If $\Delta G = 0$, the forward and reverse processes show an equal tendency to occur, and the system is at *equilibrium*. At this point, even an infinitesimal change in one of the system variables (such as temperature or pressure) will cause a net change to occur. But if a system at equilibrium is left undisturbed, no net change occurs with time.

Figure 19-9 illustrates a hypothetical process in which ΔH and ΔS are independent of temperature, and both are positive. This corresponds to case 3

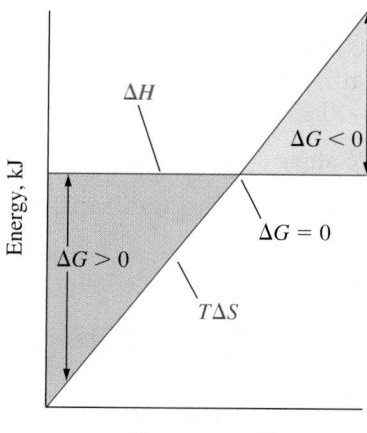

Temperature, K

◀ FIGURE 19-9
Gibbs energy change as a function of temperature
The green shaded region corresponds to ΔG at temperatures in which the process is *nonspontaneous*. The orange shaded region corresponds to ΔG at temperatures in which the process is *spontaneous*.

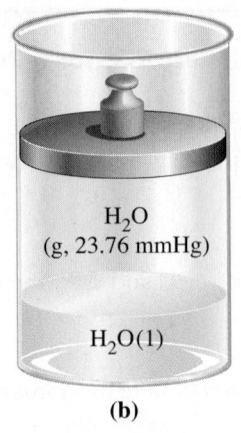
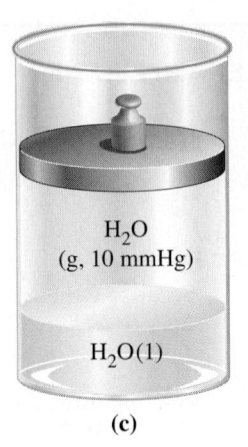

$H_2O(g, 1\text{ atm})$

$H_2O(l)$

(a)

H_2O
$(g, 23.76\text{ mmHg})$

$H_2O(l)$

(b)

H_2O
$(g, 10\text{ mmHg})$

$H_2O(l)$

(c)

▲ FIGURE 19-10
Liquid–vapor equilibrium and the direction of spontaneous change
(a) For the vaporization of water at 298.15 K and 1 atm, $H_2O(l, 1\text{ atm}) \longrightarrow$ $H_2O(g, 1\text{ atm})$, $\Delta G° = 8.590$ kJ. The direction of spontaneous change is the *condensation* of $H_2O(g)$. **(b)** At 298.15 K and 23.76 mmHg, the liquid and vapor are in equilibrium and $\Delta G° = 0$. **(c)** At 298.15 K and 10 mmHg, the vaporization of $H_2O(l)$ occurs spontaneously: $H_2O(l, 10\text{ mmHg}) \longrightarrow H_2O(g, 10\text{ mmHg})$, and $\Delta G < 0$.

in Table 19.1. If we start at the left side of the figure (i.e., at a relatively low temperature), we see that the magnitude of ΔH exceeds the magnitude of $T\,\Delta S$ and that $\Delta H - T\,\Delta S$, which was defined as ΔG, is positive; the process is *nonspontaneous*. As the temperature increases, the magnitude of ΔG decreases. At the right side of the figure (i.e., at a relatively high temperature), the magnitude of $T\,\Delta S$ exceeds the magnitude of ΔH; therefore, ΔG is negative; the process is *spontaneous*. The temperature at which the two lines (ΔH and $T\,\Delta S$) intersect, $\Delta G = 0$, and the system is at equilibrium.

For the vaporization of water, *with both liquid and vapor in their standard states* (which means that $\Delta G = \Delta G°$), the intersection of the two lines in Figure 19-9 is at $T = 373.15$ K (100.00 °C). That is, for the vaporization of water at 1 atm,

$$H_2O(l, 1\text{ atm}) \rightleftharpoons H_2O(g, 1\text{ atm}) \qquad \Delta G° = 0 \text{ at } 373.15 \text{ K}$$

At 25 °C, the $\Delta H°$ line lies above the $T\,\Delta S$ line in Figure 19-9. This means that $\Delta G° > 0$.

$$H_2O(l, 1\text{ atm}) \longrightarrow H_2O(g, 1\text{ atm}) \qquad \Delta G° = +8.590 \text{ kJ at } 298.15 \text{ K}$$

The positive value of $\Delta G°$ does not mean that vaporization of water will not occur. From common experience, we know that water spontaneously evaporates at room temperature. What the positive value means is that liquid water will not spontaneously produce $H_2O(g)$ at 1 atm pressure at 25 °C. Instead, $H_2O(g)$ is produced with a vapor pressure that is less than 1 atm pressure. The equilibrium vapor pressure of water at 25 °C is 23.76 mmHg = 0.03126 atm; that is,

$$H_2O(l, 0.03126\text{ atm}) \rightleftharpoons H_2O(g, 0.03126\text{ atm}) \qquad \Delta G = 0$$

Figure 19-10 offers a schematic summary of ideas concerning the transition between liquid and gaseous water at 25 °C.

▶ Here, we have used a standard state pressure of 1 atm. See Feature Problem 93 for an appraisal of this matter using a standard state pressure of 1 bar.

▶ The liquid–vapor equilibrium represented here is out of contact with the atmosphere. In the presence of the atmosphere, the pressure on the liquid would be barometric pressure, whereas that of the vapor would remain essentially unchanged at 0.03126 atm.

🔍 **19-3 CONCEPT ASSESSMENT**

Redraw Figure 19-9 for case 2 in Table 19.1. How does your drawing compare with Figure 19-9? State the similarities and differences.

Relationship of $\Delta G°$ To ΔG for Nonstandard Conditions

If you think about the situation just described for the vaporization of water, there is not much value in describing equilibrium in a process in terms of its $\Delta G°$ value. At only one temperature are the reactants in their standard states in equilibrium with products in their standard states; that is, at only one temperature does $\Delta G° = 0$. We want to be able to describe equilibrium for a variety of conditions, typically *nonstandard* conditions. Many reactions, such as processes occurring under physiological conditions, take place under nonstandard conditions. How, under such circumstances, can a biochemist decide which processes are spontaneous? For this we need to work with ΔG, not $\Delta G°$.

To obtain the relationship between ΔG and $\Delta G°$, we will consider a reaction between ideal gas molecules. We assume this is the case in the reaction of nitrogen with hydrogen to produce ammonia.

$$N_2(g) + 3\,H_2(g) \rightleftharpoons 2\,NH_3(g)$$

The expressions for ΔG and $\Delta G°$ are $\Delta G = \Delta H - T\,\Delta S$ and $\Delta G° = \Delta H° - T\,\Delta S°$, respectively. First consider how the enthalpy terms ΔH and $\Delta H°$ are related for an ideal gas. The enthalpy of an ideal gas is a function of temperature only; it is independent of pressure. Thus, under any mixing conditions for an ideal gas, $\Delta H = \Delta H°$. We can write

$$\Delta G = \Delta H° - T\,\Delta S \qquad \textit{(ideal gas)}$$

We now need to obtain a relationship between ΔS and $\Delta S°$. To do so, we consider the isothermal expansion of an ideal gas for which $q = -w$ and $\Delta U = 0$. If the expansion occurs reversibly (recall Figure 7-12), the work of expansion for one mole of an ideal gas is given by an equation derived in Feature Problem 125 in Chapter 7.

$$w = -RT \ln\frac{V_f}{V_i} \qquad \textit{(reversible, isothermal)}$$

The reversible, isothermal heat of expansion is

$$q_{rev} = -w = RT \ln\frac{V_f}{V_i}$$

From equation (19.2), we obtain the entropy change for the isothermal expansion of one mole of an ideal gas.

$$\Delta S = \frac{q_{rev}}{T} = R \ln\frac{V_f}{V_i} \qquad \textbf{(19.10)}$$

Using equation (19.10), we can now evaluate the entropy of an ideal gas under any conditions of pressure. From the ideal gas equation, we know that the volume of an ideal gas is inversely proportional to the pressure, so we can recast equation (19.10) as

$$\Delta S = S_f - S_i = R \ln\frac{V_f}{V_i} = R \ln\frac{P_i}{P_f} = -R \ln\frac{P_f}{P_i}$$

where P_i and P_f are the initial and final pressures, respectively. If we set $P_i = 1$ bar and designate P_i as $P°$ and S_i as $S°$, we obtain for the entropy at any pressure P

$$S = S° - R \ln\frac{P}{P°} = S° - R \ln\frac{P}{1} = S° - R \ln P \qquad \textbf{(19.11)}$$

Now, let's return to the ammonia synthesis reaction and calculate the entropy change for that reaction. We begin by applying equation (19.11) to each of the three gases.

$$S_{NH_3} = S°_{NH_3} - R \ln P_{NH_3} \qquad S_{N_2} = S°_{N_2} - R \ln P_{N_2} \qquad S_{H_2} = S°_{H_2} - R \ln P_{H_2}$$

Then we substitute the above values into the equation $\Delta S = 2S_{NH_3} - S_{N_2} - 3S_{H_2}$ to obtain

$$\Delta S = 2S^\circ_{NH_3} - 2R \ln P_{NH_3} - S^\circ_{N_2} + R \ln P_{N_2} - 3S^\circ_{H_2} + 3R \ln P_{H_2}$$

By rearranging the terms, we get

$$\Delta S = 2S^\circ_{NH_3} - S^\circ_{N_2} - 3S^\circ_{H_2} - 2R \ln P_{NH_3} + R \ln P_{N_2} + 3R \ln P_{H_2}$$

and, since the first three terms on the right-hand side of the above equation represent ΔS°, we have

$$\Delta S = \Delta S^\circ - 2R \ln P_{NH_3} + R \ln P_{N_2} + 3R \ln P_{H_2}$$

$$\Delta S = \Delta S^\circ - R \ln P^2_{NH_3} + R \ln P_{N_2} + R \ln P^3_{H_2}$$

$$\Delta S = \Delta S^\circ + R \ln \frac{P_{N_2} P^3_{H_2}}{P^2_{NH_3}}$$

Finally, we can write the equation for ΔG by substituting the expression for ΔS into the equation

$$\Delta G = \Delta H^\circ - T \Delta S \qquad\qquad (ideal\ gas)$$

▶ It is commonly stated that we are headed for a state of maximum entropy. Although this is true, it gives the incorrect notion that entropy, as defined here, and time are related. They are not. Entropy is a property of an equilibrium state. Here we calculate the difference between the entropies of two different equilibrium states.

19-4 ARE YOU WONDERING...

If there is a microscopic approach to obtaining equation (19.10)?

To do this, we use the ideas of Ludwig Boltzmann. Consider an ideal gas at an initial volume V_i and allow the gas to expand isothermally to a final volume V_f. By using the Boltzmann equation, we find that for the change in entropy,

$$\Delta S = S_f - S_i = k \ln W_f - k \ln W_i$$

$$\Delta S = k \ln \frac{W_f}{W_i}$$

where k is the Boltzmann constant, S_i and S_f are the initial and final entropies, respectively, and W_i and W_f are the number of microstates for the initial and final macroscopic states of the gas, respectively. We must now obtain a value for the ratio W_f/W_i. To do that, suppose that there is only a single gas molecule in a container. The number of microstates available to this single molecule should be proportional to the number of positions where the molecule can be and, hence, to the volume of the container. That is also true for each molecule in a system of N_A particles—Avogadro's number of particles. The number of microstates available to the whole system is

$$W_{total} = W_{particle\ 1} \times W_{particle\ 2} \times W_{particle\ 3} \times \cdots$$

Because the number of microstates for each particle is proportional to the volume V of the container, the number of microstates for N_A (Avogadro's number) ideal gas molecules is

$$W \propto V^{N_A}$$

Thus, the ratio of the microstates for isothermal expansion is

$$\frac{W_f}{W_i} = \left(\frac{V_f}{V_i}\right)^{N_A}$$

We can now calculate ΔS as follows:

$$\Delta S = k \ln \frac{W_f}{W_i} = k \ln \left(\frac{V_f}{V_i}\right)^{N_A} = N_A k \ln \left(\frac{V_f}{V_i}\right) = R \ln \left(\frac{V_f}{V_i}\right)$$

where R is the ideal gas constant. This equation, which gives the entropy change for the expansion of one mole of ideal gas, is simply equation (19.10).

This leads to

$$\Delta G = \Delta H^\circ - T\,\Delta S^\circ - RT\ln\frac{P_{N_2}P_{H_2}^3}{P_{NH_3}^2}$$

$$\Delta G = \underbrace{\Delta H^\circ - T\,\Delta S^\circ} + RT\ln\frac{P_{NH_3}^2}{P_{N_2}P_{H_2}^3}$$

$$\Delta G = \Delta G^\circ + RT\ln\frac{P_{NH_3}^2}{P_{N_2}P_{H_2}^3}$$

To simplify, we designate the quotient in the logarithmic term as the *reaction quotient Q* (recall page 670).

$$\Delta G = \Delta G^\circ + RT\ln Q \qquad\qquad \textbf{(19.12)}$$

Equation (19.12) is the relationship between ΔG and ΔG° that we have been seeking in this section, and we see that the key term in the equation is the reaction quotient formulated for the actual, nonstandard conditions. We can use equation (19.12) to decide on the spontaneity of a reaction under any conditions of composition, provided that the temperature and pressure at which we observe the reaction are constant. We turn now to describing how the standard Gibbs energy change is related to the equilibrium constant.

◀ Equation (19.12) shows that the value of the reaction quotient affects whether the forward or reverse reaction is favored under a particular set of conditions. Recall that all reactions proceed toward equilibrium where Gibbs energy change is equal to zero.

Relationship of ΔG° to the Equilibrium Constant K

We encounter an interesting situation when we apply equation (19.12) to a reaction at equilibrium. We have learned that at equilibrium $\Delta G = 0$, and in Chapter 15 we saw that if a system is at equilibrium $Q = K$. So, we can write that, *at equilibrium*,

$$\Delta G = \Delta G^\circ + RT\ln K = 0$$

which means that

$$\Delta G^\circ = -RT\ln K \qquad\qquad \textbf{(19.13)}$$

If we have a value of ΔG° at a given temperature, we can use equation (19.13) to calculate an equilibrium constant K. This means that the tabulation of thermodynamic data in Appendix D can serve as a direct source of countless equilibrium constant values at 298.15 K.

We need to say a few words about the units required in equation (19.13). Because logarithms can be taken of dimensionless numbers only, K has no units; neither does $\ln K$. The right-hand side of equation (19.13) has the unit of "RT": $J\,mol^{-1}\,K^{-1} \times K = J\,mol^{-1}$. ΔG°, on the left-hand side of the equation, must have the same unit: $J\,mol^{-1}$. The "mol^{-1}" part of this unit means "per mole of reaction." One mole of reaction is simply the reaction based on the stoichiometric coefficients chosen for the balanced equation. Strictly speaking the "mol^{-1}" should be included. Remember that the "mol^{-1}" is of reaction not per mole of substance.

◀ When a ΔG° value is accompanied by a chemical equation, the "mol^{-1}" portion of the unit is often dropped.

🔍 19-4 CONCEPT ASSESSMENT

For the reaction below, $\Delta G^\circ = 326.4\ kJ\,mol^{-1}$:

$$3\,O_2(g) \longrightarrow 2\,O_3(g)$$

What is the Gibbs energy change for the system when 1.75 mol $O_2(g)$ at 1 bar reacts completely to give $O_3(g)$ at 1 bar?

19-5 ARE YOU WONDERING...

Whether there is a relationship between thermodynamics and the reaction rates?

Recall that when a reaction achieves equilibrium, the forward rate and reverse rate are equal. When the elementary reaction $A \underset{k_{-1}}{\overset{k_1}{\rightleftharpoons}} B$ reaches equilibrium we can write $k_1[A] = k_{-1}[B]$. Let us rearrange this expression such that the rate constants are on one side, and the concentrations are on the other.

$$\frac{k_1}{k_{-1}} = \frac{[B]}{[A]}$$

The right side of the above equation is what we have defined as the equilibrium constant expression, $K = [B]/[A]$. The equilibrium constant is related to Gibbs energy by the expression $K = e^{-\Delta G°/RT}$, showing that $\frac{k_1}{k_{-1}} = e^{-\Delta G°/RT}$. This expression is valid only when the overall reaction mechanism is one step.

▶ From a theoretical standpoint, as can be seen from Figure 19-11, all chemical reactions reach equilibrium and no chemical reaction goes totally to completion.

Criteria for Spontaneous Change: Our Search Concluded

The graphs plotted in Figure 19-11 represent the culmination of our quest for criteria for spontaneous change. Unfortunately, to construct these plots in all their detail is beyond the scope of this text. However, we can rationalize their

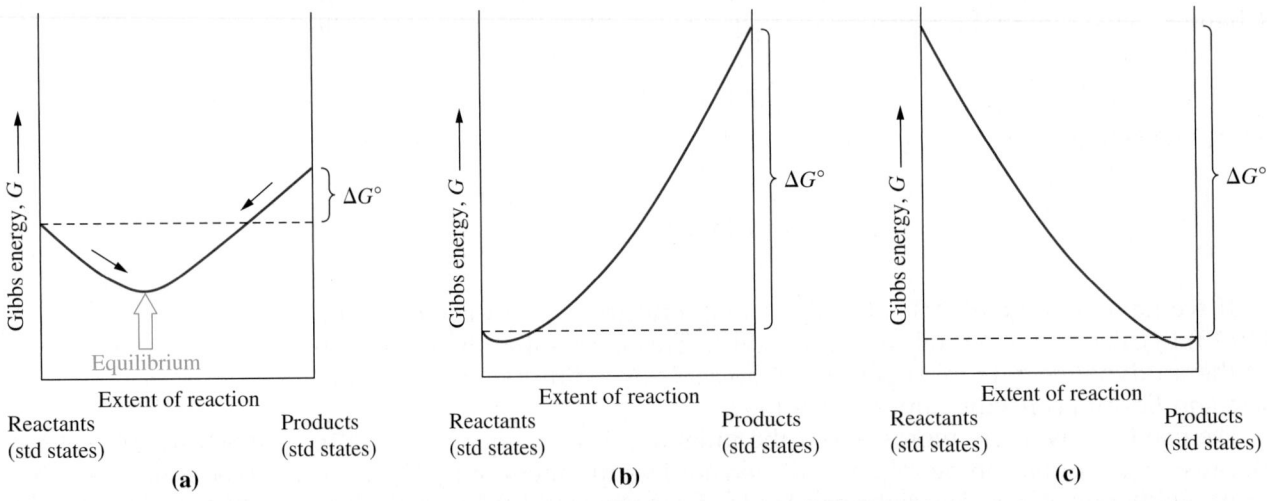

(a)   (b)   (c)

▲ FIGURE 19-11
Gibbs energy change, equilibrium, and the direction of spontaneous change
Gibbs energy is plotted against the extent of reaction for a hypothetical reaction. $\Delta G°$ is the difference between the standard Gibbs energies of formation of products and reactants. The equilibrium point lies somewhere between pure reactants and pure products. **(a)** $\Delta G°$ is small, so the equilibrium mixture lies about midway between the two extremes of pure products or reactants in their standard states. The effect of nonstandard conditions can be deduced from the slope of the curve. Mixtures with $Q > K$ are to the right of the equilibrium point and undergo spontaneous change in the direction of lower Gibbs energy, eventually coming to equilibrium. Similarly, mixtures with $Q < K$ are to the left of the equilibrium point and spontaneously yield more products before reaching equilibrium. **(b)** $\Delta G°$ is large and positive, so the equilibrium point lies close to the extreme of pure reactants in their standard states. Consequently, very little reaction takes place before equilibrium is reached. **(c)** $\Delta G°$ is large and negative, so the equilibrium point lies close to the extreme of pure products in their standard states; the reaction goes essentially to completion.

▶ Here the system starts at standard conditions and runs down to equilibrium because the reactants are not replenished. At equilibrium there is no driving force for the reaction, so there is no further reaction and the Gibbs energy must be at a minimum.

TABLE 19.2   Significance of the Magnitude of $\Delta G°$ (at 298 K)

$\Delta G°$	K	Significance
+200 kJ/mol	9.1×10^{-36}	Equilibrium favors reactants
+100	3.0×10^{-18}	
+50	1.7×10^{-9}	
+10	1.8×10^{-2}	
+1.0	6.7×10^{-1}	Equilibrium calculation is necessary
0	1.0	
−1.0	1.5	
−10	5.6×10^{1}	
−50	5.8×10^{8}	Equilibrium favors products
−100	3.3×10^{17}	
−200	1.1×10^{35}	

general shape on the basis of two ideas: (1) Every chemical reaction consists of both a forward and a reverse reaction, even if one of these occurs only to a very slight extent. (2) The direction of spontaneous change in both the forward and reverse reactions is the direction in which Gibbs energy decreases ($\Delta G < 0$). As a consequence, Gibbs energy reaches a minimum at some point between the left-hand and right-hand sides of the graph. This minimum is the equilibrium point in the reaction.

Now consider the vertical distance between the two end points of the graph; this distance represents $\Delta G°$ of the reaction. If, as in Figure 19-11(a), $\Delta G°$ of a reaction is small, either positive or negative, the equilibrium condition is one in which significant amounts of both reactants and products will be found. If $\Delta G°$ is a *large, positive* quantity, as in Figure 19-11(b), the equilibrium point lies far to the left—that is, very close to the reactants side. If $\Delta G°$ is a *large, negative* quantity, as in Figure 19-11(c), the equilibrium point lies far to the right—that is, very close to the products side. Table 19.2 summarizes the conclusions of this discussion, gives approximate magnitudes to the terms *small* and *large*, and relates $\Delta G°$ values to values of K.

$\Delta G°$ and ΔG: Predicting the Direction of Chemical Change

We have considered both $\Delta G°$ and ΔG in relation to the spontaneity of chemical reactions, and this is a good time to summarize some ideas about them.

$\Delta G < 0$ signifies that a reaction or process is spontaneous in the forward direction (to the right) for the stated conditions.

$\Delta G° < 0$ signifies that the forward reaction is spontaneous when reactants and products are in their standard states. It further signifies that $K > 1$, whatever the initial concentrations or pressures of reactants and products.

$\Delta G = 0$ signifies that the reaction is at equilibrium under the stated conditions.

$\Delta G° = 0$ signifies that the reaction is at equilibrium when reactants and products are in their standard states. It further signifies that $K = 1$, which can occur only at a particular temperature.

$\Delta G > 0$ signifies that the reaction or process is nonspontaneous in the forward direction under the stated conditions.

$\Delta G° > 0$ signifies that the forward reaction is nonspontaneous when reactants and products are in their standard states. It further signifies that $K < 1$, whatever the initial concentrations or pressures of reactants and products.

$\Delta G = \Delta G°$ only when all reactants and products are in their standard states. Otherwise, $\Delta G = \Delta G° + RT \ln Q$.

The Thermodynamic Equilibrium Constant: Activities

When we derived the equation $\Delta G = \Delta G° + RT \ln Q$, we used the relationship given in equation (19.11),

$$S = S° - R \ln \frac{P}{P°} = S° - R \ln \frac{P}{1}$$

where for a gas, we defined a standard state of 1 bar, the reference state for values of entropy. The ratio $P/P°$ is dimensionless, which is essential for a term appearing in a logarithm. At the time, we were considering a gas-phase reaction, but we must be able to discuss reactions in solutions also, so we need a more general approach. For this, we need to return to the concept of *activity* introduced in Section 15-2, which requires us to write

$$S = S° - R \ln a$$

where a is the activity, defined as

$$a = \frac{\text{the effective concentration of a substance in the system}}{\text{the effective concentration of that substance in a standard reference state}}$$

In a gas-phase reaction, we express pressure in bars and take the reference state to be 1 bar. In this way and as expected, activity is a dimensionless quantity.

In extending the activity concept to solutions, we define the reference state as a 1 M solution, so the activity of a substance is the numerical value of its molarity. Thus, the activity of protons in a 0.1 M solution of HCl in water is

$$a_{H^+} = \frac{0.1 \text{ M}}{1 \text{ M}} = 0.1$$

Another situation that we have encountered is that of a heterogeneous equilibrium, such as

$$CaCO_3(s) \rightleftharpoons CaO(s) + CO_2(g)$$

Recall that we choose the pure solids as the reference states, but the effective concentrations of the $CaCO_3(s)$ and $CaO(s)$ in the system are also those of the pure solids. Consequently, the activity of a solid is unity. This conclusion agrees with our observation in Figure 15-4 that the addition of either $CaO(s)$ or $CaCO_3(s)$ to an equilibrium mixture of $CaCO_3(s)$, $CO_2(g)$, and $CaO(s)$ has no effect on the pressure of $CO_2(g)$. The activities of $CaO(s)$ and $CaCO_3(s)$ are constant (unity).

19-6 ARE YOU WONDERING...

How to deal with activities under nonideal conditions?

For nonideal conditions, we sweep all deviations from ideality into an experimentally determined correction factor called the *activity coefficient*, γ. Thus, for real (nonideal) systems, we write

For a gas:	$a_G = \gamma P_G$
For a solute:	$a_X = \gamma[X]$
For a pure solid or liquid:	$a_L = a_S = 1$

More advanced treatments of this topic show how γ is related to the composition of the system. For dilute solutions or near ideal gases, we will assume that $\gamma = 1$. However, keep in mind that the results we obtain are limited by the validity of this assumption.

In summary, we can make these statements:

- *For pure solids and liquids:* The activity $a = 1$. The reference state is the pure solid or liquid.
- *For gases:* With ideal gas behavior assumed, the activity is replaced by the numerical value of the gas pressure in bars. The reference state is the gas at 1 bar at the temperature of interest. Thus, the activity of a gas at 0.50 bar pressure is $a = (0.50\ \text{bar})/(1\ \text{bar}) = 0.50$. (Recall also, that 1 bar of pressure is almost identical to 1 atm.)
- *For solutes in aqueous solution:* With ideal solution behavior assumed (for example, no interionic attractions), the activity is replaced by the numerical value of the molarity. The reference state is a 1 M solution. Thus, the activity of the solute in a 0.25 M solution is $a = (0.25\ \text{M})/(1\ \text{M}) = 0.25$.

When an equilibrium expression is written in terms of activities, the equilibrium constant is called the **thermodynamic equilibrium constant**. The thermodynamic equilibrium constant is dimensionless and thus appropriate for use in equation (19.13).

Thermodynamic equilibrium constants, K, are sometimes identical to K_c and K_p values, as in parts (a) and (b) of Example 19-6. In other instances, as in part (c), this is not the case. In working through Example 19-6, keep in mind that the sole reason for writing thermodynamic equilibrium constants in this text is to get the proper value to use in equation (19.13). Note that the reaction quotient Q must also be written in the same manner as K when used in equation (19.12), as demonstrated in Example 19-7.

EXAMPLE 19-6   **Writing Thermodynamic Equilibrium Constant Expressions**

For the following reversible reactions, write thermodynamic equilibrium constant expressions, making appropriate substitutions for activities. Then equate K to K_c or K_p, where this can be done.

(a) The water gas reaction

$$C(s) + H_2O(g) \rightleftharpoons CO(g) + H_2(g)$$

(b) Formation of a saturated aqueous solution of lead(II) iodide, a very slightly soluble solute

$$PbI_2(s) \rightleftharpoons Pb^{2+}(aq) + 2\,I^-(aq)$$

(c) Oxidation of sulfide ion by oxygen gas (used in removing sulfides from wastewater, as in pulp and paper mills)

$$O_2(g) + 2\,S^{2-}(aq) + 2\,H_2O(l) \rightleftharpoons 4\,OH^-(aq) + 2\,S(s)$$

Analyze

In each case, once we have made the appropriate substitutions for activities, if all terms are molarities, the thermodynamic equilibrium constant is the same as K_c. If all terms are partial pressures, $K = K_p$. If both molarities *and* partial pressures appear in the expression, however, the equilibrium constant expression can be designated only as K.

Solve

(a) The activity of solid carbon is 1. Partial pressures are substituted for the activities of the gases.

$$K = \frac{a_{CO(g)}}{a_{C(s)}a_{H_2O(g)}} = \frac{(P_{CO})(P_{H_2})}{(P_{H_2O})} = \frac{(P_{CO})(P_{H_2})}{(P_{H_2O})} = K_p$$

(b) The activity of solid lead(II) iodide is 1. Molarities are substituted for activities of the ions in aqueous solution.

$$K = \frac{a_{Pb^{2+}(aq)}a^2_{I^-(aq)}}{a_{PbI_2(s)}} = [Pb^{2+}][I^-]^2 = K_c = K_{sp}$$

(continued)

(c) The activity of both the solid sulfur and the liquid water is 1. Molarities are substituted for the activities of $OH^-(aq)$ and $S^{2-}(aq)$. The partial pressure of $O_2(g)$ is substituted for its activity. Thus, the resulting K is neither a K_c nor a K_p.

$$K = \frac{a^4_{OH^-(aq)}a^2_{S(s)}}{a_{O_2(g)}a^2_{S^{2-}(aq)}a^2_{H_2O(l)}} = \frac{[OH^-]^4 \cdot (1)^2}{P_{O_2} \cdot [S^{2-}]^2 \cdot (1)^2} = \frac{[OH^-]^4}{P_{O_2} \cdot [S^{2-}]^2}$$

Assess

These are thermodynamic equilibrium expressions since they are written in terms of their activities. The values of the thermodynamic equilibrium constant will be dimensionless. The standard pressure and concentration terms have been left out for clarity.

PRACTICE EXAMPLE A: Write thermodynamic equilibrium constant expressions for each of the following reactions. Relate these to K_c or K_p where appropriate.

(a) $Si(s) + 2\,Cl_2(g) \rightleftharpoons SiCl_4(g)$

(b) $Cl_2(g) + H_2O(l) \rightleftharpoons HOCl(aq) + H^+(aq) + Cl^-(aq)$

PRACTICE EXAMPLE B: Write a thermodynamic equilibrium constant expression to represent the reaction of solid lead(II) sulfide with aqueous nitric acid to produce solid sulfur, a solution of lead(II) nitrate, and nitrogen monoxide gas. Base the expression on the balanced net ionic equation for the reaction.

EXAMPLE 19-7   Assessing Spontaneity for Nonstandard Conditions

For the decomposition of 2-propanol to form propanone (acetone) and hydrogen,

$$(CH_3)_2CHOH(g) \rightleftharpoons (CH_3)_2CO(g) + H_2(g)$$

the equilibrium constant is 0.444 at 452 K. Is this reaction spontaneous under standard conditions? Will the reaction be spontaneous when the partial pressures of 2-propanol, propanone, and hydrogen are 0.1 bar, respectively?

Analyze

We are asked two questions. The first is whether this reaction is spontaneous under standard conditions, which means the reactants and products are at 1 bar. We can easily answer by applying the equation $\Delta G° = -RT \ln K$. The second part asks whether this reaction is spontaneous under a set of nonstandard conditions. We are asked to determine ΔG.

Solve

In each case, we first must obtain the value of $\Delta G°$, which we can get from equation (19.13).

$$\Delta G° = -RT \ln K = -8.3145 \text{ J mol}^{-1}\text{K}^{-1} \times 452\,\text{K} \times \ln(0.444)$$
$$= 3.05 \times 10^3 \text{ J mol}^{-1}$$

This result enables us to state categorically that the reaction will not proceed spontaneously if all reactants and products are in their standard states—that is, with the partial pressures of reactants and products at 1 bar.

To determine whether the reaction is spontaneous under the nonstandard-state conditions given, we must calculate ΔG. We first write Q in terms of activities and then substitute the partial pressures of the gases for the activities of the gases.

$$Q = \frac{a_{(CH_3)_2CO}a_{H_2}}{a_{(CH_3)_2CHOH}} = \frac{P_{(CH_3)_2CO}P_{H_2}}{P_{(CH_3)_2CHOH}}$$

Then use this expression in equation (19.12).

$$\Delta G = \Delta G° + RT \ln \frac{P_{(CH_3)_2CO}P_{H_2}}{P_{(CH_3)_2CHOH}}$$

$$= 3.05 \times 10^3 \text{ J mol}^{-1}$$

The value of ΔG is negative, so we can conclude that under this second set of conditions the reaction should proceed spontaneously.

$$+ 8.3145 \text{ J mol}^{-1}\text{K}^{-1} \times 452\,\text{K} \times \ln\frac{0.1 \times 0.1}{0.1}$$

$$= -5.60 \times 10^3 \text{ J mol}^{-1}$$

Assess

Remember, however, that thermodynamics says nothing about the rate of the reaction, only that the reaction will proceed in its own good time!

PRACTICE EXAMPLE A: Use the data in Appendix D to decide whether the following reaction is spontaneous under standard conditions at 298.15 K.

$$N_2O_4(g) \longrightarrow 2\,NO_2(g)$$

PRACTICE EXAMPLE B: If a gaseous mixture of N_2O_4 and NO_2, both at a pressure of 0.5 bar, is introduced into a previously evacuated vessel, which of the two gases will spontaneously convert into the other at 298.15 K?

We have now acquired all the tools with which to perform one of the most practical calculations of chemical thermodynamics: *determining the equilibrium constant for a reaction from tabulated data.* Example 19-8, which demonstrates this application, uses thermodynamic properties of ions in aqueous solution as well as of compounds. An important idea to note about the thermodynamic properties of ions is that they are relative to $H^+(aq)$, which, by convention, is assigned values of *zero* for ΔH_f°, ΔG_f°, and S°. This means that entropies listed

EXAMPLE 19-8 **Calculating the Equilibrium Constant of a Reaction from the Standard Gibbs Energy Change**

Determine the equilibrium constant at 298.15 K for the dissolution of magnesium hydroxide in an acidic solution.

$$Mg(OH)_2(s) + 2\,H^+(aq) \rightleftharpoons Mg^{2+}(aq) + 2\,H_2O(l)$$

Analyze

The key to solving this problem is to find a value of ΔG° and then to use the expression $\Delta G^\circ = -RT\ln K$.

Solve

We can obtain ΔG° from standard Gibbs energies of formation listed in Appendix D. Note that because its value is zero, the term $\Delta G_f^\circ[H^+(aq)]$ is not included.

$$\Delta G^\circ = 2\,\Delta G_f^\circ[H_2O(l)] + \Delta G_f^\circ[Mg^{2+}(aq)] - \Delta G_f^\circ[Mg(OH)_2(s)]$$
$$= 2(-237.1\ \text{kJ mol}^{-1}) + (-454.8\ \text{kJ mol}^{-1}) - (-833.5\ \text{kJ mol}^{-1})$$

Now solve for $\ln K$ and K.

$$\Delta G^\circ = -RT\ln K = -95.5\ \text{kJ mol}^{-1} = -95.5 \times 10^3\ \text{J mol}^{-1}$$

$$\ln K = \frac{-\Delta G^\circ}{RT} = \frac{-(-95.5 \times 10^3\ \text{J mol}^{-1})}{8.3145\ \text{J mol}^{-1}\ \text{K}^{-1} \times 298.15\ \text{K}} = 38.5$$

$$K = e^{38.5} = 5 \times 10^{16}$$

Assess

The value of K obtained here is the thermodynamic equilibrium constant. According to the conventions we have established, the activities of both $Mg(OH)_2(s)$ and $H_2O(l)$ are 1, and molarities can be substituted for the activities of the ions.

$$K = \frac{a_{Mg^{2+}(aq)}a^2_{H_2O(l)}}{a_{Mg(OH)_2(s)}a^2_{H^+(aq)}} = \frac{[Mg^{2+}]}{[H^+]^2} = K_c = 5 \times 10^{16}$$

PRACTICE EXAMPLE A: Determine the equilibrium constant at 298.15 K for $AgI(s) \rightleftharpoons Ag^+(aq) + I^-(aq)$. Compare your answer to the K_{sp} for AgI in Appendix D.

PRACTICE EXAMPLE B: At 298.15 K, should manganese dioxide react to an appreciable extent with 1 M HCl(aq), producing manganese(II) ion in solution and chlorine gas?

for ions are not absolute entropies, as they are for compounds. Negative values of $S°$ simply denote an entropy less than that of $H^+(aq)$.

When a question requires the use of thermodynamic properties, it is a good idea to think qualitatively about the problem before diving into calculations. The dissolution of $Mg(OH)_2(s)$ in acidic solution considered in Example 19-8 is an acid–base reaction that was used as an example in Chapter 18 to illustrate the effect of pH on solubility. It was also mentioned in Chapter 5 as the basis for using milk of magnesia as an antacid. We should certainly expect the reaction to be spontaneous. This means that the value of K should be large, which we found to be the case. If we had made an error in sign in our calculation (an easy thing to do using the expression $\Delta G° = -RT \ln K$), we would have obtained $K = 2 \times 10^{-17}$. But we would have seen immediately that this is the wrong answer. This erroneous value suggests a reaction in which the concentration of products is extremely low at equilibrium.

The data listed in Appendix D are for 25 °C. Thus, values of $\Delta G°$ and K obtained with these data are also at 25 °C. Most chemical reactions are carried out at temperatures other than 25 °C, however. In Section 19-7, we will learn how to calculate values of equilibrium constants at various temperatures.

19-7 $\Delta G°$ and K as Functions of Temperature

▶ In general, equilibrium constants for exothermic reactions will be lower at higher temperatures and higher at lower temperatures. Alternatively, equilibrium constants for endothermic reactions are lower at lower temperatures and higher at higher temperatures. This is seen by evaluating equation (19.15).

In Chapter 15, we used Le Châtelier's principle to make qualitative predictions of the effect of temperature on an equilibrium condition. We can now describe a *quantitative* relationship between the equilibrium constant and temperature. In the method illustrated in Example 19-9, we assume that $\Delta H°$ is practically independent of temperature. Although absolute entropies depend on temperature, we assume that the entropy *change* $\Delta S°$ for a reaction is also independent of temperature. Yet, the term "$T\Delta S$" is strongly temperature-dependent because of the temperature factor T. As a result, $\Delta G°$, which is equal to $\Delta H° - T\Delta S°$, is also dependent on temperature.

EXAMPLE 19-9 **Determining the Relationship Between an Equilibrium Constant and Temperature by Using Equations for Gibbs Energy Change**

At what temperature will the equilibrium constant for the formation of $NOCl(g)$ be $K = K_p = 1.00 \times 10^3$? Data for this reaction at 25 °C are

$$2\,NO(g) + Cl_2(g) \rightleftharpoons 2\,NOCl(g)$$
$$\Delta G° = -40.9 \text{ kJ mol}^{-1} \qquad \Delta H = -77.1 \text{ kJ mol}^{-1}$$
$$\Delta S° = -121.3 \text{ J mol}^{-1}\,K^{-1}$$

Analyze

To determine an unknown temperature from a known equilibrium constant, we need an equation in which both of these terms appear. The required equation is $\Delta G° = -RT \ln K$. However, to solve for the unknown temperature, we need the value of $\Delta G°$ at that temperature. We know the value of $\Delta G°$ at 25 °C $(-40.9 \text{ kJ mol}^{-1})$, but we also know that this value will be different at other temperatures. We can assume, however, that the values of $\Delta H°$ and $\Delta S°$ will not change much with temperature. This means that we can obtain a value of $\Delta G°$ from the equation $\Delta G° = \Delta H° - T\Delta S°$, where T is the *unknown* temperature and the values of $\Delta H°$ and $\Delta S°$ are those at 25 °C. Now we have two equations that we can set equal to each other.

Solve

That is,

$$\Delta G° = \Delta H° - T\,\Delta S° = -RT \ln K$$

We can gather the terms with T on the right,

$$\Delta H° = T\,\Delta S° - RT\ln K = T(\Delta S° - R\ln K)$$

and solve for T.

$$T = \frac{\Delta H°}{\Delta S° - R\ln K}$$

Now substitute values for $\Delta H°$, $\Delta S°$, R, and $\ln K$.

$$T = \frac{-77.1 \times 10^3\,\text{J mol}^{-1}}{-121.3\,\text{J mol}^{-1}\,\text{K}^{-1} - [8.3145\,\text{J mol}^{-1}\,\text{K}^{-1} \times \ln(1.00 \times 10^3)]}$$

$$= \frac{-77.1 \times 10^3\,\text{J mol}^{-1}}{-121.3\,\text{J mol}^{-1}\,\text{K}^{-1} - (8.3145 \times 6.908)\,\text{J mol}^{-1}\,\text{K}^{-1}}$$

$$= \frac{-77.1 \times 10^3\,\text{J mol}^{-1}}{-178.7\,\text{J mol}^{-1}\,\text{K}^{-1}} = 431\,\text{K}$$

Assess

Although the answer shows three significant figures, the final result should probably be rounded to just two significant figures. The assumption we made about the constancy of $\Delta H°$ and $\Delta S°$ is probably no more valid than that.

PRACTICE EXAMPLE A: At what temperature will the formation of $NO_2(g)$ from $NO(g)$ and $O_2(g)$ have $K_p = 1.50 \times 10^2$? For the reaction $2\,NO(g) + O_2(g) \rightleftharpoons 2\,NO_2(g)$ at 25 °C, $\Delta H° = -114.1\,\text{kJ mol}^{-1}$ and $\Delta S° = -146.5\,\text{J mol}^{-1}\text{K}^{-1}$.

PRACTICE EXAMPLE B: For the reaction $2\,NO(g) + Cl_2(g) \rightleftharpoons 2\,NOCl(g)$, what is the value of K at **(a)** 25 °C; **(b)** 75 °C? Use data from Example 19-9. [*Hint:* The solution to part (a) can be done somewhat more simply than that for (b).]

An alternative to the method outlined in Example 19-9 is to relate the equilibrium constant and temperature directly, without specific reference to a Gibbs energy change. We start with the same two expressions as in Example 19-9,

$$-RT\ln K = \Delta G° = \Delta H° - T\,\Delta S°$$

and divide by $-RT$.

$$\ln K = \frac{-\Delta H°}{RT} + \frac{\Delta S°}{R} \tag{19.14}$$

If we assume that $\Delta H°$ and $\Delta S°$ are constant, equation (19.14) describes a straight line with a slope of $-\Delta H°/R$ and a y-intercept of $\Delta S°/R$. Table 19.3 lists equilibrium constants as a function of the reciprocal of Kelvin temperature for

TABLE 19.3 Equilibrium Constants, K_p, for the Reaction $2\,SO_2(g) + O_2(g) \rightleftharpoons 2\,SO_3(g)$ at Several Temperatures

T, K	$1/T$, K^{-1}	K_p	$\ln K_p$
800	12.5×10^{-4}	9.1×10^2	6.81
850	11.8×10^{-4}	1.7×10^2	5.14
900	11.1×10^{-4}	4.2×10^1	3.74
950	10.5×10^{-4}	1.0×10^1	2.30
1000	10.0×10^{-4}	3.2×10^0	1.16
1050	9.52×10^{-4}	1.0×10^0	0.00
1100	9.09×10^{-4}	3.9×10^{-1}	-0.94
1170	8.5×10^{-4}	1.2×10^{-1}	-2.12

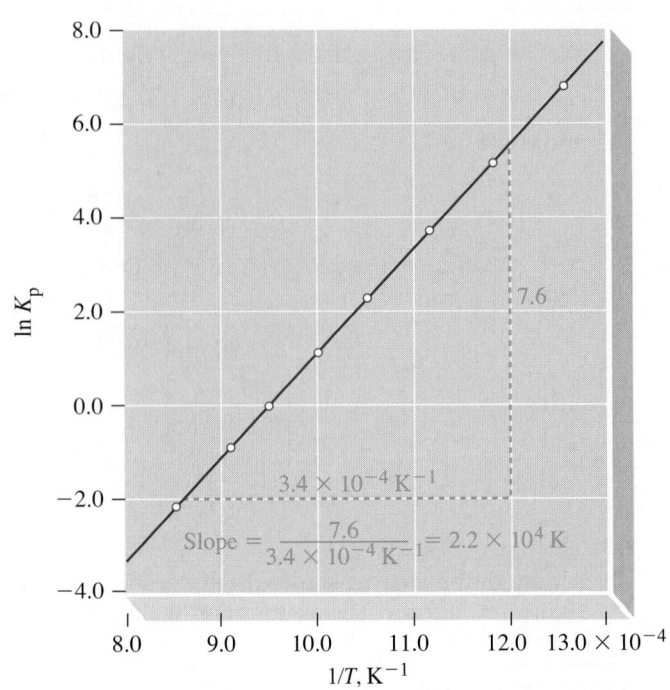

▶ FIGURE 19-12
Temperature dependence of the equilibrium constant K_p for the reaction

$$2\,SO_2(g) + O_2(g) \rightleftharpoons 2\,SO_3(g)$$

This graph can be used to establish the enthalpy of reaction, $\Delta H°$ (see equation 19.14).

$$\text{slope} = \Delta H°/R = 2.2 \times 10^4 \text{ K}$$
$$\Delta H° = -8.3145 \text{ J mol}^{-1} \text{ K}^{-1} \times 2.2 \times 10^4 \text{ K}$$
$$= -1.8 \times 10^5 \text{ J mol}^{-1}$$
$$= -1.8 \times 10^2 \text{ kJ mol}^{-1}$$

the reaction of $SO_2(g)$ and $O_2(g)$ that forms $SO_3(g)$. The ln K_p and $1/T$ data from Table 19.3 are plotted in Figure 19-12 and yield the expected straight line.

Now we can follow the procedure used in Appendix A-4 to derive the Clausius–Clapeyron equation. We can write equation (19.14) twice, for two different temperatures and with the corresponding equilibrium constants. Then, if we subtract one equation from the other, we obtain the result shown here,

$$\ln\frac{K_2}{K_1} = -\frac{\Delta H°}{R}\left(\frac{1}{T_2} - \frac{1}{T_1}\right) \qquad \text{(19.15)}$$

where T_2 and T_1 are two Kelvin temperatures; K_2 and K_1 are the equilibrium constants at those temperatures; $\Delta H°$ is the enthalpy of reaction, expressed in J mol^{-1}; and R is the gas constant, expressed as 8.3145 J mol^{-1} K^{-1}. Jacobus van't Hoff (1852–1911) derived equation (19.15), which is often referred to as *the van't Hoff equation*.

🔍 **19-5 CONCEPT ASSESSMENT**

The normal boiling point of water is 100 °C. At 120 °C and 1 atm, is ΔH or $T\,\Delta S$ greater for the vaporization of water?

EXAMPLE 19-10 Relating Equilibrium Constants and Temperature Through the Van't Hoff Equation

Use data from Table 19.3 and Figure 19-12 to estimate the temperature at which $K_p = 1.0 \times 10^6$ for the reaction

$$2\,SO_2(g) + O_2(g) \rightleftharpoons 2\,SO_3(g)$$

Analyze

To use the van't Hoff equation select one known temperature and equilibrium constant from Table 19.3 and the enthalpy change of the reaction, $\Delta H°$, from Figure 19-12.

Solve

The data to be substituted into equation (19.15) are $T_1 = ?$, $K_1 = 1.0 \times 10^6$; $T_2 = 800$ K, $K_2 = 9.1 \times 10^2$; and $\Delta H° = -1.8 \times 10^5$ J mol^{-1}.

In the following setup, we have dropped the units for simplicity. However, you should be able to show that units cancel properly.

$$\ln\frac{K_2}{K_1} = \frac{\Delta H°}{R}\left(\frac{1}{T_1} - \frac{1}{T_2}\right)$$

$$\ln\frac{9.1 \times 10^2}{1.0 \times 10^6} = \frac{-1.8 \times 10^5}{8.3145}\left(\frac{1}{T_1} - \frac{1}{800}\right)$$

$$-7.00 = -2.2 \times 10^4\left(\frac{1}{T_1} - \frac{1}{800}\right)$$

$$\frac{-7.00}{-2.2 \times 10^4} + \frac{1}{800} = \frac{1}{T_1}$$

$$\frac{1}{T_1} = (3.2 \times 10^{-4}) + (1.25 \times 10^{-3}) = 1.57 \times 10^{-3}$$

$$T_1 = \frac{1}{1.57 \times 10^{-3}} = 6.37 \times 10^2 \text{ K}$$

Assess

A common error in this type of problem is the use of incorrect temperature units. It should be Kelvin (K).

PRACTICE EXAMPLE A: Estimate the temperature at which $K_p = 5.8 \times 10^{-2}$ for the reaction in Example 19-10. Use data from Table 19.3 and Figure 19-12.

PRACTICE EXAMPLE B: What is the value of K_p for the reaction $2 SO_2(g) + O_2(g) \rightleftharpoons 2 SO_3(g)$ at 235 °C? Use data from Table 19.3, Figure 19-12 and the van't Hoff equation (19.15).

19-8   Coupled Reactions

We have seen two ways to obtain product from a nonspontaneous reaction: (1) Change the reaction conditions to ones that make the reaction spontaneous (mostly by changing the temperature), and (2) carry out the reaction by electrolysis. But there is a third way also. Combine a pair of reactions, one with a positive ΔG and one with a negative ΔG, to obtain a spontaneous overall reaction. Such paired reactions are called **coupled reactions**. Consider the extraction of a metal from its oxide.

When copper(I) oxide is heated to 673 K, no copper metal is obtained. The decomposition of Cu_2O to form products in their standard states (for instance, $P_{O_2} = 1.00$ bar) is nonspontaneous at 673 K.

◀ In general chemistry, simple examples of reactions are generally used. In fact, in almost all interesting cases, one reaction is coupled to another, and so forth. No better example exists than the complex cycles of coupled chemical reactions in biological processes.

$$Cu_2O(s) \xrightarrow{\Delta} 2\,Cu(s) + \frac{1}{2}O_2(g) \qquad \Delta G°_{673K} = +125 \text{ kJ} \qquad \textbf{(19.16)}$$

Suppose this nonspontaneous decomposition reaction is coupled with the partial oxidation of carbon to carbon monoxide—a spontaneous reaction. The overall reaction (19.17), because it has a negative value of $\Delta G°$, is spontaneous when reactants and products are in their standard states.

$$Cu_2O(s) \longrightarrow 2\,Cu(s) + \frac{1}{2}O_2(g) \qquad \Delta G°_{673\,K} = +125 \text{ kJ}$$

$$C(s) + \frac{1}{2}O_2(g) \longrightarrow CO(g) \qquad \Delta G°_{673\,K} = -175 \text{ kJ}$$

$$Cu_2O(s) + C(s) \longrightarrow 2\,Cu(s) + CO(g) \qquad \Delta G°_{673\,K} = -50 \text{ kJ} \qquad \textbf{(19.17)}$$

Note that reactions (19.16) and (19.17) are not the same, even though each has Cu(s) as a product. The purpose of coupled reactions, then, is to produce a spontaneous overall reaction by combining two other processes: one nonspontaneous and one spontaneous. Many metallurgical processes employ coupled reactions, especially those that use carbon or hydrogen as reducing agents.

To sustain life, organisms must synthesize complex molecules from simpler ones. If carried out as single-step reactions, these syntheses would generally be accompanied by increases in enthalpy, decreases in entropy, and increases in Gibbs energy—in short, they would be nonspontaneous and would not occur. In living organisms, changes in temperature and electrolysis are not viable options for dealing with nonspontaneous processes. Here, coupled reactions are crucial. See Focus On 19-1 on the MasteringChemistry website for an example.

Mastering**CHEMISTRY**   www.masteringchemistry.com

Adenosine triphosphate (ATP) is the energy currency used in nature to drive reactions. The conversion of adenosine diphosphate (ADP) to ATP is a nonspontaneous process. For a discussion of how Nature is able to convert ADP to ATP, go to the Focus On feature for Chapter 19, Coupled Reactions in Biological Systems, on the MasteringChemistry site.

Summary

19-1 Spontaneity: The Meaning of Spontaneous Change—A process that proceeds without need of external intervention is said to be a **spontaneous process**. A **nonspontaneous process** cannot occur without external intervention. A process that is spontaneous in one direction is nonspontaneous in the reverse direction. Some spontaneous processes are exothermic, others are endothermic so that criteria other than enthalpy change are needed to define spontaneity.

19-2 The Concept of Entropy—**Entropy** is a thermodynamic property related to the distribution of a system's energy among the available microscopic energy levels. Boltzmann's formula (equation 19.1) illustrates the relationship between entropy and the number of microstates of a system. The thermodynamic definition of an **entropy change**, ΔS, is of a quantity of heat (q_{rev}) divided by a Kelvin temperature (equation 19.2), and having the units $J\,K^{-1}$. The quantity of heat released (or absorbed) during the process must be from a reversible process in order for the entropy change to be path-independent.

19-3 Evaluating Entropy and Entropy Changes—The **third law of thermodynamics** states that the entropy of a pure, perfect crystal at 0 K is zero. Thus there are absolute entropies, unlike internal energy and enthalpy. The entropy of a substance in its standard state is called **standard molar entropy**, S°. Standard molar entropies of **reactants** and products can be used to calculate standard entropy changes in chemical reactions (equation 19.5). Another important entropy-related relationship is **Trouton's rule**, which states that the standard entropy of vaporization at the normal boiling point is approximately constant at $87\,J\,mol^{-1}\,K^{-1}$ (equation 19.4).

19-4 Criteria for Spontaneous Change: The Second Law of Thermodynamics—The basic criterion for spontaneous change is that the entropy change of the universe, which is the sum of the entropy change of the system plus that of the surroundings (equation 19.6),

must be greater than zero (equation 19.7). This statement is known as the **second law of thermodynamics**. An equivalent criterion applied to the system alone is based on a thermodynamic function known as the **Gibbs energy**. The **Gibbs energy change**, ΔG, is the enthalpy change for the system (ΔH) minus the product of the temperature and entropy change for the system ($T\,\Delta S$) (equation 19.9). Table 19.1 summarizes the criteria for spontaneous change based on Gibbs energy change.

19-5 Standard Gibbs Energy Change, ΔG°—The **standard Gibbs energy change**, ΔG°, is based on the conversion of reactants in their standard states to products in their standard states. Tabulated Gibbs energy data are usually **standard Gibbs energies of formation**, ΔG_f°, and usually at 298.15 K.

19-6 Gibbs Energy Change and Equilibrium—The relationship between the standard Gibbs energy change and the equilibrium constant for a reaction is $\Delta G^\circ = -RT \ln K$. The constant, K, is called a **thermodynamic equilibrium constant**. It is based on the activities of reactants and products, but these activities can be related to solution molarities and gas partial pressures by means of a few simple conventions.

19-7 ΔG° and K as Functions of Temperature—By starting with the relationship between standard Gibbs energy change and the equilibrium constant, the van't Hoff equation—relating the equilibrium constant and temperature—can be written. With this equation, tabulated data at 25 °C can be used to determine equilibrium constants not just at 25 °C but at other temperatures as well.

19-8 Coupled Reactions—Nonspontaneous processes can be made spontaneous by coupling them with spontaneous reactions and by taking advantage of the state function property of G. **Coupled reactions**, that is, paired reactions that yield a spontaneous overall reaction, occur in metallurgical processes and in biochemical transformations.

Integrative Example

The synthesis of methanol is of great importance because methanol can be used directly as a motor fuel, mixed with gasoline for fuel use, or converted to other organic compounds. The synthesis reaction, carried out at about 500 K, is

$$CO(g) + 2 H_2(g) \rightleftharpoons CH_3OH(g)$$

What is the value of K_p at 500 K?

Analyze

Our approach to this problem begins with determining $\Delta G°$ from Gibbs energy of formation data and using $\Delta G°$ to find K_p at 298 K. The next step is to calculate $\Delta H°$ from enthalpy of formation data, and use this value together with K_p at 298 K in expression (19.15) to find K_p at 500 K.

Solve

Write the equation for methanol synthesis; place Gibbs energy of formation data from Appendix D under formulas in the equation, and use these data to calculate $\Delta G°$ at 298 K.

$$CO(g) + 2 H_2(g) \rightleftharpoons CH_3OH(g)$$

$$\Delta G_f°, \text{kJ mol}^{-1} \qquad -137.2 \qquad 0 \qquad -162.0$$

$$\Delta G° = 1 \text{ mol CH}_3\text{OH} \times (-162.0 \text{ kJ/mol CH}_3\text{OH})$$
$$-1 \text{ mol CO} \times (-137.2 \text{ kJ/mol CO}) = -24.8 \text{ kJ}$$

To calculate K_p at 298 K, use $\Delta G°$ at 298 K, written as -24.8×10^3 J mol^{-1}, in the expression $\Delta G° = -RT \ln K_p$.

$$\ln K_p = -\Delta G°/RT = \frac{-(-24.8 \times 10^3 \text{ J mol}^{-1})}{8.3145 \text{ J mol}^{-1} \text{ K}^{-1} \times 298 \text{ K}} = 10.0$$

$$K_p = e^{10.0} = 2.2 \times 10^4$$

To determine $\Delta H°$ at 298 K, use standard enthalpy of formation data from Appendix D, applied in the same manner as was previously used for $\Delta G°$.

$$CO(g) + 2 H_2(g) \rightleftharpoons CH_3OH(g)$$

$$\Delta H_f°, \text{kJ mol}^{-1} \qquad -110.5 \qquad 0 \qquad -200.7$$

$$\Delta H° = 1 \text{ mol CH}_3\text{OH} \times (-200.7 \text{ kJ/mol CH}_3\text{OH})$$
$$-1 \text{ mol CO} \times (-110.5 \text{ kJ/mol CO}) = -90.2 \text{ kJ}$$

Use the van't Hoff equation with $K_p = 2.2 \times 10^4$ at 298 K and $\Delta H° = -90.2 \times 10^3$ J mol^{-1}. Solve for K_p at 500 K.

$$\ln \frac{K_p}{2.2 \times 10^4} = \frac{-90.2 \times 10^3 \text{ J mol}^{-1}}{8.3145 \text{ J mol}^{-1} \text{ K}^{-1}} \left(\frac{1}{298 \text{ K}} - \frac{1}{500 \text{ K}} \right)$$

$$= -14.7$$

$$\frac{K_p}{2.2 \times 10^4} = e^{-14.7} = 4 \times 10^{-7} \qquad K_p = 9 \times 10^{-3}$$

Assess

We have been successful in determining the equilibrium constant for the synthesis of methanol at 500 K by using tabulated thermodynamic data and the van't Hoff equation. Note that increasing the temperature favors the reverse reaction, as we should expect from Le Châtelier's principle. That is, an increase in temperature in a reversible reaction favors the heat absorbing (endothermic) reaction. Here, the forward reaction is exothermic and the reverse reaction is endothermic.

PRACTICE EXAMPLE A: Dinitrogen pentoxide, N_2O_5, is a solid with a high vapor pressure. Its vapor pressure at 7.5 °C is 100 mmHg, and the solid sublimes at a pressure of 1.00 atm at 32.4 °C. What is the standard Gibbs energy change for the process at $N_2O_5(s) \longrightarrow N_2O_5(g)$ at 25 °C?

PRACTICE EXAMPLE B: A plausible reaction for the production of ethylene glycol (used as antifreeze) is

$$2 CO(g) + 3H_2(g) \longrightarrow CH_2OHCH_2OH(l)$$

The following thermodynamic properties of $CH_2OHCH_2OH(l)$ at 25 °C are given: $\Delta H_f° = -454.8$ kJ mol^{-1} and $\Delta G_f° = -323.1$ kJ mol^{-1}. Use these data, together with values from Appendix D, to obtain a value of $S°$, the standard molar entropy of $CH_2OHCH_2OH(l)$ at 25 °C.

Mastering**CHEMISTRY**

**You'll find a link to additional self study questions in the study area on
www.masteringchemistry.com**

Exercises

Spontaneous Change and Entropy

1. Indicate whether each of the following changes represents an increase or a decrease in entropy in a system, and explain your reasoning: **(a)** the freezing of ethanol; **(b)** the sublimation of dry ice; **(c)** the burning of a rocket fuel.

2. Arrange the entropy changes of the following processes, all at 25 °C, in the expected order of increasing ΔS, and explain your reasoning:
    (a) $H_2O(l, 1 \text{ atm}) \longrightarrow H_2O(g, 1 \text{ atm})$
    (b) $CO_2(s, 1 \text{ atm}) \longrightarrow CO_2(g, 10 \text{ mmHg})$
    (c) $H_2O(l, 1 \text{ atm}) \longrightarrow H_2O(g, 10 \text{ mmHg})$

3. Use ideas from this chapter to explain this famous remark attributed to Rudolf Clausius (1865): "Die Energie der Welt ist konstant; die Entropie der Welt strebt einem Maximum zu." ("The energy of the world is constant; the entropy of the world increases toward a maximum.")

4. Comment on the difficulties of solving environmental pollution problems from the standpoint of entropy changes associated with the formation of pollutants and with their removal from the environment.

5. Indicate whether the entropy of the system would increase or decrease in each of the following reactions. If you cannot be certain simply by inspecting the equation, explain why.
    (a) $CCl_4(l) \longrightarrow CCl_4(g)$
    (b) $CuSO_4 \cdot 3\,H_2O(s) + 2\,H_2O(g) \longrightarrow$
    $\qquad\qquad\qquad\qquad\qquad CuSO_4 \cdot 5\,H_2O(s)$

6. **(c)** $SO_3(g) + H_2(g) \longrightarrow SO_2(g) + H_2O(g)$
    (d) $H_2S(g) + O_2(g) \longrightarrow H_2O(g) + SO_2(g)$
    $\qquad\qquad\qquad\qquad\qquad\text{(not balanced)}$

6. Which substance in each of the following pairs would have the greater entropy? Explain.
    (a) at 75 °C and 1 atm: 1 mol $H_2O(l)$ or 1 mol $H_2O(g)$
    (b) at 5 °C and 1 atm: 50.0 g Fe(s) or 0.80 mol Fe(s)
    (c) 1 mol Br_2 (l, 1 atm, 8 °C) or 1 mol Br_2 (s, 1 atm, −8 °C)
    (d) 0.312 mol SO_2 (g, 0.110 atm, 32.5 °C) or 0.284 mol O_2 (g, 15.0 atm, 22.3 °C)

7. For each of the following reactions, indicate whether ΔS for the reaction should be positive or negative. If it is not possible to determine the sign of ΔS from the information given, indicate why.
    (a) $CaO(s) + H_2O(l) \longrightarrow Ca(OH)_2(s)$
    (b) $2\,HgO(s) \longrightarrow 2\,Hg(l) + O_2(g)$
    (c) $2\,NaCl(l) \longrightarrow 2\,Na(l) + Cl_2(g)$
    (d) $Fe_2O_3(s) + 3\,CO(g) \longrightarrow 2\,Fe(s) + 3\,CO_2(g)$
    (e) $Si(s) + 2\,Cl_2(g) \longrightarrow SiCl_4(g)$

8. By analogy to ΔH_f° and ΔG_f° how would you define entropy of formation? Which would have the largest entropy of formation: $CH_4(g)$, $CH_3CH_2OH(l)$, or $CS_2(l)$? First make a qualitative prediction; then test your prediction with data from Appendix D.

Phase Transitions

9. In Example 19-2, we dealt with ΔH_{vap}° and ΔS_{vap}° for water at 100 °C.
    (a) Use data from Appendix D to determine values for these two quantities at 25 °C.
    (b) From your knowledge of the structure of liquid water, explain the differences in ΔH_{vap}° values and in ΔS_{vap}° values between 25 °C and 100 °C.

10. Pentane is one of the most volatile of the hydrocarbons in gasoline. At 298.15 K, the following enthalpies of formation are given for pentane: $\Delta H_f^\circ C_5H_{12}(l) = -173.5 \text{ kJ mol}^{-1}$; $\Delta H_f^\circ[C_5H_{12}(g)] = -146.9 \text{ kJ mol}^{-1}$.
    (a) Estimate the normal boiling point of pentane.
    (b) Estimate ΔG° for the vaporization of pentane at 298 K.
    (c) Comment on the significance of the sign of ΔG° at 298 K.

11. Which of the following substances would obey Trouton's rule most closely: HF, $C_6H_5CH_3$ (toluene), or CH_3OH (methanol)? Explain your reasoning.

12. Estimate the normal boiling point of bromine, Br_2, in the following way: Determine ΔH_{vap}° for Br_2 from data in Appendix D. Assume that ΔH_{vap}° remains constant and that Trouton's rule is obeyed.

13. In what temperature range can the following equilibrium be established? Explain.

    $$H_2O(l, 0.50 \text{ atm}) \rightleftharpoons H_2O(g, 0.50 \text{ atm})$$

14. Refer to Figures 12-28 and 19-9. Which has the lowest Gibbs energy at 1 atm and −60 °C: solid, liquid, or gaseous carbon dioxide? Explain.

Gibbs Energy and Spontaneous Change

15. Which of the following changes in a thermodynamic property would you expect to find for the reaction $Br_2(g) \longrightarrow 2\,Br(g)$ *at all temperatures:* **(a)** $\Delta H < 0$; **(b)** $\Delta S > 0$; **(c)** $\Delta G < 0$; **(d)** $\Delta S < 0$? Explain.

16. If a reaction can be carried out only by electrolysis, which of the following changes in a thermodynamic property *must* apply: **(a)** $\Delta H > 0$; **(b)** $\Delta S > 0$; **(c)** $\Delta G = \Delta H$; **(d)** $\Delta G > 0$? Explain.

17. Indicate which of the four cases in Table 19.1 applies to each of the following reactions. If you are unable to decide from only the information given, state why.
    (a) $PCl_3(g) + Cl_2(g) \longrightarrow PCl_5(g)$ $\Delta H^\circ = -87.9 \text{ kJ}$
    (b) $CO_2(g) + H_2(g) \longrightarrow CO(g) + H_2O(g)$
    $\qquad\qquad\qquad\qquad\qquad\Delta H^\circ = +41.2 \text{ kJ}$
    (c) $NH_4CO_2NH_2(s) \longrightarrow 2\,NH_3(g) + CO_2(g)$
    $\qquad\qquad\qquad\qquad\qquad\Delta H^\circ = +159.2 \text{ kJ}$

18. Indicate which of the four cases in Table 19.1 applies to each of the following reactions. If you are unable to decide from only the information given, state why.

(a) $H_2O(g) + \frac{1}{2}O_2(g) \longrightarrow H_2O_2(g)$

$$\Delta H° = +105.5 \text{ kJ}$$

(b) $C_6H_6(l) + \frac{15}{2}O_2(g) \longrightarrow 6\,CO_2(g) + 3\,H_2O(g)$

$$\Delta H° = -3135 \text{ kJ}$$

(c) $NO(g) + \frac{1}{2}Cl_2(g) \longrightarrow NOCl(g)$

$$\Delta H° = -38.54 \text{ kJ}$$

Standard Gibbs Energy Change

23. From the data given in the following table, determine $\Delta S°$ for the reaction $NH_3(g) + HCl(g) \longrightarrow NH_4Cl(s)$. All data are at 298 K.

	$\Delta H_f°$, kJ mol^{-1}	$\Delta G_f°$, kJ mol^{-1}
$NH_3(g)$	−46.11	−16.48
$HCl(g)$	−92.31	−95.30
$NH_4Cl(s)$	−314.4	−202.9

24. Use data from Appendix D to determine values of $\Delta G°$ for the following reactions at 25 °C.

(a) $C_2H_2(g) + 2\,H_2(g) \longrightarrow C_2H_6(g)$
(b) $2\,SO_3(g) \longrightarrow 2\,SO_2(g) + O_2(g)$
(c) $Fe_3O_4(s) + 4\,H_2(g) \longrightarrow 3\,Fe(s) + 4\,H_2O(g)$
(d) $2\,Al(s) + 6\,H^+(aq) \longrightarrow 2\,Al^{3+}(aq) + 3\,H_2(g)$

25. At 298 K, for the reaction $2\,PCl_3(g) + O_2(g) \longrightarrow 2\,POCl_3(l)$, $\Delta H° = -620.2$ kJ and the standard molar entropies are $PCl_3(g)$, 311.8 J K^{-1}; $O_2(g)$, 205.1 J K^{-1}; and $POCl_3(l)$, 222.4 J K^{-1}. Determine **(a)** $\Delta G°$ at 298 K and **(b)** whether the reaction proceeds spontaneously in the forward or the reverse direction when reactants and products are in their standard states.

26. At 298 K, for the reaction $2\,H^+(aq) + 2\,Br^-(aq) + 2\,NO_2(g) \longrightarrow Br_2(l) + 2\,HNO_2(aq)$, $\Delta H° = -61.6$ kJ and the standard molar entropies are $H^+(aq)$, 0 J K^{-1}; $Br^-(aq)$, 82.4 J K^{-1}; $NO_2(g)$, 240.1 J K^{-1}; $Br_2(l)$, 152.2 J K^{-1}; $HNO_2(aq)$, 135.6 J K^{-1}. Determine **(a)** $\Delta G°$ at 298 K and **(b)** whether the reaction proceeds spontaneously in the forward or the reverse direction when reactants and products are in their standard states.

27. The following standard Gibbs energy changes are given for 25 °C.

(1) $N_2(g) + 3\,H_2(g) \longrightarrow 2\,NH_3(g)$

$$\Delta G° = -33.0 \text{ kJ}$$

(2) $4\,NH_3(g) + 5\,O_2(g) \longrightarrow 4\,NO(g) + 6\,H_2O(l)$

$$\Delta G° = -1010.5 \text{ kJ}$$

(3) $N_2(g) + O_2(g) \longrightarrow 2\,NO(g)$

$$\Delta G° = +173.1 \text{ kJ}$$

(4) $N_2(g) + 2\,O_2(g) \longrightarrow 2\,NO_2(g)$

$$\Delta G° = +102.6 \text{ kJ}$$

(5) $2\,N_2(g) + O_2(g) \longrightarrow 2\,N_2O(g)$

$$\Delta G° = +208.4 \text{ kJ}$$

Combine the preceding equations, as necessary, to obtain $\Delta G°$ values for each of the following reactions.

19. For the mixing of ideal gases (see Figure 19-2), explain whether a positive, negative, or zero value is expected for ΔH, ΔS, and ΔG.

20. What values of ΔH, ΔS, and ΔG would you expect for the formation of an ideal solution of liquid components? (Is each value positive, negative, or zero?)

21. Explain why **(a)** some exothermic reactions do not occur spontaneously, and **(b)** some reactions in which the entropy of the system increases do not occur spontaneously.

22. Explain why you would expect a reaction of the type $AB(g) \longrightarrow A(g) + B(g)$ always to be spontaneous at *high* rather than at low temperatures.

(a) $N_2O(g) + \frac{3}{2}O_2(g) \longrightarrow 2\,NO_2(g)$ $\qquad \Delta G° = \,?$
(b) $2\,H_2(g) + O_2(g) \longrightarrow 2\,H_2O(l)$ $\qquad \Delta G° = \,?$
(c) $2\,NH_3(g) + 2\,O_2(g) \longrightarrow N_2O(g) + 3\,H_2O(l)$

$$\Delta G° = \,?$$

Of reactions (a), (b), and (c), which would tend to go to completion at 25 °C, and which would reach an equilibrium condition with significant amounts of all reactants and products present?

28. The following standard Gibbs energy changes are given for 25 °C.

(1) $SO_2(g) + 3\,CO(g) \longrightarrow COS(g) + 2\,CO_2(g)$

$$\Delta G° = -246.4 \text{ kJ}$$

(2) $CS_2(g) + H_2O(g) \longrightarrow COS(g) + H_2S(g)$

$$\Delta G° = -41.5 \text{ kJ}$$

(3) $CO(g) + H_2S(g) \longrightarrow COS(g) + H_2(g)$

$$\Delta G° = +1.4 \text{ kJ}$$

(4) $CO(g) + H_2O(g) \longrightarrow CO_2(g) + H_2(g)$

$$\Delta G° = -28.6 \text{ kJ}$$

Combine the preceding equations, as necessary, to obtain $\Delta G°$ values for the following reactions.

(a) $COS(g) + 2\,H_2O(g) \longrightarrow$
$\qquad SO_2(g) + CO(g) + 2\,H_2(g)$ $\qquad \Delta G° = \,?$
(b) $COS(g) + 3\,H_2O(g) \longrightarrow$
$\qquad SO_2(g) + CO_2(g) + 3\,H_2(g)$ $\qquad \Delta G° = \,?$
(c) $COS(g) + H_2O(g) \longrightarrow CO_2(g) + H_2S(g)$

$$\Delta G° = \,?$$

Of reactions (a), (b), and (c), which is spontaneous in the forward direction when reactants and products are present in their standard states?

29. Write an equation for the combustion of one mole of benzene, $C_6H_6(l)$, and use data from Appendix D to determine $\Delta G°$ at 298 K if the products of the combustion are **(a)** $CO_2(g)$ and $H_2O(l)$, and **(b)** $CO_2(g)$ and $H_2O(g)$. Describe how you might determine the *difference* between the values obtained in (a) and (b) without having either to write the combustion equation or to determine $\Delta G°$ values for the combustion reactions.

30. Use molar entropies from Appendix D, together with the following data, to estimate the bond-dissociation energy of the F_2 molecule.

$$F_2(g) \longrightarrow 2\,F(g) \quad \Delta G° = 123.9 \text{ kJ}$$

Compare your result with the value listed in Table 10.3.

31. Assess the feasibility of the reaction

$$N_2H_4(g) + 2\,OF_2(g) \longrightarrow N_2F_4(g) + 2\,H_2O(g)$$

by determining each of the following quantities for this reaction at 25 °C.
(a) $\Delta S°$ (The standard molar entropy of $N_2F_4(g)$ is 301.2 J K^{-1}.)
(b) $\Delta H°$ (Use data from Table 10.3 and F—O and N—F bond energies of 222 and 301 kJ mol^{-1}, respectively.)
(c) $\Delta G°$

Is the reaction feasible? If so, is it favored at high or low temperatures?

32. Solid ammonium nitrate can decompose to dinitrogen oxide gas and liquid water. What is $\Delta G°$ at 298 K? Is the decomposition reaction favored at temperatures above or below 298 K?

The Thermodynamic Equilibrium Constant

33. For one of the following reactions, $K_cK_p = K$. Identify that reaction. For the other two reactions, what is the relationship between K_c, K_p, and K? Explain.
(a) $2\,SO_2(g) + O_2(g) \rightleftharpoons 2\,SO_3(g)$

(b) $HI(g) \rightleftharpoons \frac{1}{2}H_2(g) + \frac{1}{2}I_2(g)$

(c) $NH_4HCO_3(s) \rightleftharpoons NH_3(g) + CO_2(g) + H_2O(l)$
34. $H_2(g)$ can be prepared by passing steam over hot iron: $3\,Fe(s) + 4\,H_2O(g) \rightleftharpoons Fe_3O_4(s) + 4\,H_2(g)$.
(a) Write an expression for the thermodynamic equilibrium constant for this reaction.
(b) Explain why the partial pressure of $H_2(g)$ is independent of the amounts of $Fe(s)$ and $Fe_3O_4(s)$ present.
(c) Can we conclude that the production of $H_2(g)$ from $H_2O(g)$ could be accomplished regardless of

the proportions of $Fe(s)$ and $Fe_3O_4(s)$ present? Explain.

35. In the synthesis of gaseous methanol from carbon monoxide gas and hydrogen gas, the following equilibrium concentrations were determined at 483 K: $[CO(g)] = 0.0911$ M, $[H_2O(g)] = 0.0822$ M, and $[CH_3OH(g)] = 0.00892$ M. Calculate the equilibrium constant and Gibbs energy for this reaction.
36. Calculate the equilibrium constant and Gibbs energy for the reaction $CO(g) + 2\,H_2(g) \longrightarrow CH_3OH(g)$ at 483 K by using the data tables from Appendix D. Are the values determined here different from or the same as those in exercise 35? Explain.

Relationships Involving ΔG, $\Delta G°$, Q, and K

37. Use data from Appendix D to determine K_p at 298 K for the reaction $N_2O(g) + \frac{1}{2}O_2(g) \rightleftharpoons 2\,NO(g)$.
38. Use data from Appendix D to establish for the reaction $2\,N_2O_4(g) + O_2(g) \rightleftharpoons 2\,N_2O_5(g)$:
(a) $\Delta G°$ at 298 K for the reaction as written;
(b) K_p at 298 K.
39. Use data from Appendix D to determine values at 298 K of $\Delta G°$ and K for the following reactions. (Note: The equations are not balanced.)
(a) $HCl(g) + O_2(g) \rightleftharpoons H_2O(g) + Cl_2(g)$
(b) $Fe_2O_3(s) + H_2(g) \rightleftharpoons Fe_3O_4(s) + H_2O(g)$
(c) $Ag^+(aq) + SO_4^{2-}(aq) \rightleftharpoons Ag_2SO_4(s)$
40. In Example 19-1, we were unable to conclude by inspection whether $\Delta S°$ for the reaction $CO(g) + H_2O(g) \longrightarrow CO_2(g) + H_2(g)$ should be positive or negative. Use data from Appendix D to obtain $\Delta S°$ at 298 K.
41. Use thermodynamic data at 298 K to decide in which direction the reaction

$$2\,SO_2(g) + O_2(g) \rightleftharpoons 2\,SO_3(g)$$

is spontaneous when the partial pressures of SO_2, O_2, and SO_3 are 1.0×10^{-4}, 0.20, and 0.10 atm, respectively.
42. Use thermodynamic data at 298 K to decide in which direction the reaction

$$H_2(g) + Cl_2(g) \rightleftharpoons 2\,HCl(g)$$

is spontaneous when the partial pressures of H_2, Cl_2, and HCl are all 0.5 atm.

43. The standard Gibbs energy change for the reaction

$$CH_3CO_2H(aq) + H_2O(l) \rightleftharpoons$$
$$CH_3CO_2^-(aq) + H_3O^+(aq)$$

is 27.07 kJ mol^{-1} at 298 K. Use this thermodynamic quantity to decide in which direction the reaction is spontaneous when the concentrations of $CH_3CO_2H(aq)$, $CH_3CO_2^-(aq)$, and $H_3O^+(aq)$ are 0.10 M, 1.0×10^{-3} M, and 1.0×10^{-3} M, respectively.
44. The standard Gibbs energy change for the reaction

$$NH_3(aq) + H_2O(l) \rightleftharpoons NH_4^+(aq) + OH^-(aq)$$

is 29.05 kJ mol^{-1} at 298 K. Use this thermodynamic quantity to decide in which direction the reaction is spontaneous when the concentrations of $NH_3(aq)$, $NH_4^+(aq)$, and $OH^-(aq)$ are 0.10 M, 1.0×10^{-3} M, and 1.0×10^{-3} M, respectively.
45. For the reaction $2\,NO(g) + O_2(g) \longrightarrow 2\,NO_2(g)$ all but one of the following equations is correct. Which is *incorrect*, and why? (a) $K = K_p$; (b) $\Delta S° = (\Delta G° - \Delta H°)/T$; (c) $K_p = e^{-\Delta G°/RT}$; (d) $\Delta G = \Delta G° + RT \ln Q$.
46. Why is $\Delta G°$ such an important property of a chemical reaction, even though the reaction is generally carried out under *nonstandard* conditions?
47. At 1000 K, an equilibrium mixture in the reaction $CO_2(g) + H_2(g) \rightleftharpoons CO(g) + H_2O(g)$ contains 0.276 mol H_2, 0.276 mol CO_2, 0.224 mol CO, and 0.224 mol H_2O.
(a) What is K_p at 1000 K?

(b) Calculate $\Delta G°$ at 1000 K.
(c) In which direction would a spontaneous reaction occur if the following were brought together at 1000 K: 0.0750 mol CO_2, 0.095 mol H_2, 0.0340 mol CO, and 0.0650 mol H_2O?

48. For the reaction $2\,SO_2(g) + O_2(g) \rightleftharpoons 2\,SO_3(g)$, $K_c = 2.8 \times 10^2$ at 1000 K.
    (a) What is $\Delta G°$ at 1000 K? [*Hint:* What is K_p?]
    (b) If 0.40 mol SO_2, 0.18 mol O_2, and 0.72 mol SO_3 are mixed in a 2.50 L flask at 1000 K, in what direction will a net reaction occur?

49. For the following equilibrium reactions, calculate $\Delta G°$ at the indicated temperature. [*Hint:* How is each equilibrium constant related to a thermodynamic equilibrium constant, K?]
    (a) $H_2(g) + I_2(g) \rightleftharpoons 2\,HI(g)$ $K_c = 50.2$ at 445 °C
    (b) $N_2O(g) + \frac{1}{2}O_2(g) \rightleftharpoons 2NO(g)$
    $$K_c = 1.7 \times 10^{-13} \text{ at 25 °C}$$
    (c) $N_2O_4(g) \rightleftharpoons 2NO_2(g)$
    $$K_c = 4.61 \times 10^{-3} \text{ at 25 °C}$$
    (d) $2\,Fe^{3+}(aq) + Hg_2^{2+}(aq) \rightleftharpoons$
    $$2\,Fe^{2+}(aq) + 2\,Hg^{2+}(aq)$$
    $$K_c = 9.14 \times 10^{-6} \text{ at 25 °C}$$

50. Two equations can be written for the dissolution of $Mg(OH)_2(s)$ in acidic solution.

    $$Mg(OH)_2(s) + 2\,H^+(aq) \rightleftharpoons Mg^{2+}(aq) + 2\,H_2O(l)$$
    $$\Delta G° = -95.5 \text{ kJ mol}^{-1}$$

    $$\frac{1}{2}Mg(OH)_2(s) + H^+(aq) \rightleftharpoons \frac{1}{2}Mg^{2+}(aq) + H_2O(l)$$
    $$\Delta G° = -47.8 \text{ kJ mol}^{-1}$$

    (a) Explain why these two equations have different $\Delta G°$ values.
    (b) Will K for these two equations be the same or different? Explain.

$\Delta G°$ and K as Functions of Temperature

55. Use data from Appendix D to establish at 298 K for the reaction:

    $$2\,NaHCO_3(s) \longrightarrow Na_2CO_3(s) + H_2O(l) + CO_2(g)$$

    (a) $\Delta S°$;   **(b)** $\Delta H°$;   **(c)** $\Delta G°$;   **(d)** K.

56. A possible reaction for converting methanol to ethanol is

    $$CO(g) + 2\,H_2(g) + CH_3OH(g) \longrightarrow$$
    $$C_2H_5OH(g) + H_2O(g)$$

    (a) Use data from Appendix D to calculate $\Delta H°$, $\Delta S°$, and $\Delta G°$ for this reaction at 25 °C.
    (b) Is this reaction thermodynamically favored at high or low temperatures? At high or low pressures? Explain.
    (c) Estimate K_p for the reaction at 750 K.

57. What must be the temperature if the following reaction has $\Delta G° = -45.5$ kJ, $\Delta H° = -24.8$ kJ, and $\Delta S° = 15.2$ J K^{-1}?

    $$Fe_2O_3(s) + 3\,CO(g) \longrightarrow 2\,Fe(s) + 3\,CO_2(g)$$

(c) Will the solubilities of $Mg(OH)_2(s)$ in a buffer solution at pH = 8.5 depend on which of the two equations is used as the basis of the calculation? Explain.

51. At 298 K, $\Delta G_f°[CO(g)] = -137.2$ kJ/mol and $K_p = 6.5 \times 10^{11}$ for the reaction $CO(g) + Cl_2(g) \rightleftharpoons COCl_2(g)$. Use these data to determine $\Delta G_f°[COCl_2(g)]$, and compare your result with the value in Appendix D.

52. Use thermodynamic data from Appendix D to calculate values of K_{sp} for the following sparingly soluble solutes: **(a)** AgBr; **(b)** $CaSO_4$; **(c)** $Fe(OH)_3$. [*Hint:* Begin by writing solubility equilibrium expressions.]

53. To establish the law of conservation of mass, Lavoisier carefully studied the decomposition of mercury(II) oxide:

    $$HgO(s) \longrightarrow Hg(l) + \frac{1}{2}O_2(g)$$

    At 25 °C, $\Delta H° = +90.83$ kJ and $\Delta G° = +58.54$ kJ.
    (a) Show that the partial pressure of $O_2(g)$ in equilibrium with HgO(s) and Hg(l) at 25 °C is extremely low.
    (b) What conditions do you suppose Lavoisier used to obtain significant quantities of oxygen?

54. Currently, CO_2 is being studied as a source of carbon atoms for synthesizing organic compounds. One possible reaction involves the conversion of CO_2 to methanol, CH_3OH.

    $$CO_2(g) + 3\,H_2(g) \longrightarrow CH_3OH(g) + H_2O(g)$$

    With the aid of data from Appendix D, determine
    (a) if this reaction proceeds to any significant extent at 25 °C;
    (b) if the production of $CH_3OH(g)$ is favored by raising or lowering the temperature from 25 °C;
    (c) K_p for this reaction at 500 K;
    (d) the partial pressure of $CH_3OH(g)$ at equilibrium if $CO_2(g)$ and $H_2(g)$, each initially at a partial pressure of 1 atm, react at 500 K.

58. Estimate K_p at 100 °C for the reaction $2\,SO_2(g) + O_2(g) \rightleftharpoons 2\,SO_3(g)$. Use data from Table 19.3 and Figure 19-12.

59. The synthesis of ammonia by the Haber process occurs by the reaction $N_2(g) + 3\,H_2(g) \rightleftharpoons 2\,NH_3(g)$ at 400 °C. Using data from Appendix D and assuming that $\Delta H°$ and $\Delta S°$ are essentially unchanged in the temperature interval from 25 to 400 °C, estimate K_p at 400 °C.

60. Use data from Appendix D to determine **(a)** $\Delta H°$, $\Delta S°$, and $\Delta G°$ at 298 K and **(b)** K_p at 875 K for the water gas shift reaction, used commercially to produce $H_2(g)$:
    $CO(g) + H_2O(g) \rightleftharpoons CO_2(g) + H_2(g)$.
    [*Hint:* Assume that $\Delta H°$ and $\Delta S°$ are essentially unchanged in this temperature interval.]

61. In Example 19-10, we used the van't Hoff equation to determine the temperature at which $K_p = 1.0 \times 10^6$ for the reaction $2\,SO_2(g) + O_2(g) \rightleftharpoons 2\,SO_3(g)$. Obtain another estimate of this temperature with data from Appendix D and equations (19.9) and (19.13). Compare your result with that obtained in Example 19-10.

62. The following equilibrium constants have been determined for the reaction $H_2(g) + I_2(g) \rightleftharpoons 2 HI(g)$: $K_p = 50.0$ at 448 °C and 66.9 at 350 °C. Use these data to estimate $\Delta H°$ for the reaction.

63. For the reaction $N_2O_4(g) \rightleftharpoons 2 NO_2(g)$, $\Delta H° = +57.2$ kJ mol^{-1} and $K_p = 0.113$ at 298 K.
    (a) What is K_p at 0 °C?
    (b) At what temperature will $K_p = 1.00$?

64. Use data from Appendix D and the van't Hoff equation (19.15) to estimate a value of K_p at 100 °C for the reaction $2 NO(g) + O_2(g) \rightleftharpoons 2 NO_2(g)$. [*Hint:* First determine K_p at 25 °C. What is $\Delta H°$ for the reaction?]

65. For the reaction

    $$CO(g) + 3 H_2(g) \rightleftharpoons CH_4(g) + H_2O(g),$$
    $$K_p = 2.15 \times 10^{11} \text{ at } 200 °C$$
    $$K_p = 4.56 \times 10^8 \text{ at } 260 °C$$

Coupled Reactions

67. Titanium is obtained by the reduction of $TiCl_4(l)$, which in turn is produced from the mineral rutile (TiO_2).
    (a) With data from Appendix D, determine $\Delta G°$ at 298 K for this reaction.

    $$TiO_2(s) + 2 Cl_2(g) \longrightarrow TiCl_4(l) + O_2(g)$$

    (b) Show that the conversion of $TiO_2(s)$ to $TiCl_4(l)$, with reactants and products in their standard states, is spontaneous at 298 K if the reaction in (a) is coupled with the reaction

    $$2 CO(g) + O_2(g) \longrightarrow 2 CO_2(g)$$

68. Following are some standard Gibbs energies of formation, $\Delta G_f°$, per mole of metal oxide at 1000 K: NiO, -115 kJ; MnO, -280 kJ; TiO_2, -630 kJ. The standard Gibbs energy of formation of CO at 1000 K is -250 kJ per mol CO. Use the method of coupled reactions (page 851) to determine which of these metal oxides

determine $\Delta H°$ by using the van't Hoff equation (19.15) and by using tabulated data in Appendix D. Compare the two results, and comment on how good the assumption is that $\Delta H°$ is essentially independent of temperature in this case.

66. Sodium carbonate, an important chemical used in the production of glass, is made from sodium hydrogen carbonate by the reaction

    $$2 NaHCO_3(s) \rightleftharpoons Na_2CO_3(s) + CO_2(g) + H_2O(g)$$

    Data for the temperature variation of K_p for this reaction are $K_p = 1.66 \times 10^{-5}$ at 30 °C; 3.90×10^{-4} at 50 °C; 6.27×10^{-3} at 70 °C; and 2.31×10^{-1} at 100 °C.
    (a) Plot a graph similar to Figure 19-12, and determine $\Delta H°$ for the reaction.
    (b) Calculate the temperature at which the total gas pressure above a mixture of $NaHCO_3(s)$ and $Na_2CO_3(s)$ is 2.00 atm.

can be reduced to the metal by a spontaneous reaction with carbon at 1000 K and with all reactants and products in their standard states.

69. In biochemical reactions the phosphorylation of amino acids is an important step. Consider the following two reactions and determine whether the phosphorylation of arginine with ATP is spontaneous.

    $$ATP + H_2O \longrightarrow ADP + P \quad \Delta G°' = -31.5 \text{ kJ mol}^{-1}$$
    $$arginine + P \longrightarrow phosphorarginine + H_2O$$
    $$\Delta G°' = 33.2 \text{ kJ mol}^{-1}$$

70. The synthesis of glutamine from glutamic acid is given by $Glu^- + NH_4^+ \longrightarrow Gln + H_2O$. The Gibbs energy for this reaction at pH = 7 and $T = 310$ K is $\Delta G°' = 14.8$ kJ mol^{-1}. Will this reaction be spontaneous if coupled with the hydrolysis of ATP?

    $$ATP + H_2O \longrightarrow ADP + P$$
    $$\Delta G°' = -31.5 \text{ kJ mol}-1$$

Integrative and Advanced Exercises

71. Use data from Appendix D to estimate (a) the normal boiling point of mercury and (b) the vapor pressure of mercury at 25 °C.

72. Consider the vaporization of water: $H_2O(l) \longrightarrow H_2O(g)$ at 100 °C, with $H_2O(l)$ in its standard state, but with the partial pressure of $H_2O(g)$ at 2.0 atm. Which of the following statements about this vaporization at 100 °C are true? (a) $\Delta G° = 0$, (b) $\Delta G = 0$, (c) $\Delta G° > 0$, (d) $\Delta G > 0$? Explain.

73. At 298 K, 1.00 mol BrCl(g) is introduced into a 10.0 L vessel, and equilibrium is established in the reaction $BrCl(g) \rightleftharpoons \frac{1}{2} Br_2(g) + \frac{1}{2} Cl_2(g)$. Calculate the amounts of each of the three gases present when equilibrium is established. [*Hint:* Use data from Appendix D as necessary.]

74. Use data from Appendix D and other information from this chapter to estimate the temperature at which the dissociation of $I_2(g)$ becomes appreciable

[for example, with the $I_2(g)$ 50% dissociated into I(g) at 1 atm total pressure].

75. The following table shows the enthalpies and Gibbs energies of formation of three metal oxides at 25 °C.
    (a) Which of these oxides can be most readily decomposed to the free metal and $O_2(g)$?
    (b) For the oxide that is most easily decomposed, to what temperature must it be heated to produce $O_2(g)$ at 1.00 atm pressure?

	$\Delta H_f°$, kJ mol^{-1}	$\Delta G_f°$, kJ mol^{-1}
PbO(red)	-219.0	-188.9
Ag_2O	-31.05	-11.20
ZnO	-348.3	-318.3

76. The following data are given for the two solid forms of HgI_2 at 298 K.

	ΔH_f°, kJ mol^{-1}	ΔG_f°, kJ mol^{-1}	S°, J mol^{-1} K^{-1}
HgI_2(red)	-105.4	-101.7	180
HgI_2(yellow)	-102.9	(?)	(?)

Estimate values for the two missing entries. To do this, assume that for the transition HgI_2(red) $\longrightarrow$ HgI_2(yellow), the values of ΔH° and ΔS° at 25 °C have the same values that they do at the equilibrium temperature of 127 °C.

77. Oxides of nitrogen are produced in high-temperature combustion processes. The essential reaction is $N_2(g) + O_2(g) \rightleftharpoons 2\,NO(g)$. At what approximate temperature will an *equimolar* mixture of $N_2(g)$ and $O_2(g)$ be 1.0% converted to $NO(g)$? [*Hint:* Use data from Appendix D as necessary.]

78. Use the following data, as appropriate, to estimate the molarity of a saturated aqueous solution of $Sr(IO_3)_2$.

	ΔH_f°, kJ mol^{-1}	ΔG_f°, kJ mol^{-1}	S°, J mol^{-1} K^{-1}
$Sr(IO_3)_2$(s)	-1019.2	-855.1	234
Sr^{2+}(aq)	-545.8	-599.5	-32.6
IO_3^-(aq)	-221.3	-128.0	118.4

79. Use the following data together with other data from the text to determine the temperature at which the equilibrium pressure of water vapor above the two solids in the following reaction is 75 Torr.

$$CuSO_4 \cdot 3\,H_2O(s) \rightleftharpoons CuSO_4 \cdot H_2O(s) + 2\,H_2O(g)$$

	ΔH_f°, kJ mol^{-1}	ΔG_f°, kJ mol^{-1}	S°, J mol^{-1} K^{-1}
$CuSO_4 \cdot 3\,H_2O$(s)	-1684.3	-1400.0	221.3
$CuSO_4 \cdot H_2O$(s)	-1085.8	-918.1	146.0

80. For the dissociation of $CaCO_3(s)$ at 25 °C, $CaCO_3(s) \rightleftharpoons CaO(s) + CO_2(g)$ $\Delta G^\circ = +131$ kJ mol^{-1}. A sample of pure $CaCO_3(s)$ is placed in a flask and connected to an ultrahigh vacuum system capable of reducing the pressure to 10^{-9} mmHg.
(a) Would $CO_2(g)$ produced by the decomposition of $CaCO_3(s)$ at 25 °C be detectable in the vacuum system at 25 °C?
(b) What additional information do you need to determine P_{CO_2} as a function of temperature?
(c) With necessary data from Appendix D, determine the minimum temperature to which $CaCO_3(s)$ would have to be heated for $CO_2(g)$ to become detectable in the vacuum system.

81. Introduced into a 1.50 L flask is 0.100 mol of $PCl_5(g)$; the flask is held at a temperature of 227 °C until equilibrium is established. What is the total pressure of the gases in the flask at this point?

$$PCl_5(g) \rightleftharpoons PCl_3(g) + Cl_2(g)$$

[*Hint:* Use data from Appendix D and appropriate relationships from this chapter.]

82. From the data given in Exercise 66, estimate a value of ΔS° at 298 K for the reaction

$$2\,NaHCO_3(s) \longrightarrow Na_2CO_3(s) + H_2O(g) + CO_2(g)$$

83. The normal boiling point of cyclohexane, C_6H_{12}, is 80.7 °C. Estimate the temperature at which the vapor pressure of cyclohexane is 100.0 mmHg.

84. The term *thermodynamic stability* refers to the sign of ΔG_f°. If ΔG_f° is negative, the compound is stable with respect to decomposition into its elements. Use the data in Appendix D to determine whether $Ag_2O(s)$ is thermodynamically stable at (a) 25 °C and (b) 200 °C.

85. At 0 °C, ice has a density of 0.917 g mL^{-1} and an absolute entropy of 37.95 J mol^{-1} K^{-1}. At this temperature, liquid water has a density of 1.000 g mL^{-1} and an absolute entropy of 59.94 J mol^{-1} K^{-1}. The pressure corresponding to these values is 1 bar. Calculate ΔG, ΔG°, ΔS°, and ΔH° for the melting of two moles of ice at its normal melting point.

86. The decomposition of the poisonous gas phosgene is represented by the equation $COCl_2(g) \rightleftharpoons CO(g) + Cl_2(g)$. Values of K_p for this reaction are $K_p = 6.7 \times 10^{-9}$ at 99.8 °C and $K_p = 4.44 \times 10^{-2}$ at 395 °C. At what temperature is $COCl_2$ 15% dissociated when the total gas pressure is maintained at 1.00 atm?

87. Use data from Appendix D to estimate the aqueous solubility, in milligrams per liter, of $AgBr(s)$ at 100 °C.

88. The standard molar entropy of solid hydrazine at its melting point of 1.53 °C is 67.15 J mol^{-1} K^{-1}. The enthalpy of fusion is 12.66 kJ mol^{-1}. For N_2H_4(l) in the interval from 1.53 °C to 298.15 K, the molar heat capacity at constant pressure is given by the expression $C_p = 97.78 + 0.0586(T - 280)$. Determine the standard molar entropy of N_2H_4(l) at 298.15 K. [*Hint:* The heat absorbed to produce an infinitesimal change in the temperature of a substance is $dq_{rev} = C_p\,dT$.]

89. Use the following data to estimate the standard molar entropy of gaseous benzene at 298.15 K; that is, $S^\circ[C_6H_6(g, 1\,atm)]$. For C_6H_6 (s, 1 atm) at its melting point of 5.53 °C, S° is 128.82 J mol^{-1} K^{-1}. The enthalpy of fusion is 9.866 kJ mol^{-1}. From the melting point to 298.15 K, the average heat capacity of liquid benzene is 134.0 J K^{-1} mol^{-1}. The enthalpy of vaporization of C_6H_6(l) at 298.15 K is 33.85 kJ mol^{-1}, and in the vaporization, C_6H_6(g) is produced at a pressure of 95.13 Torr. Imagine that this vapor could be compressed to 1 atm pressure without condensing and while behaving as an ideal gas. Calculate $S^\circ[C_6H_6(g, 1\,atm)]$. [*Hint:* Refer to Exercise 88, and note the following: For infinitesimal quantities, $dS = dq/dT$; for the compression of an ideal gas, $dq = -dw$; and for pressure–volume work, $dw = -P\,dV$.]

90. On page 822 the terms *states* and *microstates* were introduced. Consider a system that has four states (i.e., energy levels), with energy $\varepsilon = 0, 1, 2$, and 3 energy units, and three particles labeled A, B, and C. The total energy of the system, in energy units, is 3. How many microstates can be generated?

91. In Figure 19-7, page 830, the temperature dependence of the standard molar entropy for chloroform is plotted. (a) Explain why the slope for the standard molar entropy of the solid is greater than the slope for the standard molar entropy of the liquid, which is greater

than the slope for the standard molar entropy of the gas. **(b)** Explain why the change in the standard molar entropy from solid to liquid is smaller than that for the liquid to gas.

92. The following data are from a laboratory experiment that examines the relationship between solubility and thermodynamics. In this experiment $KNO_3(s)$ is placed in a test tube containing some water. The solution is heated until all the $KNO_3(s)$ is dissolved and then allowed to cool. The temperature at which crystals appear is then measured. From this experiment we can determine the equilibrium constant, Gibbs energy, enthalpy, and entropy for the reaction. Use the following data to calculate ΔG, ΔH, and ΔS for the dissolution of $KNO_3(s)$. (Assume the initial mass of $KNO_3(s)$ was 20.2 g.)

Total Volume, mL	Temperature Crystals Formed, K
25.0	340
29.2	329
33.4	320
37.6	313
41.8	310
46.0	306
51.0	303

Data reported by J. O. Schreck, *J. Chem. Educ.*, **73**, 426 (1996).

Feature Problems

93. A tabulation of more precise thermodynamic data than are presented in Appendix D lists the following values for $H_2O(l)$ and $H_2O(g)$ at 298.15 K, at a standard state pressure of 1 bar.

	ΔH_f°, kJ mol^{-1}	ΔG_f°, kJ mol^{-1}	S°, J mol^{-1} K^{-1}
$H_2O(l)$	−285.830	−237.129	69.91
$H_2O(g)$	−241.818	−228.572	188.825

(a) Use these data to determine, in two different ways, ΔG° at 298.15 K for the vaporization: $H_2O\,(l, 1\,bar) \rightleftharpoons H_2O(g, 1\,bar)$. The value you obtain will differ slightly from that on page 838, because here, the standard state pressure is 1 bar, and there, it is 1 atm.
(b) Use the result of part (a) to obtain the value of K for this vaporization and, hence, the vapor pressure of water at 298.15 K.
(c) The vapor pressure in part (b) is in the unit bar. Convert the pressure to millimeters of mercury.
(d) Start with the value $\Delta G^\circ = 8.590$ kJ, given on page 838 and calculate the vapor pressure of water at 298.15 K in a fashion similar to that in parts (b) and (c). In this way, demonstrate that the results obtained in a thermodynamic calculation do not depend on the convention we choose for the standard state pressure, as long as we use standard state thermodynamic data consistent with that choice.

94. The graph shows how ΔG° varies with temperature for three different oxidation reactions: the oxidations of C(graphite), Zn, and Mg to CO, ZnO, and MgO, respectively. Such graphs as these can be used to show the temperatures at which carbon is an effective reducing agent to reduce metal oxides to the free metals. As a result, such graphs are important to metallurgists. Use these graphs to answer the following questions.
(a) Why can Mg be used to reduce ZnO to Zn at all temperatures, but Zn cannot be used to reduce MgO to Mg at any temperature?

(b) Why can C be used to reduce ZnO to Zn at some temperatures but not at others? At what temperatures can carbon be used to reduce zinc oxide?
(c) Is it possible to produce Zn from ZnO by its direct decomposition without requiring a coupled reaction? If so, at what approximate temperatures might this occur?
(d) Is it possible to decompose CO to C and O_2 in a spontaneous reaction? Explain.

▲ ΔG° for three reactions as a function of temperature. The reactions are indicated by the equations written above the graphs. The points noted by arrows are the melting points (mp) and boiling points (bp) of zinc and magnesium.

(e) To the set of graphs, add straight lines representing the reactions

$$C(graphite) + O_2(g) \longrightarrow CO_2(g)$$
$$2\,CO(g) + O_2(g) \longrightarrow 2\,CO_2(g)$$

given that the three lines representing the formation of oxides of carbon intersect at about 800 °C. [*Hint:* At what other temperature can you relate ΔG° and temperature?]

The slopes of the three lines described above differ sharply. Explain why this is so—that is, explain the slope of each line in terms of principles governing Gibbs energy change.

(f) The graphs for the formation of oxides of other metals are similar to the ones shown for Zn and Mg; that is, they all have positive slopes. Explain why carbon is such a good reducing agent for the reduction of metal oxides.

95. In a heat engine, heat (q_h) is absorbed by a working substance (such as water) at a high temperature (T_h). Part of this heat is converted to work (w), and the rest (q_l) is released to the surroundings at the lower temperature (T_l). The *efficiency* of a heat engine is the ratio w/q_h. The second law of thermodynamics establishes the following equation for the maximum efficiency of a heat engine, expressed on a percentage basis.

$$\text{efficiency} = \frac{w}{q_h} \times 100\% = \frac{T_h - T_l}{T_h} \times 100\%$$

In a particular electric power plant, the steam leaving a steam turbine is condensed to liquid water at 41 °C(T_l) and the water is returned to the boiler to be regenerated as steam. If the system operates at 36% efficiency,

(a) What is the minimum temperature of the steam [$H_2O(g)$] used in the plant?
(b) Why is the actual steam temperature probably higher than that calculated in part (a)?
(c) Assume that at T_h the $H_2O(g)$ is in equilibrium with $H_2O(l)$. Estimate the steam pressure at the temperature calculated in part (a).
(d) Is it possible to devise a heat engine with greater than 100 percent efficiency? With 100 percent efficiency? Explain.

96. The Gibbs energy available from the complete combustion of 1 mol of glucose to carbon dioxide and water is

$$C_6H_{12}O_6(aq) + 6\,O_2(g) \longrightarrow 6\,CO_2(g) + 6\,H_2O(l)$$
$$\Delta G° = -2870 \text{ kJ mol}^{-1}$$

(a) Under biological standard conditions, compute the maximum number of moles of ATP that could form from ADP and phosphate if all the energy of combustion of 1 mol of glucose could be utilized.
(b) The actual number of moles of ATP formed by a cell under aerobic conditions (that is, in the presence of oxygen) is about 38. Calculate the efficiency of energy conversion of the cell.
(c) Consider these typical physiological conditions.

$$P_{CO_2} = 0.050 \text{ bar}; P_{O_2} = 0.132 \text{ bar};$$
$$[\text{glucose}] = 1.0 \text{ mg/mL}; \text{pH} = 7.0;$$
$$[\text{ATP}] = [\text{ADP}] = [P_i] = 0.00010 \text{ M}.$$

Calculate ΔG for the conversion of 1 mol ADP to ATP and ΔG for the oxidation of 1 mol glucose under these conditions.
(d) Calculate the efficiency of energy conversion for the cell under the conditions given in part (c). Compare this efficiency with that of a diesel engine that attains 78% of the theoretical efficiency operating with $T_h = 1923$ K and $T_l = 873$ K. Suggest a reason for your result. [*Hint:* See Feature Problem 95.]

97. The entropy of materials at $T = 0$ K should be zero; however, for some substances, such as CO and H_2O, this is not true. The difference between the measured value and expected value of zero is known as residual entropy. **(a)** Calculate the residual entropy for one mole of CO by using the Boltzmann equation for entropy. **(b)** Calculate the residual entropy for one mole of H_2O in the same manner.

Self-Assessment Exercises

98. In your own words, define the following symbols: **(a)** ΔS_{univ}; **(b)** $\Delta G_f°$; **(c)** K.

99. Briefly describe each of the following ideas, methods, or phenomena: **(a)** absolute molar entropy; **(b)** coupled reactions; **(c)** Trouton's rule; **(d)** evaluation of an equilibrium constant from tabulated thermodynamic data.

100. Explain the important distinctions between each of the following pairs: **(a)** spontaneous and nonspontaneous processes; **(b)** the second and third laws of thermodynamics; **(c)** ΔG and $\Delta G°$.

101. For a process to occur spontaneously, **(a)** the entropy of the system must increase; **(b)** the entropy of the surroundings must increase; **(c)** both the entropy of the system and the entropy of the surroundings must increase; **(d)** the net change in entropy of the system and surroundings considered together must be a positive quantity; **(e)** the entropy of the universe must remain constant.

102. The Gibbs energy change of a reaction can be used to assess **(a)** how much heat is absorbed from the surroundings; **(b)** how much work the system does on the surroundings; **(c)** the net direction in which the reaction occurs to reach equilibrium; **(d)** the proportion of the heat evolved in an exothermic reaction that can be converted to various forms of work.

103. The reaction, $2\,Cl_2O(g) \longrightarrow 2\,Cl_2(g) + O_2(g)$ $\Delta H = -161$ kJ, is expected to be **(a)** spontaneous at all temperatures; **(b)** spontaneous at low temperatures, but nonspontaneous at high temperatures; **(c)** nonspontaneous at all temperatures; **(d)** spontaneous at high temperatures only.

104. If $\Delta G° = 0$ for a reaction, it must also be true that **(a)** $K = 0$; **(b)** $K = 1$; **(c)** $\Delta H° = 0$; **(d)** $\Delta S° = 0$; **(e)** the equilibrium activities of the reactants and products do not depend on the initial conditions.

105. Two correct statements about the reversible reaction $N_2(g) + O_2(g) \rightleftharpoons 2\,NO(g)$ are **(a)** $K = K_p$; **(b)** the equilibrium amount of NO increases with an increased total gas pressure; **(c)** the equilibrium amount of NO increases if an equilibrium mixture is transferred from a 10.0 L container to a 20.0 L container; **(d)** $K = K_c$; **(e)** the composition of an equilibrium mixture of the gases is independent of the temperature.

106. Suppose a graph similar to (Figure 19-9) were drawn for the process $I_2(s) \longrightarrow I_2(l)$ at 1 atm.
(a) Refer to Figure 12-27 and determine the temperature at which the two lines would intersect.
(b) What would be the value of $\Delta G°$ at this temperature? Explain.

107. *Without performing detailed calculations,* indicate whether any of the following reactions would occur to a measurable extent at 298 K.
(a) Conversion of dioxygen to ozone:
$$3 O_2(g) \longrightarrow 2 O_3(g)$$
(b) Dissociation of N_2O_4 to NO_2:
$$N_2O_4(g) \longrightarrow 2 NO_2(g)$$
(c) Formation of BrCl:
$$Br_2(l) + Cl_2(g) \longrightarrow 2 BrCl(g)$$

108. Explain briefly why
(a) the change in entropy in a system is not always a suitable criterion for spontaneous change;

(b) $\Delta G°$ is so important in dealing with the question of spontaneous change, even though the conditions employed in a reaction are very often nonstandard.

109. A handbook lists the following standard enthalpies of formation at 298 K for cyclopentane, C_5H_{10}: $\Delta H_f°[C_5H_{10}(l)] = -105.9$ kJ/mol and $\Delta H_f°[C_5H_{10}(g)] = -77.2$ kJ/mol.
(a) Estimate the normal boiling point of cyclopentane.
(b) Estimate $\Delta G°$ for the vaporization of cyclopentane at 298 K.
(c) Comment on the significance of the sign of $\Delta G°$ at 298 K.

110. Consider the reaction $NH_4NO_3(s) \longrightarrow N_2O(g) + 2 H_2O(l)$ at 298 K.
(a) Is the forward reaction endothermic or exothermic?
(b) What is the value of $\Delta G°$ at 298 K?
(c) What is the value of K at 298 K?
(d) Does the reaction tend to occur spontaneously at temperatures above 298 K, below 298 K, both, or neither?

111. Which of the following diagrams represents an equilibrium constant closest to 1?

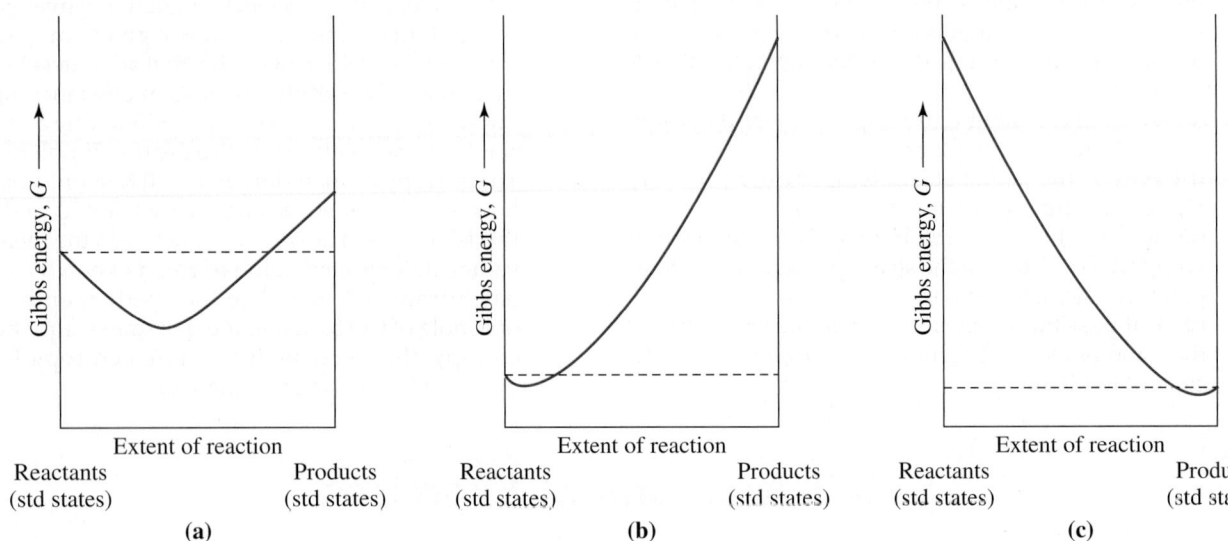

(a) (b) (c)

112. At room temperature and normal atmospheric pressure, is the entropy of the universe positive, negative, or zero for the transition of carbon dioxide solid to liquid?

Electrochemistry

A transit bus fitted with hydrogen–oxygen fuel cells. The use of fuel cells could dramatically reduce urban air pollution. The conversion of chemical energy into electrical energy is one of the main subjects of this chapter.

A conventional gasoline-powered automobile is only about 25% efficient in converting chemical energy into kinetic energy (energy of motion). An electric-powered auto is about three times as efficient. Unfortunately, when automotive technology was first being developed, devices for converting chemical energy to electrical energy did not perform at their intrinsic efficiencies. This fact, together with the availability of high-quality gasoline at a low cost, resulted in the preeminence of the internal combustion automobile. Now, with concern about long-term energy supplies and environmental pollution, there is a renewed interest in electric-powered buses and automobiles.

In this chapter, we will see how chemical reactions can be used to produce electricity and how electricity can be used to cause chemical reactions. The practical applications of electrochemistry are countless, ranging from batteries and fuel cells as electric power sources to the manufacture of key chemicals, the refining of metals, and methods for controlling corrosion. Also important, however, are the theoretical implications. Because electricity involves a flow of electric charge, a study of the relationship between chemistry and electricity gives us additional insight into reactions in which electrons are transferred—*oxidation–reduction reactions.*

20-1 Electrode Potentials and Their Measurement

The criteria for spontaneous change developed in Chapter 19 apply to reactions of all types—precipitation, acid–base, and oxidation–reduction (redox). We can devise an additional useful criterion for redox reactions, however.

Figure 20-1 shows that a redox reaction occurs between Cu(s) and $Ag^+(aq)$, but not between Cu(s) and $Zn^{2+}(aq)$. Specifically, we see that silver ions are reduced to silver atoms on a copper surface, whereas zinc ions are *not* reduced to zinc atoms on a copper surface. We can say that Ag^+ is more readily reduced than is Zn^{2+}. In this section, we will introduce the *electrode potential*, a property related to these reduction tendencies.

▶ The term *electrode* is sometimes used for the entire half-cell assembly.

When used in electrochemical studies, a strip of metal, M, is called an **electrode**. An electrode immersed in a solution containing ions of the same metal, M^{n+}, is called a **half-cell**. Two kinds of interactions are possible between metal atoms on the electrode and metal ions in solution (Fig. 20-2):

1. A metal ion M^{n+} from solution may collide with the electrode, gain n electrons from it, and be converted to a metal atom M. *The ion is reduced.*

2. A metal atom M on the surface may lose n electrons to the electrode and enter the solution as the ion M^{n+}. *The metal atom is oxidized.*

An equilibrium is quickly established between the metal and the solution, which can be represented as

KEEP IN MIND

that although $M^{n+}(aq)$ and ne^- appear together on the right-hand side of this expression, only the ion M^{n+} enters the solution. The electrons remain on the electrode, M(s). Free electrons are never found in an aqueous solution.

$$M(s) \xrightleftharpoons[\text{reduction}]{\text{oxidation}} M^{n+}(aq) + ne^- \qquad (20.1)$$

However, any changes produced at the electrode or in the solution as a consequence of this equilibrium are too slight to measure. Instead, measurements must be based on a combination of *two different* half-cells. Specifically, we must measure the tendency for electrons to flow from the electrode of one half-cell to the electrode of the other. Electrodes are classified according to whether oxidation or reduction takes place there. If oxidation takes place, the electrode is called the **anode**. If reduction takes place, the electrode is called the **cathode**.

▶ FIGURE 20-1
Behavior of $Ag^+(aq)$ and $Zn^+(aq)$ in the presence of copper
(a) Copper metal displaces silver ions from colorless $AgNO_3(aq)$ as a deposit of silver metal; the copper enters the solution as blue $Cu^{2+}(aq)$.

$$Cu(s) + 2\,Ag^+(aq) \longrightarrow Cu^{2+}(aq) + 2\,Ag(s)$$

(b) Cu(s) *does not* displace colorless Zn^{2+} from $Zn(NO_3)_2(aq)$.

$$Cu(s) + Zn^{2+}(aq) \longrightarrow \text{no reaction}$$

(a)   (b)

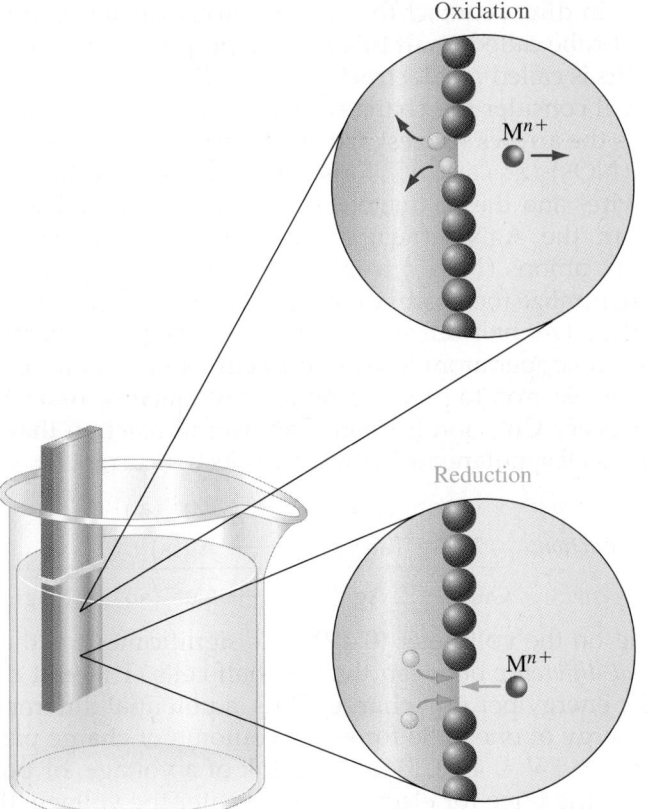

Oxidation

Reduction

◀ FIGURE 20-2
An electrochemical half-cell
The half-cell consists of a metal electrode, M, partially immersed in an aqueous solution of its ions, M^{n+}. (The anions required to maintain electrical neutrality in the solution are not shown.) The situation illustrated here is limited to metals that do not react with water.

Figure 20-3 depicts a combination of two half-cells, one with a Cu electrode in contact with $Cu^{2+}(aq)$, and the other with an Ag electrode in contact with $Ag^+(aq)$. The two electrodes are joined by wires to an electric meter—here, a *voltmeter*. To complete the electric circuit, the two solutions must also be connected electrically. However, because charge is carried through solutions by the migration of *ions*, a wire cannot be used for this connection. The solutions

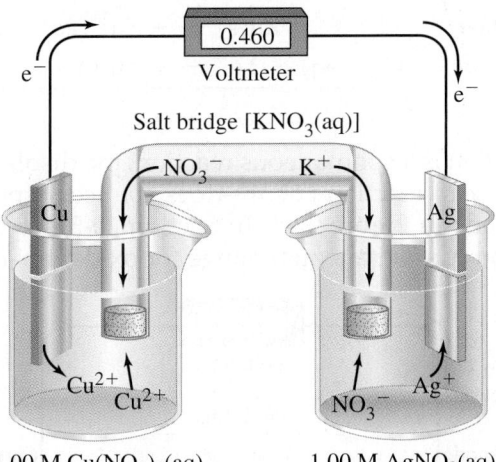

0.460
Voltmeter
Salt bridge [$KNO_3(aq)$]
NO_3^- K^+
Cu   Ag
Cu^{2+} Cu^{2+} NO_3^- Ag^+
1.00 M $Cu(NO_3)_2$(aq)   1.00 M $AgNO_3$(aq)

▲ FIGURE 20-3
Measurement of the electromotive force of an electrochemical cell
An electrochemical cell consists of two half-cells with electrodes joined by a wire and solutions joined by a salt bridge. (The ends of the salt bridge are plugged with a porous material that allows ions to migrate but prevents the bulk flow of liquid.) Electrons flow from the Cu electrode, the anode, where oxidation occurs to the Ag electrode, the cathode, where reduction occurs. For precise measurements, the amount of electric current drawn from the cell must be kept very small by means of either a specially designed voltmeter or a device called a potentiometer.

◀ <u>An</u>ions migrate toward the <u>an</u>ode, and <u>cat</u>ions toward the <u>cat</u>hode.

must either be in direct contact through a porous barrier or joined by a third solution in a U-tube called a **salt bridge**. The properly connected combination of two half-cells is called an **electrochemical cell**.

Now, we will consider the changes that occur in the electrochemical cell in Figure 20-3. As the arrows suggest, Cu atoms release electrons at the anode and enter the $Cu(NO_3)_2(aq)$ as Cu^{2+} ions. Electrons lost by the Cu atoms pass through the wires and the voltmeter to the cathode, where they are gained by Ag^+ ions from the $AgNO_3(aq)$, producing a deposit of metallic silver. Simultaneously, anions (NO_3^-) from the salt bridge migrate into the copper half-cell and neutralize the positive charge of the excess Cu^{2+} ions; cations (K^+) migrate into the silver half-cell and neutralize the negative charge of the excess NO_3^- ions. Each copper atom loses two electrons to produce Cu^{2+}; each Ag^+ ion requires one electron to produce Ag(s); consequently, two silver atoms are produced for every Cu^{2+} ion formed. The overall reaction that occurs as the electrochemical cell spontaneously produces electric current is

Oxidation: $Cu(s) \longrightarrow Cu^{2+}(aq) + 2\,e^-$

Reduction: $2\,\{Ag^+(aq) + e^- \longrightarrow Ag(s)\}$

Overall: $Cu(s) + 2\,Ag^+(aq) \longrightarrow Cu^{2+}(aq) + 2\,Ag(s)$ **(20.2)**

> **KEEP IN MIND**
>
> that the overall reaction occurring in the electrochemical cell is identical to what happens in the direct addition of Cu(s) to $Ag^+(aq)$ pictured in Figure 20-1(a).

The reading on the voltmeter (0.460 V) is significant. It is the **cell voltage**, or the *potential difference* between the two half-cells. The unit of cell voltage, **volt (V)**, is the energy per unit charge. Thus, a potential difference of one volt signifies an energy of one joule for every coulomb of charge passing through an electric circuit: $1\ V = 1\ J/C$. We can think of a voltage, or potential difference, as the driving force for electrons; the greater the voltage, the greater the driving force. The flow of water from a higher to a lower level is analogous to this situation. The greater the difference in water levels, the greater the force behind the flow of water. Cell voltage is also called **electromotive force (emf)**, or **cell potential**, and represented by the symbol E_{cell}.

> ▶ Such formulations as $Zn(s)/Zn^{2+}(aq)$ are called *couples* and are often used as abbreviations for half-cells.

Now let's return to the question raised by Figure 20-1: Why does copper *not* displace Zn^{2+} from solution? In an electrochemical cell consisting of a $Zn(s)/Zn^{2+}(aq)$ half-cell and a $Cu^{2+}(aq)/Cu(s)$ half-cell, electrons flow *from the Zn to the Cu*. The spontaneous reaction in the electrochemical cell in Figure 20-4 is

Oxidation: $Zn(s) \longrightarrow Zn^{2+}(aq) + 2\,e^-$

Reduction: $Cu^{2+}(aq) + 2\,e^- \longrightarrow Cu(s)$

Overall: $Zn(s) + Cu^{2+}(aq) \longrightarrow Zn^{2+}(aq) + Cu(s)$ **(20.3)**

Because reaction (20.3) is a spontaneous reaction, the displacement of $Zn^{2+}(aq)$ by Cu(s)—the *reverse* of reaction (20.3)—does *not* occur spontaneously. This is the observation made in Figure 20-1. In Section 20-3, we will discuss how to predict the direction of spontaneous change for oxidation–reduction reactions:

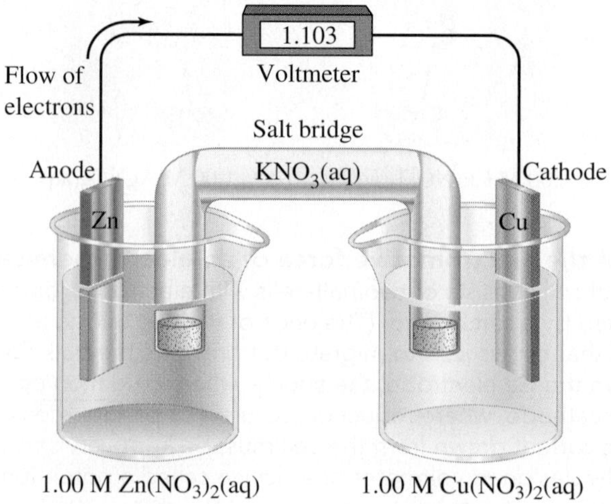

▶ FIGURE 20-4
The reaction
$Zn(s) + Cu^{2+}(aq) \longrightarrow Zn^{2+}(aq) + Cu(s)$ in
an electrochemical cell

Cell Diagrams and Terminology

Drawing sketches of electrochemical cells, as in Figures 20-3 and 20-4, is helpful, but more often a simpler representation is used. A **cell diagram** shows the components of an electrochemical cell in a symbolic way. We will use the following, generally accepted conventions in writing cell diagrams.

- The anode, the electrode at which *oxidation* occurs, is placed at the *left* side of the diagram.

- The cathode, the electrode at which *reduction* occurs, is placed at the *right* side of the diagram.

- A boundary between different phases (for example, an electrode and a solution) is represented by a *single vertical line* ($|$).

- The boundary between half-cell compartments, commonly a salt bridge, is represented by a *double vertical line* ($\|$). Species in aqueous solution are placed on either side of the double vertical line. Different species within the same solution are separated from each other by a comma. Although IUPAC recommends the use of double dashed vertical lines to represent the salt bridge, the recommendation has not yet been universally adopted by chemists. In this text, we will continue to use the double vertical line ($\|$) for a salt bridge.

◀ Several memory devices have been proposed for the oxidation/anode and reduction/cathode relationships. Perhaps the simplest is that in the oxidation/anode relationship, both terms begin with a vowel: *o/a*; in the reduction/cathode relationship, both begin with a consonant: *r/c*.

The cell diagram corresponding to both Figure 20-4 and reaction (20.3) is customarily written as

$$\text{anode} \longrightarrow \underset{\substack{\text{Half-cell} \\ \text{(oxidation)}}}{Zn(s)|Zn^{2+}(aq)} \underset{\substack{\text{Salt} \\ \text{bridge}}}{\|} \underset{\substack{\text{Half-cell} \\ \text{(reduction)}}}{Cu^{2+}(aq)|Cu(s)} \longleftarrow \text{cathode} \quad \underset{\text{Cell voltage}}{E_{cell} = 1.103\ V}$$

$$(20.4)$$

The electrochemical cells of Figures 20-3 and 20-4 produce electricity as a result of spontaneous chemical reactions; as such, they are called **voltaic**, or **galvanic**, **cells**. In Section 20-7 we will consider *electrolytic cells*—electrochemical cells in which electricity is used to accomplish a nonspontaneous chemical change.

KEEP IN MIND

that the spectator ions are not shown in a cell diagram, but they are present. They pass through the salt bridge to maintain electrical neutrality.

EXAMPLE 20-1 Representing a Redox Reaction by Means of a Cell Diagram

Aluminum metal displaces zinc(II) ion from aqueous solution.

(a) Write oxidation and reduction half-equations and an overall equation for this redox reaction.

(b) Write a cell diagram for a voltaic cell in which this reaction occurs.

Analyze

The term *displaces* means that aluminum goes into solution as $Al^{3+}(aq)$, forcing $Zn^{2+}(aq)$ out of solution as zinc metal. Al is oxidized to Al^{3+}, and Zn^{2+} is reduced to Zn. In combining the half-equations to produce the overall equation, we must take care to ensure that the *number of electrons involved in reduction equals the number involved in oxidation*. (This is the half-reaction method of balancing redox equations discussed in Section 5-5.) The cell diagram is written with the reduction half-equation as the right-hand electrode.

Solve

(a) The two half equations are

$$\textit{Oxidation:} \qquad\qquad Al(s) \longrightarrow Al^{3+}(aq) + 3\,e^-$$
$$\textit{Reduction:}\ \ Zn^{2+}(aq) + 2\,e^- \longrightarrow Zn(s)$$

On inspecting these half-equations, we see that the number of electrons involved in oxidation and reduction are different. In writing the overall equation, the coefficients must be adjusted so that equal numbers of electrons are involved in oxidation and in reduction.

$$\textit{Oxidation:} \qquad\qquad 2\,\{Al(s) \longrightarrow Al^{3+}(aq) + 3\,e^-\}$$
$$\textit{Reduction:}\ \ \underline{3\,\{Zn^{2+}(aq) + 2\,e^- \longrightarrow Zn(s)\}}$$
$$\textit{Overall:}\ \ 2\,Al(s) + 3\,Zn^{2+}(aq) \longrightarrow 2\,Al^{3+}(aq) + 3\,Zn(s)$$

(continued)

(b) Al(s) is oxidized to Al^{3+}(aq) in the anode half-cell (written on the left of the cell diagram), and Zn^{2+}(aq) is reduced to Zn(s) in the cathode half-cell (written on the right of the cell diagram).

$$Al(s)|Al^{3+}(aq)||Zn^{2+}(aq)|Zn(s)$$

Assess

Whenever balancing redox equations, it is important to ensure that the number of electrons in the oxidation step equals the number of electrons in the reduction step. This is achieved by multiplying the entire half-reaction(s) by the appropriate factor(s).

PRACTICE EXAMPLE A: Write the overall equation for the redox reaction that occurs in the voltaic cell Sc(s)|Sc^{3+}(aq)||Ag^+(aq)|Ag(s).

PRACTICE EXAMPLE B: Draw a voltaic cell in which silver ion is displaced from solution by aluminum metal. Label the cathode, the anode, and other features of the cell. Show the direction of flow of electrons. Also, indicate the direction of flow of cations and anions from a KNO_3(aq) salt bridge. Write an equation for the half-reaction occurring at each electrode, write a balanced equation for the overall cell reaction, and write a cell diagram.

EXAMPLE 20-2   Deducing the Balanced Redox Reaction from a Cell Diagram

The cell diagram for an electrochemical cell is written as

$$Ni(s)|NiCl_2(aq)||Ce(ClO_4)_4(aq), Ce(ClO_4)_3(aq)|Pt(s)$$

Write the equations for the half-reactions that occur at the electrodes. Balance the overall cell reaction.

Analyze

When inspecting a cell diagram, we first need to identify the species involved in oxidation and in reduction. Then we can write balanced half-cell equations. Finally, we can combine the half-cell equations to give the overall cell reaction. The new part to this example is the Ce^{4+}/Ce^{3+} couple in the presence of the inert platinum electrode at which the reduction takes place. Again, when balancing redox equations, it is important to ensure that the number of electrons in the oxidation step equals the number of electrons in the reduction step.

Solve

The cerium reduction reaction is

$$Ce^{4+}(aq) + e^- \longrightarrow Ce^{3+}(aq)$$

The nickel oxidation reaction is

$$Ni(s) \longrightarrow Ni^{2+}(aq) + 2\,e^-$$

In these half-equations the number of electrons involved in oxidation and reduction are different. In writing the overall equation, the coefficients must be adjusted so that equal numbers of electrons are involved in oxidation and in reduction.

Oxidation:	$Ni(s) \longrightarrow Ni^{2+}(aq) + 2\,e^-$
Reduction:	$2\,\{Ce^{4+}(aq) + e^- \longrightarrow Ce^{3+}(aq)\}$
Overall:	$Ni(s) + 2\,Ce^{4+}(aq) \longrightarrow Ni^{2+}(aq) + 2\,Ce^{3+}(aq)$

Assess

We can see the importance of balancing each half-equation with respect to charge and mass.

PRACTICE EXAMPLE A: The cell diagram for an electrochemical cell is written as

$$Sn(s)|SnCl_2(aq)||AgNO_3(aq)|Ag(s)$$

Write the equations for the half-reactions that occur at the electrodes. Balance the overall cell reaction.

PRACTICE EXAMPLE B: The cell diagram for an electrochemical cell is written as

$$In(s)|In(ClO_4)_3(aq)||CdCl_2(aq)|Cd(s)$$

Write the equations for the half-reactions that occur at the electrodes. Balance the overall cell reaction.

Add appropriate arrows to Figure 20-4 to show the direction of migration of ions through the electrochemical cell.

20-2   Standard Electrode Potentials

Cell voltages—*potential differences* between electrodes—are among the most precise scientific measurements possible. Potentials of individual electrodes, however, cannot be precisely established. If we could make such measurements, cell voltages could be obtained just by subtracting one electrode potential from another. The same result can be achieved by *arbitrarily* choosing a particular half-cell that is assigned an electrode potential of *zero*. Other half-cells can then be compared with this reference. The commonly accepted reference is the standard hydrogen electrode.

◀ This method is comparable to establishing standard enthalpies or Gibbs energies of formation on the basis of an arbitrary zero value.

The **standard hydrogen electrode (SHE)** is depicted in Figure 20-5. The SHE involves equilibrium established on the surface of an inert metal (such as platinum) between H_3O^+ ions from a solution in which they are at unit activity (that is, $a_{H_3O^+} = 1$) and H_2 molecules from the gaseous state at a pressure of 1 bar. The equilibrium reaction produces a particular potential on the metal surface, but this potential is arbitrarily taken to be *zero*.

$$2\,H^+(a = 1) + 2\,e^- \underset{\text{on Pt}}{\overset{\text{on Pt}}{\rightleftharpoons}} H_2(g, 1\text{ bar}) \qquad E° = 0 \text{ volt (V)} \qquad \textbf{(20.5)}$$

The diagram for this half-cell is

$$Pt|H_2(g, 1\text{ bar})|H^+(a = 1)$$

The two vertical lines signify that three phases are present: solid platinum, gaseous hydrogen, and aqueous hydrogen ion. For simplicity, we will usually write H^+ for H_3O^+, assume that unit activity ($a = 1$) exists at roughly $[H^+] = 1$ M, and replace a pressure of 1 bar by 1 atm.

By international agreement, a **standard electrode potential, $E°$**, measures the tendency for a *reduction* process to occur at an electrode. In all cases, the

◀ In this chapter, we shall use the units bar and atm interchangeably. The reason is that, although the current standard pressure is defined as 1 bar, the most extensive and authoritative tabulations of $E°$ values are based on the old standard of 1 atm. Fortunately, the difference between 1 bar and 1 atm is small and so too are the differences between $E°$ values based on the old and new standards.

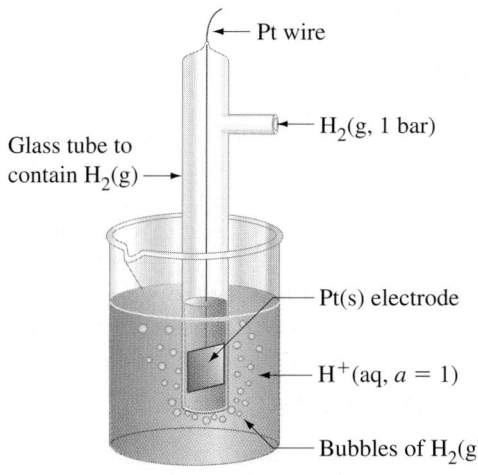

▲ FIGURE 20-5
The standard hydrogen electrode (SHE)
Because hydrogen is a gas at room temperature, electrodes cannot be constructed from it. The standard hydrogen electrode consists of a piece of platinum dipped into a solution containing 1 M H^+(aq) with a stream of hydrogen passing over its surface. The platinum does not react but provides a surface for the reduction of $H_3O(aq)^+$ to $H_2(g)$ as well as the reverse oxidation half-reaction.

ionic species are present in aqueous solution at unit activity (approximately 1 M), and gases are at 1 bar pressure (approximately 1 atm). Where no metallic substance is indicated, the potential is established on an inert metallic electrode, such as platinum.

To emphasize that $E°$ refers to a reduction, we will write a reduction couple as a subscript to $E°$, as shown in half-reaction (20.6). The substance being reduced is written on the left of the slash sign (/), and the chief reduction product on the right.

$$Cu^{2+}(1\ M) + 2\ e^- \longrightarrow Cu(s) \qquad E°_{Cu^{2+}/Cu} = ? \qquad \textbf{(20.6)}$$

To determine the value of $E°$ for a standard electrode such as that to which half-reaction (20.6) applies, we compare it with a standard hydrogen electrode (SHE). In this comparison, the SHE is always taken as the electrode on the *left* of the cell diagram—the anode—and the compared electrode is the electrode on the *right*—the cathode. In the following voltaic cell, the measured potential difference is 0.340 V, with electrons flowing from the H_2 to the Cu electrode.

$$Pt|H_2(g, 1\ atm)|H^+(1\ M)\|Cu^{2+}(1\ M)|Cu(s) \qquad E°_{cell} = 0.340\ V \qquad \textbf{(20.7)}$$
$$\underset{\text{anode}}{} \qquad\qquad\qquad \underset{\text{cathode}}{\phantom{Cu^{2+}(1M)|Cu(s)}}$$

A **standard cell potential, $E°_{cell}$,** is the potential difference, or voltage, of a cell formed from two *standard* electrodes. The *difference* is always taken in the following way:

$$E°_{cell} = E°(\text{right}) - E°(\text{left})$$
$$\phantom{E°_{cell} = }\ \ {\scriptstyle(\text{cathode})} \qquad {\scriptstyle(\text{anode})}$$

Applied to the cell diagram (20.7), we get

$$E°_{cell} = E°_{Cu^{2+}/Cu} - E°_{H^+/H_2} = 0.340\ V$$
$$= E°_{Cu^{2+}/Cu} - 0\ V = 0.340\ V$$
$$E°_{Cu^{2+}/Cu} = 0.340\ V$$

Thus, the standard *reduction* half-reaction can be written as

$$Cu^{2+}(1\ M) + 2\ e^- \longrightarrow Cu(s) \qquad E°_{Cu^{2+}/Cu} = +0.340\ V \qquad \textbf{(20.8)}$$

The overall reaction occurring in the voltaic cell diagrammed in (20.7) can be represented as

$$H_2(g, 1\ atm) + Cu^{2+}(1\ M) \longrightarrow 2\ H^+(1\ M) + Cu(s) \qquad E°_{cell} = 0.340\ V \quad \textbf{(20.9)}$$

Cell reaction (20.9) indicates that $Cu^{2+}(1\ M)$ is more easily reduced than is $H^+(1\ M)$.

Suppose the standard copper electrode in cell diagram (20.7) is replaced by a standard zinc electrode, and the potential difference between the standard hydrogen and zinc electrodes is measured by using the same voltmeter connections as in (20.7). In this case, the voltage is found to be -0.763 V. The negative sign indicates that electrons flow in the direction *opposite* that in (20.7)—that is, *from* the zinc electrode to the hydrogen electrode. Here $H^+(1\ M)$ is more easily reduced than is $Zn^{2+}(1\ M)$. These findings are represented in the following cell diagram, in which the zinc electrode appears on the *right*.

$$Pt|H_2(g, 1\ atm)|H^+(1\ M)\|Zn^{2+}(1\ M)|Zn(s) \qquad E°_{cell} = -0.763\ V \qquad \textbf{(20.10)}$$

The standard electrode potential for the Zn^{2+}/Zn couple can be written as

$$E°_{cell} = E°(\text{right}) - E°(\text{left})$$
$$= E°_{Zn^{2+}/Zn} - 0\ V = -0.763\ V$$
$$E°_{Zn^{2+}/Zn} = -0.763\ V$$

Thus, the standard *reduction* half-reaction is

$$Zn^{2+}(1\ M) + 2\ e^- \longrightarrow Zn(s) \qquad E°_{Zn^{2+}/Zn} = -0.763\ V \qquad \textbf{(20.11)}$$

In summary, the potential of the standard hydrogen electrode is set at exactly 0 V. Any electrode at which a reduction half-reaction shows a *greater* tendency to occur than does the reduction of H^+ (1 M) to H_2 (g, 1 atm) has a *positive* value for its standard electrode potential, $E°$. Any electrode at which a reduction half-reaction shows a *lesser* tendency to occur than does the reduction of H^+(1 M) to H_2 (g, 1 atm) has a *negative* value for its standard reduction potential, $E°$. Comparisons of the standard copper and zinc electrodes to the standard hydrogen electrode are illustrated in Figure 20-6. Table 20.1 on the next page lists some common reduction half-reactions and their standard electrode potentials at 25 °C.

🔍 20-2   CONCEPT ASSESSMENT

For Figure 20-6, describe any changes in mass that might be detected at the Pt, Cu, and Zn electrodes as electric current passes through the electrochemical cells.

Standard reduction potentials are used throughout this chapter for many purposes. Our first objective will be to calculate standard cell potentials for redox reactions—$E°_{cell}$ values—from standard electrode potentials for half-cell reactions—$E°$ values. The procedure used is illustrated here for reaction (20.3) and cell diagram (20.4). Note that the first three equations are alternative ways of stating the same thing; we will generally not write all of them.

◄ Reaction (20.3) is shown on page 866 and diagram (20.4) on page 867.

$$E°_{cell} = E°(\text{right}) - E°(\text{left})$$
$$= E°(\text{cathode}) - E°(\text{anode})$$
$$= E°(\text{reduction half-cell}) - E°(\text{oxidation half-cell})$$
$$= E°_{Cu^{2+}/Cu} - E°_{Zn^{2+}/Zn}$$
$$= 0.340 \text{ V} - (-0.763 \text{ V}) = 1.103 \text{ V}$$

Example 20-3 predicts $E°_{cell}$ for a new battery system. Example 20-4 uses one known electrode potential and a measured $E°_{cell}$ value to determine an unknown $E°$.

KEEP IN MIND

that the $E°$ values in this formulation are for a reduction half-reaction, regardless of whether oxidation or reduction occurs in the half-cell.

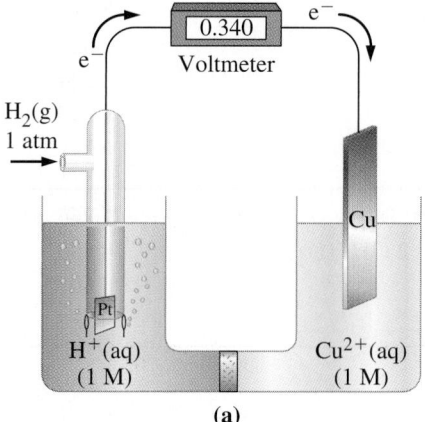 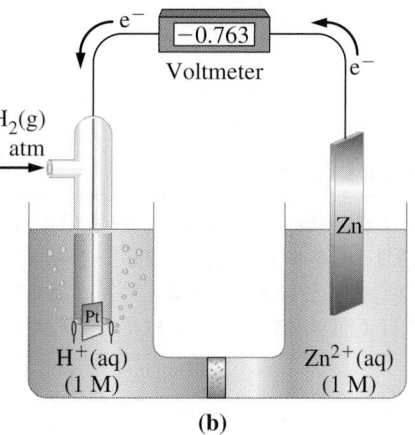

(a)   (b)

▲ FIGURE 20-6
Measuring standard electrode potentials
(a) A standard hydrogen electrode is the anode, and copper is the cathode. Contact between the half-cells occurs through a porous plate that prevents bulk flow of the solutions while allowing ions to pass. **(b)** This cell has the same connections as that in part (a), but with zinc substituting for copper. However, the electron flow is opposite that in (a), as noted by the *negative* voltage. (Zinc is the anode.)

▶ A more extensive listing of reduction half-reactions and their potentials is given in Appendix D.

▶ The placement of *oxidizing agents* is as follows: strongest oxidizing agents ($F_2, O_3, \ldots$), *left* sides, *top* of the list; weakest oxidizing agents ($Li^+, K^+, \ldots$), *left* sides, *bottom* of list. The placement of *reducing agents* is as follows: strongest reducing agents ($Li, K, \ldots$), *right* sides, *bottom* of list; weakest reducing agents ($F^-, O_2, \ldots$), *right* sides, *top* of list.

TABLE 20.1 Some Selected Standard Electrode (Reduction) Potentials at 25 °C

Reduction Half-Reaction	$E°$, V
Acidic solution	
$F_2(g) + 2\,e^- \longrightarrow 2\,F^-(aq)$	+2.866
$O_3(g) + 2\,H^+(aq) + 2\,e^- \longrightarrow O_2(g) + H_2O(l)$	+2.075
$S_2O_8^{2-}(aq) + 2\,e^- \longrightarrow 2\,SO_4^{2-}(aq)$	+2.01
$H_2O_2(aq) + 2\,H^+(aq) + 2\,e^- \longrightarrow 2\,H_2O(l)$	+1.763
$MnO_4^-(aq) + 8\,H^+(aq) + 5\,e^- \longrightarrow Mn^{2+}(aq) + 4\,H_2O(l)$	+1.51
$PbO_2(s) + 4\,H^+(aq) + 2\,e^- \longrightarrow Pb^{2+}(aq) + 2\,H_2O(l)$	+1.455
$Cl_2(g) + 2\,e^- \longrightarrow 2\,Cl^-(aq)$	+1.358
$Cr_2O_7^{2-}(aq) + 14\,H^+(aq) + 6\,e^- \longrightarrow 2\,Cr^{3+}(aq) + 7\,H_2O(l)$	+1.33
$MnO_2(s) + 4\,H^+(aq) + 2\,e^- \longrightarrow Mn^{2+}(aq) + 2\,H_2O(l)$	+1.23
$O_2(g) + 4\,H^+(aq) + 4\,e^- \longrightarrow 2\,H_2O(l)$	+1.229
$2\,IO_3^-(aq) + 12\,H^+(aq) + 10\,e^- \longrightarrow I_2(s) + 6\,H_2O(l)$	+1.20
$Br_2(l) + 2\,e^- \longrightarrow 2\,Br^-(aq)$	+1.065
$NO_3^-(aq) + 4\,H^+(aq) + 3\,e^- \longrightarrow NO(g) + 2\,H_2O(l)$	+0.956
$Ag^+(aq) + e^- \longrightarrow Ag(s)$	+0.800
$Fe^{3+}(aq) + e^- \longrightarrow Fe^{2+}(aq)$	+0.771
$O_2(g) + 2\,H^+(aq) + 2\,e^- \longrightarrow H_2O_2(aq)$	+0.695
$I_2(s) + 2\,e^- \longrightarrow 2\,I^-(aq)$	+0.535
$Cu^{2+}(aq) + 2\,e^- \longrightarrow Cu(s)$	+0.340
$SO_4^{2-}(aq) + 4\,H^+(aq) + 2\,e^- \longrightarrow 2\,H_2O(l) + SO_2(g)$	+0.17
$Sn^{4+}(aq) + 2\,e^- \longrightarrow Sn^{2+}(aq)$	+0.154
$S(s) + 2\,H^+(aq) + 2\,e^- \longrightarrow H_2S(g)$	+0.14
$2\,H^+(aq) + 2\,e^- \longrightarrow H_2(g)$	0
$Pb^{2+}(aq) + 2\,e^- \longrightarrow Pb(s)$	−0.125
$Sn^{2+}(aq) + 2\,e^- \longrightarrow Sn(s)$	−0.137
$Fe^{2+}(aq) + 2\,e^- \longrightarrow Fe(s)$	−0.440
$Zn^{2+}(aq) + 2\,e^- \longrightarrow Zn(s)$	−0.763
$Al^{3+}(aq) + 3\,e^- \longrightarrow Al(s)$	−1.676
$Mg^{2+}(aq) + 2\,e^- \longrightarrow Mg(s)$	−2.356
$Na^+(aq) + e^- \longrightarrow Na(s)$	−2.713
$Ca^{2+}(aq) + 2\,e^- \longrightarrow Ca(s)$	−2.84
$K^+(aq) + e^- \longrightarrow K(s)$	−2.924
$Li^+(aq) + e^- \longrightarrow Li(s)$	−3.040
Basic solution	
$O_3(g) + H_2O(l) + 2\,e^- \longrightarrow O_2(g) + 2\,OH^-(aq)$	+1.246
$OCl^-(aq) + H_2O(l) + 2\,e^- \longrightarrow Cl^-(aq) + 2\,OH^-(aq)$	+0.890
$O_2(g) + 2\,H_2O(l) + 4\,e^- \longrightarrow 4\,OH^-(aq)$	+0.401
$2\,H_2O(l) + 2\,e^- \longrightarrow H_2(g) + 2\,OH^-(aq)$	−0.828

EXAMPLE 20-3 Combining $E°$ Values into $E°_{cell}$ for a Reaction

A new battery system currently under study for possible use in electric vehicles is the zinc–chlorine battery. The overall reaction producing electricity in this cell is $Zn(s) + Cl_2(g) \longrightarrow ZnCl_2(aq)$. What is $E°_{cell}$ of this voltaic cell?

Analyze

First we identify the species that are oxidized and reduced. Then we obtain the standard reduction potentials for the cathode and anode from Table 20.1 or Appendix D and calculate E°_{cell}.

Solve

The oxidation state of zinc changes from 0 to +2 and therefore is oxidized; consequently, the chlorine is reduced. The half-reactions are indicated below and are combined into the overall equation (20.12).

$$
\begin{aligned}
\textit{Oxidation:} \qquad & Zn(s) \longrightarrow Zn^{2+}(aq) + 2\,e^{-} \\
\textit{Reduction:} \qquad & \underline{Cl_2(g) + 2\,e^{-} \longrightarrow 2\,Cl^{-}(aq)} \\
\textit{Overall:} \qquad & Zn(s) + Cl_2(g) \longrightarrow Zn^{2+}(aq) + 2\,Cl^{-}(aq) \qquad \textbf{(20.12)}
\end{aligned}
$$

$$
\begin{aligned}
E^{\circ}_{cell} &= E^{\circ}(\text{reduction half-cell}) - E^{\circ}(\text{oxidation half-cell}) \\
&= 1.358\ V - (-0.763\ V) = 2.121\ V
\end{aligned}
$$

Assess

Once the oxidized and reduced species are identified, we can establish E°_{cell}.

PRACTICE EXAMPLE A: What is E°_{cell} for the reaction in which $Cl_2(g)$ oxidizes $Fe^{2+}(aq)$ to $Fe^{3+}(aq)$?

$$2\,Fe^{2+}(aq) + Cl_2(g) \longrightarrow 2\,Fe^{3+}(aq) + 2\,Cl^{-}(aq) \qquad E^{\circ}_{cell} = ?$$

PRACTICE EXAMPLE B: Use data from Table 20.1 to determine E°_{cell} for the redox reaction in which $Fe^{2+}(aq)$ is oxidized to $Fe^{3+}(aq)$ by $MnO_4^{-}(aq)$ in acidic solution.

EXAMPLE 20-4   Determining an Unknown E° from an E°_{cell} Measurement

Cadmium is found in small quantities wherever zinc is found. Unlike zinc, which in trace amounts is an essential element, cadmium is an environmental poison. To determine cadmium ion concentrations by electrical measurements, we need the standard electrode potential for the Cd^{2+}/Cd electrode. The voltage of the following voltaic cell is measured.

$$Cd(s)|Cd^{2+}(1\ M)||Cu^{2+}(1\ M)|Cu(s) \qquad E^{\circ}_{cell} = 0.743\ V$$

What is the standard electrode potential for the Cd^{2+}/Cd electrode?

Analyze

We know one half-cell potential and E°_{cell} for the overall redox reaction. We can solve for the unknown standard electrode potential, $E^{\circ}_{Cd^{2+}/Cd}$.

Solve

$$
\begin{aligned}
E^{\circ}_{cell} &= E^{\circ}(\text{right}) - E^{\circ}(\text{left}) \\
0.743\ V &= E^{\circ}_{Cu^{2+}/Cu} - E^{\circ}_{Cd^{2+}/Cd} \\
&= 0.340\ V - E^{\circ}_{Cd^{2+}/Cd} \\
E^{\circ}_{Cd^{2+}/Cd} &= 0.340\ V - 0.743\ V = -0.403\ V
\end{aligned}
$$

Assess

Based on the entries in Table 20.1 we see that $Cd(s)$ is a stronger reducing agent than $Sn(s)$, but weaker than $Fe(s)$. $Cd^{2+}(aq)$ is a weaker oxidizing agent than $Sn^{2+}(aq)$.

PRACTICE EXAMPLE A: In acidic solution, dichromate ion oxidizes oxalic acid, $H_2C_2O_4(aq)$, to $CO_2(g)$ in a reaction with $E^{\circ}_{cell} = 1.81\ V$.

$$Cr_2O_7^{2-}(aq) + 3\,H_2C_2O_4(aq) + 8\,H^{+}(aq) \longrightarrow 2\,Cr^{3+}(aq) + 7\,H_2O + 6\,CO_2(g)$$

Use the value of E°_{cell} for this reaction, together with appropriate data from Table 20.1, to determine E° for the $CO_2(g)/H_2C_2O_4(aq)$ electrode.

PRACTICE EXAMPLE B: In an acidic solution, $O_2(g)$ oxidizes $Cr^{2+}(aq)$ to $Cr^{3+}(aq)$. The $O_2(g)$ is reduced to $H_2O(l)$. E°_{cell} for the reaction is 1.653 V. What is the standard electrode potential for the couple Cr^{3+}/Cr^{2+}?

For the half-cell reaction $ClO_4^-(aq) + 8\,H^+(aq) + 7\,e^- \longrightarrow \frac{1}{2}Cl_2(g) + 4\,H_2O(l)$, what are the standard-state conditions for the reactants and products?

20-3 E_{cell}, ΔG, and K

When a reaction occurs in a voltaic cell, the cell does work—electrical work. Think of this as the work of moving electric charges. The total work done is the product of three terms: (a) E_{cell}; (b) z, the number of electrons transferred between the electrodes; and (c) the electric charge per mole of electrons, called the **Faraday constant (F)**. The Faraday constant is equal to 96,485 coulombs per mole of electrons (96,485 C/mol). Because the product volt $\times$ coulomb = joule, the unit of w_{elec} is joules (J).

$$w_{elec} = zFE_{cell} \tag{20.13}$$

Expression (20.13) applies only if the cell operates reversibly.* The last paragraph in Section 19-4 on page 836 described the amount of available energy (work) that can be derived from a process as equal to $-\Delta G$. Thus,

$$\Delta G = -zFE_{cell} \tag{20.14}$$

In the special case in which the reactants and products are in their standard states,

$$\Delta G^{\circ} = -zFE_{cell}^{\circ} \tag{20.15}$$

At this point we should clear up some issues of units and definitions. The symbol z in equations (20.14) and (20.15) is properly called the *electron number* of an electrochemical reaction and is occasionally referred to as the *charge number*. The electron number has no units; that is, it is simply a number. For any given cell reaction we can write the reaction with a charge number of one or two. Thus, the hydrogen electrode reaction can be written as either

$$2\,H^+(aq) + 2\,e^- \longrightarrow H_2(g) \quad \text{or} \quad H^+(aq) + e^- \longrightarrow \frac{1}{2}H_2(g)$$

However, in considering an overall cell reaction, we must balance the electrons. Thus for the cell

$$Pt(s)|H_2(g)|H^+(aq, 1\,M)\|Cu^{2+}(aq)|Cu(s)$$

the half-cell reactions can be written as

$$2\,H^+(aq) + 2\,e^- \longrightarrow H_2(g) \quad \text{and} \quad 2\,e^- + Cu^{2+}(aq) \longrightarrow Cu(s)$$

Thus, the electron number is two and the overall electrochemical reaction is

$$H_2(g) + Cu^{2+}(aq) \longrightarrow 2\,H^+(aq) + Cu(s)$$

The standard reduction potential for this reaction is

$$E_{cell}^{\circ} = E^{\circ}(\text{right}) - E^{\circ}(\text{left})$$
$$= E_{Cu^{2+}/Cu}^{\circ} - 0\,V = 0.340\,V$$

The standard Gibbs energy is given by

$$\Delta G_{rxn}^{\circ} = -zFE_{cell}^{\circ} = 2 \times \frac{96485\,C}{mol} \times 0.340\,V$$
$$= -6.5610 \times 10^4\,J\,mol^{-1} = -65.6\,kJ\,mol^{-1}$$

▶ It is quite common to see the symbol n used for the number of electrons; however, IUPAC recommends z to avoid confusion with the use of n as the amount of substance, such as the number of moles of gas in the ideal gas equation.

▲ **Michael Faraday (1791–1867)**
Faraday, an assistant to Humphry Davy and often called "Davy's greatest discovery," made many contributions to both physics and chemistry, including systematic studies of electrolysis.

---

*The meaning of a reversible process was illustrated by Figure 7-12 on page 258. The reversible operation of a voltaic cell requires that electric current be drawn from the cell only very, very slowly.

That is, 65.6 kJ of energy is generated when 1 mole of Cu^{2+} ions is reduced or 2 moles of H^+ are produced. The process is accompanied by the passage of two moles of electrons around the outer circuit. We could also have written the reactions as

$$\text{Oxidation:} \qquad \frac{1}{2}H_2(g) \longrightarrow H^+(aq) + e^-$$

$$\text{Reduction:} \qquad \frac{1}{2}Cu^{2+}(aq) + e^- \longrightarrow \frac{1}{2}Cu(s)$$

$$\text{Overall:} \qquad \frac{1}{2}H_2(g) + \frac{1}{2}Cu^{2+}(aq) \longrightarrow \frac{1}{2}Cu(s) + H^+(aq)$$

This reaction is represented by the same cell diagram given above, but the electron number is one; consequently, the Gibbs energy is one-half of that previously calculated, but the value of E°_{cell} is the same. This result supports the fact that the standard reduction potential is an intensive property but the Gibbs energy is an extensive property. Finally, the reaction tells us that when 0.5 mole of Cu^{2+} is reduced, 32.8 kJ of energy is released and one mole of electrons passes from the anode to the cathode.

Our primary interest is not in calculating quantities of work but in using expression (20.15) as a means of evaluating Gibbs energy changes from measured cell potentials, as illustrated in Example 20-5.

EXAMPLE 20-5   **Determining a Gibbs Energy Change from a Cell Potential**

Use E° data to determine ΔG° for the reaction

$$Zn(s) + Cl_2(g, 1\text{ atm}) \longrightarrow ZnCl_2(aq, 1\text{ M})$$

Solution

This reaction is cell reaction (20.12) occurring in the voltaic cell described in Example 20-3. In this type of problem, the overall equation generally needs to be separated into two half-equations. Then the value of E°_{cell} and the number of moles of electrons (n) involved in the cell reaction can be determined. Refer to Example 20-3 to see that $E^\circ_{cell} = 2.121$ V and $z = 2$ mol e^-. Now use equation (20.15).

$$\Delta G^\circ = -zFE^\circ_{cell} = -\left(2\text{ mol } e^- \times \frac{96{,}485\text{ C}}{1\text{ mol } e^-} \times 2.121\text{ V}\right)$$

$$= -4.093 \times 10^5\text{ J} = -409.3\text{ kJ mol}^{-1}$$

PRACTICE EXAMPLE A: Use electrode potential data to determine ΔG° for the reaction

$$2\text{ Al}(s) + 3\text{ Br}_2(l) \longrightarrow 2\text{ Al}^{3+}(aq, 1\text{ M}) + 6\text{ Br}^-(aq, 1\text{ M}) \qquad \Delta G^\circ = ?$$

PRACTICE EXAMPLE B: The hydrogen–oxygen fuel cell is a voltaic cell with a cell reaction of $2\text{ H}_2(g) + O_2(g) \longrightarrow 2\text{ H}_2O(l)$. Calculate E°_{cell} for this reaction. [*Hint:* Use thermodynamic data from Appendix D (Table D-2).]

Combining Reduction Half-Equations

Not only can equation (20.15) be used to determine ΔG° from E°_{cell}, as in Example 20-5, but the calculation can be reversed and an E°_{cell} value determined from ΔG°. Moreover, equation (20.15) can be applied to half-cell reactions and half-cell potentials—that is, to standard electrode potentials, E°. That is what we must do, for example, to determine E° for the half-reaction

$$Fe^{3+}(aq) + 3\text{ e}^- \longrightarrow Fe(s)$$

Both in Table 20.1 and in Appendix D, the only entries that deal with Fe(s) and its ions are

$$Fe^{2+}(aq) + 2\text{ e}^- \longrightarrow Fe(s), E^\circ = -0.440\text{ V} \quad \text{and} \quad Fe^{3+}(aq) + e^- \longrightarrow Fe^{2+}(aq), E^\circ = 0.771\text{ V}$$

The half-equation we are seeking is simply the sum of these two half-equations, but the $E°$ value we are seeking is *not* the sum of -0.440 V and 0.771 V. What we *can* add together, though, are the $\Delta G°$ values for the two known half-reactions.

$$Fe^{2+}(aq) + 2\,e^- \longrightarrow Fe(s); \qquad \Delta G° = -2 \times F \times (-0.440\ V)$$
$$\underline{Fe^{3+}(aq) + e^- \longrightarrow Fe^{2+}(aq); \qquad \Delta G° = -1 \times F \times (0.771\ V)}$$
$$Fe^{3+}(aq) + 3\,e^- \longrightarrow Fe(s); \qquad \Delta G° = (0.880F)\ V - (0.771F)\ V = (0.109F)\ V$$

Now, to get $E°_{Fe^{3+}/Fe}$, we can again use equation (20.15) and solve for $E°_{Fe^{3+}/Fe}$.

$$\Delta G° = -zFE°_{Fe^{3+}/Fe} = -3FE°_{Fe^{3+}/Fe} = (0.109F)\ V$$
$$E°_{Fe^{3+}/Fe} = (-0.109F/3F)\ V = -0.0363\ V$$

20-1 ARE YOU WONDERING...

How the procedure for combining two $E°$ values to obtain an unknown $E°_{cell}$ relates to combining two $E°$ values to obtain an unknown $E°$?

We have just seen how to obtain an unknown $E°$ from two known values of $E°$ by working through the expression $\Delta G = -zFE°$. As shown below for a hypothetical displacement reaction, we can similarly calculate an unknown $E°_{cell}$ through the expression $\Delta G° = -zFE°_{cell}$. (Note that for the oxidation half-reaction, $\Delta G°_{ox}$ is simply the negative of the value for the reverse half-reaction, $\Delta G°_{red}$.)

Reduction: $\quad M^{z+}(aq) + z\,e^- \longrightarrow M(s) \qquad\qquad\qquad \Delta G°_{red} = -zFE°_{M^{z+}/M}$

Oxidation: $\qquad\qquad\qquad N(s) \longrightarrow N^{z+}(aq) + z\,e^-$

$$\Delta G°_{ox} = -(\Delta G°_{red}) = -(-zFE°_{N^{z+}/N}) = zFE°_{N^{z+}/N}$$

Overall: $\quad M^{z+}(aq) + N(s) \longrightarrow M(s) + N^{z+}(aq)$

$$\Delta G° = \Delta G°_{red} + \Delta G°_{ox} = -zFE°_{cell} = -zFE°_{M^{z+}/M} + zFE°_{N^{z+}/N}$$

Dividing through the above equation by the term $-zF$, we obtain $E°_{cell}$ as the familiar difference in two electrode potentials.

$$E°_{cell} = E°_{M^{z+}/M} - E°_{N^{z+}/N}$$

We have been able to skip this calculation based on $\Delta G°$ values and proceed straight to the expression

$$E°_{cell} = E°(reduction) - E°(oxidation)$$

because the term $-zF$ always cancels out. That is, z, the number of electrons, must have the same value for the oxidation and reduction half-reactions and the overall reaction. By contrast, when obtaining an unknown $E°$ from the known $E°$ values, the value for z will not be the same in all three places where it appears, and so we do have to work through the $\Delta G°$ expressions.

Spontaneous Change in Oxidation–Reduction Reactions

Our main criterion for spontaneous change is that $\Delta G < 0$. According to equation (20.14), however, redox reactions have the property that, if $\Delta G < 0$, then $E_{cell} > 0$. That is, E_{cell} must be *positive* if ΔG is to be negative. Predicting the direction of spontaneous change in a redox reaction is a relatively simple matter by using the following ideas:

- If E_{cell} is *positive*, a reaction occurs spontaneously in the *forward* direction for the stated conditions. If E_{cell} is *negative*, the reaction occurs spontaneously in the *reverse* direction for the stated conditions. If $E_{cell} = 0$, the reaction is at equilibrium for the stated conditions.

- If a cell reaction is *reversed*, E_{cell} changes sign.

EXAMPLE 20-6   Applying the Criterion for Spontaneous Change in a Redox Reaction

Will aluminum metal displace Cu^{2+} ion from aqueous solution? That is, will a spontaneous reaction occur in the forward direction for the following reaction?

$$2\,Al(s) + 3\,Cu^{2+}(1\,M) \longrightarrow 3\,Cu(s) + 2\,Al^{3+}(1\,M)$$

Analyze

We need to identify the species reduced in the reaction as it is written. We then calculate E_{cell}°. If E_{cell}° is positive, then the reaction will occur spontaneously.

Solve

The cell diagram corresponding to the reaction is $Al(s)|Al^{3+}(aq)\|Cu^{2+}(aq)|Cu(s)$, and E_{cell}° is

$$\begin{aligned}
E_{cell}^{\circ} &= E^{\circ}(\text{cathode}) - E^{\circ}(\text{anode}) \\
&= E_{Cu^{2+}/Cu}^{\circ} - E_{Al^{3+}/Al}^{\circ} \\
&= 0.340\,V - (-1.676\,V) = 2.016\,V
\end{aligned}$$

Because E_{cell}° is positive, the direction of spontaneous change is that of the forward reaction. $Al(s)$ will displace Cu^{2+} from aqueous solution under standard-state conditions.

Assess

The positive value of E_{cell}° means that the Gibbs energy change for the reaction is negative; hence, the reaction as written is spontaneous. Keep in mind that both E° and E_{cell}° are *intensive* properties. They do not depend on the quantities of materials involved, which means that their values are not affected by the choice of coefficients used to balance the equation for the cell reaction. We could just as well have written:

$$Al(s) + \frac{3}{2}Cu^{2+}(1\,M) \longrightarrow \frac{3}{2}Cu(s) + Al^{3+}(1\,M) \ \text{or}$$

$$\frac{2}{3}Al(s) + Cu^{2+}(1\,M) \longrightarrow Cu(s) + \frac{2}{3}Al^{3+}(1\,M)$$

PRACTICE EXAMPLE A: Name one metal ion that $Cu(s)$ will displace from aqueous solution, and determine E_{cell}° for the reaction.

PRACTICE EXAMPLE B: When sodium metal is added to seawater, which has $[Mg^{2+}] = 0.0512\,M$, no magnesium metal is obtained. According to E° values, should this displacement reaction occur? What reaction does occur?

In the special case in which reactants and products are in their standard states, we work with ΔG° and E_{cell}° values, as illustrated in Examples 20-6 and 20-7.

Even though we used electrode potentials and cell voltage to predict a spontaneous reaction in Example 20-6, we do not have to carry out the reaction in a voltaic cell. This is an important point to keep in mind. Thus, Cu^{2+} is displaced from aqueous solution simply by adding aluminum metal, as shown in Figure 20-7. Another point, illustrated by Example 20-7, is that qualitative answers to questions concerning redox reactions can be found without going through a complete calculation of E_{cell}°.

The Behavior of Metals Toward Acids

In the discussion of redox reactions in Chapter 5, it was noted that most metals react with an acid, such as HCl, but that a few do not. This observation can now be explained. When a metal, M, reacts with an acid, such as HCl, the metal is oxidized to the metal ion, such as M^{2+}. The reduction involves H^+ being reduced to $H_2(g)$. These ideas can be expressed as

▲ FIGURE 20-7
Reaction of Al(s) and Cu^{2+}(aq)
Notice the holes in the foil where Al(s) has dissolved. Notice also the dark deposit of Cu(s) at the bottom of the beaker.

$$\begin{array}{lrcl}
\textit{Oxidation:} & M(s) & \longrightarrow & M^{2+}(aq) + 2\,e^- \\
\textit{Reduction:} & 2\,H^+(aq) + 2\,e^- & \longrightarrow & H_2(g) \\
\hline
\textit{Overall:} & M(s) + 2\,H^+(aq) & \longrightarrow & M^{2+}(aq) + H_2(g)
\end{array}$$

$$E_{cell}^{\circ} = E_{H^+/H_2}^{\circ} - E_{M^{2+}/M}^{\circ} = 0\,V - E_{M^{2+}/M}^{\circ} = -E_{M^{2+}/M}^{\circ}$$

EXAMPLE 20-7 **Making Qualitative Predictions with Electrode Potential Data**

Peroxodisulfate salts, such as $Na_2S_2O_8$, are oxidizing agents used in bleaching. Dichromates such as $K_2Cr_2O_7$ have been used as laboratory oxidizing agents. Which is the better oxidizing agent in acidic solution under standard conditions, $S_2O_8^{2-}$ or $Cr_2O_7^{2-}$?

Analyze

In a redox reaction, the oxidizing agent is reduced; the greater the tendency for this reduction to occur, the better the oxidizing agent. The reduction tendency, in turn, is measured by the $E°$ value.

Solve

Because the $E°$ value for the reduction of $S_2O_8^{2-}(aq)$ to $SO_4^{2-}(aq)$ (2.01 V) is larger than that for the reduction of $Cr_2O_7^{2-}(aq)$ to $Cr^{3+}(aq)$ (1.33 V), $S_2O_8^{2-}(aq)$ should be the better oxidizing agent.

Assess

Inspection of standard reduction potentials enables us to qualitatively assess the spontaneity of a particular redox reaction.

PRACTICE EXAMPLE A: An inexpensive way to produce peroxodisulfates would be to pass $O_2(g)$ through an acidic solution containing sulfate ion. Is this method feasible under standard conditions? [*Hint:* What would be the reduction half-reaction?]

PRACTICE EXAMPLE B: Consider the following observations: (1) Aqueous solutions of Sn^{2+} are difficult to maintain because atmospheric oxygen easily oxidizes Sn^{2+} to Sn^{4+}. (2) One way to preserve the $Sn^{2+}(aq)$ solutions is to add some metallic tin. *Without doing detailed calculations,* explain these two statements by using $E°$ data.

Metals with *negative* standard electrode potentials yield *positive* values of $E°_{cell}$ in the above expression. These are the metals that should displace $H_2(g)$ from acidic solutions. Thus, all the metals listed *below* hydrogen in Table 20.1 (Pb through Li) should react with acids.

In acids, such as HCl, HBr, and HI, the oxidizing agent is H^+ (that is, H_3O^+). Certain metals that will not react with HCl will react with an acid in the presence of an *anion* that is a better oxidizing agent than H^+. Nitrate ion is a good oxidizing agent in acidic solution, and silver metal, which does not react with HCl(aq), readily reacts with nitric acid, $HNO_3(aq)$.

$$3\,Ag(s) + NO_3^-(aq) + 4\,H^+(aq) \longrightarrow 3\,Ag^+(aq) + NO(g) + 2\,H_2O \quad E°_{cell} = 0.156\ V$$

The Relationship Between $E°_{cell}$ and K

$\Delta G°$ and $E°_{cell}$ were related through equation (20.15). In Chapter 19, $\Delta G°$ and K were related through equation (19.12). The three quantities are thus related in this way.

$$\Delta G° = -RT \ln K = -zFE°_{cell}$$

and therefore,

$$E°_{cell} = \frac{RT}{zF} \ln K \qquad \qquad \textbf{(20.16)}$$

In equation (20.16), R has a value of 8.3145 J mol^{-1} K^{-1} and z represents the number of electrons involved in the reaction. If we then specify a temperature of 25 °C = 298.15 K (the temperature at which electrode potentials are generally determined), the combined terms "RT/F" in equation (20.16) can be replaced by a single constant. This constant has the value 0.025693 J/C = 0.025693 V.

$$E°_{cell} = \frac{RT}{zF} \ln K = \frac{8.3145\ \text{J mol}^{-1}\ \text{K}^{-1} \times 298.15\ \text{K}}{z \times 96485\ \text{C mol}^{-1}} \ln K$$

► Note that any electrochemical cell, if left in a completed circuit, will eventually die as the redox reaction goes to completion. This means that the cell potential will eventually drop to zero. In Chapter 19, the relationship between Gibbs energy and equilibrium was established in a similar way.

► Equation 20.16 gives the expected result that reactions with equilibrium constants larger than one have a positive standard cell potential.

$$E^\circ_{cell} = \frac{0.025693 \text{ V}}{z} \ln K \qquad (20.17)$$

The relationship between E°_{cell} and K is illustrated in Example 20-8. Also, Figure 20-8 summarizes several important relationships from thermodynamics, equilibrium, and electrochemistry.

EXAMPLE 20-8   **Relating K to E°_{cell} for a Redox Reaction**

What is the value of the equilibrium constant K for the reaction between copper metal and iron(III) ions in aqueous solution at 25 °C?

$$Cu(s) + 2 Fe^{3+}(aq) \longrightarrow Cu^{2+}(aq) + 2 Fe^{2+}(aq) \qquad K = ?$$

Analyze

We first identify the reactant that is reduced and the reactant that is oxidized. We then use the data in Table 20.1 to obtain the standard reduction potentials and hence the cell potential. Finally, we use equation (20.17) to obtain K from E°_{cell}.

Solve

First, we use data from Table 20.1 to determine E°_{cell}.

$$E^\circ_{cell} = E^\circ(\text{reduction half-cell}) - E^\circ(\text{oxidation half-cell})$$
$$= E^\circ_{Fe^{3+}/Fe^{2+}} - E^\circ_{Cu^{2+}/Cu}$$
$$= 0.771 \text{ V} - 0.340 \text{ V} = 0.431 \text{ V}$$

The charge number (z) for the cell reaction is 2.

$$E^\circ_{cell} = 0.431 \text{ V} = \frac{0.02569 \text{ V}}{2} \ln K$$

$$\ln K = \frac{2 \times 0.431 \text{ V}}{0.02569 \text{ V}} = 33.6$$

$$K = e^{33.6} = 4 \times 10^{14}$$

Assess

The positive value of the cell potential means that the equilibrium constant is greater than one. The value of the equilibrium constant is very large, and so we can expect this reaction to go to completion.

PRACTICE EXAMPLE A: Should the displacement of Cu^{2+} from aqueous solution by Al(s) go to completion? [*Hint:* Base your assessment on the value of K for the displacement reaction. We determined E°_{cell} for this reaction in Example 20-6.]

PRACTICE EXAMPLE B: Should the reaction of Sn(s) and $Pb^{2+}(aq)$ go to completion? Explain.

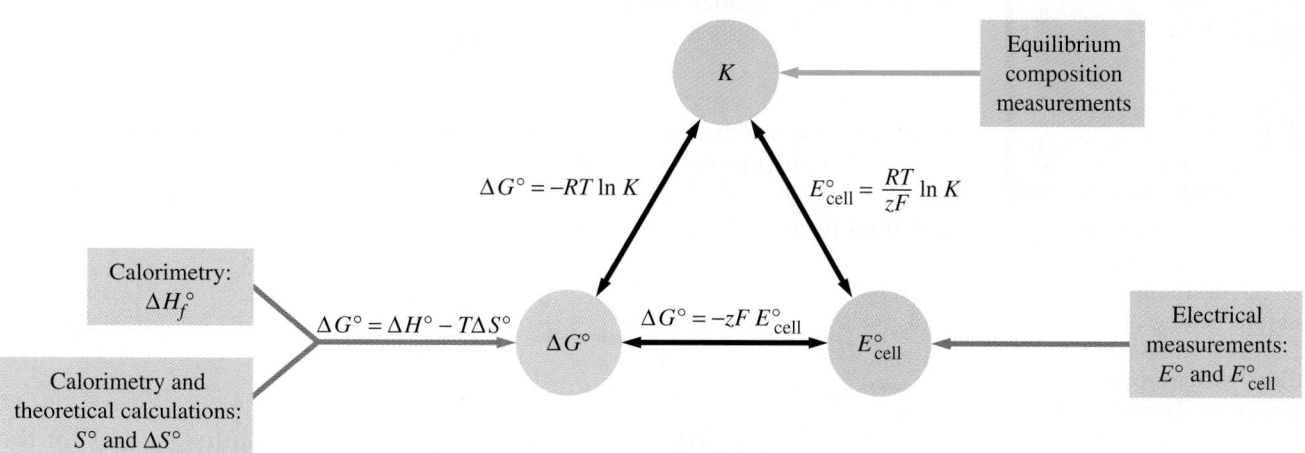

▲ FIGURE 20-8
A summary of important thermodynamic, equilibrium, and electrochemical relationships under standard conditions

20-4 E_{cell} as a Function of Concentrations

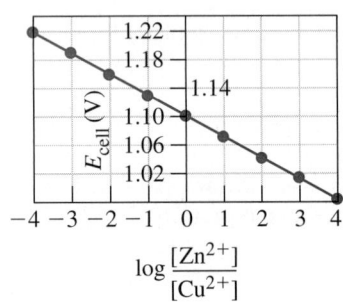

▲ FIGURE 20-9
Variation of E_{cell} with ion concentrations
The cell reaction is Zn(s) + Cu^{2+}(aq) ⟶ Zn^{2+}(aq) + Cu(s) and has E°_{cell} = 1.103 V.

When we combine standard electrode potentials, we obtain a standard E°_{cell}, such as E°_{cell} = 1.103 V for the voltaic cell of Figure 20-4. For the following cell reaction at *nonstandard* conditions, however, the measured E_{cell} is not 1.103 V.

$$Zn(s) + Cu^{2+}(2.0\,M) \longrightarrow Zn^{2+}(0.10\,M) + Cu(s) \qquad E_{cell} = 1.142\,V$$

Experimental measurements of cell potentials are often made for nonstandard conditions; these measurements have great significance, especially for performing chemical analyses.

From Le Châtelier's principle, it would seem that *increasing* the concentration of a reactant (Cu^{2+}) while *decreasing* the concentration of a product (Zn^{2+}) should favor the forward reaction. Zn(s) should displace Cu^{2+}(aq) even more readily than for standard-state conditions and E_{cell} > 1.103 V. E_{cell} is found to vary linearly with log ($[Zn^{2+}]/[Cu^{2+}]$), as illustrated in Figure 20-9.

It is not difficult to establish the relationship between the cell potential, E_{cell}, and the concentrations of reactants and products. From Chapter 19 we can write equation (19.11).

$$\Delta G = \Delta G^\circ + RT \ln Q$$

For ΔG and ΔG°, we can substitute $-zFE_{cell}$ and $-zFE^\circ_{cell}$, respectively.

$$-zFE_{cell} = -zFE^\circ_{cell} + RT \ln Q$$

Dividing through by $-zF$ gives

$$E_{cell} = E^\circ_{cell} - \frac{RT}{zF} \ln Q$$

This equation was first proposed by Walther Nernst in 1889. It is commonly used by analytical chemists in the form in which Nernst expressed it. The **Nernst equation** can be obtained by switching from natural to common logarithms ($\ln Q = 2.3026 \log Q$).

$$E_{cell} = E^\circ_{cell} - \frac{2.3026\,RT}{zF} \log Q$$

By specifying a temperature of 298.15 K and replacing RT/F by 0.025693 V, as in the development of equation (20.17), we find that the term 2.3026 RT/F = 2.3026 × 0.025693 V = 0.059161 V, usually rounded off to 0.0592 V. The final form of the Nernst equation is

$$E_{cell} = E^\circ_{cell} - \frac{0.0592\,V}{z} \log Q \qquad (20.18)$$

In the Nernst equation, we make the usual substitutions into Q: $a = 1$ for the activities of pure solids and liquids, partial pressures (atm) for the activities of gases, and molarities for the activities of solution components. Example 20-9 demonstrates that the Nernst equation makes it possible to calculate E_{cell} for any chosen concentrations, not just for standard conditions.

▲ **Walther Nernst (1864–1941)**
Nernst was only 25 years old when he formulated his equation relating cell voltages and concentrations. He is also credited with proposing the solubility product concept in the same year. In 1906, he announced his "heat theorem," which we now know as the third law of thermodynamics.

EXAMPLE 20-9   **Applying the Nernst Equation for Determining E_{cell}**

What is the value of E_{cell} for the voltaic cell pictured in Figure 20-10 and diagrammed as follows?

$$Pt|Fe^{2+}(0.10\ M),\ Fe^{3+}(0.20\ M)\|Ag^+(1.0\ M)|Ag(s) \qquad E_{cell} = ?$$

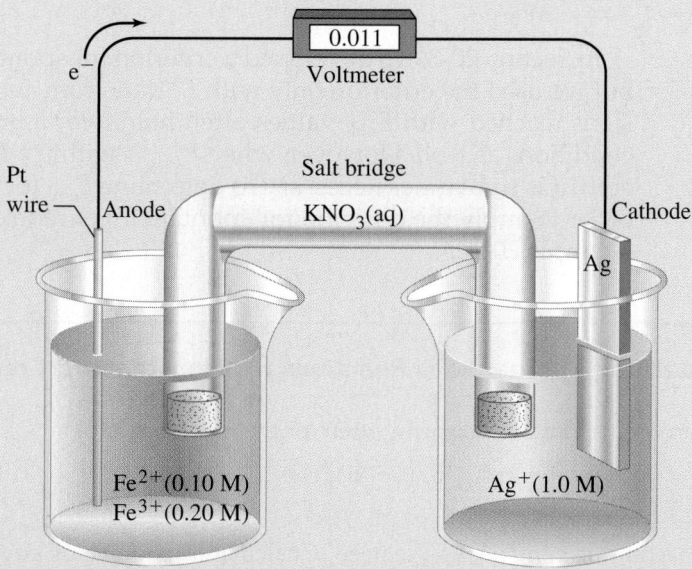

▲ FIGURE 20-10
A voltaic cell with nonstandard conditions—Example 20-9 illustrated

Analyze

To use the Nernst equation we need to establish E°_{cell} and the reaction to which the cell diagram corresponds so that the form of the reaction quotient (Q) can be revealed (see Example 20-2). Once we have determined the form of the Nernst equation, we can insert the concentration of the species.

Solve

Two steps are required when using the Nernst equation. First, to determine E°_{cell}, use data from Table 20.1 to write

$$\begin{aligned} E^\circ_{cell} &= E^\circ(\text{cathode}) - E^\circ(\text{anode}) \\ &= E^\circ_{Ag^+/Ag} - E^\circ_{Fe^{3+}/Fe^{2+}} \\ &= 0.800\ V - 0.771\ V = 0.029\ V \end{aligned}$$ **(20.19)**

Now, to determine E_{cell} for the reaction

$$Fe^{2+}(0.10\ M) + Ag^+(1.0\ M) \longrightarrow Fe^{3+}(0.20\ M) + Ag(s) \quad E_{cell} = ? \quad \textbf{(20.20)}$$

substitute appropriate values into the Nernst equation (20.18), starting with $E^\circ_{cell} = 0.029\ V$ and $z = 1$,

$$E_{cell} = 0.029\ V - \frac{0.0592\ V}{1} \log\frac{[Fe^{3+}]}{[Fe^{2+}][Ag^+]}$$

and for concentrations $[Fe^{2+}] = 0.10\ M$; $[Fe^{3+}] = 0.20\ M$; $[Ag^+] = 1.0\ M$.

$$\begin{aligned} E_{cell} &= 0.029\ V - 0.0592\ V \times \log\frac{0.20}{0.10 \times 1.0} \\ &= 0.029\ V - 0.0592\ V \times \log 2 = 0.029\ V - 0.018\ V \\ &= 0.011\ V \end{aligned}$$

Assess

The E_{cell} is positive so that the reaction is spontaneous in the direction of the reduction of silver.

PRACTICE EXAMPLE A: Calculate E_{cell} for the following voltaic cell.

$$Al(s)|Al^{3+}(0.36\ M)\|Sn^{4+}(0.086\ M),\ Sn^{2+}(0.54\ M)|Pt$$

PRACTICE EXAMPLE B: Calculate E_{cell} for the following voltaic cell.

$$Pt(s)|Cl_2(1\ atm)|Cl^-(1.0\ M)\|Pb^{2+}(0.050\ M),\ H^+(0.10\ M)|PbO_2(s)$$

Describe *two* sets of conditions under which the measured E_{cell} for a reaction is equal to E°_{cell}.

In Section 20-3, we developed a criterion for spontaneous change ($E_{cell} > 0$), but we used the criterion only with E° data from Table 20.1 Qualitative conclusions reached with E°_{cell} values often hold over a broad range of nonstandard conditions as well. However, when E°_{cell} is within a few hundredths of a volt of zero, it is sometimes necessary to determine E_{cell} for nonstandard conditions in order to apply the criterion for spontaneity of redox reactions, as illustrated in Example 20-10.

EXAMPLE 20-10   Predicting Spontaneous Reactions for Nonstandard Conditions

Will the cell reaction proceed spontaneously as written for the following cell?

$$Ag(s)|Ag^+(0.075\ M)\|Hg^{2+}(0.85\ M)|Hg(l)$$

Analyze

To decide whether a reaction is spontaneous, we need to calculate E_{cell} by using equation (20.18) with the concentrations given. We then identify the oxidized and reduced species and look up the appropriate standard half-cell potentials. Then we construct the chemical equation that corresponds to the cell diagram, choosing an appropriate electron number (z).

Solve

To determine E°_{cell} from E° data we write

$$\begin{aligned}
\textit{Oxidation:} \quad & 2\ Ag(s) \longrightarrow 2\ Ag^+(aq) + 2\ e^- \\
\textit{Reduction:} \quad & Hg^{2+}(aq) + 2\ e^- \longrightarrow Hg(l) \\
\hline
\textit{Overall:} \quad & 2\ Ag(s) + Hg^{2+}(aq) \longrightarrow 2\ Ag^+(aq) + Hg(l)
\end{aligned}$$

$$\begin{aligned}
E^\circ_{cell} &= E^\circ(\text{reduction half-cell}) - E^\circ(\text{oxidation half-cell}) \\
&= 0.854\ V - (0.800\ V) = 0.054\ V
\end{aligned}$$

The overall reaction that we have written has an electron number, $z = 2$, so that the Nernst equation is

$$E_{cell} = 0.054\ V - \frac{0.0592\ V}{2} \log \frac{[Ag^+]^2}{[Hg^{2+}]}$$

By using the concentrations $[Ag^+] = 0.075\ M$ and $[Hg^{2+}] = 0.85\ M$ provided, we obtain

$$E_{cell} = 0.054\ V - 0.0296\ V \log \frac{[0.075]^2}{[0.85]}$$

$$= 0.054\ V - 0.0296\ V \log(0.0066) = 0.054 - 0.0296 \times (-2.179)\ V$$

$$= 0.054 + 0.065\ V = -0.119\ V$$

Because $E_{cell} > 0$, we conclude that the reaction as written is spontaneous.

Assess

If we had used an electron number of $z = 1$, the overall reaction would have been

$$\textit{Overall } (z = 1)\text{: } Ag(s) + \frac{1}{2} Hg^{2+}(aq) \longrightarrow Ag^+(aq) + \frac{1}{2} Hg(l)$$

The corresponding Nernst equation is

$$E_{cell} = 0.054\ V - \frac{0.0592\ V}{1} \log \frac{[Ag^+]}{[Hg^{2+}]^{1/2}}$$

By rearranging slightly and using a property of logarithms,

$$E_{cell} = 0.054 \text{ V} - \frac{0.0592 \text{ V}}{1} \log \left(\frac{[Ag^+]^2}{[Hg^{2+}]} \right)^{1/2} = 0.054 \text{ V} - \frac{0.0592 \text{ V}}{2} \log \frac{[Ag^+]^2}{[Hg^{2+}]}$$

we have recovered the Nernst equation for the cell reaction by using an electron number (z) of 2. We conclude that as long as we balance charge and the electron number correctly, we will always get the correct result.

PRACTICE EXAMPLE A: Will the cell reaction proceed spontaneously as written for the following cell?

$$Cu(s) | Cu^{2+}(0.15 \text{ M}) \| Fe^{3+}(0.35 \text{ M}), Fe^{2+}(0.25 \text{ M}) | Pt(s)$$

PRACTICE EXAMPLE B: For what ratio of $[Ag^+]^2 / [Hg^{2+}]$ will the cell reaction in Example 20-10 not be spontaneous in either direction?

20-6   CONCEPT ASSESSMENT

The following cell is set up under standard-state conditions.

$$Pb(s) | Pb^{2+} \| Cu^{2+} | Cu(s) \qquad E_{cell}^{\circ} = 0.47 \text{ V}$$

When sodium sulfate is added to the anode half-cell, formation of a white precipitate is observed, accompanied by a change in the value of E_{cell}. Explain these observations, and predict whether the new E_{cell} is greater or less than E_{cell}°.

Concentration Cells

The voltaic cell in Figure 20-11 consists of two hydrogen electrodes. One is a standard hydrogen electrode (SHE), and the other is a hydrogen electrode immersed in a solution of unknown $[H^+]$, less than 1 M. The cell diagram is

$$Pt | H_2(g, 1 \text{ atm}) | H^+(x \text{ M}) \| H^+(1 \text{ M}) | H_2(g, 1 \text{ atm}) | Pt$$

The reaction occurring in this cell is

Reduction:	$2 H^+(1 \text{ M}) + 2e^- \longrightarrow H_2(g, 1 \text{ atm})$
Oxidation:	$H_2(g, 1 \text{ atm}) \longrightarrow 2 H^+(x \text{ M}) + 2e^-$
Overall:	$2 H^+(1 \text{ M}) \longrightarrow 2 H^+(x \text{ M})$ **(20.21)**

$$E_{cell}^{\circ} = E_{H^+/H_2}^{\circ} - E_{H^+/H_2}^{\circ} = 0 \text{ V}$$

The voltaic cell in Figure 20-11 is called a concentration cell. A **concentration cell** consists of two half-cells with *identical electrodes* but different ion concentrations. Because the electrodes are identical, the standard electrode potentials are numerically equal and subtracting one from the other leads to the value $E_{cell}^{\circ} = 0$. However, because the ion concentrations differ, there is a potential

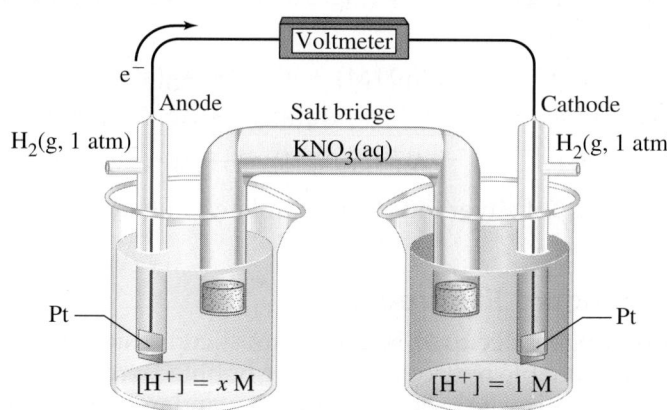

◀ FIGURE 20-11
A concentration cell
The cell consists of two hydrogen electrodes. The electrode on the right is a SHE. Oxidation occurs at the anode on the left, where $[H^+]$ is less than 1 M. The reading on the voltmeter is directly proportional to the pH of the solution in the anode compartment.

difference between the two half-cells. The spontaneous change in a concentration cell always occurs such that the concentrated solution becomes more dilute, and the dilute solution becomes more concentrated. The final result is as if the solutions were simply mixed. In a concentration cell, however, the natural tendency for entropy to increase in a mixing process is used as a means of producing electricity.

The Nernst equation for reaction (20.21) takes the form

$$E_{cell} = E_{cell}^\circ - \frac{0.0592 \text{ V}}{2} \log \frac{x^2}{1^2}$$

which simplifies to

$$E_{cell} = 0 - \frac{0.0592 \text{ V}}{2} \times 2 \log \frac{x}{1} = -0.0592 \text{ V} \log x$$

Because x is $[H^+]$ in the unknown solution and $-\log x = -\log[H^+] = pH$, the final result is

$$E_{cell} = (0.0592 \text{ pH}) \text{ V} \qquad \qquad \textbf{(20.22)}$$

where the pH is that of the unknown solution. If an unknown solution has a pH of 3.50, for example, the measured cell voltage in Figure 20-11 will be $E_{cell} = (0.0592 \times 3.50) \text{ V} = 0.207 \text{ V}$.

Constructing and using a hydrogen electrode is difficult. The Pt metal surface must be specially prepared and maintained, gas pressure must be controlled, and the electrode cannot be used in the presence of strong oxidizing or reducing agents. The solution to these problems is discussed later in this chapter.

🔍 20-7 CONCEPT ASSESSMENT

Write a cell diagram for a possible voltaic cell in which the cell reaction is $Cl^-(0.50 \text{ M}) \longrightarrow Cl^-(0.10 \text{ M})$. What would be E_{cell} for this reaction?

Measurement of K_{sp}

The difference in concentration of ions in the two half-cells of a concentration cell accounts for the observed E_{cell}. It also provides a basis for determining K_{sp} values for sparingly soluble ionic compounds. Consider the following concentration cell.

$$Ag(s)|Ag^+(\text{satd AgI})\|Ag^+(0.100 \text{ M})|Ag \qquad E_{cell} = 0.417 \text{ V}$$

At the anode, a silver electrode is placed in a saturated aqueous solution of silver iodide. At the cathode, a second silver electrode is placed in a solution with $[Ag^+] = 0.100$ M. The two half-cells are connected by a salt bridge, and the measured cell voltage is 0.417 V (Fig. 20-12). The cell reaction occurring in this *concentration cell* is

Reduction:	$Ag^+(0.100 \text{ M}) + e^- \longrightarrow Ag(s)$
Oxidation:	$Ag(s) \longrightarrow Ag^+(\text{satd AgI})$
Overall:	$Ag^+(0.100 \text{ M}) \longrightarrow Ag^+(\text{satd AgI})$ **(20.23)**

The calculation of K_{sp} of silver iodide is completed in Example 20-11.

Alternative Standard Electrodes

The standard hydrogen electrode is not the most convenient to use because it requires highly flammable hydrogen gas to be bubbled over the platinum electrode. Other electrodes can be used as secondary standard electrodes, such as the

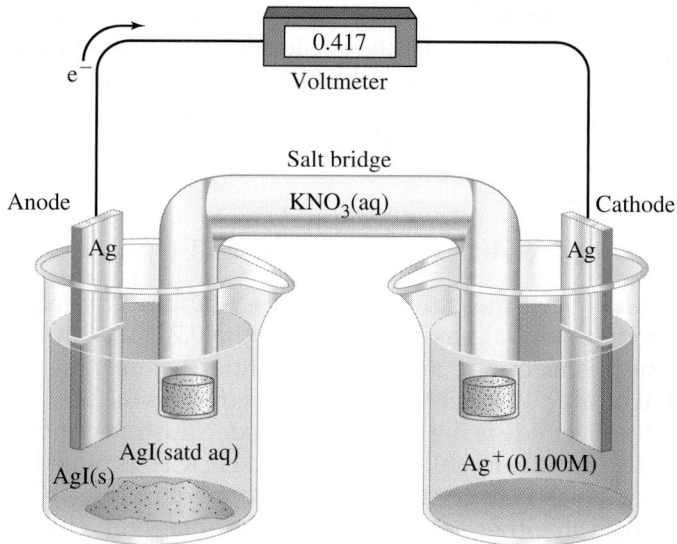

▲ FIGURE 20-12
A concentration cell for determining K_{sp} of AgI
The silver electrode in the anode compartment is in contact with a saturated solution
of AgI. In the cathode compartment, $[Ag^+] = 0.100$ M.

EXAMPLE 20-11   Using a Voltaic Cell to Determine K_{sp} of a Slightly Soluble Solute

With the data given for reaction (20.23), calculate K_{sp} for AgI.

$$AgI(s) \rightleftharpoons Ag^+(aq) + I^-(aq) \qquad K_{sp} = ?$$

Analyze

Once we have determined the concentration of Ag^+ ions from the Nernst equation for the cell, we can calculate the equilibrium constant by using the expression for the solubility product.

Solve

$[Ag^+]$ in saturated silver iodide solution can be represented as x. The Nernst equation is then applied to reaction (20.23). (To simplify the equations that follow, we have dropped the unit V, which would otherwise appear in several places.)

$$E_{cell} = E°_{cell} - \frac{0.0592}{z}\log\frac{[Ag^+]_{satd\ AgI}}{[Ag^+]_{0.100\ M\ soln}}$$

$$= E°_{cell} - \frac{0.0592}{1}\log\frac{x}{0.100}$$

$$0.417 = 0 - 0.0592(\log x - \log 0.100)$$

Divide both sides of the equation by 0.0592.

$$\frac{0.417}{0.0592} = -\log x + \log 0.100$$

$$\log x = \log 0.100 - \frac{0.417}{0.0592} = -1.00 - 7.04 = -8.04$$

$$x = [Ag^+] = 10^{-8.04} = 9.1 \times 10^{-9}\ M$$

Because in saturated AgI the concentrations of Ag^+ and I^- are equal,

$$K_{sp} = [Ag^+][I^-] = (9.1 \times 10^{-9})(9.1 \times 10^{-9}) = 8.3 \times 10^{-17}$$

Assess

Apart from the use of the Nernst equation, the other essential aspect is the realization that the only source of Ag^+ and I^- is from the AgI present; the saturated AgI(s) electrode has $[Ag^+] = [I^-]$.

PRACTICE EXAMPLE A: K_{sp} for AgCl $= 1.8 \times 10^{-10}$. What would be the measured E_{cell} for the voltaic cell in Example 20-11 if the contents of the anode half-cell were saturated AgCl(aq) and AgCl(s)?

PRACTICE EXAMPLE B: Calculate the K_{sp} for PbI_2 given the following concentration cell information.

$$Pb(s)|Pb^{2+}(satd\ PbI_2)||Pb^{2+}(0.100\ M)|Pb(s) \qquad E_{cell} = 0.0567\ V$$

silver–silver chloride electrode, in which a silver wire is covered with a layer of insoluble solid silver chloride. The silver-chloride-coated silver wire is immersed in a 1 M potassium chloride solution (see Fig. 20-13a), giving the electrode

$$Ag(s)|AgCl(s)|Cl^-(1.0 \text{ M})$$

with a half-cell reaction of

$$AgCl(s) + e^- \longrightarrow Ag(s) + Cl^-(aq)$$

This electrode has been measured against the standard hydrogen electrode, and the electrode potential has been found to be 0.22233 V at 25 °C. Since all components of this electrode are in their standard states, the standard electrode potential of the silver–silver chloride electrode is 0.22233 V at 25 °C.

An alternative electrode is the calomel electrode, illustrated in Figure 20-13(b). In this electrode, mercurous chloride (calomel, Hg_2Cl_2) is mixed with mercury to form a paste, which is in contact with liquid mercury, Hg(l), and the whole setup is immersed in either a 1.0 M solution of potassium chloride or a saturated solution of potassium chloride. The electrode is

$$Hg(l)|Hg_2Cl_2(s)|Cl^-(1.0 \text{ M})$$

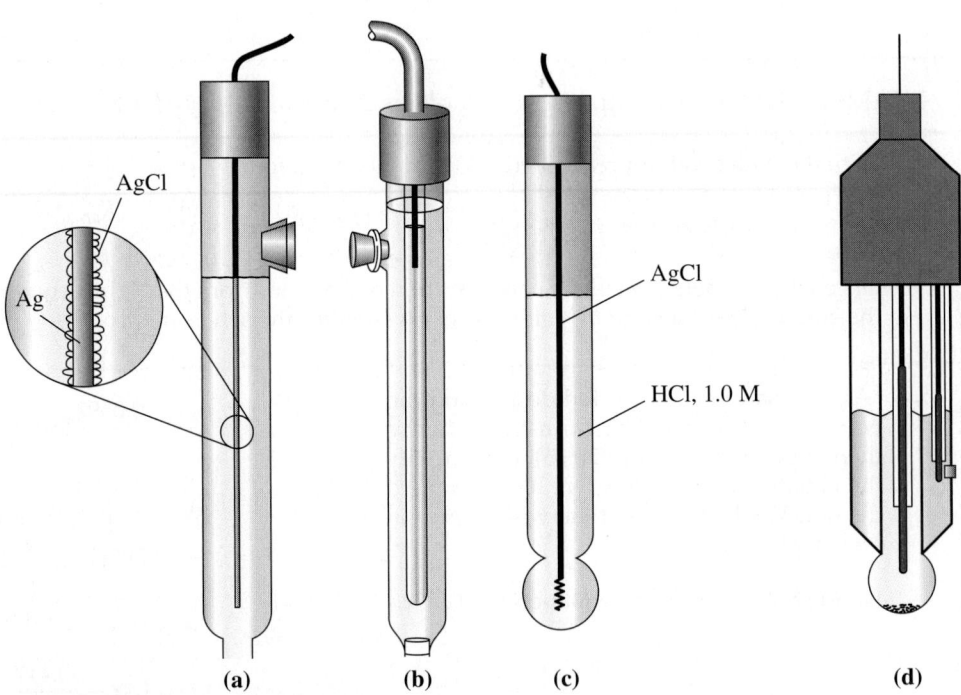

▲ FIGURE 20-13
Schematic diagrams of some common electrodes
(a) The silver–silver chloride electrode. Silver wire is coated with silver chloride and immersed in a 1 M aqueous solution of KCl. At the bottom of the tube is a fritted (porous) disc to allow contact with a solution of interest. **(b)** The standard calomel electrode is a tube containing a paste of calomel and mercury, immersed in a 1.0 M solution of KCl. Contact with an external circuit is made with a Pt wire inserted in the inner tube; the inner tube makes contact with the outer 1.0 M solution of KCl through a small hole in the bottom of the inner tube. **(c)** A glass electrode consists of tube with a very thin glass bulb at the end and a Ag–AgCl electrode immersed in a 1.0 M HCl solution. When the glass electrode is dipped into a solution, ions interact with the membrane. The potential established on the silver wire depends on the solution being tested. **(d)** A modern pH electrode consists of a glass electrode and an internal Ag–AgCl reference electrode. There is a small sintered disc in the side of the outer tube that acts as a salt bridge between the electrode and the unknown solution.

and the half-cell reaction is

$$\frac{1}{2}Hg_2Cl_2(s) + e^- \longrightarrow Hg(l) + Cl^-(aq)$$

The standard electrode potential at 25 °C is 0.2680 V. If, however, a saturated solution of KCl is used, as opposed to one of 1 M, the reduction potential is 0.2412 V. This electrode is known as the saturated calomel electrode (SCE) and is often used as a reference.

In practice a variety of reference electrodes are used; therefore, it is necessary to quote reduction potentials with respect to a specific reference.

🔍 20-8   CONCEPT ASSESSMENT

Why do the calomel standard reduction potential and the saturated calomel electrode have different potentials?

The Glass Electrode and the Electrochemical Measurement of pH

To measure the pH of a solution electrochemically, we need an electrode that responds to changes in $[H^+(aq)]$. We noted that the standard hydrogen electrode is difficult to use for this purpose, and so, for routine use, a simpler and safer electrode is needed. Such an electrode is the *glass electrode*, which consists of a very thin walled glass bulb (see Fig. 20-13c) at the end of a tube that contains a silver–silver chloride electrode and a HCl solution of known composition (e.g., 1 M). When the bulb is placed in a solution of unknown pH, a potential develops because of the concentration difference across the membrane, analogous to a concentration cell. To measure this potential difference, a reference electrode is used, which can be either a saturated calomel electrode or a second silver–silver chloride electrode, as in the combination electrode shown in Figure 20-13(d). The overall cell can be represented as

$$Ag(s)|AgCl(s)|Cl^-(1.0\ M), H^+(1.0\ M)|\text{glass membrane}|H^+(\text{unknown})\|Cl^-(1.0\ M)|AgCl(s)|Ag(s)$$

where the two electrodes are connected by a salt bridge. The half-reactions are

$$Ag(s) + Cl^-(aq) \longrightarrow AgCl(s) + e^-$$
$$H^+(1.0\ M) \longrightarrow H^+(\text{unknown})$$
$$AgCl(s) + e^- \longrightarrow Ag(s) + Cl^-(aq)$$

The half-cell potentials of the two half-reactions for the silver–silver chloride electrodes cancel each other out and make no contribution to the cell potential. The Gibbs energy change corresponding to the dilution of protons from a known concentration of 1.0 M to the unknown solution is the source of the potential difference across the glass membrane. The Gibbs energy difference across the membrane, using $G = G° + RT \ln[H^+]$, is

$$\Delta G = G(\text{unknown}) - G(1.0\ M)$$
$$= G° + RT \ln[\text{unknown}] - G° - RT \ln 1.0$$
$$= RT \ln[\text{unknown}]$$

Converting this to a potential by dividing by $-zF, z = 1$, and assuming $T = 25\ °C$, we obtain

$$E_{cell} = 0.0592\ pH$$

after converting the logarithm to base 10 and using the definition of $pH = -\log_{10}[\text{unknown}]$. The cell potential is measured with a pH meter, a voltage measuring device that electronically converts E_{cell} to pH and displays the result in pH units.

The glass electrode was devised in 1906 by German biologist Max Cremer, and it was the prototype for a large number of membrane electrodes that are selective for a particular ion, such as the ions K^+, NH_4^+, Cl^-, and many others. Such electrodes are known collectively as *ion-selective electrodes*, and they have many applications in environmental chemistry and biochemistry.

20-5 Batteries: Producing Electricity Through Chemical Reactions

▶ Batteries are vitally important to modern society. Annual production in developed nations has been estimated at more than 10 batteries per person per year.

A **battery** is a device that stores chemical energy for later release as electricity. Some batteries consist of a single voltaic cell with two electrodes and the appropriate electrolyte(s); an example is a flashlight cell. Other batteries consist of two or more voltaic cells joined in series fashion—plus to minus—to increase the total voltage; an example is an automobile battery. In this section, we will consider three types of voltaic cells and the batteries based on them.

- **Primary cells.** The cell reaction in a primary cell is not reversible. When the reactants have been mostly converted to products, no more electricity is produced and a battery employing a primary cell(s) is dead.

▶ Cell phones, laptop computers, and many other devices rely heavily on rechargeable batteries. Advances in electrochemistry and engineering are leading to the development of batteries that weigh less, last longer, and provide more power for portable electronic devices.

- **Secondary cells.** The cell reaction in a secondary cell *can* be reversed by passing electricity through the cell (charging). A battery employing secondary cells can be used through several hundred or more cycles of discharging followed by charging.

- **Flow batteries** and **fuel cells.** Materials (reactants, products, and electrolytes) pass through the battery, which is simply a converter of chemical energy to electric energy. These types of batteries can be run indefinitely as long as they are supplied by electrolytes.

The Leclanché (Dry) Cell

The most common form of voltaic cell is the *Leclanché cell*, invented by the French chemist Georges Leclanché (1839–1882) in the 1860s. Popularly called a *dry cell*, because no free liquid is present, or *flashlight battery*, the Leclanché cell is diagrammed in Figure 20-14. In this cell, oxidation occurs at a zinc anode and reduction at an inert carbon (graphite) cathode. The electrolyte is a moist paste of MnO_2, $ZnCl_2$, NH_4Cl, and carbon black (soot). The maximum cell voltage is 1.55 V. The anode (oxidation) half-reaction is simple.

$$\text{Oxidation:} \quad Zn(s) \longrightarrow Zn^{2+}(aq) + 2\,e^-$$

The reduction is more complex. Essentially, it involves the reduction of MnO_2 to compounds having Mn in a +3 oxidation state, for example,

$$\text{Reduction:} \quad 2\,MnO_2(s) + H_2O(l) + 2\,e^- \longrightarrow Mn_2O_3(s) + 2\,OH^-(aq)$$

An acid–base reaction occurs between NH_4^+ (from NH_4Cl) and OH^-.

$$NH_4^+(aq) + OH^-(aq) \longrightarrow NH_3(g) + H_2O(l)$$

A buildup of $NH_3(g)$ around the cathode would disrupt the current because the $NH_3(g)$ adheres to the cathode. That buildup is prevented by a reaction between Zn^{2+} and $NH_3(g)$ to form the complex ion $[Zn(NH_3)_2]^{2+}$, which crystallizes as the chloride salt.

$$Zn^{2+}(aq) + 2\,NH_3(g) + 2\,Cl^-(aq) \longrightarrow [Zn(NH_3)_2]Cl_2(s)$$

The Leclanché cell is a *primary cell*; it cannot be recharged. This cell is cheap to make, but it has some drawbacks. When current is drawn rapidly from the cell, products, such as NH_3, build up on the electrodes, causing the voltage to drop. Also, because the electrolyte medium is acidic, zinc metal slowly dissolves.

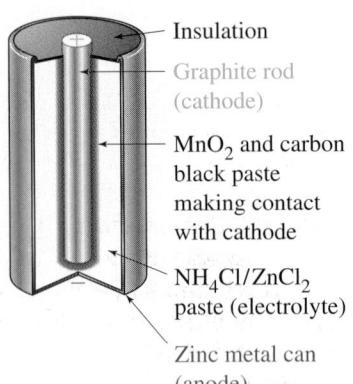

Insulation

Graphite rod (cathode)

MnO_2 and carbon black paste making contact with cathode

$NH_4Cl/ZnCl_2$ paste (electrolyte)

Zinc metal can (anode)

▲ FIGURE 20-14
The Leclanché (dry) cell
The chief components of the cell are a graphite (carbon) rod serving as the cathode, a zinc container serving as the anode, and an electrolyte.

A superior form of the Leclanché cell is the *alkaline cell,* which uses NaOH or KOH in place of NH_4Cl as the electrolyte. The reduction half-reaction is the same as that shown above, but the oxidation half-reaction involves the formation of $Zn(OH)_2(s)$, which can be thought of as occurring in two steps.

$$Zn(s) \longrightarrow Zn^{2+}(aq) + 2\,e^-$$

$$\underline{Zn^{2+}(aq) + 2\,OH^-(aq) \longrightarrow Zn(OH)_2(s)}$$

$$Zn(s) + 2\,OH^-(aq) \longrightarrow Zn(OH)_2(s) + 2\,e^-$$

The advantages of the alkaline cell are that zinc does not dissolve as readily in a basic (alkaline) medium as in an acidic medium, and the cell does a better job of maintaining its voltage as current is drawn from it.

The Lead–Acid (Storage) Battery

Secondary cells are commonly encountered joined together in series in the *lead–acid battery,* or *storage battery,* which has been used in automobiles since about 1915 (Fig. 20-15). A storage battery is capable of repeated use because its chemical reactions are reversible. That is, the discharged energy can be restored by supplying electric current to recharge the cells in the battery.

The reactants in a lead–acid cell are spongy lead packed into a lead grid at the anode, red-brown lead(IV) oxide packed into a lead grid at the cathode, and an electrolyte solution consisting of dilute sulfuric acid (about 35% H_2SO_4, by mass). In this strongly acidic medium, the ionization of H_2SO_4 does not go to completion. Both $HSO_4^-(aq)$ and $SO_4^{2-}(aq)$ are present, but $HSO_4^-(aq)$ predominates. The half-reactions and overall reaction are

▲ $E°_{cell}$ **is an intensive property** The voltage of a dry cell battery does not depend on the size of the battery—all of those pictured here are 1.5 V batteries. Although these batteries deliver the same voltage, the total energy output of each battery is different.

Reduction: $PbO_2(s) + 3\,H^+(aq) + HSO_4^-(aq) + 2\,e^- \longrightarrow PbSO_4(s) + 2\,H_2O(l)$

Oxidation: $\underline{\hspace{4cm} Pb(s) + HSO_4^-(aq) \longrightarrow PbSO_4(s) + H^+(aq) + 2\,e^-}$

Overall: $PbO_2(s) + Pb(s) + 2\,H^+(aq) + 2\,HSO_4^-(aq) \longrightarrow 2\,PbSO_4(s) + 2\,H_2O(l)$

$$E_{cell} = E_{PbO_2/PbSO_4} - E_{PbSO_4/Pb} = 1.74\,V - (-0.28\,V) = 2.02\,V \qquad \textbf{(20.24)}$$

◀ You can think of the half-reactions as occurring in two steps: (1) oxidation of Pb(s) to $Pb^{2+}(aq)$ and reduction of $PbO_2(s)$ to $Pb^{2+}(aq)$, followed by (2) precipitation of $PbSO_4(s)$ at each electrode.

▲ FIGURE 20-15
A lead–acid (storage) cell
The composition of the electrodes is described in the text. The cell reaction that occurs as the cell is discharged is given in equation (20.24). The voltage of the cell is 2.02 V. In this figure, two anode plates are connected in parallel fashion, as are two cathode plates. The battery shown in this figure is the good battery used to charge a dead battery.

▶ In spite of its usefulness and ability to deliver a strong current, the lead storage battery is also a pollution hazard. All batteries should be disposed of properly and should not be dumped in land fills or garbage disposal sites.

▶ Lead-acid storage batteries are also used to power golf carts, wheelchairs, and passenger carts in airport terminals.

▶ Rechargeable silver oxide batteries have been developed and provide alternatives to lithium-ion batteries.

When an automobile engine is started, the battery is at first discharged. Once the car is in motion, an alternator powered by the engine constantly recharges the battery. At times, the plates of the battery become coated with $PbSO_4(s)$ and the electrolyte becomes sufficiently diluted with water that the battery must be recharged by connecting it to an external electric source. This forces the reverse of reaction (20.24), a nonspontaneous reaction.

$$2\,PbSO_4(s) + 2\,H_2O(l) \longrightarrow Pb(s) + PbO_2(s) + 2\,H^+(aq) + 2\,HSO_4^-(aq)$$
$$E_{cell} = -2.02\text{ V}$$

To prevent the anode and cathode from coming into contact with each other, causing a *short circuit*, sheets of an insulating material are used to separate alternating anode and cathode plates. A group of anodes is connected together electrically, as is a group of cathodes. This parallel connection increases the electrode area in contact with the electrolyte solution and increases the current-delivering capacity of the cell. Cells are then joined in a series fashion, positive to negative, to produce a battery. The typical 12 V battery consists of six cells, each cell with a potential of about 2 V.

The Silver–Zinc Cell: A Button Battery

The cell diagram of a *silver–zinc cell* (Fig. 20-16) is

$$Zn(s), ZnO(s)|KOH(satd)|Ag_2O(s), Ag(s)$$

The half-reactions on discharging are

Reduction:	$Ag_2O(s) + H_2O(l) + 2\,e^- \longrightarrow 2\,Ag(s) + 2\,OH^-(aq)$
Oxidation:	$Zn(s) + 2\,OH^-(aq) \longrightarrow ZnO(s) + H_2O(l) + 2\,e^-$
Overall:	$Zn(s) + Ag_2O(s) \longrightarrow ZnO(s) + 2\,Ag(s)$ **(20.25)**

Because no solution species is involved in the cell reaction, the quantity of electrolyte is very small and the electrodes can be maintained very close together. The cell voltage is 1.8 V, and its storage capacity is six times greater than that of a lead–acid battery of the same size. These characteristics make batteries, such as the silver–zinc cell, useful in button batteries. These miniature batteries are used in watches, hearing aids, and cameras. In addition, silver–zinc batteries fulfill the requirements of spacecraft, satellites, missiles, rockets, space launch vehicles, torpedoes, underwater vehicles, and life-support systems. On the Mars *Pathfinder* mission, the rover and the cruise system were powered by solar cells. The energy storage requirements of the lander were met by modified silver–zinc batteries with about three times the storage capacity of the standard nickel–cadmium rechargeable battery.

The Nickel–Cadmium Cell: A Rechargeable Battery

The *nickel–cadmium cell* (or *nicad battery*) is commonly used in cordless electric devices, such as electric shavers and handheld calculators. The anode in this cell is cadmium metal, and the cathode is the Ni(III) compound NiO(OH)

▲ A rechargeable nickel–cadmium cell, or nicad battery.

▶ FIGURE 20-16
A silver–zinc button (miniature) cell

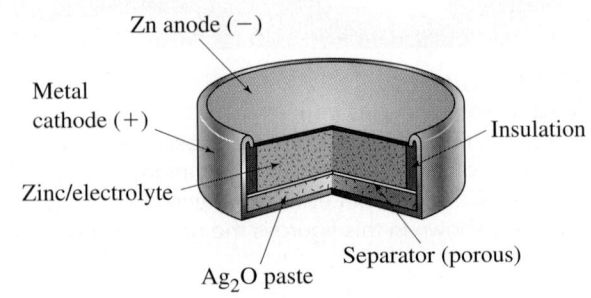

Zn anode (−)

Metal cathode (+)

Insulation

Zinc/electrolyte

Separator (porous)

Ag_2O paste

supported on nickel metal. The half-cell reactions for a nickel–cadmium battery during discharge are

Reduction: $2 NiO(OH)(s) + 2 H_2O(l) + 2 e^- \longrightarrow 2 Ni(OH)_2(s) + 2 OH^-(aq)$

Oxidation: $\underline{\qquad\qquad Cd(s) + 2 OH^-(aq) \longrightarrow Cd(OH)_2(s) + 2 e^-}$

Overall: $Cd(s) + 2 NiO(OH)(s) + 2 H_2O(l) \longrightarrow 2 Ni(OH)_2(s) + Cd(OH)_2(s)$

This cell gives a fairly constant voltage of 1.4 V. When the cell is recharged by connection to an external voltage source, the reactions above are reversed. Nickel–cadmium batteries can be recharged many times because the solid products adhere to the surface of the electrodes.

In primary cells the positive and negative electrodes are known as the cathode, where reduction takes place, and the anode, where oxidation takes place. In rechargeable systems, however, we have either a charging mode or a discharging mode, and so depending whether electrons are flowing out of the cell or flowing into the cell, the notion of the anode and the cathode changes. On the discharge of a nicad battery, the NiO(OH) electrode is the cathode because reduction is taking place, but on the charge, it is the anode because oxidation is taking place (the reverse reaction). In discharge mode the NiO(OH) electrode electrons are removed from the electrode because of the reduction process, and so this electrode is positively charged. In the charging mode electrons are being removed from this electrode by the oxidation process; this is the anode and it is positively charged. Therefore, regardless of charging or discharging, the NiO(OH) electrode is positive.

The negative electrode, the cadmium electrode in a nicad battery, is the anode on discharging (oxidation) and the cathode (reduction) on charging. In both charging and discharging, the anode is the electrode from which electrons exit the battery, and the cathode is the electrode at which electrons enter the battery.

In summary, when dealing with rechargeable batteries, it is better to speak of the positive and negative electrodes and avoid the terms *cathode* and *anode*.

The Lithium-Ion Battery

Lithium-ion batteries are a type of rechargeable battery now commonly used in consumer electronics, such as cell phones, laptop computers, and MP3 players. In a lithium-ion battery, the lithium ion moves between the positive and negative electrodes. The positive electrode consists of lithium cobalt(III) oxide, $LiCoO_2$, and the negative electrode is highly crystallized graphite. To complete the battery an electrolyte is needed, which can consist of an organic solvent and ions, such as $LiPF_6$. The structure of $LiCoO_2$ and graphite electrodes is illustrated in Figure 20-17. In the charging cycle at the positive electrode, lithium ions are released into the electrolyte solution as electrons are removed from the electrode. To maintain a charge balance, one cobalt(III) ion is oxidized to cobalt(IV) for each lithium ion released:

$$LiCoO_2(s) \longrightarrow Li_{(1-x)}CoO_2(s) + xLi^+(solvent) + x e^-$$

At the negative electrode, lithium ions enter between the graphite layers and are reduced to lithium metal. This insertion of a guest atom into a host solid is called *intercalation*, and the resulting product is called an *intercalation compound*:

$$C(s) + xLi^+(solvent) + x e^- \longrightarrow Li_xC(s)$$

In the operation of a lithium-ion battery the source of the electrons is the oxidation of the Co(III) to Co(IV). The lithium ion takes these electrons to the graphite electrode during charging and returns them to the positive electrode during discharge.

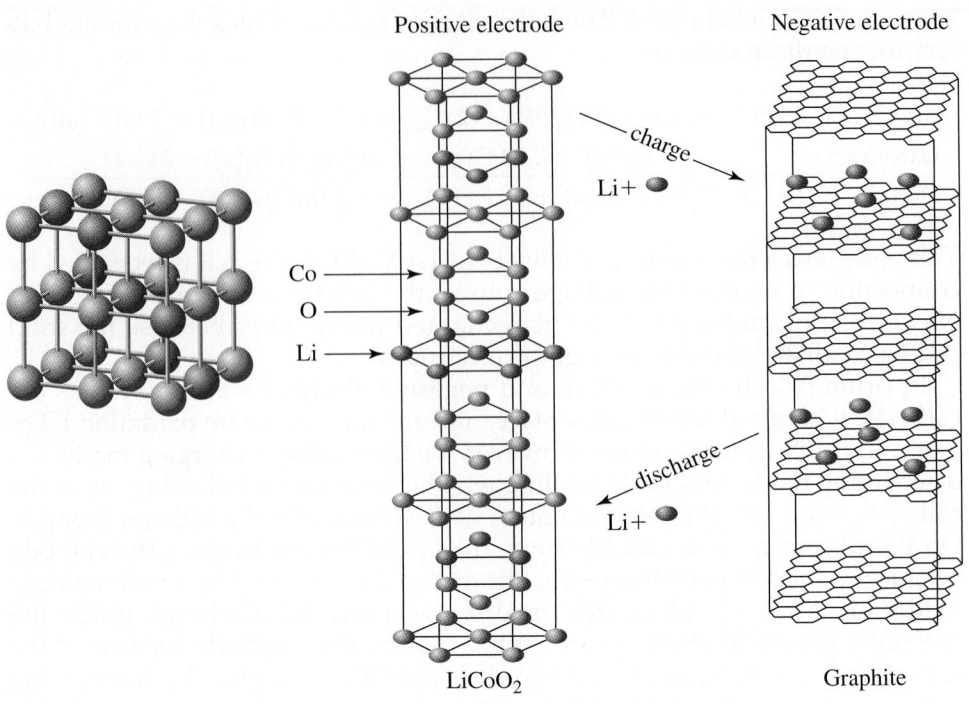

▲ FIGURE 20-17
The electrodes of a lithium-ion battery
The layered graphite electrode is shown with lithium ions (violet) intercalated. The $LiCoO_2$ is shown as a face-centered cubic lattice, with the oxygen atoms (red) occupying the corners and the faces, the cobalt atoms (pink) occupying half of the edges, and the lithium atoms occupying half of the edges and the central octahedral hole. This arrangement leads to planes of oxygen, cobalt, oxygen, lithium, oxygen, cobalt, and oxygen atoms, as indicated in the figure.

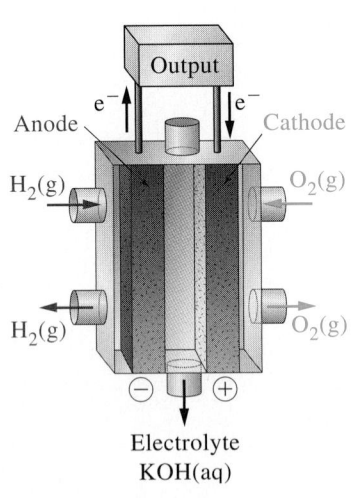

▲ FIGURE 20-18
A hydrogen–oxygen fuel cell
A key requirement in fuel cells is porous electrodes that allow for easy access of the gaseous reactants to the electrolyte. The electrodes chosen should also catalyze the electrode reactions.

Many other lithium batteries exist that use many different materials for the positive electrode, while graphite is the most common negative electrode. A major development is in the use of conducting polymers as the electrolyte, which has led to a whole range of *lithium-ion polymer batteries*. The development of new batteries based on lithium ions is currently an area of great interest.

Fuel Cells

The three types of cells considered in the remainder of this section fall into the third category mentioned on page 888; they are found in flow batteries.

For most of the twentieth century, scientists explored the possibility of converting the chemical energy of fuels directly to electricity. The essential process in a fuel cell is *fuel + oxygen ⟶ oxidation products*. The first fuel cells were based on the reaction of hydrogen and oxygen. Figure 20-18 represents such a fuel cell. The overall change is that $H_2(g)$ and $O_2(g)$ in an alkaline medium produce $H_2O(l)$.

Reduction: $O_2(g) + 2 H_2O(l) + 4 e^- \longrightarrow 4 OH^-(aq)$

Oxidation: $2 \{H_2(g) + 2 OH^-(aq) \longrightarrow 2 H_2O(l) + 2 e^-\}$

Overall: $\overline{2 H_2(g) + O_2(g) \longrightarrow 2 H_2O(l)}$ **(20.26)**

$$E^\circ_{cell} = E^\circ_{O_2/OH^-} - E^\circ_{H_2O/H_2} = 0.401 \text{ V} - (-0.828 \text{ V}) = 1.229 \text{ V}$$

The theoretical maximum energy available as electric energy in any electrochemical cell is the Gibbs energy change for the cell reaction, ΔG°. The maximum energy release when a fuel is burned is the enthalpy change, ΔH°. One of the measures used to evaluate a fuel cell is the *efficiency value*, $\varepsilon = \Delta G^\circ / \Delta H^\circ$. For the hydrogen–oxygen fuel cell, $\varepsilon = -474.4 \text{ kJ} / -571.6 \text{ kJ} = 0.83$.

The day is fast approaching when fuel cells based on the direct oxidation of common fuels will become a reality. For example, the half-reaction and cell reaction for a fuel cell using methane (natural gas) are

Reduction: $2\{O_2(g) + 4H^+ + 4e^- \longrightarrow 2H_2O(l)\}$

Oxidation: $CH_4(g) + 2H_2O(l) \longrightarrow CO_2(g) + 8H^+ + 8e^-$

Overall: $CH_4(g) + 2O_2(g) \longrightarrow CO_2(g) + 2H_2O(l)$

$$\Delta H° = -890\ kJ \qquad \Delta G° = -818\ kJ \qquad \varepsilon = 0.92 \qquad \textbf{(20.27)}$$

▲ This Toyota prototype is a fuel-cell-powered electric car producing hydrogen from gasoline.

Although methane fuel cells are still in the research stage, an automobile engine is currently under development in which (1) a liquid hydrocarbon is vaporized; (2) the vaporized fuel is partially oxidized to $CO(g)$; (3) in the presence of a catalyst, steam converts the $CO(g)$ to $CO_2(g)$ and $H_2(g)$; and (4) $H_2(g)$ and air are fed through a fuel cell, producing electric energy.

A fuel cell should actually be called an *energy converter* rather than a battery. As long as fuel and $O_2(g)$ are available, the cell will produce electricity. It does not have the limited capacity of a primary battery or the fixed storage capacity of a secondary battery. Fuel cells based on reaction (20.26) have had their most notable successes as energy sources in space vehicles. (Water produced in the cell reaction is also a valuable product of the fuel cell.)

◄ Fuel cells are environmentally friendly. Oxygen and hydrogen are readily available. Hydrogen, although dangerous, can now be transported safely by the use of special materials that can adsorb large volumes.

Air Batteries

In a fuel cell, $O_2(g)$ is the oxidizing agent that oxidizes a fuel such as $H_2(g)$ or $CH_4(g)$. Another kind of flow battery, because it uses $O_2(g)$ from air, is known as an *air battery*. The substance that is oxidized in an air battery is typically a metal.

One heavily studied battery system is the aluminum–air battery in which oxidation occurs at an aluminum anode and reduction at a carbon–air cathode. The electrolyte circulated through the battery is $NaOH(aq)$. Because it is in the presence of a high concentration of OH^-, Al^{3+} produced at the anode forms the complex ion $[Al(OH)_4]^-$. The operation of the battery is suggested by Figure 20-19. The half-reactions and the overall cell reaction are

Reduction: $3\{O_2(g) + 2H_2O(l) + 4e^- \longrightarrow 4OH^-(aq)\}$

Oxidation: $4\{Al(s) + 4OH^-(aq) \longrightarrow [Al(OH)_4]^-(aq) + 3e^-\}$

Overall: $4Al(s) + 3O_2(g) + 6H_2O(l) + 4OH^-(aq) \longrightarrow 4[Al(OH)_4]^-(aq)$ **(20.28)**

The battery is kept charged by feeding chunks of Al and water into it. A typical air battery can power an automobile several hundred miles before refueling is necessary. The electrolyte is circulated outside the battery, where $Al(OH)_3(s)$ is precipitated from the $[Al(OH)_4]^-(aq)$. This $Al(OH)_3(s)$ is collected and can then be converted back to aluminum metal at an aluminum manufacturing facility.

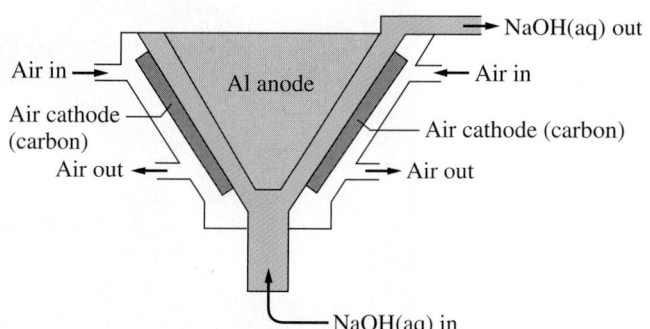

◄ FIGURE 20-19
A simplified aluminum–air battery

20-6   Corrosion: Unwanted Voltaic Cells

The reactions occurring in voltaic cells (batteries) are important sources of electricity, but similar reactions also underlie corrosion processes. First, we will consider the electrochemical basis of corrosion, and then we will see how electrochemical principles can be applied to control corrosion.

Figure 20-20(a) demonstrates the basic processes in the corrosion of an iron nail. The nail is embedded in an agar gel in water. The gel contains the common acid–base indicator phenolphthalein and potassium ferricyanide, $K_3[Fe(CN)_6]$. Within hours of starting the experiment, a deep blue precipitate forms at the head and tip of the nail. Along the body of the nail, the agar gel turns pink. The blue precipitate, Turnbull's blue, establishes the presence of iron(II). The pink color is that of phenolphthalein in basic solution. From these observations, we write two simple half-equations.

▶ This and other tests for iron(II) are described more fully in Section 23-5.

$$\textit{Reduction:} \quad O_2(g) + 2\,H_2O(l) + 4\,e^- \longrightarrow 4\,OH^-(aq)$$
$$\textit{Oxidation:} \quad 2\,Fe(s) \longrightarrow 2\,Fe^{2+}(aq) + 4\,e^-$$

The potential difference for these two half-reactions is

$$E^{\circ}_{cell} = E^{\circ}_{O_2/OH^-} - E^{\circ}_{Fe^{2+}/Fe} = 0.401\ V - (-0.440\ V) = 0.841\ V$$

indicating that the corrosion process should be spontaneous when reactants and products are in their standard states. Typically, the corrosion medium has $[OH^-] \ll 1\ M$, the reduction half-reaction is even more favorable, and E_{cell} is even greater than 0.841 V. Corrosion is especially significant in acidic solutions, in which the reduction half-reaction is

$$O_2(g) + 4\,H^+(aq) + 4\,e^- \longrightarrow 2\,H_2O(l) \qquad E^{\circ}_{O_2/H_2O} = 1.229\ V$$

In the corroding nail of Figure 20-20(a), oxidation occurs at the head and tip. Electrons given up in the oxidation move along the nail and are used to reduce dissolved O_2. The reduction product, OH^-, is detected by the phenolphthalein. In the bent nail in Figure 20-20(b), oxidation occurs at *three* points: the head and tip and also the bend. The nail is preferentially oxidized

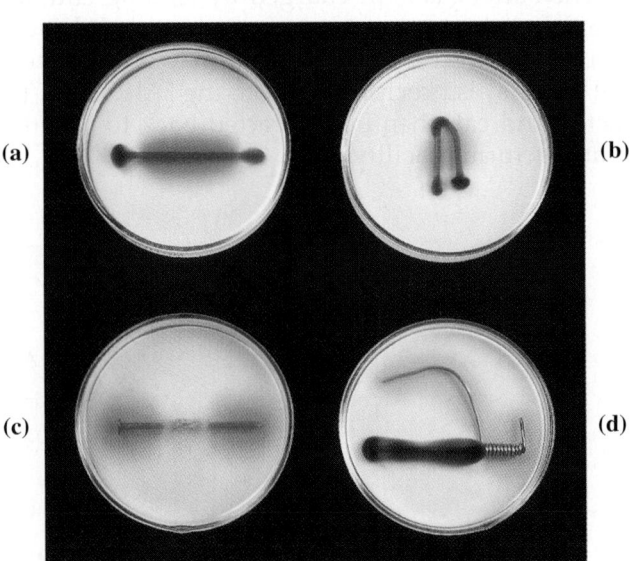

(a)

(b)

(c)

(d)

▶ FIGURE 20-20
Demonstration of corrosion and methods of corrosion protection
The pink color results from the indicator phenolphthalein in the presence of base; the dark blue color results from the formation of Turnbull's blue $KFe[Fe(CN)_6]$. Corrosion (oxidation) of the nail occurs at strained regions: **(a)** the head and tip and **(b)** a bend in the nail. **(c)** Contact with zinc protects the nail from corrosion. Zinc is oxidized instead of the iron (forming the faint white precipitate of zinc ferricyanide). **(d)** Copper does not protect the nail from corrosion. Electrons lost in the oxidation half-reaction distribute themselves along the copper wire, as seen by the pink color that extends the full length of the wire.

at these points because the strained metal is more active (more anodic) than the unstrained metal. This situation is similar to the preferential rusting of a dented automobile fender.

Some metals, such as aluminum, form corrosion products that adhere tightly to the underlying metal and protect it from further corrosion. Iron oxide (rust), however, flakes off and constantly exposes fresh surface. This difference in corrosion behavior explains why cans made of iron deteriorate rapidly in the environment, whereas aluminum cans have an almost unlimited lifetime. The simplest method of protecting a metal from corrosion is to cover it with paint or some other protective coating impervious to water, an important reactant and solvent in corrosion processes.

Another method of protecting an iron surface is to plate it with a thin layer of a second metal. Iron can be plated with copper by electroplating or with tin by dipping the iron into molten tin. In either case, the underlying metal is protected as long as the coating remains intact. If the coating is cracked, as when a "tin" can is dented, the underlying iron is exposed and begins to corrode. Iron, being more active than copper and tin, undergoes oxidation; the reduction half-reaction occurs on the plating (Figs. 20-20d and 20-21).

When iron is coated with zinc (galvanized iron), the situation is different. Zinc is more active than iron. If a break occurs in the zinc plating, the iron is still protected because the zinc is oxidized instead of the iron, and corrosion products protect the zinc from further corrosion (Figs. 20-20c and 20-21).

Still another method is used to protect large iron and steel objects in contact with water or moist soils—ships, storage tanks, pipelines, plumbing systems. This method involves connecting a chunk of magnesium or some other active metal to the object, either directly or through a wire. Oxidation occurs at the active metal, which slowly dissolves. The iron surface acquires electrons from the oxidation of the active metal; the iron acts as a cathode and supports a *reduction* half-reaction. As long as some of the active metal remains, the iron is protected. This type of protection is called **cathodic protection**, and the active metal is called, appropriately, a *sacrificial anode*. Millions of pounds of magnesium are used annually in the United States in sacrificial anodes.

▲ Galvanized nails

▲ **Magnesium sacrificial anodes**
The small cylindrical bars of magnesium attached to the steel ship provide cathodic protection against corrosion.

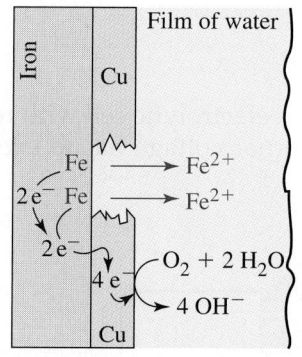
(a) Copper-plated iron

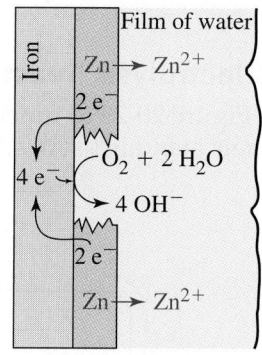
(b) Galvanized iron

▲ FIGURE 20-21
Protection of iron against electrolytic corrosion
In the *anodic* reaction, the metal that is more easily oxidized loses electrons to produce metal ions. In **(a)**, this is iron; in **(b)**, it is zinc. In the *cathodic* reaction, oxygen gas, which is dissolved in a thin film of water on the metal, is reduced to OH^-. Rusting of iron occurs in (a), but it does not in (b). When iron corrodes, Fe^{2+} and OH^- ions from the half-reactions initiate these further reactions.

$$Fe^{2+} + 2\,OH^- \longrightarrow Fe(OH)_2(s)$$

$$4\,Fe(OH)_2(s) + O_2 + 2\,H_2O \longrightarrow 4\,Fe(OH)_3(s)$$

$$2\,Fe(OH)_3(s) \longrightarrow \underset{\text{rust}}{Fe_2O_3 \cdot H_2O(s)} + H_2O(l)$$

20-7 Electrolysis: Causing Nonspontaneous Reactions to Occur

Until now, the emphasis has been on voltaic (galvanic) cells, electrochemical cells in which chemical change is used to produce electricity. Another type of electrochemical cell—the **electrolytic cell**—uses electricity to produce a *nonspontaneous* reaction. The process in which a nonspontaneous reaction is driven by the application of electric energy is called **electrolysis**.

Let's explore the relationship between voltaic and electrolytic cells by returning briefly to the cell shown in Figure 20-4. When the cell functions spontaneously, electrons flow from the zinc to the copper and the overall chemical change in the voltaic cell is

$$Zn(s) + Cu^{2+}(aq) \longrightarrow Zn^{2+}(aq) + Cu(s) \qquad E^{\circ}_{cell} = 1.103 \text{ V}$$

Now suppose the same cell is connected to an external electric source of voltage greater than 1.103 V (Fig. 20-22). That is, the connection is made so that electrons are forced into the zinc electrode (now the cathode) and removed from the copper electrode (now the anode). The overall reaction in this case is the *reverse* of the voltaic cell reaction, and E°_{cell} is *negative*.

Reduction: $\quad Zn^{2+}(aq) + 2 e^- \longrightarrow Zn(s)$

Oxidation: $\qquad\qquad\qquad Cu(s) \longrightarrow Cu^{2+}(aq) + 2 e^-$

Overall: $\quad Cu(s) + Zn^{2+}(aq) \longrightarrow Cu^{2+}(aq) + Zn(s)$

$$E^{\circ}_{cell} = E^{\circ}_{Zn^{2+}/Zn} - E^{\circ}_{Cu^{2+}/Cu} = -0.763 \text{ V} - 0.340 \text{ V} = -1.103 \text{ V}$$

Thus, reversing the direction of the electron flow changes the voltaic cell into an electrolytic cell.

Predicting Electrolysis Reactions

For the cell in Figure 20-22 to function as an electrolytic cell with reactants and products in their standard states, the external voltage has to exceed 1.103 V.

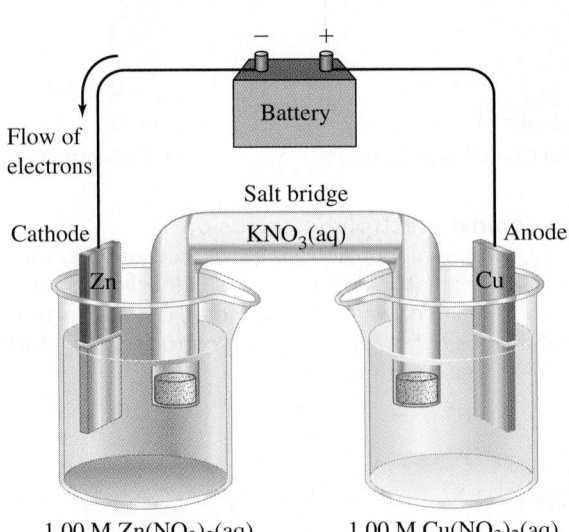

▶ FIGURE 20-22
An electrolytic cell
The direction of electron flow is the reverse of that in the voltaic cell of Figure 20-4, and so is the cell reaction. Now the zinc electrode is the *cathode* and the copper electrode, the *anode.* The battery must have a voltage in excess of 1.103 V in order to force electrons to flow in the reverse (nonspontaneous) direction.

20-2 ARE YOU WONDERING...

Why the anode is (+) in an electrolytic cell but (−) in a voltaic cell?

Assigning the terms *anode* and *cathode* is not based on the electrode charges; it is based on the half-reactions at the electrode surfaces. Specifically,

- *Oxidation* always occurs at the *anode* of an electrochemical cell. Because of the buildup of electrons freed in the oxidation half-reaction, the anode of a *voltaic* cell is (−). Because electrons are withdrawn from it, the anode in an *electrolytic* cell is (+). For both cell types, the anode is the electrode from which electrons *exit* the cell.

- *Reduction* always occurs at the cathode of an electrochemical cell. Because of the *removal* of electrons by the reduction half-reaction, the cathode of a *voltaic* cell is (+). Because of the electrons forced onto it, the cathode of an *electrolytic* cell is (−). For both cell types, the cathode is the electrode at which electrons *enter* the cell.

The following table summarizes the relationship between a voltaic cell and an electrolytic cell.

	Voltaic Cell			Electrolytic Cell	
Oxidation:	$A \longrightarrow A^+ + e^-$	Anode (negative)	*Oxidation:*	$B \longrightarrow B^+ + e^-$	Anode (positive)
Reduction:	$B^+ + e^- \longrightarrow B$	Cathode (positive)	*Reduction:*	$A^+ + e^- \longrightarrow A$	Cathode (negative)
Overall:	$A + B^+ \longrightarrow A^+ + B$ $\Delta G < 0$ Spontaneous redox reaction releases energy		*Overall:*	$A^+ + B \longrightarrow A + B^+$ $\Delta G > 0$ Nonspontaneous redox reaction absorbs energy to drive it	
	The system (the cell) does work on the surroundings			The surroundings (the source of energy) do work on the system	

Note that the sign of each electrode in an electrolytic cell is the same as the sign of the battery electrode to which it is attached.

We can make similar calculations for other electrolyses. What actually happens, however, does not always correspond to these calculations. Four complicating factors must be considered:

1. A voltage significantly in excess of the calculated value, an **overpotential**, may be necessary to cause a particular electrode reaction to occur. Overpotentials are needed to overcome interactions at the electrode surface and are particularly common when gases are involved. For example, the overpotential for the discharge of $H_2(g)$ at a mercury cathode is approximately 1.5 V; the overpotential on a platinum cathode is practically zero.

2. Competing electrode reactions may occur. In the electrolysis of *molten* sodium chloride with inert electrodes, only one oxidation and one reduction are possible.

$$\text{Reduction:} \quad 2\,Na^+ + 2\,e^- \longrightarrow 2\,Na(l)$$
$$\text{Oxidation:} \quad 2\,Cl^- \longrightarrow Cl_2(g) + 2\,e^-$$

In the electrolysis of *aqueous* sodium chloride with inert electrodes, there are *two* possible reduction half-reactions and *two* possible oxidation half-reactions.

$$\text{Reduction:}\quad 2\,Na^+(aq) + 2\,e^- \longrightarrow Na(s) \qquad\qquad E^\circ_{Na^+/Na} = -2.71\ V \qquad \textbf{(20.29)}$$

$$2\,H_2O(l) + 2\,e^- \longrightarrow H_2(g) + 2\,OH^-(aq) \qquad E^\circ_{H_2O/H_2} = (-0.83\ V) \qquad \textbf{(20.30)}$$

$$\text{Oxidation:}\quad 2\,Cl^-(aq) \longrightarrow Cl_2(g) + 2\,e^- \qquad\qquad -E^\circ_{Cl_2/Cl^-} = -(1.36\ V) \qquad \textbf{(20.31)}$$

$$2\,H_2O(l) \longrightarrow O_2(g) + 4\,H^+(aq) + 4\,e^- \qquad -E^\circ_{O_2/H_2O} = -(1.23\ V) \qquad \textbf{(20.32)}$$

▶ We have written a *minus* sign in front of the electrode potentials in (20.31) and (20.32) as a way of emphasizing the *oxidation* rather than the reduction tendency.

Reaction (20.29) can be eliminated as a possible reduction half-reaction: Unless the overpotential for $H_2(g)$ is unusually high, the reduction of Na^+ is far more difficult to accomplish than that of H_2O. This leaves two possibilities for the cell reaction.

Half-reaction (20.30) + half-reaction (20.31):

Reduction:	$2\,H_2O(l) + 2\,e^- \longrightarrow H_2(g) + 2\,OH^-(aq)$
Oxidation:	$2\,Cl^-(aq) \longrightarrow Cl_2(g) + 2\,e^-$
Overall:	$2\,Cl^-(aq) + 2\,H_2O(l) \longrightarrow Cl_2(g) + H_2(g) + 2\,OH^-(aq)$ **(20.33)**

$$E^\circ_{cell} = E^\circ_{H_2O/H_2} - E^\circ_{Cl_2/Cl^-} = -0.83\ V - (1.36\ V) = -2.19\ V$$

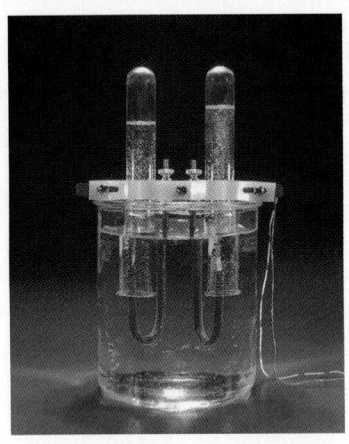

▲ The electrolysis of water into $H_2(g)$ and $O_2(g)$, shown by bubbles at the electrodes and the gases collecting in the test tubes—twice the volume of $H_2(g)$ as $O_2(g)$.

Half-reaction (20.30) + half-reaction (20.32):

Reduction:	$2\,\{2\,H_2O(l) + 2\,e^- \longrightarrow H_2(g) + 2\,OH^-(aq)\}$
Oxidation:	$2\,H_2O(l) \longrightarrow O_2(g) + 4\,H^+(aq) + 4\,e^-$
Overall:	$2\,H_2O(l) \longrightarrow 2\,H_2(g) + O_2(g)$ **(20.34)**

$$E^\circ_{cell} = E^\circ_{H_2O/H_2} - E^\circ_{O_2/H_2O} = -0.83\ V - (1.23\ V) = -2.06\ V$$

In the electrolysis of NaCl(aq), $H_2(g)$ is the product expected at the *cathode*. Because cell reactions (20.33) and (20.34) have E°_{cell} values that are so similar, a mixture of $Cl_2(g)$ and $O_2(g)$ would be the expected product at the *anode*. Actually, because of the high overpotential of $O_2(g)$ compared to $Cl_2(g)$, cell reaction (20.33) predominates; $Cl_2(g)$ is essentially the only product at the anode.

3. The reactants very often are in *nonstandard* states. In the industrial electrolysis of NaCl(aq), $[Cl^-] \approx 5.5\ M$, not the unit activity ($[Cl^-] \approx 1\ M$) implied in half-reaction (20.31); therefore $E_{Cl_2/Cl^-} = 1.31\ V$ (*not* 1.36 V). Also, the pH in the anode half-cell is adjusted to 4, not the unit activity ($[H_3O^+] \approx 1\ M$) implied in half-reaction (20.32); hence $E_{O_2/H_2O} = 0.99\ V$ (*not* 1.23 V). The net effect of these nonstandard conditions is to favor the production of O_2 at the anode. In practice, however, the $Cl_2(g)$ obtained contains less than 1% $O_2(g)$, indicating the overpowering effect of the high overpotential of $O_2(g)$. Not surprisingly, the proportion of $O_2(g)$ increases significantly in the electrolysis of very dilute NaCl(aq).

4. The nature of the electrodes matters. An *inert electrode*, such as platinum, provides a surface on which an electrolysis half-reaction occurs, but the reactants themselves must come from the electrolyte solution. An *active electrode* is one that can itself participate in the oxidation or reduction half-reaction. The distinction between inert and active electrodes is explored in Figure 20-23 and Example 20-12.

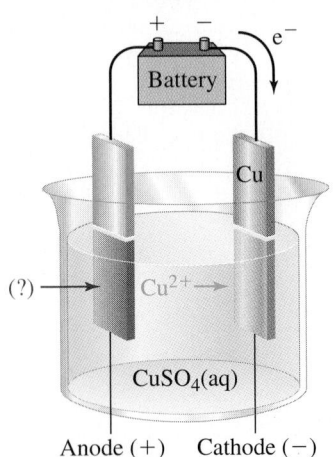

▲ FIGURE 20-23
Predicting electrode reactions in electrolysis— Example 20-11 illustrated
Electrons are forced onto the copper cathode by the external source (battery). Cu^{2+} ions are attracted to the cathode and are reduced to Cu(s). The oxidation half-reaction depends on the metal used for the anode.

Quantitative Aspects of Electrolysis

We have seen how to calculate the theoretical voltage required for electrolysis. Equally important are calculations of the quantities of reactants consumed and products formed in an electrolysis. For these calculations, we will continue to use stoichiometric factors from the chemical equation, but another

EXAMPLE 20-12   Predicting Electrode Half-Reactions and Overall Reactions in Electrolysis

Refer to Figure 20-23. Predict the electrode reactions and the overall reaction when the anode is made of **(a)** copper and **(b)** platinum.

Analyze

In both cases we have to decide on the likely oxidation and reduction processes. The low reduction potential of $Cu^{2+}(aq)$ makes this the likely reduction process in both cases. What about oxidation processes? The possibilities are in **(a)** oxidation of the copper electrode (anode) ($E° = 0.340$ V), oxidation of sulfate anion (2.01 V), and oxidation of water (1.23 V). Thus, the most easily oxidized is the copper at the anode. In **(b)**, the platinum electrode is inert and is not easily oxidized. Of the other two candidates, sulfate anion and water, water has the lower oxidation potential.

Solve

$$Reduction: \quad Cu^{2+}(aq) + 2\,e^- \longrightarrow Cu(s) \quad E°_{Cu^{2+}/Cu} = 0.340 \text{ V}$$

(a) At the cathode we have the reduction of $Cu^{2+}(aq)$. At the anode, Cu(s) can be oxidized to $Cu^{2+}(aq)$, as represented by

$$Oxidation: \quad Cu(s) \longrightarrow Cu^{2+}(aq) + 2\,e^-$$

If the oxidation and reduction half-equations are added, $Cu^{2+}(aq)$ cancels out. The electrolysis reaction is simply

$$Cu(s)[\text{anode}] \longrightarrow Cu(s)[\text{cathode}] \tag{20.35}$$

$$E°_{cell} = E°_{Cu^{2+}/Cu} - E°_{Cu^{2+}/Cu} = 0.340 \text{ V} - 0.340 \text{ V} = 0$$

(b) The oxidation that occurs most readily is that of H_2O, shown in reaction (20.32).

$$Oxidation: \quad 2\,H_2O(l) \longrightarrow O_2(g) + 4\,H^+(aq) + 4\,e^-$$

$$-E°_{O_2/H_2O} = -1.23 \text{ V}$$

The electrolysis reaction and its $E°_{cell}$ are

$$2\,Cu^{2+}(aq) + 2\,H_2O(l) \longrightarrow 2\,Cu(s) + 4\,H^+(aq) + O_2(g) \tag{20.36}$$

$$E°_{cell} = E°(\text{reduction half-cell}) - E°(\text{oxidation half-cell})$$

$$= E°_{Cu^{2+}/Cu} - E°_{O_2/H_2O}$$

$$= 0.340 \text{ V} - 1.23 \text{ V} = -0.89 \text{ V}$$

Assess

(a) Only a very small voltage is needed to overcome the resistance in the electric circuit for this electrolysis. For every Cu atom that enters the solution at the anode, an active electrode, one Cu^{2+} ion, deposits as a Cu atom at the cathode. Copper is transferred from the anode to the cathode through the solution as Cu^{2+} and the concentration of $CuSO_4(aq)$ remains unchanged.

(b) A potential greater than 0.89 V is required to electrolyze water and deposit copper. Keep in mind that when calculating $E°_{cell}$ as a *difference* between two $E°$ values, the $E°$ values are reduction potentials. Because $-E°$ corresponds to the half-cell potential for the oxidation process, the *difference* between two reduction potentials is equivalent to the *sum* of a reduction potential and an oxidation potential.

PRACTICE EXAMPLE A: Use data from Table 20.1 to predict the probable products when Pt electrodes are used in the electrolysis of KI(aq).

PRACTICE EXAMPLE B: In the electrolysis of $AgNO_3(aq)$, what are the expected electrolysis products if the anode is silver metal and the cathode is platinum?

factor enters in as well: the quantity of electric charge associated with one mole of electrons. This factor is provided by the Faraday constant, which we can write as

$$1 \text{ mol } e^- = 96{,}485 \text{ C}$$

Generally, electric charge is not measured directly; instead, it is the electric current that is measured. One *ampere* (A) of electric current represents the passage of 1 coulomb of charge per second (C/s). The product of current and time yields the total quantity of charge transferred.

$$\text{charge (C)} = \text{current (C/s)} \times \text{time (s)}$$

To determine the number of moles of electrons involved in an electrolysis reaction, we can write

$$\text{number of mol e}^- = \text{current}\left(\frac{\text{C}}{\text{s}}\right) \times \text{time (s)} \times \frac{1 \text{ mol e}^-}{96{,}485 \text{ C}}$$

As illustrated in Example 20-13, to determine the mass of a product in an electrolysis reaction, follow this conversion pathway.

$$\text{C/s} \longrightarrow \text{C} \longrightarrow \text{mol e}^- \longrightarrow \text{mol product} \longrightarrow \text{g product}$$

EXAMPLE 20-13   Calculating Quantities Associated with Electrolysis Reactions

The electrodeposition of copper can be used to determine the copper content of a sample. The sample is dissolved to produce $Cu^{2+}(aq)$, which is electrolyzed. At the cathode, the reduction half-reaction is $Cu^{2+}(aq) + 2 e^- \longrightarrow Cu(s)$. What mass of copper can be deposited in 1.00 hour by a current of 1.62 A?

Analyze

To find the mass of copper deposited, we first need to determine the number of moles of electrons generated in the given time. Because we know that for each copper(II) ion we need two electrons, we can calculate the mass by using the number of moles of electrons.

Solve

First, we determine the number of moles of electrons involved in the electrolysis in the manner outlined above:

$$1.00 \text{ h} \times \frac{60 \text{ min}}{1 \text{ h}} \times \frac{60 \text{ s}}{1 \text{ min}} \times \frac{1.62 \text{ C}}{1 \text{ s}} \times \frac{1 \text{ mol e}^-}{96{,}485 \text{ C}} = 0.0604 \text{ mol e}^-$$

The mass of Cu(s) produced at the cathode by this number of moles of electrons is calculated as follows:

$$\text{mass of Cu} = 0.0604 \text{ mol e}^- \times \frac{1 \text{ mol Cu}}{2 \text{ mol e}^-} \times \frac{63.5 \text{ g Cu}}{1 \text{ mol Cu}} = 1.92 \text{ g Cu}$$

Assess

The key factor in this calculation, relating moles of copper to moles of electrons, is printed in blue. This type of conversion is very similar to the one you learned when doing stoichiometric problems.

PRACTICE EXAMPLE A: If 12.3 g of Cu is deposited at the cathode of an electrolytic cell after 5.50 h, what was the current used?

PRACTICE EXAMPLE B: For how long would the electrolysis in Example 20-13 have to be carried out, using Pt electrodes and a current of 2.13 A, to produce 2.62 L $O_2(g)$ at 26.2 °C and 738 mmHg pressure at the anode?

20-8   Industrial Electrolysis Processes

Modern industry could not function in its present form without electrolysis reactions. A number of elements are produced almost exclusively by electrolysis—for example, aluminum, magnesium, chlorine, and fluorine. Among chemical compounds produced industrially by electrolysis are NaOH, $K_2Cr_2O_7$, $KMnO_4$, $Na_2S_2O_8$, and a number of organic compounds.

◀ The refining of copper by electrolysis.

Electrorefining

The *electrorefining* of metals involves the deposition of pure metal at a cathode, from a solution containing the metal ion. Copper produced by the smelting of copper ores is of sufficient purity for some uses, such as plumbing, but it is not pure enough for applications in which high electrical conductivity is required. For these applications, the copper must be more than 99.5% pure. The electrolysis reaction (20.35) on page 899 is used to obtain such high-purity copper. A chunk of impure copper is the anode and a thin sheet of pure copper is the cathode. During the electrolysis, Cu^{2+} produced at the anode migrates through an aqueous sulfuric acid–copper(II) sulfate solution to the cathode, where it is reduced to Cu(s). The pure copper cathode increases in size as the impure chunk of copper is consumed. As noted in Example 20-12a, the electrolysis is carried out at a low voltage—from 0.15 to 0.30 V. Under these conditions, Ag, Au, and Pt impurities are not oxidized at the anode, and they drop to the bottom of the tank as a sludge called *anode mud*. Sn, Bi, and Sb are oxidized, but they precipitate as oxides or hydroxides; Pb is oxidized but precipitates as $PbSO_4(s)$. As, Fe, Ni, Co, and Zn are oxidized but form water-soluble species. Recovery of Ag, Au, and Pt from the anode mud helps offset the cost of the electrolysis.

Electroplating

In *electroplating,* one metal is plated onto another, often less expensive, metal by electrolysis. This procedure is done for decorative purposes or to protect the underlying metal from corrosion. Silver-plated flatware, for example, consists of a thin coating of metallic silver on an underlying base of iron. In electroplating, the item to be plated is the cathode in an electrolytic cell. The electrolyte contains ions of the metal to be plated, which are attracted to the cathode, where they are reduced to metal atoms.

In copper plating, the electrolyte is usually copper sulfate. In silver plating, it is commonly $K[Ag(CN)_2](aq)$. The concentration of free silver ion in a solution of the complex ion $[Ag(CN)_2]^-(aq)$ is very low, and electroplating under these conditions promotes a strongly adherent microcrystalline deposit of the metal. Chromium plating is useful for its resistance to corrosion as well as its appearance. Steel can be chromium-plated from an aqueous solution of CrO_3 and H_2SO_4. The plating obtained, however, is thin and porous and tends to develop cracks. In practice, the steel is first plated with a thin coat of copper or nickel, and then the chromium plating is applied. Chromium plating or cadmium plating is used to weatherproof machine parts. Metal plating can even be applied to some plastics. The plastic must first be made electrically conductive—for example, by coating it with graphite powder. Copper plating of plastics has been used to improve the quality of some microelectronic circuit boards. Electroplating is even used, quite literally, to make money. The U.S. penny is no longer copper throughout. A zinc plug is electroplated with a thin coat of copper, and the copper-plated plug is stamped to create a penny.

▲ A rack of metal parts being lifted from the electrolyte solution after electroplating.

Electrosynthesis

Electrosynthesis is a method of producing substances through electrolysis reactions. It is useful for certain syntheses in which reaction conditions must be carefully controlled. Manganese dioxide occurs naturally as the mineral *pyrolusite*, but small crystal size and lattice imperfections make this material inadequate for certain modern applications, such as alkaline batteries. The electrosynthesis of MnO_2 is carried out in a solution of $MnSO_4$ in $H_2SO_4(aq)$ Pure $MnO_2(s)$ is formed by the oxidation of Mn^{2+} at an inert anode, such as graphite.

$$\text{Oxidation:} \quad Mn^{2+}(aq) + 2\,H_2O(l) \longrightarrow MnO_2(s) + 4\,H^+(aq) + 2\,e^-$$

The reaction at the cathode is the reduction of H^+ to $H_2(g)$, and the overall electrolysis reaction is

$$Mn^{2+}(aq) + 2\,H_2O(l) \longrightarrow MnO_2(s) + 2\,H^+(aq) + H_2(g)$$

An example of electrosynthesis in organic chemistry is the reduction of acrylonitrile, $CH_2{=}CH{-}C{\equiv}N$, to adiponitrile, $N{\equiv}C(CH_2)_4C{\equiv}N$, at a lead cathode (chosen because of the high overpotential of H_2 on lead). Oxygen is released at the anode.

$$\text{Reduction:} \quad 2\,CH_2{=}CH{-}C{\equiv}N + 2\,H_2O + 2\,e^- \longrightarrow$$
$$N{\equiv}C(CH_2)_4C{\equiv}N + 2\,OH^-$$

The commercial importance of this electrolysis is that adiponitrile can be readily converted to two other compounds: hexamethylenediamine, $H_2NCH_2(CH_2)_4CH_2NH_2$, and adipic acid, $HOOCCH_2(CH_2)_2CH_2COOH$. These two compounds are the monomers used to make the polymer *Nylon-66* (page 1253).

The Chlor–Alkali Process

On page 898 we described the electrolysis of NaCl(aq) through the reduction half-reaction (20.30) and the oxidation half-reaction (20.31).

$$2\,Cl^-(aq) + 2\,H_2O(l) \longrightarrow 2\,OH^-(aq) + H_2(g) + Cl_2(g) \qquad E^\circ_{cell} = -2.19\text{ V}$$

When conducted on an industrial scale, this electrolysis is called the *chlor–alkali process*, named after the two principal products: *chlorine* and the *alkali* NaOH(aq). The chlor–alkali process is one of the most important of all electrolytic processes because of the high value of these products.

In the *diaphragm cell* depicted in Figure 20-24, $Cl_2(g)$ is produced in the anode compartment, and $H_2(g)$ and NaOH(aq) in the cathode compartment. If $Cl_2(g)$ comes in contact with NaOH(aq), the Cl_2 disproportionates into $ClO^-(aq)$, $ClO_3^-(aq)$, and $Cl^-(aq)$. The purpose of the diaphragm is to prevent this contact. The NaCl(aq) in the anode compartment is kept at a slightly higher level than that in the cathode compartment. This disparity creates a gradual flow of NaCl(aq) between the compartments and reduces the backflow of NaOH(aq) into the anode compartment. The solution in the cathode compartment, about 10%–12% NaOH(aq) and 14%–16% NaCl(aq), is concentrated and purified by evaporating some water and crystallizing the NaCl(s). The final product is 50% NaOH(aq), with up to 1% NaCl(aq).

The theoretical voltage required for this electrolysis is 2.19 V. However, as a result of the internal resistance of the cell and overpotentials at the electrodes, a voltage of about 3.5 V is used. The current is kept very high, typically about 1×10^5 A.

The NaOH(aq) produced in a diaphragm cell is not pure enough for certain uses, such as rayon manufacture. A higher purity is achieved if electrolysis is carried out in a mercury cell, illustrated in Figure 20-25. This cell takes advantage of the high overpotential for the reduction of $H_2O(l)$ to $H_2(g)$ and $OH^-(aq)$ at a

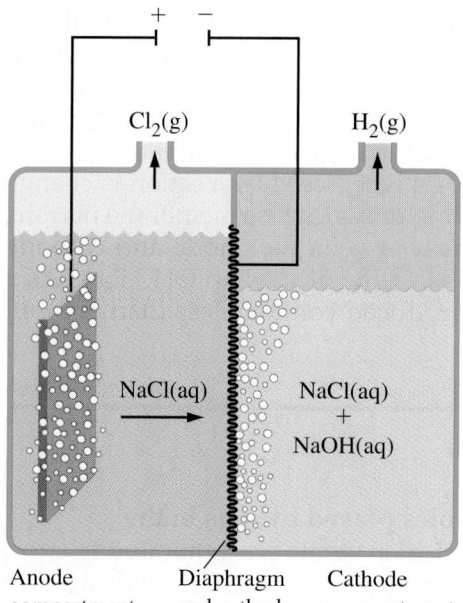

◀ FIGURE 20-24
A diaphragm chlor–alkali cell
The anode may be made of graphite or, in more modern technology, specially treated titanium metal. The diaphragm and cathode are generally fabricated as a composite unit consisting of asbestos or an asbestos–polymer mixture deposited on a steel wire mesh. To avoid the use of asbestos, a more modern development substitutes a fluorocarbon mesh for the asbestos.

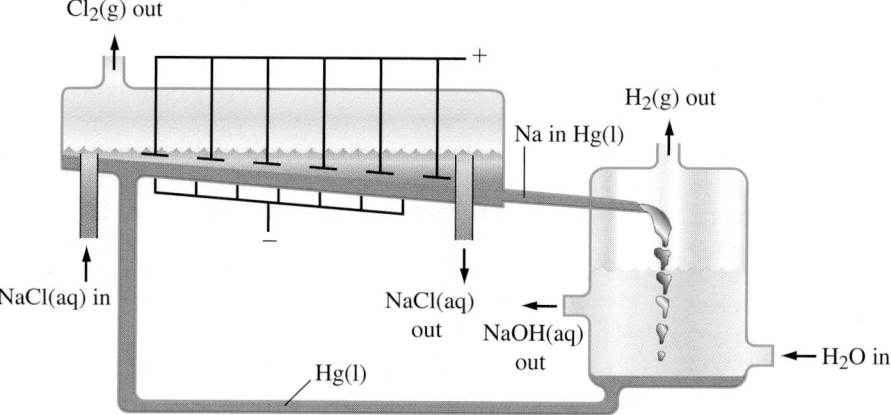

▲ FIGURE 20-25
The mercury-cell chlor–alkali process
The cathode is a layer of Hg(l) that flows along the bottom of the tank. Anodes, at which $Cl_2(g)$ forms, are situated in the NaCl(aq) just above the Hg(l). Sodium formed at the cathode dissolves in the Hg(l), and the sodium amalgam is decomposed with water to produce NaOH(aq) and $H_2(g)$. The regenerated Hg(l) is recycled.

mercury cathode. The reduction that occurs instead is that of $Na^+(aq)$ to Na, which dissolves in Hg(l) to form an amalgam (Na–Hg alloy) with about 0.5% Na by mass.

$$2\,Na^+(aq) + 2\,Cl^-(aq) \longrightarrow 2\,Na(in\ Hg) + Cl_2(g) \qquad E^\circ_{cell} = -3.20\ V$$

When the Na amalgam is removed from the cell and treated with water, NaOH(aq) forms,

$$2\,Na(in\ Hg) + 2\,H_2O(l) \longrightarrow 2\,Na^+(aq) + 2\,OH^-(aq) + H_2(g) + Hg(l)$$

and the liquid mercury is recycled back to the electrolytic cell.

Although the mercury cell has the advantage of producing concentrated high-purity NaOH(aq), it has some disadvantages. The mercury cell requires a higher voltage (about 4.5 V) than does the diaphragm cell (3.5 V) and consumes more electrical energy, about 3400 kWh/ton Cl_2 in a mercury cell, compared with 2500 kWh/ton Cl_2 in a diaphragm cell. Another serious drawback is the need to control mercury effluents to the environment. Mercury losses, which at one time were as high as 200 g Hg per ton Cl_2, have been reduced to about 0.25 g Hg per ton of Cl_2 in older plants and half this amount in new plants.

The ideal chlor–alkali process is one that is energy-efficient and does not use mercury. A type of cell offering these advantages is the *membrane cell,* in which the porous diaphragm of Figure 20-24 is replaced by a cation-exchange membrane, normally made of a fluorocarbon polymer. The membrane permits hydrated cations (Na^+ and H_3O^+) to pass between the anode and cathode compartments but severely restricts the backflow of Cl^- and OH^- ions. As a result, the sodium hydroxide solution produced contains less than 50 ppm chloride ion contaminant.

Mastering**CHEMISTRY**   www.masteringchemistry.com

The concepts presented in this chapter can be used to explain the roles played by ions in the generation of biological electric currents. Electric currents in biological systems are generated in muscle contraction and neuron activity, for example. For a discussion of the source of biological electric currents, go to the Focus On feature for Chapter 20, Membrane Potentials, on the MasteringChemistry site.

Summary

20-1 Electrode Potentials and Their Measurement—In an **electrochemical cell** electrons in an oxidation–reduction reaction are transferred at metal strips called **electrodes** and conducted through an external circuit. The oxidation and reduction half-reactions occur in separate regions called **half-cells.** In a half-cell, an electrode is immersed in a solution. The electrodes of the two half-cells are joined by a wire, and an electrical connection between the solutions is also made, as through a **salt bridge** (Fig. 20-3). The cell reaction involves oxidation at one electrode called the **anode** and reduction at the other electrode called the **cathode.** A **voltaic (galvanic) cell** produces electricity from a spontaneous oxidation–reduction reaction. The difference in electric potential between the two electrodes is the **cell voltage**; the unit of cell voltage is the **volt (V).** The cell voltage is also called the **cell potential** or **electromotive force (emf)** and designated as E_{cell}. A **cell diagram** displays the components of a cell in a symbolic way (expression 20-4).

20-2 Standard Electrode Potentials—The reduction occurring at a **standard hydrogen electrode (SHE)**, $2 H^+(a = 1) + 2 e^- \xrightarrow{\text{on Pt}} H_2(g, 1\,bar)$, is arbitrarily assigned a potential of zero. A half-cell has a **standard electrode potential**, $E°$, in which all reactants and products are at unit activity. A half-cell reaction with a *positive* standard electrode potential ($E°$) occurs more readily than does reduction of H^+ ions at the SHE. A *negative* standard electrode potential signifies a lesser tendency to undergo reduction. The **standard cell potential** ($E°_{cell}$) of a voltaic cell is the *difference* between $E°$, of the cathode and $E°$, of the anode; that is, $E°_{cell} = E°(\text{cathode}) - E°(\text{anode})$.

20-3 E_{cell}, ΔG, and K—Cell voltages based on standard electrode potentials are $E°_{cell}$ values. The electrical work that can be obtained from a cell depends on the

number of electrons involved in the cell reaction, the cell potential, and the **Faraday constant (*F*)**, which is the number of coulombs of charge per mole of electrons—96,485 C/mol e^-. An important relationship exists between $E°_{cell}$ and $\Delta G°$, namely, $\Delta G° = -zFE°_{cell}$. The equilibrium constant of the cell reaction K is related to $E°_{cell}$ through the expression $\Delta G° = -RT \ln K = -zFE°_{cell}$.

20-4 E_{cell} as a Function of Concentrations—In the **Nernst equation,** E_{cell} for nonstandard conditions is related to $E°_{cell}$ and the reaction quotient Q (equation 20.18). If $E_{cell} > 0$, the cell reaction is spontaneous in the forward direction for the stated conditions; if $E_{cell} < 0$, the forward reaction is *nonspontaneous* for those conditions. A **concentration cell** consists of two half-cells with identical electrodes but different solution concentrations.

20-5 Batteries: Producing Electricity Through Chemical Reactions—An important application of voltaic cells is found in various battery systems. A **battery** stores chemical energy so that it can be released as energy. Batteries consist of one or more voltaic cells and are divided into three major classes: **primary** (Leclanché), **secondary** (lead–acid, silver–zinc nicad, and lithium-ion), and **flow batteries** or **fuel cells** in which reactants, such as hydrogen and oxygen, are continuously fed into the battery and chemical energy is converted to electric energy.

20-6 Corrosion: Unwanted Voltaic Cells—Electrochemistry plays a key role in corrosion and its control. Oxidation half-reactions produce anodic regions and reduction half-reactions, cathodic regions. **Cathodic protection** is achieved when a more active metal is attached to the metal being protected from corrosion. The more active metal, a "sacrificial" anode, is preferentially oxidized while the protected metal is a cathode at which a harmless reduction half-reaction occurs.

20-7 Electrolysis: Causing Nonspontaneous Reactions to Occur—In **electrolysis**, a nonspontaneous chemical reaction occurs as electrons from an external source are forced to flow in a direction opposite that in which they would flow spontaneously. The electrochemical cell in which electrolysis is conducted is called an **electrolytic cell**. $E°$, values are used to establish the theoretical voltage requirements for an electrolysis. Sometimes, particularly when a gas is liberated at an electrode, the voltage requirement for the electrode reaction exceeds the value of $E°$. The additional voltage requirement is called the **overpotential**. The amounts of reactants and products involved in an electrolysis can be calculated from the amount of electric charge passing through the electrolytic cell. The Faraday constant is featured in these calculations.

20-8 Industrial Electrolysis Processes—Electrolysis has many industrial applications, including electroplating, refining of metals, and production of substances such as $NaOH(aq)$, $H_2(g)$, and $Cl_2(g)$.

Integrative Example

Two electrochemical cells are connected as shown.

Cell A

$$Zn(s)|Zn^{2+}(0.85\ M)||Cu^{2+}(1.10\ M)|Cu(s)$$

$$Zn(s)|Zn^{2+}(1.05\ M)||Cu^{2+}(0.75\ M)|Cu(s)$$

Cell B

(a) Do electrons flow in the direction of the red arrows or the blue arrows?

(b) What are the ion concentrations in the half-cells at the point at which current ceases to flow?

Analyze

The two cells differ only in their ion concentrations, which means that they have the same $E°_{cell}$ value but different E_{cell} values. In part **(a)**, use the Nernst equation (20.18) to determine which cell has the greater E_{cell} value when functioning as a voltaic cell. This will establish the direction that electrons flow. In part **(b)**, write and solve an equation relating ion concentrations to the condition where the two cells have the same voltage but, being connected in opposition to each other, produce no net electric current.

Solve

(a) In each voltaic cell zinc is the anode, copper is the cathode, and the cell reaction is

$$Zn(s) + Cu^{2+}(aq) \longrightarrow Zn^{2+}(aq) + Cu(s)$$

The E_{cell} values are given by the Nernst equation.

$$E_{cell} = E°_{cell} - (0.0592/2)\log[Zn^{2+}]/[Cu^2] \qquad \textbf{(20.37)}$$

Note that for Cell A,

$$[Zn^{2+}]/[Cu^{2+}] = 0.85\ M/1.10\ M < 1$$
$$\log[Zn^{2+}]/[Cu^{2+}] < 0, \quad \text{and} \quad E_{cell} > E°_{cell}$$

For Cell B,

$$[Zn^{2+}]/[Cu^{2+}] = 1.05\ M/0.75\ M > 1$$
$$\log[Zn^{2+}]/[Cu^{2+}] > 0, \text{ and } E_{cell} < E°_{cell}$$

The voltage of Cell A is greater than that of Cell B.

In the connection of the two cells shown in the diagram, Cell A is a *voltaic cell* and Cell B is an *electrolytic cell*. There is an emf from Cell B that resists that of Cell A, but on balance, the electron flow is in the direction of the red arrows.

(b) As electrons flow between the two cells, the overall reaction in Cell A causes $[Zn^{2+}]$ to increase, $[Cu^{2+}]$ to decrease, and E_{cell} to decrease. The overall reaction in Cell B causes $[Zn^{2+}]$ to decrease, $[Cu^{2+}]$ to increase, and the back emf to increase. When the back emf from Cell B equals E_{cell} of Cell A, electrons cease to flow.

The cell diagrams when this condition is reached, with x representing *changes* in concentrations, are

Cell A: $Zn(s)|Zn^{2+}(0.85 + x)\ M||Cu^{2+}(1.10 - x)\ M|Cu(s)$

Cell B: $Zn(s)|Zn^{2+}(1.05 - x)\ M||Cu^{2+}(0.75 + x)\ M|Cu(s)$

Use equation (20.37) to obtain E_{cell} for each cell.

Cell A: $\quad E_{cell} = E^\circ_{cell} - \dfrac{0.0592\ \text{V}}{2}\log\dfrac{(0.85 + x)}{(1.10 - x)}$

Cell B: $\quad E_{cell} = E^\circ_{cell} - \dfrac{0.0592\ \text{V}}{2}\log\dfrac{(1.05 - x)}{(0.75 + x)}$

Set the two expressions equal to one another, cancel the terms E°_{cell} and $(0.0592\ \text{V})/2$, to obtain

$$\log\dfrac{(0.85 + x)}{(1.10 - x)} = \log\dfrac{(1.05 - x)}{(0.75 + x)}$$

Because the logarithms of the quantities on the two sides are equal, so too are the quantities themselves.

$$\dfrac{(0.85 + x)}{(1.10 - x)} = \dfrac{(1.05 - x)}{(0.75 + x)}$$

The expression to be solved is a quadratic equation

$$(0.85 + x)(0.75 + x) = (1.10 - x)(1.05 - x)$$

Cancel x^2 on each side

$$0.64 + 1.60x + x^2 = 1.16 - 2.15x + x^2$$

$$0.64 + 1.60x = 1.16 - 2.15x$$

Solve to obtain

$$3.75x = 0.52 \quad \text{and} \quad x = 0.14$$

When electrons no longer flow, the ion concentrations are as follows:

Cell A: $\quad [\text{Zn}^{2+}] = 0.99\ \text{M}; [\text{Cu}^{2+}] = 0.96\ \text{M}.$

Cell B: $\quad [\text{Zn}^{2+}] = 0.91\ \text{M}; [\text{Cu}^{2+}] = 0.89\ \text{M}.$

Assess

Once the direction of electron flow had been established, it was possible to decide in which cell $[\text{Zn}^{2+}]$ would increase and in which cell it would decrease. At equilibrium the two cell potentials became equal. That the calculated equilibrium concentrations are correct can be seen in $0.99/0.96 \approx 0.91/0.89$.

PRACTICE EXAMPLE A: Current fuel cells use the reaction of $H_2(g)$ and $O_2(g)$ to form $H_2O(l)$. Often, the $H_2(g)$ is obtained by the steam reforming of a hydrocarbon, such as $C_3H_8(g) + 3\,H_2O(g) \longrightarrow 3\,CO(g) + 7\,H_2(g)$. A future possibility is a fuel cell that converts a hydrocarbon, such as propane, directly to $CO_2(g)$ and $H_2O(l)$:

$$C_3H_8(g) + 5\,O_2(g) \longrightarrow 3\,CO_2(g) + 4\,H_2O(l)$$

Based on this reaction, use data from Table 20.1 and Appendix D to determine E° for the reduction of $CO_2(g)$ to $C_3H_8(g)$ in an acidic solution.

PRACTICE EXAMPLE B: A battery system that may be used to power automobiles in the future is the aluminum–air battery. This is a *flow battery* in which oxidation occurs at an aluminum anode and reduction at a carbon–air cathode. The electrolyte circulated through the battery is $NaOH(aq)$; the ultimate reaction product is $Al(OH)_3(s)$, which is removed from the battery as it is formed. In operation the battery can be kept charged by feeding Al anode slugs and water into it; oxygen is drawn from the air (see Fig. 20-18). The battery can power an automobile several hundred miles between charges. The $Al(OH)_3(s)$ removed from the battery can be converted back to aluminum in an aluminum manufacturing facility.

(a) In actual practice Al^{3+} produced at the anode does not precipitate as $Al(OH)_3(s)$ but is obtained as the complex ion $[Al(OH)_4]^-$ in the presence of $NaOH(aq)$. $Al(OH)_3$ is precipitated from the circulating $NaOH(aq)$ electrolyte *outside* the battery. Write plausible equations for oxidation and reduction half-reactions, and for the net reaction that occurs in the battery.

(b) The theoretical voltage of the aluminum–air cell is $+2.73$ V. Use this information and data from Table 20.1 to obtain E° for the reduction

$$[Al(OH)_4]^- + 3\,e^- \longrightarrow Al(s) + 4\,OH^- \qquad E^\circ = ?$$

(c) Given that E°_{cell} for the reaction is $+2.73$ V, that $\Delta G^\circ_f[OH^-(aq)] = -157\ \text{kJ mol}^{-1}$, and that $\Delta G^\circ_f[H_2O(l)] = -237.2\ \text{kJ mol}^{-1}$, determine the Gibbs energy of formation, ΔG°_f, of the aluminate ion, $[Al(OH)_4]^-$.

(d) What mass of aluminum is consumed if 10.0 A of electric current is drawn from the battery for 4.00 h?

Mastering**CHEMISTRY**

You'll find a link to additional self study questions in the study area on
www.masteringchemistry.com

Exercises

(Use data from Table 20.1 and Appendix D as necessary.)

Standard Electrode Potentials

1. From the observations listed, estimate the value of $E°$ for the half-reaction $M^{2+}(aq) + 2 e^- \longrightarrow M(s)$.
   (a) The metal M reacts with $HNO_3(aq)$, but not with $HCl(aq)$; M displaces $Ag^+(aq)$, but not $Cu^{2+}(aq)$.
   (b) The metal M reacts with $HCl(aq)$, producing $H_2(g)$, but displaces neither $Zn^{2+}(aq)$ nor $Fe^{2+}(aq)$.

2. You must estimate $E°$ for the half-reaction $In^{3+}(aq) + 3 e^- \longrightarrow In(s)$. You have no electrical equipment, but you do have all of the metals listed in Table 20.1 and aqueous solutions of their ions, as well as $In(s)$ and $In^{3+}(aq)$. Describe the experiments you would perform and the accuracy you would expect in your result.

3. $E°_{cell} = 0.201$ V for the reaction

   $$3 Pt(s) + 12 Cl^-(aq) + 2 NO_3^-(aq) + 8 H^+(aq) \longrightarrow$$
   $$3[PtCl_4]^{2-}(aq) + 2 NO(g) + 4 H_2O(l)$$

   What is $E°$ for the reduction of $[PtCl_4]^{2-}$ to Pt in acidic solution?

4. Given that $E°_{cell} = 3.20$ V for the reaction

   $$2 Na(in\ Hg) + Cl_2(g) \longrightarrow 2 Na^+(aq) + 2 Cl^-(aq)$$

   What is $E°$ for the reduction $2 Na^+(aq) + 2 e^- \longrightarrow 2 Na(in\ Hg)$?

5. Given that $E°_{cell}$ for the aluminum-air battery is 2.71 V, what is $E°$ for the reduction half-reaction $[Al(OH)_4]^-(aq) + 3 e^- \longrightarrow Al(s) + 4 OH^-(aq)$? [*Hint:* Refer to cell reaction (20.28).]

6. The theoretical $E°_{cell}$ for the methane–oxygen fuel cell is 1.06 V. What is $E°$ for the reduction half-reaction $CO_2(g) + 8 H^+(aq) + 8 e^- \longrightarrow CH_4(g) + 2 H_2O(l)$? [*Hint:* Refer to cell reaction (20.27).]

7. The following sketch is of a voltaic cell consisting of two standard electrodes for two metals, M and N:

   $$M^{z+}(aq) + z e^- \longrightarrow M(s) \qquad E°_{M^{z+}/M}$$
   $$N^{z+}(aq) + z e^- \longrightarrow N(s) \qquad E°_{N^{z+}/N}$$

   Use the standard reduction potentials of these half-reactions to answer the questions that follow:

   $$Ag^+(aq) + e^- \longrightarrow Ag(s)$$
   $$Zn^{2+}(aq) + 2 e^- \longrightarrow Zn(s)$$

Predicting Oxidation–Reduction Reactions

9. Ni^{2+} has a more positive reduction potential than Cd^{2+}.
   (a) Which ion is more easily reduced to the metal?
   (b) Which metal, Ni or Cd, is more easily oxidized?

10. Refer to standard reduction potentials, and predict which metal in each of the following pairs is the stronger reducing agent:
    (a) sodium or potassium
    (b) magnesium or barium

11. Assume that all reactants and products are in their standard states, and use data from Table 20.1 to predict whether a spontaneous reaction will occur in the forward direction in each case.
    (a) $Sn(s) + Pb^{2+}(aq) \longrightarrow Sn^{2+}(aq) + Pb(s)$

$$Cu^{2+}(aq) + 2 e^- \longrightarrow Cu(s)$$
$$Al^{3+}(aq) + 3 e^- \longrightarrow Al(s)$$

(a) Determine which pair of these half-cell reactions leads to a cell reaction with the largest positive cell potential, and calculate its value. Which couple is at the anode and which at the cathode?
(b) Determine which pair of these half-cell reactions leads to the cell with the smallest positive cell potential, and calculate its value. Which couple is at the anode and which is at the cathode?

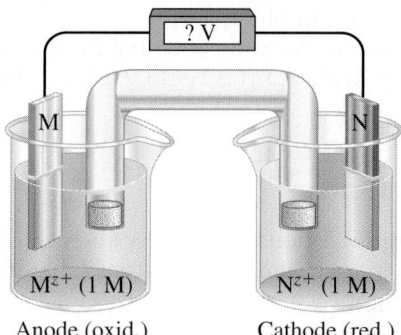

Anode (oxid.)   Cathode (red.)

8. Given these half-reactions and associated standard reduction potentials, answer the questions that follow:

   $$[Zn(NH_3)_4]^{2+}(aq) + 2 e^- \longrightarrow Zn(s) + 4 NH_3(aq)$$
   $$E° = -1.015\ V$$
   $$Ti^{3+}(aq) + e^- \longrightarrow Ti^{2+}(aq)$$
   $$E° = -0.37\ V$$
   $$VO^{2+}(aq) + 2 H^+(aq) + e^- \longrightarrow V^{3+}(aq) + H_2O(l)$$
   $$E° = 0.340\ V$$
   $$Sn^{2+}(aq) + 2 e^- \longrightarrow Sn(aq)$$
   $$E° = -0.14\ V$$

   (a) Determine which pair of half-cell reactions leads to a cell reaction with the largest positive cell potential, and calculate its value. Which couple is at the anode and which is at the cathode?
   (b) Determine which pair of these half-cell reactions leads to the cell with the smallest positive cell potential, and calculate its value. Which couple is at the anode and which is at the cathode?

(b) $Cu^{2+}(aq) + 2 I^-(aq) \longrightarrow Cu(s) + I_2(s)$
(c) $4 NO_3^-(aq) + 4 H^+(aq) \longrightarrow$
$$3 O_2(g) + 4 NO(g) + 2 H_2O(l)$$
(d) $O_3(g) + Cl^-(aq) \longrightarrow OCl^-(aq) + O_2(g)$
(basic solution)

12. For the reduction half-reaction $Hg_2^{2+}(aq) + 2 e^- \longrightarrow 2 Hg(l)$, $E° = 0.797$ V. Will $Hg(l)$ react with and dissolve in $HCl(aq)$? in $HNO_3(aq)$? Explain.

13. Use data from Table 20.1 to predict whether, to any significant extent,
    (a) $Mg(s)$ will displace Pb^{2+} from aqueous solution;
    (b) $Sn(s)$ will react with and dissolve in 1 M HCl;

(c) SO_4^{2-} will oxidize Sn^{2+} to Sn^{4+} in acidic solution;

(d) $MnO_4^-(aq)$ will oxidize $H_2O_2(aq)$ to $O_2(g)$ in acidic solution;

(e) $I_2(s)$ will displace $Br^-(aq)$ to produce $Br_2(l)$.

14. Consider the reaction $Co(s) + Ni^{2+}(aq) \longrightarrow Co^{2+}(aq) + Ni(s)$, with $E^\circ_{cell} = 0.02$ V. If $Co(s)$ is added to a solution with $[Ni^{2+}] = 1$ M, should the reaction go to completion? Explain.

15. Dichromate ion $(Cr_2O_7^{2-})$ in acidic solution is a good oxidizing agent. Which of the following oxidations can be accomplished with dichromate ion in acidic solution? Explain.

(a) $Sn^{2+}(aq)$ to $Sn^{4+}(aq)$

(b) $I_2(s)$ to $IO_3^-(aq)$

(c) $Mn^{2+}(aq)$ to $MnO_4^-(aq)$

16. The standard electrode potential for the reduction of $Eu^{3+}(aq)$ to $Eu^{2+}(aq)$ is -0.43 V. Use the data in Appendix D to determine which of the following is capable of reducing $Eu^{3+}(aq)$ to $Eu^{2+}(aq)$ under standard-state conditions: $Al(s)$, $Co(s)$, $H_2O_2(aq)$, $Ag(s)$, $H_2C_2O_4(aq)$.

17. Predict whether the following metals will react with the acid indicated. If a reaction does occur, write the net ionic equation for the reaction. Assume that reactants and products are in their standard states. (a) Ag in $HNO_3(aq)$; (b) Zn in $HI(aq)$; (c) Au in HNO_3 (for the couple Au^{3+}/Au, $E^\circ = 1.52$ V).

18. Predict whether, to any significant extent,

(a) $Fe(s)$ will displace $Zn^{2+}(aq)$;

(b) $MnO_4^-(aq)$ will oxidize $Cl^-(aq)$ to $Cl_2(g)$ in acidic solution;

(c) $Ag(s)$ will react with 1 M $HCl(aq)$;

(d) $O_2(g)$ will oxidize $Cl^-(aq)$ to $Cl_2(g)$ in acidic solution.

Galvanic Cells

19. Write cell reactions for the electrochemical cells diagrammed here, and use data from Table 20.1 to calculate E°_{cell} for each reaction.

(a) $Al(s)|Al^{3+}(aq)||Sn^{2+}(aq)|Sn(s)$

(b) $Pt(s)|Fe^{2+}(aq), Fe^{3+}(aq)||Ag^+(aq)|Ag(s)$

(c) $Cr(s)|Cr^{2+}(aq)||Au^{3+}(aq)|Au(s)$

(d) $Pt(s)|O_2(g)|H^+(aq)||OH^-(aq)|O_2(g)|Pt(s)$

20. Write the half-reactions and the balanced chemical equation for the electrochemical cells diagrammed here. Use data from Table 20.1 and Appendix D to calculate E°_{cell} for each cell.

(a) $Cu(s)|Cu^{2+}(aq)||Cu^+(aq)|Cu(s)$

(b) $Ag(s)|AgI(s)|I^-(aq)||Cl^-(aq)|AgCl(s)|Ag(s)$

(c) $Pt|Ce^{4+}(aq), Ce^{3+}(aq)||I^-(aq), I_2(s)|C(s)$

(d) $U(s)|U^{3+}(aq)||V^{2+}(aq)|V(s)$

21. Use the data in Appendix D to calculate the standard cell potential for each of the following reactions. Which reactions will occur spontaneously?

(a) $H_2(g) + F_2(g) \longrightarrow 2 H^+(aq) + 2 F^-(aq)$

(b) $Cu(s) + Ba^{2+}(aq) \longrightarrow Cu^{2+}(aq) + Ba(s)$

(c) $3 Fe^{2+}(aq) \longrightarrow Fe(s) + 2 Fe^{3+}(aq)$

(d) $Hg(l) + HgCl_2(aq) \longrightarrow Hg_2Cl_2(s)$

22. In each of the following examples, sketch a voltaic cell that uses the given reaction. Label the anode and cathode; indicate the direction of electron flow; write a balanced equation for the cell reaction; and calculate E°_{cell}.

(a) $Cu(s) + Fe^{3+}(aq) \longrightarrow Cu^{2+}(aq) + Fe^{2+}(aq)$

(b) $Pb^{2+}(aq)$ is displaced from solution by $Al(s)$.

(c) $Cl_2(g) + H_2O(l) \longrightarrow Cl^-(aq) + O_2(g) + H^+(aq)$

(d) $Zn(s) + H^+ + NO_3^- \longrightarrow Zn^{2+} + H_2O(l) + NO(g)$

23. Use the data in Appendix D to calculate the standard cell potential for each of the following reactions. Which reactions will occur spontaneously?

(a) $Fe^{3+}(aq) + Ag(s) \longrightarrow Fe^{2+}(aq) + Ag^+(aq)$

(b) $Sn(s) + Sn^{4+}(aq) \longrightarrow 2 Sn^{2+}(aq)$

(c) $2 Hg^{2+}(aq) + 2 Br^-(aq) \longrightarrow Hg_2^{2+}(aq) + Br_2(l)$

(d) $2 NO_3^-(aq) + 4 H^+(aq) + Zn(s) \longrightarrow Zn^{2+}(aq) + 2 NO_2(g) + 2 H_2O(l)$

24. Write a cell diagram and calculate the value of E°_{cell} for a voltaic cell in which

(a) $Cl_2(g)$ is reduced to $Cl^-(aq)$ and $Fe(s)$ is oxidized to $Fe^{2+}(aq)$;

(b) $Ag^+(aq)$ is displaced from solution by $Zn(s)$;

(c) The cell reaction is $2 Cu^+(aq) \longrightarrow Cu^{2+}(aq) + Cu(s)$;

(d) $MgBr_2(aq)$ is produced from $Mg(s)$ and $Br_2(l)$.

ΔG°, E°_{cell}, and K

25. Determine the values of ΔG° for the following reactions carried out in voltaic cells.

(a) $2 Al(s) + 3 Cu^{2+}(aq) \longrightarrow 2 Al^{3+}(aq) + 3 Cu(s)$

(b) $O_2(g) + 4 I^-(aq) + 4 H^+(aq) \longrightarrow 2 H_2O(l) + 2 I_2(s)$

(c) $Cr_2O_7^{2-}(aq) + 14 H^+(aq) + 6 Ag(s) \longrightarrow 2 Cr^{3+}(aq) + 6 Ag^+(aq) + 7 H_2O(l)$

26. Write the equilibrium constant expression for each of the following reactions, and determine the value of K at 25 °C. Use data from Table 20.1.

(a) $2 V^{3+}(aq) + Ni(s) \longrightarrow 2 V^{2+}(aq) + Ni^{2+}(aq)$

(b) $MnO_2(s) + 4 H^+(aq) + 2 Cl^-(aq) \longrightarrow Mn^{2+}(aq) + 2 H_2O(l) + Cl_2(g)$

(c) $2 OCl^-(aq) \longrightarrow 2 Cl^-(aq) + O_2(g)$

(basic solution)

27. For the reaction

$MnO_4^-(aq) + 8 H^+(aq) + 5 Ce^{3+}(aq) \longrightarrow 5 Ce^{4+}(aq) + Mn^{2+}(aq) + 4 H_2O(l)$

use data from Table 20.1 to determine (a) E_{cell}°; (b) ΔG°; (c) K; (d) whether the reaction goes substantially to completion when the reactants and products are initially in their standard states.

28. For the reaction that occurs in the voltaic cell

$$Pt|Pb^{4+}(aq), Pb^{2+}(aq)||Sn^{4+}(aq), Sn^{2+}(aq)|C(s)$$

use data from Appendix D to determine (a) the equation for the cell reaction; (b) E_{cell}°; (c) ΔG°; (d) K; (e) whether the reaction goes substantially to completion when the reactants and products are initially in their standard states.

29. For the reaction $2\,Cu^+(aq) + Sn^{4+}(aq) \longrightarrow 2\,Cu^{2+}(aq) + Sn^{2+}(aq)$, $E_{cell}^\circ = -0.0050$ V,
    (a) can a solution be prepared at 298 K that is 0.500 M in each of the four ions?
    (b) If not, in which direction will a reaction occur?

30. For the reaction $2\,H^+(aq) + BrO_4^-(aq) + 2\,Ce^{3+}(aq) \longrightarrow BrO_3^-(aq) + 2\,Ce^{4+}(aq) + H_2O(l)$, $E_{cell}^\circ = -0.017$ V, answer the following questions:

(a) Can a solution be prepared at 298 K that has $[BrO_4^-] = [Ce^{4+}] = 0.675$ M, $[BrO_3^-] = [Ce^{3+}] = 0.600$ M and pH $= 1$?
(b) If not, in which direction will a reaction occur?

31. Use thermodynamic data from Appendix D to calculate a theoretical voltage of the silver–zinc button cell described on page 890.

32. The theoretical voltage of the aluminum–air battery is $E_{cell}^\circ = 2.71$ V. Use data from Appendix D and equation (20.28) to determine ΔG_f° for $Al[(OH)_4]^-$.

33. By the method of combining reduction half-equations illustrated on page 875, determine $E_{IrO_2/Ir}^\circ$, given that $E_{Ir^{3+}/Ir}^\circ = 1.156$ V and $E_{IrO_2/Ir^{3+}}^\circ = 0.223$ V.

34. Determine $E_{MoO_2/Mo^{3+}}^\circ$, given that $E_{H_2MoO_4/MoO_2}^\circ = 0.646$ V and $E_{H_2MoO_4/Mo^{3+}}^\circ = 0.428$ V. (See page 875).

Concentration Dependence of E_{cell}—The Nernst Equation

35. A voltaic cell represented by the following cell diagram has $E_{cell} = 1.250$ V. What must be $[Ag^+]$ in the cell?

$$Zn(s)|Zn^{2+}(1.00\ M)||Ag^+(x\ M)|Ag(s)$$

36. For the cell pictured in Figure 20-11, what is E_{cell} if the unknown solution in the half-cell on the left (a) has pH = 5.25; (b) is 0.0103 M HCl; (c) is 0.158 M $HC_2H_3O_2$ ($K_a = 1.8 \times 10^{-5}$)?

37. Use the Nernst equation and Table 20.1 to calculate E_{cell} for each of the following cells.
    (a) $Al(s)|Al^{3+}(0.18\ M)||Fe^{2+}(0.85\ M)|Fe(s)$
    (b) $Ag(s)|Ag^+(0.34\ M)||Cl^-(0.098\ M),$
    $$Cl_2(g, 0.55\ atm)|Pt(s)$$

38. Use the Nernst equation and data from Appendix D to calculate E_{cell} for each of the following cells.
    (a) $Mn(s)|Mn^{2+}(0.40\ M)||Cr^{3+}(0.35\ M),$
    $$Cr^{2+}(0.25\ M)|Pt(s)$$
    (b) $Mg(s)|Mg^{2+}(0.016\ M)||[Al(OH)_4]^-(0.25\ M),$
    $$OH^-(0.042\ M)|Al(s)$$

39. Consider the reduction half-reactions listed in Appendix D, and give plausible explanations for the following observations:
    (a) For some half-reactions, E depends on pH; for others, it does not.
    (b) Whenever H^+ appears in a half-equation, it is on the *left* side.
    (c) Whenever OH^- appears in a half-equation, it is on the *right* side.

40. Write an equation to represent the oxidation of $Cl^-(aq)$ to $Cl_2(g)$ by $PbO_2(s)$ in an acidic solution. Will this reaction occur spontaneously in the forward direction if all other reactants and products are in their standard states and (a) $[H^+] = 6.0$ M; (b) $[H^+] = 1.2$ M; (c) pH = 4.25? Explain.

41. If $[Zn^{2+}]$ is maintained at 1.0 M,
    (a) what is the minimum $[Cu^{2+}]$ for which reaction (20.3) is spontaneous in the forward direction?

(b) Should the displacement of $Cu^{2+}(aq)$ by $Zn(s)$ go to completion? Explain.

42. Can the displacement of Pb(s) from 1.0 M $Pb(NO_3)_2$ be carried to completion by tin metal? Explain.

43. A concentration cell is constructed of two hydrogen electrodes: one immersed in a solution with $[H^+] = 1.0$ M and the other in 0.65 M KOH.
    (a) Determine E_{cell} for the reaction that occurs.
    (b) Compare this value of E_{cell} with E° for the reduction of H_2O to $H_2(g)$ in basic solution, and explain the relationship between them.

44. If the 0.65 M KOH of Exercise 43 is replaced by 0.65 M NH_3,
    (a) will E_{cell} be higher or lower than in the cell with 0.65 M KOH?
    (b) What will be the value of E_{cell}?

45. A voltaic cell is constructed as follows:

$$Ag(s)|Ag^+(satd\ Ag_2CrO_4)||Ag^+(0.125\ M)|Ag(s)$$

What is the value of E_{cell}? For Ag_2CrO_4, $K_{sp} = 1.1 \times 10^{-12}$.

46. A voltaic cell, with $E_{cell} = 0.180$ V, is constructed as follows:

$$Ag(s)|Ag^+(satd\ Ag_3PO_4)||Ag^+(0.140\ M)|Ag(s)$$

What is the K_{sp} of Ag_3PO_4?

47. For the voltaic cell,

$$Sn(s)|Sn^{2+}(0.075\ M)||Pb^{2+}(0.600\ M)|Pb(s)$$

(a) what is E_{cell} initially?
(b) If the cell is allowed to operate spontaneously, will E_{cell} increase, decrease, or remain constant with time? Explain.
(c) What will be E_{cell} when $[Pb^{2+}]$ has fallen to 0.500 M?
(d) What will be $[Sn^{2+}]$ at the point at which $E_{cell} = 0.020$ V?
(e) What are the ion concentrations when $E_{cell} = 0$?

48. For the voltaic cell,

$$Ag(s)|Ag^+(0.015 \text{ M})\|Fe^{3+}(0.055 \text{ M}),$$
$$Fe^{2+}(0.045 \text{ M})|Pt(s)$$

(a) what is E_{cell} initially?
(b) As the cell operates, will E_{cell} increase, decrease, or remain constant with time? Explain.
(c) What will be E_{cell} when $[Ag^+]$ has increased to 0.020 M?
(d) What will be $[Ag^+]$ when $E_{cell} = 0.010$ V?
(e) What are the ion concentrations when $E_{cell} = 0$?

49. Show that the oxidation of $Cl^-(aq)$ to $Cl_2(g)$ by $Cr_2O_7{}^{2-}(aq)$ in acidic solution, with reactants and products in their standard states, does not occur spontaneously. Explain why it is still possible to use this method to produce $Cl_2(g)$ in the laboratory. What experimental conditions would you use?

50. Derive a balanced equation for the reaction occurring in the cell:

$$Fe(s)|Fe^{2+}(aq)\|Fe^{3+}(aq), Fe^{2+}(aq)|Pt(s)$$

(a) If $E°_{cell} = 1.21$ V, calculate $\Delta G°$ and the equilibrium constant for the reaction.
(b) Use the Nernst equation to determine the potential for the cell:

$$Fe(s)|Fe^{2+}(aq, 1.0 \times 10^{-3} \text{ M})\|Fe^{3+}(aq, 1.0 \times 10^{-3} \text{ M}),$$
$$Fe^{2+}(aq, 0.10 \text{ M})|Pt(s)$$

(c) In light of (a) and (b), what is the likelihood of being able to observe the disproportionation of Fe^{2+} into Fe^{3+} and Fe under standard conditions?

Batteries and Fuel Cells

51. The iron–chromium redox battery makes use of the reaction

$$Cr^{2+}(aq) + Fe^{3+}(aq) \longrightarrow Cr^{3+}(aq) + Fe^{2+}(aq)$$

occurring at a chromium anode and an iron cathode.
(a) Write a cell diagram for this battery.
(b) Calculate the theoretical voltage of the battery.

52. Refer to the discussion of the Leclanché cell (page 888).
(a) Combine the several equations written for the operation of the Leclanché cell into a single overall equation.
(b) Given that the voltage of the Leclanché cell is 1.55 V, estimate the electrode potentials, E, for each of the half-reactions. Why are your values only estimates?

53. What is the theoretical standard cell voltage, $E°_{cell}$, of each of the following voltaic cells: (a) the hydrogen–oxygen fuel cell described by equation (20.26); (b) the zinc–air battery; (c) a magnesium–iodine battery?

54. For the alkaline Leclanché cell (page 888)
(a) write the overall cell reaction.
(b) Determine $E°_{cell}$ for that cell reaction.

55. One of the advantages of the aluminum-air battery over the iron–air and zinc–air batteries is the greater quantity of charge transferred per unit mass of metal consumed. Show that this is indeed the case. Assume that zinc and iron are oxidized to oxidation state +2 in air batteries.

56. Describe how you might construct batteries with each of the following voltages: (a) 0.10 V; (b) 2.5 V; (c) 10.0 V. Be as specific as you can about the electrodes and solution concentrations you would use, and indicate whether the battery would consist of a single cell or two or more cells connected in series.

57. A lithium battery, which is different from a lithium-ion battery, uses lithium metal as one electrode and carbon in contact with MnO_2 in a paste of KOH as the other electrode. The electrolyte is lithium perchlorate in a nonaqueous solvent, and the construction is similar to the silver battery. The half-cell reactions involve the oxidation of lithium and the reaction

$$MnO_2(s) + 2 H_2O(l) + e^- \longrightarrow Mn(OH)_3(s) + OH^-(aq) \qquad E° = -0.20 \text{ V}$$

Draw a cell diagram for the lithium battery, identify the negative and positive electrodes, and estimate the cell potential under standard conditions.

58. For each of the following potential battery systems, describe the electrode reactions and the net cell reaction you would expect. Determine the theoretical voltage of the battery.
(a) $Zn-Br_2$
(b) $Li-F_2$

Electrochemical Mechanism of Corrosion

59. Refer to Figure 20-20, and describe in words or with a sketch what you would expect to happen in each of the following cases.
(a) Several turns of copper wire are wrapped around the head and tip of an iron nail.
(b) A deep scratch is filed at the center of an iron nail.
(c) A galvanized nail is substituted for the iron nail.

60. When an iron pipe is partly submerged in water, the iron dissolves more readily below the waterline than

at the waterline. Explain this observation by relating it to the description of corrosion given in Figure 20-21.

61. Natural gas transmission pipes are sometimes protected against corrosion by the maintenance of a small potential difference between the pipe and an inert electrode buried in the ground. Describe how the method works.

62. In the construction of the Statue of Liberty, a framework of iron ribs was covered with thin sheets of copper less than 2.5 mm thick. A layer of asbestos separated the copper skin and iron framework. Over time, the asbestos wore away and the iron ribs corroded. Some of the ribs lost more than half their mass in the 100 years before the statue was restored. At the same time, the copper skin lost only about 4% of its thickness. Use electrochemical principles to explain these observations.

Electrolysis Reactions

63. How many grams of metal are deposited at the cathode by the passage of 2.15 A of current for 75 min in the electrolysis of an aqueous solution containing (a) Zn^{2+}; (b) Al^{3+}; (c) Ag^+; (d) Ni^{2+}?

64. A quantity of electric charge brings about the deposition of 3.28 g Cu at a cathode during the electrolysis of a solution containing $Cu^{2+}(aq)$. What volume of $H_2(g)$, measured at 28.2 °C and 763 mmHg, would be produced by this same quantity of electric charge in the reduction of $H^+(aq)$ at a cathode?

65. Which of the following reactions occur spontaneously, and which can be brought about only through electrolysis, assuming that all reactants and products are in their standard states? For those requiring electrolysis, what is the *minimum* voltage required?
    (a) $2 H_2O(l) \longrightarrow 2 H_2(g) + O_2(g)$ [in 1 M $H^+(aq)$]
    (b) $Zn(s) + Fe^{2+}(aq) \longrightarrow Zn^{2+}(aq) + Fe(s)$
    (c) $2 Fe^{2+}(aq) + I_2(s) \longrightarrow 2 Fe^{3+}(aq) + 2 I^-(aq)$
    (d) $Cu(s) + Sn^{4+}(aq) \longrightarrow Cu^{2+}(aq) + Sn^{2+}(aq)$

66. An aqueous solution of K_2SO_4 is electrolyzed by means of Pt electrodes.
    (a) Which of the following gases should form at the *anode*: O_2, H_2, SO_2, SO_3? Explain.
    (b) What product should form at the *cathode*? Explain.
    (c) What is the *minimum* voltage required? Why is the actual voltage needed likely to be higher than this value?

67. If a lead storage battery is charged at too high a voltage, gases are produced at each electrode. (It is possible to recharge a lead-storage battery only because of the high overpotential for gas formation on the electrodes.)
    (a) What are these gases?
    (b) Write a cell reaction to describe their formation.

68. A dilute aqueous solution of Na_2SO_4 is electrolyzed between Pt electrodes for 3.75 h with a current of 2.83 A. What volume of gas, saturated with water vapor at 25 °C and at a total pressure of 742 mmHg, would be collected at the *anode*? Use data from Table 12.4, as required.

69. Calculate the quantity indicated for each of the following electrolyses.
    (a) the mass of Zn deposited at the cathode in 42.5 min when 1.87 A of current is passed through an aqueous solution of Zn^{2+}
    (b) the time required to produce 2.79 g I_2 at the anode if a current of 1.75 A is passed through KI(aq)

70. Calculate the quantity indicated for each of the following electrolyses.
    (a) $[Cu^{2+}]$ *remaining* in 425 mL of a solution that was originally 0.366 M $CuSO_4$, after passage of 2.68 A for 282 s and the deposition of Cu at the cathode
    (b) the time required to reduce $[Ag^+]$ in 255 mL of $AgNO_3(aq)$ from 0.196 to 0.175 M by electrolyzing the solution between Pt electrodes with a current of 1.84 A

71. A *coulometer* is a device for measuring a quantity of electric charge. In a silver coulometer, $Ag^+(aq)$ is reduced to $Ag(s)$ at a Pt cathode. If 1.206 g Ag is deposited in 1412 s by a certain quantity of electricity, (a) how much electric charge (in C) must have passed, and (b) what was the magnitude (in A) of the electric current?

72. Electrolysis is carried out for 2.00 h in the following cell. The platinum *cathode*, which has a mass of 25.0782 g, weighs 25.8639 g after the electrolysis. The platinum *anode* weighs the same before and after the electrolysis.
    (a) Write plausible equations for the half-reactions that occur at the two electrodes.
    (b) What must have been the magnitude of the current used in the electrolysis (assuming a constant current throughout)?
    (c) A gas is collected at the anode. What is this gas, and what volume should it occupy if (when dry) it is measured at 23 °C and 755 mmHg pressure?

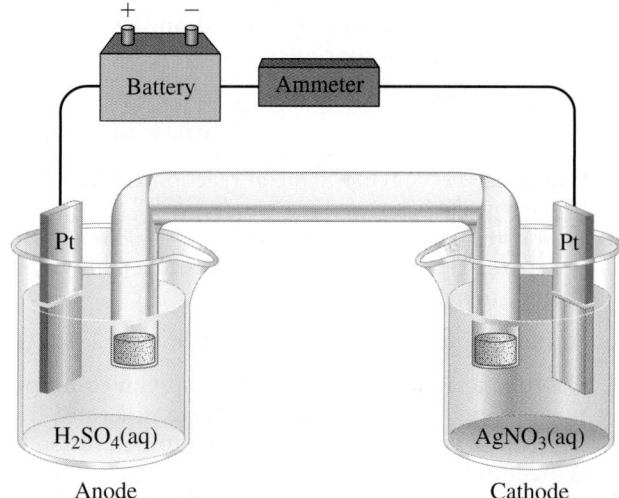

Anode   Cathode

73. A solution containing both Ag^+ and Cu^{2+} ions is subjected to electrolysis. (a) Which metal should plate out first? (b) Plating out is finished after a current of 0.75 A is passed through the solution for 2.50 hours. If the total mass of metal is 3.50 g, what is the mass percent of silver in the product?

74. A solution containing a mixture of a platinum(II) salt contaminated by approximately 10 mole % of another oxidation state is electrolyzed at 1.20 A for 32.0 minutes, at which point no more platinum is deposited.
    (a) What is the oxidation state of the contaminant?
    (b) What is the composition of the mixture?

Integrative and Advanced Exercises

75. Two voltaic cells are assembled in which the following reactions occur.

$$V^{2+}(aq) + VO^{2+}(aq) + 2H^+(aq) \longrightarrow$$
$$2V^{3+}(aq) + H_2O(l) \quad E°_{cell} = 0.616 \text{ V}$$

$$V^{3+}(aq) + Ag^+(aq) + H_2O(l) \longrightarrow$$
$$VO^{2+}(aq) + 2H^+(aq) + Ag(s) \quad E°_{cell} = 0.439 \text{ V}$$

Use these data and other values from Table 20.1 to calculate E° for the half-reaction $V^{3+} + e^- \longrightarrow V^{2+}$.

76. Suppose that a fully charged lead–acid battery contains 1.50 L of 5.00 M H_2SO_4. What will be the concentration of H_2SO_4 in the battery after 2.50 A of current is drawn from the battery for 6.0 h?

77. The energy consumption in electrolysis depends on the product of the charge and the voltage [volt × coulomb = V·C = J(joules)]. Determine the theoretical energy consumption per 1000 kg Cl_2 produced in a diaphragm chlor–alkali cell (page 902) that operates at 3.45 V. Express this energy in **(a)** kJ; **(b)** kilowatt-hours, kWh.

78. For the half-reaction $Cr^{3+} + e^- \longrightarrow Cr^{2+}$, $E° = -0.424$ V. If excess Fe(s) is added to a solution in which $[Cr^{3+}] = 1.00$ M, what will be $[Fe^{2+}]$ when equilibrium is reached at 298 K?

$$Fe(s) + 2Cr^{3+} \rightleftharpoons Fe^{2+} + 2Cr^{2+}$$

79. A voltaic cell is constructed based on the following reaction and initial concentrations:

$$Fe^{2+}(0.0050 \text{ M}) + Ag^+(2.0 \text{ M}) \rightleftharpoons$$
$$Fe^{3+}(0.0050 \text{ M}) + Ag(s)$$

Calculate $[Fe^{2+}]$ when the cell reaction reaches equilibrium.

80. To construct a voltaic cell with $E_{cell} = 0.0860$ V, what $[Cl^-]$ must be present in the cathode half-cell to achieve this result?

$$Ag(s)|Ag^+(\text{satd AgI})\|Ag^+(\text{satd AgCl}, x \text{ M } Cl^-)|Ag(s)$$

81. Describe a laboratory experiment that you could perform to evaluate the Faraday constant, F, and then show how you could use this value to determine the Avogadro constant.

82. The hydrazine fuel cell is based on the reaction

$$N_2H_4(aq) + O_2(g) \longrightarrow N_2(g) + 2H_2O(l)$$

The theoretical $E°_{cell}$ of this fuel cell is 1.559 V. Use this information and data from Appendix D to calculate a value of $\Delta G°_f$ for $[N_2H_4(aq)]$.

83. It is sometimes possible to separate two metal ions through electrolysis. One ion is reduced to the free metal at the cathode, and the other remains in solution. In which of these cases would you expect complete or nearly complete separation: **(a)** Cu^{2+} and K^+; **(b)** Cu^{2+} and Ag^+; **(c)** Pb^{2+} and Sn^{2+}? Explain.

84. Show that for some fuel cells the efficiency value, $\varepsilon = \Delta G°/\Delta H°$, can have a value greater than 1.00. Can you identify one such reaction? [*Hint:* Use data from Appendix D.]

85. In one type of Breathalyzer (alcohol meter), the quantity of ethanol in a sample is related to the amount of electric current produced by an ethanol–oxygen fuel cell. Use data from Table 20.1 and Appendix D to determine **(a)** $E°_{cell}$ and **(b)** E° for the reduction of $CO_2(g)$ to $CH_3CH_2OH(g)$.

86. You prepare 1.00 L of a buffer solution that is 1.00 M NaH_2PO_4 and 1.00 M Na_2HPO_4. The solution is divided in half between the two compartments of an electrolytic cell. Both electrodes used are Pt. Assume that the only electrolysis is that of water. If 1.25 A of current is passed for 212 min, what will be the pH in each cell compartment at the end of the electrolysis?

87. Assume that the volume of each solution in Figure 20-22 is 100.0 mL. The cell is operated as an electrolytic cell, using a current of 0.500 A. Electrolysis is stopped after 10.00 h, and the cell is allowed to function as a voltaic cell. What is E_{cell} at this point?

88. A common reference electrode consists of a silver wire coated with AgCl(s) and immersed in 1 M KCl.

$$AgCl(s) + e^- \longrightarrow Ag(s) + Cl^-(1 \text{ M}) \quad E° = 0.2223 \text{ V}$$

(a) What is $E°_{cell}$ when this electrode is a cathode in combination with a standard zinc electrode as an anode?
(b) Cite several reasons why this electrode should be easier to use than a standard hydrogen electrode.
(c) By comparing the potential of this silver–silver chloride electrode with that of the silver–silver ion electrode, determine K_{sp} for AgCl.

89. The electrodes in the following electrochemical cell are connected to a voltmeter as shown. The half-cell on the right contains a standard silver–silver chloride electrode (see Exercise 88). The half-cell on the left contains a silver electrode immersed in 100.0 mL of 1.00×10^{-3} M $AgNO_3(aq)$. A porous plug through which ions can migrate separates the half-cells.
(a) What is the initial reading on the voltmeter?
(b) What is the voltmeter reading after 10.00 mL of 0.0100 M K_2CrO_4 has been added to the half-cell on the left and the mixture has been thoroughly stirred?
(c) What is the voltmeter reading after 10.00 mL of 10.0 M NH_3 has been added to the half-cell described in part (b) and the mixture has been thoroughly stirred?

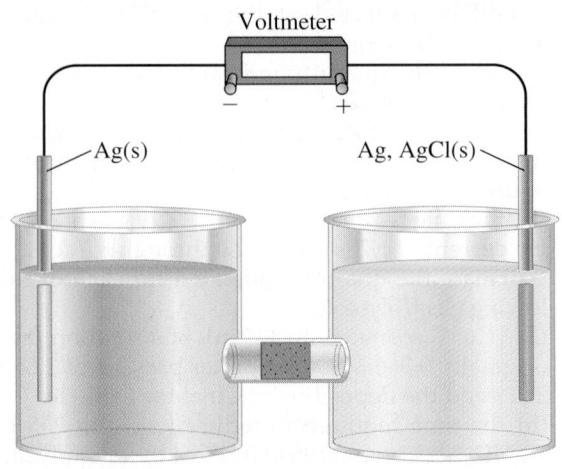

Voltmeter

Ag(s)   Ag, AgCl(s)

1.00×10^{-3} M $AgNO_3(aq)$ 1.00 M KCl

90. An important source of Ag is recovery as a by-product in the metallurgy of lead. The percentage of Ag in lead was determined as follows. A 1.050-g sample was dissolved in nitric acid to produce $Pb^{2+}(aq)$ and $Ag^+(aq)$. The solution was then diluted to 500.0 mL with water, a Ag electrode was immersed in the solution, and the potential difference between this electrode and a SHE was found to be 0.503 V. What was the percent Ag by mass in the lead metal?

91. A test for completeness of electrodeposition of Cu from a solution of $Cu^{2+}(aq)$ is to add $NH_3(aq)$. A blue color signifies the formation of the complex ion $[Cu(NH_3)_4]^{2+}$ ($K_f = 1.1 \times 10^{13}$). Let 250.0 mL of 0.1000 M $CuSO_4(aq)$ be electrolyzed with a 3.512 A current for 1368 s. At this time, add a sufficient quantity of $NH_3(aq)$ to complex any remaining Cu^{2+} and to maintain a free $[NH_3] = 0.10$ M. If $[Cu(NH_3)_4]^{2+}$ is detectable at concentrations as low as 1×10^{-5} M, should the blue color appear?

92. A solution is prepared by saturating 100.0 mL of 1.00 M $NH_3(aq)$ with AgBr. A silver electrode is immersed in this solution, which is joined by a salt bridge to a standard hydrogen electrode. What will be the measured E_{cell}? Is the standard hydrogen electrode the anode or the cathode?

93. The electrolysis of $Na_2SO_4(aq)$ is conducted in two separate half-cells joined by a salt bridge, as suggested by the cell diagram $Pt|Na_2SO_4(aq)|\,|Na_2SO_4(aq)|Pt$.
    (a) In one experiment, the solution in the anode compartment becomes more acidic and that in the cathode compartment, more basic during the electrolysis. When the electrolysis is discontinued and the two solutions are mixed, the resulting solution has pH = 7. Write half-equations and the overall electrolysis equation.
    (b) In a second experiment, a 10.00-mL sample of an unknown concentration of $H_2SO_4(aq)$ and a few drops of phenolphthalein indicator are added to the $Na_2SO_4(aq)$ in the cathode compartment. Electrolysis is carried out with a current of 21.5 mA (milliamperes) for 683 s, at which point, the solution in the cathode compartment acquires a lasting pink color. What is the molarity of the unknown $H_2SO_4(aq)$?

94. A Ni anode and an Fe cathode are placed in a solution with $[Ni^{2+}] = 1.0$ M and then connected to a battery. The Fe cathode has the shape shown. How long must electrolysis be continued with a current of 1.50 A to build a 0.050-mm-thick deposit of nickel on the iron? (Density of nickel = 8.90 g/cm^3.)

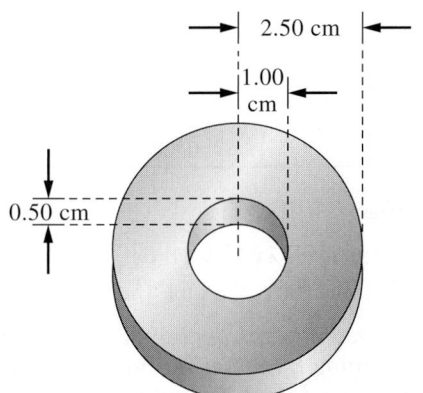

95. Initially, each of the half-cells in Figure 20-21 contained a 100.0-mL sample of solution with an ion concentration of 1.000 M. The cell was operated as an electrolytic cell, with copper as the anode and zinc as the cathode. A current of 0.500 A was used. Assume that the only electrode reactions occurring were those involving Cu/Cu^{2+} and Zn/Zn^{2+}. Electrolysis was stopped after 10.00 h, and the cell was allowed to function as a voltaic cell. What was E_{cell} at that point?

96. Silver tarnish is mainly Ag_2S:

$$Ag_2S(s) + 2\,e^- \longrightarrow 2\,Ag(s) + S^{2-}(aq)$$
$$E° = -0.691 \text{ V}$$

A tarnished silver spoon is placed in contact with a commercially available metallic product in a glass baking dish. Boiling water, to which some $NaHCO_3$ has been added, is poured into the dish, and the product and spoon are completely covered. Within a short time, the removal of tarnish from the spoon begins.
    (a) What metal or metals are in the product?
    (b) What is the probable reaction that occurs?
    (c) What do you suppose is the function of the $NaHCO_3$?
    (d) An advertisement for the product appears to make two claims: (1) No chemicals are involved, and (2) the product will never need to be replaced. How valid are these claims? Explain.

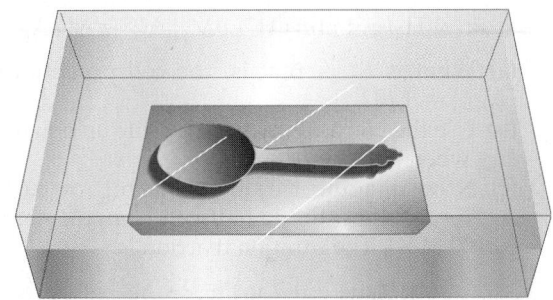

97. Your task is to determine $E°$ for the reduction of $CO_2(g)$ to $C_3H_8(g)$ in two different ways and to explain why each gives the same result. (a) Consider a fuel cell in which the cell reaction corresponds to the complete combustion of propane gas. Write the half-cell reactions and the overall reaction. Determine $\Delta G°$ and $E_{cell}°$ for the reaction, then obtain $E°_{CO_2/C_3H_8}$. (b) Without considering the oxidation that occurs simultaneously, obtain $E°_{CO_2/C_3H_8}$ directly from tabulated thermodynamic data for the reduction half-reaction.

98. Equation (20.15) gives the relationship between the standard Gibbs energy of a reaction and the standard cell potential. We know how the Gibbs energy varies with temperature.
    (a) Making the assumption that $\Delta H°$ and $\Delta S°$ do not vary significantly over a small temperature range, derive an equation for the temperature variation of $E_{cell}°$.
    (b) Calculate the cell potential of a Daniell cell at 50 °C under standard conditions. The overall cell reaction for a Daniell cell is $Zn(s) + Cu^{2+}(aq) \rightarrow Zn^{2+}(aq) + Cu(s)$.

99. Show that for nonstandard conditions the temperature variation of a cell potential is

$$E(T_1) - E(T_2) = (T_1 - T_2)\frac{(\Delta S° - R\ln Q)}{zF}$$

where $E(T_1)$ and $E(T_2)$ are the cell potentials at T_1 and T_2, respectively. We have assumed that the value of Q is maintained at a constant value. For the nonstandard cell below, the potential drops from 0.394 V at 50.0 °C to 0.370 V at 25.0 °C. Calculate Q, $\Delta H°$, and $\Delta S°$ for the reaction, and calculate K for the two temperatures.

$$Cu(s)|Cu^{2+}(aq)\|Fe^{3+}(aq), Fe^{2+}(aq)|Pt(s)$$

Choose concentrations of the species involved in the cell reaction that give the value of Q that you have calculated, and then determine the equilibrium concentrations of the species at 50.0 °C.

100. Show that for a combination of half-cell reactions that produce a standard reduction potential for a half-cell that is not directly observable, the standard reduction potential is

$$E° = \frac{\Sigma n_i E_i°}{\Sigma n_i}$$

where n_i is the number of electrons in each half-reaction of potential $E_i°$. Use the following half-reactions:

$$H_5IO_6(aq) + H^+(aq) + 2\,e^- \longrightarrow IO_3^-(aq) +$$
$$3H_2O(l) \qquad E° = 1.60\ V$$

$$IO_3^-(aq) + 6\,H^+(aq) + 5\,e^- \longrightarrow \frac{1}{2}I_2(s) + 3\,H_2O(l)$$
$$E° = 1.19\ V$$

$$2\,HIO(aq) + 2\,H^+(aq) + 2\,e^- \longrightarrow I_2(s) + 2\,H_2O(l)$$
$$E° = 1.45\ V$$

$$I_2(s) + 2\,e^- \longrightarrow 2\,I^-(aq) \qquad E° = 0.535\ V$$

Calculate the standard reduction potential for

$$H_6IO_6 + 5\,H^+ + 2\,I^- + 3\,e^- \longrightarrow$$
$$\frac{1}{2}I_2 + 4\,H_2O + 2\,HIO$$

Feature Problems

101. Consider the following electrochemical cell:

$$Pt(s)|H_2(g, 1\ atm)|H^+(1\ M)\|Ag^+(x\ M)|Ag(s)$$

(a) What is $E°_{cell}$—that is, the cell potential when $[Ag^+] = 1\ M$?
(b) Use the Nernst equation to write an equation for E_{cell} when $[Ag^+] = x$.
(c) Now imagine titrating 50.0 mL of 0.0100 M AgNO$_3$ in the cathode half-cell compartment with 0.0100 M KI. The titration reaction is

$$Ag^+(aq) + I^-(aq) \longrightarrow AgI(s)$$

Calculate $[Ag^+]$ and then E_{cell} after addition of the following volumes of 0.0100 M KI: (i) 0.0 mL; (ii) 20.0 mL; (iii) 49.0 mL; (iv) 50.0 mL; (v) 51.0 mL; (vi) 60.0 mL.
(d) Use the results of part (c) to sketch the titration curve of E_{cell} versus volume of titrant.

102. Ultimately, $\Delta G_f°$ values must be based on experimental results; in many cases, these experimental results are themselves obtained from $E°$ values. Early in the twentieth century, G. N. Lewis conceived of an experimental approach for obtaining standard potentials of the alkali metals. This approach involved using a solvent with which the alkali metals do not react. Ethylamine was the solvent chosen. In the following cell diagram, Na(amalg, 0.206%) represents a solution of 0.206% Na in liquid mercury.

1. Na(s)|Na$^+$(in ethylamine)|Na(amalg, 0.206%)
$$E_{cell} = 0.8453\ V$$

Although Na(s) reacts violently with water to produce H$_2$(g), at least for a short time, a sodium amalgam electrode does not react with water. This makes it possible to determine E_{cell} for the following voltaic cell.

2. Na(amalg, 0.206%)|Na$^+$(1 M)\|H$^+$(1 M)|
$$H_2(g, 1\ atm) \qquad E_{cell} = 1.8673\ V$$

(a) Write equations for the cell reactions that occur in the voltaic cells (1) and (2).
(b) Use equation (20.14) to establish ΔG for the cell reactions written in part (a).
(c) Write the overall equation obtained by combining the equations of part (a), and establish $\Delta G°$ for this overall reaction.
(d) Use the $\Delta G°$ value from part (c) to obtain $E°_{cell}$ for the overall reaction. From this result, obtain $E°_{Na^+/Na}$. Compare your result with the value listed in Appendix D.

103. The following sketch is called an electrode potential diagram. Such diagrams summarize electrode potential data more efficiently than do listings such as that in Appendix D. In this diagram for bromine and its ions in basic solution,

$$BrO_4^- \xrightarrow{\ 1.025\ V\ } BrO_3^-$$

signifies

$$BrO_4^-(aq) + H_2O(l) + 2\,e^- \longrightarrow$$
$$BrO_3^-(aq) + 2\,OH^-(aq),\ E°_{BrO_4^-/BrO_3^-} = 1.025\ V$$

Similarly,

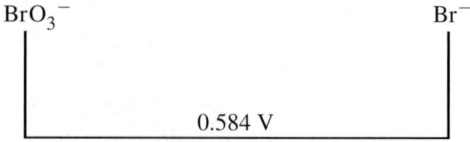

signifies

$$BrO_3^-(aq) + 3\,H_2O(l) + 6\,e^- \longrightarrow$$
$$Br^-(aq) + 6\,OH^-(aq),\ E°_{BrO_3^-/Br^-} = 0.584\ V$$

With reference to Appendix D and to the method of determining $E°$ values outlined on page 875, supply the missing data in the following diagram.

Basic solution ($[OH^-]$ = 1 M):

$$BrO_4^- \xrightarrow{\text{1.025 V}} BrO_3^- \xrightarrow{\text{(?)}} BrO^- \xrightarrow{\text{(?)}} Br_2 \xrightarrow{\text{(?)}} Br^-$$

with a bracket from BrO_3^- through Br_2 to Br^- labeled (?) and an overall bracket from BrO_3^- to Br^- labeled 0.584 V.

104. Only a tiny fraction of the diffusible ions move across a cell membrane in establishing a Nernst potential (see Focus On 20: Membrane Potentials), so there is no detectable concentration change. Consider a typical cell with a volume of 10^{-8} cm³, a surface area (A) of 10^{-6} cm², and a membrane thickness (l) of 10^{-6} cm. Suppose that $[K^+]$ = 155 mM inside the cell and $[K^+]$ = 4 mM outside the cell and that the observed Nernst potential across the cell wall is 0.085 V. The membrane acts as a charge-storing device called a *capacitor*, with a *capacitance*, C, given by

$$C = \frac{\varepsilon_0 \varepsilon A}{l}$$

where ε_0 is the *dielectric constant* of a vacuum and the product $\varepsilon_0 \varepsilon$ is the dielectric constant of the membrane, having a typical value of $3 \times 8.854 \times 10^{-12}$ $C^2 N^{-1} m^{-2}$ for a biological membrane. The SI unit of capacitance is the *farad*, 1 F = 1 coulomb per volt = 1 C V^{-1} = $1 \times C^2 N^{-1} m^{-1}$.

(a) Determine the capacitance of the membrane for the typical cell described.
(b) What is the net charge required to maintain the observed membrane potential?
(c) How many K^+ ions must flow through the cell membrane to produce the membrane potential?
(d) How many K^+ ions are in the typical cell?
(e) Show that the fraction of the intracellular K^+ ions transferred through the cell membrane to produce the membrane potential is so small that it does not change $[K^+]$ within the cell.

105. When deciding whether a particular reaction corresponds to a cell with a positive standard cell potential, which of the following thermodynamic properties would you use to get your answer without performing any calculations? Which would you not use? Explain. (a) ΔG°; (b) ΔS°; (c) ΔH°; (d) ΔU°; (e) K.

106. Consider two cells involving two metals X and Y

$$X(s)|X^+(aq)\|H^+(aq), H_2(g, 1\ bar)|Pt(s)$$
$$X(s)|X^+(aq)\|Y^{2+}(aq)|Y(s)$$

In the first cell electrons flow from the metal X to the standard hydrogen electrode. In the second cell electrons flow from metal X to metal Y. Is $E^\circ_{X^+/X}$ greater or less than zero? Is $E^\circ_{X^+/X} > E^\circ_{Y^{2+}/Y}$? Explain.

107. Describe in words how you would calculate the standard potential of the $Fe^{2+}/Fe(s)$ couple from those of Fe^{3+}/Fe^{2+} and $Fe^{3+}/Fe(s)$.

Self-Assessment Exercises

108. In your own words, define the following symbols or terms: (a) E°; (b) F; (c) anode; (d) cathode.

109. Briefly describe each of the following ideas, methods, or devices: (a) salt bridge; (b) standard hydrogen electrode (SHE); (c) cathodic protection; (d) fuel cell.

110. Explain the important distinctions between each pair of terms: (a) half-reaction and overall cell reaction; (b) voltaic cell and electrolytic cell; (c) primary battery and secondary battery; (d) E_{cell} and E°_{cell}.

111. Of the following statements concerning electrochemical cells, the correct ones are: (a) The cathode is the negative electrode in both voltaic and electrolytic cells. (b) The function of a salt bridge is to permit the migration of electrons between the half-cell compartments of an electrochemical cell. (c) The anode is the negative electrode in a voltaic cell. (d) Electrons leave the cell from either the cathode or the anode, depending on what electrodes are used. (e) Reduction occurs at the cathode in both voltaic and electrolytic cells. (f) If electric current is drawn from a voltaic cell long enough, the cell becomes an electrolytic cell. (g) The cell reaction is an oxidation-reduction reaction.

112. For the half-reaction $Hg^{2+}(aq) + 2\ e^- \longrightarrow Hg(l)$, E° = 0.854 V. This means that (a) Hg(l) is more readily oxidized than $H_2(g)$; (b) $Hg^{2+}(aq)$ is more readily reduced than $H^+(aq)$; (c) Hg(l) will dissolve in 1 M HCl; (d) Hg(l) will displace Zn(s) from an aqueous solution of Zn^{2+} ion.

113. The value of E°_{cell} for the reaction Zn(s) + $Pb^{2+}(aq) \longrightarrow Zn^{2+}(aq) + Pb(s)$ is 0.66 V. This means that for the reaction Zn(s) + $Pb^{2+}(0.01\ M)$ $\longrightarrow Zn^{2+}(0.10\ M) + Pb(s)$, E_{cell} equals (a) 0.72 V; (b) 0.69 V; (c) 0.66 V; (d) 0.63 V.

114. For the reaction Co(s) + $Ni^{2+}(aq) \longrightarrow Co^{2+}(aq) + $ Ni(s), E°_{cell} = 0.03 V. If cobalt metal is added to an aqueous solution in which $[Ni^{2+}]$ = 1.0 M, (a) the reaction will not proceed in the forward direction at all; (b) the displacement of Ni(s) from the $Ni^{2+}(aq)$ will go to completion; (c) the displacement of Ni(s) from the solution will proceed to a considerable extent, but the reaction will not go to completion; (d) there is no way to predict how far the reaction will proceed.

115. The gas evolved at the *anode* when $K_2SO_4(aq)$ is electrolyzed between Pt electrodes is most likely to be (a) O_2; (b) H_2; (c) SO_2; (d) SO_3; (e) a mixture of sulfur oxides.

116. The quantity of electric charge that will deposit 4.5 g Al at a cathode will also produce the following volume at STP of $H_2(g)$ from $H^+(aq)$ at a cathode: (a) 44.8 L; (b) 22.4 L; (c) 11.2 L; (d) 5.6 L.

117. If a chemical reaction is carried out in a fuel cell, the maximum amount of useful work that can be obtained is (a) ΔG; (b) ΔH; (c) $\Delta G/\Delta H$; (d) $T\ \Delta S$.

118. For the reaction Zn(s) + $H^+(aq)$ + $NO_3^-(aq) \longrightarrow$ $Zn^{2+}(aq) + H_2O(l) + NO(g)$, describe the voltaic cell in which it occurs, label the anode and cathode,

use a table of standard electrode potentials to evaluate E°_{cell}, and balance the equation for the cell reaction.

119. The following voltaic cell registers an $E_{cell} = 0.108$ V. What is the pH of the unknown solution?

$$Pt|H_2(g, 1\ atm)|H^+(x\ M)\|H^+(1.00\ M)|$$

$$H_2(g, 1\ atm)|Pt$$

120. $E^\circ_{cell} = -0.0050$ V for the reaction, $2\ Cu^+(aq) + Sn^{4+}(aq) \longrightarrow 2\ Cu^{2+}(aq) + Sn^{2+}(aq)$.

(a) Can a solution be prepared that is 0.500 M in each of the four ions at 298 K?

(b) If not, in what direction must a net reaction occur?

121. For each of the following combinations of electrodes (A and B) and solutions, indicate

• the overall cell reaction
• the direction in which electrons flow spontaneously (from A to B, or from B to A)
• the magnitude of the voltage read on the voltmeter, V

	A	Solution A	B	Solution B
(a)	Cu	1.0 M Cu^{2+}	Fe	1.0 M Fe^{2+}
(b)	Pt	1.0 M Sn^{2+}/1.0 M Sn^{4+}	Ag	1.0 M Ag^+
(c)	Zn	0.10 M Zn^{2+}	Fe	1.0×10^{-3} M Fe^{2+}

122. Use data from Table 20.1, as necessary, to predict the probable products when Pt electrodes are used in the electrolysis of (a) $CuCl_2$(aq); (b) Na_2SO_4(aq); (c) $BaCl_2$(l); (d) KOH(aq).

123. Using the method presented in Appendix E, construct a concept map showing the relationship between electrochemical cells and thermodynamic properties.

124. Construct a concept map illustrating the relationship between batteries and electrochemical ideas.

125. Construct a concept map illustrating the principles of electrolysis and its industrial applications.

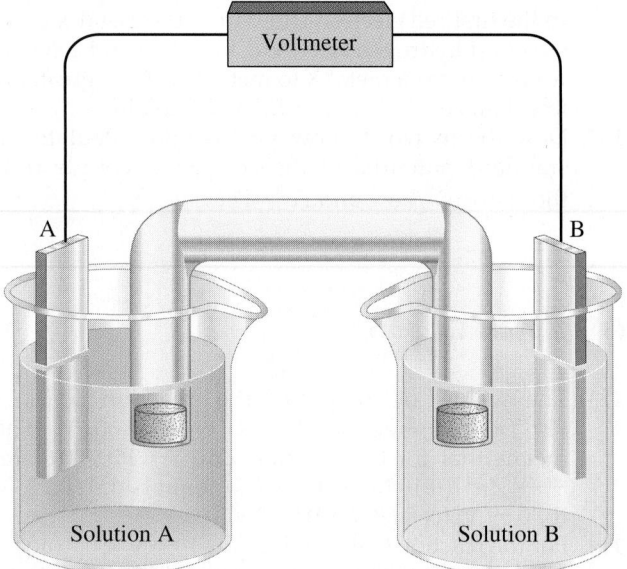

Chemistry of the Main-Group Elements I: Groups 1, 2, 13, and 14

21

CONTENTS

Some of the dramatic colors seen in fireworks displays are the flame colors of some of the groups 1 and 2 metals. These colors, as we will see, are related to the electronic structures of those metal atoms.

The chemistry of the elements is best described by using the periodic table as a basis. The trends within the groups and across the periods allow us to organize our thinking about the chemistry of the elements by using patterns. The chemistry of each group has both similarities and differences that can be understood in terms of the underlying principles that organize the periodic table. As we explore the chemistry of these groups, we will discover that the first member of a group is often markedly different from the others in its physical and chemical properties. Typically, the second member of the group exhibits properties that are most representative of the group.

In this chapter, our focus is on groups 1, 2, 13, and 14—the first four groups of the main group of elements. We will start with the chemistry of the group 1 metals: the alkali metals. They are the first members of the s-block elements. Atoms of the group 1 elements have ground state valence configurations that consist of a single electron in an s orbital.

KEEP IN MIND

that hydrogen is often placed in group 1 of the periodic table, but this element is not an alkali metal. Francium is an alkali metal, but this highly radioactive metal is so rare that few of its properties have been measured.

As discussed in Chapter 9, the group 1 atom in a given period is always the largest (having the largest atomic radius) and is the most easily ionized (having the lowest first ionization energy). As a result, the group 1 elements have low densities (some have densities of less than 1 g cm^{-3}), are easily oxidized, and are highly reactive. The reactivity of the alkali metals is evident in their violent reactions with water.

Next, we will discuss the group 2 metals: the alkaline earth metals. The group 2 metals are considered s-block elements because atoms of these metals have valence configurations that consist of two electrons in an s orbital. Like their group 1 neighbors, the metals of group 2 are highly reactive and less dense than a typical metal, though they are less reactive and denser than the alkali metals. Unlike the group 1 metals, the group 2 metals react only slowly, or not at all, with water, and all of them have densities greater than that of water.

Atoms of the p-block elements are characterized by valence configurations involving electrons in p-orbitals. In this chapter we will discuss groups 13 and 14. Atoms of these elements have the configurations ns^2np^1 and ns^2np^2, respectively. These are the first groups in which both metals and nonmetals are encountered. Boron is the only nonmetal in group 13 and has interesting chemistry because it tends to form molecules with incomplete octets around the central boron atoms. Aluminum is the most abundant of the metals and one of the most widely used. Aluminum metal is obtained from its compounds by using electrolysis. Because aluminum production requires prodigious quantities of electricity, aluminum-production plants are often located near plentiful sources of hydroelectricity.

The remaining elements of group 13—gallium, indium, and thallium—are all metals. The chemistry of group 13 is dominated by boron and aluminum, and we will mention the heavier elements only briefly in this chapter. Group 14 contains a nonmetal (carbon), two metalloids (silicon and germanium), and two metals (tin and lead). Carbon has the most important chemistry of the group since it occurs in all living systems and we devote three chapters (26, 27 and 28) to the chemistry of carbon. Silicon is found in numerous minerals, forming many different and interesting oxoanions. In contrast to aluminum, tin and lead can be obtained by chemical reduction, using methods known since ancient times.

This chapter and the remaining chapters offer many opportunities to relate new information to principles presented earlier in the text. Ideas of atomic structure, periodic trends in atomic and ionic radii, chemical bonding, and thermodynamics will help us to understand the chemical behavior of the elements.

21-1   Periodic Trends and Charge Density

The chemistry of the elements can, to some extent, be rationalized in terms of the periodic trends that we have covered previously in this text. The atoms of each group of the periodic table have similar electronic configurations and consequently, the elements in a given group have similar—but not exactly the same—chemical properties. The first member of a group is the lightest and often has features that are different from the remaining members of the group. In this section we will briefly review trends in atomic properties and introduce a new ionic property. With these ideas as a basis we will begin to understand the trends in the chemistry of the elements.

The atomic properties that are responsible for the chemistry of an element are atomic radius, ionization energy, electron affinity, and polarizability. The electronegativity of an atom is also an important consideration. Before we attempt to explain the chemistry of the elements in terms of atomic properties, it will be helpful to examine Figure 21-1, which summarizes the periodic

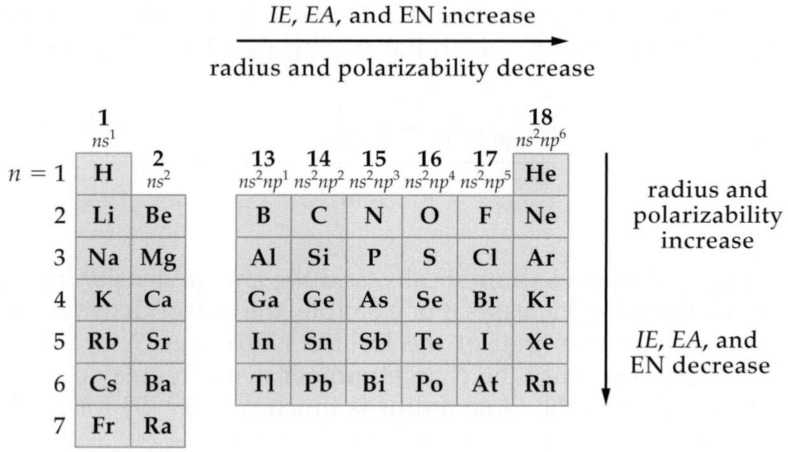

IE, EA, and EN increase
⟶
radius and polarizability decrease

	1 ns^1	**2** ns^2		**13** ns^2np^1	**14** ns^2np^2	**15** ns^2np^3	**16** ns^2np^4	**17** ns^2np^5	**18** ns^2np^6
$n=1$	H								He
2	Li	Be		B	C	N	O	F	Ne
3	Na	Mg		Al	Si	P	S	Cl	Ar
4	K	Ca		Ga	Ge	As	Se	Br	Kr
5	Rb	Sr		In	Sn	Sb	Te	I	Xe
6	Cs	Ba		Tl	Pb	Bi	Po	At	Rn
7	Fr	Ra							

radius and polarizability increase

IE, EA, and EN decrease

▲ FIGURE 21-1
A summary of trends in atomic radius, first ionization energy, electron affinity, electronegativity, and atomic polarizability
Atomic radius and polarizability decrease from left to right in a given period and increase from top to bottom in a given group. Ionization energy (IE), electron affinity (EA) and electronegativity (EN) increase from left to right in a given period and decrease from top to bottom in a given group. The shaded elements are the focus of this chapter. Atomic radii, ionization energies, and electron affinities were discussed in Chapter 9. Electronegativities were discussed in Chapter 10 and polarizabilities in Chapter 12.

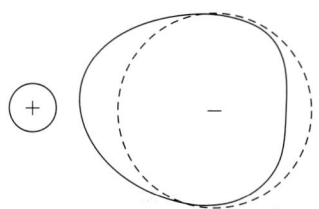

▲ FIGURE 21-2
Polarization of the electron cloud of an anion by a cation
A cation with a high charge density distorts the electron cloud around the anion. The undistorted electron cloud is shown with a dashed line and the distorted electron cloud is shown with a solid line. The greater the distortion of the electron cloud, the greater the degree of covalency of the bond.

trends we have discussed previously in this text. The figure shows that atomic radii and polarizabilities decrease from left to right in a period and increase from top to bottom in a group. First ionization energies, electron affinities, and electronegativities show the opposite trend: these quantities increase across a period and decrease down a group. It is also important to remember that ionic radii follow the same trend as atomic radii and that cations are smaller than the parent atoms while anions are larger. Explanations for these trends have been given elsewhere in the text (see Chapters 9 and 10).

Most of the elements we consider in this chapter are metals, and when a metal combines with a nonmetallic element to form a compound, a metal atom is converted into a cation. When a cation interacts with an anion, the electron cloud of the anion is distorted into the internuclear region toward the cation, as suggested by Figure 21-2. As a result of this distortion, the bond between the cation and the anion has some covalent character in addition to its ionic character; the greater the distortion, the greater the covalent character.

The polarizability of the anion and the polarizing power of the cation determine the extent to which the electron cloud of the anion is distorted. It is generally true that anions derived from atoms that are lower down in a group are larger and more polarizable. For example, the I^- ion is much more polarizable than F^-. The polarizing power of a cation is related to its **charge density**. The charge density, ρ, can be defined as charge per unit volume. For a metal cation, M^{z+}, with ionic radius r, the charge density is calculated by using the formula below:

$$\rho = \frac{(1.60 \times 10^{-19}\,\text{C})(z)}{\frac{4}{3}\pi r^3}$$
(21.1)

In equation (21.1), the numerator represents the total charge, in coulombs, of the cation; the denominator represents the volume of the cation, which is assumed to be spherical. If the ionic radius is expressed in millimeters, then

KEEP IN MIND

that a variety of different definitions have been used for charge density. For example, some authors have defined charge density as the amount of charge distributed over the *surface* of the cation and thus, they calculate the charge density as $ze/(4\pi r^2)$, with $e = 1.60 \times 10^{-19}$ C. Both definitions provide a measure of the charge-to-size ratio for ions. The use of charge-to-size ratios can be traced back to at least 1928, when G. H. Cartledge introduced the *ionic potential*, z/r, in an attempt to explain variations in the properties of compounds.

the charge density is typically a number between 1 and 1000. As an example, consider the lithium cation, which has a charge of 1+ and an ionic radius of 73 pm:

$$\rho_{Li^+} = \frac{(1.60 \times 10^{-19}\,C)(1)}{\frac{4}{3}\pi\,(73 \times 10^{-9}\,mm)^3} = 98\,C\,mm^{-3}$$

The ionic radius decreases dramatically, and the charge density increases, as the charge on the cation increases. For example, the charge density of the Al^{3+} ion is much greater than that of the Li^+ ion ($770\,C\,mm^{-3}$ for Al^{3+} versus $98\,C\,mm^{-3}$ for Li^+) because the charge number is much greater (+3 for Al^{3+} versus +1 for Li^+) and the ionic radius is much smaller (53 for Al^{3+} versus 73 pm for Li^+).

We anticipate that the higher the charge density, the greater the polarizing power of the cation and the greater the ability of a cation to distort the electron cloud of an anion toward itself. For example, because the charge density of the Al^{3+} ion is much greater than that of the Li^+ ion, we expect the interaction between Al^{3+} and I^- to have a greater degree of covalency than does the interaction between Li^+ and I^-.

Throughout this chapter, we will use the charge density concept to rationalize certain observations. For example, we will use it to help us understand why sometimes there are dramatic differences in the properties of elements in the same group and interesting similarities in the properties of elements in different groups. However, a single quantity—such as charge density—cannot be used to rationalize all things and it should never be used as a substitute for careful consideration of all contributing factors.

TABLE 21.1 Abundances of Group 1 Elements		
Elements	Abundances, ppm[a]	Rank
Li	18	35
Na	22,700	7
K	18,400	8
Rb	78	23
Cs	2.6	46
Fr	Trace	—

[a]Grams per 1000 kg of solid crust.

21-1 CONCEPT ASSESSMENT

One way of distinguishing ionic behavior from covalent behavior is by comparing melting points. Ionic compounds tend to have higher melting points than covalent compounds. Which of the two halides of aluminium, AlF_3 and AlI_3, is expected to have the lowest melting point?

21-2 Group 1: The Alkali Metals

As Table 21.1 indicates, the group 1 elements, the **alkali metals**, are relatively abundant. Some of their compounds have been known and used since prehistoric times. Yet these elements were not isolated in pure form until about 200 years ago. The compounds of the alkali metals are difficult to decompose by ordinary chemical means, so discovery of the elements had to await new scientific developments. Sodium (1807) and potassium (1807) were discovered through electrolysis. Lithium was discovered in 1817. Cesium (1860) and rubidium (1861) were identified as new elements through their emission spectra. Francium (1939) was isolated in the radioactive decay products of actinium.

Because most alkali metal compounds are water soluble, a number of Li, Na, and K compounds, including chlorides, carbonates, and sulfates, can be obtained from natural brines. A few alkali metal compounds, such as NaCl, KCl, and Na_2CO_3, can be mined as solid deposits. Sodium chloride is also obtained from seawater. An important source of lithium is the mineral *spodumene*, $LiAl(SiO_3)_2$, shown in the margin. Rubidium and cesium are obtained as by-products in the processing of lithium ores.

▲ The mineral spodumene, $LiAl(SiO_3)_2$.

Physical Properties of the Alkali Metals

By any measure, the group 1 elements are the most active metals. Table 21.2 lists several of their properties, and a few of these properties are discussed next.

Flame Colors The energy differences between the valence-shell s and p orbitals of the group 1 atoms match those of certain wavelengths of visible light. As a result, when heated in a flame, group 1 compounds produce characteristic flame colors, as was shown in Figure 8.8. For example, when NaCl is vaporized in a flame, ion pairs are converted to gaseous atoms. Sodium atoms, Na(g), are excited to higher energies, and light with a wavelength of 589 nm (yellow) is emitted as the excited atoms (Na*) revert to their ground-state electron configurations.

$$Na^+Cl^-(g) \longrightarrow Na(g) + Cl(g)$$
$$Na(g) \longrightarrow Na^*(g)$$
$$[Ne]3s^1 \qquad [Ne]3p^1$$
$$Na^*(g) \longrightarrow Na(g) + h\nu \text{ (589 nm; yellow)}$$

Alkali metal compounds are used in pyrotechnic displays—fireworks.

Densities and Melting Points Atoms of the group 1 elements are the largest in their respective periods, and the atomic radii increase from the top to the bottom within this group, as was described in Chapter 9. These large atoms make for a relatively low mass per unit volume—that is, low density. The lighter of the alkali metals (Li, Na, and K) will float on water. The large atomic sizes, together with the fact that each of these atoms has only one valence electron, leads to rather weak metallic bonding. This property, in turn, leads to soft metals with low melting points. A bar of sodium has the consistency of a stick of butter and is easily cut with a knife, as shown in the photo in the margin.

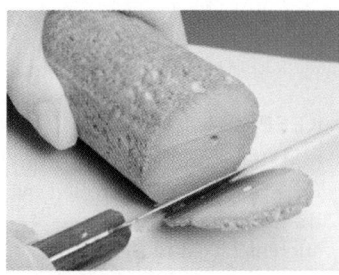

▲ **The cutting of metallic sodium**
The sodium, an active metal, is covered with a thick oxide coating.

TABLE 21.2   Some Properties of the Group 1 (Alkali) Metals

	Li	Na	K	Rb	Cs
Atomic number	3	11	19	37	55
Valence-shell electron configuration	$2s^1$	$3s^1$	$4s^1$	$5s^1$	$6s^1$
Atomic (metallic) radius, pm	152	186	227	248	265
Ionic (M^+) radius, pm[a]	73	116	152	166	181
Electronegativity	1.0	0.9	0.8	0.8	0.8
Charge density of M^+ (C mm^{-3})	98	24	11	8	6
First ionization energy, kJ mol^{-1}	520.2	495.8	418.8	403.0	375.7
Electrode potential $E°$, V[b]	−3.040	−2.713	−2.924	−2.924	−2.923
Melting point, °C	180.54	97.81	63.65	39.05	28.4
Boiling point, °C	1347	883.0	773.9	687.9	678.5
Density, g cm^{-3} at 20 °C	0.534	0.971	0.862	1.532	1.873
Hardness[c]	0.6	0.4	0.5	0.3	0.2
Electrical conductivity[d]	17.1	33.2	22.0	12.4	7.76
Flame color	Carmine	Yellow	Violet	Bluish red	Blue
Principal visible emission lines, nm	610,671	589	405,767	780,795	456,459

[a]The values given here assume a coordination number of 4 for Li$^+$ and 6 for the others.
[b]For the reduction $M^+(aq) + e^- \longrightarrow M(s)$.
[c]Hardness measures the ability of substances to scratch, abrade, or indent one another. On the Mohs scale, ten minerals are ranked by hardness, ranging from that of talc (0) to diamond (10). Other values: wax (0 °C), 0.2; asphalt, 1–2; fingernail, 2.5; copper, 2.5–3; iron, 4–5; chromium, 9. Each substance can scratch only other substances with hardness values lower than its own.
[d]On a scale relative to silver as 100.

Electrode Potentials A good indicator of the extreme metallic character of the group 1 elements is their electrode potentials, which are large, negative quantities. The ions $M^+(aq)$ are very difficult to reduce to the metals $M(s)$, and in turn, the metals are very easily oxidized to $M^+(aq)$. All the alkali metals easily displace $H_2(g)$ from water.

$$2\,M(s) + 2\,H_2O(l) \longrightarrow 2\,M^+(aq) + 2\,OH^-(aq) + H_2(g) \qquad \textbf{(21.2)}$$

$$E^\circ_{cell} = E^\circ_{H_2O/H_2} - E^\circ_{M^+/M}$$
$$= -0.828\ \text{V} - E^\circ_{M^+/M}$$

Using the standard reduction potentials given in Table 21.2, we can calculate the following E°_{cell} values.

$$E^\circ_{cell} = 2.212\ \text{V (for Li)} \qquad 1.885\ \text{V (for Na)} \qquad 2.096\ \text{V (for K)}$$
$$2.096\ \text{V (for Rb)} \qquad 2.095\ \text{V (for Cs)}$$

The E°_{cell} values indicate that of the alkali metals, lithium is the strongest reducing agent in aqueous solution. However, all the alkali metals are strong reducing agents in aqueous solution. If we convert the E°_{cell} values given above to equilibrium constants (K) by using the equation $K = nE^\circ_{cell}/(0.0592\ \text{V})$, we find that the values of K range from 7×10^{31} for the reaction of $Na(s)$ and water to 2×10^{37} for the reaction of $Li(s)$ and water. The equilibrium position for reaction (21.2) is always very far to the right, and so reaction (21.2) goes essentially to completion regardless of which alkali metal is involved.

Experiments show that lithium reacts more slowly and less vigorously with water than do any of the other alkali metals. To explain this observation, it is necessary to consider what happens to the energy that is released by the reaction as the metal is oxidized by water. As the metal reacts, energy released by the reaction is used to heat the system, including unreacted metal. For all the alkali metals except lithium, energy released by the reaction is sufficient to melt the unreacted metal. The melting of the metal increases the rate of reaction because it causes more metal atoms to come into contact with water molecules. Because lithium metal does not melt as the reaction proceeds, the reaction of lithium and water is neither as fast nor as vigorous as it is for the other alkali metals.

Production and Uses of the Alkali Metals

Lithium and sodium are produced from their molten chlorides by electrolysis, Figure 21-3. The electrolysis of $NaCl(l)$, for example, is carried out at about 600 °C.

$$2\,NaCl(l) \xrightarrow{\text{electrolysis}} 2\,Na(l) + Cl_2(g) \qquad \textbf{(21.3)}$$

The melting point of NaCl is 801 °C, which is too high a temperature to carry out this electrolysis economically. Adding $CaCl_2$ to the mixture reduces the melting point. (Calcium metal, also produced in the electrolysis, precipitates out from the $Na(l)$ as the liquid metal is cooled. The final product is 99.95% Na.)

Potassium metal is produced by the reduction of molten KCl by liquid sodium.

$$KCl(l) + Na(l) \xrightarrow{850\ °C} NaCl(l) + K(g) \qquad \textbf{(21.4)}$$

Reaction (21.4) is reversible; at low temperatures, most of the $KCl(l)$ remains unreacted. At 850 °C, however, the equilibrium is displaced far to the right as $K(g)$ escapes from the molten mixture (an application of Le Châtelier's principle). The $K(g)$ is freed of any $Na(g)$ present by condensation of the vapor, followed by fractional distillation of the liquid metals. Rb and Cs can be produced in much the same way, with Ca metal as the reducing agent.

Because sodium metal is so easily oxidized, its most important use is as a reducing agent—for example, in obtaining such metals as titanium, zirconium, and hafnium.

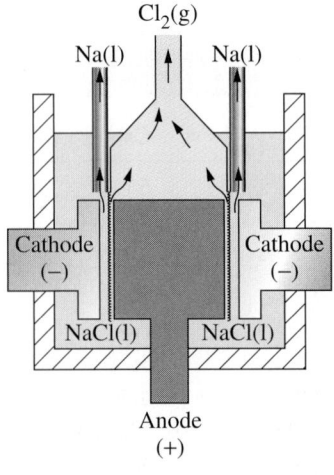

▲ FIGURE 21-3
The Downs cell used for the production of sodium
The electrolysis cell shown here is the Downs cell. The electrolyte is molten $NaCl(l)$ to which $CaCl_2$ has been added to lower the melting point of $NaCl(s)$. Liquid sodium metal forms at the steel cathode and $Cl_2(g)$ forms at the graphite anode. The chlorine and sodium are kept apart by a steel gauze diaphragm.

For example, titanium metal can be obtained from the reduction of $TiCl_4$ by Na, as shown below.

$$TiCl_4 + 4\,Na \xrightarrow{\Delta} Ti + 4\,NaCl$$

Sodium is also used as a heat-transfer medium in nuclear reactors. Liquid sodium is especially good for this purpose because it has a low melting point, a high boiling point, and a low vapor pressure. Also, it has better thermal conductivity and a higher specific heat than do most liquid metals. Finally, its low density and low viscosity make it easy to pump. Sodium is also used in sodium vapor lamps, which are very popular for outdoor lighting. Because each lamp uses only a few milligrams of Na, however, the total quantity consumed in this application is rather small.

Lithium metal is used as an alloying agent to make high-strength, low-density alloys with aluminum and with magnesium. These alloys are used in the aerospace and aircraft industries. The use of lithium as an anode material in batteries is on the increase, in part because of its ease of oxidation and in part because a small mass of lithium produces a large number of electrons. Only 6.94 g Li (1 mol) needs to be consumed to produce one mole of electrons. Lithium batteries are particularly useful where the installed battery must have high reliability and a long lifetime, as in cardiac pacemakers. An X-ray photograph of a pacemaker is shown in the margin.

◀ We introduced the symbol $\xrightarrow{\Delta}$ in Chapter 4 (page 116) to indicate that a reaction is carried out at an elevated temperature.

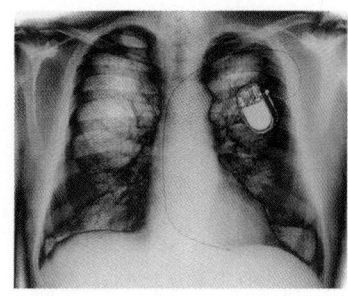

▲ An X-ray photograph showing a heart pacemaker powered by a lithium battery.

21-1 ARE YOU WONDERING...

Why lithium is the most easily oxidized of the group 1 elements on the basis of $E°$ values but not on the basis of ionization energies?

Being the smallest of the alkali metal atoms, Li has the highest first ionization energy. It is the most difficult to oxidize in the reaction $M(g) \longrightarrow M^+(g) + e^-$. However, oxidation to produce $M^+(aq)$ is a different matter. We can think of it as the overall result of a hypothetical three-step process.

Sublimation:	$M(s) \longrightarrow M(g)$
Ionization:	$M(g) \longrightarrow M^+(g) + e^-$
Hydration:	$M^+(g) \longrightarrow M^+(aq).$
Overall:	$M(s) \longrightarrow M^+(aq) + e^-$

Thus, to compare tendencies to form $M^+(aq)$ by oxidation of the metals, we must compare tendencies in each of these three steps. The data given in Exercise 59 reveal that Li^+ has an unusually large hydration energy—enough to make $Li^+(aq)$ especially hard to reduce and $Li(s)$ especially easy to oxidize. The large hydration energy is a consequence of the small size of the Li^+ ion, which allows a close approach to surrounding water molecules and strong attractive ion–dipole forces between them, as shown in Figure 21-4.

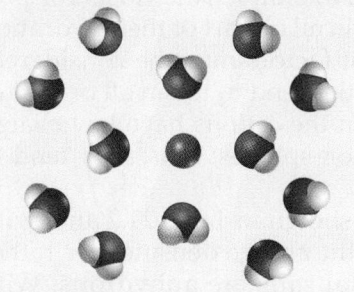

◀ FIGURE 21-4
Hydration of a Li^+ ion
Electrostatic forces hold a small number of H_2O molecules around a Li^+ ion in a primary hydration sphere. These molecules, in turn, hold other molecules, but more weakly, in a secondary hydration sphere.

KEEP IN MIND

that the more negative the value of $E°$, the more difficult is the reduction but the more easily does the reverse process—the oxidation half-reaction—occur. The value of $E°_{Li^+/Li}$, -3.040 V, comes at the very bottom of a listing of electrode potentials (recall Table 20-1), making $Li(s)$ the easiest substance to oxidize.

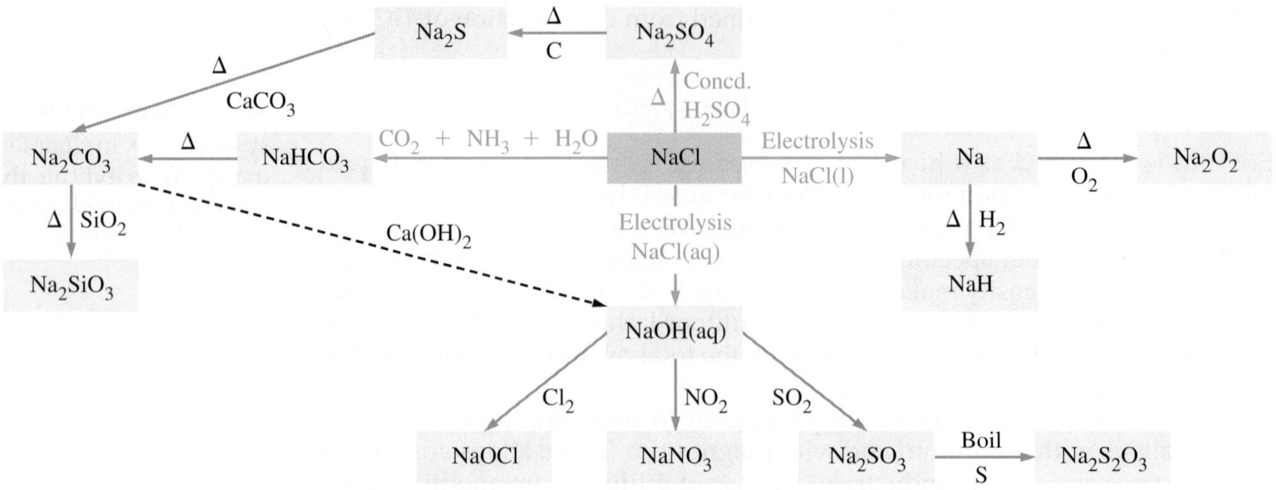

▲ FIGURE 21-5
Preparation and reactions of sodium compounds
This is a simple and common method of summarizing important reactions, starting from a compound of control importance. Most of these reactions are described in this section. A number of these compounds may be prepared by alternative methods. The conversion of Na_2CO_3 to NaOH (dashed arrow) is no longer of commercial importance.

Group 1 Compounds

Group 1 metals, Li–Fr, have a valence configuration of ns^1 and occur exclusively in the +1 oxidation state. Most of the compounds of group 1 metals are stable, ionic solids. When studying the chemistry of the elements we are faced with a large amount of information that is challenging to organize. Figure 21-5 introduces a useful diagrammatic format for summarizing reaction chemistry. Although this diagram deals with sodium compounds, some of which are discussed in this section, similar diagrams can be constructed for other compounds. Such diagrams note a compound of central importance (NaCl in Figure 21-5) and show how several other compounds can be obtained from it. Some of these conversions occur in one step, such as the reaction of NaCl with H_2SO_4 to form Na_2SO_4 (discussed on page 930). Other conversions involve two or more consecutive reactions, such as the preparation of sodium silicate, Na_2SiO_3. The principal reactants required for the conversions are written near the connecting arrows. If a reaction mixture must be heated, a Δ symbol is used to indicate this requirement. The reactions may also produce by-products that are not noted in the diagram.

Hydration of Salts When salts are dissolved in water, the cations are hydrated, as suggested by Figure 21-4. The anions are similarly hydrated but with the slightly positive hydrogen atoms of the water molecules directed toward the anion. When a salt crystallizes from an aqueous solution, the salt that is obtained may or may not contain water molecules—called water of crystallization—as part of the solid structure. No simple rule exists for predicting with certainty whether the ions will retain all or part of their hydration spheres in the solid state because a number of factors must be considered. That being said, cations with high charge densities tend to retain all or part of their hydration spheres in the solid state. When the cations have low charge densities, the cations tend to lose their hydration spheres; thus, they tend to form anhydrous salts.

The charge densities of the alkali metals are shown in Table 21.2 and, with the exception of lithium and perhaps sodium, the charge densities are rather low; consequently, the majority of alkali metal salts are anhydrous. With lithium and sodium, salts are most likely to exist as hydrated salts. The

---

EXAMPLE 21-1   Writing Chemical Equations from a Summary Diagram of Reaction Chemistry

Use the information provided in Figure 21-5 to write balanced chemical equations for the reactions involved in synthesizing sodium carbonate from sodium chloride.

Analyze

Consult Figure 21-5 to find a route from NaCl to Na_2CO_3. One possible route involves the conversions $NaCl \longrightarrow Na_2SO_4 \longrightarrow Na_2S \longrightarrow Na_2CO_3$. For each conversion, the other necessary reactants (H_2SO_4, C, $CaCO_3$) are noted. Use this information to write balanced chemical equations for each conversion, keeping in mind that other reaction products (such as HCl, CO and CaS) must be included.

Solve

First, $Na_2SO_4(s)$ is produced from NaCl(s) and concentrated sulfuric acid.

$$2\,NaCl(s) + H_2SO_4(concd\ aq) \xrightarrow{\Delta} Na_2SO_4(s) + 2\,HCl(g)$$

Next, the Na_2SO_4 is reduced to Na_2S with carbon.

$$Na_2SO_4(s) + 4\,C(s) \xrightarrow{\Delta} Na_2S(s) + 4\,CO(g)$$

The final step is a reaction between Na_2S and $CaCO_3$.

$$Na_2S(s) + CaCO_3(s) \xrightarrow{\Delta} CaS(s) + Na_2CO_3(s)$$

Assess

There are three conversions and thus three separate chemical equations. Each chemical equation is balanced and produces one of the sodium compounds on the route chosen.

PRACTICE EXAMPLE A: Write chemical equations for the reactions involved in synthesizing sodium nitrate from sodium chloride.

PRACTICE EXAMPLE B: Write chemical equations for the reactions involved in synthesizing sodium thiosulfate from sodium chloride.

---

empirical formulas of the alkali metal perchlorates illustrate nicely the greater tendency of Li and Na to form hydrated salts: $LiClO_4 \cdot 3\,H_2O$, $NaClO_4 \cdot H_2O$, $KClO_4$, $RbClO_4$, $CsClO_4$.

Halides All alkali metals react vigorously, sometimes explosively, with halogens to produce ionic halides, the most important of which are NaCl and KCl. Sodium chloride—salt—is the most used of all minerals for the production of chemicals. It is not listed among the top chemicals, however, because it is considered a raw material, not a manufactured chemical. Large quantities of NaCl can be obtained by evaporation of seawater, as shown in the photograph. Annual use of NaCl in the United States amounts to about 50 million metric tons. Salt is used to preserve meat and fish, control ice on roads, and regenerate water softeners. In the chemical industry, NaCl is a source of many chemicals, including sodium metal, chlorine gas, hydrochloric acid, and sodium hydroxide.

Potassium chloride, KCl, is obtained from naturally occurring brines (concentrated solutions of salts). It is most extensively used in plant fertilizers because potassium is a major essential element for plant growth. Potassium

◄ Sea salt (sodium chloride) stacks that have been harvested by evaporation of seawater.

chloride is also used as a raw material in the manufacture of KOH, KNO_3, and other industrially important potassium compounds.

Alkali Metal Hydrides When an alkali metal (M) is heated in the presence of hydrogen gas, an ionic hydride is formed:

$$2\,M(s) + H_2(g) \xrightarrow{\Delta} 2\,MH(s)$$

Alkali metal hydrides contain the hydride ion, H^-, and have the sodium chloride structure. All the alkali metal hydrides are very reactive. For example, they react readily with water to give a hydroxide salt and hydrogen gas:

$$MH(s) + H_2O(l) \longrightarrow MOH(aq) + H_2(g)$$

Alkali metal hydrides also react with metal halides. An important example is the reaction of lithium hydride and aluminum chloride, $AlCl_3$, which produces lithium aluminum hydride, $LiAlH_4$, and lithium chloride. Lithium aluminum hydride (LAH) is a powerful reducing agent used in organic chemistry. $LiAlH_4$ can be made using the following reaction:

$$4LiH + AlCl_3 \xrightarrow{(C_2H_5)_2O} LiAlH_4 + 3\,LiCl$$

The reaction is carried out by adding finely divided LiH(s) to a solution of $AlCl_3$ in a nonaqueous solvent, such as diethyl ether, $(C_2H_5)_2O$. A nonaqueous solvent is used because both $LiAlH_4$ and LiH react vigorously with water. $LiAlH_4$ is obtained as a white solid by careful and controlled evaporation of the solvent.

Oxides and Hydroxides The alkali metals react rapidly with oxygen to produce several different ionic oxides. Under appropriate conditions—generally by carefully controlling the supply of oxygen—the oxide M_2O can be prepared for each of the alkali metals. Lithium reacts with excess oxygen to give Li_2O and a small amount of lithium *peroxide*, Li_2O_2. Sodium reacts with excess oxygen to give mostly the peroxide Na_2O_2 and a small amount of Na_2O. Potassium, rubidium, and cesium react to form the *superoxides*, MO_2. The oxides, peroxides, and superoxides are ionic compounds. The Lewis symbol for the O^{2-} ion and the Lewis structures for the O_2^{2-} and O_2^- ions are shown below:

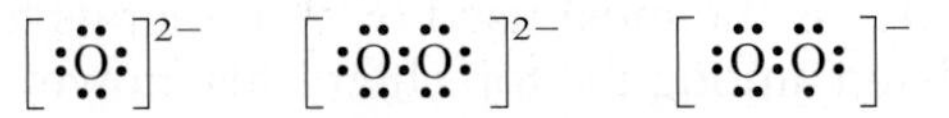

Oxide ion   Peroxide ion   Superoxide ion

The principal products of the reactions of the alkali metals with excess oxygen are summarized in the following table.

Principal Combustion Product			
Akali Metal	Oxide	Peroxide	Superoxide
Li	Li_2O		
Na		Na_2O_2	
K			KO_2
Rb			RbO_2
Cs			CsO_2

▶ These are not the only oxides formed by the alkali metals. About nine different cesium oxides are known.

Notice that the principal combustion product shifts from the oxide (M_2O) to the superoxide (MO_2) as we move down the group from Li to Cs. If we assume that the principal combustion product is the compound that is *most stable with respect to the starting materials,* then we conclude that $Li_2O(s)$ is more stable than $Li_2O_2(s)$ or $LiO_2(s)$, but for the larger alkali metal ions, $MO_2(s)$ is more stable than $M_2O(s)$ or $M_2O_2(s)$. In Figure 21-6, we use ΔG_f° values to compare the stabilities of the various oxides, relative to the elements M(s) and $O_2(g)$.

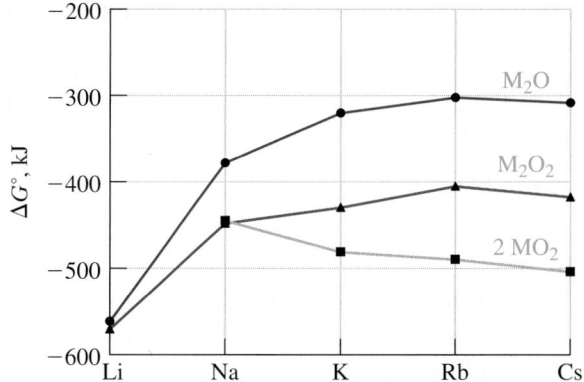

◀ **FIGURE 21-6**
Relative stabilities of M_2O, M_2O_2, and $2MO_2$ for the alkali metals
The values of $\Delta G°$ at 298 K for forming $M_2O(s)$, $M_2O_2(s)$, and 2 $MO_2(s)$ from $M(s)$ and $O_2(g)$ are plotted for the alkali metals. For K, Rb, and Cs, the superoxides are more stable relative to $M(s)$ and $O_2(g)$ than are the peroxides and monoxides. The $\Delta G°$ values for $Li_2O(s)$ and $Li_2O_2(s)$ are very similar at 298 K. At higher temperatures, $\Delta G_f°$ for $Li_2O(s)$ is more negative than that of $Li_2O_2(s)$.

For K, Rb, and Cs, the superoxide is most stable (lowest in energy) and for Li and Na, the oxides (M_2O) and peroxides (M_2O_2) are most stable.

Perhaps it comes as a surprise that the principal combustion product in these reactions is MO_2 (or M_2O_2) and not M_2O. Because the O^{2-} ion has a complete octet, it is tempting to jump to the conclusion that O^{2-} is a very stable ion. However, stability can be measured only relative to some set of reference conditions. Compared with oxygen molecules, the O^{2-} ion is not very stable. The formation of the O^{2-} ion from O_2 is a very endothermic process because the $O{=}O$ bond must be broken and two electrons must be added to an O atom. The formation of one mole of O^{2-} ions from O_2 molecules requires 852 kJ of energy, as shown below:

$$\frac{1}{2} O_2(g) \longrightarrow O(g) \qquad \Delta H = \frac{1}{2}(498 \text{ kJ})$$

$$O(g) + e^- \longrightarrow O^-(g) \qquad \Delta H = -141 \text{ kJ}$$

$$\underline{O^-(g) + e^- \longrightarrow O^{2-}(g) \qquad \Delta H = +744 \text{ kJ}}$$

$$\frac{1}{2} O_2(g) + 2e^- \longrightarrow O^{2-}(g) \qquad \Delta H = 852 \text{ kJ}$$

The formation of O_2^- or O_2^{2-} ions requires much less energy because the $O{=}O$ bond does not have to be broken. Because the energy requirement for forming O^{2-} ions is so large, it is somewhat surprising that $M_2O(s)$ is formed at all.

The energy consumed in the formation of M^+ and O^{2-} ions is offset by the energy released when M^+ and O^{2-} ions combine to give $M_2O(s)$, as shown below.

$$2 M^+(g) + O^{2-}(g) \longrightarrow M_2O(s) \qquad \Delta H < 0$$

The enthalpy change for the process above is the lattice energy (see Chapter 12). The formation of $Li_2O(s)$ starting from $Li(s)$ and $O_2(s)$ is very favorable because the lattice energy of Li_2O is very large (very negative). The lattice energy of Li_2O is large because the Li^+ ion is small and can pack around the O^{2-} ions very efficiently. The lattice energies of Na_2O, K_2O, Rb_2O, and Cs_2O are much smaller (less negative) than that of Li_2O, and so the formation of the M_2O lattice is much less favorable. The larger alkali metal ions pack more efficiently around the larger anions (O_2^{2-} or O_2^-), and thus the heavier alkali metals react with excess oxygen to give either M_2O_2 or MO_2.

The combustion of $Li(s)$ in excess oxygen produces some $Li_2O_2(s)$ as a minor product, and this suggests that $Li_2O_2(s)$ is a reasonably stable compound. Solid Li_2O_2 decomposes to Li_2O, as shown below, when heated to about 300 °C:

$$Li_2O_2(s) \xrightarrow{\Delta} Li_2O(s) + \frac{1}{2} O_2(g)$$

▲ The data in Figure 21-6 show that the $\Delta G_f°$ values for $Li_2O(s)$ and $Li_2O_2(s)$ are almost the same at 298 K. In a combustion reaction, heat is not dissipated immediately, and the system will be heated temporarily to a very high temperature. At high temperatures, $\Delta G_f°$ for $Li_2O(s)$ is more negative than it is for $Li_2O_2(s)$—see exercise 59—and the formation of $Li_2O(s)$ is thermodynamically favored.

The other alkali metal peroxides must be heated to higher temperatures ($>500 \,°C$) before decomposition occurs. The decomposition of $Li_2O_2(s)$ to $Li_2O(s)$ is thermodynamically favorable because of large differences in the lattice energies of the two solids, but the decomposition may also be assisted kinetically, as suggested below:

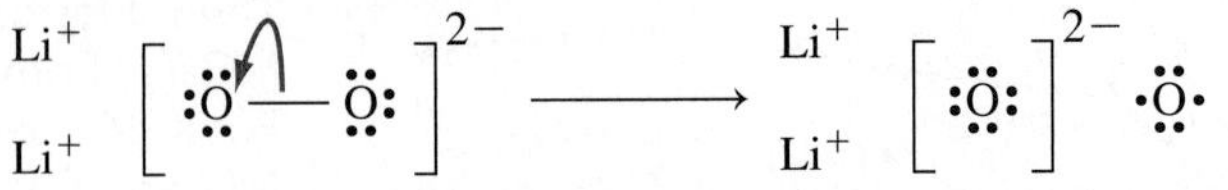

Because of the high polarizing power of the Li^+ ion, electrons in the O_2^{2-} bond are dragged from the bond to one of the oxygen atoms, yielding an O^{2-} ion and an oxygen atom. The oxygen atom combines with another oxygen atom to form an O_2 molecule.

The peroxides have many important uses. For example, sodium peroxide is used as a bleaching agent and a powerful oxidant. Lithium peroxide or sodium peroxide are sometimes used in emergency breathing devices in submarines and spacecraft because these compounds react with carbon dioxide to produce oxygen, as shown below:

$$2\,M_2O_2(s) + 2\,CO_2(s) \longrightarrow 2\,M_2CO_3(s) + O_2(s) \qquad (M = Li, Na) \quad \textbf{(21.5)}$$

Potassium superoxide, KO_2, can also be used for this purpose. The oxides, peroxides, and superoxides of the alkali metals react with water to form basic solutions. The reaction of an alkali metal oxide with water is an acid–base reaction that produces the alkali metal hydroxide.

The chemical equation and net ionic equation for the reaction are given below:

$$M_2O(s) + H_2O(l) \longrightarrow 2\,MOH(aq)$$

$$O^{2-}(aq) + H_2O(l) \longrightarrow 2\,OH^-(aq)$$

The resulting solution is quite basic because one mole of the oxide produces two moles of hydroxide ions.

The peroxide ion reacts with water in a similar manner to produce hydroxide ion and hydrogen peroxide.

$$O_2^{2-}(aq) + 2\,H_2O(l) \longrightarrow 2\,OH^-(aq) + H_2O_2(aq)$$

Hydrogen peroxide then slowly disproportionates into water and oxygen (page 174).

The superoxide ion reacts with water to give hydroxide ions, hydrogen peroxide, and oxygen.

$$2\,O_2^-(aq) + 2\,H_2O(l) \longrightarrow 2\,OH^-(aq) + H_2O_2(aq) + O_2(g)$$

The hydroxides of the group 1 metals are strong bases because they dissociate to release hydroxide ions in aqueous solution. As we learned in Section 20-8, sodium hydroxide is produced commercially by the electrolysis of $NaCl(aq)$. $Na^+(aq)$ goes through the electrolysis unchanged; $Cl^-(aq)$ is oxidized to $Cl_2(g)$; and H_2O is reduced to $H_2(g)$. Potassium hydroxide and lithium hydroxide are made in a similar fashion. Alkali metal hydroxides can also be prepared by the reaction of the group 1 metals with water (equation 21.2). Alkali hydroxides are important in the manufacture of soaps and detergents, described later in this section.

Carbonates and Sulfates Except for Li_2CO_3, all the alkali metal carbonates (M_2CO_3) are soluble in water and can be heated to very high temperatures ($>800 \,°C$) before decomposing to $M_2O(s)$ and $CO_2(g)$. The lower aqueous solubility of Li_2CO_3 and its lower stability with respect to the oxide are specific examples of a more general observation: *compounds derived from the first member of a group are often markedly different from compounds derived from the other members of the group.* Typically, the second member of the group exhibits chemistry most representative of the group. No single explanation accounts for the fact that the first member is not representative of the group. However, in many cases, the distinctive chemistry of the first member of the group can be attributed to the small atomic size, inability to expand the valence shell and especially for the *p*-block elements, extensive $\pi-$ bonding.

Lithium carbonate is used in the treatment of individuals who have bipolar disorder. Daily dosages of 1–2 g Li_2CO_3 maintain a level of Li^+ in the blood of about one millimole per liter. This treatment apparently influences the balance of Na^+ and K^+, that of Mg^{2+} and Ca^{2+}, or both, across cell membranes.

Sodium carbonate (soda ash) is used primarily in the manufacture of glass. The Na_2CO_3 produced in the United States now comes mostly from natural sources, such as the mineral *trona*, $Na_2CO_3 \cdot NaHCO_3 \cdot nH_2O$, found in dry lakes in California and in immense deposits in western Wyoming. In the past, sodium carbonate was manufactured mostly from NaCl, $CaCO_3$, and NH_3, using a process introduced by the Belgian chemist Ernest Solvay in 1863.

The great success of the *Solvay process* over the synthetic method outlined in Example 21-1 lies in the efficient use of certain raw materials through recycling. An outline of the process is shown in Figure 21-7. The key step involves the reaction of $NH_3(g)$ and $CO_2(g)$ in saturated NaCl(aq). Of the possible ionic compounds that could precipitate from such a mixture (NaCl, NH_4Cl, $NaHCO_3$, and NH_4HCO_3), the least soluble is *sodium hydrogen carbonate* (sodium bicarbonate). The following equation gives a simplified description of the process.

$$Na^+(aq) + Cl^-(aq) + NH_3(g) + CO_2(g) + H_2O(l) \longrightarrow$$
$$NaHCO_3(s) + NH_4^+(aq) + Cl^-(aq) \quad \textbf{(21.6)}$$

In industry, the reaction above is actually carried out in two stages. In the first stage, ammonia is bubbled into a concentrated brine (NaCl) solution, and in

◀ The term *ash* signifies a product obtained by heating or burning. The earliest alkaline materials (notably, potassium carbonate, *potash*) were extracted from the ashes of burned plants. The first commercial source of *soda ash* was the Leblanc process, featured in Example 21-1.

▲ The mineral *trona*, from Green River, Wyoming, is currently the principal source of Na_2CO_3 in the United States.

▲ FIGURE 21-7
The Solvay process for the manufacture of $NaHCO_3$
The main reaction sequence is traced by solid arrows. Recycling reactions are shown by dashed arrows.

the second stage, CO_2 is bubbled through the ammoniated brine. The sodium bicarbonate is isolated and sold, or it is converted to sodium carbonate by heating, as shown below in reaction (21.7).

$$2 \, NaHCO_3(s) \xrightarrow{\Delta} Na_2CO_3(s) + H_2O(g) + CO_2(g) \qquad \textbf{(21.7)}$$

One noteworthy feature of the Solvay process is that it involves only simple precipitation and acid–base reactions. Another feature is that materials produced in one step are efficiently recycled into a subsequent step. Recycling makes good economic sense: A process that recycles materials minimizes the use of raw materials (an expense in purchasing) and cuts down on the production of by-products (an expense in disposal). Thus, when limestone ($CaCO_3$) is heated to produce the reactant CO_2, the other reaction product, CaO, is also used. It is converted to $Ca(OH)_2$, which is then used to convert NH_4Cl (another reaction by-product) to $NH_3(g)$. The $NH_3(g)$ is recycled into the production of ammoniated brine.

The Solvay process has ultimately only one by-product—$CaCl_2$—for which the demand is very limited. In the past, some $CaCl_2$ was used for deicing roads in the winter (page 584) and for dust control on dirt roads in the summer (by means of the deliquescence of $CaCl_2 \cdot 6 \, H_2O$, page 578). The bulk of the $CaCl_2$, however, was dumped into local lakes and streams (notably, Onondaga Lake, near Solvay, New York), resulting in extensive environmental contamination. Environmental regulations no longer permit such dumping. Partly because of these regulations, but mainly for economic reasons, natural sources of sodium carbonate have supplanted the Solvay process in the United States. The process is still widely used elsewhere in the world, however.

Sodium sulfate, Na_2SO_4, is obtained partly from natural sources, partly from neutralization reactions, and partly through a process discovered by Johann Rudolf Glauber in 1625. Glauber's process is based on the following reactions.

▶ Equations (21.8) are often replaced by the overall equation, H_2SO_4(concd aq) + 2 NaCl(s) $\longrightarrow$ Na_2SO_4(s) + 2 HCl(g)

$$H_2SO_4(\text{concd aq}) + NaCl(s) \xrightarrow{\Delta} NaHSO_4(s) + HCl(g)$$
$$NaHSO_4(s) + NaCl(s) \xrightarrow{\Delta} Na_2SO_4(s) + HCl(g) \qquad \textbf{(21.8)}$$

The strategy behind reactions (21.8) is the production of a *volatile* acid (HCl) by heating one of its salts (NaCl) with a *nonvolatile* acid (H_2SO_4). Several other acids can be produced by similar reactions. The major use of Na_2SO_4 is in the paper industry. For instance, in the kraft process for papermaking, undesirable lignin is removed from wood by digesting the wood in an alkaline solution of Na_2S. The Na_2S required for this step is produced by the reduction of Na_2SO_4 with carbon.

$$Na_2SO_4(s) + 4 \, C(s) \xrightarrow{\Delta} Na_2S(s) + 4 \, CO(g)$$

About 45 kilograms of Na_2SO_4 is required for every metric ton of paper produced.

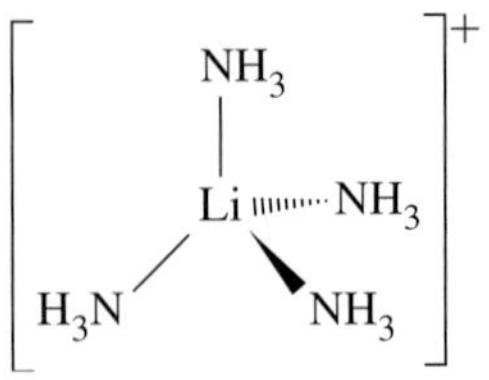

Alkali Metal Complexes Most alkali metals show little or no tendency to form complexes with neutral Lewis bases, such as NH_3, because most of them have relatively low charge densities. However, the Li^+ ion has an unusually high charge density and shows a greater tendency to form complexes with simple Lewis bases. For example, in the presence of ammonia, the lithium ion forms a tetrahedral complex, $[Li(NH_3)_4]^+$, which is shown in the margin.

In 1967, Charles J. Pedersen reported the discovery of a type of Lewis base that could coordinate the "recalcitrant alkali metal ions," a quote from his 1987 Nobel lecture. The structure of the Lewis base is that of a ring of oxygen atoms connected by $-CH_2CH_2-$ units, as shown in Figure 21-8(a). The ring structures were called *crown ethers* by Pedersen because they resemble a monarch's crown. Crown ethers are given a name that reflects its structure in

▲ The $[Li(NH_3)_4]^+$ complex

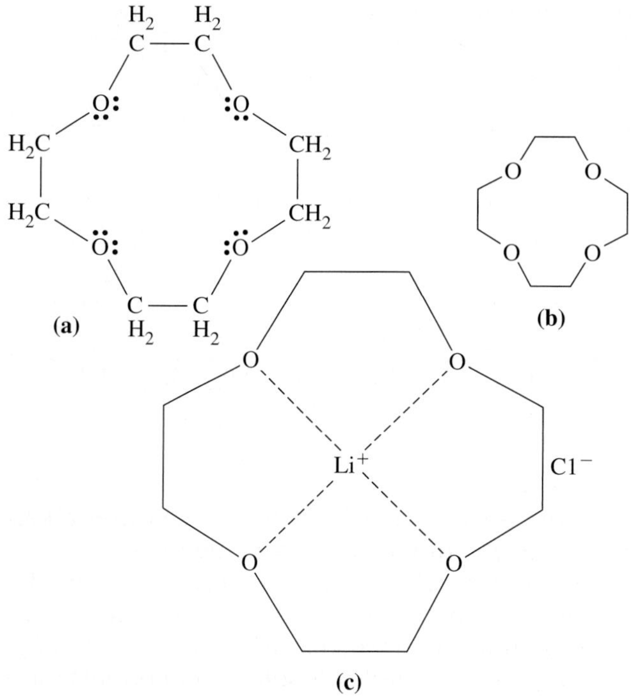

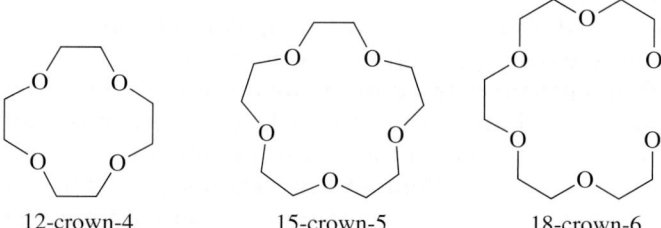

◀ FIGURE 21-8
12-crown-4 ether and the $[Li(12\text{-}crown\text{-}4)]^+$ complex
(a) The 12-crown-4 ether consists of a ring of 12 atoms, of which 4 are oxygen atoms. **(b)** A simplified illustration of 12-crown-4. **(c)** In the $[Li(12\text{-}crown\text{-}4)]^+$ complex, electron density is donated from the oxygen atoms to the Li^+ ion.

a very intuitive way. For example, the crown ether shown in Figure 21-8 is called 12-crown-4 because the ring consists of 12 carbon and oxygen atoms, and of those 12 ring atoms, 4 are oxygen. A line-angle formula of 12-crown-4 is shown in Figure 21-8(b). When a crown ether forms a complex with a metal ion, it encapsulates the metal ion, as shown in Figure 21-8(c). The metal ion is held in place by the donation of electron density from the oxygen atoms.

The cavity size of a crown ether depends on the number of $-OCH_2CH_2-$ units in the ring. This is illustrated in Figure 21-9. Cavity size is one of the factors that determine which cations will bind to a particular crown ether. The diameter of a K^+ ion is about 304 pm, and thus a K^+ ion fits nicely in the cavity of the 18-crown-6 ether. Thus, the K^+ ion binds preferentially to 18-crown-6, whereas Li^+ and Na^+ bind less effectively. A chemical equation representing the binding of a metal cation and 18-crown-6 is given below:

$$M^+ + 18\text{-}crown\text{-}6 \rightleftharpoons [M(18\text{-}crown\text{-}6)]^+$$

Equilibrium constants for the reaction of the alkali metal ions with 18-crown-6 are plotted in Figure 21-10. The figure shows that the equilibrium constant is greatest when the cation is K^+, partly because K^+ fits best into the cavity of 18-crown-6. However, the values of the equilibrium constants are not

12-crown-4   15-crown-5   18-crown-6

◀ FIGURE 21-9
Three different crown ethers
The cavity size of a crown ether increases with the number of atoms in the ring.

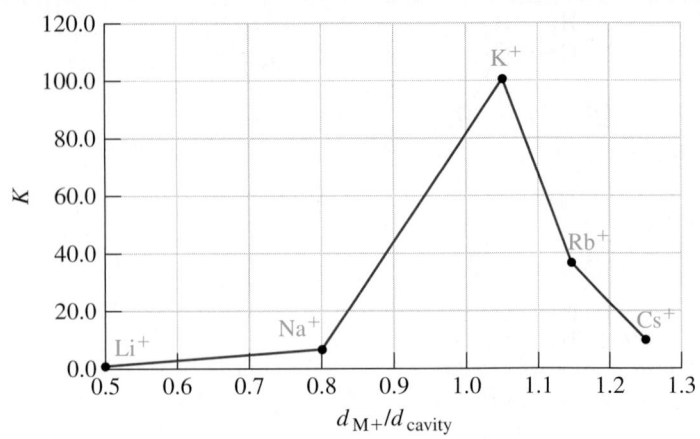

▶ FIGURE 21-10
Selectivity of 18-crown-6
The $\log_{10}K$ values for the reaction of alkali metal ions
(M^+) with 18-crown-6 in water at 25 °C are plotted
against the ratio of diameters, d_{M^+}/d_{cavity}. The
equilibrium constant is largest for K^+ because
the radius of this cation most closely matches that
of the cavity.

different enough to make 18-crown-6 selective for just K^+; the other alkali metal ions also form complexes with 18-crown-6 to varying degrees.

The formation of crown ether complexes has been exploited in synthetic organic chemistry to dissolve ionic reagents in nonpolar solvents, such as benzene. For example, $KMnO_4$ is not soluble in pure benzene but $[K(18\text{-crown-}6)]$ $[MnO_4]$ is soluble. The $[K(18\text{-crown-}6)]^+$ complex is soluble in nonpolar solvents because most of the complex is made up of a hydrocarbon skeleton, which is compatible with other nonpolar molecules. The MnO_4^- ion remains in the vicinity of the complexed cation, because of electrostatic attraction to the positive charge on the complex. However, the MnO_4^- ion does not interact strongly with the nonpolar benzene molecules, and thus it is hardly solvated by benzene molecules. Consequently, MnO_4^- is much more reactive in benzene than it is in water.

▶ An emulsion is a suspension of one liquid in another (see Table 13.4). Soap is an emulsifier because it facilitates the formation of an emulsion of oil droplets in water.

Alkali Metal Detergents and Soaps A **detergent** is a cleansing agent used primarily because it can emulsify oils. Although the term *detergent* includes common soaps, it is used primarily to describe certain synthetic products, such as sodium lauryl sulfate, whose manufacture involves the following conversions.

$$CH_3(CH_2)_{10}CH_2OH \longrightarrow CH_3(CH_2)_{10}CH_2OSO_3H \longrightarrow CH_3(CH_2)_{10}CH_2OSO_3^- Na^+$$

Lauryl alcohol · Lauryl hydrogen sulfate · Sodium lauryl sulfate

Notice that the lauryl sulfate anion has a long, nonpolar tail and a highly polar (negatively charged) head. These structural features are common to all detergents and soaps, as described in more detail below.

A **soap** is a specific kind of detergent that is the salt of a metal hydroxide and a fatty acid. An example is sodium palmitate, which is the product of the reaction of palmitic acid and NaOH.

$$CH_3(CH_2)_{14}{-}\overset{\overset{\displaystyle O}{\|}}{C}{-}O{-}H + Na^+ + OH^- \longrightarrow CH_3(CH_2)_{14}{-}\overset{\overset{\displaystyle O}{\|}}{C}{-}O^- Na^+ + H_2O \quad \textbf{(21.9)}$$

Palmitic acid · Sodium palmitate (a soap)

Again, notice that the anion of the soap has a long, nonpolar tail and a polar head. Figure 21-11 illustrates the detergent action of sodium palmitate.

Sodium soaps are the familiar hard (bar) soaps. Potassium soaps have low melting points and are soft soaps. Lithium stearate ($C_{17}H_{35}CO_2Li$), a soap not seen in ordinary household use, is used to thicken some oils into greases. These greases have excellent water-repellent and lubricating properties at both high and low temperatures. The greases remain in contact with moving metal parts under conditions in which oil by itself would run off.

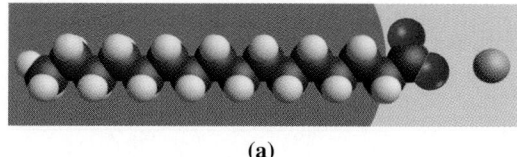

(a)

▲ FIGURE 21-11
Illustration of soap molecules and their cleaning action
(a) The sodium palmitate molecule has a long nonpolar portion buried in a droplet of oil and a polar head projecting into an aqueous medium. (b) Electrostatic attractions between these polar heads and water molecules cause the droplet to be emulsified, or solubilized.

(b)

🔍 **21-2   CONCEPT ASSESSMENT**

The nitrates MNO_3 (M = Na, K, Rb, Cs) decompose to the nitrites (MNO_2) on heating; while in contrast, $LiNO_3$ decomposes to Li_2O. Suggest a reason for this difference in behavior. Write balanced equations for both reactions.

21-3   Group 2: The Alkaline Earth Metals

As a group, the alkaline earth metals of group 2 are as common as the elements of group 1. Table 21.3 shows that calcium and magnesium are particularly abundant. Even beryllium, the least abundant member of group 2, is accessible because it occurs in deposits of the mineral *beryl*, $Be_3Al_2Si_6O_{18}$, which is shown in the margin. The other group 2 elements are found primarily as carbonates, sulfates, and silicates. Radium, like its group 1 neighbor, francium, is a radioactive element found only in trace amounts. Radium is more interesting for its radioactive properties than for its chemical similarities to the other group 2 elements.

Even though the group 2 metal oxides and hydroxides are only very slightly soluble in water, they are basic, or *alkaline*. At one time, insoluble substances that do not decompose on heating were called "earths." This term is the basis of the group 2 name: **alkaline earth metals**.

From a chemical standpoint (for example, in their abilities to react with water and acids and to form ionic compounds), the heavier group 2 metals—Ca, Sr, Ba, and Ra—are nearly as active as the group 1 metals. In terms of certain physical properties (for example, density, hardness, and melting point), all the group 2 elements are more typically metallic than the group 1 elements, as we can see by comparing Tables 21.2 and 21.4.

Table 21.4 shows that beryllium is out of step with the other group 2 elements in some of its physical properties. For instance, it has a higher melting point and is much harder than the others. Its chemical properties also differ significantly. For example,

- Be is quite unreactive with air and water;
- BeO does not react with water, whereas the other MO oxides form $M(OH)_2$;

TABLE 21.3   Abundances of Group 2 Elements

Elements	Abundances, ppm[a]	Rank
Be	2	51
Mg	27,640	6
Ca	46,600	5
Sr	384	15
Ba	390	14
Ra	Trace	—

[a]Grams per 1000 kg of solid crust.

▲ An emerald crystal, which is based on the mineral *beryl*, embedded in a calcite matrix.

TABLE 21.4 Some Properties of Group 2 (Alkaline Earth) Metals

	Be	Mg	Ca	Sr	Ba
Atomic number	4	12	20	38	56
Atomic (metallic) radius, pm	111	160	197	215	222
Ionic (M^{2+}) radius, pm[a]	41	86	114	132	149
Electronegativity	1.5	1.2	1.0	1.0	0.9
Charge Density of M^{2+}($C\ mm^{-3}$)	1108	120	52	33	23
First ionization energy, $kJ\ mol^{-1}$	899.4	737.7	589.7	549.5	502.8
Electrode potential $E°$, V[b]	−1.85	−2.356	−2.84	−2.89	−2.92
Melting point, °C	1278	648.8	839	769	729
Boiling point, °C	2970[c]	1090	1483.6	1383.9	1637
Density, $g\ cm^{-3}$ at 20 °C	1.85	1.74	1.55	2.54	3.60
Hardness[d]	~5	2.0	1.5	1.8	~2
Electrical conductivity[d]	39.7	35.6	40.6	6.90	3.20
Flame color	None	None	Orange-red	Scarlet	Green

[a]Ionic radii are for a coordination number of 6, except for Be^{2+}, for which the coordination number is 4.
[b]For the reduction $M^{2+}(aq) + 2\,e^- \longrightarrow M(s)$.
[c]Boiling point at 5 mmHg pressure.
[d]See footnotes c and d of Table 21.2.

- Be and BeO dissolve in strongly basic solutions to form the ion $[Be(OH)_4]^{2-}$;
- $BeCl_2$ and BeF_2 in the molten state are poor conductors of electricity; they are covalent substances.

The unusual chemical behavior of beryllium is related to the high charge density of the beryllium cation. Beryllium shows only a limited tendency to form ionic compounds because the small Be^{2+} ion polarizes any nearby anion, drawing electron density toward itself, creating a bond with significant covalent character. Thus, compounds of beryllium show some properties that are more typical of covalent solids. For example, while the other oxides of group 2 are basic, BeO is an *amphoteric* oxide, reacting with both strong acids and bases to yield complex ions:

$$BeO(s) + H_2O(l) + 2\,H_3O^+(aq) \longrightarrow [Be(H_2O)_4]^{2+}(aq)$$

$$BeO(s) + H_2O(l) + 2\,OH^-(aq) \longrightarrow [Be(OH)_4]^{2-}(aq)$$

In these complex ions, electron density is donated from H_2O or OH^- to Be^{2+} because of the high polarizing power of the Be^{2+} ion and the resulting complex has a well-defined structure. The figure in the margin shows that the $[Be(H_2O)_4]^{2+}$ ion, for example, has a tetrahedral structure.

The reaction of beryllium metal with aqueous acids yields hydrogen and ionic compounds, such as $BeCl_2 \cdot 4\,H_2O$. In $BeCl_2 \cdot 4\,H_2O$, water molecules are covalently bonded to Be^{2+} ions, producing the complex cations $[Be(OH_2)_4]^{2+}$ that, together with the anions Cl^-, form the crystal lattice. In covalent compounds, Be atoms appear to use hybrid orbitals—sp orbitals in $BeCl_2(g)$ and sp^3 orbitals in $BeCl_2(s)$ (Fig. 21-12).

▲ Tetrahedral shape of the $[Be(OH_2)_4]^{2+}$ ion.

21-3 CONCEPT ASSESSMENT

Why, on going from monomeric $BeCl_2$ to dimeric $(BeCl_2)_2$ to polymeric $(BeCl_2)_n$, does the atomic arrangement around the beryllium change from linear to trigonal planar to tetrahedral?

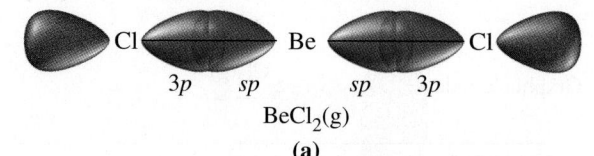

BeCl$_2$(g)
(a)

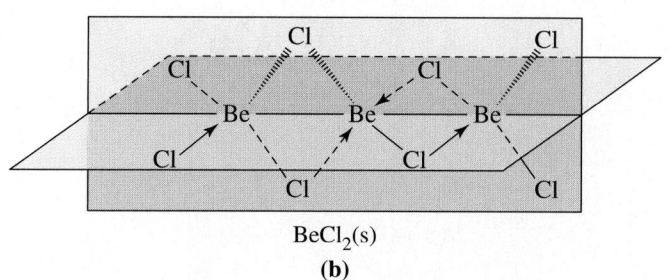

BeCl$_2$(s)
(b)

◀ FIGURE 21-12
Covalent bonds in BeCl$_2$
(a) In gaseous BeCl$_2$, discrete molecules exist with the bonding scheme shown. **(b)** In solid BeCl$_2$, two Cl atoms are bonded to one Be atom by normal covalent bonds. Two other Cl atoms are bonded by coordinate covalent bonds, using lone-pair electrons on the Cl atoms. Once formed, these two types of bonds are indistinguishable. BeCl$_2$ units are linked into long, chain-like polymeric molecules as (BeCl$_2$)$_n$.

Production and Uses of the Alkaline Earth Metals

The preferred method of producing the group 2 metals (except Mg) is by reducing their salts with other active metals. Beryl, Be$_3$Al$_2$Si$_6$O$_{18}$, is the natural source of beryllium compounds. This mineral is processed to produce BeF$_2$, which is then reduced with Mg to give Be(s). Beryllium metal is used as an alloying agent when low density is a primary requirement. Because Be can withstand metal fatigue, an alloy of copper with about 2% Be is used in springs, clips, and electrical contacts. The Be atom does not readily absorb X-rays or neutrons, so beryllium is also used to make windows for X-ray tubes and for various components in nuclear reactors. Beryllium and its compounds are limited in their use, however, because they are toxic. In addition, they are suspected of being carcinogens even at a level as low as 0.002 ppm in air.

Calcium, strontium, and barium are obtained by the reduction of their oxides with aluminum; Ca and Sr also can be obtained by electrolysis of their molten chlorides. Calcium metal is used primarily as a reducing agent to prepare, from their oxides or fluorides, other metals such as U, Pu, and most of the lanthanides. Strontium and barium have limited use in alloys, but some of their compounds (discussed later) are quite important. Some salts of Sr and Ba provide vivid colors for pyrotechnic displays.

Magnesium metal is obtained by electrolysis of the molten chloride in the *Dow process.* The Dow process is outlined in Figure 21-13, and the electrolysis of

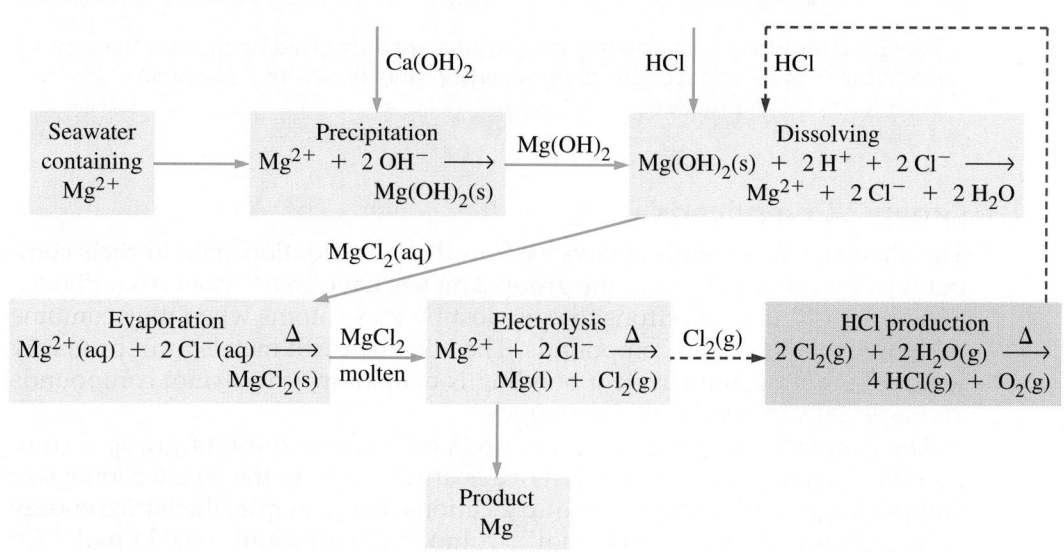

◀ FIGURE 21-13
The Dow process for the production of Mg
The main reaction sequence is traced by solid arrows. The recycling of Cl$_2$(g) is shown by dashed arrows.

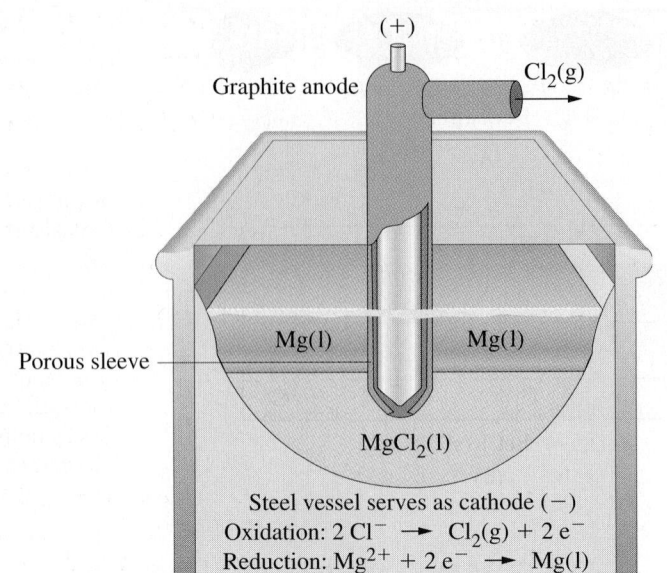

▶ FIGURE 21-14
The electrolysis of molten MgCl₂
The electrolyte is a mixture of molten NaCl, CaCl₂, and MgCl₂. This mixture has a lower melting point and higher electrical conductivity than MgCl₂ alone, but if the voltage is carefully controlled, only Mg^{2+} is reduced in the electrolysis.

▶ In Example 18-6, we demonstrated that this precipitation goes to completion.

$MgCl_2(l)$ is pictured in Figure 21-14. Like the Solvay process for making $NaHCO_3$, the Dow process takes advantage of simple chemistry and recycling.

The source of magnesium is seawater or natural brines. The abundance of Mg^{2+} in seawater is about 1350 mg/L. The first step in the Dow process is the precipitation of $Mg(OH)_2(s)$ with slaked lime, $Ca(OH)_2$, as the source of OH^-. Slaked lime is formed by the reaction of quicklime (CaO) with water. The precipitated $Mg(OH)_2(s)$ is washed, filtered, and dissolved in HCl(aq). The resulting concentrated $MgCl_2(aq)$ is dried by evaporation, melted, and electrolyzed, yielding pure Mg metal and $Cl_2(g)$. The $Cl_2(g)$ is converted to HCl, which is recycled.

Magnesium has a lower density than that of any other metal used for structural purposes. Lightweight objects, such as aircraft parts, are manufactured from magnesium alloyed with aluminum and other metals. Magnesium is a good reducing agent and is used in a number of metallurgical processes, such as the production of beryllium, as mentioned earlier. The ease with which magnesium is oxidized also underlies its use in sacrificial anodes for corrosion protection (page 895). Magnesium's most spectacular use may be in firework, as it burns in air with a brilliant white light.

🔍 21-4   CONCEPT ASSESSMENT

The good reducing properties of magnesium are illustrated by the fact that the metal burns in an atmosphere of pure carbon dioxide. Write a plausible equation(s) for this reaction.

Group 2 Compounds

The alkaline earth metals always exist in the +2 oxidation state in their compounds. Recall that atoms of the group 2 metals have an ns^2 valence configuration and it is the ns^2 electrons that are lost by these atoms when they combine with nonmetals to form compounds. The alkaline earth metals form primarily ionic compounds, but covalent bonding is evident in magnesium compounds and especially in beryllium compounds.

The properties of group 2 compounds differ from those of group 1 compounds. In some cases, this difference is attributable to the smaller ionic size and the larger ionic charge of group 2 cations. For example, the lattice energy of $Mg(OH)_2$ is about -3000 kJ mol^{-1}, compared with about -900 kJ mol^{-1} for

NaOH. This difference in lattice energy helps explain why NaOH is very soluble in water—up to about 20 M NaOH(aq)—while $Mg(OH)_2$ is only sparingly soluble, with $K_{sp} = 1.8 \times 10^{-11}$. The heavier group 2 hydroxides are somewhat more soluble than $Mg(OH)_2$. Other alkaline earth compounds that are only slightly soluble include the carbonates, fluorides, and oxides.

Halides The group 2 metals react directly with the halogens to form halides:

$$M(s) + X_2 \longrightarrow MX_2(s) \qquad \textbf{(21.10)}$$

(M = a group 2 metal, and X = F, Cl, Br, or I)

However, the standard method for preparing $MX_2(s)$ in anhydrous form is to dehydrate the hydrates obtained from the reaction of the metal and aqueous hydrohalic acid, HX(aq). However, this method of preparation cannot be used to prepare beryllium halides because the hydrates of beryllium halides decompose to $Be(OH)_2(s)$ on heating. For example, when $[Be(H_2O)_4]Cl_2(s)$ is heated, the following reaction occurs:

$$[Be(H_2O)_4]Cl_2(s) \xrightarrow{\Delta} Be(OH)_2(s) + 2H_2O(g) + 2HCl(g)$$

Instead anhydrous $BeCl_2$ is prepared from BeO and CCl_4, as shown below.

$$2\,BeO + CCl_4 \xrightarrow{1070\ K} 2\,BeCl_2 + CO_2$$

The halides have varied uses. For example, $MgCl_2$ is used in the preparation of magnesium metal, in fireproofing wood, in special cements, in ceramics, in treating fabrics, and as a refrigeration brine.

Oxides and Hydroxides All the group 2 metals burn in air to produce oxides, MO(s):

$$2\,M(s) + O_2(g) \longrightarrow 2\,MO(s)$$

In the presence of excess oxygen, the heavier group 2 metals—such as barium—form peroxides:

$$Ba(s) + O_2(g) \longrightarrow BaO_2(s)$$

The peroxides of Mg, Ca, and Sr are also known but they are less stable, presumably because the charge densities of Mg^{2+}, Ca^{2+}, and Sr^{2+} are high enough to assist the decomposition of O_2^{2-}, as suggested on page 928. On heating, the peroxides decompose to give MO(s) and $O_2(g)$:

$$MO_2(s) \xrightarrow{\Delta} MO(s) + \frac{1}{2}O_2(g)$$

Although oxides of the group 2 metals can be obtained from burning the metals in air or oxygen (or by thermal decomposition of the peroxides), the chief method of preparing them, except for BeO(s), is the thermal decomposition of the respective carbonate:

$$MCO_3(s) \xrightarrow{\Delta} MO(s) + CO_2(g) \quad (M = Mg, Ca, Sr, and Ba) \qquad \textbf{(21.11)}$$

A particularly useful oxide is calcium oxide, CaO(s), also known as **quicklime**. It is used in water treatment, in the removal of $SO_2(g)$ from the smokestack gases in electric power plants, and in the making of $Ca(OH)_2$—an important and inexpensive strong base. The conversion of CaO to $Ca(OH)_2$ is a specific example of the reaction that converts a group 2 oxide into a hydroxide:

$$MO(s) + H_2O(l) \longrightarrow M(OH)_2(aq) \qquad \textbf{(21.12)}$$

All the hydroxides of the group 2 metals are strong bases. Calcium hydroxide, $Ca(OH)_2$, also known as **slaked lime**, is the cheapest commercial strong base. It is not very soluble in water, but it is used in a variety of applications, such as the Solvay and Dow processes.

A mixture of slaked lime, sand, and water composes the mortar used in bricklaying. Excess water in the mortar is absorbed by the bricks and then lost by evaporation. In the final setting of the mortar, $CO_2(g)$ from the air reacts with $Ca(OH)_2(s)$ and converts it to $CaCO_3(s)$, as shown below:

$$Ca(OH)_2(s) + CO_2(g) \longrightarrow CaCO_3(s) + H_2O(g) \qquad \textbf{(21.13)}$$

The final form of the mortar is a complex mixture of hydrated calcium carbonate and silicate (from the sand).

Reaction (21.13) is general for all group 2 hydroxides. Art conservators have used this reaction to preserve art objects. For example, an aqueous solution of $Ba(NO_3)_2$ is sprayed onto cracking frescos (paintings embedded in plaster). When the solution has had time to fill small cracks and spaces, an aqueous ammonia solution is applied to the surface of the fresco. The ammonia raises the pH of the solution, resulting in formation of $Ba(OH)_2$. As excess water evaporates, carbon dioxide from the air reacts with the barium hydroxide. Insoluble barium carbonate is produced, binding the cracking fresco together and strengthening it without affecting the delicate colors.

🔍 21-5 CONCEPT ASSESSMENT

Comment on the statement "The best way to way to prepare BeO is to heat the compound $BeCO_3(s)$." If you disagree with this statement suggest an alternative.

Hydration of Salts Another common characteristic of alkaline earth compounds is the formation of hydrates. Typical hydrates are $MX_2 \cdot 6\,H_2O$, where M = Mg, Ca, or Sr, and X = Cl or Br. The Ba^{2+} ion has a low charge density and shows little or no tendency to retain its hydration sphere in the solid state. The formulas for the hydrates of the alkaline earth nitrates illustrate that the degree of hydration typically decreases as the charge density of the metal ion decreases: $Mg(NO_3)_2 \cdot 6\,H_2O$, $Ca(NO_3)_2 \cdot 4\,H_2O$, $Sr(NO_3)_2 \cdot 4\,H_2O$, $Ba(NO_3)_2$.

Carbonates and Sulfates The group 2 carbonates are insoluble in water, as are the sulfates of Ca, Sr, and Ba. Because of this insolubility, these compounds are the most important minerals of the group 2 metals. The most familiar is $CaCO_3$, the principal component of the rock *limestone*. If a limestone contains more than 5% $MgCO_3$, it is usually called dolomite limestone, or *dolomite*. Some clay, sand, or quartz may also be present in limestone. The primary use of limestone (about 70%) is as a building stone. Among other applications, limestone is used in the manufacture of quicklime and slaked lime, as an ingredient in glass, and as a flux in metallurgical processes. A *flux* is a material that combines with impurities and removes them as a free-flowing liquid—a *slag*—during production of a metal (page 1044).

Portland cement, another important product of limestone, is a complex mixture of calcium silicates and aluminates. It is produced in long rotary kilns like the one shown in the photograph in the margin. In the kiln, mixtures of limestone, clay, and sand are heated to progressively higher temperatures as they slowly move down the inclined kiln. First, moisture and then chemically bound water are driven off. This process is followed by decomposition (calcination) of the limestone to $CaO(s)$ and $CO_2(g)$. Finally, CaO combines with silica (SiO_2) and alumina (Al_2O_3) from the sand and clay to form silicates and aluminates. Pure cement does not have much strength. When mixed with sand, gravel, and water, however, it sets into the familiar rock-like mass called *concrete*. Portland cement is an especially valuable material for building bridge piers and other underwater structures because it hardens even when underwater—it is a *hydraulic* cement.

▲ The decomposition (calcination) of limestone is carried out in a long rotary kiln, whether for the production of quicklime, CaO, or for the manufacture of Portland cement.

Pure, white $CaCO_3$ is used in a wide variety of products. For example, it is used in papermaking to impart brightness, opacity, smoothness, and good ink-absorbing qualities to paper. It is particularly suited to newer papermaking processes that produce acid-free (alkaline) paper with an expected shelf life of 300 years or more. $CaCO_3$ is used as a filler in plastics, rubber, floor tiles, putties, and adhesives, as well as in foods and cosmetics. It is also used as an antacid and as a dietary supplement for the prevention of osteoporosis, a condition in which the bones become porous and brittle and break easily.

Three steps are required to obtain pure $CaCO_3$ from limestone: (1) thermal decomposition of limestone (called **calcination**), (2) reaction of CaO with water (slaking), and (3) conversion of an aqueous suspension of $Ca(OH)_2(s)$ to precipitated $CaCO_3$ (carbonation).

Calcination: $\qquad\qquad CaCO_3(s) \xrightarrow{\Delta} CaO(s) + CO_2(g)$ **(21.14)**

Slaking: $\qquad\quad CaO(s) + H_2O(l) \longrightarrow Ca(OH)_2(s)$ **(21.15)**

Carbonation: $\ Ca(OH)_2(s) + CO_2(aq) \xrightarrow{aq} CaCO_3(s) + H_2O(l)$ **(21.16)**

The calcination of limestone, reaction (21.14), is highly reversible at room temperature. Thus, a high temperature must be used and $CO_2(g)$ must be continuously removed from the kiln to prevent the reverse reaction.

Limestone ($CaCO_3$) is also responsible for the beautiful natural formations found in limestone caves. Natural groundwater is slightly acidic because of dissolved $CO_2(g)$ and is essentially a solution of carbonic acid, H_2CO_3.

$$CO_2(aq) + 2\,H_2O(l) \rightleftharpoons H_3O(aq)^+ + HCO_3^-(aq) \quad K_{a_1} = 4.4 \times 10^{-7} \quad \textbf{(21.17)}$$

$$HCO_3^-(aq) + H_2O(l) \rightleftharpoons H_3O^+(aq) + CO_3^{2-}(aq) \qquad K_{a_2} = 4.7 \times 10^{-11} \quad \textbf{(21.18)}$$

Although the carbonates are not very soluble in water, they are bases; thus, they dissolve readily in acidic solutions. As mildly acidic groundwater seeps through limestone beds, insoluble $CaCO_3$ is converted to soluble $Ca(HCO_3)_2$.

$$CaCO_3(s) + H_2O(l) + CO_2(aq) \rightleftharpoons Ca(HCO_3)_2(aq) \qquad K = 2.6 \times 10^{-5} \quad \textbf{(21.19)}$$

Over time, this dissolving action can produce a large cavity in the limestone bed—a limestone cave. Reaction (21.19) is reversible, however, and evaporation of the solution causes a loss of both water and CO_2 and conversion of $Ca(HCO_3)_2(aq)$ back to $CaCO_3(s)$. This process occurs very slowly. Yet over a period of many years, as $Ca(HCO_3)_2(aq)$ drips from the ceiling of a cave, $CaCO_3(s)$ remains as icicle-like deposits called **stalactites**. Some of the dripping solution may hit the floor of the cave before decomposition occurs; in this way, limestone deposits build up from the floor in formations called **stalagmites**. Eventually, some stalactites and stalagmites grow together into limestone columns, as shown in Figure 21-15.

Another important calcium-containing mineral is *gypsum*, $CaSO_4 \cdot 2\,H_2O$. In the United States, about 50 million metric tons of gypsum are consumed annually. About half of this is converted to **plaster of Paris**, $CaSO_4 \cdot \frac{1}{2}H_2O$, a hemihydrate.

$$CaSO_4 \cdot 2\,H_2O(s) \xrightarrow{\Delta} CaSO_4 \cdot \tfrac{1}{2}H_2O(s) + \tfrac{3}{2}H_2O(g) \qquad \textbf{(21.20)}$$

When mixed with water, plaster of Paris reverts to gypsum. Because it expands as it sets, a mixture of plaster of Paris and water is useful in making castings where sharp details of an object must be retained. Plaster of Paris is used extensively in jewelry making and in dental work. The most important application, though, is in producing gypsum wallboard (drywall), which has all but supplanted other interior wall coverings in the construction industry.

Barium sulfate has had important applications in medical imaging because barium is opaque to X-rays. Although barium ion is toxic, the compound $BaSO_4$ is so insoluble as to be safe to use as a "barium milkshake" to coat the stomach or upper gastrointestinal tract or in a "barium enema" for the lower tract.

▲ FIGURE 21-15
Stalactites and stalagmites
Stalactites hang from the top; stalagmites rise from the ground.

▲ Plaster of Paris castings and the molds used to form them.

Diagonal Relationship of Lithium and Magnesium

We have described trends in the periodic table either vertically (down a group) or horizontally (across periods), but an investigation of the chemistry of lithium and magnesium reveals a **diagonal relationship**. Let us look at these similarities.

The following list contains some of the chemical similarities between lithium and magnesium that lead to the diagonal relationship between them.

- Lithium and magnesium combine with O_2 to give the oxide rather than the peroxide.
- The carbonates of lithium and magnesium can be decomposed thermally to give the oxide and carbon dioxide. The carbonates of the remaining group 1 metals are stable to thermal decomposition.
- As mentioned earlier the salts of lithium and magnesium are strongly hydrated.
- The fluorides of lithium and magnesium are sparingly soluble in water, whereas the later group 1 fluorides are soluble.
- LiOH is the least soluble of the group 1 hydroxides and $Mg(OH)_2$ is only sparingly soluble in water.

The diagonal relationship between lithium and magnesium can be understood in terms of charge densities. The similarities in charge density arises from the increase in size of the Mg^{2+} relative to Li^+, which is counterbalanced by the increase in charge. Both Li^+ and Mg^{2+} have high polarizing power and their compounds exhibit a high degree of covalency. Diagonal relationships also exist between Be and Al and between B and Si also. These relationships are emphasized in Figure 21-16.

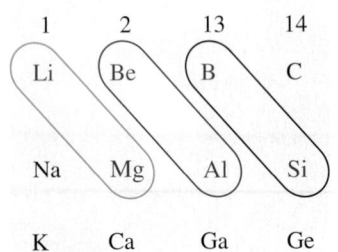

▲ FIGURE 21-16
Diagonal relationships
The two elements in each encircled pair exhibit many similar properties.

EXAMPLE 21-2 Identifying Elements and Compounds from a Description of Reaction Chemistry

In an experiment, 0.1 moles of M, a group 1 metal, react with sufficient oxygen to give 0.05 moles of compound X. Compound X is then allowed to react with water, and a hydroxide is the only product. In a separate experiment, 0.1 moles of the metal reacts with water to give 0.1 moles of a hydroxide and 0.05 moles of a gas, Y. Identify the metal and compounds X and Y. Write balanced chemical equations for the reactions described.

Analyze

Three reactions are described and some information is provided for each reaction. One way to tackle this problem is to write partial chemical equations for these reactions, and then use the information provided to complete the equations.

Solve

We are told that M is a group 1 metal, so the hydroxide has the formula MOH. The partial chemical equations are as follows:

$$M + O_2 \longrightarrow X$$

$$X + H_2O \longrightarrow MOH$$

$$M + H_2O \longrightarrow MOH + \frac{1}{2}Y$$

All the alkali metals react with water to give MOH and H_2. Thus, Y must be H_2. In the first reaction, X could be M_2O, M_2O_2, or MO_2. However, the second reaction has only one product. Because M_2O_2 and MO_2 react with water to give MOH and other products, we know that X cannot be M_2O_2 or MO_2. Thus, X must be M_2O. The only alkali metal that reacts with oxygen to give M_2O as the principal product is lithium, so M must be Li. The complete chemical equations are

$$2\,Li(s) + \frac{1}{2}O_2(g) \longrightarrow Li_2O(s)$$

$$Li_2O(s) + H_2O(l) \longrightarrow 2\,LiOH(s)$$

$$Li(s) + H_2O(l) \longrightarrow LiH(aq) + \frac{1}{2}H_2(g)$$

Assess

The key to solving this problem was in recognizing that only the normal oxide, M_2O, reacts with water to give MOH as the only product and that lithium is the only alkali metal that gives M_2O.

PRACTICE EXAMPLE A: Sodium nitrite (0.1 mol) reacts with sodium metal (0.3 mol) to give compound X (0.2 mol) and nitrogen gas (0.05 mol). Compound X (0.1 mol) reacts with oxygen (0.05 mol) to give compound Y (0.1 mol) only. Identify X and Y, and write balanced chemical equations for the reactions described.

PRACTICE EXAMPLE B: A group 2 metal (0.1 mol) is heated with carbon (0.2 mol) at 1100 °C to produce a single compound X (0.1 mol). When compound X (0.1 mol) is heated in excess $N_2(g)$ at 1100 °C solid carbon (0.1 mol) is produced along with compound Y (0.1 mol). The group 2 metal sulfate is used in plaster of Paris and the anion in compound Y is isoelectronic with CO_2. Identify compounds X and Y, write balanced chemical equations for the reactions described and describe the shape of the anion in Y.

21-4   Group 13: The Boron Family

The only group 13 element that is almost exclusively nonmetallic in its physical and chemical properties is boron. The remaining members of group 13—Al, Ga, In, and Tl—are metals and will be discussed later in this section. In group 13 we find for the first time elements possessing more than one oxidation state. All the elements of this group exhibit both the +1 and +3 oxidation states. Boron is a nonmetal and forms primarily covalent bonds. The other members of the group, despite being metals, commonly form covalent bonds. The tendency for forming covalent bonds can be attributed to the high charge densities (Table 21.5) of the group 13 ions. The high charge density means that the group 13 ions have a high polarizing power that leads to covalent bonding between cation and an anion.

TABLE 21.5 Charge Densities of Group 13 Elements in the +3 Oxidation State	
Element	Charge Density, $C\,mm^{-3}$
B	1663
Al	364
Ga	261
In	138
Tl	105

Boron and Its Compounds

Many boron compounds lack an octet of electrons about the central boron atom, which makes the compounds *electron deficient*. This deficiency also makes them strong Lewis acids. The electron deficiency of some boron compounds leads to bonding of a type that we have not previously encountered. This type of bonding occurs in the boron hydrides.

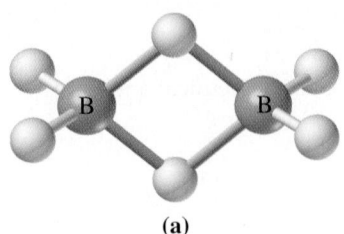

(a)

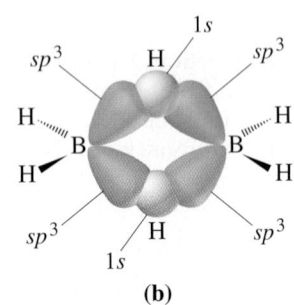

(b)

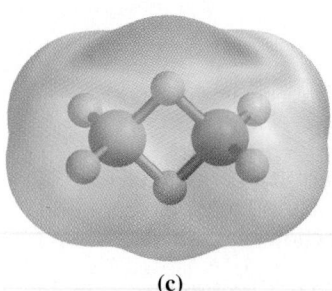

(c)

▲ FIGURE 21-17
Structure of diborane, B_2H_6
(a) The molecular structure.
(b) Bonding.
(c) Electrostatic potential map.

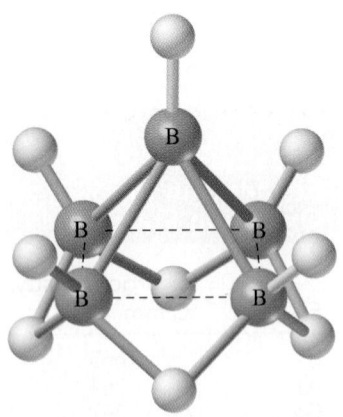

▲ FIGURE 21-18
Structure of pentaborane, B_5H_9
The boron atoms are joined by multicenter B—B—B bonds. Five of the H atoms are bonded to one B atom each. The other four H atoms bridge pairs of B atoms.

Boron Hydrides The molecule BH_3 (borane) may exist as a reaction intermediate, but it has not been isolated as a stable compound. The B atom in BH_3 lacks a complete octet—it has only six electrons in its valence shell. The simplest boron hydride that has been isolated is diborane, B_2H_6, but this molecule defies simple description. In the following structural formula, what holds the two borane units together?

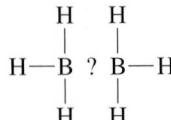

To explain the structure and bonding in B_2H_6, we need to use molecular orbital theory because simpler bonding theories fail for this molecule. The problem is this: The B_2H_6 molecule has only *12* valence electrons (three each from the two B atoms, and one each from the six H atoms). The minimum number of valence-shell atomic orbitals required to make a Lewis structure for B_2H_6 resembling that of C_2H_6, however, is *14* (four each from the two B atoms, and one each from the six H atoms).

The structure of diborane is illustrated in Figure 21-17. The two B atoms and four of the H atoms lie in the same plane (a plane perpendicular to the plane of the page). The orbitals used by the B atoms to bond these particular four H atoms can be viewed as sp^3. Eight electrons are involved in these four bonds. Four electrons are left to bond the two remaining H atoms to the two B atoms and also to bond the B atoms together. This is accomplished if each of the two H atoms simultaneously bonds to *both* B atoms.

Atom bridges are actually fairly common, although we have not had much occasion to deal with them before. (See the discussion of Al_2Cl_6, page 949.) The B—H—B bridges are unusual, however, in having only *two* electrons shared among *three* atoms. For this reason, these bonds are referred to as *three-center two-electron bonds*.

We can rationalize the bonding in these three-center bonds with molecular orbital theory. The *six* atomic orbitals shown in Figure 21-17(b)—two sp^3 orbitals from each B atom and an s orbital from each bridging H atom—are combined into *six* molecular orbitals in these two bridge bonds. Of these six molecular orbitals, *two* are bonding orbitals, and these are the orbitals into which the four electrons are placed. The concept of bridge bonds can be extended to B—B—B bonds to describe the structure of higher boranes, such as B_5H_9 (Fig. 21-18).

Boron hydrides are widely used in reactions for synthesizing organic compounds. They continue to provide new and exciting developments in chemistry.

Other Boron Compounds Boron compounds are widely distributed in Earth's crust, but concentrated ores are found in only a few locations—in Italy, Russia, Tibet, Turkey, and the desert regions of California. Typical of these ores is the hydrated borate *borax*, $Na_2B_4O_7 \cdot 10 H_2O$. Figure 21-19 illustrates how borax can be converted to a variety of boron compounds.

One useful compound that can be obtained by crystallization from a solution of borax and hydrogen peroxide is *sodium perborate*, $NaBO_3 \cdot 4 H_2O$. This formula is deceptively simple; a more precise formula is $Na_2[B_2(O_2)_2(OH)_4] \cdot 6 H_2O$. Sodium perborate contains the perborate ion, $[B_2(O_2)_2(OH)_4]^{2-}$, the structure of which is shown in Figure 21-20. Sodium perborate is a bleach alternative used in many color-safe bleaches. The key to the bleaching action is the presence of the two peroxo groups (—O—O—) bridging the boron atoms in the $[B_2(O_2)_2(OH)_4]^{2-}$ ion.

One of the key compounds from which other boron compounds can be synthesized is *boric acid*, $B(OH)_3$. The weakly acidic nature of boric acid comes about in a rather unusual way. The electron-deficient $B(OH)_3$ molecule accepts a

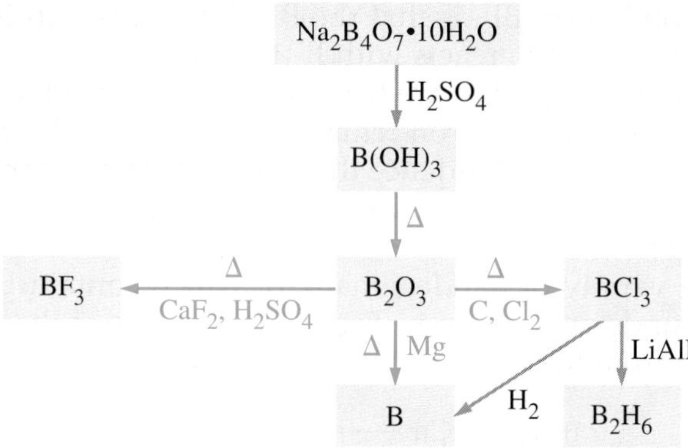

$Na_2B_4O_7 \cdot 10H_2O$

$\downarrow H_2SO_4$

$B(OH)_3$

$\downarrow \Delta$

$BF_3 \xleftarrow[CaF_2, H_2SO_4]{\Delta} B_2O_3 \xrightarrow[C, Cl_2]{\Delta} BCl_3$

$\Delta \downarrow Mg$ $\downarrow LiAlH_4$

$B \xleftarrow{H_2} B_2H_6$

▲ FIGURE 21-19
Preparation of some boron compounds
$Na_2B_4O_7 \cdot 10\ H_2O$ (borax) is converted to $B(OH)_3$ by reaction with H_2SO_4. When heated strongly, $B(OH)_3$ is converted to B_2O_3. A variety of boron-containing compounds and boron itself can be prepared from B_2O_3.

OH^- ion from the self-ionization of water, forming the complex ion $[B(OH)_4]^-$. Thus, the source of the H_3O^+ in $B(OH)_3(aq)$ is the water itself. This ionization scheme, together with the fact that $B(OH)_3$ is a *mono*protic, not *tri*protic, acid suggests that the best formula of boric acid is $B(OH)_3$, not H_3BO_3.

$$B(OH)_3(aq) + 2\,H_2O(l) \longrightarrow H_3O^+(aq) + B(OH)_4^-(aq) \qquad K_a = 5.6 \times 10^{-10}$$

Borate salts, as expected of the salts of a weak acid, produce basic solutions by hydrolysis, which accounts for their use in cleaning agents. Boric acid also acts as an insecticide, particularly to kill roaches, and as an antiseptic in eye-wash solutions. Boron compounds are used in products as varied as adhesives, cement, disinfectants, fertilizers, fire retardants, glass, herbicides, metallurgical fluxes, and textile bleaches and dyes.

The halides of boron, particularly BF_3 and BCl_3, provide examples of the Lewis acid behavior of boron compounds. For example, BF_3 can react with diethyl ether, $(C_2H_5)_2O$, to form an **adduct**:

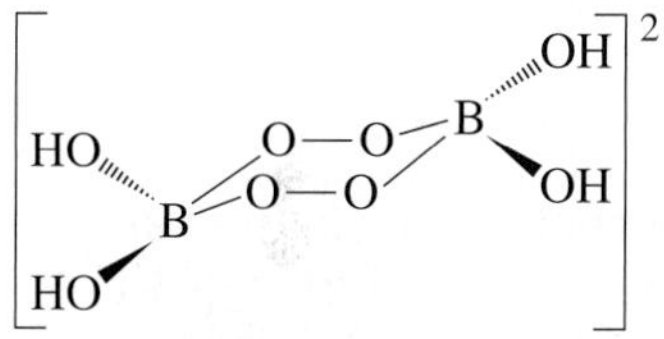

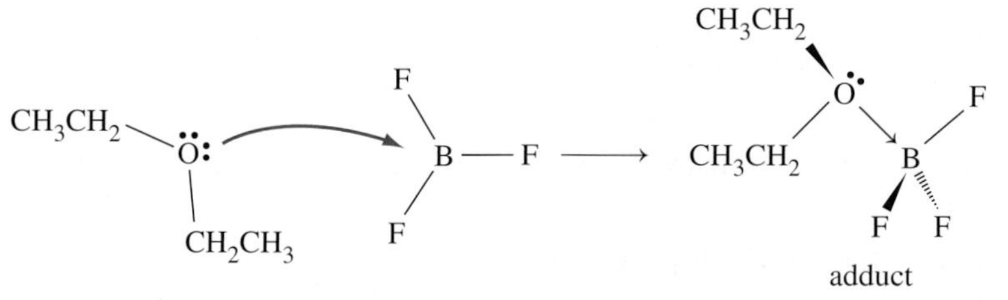

In the formation of an adduct, a coordinate covalent bond is formed between a Lewis acid and a Lewis base, with the pair of electrons for the bond coming from the Lewis base. In the reaction above, the transfer of electron density from the Lewis base to the Lewis acid is shown by the red arrow. Although a coordinate covalent bond, once formed, is indistinguishable from a regular covalent bond, we sometimes use an arrow, rather than a straight line, to identify a coordinate covalent bond. Notice that the hybridization of the boron atom changes from sp^2 to sp^3 when BF_3 and $(C_2H_5)_2O$ form an adduct.

Unlike BH_3, BF_3 does not dimerize, and the boron–fluorine bond energy in BF_3 is extremely high (646 kJ mol^{-1}) and is comparable to the bond energies of many double bonds. A possible explanation for these observations is that in a BF_3 molecule, some π bonding occurs in addition to σ bonding. A delocalized π system involving the empty $2p$ orbital on the boron atom and one full $2p$ orbital on each of the fluorine atoms can be formed (Figure 21-21).

▲ FIGURE 21-20
Perborate ion

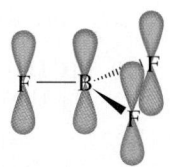

▲ FIGURE 21-21
Delocalized π system in BF_3
The $2p$ orbitals on the B and F atoms overlap to form a delocalized π system.

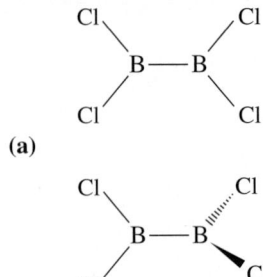

(a)

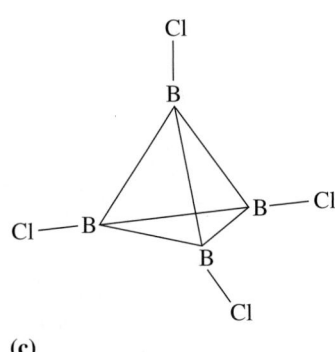

(b)

(c)

▲ FIGURE 21-22
Structures of (a) planar B₂Cl₄, (b) nonplanar B₂Cl₄, (c) tetrahedral B₄Cl₄,

Evidence for π bonding in BF_3 is that the B—F bond length increases (from 130 pm to 145 pm) when BF_3 reacts with F^- to form the BF_4^-, ion. In BF_4^-, the $2s$ and $2p$ orbitals on the B atom are used in σ bonding and are not available for π bonding. In the other boron trihalides, such as BCl_3, π bonding also occurs but to a lesser extent. Evidence that supports less π bonding in BCl_3 is that BCl_3 is a stronger Lewis acid than BF_3 and shows a greater tendency than BF_3 to form adducts.

Boron forms a variety of halides, many of which contain B—B bonds. A few such compounds are shown in Figure 21-22. One of these compounds, B_2Cl_4, is obtained from the reaction of BCl_3 and Cu:

$$2\ BCl_3 + 2\ Cu \longrightarrow B_2Cl_4 + 2\ CuCl$$

B_2Cl_4 is an interesting compound because, in the solid phase (mp $-92.6\ °C$), the B_2Cl_4 molecule adopts a planar geometry (Fig. 21-22a) but in the gas phase (bp $65.5\ °C$), it is nonplanar (Fig. 21-22b).

Another interesting chloride of boron is B_4Cl_4, which consists of a tetrahedral arrangement of boron atoms with a chlorine atom bonded to each B atom (Fig. 21-22c). The B_4Cl_4 molecule is highly electron deficient and the bonding is usually described in terms of three-center two-electron bonds.

🔍 **21-8   CONCEPT ASSESSMENT**

Boron forms a compound with the formula $B_2H_2(CH_3)_4$. Suggest a probable structure.

EXAMPLE 21-3   Writing Chemical Equations from a Summary Diagram of Reaction Chemistry

Using Figure 21-19, write chemical equations for the successive conversions of borax to (a) boric acid, (b) B_2O_3, and (c) impure boron metal.

Analyze

Figure 21-19 lists the key substances involved in each reaction. We can write an incomplete chemical equation for each reaction and then identify other plausible reactants and products.

Solve

(a) The conversion of a borax, a salt, to boric acid, $B(OH)_3$, requires H_2SO_4. An incomplete chemical equation for the reaction is as follows:

$$Na_2B_4O_7 \cdot 10\ H_2O(aq) + H_2SO_4(aq) \longrightarrow B(OH)_3(aq)$$

This is an acid–base reaction. The other products of the reaction are Na_2SO_4 and H_2O. The reaction does not involve changes in oxidation states, and thus the equation can be balanced by inspection. The balanced chemical equation is given below:

$$Na_2B_4O_7 \cdot 10\ H_2O(aq) + H_2SO_4(aq) \longrightarrow 4\ B(OH)_3(aq) + Na_2SO_4(aq) + 5\ H_2O(l)$$

(b) $B(OH)_3$ can be converted to B_2O_3 by heating. The conversion of a hydroxide to an oxide requires that H_2O be driven off. The balanced chemical equation for the conversion of $B(OH)_3$ to B_2O_3 is given below:

$$2\ B(OH)_3(s) \xrightarrow{\Delta} B_2O_3(s) + 3\ H_2O(g)$$

(c) B is obtained when B_2O_3 is heated with Mg. An incomplete chemical equation for the reaction is as follows:

$$B_2O_3(s) + Mg(s) \xrightarrow{\Delta} B(s)$$

The other product must be MgO. The balanced chemical equation is given below.

$$B_2O_3(s) + 3\ Mg(s) \xrightarrow{\Delta} 2\ B(s) + 3\ MgO(s)$$

Assess

The reaction summary diagram identifies only some of the substances involved in the various reaction pathways. Writing down an incomplete chemical equation for a reaction can often make it easier to identify other reactants and products that are involved.

PRACTICE EXAMPLE A: Using Figure 21-19, write chemical equations for the sequence of reactions by which borax is converted to diborane.

PRACTICE EXAMPLE B: Using Figure 21-19, write chemical equations for the sequence of reactions by which borax is converted to BF_3.

Properties and Uses of Group 13 Metals

In their appearance and physical properties and in most of their chemical behaviors, aluminum, gallium, indium, and thallium are metallic. Some properties of the group 13 metals are listed in Table 21.6.

The most important of the group 13 metals is aluminum, which is used mainly in lightweight alloys. Aluminum, like most of the other main-group metals, is an active metal. Because it is easily oxidized to the 3+ ion, aluminum is an excellent reducing agent—for example, reacting with acids to reduce $H^+(aq)$ to $H_2(g)$.

$$2\,Al(s) + 6\,H^+(aq) \longrightarrow 2\,Al^{3+}(aq) + 3\,H_2\,(g) \qquad \textbf{(21.21)}$$

Aluminum also reacts in basic solutions as shown below:

$$2\,Al(s) + 2\,OH^-(aq) + 6\,H_2O(l) \longrightarrow 2[Al(OH)_4]^-(aq) + 3\,H_2(g) \quad \textbf{(21.22)}$$

Air or other oxidants easily oxidize powdered aluminum in highly exothermic reactions. The reaction of Al and O_2 yields Al_2O_3:

$$2\,Al(s) + \tfrac{3}{2}O_2(g) \longrightarrow Al_2O_3(s) \qquad \Delta H = -1676 \text{ kJ} \qquad \textbf{(21.23)}$$

Aluminum is such a good reducing agent that it will extract oxygen from other metal oxides, producing aluminum oxide while liberating the other metal in its free state. The following reaction, known as the **thermite reaction**, produces liquid iron and is used in the on-site welding of large metal objects.

$$Fe_2O_3(s) + 2\,Al(s) \longrightarrow Al_2O_3(s) + 2\,Fe(l) \qquad \textbf{(21.24)}$$

◀ Certain drain cleaners are a mixture of NaOH and Al(s). When they are added to water, reaction (21.22) occurs. The evolved $H_2(g)$ helps unplug a stopped-up drain. The heat of reaction melts fats and grease, and the NaOH(aq) solubilizes them.

TABLE 21.6   Some Properties of the Group 13 Metals

	Al	Ga	In	Tl
Atomic number	13	31	49	81
Atomic (metallic) radius, pm	143	122	163	170
Ionic (M^{3+}) radius, pm	53	62	79	88
Electronegativity	1.5	1.6	1.7	1.8
First ionization energy, kJ mol^{-1}	577.6	578.8	558.3	589.3
Electrode potential $E°$, V[a]	−1.676	−0.56	−0.34	+0.72
Melting point, °C	660.37	29.78	156.17	303.55
Boiling point, °C	2467	2403	2080	1457
Density, g cm^{-3} at 20 °C	2.698	5.907	7.310	11.85
Hardness[b]	2.75	1.5	1.2	1.25
Electrical conductivity[b]	59.7	9.1	19.0	8.82

[a]For the reduction $M^{3+}(aq) + 3\,e^- \longrightarrow M(s)$.
[b]See footnotes b and c of Table 21.2.

▲ Thermite reaction.

The thermite reaction is highly exothermic and visually spectacular. It is shown in the photograph in the margin.

Gallium metal is of great importance in the electronics industry. It is used to make gallium arsenide (GaAs), a compound that can convert light directly into electricity (photoconduction). This semiconducting material is also used in light-emitting diodes (LEDs) (see Focus On 21: Gallium Arsenide) and in solid-state devices such as transistors.

Indium is a soft silvery metal used to make low-melting alloys. Like GaAs, InAs also finds use in low-temperature transistors and as a photoconductor in optical devices.

Thallium and its compounds are extremely toxic; as a result, they have few industrial uses. One possible use, however, is in high-temperature super conductors. For example, a thallium-based ceramic with the approximate formula $Tl_2Ba_2Ca_2Cu_3O_{8+x}$ exhibits superconductivity at temperatures as high as 125 K.

▶ As discussed on page 1059 a superconducting material loses its electrical resistance below a certain temperature. Metals typically become superconducting only a few degrees above 0 K.

Oxidation States of Group 13 Metals

Aluminum, at the top of the group of four metals in group 13, occurs almost exclusively in the +3 oxidation state in its compounds. Gallium also favors the +3 oxidation state. Indium compounds can be found with +3 and +1 oxidation states, though the +3 is more common. In thallium, this preference is reversed. For example, thallium forms the oxide Tl_2O, the hydroxide TlOH, and the carbonate Tl_2CO_3. These compounds are ionic and in some respects resemble group 1 compounds. Thus, TlOH is both very soluble and a strong base in aqueous solution. The higher stability of the +1 over the +3 oxidation state of thallium is often described as the **inert pair effect**. Thallium has the electron configuration $[Xe]4f^{14}5d^{10}6s^26p^1$. In forming the Tl^+ ion, a Tl atom loses the $6p$ electron and retains two electrons in its $6s$ subshell. It is this pair of electrons—$6s^2$—that is called the inert pair. The electron configuration $(n-1)s^2(n-1)p^6(n-1)d^{10}ns^2$ is commonly encountered in ions of the post-transition elements. One explanation of the inert pair effect is that the small bond energies and lattice energies associated with the large atoms and ions at the bottom of a group are not sufficiently great to offset the ionization energies of the ns^2 electrons.

Aluminum

Aluminum is the third most abundant element and composes 8.3% by mass of Earth's solid crust. On average, more than 5 million metric tons of aluminum are produced per year in the United States.

▶ Stimulated by a remark by one of his professors, Charles Martin Hall invented the electrolytic process for the production of aluminum at the age of 23, eight months after his graduation from Oberlin College. Paul Héroult, a student of Le Châtelier, and also at the age of 23, invented the identical process in the same year.

Production of Aluminum When an aluminum cap was placed atop the Washington Monument in 1884, aluminum was still a semiprecious metal. At that time it cost $1 per ounce to produce, equivalent to the daily wage of a skilled laborer working a 10-hour day. As a result, aluminum was used mainly in jewelry and artwork. But just two years later, all this changed. In 1886, Charles Martin Hall in the United States and Paul Héroult in France independently discovered an economically feasible method of producing aluminum from Al_2O_3 by electrolysis.

The manufacture of Al involves several interesting principles. The chief ore, *bauxite*, contains Fe_2O_3 as an impurity that must be removed. The principle used in the separation is that Al_2O_3 is an *amphoteric* oxide and dissolves in NaOH(aq), whereas the iron oxide is a *basic* oxide and does not. When $Al_2O_3(s)$ is added to NaOH(aq), $Al_2O_3(s)$ dissolves because the following reaction converts the solid to soluble $[Al(OH)_4]^-$.

$$Al_2O_3(s) + 2\,OH^-(aq) + 3\,H_2O(l) \longrightarrow 2[Al(OH)_4]^-(aq)$$

When the solution containing $[Al(OH)_4]^-$ is slightly acidified, $Al(OH)_3(s)$ precipitates. Pure Al_2O_3 is obtained by heating the $Al(OH)_3$.

$$[Al(OH)_4]^-(aq) + H_3O^+(aq) \longrightarrow Al(OH)_3(s) + 2H_2O$$

$$2\,Al(OH)_3(s) \xrightarrow{\Delta} Al_2O_3(s) + 3\,H_2O(g)$$

The separation of a mixture of $Fe^{3+}(aq)$ and $Al^{3+}(aq)$, which makes use of these reactions, is illustrated in Figure 21-23.

Al$_2$O$_3$ has a very high melting point (2020 °C) and produces a liquid that is a poor electrical conductor. Thus, its electrolysis is not feasible without a better conducting solvent. That was the crux of the discovery by Hall and Héroult. They found, independently, that up to 15% Al$_2$O$_3$, by mass, can be dissolved in the molten mineral *cryolite*, Na$_3$AlF$_6$, at about 1000 °C. Cryolite consists of Na$^+$ and AlF$_6^{3-}$ ions. In the molten state, it is a good electrical conductor. In the Hall–Héroult process, aluminum metal is produced by electrolysis of Al$_2$O$_3$ in molten cryolite. A typical electrolysis cell, pictured in Figure 21-24, produces aluminum of 99.6%–99.8% purity. The electrode reactions are not known with certainty, but the overall electrolysis reaction is

Oxidation: $3\,\{C(s) + 2\,O^{2-} \longrightarrow CO_2(g) + 4\,e^-\}$

Reduction: $\underline{\qquad 4\,\{Al^{3+} + 3\,e^- \longrightarrow Al(l)\} \qquad}$

Overall: $3\,C(s) + 4\,Al^{3+} + 6\,O^{2-} \longrightarrow 4\,Al(l) + 3\,CO_2(g)$ **(21.25)**

The energy consumed to produce aluminum by electrolysis is very high, about 15 kWh per kg of Al. This is more than three times the amount of energy consumed per kg of Na in the electrolysis of NaCl(aq). Because of the high energy requirements for producing Al(s), aluminum production facilities are generally located near low-cost hydroelectric power sources. The energy required to recycle Al is only about 5% of that to produce the metal from bauxite, and currently about 45% of the Al produced in the United States is obtained from the recycling of scrap aluminum.

Why is so much energy consumed in the electrolytic production of aluminum? Any process that must be carried out at a high temperature requires

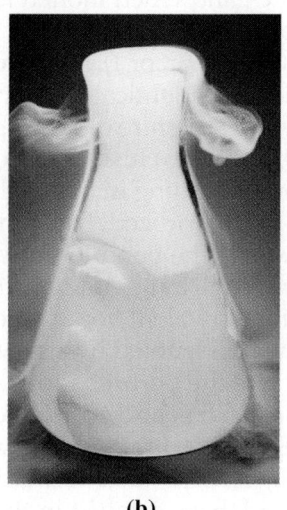

(a)   (b)   (c)

▲ FIGURE 21-23
Purifying bauxite
(a) When an excess of OH$^-$(aq) is added to a solution containing Al^{3+}(aq) and Fe^{3+}(aq), the Fe^{3+} precipitates as Fe(OH)$_3$(s) and the Al(OH)$_3$(s) first formed redissolves to produce [Al(OH)$_4$]$^-$(aq). **(b)** The Fe(OH)$_3$(s) is filtered off, and the [Al(OH)$_4$]$^-$(aq) is made slightly acidic through the action of CO$_2$, here added as dry ice. **(c)** The precipitated Al(OH)$_3$(s) collects at the bottom of a clear, colorless solution.

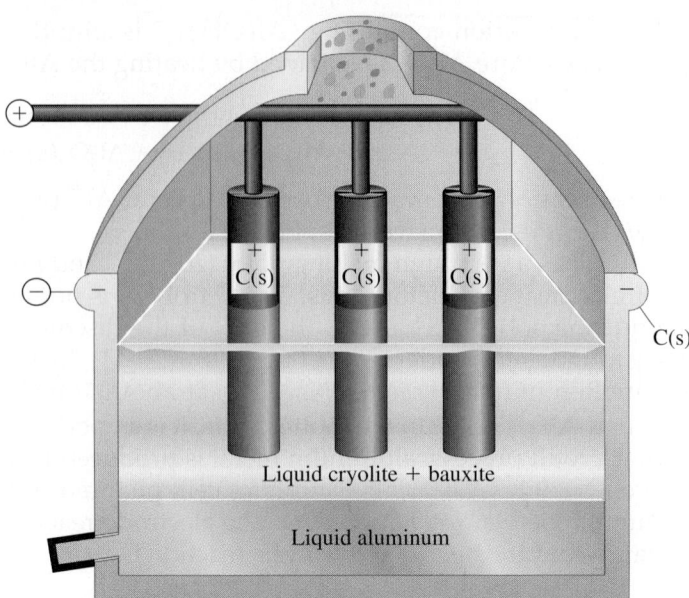

▶ FIGURE 21-24
Electrolysis cell for aluminum production
The cathode is a carbon lining in a steel tank. The anodes are also made of carbon. Liquid aluminum is denser than the electrolyte medium and collects at the bottom of the tank.

large amounts of energy for heating. In the electrolytic production of Al, the electrolysis bath must be kept at about 1000 °C, which is done by means of electric heating. Two other factors, however, are also involved in the large energy consumption. First, to produce one mole of Al, *three* moles of electrons must be transferred: $Al^{3+} + 3\,e^- \longrightarrow Al(l)$. Additionally, the molar mass of Al is relatively low, 27 g mol^{-1}. The electric current equivalent to the passage of one mole of electrons produces only 9 g Al. In contrast, one mole of electrons produces 12 g Mg, 20 g Ca, or 108 g Ag. Yet the same factors that make Al production a significant energy consumer make Al an outstanding energy producer when it is used in a battery. (Recall the aluminum–air battery described on page 893.)

Aluminum Halides Aluminum fluoride, AlF_3, has considerable ionic character. It has a high melting point (1040 °C) and when molten it is a conductor of electric current. In contrast, the other aluminum halides exist as *molecular* species with the formula Al_2X_6 (for which X = Cl, Br, or I). We can think of this molecule as comprising two AlX_3 units. When two identical units combine, the resulting molecule is called a **dimer**. The dimeric structure of Al_2Cl_6 is shown in Figure 21-25. Notice that two Cl atoms are bonded exclusively to each Al atom and two Cl atoms bridge the two metal atoms. Bonding in this molecule can be described by assuming that the Al atoms are sp^3 hybridized. Each bridging Cl atom appears to bond to two Al atoms in two ways. The bond to one Al atom is a conventional covalent bond because each atom contributes one electron to the bond. The bond to a second Al atom is a coordinate covalent bond, where the chlorine atom provides the pair of electrons for the bond, noted by arrows in Figure 21-25.

The aluminum halides, just like the boron halides, are reactive Lewis acids. They readily accept a pair of electrons and form adducts. For example, they form adducts with ethers, as does BF_3. In another example, addition of Cl$^-$ to $AlCl_3$ produces the tetrahedral $[AlCl_4]^-$. The formation of $[AlCl_4]^-$ is important in the use of $AlCl_3$ as an acid–base catalyst in the Friedel–Crafts reaction. The most common reaction of this type involves the addition of an alkyl group, such as the ethyl group, C_2H_5—, to a benzene ring, as shown below.

Bonding scheme   Electrostatic potential map

▲ FIGURE 21-25
Bonding in Al₂Cl₆
Two Cl atoms bridge the AlCl₃ units to form Al₂Cl₆. Electrons donated by these Cl atoms to Al atoms are indicated by arrows.

In this reaction, AlCl₃ acts as a Lewis acid and assists in generating the positive cation $[C_2H_5]^+$:

$$C_2H_5Cl + AlCl_3 \longrightarrow [C_2H_5]^+ + [AlCl_4]^+$$

The $[C_2H_5]^+$ cation attacks the benzene ring, liberating a proton that reacts with $[AlCl_4]^-$ to regenerate AlCl₃ and HCl.

As pointed out on page 947, a very important halide complex of aluminum is cryolite, NaAlF₆. Natural deposits of cryolite were discovered in Greenland in 1794 and occur almost nowhere else. For aluminum production, natural cryolite has been largely displaced by cryolite synthesized in a lead-clad vessel by using the following reaction:

$$6\,HF + Al(OH)_3 + 3\,NaOH \longrightarrow Na_3AlF_6 + 6\,H_2O \qquad \textbf{(21.26)}$$

Aluminum Oxide and Hydroxide Aluminum oxide is often referred to as *alumina*; when in crystalline form, it is called *corundum*.

The bonding and crystal structure of alumina account for its physical properties. The small Al^{3+} ions and small O^{2-} ions form a very stable ionic lattice. The crystal has a cubic closest packed structure of O^{2-} ions, with Al^{3+} ions occupying octahedral holes. Alumina is a very hard material and is often used as an abrasive. It is also resistant to heat (mp 2020 °C) and is used in linings for high-temperature furnaces and as a catalyst support in industrial chemical processes. Aluminum oxide is relatively unreactive except at very high temperatures. Its stability at high temperatures classifies it as a *refractory* material.

As mentioned previously, aluminum is protected against reaction with water in the pH range 4.5–8.5 by a thin, impervious coating of Al₂O₃. This coating can be purposely thickened to enhance the corrosion resistance of the metal by a process known as *anodizing*. An aluminum object is used for the anode and a graphite electrode is used for the cathode in an electrolyte bath of H₂SO₄(aq). The half-reaction occurring at the anode during electrolysis is shown below:

$$2\,Al(s) + 3\,H_2O(l) \longrightarrow Al_2O_3(s) + 6\,H^+(aq) + 6\,e^-$$

Al₂O₃ coatings of varying porosity and thickness can be obtained. Also, the oxide can be made to absorb pigments or other additives. Anodized aluminum is used to make everyday items, such as the drinking cups shown in the photograph in the margin, and is used in architectural components of buildings, such as bronze or black window frames.

Aluminum hydroxide is *amphoteric*. It reacts with acids to form $[Al(H_2O)_6]^{3+}$, as shown below:

$$Al(OH)_3(s) + 3\,H_3O^+(aq) \longrightarrow [Al(H_2O)_6]^{3+}(aq) \qquad \textbf{(21.27)}$$

Also, it reacts with bases to form $[Al(OH)_4]^-$:

$$Al(OH)_3(s) + OH^-(aq) \longrightarrow [Al(OH)_4]^-(aq) \qquad \textbf{(21.28)}$$

◀ Corundum, when pure, is also known as the gemstone *white sapphire*. Certain other gemstones consist of corundum with small amounts of transition metal ions as impurities: Cr^{3+} in ruby and Fe^{3+} and Ti^{4+} in blue sapphire, for example. Artificial gemstones are made by fusing corundum with carefully controlled amounts of other oxides.

▲ Drinking cups made of anodized aluminum.

▶ A disadvantage of the use of aluminum sulfate in sizing paper is that its acidic character contributes to deterioration of the paper. In contrast, calcium carbonate maintains an alkaline medium in paper (page 939).

Aluminum Sulfate and Alums Aluminum sulfate, $Al_2(SO_4)_3$, is the most important commercial aluminum compound. It is prepared by the reaction of hot concentrated $H_2SO_4(aq)$ on $Al_2O_3(s)$. The product that crystallizes from solution is $Al_2(SO_4)_3 \cdot 18\ H_2O$. More than 1 million metric tons of aluminum sulfate are produced annually in the United States, about half of which is used in water purification. In this application, the pH of the water is adjusted so that $Al(OH)_3(s)$ precipitates when aluminum sulfate is added. As the $Al(OH)_3(s)$ settles, it removes suspended solids in the water. Another important use is in the sizing of paper. *Sizing* refers to incorporating materials, such as waxes, glues, or synthetic resins, into paper to make the paper more water resistant. $Al(OH)_3$ precipitated from $Al_2(SO_4)_3(aq)$ helps deposit the sizing agent in the paper.

When an aqueous solution of equimolar amounts of $Al_2(SO_4)_3$ and K_2SO_4 is allowed to crystallize, the crystals obtained are of potassium aluminum sulfate, $KAl(SO_4)_2 \cdot 12\ H_2O$. This is just one of a large class of double salts called alums. **Alums** have the formula $M(I)M(III)(SO_4)_2 \cdot 12\ H_2O$, where M(I) is a unipositive cation (other than Li^+) and M(III) is a tripositive cation—Al^{3+}, Ga^{3+}, In^{3+}, Ti^{3+}, V^{3+}, Cr^{3+}, Mn^{3+}, Fe^{3+}, Co^{3+}, Re^{3+}, or Ir^{3+}. The actual ions present in the alums are $[M(H_2O)_6]^+$, $[M(H_2O)_6]^{3+}$, and SO_4^{2-}. The most common alums have M(I) = K^+, Na^+, or NH_4^+ and M(III) = Al^{3+}. Li^+ does not form alums because the ion is too small to accommodate six water molecules. Sodium aluminum sulfate is the leavening acid in baking powders, and potassium aluminum sulfate is used in dyeing. The fabric to be dyed is dipped into a solution of the alum and heated in steam. Hydrolysis of $[Al(H_2O)_6]^{3+}$ deposits $Al(OH)_3$ into the fibers of the material, and the dye is adsorbed on the $Al(OH)_3$ that deposits.

▶ In the industrial world, *alum* usually refers to simple aluminum sulfate; terms such as *potash* (potassium) *alum* and *ammonium alum* designate the double salts.

Diagonal Relationship of Beryllium and Aluminum

The charge densities of Be^{2+} and Al^{3+} are very similar and this similarity has been used to rationalize the following observations:

- The Be^{2+} ion is hydrated in aqueous solutions, forming $[Be(H_2O)_4]^{2+}$. Similarly, Al^{3+} is also hydrated as $[Al(H_2O)_6]^{3+}$. Because the Al^{3+} ion is larger than the Be^{2+} ion, it can accommodate a greater number of water molecules in its primary hydration sphere. In both cases, the cations are sufficiently polarizing that the following reactions occur in aqueous solutions:

$$[Be(H_2O)_4]^{2+}(aq) + H_2O(l) \rightleftharpoons [Be(H_2O)_3(OH)]^+(aq) + H_3O^+(aq)$$
$$[Al(H_2O)_6]^{3+}(aq) + H_2O(l) \rightleftharpoons [Al(H_2O)_5(OH)]^{2+}(aq) + H_3O^+(aq)$$

As a result, aqueous solutions of beryllium and aluminum salts, such as $Be(NO_3)_2$ or $Al(NO_3)_3$, are slightly acidic.

- The hydroxides of beryllium and aluminum are both amphoteric and form tetrahydroxo complexes, $[Be(OH)_4]^{2-}$ or $[Al(OH)_4]^-$, in highly basic solutions.

- In air, both metals form a strong oxide coating, which protects the metals from reaction.

- Both metals form carbides (Be_2C and Al_4C_3) containing the C^{4-} ion. The carbides react with water to form methane:

$$Be_2C(s) + 4\ H_2O(l) \longrightarrow 2\ Be(OH)_2(s) + CH_4(g)$$
$$Al_4C_3(s) + 12\ H_2O(l) \longrightarrow 4\ Al(OH)_3(s) + 3\ CH_4(g)$$

- Both Be and Al form halides, such as $BeCl_2$ or $AlCl_3$, that can act as Lewis acids and Friedel–Crafts catalysts.

21-5   Group 14: The Carbon Family

The properties of the group 14 elements vary dramatically within the group. Tin and lead, at the bottom of the group, have mainly metallic properties. Germanium is a *metalloid* (or *semi-metal*), and it exhibits semiconductor behavior. Silicon is mostly nonmetallic in its chemical behavior but is sometimes classified as a metalloid. Silicon also exhibits semiconductor behavior. Carbon, the first member of group 14, is a nonmetal. We will first discuss carbon and silicon and then tin and lead; germanium is mentioned only briefly.

The essential differences between carbon and silicon, as outlined in Table 21.7, are perhaps the most striking between any second- and third-period elements within a group in the periodic table. As suggested by the approximate bond energies, strong C—C and C—H bonds account for the central role of carbon–atom chains and rings in establishing the chemical behavior of carbon. A study of these chains and rings and their attached atoms is the focus of organic chemistry (Chapters 26 and 27) and biochemistry (Chapter 28). The Si—Si and Si—H bonds are much weaker than the Si—O bond. The strength of the Si—O bond accounts for the predominance of the silicates and related compounds among silicon compounds.

Carbon

Carbon is so much the central element to the study of organic and biochemistry that the rich and important inorganic chemistry of carbon is sometimes

---

TABLE 21.7   Comparison of Carbon and Silicon

Carbon	Silicon
Two principal allotropes: graphite and diamond	One stable, diamond-type crystalline structure
Forms two stable *gaseous* oxides, CO and CO$_2$, and several less stable ones, such as C$_3$O$_2$	Forms only one *solid* oxide (SiO$_2$) that is stable at room temperature; a second oxide (SiO) is stable only in the temperature range of 1180–2480 °C
Insoluble in alkaline media	Reacts in alkaline media, forming H$_2$(g) and SiO$_4{}^{4-}$(aq)
Principal oxoanion is CO$_3{}^{2-}$, which has a trigonal-planar shape	Principal oxoanion is SiO$_4{}^{2-}$, which has a tetrahedral shape
Strong tendency for catenation,[a] with straight and branched chains and rings containing up to hundreds of C atoms	Less tendency for catenation,[a] with silicon atom chains limited to about six Si atoms
Readily forms multiple bonds through use of the orbital sets $sp^2 + p$ and $sp + p^2$	Multiple bond formation much less common than with carbon
Approximate single-bond energies, kJ mol^{-1}: C—C, 347 C—H, 414 C—O, 360	Approximate single-bond energies, kJ mol^{-1}: Si—Si, 226 Si—H, 318 Si—O, 464

[a]Catenation is the joining together of like atoms into chains.

overlooked. It is the inorganic aspects of the chemistry of carbon that we emphasize in this section.

Production and Uses of Carbon Graphite is a form of carbon that is widely distributed in Earth's crust, some of it in deposits rich enough for commercial exploitation. The bulk of industrial graphite, however, is synthesized from carbon-containing materials. The key requirement is to heat the high-carbon-content material to a temperature of about 3000 °C in an electric furnace. In this process, the carbon atoms fuse into larger and larger ring systems, leading ultimately to the graphite structure shown in Figure 12-33.

Graphite has excellent lubricating properties, even when dry. The planes of carbon atoms in graphite are held together by relatively weak forces and can easily slip past one another. This property is handy in pencil "lead," which is actually a thin rod made from a mixture of graphite and clay that glides easily on paper. Graphite's principal use is based on its ability to conduct electric current; it is used for electrodes in batteries and industrial electrolysis. Graphite's use in foundry molds, furnaces, and other high-temperature environments is based on its ability to withstand high temperatures.

A newly developed use of graphite is in the manufacture of strong, lightweight composites consisting of graphite fibers, shown in the margin, and various plastics. These composites are used in products ranging from tennis rackets to lightweight aircraft. When carbon-based fibers, such as rayon, are carefully heated to a very high temperature, all volatile matter is driven off, leaving a carbon residue with the graphite structure.

As indicated in the phase diagram for carbon in Figure 21-26, graphite is the more stable form of carbon, not only at room temperature and normal atmospheric pressure but at temperatures up to 3000 °C and pressures of 10^4 atm and higher. Diamond is the more stable form of carbon at very high pressures. Diamonds can be synthesized from graphite by heating the graphite to temperatures of 1000–2000 °C and subjecting it to pressures of 10^5 atm or more. Usually the graphite is mixed with a metal, such as iron. The metal melts, and the graphite is converted to diamond within the liquid metal. Diamonds can then be picked out of the solidified metal.

According to Figure 21-26, we might expect diamond to revert to graphite at room temperature and pressure. Fortunately for the jewelry industry and for those who treasure diamonds as gems, many phase changes that require a rearrangement in bond type and crystal structure occur extremely slowly. That is very much the case with the diamond–graphite transition.

Most diamonds that are used as gemstones are natural diamonds. For industrial purposes, inferior specimens of natural diamonds or, increasingly,

▲ Fabric for use in composite materials woven from carbon fibers.

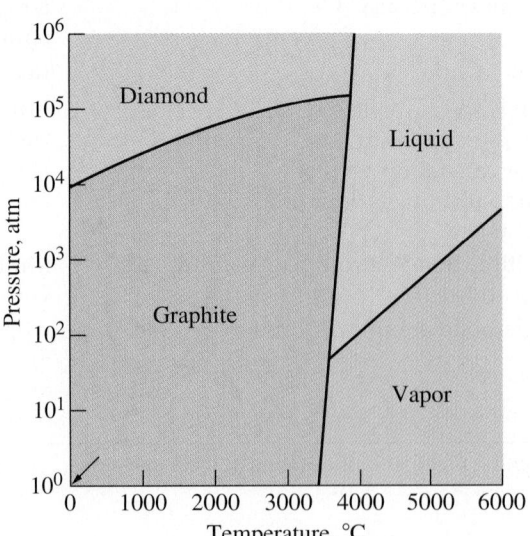

▶ FIGURE 21-26
Simplified phase diagram for carbon
Notice that pressure is plotted on a logarithmic scale. The arrow marks the point 1 atm, 25 °C.

synthetic diamonds, shown in the margin, are used. The industrial use depends on two key properties. Diamonds are used as abrasives because they are extremely hard (10 on the Mohs hardness scale). No harder substance is known. Diamonds also have a high thermal conductivity (they dissipate heat quickly), so they are used in drill bits for cutting steel and other hard materials. The rapid dissipation of heat makes the drilling process faster and increases the lifetime of the bit. A recent development has been the creation of diamond films that can be deposited directly onto metals, which imparts some of the properties of diamond to the metal. For example, when a metal is coated with a thin diamond film, the resulting material has a high thermal conductivity. Such materials have been used in heat sinks for computer chips to help dissipate the heat that computer chips generate.

◀ Because of their expected future importance, diamond films were named "Molecule of the Year" by the journal *Science* in 1990. In 1991, the fullerenes earned that honor.

Carbon can be obtained in several other forms, consisting of mixed crystalline or amorphous structures. Incomplete combustion of natural gas, such as in an improperly adjusted Bunsen burner in the laboratory, produces a smoky flame. The smoke can be deposited as finely divided soot called **carbon black**. Carbon black is used as filler in rubber tires (several kilograms per tire), as a pigment in printing inks, and as the transfer material in carbon paper, typewriter ribbons, laser printers, and photocopying machines. Recently, new allotropes of carbon have been isolated from the decomposition of graphite. These allotropes, known as fullerenes and nanotubes, were presented in Chapter 12 (page 528). Recall that the molecule C_{60} is remarkably stable and has a shape resembling a soccer ball (12 pentagonal faces, 20 hexagonal faces, and 60 vertices). Other fullerenes include C_{70}, C_{74}, and C_{82}. Fullerenes are typically produced by laser decomposition of graphite under a helium atmosphere. Because nitrogen and oxygen interfere with the process of forming fullerenes, soot, which is formed by the combustion of hydrocarbons in air, does not contain fullerenes.

In Chapter 12, we discussed various forms of carbon: graphite, diamond, fullerenes, and nanotubes. In this chapter we focus on graphene. **Graphene** is a sheet of carbon atoms only one atom thick. The layers of graphite are graphene. A carbon nanotube is a sheet of graphene rolled into a cylinder. A fullerene is obtained when an appropriate number of hexagonal rings in graphene are replaced by pentagonal rings; the presence of the pentagonal rings causes the flat sheet to pucker and form a spherical ball. The relationship between graphene and other forms of carbon is illustrated in Figure 21-27. Graphene has very interesting electronic properties because electrons in these sheets are moving very quickly—at approximately 1/300 of the speed of light. It is expected that graphene will play an important role in the development of electronic devices, possibly replacing silicon.

▲ Synthetic diamonds

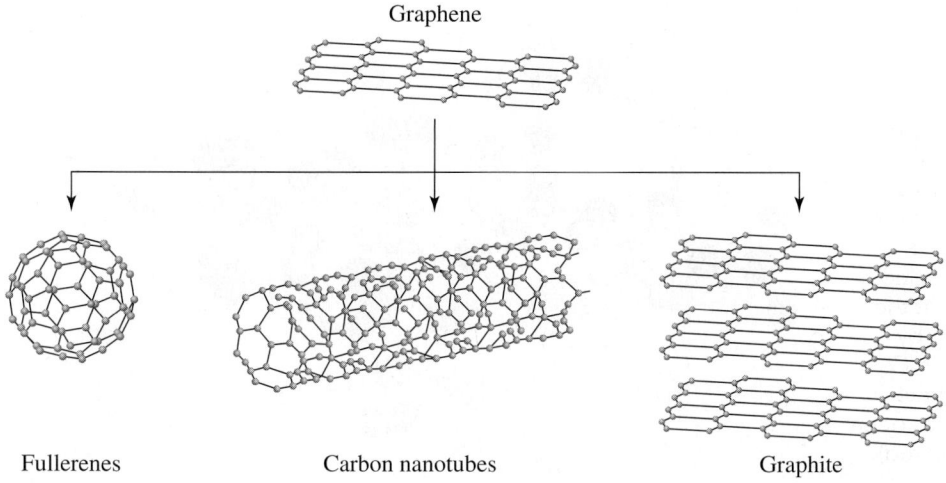

Graphene

Fullerenes   Carbon nanotubes   Graphite

◀ FIGURE 21-27
The relationship between graphene and the other forms of carbon

It was once thought impossible that a sheet of carbon, only one atom thick, could be made or isolated. However, scientists in the United Kingdom (Andre Geim and Kostya Novosalov at Manchester University) were able to isolate graphene in 2004 by using a technique called micromechanical cleavage. Essentially, the process is just like drawing with a pencil, the "lead" of which contains graphite, and looking among the traces for graphene pieces in the trail left by the pencil. Another way to isolate graphene is to use adhesive tape repeatedly to peel away layers of carbon atoms from a graphite surface in a process called exfoliation. The exfoliation of graphite is illustrated schematically in Figure 21-28. These methods produce flakes that contain 1 to 10 layers of graphene and are up to 100 μm thick. The single-layer flakes have to be found among the thousands of thicker flakes.

Two other noteworthy carbon-containing materials are coke and charcoal. When coal is heated in the absence of air, volatile substances are driven off, leaving a high-carbon residue called *coke*. This same type of destructive distillation of wood produces *charcoal*. Currently, coke is the principal metallurgical reducing agent. It is used in blast furnaces, for instance, to reduce iron oxide to iron metal.

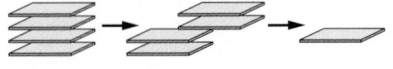

▲ FIGURE 21-28
Exfoliation of graphite

Carbon Dioxide The chief oxides of carbon are carbon monoxide, CO, and carbon dioxide, CO_2. There are about 380 ppm of CO_2 in air (0.038% by volume). CO occurs to a much lesser extent. Although they are only minor constituents of air, these two oxides are important in many ways.

Carbon dioxide is the only oxide of carbon formed when carbon or carbon-containing compounds are burned in an *excess* of air (providing an abundance of O_2). This condition exists when a fuel-lean mixture is burned in an automobile engine. Thus, for the combustion of the gasoline component octane,

$$C_8H_{18}(l) + \frac{25}{2} O_2(g) \longrightarrow 8 CO_2(g) + 9 H_2O(l) \tag{21.29}$$

If the combustion occurs in a *limited* quantity of air, carbon monoxide is also produced. This condition prevails when a fuel-rich mixture is burned in an automobile engine. One possibility for the incomplete combustion of octane is

$$C_8H_{18}(l) + 12 O_2(g) \longrightarrow 7 CO_2(g) + CO(g) + 9 H_2O(l) \tag{21.30}$$

CO as an air pollutant comes chiefly from the incomplete combustion of fossil fuels in automobile engines. CO is an inhalation poison because CO molecules bond irreversibly to Fe atoms in hemoglobin in blood and displace the O_2 molecules that the hemoglobin normally carries, as illustrated in Figure 21-29.

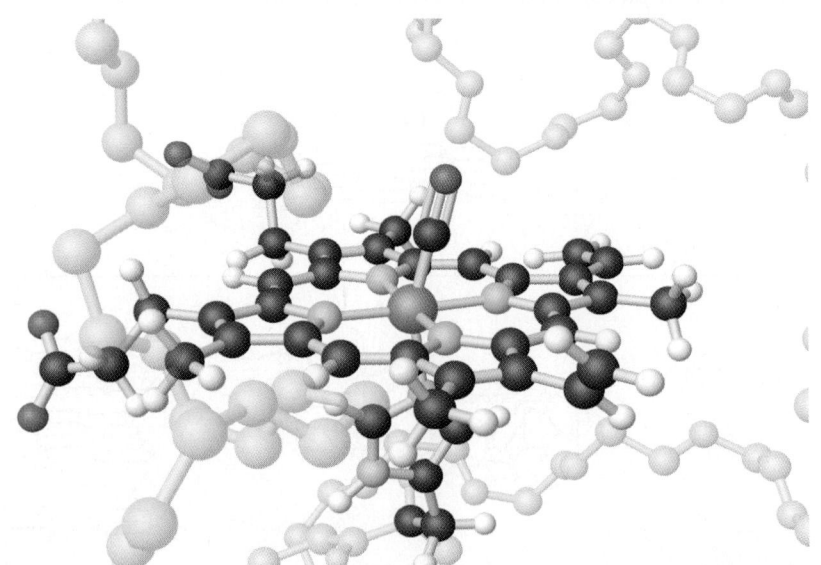

▶ FIGURE 21-29
CO bound to hemoglobin
Carbon monoxide binds to the iron atoms in hemoglobin more strongly than does oxygen. Thus, carbon monoxide's toxicity arises because it prevents hemoglobin from binding with oxygen. The portion of the hemoglobin molecule shown here is called a *heme* group. An iron atom (yellow) is at the center of the group and is surrounded by four nitrogen atoms. In hemoglobin, an O_2 molecule projects above the plane of the iron and nitrogen atoms, but here it has been replaced by a CO molecule (black and red).

TABLE 21.8 Some Industrial Methods of Preparing CO_2	
Method	Chemical Reaction
Recovery from exhaust stack gases in the combustion of carbonaceous fuels, such as the combustion of coke	$C(s) + O_2(g) \longrightarrow CO_2(g)$
Recovery in ammonia plants from steam–reforming reactions used to produce hydrogen	$CH_4(g) + 2\,H_2O(g) \longrightarrow CO_2(g) + 4\,H_2(g)$
Decomposition (calcination) of limestone at about 900 °C	$CaCO_3(s) \longrightarrow CaO(s) + CO_2(g)$
Fermentation by-product in the production of ethanol	$C_6H_{12}O_6(aq) \longrightarrow 2\,C_2H_5OH(aq) + 2\,CO_2(g)$ a sugar

Not only does incomplete combustion of gasoline contribute to air pollution, but also it represents a loss of efficiency. A given quantity of gasoline evolves less heat if $CO(g)$ is formed as a combustion product rather than $CO_2(g)$.

Although carbon dioxide can be obtained directly from the atmosphere as a by-product of the liquefaction of air, this is not an important source. Some of the principal commercial sources of CO_2 are summarized in Table 21.8.

The major use of carbon dioxide (about 50%) is as a refrigerant in the form of dry ice for freezing, preserving, and transporting food. Carbonated beverages account for about 20% of CO_2 consumption. Other important uses are in oil recovery in oil fields and in fire-extinguishing systems. Of course, the major use is not by humans but by algae and plants.

Atmospheric CO_2 is the source of all the carbon-containing compounds (organic compounds) synthesized by green plants. Here we briefly consider the *carbon cycle*—the major exchanges that occur between the atmosphere and the surface of Earth. A portion of the carbon cycle is represented in Figure 21-30. Atmospheric CO_2 is the only source of carbon available to plants for making

🔍 21-10   CONCEPT ASSESSMENT

With a minimum of calculation, determine how much less heat is produced per mol of $C_8H_{18}(l)$ burned in reaction (21.30) than in reaction (21.29).

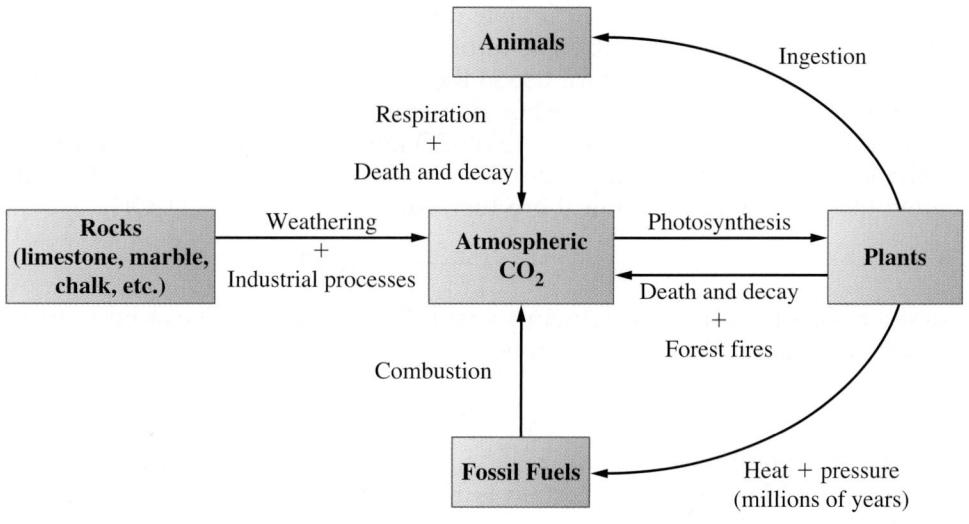

▲ FIGURE 21-30
The carbon cycle

▶ Melvin Calvin won the Nobel Prize in Chemistry in 1961 for his research on the assimilation of carbon dioxide in plants.

organic compounds by the process of *photosynthesis*. The process is extremely complicated, and its details have been known for only a few decades. It involves up to 100 sequential steps for the conversion of 6 mol CO_2 to 1 mol $C_6H_{12}O_6$ (glucose). The overall change is represented by the chemical equation below:

$$6\,CO_2(g) + 6\,H_2O(l) \xrightarrow[\text{sunlight}]{\text{chlorophyll}} C_6H_{12}O_6(s) + 6\,O_2(g) \qquad \Delta H = +2.8 \times 10^3\ \text{kJ}$$

The overall reaction is highly endothermic. The required energy comes from sunlight. Chlorophyll, a green pigment in plants, is crucial to the process. Atmospheric oxygen is a by-product of the reaction.

Here are some of the ideas illustrated in Figure 21-30. When animals consume plants, carbon atoms pass to the animals. Some carbon is returned to the atmosphere as CO_2 when the animals breathe and when they expel gas (methane). Additional CO_2 returns to the atmosphere as plants and animals die and their remains are broken down by bacteria. Some carbon in decaying organic matter is converted to coal, petroleum, and natural gas. This carbon is unavailable for photosynthesis.

Not represented in the drawing is the cycle of CO_2 through the oceans of the world. Phytoplankton (small, floating green organisms) also carry on photosynthesis, converting CO_2 to organic compounds. Phytoplankton are at the bottom of the ocean food chain, directly and indirectly supporting all the animals in the oceans.

Huge quantities of carbon have accumulated in the form of carbonate rocks (mostly $CaCO_3$). These come from the shells of decayed mollusks in ancient seas.

Human activities now play a far more significant role in the carbon cycle than in preindustrial times. The combustion of fossil fuels is replacing stored carbon by carbon dioxide to an even greater extent. We have already seen the possible consequences of this distortion of the carbon cycle: an increased level of atmospheric CO_2 and a future global warming (page 276). Human disruption of the natural carbon cycle has become a widely debated issue.

Carbon Monoxide A modern method of making carbon monoxide is the steam reforming of natural gas, which is based on the following chemical reaction:

$$CH_4(g) + H_2O(g) \longrightarrow CO(g) + 3\,H_2(g) \qquad \textbf{(21.31)}$$

Steam refers to gaseous water. *Reforming* refers to the restructuring of a carbon compound, such as CH_4 to CO. The reforming of natural gas (mostly CH_4) is an important source of $H_2(g)$ for use in the synthesis of NH_3 (page 1008).

There are three main uses of carbon monoxide. One is in synthesizing other compounds. For example, a mixture of CO and H_2 produced by reforming methane or some other hydrocarbon and known as *synthesis gas* can be converted to a new organic chemical product, such as methanol, CH_3OH:

$$CO(g) + 2\,H_2(g) \longrightarrow CH_3OH(l)$$

Another use of CO is as a reducing agent. For instance, CO can be used to reduce iron oxide to iron, as shown below:

$$Fe_2O_3(s) + 3\,CO(g) \longrightarrow 2\,Fe(l) + 3\,CO_2(g)$$

The reaction can be carried out by heating Fe_2O_3 and coke—a form of pure carbon—in a blast furnace. Carbon is first converted to CO and then CO reduces Fe_2O_3 to Fe.

A third use of CO is as a fuel, usually mixed with CH_4, H_2, and other combustible gases. This was discussed in Section 7-9.

Other Inorganic Carbon Compounds Carbon combines with metals to form *carbides*. In many cases, the carbon atoms occupy the holes or voids, also called interstitial sites, in metal structures, forming *interstitial carbides*. With active metals, the carbides are ionic. *Calcium carbide* forms in the high-temperature reaction of lime and coke:

$$CaO(s) + 3\,C(s) \xrightarrow{2000\ °C} CaC_2(s) + CO(g) \tag{21.32}$$

Calcium carbide is important because acetylene ($HC\equiv CH$) can easily be made from it, and acetylene can be used to synthesize many chemical compounds:

$$CaC_2(s) + 2\,H_2O(l) \longrightarrow Ca(OH)_2(s) + C_2H_2(g) \tag{21.33}$$

Solid CaC_2 can be regarded as a face-centered cubic array of Ca^{2+} ions with the C_2^{2-} ions in the octahedral holes, as shown in the figure in the margin.

   Carbon disulfide, CS_2, can be synthesized by the reaction of methane and sulfur vapor in the presence of a catalyst:

$$CH_4(g) + 4\,S(g) \longrightarrow CS_2(l) + 2\,H_2S(g) \tag{21.34}$$

Carbon disulfide is a highly flammable, volatile liquid that acts as a solvent for sulfur, phosphorus, bromine, iodine, fats, and oils. Its uses as a solvent are decreasing, however, because CS_2 is poisonous. Other important uses are in the manufacture of rayon and cellophane.

   Carbon tetrachloride, CCl_4, can be prepared by the direct chlorination of methane, as shown below:

$$CH_4(g) + 4\,Cl_2(g) \longrightarrow CCl_4(l) + 4\,HCl(g) \tag{21.35}$$

Although CCl_4 has been extensively used as a solvent, dry-cleaning agent, and fire extinguisher, these uses have been steadily declining because CCl_4 causes liver and kidney damage and is a known carcinogen.

   Certain groupings of atoms, several containing C atoms, have some of the characteristics of a halogen atom. They are called *pseudohalogens* and include the following groupings of atoms:

   —CN (cyanide)   —OCN (cyanate)   —SCN (thiocyanate)

The *cyanide* ion, CN^-, is similar to the halide ions, X^-, in that it forms an insoluble silver salt, AgCN, and an acid, HCN. Hydrocyanic acid, HCN, is a liquid that boils at about room temperature. It is a very weak acid, unlike HCl. Despite its extreme toxicity, HCN has important uses in the manufacture of plastics. The combination of two cyanide groups produces *cyanogen*, $(CN)_2$. This gas resembles chlorine gas in undergoing a disproportionation reaction in basic solution:

$$(CN)_2 + 2\,OH^-(aq) \longrightarrow CN^-(aq) + OCN^-(aq) + H_2O(l) \tag{21.36}$$

Cyanogen is used in organic synthesis, as a fumigant, and as a rocket propellant.

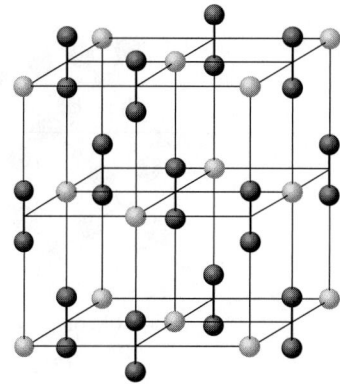

$\vert \;= C_2^{2-}$ ions

$\bigcirc = Ca^{2+}$ ions

▲ **Structure of calcium carbide**

Silicon

Among the elements, silicon is second only to oxygen in its abundance in Earth's crust. Silicon is to the mineral world what carbon is to the living world—the backbone element.

Production and Uses of Silicon Elemental silicon is produced when quartz or sand (SiO_2) is reduced by reaction with coke in an electric arc furnace. The balanced chemical equation for the reaction is given below:

$$SiO_2 + 2\,C \xrightarrow{\Delta} Si + 2\,CO(g) \tag{21.37}$$

Very high purity Si for solar cells can be made by reducing Na_2SiF_6 with metallic Na. The Na_2SiF_6 required for this process is obtained as a by-product

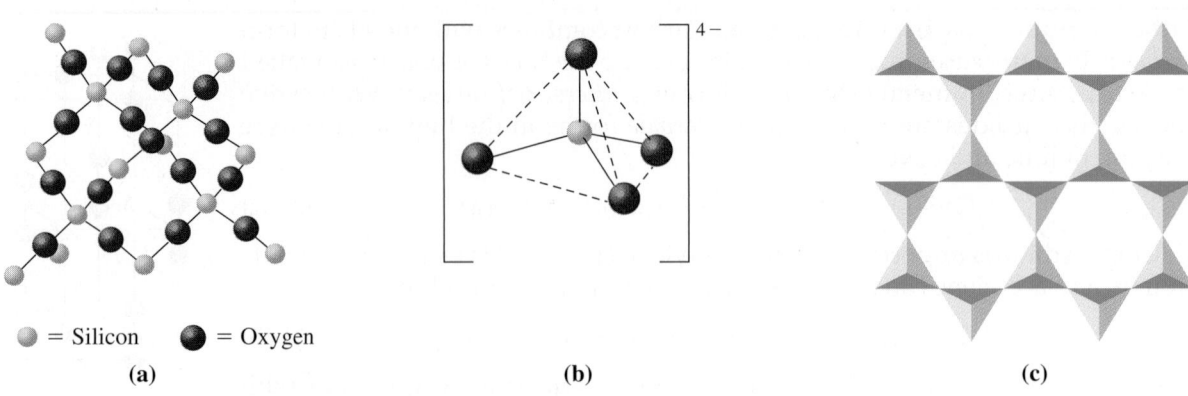

= Silicon = Oxygen

(a) **(b)** **(c)**

▲ FIGURE 21-31
Structures of silica and silicates
(a) A three-dimensional network of bonds in silica, SiO_2. (b) The silicate anion, SiO_4^{4-}, commonly found in silicate materials. The Si atom is in the center of the tetrahedron and is surrounded by four O atoms. (c) A depiction of the structure of a mica, using tetrahedra to represent the SiO_4 units. The cations K^+ and Al^{3+} are also present but are not shown. This view, looking down on top of a tetrahedron, is the usual way that materials scientists represent silicates and similar materials.

of the formation of phosphate fertilizers (page 1017). High-purity silicon is also required in the manufacture of transistors and other semiconductor devices.

Oxides of Silicon; Silicates Silica, SiO_2, is the only stable oxide of silicon. Silica is a network covalent solid (not a molecular solid, like CO_2). In silica, each Si atom is bonded to *four* O atoms and each O atom to *two* Si atoms. The structure is that of a network covalent solid, as suggested by Figure 21-31(a). This structure is reminiscent of the diamond structure, and silica has certain properties that resemble those of diamond. For example, quartz, a form of silica, is fairly hard (with a Mohs hardness of 7) but not as hard as diamond (with a Mohs hardness of 10), has a high melting point (about 1700 °C), and is a nonconductor of electricity. Silica is the basic raw material of the glass and ceramics industries.

The central feature of all *silicates* is the SiO_4^{4-} tetrahedron depicted in Figure 21-31(b). These tetrahedra may be arranged in a wide variety of ways. Here are just a few examples:

- *Simple SiO_4 tetrahedra.* Typical minerals in which the anions are simple SiO_4^{4-} tetrahedra are *thorite* ($ThSiO_4$) and *zircon* ($ZrSiO_4$).

- *Two SiO_4 tetrahedra joined end-to-end.* The silicon atoms in the two tetrahedra share an O atom between them in the anion $Si_2O_7^{6-}$, found in the mineral *thortveitite* ($Sc_2Si_2O_7$). See Figure 21-32(a).

- *SiO_4 tetrahedra joined into long chains.* Each Si atom shares an O atom with the Si atom in an adjacent tetrahedron on either side. An example is *spodumene*, the principal source of lithium and lithium compounds. Its empirical formula is $LiAl(SiO_3)_2$. See Figure 21-32(b).

- *SiO_4 tetrahedra joined into a double chain.* Half the Si atoms share three of their four O atoms with Si atoms in adjacent tetrahedra, and half share only two. In *chrysotile asbestos*, double chains are held together by cations (chiefly Mg^{2+}); this mineral has a fibrous appearance. The empirical formula is $Mg_3(Si_2O_5)(OH)_4$. See Figure 21-32(c).

- *SiO_4 tetrahedra are bonded together in two-dimensional sheets.* Each Si atom shares O atoms with the Si atoms in three adjacent tetrahedra. This structure is depicted in Figure 21-31(c). In *muscovite mica*, with the empirical

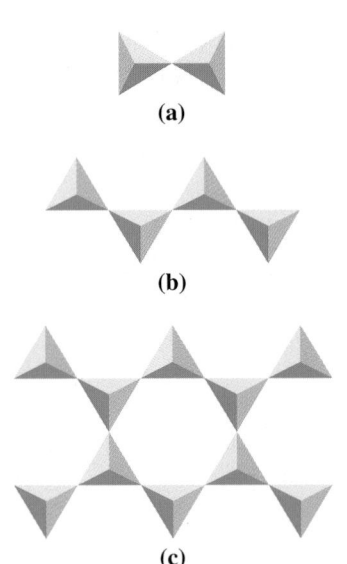

(a)

(b)

(c)

▲ FIGURE 21-32
Linking of SiO₄ tetrahedra
In (a), two tetrahedra are linked at one corner to form the $Si_2O_7^{6-}$ ion. In (b), tetrahedra are linked at two corners to form a long chain. The empirical formula of the chain is SiO_3^{2-}. In (c), SiO_4 tetrahedra are joined into a double chain.

formula, $KAl_2(AlSi_3O_{10})(OH)_2$, the counterions that bond sheets together in layers are mostly K^+ and Al^{3+}. Bonding within the sheets is stronger than that between the sheets, so muscovite mica flakes easily.

- *SiO_4 tetrahedra are bonded together in three-dimensional structures.* Three-dimensional arrays of tetrahedra result when each Si atom shares all four O atoms with the Si atoms in four adjoining tetrahedra (as shown in Figure 21-31a). This is the most common arrangement, occurring in silica (quartz) and in the majority of silicate minerals.

 **21-11   CONCEPT ASSESSMENT**

The empirical formula of the mineral *beryl* is $Be_3Al_2Si_6O_{18}$. By using the several descriptions of silicate minerals just given as a guide, describe the structure of the silicate anion in beryl.

SiO_2 is a weakly acidic oxide and slowly dissolves in strong bases. It forms a series of silicates, such as Na_2SiO_4 (sodium orthosilicate) and Na_2SiO_3 (sodium metasilicate). These compounds are somewhat soluble in water and, as a result, are sometimes referred to as "water glass."

Silicate anions are bases; when acidified, they produce silicic acids, which are unstable and decompose to silica. The silica obtained, however, is not a crystalline solid or powder. Depending on the acidity of the solution, the silica is obtained as a colloidal dispersion, a gelatinous precipitate, or a solid-like gel in which all the water is entrapped. These hydrated silicates are polymers of silica formed by the elimination of water molecules between neighboring molecules of silicic acid. The process begins with the following reaction:

$$SiO_4^{4-}(aq) + 4\,H^+(aq) \longrightarrow Si(OH)_4$$

◀ H_4SiO_4 is called *ortho*silicic acid. When a combination of one H atom and one OH group (equivalent to a molecule of H_2O) are eliminated from the *ortho* acid, the resulting acid, H_2SiO_3, is called *meta*silicic acid.

21-2 ARE YOU WONDERING...

Why silica (SiO_2) does not exist as simple molecules as CO_2 does?

Because both silicon and carbon are in group 14 of the periodic table and have four valence electrons, we might expect them to form oxides with similar properties. In CO_2, the side-to-side overlap of $2p$ orbitals of the C and O atoms is extensive (see page 461). The carbon-to-oxygen double bond in CO_2 proves to be stronger (799 kJ mol^{-1}) than two single bonds (2 × 360 kJ mol^{-1}). This results in the familiar Lewis structure of CO_2:

$$:\ddot{O}{=}C{=}\ddot{O}:$$

Silicon, being in the third period, would have to use $3p$ orbitals to form double bonds with oxygen. The side-to-side overlap of these orbitals with the $2p$ orbitals of oxygen is quite limited. In terms of energy, a stronger bonding arrangement results if the Si atoms form *four* single bonds with O atoms (bond energy: 4 × 464 kJ mol^{-1} = 1856 kJ total) rather than *two* double bonds (bond energy: 2 × 640 kJ mol^{-1} = 1280 kJ total). Because each O atom must simultaneously bond to two Si atoms, the result is a network of —Si—O—Si— bonds (see Figure 21-31a).

On page 935 we contrasted Be and Mg, the second- and third-period members of group 2. Here is another example of how the second-period member of a group (that is, carbon) differs from the higher period members.

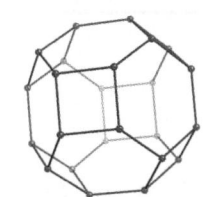

(a)

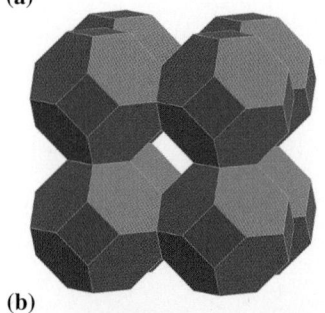

(b)

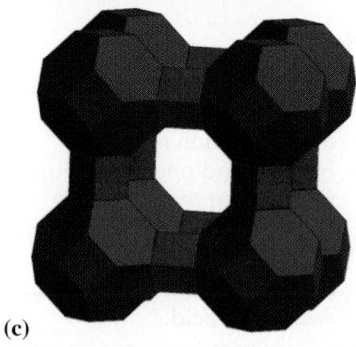

(c)

▲ **FIGURE 21-33**
Structures of zeolites
(a) The basic structural unit in many zeolites is a β-cage, which consists of six-membered and four-membered rings. **(b)** This is the structure obtained when eight β-cages are joined together by sharing faces of the four-membered rings. This structural unit is found in sodalite, a naturally occurring zeolite. **(c)** This is the structure obtained when the four-membered rings of the β-cages are joined together by bridges. The resulting structure is more open than the structure shown in (b). Such structures are found in zeolite-A, a synthetic zeolite.

▶ The dehydrated form of a zeolite is obtained by heating the zeolite under vacuum. The heating process drives off the waters of hydration from the zeolite structure, leaving behind a structure that has a high affinity for water.

It is followed by the reaction below:

$$(HO)-\underset{\underset{\textstyle OH}{|}}{\overset{\overset{\textstyle O\!H}{|}}{Si}}-O\!H + (HO)-\underset{\underset{\textstyle OH}{|}}{\overset{\overset{\textstyle O\!H}{|}}{Si}}-O\!H + \cdots \longrightarrow \longrightarrow (x\,SiO_2 \cdot y\,H_2O)$$
$$\text{colloidal silica}$$

Notice that —H and —OH combine to form water (HOH) and that Si—O—Si bridges are produced.

Zeolites: An Important Class of Aluminosilicates

A **zeolite** is a three-dimensional network of SiO_4 and AlO_4 tetrahedra. Many of the zeolite structures contain a ring based on $Si_6O_{18}^{12-}$. The ring of Si atoms, some of which can be replaced by Al atoms, is represented by a hexagon. The hexagons can be joined together to give the structure in Figure 21-33(a). Bear in mind that the hexagons shown in Figure 21-33 emphasize the positions of the Si and Al atoms. The straight lines do not represent bonds; the atoms at the vertices of the hexagons are, in fact, connected by bent Si—O—Si or Al—O—Al bridges.

The structure shown in Figure 21-33(a) is called a β-cage and it is present in a naturally occurring zeolite known as *sodalite*. In the β-cage, there are four- and six-membered rings of nonoxygen atoms. In sodalite, eight of the β-cages are joined together by sharing the faces of the four-membered rings to give the cubic structure shown in Figure 21-33(b). An alternative is to bridge oxygen at each corner of the four ring faces. This produces the structure shown in Figure 21-33(c), which is found in a synthetic zeolite, $Na_{12}(AlO_2)_{12}(SiO_2)_{12} \cdot 27\,H_2O$, known as zeolite-A. The Na^+ ions in the formula of zeolite-A are required to offset the decrease in positive charge that occurs when Al^{3+} ions replace Si^{4+} ions in the lattice.

The important consequence of the arrangements described above, and illustrated in Figure 21-33, is the presence of channels and cavities in zeolite structures. These channels and cavities give zeolites their important properties. For example, because small molecules are able to diffuse into cavities and larger molecules are excluded, zeolites have been used as molecular sieves to remove molecules of certain sizes from a mixture. Zeolites, in their dehydrated forms, have also been used for removing water from gases or organic solvents. For example, in the drying of benzene, water molecules diffuse into the zeolite lattice while benzene molecules, which are too large, are excluded. The zeolite can be separated from the benzene and can be regenerated by heating.

Another important application of zeolites is as an ion exchange material in the treatment of hard water, as illustrated in Figure 21-34. Hard water contains significant concentrations of ions, especially Ca^{2+}, Mg^{2+}, or Fe^{3+}, and a zeolite can be used to exchange these ions with Na^+ ions. It is desirable to remove Ca^{2+}, Mg^{2+}, and Fe^{3+} ions from water because these ions react with CO_3^{2-} ions or anions of soaps to form insoluble precipitates. For example, Ca^{2+}, Mg^{2+}, and Fe^{3+} ions react with CO_3^{2-} ions to form a mixed precipitate of $CaCO_3$, $MgCO_3$, and rust called *boiler scale*. The formation of boiler scale lowers the efficiency of water heaters and builds up in pipes or on the insides of containers used for boiling water. Ca^{2+} and Mg^{2+} ions in hard water combine with anions of soaps—such as the palmitate ion, equation (21.9)—to form insoluble precipitates that accumulate on the surfaces of bathtubs or showers and contribute to the formation of *soap scum* and *bathtub rings*.

The equation below represents the exchange that occurs when, for example, Ca^{2+} ions (in hard water) are exchanged with Na^+ ions (in a zeolite):

$$Na_2[zeolite](s) + Ca^{2+}(aq) \longrightarrow Ca[zeolite](s) + 2\,Na^+(aq)$$

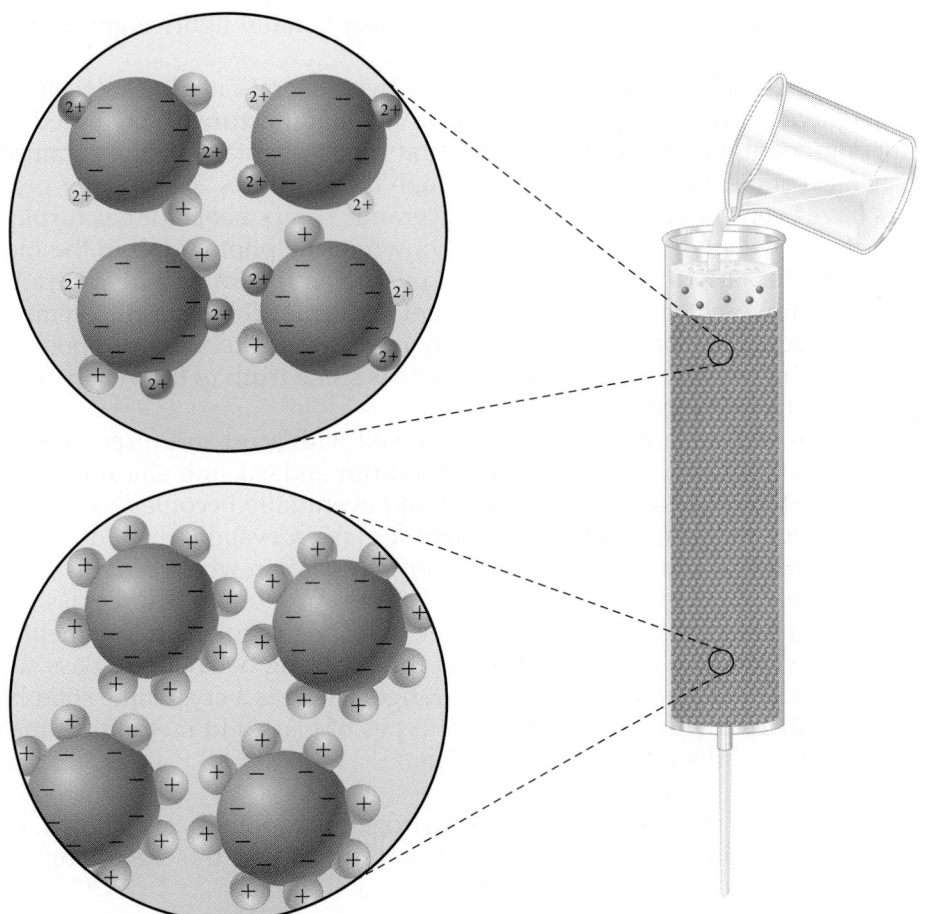

▲ FIGURE 21-34
Ion exchange
The resin shown here is a cation-exchange resin (a zeolite, for example). Multivalent cations (green, Fe^{2+}, yellow, Ca^{2+}) replace Na^+ (orange) at the top of the resin column. By the time the water has reached the bottom of the column, all the multivalent ions have been removed and only Na^+ ions remain as counterions. The exchange can be represented as $2\,NaR + M^{2+} \rightleftharpoons MR_2 + 2\,Na^+$. The reaction occurs in the forward direction during water softening. As expected from Le Châtelier's principle, in the presence of concentrated NaCl(aq), the reverse reaction is favored and the resin is recharged.

Because sodium compounds are generally soluble, the replacement of Ca^{2+} ions by Na^+ ions prevents the formation of insoluble precipitates.

Zeolites are used not only in ion exchange resins but also in detergents to help remove any Ca^{2+} and Mg^{2+} ions that might be present in water used for washing clothes. The removal of these ions helps the detergents foam better and also helps to prevent the formation of insoluble calcium and magnesium compounds.

Modern automobile engines require fuels containing short-chain hydrocarbons with low boiling points, and some require high octane fuels containing branched-chain hydrocarbons. Zeolite catalysts are used to speed up the conversion of long-chain hydrocarbons—found in crude oil—into short-chain or branched-chain hydrocarbons. Consequently, zeolites play an important role in the petroleum industry.

Silicates in Ceramics and Glass Hydrated silicate polymers are important in the ceramics industry. A colloidal dispersion of particles in a liquid is called a *sol*. The sol can be poured into a mold and, following removal of some of the liquid, is converted to a *gel*. The gel is then processed into the final ceramic

▲ Ceramic components of an automobile engine.
(Photo courtesy of Kyocera Industrial Ceramics Corp./Vancouver, WA)

product. This sol-gel process can produce exceptionally lightweight ceramic materials.

Uses of these advanced ceramics fall into two general categories: **(1)** electrical, magnetic, or optical applications (as in the manufacture of integrated circuit components) and **(2)** applications that take advantage of the ceramic's mechanical and structural properties at high temperatures. These latter properties have been explored in developing ceramic components for gas turbines and automotive engines, such as those shown in the photograph in the margin. Quite possibly the automobile engine of the future will be a ceramic engine—lightweight and more fuel-efficient because higher operating temperatures are possible. Some engines already have several ceramic components, such as engine valves and valve seats. There is some truth to the vision of this ceramic future as the new stone age.

If sodium and calcium carbonates are mixed with sand and fused at about 1500 °C, the result is a liquid mixture of sodium and calcium silicates. When cooled, the liquid becomes more viscous and eventually becomes a solid that is transparent to light; this solid is called a **glass**. Crystalline solids have a long-range order, whereas glasses are *amorphous solids* in which order is found over relatively short distances only. Think of the structural units in glass (silicate anions) as being in a jumbled rather than in a regular arrangement. A glass and a crystalline solid also differ in their melting behavior. A glass softens and melts over a broad temperature range, whereas a crystalline solid has a definite, sharp melting point. Different types of glass and methods of making them are described later in this section.

Silanes and Silicones Several silicon–hydrogen compounds are known, but because Si—Si single bonds are not particularly strong, the chain length in these compounds, called *silanes*, is limited to six.

$$\underset{\text{Monosilane}}{\overset{\displaystyle H}{\underset{\displaystyle H}{H-\overset{|}{\underset{|}{Si}}-H}}} \qquad \underset{\text{Disilane}}{\overset{\displaystyle H\;\;\;H}{\underset{\displaystyle H\;\;\;H}{H-\overset{|}{\underset{|}{Si}}-\overset{|}{\underset{|}{Si}}-H}}} \qquad \underset{\text{Trisilane}}{\overset{\displaystyle H\;\;\;H\;\;\;H}{\underset{\displaystyle H\;\;\;H\;\;\;H}{H-\overset{|}{\underset{|}{Si}}-\overset{|}{\underset{|}{Si}}-\overset{|}{\underset{|}{Si}}-H}}} \quad \ldots \quad \underset{\text{Hexasilane}}{S_6H_{14}}$$

Other atoms or groups of atoms can be substituted for H atoms in silanes to produce compounds called *organosilanes*. Typical is the direct reaction of Si and methyl chloride, CH_3Cl. The equation for the reaction is given below:

$$2\,CH_3Cl + Si \longrightarrow (CH_3)_2SiCl_2$$

The reaction of $(CH_3)_2SiCl_2$, dichlorodimethylsilane, with water produces an interesting compound, *dimethylsilanol*, $(CH_3)_2Si(OH)_2$. Dimethylsilanol undergoes a polymerization reaction in which H_2O molecules are eliminated from among large numbers of silanol molecules. The result of this polymerization is a material consisting of molecules with long silicon-oxygen chains: **silicones**.

$$HO-\overset{\displaystyle CH_3}{\underset{\displaystyle CH_3}{\overset{|}{\underset{|}{Si}}}}-O-H + HO-\overset{\displaystyle CH_3}{\underset{\displaystyle CH_3}{\overset{|}{\underset{|}{Si}}}}-OH \xrightarrow{-H_2O} HO-\overset{\displaystyle CH_3}{\underset{\displaystyle CH_3}{\overset{|}{\underset{|}{Si}}}}-O\left[\overset{\displaystyle CH_3}{\underset{\displaystyle CH_3}{\overset{|}{\underset{|}{Si}}}}-O\right]_n \overset{\displaystyle CH_3}{\underset{\displaystyle CH_3}{\overset{|}{\underset{|}{Si}}}}-OH$$

A silicone

▲ Some common applications of silicones.

Silicones are important polymers because they are versatile. They are used to make a variety of useful products, such as those shown in the photograph in the margin. Silicones can be obtained either as oils or as rubber-like materials. Silicone oils are not volatile and do not decompose when heated. Also, they can be cooled to low temperatures without solidifying or becoming viscous. Silicone oils are excellent high-temperature lubricants. In contrast, hydrocarbon oils break down at high temperatures, become very viscous, and then solidify at low temperatures. Silicone rubbers retain their elasticity at low temperatures and are chemically resistant and thermally stable. This makes them useful in caulking around windows, for instance.

Silicon Halides Silicon reacts readily with the halogens (X_2) to form the products SiX_4. As we might expect from their molecular structures and molar masses, at room temperature SiF_4 is a gas, $SiCl_4$ and $SiBr_4$ are liquids, and SiI_4 is a solid. Both SiF_4 and $SiCl_4$ are readily hydrolyzed with water, but the former is only partially hydrolyzed. The hydrolysis reactions for SiF_4 and $SiCl_4$ are as follows:

$$2\,SiF_4(g) + 4\,H_2O(l) \rightleftharpoons SiO_2(s) + 2\,H_3O^+(aq) + [SiF_6]^{2-}(aq) + 2\,HF(aq) \qquad \textbf{(21.38)}$$

$$SiCl_4(l) + 2\,H_2O(l) \longrightarrow SiO_2(s) + 4\,HCl(aq) \qquad \textbf{(21.39)}$$

The ability of the Si atom to expand its valence shell is illustrated by the formation of the ion $[SiF_6]^{2-}$. The corresponding $[SiCl_6]^{2-}$ has not been prepared, probably because the chloride ion is too big for six of them to fit around the Si atom.

$SiCl_4$ is manufactured on a large scale to produce finely divided silica (reaction 21.37), used as a reinforcing filler in silicone rubber, and very pure silicon for transistors used in computer chips.

Glass Making Glass and the art of glassmaking have been known for millennia. Beautiful stained-glass windows can be seen in medieval and modern churches; ancient glass containers for perfume and oil are displayed in many museums. Today, glass is indispensable in almost every facet of life.

Soda–lime glass is the oldest form of glass. The starting material in its manufacture is a mixture of sodium carbonate (*soda* ash), calcium carbonate (which decomposes to form quick*lime* when heated), and silicon dioxide. The mixture can be fused at a relatively low temperature (1300 °C) compared with the melting point of pure silica (1710 °C), and it is easy to form into the shapes needed. The effect of the sodium ions is to break up the crystalline lattice of the SiO_2; the calcium ions render the glass insoluble in water, so it can be used for such items as drinking glasses and windows. At the high temperatures employed, chemical reactions occur that produce a mixture of sodium and calcium silicates as the ultimate glass product.

▲ Stained-glass windows form the 20-meter-high ceiling, called the Glory Window, inside the Chapel of Thanksgiving in Dallas, Texas.

Glass containing even small amounts of Fe_2O_3 has a distinctive green color (bottle glass). Glass can be made colorless by incorporating MnO_2 into the glassmaking process. The MnO_2 oxidizes green $FeSiO_3$ to yellow $Fe_2(SiO_3)_3$ and is itself reduced to Mn_2O_3, which imparts a violet color. The yellow and violet are complementary colors, so the glass appears colorless. Where desired, color can be imparted by means of appropriate additives, such as CoO for cobalt blue glass. To produce an opaque glass, such additives as calcium phosphate are used. In Bohemian crystal, most of the Na^+ is replaced by K^+; a glass with exceptional transparency can be made by incorporating lead oxides.

A problem with soda-lime glass is its high coefficient of thermal expansion—its dimensions change significantly with temperature. The glass cannot withstand thermal shock. This limitation posed a particular problem for the lanterns used in the early days of railroads. In the rain, the hot glass in these lanterns would easily shatter. Adding B_2O_3 to the glass solved the problem. A *borosilicate* glass has a low coefficient of thermal expansion and is thus resistant to thermal shock. This is the glass more commonly known by its trade name, Pyrex. It is widely used in chemical laboratories and for cookware in the home.

Most glass has small bubbles or impurities in it that decrease its ability to transmit light without scattering—a phenomenon observed in the distorted images produced by the thick bottoms of drinking glasses. In modern *fiber-optic* cables, sound waves are converted to electrical impulses, which are transmitted as laser light beams. The light must be transmitted over long distances without distortion or loss of signal. For this purpose, a special glass made of pure silica is required. The key to making this glass is in purifying silica, which can be done by a series of chemical reactions. First, impure quartz or sand is reduced to silicon by using coke as a reducing agent. The silicon is then allowed to react with $Cl_2(g)$ to form $SiCl_4(g)$. The reactions involved in the conversion of SiO_2 to $SiCl_4$ are as follows:

$$SiO_2(s) + 2\,C(s) \longrightarrow Si(s) + 2\,CO(g)$$
$$Si(s) + 2\,Cl_2(g) \longrightarrow SiCl_4(g)$$

Finally, the $SiCl_4$ is burned in a methane–oxygen flame. SiO_2 deposits as a fine ash, and chlorocarbon compounds escape as gaseous products. The SiO_2, with impurity levels reduced to parts per billion, can then be melted and drawn into the fine filaments required in fiber-optic cable. Tens of millions of kilometers of fiber-optic cable are currently produced annually in the United States.

Diagonal Relationship of Boron and Silicon

The following similarities between B and Si are listed here:

- Boron forms a solid acidic oxide, B_2O_3, like that of silicon, SiO_2. In contrast Al_2O_3 is amphoteric and CO_2 is acidic.
- Boric acid, H_3BO_3, is a weak acid similar to silicic acid H_4SiO_4.
- There is a wide range of polymeric borates and silicates, based on shared oxygen atoms.
- Both boron and silicon form gaseous hydrides.

This diagonal relationship is not readily understood and cannot be interpreted in terms of charge density since the bonding in boron compounds and in silicon compounds is exclusively covalent. The elements are, however, both metalloids, have similar electronegativities, and have similar sizes leading to similar chemical behavior.

Properties and Uses of Tin and Lead

The data in Table 21.9 suggest that tin and lead are rather similar to each other. Both are soft and malleable and melt at low temperatures. The ionization

TABLE 21.9   Some Properties of Tin and Lead

	Sn	Pb
Atomic number	50	82
Atomic (metallic) radius, pm	141	175
Ionic (M^{2+}) radius, pm[a]	93	118
First ionization energy, kJ mol^{-1}	709	716
Electrode potential $E°$, V		
$[M^{2+}(aq) + 2\,e^- \longrightarrow M(s)]$	−0.137	−0.125
$[M^{4+}(aq) + 2\,e^- \longrightarrow M^{2+}(aq)]$	+0.154	+1.5
Melting point, °C	232	327
Boiling point, °C	2623	1751
Density, g cm^{-3} at 20 °C	5.77 (α, gray)	11.34
	7.29 (β, white)	
Hardness[b]	1.6	1.5
Electrical conductivity[b]	14.4	7.68

[a]For coordination number four.
[b]See footnotes c and d of Table 21-2.

energies and standard electrode potentials of the two metals are also about the same. This means that their tendencies to be oxidized to the +2 oxidation state are comparable.

The fact that both tin and lead can exist in two oxidation states, +2 and +4, is an example of the inert pair effect (page 946). In the +2 oxidation state, the inert pair ns^2 is not involved in bond formation, whereas in the +4 oxidation state, the pair does participate. Tin displays a stronger tendency to exist in the +4 oxidation state than does lead. That tendency is consistent with the trend observed in group 13, in which the lower oxidation state is favored farther down a group.

Another difference between tin and lead is that tin exists in two common crystalline forms (α and β), whereas lead has but a single solid form. The α (gray), or nonmetallic, form of tin is stable below 13 °C; the β (white), or metallic, form of tin is stable above 13 °C. Ordinarily, when a sample of β tin is cooled, it must be kept below 13 °C for a long time before the transition to α tin occurs. Once it does begin, however, the transformation takes place rather rapidly and with dramatic results. Because α tin is less dense than the β variety, the tin expands and crumbles to a powder. This transformation leads to the disintegration of objects made of tin. It has been a particular problem in churches in colder climates because some organ pipes are made of tin or tin alloys. The transformation is known in northern Europe as *tin disease, tin pest,* or *tin plague.*

The chief tin ore is tin(IV) oxide, SnO_2, known as *cassiterite*. After initial purification, the tin(IV) oxide is reduced with carbon (coke) to produce tin metal, as shown below:

$$SnO_2(s) + C(s) \xrightarrow{\Delta} Sn(l) + CO_2(g) \tag{21.38}$$

Nearly 50% of the tin metal produced is used in tinplate, especially in plating iron for use in cans for storing foods. The next most important use (about 25% of the total produced) is in the manufacture of **solders**—low-melting alloys used to join wires or pieces of metal. Other important alloys of tin are *bronze* (90% Cu, 10% Sn) and *pewter* (85% Sn, 7% Cu, 6% Bi, 2% Sb). Alloys of Sn and Pb are used to make organ pipes.

Lead is found chiefly as lead(II) sulfide, PbS, an ore known as *galena*. Lead(II) sulfide is first converted to lead(II) oxide by heating it strongly in air, a process called **roasting**. The oxide is then reduced with coke to produce the metal. The reactions are as follows:

$$2\,PbS(s) + 3\,O_2(g) \xrightarrow{\Delta} 2\,PbO(s) + 2\,SO_2(g) \qquad \textbf{(21.39)}$$

$$2\,PbO(s) + C(s) \xrightarrow{\Delta} 2\,Pb(l) + CO_2(g) \qquad \textbf{(21.40)}$$

More than half the lead produced is used in lead–acid (storage) batteries. Other uses include the manufacture of solder and other alloys, ammunition, and radiation shields (to protect against X-rays).

Compounds of Group 14 Metals

As mentioned earlier, both tin and lead exhibit the +2 and +4 oxidation states. The charge densities of these ions are shown in Table 21.10. The charge density of Sn^{2+} is such that many of the compounds containing tin in the +2 oxidation state are covalent; however, a few ionic solids containing the Sn^{2+} ion are known. Lead(II) exists in many ionic solids. When the metals are in the +4 oxidation state, they form covalent bonds, not ionic bonds, because these Sn^{4+} and Pb^{4+} have very large charge densities and draw electron density from surrounding anions toward themselves.

Oxides Tin forms two primary oxides: SnO and SnO_2. By heating SnO in air, it can be converted to SnO_2. One use of SnO_2 is as a jewelry abrasive.

Lead forms a number of oxides, and the chemistry of some of these oxides is not completely understood. The best known oxides of lead are yellow PbO, *litharge*; red-brown lead dioxide, PbO_2; and a mixed-valence oxide known as *red lead*, Pb_3O_4. Lead oxides are used in the manufacture of lead–acid (storage) batteries, glass, ceramic glazes, cements (PbO), metal-protecting paints (Pb_3O_4), and matches (PbO_2). Other lead compounds are generally made from the oxides.

Because lead tends to be in the +2 oxidation state, lead(IV) compounds tend to undergo reduction to compounds of lead(II) and are therefore good oxidizing agents. A case in point is PbO_2. In Chapter 20, we noted its use as the cathode in lead–acid storage cells. The reduction of $PbO_2(s)$ can be represented by the half-equation

$$PbO_2(s) + 4\,H^+(aq) + 2\,e^- \longrightarrow Pb^{2+}(aq) + 2\,H_2O(l) \qquad E° = +1.455\ V$$

$PbO_2(s)$ is a better oxidizing agent than $Cl_2(g)$ and nearly as good as $MnO_4^-(aq)$. For example, $PbO_2(s)$ can oxidize HCl(aq) to $Cl_2(g)$, as shown below:

$$PbO_2(s) + 4\,HCl(aq) \longrightarrow PbCl_2(aq) + 2\,H_2O(l) + Cl_2(g) \qquad E°_{cell} = 0.097\ V$$

Halides Tin (IV) chloride is a covalent compound. It is an oily liquid that reacts with moisture in the air as follows:

$$SnCl_4(l) + 4\,H_2O(l) \longrightarrow Sn(OH)_4(s) + 4\,HCl(g)$$

Lead(IV) chloride is also a covalent compound and a yellow oil that reacts with moisture in the air in a manner similar to $SnCl_4$.

Tin(II) and lead(II) chlorides are quite different. Lead(II) chloride is a white insoluble ionic solid, whereas tin(II) chloride is a covalent solid that is soluble in organic solvents. In the gas phase, $SnCl_2$ is a V-shaped molecule, as expected from VSEPR theory. We might expect $SnCl_2$ to act as a Lewis base, because of the presence of the lone pair. However, it acts as a Lewis acid. For example, $SnCl_2$ reacts with Cl^- to form $SnCl_3^-$.

Both chlorides of tin—$SnCl_2$ and $SnCl_4$—have important uses. Tin(II) chloride, $SnCl_2$, is a good reducing agent and is used in the quantitative analysis of iron ores to reduce iron(III) to iron(II) in aqueous solution. Tin(IV) chloride, $SnCl_4$, is formed by the direct reaction of tin and $Cl_2(g)$; it is the form in which tin is recovered from scrap tinplate. Tin(II) fluoride, SnF_2 (stannous fluoride),

---

TABLE 21.10 Charge Densities of Tin and Lead Ions

Ion	Charge Density, C mm^{-3}
Sn^{2+}	54
Pb^{2+}	32
Sn^{4+}	267
Pb^{4+}	196

▶ Another mixed-valence oxide that we have encountered on several occasions in this text is Fe_3O_4 (see, for example, page 86).

was used as an anticavity additive to toothpaste but has largely been replaced by NaF in gel toothpastes.

Other Compounds As we should expect from the solubility guidelines (Table 5.1), one of the few soluble lead compounds is lead(II) nitrate, $Pb(NO_3)_2$. It is formed in the reaction of PbO_2 with nitric acid, as shown below:

$$2\,PbO_2(s) + 4\,HNO_3(aq) \longrightarrow 2\,Pb(NO_3)_2(aq) + 2\,H_2O(l) + O_2(g)$$

The addition of a soluble chromate salt to $Pb(NO_3)_2(aq)$ produces lead(II) chromate $PbCrO_4$, a yellow pigment known as chrome yellow. Another lead-based pigment used in ceramic glazes and once extensively used in the manufacture of paint is *white lead*, $2\,PbCO_3 \cdot Pb(OH)_2$.

Lead Poisoning

Beginning with the ancient Romans and continuing to fairly recent times, lead has been used in plumbing systems, including those designed to transport drinking water. Exposure to lead has also occurred through cooking and eating utensils and pottery glazes made with lead. In colonial times, lead poisoning was clearly diagnosed as the cause of "dry bellyache" suffered by some North Carolinians who consumed rum made in New England. The distilling equipment used in the manufacture of the rum had components made of lead.

Mild forms of lead poisoning produce nervousness and depression. More severe cases can lead to permanent nerve, brain, and kidney damage. Lead interferes with the biochemical reactions that produce the iron-containing heme group in hemoglobin. As little as 10–15 μg Pb/dL in blood seems to produce physiological effects, especially in small children. The phaseout of leaded gasoline has resulted in a dramatic drop in average lead levels in blood. The drop in blood lead levels is evident in the graph shown in Figure 21-35 and parallels the decline in the use of lead in gasoline. The principal sources of lead contamination now seem to be lead-based painted surfaces found in old buildings and soldered joints in plumbing systems. Lead has been eliminated from modern plumbing solder, which is now a mixture of 95% Sn and 5% Sb. Because lead is toxic, its disposal is closely monitored. Recycling provides about three-quarters of the current lead metal production.

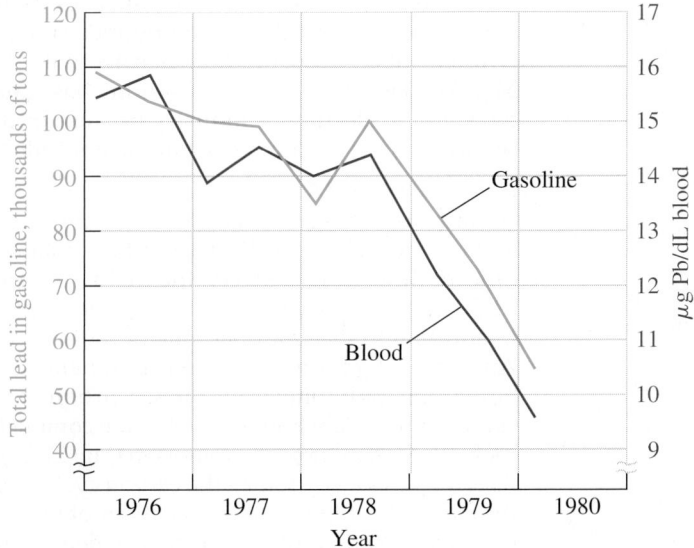

▲ FIGURE 21-35
Lead in gasoline and in blood
The level of lead in the blood of a representative human population showed a dramatic decline that paralleled the decline in the use of lead additives in gasoline in the 1970s. (Data source: Environmental Protection Agency, Office of Policy Analysis, 1984.)

21-12 CONCEPT ASSESSMENT

To prevent the air oxidation of aqueous solutions of Sn^{2+} to Sn^{4+}, metallic tin is sometimes kept in contact with the $Sn^{2+}(aq)$. Suggest how this contact helps prevent the oxidation.

Mastering**CHEMISTRY**   www.masteringchemistry.com

Gallium arsenide, GaAs, may be one of the most versatile high-tech materials of our time. For a discussion of some of its properties and uses, go to the Focus On feature for Chapter 21, Gallium Arsenide, on the MasteringChemistry site.

Summary

21-1 Periodic Trends and Charge Density—Trends in atomic or ionic radii, ionization energy, electron affinity, electronegativity, and polarizability (Fig. 21-1), and the **charge density** (equation 21.1) of an ion can be used to rationalize the chemistry of the elements. The greater the charge density, the greater is the tendency for that cation to form bonds with covalent character.

21-2 Group 1: The Alkali Metals—The **alkali metals** (group 1) are the most active of the metals, as indicated by their low ionization energies and large negative electrode potentials. Lithium exhibits some properties similar to magnesium in group 2. Most of the alkali metals are prepared by the electrolysis of their molten salts. Electrolysis of NaCl(aq) produces NaOH(aq), from which many other sodium compounds can be prepared (Fig. 21-5). Na_2CO_3 can be produced from NaCl, NH_3, and $CaCO_3$ by the Solvay process (Fig. 21-7). A **detergent** is a cleansing agent, often sodium salts with long-chain hydrocarbon anions terminating in a sulfate group. A **soap** is also a detergent, but soaps are the sodium salts of long-chain fatty acids terminating in the carboxylate group.

21-3 Group 2: The Alkaline Earth Metals—The **alkaline earth metals** (group 2), like group 1 metals, are also very active. Some group 2 metals are prepared by the electrolysis of a molten salt (Fig. 21-14) and some by chemical reduction. Among the most important of the alkaline earth compounds are the carbonates, especially $CaCO_3$. The oxide of calcium, CaO, called **quicklime**, is formed by the high-temperature decomposition (**calcination**) of limestone. The hydroxide of calcium, $Ca(OH)_2$, called **slaked lime**, is formed in the reaction of quicklime and water. Reversible reactions involving CO_3^{2-}, HCO_3^-, $CO_2(g)$, and H_2O account for the formation of limestone caves and interior features, such as **stalactites** and **stalagmites** (Fig. 21-15). **Plaster of Paris** is the hemihydrate of calcium sulfate, $CaSO_4 \cdot \frac{1}{2}H_2O$, formed by heating the mineral gypsum, $CaSO_4 \cdot 2 H_2O$. The similar properties exhibited by the pairs of elements (Li, Mg), (Be, Al), and (B, Si) are known as **diagonal relationships**.

21-4 Group 13: The Boron Family—Group 13 contains one nonmetal, B, and the metals Al, Ga, In, and Tl. Boron compounds are often electron deficient (fewer than an octet of electrons around the B atoms). Three-center, two-electron bonds are used in describing the bonding in diborane (Fig. 21-17). Aluminum occurs almost exclusively in the +3 oxidation state, whereas thallium is mostly in the +1 oxidation state. This is a manifestation of the presence of a pair of s electrons in the valence shells of certain post-transition elements. These electrons do not participate in chemical bonding, a consequence referred to as the inert **pair effect**. The principal metal of group 13 is aluminum, whose large-scale use is made possible by an effective method of production (Fig. 21-24). The amphoterism of Al_2O_3 is the basis for separating Al_2O_3 from impurities, mostly Fe_2O_3. Electrolysis is carried out in molten Na_3AlF_6 with Al_2O_3 as a solute. Aluminum is a powerful reducing agent in the **thermite reaction**, in which Fe_2O_3 is reduced to the metal. Aluminum chloride forms a **dimer**, in which two bridging chlorine atoms join together two $AlCl_3$ units (Fig. 21-25). $AlCl_3$ is very reactive. For example, a molecule of $AlCl_3$, a Lewis acid, attaches itself to a Lewis base to form an addition compound called **an adduct**. **Alums** are a class of double salts with the formula $M(I)M(III)(SO_4)_2 \cdot 12 H_2O$. Gallium has gained importance in the electronics industry because of the desirable semiconductor properties of gallium arsenide (GaAs).

21-5 Group 14: The Carbon Family—Group 14 contains one nonmetal (C), two metalloids (Si and Ge), and the metals Sn, and Pb. Group 14 is notable for the significant differences between the first two members (Table 21.7). Carbon is found in several different physical forms in addition to its allotropes; one of these is **carbon black**. An interesting form of carbon is **graphene**, a sheet of carbon atoms only one atom thick. Carbon monoxide and carbon dioxide are both formed in the combustion of fossil fuels. Carbon monoxide is a poison (Fig. 21-29) and carbon dioxide plays a key role in the carbon cycle (Fig. 21-30). All the elements of group 14 form halides of the type MX_4. All but carbon can employ expanded valence shells to form compounds of the form species, such as the anion $[MX_6]^{2-}$. Progressing down group 14, the halide MX_2 becomes more stable than MX_4 because of the inert pair effect. Silica (SiO_2) and various silicate anions are ubiquitous components of the mineral world. They are also constituents of ceramic materials and **glass**. Organosilanes, compounds in which the hydrogen atoms of silanes are replaced by organic groups, can be polymerized to form **silicones**. **Zeolites** are aluminosilicates that are used as molecular sieves and in treating hard water. Tin and lead in group 14 have some similarities (Table 21.9). They are both soft metals with low melting points. They also have some differences, including the fact that tin acquires the oxidation state +4 rather easily, whereas the +2 oxidation state is favored by lead. One of the most important uses of tin is in low-melting-point alloys known as **solders**. Both elements are obtained by the reduction of their oxides. The metallurgical method of converting a sulfide ore to an oxide, typified by the conversion of PbS to PbO on strong heating, is called **roasting**.

Integrative Example

Without performing detailed calculations, demonstrate that reaction (21.19) correctly describes the dissolving action of rainwater on limestone (for $CaCO_3$, $K_{sp} = 2.8 \times 10^{-9}$).

Analyze

Write the equations that, when combined, yield equation (21.19) for the overall reaction occurring when rainwater acts on limestone. Evaluate the sources and relative importance of the different species involved in the dissolution of the limestone. Describe the overall result.

Solve

The relevant equations to describe the dissolution of calcium carbonate in rainwater are the solubility product expression for $CaCO_3$ and two equations describing the ionization of CO_2 in water (that is, carbonic acid). The three equations are as follows:

$$CaCO_3(s) \rightleftharpoons Ca^{2+}(aq) + CO_3^{2-}(aq)$$
$$K_{sp} = 2.8 \times 10^{-9}$$

$$CO_2 + 2 H_2O \rightleftharpoons H_3O^+ + HCO_3^-$$
$$K_{a_1} = 4.4 \times 10^{-7} \quad \textbf{(21.17)}$$

$$HCO_3^- + H_2O \rightleftharpoons H_3O^+ + CO_3^{2-}$$
$$K_{a_2} = 4.7 \times 10^{-11} \quad \textbf{(21.18)}$$

Equation (21.19) is related to these three equations in this way:

$$K_{sp} \text{ expression} + \text{equation (21.17)} - \text{equation (21.18)}$$

Which leads to

$$CaCO_3(s) + H_2O + CO_2 \rightleftharpoons Ca^{2+}(aq) + 2 HCO_3^-(aq)$$
$$K = (K_{sp} \times K_{a_1})/K_{a_2} = 2.6 \times 10^{-5} \quad \textbf{(21.19)}$$

Next, consider which is the greater source of CO_3^{2-} ions in solution: a saturated aqueous solution of $CaCO_3$ or an aqueous carbonic acid solution. In saturated $CaCO_3(aq)$,

$$[Ca^{2+}] = [CO_3^{2-}] \quad K_{sp} = [Ca^{2+}][CO_3^{2-}] = [CO_3^{2-}]^2$$
$$= 2.8 \times 10^{-9}$$

$$[CO_3^{2-}] = (2.8 \times 10^{-9})^{\frac{1}{2}} = 5.3 \times 10^{-5}\, M$$

In a carbonic acid solution, H_2CO_3 is a weak *diprotic* acid with $K_{a_2} \ll K_{a_1}$. Thus,

$$[H_3O^+] = [HCO_3^-], \text{ and } [CO_3^{2-}] = K_{a_2} = 4.7 \times 10^{-11}\, M$$

Note that the carbonate ion concentration in a saturated $CaCO_3$ solution is much greater than the carbonate ion concentration observed in a carbonic acid solution. This means that when the two processes occur simultaneously, carbonate ion from the dissolution of $CaCO_3$ acts as a common ion in the carbonic acid equilibrium. This displaces reaction (21.18) *to the left*, converting CO_3^{2-} to HCO_3^- while consuming H_3O^+. Removal of H_3O^+ in reaction (21.18) stimulates reaction (21.17) to shift *to the right*, producing more H_3O^+ and, simultaneously, more HCO_3^-. The overall effect is that H_2O, CO_2, and CO_3^{2-} (from $CaCO_3$) are consumed and HCO_3^- is produced, just as suggested by equation (21.19).

Assess

We have assessed the qualitative correctness of reaction (21.19). To calculate the quantitative extent of the dissolution of $CaCO_3(s)$ in rainwater is somewhat more difficult. The calculation centers on the combined equilibrium constant expression for reaction (21.19), $K = 2.6 \times 10^{-5}$, and is affected by the partial pressure of atmospheric CO_2 in equilibrium with rainwater. Typical data are given in Chapter 16, Practice Example A, page 738.

PRACTICE EXAMPLE A: Write chemical equations for the reactions that occur when NaCN is dissolved in water and when $Al(NO_3)_3$ is dissolved in water. Then, use data from Appendix D to explain why a precipitate of $Al(OH)_3$ forms when equal volumes of 1.0 M aqueous solutions of NaCN and $Al(NO_3)_3$ are mixed.

PRACTICE EXAMPLE B: The compound $BeCl_2 \cdot 4 H_2O$ cannot be dehydrated by heating and it dissolves in water to give an acidic solution. Conversely, $CaCl_2 \cdot 6 H_2O$ can be dehydrated by heating and it dissolves in water to give a solution with neutral pH. Explain these observations and write chemical equations for the reactions that occur, if any, when the salts are heated and when they are dissolved in water.

Exercises

Group 1: The Alkali Metals

1. Use information from the chapter to write chemical equations to represent each of the following:
   (a) reaction of cesium metal with chlorine gas
   (b) formation of sodium peroxide (Na_2O_2)
   (c) thermal decomposition of lithium carbonate
   (d) reduction of sodium sulfate to sodium sulfide
   (e) combustion of potassium to form potassium superoxide
2. Use information from the chapter to write chemical equations to represent each of the following:
   (a) reaction of rubidium metal with water
   (b) thermal decomposition of aqueous $KHCO_3$
   (c) combustion of lithium metal in oxygen gas
   (d) action of concentrated aqueous H_2SO_4 on KCl(s)
   (e) reaction of lithium hydride with water
3. Describe a simple test for determining whether a pure white solid is LiCl or KCl.
4. Describe two methods for determining the identity of an unknown compound that is either Li_2CO_3 or K_2CO_3.
5. Arrange the following compounds in the expected order of increasing solubility in water, and give the basis for your arrangement: Li_2CO_3, Na_2CO_3, $MgCO_3$.
6. The first electrolytic process to produce sodium metal used molten NaOH as the electrolyte. Write probable half-equations and an overall equation for this electrolysis.
7. A 0.872 L NaCl(aq) solution is electrolyzed for 2.50 min with a current of 0.810 A.
   (a) Calculate the pH of the solution after electrolysis.
   (b) Why doesn't the result depend on the initial concentration of the NaCl(aq)?
8. A lithium battery used in a cardiac pacemaker has a voltage of 3.0 V and a capacity of 0.50 A h (ampere hour). Assume that 5.0 μW of power is needed to regulate the heartbeat. [*Hint:* See Appendix B.]

(a) How long will the implanted battery last?
(b) How many grams of lithium must be present in the battery for the lifetime calculated in part (a)?
9. An analysis of a Solvay-process plant shows that for every 1.00 kg of NaCl consumed, 1.03 kg of $NaHCO_3$ are obtained. The quantity of NH_3 consumed in the overall process is 1.5 kg.
   (a) What is the percent efficiency of this process for converting NaCl to $NaHCO_3$?
   (b) Why is so little NH_3 required?
10. Consider the reaction $Ca(OH)_2(s) + Na_2SO_4(aq) \rightleftharpoons CaSO_4(s) + 2\,NaOH(aq)$.
    (a) Write a net ionic equation for this reaction.
    (b) Will the reaction essentially go to completion?
    (c) What will be $[SO_4^{2-}]$ and $[OH^-]$ *at equilibrium* if a slurry of $Ca(OH)_2(s)$ is mixed with 1.00 M $Na_2SO_4(aq)$?
11. The Gibbs energies of formation, ΔG_f°, for $Na_2O(s)$ and $Na_2O_2(s)$ are -379.09 kJ mol^{-1} and -449.63 kJ mol^{-1}, respectively, at 298 K. Calculate the equilibrium constant for the reaction below at 298 K. Is $Na_2O_2(s)$ thermodynamically stable with respect to $Na_2O(s)$ and $O_2(g)$ at 298 K?

$$Na_2O_2(s) \longrightarrow Na_2O(s) + \frac{1}{2}O_2(g)$$

12. The Gibbs energies of formation, ΔG_f°, for $KO_2(s)$ and $K_2O(s)$ are -240.59 kJ mol^{-1} and -322.09 kJ mol^{-1}, respectively, at 298 K. Calculate the equilibrium constant for the reaction below at 298 K. Is $KO_2(s)$ thermodynamically stable with respect to $K_2O(s)$ and $O_2(g)$ at 298 K?

$$2\,KO_2(s) \longrightarrow K_2O(s) + \frac{3}{2}O_2(g)$$

Group 2: The Alkaline Earth Metals

13. In the manner used to construct Figure 21-5, complete the diagram outlined. Specifically, indicate the reactants (and conditions) you would use to produce the indicated substances from $Ca(OH)_2$.

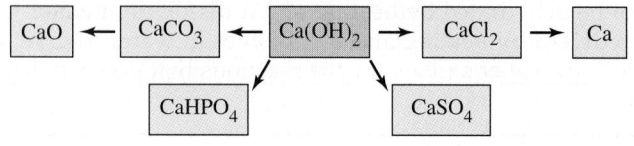

14. Replace the calcium-containing substances shown in the diagram accompanying Exercise 13 by their magnesium-containing equivalents. Then describe the reactants (and conditions) you would use to produce the indicated substances from $MgSO_4$.
15. In the Dow process (Fig. 21-13), the starting material is Mg^{2+} in seawater and the final product is Mg metal. This process seems to violate the principle of conservation of charge. Does it? Explain.

16. Which has the **(a)** higher melting point, MgO or BaO; **(b)** greater solubility in water, MgF_2 or $MgCl_2$? Explain.

17. Write chemical equations to represent the following:
    (a) reduction of BeF_2 to Be metal with Mg as a reducing agent
    (b) reaction of barium metal with $Br_2(l)$
    (c) reduction of uranium(IV) oxide to uranium metal with calcium as the reducing agent
    (d) calcination of dolomite, a mixed calcium magnesium carbonate ($MgCO_3 \cdot CaCO_3$)
    (e) complete neutralization of phosphoric acid with quicklime

18. Write chemical equations for the reactions you would expect to occur when
    (a) $Mg(HCO_3)_2(s)$ is heated to a high temperature
    (b) $BaCl_2(l)$ is electrolyzed
    (c) $Sr(s)$ is added to cold dilute $HBr(aq)$
    (d) $Ca(OH)_2(aq)$ is added to $H_2SO_4(aq)$
    (e) $CaSO_4 \cdot 2 H_2O(s)$ is heated

19. *Without performing detailed calculations*, indicate whether equilibrium is displaced either far to the left or far to the right for each of the following reactions. Use data from Appendix D as necessary.

    (a) $BaSO_4(s) + CO_3{}^{2-}(aq) \rightleftharpoons$
    $$BaCO_3(s) + SO_4{}^{2-}(aq)$$
    (b) $Mg_3(PO_4)_2(s) + 3 CO_3{}^{2-}(aq) \rightleftharpoons$
    $$3 MgCO_3(s) + 2 PO_4{}^{3-}(aq)$$
    (c) $Ca(OH)_2(s) + 2 F^-(aq) \rightleftharpoons$
    $$CaF_2(s) + 2 OH^-(aq)$$

20. *Without performing detailed calculations*, indicate why you would expect each of the following reactions to occur to a significant extent as written. Use data from Appendix D as necessary.
    (a) $BaCO_3(s) + 2 CH_3CO_2H(aq) \longrightarrow Ba^{2+}(aq) +$
    $$2 CH_3CO_2{}^-(aq) + H_2O(l) + CO_2(g)$$
    (b) $Ca(OH)_2(s) + 2 NH_4{}^+(aq) \longrightarrow$
    $$Ca^{2+}(aq) + 2 NH_3(aq) + 2 H_2O(l)$$
    (c) $BaF_2(s) + 2 H_3O^+(aq) \longrightarrow$
    $$Ba^{2+}(aq) + 2 HF(aq) + 2 H_2O(l)$$

21. With respect to decomposition to $MO(s)$ and $SO_3(g)$, which of the group 2 sulfates, $MSO_4(s)$, do you expect to be least stable? Explain your answer.

22. With respect to the decomposition to $MO(s)$ and $CO_2(g)$, which of the group 2 carbonates, $MCO_3(s)$, do you expect to be most stable? Explain your answer.

Group 13: The Boron Family

23. The molecule tetraborane has the formula B_4H_{10}.
    (a) Show that this is an electron-deficient molecule.
    (b) How many bridge bonds must occur in the molecule?
    (c) Show that butane, C_4H_{10}, is not electron deficient.

24. Write Lewis structures for the following species, both of which involve coordinate covalent bonding:
    (a) tetrafluoroborate ion, $BF_4{}^-$, used in metal cleaning and in electroplating baths
    (b) boron trifluoride ethylamine, used in curing epoxy resins (ethylamine is $C_2H_5NH_2$)

25. Write chemical equations to represent the following:
    (a) the preparation of boron from BBr_3
    (b) the formation of BF_3 from B_2O_3
    (c) the combustion of boron in hot $N_2O(g)$

26. Assign oxidation states to all the atoms in a perborate ion based on the structure on page 942.

27. Write chemical equations to represent the
    (a) reaction of $Al(s)$ with $HCl(aq)$;
    (b) reaction of $Al(s)$ with $NaOH(aq)$;
    (c) oxidation of $Al(s)$ to $Al^{3+}(aq)$ by an aqueous solution of sulfuric acid; the reduction product is $SO_2(g)$.

28. Write plausible equations for the
    (a) reaction of $Al(s)$ with $Br_2(l)$;
    (b) production of Cr from $Cr_2O_3(s)$ by the thermite reaction, with Al as the reducing agent;
    (c) separation of Fe_2O_3 impurity from bauxite ore.

29. In some foam-type fire extinguishers, the reactants are $Al_2(SO_4)_3(aq)$ and $NaHCO_3(aq)$. When the extinguisher is activated, these reactants mix, producing $Al(OH)_3(s)$ and $CO_2(g)$. The $Al(OH)_3$–CO_2 foam extinguishes the fire. Write a net ionic equation to represent this reaction.

30. Some baking powders contain the solids $NaHCO_3$ and $NaAl(SO_4)_2$. When water is added to this mixture of compounds, $CO_2(g)$ and $Al(OH)_3(s)$ are two

of the products. Write plausible net ionic equations for the formation of these two products.

31. The maximum resistance to corrosion of aluminum is between pH 4.5 and 8.5. Explain how this observation is consistent with other facts about the behavior of aluminum presented in this text.

32. Describe a series of *simple* chemical reactions that you could use to determine whether a particular metal sample is "aluminum 2S" (99.2% Al) or "magnalium" (70% Al, 30% Mg). You are permitted to destroy the metal sample in the testing.

33. In the purification of bauxite ore, a preliminary step in the production of aluminum, $[Al(OH)_4]^-(aq)$ can be converted to $Al(OH)_3(s)$ by passing $CO_2(g)$ through the solution. Write an equation for the reaction that occurs. Could $HCl(aq)$ be used instead of $CO_2(g)$? Explain.

34. In 1825, Hans Oersted produced aluminum chloride by passing chlorine over a heated mixture of carbon and aluminum oxide. In 1827, Friedrich Wöhler obtained aluminum by heating aluminum chloride with potassium. Write plausible equations for these reactions.

35. A description for preparing potassium aluminum alum calls for dissolving aluminum foil in $KOH(aq)$. The solution obtained is treated with $H_2SO_4(aq)$, and the alum is crystallized from the resulting solution. Write plausible equations for these reactions.

36. Handbooks and lists of chemicals do not contain entries under the formulas $Al(HCO_3)_3$ and $Al_2(CO_3)_2$. Explain why these compounds do not exist.

37. Digallane, Ga_2H_6, is a volatile compound and decomposes above $-10\ °C$ to gallium metal and hydrogen gas. Suggest a plausible structure for the digallane molecule and describe the bonding.

38. Draw a plausible structure for the GaH_2Cl_2 molecule and describe the bonding. [*Hint:* Each Cl atom is simultaneously bonded to two gallium atoms.]

Group 14: The Carbon Family

39. Comment on the accuracy of a jeweler's advertising that "diamonds last forever." In what sense is the statement true, and in what ways is it false?

40. A temporary fix for a "sticky" lock is to scrape a pencil point across the notches on the key and to work the key in and out of the lock a few times. What is the basis of this fix?

41. Write a chemical equation to represent
    (a) the reduction of silica to elemental silicon by aluminum;
    (b) the preparation of potassium metasilicate by the high-temperature fusion of silica and potassium carbonate;
    (c) the reaction of Al_4C_3 with water to produce methane.

42. Write a chemical equation to represent
    (a) the reaction of potassium cyanide solution with silver nitrate solution;
    (b) the combustion of Si_3H_8 in an excess of oxygen;
    (c) the reaction of dinitrogen with calcium carbide to give calcium cyanamide (CaNCN).

43. Describe what is meant by the terms *silane* and *silanol*. What is their role in the preparation of silicones?

44. Describe and explain the similarities and differences between the reaction of a silicate with an acid and that of a carbonate with an acid.

45. Methane and sulfur vapor react to form carbon disulfide and hydrogen sulfide. Carbon disulfide reacts with $Cl_2(g)$ to form carbon tetrachloride and S_2Cl_2. Further reaction of carbon disulfide and S_2Cl_2 produces additional carbon tetrachloride and sulfur. Write a series of equations for the reactions described here.

46. In a manner similar to that outlined on page 962,
    (a) write equations to represent the reaction of $(CH_3)_3SiCl$ with water, followed by the elimination of H_2O from the resulting silanol molecules.
    (b) Does a silicone polymer form from part (a)?
    (c) What would be the corresponding product obtained from CH_3SiCl_3?

47. Show that the empirical formula given for muscovite mica is consistent with the expected oxidation states of the elements present.

48. Show that the empirical formula given for crysotile asbestos is consistent with the expected oxidation states of the elements present.

49. Write plausible chemical equations for the (a) dissolving of lead(II) oxide in nitric acid; (b) heating of $SnCO_3(s)$; (c) reduction of lead(II) oxide by carbon; (d) reduction of $Fe^{3+}(aq)$ to $Fe^{2+}(aq)$ by $Sn^{2+}(aq)$; (e) formation of lead(II) sulfate during high-temperature roasting of lead(II) sulfide.

50. Write plausible chemical equations for preparing each compound from the indicated starting material: (a) $SnCl_2$ from SnO; (b) $SnCl_4$ from Sn; (c) $PbCrO_4$ from PbO_2. What reagents (acids, bases, salts) and equipment commonly available in the laboratory are needed for each reaction?

51. Lead(IV) oxide, PbO_2, is a good oxidizing agent. Use appropriate data from Appendix D to determine whether $PbO_2(s)$ in a solution with $[H_3O^+] = 1$ M is a sufficiently good oxidizing agent to carry the following oxidations to the point at which the concentration of the species being oxidized decreases to one-thousandth of its initial value.
    (a) $Fe^{2+}(1$ M$)$ to Fe^{3+}
    (b) $SO_4^{2-}(1$ M$)$ to $S_2O_8^{2-}$
    (c) $Mn^{2+}(1 \times 10^{-4}$ M$)$ to MnO_4^-

52. Tin(II) ion, Sn^{2+}, is a good reducing agent. Use data from Appendix D to determine whether Sn^{2+} is a sufficiently good reducing agent to reduce (a) I_2 to I^-; (b) Fe^{2+} to Fe(s); (c) Cu^{2+} to Cu(s); (d) $Fe^{3+}(aq)$ to $Fe^{2+}(aq)$ Assume that all reactants and products are in their standard states.

53. Would you expect the reaction of Pb(s) and $Cl_2(g)$ to yield $PbCl_2$ or $PbCl_4$?

54. Would you expect the reaction of Ge(s) and $F_2(g)$ to yield GeF_2, with germanium in the +2 oxidation state, or GeF_4, with germanium in the +4 oxidation state?

Integrative and Advanced Exercises

55. A chemical that should exist as a crystalline solid is seen to be a mixture of a solid and liquid in a container on a storeroom shelf. Give a plausible reason for that observation. Should the chemical be discarded or is it still useful for some purposes?

56. The following series of observations is made: (1) a small piece of dry ice $[CO_2(s)]$ is added to 0.005 M $Ca(OH)_2(aq)$. (2) Initially, a white precipitate forms. (3) After a short time the precipitate dissolves.
    (a) Write chemical equations to explain these observations.
    (b) If the 0.005 M $Ca(OH)_2(aq)$ is replaced by 0.005 M $CaCl_2(aq)$, would a precipitate form? Explain.
    (c) If the 0.005 M $Ca(OH)_2(aq)$ is replaced by 0.010 M $Ca(OH)_2(aq)$, a precipitate forms but does not re-dissolve. Explain why.

57. The melting point of NaCl(s) is 801 °C, much higher than that of NaOH (322 °C). More energy is consumed to melt and maintain molten NaCl than NaOH. Yet the preferred commercial process for the production of sodium is electrolysis of NaCl(l) rather than NaOH(l). Give a reason or reasons for this discrepancy.

58. Although the triiodide ion, I_3^-, is known to exist in aqueous solutions, the ion is stable in only certain ionic solids. For example, CsI_3 is stable with respect to decomposition to CsI and I_2, but LiI_3 is not stable with respect to LiI and I_2. Draw a Lewis structure for the I_3^- ion and suggest a reason why CsI_3 is stable with respect to decomposition to the iodide but LiI_3 is not.

59. At 298 K, the ΔG_f° values for $Li_2O(s)$ and $Li_2O_2(s)$ suggest that $Li_2O_2(s)$ is thermodynamically more stable than $Li_2O(s)$. At 1000 K, however, the situation is reversed. The Gibbs energies of formation, ΔG_f°, for $Li_2O(s)$ and $Li_2O_2(s)$ are -466.40 kJ mol^{-1} and -419.02 kJ mol^{-1}, respectively, at 1000 K. Calculate the equilibrium constant for the reaction below at

1000 K and the equilibrium partial pressure of $O_2(g)$ above a sample of $Li_2O_2(s)$ at 1000 K.

$$Li_2O_2(s) \longrightarrow Li_2O(s) + \frac{1}{2}O_2(g)$$

60. The chemical equation for the hydration of an alkali metal ion is $M^+(g) \rightarrow M^+(aq)$. The Gibbs energy change and the enthalpy change for the process are denoted by $\Delta G°_{hydr.}$ and $\Delta H°_{hydr.}$, respectively. $\Delta G°_{hydr.}$ and $\Delta H°_{hydr.}$ values are given below for the alkali metal ions.

M^+	Li^+	Na^+	K^+	Rb^+	Cs^+
$\Delta H°_{hydr.}$	−522	−407	−324	−299	−274 kJ mol^{-1}
$\Delta G°_{hydr.}$	−481	−375	−304	−281	−258 kJ mol^{-1}

Use the data above to calculate $\Delta S°_{hydr.}$ values for the hydration process. Explain the trend in the $\Delta S°_{hydr.}$ values.

61. Lithium superoxide, $LiO_2(s)$, has never been isolated. Use ideas from Chapter 12, together with data from this chapter and Appendix D, to estimate $\Delta H°_f$ for $LiO_2(s)$ and assess whether $LiO_2(s)$ is thermodynamically stable with respect to $Li_2O(s)$ and $O_2(g)$.
(a) Use the Kapustinskii equation, along with appropriate data below, to estimate the lattice energy, U, for $LiO_2(s)$. (See exercise 126 in Chapter 12.) The ionic radii for Li^+ and O_2^- are 73 pm and 144 pm, respectively.
(b) Use your result from part (a) in the Born–Fajans–Haber cycle to estimate $\Delta H°_f$ for $LiO_2(s)$. [*Hint:* For the process $O_2(g) + e^- \rightarrow O_2^-(g)$, $\Delta H° = -43$ kJ mol^{-1}. See Table 21.2 and Appendix D for the other data that are required.]
(c) Use your result from part (b) to calculate the enthalpy change for the decomposition of $LiO_2(s)$ to $Li_2O(s)$ and $O_2(g)$. For $Li_2O(s)$, $\Delta H°_f = -598.73$ kJ mol^{-1}.
(d) Use your result from part (c) to decide whether $LiO_2(s)$ is thermodynamically stable with respect to $Li_2O(s)$ and $O_2(g)$. Assume that entropy effects can be neglected.

62. When a 0.200 g sample of Mg is heated in air, 0.315 g of product is obtained. Assume that all the Mg appears in the product.
(a) If the product were pure MgO, what mass should have been obtained?
(b) Show that the 0.315 g product could be a mixture of MgO and Mg_3N_2.
(c) What is the mass percent of MgO in the $MgO–Mg_3N_2$ mixed product?

63. Comment on the feasibility of using a reaction similar to (21.4) to produce **(a)** lithium metal from LiCl; **(b)** cesium metal from CsCl, with Na(l) as the reducing agent in each case. [*Hint:* Consider data from Table 21.2.]

64. Concerning the thermite reaction,
(a) use data from Appendix D to calculate $\Delta H°$ at 298 K for the reaction.

$$2\,Al(s) + Fe_2O_3(s) \longrightarrow 2\,Fe(s) + Al_2O_3(s)$$

(b) Write an equation for the reaction when $MnO_2(s)$ is substituted for $Fe_2O_3(s)$, and calculate $\Delta H°$ for this reaction.

(c) Show that if MgO were substituted for Fe_2O_3, the reaction would be *endothermic*.

65. Use data from Appendix D (Table D-2) to calculate a value of $E°$ for the reduction of $Li^+(aq)$ to $Li(s)$, and compare your result with the value listed in Table 21.2.

66. The electrolysis of 0.250 L of 0.220 M $MgCl_2$ is conducted until 104 mL of gas (a mixture of H_2 and water vapor) is collected at 23 °C and 748 mmHg. Will $Mg(OH)_2(s)$ precipitate if electrolysis is carried to this point? (Use 21 mmHg as the vapor pressure of the solution.)

67. A particular water sample contains 56.9 ppm SO_4^{2-} and 176 ppm HCO_3^-, with Ca^{2+} as the only cation.
(a) How many parts per million of Ca^{2+} does the water contain?
(b) How many grams of CaO are consumed in removing HCO_3^-, from 602 kg of the water?
(c) Show that the Ca^{2+} remaining in the water after the treatment described in part (b) can be removed by adding Na_2CO_3.
(d) How many grams of Na_2CO_3 are required for the precipitation referred to in part (c)?

68. An aluminum production cell of the type pictured in Figure 21-24 operates at a current of 1.00×10^5 A and a voltage of 4.5 V. The cell is 38% efficient in using electrical energy to produce chemical change. (The rest of the electrical energy is dissipated as thermal energy in the cell.)
(a) What mass of Al can be produced by this cell in 8.00 h?
(b) If the electrical energy required to power this cell is produced by burning coal (85% C; heat of combustion of C = 32.8 kJ/g) in a power plant with 35% efficiency, what mass of coal must be burned to produce the mass of Al determined in part (a)?

69. Use data from Appendix D (Table D-2) to estimate the minimum voltage required to electrolyze Al_2O_3 in the Hall-Héroult process, reaction (21.25). Use $\Delta G°_f[Al_2O_3(l)] = -1520$ kJ mol^{-1}. Show that the oxidation of the graphite anode to $CO_2(g)$ permits the electrolysis to occur at a lower voltage than if the electrolysis reaction were $Al_2O_3(l) \longrightarrow 2\,Al(l) + \frac{3}{2}O_2(g)$.

70. At 20 °C, a saturated aqueous solution of $Pb(NO_3)_2$ maintains a relative humidity of 97%. What must be the composition of this solution, expressed as g $Pb(NO_3)_2/100.0$ g H_2O?

71. Use information from this chapter and elsewhere in this text to explain why the compounds $PbBr_4$ and PbI_4 do *not* exist.

72. The reaction of borax, calcium fluoride, and concentrated sulfuric acid yields sodium hydrogen sulfate, calcium sulfate, water, and boron trifluoride as products. Write a balanced equation for this reaction.

73. The dissolution of $MgCO_3(s)$ in $NH_4^+(aq)$ can be represented as

$$MgCO_3(s) + NH_4^+(aq) \rightleftharpoons$$
$$Mg^{2+}(aq) + HCO_3^-(aq) + NH_3(aq)$$

Calculate the molar solubility of $MgCO_3$ in each of the following solutions: **(a)** 1.00 M $NH_4Cl(aq)$; **(b)** a buffer that is 1.00 M NH_3 and 1.00 M NH_4Cl; **(c)** a buffer that is 0.100 M NH_3 and 1.00 M NH_4Cl.

74. Show that, in principle, $Na_2CO_3(aq)$ can be converted almost completely to $NaOH(aq)$ by the reaction

$$Ca(OH)_2(s) + Na_2CO_3(aq) \longrightarrow$$
$$CaCO_3(s) + 2\,NaOH(aq)$$

75. Assume that the packing of spherical atoms in crystalline metals is the same for Li, Na, and K, and explain why Na has a higher density than *both* Li and K. [*Hint:* Use data from Table 21.2.]
76. Would you expect the lattice energy of $MgS(s)$ to be less than, greater than, or about the same as that of $MgO(s)$? Use appropriate data from various locations

in this text to obtain the values of the two lattice energies. Use a value of 456 kJ for the process $S^-(g) + e^- \longrightarrow S^{2-}(g)$.
77. There has been some interest in the alkali metal fullerides, $M_nC_{60}(s)$, because at low temperatures, some of these compounds become superconducting. The alkali metal fullerides are ionic crystals comprising M^+ ions and C_{60}^{n-} ions. The value of n can be deduced from the crystal structure. If M_nC_{60} consists of a cubic closest packed array of fulleride ions, with M^+ ions occupying all the octahedral and tetrahedral holes in the fulleride lattice, then what is the value of n and what is the empirical formula of the fulleride?

Feature Problems

78. In Chapter 20, we examined the relationship of electrode potentials to thermodynamic data. In fact, electrode potentials can be calculated from tabulated thermodynamic data (many of which, in turn, were established from electrochemical measurements). To demonstrate this fact and to pursue further the discussion in Are You Wondering 21-1, combine the three steps for the oxidation of Li(s) with a corresponding set of three steps for the reduction of H^+ (1 M) to $H_2(g)$. Obtain $\Delta H°$ for the overall reaction.
    (a) Neglect entropy changes that occur (that is, assume that $\Delta G° \approx \Delta H°$), and estimate the value of $E°_{Li^+/Li}$.
    (b) Combine the calculated $\Delta H°$ value with a value of $\Delta S°$ to obtain another estimate of $E°_{Li^+/Li}$. [*Hint:* Hydration energies for $Li^+(g)$ and $H^+(g)$ to form 1 M solutions are −522 and −1094 kJ/mol, respectively. Also, use data from various locations in this text.]
79. Mono Lake in eastern California is a rather unusual salt lake. The lake has no outlets; water leaves only by evaporation. The rate of evaporation is great enough that the lake level would be lowered by three meters per year if not for fresh water entering through underwater springs and streams originating in the nearby Sierra Nevada mountains. The principal salts in the lake are the chlorides, bicarbonates, and sulfates of sodium. An approximate "recipe" for simulating the lake water is to dissolve 18 tablespoons of sodium bicarbonate, 10 tablespoons of sodium chloride, and 8 teaspoons of Epsom salt (magnesium sulfate heptahydrate) in 4.5 liters of water (although the lake water

actually contains only trace amounts of magnesium ion). Assume that 1 tablespoon of any of the salts weighs about 10 g. (1 tablespoon = 3 teaspoons.)
    (a) Expressed as grams of salt per liter, what is the approximate salinity of Mono Lake? How does this salinity compare with seawater, which is approximately 0.438 M NaCl and 0.0512 M $MgCl_2$?
    (b) Estimate an approximate pH for Mono Lake water. How does your estimate compare with the observed pH of about 9.8? Actually, the recipe for the lake water also calls for a pinch of borax. How would its presence affect the pH? [Borax is a sodium salt, $Na_2B_4O_7 \cdot 10\,H_2O$, related to the weak monoprotic boric acid ($pK_a = 9.25$).]
    (c) Mono Lake has some unusual limestone formations called *tufa*. They form at the site of underwater springs and grow only underwater, although some project above water, having formed at a time when the lake level was higher. Explain how the tufa form. [*Hint:* What chemical reaction(s) is(are) involved?]

Self-Assessment Exercises

80. In your own words, define the following terms: (a) dimer; (b) adduct; (c) calcination; (d) amphoteric oxide; (e) three-center two-electron bond.
81. Briefly describe each of the following ideas, methods, or phenomena: (a) diagonal relationship; (b) preparation of deionized water by ion exchange; (c) thermite reaction; (d) inert pair effect.
82. Explain the important distinction between each pair of terms: (a) peroxide and superoxide; (b) quicklime

and slaked lime; (c) soap and detergent; (d) silicate and silicone; (e) sol and gel.
83. Of the following oxides, the one with the highest melting point is (a) Li_2O; (b) BaO; (c) MgO; (d) SiO_2.
84. The best oxidizing agent of the following oxides is (a) Li_2O; (b) MgO; (c) Al_2O_3; (d) CO_2; (e) SnO_2; (f) PbO_2.
85. Predict the products of the following reactions:
    (a) $BCl_3(g) + NH_3(g)$
    (b) $K(s) + O_2(g)$

(c) $Li(s) + O_2(g)$

(d) $BaO_2(s) + H_2O(l)$

86. A chemist knows that aluminum is more reactive toward oxygen than is iron, but many people believe the opposite. What evidence was provided in this chapter that demonstrates that aluminum is a much stronger reducing agent (and is much more reactive) than iron? If aluminum is so reactive, then why is it that aluminum can be used for making products— pots, pans, aluminum foil, siding for houses—that last so long?

87. Listed are several pairs of substances. For some pairs, one or both members of the pair react individually with water to produce a gas. For others, neither member of the pair reacts with water. The pair for which *each* member reacts with water and yields the *same* gaseous product is **(a)** $Al(s)$ and $Ba(s)$; **(b)** $Ca(s)$ and $CaH_2(s)$; **(c)** $Na(s)$ and $Na_2O_2(s)$; **(d)** $K(s)$ and $KO_2(s)$; **(e)** $NaHCO_3(s)$ and $HCl(aq)$.

88. Complete and balance the following. Write the simplest equation possible. If no reaction occurs, so state.

(a) $Li_2CO_3(s) \xrightarrow{\Delta}$

(b) $CaCO_3(s) + HCl(aq) \longrightarrow$

(c) $Al(s) + NaOH(aq) \longrightarrow$

(d) $BaO(s) + H_2O(l) \longrightarrow$

(e) $Na_2O_2(s) + CO_2(g) \longrightarrow$

89. Assuming that water, common reagents (acids, bases, salts), and simple laboratory equipment are available, give a practical method to prepare **(a)** $MgCl_2$ from $MgCO_3(s)$; **(b)** $NaAl(OH)_4$ from $Na(s)$ and $Al(s)$; and **(c)** Na_2SO_4 from $NaCl(s)$.

90. Write the simplest chemical equation to represent the reaction of **(a)** $K_2CO_3(aq)$ and $Ba(OH)_2(aq)$; **(b)** $Mg(HCO_3)_2(aq)$ on heating; **(c)** tin(II) oxide when heated with carbon; **(d)** $CaF_2(s)$ and H_2SO_4(concd aq); **(e)** $NaHCO_3(s)$ and $HCl(aq)$; **(f)** $PbO_2(s)$ and $HBr(aq)$; and **(g)** the reduction of SiF_4 to pure Si, by using Na as the reducing agent.

91. Write an equation to represent the reaction of gypsum, $CaSO_4 \cdot 2\,H_2O$, with ammonium carbonate to produce ammonium sulfate (a fertilizer), calcium carbonate, and water.

92. Write chemical equations to represent the most probable outcome in each of the following. If no reaction is likely to occur, so state.

(a) $B(OH)_3 \xrightarrow{\Delta}$

(b) $Al_2O_3(s) \xrightarrow{\Delta}$

(c) $CaSO_4 \cdot 2\,H_2O(s) \xrightarrow{\Delta}$

93. A chemical dictionary gives the following descriptions of the production of some compounds. Write plausible chemical equations based on these descriptions.

(a) lead(II) carbonate: adding a solution of sodium bicarbonate to a solution of lead nitrate

(b) lithium carbonate: reaction of lithium oxide with ammonium carbonate solution

(c) hydrogen peroxide: by the action of dilute sulfuric acid on barium peroxide

(d) lead(IV) oxide: action of an alkaline solution of calcium hypochlorite on lead(II) oxide

94. Name the chemical compound(s) you would expect to be the *primary* constituent(s) of **(a)** stalactites; **(b)** gypsum; **(c)** "barium milkshake"; **(d)** blue sapphires.

95. How many cubic meters of $CO_2(g)$ at 102 kPa and 288 K are produced in the calcination of 5.00×10^3 kg of the mineral dolomite, $CaMg(CO_3)_2$?

22

Chemistry of the Main-Group Elements II: Groups 18, 17, 16, 15, and Hydrogen

CONTENTS

Bromine, a nonmetal, is a reactant in the synthesis of flame retardants for use in plastics.

In this chapter we continue our discussion of the *p*-block elements. We will progress from right to left through the periodic table and begin with a survey of the noble gases (group 18), the atoms of which have filled valence shells. The filled valence shells of the group 18 atoms makes these gases special—or noble—in the sense that they are particularly unreactive, though not totally so, as we will soon see. We next examine the halogens (group 17) and then proceed to groups 16 and 15. In each successive group, proceeding from right to left, the number of electrons in the *p*-subshell of an atom decreases by one and as a consequence, the tendency to form more than one covalent bond increases. Finally, in this chapter we will consider hydrogen. Hydrogen has unique chemistry and is not easily placed in any group of the periodic table. To emphasize the uniqueness of hydrogen, some chemists prefer to use a version of the periodic table, such as that shown at the top of page 977, in which hydrogen is separated from the rest of the table.

1	2												H					13	14	15	16	17	**18** He

An alternative periodic table layout:

Row with H separated, He in group 18.

- Li Be | | | | | | | | | | | | | | | | | | B C N O F Ne
- Na Mg | 3 4 5 6 7 8 9 10 11 12 | Al Si P S Cl Ar
- K Ca Sc Ti V Cr Mn Fe Co Ni Cu Zn Ga Ge As Se Br Kr
- Rb Sr Y Zr Nb Mo Tc Ru Rh Pd Ag Cd In Sn Sb Te I Xe
- Cs Ba La–Lu Hf Ta W Re Os Ir Pt Au Hg Tl Pb Bi Po At Rn
- Fr Ra Ac–Lr Rf Db Sg Bh Hs Mt Ds Rg

▲ **An alternative version of the periodic table**
This version differs slightly from the one given on the inside front cover in that
hydrogen is separated from the rest of the elements. Also, it does not explicitly show the
lanthanides (La–Lr) or the actinides (Ac–Lr). Some chemists prefer to put H on its own
because of the uniqueness of hydrogen and the difficulty in placing it in a specific group.

Among the fundamental concepts emphasized in this chapter are atomic,
physical, and thermodynamic properties; bonding and structure; acid–base
chemistry; and oxidation states, redox reactions, and electrochemistry.

22-1   Periodic Trends in Bonding

To uncover trends in bonding it is instructive to consider a series of binary
compounds with a common element, such as fluorine or oxygen. First, we will
consider the fluorides of the second- and third-row elements, with the general
formula AF_n.

As we progress from left to right through the periodic table, the metallic
character of the elements decreases (see Figure 9-12) and the bonding in the
fluorides, AF_n, changes from ionic bonding to covalent bonding. In Table 22.1
we show the formulas, bonding types, and phases at room temperature of the
binary fluorides of the second- and third-period elements. In the fluorides, we
observe a transition from ionic bonding to covalent bonding as we move left to
right across a period. Notice that BeF_2 and AlF_3 are network covalent solids. In
a network covalent solid, each atom is bonded to one or more atoms by covalent bonds to form a giant molecule. Most of the fluorides are molecular compounds, with covalent bonds holding together the atoms of the molecule and
with intermolecular forces, such as dipole-dipole or London dispersion forces,
contributing to the attraction among individual molecules. In Table 22.1, such
compounds are called *molecular* covalent compounds to distinguish them from
the *network* covalent compounds.

The diagonal relationship between beryllium and aluminum, discussed
previously in Chapter 21, is evident in Table 22.1. Both beryllium (in group 2)
and aluminum (in group 13) form network covalent solids with fluorine.

TABLE 22.1   Binary Fluorides, AF_n, of the Second- and Third-Row Elements[a]

Group	1	2	13	14	15	16	17
Formula	$LiF(s)$	$BeF_2(s)$	$BF_3(g)$	$CF_4(g)$	$NF_3(g)$	$OF_2(g)$	$F_2(g)$
Bonding	Ionic	Network Covalent	Molecular Covalent	Molecular Covalent	Molecular Covalent	Molecular Covalent	Molecular Covalent
Formula	$NaF(s)$	$MgF_2(s)$	$AlF_3(s)$	$SiF_4(g)$	$PF_5(g)$	$SF_6(g)$	$ClF_5(g)$
Bonding	Ionic	Ionic	Network Covalent	Molecular Covalent	Molecular Covalent	Molecular Covalent	Molecular Covalent

[a]For each element, the substance shown is the one that has the greatest number of fluorine atoms per atom of A. The phase
indicated is the phase at 25 °C.

The phase of the compound at room temperature is a reflection of the type of bonding in the compound. Ionic solids melt at very high temperatures because the melting process requires the breaking of strong ionic bonds in the crystal lattice. Similarly, network covalent solids have very high melting temperatures because covalent bonds have to be broken in the melting process. In the molecular covalent substances, only weak intermolecular forces contribute to the attraction between the molecules; thus, these species are gases at room temperature.

For the fluorides of the second-row elements, the formulas are relatively easy to predict and explain. The number of fluorine atoms per formula unit is equal to the number of valence electrons an atom must gain or lose to attain a noble gas configuration. For the fluorides of groups 14 to 16, the number of fluorine atoms per formula unit is equal to the number of valence electrons required to fill the valence shell of carbon, nitrogen, or oxygen. For example, a carbon atom with configuration $[He]2s^22p^2$ requires four electrons to complete its valence shell, and because each fluorine atom has only one unpaired valence electron, carbon and fluorine combine in the ratio $1:4$ to give CF_4. For the fluorides of groups 1, 2, and 13, however, the number of fluorine atoms per formula unit is equal to the number of valence electrons a lithium, beryllium, or boron atom would have to lose to attain a $2s^2$ configuration. For example, lithium must lose only one electron to attain a $2s^2$ configuration, and because a fluorine atom needs only one electron to complete its valence shell, lithium and fluorine combine $1:1$ to give LiF. The formulas of BeF_2 and BF_3 are related to the number of valence electrons a Be or B atom must lose to attain a noble gas (He) configuration, even though neither Be nor B actually loses any of its valence electrons—both Be and B form covalent bonds with fluorine.

For the fluorides of the third-row elements, the number of fluorine atoms per formula unit is more difficult to predict. For the fluorides shown in Table 22.1, the oxidation state of the element bonded to fluorine increases in steps of 1 as we move left to right across the third period, except in going from sulfur to chlorine. Chlorine has an oxidation state of $+7$ in some of its compounds, as is the case in $HClO_4$ and Cl_2O_7, but it does not form the heptafluoride (ClF_7) with Cl in the $+7$ oxidation state. Presumably, this occurs because the chlorine atom is not large enough to interact simultaneously with seven fluorine atoms. This argument is supported by the observation that iodine, also in group 17 but a much larger atom, forms the heptafluoride IF_7.

Similar trends in bonding are observed for the oxides of the second- and third-row elements, as can be seen in Table 22.2. Also shown in Table 22.2 are the acid–base properties of the oxides. Half of these oxides are acidic, only three are basic (Li_2O, Na_2O, and MgO), and two are amphoteric (BeO and Al_2O_3).

TABLE 22.2 Binary Oxides of the Second- and Third-Row Elements[a]

Group	1	2	13	14	15	16	17
Formula	$Li_2O(s)$	BeO(s)	$B_2O_3(s)$	$CO_2(g)$	$N_2O_5(g)$	$O_2(g)$	$OF_2(g)$[b]
Bonding	Ionic	Ionic	Network Covalent	Molecular	Molecular	Molecular	Molecular
Acid–Base Properties	Basic	Amphoteric	Acidic	Covalent Acidic	Covalent Acidic	Covalent Neutral	Covalent Neutral
Formula	$Na_2O(s)$	MgO(s)	$Al_2O_3(s)$	$SiO_2(s)$	$P_4O_{10}(s)$	$SO_3(l)$[c]	$Cl_2O_7(l)$
Bonding	Ionic	Ionic	Ionic	Network	Molecular	Molecular	Molecular
Acid–Base Properties	Basic	Basic	Amphoteric	Covalent Acidic	Covalent Acidic	Covalent Acidic	Covalent Acidic

[a]Some elements form more than one compound with oxygen. The substance shown is the one having the element in its most highly oxidized form. The phase indicated is the phase at 25 °C.
[b]OF_2 is included in the table even though OF_2 is not an oxide. It is a fluoride with oxygen having an oxidation state of $+2$.
[c]In the liquid phase, SO_3 consists of a trimer of SO_3 molecules, as shown in Figure 22-14.

The diagonal relationships between Be and Al and between B and Si are evident. Both Be and Al form amphoteric oxides and both B and Si form network covalent oxides that are acidic. A transition occurs from basic oxides to acidic oxides as we proceed from metallic to nonmetallic elements, or, stated another way, as we proceed from the least electronegative elements to the most electronegative elements. For example, sodium oxide reacts with water to give sodium hydroxide, while sulfur trioxide reacts with water to give sulfuric acid, as shown below:

$$Na_2O(s) + H_2O(l) \longrightarrow 2\,NaOH(aq)$$
$$SO_3(l) + H_2O(l) \longrightarrow H_2SO_4(l)$$

With the exception of OF_2, the oxides of the elements in groups 14 to 17 react with water to give oxoacids, as shown above for SO_3. OF_2 reacts with water, but the reaction gives a binary acid (HF) and not the oxoacid (HOF). That the acid–base behavior of OF_2 is different from the oxides of other p-block elements comes as no surprise; OF_2 is not an oxide. It is a fluoride with oxygen having an oxidation state of $+2$.

The oxides of Be and Al, as we saw in Chapter 21, are amphoteric because these oxides react with both acids and bases. Beryllium oxide, BeO, reacts as follows:

$$BeO(s) + 2\,H_3O^+(aq) + H_2O(l) \longrightarrow [Be(OH_2)_4]^{2+}(aq)$$
$$BeO(s) + 2\,OH^-(aq) + H_2O(l) \longrightarrow [Be(OH)_4]^{2-}(aq)$$

We should not forget the point made in Chapter 21: *as we move down a group, the metallic character of an element increases.* This has an important consequence for the oxides of the group 14 elements. The oxides of C and Si are acidic, the oxide of Sn is amphoteric, and the oxide of Pb is basic. The amphoterism of SnO is evident in the reactions of SnO with HCl(aq) and NaOH(aq), which are given below:

$$SnO(s) + 2\,HCl(aq) \longrightarrow SnCl_2(aq) + H_2O(l)$$
$$SnO(s) + NaOH(aq) + H_2O(l) \longrightarrow Na^+(aq) + [Sn(OH)_3]^-(aq)$$

As we discuss the chemistry of the elements of groups 15 through 18, we will see these trends repeated many times. Explanations of these trends are based on differences in electronegativities and differences in the sizes of the atoms involved.

22-2   Group 18: The Noble Gases

In 1785, Henry Cavendish, the discoverer of hydrogen, passed electric discharges through air to form oxides of nitrogen. (A similar process occurs during lightning storms.) He then dissolved these oxides in water to form nitric acid. Even by using excess oxygen, Cavendish was unable to get all the air to react. He suggested that air contained an unreactive gas making up "not more than 1/120 of the whole." John Rayleigh and William Ramsay isolated this gas one century later (1894) and named it argon. The name is derived from the Greek *argos*, "lazy one," meaning "inert." Its *inability* to form chemical compounds with any of the other elements—its chemical inertness—was found to be argon's most notable feature. Because argon resembled no other known element, Ramsay placed it in a separate group of the periodic table and reasoned that there should be other members of this group.

Ramsay then began a systematic search for other inert gases. In 1895, he extracted helium from a uranium mineral. A few years later, by very carefully distilling liquid argon, he was able to extract three additional inert gases: neon, krypton, and xenon. The final member of the group of inert gases, a radioactive element called radon, was discovered in 1900. In 1962, compounds of Xe were first prepared, and so the inert gases proved to be not completely

▲ **William Ramsay (1852–1916)**
This distinguished Scottish chemist received the 1904 Nobel Prize in Chemistry for his work on the noble gases.

▲ Gaseous atoms in the region of an electric discharge emit light. The light emitted by neon atoms is red-orange in color. Neon lights of other colors use other noble gases or mixtures of gases.

▶ As we learned in Chapter 13 (page 572), using helium–oxygen mixtures instead of air for deep-sea diving prevents the condition known as the *bends* because helium is more rapidly and smoothly expelled from the blood than is nitrogen.

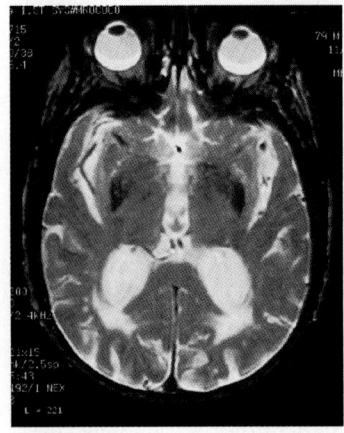

▲ An MRI image of a head.

inert after all. Since that time, this group of gases has been called the **noble gases**. They are found in group 18 of the periodic table shown on the inside front cover.

Occurrence

Air contains 0.000524% He, 0.001818% Ne, and 0.934% Ar, by volume. The proportion of Kr is about 1 ppm by volume, and that of Xe, 0.05 ppm. The atmosphere is the only source of all these gases except helium. The main source of helium is certain natural gas wells in the western United States that produce natural gas containing up to 8% He by volume. It is cost-effective to extract helium from natural gas even down to levels of about 0.3%. Underground helium accumulates as a result of α-particle emission by radioactive elements in Earth's crust. Whereas the abundance of He on Earth is very limited, it is second only to hydrogen in the universe as a whole.

Most of the noble gases have escaped from the atmosphere since Earth was formed, but Ar is an exception. The concentration of Ar remains quite high because it is constantly being formed by the radioactive decay of potassium-40, a reasonably abundant, naturally occurring radioactive isotope. Helium is also constantly produced through α-particle emissions by radioactive decay processes, but because the molar mass of He is 10 times less than that of Ar, it escapes from the atmosphere into outer space at a higher rate.

Properties and Uses

The lighter noble gases are commercially important, in part because they are chemically inert. The efficiency and life of electric lightbulbs are increased when they are filled with an argon–nitrogen mixture. Electric discharge through neon-filled glass or plastic tubes produces a distinctive red light (neon light). Krypton and xenon are used in lasers and in flashlamps in photography. Helium has several unique physical properties. Best known of these is that it exists as a liquid at temperatures approaching 0 K. All other substances freeze to solids at temperatures well above 0 K. (The melting point of solid H_2, for example, is 14 K.) Because of their inertness, both He and Ar are used to protect materials from nitrogen and oxygen in the air; consequently, He and Ar are used for this purpose in certain types of welding, in metallurgical processes, and in the preparation of ultrapure Si and Ge and other semiconductor materials. Helium mixed with oxygen is used as a breathing mixture for deep-sea diving and in certain medical applications. Large quantities of liquid helium are used to maintain low temperatures (cryogenics). Some metals essentially lose their electrical resistivity at liquid He temperatures and become *superconductors*. Powerful magnets can be made by immersing the coils of electromagnets in liquid helium. Such magnets are used in particle accelerators and in nuclear fusion research. More familiar uses of large electromagnets cooled by liquid helium are nuclear magnetic resonance (NMR) instruments in research laboratories and magnetic resonance imaging (MRI) devices in hospitals. Helium is also used to fill lighter-than-air airships (blimps). Some of these applications are highlighted in the margin.

Xenon Compounds

In this section, we focus exclusively on compounds of xenon because most of the known noble gas compounds contain xenon. A few compounds of krypton, such as KrF_2, have been synthesized and well characterized. Radon is expected to form compounds even more readily than xenon, because of its lower ionization energy, but the chemistry of radon is complicated by its radioactivity.

For a long time, the noble gases were thought to be chemically inert. This apparent inertness helped provide a theoretical framework for the Lewis theory of bonding. Subsequently it was found that compounds of xenon can be made fairly easily, and these compounds have added significantly to our knowledge of chemical bonding.

In the 1930s, Linus Pauling did theoretical calculations that suggested xenon should form oxide and fluoride compounds, but attempts to make them failed at that time. In 1962, Neil Bartlett and Derek H. Lohmann discovered that O_2 and PtF_6 would combine in a 1:1 mole ratio to form the compound O_2PtF_6. Properties of this compound suggest it to be ionic: $[O_2]^+[PtF_6]^-$. The energy required to extract an electron from O_2 is 1177 kJ mol^{-1}, almost identical to the first ionization energy of Xe, 1170 kJ mol^{-1}. The size of the Xe atom is also roughly the same as that of the dioxygen molecule. Thus, it was reasoned that the compound $XePtF_6$ might also exist. Bartlett and Lohmann were able to prepare a yellow crystalline solid with that composition.*

Soon thereafter, chemists around the world synthesized several additional noble gas compounds. In general, the conditions necessary to form noble gas compounds are as Pauling predicted. The formation of a noble gas compound requires:

- a readily ionizable (therefore, high atomic number) noble gas atom
- highly electronegative atoms (such as F or O) to bond to it

Xenon compounds have been synthesized that have Xe in four possible oxidation states, as summarized below:

	+2	+4	+6	+8
Examples:	XeF_2	$XeF_4, XeOF_2$	XeF_6, XeO_3	XeO_4, H_4XeO_6

Because it is difficult to oxidize Xe to these positive oxidation states, we would expect Xe compounds to be easily reduced, making them very strong oxidizing agents. For example, in aqueous acidic solution, XeF_2 is an excellent oxidizing agent, as indicated by the large $E°$ value for the following half-reaction:

$$XeF_2(aq) + 2\,H^+(aq) + 2\,e^- \longrightarrow Xe(g) + 2\,HF(aq) \qquad E° = +2.64\ V$$

The significance of this large $E°$ value is that XeF_2 is not very stable in aqueous solution. In aqueous solution, XeF_2 oxidizes the water, producing $O_2(g)$, as shown below:

Reduction: $2\{XeF_2(aq) + 2\,H^+(aq) + 2\,e^- \longrightarrow Xe(g) + 2\,HF(aq)\}$

Oxidation: $2\,H_2O(l) \longrightarrow 4\,H^+(aq) + O_2(g) + 4\,e^-$

Overall: $2\,XeF_2(aq) + 2\,H_2O(l) \longrightarrow 2\,Xe(g) + 4\,HF(aq) + O_2(g)$

$$E°_{cell} = E°(\text{reduction}) - E°(\text{oxidation})$$
$$= E°_{XeF_2/Xe} - E°_{O_2/H_2O}$$
$$= 2.64\ V - (1.229\ V) = 1.41\ V$$

Xenon reacts directly only with F_2. Three different xenon fluorides can be produced by heating xenon and fluorine in nickel reaction vessels, but the product obtained depends on the experimental conditions, as summarized below:

$$Xe(g) + F_2(g) \xrightarrow[400\,°C,\,1\,atm]{Xe:F_2\,=\,2:1} XeF_2(s)$$

$$Xe(g) + 2\,F_2(g) \xrightarrow[400\,°C,\,6\,atm]{Xe:F_2\,=\,1:5} XeF_4(s)$$

$$Xe(g) + 3\,F_2(g) \xrightarrow[300\,°C,\,50\,atm]{Xe:F_2\,=\,1:20} XeF_6(s)$$

All three of these xenon fluorides are colorless, volatile solids. Notice that, for each synthesis, the Xe:F_2 mole ratio is significantly different from the

◀ In 2000, chemists at the University of Helsinki reported* they had made HArF molecules that were stable at very low temperatures (below 27 K). The report has generated much discussion—and debate— among chemists about the existence of argon compounds and the possibilities for synthesizing argon compounds that are stable at higher temperatures.

*L. Khriachtchev, M. Pettersson, N. Runenburg, J. Lundell, and M. Rasanen, *Nature*, **406**, 874 (2000).

▲ Because of the explosive nature of hydrogen, helium is now used in airships.

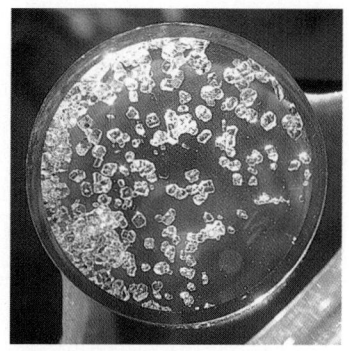

▲ Crystals of xenon tetrafluoride under magnification.

*It has since been established that this solid is more complicated than first thought. It has the formula $Xe(PtF_6)_n$, where n is between 1 and 2.

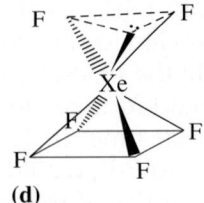

(a)

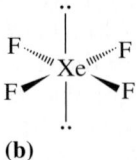

(b)

▶ FIGURE 22-1
Electron group geometries for XeF$_2$, XeF$_4$, and XeF$_6$
(a) XeF$_2$; (b) XeF$_4$; (c) XeF$_6$ pentagonal bipyramid; (d) XeF$_6$ capped trigonal prism; (e) XeF$_6$ capped octahedron.

(c)

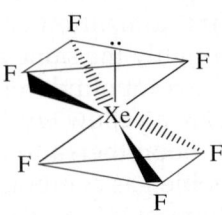

(d)

(e)

stoichiometry of the reaction. For example, the synthesis of XeF$_2$ requires an excess of Xe, and the synthesis of XeF$_6$ requires large excess of F$_2$. A major difficulty in synthesizing the xenon fluorides is that all three products tend to form simultaneously. The most difficult to prepare in pure form, starting from Xe and F$_2$, is XeF$_4$. Xenon tetrafluoride can be prepared in high purity by the reaction below, which uses O$_2$F$_2$(g) instead of F$_2$(g):

$$Xe(g) + 2\,O_2F_2(g) \longrightarrow XeF_4(g) + 2\,O_2(g)$$

The shapes of the XeF$_2$, XeF$_4$, and XeF$_6$ molecules are shown in Figure 22-1. The XeF$_2$ and XeF$_4$ molecules have shapes that are easily understood in terms of VSEPR theory, but the shape of the XeF$_6$ molecule is a little more difficult to interpret. In the XeF$_6$ molecule, the xenon atom is surrounded by seven electron pairs, and three possible arrangements of these pairs are shown in Figure 22-1. These are pentagonal bipyramid, capped trigonal prism, and capped octahedron. These structures are expected to have nearly the same energy and the preferred arrangement depends strongly on the exact conditions. In the gas phase, the XeF$_6$ molecule has a capped octahedral structure (Figure 22-1e). In this structure, the six fluorine atoms form a distorted octahedron, and the lone pair on xenon is directed toward the center of one of the triangular faces. In the solid phase, XeF$_6$ exists as XeF$_5$$^+$, which is square pyramidal, and F$^-$, with the F$^-$ ions forming bridges between XeF$_5$$^+$ ions.

To conform to the shapes indicated by VSEPR theory (or experiment, when the structure is known), bonding theory based on hybridization of orbitals requires the hybrid orbitals sp^3d for XeF$_2$, sp^3d^2 for XeF$_4$, and sp^3d^3 for XeF$_6$. However, in light of the observed bond lengths and bond energies in these fluorides and the high energy (estimated to be 1000 kJ mol^{-1}) required to promote an electron from a $5p$ to a $5d$ orbital, there is doubt as to whether d orbitals are involved in the bond formation. A molecular orbital description can be constructed that does not involve the participation of the xenon $5d$ orbitals, but in its simplest form this description cannot explain the nonoctahedral shape of XF$_6$. The situation described here reinforces the caveat that approximate theories of chemical bonding must be viewed critically.

Some properties of XeF$_2$, XeF$_4$, and XeF$_6$ are listed in Table 22.3. Because of their molecular structures, the XeF$_2$, XeF$_4$, and XeF$_6$ molecules are all nonpolar, and thus we expect that in each of these fluorides, the molecules interact with each other primarily through London dispersion forces. However, the melting points of these fluorides decrease as the molecular size increases,

▶ Actually, the XeF$_6$ molecule does possess a small dipole moment (about 0.03 debye). Compare this value with that for H$_2$O (1.84 debye) and HCl (1.03 debye).

TABLE 22.3   Some Properties[a] of XeF_2, XeF_4, and XeF_6

	XeF_2	XeF_4	XeF_6
Melting point, °C	129	117	49
$\Delta H^\circ_{f,solid}$, kJ mol^{-1}	−163	−267	−338
$\Delta H^\circ_{f,gas}$, kJ mol^{-1}	−107	−206	−279
$\Delta G^\circ_{f,gas}$, kJ mol^{-1}	−96	−138	—
Xe—F bond energy, kJ mol^{-1}	133	131	126
Xe—F bond length, pm	200	195	189

[a]The ΔH°_f and ΔG°_f values are for 25 °C. The bond energies and bond lengths are average values.

contrary to what we would expect if only London dispersion forces contributed to the attraction among molecules. The trend in melting points indicates that, in the xenon fluorides, there are additional interactions to consider. It has been suggested that, while the XeF_2 and XeF_4 molecules are nonpolar, the polar Xe—F bonds interact with each other in the manner suggested in Figure 22-2. The electrostatic potential maps shown in Figure 22-2 indicate that the Xe—F bonds in XeF_2 are more polar than they are in XeF_4; thus, the interactions of the bond dipoles are more important in $XeF_2(s)$ than in $XeF_4(s)$. The interactions of bond dipoles help to explain not only why the xenon fluorides have high melting points but also why the melting point of $XeF_2(s)$ is higher than that of $XeF_4(s)$.

Xenon forms other compounds, in which xenon is bonded to chlorine, oxygen, or nitrogen. Some of the simpler compounds include $XeCl_2$, XeO_3, XeO_4, $XeOF_2$, XeO_2F_2, and $XeOF_4$. However, many of these compounds must be prepared by using an indirect route. For example, XeO_3 can be made from XeF_6 by using the following reaction:

$$2\,XeF_6(s) + 3\,SiO_2(s) \longrightarrow 2\,XeO_3(s) + 3\,SiF_4(g)$$

Why is it that Xe(g) reacts directly with only $F_2(g)$? We can formulate an answer to this question by focusing on the following synthesis:

$$Xe(g) + F_2(g) \longrightarrow XeF_2(s)$$

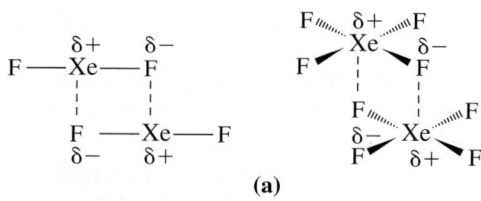

(a)

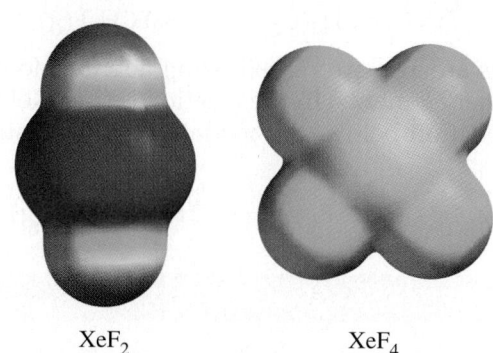

XeF_2 XeF_4

(b)

◀ FIGURE 22-2
Interactions and electrostatic potential maps of XeF_2 and XeF_4
(a) Possible interactions between bond dipoles in XeF_2 and XeF_4 lead to relatively high melting points. (b) The electrostatic potential maps of XeF_2 and XeF_4 show that the Xe—F bond dipole is smaller in XeF_4 than it is in XeF_2.

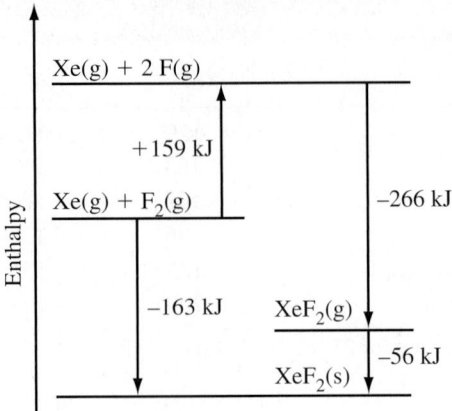

► FIGURE 22-3
Enthalpy diagram for the formation of XeF$_2$(s)

► The low bond energy of the F—F bond is probably the result of repulsions involving the lone-pair electrons on the two F atoms. The repulsions are rather strong because the atomic radius of fluorine is small, and thus the six lone pairs are in close proximity. The weakness of the F—F bond contributes to the reactivity of F$_2$.

The enthalpy diagram shown in Figure 22-3 shows that the reaction of Xe(g) and F$_2$(g) is exothermic (and energetically favorable) because the reaction involves the breaking of relatively weak F—F bonds (159 kJ mol^{-1}) and the formation of twice as many Xe—F bonds, each of which is almost as strong as an F—F bond. (For the synthesis of all the xenon fluorides, starting from Xe(g) and F$_2$(g), the number of Xe—F bonds formed is always two times the number of F—F bonds broken.) None of the other noble gases react directly with fluorine because, for the other noble gases, the bond energy of the bond formed with F is not large enough to offset the energy requirements of breaking the F—F bond. For example, the bond energy of the Kr—F bond, in KrF$_2$, is only about 50 kJ mol^{-1}.

A similar analysis can also be used to help us to understand why xenon does not react directly with Cl$_2$ or O$_2$ to form chlorides or oxides. Such an analysis (see exercise 109) reveals that the Xe—Cl or Xe—O bond energy is much too small to offset the energy requirements for breaking the Cl—Cl or O=O bonds. (The bond energies of Cl$_2$ and O$_2$ are 243 kJ mol^{-1} and 498 kJ mol^{-1}, respectively.)

All the xenon fluorides react with water to form various products. For example, in aqueous solution, xenon hexafluoride is first hydrolyzed to xenon oxide tetrafluoride, XeOF$_4$, which is further hydrolyzed to xenon trioxide. The reactions are as follows:

$$XeF_6(s) + H_2O(l) \longrightarrow XeOF_4(aq) + 2\,HF(aq)$$
$$XeOF_4(l) + 2\,H_2O(l) \longrightarrow XeO_3(aq) + 4\,HF(aq)$$

The xenon fluorides are good fluorinating agents. Xenon difluoride is sometimes used in organic chemistry to add fluorine atoms to carbon compounds. An advantage of using XeF$_2$ for this purpose is that the by-product, Xe(g), is easily separated from the desired product. For example, XeF$_2$ can be used to add F atoms to either side of a carbon–carbon double bond, as shown below:

$$XeF_2(s) + CH_2{=}CH_2(g) \longrightarrow CH_2FCH_2F(g) + Xe(g)$$

The xenon fluorides are also strong oxidizing agents and can be used to oxidize fluorides of other elements. In the reaction below, XeF$_4$ is used to oxidize SF$_4$ to SF$_6$. In the reaction below, the oxidation state of sulfur changes from +4 to +6.

$$XeF_4(s) + 2\,SF_4(g) \longrightarrow 2\,SF_6(g) + Xe(g)$$

🔍 **22-1   CONCEPT ASSESSMENT**

Which is likely to be the more stable, XeF$_2$ or XeCl$_2$? Explain.

22-3 Group 17: The Halogens

The word *halogen* was introduced in 1811 to describe the ability of chlorine to form ionic compounds with metals. The name is based on the Greek words *halos* and *gen* meaning "salt former." The name was later extended to include fluorine, bromine, and iodine as well. Mendeleev placed the halogens in Group VII of his periodic table (page 362), which is now group 17 in the IUPAC table (page 51). Current interest in the halogens extends far beyond their ability to form metallic salts.

Properties

The **halogens** exist as diatomic molecules, symbolized by X_2, where X is a generic symbol for a halogen atom. That these elements occur as nonpolar diatomic molecules accounts for their relatively low melting and boiling points (Table 22.4). As expected, melting and boiling points increase as we move down the group from the smallest and lightest member of the group, fluorine, to the largest and heaviest, iodine. Conversely, chemical reactivity toward other elements and compounds increases in the *opposite* order, with fluorine being the most reactive and iodine the least reactive.

All the halogen atoms have large electron affinities (see Table 22.4 and Figure 9-11), and show a strong tendency to gain electrons. Consequently, the halogens are rather good oxidizing agents.

We have mentioned previously, and we will see again, that the elements of period 2 have distinctly different chemistry from the rest of the group because of their small sizes and inability to expand their valence shells. However, for the halogens, the differences between the second-row element (fluorine) and the members of the group are much less dramatic. Still, fluorine differs from the other halogens in a few ways. For example, a fluorine atom almost always forms just one covalent bond, whereas chlorine, bromine, and iodine atoms typically form more than one bond and as many as seven in some of their compounds. Although all the halogens are quite reactive and are found in nature only as compounds, fluorine is considerably more reactive than the other members of the group. It reacts directly with almost all the elements, except for oxygen, nitrogen, and the lighter noble gases, and forms compounds with even the most unreactive metals. It reacts with almost all materials, especially organic compounds, to produce fluorides. The reactivity of fluorine can be attributed to the weakness of the fluorine–fluorine bond in F_2, which arises, as mentioned earlier, because of the small size of the fluorine atom and the repulsion between the lone pairs on the fluorine atoms.

◀ In the perhalic acids, HXO_4, where X = Cl, Br, or I, the halogen atom forms seven bonds, four of which are sigma bonds and three of which are pi bonds. The Lewis structure for perchloric acid is shown below.

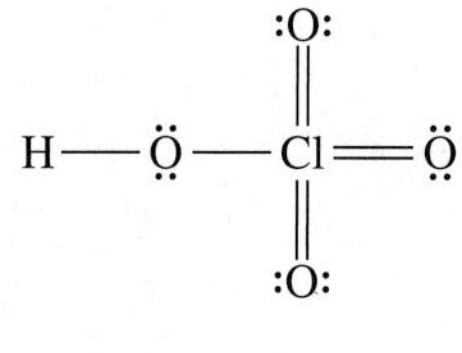

$$H - \ddot{\underset{}{O}} - Cl = \ddot{O}$$

with $:\!\ddot{O}\!:$ above and $:\!\ddot{O}\!:$ below the Cl

TABLE 22.4 Group 17 Elements: The Halogens

	Fluorine (F)	Chlorine (Cl)	Bromine (Br)	Iodine (I)
Physical form at room temperature	Pale yellow gas	Yellow-green gas	Dark red liquid	Violet-black solid
Melting point, °C	−220	−101	−7.2	114
Boiling point, °C	−188	−35	58.8	184
Electron configuration	$[He]2s^22p^5$	$[Ne]3s^23p^5$	$[Ar]3d^{10}4s^24p^5$	$[Kr]4d^{10}5s^25p^5$
Covalent radius, pm	71	99	114	133
Ionic (X^-) radius, pm	133	181	196	220
First ionization energy, kJ mol^{-1}	1681	1251	1140	1008
Electron affinity, kJ mol^{-1}	−328.0	−349.0	−324.6	−295.2
Electronegativity	4.0	3.0	2.8	2.5
Standard electrode potential, V ($X_2 + 2e^- \longrightarrow 2X^-$)	2.866	1.358	1.065	0.535

Fluorine differs from the other halogens in that it shows a much greater tendency to form ionic bonds with metals. Perhaps this is most evident when we look at the binary compounds formed by the halogens and the group 13 metals. The trifluorides of Al, Ga, and In are all ionic compounds, with very high lattice energies and very high melting points ($>1000\ °C$) whereas the trichlorides are volatile compounds with much lower melting points ($<600\ °C$). For $AlCl_3$, $GaCl_3$, and $InCl_3$, the bonding is largely covalent because chloride ions are much larger, and much more polarizable, than fluoride ions. Also, in the solid state, the chlorides of the group 13 metals contain dimers, M_2Cl_6, whereas the fluorides of the group 13 metals are all ionic lattices containing M^{3+} and F^- ions.

Another important difference between fluorine and the other halogens is that fluorine shows the ability to stabilize other elements in very high oxidation states. For example, fluorine reacts with sulfur to give SF_6, with sulfur in the $+6$ oxidation state, whereas chlorine reacts directly with molten sulfur to give S_2Cl_2, with sulfur in the $+1$ oxidation state.

Much of the reaction chemistry of the halogens involves oxidation–reduction reactions in aqueous solutions. For these reactions, standard electrode potentials are helpful for understanding the reactivity of the halogens. Among the properties of the halogens listed in Table 22.4 are potentials for the following half-reaction:

$$X_2 + 2\,e^- \longrightarrow 2\,X^-(aq)$$

By this measure, fluorine is clearly the most reactive element of the group ($E° = 2.866\ V$). Of all the elements, it shows the greatest tendency to gain electrons and is therefore the most easily reduced. Given this fact, it is not surprising that fluorine occurs naturally only in combination with other elements, and only as the fluoride ion, F^-. Although both chlorine and bromine can exist in a variety of positive oxidation states, they are found in their naturally occurring compounds only as chloride and bromide ions. There are, however, naturally occurring compounds in which iodine is in a positive oxidation state (such as the iodate ion, IO_3^-, in $NaIO_3$). In the case of iodine, the tendency for I_2 to be reduced to I^- is not particularly great ($E° = 0.535\ V$).

KEEP IN MIND

that the halogens are extremely toxic.

Electrode Potential Diagrams

When we summarize the reduction tendencies of main-group metals and their ions, generally one or, at most, a few $E°$ values tell the story, and these values are easily incorporated into tables, such as that in Appendix D. However, the oxidation–reduction chemistry of some of the nonmetals is much richer and involves a larger number of relevant $E°$ values. In these cases, *electrode potential diagrams* are particularly useful for summarizing $E°$ data. Partial diagrams for chlorine are shown in Figure 22-4. In these diagrams, a number written above a line segment is the $E°$ value for reduction of the species on the left (higher oxidation state) to the one on the right (lower oxidation state). For a reduction involving species not joined by a line segment, we generally can calculate the appropriate value of $E°$ by the method illustrated in Example 22-1.

Production and Uses

Although the existence of fluorine had been known since early in the nineteenth century, no one was able to devise a chemical reaction to extract the free element from its compounds. Finally, in 1886, H. Moissan succeeded in preparing $F_2(g)$ by an electrolysis reaction. Moissan's method, which is still the only important commercial method for fluorine extraction, involves the

Acidic solution ([H$^+$] = 1 M):

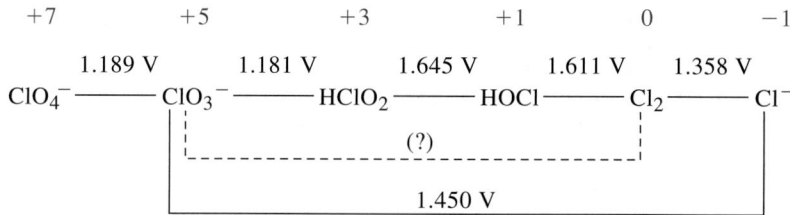

Basic solution ([OH$^-$] = 1 M):

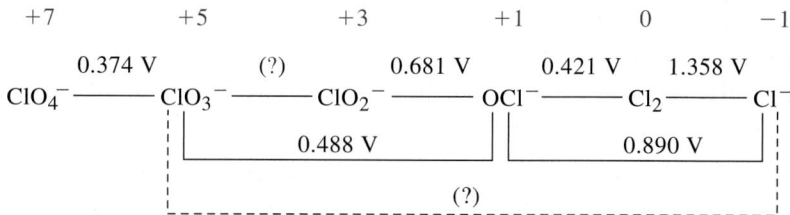

◀ FIGURE 22-4
Standard electrode potential diagrams for chlorine
The numbers in color are the oxidation states of the Cl atom. $E°$ values are written above the line segments for the reduction of the species on the left to the species on the right. All reactants and products are at unit activity. Because of the basic properties of the ClO_2^- and OCl^- ions, the weak acids $HClO_2$ and $HOCl$ form in acidic solution.

EXAMPLE 22-1   Using an Electrode Potential Diagram to Determine $E°$ for a Half-Reaction

Determine $E°$ for the reduction of ClO_3^- to ClO_2^- in a basic solution, which is marked (?) in Figure 22-4.

Analyze

One way we can approach this problem is to identify a route that (1) converts ClO_3^- to ClO_2^- and (2) involves $E°$ values that are known. One such route involves converting ClO_3^- to OCl^-, for which $E° = 0.488$ V, and then converting OCl^- to ClO_2^-, for which $E° = -0.681$ V. (The $E°$ value for converting ClO_2^- to OCl^- is given as 0.681 V in Figure 22-4. For the reverse process, the $E°$ value has the opposite sign.) First, we should write balanced equations for $ClO_3^- \longrightarrow ClO^-$ and $ClO^- \longrightarrow ClO_2^-$, and then focus on how we combine the equations and corresponding $E°$ values to obtain the equation and $E°$ value for the desired conversion, $ClO^- \longrightarrow ClO_2^-$. Because the desired process is a reduction process (and not an oxidation–reduction process), we cannot simply add or subtract the $E°$ values to obtain $E°$ for the desired process (see page 875). We must first convert the given $E°$ values to $\Delta G°$ values, then combine those $\Delta G°$ values to obtain a $\Delta G°$ value for the desired process, and finally convert this $\Delta G°$ value back to a value of $E°$.

Solve

Figure 22-4 identifies only the oxidized and reduced forms of the chlorine species. By using the method described on page 174, we obtain the following balanced equations for the conversions of ClO_3^- to ClO^- and ClO^- to ClO_2^-:

$ClO_3^-(aq) + 2 H_2O(l) + 4 e^- \longrightarrow OCl^-(aq) + 4 OH^-(aq)$ $E° = 0.488$ V $\Delta G° = -4F \times 0.488$ V

$OCl^-(aq) + 2 OH^-(aq) \longrightarrow ClO_2^-(aq) + H_2O(l) + 2 e^-$ $E° = -0.681$ V $\Delta G° = -2F \times (-0.681$ V)

The overall equation for the desired process is the sum of the two equations above, and the corresponding $\Delta G°$ value is the sum of the $\Delta G°$ values.

$ClO_3^-(aq) + H_2O(l) + 2 e^- \longrightarrow ClO_2^-(aq) + 2 OH^-(aq)$ $\Delta G° = -F (4 \times 0.488 - 2 \times 0.681)$ V

The $\Delta G°$ value for the process above is related to the $E°$ value by the expression $\Delta G° = -2FE°$, and so, the $E°$ value is calculated as follows.

$$E° = \frac{F[(2 \times 0.681) - (4 \times 0.488)] \text{ V}}{-2F} = 0.295 \text{ V}$$

Assess

When adding half-equations to give another half-equation, we must add the $\Delta G°$ values, not the $E°$ values.

PRACTICE EXAMPLE A: Determine the missing $E°$ value for the dashed line that joins ClO_3^- and Cl^- in basic solutions in Figure 22-4.

PRACTICE EXAMPLE B: Determine the missing $E°$ value for the dashed line that joins ClO_3^- and Cl_2 in acidic solutions in Figure 22-4.

▶ The hydrogen difluoride ion in KHF_2 features a strong hydrogen bond, with a H^+ midway between two F^- ions: $[F—H—F]^-$.

▲ **Salt formation in the Dead Sea**
The high concentrations of salts in the Dead Sea make it a good source of bromine and a number of other chemicals.

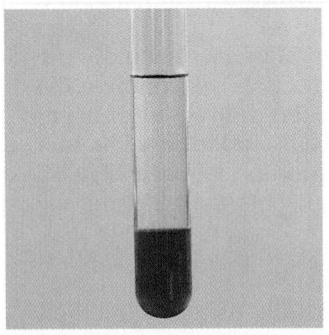

▲ **A test for Br^-(aq)**
Reaction (22.3) is used as a qualitative test in the laboratory. The liberated Br_2 is extracted from the top aqueous layer into the bottom layer, here the organic solvent chloroform, $CHCl_3$.

electrolysis of HF dissolved in molten KHF_2. The chemical equation for the reaction is given below:

$$2\,HF \xrightarrow[\text{KF} \cdot 2\,\text{HF(l)}]{\text{electrolysis}} H_2(g) + F_2(g) \tag{22.1}$$

The corresponding half-reactions are as follows:

Anode: $\quad 2\,F^- \longrightarrow F_2(g) + 2\,e^-$

Cathode: $\quad 2\,H^+ + 2\,e^- \longrightarrow H_2(g)$

Moissan also developed the electric furnace and was honored for both these achievements with the Nobel Prize in chemistry in 1906. Nevertheless, the challenge of producing fluorine by means of a chemical reaction remained. In 1986, one century after Moissan isolated fluorine, the chemical synthesis of fluorine was announced (see Exercise 14).

Although chlorine can be prepared by several chemical reactions, electrolysis of NaCl(aq) is the usual industrial method, as we have mentioned previously (see page 902). The electrolysis reaction is

$$2\,Cl^-(aq) + 2\,H_2O(l) \xrightarrow{\text{electrolysis}} 2\,OH^-(aq) + H_2(g) + Cl_2(g) \tag{22.2}$$

Bromine can be extracted from seawater, where it occurs in concentrations of about 70 ppm as Br^-, or from inland brine sources. Seawater from the Dead Sea, highlighted in the margin, is a good source of bromine. The seawater or brine solution is adjusted to pH 3.5 and treated with $Cl_2(g)$, which oxidizes Br^- to Br_2 in the following displacement reaction:

$$Cl_2(g) + 2\,Br^-(aq) \longrightarrow Br_2(l) + 2\,Cl^-(aq) \qquad E^\circ_{cell} = 0.293\,V \tag{22.3}$$

The liberated Br_2 is swept from seawater with a current of air or from brine with steam. A dilute bromine vapor forms and can be concentrated by various methods. The reaction above also forms the basis of a test for the presence of Br^-, as described in the margin.

Certain marine plants, such as seaweed, absorb and concentrate I^- selectively in the presence of Cl^- and Br^-. Iodine is obtained in small quantities from such plants. In the United States, I_2 is obtained from inland brines by a process similar to that for the production of Br_2. Another abundant natural source of iodine is $NaIO_3$, found in large deposits in Chile. Because the oxidation state of iodine must be reduced from +5 in IO_3^- to 0 in I_2, the conversion of IO_3^- to I_2 requires the use of a reducing agent. Aqueous sodium hydrogen sulfite (sodium bisulfite) is used as the reducing agent in the first part of a two-step procedure, followed by the reaction of I^- with additional IO_3^- to produce I_2. The net ionic equations for the reactions are given below:

$$IO_3^-(aq) + 3\,HSO_3^-(aq) \longrightarrow I^-(aq) + 3\,SO_4^{2-}(aq) + 3\,H^+(aq) \tag{22.4}$$

$$5\,I^-(aq) + IO_3^-(aq) + 6\,H^+(aq) \longrightarrow 3\,I_2(s) + 3\,H_2O(l) \tag{22.5}$$

The halogen elements form a variety of useful compounds, and the elements themselves are largely used to produce these compounds. All the halogens are used to make halogenated organic compounds. For example, elemental fluorine is used to produce compounds such as polytetrafluoroethylene, a plastic more commonly known as Teflon. In the past, fluorine was used to make chlorofluorocarbons (CFCs), which were used as refrigerants, but international treaties have banned the production of CFCs in most countries because they damage the stratospheric ozone layer (see Focus On 22, www.masteringchemistry.com). Now fluorine is used to make hydrochlorofluorocarbons (HCFCs), which are more environmentally benign alternatives to CFCs. Fluorinated organic compounds tend to be chemically inert, and it is this inertness that makes them useful as components in harsh chemical environments. Fluorine is a key element in a variety of useful inorganic compounds. Some of these compounds and their uses are listed in Table 22.5.

TABLE 22.5	Important Inorganic Compounds of Fluorine
Compound	Uses
Na_3AlF_6	Manufacture of aluminum
BF_3	Catalyst
CaF_2	Optical components, manufacture of HF, metallurgical flux
ClF_3	Fluorinating agent, reprocessing nuclear fuels
HF	Manufacture of F_2, AlF_3, Na_3AlF_6, and fluorocarbons
LiF	Ceramics manufacture, welding, and soldering
NaF	Fluoridating water, dental prophylasis, insecticide
SF_6	Insulating gas for high-voltage electrical equipment
SnF_2	Manufacture of toothpaste
UF_6	Manufacture of uranium fuel for nuclear reactors

With an annual production of more than 13 million metric tons, elemental chlorine ranks about eighth in quantity among manufactured chemicals in the United States. It has three main commercial uses: (1) production of chlorinated organic compounds (about 70%), chiefly ethylene dichloride, CH_2ClCH_2Cl, and vinyl chloride, $CH_2{=}CHCl$ (the monomer of polyvinyl chloride, PVC); (2) as a bleach in the paper and textile industries and for the treatment of swimming pools, municipal water, and sewage (about 20%); and (3) production of dozens of chlorine-containing inorganic chemicals (about 10%).

Bromine is used to make brominated organic compounds. Some of these are used as fire retardants and pesticides. Others are used extensively as dyes and pharmaceuticals. An important inorganic bromine compound is AgBr, the primary light-sensitive agent used in photographic film.

Iodine is of much less commercial importance than chlorine. Iodine and its compounds, however, do have applications as catalysts, antiseptics, and germicides and in the preparation of pharmaceuticals and photographic emulsions (as AgI).

◀ The major industrial source of HCl is the chlorination of organic compounds. For example, HCl is obtained as a by-product in the chlorination of methane, as shown below:

$$CH_4 + Cl_2 \longrightarrow CH_3Cl + HCl$$

Hydrogen Halides

We have encountered the hydrogen halides from time to time throughout this text. In aqueous solution, they are called the *hydrohalic acids*. Except for HF, hydrohalic acids are strong acids in water. An explanation for why HF is a weak acid was given on page 728.

One well-known property of HF is its ability to etch (and ultimately to dissolve) glass, an application that is highlighted in the margin. The reaction is similar to one between HF and silica, SiO_2.

$$SiO_2(s) + 4\,HF(aq) \longrightarrow 2\,H_2O(l) + SiF_4(g) \qquad \textbf{(22.6)}$$

Because HF reacts with glass, it must be stored in special containers coated with a lining of Teflon or polyethylene.

Hydrogen fluoride is commonly produced by a method discussed in Section 21-2. When a halide salt (such as fluorite, CaF_2) is heated with a *nonvolatile* acid, such as concentrated $H_2SO_4(aq)$, a sulfate salt and the volatile hydrogen halide are produced.

$$CaF_2(s) + H_2SO_4(\text{concd aq}) \xrightarrow{\Delta} CaSO_4(s) + 2\,HF(g) \qquad \textbf{(22.7)}$$

This method also works for preparing HCl(g) but not for HBr(g) or HI(g). Concentrated $H_2SO_4(aq)$ is a sufficiently strong oxidizing agent to oxidize Br^- to Br_2 and I^- to I_2. For example, the reaction of NaBr(s) and concentrated $H_2SO_4(aq)$ yields $Br_2(g)$ and not HBr(g).

$$2\,NaBr(s) + 2\,H_2SO_4(\text{concd aq}) \xrightarrow{\Delta} Na_2SO_4(s) + 2\,H_2O(l) + Br_2(g) + SO_2(g) \qquad \textbf{(22.8)}$$

▲ Glass etched with hydrofluoric acid.

TABLE 22.6 Free Energy of Formation of Hydrogen Halides at 298 K	
	ΔG_f°, kJ mol^{-1}
HF(g)	−273.2
HCl(g)	−95.30
HBr(g)	−53.45
HI(g)	+1.70

We can get around this difficulty by using a *nonoxidizing* nonvolatile acid, such as phosphoric acid, H_3PO_4. Also, all the hydrogen halides can be formed by the direct combination of the elements, as shown below:

$$H_2(g) + X_2(g) \longrightarrow 2\,HX(g) \qquad (22.9)$$

The reaction of $H_2(g)$ and $F_2(g)$ is very fast, however, occurring with explosive violence under some conditions. With $H_2(g)$ and $Cl_2(g)$, the reaction also proceeds rapidly (explosively) in the presence of light (photochemically initiated), although some HCl is made this way commercially. With Br_2 and I_2, the reaction occurs more slowly and a catalyst is required.

The data in Table 22.6 show that the standard free energies of formation of HF(g), HCl(g), and HBr(g) are large and negative, suggesting that for them reaction (22.9) goes to completion. For HI(g), ΔG_f° is small and positive. This suggests that even at room temperature HI(g) should dissociate to some extent into its elements. Because the dissociation of HI(g) has a high activation energy, however, the dissociation occurs only very slowly in the absence of a catalyst. As a result, HI(g) is stable at room temperature.

EXAMPLE 22-2 Determining K_p for a Dissociation Reaction from Gibbs Energies of Formation

What is the value of K_p for the dissociation of HI(g) into its elements at 298 K?

Analyze

Dissociation of HI(g) into its elements is the reverse of the formation reaction, and $\Delta G°$ for the dissociation reaction is the negative of ΔG_f° listed in Table 22.6. The value of K for the dissociation reaction can be calculated by using the expression $\Delta G° = -RT \ln K$.

Solve

The dissociation reaction and the value of $\Delta G°$ are as follows:

$$HI(g) \longrightarrow \frac{1}{2}H_2(g) + \frac{1}{2}I_2(s) \qquad \Delta G° = -1.70 \text{ kJ} \qquad (22.10)$$

Now we can use the relationship $\Delta G° = -RT \ln K$.

$$\ln K_p = \frac{-\Delta G°}{RT} = \frac{-(-1.70 \times 10^3 \text{ J mol}^{-1})}{8.3145 \text{ J mol}^{-1}\text{ K}^{-1} \times 298 \text{ K}} = 0.686$$

$$K_p = e^{0.686} = 1.99$$

Assess

The value of K is reasonable at just slightly larger than 1 because $\Delta G°$ for the dissociation reaction is just slightly less than zero. When $\Delta G° = 0$, the value of K is equal to 1. When $\Delta G°$ is a small negative number, the forward reaction is slightly favored and K is just slightly larger than one.

PRACTICE EXAMPLE A: What is the value of K_p for the dissociation of HF(g) into its elements at 298 K? Use data from Table 22.6.

PRACTICE EXAMPLE B: Use data from Table 22.6 to determine K_p and the percent dissociation of HCl(g) into its elements at 298 K.

Oxoacids and Oxoanions

Fluorine, the most electronegative element, adopts the −1 oxidation state in its compounds. The other halogens, when bonded to a more electronegative element such as oxygen, can have any one of several positive oxidation states: +1, +3, +5, or +7. This variability of oxidation states, which was illustrated through electrode potential diagrams for chlorine (Fig. 22-4), is emphasized again by the oxoacids listed in Table 22.7. The structures of some of these

TABLE 22.7   Oxoacids of the Halogens[a]

Oxidation State of Halogen	Chlorine	Bromine	Iodine
+1	HOCl	HOBr	HOI
+3	$HClO_2$	—	—
+5	$HClO_3$	$HBrO_3$	HIO_3
+7	$HClO_4$	$HBrO_4$	HIO_4; H_5IO_6

[a]In all these acids, H atoms are bonded to O atoms—as shown for HOCl, HOBr, and HOI—not to the central halogen atom. More accurate representations of the other acids would be HOClO (instead of $HClO_2$), $HOClO_2$ (instead of $HClO_3$), and so on.

oxoacids are shown in Figure 22-5. Chlorine forms a complete set of oxoacids in all these oxidation states, but bromine and iodine do not. Only a few of the oxoacids can be isolated in pure form ($HClO_4$, HIO_3, HIO_4, H_5IO_6); the rest are stable only in aqueous solution.

The shapes of the oxoacids (and the corresponding oxoanions) are based on a tetrahedral arrangement of the electron pairs around the chlorine atom. The structures of the acids and the electrostatic potential maps for the anions are displayed in Figure 22-5. Two things are noteworthy in Figure 22-5. First, in the structures for $HClO_3$ and $HClO_4$, the bonds shown with dashes are double bonds. The chlorine-oxygen double bonds, with a bond length of about 141 pm, are much shorter than a Cl—O single bond, which has a bond length of about 164 pm. Second, the electrostatic potential maps indicate that, as the number of oxygen atoms increases, the negative charge on the anion gets progressively delocalized. In OCl^-, the oxygen atom is shown in red, which indicates that the negative charge is highly localized on the oxygen atom. In the other anions, the oxygens are shown in yellow or green, which indicates that they are much less negative. (See Figure 10-5 for a description of the color scheme used in the electrostatic potential maps.) Generally speaking, the more

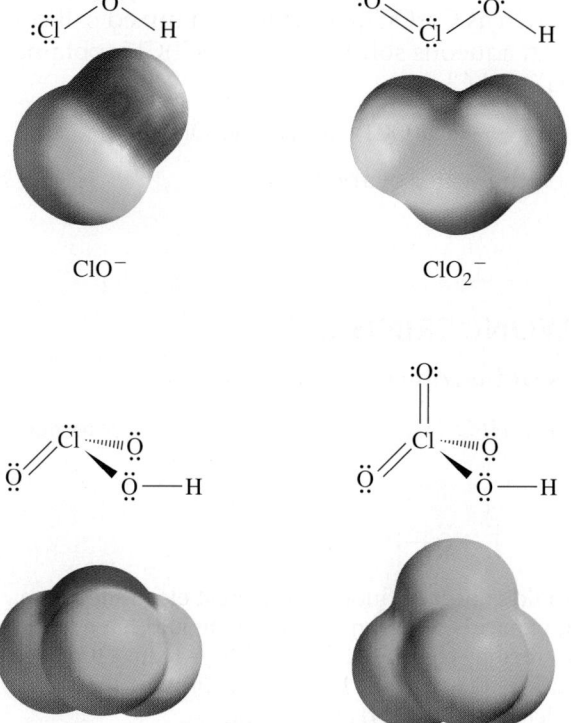

◀ FIGURE 22-5
Shapes of the oxoacids of chlorine and the electrostatic potential maps of the corresponding anions
In the structures shown for $HClO_3$ and $HClO_4$, the dotted double lines represent double bonds that point behind the plane of the page, away from the viewer.

delocalized the charge, the more stable the anion is in aqueous solution and the weaker it is as a Brønsted–Lowry base. Because the charge of the anion is delocalized to the greatest extent in ClO_4^-, we conclude that ClO_4^- is the weakest base, and $HClO_4$, its conjugate acid, is the strongest acid.

An easily prepared oxidizing agent for laboratory use is an aqueous solution of chlorine ("chlorine water"). The solution is not just one of Cl_2, however, because $Cl_2(aq)$ disproportionates into $HOCl(aq)$ and $HCl(aq)$.

▶ Never mix bleach with ammonia. The formation of toxic chlorine gas can result.

Reduction: $\quad\quad\quad\quad Cl_2(g) + 2\,e^- \longrightarrow 2\,Cl^-(aq)$

Oxidation: $\quad Cl_2(g) + 2\,H_2O(l) \longrightarrow 2\,HOCl(aq) + 2\,H^+(aq) + 2\,e^-$

Overall: $\quad\quad Cl_2(g) + H_2O(l) \longrightarrow HOCl(aq) + H^+(aq) + Cl^-(aq)$ **(22.11)**

$$E° = E°_{Cl_2/Cl^-} - E°_{HOCl/Cl_2} = 1.358\,V - 1.611\,V = -0.253\,V$$
$$\Delta G° = -nFE° = 48.8\,kJ$$

Although the disproportionation of Cl_2 in water is nonspontaneous when all reactants and products are in their standard states, the reaction does occur to a limited extent in solutions that are not strongly acidic, as the $E°$ and $\Delta G°$ values indicate.

In contrast, the disproportionation is spontaneous for standard-state conditions in basic solution, because $\Delta G°$ for the reaction is negative, as demonstrated below:

KEEP IN MIND

that E_{Cl_2/Cl^-} is independent of pH, but both E_{HOCl/Cl_2} and E_{OCl^-/Cl_2} are pH dependent. They become smaller as the pH increases, increasing E_{cell} and making ΔG less positive and eventually negative. In terms of Le Châtelier's principle, the smaller $[H^+]$ (greater $[OH^-]$), the more the disproportionation reaction is favored.

Reduction: $\quad\quad\quad\quad Cl_2(g) + 2\,e^- \longrightarrow 2\,Cl^-(aq)$

Oxidation: $\quad Cl_2(g) + 4\,OH^-(aq) \longrightarrow 2\,OCl^-(aq) + 2\,H_2O(l) + 2\,e^-$

Overall: $\quad\quad Cl_2(g) + 2\,OH^-(aq) \longrightarrow OCl^-(aq) + Cl^-(aq) + H_2O(l)$ **(22.12)**

$$E° = E°_{Cl_2/Cl^-} - E°_{OCl/Cl_2} = 1.358\,V - 0.421\,V = 0.937\,V$$
$$\Delta G° = -nFE° = -181\,kJ$$

$HOCl(aq)$ is an effective germicide used, for example, in water purification and the treatment of swimming pools. Aqueous solutions of *hypochlorite* salts, notably $NaOCl(aq)$, are used as common household bleaches. The bleaching action of $NaOCl(aq)$ is highlighted in the margin. Reaction (22.12) is used to make solid household bleaches, such as $Ca(OCl)Cl$, which is a mixed salt containing both OCl^- and Cl^- ions. An aqueous solution of $Ca(OCl)Cl$ is obtained from the reaction of $Ca(OH)_2(aq)$ and Cl_2:

$$Ca(OH)_2(aq) + Cl_2(g) \longrightarrow Ca(OCl)Cl(aq) + H_2O(l)$$

Solid $Ca(OCl)Cl$ is obtained when the water evaporates.

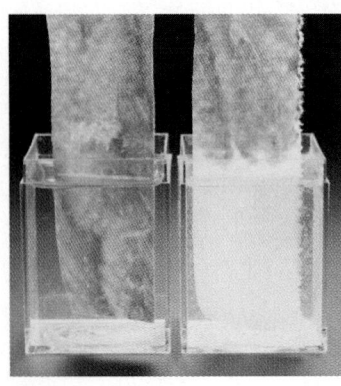

▲ **Bleaching with hypochlorite ion**
Both strips of cloth are heavily stained with tomato sauce. Pure water (left) has little ability to remove the stain. NaOCl (right) rapidly bleaches (oxidizes) the colored components of the sauce to colorless products.

22-1 ARE YOU WONDERING...

If there are oxoacids of fluorine?

The hypothetical oxoacid of fluorine, $HFO_2(HOFO)$, would have a *positive* formal charge on the central F atom.

$$\overset{\ominus}{:\!\ddot{O}}-\overset{\oplus}{\ddot{F}}-\ddot{O}-H$$

This is not something that we would expect of fluorine, the most electronegative of all the elements. Hypothetical oxoacids with more O atoms would have the F atom with even higher positive formal charges. Fluorine does form HOF, an oxoacid with no formal charges $H-\ddot{O}-\ddot{F}:$, but HOF exists only in the solid and liquid states. In water, HOF decomposes to HF, H_2O_2, and O_2.

Chlorine dioxide, ClO_2, is an important bleach for paper and fibers. Its reduction with peroxide ion in aqueous solution produces *chlorite* salts:

$$2\,ClO_2(g) + O_2^{2-}(aq) \longrightarrow 2\,ClO_2^-(aq) + O_2(g) \qquad \textbf{(22.13)}$$

Sodium chlorite is used as a bleaching agent for textiles.

Chlorate salts, which contain the ClO_3^- ion, form when $Cl_2(g)$ disproportionates in hot alkaline solutions, as shown below. (Hypochlorites form in cold alkaline solutions; recall reaction 22.12.)

$$3\,Cl_2(g) + 6\,OH^-(aq) \longrightarrow 5\,Cl^-(aq) + ClO_3^-(aq) + 3\,H_2O(l) \qquad \textbf{(22.14)}$$

Chlorates are good oxidizing agents. Also, solid chlorates produce oxygen gas when they decompose, which makes them useful in matches and fireworks. A simple laboratory method of producing $O_2(g)$ involves heating $KClO_3(s)$ in the presence of $MnO_2(s)$, a catalyst.

$$2\,KClO_3(s) \xrightarrow[\text{MnO}_2]{\Delta} 2\,KCl(s) + 3\,O_2(g) \qquad \textbf{(22.15)}$$

A similar reaction is used as a source of emergency oxygen in aircraft and submarines.

Perchlorate salts are prepared mainly by electrolyzing chlorate solutions. Oxidation of ClO_3^- occurs at a Pt anode through the half-reaction

$$ClO_3^-(aq) + H_2O(l) \longrightarrow ClO_4^-(aq) + 2\,H^+(aq) + 2\,e^- \qquad E^\circ = -1.189\ \text{V} \quad \textbf{(22.16)}$$

Compared with the other oxoacid salts, perchlorates are relatively stable. For example, they do not disproportionate, because no oxidation state higher than +7 is available to chlorine. At elevated temperatures or in the presence of a readily oxidizable compound, however, perchlorate salts may react explosively, so caution is advised when using them. Mixtures of ammonium perchlorate and powdered aluminum are used as the propellant in some solid-fuel rockets, such as those used on the space shuttle. Ammonium perchlorate is especially dangerous to handle, because an explosive reaction may occur when the oxidizing agent ClO_4^- acts on the reducing agent NH_4^+.

◄ A particularly destructive explosion of ammonium perchlorate occurred at a rocket fuel plant in Henderson, Nevada, in 1988.

Interhalogen Compounds

Interhalogen compounds contain two different halogens. The general formulas for interhalogens are XY, XY_3, XY_5, or XY_7, where X and Y represent two different halogens. Some of the more common ones are listed in Table 22.8. The molecular structures of the interhalogen compounds feature the large, less electronegative halogen as the central atom and the smaller halogen atoms as terminal atoms. Molecular shapes of the interhalogen compounds agree quite well with VSEPR theory predictions. Structures for IF_x $(x = 1, 3, 5, 7)$ are shown in Figure 22-6, including one type that we have not seen previously. In IF_7, *seven* electron pairs are distributed around the central I atom in the form of a pentagonal-bipyramid.

Most interhalogen compounds are very reactive. For example, ClF_3 and BrF_3 react explosively with water, organic materials, and some inorganic materials.

TABLE 22.8   Some Interhalogen Compounds

XY	XY_3	XY_5	XY_7
$ClF(g)^a$	$ClF_3(g)$	$ClF_5(g)$	
$BrF(g)$	$BrF_3(l)$	$BrF_5(l)$	
$BrCl(g)$	$IF_3(s)$	$IF_5(l)$	$IF_7(g)$
$ICl(s)$	$ICl_3(s)$		
$IBr(s)$			

[a]The states of matter are given for 25 °C and 1 atm, except IF_3 which decomposes above −28 °C.

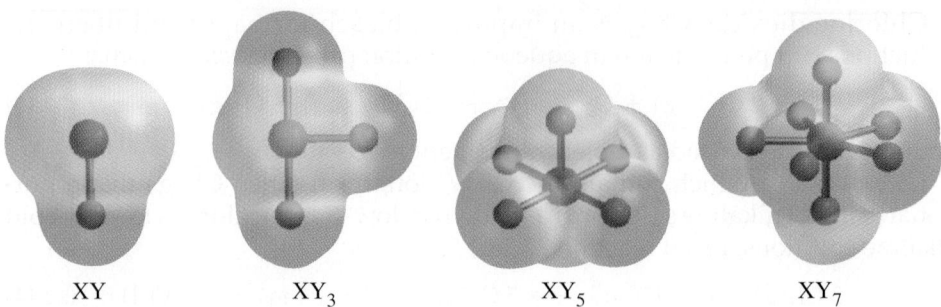

| XY | XY$_3$ | XY$_5$ | XY$_7$ |

▲ FIGURE 22-6
Structures and electrostatic potential maps of the interhalogen compounds of iodine and fluorine

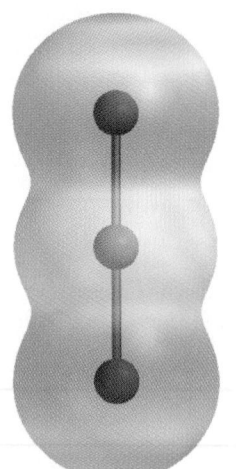

▲ FIGURE 22-7
Structure of the I$_3^-$ ion
Notice that in the electrostatic potential map for I$_3^-$, the three lone pairs around the central iodine atom lead to a neutral charge distribution at this atom, not the negative charge suggested by formal charge arguments.

These two fluorides are used to make fluorinated compounds. For example, ClF$_3$ can be used to convert U(s) to UF$_6$(g), so that the isotopes of uranium can be separated from each other by gaseous diffusion (see page 224).

$$U(s) + 3\,ClF_3(g) \longrightarrow UF_6(l) + 3\,ClF(g)$$

ICl is used as an iodination reagent in organic chemistry.

Polyhalide Ions

The triiodide ion, I$_3^-$, is one of a group of species called **polyhalide ions** that are produced by the reaction of a halide ion with a halogen molecule. In the reaction below, the I$^-$ ion acts as a *Lewis base* (an electron-pair donor) and the I$_2$ molecule as a *Lewis acid* (an electron-pair acceptor).

$$:\!\ddot{\text{I}}\!-\!\ddot{\text{I}}\!: \; + \; :\!\ddot{\text{I}}\!:^- \;\longrightarrow\; \left[:\!\ddot{\text{I}}\!-\!\ddot{\text{I}}\!-\!\ddot{\text{I}}\!:\right]^- \qquad (22.17)$$

The structure of the I$_3^-$ ion is shown in Figure 22-7. The familiar iodine solutions used as antiseptics typically contain triiodide ion, and triiodide ion solutions are widely used in analytical chemistry.

🔍 **22-2 CONCEPT ASSESSMENT**

Do you expect the ions ICl$_2^+$ and ICl$_2^-$ to have the same shape? Explain.

22-4 Group 16: The Oxygen Family

Of the group 16 elements, oxygen and sulfur are clearly nonmetallic in their behavior, but the heavier elements have some metallic properties.

Properties

On the basis of electron configurations alone, we expect oxygen and sulfur to be similar. Both elements form ionic compounds with active metals, and both form similar covalent compounds, such as H$_2$S and H$_2$O, CS$_2$ and CO$_2$, SCl$_2$ and Cl$_2$O.

Even so, there are important differences between oxygen and sulfur compounds. For example, H$_2$O has a very high boiling point (100 °C) for a compound of such low molecular mass (18 u), whereas the boiling point of H$_2$S (molecular mass, 34 u) is much lower (−61 °C). This difference in behavior can be explained in terms of the extensive hydrogen bonding that occurs in H$_2$O but not in H$_2$S (see page 503). Table 22.9 is a comparison of some properties

TABLE 22.9 Comparisons of Oxygen and Sulfur	
Oxygen	**Sulfur**
$O_2(g)$ at 298 K and 1 atm	$S_8(s)$ at 298 K and 1 atm
Two allotropes: $O_2(g)$ and $O_3(g)$	Two solid crystalline forms and many different molecular species in liquid and gaseous states
Principal oxidation states: $-2, -1, 0 \left(-\frac{1}{2} \text{ in } O_2^-\right)$	Possible oxidation states: all values from -2 to $+6$
$O_2(g)$ and $O_3(g)$ are very good oxidizing agents	$S_8(s)$ is a poor oxidizing agent
Forms, with metals, oxides that are mostly ionic in character	Forms ionic sulfides with the most active metals, but many metal sulfides have partial covalent character
O^{2-} completely hydrolyzes in water, producing OH^-	S^{2-} strongly hydrolyzes in water to HS^- (and OH^-)
O is not often the central atom in a structure and can never have more than four atoms bonded to it; more commonly it has two (as in H_2O) or occasionally three (as in H_3O^+)	S is the central atom in many structures; can easily accommodate up to six electron pairs around itself (e.g., SO_3, SO_4^{2-}, SF_6)
Can form only two-atom and three-atom chains, as in H_2O_2 and O_3; compounds with O—O bonds decompose readily	Can form molecules with up to six S atoms per chain in compounds such as H_2S_n, Na_2S_n, $H_2S_nO_6$
Forms the oxide CO_2, which reacts with NaOH(aq) to produce $Na_2CO_3(aq)$	Forms the sulfide CS_2, which reacts with NaOH(aq), producing $Na_2CS_3(aq)$ and $Na_2CO_3(aq)$
Forms, with hydrogen, the compound H_2O, which is a liquid at 298 K and 1 atm; is extensively hydrogen bonded; has a large dipole moment; is an excellent solvent for ionic solids; forms hydrates and aqua complexes; is oxidized with difficulty	Forms, with hydrogen, the compound H_2S, which is a (poisonous) gas at 298 K and 1 atm; is not hydrogen bonded; has a small dipole moment; is a poor solvent; forms no complexes; is easily oxidized

of sulfur and oxygen. In general, the differences can be attributed to the following characteristics of the oxygen atom: (1) small size, (2) high electronegativity, and (3) its inability to employ an expanded valence shell in Lewis structures.

As indicated in Table 22.9, the principal oxidation states of oxygen are -2, -1, and 0. Sulfur, conversely, can exhibit all oxidation states from -2 to $+6$, including several "mixed" oxidation states, such as $+2.5$ in the tetrathionate ion, $S_4O_6^{2-}$.

Occurrence, Production, and Uses

Oxygen Oxygen is the most abundant element in Earth's crust, making up 45.5% by mass. It is also the most abundant element in seawater, accounting for nearly 90% of the mass. In the atmosphere, it is second only to nitrogen in abundance, accounting for 23.15% by mass and 21.04% by volume. Although it is obtained to a limited extent by the decomposition of oxygen-containing compounds and the electrolysis of water, the principal commercial source of oxygen is the fractional distillation of liquid air, which also produces nitrogen, argon, and other noble gases. This process involves only physical changes and is described in Figure 22-8.

Annually, oxygen is one of the principal chemicals manufactured in the United States. The uses of oxygen gas are summarized in Table 22.10. With its commercial availability, O_2 is not commonly prepared in the laboratory. In a submarine, in spacecraft, and in an emergency breathing apparatus, however, it is necessary to generate small quantities of oxygen from solids. The reaction of potassium *superoxide* with CO_2 works well for this purpose because, as the reaction below shows, KO_2 removes CO_2 while O_2 is being formed:

$$4\,KO_2(s) + 2\,CO_2(g) \longrightarrow 2\,K_2CO_3(s) + 3\,O_2(g)$$

▶ FIGURE 22-8
The fractional distillation of liquid air—a simplified representation
Clean air is fed into a compressor and cooled by refrigeration. The cold air then expands through a nozzle and is cooled still further—enough to cause it to liquefy. The liquid air is filtered to remove solid CO_2 and hydrocarbons and then distilled. Liquid air enters the top of the column where nitrogen, the most volatile component (lowest boiling point), passes off as a gas. In the middle of the column, gaseous argon is removed. Liquid oxygen, the least volatile component, collects at the bottom. The normal boiling points of nitrogen, argon, and oxygen are 77.4, 87.5, and 90.2 K, respectively.

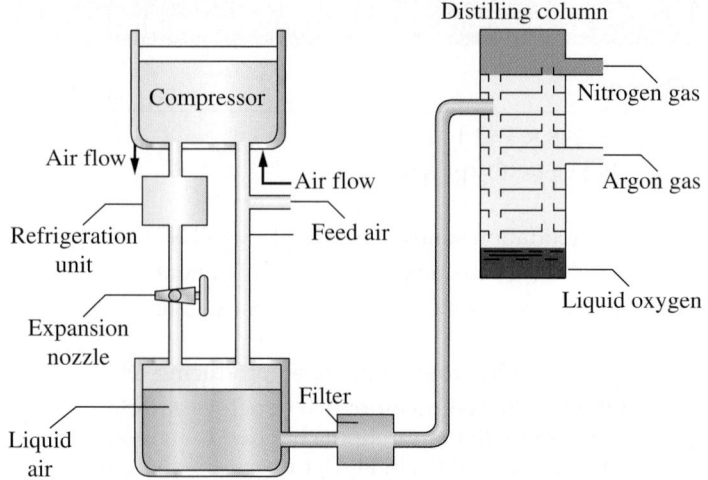

TABLE 22.10 Uses of Oxygen Gas
Manufacture of iron and steel
Manufacture and fabrication of other metals (cutting and welding)
Chemicals manufacture and other oxidation processes
Water treatment
Oxidizer of rocket fuels
Medicinal uses
Petroleum refining

🔍 **22-3 CONCEPT ASSESSMENT**

Which of the following dilute aqueous solutions can be used to obtain oxygen and hydrogen gases simultaneously through electrolysis using platinum electrodes: $H_2SO_4(aq)$, $CuSO_4(aq)$, $NaOH(aq)$, $KNO_3(aq)$, $NaI(aq)$? Explain.

▲ These vast formations of solid sulfur were formed by the solidification of liquid sulfur obtained by the Frasch process.

Sulfur Sulfur is the sixteenth most abundant element in Earth's crust, accounting for 0.0384% by mass. Sulfur occurs as elemental sulfur, as mineral sulfides and sulfates, as $H_2S(g)$ in natural gas, and as organosulfur compounds in oil and coal. Extensive deposits of elemental sulfur are found in Texas and Louisiana, some of them in offshore sites. This sulfur is mined using the **Frasch process,** which is illustrated in Figure 22-9. Superheated water (at about 160 °C and 16 atm) is forced down the outermost of three concentric pipes into an underground bed of sulfur-containing rock. The sulfur melts and forms a liquid pool. Compressed air (at 20–25 atm) is pumped down the innermost pipe and forces the liquid sulfur–water mixture up the remaining pipe. The evaporation of water from the mixture yields formations of solid sulfur, such as those shown in the margin.

Although the Frasch process was once the principal source of elemental sulfur, that is no longer the case. This change has been brought about by the need to control sulfur emissions from industrial operations. Today, most elemental sulfur is obtained from H_2S, which is a common impurity in oil and natural gas. After being removed from the fuel, H_2S is reduced to elemental sulfur in a two-step process. A stream of H_2S gas is split into two parts. One part (about one-third of the stream) is burned to convert H_2S to SO_2. The streams are rejoined in a catalytic converter at 200–300 °C, where the following reaction occurs:

$$2 H_2S(g) + SO_2(g) \longrightarrow 3 S(g) + 2 H_2O(g) \qquad \textbf{(22.18)}$$

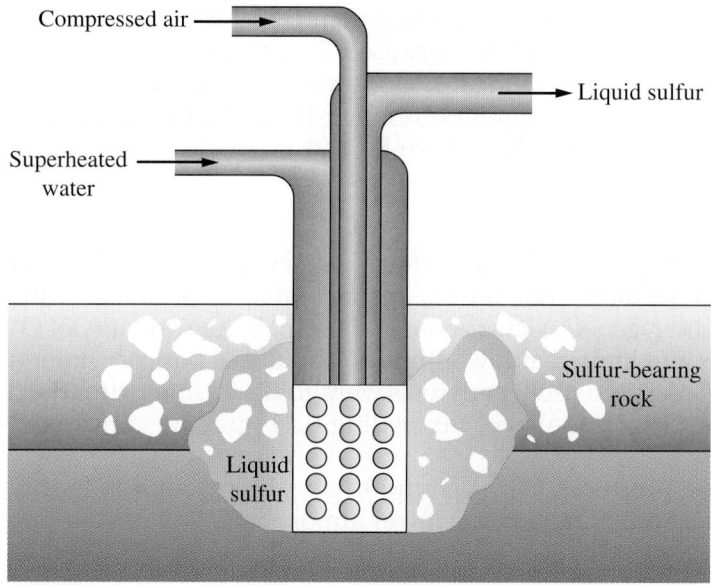

◀ FIGURE 22-9
The Frasch process
Sulfur is melted by using superheated water, and liquid sulfur is forced to the surface.

About 90% of all the sulfur produced is burned to form $SO_2(g)$, and in turn, most $SO_2(g)$ is converted to sulfuric acid, H_2SO_4. The conversion of S to H_2SO_4 is only one of several possibilities identified in Figure 22-10. Elemental sulfur does have a few uses of its own, however. One of these is in vulcanizing rubber (page 1251); another is as a fungicide used for dusting grapevines.

Selenium and Tellurium Selenium and tellurium have properties similar to those of sulfur, but they are more metallic. For example, sulfur is an electrical insulator, whereas selenium and tellurium are semiconductors. Selenium and tellurium are obtained mostly as by-products of metallurgical processes, such as in the anode mud deposited in the electrolytic refining of copper (page 901). Although there is not much use for tellurium compounds, selenium is used in the manufacture of rectifiers (devices that convert alternating to direct electric current). Both Se and Te are employed in the preparation of alloys, and their compounds are used as additives to control the color of glass.

Selenium also displays the property of *photoconductivity*: The electrical conductivity of selenium increases in the presence of light. This property is used, for example, in photocells in cameras. In some modem photocopying machines, the light-sensitive element is a thin film of Se deposited on aluminum. The light and dark areas of the image being copied are converted into a distribution of charge on the light-sensitive element. A dry black powder (toner) coats the charged portions of the light-sensitive element, and this image is transferred to a sheet of paper. Next, the dry powder is fused to the paper. In the final step, the electrostatic charge on the light-sensitive element is neutralized to prepare it for the next cycle.

▲ A selenium-coated light-sensitive element from a photocopier.

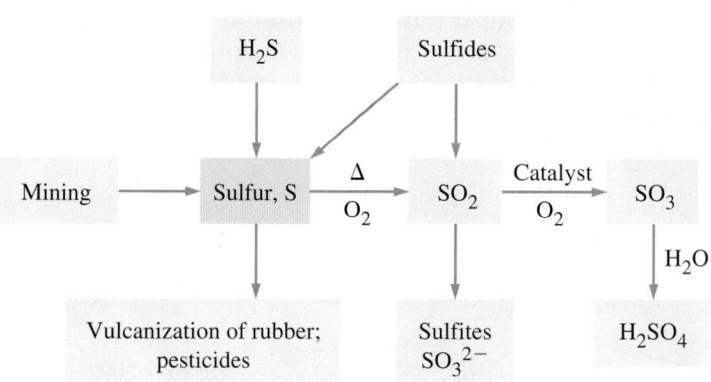

◀ FIGURE 22-10
Sources and uses of sulfur and its oxides

Polonium Polonium is a very rare, radioactive metal, and the only element that crystallizes in a simple cubic lattice. Because it is extremely low in abundance, it has not found much practical use. Polonium was the first new radioactive element isolated from uranium ore by Marie and Pierre Curie in 1898. Madame Curie named it after her native Poland.

Allotropy of Oxygen: Ozone

As we first learned in Chapter 3, allotropy refers to the existence of an element in two or more different molecular forms, and we were introduced to the allotrope of ordinary dioxygen, O_2; it is ozone, O_3. Normally, the quantity of O_3 in the atmosphere is quite limited at low altitudes, about 0.04 part per million (ppm). However, its level increases (perhaps severalfold), in smog situations as described on page 635. Ozone levels exceeding 0.12 ppm are considered unhealthful.

The reaction below produces $O_3(g)$ from $O_2(g)$. The reaction is highly endothermic and occurs only rarely in the lower atmosphere.

$$3\,O_2(g) \longrightarrow 2\,O_3(g) \qquad \Delta H° = +285 \text{ kJ}$$

This reaction does occur in high-energy environments, such as electrical storms. The pungent odor you may at times smell around heavy-duty electrical equipment or xerographic office copiers is probably O_3. The chief method of producing ozone in the laboratory, in fact, is to pass an electric discharge (high-energy electrons) through $O_2(g)$. Because ozone is unstable and decomposes back to $O_2(g)$, it is always generated at the point of use.

Ozone is an excellent oxidizing agent. Its oxidizing ability is surpassed by few other substances (two are F_2 and OF_2). The following standard reduction potentials show that, in acid solution, $O_3(g)$ is a much stronger oxidizing agent than $O_2(g)$, but not as strong as $F_2(g)$ or $OF_2(g)$.

Half-Equation	$E°$
$F_2(g) + 2\,e^- \longrightarrow 2\,F^-(aq)$	+2.87 V
$OF_2(g) + 2\,H^+(aq) + 4\,e^- \longrightarrow H_2O(l) + 2\,F^-(aq)$	+2.0 V
$O_3(g) + 2\,H^+(aq) + 2\,e^- \longrightarrow O_2(g) + H_2O(l)$	+2.07 V
$O_2(g) + 4\,H^+(aq) + 4\,e^- \longrightarrow 2\,H_2O(l)$	+1.23 V

Christian F. Schönbein, the discoverer of ozone, was a German-born chemist working at the University of Basel in Switzerland. In 1866, he noted that when ozone is passed through a concentrated aqueous KOH solution, a red color forms. Subsequently, it was established that an ozonide, containing O_3^- ions, is formed by the following reaction:

$$2\,KOH(aq) + 5\,O_3(g) \longrightarrow 2\,KO_3(aq) + 5\,O_2(g) + H_2O(l)$$

The ozonide subsequently reacts with water to give KOH(aq) and $O_2(g)$. By using other methods of preparation, it has been possible to prepare other alkali metal ozonides. However, the stability of these ozonides decreases as the size of the metal cation decreases because, as discussed in Chapter 21, a large polarizable anion, such as O_3^-, is not particularly stable in the presence of small, highly polarizing cations.

▶ FIGURE 22-11
Some molecular forms of sulfur

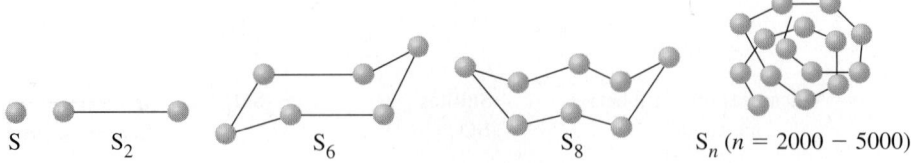

S S_2 S_6 S_8 $S_n\ (n = 2000 - 5000)$

The most important use of ozone is as a substitute for chlorine in purifying drinking water. Its advantages are that it does not impart a taste to the water and it does not form the potentially carcinogenic chlorination products that chlorine can. Its main disadvantage is that O_3 decomposes quickly and disappears from water fairly soon after it is treated. Over time, water that has been treated with ozone is not as well protected against bacterial contamination as is water treated with chlorine.

An important environmental problem centered on atmospheric ozone is discussed in Focus On 22-1, available on the MasteringChemistry website at www.masteringchemistry.com.

> ### 22-4   CONCEPT ASSESSMENT
>
> Would you expect the shape of the ozonide ion to be the same as that of ozone? Explain.

Allotropy and Polymorphism of Sulfur

Sulfur has more allotropes than most elements. The most common structural unit in the solid state is the S_8 ring, although an additional half-dozen cyclic structures are known with up to 20 S atoms per ring. In sulfur vapor, S, S_2, S_4, S_6, and S_8 can all exist under the appropriate conditions. Long-chain molecules of sulfur atoms are found in liquid sulfur. *Rhombic sulfur* (S_α), the stable solid at room temperature, is made up of cyclic S_8 molecules. At 95.5 °C, it converts to *monoclinic sulfur* (S_β), which is also made up of S_8 molecules but has a different crystal structure than S_α. At 119 °C, S_β melts, yielding *liquid sulfur* (S_λ), a straw-colored liquid comprising mostly S_8 molecules but with other cyclic molecules containing from 6 to 20 atoms. At 160 °C, the cyclic molecules open up and recombine into long spiral-chain molecules, producing another form of *liquid sulfur* (S_μ), a dark, viscous liquid. The chain length and viscosity reach a maximum at about 180 °C. At higher temperatures, the chains break up and the viscosity decreases. At 445 °C, the liquid boils, producing *sulfur vapor*. At the boiling point, S_8 molecules predominate in the vapor but they break down into smaller molecules at higher temperatures. *Plastic sulfur* forms if liquid S_μ is poured into cold water. Plastic sulfur consists of long, spiral-chain molecules and has rubberlike properties. On standing, it reverts to rhombic sulfur, a brittle solid.

The following sequence summarizes the phase transitions that occur in sulfur as the temperature increases:

$$S_\alpha \xrightarrow{95.5\,°C} S_\beta \xrightarrow{119} S_\lambda \xrightarrow{160} S_\mu \xrightarrow{445} S_8(g) \longrightarrow S_6 \longrightarrow S_4 \xrightarrow{1000} S_2 \xrightarrow{2000} S$$

Some of these forms of sulfur are shown in Figure 22-12.

Because some of these transitions, especially those in the solid state, are sluggish, additional phenomena are occasionally seen. For example, if rhombic sulfur is heated rapidly, it may melt at 113 °C and fail to convert to monoclinic sulfur. However, monoclinic sulfur may then freeze from this liquid, only to melt again at 119 °C.

Oxygen Compounds

Oxygen is so central to the study of chemistry that we constantly refer to its physical and chemical properties in developing a framework of chemical principles. For instance, our discussion of stoichiometry began with combustion reactions—reactions of substances with $O_2(g)$ to form products such as $CO_2(g)$, $H_2O(l)$, and $SO_2(g)$. Combustion reactions also figured prominently in the presentation of thermochemistry. Many of the molecules and polyatomic anions described in the chapters on chemical bonding were oxygen-containing species. Water was a primary subject in the discussion of liquids, solids, and

(a)　　　　　　　(b)　　　　　　　(c)　　　　　　　(d)

▲ FIGURE 22-12
Several macroscopic forms of sulfur
(a) Rhombic sulfur. (b) Monoclinic sulfur. (c) On the left, monoclinic sulfur has just melted to form an orange liquid. On the right, after continued heating, the liquid becomes red and more viscous. (d) Liquid sulfur is poured into water to produce plastic sulfur.

intermolecular forces as well as in the study of acid–base and other solution equilibria. The dual roles of hydrogen peroxide as an oxidizing agent and a reducing agent were described in Section 5-6; and the kinetics of the decomposition of H_2O_2 was examined in detail in Chapter 14. The acidic, basic, and amphoteric properties of element oxides were outlined in Chapter 9.

A systematic study of oxygen compounds is generally done in conjunction with the study of the other elements. Thus, the oxides of boron were considered in the discussion of boron chemistry in Chapter 21; the oxides of carbon were also considered there. The survey of the chemistry of the alkali and alkaline earth metals in Chapter 21 provided an opportunity to describe normal oxides, peroxides, and superoxides; and the important oxides of sulfur, nitrogen, and phosphorus are discussed in this chapter.

Sulfur Compounds

As in the corresponding discussion for the halogens (Section 22-3), oxidation–reduction chemistry is a primary concern here. To assist in this discussion, we provide electrode potential diagrams for some important sulfur-containing species in Figure 22-13.

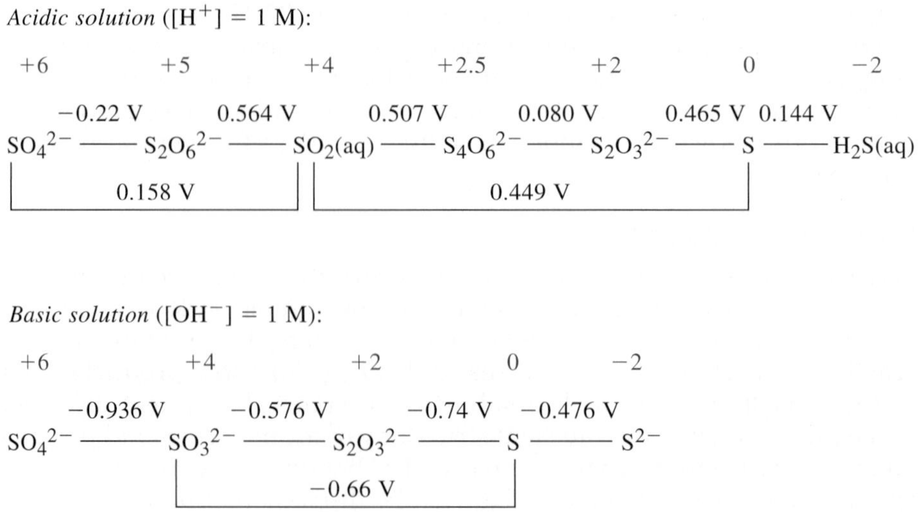

Acidic solution ($[H^+] = 1$ M):

| +6 | | +5 | | +4 | | +2.5 | | +2 | | 0 | | −2 |

$$SO_4^{2-} \overset{-0.22\ V}{\rule{1.5cm}{0.4pt}} S_2O_6^{2-} \overset{0.564\ V}{\rule{1.5cm}{0.4pt}} SO_2(aq) \overset{0.507\ V}{\rule{1.5cm}{0.4pt}} S_4O_6^{2-} \overset{0.080\ V}{\rule{1.5cm}{0.4pt}} S_2O_3^{2-} \overset{0.465\ V}{\rule{1.5cm}{0.4pt}} S \overset{0.144\ V}{\rule{1.5cm}{0.4pt}} H_2S(aq)$$

0.158 V　　　　　　　　　　　　0.449 V

Basic solution ($[OH^-] = 1$ M):

| +6 | | +4 | | +2 | | 0 | | −2 |

$$SO_4^{2-} \overset{-0.936\ V}{\rule{1.5cm}{0.4pt}} SO_3^{2-} \overset{-0.576\ V}{\rule{1.5cm}{0.4pt}} S_2O_3^{2-} \overset{-0.74\ V}{\rule{1.5cm}{0.4pt}} S \overset{-0.476\ V}{\rule{1.5cm}{0.4pt}} S^{2-}$$

−0.66 V

▶ FIGURE 22-13
Electrode potential diagrams for sulfur

Sulfur Dioxide and Sulfur Trioxide More than a dozen oxides of sulfur have been reported, but only sulfur dioxide, SO_2, and sulfur trioxide, SO_3, are commonly encountered. Typical structures are shown in Figure 22-14. The main commercial methods of producing $SO_2(g)$ are the direct combustion of sulfur, reaction (22.19), and the roasting of metal sulfides, reaction (22.20):

$$S(s) + O_2(g) \xrightarrow{\Delta} SO_2(g) \qquad \textbf{(22.19)}$$

$$2\,ZnS(s) + 3\,O_2(g) \xrightarrow{\Delta} 2\,ZnO(s) + 2\,SO_2(g) \qquad \textbf{(22.20)}$$

The main use of SO_2 is in the synthesis of SO_3 to make sulfuric acid, H_2SO_4. In the *contact process*, $SO_2(g)$ is formed by reaction (22.19) or (22.20). Then sulfur trioxide is produced by oxidizing $SO_2(g)$ in an exothermic, reversible reaction:

$$2\,SO_2(g) + O_2(g) \rightleftharpoons 2\,SO_3(g) \qquad \textbf{(22.21)}$$

Reaction (22.21) is the key step in the process, but it occurs very slowly unless catalyzed. The principal catalyst is V_2O_5 mixed with alkali metal sulfates. The catalysis involves adsorption of the $SO_2(g)$ and $O_2(g)$ on the catalyst, followed by reaction at active sites and desorption of SO_3.

Sulfuric Acid SO_3 reacts with water to form H_2SO_4, but the direct reaction of $SO_3(g)$ and water produces a fine mist of $H_2SO_4(aq)$ droplets with unreacted $SO_3(g)$ trapped inside the droplets. This misting would result in a great loss of product and a tremendous pollution problem. To avoid these outcomes, $SO_3(g)$ is instead bubbled through 98% H_2SO_4 in towers packed with a ceramic material. The $SO_3(g)$ readily dissolves in the sulfuric acid and reacts with the small amount of water present to increase the concentration of the sulfuric acid. The result is a form of sulfuric acid sometimes called *oleum* but more commonly called *fuming sulfuric acid*. In a sense, the product is greater than 100% H_2SO_4. Sufficient water is added to the circulating acid in the tower to maintain the required concentration. Later, sulfuric acid of the strength desired is produced by dilution with water. If we use the formula $H_2S_2O_7$ (disulfuric acid) as an example of a particular oleum, the reactions are

$$SO_3(g) + H_2SO_4(l) \longrightarrow H_2S_2O_7(l) \qquad \textbf{(22.22)}$$

$$H_2S_2O_7(l) + H_2O(l) \longrightarrow 2\,H_2SO_4(l) \qquad \textbf{(22.23)}$$

$$H_2SO_4(l) \xrightarrow{H_2O} H_2SO_4(aq) \qquad \textbf{(22.24)}$$

Dilute sulfuric acid, $H_2SO_4(aq)$, enters into all the common reactions of a strong acid, such as neutralizing bases. It reacts with metals to produce $H_2(g)$ and dissolves carbonates to liberate $CO_2(g)$.

Concentrated sulfuric acid has some distinctive properties. It has a very strong affinity for water, strong enough that it will even remove H and O atoms (in the proportion H_2O) from some compounds. In the reaction of concentrated sulfuric acid with a carbohydrate like sucrose, all the H and O atoms are removed and a residue of pure carbon is left, as shown in the photo. The chemical equation for the reaction that occurs is given below:

$$C_{12}H_{22}O_{11}(s) \xrightarrow{H_2SO_4(concd)} 12\,C(s) + 11\,H_2O(l) \qquad \textbf{(22.25)}$$

Concentrated sulfuric acid is a moderately good oxidizing agent and is able, for example, to react with copper.

$$Cu(s) + 2\,H_2SO_4(concd) \longrightarrow Cu^{2+}(aq) + SO_4^{2-}(aq) + 2\,H_2O(l) + SO_2(g) \qquad \textbf{(22.26)}$$

For a very long time, sulfuric acid has ranked among the top manufactured chemicals, with annual production in the United States of about 45 million metric tons. Sulfuric acid continues to have many uses, but the bulk of H_2SO_4 is used in the manufacture of fertilizers. Other uses are found in various metallurgical processes, oil refining, and the manufacture of the white pigment

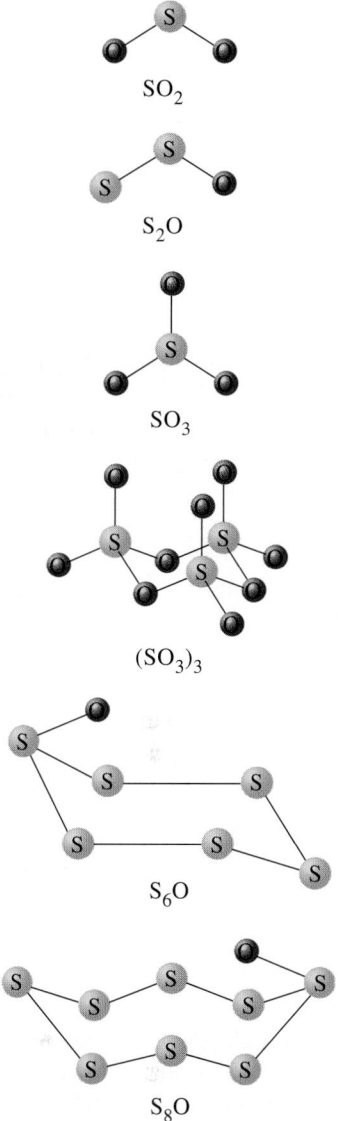

▲ FIGURE 22-14
Structures of some sulfur oxides
To conform to the observed structures of SO_2, S_2O, and SO_3, the hybridization scheme proposed for the central S atom is sp^2. S_2O has a similar structure to SO_2 but with a S atom substituted for one O atom. The SO_3 exists in equilibrium with the trimer $(SO_3)_3$, in which the O—S—O angle is approximately tetrahedral and the hybridization of the S should be sp^3. The oxides S_6O and S_8O illustrate sulfur's ability to form ring compounds.

▶ **(a)** Concentrated sulfuric acid is added to cane sugar. **(b)** Carbon is produced in the reaction.

(a)   (b)

titanium dioxide. Also familiar is its use as the electrolyte in storage batteries for automobiles and emergency power supplies. We might say that sulfuric acid was the workhorse of the "old economy" but that it has a lesser role in the "new economy."

When $SO_2(g)$ reacts with water, it produces $H_2SO_3(aq)$, but this acid, *sulfurous acid*, has never been isolated in pure form. Salts of sulfurous acid, *sulfites*, are good reducing agents and are easily oxidized by $O_2(g)$. In aqueous solution, sulfite ions, SO_3^{2-}, are oxidized by O_2 to SO_4^{2-} ions, as shown below:

$$O_2(g) + 2\,SO_3^{2-}(aq) \longrightarrow 2\,SO_4^{2-}(aq) \qquad (22.27)$$

Interestingly, sulfite ion can also act as an oxidizing agent, as in this reaction with H_2S.

$$2\,H_2S(g) + 2\,H^+(aq) + SO_3^{2-}(aq) \longrightarrow 3\,H_2O(l) + 3\,S(s) \qquad (22.28)$$

Both H_2SO_3 and H_2SO_4 are diprotic acids. They ionize in two steps and produce two types of salts, one in each ionization step. The term **acid salt is** sometimes used for salts such as $NaHSO_3$ and $NaHSO_4$ because their anions undergo a further *acid* ionization. H_2SO_3 is a weak acid in both ionization steps, whereas H_2SO_4 is strong in the first step and somewhat weak in the second. If a solution of H_2SO_4 is sufficiently dilute, however (less than about 0.001 M), we can treat the acid as if both ionization steps go to completion.

▶ Actually, $NaHSO_3$ cannot be isolated as a solid. When we attempt to crystallize this salt from an aqueous solution containing HSO_3^-, the reaction $2\,HSO_3^- \longrightarrow S_2O_5^{2-} + H_2O$ occurs. The product obtained is sodium *metabisulfite*, $Na_2S_2O_5$.

Sulfates and Sulfites Sulfate and sulfite salts have a number of important uses. Calcium sulfate dihydrate (gypsum) is used to make the hemihydrate (plaster of Paris) for the building industry (page 939). Aluminum sulfate is used in water treatment and in sizing paper (page 950). Copper(II) sulfate is employed as a fungicide and an algicide and in electroplating. The chief application of sulfites is in the pulp and paper industry. Sulfites solubilize *lignin*, a polymeric substance that coats the cellulose fibers in wood. This treatment frees the fibers for processing into wood pulp and then paper. Sulfites are also used as reducing agents in photography and as scavengers of $O_2(aq)$ in treating boiler water (reaction 22.27). Compounds of sulfur(IV) have long been used as food preservatives and antioxidants. For example, exposure to $SO_2(g)$ prevents the discoloration of dried fruits, and soluble sulfites act as antimicrobial agents in winemaking.

Thiosulfates In addition to sulfite and sulfate ions, another important sulfur–oxygen ion is thiosulfate ion, $S_2O_3^{2-}$. The prefix *thio* signifies that a S atom replaces an O atom in a compound. Thus, the thiosulfate ion can be viewed as a sulfate ion, SO_4^{2-}, in which a S atom replaces one of the O atoms. The formal oxidation state of S in $S_2O_3^{2-}$ is $+2$, but as Figure 22-15 indicates, the two S atoms are not equivalent: The central S atom is in the oxidation state $+6$, and the terminal S atom, -2. The structures of several other thio anions are also shown in Figure 22-15.

Thiosulfates can be prepared by boiling elemental sulfur in an alkaline solution of sodium sulfite. The sulfur is oxidized and the sulfite ion is reduced, both to thiosulfate ion.

$$SO_3^{2-}(aq) + S(s) \longrightarrow S_2O_3^{2-}(aq) \qquad (22.29)$$

Thiosulfate solutions are important in photographic film processing (see page 1099). They are also common analytical reagents, often used in conjunction with iodine. For example, in one method of analysis for copper, an excess of iodide ion is added to $Cu^{2+}(aq)$, producing $CuI(s)$ and triiodide ion, I_3^-.

$$2\,Cu^{2+}(aq) + 5\,I^-(aq) \longrightarrow 2\,CuI(s) + I_3^-(aq) \qquad (22.30)$$

The excess triiodide ion is then titrated with a standard solution of $Na_2S_2O_3(aq)$, forming I^- and $S_4O_6^{2-}$, the *tetrathionate* ion.

$$I_3^-(aq) + 2\,S_2O_3^{2-}(aq) \longrightarrow 3\,I^-(aq) + S_4O_6^{2-}(aq) \qquad (22.31)$$

Oxygen and Sulfur Halides

Both oxygen and sulfur form a number of interesting compounds with the halogens. Oxygen, for example, forms the fluorides OF_2 and O_2F_2, which have structures similar to those of water and hydrogen peroxide but which are much more reactive. Sulfur also forms compounds with the halogens; the analogous compounds SF_2 and S_2F_2 are known, as are the compounds SF_4 and SF_6. The reactivities of SF_4 and SF_6 are quite different. SF_6 is a colorless, odorless, and unreactive gas. It is so unreactive that it can be safely inhaled, in small quantities, resulting in a very deep voice. (As you may already know, helium gas has the opposite effect when inhaled in small quantities.) Conversely, SF_4 is a very reactive gas and a powerful fluorinating agent. In the following reaction, SF_4 converts BCl_3 to BF_3.

$$3\,SF_4 + 4\,BCl_3 \longrightarrow 4\,BF_3 + 3\,SCl_2 + 3\,Cl_2$$

Sulfur and chlorine also form the compounds S_2Cl_2 and SCl_4, but the best-known halide is SCl_2. It is a foul-smelling, red liquid (melting point, $-122\ °C$; boiling point, $59\ °C$) that has been used in the production of the notorious, and highly poisonous, mustard gas, $S(CH_2CH_2Cl)_2$. The production of mustard gas is based on the following reactions:

$$SCl_2 + 2\,CH_2CH_2 \longrightarrow S(CH_2CH_2Cl)_2$$

Mustard gas is not actually a gas, but a volatile liquid (melting point, $13\ °C$; boiling point, $235\ °C$). During World War I it was sprayed as a mist that stayed close to the ground and was blown by wind onto the enemy. Exposure to mustard gas causes blistering of the skin, internal and external bleeding, blindness, and, after four or five weeks, death.

SO_2 Emissions and the Environment

Industrial smog consists primarily of particles (ash and smoke), $SO_2(g)$, and H_2SO_4 mist. A variety of industrial operations produce significant quantities of $SO_2(g)$, but the main contributors to atmospheric releases of $SO_2(g)$ are power plants burning coal or high-sulfur fuel oils. SO_2 can oxidize to SO_3, especially when the reaction is catalyzed on the surfaces of airborne particles or through reaction with NO_2:

$$SO_2(g) + NO_2(g) \longrightarrow SO_3(g) + NO(g) \qquad (22.32)$$

SO_3^{2-}, Sulfite

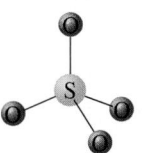
SO_4^{2-}, Sulfate

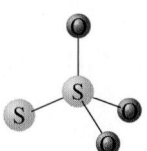
$S_2O_3^{2-}$, Thiosulfate

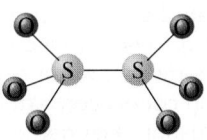
$S_2O_6^{2-}$, Dithionate

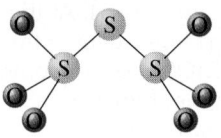
$S_3O_6^{2-}$, Trithionate

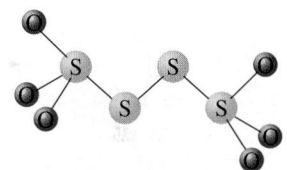
$S_4O_6^{2-}$, Tetrathionate

▲ FIGURE 22-15
Structures of some oxoanions of sulfur

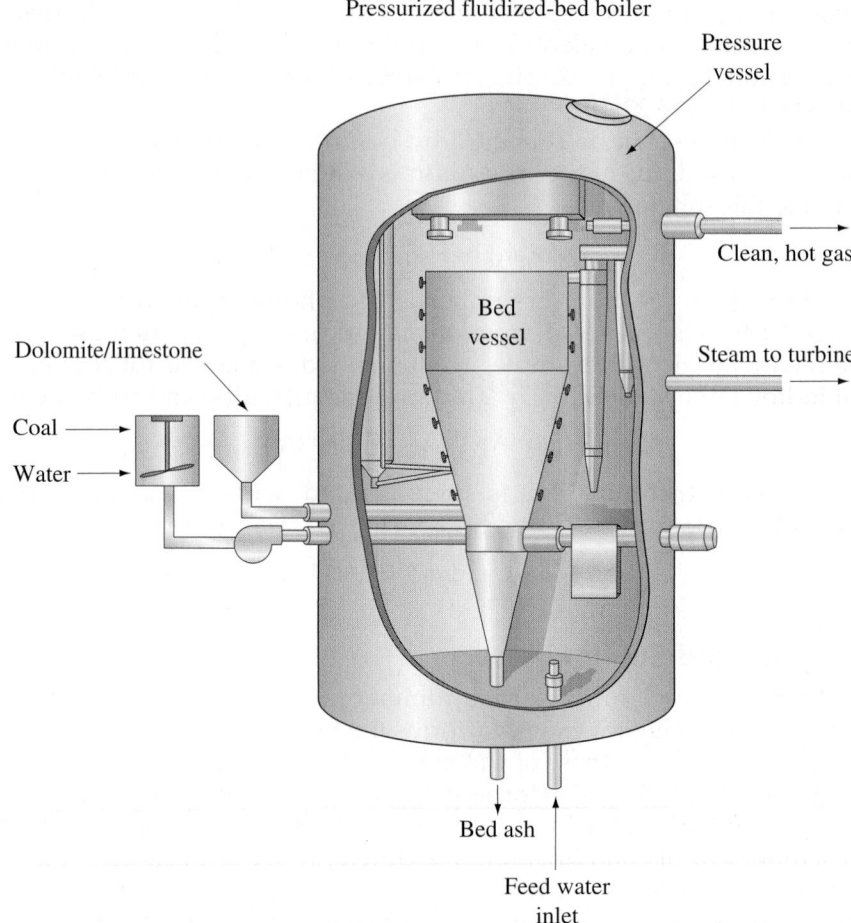

Pressurized fluidized-bed boiler

▶ FIGURE 22-16
Fluidized-bed combustion
Powdered coal, limestone, and air are introduced into a combustion chamber where water circulating through coils is converted to steam. Combustion is carried out at a relatively low temperature (760–860 °C), which minimizes the production of NO(g) from $N_2(g)$ and $O_2(g)$. At the same time, $SO_2(g)$ from sulfur in the coal reacts with CaO(s) from decomposition of the limestone, forming $CaSO_3(s)$ in a Lewis acid–base reaction.

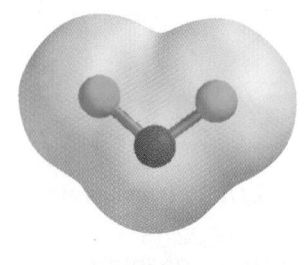

(a)

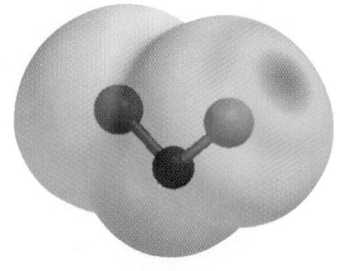

(b)

▲ Electrostatic potential maps of **(a)** OF_2 and **(b)** OCl_2.

In turn, SO_3 can react with water vapor in the atmosphere to produce H_2SO_4 mist, a component of acid rain. Also, the reaction of H_2SO_4 with airborne NH_3 produces particles of $(NH_4)_2SO_4$. The details of the effect of low concentrations of SO_2 and H_2SO_4 on the body are not well understood, but it is clear that these substances are respiratory irritants. Levels above 0.10 ppm are considered potentially harmful.

The control of industrial smog and acid rain hinges on the removal of sulfur from fuels and the control of $SO_2(g)$ emissions. Dozens of processes have been proposed for removing SO_2 from smokestack gases, one of which is illustrated in Figure 22-16. In this process, $SO_2(g)$ from the coal reacts with CaO(s) to form deposits of $CaSO_3(s)$.

🔍   **22-5   CONCEPT ASSESSMENT**

Electrostatic potential maps for OF_2 and OCl_2 are shown in the margin. Explain the differences.

22-5   Group 15: The Nitrogen Family

The chemistry of the group 15 elements is an extensive subject, especially that of nitrogen and phosphorus. We will discuss the special significance of these two elements to living matter later in the text, but even here you should get a sense of the richness of their chemistry. For example, nitrogen atoms can exist in many oxidation states, which is evident from the variety of nitrogen–containing species identified in Figure 22-17.

Acidic solution ([H$^+$] = 1 M):

+5	+4	+3	+2	+1	0	−1	−2	−3

$$\underset{NO_3^-}{} \overset{0.803\ V}{\rule{1cm}{0.4pt}} \underset{N_2O_4}{} \overset{1.065\ V}{\rule{1cm}{0.4pt}} \underset{HNO_2}{} \overset{0.996\ V}{\rule{1cm}{0.4pt}} \underset{NO}{} \overset{1.591\ V}{\rule{1cm}{0.4pt}} \underset{N_2O}{} \overset{1.766\ V}{\rule{1cm}{0.4pt}} \underset{N_2}{} \overset{-1.87\ V}{\rule{1cm}{0.4pt}} \underset{NH_3OH^+}{} \overset{1.42\ V}{\rule{1cm}{0.4pt}} \underset{N_2H_5^+}{} \overset{1.275\ V}{\rule{1cm}{0.4pt}} \underset{NH_4^+}{}$$

Basic solution ([OH$^-$] = 1 M):

+5	+4	+3	+2	+1	0	−1	−2	−3

$$\underset{NO_3^-}{} \overset{-0.86\ V}{\rule{1cm}{0.4pt}} \underset{N_2O_4}{} \overset{0.867\ V}{\rule{1cm}{0.4pt}} \underset{NO_2^-}{} \overset{-0.46\ V}{\rule{1cm}{0.4pt}} \underset{NO}{} \overset{0.76\ V}{\rule{1cm}{0.4pt}} \underset{N_2O}{} \overset{0.94\ V}{\rule{1cm}{0.4pt}} \underset{N_2}{} \overset{-3.04\ V}{\rule{1cm}{0.4pt}} \underset{NH_2OH}{} \overset{0.73\ V}{\rule{1cm}{0.4pt}} \underset{N_2H_4}{} \overset{0.10\ V}{\rule{1cm}{0.4pt}} \underset{NH_3}{}$$

◀ FIGURE 22-17
Electrode potential diagrams for nitrogen

Metallic–Nonmetallic Character in Group 15

All the elements in group 15 have the valence-shell electron configuration ns^2np^3. This configuration suggests nonmetallic behavior and doesn't give any clues as to metallic character that might exist. Table 22.11, however, indicates the usual decrease of ionization energy with increasing atomic number. These values, taken together with physical properties from the table, do suggest the order of metallic character within the group. Nitrogen and phosphorus are nonmetallic, arsenic and antimony are metalloids, and bismuth is metallic. The first ionization energy of bismuth is actually somewhat less than that of magnesium, and its third ionization energy $(2466\ kJ\ mol^{-1})$ is less than the third ionization energy of aluminum $(2745\ kJ\ mol^{-1})$. The electronegativities indicate a high degree of nonmetallic character for nitrogen and less so for the remaining members of the group.

Three group 15 elements—phosphorus, arsenic, and antimony—exhibit allotropy. The common forms of phosphorus at room temperature, both nonmetallic, are white and red phosphorus. For arsenic and antimony, the more stable allotropic forms are the metallic ones. These forms have high densities, moderate thermal conductivities, and limited abilities to conduct electricity. Bismuth is a metal despite its low electrical conductivity, which, nevertheless, is better than that of manganese and almost as good as that of mercury. The nonmetals and metals in group 15 are also distinguishable by their oxides. The oxides of nitrogen and phosphorus (for example, N_2O_3 and P_4O_6) are acidic

TABLE 22.11 Selected Properties of Group 15 Elements

Element	Covalent Radius, pm	Electronegativity	First Ionization Energy kJ mol^{-1}	Common Physical Form(s)	Density of Solid, g cm^{-3}	Comparative Electrical Conductivity[a]
N	75	3.0	1402	Gas	1.03 (−252 °C)	—
P	110	2.1	1012	Wax-like white solid;	1.82	—
				Red solid	2.20	10^{-17}
As	121	2.0	947	Yellow solid;	2.03	—
				Gray solid with metallic luster	5.78	6.1
Sb	140	1.9	834	Yellow solid;	5.3	—
				Silvery white metallic solid	6.69	4.0
Bi	155	1.9	703	Pinkish white metallic solid	9.75	1.5

[a]These values are relative to an assigned value of 100 for silver.

when they react with water. As with the nonmetals in groups 16 and 17, this is typical for nonmetal oxides. Arsenic(III) oxide and antimony(III) oxide are amphoteric, whereas bismuth(III) oxide acts only as a base, a property typical of metal oxides.

Occurrence, Production, and Uses

▲ Viruses for use in medical research are frozen in liquid nitrogen.

Nitrogen Nitrogen occurs mainly in the atmosphere. Its abundance in Earth's solid crust is only 0.002% by mass. The only important nitrogen-containing minerals are KNO_3 (niter, or saltpeter) and $NaNO_3$ (soda niter, or Chile saltpeter), found in a few desert regions. Other natural sources of nitrogen-containing compounds are plant and animal protein and the fossilized remains of ancient plant life, such as coal.

Until about 100 years ago, sources of pure nitrogen and its compounds were quite limited. This all changed with the invention of a process for the liquefaction of air in 1895 (see Figure 22-8) and the development of the Haber–Bosch process for converting nitrogen to ammonia in 1908 (page 1008). A host of nitrogen compounds can be made from ammonia. Nitrogen has many important uses of its own in addition to being a precursor of manufactured nitrogen compounds. Some of these uses are listed in Table 22.12.

Phosphorus Although phosphorus is the eleventh most abundant element and makes up about 0.11% of Earth's crust by mass, it was not discovered until 1669. It was originally isolated from putrefied urine—an effective if not particularly pleasant source. Today the principal source of phosphorus compounds is phosphate rock, a class of minerals known as *apatites*, such as fluorapatite $Ca_5(PO_4)_3F$ or $3 Ca_3(PO_4)_2 \cdot CaF_2$. Elemental phosphorus is prepared by heating phosphate rock, silica (SiO_2), and coke (C) in an electric furnace. The overall change that occurs is

$$2 Ca_3(PO_4)_2(s) + 10 C(s) + 6 SiO_2(s) \xrightarrow{\Delta} 6 CaSiO_3(l) + 10 CO(g) + P_4(g) \quad \textbf{(22.33)}$$

The $P_4(g)$ is condensed, collected, and stored under water as white phosphorus.

Although compounds of phosphorus are vitally important to living organisms (DNA and phosphates in bones and teeth, for example), the element itself is not widely used. Almost all the elemental phosphorus produced is reoxidized to give P_4O_{10} for the manufacture of high-purity phosphoric acid. The rest is used to make organophosphorus compounds and phosphorus sulfides (P_4S_3) in match heads.

Arsenic, Antimony, and Bismuth Arsenic is obtained by heating arsenic-containing metal sulfides. For example, FeAsS yields FeS and As(g). The As(g) deposits as As(s), which can be used to make other compounds. Some arsenic is also obtained by the reduction of arsenic(III) oxide with CO(g). Antimony is obtained mainly from its sulfide ores. Bismuth is obtained as a by-product of the refining of other metals.

Both As and Sb are used in making alloys of other metals. For example, the addition of As and Sb to Pb produces an alloy that has desirable properties for use as electrodes in lead–acid batteries. Arsenic and antimony are used to produce semiconductor materials, such as GaAs, GaSb, and InSb, in electronic devices.

TABLE 22.12   Uses of Nitrogen Gas
Provide a blanketing (inert) atmosphere for the production of chemicals and electronic components
Pressurized gas for enhanced oil recovery
Metals treatment
Refrigerant (e.g., fast freezing of foods)

Nitrogen Compounds

The substance from which all nitrogen compounds are ultimately derived, $N_2(g)$, is unusually stable. One explanation of the limited reactivity of the N_2 molecule is based on its electronic structure. As discussed in Chapter 10, the bond between the two N atoms in N_2 is a *triple* covalent bond, which is unusually strong and difficult to break. In thermochemical terms, the enthalpy change associated with breaking the bonds in one mole of N_2 molecules is very high—the dissociation reaction is highly endothermic.

$$N{\equiv}N(g) \longrightarrow 2\,N(g) \quad \Delta H^\circ = +945.4 \text{ kJ}$$

Also, the Gibbs energies of formation of many nitrogen compounds are positive, which means that their formation reactions are not spontaneous. For $NO(g)$,

$$\frac{1}{2}N_2(g) + \frac{1}{2}O_2 \longrightarrow NO(g) \quad \Delta G_f^\circ = 86.55 \text{ kJ}$$

Reactions with a positive Gibbs energy, such as the formation of $NO(g)$ from its elements, do not occur to any significant extent at normal temperatures and atmospheric conditions. Just imagine the situation if $\Delta G_f^\circ[NO(g)] = -86.55 \text{ kJ mol}^{-1}$ instead of $+86.55 \text{ kJ mol}^{-1}$. The reaction of $N_2(g)$ and $O_2(g)$ to form $NO(g)$ would proceed to a far greater extent. With an atmosphere depleted in $O_2(g)$ and rich in noxious $NO(g)$, life as we know it would not be possible.

Nitrides Nitrogen forms binary compounds with most other elements, and these compounds can be grouped into four categories. In ionic (salt-like) nitrides, the nitrogen is present as the N^{3-} ion. These compounds form with lithium and the group 2 metals. Thus, when magnesium is burned in air (see Figure 2-1), a small quantity of magnesium nitride forms, together with the principal product, magnesium oxide.

$$3\,Mg(s) + N_2(g) \xrightarrow{\Delta} Mg_3N_2(s) \qquad \text{(22.34)}$$

The nitride ion is a very strong base. In aqueous solution, it accepts protons from water molecules to form ammonia molecules and hydroxide ions.

$$N^{3-}(aq) + 3\,H_2O(l) \longrightarrow NH_3(aq) + 3\,OH^-(aq)$$

In the reaction of magnesium nitride with water, magnesium and hydroxide ions combine to form insoluble $Mg(OH)_2$, and the ammonia is released as a gas, which is easily detectable by its odor.

$$Mg_3N_2(s) + 6\,H_2O(l) \longrightarrow 3\,Mg(OH)_2(s) + 2\,NH_3(g)$$

When nitrogen combines with other typical nonmetals, it does so by forming covalent bonds, yielding covalent nitrides. Bonding in these nitrides can be described in terms of the general principles presented in Chapters 10 and 11. Some binary covalent nitrides are $(CN)_2$, P_3N_5, As_4N_4, S_2N_2, and S_4N_4. Nitrogen combines with elements of group 13, producing compounds of the form MN (where M = B, Al, Ga, In, or Tl). These compounds have solid structures that resemble those of graphite or diamond, with M and N atoms bonded to form planes of hexagonal rings (the graphite-like form) or a diamond-like lattice. A fourth type of binary nitrides are the metallic nitrides with formulas such as MN, M_3N, and M_4N. These are *interstitial compounds*, in which N atoms occupy some or all of the interstices (voids) in the structure of the metal. They are hard, chemically inert, high-melting-point solids often used to harden and protect surfaces. Typical examples are TiN, VN, and UN, with melting points of 2950 °C, 2050 °C, and 2800 °C, respectively.

▲ **Fritz Haber (1868–1934)**
Haber's perfection of the ammonia synthesis reaction, which made the manufacture of inexpensive explosives possible, was of critical importance to Germany during World War I. After the war, Haber again applied his chemical knowledge for his country's benefit by attempting, unsuccessfully, to extract gold from seawater for use in paying war reparations. Despite his past services, this Jewish scientist was driven from his academic post by the Nazi regime in 1933.

▲ Anhydrous liquid ammonia being applied directly to the soil.

Ammonia and Related Compounds The Haber–Bosch process for the synthesis of ammonia involves the reaction below:

$$N_2(g) + 3 H_2(g) \rightleftharpoons 2 NH_3(g) \qquad (22.35)$$

As we noted in an earlier encounter with this reaction (see Focus On 15, www.masteringchemistry.com), a high yield of ammonia requires (1) a high temperature (400 °C), (2) a catalyst to speed up the reaction, and (3) a high pressure (about 200 atm). The key to achieving essentially 100% yield is continuous removal of NH_3 and recycling of the unreacted $N_2(g)$ and $H_2(g)$. The NH_3 is removed by liquefaction. The Haber–Bosch process is outlined in Figure 22-18. A critical aspect of the process is having a source of $H_2(g)$. Mostly, this is made by the re-forming of natural gas (see page 956).

Ammonia is the starting material in the manufacture of most other nitrogen compounds, but it has some direct uses of its own. Its most important use is as a fertilizer. The highest concentration in which nitrogen fertilizer can be applied to fields is as pure liquid NH_3, known as "anhydrous ammonia." $NH_3(aq)$ is also applied in a variety of household cleaning products, such as commercial glass cleaners. In these products, the ammonia acts as an inexpensive base to produce $OH^-(aq)$. The $OH^-(aq)$ reacts with grease and oil molecules to convert them into compounds that are more soluble in water. In addition, the aqueous ammonia solution dries quickly, leaving few streaks on glass.

Because ammonia is a base, a simple approach to producing certain nitrogen compounds is to neutralize ammonia with an appropriate acid. The acid–base reaction that forms ammonium sulfate, an important solid fertilizer, is

$$2 NH_3(aq) + H_2SO_4(aq) \longrightarrow (NH_4)_2SO_4(aq) \qquad (22.36)$$

Ammonium chloride, made by the reaction of $NH_3(aq)$ and $HCl(aq)$, is used in the manufacture of dry-cell batteries, in cleaning metals, and as an agent to help solder flow smoothly when soldering metals. Ammonium nitrate, made by the reaction of $NH_3(aq)$ and $HNO_3(aq)$, is used both as a fertilizer and as an explosive. The explosive power of ammonium nitrate was not fully appreciated until a shipload of this material exploded in Texas City, Texas, in 1947, killing many people. More recently, mixtures of ammonium

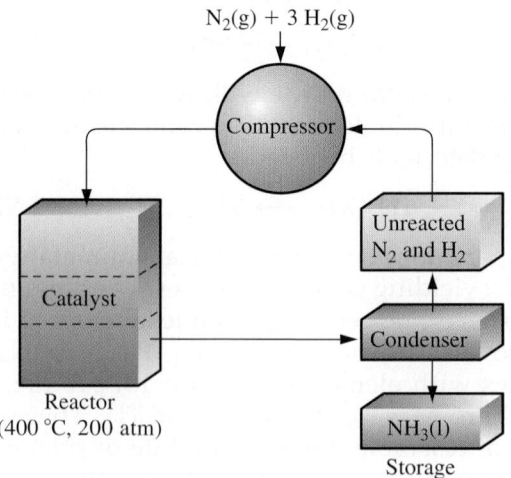

▲ FIGURE 22-18
Ammonia synthesis reaction—the Haber–Bosch process
The gaseous N_2–H_2 mixture is introduced into a reactor at high temperature and pressure in the presence of a catalyst. The gaseous N_2–H_2–NH_3 mixture leaves the reactor and is cooled as it passes through a condenser. Liquefied NH_3 is removed, and the remaining N_2–H_2 mixture is compressed and returned to the reactor. The yield is essentially 100%.

nitrate and fuel oil were used as explosives in the terrorist attacks on the World Trade Center in New York City in 1993 and the Murrah Federal Building in Oklahoma City in 1995. The reaction of $NH_3(aq)$ and $H_3PO_4(aq)$ yields ammonium phosphates, such as $NH_4H_2PO_4$ and $(NH_4)_2HPO_4$. These compounds are good fertilizers because they supply two vital plant nutrients, N and P; they are also used as fire retardants.

Urea, which contains 46% nitrogen by mass, is often manufactured at ammonia synthesis plants by using the following reaction:

$$2 NH_3 + CO_2 \longrightarrow CO(NH_2)_2 + H_2O \quad \text{(22.37)}$$

The structure of the urea molecule is shown in the margin. Urea is an excellent fertilizer, either as a pure solid, as a solid mixed with ammonium salts, or in a very concentrated aqueous solution mixed with NH_4NO_3 or NH_3 (or both). Urea is also used as a feed supplement for cattle and in the production of polymers and pesticides.

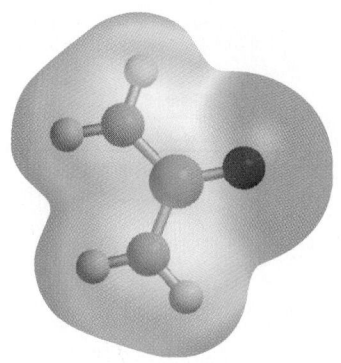

▲ Urea.

Other Hydrides of Nitrogen A great deal has been said in this text about the principal hydride of nitrogen: ammonia, NH_3. Here we describe some lesser-known hydrides. If a H atom in NH_3 is replaced by the group $—NH_2$, the resulting molecule is $H_2N—NH_2$ or N_2H_4, *hydrazine* ($pK_{b_1} = 6.07$; $pK_{b_2} = 15.05$). Replacement of a H atom in NH_3 by $—OH$ produces NH_2OH, *hydroxylamine* ($pK_b = 8.04$). Hydrazine and hydroxylamine are weak bases. Because it has two N atoms, N_2H_4 ionizes in two steps. Hydrazine and hydroxylamine form salts analogous to ammonium salts, such as $N_2H_5^+NO_3^-$, $N_2H_6^{2+}SO_4^{2-}$, and $NH_3OH^+Cl^-$. As expected, these salts hydrolyze in water to yield acidic solutions.

Hydrazine and some of its derivatives burn in air with the evolution of large quantities of heat; they are used as rocket fuels (see margin). For the combustion of hydrazine,

$$N_2H_4(l) + O_2(g) \longrightarrow N_2(g) + 2 H_2O(l) \quad \Delta H° = -622.2 \text{ kJ} \quad \text{(22.38)}$$

Reaction (22.38) can also be used to remove dissolved $O_2(g)$ from boiler water. Hydrazine is particularly valued for this purpose because no salts (ionic compounds) form that would be objectionable in the water. The industrial preparation of hydrazine is described in Focus On 4, www.masteringchemistry.com.

Both hydrazine and hydroxylamine can act as either oxidizing or reducing agents (usually the latter), depending on the pH and the substances with which they react. The oxidation of hydrazine in acidic solution by nitrite ion produces *hydrazoic acid*, $HN_3(aq)$.

$$N_2H_5^+(aq) + NO_2^-(aq) \longrightarrow HN_3(aq) + 2 H_2O(l) \quad \text{(22.39)}$$

Pure HN_3 is a colorless liquid that boils at 37 °C. It is very unstable and will detonate when subjected to shock. Resonance structures for the HN_3 molecule are shown below:

$$:\ddot{N}=N=\ddot{N} \longleftrightarrow :N\equiv N-\ddot{N}$$

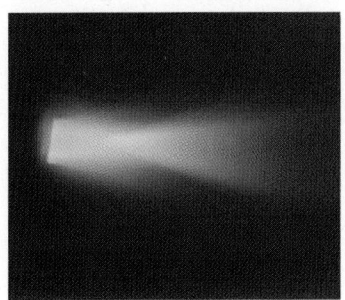

▲ The lifting thrusters of a space shuttle, shown here in an in-flight test, use methylhydrazine, CH_3NHNH_2, as a fuel.

In aqueous solution, HN_3 is a weak acid; its salts are called *azides*. Azides contain the azide ion, N_3^-, and resemble chlorides in some properties (for instance, AgN_3 is insoluble in water), but they are unstable. Some azides (such as lead azide) are used to make detonators. The release of $N_2(g)$ by the decomposition of sodium azide, NaN_3, is the basis of the air-bag safety system in automobiles (page 210).

Oxides of Nitrogen

Nitrogen forms a series of oxides in which the oxidation state of N can have every value ranging from +1 to +5 (Table 22.13). All these oxides are gases at

TABLE 22.13 Oxides of Nitrogen

O.S. of N	Formula
+1	N_2O
+2	NO
+3	N_2O_3
+4	NO_2
+4	N_2O_4
+5	N_2O_5

TABLE 22.14 Preparation of Oxides of Nitrogen	
Oxide	A Method of Preparation
N_2O	$NH_4NO_3(s) \xrightarrow{\Delta} N_2O(g) + 2\,H_2O(g)$
NO	$3\,Cu(s) + 8\,H^+(aq) + 2\,NO_3^-(aq) \longrightarrow 3\,Cu^{2+}(aq) + 2\,NO(g) + 4\,H_2O(l)$
N_2O_3	$2\,NO(g) + N_2O_4(g) \xrightarrow{-20\,°C} 2\,N_2O_3(l)$
NO_2	$2\,Pb(NO_3)_2(s) \xrightarrow{\Delta} 2\,PbO(s) + 4\,NO_2(g) + O_2(g)$
	$2\,NO(g) + O_2(g) \rightleftharpoons 2\,NO_2(g) \qquad K_p = 1.6 \times 10^{12}$ (at 298 K)
N_2O_4	$2\,NO_2(g) \rightleftharpoons N_2O_4(g) \qquad K_p = 8.84$ (at 298 K)
N_2O_5	$4\,HNO_3(l) + P_4O_{10}(s) \xrightarrow{-10\,°C} 4\,HPO_3(s) + 2\,N_2O_5(s)$

25 °C, except N_2O_5, which is a solid with a sublimation pressure of 1 atm at 32.5 °C. It is impossible to obtain either brown $NO_2(g)$ or its colorless dimer, $N_2O_4(g)$, as a pure gas at temperatures between about -10 °C and 140 °C because of the equilibrium between them (review Example 15-9). At lower temperatures, N_2O_4 can be obtained as a pure solid, and above 140 °C, the gas-phase equilibrium strongly favors $NO_2(g)$. In the solid state, N_2O_3 is pale blue; in the liquid state, it is bright blue.

All nitrates decompose on heating, but only NH_4NO_3 yields $N_2O(g)$. Nitrates of active metals, such as $NaNO_3$, yield the corresponding nitrite, such as $NaNO_2$, and $O_2(g)$. Nitrates of less active metals, such as $Pb(NO_3)_2$, yield the metal oxide, $NO(g)$ and $O_2(g)$. These methods of preparing oxides of nitrogen are outlined in Table 22.14.

As mentioned in our assessment of the metallic–nonmetallic character of the group 15 elements, the oxides of nitrogen are acidic, and they react with water to give acidic solutions. For example, the reaction of N_2O_5 and H_2O yields HNO_3, as shown below, and thus N_2O_5 is the acid anhydride of HNO_3.

$$N_2O_5(s) + H_2O(l) \longrightarrow 2\,HNO_3(aq)$$

The acid anhydride of nitrous acid, HNO_2, is N_2O_3. Nitrogen dioxide, NO_2, produces both HNO_3 and NO when it reacts with water. But N_2O is not an acid anhydride in the usual sense; it is related to *hyponitrous* acid, $H_2N_2O_2$ (HON=NOH), which yields N_2O and H_2O on decomposition.

$$H_2N_2O_2 \longrightarrow N_2O + H_2O$$

Among the oxides of nitrogen, N_2O (laughing gas) has anesthetic properties and finds some use in dentistry and in providing pain relief during childbirth. N_2O has also been used as a propellant in pressured cans of fats (such as whipped cream) and as a power booster in combustion engines. The power-boosting capability of N_2O arises because at high temperatures—such as those in a combustion chamber—two moles of N_2O decompose to give a total of three moles of N_2 and O_2. The increase in the number of moles of gas increases the force on the pistons in the engine, leading ultimately to increased acceleration.

Other oxides of nitrogen include NO_2, which is employed in the manufacture of nitric acid; N_2O_4, which is used extensively as an oxidizer in rocket fuels; and NO, which is arguably the most important oxide of nitrogen, at least from a biological standpoint. Among its many biological functions, NO helps to protect the heart, stimulate the brain, and kill bacteria. It has been called the "miracle molecule" because it helps to regulate the health of almost every cell in the body.

The discovery of the role of NO in biological systems followed a somewhat tortuous route. In 1829, it was discovered that the compound trinitroglycerine, $C_3H_5(NO_3)_3$, a highly explosive compound, helped dilate (enlarge) blood vessels and relieved symptoms of a heart attack. Later acetylcholine, $C_7H_{16}NO_2^+$,

▲ Lewis structures of **(a)** trinitroglycerine, and **(b)** acetylcholine.

was discovered to have a similar effect on blood vessels. Lewis structures of these two dissimilar molecular species are shown above. Surprisingly, the link between them and their action on blood vessels was not explained until 1987. In that year, two reports were published by two research groups, one lead by Louis Ignarro in the United States and the other lead by Salvador Moncada in the United Kingdom, which demonstrated that the link is the nitric oxide molecule, NO. It is now known that acetylcholine causes enzymes in the blood vessel to release NO, which in turn causes other enzymes to relax the muscle of the vessel. Trinitroglycerine is converted to NO(g) by metabolic processes and NO(g) acts as a signaling agent in blood vessels and causes the muscle in the vessels to relax.

The NO molecule has also been implicated in the human response to infection. When infection occurs, the immune system produces both NO and O_2^-. These two radicals react as follows to produce the peroxynitrite anion, ONO_2^-:

$$NO + O_2^- \longrightarrow O{=}N{-}O{-}O^-$$

The peroxynitrite anion is a strong and versatile oxidant that can break down the structures of, and thus kill, cells of invading species. Over time, excess peroxynitrite anions isomerizes to harmless nitrate anions.

One of the interesting features of the oxides of nitrogen is that their free energies of formation are all *positive* quantities. This feature suggests that the oxides, such as $N_2O(g)$, are thermodynamically unstable, and that decomposition to the elements is spontaneous under standard conditions.

$$2\,N_2O(g) \longrightarrow 2\,N_2(g) + O_2(g) \qquad \Delta G^\circ = -208 \text{ kJ (at 298 K)} \qquad \textbf{(22.40)}$$

Actually, $N_2O(g)$ is quite stable at room temperature because the activation energy for its decomposition is very high—about 250 kJ mol^{-1}. At higher temperatures (about 600 °C), its rate of decomposition becomes appreciable. Reaction (22.40) accounts for the ability of N_2O to support combustion because the $O_2(g)$ necessary for the combustion is produced by the decomposition of N_2O. Overall reactions of the following sort occur.

$$H_2(g) + N_2O(g) \longrightarrow H_2O(l) + N_2(g)$$
$$Cu(s) + N_2O(g) \longrightarrow CuO(s) + N_2(g)$$

Notice that one of the combustion products is the normal combustion product and the other is N_2.

Nitrogen monoxide (nitric oxide), NO(g), is produced commercially by the Ostwald process, in which $NH_3(g)$ is oxidized in the presence of a catalyst:

$$4\,NH_3(g) + 5\,O_2(g) \xrightarrow[850\,^\circ\text{C}]{\text{Pt}} 4\,NO(g) + 6\,H_2O(g) \qquad \textbf{(22.41)}$$

▲ The combustion of copper gauze in $N_2O(g)$.

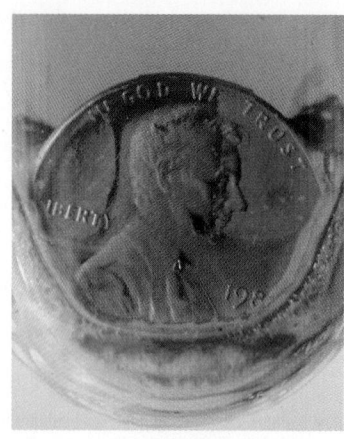

▲ A copper penny reacting with nitric acid. The reaction is that given by equation (22.43). The blue-green color of the solution is due to $Cu^{2+}(aq)$, and the reddish brown color is that of nitrogen dioxide, $NO_2(g)$.

▲ Flash paper is used by magicians for dramatic effect. It can be made by treating paper with nitric and sulfuric acids. This process converts the cellulose fibers into nitrocellulose, which burns cleanly and rapidly.

The oxidation of NH_3 to NO is the first step in converting NH_3 to a number of other nitrogen compounds.

Another source of NO, usually unwanted, is in high-temperature combustion processes, such as those that occur in automobile engines and in electric power plants. At the same time that the fuel combines with oxygen from air to produce a high temperature, $N_2(g)$ and $O_2(g)$ in the hot air combine to a limited extent to form $NO(g)$.

$$N_2(g) + O_2(g) \xrightarrow{\Delta} 2\,NO(g) \tag{22.42}$$

Brown nitrogen dioxide, $NO_2(g)$, is often seen in reactions involving nitric acid. An example is the reaction of $Cu(s)$ with warm concentrated $HNO_3(aq)$.

$$Cu(s) + 4\,H^+(aq) + 2\,NO_3^-(aq) \longrightarrow Cu^{2+}(aq) + 2\,H_2O(l) + 2\,NO_2(g) \tag{22.43}$$

Of particular interest to atmospheric chemists is the key role of $NO_2(g)$ in the formation of photochemical smog (page 636).

Nitric Acid and Nitrates The commercial synthesis of nitric acid does not use N_2O_5, as might be expected. It involves the following three reactions, the first of which—the Ostwald process—was described previously. $NO(g)$ from reaction (22.45) is recycled into reaction (22.44).

$$4\,NH_3(g) + 5\,O_2(g) \xrightarrow[850\,°C]{Pt} 4\,NO(g) + 6\,H_2O(g) \tag{22.41}$$

$$2\,NO(g) + O_2(g) \longrightarrow 2\,NO_2(g) \tag{22.44}$$

$$3\,NO_2(g) + H_2O(l) \longrightarrow 2\,HNO_3(aq) + NO(g) \tag{22.45}$$

Nitric acid is used in the preparation of various dyes; drugs; fertilizers (ammonium nitrate); and explosives, such as nitroglycerin, nitrocellulose, and trinitrotoluene (TNT). It is also used in metallurgy and in reprocessing spent nuclear fuels. Nitric acid is about twelfth, by mass, among the top chemicals produced in the United States.

Nitric acid is also a good oxidizing agent. For example, copper reacts with dilute $HNO_3(aq)$, producing primarily NO or, with concentrated $HNO_3(aq)$, NO_2 (reaction 22.43). With a more active metal, such as Zn, the reduction product has N in one of its lower oxidation states, for example, NH_4^+. Nitrates can be made by neutralizing nitric acid with appropriate bases.

Nitrogen Halides

Nitrogen forms halides with the elements of group 17. Nitrogen trifluoride, NF_3, can be made by the fluorination of ammonia in the presence of a Cu catalyst.

$$4\,NH_3(g) + 3\,F_2(g) \xrightarrow{Cu} NF_3(g) + 3\,NH_4F(s)$$

Nitrogen trifluoride is a colorless, odorless gas and is one of the few nitrogen compounds that is thermodynamically stable with respect to the elements ($\Delta G_f^\circ = -83.3\ \text{kJ mol}^{-1}$). Although the nitrogen atom in NF_3 has a lone pair of electrons, NF_3 has very little tendency to act as a Lewis base, unlike ammonia, but it can be made to react, in the gas phase, with oxygen to give NF_3O, a stable but somewhat unusual molecule. In the NF_3O molecule, the nitrogen atom is the central atom, and the nitrogen–oxygen bond is much shorter than the nitrogen-fluorine bonds. The following resonance structures have been proposed, and the structure on the right appears to be an important contributor.

In contrast to nitrogen trifluoride, nitrogen trichloride, NCl_3, is neither a gas nor stable. It is an oily, yellow, highly explosive liquid and care must be taken in handling or making this compound. Its explosive nature can be attributed to its endothermic enthalpy of formation ($\Delta H_f^\circ = 230.0$ kJ mol^{-1}), which indicates that the decomposition to $N_2(g)$ and $Cl_2(g)$ is highly exothermic and thermodynamically favorable. Nitrogen trichloride is not prepared from the direct reaction of $N_2(g)$ and $Cl_2(g)$ but from the reaction of ammonium chloride with chlorine, as shown below:

$$NH_4Cl(s) + 3\,Cl_2(g) \rightleftharpoons NCl_3(l) + 4\,HCl(g)$$

The equilibrium is shifted to the right-hand side by dissolving the NCl_3 in an organic solvent. In contrast to NF_3, nitrogen trichloride reacts with water to give ammonia.

$$NCl_3(aq) + 3\,H_2O(l) \longrightarrow NH_3(aq) + 3\,HOCl(aq)$$

This reaction produces HOCl, a bleaching agent, and thus nitrogen trichloride, diluted in air, has been used to bleach flour. The compounds NBr_3 and NI_3 are also known, and are even more reactive and explosive (and thus, more dangerous to handle) than NCl_3. NI_3 is so unstable that it detonates with the slightest contact, even if touched with a feather or breathed on.

Two other fluorides of nitrogen are known: N_2F_4 and N_2F_2. The structures of the N_2F_4 and N_2F_2 molecules are shown in Figure 22-19. Dinitrogen tetrafluoride, N_2F_4, interconverts between two conformations—staggered and gauche—because the two NF_2 units can rotate independently about the N—N bond. (We will encounter these conformations again in organic chemistry in Chapters 26 and 27.) Dinitrogen difluoride, N_2F_2, exists in two geometrical forms, called geometrical isomers, that are not easily interconverted. Because the double bond prevents twisting of the molecule about the nitrogen–nitrogen bond axis, the N_2F_2, molecule exists as either the *cis* isomer or the *trans* isomer. The *cis* isomer has both fluorine atoms on the same side of the double bond, whereas the *trans* isomer has the fluorine atoms on opposite sides of the double bond. (We will meet this type of isomerism again in Chapter 26.)

Allotropes of Phosphorus

White phosphorus is a white, waxy, phosphorescent solid that can be cut with a knife. (A phosphorescent material glows in the dark.) It is a nonconductor of electricity, can ignite spontaneously in air (hence it is stored under water), and is insoluble in water but soluble in some nonpolar solvents, such as CS_2. The solid has P_4 molecules as its basic structural units (Fig. 22-20a). The P_4 molecule is tetrahedral, with a P atom at each corner. The phosphorus-to-phosphorus bonds in P_4 appear to involve the overlap of $3p$ orbitals almost exclusively. Such overlap normally produces 90° bond angles, but in P_4 the P—P—P bond angles are 60°. The bonds are strained, and as might be expected, species with strained bonds are reactive.

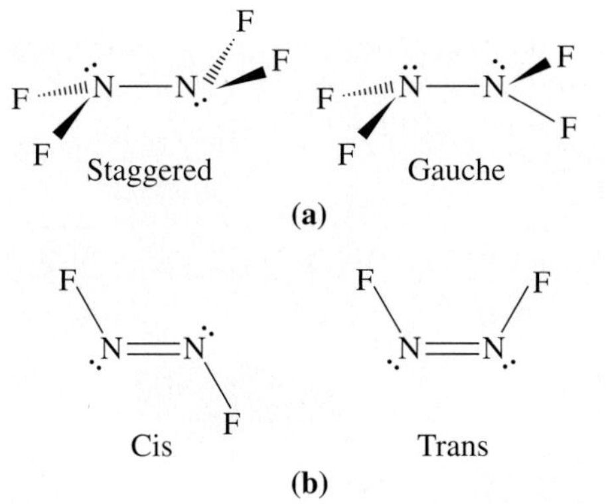

(a) Staggered   Gauche

(b) Cis   Trans

◄ FIGURE 22-19
Structures of N_2F_4 and N_2F_2
(a) In the staggered conformation of N_2F_4, the lone pairs on the nitrogen atoms are diametrically opposed to each other. The gauche conformation is obtained from the staggered conformation by rotating one NF_2 group by 60° with respect to the other. **(b)** Because the double bond prevents the molecule from twisting about the nitrogen–nitrogen bond axis, there are two geometrical isomers for the F—N≡N—F molecule.

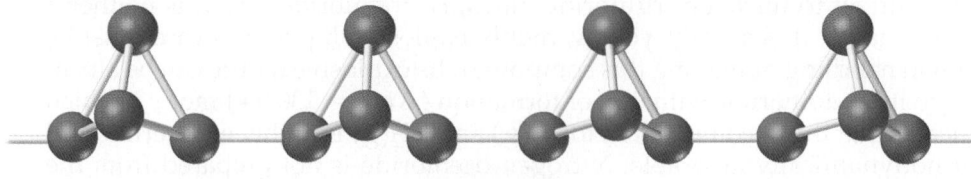

(a) White phosphorus 60°   (b) Red phosphorus

▲ FIGURE 22-20
Two forms of phosphorus
(a) Structure of white phosphorus: the P_4 molecule. (b) Structure of red phosphorus.

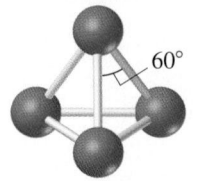

▲ The glow of white phosphorus gave the element its name—*phos*, light, and *phorus*, bringing. The solid has a relatively high vapor pressure, and the glow results from the slow reaction between phosphorus vapor and oxygen in air.

When white phosphorus is heated to about 300 °C out of contact with air, it transforms to *red phosphorus*. What appears to happen is that one P—P bond per P_4 molecule breaks, and the resulting fragments join together into long chains, as suggested in Figure 22-20(b). Red P is less reactive than white P. Because they have a different atomic arrangement in their basic structural units, red and white phosphorus are allotropic forms of phosphorus rather than just different solid phases. These two allotropes of phosphorus are shown in the photograph below. The triple point of red phosphorus is 590 °C and 43 atm. Thus, red phosphorus sublimes without melting (at about 420 °C).

Despite the fact that white P is the form obtained by condensing $P_4(g)$ and that the conversion of white P to red P is a very slow process at ambient temperatures, red P is actually the more thermodynamically stable of the two forms at 298.15 K. Nevertheless, white P is assigned values of 0 for ΔH_f° and ΔG_f°, and for red P, these values are negative.

Phosphorus Compounds

In Chapter 28 we will discover that compounds of phosphorus are essential to all living organisms. However, there is also a significant "inorganic" chemistry of phosphorus, as we will see in this section.

▶ Phosphine is extremely poisonous and has been used as a fumigant against rodents and insects.

Phosphine The most important compound of phosphorus and hydrogen is phosphine, PH_3. This compound is analogous to ammonia in that it acts as a base and forms phosphonium (PH_4^+) compounds. Unlike ammonia, PH_3 is thermally unstable. Phosphine is produced by the disproportionation of P_4 in aqueous base.

$$P_4(s) + 3\,OH^-(aq) + 3\,H_2O(l) \longrightarrow 3\,H_2PO_2^-(aq) + PH_3(g)$$

Phosphorus Trichloride Another phosphorus(III) compound is phosphorus trichloride, PCl_3. One typical reaction of PCl_3 is its hydrolysis, which produces hydrochloric and phosphorous acids.

$$PCl_3(l) + 6\,H_2O(l) \longrightarrow H_3PO_3(aq) + 3\,H_3O^+(aq) + 3\,Cl^-(aq)$$

▶ The white and red allotropes of phosphorus.

PCl_3 is the most important phosphorus halide, and a variety of phosphorus(III) compounds are made from it. It is produced by the direct action of $Cl_2(g)$ on elemental phosphorus. Although you may never see PCl_3, chemicals made from it are everywhere—soaps and detergents, plastics and synthetic rubber, nylon, motor oils, and insecticides and herbicides. A variety of organic groups can replace one or more chlorine atoms of PCl_3 to give a family of phosphine-like compounds. These are good Lewis bases and can act as ligands in complex-ion formation.

The halide PCl_5 is obtained by the reaction of Cl_2 with PCl_3 in tetrachloromethane (CCl_4). In the gas phase, PCl_5 exists as discrete trigonal bipyramidal molecules. In the solid state, it exists as $[PCl_4]^+[PCl_6]^-$, in which the ions are tetrahedral and octahedral, respectively.

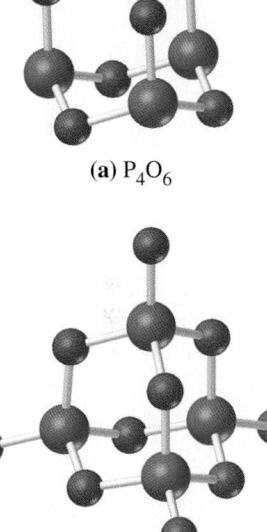

(a) P_4O_6

(b) P_4O_{10}

⬤ = P   ⬤ = O

▲ FIGURE 22-21
Molecular structures of P_4O_6 and P_4O_{10}

🔍 22-6   CONCEPT ASSESSMENT

In the solid phase, PCl_5 forms PCl_4^+ and PCl_6^+. However, PBr_5 forms $PBr_4^+Br^-$. Suggest a reason for this difference in structure.

Oxides of Phosphorus The simplest formulas we can write for the oxides that have phosphorus in the oxidation states +3 and +5 are P_2O_3 and P_2O_5, respectively. The corresponding names are "phosphorus trioxide" and "phosphorus pentoxide." P_2O_3 and P_2O_5 are only empirical formulas, however. The true molecular formulas of the oxides are double those—that is, P_4O_6 and P_4O_{10}.

The structure of each oxide molecule is based on the P_4 tetrahedron and so must have *four* P atoms, not two. As shown in Figure 22-21(a), in P_4O_6 one O atom bridges each pair of P atoms in the P_4 tetrahedron, which means that there are *six* O atoms per P_4 tetrahedron. The structure of P_4O_{10} is shown in Figure 22-21(b). In addition to the six bridging O atoms, one O atom is bonded to each corner P atom. This means that there are a total of *ten* O atoms per P_4 tetrahedron.

The reaction of P_4 with a limited quantity of $O_2(g)$ produces P_4O_6. If an excess of $O_2(g)$ is used, P_4O_{10} is obtained. Both oxides react with water to form oxoacids—both are *acid anhydrides*.

$$P_4O_6(l) + 6\,H_2O(l) \longrightarrow 4\,H_3PO_3(aq)$$
phosphorous acid

$$P_4O_{10}(s) + 6\,H_2O(l) \longrightarrow 4\,H_3PO_4(aq) \qquad \textbf{(22.46)}$$
phosphoric acid

Because the formulas of phosphoric acid (H_3PO_4) and phosphorous acid (H_3PO_3) are very similar, it is tempting to think that the acids are somewhat similar. It turns out that these two acids have structures that are different in a very significant way. The structures and electrostatic potential maps of H_3PO_4 and H_3PO_3 are shown on page 1016. In H_3PO_4, the P atom is bonded to four oxygen atoms and each of the H atoms is bonded to an oxygen atom, but in H_3PO_3, the P atom is bonded to three oxygen atoms and one H atom. All three H atoms in phosphoric acid are ionizable, and so H_3PO_4 is a triprotic acid. Only two of the H atoms in H_3PO_3 are ionizable and so H_3PO_3 is a diprotic acid.

◀ Ionizable H atoms in oxoacids are associated with the linkage E—O—H. See page 729.

🔍 22-7   CONCEPT ASSESSMENT

Write condensed structural formulas for phosphoric acid and phosphorous acid. Condensed structural formulas are discussed on page 70.

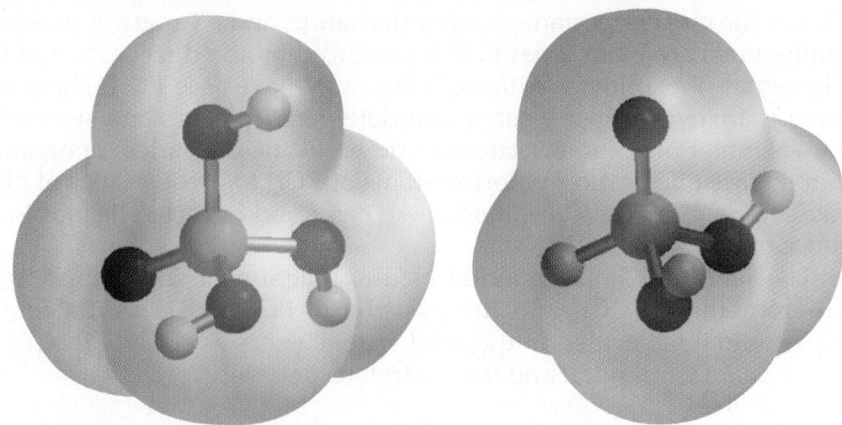

▶ Phosphoric and phosphorous acids.

Phosphoric Acid Phosphoric acid, H_3PO_4, ranks about seventh among the chemicals manufactured in the United States, with an annual production of more than 13 million tons. It is used mainly to make fertilizers, but it is also used to treat metals to make them more corrosion-resistant. Phosphoric acid has many uses in the food industry: It is used to make baking powders and instant cereals, in cheese making, in curing hams, and in soft drinks to impart tartness.

If P_4O_{10} and H_2O are combined in a 1:6 mole ratio in reaction (22.46), the liquid product should be pure H_3PO_4, that is, 100% H_3PO_4, a compound called *orthophosphoric* acid. An analysis of the liquid, however, shows it to be only about 87.3% H_3PO_4. The "missing" phosphorus is still present in the liquid, but as $H_4P_2O_7$, a compound called either *diphosphoric* or *pyrophosphoric acid*. A molecule of diphosphoric acid forms when a molecule of H_2O is eliminated from between two molecules of orthophosphoric acid, as shown in Figure 22-22. If a third molecule of orthophosphoric acid joins in by the elimination of another H_2O molecule, the product is $H_5P_3O_{10}$, triphosphoric acid, and so on. As a class, the chain-like phosphoric acid structures are called *polyphosphoric* acids, and their salts are called *polyphosphates*. Two especially important derivatives of the polyphosphoric acids present in living organisms are the substances known as ADP and ATP. The "A" portion of the acronym stands for adenosine, a combination of an organic base called adenine and a

Orthophosphoric acid
H_3PO_4

Diphosphoric acid
(pyrophosphoric acid)
$H_4P_2O_7$

Triphosphoric acid
$H_5P_3O_{10}$

▲ FIGURE 22-22
Formation of polyphosphoric acids
Removal of H_2O molecules results in P—O—P bridges.

five-carbon sugar called ribose. If this adenosine combination is linked to a diphosphate ion, the product is ADP, adenosine diphosphate. Addition of a phosphate ion to ADP yields ATP; adenosine triphosphate. These polyphosphates are described in more detail in Chapter 28.

Most phosphoric acid is made by the action of sulfuric acid on phosphate rock. The chemical equation for the reaction is given below:

$$3\,Ca_3(PO_4)_2 \cdot CaF_2(s) + 10\,H_2SO_4(concd\ aq) + 20\,H_2O(l) \longrightarrow$$
Fluorapatite

$$6\,H_3PO_4(aq) + 10\,CaSO_4 \cdot 2\,H_2O(s) + 2\,HF(aq) \qquad \textbf{(22.47)}$$
Gypsum

The HF is converted to insoluble Na_2SiF_6, and the gypsum is filtered off along with other insoluble impurities. The phosphoric acid is concentrated by evaporation. Phosphoric acid obtained by this "wet process" contains a variety of metal ions as impurities and is dark green or brown. Nevertheless, it is satisfactory for the manufacture of fertilizers and for metallurgical operations.

If H_3PO_4 from reaction (22.47) is used in place of H_2SO_4 to treat phosphate rock, the principal product is calcium dihydrogen phosphate. This compound, a fertilizer containing 20–21% P, is marketed under the name *triple superphosphate*:

$$3\,Ca_3(PO_4)_2 \cdot CaF_2(s) + 14\,H_3PO_4(concd\ aq) + 10\,H_2O(l) \longrightarrow$$

$$10\,Ca(H_2PO_4)_2 \cdot H_2O(s) + 2\,HF(aq) \qquad \textbf{(22.48)}$$
Triple superphosphate

Phosphorus and the Environment

Phosphates are widely used as fertilizers because phosphorus is an essential nutrient for plant growth. However, heavy fertilizer use may lead to phosphate pollution of lakes, ponds, and streams, causing an explosion of plant growth, particularly algae. The algae deplete the oxygen content of the water, eventually killing fish. This type of change, occurring in freshwater bodies as a result of their enrichment by nutrients, is called **eutrophication**. Eutrophication is a natural process that occurs over geological time periods, but it can be greatly accelerated by human activities, as shown in the photograph in the margin.

Natural sources of plant nutrients include animal wastes, decomposition of dead organic matter, and natural nitrogen fixation. Human sources include industrial wastes and municipal sewage plant effluents, in addition to fertilizer runoff. One way to reduce phosphate discharges into the environment is to remove them from the wastewater in sewage treatment plants. In the processing of sewage, polyphosphates are degraded to orthophosphates by bacterial action. The orthophosphates can then be precipitated, either as iron(III) phosphates, aluminum phosphates, or as calcium phosphate or hydroxyapatite $[Ca_5(OH)(PO_4)_3]$. The precipitating agents are generally aluminum sulfate, iron(III) chloride, or calcium hydroxide (slaked lime). In a fully equipped modern sewage treatment plant, up to 98% of the phosphates in sewage can be removed.

▲ The natural eutrophication of a lake is greatly accelerated by phosphates in wastewater and the agricultural runoff of fertilizers.

22-6 Hydrogen: A Unique Element

The first period has only two elements: hydrogen and helium. Hydrogen is quite reactive, but helium is inert. In the case of helium there is no difficulty relating its electronic structure and chemical properties to those of the other noble gases in group 18. Conversely, the chemical and physical properties of hydrogen cannot be correlated with any of the main groups in the periodic table. Hydrogen is truly unique and is best considered on its own.

The ground state electron configuration of a hydrogen atom ($1s^1$) is similar to that of the alkali metal atoms (ns^1), so it seems logical to place hydrogen in group 1. However, such a placement suggests that hydrogen will have properties similar to those of the alkali metals. This is not the case. Alkali metal atoms show a tendency to form M^+ ions. Although H^+ is known in acid–base chemistry, a hydrogen atom has a much greater tendency to form a covalent bond through the sharing of a pair of electrons. Hydrogen's electron configuration also resembles that of the halogens in being just one electron short of that of a noble gas. But unlike the halogens hydrogen rarely form H^- ions, except with the most active metals.

In still other ways hydrogen is like the group 14 elements—both have half-filled valence shells and similar electronegativity values. Thus, the groups H— and CH_3— have one unpaired electron and can form compounds such as LiH and $LiCH_3$. In spite of all of this, hydrogen is best treated as a group on its own.

Think how important hydrogen has been in the study of chemistry. John Dalton based atomic masses on a value of 1 for the H atom. Humphry Davy (1810) proposed that hydrogen is the key element in acids. We saw in Chapter 8 that theoretical studies of the H atom provided us with our modern view of atomic structure. In Chapter 11 we found that the H_2 molecule was the starting point for modern theories of chemical bonding. For all of its theoretical significance, though, hydrogen is also of great practical importance, as we will emphasize in this section.

Occurrence and Preparation

Hydrogen is a very minor component of the atmosphere, about 0.5 ppm at Earth's surface. At altitudes above 2500 km, the atmosphere is mostly atomic hydrogen at extremely low pressures. In the universe as a whole, hydrogen accounts for about 90% of the atoms and 75% of the mass. On Earth, *hydrogen occurs in more compounds than does any other element.*

The free element can easily be produced, but from only a few of its compounds. Our first choice might be H_2O—the most abundant hydrogen compound. To extract hydrogen from water means reducing the oxidation state of H from +1 in H_2O to 0 in H_2. This requires an appropriate *reducing* agent, such as carbon (coal or coke), carbon monoxide, or a hydrocarbon—particularly methane (natural gas). The first pair of reactions that follow are called the **water gas** reactions; they represent a way of making combustible gases—CO and H_2—from steam.

Water gas reactions: $C(s) + H_2O(g) \longrightarrow CO(g) + H_2(g)$ **(22.49)**

$CO(g) + H_2O(g) \longrightarrow CO_2(g) + H_2(g)$ **(22.50)**

Re-forming of methane: $CH_4(g) + H_2O(g) \longrightarrow CO(g) + 3\,H_2(g)$

Another source of $H_2(g)$ is as a by-product in petroleum refining.

Often we use methods in the chemical laboratory that are not commercially feasible. Electrolysis of water is one useful laboratory method for producing small quantities of $H_2(g)$. Another involves the reaction of active metals (recall the list in Table 5.3) in acidic solutions, an example of which is given below.

$$Zn(s) + 2\,H^+(aq) \longrightarrow Zn^{2+}(aq) + H_2(g)$$

Hydrogen Compounds

Hydrogen forms binary compounds, called **hydrides**, with most of the other elements. Binary hydrides are usually grouped into three broad categories: covalent, ionic, and metallic. *Covalent hydrides* are those formed between hydrogen and nonmetals. Some of these hydrides are simple molecules that can be formed by the direct union of hydrogen and the second element. Two examples are given below:

$$H_2(g) + Cl_2(g) \longrightarrow 2\,HCl(g)$$
$$3\,H_2(g) + N_2(g) \longrightarrow 2\,NH_3(g)$$

Ionic hydrides form between hydrogen and the most active metals, particularly those of groups 1 and 2. In these compounds hydrogen exists as the hydride *ion*, H^-.

$$2\,M(s) + H_2(g) \longrightarrow 2\,MH(s) \qquad M(s) + H_2(g) \longrightarrow MH_2(s)$$

(M is any group 1 metal)   (M is Ca, Sr, or Ba)

Ionic hydrides react vigorously with water to produce $H_2(g)$. CaH_2, a gray solid, has been used as a portable source of $H_2(g)$ for filling weather observation balloons.

$$CaH_2(s) + 2\,H_2O(l) \longrightarrow Ca(OH)_2(s) + 2\,H_2(g) \qquad \textbf{(22.51)}$$

The reaction between CaH_2 and water is highlighted in the margin.

Metal hydrides feature prominently in organic chemistry. For example, because CaH_2 reacts readily with water, it is often used to remove water from organic solvents. Sodium hydride (NaH) is used as a strong base in the synthesis of many organic compounds. Lithium hydride (LiH) is used to make lithium aluminum hydride ($LiAlH_4$), a powerful reducing agent used in organic chemistry (see page 1282).

Metallic hydrides are commonly formed with the transition elements—groups 3 to 12. A distinctive feature of these hydrides is that in many cases they are *nonstoichiometric*—the ratio of H atoms to metal atoms is variable, not fixed. This is because H atoms can enter the voids or holes among the metal atoms in a crystalline lattice and fill some but not others.

▲ **Reaction of CaH₂ with water**
The pink color of phenolphthalein indicator added to the water signals the production of $Ca(OH)_2$ in reaction (22.51).

Uses of Hydrogen

Hydrogen is not listed among the top chemicals produced, because only a small percentage is ever sold to customers. Most hydrogen is produced and used on the spot. In these terms, its most important use (about 42%) is in the manufacture of NH_3 (reaction 22.35). The next most important use of H_2 (about 38%) is in petroleum refining, where it is produced in some operations and consumed in others, such as in the production of the high-octane gasoline component, isooctane, from diisobutylene.

$$
\begin{array}{c}
\underset{\text{Diisobutylene}}{
\overset{\displaystyle CH_3}{\underset{\displaystyle CH_3\ \ H}{CH_3{-}\overset{|}{\underset{|}{C}}{-}C{=}C{-}CH_3}}
}
\ +\ H_2(g)\ \xrightarrow{\text{catalyst}}\
\underset{\text{Isooctane}}{
\overset{\displaystyle CH_3\ \ H\ \ H}{\underset{\displaystyle CH_3\ \ H\ \ CH_3}{CH_3{-}\overset{|}{\underset{|}{C}}{-}\overset{|}{\underset{|}{C}}{-}\overset{|}{\underset{|}{C}}{-}CH_3}}
}
$$

In similar reactions, called **hydrogenation reactions**, hydrogen atoms, in the presence of a catalyst, can be added to double or triple bonds in other molecules. This type of reaction, for example, converts liquid oleic acid, $C_{17}H_{33}COOH$, to solid stearic acid, $C_{17}H_{35}COOH$.

$$CH_3(CH_2)_7CH{=}CH(CH_2)_7COOH + H_2(g) \xrightarrow{\text{Ni}} CH_3(CH_2)_{16}COOH \qquad \textbf{(22.52)}$$

Oleic acid   Stearic acid

Similar reactions serve as the basis for converting oils that contain carbon-carbon double bonds, such as vegetable oils, into solid or semisolid fats, such as shortening, as described in the margin.

▲ Liquid vegetable oils contain long molecules with some carbon-to-carbon double bonds. When some of these double bonds in the molecules are hydrogenated to give carbon-to-carbon single bonds, the result is conversion of the liquid to a solid "partially hydrogenated vegetable oil."

TABLE 22.15   Some Uses of Hydrogen

Synthesis of
  ammonia, NH_3
  hydrogen chloride, HCl
  methanol, CH_3OH

Hydrogenation reactions in
  petroleum refining
  converting oils to fats

Reduction of metal oxides, such as those of iron, cobalt, nickel, copper, tungsten,   molybdenum
Metal cutting and welding with atomic and oxyhydrogen torches
Rocket fuel, usually $H_2(l)$ in combination with $O_2(l)$
Fuel cells for generating electricity, in combination with $O_2(g)$

Another important chemical manufacturing process that uses hydrogen is the synthesis of methyl alcohol (methanol), an alternative fuel.

$$CO(g) + 2\,H_2(g) \xrightarrow{\text{catalyst}} CH_3OH(g)$$

Hydrogen gas is an excellent reducing agent and in some cases is used to produce metals from their oxide ores. For example, the following reaction is used, at 850 °C, to produce tungsten metal from its oxide:

$$WO_3(s) + 3\,H_2(g) \longrightarrow W(s) + 3\,H_2O(g)$$

The uses of hydrogen described here, together with several others, are listed in Table 22.15.

Hydrogen and the Environment—A Hydrogen Economy

As we contemplate the eventual decline of the world's supplies of fossil fuels, hydrogen emerges as an attractive means of storing, transporting, and using energy. For example, when an automobile engine burns hydrogen rather than gasoline, its exhaust is essentially pollution free. The range of supersonic aircraft could be increased if they used liquid hydrogen as a fuel. A hypersonic airplane (the space plane) might also become possible. As we learned in Chapter 20, one method of using hydrogen that is already available combines H_2 and O_2 to form H_2O in an electrochemical fuel cell, which converts chemical energy directly to electricity. The subsequent conversion of electrical energy to mechanical energy (work) can be carried out much more efficiently than can the conversion of heat to mechanical energy.

The basic problems are in finding a cheap source of hydrogen and an effective means of storing it. One possibility is to use hydrogen made by the electrolysis of seawater. This possibility requires an abundant energy source, however—perhaps nuclear fusion energy if it can be developed. Another alternative is the thermal decomposition of water. The problem here is that even at 2000 °C, water is only about 1% decomposed. What is needed is a thermochemical cycle, a series of reactions that have as their overall reaction: $2\,H_2O(l) \longrightarrow 2\,H_2(g) + O_2(g)$. Ideally, no single reaction in the cycle would require a very high temperature. Still another alternative being studied involves the use of solar energy to decompose water—photodecomposition.

Storage of gaseous hydrogen is difficult because of the bulk of the gas. When liquefied, hydrogen occupies a much smaller volume, but because of its very low boiling point (-253 °C), the $H_2(l)$ must be stored at very low temperatures. Also, hydrogen must be maintained out of contact with oxygen or air, with which it forms explosive mixtures. One approach may be to dissolve

$H_2(g)$ in a metal or metal alloy, such as an iron–titanium alloy. The gas can be released by mild heating. In an automobile, this storage system would replace the gasoline tank. The heat required to release hydrogen from the metal would come from the engine exhaust.

If the problems described here can be solved, not only could hydrogen be used to supplant gasoline as a fuel for transportation but it could also replace natural gas for space heating. Because H_2 is a good reducing agent, it could replace carbon (as coal or coke) in metallurgical processes and, of course, it would be abundantly available for reaction with N_2 to produce NH_3 for the manufacture of fertilizers. The combination of all these potential uses of hydrogen could lead to a fundamental change in our way of life and give rise to what is called a **hydrogen economy**.

Mastering**CHEMISTRY**   www.masteringchemistry.com

Ozone plays a vital role in protecting life on Earth because it absorbs potentially harmful ultraviolet radiation and also helps to maintain a heat balance in the atmosphere. For a discussion of some ozone-producing and ozone-destroying reactions occurring in the atmosphere, and the impact made by human activities, go to the Focus On feature for Chapter 22, The Ozone Layer and Its Environmental Role, on the MasteringChemistry site.

Summary

22-1 Periodic Trends in Bonding—The bonding in the fluorides changes from ionic bonding to covalent bonding as we move from left to right across the periodic table. In the transition from ionic bonding to covalent bonding, we pass through a group of elements that react with fluorine to give network covalent compounds. The bonding in the oxides exhibits a similar transition. The acid–base character of the oxides also changes as we move left to right across the periodic table. Oxides of the metallic elements (on the left of the periodic table) are generally basic, and oxides of nonmetallic elements are generally acidic. Between these two extremes are the amphoteric oxides derived from some of the elements in groups 2 and 13.

22-2 Group 18: The Noble Gases—The **noble gases** (group 18) form very few compounds. The chemistry of this group is concerned mainly with compounds of xenon and the two most electronegative elements, F and O.

22-3 Group 17: The Halogens—The **halogens** (group 17) are among the most reactive elements, forming compounds with all elements in the periodic table. Electrode potential diagrams are a way to summarize the oxidation–reduction chemistry of an element, and are introduced for the oxidation states of chlorine. Often electrode potentials for reduction processes not specifically represented in a diagram can be obtained by means of calculations based on the relationship $\Delta G° = -nFE°$. The oxoacids and oxoanions of chlorine are described in terms of the methods required to prepare them, their acid–base properties, their strengths as oxidizing or reducing agents, and their structures. In this group we observe several differences between the first and higher members of a group of the periodic table. Thus, for example, we note the failure

of fluorine to form stable oxoacids. The later members of the group form **interhalogens**, such as ICl, and **polyhalide ions**, such as I_3^-.

22-4 Group 16: The Oxygen Family—The oxygen family (group 16) includes two reactive nonmetals (O and S), two metalloids (Se and Te), and a metal (Po). Again note the difference in properties between the first and second member of this group. The ability to form strong π bonds with $2p$ orbitals by oxygen contrasts with the single bond chain-like structures favored by sulfur. Sulfur is extracted from underground sources by the **Frasch process**. Sulfur forms the important acids H_2SO_4 and H_2SO_3. The salt $NaHSO_4$ is an example of an **acid salt** since the anion HSO_4^- possesses an ionizable proton.

22-5 Group 15: The Nitrogen Family—The nitrogen family (group 15) shows the progression of properties from nonmetallic to metallic within a group. The chemistry of nitrogen and phosphorus is wide and diverse. The ammonium salts of nitric acid are used in fertilizers. Phosphate ions are also important in fertilizers and in living systems as ADP or ATP. Also, phosphates are implicated in the pollution process of **eutrophication**.

22-6 Hydrogen: A Unique Element—Hydrogen is a unique element that does not fit into any group. Hydrogen can occur as $H^+(aq)$ in aqueous solutions and as H^- in **hydrides,** for example, $NaH(s)$. Hydrogen is a good reducing agent and can be used in **hydrogenation reactions** (equation 22.52). Hydrogen can be obtained from water by electrolysis and also by the **water gas** reaction (equation 22.49). Hydrogen gas is potentially a very promising nonpolluting means of storing, transporting, and using energy and could lead to the so-called **hydrogen economy**.

Integrative Example

For use in analytical chemistry, sodium thiosulfate solutions must be carefully prepared. In particular, the solutions must be kept from becoming acidic. In strongly acidic solutions, thiosulfate ion disproportionates into $SO_2(g)$ and $S_8(s)$.

▲ **Decomposition of thlosulfate lon**
When an aqueous solution of $Na_2S_2O_3$ is acidified, the sulfur is in the colloidal state when first formed (right).

Show that the disproportionation of $S_2O_3^{2-}(aq)$ is spontaneous for standard-state conditions in acidic solution, but not in basic solution.

Analyze

Begin by writing the half-equations and an overall equation for the disproportionation reaction. Determine $E°_{cell}$ for the reaction and thus whether the reaction is spontaneous for standard-state conditions in acidic solution. Then make a qualitative assessment of whether the reaction is likely to be more spontaneous or less spontaneous in basic solution.

Solve

Base the overall equation on the verbal description of the reaction.

Reduction:
$$4 S_2O_3^{2-}(aq) + 24 H^+(aq) + 16 e^- \longrightarrow S_8(s) + 12 H_2O(l)$$

Oxidation:
$$4\{S_2O_3^{2-}(aq) + H_2O(l) \longrightarrow 2 SO_2(g) + 2 H^+(aq) + 4 e^-\}$$

Overall:
$$8 S_2O_3^{2-}(aq) + 16 H^+(aq) \longrightarrow$$
$$S_8(s) + 8 SO_2(g) + 8 H_2O(l) \quad \textbf{(22.53)}$$

To determine $E°_{cell}$ for the reaction (22.53), use data from Figure 22-13. That figure gives an $E°$ value for the reduction half-reaction (0.465 V) but no value for the oxidation. To obtain this missing $E°$, use additional data from Figure 22-13 together with the method of Example 22-1. That is, the sum of the half-equation

$$4 SO_2(g) + 4 H^+(aq) + 6 e^- \longrightarrow S_4O_6^{2-}(aq) + 2 H_2O(l)$$
$$\Delta G° = -6FE° = -6F \times 0.507 \text{ V}$$

and the half-equation

$$S_4O_6^{2-}(aq) + 2 e^- \longrightarrow 2 S_2O_3^{2-}(aq)$$
$$\Delta G° = -2FE° = -2F \times 0.080 \text{ V}$$

yields the desired new half-equation and its $E°$ value.

$$4 SO_2(g) + 4 H^+(aq) + 8 e^- \longrightarrow$$
$$2 S_2O_3^{2-}(aq) + 2 H_2O(l)$$
$$\Delta G° = -F[(6 \times 0.507) + (2 \times 0.080)] \text{ V}$$
$$\Delta G° = -8FE° = -F(3.202) \text{ V}$$
$$E° = (3.202/8) \text{ V} = 0.400 \text{ V}$$

Now we can calculate $E°_{cell}$ for reaction (22.53).

$$E°_{cell} = E°(\text{reduction}) - E°(\text{oxidation})$$
$$= 0.465 \text{ V} - 0.400 \text{ V} = 0.065 \text{ V}$$

The disproportionation is spontaneous for standard-state conditions in acidic solution.

Increasing $[OH^-]$, as would be the case in making the solution basic, means decreasing $[H^+]$. In fact, $OH^- = 1$ M corresponds to $[H^+] = 1 \times 10^{-14}$ M. Because equation (22.53) has $H^+(aq)$ on the *left* side of the equation, a decrease in $[H^+]$ favors the *reverse* reaction (by Le Châtelier's principle). At some point before the solution becomes basic, the forward reaction is no longer spontaneous.

Assess

This calculation demonstrated in a qualitative way that $S_2O_3^{2-}(aq)$ is stable in basic solutions and spontaneously disproportionates in acidic solutions. To determine the pH at which the disproportionation becomes spontaneous, one can use the Nernst equation, as seen in Exercise 100.

PRACTICE EXAMPLE A: Use information from Figure 22-17 to decide whether the nitrite anion, NO_2^-, disproportionates spontaneously in basic solution to NO_3^- and NO. Assume standard-state conditions.

PRACTICE EXAMPLE B: Does HNO_2 spontaneously disproportionate to NO_3^- and NO in acidic solution? Assume standard-state conditions. [*Hint*: Use data from Figure 22-17.]

Exercises

Periodic Trends in Bonding and Acid–Base Character of Oxides

1. Give the formula of the stable fluoride formed by Li, Be, B, C, N, and O. For these fluorides, describe the variation in the bonding that occurs as we move from left to right across the period.

2. Fluorine is able to stabilize elements in very high oxidation states. For each of the elements Na, Mg, Al, Si, P, S, and Cl, give the formula of the highest-order fluoride that is known to exist. Then, describe the variation in bonding that occurs as we move from left to right across the period.

3. The oxides of the phosphorus(III), antimony(III), and bismuth(III) are P_4O_6, Sb_4O_6, and Bi_2O_3. Only one of these oxides is amphoteric. Which one? Which of these oxides is most acidic? Which is most basic?

4. The oxides of the selenium(IV) and tellurium(IV) are SeO_2 and TeO_2. One of these oxides is amphoteric and one is acidic. Which is which?

The Noble Gases

5. A 55 L cylinder contains Ar at 145 atm and 26 °C. What minimum volume of air at STP must have been liquefied and distilled to produce this Ar? Air contains 0.934% Ar, by volume.

6. Some sources of natural gas contain 8% He by volume. How many liters of such a natural gas must be processed at STP to produce 5.00 g of He? [*Hint:* What are the apparent molar masses of air and of the $He-O_2$ mixture?]

7. Use VSEPR theory to predict the probable geometric structures of **(a)** XeO_3; **(b)** XeO_4; **(c)** XeF_5^+.

8. Use VSEPR theory to predict the probable geometric structures of the molecules **(a)** O_2XeF_2; **(b)** O_3XeF_2; **(c)** $OXeF_4$.

9. Write a chemical equation for the hydrolysis of XeF_4 that yields XeO_3, Xe, O_2, and HF as products.

10. Write a chemical equation for the hydrolysis in alkaline solution of XeF_6 that yields XeO_6^{4-}, Xe, O_2, F^-, and H_2O as products.

11. Provide an explanation for the observation that helium, neon, and argon do not react directly with fluorine.

12. Provide an explanation for the inability of O_2 to react directly with xenon.

The Halogens

13. Freshly prepared solutions containing iodide ion are colorless, but over time they usually turn yellow. Describe a plausible chemical reaction (or reactions) to account for this observation.

14. Fluorine can be prepared by the reaction of hexafluoromanganate(IV) ion, MnF_6^{2-}, with antimony pentafluoride to produce manganese(IV) fluoride and SbF_6^-, followed by the disproportionation of manganese(IV) fluoride to manganese(III) fluoride and $F_2(g)$. Write chemical equations for these two reactions.

15. Make a general prediction about which of the halogen elements, F_2, Cl_2, Br_2, or I_2, displaces other halogens from a solution of halide ions. Which of the halogens is able to displace $O_2(g)$ from water? Which is able to displace $H_2(g)$ from water?

16. The following properties of astatine have been measured or estimated: **(a)** covalent radius; **(b)** ionic radius (At^-); **(c)** first ionization energy; **(d)** electron affinity; **(e)** electronegativity; **(f)** standard reduction potential. Based on periodic relationships and data in Table 22.4, what values would you expect for these properties?

17. The abundance of F^- in seawater is 1 g F^- per ton of seawater. Suppose that a commercially feasible method could be found to extract fluorine from seawater.
(a) What mass of F_2 could be obtained from 1 km^3 of seawater ($d = 1.03$ g cm^{-3})?
(b) Would the process resemble that for extracting bromine from seawater? Explain.

18. Fluorine is produced chiefly from fluorite, CaF_2. Fluorine can also be obtained as a by-product of the production of phosphate fertilizers, derived from phosphate rock [$3 Ca_3(PO_4)_2 \cdot CaF_2$]. What is the maximum mass of fluorine that could be extracted as a by-product from 1.00×10^3 kg of phosphate rock?

19. Show by calculation whether the disproportionation of chlorine gas to chlorate and chloride ions will occur under standard-state conditions in an acidic solution.

20. Show by calculation whether the reaction
$2 HOCl(aq) \longrightarrow HClO_2(aq) + H^+(aq) + Cl^-(aq)$
will go essentially to completion as written for standard-state conditions.

21. Predict the geometric structures of **(a)** $\underline{Br}F_3$; **(b)** $\underline{I}F_5$; **(c)** $Cl_3\underline{I}F^-$. (Central atom underlined.)

22. Which of the following species has a linear structure: $\underline{Cl}F_2^+$, $IBr\underline{F}^-$, $O\underline{Cl}_2$, $\underline{Cl}F_3$, or $\underline{S}F_4$? (Central atom underlined.) Do any two of these species have the same structure?

23. When iodine is added to an aqueous solution of iodide ion, the I_3^- ion is formed, according to the reaction below:

$$I_2(aq) + I^-(aq) \rightleftharpoons I_3^-(aq)$$

The equilibrium constant for the reaction above is $K = 7.7 \times 10^2$ at 25 °C.

(a) What is $E°$ for the reaction above?

(b) If a 0.0010 mol sample of I_2 is added to 1.0 L of 0.0050 M NaI(aq) at 25 °C, then what fraction of the I_2 remains unreacted at equilibrium?

24. The trichloride ion, Cl_3^-, is not very stable in aqueous solution. The equilibrium constant for the following dissociation reaction is 5.5 at 25 °C:

$$Cl_3^-(aq) \rightleftharpoons Cl^-(aq) + Cl_2(aq)$$

(a) Draw a Lewis structure for the Cl_3^- ion and predict the geometry.

(b) Calculate the equilibrium concentration of Cl_3^- if 0.0010 moles each of KCl and Cl_2 are dissolved in water at 25 °C to make 1.0 L of solution.

Oxygen

25. Each of the following compounds decomposes to produce $O_2(g)$ when heated: (a) HgO(s); (b) $KClO_4(s)$. Write plausible equations for these reactions.

26. $O_3(g)$ is a powerful oxidizing agent. Write equations to represent oxidation of (a) I^- to I_2 in acidic solution; (b) sulfur in the presence of moisture to sulfuric acid; (c) $[Fe(CN)_6]^{4-}$ to $[Fe(CN)_6]^{3-}$ in basic solution. In each case $O_3(g)$ is reduced to $O_2(g)$.

27. *Without performing detailed calculations*, determine which of the following compounds has the greatest percent oxygen by mass: dinitrogen tetroxide, aluminum oxide, tetraphosphorus hexoxide, or carbon dioxide.

28. *Without performing detailed calculations*, determine which decomposition yields the most $O_2(g)$ (a) per mole and (b) per gram of substance.
    (1) ammonium nitrate $\longrightarrow$
    $\qquad$ nitrogen + oxygen + water
    (2) hydrogen peroxide $\longrightarrow$ oxygen + water
    (3) potassium chlorate $\longrightarrow$
    $\qquad$ potassium chloride + oxygen

29. The natural abundance of O_3 in unpolluted air at ground level is about 0.04 parts per million (ppm) by volume. What is the approximate partial pressure of O_3 under these conditions, expressed in millimeters of mercury?

30. A typical concentration of O_3 in the ozone layer is 5×10^{12} O_3 molecules cm^{-3}. What is the partial pressure of O_3, expressed in millimeters of mercury, in that layer? Assume a temperature of 220 K.

31. Explain why the volumes of $H_2(g)$ and $O_2(g)$ obtained in the electrolysis of water are not the same.

32. In the electrolysis of a sample of water 22.83 mL of $O_2(g)$ was collected at 25.0 °C at an oxygen partial pressure of 736.7 mmHg. Determine the mass of water that was decomposed.

33. Hydrogen peroxide is a somewhat stronger acid than water. For the ionization

$$H_2O_2(aq) + H_2O(l) \longrightarrow H_3O^+(aq) + HO_2^-(aq)$$

$pK_a = 11.75$. Calculate the expected pH of a typical antiseptic solution that is 3.0% H_2O_2 by mass.

34. In water, O^{2-} is a strong base. If 50.0 mg of Li_2O is dissolved in 750.0 mL of aqueous solution, what will be the pH of the solution?

35. The conversion of $O_2(g)$ to $O_3(g)$ can be accomplished in an electric discharge, $3 O_2(g) \longrightarrow 2 O_3(g)$. Use a bond dissociation energy of 498 kJ mol^{-1} for $O_2(g)$ and data from Appendix D to calculate an average oxygen-to-oxygen bond energy in $O_3(g)$.

36. Estimate the average bond energy in $O_3(g)$ from the structure on page 416 and data in Table 10.3. Compare this result with that of Exercise 35.

37. Use Lewis structures and other information to explain the observation that
    (a) H_2S is a gas at room temperature, whereas H_2O is a liquid.
    (b) O_3 is diamagnetic.

38. Use Lewis structures and other information to explain the observation that
    (a) the oxygen-to-oxygen bond lengths in O_2, O_3, and H_2O_2 are 121, 128, and 148 pm, respectively.
    (b) the oxygen-to-oxygen bond length of O_2 is 121 pm and for O_2^+ is 112 pm. Why is the bond length for O_2^+ so much shorter than for O_2?

39. Which of the following reactions are likely to go to completion or very nearly so?
    (a) $H_2O_2(aq) + 2 I^-(aq) + 2 H^+(aq) \longrightarrow$
    $\qquad I_2(s) + 2 H_2O(l)$
    (b) $O_2(g) + 2 H_2O(l) + 4 Cl^-(aq) \longrightarrow$
    $\qquad 2 Cl_2(g) + 4 OH^-(aq)$
    (c) $O_3(g) + Pb^{2+}(aq) + H_2O(l) \longrightarrow$
    $\qquad PbO_2(s) + 2 H^+(aq) + O_2(g)$
    (d) $HO_2^-(aq) + 2 Br^-(aq) + H_2O(l) \longrightarrow$
    $\qquad 3 OH^-(aq) + Br_2(l)$

40. Each of the following compounds produces $O_2(g)$ when strongly heated: (a) HgO(s); (b) $KClO_4(s)$; (c) $Hg(NO_3)_2(s)$; (d) $H_2O_2(aq)$. Write a plausible equation for the reaction that occurs in each instance.

41. In the laboratory, small quantities of oxygen gas can be prepared by heating potassium chlorate, $KClO_3(s)$, in the presence of $MnO_2(s)$, a catalyst. What volume of oxygen, measured at 25 °C and 101 kPa, is obtained from the decomposition of 1.0 g $KClO_3(s)$? [*Hint*: The other product of the reaction is KCl(s).]

42. Joseph Priestley, a British chemist, was credited with the discovering oxygen in 1774. In his experiments, he generated oxygen gas by heating HgO(s). The other product of the decomposition reaction is Hg(l). What volume of wet $O_2(g)$ is obtained from the decomposition of 1.0 g HgO(s), if the gas is collected over water at 25 °C and a barometric pressure of 756 mmHg? The vapor pressure of water is 23.76 mmHg at 25 °C.

Sulfur

43. Give an appropriate name to each of the following compounds: **(a)** ZnS; **(b)** $KHSO_3$; **(c)** $K_2S_2O_3$; **(d)** SF_4.

44. Give an appropriate formula for each of the following compounds: **(a)** calcium sulfate dihydrate; **(b)** hydrosulfuric acid; **(c)** sodium hydrogen sulfate; **(d)** disulfuric acid.

45. Give a specific example of a chemical equation that illustrates the
(a) reaction of a metal sulfide with HCl(aq);
(b) action of a *nonoxidizing* acid on a metal sulfite;
(c) oxidation of $SO_2(aq)$ to $SO_4^{2-}(aq)$ by $MnO_2(s)$ in acidic solution;
(d) disproportionation of $S_2O_3^{2-}$ in acidic solution.

46. Show how you would use elemental sulfur, chlorine gas, metallic sodium, water, and air to produce aqueous solutions containing **(a)** Na_2SO_3; **(b)** Na_2SO_4; **(c)** $Na_2S_2O_3$. [*Hint:* You will have to use information from other chapters as well as this one.]

47. Describe a chemical test you could use to determine whether a white solid is Na_2SO_4 or $Na_2S_2O_3$. Explain the basis of this test using a chemical equation or equations.

48. Explain why sulfur can occur naturally as sulfates, but not as sulfites.

49. Salts like $NaHSO_4$ are called *acid* salts because their anions undergo further ionization. What is the pH of 250 mL of water solution containing 12.5 g $NaHSO_4$? [*Hint:* Use data from Chapter 16, as necessary.]

50. What mass of Na_2SO_3 was present in a sample that required 26.50 mL of 0.0510 M $KMnO_4$ for its oxidation to Na_2SO_4 in an acidic solution? MnO_4^- is reduced to Mn^{2+}.

51. A 1.100 g sample of copper ore is dissolved, and the $Cu^{2+}(aq)$ is treated with excess KI. The liberated I_3^- requires 12.12 mL of 0.1000 M $Na_2S_2O_3$ for its titration. What is the mass percent copper in the ore?

52. A 25.0 L sample of a natural gas, measured at 25 °C. and 740.0 Torr, is bubbled through $Pb^{2+}(aq)$, yielding 0.535 g of PbS(s). What mass of sulfur can be recovered per cubic meter of this natural gas?

53. What is the oxidation state of sulfur in the following compounds? **(a)** SF_4; **(b)** S_2F_{10}; **(c)** H_2S; **(d)** $CaSO_3$.

54. What is the oxidation state of sulfur in the following compounds? **(a)** S_2Br_2; **(b)** SCl_2; **(c)** $Na_2S_2O_3$; **(d)** $(NH_4)_2S_4O_6$.

Nitrogen Family

55. Write balanced equations for the following important commercial reactions involving nitrogen and its compounds.
(a) the principal artificial method of fixing atmospheric N_2
(b) oxidation of ammonia to NO
(c) preparation of nitric acid from NO

56. When heated, each of the following substances decomposes to the products indicated. Write balanced equations for these reactions.
(a) $NH_4NO_3(s)$ to $N_2(g)$, $O_2(g)$, and $H_2O(g)$
(b) $NaNO_3(s)$ to sodium nitrite and oxygen gas
(c) $Pb(NO_3)_2(s)$ to lead(II) oxide, nitrogen dioxide, and oxygen

57. Sodium nitrite can be made by passing oxygen and nitrogen monoxide gases into an aqueous solution of sodium carbonate. Write a balanced equation for this reaction.

58. Concentrated $HNO_3(aq)$ used in laboratories is usually 15 M HNO_3 and has a density of 1.41 g mL^{-1}. What is the percent by mass of HNO_3 in this concentrated acid?

59. In 1968, before pollution controls were introduced, over 75 billion gallons of gasoline were used in the United States as a motor fuel. Assume an emission of oxides of nitrogen of 5 grams per vehicle mile and an average mileage of 15 miles per gallon of gasoline. How many kilograms of nitrogen oxides were released into the atmosphere in the United States in 1968?

60. One reaction that competes with reaction (22.41), the Ostwald process, is the reaction of gaseous ammonia and nitrogen monoxide to produce gaseous nitrogen and gaseous water. Use data from Appendix D to determine $\Delta H°$ for this reaction, per mole of ammonia consumed.

61. Use information from this chapter and previous chapters to write chemical equations to represent the following:
(a) equilibrium between nitrogen dioxide and dinitrogen tetroxide in the gaseous state
(b) the reduction of nitrous acid by $N_2H_5^+$ forming hydrazoic acid, followed by the reduction of additional nitrous acid by the hydrazoic acid, yielding nitrogen and dinitrogen monoxide
(c) the neutralization of $H_3PO_4(aq)$ to the second equivalence point by $NH_3(aq)$

62. Use information from this chapter and previous chapters to write plausible chemical equations to represent the following:
(a) the reaction of silver metal with $HNO_3(aq)$
(b) the complete combustion of the rocket fuel, unsymmetrical dimethylhydrazine, $(CH_3)_2NNH_2$
(c) the preparation of sodium triphosphate by heating a mixture of sodium dihydrogen phosphate and sodium hydrogen phosphate.

63. Draw plausible Lewis structures for
(a) dimethylhydrazine, $(CH_3)_2NNH_2$
(b) nitryl chloride, $ClNO_2$ (Central atom is N)
(c) phosphorous acid, a *diprotic* acid with the empirical formula H_3PO_3

64. Both nitramide and hyponitrous acid have the formula $H_2N_2O_2$. Hyponitrous acid is a weak diprotic acid; nitramide contains the amide group ($-NH_2$). Draw plausible Lewis structures for these two substances.

65. Supply an appropriate name for each of the following: **(a)** HPO_4^{2-}; **(b)** $Ca_2P_2O_7$; **(c)** $H_6P_4O_{13}$.

66. Write an appropriate formula for each of the following: **(a)** hydroxylamine; **(b)** calcium hydrogen phosphate; **(c)** lithium nitride.

67. Use Figure 22-17 to establish $E°$ for the reduction of N_2O_4 to NO in an acidic solution.

68. Use Figure 22-17 to establish $E°$ for the reduction of NO_3^- to NO_2^- in a basic solution.

69. All the group 15 elements form trifluorides, but nitrogen is the only group 15 element that does not form a pentafluoride.

(a) Suggest a reason why nitrogen does not form a pentafluoride.
(b) The observed bond angle in NF_3 is approximately 102.5 °C. Use VSEPR theory to rationalize the structure of the NF_3 molecule.

70. The structures of the NH_3 and NF_3 molecules are similar, yet the dipole moment for the NH_3 molecule is rather large (1.47 debye) and that of the NF_3 molecule is rather small (0.24 debye). Provide an explanation for this difference in the dipole moments.

Hydrogen

71. Use data from Table 7.2 (page 269) to calculate the standard enthalpies of combustion of the four alkane hydrocarbons listed there.

72. Based on the results of Exercise 71, which alkane evolves the greatest amount of heat upon combustion on **(a)** a *per mole* basis and **(b)** a *per gram* basis? Which is the most desirable alkane from the standpoint of reducing the emission of carbon dioxide to the atmosphere? Explain.

73. Write chemical equations for the following reactions:
    (a) the displacement of $H_2(g)$ from HCl(aq) by Al(s)
    (b) the re-forming of propane gas (C_3H_8) with steam
    (c) the reduction of $MnO_2(s)$ to Mn(s) with $H_2(g)$

74. Write equations to show how to prepare $H_2(g)$ from each of the following substances: **(a)** H_2O; **(b)** HI(aq); **(c)** Mg(s); **(d)** CO(g). Use other common laboratory reactants as necessary, that is, water, acids or bases, metals, and so on.

75. $CaH_2(s)$ reacts with water to produce $Ca(OH)_2$ and $H_2(g)$. Ca(s) reacts with water to produce the same products. Na(s) reacts with water to form NaOH and $H_2(g)$. *Without doing detailed calculations, determine*

(a) which of these reactions produces the most H_2 per liter of water used, and **(b)** which solid—CaH_2, Ca, or Na—produces the most $H_2(g)$ *per gram* of the solid.

76. What volume of $H_2(g)$ at 25 °C and 752 mmHg is required to hydrogenate oleic acid, $C_{17}H_{33}COOH(l)$, to produce one mole of stearic acid, $C_{17}H_{35}COOH(s)$? Assume reaction (22.52) proceeds with a 95% yield.

77. *Without doing detailed calculations*, explain in which of the following materials you would expect to find the greatest mass percent of hydrogen: seawater, the atmosphere, natural gas (CH_4), ammonia.

78. How many grams of $CaH_2(s)$ are required to generate sufficient $H_2(g)$ to fill a 235 L weather observation balloon at 722 mmHg and 19.7 °C?

$$CaH_2(s) + 2 H_2O(l) \longrightarrow Ca(OH)_2(aq) + 2 H_2(g)$$

79. The amide anion NH_2^- is a very strong base. On the basis of molecular orbital theory, would you expect NH_2^- to be linear or bent?

80. On the basis of molecular orbital theory, would you expect NH_2^+ to be linear or bent?

Integrative and Advanced Exercises

81. The boiling points of oxygen and argon are −183 °C and −189 °C, respectively. Because the boiling points are so similar, argon obtained from the fractional distillation of liquid air is contaminated with oxygen. The following three-step procedure can be used to obtain pure argon from the oxygen-contaminated sample:
    (1) Excess hydrogen is added to the mixture and then the mixture is ignited.
    (2) The mixture from step (1) is then passed over hot copper(II) oxide.
    (3) The mixture from step (2) is passed over a dehydrated zeolite material (see Chapter 21).
    Explain the purpose of each step, writing chemical equations for any reactions that occur.

82. In 1988, G. J. Schrobilgen, professor of chemistry at McMaster University in Canada, reported the synthesis of an ionic compound, $[HCNKrF][AsF_6]$, which consists of $HCNKrF^+$ and AsF_6^- ions. In the $HCNKrF^+$ ion, the krypton is covalently bonded to both fluorine and nitrogen. Draw Lewis structures for these ions, and estimate the bond angles.

83. Suppose that no attempt is made to separate the $H_2(g)$ and $O_2(g)$ produced by the electrolysis of water. What volume of a H_2—O_2 mixture, saturated

with $H_2O(g)$ and collected at 23 °C and 755 mmHg, would be produced by electrolyzing 17.3 g water? Assume that the water vapor pressure of the dilute electrolyte solution is 20.5 mmHg.

84. The photograph was taken after a few drops of a deep-purple acidic solution of $KMnO_4(aq)$ were added to $NaNO_3(aq)$ (left) and to $NaNO_2(aq)$ (right). Explain the difference in the results shown.

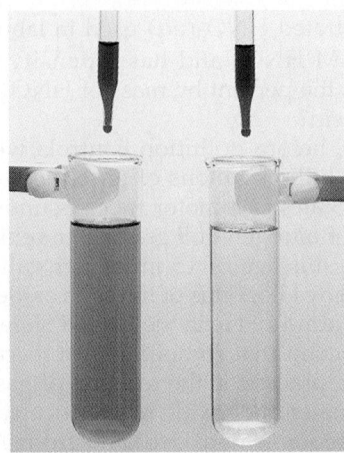

85. Zn can reduce NO_3^- to $NH_3(g)$ in basic solution. (The following equation is not balanced.)

$$NO_3^-(aq) + Zn(s) + OH^-(aq) + H_2O(l) \longrightarrow$$
$$[Zn(OH)_4]^{2-}(aq) + NH_3(g)$$

The NH_3 can be neutralized with an excess of $HCl(aq)$. Then, the unreacted HCl can be titrated with NaOH. In this way a quantitative determination of NO_3^- can be achieved. A 25.00 mL sample of nitrate solution was treated with zinc in basic solution. The $NH_3(g)$ was passed into 50.00 mL of 0.1500 M HCl. The excess HCl required 32.10 mL of 0.1000 M NaOH for its titration. What was the $[NO_3^-]$ in the original sample?

86. Oxygen atoms are an important constituent of the thermosphere, a layer of the atmosphere with temperatures up to 1500 K. Calculate the average translational kinetic energy of O atoms at 1500 K.

87. One reaction for the production of adipic acid, $HOOC(CH_2)_4COOH$, used in the manufacture of nylon, involves the oxidation of cyclohexanone, $C_6H_{10}O$, in a nitric acid solution. Assume that dinitrogen monoxide is also formed, and write a balanced equation for this reaction.

88. Despite the fact that it has the higher molecular mass, XeO_4 exists as a gas at 298 K, whereas XeO_3 is a solid. Give a plausible explanation for this observation.

89. The text mentions that ammonium perchlorate is an explosion hazard. Assuming that NH_4ClO_4 is the sole reactant in the explosion, write a plausible equation(s) to represent the reaction that occurs.

90. The following bond energies are given for 298 K: O_2, 498; N_2, 946; F_2, 159; Cl_2, 243; ClF, 251; OF (in OF_2), 213; ClO (in Cl_2O), 205; and NF (in NF_3), 280 kJ mol^{-1}. Calculate ΔH_f° at 298 K for 1 mol of (a) ClF(g); (b) $OF_2(g)$; (c) $Cl_2O(g)$; (d) $NF_3(g)$.

91. The standard electrode potential of fluorine cannot be measured directly because F_2 reacts with water, displacing O_2. Using thermodynamic data from Appendix D, obtain a value of $E^\circ_{F_2/F^-}$.

92. Polonium is the only element known to crystallize in the simple cubic form. In this structure, the interatomic distance between a Po atom and each of its six nearest neighbors is 335 pm. Use this description of the crystal structure to estimate the density of polonium.

93. Refer to Figure 11-26 to arrange the following species in the expected order of increasing (a) bond length and (b) bond strength (energy): $O_2, O_2^+, O_2^-, O_2^{2-}$. State the basis of your expectation.

94. One reaction of a chlorofluorocarbon implicated in the destruction of stratospheric ozone is $CFCl_3 + hv \longrightarrow CFCl_2 + Cl$.
    (a) What is the energy of the photons (hv) required to bring about this reaction, expressed in kilojoules per mole?
    (b) What is the frequency and wavelength of the light necessary to produce the reaction? In what portion of the electromagnetic spectrum is this light found?

95. The composition of a phosphate mineral can be expressed as % P, % P_4O_{10}, or % BPL [bone phosphate of lime, $Ca_3(PO_4)_2$].
    (a) Show that % P = $0.436 \times$ (% P_4O_{10}) and % BPL = $2.185 \times$ (% P_4O_{10}).

(b) What is the significance of a % BPL greater than 100?
(c) What is the % BPL of a typical phosphate rock?

96. Estimate the percent dissociation of $Cl_2(g)$ into Cl(g) at 1 atm total pressure and 1000 K. Use data from Appendix D and equations found elsewhere in this text, as necessary.

97. *Peroxonitrous acid* is an unstable intermediate formed in the oxidation of HNO_2 by H_2O_2. It has the same formula as *nitric acid*, HNO_3. Show how you would expect peroxonitrous and nitric acids to differ in structure.

98. The structure of $N(SiH_3)_3$ involves a planar arrangement of N and Si atoms, whereas that of the related compound $N(CH_3)_3$ has a pyramidal arrangement of N and C atoms. Propose bonding schemes for these molecules that are consistent with this observation.

99. In the extraction of bromine from seawater (reaction 22.3), seawater is first brought to a pH of 3.5 and then treated with $Cl_2(g)$. In practice, the pH of the seawater is adjusted with H_2SO_4, and the mass of chlorine used is 15% in excess of the theoretical. Assuming a seawater sample with an initial pH of 7.0, a density of 1.03 g cm^{-3}, and a bromine content of 70 ppm by mass, what masses of H_2SO_4 and Cl_2 would be used in the extraction of bromine from 1.00×10^3 L of seawater?

100. Refer to the Integrative Example on page 1022. Assume that the disproportionation of $S_2O_3^{2-}$ is no longer spontaneous when the partial pressure of $SO_2(g)$ above a solution with $[S_2O_3^{2-}] = 1$ M has dropped to 1×10^{-6} atm. Show that this condition is reached while the solution is still acidic.

101. The bond energies of Cl_2 and F_2 are 243 and 159 kJ mol^{-1}, respectively. Use these data to explain why XeF_2 is a much more stable compound than $XeCl_2$. [*Hint:* Recall that Xe exists as a monatomic gas.]

102. Write plausible half-equations and a balanced oxidation–reduction equation for the disproportionation of XeF_4 to Xe and XeO_3 in aqueous acidic solution. Xe and XeO_3 are produced in a 2:1 mol ratio, and $O_2(g)$ is also produced.

103. A handbook gives the value $E^\circ = 0.174$ V for the reduction half-reaction $S + 2 H^+ + 2 e^- \longrightarrow H_2S(g)$. In Figure 22-13, the value given for the segment $S—H_2S(aq)$ is 0.144 V. Why are these E° values different? Can both be correct?

104. The solubility of $Cl_2(g)$ in water is 6.4 g L^{-1} at 25 °C. Some of this chlorine is present as Cl_2, and some is found as HOCl or Cl^-. For the hydrolysis reaction

$$Cl_2(aq) + 2 H_2O(l) \longrightarrow$$
$$HOCl(aq) + H_3O^+(aq) + Cl^-(aq)$$
$$K_c = 4.4 \times 10^{-4}$$

For a saturated solution of Cl_2 in water, calculate $[Cl_2]$, $[HOCl]$, $[H_3O^+]$, and $[Cl^-]$.

105. Not shown in Figure 22-17 are electrode potential data involving hydrazoic acid. Given that $E^\circ = -3.09$ V for the reduction of HN_3 to N_2 in acidic solution, what is E° for the reduction of HN_3 to NH_4^+ in acidic solution?

106. The heavier halogens (Cl, Br, and I) form compounds in which the central halogen atom, X, is bonded directly to oxygen and to fluorine. Several examples are known, including those with formulas of the type FXO_2, FXO_3, and F_3XO. The structures of these molecules are all consistent with the VSEPR model. Draw Lewis structures and predict the geometries of **(a)** chloryl fluoride, $FClO_2$; **(b)** perchloryl fluoride, $FClO_3$; **(c)** F_3ClO.

107. Draw Lewis structures for O_3 and SO_2. In what ways are the structures similar? In what ways do they differ?

108. Chemists have successfully synthesized the ionic compound $[N_5][SbF_6]$, which consists of N_5^+ and SbF_6^- ions. Draw Lewis structures for these ions and assign formal charges to the atoms in your structures. Describe the structures of these ions. [*Hint*: The skeleton structure for N_5^+ is N—N—N—N—N and several resonance structures can be drawn.]

109. Refer to Figure 22-3 and then construct an enthalpy diagram for forming $XeO_3(g)$ from $Xe(g)$ and $O_2(g)$. For XeO_3, the average bond enthalpy is about 36 kJ mol^{-1}, and for O_2, the bond enthalpy is 498 kJ mol^{-1}. What is ΔH_f° for $XeO_3(g)$? Does your result support the observation that $Xe(g)$ does not react directly with $O_2(g)$ to form $XeO_3(g)$?

Feature Problems

110. Various thermochemical cycles are being explored as possible sources of $H_2(g)$. The object is to find a series of reactions that can be conducted at moderate temperatures (about 500 °C) and that results in the decomposition of water into H_2 and O_2. Show that the following series of reactions meets these requirements.

$$FeCl_2 + H_2O \longrightarrow Fe_3O_4 + HCl + H_2$$
$$Fe_3O_4 + HCl + Cl_2 \longrightarrow FeCl_3 + H_2O + O_2$$
$$FeCl_3 \longrightarrow FeCl_2 + Cl_2$$

111. The decomposition of aqueous hydrogen peroxide is catalyzed by $Fe^{3+}(aq)$. A proposed mechanism for this catalysis involves two reactions. In the first reaction, Fe^{3+} is reduced by H_2O_2. In the second, the iron is oxidized back to its original form, while hydrogen peroxide is reduced. Write an equation for the overall reaction and show that the overall reaction is indeed spontaneous. What are the minimum and maximum values of E° for a catalyst to function in this way? Which of the following should be able to catalyze the decomposition of hydrogen peroxide by the mechanism outlined here: **(a)** Cu^{2+}; **(b)** Br_2; **(c)** Al^{3+}; **(d)** Au^{3+}? In the reaction between iodic acid and hydrogen peroxide in the presence of starch indicator, the color of the reaction mixture oscillates between deep blue and colorless. What is the basis of these changes in color? Will this oscillation of color continue indefinitely? Explain. [*Hint*: Refer to Problem 97 in Chapter 14, particularly to equation (c).]

112. Both in this chapter and in Chapter 20, we have stressed the relationship between E° values and thermodynamic properties. We can use this relationship to add some missing features to an electrode potential diagram. For example, note that $ClO_2(g)$, which has Cl in the oxidation state +4, is not included in Figure 22-4. Using data from Figure 22-4 and Appendix D, add $ClO_2(g)$ to the electrode potential diagram for acidic solutions, and indicate the E° values that link $ClO_2(g)$ to $ClO_3^-(aq)$ and to $HClO_2(aq)$.

113. Figure 15-1 (page 656) shows that I_2 is considerably more soluble in $CCl_4(l)$ than it is in $H_2O(l)$. The concentration of I_2 in its saturated aqueous solution is 1.33×10^{-3} M, and the equilibrium achieved when I_2 distributes itself between H_2O and CCl_4 is

$$I_2(aq) \rightleftharpoons I_2(CCl_4) \qquad K_c = 85.5$$

(a) A 10.0 mL sample of saturated $I_2(aq)$ is shaken with 10.0 mL CCl_4. After equilibrium is established, the two liquid layers are separated. How many milligrams of I_2 will be in the aqueous layer?
(b) If the 10.0 mL of aqueous layer from part (a) is extracted with a second 10.0 mL portion of CCl_4, how many milligrams of I_2 will remain in the aqueous layer when equilibrium is reestablished?
(c) If the 10.0 mL sample of saturated $I_2(aq)$ in part (a) had originally been extracted with 20.0 mL CCl_4, would the mass of I_2 remaining in the aqueous layer have been less than, equal to, or greater than that in part (b)? Explain.

114. The so-called pyroanions, $X_2O_7^{n-}$, form a series of structurally similar polyatomic anions for the elements Si, P, and S.
(a) Draw the Lewis structures of these anions, and predict the geometry of the anions. What is the maximum number of atoms that can lie in a plane?
(b) Each pyroanion in part (a) corresponds to a pyroacid, $X_2O_7H_n$. Compare each pyroacid to the acid containing only one atom of the element in its maximum oxidation state. From this comparison, suggest a strategy for the preparation of these pyroacids.
(c) What is the chlorine analogue of the pyroanions? For which acid is this species the anhydride?

115. A description of bonding in XeF_2 based on the valence bond model requires the $5d$ orbitals of Xe. A more satisfactory description uses a molecular orbital approach involving three-center bonds. Assume that bonding involves the $5p_z$ orbital of Xe and the $2p_z$ orbitals of the two F atoms. These three atomic orbitals combine to give three molecular orbitals: one bonding, one nonbonding, and one antibonding. Recall that for bonding to occur, atomic orbitals with the same phase must overlap to form bonding molecular orbitals (see Chapter 11).
(a) Construct diagrams similar to Figure 11-33 to indicate the overlap of the three atomic orbitals in

forming the three molecular orbitals. Assume that the order of energy of the molecular orbitals is bonding MO < nonbonding MO < antibonding MO.

(b) Sketch a molecular orbital energy-level diagram, and assign the appropriate number of electrons from fluorine and xenon to the molecular orbitals. What is the bond order?

(c) With the aid of VSEPR theory, show that this molecular orbital description based on three-center bonds works well for XeF$_4$ but not for XeF$_6$.

116. The sketch is a portion of the phase diagram for the element sulfur. The transition between solid orthorhombic (S$_\alpha$) and solid monoclinic (S$_\beta$) sulfur, in the presence of sulfur vapor, is at 95.3 °C. The triple point involving monoclinic sulfur, liquid sulfur, and sulfur vapor is at 119 °C.

(a) How would you modify the phase diagram to represent the melting of orthorhombic sulfur that is sometimes observed at 113 °C? [*Hint:* What would the phase diagram look like if the monoclinic sulfur did not form?]

(b) Account for the observation that if a sample of rhombic sulfur is melted at 113 °C and then heated, the liquid sulfur freezes at 119 °C upon cooling.

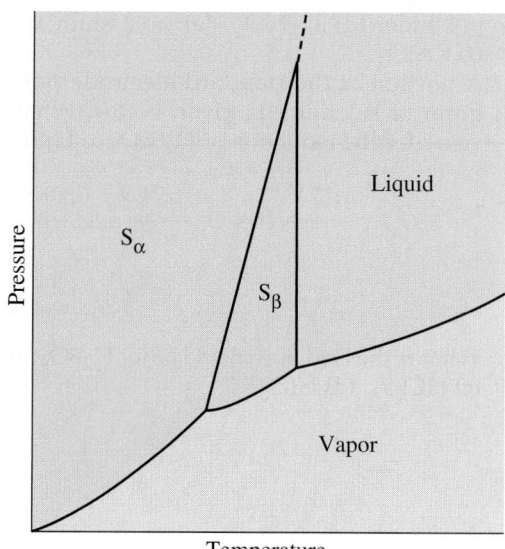

Self-Assessment Exercises

117. In your own words, define the following terms: **(a)** polyhalide ion; **(b)** polyphosphate; **(c)** interhalogen; **(d)** disproportionation.

118. Briefly describe each of the following terms: **(a)** Frasch process; **(b)** water gas reactions; **(c)** eutrophication; **(d)** electrode potential diagram.

119. Explain the important distinctions between each pair of terms: **(a)** acid salt and acid anhydride; **(b)** azide and nitride; **(c)** white phosphorus and red phosphorus; **(d)** ionic hydride and metallic hydride.

120. To displace Br$_2$ from an aqueous solution of Br$^-$, add **(a)** I$_2$(aq); **(b)** Cl$_2$(aq); **(c)** H$_2$(g); **(d)** Cl$^-$(aq); **(e)** I$_3^-$(aq).

121. All of the following compounds yield O$_2$(g) when heated to about 1000 K except **(a)** KClO$_3$; **(b)** KClO$_4$; **(c)** N$_2$O; **(d)** CaCO$_3$; **(e)** Pb(NO$_3$)$_2$.

122. All of the following substances are bases except for **(a)** H$_2$NNH$_2$; **(b)** NH$_3$; **(c)** HN$_3$; **(d)** NH$_2$OH; **(e)** CH$_3$NH$_2$.

123. The best reducing agent of the following substances is **(a)** H$_2$S; **(b)** O$_3$; **(c)** H$_2$SO$_4$; **(d)** NaF; **(e)** H$_2$O.

124. Of the following substances the one that is unimportant is the production of fertilizers is **(a)** NH$_3$; **(b)** phosphate rock; **(c)** HNO$_3$; **(d)** Na$_2$CO$_3$; **(e)** H$_2$SO$_4$.

125. All of the following have a tetrahedral shape except **(a)** SO$_4^{2-}$; **(b)** XeF$_4$; **(c)** CCl$_4$; **(d)** XeO$_4$; **(e)** NH$_4^+$.

126. Two of the following, through a reaction occurring in a weakly acidic solution, produce the same gaseous product. They are **(a)** CaH$_2$(s); **(b)** Na$_2$O$_2$(s); **(c)** NaOH(s); **(d)** Al(s); **(e)** NaHCO$_3$(s); **(f)** N$_2$H$_4$(l)

127. Write a plausible chemical equation to represent the reaction of **(a)** Cl$_2$(g) with cold NaOH(aq); **(b)** NaI(s) with hot H$_2$SO$_4$(concd aq); **(c)** Cl$_2$(g) with KI$_3$(aq); **(d)** NaBr(s) with hot H$_3$PO$_4$ (concd aq); **(e)** NaHSO$_3$(aq) with MnO$_4^-$(aq) in dilute H$_2$SO$_4$(aq).

128. Give a practical laboratory method that you might use to produce small quantities of the following gases and comment on any difficulties that might arise: **(a)** O$_2$; **(b)** NO; **(c)** H$_2$; **(d)** NH$_3$; **(e)** CO$_2$.

129. Complete and balance equations for these reactions.

(a) LiH(s) + H$_2$O(l) $\longrightarrow$

(b) C(s) + H$_2$O(g) $\xrightarrow{\Delta}$

(c) NO$_2$(g) + H$_2$O(l) $\longrightarrow$

130. If Br$^-$ and I$^-$ occur together in an aqueous solution, I$^-$ can be oxidized to IO$_3^-$ with an excess of Cl$_2$(aq). Simultaneously, Br$^-$ is oxidized to Br$_2$, which is extracted with CS$_2$(l). Write chemical equations for the reactions that occur.

131. Suppose that the sulfur present in seawater as SO$_4^{2-}$ (2650 mg L^{-1}) could be recovered as elemental sulfur. If this sulfur were then converted to H$_2$SO$_4$, how many cubic kilometers of seawater would have to be processed to yield the average U.S. annual consumption of about 45 million tons of H$_2$SO$_4$?

132. Although relatively rare, all of the following compounds exist. Based on what you know about related compounds (for example, from the periodic table), propose a plausible name or formula for each compound: **(a)** silver(I) astatide; **(b)** Na$_4$XeO$_6$; **(c)** magnesium polonide; **(d)** H_2TeO_3; **(e)** potassium thioselenate; **(f)** $KAtO_4$.

133. A portion of the standard electrode potential diagram of selenium is given below. What is the $E°$ value for the reduction of H_2SeO_3 to H_2Se?

$$SeO_4^{2-} \xrightarrow{\text{1.15 V}} H_2SeO_3 \xrightarrow{\text{0.74 V}} Se \xrightarrow{\text{-0.35 V}} H_2Se$$

$$\text{(?)}$$

134. What is the acid anhydride of **(a)** H_2SO_4; **(b)** H_2SO_3; **(c)** $HClO_4$; **(d)** HIO_3?

135. Use the following electrode potential diagram for basic solutions to classify each of the statements below as true or false. Assume standard conditions.

$$SO_4^{2-} \xrightarrow{\text{-0.936 V}} SO_3^{2-} \xrightarrow{\text{-0.576 V}}$$

$$S_2O_3^{2-} \xrightarrow{\text{-0.74 V}} S \xrightarrow{\text{-0.476 V}} S^{2-}$$

(a) Sulfate (SO_4^{2-}) is a stronger oxidant than thiosulfate ($S_2O_3^{2-}$) in basic solution.

(b) S^{2-} can be used as a reducing agent in basic solutions.

(c) $S_2O_3^{2-}$ is stable with respect to disproportionation to SO_3^{2-} and S in basic solution.

The Transition Elements

Whiskers of rutile, TiO_2, in quartz (left) and titanium ore (right), a source of rutile. Titanium metal, obtained from rutile, is used in industry because it has low density and high strength. Pure TiO_2 is a bright white pigment used in paints and specialty papers.

There are more transition elements—members of the d and f blocks—than main-group elements. Although some of the transition elements are rare and of limited use, others play crucial roles in many aspects of modern life. All the transition elements are metals; among them are both the chief structural metal, iron (Fe), and important alloying metals in the manufacture of steel (V, Cr, Mn, Co, Ni, Mo, W). The best electrical conductors (Ag, Cu) are transition metals. The compounds of several transition metals (Ti, Fe, Cr) are the primary constituents of paint pigments. Compounds of silver (Ag) provide the essential material for photographic film. Specialized materials for modern applications, such as color television screens, use compounds of the f-block elements (lanthanide oxides). Nine of the transition metals are essential elements for living organisms.

The chemistry of the d-block and f-block elements has both theoretical and practical significance. These elements and their compounds provide insight into fundamental aspects of bonding, magnetism, and reaction chemistry.

23-1 General Properties

The high melting points, good electrical conductivity, and moderate-to-extreme hardness of the transition elements result from the ready availability of electrons and orbitals for metallic bonding (see Section 11-7). Some similarities are found among the transition elements, but each element also has some unique properties that make it and its compounds useful in particular ways. Table 23.1 lists properties of the fourth-period transition elements—the first transition series.

Atomic (Metallic) Radii

In Table 23.1, with the exception of Sc and Ti, we find little variation among the atomic radii across the first transition series. The chief difference in atomic structure between successive elements involves one unit of positive charge on the nucleus and one electron in an orbital of an *inner* electron shell. This is not a major difference and does not cause much of a change in atomic radius, especially in the middle of the series.

When an element of the first transition series is compared with elements of the second and third series within the same group, important differences appear. Consider the members of group 6—Cr, Mo, and W. As we might expect, the atomic radius of Mo is larger than that of Cr; but contrary to our expectation, the atomic radius of W is the same as that of Mo, not larger. In the aufbau process, 18 electrons are added in progressing from Cr to Mo, and all of them enter *s*, *p*, and *d* subshells. Between Mo and W, however, 32 electrons must be added, and 14 of them enter the 4*f* subshell. Electrons in an *f* subshell are not very effective in screening outer-shell electrons from the nucleus. As a result, the outer-shell electrons are held more tightly by the nucleus than we would otherwise expect. Atomic radii do not increase. In fact, in the series of elements in which the 4*f* subshell is filled, atomic radii decrease somewhat. This phenomenon occurs in the lanthanide series (Z = 58 to 71) and is called the **lanthanide contraction**. The lanthanide contraction is made more apparent in the graphs in Figure 23-1.

TABLE 23.1   Selected Properties of Elements of the First Transition Series

	Sc	Ti	V	Cr	Mn	Fe	Co	Ni	Cu	Zn
Atomic number	21	22	23	24	25	26	27	28	29	30
Electron config.[a]	$3d^14s^2$	$3d^24s^2$	$3d^34s^2$	$3d^54s^1$	$3d^54s^2$	$3d^64s^2$	$3d^74s^2$	$3d^84s^2$	$3d^{10}4s^1$	$3d^{10}4s^2$
Metallic radius, pm	161	145	132	125	124	124	125	125	128	133
Ioniz. energy, kJ mol^{-1}										
First	631	658	650	653	717	759	758	737	745	906
Second	1235	1310	1414	1592	1509	1561	1646	1753	1958	1733
Third	2389	2653	2828	2987	3248	2957	3232	3393	3554	3833
$E°$, V[b]	−2.03	−1.63	−1.13	−0.90	−1.18	−0.440	−0.277	−0.257	+0.340	−0.763
Common positive oxidation states[c]	3	2, 3, 4	2, 3, 4, 5	2, 3, 6	2, 3, 4, 7	2, 3, 6	2, 3	2, 3	1, 2	2
mp, °C	1397	1672	1710	1900	1244	1530	1495	1455	1083	420
Density, g cm^{-3}	3.00	4.50	6.11	7.14	7.43	7.87	8.90	8.91	8.95	7.14
Hardness[d]	—	—	—	9.0	5.0	4.5	—	—	2.8	2.5
Electrical conductivity[e]	3	4	6	12	1	16	25	23	93	27

[a]Each atom has an argon inner-core configuration.
[b]For the reduction process, $M^{2+}(aq) + 2 e^- \longrightarrow M(s)$ [except for scandium, where the ion is $Sc^{3+}(aq)$].
[c]The most important oxidation states are printed in red.
[d]Hardness values are on the Mohs scale (see Table 21.2).
[e]Electrical conductivity compared with an arbitrarily assigned value of 100 for silver.

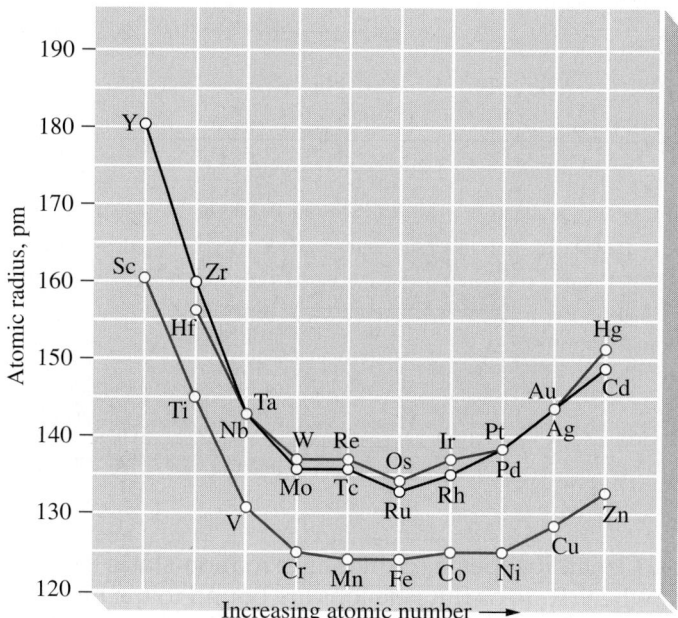

◀ FIGURE 23-1
Atomic radii of the *d*-block elements
The radii of the fourth-period transition elements (blue) are smaller than those of the corresponding group members in the succeeding periods. This trend is not seen between the fifth-period (black) and sixth-period (red) members, illustrating the contraction in atomic radii associated with the lanthanide series.

Electron Configurations and Oxidation States

The elements of the first transition series have electron configurations with the following characteristics:

- an inner core of electrons in the argon configuration
- two electrons in the 4s orbital for eight members and one 4s electron for the remaining two (Cr and Cu)
- a number of 3*d* electrons, ranging from one in Sc to ten in Cu and Zn

As we have seen for some of the main-group elements in Chapters 21 and 22, an element may display several oxidation states. Often, however, one particular oxidation state is the most common for an element. Ti atoms, with the electron configuration $[Ar]3d^2 4s^2$, tend to use all four electrons beyond the argon core in compound formation and display the oxidation state +4. It is also possible, however, for Ti atoms to use fewer electrons, as through the loss of the $4s^2$ electrons to form the ion Ti^{2+}. With Ti, then, we note two features: (1) several possible oxidation states, as shown in Figure 23-2, and (2) a maximum oxidation state corresponding to the group number, 4. These two features continue with V, Cr, and Mn, for which the maximum oxidation states are +5, +6 and +7, respectively. A shift in behavior occurs in groups 8–12, however. Thus, although Fe, Co, and Ni can all exist in more than one oxidation state, they do not display the wide variety found in the earlier members of the first transition series. Nor do they exhibit a maximum oxidation state corresponding to their group number. In crossing the first transition series, the nuclear charge, number of *d* electrons, and energy requirement for the successive ionization of d electrons increase. Involvement of a large number of *d* electrons in bond formation becomes increasingly unfavorable energetically, and only the lower oxidation states are commonly encountered for these later elements of the first transition series.

Although the transition elements display a variety of oxidation states, they differ in the ease with which these oxidation states can be attained and in their stabilities. The stability of an oxidation state for a given transition metal depends on a number of factors—other atoms to which the transition metal atom is bonded; whether the compound is in solid form or in solution; and the pH of the solution. For example, $TiCl_2$ is a well-characterized compound as a solid, but the Ti^{2+} ion is oxidized to Ti^{3+} by dissolved oxygen in aqueous solutions and even by the water itself. Conversely, $Co^{3+}(aq)$ readily oxidizes water to $O_2(g)$ and is itself reduced to $Co^{2+}(aq)$. $Co^{3+}(aq)$ can be stabilized, however, in certain

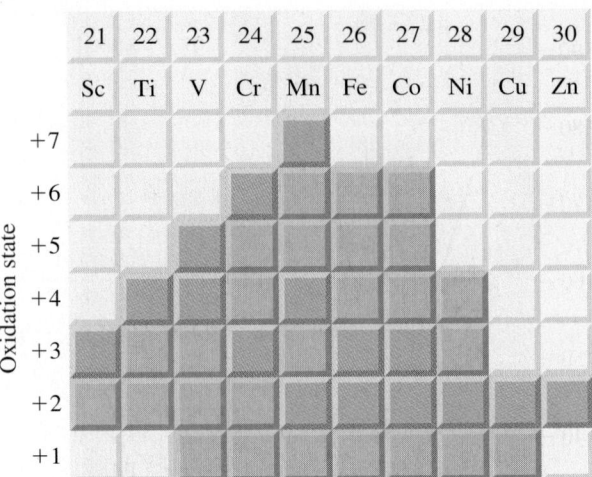

▶ FIGURE 23-2
Positive oxidation states of the elements of the first transition series
Common oxidation states are shown in red and less common ones in gray. Some oxidation states are rather rare, and a zero or negative oxidation state is occasionally found, chiefly in transition metal complexes. For example, the oxidation state of Cr is -2 in $Na_2[Cr(CO)_5]$, -1 in $Na_2[Cr_2(CO)_{10}]$, and 0 in $Cr(CO)_6$.

complex ions. Generally speaking, higher oxidation states in transition metals are stabilized when oxide or fluoride ions are bound to the metal. In Figure 23-2 and elsewhere, the term *common* oxidation state refers to an oxidation state often found in aqueous solution.

Another feature of the transition metals is the progressively increasing stability of higher oxidation states in descending a group of the periodic table, the reverse of the trend often seen for main-group elements. Consider Cr, Mo, and W in group 6, all of which can exhibit oxidation states ranging from $+6$ to -2. The number of compounds with Cr in the $+6$ oxidation state is rather limited, whereas those of Mo and W abound. Chromium is encountered in the oxidation states $+5$ and $+4$ mostly in unstable intermediates, whereas Mo and W exhibit a rich chemistry in these states. The most stable oxidation state of chromium is $+3$. Although it is a strong reducing agent, $Cr^{2+}(aq)$ nevertheless is readily obtainable, whereas Mo and W are not obtainable as the simple $+2$ cation. This trend favoring lower oxidation states for the first group member and higher oxidation states for the later members is also found in other groups of transition metals. For instance, although Fe does not exhibit an oxidation state corresponding to the group number, Os does form the stable oxide OsO_4 with Os in the $+8$ oxidation state.

Ionization Energies and Electrode Potentials

Ionization energies are fairly constant across the first transition series. Values of the first ionization energies are about the same as for the group 2 metals. Standard electrode potentials gradually increase in value across the series. With the exception of the oxidation of Cu to Cu^{2+}, however, all these elements are more readily oxidized than hydrogen. This means that the metals displace $H_2(g)$ from $H^+(aq)$. Additional comments on electrode potentials, some supported by electrode potential diagrams, are found throughout the chapter.

Ionic and Covalent Compounds

▶ The oxidation state of manganese is $+7$ in Mn_2O_7. The bonding in this compound is not ionic, because an ion with a charge of $+7$ would have a very high charge density and would be strongly polarizing. Mn_2O_7 is a molecular compound with covalent bonds between Mn and O atoms.

We tend to think of metals as forming ionic compounds with nonmetals. This is certainly the case with group 1 and most group 2 metal compounds. However, some metal compounds have significant covalent character; $BeCl_2$ and $AlCl_3(Al_2Cl_6)$, for example, are molecular compounds. Transition metal compounds display both ionic and covalent character. In general, compounds with the transition metal in lower oxidation states are essentially ionic, while those in higher oxidation states have covalent character. As an example, MnO is a green ionic solid with a melting point of 1785 °C, whereas Mn_2O_7 is a dark red, oily, molecular liquid that boils at room temperature and is highly explosive. Another feature of ionic compounds of the transition metals is that the metal atoms often occur in polyatomic cations or anions rather than as the simple monatomic ion. Some common examples are VO_2^+, MnO_4^-, and $Cr_2O_7^{2-}$.

Catalytic Activity

An unusual ability to adsorb gaseous species makes some transition metals, such as Ni and Pt, good heterogeneous catalysts. The possibility of multiple oxidation states seems to account for the ability of some transition metal ions to serve as catalysts in certain oxidation–reduction reactions. In still other types of catalysis, complex-ion formation may play an important role. As we saw in a limited way in Chapter 18 and will explore more fully in Chapter 24, complex-ion formation is a particularly distinctive feature of transition metal chemistry.

Catalysis is an essential aspect of about 90% of all chemical manufacturing processes, and the transition metals are often the key elements in the catalysts used. For example, Ni is used in the hydrogenation of oils (page 1019); Pt, Pd, and Rh are used in catalytic converters in automobiles (page 636); Fe_3O_4 is the main component of the catalyst used in the synthesis of ammonia (Focus On 15, www.masteringchemistry.com) and V_2O_5 is used in the conversion of $SO_2(g)$ to $SO_3(g)$ in the manufacture of sulfuric acid (page 1001).

Transition metal catalysts are used in both homogeneous and heterogeneous catalysis. In homogeneous catalysis (page 637), the reactants, products, and catalyst are all in the same phase (often liquid or gas) and the transition metal is part of a compound or a complex. In homogeneous catalysis, the transition metal atoms or ions serve as electron banks that lend out electrons at the appropriate time or store them for later use. In heterogeneous catalysis (page 638), the catalyst is in a different phase from the reactants or the products, and, typically, the catalyst provides a surface on which the reaction occurs. The hydrogenation of oils (page 1019) makes use of heterogeneous catalysis. The details concerning how a catalyst functions in a hydrogenation reaction depend not only on the metal used but also on the experimental conditions (for example, the amounts of reactants used and the temperature). In Figure 23-3, the events that are believed to occur in the hydrogenation of C_2H_4 are illustrated. A C_2H_4 molecule is adsorbed onto the metal surface, with electron density being transferred from the π bond in C_2H_4 to metal atoms in the surface (a Lewis acid–base reaction). Hydrogen molecules may also be adsorbed onto the surface near the C_2H_4 molecule, causing a weakening of the H—H bond. Hydrogen atoms are then transferred to the carbon atoms in the C_2H_4 molecule, producing C_2H_6 molecules, which detach from the metal surface.

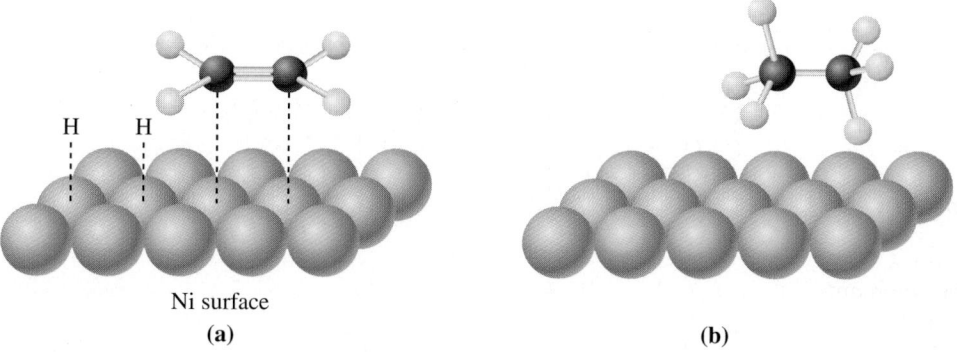

H H

Ni surface
(a)

(b)

▲ FIGURE 23-3
Schematic representation of the metal-catalyzed hydrogenation of C_2H_4
(a) H_2 and C_2H_4 molecules adsorb onto the metal surface, causing weakening of the H—H bond and the π bond in C_2H_4. **(b)** H atoms become bonded to carbon atoms, converting C_2H_4 to C_2H_6, which desorbs from the metal surface. The exact details of the mechanism are a matter of debate.

▲ Color-enhanced image of magnetic domains in a ferromagnetic garnet film.

Color and Magnetism

As we will explain in Section 24-6, the five d orbitals in a transition metal atom (or ion) do not all have the same energy when the atom is part of a compound. Electronic transitions that occur within the d orbitals impart color to transition compounds and their solutions. Consequently, transition metal compounds and their solutions exhibit a wide variety of colors.

Because most transition elements have partially filled d subshells, many transition metals and their compounds are paramagnetic—that is, they have unpaired electrons. This description certainly fits Fe, Co, and Ni, but these three metals are unique among the elements in displaying a special magnetic property: the ability to be made into permanent magnets, a property known as **ferromagnetism**. A key feature of ferromagnetism is that in the solid state, the metal atoms are thought to be grouped together into small regions—called *domains*—containing rather large numbers of atoms. Instead of the individual magnetic moments of the atoms within a domain being randomly oriented, all the magnetic moments are directed in the same way. In an unmagnetized piece of iron, the domains are oriented in several directions and their magnetic effects cancel. When the metal is placed in a magnetic field, however, the domains line up and a strong resultant magnetic effect is produced. This alignment of domains may actually involve the growth of domains with favorable orientations at the expense of those with unfavorable orientations (rather like a recrystallization of the material). The ordering of domains can persist when the object is removed from the magnetic field, and thus permanent magnetism results. Paramagnetism and ferromagnetism are compared in Figure 23-4.

The key factors in ferromagnetism are that (1) the atoms involved have unpaired electrons (a property possessed by many atoms), and (2) interatomic distances are of just the right magnitude to make possible the ordering of atoms into domains. If atoms are too large, interactions among them are too weak to produce this ordering. With small atoms, the tendency is for atoms to pair and their magnetic moments to cancel. This critical factor of atomic size is just met in Fe, Co, and Ni. It is possible, however, to prepare *alloys* of other metals in which this condition is also met. Some examples are Al–Cu–Mn, Ag–Al–Mn, and Bi–Mn.

Comparison of Transition and Main-Group Elements

With the main-group elements, the s and p orbitals of the outermost electron shell are the most important in determining the nature of the chemical bonding that occurs. Participation in bonding by d orbitals is essentially nonexistent for

▶ FIGURE 23-4
Ferromagnetism and paramagnetism compared
In a paramagnetic material, the effect of a magnetic field is to align the magnetic moments of the individual atoms. In a ferromagnetic material, the magnetic moments are aligned within domains even in the absence of a magnetic field, but the direction of the alignment varies from one domain to another. The effect of the magnetic field is to change the orientation of these varied alignments into a single direction—the direction of the magnetic field.

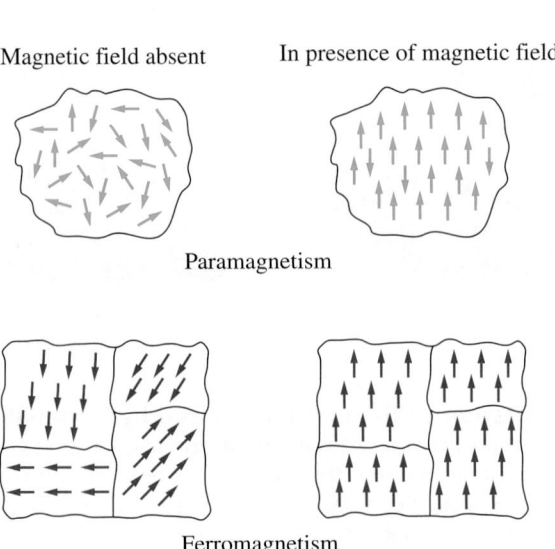

second-period elements and for group 1 and 2 metals. With the transition elements, d orbitals are as important as s and p orbitals. Most of the observed behavioral differences between the transition and main-group elements—multiple versus single oxidation states, complex-ion formation, color, magnetic properties, and catalytic activity—can be traced to the orbitals that are most involved in bond formation.

🔍 23-1 CONCEPT ASSESSMENT

Why is +3 the most stable oxidation state for Fe, while it is +2 for Co and Ni?

23-2 Principles of Extractive Metallurgy

Many of the transition elements have important uses related to their metallic properties—iron for its structural strength and copper for its excellent electrical conductivity, for example. Unlike the more chemically reactive metals of groups 1 and 2 and aluminum in group 3, which are produced mainly by modern methods of electrolysis, the transition metals are obtained by procedures developed over many centuries.

The term *metallurgy* describes the general study of metals. **Extractive metallurgy** describes the winning of metals from their ores. There is no single method of extractive metallurgy, but a few basic operations generally apply. Let us illustrate them with the extractive metallurgy of zinc.

Concentration In mining operations, the desired mineral from which a metal is to be extracted often constitutes only a small percentage (or occasionally just a fraction of a percent) of the material mined. It is necessary to separate the desired ore from waste rock before proceeding with other metallurgical operations. One useful method, *flotation*, is described in Figure 23-5.

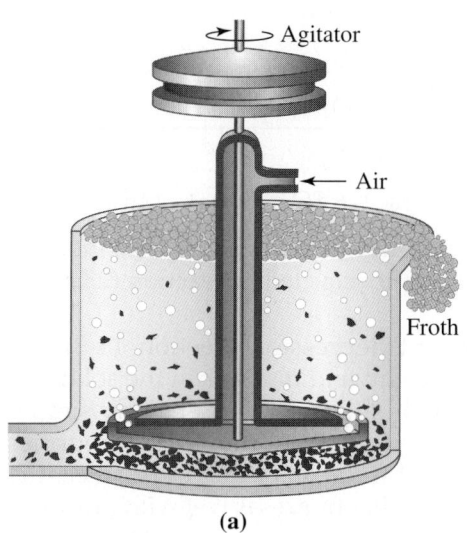
(a)

(b)

▲ FIGURE 23-5
Concentration of an ore by flotation
(a) Powdered ore is suspended in water in a large vat, together with suitable additives, and the mixture is agitated with air. Particles of ore become attached to air bubbles, rise to the top of the vat, and are collected in the overflow froth. Particles of undesired waste rock (gangue) fall to the bottom. (b) The froth formed in the flotation process.

Roasting An ore is roasted (heated to a high temperature) to convert a metal compound to its oxide, which can then be reduced. For zinc, the commercially important ores are $ZnCO_3$ (smithsonite) and ZnS (sphalerite). $ZnCO_3(s)$, like the carbonates of the group 2 metals, decomposes to $ZnO(s)$ and $CO_2(g)$ when it is strongly heated. When strongly heated in air, $ZnS(s)$ reacts with $O_2(g)$, producing $ZnO(s)$ and $SO_2(g)$. In modern smelting operations, $SO_2(g)$ is converted to sulfuric acid rather than being vented to the atmosphere.

$$ZnCO_3(s) \xrightarrow{\Delta} ZnO(s) + CO_2(g) \tag{23.1}$$

$$2\,ZnS(s) + 3\,O_2(g) \xrightarrow{\Delta} 2\,ZnO(s) + 2\,SO_2(g) \tag{23.2}$$

Reduction Because it is inexpensive and easy to handle, carbon, in the form of coke or powdered coal, is used as the reducing agent whenever possible. Several reactions occur simultaneously in which both $C(s)$ and $CO(g)$ act as reducing agents. The reduction of ZnO is carried out at about 1100 °C, a temperature above the boiling point of zinc. The zinc is obtained as a vapor and condensed to the liquid.

$$ZnO(s) + C(s) \xrightarrow{\Delta} Zn(g) + CO(g) \tag{23.3}$$

$$ZnO(s) + CO(g) \xrightarrow{\Delta} Zn(g) + CO_2(g) \tag{23.4}$$

Refining The metal produced by chemical reduction is usually not pure enough for its intended uses. Impurities must be removed; that is, the metal must be refined. The refining process chosen depends on the nature of the impurities. The impurities in zinc are mostly Cd and Pb, which can be removed by the fractional distillation of liquid zinc.

Most of the zinc produced worldwide, however, is refined electrolytically, usually in a process that combines reduction and refining. ZnO from the roasting step is dissolved in $H_2SO_4(aq)$. This is represented by the ionic equation

$$ZnO(s) + 2\,H^+(aq) + SO_4{}^{2-}(aq) \longrightarrow Zn^{2+}(aq) + SO_4{}^{2-}(aq) + H_2O(l) \tag{23.5}$$

Powdered Zn is added to the solution to displace less active metals, such as Cd. Then the solution is electrolyzed. The electrode reactions are

▶ The sulfate anion is a spectator ion in this reaction and could be cancelled.

Cathode: $\qquad Zn^{2+}(aq) + 2\,e^- \longrightarrow Zn(s)$

Anode: $\qquad H_2O \longrightarrow \frac{1}{2}O_2(g) + 2\,H^+(aq) + 2\,e^-$

Unchanged: $\qquad \underline{\quad SO_4{}^{2-}(aq) \longrightarrow SO_4{}^{2-}(aq) \quad}$

Overall: $\quad Zn^{2+}(aq) + SO_4{}^{2-}(aq) + H_2O(l) \longrightarrow$
$$Zn(s) + 2\,H^+(aq) + SO_4{}^{2-}(aq) + \tfrac{1}{2}O_2(g) \tag{23.6}$$

Note that in the overall electrolysis reaction, Zn^{2+} is reduced to pure metallic zinc and sulfuric acid is regenerated. The acid is recycled in reaction (23.5).

Zone Refining In discussing freezing-point depression (Section 13-8), we assumed that a solute is *soluble* in a liquid solvent and *insoluble* in the solid solvent that freezes from solution. This behavior suggests a particularly simple way to purify a solid: Melt the solid and then refreeze a portion of it. Impurities remain in the liquid phase, and the solid that freezes is pure. In practice, the method is not quite so simple because the solid that freezes is wet with unfrozen liquid and thereby retains some impurities. Also, one or more solutes (impurities) might be slightly soluble in the solid solvent. In any case, the impurities do distribute themselves between the solid and liquid, concentrating in the liquid phase. If the solid that freezes from a liquid is remelted and the molten material refrozen, the solid obtained in the second freezing is purer than that in the first. Repeating the melting and refreezing procedure hundreds of times produces a very pure solid product.

EXAMPLE 23-1 Writing Chemical Equations for Metallurgical Processes

Write chemical equations to represent the **(a)** roasting of galena, PbS; **(b)** reduction of $Cu_2O(s)$ with charcoal as a reducing agent; **(c)** deposition of pure silver from an aqueous solution of Ag^+.

Analyze

When answering these questions we have to make sure we understand the meaning of the terms *roasting*, *reduction*, and *deposition*; these terms have all been discussed in the text.

Solve

(a) We expect this process to be essentially the same as reaction (23.2).

$$2\,PbS(s) + 3\,O_2(g) \xrightarrow{\Delta} 2\,PbO(s) + 2\,SO_2(g)$$

(b) The simplest possible equation is

$$Cu_2O(s) + C(s) \xrightarrow{\Delta} 2\,Cu(l) + CO(g)$$

(c) The reduction half-reaction is

$$Ag^+(aq) + e^- \longrightarrow Ag(s)$$

Assess

Roasting converts a metal sulfide to a metal oxide. When carbon is used as a reducing agent, $CO(g)$ is produced, not $CO_2(g)$. The deposition of $Ag(s)$ involves a reduction half-reaction. The accompanying oxidation half-reaction is not specified and neither is it specified whether this is an electrolysis process or whether silver is displaced by a more active metal, but these distinctions are not important in answering the question.

PRACTICE EXAMPLE A: Write plausible chemical equations to represent the **(a)** roasting of Cu_2S; **(b)** reduction of WO_3 with $H_2(g)$; **(c)** thermal decomposition of HgO to its elements.

PRACTICE EXAMPLE B: Write chemical equations to represent the **(a)** reduction of Cr_2O_3 to chromium with silicon as the reducing agent; **(b)** conversion of $Co(OH)_3(s)$ to $Co_2O_3(s)$ by roasting; **(c)** production of pure $MnO_2(s)$ from $MnSO_4(aq)$ at the anode in an electrolysis cell. [*Hint:* A few simple products are not specifically mentioned; propose plausible ones.]

The purification procedure just described implies that the melting and refreezing is done in batches, but in practice it is done *continuously*. In the method known as **zone refining**, a cylindrical rod of material is alternately melted and refrozen as a series of heating coils passes along the rod (Fig. 23-6). Impurities concentrate in the molten zones, and the portions of the rod behind

◄ FIGURE 23-6
Zone refining
As a heating coil moves up the rod of material, melting occurs. Impurities concentrate in the molten zone. The portion of the rod below the molten zone is purer than the portion in or above the molten zone. With each successive passage of the heating coil, the rod becomes purer.

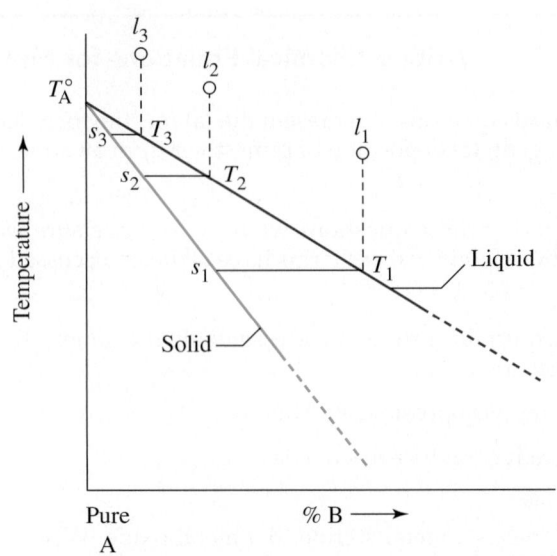

▲ FIGURE 23-7
The principle of zone refining
The red line shows the freezing points of solutions of impurity B in substance A.
The blue line gives the composition of the solid that freezes from these solutions.
In some cases, the blue line is nearly coincident with the temperature axis. When a
solution of composition l_1 is cooled to temperature T_1, it freezes to produce a solid of
composition s_1. If a small quantity of this solid is removed from the solution and melted,
it produces a new liquid, l_2. The freezing point of l_2 is T_2, and the composition of the
solid freezing from the solution is s_2. When removed from the solution, a small portion
of this solid produces liquid l_3 and so on. With each melting/freezing cycle, the melting
point increases and the point representing the composition of the solid moves closer
to pure A. In zone refining the melting/freezing cycles are conducted continuously,
not in batches as described here.

these zones are somewhat purer than the portions in front of the zones.
Eventually, impurities are swept to the end of the rod, which is cut off. The
principle of zone refining is shown graphically in Figure 23-7. This process
is capable of producing materials in which the impurity levels are as low as
10 parts per billion (ppb), a common requirement of substances used in
semiconductors.

Thermodynamics of Extractive Metallurgy

It is interesting to think of the reduction of zinc oxide by carbon, as shown in
reaction (23.3), as a competition between zinc and carbon for O atoms. Zinc
has them initially in ZnO, and carbon acquires them in forming CO. To establish the conditions under which carbon will reduce zinc oxide to zinc, we start
by comparing the relative tendencies for zinc and carbon to undergo oxidation. We can assess these tendencies through Gibbs energy changes.

$$\text{(a)} \quad 2\,C(s) + O_2(g) \longrightarrow 2\,CO(g) \qquad \Delta G^\circ_{(a)}$$
$$\text{(b)} \quad 2\,Zn(s) + O_2(g) \longrightarrow 2\,ZnO(s) \qquad \Delta G^\circ_{(b)}$$

To determine if the reduction of zinc oxide to zinc by carbon is a spontaneous reaction, we need the value of ΔG° for the overall reaction, represented
below by reversing equation (b) and adding it to equation (a).

(a)	$2\,C(s) + O_2(g) \longrightarrow 2\,CO(g)$	$\Delta G^\circ_{(a)}$
$-$ (b)	$2\,ZnO(s) \longrightarrow 2\,Zn(s) + O_2(g)$	$-\Delta G^\circ_{(b)}$

Overall: $\quad 2\,ZnO(s) + 2\,C(s) \longrightarrow 2\,Zn(s) + 2\,CO(g) \qquad \Delta G^\circ = \Delta G^\circ_{(a)} - \Delta G^\circ_{(b)}$

KEEP IN MIND

that a nonspontaneous
process can sometimes be
achieved by *coupling* with a
spontaneous process. In the
reduction of ZnO with C,
the nonspontaneous
$2\,ZnO \longrightarrow 2\,Zn + O_2$
is coupled with the
spontaneous
$2\,C + O_2 \longrightarrow 2\,CO$.

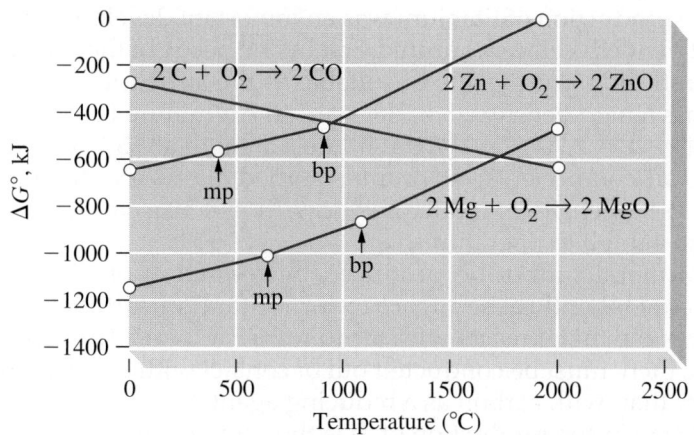

◀ FIGURE 23-8
$\Delta G°$ as a function of temperature for some reactions of extractive metallurgy
The points on the lines marked by arrows indicate the melting points and boiling points of Zn and Mg. At these points, the states of matter in which the metal exists change from (s) to (l) and from (l) to (g), respectively.

We still need some numerical data to complete our assessment. In Figure 23-8, the blue line gives $\Delta G°_{(a)}$ as a function of temperature, and the top red line gives $\Delta G°_{(b)}$.

Figure 23-8 shows that at low temperatures, $\Delta G°_{(b)}$ is much more negative than $\Delta G°_{(a)}$. This makes $\Delta G°$ for the overall reaction positive and the reaction *nonspontaneous*. At high temperatures, the situation is reversed: $\Delta G°_{(a)}$ is more negative than $\Delta G°_{(b)}$, and the overall reaction is *spontaneous*. The switchover from nonspontaneous to spontaneous occurs at the point of intersection of the blue line and the red line—about 950 °C. There, $\Delta G°$ for the overall reaction is *zero*.

When we make a similar assessment for the reduction $2\,MgO(s) + 2\,C(s) \longrightarrow 2\,Mg(g) + 2\,CO(g)$, we conclude that the reaction does not become spontaneous until a temperature in excess of 1700 °C is reached. This is an exceedingly high temperature at which to carry out a chemical reaction, and it is not used in the metallurgy of magnesium.

Alternative Methods in Extractive Metallurgy Some common variations of the methods previously discussed are worth mentioning. First, many ores contain several metals, and it is not always necessary to separate them. For example, a major use of vanadium, chromium, and manganese is in making alloys with iron. Obtaining each metal by itself is not commercially important. Thus, the principal chromium ore *chromite*, $Fe(CrO_2)_2$, can be reduced to give an alloy of Fe and Cr called *ferrochrome*. Ferrochrome may be added directly to iron, together with other metals, to produce one type of steel. Vanadium and manganese can be isolated as the oxides V_2O_5 and MnO_2, respectively. When iron-containing compounds are added to these oxides and the mixtures are reduced, ferrovanadium and ferromanganese alloys form.

23-1 ARE YOU WONDERING...

Why the slope of the blue line in Figure 23-8 is negative, whereas the other slopes are positive?

Recall from Chapter 19 that $\Delta G° = \Delta H° - T\,\Delta S°$ and that $\Delta H°$ does not change appreciably with temperature. Thus, the temperature variation of $\Delta G°$ is determined primarily by the $-T\,\Delta S°$ term. We expect $\Delta S°$ to be negative for the reaction $2\,Zn(s) + O_2(g) \longrightarrow 2\,ZnO(s)$, because a mole of gas is lost. The term $-T\,\Delta S°$ is *positive*, and $\Delta G°$ *increases* with temperature. Conversely, in the reaction $2\,C(s) + O_2(g) \longrightarrow 2\,CO(g)$, an additional mole of gas forms, resulting in a positive value of $\Delta S°$. As a consequence, $-T\,\Delta S°$ is *negative* and $\Delta G°$ *decreases* with temperature.

▲ Vacuum-distilled metallic titanium sponge produced by the Kroll process.

Extensive production of titanium was an important development of the latter half of the twentieth century, spurred first by the needs of the military and then by the aircraft industry. Steel is unsuitable as the structural metal for aircraft because it has a high density ($d = 7.8 \text{ g cm}^{-3}$). Aluminum has the advantage of a low density ($d = 2.70 \text{ g cm}^{-3}$), but it loses strength at high temperatures. For certain aircraft components, titanium is a good alternative to aluminum and steel because its density is moderately low ($d = 4.50 \text{ g cm}^{-3}$) and it does not lose strength at high temperatures.

Titanium metal cannot be produced by reduction of TiO_2 with carbon because the metal and carbon react to form titanium carbides. Also, at high temperatures the metal reacts with air to form TiO_2 and TiN. The metallurgy of titanium, then, must be conducted out of contact with air and with an active metal rather than with carbon as a reducing agent.

The first step in the production of Ti is the conversion of *rutile* ore (TiO_2) to $TiCl_4$ by reaction with carbon and $Cl_2(g)$.

$$TiO_2(s) + 2\,C(s) + 2\,Cl_2(g) \xrightarrow{800\,°C} TiCl_4(g) + 2\,CO(g) \qquad \textbf{(23.7)}$$

The purified $TiCl_4$ is next reduced to Ti with a good reducing agent. The *Kroll process* uses Mg.

$$TiCl_4(g) + 2\,Mg(l) \xrightarrow{1000\,°C} Ti(s) + 2\,MgCl_2(l) \qquad \textbf{(23.8)}$$

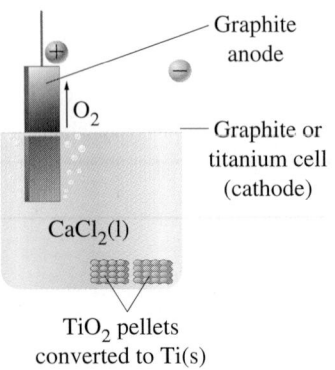

▲ Electrolytic production of Ti(s) from $TiO_2(s)$.

The $MgCl_2(l)$ is removed and electrolyzed to produce Cl_2 and Mg, which are recycled in reactions (23.7) and (23.8), respectively. The Ti is obtained as a sintered (fused) mass called *titanium sponge*. This sponge must be subjected to further treatment and alloying with other metals before it can be used.

The Kroll process is slow; it takes a week to produce a few tons of Ti. It is also demanding in health and safety terms because it requires high-temperature vacuum distillation to remove the Mg and $MgCl_2$ from the titanium. Recently, an electrolytic process has been suggested for the production of Ti from rutile. Porous pellets of TiO_2 are placed at the cathode of an electrolytic cell containing molten calcium chloride. The pellets dissolve in the electrolyte and oxide ions (O^{2-}) are discharged as oxygen at a graphite anode. The Ti(IV) is reduced at the cathode, which is the vessel containing the electrolytic cell and is made of either graphite or titanium. The titanium metal is obtained as sponge. This method, devised by Derek Fray, George Chen, and Tom Farthing in the United Kingdom, is being developed as a commercial process for the production of Ti(s) at a substantially lower cost than the Kroll process.

Metallurgy of Copper The extraction of copper from its ores (generally sulfides) is rather complicated, chiefly because the copper ores usually contain iron sulfides. The scheme of extractive metallurgy previously discussed produces copper contaminated with iron. For some metals, such as V, Cr, and Mn, contamination with iron is not a problem because the metals are mostly used in the manufacture of steel. Copper, however, is prized commercially for the properties of the pure metal. To avoid contamination with iron, several changes to the usual metallurgical methods are necessary.

Concentration of copper is done by flotation, and roasting converts iron sulfides to iron oxides. The copper remains as the sulfide if the temperature is kept below 800 °C. Smelting of the roasted ore in a furnace at 1400 °C causes the material to melt and separate into two layers. The bottom layer, called *copper matte*, consists chiefly of the molten sulfides of copper and iron. The top layer is a silicate slag formed by the reaction of oxides of Fe, Ca, and Al with SiO_2 (which typically is present in the ore or can be added). For example,

$$FeO(s) + SiO_2(s) \xrightarrow{\Delta} FeSiO_3(l) \qquad \textbf{(23.9)}$$

▲ Slag formed during the smelting of copper ore.

A process called *conversion* is carried out in another furnace, where air is blown through the molten copper matte. First, the remaining iron sulfide is converted to the oxide, followed by formation of slag [$FeSiO_3(l)$]. The slag is poured off, and air is again blown through the furnace. The following reactions occur and yield a product that is about 98%–99% Cu:

$$2\,Cu_2S(l) + 3\,O_2(g) \xrightarrow{\Delta} 2\,Cu_2O(l) + 2\,SO_2(g) \tag{23.10}$$

$$2\,Cu_2O(l) + Cu_2S(l) \xrightarrow{\Delta} 6\,Cu(l) + SO_2(g) \tag{23.11}$$

The product of reaction (23.11) is called *blister copper* because frozen bubbles of $SO_2(g)$ are present. Blister copper can be used where high purity is not required (as in plumbing).

Refining blister copper to obtain high-purity copper is done electrolytically by the method outlined on page 901. High-purity copper is essential in electrical applications.

Pyrometallurgical Processes The metallurgical method based on roasting an ore, followed by reduction of the oxide to the metal, is called **pyrometallurgy**, the prefix *pyro-* suggesting that high temperatures are involved. Some of the characteristics of pyrometallurgy are as follows:

- large quantities of waste materials produced in concentrating low-grade ores
- high energy consumption to maintain high temperatures necessary for roasting and reduction of ores
- gaseous emissions that must be controlled, such as $SO_2(g)$ in roasting

Many of the metallurgical processes described earlier fall into this category.

Hydrometallurgical Processes In **hydrometallurgy**, the materials handled are water and aqueous solutions at moderate temperatures rather than dry solids at high temperatures. Generally, three steps are involved in hydrometallurgy:

1. *Leaching:* Metal ions are extracted (leached) from the ore by a liquid. Leaching agents include water, acids, bases, and salt solutions. Oxidation–reduction reactions may also be involved.

2. *Purification and concentration:* Impurities are separated, and the solution produced by leaching may be made more concentrated. Methods include the adsorption of impurities on the surface of activated charcoal, ion exchange, and the evaporation of water.

3. *Precipitation:* The desired metal ions are either precipitated in an ionic solid or reduced to the free metal, often electrolytically.

Hydrometallurgy has long been used in obtaining silver and gold from natural sources. A typical gold ore currently being processed in the United States has only about 10 g Au per metric ton of ore. The leaching step in gold processing is known as *cyanidation*. The process is based on the following reaction:

$$4\,Au(s) + 8\,CN^-(aq) + O_2(g) + 2\,H_2O(l) \longrightarrow 4[Au(CN)_2]^-(aq) + 4\,OH^-(aq) \tag{23.12}$$

The $[Au(CN)_2]^-(aq)$ is then filtered and concentrated. This is followed by displacement of Au(s) from solution by an active metal, such as zinc.

$$2[Au(CN)_2]^-(aq) + Zn(s) \longrightarrow 2\,Au(s) + [Zn(CN)_4]^{2-}(aq) \tag{23.13}$$

In one hydrometallurgical process for zinc, a zinc sulfide ore is leached with a sulfuric acid solution at 150 °C and an oxygen pressure of about 7 atm. The overall reaction is

$$ZnS(s) + H_2SO_4(aq) + \tfrac{1}{2}O_2(g) \longrightarrow ZnSO_4(aq) + S(s) + H_2O(l) \tag{23.14}$$

In this process, there is no $SO_2(g)$ emission. Also, mercury impurities in the ZnS ore are retained in the leaching solution rather than being emitted with $SO_2(g)$ as in the traditional roasting process. Following the leaching process,

ZnSO$_4$(aq) is electrolyzed to produce pure Zn at the cathode, and H$_2$SO$_4$(aq) is regenerated; see reaction (23.6). The H$_2$SO$_4$(aq) is recycled into the leaching operation.

> **23-2 CONCEPT ASSESSMENT**
>
> The text describes the production of pure zinc from zinc sulfide ore both by a pyrometallurgical and by a hydrometallurgical process. Which aspects of the two processes are different and which are similar?

23-3 Metallurgy of Iron and Steel

▶ In the United States, slightly more than half of all iron and steel production comes from recycled iron and steel.

Iron is the most widely used metal from Earth's crust, and for this reason, we use this section to explore the metallurgy of iron and its principal alloy—steel—somewhat more fully. A type of steel called *wootz steel* was first produced in India about 3000 years ago; that same steel became famous in ancient times as Damascus steel, prized for making swords because of its suppleness and ability to hold a cutting edge. Many technological advances have been made since ancient times. These include introduction of the blast furnace around AD 1300, the Bessemer converter in 1856, the open-hearth furnace in the 1860s, and the basic oxygen furnace in the 1950s. A true understanding of the iron- and steelmaking processes has developed only within the past few decades, however. This understanding is based on concepts of thermodynamics, equilibrium, and kinetics.

Pig Iron

The reactions that occur in a blast furnace are complex. A highly simplified representation of the reduction of iron ore to impure iron is

$$\text{Fe}_2\text{O}_3(\text{s}) + 3\,\text{CO}(\text{g}) \longrightarrow 2\,\text{Fe}(\text{l}) + 3\,\text{CO}_2(\text{g}) \tag{23.15}$$

A more complete description of the blast furnace reactions, including the removal of impurities as slag, is given in Table 23.2. Approximate temperatures are given for these reactions so that you can key them to regions of the blast furnace pictured in Figure 23-9.

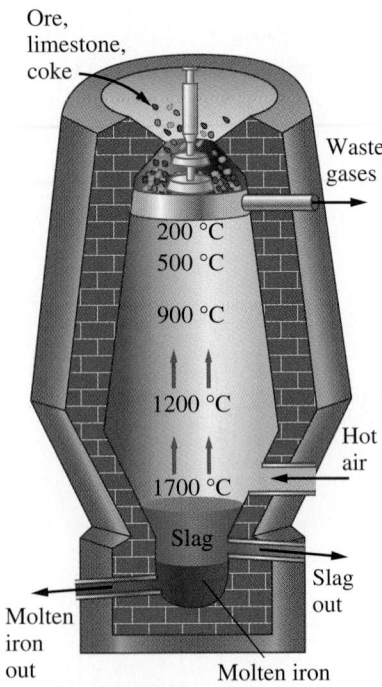

▲ FIGURE 23-9
Typical blast furnace
Iron ore, coke, and limestone are added at the top of the furnace, and hot air is introduced through the bottom. Maximum temperatures are attained near the bottom of the furnace where molten iron and slag are drained off. The principal reactions occurring in the blast furnace are outlined in Table 23.2.

TABLE 23.2 Some Blast Furnace Reactions

Formation of gaseous reducing agents CO(g) and H$_2$(g):
$$\text{C} + \text{H}_2\text{O} \longrightarrow \text{CO} + \text{H}_2 \ (> 600\ °\text{C})$$
$$\text{C} + \text{CO}_2 \longrightarrow 2\,\text{CO}\ (1700\ °\text{C})$$
$$2\,\text{C} + \text{O}_2 \longrightarrow 2\,\text{CO}\ (1700\ °\text{C})$$

Reduction of iron oxide:
$$3\,\text{CO} + \text{Fe}_2\text{O}_3 \longrightarrow 2\,\text{Fe} + 3\,\text{CO}_2\ (900\ °\text{C})$$
$$3\,\text{H}_2 + \text{Fe}_2\text{O}_3 \longrightarrow 2\,\text{Fe} + 3\,\text{H}_2\text{O}\ (900\ °\text{C})$$

Slag formation to remove impurities from ore:
$$\text{CaCO}_3 \longrightarrow \text{CaO} + \text{CO}_2\ (800–900\ °\text{C})$$
$$\text{CaO} + \text{SiO}_2 \longrightarrow \text{CaSiO}_3(\text{l})\ (1200\ °\text{C})$$
$$6\,\text{CaO} + \text{P}_4\text{O}_{10} \longrightarrow 2\,\text{Ca}_3(\text{PO}_4)_2(\text{l})\ (1200\ °\text{C})$$

Impurity formation in the iron:
$$\text{MnO} + \text{C} \longrightarrow \text{Mn} + \text{CO}\ (1400\ °\text{C})$$
$$\text{SiO}_2 + 2\,\text{C} \longrightarrow \text{Si} + 2\,\text{CO}\ (1400\ °\text{C})$$
$$\text{P}_4\text{O}_{10} + 10\,\text{C} \longrightarrow 4\,\text{P} + 10\,\text{CO}\ (1400\ °\text{C})$$

The blast furnace charge—that is, the solid reactants—consists of iron ore, coke, a slag-forming flux, and perhaps some scrap iron. The exact proportions depend on the composition of the iron ore and its impurities. The common ores of iron are the oxides and carbonate: hematite (Fe_2O_3), magnetite (Fe_3O_4), limonite ($2\ Fe_2O_3 \cdot 3\ H_2O$), and siderite ($FeCO_3$). The purpose of the flux is to maintain the proper ratio of acidic oxides (SiO_2, Al_2O_3, and P_4O_{10}) to basic oxides (CaO, MgO, and MnO) to obtain an easily liquefied silicate, aluminate, or phosphate slag. Because acidic oxides predominate in most ores, the flux generally employed is limestone, $CaCO_3$, or dolomite, $CaCO_3 \cdot MgCO_3$.

The iron obtained from a blast furnace is called **pig iron**. It contains about 95% Fe, 3%–4% C, and varying quantities of other impurities. *Cast iron* can be obtained by pouring pig iron directly into molds of the desired shape. Cast iron is very hard and brittle and is used only where it is not subjected to mechanical or thermal shock, such as in engine blocks, brake drums, and transmission housings in automobiles.

▲ Pouring pig iron.

◀ Thermal shock occurs when an object undergoes a rapid change of temperature. Engine blocks get quite hot, but because they cool slowly, they are not usually subject to thermal shock.

Steel

Three fundamental changes must be made to convert pig iron to **steel**:

1. reduction of the carbon content from 3%–4% in pig iron to 0%–1.5% in steel

2. removal, through slag formation, of Si, Mn, and P (each present in pig iron to the extent of 1% or so), together with other minor impurities

3. addition of alloying elements (such as Cr, Ni, Mn, V, Mo, and W) to give the steel its desired end properties

◀ Steel made with 18% Cr and 8% Ni resists corrosion and is commonly known as *stainless steel*.

The most important method of steelmaking today is the **basic oxygen process**. Oxygen gas at about 10 atm pressure and a stream of powdered limestone are fed through a water-cooled tube (called a *lance*) and discharged above the molten pig iron (Fig. 23-10). The reactions that occur (Table 23.3) accomplish the first two objectives. A typical reaction time is 22 minutes. The reaction vessel is tilted to pour off the liquid slag floating on top of the iron, and then the desired alloying elements are added.

Steelmaking has been undergoing rapid technological changes. It is now possible to make iron and steel directly from iron ore in a single-step, continuous process at temperatures below the melting point of any of the materials used in the process. In the *direct reduction of iron* (DRI), CO(g) and H_2(g), obtained in the reaction of steam with natural gas, are used as reducing agents. The economic viability of the DRI process depends on an abundant supply of natural gas. Currently, only a small percentage of the world's iron production is by direct reduction, but this is a fast-growing component of the iron and steel industry, particularly in the Middle East and South America.

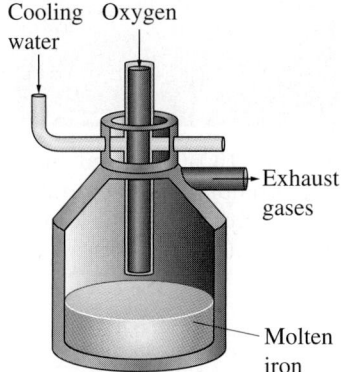

Cooling water   Oxygen

Exhaust gases

Molten iron

▲ FIGURE 23-10
A basic oxygen furnace

TABLE 23.3   Some Reactions Occurring in Steelmaking Processes

$$2\ C + O_2 \longrightarrow 2\ CO$$
$$2\ FeO + Si \longrightarrow 2\ Fe + SiO_2$$
$$FeO + Mn \longrightarrow Fe + MnO$$
$$FeO + SiO_2 \longrightarrow \underset{\text{slag}}{FeSiO_3}$$
$$MnO + SiO_2 \longrightarrow \underset{\text{slag}}{MnSiO_3}$$
$$4\ P + 5\ O_2 \longrightarrow P_4O_{10}$$
$$6\ CaO + P_4O_{10} \longrightarrow \underset{\text{slag}}{2\ Ca_3(PO_4)_2}$$

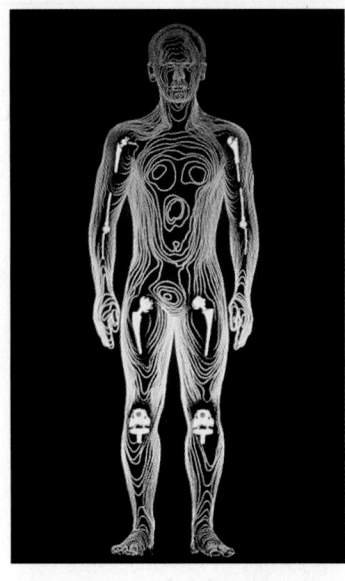

▲ A computer-generated representation of titanium joint implants at the shoulders, elbows, hips, and knees.

▲ White $TiO_2(s)$, mixed with other components to produce the desired color, is the leading pigment used in paints.

> 🔍 **23-3 CONCEPT ASSESSMENT**
>
> Write a balanced chemical equation for the direct reduction of iron(III) oxide by hydrogen gas.

23-4 First-Row Transition Metal Elements: Scandium to Manganese

The properties and uses of the first-row transition metals span a wide range, strikingly illustrating periodic behavior despite the small variation in some of the atomic properties listed in Table 23.1. The preparation, uses, and reactions of the compounds of these metals illustrate concepts we have previously discussed, including the variability of oxidation states.

Scandium

Scandium is a rather obscure metal, though not especially rare. It constitutes about 0.0025% of Earth's crust, which makes it more abundant than many better-known metals, including lead, uranium, molybdenum, tungsten, antimony, silver, mercury, and gold. Its principal mineral form is *thortveitite*, $Sc_2Si_2O_7$. Most scandium is obtained from uranium ores, however, in which it occurs only to the extent of about 0.01% Sc by mass. The commercial uses of scandium are limited, and its production is measured in gram or kilogram quantities, not in tons. One application is in high-intensity lamps. The pure metal is usually prepared by the electrolysis of a fused mixture of $ScCl_3$ with other chlorides.

Because of its noble-gas electron configuration, the Sc^{3+} ion lacks some of the characteristic properties of transition metal ions. For instance, the ion is colorless and diamagnetic, as are most of its salts. In its chemical behavior, Sc^{3+} most closely resembles Al^{3+}, as in the hydrolysis of $[Sc(H_2O)_6]^{3+}(aq)$ to yield acidic solutions and in the formation of an amphoteric gelatinous hydroxide, $Sc(OH)_3$.

Titanium

Titanium is the ninth most abundant element, constituting 0.6% of Earth's solid crust. The metal is greatly valued for its low density, high structural strength, and corrosion resistance. The first two properties account for its extensive use in the aircraft industry and the third for its uses in the chemical industry: in pipes, component parts of pumps, and reaction vessels. Titanium is also used in dental and other bone implants. The metal provides a strong support and bone bonds directly to a titanium implant, making it a part of the body.

Several compounds of titanium are of particular commercial importance. *Titanium tetrachloride*, $TiCl_4$, is the starting material for preparing other Ti compounds and plays a central role in the metallurgy of titanium. $TiCl_4$ is also used to formulate catalysts for the production of plastics. The usual method of preparing $TiCl_4$ involves the reaction of naturally occurring rutile (TiO_2) with carbon and $Cl_2(g)$ [reaction (23.7)].

$TiCl_4$ is a colorless liquid (mp -24 °C; bp 136 °C). In the $+4$ oxidation state, all the valence-shell electrons of Ti atoms are employed in bond formation. In this oxidation state, Ti bears a strong resemblance to the group 14 elements, with some properties and a molecular shape (tetrahedral) similar to those of CCl_4 and $SiCl_4$. The hydrolysis of $TiCl_4$, when carried out in moist air, is the basis for a type of smoke grenade in which $TiO_2(s)$ is the smoke.

$$TiCl_4(l) + 2\,H_2O(l) \longrightarrow TiO_2(s) + 4\,HCl(g)$$

$SiCl_4$ also fumes in moist air in a similar reaction.

Titanium dioxide, TiO_2, is bright white, opaque, inert, and nontoxic. Because of these properties and its relative low cost, it is now the most widely used white pigment for paints. In this application, TiO_2 has displaced toxic basic lead carbonate—so-called white lead. TiO_2 is also used as a paper whitener and in glass, ceramics, floor coverings, and cosmetics.

To produce pure TiO_2 for these and other uses, a gaseous mixture of $TiCl_4$ and O_2 is passed through a silica tube at about 700 °C:

$$TiCl_4(g) + O_2(g) \xrightarrow{\Delta} TiO_2(s) + 2\,Cl_2(g)$$

Vanadium

Vanadium is a fairly abundant element (0.02% of Earth's crust) found in several dozen ores. Its principal ores are rather complex, such as *vanadinite*, $3\,Pb_3(VO_4)_2 \cdot PbCl_2$. The metallurgy of vanadium is not simple, but vanadium of high purity (99.99%) is obtainable. For most of its applications, though, vanadium is prepared as an iron–vanadium alloy, *ferrovanadium*, containing from 35% to 95% V. About 80% of the vanadium produced is for the manufacture of steel. Vanadium-containing steels are used in applications requiring strength and toughness, such as in springs and high-speed machine tools.

The most important compound of vanadium is the pentoxide, V_2O_5, used mainly as a catalyst, as in the conversion of $SO_2(g)$ to $SO_3(g)$ in the contact method for the manufacture of sulfuric acid. The activity of V_2O_5 as an oxidation catalyst may be linked to its reversible loss of oxygen occurring from 700 to 1100 °C.

In its compounds, vanadium can exist in a variety of oxidation states. In each of these oxidation states, vanadium forms an oxide or ion. The ions display distinctive colors in aqueous solution (Fig. 23-11). The acid–base properties of vanadium oxides are in accord with factors established earlier in the text: If the central metal atom is in a low oxidation state, the oxide acts as a base; in higher oxidation states for the central atom, acidic properties become important. Vanadium oxides with V in the +2 and +3 oxidation states are basic, whereas those in the +4 and +5 oxidation states are amphoteric.

Most compounds with vanadium in its highest oxidation state (+5) are good oxidizing agents. In its +2 oxidation state, vanadium (as V^{2+}) is a good reducing agent. The oxidation–reduction relationships between the ionic species pictured in Figure 23-11 are summarized in Table 23.4.

Chromium

Although it is found only to the extent of 122 parts per million (0.0122%) in Earth's crust, chromium is one of the most important industrial metals. The production of ferrochrome from *chromite*, $Fe(CrO_2)_2$, was discussed in Section 23-2. Chromium metal is hard and maintains a bright surface through the protective action of an invisible oxide coating. Because it is resistant to corrosion, chromium is extensively used in plating other metals.

▲ FIGURE 23-11
Some vanadium species in solution
The yellow solution has vanadium in the +5 oxidation state, as VO_2^+. In the blue solution, the oxidation state is +4, in VO^{2+}. The green solution contains V^{3+}, and the violet solution contains V^{2+}.

◄ The word *chromium* is derived from the Greek *chroma*, meaning "color"—an apt name, given the range of colors found in chromium compounds.

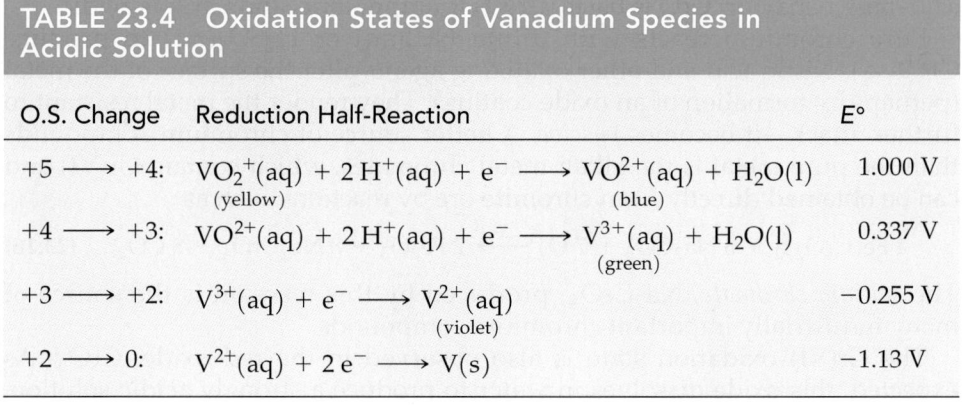

TABLE 23.4	Oxidation States of Vanadium Species in Acidic Solution	
O.S. Change	**Reduction Half-Reaction**	**$E°$**
+5 ⟶ +4:	$VO_2^+(aq) + 2\,H^+(aq) + e^- \longrightarrow VO^{2+}(aq) + H_2O(l)$ (yellow) (blue)	1.000 V
+4 ⟶ +3:	$VO^{2+}(aq) + 2\,H^+(aq) + e^- \longrightarrow V^{3+}(aq) + H_2O(l)$ (green)	0.337 V
+3 ⟶ +2:	$V^{3+}(aq) + e^- \longrightarrow V^{2+}(aq)$ (violet)	−0.255 V
+2 ⟶ 0:	$V^{2+}(aq) + 2\,e^- \longrightarrow V(s)$	−1.13 V

EXAMPLE 23-2   **Using Electrode Potential Data to Predict an Oxidation–Reduction Reaction**

Can $MnO_4^-(aq)$ be used to oxidize $VO^{2+}(aq)$ to $VO_2^+(aq)$ for standard-state conditions in an acidic solution? If so, write a balanced equation for the redox reaction.

Analyze

We need to start by writing two half-equations, one for the reduction of MnO_4^- to Mn^{2+} the other for the oxidation of VO^{2+} to VO_2^+. Both occur in acidic solution. We find one $E°$ value in Table 23.4 and the other in Table 20.1. We then combine the $E°$ values to obtain a value for $E°_{cell}$.

Reduction: $\qquad\qquad\qquad MnO_4^- + 8\,H^+ + 5\,e^- \longrightarrow Mn^{2+} + 4\,H_2O$

Oxidation: $\qquad\qquad\qquad 5(VO^{2+} + H_2O \longrightarrow VO_2^+ + 2\,H^+ + e^-)$

Overall: $\quad 5\,VO^{2+}(aq) + MnO_4^-(aq) + H_2O(l) \longrightarrow 5\,VO_2^+(aq) + Mn^{2+}(aq) + 2\,H^+(aq)$

$$E°_{cell} = E°_{MnO_4^-/Mn^{2+}} - E°_{VO_2^+/VO^{2+}} = 1.51\ V - 1.000\ V = 0.51\ V$$

Because $E°_{cell}$ is positive, we predict that MnO_4^- should oxidize VO^{2+} to VO_2^+ for standard-state conditions in acidic solution.

Assess

By using equation (20.17), we can easily verify that K for the overall reaction is very large (greater than 1×10^{43}), which is another indication that $MnO_4^-(aq)$ could be effectively used for oxidizing $VO^{2+}(aq)$ to $VO_2^+(aq)$.

PRACTICE EXAMPLE A: Use data from Tables 20.1 and 23.4 to determine whether nitric acid can be used to oxidize $V^{3+}(aq)$ to $VO^{2+}(aq)$ for standard-state conditions. If so, write a balanced equation for the reaction.

PRACTICE EXAMPLE B: Select a *reducing agent* from Table 20.1 that can be used to reduce VO^{2+} to V^{2+} for standard-state conditions in acidic solution. Consider that the reduction occurs in two stages: $VO^{2+} \longrightarrow V^{3+} \longrightarrow V^{2+}$, but note that the V^{2+} must not be reduced to $V(s)$.

Steel is chrome-plated from an aqueous solution containing CrO_3 and H_2SO_4. The plating obtained is thin and porous. It tends to develop cracks unless the steel is first plated with copper or nickel, which provides the true protective coating. Then chromium is plated over this layer for extra protection and decorative purposes. The efficiency of chrome-plating is limited by the fact that reduction of Cr(VI) to Cr(0) produces only $\frac{1}{6}$ mol Cr per mole of electrons. In other words, large quantities of electric energy are required for chrome-plating relative to other types of metal plating.

Chromium, like vanadium, has a variety of oxidation states in aqueous solution, each having a different color.

▶ The colors may also depend on other species present in solution. For example, if $[Cl^-]$ is high, $[Cr(H_2O)_6]^{3+}$ is converted to $[CrCl_2(H_2O)_4]^+$ and the violet color changes to green.

O.S. +2:	$[Cr(H_2O)_6]^{2+}$, blue	
O.S. +3:	(acidic) $[Cr(H_2O)_6]^{3+}$, violet	(basic) $[Cr(OH)_4]^-$, green
O.S. +6:	(acidic) $Cr_2O_7^{2-}$, orange	(basic) CrO_4^{2-}, yellow

The oxides and hydroxides of chromium conform to the general principles of acid–base behavior: CrO is basic, Cr_2O_3 is amphoteric, and CrO_3 is acidic.

Pure chromium reacts with dilute HCl(aq) or H_2SO_4(aq) to produce Cr^{2+}(aq). Nitric acid and other oxidizing agents alter the surface of the metal (perhaps by formation of an oxide coating). They render the metal resistant to further attack—it becomes *passive*. A better source of chromium compounds than the pure metal is the alkali metal chromates, which contain Cr(VI) and can be obtained directly from chromite ore by reactions such as

$$4\,Fe(CrO_2)_2 + 8\,Na_2CO_3 + 7\,O_2 \xrightarrow{\Delta} 2\,Fe_2O_3 + 8\,Na_2CrO_4 + 8\,CO_2 \qquad \textbf{(23.16)}$$

The *sodium chromate*, Na_2CrO_4, produced by this reaction is the source of many industrially important chromium compounds.

The Cr(VI) oxidation state is also observed in the red oxide, CrO_3. As expected, this oxide dissolves in water to produce a strongly acidic solution.

The product of the reaction is not the expected chromic acid, H_2CrO_4, however, which has never been isolated in the pure state. Instead, the observed reaction is

$$2\,CrO_3(s) + H_2O(l) \longrightarrow 2\,H^+(aq) + Cr_2O_7^{2-}(aq) \qquad \textbf{(23.17)}$$

It is possible to crystallize a *dichromate* salt from an aqueous solution of CrO_3. If the solution is made basic, the color turns from orange to yellow. From basic solutions, only *chromate* salts can be crystallized. Thus, whether a solution contains Cr(VI) as $Cr_2O_7^{2-}$ or CrO_4^{2-} or a mixture of the two depends on the pH. The relevant equations follow.

$$2\,CrO_4^{2-}(aq) + 2\,H^+(aq) \rightleftharpoons Cr_2O_7^{2-}(aq) + H_2O(l) \qquad \textbf{(23.18)}$$

$$K_c = \frac{[Cr_2O_7^{2-}]}{[CrO_4^{2-}]^2[H^+]^2} = 3.2 \times 10^{14} \qquad \textbf{(23.19)}$$

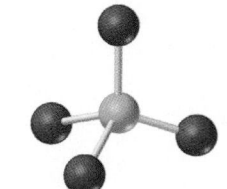

Chromate ion (CrO_4^{2-})

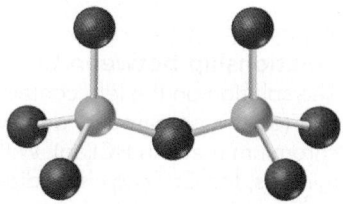

Dichromate ion ($Cr_2O_7^{2-}$)

Le Châtelier's principle predicts that the forward reaction of (23.18) is favored in acidic solutions and that the predominant Cr(VI) species is $Cr_2O_7^{2-}$. In basic solution, H^+ ions are removed and the reverse reaction is favored, forming CrO_4^{2-} as the principal species. Careful control of the pH is necessary when $Cr_2O_7^{2-}$ is used as an oxidizing agent or CrO_4^{2-} as a precipitating agent. In addition, equation (23.19) can be used to calculate the relative amounts of the two ions as a function of $[H^+]$.

Chromate ion in basic solution can be used to precipitate metal chromates such as $BaCrO_4(s)$ and $PbCrO_4(s)$. It is not a good oxidizing agent, however; it is not readily reduced.

$$CrO_4^{2-}(aq) + 4\,H_2O(l) + 3\,e^- \longrightarrow [Cr(OH)_4]^-(aq) + 4\,OH^-(aq) \quad E^\circ = -0.13\,V$$

Dichromates are poor precipitating agents but excellent oxidizing agents, which are used in a variety of industrial processes. In the chrome leather tanning process, for example, animal hides are immersed in $Na_2Cr_2O_7(aq)$, which is then reduced by $SO_2(g)$ to soluble basic chromic sulfate, $Cr(OH)SO_4$. Collagen, a protein in hides, reacts to form an insoluble chromium complex. The hides become *leather*, a tough, pliable material resistant to biological attack.

Dichromates are easily reduced to Cr_2O_3. In the case of ammonium dichromate, simply heating the compound produces Cr_2O_3 in a dramatic reaction (Fig. 23-12).

▲ FIGURE 23-12
Decomposition of $(NH_4)_2Cr_2O_7$
Ammonium dichromate (left) contains both an oxidizing agent, $Cr_2O_7^{2-}$, and a reducing agent, NH_4^+. The products of the reaction between these two ions are $Cr_2O_3(s)$ (right), $N_2(g)$, and $H_2O(g)$. Considerable heat and light are also evolved (center).

▶ FIGURE 23-13
Relationship between Cr²⁺ and Cr³⁺
The solution on the left, containing blue $Cr^{2+}(aq)$, is prepared by dissolving chromium metal in HCl(aq). Within minutes, the $Cr^{2+}(aq)$ is oxidized to green $Cr^{3+}(aq)$ by atmospheric oxygen (right). The green color is that of the complex ion $[CrCl_2(H_2O)_4]^+(aq)$.

Chromium(II) compounds can be prepared by the reduction of Cr(III) compounds with zinc in acidic solution or electrolytically at a lead cathode. The most distinctive feature of Cr(II) compounds is their reducing power.

$$Cr^{3+}(aq) + e^- \longrightarrow Cr^{2+}(aq) \qquad E° = -0.424 \text{ V}$$

That is, the oxidation of $Cr^{2+}(aq)$ occurs readily. In fact, Cr(II) solutions can be used to purge gases of trace amounts of $O_2(g)$, through the following reaction, illustrated in Figure 23-13.

$$4\,Cr^{2+}(aq) + O_2(g) + 4\,H^+(aq) \longrightarrow 4\,Cr^{3+}(aq) + 2\,H_2O(l) \qquad E°_{cell} = +1.653 \text{ V}$$

Pure Cr can be obtained in small amounts by reducing Cr_2O_3 with Al in a reaction similar to the thermite reaction.

$$Cr_2O_3(s) + 2\,Al(s) \longrightarrow Al_2O_3(s) + 2\,Cr(l) \qquad \textbf{(23.20)}$$

Manganese

Manganese is a fairly abundant element, constituting about 1% of Earth's crust. Its principal ore is *pyrolusite*, MnO_2. Like V and Cr, Mn is most important in steel production, generally as the iron–manganese alloy, *ferromanganese*. Ferromanganese can be obtained by the reduction of a mixture of pyrolusite and hematite iron ores with carbon.

$$MnO_2 + Fe_2O_3 + 5\,C \xrightarrow{\Delta} \underset{\text{Ferromanganese}}{Mn + 2\,Fe} + 5\,CO$$

Mn participates in the purification of iron by reacting with sulfur and oxygen and removing them through slag formation. In addition, Mn increases the hardness of steel. Steel containing high proportions of Mn is extremely tough and wear-resistant in such applications as railroad rails, bulldozers, and road scrapers.

The electron configuration of Mn is $[Ar]3d^54s^2$. By employing first the two 4s electrons and then, consecutively, up to all five of its unpaired 3d electrons, manganese exhibits all oxidation states from +2 to +7. The most important reactions of manganese compounds are oxidation–reduction reactions. Standard electrode potential diagrams are given in Figure 23-14. These diagrams help explain the following observations:

- $Mn^{3+}(aq)$ is unstable; that is, its disproportionation is spontaneous.

$$2\,Mn^{3+}(aq) + 2\,H_2O(l) \longrightarrow Mn^{2+}(aq) + MnO_2(s) + 4\,H^+(aq)$$

$$E°_{cell} = 0.54 \text{ V} \qquad \textbf{(23.21)}$$

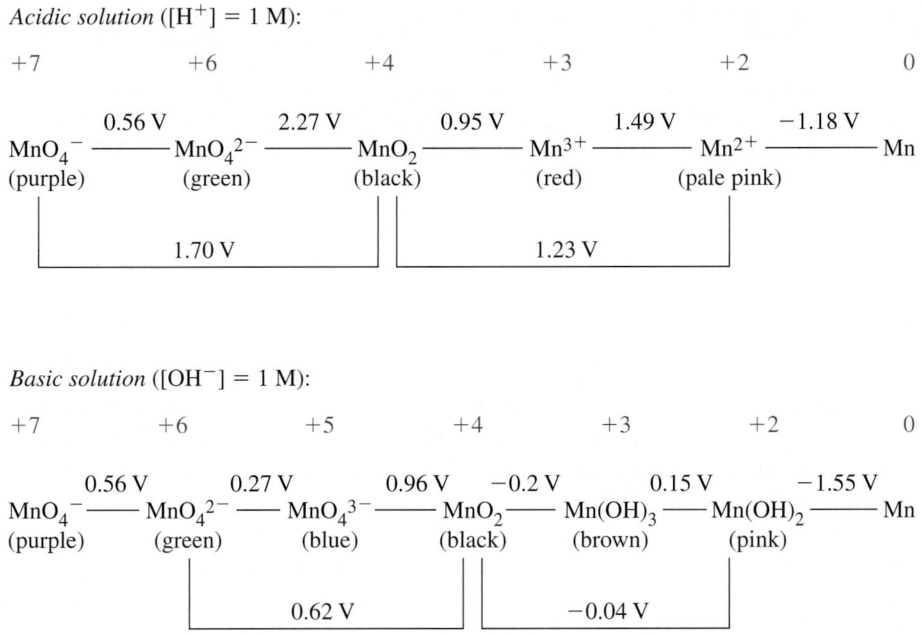

Acidic solution ([H$^+$] = 1 M):

| +7 | +6 | +4 | +3 | +2 | 0 |

$$\underset{\substack{(\text{purple})}}{\text{MnO}_4^-} \xrightarrow{0.56\text{ V}} \underset{\substack{(\text{green})}}{\text{MnO}_4^{2-}} \xrightarrow{2.27\text{ V}} \underset{\substack{(\text{black})}}{\text{MnO}_2} \xrightarrow{0.95\text{ V}} \underset{\substack{(\text{red})}}{\text{Mn}^{3+}} \xrightarrow{1.49\text{ V}} \underset{\substack{(\text{pale pink})}}{\text{Mn}^{2+}} \xrightarrow{-1.18\text{ V}} \text{Mn}$$

1.70 V   1.23 V

Basic solution ([OH$^-$] = 1 M):

| +7 | +6 | +5 | +4 | +3 | +2 | 0 |

$$\underset{\substack{(\text{purple})}}{\text{MnO}_4^-} \xrightarrow{0.56\text{ V}} \underset{\substack{(\text{green})}}{\text{MnO}_4^{2-}} \xrightarrow{0.27\text{ V}} \underset{\substack{(\text{blue})}}{\text{MnO}_4^{3-}} \xrightarrow{0.96\text{ V}} \underset{\substack{(\text{black})}}{\text{MnO}_2} \xrightarrow{-0.2\text{ V}} \underset{\substack{(\text{brown})}}{\text{Mn(OH)}_3} \xrightarrow{0.15\text{ V}} \underset{\substack{(\text{pink})}}{\text{Mn(OH)}_2} \xrightarrow{-1.55\text{ V}} \text{Mn}$$

0.62 V   −0.04 V

◀ FIGURE 23-14
Electrode potential diagrams for manganese

- Manganate ion, MnO_4^{2-}, is also unstable in acidic solution; its disproportionation is spontaneous.

  $$3\,MnO_4^{2-}(aq) + 4\,H^+(aq) \longrightarrow MnO_2(s) + 2\,MnO_4^-(aq) + 2\,H_2O(l)$$
  $$E^\circ_{cell} = 1.70\text{ V} \quad \textbf{(23.22)}$$

- If $[OH^-]$ is kept sufficiently high, the following reaction can be reversed; thus, manganate ion can be maintained as a stable species in a strongly basic medium.

  $$3\,MnO_4^{2-}(aq) + 2\,H_2O(l) \longrightarrow MnO_2(s) + 2\,MnO_4^-(aq) + 4\,OH^-(aq)$$
  $$E^\circ_{cell} = 0.04\text{ V} \quad \textbf{(23.23)}$$

Manganese dioxide is used in dry-cell batteries, in glass and ceramic glazes, and as a catalyst; it is also the principal source of manganese compounds. When MnO_2 is heated in the presence of an alkali and an oxidizing agent, a manganate salt is produced.

$$3\,MnO_2 + 6\,KOH + KClO_3 \xrightarrow{\Delta} 3\,K_2MnO_4 + KCl + 3\,H_2O$$

K_2MnO_4 is extracted from the fused mass with water and can then be oxidized to $KMnO_4$, *potassium permanganate* (with Cl_2 as an oxidizing agent, for instance). Potassium permanganate, $KMnO_4$, is an important laboratory oxidizing agent. For chemical analyses, it is generally used in acidic solutions in which it is reduced to $Mn^{2+}(aq)$. In the analysis of iron by MnO_4^-, a sample of Fe^{2+} is prepared by dissolving iron in an acid and reducing any Fe^{3+} back to Fe^{2+}. Then the sample is titrated with $MnO_4^-(aq)$.

$$5\,Fe^{2+}(aq) + MnO_4^-(aq) + 8\,H^+(aq) \longrightarrow$$
$$5\,Fe^{3+}(aq) + Mn^{2+}(aq) + 4\,H_2O(l) \quad \textbf{(23.24)}$$

$Mn^{2+}(aq)$ has a barely discernible pale pink color. $MnO_4^-(aq)$ is an intense purple color. At the end point of the titration reaction (23.24), the solution acquires a lasting light purple color with just one drop of excess $MnO_4^-(aq)$ (recall Figure 5-18). $MnO_4^-(aq)$ is less satisfactory for titrations in alkaline solutions because the insoluble reduction product, brown $MnO_2(s)$, obscures the end point.

▲ Cobalt–samarium magnets are used in high-efficiency motors.

23-5   The Iron Triad: Iron, Cobalt, and Nickel

The transition elements iron, cobalt, and nickel comprise the *iron triad*. Iron, with an annual worldwide production of more than 500 million metric tons, is the most important metal in modern civilization. It is widely distributed in Earth's crust at an abundance of 4.7%. The major commercial use of iron is to make steel (see Section 23-3).

Cobalt is among the rarer elements. It comprises only about 0.0020% of Earth's crust, but it occurs in sufficiently concentrated deposits (ores) so that its annual production runs into the millions of pounds. Cobalt is used primarily in alloys with other metals. Like iron, cobalt is ferromagnetic. One alloy of cobalt, Co_5Sm, makes a particularly strong and lightweight permanent magnet. Because of the strength of its magnetic field, magnets of this alloy are used in the manufacture of miniature electronic devices.

Nickel ranks twenty-fourth in abundance among the elements in Earth's crust. Its ores are mainly the sulfides, oxides, silicates, and arsenides. Particularly large deposits are found in Canada. Of the 136 million kilograms of nickel consumed annually in the United States, about 80% goes to the production of alloys. Another 15% is used in electroplating, and the remainder for miscellaneous purposes (for example, as catalysts).

Oxidation States

Variability of oxidation state is seen in the iron triad, even if not to the same degree as with vanadium, chromium, and manganese. The +2 oxidation state is commonly encountered in all three metals.

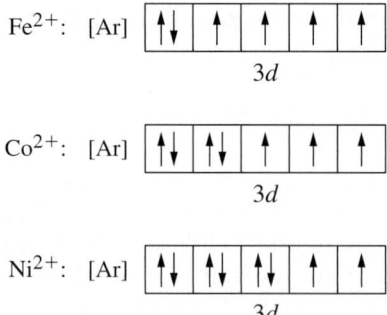

For cobalt and nickel, the +2 oxidation state is the most stable, but for iron, the most stable is +3.

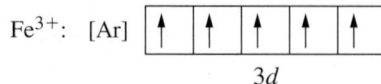

The electron configuration of Fe^{3+} has a half-filled d subshell with all the d electrons unpaired. This type of electron configuration has a special stability. As we can see from the orbital diagrams of Co^{2+} and Ni^{2+}, neither would have a half-filled $3d$ subshell following the loss of one additional electron. Co^{3+} and

Ni^{3+} are not readily formed. To illustrate the contrast between Fe^{3+} on the one hand, and Co^{3+} and Ni^{3+}, on the other, consider the following observations.

- $Fe^{2+}(aq)$ is spontaneously oxidized to $Fe^{3+}(aq)$ by $O_2(g)$ at 1 atm in a solution with $[H^+] = 1$ M.

$$4\,Fe^{2+}(aq) + O_2(g) + 4\,H^+(aq) \longrightarrow 4\,Fe^{3+}(aq) + 2\,H_2O(l)$$

$$E^\circ_{cell} = 0.44\ \text{V} \quad \textbf{(23.25)}$$

  This reaction is still spontaneous with lower O_2 partial pressures and less acidic solutions.

- For the half-reaction $Co^{3+}(aq) + e^- \longrightarrow Co^{2+}(aq)$, $E^\circ = 1.82$ V. The reduction of $Co^{3+}(aq)$ to $Co^{2+}(aq)$ occurs readily; conversely, the oxidation of $Co^{2+}(aq)$ to Co^{3+} occurs only with difficulty. As will be noted in Chapter 24, however, the oxidation state +3 can be attained when Co^{3+} is the central metal ion in very stable complex ions.

- Nickel(III) compounds are used in batteries. For example, in the nickel–cadmium (Nicad) cell, the cathode half-reaction is $NiO(OH) + H^+ + e^- \longrightarrow Ni(OH)_2$. It is the ease of this reduction, when combined with the oxidation half-reaction, $Cd(s) + 2\,OH^-(aq) \longrightarrow Cd(OH)_2(s) + 2\,e^-$, that produces a cell with a voltage of about 1.5 V.

Some Reactions of the Iron Triad Elements

The reactions of the iron triad elements are many and varied. The metals are more active than hydrogen and liberate $H_2(g)$ from an acidic solution. Hydrated Co^{2+} and Ni^{2+} are red and green, respectively. In aqueous solution, Fe^{2+} is pale green and fully hydrated Fe^{3+} is purple. Generally, however, solutions of $Fe^{2+}(aq)$ are yellow to brown, but this color probably results from the presence of species formed in the hydrolysis of $Fe^{3+}(aq)$. Like the hydrolysis of $Al^{3+}(aq)$, described on page 733, that of $Fe^{3+}(aq)$ produces an acidic solution.

$$[Fe(H_2O)_6]^{3+}(aq) + H_2O(l) \rightleftharpoons [FeOH(H_2O)_5]^{2+}(aq) + H_3O^+(aq)$$

$$K_a = 8.9 \times 10^{-4} \quad \textbf{(23.26)}$$

Some reactions that can be used to identify and distinguish between $Fe^{2+}(aq)$ and $Fe^{3+}(aq)$ are summarized in Table 23.5.

An interesting set of reactions involves the complex ions $[Fe(CN)_6]^{4-}$ and $[Fe(CN)_6]^{3-}$. These ions are commonly called *ferrocyanide* and *ferricyanide*, respectively. $Fe^{3+}(aq)$ yields a dark blue precipitate called *Prussian blue* when treated with potassium ferrocyanide, $K_4[Fe(CN)_6](aq)$, whereas $Fe^{2+}(aq)$ yields a similar blue precipitate, called *Turnbull's blue*, when treated with potassium ferricyanide, $K_3[Fe(CN)_6](aq)$. Together, these two and similar related precipitates are known commercially as *iron blue*. Iron blue is used as a pigment for paints, printing inks, laundry bluing, art colors, cosmetics (eye shadow), and blueprinting. An additional sensitive test for $Fe^{3+}(aq)$ is the formation of a blood-red complex ion with thiocyanate ion, $SCN^-(aq)$.

$$[Fe(H_2O)_6]^{3+} + SCN^-(aq) \longrightarrow [FeSCN(H_2O)_5]^{2+} + H_2O(l)$$

◀ The blue precipitate that establishes the presence of $Fe^{2+}(aq)$ from the corroding nails in Figure 20-20 is Turnbull's blue.

◀ In Chapter 24, we will learn to write the systematic names for ferrocyanide and ferricyanide ions; they are hexacyanoferrate(II) and hexacyanoferrate(III), respectively.

TABLE 23.5   Some Qualitative Tests for $Fe^{2+}(aq)$ and $Fe^{3+}(aq)$

Reagent	$Fe^{2+}(aq)$	$Fe^{3+}(aq)$
$NaOH(aq)$	Green precipitate	Red-brown precipitate
$K_4[Fe(CN)_6]$	White precipitate, turning blue rapidly	Prussian blue precipitate
$K_3[Fe(CN)_6]$	Turnbull's blue precipitate	Red-brown (no precipitate)
$KSCN(aq)$	No color	Deep red

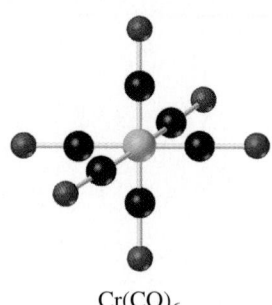

$Cr(CO)_6$

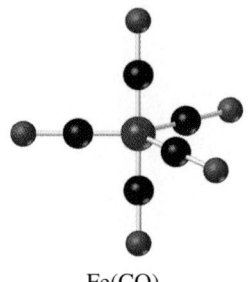

$Fe(CO)_5$

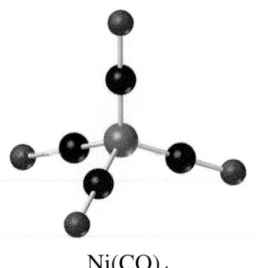

$Ni(CO)_4$

▲ FIGURE 23-15
Structure of some simple carbonyls

TABLE 23.6 Three Metal Carbonyls

	Number of e^-		
	From Metal	From CO	Total
$Cr(CO)_6$	24	12	36
$Fe(CO)_5$	26	10	36
$Ni(CO)_4$	28	8	36

With only a few exceptions, the transition metals form compounds with carbon monoxide (CO), called **metal carbonyls**. In the simple metal carbonyls listed in Table 23.6,

- each CO molecule contributes an electron pair to an empty orbital of the metal atom
- all electrons are paired (most metal carbonyls are diamagnetic)
- the metal atom acquires the electron configuration of the noble gas Kr

The structures of the simple carbonyls in Figure 23-15 are those that we would predict from VSEPR theory (based on a number of electron pairs around the central atom equal to the number of CO molecules).

Metal carbonyls are produced in several ways. Nickel combines with CO(g) at ordinary temperatures and pressures in a reversible reaction.

$$Ni(s) + 4\,CO(g) \rightleftharpoons Ni(CO)_4(l)$$

The reaction above forms the basis of an important industrial process called the Mond process for obtaining nickel metal from its oxides. In the Mond process, CO(g) is passed over a mixture of metal oxides. Nickel is carried off as $Ni(CO)_4(g)$, while the other oxides are reduced to the metals. When $Ni(CO)_4(g)$ is subsequently heated to about 250 °C, the carbonyl complex decomposes, yielding Ni(s). With iron, it is necessary to use higher temperatures (200 °C) and CO(g) pressures (100 atm).

$$Fe(s) + 5\,CO(g) \rightleftharpoons Fe(CO)_5(g)$$

In other cases, the carbonyl is obtained by reducing a metal compound in the presence of CO(g).

Carbon monoxide poisoning results from a reaction similar to carbonyl formation. CO molecules coordinate with Fe atoms in hemoglobin in the blood, displacing the O_2 molecules normally carried by hemoglobin. The metal carbonyls themselves are also very poisonous.

🔍 23-5 CONCEPT ASSESSMENT

What is the oxidation state of iron in the compound $Fe(CO)_5$?

23-6 Group 11: Copper, Silver, and Gold

Throughout the ages, Cu, Ag, and Au have been the preferred metals for coins because they are so durable and resistant to corrosion. The data in Table 23.7 help us understand why this is so. The metal ions are easy to reduce to the free metals, which means that the metals are difficult to oxidize.

In Mendeleev's periodic table, the alkali metals (group 1) and the coinage metals (group 11) appear together as group I. The only similarity between the two subgroups, however, is that both have a single s electron in the valence

TABLE 23.7   Some Properties of Copper, Silver, and Gold

	Cu	Ag	Au
Electron configuration	$[Ar]3d^{10}4s^1$	$[Kr]4d^{10}5s^1$	$[Xe]4f^{14}5d^{10}6s^1$
Metallic radius, pm	128	144	144
First ioniz. energy, kJ mol^{-1}	745	731	890
Electrode potential, V			
$M^+(aq) + e^- \longrightarrow M(s)$	+0.520	+0.800	+1.83
$M^{2+}(aq) + 2\,e^- \longrightarrow M(s)$	+0.340	+1.39	—
$M^{3+}(aq) + 3\,e^- \longrightarrow M(s)$	—	—	+1.52
Oxidation states[a]	+1, +2	+1, +2	+1, +3

[a]The most common oxidation states are shown in red.

shells of their atoms. More significant are the differences between the group 1 and group 11 metals. For example, the first ionization energies for the group 11 metals are much larger than for the group 1 metals, and the standard electrode potentials, $E°$, are positive for the group 11 metals and negative for the group 1 metals.

Like the other transition elements that precede them in the periodic table, the group 11 metals can use d electrons in chemical bonding. Thus they can exist in different oxidation states, exhibit paramagnetism and color in some of their compounds, and form complex ions. They also possess to a high degree some of the distinctive physical properties of metals—malleability, ductility, and excellent electrical and thermal conductivity.

Copper, silver, and gold—the coinage metals—are used in jewelry making and the decorative arts. Gold, for instance, is extraordinarily malleable and can be pounded into thin translucent sheets known as gold leaf. The coinage metals are valued by the electronics industry for their ability to conduct electricity. Silver has the highest electrical conductivity of any pure element, but both copper and gold are more often used as electrical conductors because copper is inexpensive and gold does not readily corrode. The most important use of gold is as the monetary reserve of nations throughout the world.

Generally, the coinage metals are resistant to air oxidation, although silver will tarnish through reactions with sulfur compounds in air to produce black Ag_2S. In moist air, copper corrodes to produce green basic copper carbonate. This is the green color associated with copper roofing and gutters and bronze statues. (Bronze is an alloy of Cu and Sn.) Fortunately, this corrosion product forms a tough adherent coating that protects the underlying metal. The corrosion reaction is complex but may be summarized as

$$2\,Cu(s) + H_2O(g) + CO_2(g) + O_2(g) \longrightarrow Cu_2(OH)_2CO_3(s) \qquad \text{(23.27)}$$
Basic copper carbonate

The group 11 metals do not react with HCl(aq), but both Cu and Ag react with concentrated $H_2SO_4(aq)$ or $HNO_3(aq)$. The metals are oxidized to Cu^{2+} and Ag^+, respectively, and the reduction products are $SO_2(g)$ in $H_2SO_4(aq)$ and either $NO(g)$ or $NO_2(g)$ in $HNO_3(aq)$.

Au does not react with either acid, but it will react with "royal water"—*aqua regia* (1 part HNO_3 and 3 parts HCl). The $HNO_3(aq)$ oxidizes the metal and Cl^- from the HCl(aq) promotes the formation of the stable complex ion $[AuCl_4]^-$.

$$Au(s) + 4\,H^+(aq) + NO_3{}^-(aq) + 4\,Cl^-(aq) \longrightarrow$$
$$[AuCl_4]^-(aq) + 2\,H_2O(l) + NO(g) \qquad \text{(23.28)}$$

Trace amounts of Cu are essential to life, but larger quantities are toxic, especially to bacteria, algae, and fungi. Among the many copper compounds used

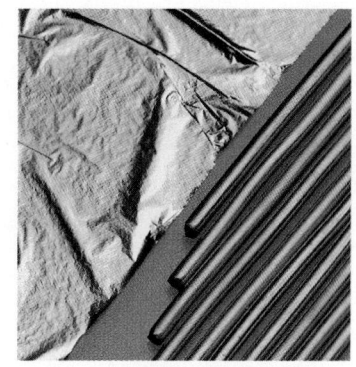

▲ Gold leaf and copper wire

as pesticides are the basic acetate, carbonate, chloride, hydroxide, and sulfate. Commercially, the most important copper compound is $CuSO_4 \cdot 5\,H_2O$. In addition to its agricultural applications, $CuSO_4$ is employed in batteries and electroplating, in the preparation of other copper salts, and in a variety of industrial processes.

Silver(I) nitrate is the principal silver compound of commerce and is also an important laboratory reagent for the precipitation of anions, most of which form insoluble silver salts. These precipitation reactions can be used for the quantitative determination of anions, either gravimetrically (by weighing precipitates) or volumetrically (by titration). Most other Ag compounds are derived from $AgNO_3$. Ag compounds are used in electroplating, in the manufacture of batteries, as catalysts, and in cloud seeding (AgI). Silver halides, such as AgBr, are used in photography (see Section 24-10), although this is diminishing now because of the advent of digital cameras.

Gold compounds are used in electroplating, photography, medicinal chemistry (as anti-inflammatory agents for severe rheumatoid arthritis, for example), and the manufacture of special glasses and ceramics (such as ruby glass).

▲ Gold plating on a space antenna.

23-6 CONCEPT ASSESSMENT

What is the electron configuration of gold(III) ion?

23-7 Group 12: Zinc, Cadmium, and Mercury

The properties of the group 12 elements are consistent with elements having full subshells, $(n-1)d^{10}ns^2$; some of those properties are summarized in Table 23.8. The low melting and boiling points of the group 12 metals can probably be attributed to the fact that, with only the ns^2 electrons participating, metallic bonding is weak. Mercury is the only metal that exists as a liquid at room temperature and below (although liquid gallium can easily be supercooled to room temperature). Mercury differs from Zn and Cd in a number of ways in addition to physical appearance.

- Mercury has little tendency to combine with oxygen. Its oxide, HgO, is thermally unstable.

- Very few mercury compounds are water soluble, and most are not hydrated.

TABLE 23.8 Some Properties of the Group 12 Metals

	Zn	Cd	Hg
Density, g cm^{-3}	7.14	8.64	13.59 (liquid)
Melting point, °C	419.6	320.9	−38.87
Boiling point, °C	907	765	357
Electron configuration	$[Ar]3d^{10}4s^2$	$[Kr]4d^{10}5s^2$	$[Xe]4f^{14}5d^{10}6s^2$
Atomic radius, pm	133	149	160
Ionization energy, kJ mol^{-1}			
First	906	867	1006
Second	1733	1631	1809
Principal oxidation state(s)	+2	+2	+1, +2
Electrode potential $E°$, V			
$[M^{2+}(aq) + 2\,e^- \longrightarrow M]$	−0.763	−0.403	+0.854
$[M_2^{2+}(aq) + 2\,e^- \longrightarrow 2\,M]$	—	—	+0.796

- Many mercury compounds are covalent. Except for HgF_2, mercury halides are only slightly ionized in aqueous solution.
- Mercury(I) forms a common diatomic ion with a metal–metal covalent bond, Hg_2^{2+}.
- Mercury will not displace $H_2(g)$ from $H^+(aq)$.

Some of these differences exhibited by mercury can probably be attributed to the *relativistic effect* discussed in Chapter 9 (Focus On 9, www.masteringchemistry.com). Because the speed of the $6s$ electrons reaches a significant fraction of the speed of light, as they approach the high positive charge of the mercury nucleus, their masses increase (as predicted by Einstein's theory of relativity) and the $6s$ orbital shrinks in size. The closer approach of the $6s$ electrons to the nucleus subjects them to a greater attractive force than that experienced by the ns^2 electrons in Zn and Cd. As a result, for example, the first ionization energy of Hg is greater than that of Zn or Cd.

Uses of the Group 12 Metals and Their Compounds

About one-third of the zinc produced is used in coating iron to give it corrosion protection (Section 20-6). The product is called *galvanized iron*. Large quantities of Zn are consumed in the manufacture of alloys. For example, about 20% of the production of Zn is used in *brass*, a copper alloy having 20%–45% Zn and small quantities of Sn, Pb, and Fe. Brass is a good electrical conductor and is corrosion resistant. Zinc is also employed in the manufacture of dry-cell batteries, in printing (lithography), in the construction industry (roofing materials), and as sacrificial anodes in corrosion protection (Section 20-6).

Although poisonous, cadmium is substituted for zinc as a shiny and protective plating on iron in special applications. It is used in bearing alloys, in low-melting solders, in aluminum solders, and as an additive to impart strength to copper. Another application, based on its neutron-absorbing capacity, is in control rods and shielding for nuclear reactors.

The principal uses of mercury take advantage of its metallic and liquid properties and its high density. It is used in thermometers, barometers, gas-pressure regulators, and electrical relays and switches, and as electrodes, as in the chlor-alkali process (Section 20-8). Mercury vapor is used in fluorescent tubes and street lamps. Mercury alloys, called **amalgams**, can be made with most metals, and some of these amalgams are of commercial importance. A silver dental filling is an amalgam of mercury with an alloy containing about 70% Ag, 26% Sn, 3% Cu, and 1% Zn.

Table 23.9 lists a few important compounds of the group 12 metals and some of their uses. Some of the most interesting compounds are the

▲ A brass seagoing chronometer made by John Harrison in the eighteenth century.

◀ Iron is one of the few metals that does not form an amalgam. Mercury is generally stored and shipped in iron containers.

Compound	Uses
TABLE 23.9 Some Important Compounds of the Group 12 Metals	
ZnO	Reinforcing agent in rubber; pigment; cosmetics; dietary supplement; photoconductors in copying machines
ZnS	Phosphors in X-ray and television screens; pigment; luminous paints
$ZnSO_4$	Rayon manufacture; animal feeds; wood preservative
CdO	Electroplating; batteries; catalyst
CdS	Solar cells; photoconductor in photocopying; phosphors; pigment
$CdSO_4$	Electroplating; standard voltaic cells (Weston cell)
HgO	Polishing compounds; dry cells; antifouling paints; fungicide; pigment
$HgCl_2$	Manufacture of Hg compounds; disinfectant; fungicide; insecticide; wood preservative
Hg_2Cl_2	Electrodes; pharmaceuticals; fungicide

semiconductors ZnO, CdS, and HgS, which are also the artist's pigments zinc white, cadmium yellow, and vermilion (red), respectively. Like all semiconductor materials, these compounds have an electronic structure consisting of a valence band and a conduction band (see Section 11-7). When light interacts with these compounds, electrons from the valence band may absorb photons and be excited into the conduction band. The energy of the light absorbed must equal or exceed the energy difference between the bands, called the *band gap*. The characteristic colors of these materials depend on the widths of the band gaps, as described in Focus On 21, www.masteringchemistry.com.

Mercury and Cadmium Poisoning

Accumulations of mercury in the body affect the nervous system and cause brain damage. One form of chronic mercury poisoning, "hatter's disease," was fairly common in the nineteenth century. Mercury compounds were used to convert fur to felt for making hats. Many hat makers of the time worked in hot, cramped spaces and used these compounds without special precautions. The hatters inadvertently ingested or inhaled the toxic mercury compounds while they worked.

One proposed mechanism of mercury poisoning, based on the fact that Hg has a high affinity for sulfur, involves interference with the functioning of sulfur-containing enzymes. Organic mercury compounds are generally more poisonous than inorganic ones and much more toxic than the element itself. An insidious aspect of mercury poisoning is that certain microorganisms have the ability to convert mercury(II) compounds to methylmercury (CH_3Hg^+) compounds, which then concentrate in the food chains of fish and other aquatic life. An early discovery of the environmental hazard of mercury was in Japan in the 1950s. Dozens of cases of mercury poisoning, including more than 40 deaths, occurred among residents of the shores of Minamata Bay. Local seafood with up to 20 ppm of mercury was a major component of the victims' diet. The source of contamination was traced to a chemical plant discharging mercury waste into the bay.

In the free state, mercury is most poisonous as a vapor. Levels of mercury that exceed 0.05 mg Hg/m^3 air are considered unsafe. Although we think of mercury as having a low vapor pressure, the concentration of Hg in its saturated vapor far exceeds this limit, and mercury vapor levels sometimes exceed safe limits where mercury is used—as in chlor–alkali plants, thermometer factories, and smelters.

Although zinc is an essential element in trace amounts, cadmium, which so closely resembles zinc, is a poison. One effect of cadmium poisoning is an extremely painful skeletal disorder known as *itai-itai kyo* (Japanese for "ouch-ouch" disease). This disorder was discovered in an area of Japan where effluents from a zinc mine became mixed with irrigation water used in rice fields, and cadmium poisoning was discovered in people who ate the rice. Cadmium poisoning can also cause liver damage, kidney failure, and pulmonary disease. The mechanism of cadmium poisoning may involve substitution in certain enzymes of Cd (a poison) for Zn (an essential element). Concern over cadmium poisoning has increased with an awareness that some cadmium is almost always found in zinc and zinc compounds, materials that have many commercial applications.

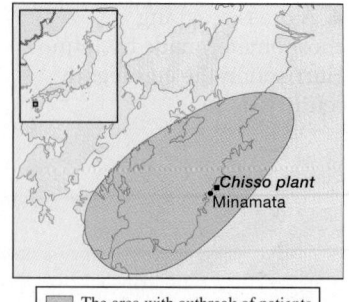

Chisso plant
Minamata

☐ The area with outbreak of patients

▲ The area of contamination in Japan where mercury poisoning was observed.

🔍   **23-7   CONCEPT ASSESSMENT**

Suggest the geometry of the cadmium anion $[CdCl_5]^{3-}$.

23-8   Lanthanides

The elements from lanthanum ($Z = 57$) through lutetium ($Z = 71$) are variously called the *lanthanide, lanthanoid,* or *rare earth elements.* The rare earth elements are "rare" only relative to the alkaline earth metals (group 2). Otherwise, they are not particularly rare. Ce, Nd, and La, for example, are more abundant than lead, and Tm is about as abundant as iodine. The lanthanides occur primarily as oxides, and mineral deposits containing them are found in various locations. Large deposits near the California-Nevada border are being developed to provide oxides of the lanthanides for use as phosphors in color monitors and television sets. Cobalt–samarium (Co_5Sm) magnets are used in high-efficiency motors.

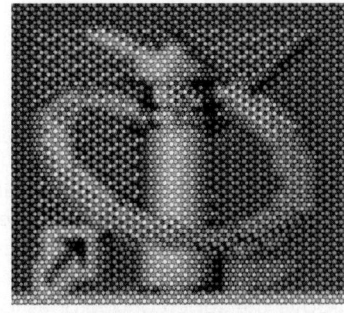

▲ The individual elements of a CRT computer screen are phosphors that contain the lanthanide elements. The phosphors glow red, blue, or green depending on their composition. (^{TM} Netscape ®)

Because the differences in electron configuration among the lanthanides are mainly in $4f$ orbitals, and because $4f$ electrons play a minor role in chemical bonding, strong similarities are found among these elements. For example, $E°$ values for the reduction process $M^{3+}(aq) + 3e^- \longrightarrow M(s)$ do not show much variation. All fall between -2.38 V (La) and -1.99 V (Eu). The differences in properties that do exist among the lanthanides arise mostly from the lanthanide contraction discussed in Section 23-1. This contraction is best illustrated in the radii of the ions M^{3+}. These radii decrease regularly by about 1–2 pm for each unit increase in atomic number, from a radius of 106 pm for La^{3+} to 85 pm for Lu^{3+}.

The lanthanides, which we can represent by the general symbol Ln, are reactive metals that liberate $H_2(g)$ from hot water and from dilute acids by undergoing oxidation to $Ln^{3+}(aq)$. The lanthanides combine with $O_2(g)$, sulfur, the halogens, $N_2(g)$, $H_2(g)$, and carbon in much the same way as expected for metals about as active as the alkaline earths. The pure metals can be prepared by electrolytic reduction of Ln^{3+} in a molten salt.

The most common oxidation state for the lanthanides is $+3$. About half the lanthanides can also be obtained in the oxidation state $+2$; the other half, $+4$. The special stability associated with an electron configuration involving half-filled f orbitals (f^7) may account for some of the observed oxidation states, but the reason for the predominance of the $+3$ oxidation state is less clear. Most of the lanthanide ions are paramagnetic and colored in aqueous solution.

The lanthanide elements are extremely difficult to extract from their natural sources and to separate from one another. All the methods for doing so are based on the following principle: Species that are strongly dissimilar can often be completely separated in a one-step process, such as separating $Ag^+(aq)$ and $Cu^{2+}(aq)$ by adding $Cl^-(aq)$; AgCl is insoluble. At best, species that are very similar can only be fractionated in a one-step process. That is, the ratio of the concentration of one species to that of another is altered slightly. To achieve a complete separation means repeating the same basic step hundreds or even thousands of times. To achieve the separation of the lanthanides, the methods of fractional crystallization, fractional precipitation, solvent extraction, and ion exchange were brought to their highest level of performance.

23-9   High-Temperature Superconductors

Magnetically levitated trains, magnetic resonance imaging (MRI) for medical diagnoses, and particle accelerators used in high-energy physics all require high magnetic fields generated by superconducting electromagnets. Superconductors offer no resistance to an electric current, so electricity is conducted with no loss of energy (Fig. 23-16).

If cooled to near absolute zero, all metals become superconducting. Several metals and alloys superconduct even at marginally higher temperatures of

▶ FIGURE 23-16
The small magnet induces an electric current in the superconductor below it. Associated with this current is another magnetic field that opposes the field of the small magnet, causing it to be repelled. The magnet remains suspended above the superconductor as long as the superconducting current is present, and the current persists as long as the temperature of the superconductor is maintained at the boiling point of liquid nitrogen (77 K).

10–15 K. To maintain a superconductor at these extremely low temperatures requires liquid helium (bp, 4 K) as a coolant.

In the mid-1980s, materials made of lanthanum, strontium, copper, and oxygen were found to become superconducting at 30 K. This was a much higher temperature for superconductivity than had been previously achieved. More surprising, the new materials were not metals but *ceramics*! In short order, other types of ceramic superconductors were discovered.

One of these new types was particularly easy to make. When a stoichiometric mixture of yttrium oxide (Y_2O_3), barium carbonate ($BaCO_3$), and copper(II) oxide (CuO) is heated in a stream of $O_2(g)$, a ceramic is produced with the approximate formula $YBa_2Cu_3O_x$ (where x is slightly less than 7), which is an oxygen-deficient version of the ceramic $YBa_2Cu_3O_7$ shown in Figure 23-17). This so-called YBCO ceramic becomes superconducting at the remarkably high temperature of 92 K. Although a temperature of 92 K is still quite low, it is far above the boiling point of helium. In fact, it is above the boiling point of nitrogen (77 K). Thus, inexpensive liquid nitrogen can be used as the coolant.

Many variations of the basic YBCO formula are possible. Almost any lanthanide element can be substituted for yttrium, and combinations of group 2 elements can be substituted for barium. All these variations yield materials that are superconducting at relatively high temperatures, but of the group, the yttrium compound is superconducting at the highest temperature.

▶ FIGURE 23-17
Structures of $YBa_2Cu_3O_7$ and LaOFeAs
Two important classes of superconductors are based on the structures of **(a)** $YBa_2Cu_3O_7$ and **(b)** LaOFeAs.

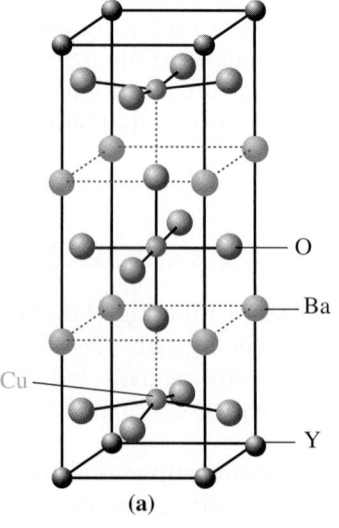

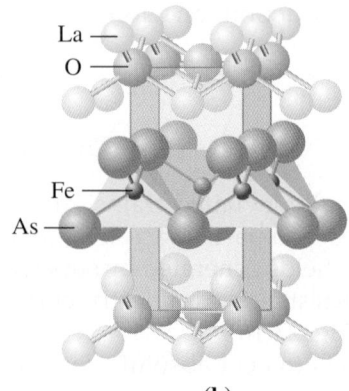

(a)   **(b)**

The record high temperature for superconductivity set by the YBCO ceramics was soon eclipsed by another group of ceramics containing bismuth and copper, such as $Bi_2Sr_2CaCu_2O_8$. One of these is superconducting at 110 K, but this record was also short-lived. A ceramic containing thallium and copper, with the approximate formula $TlBa_2Ca_3Cu_4O_y$ (where y is slightly larger than 10), was found to become superconducting at 125 K. Now, the search continues for materials that might become superconducting at room temperature (about 293 K).

To date, many of the ceramic superconductors contain copper and share a common structural feature: carbon and oxygen atoms bonded together in planar sheets. In YBCO superconductors, the Cu—O planes are widely separated. In bismuth superconductors, the Cu—O planes occur in "sandwiches" consisting of two closely spaced sheets separated by a layer of group 2 ions. These sandwiches are separated from one another by several layers of bismuth oxide. In the thallium superconductors, the Cu—O planes are stacked in groups of three, like triple-decker sandwiches.

Recently,[*] the first copper-free ceramic superconductors were discovered. This new class of superconducting ceramics, with approximate formula $LaO_{1-x}F_xFeAs$, are based on iron, a metal that is much more abundant than copper. These superconducting materials are based on LaOFeAs, shown in Figure 23-17(b), but some of the O^{2-} ions are replaced by F^- ions. LaOFeAs has a stack of alternating layers of LaO and FeAs, as illustrated in Figure 23-17(b). The LaO layer is positively charged, consisting of La^{3+} and O^{2-} ions, and the FeAs layer is negatively charged, with the bonding between Fe and As being predominantly covalent. Replacing some of the O^{2-} ions with F^- ions gives a material that superconducts up to about 26 K. Although this temperature is not particularly high, the discovery has generated much excitement because it has opened up new possibilities for developing high-temperature superconductors.

The current theory of superconductivity, developed in the 1950s, explains the superconducting behavior of metals at very low temperatures but not the higher-temperature superconductivity of ceramics. It seems that the electrons in all known superconductors move through the material in pairs—a sort of buddy system that allows the electrons to move without resistance. The mechanism by which electron pairs form in high-temperature superconductors, however, is clearly different from that in low-temperature superconductors. Lack of a suitable theory complicates the search for higher-temperature superconductors. When the mechanism for high-temperature superconductors is better understood, new breakthroughs might be easier to accomplish. Perhaps a room-temperature superconducting material will be possible.

Despite this less-than-complete understanding of high-temperature superconductors, engineers are already building devices that use the new materials. Wires have been made that are superconducting at liquid nitrogen temperatures, and new devices for precise magnetic field measurements using ceramic superconductors are now being produced. Ultimately, ceramic superconductors may find application in low-cost, energy-efficient electric power transmission.

Mastering**CHEMISTRY**   www.**masteringchemistry.com**

Nanoparticles are chemical structures whose dimensions are in the range 1–100 nm. One example of a nanoparticle is a quantum dot, a cluster of atoms with a diameter of tens of nanometers. For a discussion of the interesting and unusual properties of quantum dots, go to the Focus On feature for Chapter 23, Quantum Dots, on the MasteringChemistry site.

*Y. Kamihara, T. Watanabe, M. Hirano, and H. Hosono, *J. Am. Chem. Soc.*, **130**, 11 (2008).

Summary

23-1 General Properties—More than half the elements are the metals known as transition elements. Most are more reactive than hydrogen. Transition metals tend to exist in several different oxidation states in their compounds (see Table 23.1), and they readily form complex ions (discussed in Chapter 24). Many transition metals and their compounds are paramagnetic, and certain of the metals (Fe, Co, and Ni) and their alloys exhibit **ferromagnetism**. Within a group of *d*-block elements, the members of the second and third transition series resemble one another more than they do the group member in the first transition series. This is a consequence of the phenomenon known as the **lanthanide contraction** occurring in the sixth period.

23-2 Principles of Extractive Metallurgy—The extraction of metals from their ores is called **extractive metallurgy**. **Pyrometallurgy** relies on roasting an ore with subsequent reduction of the oxide to the metal. In **hydrometallurgy** metal ions are leached from ores using aqueous solutions of acid or bases. Various methods can be used to refine impure metals. One technique for producing an ultrapure metal is **zone refining**, which involves the continuous melting and refreezing of the metal to concentrate the impurities in one region of the sample, which is then discarded (Fig. 23-6). The thermodynamics of metallurgy can be viewed as the appropriate coupling of reactions into a spontaneous process (Fig. 23-8).

23-3 Metallurgy of Iron and Steel—The conversion of iron(III) oxide into iron by reduction with carbon monoxide in a blast furnace (Fig. 23-9) leads to the formation of **pig iron**. Pig iron is converted to **steel** through the **basic oxygen process** (Fig. 23-10).

23-4 First-Row Transition Elements: Scandium to Manganese—Oxidation–reduction reactions are commonly encountered with transition metal compounds of Sc, Ti, V, Cr, and Mn. Two common types of oxidizing agents are the dichromates and permanganates. In aqueous solution, dichromate ion is in equilibrium with chromate ion,

which is a good precipitating agent for a number of metal ions. Most oxides and hydroxides of the transition metals are basic if the metal is in one of its lower oxidation states. In higher oxidation states, some transition metal oxides and hydroxy compounds are amphoteric, and in the highest oxidation states, a few are acidic (as is CrO_3, for example).

23-5 The Iron Triad: Iron, Cobalt, and Nickel—The iron triad elements (Fe, Co, and Ni) exhibit a variability in oxidation state with +2 the most common. Like other transition elements these metals form compounds with carbon monoxide called metal **carbonyls**.

23-6 Group 11: Copper, Silver, and Gold—The metals Cu, Ag, and Au are resistant to corrosion and are widely used in coins and jewelry. These so-called coinage metals are excellent conductors of heat and electricity and are highly malleable. They also find many uses in the electronics industry.

23-7 Group 12: Zinc, Cadmium, and Mercury—These group 12 elements have chemical properties consistent with a filled *d* subshell. Zn is used in an alloy with copper to produce brass, which is a good electrical conductor and is resistant to corrosion. Most metals form alloys with mercury called **amalgams**.

23-8 Lanthanides—This first series of *f*-block elements have long been known as the rare earth elements, not because of their scarcity but because they are difficult to separate from one another. Their chemical behavior is strongly influenced by the lanthanide contraction, which produces similarities in their atomic and ionic sizes.

23-9 High-Temperature Superconductors—High-temperature superconducting materials have been made using some of the transition elements. In particular, a ceramic material that is superconducting at 92 K is made from yttrium, barium, copper, and oxygen. These ceramic materials have the potential for many applications, for example, more efficient electric power transmission.

Integrative Example

Although a number of slightly soluble copper(I) compounds (such as CuCN) can exist in contact with water, it is not possible to prepare a solution with a high concentration of Cu^+ ion.

Show that $Cu^+(aq)$ disproportionates to $Cu^{2+}(aq)$ and $Cu(s)$, and explain why a high $[Cu^+]$ cannot be maintained in aqueous solution.

Analyze

Write a plausible equation describing the disproportionation reaction. Then determine E°_{cell} for the reaction, followed by the equilibrium constant K, and see what conclusions can be drawn from the numerical value of K.

Solve

The half-equations and overall equation for the disproportionation are

Reduction:	$Cu^+(aq) + e^- \longrightarrow Cu(s)$
Oxidation:	$Cu^+(aq) \longrightarrow Cu^{2+}(aq) + e^-$
Overall:	$2\,Cu^+(aq) \longrightarrow Cu^{2+}(aq) + Cu(s)$

Find E° values for the couples $Cu^+/Cu(s)$ and Cu^{2+}/Cu^+ in Appendix D and combine them to obtain E°_{cell}.

$E^\circ_{cell} = E^\circ(\text{reduction}) - E^\circ(\text{oxidation})$
$= E^\circ_{Cu^+/Cu(s)} - E^\circ_{Cu^{2+}/Cu^+} = 0.520\ V - 0.159\ V = 0.361\ V$

Use equation (20.17) to obtain the value of K. In the equation $n = 1$,

$$E^\circ_{cell} = \frac{0.025693 \text{ V}}{n} \ln K$$

$$\ln K = \frac{n \times E^\circ_{cell}}{0.025693 \text{ V}} = \frac{1 \times 0.361 \text{ V}}{0.025693 \text{ V}} = 14.1$$

$$K = e^{14.1} = 1.3 \times 10^6$$

Thus, for the disproportionation reaction,

$$K = \frac{[Cu^{2+}]}{[Cu^+]^2} = 1.3 \times 10^6$$

and

$$[Cu^{2+}] = 1.3 \times 10^6 \times [Cu^+]^2$$

To maintain $[Cu^+] = 1$ M in solution, $[Cu^{2+}]$ would have to be more than 1×10^6 M—a clear impossibility.

Assess
As a practical matter, we could not maintain $[Cu^+]$ at much more than 0.002 M, for even this would require that $[Cu^{2+}] \approx 5$ M.

PRACTICE EXAMPLE A: Consider a galvanic cell based on the following half-reactions. Assuming the cell operates under standard conditions at 25 °C, what is the spontaneous cell reaction? Under what nonstandard conditions is the spontaneous formation of $[PtCl_6]^{2-}$ favored? Do you think that, in practical terms, a significant amount of $PtCl_6{}^{2-}$ can be obtained by altering the concentrations in a galvanic cell?

$$PtCl_6{}^{2-} + 2 e^- \longrightarrow PtCl_4{}^{2-} + 2 Cl^- \qquad E^\circ = 0.68 \text{ V}$$
$$V^{3+} + e^- \longrightarrow V^{2+} \qquad E^\circ = -0.255 \text{ V}$$

PRACTICE EXAMPLE B: Because metallic titanium exhibits excellent corrosion resistance, it is often desirable to coat iron objects with a thin coating of titanium metal. One approach involves production of Ti(s) from electrolysis of molten mixtures of NaCl and $TiCl_2$. The production of Ti(s) involves the disproportionation of Ti^{2+} to Ti^{3+} and Ti. Write a balanced chemical equation for the disproportionation reaction and use the following half-reactions to decide whether the disproportionation reaction is spontaneous under standard conditions. $Ti^{2+} + 2 e^- \longrightarrow Ti, \ E^\circ = -1.630 \text{ V}$; $Ti^{3+} + e^- \longrightarrow Ti^{2+}, \ E^\circ = -0.369 \text{ V}$.

Mastering **CHEMISTRY**

You'll find a link to additional self study questions in the study area on www.masteringchemistry.com

Exercises

Properties of the Transition Elements

1. By means of orbital diagrams, write electron configurations for the following transition element atom and ions: **(a)** Ti; **(b)** V^{3+}; **(c)** Cr^{2+}; **(d)** Mn^{4+}; **(e)** Mn^{2+}; **(f)** Fe^{3+}.

2. Arrange the following species according to the number of unpaired electrons they contain, starting with the one that has the greatest number: Fe, Sc^{3+}, Ti^{2+}, Mn^{4+}, Cr, Cu^{2+}.

3. Describe how the transition elements compare with main-group metals (such as group 2) with respect to oxidation states, formation of complexes, colors of compounds, and magnetic properties.

4. With only minor irregularities, the melting points of the first series of transition metals rise from that of Sc to that of Cr and then fall to that of Zn. Give a plausible explanation for this phenomenon based on atomic structure.

5. Why do the atomic radii vary so much more for two main-group elements that differ by one unit in atomic number than they do for two transition elements that differ by one unit?

6. The metallic radii of Ni, Pd, and Pt are 125, 138, and 139 pm, respectively. Why is the difference in radius between Pt and Pd so much less than between Pd and Ni?

7. Which of the first transition series elements exhibits the greatest number of different oxidation states in its compounds? Explain.

8. Why is the number of common oxidation states for the elements at the beginning and those at the end of the first transition series less than for elements in the middle of the series?

9. As a group, the lanthanides are more reactive metals than are those in the first transition series. How do you account for this difference?

10. The maximum difference in standard reduction potential, $E^\circ_{M^{2+}/M(s)}$, among members of the first transition series is about 2.4 V. For the lanthanides, the maximum difference in $E^\circ_{M^{3+}/M(s)}$ is only about 0.4 V. How do you account for this fact?

Reactions of Transition Metals and Their Compounds

11. Complete and balance the following equations. If no reaction occurs, so state.

(a) $TiCl_4(g) + Na(l) \xrightarrow{\Delta}$

(b) $Cr_2O_3(s) + Al(s) \xrightarrow{\Delta}$

(c) $Ag(s) + HCl(aq) \longrightarrow$

(d) $K_2Cr_2O_7(aq) + KOH(aq) \longrightarrow$

(e) $MnO_2(s) + C(s) \xrightarrow{\Delta}$

12. By means of a chemical equation, give an example to represent the reaction of **(a)** a transition metal with a nonoxidizing acid; **(b)** a transition metal oxide with $NaOH(aq)$; **(c)** an inner transition metal with $HCl(aq)$.

13. Write balanced chemical equations for the following reactions described in the chapter.

(a) the reaction of $Sc(OH)_3(s)$ with $HCl(aq)$

(b) oxidation of $Fe^{2+}(aq)$ by $MnO_4^-(aq)$ in basic solution to give $Fe^{3+}(aq)$ and $MnO_2(s)$

(c) the reaction of $TiO_2(s)$ with molten KOH to form K_2TiO_3

(d) oxidation of $Cu(s)$ to $Cu^{2+}(aq)$ with H_2SO_4 (concd aq) to form $SO_2(g)$

14. Write balanced equations for the following reactions described in the chapter.

(a) $Sc(l)$ is produced by the electrolysis of Sc_2O_3 dissolved in $Na_3ScF_6(l)$.

(b) $Cr(s)$ reacts with $HCl(aq)$ to produce a blue solution containing $Cr^{2+}(aq)$.

(c) $Cr^{2+}(aq)$ is readily oxidized by $O_2(g)$ to $Cr^{3+}(aq)$.

(d) $Ag(s)$ reacts with concentrated $HNO_3(aq)$, and $NO_2(g)$ is evolved.

15. Suggest a series of reactions, using common chemicals, by which each of the following syntheses can be performed.

(a) $Fe(OH)_3(s)$ from $FeS(s)$

(b) $BaCrO_4(s)$ from $BaCO_3(s)$ and $K_2Cr_2O_7(aq)$

16. Suggest a series of reactions, using common chemicals, by which each of the following syntheses can be performed.

(a) $Cu(OH)_2(s)$ from $CuO(s)$

(b) $CrCl_3(aq)$ from $(NH_4)_2Cr_2O_7(s)$

Extractive Metallurgy

17. One of the simplest metals to extract from its ores is mercury. Mercury vapor is produced by roasting cinnabar ore (HgS) in air. Alternatives to this simple roasting, designed to reduce or eliminate SO_2 emissions, is to roast the ore in the presence of a second substance. For example, when cinnabar is roasted with quicklime, the products are mercury vapor and calcium sulfide and calcium sulfate. Write equations for the two reactions described here.

18. According to Figure 23-8, $\Delta G°$ decreases with temperature for the reaction $2 C(s) + O_2(g) \longrightarrow 2 CO(g)$. How would you expect $\Delta G°$ to vary with temperature for the following reactions?

(a) $C(s) + O_2(g) \longrightarrow CO_2(g)$

(b) $2 CO(g) + O_2(g) \longrightarrow 2 CO_2(g)$

19. Calcium will reduce $MgO(s)$ to $Mg(s)$ at all temperatures from 0 to 2000 °C. Use this fact, together with the melting point (839 °C) and boiling point (1484 °C) of calcium, to sketch a plausible graph of $\Delta G°$ as a function of temperature for the reaction $2 Ca(s) + O_2(g) \longrightarrow 2 CaO(s)$.

20. One method of obtaining chromium metal from chromite ore is as follows. After reaction (23.16), sodium chromate is reduced to chromium(III) oxide by carbon. Then the chromium(III) oxide is reduced to chromium metal by silicon. Write plausible equations to describe these two reactions.

Oxidation–Reduction

21. Write plausible half-equations to represent each of the following in acidic solution.

(a) $VO^{2+}(aq)$ as an oxidizing agent

(b) $Cr^{2+}(aq)$ as a reducing agent

22. Write plausible half-equations to represent each of the following in basic solution.

(a) oxidation of $Fe(OH)_3(s)$ to FeO_4^{2-}

(b) reduction of $[Ag(CN)_2]^-$ to silver metal

23. Use electrode potential data from this chapter or Appendix D to predict whether each of the following reactions will occur to any significant extent under standard-state conditions.

(a) $2 VO_2^+ + 6 Br^- + 8 H^+ \longrightarrow$
$$2 V^{2+} + 3 Br_2(l) + 4 H_2O$$

(b) $VO_2^+ + Fe^{2+} + 2 H^+ \longrightarrow VO^{2+} + Fe^{3+} + H_2O$

(c) $MnO_2(s) + H_2O_2 + 2 H^+ \longrightarrow$
$$Mn^{2+} + 2 H_2O + O_2(g)$$

24. You are given these three reducing agents: $Zn(s)$, $Sn^{2+}(aq)$, and $I^-(aq)$. Use data from Appendix D to determine which of them can, under standard-state conditions in acidic solution, reduce

(a) $Cr_2O_7^{2-}(aq)$ to $Cr^{3+}(aq)$

(b) $Cr^{3+}(aq)$ to $Cr^{2+}(aq)$

(c) $SO_4^{2-}(aq)$ to $SO_2(g)$

25. Refer to Example 23-2. Select a reducing agent (from Table 23.1 or Appendix D) that will reduce VO^{2+} to V^{3+} and no further in acidic solution.

26. The electrode potential diagram for manganese in acidic solutions in Figure 23-14 does not include a value of $E°$ for the reduction of MnO_4^- to Mn^{2+}. Use other data in the figure to establish this $E°$, and compare your result with the value found in Table 20.1.

27. Use data from the text to construct a standard electrode potential diagram relating the following chromium species in acidic solution.

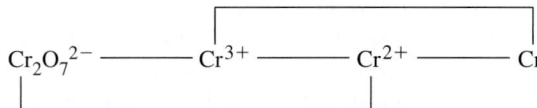

28. Use data from the text to construct a standard electrode potential diagram relating the following vanadium species in acidic solution.

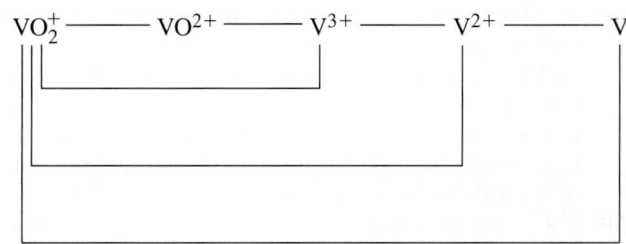

Chromium and Chromium Compounds

29. When a soluble lead compound is added to a solution containing primarily *orange* dichromate ion, *yellow* lead chromate precipitates. Describe the equilibria involved.

30. When *yellow* $BaCrO_4$ is dissolved in $HCl(aq)$, a *green* solution is obtained. Write a chemical equation to account for the color change.

31. When $Zn(s)$ is added to $K_2Cr_2O_7$ dissolved in $HCl(aq)$, the color of the solution changes from orange to green, then to blue, and, over a period of time, back to green. Write equations for this series of reactions.

32. If $CO_2(g)$ under pressure is passed into $Na_2CrO_4(aq)$, $Na_2Cr_2O_7(aq)$ is formed. What is the function of the $CO_2(g)$? Write a plausible equation for the net reaction.

33. Use equation (23.19) to determine $[Cr_2O_7^{2-}]$ in a solution that has $[CrO_4^{2-}] = 0.20 \text{ M}$ and a pH of **(a)** 6.62 and **(b)** 8.85.

34. If a solution is prepared by dissolving 1.505 g Na_2CrO_4 in 345 mL of a buffer solution with pH = 7.55, what will be $[CrO_4^{2-}]$ and $[Cr_2O_7^{2-}]$?

35. How many grams of chromium would be deposited on an object in a chrome-plating bath (see page 1048) after 1.00 h at a current of 3.4 A?

36. How long would an electric current of 3.5 A have to pass through a chrome-plating bath (see page 1048) to produce a chromium deposit 0.0010 mm thick on an object with a surface area of 0.375 m^2? (The density of Cr is 7.14 g cm^{-3}.)

37. Why is it reasonable to expect the chemistry of dichromate ion to involve mainly oxidation–reduction reactions and that of chromate ion to involve mainly precipitation reactions?

38. What products are obtained when $Mg^{2+}(aq)$ and $Cr^{3+}(aq)$ are each treated with a limited amount of $NaOH(aq)$? With an excess of $NaOH(aq)$? Why are the results different in these two cases?

The Iron Triad

39. Will reaction (23.25) still be spontaneous in the forward direction in a solution containing equal concentrations of Fe^{2+} and Fe^{3+}, a pH of 3.25, and under an $O_2(g)$ partial pressure of 0.20 atm?

40. Based on the description of the nickel–cadmium cell on page 1053, and with appropriate data from Appendix D, estimate $E°$ for the reduction of $NiO(OH)$ to $Ni(OH)_2$.

41. Write a net ionic equation to represent the precipitation of Prussian blue, described on page 1053.

42. The reaction to form Turnbull's blue (page 1053) appears to occur in two stages. First, $Fe^{2+}(aq)$ is oxidized to $Fe^{3+}(aq)$ and ferricyanide ion is reduced to ferrocyanide ion. Then, the $Fe^{3+}(aq)$ and ferrocyanide ion combine. Write equations for these reactions.

Group 11 Metals

43. Write plausible equations for the following reactions occurring in the hydrometallurgy of the coinage metals.
    (a) Copper is precipitated from a solution of copper(II) sulfate by treatment with $H_2(g)$.
    (b) Gold is precipitated from a solution of Au^+ by adding iron(II) sulfate.
    (c) Copper(II) chloride solution is reduced to copper(I) chloride when treated with $SO_2(g)$ in acidic solution.

44. In the metallurgical extraction of silver and gold, an alloy of the two metals is often obtained. The alloy can be separated into Ag and Au either with concentrated HNO_3 or boiling concentrated H_2SO_4, in a process called *parting*. Write chemical equations to show how these separations work.

45. Use the result of the Integrative Example to determine whether a solution can be prepared with $[Cu^+]$ equal to **(a)** 0.20 M; **(b)** 1.0×10^{-10} M.

46. Show that the corrosion reaction in which Cu is converted to its basic carbonate (reaction 23.27) can be thought of in terms of a combination of oxidation–reduction, acid–base, and precipitation reactions.

Group 12 Metals

47. Use data from Table 23.8 to determine $E°$ for the reduction of Hg^{2+} to Hg_2^{2+} in aqueous solution.

48. At $400 °C, \Delta G° = -25$ kJ for the reaction $2 Hg(l) + O_2(g) \longrightarrow 2 HgO(s)$. If a sample of $HgO(s)$ is heated to $400 °C$, what will be the equilibrium partial pressure of $O_2(g)$?

49. Use Figure 23-8 to estimate for the reaction $ZnO(s) + C(s) \rightleftharpoons Zn(l) + CO(g)$, at about $800 °C$, **(a)** a value of K_p and **(b)** the equilibrium pressure of $CO(g)$.

50. The vapor pressure of $Hg(l)$ as a function of temperature is $\log P(mmHg) = (-0.05223a/T) + b$, where $a = 61,960$ and $b = 8.118$; T is the Kelvin temperature.

Show that at $25 °C$, the concentration of $Hg(g)$ in equilibrium with $Hg(l)$ greatly exceeds the maximum permissible level of 0.05 mg Hg/m^3 air.

51. In ZnO, the band gap between the valence and conduction bands is 290 kJ mol^{-1}, and in CdS it is 250 kJ mol^{-1}. Show that CdS absorbs some visible light but ZnO does not. Explain the observed colors: ZnO is white and CdS is yellow.

52. CdS is yellow, HgS is red, and CdSe is black. Which of these materials has the largest band gap? the smallest? How does the band gap relate to the observed color?

Integrative and Advanced Exercises

53. Although Au reacts with and dissolves in aqua regia (3 parts HCl + 1 part HNO_3), Ag does not dissolve. What is (are) the likely reason(s) for this difference?

54. The text mentions that scandium metal is obtained from its molten chloride by electrolysis, and that titanium is obtained from its chloride by reduction with magnesium. Why are these metals not obtained by the reduction of their oxides with carbon (coke), as are metals such as zinc and iron?

55. The text notes that in small quantities, zinc is an essential element (though it is toxic in higher concentrations). Tin is considered to be a toxic metal. Can you think of reasons why, for food storage, tinplate instead of galvanized iron is used in cans?

56. In an atmosphere polluted with industrial smog, Cu corrodes to a basic sulfate, $Cu_2(OH)_2SO_4$. Propose a series of chemical reactions to describe this corrosion.

57. What formulas would you expect for the metal carbonyls of **(a)** molybdenum, **(b)** osmium; **(c)** rhenium? Note that the simple carbonyls shown in Figure 23-15 have one metal atom per molecule. Some metal carbonyls are *binuclear*; that is, they have two metal atoms bonded together in the carbonyl structure. Also, **(d)** explain why iron and nickel carbonyls are liquids at room temperature, whereas that of cobalt is a solid, and **(e)** describe the probable nature of the bonding in the compound $Na[V(CO)_6]$.

58. For the straight-line graphs in Figure 23-8, explain why **(a)** breaks occur at the melting points and boiling points of the metals; **(b)** the slopes of the lines become more positive at these breaks; **(c)** the break at the boiling point is sharper than at the melting point.

59. Attempts to make CuI_2 by the reaction of $Cu^{2+}(aq)$ and $I^-(aq)$ produce $CuI(s)$ and $I_3^-(aq)$ instead. Without performing detailed calculations, show why this reaction should occur.

$$2 Cu^{2+}(aq) + 5 I^-(aq) \longrightarrow 2 CuI(s) + I_3^-(aq)$$

60. Without performing detailed calculations, show that significant disproportionation of AuCl occurs if you attempt to make a saturated aqueous solution. Use data from Table 23.7 and $K_{sp}(AuCl) = 2.0 \times 10^{-13}$.

61. In acidic solution, silver(II) oxide first dissolves to produce $Ag^{2+}(aq)$. This is followed by the oxidation of $H_2O(l)$ to $O_2(g)$ and the reduction of Ag^{2+} to Ag^+.
(a) Write equations for the dissolution and oxidation–reduction reactions.
(b) Show that the oxidation–reduction reaction is indeed spontaneous.

62. Equation (23.18), which represents the chromate–dichromate equilibrium, is actually the sum of two equilibrium expressions. The first is an acid–base reaction, $H^+ + CrO_4^{2-} \rightleftharpoons HCrO_4^-$. The second reaction involves elimination of a water molecule between two $HCrO_4^-$ ions (a dehydration reaction), $2 HCrO_4^- \rightleftharpoons Cr_2O_7^{2-} + H_2O$. If the ionization constant, K_a, for $HCrO_4^-$ is 3.2×10^{-7}, what is the value of K for the dehydration reaction?

63. Show that under the following conditions, $Ba^{2+}(aq)$ can be separated from $Sr^{2+}(aq)$ and $Ca^{2+}(aq)$ by precipitating $BaCrO_4(s)$ with the other ions remaining in solution:

$$[Ba^{2+}] = [Sr^{2+}] = [Ca^{2+}] = 0.10 \text{ M}$$
$$[HC_2H_3O_2] = [C_2H_3O_2^-] = 1.0 \text{ M}$$
$$[Cr_2O_7^{2-}] = 0.0010 \text{ M}$$
$$K_{sp}(BaCrO_4) = 1.2 \times 10^{-10}$$
$$K_{sp}(SrCrO_4) = 2.2 \times 10^{-5}$$

Use data from this and previous chapters, as necessary.

64. A 0.589 g sample of pyrolusite ore (impure MnO_2) is treated with 1.651 g of oxalic acid ($H_2C_2O_4 \cdot 2 H_2O$) in an acidic medium (reaction 1). Following this, the excess oxalic acid is titrated with 30.06 mL of 0.1000 M $KMnO_4$ (reaction 2). What is the mass percent of MnO_2 in the pyrolusite? The following equations are neither complete nor balanced.

(1) $\quad H_2C_2O_4(aq) + MnO_2(s) \longrightarrow Mn^{2+}(aq) + CO_2(g)$
(2) $\quad H_2C_2O_4(aq) + MnO_4^-(aq) \longrightarrow Mn^{2+}(aq) + CO_2(g)$

65. Both $Cr_2O_7^{2-}(aq)$ and $MnO_4^-(aq)$ can be used to titrate $Fe^{2+}(aq)$ to $Fe^{3+}(aq)$. Suppose you have available as titrants two solutions: 0.1000 M $Cr_2O_7^{2-}(aq)$ and 0.1000 M $MnO_4^-(aq)$.

(a) For which solution would the greater volume of titrant be required for the titration of a particular sample of $Fe^{2+}(aq)$? Explain.

(b) How many mL of 0.1000 M $MnO_4^-(aq)$ would be required for a titration if the same titration requires 24.50 mL of 0.1000 M $Cr_2O_7^{2-}(aq)$?

66. The only important compounds of Ag(II) are AgF_2 and AgO. Why would you expect these two compounds to be stable, but not other silver(II) compounds such as $AgCl_2$, $AgBr_2$, and AgS?

67. A certain steel is to be analyzed for Cr and Mn. By suitable treatment, the Cr in the steel is oxidized to $Cr_2O_7^{2-}(aq)$ and the Mn to $MnO_4^-(aq)$. A 10.000 g sample of steel is used to produce 250.0 mL of a solution containing $Cr_2O_7^{2-}(aq)$ and $MnO_4^-(aq)$. A 10.00 mL portion of this solution is added to $BaCl_2(aq)$, and by proper adjustment of the pH, the chromium is completely precipitated as $BaCrO_4(s)$; 0.549 g is obtained. A second 10.00 mL portion of this solution requires exactly 15.95 mL of 0.0750 M $Fe^{2+}(aq)$ for its titration in acidic solution. Calculate the % Cr and % Mn in the steel sample. [*Hint:* In the titration $MnO_4^-(aq)$ is reduced to $Mn^{2+}(aq)$ and $Cr_2O_7^{2-}(aq)$ is reduced to $Cr^{3+}(aq)$; the $Fe^{2+}(aq)$ is oxidized to $Fe^{3+}(aq)$.]

68. Nickel can be determined as nickel dimethylglyoximate, a brilliant scarlet precipitate that has the composition 20.31% Ni, 33.26% C, 4.88% H, 22.15% O, and 19.39% N. A 15.020 g steel sample is dissolved in concentrated HCl(aq). The solution obtained is suitably treated to remove interfering ions, to establish the proper pH, and to obtain a final solution volume of 250.0 mL. A 10.00 mL sample of this solution is then treated with dimethylglyoxime. The mass of purified, dry nickel dimethylglyoximate obtained is 0.104 g.

(a) What is the empirical formula of nickel dimethylglyoximate?

(b) What is the mass percent nickel in the steel sample?

69. A solution is believed to contain one or more of the following ions: Cr^{3+}, Zn^{2+}, Fe^{3+}, Ni^{2+}. When the solution is treated with excess NaOH(aq), a precipitate forms. The solution in contact with the precipitate is colorless. The precipitate is dissolved in HCl(aq), and the resulting solution is treated with $NH_3(aq)$. No precipitation occurs. Based solely on these observations, what conclusions can you draw about the ions present in the original solution? That is, which ion(s) are likely present, which are most likely not present, and about which can we not be certain? [*Hint:* Refer to Appendix D for solubility product and complex-ion formation data.]

70. Nearly all mercury(II) compounds exhibit covalent bonding. Mercury(II) chloride is a covalent molecule that dissolves in warm water. The stability of this compound is exploited in the determination of the levels of chloride ion in blood serum. Typical human blood serum levels range from 90 to 115 mmol L^{-1}. The chloride concentration is determined by titration with $Hg(NO_3)_2$. The indicator used in the titration is diphenylcarbazone, $C_6H_5N{=}NCONHNHC_6H_5$, which complexes with the mercury(II) ion after all the chloride has reacted with the mercury(II). Free diphenylcarbazone is pink in solution, and when it is complexed with mercury(II), it is blue. Thus, the diphenylcarbazone acts as an indicator, changing from pink to blue when the first excess of mercury(II) appears. In an experiment, $Hg(NO_3)_2(aq)$ solution is standardized by titrating 2.00 mL of 0.0108 M NaCl solution. It takes 1.12 mL of $Hg(NO_3)_2(aq)$ to reach the diphenylcarbazone end point. A 0.500 mL serum sample is treated with 3.50 mL water, 0.50 mL of 10% sodium tungstate solution, and 0.50 mL of 0.33 M $H_2SO_4(aq)$ to precipitate proteins. After the proteins are precipitated, the sample is filtered and a 2.00 mL aliquot of the filtrate is titrated with $Hg(NO_3)_2$ solution, requiring 1.23 mL. Calculate the concentration of Cl^-. Express your answer in mmol L^{-1}. Does this concentration fall in the normal range?

71. Covalent bonding is involved in many transition metal compounds. Draw Lewis structures, showing any nonzero formal charges, for the following molecules or ions: (a) Hg_2^{2+}; (b) Mn_2O_7; (c) OsO_4. [*Hint:* In (b), there is one $Mn{-}O{-}Mn$ linkage in the molecule.]

72. For a coordination number of four, the radius of Mn^{7+} has been estimated to be 39 pm. Estimate the charge density for the Mn^{7+} ion. Express your answer in C mm^{-3}. How does this compare with the charge density of Be^{2+} given in Table 21.4? Would you expect the bonding in Mn_2O_7 to be primarily ionic or primarily covalent? Explain.

73. Nitinol is a nickel–titanium alloy known as *memory metal*. The name nitinol is derived from the symbols for nickel (Ni), titanium (Ti), and the acronym for the Naval Ordinance Laboratory (NOL), where it was discovered. If an object made out of nitinol is heated to about 500 °C for about an hour and then allowed to cool, the original shape of the object is "remembered," even if the object is deformed into a different shape. The original shape can be restored by heating the metal. Because of this property, nitinol has found many uses, especially in medicine and orthodontics (for braces). Nitinol exists in a number of different solid phases. In the so-called austerite phase, the metal is relatively soft and elastic. The crystal structure for the austerite phase can be described as a simple cubic lattice of Ti atoms with Ni atoms occupying cubic holes in the lattice of Ti atoms. What is the empirical formula of nitinol and what is the percent by mass of titanium in the alloy?

Feature Problems

74. As a continuation of Problem 94 of Chapter 19 and the discussion on page 1040, consider the three graphs of $\Delta G°$ as a function of temperature shown in the following figure.

(a) Explain the shapes of the three graphs. Specifically, why is one line essentially parallel to the temperature axis, why does one have a positive slope, and why does one have a negative slope?

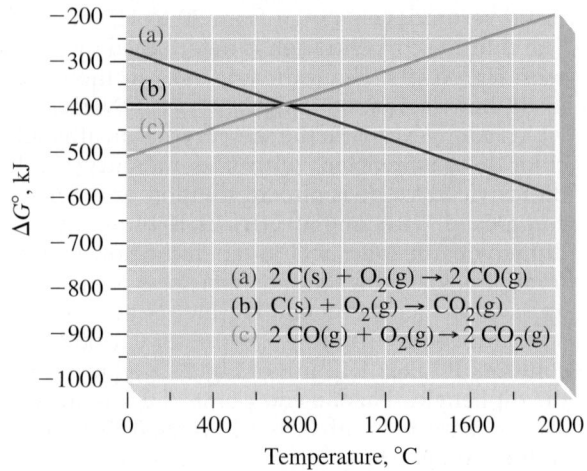

(a) $2\,C(s) + O_2(g) \rightarrow 2\,CO(g)$
(b) $C(s) + O_2(g) \rightarrow CO_2(g)$
(c) $2\,CO(g) + O_2(g) \rightarrow 2\,CO_2(g)$

(b) Table 23.2 lists as an additional blast furnace reaction, $C(s) + CO_2(g) \longrightarrow 2\,CO(g)$. Determine how ΔG° for this reaction is related to the three reactions shown in the figure, and plot ΔG° for this reaction as a function of temperature. If an equilibrium is established in this reaction at 1000 °C and the partial pressure of $CO_2(g)$ is 0.25 atm, what should be the equilibrium partial pressure of $CO(g)$?

75. Several transition metal ions are found in cation group 3 of the qualitative analysis scheme outlined in Figure 18-7. At one point in the separation and testing of this group, a solution containing Fe^{3+}, Co^{2+}, Ni^{2+},

Al^{3+}, Cr^{3+}, and Zn^{2+} is treated with an excess of NaOH(aq), together with H_2O_2(aq).

(1) The excess NaOH(aq) causes *three* of the cations to precipitate as hydroxides and *three* to form hydroxo complex ions.
(2) In the presence of H_2O_2(aq), the cation in one of the insoluble hydroxides is oxidized from the +2 to the +3 oxidation state, and one of the hydroxo complex ions is also oxidized.
(3) The three insoluble hydroxides are found as a dark precipitate.
(4) The solution above the precipitate has a yellow color.
(5) The dark precipitate from (3) reacts with HCl(aq), and all the cations return to solution; one of the cations is reduced from the +3 to the +2 oxidation state.
(6) The solution from (5) is treated with 6 M NH_3(aq), and a precipitate containing one of the cations forms.

(a) Write equations for the reactions referred to in item (1).
(b) Write an equation for the most likely reaction in which a hydroxide precipitate is oxidized in item (2).
(c) What is the ion responsible for the yellow color of the solution in item (4)? Write an equation for its formation.
(d) Write equations for the dissolution of the precipitate and the reduction of the cation in item (5).
(e) Write an equation for the precipitate formation in item (6). [*Hint:* You may need solubility product and complex-ion formation data from Appendix D, together with descriptive information from this chapter and from elsewhere in the text.]

Self-Assessment Exercises

76. In your own words, define the following terms: **(a)** domain; **(b)** flotation; **(c)** leaching; **(d)** amalgam.
77. Briefly describe each of the following ideas, phenomena, or methods: **(a)** lanthanide contraction; **(b)** zone refining; **(c)** basic oxygen process; **(d)** slag formation.
78. Explain the important distinctions between each pair of terms: **(a)** ferromagnetism and paramagnetism; **(b)** roasting and reduction; **(c)** hydrometallurgy and pyrometallurgy; **(d)** chromate and dichromate.
79. Describe the chemical composition of the material called **(a)** pig iron; **(b)** ferromanganese alloy; **(c)** chromite ore; **(d)** brass; **(e)** aqua regia; **(f)** blister copper; **(g)** stainless steel.
80. Three properties expected for transition elements are **(a)** low melting points; **(b)** high ionization energies; **(c)** colored ions in solution; **(d)** positive standard electrode (reduction) potentials; **(e)** diamagnetism; **(f)** complex ion formation; **(g)** catalytic activity.
81. The only diamagnetic ion of the following group is **(a)** Cr^{2+}; **(b)** Zn^{2+}; **(c)** Fe^{3+}; **(d)** Ag^{2+}; **(e)** Ti^{3+}.
82. All of the following elements have an ion displaying the +6 oxidation state except **(a)** Mo; **(b)** Cr; **(c)** Mn; **(d)** V; **(e)** S.
83. The best oxidizing agent of the following group of ions is **(a)** Ag^+(aq); **(b)** Cl^-(aq); **(c)** H^+(aq); **(d)** Na^+(aq); **(e)** OH^-(aq).
84. To separate Fe^{3+} and Ni^{2+} from an aqueous solution containing both ions, with one cation forming a

precipitate and the other remaining in solution, add to the solution **(a)** NaOH(aq); **(b)** H_2S(g); **(c)** HCl(aq); **(d)** NH_3(aq).
85. Of the following, the two solids that will liberate Cl_2(g) when heated with HCl(aq) are **(a)** NaCl(s); **(b)** $ZnCl_2$(s); **(c)** MnO_2(s); **(d)** CuO(s); **(e)** $K_2Cr_2O_7$(s); **(f)** NaOH(s).
86. Provide the missing name or formula for the following:
(a) chromium(VI) oxide _____
(b) _____ K_2MnO_4
(c) _____ $Cr(CO)_6$
(d) barium dichromate _____
(e) _____ $La_2(SO_4)_3 \cdot 9\,H_2O$
(f) gold(III) cyanide trihydrate _____
87. Balance the following oxidation–reduction equations.
(a) $Fe_2S_3(s) + H_2O + O_2(g) \longrightarrow Fe(OH)_3(s) + S(s)$
(b) $Mn^{2+}(aq) + S_2O_8^{2-}(aq) + H_2O \longrightarrow$
$\qquad MnO_4^-(aq) + SO_4^{2-}(aq) + H^+(aq)$
(c) $Ag(s) + CN^-(aq) + O_2(g) + H_2O \longrightarrow$
$\qquad [Ag(CN)_2]^-(aq) + OH^-(aq)$
88. Explain why Zn, Cd, and Hg resemble the group 2 metals in some of their properties.
89. Explain why gold dissolves in aqua regia but not in HNO_3(aq).
90. Explain why 1.0 M $Fe(NO_3)_3$(aq) is acidic.

Complex Ions and Coordination Compounds

24

CONTENTS

Turquoise is a mineral of copper, $CuAl_6(PO_4)_4(OH)_8 \cdot 4\,H_2O$. The distinctive color of this gemstone and many others is a consequence of the nature of metal–ligand bonding in complex ions, a central topic of this chapter.

Chapter 23 included discussions of several situations involving a succession of color changes that were attributed to changes in oxidation state. The color changes discussed in this chapter are not generally caused by oxidation–reduction reactions. Instead, they are observed with changes in the groups (ligands) bound to a metal center, even though the oxidation state of the metal remains unchanged. Explanation of this observation requires a fuller exploration of the nature of complex ions and coordination compounds, a subject that was briefly introduced in Chapter 18. One topic we will consider is the geometric structure of complex ions, which reveals new possibilities for isomerism: the existence of compounds having identical compositions but different structures and properties. We will also examine the nature of the bonding between ligands and the metal centers to which they are attached. It is through an understanding of bonding in complex ions that we can gain some insight into the origin of their colors.

24-1 Werner's Theory of Coordination Compounds: An Overview

▲ **Alfred Werner (1866–1919)**
Werner's success in explaining coordination compounds came in large part through his application of new ideas: the theory of electrolytic dissociation and principles of structural chemistry.

Prussian blue (page 1053), accidentally discovered early in the eighteenth century, was perhaps the first known coordination compound—the type explored in this chapter. However, nearly a century passed before the uniqueness of these compounds came to be appreciated. In 1798, B. M. Tassaert obtained yellow crystals of a compound having the formula $CoCl_3 \cdot 6 NH_3$ from a mixture of $CoCl_3$ and $NH_3(aq)$. What seemed unusual was that both $CoCl_3$ and NH_3 are stable compounds capable of independent existence, yet they combine to form still another stable compound. Such compounds made up of two simpler compounds came to be called **coordination compounds**.

In 1851, another coordination compound of $CoCl_3$ and NH_3 was discovered. This one had the formula $CoCl_3 \cdot 5 NH_3$ and formed purple crystals. The two compounds are shown in Figure 24-1. Their formulas follow.

$$CoCl_3 \cdot 6 NH_3 \qquad CoCl_3 \cdot 5 NH_3$$

$$\text{(yellow)} \qquad \text{(purple)}$$

$$\text{(a)} \qquad \text{(b)}$$

The mystery of coordination compounds deepened as more were discovered and studied. For example, when treated with $AgNO_3(aq)$, compound (a) formed *three* moles of $AgCl(s)$, as expected, but compound (b) formed only *two* moles of $AgCl(s)$.

Inorganic coordination chemistry was a hot field of research in the second half of the nineteenth century, and all the pieces started to fall into place with the work of the Swiss chemist Alfred Werner. Werner's theory of coordination compounds explained the reactions of compounds (a) and (b) with $AgNO_3(aq)$ by considering that, in aqueous solutions, these two compounds ionize in the following way:

(a) $[Co(NH_3)_6]Cl_3(s) \xrightarrow{H_2O} [Co(NH_3)_6]^{3+}(aq) + 3\,Cl^-(aq)$

(b) $[CoCl(NH_3)_5]Cl_2(s) \xrightarrow{H_2O} [CoCl(NH_3)_5]^{2+}(aq) + 2\,Cl^-(aq)$

Thus, compound (a) produces the three moles of Cl^- per mole of compound necessary to precipitate three moles of $AgCl(s)$, while compound (b) produces only two moles of Cl^-. Werner's proposal of this ionization scheme was based on extensive studies of the electrical conductivity of coordination compounds. Compound (a) is a better conductor than compound (b), consistent with producing four ions per formula unit compared to three for compound (b). $CoCl_3 \cdot 4 NH_3$ is a still poorer conductor, corresponding to the formula $[CoCl_2(NH_3)_4]Cl$. $CoCl_3 \cdot 3 NH_3$ is a *nonelectrolyte*, corresponding to the formula $[CoCl_3(NH_3)_3]$.

▲ **FIGURE 24-1**
Two coordination compounds
The compound on the left is $[Co(NH_3)_6]Cl_3$. The compound on the right is $[CoCl(NH_3)_5]Cl_2$.

The heart of Werner's theory, proposed in 1893, was that certain metal atoms, primarily those of transition metals, have two types of valence or bonding capacity. One, the *primary* valence, is based on the number of electrons the atom loses in forming the metal ion. A *secondary* valence is responsible for the bonding of other groups, called **ligands**, to the central metal ion.

In modern usage, a **complex** is any species involving coordination of ligands to a metal center. The metal center can be an atom or an ion, and the complex can be a cation, an anion, or a neutral molecule. In a chemical formula, a complex—a metal center and attached ligands—is set off by square brackets, [ ]. Compounds that are complexes or contain complex ions are known as coordination compounds.

$$[Co(NH_3)_6]^{3+} \qquad [CoCl_4(NH_3)_2]^{-} \qquad [CoCl_3(NH_3)_3] \qquad K_4[Fe(CN)_6]$$

| Complex cation | Complex anion | Neutral complex | Coordination compound |

The **coordination number** of a complex is the number of points around the metal center at which bonds to ligands can form. Coordination numbers ranging from 2 to 12 have been observed, although 6 is by far the most common number, followed by 4. Coordination number 2 is limited mostly to complexes of Cu(I), Ag(I), and Au(I). Coordination numbers greater than 6 are not often found in members of the first transition series, but are more common in those of the second and third series. Stable complexes with coordination numbers 3 and 5 are rare. The coordination number observed in a complex depends on a number of factors, such as the ratio of the radius of the central metal atom or ion to the radii of the attached ligands.

Coordination numbers of some common ions are listed in Table 24.1. The four most commonly observed geometric shapes of complex ions are shown in Figure 24-2. One practical use of the coordination number is to assist in writing and interpreting formulas of complexes, as illustrated in Example 24-1.

TABLE 24.1 Some Common Coordination Numbers of Metal Ions

Cu^+	2, 4		
Ag^+	2		
Au^+	2, 4	Al^{3+}	4, 6
		Sc^{3+}	6
		Cr^{3+}	6
Fe^{2+}	6	Fe^{3+}	6
Co^{2+}	4, 6	Co^{3+}	6
Ni^{2+}	4, 6	Au^{3+}	4
Cu^{2+}	4, 6	Pt^{4+}	6
Zn^{2+}	4		
Pt^{2+}	4		

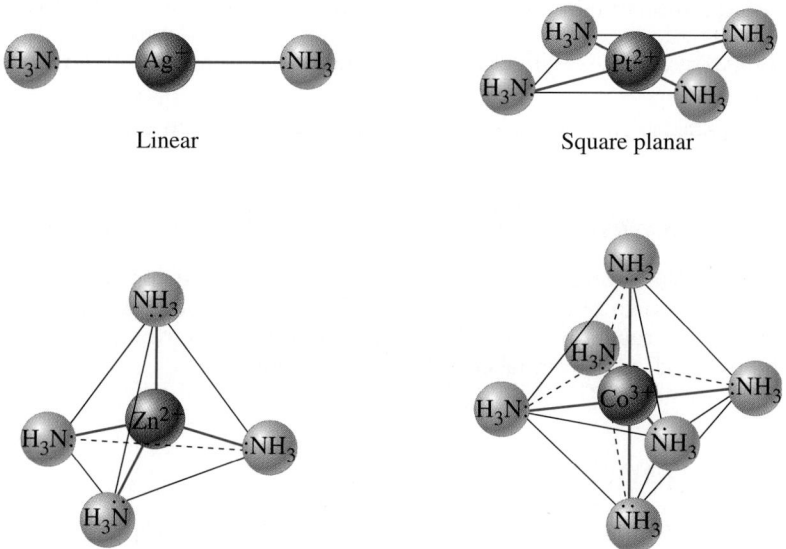

Linear   Square planar

Tetrahedral   Octahedral

▲ FIGURE 24-2
Structures of some complex ions
Attachment of the NH_3 molecules occurs through the lone-pair electrons on the N atoms. In each complex, all the ligands are the same and hence no distortions of these shapes are observed.

EXAMPLE 24-1   Relating the Formula of a Complex to the Coordination Number and Oxidation State of the Central Metal

What are the coordination number and oxidation state of Co in the complex ion $[CoCl(NO_2)(NH_3)_4]^+$?

Analyze

In determining the oxidation state of the metal ion in a complex, it is important to recognize which ligands are charged (invariably negatively charged) and which are neutral.

Solve

The complex ion has as ligands *one* Cl^- ion, *one* NO_2^- ion, and *four* NH_3 molecules. The coordination number is 6. Of these six ligands, *two* carry a charge of $1-$ each (the Cl^- and NO_2^- ions) and four are *neutral* (the NH_3 molecules). The total contribution of the anions to the net charge on the complex ion is $2-$. Because the net charge on the complex ion is $1+$, the oxidation state of the central cobalt ion is $+3$. Diagrammatically, we can write

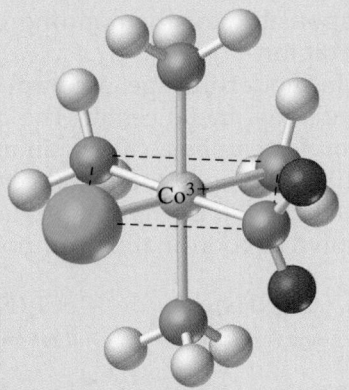

▲ The complex ion $[CoCl(NO_2)(NH_3)_4]^+$

Oxidation state = x ⟶ Charge of $1-$ on Cl^-

Charge of $1-$ on NO_2^-

$\left.\begin{array}{c}\text{Total negative}\\\text{charge: }2-\end{array}\right\}$

$[CoCl(NO_2)(NH_3)_4]^+$

Net charge on complex ion

Coordination number = 6

$x - 2 = +1$
$x = +3$

Assess

The oxidation state of the metal can be determined by using the following relationship:

charge on the complex ion = oxidation state of the metal + sum of charges on the ligands

In this case,

$$(1+) = \text{oxidation state of the metal} + (1-) + (1-)$$
$$\text{oxidation state of the metal} = +3$$

PRACTICE EXAMPLE A: What are the coordination number and oxidation state of nickel in the ion $[Ni(CN)_4I]^{3-}$?

PRACTICE EXAMPLE B: Write the formula of a complex with cyanide ion ligands, an iron ion with an oxidation state of $+3$, and a coordination number of 6.

🔍 **24-1   CONCEPT ASSESSMENT**

A complex of Al(III) can be formulated as $AlCl_3 \cdot 3\,H_2O$. The coordination number is not known but is expected to be 4 or 6. Describe how Werner's methods, that is, reaction with $AgNO_3(aq)$ or conductivity measurements, could help elucidate the actual coordination number.

KEEP IN MIND

that the covalent bond formed in the Lewis acid–base reaction is a *coordinate* covalent bond. Thus we can think of the coordination number of a transition metal ion in a complex as the number of coordinate covalent bonds in the complex.

24-2   Ligands

A common feature shared by the ligands in coordination complexes is the ability to donate electron pairs to central metal atoms or ions. Ligands are *Lewis bases*. In accepting electron pairs, central metal atoms or ions act as *Lewis acids*. A ligand that uses one pair of electrons to form one point of attachment to the central metal atom or ion is called a **monodentate** ligand. Some examples of monodentate ligands are monatomic anions such as the halide ions, polyatomic anions such as hydroxide ion, simple molecules, such as ammonia (called *ammine* when it is a ligand), and more complex molecules, such as methylamine, CH_3NH_2 (Table 24.2).

TABLE 24.2 Some Common Monodentate Ligands

Formula	Name as Ligand	Formula	Name as Ligand	Formula	Name as Ligand
Neutral molecules		**Anions**		**Anions**	
H_2O	Aqua	F^-	Fluoro	SO_4^{2-}	Sulfato
NH_3	Ammine	Cl^-	Chloro	$S_2O_3^{2-}$	Thiosulfato
CO	Carbonyl	Br^-	Bromo	NO_2^-	Nitrito-N-[a]
NO	Nitrosyl	I^-	Iodo	ONO^-	Nitrito-O-[a]
CH_3NH_2	Methylamine	O^{2-}	Oxo	SCN^-	Thiocyanato-S-[b]
C_5H_5N	Pyridine	OH^-	Hydroxo	NCS^-	Thiocyanato-N-[b]
		CN^-	Cyano		

[a]If the nitrite ion is attached through the N atom ($-NO_2$), the designation *nitrito-N-* is used; if attached through an O atom ($-ONO$), *nitrito-O-*.
[b]If the thiocyanate ion is attached through the S atom ($-SCN$), the name *thiocyanato-S-* is used; if attachment is through the N atom ($-NCS$), *thiocyanato-N-*.

KEEP IN MIND

that the lone pairs of electrons of a polydentate ligand must be far enough apart to attach to the metal center at two or more points; the donated pairs of electrons must be on different atoms.

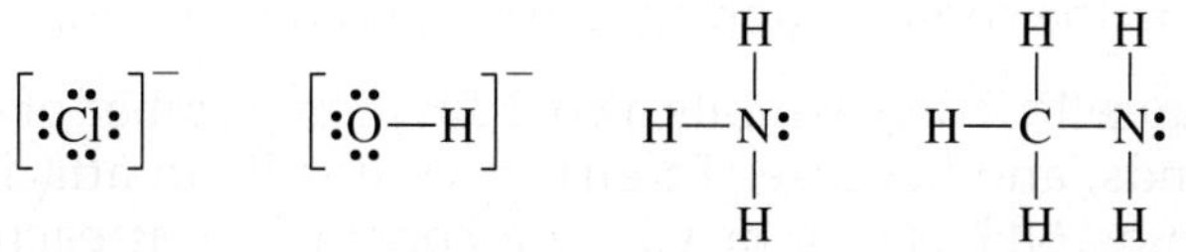

Ligand name: Chloro   Hydroxo   Ammine   Methylamine

Some ligands are capable of donating more than a single electron pair from *different* atoms in the ligand and to *different* sites in the geometric structure of a complex. These are called **polydentate** ligands. The molecule *ethylenediamine* (en) can donate two electron pairs, one from each N atom. Since en attaches to the metal center at two points, it is called a **bidentate** ligand.

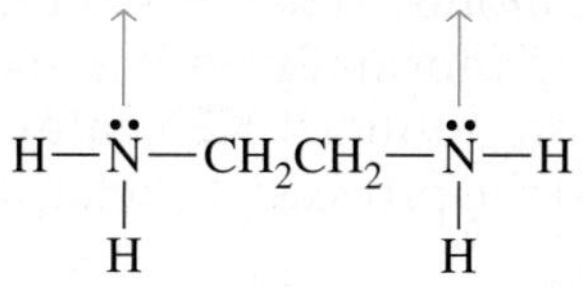

Three common polydentate ligands are shown in Table 24.3.

KEEP IN MIND

that the $EDTA^{4-}$ ligand is not planar as suggested in Table 24.3. Use VSEPR theory to gain an insight into the shape of this polyatomic anion. Also, refer to Figure 24-23.

TABLE 24.3 Some Common Polydentate Ligands (Chelating Agents)

Abbreviation	Name	Formula
en	Ethylenediamine	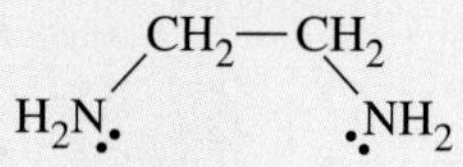
ox^{2-}[a]	Oxalato	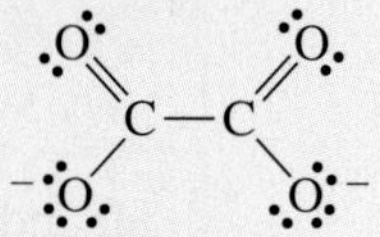
$EDTA^{4-}$[b]	Ethylenediaminetetraacetato	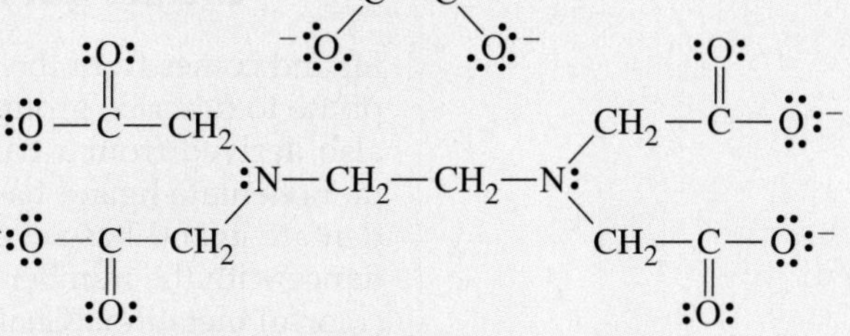

[a]Oxalic acid is a diprotic acid denoted H_2ox. It is the ox^{2-} anion that binds as a bidentate ligand.
[b]Ethylenediaminetetraacetic acid, a tetraprotic acid, is denoted H_4EDTA.

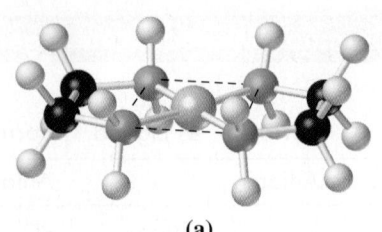

(a)

► FIGURE 24-3
Three representations of the chelate [Pt(en)₂]²⁺
(a) Overall structure. **(b)** The ligands attach at adjacent corners along an edge of the square. They do *not* bridge the square by attaching to opposite corners. Bonds are shown in red, and the square-planar shape is indicated by the black parallelogram.

(b)

Figure 24-3 represents the attachment of two ethylenediamine (en) ligands to a Pt^{2+} ion. Here is how we can establish that each ligand is attached to two positions in the coordination sphere around the Pt^{2+} ion.

- Because Pt^{2+} ion exhibits a coordination number of 4 with monodentate ligands, and because $[Pt(en)_2]^{2+}$ is unable to attach additional ligands such as NH_3, H_2O, or Cl^-, we conclude that each en group must be attached at two points.

- The en ligands in the complex ion exhibit no further basic properties. They cannot accept protons from water to produce OH^-, as they would if they had an available lone pair of electrons. Both $—NH_2$ groups of each en molecule must be tied up in the complex ion.

Note the two five-member rings (pentagons) outlined in Figure 24-3(b). They consist of Pt, N, and C atoms. When the bonding of a polydentate ligand to a metal ion produces a ring (normally with five or six members), we refer to the complex as a **chelate** (pronounced KEY-late). The polydentate ligand is called a **chelating agent**, and the process of chelate formation is called *chelation*.

🔍 **24-2   CONCEPT ASSESSMENT**

The ligand diethylenetriamine (abbreviated det), $H_2NCH_2CH_2NHCH_2CH_2NH_2$, can form complexes. Classify this ligand as mono-, bi-, tri-, or quadridentate.

24-1 ARE YOU WONDERING...

How such strange terms as *ligand*, *monodentate*, and *chelate* got into the vocabulary of chemistry?

Ligand comes from the Latin word *ligare*, which means to bind. It is quite appropriate to describe groups that are bound to a metal center as ligands. Dentate is also derived from a Latin word, *dens*, meaning tooth. Figuratively speaking, a monodentate ligand has one tooth; a bidentate ligand has two teeth; and a polydentate ligand has several. A ligand attaches itself to the metal center in accordance with the number of "teeth" it possesses. This is an easily remembered and colorful metaphor. Chelate is derived from the Greek word *chela*, which means a crab's claw. The way in which a chelating agent attaches itself to a metal ion resembles a crab's claw—another colorful metaphor.

24-3 Nomenclature

The system of naming complexes originated with Werner, but it has been modified several times over the years. Even today, usage varies somewhat. Our approach will be to consider a few general rules that permit us to relate names and formulas of simple complexes. We will not consider any of the complicated cases for which writing names and formulas is more challenging.

1. *Anions as ligands are named by using the ending* -o. As implied by Table 24.2, normally -*ide* endings change to -*o*, -*ite* to -*ito*, and -*ate* to -*ato*.

2. *Neutral molecules as ligands generally carry the unmodified name.* For example, the name ethylenediamine is used both for the free molecule and for the molecule as a ligand. Aqua, ammine, carbonyl, and nitrosyl are important exceptions (see Table 24.2).

3. *The number of ligands of a given type is denoted by a prefix.* The usual prefixes are *mono* = 1, *di* = 2, *tri* = 3, *tetra* = 4, *penta* = 5, and *hexa* = 6. As in many other cases, the prefix *mono-* is often not used. If the ligand name is a composite name that itself contains a numerical prefix, such as ethylene*di*amine, place parentheses around the name and precede it with *bis* = 2, *tris* = 3, *tetrakis* = 4, and so on. Thus, dichloro signifies *two* Cl^- ions as ligands, and pentaaqua signifies *five* H_2O molecules. To indicate the presence of *two* ethylenediamine (en) ligands, we write *bis*(ethylenediamine).

4. *When we name a complex, ligands are named first, in alphabetical order, followed by the name of the metal center. The oxidation state of the metal center is denoted by a Roman numeral. If the complex is an anion, the ending* -*ate* *is attached to the name of the metal.* Prefixes (*di*, *tri*, *bis*, *tris*,...) are ignored in establishing the alphabetical order. Thus, $[CrCl_2(H_2O)_4]^+$ is called tetraaquadichlorochromium(III) ion; $[CoCl_2(en)_2]^+$ is dichlorobis(ethylenediamine)cobalt(III) ion; and $[Cr(OH)_4]^-$ is tetrahydroxochromate(III) ion. For complex anions of a few of the metals, the English name is replaced by the Latin name given in Table 24.4. Thus, $[CuCl_4]^{2-}$ is the tetrachlorocuprate(II) ion.

5. *When we write the formula of a complex, the chemical symbol of the metal center is written first, followed by the formulas of anions and then neutral molecules.* If there are two or more different anions or neutral molecules as ligands, they are written in alphabetical order according to the first chemical symbols of their formulas. Thus, in the formula of the tetraaminechloronitrito-*N*-cobalt(III) ion, Cl^- precedes NO_2^-, and both are placed ahead of the neutral NH_3 molecules: $[CoCl(NO_2)(NH_3)_4]^+$.

6. *In names and formulas of coordination compounds, cations come first followed by anions.* This is the same order as in simple ionic compounds like NaCl for sodium chloride. For example, the formula $[Pt(NH_3)_4][PtCl_4]$ represents the coordination compound tetraammineplatinum(II) tetrachloroplatinate(II).

◄ Occasionally, the metal center will be in the oxidation state 0, as in $[W(CO)_6]$ hexacarbonyltungsten(0).

TABLE 24.4 Names for Some Metals in Complex Anions

Iron	⟶	Ferrate
Copper	⟶	Cuprate
Tin	⟶	Stannate
Silver	⟶	Argentate
Lead	⟶	Plumbate
Gold	⟶	Aurate

EXAMPLE 24-2   Relating Names and Formulas of Complexes

(a) What is the name of the complex $[CoCl_3(NH_3)_3]$? **(b)** What is the formula of the compound pentaaquachlorochromium(III) chloride? **(c)** What is the name of the compound $K_3[Fe(CN)_6]$?

Analyze

To translate a chemical formula into the appropriate IUPAC name, we need to determine the oxidation state of the metal center (as done in Example 24-1), and then focus on naming the ligands. Translating an IUPAC name to a chemical formula is a simpler exercise, provided we know the names of common ligands and understand the meanings of the various prefixes.

(continued)

Solve

(a) $[CoCl_3(NH_3)_3]$ consists of *three* ammonia molecules and *three* chloride ions attached to a central Co^{3+} ion; it is electrically neutral. The name of this neutral complex is triamminetrichlorocobalt(III).

(b) The central metal ion is Cr^{3+}. There are five H_2O molecules and one Cl^- ion as ligands. The complex ion carries a net charge of 2+. Two Cl^- ions are required to neutralize the charge on this complex cation. The formula of the coordination compound is $[CrCl(H_2O)_5]Cl_2$.

(c) This compound consists of K^+ cations and complex anions having the formula $[Fe(CN)_6]^{3-}$. Each cyanide ion carries a charge of $1-$, so the oxidation state of the iron must be $+3$. The Latin-based name "ferrate" is used because the complex ion is an anion. The name of the anion is hexacyanoferrate(III) ion. The coordination compound is potassium hexacyanoferrate(III).

Assess

The number of counter ions is never stated explicitly but is derived from the name. In part **(b)**, for example, the number of chloride ions is not stated.

PRACTICE EXAMPLE A: What is the formula of the compound potassium hexachloroplatinate(IV)?

PRACTICE EXAMPLE B: What is the name of the compound $[Co(SCN)(NH_3)_5]Cl_2$?

Although most complexes are named in the manner just outlined, some common, or trivial, names are still in use. Two such trivial names are ferrocyanide for $[Fe(CN)_6]^{4-}$ and ferricyanide for $[Fe(CN)_6]^{3-}$. These common names suggest the oxidation state of the central metal ions through the *o* and *i* designations (*o* for the ferrous ion, Fe^{2+}, in $[Fe(CN)_6]^{4-}$ and *i* for the ferric ion, Fe^{3+}, in $[Fe(CN)_6]^{3-}$). These trivial names do not indicate that the metal ions have a coordination number of 6, however. The systematic names—hexacyanoferrate(II) and hexacyanoferrate(III)—are more informative.

24-3 CONCEPT ASSESSMENT

A student named a coordination complex tripotassium dichlorodibromodihydroxoiron. The student's instructor pointed out that although the correct formula for the compound could be deduced, this name violates the IUPAC convention. How does this name violate IUPAC rules, and what is the correct name?

24-4 Isomerism

As previously noted (page 95), *isomers* are substances that have the same formulas but differ in their structures and properties. Several kinds of isomerism are found among complex ions and coordination compounds. These can be lumped into two broad categories: **Structural isomers** differ in basic structure or bond type—what ligands are bonded to the metal center and through which atoms. **Stereoisomers** have the same number and types of ligands and the same mode of attachment, but they differ in the way in which the ligands occupy the space around the metal center. Of the following five examples, the first three are types of structural isomerism (ionization, coordination, and linkage isomerism), and the remaining two are types of stereoisomerism (geometric and optical isomerism).

▶ Structural isomers are also known as constitutional isomers.

Ionization Isomerism

The two coordination compounds whose formulas are shown here have the same central ion (Cr^{3+}), and five of the six ligands (NH_3 molecules) are the same. The compounds differ in that one has SO_4^{2-} ion as the sixth ligand, with a Cl^- ion to neutralize the charge of the complex ion, whereas the other has Cl^- as the sixth ligand and SO_4^{2-} to neutralize the charge of the complex ion.

$$[CrSO_4(NH_3)_5]Cl$$
Pentaamminesulfatochromium(III) chloride
(a)

$$[CrCl(NH_3)_5]SO_4$$
Pentaamminechlorochromium(III) sulfate
(b)

Coordination Isomerism

A situation somewhat similar to that just described can arise when a coordination compound is composed of both complex cations and complex anions. The ligands can be distributed differently between the two complex ions, as are $NH_3(aq)$ and CN^- in these two compounds.

$$[Co(NH_3)_6][Cr(CN)_6]$$
Hexaamminecobalt(III) hexacyanochromate(III)
(a)

$$[Cr(NH_3)_6][Co(CN)_6]$$
Hexaaminechromium(III) hexacyanocobaltate(III)
(b)

Linkage Isomerism

Some ligands may attach to the central metal ion of a complex ion in different ways. For example, the nitrite ion, a monodentate ligand, has electron pairs available for coordination both on the N and O atoms.

$$\left[\begin{matrix} \ddot{N} \\ :\ddot{O}: \quad \ddot{O}: \end{matrix} \right]^-$$

Whether attachment of this ligand is through the N or an O atom, the formula of the complex ion is unaffected. The properties of the complex ion, however, may be affected. When attachment occurs through the N atom, the ligand can be referred to as nitro or, more properly, nitrito-*N*- when naming the complex. Coordination through an O atom can be referred to as nitrito or, more properly, nitrito-*O*- when naming the complex.

$$[Co(NO_2)(NH_3)_5]^{2+}$$
Pentaamminenitrito-*N*-cobalt(III) ion
(a)

$$[Co(ONO)(NH_3)_5]^{2+}$$
Pentaamminenitrito-*O*-cobalt(III) ion
(b)

The structures of these compounds are illustrated in Figure 24-4 on page 1078.

Geometric Isomerism

If a single Cl^- ion is substituted for an NH_3 molecule in the square-planar complex ion $[Pt(NH_3)_4]^{2+}$ in Figure 24-2, it does not matter at which corner of the square this substitution is made. All four possibilities are alike. If a *second* Cl^- is substituted for an NH_3, there are two distinct possibilities (Fig. 24-5b). The two Cl^- ions can either be along the same edge of the square (*cis*) or on opposite corners, across from each other (*trans*). To distinguish clearly between these two possibilities, we must either draw a structure or refer to the appropriate name. The formula alone will not distinguish between them. (Note that this complex is a neutral species, not an ion.)

$$[PtCl_2(NH_3)_2]$$
cis-diamminedichloroplatinum(II)

or

trans-diamminedichloroplatinum(II)

The type of isomerism described here in which the positions of the ligands produce distinct isomers is called **geometric isomerism**. Interestingly, if a *third*

◀ The term *cis* means "on this side" in Latin, and *trans* means "across." Domestic airline flights in the United States are *cis*atlantic, whereas flights to Europe are *trans*atlantic.

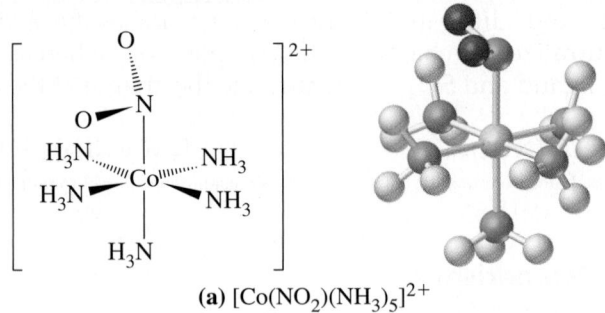

(a) $[Co(NO_2)(NH_3)_5]^{2+}$

▶ FIGURE 24-4
Linkage isomerism—illustrated
(a) Pentaamminenitrito-*N*-cobalt(III) cation.
(b) Pentaamminenitrito-*O*-cobalt(III) cation.

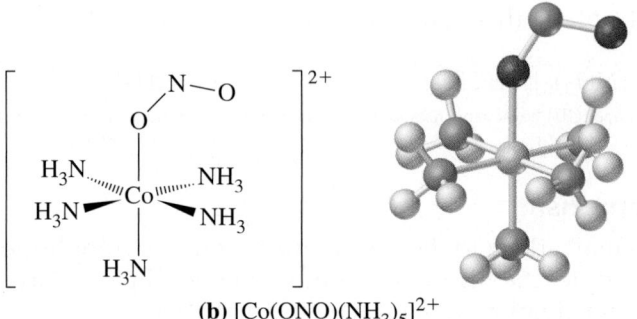

(b) $[Co(ONO)(NH_3)_5]^{2+}$

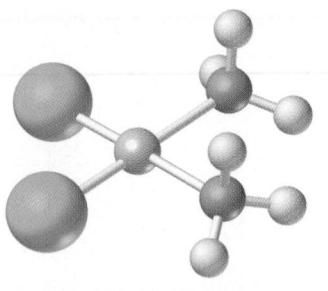

cis-$[PtCl_2(NH_3)_2]$

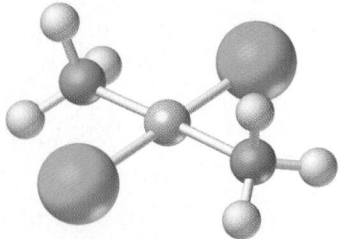

trans-$[PtCl_2(NH_3)_2]$

▲ The geometric isomers of
$[PtCl_2(NH_3)_2]$

▶ FIGURE 24-5
Geometric isomerism—illustrated
For the square-planar complexes shown
here, isomerism exists only when two
Cl⁻ ions have replaced NH₃ molecules.

Cl⁻ ion is substituted, isomerism disappears (Fig. 24-5c). There is only one complex ion with the formula $[PtCl_3(NH_3)]^-$.

With an octahedral complex, the situation is a bit more complicated. Take the complex ion $[Co(NH_3)_6]^{3+}$ of Figure 24-2 as an example. If one Cl⁻ is substituted for an NH₃, a single structure results. With the substitution of *two* Cl⁻ ions for NH₃ molecules, *cis* and *trans* isomers result. The *cis* isomer has two Cl⁻ ions along the same edge of the octahedron (Fig. 24-6a). The *trans* isomer has two Cl⁻ ions on opposite corners, that is, at opposite ends of a line drawn through the central metal ion (Fig. 24-6b). One difference between the two is that the *cis* isomer has a purple color and the *trans* has a bright green color.

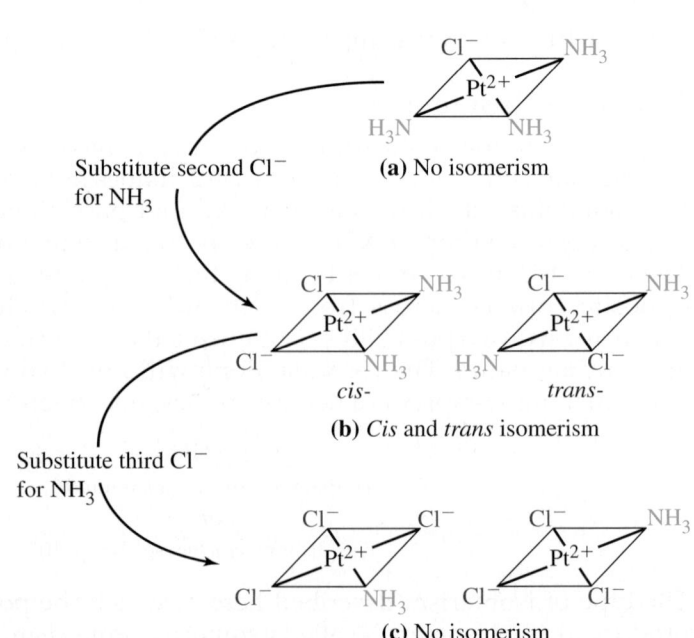

Substitute second Cl⁻
for NH₃

(a) No isomerism

cis-   *trans-*

(b) *Cis* and *trans* isomerism

Substitute third Cl⁻
for NH₃

(c) No isomerism

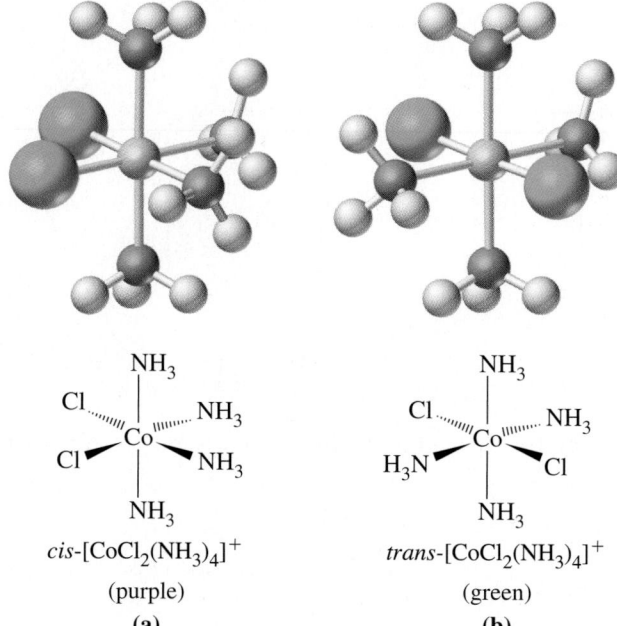

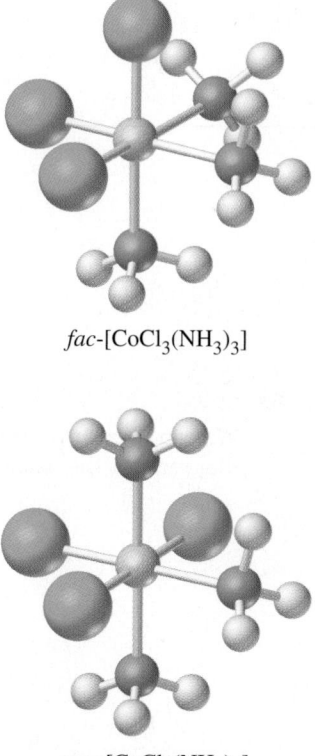

fac-[CoCl$_3$(NH$_3$)$_3$]

mer-[CoCl$_3$(NH$_3$)$_3$]

▲ The geometric isomers of [CoCl$_3$(NH$_3$)$_3$]

▲ FIGURE 24-6
Cis and trans isomers of an octahedral complex
The Co^{3+} ion is at the center of the octahedron, and NH$_3$ and Cl$^-$ ligands are at the vertices.

cis-[CoCl$_2$(NH$_3$)$_4$]$^+$
(purple)
(a)

trans-[CoCl$_2$(NH$_3$)$_4$]$^+$
(green)
(b)

If a *third* Cl$^-$ is substituted for an NH$_3$ in Figure 24-6(a), two possibilities exist. If this third substitution is at either the top or bottom of the structure, the result is that three Cl$^-$ ions appear on the same face of the octahedron. This is called a *fac* (facial) isomer. If the third substitution is at either of the other two positions, the result is three Cl$^-$ ions around a perimeter or meridian of the octahedron. This is a *mer* (meridional) isomer.

🔍 **24-4   CONCEPT ASSESSMENT**

Explain why the substitution of a *fourth* Cl$^-$ for an NH$_3$ in *mer*-[CoCl$_3$(NH$_3$)$_3$] may produce some of the same product as the substitution of a fourth Cl$^-$ for an NH$_3$ in *fac*-[CoCl$_3$(NH$_3$)$_3$], or it may also produce a different product.

EXAMPLE 24-3   Identifying Geometric Isomers

Sketch structures of all the possible isomers of [CoCl(ox)(NH$_3$)$_3$].

Analyze

The Co^{3+} ion exhibits a coordination number of 6. The structure is octahedral. Recall that ox (oxalate ion) is a bidentate ligand carrying a charge of 2– (see Table 24.3). Also, as shown in Figure 24-3, a bidentate ligand must be attached in *cis* positions, not *trans*.

Solve

Once the ox ligand is placed, any position is available to the Cl$^-$. This leaves two possibilities for the three NH$_3$ molecules. They can be situated (1) on the same face of the octahedron (*fac* isomer) or (2) around a perimeter of the octahedron (*mer* isomer).

(continued)

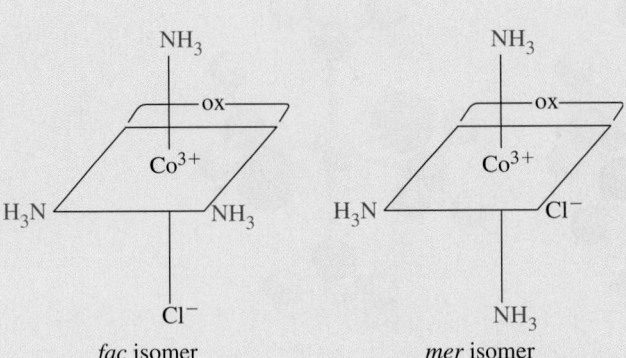

fac isomer   *mer* isomer

Assess

We can also draw these stereoisomers by using the dashed and solid wedge symbols, together with a line diagram for the oxalate anion.

fac isomer   *mer* isomer

PRACTICE EXAMPLE A: Sketch the geometric isomers of $[CoCl_2(ox)(NH_3)_2]^-$.

PRACTICE EXAMPLE B: Sketch the geometric isomers of $[MoCl_2(C_5H_5N)_2(CO)_2]^+$.

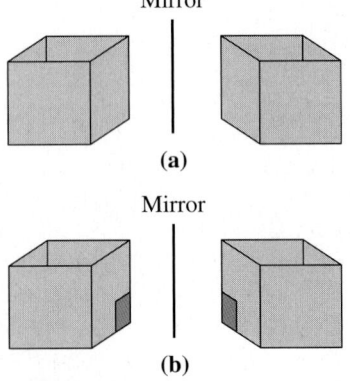

▲ FIGURE 24-7
Superimposable and nonsuperimposable objects—an open-top box
(a) You can place the box into its mirror image (hypothetically) in several ways. The box and its mirror image are superimposable. (b) No matter how you place the box in its mirror image, the stickers will not appear in the same position. This box and its mirror image are nonsuperimposable.

Optical Isomerism

To understand optical isomerism, we need to understand the relationship between an object and its mirror image. Features on the right side of the object appear on the left side of its image in a mirror, and vice versa. Certain objects can be rotated in such a way as to be *superimposable* on their mirror images, but other objects are *nonsuperimposable* on their mirror images. An unmarked tennis ball is superimposable on its mirror image, but a left hand is nonsuperimposable on its mirror image (a right hand).

Consider the open-top cardboard box pictured in Figure 24-7(a). There are a number of hypothetical ways in which the box can be superimposed on its mirror image. Now imagine that a distinctive sticker is placed at a corner of one side of the box (Fig. 24-7b). In this case, there is no way that the box and its mirror image can be superimposed; they are clearly different. This is equivalent to saying that there is no way that a formfitting left glove can be worn on a right hand (turning it inside out is not allowed).

The two structures of $[Co(en)_3]^{3+}$ depicted in Figure 24-8 are related to each other as are an object and its image in a mirror. Furthermore, the two structures are *nonsuperimposable*, like a left and a right hand. The two structures represent two different complex ions; they are isomers.

Structures that are nonsuperimposable mirror images of each other are called **enantiomers** and are said to be **chiral** (pronounced KYE-rull). (Structures that are superimposable are *achiral*.) Whereas other types of isomers may differ significantly in their physical and chemical properties,

◀ The prefixes *dextro* and *levo* are derived from the Latin words *dexter*, "right," and *levo*, "left."

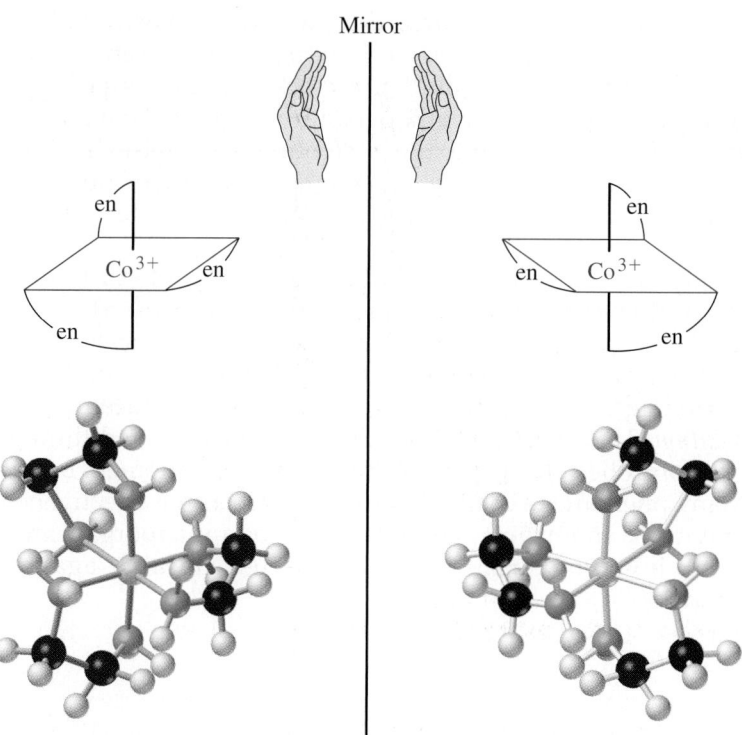

▲ FIGURE 24-8
Optical isomers
The two structures are nonsuperimposable mirror images. Like a right hand and a left hand, one structure cannot be superimposed onto the other.

enantiomers have identical properties except in a few specialized situations. These exceptions involve phenomena that are directly linked to chirality, or handedness, at the molecular level. An example is *optical activity*, pictured in Figure 24-9.

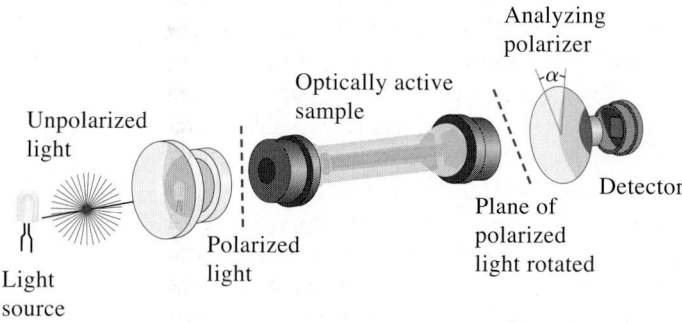

▲ FIGURE 24-9
Optical activity
Light from an ordinary source consists of electromagnetic waves vibrating in all planes; it is unpolarized. This light is passed through a polarizer, a material that screens out all waves except those vibrating in a particular plane. The plane of polarization of transmitted polarized light is then changed by passage through an optically active substance. The angle through which the plane of polarization has been rotated is determined by rotating an analyzer (a second polarizer) to the extent that all the polarized light is absorbed.

◀ Rotation of plane polarized light is a physical property. Chemical activity, or lack of it, is not a physical property. The physical and chemical properties, however, both depend on the existence of a center of asymmetry.

▶ A process based on chirality would be creating the maximum number of matched pairs of gloves from a bin of gloves of identical sizes of which 90 are for the right hand, but only 10 are for the left hand.

Interactions between a beam of polarized light and the electrons in an enantiomer cause the plane of the polarized light to rotate. One enantiomer rotates the plane of polarized light to the right (clockwise) and is said to be *dextrorotatory* (designated + or *d*). The other enantiomer rotates the plane of polarized light to the same extent, but to the left (counterclockwise). It is said to be *levorotatory* (− or *l*). Because they can rotate the plane of polarized light, the enantiomers are said to be *optically active* and are referred to as **optical isomers**.

When an optically active complex is synthesized, a mixture of the two optical isomers (enantiomers) is obtained, such as the two $[Co(en)_3]^{3+}$ isomers in Figure 24-8. The optical rotation of one isomer just cancels that of the other. The mixture, called a *racemic mixture*, produces no net rotation of the plane of polarized light. Separating the *d* and *l* isomers of a racemic mixture is called *resolution*. This separation can sometimes be achieved through chemical reactions controlled by chirality, with the two enantiomers behaving differently. Many phenomena of the living state, such as the effectiveness of a drug, the activity of an enzyme, and the ability of a microorganism to promote a reaction, involve chirality. We will discuss chirality again in Chapters 26, 27, and 28.

24-2 ARE YOU WONDERING...

How to tell if a molecule is superimposable on its mirror image?

The complex ion $[CrBr_2(H_2O)_2(NH_3)_2]^+$ has five isomers; one is shown below with its mirror image. To test if the mirror image is superimposable on the original, imagine rotating the mirror image about the vertical axis (H_2O—Cr—NH_3) by 180° so that the two Br^- ligands are in the same position as in the original molecule. We see that the NH_3 and H_2O ligands that are in the same plane as the two Br^- ligands are in reversed positions when compared with the original molecule. Thus the molecule and its mirror image are not superimposable. The molecule is potentially optically active.

▲ FIGURE 24-10
Hypothetical structures for $[CoCl_2(NH_3)_4]^+$
If the complex ion had this hexagonal structure, there should be *three* distinct isomers, but only two isomers exist—*cis* and *trans*.

Isomerism and Werner's Theory

The study of isomerism played a crucial role in the development of Werner's theory of coordination chemistry. Werner proposed that complexes with coordination number 6 have an octahedral structure, but other possibilities were also proposed. For example, Figure 24-10 shows a hypothetical hexagonal structure for $[CoCl_2(NH_3)_4]^+$. However, this hexagonal structure would require the existence of *three* isomers, but a third isomer was never found; the only two isomers are those pictured in Figure 24-6. Additional direct evidence came with the discovery of optical isomerism in the tris(ethylenediamine) cobalt(III) ion (Fig. 24-8). Neither the hexagonal structure nor alternative structures can account for this isomerism—but, as we have seen, the octahedral structure does. Werner even succeeded in preparing an optically active octahedral complex with only inorganic ligands, to overcome objections that the optical activity of tris(ethylenediamine)cobalt(III) ion owed its optical activity to its carbon atoms and not to its geometric structure.

🔍 24-5   CONCEPT ASSESSMENT

Of the following complex cations, which are identical, which are geometrical isomers, which are enantiomers?

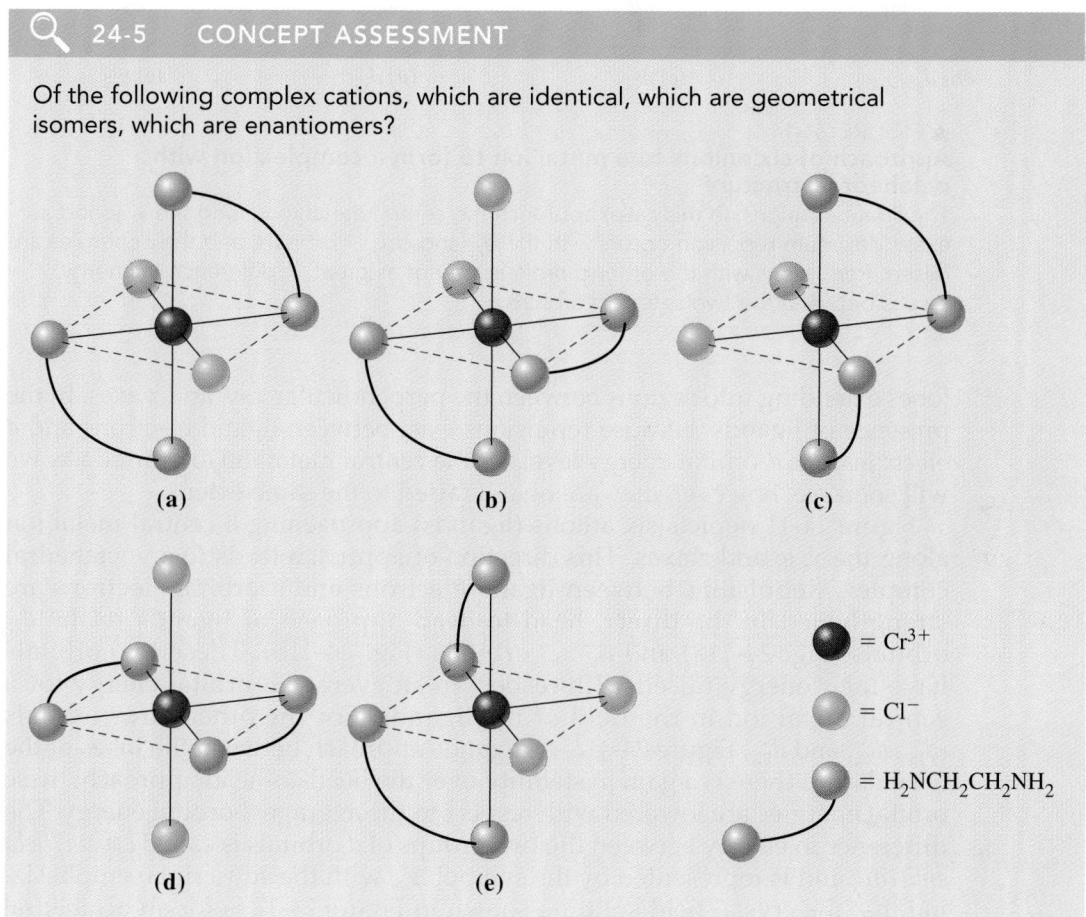

(a)   (b)   (c)

(d)   (e)

● = Cr^{3+}

○ = Cl^-

○ = $H_2NCH_2CH_2NH_2$

24-5   Bonding in Complex Ions: Crystal Field Theory

The theories of chemical bonding that were so useful in earlier chapters do not help much in explaining the characteristic colors and magnetic properties of complex ions. In transition metal ions, we need to focus on how the electrons in the d orbitals of a metal ion are affected when they are in a complex. A theory that provides that focus and an explanation of these properties is crystal field theory.

In the **crystal field theory**, bonding in a complex ion is considered to be an electrostatic attraction between the positively charged nucleus of the central metal ion and electrons in the ligands. Repulsions also occur between the ligand electrons and electrons in the central ion. In particular, the crystal field theory focuses on the repulsions between ligand electrons and d electrons of the central ion.

First, a reminder about the d orbitals introduced in Figure 8-30: All five of the orbitals are alike in energy when in an isolated atom or ion, but they are *unlike* in their spatial orientations. One of them, d_{z^2}, is directed along the z axis, and another, $d_{x^2-y^2}$, has lobes along the x and y axes. The remaining three have

◄ Modifications of the simple crystal field theory that take into account such factors as the partial covalency of the metal–ligand bond are called *ligand field theory*. This term is often used to signify both the purely electrostatic crystal field theory and its modifications.

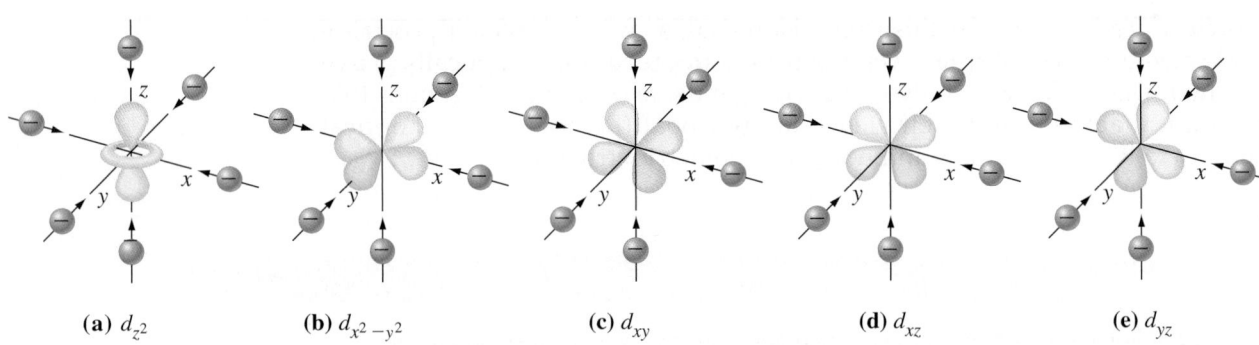

(a) d_{z^2} **(b)** $d_{x^2-y^2}$ **(c)** d_{xy} **(d)** d_{xz} **(e)** d_{yz}

▲ FIGURE 24-11
Approach of six anions to a metal ion to form a complex ion with octahedral structure
The ligands (anions, in this case) approach the central metal ion along the x, y, and z axes. Maximum repulsion occurs with the d_{z^2} and $d_{x^2-y^2}$ orbitals, and their energies are raised. Repulsions with the other d orbitals are not as great. A difference in energy results between the two sets of d orbitals.

lobes extending into regions between the perpendicular x, y, and z axes. In the presence of ligands, because repulsions exist between ligand electrons and d electrons, the d-orbital energy levels of the central metal ion are raised. As we will soon see, however, they are not all raised to the same extent.

Figure 24-11 depicts six anions (ligands) approaching a central metal ion along the x, y, and z axes. This direction of approach leads to an octahedral complex. Repulsions between ligand electrons and d-orbital electrons are strengthened in the direct, head-to-head approach of ligands to the d_{z^2} orbitals (Fig. 24-11a) and $d_{x^2-y^2}$ orbitals (Fig. 24-11b). These two orbitals have their energy raised with respect to an average d-orbital energy for a central metal ion in the field of the ligands. For the other three orbitals (d_{xy}, d_{xz}, and d_{yz}, Figure 24-11c–e), ligands approach between the lobes of the orbitals and there is a gain in stability over the head-to-head approach; these orbital energies are lowered with respect to the average d-orbital energy. The difference in energy between the two groups of d orbitals is called *crystal field splitting* and is represented by the symbol Δ_o, with the subscript o emphasizing that the crystal field splitting shown in Figure 24-12 is for an octahedral complex.

The removal of the degeneracy of the d orbitals by the crystal field has important consequences for the electron configurations of transition metal ions having between 4 and 7 d electrons. Consider the transition metal ion Cr^{2+} with a d^4 configuration. If the four d electrons are assigned to the orbitals of lowest energy first, the first three electrons go into the d_{xy}, d_{xz}, and d_{yz} orbitals according to Hund's rule (page 339)—but what about the fourth

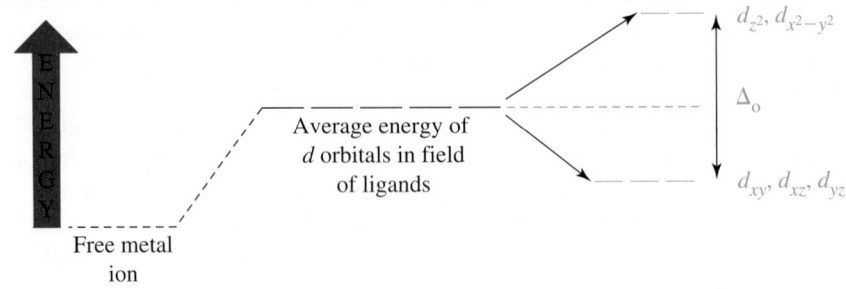

ENERGY

Free metal ion

Average energy of d orbitals in field of ligands

$d_{z^2}, d_{x^2-y^2}$

Δ_o

d_{xy}, d_{xz}, d_{yz}

▲ FIGURE 24-12
Splitting of d energy levels in the formation of an octahedral complex ion
The d-orbital energy levels of the free central ion are raised in the presence of ligands to the average level shown, but the five levels are split into two groups.

electron? The aufbau process (page 340) suggests that the fourth electron should pair up with any one of the three electrons already in the d_{xy}, d_{xz}, and d_{yz} orbitals.

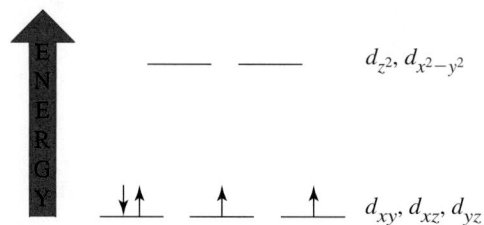

Placing the fourth electron in the lower level confers extra stability (lower energy) on the complex, but some of this stability is offset because it requires energy, called the **pairing energy** (P), to force an electron into an orbital that is already occupied by an electron. Alternatively, the electron could be assigned to either the $d_{x^2-y^2}$ or d_{z^2} orbital, avoiding the pairing energy.

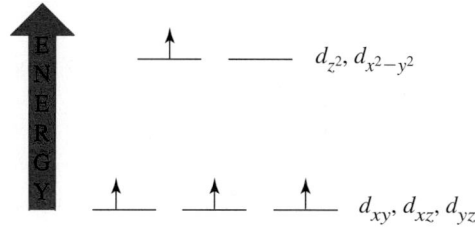

Placing the fourth electron in the upper level requires energy and offsets the extra stability acquired by placing the first three electrons in the lower level. To pair or not to pair, that is the question.

Whether the fourth electron enters the lowest level and becomes paired or, instead, enters the upper level with the same spin as the first three electrons depends on the magnitude of Δ_o. If Δ_o is greater than the pairing energy, P, greater stability is obtained if the fourth electron is paired with one in the lower level. If Δ_o is less than the pairing energy, greater stability is obtained by keeping the electrons unpaired. Thus, for octahedral chromium(II) complexes, there are two possibilities for the number of unpaired electrons. In one case, there are four unpaired electrons when $\Delta_o < P$; this situation corresponds to the maximum number of unpaired electrons and is referred to as **high spin**. Ligands such as H_2O and F^- produce only a small crystal field splitting, leading to high-spin complexes; such ligands are said to be *weak-field ligands*. As an example, $[Cr(H_2O)_6]^{2+}$ is a weak-field complex. In the other case, when $\Delta_o > P$, there are two unpaired electrons; this corresponds to the minimum number of unpaired electrons and is referred to as **low spin**. Ligands, such as NH_3 and CN^-, produce large crystal field splitting, leading to low-spin complexes; such ligands are said to be *strong-field ligands*. $[Cr(CN)_6]^{4-}$ is a strong-field complex.

Different ligands can be arranged in order of their abilities to produce a splitting of the d energy levels. This arrangement is known as the **spectrochemical series**.

◄ These two possibilities exist for complex ions because the crystal field splitting and pairing energies are small and of comparable value. In considering the electron configurations in atoms, the spacing between energy levels is much greater than the pairing energy.

Strong field

(large Δ_o)

$CN^- > NO_2^- > en > py \approx NH_3 > EDTA^{4-} > SCN^- > H_2O >$

$\qquad\qquad ONO^- > ox^{2-} > OH^- > F^- > SCN^- > Cl^- > Br^- > I^-$

$\qquad\qquad\qquad\qquad\qquad\qquad\qquad\qquad$ (small Δ_o)

The red color indicates the donor atom.   *Weak field*

To summarize, consider these two complexes of Co(III): $[CoF_6]^{3-}$ and $[Co(NH_3)_6]^{3+}$. The F^- ion is a weak-field ligand, whereas NH_3 is a strong-field

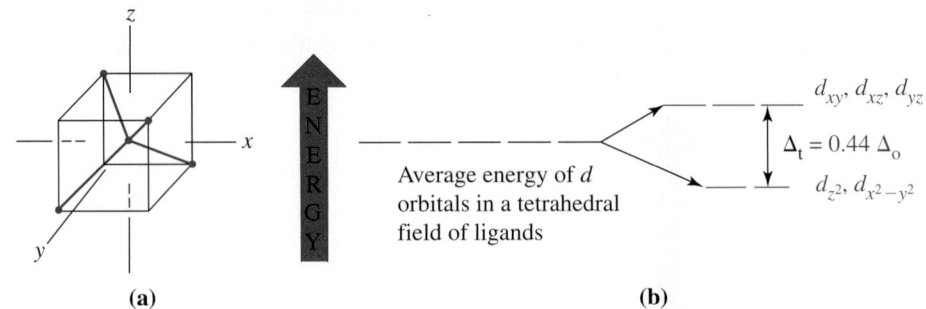

▲ FIGURE 24-13
Crystal field splitting in a tetrahedral complex ion
(a) The positions of attachment of ligands to a metal ion leading to the formation of a tetrahedral complex ion. **(b)** Interference with the d orbitals directed along the x, y, and z axes is not as great as with those that lie between the axes (see Figure 24-11). As a result, the pattern of crystal field splitting is reversed from that of an octahedral complex. Δ_t denotes the energy separation for a tetrahedral complex.

ligand. Because the crystal field splitting for NH_3 is greater than the pairing energy for Co^{3+}, we have the following situations.

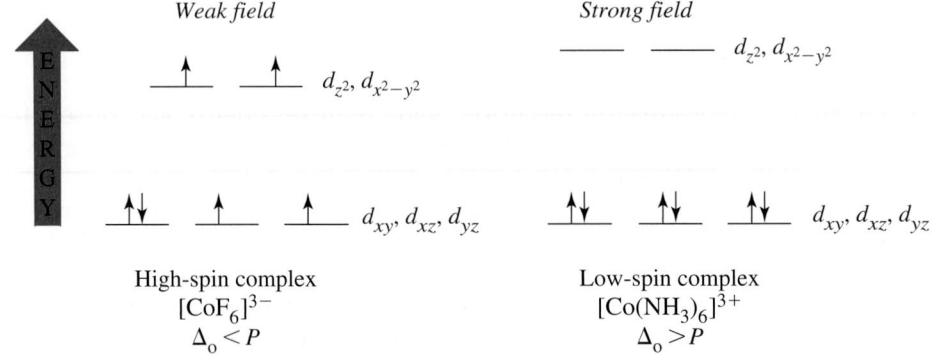

So far, we have considered just octahedral complexes. In the formation of complex ions of other geometric structures, ligands approach from different directions and produce different patterns of splitting of the d energy level. Figure 24-13 shows the pattern for tetrahedral complexes, and Figure 24-14 shows the pattern for square-planar complexes. In comparisons of hypothetical complexes of different structures having the same combinations of ligands, metal ions, and metal–ligand distances, we find the greatest energy separation of the d levels for the square-planar complex and the smallest energy separation for the tetrahedral complex. Because the tetrahedral splitting is small, almost all tetrahedral complexes are high spin.

24-3   ARE YOU WONDERING...

If the two groups of orbitals split equally with respect to the average energy of the d orbitals?

The answer is no. The reason is that the total energy must be constant. If we consider a d^{10} ion such as Zn^{2+} in an octahedral field, the destabilization caused by the four electrons in the $d_{x^2-y^2}$ and d_{z^2} orbitals must be offset by the stabilization gained by the six electrons in the d_{xy}, d_{xz}, and d_{yz}, orbitals. This requires that

$$(6 \times \text{the energy of } d_{xy}, d_{xy}, d_{yz} \text{ orbitals}) + (4 \times \text{the energy of } d_{x^2-y^2}, d_{z^2}) = 0$$

That is, the energy gained equals the energy lost. Also, the energy difference between the orbitals is

(the energy of $d_{x^2-y^2}, d_{z^2}$) − (the energy of d_{xy}, d_{xy}, d_{yz} orbitals) = Δ_o

To satisfy these two relationships, it is necessary that

the energy of d_{xy}, d_{xz}, d_{yz} orbitals = $-0.4\,\Delta_o$
the energy of $d_{x^2-y^2}, d_{z^2}$ = $0.6\,\Delta_o$

The splitting is as shown.

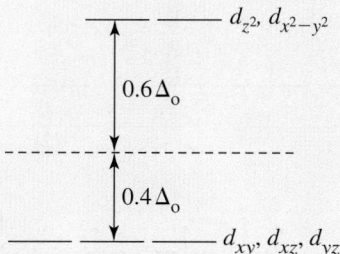

The splitting of the two groups of orbitals is not equal with respect to the average energy of the d orbitals.

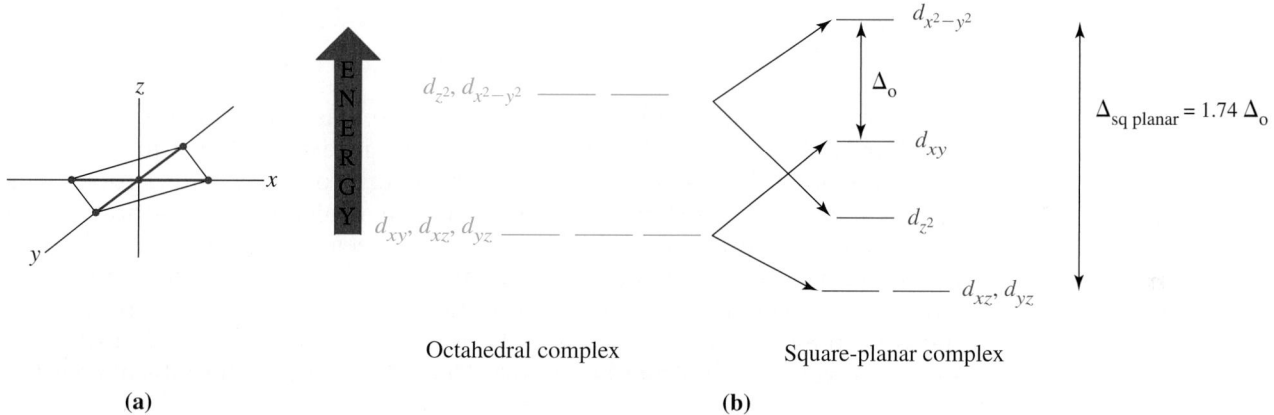

Octahedral complex   Square-planar complex

(a)   (b)

▲ FIGURE 24-14
Comparison of crystal field splitting in a square-planar and an octahedral complex
(a) The positions of attachment of ligands to a metal ion leading to the formation of a square-planar complex. (b) Splitting of the d energy level in a square-planar complex can be related to that of the octahedral complex. There are no ligands along the z axis in a square-planar complex, so we expect the repulsion between ligands and d_{z^2} electrons to be much less than in an octahedral complex. The d_{z^2} energy level is lowered considerably from that in an octahedral complex. Similarly, the energy levels of the d_{xz} and d_{yz} orbitals are lowered slightly because the electrons in these orbitals are concentrated in planes perpendicular to that of the square-planar complex. The energy of the $d_{x^2-y^2}$ orbital is raised because the x and y axes represent the direction of approach of four ligands to the central ion. The energy of the d_{xy} orbital is also raised because this orbital lies in the plane of the ligands in the square-planar complex. The energy difference between the d_{xy} and $d_{x^2-y^2}$ orbitals in a square-planar complex is the same as in an octahedral complex because these orbitals are equally affected by ligand repulsions in both complexes. The maximum energy difference between d orbitals in a square-planar complex is $\Delta_{\text{sq planar}} = 1.74\Delta_o$.

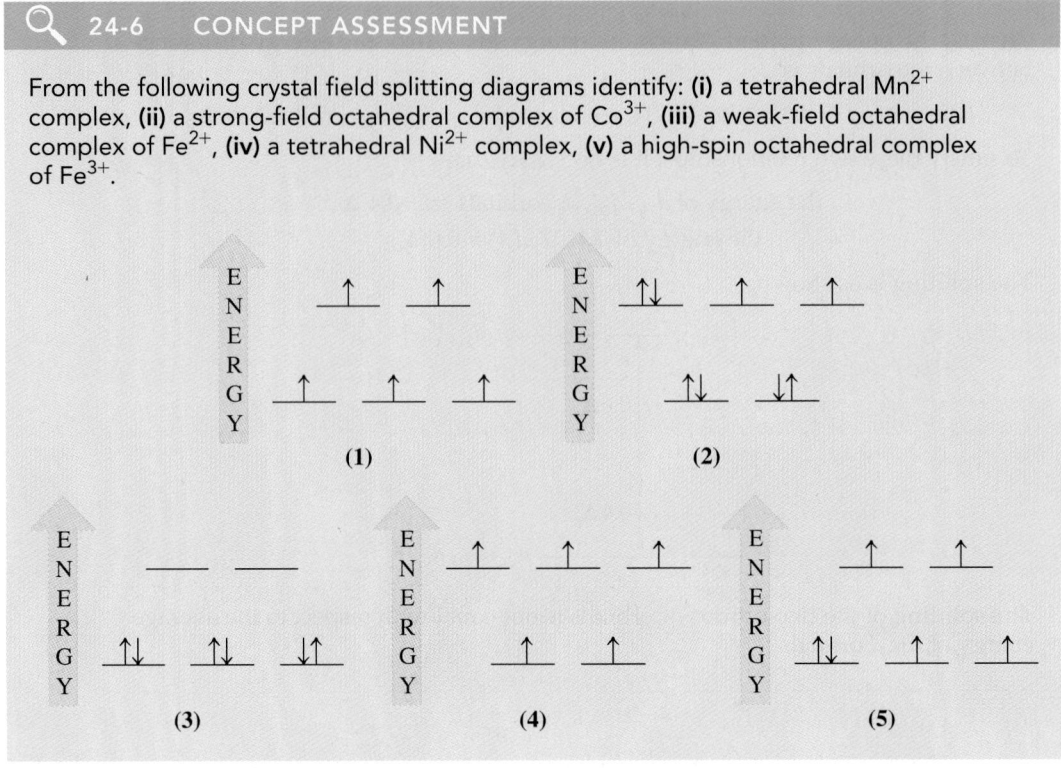

24-6 CONCEPT ASSESSMENT

From the following crystal field splitting diagrams identify: **(i)** a tetrahedral Mn^{2+} complex, **(ii)** a strong-field octahedral complex of Co^{3+}, **(iii)** a weak-field octahedral complex of Fe^{2+}, **(iv)** a tetrahedral Ni^{2+} complex, **(v)** a high-spin octahedral complex of Fe^{3+}.

24-6 Magnetic Properties of Coordination Compounds and Crystal Field Theory

The paramagnetism of the dioxygen molecule was dramatically illustrated in Chapter 10 (Figure 10-3, page 401) by the interaction of liquid oxygen with the poles of a strong magnet. The origin of the paramagnetism is the existence of unpaired electrons in the molecule. Transition metal coordination compounds exhibit varying degrees of paramagnetism. Some are diamagnetic. A paramagnetic substance is pulled into, and a diamagnetic substance is pushed out of, a magnetic field. A straightforward way to measure magnetic properties is to weigh a substance "in" and "out" of a magnetic field, as illustrated in Figure 24-15. The mass of the substance is the

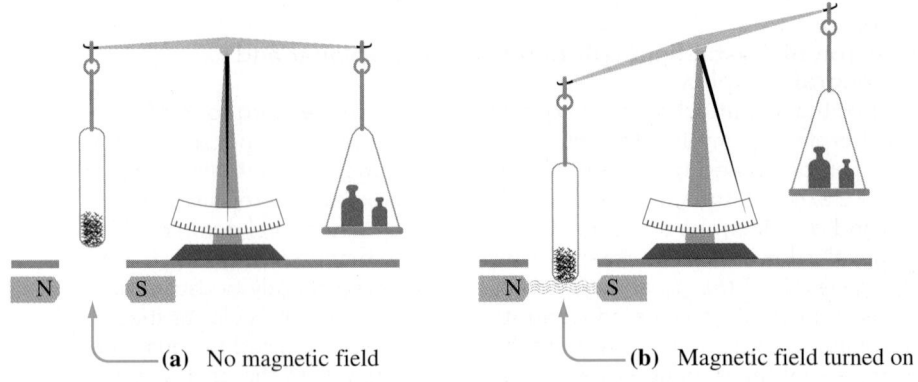

(a) No magnetic field   **(b)** Magnetic field turned on

▲ FIGURE 24-15
Paramagnetism—illustrated
(a) A sample is weighed in the absence of a magnetic field. **(b)** When the field is turned on, the balanced condition is upset. The sample gains weight because it is now subjected to two attractive forces: the force of gravity *and* the force of interaction of the external magnetic field and the unpaired electrons.

same whatever magnetic property the substance possesses. However, if the substance is diamagnetic, it is slightly repelled by a magnetic field and *weighs* less within the field. If the substance is paramagnetic, it *weighs* more within the field.

The degree to which a substance weighs more in the magnetic field depends on the number of unpaired electrons. In the previous section, we saw that a high-spin d^n complex has more unpaired electrons than a low-spin d^n complex. Thus, measuring the change in weight of the complex in a magnetic field allows us to determine whether a complex is high or low spin. The magnetic properties of a complex depend on the magnitude of the crystal field splitting. Strong-field ligands tend to form low-spin, weakly paramagnetic, or even diamagnetic, complexes. Weak-field ligands tend to form high-spin, strongly paramagnetic complexes. The results of measuring the magnetic properties of coordination compounds can therefore be interpreted from crystal field theory, as demonstrated in Examples 24-4 and 24-5.

EXAMPLE 24-4   Using the Spectrochemical Series to Predict Magnetic Properties

How many unpaired electrons would you expect to find in the octahedral complex $[Fe(CN)_6]^{3-}$?

Analyze

We expect complexes with strong field ligands (that is, ligands that are high in the spectrochemical series) to be low spin.

Solve

The Fe atom has the electron configuration $[Ar]3d^64s^2$. The Fe^{3+} ion has the configuration $[Ar]3d^5$. CN^- is a strong-field ligand. Because of the large energy separation in the d levels of the metal ion produced by this ligand, we expect all the electrons to be in the lowest energy level. There should be only one unpaired electron.

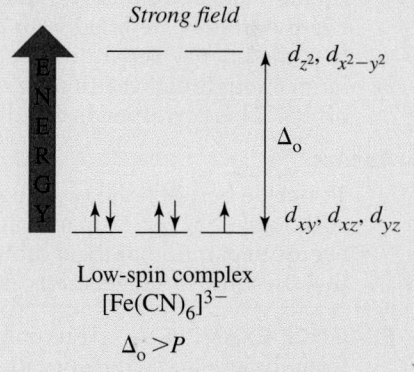

Low-spin complex
$[Fe(CN)_6]^{3-}$

$\Delta_o > P$

Assess

When dealing with ligands at the extremes of the spectrochemical series, it is easier to determine whether the complex is high or low spin. Occasionally, the crystal field splitting is comparable to the pairing energy; in such cases, changes in pressure or temperature can cause a crossover from low spin to high spin. Such a phenomenon is called *spin crossover*. A complex that displays spin crossover is tris(2-picolylamine)iron(II).

Tris(2-picolylamine)iron(II)

PRACTICE EXAMPLE A: How many unpaired electrons would you expect to find in the octahedral complex $[MnF_6]^{2-}$?

PRACTICE EXAMPLE B: How many unpaired electrons would you expect to find in the tetrahedral complex $[CoCl_4]^{2-}$? Would you expect more, fewer, or the same number of unpaired electrons as in the octahedral complex $[Co(H_2O)_6]^{2+}$?

EXAMPLE 24-5 Using the Crystal Field Theory to Predict the Structure of a Complex from Its Magnetic Properties

The complex ion $[Ni(CN)_4]^{2-}$ is diamagnetic. Use ideas from the crystal field theory to speculate on its probable structure.

Analyze

We can eliminate an octahedral structure because the coordination number is 4 (not 6). Our choice is between tetrahedral and square-planar geometries. We compare the crystal field splitting diagrams for each.

Solve

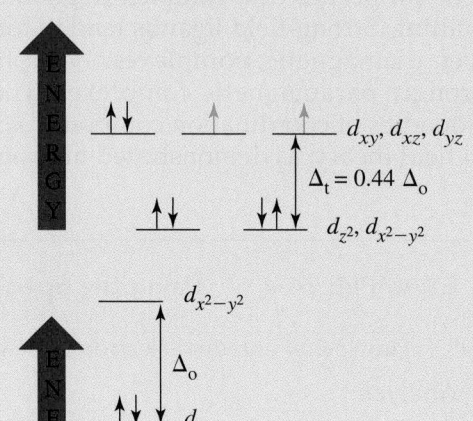

The electron configuration of Ni is $[Ar]3d^84s^2$, and that of Ni(II) is $[Ar]3d^8$. Because the complex ion is diamagnetic, all $3d$ electrons must be paired. Let us see how we would distribute these $3d$ electrons if the structure were tetrahedral (recall Figure 24-13). We would place four electrons (all paired) into the two lowest d levels. We would then distribute the remaining four electrons among the three higher-level d orbitals. Two of the electrons would be unpaired, and the complex ion would be paramagnetic.

Because the tetrahedral structure would be paramagnetic, we can conclude that the structure of the diamagnetic $[Ni(CN)_4]^{2-}$ ion must be square-planar. Let's also demonstrate that this is a reasonable conclusion based on the d-orbital energy-level diagram for a square-planar complex (recall Figure 24-14). First, the three lowest-energy orbitals are filled with electrons (six), and then we assume that the energy separation between the d_{xy} and $d_{x^2-y^2}$ orbitals is large enough that the final two electrons remain paired in the d_{xy} orbital. This corresponds to a diamagnetic complex ion.

Assess

To decide between the two possible geometries, we needed to know the magnetic properties of the complex. Although $[NiCl_4]^{2-}$ is paramagnetic and tetrahedral, the stronger field ligands in $[Ni(CN)_4]^{2-}$ increase the energy separation of the d orbitals, making the square-planar geometry more stable. (VSEPR theory predicts that the tetrahedral geometry is more stable.)

PRACTICE EXAMPLE A: The complex ion $[Co(CN)_4]^{2-}$ is paramagnetic with three unpaired electrons. Use ideas from the crystal field theory to speculate on its probable structure.

PRACTICE EXAMPLE B: Would you expect $[Cu(NH_3)_4]^{2+}$ to be diamagnetic or paramagnetic? Can you use this information about the magnetic properties of $[Cu(NH_3)_4]^{2+}$ to help you determine whether the structure of $[Cu(NH_3)_4]^{2+}$ is tetrahedral or square-planar? Explain.

24-7 Color and the Colors of Complexes

Figure 24-16 uses two situations for mixing color to help clarify the nature of colors. Figure 24-16(a) represents *additive mixing*, the type of color mixing that occurs when colored spotlights are superimposed. Figure 24-16(b), conversely, represents *subtractive mixing*, the type of color mixing that occurs when paint pigments or colored solutions are mixed.

Primary, Secondary, and Complementary Colors

For the additive mixing of light beams in Figure 24-16(a), the **primary colors** are defined as any three colors that, when combined, yield white light (W). The colors used in the figure are *red* (R), *green* (G), and *blue* (B); their sum can be represented as R + G + B = W. The **secondary colors** are those produced by combining two primary colors. Figure 24-16(a) indicates that the secondary colors are *yellow* (Y = R + G), *cyan* (C = G + B), and *magenta* (M = B + R).

Each secondary color is a **complementary color** of one of the primary colors. Again, from Figure 24-16(a), *cyan* (C) is the complementary color of red (R);

▶ Additive color mixing is used in color monitors and color television screens. Subtractive color mixing is used for all the photographs and art in this text.

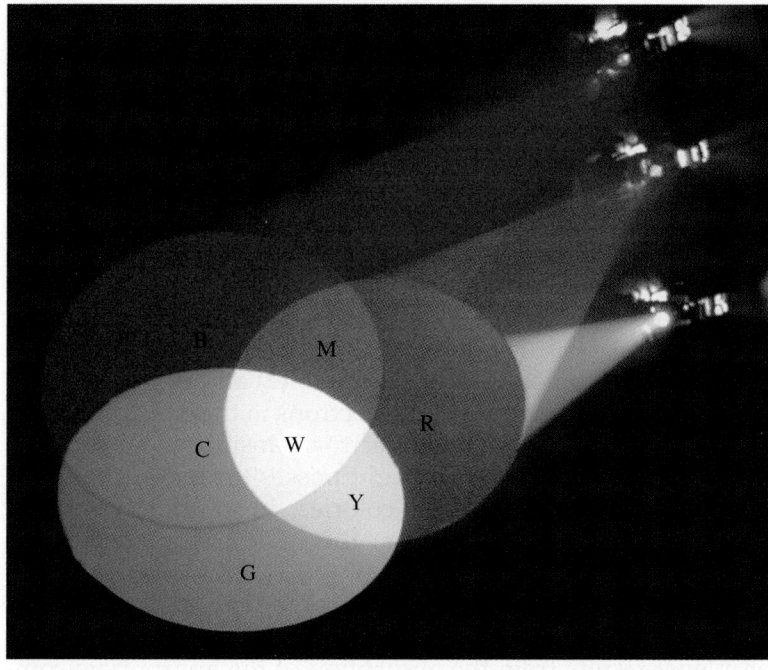

(a) Additive color mixing

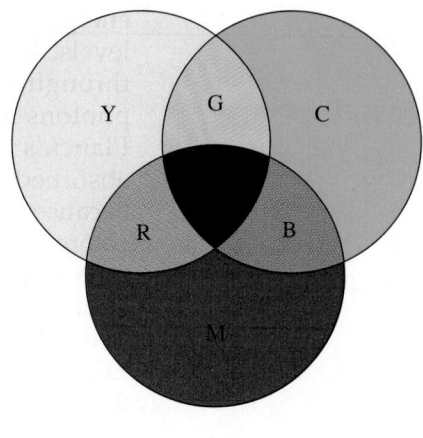

(b) Subtractive color mixing

▲ FIGURE 24-16
The mixing of colors
(a) The additive mixing of three beams of colored light: red (R), green (G), and blue (B). The secondary colors—yellow (Y), cyan (C), and magenta (M)—are produced in regions where two of the beams overlap. The overlap of all three beams produces white light (W). **(b)** The subtractive mixing of three pigments with the primary colors magenta (M), yellow (Y), and cyan (C). Here, the secondary colors—red (R), green (G), and blue (B)—form when two of the primary colors are mixed. A mixture of all three primary colors produces a very dark brown to black color.

   In four-color printing, such as in this book and in color printers in computer systems, the basic inks used are magenta, yellow, cyan, and black. In photographs and drawings in this book, very small dots of these four basic colors, printed singly and in various combinations, produce the colored images you see. (With a magnifying glass, you can see individual dots of color.)

magenta (M), of green (G); and yellow (Y), of blue (B). Figure 24-16(a) shows that when a primary color and its complementary color are mixed, the result is white light. This must be the case: Because cyan, for example, is itself the combination of two of the primary colors—green and blue (C = G + B)—the combination of cyan and red (C + R) is the same as the combination of the three primary colors: G + B + R = W.

   In subtractive color mixing, some of the wavelength components of white light are removed by absorption, and the reflected light (for example, from a painted surface or colored fabric) or transmitted light (as seen through glass or a solution) is deficient in some wavelength components. The reflected or transmitted light is colored. In subtractive color mixing (Fig. 24-16b), there are also primary, secondary, and complementary colors. In this case, *magenta* (M), *yellow* (Y), and *cyan* (C) are the primary colors and *green* (G), *blue* (B), and *red* (R) are the secondary colors. If a material absorbs all three primary colors, there is essentially no light left to be reflected or transmitted; the material appears black, or nearly so. If a material absorbs one color, primary or secondary, the reflected or transmitted light is the complementary color. Thus, a magenta sweater has that color because the dye it contains strongly absorbs green light and reflects magenta, the complement of green. A solution of red food dye has that color because the dye absorbs cyan light and transmits red.

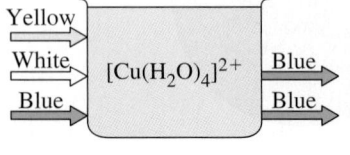

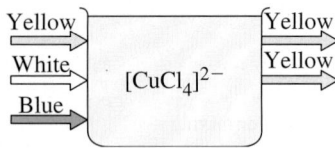

▲ FIGURE 24-17
Light absorption and transmission
$[Cu(H_2O)_4]^{2+}$ absorbs in the yellow region of the spectrum and transmits blue light. $[CuCl_4]^{2-}$ absorbs in the blue region of the spectrum and transmits yellow light.

Colored Solutions

Now let's direct our attention specifically to colored solutions. Colored solutions contain species that can absorb photons of visible light and use the energy of those photons to promote electrons in the species to higher energy levels. The energies of the photons must just match the energy differences through which the electrons are to be promoted. Because the energies of photons are related to the frequencies (and wavelengths) of light (recall Planck's equation, $E = hv$), only certain wavelength components are absorbed as white light passes through the solution. The emerging light, because it is lacking some wavelength components, is no longer white; it is colored.

Ions having (1) a noble-gas electron configuration, (2) an outer shell of 18 electrons, or (3) the "18 + 2" configuration (18 electrons in the $n - 1$ shell and two in the n, or outermost, shell) do not have electron transitions in the energy range corresponding to visible light. White light passes through these solutions without being absorbed; these ions are colorless in solution. Examples are the alkali and alkaline earth metal ions, the halide ions, Zn^{2+}, Al^{3+}, and Bi^{3+}.

Crystal field splitting of the d energy levels produces the energy difference, Δ, that accounts for the colors of complex ions. Promotion of an electron from a lower to a higher d level results from the absorption of the appropriate components of white light; the transmitted light is colored. Subtracting one color from white light leaves the complementary color. A solution containing $[Cu(H_2O)_4]^{2+}$ absorbs most strongly in the yellow region of the spectrum (about 580 nm). The wavelength components of the light transmitted combine to produce the color *blue*. Thus, aqueous solutions of copper(II) compounds usually have a characteristic blue color. In the presence of high concentrations of Cl^-, copper(II) forms the complex ion $[CuCl_4]^{2-}$. This species absorbs strongly in the blue region of the spectrum. The transmitted light is *yellow*, as is the color of the solution. Figure 24-17 suggests light absorption by these solutions. The colors of some complex ions of chromium are given in Table 24.5. The colors of six related coordination compounds of cobalt(III) are shown in Figure 24-18.

TABLE 24.5 Some Coordination Compounds of Cr^{3+} and Their Colors	
Isomer	Color
$[Cr(H_2O)_6]Cl_3$	Violet
$[CrCl(H_2O)_5]Cl_2$	Blue-green
$[Cr(NH_3)_6]Cl_3$	Yellow
$[CrCl(NH_3)_5]Cl_2$	Purple

▶ FIGURE 24-18
Effect of ligands on the colors of coordination compounds
These compounds all consist of a six-coordinate cobalt complex ion in combination with nitrate ions. In each case, the complex ion has five NH_3 molecules and one other group as ligands.

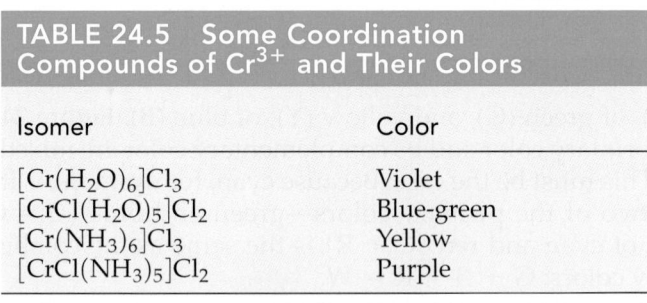

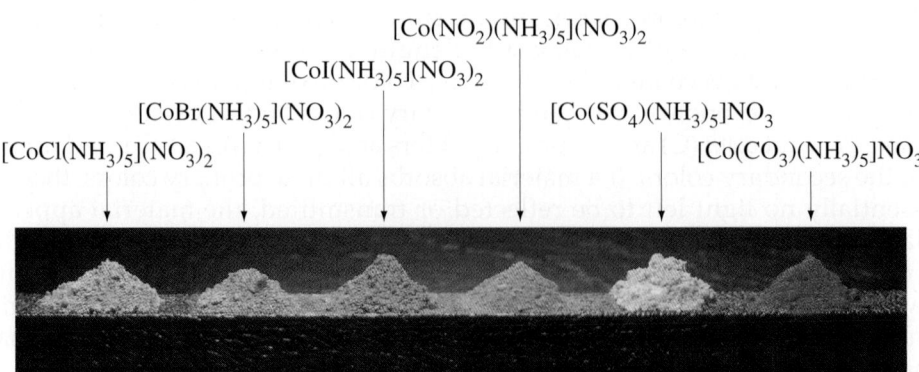

EXAMPLE 24-6   Relating the Colors of Complexes to the Spectrochemical Series

Table 24.5 lists the color of $[Cr(H_2O)_6]Cl_3$ as violet, whereas that of $[Cr(NH_3)_6]Cl_3$ is yellow. Explain this difference in color.

Analyze

Both chromium(III) complexes are octahedral, and the electron configuration of Cr^{3+} is $[Ar]3d^3$. From these facts, we can construct the energy-level diagram shown here. The three unpaired electrons go into the three lower-energy d orbitals. When a photon of light is absorbed, an electron is promoted from a d orbital of lower energy to a d orbital of higher energy. The quantity of energy required for this promotion depends on the energy level separation, Δ_o.

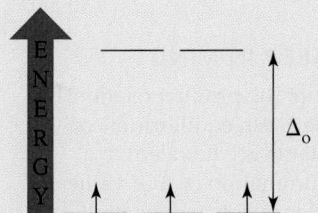

Solve

According to the spectrochemical series, NH_3 produces a greater splitting of the d energy level than does H_2O. We should expect $[Cr(NH_3)_6]^{3+}$ to absorb light of a *shorter* wavelength (higher energy) than does $[Cr(H_2O)_6]^{3+}$. Thus, $[Cr(NH_3)_6]^{3+}$ absorbs in the violet region of the spectrum, and the transmitted light is *yellow*. $[Cr(H_2O)_6]^{3+}$ absorbs in the yellow region of the spectrum, and the transmitted light is *violet*.

Assess

If we measure the wavelength, λ, of the photon absorbed, we can then calculate the splitting energy, $\Delta_o = hc/\lambda$. Stronger field ligands increase the magnitude of Δ_o; therefore, shorter wavelengths of light are absorbed and longer wavelengths are transmitted.

PRACTICE EXAMPLE A: The color of $[Co(H_2O)_6]^{2+}$ is pink, whereas that of tetrahedral $[CoCl_4]^{2-}$ is blue. Explain this difference in color.

PRACTICE EXAMPLE B: One of the following solids is yellow, and the other is green: $Fe(NO_3)_2 \cdot 6\,H_2O$; $K_4[Fe(CN)_6] \cdot 3\,H_2O$. Indicate which is which, and explain your reasoning.

24-8   Aspects of Complex-Ion Equilibria

In Chapter 18, we learned that complex-ion formation can have a great effect on the solubilities of substances, as in the ability of $NH_3(aq)$ to dissolve rather large quantities of $AgCl(s)$. To calculate the solubility of $AgCl(s)$ in $NH_3(aq)$, we had to use the formation constant, K_f, of $[Ag(NH_3)_2]^+$. Equations 24.1 and 24.2 illustrate how we dealt with formation constants in Chapter 18, with $[Zn(NH_3)_4]^{2+}$ as an example.

$$Zn^{2+}(aq) + 4\,NH_3(aq) \rightleftharpoons [Zn(NH_3)_4]^{2+}(aq) \tag{24.1}$$

$$K_f = \frac{[[Zn(NH_3)_4]^{2+}]}{[Zn^{2+}][NH_3]^4} = 4.1 \times 10^8 \tag{24.2}$$

In fact, cations in aqueous solution exist mostly in *hydrated* form. That is, $Zn^{2+}(aq)$ is actually $[Zn(H_2O)_4]^{2+}$. As a result, when NH_3 molecules bond to Zn^{2+} and form an ammine complex ion, they do not enter an empty coordination sphere. They must displace H_2O molecules. This displacement occurs in a stepwise fashion. The reaction

$$[Zn(H_2O)_4]^{2+} + NH_3 \rightleftharpoons [Zn(H_2O)_3NH_3]^{2+} + H_2O \tag{24.3}$$

for which

$$K_1 = \frac{[[Zn(H_2O)_3NH_3]^{2+}]}{[[Zn(H_2O)_4]^{2+}][NH_3]} = 3.9 \times 10^2 \tag{24.4}$$

is followed by

$$[Zn(H_2O)_3NH_3]^{2+} + NH_3 \rightleftharpoons [Zn(H_2O)_2(NH_3)_2]^{2+} + H_2O \tag{24.5}$$

for which

$$K_2 = \frac{[[Zn(H_2O)_2(NH_3)_2]^{2+}]}{[[Zn(H_2O)_3NH_3]^{2+}][NH_3]} = 2.1 \times 10^2 \qquad (24.6)$$

and so on.

KEEP IN MIND

that the product relationship between equilibrium constants is equivalent to a summation of $\Delta G°$ values. Hence, when an overall $\Delta G°$ is used to evaluate K_n

$\Delta G°$ (overall) =

$\Delta G_1° + \Delta G_2° + \Delta G_3° + \cdots$

$= -RT \ln K_1 - RT \ln K_2 -$

$RT \ln K_3 - \cdots$

$= -RT \ln(K_1 \times K_2 \times K_3 \cdots)$

$= -RT \ln \beta_n = -RT \ln K_f$

The value of K_1 in equation (24.4) is often designated as β_1 and called the formation constant for the complex ion $[Zn(H_2O)_3NH_3]^{2+}$. The formation of $[Zn(H_2O)_2(NH_3)_2]^{2+}$ is represented by the *sum* of equations (24.3) and (24.5),

$$[Zn(H_2O)_4]^{2+} + 2\,NH_3 \rightleftharpoons [Zn(H_2O)_2(NH_3)_2]^{2+} + 2\,H_2O \qquad (24.7)$$

The formation constant β_2, in turn, is given by the *product* of equations (24.4) and (24.6).

$$\beta_2 = \frac{[[Zn(H_2O)_2(NH_3)_2]^{2+}]}{[[Zn(H_2O)_4]^{2+}][NH_3]^2} = K_1 \times K_2 = 8.2 \times 10^4 \qquad (24.8)$$

For the next ion in the series, $[Zn(H_2O)(NH_3)_3]^{2+}$, $\beta_3 = K_1 \times K_2 \times K_3$. For the final member, $[Zn(NH_3)_4]^{2+}$, $\beta_4 = K_1 \times K_2 \times K_3 \times K_4$, and it is this product of terms that we called the formation constant in Section 18-8 and listed as K_f in Table 18.2. Additional stepwise formation constants are presented in Table 24.6.

The large numerical value of K_1 for reaction (24.3) indicates that Zn^{2+} has a greater affinity for NH_3 (a stronger Lewis base) than it does for H_2O. Displacement of ligand H_2O molecules by NH_3 occurs even if the number of NH_3 molecules present in aqueous solution is much smaller than the number of H_2O molecules, as in dilute $NH_3(aq)$. The fact that the successive K values decrease regularly in the displacement process, at least for displacements involving neutral molecules as ligands, is due in part to statistical factors: An NH_3 molecule has a better chance of replacing a H_2O molecule in $[Zn(H_2O)_4]^{2+}$, in which each coordination position is occupied by H_2O, than in $[Zn(H_2O)_3NH_3]^{2+}$, in which one of the positions is already occupied by NH_3. Also, once the degree of substitution of NH_3 for H_2O has become large, the chances improve for H_2O molecules to replace NH_3 molecules in a reverse reaction. Again, this tends to reduce the value of K. If irregularities arise in the succession of K values, it is often because of a change in structure of the complex ion at some point in the series of displacement reactions.

If the ligand in a substitution process is polydentate, it displaces as many H_2O molecules as there are points of attachment. Thus, ethylenediamine (en)

TABLE 24.6   Stepwise and Overall Formation (Stability) Constants for Several Complex Ions

Metal[a] Ion	Ligand	K_1	K_2	K_3	K_4	K_5	K_6	β_n (or K_f)[b]
Ag^+	NH_3	2.0×10^3	7.9×10^3					1.6×10^7
Zn^{2+}	NH_3	3.9×10^2	2.1×10^2	1.0×10^2	5.0×10^1			4.1×10^8
Cu^{2+}	NH_3	1.9×10^4	3.9×10^3	1.0×10^3	1.5×10^2			1.1×10^{13}
Ni^{2+}	NH_3	6.3×10^2	1.7×10^2	5.4×10^1	1.5×10^1	5.6	1.1	5.3×10^8
Cu^{2+}	en	5.2×10^{10}	2.0×10^9					1.0×10^{20}
Ni^{2+}	en	3.3×10^7	1.9×10^6	1.8×10^4				1.1×10^{18}
Ni^{2+}	EDTA	4.2×10^{18}						4.2×10^{18}

[a]In many tabulations in the chemical literature, formation-constant data are presented as logarithms: that is, $\log K_1, \log K_2, \ldots,$ and $\log \beta_n$.
[b]The β_n listed is for the number of steps shown: e.g., for $[Ag(NH_3)_2]^+$, $\beta_2 = K_f = K_1 \times K_2$; for $[Ni(en)_3]^{2+}$, $\beta_3 = K_f = K_1 \times K_2 \times K_3$; and for $[Ni(EDTA)]^{2-}$, $\beta_1 = K_f = K_1$.

displaces H_2O molecules in $[Ni(H_2O)_6]^{2+}$ two at a time, in three steps. The first step is

$$[Ni(H_2O)_6]^{2+} + en \rightleftharpoons [Ni(en)(H_2O)_4]^{2+} + 2\,H_2O \qquad K_1(\beta_1) = 3.3 \times 10^7 \quad \textbf{(24.9)}$$

Note from Table 24.6 that the complex ions with polydentate ligands have much larger formation constants than do those with monodentate ligands. For example, K_f (that is, β_3) for $[Ni(en)_3]^{2+}$ is 1.1×10^{18}, whereas K_f (that is, β_6) for $[Ni(NH_3)_6]^{2+}$ is 5.3×10^8. The additional stability of chelates over complexes with monodentate ligands is known as the **chelation (or chelate) effect**. This effect can be partly attributed to the increase in entropy associated with chelation. In the displacement of H_2O by NH_3, the entropy change is small [two particles on each side of an equation such as (24.3)]. An ethylenediamine molecule, conversely, displaces *two* H_2O molecules [two particles on the left and three on the right of equation (24.9)]. The larger, positive value of $\Delta S°$ for the displacement by ethylenediamine means a more negative $\Delta G°$ and a larger K.

🔍 24-7   CONCEPT ASSESSMENT

The Cr(III) cation in aqueous solution can form 6-coordinate complexes with ethylenediamine and $EDTA^{4-}$. Which complex do you expect to be the most stable if all water molecules in the six-coordinate $Cr^{3+}(aq)$ ion are replaced?

24-9   Acid–Base Reactions of Complex Ions

We have described complex-ion formation in terms of Lewis acids and bases. Complex ions may also exhibit acid–base properties in the Brønsted–Lowry sense; that is, they may act as proton donors or acceptors. Figure 24-19 represents the ionization of $[Fe(H_2O)_6]^{3+}$ as an acid. A proton from a *ligand* water molecule in hexaaquairon(III) ion is transferred to a *solvent* water molecule. The H_2O ligand is converted to OH^-.

$$[Fe(H_2O)_6]^{3+} + H_2O \rightleftharpoons [FeOH(H_2O)_5]^{2+} + H_3O^+ \quad K_{a_1} = 9 \times 10^{-4} \quad \textbf{(24.10)}$$

The second ionization step is

$$[FeOH(H_2O)_5]^{2+} + H_2O \rightleftharpoons [Fe(OH)_2(H_2O)_4]^+ + H_3O^+ \quad K_{a_2} = 5 \times 10^{-4} \quad \textbf{(24.11)}$$

From these K_a values, we see that $Fe^{3+}(aq)$ is fairly acidic (compared, for example, with acetic acid, with $K_a = 1.8 \times 10^{-5}$). To suppress ionization (hydrolysis) of $[Fe(H_2O)_6]^{3+}$, we need to maintain a low pH by the addition of acids such as HNO_3 or $HClO_4$. The ion $[Fe(H_2O)_6]^{3+}$ is violet in color, but aqueous solutions of $Fe^{3+}(aq)$ are generally yellow because of the presence of hydroxo complex ions.

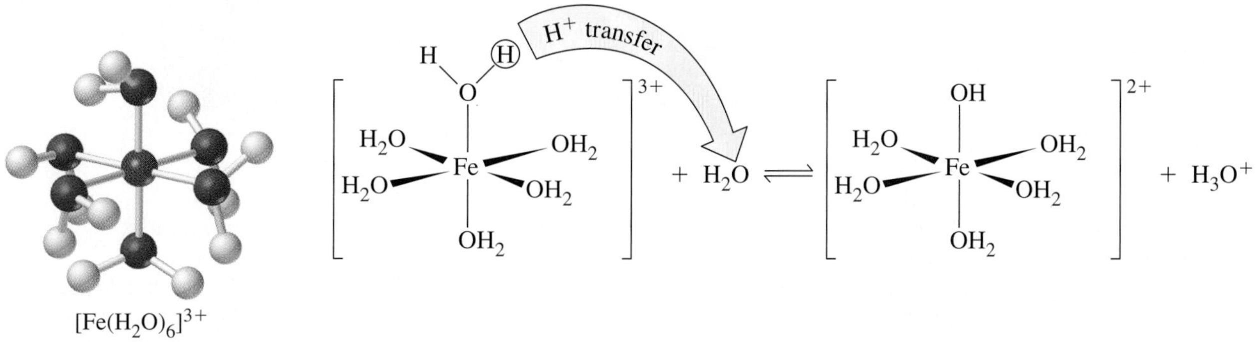

▲ FIGURE 24-19
Ionization of $[Fe(H_2O)_6]^{3+}$

The ions Cr^{3+} and Al^{3+} behave in a manner similar to Fe^{3+} except that, with them, hydroxo complex-ion formation can continue until complex anions are produced. $Cr(OH)_3$ and $Al(OH)_3$, as we previously noted, are soluble in alkaline as well as acidic solutions; they are amphoteric.

Regarding the acid strengths of aqua complex ions, a critical factor is the charge-to-radius ratio of the central metal ion. Thus, the small, highly charged Fe^{3+} attracts electrons away from an O—H bond in a ligand water molecule more strongly than does Fe^{2+}. Hence, $[Fe(H_2O)_6]^{3+}$ is a stronger acid ($K_{a_1} = 9 \times 10^{-4}$) than is $[Fe(H_2O)_6]^{2+}$ ($K_{a_1} = 1 \times 10^{-7}$).

24-10 Some Kinetic Considerations

When $NH_3(aq)$ is added to an aqueous solution containing Cu^{2+}, there is a change in color from pale blue to very deep blue. The reaction involves NH_3 molecules displacing H_2O molecules as ligands.

$$\underset{\text{(pale blue)}}{[Cu(H_2O)_4]^{2+}} + 4\,NH_3 \longrightarrow \underset{\text{(very deep blue)}}{[Cu(NH_3)_4]^{2+}} + 4\,H_2O$$

This reaction occurs very rapidly—as rapidly as the two reactants can be brought together. The addition of $HCl(aq)$ to an aqueous solution of Cu^{2+} produces an immediate color change from pale blue to green, or even yellow if the $HCl(aq)$ is sufficiently concentrated.

$$\underset{\text{(pale blue)}}{[Cu(H_2O)_4]^{2+}} + 4\,Cl^- \longrightarrow \underset{\text{(yellow)}}{[CuCl_4]^{2-}} + 4\,H_2O$$

Complex ions in which ligands can be interchanged rapidly are said to be **labile**. $[Cu(H_2O)_4]^{2+}$, $[Cu(NH_3)_4]^{2+}$, and $[CuCl_4]^{2-}$ are all labile (Fig. 24-20).

In freshly prepared $CrCl_3(aq)$, the ion *trans*-$[CrCl_2(H_2O)_4]^+$ produces a green color, but the color gradually turns to violet (Fig. 24-21). This color change results from the very slow exchange of H_2O for Cl^- ligands. A complex ion that exchanges ligands slowly is said to be *nonlabile*, or **inert**. In general, complex ions of the first transition series, except for those of Cr(III) and Co(III), are kinetically labile. Those of the second and third transition series are generally kinetically inert. Whether a complex ion is labile or inert affects the ease with which it can be studied. The inert ones are easiest to isolate and characterize, which may explain why so many of the early studies of complex ions were based on Cr(III) and Co(III).

▶ The terms *labile* and *inert* are not related to the thermodynamic stabilities of complex ions or to the equilibrium constants for ligand-substitution reactions. The terms are *kinetic* terms, referring to the rates at which ligands are exchanged.

▲ FIGURE 24-21
Inert complex ions
The green solid $CrCl_3 \cdot 6\,H_2O$ produces the green aqueous solution on the left. The color is due to *trans*-$[CrCl_2(H_2O)_4]^+$. A slow exchange of H_2O for Cl^- ligands leads to a violet solution of $[Cr(H_2O)_6]^{3+}$ in one or two days (right).

▲ FIGURE 24-20
Labile complex ions
The exchange of ligands in the coordination sphere of Cu^{2+} occurs very rapidly. The solution at the extreme left is formed by dissolving $CuSO_4$ in concentrated $HCl(aq)$. Its yellow color is due to $[CuCl_4]^{2-}$. When a small amount of water is added, the mixture of $[Cu(H_2O)_4]^{2+}$ and $[CuCl_4]^{2-}$ ions produces a yellow-green color. When $CuSO_4$ is dissolved in water, a light blue solution of $[Cu(H_2O)_4]^{2+}$ forms. NH_3 molecules readily displace H_2O molecules as ligands and produce deep blue $[Cu(NH_3)_4]^{2+}$ (extreme right).

24-11 Applications of Coordination Chemistry

The applications of coordination chemistry are numerous and varied. They range from analytical chemistry to biochemistry. The several brief examples in this section give some idea of this diversity.

Cisplatin: A Cancer-Fighting Drug

Chemotherapy is a treatment used for some types of cancer. The treatment employs anticancer drugs to destroy cancer cells. An important cancer-fighting drug is cisplatin, which is commonly used to treat testicular, bladder, lung, esophagus, stomach, and ovarian cancers.

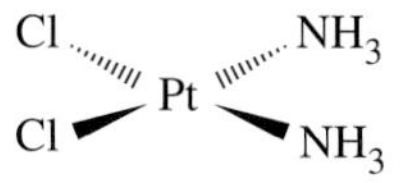

Cisplatin was first synthesized in 1845 by Michel Peyrone, an Italian doctor, but its structure was not elucidated until almost fifty years later, in 1893, by Werner. The anticancer activity of cisplatin was discovered in the 1960s by Barrett Rosenberg, a professor of chemistry at Michigan State University. Not only did Rosenberg discover the anticancer activity of cisplatin, but he was also the first to report that the *trans* isomer (transplatin) was ineffective in killing cancer cells.

Before examining the anticancer activity of cisplatin, let's first consider its synthesis. One method for making cisplatin starts from $K_2[PtCl_4]$, which is converted to $K_2[PtI_4]$, by treatment with an aqueous solution of KI:

$$K_2[PtCl_4] + KI \longrightarrow K_2[PtI_4] + 4\ KCl$$

In the next step, NH_3 is added, forming a yellow compound, *cis*-$[PtI_2(NH_3)_2]$. The formation of *cis*-$[PtI_2(NH_3)_2]$ is a key step, which occurs in two stages.

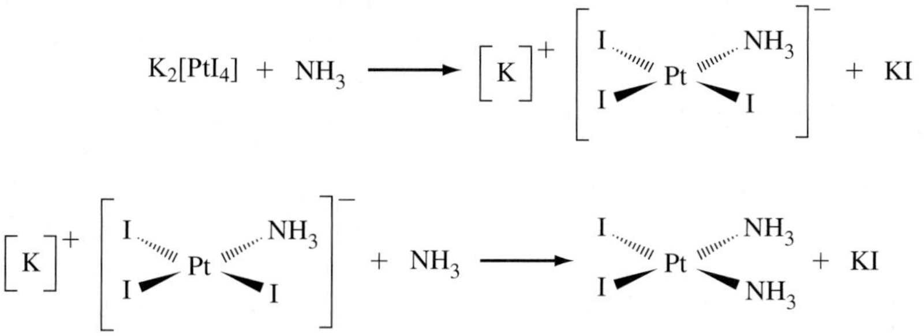

It might seem strange that only one isomer (the *cis* isomer) is obtained. One way to rationalize why this happens is to consider the ligands in $[Pt(NH_3)I_3]^-$ in terms of their tendencies to direct an incoming ligand toward the *trans* position. Empirical studies indicate that NH_3 is a weaker *trans* director than is I^-, and so the second NH_3 molecule is preferentially directed to a position that is *trans* to I^-, rather than being directed to a position that is *trans* to NH_3. As pointed out in the margin note, Cl^- is a weaker *trans* director than is I^-, so treatment of $K_2[PtCl_4]$ with NH_3 will give a lower yield of the *cis* isomer of $Pt(NH_3)_2Cl_2$. Consequently, the conversion of $K_2[PtCl_4]$ to $K_2[PtI_4]$ is an important step in the synthesis.

The remaining steps in preparing cisplatin are as follows. When *cis*-$[PtI_2(NH_3)_2]$ is treated with $AgNO_3$, insoluble AgI precipitates, leaving behind a solution of *cis*-$[Pt(OH_2)_2(NH_3)_2]^{2+}$. Finally, treatment with KCl gives *cis*-$[PtCl_2(NH_3)_2]$ as a yellow precipitate.

Let's now consider the essential features of the anticancer activity of cisplatin. Cisplatin enters cancer cells mainly by diffusion. Once inside the cell, one of the chloride ions in cisplatin is replaced by a water molecule.

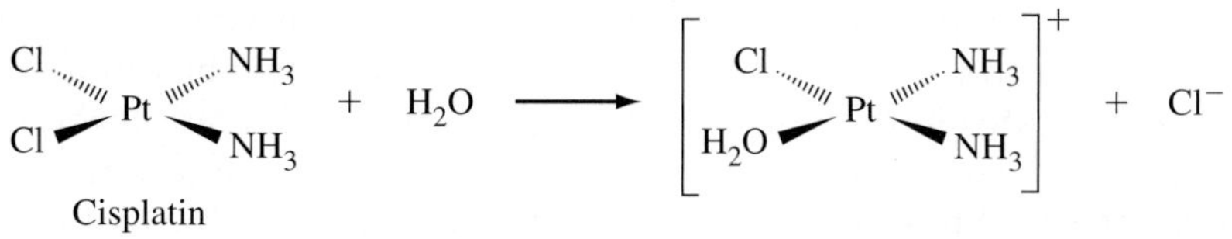

Cisplatin

(margin, right side)

cis-$[Pt(NH_3)_2Cl_2]$
(cisplatin)

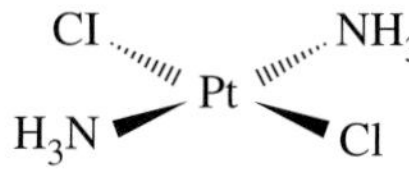

trans-$[Pt(NH_3)_2Cl_2]$
(transplatin)

◄ Empirically, certain ligands are found to have a greater tendency than others to direct an incoming ligand toward a *trans* position. For the monodentate ligands we have encountered, the approximate ordering from strongest *trans* directors to weakest *trans* directors is $CN^-, CO > NO_2^-$, $I^- > Br^- > Cl^- > NH_3 >$ $OH^- > H_2O$.

The anticancer activity of cisplatin is associated with the binding of $[Pt(NH_3)_2(Cl)(H_2O)]^+$ to a cellular DNA molecule. (We will discuss DNA in more detail in Chapter 28.) When $[Pt(NH_3)_2(Cl)(H_2O)]^+$ binds to a DNA molecule, structural deformations occur in the DNA molecule and these deformations, if not repaired by proteins in the cell, lead ultimately to cell death.

It may seem surprising that the *cis* and *trans* isomers of $Pt(NH_3)_2Cl_2$ exhibit a dramatic difference in anticancer activity, given their structural similarities. Overall, the *trans* isomer (transplatin) is more reactive and potentially more potent than cisplatin but the higher reactivity leads ultimately to lower anticancer activity. Because of its increased reactivity, transplatin can undergo many side reactions before it reaches its target; thus, transplatin is less effective in killing cancer cells.

The discovery of the anticancer activity of cisplatin was nothing less than monumental. To date, thousands of platinum-containing compounds have been investigated as potential chemotherapy drugs. Worldwide annual sales of platinum-based anticancer drugs are currently in excess of $2 billion.

Hydrates

When a compound is crystallized from an aqueous solution of its ions, the crystals obtained are often hydrated. As originally described in Chapter 3, a hydrate is a substance that has a fixed number of water molecules associated with each formula unit. In some cases, the water molecules are ligands bonded directly to a metal ion. The coordination compound $[Co(H_2O)_6](ClO_4)_2$ may be represented as the hexahydrate, $Co(ClO_4)_2 \cdot 6\,H_2O$. In the hydrate $CuSO_4 \cdot 5\,H_2O$, four H_2O molecules are associated with copper in the complex ion $[Cu(H_2O)_4]^{2+}$, and the fifth with the SO_4^{2-} anion by hydrogen bonding. Another possibility for hydrate formation is that the water molecules may be incorporated into definite positions in the solid crystal but not associated with any particular cations or anions, as in $BaCl_2 \cdot 2\,H_2O$. This is called *lattice water*. Finally, part of the water may be coordinated to an ion and part of it may be lattice water, as appears to be the case with alums, such as $KAl(SO_4)_2 \cdot 12\,H_2O$.

Stabilization of Oxidation States

The standard electrode potential for the reduction of Co(III) to Co(II) is

$$Co^{3+}(aq) + e^- \longrightarrow Co^{2+}(aq) \quad E° = +1.82\ \text{V}$$

This large positive value suggests that $Co^{3+}(aq)$ is a strong oxidizing agent, strong enough to oxidize water to $O_2(g)$.

$$4\,Co^{3+}(aq) + 2\,H_2O(l) \longrightarrow 4\,Co^{2+}(aq) + 4\,H^+(aq) + O_2(g) \quad E°_{cell} = +0.59\ \text{V}$$

$$(24.12)$$

Yet one of the complex ions featured in this chapter has been $[Co(NH_3)_6]^{3+}$. This ion is stable in water solution, even though it contains cobalt in the +3 oxidation state. Reaction (24.12) will not occur if the concentration of Co^{3+} is sufficiently low and $[Co^{3+}]$ is kept very low because of the great stability of the complex ion.

$$Co^{3+}(aq) + 6\,NH_3(aq) \rightleftharpoons [Co(NH_3)_6]^{3+}(aq) \quad \beta_6 = K_f = 4.5 \times 10^{33}$$

In fact, the concentration of free Co^{3+} is so low that for the half-reaction

$$[Co(NH_3)_6]^{3+} + e^- \longrightarrow [Co(NH_3)_6]^{2+}$$

$E°$ is only $+0.10$ V. As a consequence, not only is $[Co(NH_3)_6]^{3+}$ stable but also $[Co(NH_3)_6]^{2+}$ is rather easily oxidized to the Co(III) complex.

The ability of strong electron-pair donors (strong Lewis bases) to stabilize high oxidation states in the way that NH_3 does in Co(III) complexes and O^{2-} in Mn(VII) complexes (such as MnO_4^-) affords a means of attaining certain oxidation states that might otherwise be difficult or impossible to attain.

Photography: Fixing a Photographic Film

A black-and-white photographic film is an emulsion of a finely divided silver halide (typically AgBr) coated on a strip of polymer, such as a modified cellulose. In the *exposure* step, the film is exposed to light and some of the tiny granules of AgBr(s) absorb photons. The photons promote the oxidation of Br^- to Br and the reduction of Ag^+ to Ag. The Ag and Br atoms remain in the crystalline lattice of AgBr(s) as "defects," in numbers that depend on the intensity of the light absorbed: The brighter the light, the more Ag atoms. Because the actual number of Ag atoms produced in the exposure is not large, the silver is invisible to the eye. The pattern of distribution of the Ag atoms, however, creates a *latent* image of the object photographed. To obtain a visible image, the film is developed.

In the *developing* step, the exposed film is placed in a solution of a mild reducing agent such as hydroquinone, $C_6H_4(OH)_2$. An oxidation–reduction reaction occurs in which Ag^+ ions are reduced to Ag and the hydroquinone is oxidized. The action of the developer is such that reduction of Ag^+ to Ag occurs just in those granules of AgBr(s) that contain Ag atoms of the latent image. As a result, the number of Ag atoms in the film is greatly increased, and the latent image becomes visible. Bright regions of the photographed object appear as dark regions in the photographic image. At this stage, the film is a photographic *negative*. This negative cannot be exposed to light, however, because reduction of Ag^+ to Ag could still occur in the previously unexposed granules of AgBr(s). The negative must be fixed.

The *fixing* step requires that the black metallic silver of the negative remain on the film and the unexposed AgBr(s) be removed. A common fixer is an aqueous solution of sodium thiosulfate (also known as sodium hyposulfite or hypo). Because the complex ion $[Ag(S_2O_3)_2]^{3-}$ has a large formation constant, the following reaction is driven to completion—the AgBr(s) dissolves.

$$AgBr(s) + 2\,S_2O_3{}^{2-}(aq) \longrightarrow [Ag(S_2O_3)_2]^{3-}(aq) + Br^-(aq) \qquad \textbf{(24.13)}$$

Once the negative has been fixed, it is used to produce a *positive* image, the final photograph. This is done by projecting light through the negative onto a piece of photographic paper. Regions of the negative that are dark transmit little light to the photographic paper and will appear light when the photographic paper is subsequently developed and fixed. Conversely, light areas of the negative will appear dark in the final print. In this way, the areas of light and dark in the final print are the same as in the photographed object.

Qualitative Analysis

In discussing qualitative cation analysis in Section 18-9 (page 805), we showed how the group 1 precipitate—AgCl(s), $PbCl_2(s)$, or $Hg_2Cl_2(s)$—is separated by taking advantage of the stable complex ion formed by $Ag^+(aq)$ and $NH_3(aq)$.

$$AgCl(s) + 2\,NH_3(aq) \longrightarrow [Ag(NH_3)_2]^+(aq) + Cl^-(aq)$$

The qualitative analysis scheme abounds in other examples of complexion formation. For example, at a point in the procedure for cation group 3, a test is needed for Co^{2+}. In the presence of SCN^- ion, Co^{2+} forms a blue thiocyanato complex ion, $[Co(SCN)_4]^{2-}$ (Fig. 24-22a). A problem develops, however, if even a trace amount of Fe^{3+} is present in the solution. Fe^{3+} reacts with SCN^- to produce $[FeSCN(H_2O)_5]^{2+}$, a strongly colored, blood-red complex ion (Fig. 24-22b.) Fortunately, this complication can be resolved by treating a solution containing both Co^{2+} and Fe^{3+} with an excess of F^-. The Fe^{3+} is converted to the extremely stable, pale yellow $[FeF_6]^{3-}$. The complex ion $[CoF_4]^{2-}$, being much less stable than $[Co(SCN)_4]^{2-}$, does not form. As a result, the $[Co(SCN)_4]^{2-}(aq)$ can be detected via the blue-green solution color (Fig. 24-22c).

► FIGURE 24-22
Qualitative tests for Co^{2+} and Fe^{3+}

(a) $[Co(SCN)_4]^{2-}$ complex ion.
(b) $[FeSCN(H_2O)_5]^{2+}$ complex ion.
(c) Mixture of $[FeF_6]^{3-}$ and $[Co(SCN)_4]^{2-}$.

Sequestering Metal Ions

Metal ions can act as unintended catalysts in promoting undesirable chemical reactions in a manufacturing process, or they may alter the properties of the material being manufactured. Thus, for many industrial purposes, it is imperative to remove mineral impurities from water. Often these impurities, such as Cu^{2+}, are present only in trace amounts, and precipitation of metal ions is feasible only if K_{sp} for the precipitate is very small. An alternative is to treat the water with a chelating agent. This reduces the free cation concentrations to the point at which the cations can no longer enter into objectionable reactions. The cations are said to be *sequestered*. Among the chelating agents widely employed are the salts of *ethylenediaminetetraacetic acid* (H_4EDTA), usually as the sodium salt.

$$4\,Na^+ \left[\begin{array}{c} {}^-OOCCH_2 \\ \\ {}^-OOCCH_2 \end{array} NCH_2CH_2N \begin{array}{c} CH_2COO^- \\ \\ CH_2COO^- \end{array} \right]$$

A representative complex ion formed by a metal ion, M^{n+}, with the hexadentate $EDTA^{4-}$ anion is depicted in Figure 24-23. The high stability of such complexes can be attributed to the presence of five, five-member chelate rings.

► FIGURE 24-23
Structure of a metal–EDTA complex
The central metal ion M^{n+} (pale green) can be Ca^{2+}, Mg^{2+}, Fe^{2+}, Fe^{3+}, and so on. The ligand is $EDTA^{4-}$, and the net charge on the complex is $+n-4$. Structural diagrams and ball-and-stick models are shown for the two optical isomers of an $[MEDTA]^{n-4}$ complex.

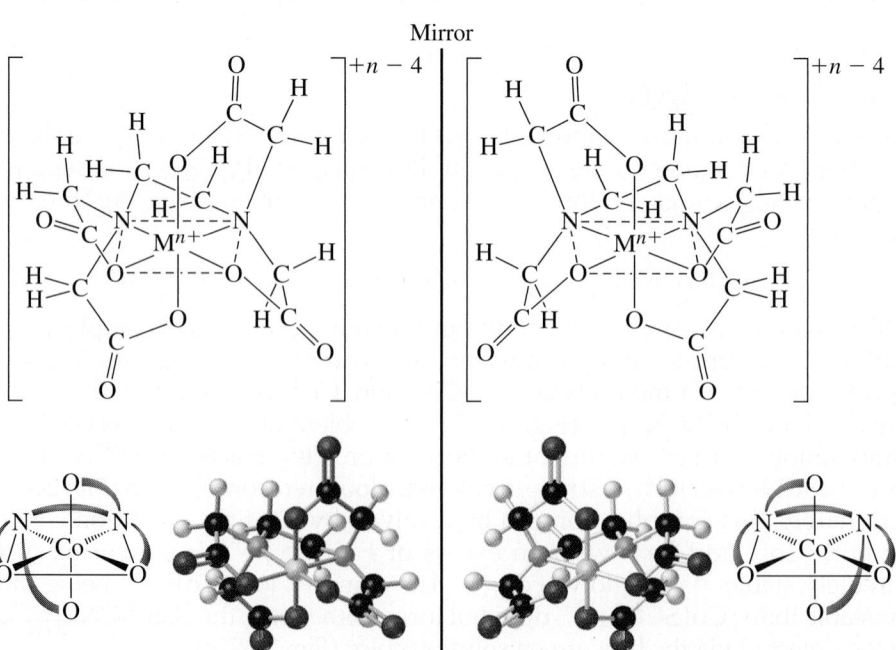

In the presence of EDTA^{4-}(aq), Ca^{2+}, Mg^{2+}, and Fe^{3+} in hard water are unable to form boiler scale or to precipitate as insoluble soaps. The cations are sequestered in the complex ions: [Ca(EDTA)]$^{2-}$, [Mg(EDTA)]$^{2-}$, and [Fe(EDTA)]$^-$, having K_f values of 4.0×10^{10}, 4.0×10^8, and 1.7×10^{24}, respectively.

Chelation with EDTA can be used in treating some cases of metal poisoning. If a person with lead poisoning is fed [Ca(EDTA)]$^{2-}$, the following exchange occurs because [Pb(EDTA)]$^{2-}$ ($K_f = 2 \times 10^{18}$) is even more stable than [Ca(EDTA)]$^{2-}$ ($K_f = 4 \times 10^{10}$).

$$Pb^{2+} + [Ca(EDTA)]^{2-} \longrightarrow [Pb(EDTA)]^{2-} + Ca^{2+}$$

The body excretes the lead complex, and the Ca^{2+} remains as a nutrient. A similar method can be used to rid the body of radioactive isotopes, as in the treatment of plutonium poisoning.

Some plant fertilizers contain EDTA chelates of metals such as Cu^{2+} as a soluble form of the metal ion for the plant to use. Metal ions can catalyze reactions that cause mayonnaise and salad dressings to spoil, and the addition of EDTA reduces the concentration of metal ions by chelation.

▲ Some products containing EDTA.

Biological Applications: Porphyrins

The structure in Figure 24-24 is commonly found in both plant and animal matter. If the eight R groups are all H atoms, the molecule is called *porphin*. The central N atoms can give up their H atoms, and a metal atom can coordinate simultaneously with all four N atoms. The porphin is a tetradentate ligand for the central metal, and the metal–porphin complex is called a *porphyrin*. Specific porphyrins differ in their central metals and in the R groups on the porphin rings.

In *photosynthesis*, carbon dioxide and water, in the presence of inorganic salts, a catalytic agent called *chlorophyll*, and sunlight, combine to form carbohydrates.

$$n\, CO_2 + n\, H_2O \xrightarrow[\text{chlorophyll}]{\text{sunlight}} \underset{\text{carbohydrates}}{(CH_2O)_n} + n\, O_2 \qquad \textbf{(24.14)}$$

Carbohydrates are the main structural materials of plants. Chlorophyll is a green pigment that absorbs sunlight and directs the storage of this energy into the chemical bonds of the carbohydrates. The structure of one type of chlorophyll is shown in Figure 24-25; it is a porphyrin. The central metal ion is Mg^{2+}.

Green is the complementary color of magenta—a purplish red—so we should expect chlorophyll to absorb light in the red region of the spectrum (about 670–680 nm). This suggests that green plants should grow more readily in red light than in light of other colors, and some experimental evidence indicates that this is the case. For example, the maximum rate of formation of O$_2$(g) by reaction (24.14) occurs with red light.

In Chapter 28, we will consider another porphyrin structure that is essential to life—hemoglobin.

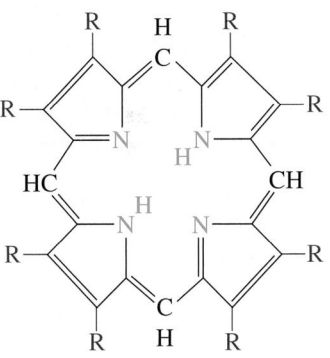

▲ FIGURE 24-24
The porphyrin structure

KEEP IN MIND

magnesium is not a transition metal.

◀ FIGURE 24-25
Structure of chlorophyll *a*

Summary

24-1 Werner's Theory of Coordination Compounds: An Overview—Many metal atoms or ions, particularly among the transition elements, have the ability to bond with **ligands** (electron-pair donors) to form a **complex**, a species in which there are coordinate covalent bonds between ligands and metal centers. The number of electron pairs donated to the central metal atom or ion by the ligands is the **coordination number**. A compound having one or more complexes as a constituent is called a **coordination compound**.

24-2 Ligands—A ligand that attaches to a central metal atom or ion by donating a single electron pair is said to be **monodentate**. A **bidentate** ligand binds to the central metal atom or ion with two pairs of electrons (Fig. 24-3). **Polydentate** ligands are able to attach simultaneously to two or more positions at the metal center. This multiple attachment produces complexes with five- or six-member rings of atoms—**chelates**. A polydentate ligand is also referred to as a **chelating agent**.

24-3 Nomenclature—The name of a complex conveys information about the number and kinds of ligands; the oxidation state of the metal center; and whether the complex is a neutral species, a cation, or an anion.

24-4 Isomerism—The positions at which ligands are attached to the metal center are not always equivalent, and as a consequence isomerism occurs in coordination complexes. **Structural isomers** differ in what ligands are attached to the metal ion and through which atoms, one example being linkage isomerism (Fig. 24-4). In **geometric isomerism**, different structures with different properties result, depending on where the attachment of ligands occurs. In coordination complexes in which there are two ligands of one type and the remainder of another type, the *cis* isomer has the two identical ligands on the same side of a square in a planar structure (Fig. 24-5) or the same edge of an octahedral structure (Fig. 24-6). The *trans* isomer has the two ligands on opposite corners of a square (Fig. 24-5) or diametrically opposed in an octahedral structure (Fig. 24-6). **Optical isomers** differ by being nonsuperimposable mirror images of each other (Fig. 24-8). The two isomers so related are called **enantiomers** and they are said to be **chiral**. Optical isomerism and geometrical isomerism are forms of **stereoisomerism**, that is, isomerism based on the way ligands occupy the three-dimensional space around the metal center.

24-5 Bonding in Complex Ions: Crystal Field Theory—**Crystal field theory** is a bonding theory useful in explaining the magnetic properties and characteristic colors of complex ions (Fig. 24-11). This theory emphasizes the splitting of the *d* energy level of the central metal ion as a result of repulsions between *d*-orbital electrons of the central ion and electrons of the ligands (Fig. 24-12). Whether or not electrons are accommodated in the upper *d* orbitals depends on whether the crystal field splitting, Δ, is greater than the **pairing energy** (P). In d^n complexes, where $4 \leq n \leq 7$, **low spin** complexes occur when $P > \Delta$, and **high spin** complexes occur when $P < \Delta$. A prediction of the magnitude of *d*-level splitting produced by a ligand can be made via the ranking known as the **spectrochemical series**.

24-6 Magnetic Properties of Coordination Compounds and Crystal Field Theory—The magnetic properties of coordination compounds—whether diamagnetic or paramagnetic, and to what degree—can be assessed by measuring the change in weight of the substance when placed in a magnetic field (Fig. 24-15).

24-7 Color and the Colors of Complexes—The colors displayed by coordination complexes (Fig. 24-18) arise from the absorption of certain wavelength components of white light as an electron is excited from a lower energy *d* orbital to a higher energy *d* orbital in a crystal field. The transmitted light is deficient in the absorbed wavelengths and hence of a different color. The relationship of **primary, secondary**, and **complementary colors** is described and presented visually in Figure 24-16.

24-8 Aspects of Complex-Ion Equilibria—Formation of a coordination complex can be viewed as a stepwise equilibrium process in which other ligands displace H_2O molecules from aqua complex ions. The stepwise constants can be combined into an overall formation constant for the complex ion, K_f. When a chelating agent binds to a metal ion, an increased stability is observed because of the entropy changes produced when the chelating agent displaces water molecules from binding sites in the coordination sphere. This greater stability achieved is called the **chelation (chelate) effect**.

24-9 Acid–Base Reactions of Complex Ions—The ability of ligand water molecules to ionize causes some aqua complexes to exhibit acidic properties and helps explain amphoterism.

24-10 Some Kinetic Considerations—Also important in determining properties of a complex ion is the rate at which the ion exchanges ligands between its coordination sphere and the solution. Exchange is rapid in a **labile** complex and slow in an **inert** complex.

24-11 Applications of Coordination Chemistry—
Complex-ion formation can be used to stabilize certain oxidation states, such as Co(III). Other applications include dissolving precipitates, such as AgCl by $NH_3(aq)$ in the qualitative analysis scheme and AgBr by $Na_2S_2O_3(aq)$ in the photographic process, and sequestering ions by chelation, as with EDTA.

Integrative Example

Absorbance is a measure of the proportion of monochromatic (single-color) light that is absorbed as the light passes through a solution. An *absorption spectrum* is a graph of absorbance as a function of wavelength. High absorbances correspond to large proportions of the light entering a solution being absorbed. Low absorbances signify that large proportions of the light are transmitted. The absorption spectrum of $[Ti(H_2O)_6]^{3+}(aq)$ is shown in Figure 24-26(a).

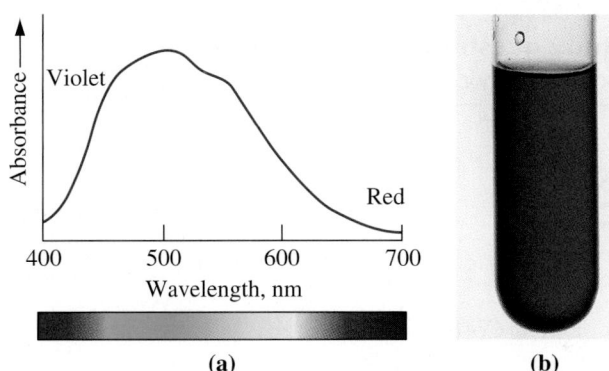

◀ FIGURE 24-26
The color of $[Ti(H_2O)_6]^{3+}(aq)$
(a) The absorption spectrum of $[Ti(H_2O)_6]^{3+}(aq)$.
(b) A solution containing $[Ti(H_2O)_6]^{3+}(aq)$.

(a) Describe the color of light that $[Ti(H_2O)_6]^{3+}(aq)$ absorbs most strongly, and the color of the solution. (b) Describe the electron transition responsible for the absorption peak, and determine the energy associated with this absorption.

Analyze
The highest absorbances in the spectrum shown in Fig. 24-26(a)) come at about 500 nm. We use the electromagnetic spectrum in Figure 8-3 to determine the color of the absorbed light. We determine that the electron configuration of Ti^{3+}, the central ion in the complex ion, is $[Ar]3d^1$. In the ground state, the $3d$ electron is in one of the three degenerate lower levels in the d-orbital splitting diagram for an octahedral complex (Fig. 24-12).

Solve
(a) The electromagnetic spectrum in Figure 8-3 indicates that the absorbed light should be dark green. Fig. 24-16(b) indicates that, in subtractive color mixing, the complementary color of green is magenta. Thus the color of the transmitted light (and hence the observed color of the solution) is a blend of red and blue.

(b) The quantity of energy we seek is that corresponding to electromagnetic radiation at the peak of the absorption spectrum: 500 nm. First we can establish the frequency of this light from $c = \nu \times \lambda$.

$$\nu = \frac{c}{\lambda} = \frac{2.998 \times 10^8 \text{ m s}^{-1}}{500 \times 10^{-9} \text{ m}} = 6.00 \times 10^{14} \text{ s}^{-1}$$

Then, we use Planck's equation to determine E:

$$E = h\nu = (6.626 \times 10^{-34} \text{ J s}) \times (6.00 \times 10^{14} \text{ s}^{-1}) = 3.98 \times 10^{-19} \text{ J}$$

This is the energy per photon. If we want the energy on a per-mole basis, we can write

$$E = (3.98 \times 10^{-19} \text{ J photon}^{-1}) \times (6.022 \times 10^{23} \text{ photon mol}^{-1}) \times \frac{1 \text{ kJ}}{1000 \text{ J}}$$

$$= 240 \text{ kJ mol}^{-1}$$

Assess
Using ideas of subtractive color mixing, we ascertained the wavelength components of white light that the complex cation absorbs. The broadness of the absorption band is due to vibrations in the metal to oxygen bonds. By using the maximum of the absorption curve, we calculated the most probable energy of the absorbed photons.

PRACTICE EXAMPLE A: The compound $Co(en)_2(NO_2)_2Cl$ has been prepared in a number of isomeric forms. One form undergoes no reaction with either $AgNO_3$ or en and is optically inactive. A second form reacts with $AgNO_3$ to form a white precipitate, does not react readily with en, and is optically inactive. A third form is optically active and reacts both with en and $AgNO_3$. Assuming a coordination number of 6 for the cobalt ion, identify each of the three isomeric forms by name, and sketch each of the structures.

PRACTICE EXAMPLE B: A compound is analyzed and found to contain 46.2% Pt, 33.6% Cl, 16.6% N, and 3.6% H. The freezing point of a 0.1 M aqueous solution of the compound is $-0.74\,°C$. What is the structural formula of the compound? What possible isomeric forms are there for this compound?

Mastering**CHEMISTRY**

You'll find a link to additional self study questions in the study area on www.masteringchemistry.com

---

Exercises

Nomenclature

1. Write the formula and name of
   (a) a complex ion having Cr^{3+} as the central ion and two NH_3 molecules and four Cl^- ions as ligands
   (b) a complex ion of iron(III) having a coordination number of 6 and CN^- as ligands
   (c) a coordination compound comprising two types of complex ions: one a complex of Cr(III) with ethylenediamine (en), having a coordination number of 6; the other, a complex of Ni(II) with CN^-, having a coordination number of 4

2. What are the coordination number and the oxidation state of the central metal ion in each of the following complexes? Name each complex.
   (a) $[Co(NH_3)_6]^{2+}$
   (b) $[AlF_6]^{3-}$
   (c) $[Cu(CN)_4]^{2-}$
   (d) $[CrBr_2(NH_3)_4]^+$

   (e) $[Co(ox)_3]^{4-}$
   (f) $[Ag(S_2O_3)_2]^{3-}$

3. Supply acceptable names for the following.
   (a) $[Co(OH)(H_2O)_4(NH_3)]^{2+}$
   (b) $[Co(ONO)_3(NH_3)_3]$
   (c) $[Pt(H_2O)_4][PtCl_6]$
   (d) $[Fe(ox)_2(H_2O)_2]^-$
   (e) $Ag_2[HgI_4]$

4. Write appropriate formulas for the following.
   (a) potassium hexacyanoferrate(III)
   (b) bis(ethylenediamine)copper(II) ion
   (c) pentaaquahydroxoaluminum(III) chloride
   (d) amminechlorobis(ethylenediamine) chromium(III) sulfate
   (e) tris(ethylenediamine)iron(III) hexacyanoferrate(II)

Bonding and Structure in Complex Ions

5. Draw Lewis structures for the following ligands: **(a)** H_2O; **(b)** CH_3NH_2; **(c)** ONO^-; **(d)** SCN^-.
6. Draw Lewis structures for the following ligands: **(a)** hydroxo; **(b)** sulfato; **(c)** oxalato; **(d)** thiocyanato-N-.
7. Draw a plausible structure to represent:
   (a) $[PtCl_4]^{2-}$
   (b) fac-$[Co(H_2O)_3(NH_3)_3]^{2+}$
   (c) $[CrCl(H_2O)_5]^{2+}$
8. Draw plausible structures of the following chelate complexes.
   (a) $[Pt(ox)_2]^{2-}$
   (b) $[Cr(ox)_3]^{3-}$
   (c) $[Fe(EDTA)]^{2-}$

9. Draw plausible structures corresponding to each of the following names.
   (a) pentamminesulfatochromium(III) ion
   (b) trioxalatocobaltate(III) ion
   (c) triamminedichloronitrito-O-cobalt(III)
10. Draw plausible structures corresponding to each of the following names.
   (a) pentamminenitrito-N-cobalt(III) ion
   (b) ethylenediaminedithiocyanato-S-copper(II)
   (c) hexaaquanickel(II) ion

Isomerism

11. Which of these general structures for a complex ion would you expect to exhibit *cis* and *trans* isomerism? Explain.
    (a) tetrahedral
    (b) square-planar
    (c) linear

12. Which of these octahedral complexes would you expect to exhibit *geometric* isomerism? Explain.
    (a) $[CrOH(NH_3)_5]^{2+}$
    (b) $[CrCl_2(H_2O)(NH_3)_3]^+$
    (c) $[CrCl_2(en)_2]^+$
    (d) $[CrCl_4(en)]^-$
    (e) $[Cr(en)_3]^{3+}$

13. If A, B, C, and D are four different ligands,
    (a) how many geometric isomers will be found for square-planar $[PtABCD]^{2+}$?
    (b) Will tetrahedral $[ZnABCD]^{2+}$ display optical isomerism?

14. Write the names and formulas of three coordination isomers of $[Co(en)_3][Cr(ox)_3]$.

15. Draw a structure for *cis*-dichlorobis(ethylenediamine)cobalt(III) ion. Is this ion chiral? Is the *trans* isomer chiral? Explain.

16. The structures of four complex ions are given. Each has Co^{3+} as the central ion. The ligands are H_2O, NH_3, and oxalate ion, $C_2O_4^{2-}$. Determine which, *if any*, of these complex ions are isomers (geometric or optical);

which, *if any*, are identical (that is, have identical structures); and which, *if any*, are distinctly different.

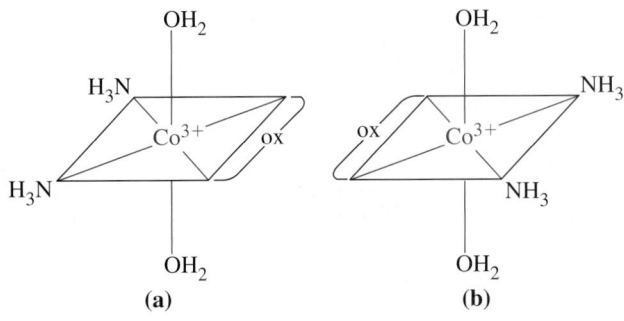

(a)   (b)

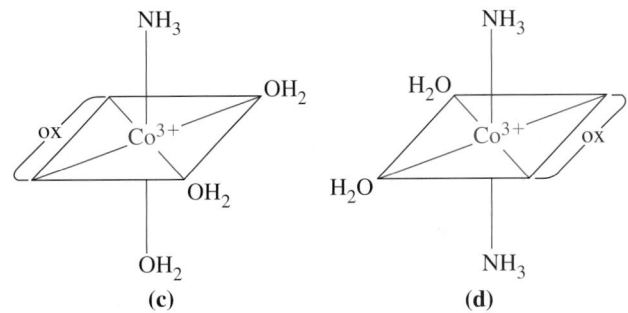

(c)   (d)

Crystal Field Theory

17. Describe how the crystal field theory explains the fact that so many transition metal compounds are colored.

18. Cyano complexes of transition metal ions (such as Fe^{2+} and Cu^{2+}) are often yellow, whereas aqua complexes are often green or blue. Explain the basis for this difference in color.

19. If the ion Co^{2+} is linked with strong-field ligands to produce an octahedral complex, the complex has *one* unpaired electron. If Co^{2+} is linked with weak-field ligands, the complex has *three* unpaired electrons. How do you account for this difference?

20. In contrast to the case of Co^{2+} considered in Exercise 19, no matter what ligand is linked to Ni^{2+} to form an octahedral complex, the complex always has *two* unpaired electrons. Explain this fact.

21. Predict:
    (a) which of the complex ions, $[MoCl_6]^{3-}$ and $[Co(en)_3]^{3+}$, is diamagnetic and which is paramagnetic;

    (b) the number of unpaired electrons expected for the tetrahedral complex ion $[CoCl_4]^{2-}$.

22. Predict:
    (a) whether the square-planar complex ion $[Cu(py)_4]^{2+}$ is diamagnetic or paramagnetic
    (b) whether octahedral $[Mn(CN)_6]^{3-}$ or tetrahedral $[FeCl_4]^-$ has the greater number of unpaired electrons.

23. In Example 24-5, we chose between a tetrahedral and a square-planar structure for $[Ni(CN)_4]^{2-}$ based on magnetic properties. Could we similarly use magnetic properties to establish whether the ammine complex of Ni(II) is octahedral $[Ni(NH_3)_6]^{2+}$ or tetrahedral $[Ni(NH_3)_4]^{2+}$? Explain.

24. In both $[Fe(H_2O)_6]^{2+}$ and $[Fe(CN)_6]^{4-}$ ions, the iron is present as Fe(II); however, $[Fe(H_2O)_6]^{2+}$ is paramagnetic, whereas $[Fe(CN)_6]^{4-}$ is diamagnetic. Explain this difference.

Complex-Ion Equilibria

25. Write equations to represent the following observations.
    (a) A mixture of $Mg(OH)_2(s)$ and $Zn(OH)_2(s)$ is treated with $NH_3(aq)$. The $Zn(OH)_2$ dissolves, but the $Mg(OH)_2(s)$ is left behind.
    (b) When NaOH(aq) is added to $CuSO_4(aq)$, a pale blue precipitate forms. If $NH_3(aq)$ is added, the precipitate redissolves, producing a solution with an intense deep blue color. If this deep blue solution is

made acidic with $HNO_3(aq)$, the color is converted back to pale blue.

26. Write equations to represent the following observations.
    (a) A quantity of $CuCl_2(s)$ is dissolved in concentrated HCl(aq) and produces a yellow solution. The solution is diluted to twice its volume with water and assumes a green color. On dilution to ten times its original volume, the solution becomes pale blue.

(b) When chromium metal is dissolved in HCl(aq), a blue solution is produced that quickly turns green. Later the green solution becomes blue-green and then violet.

27. Which of the following complex ions would you expect to have the largest overall K_f, and why? $[Co(NH_3)_6]^{3+}$, $[Co(en)_3]^{3+}$, $[Co(H_2O)_6]^{3+}$, $[Co(H_2O)_4(en)]^{3+}$.

28. Use data from Table 24.6 to determine values of **(a)** β_4 for the formation of $[Zn(NH_3)_4]^{2+}$; **(b)** β_4 for the formation of $[Ni(H_2O)_2(NH_3)_4]^{2+}$.

29. Write a series of equations to show the stepwise displacement of H_2O ligands in $[Fe(H_2O)_6]^{3+}$ by ethylenediamine, for which $\log K_1 = 4.34$, $\log K_2 = 3.31$, and $\log K_3 = 2.05$. What is the overall formation constant, $\beta_3 = K_f$, for $[Fe(en)_3]^{3+}$?

30. A tabulation of formation constant data lists the following $\log K$ values for the formation of $[CuCl_4]^{2-}$: $\log K_1 = 2.80$, $\log K_2 = 1.60$, $\log K_3 = 0.49$, and $\log K_4 = 0.73$. What is the overall formation constant $\beta_4 = K_f$ for $[CuCl_4]^{2-}$?

31. Explain the following observations in terms of complex-ion formation.
    (a) $Al(OH)_3(s)$ is soluble in NaOH(aq) but insoluble in NH_3(aq).
    (b) $ZnCO_3(s)$ is soluble in NH_3(aq), but ZnS(s) is not.
    (c) The molar solubility of AgCl in pure water is about 1×10^{-5} M; in 0.04 M NaCl(aq), it is about 2×10^{-6} M; but in 1 M NaCl(aq), it is about 8×10^{-5} M.

32. Explain the following observations in terms of complex-ion formation.
    (a) $CoCl_3$ is unstable in aqueous solution, being reduced to $CoCl_2$ and liberating $O_2(g)$. Yet, $[Co(NH_3)_6]Cl_3$ can be easily maintained in aqueous solution.
    (b) AgI is insoluble in water and in dilute NH_3(aq), but AgI will dissolve in an aqueous solution of sodium thiosulfate.

Acid–Base Properties

33. Which of the following would you expect to react as a Brønsted–Lowry acid: $[Cu(NH_3)_4]^{2+}$, $[FeCl_4]^-$, $[Al(H_2O)_6]^{3+}$, or $[Zn(OH)_4]^{2-}$? Why?

34. Write simple chemical equations to show how the complex ion $[CrOH(H_2O)_5]^{2+}$ acts as **(a)** an acid; **(b)** a base.

Applications

35. From data in Chapter 18,
    (a) Derive an equilibrium constant for reaction (24.13), and explain why this reaction (the fixing of photographic film) is expected to go essentially to completion.
    (b) Explain why NH_3(aq) cannot be used in the fixing of photographic film.

36. Show that the oxidation of $[Co(NH_3)_6]^{2+}$ to $[Co(NH_3)_6]^{3+}$ referred to on page 1098 should occur spontaneously in alkaline solution with H_2O_2 as an oxidizing agent.

37. Explain why $K_2[PtCl_4]$ is first converted to $K_2[PtI_4]$ in the synthesis of the anticancer drug cisplatin, *cis*-$Pt(NH_3)_2Cl_2$.

38. Draw dashed and solid wedge diagrams of transplatin, *trans*-$Pt(NH_3)_2Cl_2$, and cisplatin, *cis*-$Pt(NH_3)_2Cl_2$. Then, explain how transplatin can be more reactive yet less effective at killing cancer cells than is cisplatin.

Integrative and Advanced Exercises

39. From each of the following names, you should be able to deduce the formula of the complex ion or coordination compound intended. Yet, these are not the best systematic names that can be written. Replace each name with one that is more acceptable: **(a)** cupric tetraammine ion; **(b)** dichlorotetraammine cobaltic chloride; **(c)** platinic(IV) hexachloride ion; **(d)** disodium copper tetrachloride; **(e)** dipotassium antimony(III) pentachloride.

40. Magnus's green salt has the empirical formula $PtCl_2 \cdot 2 NH_3$. It is a coordination compound consisting of both complex cations and complex anions. Write the probable formula of this coordination compound according to Werner's theory, and assign it a systematic name.

41. How many isomers are there of the complex ion $[CoCl_2(en)(NH_3)_2]^+$? Sketch their structures.

42. Explain the following observations through a series of equations. The green solid $CrCl_3 \cdot 6 H_2O$ dissolves in water to form a green solution. The solution slowly turns blue-green; after a day or two, the solution is violet. When the violet solution evaporates to dryness, a green solid remains.

43. The *cis* and *trans* isomers of $[CoCl_2(en)_2]^+$ can be distinguished via a displacement reaction with oxalate ion. What difference in reactivity toward oxalate ion would you expect between the *cis* and *trans* isomers? Explain.

44. Write half-equations and an overall equation to represent the oxidation of tetraammineplatinum(II) ion to *trans*-tetraamminedichloroplatinum(IV) ion by Cl_2. Then make sketches of the two complex ions.

45. We learned in Chapter 16 that for polyprotic acids, ionization constants for successive ionization steps decrease rapidly. That is, $K_{a_1} \gg K_{a_2} \gg K_{a_3}$. The ionization constants for the first two steps in the ionization of $[Fe(H_2O)_6]^{3+}$ (reactions 24.10 and 24.11) are more nearly equal in magnitude. Why does this multistep ionization seem not to follow the pattern for polyprotic acids?

46. Following are the names of five coordination compounds containing complexes with platinum(II) as the central metal ion and ammonia molecules and/or chloride ions as ligands: **(a)** potassium amminetrichloroplatinate(II), **(b)** diamminedichloroplatinum(II), **(c)** triamminechloroplatinum(II) chloride, **(d)** tetraammineplatinum(II) chloride, **(e)** potassium tetrachloroplatinate(II). Make a rough sketch of the expected graph when electric conductivity is plotted as a function of the chlorine content of the compounds. [*Hint:* Your graph should be based on five points, but no quantitative data are given.]

47. For a solution that is 0.100 M in $[Fe(H_2O)_6]^{3+}$,
   (a) assuming that ionization of the aqua complex ion proceeds only through the first step, equation (24.10), calculate the pH of the solution.
   (b) Calculate $[[FeOH(H_2O)_5]^{2+}]$ if the solution is also 0.100 M $HClO_4$. (ClO_4^- does not complex with Fe^{3+}.)
   (c) Can the pH of the solution be maintained so that $[[FeOH(H_2O)_5]^{2+}]$ does not exceed 1×10^{-6} M? Explain.

48. A solution that is 0.010 M in Pb^{2+} is also made to be 0.20 M in a salt of EDTA (that is, having a concentration of the $EDTA^{4-}$ ion of 0.20 M). If this solution is now made 0.10 M in H_2S and 0.10 M in H_3O^+, will $PbS(s)$ precipitate?

49. Without performing detailed calculations, show why you would expect the concentrations of the various ammine–aqua complex ions to be negligible compared with that of $[Cu(NH_3)_4]^{2+}$ in a solution having a total Cu(II) concentration of 0.10 M and a total concentration of NH_3 of 1.0 M. Under what conditions would the concentrations of these ammine–aqua complex ions (such as $[Cu(H_2O)_3NH_3]^{2+}$) become more significant relative to the concentration of $[Cu(NH_3)_4]^{2+}$? Explain.

50. Verify the statement on page 1101 that neither Ca^{2+} nor Mg^{2+} found in natural waters is likely to precipitate from the water on the addition of other reagents if the ions are complexed with EDTA. Assume reasonable values for the total metal ion concentration and that of *free* EDTA, such as 0.10 M each.

51. Estimate the total $[Cl^-]$ required in a solution that is initially 0.10 M $CuSO_4$ to produce a visible yellow color.

$$[Cu(H_2O)_4]^{2+} + 4\,Cl^- \rightleftharpoons [CuCl_4]^{2-} + 4\,H_2O$$
   (blue)   (yellow)

$$K_f = 4.2 \times 10^5$$

Assume that 99% conversion of $[Cu(H_2O)_4]^{2+}$ to $[CuCl_4]^{2-}$ is sufficient for this to happen, and ignore the presence of any mixed aqua–chloro complex ions.

52. Refer to the stability of $[Co(NH_3)_6]^{3+}(aq)$ on page 1098, and
   (a) verify that E°_{cell} reaction (24.12) is +0.59 V.
   (b) Calculate $[Co^{3+}]$ in a solution that has a total concentration of cobalt of 0.1 M and $[NH_3] = 0.1$ M.
   (c) Show that for the value of $[Co^{3+}]$ calculated in part (b), reaction (24.12) will not occur. [*Hint:* Assume a low, but reasonable, concentration of Co^{2+} (say, 1×10^{-4} M) and a partial pressure of $O_2(g)$ of 0.2 atm.]

53. A Cu electrode is immersed in a solution that is 1.00 M NH_3 and 1.00 M in $[Cu(NH_3)_4]^{2+}$. If a standard hydrogen electrode is the cathode, E_{cell} is +0.08 V. What is the value obtained by this method for the formation constant, K_f, of $[Cu(NH_3)_4]^{2+}$?

54. The following concentration cell is constructed.

$$Ag|Ag^+(0.10\ M[Ag(CN)_2]^-, 0.10\ M\ CN^-)$$
$$\|Ag^+(0.10\ M)|Ag$$

If K_f for $[Ag(CN)_2]^-$ is 5.6×10^{18}, what value would you expect for E_{cell}? [*Hint:* Recall that the anode is on the left.]

55. The compound $CoCl_2 \cdot 2\,H_2O \cdot 4\,NH_3$ may be one of the hydrate isomers $[CoCl(H_2O)(NH_3)_4]Cl \cdot H_2O$ or $[Co(H_2O)_2(NH_3)_4]Cl_2$. A 0.10 M aqueous solution of the compound is found to have a freezing point of $-0.56\,°C$. Determine the correct formula of the compound. The freezing-point depression constant for water is $1.86\ mol\ kg^{-1}\,°C$, and for aqueous solutions, molarity and molality can be taken as approximately equal.

56. Explain why aqueous solutions of $[Sc(H_2O)_6]Cl_3$ and $[Zn(H_2O)_4]Cl_2$ are colorless, but an aqueous solution of $[Fe(H_2O)_6]Cl_3$ is not.

57. Provide a valence bond description of the bonding in the $Cr(NH_3)_6^{3+}$ ion. According to the valence bond description, how many unpaired electrons are there in the $Cr(NH_3)_6^{3+}$ complex? How does this prediction compare with that of crystal field theory?

58. The formation constants of the ions hexamminenickel(II), tris(ethylenediamine)nickel(II) and pentaethylenehexaminenickel(II) are $\beta_6 = 3.2 \times 10^8$, $\beta_3 = 1.6 \times 10^{18}$, and $\beta_1 = 2.0 \times 10^{19}$, respectively. Make reasonable assumptions concerning the enthalpy changes, and show that the formation constants illustrate the chelate effect. Pentaethylenehexamine (penten) is the hexadentate ligand shown below.

$$H_2NCH_2CH_2\ CH_2CH_2\ CH_2CH_2\ CH_2CH_2\ CH_2CH_2NH_2$$
$$NH \quad\quad NH \quad\quad NH \quad\quad NH$$

59. Acetyl acetone undergoes an isomerization to form a type of alcohol called an *enol*.

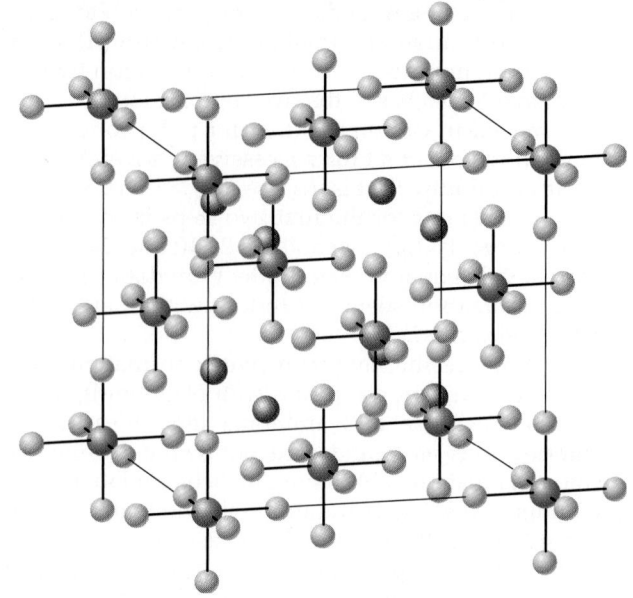

The enol, abbreviated acacH, can act as a bidentate ligand as the anion acac⁻. Which of the following compounds are optically active: Co(acac)₃; *trans*-[Co(acac)₂(H₂O)₂]Cl₂; *cis*-[Co(acac)₂(H₂O)₂]Cl₂?

60. We have seen that complex formation can stabilize oxidation states. An important illustration of this fact is the oxidation of water in acidic solutions by $Co^{3+}(aq)$ but not by $[Co(en)_3]^{3+}$. Use the following data.

$$[Co(H_2O)_6]^{3+} + e^- \longrightarrow [Co(H_2O)_6]^{2+}$$
$$E° = 1.82 \text{ V}$$

$$[Co(H_2O)_6]^{2+} + 3 \text{ en} \longrightarrow [Co(en)_3]^{2+} + 6 H_2O(l)$$
$$\log \beta_3 = 12.18$$

$$[Co(H_2O)_6]^{3+} + 3 \text{ en} \longrightarrow [Co(en)_3]^{3+} + 6 H_2O(l)$$
$$\log \beta_3 = 47.30$$

Calculate $E°$ for the reaction

$$[Co(en)_3]^{3+} + e^- \longrightarrow [Co(en)_3]^{2+}$$

Show that $[Co(en)_3]^{3+}$ is stable in water but $Co^{3+}(aq)$ is not.

61. The amino acid glycine ($NH_2CH_2CO_2H$, denoted Hgly) binds as an anion and is a bidentate ligand. Draw and name all possible isomers of $[Co(gly)_3]$. How many isomers are possible for the compound $[Co(gly)_2Cl(NH_3)]$ [*Hint:* $NH_2CH_2CO_2^-$ is the glycinate anion.]

62. The structure of $K_2[PtCl_6]$ in the solid state is shown here. Identify the type of cubic unit cell, and describe the structure in terms of the holes occupied by the various ions.

63. The graph below represents the molar conductivity of some Pt(IV) complexes. The ligands in these complexes are NH_3 molecules or Cl^- ions, the coordination number of Pt(IV) is 6, and the counter ions (for balancing charge) are K^+ or Cl^-. Write formulas for the coordination compounds corresponding to each point in the graph. (Molar conductivity is the electrical conductivity, under precisely defined conditions, of an aqueous solution containing one mole of a compound.)

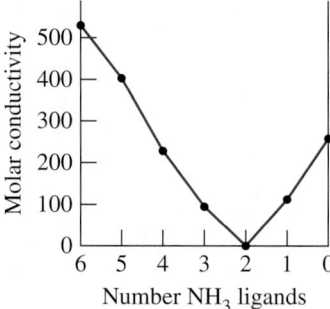

Feature Problems

64. A structure that Werner examined as a possible alternative to the octahedron is the trigonal prism.

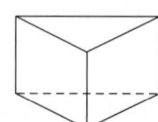

   (a) Does this structure predict the correct number of isomers for the complex ion $[CoCl_2(NH_3)_4]^+$? If not, why not?

   (b) Does this structure account for optical isomerism in $[Co(en)_3]^{3+}$? Explain.

65. Werner demonstrated that octahedral complexes can exhibit optical isomerism, and to Werner's satisfaction, this confirmed the octahedral arrangement of

ligands. However, skeptics of his theory said that because the ligands contained carbon atoms, he could not rule out carbon as the source of the optical activity. Werner devised and prepared the following compound in which the OH⁻ groups act as bridging groups.

$$\left[Co \left(\begin{array}{c} H \\ O \\ \diagup \diagdown \\ \diagdown \diagup \\ O \\ H \end{array} Co(NH_3)_4 \right)_3 \right]^{6+}$$

Werner resolved this compound into its optical isomers, confirming his theory and confounding his critics. What are the oxidation states of the Co ions? If the complex is low spin, what is the number of unpaired electrons in the molecule? Draw the structures of the two optical isomers.

66. The crystal field model discussed in the text describes how the degeneracy of the d orbitals is removed by an octahedral field of ligands. We have seen that the d_{xy}, d_{xz}, and d_{yz} orbitals are stabilized (lower energy) with respect to the average energy of the d orbitals and that the $d_{x^2-y^2}$ and d_{z^2} orbitals are destabilized. As described in Are You Wondering 24-3, the stabilization is $-0.4 \Delta_o$ and destabilization is $0.6 \Delta_o$. The crystal field stabilization energy (CFSE) can be defined as CFSE = [(number of electrons in the d_{xy}, d_{xz}, and d_{yz} orbitals) $\times (-0.4 \Delta_o)$] + [(number of electrons in the $d_{x^2-y^2}$ and d_{z^2} orbitals) $\times (0.6 \Delta_o)$]. The following table contains the enthalpy of hydration for the reaction

$$M^{2+}(g) + 6 H_2O(l) \longrightarrow [M(H_2O)_6]^{2+}(aq)$$

(a) Plot the hydration energies as a function of the atomic number of the metals shown.
(b) Assuming that all the hexaaqua complexes are high spin, which ions have zero CFSE?
(c) If lines are drawn between those ions with CFSE = 0, a line of negative slope is obtained. Can you explain this finding?

Dipositive Metal Ion	Hydration Energy, kJ mol^{-1}
Ca	−2468
Sc	−2673
Ti	−2750
V	−2814
Cr	−2799
Mn	−2743
Fe	−2843
Co	−2904
Ni	−2986
Cu	−2989
Zn	−2939

(d) The ions that do not have CFSE = 0 have heats of hydration that are more negative than the lines drawn in part (c). What is the explanation for this?
(e) Estimate the value of Δ_o for the Fe(II) ion in an octahedral field of water molecules.
(f) What wavelength of light would the $[Fe(H_2O)_6]^{2+}$ ion absorb?

Self-Assessment Exercises

67. In your own words, describe the following terms or symbols: **(a)** coordination number; **(b)** Δ_o; **(c)** ammine complex; **(d)** enantiomer.

68. Briefly describe each of the following ideas, phenomena, or methods: **(a)** spectrochemical series; **(b)** crystal field theory; **(c)** optical isomer; **(d)** structural isomerism.

69. Explain the important distinction between each of the following pairs: **(a)** coordination number and oxidation number; **(b)** monodentate and polydentate ligands; **(c)** cis and trans isomers; **(d)** dextrorotatory and levorotatory compounds; **(e)** low-spin and high-spin complexes.

70. The oxidation state of Ni in the complex ion $[Ni(CN)_4I]^{3-}$ is **(a)** −3; **(b)** −2; **(c)** 0; **(d)** +2; **(e)** +3.

71. The coordination number of Pt in the complex ion $[PtCl_2(en)_2]^{2+}$ is **(a)** 2; **(b)** 3; **(c)** 4; **(d)** 5; **(e)** 6.

72. Of the following complex ions, the one that exhibits isomerism is **(a)** $[Ag(NH_3)_2]^+$; **(b)** $[CoNO_2(NH_3)_5]^{2+}$; **(c)** $[Pt(en)(NH_3)_2]^{2+}$; **(d)** $[CoCl(NH_3)_5]^{2+}$; **(e)** $[PtCl_6]^{2-}$.

73. Of the following complex ions, the one that is optically active is **(a)** cis-$[CoCl_2(en)_2]^+$; **(b)** $[CoCl_2(NH_3)_4]^+$; **(c)** $[CoCl_4(NH_3)_2]^-$; **(d)** $[CuCl_4]^-$.

74. The number of unpaired electrons in the complex ion $[Cr(NH_3)_6]^{2+}$ is **(a)** 5; **(b)** 4; **(c)** 3; **(d)** 2; **(e)** 1.

75. Of the following, the one that is a Brønsted–Lowry acid is **(a)** $[Cu(NH_3)_4]^{2+}$; **(b)** $[FeCl_4]^-$; **(c)** $[Fe(H_2O)_6]^{3+}$; **(d)** $[Zn(OH)_4]^-$.

76. The most soluble of the following solids in NH$_3$(aq) is **(a)** Ca(OH)$_2$; **(b)** Cu(OH)$_2$; **(c)** BaSO$_4$; **(d)** MgCO$_3$; **(e)** Fe$_2$O$_3$.

77. Name the following coordination compounds. Which one(s) exhibit(s) isomerism? Explain.
(a) $[CoBr(NH_3)_5]SO_4$
(b) $[Cr(NH_3)_6][Co(CN)_6]$
(c) $Na_3[Co(NO_2)_6]$
(d) $[Co(en)_3]Cl_3$

78. Write appropriate formulas for the following species.
(a) dicyanoargentate(I) ion
(b) triamminenitrito-N-platinum(II) ion
(c) aquachlorobis(ethylenediamine)cobalt(III) ion
(d) potassium hexacyanochromate(II)

79. Draw structures to represent these four complex ions:
(a) $[PtCl_4]^{2-}$; **(b)** $[FeCl_4(en)]^-$; **(c)** cis-$[FeCl_2(ox)(en)]^-$; **(d)** trans-$[CrCl(OH)(NH_3)_4]^+$.

80. How many different structures are possible for each of the following complex ions?
(a) $[Co(H_2O)(NH_3)_5]^{3+}$
(b) $[Co(H_2O)_2(NH_3)_4]^{3+}$
(c) $[Co(H_2O)_3(NH_3)_3]^{3+}$
(d) $[Co(H_2O)_4(NH_3)_2]^{3+}$

81. Indicate what type of isomerism may be found in each of the following cases. If no isomerism is possible, so indicate.
(a) $[Zn(NH_3)_4][CuCl_4]$
(b) $[Fe(CN)_5SCN]^{4-}$
(c) $[NiCl(NH_3)_5]^+$
(d) $[PtBrCl_2(py)]^-$
(e) $[Cr(OH)_3(NH_3)_3]^-$

82. Indicate what type of isomerism may be found in each of the following cases. If no isomerism is possible, so indicate.
(a) $[Cr(en)_2Br_2]^+$
(b) $[Co(ox)_2Br(SCN)]^{3-}$
(c) $[NiCl_4(en)]^{2-}$
(d) $[PtBrCl(ox)]^-$
(e) $[Cr(Cl)_3(det)]$, det is $H_2N(CH_2)_2NH(CH_2)_2NH_2$

83. Of the complex ions $[Co(H_2O)_6]^{3+}$ and $[Co(en)_3]^{3+}$, one has a yellow color in aqueous solution; the other, blue. Match each ion with its expected color, and state your reason for doing so.

84. Using the method presented in Appendix E, construct a concept map depicting the essential ideas of crystal field theory, and how the theory explains the colors and magnetic properties of transition metal complexes.

25

Nuclear Chemistry

The thin strands of nebulosity are the remains of a star that ended its life in an enormous supernova explosion around 11,000 years ago. Near the center of this large shell of gas is a rapidly spinning neutron star believed to be the surviving core of the original star. Elementary particles, such as neutrons, and nuclear reactions, such as those occurring in stars, are discussed in this chapter.

CONTENTS

◀ When radioactivity was first discovered, its dangers were not appreciated. The first workers in the area did not take the precautions that are routine today.

The origin of the elements is the stars, including our sun. Nuclear fusion in stars creates heavier elements from lighter ones. The heaviest elements, those with atomic numbers greater than 83, have unstable nuclei—that is, they are *radioactive*.

Certain isotopes of the lighter elements are also radioactive. The isotope carbon-14, for example, has chemical and physical properties that are essentially identical to those of the much more abundant isotopes carbon-12 and carbon-13. Carbon-14, however, is radioactive, and it is this property that is used in the technique known as radiocarbon dating.

In this chapter, we will consider a variety of phenomena that originate within the nuclei of atoms. Collectively, we refer to these phenomena as *nuclear chemistry*. Although stars are the natural source of all the elements, we will discuss how new heavy elements and radioactive isotopes of existing lighter elements can be made artificially. We will also discuss the effects of ionizing radiation on matter. These effects can have both positive and negative outcomes and are a subject of society's continuing nuclear debate.

25-1 Radioactivity

The term *radioactivity* was proposed by Marie Curie to describe the emission of ionizing radiation by some of the heavier elements. Ionizing radiation, as the name implies, interacts with matter to produce ions. This means that the radiation is sufficiently energetic to break chemical bonds. Some ionizing radiation is particulate (consisting of particles), and some is nonparticulate. We introduced α, β, and γ radiation in Section 2-2. Let's describe them again in more detail, together with two other nuclear processes.

Alpha Particles

Alpha (α) particles are the nuclei of helium-4 atoms, $_2^4\text{He}^{2+}$, ejected spontaneously from the nuclei of certain radioactive atoms. We can think of α-particle emission as a process in which a bundle of two protons and two neutrons is emitted by a radioactive nucleus, resulting in a lighter nucleus. Alpha particles produce large numbers of ions via their collisions and near collisions with atoms as they travel through matter, but their penetrating power is low. (Generally, a few sheets of paper can stop them.) Because they have a positive charge, α particles are deflected by electric and magnetic fields (recall Figure 2-10).

We can represent the production of α particles by means of a nuclear equation. A **nuclear equation is** written to conform to two rules:

> 1. The sum of mass numbers must be the same on both sides.
> 2. The sum of atomic numbers must be the same on both sides.

In equation (25.1), the alpha particle is represented as $_2^4\text{He}$.

$$_{92}^{238}\text{U} \longrightarrow {_{90}^{234}}\text{Th} + {_2^4}\text{He} \qquad (25.1)$$

Mass numbers total 238, and atomic numbers total 92. The loss of an α particle results in a *decrease* of 2 in the atomic number and 4 in the mass number of the nucleus.

Beta Particles

Beta (β^-) particles are deflected by electric and magnetic fields in the *opposite* direction from α particles. They are less massive than α particles, so they are deflected more strongly than α particles (recall Figure 2-10). Also, they have a greater penetrating power through matter than do α particles (a book, rather than just a few sheets of paper, may be required to stop them). Beta ($-$) particles are electrons, but they are electrons that originate from the nuclei of atoms in nuclear decay processes and are therefore extremely energetic. Electrons that surround the nucleus are given the familiar symbol, e^-.

The simplest decay process producing a β^- particle is the decay of a free neutron, which is unstable outside the nucleus of an atom.

$$_0^1\text{n} \longrightarrow {_1^1}\text{p} + {_{-1}^0}\beta + \nu \qquad (25.2)$$

A β^- particle does not have an atomic number, but its $1-$ charge is equivalent to an atomic number of -1. In nuclear equations, the β^- particle is represented as $_{-1}^0\beta$. Also, a β^- particle is small enough, compared to protons and neutrons, that its mass can be ignored in most calculations. Equation (25.2) introduces the symbol ν to represent an entity called a *neutrino*. This particle was first postulated in the 1930s as necessary for the conservation of certain properties during the β^- decay process. Because they interact so weakly with matter, neutrinos were not detected until the 1950s. Even today, little is known of their properties, including their rest mass. (Rest mass is discussed on page 305).

▲ Alpha particles leave a trail of liquid droplets, artificially colored green in this photograph, as they pass through a supersaturated vapor in a detector known as a cloud chamber. The chamber also contains He(g), and the trail of one α particle (yellow) is striking the nucleus of a He atom. Following the collision, the α particle and the He atom move apart along lines at about a 90° angle.

KEEP IN MIND

that in a *nuclear* equation we represent only the nuclei of atoms, not the atoms as a whole. Although we do not keep track of electrons, electric charge is conserved by the requirement that the sums of the atomic numbers be the same on the two sides of the equation.

▶ A neutrino has no charge and its detection was not easy. Theoretically, to balance nuclear equations, it was known that such a particle must exist. Many elementary particles have been found through experiments based on symmetry arguments.

For a typical β^- decay process, as represented by equation (25.3), we can think of a neutron *within* the nucleus of an atom spontaneously converting to a proton and an electron. This proton remains in the nucleus, whereas the electron is emitted as a β^- particle. Because of the extra proton, the atomic number *increases* by one unit, while the mass number is unchanged. The elusive neutrino is generally not included in the nuclear equation.

$$^{234}_{90}\text{Th} \longrightarrow {}^{234}_{91}\text{Pa} + {}^{0}_{-1}\beta \qquad (25.3)$$

In a similar manner, in some decay processes a *proton* within the nucleus is converted to a neutron, and a β^+ particle and a neutrino* are emitted.

$$^{1}_{1}\text{p} \longrightarrow {}^{1}_{0}\text{n} + {}^{0}_{+1}\beta + \nu \qquad (25.4)$$

The **β^+** particle, also called a **positron**, has properties similar to the β^- particle, except that it carries a *positive* charge. (See the photograph in the margin.) This particle is also known as a *positive electron* and is designated ${}^{0}_{+1}\beta$ in nuclear equations. Positron emission is commonly encountered with artificially produced radioactive nuclei of the lighter elements. For example,

$$^{30}_{15}\text{P} \longrightarrow {}^{30}_{14}\text{Si} + {}^{0}_{+1}\beta \qquad (25.5)$$

Electron Capture

Another process that achieves the same effect as positron emission is **electron capture (EC)**. In this case, an electron from an inner electron shell (usually the shell $n = 1$) is absorbed by the nucleus, where it converts a proton to a neutron. When an electron from a higher quantum level drops to the energy level vacated by the captured electron, X radiation is emitted. For example,

$$^{202}_{81}\text{Tl} + {}^{0}_{-1}\text{e} \longrightarrow {}^{202}_{80}\text{Hg} \text{ (followed by X radiation)} \qquad (25.6)$$

Gamma Rays

Some radioactive decay processes that yield α or β^- particles leave the nucleus in an excited state. The nucleus then loses energy in the form of electromagnetic radiation called gamma rays. **Gamma (γ) rays** are a highly penetrating form of radiation that are *undeflected* by electric and magnetic fields (recall Figure 2-10). (Lead bricks more than several centimeters thick may be required to stop them.) In the radioactive decay of $^{234}_{92}\text{U}$, 77% of the nuclei emit α particles having an energy of 4.18 MeV. The remaining 23% of the $^{234}_{92}\text{U}$ nuclei produce α particles with energies of 4.13 MeV. In the latter case, the $^{230}_{90}\text{Th}$ nuclei are left with an excess energy of 0.05 MeV. This energy is released as γ rays. If the unstable excited Th nucleus is denoted as $^{230}_{90}\text{Th}^{\ddagger}$, we can write

$$^{234}_{92}\text{U} \longrightarrow {}^{230}_{90}\text{Th}^{\ddagger} + {}^{4}_{2}\text{He} \qquad (25.7)$$

$$^{230}_{90}\text{Th}^{\ddagger} \longrightarrow {}^{230}_{90}\text{Th} + \gamma \qquad (25.8)$$

This γ-emission process is represented diagrammatically in Figure 25-1.

25-1 ARE YOU WONDERING...

How an α particle gets out of a nucleus?

The answer has to do with quantum theory and the nature of the forces involved. Consider the potential energy diagram shown below. The blue line represents the potential energy, where we imagine an α particle as a separate particle within a nucleus such as $^{238}_{92}\text{U}$. Region A represents the potential energy of the α particle when it is held within the nucleus by the forces inside the uranium nucleus.

(continued)

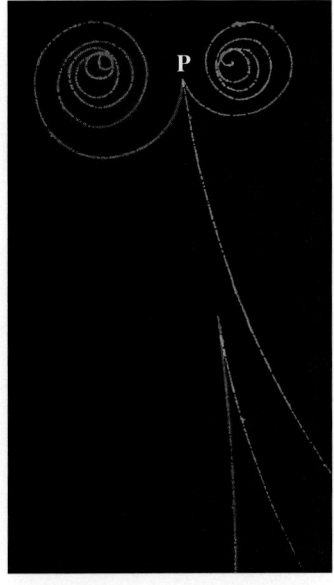

▲ **A colorized cloud chamber photograph**
Point P marks an atomic nucleus that interacts with a γ-ray photon (not visible), producing a β^- particle and a positron (spiral green and red tracks, respectively). The photon also dislodges an orbital electron (vertical green track).

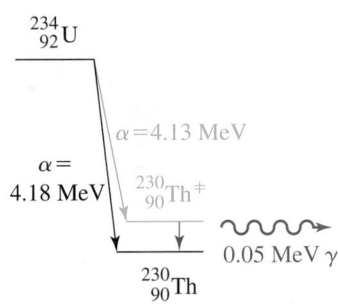

▲ FIGURE 25-1
Production of γ rays
The transition of a $^{230}_{90}\text{Th}$ nucleus between the two energy states shown results in the emission of 0.05 MeV of energy in the form of γ rays.

An electronvolt (eV) is the energy acquired by an electron when it falls through an electric potential difference of 1 volt:

$$1 \text{ eV} = 1.6022 \times 10^{-19} \text{ J}$$
$$1 \text{ MeV} = 1 \times 10^{6} \text{ eV}$$

---

*There appear to be two related entities: the neutrino and antineutrino. Neutrinos accompany positron emission and electron capture; antineutrinos are associated with β^- emission.

Region C represents the potential energy of the α particle when it is free of the nucleus. The potential energy along the downward curving portion of the blue line represents the Coulomb (electrostatic) repulsion between the positively charged α particle and the nucleus remaining after the α particle has escaped ($^{234}_{90}$Th).

To get to region C, the α particle must get past the barrier in region B. The potential energy just beyond A, the radius of the nucleus, is greater than the energy of the α particle. (We know this from the measured energy of the α particle.) The α particle could not escape the nucleus if it were governed by classical physics because this would require an input of energy equal to the height of the barrier. Radioactive nuclei decay spontaneously, however, without an input of energy. How can the α particle get from region A to region C?

It passes through the barrier in a process known as *tunneling*. Classically, to go from A to C, the α particle would violate the principle of the conservation of energy. The α particle, however, possesses wave-like properties, as seen through the wave function at the bottom of the figure. Quantum mechanics predicts a finite probability of finding the α particle in a classically forbidden region. The wave function for the α particle trails off in the barrier region (B) and then reaches the outside, where it appears as a wave with much smaller amplitude. There is a finite probability (ψ^2) of finding the α particle outside the nucleus: The α particle has tunneled through the barrier.

Moreover, an alternative form of the uncertainty principle tells us that energy conservation can be violated by an amount ΔE for the length of time Δt given by

$$\Delta E \times \Delta t = \frac{h}{4\pi}$$

That is, the wave–particle duality of quantum theory allows the conservation of energy to be violated for brief periods—long enough for an α particle to tunnel through the barrier. ΔE corresponds to the energy difference between the barrier height and the α particle's energy, and Δt corresponds to the time required to pass through the barrier. The higher and wider the potential energy barrier, the less time the particle has to escape and the less likely it will do so. Thus, the height and width of the barrier control the rate of decay.

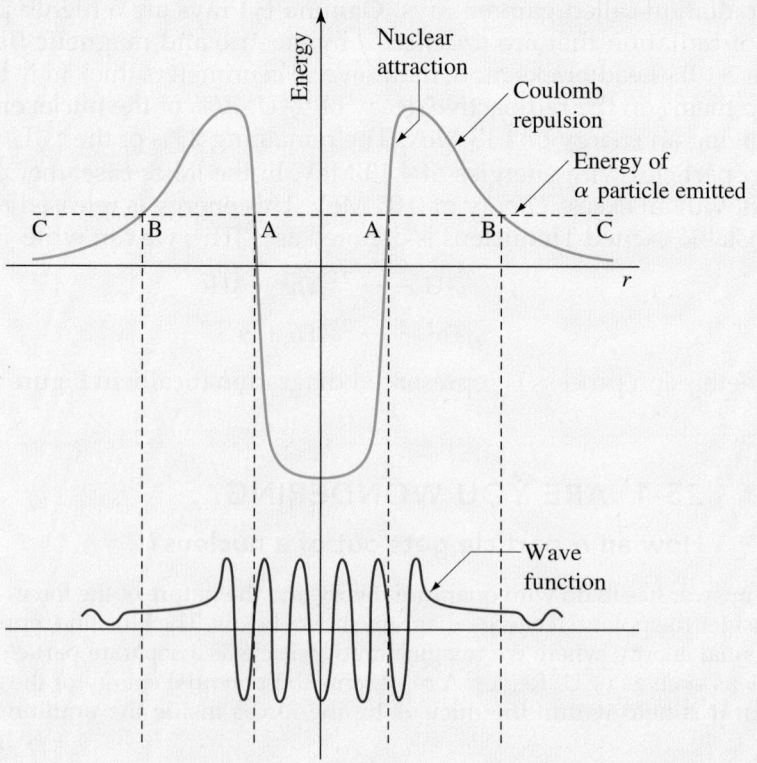

▲ Tunneling of an α particle out of the nucleus.

EXAMPLE 25-1   Writing Nuclear Equations for Radioactive Decay Processes

Write nuclear equations to represent **(a)** α-particle emission by ^{222}Rn and **(b)** radioactive decay of bismuth-215 to polonium-215.

Analyze

In part **(a)** we can identify two of the species involved in this process from the information given and we can write an incomplete equation. The unknown element in the incomplete equation is identified by determining the atomic number (Z) and mass number (A) that will balance the incomplete equation. Part **(b)** is done in a way very similar to that in part **(a)**.

Solve

(a) Since the $^{222}_{86}$Rn nucleus ejects an α particle, $^{4}_{2}$He, as shown in the following incomplete nuclear equation.

$$^{222}_{86}\text{Rn} \longrightarrow \text{?} + {}^{4}_{2}\text{He}$$

Because the ejected α particle contains two protons, the unknown product must contain two fewer protons than $^{222}_{86}$Rn: $Z = 86 - 2 = 84$. This atomic number identifies the element as polonium, $_{84}$Po. The mass number (A) of the product can be obtained by subtracting the mass number of the α particle from that of the radon isotope: $A = 222 - 4 = 218$. The completed nuclear equation is

$$^{222}_{86}\text{Rn} \longrightarrow {}^{218}_{84}\text{Po} + {}^{4}_{2}\text{He}$$

(b) The atomic number of bismuth is 83 and that of polonium is 84. We can approach this problem as we did part **(a)**.

$$^{215}_{83}\text{Bi} \longrightarrow {}^{215}_{84}\text{Po} + \text{?}$$

There is no change in mass number, so the particle has a zero mass number. Its atomic number is $Z = 83 - 84 = -1$. Only a $^{0}_{-1}\beta$ particle fits these parameters: Beta ($-$) decay is the only type of emission leading to an increase of one unit in atomic number without a change in the mass number.

$$^{215}_{83}\text{Bi} \longrightarrow {}^{215}_{84}\text{Po} + {}^{0}_{-1}\beta$$

Assess

It is interesting that when starting from different elements, we can arrive at the same element through different types of particle emission. Note that even though we came to the same decayed element, we ended with different isotopes of that element.

PRACTICE EXAMPLE A: Write a nuclear equation to represent β^--particle emission by $^{241}_{94}$Pu.

PRACTICE EXAMPLE B: Write a nuclear equation to represent the decay of a radioactive nucleus to produce ^{58}Ni and a positron.

25-1   CONCEPT ASSESSMENT

Which type(s) of radioactive decay transform(s) the nucleus of an atom to that of a different element, and which type(s) do not?

25-2   Naturally Occurring Radioactive Isotopes

Of the stable nuclides, $^{209}_{83}$Bi has the highest atomic number and mass number. All known nuclides beyond it in atomic and mass numbers are radioactive. Naturally occurring $^{238}_{92}$U is radioactive and disintegrates by the loss of α particles.

$$^{238}_{92}\text{U} \longrightarrow {}^{234}_{90}\text{Th} + {}^{4}_{2}\text{He}$$

KEEP IN MIND

that *nuclide* is the general term for an atom with a particular atomic number and mass number. Different nuclides of an element are referred to as *isotopes*.

$^{234}_{90}$Th is also radioactive; it decays by β^- emission.

$$^{234}_{90}\text{Th} \longrightarrow\ ^{234}_{91}\text{Pa} +\ ^{\ 0}_{-1}\beta$$

$^{234}_{91}$Pa also decays by β^- emission to produce $^{234}_{92}$U, which is also radioactive.

$$^{234}_{91}\text{Pa} \longrightarrow\ ^{234}_{92}\text{U} +\ ^{\ 0}_{-1}\beta$$

The term *daughter* is commonly used to describe the new nuclide produced in a radioactive decay. Thus, ^{234}Th is the daughter of ^{238}U, ^{234}Pa is the daughter of ^{234}Th, and so on.

The chain of radioactive decay that begins with $^{238}_{92}$U continues through a number of steps of α and β^- emission until it eventually terminates with a stable isotope of lead—$^{206}_{82}$Pb. The entire scheme is outlined in Figure 25-2. All naturally occurring radioactive nuclides of high atomic number belong to one of three **radioactive decay series**: the *uranium* series just described, the *thorium* series, or the *actinium* series. (The actinium series actually begins with uranium-235, which was once called actino-uranium.)

Even though some of the daughters in natural radioactive decay schemes have very short half-lives, all are present because they are constantly forming as well as decaying. It is likely that only about one gram of radium-226 was present in several tons of uranium ore processed by Marie Curie in her discovery of radium in 1898. Nevertheless, she was successful in isolating it. The ore also contained only a fraction of a milligram of polonium, which she was able to detect but not isolate.

Radioactive decay schemes can be used to determine the ages of rocks and thereby the age of Earth (see Section 25-5). The appearance of certain radioactive substances in the environment can also be explained through radioactive decay series. The nuclides ^{210}Po and ^{210}Pb have been detected in cigarette smoke. These radioactive isotopes are derived from ^{238}U, found in trace amounts in the phosphate fertilizers used in tobacco fields. These α-emitting isotopes have been implicated in the link between cigarette smoking and cancer and heart disease.

▲ Marie Sklodowska Curie (1867–1934)
Marie Curie shared the 1903 Nobel Prize in physics for studies on radiation phenomena. In 1911, she won the Nobel Prize in chemistry for her discovery of polonium and radium.

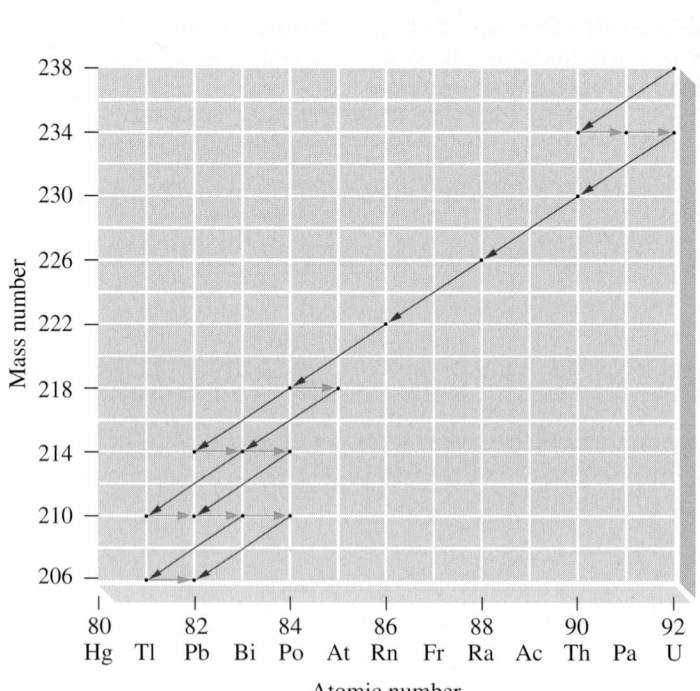

▲ FIGURE 25-2
The natural radioactive decay series for $^{238}_{92}$U (uranium series)
The long arrows pointing down and to the left correspond to α-particle emissions. The short horizontal arrows represent β^- emissions. Other natural decay series originate with $^{232}_{90}$Th (thorium series) and $^{235}_{92}$U (actinium series).

Radioactivity, which is so common among isotopes of high atomic number, is a relatively rare phenomenon among the naturally occurring lighter isotopes. Even so, ^{40}K is a radioactive isotope, as are ^{50}V and ^{138}La. ^{40}K decays by β^- emission and by electron capture.

$$^{40}_{19}\text{K} \longrightarrow {}^{40}_{20}\text{Ca} + {}^{0}_{-1}\beta \quad \text{and} \quad {}^{40}_{19}\text{K} + {}^{0}_{-1}\text{e} \longrightarrow {}^{40}_{18}\text{Ar}$$

At the time Earth was formed ^{40}K was much more abundant than it is now. It is believed that the high argon content of the atmosphere (0.934% by volume and almost all of it as ^{40}Ar) is derived from the radioactive decay of ^{40}K. Aside from ^{40}K and ^{14}C (produced by cosmic radiation), the most important radioactive isotopes of the lighter elements are produced *artificially*.

25-2   CONCEPT ASSESSMENT

Explain why francium, the heaviest of the alkali metals (group 1), is not found in minerals containing the other alkali metals and is also one of the rarest elements.

25-3   Nuclear Reactions and Artificially Induced Radioactivity

Ernest Rutherford discovered that atoms of one element can be transformed into atoms of another element. He did this in 1919 by bombarding $^{14}_{7}\text{N}$ nuclei with α particles, producing $^{17}_{8}\text{O}$ and protons. In this way, he was able to obtain protons outside atomic nuclei. The process can be represented as

$$^{14}_{7}\text{N} + {}^{4}_{2}\text{He} \longrightarrow {}^{17}_{8}\text{O} + {}^{1}_{1}\text{H} \tag{25.9}$$

In reaction (25.9), instead of a nucleus disintegrating spontaneously, it must be struck by another small particle to induce a nuclear reaction. $^{17}_{8}\text{O}$ is a naturally occurring *nonradioactive* isotope of oxygen (0.037% natural abundance). The situation with $^{30}_{15}\text{P}$, which can also be produced by a nuclear reaction, is somewhat different.

In 1934, when bombarding aluminum with α particles, Iréne Joliot-Curie (daughter of Marie and Pierre Curie) and her husband, Frédéric Joliot, observed the emission of two types of particles: neutrons and positrons. The Joliots observed that when bombardment by α particles was stopped, the emission of neutrons also stopped; the emission of positrons continued, however. Their conclusion was that the nuclear bombardment produces $^{30}_{15}\text{P}$, which undergoes radioactive decay by the emission of positrons.

$$^{27}_{13}\text{Al} + {}^{4}_{2}\text{He} \longrightarrow {}^{30}_{15}\text{P} + {}^{1}_{0}\text{n}$$
$$^{30}_{15}\text{P} \longrightarrow {}^{30}_{14}\text{Si} + {}^{0}_{+1}\beta$$

The first radioactive nuclide obtained by artificial means was $^{30}_{15}\text{P}$. Now, over 1000 artificially radioactive nuclides have been produced, and their number considerably exceeds the number of nonradioactive ones (about 280).

▲ **Iréne Joliot-Curie (1897–1956)**
Iréne Joliot-Curie and her husband, Frédéric Joliot, shared the 1935 Nobel Prize in chemistry for the artificial production of radioactive nuclides.

EXAMPLE 25-2   Writing Equations for Nuclear Bombardment Reactions

Write a nuclear equation for the production of ^{56}Mn by bombardment of ^{59}Co with neutrons.

Analyze

In this example we proceed in a manner similar to that in Example 25-1: We first write an incomplete nuclear equation from the given information. We must also realize that a particle is produced along with ^{56}Mn. To find the mass number (A) of the unknown particle, we must subtract the mass number of the Mn atom from that of

(continued)

the Co atom plus the neutron that initiates the reaction. Thus, for the unknown particle, $A = 59 + 1 - 56 = 4$. Subtracting the atomic number of Mn from that of Co provides the atomic number of the unknown particle: $Z = 27 - 25 = 2$.

Solve

The unknown particle must have $A = 4$ and $Z = 2$; it is an α particle.

$$^{59}_{27}\text{Co} + {}^{1}_{0}\text{n} \longrightarrow {}^{56}_{25}\text{Mn} + {}^{4}_{2}\text{He}$$

Assess

By using the concept that a balanced nuclear equation has the same overall atomic number and mass number on both sides of the equation, we can identify the unknown particle that has been ejected during bombardment.

PRACTICE EXAMPLE A: Write a nuclear equation for the production of ^{147}Eu by bombardment of ^{139}La with ^{12}C.

PRACTICE EXAMPLE B: Write a nuclear equation for the production of ^{124}I by bombardment of ^{121}Sb with α particles. Also, write an equation for the subsequent decay of ^{124}I by positron emission.

25-4 Transuranium Elements

Until 1940, the only known elements were those that occur naturally. In 1940, bombardment of $^{238}_{92}\text{U}$ atoms with neutrons produced the first synthetic element. First, the unstable nucleus $^{239}_{92}\text{U}$ forms. This nucleus then undergoes β^- decay, yielding the element neptunium, with $Z = 93$.

$$^{238}_{92}\text{U} + {}^{1}_{0}\text{n} \longrightarrow {}^{239}_{92}\text{U} + \gamma$$

$$^{239}_{92}\text{U} \longrightarrow {}^{239}_{93}\text{Np} + {}^{0}_{-1}\beta$$

Bombardment by neutrons is an effective way to produce nuclear reactions because these heavy uncharged particles are not repelled as they approach a nucleus.

Since 1940, all the elements from $Z = 93$ to 112, as well as elements 114 and 116, have been synthesized. Many of the new elements of high atomic number have been formed by bombarding transuranium atoms with the nuclei of lighter elements. For example, an isotope of the element $Z = 105$ can be produced by bombarding atoms of $^{249}_{98}\text{Cf}$ with $^{15}_{7}\text{N}$ nuclei.

$$^{249}_{98}\text{Cf} + {}^{15}_{7}\text{N} \longrightarrow {}^{260}_{105}\text{Db} + 4\,{}^{1}_{0}\text{n} \tag{25.10}$$

To bring about nuclear reactions such as (25.10) requires bombarding atomic nuclei with energetic particles. Such energetic particles can be obtained in an accelerator. A type of accelerator known as a cyclotron is illustrated in Figure 25-3.

A *charged-particle accelerator*, as the name implies, can produce only beams of charged particles (such as $^{1}_{1}\text{H}^+$) as projectiles. In many cases, neutrons are most effective as projectiles for nuclear bombardment. The neutrons required can be generated through a nuclear reaction produced by a charged-particle beam. In the following reaction, $^{2}_{1}\text{H}$ represents a beam of deuterons (actually, $^{2}_{1}\text{H}^+$) from an accelerator.

$$^{9}_{4}\text{Be} + {}^{2}_{1}\text{H} \longrightarrow {}^{10}_{5}\text{B} + {}^{1}_{0}\text{n}$$

Another important source of neutrons for nuclear reactions is a nuclear reactor (as we will see in Section 25-8).

▲ Stanford Linear Accelerator Center (SLAC0 PEP-II collider). Electrons (blue) and positrons (pink) circulate in opposite directions along the two rings before being forced to collide. These collisions produce large quantities of subatomic particles (e.g., "B mesons and anti-B mesons").

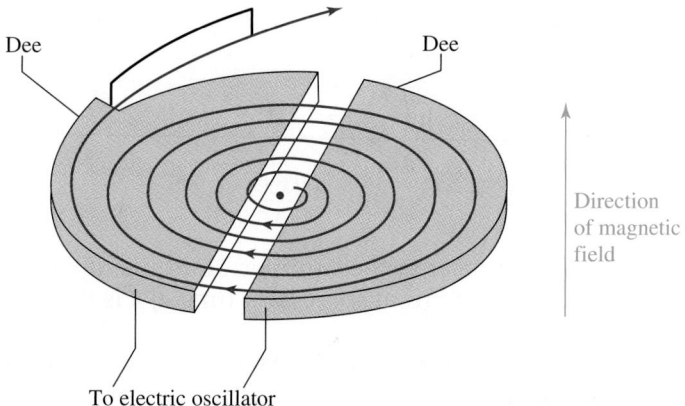

Dee   Dee

Direction
of magnetic
field

To electric oscillator

▲ FIGURE 25-3
The cyclotron
This type of accelerator consists of two hollow, flat, semicircular boxes, called
dees, that are kept electrically charged. The entire assembly is maintained within a
magnetic field. The particles to be accelerated, in the form of positive ions, are
produced at the center of the opening between the dees. They are then attracted
into the negatively charged dee and forced into a circular path by the magnetic
field. When the particles leave the dee and enter the gap, the electric charges on
the dees are reversed, so that the particles are attracted into the opposite dee.
The particles are accelerated as they pass the gap and travel a wider circular path
in the new dee. This process is repeated many times until the particles are brought
to the required energy.

25-5   Rate of Radioactive Decay

In time, we can expect every atomic nucleus of a radioactive nuclide to disintegrate, but it is impossible to predict when any one nucleus will do so. Although
we cannot make predictions for a particular atom, we can use statistical methods to make predictions for a collection of atoms. Based on experimental observations, a **radioactive decay law** has been established.

> *The rate of disintegration of a radioactive material—called the
> activity, A, or the decay rate—is directly proportional to the
> number of atoms present.*

In mathematical terms,

$$\text{rate of decay} \propto N \quad \text{and} \quad \text{rate of decay} = A = \lambda N \qquad (25.11)$$

The activity is expressed in atoms per unit time, such as atoms per second. N is
the number of atoms in the sample being observed; λ is the **decay constant**,
which has units of time^{-1}. Consider the case of a 1,000,000-atom sample disintegrating at the rate of 100 atoms per second. In such a case, $N = 1.0 \times 10^6$ and

$$\lambda = A/N = 100 \text{ atom s}^{-1}/1.0 \times 10^6 \text{ atom} = 1.0 \times 10^{-4} \text{ s}^{-1}$$

Radioactive decay is a *first-order* process. To relate it to the first-order kinetics
that we studied in Chapter 14, think of the activity as corresponding to a rate
of reaction; the number of atoms as corresponding to the concentration of a
reactant; and the decay constant, λ, as corresponding to a rate constant, k. This
correspondence can be carried further by writing an integrated radioactive
decay law and a relationship between the decay constant and the **half-life** of

the process—the length of time required for half of a radioactive sample to disintegrate.

$$\ln\left(\frac{N_t}{N_0}\right) = -\lambda t \tag{25.12}$$

$$t_{1/2} = \frac{0.693}{\lambda} \tag{25.13}$$

In these equations, N_0 represents the number of atoms at some initial time ($t = 0$); N_t is the number of atoms at some later time, t; λ is the decay constant; and $t_{1/2}$ is the half-life.

Recall from Chapter 14 (page 615) that the half-life of a first-order process is a *constant*. Thus, if half the atoms of a radioactive sample disintegrate in 2.5 min, the number of atoms remaining will be reduced to one-fourth the original number in 5.0 min, one-eighth in 7.5 min, and so on. The shorter the half-life, the larger the value of λ and the faster the decay process. Half-lives of radioactive nuclides range from extremely short to very long, as suggested by the representative data in Table 25.1.

You may be wondering whether, like first-order chemical reactions, radioactive decay is temperature dependent. We learned in Chapter 14 that an important determinant of the rate of a chemical reaction is the height of the energy barrier between the reactants and products (the activation energy). The higher the temperature, the greater the number of molecules that can surmount the barrier as a result of collisions and the faster the reaction proceeds. Although there is also an energy barrier that confines nuclear particles to the nucleus, molecular collisions do not invest any energy in nuclear particles. Moreover, in radioactive decay, nuclear particles do not escape the nucleus by surmounting an energy barrier—they tunnel through it. Thus, the rates of radioactive decay processes are independent of temperature.

TABLE 25.1   Some Representative Half-Lives

Nuclide	Half-Life[a]	Nuclide	Half-Life[a]	Nuclide	Half-Life[a]
$^{3}_{1}\text{H}$	12.26 y	$^{40}_{19}\text{K}$	1.25×10^{9} y	$^{214}_{84}\text{Po}$	1.64×10^{-4} s
$^{14}_{6}\text{C}$	5730 y	$^{80}_{35}\text{Br}$	17.6 min	$^{222}_{86}\text{Rn}$	3.823 d
$^{13}_{8}\text{O}$	8.7×10^{-3} s	$^{90}_{38}\text{Sr}$	27.7 y	$^{226}_{88}\text{Ra}$	1.60×10^{3} y
$^{28}_{12}\text{Mg}$	21 h	$^{131}_{53}\text{I}$	8.040 d	$^{234}_{90}\text{Th}$	24.1 d
$^{32}_{15}\text{P}$	14.3 d	$^{137}_{55}\text{Cs}$	30.23 y	$^{238}_{92}\text{U}$	4.51×10^{9} y
$^{35}_{16}\text{S}$	88 d				

[a]s, second; min, minute; h, hour; d, day; y, year.

EXAMPLE 25-3   Using the Half-Life Concept and the Radioactive Decay Law to Describe the Rate of Radioactive Decay

The phosphorus isotope ^{32}P is used in biochemical studies to determine the pathways of phosphorus atoms in living organisms. Its presence is detected through its emission of β^- particles. **(a)** What is the decay constant for ^{32}P, expressed in the unit s^{-1}? **(b)** What is the activity of a 1.00 mg sample of ^{32}P (that is, how many atoms disintegrate per second)? **(c)** Approximately what mass of ^{32}P will remain in the original 1.00 mg sample after 57 days? (See Table 25.1.) **(d)** What will be the rate of radioactive decay after 57 days?

Analyze

To solve these types of problems, we need to determine the decay constant, λ, of the radioactive species, which is related to the concept of half-life through equation (25.13). After determining the decay constant from the half-life of the sample, we can then use it to determine the activity for part **(b)**.

Solve

(a) We can determine λ from $t_{1/2}$ with equation (25.13). The first result we get has the unit d^{-1}. We must convert this unit to h^{-1}, min^{-1}, and s^{-1}.

$$\lambda = \frac{0.693}{14.3 \text{ d}} \times \frac{1 \text{ d}}{24 \text{ h}} \times \frac{1 \text{ h}}{60 \text{ min}} \times \frac{1 \text{ min}}{60 \text{ s}} = 5.61 \times 10^{-7} \text{ s}^{-1}$$

(b) First, let us find the number of atoms, N, in 1.00 mg of ^{32}P.

$$N\left(^{32}\text{P atoms}\right) = 0.00100 \text{ g} \times \frac{1 \text{ mol } ^{32}\text{P}}{32.0 \text{ g}} \times \frac{6.022 \times 10^{23}\ ^{32}\text{P atoms}}{1 \text{ mol } ^{32}\text{P}}$$
$$= 1.88 \times 10^{19}\ ^{32}\text{P atoms}$$

Then, we can multiply this number by the decay constant to get the activity or decay rate.

$$\text{activity} = \lambda N = 5.61 \times 10^{-7} \text{ s}^{-1} \times 1.88 \times 10^{19} \text{ atoms}$$
$$= 1.05 \times 10^{13} \text{ atoms/s}$$

(c) A period of 57 days is $57/14.3 = 4.0$ half-lives. As shown in Figure 25-4, the quantity of radioactive material decreases by one-half for every half-life. The quantity remaining is $\left(\frac{1}{2}\right)^4$ of the original quantity.

$$? \text{ mg } ^{32}\text{P} = 1.00 \text{ mg} \times \left(\frac{1}{2}\right)^4 = 1.00 \text{ mg} \times \frac{1}{16} = 0.063 \text{ mg } ^{32}\text{P}$$

(d) The activity is directly proportional to the number of radioactive atoms remaining ($\text{activity} = \lambda N$), and the number of atoms is directly proportional to the mass of ^{32}P. When the mass of ^{32}P has dropped to one-sixteenth its original mass, the number of ^{32}P atoms also falls to one-sixteenth the original number, and the rate of decay is one-sixteenth the original activity.

$$\text{rate of decay} = \frac{1}{16} \times 1.05 \times 10^{13} \text{ atoms/s} = 6.56 \times 10^{11} \text{ atoms/s}$$

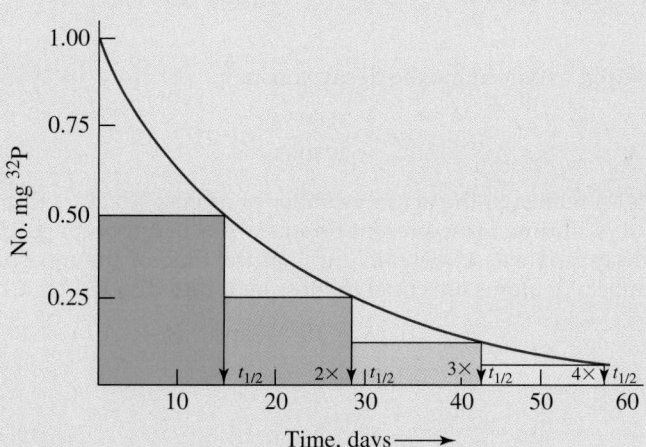

▲ FIGURE 25-4
Radioactive decay of a hypothetical ^{32}P sample—Example 25-3 illustrated

Assess

Phosphorus-32 isotope is an ideal radionuclide to use because of its short half-life. This time is long enough to carry out experiments, yet short enough that disposal is not a major problem. We see in this example that after 57 days, approximately 6% of the original mass of phosphorus remains.

PRACTICE EXAMPLE A: ^{131}I is a β^- emitter used as a tracer for radioimmunoassays in biological systems. Use information in Table 25.1 to determine **(a)** the decay constant in s^{-1}; **(b)** the activity of a 2.05 mg sample of ^{131}I; **(c)** the percentage of ^{131}I remaining after 16 days; and **(d)** the rate of β^- emission after 16 days.

PRACTICE EXAMPLE B: ^{223}Ra has a half-life of 11.4 days. How long would it take for the activity associated with a sample of ^{223}Ra to decrease to 1.0% of its current value?

25-3   CONCEPT ASSESSMENT

Why are radioactive nuclides with intermediate half-lives generally more hazardous than those with either extremely short or extremely long half-lives?

Radiocarbon Dating

In the upper atmosphere, $^{14}_{6}C$ is formed at a constant rate by the bombardment of $^{14}_{7}N$ with neutrons.

$$^{14}_{7}N + ^{1}_{0}n \longrightarrow ^{14}_{6}C + ^{1}_{1}H$$

The neutrons are produced by cosmic rays. $^{14}_{6}C$ disintegrates by β^{-} emission.

Carbon-containing compounds in living organisms are in equilibrium with ^{14}C in the atmosphere—that is, these organisms replace ^{14}C atoms that have undergone radioactive decay with "fresh" ^{14}C atoms through interactions with their environment. The ^{14}C isotope is radioactive and has a half-life of 5730 years. The activity associated with ^{14}C that is in equilibrium with its

---

EXAMPLE 25-4   Applying the Integrated Rate Law for Radioactive Decay: Radiocarbon Dating

A wooden object found in an Indian burial mound is subjected to radiocarbon dating. The activity associated with its ^{14}C content is 10 dis min^{-1} g^{-1}. What is the age of the object? In other words, how much time has elapsed since the tree from which the wood came was cut down?

Analyze

The solution requires three equations: (25.11), (25.12), and (25.13).

Solve

Equation (25.13) is used to determine the decay constant.

$$\lambda = \frac{0.693}{5730 \text{ y}} = 1.21 \times 10^{-4} \text{ y}^{-1}$$

Next, equation (25.11) relates to the actual number of atoms: N at $t = 0$ (the time when the ^{14}C equilibrium was destroyed) and N_t at time t (the present time). As discussed on page 1123, the activity just before the ^{14}C equilibrium was destroyed was 15 dis min^{-1} g^{-1}; at the time of the measurement, it is 10 dis min^{-1} g^{-1}. The corresponding numbers of atoms are equal to these activities divided by λ.

$$N_0 = A_0/\lambda = 15/\lambda \quad \text{and} \quad N_t = A_t/\lambda = 10/\lambda$$

Finally, we substitute into equation (25.12).

$$\ln\frac{N_t}{N_0} = \ln\frac{10/\lambda}{15/\lambda} = \ln\frac{10}{15} = -(1.21 \times 10^{-4} \text{ y}^{-1})t$$

$$-0.41 = -(1.21 \times 10^{-4} \text{ y}^{-1})t$$

$$t = \frac{0.41}{1.21 \times 10^{-4} \text{ y}^{-1}} = 3.4 \times 10^{3} \text{ y}$$

Assess

In the previous example, we observed that the activity depends on the amount of material. Therefore, the results of radiocarbon dating depend on knowing the activity at the time the equilibrium between ^{14}C and the other nonradioactive carbon isotopes ceases. If at the time equilibrium was destroyed the activity was 14 dis min^{-1} g^{-1}, we would have determined the object to be 2.8×10^{3} years old, which is approximately a 17% error.

PRACTICE EXAMPLE A: What is the age of a mummy, given a ^{14}C activity of 8.5 dis min^{-1} g^{-1}?

PRACTICE EXAMPLE B: What should be the current activity, in dis min^{-1} g^{-1}, of a wooden object believed to be 1100 years old?

◀ The remains of a man frozen in a glacier in the Austrian Alps have been dated by the radiocarbon method as 5300 years old.

environment is about 15 disintegrations per minute (dis min^{-1}) per gram of carbon. When an organism dies (for instance, when a tree is cut down), this equilibrium is destroyed and the disintegration rate falls off because the dead organism no longer absorbs new ^{14}C. From the measured disintegration rate at some later time, the age can be estimated (that is, the elapsed time since the ^{14}C equilibrium was disrupted).

The Age of Earth

The natural radioactive decay scheme of Figure 25-2 suggests the eventual fate of all $^{238}_{92}U$ in nature—conversion to lead. Naturally occurring uranium minerals always have associated with them some nonradioactive lead formed by radioactive decay. From the mass ratio of $^{206}_{82}Pb$ to $^{238}_{92}U$ in such a mineral, it is possible to estimate the age of the igneous rock containing the mineral. The age of the rock refers to the time elapsed since molten magma solidified to form the rock. One assumption of this method is that the initial radioactive nuclide, the final stable nuclides, and all the products of a decay series remain in the rock. Another assumption is that any lead present in the rock initially consisted of the several isotopes of lead in their present, naturally occurring abundances.

▲ A lunar rock that has been radiometrically dated to be about 4.6 billion years old.

The half-life of $^{238}_{92}U$ is 4.5×10^9 years. According to the natural decay scheme of Figure 25-2, the basic change that occurs as atoms of $^{238}_{92}U$ and its daughters pass through the entire sequence of steps is

$$^{238}_{92}U \longrightarrow {}^{206}_{82}Pb + 8\,{}^{4}_{2}He + 6\,{}^{0}_{-1}\beta$$

The decay sequence for $^{238}U \longrightarrow {}^{206}Pb$ has 14 steps. The first step, however, has a much longer half-life than any of the other steps in the series and can thus be thought of as the rate-determining step, with the subsequent steps being "fast." As discussed in Chapter 14, we can ignore the effect on the overall rate of fast steps that occur after the slow, rate-determining step. Thus, the half-life for ^{238}U is essentially equal to the time it takes to convert half the initial ^{238}U to the ^{206}Pb isotope. Discounting the mass associated with the β^- particles, for every 238 g of uranium that undergoes complete decay, 206 g of lead and 32 g of helium are produced.

Suppose that in a hypothetical rock containing no lead initially, 1.000 g of $^{238}_{92}U$ had disintegrated through one half-life, 4.51×10^9 years. At the end of that time, 0.500 g $^{238}_{92}U$ would have disintegrated and another 0.500 g would remain. The quantity of $^{206}_{82}Pb$ now present in the rock would be

$$0.500 \text{ g } {}^{238}_{92}U \times \frac{206 \text{ g } {}^{206}_{82}Pb}{238 \text{ g } {}^{238}_{92}U} = 0.433 \text{ g } {}^{206}_{82}Pb$$

The ratio of lead-206 to uranium-238 in the rock would be

$$^{206}_{82}Pb/^{238}_{92}U = 0.433/0.500 = 0.866$$

If the $^{206}_{82}Pb/^{238}_{92}U$ mass ratio is less than 0.866, the age of the rock is less than one half-life of $^{238}_{92}U$. A higher ratio indicates a greater age for the rock. The best estimates of the age of the oldest rocks, and presumably of Earth itself, are about 4.5×10^9 years. These estimates are based on the $^{206}_{82}\text{Pb}/^{238}_{92}\text{U}$ ratio and on ratios for other pairs of isotopes from natural radioactive decay series.

Modern Radioactive Dating (Geochronology)

Modern radioactive dating techniques use mass spectrometry to analyze parent or daughter nuclides, or both. A mass spectrometer (described in Chapter 2) can be used to determine the amount of the various isotopes found in geological material (e.g., rocks), and that data can be used to calculate the age of the material. One particular pair of isotopes currently used in dating geological material is potassium-40 and argon-40. Potassium-40 is radioactive and mainly undergoes β-decay to calcium-40; however, some potassium-40 undergoes electron capture and decays to argon-40. Attempts have been made to perform radiodating for the potassium-40–calcium-40 pair, but this is not currently possible because of the large natural abundance of calcium.

Let us look at how isotope pairs can be used to determine the age of geological material. We begin by rewriting equation (25.12) as

$$N = N_0 e^{-\lambda t} \qquad (25.14)$$

where N is the number of atoms currently present, N_0 is the initial number of atoms at $t = 0$, and λ is the decay constant for ^{40}K. Since we don't know how many atoms were present at $t = 0$, we can determine that by recognizing that

$$N_0 = N + D \qquad (25.15)$$

where D is the number of daughter atoms, that is, the atoms resulting from the decay of an isotope. An example of a daughter atom is ^{40}Ar, which is a daughter atom of ^{40}K. Combining equations (25.14) and (25.15) and solving for t we have

$$t = \frac{1}{\lambda} \ln\left(\frac{D}{N} + 1\right) \qquad (25.16)$$

Equation (25.16) neglects the presence of daughter atoms, D_0, that may have been present at $t = 0$. To account for this we rewrite equation (25.16) as

$$t = \frac{1}{\lambda} \ln\left(\frac{D_t - D_0}{N} + 1\right) \qquad (25.17)$$

where D_t are the total daughter atoms at the current time. Equation (25.17) cannot be used in its current form to accurately determine the age of geological material. To accurately determine the geological age, we must consider the branching decay. Branching decay occurs when an isotope can decay through different pathways to other nuclides. The figure below illustrates branching for potassium-40. When branching is taken into account, we obtain

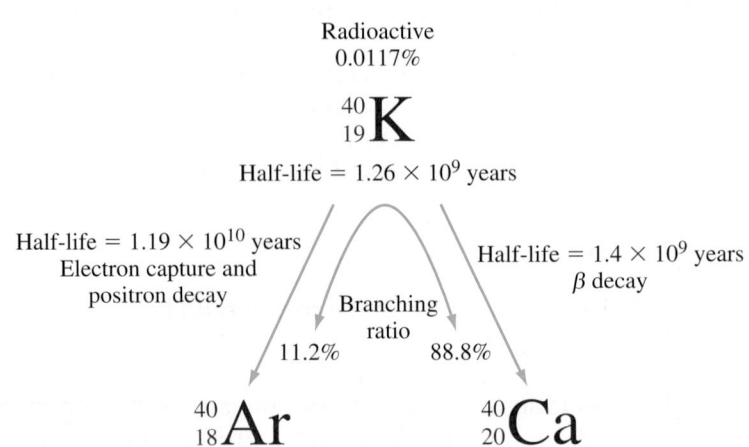

$$t = \frac{1}{\lambda_\varepsilon + \lambda_\beta} \ln\left[\frac{^{40}\text{Ar}_{\text{rad}}}{^{40}\text{K}}\left(\frac{\lambda_\varepsilon + \lambda_\beta}{\lambda_\varepsilon} + 1\right)\right] \qquad \textbf{(25.18)}$$

where λ_ε is the decay constant for the electron capture branch of ^{40}K decay and λ_β is the decay constant for the β-decay branch. $^{40}\text{Ar}_{\text{rad}}$ is the argon-40 produced by the decay of potassium-40 and corrected for the presence of trapped atmospheric argon-40. Several assumptions are made in deriving equation (25.18). The assumption most likely to cause problems is what the source of the argon is in the sample. The argon may be from the potassium-40 decay, it may be atmospheric argon, or it may be a mixture in which it is possible to identify and subtract any atmospheric argon-40 that contaminates the sample.

25-6   Energetics of Nuclear Reactions

The energy change accompanying a nuclear reaction can be described by using the mass–energy equivalence derived by Albert Einstein:

$$E = mc^2 \qquad \textbf{(25.19)}$$

KEEP IN MIND

that the mass in equation (25.19) is the rest mass of the particle, as discussed on page 305.

An energy change in a process is always accompanied by a mass change, and the constant that relates them is the square of the speed of light. In chemical reactions, energy changes are so small that the equivalent mass changes are undetectable (though real nevertheless). In fact, we base the balancing of equations and stoichiometric calculations on the principle that mass is conserved (unchanged) in a chemical reaction. In nuclear reactions, energies are orders of magnitude greater than in chemical reactions. Perceptible changes in mass do occur.

If the exact masses of atoms are known, the energy of a nuclear reaction can be calculated with equation (25.19). The term m is the net change in mass, in kilograms, and c is expressed in meters per second. The resulting energy is in joules. Another common unit for expressing nuclear energy is the megaelectronvolt (MeV). (Recall from page 1113 that $1\text{ eV} = 1.6022 \times 10^{-19}\text{ J}$.)

$$1\text{ MeV} = 1.6022 \times 10^{-13}\text{ J} \qquad \textbf{(25.20)}$$

Equation (25.20) is a conversion factor between megaelectronvolts and joules. A conversion factor between atomic mass units (u) and joules (J) is also helpful. This relationship can be established by determining the energy associated with a mass of 1 u. This calculation is based on carbon-12, and note that 1 u is exactly $\frac{1}{12}$ of the mass of a carbon-12 atom. The mass, in grams, corresponding to 1 u can be calculated as follows:

$$1\text{ u} \times \frac{1\ ^{12}\text{C atom}}{12\text{ u}} \times \frac{1\text{ mol }^{12}\text{C}}{6.0221 \times 10^{23}\text{ atoms }^{12}\text{C}} \times \frac{12\text{ g}}{1\text{ mol }^{12}\text{C}} = 1.6606 \times 10^{-24}\text{ g}$$

Converting this value of m to kilograms and using it in equation (25.19) gives

$$E = 1\text{ u} \times \frac{1.6606 \times 10^{-24}\text{ g}}{\text{u}} \times \frac{1\text{ kg}}{1000\text{ g}} \times (2.9979 \times 10^8)^2\frac{\text{m}^2}{\text{s}^2}$$

$$= 1.4924 \times 10^{-10}\text{ J}$$

Thus, the energy equivalent of 1 u is

$$1\text{ atomic mass unit (u)} = 1.4924 \times 10^{-10}\text{ J} \qquad \textbf{(25.21)}$$

Finally, to express this energy in MeV,

$$1\text{ atomic mass unit (u)} = 1.4924 \times 10^{-10}\text{ J} \times \frac{1\text{ MeV}}{1.6022 \times 10^{-13}\text{ J}} = 931.5\text{ MeV} \qquad \textbf{(25.22)}$$

These conversion factors are used in Example 25-5, together with the principle that the total mass–energy of the products of a nuclear reaction is equal to the total mass–energy of the reactants. The masses required in calculations based on nuclear reactions are *nuclear* masses. The relationship of a nuclear mass to a nuclidic (atomic) mass is

$$\text{nuclear mass} = \text{nuclidic (atomic) mass} - \text{mass of extranuclear electrons}$$

EXAMPLE 25-5 **Calculating the Energy of a Nuclear Reaction with the Mass–Energy Relationship**

What is the energy, in joules and in megaelectronvolts, associated with the α decay of ^{238}U?

$$^{238}_{92}\text{U} \longrightarrow {}^{234}_{90}\text{Th} + {}^{4}_{2}\text{He}$$

The nuclidic (atomic) masses in atomic mass units (u) are from Table D.5 in Appendix D:

$$^{238}_{92}\text{U} = 238.0508 \text{ u} \qquad ^{234}_{90}\text{Th} = 234.0437 \text{ u} \qquad ^{4}_{2}\text{He} = 4.0026 \text{ u}$$

Analyze

The key concept here is the fact that, during a nuclear reaction, a loss or gain of mass is balanced by a gain or loss of energy. We need to determine the loss or gain of mass and then use the conversion factors (25.21) and (25.22) to convert the loss or gain of mass to the corresponding amount of energy.

Solve

The net change in mass that accompanies the decay of a single nucleus of ^{238}U is shown below. Note that the masses of the extranuclear electrons do not enter into the calculation of the net change in mass.

change in mass
$$= \text{nuclear mass of } ^{234}_{90}\text{Th} + \text{nuclear mass of } ^{4}_{2}\text{He} - \text{nuclear mass of } ^{238}_{92}\text{U}$$
$$= [234.0437 \text{ u} - (90 \times \text{mass e}^-)] + [4.0026 \text{ u} - (2 \times \text{mass e}^-)] - [238.0508 \text{ u} - (92 \times \text{mass e}^-)]$$
$$= 234.0437 \text{ u} + 4.0026 \text{ u} - 238.0508 \text{ u} - 92 \times \text{mass e}^- + 92 \times \text{mass e}^-$$
$$= -0.0045 \text{ u}$$

We can use this loss of mass and conversion factors (25.21) and (25.22) to write

$$E = -0.0045 \text{ u} \times \frac{1.49 \times 10^{-10} \text{ J}}{\text{u}} = -6.7 \times 10^{-13} \text{ J}$$

or

$$E = -0.0045 \text{ u} \times \left(\frac{931.5 \text{ MeV}}{\text{u}}\right) = -4.2 \text{ MeV}$$

Assess

The negative sign denotes that energy is lost in the nuclear reaction. This is the kinetic energy of the departing α particle. Note that you can use either nuclear or nuclidic (atomic) masses in calculations based on a nuclear equation. The change in mass will be the same in either case because the masses of the electrons will cancel out.

PRACTICE EXAMPLE A: What is the energy associated with the α decay of ^{146}Sm (145.913053 u) to ^{142}Nd (141.907719 u)? Use 4.002603 u as the mass of ^{4}He.

PRACTICE EXAMPLE B: The decay of ^{222}Rn by α-particle emission is accompanied by a loss of 5.590 MeV of energy. What quantity of mass, in atomic mass units (u), is converted to energy in this process?

Figure 25-5 suggests formation of the nucleus of a $^{4}_{2}$He atom from two protons and two neutrons. In this process, there is a *mass defect* of 0.0305 u. That is, the experimentally determined mass of a $^{4}_{2}$He nucleus is 0.0305 u *less* than the combined mass of two protons and two neutrons. This "lost" mass is liberated as energy. With expression (25.22), we can show that 0.0305 u of mass is equivalent to an energy of 28.4 MeV. Because this is the energy released in forming a $^{4}_{2}$He nucleus, it is called the **nuclear binding energy**. Viewed another way, a $^{4}_{2}$He nucleus would have to absorb 28.4 MeV to cause its protons and neutrons

▶ Nucleons are nuclear particles: protons and neutrons.

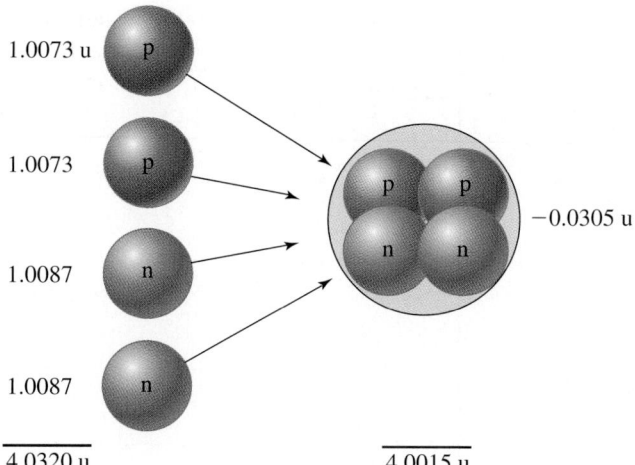

▲ FIGURE 25-5
Nuclear binding energy in $^4_2$He
The mass of a helium nucleus ($^4_2$He) is 0.0305 u (atomic mass unit) less than the combined masses of two protons and two neutrons. The energy equivalent to this loss of mass (called the mass defect) is the nuclear energy that binds the nuclear particles together.

◀ Only four forces are known: gravity, electromagnetic forces, weak nuclear force, and strong nuclear force. The nuclear forces act only within a nucleus, whereas gravity and EM can operate over essentially infinite distances.

to become separated. If the binding energy is considered to be apportioned equally among the two protons and two neutrons in $^4_2$He, the binding energy is 7.10 MeV per nucleon.

Let us calculate the binding energy per nucleon on the following elements: carbon-12, iron-56, and uranium-235. For carbon-12 we first calculate the change in mass:

$$\text{change in mass} = [(6 \times \text{mass } ^1\text{H atom}) + (6 \times \text{mass neutron})] - \text{mass } ^{12}\text{C atom}$$
$$= [6 \times 1.007825 \text{ u} + 6 \times 1.008665 \text{ u}] - 12.0000 \text{ u}$$
$$= 0.09894 \text{ u}$$

The mass for the proton and neutron were obtained from Appendix D. The next step is to convert this to the binding energy per nucleon:

$$\text{binding energy per nucleon} = \frac{0.09894 \text{ u} \times \dfrac{931.5 \text{ MeV}}{\text{u}}}{12 \text{ nucleons}} = 7.680 \text{ MeV/nucleon}$$

For iron-56 we find the binding energy per nucleon is

$$\text{binding energy per nucleon} = \frac{0.52846 \text{ u} \times \dfrac{931.5 \text{ MeV}}{\text{u}}}{56 \text{ nucleons}} = 8.790 \text{ MeV/nucleon}$$

and for uranium-235

$$\text{binding energy per nucleon} = \frac{1.921507 \text{ u} \times \dfrac{931.5 \text{ MeV}}{\text{u}}}{235 \text{ nucleons}} = 7.591 \text{ MeV/nucleon}$$

At this point we notice that iron-56 has the larger binding energy per nucleon, meaning that more energy is required to separate the iron-56 nucleons than those of carbon-12 and uranium-235. In other words, iron-56 is more stable than the other two elements.

Figure 25-6 indicates that the maximum binding energy per nucleon is found in a nucleus with a mass number of approximately 60. This finding leads to two interesting conclusions: (1) If small nuclei are combined into a heavier one (up to about $A = 60$), the binding energy per nucleon increases and a

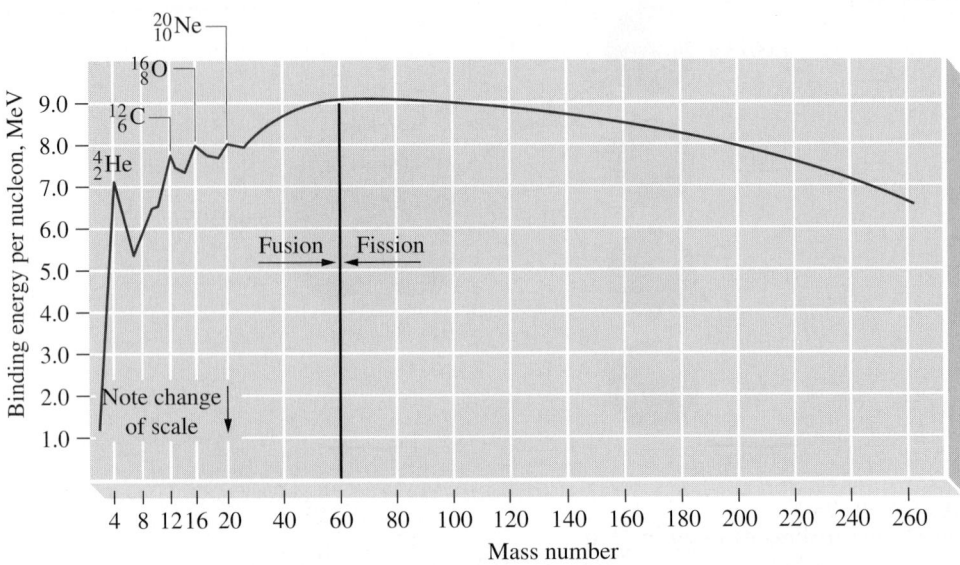

▲ FIGURE 25-6
Average binding energy per nucleon as a function of mass number

certain quantity of mass must be converted to energy. The nuclear reaction is highly exothermic. This *fusion* process serves as the basis of the hydrogen bomb. (2) For nuclei with mass numbers above 60, the addition of extra nucleons to the nucleus would require the expenditure of energy (since the binding energy per nucleon decreases). However, the *disintegration* of heavier nuclei into lighter ones is accompanied by the release of energy. This nuclear *fission* process serves as the basis of the atomic bomb and conventional nuclear power reactors. Nuclear fission and fusion will be discussed after we examine Figure 25-6 and consider the question of nuclear stability.

25-7 Nuclear Stability

A number of basic questions have probably occurred to you as nuclear decay processes have been described: Why do some radioactive nuclei decay by α emission, some by β^- emission, and so on? Why do the lighter elements have so few naturally occurring radioactive nuclides, whereas all those of the heavier elements seem to be radioactive?

Our first clue to answers for such questions comes from Figure 25-6, in which several nuclides are specifically noted. These nuclides have higher binding energies per nucleon than those of their neighbors. Their nuclei are especially stable. This observation is consistent with a theory of nuclear structure known as the *shell theory*. In the formation of a nucleus, protons and neutrons are believed to occupy a series of nuclear shells. This process is analogous to building up the electronic structure of an atom by the successive addition of electrons to electronic shells. Just as the aufbau process periodically produces electron configurations of exceptional stability, so do certain nuclei acquire a special stability as nuclear shells are closed. This condition of special stability of an atomic nucleus occurs for certain numbers of protons or neutrons known as **magic numbers** (Table 25.2).

Another observation concerning nuclei is that among stable nuclei, the number of protons and the number of neutrons is most commonly *even*. There are fewer stable nuclei with odd numbers of protons and of neutrons. The relationship between numbers of protons (Z), numbers of neutrons (N), and the stability of isotopes is summarized in Table 25.3. Note particularly that stable atoms with the combination Z odd–N odd are very rare. This combination is found only in

TABLE 25.2 Magic Numbers for Nuclear Stability

Number of Protons	Number of Neutrons
2	2
8	8
20	20
28	28
50	50
82	82
114	126
	184

TABLE 25.3 Distribution of Naturally Occurring Stable Nuclides

Combination	Number of Nuclides
Z even–N even	163
Z even–N odd	55
Z odd–N even	50
Z odd–N odd	4

the nuclides ^2_1H, ^6_3Li, $^{10}_5B$, and $^{14}_7N$. Still another observation is that elements of *odd* atomic number generally have only one or two stable isotopes, whereas those of *even* atomic number have several. Thus, F ($Z = 9$) and I ($Z = 53$) each have only one stable nuclide, and Cl ($Z = 17$) and Cu ($Z = 29$) each have two. On the other hand, O ($Z = 8$) has three and Ca ($Z = 20$) has six.

Neutrons are thought to provide a nuclear force to bind protons and neutrons together into a stable unit. Without neutrons, the electrostatic forces of repulsion between positively charged protons would cause the nucleus to fly apart. For the elements of lower atomic numbers (up to about $Z = 20$), the required number of neutrons for a stable nucleus is about equal to the number of protons, for example, ^4_2He, $^{12}_6C$, $^{16}_8O$, $^{28}_{14}Si$, $^{40}_{20}Ca$. For higher atomic numbers, because of increasing repulsive forces between protons, larger numbers of neutrons are required and the neutron–proton (N/Z) ratio increases. For bismuth, the ratio is about $1.5:1$. Above atomic number 83, no matter how many neutrons are present, the nucleus is unstable. Thus, all isotopes of the known elements with $Z > 83$ are radioactive. Figure 25-7 indicates roughly the range of N/Z ratios as a function of atomic number for stable atoms.

Using the ideas outlined here, nuclear scientists have predicted the possible existence of atoms of high atomic number that should have very long half-lives, a belt of stability in Figure 25-7. After a search of many years, such atoms have been created. In 1999, the bombardment of a plutonium-242 target with calcium-48 ions produced the as-yet unnamed isotopes $^{287}114$ and $^{289}114$, with half-lives of 5 s and 30 s, respectively. Although these half-lives may appear short, they are practically an eternity compared with the half-lives of many other superheavy atoms, which are in the microsecond range.

25-4 CONCEPT ASSESSMENT

What proton number is found in a greater number of isotopes than any other proton number? What is the corresponding situation for neutron numbers?

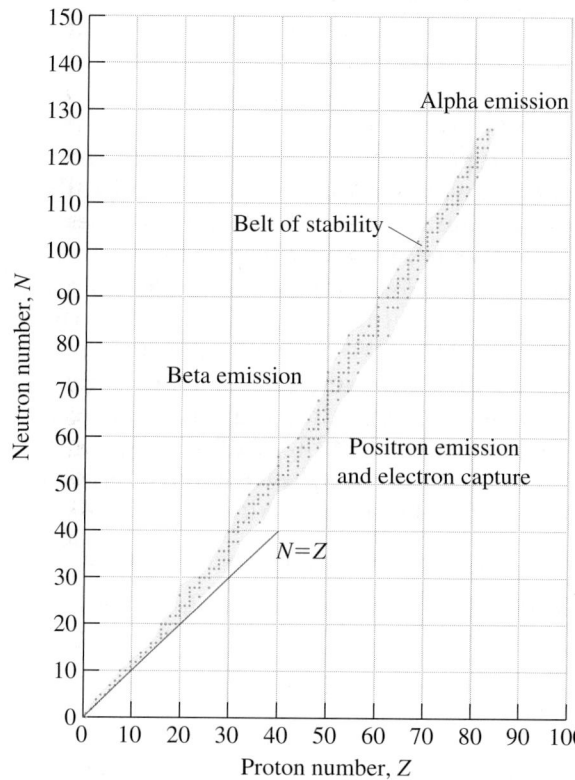

◀ FIGURE 25-7
Neutron-to-proton ratio and the stable nuclides up to $Z = 83$
The points within the belt of stability identify the stable nuclides. Some radioactive nuclides are also in this belt, but most lie outside, and the mode of their radioactive decay is indicated. The stable nuclides of low atomic numbers lie on or near the line $N = Z$; they have a neutron-to-proton ratio of one, or nearly so. At higher atomic numbers, the neutron-to-proton ratios increase to about 1.5.

EXAMPLE 25-6   Predicting Which Nuclei Are Radioactive

Which of the following nuclides would you expect to be stable, and which radioactive? **(a)** ^{82}As; **(b)** ^{118}Sn; **(c)** ^{214}Po.

Analyze

This question requires us to apply the rules of nuclear stability. We may also use Figure 25-7 to determine whether the nuclide lies within the belt of stability.

Solve

(a) Arsenic-82 has $Z = 33$ and $N = 49$. This is an odd–odd combination that is found in only four of the lighter elements. ^{82}As is radioactive. (Note also that this nuclide is outside the belt of stability in Figure 25-7.)

(b) Tin has an atomic number of 50—a magic number. The neutron number is 68 in the nuclide ^{118}Sn. This is an even–even combination, and we should expect the nucleus to be stable. Moreover, Figure 25-7 shows that this nuclide is within the belt of stability. ^{118}Sn is a stable nuclide.

(c) ^{214}Po has an atomic number of 84. All known atoms with $Z > 83$ are radioactive. ^{214}Po is radioactive.

Assess

When considering whether a nuclide is radioactive, we must remember that all elements with $Z > 83$ are radioactive, and those with $Z < 84$ are radioactive if their Z and N form odd–odd combinations. A few exceptions exist to these rules, specifically when nuclides have atomic numbers that are one of the magic numbers.

PRACTICE EXAMPLE A: Which of the following nuclides would you expect to be stable, and which radioactive? **(a)** ^{88}Sr; **(b)** ^{118}Cs; **(c)** ^{30}S.

PRACTICE EXAMPLE B: Write plausible nuclear equations to represent the radioactive decay of the fluorine isotopes ^{17}F and ^{22}F.

🔍 25-5   CONCEPT ASSESSMENT

Of the radioactive nuclides among the following, which one is most likely to decay by β^- emission, and which one by β^+ emission: ^{44}Ca, ^{57}Cu, ^{100}Zr, ^{235}U? Explain.

25-8   Nuclear Fission

▶ Ida Tacke Nodack, codiscoverer of the element rhenium, was the first person to suggest that Fermi's proposed experiments had produced fission. Her explanation was not generally accepted, however, until several years later.

In 1934, Enrico Fermi proposed that transuranium elements might be produced by bombarding uranium with neutrons. He reasoned that the successive loss of β^- particles would cause the atomic number to increase, perhaps to as high as 96. When such experiments were carried out, it was found that, in fact, the product did emit β^- particles. But in 1938, Otto Hahn, Lise Meitner, and Fritz Strassman found by chemical analysis that the products did not correspond to elements with $Z > 92$. Neither were they the neighboring elements of uranium—Ra, Ac, Th, and Pa. Instead, the products were radioisotopes of much lighter elements, such as Sr and Ba. Neutron bombardment of uranium nuclei causes certain of them to undergo **fission** into smaller fragments, as suggested by Figure 25-8.

The energy equivalent of the mass destroyed in a fission event is somewhat variable, but the average energy is approximately 3.20×10^{-11} J (200 MeV).

$$^{235}_{92}\text{U} + ^{1}_{0}\text{n} \longrightarrow ^{236}_{92}\text{U} \longrightarrow \text{fission fragments} + \text{neutrons} + 3.20 \times 10^{-11}\text{ J}$$

An energy of 3.20×10^{-11} J may seem small, but this energy is for the fission of a *single* $^{235}_{92}\text{U}$ nucleus. What if 1.00 g $^{235}_{92}\text{U}$ were to undergo fission?

$$? \text{kJ} = 1.00 \text{ g} \ ^{235}\text{U} \times \frac{1 \text{ mol} \ ^{235}\text{U}}{235 \text{ g} \ ^{235}\text{U}} \times \frac{6.022 \times 10^{23} \text{ atoms} \ ^{235}\text{U}}{1 \text{ mol} \ ^{235}\text{U}} \times \frac{3.20 \times 10^{-11}\text{ J}}{1 \text{ atoms} \ ^{235}\text{U}}$$

$$= 8.20 \times 10^{10} \text{ J} = 8.20 \times 10^{7} \text{ kJ}$$

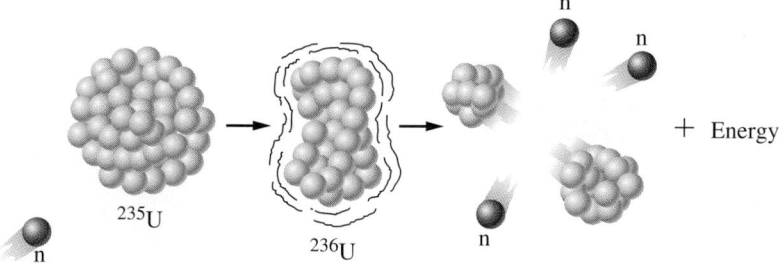

▲ FIGURE 25-8
Nuclear fission of $^{235}_{92}$U with thermal neutrons
A neutron possessing ordinary thermal energy strikes a $^{235}_{92}$U nucleus. First, the unstable nucleus $^{236}_{92}$U is produced; this then breaks up into a light fragment, a heavy fragment, and several neutrons. Various nuclear fragments are possible, but the most probable mass numbers are 97 for the light fragment and 137 for the heavy one.

(a)

This is an enormous quantity of energy! To release that quantity of energy would require the complete combustion of nearly three tons of coal.

Nuclear Reactors

In the fission of $^{235}_{92}$U, on average, 2.5 neutrons are released per fission event. These neutrons, on average, produce two or more fission events. The neutrons produced by the second round of fission produce another four or five events, and so on. The result is a *chain reaction*. If the reaction is uncontrolled, the released energy causes an explosion; this is the basis of the atomic bomb. Fission leading to an uncontrolled explosion occurs only if the quantity of ^{235}U exceeds the critical mass. The *critical mass* is the quantity of ^{235}U sufficiently large to retain enough neutrons to sustain a chain reaction. Quantities smaller than this are subcritical; neutrons escape at too great a rate to produce a chain reaction.

In a nuclear reactor, the release of fission energy is controlled. One common design, called the *pressurized water reactor* (PWR), is pictured in Figure 25-9. In the core of the reactor, rods of uranium-rich fuel are suspended in water maintained under a pressure of 70 to 150 atm. The water serves a dual purpose. First, it slows down the neutrons from the fission process so that they possess only normal thermal energy. These thermal neutrons are better able to induce fission than highly energetic ones. In this capacity, the water acts as a **moderator**. Water

(b)

◀ **The core of a nuclear reactor**
The characteristic blue glow in the water surrounding the core is called *Cerenkov radiation*. It results when charged particles pass through a transparent medium faster than does light in the same medium. The charged particles are produced by nuclear fission. The radiation is analogous to the shock wave produced in a sonic boom.

(c)

▲ **(a)** The Three Mile Island nuclear reactor, site of a small nuclear accident in 1979. **(b)** The Chernobyl nuclear reactor, site of a major nuclear accident in 1986. **(c)** A nuclear power plant in France that has operated for decades without mishap.

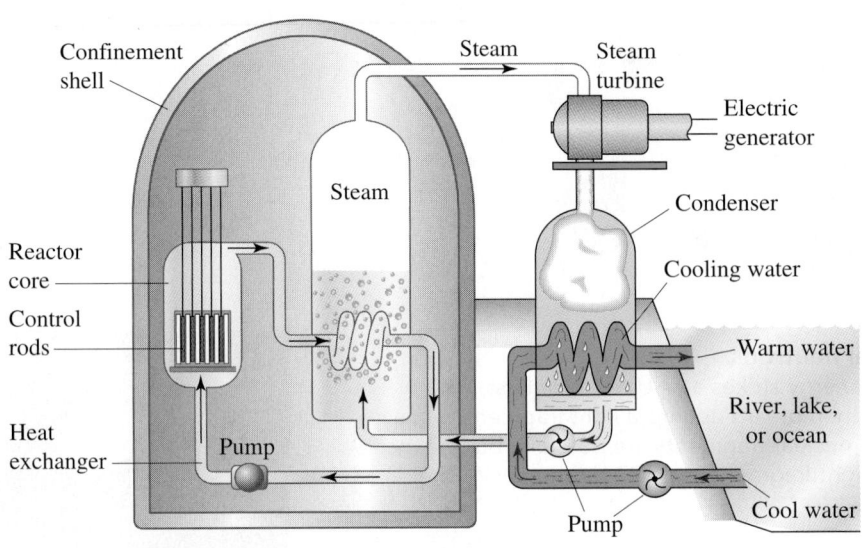

▲ FIGURE 25-9
Pressurized water nuclear reactor
(a) Schematic of a nuclear power plant. **(b)** A nuclear power plant in New York State.

also functions as a heat transfer medium. Fission energy maintains the water at a high temperature (about 300 °C). The high-temperature water is brought in contact with colder water in a heat exchanger. The colder water is converted to steam, which drives a turbine, which in turn drives an electric generator. A final component of the nuclear reactor is a set of **control rods**, usually cadmium

25-2   ARE YOU WONDERING...

About nuclear reactor safety?

There have been two notorious accidents at nuclear power stations. The first occurred at the Three Mile Island (TMI) generating plant near Middleton, Pennsylvania, in 1979. The TMI reactor is of the light-water type, in which water is used as the moderator and coolant. In this accident, some coolant (also the moderator) was lost; the chain reaction in the reactor stopped because there were too few slow neutrons. However, radioactive decay of the fission fragments continued, causing the fuel rods to get very hot. A partial meltdown resulted and, in turn, caused a fracture in one of the reactors. The fracture permitted the venting of a small amount of radioactive steam into the atmosphere. The reactor is now sealed, but electronic robots have discovered substantial damage to the fuel rods.

The second incident occurred at Chernobyl, in the Ukraine, in 1986. Graphite was used as the moderator there. When the coolant was turned off because of human error, the chain reaction went out of control. A tremendous rise in temperature followed, leading to a meltdown. During the meltdown, the graphite moderator surrounding the rods burned and radioactive smoke spewed out of the reactor. Radioactive materials were dispersed over much of Europe, Canada, and the United States. Although only a few dozen people were killed in the Chernobyl accident, many more will eventually die of cancer because of the related radiation. The type of accident observed at Chernobyl cannot happen in a light-water reactor, in which the coolant is the moderator.

Other nations have used nuclear power without experiencing accidents. In France and Japan, two-thirds of the electrical energy is produced by nuclear power stations. The safety record of nuclear reactors is particularly high, but the need for the storage of the nuclear waste produced is a vexing problem. Ways of dealing with these problems are discussed in the Focus On feature for Chapter 25 on the Mastering Chemistry website.

metal, whose function is to absorb neutrons. When the rods are lowered into the reactor, the fission process is slowed down. When the rods are raised, the density of neutrons and the rate of fission increase.

Breeder Reactors

All that is required to initiate the fission of $^{235}_{92}U$ are neutrons of ordinary thermal energies. Conversely, nuclei of $^{238}_{92}U$, the abundant nuclide of uranium (99.28%), undergo the following reactions only when struck by energetic neutrons.

$$^{238}_{92}U + ^{1}_{0}n \longrightarrow {}^{239}_{92}U$$

$$^{239}_{92}U \longrightarrow {}^{239}_{93}Np + {}^{0}_{-1}\beta$$

$$^{239}_{93}Np \longrightarrow {}^{239}_{94}Pu + {}^{0}_{-1}\beta$$

A fissionable nuclide such as $^{235}_{92}U$ is called *fissile*; $^{239}_{94}Pu$ is also fissile. A nuclide such as $^{238}_{92}U$, which can be converted into a fissile nuclide, is said to be *fertile*. In a *breeder nuclear reactor*, a small quantity of fissile nuclide provides the neutrons that convert a large quantity of a fertile nuclide into a fissile one. (The newly formed fissile nuclide then participates in a self-sustaining chain reaction.)

An obvious advantage of the breeder reactor is that the amount of uranium fuel available immediately jumps by a factor of about 100. This is the ratio of naturally occurring $^{238}_{92}U$ to $^{235}_{92}U$. But the potential advantage is even greater than this. Breeder reactors might use as nuclear fuels materials that have even very low uranium contents, such as shale deposits with about 0.006% U by mass.

There are, however, important disadvantages to breeder reactors. This is especially true of the type known as the *liquid-metal-cooled fast breeder reactor* (LMFBR). Systems must be designed to handle a liquid metal, such as sodium, which becomes highly radioactive in the reactor. Also, both the rates of heat and neutron production are greater in the LMFBR than in the PWR, so materials deteriorate more rapidly. Perhaps the greatest unsolved problems are those of handling radioactive wastes and reprocessing plutonium fuel. Plutonium is one of the most toxic substances known. It can cause lung cancer when inhaled in even microgram (10^{-6} g) amounts. Furthermore, because plutonium has a long half-life (24,000 y), any accident involving it could leave an affected area almost permanently contaminated.

25-9   Nuclear Fusion

The **fusion** of atomic nuclei is the process that produces energy in the sun. An uncontrolled fusion reaction is the basis of the hydrogen bomb. A controlled fusion reaction could provide an almost unlimited source of energy. The nuclear reaction that holds the most immediate promise is the deuterium–tritium reaction.

$$^{2}_{1}H + ^{3}_{1}H \longrightarrow {}^{4}_{2}He + {}^{1}_{0}n + energy$$

The difficulties in developing a fusion energy source are probably without parallel in the history of technology. In fact, the feasibility of a controlled fusion reaction has yet to be fully demonstrated. There are a number of problems. To permit their fusion, the nuclei of deuterium and tritium must be forced into close proximity. Because atomic nuclei repel one another, this close approach requires the nuclei to have very high thermal energies. At the temperatures necessary to initiate a fusion reaction, gases are completely ionized into a mixture of atomic nuclei and electrons known as a *plasma*. Still higher plasma temperatures—more than 40,000,000 K—are required to initiate a self-sustaining reaction (one that releases more energy than is required to get it started). A method must be devised to confine the plasma out of contact with other materials. The plasma loses thermal energy to any material it strikes. Also, a plasma must be at a sufficiently high density for a sufficient time to

◄ In the hydrogen bomb, these high temperatures are attained by exploding an atomic (fission) bomb, which triggers the fusion reaction.

KEEP IN MIND

that *both* nuclear fusion and nuclear fission reactions release energy.

permit the fusion reaction to occur. The two methods receiving greatest attention are confinement in a magnetic field and heating of a frozen deuterium-tritium pellet with laser beams. Another series of technical problems involves the handling of liquid lithium, which is the anticipated heat-transfer medium and tritium (^3_1H) source.

$$^7_3Li + ^1_0n \longrightarrow ^4_2He + ^3_1H + ^1_0n$$
(fast)   (slow)

Finally, for the magnetic containment method, the magnetic field must be produced by superconducting magnets, which currently are very expensive to operate.

A fusion reactor is currently being developed in France by an international consortium consisting of the European Union (EU), India, Japan, People's Republic of China, Russia, South Korea, the United States, and Portugal. The project, titled ITER (International Thermonuclear Experimental Reactor), is based on the tokamak design shown in the photo in the margin. The goal of the project is to generate and sustain 500 MW of fusion power for 1000 seconds through the fusion of 0.5 g of a deuterium–tritium mixture. ITER is expected to generate 5 to 10 times as much energy as is needed to initiate the reaction at its fusion temperature. The first operation of the fusion reactor should take place in 2018.

The advantages of fusion over fission could be enormous. Since deuterium constitutes about one in every 6500 H atoms, the oceans of the world can supply an almost limitless amount of nuclear fuel. It is estimated that there is sufficient lithium on Earth to provide a source of tritium for about 1 million years. Also, nuclear fusion would not pose the vexing problems of radioactive waste storage and disposal associated with nuclear fission.

▲ The plasma chamber of a fusion reactor of the magnetic confinement type (called a *tokamak*). The chamber walls are lined with carbon-fiber composite tiles to protect against the high-temperature plasma.

25-10 Effect of Radiation on Matter

Although there are substantial differences in the way in which α particles, β^- and β^+ particles, and γ rays interact with matter, they share an important feature: They dislodge electrons from atoms and molecules to produce ions. The ionizing power of radiation may be described in terms of the number of ion pairs formed per centimeter of path through a material. An *ion pair* consists of an ionized electron and the resulting positive ion. Alpha particles have the greatest ionizing power, followed by β particles and then γ rays. The ionized electrons produced directly by the collisions of particles of radiation with atoms are called *primary* electrons. These electrons may themselves possess sufficient energies to cause *secondary* ionizations.

Not all interactions between radiation and matter produce ion pairs. In some cases, electrons are simply raised to higher atomic or molecular energy levels. The return of these electrons to their normal states is then accompanied by radiation—X rays, ultraviolet light, or visible light, depending on the energies involved.

Some of the possibilities described here are depicted in Figure 25-10.

Radiation Detectors

The interactions of radiation with matter can serve as bases for the detection of radiation and the measurement of its intensity. One of the simplest methods is that used by Henri Becquerel in his discovery of radioactivity—the exposure of a photographic film, as in film badge radiation detectors. The effect of α and β particles, and γ rays on a photographic emulsion is similar to that of X rays.

One type of detector used to study high-energy radiation such as γ rays is the *bubble chamber*. In this device, a liquid, usually hydrogen, is kept just at its boiling point. As ion pairs are produced by the transit of an ionizing ray,

▲ A film badge used for detecting radiation.

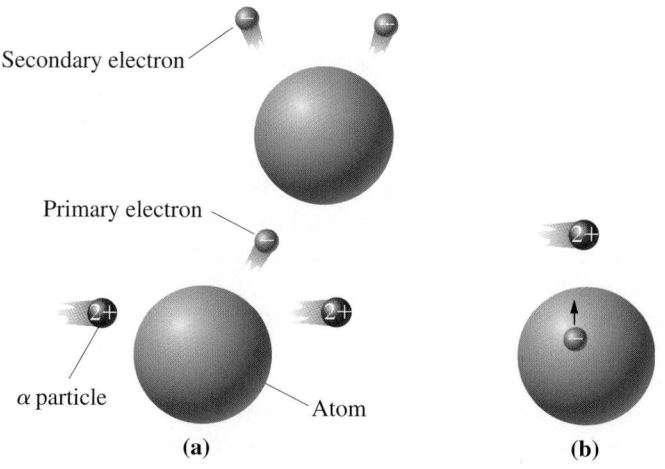

▲ FIGURE 25-10
Some interactions of radiation with matter
(a) The production of primary and secondary electrons by collisions. (b) The excitation of an atom by the passage of an α particle. An electron is raised to a higher energy level within the atom. The excited atom reverts to its normal state by emitting radiation.

bubbles of vapor form around the ions. The tracks of bubbles can be photographed and analyzed, and different types of radiation produce different tracks. Charged particles, for example, can be detected by their deflection in a magnetic field.

The most common device for detecting and measuring ionizing radiation is the *Geiger–Müller (G–M) counter*, depicted in Figure 25-11. The G–M counter consists of a cylindrical cathode with a wire anode running along its axis. The anode and cathode are sealed in a gas-filled glass tube. Ionizing radiation passing through the tube produces primary ions, and this is followed by secondary ionization. The positive ions are attracted to the cathode and the electrons to the anode, leading to a pulse of electric current. The tube is quickly recharged in preparation of the next ionizing event. The pulses of electric current are counted.

A detector widely used in biological studies is the *scintillation counter*. It is especially useful in detecting radiation that is not energetic enough to cause ionization. The radiation excites certain atoms in the detecting medium; when these atoms revert to their ground state, they emit pulses of light that can be counted. The light emission is similar to that produced when phosphors in a television screen are struck by cathode rays.

Effect of Ionizing Radiation on Living Matter

All life exists against a background of naturally occurring ionizing radiation—cosmic rays, ultraviolet light, and emanations from radioactive elements, such as uranium in rocks. The level of this radiation varies from point to point on Earth, being greater, for instance, at higher elevations. Only in recent times have humans been able to create situations in which organisms might be exposed to radiation at levels significantly higher than natural background radiation.

The interactions of radiation with living matter are the same as with other forms of matter—ionization, excitation, and dissociation of molecules. There is no question of the effect of large doses of ionizing radiation on organisms—the organisms are killed. But even slight exposures to ionizing radiation can cause changes in cell chromosomes. Thus it is believed that, even at low dosage rates, ionizing radiation can result in birth defects, leukemia, bone cancer, and other forms of cancer. The nagging question that has eluded any definitive answers is how great an increase in the incidence of birth defects and cancers might be caused by certain levels of radiation.

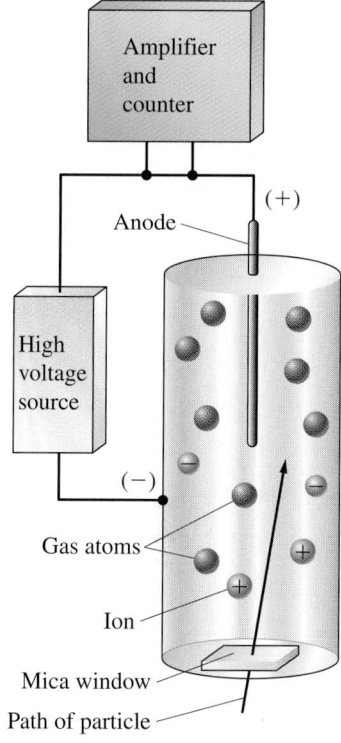

▲ FIGURE 25-11
A Geiger–Müller counter
Radiation enters the G–M tube through the mica window. The radiation ionizes some of the gas (usually argon) in the tube. A pulse of electric current passes through the electric circuit and is counted.

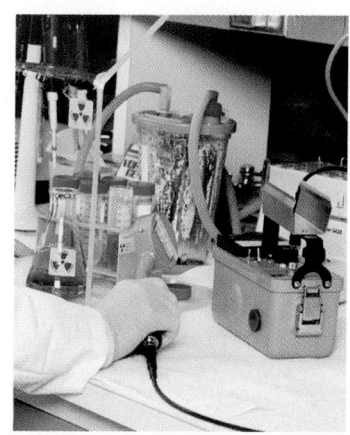

▲ Laboratory technician checking for radioactive isotopes.

Radiation Dosage

One unit long used to describe exposure to radiation is the rad. One **rad** (*radiation absorbed dose*) corresponds to the absorption of 1×10^{-2} J of energy per kilogram of matter. The effect of a dose of one rad on living matter is variable, however, and a better unit is one that takes this variability into account. The **rem** (*radiation equivalent for man*) is the rad multiplied by the *relative biological effectiveness* (*Q*). The factor *Q* takes into account that equal doses of radiation of different types may have differing effects. Table 25.4 summarizes several radiation units.

It is thought that a dose of 1000 rem absorbed in a short time interval will kill 100% of the population. A short-term dose of 450 rem would probably result in death within 30 days of about 50% of the population. A single dose of 1 rem delivered to 1 million people would probably produce about 100 cases of cancer within 20–30 years. The total body radiation received by most of the world's population from normal background sources is about 0.13 rem [130 millirem (mrem)] per year. The dose delivered in a chest X-ray examination is about 20 mrem.

Some of the foregoing statements about radiation dosages and their anticipated effects are based on (1) medical histories of the survivors of the Hiroshima and Nagasaki atomic blasts, (2) the incidence of leukemia and other cancers in children whose mothers received diagnostic radiation during pregnancy, and (3) the occurrence of lung cancers among uranium miners in the United States. What does all of this tell us about a safe level of radiation exposure? One approach has been to extrapolate from these high doses to the lower doses affecting the general population. This has led the U.S. National Council on Radiation Protection and Measurements to recommend that the dosage for the general population be limited to 0.17 rem (170 mrem) per year from all sources above background level. Experts disagree, however, on how the data observed for high dosages should be extrapolated to low doses. Some experts believe that the 0.17 rem/y figure is too high. If they are right,

TABLE 25.4 Radiation Units[a]

Unit	Definition
Radioactive decay:	
Becquerel, Bq	s^{-1} (disintegrations per second)
Curie, Ci	An amount of radioactive material decaying at the same rate as 1 g of radium (3.70×10^{10} dis s^{-1})
	1 Ci = 3.70×10^{10} Bq
Absorbed dose:	
Gray, Gy	One gray of radiation deposits one joule of energy per kilogram of matter
Rad	1 rad = 0.01 Gy
Equivalent dose:	
Sievert, Sv	1 Sv = 100 rem
Rem	1 rem = 1 rad $\times$ Q
	The quality factor, Q, is about 1 for X rays, γ rays, and β^- particles; 3 for slow neutrons; 10 for protons and fast neutrons; and 20 for α particles

[a]SI units are shown in blue. Sources of α radiation are relatively harmless when external to the body and extremely hazardous when taken internally, as in the lungs or stomach. Other forms of radiation (X rays, γ rays), because they are highly penetrating, are hazardous even when external to the body.

an additional dosage of 0.17 rem/y above normal background levels might cause statistically significant increases in the incidence of birth defects and cancers.

An Environmental Issue Involving Radon

All atoms with an atomic number greater than 83 are radioactive. The nuclei of these atoms are unstable and emit α, β, and γ radiation, eventually breaking down to more stable elements with lower atomic numbers. Radon-222, a colorless, odorless gas, is produced by the loss of α particles from radium-226, which in turn results from the radioactive decay, through several steps, of uranium-238.

In December 1984, a worker at a nuclear power plant in New Jersey registered high readings on a radiation detector during a routine safety check. But the radiation he had been exposed to didn't come from within the plant—it came from his own home. This incident led to the recognition that people can be exposed to high levels of radioactivity from radon. An estimated 21,000 lung cancer deaths per year are attributable to residential radon exposure. It is also estimated that 7 million homes have radon limits in excess of the U.S. Environmental Protection Agency's reference level of 150 Bq m^{-3}. The possible harmful effects of this exposure, primarily an increased risk of lung cancer, are fairly well documented, but the topic remains one of continuing research and debate.

In some instances, the source of radon is in wastes from uranium mining or phosphate production. In most cases, it is emitted by the radioactive decay of ^{238}U present in small amounts in rocks and soils. Because radon is a gas, it readily passes through air passages in the body and is breathed in and out. The product formed when a ^{222}Rn atom gives up an α particle is the isotope polonium-218, which also emits α particles. Unlike radon, polonium is a solid. Health hazards posed by radon seem to be from ^{218}Po and other radioactive decay products becoming attached to dust particles in the air and then being breathed into the lungs.

Fortunately, indoor radon can be rather easily detected because of its radioactivity. The chief method of reducing radon levels is through improved ventilation and by venting subsoil radon to keep it from concentrating within a building. Minimizing indoor radon should be a part of building construction.

25-11   Applications of Radioisotopes

We have described both the destructive capacity of nuclear reactions and the potential of these reactions to provide new sources of energy. Less heralded but equally important are a variety of practical applications of radioactivity. We close this chapter with a brief survey of some uses of radioisotopes.

Cancer Therapy

Ionizing radiation in low doses can induce cancers, but this same radiation, particularly γ rays, is also used in the treatment of cancer. Although ionizing radiation tends to destroy all cells, cancerous cells are more easily destroyed than normal ones. Thus, a carefully directed beam of γ rays or high-energy X rays of the appropriate dosage may be used to arrest the growth of cancerous cells. Also coming into use for some forms of cancer is radiation therapy that employs beams of protons or neutrons.

Radioactive Tracers

The small mass differences between the isotopes of an element may produce small differences in physical properties. The differing rates of diffusion of ^{238}UF$_6$(g) and ^{235}UF$_6$(g) result from such a mass difference (page 224). Generally speaking, though, the physical and chemical properties of the various isotopes

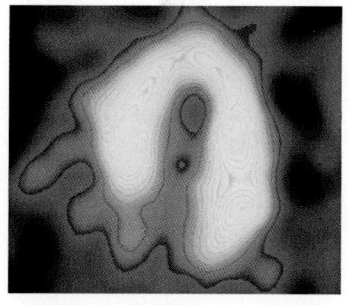

(a)

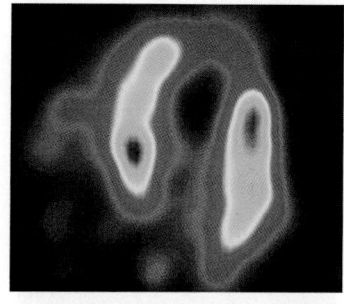

(b)

▲ **Nuclear medicine provides methods for diagnosing life-threatening conditions** In these thallium-201 scans of a heart, γ rays released in the radioactive decay of ^{201}Tl are detected and used to provide an image of the blood flow to the heart wall (heart muscle). **(a)** Normal heart wall; **(b)** heart with deficient blood flow.

of an element are practically identical. However, if one of the isotopes is radioactive, its actions can be easily followed with radiation detectors. This is the principle behind the use of *radioactive tracers*, or tagged atoms. For example, if a small quantity of radioactive ^{32}P (as a phosphate) is added to a nutrient solution fed to plants, the uptake of all the phosphorus atoms—radioactive and nonradioactive alike—can be followed by charting the regions of the plant that become radioactive. Similarly, the fate of iodine (as iodide ion) in the body can be determined by having a person drink a solution of dissolved iodides containing a small quantity of a radioactive iodide as a tracer. Abnormalities in the thyroid gland can be detected in this way. Because I^- concentrates in the thyroid, people can protect themselves by ingesting nonradioactive iodides before being exposed to a radioactive iodide. The thyroid becomes saturated with the nonradioactive iodide and rejects the radioactive iodide. Iodide tablets were distributed to the general population in some regions of Europe before the arrival of fallout from the Chernobyl nuclear accident in 1986.

Industrial applications of tracers are also numerous. The fate of a catalyst in a chemical plant can be followed by incorporating a radioactive tracer in the catalyst, such as ^{192}Ir in a Pt-Ir catalyst. Monitoring the activity of the ^{192}Ir makes it possible to determine the rate at which the catalyst is being carried away and to which parts of the plant it is being carried.

Structures and Mechanisms

Often the mechanism of a chemical reaction or the structure of a species can be inferred from experiments using radioisotopes as tracers. Consider the following experimental proof that the two S atoms in the thiosulfate ion, $S_2O_3^{2-}$, are not equivalent.

$S_2O_3^{2-}$ is prepared from radioactive sulfur (^{35}S) and sulfite ion containing the nonradioactive isotope ^{32}S.

$$^{35}S + {}^{32}SO_3^{2-} \longrightarrow {}^{35}S^{32}SO_3^{2-} \tag{25.23}$$

When the thiosulfate ion is decomposed by acidification, all the radioactivity appears in the precipitated sulfur and none in the $SO_2(g)$. The bonds of the ^{35}S atoms must be different from those of the ^{32}S atoms (Fig. 25-12).

$$^{35}S^{32}SO_3^{2-} + 2H^+ \longrightarrow H_2O + {}^{32}SO_2(g) + {}^{35}S(s) \tag{25.24}$$

In reaction (25.25), nonradioactive KIO_4 is added to a solution containing iodide ion labeled with the radioisotope ^{128}I. All the radioactivity appears in the I_2 and none in the IO_3^-. This proves that all the IO_3^- is produced by reduction of IO_4^- and none by oxidation of I^-.

$$IO_4^- + 2\,{}^{128}I^- + H_2O \longrightarrow {}^{128}I_2 + IO_3^- + 2\,OH^- \tag{25.25}$$

Analytical Chemistry

The usual procedure of analyzing a substance by precipitation involves filtering, washing, drying, and weighing a pure precipitate. An alternative is to incorporate a radioactive isotope in the precipitating reagent. By measuring the activity of the precipitate and comparing it with that of the original solution, the amount of precipitate can be calculated without having to purify, dry, and weigh it.

Another method of importance in analytical chemistry is *neutron activation analysis*. In this procedure, the sample to be analyzed, normally nonradioactive, is bombarded with neutrons; the element of interest is converted to a radioisotope. The conversion of a stable isotope (X) to an unstable isotope (X*) by neutron capture can be represented as

$$^AX + {}^1_0n \longrightarrow {}^{(A+1)}X^* \longrightarrow {}^{(A+1)}X + \gamma$$

In this example, the excited nucleus that is formed decays with the emission of a γ ray with a distinctive energy. Neutron-activated nuclei may decay by other

▲ FIGURE 25-12
Structure of thiosulfate ion, $S_2O_3^{2-}$
The central S atom is in the +6 oxidation state. The terminal S atom is in the −2 oxidation state.

▲ Neutron activation was used to establish that this painting was not painted by Rembrandt, but by an artist in the school of Rembrandt.

modes, however, such as β decay. The activity of the radioisotope formed is measured. This measurement, together with such factors as the rate of neutron bombardment, the half-life of the radioisotope, and the efficiency of the radiation detector, can be used to calculate the quantity of the element in the sample. The method is especially attractive because (1) trace quantities of elements can be determined (sometimes in parts per billion or less); (2) a sample can be tested without destroying it; and (3) the sample can be in any state of matter, including biological materials. Neutron activation analysis has been used, for instance, to determine the authenticity of old paintings. Old masters formulated their own paints. Differences between formulations are easily detected through the trace elements they contain.

▲ The strawberries on the right have been preserved by irradiation.

Radiation Processing

Radiation processing describes industrial applications of ionizing radiation—γ rays from ^{60}Co or electron beams from electron accelerators. The ionizing radiation is used in the production of certain materials or to modify their properties. Its most extensive current use is in breaking, re-forming, and cross-linking polymer chains to affect the physical and mechanical properties of plastics used in foamed products, electrical insulation, and packaging materials. Ionizing radiation is used to sterilize medical supplies, such as sutures, syringes, and hospital garb. In sewage treatment plants, radiation processing has been used to decrease the settling time of sewage sludge and to kill pathogens. Radiation is now being used in some instances in the preservation of foods as an alternative to canning, freeze-drying, or refrigeration. In radiation processing, the irradiated material is not rendered radioactive, although the ionizing radiation may produce some chemical changes.

Mastering**CHEMISTRY**   **www.masteringchemistry.com**

The use of nuclear reactions is an attractive "clean" approach to meeting today's energy consumption. However, the waste products created by nuclear power plants are not easily disposed. For a discussion of different methods of handling nuclear waste, go to the Focus On feature for Chapter 25, Radioactive Waste Disposal, on the Mastering Chemistry site.

Summary

25-1 Radioactivity—All the heavier elements ($Z > 83$) and a few of the lighter ones have naturally occurring nuclides that produce ionizing radiation. **Alpha (α) particles** emanating from radioactive nuclei are identical to the nuclei of helium-4 atoms, $^4_2\text{He}^{2+}$. **Beta (β^-) particles** are electrons that originate from the conversion of a neutron to a proton in the radioactive nucleus; **positrons (β^+)** are positively charged particles otherwise identical to electrons that originate in the nucleus in the conversion of a proton to a neutron. Beta ($-$) particles and positrons have a greater penetrating power through matter than do α particles. **Electron capture (EC)** occurs when an electron from a low-lying energy level is absorbed by the nucleus. The electron and a proton combine to form a neutron and liberate X radiation. **Gamma (γ) rays** are a nonparticulate form of electromagnetic radiation released when a nucleus in an excited state returns to the ground state. **Nuclear equations** to represent nuclear processes must be balanced, which requires that (1) the sum of the mass numbers and (2) the sum of the atomic numbers must be the same on both sides of the nuclear equation (equation 25.1).

25-2 Naturally Occurring Radioactive Isotopes—All naturally occurring radioactive nuclides of high atomic number are members of a **radioactive decay series** that originates with a long-lived isotope of high atomic number and terminates with a stable isotope, such as ^{206}Pb (Fig. 25-2).

25-3 Nuclear Reactions and Artificially Induced Radioactivity—Radioactive nuclides can be synthesized by nuclear reactions in which target nuclei are bombarded with energetic particles such as α particles or neutrons. More than 1000 radioactive nuclides have been produced in this way. Many of them have important practical applications.

25-4 Transuranium Elements—Elements beyond $Z = 92$, commonly referred to as the transuranium elements, have been synthesized by bombarding naturally occurring isotopes with charged particles produced in a charged-particle accelerator (Fig. 25-3).

25-5 Rate of Radioactive Decay—The **radioactive decay law**, based on experimental observations, states that the decay rate is directly proportional to the number of atoms present; hence it is a first-order process (equation 25.11). The proportionality constant between the decay rate and the number of atoms is the **decay constant**, λ. The **half-life**, $t_{1/2}$, is the time it takes for half of the radioactive atoms in a sample to decay. Important applications of the radioactive decay law are radiocarbon dating to determine the age of materials derived from once-living matter and determining the age of rocks and of Earth itself.

25-6 Energetics of Nuclear Reactions—The energy changes in nuclear reactions are a consequence of Einstein's discovery of a mass–energy equivalence (equation 25.19). **Nuclear binding energy** is the energy released when nucleons fuse together into a nucleus; the "lost" mass is known as the mass defect. A plot of binding energy per nucleon versus mass number passes through a maximum at about $A = 60$. Thus, the *fusion* of nucleons is favored at lower mass numbers, and the *fission* of nuclei into lighter fragments is favored at higher mass numbers (Fig. 25-6).

25-7 Nuclear Stability—The stability of a nuclide depends on the ratio N/Z, on whether the numbers of neutrons (N) and protons (Z) are odd or even, and whether either is a **magic number** arising from nuclear shell theory (Table 25.2). A plot of N versus Z shows that all stable nuclides lie within a belt of stability originating along the line $N = Z$, and expanding above the line at higher values of Z (Figure 25-7). Most radioactive nuclides lie outside the belt and undergo radioactive decay of a type that moves daughter nuclides into the belt.

25-8 Nuclear Fission—In the process of **fission**, the nucleus of a heavy nuclide, such as ^{235}U, splits into two smaller fragments after being struck by a thermal neutron (Fig. 25-8). Also released are two or three neutrons that can trigger the fission of other nuclei in a chain reaction. Essential components of a nuclear power reactor (Fig. 25-9) are the fissionable (*fissile*) nuclide, a **moderator** (such as water) to slow down the neutrons released during fission, **control rods** (such as cadmium) to control the fission by absorbing neutrons, and a heat-transfer medium (such as water).

25-9 Nuclear Fusion—The **fusion** of lighter nuclei into heavier ones converts small quantities of mass into enormous amounts of energy. This process occurs continuously in stars and in the hydrogen bomb. A controlled fusion reaction has yet to be achieved, but fusion research is being actively pursued because of the enormous potential of fusion as an energy source.

25-10 Effect of Radiation on Matter—The ionizing power of radiation (Fig. 25-10) is the basis both of radiation's effects on matter and methods used to detect radiation. Radiation detectors include simple film badges, Geiger–Müller (G–M) counters for routine measurements (Fig. 25-11), and scintillation counters in biomedical studies. Two units of measure are used to quantify exposure to radioactivity. One—the **rad** (radiation absorbed dose)—is related to the amount of radiation energy absorbed, while the other—the **rem** (radiation equivalent for man)—takes into account the differing effects of the various types of radiation (Table 25.4).

25-11 Applications of Radioisotopes—Radioactive nuclides have important uses in diagnostic medicine, in cancer therapy, in studying the mechanisms of chemical and biochemical reactions, and as tracers in various scientific and industrial settings. Additionally, radiation processing is used in processing and preserving food.

Integrative Example

On April 26, 1986, an explosion at the nuclear power plant at Chernobyl, Ukraine, released the greatest quantity of radioactive material ever associated with an industrial accident. (See the discussion and photo on pages 1131–1132.) One of the radioisotopes in this emission was ^{131}I, a β^- emitter with a half-life of 8.04 days.

Assume that the total quantity of ^{131}I released was 250 g, and determine the number of *curies* associated with this ^{131}I 30.0 days after the accident.

Analyze

From the initial mass of ^{131}I and the known half-life of ^{131}I determine how many grams of ^{131}I remain after 30 days. Convert this mass to a number of atoms using the molar mass of ^{131}I and Avogadro's number. Finally, the product of the number of atoms and the decay constant yields the activity in disintegrations per second, which can be converted to curies.

Solve

In equation (25.12), use $\lambda = 0.693/8.04$ d and $t = 30.0$ d. Because the mass of ^{131}I is directly proportional to the number of ^{131}I atoms, we can replace N_0 by $mass_0 = 250$ g, and N_t by $mass_t$, the mass we seek.

$$\ln\frac{mass_t}{mass_0} = \ln\frac{mass_t}{250\text{ g}} = \frac{-0.693}{8.04\text{ d}} \times 30.0\text{ d} = -2.59$$

$$mass_t = 250\text{ g} \times e^{-2.59} = 250\text{ g} \times 0.0750 = 18.8\text{ g}$$

To determine the number of iodine-131 atoms, use the molar mass, 131 g ^{131}I/mol ^{131}I, and the Avogadro constant.

$$N_t = 18.8\text{ g }^{131}I \times \frac{1\text{ mol }^{131}I}{131\text{ g }^{131}I} \times \frac{6.022 \times 10^{23}\ ^{131}I\text{ atoms}}{1\text{ mol }^{131}I}$$

$$= 8.64 \times 10^{22}\ ^{131}I\text{ atoms}$$

To establish the decay constant in s^{-1} and to obtain λ from $t_{1/2}$, use the method outlined in Example 25-3a.

$$\lambda = 0.693/t_{1/2} = \frac{0.693}{8.04\ d} \times \frac{1\ d}{24\ h} \times \frac{1\ h}{60\ min} \times \frac{1\ min}{60\ s}$$

$$= 9.98 \times 10^{-7}\ s^{-1}$$

To determine the decay rate (activity) after 30.0 days, use the two previous results in equation (25.11).

$$A = \lambda N = 9.98 \times 10^{-7}\ s^{-1} \times 8.64 \times 10^{22}\ atoms$$

$$= 8.62 \times 10^{16}\ dis\ s^{-1}$$

Finally, to express the activity in curies use the definition of a curie given in Table 25.4.

$$A = 8.62 \times 10^{16}\ dis\ s^{-1} \times \frac{1\ Ci}{3.70 \times 10^{10}\ dis\ s^{-1}}$$

$$= 2.33 \times 10^{6}\ Ci$$

Assess

In this calculation we could also have shown that in 30 days (about $4 \times t_{1/2}$) approximately 93% of the ^{131}I had disintegrated. In the process, approximately 2×10^{6} Ci of radiation was emitted. If we also knew the energy of the β^- particles and something about the exposed population, we could also have estimated radiation doses in rads.

PRACTICE EXAMPLE A: ^{40}K undergoes radioactive decay by electron capture to ^{40}Ar and by β^- emission to ^{40}Ca. The fraction of the decay that occurs by electron capture is 0.110. The half-life of ^{40}K is 1.25×10^9 years. Assuming that a rock in which ^{40}K has undergone decay retains all the ^{40}Ar produced, what would be the $^{40}Ar/^{40}K$ mass ratio in a rock that is 1.5×10^9 years old?

PRACTICE EXAMPLE B: Most nuclear power plants use zirconium in the fuel rods because zirconium maintains its structural integrity under exposure to the radiation in the nuclear reactor. The nuclear accidents at both Three Mile Island and Chernobyl involved the evolution of hydrogen gas from the reduction of water. The half-cell reduction potential for Zr is

$$ZrO_2(s) + 4\ H_3O^+(aq) + 4\ e^- \longrightarrow Zr(s) + 6\ H_2O(l) \qquad E° = -1.43\ V$$

(a) Can Zr reduce water under standard-state conditions?
(b) Calculate the equilibrium constant for the reduction of water by zirconium.
(c) Is the reaction spontaneous if the pH = 7 and Zr, ZrO_2, and water are in their standard states?
(d) Was the reduction of water by Zr the culprit in the reactor accidents mentioned above?

Mastering**CHEMISTRY**

You'll find a link to additional self study questions in the study area on www.masteringchemistry.com

Exercises

Radioactive Processes

1. What nucleus is obtained in each process?
   (a) $^{234}_{94}Pu$ decays by α emission.
   (b) $^{248}_{97}Bk$ decays by β^- emission.
   (c) $^{196}_{82}Pb$ goes through two successive EC processes.
2. What nucleus is obtained in each process?
   (a) $^{214}_{82}Pb$ decays through two successive β^- emissions.
   (b) $^{226}_{88}Ra$ decays through three successive α emissions.
   (c) $^{69}_{33}As$ decays by β^+ emission.

3. Based on a favorable $N–Z$ ratio for the product nucleus, write the most plausible equation for the decay of $^{14}_{6}C$.
4. Write a plausible equation for the decay of tritium, $^{3}_{1}H$, the radioactive isotope of hydrogen.

Radioactive Decay Series

5. The natural decay series starting with the radionuclide $^{232}_{90}\text{Th}$ follows the sequence represented here. Construct a graph of this series, similar to Figure 25-2.

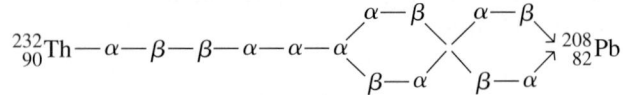

6. The natural decay series starting with the radionuclide $^{235}_{92}\text{U}$ follows the sequence represented here. Construct a graph of this series, similar to Figure 25-2.

$$^{235}_{92}\text{U}-\alpha-\beta-\alpha \underset{\beta-\alpha}{\overset{\alpha-\beta}{\diamond}} -\alpha-\alpha-\alpha-\beta \underset{\beta-\alpha}{\overset{\alpha-\beta}{\diamond}} \nearrow ^{207}_{82}\text{Pb}$$

7. The uranium series described in Figure 25-2 is also known as the "4n + 2" series because the mass number of each nuclide in the series can be expressed by the equation $A = 4n + 2$, where n is an integer. Show that this equation is indeed applicable to the uranium series.

8. Just as the uranium series is called the "4n + 2" series, the thorium series can be called the "4n" series and the actinium series the "4n + 3" series. A "4n + 1" series has also been established, with $^{241}_{94}\text{Pu}$ as the parent nuclide. To which series does each of the following belong: (a) $^{214}_{83}\text{Bi}$; (b) $^{216}_{84}\text{Po}$; (c) $^{215}_{85}\text{At}$; (d) $^{235}_{92}\text{U}$?

Nuclear Reactions

9. Supply the missing information in each of the following nuclear equations representing a radioactive decay process.
   (a) $^{160}_{?}\text{W} \longrightarrow ^{?}_{?}\text{Hf} + ?$
   (b) $^{38}_{?}\text{Cl} \longrightarrow ^{?}_{?}\text{Ar} + ?$
   (c) $^{214}_{?}? \longrightarrow ^{?}_{?}\text{Po} + ^{0}_{-1}\beta$
   (d) $^{32}_{17}\text{Cl} \longrightarrow ^{?}_{16}? + ?$

10. Complete the following nuclear equations.
   (a) $^{23}_{11}\text{Na} + ? \longrightarrow ^{24}_{11}\text{Na} + ^{1}_{1}\text{H}$
   (b) $^{59}_{27}\text{Co} + ^{1}_{0}\text{n} \longrightarrow ^{56}_{25}\text{Mn} + ?$
   (c) $? + ^{2}_{1}\text{H} \longrightarrow ^{240}_{94}\text{Pu} + ^{0}_{-1}\beta$
   (d) $^{246}_{96}\text{Cm} + ? \longrightarrow ^{254}_{102}\text{No} + 5\,^{1}_{0}\text{n}$
   (e) $^{238}_{92}\text{U} + ? \longrightarrow ^{246}_{99}\text{Es} + 6\,^{1}_{0}\text{n}$

11. Write equations for the following nuclear reactions.
   (a) bombardment of ^{7}Li with protons to produce ^{8}Be and γ rays
   (b) bombardment of ^{9}Be with $^{2}_{1}\text{H}$ to produce ^{10}B
   (c) bombardment of ^{14}N with neutrons to produce ^{14}C

12. Write equations for the following nuclear reactions.
   (a) bombardment of ^{238}U with α particles to produce ^{239}Pu
   (b) bombardment of tritium ($^{3}_{1}\text{H}$) with $^{2}_{1}\text{H}$ to produce ^{4}He
   (c) bombardment of ^{33}S with neutrons to produce ^{33}P

13. Write nuclear equations to represent the formation of an isotope of element 111 with a mass number of 272 by the bombardment of bismuth-209 by nickel-64 nuclei, followed by a succession of five α-particle emissions.

14. Write nuclear equations to represent the formation of a hypothetical isotope of element 118 with a mass number of 293 by the bombardment of lead-208 by krypton-86 nuclei, followed by a chain of α-particle emissions to the element seaborgium.

15. Scientists from Dubna, Russia, observed the existence of elements 118 and 116 at the Joint Institute for Nuclear Research U400 cyclotron in 2005. This was the result of bombarding calcium-48 ions on a californium-249 target. Write a complete nuclear equation for this reaction.

16. The immediate decay product of element 118 is thought to be element 116. Write a complete nuclear equation for this reaction.

17. Element-120 is located in a region of the neutron versus proton map known as the island of stability. Write a nuclear equation for the generation of element-120 by bombarding iron isotopes on a plutonium target.

18. Another possible nuclear reaction leading to the formation of element-120 is between uranium-238 and nickel-64. Write a nuclear equation for this nuclear reaction.

Rate of Radioactive Decay

19. Of the radioactive nuclides in Table 25.1,
   (a) which one has the largest value for the decay constant, λ?
   (b) Which one loses 75% of its radioactivity in approximately one month?
   (c) Which ones lose more than 99% of their radioactivity in one month?

20. In a comparison of two radioisotopes, isotope A requires 18.0 hours for its decay rate to fall to $\frac{1}{16}$ its initial value, while isotope B has a half-life that is 2.5 times that of A. How long does it take for the decay rate of isotope B to decrease to $\frac{1}{32}$ of its initial value?

21. The disintegration rate for a sample containing $^{60}_{27}\text{Co}$ as the only radioactive nuclide is 6740 dis h^{-1}. The half-life of $^{60}_{27}\text{Co}$ is 5.2 years. Estimate the number of atoms of $^{60}_{27}\text{Co}$ in the sample.

22. How many years must the radioactive sample of Exercise 21 be maintained before the disintegration rate falls to 101 dis min^{-1}?

23. A sample containing $^{224}_{88}\text{Ra}$, which decays by α-particle emission, disintegrates at the following rate, expressed as disintegrations per minute or counts per minute (cpm): $t = 0$, 1000 cpm; $t = 1$ h, 992 cpm; $t = 10$ h, 924 cpm; $t = 100$ h, 452 cpm;

$t = 250$ h, 138 cpm. What is the half-life of this nuclide?

24. Iodine-129 is a product of nuclear fission, whether from an atomic bomb or a nuclear power plant. It is a β^- emitter with a half-life of 1.7×10^7 years. How many disintegrations per second would occur in a sample containing 1.00 mg ^{129}I?

25. Suppose that a sample containing ^{32}P has an activity 1000 times the detectable limit. How long would an experiment have to be run with this sample before the radioactivity could no longer be detected?

26. What mass of carbon-14 must be present in a sample to have an activity of 1.00 mCi?

Age Determinations with Radioisotopes

27. A wooden object is claimed to have been found in an Egyptian pyramid and is offered for sale to an art museum. Radiocarbon dating of the object reveals a disintegration rate of 10.0 dis $\min^{-1}$ g^{-1}. Do you think the object is authentic? Explain.

28. The lowest level of ^{14}C activity that seems possible for experimental detection is 0.03 dis $\min^{-1}$ g^{-1}. What is the maximum age of an object that can be determined by the carbon-14 method?

29. What should be the mass ratio ^{208}Pb/^{232}Th in a meteorite that is approximately 2.7×10^9 years old? The half-life of ^{232}Th is 1.39×10^{10} years. [*Hint:* One ^{208}Pb atom is the final decay product of one ^{232}Th atom.]

30. Concerning the decay of ^{232}Th described in Exercise 29, a certain rock has a ^{208}Pb/^{232}Th mass ratio of 0.25/1.00. Estimate the age of the rock.

31. A lunar rock was analyzed for argon by mass spectrometry and for potassium by atomic absorption. The results of these analyses showed that the sample contained 3.02×10^{-5} mL g^{-1} of argon and 0.083% of potassium. The half-life of potassium-40 is 1.248×10^9 y. Calculate the age of the lunar rock.

32. What is the age of a piece of volcanic rock that has a mass ratio of argon-40 to potassium-40 of 2.9? The half-life of potassium-40 by β decay is 1.248×10^9 y and by electron capture $t_{1/2} = 1.4 \times 10^9$ y.

Energetics of Nuclear Reactions

33. Using appropriate equations in the text, determine
   (a) the energy in joules corresponding to the destruction of 6.02×10^{-23} g of matter;
   (b) the energy in megaelectronvolts that would be released if one α particle were completely destroyed.

34. The measured mass of the nucleus of an atom of silver-107 is 106.879289 u. For this atom, determine the binding energy per nucleon in megaelectronvolts.

35. Use the electron mass from Table 2.1 and the measured mass of the nuclide $^{19}_{9}$F, 18.998403 u, to determine the binding energy per nucleon (in megaelectronvolts) of this atom.

36. Use the electron mass from Table 2.1 and the measured mass of the nuclide $^{56}_{26}$Fe, 55.934939 u, to determine the binding energy per nucleon (in megaelectronvolts) of this atom.

37. Calculate the energy, in megaelectronvolts, released in the nuclear reaction

$$^{10}_{5}\text{B} + ^{4}_{2}\text{He} \longrightarrow ^{13}_{6}\text{C} + ^{1}_{1}\text{H}$$

The nuclidic masses are $^{10}_{5}$B = 10.01294 u; $^{4}_{2}$He = 4.00260 u; $^{13}_{6}$C = 13.00335 u; $^{1}_{1}$H = 1.00783 u.

38. You are given the following nuclidic masses: $^{6}_{3}$Li = 6.01513 u; $^{4}_{2}$He = 4.00260 u; $^{3}_{1}$H = 3.01604 u; $^{1}_{0}$n = 1.008665 u. How much energy, in megaelectronvolts, is released in the following nuclear reaction?

$$^{6}_{3}\text{Li} + ^{1}_{0}\text{n} \longrightarrow ^{4}_{2}\text{He} + ^{3}_{1}\text{H}$$

39. Calculate the number of neutrons that could be created with 6.75×10^6 MeV of energy.

40. When β^+ and β^- particles collide, they annihilate each other, producing two γ rays that move away from each other along a straight line. What is the approximate energies of these two γ rays, in MeV?

Nuclear Stability

41. Which member of the following pairs of nuclides would you expect to be most abundant in natural sources: (a) $^{20}_{10}$Ne or $^{22}_{10}$Ne; (b) $^{17}_{8}$O or $^{18}_{8}$O; (c) $^{6}_{3}$Li or $^{7}_{3}$Li? Explain your reasoning.

42. Which member of the following pairs of nuclides would you expect to be most abundant in natural sources: (a) $^{40}_{20}$Ca or $^{42}_{20}$Ca; (b) $^{31}_{15}$P or $^{32}_{15}$P; (c) $^{63}_{30}$Zn or $^{64}_{30}$Zn? Explain your reasoning.

43. One member each of the following pairs of radioisotopes decays by β^- emission, and the other by positron (β^+) emission: (a) $^{29}_{15}$P and $^{33}_{15}$P; (b) $^{120}_{53}$I and $^{134}_{53}$I. Which is which? Explain your reasoning.

44. Each of the following isotopes is radioactive: (a) $^{28}_{15}$P; (b) $^{45}_{19}$K; (c) $^{73}_{30}$Zn. Which would you expect to decay by β^+ emission?

45. Some nuclides are said to be doubly magic. What do you suppose this term means? Postulate some nuclides that might be doubly magic, and locate them in Figure 25-7.

46. Both β^- and β^+ emissions are observed for artificially produced radioisotopes of low atomic numbers, but only β^- emission is observed with naturally occurring radioisotopes of high atomic number. Why do you suppose this is so?

Fission and Fusion

47. Refer to the Integrative Example. In contrast to the Chernobyl accident, the 1979 nuclear accident at Three Mile Island released only 170 curies of ^{131}I. How many milligrams of ^{131}I does this represent?

48. Explain why more energy is released in a fusion process than in a fission process.

Effect of Radiation on Matter

49. Explain why the rem is more satisfactory than the rad as a unit for measuring radiation dosage.
50. Discuss briefly the basic difficulties in establishing the physiological effects of low-level radiation.
51. ^{90}Sr is both a product of radioactive fallout and a radioactive waste in a nuclear reactor. This radioisotope is a β^- emitter with a half-life of 27.7 years.

Suggest reasons why ^{90}Sr is such a potentially hazardous substance.
52. ^{222}Rn is an α-particle emitter with a half-life of 3.82 days. Is it hazardous to be near a flask containing this isotope? Under what conditions might ^{222}Rn be hazardous?

Applications of Radioisotopes

53. Describe how you might use radioactive materials to find a leak in the $H_2(g)$ supply line in an ammonia synthesis plant.
54. Explain why neutron activation analysis is so useful in identifying trace elements in a sample, in contrast to ordinary methods of quantitative analysis, such as precipitation or titration.
55. A small quantity of NaCl containing radioactive $^{24}_{11}Na$ is added to an aqueous solution of $NaNO_3$. The solution is cooled, and $NaNO_3$ is crystallized from the solution. Would you expect the $NaNO_3(s)$ to be radioactive? Explain.

56. The following reactions are carried out with HCl(aq) containing some tritium (^3_1H) as a tracer. Would you expect any of the tritium radioactivity to appear in the $NH_3(g)$? In the H_2O? Explain.

$$NH_3(aq) + HCl(aq) \longrightarrow NH_4Cl(aq)$$

$$NH_4Cl(aq) + NaOH(aq) \longrightarrow$$
$$NaCl(aq) + H_2O(l) + NH_3(g)$$

Integrative and Advanced Exercises

57. In some cases, the most abundant isotope of an element can be established by rounding off the atomic mass to the nearest whole number, as in ^{39}K, ^{85}Rb, and ^{88}Sr. But in other cases, the isotope corresponding to the rounded-off atomic mass does not even occur naturally, as in ^{64}Cu. Explain the basis of this observation.
58. The overall change in the radioactive decay of $^{238}_{92}U$ to $^{206}_{82}Pb$ is the emission of eight α particles. Show that if this loss of eight α particles were not also accompanied by six β^- emissions, the product nucleus would still be radioactive.
59. Use data from the text to determine how many metric tons (1 metric ton = 1000 kg) of bituminous coal (85% C) would have to be burned to release as much energy as is produced by the fission of 1.00 kg $^{235}_{92}U$.
60. One method of dating rocks is based on their $^{87}Sr/^{87}Rb$ ratio. ^{87}Rb is a β^- emitter with a half-life of 5×10^{11} years. A certain rock has a mass ratio $^{87}Sr/^{87}Rb$ of 0.004/1.00. What is the age of the rock?
61. How many millicuries of radioactivity are associated with a sample containing 5.10 mg ^{229}Th, which has a half-life of 7340 years?
62. What mass of ^{90}Sr, with a half-life of 27.7 years, is required to produce 1.00 millicurie of radioactivity?
63. Refer to the Integrative Example. Another radioisotope produced in the Chernobyl accident was ^{137}Cs. If a 1.00 mg sample of ^{137}Cs is equivalent to 89.8 millicuries, what must be the half-life (in years) of ^{137}Cs?

64. The percent natural abundance of ^{40}K is 0.0117%. The radioactive decay of ^{40}K atoms occurs 89% by β^- emission; the rest is by electron capture and β^+ emission. The half-life of ^{40}K is 1.25×10^9 years. Calculate the number of β^- particles produced per second by the ^{40}K present in a 1.00 g sample of the mineral *microcline*, $KAlSi_3O_8$.
65. The carbon-14 dating method is based on the assumption that the rate of production of ^{14}C by cosmic ray bombardment has remained constant for thousands of years and that the ratio of ^{14}C to ^{12}C has also remained constant. Can you think of any effects of human activities that could invalidate this assumption in the future?
66. Calculate the minimum kinetic energy (in megaelectronvolts) that α particles must possess to produce the nuclear reaction

$$^4_2He + ^{14}_7N \longrightarrow ^{17}_8O + ^1_1H$$

The nuclidic masses are $^4_2He = 4.00260$ u; $^{14}_7N = 14.00307$ u; $^1_1H = 1.00783$ u; $^{17}_8O = 16.99913$ u.
67. Hydrogen gas is spiked with tritium to the extent of 5.00% by mass. What is the activity in curies of a 4.65 L sample of this gas at 25.0 °C and 1.05 atm pressure? [*Hint:* Use 3.02 u as the atomic mass of tritium and data from elsewhere in the text, as necessary.]
68. A certain shale deposit containing 0.006% U by mass is being considered for use as a potential fuel in a

breeder reactor. Assuming a density of 2.5 g/cm^3, how much energy could be released from 1.00×10^3 cm^3 of this material? Assume a fission energy of 3.20×10^{-11} J per fission event (that is, per U atom).

69. An ester forms from a carboxylic acid and an alcohol.

$$RCO_2H + HOR' \longrightarrow RCO_2R' + H_2O$$

This reaction is superficially similar to the reaction of an acid with a base such as sodium hydroxide. The mechanism of the reaction can be followed by using the tracer ^{18}O. This isotope is not radioactive, but other physical measurements can be used to detect its presence. When the *esterification* reaction is carried out with the alcohol containing oxygen-18 atoms, no oxygen-18 beyond its naturally occurring abundance is found in the water produced. How does this result affect the perception that this reaction is like an acid–base reaction?

70. The conversion of CO_2 into carbohydrates by plants via photosynthesis can be represented by the reaction

$$6\,CO_2(g) + 6\,H_2O \xrightarrow{\text{light}} C_6H_{12}O_6 + 6\,O_2(g)$$

To study the mechanism of photosynthesis, algae were grown in water containing ^{18}O, that is, $H_2{}^{18}O$. The oxygen evolved contained oxygen-18 in the same ratio to the other oxygen isotopes as the water in which the reaction was carried out. In another experiment, algae were grown in water containing only ^{16}O, but with oxygen-18 present in the CO_2. The oxygen evolved in this experiment contained no oxygen-18. What conclusion can you draw about the mechanism of photosynthesis from these experiments?

71. Assume that when Earth formed, uranium-238 and uranium-235 were equally abundant. Their current percent natural abundances are 99.28% uranium-238 and 0.72% uranium-235. Given half-lives of 4.5×10^9 years for uranium-238 and 7.1×10^8 years for uranium-235, determine the age of Earth corresponding to this assumption.

Feature Problems

72. The *packing fraction* of a nuclide is related to the fraction of the total mass of a nuclide that is converted to nuclear binding energy. It is defined as the fraction $(M - A)/A$, where M is the actual nuclidic mass and A is the mass number. Use data from a handbook (such as the *Handbook of Chemistry and Physics*, published by the CRC Press) to determine the packing fractions of some representative nuclides. Plot a graph of packing fraction versus mass number, and compare it with Figure 25-6. Explain the relationship between the two.

73. For medical uses, radon-222 formed in the radioactive decay of radium-226 is allowed to collect over the radium metal. Then, the gas is withdrawn and sealed into a glass vial. Following this, the radium is allowed to disintegrate for another period, when a new sample of radon-222 can be withdrawn. The procedure can be continued indefinitely. The process is somewhat complicated by the fact that radon-222 itself undergoes radioactive decay to polonium-218, and so on. The half-lives of radium-226 and radon-222 are 1.60×10^3 years and 3.82 days, respectively.

(a) Beginning with pure radium-226, the number of radon-222 atoms present starts at zero, increases for a time, and then falls off again. Explain this behavior. That is, because the half-life of radon-222 is so much shorter than that of radium-226, why doesn't the radon-222 simply decay as fast as it is produced, without ever building up to a maximum concentration?

(b) Write an expression for the rate of change (dD/dt) in the number of atoms (D) of the radon-222 daughter in terms of the number of radium-226 atoms present initially (P_0) and the decay constants of the parent (λ_p) and daughter (λ_d).

(c) Integration of the expression obtained in part (b) yields the following expression for the number of atoms of the radon-222 daughter (D) present at a time t.

$$D = \frac{P_0\lambda_p(e^{-\lambda_p \times t} - e^{-\lambda_d \times t})}{\lambda_d - \lambda_p}$$

Starting with 1.00 g of pure radium-226, approximately how long will it take for the amount of radon-222 to reach its maximum value: one day, one week, one year, one century, or one millennium?

74. Radioactive decay and mass spectrometry are often used to date rocks after they have cooled from a magma. ^{87}Rb has a half-life of 4.8×10^{10} years and follows the radioactive decay

$$^{87}Rb \longrightarrow {}^{87}Sr + \beta^-$$

A rock was dated by assaying the product of this decay. The mass spectrum of a homogenized sample of rock showed the $^{87}Sr/^{86}Sr$ ratio to be 2.25. Assume that the original $^{87}Sr/^{86}Sr$ ratio was 0.700 when the rock cooled. Chemical analysis of the rock gave 15.5 ppm Sr and 265.4 ppm Rb, using the average atomic masses from a periodic table. The other isotope ratios were $^{86}Sr/^{88}Sr = 0.119$ and $^{84}Sr/^{88}Sr = 0.007$. The isotopic ratio for $^{87}Rb/^{85}Rb$ is 0.330. The isotopic masses are as follows:

Isotope	Atomic Mass, u
^{87}Rb	86.909
^{85}Rb	84.912
^{88}Sr	87.906
^{86}Sr	85.909
^{84}Sr	83.913
^{87}Sr	86.909

Calculate the following:
(a) the average atomic mass of Sr in the rock
(b) the original concentration of Rb in the rock in ppm
(c) the percentage of rubidium-87 decayed in the rock
(d) the time since the rock cooled

Self-Assessment Exercises

75. In your own words, define the following symbols: **(a)** α; **(b)** β^-; **(c)** β^+; **(d)** γ; **(e)** $t_{1/2}$.

76. Briefly describe each of the following ideas, phenomena, or methods: **(a)** radioactive decay series; **(b)** charged-particle accelerator; **(c)** neutron-to-proton ratio; **(d)** mass–energy relationship; **(e)** background radiation.

77. Explain the important distinctions between each pair of terms: **(a)** electron and positron; **(b)** half-life and decay constant; **(c)** mass defect and nuclear binding energy; **(d)** nuclear fission and nuclear fusion; **(e)** primary and secondary ionization.

78. Which of the following types of radiation is deflected in a magnetic field? **(a)** X ray; **(b)** γ ray; **(c)** β ray; **(d)** neutrons.

79. A process that produces a one-unit increase in atomic number is **(a)** electron capture; **(b)** β^- emission; **(c)** α emission; **(d)** γ-ray emission.

80. Of the following nuclides, the one most likely to be radioactive is **(a)** ^{31}P; **(b)** ^{66}Zn; **(c)** ^{35}Cl; **(d)** ^{108}Ag.

81. One of the following elements has eight naturally occurring *stable* isotopes. We should expect that one to be **(a)** Ra; **(b)** Au; **(c)** Cd; **(d)** Br.

82. Of the following nuclides, the highest nuclear binding energy per nucleon is found in **(a)** $^3_1$H; **(b)** $^{16}_8$O; **(c)** $^{56}_{26}$Fe; **(d)** $^{235}_{92}$U.

83. The most radioactive of the isotopes of an element is the one with the largest value of its **(a)** half-life, $t_{1/2}$; **(b)** neutron number, N; **(c)** mass number, Z; **(d)** radioactive decay constant, λ.

84. Given a radioactive nuclide with $t_{1/2} = 1.00$ h and a current disintegration rate of 1000 atoms s^{-1}, three hours from now the disintegration rate will be **(a)** 1000 atoms s^{-1}; **(b)** 333 atoms s^{-1}; **(c)** 250 atoms s^{-1}; **(d)** 125 atoms s^{-1}.

85. Write nuclear equations to represent
    (a) the decay of ^{214}Ra by α-particle emission
    (b) the decay of ^{205}At by positron emission

(c) the decay of ^{212}Fr by electron capture

(d) the reaction of two deuterium nuclei (deuterons) to produce a nucleus of $^3_2$He

(e) the production of $^{243}_{97}$Bk by the α-particle bombardment of $^{241}_{95}$Am

(f) a nuclear reaction in which thorium-232 is bombarded with α particles, producing a new nuclide and four neutrons.

86. ^{232}Ra has a half-life of 11.4 d. How long would it take for the radioactivity associated with a sample of ^{232}Ra to decrease to 1% of its current value?

87. A sample of radioactive $^{35}_{16}$S disintegrates at a rate of 1.00×10^3 atoms min^{-1}. The half-life of $^{35}_{16}$S is 87.9 d. How long will it take for the activity of this sample to decrease to the point of producing **(a)** 253; **(b)** 104; and **(c)** 52 dis min^{-1}?

88. Neutron bombardment of ^{23}Na produces an isotope that is a β emitter. After β emission, the final product is **(a)** ^{24}Na; **(b)** ^{23}Mg; **(c)** ^{23}Ar; **(d)** ^{24}Ar; **(e)** none of these.

89. A nuclide has a decay rate of 2.00×10^{10} s^{-1}. After 25.0 days, its decay rate is 6.25×10^8 s^{-1}. What is the nuclide's half-life? **(a)** 25.0 d; **(b)** 12.5 d; **(c)** 50.0 d; **(d)** 5.00 d; **(e)** none of these.

90. A nuclide has a half-life of 1.91 y. Its decay constant has a numerical value of **(a)** 1.32; **(b)** 2.76; **(c)** 0.363; **(d)** 0.524; **(e)** none of these.

91. A nuclide has a decay constant of 4.28×10^{-4} h^{-1}. If the activity of a sample is 3.14×10^5 s^{-1}, how many atoms of the nuclide are present in the sample? **(a)** 2.64×10^{12}; **(b)** 7.34×10^8; **(c)** 2.04×10^5; **(d)** 4.40×10^{10}; **(e)** none of these.

92. Based on magic numbers, which nuclide is the least stable? **(a)** ^{59}Ni; **(b)** ^{51}V; **(c)** ^{122}Sb; **(d)** ^{16}O; **(e)** ^{12}C.

Structures of Organic Compounds

Coffee beans contain an alkaloid compound commonly known as caffeine. Caffeine is a stimulant of the central nervous system producing alertness and heightened concentration. (Greg Vaughn/PacificStock.com)

To early nineteenth-century chemists, organic chemistry meant the study of compounds obtainable only from living matter, which was thought to have the "vital force" needed to make these compounds. In 1828, Friedrich Wöhler set out to synthesize ammonium cyanate, NH_4OCN, using the reaction below:

$$AgOCN(s) + NH_4Cl(aq) \xrightarrow{\Delta} AgCl(s) + NH_4OCN(aq)$$

The white crystalline solid he obtained from the solution had none of the properties of ammonium cyanate, even though it had the same composition. The compound was not NH_4OCN but $(NH_2)_2CO$—*urea*, an organic compound. As Wöhler excitedly reported to J. J. Berzelius, "I must tell you that I can make urea without the use of kidneys, either man or dog. Ammonium cyanate is urea."

Since that time, chemists have synthesized millions of organic compounds, and today organic compounds represent about 98% of all known chemical substances. In this chapter, we build on the introduction to organic compounds in Chapter 3 by exploring some of the principal types

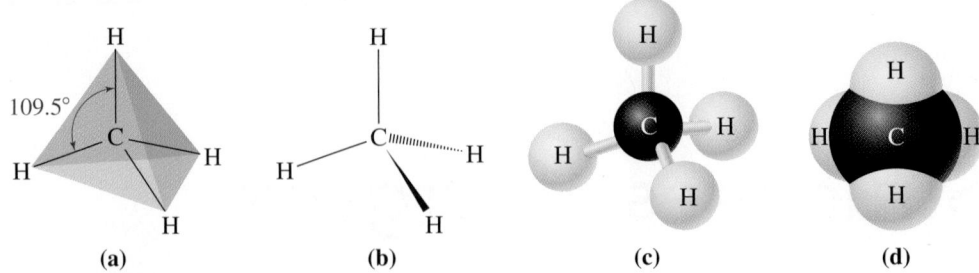

▲ FIGURE 26-1
Representation of the methane molecule
(a) Tetrahedral structure showing bond angle. (b) Dashed-wedged line structure convention used to suggest a three-dimensional structure through a structural formula. The solid lines represent bonds in the plane of the page. The dashed wedge projects *away* from the viewer (behind the plane of the page), and the heavy wedge projects *toward* the viewer (out of the page). (c) Ball-and-stick model. (d) Space-filling model.

KEEP IN MIND

that the hybridization of the C atoms in ethane and propane is sp^3, as in methane. The tetrahedral geometry at the C atoms in alkanes means that the propane chain is not linear.

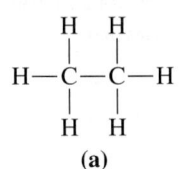

(a)

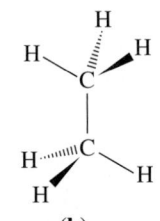

(b)

(c)

▲ FIGURE 26-2
The ethane molecule, C_2H_6
(a) Structural formula.
(b) Dashed-wedged line structure. (c) Space-filling model.

of organic compounds. In this chapter, we will focus on the structures and properties of organic compounds, and we will consider the preparation and some uses of these compounds. In the next chapter, we will turn our attention to reactions that interconvert these compounds.

26-1   Organic Compounds and Structures: An Overview

As we learned in Chapter 3, organic compounds contain carbon and hydrogen atoms or carbon and hydrogen in combination with a few other types of atoms, such as oxygen, nitrogen, and sulfur. Carbon is singled out for special study because the ability of C atoms to form strong covalent bonds with one another allows them to join together into straight chains, branched chains, and rings. The nearly infinite number of possible bonding arrangements of C atoms accounts for the vast number and variety of organic compounds.

The simplest organic compounds are those of carbon and hydrogen—**hydrocarbons**—and the simplest hydrocarbon is methane, CH_4, the chief constituent of natural gas.

From VSEPR theory we expect the electron group geometry around the central C atom in CH_4 to be tetrahedral, as illustrated in Figure 26-1(a). The four H atoms are equivalent: They are equidistant from the C atom and attached to it by covalent bonds of equal strength. The angle between any two C—H bonds is 109.5°. In Figure 26-1(b), the three-dimensional structure of CH_4 is represented by using the **dashed and solid wedge line notation** we introduced in Chapter 10. Other commonly employed depictions of methane are also shown in Figure 26-1.

The bonding in CH_4 is most easily described in terms of valence bond theory. As shown in Figure 11-7 (page 454), each of the carbon–hydrogen bonds is a σ bond formed by the overlap of a 1s orbital on hydrogen with an sp^3 orbital on carbon.

The removal of one H atom from a CH_4 molecule leaves the —CH_3 group. Now imagine forming a covalent bond between two —CH_3 groups. The resulting molecule is *ethane*, C_2H_6 (Fig. 26-2). By increasing the number of C atoms in the chain, we can obtain still more hydrocarbons. The three-carbon molecule propane, C_3H_8, is pictured in Figure 26-3.

Constitutional Isomerism in Organic Compounds

As we have previously learned, compounds that have the same molecular formula but different structural formulas are called *isomers*. Forms of isomerism

H—C—C—C—H $\quad$ CH$_3$CH$_2$CH$_3$

(a) $\qquad$ (b) $\qquad$ (c) $\qquad$ (d) $\qquad$ (e)

▲ FIGURE 26-3
The propane molecule, C$_3$H$_8$
(a) Structural formula. (b) Condensed structural formula. (c) Ball-and-stick model.
(d) Space-filling model. (e) Dashed-wedged line notation.

abound in organic chemistry. We will encounter two types of isomers in this chapter—constitutional isomers and stereoisomers—but our focus in this section is on constitutional isomers. **Constitutional isomers** have different bond connectivities and thus different skeletal structures. For example, C$_4$H$_{10}$ has two constitutional isomers, as shown below:

◀ Because constitutional isomers have different structural formulas, they are sometimes called structural isomers. However, the term is not recommended by IUPAC.

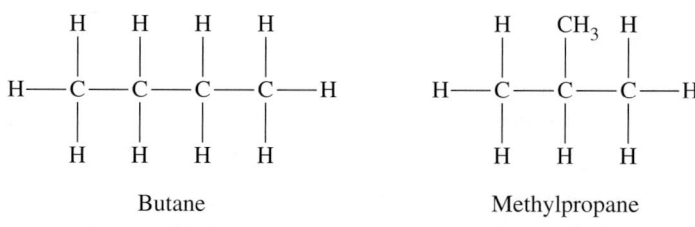

Butane $\qquad\qquad$ Methylpropane

Butane has a single chain of four carbon atoms. Methylpropane has a three-carbon chain with a —CH$_3$ group bonded to the second carbon. Butane is called a straight-chain hydrocarbon—although the molecule does not have a straight shape—and methylpropane is an example of a branched-chain hydrocarbon. Because butane and methylpropane have different structural formulas, they are different compounds and have different physical properties. For example, the boiling point of butane is $-0.5\ °C$ and that of methylpropane is $-11.7\ °C$.

The number of constitutional isomers increases rapidly with the number of carbon atoms. For example, C$_5$H$_{12}$ has only three isomers (see Example 26-1) whereas C$_{10}$H$_{22}$ has 75 and C$_{20}$H$_{42}$ has more than 300,000!

In Chapter 3, we mentioned a way of greatly simplifying the writing of organic structures. We draw lines to represent chemical bonds, and wherever a line ends or meets another line, there is a C atom. We then assume that enough H atoms are bonded to the C atoms to satisfy the need for each C atom to form *four* bonds. Such structural formulas are called *line-angle formulas*, or *line structures*. Notice that in the line-angle formulas written below for the three isomers in Example 26-1, we arrange the lines in a zigzag fashion, which is consistent with the three-dimensional structures of these molecules.

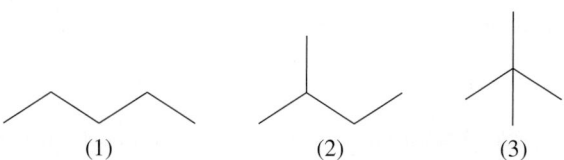

(1) $\qquad\qquad$ (2) $\qquad\qquad$ (3)

Nomenclature

Early organic chemists often assigned names related to the origin or properties of new compounds. Some of these names are still in common use. Citric acid is found in citrus fruit; uric acid is present in urine; formic acid is found in ants (from the Latin word for ant, *formica*); and morphine induces sleep (from *Morpheus*, the ancient Greek god of sleep). As thousands upon thousands of new compounds were synthesized, it became apparent that a system of common names was unworkable. Following several interim systems, one recommended by the International Union of Pure and Applied Chemistry (IUPAC) was adopted.

🔍 26-1   CONCEPT ASSESSMENT

In 1874, van't Hoff and Le Bel published separate papers advancing the hypothesis that the four bonds from a central carbon extend tetrahedrally. This marked the beginning of the field of stereochemistry. Part of their justification was that there was only one known compound with the formula CH_2X_2. Using the compound CH_2F_2 as an example, determine the expected number of isomers if the orientation of the bonds to a carbon atom is tetrahedral and the expected number if the orientation is square-planar.

EXAMPLE 26-1   Identifying Isomers

Write structural formulas for all the constitutional isomers with the molecular formula C_5H_{12}.

Analyze

First we write the longest chain of C atoms and add an appropriate number of H atoms (12, in this case) to give each C atom four bonds. Next, we look for isomers that have fewer carbon atoms in the longest chain.

Solve

The longest chain of carbon atoms that we can draw has five carbon atoms in it. When we add hydrogen atoms to give each carbon atom four bonds, we obtain structure (1).

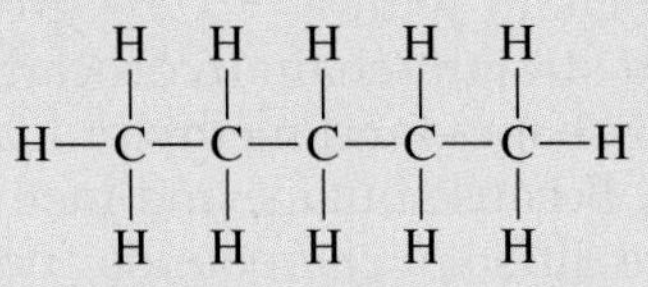

Now, let's look for isomers with four C atoms in the longest chain and one C atom as a branch (five C atoms in all). There is only *one* possibility. Notice that if structure (2) is flipped from left to right, the identical structure (2') is obtained.

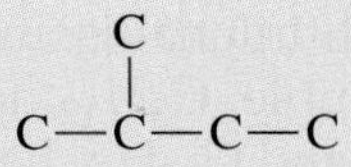

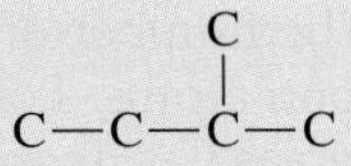

Again, we complete the structure of this isomer by adding H atoms.

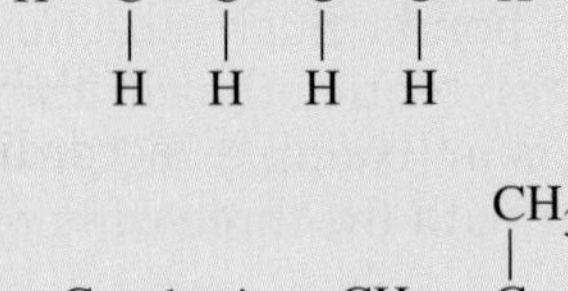

Finally, let's consider a three-carbon chain with two one-carbon branches. Again, there is only *one* possibility.

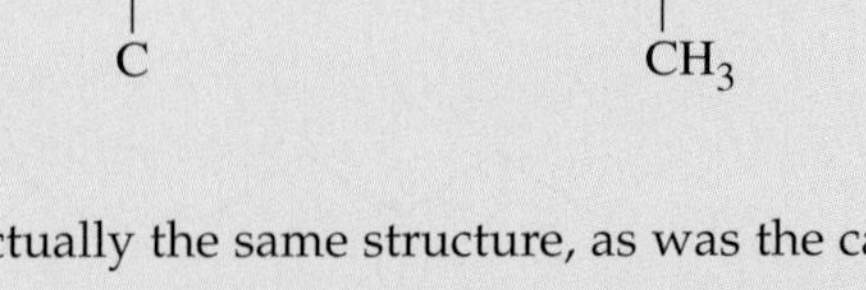

The number of isomers with the formula C_5H_{12} is three.

Assess

We must be careful to recognize when two possible structures are actually the same structure, as was the case for structures (2) and (2').

PRACTICE EXAMPLE A: Write condensed structural formulas for the five constitutional isomers with the formula C_6H_{14}.

PRACTICE EXAMPLE B: Write condensed structural formulas for the nine constitutional isomers with the formula C_7H_{16}.

In this introduction to nomenclature, we will consider only hydrocarbons with all carbon-to-carbon bonds as single bonds. These are known as **saturated hydrocarbons**, or alkanes. We have no trouble naming the first few: CH_4, methane; C_2H_6, ethane; C_3H_8, propane. We encounter our first difficulty with C_4H_{10}, which has two constitutional isomers. This problem is resolved by assigning the name *butane* to the straight-chain isomer, $CH_3CH_2CH_2CH_3$, and *isobutane* to the branched-chain isomer, $CH_3CH(CH_3)_2$. This method is inadequate for the C_5H_{12} alkanes, for which there are three structural isomers (Example 26-1), and it is even less satisfactory for alkanes with greater numbers of C atoms.

To be able to name molecules that have even greater complexity, we must first consider the nature of some of the possible side chains. A side chain is an alkane with one hydrogen atom removed. The resulting group of atoms is called an **alkyl** group, which is named by replacing the ending *-ane* in the corresponding alkane with *-yl*. For example, $—CH_3$ is the methyl group and $—CH_2CH_2CH_3$ is the propyl group. An alkyl side chain is also called a substituent alkyl group because it replaces (substitutes for) a hydrogen atom in the main chain. Table 26.1 shows some common alkyl groups. We see that some of the names incorporate prefixes, such as *sec-*, an abbreviation for secondary, or *tert-*, an abbreviation for tertiary. We use the terms *primary, secondary*, and *tertiary* to provide information about the nature of carbon atoms in organic molecules. A **primary carbon** is attached to one other carbon atom. Carbon atoms at the ends of an alkane chain are always primary carbons. The hydrogen atoms attached to a primary carbon atom are called **primary hydrogen atoms**, and an alkyl group formed by removing a primary hydrogen atom is a primary group. A **secondary carbon** is attached to two other carbon atoms, and a **tertiary carbon**, to three others. Their hydrogen atoms are labeled similarly. As we can see from Table 26.1, the removal of a secondary hydrogen results in the formation of a secondary alkyl group, and the removal of a tertiary hydrogen results in a tertiary alkyl group. Finally, a carbon attached to four carbon atoms is called **quaternary**. The classification of carbon and hydrogen atoms is illustrated in Figure 26-4.

The following rules enable us to name branched-chain hydrocarbons unambiguously as long as we apply the rules in sequence.

1. Select the *longest* continuous carbon chain in the molecule and use the hydrocarbon name of this chain as the base name. Except for the common names methane, ethane, propane, and butane, standard Greek prefixes relate the name to the number of C atoms in the chain, as in *pent*ane (C_5), *hex*ane (C_6), *hept*ane (C_7), *oct*ane (C_8), ...

2. Consider every branch of the main branch to be a substituent alkyl group. Table 26.1 gives the names of common alkyl substituents. When the substituent is more complex, we use the rules given previously to name the side group, bearing in mind that we would change the *-ane* ending to *-yl*.

3. Number the C atoms of the continuous base chain so that the substituents appear *at the lowest numbers* possible.

4. Name each substituent according to its chemical identity and the number of the C atom to which it is attached. For identical substituents use *di, tri, tetra*, and so on, and write the appropriate carbon number for each substituent.

5. Separate numbers from one another with commas but no spaces, and separate numbers from letters with hyphens.

6. List the substituents *alphabetically* by name. When determining alphabetical order, the prefixes *di-, tri-, sec-*, and *tert-* are ignored. Thus, *tert-*butyl, precedes methyl in the name 4-*tert*-butyl-2-methylheptane. However, the prefix *iso-* is not ignored when deciding the alphabetical order.

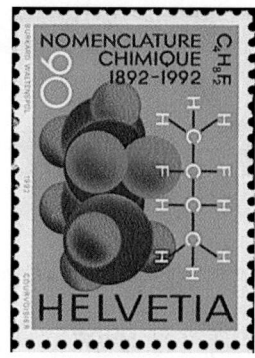

▲ A Swiss stamp commemorating the 100th anniversary of an international congress in Geneva, at which a systematic nomenclature of organic compounds was adopted.

◀ When distinguishing the different types of H and C atoms, the symbols 1°, 2°, 3°, and 4° are often used in place of the words primary, secondary, tertiary, and quaternary.

◀ Do not try to memorize these rules at the outset. Refer to them as you proceed through the examples, and they will become part of your vocabulary of organic chemistry.

26-2   CONCEPT ASSESSMENT

Is it possible that an organic molecule contains a quaternary hydrogen atom?

TABLE 26.1 Some Common Alkyl Groups

Common Name	IUPAC Name	Structural Formula
Methyl	Methyl	$-CH_3$
Ethyl	Ethyl	$-CH_2CH_3$
Propyl[a]	Propyl	$-CH_2CH_2CH_3$
Isopropyl	1-Methylethyl	CH_3CHCH_3 with lower bond
Butyl[a]	Butyl	$-CH_2CH_2CH_2CH_3$
Isobutyl	2-Methylpropyl	$-CH_2CHCH_3$ with CH_3 above
sec-Butyl[b]	1-Methylpropyl	$CH_3CHCH_2CH_3$ with lower bond
tert-Butyl[c]	1,1-Dimethylethyl	CH_3 above and below CH_3CCH_3

[a]In the past, the prefix *normal* or *n-* was used for a straight-chain alkyl group, such as *n*-propyl or *n*-butyl.
[b]*sec* = secondary
[c]*tert* = tertiary

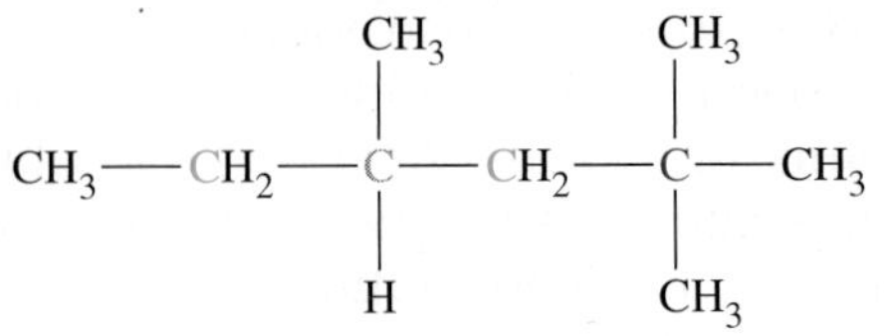

CH$_3$—CH$_2$—C—CH$_2$—C—CH$_3$ with CH$_3$ above on two carbons, H below third carbon, CH$_3$ below fifth carbon

2,2,4-trimethylhexane

C = primary carbon   C = tertiary carbon
C = secondary carbon   C = quaternary carbon

▲ FIGURE 26-4
Classification of carbon and hydrogen atoms
In 2,2,4-trimethylhexane, there are five primary carbons (shown in black), two secondary carbon atoms (shown in blue), one tertiary carbon atom (shown in gray), and one quaternary carbon atom (shown in red). The hydrogen atoms bonded to a primary carbon atom are called primary hydrogen atoms. Similarly, secondary or tertiary hydrogens are bonded, respectively, to secondary or tertiary carbon atoms.

Functional Groups

Organic compounds typically contain elements in addition to carbon and hydrogen. These elements occur as distinctive groupings of one or several atoms called **functional groups**. We have already encountered (in Chapter 3) a

---

EXAMPLE 26-2   Naming an Alkane Hydrocarbon

Give an appropriate IUPAC name for the following compound, an important constituent of gasoline.

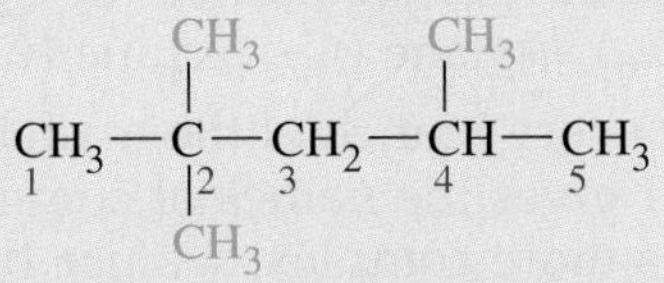

$$CH_3-\underset{2}{\overset{CH_3}{C}}-\underset{3}{CH_2}-\underset{4}{\overset{CH_3}{CH}}-\underset{5}{CH_3}$$
with CH_3 below carbon 2

Analyze

We apply the rules listed above. With practice, you will be able to apply the rules without referring back to the list.

Solve

The C atoms are numbered in red, and the side-chain substituents to be named are shown in blue. The longest chain of C atoms is five, and the carbons are numbered so that the one with two substituent groups is number 2 instead of number 4. Each substituent is a methyl group, $-CH_3$. Two methyl groups are on the second C atom, and one methyl group is on the fourth C atom. The correct name is 2,2,4-trimethylpentane.

Assess

If we had numbered the C atoms from right to left, we would have obtained the name 2,4,4-trimethylpentane. This is *not* an acceptable name because it does not use the *smallest* numbers possible.

PRACTICE EXAMPLE A: Give an IUPAC name for $CH_3CH_2CH(CH_3)CH_2CH_2C(CH_3)_2CH_2CH_3$. (*Hint:* Any CH_3 group enclosed in parentheses is bonded only to the C atom preceding it.)

PRACTICE EXAMPLE B: Give an IUPAC name for $CH_3CH_2CH(CH_3)CH_2CH_2CH(CH_3)CH_2CH_3$. (*Hint:* See the hint given in Practice Example A.)

EXAMPLE 26-3   Writing the Formula to Correspond to the Name of an Alkane Hydrocarbon

Write a condensed structural formula for 4-*tert*-butyl-2-methylheptane.

Analyze

The name tells us that the compound is a heptane with two substituent groups, *tert*-butyl and methyl, positioned at carbons 4 and 2, respectively.

Solve

Because the compound is a heptane, the longest chain of C atoms is seven.

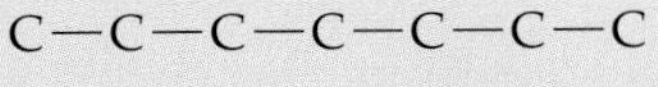

$$C-C-C-C-C-C-C$$

Starting on the left, we attach a methyl group to the second C atom.

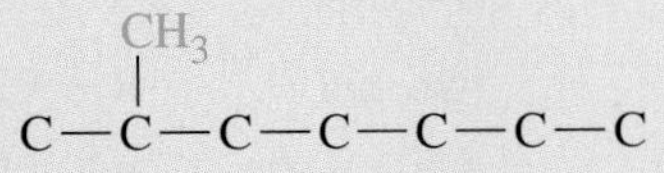

$$\begin{array}{c} CH_3 \\ | \\ C-C-C-C-C-C-C \end{array}$$

Next, we attach a *tert*-butyl group to the fourth C atom.

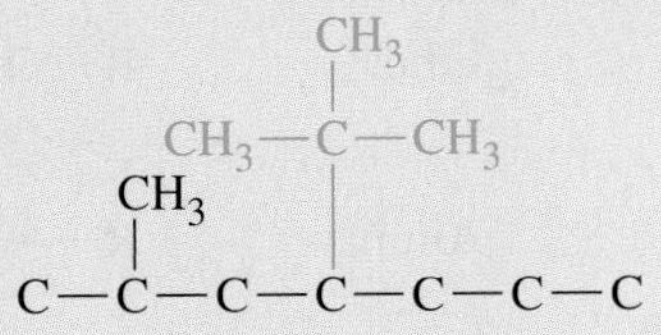

$$\begin{array}{c} CH_3 \\ | \\ CH_3-C-CH_3 \\ CH_3 | \\ | \\ C-C-C-C-C-C-C \end{array}$$

Finally, we add the remaining hydrogen atoms to give each C atom four bonds.

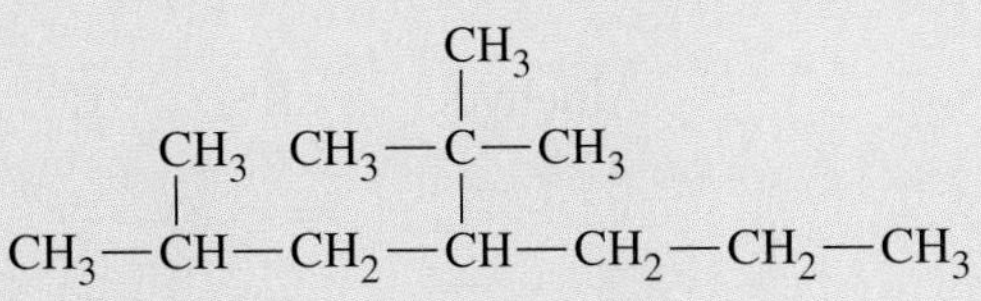

$$\begin{array}{c} CH_3 \\ | \\ CH_3 CH_3-C-CH_3 \\ | | \\ CH_3-CH-CH_2-CH-CH_2-CH_2-CH_3 \end{array}$$

Assess

To check the answer, we use the nomenclature rules given on page 1152 to name the structure we've drawn. We should obtain the name that was given.

PRACTICE EXAMPLE A: Write a condensed structural formula for 3-ethyl-2,6-dimethylheptane.

PRACTICE EXAMPLE B: Write a condensed structural formula for 3-isopropyl-2-methylpentane.

few functional groups, such as the —OH group in alcohols and the —COOH group in carboxylic acids. Table 26.2 lists the major types of organic compounds with their distinctive functional groups shown in red. The physical and chemical properties of organic molecules generally depend on the particular functional groups present. Compounds with the same functional group generally have similar chemical properties. Thus, a convenient way to study organic chemistry is to consider the properties associated with specific functional groups.

In some cases, a functional group simply takes the place of an H atom in a hydrocarbon chain or ring. Such is the case with alcohols and alkyl halides (also called haloalkanes). When naming alcohols and alkyl halides, we must identify the position of the —OH group or halogen atom in the molecule. For example, consider the following possibilities that arise when a Br atom takes the place of a hydrogen atom in pentane:

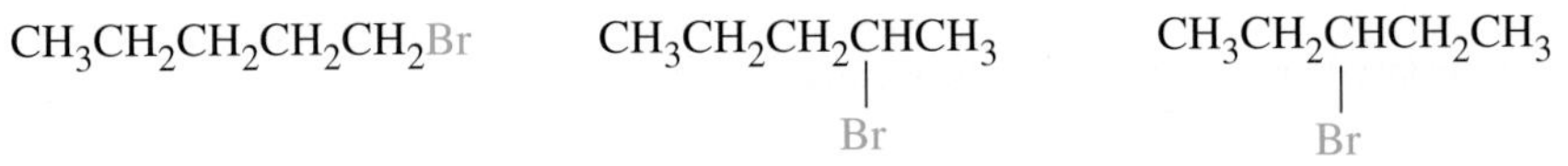

$$CH_3CH_2CH_2CH_2CH_2Br$$

1-Bromopentane

$$\begin{array}{c} CH_3CH_2CH_2CHCH_3 \\ | \\ Br \end{array}$$

2-Bromopentane

$$\begin{array}{c} CH_3CH_2CHCH_2CH_3 \\ | \\ Br \end{array}$$

3-Bromopentane

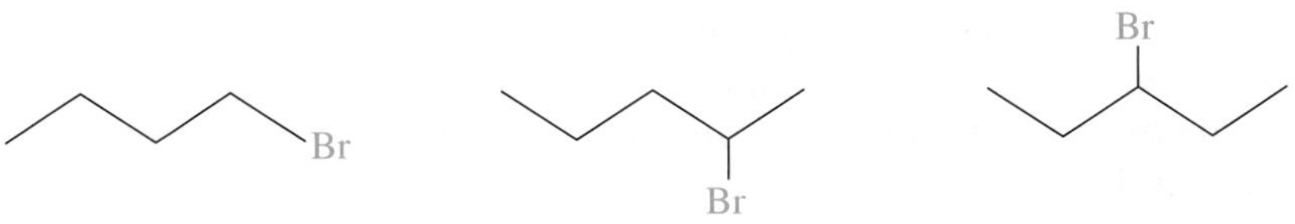

TABLE 26.2 Some Classes of Organic Compounds and Their Functional Groups

Class	General Structural Formula[a]	Example	Name of Example
Alkane	R—H	$CH_3CH_2CH_2CH_2CH_2CH_3$	Hexane
Alkene	$\diagdown C=C \diagup$	$CH_2=CHCH_2CH_2CH_3$	1-Pentene
Alkyne	—C≡C—	$CH_3C≡CCH_2CH_2CH_2CH_2CH_3$	2-Octyne
Alcohol	R—OH	$CH_3CH_2CH_2CH_2OH$	1-Butanol
Alkyl halide	R—X[b]	$CH_3CH_2CH_2CH_2CH_2CH_2Br$	1-Bromohexane
Ether	R—O—R′	$CH_3—O—CH_2CH_2CH_3$	1-Methoxypropane (methyl propyl ether)[c]
Amine	R—NH$_2$	$CH_3CH_2CH_2—NH_2$	1-Aminopropane (propylamine)[c]
Aldehyde	$R-\overset{\overset{\displaystyle O}{\|\|}}{C}-H$	$CH_3CH_2CH_2\overset{\overset{\displaystyle O}{\|\|}}{C}-H$	Butanal (butyraldehyde)[c]
Ketone	$R-\overset{\overset{\displaystyle O}{\|\|}}{C}-R'$	$CH_3CH_2\overset{\overset{\displaystyle O}{\|\|}}{C}CH_2CH_2CH_3$	3-Hexanone (ethyl propyl ketone)[c]
Carboxylic acid	$R-\overset{\overset{\displaystyle O}{\|\|}}{C}-OH$	$CH_3CH_2CH_2\overset{\overset{\displaystyle O}{\|\|}}{C}-OH$	Butanoic acid (butyric acid)[c]
Ester	$R-\overset{\overset{\displaystyle O}{\|\|}}{C}-OR'$	$CH_3CH_2CH_2\overset{\overset{\displaystyle O}{\|\|}}{C}-OCH_3$	Methyl butanoate (methyl butyrate)[c]
Amide	$R-\overset{\overset{\displaystyle O}{\|\|}}{C}-NH_2$	$CH_3CH_2CH_2\overset{\overset{\displaystyle O}{\|\|}}{C}-NH_2$	Butanamide (butyramide)[c]
Arene	Ar—H[d]	⬡—CH_2CH_3	Ethylbenzene
Aryl halide	Ar—X[b]	⬡—Br	Bromobenzene
Phenol	Ar—OH	Cl—⬡—OH	4-Chlorophenol (p-chlorophenol)[c]

[a]The functional group is shown in red. R and R′ represent alkyl groups.
[b]X stands for a halogen atom: F, Cl, Br, or I.
[c]Common name.
[d]Ar stands for an aromatic (*aryl*) group such as the benzene ring.

These three monobromopentanes possess the same carbon skeleton and differ only in the position of the bromine atom on the carbon chain. Notice that we include the appropriate carbon number for the Br substituent when naming the compound.

We will discuss several classes of organic compounds in this chapter, focusing primarily on their structures and properties. In the next chapter, we will focus on some of their characteristic reactions.

 26-3   CONCEPT ASSESSMENT

What is the IUPAC name for $CH_3CH_2CH_2CH(OH)CH_3$? You may find it useful to review Section 3-7.

26-2   Alkanes

In this section we will explore some properties of the alkanes. The essential characteristic of **alkane** hydrocarbon molecules is that they have only single covalent bonds. The bonds in these compounds are said to be *saturated*.

The alkanes range in complexity from methane, CH_4, to molecules containing fifty C atoms or more. Most have the formula C_nH_{2n+2}, and each alkane differs from the preceding one in a sequence by a $-CH_2-$, or *methylene* group. Substances whose molecules differ only by a constant unit such as $-CH_2-$ are said to form a **homologous series**. Members of such a series usually have closely related chemical and physical properties. The data in Table 26.3 indicate that boiling points of alkanes are related to polarizabilities and shapes in the ways discussed in Section 12-1. Intermolecular attractions between the straight-chain molecules are strongest, and these molecules have the highest boiling points. Isomers with more compact structures have lower boiling points.

◀ Alkanes are nonpolar, water-insoluble compounds with relatively low melting points and boiling points.

26-4   CONCEPT ASSESSMENT

Illustrate the meaning of homologous series by writing a homologous series for alkyl halides.

Conformations

With ball-and-stick models, we can visualize an important type of motion in alkane molecules—rotation of groups with respect to one another about the σ bond connecting them. **Conformations** are different spatial arrangements that are possible in a molecule. One conformation can be converted into another by rotations about σ bonds. In Figure 26-5, we focus on one of the many possible conformations of the CH_3CH_3 molecule. The spatial arrangement of the H atoms can be seen more clearly if we view the molecule along the $C-C$ axis, as suggested in Figure 26-5(a). When the molecule is viewed in this way, the carbon atom toward the rear is obscured by the one in front, as shown in Figure 26-5(b), but the bonds to the hydrogen atoms are clearly seen. The view along the $C-C$ bond is shown in a slightly different way in Figure 26-5(c). In this

TABLE 26.3   Boiling Points of Some Isomeric Alkanes

Formula	Isomer	Boiling Point, °C	Formula	Isomer	Boiling Point, °C
C_4H_{10}	Butane	−0.5	C_6H_{14}	Hexane	68.7
	Methylpropane	−11.7		3-Methylpentane	63.3
C_5H_{12}	Pentane	36.1		2-Methylpentane	60.3
	2-Methylbutane	27.9		2,3-Dimethylbutane	58.0
	2,2-Dimethylpropane	9.5		2,2-Dimethylbutane	49.7

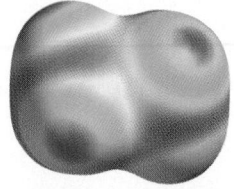

▲ FIGURE 26-5
Staggered conformation of ethane
When the ethane molecule is viewed along the C—C axis, as suggested in **(a)**, the rear carbon atom is obscured by the carbon atom in front, as shown in **(b)**. In the Newman projection, shown in **(c)**, the front carbon is located at the intersection of the three arms of the inverted Y and the rear carbon is represented by a circle.

▶ Newman projections are named after an organic chemistry professor, Melvin S. Newman, from Ohio State University. He introduced these representations in 1952 to help students understand conformations, stereochemistry, and symmetry of organic molecules.

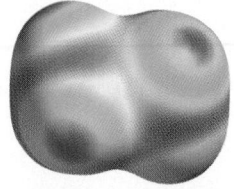

Staggered conformation

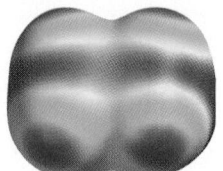

Eclipsed conformation

KEEP IN MIND

that the barrier to rotation in ethane (12 kJ mol⁻¹) is comparable to the strengths of various intermolecular forces we discussed in Chapter 12.

representation, called a *Newman projection,* the carbon atom toward the front is located at the point where the lines representing the three carbon–hydrogen bonds intersect with each other. The carbon atom toward the rear is depicted by a circle and its bonds project from the outer edge of the circle. Newman projections are used to represent the many different spatial arrangements of atoms that result from rotations about a σ bond. The arrangement (or conformation) shown in Figure 26-5 is called the *staggered conformation*. In this conformation, the carbon–hydrogen bonds in one —CH$_3$ group are positioned exactly halfway between those of the other —CH$_3$ group and the H atoms are located a maximum distance apart.

A second conformation can be obtained by rotating one of the methyl groups in the staggered conformation by 60° about the C—C axis, as suggested in Figure 26-6. When the resulting conformation is viewed along the C—C axis, all the hydrogen atoms on the first carbon atom are directly in front of those on the second carbon atom. This is called the *eclipsed* conformation. To make the three rear hydrogen atoms more visible in the Newman projection, they are drawn slightly out of the perfectly eclipsed position. The eclipsed and staggered conformations represent two extremes and all the possible conformations in between are called collectively *skew conformations*.

Electrostatic potential maps for the staggered and eclipsed conformations of ethane are shown in the margin. In the staggered conformation, the hydrogen atoms (in the blue regions) are as far apart from one another as possible, and in the eclipsed conformation, they are as close as possible.

Energy is required to convert from the staggered conformation to the eclipsed conformation, and for ethane, the energy required is about 12.0 kJ mol⁻¹. Because of this energy requirement, there is a *barrier to internal rotation*, and thus the —CH$_3$ groups in ethane do not rotate entirely freely about the C—C bond. However, the barrier to internal rotation is small enough that, at room

▶ FIGURE 26-6
Staggered and eclipsed conformations of ethane
When the methyl group on the right is rotated by 60° about the C—C bond, the staggered conformation of ethane is converted into the eclipsed conformation.

Staggered   Eclipsed

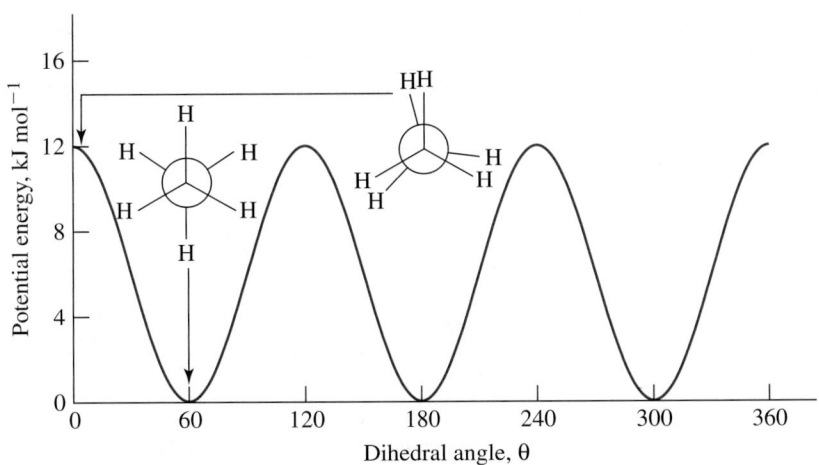

The dihedral angle, θ, refers to the angle of rotation about a carbon–carbon bond.

▲ FIGURE 26-7
Potential energy diagram for the internal rotation of the methyl groups in ethane

temperature, molecules interconvert rapidly from one conformation to another. For this reason, rotation about the C—C bond is sometimes called *free rotation*. The barrier to internal rotation arises because as the molecule converts from the staggered conformation through all the skew conformations to the eclipsed conformation, the C—H bonds of one —CH$_3$ group draw closer to the C—H bonds of the other —CH$_3$ group and the electrons in these bonds experience increased repulsion. The eclipsed conformation is highest in energy because in this conformation, the electrons in the C—H bonds of one methyl group are close to the electrons in the C—H bonds of the other methyl group.

The conversion from one conformation to another can be followed by using the dihedral angle, θ, which refers to the angle of rotation about the carbon–carbon bond, as shown in the margin. When $\theta = 0°$, the molecule is in the eclipsed conformation, and when $\theta = 60°$, it is in the staggered conformation. The energy changes that occur as the ethane molecule converts from one conformation to another are shown graphically in Figure 26-7. The difference in energy between the eclipsed and staggered conformations is called the rotational or **torsional energy**. In the eclipsed conformation, there are three C—H bond interactions, and thus each C—H bond interaction contributes 4.0 kJ mol^{-1}.

Let's consider rotation about a C—C bond in propane, CH$_3$CH$_2$CH$_3$, which is the next member of the homologous series. The potential energy diagram of propane is similar to that of ethane. However, an important difference is that the energy difference between the eclipsed and staggered conformations, and thus the barrier to rotation about a C—C bond, is slightly greater in propane (13.6 kJ mol^{-1}) than it is in ethane (12.0 kJ mol^{-1}). Newman projections for the conformations of propane can help us understand this difference. The Newman projections for the staggered and eclipsed conformations of propane, shown in the margin, are similar to those of ethane except that one of the H atoms has been replaced by a methyl group. The eclipsed conformation of propane has two C—H bond interactions and an interaction involving a C—H bond and a carbon–methyl bond. If we assume that each C—H bond interaction contributes 4.0 kJ mol^{-1} to the barrier to rotation, as was the case in ethane, then the interaction between the carbon–hydrogen bond and the carbon–methyl bond contributes $(13.6 - 2 \times 4.0) = 5.6$ kJ mol^{-1}. Thus, the interaction between a carbon–hydrogen bond and a carbon-methyl bond is slightly more repulsive than the interaction of two C—H bonds.

The next member in the homologous series is butane, CH$_3$CH$_2$CH$_2$CH$_3$. If we number the carbon atoms 1 through 4, then we have two distinct C—C bonds, C1—C2 and C2—C3, about which conformers can be formed. (The C3—C4 bond axis is equivalent to the C1—C2 bond axis because it does not

Staggered

Eclipsed

matter whether we number the carbon atoms from left to right or right to left.) When the butane molecule is viewed along the C1—C2 bond, we obtain the following Newman projections for the staggered and eclipsed conformations:

▶ When drawing Newman projections for molecules, it is customary to put the lower-numbered carbon atom at the front of the Newman projection.

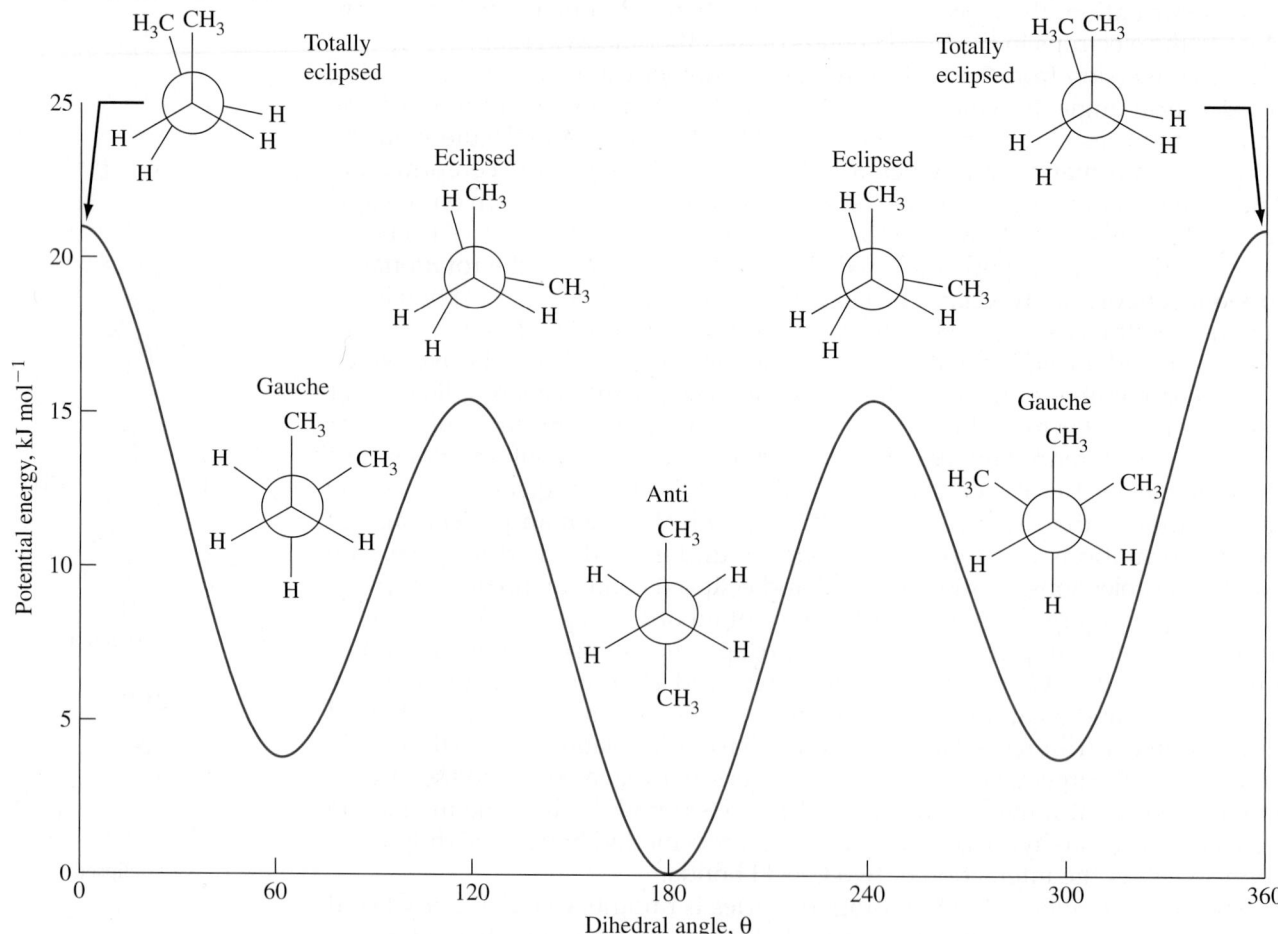

Staggered   Eclipsed

🔍 **26-5   CONCEPT ASSESSMENT**

Will all the eclipsed conformers for rotation about the C1—C2 axis in butane have the same energy?

When the butane molecule is viewed along the C2—C3 bond, several eclipsed and staggered conformations can be identified (Figure 26-8). There

▲ FIGURE 26-8
Conformations and their potential energies of butane for rotation about the C2—C3 bond

are two distinct eclipsed conformations, eclipsed and totally eclipsed, and two distinct staggered conformations, anti and gauche. The totally eclipsed is highest in energy because in this conformation, the two largest constituents are eclipsed and interfere with each other, as suggested in Figure 26-9. The interference or crowding of large substituents is called *steric hindrance*.

Of the conformations shown in Figure 26-8, the anti conformation is the lowest in energy because the methyl groups are as far apart as possible. This conformation is called the **anti** conformation because the methyl groups are diagonally opposite each other. The other staggered conformation has the methyl groups to the left and right of each other and is called the **gauche** conformation. (*Gauche* is a French word meaning "left" or "awkward.")

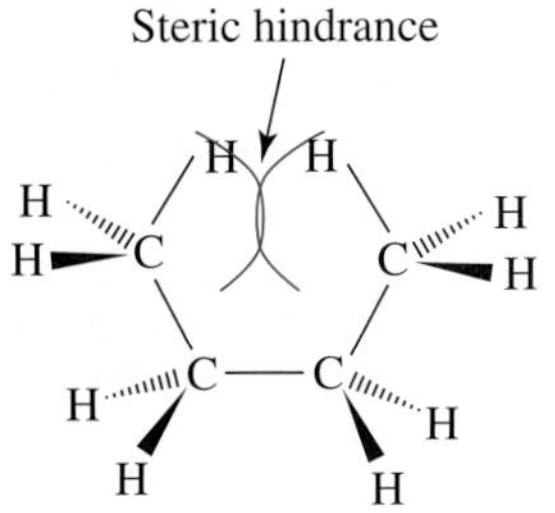

Steric hindrance

▲ FIGURE 26-9
Steric hindrance in a totally eclipsed conformation of butane
In the totally eclipsed conformation, the two methyl groups interfere with each other.

EXAMPLE 26-4   Choosing the Most Stable Conformer of an Alkane

Draw the conformation of 2,3-dimethylpentane that is lowest in energy when the molecule is viewed along C2—C3 axis.

Analyze

We must first draw a structural diagram for the molecule and identify the C2—C3 axis and then draw a Newman projection corresponding to a view along the C2—C3 bond. The lowest energy structure is the one that minimizes the repulsions among the substituents.

Solve

A structural diagram for 2,3-dimethylpentane is given here.

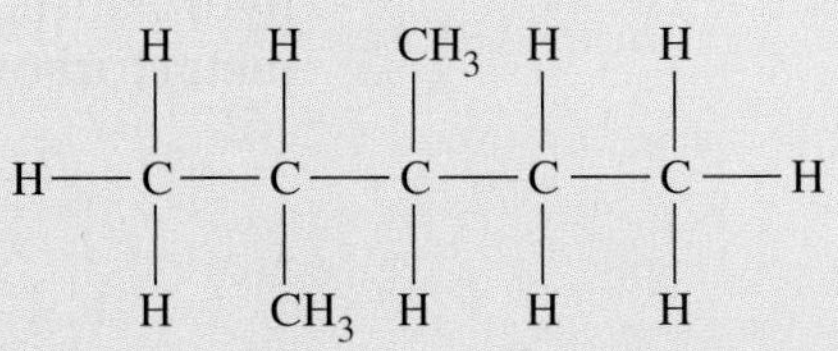

A hydrogen atom and two methyl groups are bonded to C2. Three different groups are bonded to C3: a hydrogen atom, a methyl group, and an ethyl group. The conformation of lowest energy is the one in which the alkyl groups are staggered. In addition to this, we should minimize the number of gauche interactions between the larger groups.

We must now construct a Newman projection corresponding to the view along the C2—C3 bond. First, we draw a circle to represent the rear carbon, and then we add lines for the bonds formed by the front carbon atom. Finally, we add lines for the bonds formed by the rear carbon atom.

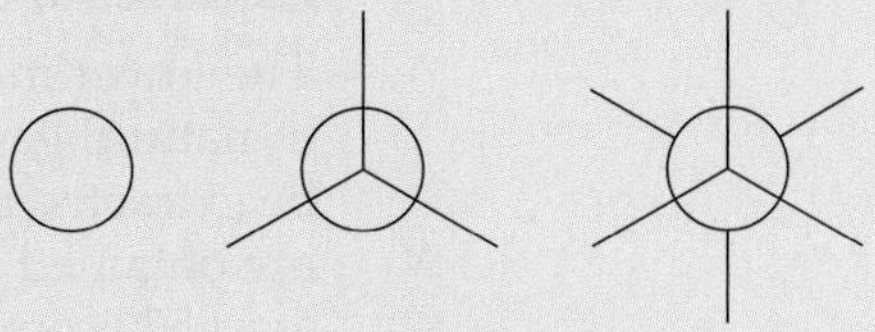

We are now ready to attach the groups to construct 2,3-dimethylpentane. Let us add two methyl groups and a hydrogen atom to the second carbon atom. Note it does not matter where we place the groups.

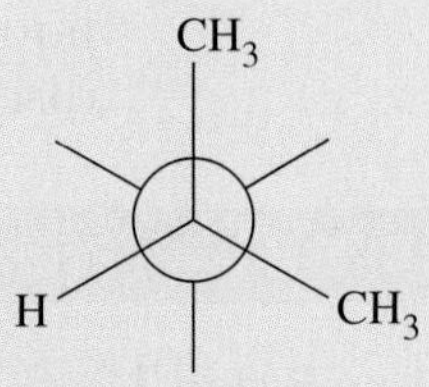

We must now add a hydrogen atom, as well as ethyl and methyl groups to the rear carbon. There are three possible staggered conformations:

(a)   (b)   (c)

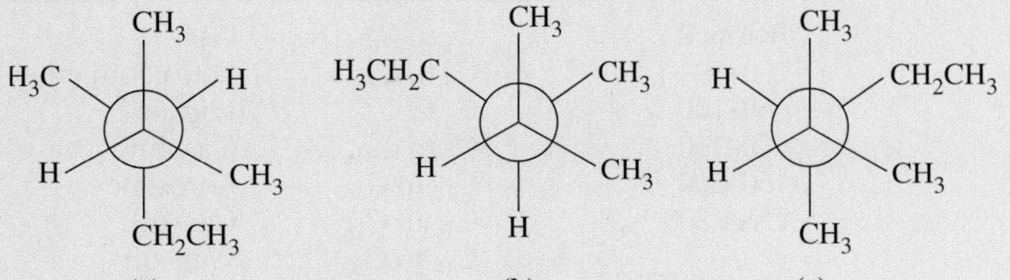

(continued)

Which one of these conformations is lowest in energy? Conformation (a) has two gauche interactions, but conformations (b) and (c) have three such interactions. Therefore, conformation (a) has the lowest energy.

Assess

In solving this problem, we considered rotation about a particular carbon–carbon bond. Keep in mind that rotations about other carbon–carbon bonds also occur simultaneously, a complication that we will not consider.

PRACTICE EXAMPLE A: Draw Newman projections for the staggered conformations of 2-methylpentane when the molecule is viewed along the C1—C2 bond. Rank the conformations in order of increasing energy (from lowest to highest).

PRACTICE EXAMPLE B: Sketch a potential energy diagram for rotation about the C1—C2 axis in 1-chloropropane.

Preparation of Alkanes

The chief source of alkanes is petroleum, as we describe at the end of this section, but several laboratory methods are also available for their preparation. In the presence of a metal catalyst such as Pt, Pb, or Ni, unsaturated hydrocarbons, whether containing double or triple bonds, may be converted to alkanes by the addition of H atoms to the multiple bond systems.

► Figure 23-3 gives a schematic representation of the role played by the catalyst in equation (26.1).

$$CH_2{=}CH_2 + H_2 \xrightarrow{\text{Pt, Pd, or Ni}} CH_3{-}CH_3 \qquad (26.1)$$

In another type of reaction, halogenated hydrocarbons react with alkali metals to produce alkanes of double the carbon content (equation 26.2). Finally, alkali metal salts of carboxylic acids can be fused with alkali metal hydroxides. Sodium carbonate and an alkane with one carbon fewer than the metal carboxylate are formed (equation 26.3).

$$2\,CH_3CH_2Br + 2\,Na \xrightarrow{\text{heat/pressure}} 2\,NaBr + CH_3CH_2CH_2CH_3 \qquad (26.2)$$

$$CH_3{-}\overset{\overset{\displaystyle O}{\|}}{C}{-}ONa + NaOH \xrightarrow{\Delta} Na_2CO_3 + CH_4 \qquad (26.3)$$

Alkanes from Petroleum

The lower molecular mass alkanes, methane and ethane, are found principally in natural gas. Propane and butane are found dissolved in petroleum; they can be extracted as gases and sold as liquefied petroleum gas (LPG). Higher alkanes are obtained by the fractional distillation of petroleum, a complex mixture of at least 500 compounds. The main petroleum fractions are listed in Table 26.4. (The liquid–vapor equilibrium principles that underlie fractional distillation were discussed on page 575.)

TABLE 26.4 Principal Petroleum Fractions

Boiling Range, °C	Composition	Fraction	Uses
Below 0	C_1 to C_4	Gas	Gaseous fuel
0–50	C_5 to C_7	Petroleum ether	Solvents
50–100	C_6 to C_8	Ligroin	Solvents
70–150	C_6 to C_9	Gasoline	Motor fuel
150–300	C_{10} to C_{16}	Kerosene	Jet fuel, diesel oil
Over 300	C_{16} to C_{18}	Gas–oil	Diesel oil, cracking stock
—	C_{18} to C_{20}	Wax–oil	Lubricating oil, mineral oil, cracking stock
—	C_{21} to C_{40}	Paraffin wax	Candles, wax paper
—	above C_{40}	Residuum	Roofing tar, road materials, waterproofing

Not all the hydrocarbons found in gasoline are equally desirable because some hydrocarbons burn more smoothly than others. (Explosive burning results in engine knocking.) The hydrocarbon *2,2,4-trimethylpentane*, a structural isomer of octane, has excellent engine performance, and it is given an octane rating of 100. *Heptane* has poor engine performance; its octane rating is 0. These two hydrocarbons serve as a reference system for rating gasoline. By identifying the composition of the *n*-heptane and 2,2,4-trimethylpentane (isooctane) mixture that matches the performance characteristics of the gasoline being tested, an octane rating can be assigned. For example, a gasoline that gives the same performance as a mixture of 87% 2,2,4-trimethylpentane and 13% heptane is assigned an octane number of 87. In general, branched-chain hydrocarbons have higher octane ratings (burn more smoothly) than their straight-chain counterparts.

▲ A catalytic cracking unit (cat cracker) at a petroleum refinery.

$$CH_3-\underset{\underset{CH_3}{|}}{\overset{\overset{CH_3}{|}}{C}}-CH_2-\underset{\underset{H}{|}}{\overset{\overset{CH_3}{|}}{C}}-CH_3 \qquad\qquad CH_3-CH_2-CH_2-CH_2-CH_2-CH_2-CH_3$$

2,2,4-Trimethylpentane
Octane rating: 100

n-Heptane
Octane rating: 0

Gasoline obtained by fractional distillation of petroleum has an octane number of 50–55 and is not acceptable for use in automobiles, which require fuels with octane numbers near 90. Extensive modifications of the gasoline fraction are required. In *thermal cracking*, large hydrocarbon molecules (called *cracking stock*) are broken down into molecules in the gasoline range, and the presence of special catalysts promotes the production of branched-chain hydrocarbons. For example, the molecule $C_{15}H_{32}$ might be broken down into C_8H_{18} and C_7H_{14}. The process of *re-forming*, or isomerization, converts straight-chain to branched-chain hydrocarbons and other types of hydrocarbons having higher octane numbers. In thermal and catalytic cracking, some low-molecular mass hydrocarbons are rejoined into higher molecular mass hydrocarbons by a process known as *alkylation*.

◀ Not only do the processes of cracking, re-forming, and alkylation produce a higher grade of gasoline but they also increase the yield of gasoline obtained from crude oil.

The octane rating of gasoline can be further improved by adding antiknock compounds to prevent premature combustion. At one time, the preferred additive was tetraethyllead, $(C_2H_5)_4Pb$. Lead additives have been phased out of gasoline in most countries because lead is toxic, and substitutes such as the oxygenated hydrocarbons methanol and ethanol are used instead.

26-3   Cycloalkanes

Alkanes in chain structures have the formula C_nH_{2n+2}. However, alkanes can also exist in ring, or cyclic, structures; such alkanes are called *cycloalkanes*. For simple cycloalkanes, we can think of the rings as having formed by the joining of the two ends of a straight-chain alkane after the elimination of an H atom from each end. Simple cycloalkanes have the formula C_nH_{2n}. Here are a few examples:

Cyclopropane
C_3H_6

Cyclobutane
C_4H_8

Cyclopentane
C_5H_{10}

Cyclohexane
C_6H_{12}

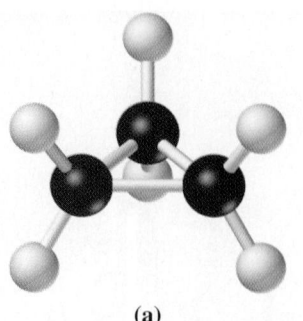

(a)

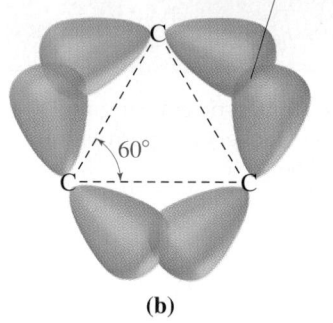

No "head-on" overlap of atomic orbitals

(b)

▲ FIGURE 26-10
Ring strain in cyclopropane
(a) Ball-and-stick model.
(b) Valence-bond picture of cyclopropane using sp^3 hybrid orbitals, which leads to poor overlap of the orbitals and hence weak bonds.

▶ The C—C bond energy in cyclopropane is about 289 kJ mol^{-1}. This is substantially smaller than the C—C bond energy of 347 kJ mol^{-1} given in Table 10.3.

The line-angle representations shown at the bottom of page 1161 might lead you to believe that the carbon atoms in cycloalkanes all lie in the same plane. However, this is not generally the case. As we will soon see, cyclopropane is the only cycloalkane in which the carbon atoms form a planar ring.

We can use the nomenclature rules on page 1151 to name a cycloalkane having substituent groups. For example, we would name the following compound 1,1,3-trimethylcyclopentane:

Ring Strain in Cycloalkanes

Some discussion of the bonding in cycloalkanes is warranted because, in these molecules, the sp^3-hybridized carbons do not necessarily form bonds that are 109.5° apart. For example, the C—C—C bond angles in cyclopropane are only 60° (Fig. 26-10). Because the bond angle in cyclopropane is much smaller than the ideal bond angle of 109.5°, the cyclopropane molecule is "strained." As illustrated in Figure 26-10(b), the carbon sp^3 orbitals in propane do not overlap as extensively as they do in straight-chain alkanes. As a result of the poor orbital overlap, the C—C bonds in cyclopropane are substantially weaker than they are in propane and other straight-chain alkanes, and cyclopropane is considerably more reactive than a straight-chain alkane.

We can use heats of combustion to estimate the amount of ring strain in cycloalkanes. The heats of combustion of propane, butane, pentane, and hexane are given in Table 26.5 along with those of the corresponding cycloalkanes. For the straight-chain alkanes, the heat of combustion changes by about 658 kJ mol^{-1} with each CH_2 group added. This suggests that each CH_2 group contributes, on average, 658 kJ mol^{-1} to the heat of combustion. The formula of a cycloalkane may be written as $(CH_2)_n$ and, if we assume that the C—C bonds that link together CH_2 groups in cycloalkanes are the same as they are in straight-chain alkanes, then the heat of combustion of a cycloalkane should be about $-n \times 658 \text{ kJ mol}^{-1}$. The energy associated with ring strain is released as heat when the compound is burned and thus, the heat of combustion is more negative than expected. The estimated and experimental heats of combustion for a few cycloalkanes are given in Table 26.5. The difference between the estimated and experimental values provides a measure of the ring strain in the cycloalkanes, and these values are also shown in the table. The data in Table 26.5 show that, as we proceed from cyclopropane to cyclohexane, the amount of ring strain decreases. Interestingly, the cyclohexane ring is essentially free of ring strain. We will soon see why.

The data in Table 26.5 show that cyclopropane and cyclobutane experience a significant amount of ring strain. However, the *ring strain per CH_2 group* is

TABLE 26.5 Heats of Combustion and Ring Strain in Some Cycloalkanes (All Values in kJ mol^{-1})

Alkane	Experimental ΔH°_{comb}	Contribution to ΔH°_{comb} from Each Additional CH_2 Group	Cycloalkane	Experimental ΔH°_{comb}	Estimated ΔH°_{comb}[b]	Ring Strain[c]
propane	−2220		cyclopropane	−2092	−1974	118
butane	−2877	−657[a]	cyclobutane	−2744	−2632	112
pentane	−3536	−659	cyclopentane	−3320	−3290	30
hexane	−4194	−658	cyclohexane	−3948	−3948	0

[a]The difference between successive values of ΔH°_{comb} is the contribution to ΔH°_{comb} made by adding another CH_2 group. For example, ΔH°_{comb} (butane) − ΔH°_{comb} (propane) = -657 kJ mol^{-1}.
[b]If there is no ring strain, then the heat of combustion of a cycloalkane should be $-n \times 658 \text{ kJ mol}^{-1}$.
[c]The ring strain is equal to the difference between the estimate and experimental values of ΔH°_{comb}.

substantially higher in cyclopropane (39 kJ mol^{-1}) than it is in cyclobutane (28 kJ mol^{-1}). The cyclopropane ring is strained not only because of poor orbital overlap but also because the carbon–hydrogen bonds in this molecule are eclipsed. This is most evident from the Newman projection of cyclopropane, which is shown in Figure 26-11.

In cyclopropane, the carbon atoms lie in the same plane, but in other cycloalkanes, they do not. For example, both the cyclobutane and the cyclopentane molecules buckle slightly to give rings that are puckered rather than planar, as shown below.

Cyclobutane   Cyclopentane

▲ FIGURE 26-11
Newman projection for cyclopropane
The carbon atom in front (C1) and the carbon atom in the rear (C2) are both bonded to the other carbon atom (C3). The H—C1 and H—C2 bonds are eclipsed.

When these molecules buckle, some or all of the hydrogen atoms move slightly out of the totally eclipsed conformation, relieving repulsion associated with the eclipsing of carbon–hydrogen bonds but at the expense of decreasing the already strained C—C—C bond angles even further.

For cyclohexane, two conformations of the molecules are important. They are shown as ball-and-stick models in Figure 26-12. These are called the *boat* and the *chair conformations*. The boat conformation is less stable (29 kJ mol^{-1} higher in energy) than the chair conformation. There are other conformations of cyclohexane, but we will not discuss them. You may encounter them in more advanced organic chemistry courses.

Cis–Trans Isomerism in Disubstituted Cycloalkanes

The ring in a cycloalkane has two distinct faces, and the substituents bonded to a particular carbon atom are adjacent to opposite faces of the ring. Let's consider, for example, the ball-and-stick model of cyclopropane, which was shown in Figure 26-10. If we focus on the hydrogen atoms bonded to the rearmost carbon, for example, then we see that one of the hydrogen atoms is above the plane of the three carbon atoms and the other is below the plane. Stated another way, one hydrogen atom is adjacent to the upper face and the other is adjacent to the lower face. If hydrogen atoms

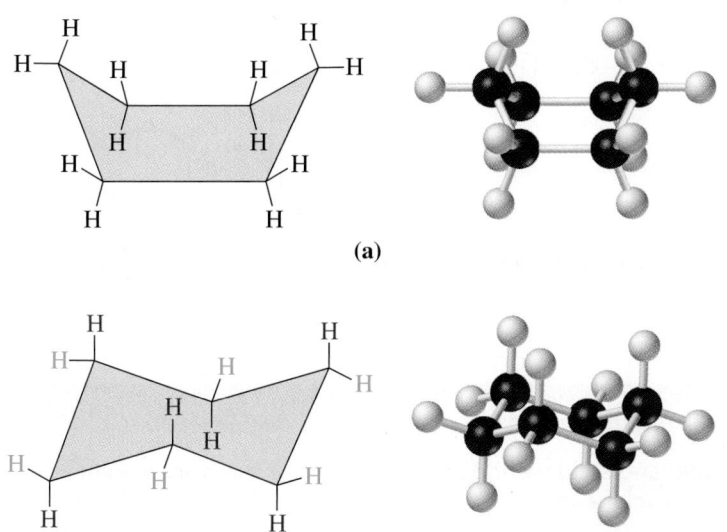

(a)

(b)

◄ FIGURE 26-12
Two important conformations of cyclohexane
(a) Boat form. (b) Chair form. The H atoms extending laterally from the ring—equatorial H atoms—are shown in blue. The H atoms projecting above and below the ring—axial H atoms—are in red.

on different carbon atoms are replaced by other substituents, as is the case in 1,2-dimethylcyclopropane or 1,2-dimethylhexane, then various isomers are possible. The *cis* and *trans* isomers of 1,2-dimethylcyclopropane are shown below. (*cis* is a Latin word meaning "on this side"; *trans* is Latin for "across.")

Side view

Top view

cis-1,2-dimethylcyclopropane   *trans*-1,2-dimethylcyclopropane

The *cis* and *trans* isomers of 1,2-dimethylcyclohexane are

trans-1,2-dimethylcyclohexane   *cis*-1,2-dimethylcyclohexane

► Conformers are interconverted by rotation about bonds. However, stereoisomers can be interconverted only by breaking and reforming bonds.

In the *cis* isomers, the two substituents are adjacent to the same face of the ring and in the *trans* isomers they are adjacent to opposite faces. To convert the *cis* isomer into the *trans* isomer (or vice versa), bonds must be broken and reformed. **cis–trans isomerism** is only one type of a more general kind of isomerism known as **stereoisomerism**. (This term is derived from the Greek word *stereos*, meaning "solid, or three-dimensional, in nature"). For stereoisomers, the number and types of atoms and bonds are the same, but certain atoms are oriented differently in space.

For disubstituted cycloalkanes, many isomers are possible. Consider chloromethylcyclohexane, for example. It has eight isomers in total, three of which are shown below:

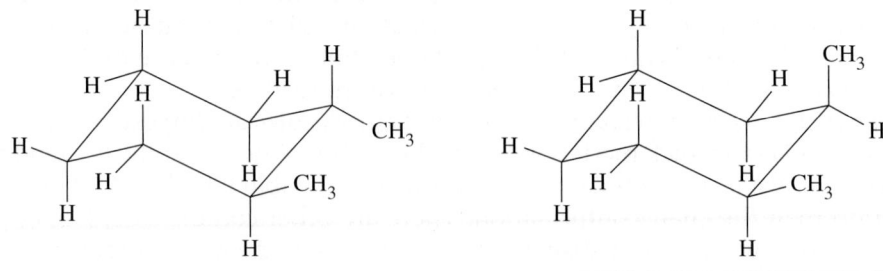

Chloromethylcyclohexane   *cis*-1-chloro-2-   1-chloro-1-
                             methylcyclohexane   methylcyclohexane

🔍 **26-6   CONCEPT ASSESSMENT**

Draw dashed and solid wedge line structures for the *trans* isomers of chloromethylcyclohexane.

A Closer Look at Cyclohexane

The bond angles are approximately 109.5° in both the chair and the boat forms of cyclohexane. However, the chair form of cyclohexane is slightly lower in energy. Why is this? To answer this question, we will first describe how to draw the chair form of cyclohexane and then use a Newman projection to provide the insight needed.

1. Draw two parallel lines that are slightly tilted.

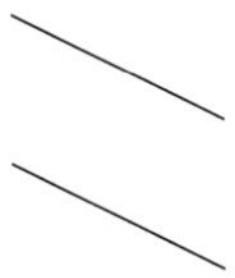

2. Connect the lower ends with a cap that points upward.

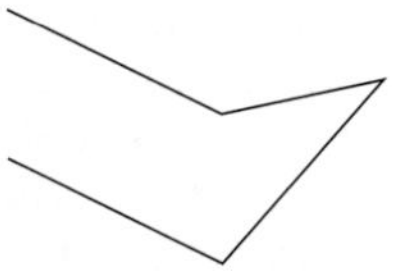

3. Connect the upper ends with a cap that points downward.

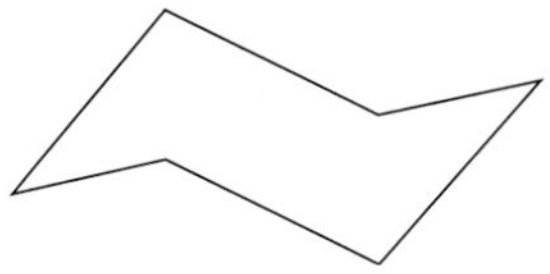

4. Add an axial hydrogen atom to each carbon atom. The bonds to the axial hydrogen atoms point up when the carbon atom points upward, and down when the carbon atom points downward. They are parallel to an imaginary line that passes through the center of the cyclohexane ring.

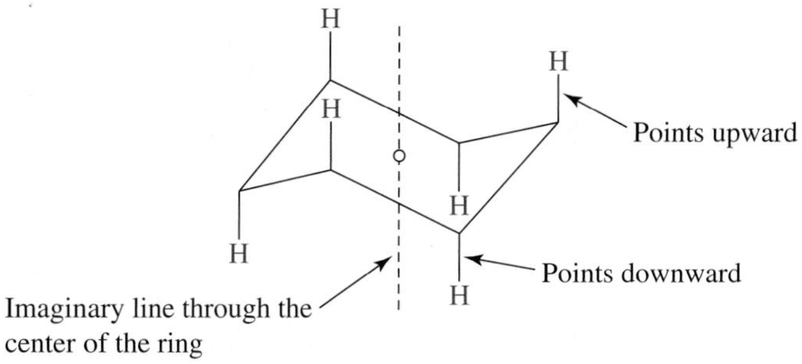

Points upward

Points downward

Imaginary line through the center of the ring

5. Add an equatorial hydrogen atom to each carbon atom. The bonds to the equatorial hydrogen atoms point sideways and complete the tetrahedral arrangement of bonds around each carbon atom. In this diagram, the axial hydrogens are shown in red and the equatorial hydrogens are shown in blue.

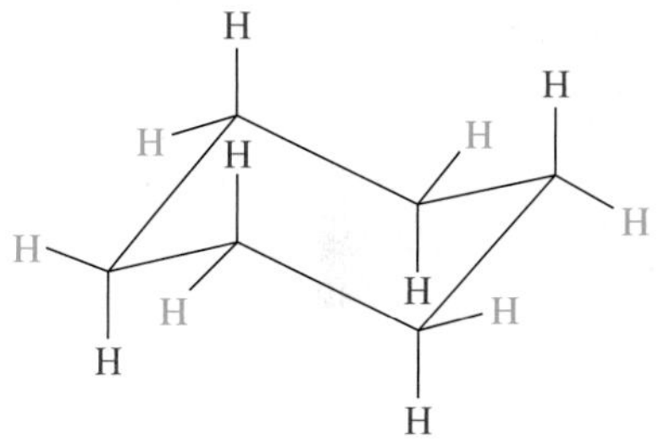

As we can see from these drawings, there are two types of hydrogen atoms: *axial* and *equatorial*. The bonds to the axial hydrogen atoms are directed vertically up or down and are parallel to an imaginary axis that passes through the center of the ring. The bonds to the equatorial hydrogen atoms are directed sideways from the ring. As we move around the ring, the axial hydrogen atoms alternate such that one points upward and the next points downward. There is a similar alternation for the equatorial hydrogen atoms. For the equatorial hydrogen atoms, it is instructive to note which pairs of C—H and C—C bonds are parallel.

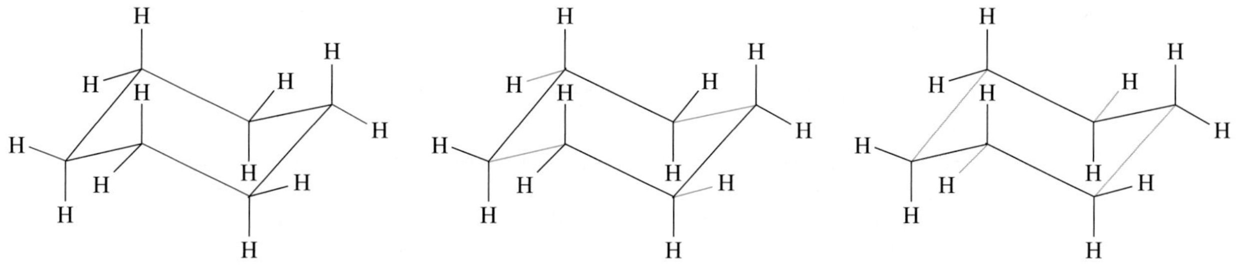

It is possible to construct a Newman projection for the chair form of cyclohexane if we view the molecule along a pair of C—C bonds that are parallel to each other, as suggested below:

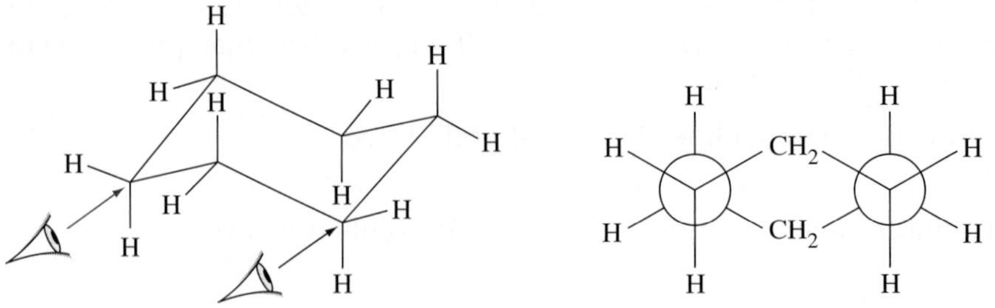

▶ Building a molecular model will help you visualize many of the concepts being discussed in this section. Build a model of cyclohexane to convince yourself that the same Newman projection is obtained when the molecule is viewed along the other two pairs of parallel C—C bonds.

Line structure

Newman projection

The Newman projection above shows that the carbon–hydrogen bonds are staggered. Some of the carbon–hydrogen bonds in the boat form are eclipsed, and consequently the boat form is higher in energy (and is less stable) than the chair form.

The cyclohexane molecule interconverts rapidly between two stable chair conformations, as suggested in Figure 26-13. This interconversion is called a ring flip. When the ring flips from one chair conformation to another, every axial hydrogen in one conformation becomes an equatorial hydrogen in the other conformation, and vice versa. At room temperature, the cyclohexane ring undergoes approximately 100,000 ring flips per second.

The two chair conformations of cyclohexane shown in Figure 26-13 are equivalent and have exactly the same energy. However, when H atoms in cyclohexane are replaced by substituents, the chair conformations no longer have the same energy. Consider, for example, the following chair conformations for methylcyclohexane:

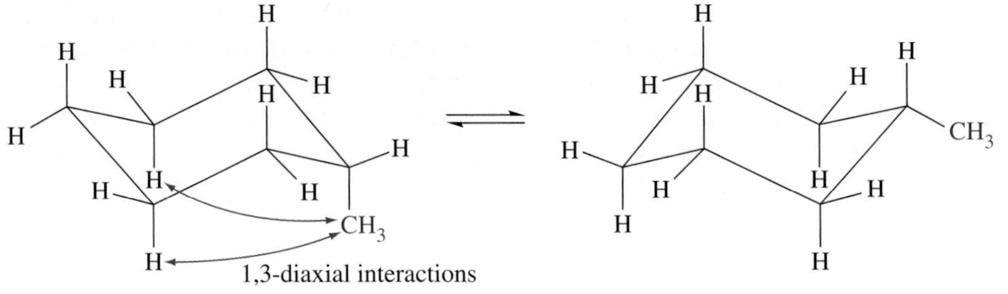

1,3-diaxial interactions

Less stable

More stable

The conformer with the methyl group in the equatorial position is of lower energy than the one with the methyl group in the axial position because, when

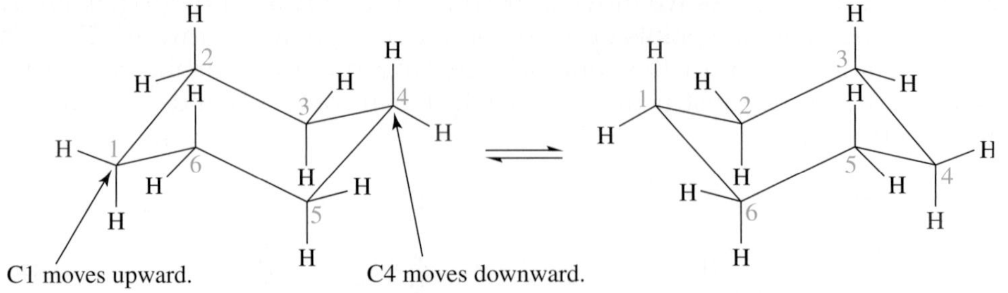

C1 moves upward.

C4 moves downward.

▲ FIGURE 26-13
The interconversion of two chair conformations in cyclohexane
In the figure, two of the H atoms are shown in red to emphasize that when the cyclohexane ring converts from one chair conformation to another, the equatorial hydrogen atoms are converted into axial hydrogen atoms and vice versa. The interconversion of the two chair forms proceeds through other conformations, including the boat form.

the methyl group is in an axial position, it interacts simultaneously (blue arrows) with two axial H atoms. Each interaction is called a *1,3-diaxial interaction* because it involves a substituent on the carbon atom that is numbered C3, assuming the carbon atom bonded to the methyl group is C1.

26-7 CONCEPT ASSESSMENT

If you could measure the individual heats of combustion of axial and equatorial methylcyclohexane, which conformer would burn more exothermically?

The energy differences between the axial and the equatorial forms of several mono-substituted cyclohexanes have been measured. Some of these energy differences are given in Table 26.6. Table 26.6 shows that for alkyl groups the energy difference between the axial and equatorial forms increases with the size of the group, a direct consequence of increasingly unfavorable steric interactions. This effect is particularly pronounced in *tert*-butylcyclohexane. Only about 0.01% of these molecules exist as the axial conformer at room temperature. The molecule is effectively locked with the *tert*-butyl group in the equatorial position!

If there are two substituents, they compete for the equatorial position. In general, the conformation of lowest energy has the bulkier group in the equatorial position.

Let us compare some isomers of dimethylcyclohexane. In 1,1-dimethylcyclohexane, one methyl group is axial and the other equatorial. A ring flip produces a conformation of equal energy, as shown below:

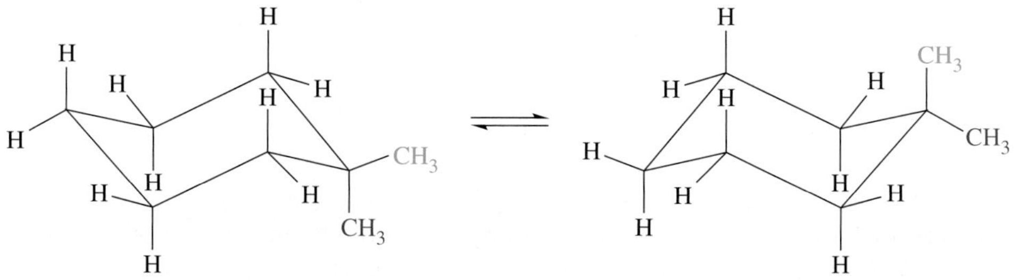

TABLE 26.6 Gibbs Energy Differences between Axial and Equatorial Conformers of Mono-substituted Cyclohexanes

Substituent	$\Delta G°$, kJ mol^{-1}
—H	0
—CH$_3$	7.1
—CH$_2$CH$_3$	7.3
—CH(CH$_3$)$_2$	9.2
—C(CH$_3$)$_3$	21

Now consider *cis*-1,4-dimethylcyclohexane. Both of the chair conformations shown below have one axial methyl group and one equatorial group. The two conformations have the same energy:

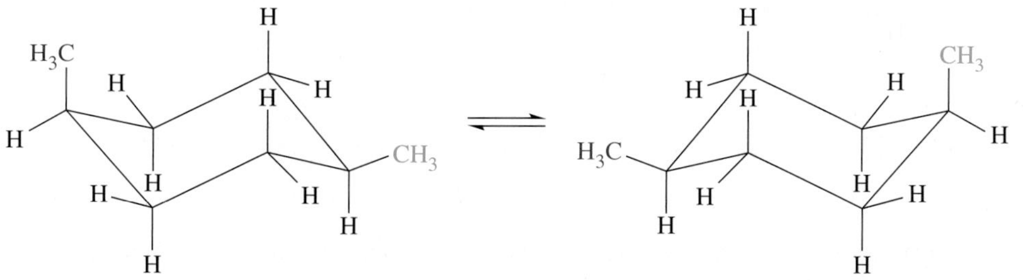

Unlike the *cis* isomer, the *trans* isomer can exist in one of two different chair conformations. One conformation has two axial methyl groups and the other has two equatorial groups, as shown below:

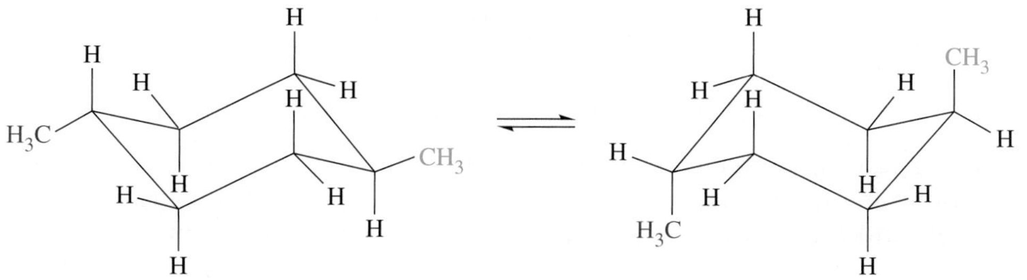

The conformation that has both methyl groups in equatorial positions is 14.2 kJ mol^{-1} lower in energy than the conformation that has the methyl groups in axial positions. This energy difference is twice that given in Table 26.6 for methylcyclohexane.

🔍 **26-8   CONCEPT ASSESSMENT**

Which chair conformation of *cis*-1-fluoro-4-methylcyclohexane is lower in energy?

EXAMPLE 26-5   Predicting Which Conformer of a Disubstituted Cycloalkane Is Lowest in Energy

Draw the lowest energy conformation of *cis*-1,3-dimethylcyclohexane.

Analyze

First, we draw a cyclohexane ring showing both the axial and the equatorial bonds but without the substituents added to the ring. Then, we consider the placement of the substituents. A useful tip is to realize that when the carbon atoms in a cyclohexane ring are numbered, there is a relationship between the bonds on the odd- and even-numbered carbon atoms. When the odd-numbered carbons have their *up* bonds axial, the even-numbered carbon atoms have their *down* bonds axial. When the odd-numbered carbons have their *up* bonds equatorial, the even-numbered carbons have their *down* bonds equatorial. The relationship is illustrated in the diagram below:

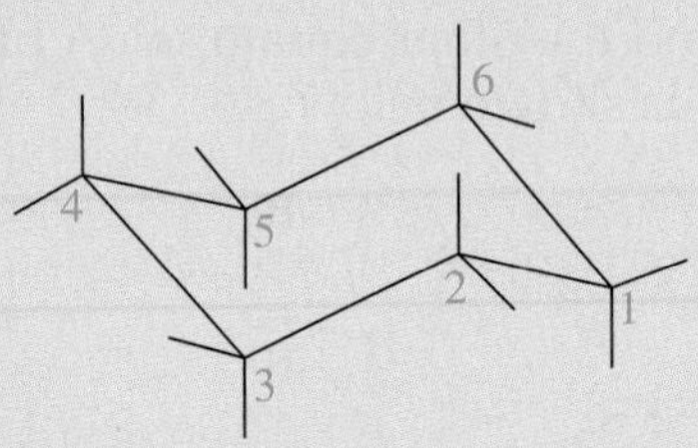

Solve

We are dealing with a *cis* isomer, and so both methyl groups are adjacent to the same face of the ring. This is possible only if the two methyl groups are bonded to the two carbon atoms with both bonds up or both bonds down. The conformation of lowest energy will be the one that has the methyl groups in the equatorial positions.

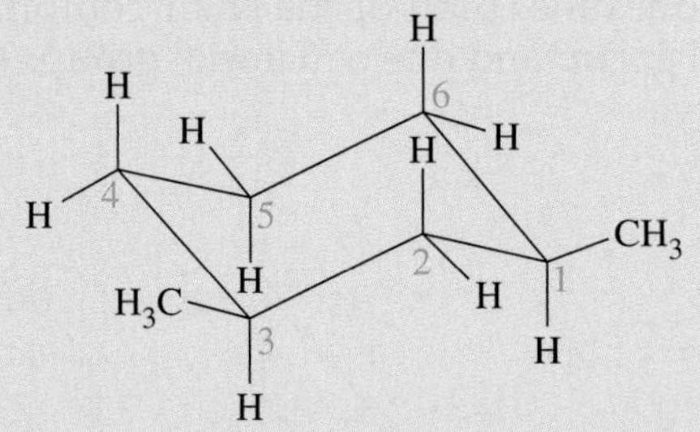

Assess

If the molecule above undergoes a ring flip, the methyl groups will be in the axial positions and the resulting conformation will be higher in energy.

PRACTICE EXAMPLE A: Draw the lower energy conformation of *trans*-1,4-dimethylcyclohexane.

PRACTICE EXAMPLE B: Draw the lower energy conformation of *cis*-1-*tert*-butyl-2-methylcyclohexane.

26-4   Stereoisomerism in Organic Compounds

In Figure 26-14, we have summarized different forms of isomerism that we encounter in organic chemistry. We have already discussed constitutional isomerism and learned about one form of stereoisomerism, namely *cis–trans*

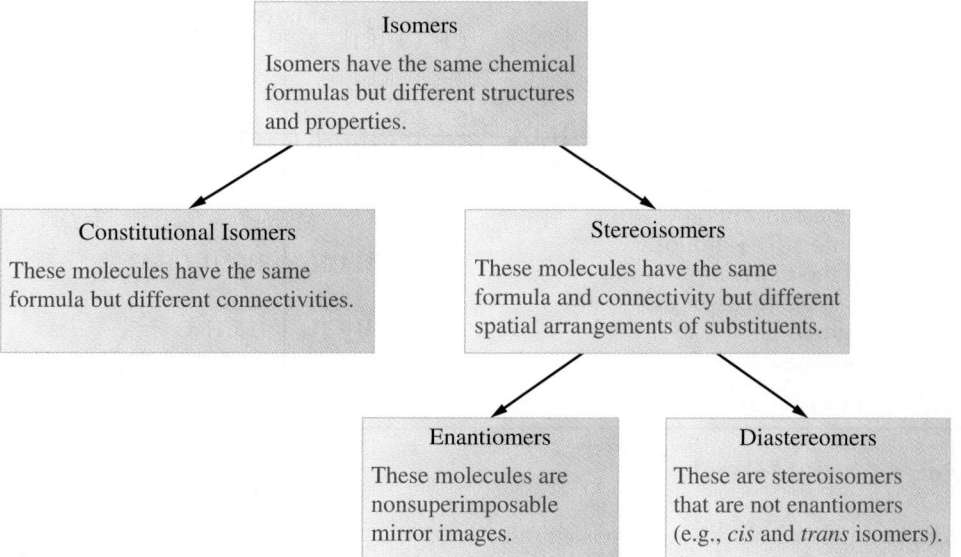

Isomers
Isomers have the same chemical formulas but different structures and properties.

Constitutional Isomers
These molecules have the same formula but different connectivities.

Stereoisomers
These molecules have the same formula and connectivity but different spatial arrangements of substituents.

Enantiomers
These molecules are nonsuperimposable mirror images.

Diastereomers
These are stereoisomers that are not enantiomers (e.g., *cis* and *trans* isomers).

◀ FIGURE 26-14
Isomerism summarized

isomerism in cycloalkanes. As illustrated in Figure 26-14, *cis–trans* isomers are also known as **diastereomers**. In this section we will focus on another form of stereoisomerism—**enantiomerism**—which arises when an organic compound has an asymmetric carbon. An **asymmetric** carbon (also called a **chiral** carbon) is one that is bonded to four different groups. Molecules with an asymmetric carbon exist as stereoisomers that cannot be interconverted without breaking and reforming bonds. We will see (in Chapter 28) that molecules with asymmetric carbon atoms figure prominently in biochemistry.

Chirality

We saw in Chapter 24 that a solution of an optically active compound can rotate the plane of polarized light. The requirement for optical activity is that the molecule be asymmetric—that is, its mirror image cannot be superimposed on the original molecule. This situation can arise at a tetrahedral C atom when all four groups attached to the C atom are different. Consider the molecule 3-methylhexane shown in Figure 26-15. The illustration shows that there are two nonsuperimposable isomers of 3-methylhexane related as mirror images. The two isomers are said to be *enantiomers*. A molecule that is not superimposable on its mirror image is said to be *chiral*. Compounds whose structures are superimposable on their mirror images are **achiral**. Examples of chiral and achiral molecules are shown here, using the dashed and solid wedge line notation:

$$\underset{\text{Chiral}}{\overset{\displaystyle H}{\underset{\displaystyle Cl}{F\cdots C^* Br}}} \qquad \underset{\text{Chiral}}{\overset{\displaystyle F}{\underset{\displaystyle Br}{Cl\cdots C^* I}}} \qquad \underset{\text{Achiral}}{\overset{\displaystyle H}{\underset{\displaystyle CH_3}{H C\cdots Cl}}} \qquad \underset{\text{Chiral}}{\overset{\displaystyle OH}{\underset{\displaystyle CH_3}{CH_3CH_2 C^* \cdots H}}}$$

All the chiral molecules shown contain an atom that is connected to four different substituent groups. The C atom to which the four different groups are attached is said to be **asymmetric**, or a **stereocenter**. Centers of this type are sometimes denoted by an asterisk. Molecules with one stereocenter are always chiral. As we will see in Chapter 28, molecules incorporating more than one stereocenter need not be chiral, and many chiral molecules occur in nature. As described in Chapter 27, the existence of chirality can play an important role in establishing some reaction mechanisms.

KEEP IN MIND

that a solution of one enantiomer rotates the plane of polarized light in one direction whereas a solution of the other enantiomer rotates the light in the opposite direction. With a 50:50 mixture of the two enantiomers—a racemic mixture—no rotation of the plane of polarized light is observed.

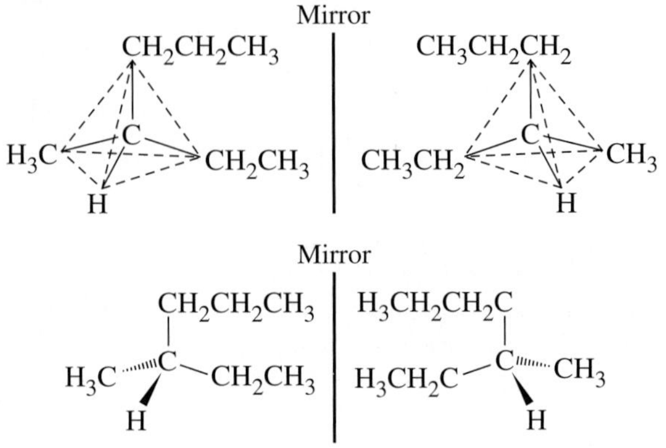

► FIGURE 26-15
Nonsuperimposable mirror images of 3-methylhexane
Notice in the diagram that we have adopted the convention used by organic chemists in drawing chiral centers: The groups attached to the central carbon in the plane of the paper are connected with solid lines, the group in front of the plane of the paper is attached by a solid wedge, and the group behind the plane of the paper is indicated by a dashed wedge.

EXAMPLE 26-6   Identifying a Chiral Molecule

Predict whether either 2-chloropentane or 3-chloropentane is chiral.

Analyze

To decide whether a molecule is chiral, we look for a C atom that has four different groups attached.

Solve

The two compounds are shown below.

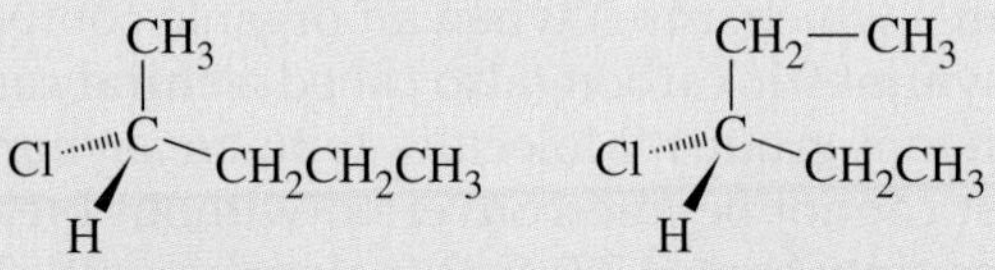

We see that 2-chloropentane contains a C atom that has four different groups attached; hence, 2-chloropentane is chiral. However, 3-chloropentane does not have such a C atom; its structure is identical to its mirror image and thus, 3-chloropentane is achiral.

Assess

In drawing the structures, we focused only on the carbons to which the Cl atoms were attached. All the other carbons are bonded to at least two H atoms (two groups are the same) and cannot possibly be chiral

---

PRACTICE EXAMPLE A: Which of the following chlorofluorohydrocarbons is chiral: (a) $CF_3CH_2CCl_3$; (b) $CF_2HCHFCCl_3$; (c) $CClFHCHHCCl_2F$?

PRACTICE EXAMPLE B: Which of the following chloroalcohols is chiral: (a) $CH_2ClCH_2CH_2OH$; (b) $CH_2ClCH(OH)CH_3$; (c) $CH(OH)ClCH_2CH_3$?

🔍 **26-9   CONCEPT ASSESSMENT**

How many chiral centers are there in 2,3-bromopentane? How many different stereoisomers are there?

EXAMPLE 26-7   Identifying Chiral Carbon Atoms in Cycloalkanes

Identify any chiral carbon atoms in the molecules to the right.

Analyze

A carbon atom is chiral if it is bonded to four different groups. To determine whether a carbon

Methylcyclohexane   cis-1,3-dimethylcyclohexane

atom in a ring structure is chiral, we must go around the ring in each direction (clockwise and counterclockwise) to see if there is a point of difference. If there is a point of difference, then the carbon atom is chiral.

Solve

When considering carbon atoms in the rings of these molecules, we need only focus on carbon atoms where hydrogen atoms have been replaced by other substituents. If there is no substituent, then the carbon atom has two identical groups (two hydrogen atoms) and thus, cannot be chiral. Thus, for methylcyclohexane, we focus only on C1, the carbon bonded to the methyl substituent. The clockwise and counterclockwise paths starting from C1 have identical constitutions ($-CH_2CH_2CH_2CH_2CH_2-$) and so, we conclude that C1 in methylcyclohexane is not chiral.

Clockwise path   Counterclockwise path

When we compare the clockwise and counterclockwise paths, starting from C1, in 1,3-dimethylcyclohexane, we notice that there is a point of difference. The constitution of the clockwise path (blue) is $-CH_2CH_2CH_2CH(CH_3)CH_2-$ and that of the counterclockwise path (red) is $-CH_2CH(CH_3)CH_2CH_2CH_2CH_2-$. The two paths are different; thus, C1 is chiral. Following similar reasoning, we find that C3 is also chiral. Thus, in cis-1,3-dimethylcyclohexane, C1 and C3 are chiral.

Clockwise path   Counterclockwise path

Assess

In most of the examples we have considered so far, there has been only one chiral atom per molecule. Such molecules are optically active. In this example, we see that 1,3-dimethylcyclohexane has two chiral carbon atoms. A molecule with two or more chiral carbon atoms may or may not be optically active. Optical activity of molecules containing two or more chiral atoms is discussed in advanced organic chemistry courses.

PRACTICE EXAMPLE A: Identify the chiral carbon atoms in the molecule shown in the diagram on the right.

PRACTICE EXAMPLE B: How many chiral atoms are there in 1,1,4-trimethylcyclohexane?

Naming Enantiomers: The *R, S* System of Nomenclature

To differentiate two enantiomers, we need a system of nomenclature that indicates the arrangement or configuration of the four groups about the chiral center. Such a system was developed by three chemists, R. S. Cahn, C. Ingold, and V. Prelog. The first step is to rank the four substituents in order of decreasing priority. The rules for assigning priorities will be described shortly.

Suppose that in the hypothetical compound *Cabcd,* substituent *a* has the highest priority, *b* the second highest, *c* the third highest, and *d* the lowest. Now position the molecule (mentally, on paper, or with a model set) so that the lowest priority substituent is as far away from the viewer as possible (Fig. 26-16). This procedure results in two possible arrangements for the remaining substituents (one for each enantiomer). In the *R, S* system, if the sequence from *a* to *b* to *c*, as viewed toward the substituent of lowest priority, is clockwise, the configuration of the stereocenter is named *R* (*rectus,* Latin, meaning right). Conversely, if the sequence is counterclockwise, the stereocenter is named *S* (*sinister,* Latin, meaning left). The symbol (*R*) or (*S*) is added as a prefix to the name of the chiral compound, as in (*R*)-2-chlorobutane

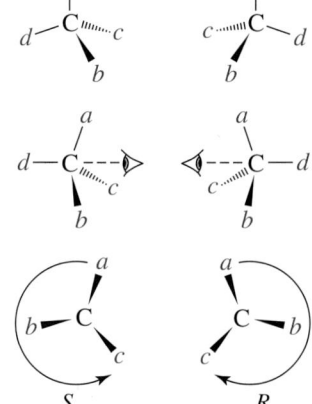

▲ FIGURE 26-16
Assignment of *R* and *S* configuration at a tetrahedral stereocenter
The group of lowest priority is placed as far away from the viewer as possible.

and (*S*)-2-chlorobutane, shown below. A racemic mixture of the enantiomers is designated (*R, S*) as in (*R, S*)-2-chlorobutane.

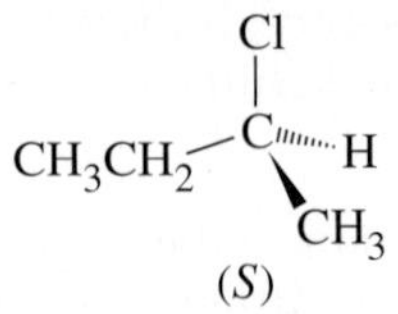

$$CH_3CH_2-\overset{\overset{\displaystyle Cl}{|}}{\underset{\underset{\displaystyle CH_3}{}}{C}}\cdots\cdots H$$

(*S*)

$$\underset{\underset{\displaystyle H_3C}{}}{\overset{\overset{\displaystyle Cl}{|}}{H\cdots\cdots C}}-CH_2CH_3$$

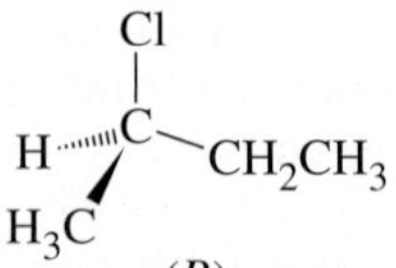

(*R*)

Rules for Assigning Priorities to Substituents In order to apply the *R, S* nomenclature to a stereocenter, we must first describe how the priorities are assigned to substituents. When looking at the atoms attached directly to the stereocenter, the rules devised by Cahn, Ingold, and Prelog are as follows:

▶ Strictly speaking, the parameter that decides priority is atomic mass, not atomic number. However, atomic mass increases as the atomic number increases, so unless we are comparing isotopes of the same element (same atomic number, different atomic masses), we can assign priorities by focusing on atomic number.

Rule 1. A substituent atom of higher atomic number takes precedence over one of lower atomic number. Consider the enantiomer of 1-chloro-1-iodoethane shown below.

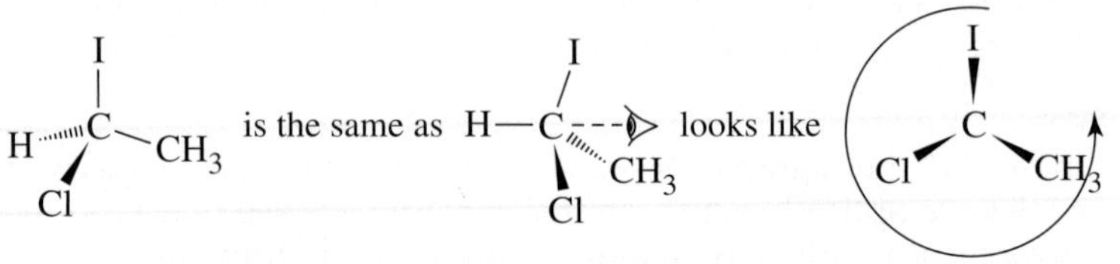

$$H\cdots\cdots\overset{\overset{\displaystyle I}{|}}{\underset{\underset{\displaystyle Cl}{}}{C}}-CH_3 \quad \text{is the same as} \quad H-\overset{\overset{\displaystyle I}{|}}{\underset{\underset{\displaystyle Cl}{}}{C}}\cdots\cdots\blacktriangleright\text{CH}_3 \quad \text{looks like}$$

Priority I > Priority Cl > Priority C
(*S*)-1-chloro-1-iodoethane

The order of priority, as viewed toward the H atom (which has the lowest priority), is counterclockwise. Therefore the enantiomer is (*S*)-1-chloro-1-iodoethane.

Rule 2. If two substituent atoms attached to the stereocenter have the same priority, proceed along the two substituent chains until a point of difference is reached. The atom of higher atomic number at this point establishes the priority. Thus, an ethyl group takes priority over a methyl group for the following reason. Although at the point of attachment to the stereocenter each substituent has a C atom, equal in priority, beyond these C atoms the methyl group has an H atom and the ethyl group has a higher-priority C atom.

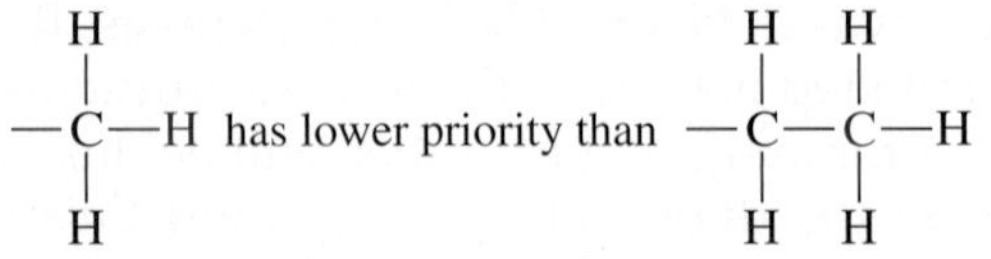

$$-\overset{\overset{\displaystyle H}{|}}{\underset{\underset{\displaystyle H}{|}}{C}}-H \quad \text{has lower priority than} \quad -\overset{\overset{\displaystyle H}{|}}{\underset{\underset{\displaystyle H}{|}}{C}}-\overset{\overset{\displaystyle H}{|}}{\underset{\underset{\displaystyle H}{|}}{C}}-H$$

It is important to understand that the decision on priority is made at the *first* point of difference along otherwise similar substituent chains. When that point has been reached, the constitution of the remainder of the chain is immaterial.

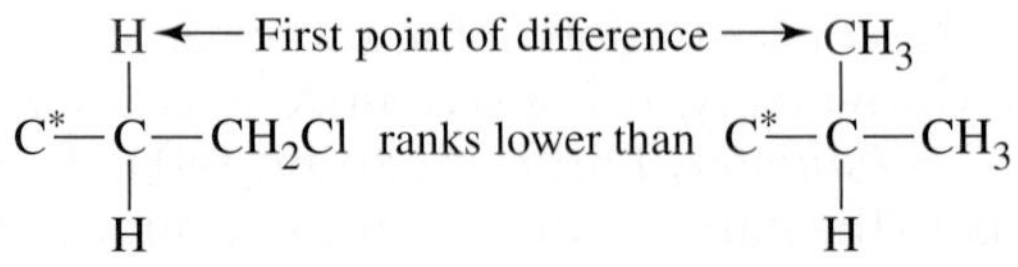

$$C^*-\overset{\overset{\displaystyle H}{|}}{\underset{\underset{\displaystyle H}{|}}{C}}-CH_2Cl \quad \text{ranks lower than} \quad C^*-\overset{\overset{\displaystyle CH_3}{|}}{\underset{\underset{\displaystyle H}{|}}{C}}-CH_3$$

H ◀── First point of difference ──▶ CH₃

Rule 3. Double bonds and triple bonds are treated as if they were single, and the atoms in them are duplicated or triplicated at each end by the particular atoms at the *other end of the multiple bond*. For example,

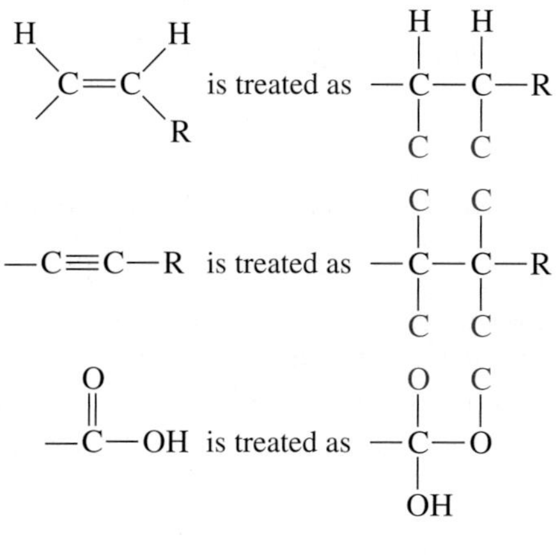

Note that the atoms shown in red are added simply for the purpose of assigning priority to the groups containing a multiple bond; they are *not* really there! To illustrate, the —CH_2OH group has lower priority than —CHO.

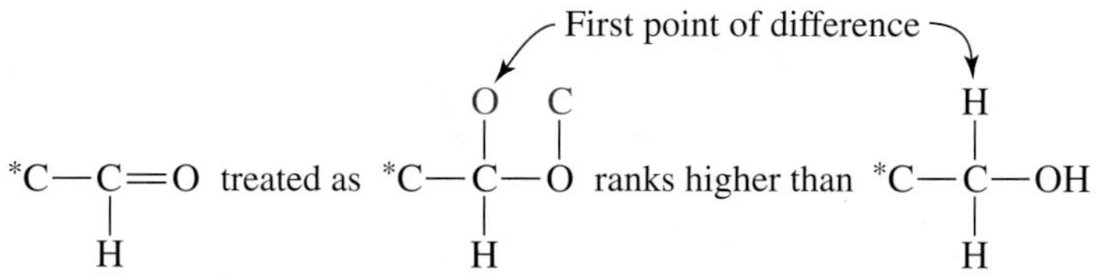

The assignment of priorities and configuration of a chiral center is illustrated in Example 26-8.

EXAMPLE 26-8   Assignment of Priorities and Configuration of a Chiral Center

Name each of the following compounds, including the assignment of configuration.

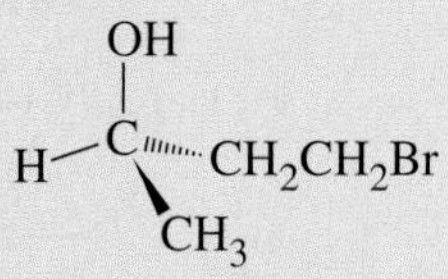

Analyze

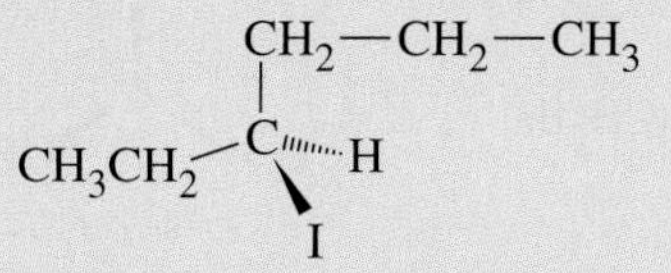

To assign the configuration at the stereocenter, we must first assign the priorities of the substituents. Then we determine the *R* or *S* configuration by viewing the molecule toward the atom of lowest priority.

Solve

   (a) This molecule is 3-iodohexane with C3 as the chiral center. The order of priorities of atoms attached to C3 is

$$I > C \text{ (ethyl)} = C \text{ (propyl)} > H$$

(continued)

In order to decide the ranking of the ethyl group relative to the propyl group, we go to the first point of difference in the chains of these substituents.

First point of difference

$-\overset{\overset{\displaystyle H}{|}}{\underset{\underset{\displaystyle H}{|}}{C}}-\overset{\overset{\displaystyle H}{|}}{\underset{\underset{\displaystyle H}{|}}{C}}-H$ ranks lower than $-\overset{\overset{\displaystyle H}{|}}{\underset{\underset{\displaystyle H}{|}}{C}}-\overset{\overset{\displaystyle H}{|}}{\underset{\underset{\displaystyle H}{|}}{C}}-\overset{\overset{\displaystyle H}{|}}{\underset{\underset{\displaystyle H}{|}}{C}}-H$

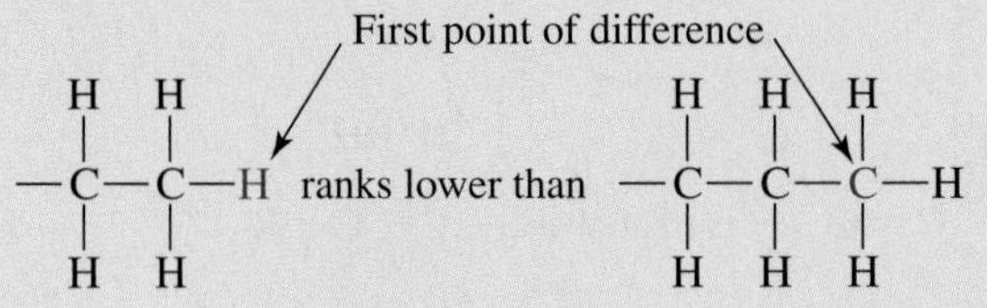

We see that the propyl group ranks higher than the ethyl group, so the order of priorities is

$$I > C \text{ (propyl)} > C \text{ (ethyl)} > H$$

Viewing the molecule toward the lowest priority H atom, we see that the priorities decrease in a counterclockwise manner, so the configuration at the stereo center is *S*. The complete name of the molecule is (*S*)-3-iodohexane.

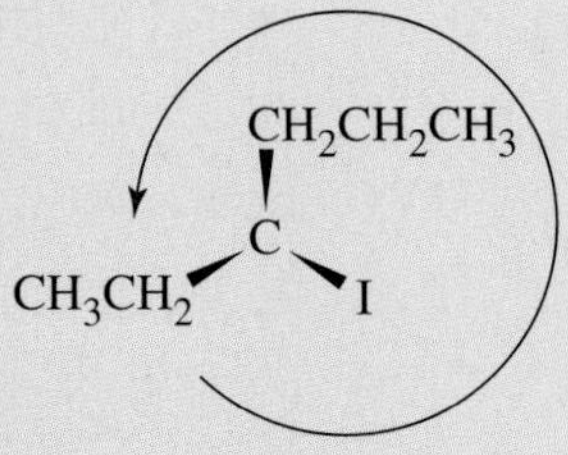

(b) This molecule is 4-bromo-2-butanol, with C2 as the chiral center. The order of priorities of atoms attached to C2 is

$$O > C \text{ (bromoethyl)} = C \text{ (methyl)} > H$$

In order to decide the ranking of the bromoethyl group relative to the methyl group, we go to the first point of difference in the chains of these substituents.

First point of difference

$-\overset{\overset{\displaystyle H}{|}}{\underset{\underset{\displaystyle H}{|}}{C}}-H$ ranks lower than $-\overset{\overset{\displaystyle H}{|}}{\underset{\underset{\displaystyle H}{|}}{C}}-\overset{\overset{\displaystyle H}{|}}{\underset{\underset{\displaystyle H}{|}}{C}}-Br$

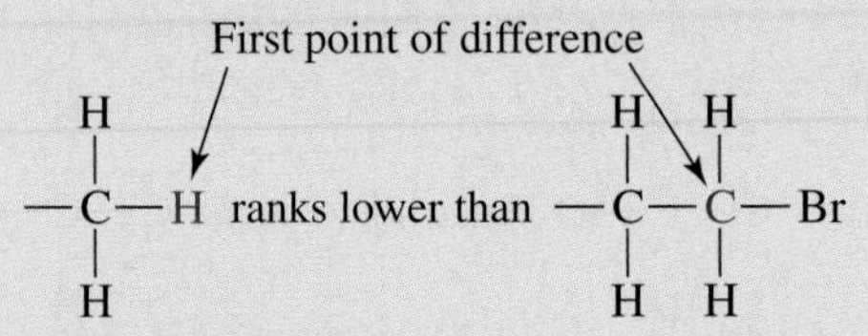

We see that the bromoethyl group ranks higher than the methyl group, so the order of priorities is

$$I > C \text{ (bromoethyl)} > C \text{ (methyl)} > H$$

Viewing the molecule toward the lowest priority H atom is not quite as straightforward as in part (a). This is because the H atom is in the plane of the paper and thus not as far away from the viewers as possible. We can tackle this problem in one of two ways. The first requires some three-dimensional "vision," or "stereoperception," in that we imagine picking up the molecule by the methyl group and bromoethyl group and reorienting the molecule so that the H atom points away from us, to give

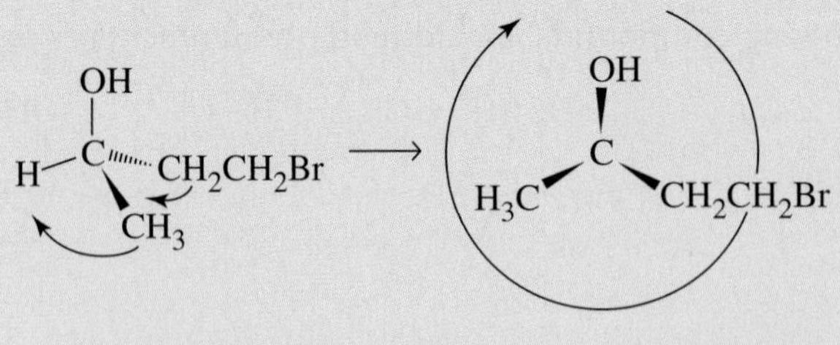

where we can now discern the sequence of priorities as clockwise, so the compound is (*R*)-4-bromo-2-butanol.

Alternatively, we can switch a pair of groups so that the group of lowest priority is bonded by a dashed wedge.

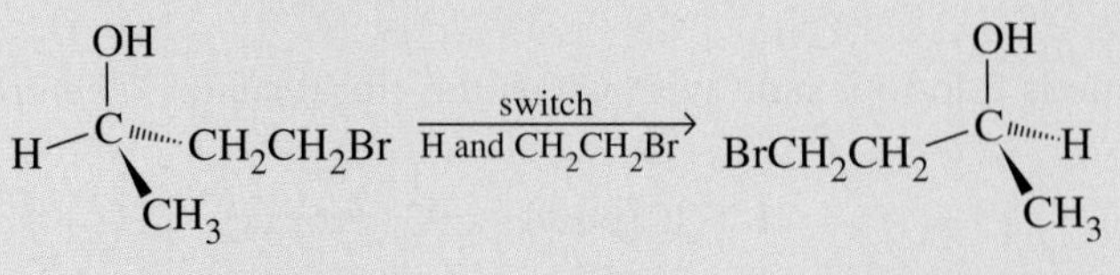

Now draw the view of the molecule toward the group of lowest priority. Here the sequence of priorities is counterclockwise. Because we switched two groups, we created the enantiomer of the molecule whose configuration we are actually seeking. Thus, although in the switched molecule the order of priorities corresponds to an S configuration, the *original* molecule corresponds to the R configuration. Therefore, as determined previously, the molecule is (R)-4-bromo-2-butanol.

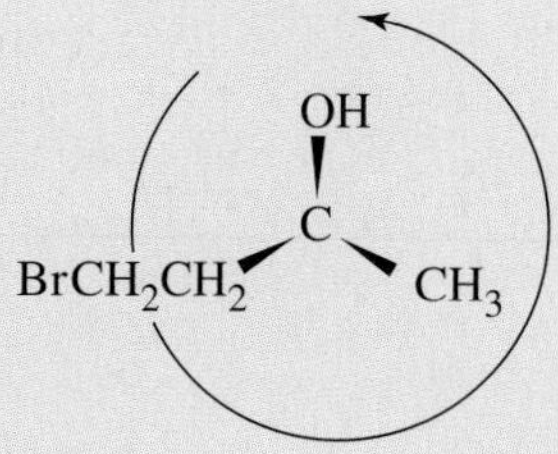

Assess

When assigning priorities to groups, it may be necessary, as it was in **(b)**, to work backward along the chain to find the first point of difference and then compare the atomic numbers of the atoms at that point.

PRACTICE EXAMPLE A: Indicate whether each of the following structures has the R configuration or the S configuration.

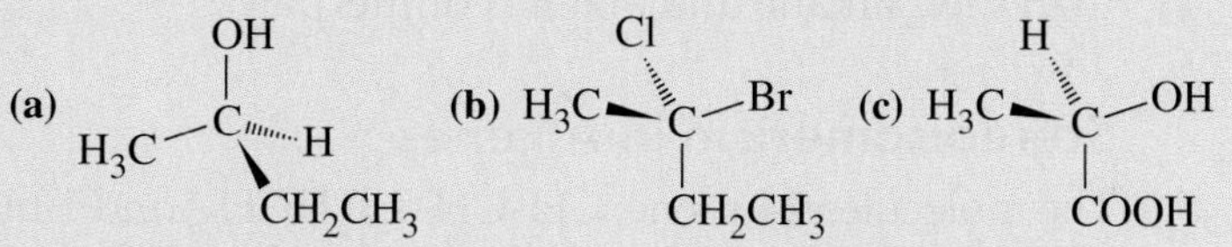

PRACTICE EXAMPLE B: Do the structures in each of the following pairs represent identical molecules or pairs of enantiomers?

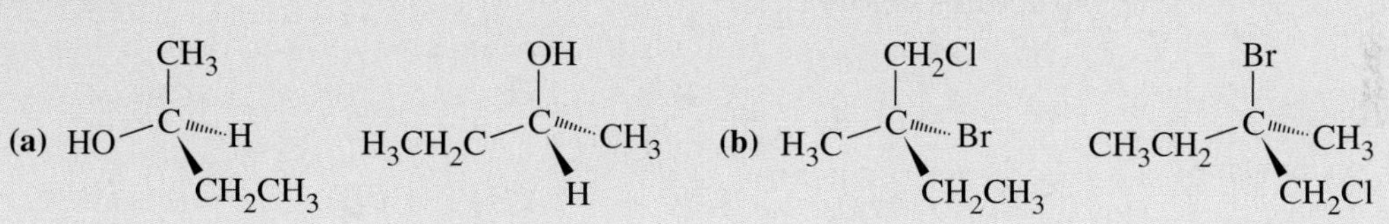

26-5   Alkenes and Alkynes

A straight- or branched-chain alkane, with formula C_nH_{2n+2}, has the maximum number of H atoms possible for its number of C atoms. In other classes of hydrocarbons, compounds with the same number of C atoms but fewer H atoms, the C atoms must join into rings, form carbon-to-carbon multiple bonds, or do both to ensure that each C atom forms a total of four bonds. We have already discussed some aspects of ring structures (in Section 26-3). In this section, we focus on hydrocarbons whose molecules contain some double or triple bonds between C atoms. Such molecules are said to be **unsaturated**. If the molecule has *one double bond*, the hydrocarbons are the simple **alkenes**, or **olefins**; they have the general formula C_nH_{2n}. Simple alkynes have *one triple bond* in their molecules and have the general formula C_nH_{2n-2}.

In the examples that follow, systematic names are shown in blue. The names given in parentheses are also commonly used.

$$CH_3CH_2CH_2C{=}CH_2$$
$$\underset{\text{2-Ethyl-1-pentene}}{\overset{|}{CH_2CH_3}}$$

$$\underset{\substack{\text{Ethene} \\ \text{(ethylene)}}}{CH_2{=}CH_2} \qquad \underset{\text{1-Butene}}{CH_3CH_2CH{=}CH_2}$$

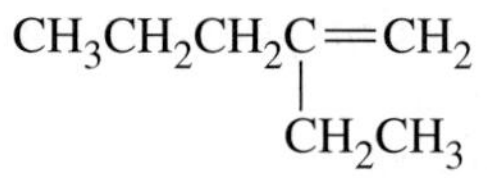

$$\underset{\substack{\text{Ethyne} \\ \text{(acetylene)}}}{HC{\equiv}CH} \qquad \underset{\substack{\text{1-Butyne} \\ \text{(ethylacetylene)}}}{CH_3CH_2C{\equiv}CH} \qquad \underset{\substack{\text{4-Methyl-2-pentyne} \\ \text{(isopropylmethylacetylene)}}}{CH_3CHC{\equiv}CCH_3}$$
$$\overset{|}{CH_3}$$

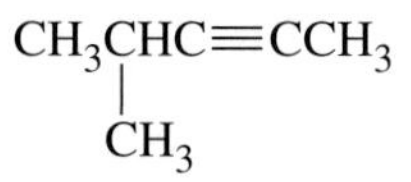

The following are the modifications of the rules on page 1151 required to name alkenes and alkynes:

1. Select as the base chain the longest chain containing the multiple bond.
2. Number the C atoms of the chain to place the multiple bond at the lowest possible number.
3. Use the ending -*ene* for alkenes and -*yne* for alkynes.

Thus, in the name 2-ethyl-1-pentene, the longest chain *containing the multiple bond* is a five-carbon chain (*pent-*). This is *not* a substituted hexane, and it is *not* a hexene. The double bond makes the molecule an alkene, and its location between the first and second carbon makes it a 1-pentene. The ethyl group is attached to the second C atom, and so the alkene is named 2-ethyl-1-pentene.

The alkenes are similar to the alkanes in physical properties. At room temperature, those containing 2 to 4 C atoms are gases; those with 5 to 18 are liquids; those with more than 18 are solids. In general, alkynes have higher boiling points than their alkane and alkene counterparts.

Stereoisomerism in Alkenes

The molecules 2-butene, $CH_3CH=CHCH_3$, and 1-butene, $CH_2=CHCH_2CH_3$, differ in the position of the double bond and are constitutional isomers. However, another type of isomerism is possible in 2-butene, as shown in the two structures below:

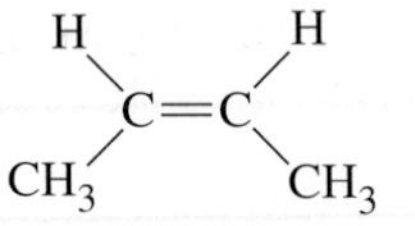

(a) *cis*-2-Butene 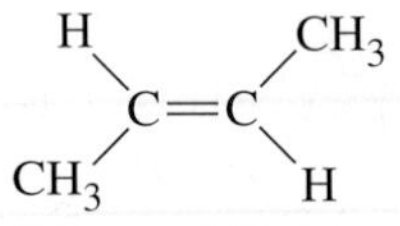   (b) *trans*-2-Butene

The two isomers of 2-butene are stereoisomers because they have exactly the same *constitution* but a different spatial arrangement of their atoms in space. As we learned in Section 11-4, a double bond between C atoms consists of the overlap of hybrid orbitals to form a σ bond *and* the sideways overlap of p orbitals to form a π bond. Because of the π bond, rotation about a double bond is severely restricted. Molecule (a) cannot be converted into molecule (b) simply by twisting one end of the molecule through 180°, so the two molecules are distinctly different above. To differentiate these two molecules, we call molecule (a) *cis*-2-butene and we call molecule (b) *trans*-2-butene. Because of differences in their molecular structures, the compounds have different physical properties. For example, the melting points are −139 °C for *cis*-2-butene and −106 °C for *trans*-2-butene; the boiling points are 3.7 °C for *cis*-2-butene and 0.9 °C for *trans*-2-butene.

Preparation and Uses of Alkenes and Alkynes

The general laboratory preparation of alkenes uses an *elimination reaction*, a reaction in which atoms are removed from adjacent positions on a carbon chain. Elimination reactions will be examined in more detail in Chapter 27. For now, it is enough to know that, in an elimination reaction, a small molecule is produced, and an additional bond is formed between the C atoms. For example, H_2O is eliminated in the following reaction:

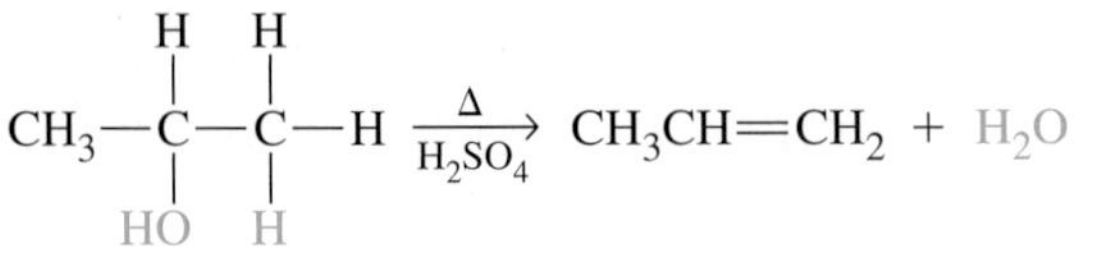

$$CH_3-\underset{\underset{HO}{|}}{\overset{\overset{H}{|}}{C}}-\underset{\underset{H}{|}}{\overset{\overset{H}{|}}{C}}-H \xrightarrow[H_2SO_4]{\Delta} CH_3CH=CH_2 + H_2O \qquad (26.4)$$

The principal alkene of the chemical industry is ethene (ethylene). Its chief use is in the manufacture of polymers (Chapter 27), although it is also used to manufacture other organic chemicals. Reaction (26.4) is relatively unimportant in the commercial production of ethylene, which is obtained mainly by thermal cracking of other hydrocarbons.

The simplest alkyne is ethyne (acetylene), which can be prepared from coal, water, and limestone in a three-step process:

$$CaCO_3 \xrightarrow{\Delta} CaO + CO_2$$

$$CaO + 3\,C \xrightarrow[2000\,°C]{\text{electric furnace}} \underset{\substack{\text{calcium acetylide}\\ \text{(calcium carbide)}}}{CaC_2} + CO$$

$$CaC_2 + 2\,H_2O \longrightarrow \underset{\text{acetylene}}{HC{\equiv}CH} + Ca(OH)_2 \qquad \textbf{(26.5)}$$

Most other alkynes are prepared from acetylene by taking advantage of the acidity of the C—H bond. In the presence of a very strong base, such as sodium amide ($NaNH_2$), the amide anion removes the proton from acetylene to form ammonia and the salt sodium acetylide. The acetylide can then react with an alkyl halide, such as CH_3Br:

$$H{-}C{\equiv}C{-}H + Na^+NH_2^- \longrightarrow H{-}C{\equiv}C^-Na^+ + NH_3$$

$$H{-}C{\equiv}C^-Na^+ + CH_3Br \longrightarrow H{-}C{\equiv}C{-}CH_3 + Na^+Br^- \qquad \textbf{(26.6)}$$

By continuing this reaction, the triple bond can be positioned as desired in the chain, as in the synthesis of 2-pentyne:

$$H{-}C{\equiv}C{-}CH_3 + NaNH_2 \longrightarrow Na^{+\,-}C{\equiv}C{-}CH_3 + NH_3$$

$$Na^{+\,-}C{\equiv}C{-}CH_3 + CH_3CH_2Br \longrightarrow CH_3CH_2C{\equiv}CCH_3 + Na^+Br^- \qquad \textbf{(26.7)}$$

At one time, acetylene was one of the most important organic raw materials in the chemical industry. At present, the chief use of acetylene is in the manufacture of other chemicals for polymer production, such as vinyl chloride, $H_2C{=}CHCl$, which is polymerized to polyvinyl chloride (PVC). We will discuss polymers and polymerization reactions in the next chapter.

Acetylene is also used to produce high-temperature flames that are used in a variety of applications. For example, the combustion of acetylene in excess oxygen is the basis of oxyacetylene torches used for cutting and welding metals.

$$HC{\equiv}CH(g) + \frac{5}{2}\,O_2(g) \longrightarrow 2\,CO_2(g) + H_2O(l) \qquad \Delta H = -1300\ \text{kJ}$$

The large negative enthalpy of combustion of acetylene results from acetylene's large *positive* enthalpy of formation: $\Delta H_f^\circ[C_2H_2(g)] = +226.7\ \text{kJ mol}^{-1}$.

Alkenes and alkynes are used by chemists to make other compounds. The reactive part of these compounds are the π bonds and the characteristic reaction is an addition reaction, in which atoms or grouping of atoms add to the carbon atoms on either side of a double or triple bond. We have already seen one example of an addition reaction, reaction (26.1), in which hydrogen atoms add across a carbon–carbon bond of an alkene to give an alkane. Although we will examine reactions of alkenes in more detail in Chapter 27, it is worth mentioning now that certain addition reactions form the basis of simple qualitative tests that can be used to determine whether a compound is an alkene or an alkyne. For example, $H_2C{=}CH_2$ will absorb H_2 in the presence of a metal catalyst, reaction (26.1), and it will decolorize a solution of bromine water, $Br_2(aq)$, because of the following reaction:

▲ Cutting steel with an oxyacetylene torch.

$$
\begin{array}{c}
\underset{\substack{H \\ | }}{\overset{\substack{H}}{}}C{=}C\underset{\substack{| \\ H}}{\overset{\substack{H}}{}} + \underset{\text{red-brown}}{Br_2(aq)} \longrightarrow H{-}\underset{\substack{| \\ H}}{\overset{\substack{Br \\ |}}{C}}{-}\underset{\substack{| \\ Br}}{\overset{\substack{H \\ |}}{C}}{-}H \qquad \textbf{(26.8)}
\end{array}
$$

The decolorization of bromine by cyclohexene can be seen in the photo in the margin on the next page.

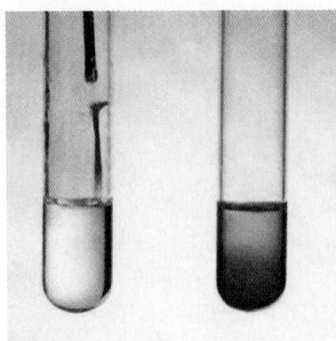

▲ The test tube on the left contains cyclohexene and the one on the right contains cyclohexane. When bromine Br₂, is added to the tube containing cyclohexene, the red-brown color disappears because Br₂ adds across the double bond. The red-brown color persists in the tube filled with cyclohexane.

Naming the Stereoisomers of Highly Substituted Alkenes: The *E, Z* System of Nomenclature

Unfortunately, the *cis–trans* nomenclature is not useful for naming highly substituted alkenes. Consider for example the following stereoisomers of 1-bromo-1-chloro-2-fluoroethene:

$$\underset{Cl}{\overset{Br}{\diagdown}}C=C\underset{H}{\overset{F}{\diagup}} \qquad \underset{Cl}{\overset{Br}{\diagdown}}C=C\underset{F}{\overset{H}{\diagup}}$$

Which is *cis*? Which is *trans*?

In the structure on the left, the F atom is *cis* to Br and *trans* to Cl; in the structure on the right, F is *cis* to Cl and *trans* to Br. Clearly, the *cis–trans* descriptors are not very helpful for naming highly substituted alkenes. An alternative system for naming such alkenes has been adopted by IUPAC: the *E, Z* **system**. The Cahn–Ingold–Prelog rules, discussed previously, are used systematically to assign priorities to the substituents on the carbon atoms of the double bond. The stereochemistry about the double bond is assigned Z (from the German word *zusammen*, meaning "together") if the two groups of higher priority *at each end of the double bond* are on the same side of the molecule. If the two groups of higher priority are on opposite sides of the double bond, the configuration is denoted by an *E* (from the German word *entgegen*, meaning "opposite"). These ideas are summarized in the diagram below:

$$\underset{\text{Low priority}}{\overset{\text{High priority}}{\diagdown}}C=C\underset{\text{Low priority}}{\overset{\text{High priority}}{\diagup}} \qquad \underset{\text{Low priority}}{\overset{\text{High priority}}{\diagdown}}C=C\underset{\text{High priority}}{\overset{\text{Low priority}}{\diagup}}$$

The *Z* isomer   The *E* isomer

We can use the *E, Z* system to name the two stereoisomers of 1-bromo-1-chloro-2-fluoroethene:

Higher priority on C1   Higher priority on C2

$$\underset{Cl}{\overset{Br}{\diagdown}}C=C\underset{H}{\overset{F}{\diagup}} \qquad \underset{Cl}{\overset{Br}{\diagdown}}C=C\underset{F}{\overset{H}{\diagup}}$$

(*Z*)-1-bromo-1-chloro-   (*E*)-1-bromo-1-chloro-
2-fluoroethene   2-fluoroethene

Note that, as in *R, S* nomenclature, the *E* or *Z* is placed in parentheses. The *E, Z* system of nomenclature can be used for all alkene stereoisomers; consequently, the IUPAC recommends that this system be used exclusively. However, many chemists continue to use the *cis* and *trans* designations for simple alkenes.

EXAMPLE 26-9   Assignment of Configurations of Alkenes

Name each of the following compounds, including the assignment of configuration.

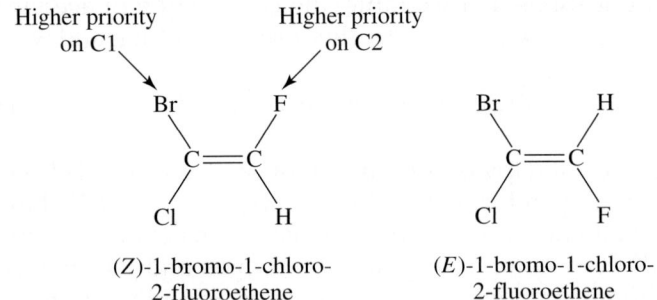

Analyze

To assign the configuration of the alkene, we must first assign the priorities to the substituents attached to each carbon in the double bond. Once this is done, we can assign the E or Z designation.

Solve

(a) One of the sp^2 carbon atoms is bonded to a Cl atom and a C atom; thus, the chlorine atom has highest priority. The other sp^2 carbon is bonded to an ethyl group and an isopropyl group. The carbon of the isopropyl group is bonded to C, C, and H, and the carbon of the ethyl group to C, H, and H. A carbon can be canceled in each group. Of the remaining atoms, the carbon has the highest priority. Thus, the isopropyl group takes precedence. The groups of highest priority are on the same side of the double bond. The complete name of the molecule is (Z)-1,2-dichloro-3-ethyl-4-methyl-2-pentene.

(b) One of the sp^2 carbon atoms is bonded to a fluoromethyl group and an isopropyl group. The carbon of the fluoromethyl group is bonded to F, H, and H; the carbon of the isopropyl group to C, C, and H. Of these six atoms, the fluorine has the highest priority. Thus, the fluoromethyl group takes precedence. The other sp^2 carbon atom is bonded to an ethyl group and a chloroethyl group. Because the first point of difference on these substituent chains is a Cl atom, the chloroethyl group takes precedence. The groups of highest priority are on opposite sides of the double bond. The complete name of the molecule is (E)-2-chloro-3-ethyl-4-fluoromethyl-5-methyl-3-hexene.

Assess

For **(b)**, the carbon atom bonded to chlorine is chiral and so, the molecule exists as either the R or S enantiomer. It takes quite a bit of practice to master organic nomenclature and, admittedly, it is not the most exciting of topics. However, it is an important part of organic chemistry. There could very well be someone else who shares your name, but there is only one (Z)-1,2-dichloro-3-ethyl-4-methyl-2-pentene!

PRACTICE EXAMPLE A: Assign a configuration to each of the following alkenes:

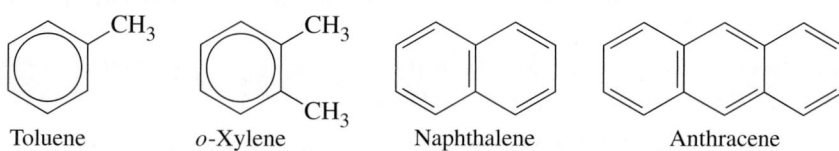

PRACTICE EXAMPLE B: Draw and label the E and Z isomers of the following compounds:

(a) $CH_3CH_2CH{=}CHCH_3$ **(b)** $CH_3CH_2C{=}CHCH_2CH_3$ **(c)** $CH_3CH_2C{=}CHCH_3$
                                                           $|$ $|$
                                                          Cl CH_3

26-6   Aromatic Hydrocarbons

Aromatic hydrocarbons have ring structures with unsaturation (multiple-bond character) in the carbon-to-carbon bonds in the rings. Most aromatic hydrocarbons are based on the molecule benzene, C_6H_6. In Section 11-6, bonding in the benzene molecule was discussed in some detail. Several ways of representing the molecule were shown, including those in the margin.

Other examples of aromatic hydrocarbons include

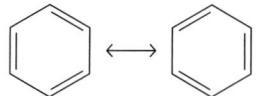

Kekulé structures

Simplified molecular orbital representation

▲ As in the case of alicyclic compounds, neither the C atoms at the vertices nor the H atoms bonded to them are shown in these structures.

CH₃ structures — Toluene   o-Xylene   Naphthalene   Anthracene

Toluene and o-xylene are *substituted benzenes*, and naphthalene and anthracene feature *fused benzene rings*. When rings are fused together, the resultant structure has two C and four H atoms fewer than the starting structures. Thus, the formula for naphthalene is $C_6H_6 + C_6H_6 - 2\,C - 4\,H = C_{10}H_8$; for anthracene: $C_{10}H_8 + C_6H_6 - 2\,C - 4\,H = C_{14}H_{10}$. In this text, we use the inscribed circle for the simple benzene ring and alternating single and double bonds for fused rings. For the fused-ring systems, the bond arrangement represents one of the possible resonance structures for the molecule.

▲ August Kekulé (1829–1896), who proposed the hexagonal ring structure for benzene in 1865. His representation of this molecule is still widely used.

$$CH_2{=}CH{-}CH{=}CH{-}CH{=}CH_2$$

1,3,5-Hexatriene

1,3-Cyclopentadiene

Biphenyl

Phenylhydrazine

Characteristics of Aromatic Hydrocarbons

Aromatic hydrocarbons are highly flammable and should always be handled with care. Prolonged inhalation of benzene vapor results in a decreased production of both red blood cells and white blood cells, which can be fatal. Also, benzene is a carcinogen. Benzene and some other toxic aromatic compounds have been isolated in the tar formed by burning cigarettes, in polluted air, and as a decomposition product of grease in the charcoal grilling of meat.

A close examination of the structures of aromatic molecules shows that they all share two common features.

- They are *planar* (flat), *cyclic* molecules.
- They have a *conjugated* bonding system—a bonding scheme among the ring atoms that consists of alternating single and double bonds. The system must extend throughout the ring, and the π electron clouds associated with the double bonds must involve $(4n + 2)$ electrons, where $n = 1, 2, \ldots$.

Thus, the benzene molecule has six electrons in the π electron clouds: $(4 \times 1) + 2 = 6$. The naphthalene molecule has 10: $(4 \times 2) + 2 = 10$. And the anthracene molecule has 14: $(4 \times 3) + 2 = 14$.

Neither of the two molecules depicted in the margin is aromatic. The 1,3,5-hexatriene molecule has six π electrons in its conjugated bonding system, but it is not cyclic. The 1,3-cyclopentadiene molecule is cyclic, but has only four π electrons in a conjugated bonding system that does not extend completely around the ring.

Benzene and its homologues are similar to other hydrocarbons in being insoluble in water but soluble in organic solvents. The boiling points of the aromatic hydrocarbons are slightly higher than those of the alkanes of similar carbon content. For example, hexane, C_6H_{14}, boils at 69 °C, whereas benzene boils at 80 °C. This can be explained by the planar structure and delocalized electron charge density of benzene, which increases the attractive forces between molecules. The symmetrical structure of benzene permits closer packing of molecules in the crystalline state and results in a higher melting point than for hexane. Benzene melts at 5.5 °C and hexane melts at −95 °C.

Two important aromatic functional groups are the phenyl and benzyl groups. A **phenyl group** is obtained when one of the six equivalent H atoms of a benzene molecule is removed. A **benzyl group** is obtained by replacing one of the H atoms in a methyl group with a phenyl group.

Phenyl group   Benzyl group

Two phenyl groups may bond together, as in biphenyl, or phenyl groups may be substituents in other molecules, as in phenylhydrazine, used in the detection of sugars. The structures of biphenyl and phenylhydrazine are shown in the margin.

> 🔍 **26-10   CONCEPT ASSESSMENT**
>
> What is the structure of the molecule with the name (*E*)-3-benzyl-2,5-dichloro-4-methyl-3-hexene?

Naming Aromatic Hydrocarbons

Other atoms or groups may be substituted for H atoms on the benzene molecule, and to name these compounds, we use a numbering system for the C atoms in the ring. If the name of an aromatic compound is based on a common name other than benzene (such as toluene), the characteristic substituent group (for example, the —CH_3 in toluene) is assigned position "1" on the benzene ring. Otherwise, the substituents are listed alphabetically and the carbon atoms in the ring are numbered so that the substituents appear at the lowest numbers possible, as shown below for 1-bromo-2-chlorobenzene.

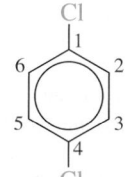

3-Bromotoluene
(*m*-bromotoluene)

1-bromo-2-chlorobenzene
(*o*-bromochlorobenzene)

1,4-Dichlorobenzene
(*p*-dichlorobenzene)

2-Chlorotoluene
(*o*-chlorotoluene)

The terms *ortho, meta,* and *para* (*o-, m-, p-*) can be used when there are two substituents on the benzene ring. **Ortho** refers to substituents on adjacent carbon atoms, **meta** to substituents with one carbon atom between them, and **para** to substituents opposite one another on the ring.

◀ *p*-dichlorobenzene is used in mothballs. Naphthalene has also been used in mothballs but is being replaced by *p*-dichlorobenzene because *p*-dichlorobenzene has a less intense smell.

Uses of Aromatic Hydrocarbons

Over 90% of the billions of pounds of benzene produced annually in the United States is derived from petroleum. The process involves dehydrogenation and cyclization of hexane to the aromatic hydrocarbon. The most important use of the petroleum-produced benzene is in manufacturing ethylbenzene for the production of styrene plastics. Other applications include the manufacture of phenol, the synthesis of dodecylbenzene (for detergents), and as an octane enhancer in gasoline. The production of aromatic compounds by dehydrogenation (removal of hydrogen) of alkanes yields large amounts of hydrogen gas, which is an important reactant in the synthesis of ammonia (page 1008).

26-7   Organic Compounds Containing Functional Groups

In this section we will describe the structure and nomenclature of organic compounds containing functional groups. We will defer a discussion of the chemical transformations between some of them until the next chapter. As we proceed you will see how all the stereochemistry that we have discussed is also applied to these molecules.

Alcohols and Phenols

Alcohols and **phenols** are characterized by the **hydroxyl** group, —OH. In alcohols, the hydroxyl group is bound to an sp^3 hybridized (aliphatic) carbon atom. If this C atom also has *one* R group (and two H atoms) bonded to it, the alcohol is a *primary* alcohol. If the C atom has *two* R groups (and one H atom), the alcohol is a *secondary* alcohol. Finally, if there are *three* R groups on the C atom (and no H atoms), the alcohol is a *tertiary* alcohol. The systematic naming of alcohols uses the suffix *-ol* and was discussed in Chapter 3. In phenols, the hydroxyl group is attached to a benzene ring.

1-Butanol
(butyl alcohol)
(a *primary* alcohol)

2-Butanol
(*sec*-butyl alcohol)
(a *secondary* alcohol)

Methyl-2-propanol
(*tert*-butyl alcohol)
(a *tertiary* alcohol)

Phenol
(carbolic acid)

2,4,6-Trinitrophenol
(picric acid)

A molecule may have more than one —OH group. Those with two —OH groups are known as *diols* (or glycols), and those with more than two —OH groups are called *polyols*. Ethylene glycol, a diol, is used in automobile antifreeze solutions; glycerol, a polyol, is an important biological molecule used as part of the body's mechanism for fat storage.

$$CH_2-CH_2 \qquad CH_2-CH-CH_2$$
$$\begin{array}{cc} | & | \\ OH & OH \end{array} \qquad \begin{array}{ccc} | & | & | \\ OH & OH & OH \end{array}$$

1,2-Ethanediol   1,2,3-Propanetriol
(ethylene glycol)   (glycerol)

Physical properties of aliphatic alcohols are strongly influenced by hydrogen bonding. As the chain length increases, however, the influence of the polar hydroxyl group on the properties of the molecule diminishes. The molecule becomes less like water and more like a hydrocarbon. As a consequence, low-molecular mass alcohols tend to be water soluble, whereas high-molecular mass alcohols are not. The boiling points and solubilities of the phenols vary widely, depending on the nature of the other substituents on the benzene ring.

26-11 CONCEPT ASSESSMENT

Explain why the name *sec*-pentyl alcohol does not unambiguously identify a compound whereas the name *sec*-butyl alcohol does.

Preparation and Uses of Alcohols Two methods by which alcohols can be synthesized are by the hydration of alkenes and the hydrolysis of alkyl halides.

$$CH_3CH{=}CH_2 + H_2O \xrightarrow{H_2SO_4} \overset{\displaystyle OH}{\underset{}{CH_3CHCH_3}} \qquad \textbf{(26.9)}$$
     propene   2-propanol
    (propylene)   (isopropyl alcohol)

$$CH_3CH_2CH_2Br + OH^- \longrightarrow CH_3CH_2CH_2OH + Br^- \qquad \textbf{(26.10)}$$
1-bromopropane   1-propanol

Reaction (26.9) is an example of an *addition reaction* and reaction (26.10) is an example of a *substitution reaction*. In an addition reaction, one or more atoms add to a molecule. In a substitution reaction, an atom or a grouping of atoms is replaced by another atom or group of atoms. We will have a closer look at these types of reactions in Chapter 27.

Methanol (wood alcohol) is the simplest alcohol. It is a highly toxic substance that can lead to blindness or death if ingested. Most methanol is manufactured from carbon monoxide and hydrogen.

$$CO(g) + 2\,H_2(g) \xrightarrow[\substack{200\ atm \\ ZnO,\ Cr_2O_3}]{350\ °C} CH_3OH(g)$$

Methanol is the most extensively produced alcohol. It is used in the synthesis of other organic chemicals and as a solvent, but potentially its most important use may be as a motor fuel (recall Section 7-9).

Ethanol, CH_3CH_2OH, is grain alcohol, which is found in alcoholic beverages. It is easily produced by the fermentation of the juices of sugarcane or other materials that contain natural sugars. The industrial method involves the hydration of ethylene with sulfuric acid catalyst (similar to reaction 26.9).

Ethylene glycol, $HOCH_2CH_2OH$, is water soluble and has a higher boiling point (197 °C) than water. These properties make it an excellent permanent, nonvolatile antifreeze for automobile radiators. It is also used in the manufacture of solvents, paint removers, and plasticizers (softeners).

Glycerol (glycerin), $HOCH_2CH(OH)CH_2OH$, is obtained commercially as a by-product in the manufacture of soap. It is a sweet, syrupy liquid that is miscible with water in all proportions. Because it takes up moisture from the air, glycerol can be used to keep skin moist and soft and is found in lotions and cosmetics.

An interesting and useful derivative of an alcohol is the alkoxide ion, RO^-, a deprotonated form of the alcohol. The alkoxide ion is formed by the reaction of sodium or potassium metal with the alcohol. For example, the methoxide ion, CH_3O^-, is produced in the following reaction:

$$CH_3OH + Na \longrightarrow CH_3O^- + Na^+ + \frac{1}{2}H_2(g) \qquad \textbf{(26.11)}$$

Reaction (26.11) provides the basis of a simple qualitative test that is sometimes used to confirm whether a compound is an alcohol. For alcohols, the addition of sodium metal results in the formation of $H_2(g)$ bubbles. Reaction (26.11) is more commonly used to produce alkoxide ions that can be used in other reactions. For example, an alkoxide ion can react with a haloalkane, as suggested below:

$$CH_3O^- + CH_3CH_2Cl \longrightarrow CH_3OCH_2CH_3 + Cl^-$$

This reaction is yet another example of a substitution reaction (Chapter 27). The compound formed is an ether, discussed next.

Ethers

An **ether** is a compound with the general formula R—O—R'. Structurally, ethers can be pure aliphatic, pure aromatic, or mixed.

▲ **Giant soap bubble**
A soap bubble consists of an air pocket enclosed in a thin film of soap in water. As the water evaporates, the film breaks and the bubble bursts. Glycerol added to the soap–water mixture forms hydrogen bonds to both the soap and water molecules. This slows the rate of evaporation of the water and increases the strength of the film, allowing the production of very large bubbles.

$$CH_3-O-CH_3 \qquad \text{(Diphenyl ether structure)} \qquad \text{(Methyl phenyl ether structure)}$$

Dimethyl ether   Diphenyl ether   Methyl phenyl ether
(anisole)

Dimethyl ether has the same formula as ethanol, but the two substances have quite different physical and chemical properties. They have different properties because each has a different functional group (see Table 26.2). Dimethyl ether and ethanol are constitutional isomers.

Ethers can be viewed as alkane or aromatic compounds containing the group RO—; this group is known as the *alkoxy group*. The IUPAC system for naming ethers treats them as alkanes that bear an alkoxy substituent, that is, as alkoxyalkanes. The smaller substituent is considered part of the alkoxy group and the larger substituent defines the stem. Thus, for example, the IUPAC name for ethylmethyl ether is methoxyethane, and for anisole, it is methoxybenzene.

Ethers can also be cyclic. For example, if an oxygen atom takes the place of a carbon atom in a cyclohexane molecule, we obtain the cyclic ether shown in the margin. The oxygen atom in this structure is called a *heteroatom* (because it is not a carbon atom) and the compound is called a **heterocyclic compound**. The simplest system for naming cyclic ethers is based on using the prefix *oxa-* in front of the name of the cycloalkane. The prefix *oxa-* indicates that a carbon atom has been replaced by an oxygen atom. Thus, the name of the compound above is oxacyclohexane.

Preparation and Uses of Ethers Symmetrical ethers, such as diethyl ether, can be prepared by the elimination of a water molecule from between two alcohol molecules with a strong dehydrating agent, such as concentrated H_2SO_4.

$$CH_3CH_2OH + HOCH_2CH_3 \xrightarrow[140\ °C]{\text{conc. } H_2SO_4} CH_3CH_2OCH_2CH_3 + H_2O \qquad \textbf{(26.12)}$$

▶ The molecule below should be named *tert*-butyl methyl ether (TBME), but has been referred to by industrial chemists as MTBE for so long that methyl *tert*-butyl ether has become the commonly used name for this compound.

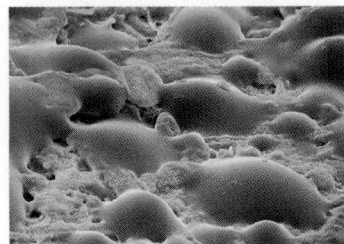

Methyl *tert*-butyl ether (MTBE)

Chemically, ethers are comparatively unreactive. The ether linkage is stable in the presence of most oxidizing and reducing agents, as well as dilute acids and alkalis.

Diethyl ether has been used extensively as a general anesthetic. It is easy to administer and produces excellent relaxation of the muscles. Also, it affects the pulse rate, rate of respiration, and blood pressure only slightly. However, it is somewhat irritating to the respiratory passages and produces nausea. Methyl propyl ether (neothyl) is also used as an anesthetic and is less irritating to the respiratory passages. Dimethyl ether, a gas at room temperatures, is used as a propellant for aerosol sprays. Higher molecular mass ethers are used as solvents for varnishes and lacquers. Methyl *tert*-butyl ether, an unsymmetrical ether, marketed under the name MTBE, has been used as an octane enhancer in gasoline. However, because of its relatively high solubility in water (≈ 5 g/100 g H_2O), in some localities it has caused rather extensive groundwater pollution through leakage from underground storage tanks; it is currently being phased out of use.

Aldehydes and Ketones

Aldehydes and ketones contain the **carbonyl group**.

$$\begin{matrix} R \\ \diagdown \\ C{=}O \\ \diagup \\ R' \end{matrix}$$

▲ False-color scanning electron micrograph of the sticky surface of a 3M Post-it note. The bubbles are 15–40 μm in diameter and consist of a urea–formaldehyde adhesive. Each time the note is pressed to a surface, fresh adhesive is released. The note can be reattached to a surface as long as some of the bubbles remain.

If either the R or R' group is an H atom, the compound is an **aldehyde**. If the R and R' groups are alkyl or aromatic (aryl) groups, the compound is a **ketone**.

Methanal (formaldehyde)

3-Chlorobutanal (β-chlorobutyraldehyde)

Benzaldehyde

Propanone (acetone)

3-Pentanone (diethyl ketone)

1-Phenylethanone (acetophenone)

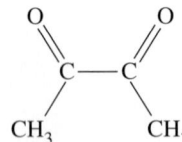

Butanedione, a diketone

Aldehydes and ketones often have characteristic and recognizable odors. For example, 2-heptanone is a liquid with a clove-like odor that accounts for the odors of many fruits and dairy products. Some aldehydes and ketones find use as flavoring agents. For example, vanillin, the compound responsible for vanilla flavor, is an aldehyde. Alpha-demascone and 2-octanone are ketones responsible for berry and mushroom flavors, respectively. Butanedione, shown in the margin, is a yellow liquid with a cheese-like smell that gives butter its flavor.

The IUPAC naming system for aldehydes uses the suffix -*al*. The parent chain is the longest chain containing the aldehyde group. Accordingly, the four-carbon aldehyde is called butanal because the name is derived from the alkane (butane) with the -*e* replaced by -*al*. The numbering of the chain starts at the carbon of the aldehyde group; it is always at position one and need not be specified. Thus 3-chloro-2-methylbutanal is

KEEP IN MIND

that the CHO functional group can be only at the terminus of a carbon chain, whereas the CO functional group cannot be at the end of a carbon chain.

$$CH_3{-}\underset{\underset{Cl}{|}}{CH}{-}\underset{\underset{CH_3}{|}}{CH}{-}\overset{\overset{O}{\|}}{C}{-}H$$

The IUPAC naming system for ketones uses the suffix -*one*. The parent chain must include the carbonyl group and be numbered from the end of the chain that reaches the carbonyl group first; this ensures, as required by IUPAC rules, that the number for the carbonyl group is as low as possible. For example, the molecule

$$CH_3-\underset{\underset{Cl}{|}}{CH}-\underset{\underset{CH_3}{|}}{CH}-\underset{\overset{O}{\|}}{C}-CH_3$$

is 4-chloro-3-methyl-2-pentanone.

Preparation and Uses of Aldehydes and Ketones Aldehydes can be produced by the oxidation of a *primary* alcohol with an oxidizing agent such as dichromate ion in acidic solution. However, the aldehyde so formed is readily oxidized further to a carboxylic acid.

$$CH_3CH_2OH \xrightarrow[H^+]{Cr_2O_7^{2-}} CH_3CHO \xrightarrow[H^+]{Cr_2O_7^{2-}} CH_3CO_2H \qquad (26.13)$$

ethanol   acetaldehyde   acetic acid
(a primary alcohol)   (an aldehyde)   (an acid)

A milder oxidizing agent in a nonaqueous medium is required to stop the oxidation at the aldehyde. This partial oxidation can be done with the reagent pyridinium chlorochromate (PCC) in an organic solvent such as dichloromethane. For example,

$$CH_3(CH_2)_8CH_2OH \xrightarrow{PCC, CH_2Cl_2} CH_3(CH_2)_8CHO$$

Oxidation of a *secondary* alcohol produces a ketone.

$$CH_3CH(OH)CH_3 \xrightarrow[H^+]{Cr_2O_7^{2-}} CH_3\overset{\overset{O}{\|}}{C}CH_3 \qquad (26.14)$$

2-propanol   acetone
(a secondary alcohol)   (a ketone)

Ketones are much more resistant to oxidation than are alcohols and aldehydes. Aldehydes and ketones occur widely in nature. Typical natural sources are

Benzaldehyde   Cinnamaldehyde   Camphor
(almonds)   (cinnamon)   (obtained from
                                              camphor tree)

Aldehydes and ketones can be reduced to primary and secondary alcohols, respectively, by sodium borohydride, $NaBH_4$. For example, 2-hexanone is reduced to 2-hexanol by $NaBH_4$:

$$CH_3-CH_2-CH_2-CH_2-\overset{\overset{O}{\|}}{C}-CH_3 \xrightarrow[H_2O]{NaBH_4} CH_3-CH_2-CH_2-CH_2-\underset{\underset{H}{|}}{\overset{\overset{OH}{|}}{C}}-CH_3$$

The reaction effectively adds two H atoms across the carbon–oxygen double bond. However, H atoms are not directly involved. The reaction involves two distinct steps (1) attack of the carbonyl carbon atom by a hydride (H^-) ion; (2) protonation of the carbonyl oxygen atom.

◀ PCC is formed as yellow-orange crystals by the reaction between CrO_3 and HCl in pyridine.

$$\text{ClCrO}_3^-$$

KEEP IN MIND

that tertiary alcohols cannot be oxidized to aldehydes or ketones because a tertiary carbon is bonded to three alkyl groups and cannot form a double bond with an oxygen atom unless one of the carbon–carbon bonds is broken. Oxidation of a tertiary alcohol is typically difficult and requires a very strong oxidizing agent.

Aldehydes and ketones find use as starting materials and reagents for the synthesis of other organic compounds, and, generally speaking, aldehydes are more reactive than ketones. The carbon atom in the carbonyl group is slightly positive and is prone to attack by species that are attracted to centers of positive charge.

The simplest aldehyde is formaldehyde ($H_2C=O$), a colorless gas that dissolves readily in water. Billions of kilograms of formaldehyde are used each year in the manufacture of synthetic resins. A polymer of formaldehyde called paraformaldehyde is used as an antiseptic and an insecticide.

Acetone is the most important of the ketones. It is a volatile liquid (boiling point, 56 °C) and highly flammable. Acetone is a good solvent for a variety of organic compounds and is widely used in solvents for varnishes, lacquers, and plastics. Unlike many common organic solvents, acetone is miscible with water in all proportions.

Carboxylic Acids

▶ The carboxyl group is represented either as —COOH or —CO₂H.

As we learned in Chapter 3, **carboxylic acids** contain the **carboxyl group** (*car*bonyl and hydr*oxyl*).

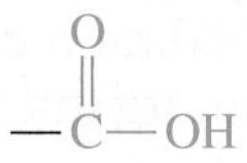

These acids have the general formula R—COOH. In many compounds, R is an aliphatic residue. Such compounds are called fatty acids because high-molecular mass compounds of this type are readily available from naturally occurring fats and oils. If two carboxyl groups are found on the same molecule, the acid is called a dicarboxylic acid. The carboxyl group can also be found attached to the benzene ring.

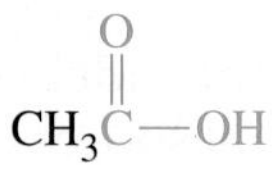

| Acetic acid (an aliphatic acid) | Benzoic acid (an aromatic acid) | Oxalic acid (an aliphatic dicarboxylic acid) | Phthalic acid (an aromatic dicarboxylic acid) |

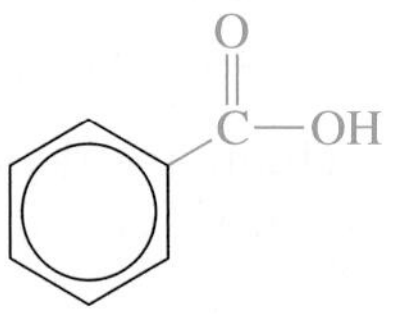

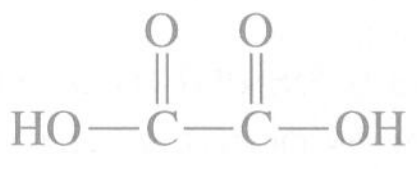

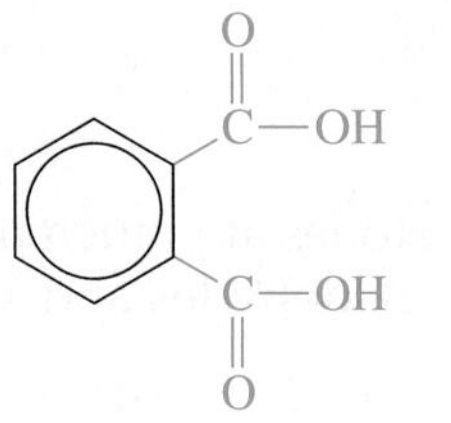

Straight- or branched-chain acids can be named either by their IUPAC names or by using Greek letters in conjunction with common names. Cycloalkanes with a —COOH substituent are named as *cycloalkanecarboxylic acids*; that is, the ending "carboxylic acid" is combined with the name of the cycloalkane. Aromatic acids are named as derivatives of benzoic acid. Here are a few examples:

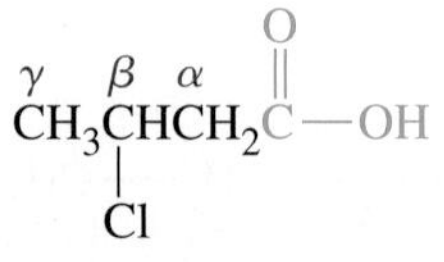

3-Chlorobutanoic acid
β-chlorobutyric acid

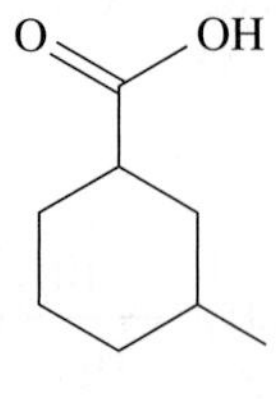

3-methylcyclohexanecarboxylic acid

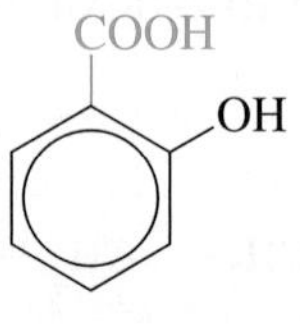

2-Hydroxybenzoic acid
o-hydroxybenzoic acid
(also salicylic acid)

Notice that in straight- or branched-chain carboxylic acids, the C atom in the —COOH group is C1, but in cyclic and aromatic acids, the C atom bonded to the —COOH group is C1.

Carboxylic acids are found widely in nature. Spinach, rhubarb, and other green leafy vegetables are rich in *oxalic acid* (see previous page). Sour milk and sore muscles have elevated levels of *lactic acid*. Citrus fruits are rich in *citric acid*. Line-angle formulas for these acids are shown in the margin.

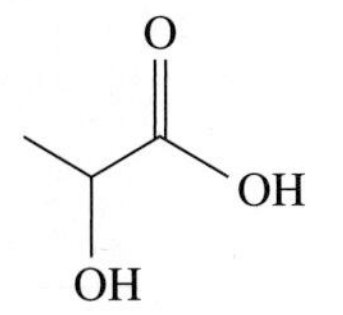

Carboxylic acids, especially those of low molecular weight, exhibit characteristic odors. Ethanoic acid (acetic acid) is the acid that gives vinegar its characteristic smell. The presence of butanoic acid contributes to the strong flavor and aroma of many cheeses, and (*E*)-3-methyl-2-hexenoic acid has been identified as the principal compound responsible for the smell of human sweat.

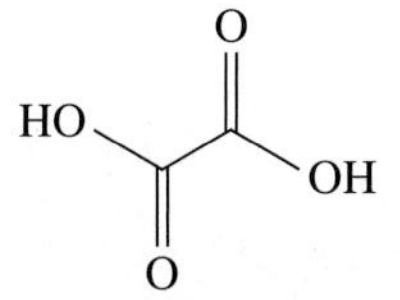

Lactic acid

Oxalic acid

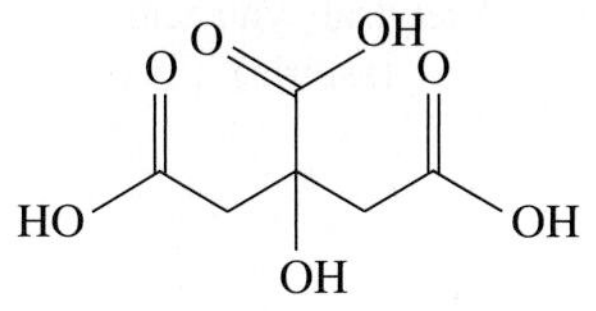

26-12   CONCEPT ASSESSMENT

Draw the structure of (*E*)-3-methyl-2-hexenoic acid.

Citric acid

Some properties of carboxylic acids can be rationalized in terms of the ability of carboxylic acid molecules to form hydrogen bonds among themselves or with other molecules (such as water). For example, low-molecular-weight carboxylic acids are soluble in water because of their ability to hydrogen bond with water. Also, carboxylic acids also have relatively high melting and boiling points because of hydrogen bonding.

As discussed in Chapter 16, soluble carboxylic acids behave as weak acids when they dissolve in water. A simple way to test whether a compound is a carboxylic acid is to add it to an aqueous solution of sodium hydrogen carbonate, $NaHCO_3(aq)$, or sodium carbonate, $Na_2CO_3(aq)$. If the compound is an acid, bubbles of $CO_2(g)$ will be visible because of the following reaction. For example,

$$2\,RCOOH(aq) + Na_2CO_3(aq) \longrightarrow 2\,RCOONa(aq) + CO_2(g) + H_2O(l) \quad \textbf{(26.15)}$$

Preparation and Uses of Carboxylic Acids   Carboxylic acids can be obtained in the laboratory by the oxidation of a primary alcohol or an aldehyde. For this purpose, the oxidizing agent is generally $KMnO_4(aq)$ in an alkaline medium. Because the medium is alkaline, the product is the potassium salt, but the free carboxylic acid can be regenerated by making the medium acidic.

$$CH_3CH_2OH \xrightarrow[\text{OH}^-,\text{ heat}]{KMnO_4} CH_3COO^-K^+ \xrightarrow{H^+} CH_3COOH + K^+ \quad \textbf{(26.16)}$$

$$CH_3CH_2CH_2OH \xrightarrow[\text{OH}^-,\text{ heat}]{KMnO_4} CH_3CH_2COO^-K^+ \xrightarrow{H^+} CH_3CH_2COOH + K^+ \quad \textbf{(26.17)}$$

We have seen (earlier in this section) that primary alcohols and aldehydes can be oxidized to carboxylic acids. Carboxylic acids can also be prepared by the hydrolysis of nitriles. The hydrolysis of a nitrile can be carried out in either acidic or basic solution, as suggested below:

$$R-C{\equiv}N + 2\,H_2O \xrightarrow[\Delta]{H^+ \text{ or } OH^-} RCOOH + NH_3$$

If the reaction is carried out in basic solution, the carboxylate anion, $RCOO^-$, is produced. To obtain RCOOH, the solution must be acidified.

Carboxylic acids are used extensively in organic chemistry to make other compounds and we will examine some of these reactions next (and in Chapter 27). Because many derivatives of carboxylic acids involve simple replacement of the hydroxyl groups, the following grouping of atoms is encountered frequently:

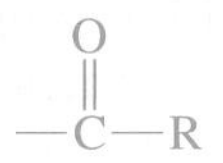

The grouping above is called an *acyl group*. The names of acyl groups are derived from the acid names by replacing *-ic acid* with the suffix *-yl*, as shown below:

$\overset{\displaystyle O}{\underset{\displaystyle \parallel}{}}$		
—C—H	—C—CH$_3$	—C— ⬡
IUPAC name:		
Methanoyl	Ethanoyl	Benzoyl
(from methanoic acid)	(from ethanoic acid)	(from benzoic acid)
Common name:		
Formyl	Acetyl	Benzoyl
(from formic acid)	(from acetic acid)	

Acetylsalicylic acid
(aspirin)

The IUPAC names for these acyl groups are almost never used. The acetyl derivative of *o*-hydroxybenzoic acid (salicylic acid) is what we know as ordinary aspirin, acetylsalicylic acid (ASA), the structure of which is shown in the margin.

Esters

As shown in Table 26.2, the general formula of an ester is RCOOR′. Comparing the general formula of an ester with that of a carboxylic acid, we see that an **ester** is a molecule in which the hydroxyl group (—OH) of a carboxylic acid functional group is replaced by an alkoxyl group (—OR′). In the lab, esters can be prepared by the reaction of a carboxylic acid and an alcohol. The products of the reaction are an ester and a water molecule.

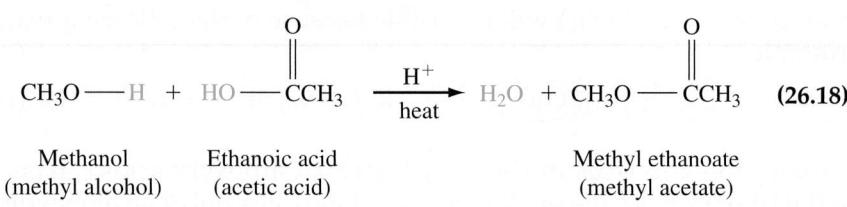

$$CH_3O\text{---H} + HO\text{---}\overset{O}{\overset{\parallel}{C}}CH_3 \xrightarrow[\text{heat}]{H^+} H_2O + CH_3O\text{---}\overset{O}{\overset{\parallel}{C}}CH_3 \qquad (26.18)$$

Methanol	Ethanoic acid	Methyl ethanoate
(methyl alcohol)	(acetic acid)	(methyl acetate)

Because the reaction above is reversible, an excess of alcohol is often used to ensure a high yield of the ester.

Esters have two-part names. The first part is the alkyl designation of the —OR′ group. The second part is the carboxylic acid name with *-ic acid* changed to *-ate*. For example, the combination based on ethanoic acid and the methoxy group is *methyl ethanoate*. A few more examples are shown here.

▲ The distinctive aroma and flavor of oranges are due in part to the ester octyl acetate, $CH_3(CH_2)_6CH_2OOCCH_3$.

Ethyl ethanoate	Phenyl propanoate	Pentyl butanoate
(ethyl acetate)	(phenyl propionate)	(pentyl butyrate)

Unlike the pungent odors of the carboxylic acids from which they are derived, esters have very pleasant aromas. The characteristic fragrances of many flowers and fruits can be traced to the esters they contain. Esters are used in perfumes and in the manufacture of flavoring agents for the confectionery and soft drink industries. Most esters are colorless liquids that are insoluble in water. Their melting points and boiling points are generally lower than those of alcohols and acids of comparable carbon content. This is because of the absence of hydrogen bonding in esters.

Amides

The replacement of the hydroxyl group in the carboxylic acid functional group with an —NH_2 group produces an **amide**. For example, the molecule butanamide (shown in margin) is formed by replacing the hydroxyl group in ethanoic acid by the —NH_2 group.

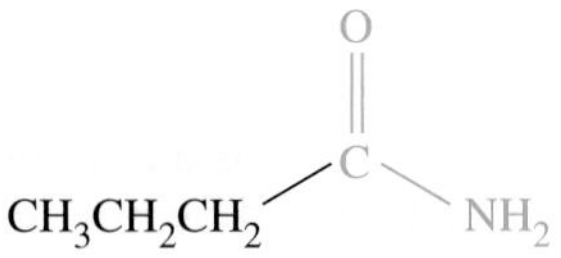

The name of an amide is constructed from the alkane part of the acid name and the suffix *-amide* is added. If hydrogen atoms on the nitrogen atom are replaced with other groups, then a substituted amide is obtained. For example, in the molecules below, one or both of the H atoms in the —NH_2 group of ethanamide have been replaced by other groups. Here are a few examples.

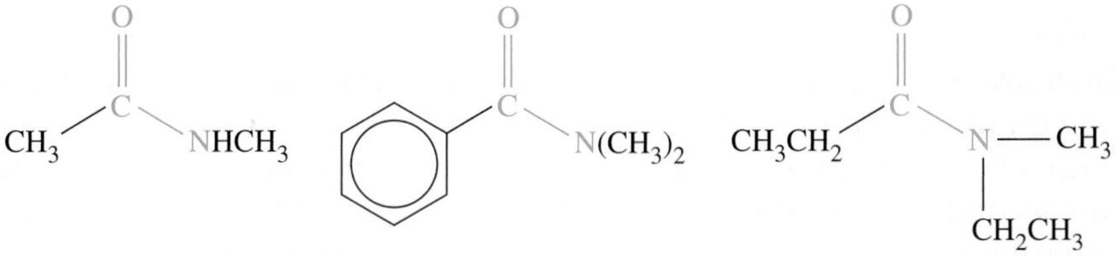

Butanamide

N-methylethanamide   *N,N*-dimethylbenzamide   *N*-ethyl-*N*-methylpropanamide

As the examples above show, we name a substituted amide by using the prefix *N-* in front of each nitrogen substituent. The replacement of one H atom gives a *N*-substituted amide. The active ingredient in Tylenol® is an *N*-substituted amide (see margin). The —CO—NH— linkage present in *N*-substituted amides is called a *peptide bond*, and it is a key bond type in proteins, as we will discover in Chapter 28.

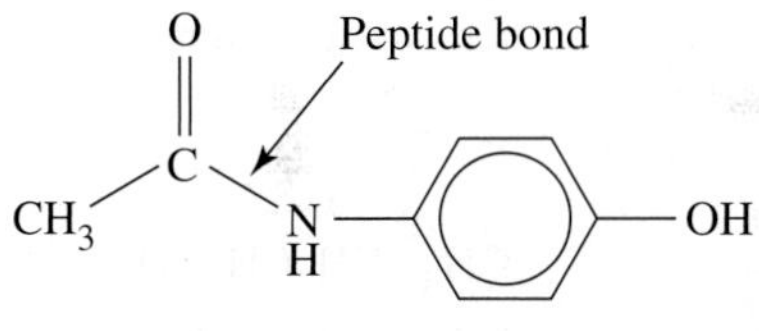

Despite the presence of the —NH_2 group in simple amides, they are not Brønsted bases like amines or ammonia because the carbonyl group promotes the resonance structures shown below.

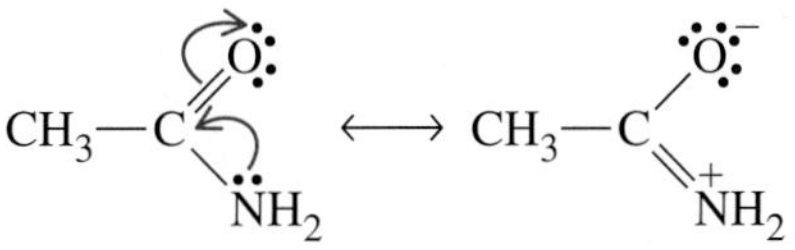

Peptide bond

N-(4-hydroxyphenyl)ethanamide
(Tylenol®)

◀ Resonance in an amide.

The lone pair of electrons on the nitrogen atom is delocalized over the carbonyl group toward the more electronegative oxygen atom. In fact, as seen from the resonance structures, it is the carbonyl group that is most likely to be protonated in acidic conditions. The protonation of the oxygen in a carbonyl group is often an important first step in reactions of esters and amides under acidic conditions.

Preparation and Uses of Amides One way of making an amide is to treat a carboxylic acid with ammonia, forming an ammonium salt, followed by heating. The following sequence summarizes a pathway for making ethanamide.

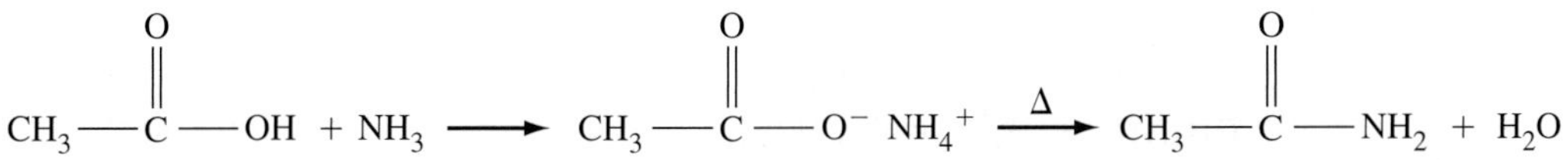

$$CH_3-\overset{\overset{\textstyle O}{\|}}{C}-OH + NH_3 \longrightarrow CH_3-\overset{\overset{\textstyle O}{\|}}{C}-O^-\ NH_4^+ \xrightarrow{\ \Delta\ } CH_3-\overset{\overset{\textstyle O}{\|}}{C}-NH_2 + H_2O$$

(26.19)

However, amides are seldom prepared as described above. As we will see in Chapter 27, a more effective way of converting CH_3COOH to CH_3CONH_2 is to treat the acid with $SOCl_2$ first, which converts the acid to a more reactive *acid chloride* (CH_3COCl), and then treat the acid chloride with ammonia.

▶ A polyamide is a polymer of amides joined by peptide bonds. Polymers are discussed in Chapter 27.

Amides can be converted into other compounds, but compared with other carboxylic acid derivatives, amides are relatively unreactive. The stability of biological structures and the properties of *polyamides*, such as silk and nylon, depend, in part, on the presence of amide linkages. Amides can, however, be converted into nitriles, $R-C\equiv N$, using a strong dehydrating agent such as P_4O_{10}.

$$CH_3CH_2-\overset{\overset{\displaystyle O}{\|}}{C}-NH_2 \xrightarrow{P_4O_{10}} CH_3CH_2-C\equiv N + H_2O$$

Amides are also routinely used to make amines, a class of compounds we discuss next.

Amines

$$CH_3CH_3NH_2$$

Ethylamine
(a primary amine)

Diphenylamine
(a secondary amine)

$$\underset{\displaystyle CH_3NCH_2CH_2CH_3}{\overset{\displaystyle CH_2CH_3}{|}}$$

Ethylmethylpropylamine
(a tertiary amine)

Amines are organic derivatives of ammonia, NH_3, in which one or more organic groups (R) are substituted for H atoms. Their classification is based on the number of R groups bonded to the nitrogen atom—one for primary amines, two for secondary, and three for tertiary (see margin). One of the chief methods of preparing an amine is by the reduction of a nitro compound.

$$\bigcirc-NO_2 \xrightarrow[HCl]{Fe} \bigcirc-NH_3{}^+Cl^- \xrightarrow{NaOH} \bigcirc-NH_2 \qquad \textbf{(26.20)}$$

An alternative method of preparing an amine involves the reaction of ammonia with an alkyl halide. The methylammonium bromide formed reacts with additional ammonia to give the amine.

$$NH_3 + CH_3Br \longrightarrow CH_3NH_3{}^+Br^-$$
$$CH_3NH_3{}^+Br^- + NH_3 \longrightarrow CH_3NH_2 + NH_4{}^+Br^-$$

The formation of pure primary amines is difficult by this method because when the primary amine is formed, it can take part in a further substitution reaction with the alkyl halide to form a secondary amine, for example,

$$CH_3NH_2 + CH_3Br \longrightarrow \underset{\displaystyle CH_3N H}{\overset{\displaystyle CH_3}{\overset{\displaystyle |}{}}} + HBr \qquad \textbf{(26.21)}$$

The compound formed here is dimethylamine. Further reaction with bromomethane produces the tertiary amine, trimethylamine.

Amines of low molecular mass are gases that are readily soluble in water, yielding basic solutions. The volatile members have odors similar to ammonia, but more fish-like. Primary and secondary amines form hydrogen bonds, but these bonds are weaker than those in water because nitrogen is less electronegative than oxygen. Like ammonia, amines have trigonal pyramidal structures with a lone pair of electrons on the N atom. Also like ammonia, amines owe their basicity to these lone-pair electrons. In aromatic amines, because of unsaturation in the benzene ring, electrons are drawn into the ring, which reduces the electron charge density on the nitrogen atom. As a result, aromatic amines are *weaker* bases than ammonia. Aliphatic amines are somewhat stronger bases than ammonia.

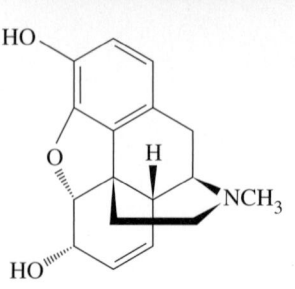

Morphine

▲ Morphine, a very powerful and addictive painkiller, can be isolated from the opium poppy (*Papaver somniferum*).

$$NH_3 + H_2O \rightleftharpoons NH_4{}^+ + OH^- \qquad K_b = 1.8 \times 10^{-5}$$
$$\underset{\text{methylamine}}{CH_3NH_2} + H_2O \rightleftharpoons CH_3NH_3{}^+ + OH^- \qquad K_b = 42 \times 10^{-5}$$

$$\underset{\text{aniline}}{\bigcirc-NH_2} + H_2O \rightleftharpoons \bigcirc-NH_3{}^+ + OH^- \qquad K_b = 7.4 \times 10^{-10}$$

Dimethylamine is an accelerator for the removal of hair from hides in the processing of leather. Butyl- and pentylamines are used as antioxidants, corrosion inhibitors, and in the manufacture of oil-soluble soaps. Dimethylamine and trimethylamine are used in the manufacture of ion-exchange resins. Additional applications are found in the manufacture of disinfectants, insecticides, herbicides, drugs, dyes, fungicides, soaps, cosmetics, and photographic developers. Many amines demonstrate biological activity, including cocaine, nicotine, morphine (see margin, previous page), quinine, and vitamin B6, to name just a few.

Important derivatives of ammonia are the *tetraalkyl ammonium salts*. The cation in these salts has four organic groups attached to the nitrogen atom and is called a *quaternary ammonium ion*. In a quaternary ammonium ion, the nitrogen atom has a formal charge of 1+. Two examples of such cations are shown in the margin.

The acetylcholine cation is similar to the much simpler tetramethylammonium cation, except that one of the methyl groups has been replaced by a $CH_3COOCH_2CH_2-$ group. Quaternary ammonium ions, such as acetylcholine, are involved in the systems that transmit nerve impulses in the human body. Also, many poisons that affect the central nervous system contain a quaternary ammonium functional group.

Heterocyclic Compounds

In most of the ring structures considered to this point, all the ring atoms have been carbon; these structures are said to be carbocyclic. However, in many compounds, both natural and synthetic, one or more of the atoms in a ring structure is not carbon. These ring structures are said to be **heterocyclic**. The heterocyclic systems most commonly encountered contain N, O, and S atoms, and the rings are of various sizes.

Pyridine is a nitrogen analogue of benzene (Fig. 26-17). Unlike benzene, it is water soluble and has basic properties (the unshared pair of electrons on the N atom is not part of the π electron cloud of the ring system). Pyridine, a liquid with a disagreeable odor, was once obtained exclusively from coal tar, but it is now used so extensively that several synthetic methods have been developed for its production. It is used in the production of pharmaceuticals such as sulfa drugs and antihistamines, as a denaturant for ethyl alcohol, as a solvent for organic chemicals, and in the preparation of waterproofing agents for textiles. A number of other examples of heterocyclic compounds are discussed in Chapter 28.

Tetramethylammonium ion

Acetylcholine

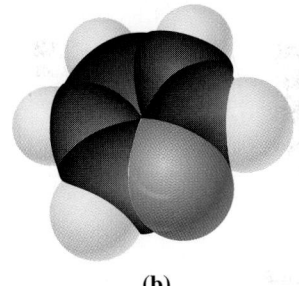

(b)

▲ FIGURE 26-17
Pyridine
In the pyridine molecule, a nitrogen atom replaces one of the CH units of benzene. Its formula is C_5H_5N
(a) Structural formula
(b) Space-filling model

26-1 ARE YOU WONDERING...

How to name compounds with more than one functional group?

In our discussion of organic compounds we restricted our discussion to molecules with only one functional group. How do we name the following compounds, each of which has more than one functional group?

(a)

(b)

(continued)

The IUPAC rules provide a solution to this problem by assigning priorities to functional groups. The priorities of some of the common functional are as follows:

carboxylic acid > ester > amide > aldehyde >
Highest priority   ketone > alcohol > amine > alkoxy
                                                                                    Lowest priority

Notice that functional groups with two heteroatoms (carboxylic acids, esters, and amides) have higher priorities than do those with only one heteroatom.

The priority order enables us to decide that molecule (a) is an acid, in fact a pentanoic acid. But what prefix do we use for the carbonyl group? The prefix *oxo-* is used to indicate either an aldehyde or a ketone as a substituent. We name the molecule 3-methyl-4-oxopentanoic acid.

The prefixes for the —OH and —NH$_2$ substituents are *hydroxy* and *amino* respectively. Thus, the name of molecule (b) is 5-hydroxy-6-methyl-2-heptanone.

A more complete discussion of nomenclature is normally covered in more advanced organic chemistry courses.

26-8 From Molecular Formula to Molecular Structure

Given only a molecular formula (perhaps derived from an experimentally determined percentage composition and a molecular weight determination) and some chemical evidence that suggests what functional groups might be present, how might we proceed to propose a possible structure?

First, it is helpful to determine how many elements of unsaturation are in a molecule. An element of unsaturation is a structural feature, such as a carbon–carbon double or triple bond or a ring structure, that causes the number of hydrogen atoms in a molecule to be less than the maximum number possible. We have learned that a saturated hydrocarbon with n carbon atoms has $2n + 2$ hydrogen atoms, the maximum number possible. We have also learned that the general formula is C_nH_{2n} for a simple alkene (a hydrocarbon having only one double bond) or a cycloalkane. Thus, a simple alkene or a cycloalkane has two fewer H atoms than the maximum number possible. We conclude that the number of H atoms decreases by two for each carbon–carbon π bond or ring structure that is present. Consider, for example, a molecule with the formula C_5H_{10}. The molecule has two fewer hydrogen atoms than the maximum possible and it could be one of the following molecules. (There are other isomers too. Try to draw structures for them.)

Each of these molecules contains one structural element (a π bond or a ring structure) that contributes to it being unsaturated. For each of these structures, the **degree of unsaturation** is equal to one. When the degree of unsaturation is equal to one, there will be two fewer H atoms than the maximum possible. When there are four hydrogen atoms less than the maximum, the degree of unsaturation is two.

> If an organic molecule with n carbon atoms has fewer than $2n + 2$ hydrogen atoms, then it contains elements of unsaturation, such as π bonds or ring structures.

Up to this point, we have focused on molecules containing only carbon and hydrogen. The determination of the degree of unsaturation in molecules

KEEP IN MIND

that benzene, C$_6$H$_6$, has eight fewer H atoms than the maximum possible. Thus, the degree of unsaturation associated with a benzene ring is four, a result which is consistent with one ring plus the equivalent of three delocalized π bonds. A high degree of unsaturation is often accounted for by a benzene ring.

containing O, N, or Cl atoms requires some explanation. In organic compounds, halogen atoms, such as chlorine, are typically terminal atoms, not central atoms. In this regard, halogen atoms are similar to hydrogen atoms. When determining the degree of unsaturation, halogen atoms are counted as if they were hydrogen atoms. Stated another way, each halogen atom effectively adds one hydrogen to the hydrogen count. Thus, C_5H_9Cl has one degree of unsaturation because $9 + 1$ is two less than the maximum number of hydrogen atoms possible. Some possible structures for C_5H_9Cl are shown below. (Again, there are other possible structures. Draw structures for the other isomers.)

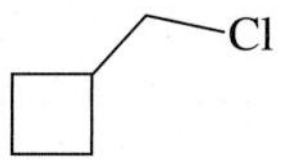

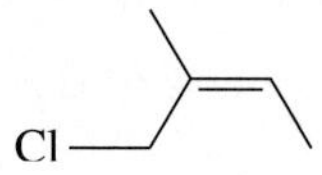

Now let us consider the situations that can arise if oxygen atoms are present in a molecule. It turns out that the degree of unsaturation in a carbon–hydrogen–oxygen compound can be established by considering only the number of carbon and hydrogen atoms in the molecule; that is, we can ignore the oxygen atoms. To understand why, let's consider a molecule that contains only one oxygen atom. The following linkages are possible:

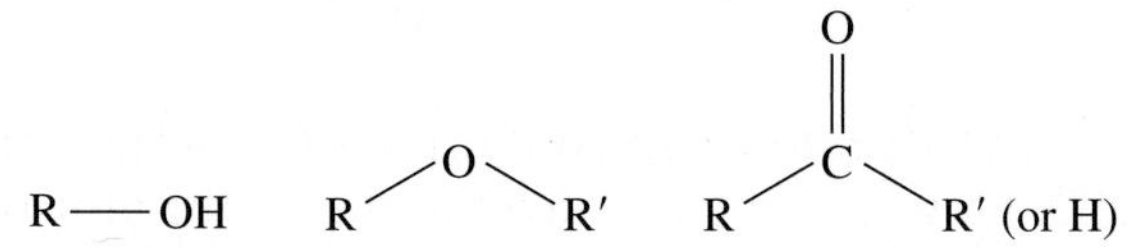

For an alcohol or ether, we see that the oxygen atom is inserted either between a carbon atom and a hydrogen atom or between two carbon atoms. Neither of these linkages causes a decrease in the number of hydrogen atoms and thus, neither is considered an element of unsaturation. Turning this idea around, we can say that if the formula of a molecule is $C_nH_{2n+2}O$, then it must be an alcohol or an ether with no π bonds or ring structures. In a carbonyl compound, however, the oxygen atom takes the place of two hydrogen atoms. Thus, a carbonyl compound has one element of unsaturation and a general formula of $C_nH_{2n}O$.

To illustrate these ideas, consider a molecule having the formula C_3H_6O. For three carbon atoms, the maximum number of hydrogen atoms is eight. Ignoring the oxygen atom, we see that the number of hydrogen atoms is two less than the maximum, and so there is one degree of unsaturation. Three possible isomers are shown below. (Try to identify and draw structures for the other isomers.)

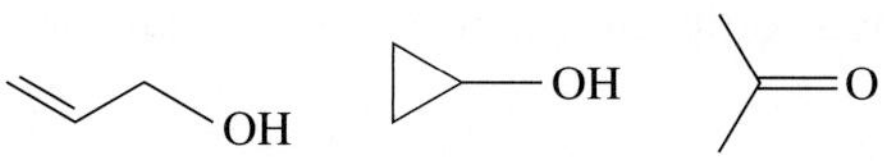

Now consider the molecule C_3H_8O. Ignoring the oxygen atoms, we determine that there are no elements of unsaturation. Consequently, there are no π bonds or rings in the molecule. Two possible isomers are shown below:

The ideas discussed above are easily extended to molecules containing more than one oxygen atom. For example, let us consider a molecule with the formula $C_3H_6O_2$. Again, in determining the degree of unsaturation, we ignore the oxygen atoms. Because there are only six hydrogen atoms and not eight, we conclude that the degree of unsaturation is one. Two possible isomers are a carboxylic acid or an ester, which are shown below:

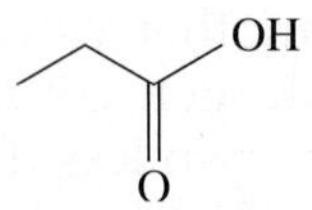

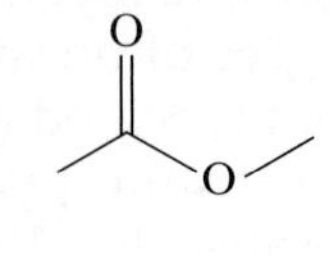

Is it possible one of the isomers of $C_3H_6O_2$ is a dialdehyde?

Nitrogen atoms also appear in organic compounds. How do we deal with a nitrogen atom in determining the degree of unsaturation? Let's consider the molecule C_3H_9N. Two isomers are shown below:

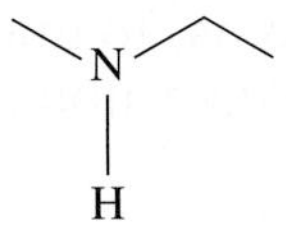

Both molecules are saturated and contain the maximum number of hydrogen atoms possible. How can we deduce the degree of unsaturation from the formula of this molecule? The key idea is that for each nitrogen atom present in the formula, the hydrogen atom count is diminished by one. Thus, the number of H atoms in C_3H_9N is considered to be 8 (not 9), which is equal to the number of H atoms expected in a saturated molecule containing three carbon atoms. For a compound with the formula C_4H_9N, the number of H atoms is effectively 8. A saturated molecule containing 4 carbon atoms should contain $2 \times 4 + 2 = 10$ hydrogen atoms. Because the molecule effectively contains only 8 hydrogen atoms (two less than the maximum), we know the degree of unsaturation is one. Two possible isomers for C_4H_9N are shown below:

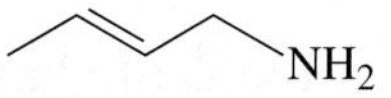

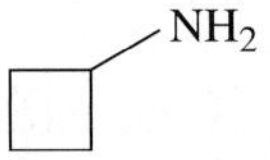

EXAMPLE 26-10 Determining the Degree of Unsaturation and Suggesting Possible Structures for a Molecule

What different types of compounds are possible for a compound with the molecular formula $C_6H_{10}O$?

Analyze

We need to establish the degree of unsaturation in the molecule and then construct one example of each type of molecule that can be formed.

Solve

As we described above, we can ignore the oxygen atom in determining the degree of unsaturation. The maximum number of hydrogen atoms is $2 \times 6 + 2 = 14$. The molecular formula has only 10 hydrogen atoms, so we know the degree of unsaturation is two. What types of molecules can be formed? We might start by considering dienes with either an alcohol or an ether functional group.

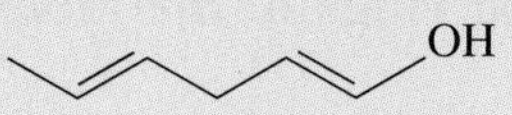

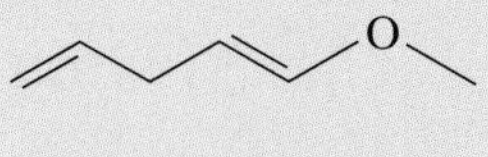

Alternatively, we can have a cyclic alkane with a carbonyl group or a cyclic alkene with an alcohol functional group. Two possible structures are shown here.

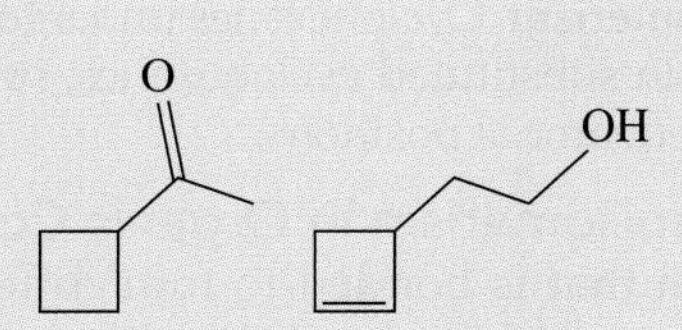

Two other possibilities are a ketone with an alkene side chain or an aldehyde with an alkene side chain. Two possible structures are shown here.

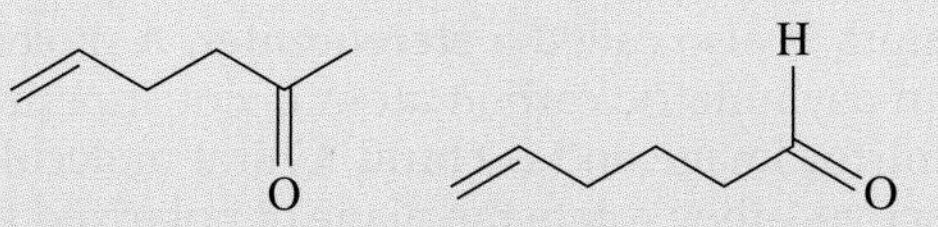

Assess

We have provided an example of an ether (with two degrees of unsaturation in the alkane chain), an aldehyde (with one degree of unsaturation in a side chain), a ketone, and some alcohols. This list is not exhaustive and you should be able to find other isomers.

PRACTICE EXAMPLE A: What different types of compounds are obtainable for a compound with the molecular formula $C_5H_{11}N$?

PRACTICE EXAMPLE B: What different types of compounds are obtainable for a compound with the molecular formula $C_5H_{10}O_2$?

Mastering**CHEMISTRY**   **www.masteringchemistry.com**

Obtaining enantiomerically pure samples of chiral molecules is extremely important when synthesizing pharmaceuticals. When a synthesis yields a mixture of enantiomers, the enantiomers may be separated from each other using a process called chemical resolution. For a discussion of this method, go to the Focus On feature for Chapter 26, Chemical Resolution of Enantiomers, on the MasteringChemistry site.

Summary

26-1 Organic Compounds and Structures: An Overview—Organic chemistry deals with compounds of carbon, the simplest of which, called **hydrocarbons**, contain only carbon and hydrogen atoms. The carbon atoms are bonded to one another in straight- or branched-chains or in rings. In **saturated hydrocarbons** all chemical bonds are single bonds. Isomers that differ in their structural skeletons are **constitutional isomers**. Alkane substituents that branch off a main hydrocarbon chain are called **alkyl groups** (Table 26.1). Carbon atoms in alkane chains are classified as **primary**, **secondary**, **tertiary**, and **quaternary** depending on the number of carbon atoms attached to them. Hydrogen atoms are similarly classified depending on the type of carbon atom to which they are attached; for example, hydrogen atoms attached to a primary carbon atom are called **primary hydrogen atoms**. A systematic nomenclature based on IUPAC rules is used to name alkanes. Distinctive groupings of atoms are referred to as **functional groups** (Table 26.2). Organic compounds are typically classified according to their functional groups. A three-dimensional representation of hydrocarbons can be accomplished using the **dashed and solid wedge line notation**. In this notation, solid lines represent bonds in the plane of the paper, solid wedges show bonds sticking out toward the viewer, and dashed wedges show bonds directed away from the viewer.

26-2 Alkanes—Among saturated hydrocarbons, also known as **alkanes**, those based on a straight- or branched-chains of carbon atoms have the formula C_nH_{2n+2} and are called **aliphatic** hydrocarbons. Aliphatic hydrocarbons that differ in sequence by a constant unit form a **homologous series**. Hydrocarbons can adopt different **conformations**, or arrangements, of their atoms. Aliphatic hydrocarbons have eclipsed and staggered conformations. Newman projections are used to depict these conformations (Figure 26-6). The energy difference between the eclipsed and staggered forms of an alkane is called the **torsional energy**. The variation in potential energy as one end of the molecule is rotated with respect to the other can be displayed in the form of a potential energy diagram (Fig. 26-8). The preference for certain conformations is driven in part by steric interactions involving bulky groups. In butane, for example, the **anti** conformation is lower in energy than the **gauche** conformation. The principal natural sources of alkanes are natural gas and petroleum.

26-3 Cycloalkanes—Hydrocarbons that form ring structures are termed **alicyclic**. Ring strain exists in cycloalkanes containing 3, 4, and 5 carbon atoms. A study of heats of combustion shows that cyclohexane is strain free. Disubstituted cycloalkanes exhibit a form of **stereoisomerism** known as *cis–trans* **isomerism.** Cyclohexanes undergo rapid ring flips (Fig. 26-13). In substituted cyclohexanes, substituents compete for the equatorial positions.

26-4 Stereoisomerism in Organic Compounds—A

carbon atom that is bonded to four different groups is called an **asymmetric** carbon atom. An asymmetric carbon atom is also called a **stereocenter**. A molecule containing an asymmetric carbon atom is not superimposable on its mirror image and is **chiral**. Chiral molecules are optically active—they rotate the plane of polarized light. Molecules with superimposable mirror images are **achiral**. Molecules with sp^3 stereocenters are named by using the *R, S* system. The rules for an *R, S* designation require assigning priorities to the substituents on a chiral C atom and "reading" the descending priority order of the substituents when the molecule is viewed in a particular orientation. If the descending priority order is clockwise, the molecule is *R*; if it is counterclockwise, the molecule is *S*. The two molecules with *R* and *S* configurations are known as **enantiomers**. Stereoisomers that are not enantiomers are called **diastereomers**. *Cis* and *trans* isomers are examples of diastereomers.

26-5 Alkenes and Alkynes—Hydrocarbons whose

molecules contain fewer hydrogen atoms than found in the corresponding alkane are **unsaturated** hydrocarbons. The unsaturation arises because of the presence of one or more carbon-to-carbon double or triple bonds. Hydrocarbons containing one or more double bonds are **alkenes**, or **olefins**, whereas those containing one or more triple bonds are **alkynes**. The placement of double or triple bonds in an unsaturated hydrocarbon can lead to different constitutional isomers, as in 2-butene versus 1-butene, for example. In 2-butene the orientation of the two end methyl groups leads to *cis* and *trans* isomers, a type of isomerism called *cis–trans* isomerism. Highly substituted alkenes are named by the *E, Z* system. Molecules are given the *Z* designation when higher priority substituents at each end of a double bond are on the same side of the double bond and the *E* designation when they are on opposite sides of the double bond.

26-6 Aromatic Hydrocarbons—Many unsaturated

hydrocarbons with ring structures are classified as **aromatic hydrocarbons**; most are based on the benzene molecule, C_6H_6. Two systems are commonly used to name substituents on a benzene ring. One uses a numbering system, such as 1,2-dichlorobenzene. The other uses the relative positions of substituents. Substituents on adjacent carbon atoms are referred to as **ortho** (*o*); substituents separated by one carbon are **meta** (*m*); and substituents opposite each other on the ring are **para** (*p*). A **phenyl** group is a benzene ring with one hydrogen atom removed. A **benzyl** group is a methyl group with one hydrogen atom replaced by a phenyl group.

26-7 Organic Compounds Containing Functional

Groups—*Alcohols and phenols:* Hydrocarbons that contain an —OH (**hydroxyl**) group are classified as **alcohols** in aliphatic compounds and **phenols** in aromatic compounds. Aliphatic alcohols are also classified as primary, secondary, or tertiary, depending on how many other substituents (excluding H atoms) are attached to the same carbon atom as the hydroxyl group: one for primary, two for secondary, and three for tertiary. Diols and polyols are molecules with two or more hydroxyl groups, respectively.

Ethers: The general formula of an **ether** is R—O—R′. The ether linkage is very stable and ethers are generally unreactive, resistant to both oxidation and reduction. One of the major uses of ethers is as solvents.

Aldehydes and ketones: The **carbonyl group** has a carbon atom attached by a double bond to an oxygen atom. If one of the two substituents attached to the carbonyl group is an H atom, the compound is an **aldehyde**, otherwise it is a **ketone**. In IUPAC nomenclature, the suffix *-al* is used for aldehydes and *-one* for ketones.

Carboxylic acids: The **carboxyl group** is a combination of the carbonyl and hydroxyl groups. **Carboxylic acids** have the general formula RCOOH and can be prepared by the oxidation of aldehydes. An acetyl group results when the —OH of a carboxyl group is replaced by a methyl group.

Esters: An **ester** is formed when an alkoxy group replaces the —OH of a carboxyl group. The general formula of an ester is RCOOR′. Most esters are colorless liquids that are insoluble in water and have pleasant fragrances.

Amides: The general formula of a primary **amide** is $RCONH_2$. A primary amide has two H atoms bonded to N and results when the hydroxyl group of a carboxylic acid is replaced by —NH_2. In secondary or tertiary amides, one or two H atoms in the —NH_2 are replaced by alkyl or aryl groups. Because of resonance, amides are much weaker bases than are ammonia or amines.

Amines: **Amines** are organic derivatives of ammonia. They are classified according to the number of organic groups (R) substituted for hydrogen atoms in NH_3. A primary amine has one substituent; a secondary amine, two substituents; and a tertiary amine, three substituents. Amines are weak bases, with the aliphatic amines being stronger bases than ammonia and the aromatic amines being weaker. Tetraalkyl ammonium salts are organic derivatives of the ammonium cation with a suitable anion.

Heterocyclic compounds: In a **heterocyclic compound**, at least one of the ring atoms is a heteroatom (not carbon). The most common heteroatoms are N, O, and S. Heterocyclic compounds tend to be more soluble in water than the parent compound because of the polarity of the heterocycle.

26-8 From Molecular Formula to Molecular

Structure—A useful idea in determining a suitable structure is the **degree of unsaturation**, which is equal to the total number of π bonds and ring structures in a molecule. The degree of unsaturation is determined by comparing the number of H atoms in the molecule to the maximum number possible for the number of carbon atoms present. In counting the number of H atoms in a molecule, each halogen atom effectively increases the hydrogen count by one; oxygen atoms can be ignored; and each nitrogen atom, in effect, decreases the hydrogen atom count by one.

Integrative Example

An acyclic organic compound with the formula $C_6H_{12}O$ is found to be optically active, does not decolorize Br_2 (in CCl_4), and does not undergo reaction when treated with a mixture of $Na_2Cr_2O_7$ and H_2SO_4. However, a reaction does occur when the compound is treated with $NaBH_4$. Identify a compound having these physical and chemical properties. Then draw structures for the two stereoisomers, using dashed and solid wedge symbols. Finally, give acceptable names for the two stereoisomers.

Analyze

There is only one oxygen atom per molecule, so the compound must be an alcohol, ether, aldehyde, or ketone. Also, because there are only 12 hydrogen atoms per molecule, rather than $2 \times 6 + 2 = 14$, we know that the molecule must contain either a π bond or a ring. We are told the compound is acyclic, and thus the molecule must contain a π bond and not a ring. Alcohols (except tertiary alcohols) and aldehydes are rather easily oxidized, so we suspect the molecule is neither an alcohol (though possibly a tertiary alcohol) nor an aldehyde. The ether linkage is stable in the presence of most oxidizing and reducing agents, and because the compound we need can be reduced, we suspect that that the compound is not an ether. At this point, we must decide whether the molecule is a ketone or a tertiary alcohol containing a carbon–carbon double bond. We can rule out a carbon–carbon double bond because the compound does not react with Br_2, a reaction characteristic of alkenes. The compound we seek is a ketone.

Solve

We have reasoned that the compound is a ketone. The linkage associated with a ketone is

$$\underset{R}{}\overset{\displaystyle O}{\underset{}{\overset{\|}{C}}}\underset{R'}{}$$

The five remaining carbon atoms are distributed among the two alkyl groups. The carbon of a carbonyl group is bonded to only three groups, and thus it cannot be chiral. The chiral carbon is part of one of the alkyl groups. Both R and R' must contain at least one carbon atom and so, the maximum number of carbon atoms in R or R' is four.

Let's choose R to be the alkyl group that contains the chiral carbon. A chiral carbon atom must be bonded to four different groups and yet R contains no more than four carbon atoms. The only possibility for R is

$$H_3C - \underset{CH_2CH_3}{\overset{H}{\underset{|}{\overset{|}{C}}}} -$$

We have accounted for five of the six carbon atoms (R contains four carbon atoms and the carbonyl group contains one). Thus, R' represents a methyl group. A skeletal structure for the molecule is

$$H_3C - \overset{}{\underset{\underset{CH_2CH_3}{|}}{\overset{H}{\underset{|}{\overset{|}{C}}}}}{}^3 - \overset{O}{\underset{}{\overset{\|}{C}}}{}^2 - {}^1CH_3$$

3-methyl-2-pentanone

The longest carbon chain has five carbon atoms with a carbonyl group bonded to C2 and a methyl group bonded to C3. The molecule is 3-methyl-2-pentanone. Because C3 is chiral, two enantiomers are possible.

(R)-3-methyl-2-pentanone (S)-3-methyl-2-pentanone

Assess

It may not be obvious that there is only one possibility for R. Try drawing structures of other ketones having the formula $C_6H_{12}O$ to convince yourself that there is only one possibility. Had we not been told the compound was acyclic, a unique identification would not have been possible.

PRACTICE EXAMPLE A: Compound A with the formula C_3H_8O is soluble in water and reacts with sodium metal, producing bubbles of gas. When compound A is treated with chromic acid (a mixture of $Na_2Cr_2O_7$ and H_2SO_4), compound B is formed. Compound B dissolves readily in $Na_2CO_3(aq)$ and reacts with ethanol, yielding compound C which has a fruity fragrance. Identify compounds A, B, and C.

PRACTICE EXAMPLE B: The following molecules all have the molecular formula $C_5H_{10}O$. You suspect that you have a sample of one of these compounds. What tests could you perform to ascertain which of these compounds you have?

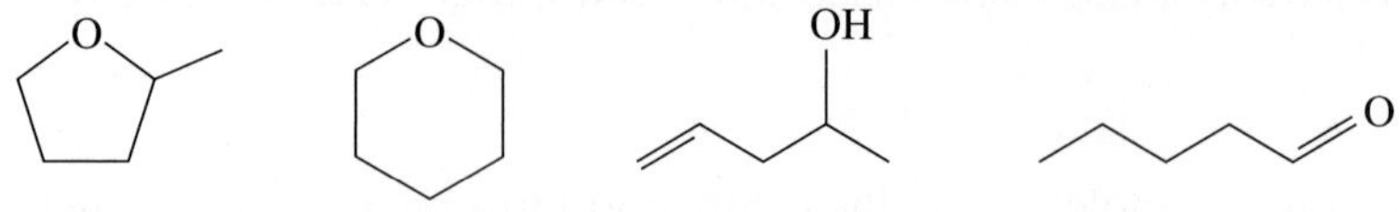

(a)   (b)   (c)   (d)

Mastering**CHEMISTRY**

**You'll find a link to additional self study questions in the study area on
www.masteringchemistry.com**

Exercises

Organic Structures

1. Write structural formulas corresponding to these condensed formulas.
   (a) $CH_3CH_2CHBrCHBrCH_3$
   (b) $(CH_3)_3CCH_2C(CH_3)_2CH_2CH_2CH_3$
   (c) $(C_2H_5)_2CHCH=CHCH_2CH_3$
2. Draw a structural formula for each of the following compounds.
   (a) 3-isopropyloctane; (b) 2-chloro-3-methylpentane;
   (c) 2-pentene; (d) dipropyl ether.
3. Supply a structural formula for each of the following compounds.
   (a) 1,3,5-trimethylbenzene; (b) p-nitrophenol;
   (c) 3-amino-2,5-dichlorobenzoic acid (a plant-growth regulator).
4. Write structural formulas corresponding to these condensed formulas.
   (a) $(CH_3)_3CCH_2CH(CH_3)CH_2CH_2CH_3$
   (b) $(CH_3)_2CHCH_2C(CH_3)_2CH_2Br$
   (c) $Cl_3CCH_2CH(CH_3)CH_2Cl$

5. Draw Lewis structures of the following simple organic molecules: (a) $CH_3CHClCH_3$; (b) $HOCH_2CH_2OH$; (c) CH_3CHO.
6. Draw Lewis structures of the following simple organic molecules: (a) CH_3CH_2COOH; (b) H_3CCN; (c) $CH_3CH_2NH_2$.
7. With appropriate sketches, represent chemical bonding in terms of the overlap of hybridized and unhybridized atomic orbitals in the following molecules.
   (a) C_4H_{10}; (b) $H_2C=CHCl$; (c) $CH_3C\equiv CH$
8. With appropriate sketches, represent chemical bonding in terms of the overlap of hybridized and unhybridized atomic orbitals in the following molecules.

$$\text{(a) } CH_3\overset{\overset{\displaystyle O}{\|}}{C}CH_3; \text{(b) } CH_3\overset{\overset{\displaystyle O}{\|}}{C}-OH; \text{(c) } H_2C=C=CH_2$$

Isomers

9. What is the relationship, if any, between the molecules in each of the following pairs? The relationship may be any of identical structures, constitutional isomers, stereoisomers, or no relationship.

   (a)

   $CH_3CH_2CH_2Cl$ and $ClCH_2CH_2CH_3$

   (b)

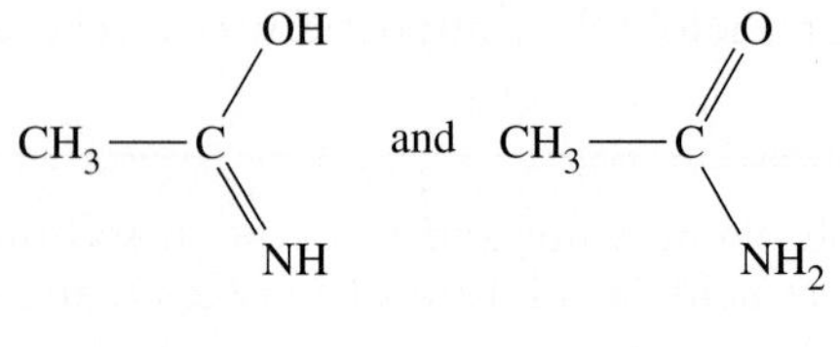

   (c)

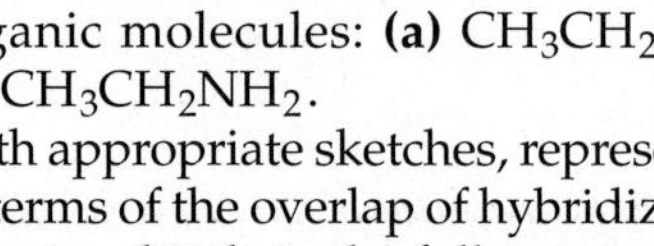

   (d)

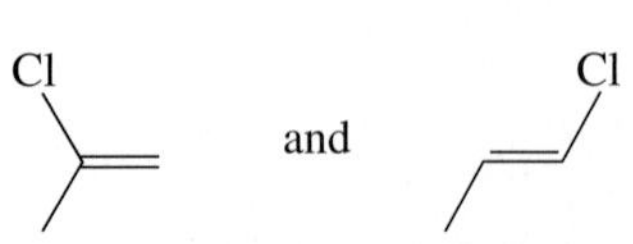

   (e)

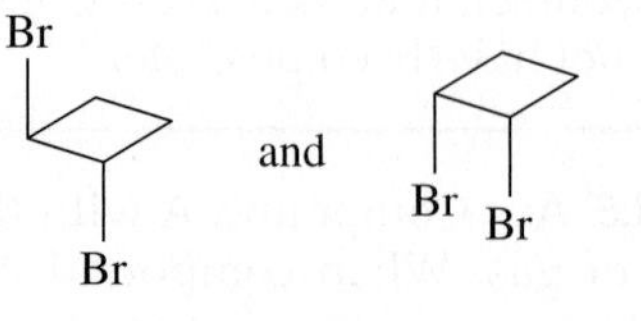

10. What is the relationship, if any, between the molecules in each of the following pairs? The relationship may be any of identical structures, constitutional isomers, stereoisomers, or no relationship.

(a) 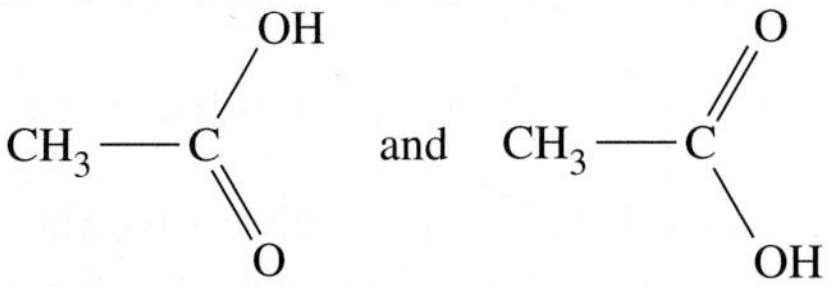 CH₃—C(OH)=O and CH₃—C(=O)—OH

(b)

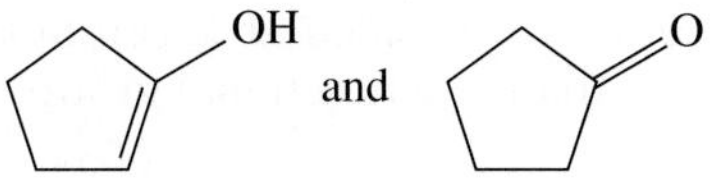

(c)

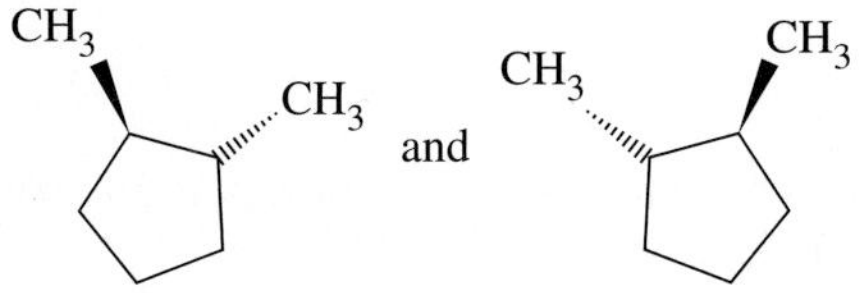

(d)

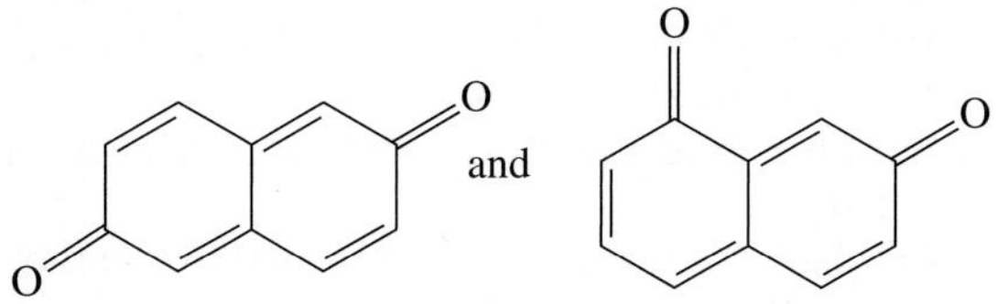

(e)

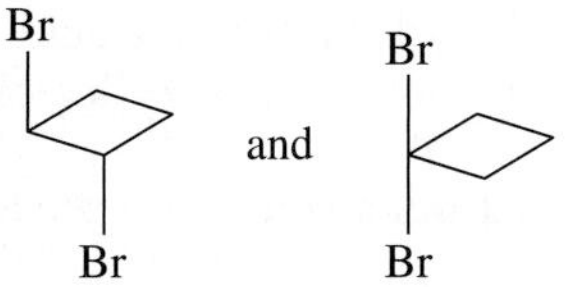

11. Draw structural formulas for all the isomers of C_7H_{16}.

12. Draw and name all the isomers of **(a)** C_6H_{14}; **(b)** C_4H_8; **(c)** C_4H_6. [*Hint:* Do not forget double bonds, rings, and combinations of these.]

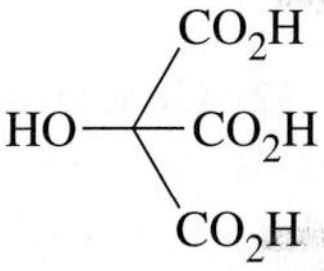

Functional Groups

17. Identify the functional group in each compound (i.e., alcohol, amine, etc.).
  (a) $CH_3CHBrCH_2CH_3$
  (b) $C_6H_5CH_2CHO$
  (c) $CH_3COCH_2CH_3$
  (d) $C_6H_4(OH)_2$

18. Identify the functional group in each compound (i.e., alcohol, amine, etc.).
  (a) CH_3CH_2COOH
  (b) $(CH_3)_2CHCH_2OCH_3$
  (c) $CH_3CH(NH_2)CH_2CH_3$
  (d) CH_3COOCH_3

19. The functional groups in each of the following pairs have certain features in common, but what is the essential difference between them?
  (a) carbonyl and carboxyl
  (b) aldehyde and ketone
  (c) acetic acid and acetyl group

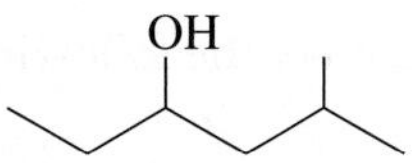

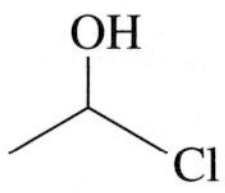

20. By name or formula, give one example of each of the following types of compounds: **(a)** aromatic nitro compound; **(b)** aliphatic amine; **(c)** chlorophenol; **(d)** aliphatic diol; **(e)** unsaturated aliphatic alcohol; **(f)** alicyclic ketone; **(g)** halogenated alkane; **(h)** aromatic dicarboxylic acid.

13. Identify the chiral carbon atoms, if any, in the following compounds.

(a)   **(b)**

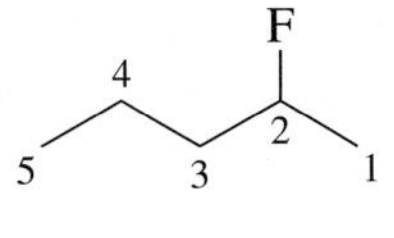

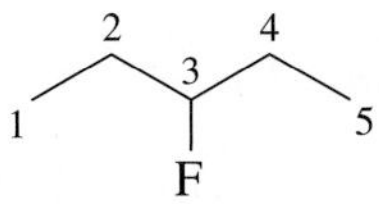

(c)

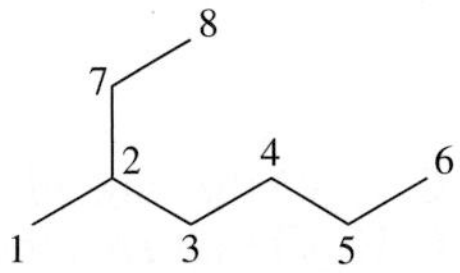

14. Identify the chiral carbon atoms, if any, in the following compounds.

(a)   **(b)**   **(c)**

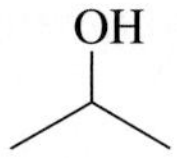

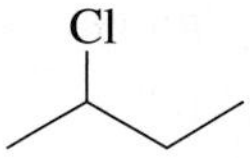

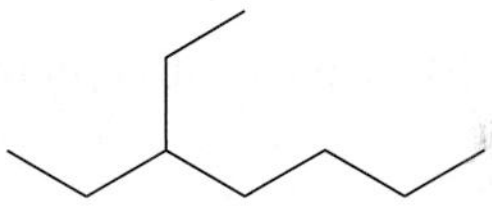

15. Identify the chiral carbon atoms, if any, in the following compounds.

(a)   **(b)**   **(c)**

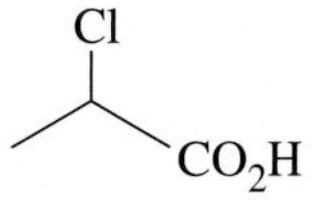

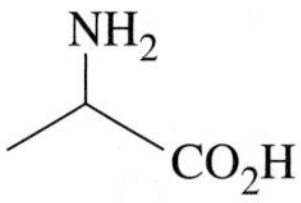

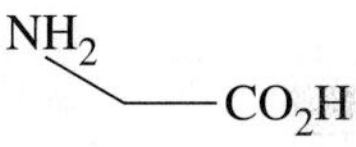

16. Identify the chiral carbon atoms, if any, in the following compounds.

(a)   **(b)**   **(c)**

21. Identify and name the functional groups in each of the following.

(a) 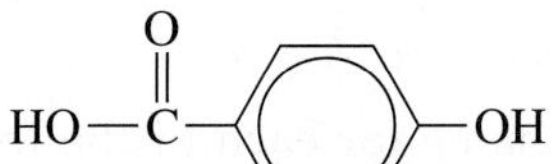 HO—C(=O)—⟨benzene⟩—OH

(b) CH_3—C(=O)—OCH_2CH_3

(c) CH_3—C(=O)—CH_2CH_2—CO_2H

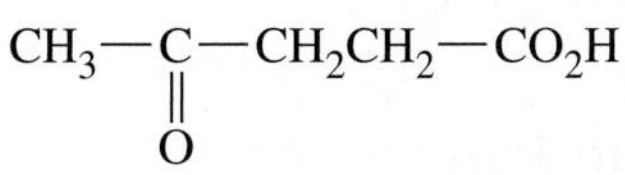

(d)

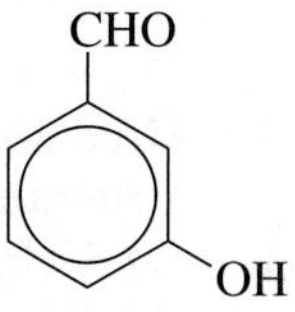

22. Identify and name the functional groups in each of the following.

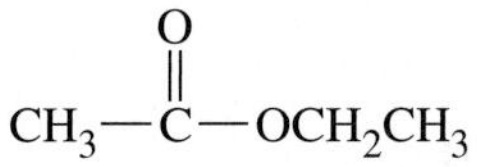

(a) 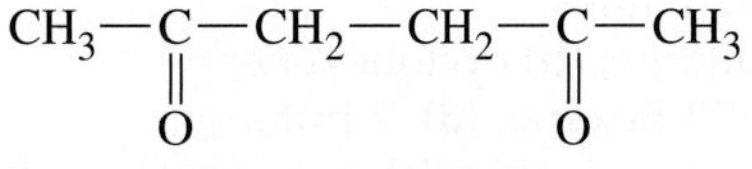 CH_3—C(=O)—CH_2—CH_2—C(=O)—CH_3

(b) CH_3—CH_2—$CONHCH_3$

(c) 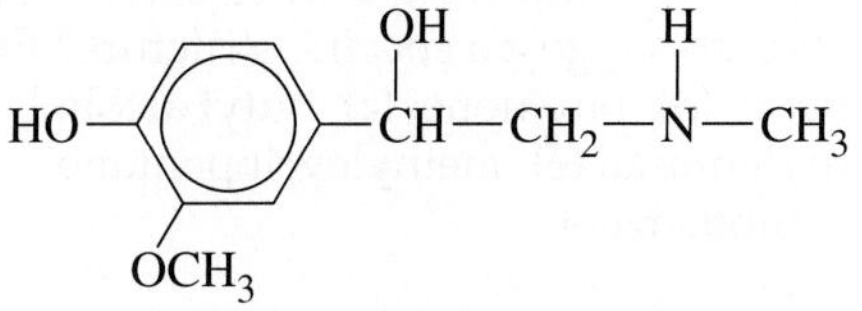 HO—⟨benzene, OCH₃⟩—CH(OH)—CH_2—N(H)—CH_3

(d)

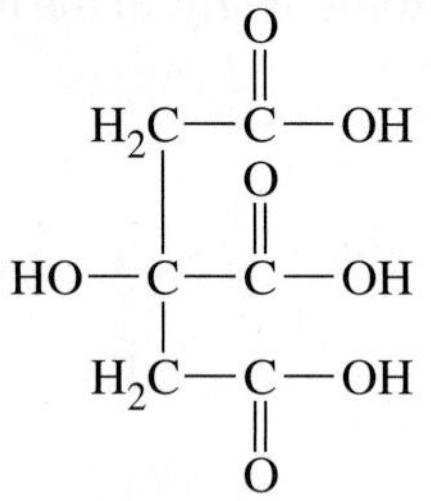

$$H_2C-\overset{\overset{\displaystyle O}{\|}}{C}-OH$$
$$HO-\overset{\overset{\displaystyle |}{}}{C}-\overset{\overset{\displaystyle O}{}}{C}-OH$$
$$H_2C-\overset{\overset{\displaystyle |}{}}{\underset{\underset{\displaystyle O}{\|}}{C}}-OH$$

23. Give the isomers of $C_4H_{10}O$ that are ethers.

24. Give the isomers of $C_5H_{12}O$ that are ethers.

Nomenclature and Formulas

31. Give an acceptable name for each of the following.

(a)
$$\overset{\overset{\displaystyle CH_3}{|}}{}\quad\overset{\overset{\displaystyle CH_3}{|}}{}$$
$$CH_3CH_2CH_2CHCH_2CHCH_2CH_3$$

(b)
$$\overset{\overset{\displaystyle CH_3}{|}}{CH_3-\underset{\underset{\displaystyle CH_3}{|}}{C}-CH_3}$$

(c)
$$\overset{\overset{\displaystyle Cl}{|}}{CH_3CH_2CHCH_2-\underset{\underset{\displaystyle Cl}{|}}{C}-CH_2CH_3}$$
$$\quad\quad\overset{\overset{\displaystyle |}{C_2H_5}}{}$$

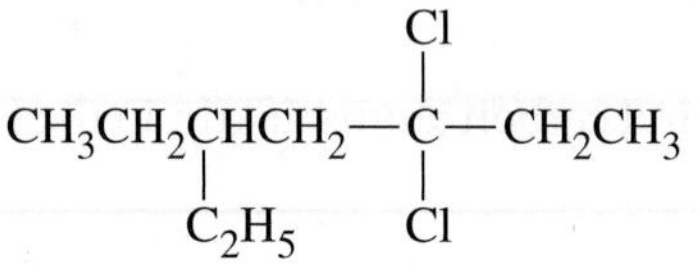

32. Give an acceptable name for each of the following.

(a) [benzene ring with Cl at top and Cl at bottom right]

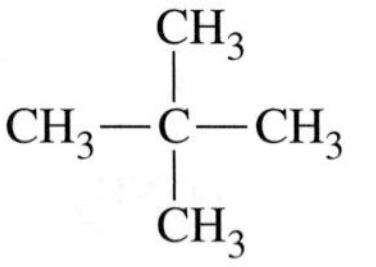

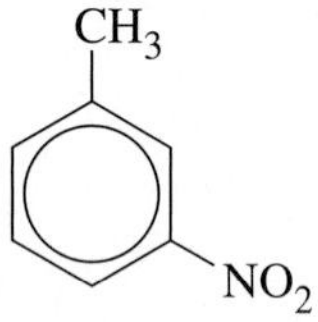

(b) [benzene ring with CH₃ at top and NO₂ at bottom right]

(c) NH_2-[benzene ring]$-COOH$

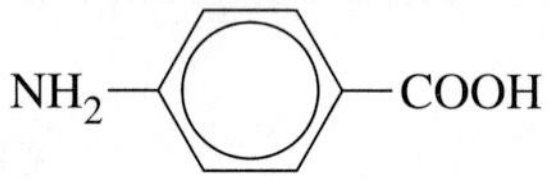

33. Give an acceptable name for each of the following structures.
(a) $CH_3CH_2C(CH_3)_3$
(b) $(CH_3)_2C=CH_2$
(c) $CH_3-CH-CH-CH_3$ with CH_2 bridging
(d) $CH_3C\equiv CCH(CH_3)_2$
(e) $CH_3CH(C_2H_5)CH(CH_3)CH_2CH_3$
(f) $CH_3CH(CH_3)CH(CH_3)CCH_2CH_2CH_3=CH_2$

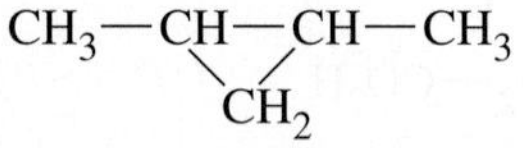

34. Draw a condensed structure to correspond to each of the following names.
(a) methylbutane; (b) cyclohexene;
(c) 2-methyl-3-hexyne; (d) 2-butanol;
(e) ethyl isopropyl ether; (f) propanal

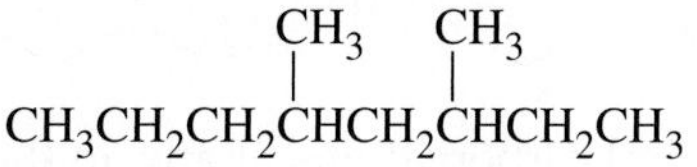

35. Does each of the following names convey sufficient information to suggest a specific structure? Explain.
(a) pentene; (b) butanone; (c) butyl alcohol;
(d) methylaniline; (e) methylcyclopentane;
(f) dibromobenzene

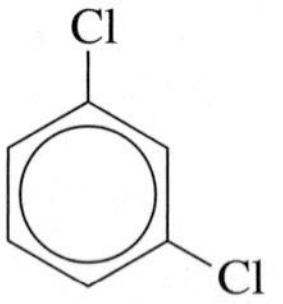

25. Give the isomers of the carboxylic acid with molecular formula $C_5H_{10}O_2$.

26. Give the isomers of the carboxylic acid with the molecular formula $C_4H_8O_2$.

27. Give the isomers of the esters having the molecular formula $C_5H_{10}O_2$.

28. Give the isomers of the esters having the molecular formula $C_4H_8O_2$.

29. Give the noncyclic isomers with molecular formula $C_4H_8O_2$ that contain more than one functional group.

30. Give the isomers with molecular formula $C_5H_{10}O_2$ that contain more than one functional group.

36. Indicate why each of these names is incorrect, and give a correct name.
(a) 3-pentene; (b) pentadiene; (c) 1-propanone;
(d) bromopropane; (e) 2,6-dichlorobenzene;
(f) 2-methyl-3-pentyne

37. Supply condensed structural formulas for the following substances.
(a) 2,4,6-trinitrotoluene (TNT—an explosive)
(b) methyl salicylate (oil of wintergreen)
[*Hint:* Salicylic acid is *o*-hydroxybenzoic acid.]
(c) 2-hydroxy-1,2,3-propanetricarboxylic acid (citric acid, $C_6H_8O_7$)

38. Supply condensed structural formulas for the following substances.
(a) *o-tert*-butylphenol (an antioxidant in aviation gasoline)
(b) 1-phenyl-2-aminopropane (benzedrine—an amphetamine, ingredient in "pep pills")
(c) 2-methylheptadecane (a sex pheromone of tiger moths—a chemical used for communication among members of the species) [*Hint:* Heptadeca means 17.]

39. Name the following amines.

(a) 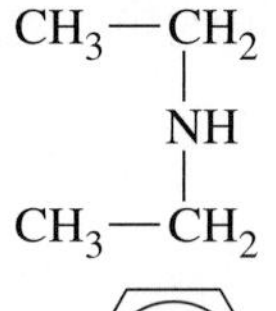
$$CH_3-CH_2$$
$$\quad\quad|$$
$$\quad\quad NH$$
$$\quad\quad|$$
$$CH_3-CH_2$$

(b) 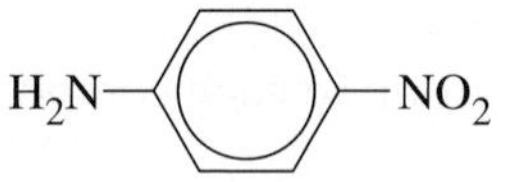 H_2N-[benzene ring]$-NO_2$

(c) 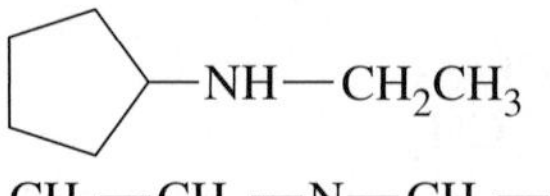 [cyclopentane ring]$-NH-CH_2CH_3$

(d) 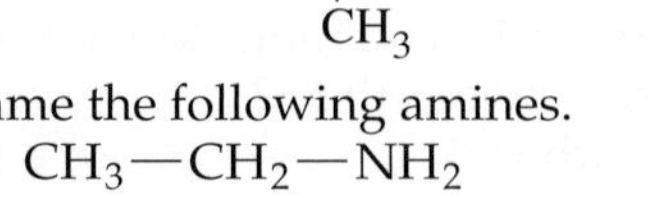
$$CH_3-CH_2-\underset{\underset{\displaystyle CH_3}{|}}{N}-CH_2-CH_3$$

40. Name the following amines.
(a) $CH_3-CH_2-NH_2$

(b) 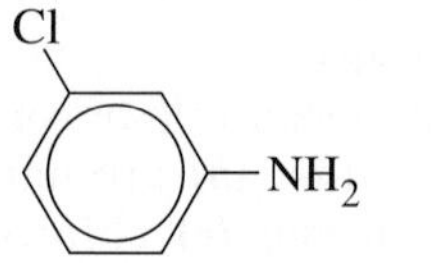 [benzene ring with Cl at top left]$-NH_2$

(c) 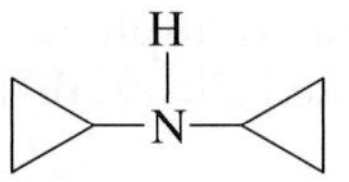 [cyclopropane ring]$-\overset{\overset{\displaystyle H}{|}}{N}-$[cyclopropane ring]

(d) $Cl-CH_2-CH_2-NH_2$

Alkanes and Cycloalkanes

41. Classify the carbon atoms in **(a)** methylbutane, and **(b)** 2,2-dimethylpropane as methyl, primary (1°), secondary (2°), tertiary (3°), or quaternary (4°).

42. Classify the carbon atoms in **(a)** 2,4-dimethylpentane, and **(b)** ethylcyclobutane as methyl, primary (1°), secondary (2°), tertiary (3°), or quaternary (4°).

43. Draw Newman projections for the staggered and eclipsed conformations of pentane for rotation about the C2—C3 bond. Which conformation is lowest in energy?

44. Draw Newman projections for the staggered and eclipsed conformations of 2-methylpentane for rotation about the C2—C3 bond. Which conformation is lowest in energy?

45. Draw the most stable conformation for the molecule below:

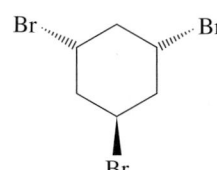

46. Draw the most stable conformation for the molecule below:

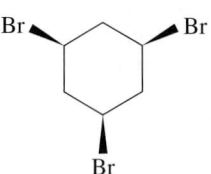

47. For each of the following substituted cyclohexanes, draw the two possible chair conformations, label each substituent as axial or equatorial, and identify the more stable conformer.
(a) cyclohexanol
(b) *trans*-3-methylcyclohexanol

48. For each of the following substituted cyclohexanes, draw the two possible chair conformations, label each substituent as axial or equatorial, and identify the more stable conformer.
(a) *cis*-1-isopropyl-3-methylcyclohexane
(b) *cis*-4-*tert*-butylcyclohexanol

Alkenes

49. Why is it not necessary to refer to ethene and propene as 1-ethene and 1-propene? Can the same be said for butene?

50. Alkenes (olefins) and cyclic alkanes (alicyclics) each have the generic formula C_nH_{2n}. In what important ways do these types of compounds differ structurally?

51. Assign a configuration (*E* or *Z*) to each of the following molecules.
(a)

(b)

(c)

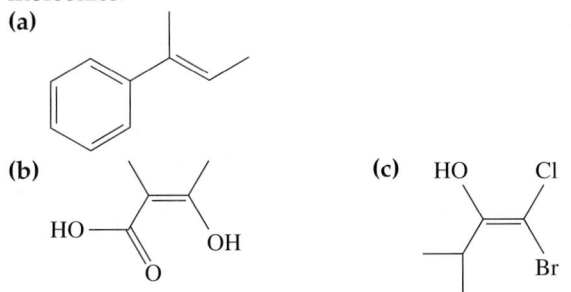

52. Assign a configuration (*E* or *Z*) to each of the following molecules.
(a)

(b)

(c)

53. Draw the *E* and *Z* isomers of **(a)** 2-chloro-2-butene; **(b)** 3-methyl-2-pentene

54. Draw the *E* and *Z* isomers of **(a)** 3-methyl-3-hexene; **(b)** 3-fluoro-2-methyl-3-hexene

Aromatic Compounds

55. Supply a name or structural formula for each of the following.
(a) phenylacetylene
(b) *m*-dichlorobenzene

(c)

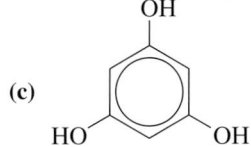

56. Supply a name or structural formula for each of the following.
(a) *p*-phenylphenol
(b) 3-hydroxy-4-isopropyltoluene (thymol—flavor constituent of the herb thyme)

(c)

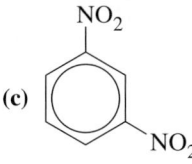

Organic Stereochemistry

57. Draw suitable structural formulas to show that there are *four* constitutional isomers of $C_3H_6Cl_2$.

58. Which of the following pairs of molecules are constitutional isomers and which are not? Explain.

(a) $CH_3CH_2CH_2CH_3$ and $CH_3CH=CHCH_3$

(b) $CH_3(CH_2)_5CH(CH_3)_2$ and
$$CH_3(CH_2)_4CH(CH_3)CH_2CH_3$$

(c) $CH_3CHClCH_2CH_3$ and $CH_3CH_2CH_2CH_2CH_2Cl$

(d) and

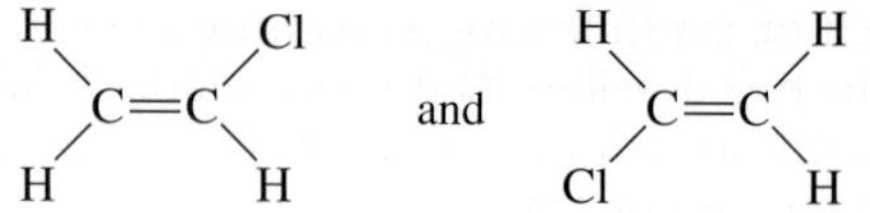

(e) and

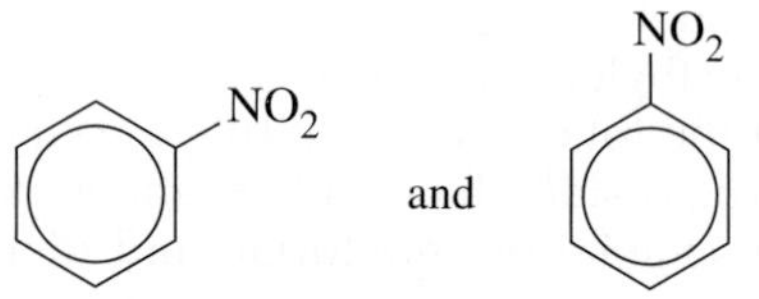

(f) and

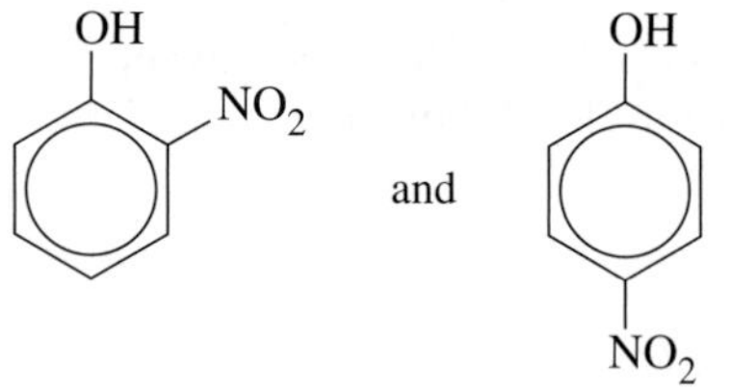

59. For each pair of structures shown below, indicate whether the two species are identical molecules, enantiomers, or isomers of some other sort.

(a) and   **(b)** and

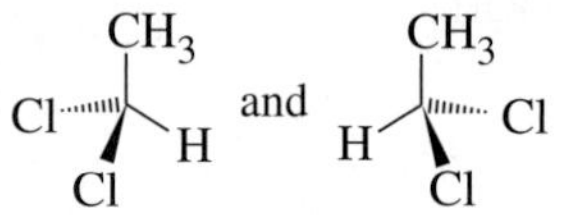

(c) and   **(d)** and

(e) and   **(f)** and

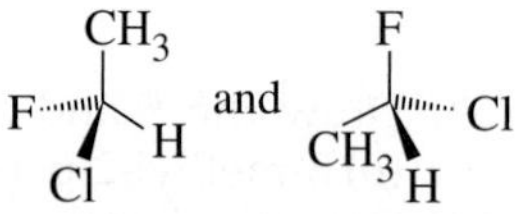

60. For each pair of structures shown below, indicate whether the two species are identical molecules, enantiomers, or isomers of some other sort.

(a)

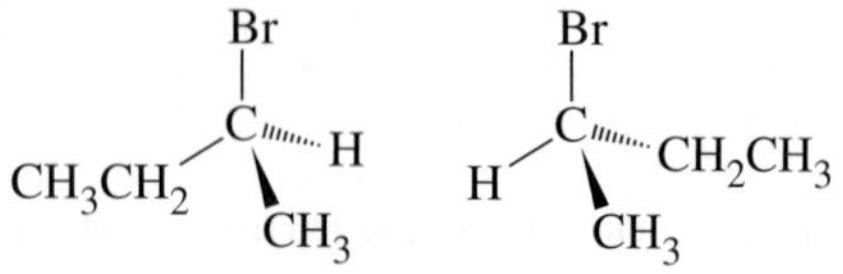

(b)

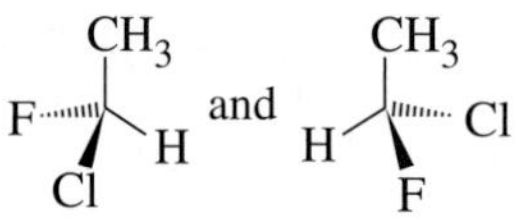

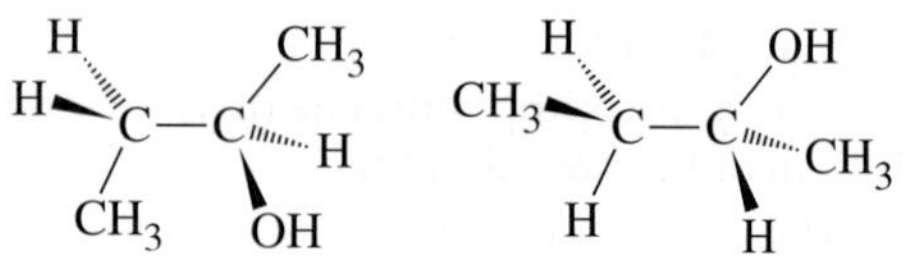

(c)

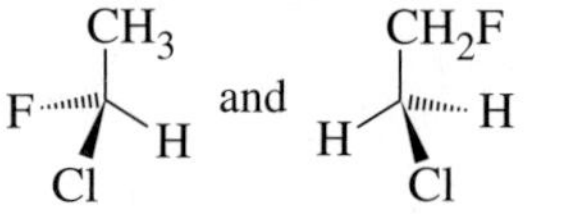

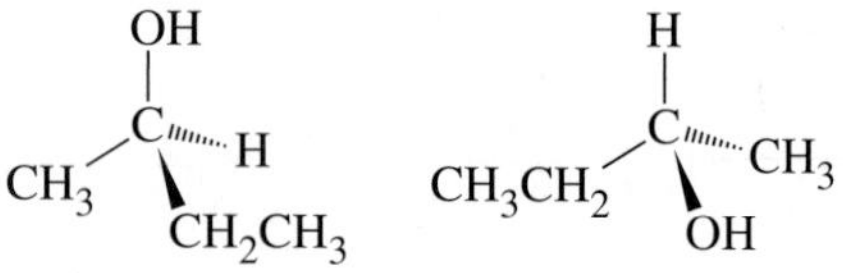

(d)

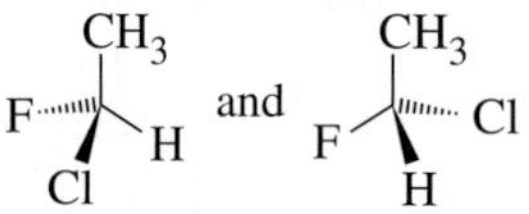

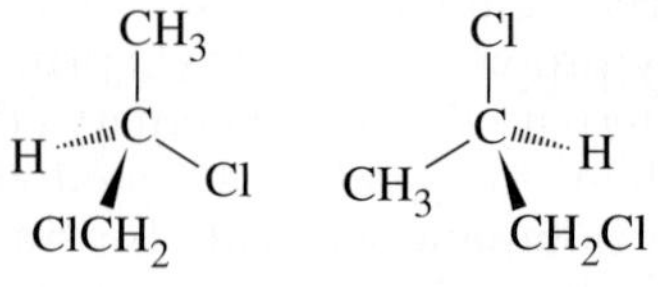

(e)

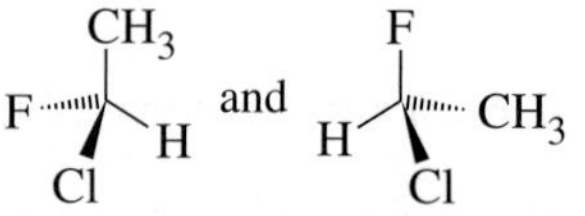

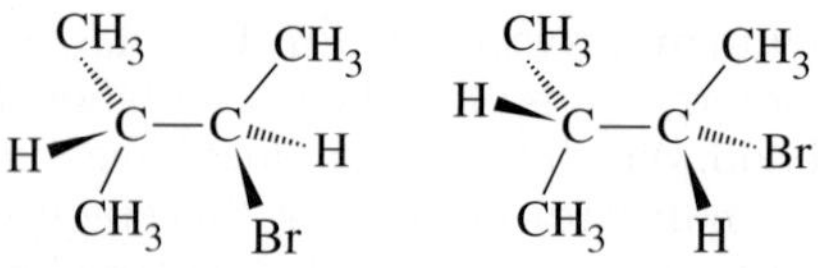

(f)

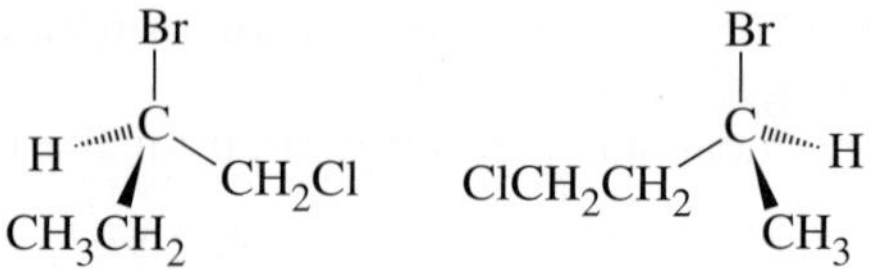

61. Name the following molecules with the appropriate stereochemical designation.

(a)   **(b)**

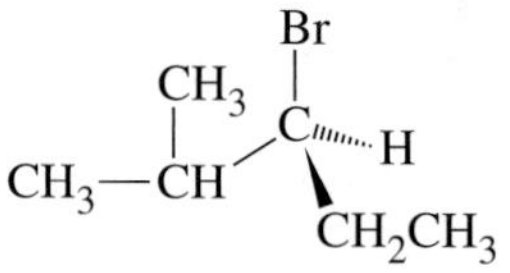

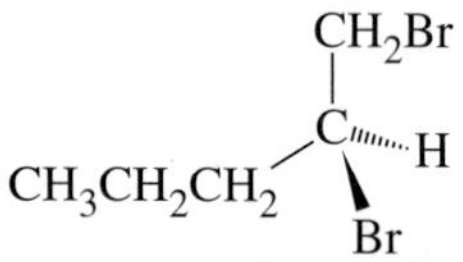

(c) 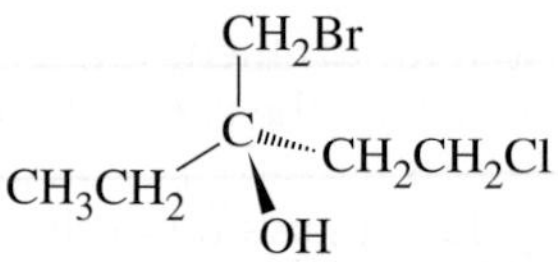   **(d)**

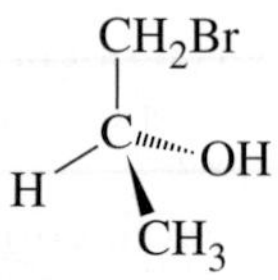

62. Name the following molecules with the appropriate stereochemical designation.

(a)   **(b)**

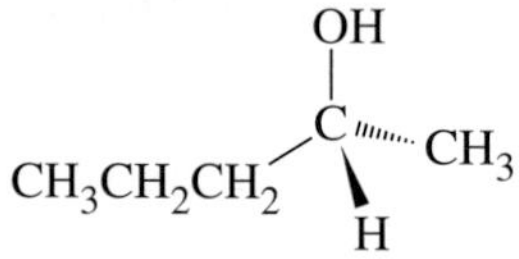

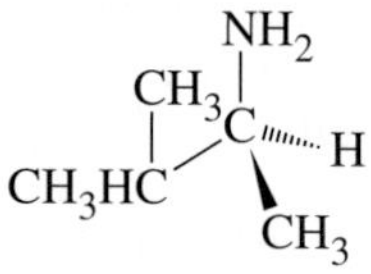

(c) 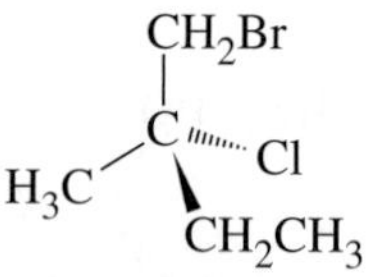   **(d)**

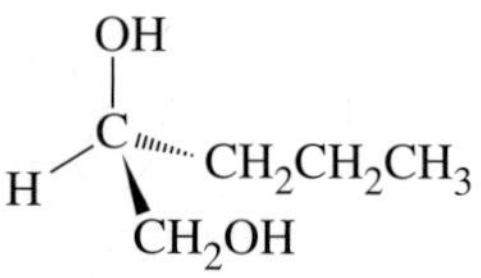

63. Name the following molecules with the appropriate stereochemical designation.

(a)   **(b)**

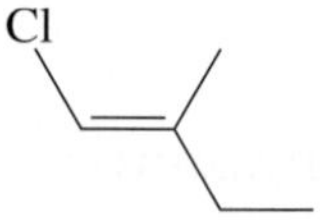

(c)

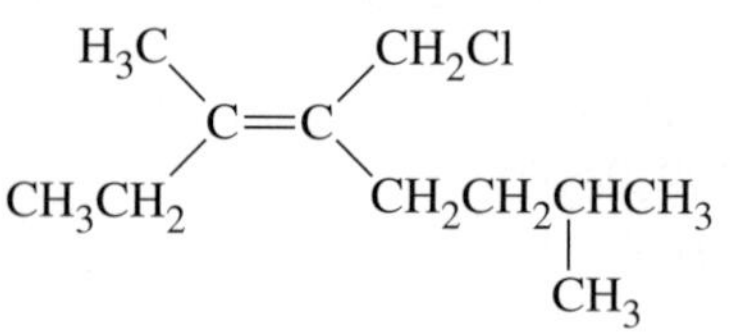

(d)

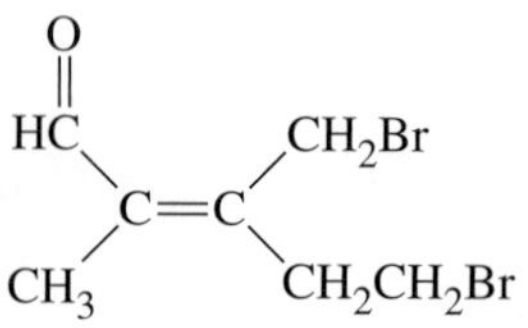

64. Name the following molecules with the appropriate stereochemical designation.

(a)

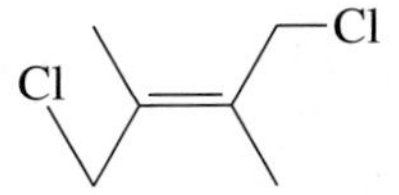

(b)

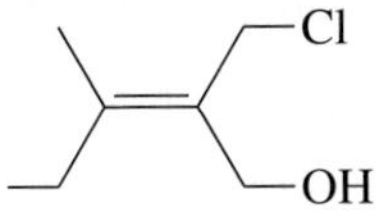

(c)

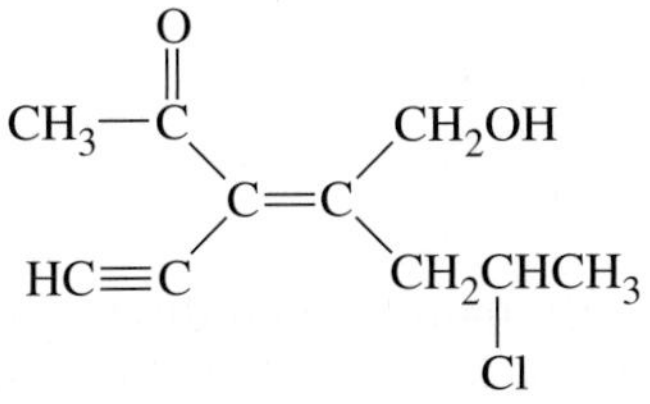

(d)

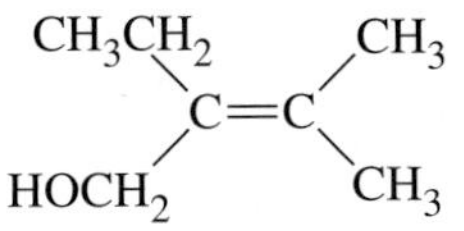

65. Draw the structure for each of the following.
(a) (Z)-1,3,5-tribromo-2-pentene
(b) (E)-1,2-dibromo-3-methyl-2-hexene
(c) (S)-1-bromo-1-chlorobutane
(d) (R)-1,3-dibromohexane
(e) (S)-1-chloro-2-propanol

66. Draw the structure for each of the following.
(a) (R)-1-bromo-1-chloroethane
(b) (E)-2-bromo-2-pentene
(c) (Z)-1-chloro-3-ethyl-3-heptene
(d) (R)-2-hydroxypropanoic acid
(e) (S)-2-aminopropanoate anion

Structures and Properties of Organic Compounds

67. Consider the following molecular formulas. How many elements of unsaturation are there in each case? (a) $C_4H_{11}N$; (b) C_4H_6O; (c) $C_9H_{15}ClO$.

68. Consider the following molecular formulas. How many elements of unsaturation are there in each case? (a) C_5H_9NO (b) $C_5H_8O_3$; (c) C_5H_9ClO.

69. How many elements of unsaturation are there in the molecule below? What is the molecular formula?

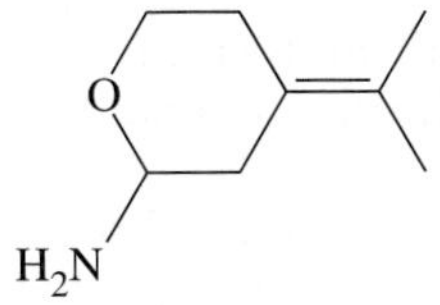

70. How many elements of unsaturation are there in the molecule below? What is the molecular formula?

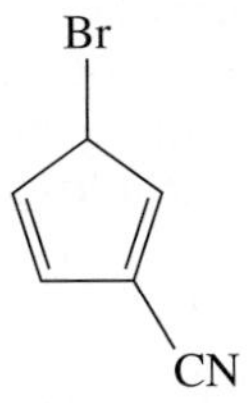

71. Match the following compounds with the chemical properties given below. Write a chemical equation for the reactions described in (a)−(d).

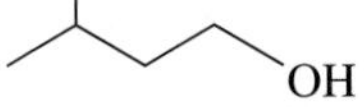

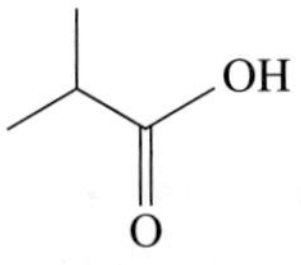

(a) is easily oxidized
(b) decolorizes bromine water
(c) generates bubbles of gas when treated with $Na_2CO_3(aq)$
(d) generates bubbles of gas when sodium metal is added

72. Match the following compounds with the chemical properties given below. Write a chemical equation for the reactions described in (a)−(d).

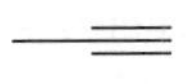

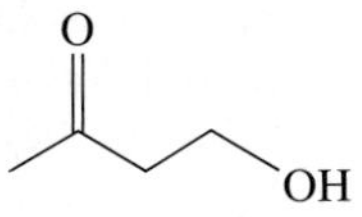

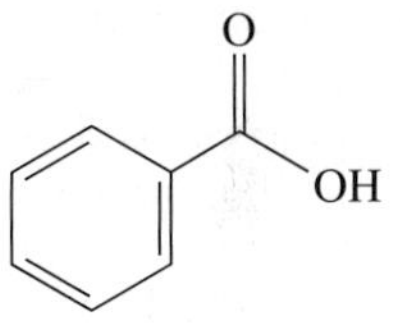

(a) forms an ester with ethanol
(b) absorbs H_2 in the presence of a metal catalyst
(c) neutralizes dilute NaOH
(d) forms an ether when heated strongly with H_2SO_4

73. Draw as many structural isomers as you can for cyclic ethers (no —OH groups) having the formula C_4H_6O. Try to draw at least six. (There are more than six.)

74. Draw as many structural isomers as you can for cyclic alcohols having the formula C_4H_6O. Try to draw at least five. (There are more than five.)

Integrative and Advanced Exercises

75. Supply condensed or structural formulas for the following substances.
  (a) 1,5-cyclooctadiene (an intermediate in the manufacture of resins)
  (b) 3,7,11-trimethyl-2,6,10-dodecatriene-1-ol (farnesol—odor of lily of the valley) [*Hint:* Dodeca means 12.]
  (c) 2,6-dimethyl-5-hepten-1-al (used in the manufacture of perfume)
76. Draw structural formulas for all the isomers listed in Table 26.3, and show that, indeed, the substances with more compact structures have lower boiling points.
77. By drawing suitable structural formulas, establish that there are *17* isomers of $C_6H_{13}Cl$. [*Hint:* Refer to Example 26-1.]
78. The symbol:

is often used to represent benzene. It is also the structural formula of cyclohexatriene. Are benzene and cyclohexatriene the same substance? Explain.
79. Use the half-reaction method to balance the following redox equations.
  (a) $C_6H_5NO_2 + Fe + H^+ \longrightarrow$
  $$C_6H_5NH_3^+ + Fe^{3+} + H_2O$$
  (b) $C_6H_5CH_2OH + Cr_2O_7^{2-} + H^+ \longrightarrow$
  $$C_6H_5CO_2H + Cr^{3+} + H_2O$$
  (c) $CH_3CH{=}CH_2 + MnO_4^- + H_2O \longrightarrow$
  $$CH_3CHOHCH_2OH + MnO_2 + OH^-$$
80. A 10.6 g sample of benzaldehyde was allowed to react with 5.9 g $KMnO_4$ in an excess of KOH(aq). After filtration of the MnO_2(s) and acidification of the solution, 6.1 g of benzoic acid was isolated. What was the percent yield of the reaction? [*Hint:* Write half-equations for the oxidation and reduction half-reactions.]
81. Combustion of a 0.1908 g sample of a compound gave 0.2895 g CO_2 and 0.1192 g H_2O. Combustion of a second sample weighing 0.1825 g yielded 40.2 mL of N_2(g), collected over 50% KOH(aq) (vapor pressure = 9 mmHg) at 25 °C and 735 mmHg barometric pressure. When 1.082 g of compound was dissolved in 26.00 g benzene (mp 5.50 °C, $K_f = 5.12$ °C m^{-1}), the solution had a freezing point of 3.66 °C. What is the molecular formula of this compound?
82. Draw and name all derivatives of benzene having the formula **(a)** C_8H_{10}; **(b)** C_9H_{12}.
83. In the monochlorination of hydrocarbons, a hydrogen atom is replaced by a chlorine atom. How many different monochloro derivatives of 2-methylbutane are possible?
84. A particular colorless organic liquid is known to be one of the following compounds: 1-butanol, diethyl ether, methyl propyl ether, butyraldehyde, or propionic acid. Can you identify which it is, based on the following tests? If not, what additional test would you perform? (1) A 2.50 g sample dissolved in 100.0 g water has a freezing point of −0.7 °C. (2) An aqueous

solution of the liquid does not change the color of blue litmus paper. (3) When alkaline $KMnO_4$(aq) is added to the liquid and the mixture is heated, the purple color of the MnO_4^- disappears.
85. Write structural formulas for the following.
  (a) 2,4-dimethyl-1,4-pentadiene
  (b) 2,3-dimethylpentane
  (c) 1,2,4-tribromobenzene
  (d) methylethanoate
  (e) butanone
86. Give the systematic names, including any stereochemical designations, for each of the following.

  (a) $CH_3{-}CH{-}CH_2{-}CH{-}CH_3$
  with CH_2, CH_2, CH_2, CH_3 chain below the first CH, and Cl below the second CH.

  (b) $CH_3{-}CH_2{-}CH{-}CH{-}CO_2H$
  with Cl and NH_2 below the two CH carbons.

  (c) $CH_3OCH_2CH_3$

  (d) $CH_2{-}CH_2{-}CH_2{-}NH_2$ with Cl below the first CH_2.

  (e) $\begin{array}{c} CH_3 \\ {} \\ ClH_2C \end{array} C{=}C \begin{array}{c} CH_3 \\ {} \\ Cl \end{array}$

87. Write structural formulas for all the isomers of C_4H_7Cl. Indicate any enantiomers or diastereomers that occur.
88. Compound A is an alcohol of formula $C_5H_{12}O$ that can be resolved into enantiomers.
  (a) Draw *three* possible structures of compound A.
  (b) Treatment of A with CrO_3/pyridine gives compound B, which also exhibits optical activity. What are the structural formulas of A and of B? Name and draw the enantiomers of A and B.
89. Levomethadyl acetate (shown below) is used in the treatment of narcotic addiction.

(a) Name the functional groups in levomethadyl acetate. **(b)** What is the hybridization of the numbered carbon atoms and the nitrogen atom? **(c)** Which, if any, of the numbered carbon atoms are chiral?

90. Thiamphenicol (shown below) is an antibacterial agent.

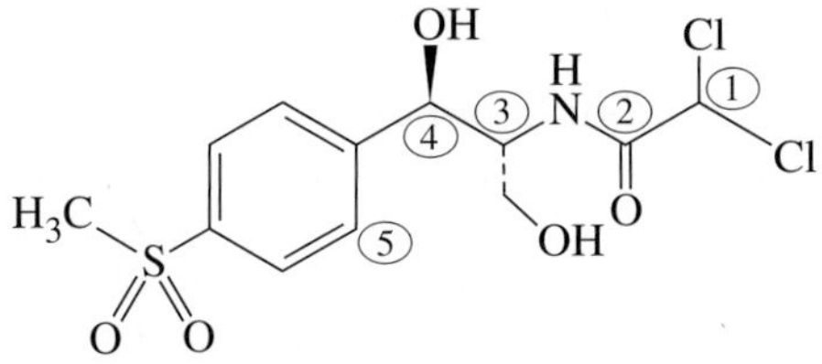

(a) Name the functional groups of thiamphenicol.
(b) What is the hybridization of the numbered carbon atoms and the nitrogen atom?
(c) Which, if any, of the numbered carbon atoms are chiral?

91. Ephedrine (shown below) is used as a decongestant in cold remedies.

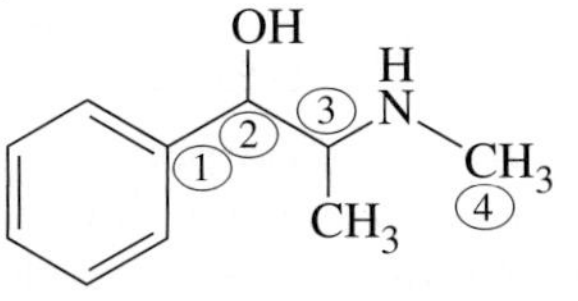

(a) Name the functional groups of ephedrine. (b) What is the hybridization of the numbered carbon atoms and the nitrogen atom? (c) Which, if any, of the numbered carbon atoms are chiral? (d) The pH of a solution of 1 g of ephedrine in 200 g of water is 10.8. What is the pK_b of ephedrine?

92. For each of the following molecules, determine the hybridization of each carbon atom, the total number of σ bonds and the total number of π bonds. Also, among the sp^3-hybridized carbon atoms, which ones are primary (1°), secondary (2°), tertiary (3°), or quaternary (4°)?

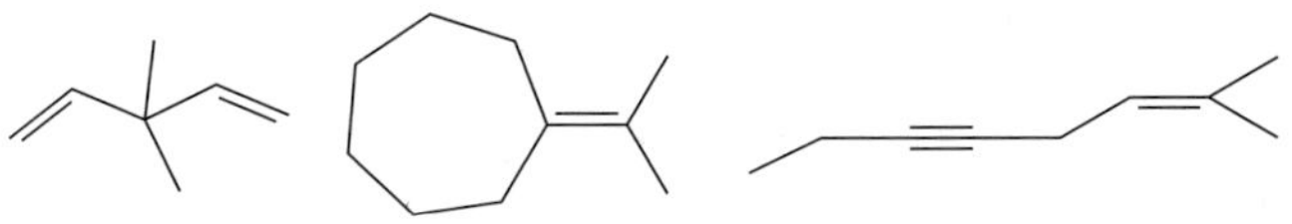

    (a)   **(b)**   **(c)**

93. Determine the configuration, R or S, of each chiral carbon atom in the molecules below. (Ph represents a phenyl group.)

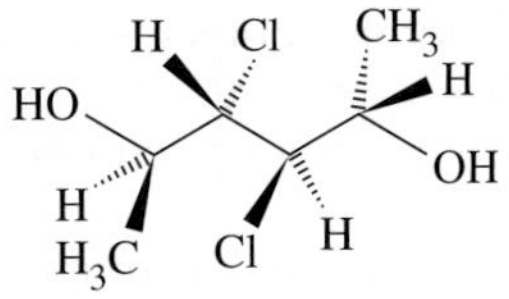

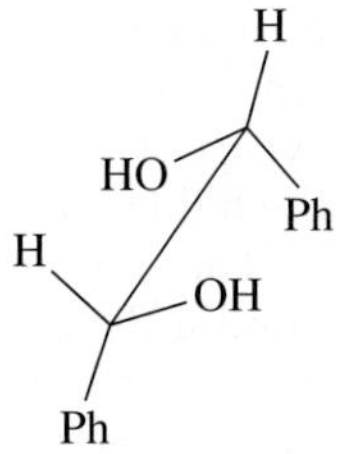

94. Among all ethers with the formula C_4H_6O, draw structures for
(a) two ethers with two sp^2 and two sp^3 carbon atoms
(b) an ether with four sp^2 carbon atoms
(c) an ether with two sp and two sp^3 carbon atoms

95. A structural formula for cholesterol is shown below. How many chiral carbon atoms are there in the cholesterol molecule? What is the configuration, R or S, of the carbon atom bonded to the —OH group? What is the configuration, E or Z, of the double bond?

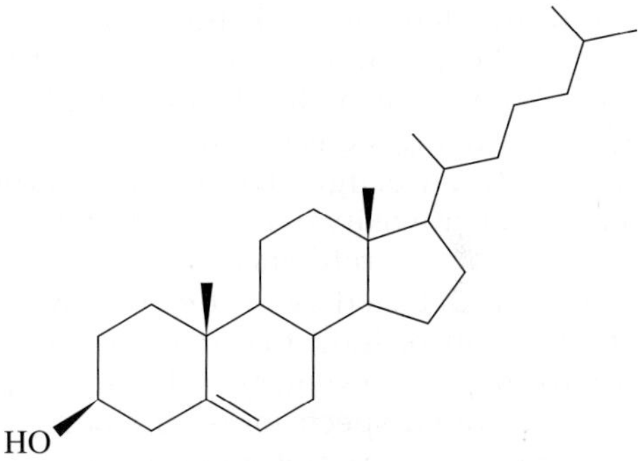

Cholesterol

96. For each of the following molecules (a) draw the two chair conformations and specify which conformation is more stable; (b) determine the number of chiral carbon atoms and for each chiral carbon, determine its configuration (R or S).

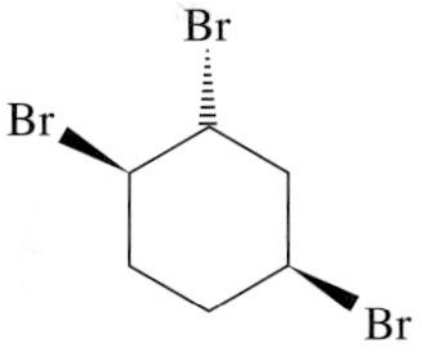

Feature Problem

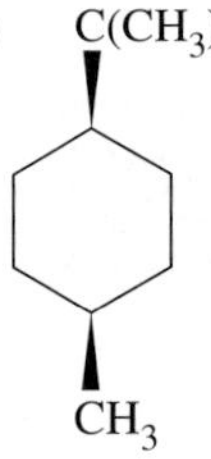

97. Organic chemists use a variety of methods to help them identify the functional groups in a molecule. In this chapter, we mentioned a few simple chemical ways to test for alkenes, alcohols, carboxylic acids, and so on. Such tests are quick and easy, but today organic chemists rely heavily on instrumental techniques. *Infrared (IR) spectroscopy* is an instrumental technique for identifying functional groups in a molecule. When infrared radiation is absorbed by a molecule, it causes atoms in bonds to vibrate back and forth with increased amplitude. We saw (in Chapter 8) that the energies of electrons in atoms are quantized and so too are the vibrations of atoms in molecules. Because each functional group has a particular grouping of atoms, there is a characteristic infrared absorption associated with each type of functional group. Some characteristic infrared absorptions are summarized in the following table. Infrared absorptions of molecules are identified by specifying the *wavenumber* of the light that is absorbed. The wavenumber is simply the reciprocal of wavelength: wavenumber = $1/\lambda = v/c$. (The definitions of λ, v, and c were given in Chapter 8.) The SI unit

for wavenumber is m^{-1} but it is often given in units of cm^{-1}. The wavenumber represents the number of cycles of the wave in each meter or centimeter along the light beam.

The data in the table indicate that molecules containing the carbonyl group (C=O) absorb light with wavenumbers between 1680 and 1750 cm^{-1}. Molecules containing a carbon–carbon triple bond (C≡C) absorb light with wavenumbers between 2100 and 2200 cm^{-1}. To identify the functional groups present in a molecule, the *infrared spectrum* of the molecule is obtained by using an instrument called an *infrared spectrometer*. A schematic diagram of an infrared spectrometer is shown below.

The light from the infrared light source is directed through a monochromator, which can be set to select a specific wavelength of light. The light then passes through a beam splitter, which splits the light into two separate beams, a reference beam and an incident beam. If the wavenumber of the incident beam matches one of the characteristic absorptions of the molecule, the sample absorbs the light, producing molecules that vibrate with greater energy. Because light has been absorbed by the sample, the intensity of the transmitted beam is less than that of the reference beam. The decrease in intensity is detected by the detector. By varying the wavenumber of light that reaches the sample and monitoring the percentage of light that is transmitted, an infrared spectrum is obtained (see graphs). An infrared spectrum is a plot of percent transmittance versus wavenumber. One hundred percent transmittance means none of the incident light was absorbed and 0% transmittance means all of the incident light was absorbed.

Type of bond	Wavenumber, cm^{-1}
Single bonds	
—C—H	2850–3300
=C—H	3000–3100
≡C—H	≈3300
N—H	3300–3500
O—H	3200–3600
Double bonds	
C=C	1620–1680
C=N	1500–1650
C=O	1680–1750
Triple bonds	
C≡C	2100–2200
C≡N	2200–2300

(a) The infrared absorptions given in the table above range from 1500 cm^{-1}. to 3600 cm^{-1}. Calculate the corresponding ranges of wavelength and frequency to verify that these absorptions correspond to the infrared region of the electromagnetic spectrum. (See Figure 8.3.)

(b) Identify the bonds responsible for the absorptions labeled A, B, C, and D in the two infrared spectra shown here. One spectrum is for acetone and the other is for 1-propanol, both of which are colorless liquids. Which of the two spectra is that of acetone?

(c) An isomer of acetone exhibits a strong IR absorption at 1645 cm^{-1} and also absorbs strongly from 2860 through 3600 cm^{-1}. The compound decolorizes bromine water, Br$_2$(aq), and produces bubbles of gas when sodium metal is added to it. What is the structure of this compound?

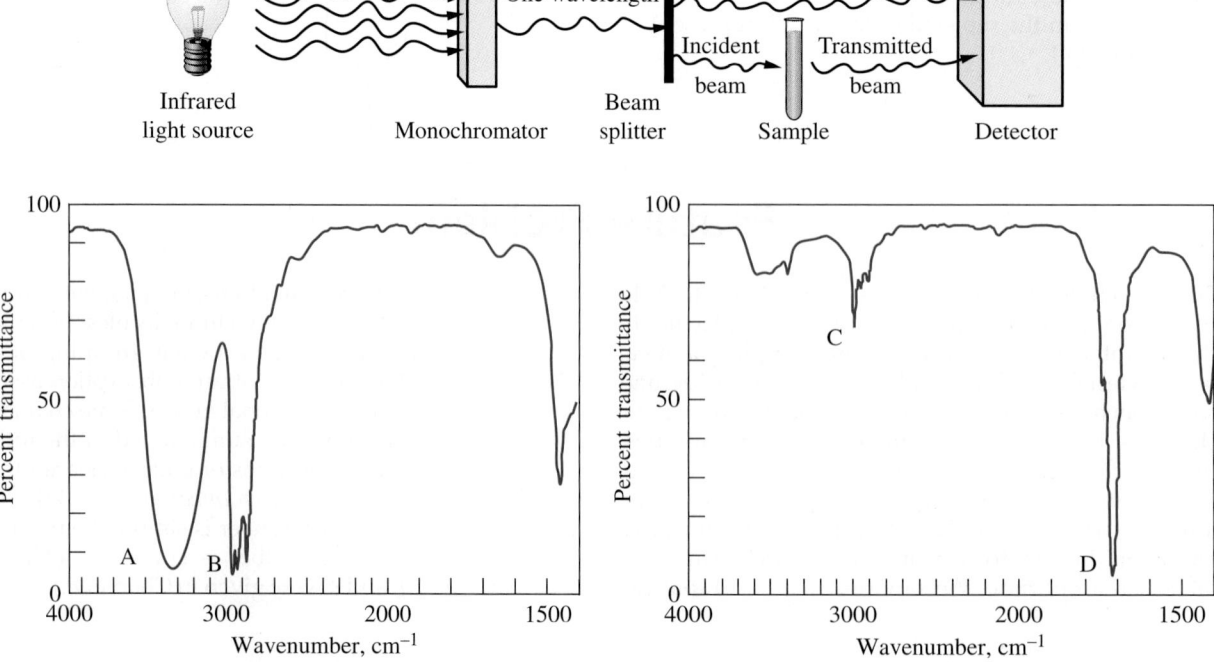

Infrared spectrometer

Self-Assessment Exercises

98. In your own words, define the following terms or symbols:
(a) *tert-*
(b) R—
(c) 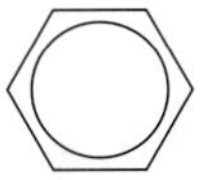
(d) carbonyl group
(e) primary amine

99. Explain the important distinctions between each pair of terms: (a) alkane and alkene; (b) aliphatic and aromatic compound; (c) alcohol and phenol; (d) ether and ester; (e) amine and ammonia.

100. Describe the characteristics of each of the following types of isomers: (a) constitutional; (b) stereoisomer (c) cis; (d) ortho.

101. The compound isoheptane is best represented by the formula (a) C_7H_{14}; (b) $CH_3(CH_2)_5CH_3$; (c) $(CH_3)_2CH(CH_2)_3CH_3$; (d) $C_6H_{11}CH_3$.

102. A compound with the same hydrogen-to-carbon ratio as cyclobutane is (a) C_4H_{10}; (b) $CH_3CH=CHCH_3$; (c) $CH_3C \equiv CCH_3$; (d) C_6H_6.

103. Three isomers exist of the hydrocarbon (a) C_3H_8; (b) C_4H_8; (c) C_4H_{10}; (d) C_6H_6; (e) C_5H_{12}.

104. Give names for the following molecules.

(a) $(CH_3)_2CBrCH_2CHClCH_2CH(CH_3)_2$

(b) 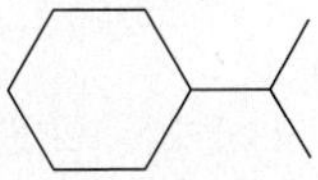
(c)

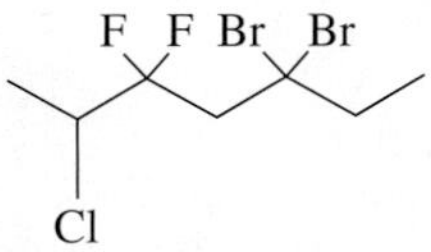

105. Assign configurations, *R* or *S*, to the chiral carbons in the molecules below. Then identify (a) any two identical structures; (b) any two constitutional isomers; (c) any two diastereomers; (d) a pair of enantiomers.

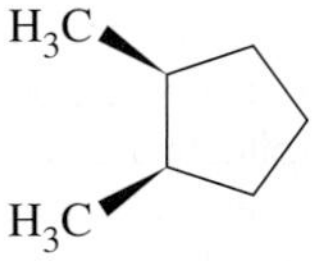

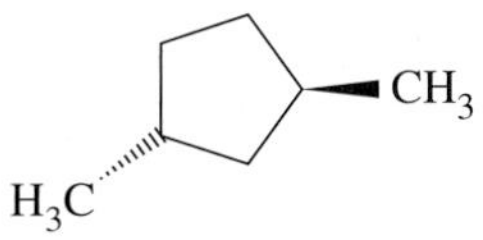

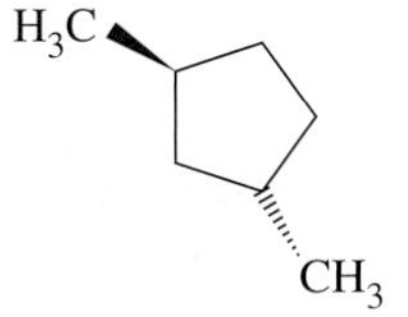

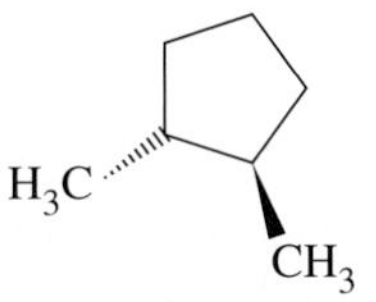

106. Consider the following pairs of structures. In each case, are the structures different conformers or are they isomers? If they are isomers, then state whether they are constitutional isomers, diastereomers, or enantiomers.

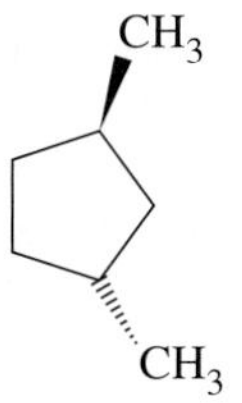

(a)

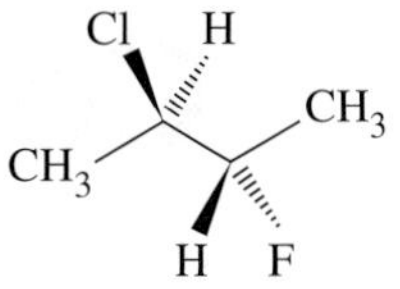

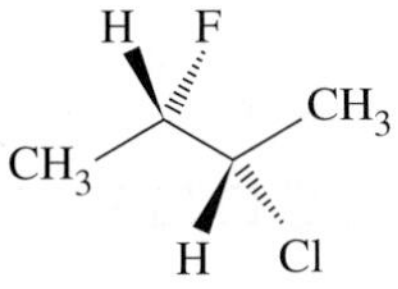

(b)

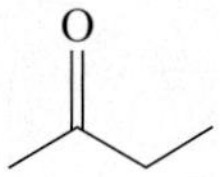

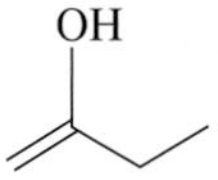

(c)

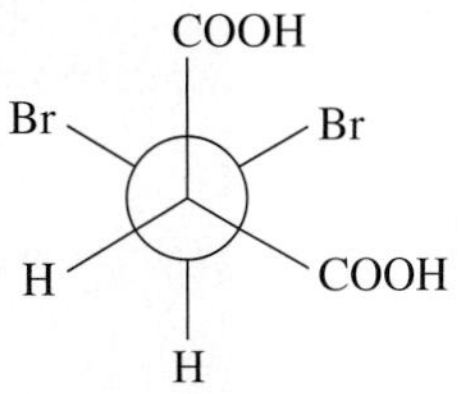

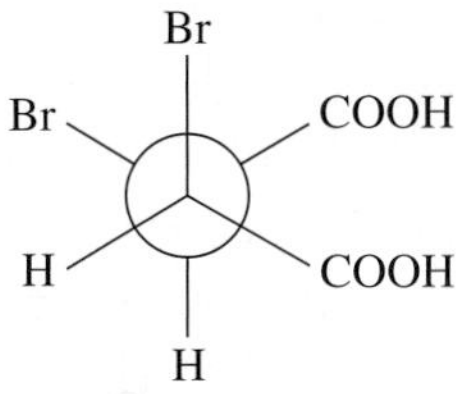

(d)

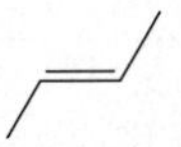

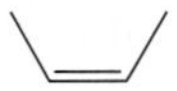

107. Draw a Newman projection for the conformation of lowest energy for viewing 2-methylhexane along the C2—C3 bond.

108. To prepare methyl ethyl ketone, one should oxidize (a) 2-propanol; (a) 1-butanol; (c) 2-butanol; (d) *tert*-butyl alcohol.

109. Which hydrocarbon has the greater number of isomers, C_4H_8 or C_4H_{10}? Explain your choice.

110. For each of the following pairs, indicate which substance has
(a) the higher boiling point, C_6H_{12} or C_6H_6
(b) the greater solubility in water, C_3H_7OH or $C_7H_{15}OH$
(c) the greater acidity in aqueous solution, C_6H_5CHO or C_6H_5COOH

111. Draw the structures of
(a) (*E*)-3-benzyl-2,5-dichloro-4-methyl-3-hexene
(b) 1-ethenyl-4-nitrobenzene
(c) *trans*-1-(4-bromophenyl)-2-methylcyclohexane

27

Reactions of Organic Compounds

CONTENTS

A branch of the Pacific Yew tree (*Taxus brevifolia*), which grows in forests along the Pacific coast of British Columbia and Washington. The needles and seeds of the Pacific Yew are poisonous but the bark contains a compound called paclitaxel (formerly known as taxol), an effective chemotherapy drug. Unfortunately, the bark of a 100-year-old Pacific yew tree yields about 3 kilograms of bark and only 300 mg of paclitaxel—barely enough for a single dose. Paclitaxel is a complicated organic molecule, the structure of which was determined in 1971. In 1994, two different groups of organic chemists, one lead by Professor Robert A. Holton of Florida State University and another lead by Professor Kyriacos C. Nicolau of the University of California (San Diego), announced independently that they were able to synthesize paclitaxel from simple starting materials.

In Chapter 26, we focused primarily on the structures of organic compounds and learned that organic compounds can be categorized into a relatively small number of families: alkanes, haloalkanes, alcohols, alkenes, and so on. In this chapter, we are interested in the characteristic reactions of some of these compounds and the mechanisms that describe how these reactions occur. Given that millions of organic compounds are known, and countless reactions involving them are possible, it may seem surprising that the reactions of organic compounds can be rationalized in terms of a relatively small number of reaction mechanisms. In this chapter, we will explore a few of these mechanisms.

We cannot possibly provide a complete perspective of organic reactions and their mechanisms in one chapter. We will focus instead on core concepts that will not only help us understand the reactions but also serve us well in more advanced organic chemistry courses. Throughout this chapter, we will emphasize the molecular aspects of chemical reactions—molecules coming together, with electron-rich regions being drawn naturally toward electron-poor regions; old bonds breaking and new bonds forming. We will use ideas about the structures and properties of molecules—gained in earlier chapters—to help us understand some important ideas about reactions of organic compounds.

We begin with a brief overview of the different types of organic reactions and then discuss these reaction types in terms of the structures of the molecules involved, the acidic or basic character of the reacting species, and the relative stabilities of reactants, products, and reaction intermediates. We will rely heavily on concepts from earlier chapters but will apply them in new ways. Also, we will make much more use of line-angle structures in this chapter to emphasize the functional groups and the prominent role they play in organic reactions.

27-1 Organic Reactions: An Introduction

Organic compounds undergo a variety of different reactions, including substitution, addition, elimination, and rearrangement reactions. In a **substitution reaction**, an atom, an ion, or a group in one molecule is replaced by (substituted with) another. Here are a few examples of substitution reactions:

$$CH_3Cl + NaOH \xrightarrow{H_2O} CH_3OH + NaCl$$

$$(CH_3)_3CBr + H_2O \xrightarrow[\Delta]{(CH_3)_2CO} (CH_3)_3COH + HBr$$

$$C_6H_6 + HNO_3 \xrightarrow{H_2SO_4} C_6H_5NO_2 + H_2O$$

In each equation, the solvent used in the reaction is written above the arrow. In the first reaction, the Cl atom in a CH_3Cl molecule is substituted with an —OH group. In the second reaction, the Br atom in a $(CH_3)_3CBr$ molecule is substituted with an —OH group, and in the third reaction, an H atom in a C_6H_6 molecule is substituted with an —NO$_2$ group. Although each reaction is a substitution reaction, we will soon see that they occur via different mechanisms.

In an **addition reaction**, a molecule adds across a double or triple bond in another molecule. Here are two examples:

$$H_2C{=}CH_2 + Br_2 \xrightarrow{CCl_4} \begin{array}{c} H \quad\ H \\ | \quad\ \ | \\ H-C-C-H \\ | \quad\ \ | \\ Br \ \ \ Br \end{array}$$

$$HC{\equiv}CH + HBr \longrightarrow H_2C{=}CHBr$$

In an **elimination reaction**, atoms or groups that are bonded to adjacent atoms are eliminated as a small molecule. Typically, the order of the bond between the two adjacent atoms increases as a result of the elimination, as shown in the examples below:

$$\begin{array}{c} H \quad\ H \\ | \quad\ \ | \\ H-C-C-H \\ | \quad\ \ | \\ OH \ \ H \end{array} \xrightarrow{H_2SO_4} H_2C{=}CH_2 + H_2O$$

$$\begin{array}{c} H \quad\ Br \\ | \quad\ \ | \\ H-C-C-H \\ | \quad\ \ | \\ Br \ \ \ H \end{array} \longrightarrow H-C{\equiv}C-H + 2\,HBr$$

The first of the two reactions above illustrates that alkenes can be obtained from alcohols (ROH). The second, in which two molecules of HBr are eliminated to yield an alkyne, is an example of a double elimination reaction.

When an organic compound undergoes a **rearrangement reaction** (or an isomerization reaction), the carbon skeleton of a molecule is rearranged.

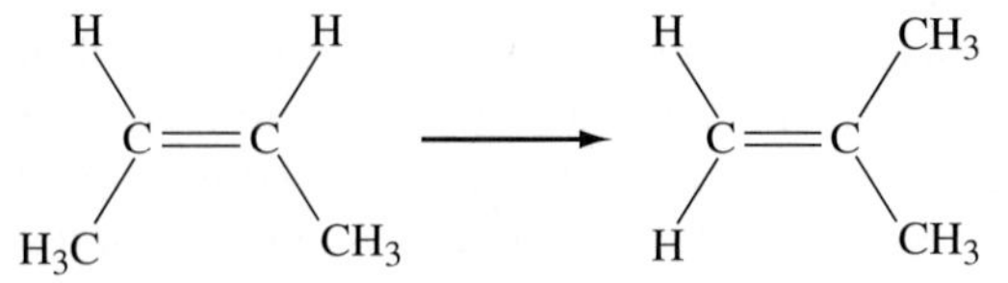

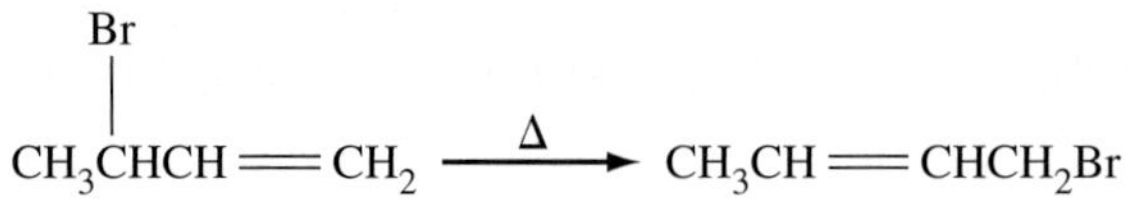

EXAMPLE 27-1 Identifying Types of Organic Reactions

State whether each of the following reactions is a substitution, an addition, an elimination, or a rearrangement reaction. (In the equations below, H^+ and Pb are catalysts.)

(a)
$$(CH_3)_2CHCONa + CH_3Br \longrightarrow (CH_3)_2CHCOCH_3 + NaBr$$

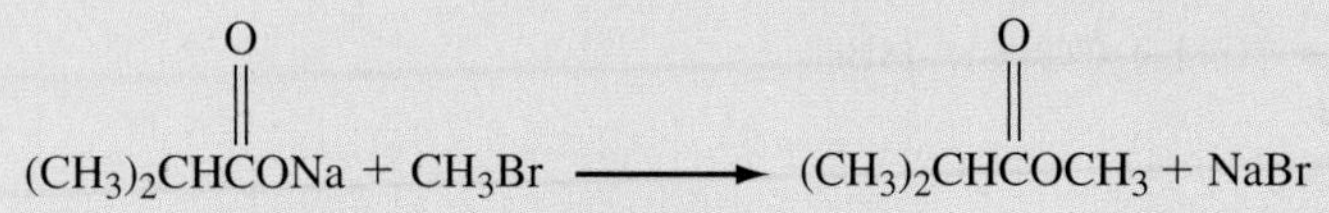

(b)

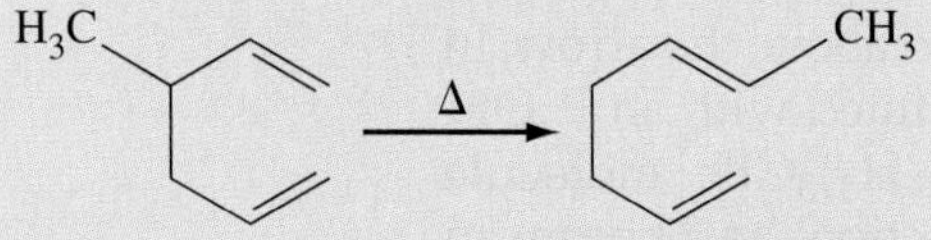

(c) $2\,CH_3OH \xrightarrow[\Delta]{H^+} CH_3OCH_3 + H_2O$

(d) cyclohexanol $\xrightarrow[\Delta]{H^+}$ cyclohexene + water

(e) 2-butyne $+ 2\,H_2 \xrightarrow{Pb}$ butane

Analyze

In a substitution reaction, an atom or a group of atoms is replaced with another atom or group of atoms. In an addition reaction, atoms add across a double or triple bond, causing a reduction in bond order between two atoms. In an elimination reaction, atoms or groups of atoms that are bonded to adjacent atoms are eliminated, causing an increase in bond order between two atoms. Typically, a rearrangement reaction involves a change in constitution or stereochemical configuration.

Solve

(a) Because a Br atom in CH_3Br is replaced by a $(CH_3)_2CHCOO$ group, the reaction is a substitution reaction.

(b) The reaction involves only a change in the skeletal structure (constitution), and so the reaction is a rearrangement reaction.

(c) When we write the reaction out as $CH_3OH + CH_3OH \longrightarrow CH_3OCH_3 + H_2O$, we see that an $-OH$ group in one molecule is replaced by a $-CH_3O$ group. The reaction is a substitution reaction.

(d) Here, we need to convert the names of the compounds into structural formulas.

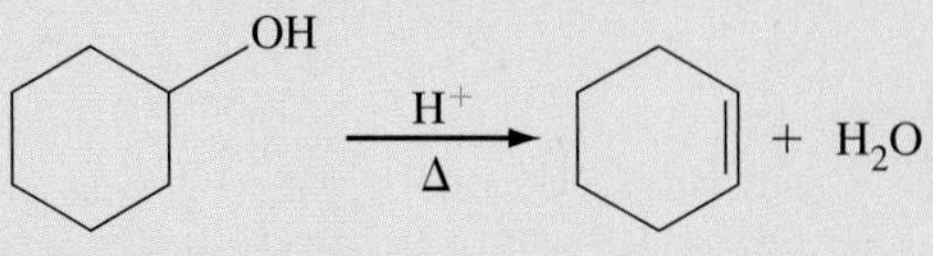

The reaction involves elimination of OH and H from adjacent carbon atoms in cyclohexanol. Thus, the reaction is an elimination reaction.

(e) In this reaction, an alkyne is converted into an alkane. Because there is a reduction in bond order, we know that the reaction is an addition reaction involving the addition of H atoms across the triple bond. Our prediction is confirmed by writing the reaction in terms of structural formulas:

$$CH_3C \equiv CCH_3 + 2H_2 \longrightarrow CH_3CH_2CH_2CH_3$$

Assess

To identify the reaction type correctly, it is often helpful to write out structural formulas for all molecules involved.

PRACTICE EXAMPLE A: Classify the following reactions as substitution, addition, elimination, or rearrangement reactions.

(a)

⬡ + Br₂ —FeBr₃→ ⬡—Br + HBr

(b)

(c)

     + CH₃CH₂OH —H₃O⁺→

PRACTICE EXAMPLE B: Classify each of the following reactions as a substitution, an addition, an elimination, or a rearrangement reaction.

(a) $CH_3CH_2CH_2CH_2Br + (CH_3)_3O^-K^+ \longrightarrow CH_3CH_2CH = CH_2 + KBr + (CH_3)_3OH$

(b)

⬡—CH₂Cl + N(CH₂CH₃)₃ ⟶ ⬡—CH₂N⁺(CH₂CH₃)₃ + Cl⁻

(c) $CH_3CH_2C \equiv C^-Na^+ + CH_3CH_2Br \longrightarrow CH_3CH_2C \equiv CCH_2CH_3 + NaBr$

27-2   Introduction to Nucleophilic Substitution Reactions

In this section, we consider the essential features of substitution reactions involving compounds in which the functional group is bonded to an sp^3-hybridized carbon atom. We will focus primarily on substitution reactions of haloalkanes. In a haloalkane, the carbon atom bonded to the halogen is sp^3 hybridized. With appropriate modifications, the concepts learned in this section can be used to understand substitution reactions of other compounds.

An example of a substitution reaction involving an sp^3 hybridized carbon is the replacement of a Cl atom in chloromethane by a hydroxide group to give methanol:

$$CH_3Cl + OH^- \longrightarrow CH_3OH + Cl^- \qquad (27.1)$$

This type of chemical reaction is called a **nucleophilic substitution reaction** because the hydroxide ion is a **nucleophile**—a reactant that seeks sites of low electron density in a molecule. The nucleophile is electron rich, and it is the species that donates a pair of electrons to another molecule. The C atom in CH_3Cl is considered electron deficient because, as discussed in Chapter 11, the Cl atom draws electron density away from the C atom; the C atom is said to be electrophilic (electron attracting). The **electrophile**, also known as the **substrate**, is the species that is being attacked by the nucleophile and accepts a pair of electrons. Electrophiles contain atoms that are electron deficient. In reaction (27.1), the hydroxide ion attacks at the electron deficient C atom, displacing the chloride ion, which is called the **leaving group**.

▲ Volcanic eruptions produce significant quantities of chloromethane, CH_3Cl.

Nucleophiles, often denoted by the abbreviation Nu, can be negatively charged or neutral, but every nucleophile contains at least one pair of unshared electrons. Nucleophiles are, in fact, Lewis bases. The nucleophilic substitution of a haloalkane is described by either of two general equations:

$$\text{Nu:}^- + \overset{\displaystyle|}{\underset{\displaystyle|}{\text{C}}}\overset{\delta+}{-}\overset{\delta-}{\text{X:}} \longrightarrow \text{Nu}-\overset{\displaystyle|}{\underset{\displaystyle|}{\text{C}}}- + \text{:X:}^- \qquad (27.2a)$$

Nucleophile   Electrophile (substrate)   Leaving group

$$\text{Nu:} + \overset{\displaystyle|}{\underset{\displaystyle|}{\text{C}}}\overset{\delta+}{-}\overset{\delta-}{\text{X:}} \longrightarrow \left[\text{Nu}-\overset{\displaystyle|}{\underset{\displaystyle|}{\text{C}}}-\right]^+ + \text{:X:}^- \qquad (27.2b)$$

Nucleophile   Electrophile (substrate)

In a nucleophilic substitution reaction, the bond (in blue) between carbon and the leaving group is broken and a new bond (in red) is formed by using a lone pair from the nucleophile. The electron pair from the bond that is broken ends up as a lone pair on the leaving group. In light of the discussion of acid–base chemistry in Section 27A on www.masteringchemistry.com, equilibrium favors forming the substitution product if the leaving group is a weaker base than the nucleophile.

As we will soon see, nucleophilic substitution of haloalkanes occurs via one of two mechanisms. Before examining the mechanisms, let's consider a few specific examples of nucleophilic substitution reactions.

In the reaction below, the $CH_3C{\equiv}C^-$ ion is the nucleophile and CH_3CH_2Br is the electrophile:

$$Na^+[CH_3C{\equiv}C:]^- + \underset{\beta\quad\alpha}{CH_3CH_2}\overset{\delta+}{-}\overset{\delta-}{Br:} \longrightarrow CH_3CH_2-C{\equiv}CCH_3 + \text{:Br:}^-$$

▶ In a haloalkane, the carbon bonded to the halogen atom is called the alpha (α) carbon. A carbon bonded to the α carbon is called a beta (β) carbon. A carbon bonded to a β carbon is called a gamma (γ) carbon, and so on. This system of labeling carbon atoms is important when discussing reactions.

The α carbon is electron deficient (d+); it is the atom of the electrophile that is attacked by the nucleophile. Because both the nucleophile and the leaving group are Lewis bases, we can use acid–base concepts to establish that the reaction above will occur as written. The $CH_3C{\equiv}C^-$ ion has the negative charge localized on C, a very small atom compared with Br, and so the $CH_3C{\equiv}C^-$ is a much stronger base than Br^-. Because the reaction involves the formation of a weaker base, the reaction will occur as written.

In contrast to the example above, the following substitution reactions will not occur as written because they both involve the formation of a base that is stronger than the nucleophile:

$$HO^- + H_3C-H \rightleftharpoons H_3C-OH + H^- \quad (H^- \text{ is a stronger base than } OH^-)$$

$$Br^- + CH_3CH_2-OH \rightleftharpoons CH_3CH_2-Br + OH^- \quad (OH^- \text{ is a stronger base than } Br^-)$$

The S$_N$1 and S$_N$2 Mechanisms of Nucleophilic Substitution Reactions

Chemical research, starting in the 1890s, has shown that nucleophilic substitution reactions can involve two types of mechanisms. The reaction between chloromethane and the hydroxide ion (27.2) has a rate law that is first order in both the nucleophile and the electrophile. That is,

$$\text{rate} = k[OH^-][CH_3Cl]$$

The mechanism for this reaction involves a bimolecular rate-determining step in which the nucleophilic group approaches the electrophilic carbon atom and, simultaneously, the leaving group departs. The entering and leaving groups act at the same time in what is known as a *concerted step*, as depicted in this schematic representation:

Acid–base concepts are among the most important of all concepts in chemistry because all chemical reactions—including those involving organic molecules—can be characterized as some form of acid–base reaction. For a discussion of acid–base concepts, with a focus on organic reactions, see Section 27A, *Organic Acids and Bases*, on the MasteringChemistry site (www.masteringchemistry.com).

Mastering**CHEMISTRY**

▲ Sir Robert Robinson (1886–1975) was awarded the 1947 Nobel Prize in Chemistry for his contributions to the synthesis of natural products. He is also credited with introducing the use of curved arrows to show the movement of electron pairs in chemical reactions.

27-1 ARE YOU WONDERING...

What the difference is between a base and a nucleophile?

We have seen that a molecule or an ion having a lone pair of electrons can act as an electron pair donor in a reaction with either a proton or an electrophilic carbon atom. In both cases, a new bond is formed. When the bond is to a proton, the electron pair donor has reacted as a base. When the bond is to a carbon atom, the electron pair donor has reacted as a nucleophile. Chemists use the terms *basicity* and *nucleophilicity* when referring to the tendency of an electron pair donor to act as a base or a nucleophile in a reaction. **Basicity** is a measure of the tendency of an electron pair donor to react with a proton. **Nucleophilicity** is a measure of how readily (how fast) a nucleophile attacks an electrophilic carbon atom. We use equilibrium constants (K_b) when comparing molecules or ions in terms of their basicity, and thus basicity is an equilibrium (thermodynamic) property. When comparing molecules or ions in terms of their nucleophilicity, we are comparing them in terms of the rates at which they attack an electrophilic carbon, and thus nucleophilicity is a kinetic property.

What can make things a little confusing is that a molecule or an ion with a lone pair can act as either a base or a nucleophile. For example, the methoxide ion, CH_3O^-, is both a strong base and a good nucleophile.

$$CH_3\ddot{O}:^- + CH_3CH_2 \overset{\delta^+}{-} \ddot{Br}:^{\delta^-} \longrightarrow CH_3O-CH_2CH_3 + :\ddot{Br}:^-$$

Nucleophile   Electrophile

Base   Acid

Thus, it is important to understand not only factors affecting basicity but also those affecting nucleophilicity. Factors affecting basicity are discussed in Section 27A on www.masteringchemistry.com. Factors affecting nucleophilicity are discussed on page 1221.

EXAMPLE 27-2 Identifying Electrophiles, Nucleophiles, Leaving Groups, Acids, and Bases

It is important to be able to distinguish when an electron pair donor in a reaction is acting as a base or as a nucleophile. The following reactions are elementary reactions. In each case, determine whether the electron pair donor is acting as a Brønsted–Lowry base or as a nucleophile. Then, identify the acid, electrophile, and leaving group, as appropriate. Finally, use arrows to indicate the movement of electrons.

(a)

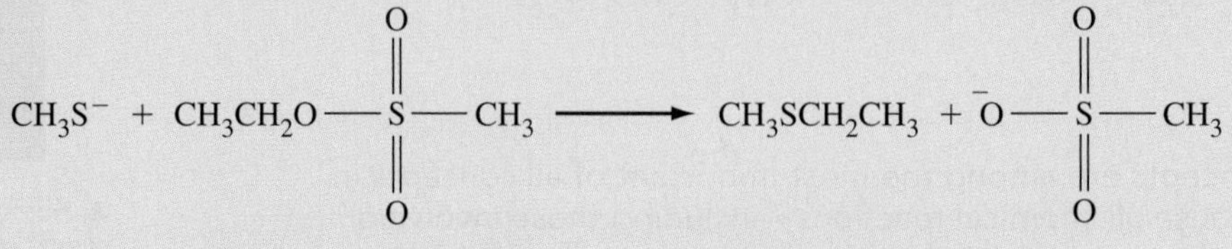

(b)

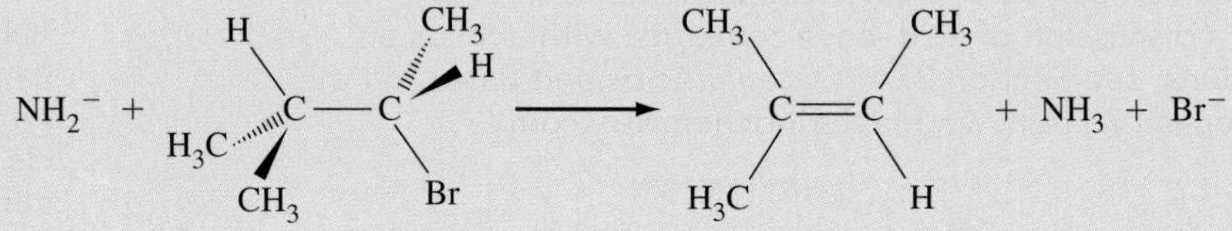

(c)

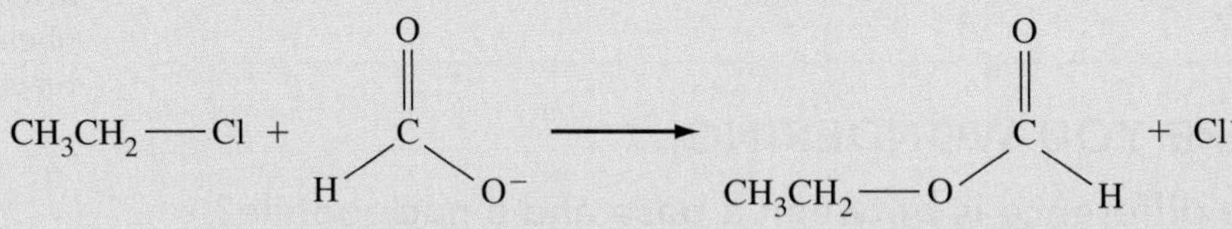

Analyze

First, we put lone pairs on the atoms, and then compare reactants and products to determine which bonds are formed and which are broken. We identify the electron pair donor and the electron pair acceptor. If the electron pair donor (the attacking species) forms a new bond with a proton, it is acting as Brønsted–Lowry base. If the attacking species forms a new bond with a carbon atom, then it is acting a nucleophile.

Solve

(a) In the equation below, reactants and products are shown with all lone pairs. We see that a new bond is formed between sulfur and carbon, and a bond between carbon and oxygen is broken. CH_3S^- is acting as a nucleophile, $CH_3CH_2OSO_2CH_3$ is the electrophile (or substrate), and $^-OSO_2CH_3$ is the leaving group. The red arrows indicate the movement of electrons.

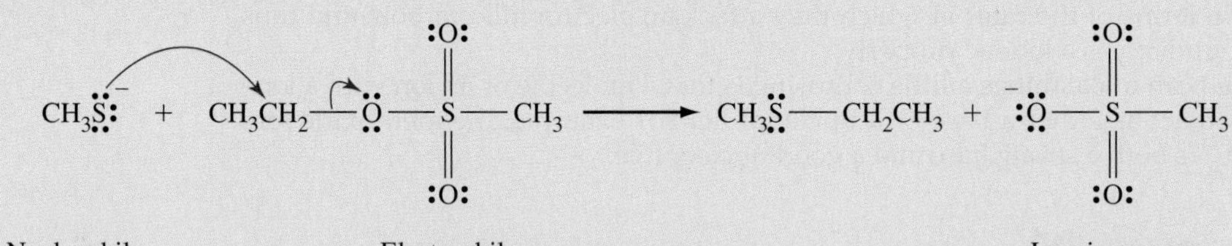

Nucleophile   Electrophile   Leaving group

(b) In this reaction, NH_2^- reacts to form NH_3. Because the chemical formulas of these species differ by a single proton, we know that NH_3 and NH_2^- are a conjugate acid–base pair. NH_2^- is acting a Brønsted–Lowry base and $(CH_3)_2CHCHBrCH_3$ is acting as an acid.

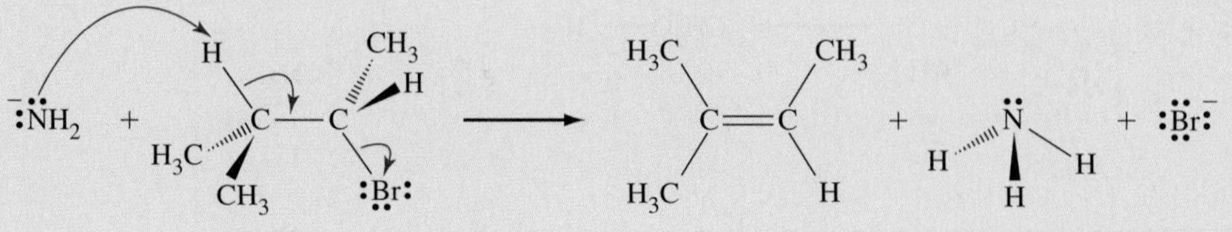

Base   Acid

(c) We see from the equation below that a new carbon–oxygen bond is formed and a carbon–chlorine bond is broken. The HCOOH molecule is acting as a nucleophile, CH_3CH_2Cl is the electrophile, and Cl^- is the leaving group.

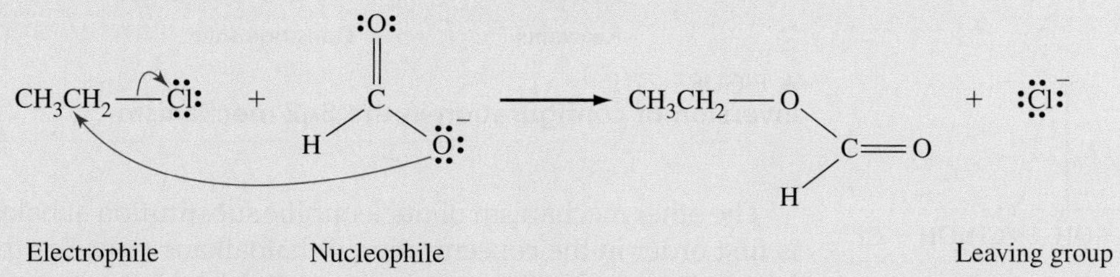

Electrophile   Nucleophile   Leaving group

Assess

We can assess whether or not the substitution reactions in this example are feasible. In **(a)**, we predict that $^-OSO_2CH_3$ is a weaker (more stable) base than CH_3S^- because the negative charge in $^-OSO_2CH_3$ is highly delocalized; it is shared equally by three electronegative oxygen atoms. Thus, equilibrium favors products and we predict that the substitution will occur to a significant extent. For **(b)**, we must compare the stabilities of the bases $HCOO^-$ and Cl^-. The conjugate acid of $HCOO^-$ is $HCOOH$, a weak acid, and so $HCOO^-$ is a weak base. Cl^- is an extremely weak base ($pK_b \approx 21$). Because the weaker base appears on the right side of the equation, we expect reaction **(c)** will occur as written.

PRACTICE EXAMPLE A: In each of the following reactions, determine whether the electron pair donor is acting as a Brønsted–Lowry base or a nucleophile. Identify the acid, electrophile, and leaving group, as appropriate, and use arrows to indicate the movement of electrons.

(a) $CH_3I + NaOH \longrightarrow CH_3OH + NaI$

(b) $NaOH + CH_3CH_2CH_2CO_2H \longrightarrow CH_3CH_2CH_2CO_2Na + H_2O$

PRACTICE EXAMPLE B: In each of the following reactions, determine whether the electron pair donor is acting as a Brønsted–Lowry base or a nucleophile. Identify the acid, electrophile, and leaving group, as appropriate, and use arrows to indicate the movement of electrons.

(a) $NH_2Na + CH_3CH_2OH \longrightarrow CH_3CH_2ONa + NH_3$

(b) $(CH_3)_2CHCH_2CH_2Br + CH_3SNa \longrightarrow (CH_3)_2CHCH_2CH_2SCH_3 + NaBr$

The pair of curved arrows at the top of page 1213 summarizes our visualization of how electrons flow to form and break bonds. Thus, the arrow starts from an electron-rich center (the electrons on the hydroxyl group in this case) to an electron-poor region (the C atom in the polar carbon-to-chlorine bond).

This mechanism is designated S_N2, with the S indicating substitution, the N indicating nucleophilic, and the 2 indicating that the rate-determining step is bimolecular. The reaction profile for an S_N2 reaction is shown in Figure 27-1, in which the transition state formed by the HO^- ion and the chloromethane is shown. The hydroxide attacks on the side opposite the Cl atom, and the C—Cl bond starts to break as the C—O bond simultaneously starts to form, producing the transition state shown. In the S_N2 mechanism, the nucleophile donates a pair of electrons to the electrophile to form a covalent bond.

Is there any experimental evidence to support this mechanism and postulated transition state? First, the rate law is suggestive of a bimolecular step; second, an observation by Paul Walden in 1893 confirmed the formation of the transition state as shown in Figure 27-2. Walden discovered that if the C atom bonded to the halogen is stereogenic, or chiral, the configuration at the chiral carbon is inverted—that is, the molecular structure is inverted and an enantiomer of the opposite configuration is formed. Thus, when (S)-2-iodobutane undergoes nucleophilic substitution by the hydroxide group, the compound (R)-2-butanol is formed (Fig. 27-2). This *inversion of configuration* is taken as confirmation of the proposed bimolecular mechanism and the five-coordinate transition state.

KEEP IN MIND

that the electrophile accepts the pair of electrons. Thus, the nucleophile acts as a Lewis base, and the electrophile as a Lewis acid.

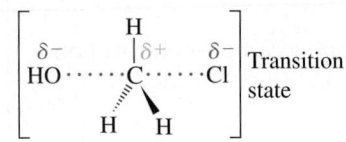

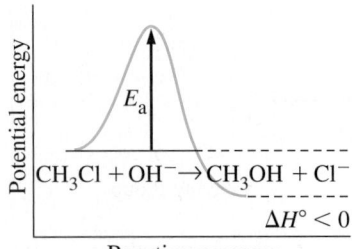

▲ FIGURE 27-1
Reaction profile for an S_N2 reaction
The bonds that are forming and breaking in the transition state are shown by dotted lines.

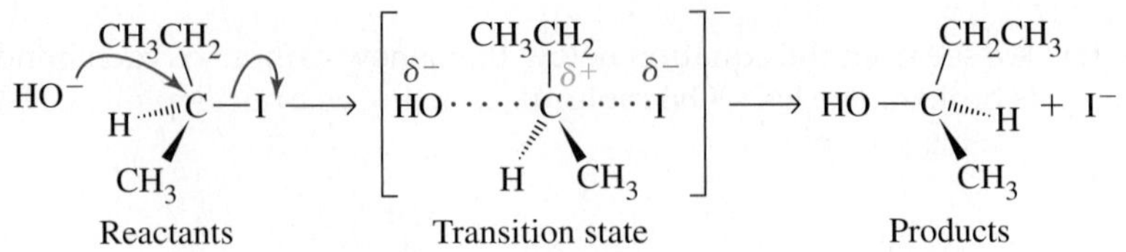

$$HO^- \quad \underset{Reactants}{\overset{CH_3CH_2}{\underset{CH_3}{\overset{|}{\underset{H}{\overset{\cdots}{C}}}}}} \!\! -I \longrightarrow \left[HO \cdots \overset{CH_3CH_2}{\underset{H \quad CH_3}{\overset{\delta-}{\underset{}{C}}}} \cdots I \right]^- \longrightarrow HO - \overset{CH_2CH_3}{\underset{CH_3}{\overset{|}{\underset{}{C}}}} H + I^-$$

Reactants   Transition state   Products

▲ FIGURE 27-2
Inversion of configuration in the S_N2 mechanism

The other mechanism of nucleophilic substitution at haloalkanes has a rate that is first order in the concentration of haloalkane only. For the reaction between 2-bromo-2-methylpropane and water, in which H_2O acts as the nucleophile,

$$(CH_3)_3CBr + H_2O \longrightarrow (CH_3)_3COH + HBr$$

the rate law is

$$\text{rate} = k[(CH_3)_3CBr]$$

This rate law suggests that the rate-determining step is unimolecular. The mechanism for this reaction is shown in Figure 27-3. The first step is a slow unimolecular step in which the haloalkane ionizes to form a bromide ion and a *carbocation*. The carbocation has a planar geometry and the positively charged carbon atom is sp^2 hybridized. In the second step, the carbocation immediately reacts with the nucleophile, a water molecule in this case, producing the conjugate acid of an alcohol (a protonated alcohol). The protonated alcohol dissociates immediately in the presence of excess water, forming the neutral alcohol and a hydronium ion.

This mechanism is designated S_N1. Again, the S indicates substitution and the N nucleophilic; in this case, the 1 indicates that the rate-determining step is unimolecular. The reaction profile for an S_N1 reaction is shown in Figure 27-4, which includes three transition states and two intermediates.

Apart from the rate law, what other evidence is there for the planar carbocation? Again, we can use a chiral haloalkane and investigate the chirality of the product. When one of the enantiomers of 3-bromo-3-methylhexane reacts

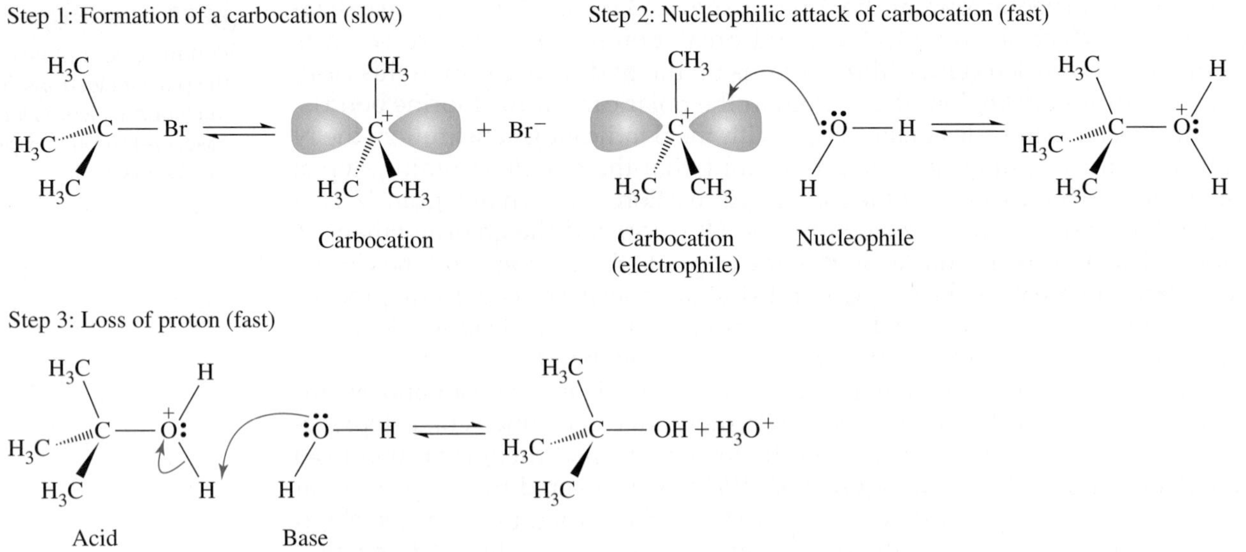

Step 1: Formation of a carbocation (slow)   Step 2: Nucleophilic attack of carbocation (fast)

Step 3: Loss of proton (fast)

▲ FIGURE 27-3
The S_N1 mechanism
The first step is the formation of a carbocation. It is slow and rate determining. In the second step, the electron pair donor acts as a nucleophile and attacks the electrophilic carbon atom of the carbocation, forming a σ bond.

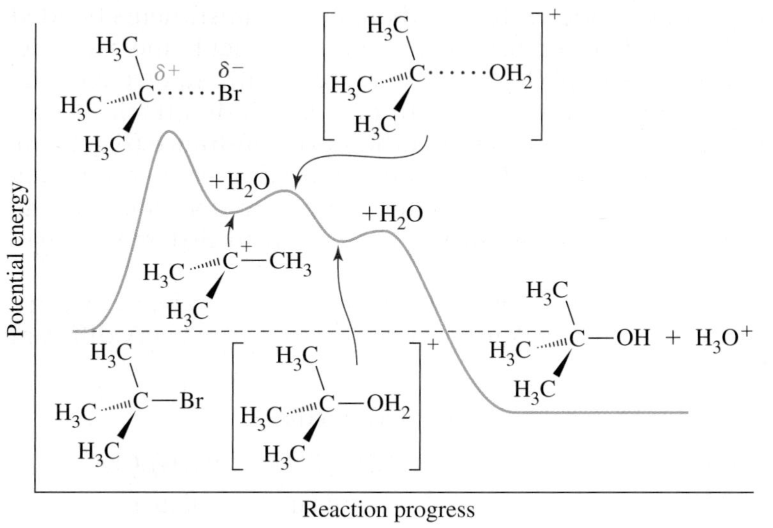

◀ FIGURE 27-4
Reaction profile for the reaction between t-butyl bromide and water
The carbocation formed in the rate-determining step is planar. The central C atom in the carbocation intermediate is hybridized. The remaining p orbital is used in the formation of the new chemical bond. The central C atom of the 2,2-dimethylethyloxonium ion (the second intermediate) becomes hybridized subsequently.

with water, the product is a racemic mixture of the enantiomers of 3-methyl-3-hexanol. The carbocation intermediate formed in the rate-determining step is planar. The water molecule (nucleophile) can form a new bond on either face of the carbocation intermediate. This results in a mixture of the product enantiomers (Fig. 27-5). The formation of a racemic mixture provides confirmation of the unimolecular rate-determining step in the S_N1 mechanism.

When is a mechanism S_N1 and when is it S_N2? To answer this question, we need to briefly discuss some kinetic studies carried out by Christopher Ingold and Edward Hughes in 1937. Measurement of rates of reaction under different sets of experimental conditions can provide insight into reaction mechanisms. For example, consider the rate of the following second-order reaction between a bromoalkane and the nucleophile Cl^-:

$$R\text{—}Br + Cl^- \xrightarrow{S_N2} R\text{—}Cl + Br^-$$

From a thermodynamic point of view, the substitution of Br^- by Cl is favorable because Br^- is a weaker base than the Cl^- ion. The *rate* of reaction shows a dramatic dependence on the degree of branching of the α carbon.

◀ In the same manner that alcohols are classified as primary, secondary, and tertiary, so are haloalkanes. Thus, 2-bromopropane is a secondary haloalkane.

$$CH_3Br > CH_3CH_2Br > CH_3CH_2CH_2Br > CH_3\underset{\underset{CH_3}{|}}{C}HBr > CH_3\underset{\underset{CH_3}{|}}{\overset{\overset{CH_3}{|}}{C}}Br$$

Relative rate   1200   40   16   1   Too slow to measure
(for S_N2)

In the S_N2 mechanism, the nucleophile attacks the electrophilic center on the side *opposite* the leaving group, often called *backside attack*. Because the

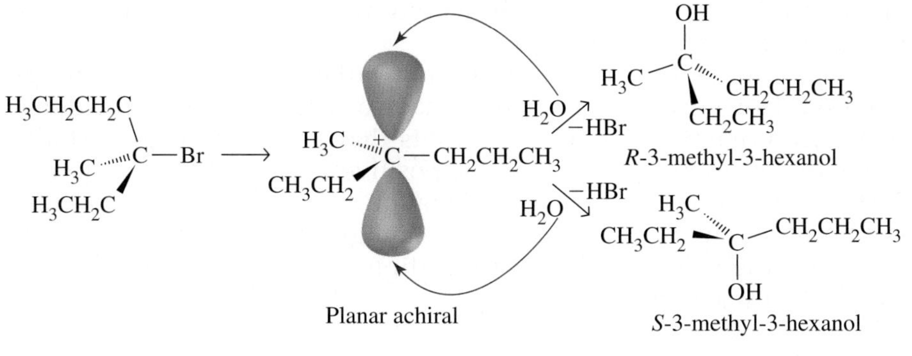

Planar achiral

R-3-methyl-3-hexanol

S-3-methyl-3-hexanol

◀ FIGURE 27-5
Formation of a racemic mixture in an S_N1 reaction

Bromomethane

Bromoethane

1-methylbromoethane

1,1-dimethylbromoethane

▲ FIGURE 27-6
Space-filling models of bromomethane, bromoethane, 1-methylbromoethane, and 1,1-dimethylbromoethane
The models are viewed from the vantage point of a nucleophile attacking the α carbon from the backside. As the number of methyl groups bonded to the α carbon increases, access to the partial positive charge on the α carbon decreases.

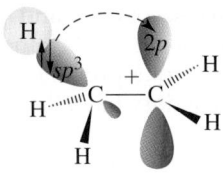

▲ FIGURE 27-7
Stabilization of a carbocation through hyperconjugation
The carbocation is stabilized through the donation of electron density from a C—H bond to an empty $2p$ orbital on the positively charged carbon atom.

nucleophile attacks the backside of the α carbon, bulky substituents bonded to the α carbon will make it harder for the nucleophile to get to the backside. The obstruction of the nucleophile from interacting with an electrophilic carbon atom is an example of steric hindrance. Backside attack of bromomethane is fastest because there is very little steric hindrance (Fig. 27-6). However, the backside of 1,1-dimethylbromoethane is almost completely blocked from nucleophilic attack, and thus this compound does not undergo an S_N2 reaction; nucleophilic substitution occurs by an S_N1 mechanism instead.

What effect does alkyl substitution have on the rate of an S_N1 reaction? When we look at a series of bromoalkanes reacting with water, a very different order of reactivity is observed:

$$R-Br + H_2O \longrightarrow R-OH + HBr$$

S_N1 reactivity: $CH_3Br < CH_3CH_2Br < (CH_3)_2CHBr < (CH_3)_3CBr$

α carbon: methyl   primary   secondary   tertiary

The order of reactivity, which is the opposite of that observed for the S_N2 reaction, follows the order of carbocation stability. The reactivity is related to the stability of the carbocation because the rate-determining step in the S_N1 reaction is the formation of the carbocation (Fig. 27-4). The order of carbocation stability is as follows:

Relative stability: $CH_3^+ < CH_3CH_2^+ < (CH_3)_2CH^+ < (CH_3)_3C^+$

Least stable   Most stable

Thus, reagents or reaction conditions that favor the formation of a carbocation will increase the rate of the S_N1 reaction.

It is important to emphasize, however, that carbocations are not particularly stable, at least not in the usual sense. Carbocations are reactive intermediates with rather short lifetimes. A tertiary carbocation, such as $(CH_3)_3C^+$, has a lifetime of about 10^{-10} s in water whereas a secondary carbocation, such as $(CH_3)_2CH^+$, has a lifetime of about 10^{-12} s in water. What is the explanation for the relative stabilities of methyl, primary, secondary, and tertiary carbocations? Alkyl groups stabilize carbocations and, in that role, they appear to be electron donating. However, the explanation of how alkyl groups stabilize a carbocation is a matter of some debate. One explanation is based on the concept of *hyperconjugation*, which is illustrated in Figure 27-7 for $CH_3CH_2^+$, a primary carbocation. The positive carbon is sp^2 hybridized. Around the positively charged carbon atom is a trigonal planar arrangement of atoms and an empty $2p$ orbital perpendicular to the plane of hybridization. The adjacent carbon atom is sp^3 hybridized and forms a σ bond with a hydrogen atom. The C—H bond can donate electron density to the empty $2p$ orbital, as suggested by the dashed arrow in Figure 27-7. If more alkyl groups are present in a carbocation, then more interactions of this type occur, and a greater degree of stabilization is the result. Thus, tertiary carbocations are more stable than secondary carbocations, which are, in turn, more stable than primary carbocations. Because no alkyl groups are present in a methyl carbocation, methyl carbocations are least stable; in fact, it is assumed they are never formed.

Other explanations for the stabilization of carbocations by alkyl groups are also used. However, all explanations have a common theme: the donation of electron density from filled orbitals that are aligned (or partially aligned) with the empty $2p$ orbital on the positively charged carbon atom of the carbocation. Irrespective of the explanation, alkyl groups help to stabilize a carbocation; thus, as the number of alkyl groups bonded to the α carbon increases, the stability of the carbocation also increases, as does the speed of the S_N1 reaction.

EXAMPLE 27-3   Recognizing S$_N$1 and S$_N$2 Reactions and Predicting the Products

Consider the following combinations of reactants. In each case, predict whether a substitution reaction will occur. If so, identify the products and suggest the likely mechanism.

(a) $CN^- + CH_3CH_2CH_2Cl \longrightarrow$

(b) $Br^- + CH_3CH_2OH \longrightarrow$

(c) $CH_3OH + (CH_3)_3CCl$ (in methanol) $\longrightarrow$

Analyze

To decide whether a reaction of this type will take place, we first identify the electrophile, nucleophile, and leaving group. Keep in mind that the order of reactivity for S$_N$2 is methyl > primary > secondary > tertiary and that the order is the opposite for S$_N$1.

Solve

(a) The nucleophile is the CN^- ion, the electrophile is the chloropropane, and the leaving group is the Cl^- ion. The CN^- is a much stronger base than the Cl^- ion, and so the equilibrium constant for the reaction should be large. The chloropropane is a primary haloalkane, so the likely mechanism is S$_N$2 and the product will be 1-cyanopropane, $CH_3CH_2CH_2CN$.

(b) The nucleophile is the Br^- ion, the electrophile is the ethanol, and the potential leaving group is the OH^- ion. The OH^- is a much stronger base than the Br^- ion, so the equilibrium constant for the reaction is much *less* than one. No reaction is expected.

(c) The nucleophile is the CH_3OH molecule, the electrophile is the *t*-butyl chloride, and the leaving group is the Cl^- ion. In this case we have to know the relative basicities of CH_3OH and Cl^- to decide in which direction the reaction will proceed. If we assume that the basicities of methanol and chloride ion are about the same (neither likes to accept a proton), we expect an equilibrium to be established. However, the fact that we are using a large excess of methanol (methanol is the solvent) shifts the equilibrium toward the product (Le Châtelier's principle). The product is *tert*-butyl methyl ether, $(CH_3)_3COCH_3$, and the likely mechanism is S$_N$1 since the electrophile is a tertiary haloalkane.

Assess

In Section 27-3 we will see that other products, not just substitution products, are possible in some of these reactions.

PRACTICE EXAMPLE A: Predict whether the following reactions will take place, and suggest the likely mechanism:

(a) $CH_3CC^- + CH_3Br \longrightarrow$

(b) $Cl^- + CH_3CH_2CN \longrightarrow$

(c) $CH_3NH_2 + (CH_3)_3CCl \longrightarrow$

PRACTICE EXAMPLE B: From the following information, identify and provide chemical equations for the mechanism of the reaction.

(a) The following reaction is carried out in acetone (propanone):

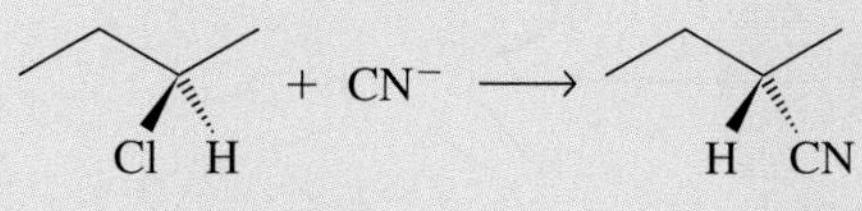

(b) A sample that contains only one enantiomer (either *R* or *S*) of an optically active compound is said to be *optically pure*. An optically pure sample of 2-iodobutane is dissolved in methanol. The resultant solution of *sec*-butyl methyl ether is optically inactive.

Solvent Effects in S$_N$1 and S$_N$2 Reactions

The nucleophile in a substitution reaction is typically a negatively charged ion or a polar molecule. Thus, to dissolve the starting materials, we must generally use polar solvents. (Recall from Chapter 13 that like dissolves like.)

Some polar aprotic solvents

Propanone
(acetone)

Diethylether

Dimethylsulfoxide
(DMSO)

Dimethoxyethane
(DME)

Dimethylformamide
(DMF)

Dimethylacetamide
(DMA)

However, the nature of the solvent—in particular, the way in which solvent molecules interact with the nucleophile—plays a critical role in determining whether a reaction will follow an S_N1 or an S_N2 mechanism. To understand solvent effects, it is helpful to differentiate between protic and aprotic solvents. **Protic solvents** are solvents whose molecules have hydrogen atoms bonded to electronegative atoms, such as oxygen and nitrogen. Water, methanol, ethanol, ethanoic acid (acetic acid), and methylamine are examples of *polar* protic solvents. (Not only do these molecules have H atoms bonded to O or N, but they are also polar.) **Aprotic solvents** are solvents whose molecules do not have a hydrogen atom bonded to an electronegative element. Aprotic solvents can be polar or nonpolar, depending on whether the solvent molecules are polar or nonpolar. A few examples of polar aprotic solvents are shown in the margin. Another example (not shown) is hexamethylphosphoric triamide (HMPT), $[(CH_3)_2N]_3P{=}O$. Examples of nonpolar aprotic solvents are hexane and benzene.

What type of solvent favors the S_N1 pathway? The rate-determining step in an S_N1 reaction is the formation of a carbocation and an anion. If the solvent does not stabilize the ions formed in this step, an S_N1 reaction will not occur. Highly polar protic solvents, such as water and methanol, promote S_N1 reactions because the molecules of such solvents stabilize both the carbocation and the anion, in much the same way that water molecules stabilize the ions of a salt when a salt dissolves in water. (See Figure 13-6.) Molecules of a polar protic solvent stabilize carbocations through the donation of lone pairs from the oxygen or nitrogen atom, and they stabilize anions through the formation of hydrogen bonds.

What type of solvent promotes an S_N2 reaction? Polar aprotic solvents work best for S_N2 reactions. The S_N2 pathway is a concerted reaction in which the nucleophile attacks an electrophile. The reactivity of the nucleophile will be reduced if the solvent molecules are able to form hydrogen bonds with the nucleophile. When the solvent molecules form hydrogen bonds with the nucleophile, each nucleophile is surrounded by a shell of solvent molecules that impedes the reaction of the nucleophile and electrophile (Fig. 27-8). The formation of hydrogen bonds is avoided by using aprotic solvents, and thus aprotic solvents are used to promote S_N2 reactions.

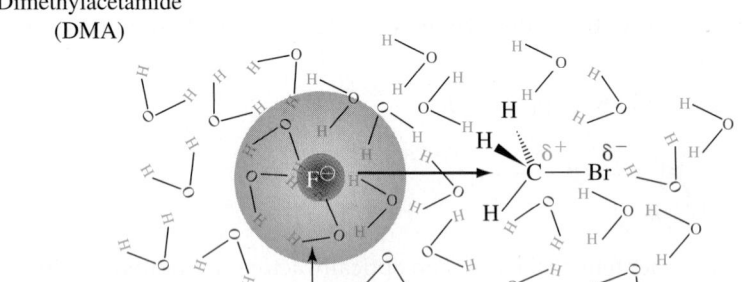

Solvent shell
Strong ion-dipole interactions.

(a)

Solvent shell
Weak ion-dipole interactions.

(b)

▲ FIGURE 27-8
Nucleophile-solvent interactions and their effect on nucleophilicity
(a) When the ion–dipole interactions are strong, nucleophilicity is decreased. The nucleophile is strongly solvated by solvent molecules and is hindered from attacking the electrophilic carbon. (b) When the ion–dipole interactions are weak, the nucleophile is not strongly solvated and is less hindered by solvent molecules.

Factors Affecting Nucleophilicity

Nucleophilicity is a measure of how readily (how fast) a nucleophile attacks an electrophilic carbon atom bearing a leaving group. But what makes a good nucleophile? It is tempting to think that there is a simple relationship between nucleophilicity and basicity because both involve the donation of an electron pair to an electrophile. No simple relationship exists because, as pointed out in Are You Wondering 27-1, nucleophilicity and basicity are fundamentally different properties. Nucleophilicity is a kinetic property (related to a rate of reaction) and basicity is a thermodynamic property (related to the thermodynamic stability of a base). Many texts correlate trends in nucleophilicity with trends in basicity. It is true that such correlations exist, but the correlations are of little use without an understanding of the underlying reasons for these correlations. The correlations exist because the explanation of trends in nucleophilicity is based, in part, on factors we use for explaining trends in basicities. For example, both nucleophilicity and basicity depend on the electronegativity, size, and hybridization of the atom bearing the lone pair; charge delocalization; and the effect of electron-withdrawing or electron-donating groups. However, trends in nucleophilicity depend on other factors too, such as the interactions between the nucleophile and solvent molecules, steric effects, and the nature of the electrophile. Therefore, trends in nucleophilicity do not always follow trends in basicity.

Let us consider a few principles that we can use for understanding trends in nucleophilicity. When applying these principles, our point of reference is the nucleophilic atom, which possesses the lone pair that is used to form a bond with the electrophile. As we work our way through these principles, we will see similarities to explanations given earlier (Section 27A on www.masteringchemistry.com), but it is important that we take careful note of the differences.

1. All else being equal, a negatively charged nucleophile will react faster than an uncharged nucleophile. For example, HO^- is a stronger nucleophile than H_2O, and CH_3O^- is a stronger nucleophile than CH_3OH. A negatively charged atom is more strongly attracted to an electrophilic center than is an atom having no charge or a partial negative charge.

2. When comparing molecules or ions in which the nucleophilic atoms are from the *same row of the periodic table*, the electronegativity of the nucleophilic atom is an important factor because it affects how available the lone pair will be for bonding to the electrophilic atom. In the following anions, the nucleophilic atoms are from the second period.

Increasing electronegativity of
the nucleophilic atom
→

$H_3C^- > H_2N^- > HO^- > F^-$

←
Increasing nucleophilicity

F^- is not as good a nucleophile as H_3C^- because F is more electronegative than C, and thus a lone pair on F^- is less available for bonding than is the lone pair on H_3C^-.

It is sometimes possible to predict the effect of other factors, such as charge delocalization or the presence of electron-withdrawing or electron-donating groups. Consider, for example, the following series of nucleophiles, arranged in order of decreasing nucleophilicity (from strongest to weakest). All the nucleophilic atoms in these ions are from the second period.

$$H_2N^- > CH_3O^- > HO^- > HCOO^-$$

Increasing nucleophilicity

H_2N^- is the strongest nucleophile because N is less electronegative than O, and thus the lone pair on N is more available for bonding than is a lone pair on O. CH_3O^- is a stronger nucleophile than HO^- because, in solution, the CH_3 group is electron donating and this helps to make the lone pairs on O less stable and more reactive. $HCOO^-$ is a weaker electrophile than HO^- because the negative charge is delocalized (shared equally by both oxygens), and thus $HCOO^-$ will be less reactive than HO^-.

3. Negatively charged nucleophiles are more reactive in polar aprotic solvents than they are in polar protic solvents. In a polar protic solvent, anions are more highly solvated and thus less reactive (Fig. 27-8).

The importance of considering how the solvent affects nucleophilicity cannot be overstated. The trend in reactivities of a series of nucleophiles can, in some instances, be completely reversed by changing the solvent.

Protic solvent: $I^- > Br^- > Cl^- > F^-$

Aprotic solvent: $F^- > Cl^- > Br^- > I^-$

Increasing nucleophilicity

4. When comparing nucleophiles in which the nucleophilic atoms are from *the same group*, nucleophilicity *usually* increases as the size (and thus, polarizability) of the nucleophilic atom increases. The larger and more polarizable the nucleophilic atom, the easier it is for the charge cloud of the nucleophile to be distorted toward the electrophilic carbon atom. The distortion of the charge cloud toward the electrophilic carbon atom helps to lower the energy of the transition state and increases the rate of reaction. (See Figure 27-9.)

5. Bulky groups adjacent to the nucleophilic atom reduce the reactivity of a nucleophile because these groups hinder the approach of the nucleophile toward the electrophilic atom. For example, consider the relative reactivities of the methoxide and *tert*-butoxide ions:

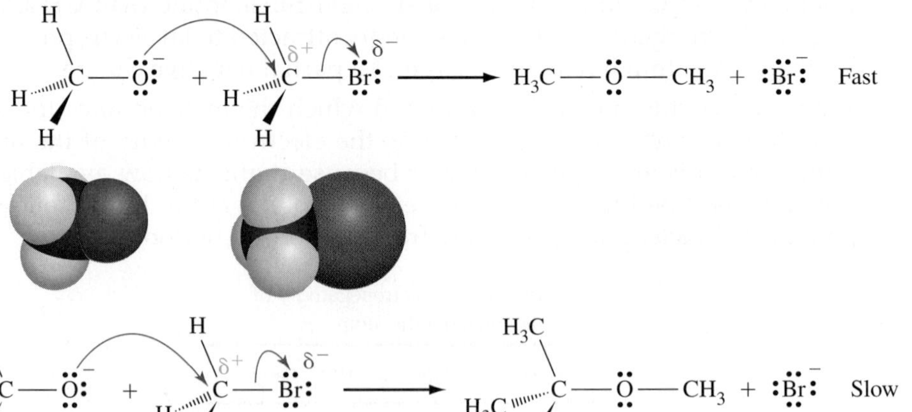

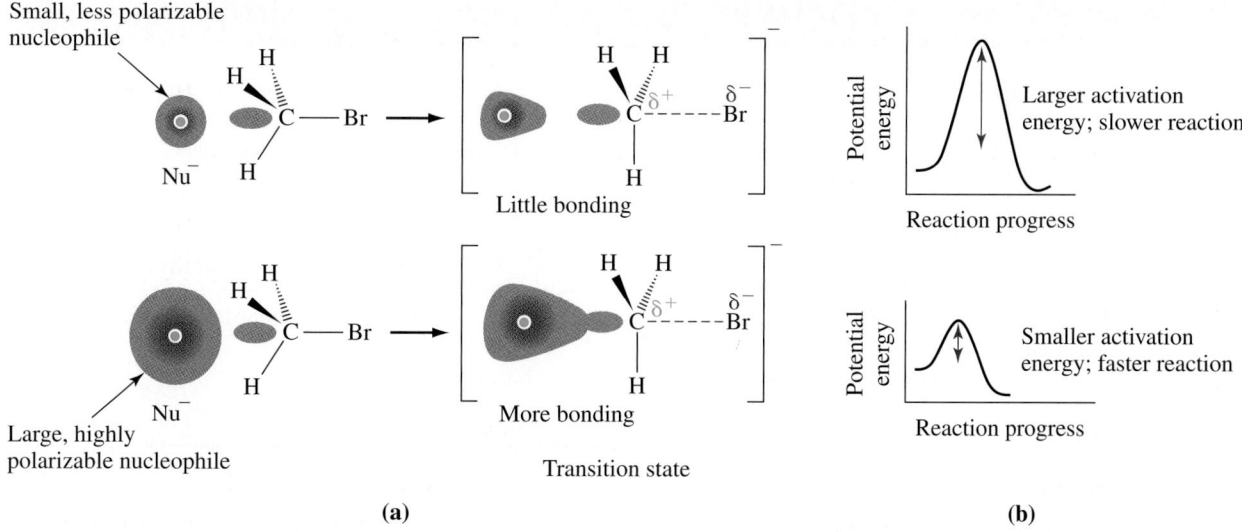

▲ FIGURE 27-9
The effect of the polarizability of the nucleophile in an S$_N$2 reaction
For simplicity, the nucleophile is represented by a spherical charge distribution.
(a) The formation of the transition state in the S$_N$2 mechanism involves transfer of electron density from the nucleophile to a vacant orbital localized on C. For the larger and more polarizable nucleophile, the transfer of electron density occurs sooner (at larger distances) and the resulting transition state is of lower energy, as shown in **(b)**. Because the rate of reaction increases as the activation energy decreases, the more polarizable nucleophile reacts faster. Keep in mind that solvent-nucleophile interactions (not shown) also affect the ability of the nucleophile to bond with the electrophilic carbon atom.

As indicated by the space-filling models of the $(CH_3)_3CO^-$ and CH_3O^- ions, the nucleophilic oxygen atom of $(CH_3)_3CO^-$ is significantly hindered from closely approaching the electrophilic carbon atom of CH_3Br but the oxygen atom of CH_3CO^- is not. Consequently, $(CH_3)_3CO^-$ is a weaker nucleophile than CH_3O^-.

In Table 27.1 we classify some commonly encountered nucleophiles as excellent (strong), good, and fair. However, it is important that we use the information in this table cautiously. The preceding discussion makes it clear that the nucleophilicity of an electron pair donor is influenced by several factors, only some of which are related to the structure of the nucleophile itself. Nucleophilicity depends also on the solvent used and the nature of the electrophile. A given electron pair donor may behave as a strong nucleophile in a particular situation, but it may not be a strong nucleophile in another.

A Summary of S$_N$1 and S$_N$2 Reactions

In this section, we considered two different mechanisms by which a substitution reaction at an sp^3-hybridized carbon atom occurs. It is important to remember that (1) S$_N$1 and S$_N$2 reactions compete with each other, and (2) the exact details of any particular reaction might place it somewhere between these two mechanistic extremes. With an appropriate selection of reactants and reaction conditions, we might be able to push a particular reaction toward one mechanism or the other. Ultimately, the molecules in a system will always react by the lowest energy pathway. Depending on the reactants and reaction conditions, the lowest energy pathway for a substitution reaction may be the S$_N$1 reaction or the S$_N$2 reaction or something in between these two extremes. Table 27.2 summarizes some key ideas from this section and identifies the combinations of reactants and reaction conditions that push a substitution reaction toward either the S$_N$1 or S$_N$2 mechanism.

TABLE 27.1   Classification of Common Nucleophiles

Excellent Nucleophiles		
		Rate[a]
NC$^-$	Cyanide	126,000
HS$^-$	Thiolate	126,000
I$^-$	Iodide	80,000
Good Nucleophiles		
		Rate
HO$^-$	Hydroxide	16,000
Br$^-$	Bromide	10,000
N$_3^-$	Azide	8,000
NH$_3$	Ammonia	8,000
NO$_2^-$	Nitrite	5,000
Fair Nucleophiles		
		Rate
Cl$^-$	Chloride	1,000
CH$_3$COO$^-$	Acetate	630
F$^-$	Fluoride	80
CH$_3$OH	Methanol	1
H$_2$O	Water	1

[a]The rate is a relative rate. A rate of 100 means that the nucleophile reacts 100 times as fast as water does.

TABLE 27.2 Relative Reactivities of Haloalkanes

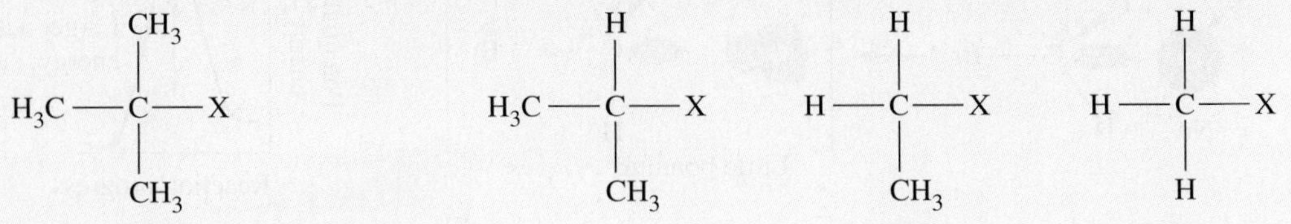

	$3°$	$2°$	$1°$	Methyl
Electrophile	$H_3C - \underset{\underset{CH_3}{\mid}}{\overset{\overset{CH_3}{\mid}}{C}} - X$	$H_3C - \underset{\underset{CH_3}{\mid}}{\overset{\overset{H}{\mid}}{C}} - X$	$H - \underset{\underset{CH_3}{\mid}}{\overset{\overset{H}{\mid}}{C}} - X$	$H - \underset{\underset{H}{\mid}}{\overset{\overset{H}{\mid}}{C}} - X$

Stability of Carbocation	Forms a relatively stable carbocation		Form relatively unstable carbocations	
S_N1 Reactivity	← increasing S_N1 reactivity		No S_N1	
S_N2 Reactivity	No S_N2		increasing S_N2 reactivity →	
α Carbon	Sterically hindered		Not sterically hindered	
Solvent	Use a polar protic solvent to promote the S_N1 reaction		Use a polar aprotic solvent to promote the S_N2 reaction	

▲ The symbols 1°, 2°, and 3° stand for primary, secondary, and tertiary, respectively.

EXAMPLE 27-4   Predicting Whether a Substitution Reaction Proceeds by an S_N1 or S_N2 Mechanism

Does the substitution reaction below follow an S_N1 or an S_N2 mechanism? What are the products? Write the steps of the mechanism and use arrows to show the movement of electrons.

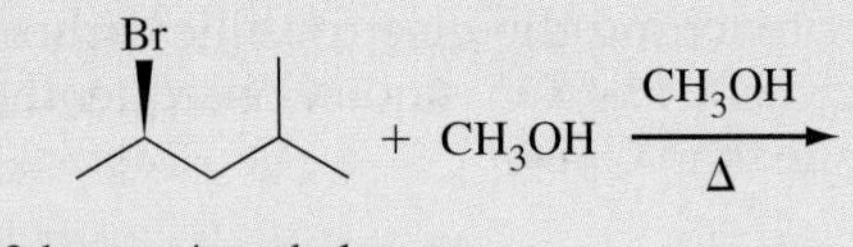

(R)-2-bromo-4-methylpentane

Analyze

The haloalkane is a 2° haloalkane. Secondary haloalkanes undergo substitution reactions by either the S_N1 or S_N2 depending on the nucleophile and solvent. (See Table 27.2.)

Solve

Methanol is the nucleophile and the haloalkane is the electrophile. Because the nucleophile is uncharged, its nucleophilicity is determined primarily by the polarizability of the nucleophilic atom (O). The O atom is relatively small and not very polarizable; thus, CH_3OH is a weak nucleophile. A weak nucleophile disfavors an S_N2 reaction. Also, the solvent is polar protic and will help to stabilize a carbocation. With a weak nucleophile and a polar protic solvent, we expect the substitution reaction to occur by an S_N1 mechanism. The carbocation that is formed reacts with a solvent molecule CH_3OH (a *solvolysis reaction*) to form a protonated ether. The final product is an ether, which is obtained when a proton is transferred from the protonated ether to a CH_3OH molecule from the solvent. We will obtain two products, the R and S stereoisomers, because CH_3OH can attack the carbocation from either side. The steps are as follows:

Step 1: Formation of a carbocation

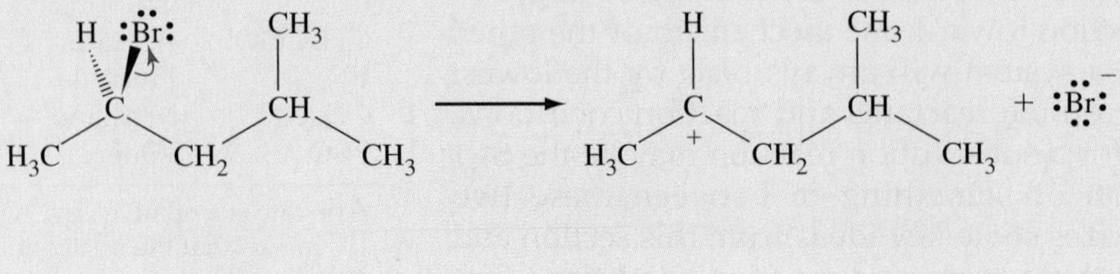

Step 2: Nucleophilic attack by CH_3OH

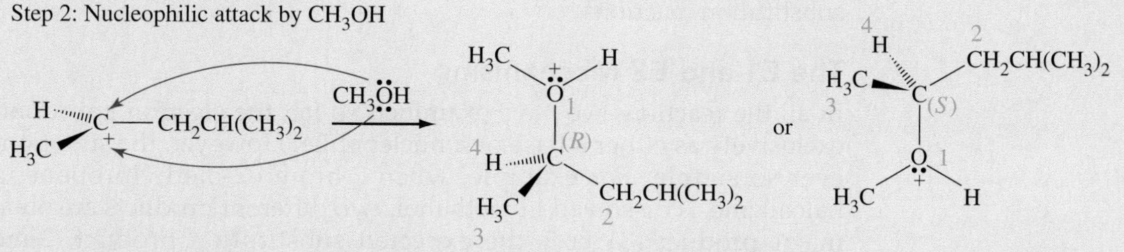

Step 3: Loss of proton to solvent (ignoring stereochemistry)

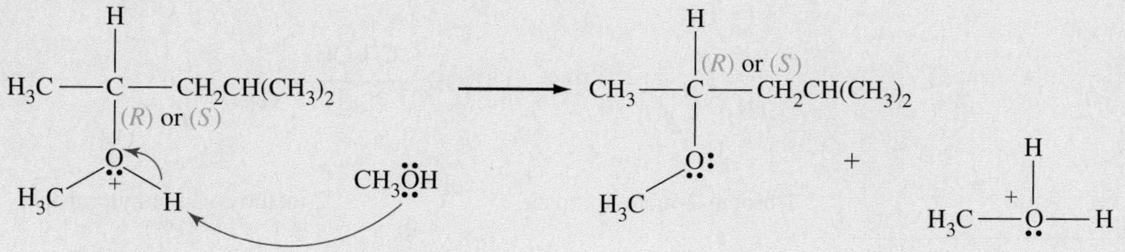

Thus, the reaction will produce a racemic mixture consisting of the (R) and (S) stereoisomers of 2-methoxy-4-methylpentane.

Assess

To name the products, you may find it helpful to review the nomenclature rules given in Chapter 26. The reaction considered in this example is also called a solvolysis reaction, because the solvent acts as the nucleophile.

PRACTICE EXAMPLE A: Predict whether the substitution product obtained in the reaction below follows an S_N1 or an S_N2 mechanism. Show all steps in the mechanism, using arrows to indicate the movement of electrons. Name the product.

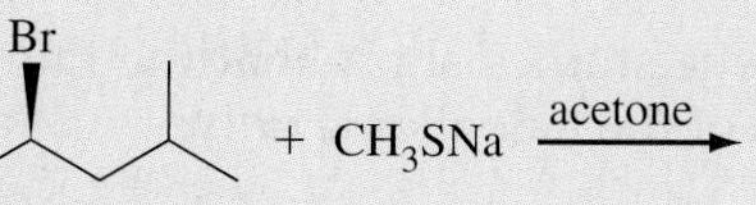

(R)-2-bromo-4-methylpentane

PRACTICE EXAMPLE B: Does the substitution reaction below occur by an S_N1 or an S_N2 mechanism?

$$\underset{Br}{CH_3CH_2CHCH_2CH_3} + H_2O \xrightarrow[80\,°C]{} \underset{OH}{CH_3CH_2CHCH_2CH_3}$$

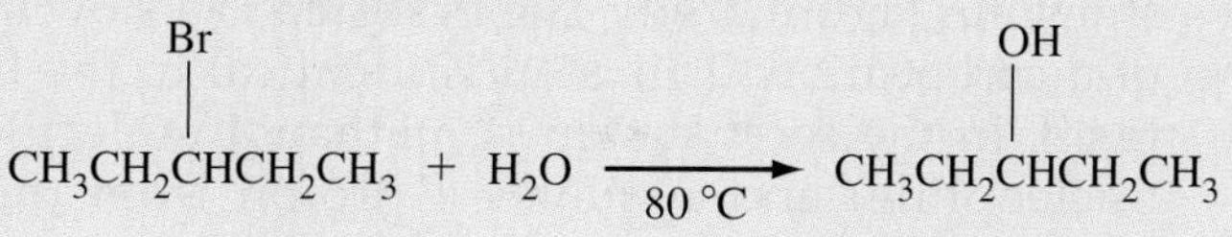

27-3 Introduction to Elimination Reactions

In the previous section, we saw that a haloalkane can undergo a substitution reaction in which the halogen atom is replaced by another group. Haloalkanes can also undergo elimination reactions, in which the halogen atom and a hydrogen atom bonded to the β carbon are removed from the molecule.

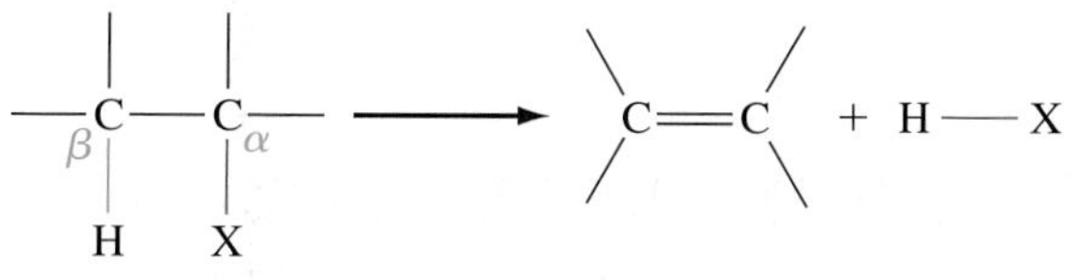

(27.3)

As is true for substitution reactions, elimination reactions of haloalkanes can occur by different mechanisms. If the rate-determining step is unimolecular, then the mechanism is called **E1**. If the rate-determining step is bimolecular, the mechanism is called **E2**. In this section, we will explore these mechanisms in some detail and discuss the competition that occurs between elimination and substitution reactions.

The E1 and E2 Mechanisms

In all the reactions we have examined so far, the electron pair donor reacted exclusively as either a base or a nucleophile. However, the situation is hardly ever so simple. For example, when 2-bromo-2-methylpropane, a tertiary haloalkane, is dissolved in methanol, two different products are obtained. The major product (81%) is the expected substitution product, 2-methoxy-2-methylpropane. The reaction proceeds via the S_N1 mechanism.

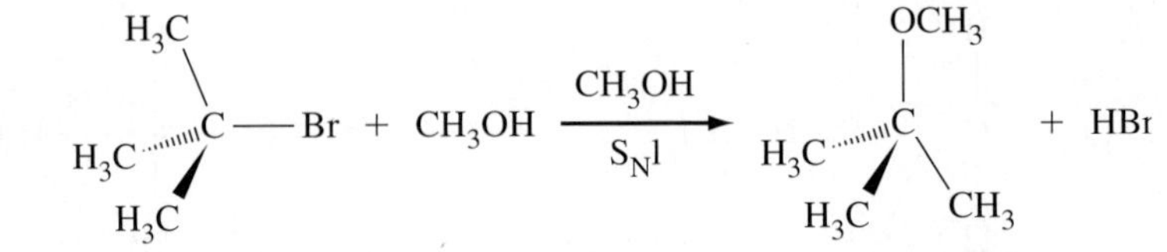

H₃C
|
H₃C......C—Br + CH₃OH —CH₃OH→ H₃C......C + HBr
| S_N1
H₃C

2-bromo-2-methylpropane   2-methoxy-2-methylpropane
(81%)

The minor product (19%) is 2-methylpropene. A balanced chemical equation for the formation of the minor product is given below:

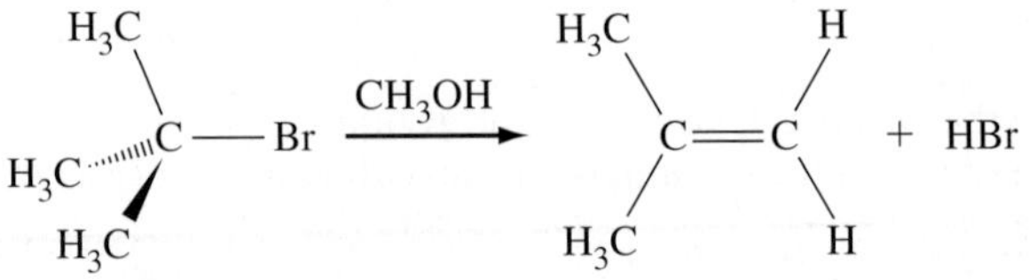

2-bromo-2-methylpropane   2-methylpropene
(19%)

Kinetic studies show that the rate of this reaction depends on the concentration of the haloalkane only:

$$rate = k[(CH_3)_3CBr]$$

Because the reaction is an elimination reaction following first-order kinetics, it is called an **E1 reaction**. The mechanism for this E1 reaction is shown in Figure 27-10. The first step is a slow step involving the formation of a carbocation. This step is identical to the first step in the S_N1 mechanism. In the second step, a methanol molecule reacts as a base (not as a nucleophile) and removes a proton from the carbocation, yielding an alkene. Notice that a proton is removed from a carbon atom adjacent to the positively charged carbon atom of the carbocation, a β carbon. The reaction profile is shown in Figure 27-11.

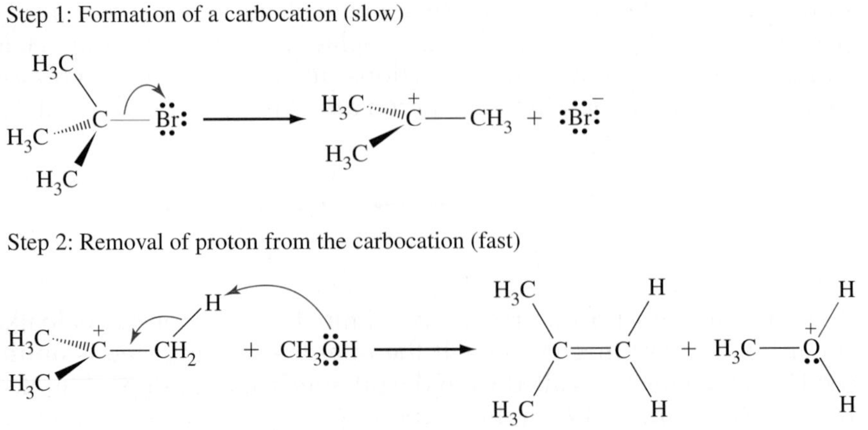

Step 1: Formation of a carbocation (slow)

▶ FIGURE 27-10
The E1 mechanism
The first step is the formation of a carbocation. It is slow and rate determining. The second step is an acid–base reaction in which a proton is removed from a carbon atom adjacent to the positively charged carbon atom of the carbocation.

Step 2: Removal of proton from the carbocation (fast)

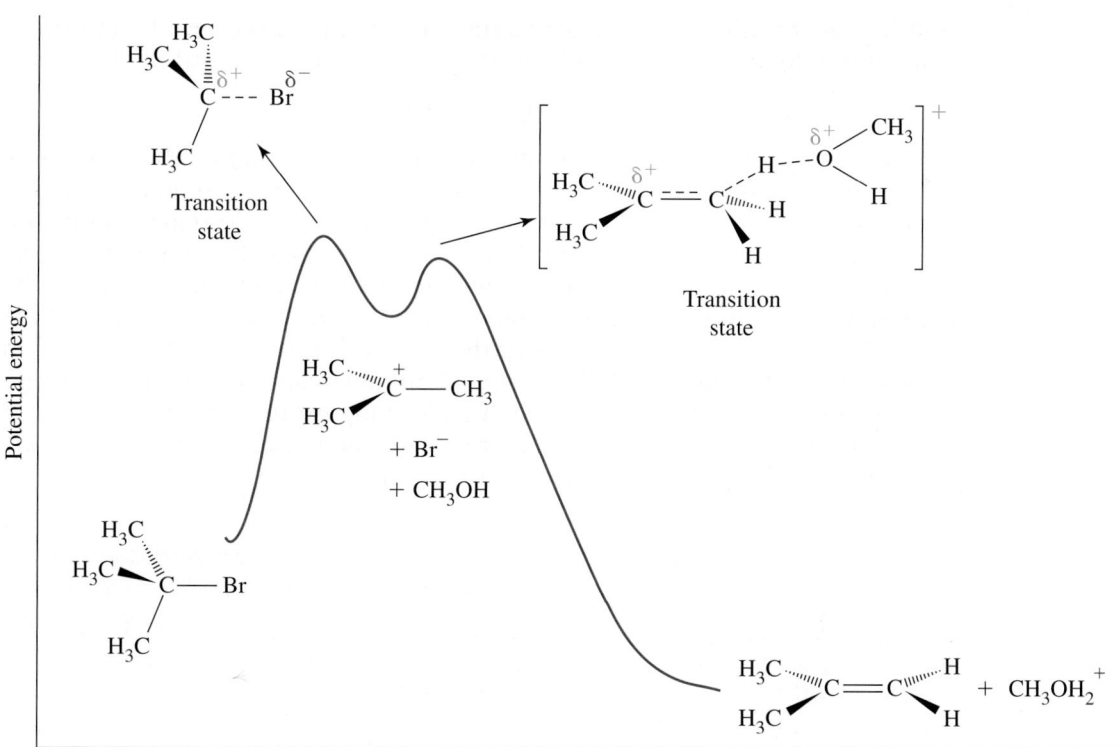

▲ FIGURE 27-11
Reaction profile for the E1 reaction between *tert*-butylbromide and methanol
The first step is a highly endothermic process involving the formation of a carbocation. In the transition state for the first step, the C—Br bond is partially broken. In the second step, CH_3OH acts as a base and removes a proton from the β carbon. In the transition state for the second step, a new π bond is forming between two C atoms. In addition, a C—H bond is breaking and a new H—O bond is forming.

When we compare the E1 mechanism to the S_N1 mechanism (Fig. 27-4), we see why the reaction of $(CH_3)_3CBr$ and CH_3OH yields a mixture of two products. After the formation of the carbocation, methanol can act like a nucleophile and attack the electrophilic carbon atom (producing the S_N1 product) or it can act as a base and remove a proton to produce an alkene. The essential difference between the S_N1 and E1 mechanism is the role played by the electron pair donor in the second step. In the S_N1 mechanism, the electron pair donor acts as a nucleophile and forms a σ bond with the electrophilic carbon atom of the carbocation. In the E1 mechanism, the electron pair donor acts as a base and forms a σ bond with a proton.

In the example we just considered, the nucleophile was CH_3OH. The CH_3OH molecule is a weak nucleophile (it is a neutral molecule and the nucleophilic atom is relatively small and not very polarizable) and also a weak base. Let us consider what would happen in the reaction of $(CH_3)_3CBr$ and CH_3CH_2ONa in ethanol, bearing in mind that the $CH_3CH_2O^-$ ion is a stronger base than CH_3OH. We anticipate that the amount of alkene in the product mixture will increase, and as the equation below shows, this is in fact what is observed:

97%

A study of the kinetics of this reaction reveals, however, that the rate of elimination depends on the concentrations of the substrate and the base:

$$rate = k[(CH_3)_3CBr][CH_3CH_2O^-]$$

Whereas the use of a stronger base promotes elimination at the expense of substitution, we can see that the mechanism of the elimination reaction cannot be the E1 mechanism. (Recall that the rate of an E1 reaction depends only on the concentration of the substrate.) Because the kinetics of this elimination reaction are second order, the rate-determining step must be bimolecular. Thus, the reaction is called an **E2 reaction**. Detailed studies of E2 reactions reveal that the E2 mechanism consists of a single step that proceeds through a single transition state where three changes are occurring simultaneously: (1) removal of a proton from the β carbon; (2) departure of the leaving group; (3) formation of a π bond between the α and β carbon atoms. The reaction profile for the E2 reaction between $(CH_3)_3CBr$ and $CH_3CH_2O^-$ is illustrated in Figure 27-12.

27-2 CONCEPT ASSESSMENT

The principal product in the reaction of $(CH_3)_3CBr$ and $NaCH_3CH_2O$ in ethanol is an alkene. What is the minor product and by what mechanism is it formed?

Predicting the Major Elimination Product

In the elimination reactions that we have studied so far, only one elimination product was possible. However, it is often the case that more than one elimination product can be produced. For example, in the reaction of 2-bromobutane with sodium methoxide in methanol, three E2 products are possible:

1-butene (18%)   (27.4a)

trans-2-butene (67%)   cis-2-butene (15%)

$$+ \ HOCH_3 \ + \ :\ddot{B}r:^-$$

(27.4b)

KEEP IN MIND

that the product mixture for an elimination reaction will be richer in the product that is formed the fastest. The relative rates of the elimination reactions are determined by the energies of the corresponding transition states. For an elimination reaction, the transition state has some double bond character, and so the factors known to stabilize an alkene will also stabilize the transition state. As a consequence, the pathway that leads to the more stable alkene usually also passes through the transition state of lowest energy; thus, the reaction leading to the most stable alkene is not only the most exothermic but also the fastest.

In 1-butene, the double bond is at the end of the carbon chain (a terminal C=C bond) and in the *cis* and *trans* isomers of 2-butene, the C=C bond is an internal double bond. Notice that there is a strong preference or selectivity for forming the internal double bond (82% versus 18%).

When more than one elimination product is possible, can we reliably predict which will be the major product? Usually, the major product of an elimination reaction is the one that is most stable. But what can we say about the relative

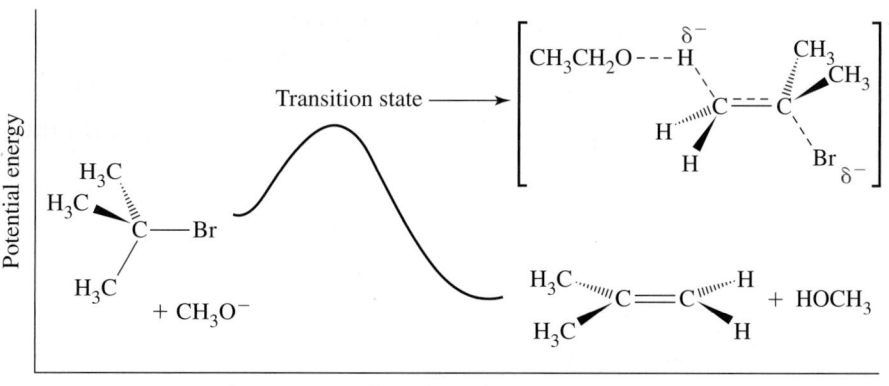

◀ FIGURE 27-12
Reaction profile for the E2 reaction of (CH$_3$)$_3$CBr and CH$_3$CH$_2$O$^-$ in methanol

stabilities of alkenes? In the diagram below, alkenes are ranked in order of increasing stability:

$$CH_2\!=\!CH_2 \; < \; RCH\!=\!CH_2 \; < \; \underset{cis}{\overset{\displaystyle \text{H}\;\;\;\;\text{H}}{\underset{\displaystyle \text{R}\;\;\;\;\text{R}}{C\!=\!C}}} \; < \; \underset{trans}{\overset{\displaystyle \text{H}\;\;\;\;\text{R}}{\underset{\displaystyle \text{R}\;\;\;\;\text{H}}{C\!=\!C}}} \; < \; \overset{\displaystyle \text{H}\;\;\;\;\text{R}}{\underset{\displaystyle \text{R}\;\;\;\;\text{R}}{C\!=\!C}} \; < \; \overset{\displaystyle \text{R}\;\;\;\;\text{R}}{\underset{\displaystyle \text{R}\;\;\;\;\text{R}}{C\!=\!C}}$$

Increasing stability of alkenes →

This order of stability is based on experimental data (heats of hydrogenation). The following statement summarizes the trend in stabilities:

> The greater the number of alkyl substituents bonded to the sp^2 carbons of an alkene, the greater its stability.

In general, more highly substituted alkenes are more stable than less highly substituted alkenes. Also, bulky alkyl groups like to be as far as apart as possible, and so, for example, the *trans* isomer is more stable than the *cis* isomer.

Let's return to reaction (27.4) and rationalize the product distribution that is obtained. The alkenes having an internal double bond are the major products because they are more highly substituted and the transition states leading to these alkenes are lower in energy than the transition state leading to the less highly substituted alkene. Thus, the alkenes having the internal double bonds are formed the fastest and are the major components of the product mixture. Of the two alkenes having an internal double bond, the *trans* stereoisomer is the more stable (and the transition state leading to this stereoisomer is lower in energy) and thus, the *trans* isomer is the major product.

A word of caution is in order. Often, the major product in an elimination reaction is the most highly substituted alkene but important exceptions exist to this rule. The formation of the most highly substituted alkene requires that the base remove a proton from a secondary or tertiary carbon atom. Access to the hydrogen atoms on a secondary or tertiary carbon atom is hindered by the adjacent groups. Small bases, such as OH$^-$, CH$_3$O$^-$ or CH$_3$CH$_2$O$^-$, can get past these groups fairly easily but larger bases, such as (CH$_3$)$_3$CO$^-$, cannot. A large base will preferentially attack the more exposed hydrogen atoms, yielding the least highly substituted alkene. In the following reactions, the same substrate reacts with different bases. In the first reaction, a smaller, less hindered base is used and the major elimination product is the more highly substituted alkene. In the second reaction, a larger, more hindered base is used and the major elimination product is the less highly substituted alkene.

◀ The explanation of why the stability of an alkene increases with the degree of alkyl substitution is a matter of some conjecture. A common explanation is based on the concept of *hyperconjugation*. According to IUPAC, in the formalism that separates bonds into σ and π types, hyperconjugation is the interaction of σ bonds (e.g., C—H, C—C, etc.) with a π network. Other explanations are based on the interactions of filled and empty molecular orbitals and call into question the appropriateness of using hyperconjugation as an explanation. Even in the absence of an explanation, the experimental facts are unambiguous: The stability of an alkene increases with the number of alkyl groups on the carbon atoms of the double bond.

CH$_3$CH$_2$CCH$_3$ (with CH$_3$ and Br) + CH$_3$CH$_2$O$^-$Na$^+$ $\xrightarrow[70\,°C]{CH_3CH_2OH}$ products

2-bromo-2-methylbutane

70%
2-methyl-2-butene

30%
2-methyl-1-butene

+ NaBr + CH$_3$CH$_2$OH

CH$_3$CH$_2$CCH$_3$ (with CH$_3$ and Br) + (CH$_3$)$_3$CO$^-$K$^+$ $\xrightarrow{(CH_3)_3COH}$ products

2-bromo-2-methylbutane

27%
2-methyl-2-butene

73%
2-methyl-1-butene

+ NaBr + CH$_3$CH$_2$OH

> For an E2 reaction, the H atom on the β carbon must be "anti" to the leaving group. This requirement places a restriction on the elimination products that can be formed from a 2° haloalkane. For a more detailed discussion and a worked example involving an elimination reaction, see Section 27B, *A Closer Look at the E2 Mechanism*, on the MasteringChemistry site (www.masteringchemistry.com).
>
> Mastering**CHEMISTRY**

Substitution and Elimination Reactions: A Summary

We have seen that haloalkanes can undergo a variety of reactions: S_N2, S_N1, E2, or E1. In principle, all these reactions compete with one another, and a variety of products are possible. When determining whether a reaction will proceed through an S_N2, S_N1, E2, or E1 mechanism, we must consider many factors. What is the nature of the electron pair donor? Is it a good nucleophile? Is it a strong base? Is it sterically hindered? What is the nature of the electrophile? Is it sterically hindered? And what about the solvent? Does it promote carbocation formation (S_N1/E1 reactions)? Does it increase or decrease the nucleophilicity of the electron pair donor?

With all these factors to consider, it is natural to ask: Is it possible to predict reliably the outcome of a reaction and the mechanisms that are involved? The answer to this question is a tentative yes. There are some guiding principles, which we present in Figure 27-13, but remember that exceptions to these guidelines exist. To make the best use of these guidelines, the following stepwise approach is recommended:

1. Show all formal charges and partial charges on atoms that have them.

2. Identify the electron pair donor and the electron pair acceptor (the electrophile).

3. Consider the electrophile. Is it primary, secondary, or tertiary?

4. Consider the electron pair donor. Is it a strong or weak nucleophile? Is it a strong or weak base? Is it sterically hindered?

5. Is the solvent protic or aprotic?

6. Determine whether the dominant reaction will be S_N2, E2, or S_N1 and E1. (Remember that S_N1 and E1 always occur at the same time.)

Before we attempt to apply this approach to specific cases, let's have a look at the guidelines summarized in Figure 27-13. The guidelines are presented in the form of a decision tree, with the first consideration being the base strength of the electron pair donor, :B. Additional considerations include steric hindrance, nucleophilicity, and solvent effects.

For primary alkanes (Figure 27-13a), the possible reactions are S_N2 and E2. An S_N1 reaction is not a likely possibility because primary carbocations are not

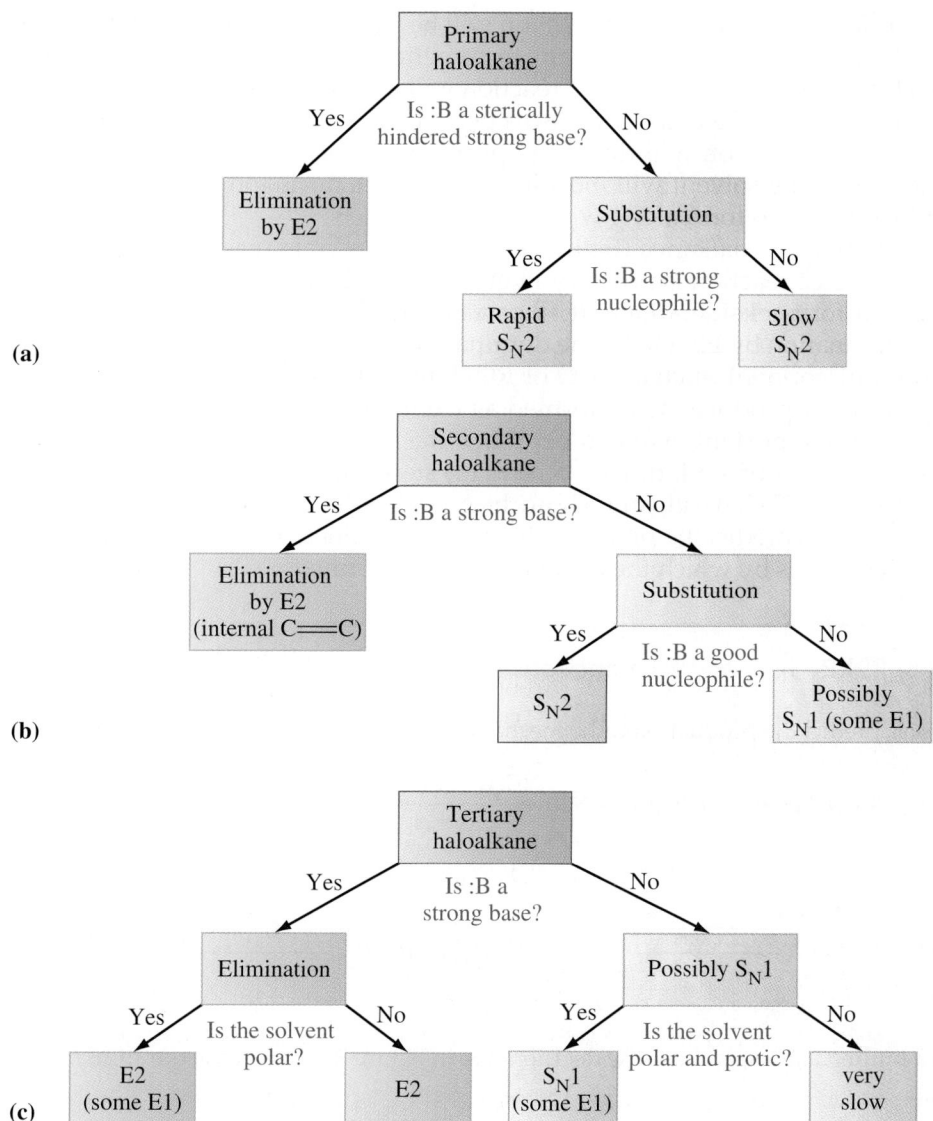

▲ FIGURE 27-13
Summary of S_N2, E2, S_N1, and E1 reactions for primary, secondary, and tertiary haloalkanes

(a) Primary haloalkanes undergo S_N2 or E2 reactions. **(b)** Secondary haloalkanes undergo S_N2, E2, S_N1, and E1 reactions. The red boxes indicate there are additional considerations. For an alkene with internal C=C bond, the possibility of (E) and (Z) stereoisomers exists. When a secondary haloalkane undergoes an S_N2 reaction, watch for an inversion of configuration, (R) ↔ (S). In polar protic solvents, a secondary haloalkane can react with a weak nucleophile to give S_N1 and E1 products. **(c)** Tertiary haloalkanes undergo S_N1, E1, and E2 reactions. Methyl halides (CH_3X) are not included because the methyl halides undergo only S_N2 reactions.

stable. If the electron donor, :B, is a sterically hindered strong base, such as R_3CO^-, then the E2 reaction will dominate because the electron donor will not be able to attack the electrophilic carbon atom from the backside. Instead, a proton will be removed from a β carbon. If :B is not sterically hindered, then it will act as a nucleophile and attack the electrophilic carbon from the backside in an S_N2 reaction. If :B is a strong nucleophile, such as I^-, ^-CN, RS^-, or RO^-, then the S_N2 reaction will occur quite rapidly. If :B is a weak nucleophile, then the S_N2 reaction will occur rather slowly.

The greatest range of possibilities occurs for secondary haloalkanes (Fig. 27-13b). If the electron donor is a strong base, then elimination by E2 is the main reaction and the major product is an alkene with an internal C=C bond. If :B is a weak base but a strong or good nucleophile, such as I^-, ^-CN,

RS⁻, or RCO₂⁻, then the main reaction will be S$_N$2, particularly if a polar aprotic solvent is used. It is important to remember that if the α carbon in the haloalkane is chiral, then the S$_N$2 reaction will invert the configuration of the α carbon. If :B is a weak base and also a poor nucleophile, such as H$_2$O or ROH, then the S$_N$1 reaction will dominate provided a polar protic solvent is used. Molecules of the solvent will most likely act as the nucleophile in this case. E1 products will also form along with the S$_N$1 products.

For tertiary haloalkanes (Figure 27-13c), the possible reactions are S$_N$1 E1, or E2. An S$_N$2 reaction is not very likely because the α carbon is too sterically hindered for backside attack. If :B is a strong base, such as R$_3$CO⁻ or RO⁻, then elimination by E2 will be the dominant reaction. If :B is a weak base (and a poor nucleophile), such as H$_2$O or ROH, then the main reaction will be S$_N$1 with some E1 products being formed as well, provided a polar protic solvent is used. It is important to remember that if the α carbon is chiral, the S$_N$1 products will consist of a mixture of (R) and (S) stereoisomers.

In Example 27-5, we illustrate how the guidelines summarized in Figure 27-13 can be used to predict the products that will be obtained in a given situation and the mechanisms by which these products will be formed.

EXAMPLE 27-5   Predicting S$_N$2, E2, S$_N$1, and E1 Reactions

For each of the following reactions, predict the products and the mechanisms by which the products are formed.

(a)
$$CH_3CH_2CH_2CH_2Br + (CH_3)_3CO^-Na^+ \xrightarrow{(CH_3)_3COH}$$

(b)

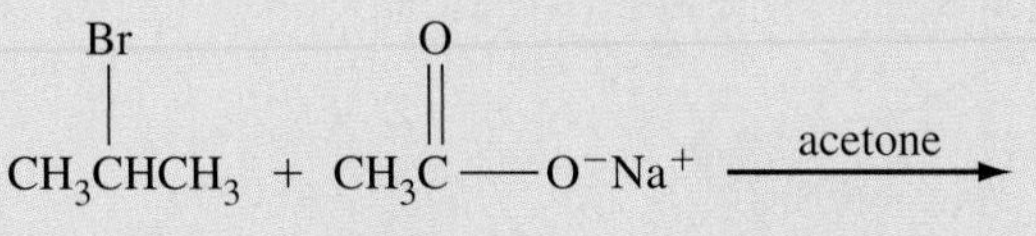

(c)

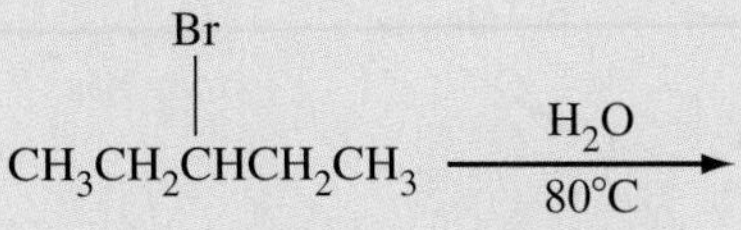

Analyze

We follow the six-step approach outlined above and use the decision tree in Figure 27-13 to guide our thinking.

Solve

(a) We identify the electrophile as CH$_3$CH$_3$CH$_2$Br, which is a primary haloalkane. The nucleophile is (CH$_3$)$_3$CO⁻, a sterically hindered strong base.

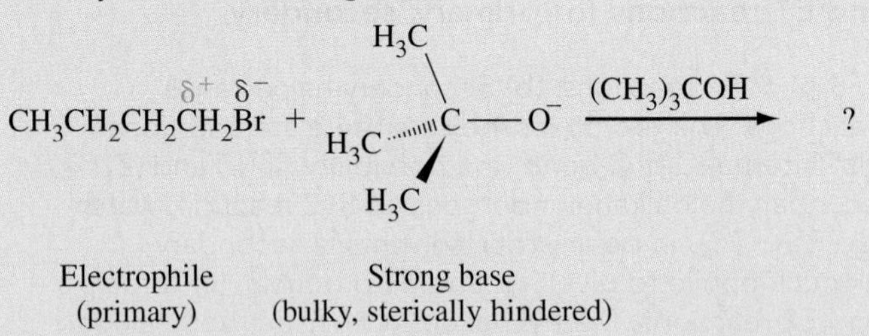

Although a strong base is usually a strong nucleophile, the bulkiness of this base disfavors substitution, and we expect that elimination by E2 will provide the major product.

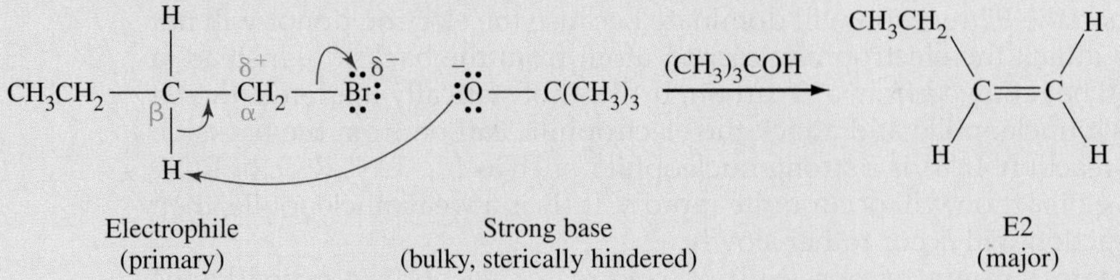

Substitution by S$_N$2 will yield CH$_3$CH$_2$CH$_3$CH$_2$OC(CH$_3$)$_3$, but the substitution product will be the minor product.

(b) The electrophile is a secondary haloalkane. The electron pair donor is a weak base (the conjugate base of a weak acid) and, because it is negatively charged, it is a reasonably good nucleophile, especially in a polar aprotic solvent. We expect that the main reaction will be S_N2.

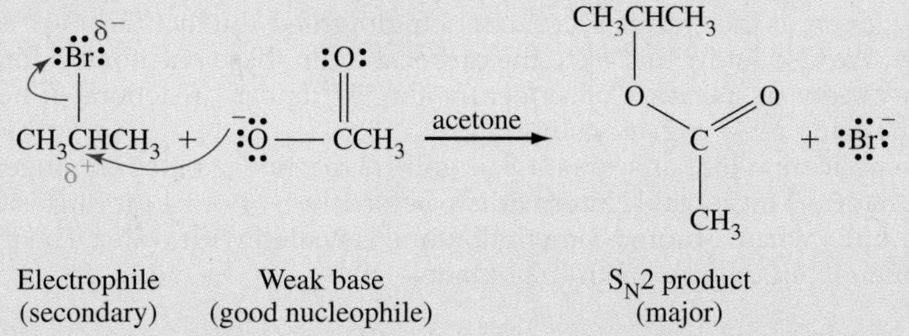

Electrophile	Weak base	S_N2 product
(secondary)	(good nucleophile)	(major)

(c) Molecules of the solvent also serve as the electron pair donor. The reaction involves a secondary haloalkane with a weak base/weak nucleophile in a polar protic solvent. These conditions do not favor S_N2 or E2 but rather S_N1 and E1. S_N1 and E1 always occur together. Secondary carbocations are relatively stable, especially in a polar protic solvent, such as water. We expect both S_N1 and E1 products. Because H_2O is a very weak base and a fair nucleophile, substitution will dominate over elimination.

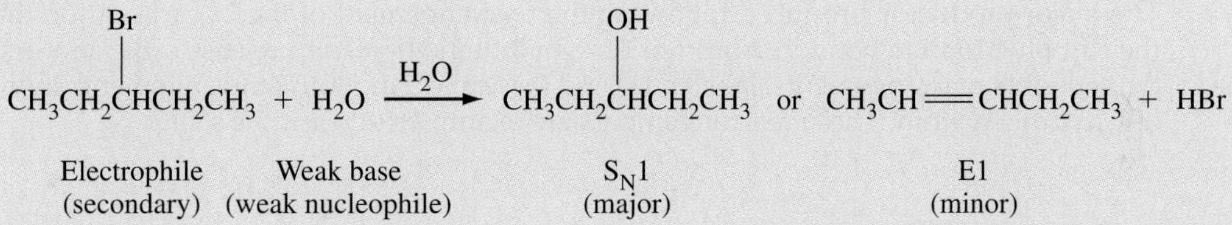

Electrophile	Weak base	S_N1	E1
(secondary)	(weak nucleophile)	(major)	(minor)

Assess

The mechanism for the reaction in **(c)** was not explicitly shown. We must make sure we are able to show the steps involved in forming the S_N1 and E1 products, including using arrows to show the movement of electrons. Also in **(c)**, (E) and (Z) isomers of 2-pentene are possible.

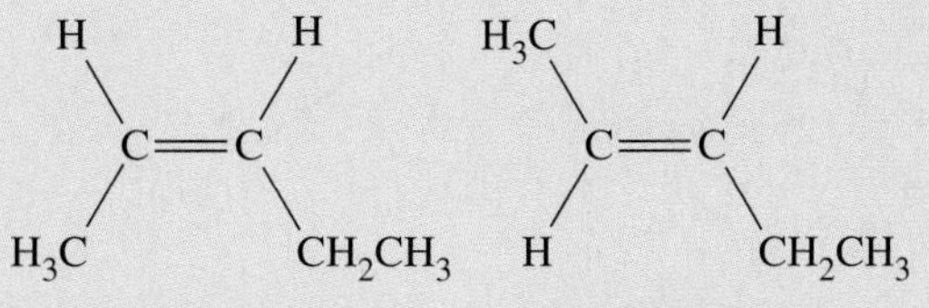

(Z)-2-pentene   (E)-2-pentene

PRACTICE EXAMPLE A: For the following reaction, predict the major product and the mechanism by which it is formed:

$$CH_3CH_2CH_2Br + CH_3S^-Na^+ \xrightarrow{\text{acetone}}$$

PRACTICE EXAMPLE B: Predict the substitution and elimination products that are possible in the following reaction. Which is the major product?

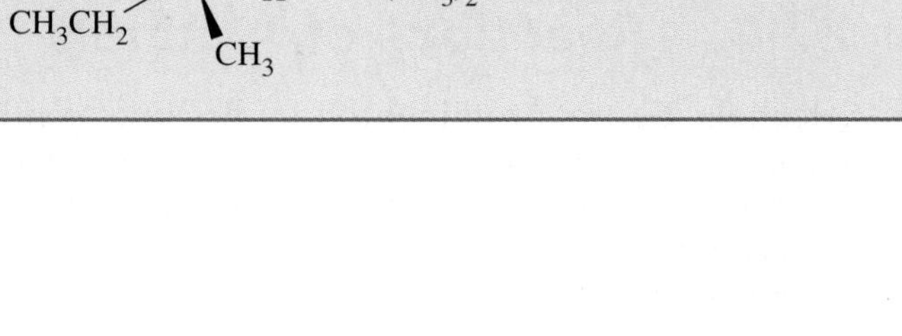

27-2 ARE YOU WONDERING...

Whether S$_N$1/E1 reactions of haloalkanes are used in organic synthesis?

The summary given at the end of Section 27-3 highlighted the fact that S$_N$1 and E1 reactions always occur together. We also know that once the carbocation in these reactions is formed, it is susceptible to attack from a variety of species. Consequently, the S$_N$1 and E1 reactions of haloalkanes are not often used for synthesizing other organic compounds.

Another complication that arises with S$_N$1 and E1 reactions is the rearrangement of the carbocation intermediate to form a more stable intermediate before the S$_N$1 or E1 product is formed. Such a situation arises, for example, when 2-bromo-3-methylbutane is heated with water. The major S$_N$1 product is not 3-methyl-2-butanol but rather 2-methyl-2-butanol.

The major product is produced following the rearrangement of the 2° carbocation that is formed when the carbon–bromine bond in 2-bromo-3-methylbutane breaks to release a Br⁻ ion. As suggested below, two possible rearrangements, labeled (a) and (b), can occur. Both rearrangements involve the movement of a hydrogen atom. These rearrangements are examples of a *hydride shift*.

Rearrangement (b) does not occur because it results in a 1° carbocation, which is highly unstable. Rearrangement (a) converts a 2° carbocation into a more stable 3° carbocation. The 3° carbocation is the one that leads to the major substitution product.

Anytime a carbocation ion forms, rearrangement of the carbocation is always a possibility. To make matters worse, the rearrangement can also involve the movement of methyl groups or alkyl chains.

Given the myriad possibilities that can arise in S$_N$1 and E1 reactions, and the difficulty in predicting or controlling what will happen, it comes as no surprise that S$_N$1 and E1 reactions are not used as often as other reactions for organic synthesis.

27-4 Reactions of Alcohols

Alcohols figure prominently in organic synthesis because they can be readily converted into other compounds. In Chapter 26, we saw that primary alcohols could be oxidized to aldehydes or carboxylic acids and secondary alcohols could be oxidized to ketones. We also mentioned that alcohols react with carboxylic acids to form esters. In Section 27-2, we saw examples of reactions in

TABLE 27.3   Some Reactions of Alcohols

Type of Reaction	Equation
Deprotonation	$ROH + Na \longrightarrow RO^-Na^+ + 1/2\,H_2$
Oxidation[a]	$ROH \xrightarrow{[O]}$ aldehyde, ketone or carboxylic acid
Esterification	$ROH + R'COOH \xrightarrow[\Delta]{H^+} R'COOR + H_2O$
Substitution	$ROH^b + R'X \longrightarrow ROR' + HX$
	$ROH^c + HX \longrightarrow RX + H_2O$
Elimination	$ROH \xrightarrow[\Delta]{H^+}$ alkenes

[a]In the equation for oxidation, [O] represents the oxidizing agent, such as $Na_2Cr_2O_7/H_2SO_4$ or PCC/CH_2Cl_2. PCC is pyridinium chlorochromate. See page 1185.
[b]ROH is acting as a nucleophile in this reaction. See Section 27-2.
[c]ROH is acting as an electrophile in this reaction.

which an alcohol (ROH) reacted as a nucleophile in a substitution reaction with a haloalkane (R′X) to form an ether (ROR′). Some of the reactions of alcohols are summarized in Table 27.3.

In this section, we focus on reactions in which the —OH group is replaced by a halogen atom (in a substitution reaction) or eliminated as H_2O (in an elimination reaction).

Substitution and Elimination Reactions of Alcohols

The —OH group of an alcohol can be replaced by a halogen atom or eliminated as H_2O under appropriate conditions. Let's consider the feasibility of the following substitution reactions by using concepts we've discussed previously in this chapter.

$$CH_3CH_2CH_2OH + NaI \longrightarrow CH_3CH_2CH_2I + NaOH \quad \text{(does not occur)}$$

$$CH_3CH_2CH_2OH + HI \xrightarrow{\Delta} CH_3CH_2CH_2I + H_2O \quad \text{(occurs slowly)} \quad \textbf{(27.5)}$$

In both of these reactions, I^- is the nucleophile and OH^- is the leaving group. Although the I^- ion is a good nucleophile, the OH^- ion is strongly basic and a poor leaving group. Consequently, the first reaction does not occur. Why does the second reaction occur? In the presence of strong acid, the oxygen atom of the —OH group is protonated, forming $R{-}^+OH_2$. For the protonated alcohol, the leaving group is H_2O instead of OH^-. Because H_2O is a much weaker base than OH^-, it is a much better leaving group. In reaction (27.5), the conditions are just right for an S_N2 reaction: the α carbon atom in $CH_3CH_2CH_2OH$ is primary (not sterically hindered) and I^- is a very good nucleophile. The mechanism of the reaction is shown in Figure 27-14. It involves a reversible protonation step followed by an S_N2 reaction. Recall that an S_N2 reaction is a concerted reaction involving backside attack at the α carbon.

KEEP IN MIND

that the weaker the base, the better it is as a leaving group. I^- is a very weak base and a good leaving group. OH^- is a strong base and a poor leaving group.

27-3   CONCEPT ASSESSMENT

Draw the structure of the transition state for the second step shown in Figure 27-14.

If a tertiary alcohol, such as $(CH_3)_3COH$, is used in place of the primary alcohol in reaction (27.5), the substitution occurring in the second step occurs by an S_N1 reaction, not by S_N2. Substitution occurs by S_N1 because the backside of

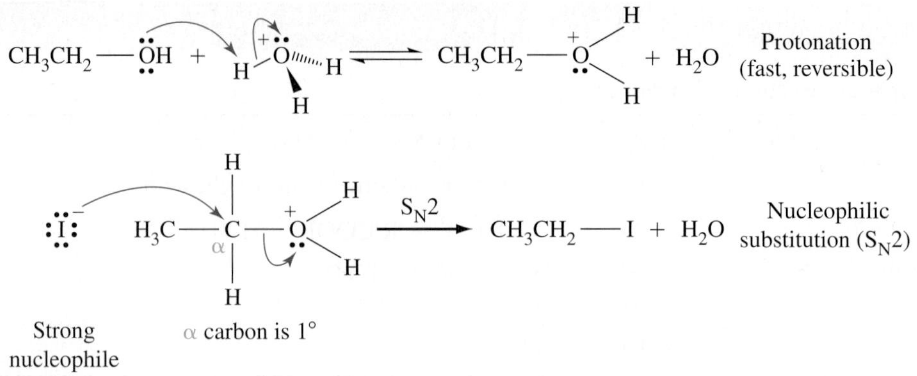

▲ FIGURE 27-14
Mechanism for the reaction of CH_3CH_2OH and HI
The first step is a reversible step in which the oxygen atom of the alcohol functional group is protonated. The second step is a nucleophilic attack of the α carbon. Because the α carbon is primary and I^- is a strong nucleophile, the substitution occurs by the S_N2 mechanism.

the α carbon in a tertiary alcohol is sterically hindered and inaccessible to a nucleophile. As shown in Figure 27-15, the substitution involves the formation of a carbocation followed by nucleophilic attack by I^-. The carbocation is stabilized by the electron-donating alkyl groups.

As we saw in the previous sections, elimination reactions compete with substitution reactions. So, it should come as no surprise that alcohols can also undergo elimination reactions. The elimination of water (H—OH) from an alcohol is an important method of synthesizing alkenes. The elimination of water from an alcohol is also called a **dehydration reaction**. The general form of a dehydration reaction is

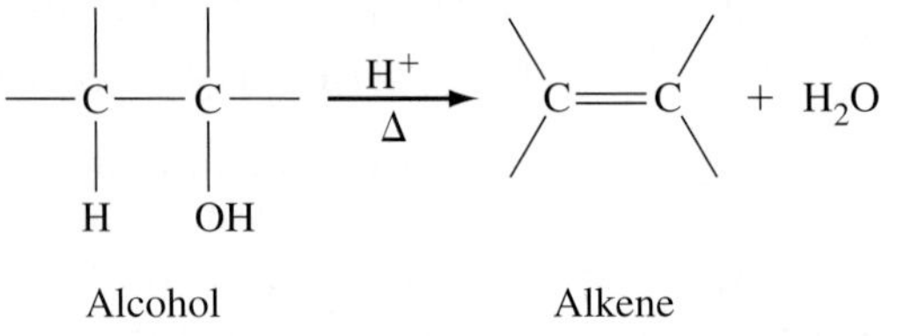

$$\tag{27.6}$$

The dehydration of an alcohol requires an acid catalyst. The acid catalyst protonates the alcohol so that the leaving group will be H_2O (a good leaving group) rather than OH^- (a poor leaving group). To promote elimination (dehydration)

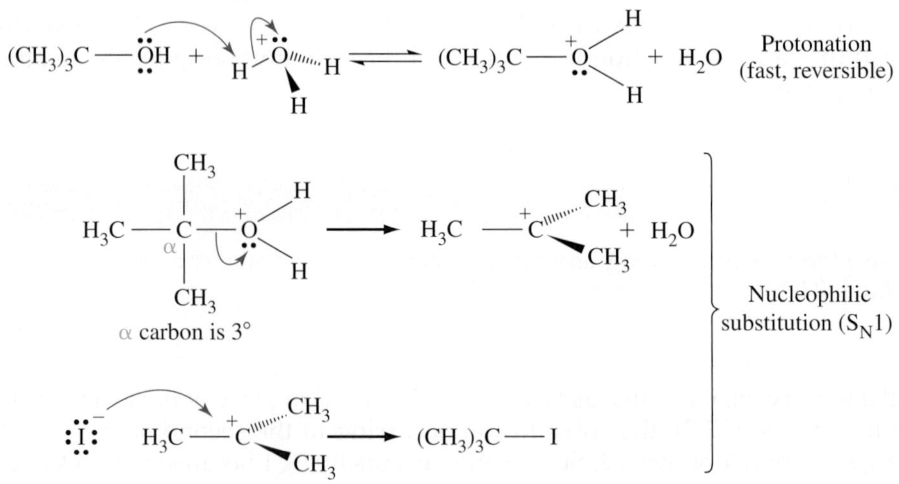

▶ FIGURE 27-15
Mechanism for the reaction of $(CH_3)_3COH$ and HI
The first step is a reversible step in which the oxygen atom of the alcohol functional group is protonated. Because the α carbon is tertiary, the backside of the α carbon is not susceptible to backside attack and so substitution by S_N2 does not occur. Instead, substitution occurs by S_N1.

over substitution, concentrated H_2SO_4 or H_3PO_4 is used rather than HI or HBr. (Remember, I^- and Br^- are strong nucleophiles, and if HI or HBr is used in place of H_2SO_4 or H_3PO_4, substitution will dominate over elimination.) Dehydration of tertiary alcohols occurs by an E1 reaction. Consider, for example, the dehydration of 2-methyl-2-butanol, a tertiary alcohol:

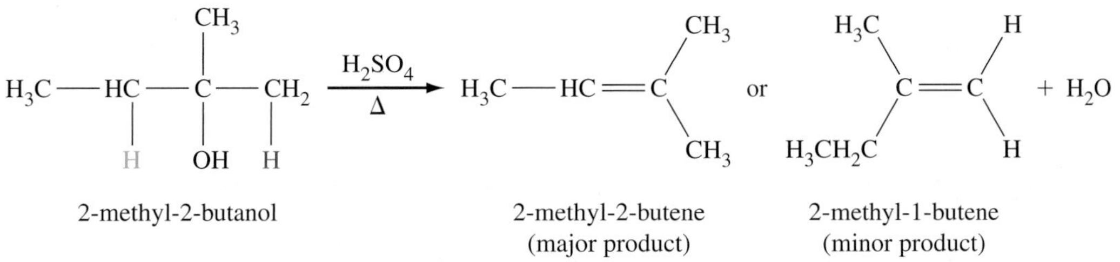

2-methyl-2-butanol → 2-methyl-2-butene (major product) or 2-methyl-1-butene (minor product) + H_2O

The mechanism for this reaction is shown in Figure 27-16. The first step (Fig. 27-16a) involves protonation of the alcohol. After protonation, dehydration occurs by E1 (Fig. 27-16b). Elimination of H—OH yields 2-methyl-2-butene and elimination of HO—H yields 2-methyl-1-butene. As discussed in Section 27-3, the major product in an elimination reaction is usually the more highly substituted alkene.

(a) Protonation of the alcohol

(b) Elimination by E1

Major elimination product (more highly substituted alkene)   Minor elimination product

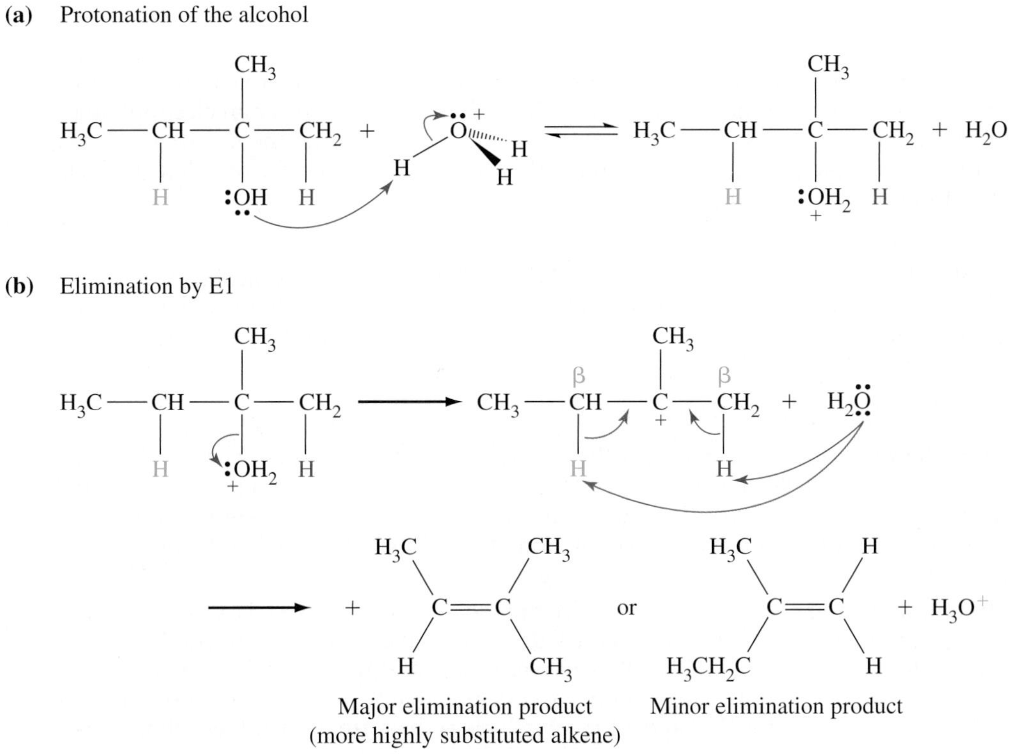

▲ FIGURE 27-16
Mechanism for the acid-catalyzed dehydration of 2-methyl-2-butanol
(a) In the presence of strong acid, the alcohol is protonated. (b) Elimination involves the formation of a carbocation, followed by the removal of a proton from one of the β carbons. The blue arrows show the movement of electrons when H is attacked. The red arrows show the movement of electrons when H is attacked. The major product is the more highly substituted alkene. Notice that the H_3O^+ ion consumed in the protonation step is regenerated in the last step.

Dehydration of secondary alcohols usually occurs by E1, but it can occur by E2. Ethanol, a primary alcohol, undergoes dehydration by a concerted E2 reaction, as shown below:

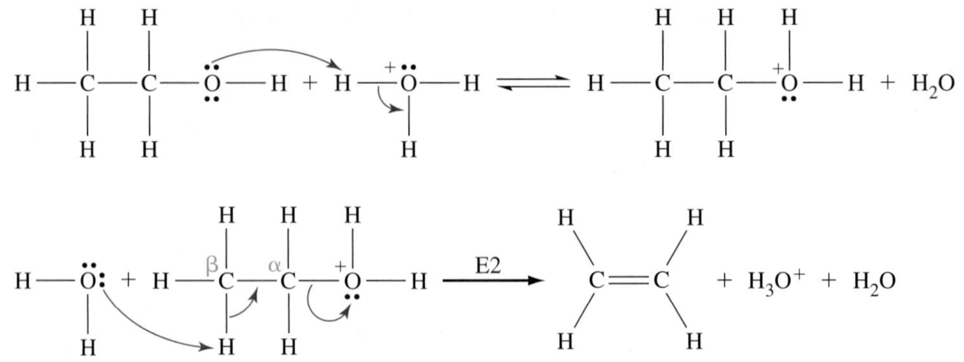

EXAMPLE 27-6 Predicting the Products of a Reaction Involving an Alcohol

Predict the products of the following reactions. If appropriate, suggest a mechanism by which the reaction occurs by using arrows to show the movement of electrons.

(a) 1-propanol + sodium hydroxide $\xrightarrow{H_2O}$

(b) (R)-2-bromo-3-methylbutane + ethanol $\xrightarrow{ethanol}$

Analyze

First, we write condensed structural formulas for the reactants, and then we consider the role of the alcohol in each case. Alcohols can be oxidized or deprotonated. They can act as either a nucleophile or an electrophile in a substitution reaction or they can be dehydrated. To make a decision, we must consider the α carbon atoms in the reactants and classify them as primary, secondary, or tertiary. We must also consider the other reactants, the solvent, and the reaction conditions.

Solve

(a) The condensed structural formula for 1-propanol is $CH_3CH_2CH_2OH$. It is a primary alcohol. Neither substitution nor elimination (dehydration) is possible under the conditions specified. We know that sodium hydroxide is a relatively strong base, and so we should consider the feasibility of the following acid–base reaction:

$$CH_3CH_2CH_2OH + NaOH \longrightarrow CH_3CH_2CH_2O^-Na^+ + H_2O$$

We must compare the strengths of the acids (or the bases) on either side of the equation. We know that ^-OH and $CH_3CH_2CH_2O^-$ are both considered strong bases, but which is stronger? Because both bases have the negative charge localized on oxygen, we expect them to be fairly similar in strength. Thus, we predict that the reaction will not go to completion. At equilibrium, significant amounts of all species will be present.

(b) The condensed structural formulas for 2-bromo-3-methylbutane and ethanol are $CH_3CHBrCH(CH_3)_2$ and CH_3CH_2OH. The carbon bonded to bromine in the haloalkane is a 2° carbon atom and Br^- is a very weak base and a good leaving group. CH_3CH_2OH is a weak base and also a weak nucleophile. The solvent is polar protic. On the basis of this information, we conclude that S_N1 and E1 reactions will occur (see Fig. 27-13). The first step in both the S_N1 and E1 mechanisms is the formation of the carbocation:

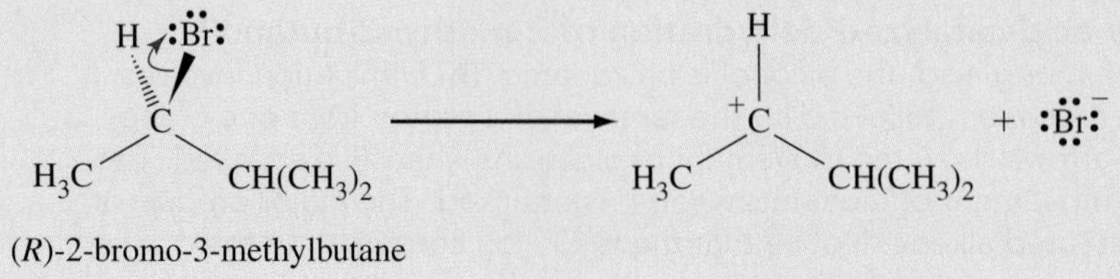

(R)-2-bromo-3-methylbutane

Once the carbocation is formed, either substitution or elimination can occur. The steps involved in the substitution are as follows:

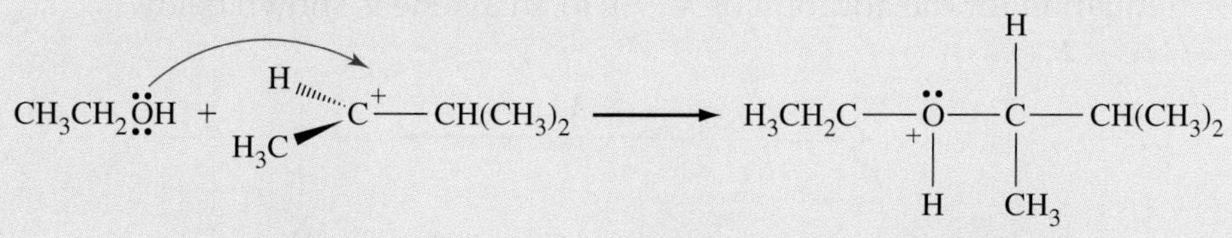

(R) and (S)

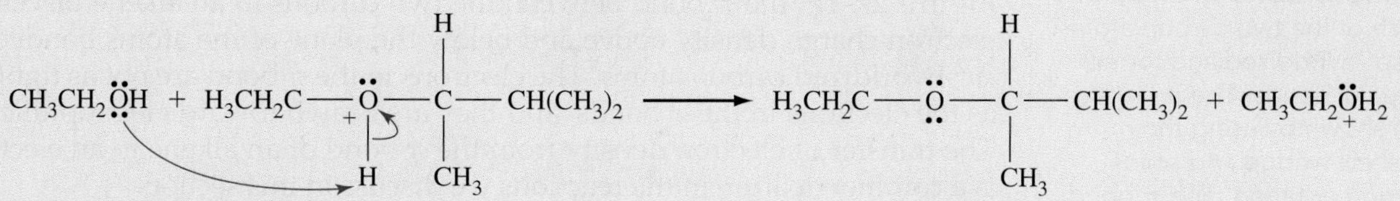

(R) and (S)

In the diagram above, the nucleophile is shown attacking the carbocation from above, but it can attack from either above or below. Consequently, both (R) and (S) stereoisomers are obtained.

The steps involved in the elimination are shown below. The elimination reaction produces two alkenes. The major product is the most highly substituted alkene.

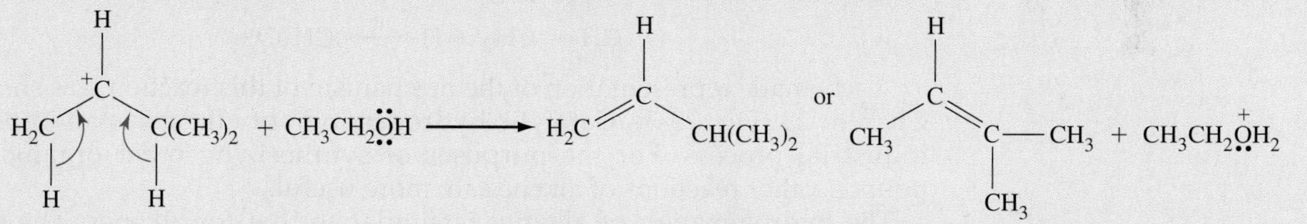

3-methyl-1-butene
(minor elimination product)

2-methyl-2-butene
(major elimination product)

Assess

As shown in **(a)**, the deprotonation of CH_3CH_2OH cannot be accomplished effectively by using NaOH. However, the deprotonation can be accomplished by using $NaNH_2$. The pK_a for NH_3 is about 34, and so pK_b for NH_2^- is about $14 - 34 = -20$. For part **(b)**, keep in mind that a carbocation can undergo a rearrangement (see Are You Wondering 27-2). Consequently, a variety of products are obtained in substitution and elimination reactions involving secondary and tertiary alcohols. We will not explore this complication.

PRACTICE EXAMPLE A: Predict the products of the following reactions. If appropriate, suggest a mechanism by which the reaction occurs, using arrows to show the movement of electrons.

(a) (R)-2-butanol $\xrightarrow[H_2SO_4]{Na_2Cr_2O_7}$

(b) 1-propanol $\xrightarrow[\Delta]{\text{conc. } H_2SO_4}$

PRACTICE EXAMPLE B: Predict the products of the following reactions. If appropriate, suggest a mechanism by which the reaction occurs, using arrows to show the movement of electrons.

(a) 3-methyl-1-butanol + hydrogen iodide $\longrightarrow$

(b) 2-methyl-2-propanol + sodium $\longrightarrow$

27-5 Introduction to Addition Reactions: Reactions of Alkenes

In the previous sections, we focused on reactions involving Lewis bases or nucleophiles in which the electron pair to be donated was a lone pair of electrons. In this section, we focus on a few reactions of alkenes, such as $(CH_3)_2C{=}CH_2$, and will see that the π bond in these molecules acts as the Lewis base or nucleophile in those reactions.

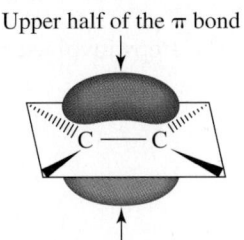

Upper half of the π bond

Lower half of the π bond

▲ FIGURE 27-17
Schematic representation of the π bond in an alkene
Each of the two carbon atoms is sp^2 hybridized and forms three σ bonds. The σ bonds are shown by using the dashed-wedge and solid-wedge notation. The π bond in an alkene places electron density above and below the plane of the atoms bonded to the carbon atoms by σ bonds.

The characteristic reaction of alkenes (and alkynes) is addition of two substituents, X and Y, to each of the carbons in the double or triple bond. In this section, we will focus primarily on addition reactions of alkenes. The general equation for the addition of X—Y to an alkene is shown below.

$$\ce{C=C} + X-Y \longrightarrow -\overset{|}{\underset{X}{C}}-\overset{|}{\underset{Y}{C}}- \qquad (27.7)$$

The region of reactivity in an alkene is the π bond. As illustrated in Figure 27-17, the π bond between the two carbons in an alkene places a high electron charge density above and below the plane of the atoms bonded to the sp^2-hybridized carbon atoms. The electrons in the π bond are not as tightly held as the electrons in the σ bonds, and they are drawn toward electrophilic atoms. The transfer of electron density from the π bond of an alkene to an electrophile is a common feature in the reactions we discuss in this section.

Addition of Hydrogen: Hydrogenation

The addition of H_2 across the double bond of an alkene yields an alkane. The reaction is very slow unless a finely divided metal catalyst, such as Ni, Pd, Pt, or Rh, is used. The equation for the hydrogenation of propene, $CH_3CH=CH_2$, by using Pt metal as a catalyst is given below:

$$CH_2=CH_2 + H_2 \xrightarrow{\text{Pt}} CH_3CH_3$$

A schematic representation of the mechanism of this reaction was shown in Chapter 23 (Fig. 23-3). Catalytic hydrogenation of alkenes is an important industrial process. For the purposes of synthesizing other organic compounds, other reactions of alkenes are more useful.

The hydrogenation of alkynes is similar to that for alkenes. The metal-catalyzed hydrogenation of an alkyne also yields an alkane, unless a special catalyst called Lindlar's catalyst is used. Lindlar's catalyst consists of barium sulfate coated with palladium metal and "poisoned" with quinoline, a heterocyclic aromatic compound shown in the margin. In Lindlar's catalyst, the catalytic activity of palladium is decreased and the hydrogenation of an alkyne stops at the alkene. The chemical equations are given below for the hydrogenation of 2-butyne by using Pt as a catalyst and Lindlar's catalyst. Notice that when a "pure" metal catalyst is used, the hydrogenation of an alkyne consumes two moles of hydrogen per mole of alkyne. When Lindlar's catalyst is used, only the *cis* (Z) stereoisomer is obtained. The two H atoms add from the same side.

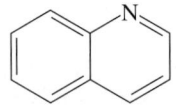

▲ Quinoline is used in Lindlar's catalyst to lower the catalytic activity of palladium metal.

$$H_3C-C\equiv C-CH_3 + 2\,H_2 \xrightarrow{\text{Pt}} CH_3CH_2CH_2CH_3$$

$$H_3C-C\equiv C-CH_3 + H_2 \xrightarrow[\text{catalyst}]{\text{Lindlar's}} \underset{\underset{H}{|}}{\overset{\overset{H_3C}{\diagdown}}{C}}=\underset{\underset{H}{|}}{\overset{\overset{CH_3}{\diagup}}{C}}$$

cis (z) isomer only

Addition of Hydrogen Halides (HX)

The addition of HX to an alkene can be represented by the following general equation:

$$\ce{C=C} + \underset{\delta^+ \quad \delta^-}{H-X} \longrightarrow H-\overset{|}{\underset{|}{C}}-\overset{|}{\underset{|}{C}}-X \qquad (27.8)$$

Alkene   Hydrogen halide   Haloalkane

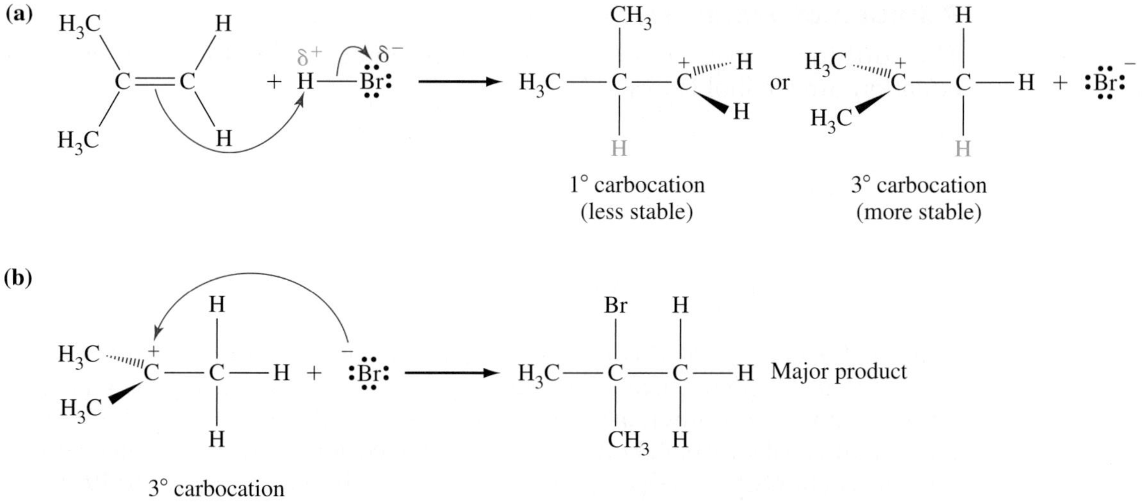

(a)

1° carbocation
(less stable)

3° carbocation
(more stable)

(b)

3° carbocation

▲ FIGURE 27-18
Mechanism for electrophilic addition of HBr to 2-methyl-1-propene
(a) The alkene acts a nucleophile and attacks the partially positive hydrogen atom of HBr. In principle, the H atom can add to either carbon atom of the double bond, but there is a strong preference for adding to the least-substituted carbon atom. When H adds to the least substituted carbon atom, the resulting carbocation is more stable.
(b) The Br⁻ ion attacks the positive carbon atom of the carbocation.

The mechanism for the reaction consists of two steps, as illustrated in Figure 27-18 for the reaction of $(CH_3)_2C{=}CH_2$ and HBr.

In the first step (Fig. 27-18a), the partially positive H atom adds to one of the carbon atoms of the double bond. This step produces a carbocation. In principle, the H atom can add to either one of the carbon atoms but, in practice, it adds preferentially to the least-substituted carbon atom. When H adds to the least substituted carbon, the positive charge of the carbocation ends up on the carbon atom most able to stabilize the positive charge, that is, on the most highly substituted carbon. (Recall that alkyl groups help to stabilize a carbocation.) In the second step (Fig. 27-18b), the Br⁻ ion acts as a nucleophile and attacks the positively charged carbon atom of the carbocation.

> When H—X adds to the double bond of an alkene, the H atom adds to the carbon atom having the smallest number of alkyl groups.

If the reaction is carried out in an aqueous solution, a mixture of products is obtained. To avoid this complication, the reaction can be carried out by bubbling HCl, HBr, or HI through the pure alkene or by carrying out the reaction in a solvent whose molecules are not nucleophilic.

27-4   CONCEPT ASSESSMENT

What are the major and minor products obtained in the reaction of (E)-3-methyl-2-pentene and HBr?

Addition of Water: Hydration

The addition of H_2O to an alkene is represented by the following general equation. An alcohol is formed.

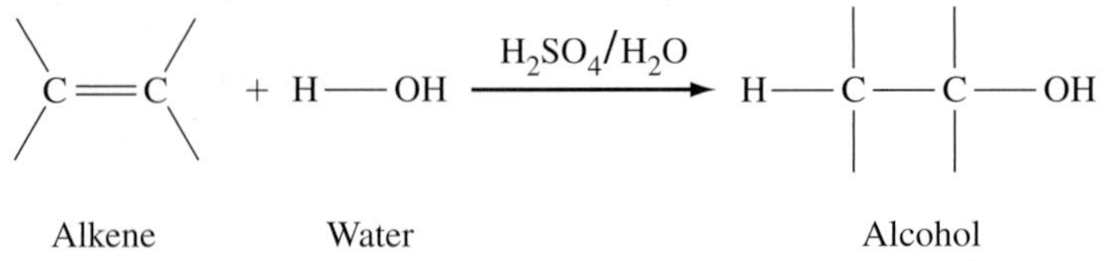

$$\ce{C=C} + \ce{H-OH} \xrightarrow{\ce{H_2SO_4/H_2O}} \ce{H-C-C-OH} \qquad (27.9)$$

Alkene   Water   Alcohol

The reaction occurs only in an acid solution and can be carried out in a mixture of H_2SO_4 and H_2O (typically 50% H_2SO_4 by volume). The mechanism is similar to the one given previously for the addition of HX to an alkene. It involves the formation of a carbocation followed by nucleophilic attack. The mechanism for the reaction of $(CH_3)_2C{=}CH_2$ and H_2O is shown in Figure 27-19. In an acidic solution, there is an excess of H_3O^+ ions, and these ions are the source of the electrophile, H^+. In the first step, an H atom from H_3O^+ adds to the carbon that has the fewest number of H atoms, yielding a carbocation (Fig. 27-19a). In the second step, a water molecule acts as a nucleophile and attacks the positively charged carbon of the carbocation (Fig. 27-19b). In the third step, a proton is removed, yielding an alcohol (Fig. 27-19c).

🔍 27-5 CONCEPT ASSESSMENT

Would hydration of (E)-2-butene yield exclusively (R)-2-butanol, (S)-2-butanol, or a mixture of the (R) and (S) enantiomers? Explain.

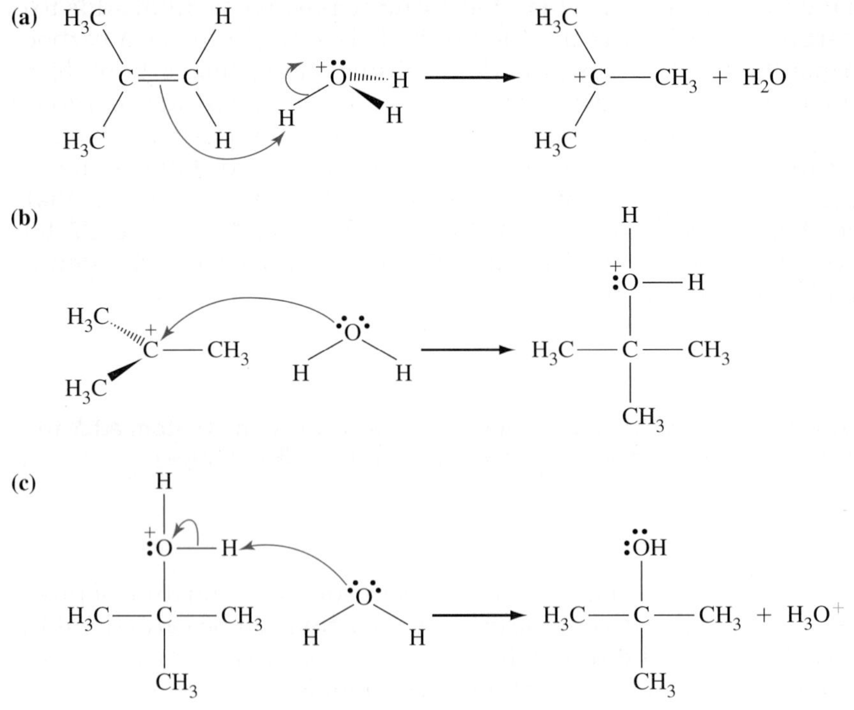

▶ FIGURE 27-19
Mechanism for the acid-catalyzed addition of H_2O to 2-methyl-1-propene
(a) In an acidic solution, an H atom from an H_3O^+ ion can add to a carbon in the double bond of an alkene. As discussed in the text, there is a preference for the H atom to add to the least-substituted carbon atom. The other carbocation that can be formed in this step is less stable and is not shown.
(b) A water molecule attacks the carbocation, forming a protonated alcohol.
(c) The protonated alcohol transfers a proton to a water molecule.

Addition of Halogens: Halogenation

When a halogen, represented by X_2, adds across the double bond of an alkene, the product is a dihalide. The halogen atoms are bonded to adjacent carbons. Such a dihalide is called a **vicinal dihalide**. (When two halogen atoms are bonded to the same carbon, the compound is a **geminal**

dihalide.) The general equation for the addition of X_2 to an alkene is shown in equation (27.10):

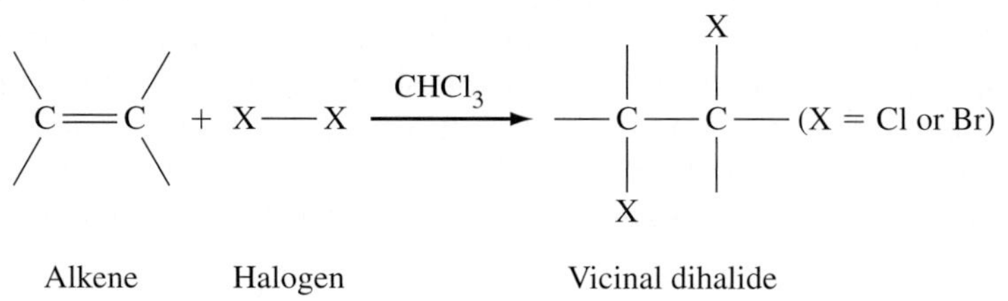

$$\underset{\text{Alkene}}{\overset{\displaystyle\mathrm{C}=\mathrm{C}}{}} + \underset{\text{Halogen}}{\mathrm{X-X}} \xrightarrow{\mathrm{CHCl_3}} \underset{\text{Vicinal dihalide}}{\overset{\displaystyle \mathrm{-C-C-}}{}} (\mathrm{X = Cl\ or\ Br}) \qquad \textbf{(27.10)}$$

The reaction above occurs readily at room temperature when the halogen is Cl_2 or Br_2. A nonaqueous solvent, such as $CHCl_3$ or CCl_4, must be used to prevent the formation of other products, including alcohols. The mechanism for the addition of X_2 to an alkene is significantly different from the mechanisms considered previously for the addition of HX and H_2O. The first step of the mechanism is the formation of a **bridged halonium ion**, rather than a carbocation (Fig. 27-20a). In the halonium ion, the halogen atom is bonded to two carbon atoms and it is the atom that carries the positive charge. Notice that the carbon atoms and the halogen atom in the halonium ion all have complete octets. Take careful note of the movement of electrons that occurs in the formation of the halonium ion. The electron pair in the π bond of the alkene is directed toward a chlorine atom. As this occurs, an electron pair from a chlorine atom is directed toward one of the carbon atoms and the Cl—Cl bond is breaking. In the next step (Fig. 27-20b), a Cl^- ion attacks a carbon atom in the bridge from the backside because the bridge blocks the carbon atoms from a frontside attack.

🔍 27-6 CONCEPT ASSESSMENT

Apply the ideas discussed above to the bromination of cyclopentene. What is the product obtained in this reaction? Specify stereochemistry, if relevant.

Carboxylic acids and their derivatives are important organic compounds. They are found widely in nature, and biological systems, and are used extensively in organic synthesis. The reactions of these compounds, and the mechanisms by which they occur, are discussed in Section 27C, *Carboxylic Acids and Their Derivatives: The Addition-Elimination Mechanism*, on the MasteringChemistry site (www.masteringchemistry.com).

Mastering**CHEMISTRY**

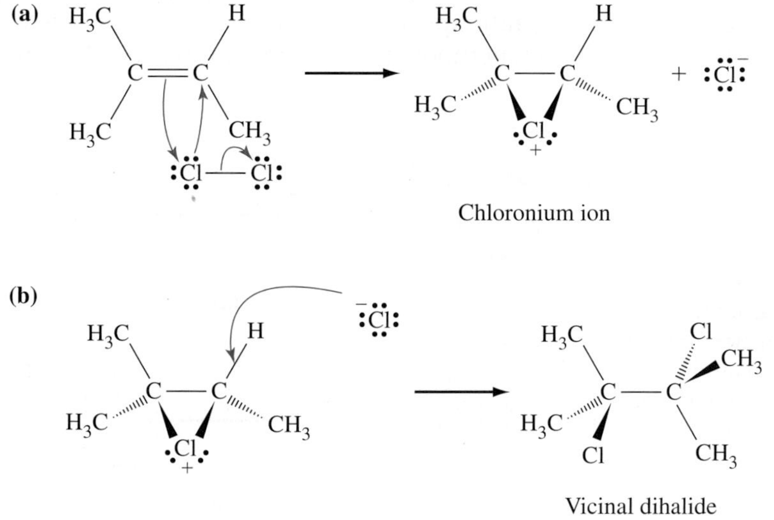

◀ FIGURE 27-20
Mechanism for chlorination of 2-methyl-1-propene
(a) A chlorine atom adds to the alkene by bonding simultaneously to both carbon atoms of the double bond. The structure formed in this step is called a chloronium ion. Notice that the chlorine atom bridges two carbon atoms and is the atom carrying the positive charge. **(b)** In the second step, a Cl^- attacks one of the carbon atoms in the bridge from the backside. A vicinal dihalide is formed.

(a) Chloronium ion

(b) Vicinal dihalide

27-6 Electrophilic Aromatic Substitution

We saw in the previous section that alkenes typically undergo addition reactions, in which substituents add across a carbon–carbon double bond. Because we often represent the structure of benzene as a six-membered ring with alternating single and double bonds, it is tempting to think that benzene would also undergo addition reactions in which substituents add across one of the carbon–carbon double bonds. As illustrated below, benzene does not undergo addition reactions:

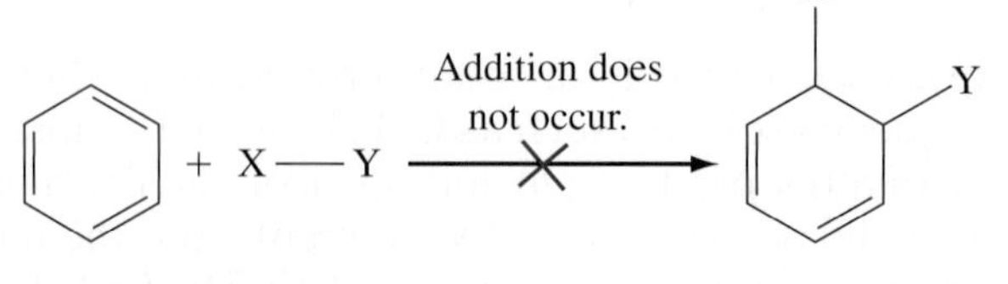

Benzene
(aromatic)

Addition does
not occur.

Hypothetical addition
product (not aromatic)

Benzene and its derivatives typically react with electrophiles in substitution reactions. The general equation for the reaction of benzene and an electrophilic species, E—Y, is shown below:

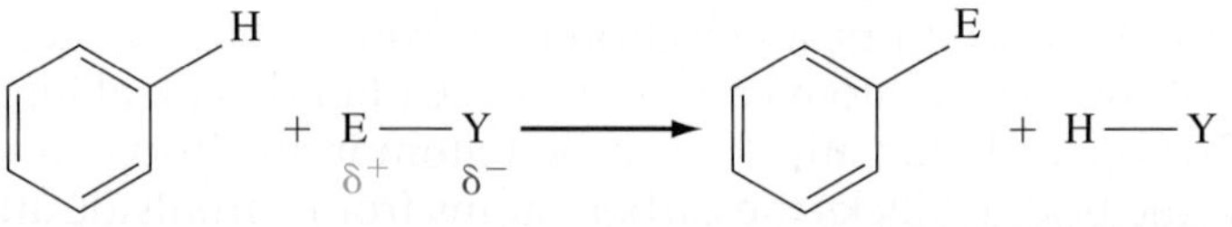

$$+ \; E \underset{\delta^+}{\text{—}} \underset{\delta^-}{Y} \longrightarrow \qquad + \; H \text{—} Y \qquad \textbf{(27.11)}$$

The reaction is an **electrophilic substitution reaction** because one of the hydrogen atoms in benzene is replaced by, or substituted with, an electrophile, E. The mechanism is shown Figure 27-21 and it involves two steps. In the first step, the electrophile accepts an electron pair from the π system of the benzene ring to form a carbocation, $C_6H_6E^+$, called an **arenium ion**. In the second step, the arenium ion loses a proton. In essence, electrophilic aromatic substitution follows an addition-elimination sequence. Notice that in the second step, Y^- acts as a base and attacks the H atom on the carbon that is bonded to E. It is tempting (but wrong) to think that Y^- would act as a nucleophile and attack the positively charged carbon atom of the arenium ion. Y^- does not attack the positively charged carbon for two reasons. First, the positive charge of the arenium ion is not localized on a particular carbon atom. It is shared equally by three carbons, as suggested in Figure 27-22, and thus these carbon atoms are only partially positive. Second, and more important, if Y^- attacks and forms a bond with one of the partially positive carbon atoms, the resulting product is not aromatic and thus is much less stable.

Various electrophilic aromatic substitutions are possible, each of which involves a different electrophilic species. We will examine a few cases, placing special consideration on the methods used for generating the electrophile.

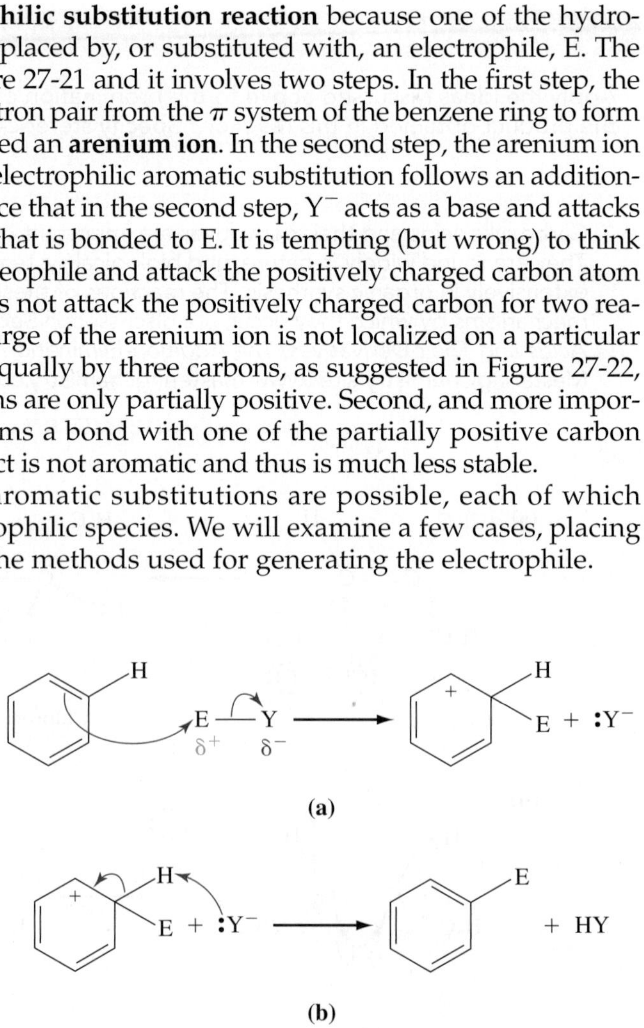

▶ FIGURE 27-21
General mechanism for electrophilic aromatic substitution of benzene
(a) The electrophile accepts an electron pair and forms a bond with a carbon atom in the benzene ring. The aromaticity of the ring is lost. (b) A proton is removed from the carbocation, restoring the aromaticity of the ring.

(a)

(b)

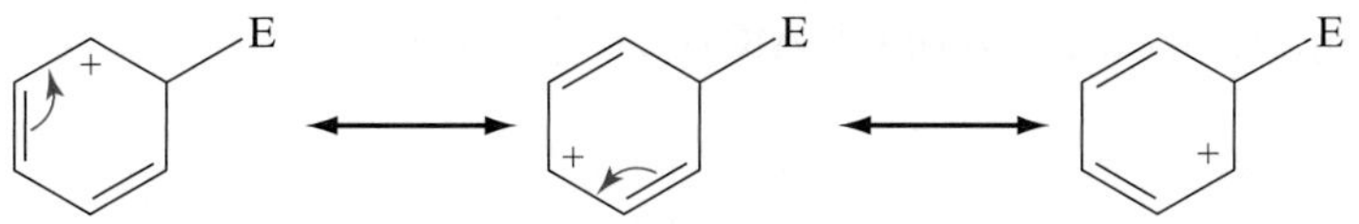

▲ FIGURE 27-22
Three equivalent contributing structures for $C_6H_6E^+$
There are three equivalent contributing structures for the $C_6H_6E^+$ carbocation. The positive charge is shared equally by three carbon atoms in the ring.

Nitration: Substitution of —H with —NO$_2$

To replace an H atom in benzene with a nitro group, —NO$_2$, benzene is treated with a mixture of sulfuric acid (H_2SO_4) and nitric acids (HNO_3). The reaction of H_2SO_4 and HNO_3 produces a nitronium ion, NO_2^+:

$$HNO_3 + 2\,H_2SO_4 \longrightarrow O{=}N^+{=}O + H_3O^+ + 2\,HSO_4^-$$

The mechanism for the nitration of benzene involves two steps. In the first step, the nitrogen atom of the nitronium ion accepts a pair of electrons from the π system of the benzene ring. A carbon–nitrogen bond is formed:

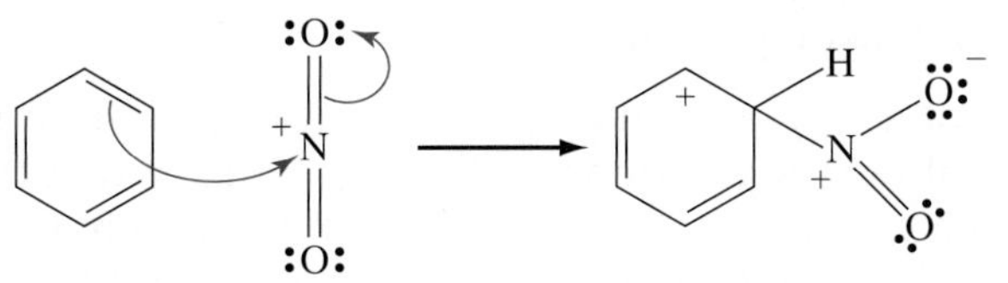

In the second step, the arenium ion is deprotonated by HSO_4^-, yielding nitrobenzene:

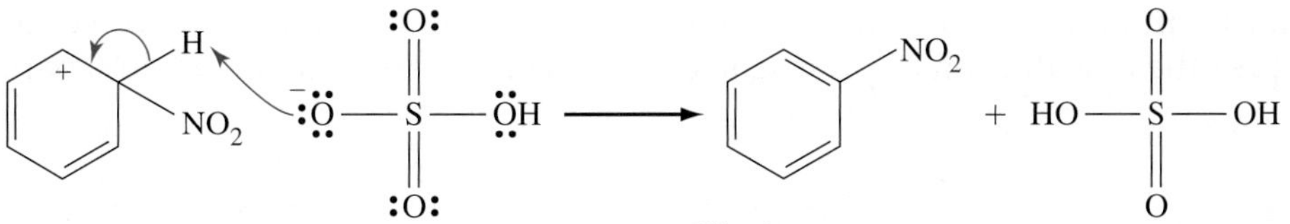

Nitrobenzene

Halogenation

Benzene can be converted into chlorobenzene or bromobenzene by treating benzene with either Cl_2 or Br_2 in the presence of an appropriate catalyst. The chemical equations for these conversions are shown below:

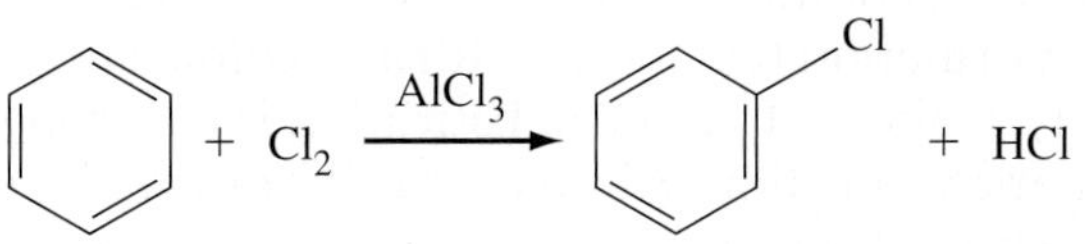

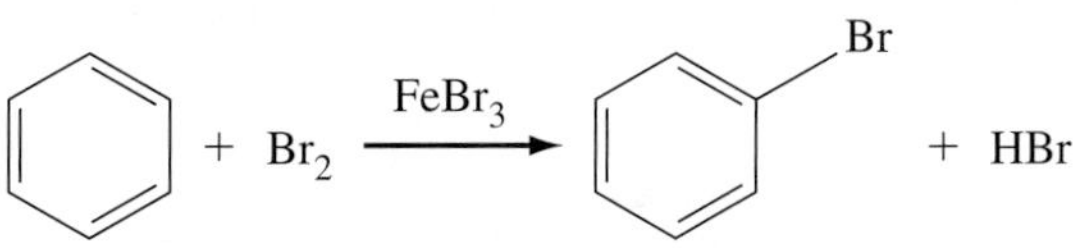

In these reactions, the catalyst reacts with Cl_2 or Br_2 to form an intermediate species that reacts with benzene. For example, Cl_2 and $AlCl_3$ react as follows:

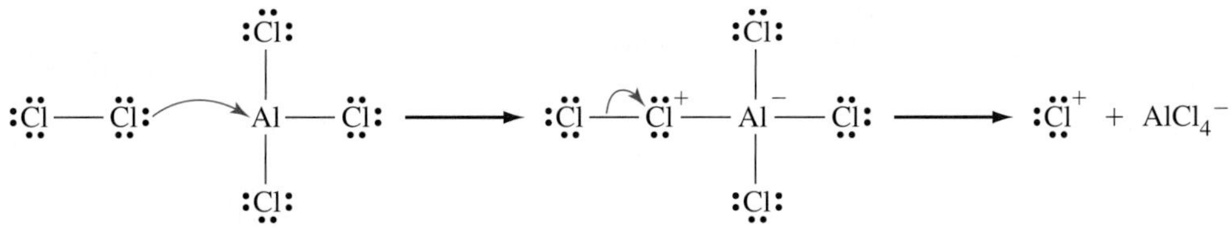

Following the formation of $Cl^+AlCl_4^-$, the chlorination of benzene proceeds as follows.

Many other substitutions are possible. For example, treating benzene with either concentrated sulfuric acid or *fuming sulfuric acid* (SO_3 in concentrated H_2SO_4) replaces a hydrogen atom with the —SO_3H group to make benzene-sulfonic acid (see Exercise 37). Methods and reagents are available for replacing an H atom with an alkyl group, R (see Exercise 38), or an acyl group, RC=O.

Benzenesulfonic acid

Ortho, Para-directing Substituents and Meta-directing Substituents

The substitution of a single atom or group, X, for an H atom in benzene can occur at any one of the six positions on the ring. We say that the six positions are equivalent. If a group Y is substituted for an H atom in C_6H_5X, this question arises: To which of the remaining five positions does the Y group go? If all the sites on the benzene ring were equally preferred, the distribution of the products would be a purely statistical one. That is, Y can be substituted in five possible positions, and we should get 20% of each one. Since two possibilities lead to an *ortho* isomer and two lead to a *meta* isomer, however, we should expect the distribution of products to be 40% ortho, 40% meta, and 20% para:

40% ortho ($\frac{2}{5}$)   40% meta ($\frac{2}{5}$)   20% para ($\frac{1}{5}$)

The following scheme describes the products resulting from nitration followed by chlorination (reaction 27.10) and chlorination followed by nitration (reaction 27.11). It shows that the substitution is *not random*. The —NO_2 group directs Cl to a meta position. Almost no *ortho* or *para* isomer is formed in reaction (27.12). The Cl group, on the other hand, is an *ortho, para* director. Essentially no *meta* isomer is produced in reaction (27.13).

$$\text{(27.12)}$$

$$\text{(27.13)}$$

▲ A Chinese rose has the aroma of black tea, primarily because of 1,3-dimethoxy-5-methylbenzene, a derivative of benzene.

Whether a group is an *ortho, para,* or *meta* director depends on how the presence of one substituent alters the electron distribution in the benzene ring. As a result, attack by a second group is more likely at one type of position than another. Examination of many reactions leads to the following order:

Ortho, para directors: —NH$_2$, —OR, —OH, —OCOR, —R, —X (X = halogen)

(from strongest to weakest)

Meta directors: —NO$_2$, —CN, —SO$_3$H, —CHO, —COR, —COOH, —COOR

(from strongest to weakest)

When two groups of the same type (both *o*-, *p*-directing or both *m*-directing) are present, the stronger director wins out. When two groups of different type (one *o*-, *p*-directing and one *m*-directing) are present, then the *o*-, *p*-directing group guides the reaction.

EXAMPLE 27-7   **Predicting the Products of an Aromatic Substitution Reaction**

Predict the products of the mononitration of

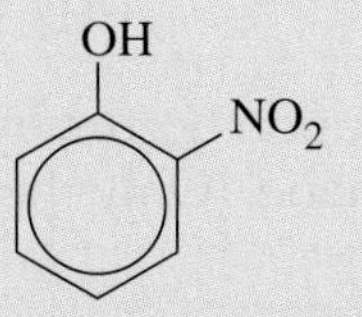

Analyze

The groups are different, and so the *o*-, *p*-directing group (—OH) guides the reaction.

Solve

Because the —OH group is an *o*-, *p*-directing group and guides the reaction, we expect the following products:

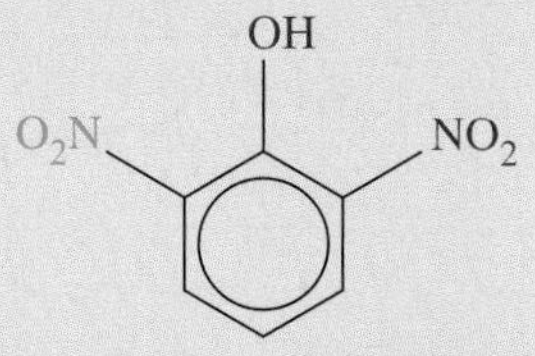

2,6-dinitrophenol 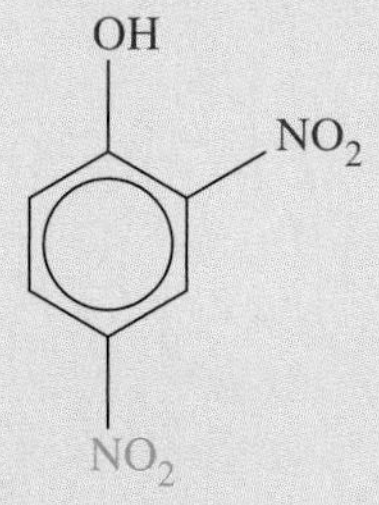   2,4-dinitrophenol

Assess

In the diagrams below, we show the flow of electron density that occurs during the formation of 2,4-dinitrophenol:

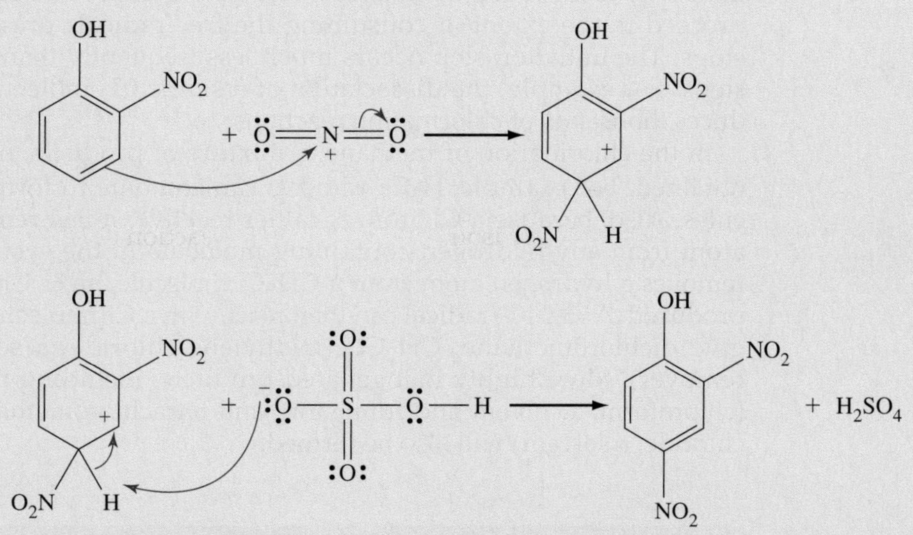

PRACTICE EXAMPLE A: Predict the major product(s) of the mononitration of benzaldehyde, C$_6$H$_5$CHO.

PRACTICE EXAMPLE B: Predict the major product(s) of the mononitration of 1,3-dichlorobenzene.

27-7 Reactions of Alkanes

Saturated hydrocarbons have little affinity for most chemical reactants. They are nonpolar substances that are insoluble in water and unreactive toward acids, bases, or oxidizing agents. The most commonly encountered reaction of alkanes is with oxygen. Alkanes burn; the *oxidation* of hydrocarbons underlies their important use as fuels. For example, octane reacts with oxygen as follows:

$$C_8H_{18}(l) + \frac{25}{2} O_2(g) \longrightarrow 8\,CO_2(g) + 9\,H_2O(l) \qquad \Delta H° = -5.48 \times 10^3 \text{ kJ} \quad \textbf{(27.14)}$$

Alkanes, however, will also react with halogens under the right conditions. For example, alkanes react only slowly with halogens at room temperature, but at higher temperatures, particularly in the presence of light, halogenation occurs. The reaction between an alkane and a halogen is a substitution reaction, with the halogen atom replacing a hydrogen atom. For example, in the reaction below, a chlorine atom replaces a hydrogen atom in methane:

$$CH_4 + Cl_2 \xrightarrow[\text{light}]{\text{heat or}} CH_3Cl + HCl \qquad \textbf{(27.15)}$$

The substitution occurs by a *chain reaction*, written as follows for the chlorination of methane. (Only the electrons involved in bond breakage or formation are shown.)

Initiation: $\quad\quad\quad\quad Cl\!:\!Cl \xrightarrow[\text{light}]{\text{heat or}} 2\,Cl\cdot$

Propagation: $\quad H_3C\!:\!H + Cl\cdot \longrightarrow H_3C\cdot + H\!:\!Cl$

$\quad\quad\quad\quad\quad\quad H_3C\cdot + Cl\!:\!Cl \longrightarrow H_3C\!:\!Cl + Cl\cdot$

Termination: $\quad\quad Cl\cdot + Cl\cdot \longrightarrow Cl\!:\!Cl$

$\quad\quad\quad\quad\quad\quad H_3C\cdot + Cl\cdot \longrightarrow H_3C\!:\!Cl$

$\quad\quad\quad\quad\quad\quad H_3C\cdot + H_3C\cdot \longrightarrow H_3C\!:\!CH_3$

The reaction is initiated when some Cl_2 molecules absorb sufficient energy to dissociate into Cl atoms (represented above as $Cl\cdot$). Cl atoms collide with CH_4 molecules to produce methyl *free radicals* ($H_3C\cdot$), which combine with Cl_2 molecules to form CH_3Cl molecules. When any or all of the last three reactions proceed to the extent of consuming the free radicals present, the reaction stops. The initiation step occurs much less frequently than the propagation steps. For example, the dissociation of a single Cl_2 molecule probably produces thousands of chlorination reactions.

In the chlorination of methane, a mixture of products, not just CH_3Cl, is obtained. For example, $H_3C\cdot$ radicals can combine to form CH_3CH_3 molecules. Also, because a Cl atom is rather reactive, it can remove a hydrogen atom from any hydrogen-containing molecule in the system. If a Cl atom removes a hydrogen atom from a CH_3Cl molecule, an $H_2ClC\cdot$ radical will be produced. A $H_2ClC\cdot$ radical can then react with a Cl_2 molecule or a Cl atom to give dichloromethane, CH_2Cl_2 (methylene chloride, a solvent and paint remover). More highly halogenated products, including trichloromethane (chloroform, a solvent and fumigant), and tetrachloromethane (carbon tetrachloride, a solvent) will also be formed.

🔍 27-7 CONCEPT ASSESSMENT

In the chlorination of methane, small amounts of chloroethane will also be produced. Suggest how the formation of chloroethane might occur.

Not all the halogens show the same reactivity with methane. Iodine is not very reactive and fluorine reacts explosively with methane unless special precautions are taken.

$$I_2 \ll Br_2 < Cl_2 \ll F_2$$
Increasing reactivity toward CH_4

How can we rationalize this trend in reactivities? Let's compare the energy changes involved in the propagation steps (Table 27.4).

The activation energy for hydrogen removal is rather small for fluorine (only 5 kJ/mol) and quite large for iodine (140 kJ mol^{-1}). Thus, the rate of hydrogen removal is highest for fluorine and lowest for iodine. However, activation energies don't tell the whole story. The enthalpy change for the halogenation process also plays a role. For example, heat released by the propagation steps is not immediately dissipated to the surroundings and is absorbed by the reacting system, causing the temperature to rise. The greater the amount of heat released, the greater the temperature rise and the greater the increase in the rate of halogenation. For the fluorination of methane, the propagation steps release a large quantity of heat ($\Delta H° = -432$ kJ mol^{-1}), leading ultimately to a large increase in the rate of halogenation, whereas in the iodination of methane, the propagation steps absorb heat.

Just as the halogens show different reactivities, so too do the hydrogen atoms in a molecule. Consider, for example, the bromination of methylpropane:

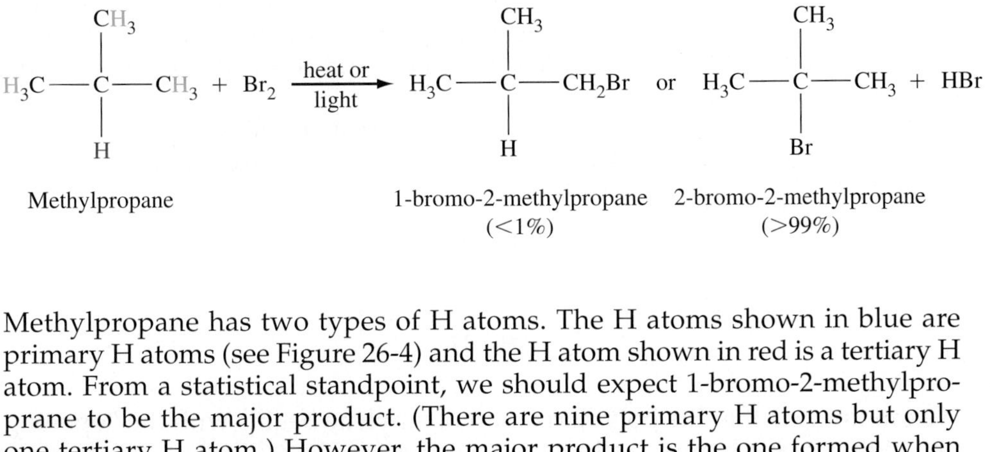

Methylpropane   1-bromo-2-methylpropane   2-bromo-2-methylpropane
                                   (<1%)   (>99%)

Methylpropane has two types of H atoms. The H atoms shown in blue are primary H atoms (see Figure 26-4) and the H atom shown in red is a tertiary H atom. From a statistical standpoint, we should expect 1-bromo-2-methylpro-prane to be the major product. (There are nine primary H atoms but only one tertiary H atom.) However, the major product is the one formed when the tertiary H atom is replaced. The selectivity of bromine for different H atoms is summarized on the next page. (Chlorine is more reactive and not as selective as bromine.)

TABLE 27.4   Activation Energy and Enthalpy Change Involved in the Halogenation of Methane

	Fluorination		Chlorination		Bromination		Iodination	
	E_a	$\Delta H°$	E_a	$\Delta H°$	E_a	$\Delta H°$	E_a	$\Delta H°$
$CH_4 + X• \longrightarrow CH_3• + HX$	5.0	-130	16	8	78	74	140	142
$CH_3• + X_2 \longrightarrow CH_3X + X•$	≈0	-302	≈0	-109	≈0	-100	≈0	-89
$CH_4 + X_2 \longrightarrow CH_3X + HX$		-432		-101		-26		53

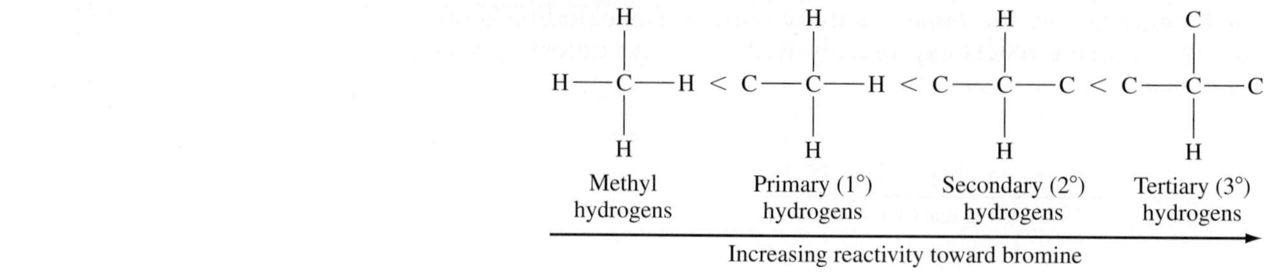

Methyl hydrogens < Primary (1°) hydrogens < Secondary (2°) hydrogens < Tertiary (3°) hydrogens

Increasing reactivity toward bromine

The trend in reactivities can be rationalized in terms of the stabilities of the radicals that are formed when a particular H atom is removed. For example, the removal of a tertiary H atom yields a 3° radical, whereas the removal of a primary H atom yields a 1° radical. Like carbocations, radicals are stabilized by alkyl groups, and so the relative stabilities of radicals follow those of carbocations: 3° > 2° > 1° > methyl.

Q 27-8   CONCEPT ASSESSMENT

What is the major product obtained in the bromination of $(CH_3)_2CHCH_2CH_3$?

27-8   Polymers and Polymerization Reactions

An Overview

Polymers are made up of simple molecules with low molecular masses joined together into extremely large molecules. Polymers with molecular masses below about 20,000 u are called low polymers and those above 20,000 u are called high polymers.

One familiar polymer is *polyethylene*. As its name implies, its basic unit, or *monomer*, is the ethylene molecule, which has the Lewis structure

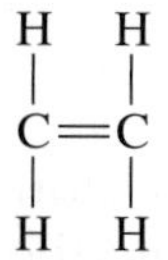

We can imagine that the polymerization of ethylene begins with the "opening up" of the double bonds in ethylene molecules.

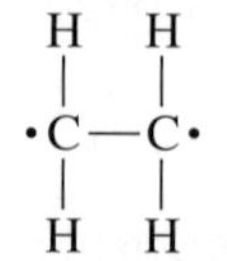

Then each C atom in the resulting molecular fragment (radical) forms an additional single covalent bond with a C atom in another molecular fragment, and so on, producing the structure shown below:

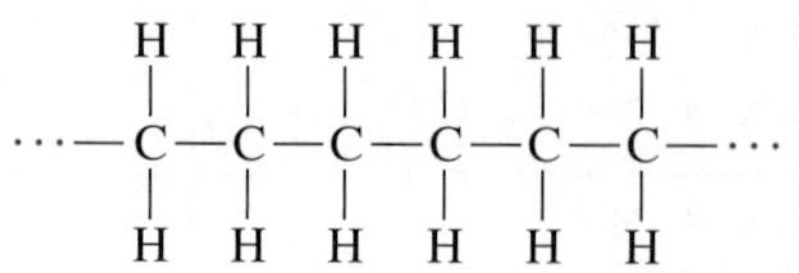

In the following notation, the monomer unit is enclosed in square brackets and the subscript n signifies the number of monomers present in the final *macromolecule*. Typically, n might range from several hundred to several thousand.

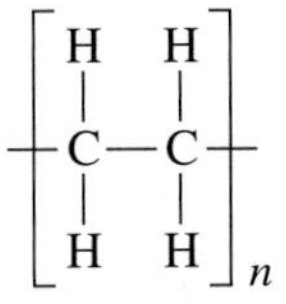

$$\left[\begin{array}{c} H \quad H \\ | \quad | \\ C - C \\ | \quad | \\ H \quad H \end{array}\right]_n$$

Another polymer in which monomer units join end to end is *latex*—natural rubber:

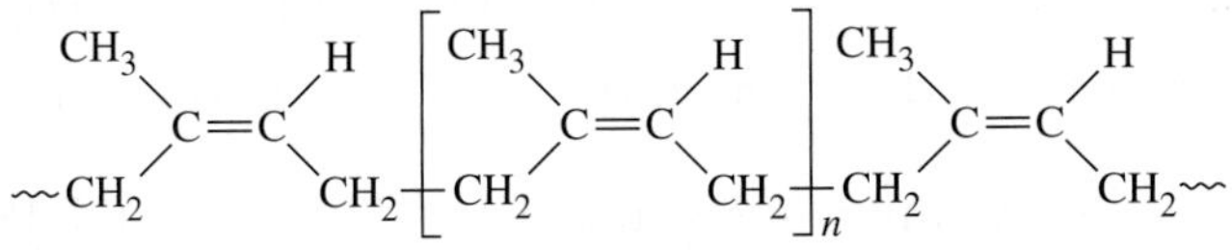

Rubber

Early rubber products were of limited use because they were sticky in hot weather and stiff in cold weather. In 1839, Charles Goodyear accidentally discovered that by heating a sulfur–rubber mixture, a product could be made that was stronger, more elastic, and more resistant to heat and cold than natural rubber. This process is now called vulcanization (after *Vulcan*, the Roman god of fire). The purpose of vulcanization is to form *cross-links* between long polymer chains. An example of a cross-link through two sulfur atoms is shown here:

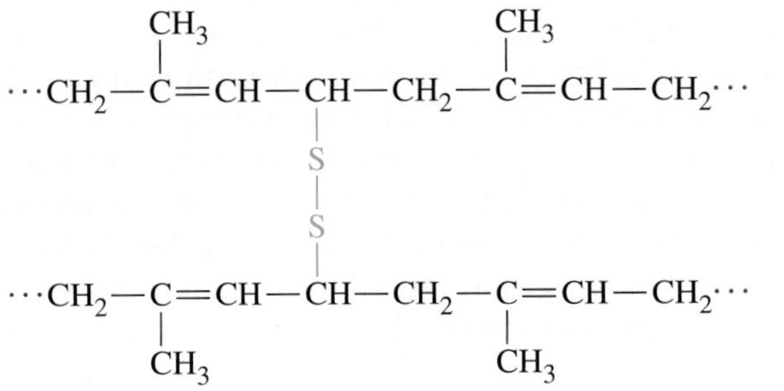

Polymers are familiar products in the modern world. Nylon, one of the first polymers developed, is like an artificial silk and is used in making clothing, ropes, and sails. The fluorine-containing polymer Teflon (polytetrafluoroethylene) is used in nonstick frying and baking pans. Polyvinyl chloride (PVC) is used in food wrap, hoses, pipes, and floor tile. In all, the polymer industry is huge. It has been estimated that about half of all chemists work with polymers. Of particular interest to them are the reactions that can be used to make polymers. We will briefly survey the main types of polymerization reactions.

Chain-Reaction Polymerization

Monomers with carbon-to-carbon double bonds typically undergo **chain-reaction polymerization**. The net result is that the double bonds open up and monomer units add to growing chains. As with other chain reactions, the mechanism involves three characteristic steps: initiation, propagation, and termination. Let us illustrate this mechanism for the formation of the polymer polyethylene from the monomer ethylene (ethene). The key to the polymerization reaction is the free-radical initiator. In reaction (27.16), an organic peroxide dissociates into two peroxy radicals. The radicals add to the double bonds of ethylene molecules to form radical intermediates that attack more ethylene molecules and form new

◀ Molecules of a *polymer* are formed by the joining together of many simple molecules called *monomers*.

▲ Extruding polyethylene film.

intermediates of longer and longer length, as in reaction (27.17). The chains terminate as a result of reactions such as (27.18) and (27.19).

Initiation:
$$R{-}O{:}O{-}R \longrightarrow 2\,R{-}O\cdot \qquad (27.16)$$
an organic peroxide

followed by

$$CH_2{=}CH_2 + RO\cdot \longrightarrow R{-}O{-}CH_2{-}CH_2\cdot$$

Propagation: $\quad ROCH_2CH_2\cdot + CH_2{=}CH_2 \longrightarrow ROCH_2CH_2CH_2CH_2\cdot \qquad (27.17)$

$$RO(CH_2)_3CH_2\cdot + CH_2{=}H_2 \longrightarrow RO(CH_2)_5CH_2\cdot$$

Termination: $\quad RO(CH_2)_xCH_2\cdot + RO\cdot \longrightarrow RO(CH_2)_xCH_2OR \qquad (27.18)$

or

$$RO(CH_2)_xCH_2\cdot + RO(CH_2)_yCH_2\cdot \longrightarrow RO(CH_2)_xCH_2CH_2(CH_2)_yOR \qquad (27.19)$$

Several polymers formed by chain-reaction polymerization are listed in Table 27.5.

Notice that the termination steps, reactions (27.18) and (27.19), produce a polymer with —OR groups at either end of the chain. The —OR groups have no effect on the properties of the polymer because the polymer has hundreds, possibly thousands, of monomer units, and the —OR groups are only at the ends of each polymer strand. The long chain of monomer units gives the polymer its particular properties, and the group serves only to initiate the reaction and to terminate the polymer chains.

Step-Reaction Polymerization

In **step-reaction polymerization**, also called *condensation polymerization*, the monomers typically have two or more functional groups that react to join the two molecules together. Usually, this involves the elimination of a small molecule, such as H_2O. In chain-reaction polymerization, the reaction of a monomer can occur only at the end of a growing polymer chain, but in step-reaction polymerization, any pair of monomers is free to join into a *dimer*; the dimer can join with a monomer to form a *trimer*; two dimers can join to form a *tetramer*; and so on. Step-reaction polymerization tends to occur slowly and produces polymers of only moderately high molecular masses (less than 10^5 u). The formation of polyethylene glycol terephthalate, Dacron, is illustrated in

TABLE 27.5 Some Polymers Produced by Chain-Reaction Polymerization

Name	Monomer	Polymer	Uses
Polyethylene	$CH_2{=}CH_2$	$-(CH_2{-}CH_2)_n-$	Bags, bottles, tubing, packaging film
Polypropylene	$CH_2{=}CHCH_3$	$\left(CH_2{-}\underset{\underset{CH_3}{\mid}}{CH}\right)_n$	Laboratory and household ware, artificial turf, surgical casts, toys
Poly(vinyl chloride) PVC	$CH_2{=}CHCl$	$\left(CH_2{-}\underset{\underset{Cl}{\mid}}{CH}\right)_n$	Bottles, floor tile, food wrap, piping, hoses
Poly(tetrafluoroethylene), Teflon	$CF_2{=}CF_2$	$-(CF_2{-}CF_2)_n-$	Bearings, insulation, nonstick surfaces, gaskets, industrial ware
Polystyrene	$CH_2{=}CH$ (phenyl)	$\left(CH_2{-}CH\text{(phenyl)}\right)_n$	Packaging, refrigerator doors, cups, ice buckets, and coolers (as foam)

TABLE 27.6   Some Polymers Produced by Step-Reaction Polymerization

Name	Monomer	Polymer	Uses
Poly(ethylene glycol terephthalate) (Dacron)	$HOCH_2CH_2OH$ and 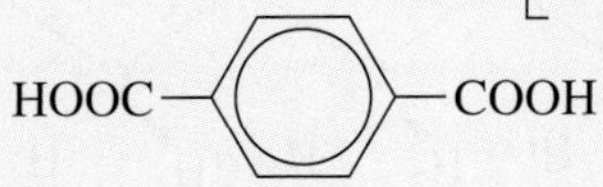 $HOOC-\bigcirc-COOH$	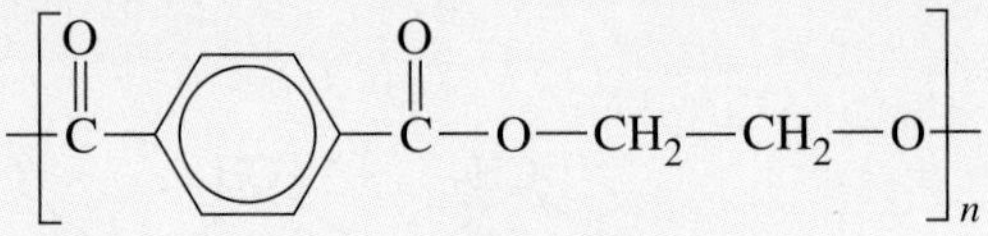	Textile fabrics, twine and rope, fire hoses, plastic containers
Poly(hexamethyleneadipamide) nylon 66	$H_2N(CH_2)_6NH_2$ and $HOOC(CH_2)_4COOH$	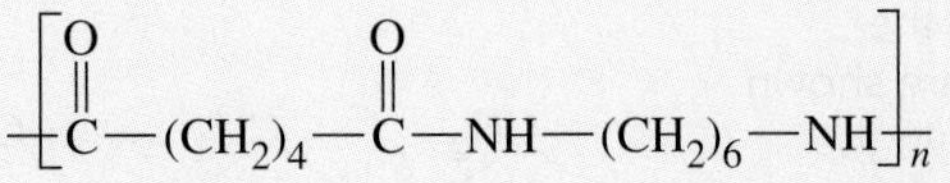	Hosiery, rope, tire cord, parachutes, artificial blood vessels
Polyurethane	$HO(CH_2)_4OH$ and $OCN(CH_2)_6NCO$	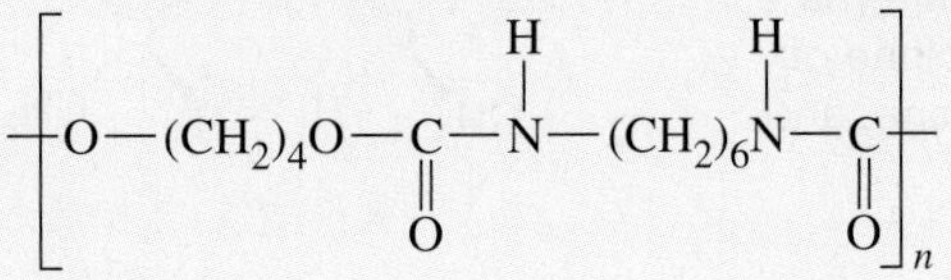	Spandex fibers, bristles for brushes, cushions and mattresses (as foam)

the following reaction, and several other polymers formed by this method are listed in Table 27.6.

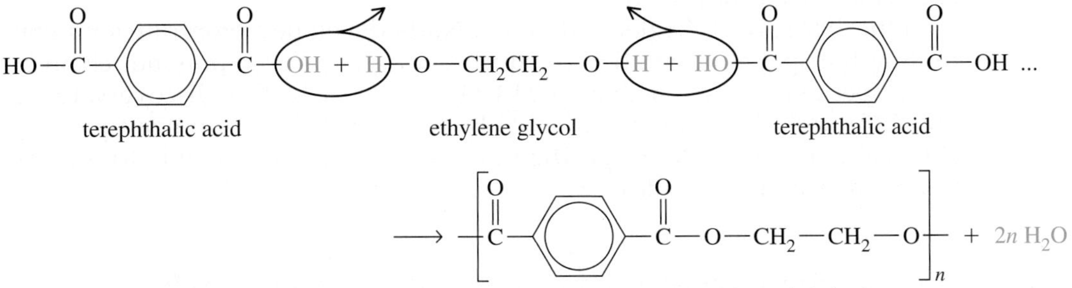

Stereospecific Polymers

The physical properties of a polymer are determined by a number of factors, such as the average length (average molecular mass) of the polymer chains and the strength of intermolecular forces between chains. Another important factor is whether the polymer chains display any crystallinity—that is, an ordered geometry and spacing of the atoms between polymer chains. In general, amorphous polymers are glass-like or rubbery. A high-strength fiber, on the other hand, must possess some crystallinity. Many polymers have both crystalline and amorphous regions. The relative amount of each type of region affects the physical properties of the polymer.

The usual designation of a polymer, as applied to polypropylene, for example, is not very revealing about the structure of the polymer:

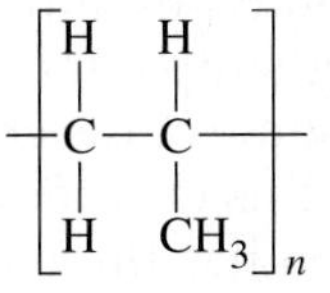

It does not indicate the orientation of the groups along the polymer chain. If propylene is polymerized by the method shown for ethylene on page 1252, the orientation of the groups is random (Fig. 27-23). A polymer of this type is called *atactic*. Because there is no regularity to the structure, atactic polymers are amorphous. In an *isotactic* polymer, all —CH_3 groups have the same orientation, and

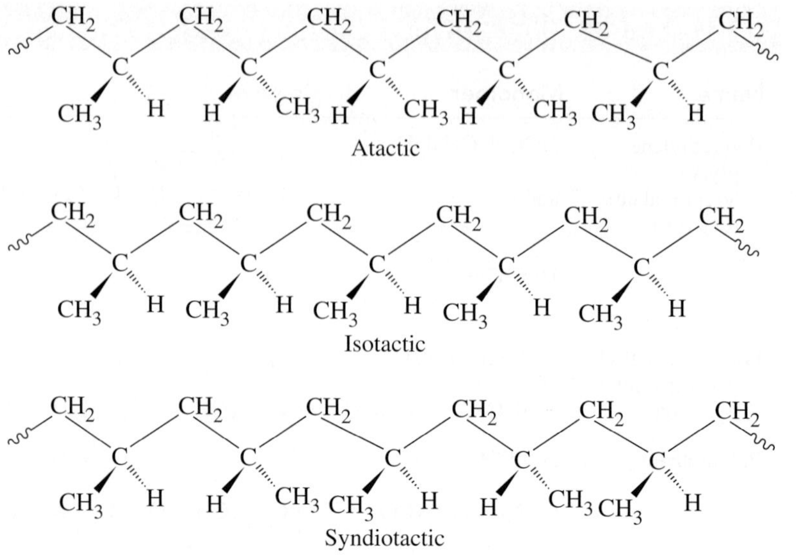

Three representations of a polypropylene chain
In the *atactic* polymer, the —CH₃ groups are randomly distributed on the chain. In the *isotactic* polymer, all —CH₃ groups are shown coming out of the plane of the page. In the *syndiotactic* polymer, —CH₃ groups alternate along the chain—one in front of the plane of the page, the next one behind the plane, and so on.

in a *syndiotactic* polymer, the —CH₃ groups alternate back and forth along the chain. Because of their structural regularity, isotactic and syndiotactic polymers possess crystallinity, which makes them stronger and more resistant to chemical attack than an atactic polymer.

In the 1950s, Karl Ziegler and Giulio Natta developed procedures for controlling the spatial orientation of substituent groups on a polymer chain by using special catalysts, such as $(CH_3CH_2)_3Al + TiCl_4$. This discovery, recognized through the award of a Nobel Prize in 1963, revolutionized polymer chemistry. Through stereospecific polymerization, it is possible to literally tailor-make large molecules.

27-9 Synthesis of Organic Compounds

Originally, all organic compounds were isolated from natural sources. However, as chemists developed an understanding of the chemical behavior of organic compounds, they began to devise methods of synthesizing compounds from simple starting materials. Moreover, some of the compounds synthesized by the newly developed methods had never been observed to occur naturally.

In organic synthesis, chemists attempt to transform simple, readily available compounds into more complex molecules with desirable physical and chemical properties. Some syntheses are designed to make biologically active compounds that are otherwise available only from natural sources and in only small quantities at high cost. Other synthetic approaches are designed to make new compounds, similar to naturally occurring ones but even more powerful in biological activity, such as medications to fight disease.

The approach that a synthetic organic chemist takes is to apply a knowledge of a wide variety of reaction types and reaction mechanisms to devise a synthetic scheme for assembling simple molecules into more complex structures. Figure 27-24 summarizes some of the chemical transformations between functional groups considered in this chapter (and Chapter 26). This summary stresses that each type of reaction has three components: the starting materials, the products, and the required reagents. For any reaction, we can complete a chemical equation if we have information about two of the three components of the reaction and if we know the type of reaction that converts the starting materials to the products. For example, in the following nucleophilic substitution reaction,

$$CH_3Br + ? \longrightarrow CH_3CN + ?$$

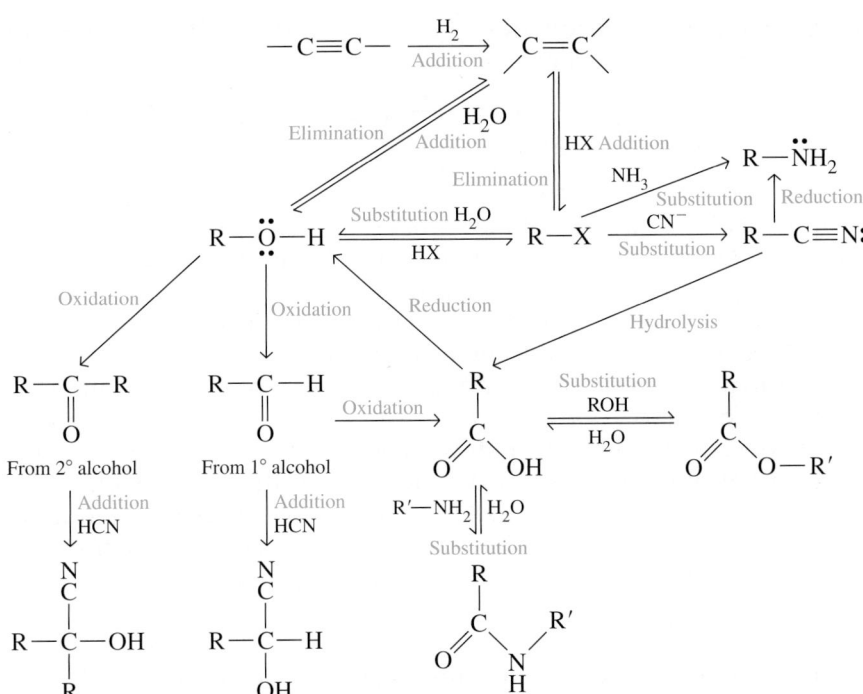

Some functional group transformations
A summary of some important functional group transformations, each of which requires a specific reagent as described in the text or on www.masteringchemistry.com. When using this diagram, remember that changes in the R group are not indicated because this depends on the particular group involved. Recall that the symbols 1°, 2°, and 3° stand for primary, secondary, and tertiary, respectively.

we see that CN has been substituted for Br as a functional group. To achieve this transformation, the nucleophile CN^- is required, and we complete the equation:

$$CH_3Br + CN^- \longrightarrow CH_3CN + Br^-$$

A somewhat different, and more common, situation is that represented by

$$? + Br^- \longrightarrow CH_3Br + ?$$

where we need to work backward to decide on the electrophile in a nucleophilic substitution reaction to get the desired product, CH_3Br. The electrophile will need to be connected to the leaving group (for example, another halogen atom) in the same manner and place as the Br atom will be in the product. An appropriate complete equation is

$$CH_3I + Br^- \longrightarrow CH_3Br + I^-$$

Let's apply the approach outlined above to the synthesis of ethyl ethanoate, starting only from inorganic substances and carbon (coke). First, we break the desired compound down into its constituent molecules, namely ethanol and ethanoic acid, which can be combined to give the ester:

$$CH_3COOH + HOCH_2CH_3 \longrightarrow H_2O + CH_3COOCH_2CH_3$$

We therefore need to be able to produce ethanol, which can, in turn, be converted into ethanoic acid:

$$HOCH_2CH_3 \xrightarrow{\ ?\ } H_3COOH$$

The reagent required to oxidize ethanol is $K_2Cr_2O_7$ in acidic solution. We now need a route to the synthesis of ethanol from carbon. One inorganic reagent that we have at our disposal that contains a —OH group is water. We therefore write the equation

$$? + H_2O \longrightarrow HOCH_2CH_3$$

A consideration of the reactions in Figure 27-24 indicates that one possibility is the addition of water to the double bond in ethene in the presence of sulfuric acid as a catalyst. Thus, we write

$$H_2C{=}CH_2 + H_2O \xrightarrow{H_2SO_4} HOCH_2CH_3$$

To continue, the ethene molecule has to be generated. Ethene is related to ethyne by a hydrogenation reaction:

$$H-C\equiv C-H + H_2 \xrightarrow[\text{heat and pressure}]{\text{Pt}} H_2C=CH_2$$

Our source of carbon is the element carbon and we need a way of producing a molecule with a carbon-chain length of two. The production of calcium carbide and its subsequent hydrolysis to give ethyne is described on page 1177. We have now completed the description of the desired synthesis. The underlying strategy is to break the desired molecule down into its constituent molecules and work backward, using the chemical transformations available. This approach is known as *retrosynthesis* and is the method of choice in modern synthetic organic chemistry.

Often, in the design of a synthetic pathway, one or more routes are possible. For example, in the synthesis described above, one of the steps required the synthesis of ethanol. We chose to make it by adding H_2O across the double bond of ethene. However, we could have taken another approach. For example, we could have used the following substitution reaction:

$$CH_3CH_2Cl + OH^- \longrightarrow CH_3CH_2OH + Cl^-$$

To obtain the chloroethane required for this reaction, we could substitute a Cl atom for an H atom in CH_3CH_3 which, in turn, can be produced by the hydrogenation of ethyne.

How do we decide which synthetic pathway is best? Frequently, the chosen pathway is the one that produces the greatest yield of the desired product. However, other factors may need to be considered. What are the yields of the various reactions involved? Are the required reagents readily available and what is their cost? Are the required reagents or intermediate compounds toxic or dangerous, and, if so, are there methods or equipment in place to be able to work with these compounds? Clearly, knowledge of the physical and chemical properties of organic compounds figure prominently in the design of synthetic pathways.

Mastering **CHEMISTRY**   **www.masteringchemistry.com**

Ionic liquids are organic salts with room temperature melting points. They are environmentally friendly, are able to dissolve most organic molecules, have high thermal stabilities and are nonflammable—properties that make these liquids attractive as solvents for organic synthesis. For a discussion of ionic liquids, go to the Focus On feature for Chapter 27, Green Chemistry and Ionic Liquids, on the MasteringChemistry site.

Summary

27-1 Organic Reactions: An Introduction— Organic compounds undergo a variety of reactions. In a **substitution reaction**, an atom, an ion, or a group in one molecule is replaced by another. In an **elimination reaction**, atoms or groups that are bonded to adjacent atoms are removed as a small molecule. In an **addition reaction**, a molecule adds across a double or triple bond in another molecule. In a **rearrangement reaction**, the carbon skeleton (or constitution) of a molecule is rearranged.

27-2 Introduction to Nucleophilic Substitution Reactions— A common reaction involving sp^3-hybridized carbon atoms is a **nucleophilic substitution reaction**, in which a **nucleophile** attacks the **electrophilic**

carbon atom of the **substrate** yielding the product and a **leaving group**. The leaving group is the substituent that is displaced from the electrophilic carbon atom during the reaction. The nucleophile contains a pair of electrons that seek the positive carbon atom of the other reactant. **Basicity** is a measure of the tendency of an electron pair donor to react with a proton. The **nucleophilicity** of an electron pair donor depends on a number of factors, including whether the nucleophile is neutral or negatively charged; the electronegativity and size of the nucleophilic atom; whether the nucleophilic atom is sterically hindered; and the solvent. Two mechanisms for nucleophilic substitution reactions are S_N1 and S_N2. The S_N1 mechanism involves two steps: (1) a unimolecular rate-determining step involving the formation of a carbocation, followed by (2) nucleophilic attack of the carbocation (Figs. 27-3 and 27-4). S_N1 reactions occur most rapidly for reactions involving tertiary substrates and weak nucleophiles reacting in a polar **protic solvent**. Methyl and primary substrates do not undergo S_N1 reactions (Table 27.2). The S_N2 mechanism is a concerted bimolecular reaction in which a nucleophile attacks the electrophilic carbon atom from the backside (Figs. 27-1 and 27-2). S_N2 reactions occur most rapidly for reactions involving methyl and primary substrates and good or strong nucleophiles reacting in a polar **aprotic solvent**. If the electrophilic carbon atom is chiral, and if both the nucleophile and leaving group are high in the R/S priority order, then backside attack (S_N2) leads to an inversion of configuration ($R \leftrightarrow S$) at the chiral carbon.

27-3 Introduction to Elimination Reactions—Under appropriate conditions, haloalkanes undergo elimination reactions to form alkenes. Usually, the major elimination product is the one that is the most highly substituted. An **E1 reaction** is an elimination reaction with a unimolecular rate-determining step, involving carbocation formation, followed by removal of a proton from a β carbon atom (Figs. 27-10 and 27-11). An **E2 reaction** is a concerted, bimolecular reaction (Fig. 27-12). Elimination reactions compete with nucleophilic substitution reactions. S_N1 and E1 reactions always occur together. The use of a strong base favors elimination by E2 and the use of a weak base favors substitution by either S_N1 or S_N2 (Fig. 27-13).

27-4 Reactions of Alcohols—Alcohols undergo a variety of reactions (Table 27.3). Like haloalkanes, alcohols can undergo substitution or elimination reactions. These reactions require the use of an acid catalyst (Figs. 27-14 and 27-15). When an alcohol acts as the substrate in a substitution reaction, the —OH group is replaced by another group. When an alcohol acts as the substrate in an elimination reaction, an alkene and a water molecule are produced (reaction 27.6). The elimination of water from an alcohol is also called a **dehydration reaction**.

27-5 Introduction to Addition Reactions: Reactions of Alkenes—The characteristic reaction of alkenes is the addition of two substituents to each of the carbon atoms in the double bond (reaction 27.7). A variety of addition reactions are possible: hydrogenation; addition of hydrogen halides (reaction 27.8); hydration (reaction 27.9), producing a **vicinal dihalide** or a **geminal dihalide**; and halogenation (reaction 27.10). In these reactions, a transfer of electron density occurs from the π bond of the alkene to an electrophilic atom. In all these reactions—except halogenation—the electrophilic atom tends to add to the least-substituted carbon atom of the double bond, yielding a carbocation that has the positive charge localized on the more highly substituted carbon atom. In a halogenation reaction, a **bridged halonium ion** is formed in which a halogen atom is bonded to both carbon atoms (Fig. 27-20). The halonium ion is then attacked from the backside by a halide ion.

27-6 Electrophilic Aromatic Substitution—Benzene reacts with electrophiles in substitution reactions in which any of the six H atoms is replaced by another atom or group. In a nitration reaction, —H is replaced by —NO_2; in a halogenation reaction, —H is replaced by a halogen atom (—X). Nitration and halogenation of benzene are examples of an **electrophilic substitution reaction** (reaction 27.11). An electrophilic substitution reaction involves two steps: (1) an electrophilic attack of a carbon atom in the benzene ring, forming a carbocation called an **arenium ion**; and (2) the loss of a proton to a base (Fig. 27-21). Derivatives of benzene can also undergo substitution reactions. A substituent that is bonded to the carbon atom in a benzene ring can be *ortho, para* directing, or *meta* directing.

27-7 Reactions of Alkanes—Alkanes and other hydrocarbons react with oxygen during combustion to form carbon dioxide and water (reaction 27.14). Alkanes are relatively unreactive except with the halogens, which can substitute for hydrogen atoms in a substitution reaction (reaction 27.15). The halogenation of alkanes occurs by a chain reaction that involves initiation, propagation, and termination steps. Among the halogens, bromine shows the greatest selectivity for which hydrogen atom in an alkane is replaced: $3° > 2° > 1° > $ methyl.

27-8 Polymers and Polymerization Reactions—Polymers are a very important class of substances in modern life. One method of polymerization, the joining of monomers to form polymers, is called **step-reaction polymerization**, or condensation polymerization. It produces polymers of moderate molecular mass. Another type of polymerization mechanism is **chain-reaction polymerization**, a three-step process involving initiation (reaction 27.16), propagation (reaction 27.17), and termination (reaction 27.18), producing polymers of high molecular mass.

27-9 Synthesis of Organic Compounds—The synthesis of larger and more complex organic compounds from simpler precursors can be achieved through a sequence of reactions of different types and varied mechanisms. Figure 27-24 summarizes a number of functional group transformations that can be used in designing synthetic strategies. One powerful strategy is to work backward from the desired product to the starting materials, a method known as retrosynthesis.

Integrative Example

The female of a species of worm produces the sex attractant *spodoptol*, which has the following structure.

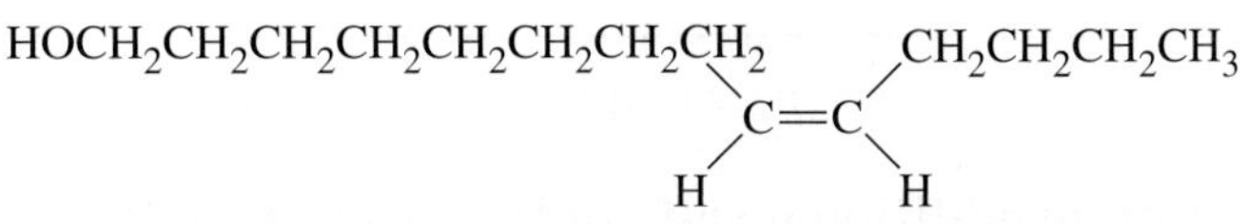

$$\text{HOCH}_2\text{CH}_2\text{CH}_2\text{CH}_2\text{CH}_2\text{CH}_2\text{CH}_2\text{CH}_2 \qquad \text{CH}_2\text{CH}_2\text{CH}_2\text{CH}_3$$
$$\underset{\text{H}}{\overset{}{\text{C}}} = \underset{\text{H}}{\overset{}{\text{C}}}$$

Spodoptol

Spodoptol is a pheromone that attracts the male worms of the species. Synthetic spodoptol in traps can be used to control the population of worms. Devise a synthesis of spodoptol starting from the alcohol shown below.

$$HOCH_2CH_2CH_2CH_2CH_2CH_2CH_2CH_2C\equiv CH$$

Analyze

In designing the synthesis, we break down the spodoptol molecule into the starting material and the necessary alkyl group that has to be attached to a double bond, namely a *n*-butyl group. In order to produce the alkene skeleton structure, we will have to convert the triple bond in the starting material to a double bond with *cis* stereochemistry.

Solve

Working backward, this *cis* isomer is obtained by using Lindlar's catalyst (recall page 1240).

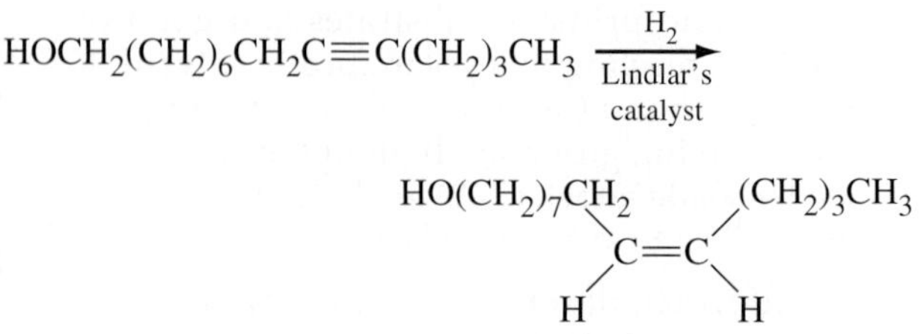

$$HOCH_2(CH_2)_6CH_2C\equiv C(CH_2)_3CH_3 \xrightarrow[\text{Lindlar's catalyst}]{H_2}$$

$$HO(CH_2)_7CH_2 \qquad (CH_2)_3CH_3$$
$$\diagdown C=C \diagup$$
$$H \qquad H$$

Now we need to make the alkyne needed in the above hydrogenation. The starting material provided lacks the butyl group at the triple bond. To attach the alkyl group, we can perform a nucleophilic substitution at 1-bromobutane with the acetylide ion generated from the starting material, by using sodium amide:

$$^-O(CH_2)_8C\equiv C:^- + CH_3(CH_2)_3Br \longrightarrow$$
$$^-O(CH_2)_8C\equiv C(CH_2)_3CH_3 + Br^-$$

Because the alcohol proton of the acetylide is also acidic, we need to add an *excess* of sodium amide to give the combined alkoxide and acetylide. (Ammonia is also produced in the reaction.)

$$HO(CH_2)_8C\equiv CH \xrightarrow[\text{NaNH}_2]{\text{excess}} {}^-O(CH_2)_8C\equiv C:^-$$

We can now combine these steps into a complete synthesis where we have converted the alkoxide to the alcohol in the next-to-last step by adding ethanol. The ethanol gives up a proton to the alkoxide we have synthesized (the ethoxide ion is the weaker base).

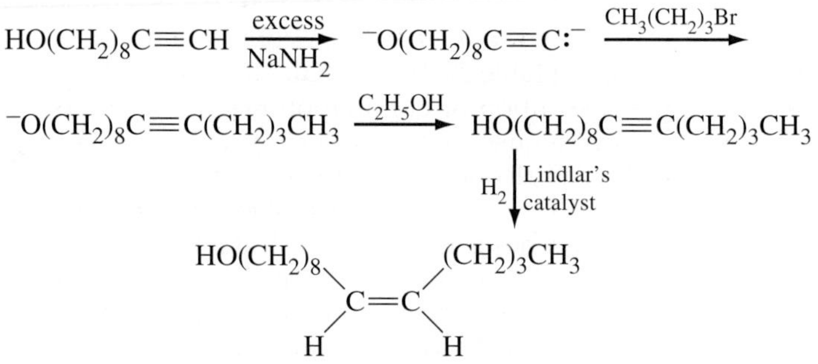

$$HO(CH_2)_8C\equiv CH \xrightarrow[\text{NaNH}_2]{\text{excess}} {}^-O(CH_2)_8C\equiv C:^- \xrightarrow{CH_3(CH_2)_3Br}$$

$$^-O(CH_2)_8C\equiv C(CH_2)_3CH_3 \xrightarrow{C_2H_5OH} HO(CH_2)_8C\equiv C(CH_2)_3CH_3$$
$$\Big\downarrow {}^{H_2}\ \text{Lindlar's catalyst}$$

$$HO(CH_2)_8 \qquad (CH_2)_3CH_3$$
$$\diagdown C=C \diagup$$
$$H \qquad H$$

Assess

By using the backward synthetic approach—retrosynthesis—we have established a possible synthetic pathway for the synthesis of spodoptol from the starting material. This is a common approach in the development of a synthetic route to a target compound. A proposed synthesis should always be carefully examined to assess whether side reactions are possible because they can decrease the yield of the desired product. In the present case, a side reaction is possible in the second step above. The $—O(CH_2)_8CC—$ ion could also react with $CH_3(CH_2)_3Br$ to give an ether.

PRACTICE EXAMPLE A: How could 1,3-cyclohexadiene be synthesized from cyclohexane?

PRACTICE EXAMPLE B: How could the compound at right be prepared from benzene by using electrophilic aromatic substitution reactions discussed in this chapter?

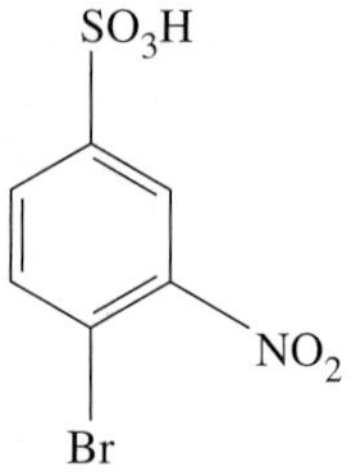

(structure: benzene ring with SO_3H, NO_2, and Br substituents)

Mastering**CHEMISTRY**

You'll find a link to additional self study questions in the study area on
www.masteringchemistry.com

Exercises

Types of Organic Reactions

1. Describe what is meant by each of the following reaction types, and illustrate with an example: **(a)** nucleophilic substitution reaction; **(b)** electrophilic substitution reaction; **(c)** addition reaction; **(d)** elimination reaction; **(e)** rearrangement reaction.

2. Describe what is meant by each of the following reaction types, and illustrate with an example from the text: **(a)** dehydration; **(b)** hydrolysis; **(c)** solvolysis; **(d)** hydration of an alkene.

3. Identify the following types of reactions.

(a)

$$\text{(structure) Cl} + \text{CH}_3\text{S}^- \longrightarrow \text{(structure) S}$$

(b)

$$\text{(structure)} + \text{H}_2 \xrightarrow{\text{Pt}} \text{(structure)}$$

(c)

$$\text{(structure) CO}_2\text{H} + \text{(structure) OH} \longrightarrow \text{(structure) CO}_2$$

4. Identify the following types of reactions.

(a)

$$\text{(structure) NH}_2 + \text{(structure) Cl} \longrightarrow \text{(structure) } \overset{\text{H}}{\underset{}{\text{N}}}$$

(b)

$$\text{(structure) OH} + \text{(structure) Cl} \longrightarrow \text{(structure) O}$$

(c)

$$\text{(structure) OH} \xrightarrow{\text{PCC, CH}_2\text{Cl}_2} \text{(structure) CHO}$$

Note: PCC is pyridinium chlorochromate (see page 1185).

5. Write a balanced chemical equation for the reaction that is described and then classify the reaction as a substitution, an elimination, an addition, or a rearrangement reaction:
   (a) Ethene and Br_2 react in carbon tetrachloride to give 1,2-dibromoethane.
   (b) Iodoethane reacts with KOH(aq) yielding ethene, water, and potassium bromide.
   (c) Chloromethane reacts with NaOH(aq) to give methanol and sodium chloride.

6. Write a balanced chemical equation for the reaction that is described and then classify the reaction as a substitution, an elimination, an addition, or a rearrangement reaction.
   (a) 3,3-dimethyl-1-butene reacts in acid solution to yield 2,3-dimethyl-2-butene.
   (b) 1-iodo-2,2-dimethylpropane reacts with water, 2,2-dimethyl-1-propanol, and HI(aq).
   (c) 2-chloro-2-methylpropane reacts with NaOH(aq) to give 2-methyl-1-propene, sodium chloride, and water.

Substitution and Elimination Reactions

7. Write equations for the substitution reaction of *n*-bromobutane, a typical primary haloalkane, with the following reagents: **(a)** NaOH; **(b)** NH_3; **(c)** NaCN; **(d)** CH_3CH_2ONa.

8. Write equations for the substitution reaction of *n*-bromopentane, a typical primary haloalkane with the following reagents: **(a)** NaN_3; **(b)** $N(CH_3)_3$; **(c)** $CH_3CH_2C\equiv CNa$; **(d)** CH_3CH_2SNa.

9. Answer the following questions for this S_N2 reaction:

$$CH_3CH_2CH_2CH_2Br + NaOH \longrightarrow$$
$$CH_3CH_2CH_2CH_2OH + NaBr$$

   (a) What is the rate expression for the reaction?
   (b) Draw the reaction profile for the reaction. Label all parts. Assume that the products are lower in energy than the reactants.
   (c) What is the effect on the rate of the reaction of doubling the concentration of *n*-butyl bromide?
   (d) What is the effect on the rate of the reaction of halving the concentration of sodium hydroxide?

10. Answer the following questions for this S_N1 reaction:

$$CH_3CH_2CH_2\overset{\overset{\textstyle CH_3}{|}}{\underset{\underset{\textstyle CH_3}{|}}{C}}Br + CH_3CH_2OH \longrightarrow$$

$$CH_3CH_2CH_2\overset{\overset{\textstyle CH_3}{|}}{\underset{\underset{\textstyle CH_3}{|}}{C}}OCH_2CH_3 + HBr$$

   (a) What is the rate expression for the reaction?
   (b) Draw the reaction profile for the reaction. Label all parts. Assume that the products are lower in energy than the reactants.
   (c) What is the effect on the rate of the reaction of doubling the concentration of 1-bromo-1-methylpentane?
   (d) The solvent for the reaction is ethanol. What is the effect on the rate of the reaction of adding more ethanol?

11. Answer the following questions for this E2 reaction:

$$CH_3CH_2CH_2Br + NaOH \longrightarrow$$
$$CH_3CH=CH_2 + NaBr + H_2O$$

(a) What is the rate expression for the reaction?
(b) Draw the reaction profile for the reaction. Label all parts. Assume that the products are lower in energy than the reactants.
(c) What is the effect on the rate of the reaction of doubling the concentration of $CH_3CH_2CH_2Br$?
(d) What is the effect on the rate of the reaction of halving the concentration of NaOH?

12. Answer the following questions for this E1 reaction:

$$CH_3CH_2C(CH_3)_2Br + CH_3OH \longrightarrow$$
$$CH_3CH=C(CH_3)_2 + Br^- + CH_3OH_2^+$$

(a) What is the rate expression for the reaction?
(b) Draw the reaction profile for the reaction. Label all parts. Assume that the products are lower in energy than the reactants.
(c) What is the effect on the rate of the reaction of doubling the concentration of $CH_3CH_2C(CH_3)_2Br$?
(d) What is the effect on the rate of the reaction of doubling the concentration of CH_3OH?

13. Identify the nucleophile, electrophile, and leaving group in each of the following substitution reactions. Predict whether equilibrium favors the reactants or products:
(a) $CH_3CH_2ONa + CH_3CH_2CH_2I \rightleftharpoons$
$$CH_3CH_2CH_2OCH_2CH_3 + NaI$$
(b) $CH_3CH_2NH_3^+ + KI \rightleftharpoons$
$$NH_3 + CH_3CH_2I + K^+$$

14. Identify the nucleophile, electrophile, and leaving group in each of the following substitution reactions. Predict whether equilibrium favors the reactants or products:
(a) $CH_3OH + I^- \rightleftharpoons CH_3OH + NaOH$
(b) $(CH_3)_2CHCl + KCN \rightleftharpoons (CH_3)_2CHCN + KCl$

15. Molecule (a) below reacts faster in the S_N2 reaction than does molecule (b). Explain this observation. [*Hint*: Draw chair structures for the most stable conformations of the two molecules.]

H₃C⋯ ⋯CH₃   H₃C⋯ CH₃

I   I
(a)   (b)

16. Molecule (a) below reacts faster in the S_N1 reaction than does molecule (b). Explain this observation. [*Hint*: Draw chair structures for the two molecules.]

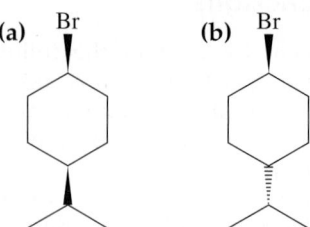

(a) Br   (b) Br

17. A sample of (S)-$CH_3CH_2CH(CH_3)Cl$ is hydrolyzed by water, and the resulting solution is optically inactive. (a) Write the formula of the product. (b) By which nucleophilic substitution reaction mechanism does this reaction occur?

18. A sample of (R)-$CH_3CH_2CH(CH_3)Cl$ reacts with CH_3O^- in dimethyl sulfoxide, $(CH_3)_2SO$, a convenient solvent for organic reactions. The resulting solution is optically active. (a) Write the formula of the product. (b) By which mechanism does this nucleophilic substitution reaction occur?

19. A sample of (S)-$CH_3CH(Cl)CH_2CH_3$ is dissolved in ethanol, and the resulting solution is optically inactive. (a) Write the formula of the product. (b) By which mechanism does this nucleophilic substitution reaction occur?

20. A sample of (R)-$CH_3CH(Cl)CH_2CH_3$ reacts with CH_3S^- in dimethyl sulfoxide, and the resulting solution is optically active. (a) Write the formula of the product. (b) By which mechanism does this nucleophilic substitution reaction occur?

21. Give the major products obtained in each of the following reactions and indicate which mechanisms are involved:
(a)

CH₂Cl + KOC(CH₃)₃ $\xrightarrow{(CH_3)_3COH}$

H

(b)

Cl + NaI $\xrightarrow{propanone}$

(c) (S)-$(CH_3)_3CCHBrCH_3 + CH_3CH_2OH \longrightarrow$

22. Give the major products obtained in each of the following reactions and indicate which mechanisms are involved:
(a)

Br + NaNH₂ $\xrightarrow{NH_3(l)}$

(b) $(CH_3)_2CHCH_2CH_2CH_2Br +$
$$NaOCH_2CH_3 \xrightarrow{CH_3CH_2OH}$$
(c) (R)-2-iodobutane + NaHS $\xrightarrow{DMSO}$

Alcohols and Alkenes

23. Predict the major organic product obtained in each of the following reactions. Assume that [O] represents $Na_2Cr_2O_7$ in H_2SO_4:
(a) $(CH_3)_2CHCH_2OH + HBr \longrightarrow$

(b) $(CH_3)_3COH + K \longrightarrow$

(c) $(CH_3)_2CHOH \xrightarrow{[O]}$

(d) $CH_3CH_2CH_2OH + CH_3CH_2I \longrightarrow$

24. Predict the products of the following reactions. Assume that [O] represents $Na_2Cr_2O_7$ in H_2SO_4:
    (a) $(CH_3)_2CHCH_2OH \xrightarrow{[O]}$
    (b) $CH_3CH_2OH + (CH_3)_2CHCOOH \xrightarrow[\Delta]{H^+}$
    (c) 3-methyl-2-butanol $+ NaCl \xrightarrow{H_2SO_4}$
    (d) $CH_3CH_2CH_2OH \xrightarrow[\Delta]{H_2SO_4}$

25. Explain how you could carry out the following conversion. Write a mechanism for the reaction:

26. Explain how you could carry out the following conversion. Write a mechanism for the reaction:

27. As discussed in Are You Wondering 27-2, carbocation rearrangements occur in some substitution and elimination reactions. What elimination product would you expect in the following reaction *if no rearrangement of the carbocation occurs*? Experiment reveals that the main product is 2,3-dimethyl-2-pentene. Suggest a mechanism that shows how 2,3-dimethyl-2-pentene is formed. [*Hint:* Consider the formation and rearrangement of a carbocation.]

28. The molecule below is 2,2-dimethyl-1-propanol, a primary alcohol. Because no H atoms are bonded to the β carbon atom in this molecule, dehydration seems unlikely. However, when 2,2-dimethyl-1-propanol is heated with an acid catalyst, 2-methyl-2-butene is obtained. Suggest a mechanism that shows how 2-methyl-2-butene is formed. [*Hint:* Consider the formation and rearrangement of a carbocation. See Are You Wondering 27-2.]

29. Draw the structures of the products of each of the following reactions:
    (a) propene + hydrogen (Pt, heat)
    (b) 2-butanol + heat (in the presence of sulfuric acid)

30. Predict the product(s) of the reaction of:
    (a) HCl with 2-chloro-1-propene
    (b) HCN with $CH_3CH{=}CH_2$
    (c) HCl with $CH_3CH{=}C(CH_3)_2$

31. Give the structures of the main organic product(s) in each of the following reactions.
    (a)

    (b) (Z)-2-butene $+ HI \longrightarrow$
    (c) 2-methyl-1-butene $+ H_2O \xrightarrow{H_2SO_4}$

32. Give the structure of the main organic product(s) in each of the following reactions:
    (a) (E)-2-butene $+ HBr \longrightarrow$
    (b) 2-methyl-2-pentene $+ H_2O \xrightarrow{H_2SO_4}$
    (c)

33. Give the major product that forms when (Z)-3-methyl-2-pentene reacts with each of the following reagents:
    (a) HI; (b) H_2 in the presence of a platinum catalyst; (c) H_2O in H_2SO_4; (d) Br_2 in CCl_4.

34. Give the major product that forms when 1-ethylcyclohexene reacts with each of the following reagents:
    (a) HI; (b) H_2 in the presence of a platinum catalyst; (c) H_2O in H_2SO_4; (d) Br_2 in CCl_4.

Electrophilic Aromatic Substitution

35. Sketch the reaction profile for the mechanism depicted in Figure 27-21, by assuming the overall reaction is exothermic. Also, experimental evidence suggests that the activation energy for the first step is greater than the activation energy for the second step. Draw structures for the intermediate and transition states.

36. The chlorination of benzene can be carried out by allowing benzene to react with Cl_2 in the presence of $AlCl_3$. As discussed on page 1245, Cl_2 and $AlCl_3$ react to form $Cl^+AlCl_4^-$. Sketch the reaction profile for the reaction of benzene and $Cl^+AlCl_4^-$. Use the same assumptions as in Exercise 35.

37. Alkylation of benzene can be accomplished by treating benzene with haloalkane (RX) in the presence of $AlCl_3$. The reaction is known as a Friedel–Crafts alkylation reaction. (The reaction is named after Charles Friedel, a French chemist, and James M. Crafts, an American chemist, who discovered this method of making alkylbenzenes in 1877.) An example of a Friedel–Crafts alkylation reaction is shown below:

1-chloro-1-methylethane

1-methyl-1-phenylethane

The mechanism for this reaction involves the following steps. First, $(CH_3)_2CHCl$ and $AlCl_3$ react in a Lewis acid–base reaction to form an adduct, $(CH_3)_2CH-Cl-AlCl_3$, in which a chlorine atom is bonded to both carbon and aluminum. The adduct then dissociates to $(CH_3)_2CH^+$, a carbocation, and $AlCl_4^-$. The carbocation acts as an electrophile in a reaction with benzene, forming an arenium ion. Finally, a proton is removed from the arenium ion by $AlCl_4^-$, yielding an alkylbenzene, HCl, and $AlCl_3$. Write chemical equations for the elementary processes involved in forming 1-methyl-1-phenylethane and HCl from $(CH_3)_2CHCl$ and benzene. Use curved arrows to show the movement of electrons.

38. Treating benzene with fuming sulfuric acid yields benzenesulfonic acid, which is formed by the following reaction:

Benzenesulfonic acid

The SO_3 that participates in the reaction above is formed by the reaction of sulfuric acid molecules:

$$2\,H_2SO_4 \longrightarrow SO_3 + HSO_4^- + H_3O^+$$

The reaction of benzene and SO_3 to give benzenesulfonic acid involves the following elementary processes. SO_3 acts as an electrophile in a reaction with benzene, forming an arenium ion. A proton is then transferred from the arenium ion to HSO_4^-, forming the benzenesulfonate ion, $C_6H_5SO_3^-$, and H_2SO_4. Finally, $C_6H_5SO_3^-$ is protonated by H_3O^+ to give benzenesulfonic acid and a water molecule. Write balanced chemical equations for these elementary processes, using curved arrows to show the movement of electrons.

39. Predict the main product(s) of (a) the mononitration of chlorobenzene; (b) the monosulfonation of nitrobenzene; (c) the monochlorination of 1-methyl-2-nitrobenzene

40. Predict the main product(s) of (a) the mononitration of benzoic acid; (b) the monosulfonation of phenol; (c) the monobromination of 2-nitrobenzaldehyde.

Benzoic acid   Phenol   2-nitrobenzaldehyde

Reactions of Alkanes

41. What is major product expected from the monobromination of 2,2,3-trimethylpentane?

42. What is the major product expected from the monobromination of methylcyclohexane?

43. (a) Write the initiation, propagation, and termination steps involved in the monofluorination of 2,3-dimethylbutane to give 1-fluoro-2,3-dimethylbutane.

(b) Explain why in the monofluorination of 2,3-dimethylbutane the major product is 1-fluoro-2,3-dimethylbutane, not 2-fluoro-2,3-dimethylbutane.

44. Write the initiation, propagation, and termination steps involved in the monobromination of 2,3-dimethylbutane to give 2-bromo-2,3-dimethylbutane.

Polymerization Reactions

45. In referring to the molecular mass of a polymer, we can speak only of the average molecular mass. Explain why the molecular mass of a polymer is not a unique quantity, as it is for a substance like benzene.

46. Explain why Dacron is called a polyester. What is the percent oxygen, by mass, in Dacron?

47. Nylon 66 is produced by the reaction of 1,6-hexanediamine with adipic acid. A different nylon polymer is obtained if sebacyl chloride is substituted for the

adipic acid. What is the basic repeating unit of this nylon structure?

48. Would you expect a polymer to be formed by the reaction of terephthalic acid with ethyl alcohol in place of ethylene glycol? With glycerol in place of ethylene glycol? Explain.

Synthesis of Organic Compounds

49. Starting with acetylene as the only source of carbon, together with any inorganic reagents required, devise a method to synthesize acetaldehyde.

50. Starting with acetylene as the only source of carbon, together with any inorganic reagents required, devise a method to synthesize 1,1,2,2-tetrabromoethane.

51. How would you synthesize (E)- and (Z)-3-heptene from acetylene and any other chemicals?

52. How would you synthesize (R)-2-butanamine from (S)-2-butanol?

53. The azide anion is a nucleophile and when attached to a carbon atom, undergoes reduction to the amino group and free nitrogen. Suggest a method of preparation of the primary amine propanamine.

54. The cyanide anion is a nucleophile and when attached to a carbon atom, is reduced to a primary amine. Suggest a method of preparing propanamine from chloroethane.

Integrative and Advanced Exercises

55. Draw a structure to represent the principal product of each of the following reactions:
(a) 1-pentanol + dichromate ion (acid solution)
(b) butyric acid + ethanol (acid solution)
(c) 2-methyl-1-butene + HBr

56. Predict the products of the monobromination of **(a)** *m*-dinitrobenzene; **(b)** aniline; **(c)** *p*-bromoanisole.

57. Write the formulas of the products formed from the reaction of propene with each of the following substances: **(a)** H_2; **(b)** Cl_2; **(c)** HCl; **(d)** H_2O (in acid).

58. Write the formulas of the products formed from the reaction of 2-butene with each of the following substances: **(a)** H_2; **(b)** Cl_2; **(c)** HCl; **(d)** H_2O (in acid).

59. Which of the following species gives the reaction indicated? Write the structures of the reaction products.

(1) $CH_3CH_2\overset{\overset{O}{\|}}{C}{-}OCH_3$ **(2)** $CH_3CH_2\overset{\overset{O}{\|}}{C}{-}OH$ **(3)** $CH_3\overset{\overset{O}{\|}}{C}{-}O^-$

(a) reacts with dilute HCl
(b) hydrolyzes
(c) neutralizes dilute NaOH

60. Which of the following species gives the reaction indicated? Write the structures of the reaction products.

(1) $CH_3CH_2\overset{\overset{O}{\|}}{C}{-}NH_2$

(2) $CH_3CH_2NH_2$ **(3)** $CH_3CH_2CH_2NH_3{}^+Cl^-$

(a) neutralizes dilute HCl
(b) hydrolyzes
(c) neutralizes dilute NaOH

61. Write the formulas of the products expected to form in the following situations. If no reaction occurs, write N.R.
(a) $CH_3CH_2NH_2(aq) + HCl(aq) \longrightarrow$
(b) $(CH_3)_3N(aq) + HBr(aq) \longrightarrow$
(c) $CH_3CH_2NH_3{}^+(aq) + H_3O^+(aq) \longrightarrow$
(d) $CH_3CH_2NH_3{}^+(aq) + OH^-(aq) \longrightarrow$

62. Write the formulas of the products expected to form in the following situations. If no reaction occurs, write N.R.

(b) $(CH_3)_4N^+(aq) + HCl(aq) \longrightarrow$

(a) [cyclopentane ring]NH(aq) + HCl(aq) $\longrightarrow$

(c) $CH_3CH_2NH_2(aq) + OH^-(aq) \longrightarrow$
(d) $(CH_3)_3NH^+ + OH^- \longrightarrow$

63. To prepare methyl ethyl ketone, which of these compounds would you oxidize: 2-propanol, 1-butanol, 2-butanol, or *tert*-butyl alcohol? Explain.

64. Indicate the principal product(s) you would expect in **(a)** treating $CH_3CH_2CH{=}CH_2$ with *dilute* $H_2SO_4(aq)$; **(b)** exposing a mixture of chlorine and propane gases to ultraviolet light; **(c)** heating a mixture of isopropyl alcohol and benzoic acid; **(d)** oxidizing *sec*-butyl alcohol with $Cr_2O_7{}^{2-}$ in acidic solution.

65. Match the following compounds with the chemical properties given below. Write the structure of the products of the reactions described in (a) to (e).

(1) [structure: propyl group]—OH **(2)** [structure]—$\overset{\overset{O}{\|}}{C}$—OH

(3) [structure]—$\overset{\overset{O}{\|}}{C}$—H

(a) is easily oxidized
(b) neutralizes NaOH(aq)
(c) forms an ester with ethanol
(d) can be oxidized to a carboxylic acid
(e) can be dehydrated to an alcohol

66. Match the following compounds with the chemical properties given below. Write the structure of the products of the reactions described in (a) to (e).

(1) [structure]—NH_2 **(2)** [structure]—Cl **(3)** [structure]—O—[structure]

(a) neutralizes HCl(aq)
(b) neutralizes NaOH(aq)
(c) forms an amide with ethanoic acid
(d) reacts with ammonia
(e) reacts with $CN^-(aq)$

67. Write the structures of the isomers you would expect to obtain in the mononitration of *m*-methoxybenzaldehyde:

[structure: benzene ring with CHO at top and OCH_3 at meta position]

68. In the chlorination of CH_4, some CH_3CH_2Cl is obtained as a product. Explain why this should be so.

69. The three isomeric tribromobenzenes, I, II, and III, when nitrated, form three, two, and one mononitrotribromobenzenes, respectively. Assign correct structures to I, II, and III.

70. Write the name and structure of the benzene derivatives described below.
(a) Formula: C_8H_{10}; forms three monochlorination products when treated with Cl_2 and $FeCl_3$
(b) Formula: C_9H_{12}; forms one monochlorination product when treated with Cl_2 and $FeCl_3$
(c) Formula: C_9H_{12}; forms four monochlorination products when treated with Cl_2 and $FeCl_3$

71. For the monochlorination of hydrocarbons, the following ratio of reactivities has been found: $3° > 2° > 1°$, 4.3:3:1. How many different monochloro derivatives of 2-methylbutane are possible, and what percentage of each would you expect to find?

72. The cyanide anion is a nucleophile and when attached to a carbon atom, undergoes hydrolysis under basic conditions to the carboxylate anion. Suggest a method of preparing sodium butanoate from chloropropane. How can sodium butanoate be converted into butanoic acid?

73. The iodide ion cannot displace the —OH group in ethanol, but excess HI will react to produce ethyl iodide. Explain.

74. Starting with the compounds chloromethane, chloroethane, sodium azide, sodium cyanide, and a reducing agent, suggest how the following compounds could be synthesized.
    (a) N-methylpropanamide
    (b) ethylethanoate
    (c) methylethylamine
    (d) tetramethylammonium chloride

75. Predict and name the product(s) obtained from the following reaction. Write out the mechanism for the reaction and use curved arrows to show the movement of electrons.

$$(CH_3)_2CHCH{=}CH_2 \xrightarrow{H_2SO_4, H_2O}$$

76. Predict and name the product(s) obtained from the following reaction. Write out the mechanism for the reaction and use curved arrows to show the movement of electrons.

77. Starting with benzene and methane, and suitable inorganic reagents, suggest how the following compound might be synthesized. [*Hint:* See Exercise 37 for a description of how an alkyl group can be added to the benzene ring.]

Feature Problem

78. The reduction of aldehydes and ketones with a suitable hydride-containing reducing agent is a good way of synthesizing alcohols. This approach would be even more effective if, instead of a hydride, we could use a source of nucleophilic carbon. Attack by a carbon atom on a carbonyl group would give an alcohol and simultaneously form a carbon-to-carbon bond. How can we make a C atom in an alkane nucleophilic? This was achieved by Victor Grignard, who created the organometallic reagent R—MgBr, with the following reaction in diethyl ether:

$$R{-}Br + Mg \longrightarrow R{-}MgBr$$

The Grignard reagent is rarely isolated. It is formed in solution and used immediately in the desired reaction. The alkylmetal bond is highly polar, with the partial negative charge on the C atom, which makes the C atom highly nucleophilic. The Grignard reagent (R—MgBr) can attack a carbonyl group in an aldehyde or ketone as follows:

metal alkoxide

Addition of dilute aqueous acid solution to the metal alkoxide furnishes the alcohol. The important synthetic consequence of this procedure is that we have prepared a product with more carbon atoms than present in the starting material. A simple starting material can be transformed into a more complex molecule.

(a) What is the product of the reaction between methanal and the Grignard reagent formed from 1-bromobutane after the addition of dilute acid?

(b) By using a Grignard reagent, devise a synthesis for 2-hexanol.

(c) By using a Grignard reagent, devise a synthesis for 2-methyl-2-hexanol.

(d) Grignard reagents can also be formed with aryl halides, such as chlorobenzene. What would be the product of the reaction between the Grignard reagent of chlorobenzene and propanone? Can you think of an alternative synthesis of this product, again using a Grignard reagent?

(e) The basicity of the C atom bound to the magnesium in the Grignard reagent can be used to make Grignard reagents of terminal alkynes. Write the equation of the reaction between ethylmagnesium bromide and 1-hexyne. [*Hint:* Ethane is evolved.]

(f) By using a Grignard reagent, suggest a synthesis for 2-heptyn-1-ol.

Self-Assessment Exercises

79. Explain the important distinctions between each pair of terms: **(a)** nucleophilic substitution and electrophilic aromatic substitution; **(b)** addition and elimination; **(c)** S_N1 and S_N2; **(d)** E1 and E2

80. Explain the important distinctions between each pair of terms: **(a)** base and nucleophile; **(b)** α carbon and β carbon; **(c)** polar protic solvent and polar aprotic solvent; **(d)** carbocation and radical.

81. **(a)** Which of the nucleophiles CN^- or Cl^- reacts faster with CH_3CH_2I in an S_N2 reaction?
(b) Which of the substrates, $CH_3CH_2CH(CH_3)CH_2I$ or CH_3I, reacts faster with OH^- in an S_N2 reaction?

82. Which of the following is the strongest nucleophile for an S_N2 reaction? **(a)** H_2O; **(b)** CH_3CH_2OH; **(c)** $CH_3CH_2O^-$; **(d)** $CH_3CO_2^-$; **(e)** CH_3S^-.

83. Which of the following reactions would give a better yield of $CH_3OCH(CH_3)_2$? Explain.

$$CH_3ONa + (CH_3)_2CHI \longrightarrow CH_3OCH(CH_3)_2 + NaI$$

$$(CH_3)_2CHONa + CH_3I \longrightarrow CH_3OCH(CH_3)_2 + NaI$$

84. What is the major elimination product obtained in the following reactions?
(a)

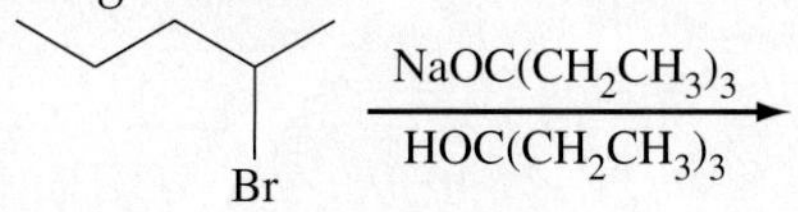

$$\xrightarrow[\text{HOC(CH}_2\text{CH}_3)_3]{\text{NaOC(CH}_2\text{CH}_3)_3}$$

(b)

$$\xrightarrow[\text{HOCH}_3]{\text{NaOCH}_3}$$

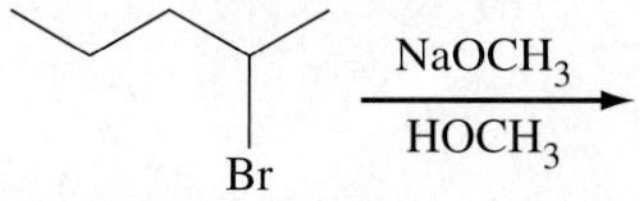

85. What is the major organic product obtained in the following reactions?
(a)

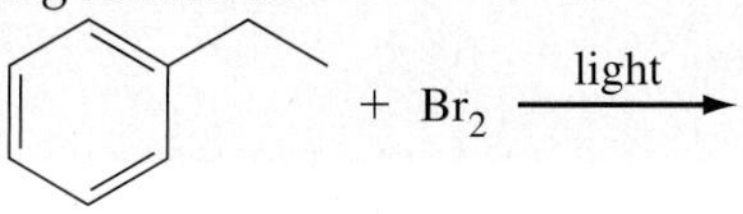

$$+ \; Br_2 \xrightarrow{\text{light}}$$

(b)

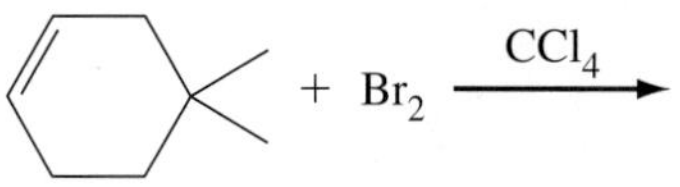

$$+ \; Br_2 \xrightarrow{\text{CCl}_4}$$

(c)

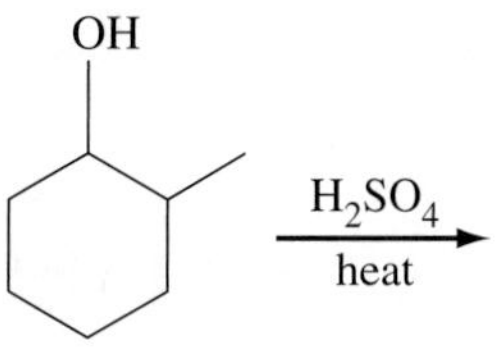

$$\xrightarrow[\text{heat}]{\text{H}_2\text{SO}_4}$$

(d)

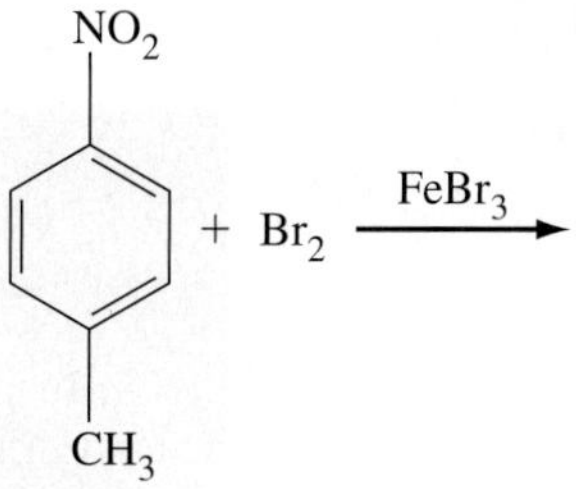

$$+ \; Br_2 \xrightarrow{\text{FeBr}_3}$$

(e)

$$+ \; CH_3OH \xrightarrow{\text{H}_2\text{SO}_4}$$

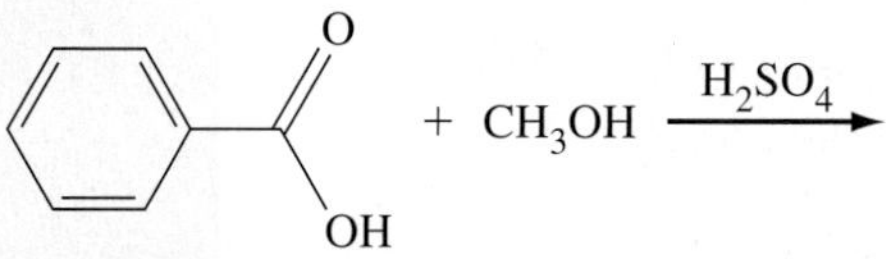

86. When $(CH_3CH_2)_3CBr$ is added to CH_3OH at room temperature, the major product is $CH_3O(CH_2CH_3)_3$ and a minor product is $CH_3CH{=}C(CH_2CH_3)_2$. Write out the mechanisms for the reactions leading to these products and use curved arrows to show the movement of electrons.

28

Chemistry of the Living State

CONTENTS

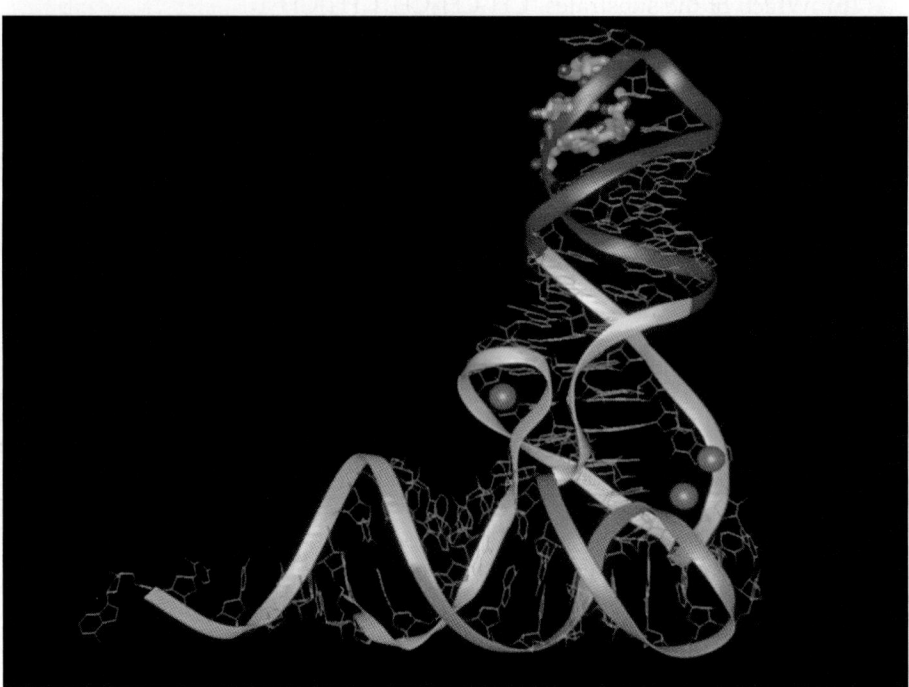

Transfer ribonucleic acid, tRNA, shown as a ribbon model. A wire frame model of tRNA is superimposed on the ribbon model. The red spheres represent magnesium ions, and the red and green spheres at the top highlight the anticodon triplet. As we learn in this chapter, nucleic acids are important components in the chemistry of life.

The planet Earth teems with life—from the tiny bacterium to the enormous whale. In spite of extraordinary differences in outward appearance, all living organisms share similar needs and chemical structures. Raw materials (food) are required for building cells and providing the energy of metabolism. The cells of almost all forms of living things on Earth use the same types of complicated structures to perform specialized functions. In this chapter, we will stress the structures of the macromolecules common to all organisms. Our discussion will emphasize how fundamental principles of chemistry as presented throughout this textbook can contribute to a knowledge of the living state.

28-1   Chemical Structure of Living Matter: An Overview

From one-celled organisms to humans, the living state includes some highly complex forms of matter. About 30 of the known elements occur in measurable concentrations in living matter. Of these, about 25 have functions that are definitely known. Four elements together—oxygen, carbon, hydrogen, and nitrogen—account for 96% of human body mass. Figure 28-1 identifies the elements found in living matter. As shown in Figure 28-2, the complex structures of living matter are synthesized in several stages from a few simple environmental precursors—N_2, H_2O, and CO_2.

In addition to water, which is the most abundant compound in the majority of living organisms, other important constituents are lipids, carbohydrates, proteins, and nucleic acids. These four types of macromolecular substances are the principal topics of this chapter.

The **cell is** the fundamental unit of all life. Cells contain a variety of substructures, such as a nucleus, mitochondria, and chloroplasts (plant cells). Cells combine to form tissues; tissues can be grouped into organs; organs combine into organ systems, and organ systems together form organisms.

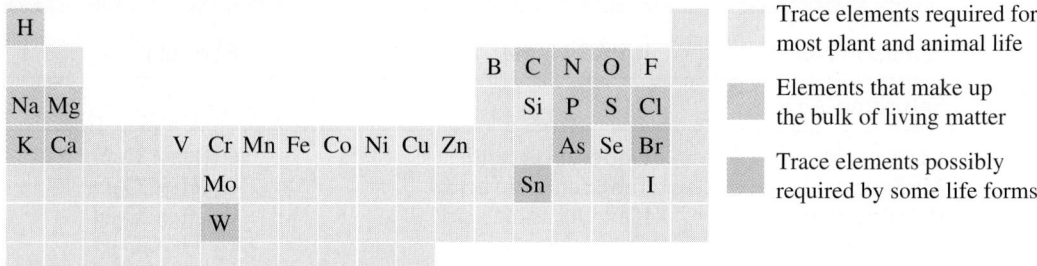

▲ FIGURE 28-1
Elements in living matter
For the most part, the elements essential to living matter are also among the more abundant elements in Earth's crust and in seawater. It is likely that lifeforms developed from the elements available to them.

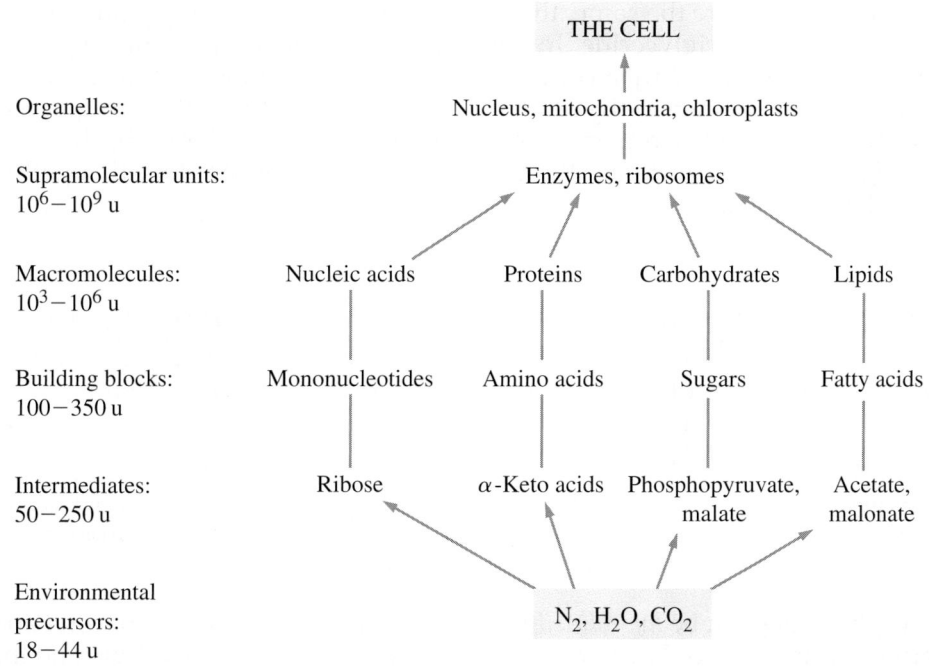

◀ FIGURE 28-2
Cellular organization

28-2   Lipids

Lipids are best described through their physical properties rather than in precise structural terms. **Lipids** are those constituents of plant and animal tissue that are soluble in low-polarity solvents, such as chloroform, carbon tetrachloride, diethyl ether, and benzene. Many compounds fit this description, but we will discuss only a few of them.

Triglycerides are esters of glycerol (1,2,3-propanetriol) and long-chain monocarboxylic acids (fatty acids). Some common fatty acids are listed in Table 28.1. *Triglyceride* is a common name; the systematic name of the triglycerides is triacylglycerols. *Glycerol* provides the three–carbon backbone, and the fatty acids provide *acyl* groups (in blue below).

KEEP IN MIND

that in forming an ester, a carboxylic acid loses its —OH group. The group that remains,

$$\begin{array}{c} O \\ \parallel \\ RC- \end{array}$$

is the *acyl* group.

$$\underset{\text{Glycerol}}{HO-\overset{\overset{\displaystyle H}{|}}{\underset{\underset{\displaystyle H}{|}}{C}}-\overset{\overset{\displaystyle OH}{|}}{\underset{\underset{\displaystyle H}{|}}{C}}-\overset{\overset{\displaystyle H}{|}}{\underset{\underset{\displaystyle H}{|}}{C}}-OH} \qquad \underset{\text{Fatty acid}}{HO-\overset{\overset{\displaystyle O}{\parallel}}{C}-R}$$

TABLE 28.1   Some Common Fatty Acids

Common Name	IUPAC Name	Formula
Saturated Acids		
Lauric acid	Dodecanoic acid	$C_{11}H_{23}CO_2H$
Myristic acid	Tetradecanoic acid	$C_{13}H_{27}CO_2H$
Palmitic acid	Hexadecanoic acid	$C_{15}H_{31}CO_2H$
Stearic acid	Octadecanoic acid	$C_{17}H_{35}CO_2H$
Unsaturated Acids		
Oleic acid	9-Octadecenoic acid	$C_{17}H_{33}CO_2H$
Linoleic acid	9,12-Octadecadienoic acid	$C_{17}H_{31}CO_2H$
Linolenic acid	9,12,15-Octadecatrienoic acid	$C_{17}H_{29}CO_2H$
Eleostearic acid	9,11,13-Octadecatrienoic acid	$C_{17}H_{29}CO_2H$

If all acyl groups are the same, the triglyceride is a *simple* triglyceride; otherwise it is a *mixed* triglyceride. In naming a triglyceride, the name *glyceryl* is written first, followed by a compound name for the three acyl groups. The acyl groups are named in the order in which they are attached to the glyceryl backbone. The first two names are given an *o* ending and the third, an *ate* ending. If all acyl groups are the same, only the ending *ate* is used, together with the prefix *tri*.

$$\begin{array}{l} CH_2OH \\ | \\ CHOH \\ | \\ CH_2OH \end{array}$$
Glycerol

$$\begin{array}{l} CH_2O\overset{\overset{\displaystyle O}{\parallel}}{C}(CH_2)_{14}CH_3 \\ | \quad\; \overset{\displaystyle O}{\parallel} \\ CHO\overset{}{C}(CH_2)_{14}CH_3 \\ | \quad\; \overset{\displaystyle O}{\parallel} \\ CH_2O\overset{}{C}(CH_2)_{14}CH_3 \end{array}$$
Glyceryl tripalmitate
Tripalmitin
(a simple triglyceride; a fat)

$$\begin{array}{l} CH_2O\overset{\overset{\displaystyle O}{\parallel}}{C}(CH_2)_7CH{=}CH(CH_2)_7CH_3 \\ | \quad\; \overset{\displaystyle O}{\parallel} \\ CHO\overset{}{C}(CH_2)_7CH{=}CH(CH_2)_7CH_3 \\ | \quad\; \overset{\displaystyle O}{\parallel} \\ CH_2O\overset{}{C}(CH_2)_7CH{=}CH(CH_2)_7CH_3 \end{array}$$
Glyceryl trioleate
Triolein
(a simple triglyceride; an oil)

$$\begin{array}{l} CH_2O\overset{\overset{\displaystyle O}{\parallel}}{C}(CH_2)_{10}CH_3 \\ | \quad\; \overset{\displaystyle O}{\parallel} \\ CHO\overset{}{C}(CH_2)_{14}CH_3 \\ | \quad\; \overset{\displaystyle O}{\parallel} \\ CH_2O\overset{}{C}(CH_2)_{16}CH_3 \end{array}$$
Glyceryl lauropalmitostearate
(a mixed glyceride)

Triglycerides can be *hydrolyzed* in alkaline solution to produce glycerol and salts of the fatty acids. The hydrolysis process is called **saponification**, and the salts are commonly known as **soaps**. For example, the hydrolysis of tristearin with aqueous KOH gives glycerol and the soap potassium stearate

$$
\begin{array}{c}
\text{CH}_2\text{OC(CH}_2)_{16}\text{CH}_3 \\
\overset{\text{O}}{\underset{|}{||}} \\
\text{CHOC(CH}_2)_{16}\text{CH}_3 \quad + \quad 3\ \text{KOH} \longrightarrow \\
\overset{\text{O}}{\underset{|}{||}} \\
\text{CH}_2\text{OC(CH}_2)_{16}\text{CH}_3
\end{array}
\quad
\begin{array}{c}
\text{CH}_2\text{OH} \\
| \\
\text{CHOH} \quad + \quad 3\ \text{CH}_3(\text{CH}_2)_{16}\overset{\text{O}}{\overset{||}{\text{C}}}\text{O}^-\text{K}^+ \\
| \\
\text{CH}_2\text{OH}
\end{array}
\qquad (28.1)
$$

Tristearin · Glycerol · Potassium stearate (a soap)

The cleansing action of soaps was described in Section 21-2 (page 932).

Fats and oils are both triglycerides (*glyceryl esters*), but differ from each other by the nature of the acid components in the triglycerides. **Fats** are glyceryl esters in which *saturated* fatty acid components predominate; they are solids at room temperature. **Oils** have a predominance of *unsaturated* fatty acids and are liquids at room temperature. The compositions of fats and oils are variable and depend not only on the particular plant or animal species involved but also on dietary and climatic factors. Some common fats and oils are listed in Table 28.2.

When pure, fats and oils are colorless, odorless, and tasteless. The characteristic colors, odors, and flavors commonly associated with fats and oils come from other organic substances present as impurities. The yellow color of butter is that of β-carotene (a yellow pigment also found in carrots and marigolds). The taste of butter is attributed to 3-hydroxy-2-butanone and diacetyl.

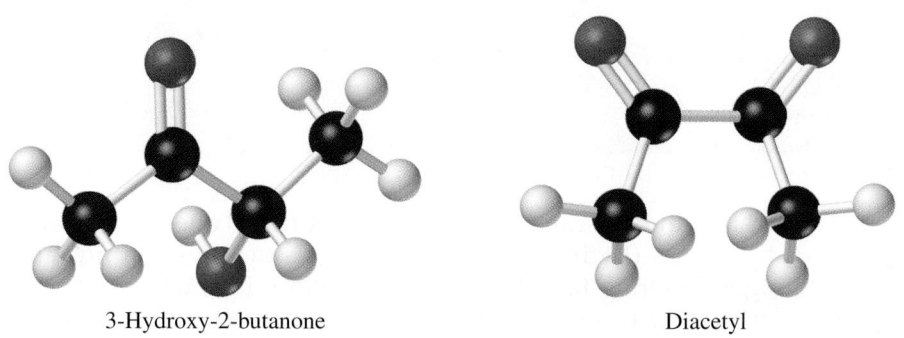

3-Hydroxy-2-butanone · Diacetyl

TABLE 28.2 Some Common Fats and Oils

| Lipid | Component Acids,[a] % by Mass | | | | | |
| | Saturated | | | Unsaturated | | |
	Myristic	Palmitic	Stearic	Oleic	Linoleic	Linolenic
Fats						
Butter	7–10	24–26	10–13	28–31	1–3	0.2–0.5
Lard	1–2	28–30	12–18	40–50	7–13	0–1
Edible Oils						
Corn	1–2	8–12	2–5	19–49	34–62	—
Safflower	—	6–7	2–3	12–14	75–80	0.5–1.5

[a]The formulas of the individual acids are listed in Table 28.1.

28-1 ARE YOU WONDERING...

What "calorie-free" fats are?

In olestra, an ester of fatty acids and sucrose, the sucrose molecule takes the place of glycerol as the backbone of the ester. When olestra is burned in a bomb calorimeter, the liberated heat is comparable to that obtained in the combustion of other oils and fats. So, in this sense, olestra is *not* calorie free. However, the sucrose backbone of the olestra molecule can bond with six, seven, or eight acyl groups instead of the three acyl groups in a triglyceride, as shown below. Human enzymes cannot break down this bulky molecule, and so it is not digested. It is in this sense that it is calorie free.

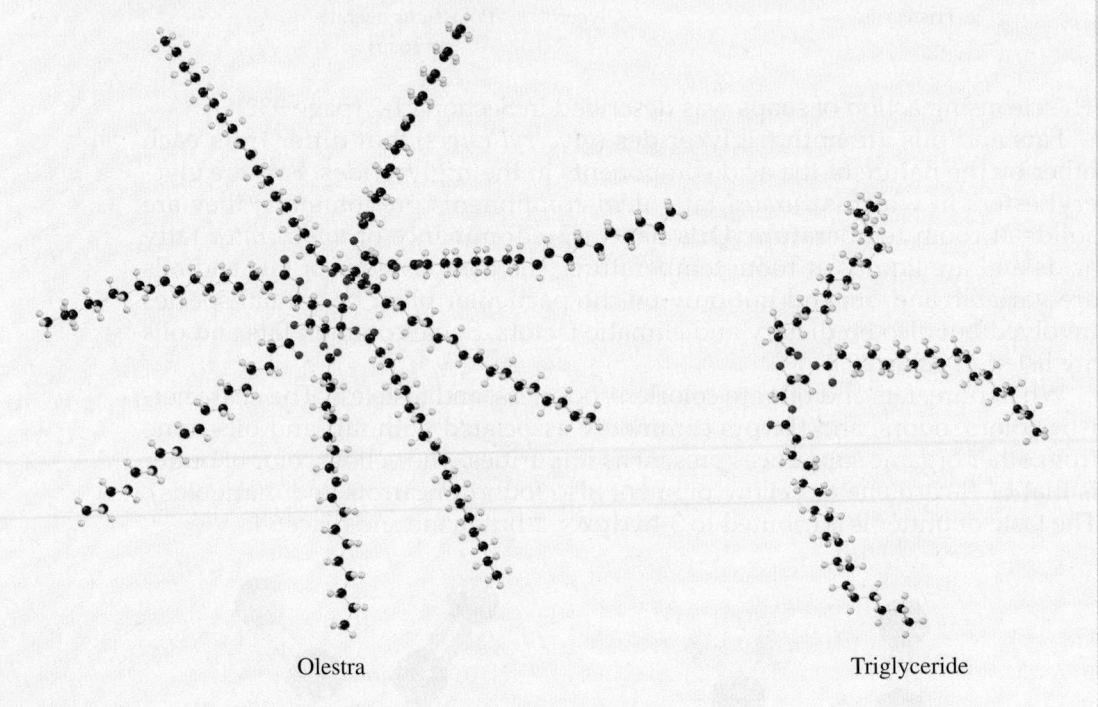

Olestra   Triglyceride

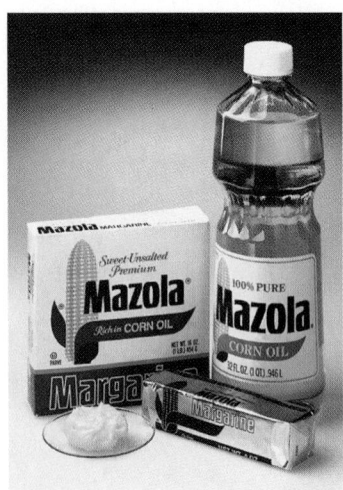

▲ Corn oil has the approximate composition listed in Table 28.2. Hydrogenation converts some of the unsaturated to saturated fatty acid components—the primary change required in making margarine from vegetable oils.

Unsaturated fats and oils can be converted to saturated ones by the catalytic addition of hydrogen (hydrogenation). Thus, oils or low-melting fats are changed to higher-melting fats. These higher-melting fats, when mixed with skim milk, fortified with vitamin A, and artificially colored, are known as *margarines*. Unsaturation in a fat or an oil is also removed when the fat or an oil decomposes. Edible fats and oils both hydrolyze and cleave at the double bonds by oxidation on exposure to heat, air, and light. When this happens, the fat becomes rancid. The low-molecular-mass fatty acids produced by this cleavage have offensive odors, as exhibited by butyric acid in rancid butter. Antioxidants are commonly added to oils used in the high-temperature cooking of potato chips and other foods to retard this oxidative rancidity.

Medical evidence suggests a relationship between a high intake of saturated fats and the incidence of coronary heart disease. For this reason, many diets call for the substitution of unsaturated fatty acids for saturated fatty acids in foods. In general, most mammal fats are saturated, whereas those derived from vegetables and seafood are unsaturated.

🔍 28-1   CONCEPT ASSESSMENT

What are the products of the saponification of one mole of glyceryl palmitooleolinolenate with three moles of sodium hydroxide?

Phospholipids

Phospholipids (phosphatides) occur in all animal cells and are especially prevalent in nerve tissue. They are derived from glycerol, fatty acids, phosphoric acid, and a nitrogen-containing base—ethanolamine in a class of phospholipids called *cephalins* and choline in class called *lecithins*. In the following structures, R and R′ are long-chain alkyl groups.

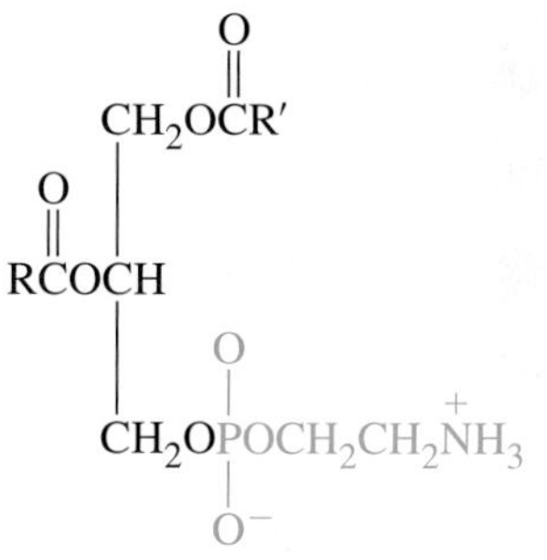

$$\underset{\substack{\text{Phosphatidylethanolamine}\\\text{(a cephalin)}}}{\begin{array}{c}\text{CH}_2\text{OCR}'\\[2pt]\text{RCOCH}\\[2pt]\text{CH}_2\text{OPOCH}_2\text{CH}_2\overset{+}{\text{N}}\text{H}_3\\[2pt]\text{O}^-\end{array}} \qquad \underset{\substack{\text{Phosphatidylcholine}\\\text{(a lecithin)}}}{\begin{array}{c}\text{CH}_2\text{OCR}'\\[2pt]\text{RCOCH}\\[2pt]\text{CH}_2\text{OPOCH}_2\text{CH}_2\overset{+}{\text{N}}(\text{CH}_3)_3\\[2pt]\text{O}^-\end{array}}$$

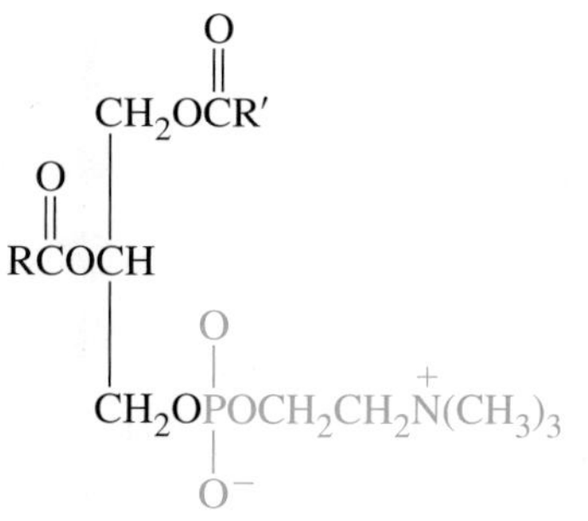

Like soap molecules, the phospholipids have a hydrophilic head (the phosphate-ethanolamine or phosphate-choline portions shown in blue above) and hydrophobic tails (the two alkyl chains). This enables phospholipids to solubilize and transport fats and oils in aqueous medium, whether this occurs in transporting lipids in the bloodstream or in emulsifying fats and oils in salad dressings.

Cell membranes (the outer boundary of all living cells) consist of a bilayer of phospholipids having their hydrophilic heads in an aqueous medium and their hydrophobic portions turned inward into a medium of cholesterol and proteins.

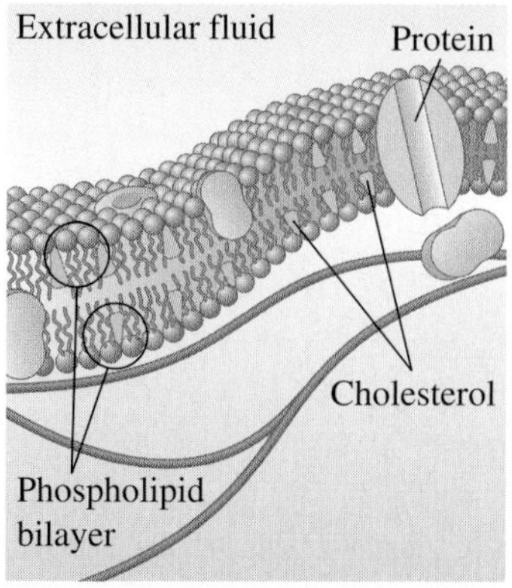

▲ A cell membrane is a bilayer (double layer) of phospholipid molecules with the polar head groups oriented toward the aqueous phase. Other lipid molecules, such as cholesterol, can be embedded in the bilayer. Some proteins are also found in the bilayer; membrane-bound proteins often act as ion pumps by providing a channel for certain ions to pass through.

28-3 Carbohydrates

The literal meaning of "carbohydrate" is hydrate of carbon: $C_x(H_2O)_y$. Thus, sucrose, or cane sugar, $C_{12}H_{22}O_{11}$, is equivalent to $C_{12}(H_2O)_{11}$. A more useful definition is that **carbohydrates** are polyhydroxy aldehydes, polyhydroxy ketones, their derivatives, and substances that yield them on hydrolysis. Carbohydrates that are aldehydes are called aldoses; those that are ketones are called ketoses. A five-carbon carbohydrate is a pentose, a six-carbon one, a hexose, and so on. The structures in the margin are those of two familiar hexoses—glucose and fructose, an aldose and a ketose, respectively.

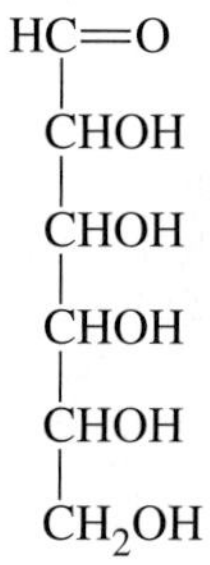

The general term for all carbohydrates is *glycoses*. The simplest carbohydrates are collections of individual small molecules of the formula $C_x(H_2O)_y$ and are called **monosaccharides. Oligosaccharides** are larger molecules composed of two to ten monosaccharide units bonded together. Names can be assigned to reflect the actual number of such units present, such as *di*saccharide and *tri*saccharide. Mono- and oligosaccharides are also called **sugars. Polysaccharides** contain more than ten monosaccharide units in their basic molecular structure, and many are in the *macromolecular* range. In summary,

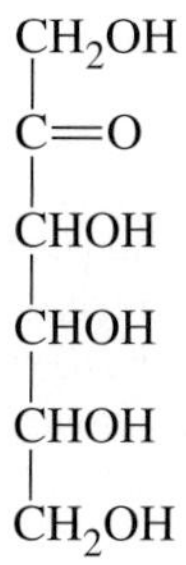

$$\underset{\text{Glucose}}{\begin{array}{c}\text{HC}=\text{O}\\[2pt]\text{CHOH}\\[2pt]\text{CHOH}\\[2pt]\text{CHOH}\\[2pt]\text{CHOH}\\[2pt]\text{CH}_2\text{OH}\end{array}} \qquad \underset{\text{Fructose}}{\begin{array}{c}\text{CH}_2\text{OH}\\[2pt]\text{C}=\text{O}\\[2pt]\text{CHOH}\\[2pt]\text{CHOH}\\[2pt]\text{CHOH}\\[2pt]\text{CH}_2\text{OH}\end{array}}$$

Glycoses

- *Monosaccharides*
  - aldoses (aldotriose, aldotetrose, . . .)
  - ketoses (ketotriose, ketotetrose, . . .)

- *Oligosaccharides* (from two to ten monosaccharide units)
  - disaccharides (e.g., sucrose)
  - trisaccharides (e.g., raffinose)
  - and so on.

- *Polysaccharides* (more than ten monosaccharide units)
  - (e.g., starch and cellulose)

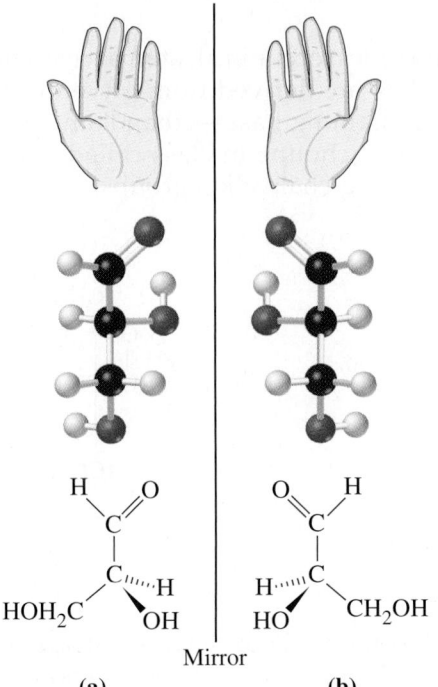

▶ FIGURE 28-3
Optical isomers of glyceraldehyde
The structure in **(a)** is not superimposable on **(b)**, just as a right hand and a left hand are not superimposable.

The simplest glycose is 2,3-dihydroxypropanal (glyceraldehyde), an *aldotriose*. As Figure 28-3 illustrates, the central C atom in glyceraldehyde has four different groups attached to it and is therefore chiral. As we have seen in Chapters 24 and 27, such molecules exhibit an interesting form of stereoisomerism—*optical isomerism*. There are *two* nonsuperimposable structures for glyceraldehyde. Such structures are related to each other like a right and a left hand, or like an object and its nonsuperimposable mirror image; they are *enantiomers* (see Section 24-4 and Section 26-4).

Optically active molecules affect plane-polarized light (Fig. 28-4). Interactions between a beam of polarized light and the electrons in an enantiomer cause a rotation of the plane of the polarized light. One enantiomer rotates the plane of polarized light to the right (clockwise) and is said to be **dextrorotatory** (designated +). The other enantiomer rotates the plane of polarized light to the same extent, but to the left (counterclockwise). It is said to be **levorotatory** (designated −). Because of their ability to rotate the plane of polarized light, isomers of these types are said to be *optically active*, and they are called *optical isomers*. Almost all molecules exhibiting optical isomerism possess at least one asymmetric, or chiral, C atom.

▶ The prefixes *dextro* and *levo* are derived from the Latin words *dexter*, meaning right, and *laevus*, meaning left.

The arrangement of groups at an asymmetric C atom is called the **absolute configuration**. We have already described the *R*, *S* system of nomenclature for describing the absolute configuration of chiral centers. The configuration

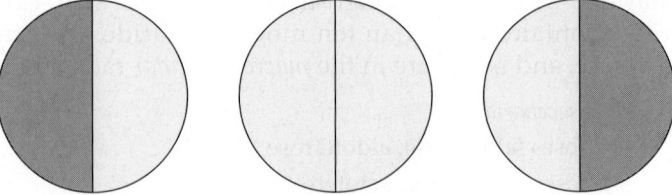

▲ FIGURE 28-4
View through the analyzer prism of a polarimeter (see also Figure 24-9)
The field of view of polarized light from a sodium vapor lamp is split in half. The analyzer prism in the eyepiece has been rotated in the correct direction—clockwise (to the right) or counterclockwise (to the left)—and through the appropriate angle, α (shown in Figure 24-9), when the two halves transmit light of equal intensity (center). For incorrect angles of rotation, one semicircle is darker than the other.

of glyceraldehyde shown in Figure 28-3(a) is (*S*)-glyceraldehyde. The (*R*)-glyceraldehyde is shown in Figure 28-3(b). But which configuration rotates the plane of polarized light in a dextrorotatory (+) sense? Optical rotation studies alone cannot tell us which of the enantiomers rotates the plane of plane-polarized light in a positive sense or a negative sense, because there is no way to assign the *R* or *S* configuration to a particular one of a pair of enantiomers. That is, there is no relationship between the absolute configuration and the sign of the rotation; both must be determined experimentally by independent methods. The assignment of which configuration, *R* or *S*, is the dextrorotary (+) isomer has been achieved using X-ray studies. These methods show that the enantiomer that corresponds to the dextrorotatory (+) isomer has the *R* configuration and is named (*R*)-(+)-glyceraldehyde (Fig. 28-3b). Correspondingly, the levorotatory (−) isomer is found to correspond to the *S* configuration and is named (*S*)-(−)-glyceraldehydes (Fig. 28-3a). It is purely fortuitous that the positive rotation corresponds to the *R* configuration.

The German chemist Emil Fischer studied carbohydrates in the late nineteenth century when techniques for determining the configuration of compounds were not available. Fischer arbitrarily assigned the configuration shown in Figure 28-3b to the dextrorotatory isomer of glyceraldehyde. Fischer designated this configuration D and named the isomer D-(+)-glyceraldehyde; D-(+)-glyceraldehyde is (*R*)-(+)-glyceraldehyde, and L-(−)-glyceraldehyde is (*S*)-(−)-glyceraldehyde. This system, extended to other chiral structures, is called the D, L convention.

We have already seen how to use a dashed and solid wedge line diagram to represent the three-dimensional structure of a molecule such as glyceraldehyde. Drawing these diagrams is not too difficult for compounds containing one or two stereocenters, but the process becomes increasingly cumbersome for compounds having several chiral centers, which carbohydrates often do. In addition to assigning the configuration to (*R*)-(+)-glyceraldehyde, Emil Fischer introduced a convention for representing three-dimensional structures in two-dimensional drawings, now called **Fischer projection formulas**.

The Fischer projections have two aspects to them: The first is how the stereochemistry at a chiral carbon atom is represented in two dimensions, and the second is how the carbon-chain backbone is arranged on the page. First, the bonds between each C atom and its four substituents are drawn in the form of a cross, the central carbon being at the point of intersection. The horizontal lines signify bonds directed toward the viewer; the vertical lines point away. Dashed and solid wedge line structures have to be arranged in this way to allow their conversion into Fischer projections. Consider (*R*)-(+)-glyceraldehyde.

$$\underset{\text{HO}}{\overset{\text{CHO}}{\underset{\text{H}}{\text{C}}}}\text{CH}_2\text{OH} \;\longrightarrow\; \text{H}-\underset{\text{CH}_2\text{OH}}{\overset{\text{CHO}}{\text{C}}}-\text{OH} \;\longrightarrow\; \text{H}-\underset{\text{CH}_2\text{OH}}{\overset{\text{CHO}}{\rule{0pt}{0pt}}}-\text{OH}$$

In order to view the molecule with two groups pointing toward us, we imagine the molecule as being picked up by the hydroxyl group and the hydrogen atom and then turned to bring these groups toward us, as shown. The CHO and CH_2OH groups naturally take up positions away from us. The Fischer projection is drawn with the H and OH groups connected by a horizontal line and the CHO and CH_2OH groups joined by a vertical line. Note that the central C atom is not drawn in, but is understood to be at the point of intersection of the lines; including this C atom would make the Fischer projection indistinguishable from a Lewis structure, which contains no stereochemical information.

In the Fischer notation, the structural formula is drawn so that the backbone of the molecule is arranged from top to bottom, with the most oxidized portion of the molecule (—CHO) at the top and the least oxidized (—CH_2OH) at the bottom. Attached groups (—H and —OH) are written to the sides. The end

▲ **Emil Fischer (1852–1919)** Fischer was awarded the Nobel Prize in 1902 for his research on the structures of sugars. Later he also elucidated how amino acid molecules join to form proteins.

groups of the backbone are considered to extend *behind* the plane of the page, *away* from the viewer. The glyceraldehyde enantiomers are written as

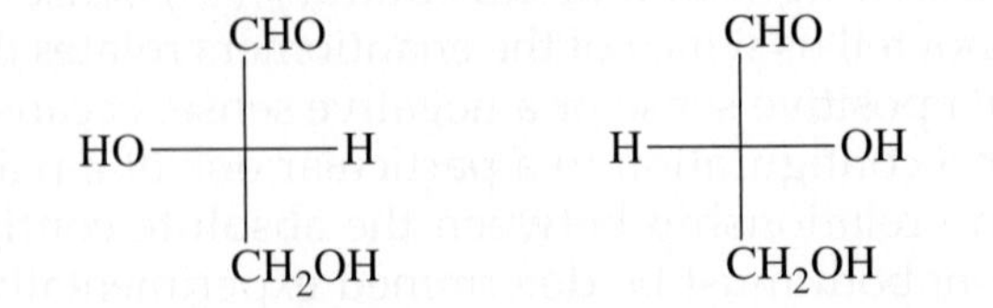

(S)-$(-)$ Glyceraldehyde
(L)-$(-)$ Glyceraldehyde

(R)-$(+)$-Glyceraldehyde
(D)-$(+)$-Glyceraldehyde

KEEP IN MIND

that D and L are like R and S in that they indicate the configuration of a chiral carbon atom, but they do not indicate whether the compound rotates plane-polarized light to the right $(+)$ or to the left $(-)$.

and are used to establish the D, L configuration for other sugars.

The —H and —OH groups on the next-to-last (penultimate) C atom extend in *front* of the page, *toward* the viewer. If the —OH group on this penultimate C atom is to the *right*, the configuration is D. If the —OH is to the left, the configuration is L. This convention is applied below to the four-carbon aldoses, where the penultimate C atom and attached groups are shaded with blue. All D sugars have the same configuration at this penultimate carbon. Figure 28-5 may help you to picture the relationship between a three-dimensional structure and its two-dimensional representation.

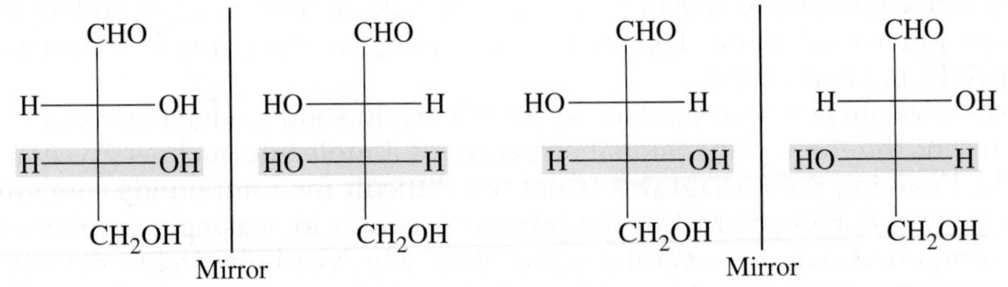

D-$(-)$-Erythrose   L-$(+)$-Erythrose   D-$(-)$-Threose   L-$(+)$-Threose

D-Erythrose and L-erythrose are enantiomers, as are D-threose and L-threose. If we compare the configurations of D-erythrose and D-threose, we note that these two molecules are *not* mirror images (see below). On the other hand, they are isomers of each other and both are optically active. Optical isomers that are *not* mirror images of each other are called **diastereomers**.

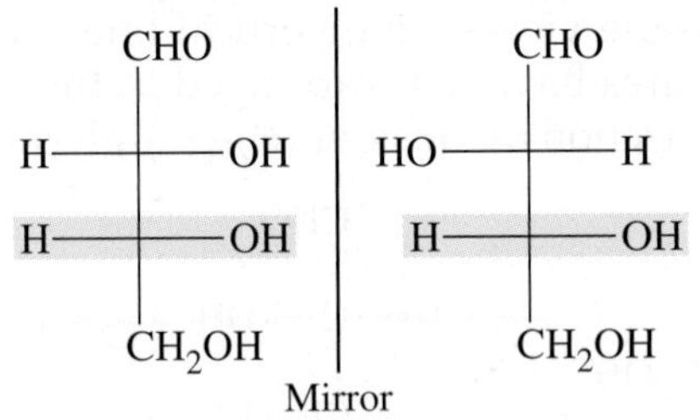

D-Erythrose   D-Threose

Not mirror images

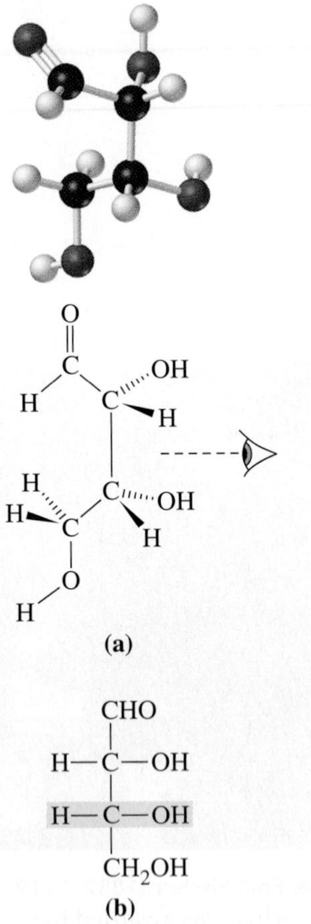

Enantiomers differ only in the *direction*, not the extent of their rotation of plane-polarized light. Diastereomers do differ in the extent to which they rotate plane-polarized light. They also differ in physical and chemical properties; for example, they have different solubilities in a particular solvent and react with chemical reagents at different rates.

A mixture of equal amounts of the D and L configurations of a particular substance, called a **racemic mixture**, does not rotate the plane of polarized light either to the left or to the right. The designation DL-erythrose, for example, signifies a racemic mixture. Often, when molecules with chiral centers are synthesized, the product is a racemic mixture. This is because the creation of these centers is a random process, like flipping a coin (an equal probability for heads or tails).

(a)

(b)

▲ FIGURE 28-5
The structure of D-$(-)$-erythrose
The three-dimensional structure **(a)** is represented in two dimensions by **(b)**.

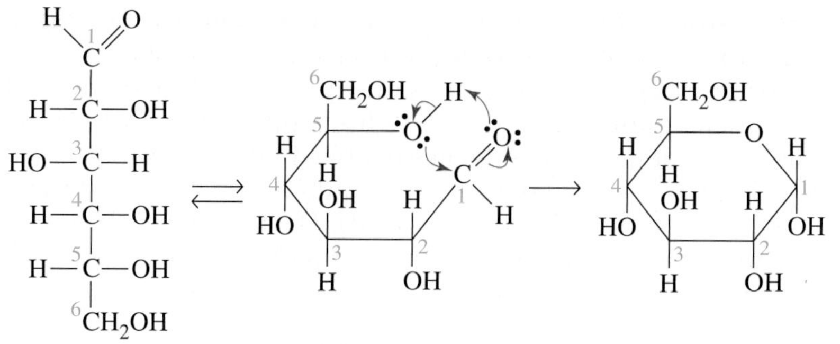

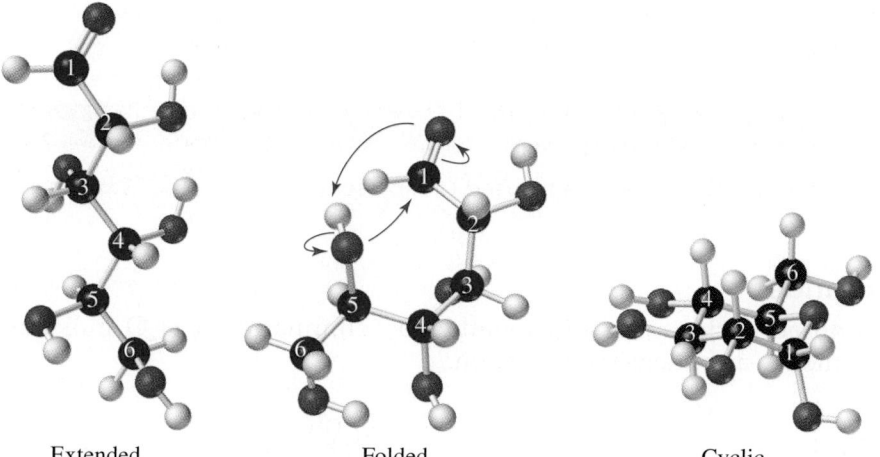

D-(+)-Glucose   α-D-(+)-Glucose

Extended   **Folded**   **Cyclic**

◀ FIGURE 28-6
Representation of ring closure in glucose molecule
The straight chain D-(+)-glucose molecule folds back on itself to bring the —OH group on the C-5 atom close to the aldehyde group of the C-1 atom. A proton is transferred, and this is followed by the formation of a six-member ring with an O atom joining the C-1 and C-5 atoms.

If optically pure isomers are desired, the racemic mixture must be separated into the component enantiomers by a process called *resolution*. Sometimes this is carried out using an enzyme that reacts with one enantiomer but not the other.

Monosaccharides

Of the 16 possible aldohexoses, only three occur widely in nature: D-glucose, D-mannose, and D-galactose. These three sugar molecules exist in a straight-chain form only to a very small extent (less than 0.5% for glucose). The predominant form for each is *cyclic*. In this ring formation, the —OH group of the fifth C atom (C-5) adds to the carbonyl of the C-1 atom and produces a ring composed of five C atoms and one O atom, as illustrated in Figure 28-6. The conformation of the six-member ring is of the chair type (recall Figure 26-12).

When the chain form of a sugar is converted to the ring form, a new chiral (asymmetrical) center is produced at the C-1 atom. There are two possible orientations at this center. In the **a** form, the —OH at C-1 is *axial* (directed down); in the **b** form, it is *equatorial* (extends out from the ring). The **a** and **b** forms of glucose are pictured in Figure 28-7.

The naming of monosaccharides in the ring form is complicated, but each item in a name conveys precise information. Thus, D-(+)-glucose refers to the straight-chain form of glucose in the D configuration; the (+) indicates this

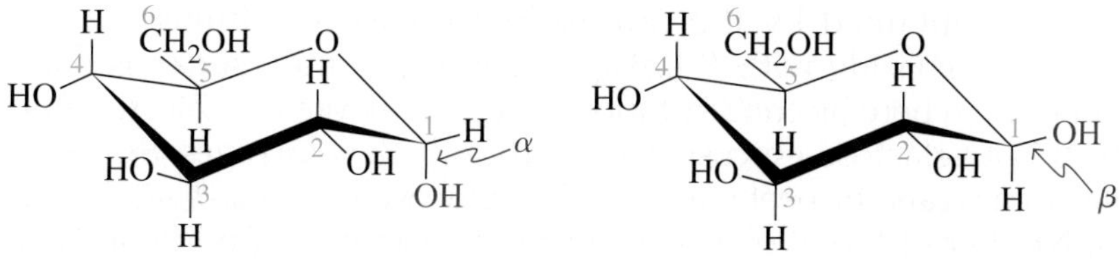

α-D-(+)-Glucose   β-D-(+)-Glucose

◀ FIGURE 28-7
α and β forms of D-glucose
In the α form, the —OH group at C-1 (in red) is axial, extending below the chair-like ring. In the β form, the —OH group is equatorial and extends out from the ring.

form is dextrorotatory. The name α-D-(+)-glucose denotes the ring form derived from D-glucose with the α configuration at the C-1 atom.

Monosaccharides such as glucose are known as **reducing sugars**, which means that a sufficient amount of the straight-chain form is in equilibrium with the cyclic form that the sugar engages in an oxidation–reduction reaction with $Cu^{2+}(aq)$. The $Cu^{2+}(aq)$ is reduced to insoluble red Cu_2O, and the aldehyde portion of the sugar is oxidized (to an acid). The test for a reducing sugar is conducted with alkaline copper ion complexed with tartrate (Fehling's solution) or citrate ion (Benedict's solution).

$$\text{Certain cyclic sugars} \rightleftharpoons \underset{\text{straight-chain form}}{\begin{array}{c}\text{CHO}\\|\\\text{CHOH}\\|\end{array}} \xrightarrow{Cu^{2+}} \underset{\text{red ppt}}{\begin{array}{c}\text{COOH}\\|\\\text{CHOH}\\|\end{array} + Cu_2O(s)} \qquad \textbf{(28.2)}$$

28-2 CONCEPT ASSESSMENT

Can a Fischer projection formula be used to represent D-(−)-glucose? Explain.

Disaccharides

Two monosaccharides can join together by eliminating a H_2O molecule between them—a condensation reaction.

The new bond formed between the two monosaccharides is call a glycosidic bond. This combination of monosaccharides is called a *disaccharide*. In describing a disaccharide, we must consider the identity of the component monosaccharides and whether the configuration of the linkage between the monosaccharides is α or β. The important, naturally occurring disaccharides—maltose, cellobiose, lactose, and sucrose—are presented in Figure 28-8.

In *maltose*, a H atom on the C-1 hydroxyl group of one glucose unit reacts with the hydroxyl group on the C-4 atom of a second glucose unit. The two units are linked in the α manner. Equilibrium is possible between the cyclic and straight-chain forms of maltose, so it is a reducing sugar. Maltose is produced by the action of malt enzyme on starch. In the presence of yeast, maltose undergoes fermentation—first to glucose and then to ethanol and $CO_2(g)$.

Cellobiose can be obtained by the careful hydrolysis of cellulose. It is a glucose-glucose disaccharide with β linkages. *Lactose*, or milk sugar, is naturally present in milk, where its concentration may range from 0 to 7% in different mammals. It is a galactose-glucose disaccharide with β linkages. *Sucrose* is ordinary table sugar (cane or beet sugar). It is a glucose-fructose disaccharide linked $1\alpha, 2\beta$. Neither of the two cyclic sugar units can open up into a chain form, and so sucrose is *not* a reducing sugar.

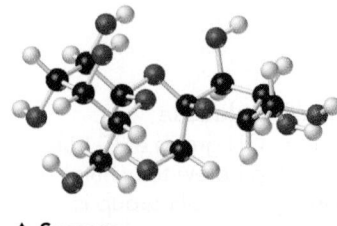

▲ Sucrose

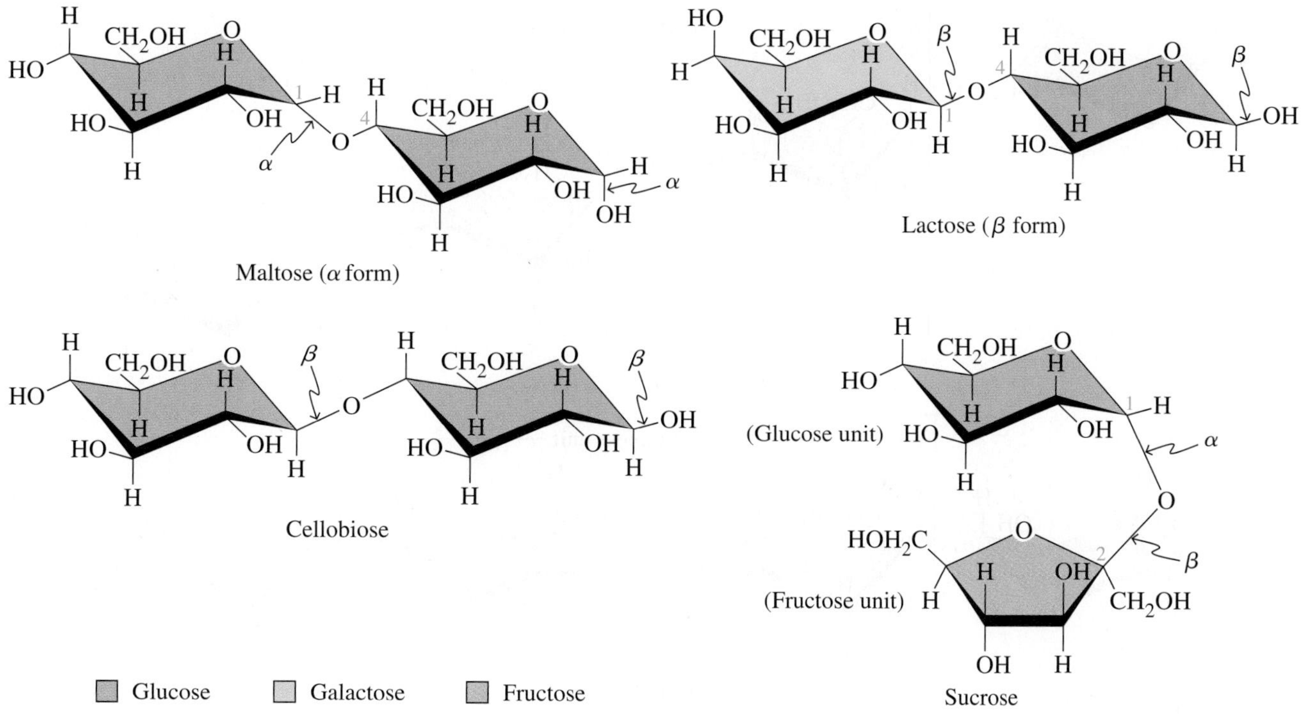

Maltose (α form)

Lactose (β form)

Cellobiose

(Glucose unit)

(Fructose unit)

Sucrose

☐ Glucose   ☐ Galactose   ☐ Fructose

▲ FIGURE 28-8
Some common disaccharides

Polysaccharides

Polysaccharides are composed of monosaccharide units joined into long chains by oxygen linkages. *Starch*, with a molecular mass between 20,000 and 1,000,000 u, is the reserve carbohydrate of many plants and is the bulk constituent of cereals and potatoes. Its structural features are shown in Figure 28-9. *Glycogen* is the reserve carbohydrate of animals; it is stored in the liver and in muscle tissue. It has a higher molecular mass than starch, and the polysaccharide chains are more branched. *Cellulose* is the main structural material of plants. It is the chief component of wood pulp, cotton, and straw. Complete hydrolysis of cellulose produces glucose. Cellulose has a molecular mass between 300,000 and 500,000 u, corresponding to 1800–3000 glucose units. Most animals, including human beings, do not possess the necessary enzymes to hydrolyze β linkages. As a result, they cannot digest cellulose. Certain bacteria in ruminants (cows, deer, camels) and termites can hydrolyze cellulose, allowing them to use it as a food. Termites, as we know, subsist on a diet of wood.

Photosynthesis

As we have noted in Sections 7-9 and 24-11, the process of photosynthesis involves the conversion by plants of carbon dioxide and water into carbohydrates. Photosynthesis requires the catalyst chlorophyll (see Figure 24-25) and sunlight.

$$n \, CO_2 + n \, H_2O \xrightarrow[\text{chlorophyll}]{\text{sunlight}} C_n(H_2O)_n + n \, O_2 \qquad \textbf{(28.3)}$$

Equation (28.3) is greatly oversimplified. The currently accepted mechanism, proposed by Melvin Calvin (Nobel Prize, 1961), involves as many as 100 sequential steps for the conversion of six moles of carbon dioxide to one mole of glucose. The elucidation of this mechanism was greatly aided by the use of carbon-14 as a radioactive tracer. For simplicity, the overall photosynthetic process is divided into two phases: (1) the conversion of solar energy to chemical energy—the light

▲ FIGURE 28-9
Two common polysaccharides
The linkage between carbon 1 on one ring and carbon 4 on the other ring is commonly designated as $1 \rightarrow 4$. The $1 \rightarrow 4$ linkage between the monosaccharides of starch (α) and cellulose (β) are different. This difference results in cellulose being able to adopt a straighter polymer chain than starch can. This means that more hydrogen bonding occurs between the long cellulose polymer chains, making it stronger for plant structure.

reactions; and (2) the synthesis, promoted by enzymes, of carbohydrates. This latter phase, called the dark reactions, can occur in the absence of light.

Biomass

Biomass is any living matter. An important component of biomass is the organic material produced by photosynthesis, that is, plants or their principal constituents—cellulose, starch, sugars. Plant biomass can be used directly as a fuel, or it may be converted to other gaseous, liquid, or solid materials, which can be used as fuels or chemical raw materials.

Perhaps the best known and most widely used conversion method is the fermentation of sugars to ethanol. Fermentation involves the decomposition of organic matter in the absence of air and through the action of a microorganism.

$$\text{hexose sugar} \xrightarrow{\text{yeast}} 2\,CH_3CH_2OH + 2\,CO_2$$

In North America, the main raw material for the production of ethanol by fermentation is corn.

The conversion of plant material to fossil fuels requires geologic processes and geologic time scales, thereby limiting the future availability of these fuels as energy sources and as raw materials. In principle, most of the compounds now being produced from fossil fuels could be made directly from cellulose. Methanol (wood alcohol) is formed by the destructive distillation (pyrolysis) of wood. Cellulose can be hydrolyzed to glucose and then converted to ethanol by fermentation. Fermentation processes can also be used to produce a series of oxygenated compounds—alcohols and ketones—which can be converted to hydrocarbons. Thus, the entire spectrum of organic chemicals could be produced from the simple molecules CO_2 and H_2O. The required energy would be mostly solar.

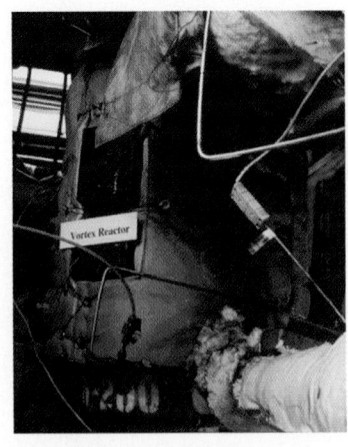

▲ A converter for the pyrolysis of biomass

28-4 Proteins

When a protein is hydrolyzed by dilute acids, bases, or hydrolytic enzymes, the result is a mixture of α-amino acids. An *amino acid* is a carboxylic acid that also contains an amine group, $-NH_2$; an α-**amino acid** has the amino group on the α carbon atom—the C atom next to the carboxyl group. Thus, **proteins** are high-molecular-mass polymers composed of α-amino acids. Of the known α-amino acids, about 20 have been identified as the building blocks of plant and animal proteins. Some of these amino acids are listed in Table 28.3.

Proteins are the basis of protoplasm and are found in all living organisms. In animals, proteins—as muscle, skin, hair, and other tissue—make up the bulk of

◀ The word *protein* is derived from the Greek word *proteios*, meaning "of first importance" (similar to the derivation of "proton").

TABLE 28.3   Some Common Amino Acids

Name	Symbol	Formula	pI[a]
Neutral Amino Acids			
Glycine	Gly	$HCH(NH_2)CO_2H$	6.03
Alanine	Ala	$CH_3CH(NH_2)CO_2H$	6.10
Valine[b]	Val	$(CH_3)_2CHCH(NH_2)CO_2H$	6.04
Leucine[b]	Leu	$(CH_3)_2CHCH_2CH(NH_2)CO_2H$	6.04
Isoleucine[b]	Ileu or Ile	$CH_3CH_2CH(CH_3)CH(NH_2)CO_2H$	6.04
Serine	Ser	$HOCH_2CH(NH_2)CO_2H$	5.70
Threonine[b]	Thr	$CH_3CH(OH)CH(NH_2)CO_2H$	5.6
Phenylalanine[b]	Phe	$C_6H_5CH_2CH(NH_2)CO_2H$	5.74
Methionine[b]	Met	$CH_3SCH_2CH_2CH(NH_2)CO_2H$	5.71
Cysteine	Cys	$HSCH_2CH(NH_2)CO_2H$	5.05
Cystine	(Cys)$_2$	$[SCH_2CH(NH_2)CO_2H]_2$	5.1
Tyrosine	Tyr	$4\text{-}HOC_6H_4CH_2CH(NH_2)CO_2H$	5.70
Tryptophan[b]	Trp	(indole ring)—$CH_2CH(NH_2)CO_2H$	5.89
Proline[c]	Pro	(pyrrolidine ring)—CO_2H	6.21
Acidic Amino Acids			
Aspartic acid	Asp	$HO_2CCH_2CH(NH_2)CO_2H$	2.96
Glutamic acid	Glu	$HO_2CCH_2CH_2CH(NH_2)CO_2H$	3.22
Basic Amino Acids			
Lysine[b]	Lys	$H_2N(CH_2)_4CH(NH_2)CO_2H$	9.74
Arginine	Arg	$H_2NC(=NH)NH(CH_2)_3CH(NH_2)CO_2H$	10.73
Histidine	His	(imidazole ring)—$CH_2CH(NH_2)CO_2H$	7.58

[a]pH of *isoelectric point.*
[b]Essential amino acids. In addition, arginine and glycine are required by the chick, arginine by the rat, and histidine by human infants.
[c]The secondary amino group makes proline an *α-imino* acid. Nevertheless, it is commonly listed with amino acids.

the body's nonskeletal structure. As enzymes, proteins catalyze biochemical reactions; as hormones, they regulate metabolic processes; and as antibodies, they counteract the effect of invading organisms.

Amino Acids Other than glycine ($H_2NCH_2CO_2H$), naturally occurring amino acids are optically active, mostly with an L configuration.

An L-amino acid

The reference structure for establishing the absolute configurations of amino acids is again glyceraldehyde, with the —NH_2 group substituting for —OH and —CO_2H, for —CHO. The molecule shown above has an L configuration because the —NH_2 group appears on the *left*.

Certain amino acids that are required for proper health and growth in human beings cannot be synthesized by the body. These amino acids, which are called the *essential amino acids,* must be ingested as food. Eight amino acids are known to be essential; the case of three others is less certain (see Table 28.3).

Amino acids are colorless, crystalline, high-melting-point solids that are moderately soluble in water. In a strongly acidic solution (low pH), the amino acid exists as a cation: A proton from solution attaches itself to the unshared pair of electrons on the nitrogen atom in the —NH_2 group. In a strongly basic solution (high pH), an anion forms through the loss of protons by the —CO_2H and —NH_3^+ groups. At an intermediate isoelectric point, a proton is lost from the —CO_2H but retained by the —NH_3^+ group. The product is a dipolar ion, or a **zwitterion**.

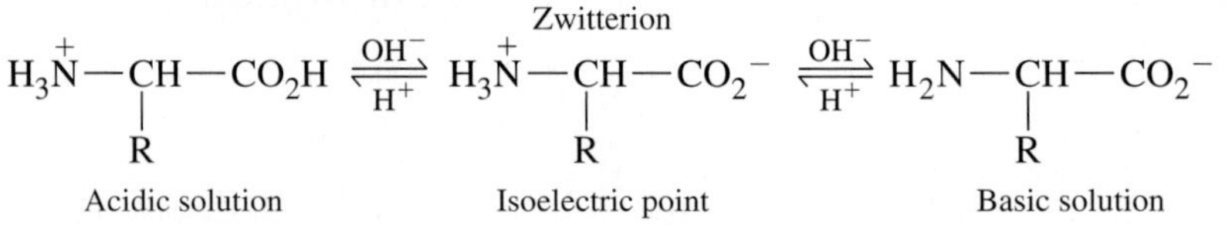

$$\underset{\text{Acidic solution}}{\overset{+}{H_3}N-\underset{\underset{R}{|}}{CH}-CO_2H} \underset{H^+}{\overset{OH^-}{\rightleftharpoons}} \underset{\text{Isoelectric point}}{\overset{\text{Zwitterion}}{\overset{+}{H_3}N-\underset{\underset{R}{|}}{CH}-CO_2^-}} \underset{H^+}{\overset{OH^-}{\rightleftharpoons}} \underset{\text{Basic solution}}{H_2N-\underset{\underset{R}{|}}{CH}-CO_2^-} \quad (28.4)$$

Amino acids are amphoteric. In most amino acids, the acidity of the —NH_3^+ group is slightly greater than the basicity of the —CO_2^- group. The largest group of amino acids are close to pH neutral. The pH at which the dipolar structure predominates is called the **isoelectric point**, or pI (see Table 28.3). At this pH, the molecule does not migrate in an electric field. At a pH above the pI, the molecule migrates to the anode (positive electrode); and below the pI, to the cathode (negative electrode). Most basic amino acids have a pI well above 7, acidic ones well below 7, and most neutral ones slightly less than 7 (5.7–6.1).

Peptides Two amino acid molecules can be joined by the elimination of a water molecule between them. The amino acids thus joined form a dipeptide. The bond between the two amino acid units is called a **peptide bond**.

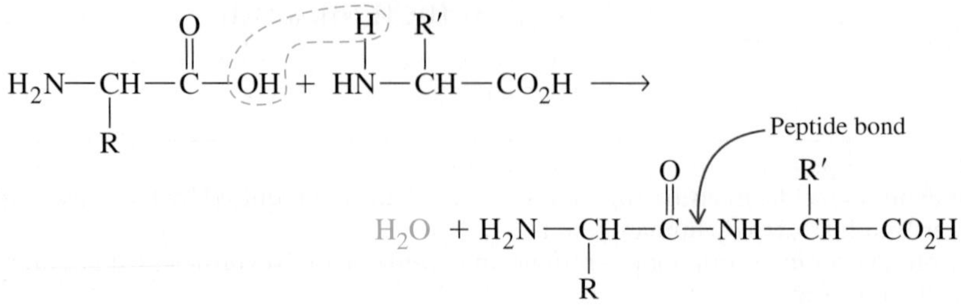

A dipeptide

A tripeptide has three amino acid residues and two peptide linkages. A large number of amino acid units may join to form a **polypeptide**.

The amino acid unit present at one end of a polypeptide chain has a free —NH₂ group; this is the N-terminal end. The other end of the chain has a free —CO₂H group; this is the C-terminal end. The polypeptide structure is written with the N-terminal end to the left and the C-terminal end to the right. The base name of the polypeptide is that of the C-terminal amino acid. The other amino acid units in the chain are named as substituents of this acid. Their names change from an *ine* to the *yl* ending. Abbreviations are also commonly used in writing polypeptide names, as illustrated in Example 28-1.

EXAMPLE 28-1   Naming a Polypeptide

What is the name of the polypeptide whose structure is shown below?

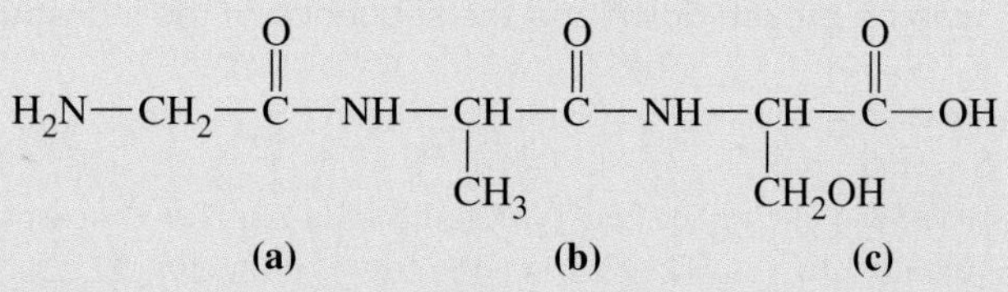

$$H_2N-CH_2-\overset{\overset{\displaystyle O}{\|}}{C}-NH-\underset{\underset{\displaystyle CH_3}{|}}{CH}-\overset{\overset{\displaystyle O}{\|}}{C}-NH-\underset{\underset{\displaystyle CH_2OH}{|}}{CH}-\overset{\overset{\displaystyle O}{\|}}{C}-OH$$

(a)   (b)   (c)

Analyze

We can identify the three amino acids in this tripeptide by using Table 28.3: (a) = glycine; (b) = alanine; and (c) = serine. The C-terminal amino acid is serine.

Solve

The name is glycylalanylserine (Gly-Ala-Ser).

Assess

When naming polypeptides, we always begin from the N-terminal and finish at the C-terminal. Remember to replace the *-ine* with *-yl* for all the amino acids except the last one, which retains the *-ine*.

PRACTICE EXAMPLE A: What is the name of the polypeptide shown below?

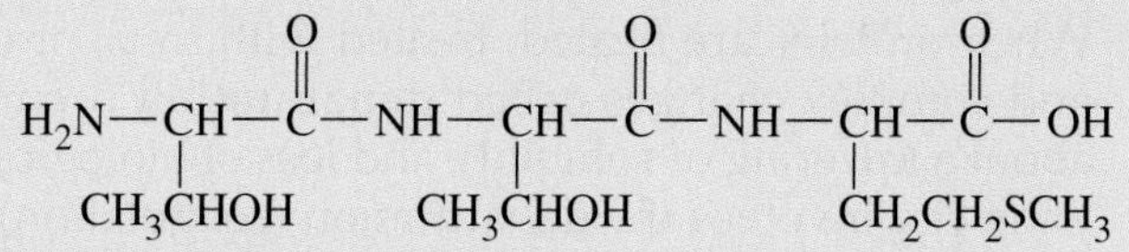

$$H_2N-\underset{\underset{\displaystyle CH_3CHOH}{|}}{CH}-\overset{\overset{\displaystyle O}{\|}}{C}-NH-\underset{\underset{\displaystyle CH_3CHOH}{|}}{CH}-\overset{\overset{\displaystyle O}{\|}}{C}-NH-\underset{\underset{\displaystyle CH_2CH_2SCH_3}{|}}{CH}-\overset{\overset{\displaystyle O}{\|}}{C}-OH$$

PRACTICE EXAMPLE B: Write the structural formula of the polypeptide serylglycylvaline.

28-3   CONCEPT ASSESSMENT

At pH = 7, what is the most probable charge on the tripeptide molecule, Gly-Ala-Ser, described in Example 28-1?

Amino Acid Sequencing Suppose that a tripeptide is known to consist of the three amino acids: A, B, and C. What is the correct structure: ABC, BAC, . . . ? Can you see that there are six possibilities? For longer chains, of course, the number of possibilities is enormous. Determining the sequence of amino acids in a polypeptide chain is one of the most significant problems in all of biochemistry. The method employed is outlined in Figure 28-10, and the structure of a typical polypeptide, beef insulin, is shown in Figure 28-11.

The distinction between large polypeptides and proteins is arbitrary. It is generally accepted that if the molecular mass is over 10,000 u (roughly 50–75 amino acid units), the substance is a protein. Proteins possess characteristic isoelectric

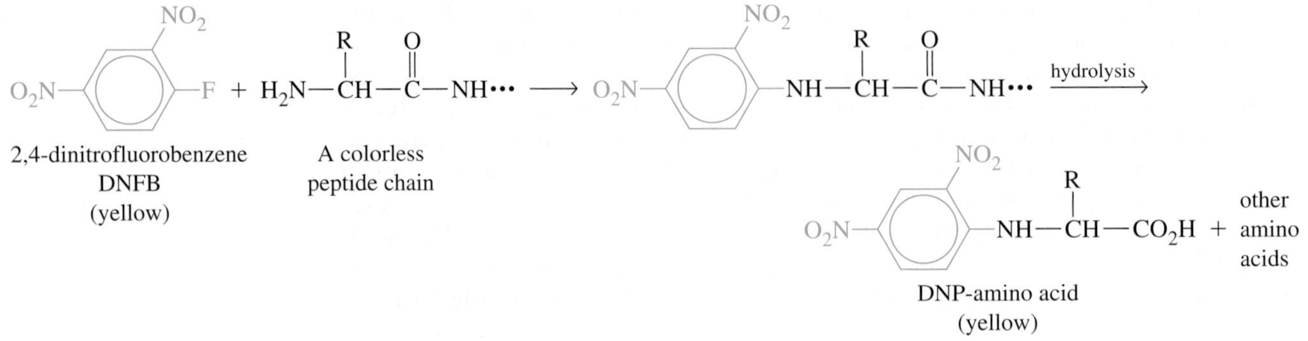

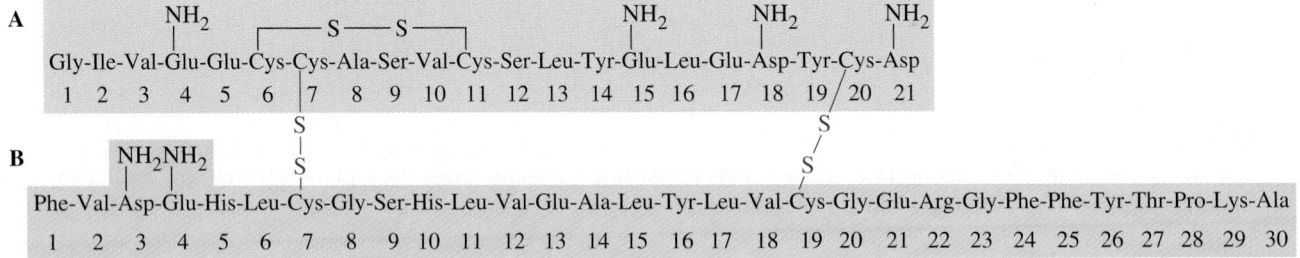

▲ FIGURE 28-10
Experimental determination of amino acid sequence
In the reaction between DNFB and a polypeptide, the N-terminal amino acid
ends up with the yellow marker (a dinitrophenyl group, DNP) attached to it.
By gentle hydrolysis and repeated use of the marker, a polypeptide chain
can be broken down and the sequence of the individual units determined.

A

$$\overset{\overset{\displaystyle NH_2}{|}}{Gly}\text{-Ile-Val-Glu-Glu-Cys-Cys-Ala-Ser-Val-Cys-Ser-Leu-Tyr-Glu-Leu-Glu-Asp-Tyr-}\overset{\overset{\displaystyle NH_2}{|}}{Cys}\text{-Asp}$$

1　2　3　4　5　6 | 7　8　9　10　11　12　13　14　15　16　17　18　19 / 20　21

B

Phe-Val-Asp-Glu-His-Leu-Cys-Gly-Ser-His-Leu-Val-Glu-Ala-Leu-Tyr-Leu-Val-Cys-Gly-Glu-Arg-Gly-Phe-Phe-Tyr-Thr-Pro-Lys-Ala

1　2　3　4　5　6　7　8　9　10　11　12　13　14　15　16　17　18　19　20　21　22　23　24　25　26　27　28　29　30

▲ FIGURE 28-11
Amino acid sequence in beef insulin—primary structure of a protein
There are two polypeptide chains joined by disulfide ($-S-S-$) linkages. One chain
has 21 amino acids, and the other 30. In chain A, the Gly at the left end is N-terminal
and the Asp is C-terminal. In chain B, Phe is N-terminal and Ala is C-terminal.

points, and their acidity or basicity depends on their amino acid composition.
When proteins are heated, treated with salts, or exposed to UV light, profound
and complex changes called **denaturation** occur. Denaturation usually brings
about a lowering of solubility and loss of biological activity. The frying or boiling
of an egg involves the denaturation (coagulation) of the egg albumin, a protein.
A hair perm takes advantage of a denaturation process that is reversible. The proteins found in hair (e.g., keratin) contain disulfide linkages ($-S-S-$). When
hair is treated with a reducing agent, these linkages break—a denaturation
process. Following this step, the hair is set into the desired style. Next, the hair is
treated with a mild oxidizing agent. The disulfide linkages are reestablished, and
the hair remains in the style in which it was set.

The Structure of Proteins

Proteins exhibit four levels of structure commonly referred to as primary, secondary, tertiary and quaternary structure. The **primary structure** of a protein
refers to the exact sequence of amino acids in the polypeptide chains that make
up the protein. But what are the shapes of the long polymeric chains themselves? Are they simply limp and entangled like a plate of spaghetti or is there
some order within chains and among chains? The structure or shape of protein
segments is referred to as **secondary structure**. In 1951, based on X-ray diffraction studies on polylysine, a synthetic polypeptide, Linus Pauling and R. B.
Corey proposed that the orientation of this polypeptide and thus of the protein
chain *is helical*. A spiral, helical, or spring-like shape can be either left- or right-
handed, but because proteins are composed of l amino acids, their helical structure is right-handed (Fig. 28-12).

EXAMPLE 28-2   Determining the Sequence of Amino Acids in a Polypeptide

A polypeptide, on complete hydrolysis, yielded the amino acids A, B, C, D, and E. Partial hydrolysis and sequence proof gave single amino acids, together with the following larger fragments: AD, CD, DCB, BE, and BC. What must be the sequence of amino acids in the polypeptide?

Analyze

Begin by arranging the fragments in the following manner,

$$AD$$
$$DC$$
$$DCB$$
$$BE$$
$$CB$$

Solve

We see that the sequence ADCBE is consistent with the fragments observed.

Assess

Typically, we write a sequence starting from the N-terminal amino acid and ending at the C-terminal amino acid. Therefore, the sequence EBCDA is also possible because we do not know whether A or E is the N-terminal end of the polypeptide.

PRACTICE EXAMPLE A: On complete hydrolysis, a pentapeptide yields the amino acids Val, Phe, Gly, Cys, and Tyr. Partial hydrolysis yields the fragments Val-Phe, Gly-Cys, Cys-Val-Phe, and Phe-Tyr. Glycine (Gly) is the N-terminal acid. What is the sequence of amino acids in the polypeptide?

PRACTICE EXAMPLE B: On complete hydrolysis, a hexapeptide yields the amino acids Ala, Gly, Ser, Trp, and Val. Partial hydrolysis yields the fragments Val-Trp, Gly-Gly-Ala, Ser-Gly-Gly, and Ala-Val-Trp. Serine (Ser) is the N-terminal acid. What is the sequence of amino acids in the polypeptide?

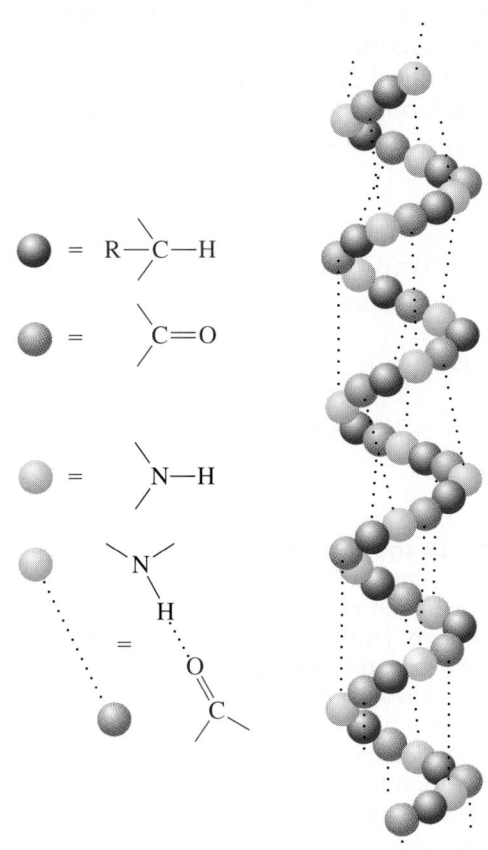

◀ FIGURE 28-12
Secondary structure of a protein—an α helix
The helical structure is stabilized by hydrogen bonds between

$$\overset{O}{\underset{\|}{-C-}}\text{groups in one turn and} \quad \overset{H}{\underset{|}{-N---}}$$

groups in the next turn above. The bulky R groups are directed outward from the atoms in the spiral.

▶ FIGURE 28-13
Pleated-sheet model of β-keratin
(a) A polypeptide chain showing the direction of interpolypeptide hydrogen bonds (other polypeptide chains lie to the left and to the right of the chain shown). Bulky R groups extend above and below the pleated sheet. (b) The stacking of pleated sheets.

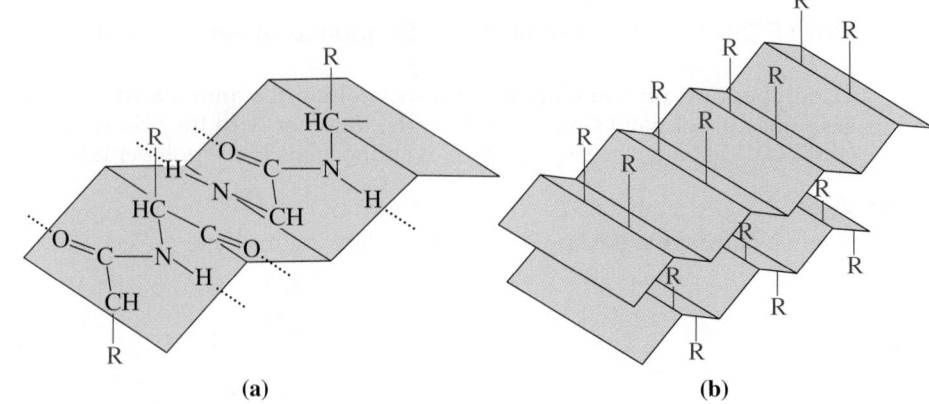

(a)   (b)

Other types of orientations are also possible. For example, β-keratin and silk fibroin are arranged in pleated sheets. In these proteins, the side chains extend above and below the pleated sheets, and there is hydrogen bonding between *different* molecules (interpeptide bonding) lying next to each other and about 0.47 nm apart in the same sheet. These sheets are stacked on top of one another about 1.0 nm apart, rather like a pile of sheets of corrugated roofing (Fig. 28-13). Some proteins, such as gamma globulin, are amorphous: They do not have a definite secondary structure.

Many proteins possess additional structural features. For example, rather than being elongated, the coils may be twisted, knotted, and so forth. The final statement regarding the shape of a protein molecule lies in a description of its **tertiary structure**. Because the internal hydrogen bonding between atoms in successive turns of a protein helix is weak, these hydrogen bonds ought to be easily broken. In particular, we should expect them to be replaced by hydrogen bonds to water molecules when the protein is placed in water. That is, the α helix should open up and become a randomized structure when placed in water. But experimental evidence indicates that this does not happen, which leads to the conclusion that other forces must be involved in entwining the long α-helical chains into definite geometric shapes. The tertiary structure of myoglobin is shown in Figure 28-14. Three types of linkages are involved in tertiary structures as described in Figure 28-15.

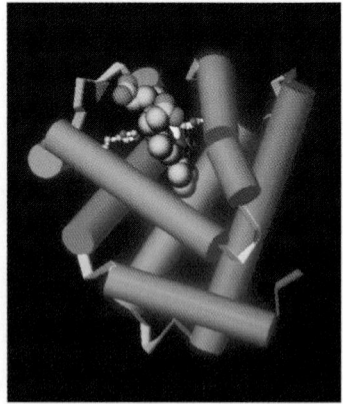

▲ FIGURE 28-14
Representation of the tertiary structure of myoglobin
The primary structure is that of a peptide of 153 units in a single chain. Secondary structure involves coiling of 70% of the chain into an α helix.

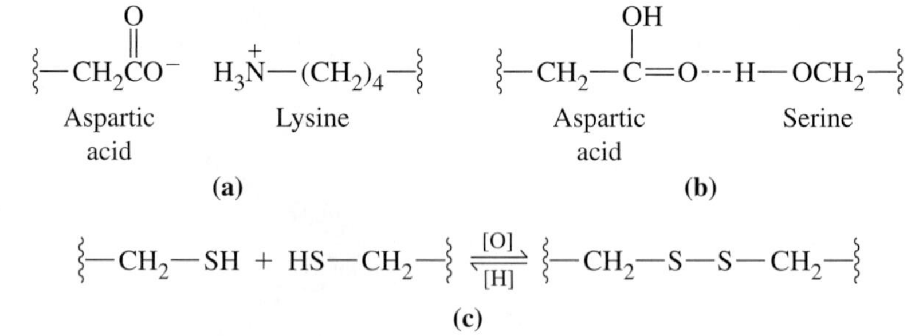

$$\{-CH_2CO^- \quad H_3\overset{+}{N}-(CH_2)_4-\} \qquad \{-CH_2-\overset{OH}{\underset{|}{C}}=O\text{---}H-OCH_2-\}$$

Aspartic   Lysine   Aspartic   Serine
acid   acid

(a)   (b)

$$\{-CH_2-SH + HS-CH_2-\} \underset{[H]}{\overset{[O]}{\rightleftharpoons}} \{-CH_2-S-S-CH_2-\}$$

(c)

▲ FIGURE 28-15
Linkages contributing to the tertiary structure of proteins
(a) *Salt linkages.* Acid–base interactions between different coils. Here, the carboxyl group of an aspartic acid unit on one coil donates a proton to the free amine group of a lysine unit on another. (b) *Hydrogen bonding.* Interactions between side chains of certain amino acids, for example, aspartic acid and serine. (c) *Disulfide linkages.* Oxidation of the highly reactive thioalcohol group (—SH) of cysteine to a disulfide (—S—S—) can occur (as in beef insulin).

The folding of a polypeptide chain into a tertiary structure is influenced by an additional factor. The hydrophobic hydrocarbon portions of the chains (R groups) tend to pull away from the aqueous medium and retreat to the interior of the structure, leaving ionic groups at the exterior.

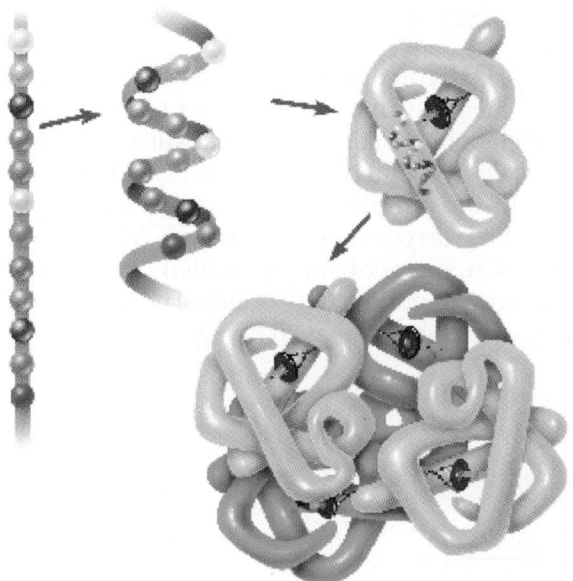

◀ FIGURE 28-16
The four levels of protein structure
Hemoglobin is the oxygen-carrying protein in red blood cells. The iron-containing heme units are shown as red discs. The primary structure of the protein is determined by its amino acid sequence. The secondary structure (α helix) is stabilized through formation of hydrogen bonds, as illustrated in Figure 28-12. The tertiary structure is determined by interactions between the R groups and their surroundings, causing the polypeptide chains to fold in a particular manner. Finally, the quaternary structure is an aggregation of two or more folded chains—four chains in the case of hemoglobin. Not all proteins have a quaternary structure.

The hemoglobin molecule consists of four separate polypeptide chains or subunits. The arrangement of these four subunits constitutes a still higher order of structure referred to as the **quaternary structure**. The levels of protein structure in hemoglobin, from primary to quaternary, are depicted in Figure 28-16.

Even minor changes in the structure of a protein can have profound effects. Hemoglobin contains four polypeptide chains; two chains are labeled α with 141 residues and two are labeled β with 146 residues. The substitution of valine for glutamic acid at one site in two of these chains gives rise to the sometimes fatal blood disease known as sickle-cell anemia. The altered hemoglobin has a reduced ability to transport oxygen through the blood, although it does seem to provide some defense against malaria.

Q   28-4   CONCEPT ASSESSMENT

Polypeptides can fold into helical coils. Would you also expect polysaccharides to form helical coils? Explain.

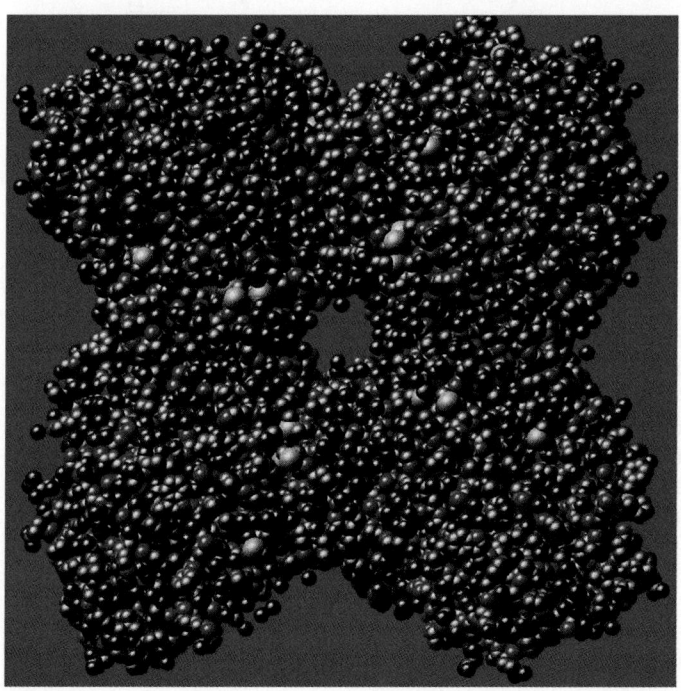

◀ A computer image of the three-dimensional structure of ribulose-1, 5-biphosphate carboxylase-oxygenase (RuBisCo), which consists of 37,792 atoms. This structure was elucidated in 1988, after an 18-year effort. RuBisCo is the most abundant protein on Earth. Its estimated annual worldwide production by plants is 4×10^{13} g (40 million tons). RuBisCo is the enzyme that initiates the process of photosynthesis.

28-5 Aspects of Metabolism

Although living organisms differ markedly in appearance, there is a striking similarity in the chemical reactions of their life processes. We refer to the totality of these reactions as **metabolism**. Metabolic reactions in which substances are broken down are referred to as *catabolism*. Metabolic reactions in which more complex substances are synthesized from simpler ones are called *anabolism*. Reactions having $\Delta G > 0$ are said to be *endergonic*, and those with $\Delta G < 0$ are *exergonic*. The substances involved in metabolism are called *metabolites*. Metabolism is a complex subject that we can outline only briefly. The discussion that follows centers on the summary of metabolism given in Figure 28-17.

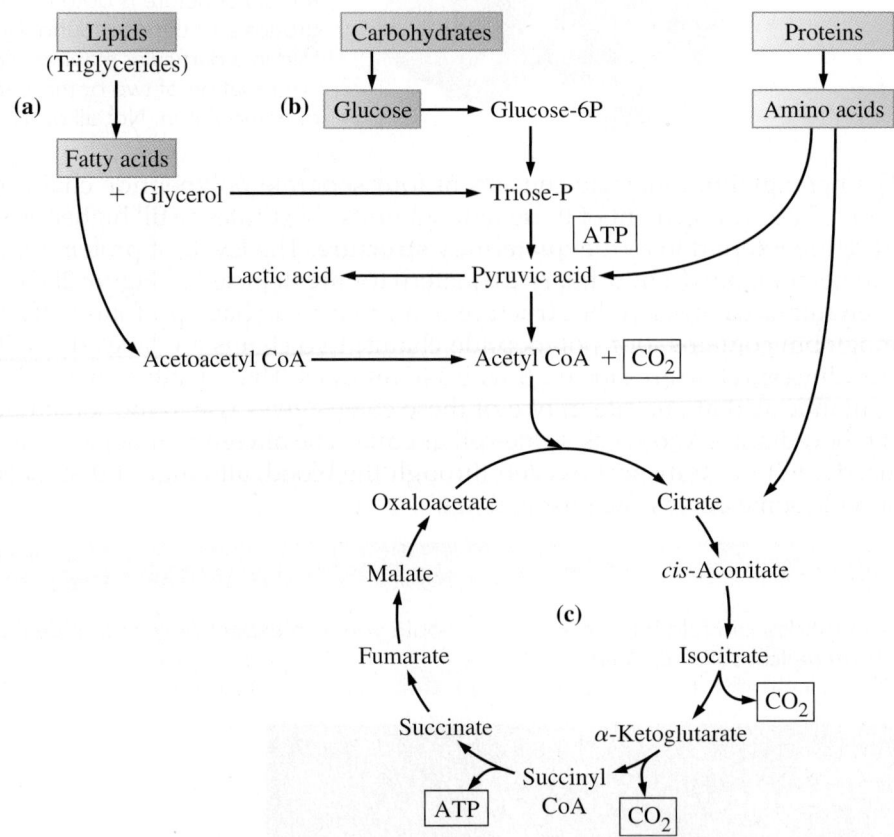

▲ FIGURE 28-17
Metabolism outline
(a) *Fatty acid section.* Fatty acids are degraded two C atoms at a time. Acetyl units enter into the citric acid cycle (c) as acetyl CoA. **(b)** *Glycolysis section (Embden–Meyerhof pathway).* These reactions are anaerobic (no oxygen required). Carbohydrates are degraded to the six-carbon sugar glucose, and then to the three-carbon triose-P (glyceraldehyde-3-phosphate). Next, the three-carbon acid pyruvic acid is formed from triose-P. Pyruvic acid loses a molecule of CO_2, yielding the two-carbon acetyl unit, which combines with coenzyme A (CoA) to form acetyl CoA. **(c)** *Citric acid cycle (Krebs cycle).* A two-carbon acetyl unit from acetyl CoA joins with the four-carbon oxalacetate unit to produce the six-carbon tricarboxylic acid citric acid (designated here as citrate). A two-step conversion to isocitrate occurs, followed by the loss of a molecule of CO_2 and formation of the five-carbon α-ketoglutarate. Another CO_2 molecule is lost in the formation of succinyl CoA. The remainder of the cycle involves a succession of four-carbon acids leading to oxaloacetate. The oxaloacetate regenerated at the end of the cycle now joins with another acetyl unit, and the cycle is repeated. The overall change occurring in the cycle is that a two-carbon acetyl unit enters the cycle and two molecules of CO_2 leave.

Carbohydrate Metabolism

Starch is the principal source of energy for humans and other animals. Digestion of starch begins in the mouth with the action of salivary enzymes, the amylases. Starch is converted to maltose and polysaccharides known as dextrins. This process continues briefly in the stomach (until the enzymes are denatured by the acid present), and the maltose and polysaccharides pass into the small intestine. Here, amylase from the pancreas completes the conversion of polysaccharides to maltose, and the enzyme *maltase* converts the maltose to glucose. Glucose is absorbed through the wall of the small intestine into the bloodstream and distributed to other organs.

Glucose is ultimately oxidized to carbon dioxide and water, with the liberation of energy. The principal intermediate in this process is glucose-6-phosphate (glucose-6P). Once formed, glucose-6P may be converted to glycogen (a polysaccharide stored in the liver), revert back to glucose, or be metabolized. The major metabolic route (Fig. 28-17) involves the anaerobic (absence of air) glycolysis pathway, followed by an aerobic cycle, the citric acid cycle.

◀ In glucose-6-phosphate, a phosphate group replaces the —OH group on the C-6 atom of the cyclic glucose molecule (see Figure 28-7).

$$H_2C-O-\overset{\displaystyle O^-}{\underset{\displaystyle O}{\overset{\displaystyle |}{\underset{\displaystyle |}{P}}}}-OH$$

Lipid Metabolism

The digestion of fats and oils occurs in the small intestine through the action of a combination of lipase enzymes. The products of this enzyme hydrolysis are glycerol, mixtures of mono- and diglycerides, and fatty acids. These are absorbed into the bloodstream through the wall of the intestine. Glycerol is converted to glyceraldehyde-3-phosphate (triose phosphate) and joins into the glucose metabolic route previously described. Fatty acids are oxidized to carbon dioxide and water, with the release of energy, in a series of reactions known as β oxidation. In this process, oxidation occurs at the β carbon atom of a fatty acid, followed by cleavage. Thus, two-carbon pieces (acetic acid) are split off. The process also requires coenzyme A (CoA). For example, with palmitic acid, $C_{15}H_{31}COOH$, the process is repeated seven times, with the formation of eight molecules of acetyl CoA, which enter the Krebs cycle (Fig. 28-17).

◀ The structure of glyceraldehyde-3-phosphate is

Protein Metabolism

In the stomach, HCl(aq) and the enzyme *pepsin* hydrolyze about 10% of the amide linkages in proteins and produce polypeptides in the molecular mass range of 500 to several thousand atomic mass units. In the small intestine, peptidases such as trypsin and chymotrypsin (from the pancreas) cleave the polypeptides into very small fragments. These fragments are then acted on by aminopeptidase and carboxypeptidase. The resulting free amino acids pass through the wall of the intestine, into the bloodstream, and to all the cells of the body, where they are the building blocks of proteins. Protein synthesis is directed by the nucleic acids DNA and RNA (page 1292).

Energy Relationships in Metabolism

The fundamental agents involved in energy transfers from exergonic to endergonic reactions are **adenosine *di*phosphate (ADP)** and **adenosine *tri*phosphate (ATP)**. The following equation represents the conversion of one mole of ATP to one mole of ADP.

$$ATP^{4-} + H_2O \longrightarrow ADP^{3-} + HPO_4^{2-} + H^+ \qquad \Delta G^{\circ\prime} = -32.4 \text{ kJ} \qquad \textbf{(28.6)}$$

Figure 28-18 represents the reverse of reaction (28.6).

The energy released in the oxidation of foods is picked up by ADP, which is converted to ATP. Enzymes catalyze each step in the overall conversion. The oxidation of one mole of glucose to CO_2 and H_2O is accompanied by the conversion of 38 mol ADP to 38 mol ATP. These two processes are represented below.

◀ Additional thermodynamic aspects of ADP/ATP conversions are discussed in the Focus On feature for Chapter 19, *Coupled Reactions in Biological Systems*, on the MasteringChemistry site (www.mastering chemistry.com).

$$C_6H_{12}O_6 + 6 O_2 \longrightarrow 6 CO_2 + 6 H_2O \quad \Delta G^{\circ} = -2880 \text{ kJ}$$

$$38 \times \{ADP^{3-} + HPO_4^{2-} + H^+ \longrightarrow ATP^{4-} + H_2O\} \quad \Delta G^{\circ\prime} = 38 \times 32.4 \text{ kJ} = 1230 \text{ kJ}$$

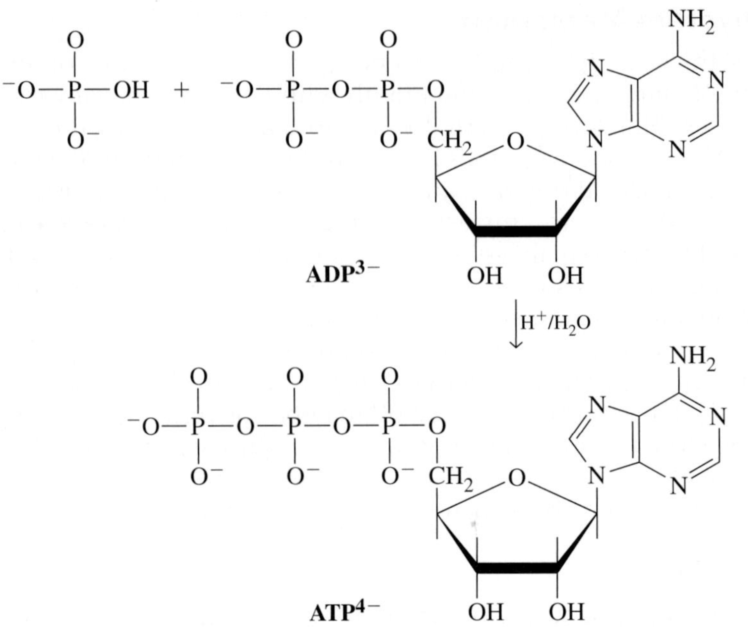

► Based on the discussion of the expanded valence shell (page 419), we represent all the P-to-O bonds in Figure 28-18 as single bonds. Sometimes these structures are written with one P-to-O double bond in each phosphate unit.

▲ FIGURE 28-18
Conversion of ADP to ATP
The H$^+$ ion entering into the reaction and the H$_2$O molecule produced are shown beside the main reaction arrow.

Thus, of the 2880 kJ of energy released in the oxidation of one mole of glucose, 1230 kJ of energy is stored in the high-energy bonds of ATP. The efficiency of this energy storage is (1230/2880) × 100% = 43%. That is, nearly half of the energy released in carbohydrate metabolism is stored in the human body for later use. This is a much more efficient use of energy than in an internal combustion engine, for example. If the metabolism of glucose occurred in a single step, with just one mole of ADP converted to one mole of ATP, the efficiency would drop to (32.4/2880) × 100% = 1.1%. So we can see why the metabolism of sucrose is such a complex, multistep process.

Enzymes

An **enzyme** is a biological catalyst that contains protein. Enzymes are specific for each biological transformation and catalyze a reaction without requiring a change in temperature or pH. Originally, enzymes were assigned common or trivial names, such as pepsin and catalase. Present practice, however, is to name them after the processes they catalyze, usually employing an *ase* ending.

A model for enzyme action was introduced in the discussion of enzyme kinetics in Section 14-11. According to this model, an enzyme can exert its catalytic activity only after combining with the reacting substance, the *substrate*, to form a complex. The site on the enzyme where the substrate bonds is called the *active site*; some enzymes have more than one active site. Reaction of the substrate (S) with the enzyme (E) to form a complex (ES) permits the reaction to proceed via a path of lower activation energy than the noncatalyzed path. When the complex decomposes, products (P) are formed and the enzyme is regenerated.

$$E + S \rightleftharpoons ES \longrightarrow E + P$$

One example is the hydrolysis of sucrose. Recall from Figure 28-8 that sucrose is a disaccharide of glucose and fructose.

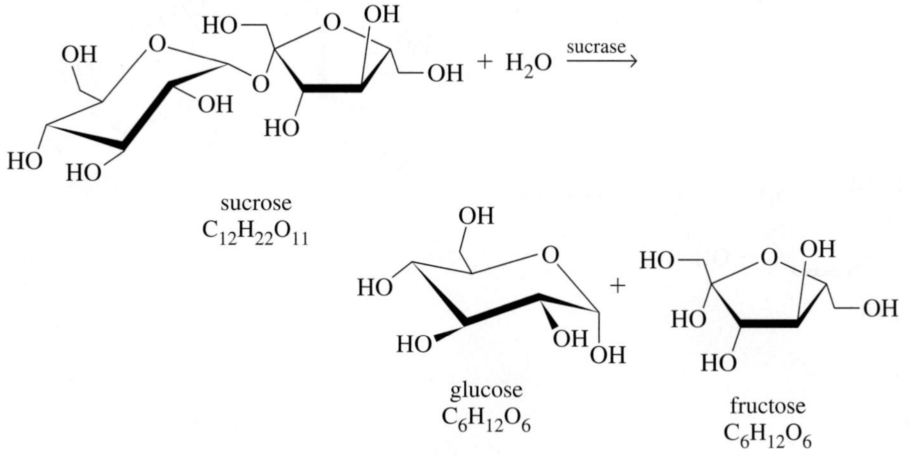

sucrose
$C_{12}H_{22}O_{11}$

glucose
$C_6H_{12}O_6$

fructose
$C_6H_{12}O_6$

The first step is the binding of sucrose to the active site of the enzyme sucrase, as illustrated in Figure 28-19. When the sucrose is in the active site of the enzyme, a number of chemical processes can take place. For example, a proton can be transferred from the sucrase to the glucose, which causes the active site of the sucrase to open up, causing the C—O bond joining fructose and glucose to weaken. The transfer of a proton from an amino acid side chain of the sucrase molecule may destroy hydrogen bonding within the enzyme and cause it to open up; this process stretches the sucrose molecule by pulling at each end, as suggested in Figure 28-20. In this distorted form, the C—O of the disaccharide is highly susceptible to attack by a water molecule. Figure 28-21 shows that after this reaction, glucose and fructose are formed and the proton is transferred back to the sucrase, which causes the active site to return to its original

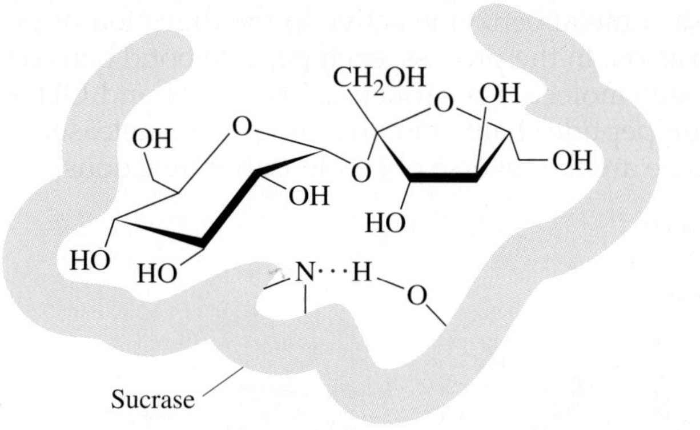

◄ FIGURE 28-19
Sucrase binding a sucrose molecule
The figure illustrates the "pocket" in the enzyme sucrase that the sucrose fits into. The globular protein sucrase (in blue) is much larger than sucrose.

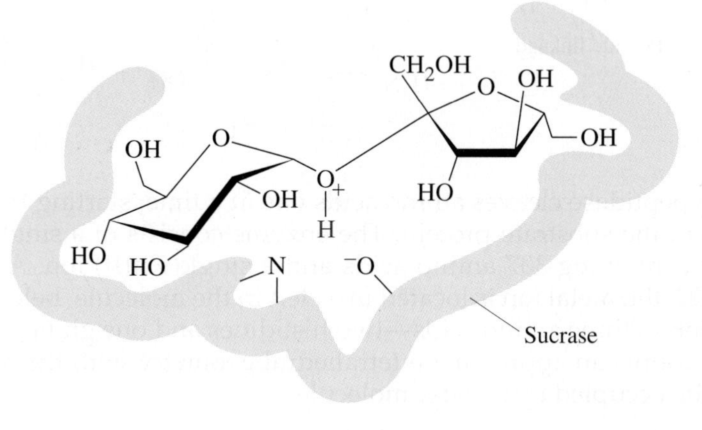

◄ FIGURE 28-20
Change in enzyme conformation after binding of substrate
The transfer of a proton (in orange) from the sucrase to the sucrose is thought to open up the enzyme structure, leading to a weakening of the C—O bond in sucrose.

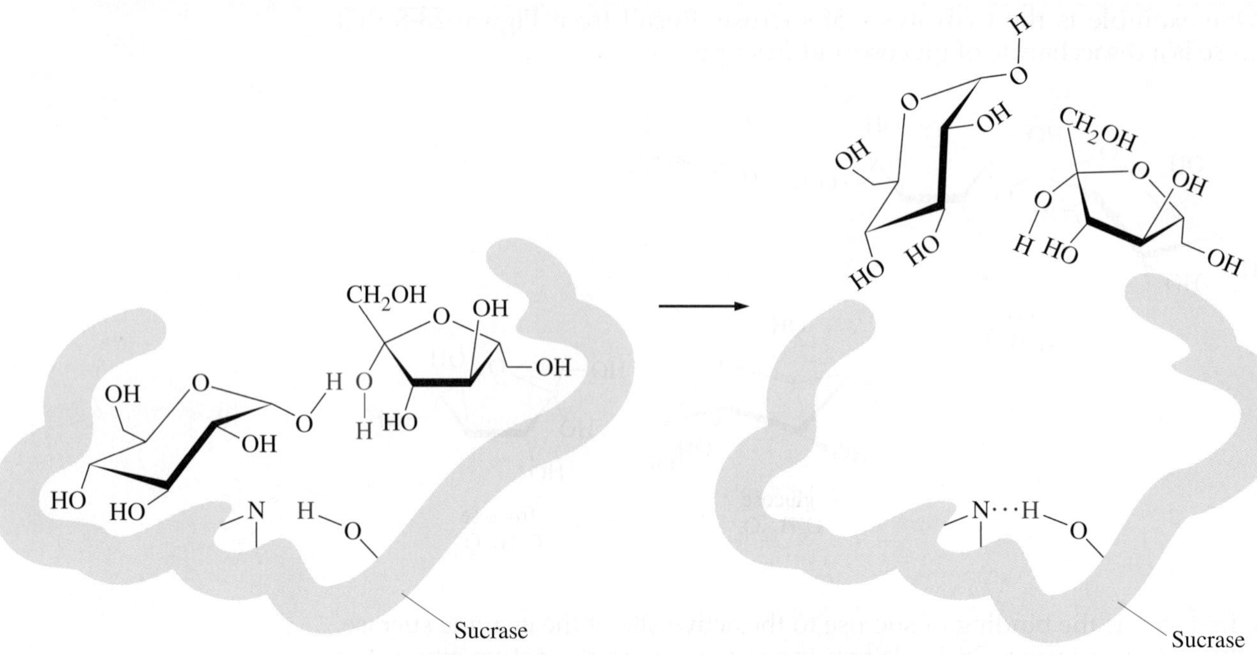

▲ FIGURE 28-21
Completion of the enzyme-catalyzed reaction
The weakened C—O bond in sucrose is susceptible to nucleophilic attack by a water molecule (in red) to produce glucose and fructose. These smaller molecules escape the pocket designed to fit sucrose. The sucrase returns to its original state with the proton (orange) returned to the hydroxyl group.

state. The smaller fructose and glucose molecules leave the active site, and the enzyme is ready to go again.

Enzymes such as sucrase do not contain metal ions, but almost a third of known enzymes do. One enzyme with a metal ion in the active site is carboxypeptidase, a metalloenzyme active in the digestion of proteins into individual amino acids. In the process, each peptide bond between amino acids is cleaved. A water molecule is broken apart, the H and OH are added to the ends of the polypeptide chain, and an amino acid is released. Hydrolysis reactions of this type are the reverse of condensation reactions.

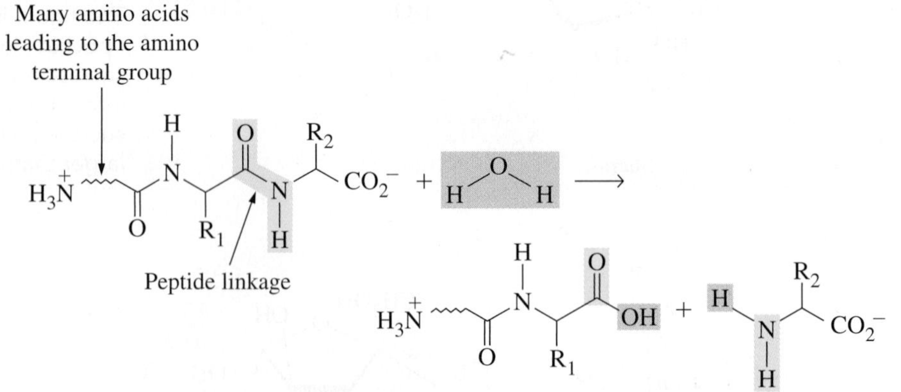

Carboxypeptidase cleaves amino acids one at a time, starting from the carboxyl end of the substrate protein. The enzyme consists of a single polypeptide chain containing 307 amino acids and a single Zn(II) ion. As shown in Figure 28-22, the metal ion is located in a cleft in the molecule, held in place by coordination to three amino acids—two histidines and one glutamic acid. The Zn(II) ion adopts an approximate tetrahedral geometry with the fourth coordination site occupied by a water molecule.

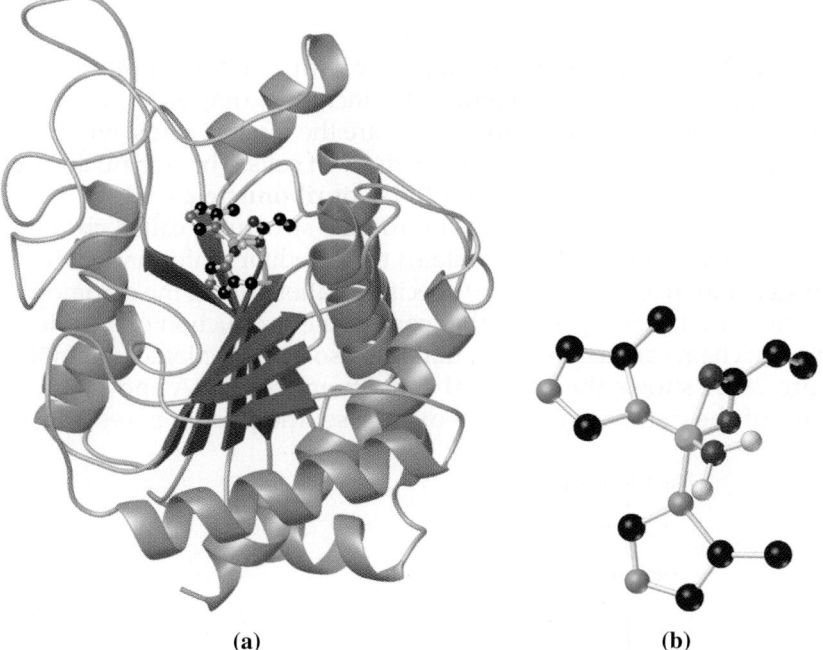

◀ FIGURE 28-22
Structure of carboxypeptidase
(a) In this illustration, the secondary structure of the helices (blue), pleated sheets (purple), and nonspecific secondary structure (yellow) combine to give a visualization of the overall tertiary structure of carboxypeptidase. (b) Coordination of the Zn(II) ion, dark green sphere, in the active site.

During the hydrolysis of a protein by carboxypeptidase, the protein substrate binds in the active site of the enzyme, as suggested by Figure 28-23. The substrate is held near the zinc center by hydrogen bonding to several amino acids in the cleft. The details of the mechanism have not been conclusively determined, except that it is known that it is the water coordinated to the Zn(II) that is used in the cleavage of the peptide bond. The proximity of the peptide linkage to the highly polarized water molecule coordinated to the Zn(II) ion permits the cleavage of the peptide bond and release of an amino acid.

The optimum temperature for enzyme activity is about 37 °C (98 °F) for enzymes present in warm-blooded animals. Above this temperature, enzyme activity declines as the secondary and tertiary structures are disrupted and the active site is distorted.

The two enzymes described here provide only a glimpse of the fascinating chemistry involved in the action of enzymes, which is currently a very active area of research.

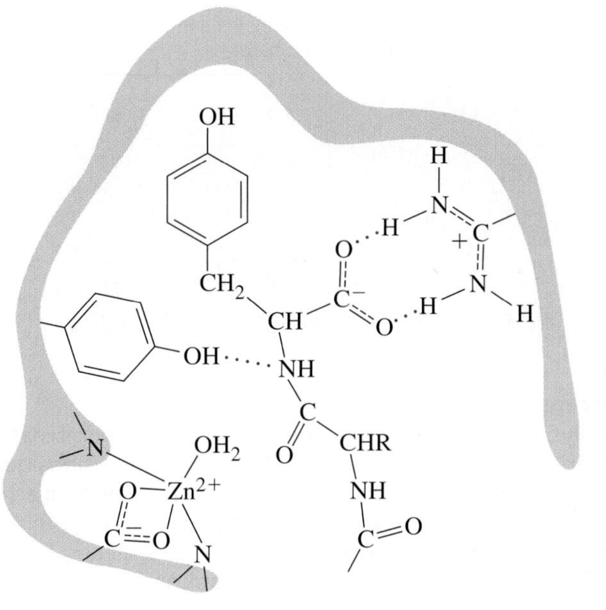

◀ FIGURE 28-23
Binding of protein to the active site of carboxypeptidase
A schematic of the C-terminus of a protein bound at the active site of carboxypeptidase. Hydrogen bonding to enzyme side chains holds the protein in place to facilitate nucleophilic attack by the water molecule.

28-6   Nucleic Acids

Lipids, carbohydrates, and proteins together with water constitute about 99% of most living organisms. The remaining 1% includes compounds of vital importance to the existence of life. Among these are the **nucleic acids**, which carry the information that directs the metabolic activity of cells. The two nucleic acids are **DNA**, or **deoxyribonucleic acid**, and **RNA**, or **ribonucleic acid**.

DNA is found in the cell nucleus on rodlike structures called chromosomes, which also contain protein. DNA contains the hereditary information that is passed from generation to generation. At specific locations along the chromosomes are *genes*, which are DNA sequences that control the production of a protein chain. This protein chain can manifest as a trait but also as a functional unit in the cell.

Figure 28-24 shows the chemical constituents of DNA and RNA. DNA is made up of the pentose sugar 2-deoxyribose, the purine bases adenine and

Nucleic acids

↓ Hydrolysis

Nucleotides

↓ Hydrolysis

Phosphoric acid   Nucleosides

↓ Hydrolysis

Pentose sugars   Purine bases   Pyrimidine bases

Pentose sugars

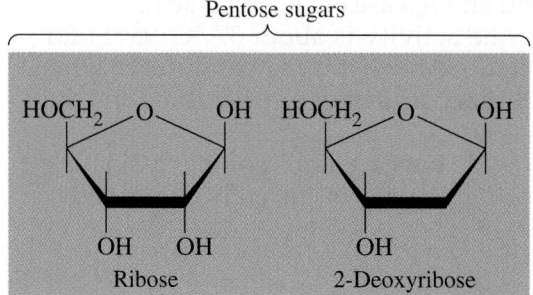

Ribose   2-Deoxyribose

Purine bases

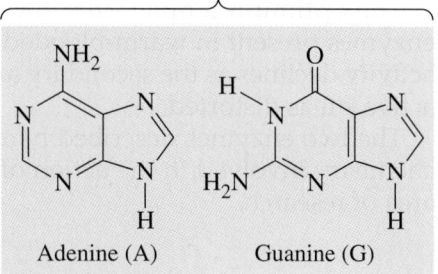

Adenine (A)   Guanine (G)

Pyrimidine bases

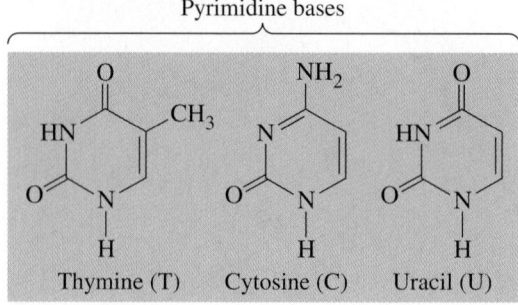

Thymine (T)   Cytosine (C)   Uracil (U)

▲ FIGURE 28-24
Constituents of nucleic acids
Tracing the hydrolysis reactions in the reverse direction, the combination of a pentose sugar and a purine or a pyrimidine base yields a nucleoside. A nucleoside in combination with phosphoric acid yields a nucleotide. A nucleic acid is a polymer of nucleotides. If the sugar is 2-deoxyribose and the bases are A, G, T, and C, the nucleic acid is DNA. If the sugar is ribose and the bases are A, G, U, and C, the nucleic acid is RNA. (The term "2-deoxy" means without an O atom on the second C atom.)

guanine, and the pyrimidine bases cytosine and thymine. RNA differs in that it contains ribose instead of deoxyribose, and the base uracil instead of thymine. Both DNA and RNA contain phosphate groups. The combination of a pentose sugar and a purine or pyrimidine base is called a *nucleoside*. The combination of a nucleoside and a phosphate group is called a *nucleotide*. A DNA molecule is made up of two strands of nucleotides held together by hydrogen bonds and twisted into a helix. The figure in the margin illustrates the hydrogen bonding between purine and pyrimidine bases. RNA is generally single stranded but folds back on itself to give helical structures or other hydrogen bonded structures. Figure 28-25 shows a portion of a nucleic acid chain and how the constituents are joined by phosphate groups. If the sugar is 2-deoxyribose and the bases are adenine (A), guanine (G), thymine (T), and cytosine (C), the nucleic acid is DNA. If the sugar is ribose and the bases are adenine (A), guanine (G), uracil (U), and cytosine (C), the nucleic acid is RNA.

The usual *double helix* form of DNA is shown in Figure 28-26. The postulation of this structure by Francis Crick and James Watson in 1953 was one of the great scientific breakthroughs of modern times. Their work was critically dependent on precise X-ray diffraction studies of DNA done by Maurice Wilkins and Rosalind Franklin and on a set of regularities regarding the purine and pyrimidine bases in nucleic acids discovered by Erwin Chargaff. These regularities, known as the *base-pairing rules*, require the following:

1. The amount of adenine is equal to the amount of thymine (A = T).
2. The amount of guanine is equal to the amount of cytosine (G = C).
3. The total amount of purine bases is equal to the total amount of pyrimidine bases (G + A = C + T).

To maintain the structure of a double helix, hydrogen bonding must occur between the two single strands. The necessary conditions for hydrogen bonding exist only if an A on one strand appears opposite a T on the other, or if a G is

▲ Hydrogen bonding between a purine and a pyrimidine base.

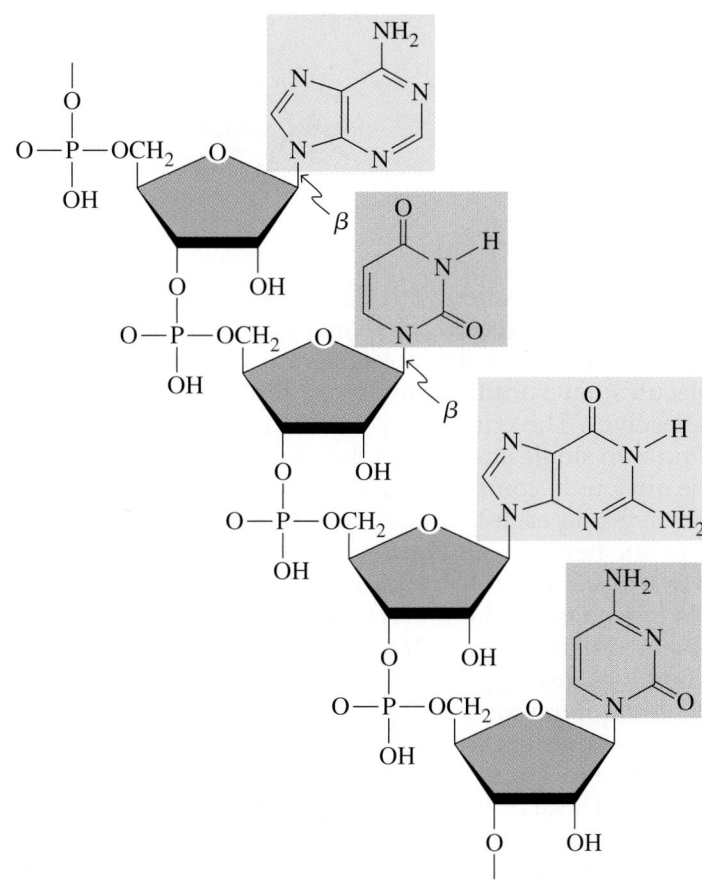

◀ FIGURE 28-25
A portion of a nucleic acid chain

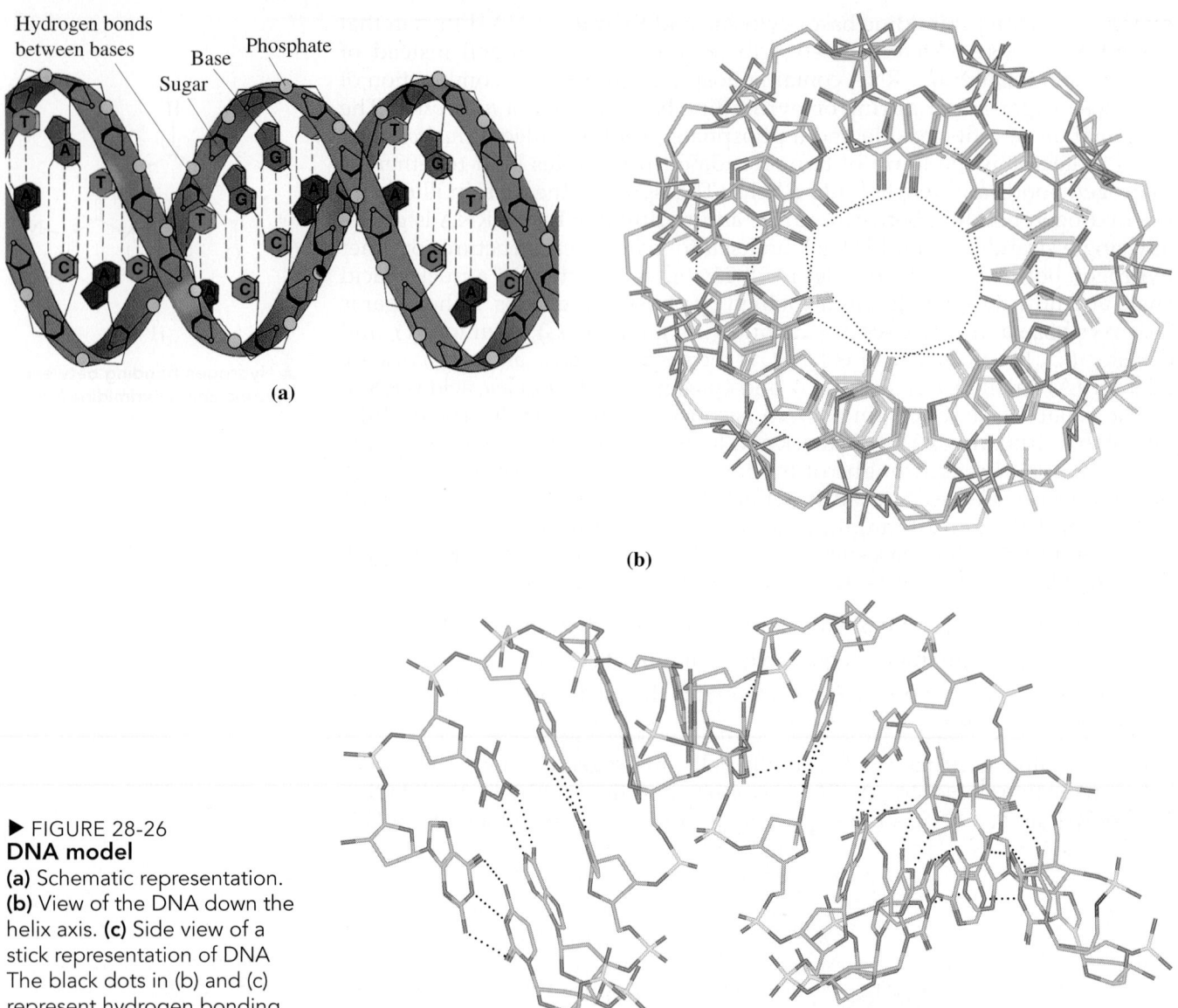

Hydrogen bonds
between bases
Base Phosphate
Sugar

(a)

(b)

(c)

▶ FIGURE 28-26
DNA model
(a) Schematic representation.
(b) View of the DNA down the
helix axis. **(c)** Side view of a
stick representation of DNA
The black dots in (b) and (c)
represent hydrogen bonding
between base pairs.

opposite a C. C cannot be paired with T because the relatively small molecules
(single ring) would not approach each other closely enough. The combination of
G and A cannot occur because the molecules are too large (double rings).

DNA molecules have a unique ability to replicate—that is, to make exact
copies of themselves. The critical step in DNA replication requires the molecule to unwind into single strands. As the unwinding occurs, free nucleotides
present in the nucleus become attached to the exposed portions of the two single strands, converting each to a new double helix of DNA consisting of one
old strand and one new strand (Fig. 28-27).

There are two pieces of evidence, each very convincing, that the process just
outlined does indeed occur. First, electron micrographs of the DNA molecule
have been obtained that capture DNA in the act of replication. Another elegant
experiment involves growing bacteria in a medium containing ^{15}N atoms so
that all the N atoms of the bases of the DNA molecules are ^{15}N. The bacteria are
then transferred to a nutrient with nucleotides containing normal ^{14}N. Here, the
bacteria are allowed to divide and reproduce. The DNA of the offspring cells
are then analyzed. Those of the first generation, for example, consist of DNA
molecules with one strand having ^{15}N and the other ^{14}N atoms. This is exactly
the result to be expected if replication occurs by the unzipping process

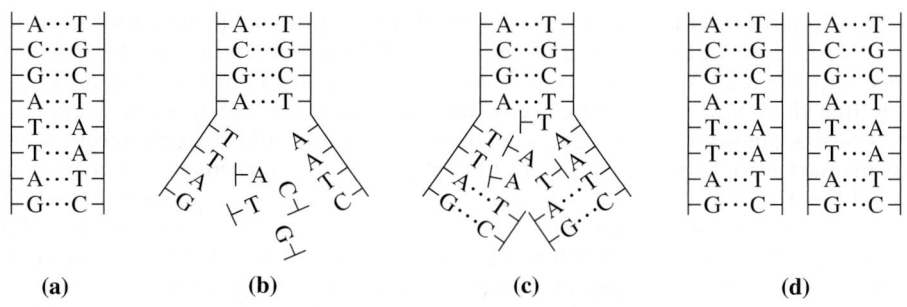

▲ FIGURE 28-27
Replication of a DNA molecule
(a) A DNA double helix. (b) The two strands begin to unzip. (c) Nucleotides from the cellular fluid crowd into the region between the two strands and attach to the exposed strands through hydrogen bonds. (d) Two DNA double helices identical to the original one.

described in Figure 28-27. The central importance of the base sequences in DNA is that they act as a chemical template for the formation of an organism's proteins; that is, they act as genes. This, together with a discussion of RNA, is outlined in Focus On 28 on the Mastering Chemistry website.

Mastering**CHEMISTRY**   www.masteringchemistry.com

How is the genetic code contained within DNA transmitted to the sites of protein synthesis? The process involves molecules acting as initiators, messengers, movers, and terminators. For a discussion of the process and the roles that various molecules play in it, go to the Focus On feature for Chapter 28, Protein Synthesis and the Genetic Code, on the MasteringChemistry site.

Summary

28-1 Chemical Structure of Living Matter: An Overview—Four general categories of substances found in living organisms are lipids, carbohydrates, proteins, and nucleic acids. **Cells**, which are the fundamental structural units of living things, contain macromolecules from each of these categories (Fig. 28-2).

28-2 Lipids—A range of substances of biological origin that are soluble in nonpolar or low-polarity solvents are classified as **lipids**. One familiar group of lipids is the **triglycerides**, esters of glycerol with long-chain monocarboxylic (fatty) acids. If saturated fatty acids predominate, the triglyceride is a **fat**. If unsaturation (in the form of double bonds) occurs in some of the fatty acid components, the triglyceride is an **oil**. **Saponification**, the hydrolysis of a triglyceride with a strong base, yields glycerol and salts of the fatty acid components (equation 28.1). The salt of a fatty acid is called a **soap**.

28-3 Carbohydrates—The most common among simple **carbohydrate** molecules are five- and six-carbon-chain polyhydroxy aldehydes and ketones. These molecules can be represented by **Fischer projection formulas**, which project the three-dimensional structures in two dimensions (Fig. 28-5). Carbohydrates contain chiral carbon atoms, existing therefore as optical isomers. One isomer, designated *d* or (+), rotates the plane of polarized light

"to the right" (clockwise) and is said to be **dextrorotatory**. Its enantiomer, designated *l* or (−), rotates polarized light "to the left" (counterclockwise) and is **levorotatory** (Fig. 28-4). Commonly encountered for carbohydrates with more than one chiral center are **diastereomers**, optical isomers that are not mirror images and hence not enantiomers. The arrangement of groups at an asymmetric carbon atom is called the **absolute configuration** and is designated by the *R*, *S* system of nomenclature. Syntheses of carbohydrates generally lead to equal amounts of enantiomers in mixtures referred to as **racemic mixtures**.

Monosaccharides are the simplest carbohydrate structures (Fig. 28-7). Two to ten monosaccharide units can join to form an **oligosaccharide**, for example, two monosaccharide units join to form a disaccharide (Fig. 28-8). Together, monosaccharides and oligosaccharides are also known as **sugars**. **Polysaccharides** contain from a few to a few thousand monosaccharide units. The most common of these are starch, glycogen, and cellulose (Fig. 28-9). A **reducing sugar** is one that can reduce Cu^{2+} to copper(I) oxide.

28-4 Proteins—Some 20 different **α-amino acids** (Table 28.3) are the basic building blocks of **proteins**, which consist of a long chain of amino acids. An α-amino acid has an amino group and a carboxyl group connected to the rest of the molecule at the α carbon atom. The pH at which the amino acid exists predominantly as a dipolar ion, a **zwitterion**, is the **isoelectric point**. Two amino acid molecules may join by eliminating a H_2O molecule between them to form a **peptide bond** (equation 28.5). When several peptides are linked together, the chain is referred to as a **polypeptide** chain. The linear sequence of amino acids of a protein is known as its **primary structure** (Fig. 28-11). The structure or shape of the entire amino acid chain in a protein is referred to as its **secondary structure**. Types of secondary structure include helical coils (Fig. 28-12) and pleated sheets (Fig. 28-13). **Tertiary structure** is the three-dimensional arrangement of the amino acid chains with respect to each other (Fig. 28-14), while **quaternary structure** is the arrangement of different polypeptide chains to form a larger biomolecule, such as hemoglobin (Fig. 28-16). **Denaturation** occurs when a protein loses it biological activity due to the unfolding of its tertiary structure.

28-5 Aspects of Metabolism—Lipids, carbohydrates, and proteins are complex molecules that are broken

down into their simplest units as part of the process of **metabolism** (Fig. 28-17). Carbohydrates are degraded into monosaccharides, proteins into amino acids, and lipids into glycerol and fatty acids. Ultimately, these degradation products are decomposed into still smaller molecules, such as CO_2, H_2O, NH_3, and urea. **Enzymes** are proteins that catalyze metabolic reactions. Energy released in catabolic processes is used to convert **adenosine diphosphate (ADP)** to **adenosine triphosphate (ATP)**, which can then supply the energy needs of other processes.

28-6 Nucleic Acids—Nucleic acids are made up of

pentose sugars, purine and pyrimidine bases, and phosphate groups (Fig. 28-24). Two classes of nucleic acids are **ribonucleic acid (RNA)** and **deoxyribonucleic acid (DNA)** (Figs. 28-25 and 28-26). RNA contains the sugar ribose, while DNA contains the sugar 2-deoxyribose. Nucleic acids carry the information that directs the metabolic activity of cells and the hereditary information that is passed from one generation to the next.

Integrative Example

L-Threonine is one of the essential amino acids. It is found in animal protein (for example, eggs and milk) but is missing in some grains, such as rice. It is now common practice to fortify grains with the essential amino acids that they lack. The structure of L-threonine is shown below.

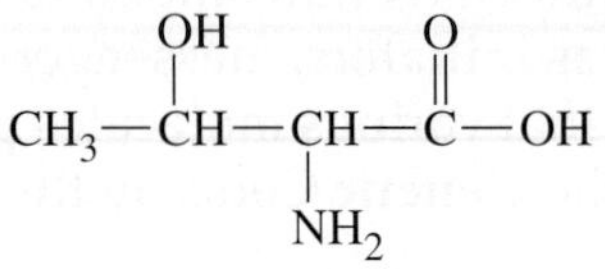

$$CH_3—\underset{\underset{NH_2}{|}}{CH}—\overset{\overset{OH}{|}}{CH}—\overset{\overset{O}{||}}{C}—OH$$

$$pK_{a_1} = 2.15 \quad pK_{a_2} = 9.12$$

Use the above data to determine the principal form of L-threonine and the direction that it will migrate under the influence of an electric field in a gel that is 0.25 M NaH_2PO_4 and 0.50 M Na_2HPO_4.

Analyze

The key to this problem is to recognize that the NaH_2PO_4/Na_2HPO_4 mixture is a buffer that controls the pH of the gel. Once the pH of the gel is established we can identify the ionization state for the bulk of the L-threonine and hence the direction it will migrate.

Solve

Use pK_{a_2} from Table 16.4 and the Henderson–Hasselbalch equation to calculate the pH of the gel.

$$pH = pK_{a_2} + \log([HPO_4^{2-}]/[H_2PO_4^-])$$

and

$$pH = 7.20 + \log(0.50/0.25) = 7.50$$

The structures of the three ionic forms in which L-threonine can appear are

Acidic solution:

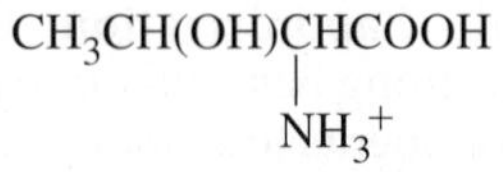

$$CH_3CH(OH)\underset{\underset{NH_3^+}{|}}{CH}COOH$$

Isoelectric point:

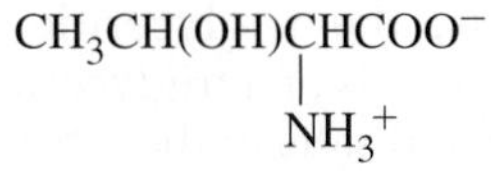

$$CH_3CH(OH)\underset{\underset{NH_3^+}{|}}{CH}COO^-$$

Basic solution:

$$CH_3CH(OH)\underset{\underset{NH_2}{|}}{CH}COO^-$$

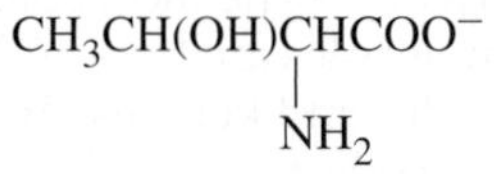

The pH at the first equivalence point in the titration of a weak diprotic acid is given by equation (17.5). Applied to L-threonine, the result is

$$pH = \frac{pK_{a_1} + pK_{a_2}}{2} = \frac{2.15 + 9.12}{2} = 5.64$$

The predominant species at this first equivalence point of pH = 5.64 is the dipolar zwitterion, which would not migrate at all in an electric field. However, in the gel medium of pH = 7.50 the hydroxide ion concentration is 100 times as great. The predominant species is the anion with charge −1. This form of the L-threonine migrates toward the positive electrode, the anode.

Assess

This problem suggests two ways to establish the isoelectric point of an amino acid. The first, used here, is to calculate pI by combining pK_a values of a polyprotic amino acid. Another way is to find the pH of a gel medium in which the amino acid does not migrate in the presence of an applied electric field, hence pH = pI. Additionally, migration through an electric field in a gel (gel electrophoresis) can be used to separate amino acid mixtures.

PRACTICE EXAMPLE A: Osmotic pressure experiments are one method to determine the molar mass of proteins. Myoglobin from beef heart muscle has a molar mass of 16.9 kg per mole. If you dissolve 0.500 g of beef heart myoglobin in 25.0 mL of water at T = 298 K, what is the height, h, of the solution generated in an osmotic pressure experiment? [*Hint*: The osmotic pressure is equivalent to the hydrostatic pressure given in equation (6.2).]

PRACTICE EXAMPLE B: Carbonic anhydrase is a metalloenzyme that reversibly catalyzes the reaction of carbon dioxide to bicarbonate ion:

$$CO_2(aq) + H_2O(l) \rightleftharpoons HCO_3^-(aq) + H^+(aq)$$

This is one mechanism of moving carbon dioxide from the cell to the lungs in which the bicarbonate ion is converted to carbon dioxide and then exhaled. **(a)** Draw a possible mechanism for the hydration of carbon dioxide to bicarbonate ion in which you illustrate the movement of electrons. **(b)** In Chapter 14, equation (14.36) represents the rate of catalysis versus substrate concentration. Equation (14.36) can be transformed to a linear form known as the Lineweaver-Burk equation:

$$\frac{1}{V} = \frac{1}{V_{max}} + \frac{K_M}{V_{max}} \times \frac{1}{[S]}$$

where $V_{max} = k_2[E]_0$. Determine K_M and k_2, given the following data taken at 0.5 °C with an initial enzyme concentration of 2.3 nM.

Rate[a], M s^{-1}	[CO$_2$], mM
9.02 × 10^{-5}	1.25
1.67 × 10^{-4}	2.5
2.90 × 10^{-4}	5.0
6.47 × 10^{-4}	20.0

[a]Data from H. DeVoe and G. B. Kistiakowsky, *J. Amer. Chem. Soc.*, **83**, 274 (1961).

Mastering**CHEMISTRY**

You'll find a link to additional self study questions in the study area on www.masteringchemistry.com

Exercises

Structure and Composition of the Cell

Exercises 1 to 4 refer to a typical Escherichia coli *bacterium. This is a cylindrical cell about 2 μm long and 1 μm in diameter, weighing about 2 × 10^{-12} g and containing about 80% water by volume.*

1. The intracellular pH is 6.4 and [K$^+$] = 1.5 × 10^{-4} M. Determine the number of **(a)** H$_3$O$^+$ ions and **(b)** K$^+$ ions in a typical cell.
2. Calculate the number of lipid molecules present, assuming their average molecular mass to be 700 u and the lipid content to be 2% by mass.

3. The cell is about 15% protein by mass with 90% of the protein in the cytoplasm. Assuming an average molecular mass of 3 × 10^4 u, how many protein molecules are present in the cytoplasm?
4. A single chromosomal DNA molecule contains about 4.5 million nucleotide units. If this molecule were extended so that the nucleotide units were 450 pm apart, what would be the length of the molecule? How does this compare with the length of the cell itself? What does this result suggest about the shape of the DNA molecule?

Lipids

5. Name the following compounds.

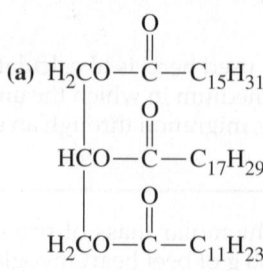

(a) H₂CO—C—C₁₅H₃₁
$$\text{H}_2\text{CO}-\overset{\overset{\displaystyle O}{\|}}{\text{C}}-\text{C}_{15}\text{H}_{31}$$
$$\text{HCO}-\overset{\overset{\displaystyle O}{\|}}{\text{C}}-\text{C}_{17}\text{H}_{29}$$
$$\text{H}_2\text{CO}-\overset{\overset{\displaystyle O}{\|}}{\text{C}}-\text{C}_{11}\text{H}_{23}$$

(b)
$$\text{H}_2\text{CO}-\overset{\overset{\displaystyle O}{\|}}{\text{C}}-\text{C}_{17}\text{H}_{33}$$
$$\text{HCO}-\overset{\overset{\displaystyle O}{\|}}{\text{C}}-\text{C}_{17}\text{H}_{33}$$
$$\text{H}_2\text{CO}-\overset{\overset{\displaystyle O}{\|}}{\text{C}}-\text{C}_{17}\text{H}_{33}$$

(c) $C_{13}H_{27}CO_2^-Na^+$

6. Write structural formulas for the following.
   (a) glyceryl palmitolauroeleostearate
   (b) tripalmitin
   (c) potassium myristate
   (d) butyl oleate

7. Describe the similarities and differences between **(a)** trilaurin and trilinolein, **(b)** a soap and a phospholipid such as a lecithin.

8. Write a structural formula for a generic phosphatidic acid—an acid that produces a cephalin when esterified with ethanolamine and a lecithin when esterified with choline. Mono- and diglycerides are found in many processed foods. Write structural formulas for a generic monoglyceride and a generic diglyceride.

9. Oleic acid is a moderately unsaturated fatty acid. Linoleic acid is *polyunsaturated*. What structural feature characterizes polyunsaturated fatty acids? Is stearic acid polyunsaturated? Is eleostearic acid? Why do you suppose safflower oil is so highly recommended in dietary programs?

10. Corn oil and safflower oil are popular cooking oils (Table 28.2). They can be converted to solid fats by hydrogenation. Which would consume the greater amount of $H_2(g)$ in its hydrogenation to a solid fat, 1 kg of corn oil or 1 kg of safflower oil? Explain.

11. Write structural formulas to represent the products of the saponification of tripalmitin with NaOH(aq).

12. Calculate the maximum mass of the sodium soap that can be prepared from 105 g of glyceryl trimyristate.

Carbohydrates

13. Characterize the two sugars represented below using the terminology from page 1273. Further indicate whether the sugar has a D or L configuration.

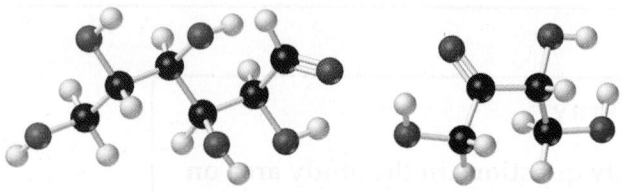

14. Write the structure for the straight-chain form of L-glucose. Does the structure determine if this isomer is levorotatory? Explain.

15. From the given structure of L-(+)-arabinose below, derive the structure of **(a)** D-(−)-arabinose; **(b)** a diastereomer of L-(+)-arabinose.

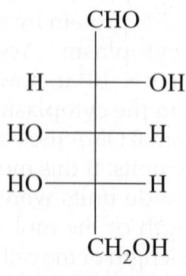

CHO

H——OH

HO——H

HO——H

CH₂OH

L-(+)-Arabinose

16. Describe the similarities and differences in the structures of the following three compounds. (a) β-D-(+)-glucose; **(b)** D-(−)-arabinose; **(c)** D-(+)-glucose.

17. The following terms are all related to stereoisomers and their optical activity. Explain the meaning of each: **(a)** dextrorotatory; **(b)** levorotatory; **(c)** racemic mixture; **(d)** (R).

18. The following terms are all related to optical isomers. Explain the meaning of each: **(a)** diastereomers; **(b)** enantiomers; **(c)** (−); **(d)** D configuration.

19. Explain the meaning of the term *reducing sugar*. What structural feature characterizes a reducing sugar, and what is the test for a reducing sugar? What mass of $Cu_2O(s)$ should be produced when 0.500 g glucose is subjected to this test?

20. There are eight aldopentoses. Draw their structures, and indicate which are enantiomers.

21. The pure α and β forms of D-glucose rotate the plane of polarized light to the right by 112° and 18.7°, respectively (denoted as +112 and +18.7). Are these two forms of glucose enantiomers or diastereomers?

22. When a mixture of the pure α and β forms of D-glucose is allowed to reach equilibrium in solution, the rotation changes to +52.7 (a phenomenon known as mutarotation). What are the percentages of the α and β forms in the equilibrium mixture? [*Hint:* Refer to Exercise 21.]

Fischer Projections and *R, S* Nomenclature

23. For each pair of Fischer projections below, decide whether they represent the same compound, enantiomers, or diastereomers. Check your conclusion by designating the configuration at each stereocenter as *R* or *S*.

(a)

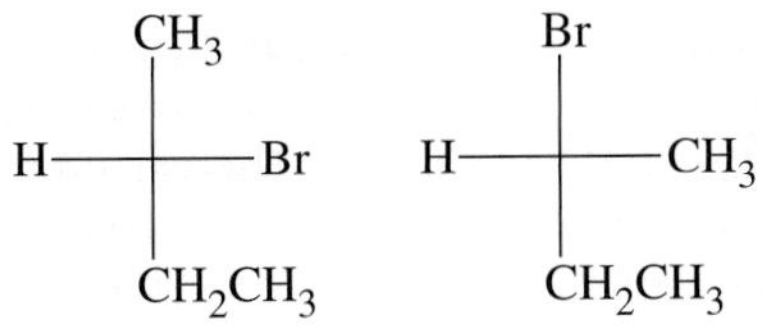

(b)

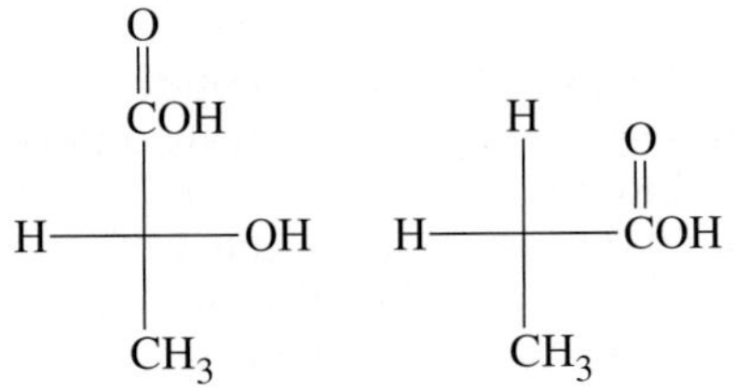

(c)

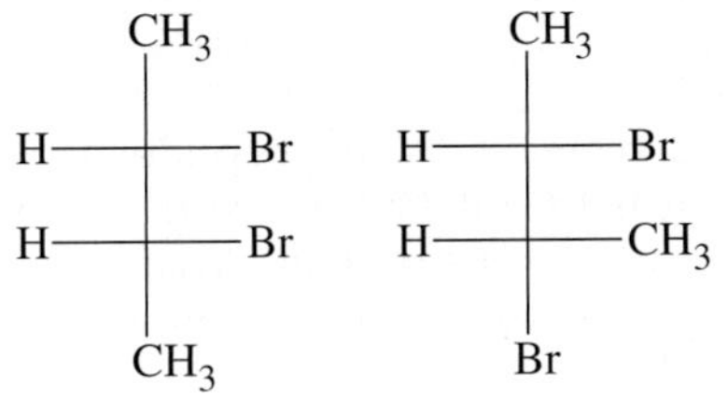

(d)

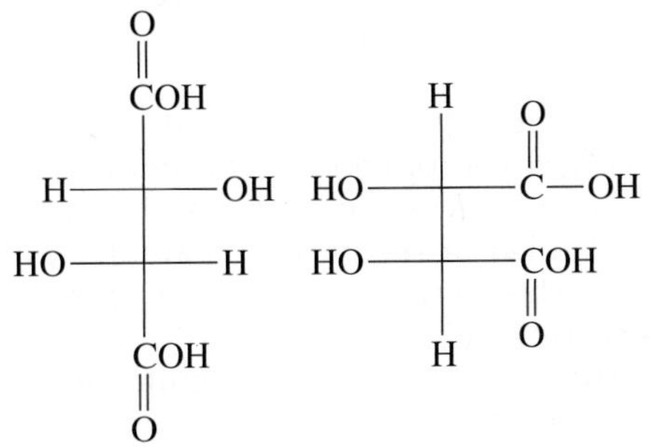

24. For each pair of Fischer projections below, decide whether they represent the same compound, enantiomers, or diastereomers. Check your conclusion by designating the configuration at each stereocenter as *R* or *S*.

(a)

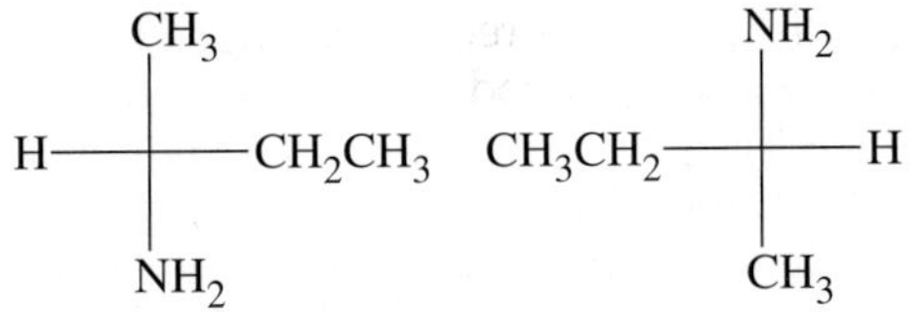

(b)

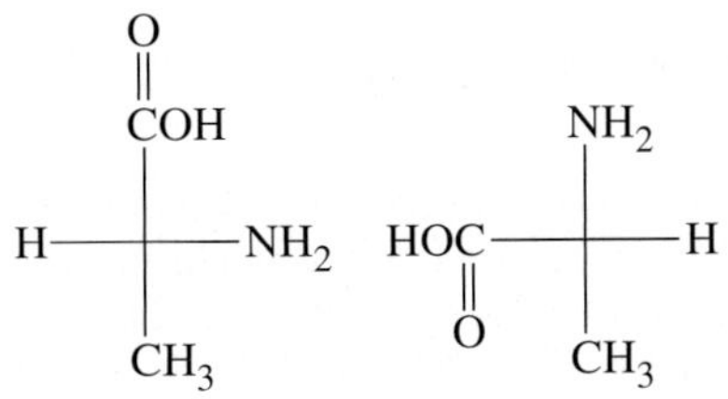

(c)

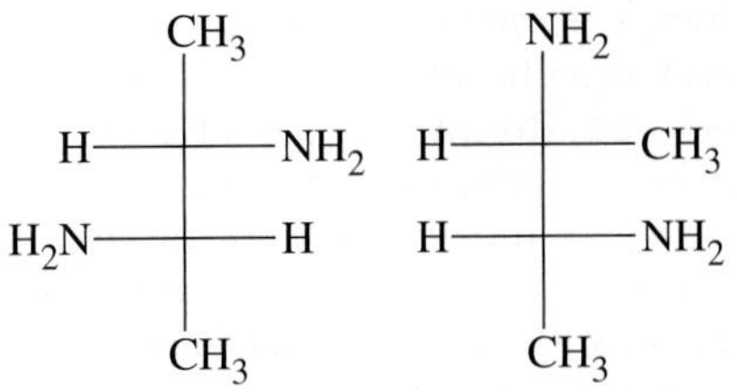

(d)

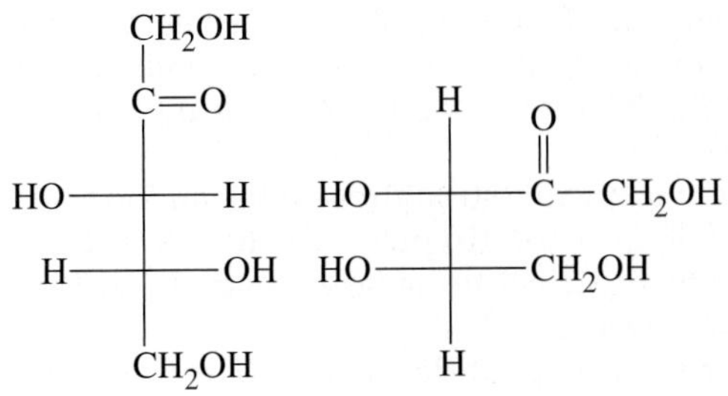

25. Redraw each of the following molecules as a Fischer projection; then assign *R* or *S* to each stereocenter.

(a)

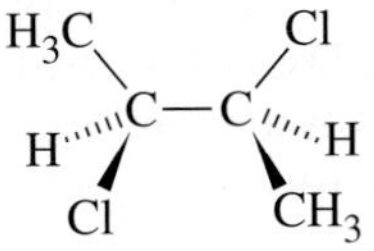

(b)

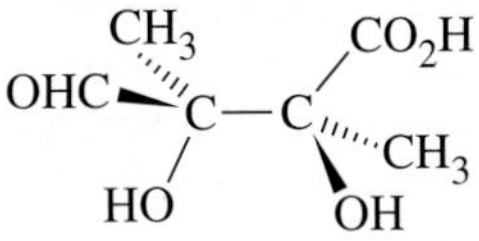

(c)

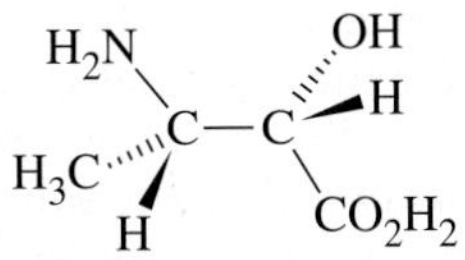

(d)

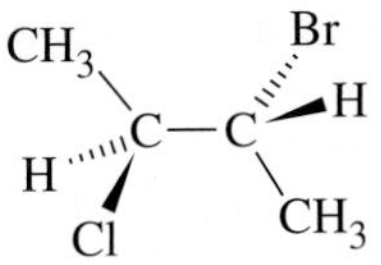

26. Redraw each of the following molecules as a Fischer projection; then assign *R* or *S* to each stereocenter.

(a) **(b)**

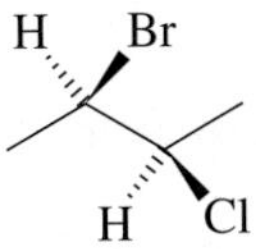

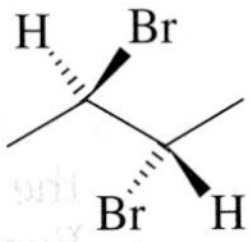

(c)

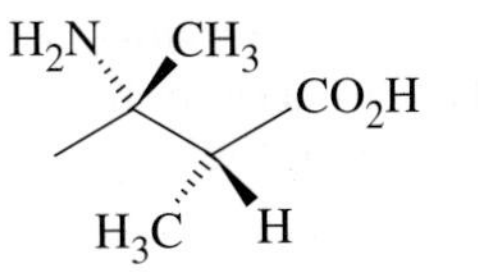

(d)

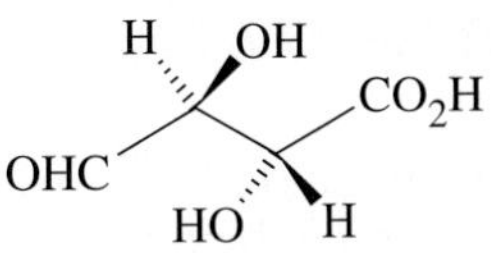

(e)

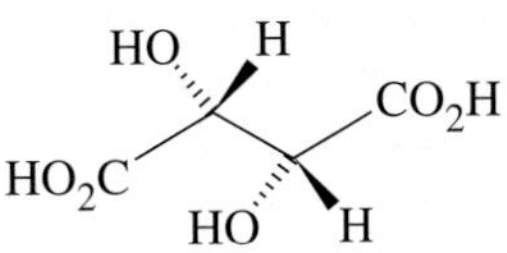

Amino Acids, Polypeptides, and Proteins

27. Describe what is meant by each of the following terms, using specific examples where appropriate: **(a)** α-amino acid; **(b)** zwitterion; **(c)** isoelectric point; **(d)** peptide bond; **(e)** tertiary structure.

28. Describe what is meant by each of the following terms, using specific examples where appropriate: **(a)** polypeptide; **(b)** protein; **(c)** N-terminal amino acid; **(d)** α helix; **(e)** denaturation.

29. Write the formulas of the species expected if the amino acid phenylalanine is maintained in **(a)** 1.0 M HCl; **(b)** 1.0 M NaOH; **(c)** a buffer solution with pH = 5.7.

30. Draw plausible structures for the amino acid histidine at **(a)** pH = 3.0; **(b)** pH = 7.6; **(c)** pH = 12.0.

31. Write the structures of **(a)** alanylcysteine; **(b)** threonyl-valylglycine.

32. For the polypeptide Met-Val-Thr-Cys, **(a)** write the structural formula; **(b)** name the polypeptide. [*Hint:* Which is the N-terminal, and which is the C-terminal amino acid?]

33. A mixture of the amino acids lysine, proline, and aspartic acid is placed in a gel buffered at pH 6.3. An electric current is applied between an anode and a cathode immersed in the gel. Toward which electrode will each amino acid migrate?

34. A mixture of the amino acids histidine, glutamic acid, and phenylalanine is placed in a gel buffered at pH 5.7. An electric current is applied between an anode and a cathode immersed in the gel. Toward which electrode will each amino acid migrate?

35. Draw condensed structural formulas showing what form you would expect for the essential amino acid threonine **(a)** in strongly acidic solutions; **(b)** at the isoelectric point; and **(c)** in strongly basic solutions.

36. Draw condensed structural formulas for the following amino acids buffered at pH 6.0: **(a)** aspartic acid; **(b)** lysine; and **(c)** alanine.

37. Write the structures of **(a)** the different tripeptides that can be obtained from a combination of alanine, serine, and lysine; **(b)** the tetrapeptides containing two serine and two alanine amino acid units.

38. Write the structures of the different tetrapeptides that can be obtained from a combination of alanine, lysine, serine, and phenylalanine. Give the abbreviated formula of each (such as Ala-Lys-Ser-Phe), starting at the N-terminal end.

39. After undergoing complete hydrolysis, a polypeptide yields the following amino acids: Gly, Leu, Ala, Val, Ser, Thr. Partial hydrolysis yields the following fragments: Ser-Gly-Val, Thr-Val, Ala-Ser, Leu-Thr-Val, Gly-Val-Thr. An experiment using a marker establishes that Ala is the N-terminal amino acid.
    (a) Establish the amino acid sequence in this polypeptide.
    (b) What is the name of the polypeptide?

40. After undergoing complete hydrolysis, a polypeptide yields the following amino acids: Ala, Gly, Lys, Ser, Phe. Partial hydrolysis yields the following fragments: Ala-Lys-Ser, Gly-Phe-Gly, Ser-Gly, Gly-Phe, Lys-Ser-Gly. An experiment using a marker establishes that Ala is the N-terminal amino acid.
    (a) Establish the amino acid sequence in this polypeptide.
    (b) What is the name of the polypeptide?

41. Describe what is meant by the primary, secondary, and tertiary structure of a protein. What is the quaternary structure? Do all proteins have a quaternary structure? Explain.

42. Sickle-cell anemia is sometimes referred to as a "molecular" disease. Comment on the appropriateness of this term.

43. The amino acid (R)-alanine is found in insect larvae. Draw the Fischer projection of this amino acid.

44. The amino acid (R)-serine is found in earthworms. Draw the Fischer projection of this amino acid.

45. Draw the dashed-wedged line structure for (S)-alanine and (S)-phenylalanine.

46. Draw the dashed-wedged line structure for (R)-proline and (S)-valine.

Nucleic Acids

47. What are the two major types of nucleic acids? List their principal components.

48. DNA has been called the "thread of life." Comment on the appropriateness of this expression.

49. If one strand of a DNA molecule has the base sequence of AGC, what must be the sequence on the opposite strand? Draw a structure of this portion of the double helix, showing all hydrogen bonds.

50. If one strand of a DNA molecule has the base sequence TCT, what must be the sequence on the opposite strand? Draw a structure of this portion of the double helix, showing all hydrogen bonds.

Integrative and Advanced Exercises

51. The protein molecule hemoglobin contains four iron atoms. The mass percent iron in hemoglobin is 0.34%. What is the molecular mass of hemoglobin?

52. A 1.00 mL solution containing 1.00 mg of an enzyme was deactivated by the addition of 0.346 μmol AgNO₃. What is the *minimum* molecular mass of the enzyme? Why does this calculation yield only a minimum value?

53. What minimum volume of natural gas is needed to produce the hydrogen consumed in converting 15.5 kg of the oil glyceryl trioleate (triolein) to the fat glyceryl tristearate (tristearin)? The natural gas is 92.0% CH_4 and 8.0% C_2H_6 by volume and is measured at 25.5 °C and 756 mmHg pressure. The hydrogen is produced by steam re-forming the natural gas. Assume that

natural gas and steam are the only reactants and that carbon monoxide is the only product other than hydrogen.

54. The text states that there are 16 possible aldohexoses. Draw their structures, and indicate which are enantiomers.

55. The term "epimer" is used to describe diastereomers that differ in the configuration about a *single* carbon atom. Which pairs of the eight possible aldopentoses are epimers?

56. The amino acid ornithine, not normally found in proteins, has the following structure at a pH of 1.0.

$$H_3\overset{+}{N}—CH_2—CH_2—CH_2—\underset{\underset{+NH_3}{|}}{CH}—\overset{\overset{O}{||}}{C}—OH$$

$pK_{a_1} = 1.94; \quad pK_{a_2} = 8.65; \quad pK_{a_3} = 10.76$

What is the p*I* value of this amino acid?

57. In the experiment described on page 1294, the first-generation offspring of DNA molecules each contained one strand with ^{15}N atoms and one with ^{14}N. If the experiment were carried through a second, third, and fourth generation, what fractions of the DNA molecules would still have strands with ^{15}N atoms?

58. Bradykinin is a nonapeptide obtained by the partial hydrolysis of blood serum protein. It causes a lowering of blood pressure and an increase in capillary permeability. Complete hydrolysis of bradykinin yields three proline (Pro), two arginine (Arg), two phenylalanine (Phe), one glycine (Gly), and one serine (Ser) amino acid units. The N-terminal and C-terminal units are both arginine (Arg). In a hypothetical experiment, partial hydrolysis and sequence proof reveals the following fragments: Gly-Phe-Ser-Pro; Pro-Phe-Arg; Ser-Pro-Phe; Pro-Pro-Gly; Pro-Gly-Phe; Arg-Pro-Pro; Phe-Arg. Deduce the sequence of amino acid units in bradykinin.

59. With the aid of the table in Focus On 28 on the Mastering Chemistry website, propose a plausible sequence on a DNA strand that would code for the synthesis of the polypeptide Ser-Gly-Val-Ala. Why is there more than one possible sequence for the DNA strand?

60. Refer to the Integrative Example. A 1.00 g sample of threonine is dissolved in 10.0 mL of 1.00 M HCl, and the solution is titrated with 1.00 M NaOH. Sketch a titration curve for the titration, indicating the approximate pH at representative points on the curve [such as the initial pH, the half-neutralization point(s) and the equivalence point(s)].

61. Projection formulas impose certain limitations and care must be taken in drawing and manipulating them. Use (*R*)-(+)-glyceraldehyde as an example and show that:
(a) Rotating the Fischer projection by 90° in the plane of the paper produces the opposite absolute configuration. What is the effect of rotation by 180° and 270° in the plane of the paper?
(b) Interchanging two substituents converts one enantiomer into the other. What is the effect of two such interchanges?

62. A pentapeptide was isolated from a cell extract and purified. A portion of the compound was treated with 2,4-dinitrofluorobenzene (DNFB), and the resulting material hydrolyzed (recall Figure 28-10). Analysis of the hydrolysis products revealed one mole of DNP-methionine, two moles of methionine, and one mole each of serine and glycine. A second portion of the original compound was partially hydrolyzed and separated into four products. Separately, the four products were hydrolyzed further, giving the following four sets of compounds: (1) one mole of DNP-methionine, one mole of methionine, and one mole of glycine; (2) one mole of DNP-methionine and one mole of methionine; (3) one mole of DNP-serine and one mole of methionine; (4) one mole of DNP-methionine, one mole of methionine, and one mole of serine. What is the amino acid sequence of the pentapeptide?

63. Eighteen of the nineteen 1-amino acids have the *S* configuration at the α carbon (the first carbon after the carboxyl carbon). Cysteine is the only 1-amino acid that has an *R* configuration. Explain.

64. The systematic (IUPAC) name for the amino acid threonine is (2*S*,3*R*)-2-amino-3-hydroxybutanoic acid. The systematic name indicates that threonine has two stereocenters, at positions 2 and 3, with *S* and *R* configurations, respectively. Draw the Fischer projection of threonine. How many other possible stereoisomers of threonine are there?

65. Coupling ATP hydrolysis to a thermodynamically unfavorable reaction can shift the equilibrium of a reaction (see Chapter 19).
(a) Calculate *K* for the hypothetical reaction A $\longrightarrow$ B when $\Delta G^{\circ\prime}$ is 23 kJ mol^{-1} at 25 °C.
(b) Calculate *K* for the same reaction when it is coupled to the hydrolysis of ATP ($\Delta G^{\circ\prime} = -30$ kJ mol^{-1}). Compare with the value obtained in part (a).
(c) Many cells maintain [ATP] to [ADP] ratios of 400 or more. Calculate the ratio of [B] to [A] when [ATP]/[ADP] = 400 and P_i = 5 mM. Compare this ratio to that for the uncoupled reaction.

Feature Problems

66. Useful information about the structures of triglycerides can be obtained through saponification reactions. *The saponification value of a triglyceride is the number of milligrams of KOH required to saponify 1.00 g of the triglyceride.* Unsaturation in the fatty acids in triglycerides can be discovered by studying reactions in which I_2 is added across double bonds. *The iodine number is the number of grams of I_2 that reacts with 100 g of a triglyceride.*

(a) Determine the saponification value of glyceryl tristearate and the iodine number of glyceryl trioleate.
(b) Castor oil is a mixture of triglycerides with about 90% of its fatty acid content in ricinoleic acid, $CH_3(CH_2)_5CH(OH)CH=CH(CH_2)_7COOH$. Estimate the saponification value and the iodine number of castor oil.

(c) Based on the composition given in Table 28.2, estimate the range of values for the saponification value and iodine number of safflower oil.

67. The so-called Ruff degradation is a chain-shortening reaction in which an aldose chain is shortened by one C atom, hexoses, for example, being converted into pentoses. In the Ruff degradation, the calcium salt of an aldonic acid (the corresponding carboxylic acid of an aldose) is oxidized with hydrogen peroxide. Ferric ion catalyzes the reaction. The calcium salt of the aldonic acid necessary for the Ruff degradation is obtained by oxidizing an aldose with an aqueous solution of bromine and then adding calcium hydroxide. The reaction scheme is as follows

$$RCH(OH)CHO \xrightarrow[\text{2. Ca (OH)}_2]{\text{1. Br}_2, \text{H}_2\text{O}} RCH(OH)CO_2^-(Ca^{2+})_{\frac{1}{2}}$$

$$\downarrow \begin{matrix} H_2O_2 \\ Fe^{3+}(aq) \end{matrix}$$

$$RCHO + CO_2$$

where R represents the rest of the chain of the aldose.
(a) Show that D-glucose can be degraded into D-arabinose. Which other aldose can be degraded into D-arabinose?
(b) Which two monosaccharides can be degraded into D-glyceraldehyde by employing the Ruff degradation only once?

68. If D-(+)-glyceraldehyde is treated with HCN in aqueous solution under basic conditions for three days at room temperature, cyanohydrins are formed (see Chapter 27). The cyanohydrins are not isolated, but are hydrolyzed to hydroxyacids in the same reaction mixture using dilute sulfuric acid. In this process, a new stereocenter is formed in the molecule. The products are diastereomers, formed in unequal amounts, and separable from each other by recrystallization because of their different physical properties, including solubilities. The trihydroxybutanoic acids are separated and then oxidized to tartaric acid with dilute nitric acid, which oxidizes only the primary alcohol group.
(a) Ignoring stereochemistry, draw the reaction sequence for the transformations described above and hence deduce the structure of tartaric acid.
(b) Starting from the Fischer projection of D-(+)-glyceraldehyde and using the reaction scheme from part (a), draw Fischer projections of the two trihydroxybutanoic acids formed and designate the chiral centers as R or S.
(c) Starting from the Fischer projection of D-(+)-glyceraldehyde and using the reaction scheme from part (a), draw Fischer projections of the two forms of tartaric acid formed and designate the chiral centers as R or S.
(d) One form of tartaric acid obtained is optically active, rotating the plane of polarized light in a negative sense (−). The other isomer formed, called *meso*-tartaric acid, is not optically active. Explain why the other isomer is not optically active. Draw the dashed-wedged line structure that corresponds to the Fischer projection of *meso*-tartaric acid. Can you describe how the two halves of the molecule are related? Using Fischer projections, write equations for the conversion of L-(−)-glyceraldehyde to tartaric acid. Show clearly the stereochemistry of the tartaric acids that are formed, and indicate whether you expect them to be optically active.

Self-Assessment Exercises

69. In your own words, define the following terms or symbols: (a) (+); (b) L; (c) sugar; (d) α-amino acid; (e) isoelectric point.
70. Briefly describe each of the following ideas, phenomena, or methods: (a) saponification; (b) chiral carbon atom; (c) racemic mixture; (d) denaturation of a protein.
71. Explain the important distinctions between each pair of terms: (a) fat and oil; (b) enantiomer and diastereomer; (c) primary and secondary structure of a protein; (d) DNA and RNA; (e) ADP and ATP.
72. The substance *glyceryl trilinoleate* (linoleic acid: $C_{17}H_{31}COOH$) is best described as a (a) fat; (b) oil; (c) wax; (d) fatty acid; (e) phospholipid.
73. The best description of the observed rotation of plane polarized light by a sugar sample labeled DL-erythrose is (a) to the left; (b) to the right; (c) alternately to the right and then to the left; (d) neither to the left nor the right; (e) clockwise.
74. Of the following names, the one that refers to a simple sugar in its cyclic (ring) form is (a) β-galactose; (b) L-(−)-glyceraldehyde; (c) D-(+)-glucose; (d) DL-erythrose; (e) (R)-(+)-glyceraldehyde.
75. The coagulation of egg whites by boiling is an example of (a) saponification; (b) inversion of a sugar; (c) hydrolysis of a protein; (d) denaturation of a protein; (e) condensation of amino acids.

76. A molecule in which the energy of metabolism is stored is (a) glucose; (b) DNA; (c) RNA; (d) glycerol; (e) ATP.
77. Of the following, the one that is not a constituent of a nucleic acid chain is (a) purine base; (b) phosphate group; (c) glycerol; (d) pentose sugar; (e) pyrimidine base.
78. The structure of the DNA molecule is best described as (a) a random coil; (b) a double helix; (c) a pleated sheet; (d) partly coiled; (e) a branched chain.
79. What is the maximum mass of a sodium soap that can be prepared from 125 g of glyceryl tripalmitate?
80. With reference to Figure 28-25, identify the purine bases, the pyrimidine bases, the pentose sugars, and the phosphate groups. Is this a chain of DNA or RNA? Explain.
81. The R group $—CH_2CH_2NH_2$ is that of what amino acid? (a) alanine; (b) serine; (c) threonine; (d) arginine; (e) none of these.
82. Which molecule is not optically active?
(a) 1,2-dichlorobutane; (b) 1,4-dichlorobutane; (c) 1,3-dichlorobutane; (d) 1,2-dichloropropane; (e) none of these.
83. Which of the following amino acids does not have a chiral carbon? (a) glycine; (b) alanine; (c) threonine; (d) lysine; (e) none of these.

84. Enantiomers always **(a)** have an asymmetric carbon; **(b)** have different physical properties; **(c)** change the color of light; **(d)** rotate polarized light; **(e)** none of these.

85. The R group of valine is **(a)** $-CH_2OH$; **(b)** $-CH_2COOH$; **(c)** $-CH(CH_3)_2$; **(d)** $-CH_2CH_2CH_2CH_2NH_2$; **(e)** none of these.

86. Amino acids are joined into proteins by **(a)** $3'-5'$ linkage; **(b)** a glycosidic bond; **(c)** $\beta(1 \longrightarrow 4)$ linkage; **(d)** a condensation reaction; **(e)** none of these.

87. Which of the following fatty acids is unsaturated? **(a)** palmitic acid; **(b)** oleic acid; **(c)** lauric acid; **(d)** stearic acid; **(e)** none of these.

88. Upon complete hydrolysis of a pentapeptide, the following amino acids are obtained: valine, phenylalanine, glycine, cysteine, and tyrosine. Partial hydrolysis yields the following fragments: Val-Phe; Cys-Gly; Cys-Val-Phe; Tyr-Phe. If glycine is the N-terminal acid, what is the sequence of amino acids in the polypeptide?

89. Appendix E describes a useful study aid known as concept mapping. Using the method presented in Appendix E, construct a concept map illustrating the concepts in Section 28-4.

88. Upon complete hydrolysis of a pentapeptide the following amino acids are obtained: valine, phenylalanine, glycine, cysteine, and tyrosine. Partial hydrolysis yields the following fragments: Val-Phe, Cys-Gly, Cys-Val-Phe, Tyr-Phe. If glycine is the N-terminal acid, what is the sequence of amino acids in the polypeptide?

89. Appendix A describes a useful study skill known as concept mapping. Using the method presented in Appendix E, construct a concept map illustrating the concepts in Section 28.4.

83. Enantiomers always (a) have an asymmetric carbon (b) have different physical properties (c) change the color of light (d) rotate polarized light (e) none of these.

85. The R group of valine is (a) —CH$_2$OH (b) —CH$_2$COOH (c) —CH$_2$SH (d) —CH—CH$_3$CH$_2$NH$_2$ (e) none of these.

86. Amino acids are joined into proteins by (a) 2—5 linkage (b) a glycosidic bond (c) PO (d) a condensation reaction (e) none of these.

87. Which of the following fatty acids is unsaturated? (a) palmitic acid (b) oleic acid (c) lauric acid (d) stearic acid (e) none of these.

A

Appendix

Mathematical Operations

A-1 Exponential Arithmetic

Measured quantities in this text range from very small to very large. For example, the mass of an individual hydrogen atom is 0.00000000000000000000000167 g, and the number of molecules in 18.0153 g of the substance water is 602,214,000,000,000,000,000,000. These numbers are difficult to write in conventional form and are even more cumbersome to handle in numerical calculations. We can greatly simplify them by expressing them in exponential form. The *exponential form* of a number consists of a coefficient (a number with value between 1 and 10) multiplied by a power of 10.

The number 10^n is the *nth power* of 10. If n is a *positive* quantity, 10^n is *greater than 1*. If n is a *negative* quantity, 10^n is *between 0 and 1*. The value of $10^0 = 1$.

Positive powers

$10^0 = 1$

$10^1 = 10$

$10^2 = 10 \times 10 = 100$

$10^3 = 10 \times 10 \times 10 = 1000$

Negative powers

$10^0 = 1$

$10^{-1} = \dfrac{1}{10} = 0.1$

$10^{-2} = \dfrac{1}{10 \times 10} = \dfrac{1}{10^2} = 0.01$

$10^{-3} = \dfrac{1}{10 \times 10 \times 10} = \dfrac{1}{10^3} = 0.001$

To express the number 3170 in exponential form, we write

$$3170 = 3.17 \times 1000 = 3.17 \times 10^3$$

For the number 0.00046 we write

$$0.00046 = 4.6 \times 0.0001 = 4.6 \times 10^{-4}$$

A simpler method of converting a number to exponential form that avoids intermediate steps is illustrated below.

$$\underset{3\;2\;1}{3\,1\,7\,0} = 3.17 \times 10^3$$

$$\underset{1\;2\;3\;4}{0.0\,0\,0\,4\,6} = 4.6 \times 10^{-4}$$

That is, to convert a number to exponential form,

- Move the decimal point to obtain a coefficient with value between 1 and 10.
- The exponent (power) of 10 is equal to the number of places the decimal point is moved.
- If the decimal point is moved *to the left*, the exponent of 10 is *positive*.
- If the decimal point is moved *to the right*, the exponent of 10 is *negative*.

To convert a number from exponential form to conventional form, move the decimal point the number of places indicated by the power of 10. That is,

$$6.1 \times 10^6 = 6.\underset{1\ 2\ 3\ 4\ 5\ 6}{\underbrace{1\ 0\ 0\ 0\ 0\ 0}} = 6{,}100{,}000$$

$$8.2 \times 10^{-5} = \underset{5\ 4\ 3\ 2\ 1}{\underbrace{0\ 0\ 0\ 0\ 0}} 8.2 = 0.000082$$

► The instructions given here are for a typical electronic calculator. The keystrokes required with your calculator may be somewhat different. Look for specific instructions in the instruction manual supplied with the calculator.

Electronic calculators designed for scientific and engineering work easily accommodate exponential numbers. A typical procedure is to key in the number, followed by the key "EXP" or "EE." Thus, the keystrokes required for the number 6.57×10^3 are

$$\boxed{6}\ \boxed{.}\ \boxed{5}\ \boxed{7}\ \boxed{\text{EXP}}\ \boxed{3}$$

and the result displayed is $\boxed{6.57^{03}}$

For the number 6.25×10^{-4}, the keystrokes are

$$\boxed{6}\ \boxed{.}\ \boxed{2}\ \boxed{5}\ \boxed{\text{EXP}}\ \boxed{4}\ \boxed{\pm}$$

and the result displayed is $\boxed{6.25^{-04}}$

Some calculators have a mode setting that automatically converts all numbers and calculated results to the exponential form, regardless of the form in which numbers are entered. In this mode setting you can generally also set the number of significant figures to be carried in displayed results.

Addition and Subtraction

To add or subtract numbers written in exponential form, first express each quantity as *the same power of 10*. Then add and/or subtract the coefficients as indicated. That is, treat the power of 10 as you would a unit common to the terms being added and/or subtracted. In the example that follows, convert 3.8×10^{-3} to 0.38×10^{-2} and use 10^{-2} as the common power of 10.

$$(5.60 \times 10^{-2}) + (3.8 \times 10^{-3}) - (1.52 \times 10^{-2}) = (5.60 + 0.38 - 1.52) \times 10^{-2}$$
$$= 4.46 \times 10^{-2}$$

Multiplication

Consider the numbers $a \times 10^y$ and $b \times 10^z$. Their product is $a \times b \times 10^{(y+z)}$. *Coefficients are multiplied, and exponents are added.*

$$0.0220 \times 0.0040 \times 750 = (2.20 \times 10^{-2})(4.0 \times 10^{-3})(7.5 \times 10^2)$$
$$= (2.20 \times 4.0 \times 7.5) \times 10^{(-2-3+2)} = 66 \times 10^{-3}$$
$$= 6.6 \times 10^1 \times 10^{-3} = 6.6 \times 10^{-2}$$

Division

Consider the numbers $a \times 10^y$ and $b \times 10^z$. Their quotient is

$$\frac{a \times 10^y}{b \times 10^z} = (a/b) \times 10^{(y-z)}$$

Coefficients are divided, and the exponent of the denominator is subtracted from the exponent of the numerator.

$$\frac{20.0 \times 636 \times 0.150}{0.0400 \times 1.80} = \frac{(2.00 \times 10^1)(6.36 \times 10^2)(1.50 \times 10^{-1})}{(4.00 \times 10^{-2}) \times 1.80}$$
$$= \frac{2.00 \times 6.36 \times 1.50 \times 10^{(1+2-1)}}{(4.00 \times 1.80) \times 10^{-2}} = \frac{19.1 \times 10^2}{7.20 \times 10^{-2}}$$
$$= 2.65 \times 10^{(2-(-2))} = 2.65 \times 10^4$$

Raising a Number to a Power

To "square" the number $a \times 10^y$ means to determine the value $(a \times 10^y)^2$, or the product $(a \times 10^y)(a \times 10^y)$. According to the rule for multiplication, this product is $(a \times a) \times 10^{(y+y)} = a^2 \times 10^{2y}$. When an exponential number is raised to a power, *the coefficient is raised to that power and the exponent is multiplied by the power*. For example,

$$(0.0034)^3 = (3.4 \times 10^{-3})^3 = (3.4)^3 \times 10^{(3)(-3)} = 39 \times 10^{-9} = 3.9 \times 10^{-8}$$

Extracting the Root of an Exponential Number

To extract the root of a number is the same as raising the number to a fractional power. This means that the square root of a number is the number to the one-half power; the cube root is the number to the one-third power; and so on. Thus,

$$\sqrt{a \times 10^y} = (a \times 10^y)^{1/2} = a^{1/2} \times 10^{y/2}$$

$$\sqrt{156} = \sqrt{1.56 \times 10^2} = (1.56)^{1/2} \times 10^{2/2} = 1.25 \times 10^1 = 12.5$$

In the following example, where the cube root is sought, the exponent (-5) is not divisible by 3; the number is rewritten with an exponent (-6) that is divisible by 3.

$$(3.52 \times 10^{-5})^{1/3} = (35.2 \times 10^{-6})^{1/3} = (35.2)^{1/3} \times 10^{-6/3} = 3.28 \times 10^{-2}$$

A-2   Logarithms

The *common* logarithm (log) of a number (N) is the exponent (x) to which the base 10 must be raised to yield the number N. That is, $\log N = x$ means that $N = 10^x = 10^{\log N}$. For simple powers of ten, for example,

$$\log 1 = \log 10^0 = 0$$
$$\log 10 = \log 10^1 = 1 \qquad \log 0.10 = \log 10^{-1} = -1$$
$$\log 100 = \log 10^2 = 2 \qquad \log 0.01 = \log 10^{-2} = -2$$

Most of the numbers that result from measurements and appear in calculations are not simple powers of 10, but it is not difficult to obtain logarithms of these numbers with an electronic calculator. To find the logarithm of a number, enter the number, followed by the "LOG" key.

$$\log 734 = 2.866$$
$$\log 0.0150 = -1.824$$

Another common example requires us to find the number having a certain logarithm. This number is often called the *antilogarithm* or the *inverse* logarithm. For example, if $\log N = 4.350$, what is N? N, the antilogarithm, is simply $10^{4.350}$, and to find its value we enter 4.350, followed by the key "10^x". Depending on the calculator used, it is usually necessary to press the key "INV" or "2nd F" before the log key.

$$\log N = 4.350$$
$$N = 10^{4.350}$$
$$N = 2.24 \times 10^4$$

If the task is to find the antilogarithm of -4.350, we again note that $N = 10^{-4.350}$, and $N = 4.47 \times 10^{-5}$. The required keystrokes on a typical electronic calculator are

$$\boxed{4}\ \boxed{.}\ \boxed{3}\ \boxed{5}\ \boxed{0}\ \boxed{\pm}\ \boxed{\text{INV}}\ \boxed{\log}$$

and the display, to three significant figures, is

$$\boxed{4.47^{-05}}$$

Some Useful Relationships

From the definition of a logarithm we can write $M = 10^{\log M}$, $N = 10^{\log N}$, and $M \times N = 10^{\log(M \times N)}$. This means that

$$\log(M \times N) = \log M + \log N$$

Similarly, it is not difficult to show that

$$\log \frac{M}{N} = \log M - \log N$$

Finally, because $N^2 = N \times N$, $10^{\log N^2} = 10^{\log N} \times 10^{\log N}$, and

$$\log N^2 = \log N + \log N = 2 \log N$$

Or, in more general terms,

$$\log N^a = a \log N$$

This expression is especially useful for extracting the roots of numbers. Thus, to determine $(2.5 \times 10^{-8})^{1/5}$, we write

$$\log(2.5 \times 10^{-8})^{1/5} = \tfrac{1}{5} \log(2.5 \times 10^{-8}) = \tfrac{1}{5}(-7.60) = -1.52$$
$$(2.5 \times 10^{-8})^{1/5} = 10^{-1.52} = 0.030$$

Significant Figures in Logarithms

To establish the number of significant figures to use in a logarithm or antilogarithm, use this fundamental rule: All digits to the *right* of the decimal point in a logarithm are significant. Digits to the *left* are used to establish the power of 10. Thus, the logarithm -2.08 is expressed to *two* significant figures. The antilogarithm of -2.08 should also be expressed to *two* significant figures; it is 8.3×10^{-3}. To help settle this point, take the antilogarithms of -2.07, -2.08, and -2.09. You will find these antilogs to be 8.5×10^{-3}, 8.3×10^{-3}, and 8.1×10^{-3}, respectively. Only *two* significant figures are justified.

Natural Logarithms

Logarithms can be expressed to a base other than 10. For instance, because $2^3 = 8$, $\log_2 8 = 3$ (read as, "the logarithm of 8 to the base 2 is equal to 3"). Similarly, $\log_2 10 = 3.322$. Several equations in this text are derived by the methods of calculus and involve logarithms. These equations require that the logarithm be a "natural" one. A *natural* logarithm has the base $e = 2.71828\ldots$. A logarithm to the base "e" is usually denoted as *ln*.

The relationship between a "natural" and "common" logarithm simply involves the factor $\log_e 10 = 2.303$. That is, for the number N, $\ln N = 2.303 \log N$. The methods and relationships described for logarithms and antilogarithms to the base 10 all apply to the base e as well, except that the relevant keys on an electronic calculator are "ln" and "e^x" rather than "LOG" and "10^x."

A-3   Algebraic Operations

An algebraic equation is solved when one of the quantities, the unknown, is expressed in terms of all the other quantities in the equation. This effect is achieved when the unknown is present, *alone*, on one side of the equation, and the rest of the terms are on the other side. To solve an equation, a rearrangement of terms may be necessary. The basic principle governing these rearrangements is quite simple. *Whatever is done to one side of the equation must be done to the other as well.*

$$3x^2 + 6 = 33 \qquad \textit{Solve for x.}$$

$$3x^2 + 6 - 6 = 33 - 6 \qquad \text{(1) Subtract 6 from each side.}$$

$$3x^2 = 27$$

$$\frac{3x^2}{3} = \frac{27}{3} \qquad \text{(2) Divide each side by 3.}$$

$$x^2 = 9$$

$$\sqrt{x^2} = \sqrt{9} \qquad \text{(3) Extract the square root of each side.}$$

$$x = 3 \qquad \text{(4) Simplify. The square root of 9 is 3.}$$

Quadratic Equations

A quadratic equation has the form $ax^2 + bx + c = 0$, where a, b, and c are constants (a cannot be equal to 0). A number of calculations in the text require us to solve a quadratic equation. At times, quadratic equations are of the form

$$(x + n)^2 = m^2$$

Such equations can be solved by extracting the square root of each side.

$$x + n = \pm m \quad \text{and} \quad x = m - n \quad \text{or} \quad x = -m - n$$

More likely, however, the *quadratic formula* will be needed.

$$x = \frac{-b \pm \sqrt{b^2 - 4ac}}{2a}$$

In Example 15-13 on page 685, the following equation must be solved.

$$\frac{(0.300 - x)}{(0.200 + x)(0.100 + x)} = 2.98$$

This is a quadratic equation, but before the quadratic formula can be applied, the equation must be rearranged to the standard form: $ax^2 + bx + c = 0$. This is accomplished in the steps that follow.

$$(0.300 - x) = 2.98(0.200 + x)(0.100 + x)$$

$$0.300 - x = 2.98(0.0200 + 0.300x + x^2)$$

$$0.300 - x = 0.0596 + 0.894x + 2.98x^2$$

$$2.98x^2 + 1.894x - 0.240 = 0 \qquad\qquad \textbf{(A.1)}$$

Now we can apply the quadratic formula.

$$x = \frac{-1.894 \pm \sqrt{(1.894)^2 + (4 \times 2.98 \times 0.240)}}{2 \times 2.98}$$

$$= \frac{-1.894 \pm \sqrt{3.587 + 2.86}}{2 \times 2.98}$$

$$= \frac{-1.894 \pm \sqrt{6.45}}{2 \times 2.98} = \frac{-1.894 \pm 2.54}{2 \times 2.98}$$

$$= \frac{-1.894 + 2.54}{2 \times 2.98} = \frac{0.65}{5.96} = 0.11$$

Note that only the ($+$) value of the ($\pm$) sign was used in solving for x. If the ($-$) value had been used, a negative value of x would have resulted. However, for the given situation a negative value of x is meaningless.

The Method of Successive Approximations

The quadratic equation that was just solved using the quadratic formula can be solved by an alternative method that can be extended to equations of higher order, such as the cubic, quartic, and quintic equations often encountered in solving equilibrium problems. To illustrate the method suppose we wish to solve expression (A.1) without recourse to the quadratic formula. We can rewrite the equation as follows

$$x = \frac{2.98x^2 - 0.240}{-1.894}$$

and make a guess at the value of x, which we substitute into the right hand side of the equation to calculate a new value of x. If we guess 0.15, which is reasonable given the starting concentrations involved in Example 15-13, we calculate

$$x = \frac{2.98 \times (0.15)^2 - 0.240}{-1.894} = 0.091$$

We can now use this value of x to calculate a new one.

$$x = \frac{2.98 \times (0.091)^2 - 0.240}{-1.894} = 0.114$$

Repeating this procedure one more time, we get

$$x = \frac{2.98 \times (0.114)^2 - 0.240}{-1.894} = 0.106$$

One more attempt gives a value of 0.11, which is in agreement with the answer previously obtained. The method that we have just used is called the method of successive approximations.

Let us now apply the method of successive approximations to the equation obtained in the Integrative Example of Chapter 15, namely

$$256x^5 - 0.74[(1.00 - x)(1.00 - 2x)^2] = 0 \tag{A.2}$$

The approach we can take here is to guess a value of x; evaluate the expression to see how close to zero it comes; and then adjust the value of x accordingly. Let us start with a guess of 0.40. The result is

$$256(0.40)^5 - 0.74[(1.00 - 0.40)(1.00 - 2 \times 0.40)^2] = 2.60$$

Clearly the value of 0.40 is too large. If we now try 0.10 we obtain a value of -0.42. We have overshot the value of x. We can now try a value of 0.25 (halfway between our two previous guesses) and obtain 0.11. We realize now that we have to reduce the guessed value of x slightly to get closer to zero. If we try 0.20 we obtain -0.13, and if we next try the value $x = 0.225$ we obtain -0.03. We are now very close to our goal of finding the value of x that satisfies the expression. One final guess of 0.23 gives a value of -0.001, a very satisfactory result.

An alternative approach is to rewrite expression (A.2) as

$$x^5 = \frac{0.74[(1.00 - x)(1.00 - 2x)^2]}{256}$$

and evaluate x as the fifth root of this new expression. If we substitute a value of $x = 0.40$ on the right side, we obtain $x = 0.15$ on the left side. Now by using this value on the right we calculate a new value of $x = 0.26$ on the left. By using this value we obtain $x = 0.22$ on the left, and finally with this last value we obtain $x = 0.23$, in agreement with our previous procedure. Which method we use is a matter of convenience, but the second method provides a new value of x whereas the first method may require more trial and error. When using the method of successive approximations, it is often a useful strategy to take the average of two results in order to speed up the convergence. The method of successive approximations can be very useful, but sometimes, depending how the equation is set up, the convergence to the correct answer may be slow or the process may even diverge. In such circumstances, the expression can be graphed as a function of x to ascertain where the solutions occur. In any event, we must always make sure that any answer obtained is reasonable from a chemical or physical point of view.

A-4   Graphs

Suppose the following sets of numbers are obtained for two quantities x and y by laboratory measurement.

$$x = 0, 1, 2, 3, 4, \ldots$$

$$y = 2, 4, 6, 8, 10, \ldots$$

The relationship between these sets of numbers is not difficult to establish.

$$y = 2x + 2$$

Ideally, the results of experimental measurements are best expressed through a mathematical equation. Sometimes, however, an exact equation cannot be written, or its form is not clear from the experimental data. The graphing of data is very useful in such cases. In Figure A-1 the points listed above are located on a coordinate grid in which x values are placed along the horizontal axis (abscissa) and y values along the vertical axis (ordinate). For each point the x and y values are indicated in parentheses.

The data points are seen to define a straight line. A mathematical equation of a straight line has the form

$$y = mx + b$$

Values of m, the *slope* of the line, and b, the *intercept*, can be obtained from the straight-line graph.

When $x = 0$, $y = b$. The intercept is the point where the straight line intersects the y-axis. The slope can be obtained from two points on the graph.

$$y_2 = mx_2 + b \quad \text{and} \quad y_1 = mx_1 + b$$

$$y_2 - y_1 = m(x_2 - x_1) + \cancel{b} - \cancel{b}$$

$$m = \frac{y_2 - y_1}{x_2 - x_1}$$

From the straight line in Figure A-1, can you establish that $m = b = 2$?

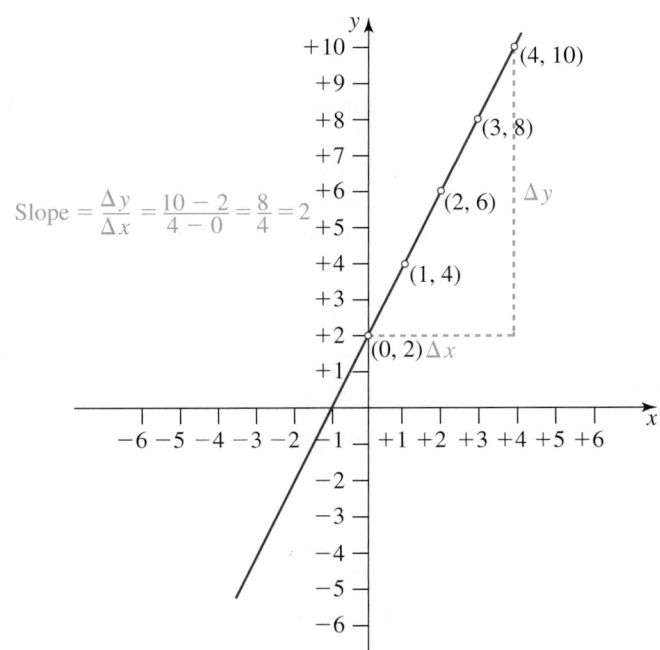

$$\text{Slope} = \frac{\Delta y}{\Delta x} = \frac{10 - 2}{4 - 0} = \frac{8}{4} = 2$$

▲ FIGURE A-1
A straight-line graph: *y* = *mx* + *b*

The technique used above to eliminate the constant b is applied to logarithmic functions in several places in the text. For example, the expression written below is from page 516. In this expression P is a pressure, T is a Kelvin temperature, and A and B are constants. The equation is that of a straight line.

$$\ln P = -A\left(\frac{1}{T}\right) + B$$

equation of straight line: $\qquad y = m \times x + b$

We can write this equation twice, for the point (P_1, T_1) and the point (P_2, T_2).

$$\ln P_1 = -A\left(\frac{1}{T_1}\right) + B \quad \text{and} \quad \ln P_2 = -A\left(\frac{1}{T_2}\right) + B$$

The difference between these equations is

$$\ln P_2 - \ln P_1 = -A\left(\frac{1}{T_2}\right) + B + A\left(\frac{1}{T_1}\right) - B$$

$$\ln \frac{P_2}{P_1} = A\left(\frac{1}{T_1} - \frac{1}{T_2}\right)$$

A-5 Using Conversion Factors (Dimensional Analysis)

Some calculations in general chemistry require that a quantity measured in one set of units be converted to another set of units. Consider this fact.

$$1 \text{ m} = 100 \text{ cm}$$

Divide each side of the equation by 1 m.

$$\frac{1 \text{ m}}{1 \text{ m}} = \frac{100 \text{ cm}}{1 \text{ m}}$$

On the left side of the equation, the numerator and denominator are identical; they cancel.

$$1 = \frac{100 \text{ cm}}{1 \text{ m}} \qquad \text{(A.3)}$$

On the right side they are not identical, but they are equal because they do represent the *same length*. The ratio 100 cm/1 m, when multiplied by a length in meters, converts that length to centimeters. The ratio is called a *conversion factor*.

Consider the question, "how many centimeters are there in 6.22 m?" The measured quantity is 6.22 m, and multiplying this quantity by 1 does not change its value.

$$6.22 \text{ m} \times 1 = 6.22 \text{ m}$$

Now replace the factor "1" by its equivalent—the conversion factor (A.3). Cancel the unit, m, and carry out the multiplication.

$$6.22 \text{ m} \times \underbrace{\frac{100 \text{ cm}}{1 \text{ m}}}_{\substack{\text{this factor} \\ \text{converts} \\ \text{m to cm}}} = 622 \text{ cm}$$

Next consider the question, "how many meters are there in 576 cm?" If we use the same factor (A.3) as before, the result is nonsensical.

$$576 \text{ cm} \times \frac{100 \text{ cm}}{1 \text{ m}} = 5.76 \times 10^4 \text{ cm}^2/\text{m}$$

Factor (A.3) must be rearranged to 1 m/100 cm.

$$576 \text{ cm} \times \underbrace{\frac{1 \text{ m}}{100 \text{ cm}}}_{\substack{\text{this factor} \\ \text{converts} \\ \text{cm to m}}} = 5.76 \text{ m}$$

This second example emphasizes two points.

1. There are two ways to write a conversion factor—in one form or its reciprocal (inverse). Because a conversion factor is equal to 1, its value is not changed by the inversion, but:

2. A conversion factor must be used in such a way as to produce the necessary cancellation of units.

Calculations based on conversion factors are always of the form

> *information sought = information given × conversion factor(s)*   **(A.4)**

◀ Because of the importance of the cancellation of units, this problem-solving method is often called *unit analysis* or *dimensional analysis*.

Often several conversions must be made in sequence in order to get to the desired result. For example, if we want to know how many yards (yd) there are in 576 cm, we find that there is no direct cm → yd conversion factor available. From the inside back cover of the text, however, we do find a conversion factor for cm → in. Thus, we can develop a *conversion pathway*, that is, a series of conversion factors that will take us from centimeters to yards:

$$\text{cm} \quad \rightarrow \quad \text{in.} \quad \rightarrow \quad \text{ft} \quad \rightarrow \quad \text{yd}$$

$$? \text{ yards} = 576 \text{ cm} \times \frac{1 \text{ in.}}{2.54 \text{ cm}} \times \frac{1 \text{ ft}}{12 \text{ in.}} \times \frac{1 \text{ yd}}{3 \text{ ft}}$$

$$= 6.30 \text{ yd}$$

We can use the same idea of a conversion pathway to deal with the somewhat more challenging situation faced when the units are squared (or cubed). Consider the question, "how many square feet (ft²) correspond to an area of 1.00 square meter (m²), given that 1 m = 39.37 in and 12 in = 1 ft?" Here, it may be helpful to begin by drawing a sketch or outline of the situation. Figure A-2 represents an area of 1.00 m². Think of it as a square with sides 1 m long. Figure A-2 also represents the length 1 ft and an area of 1.00 ft². Do you see that there is somewhat more than 9 ft² in 1 m²?

We can write expression (A.4) as follows:

$$? \text{ ft}^2 = 1.00 \text{ m}^2 \times \underbrace{\left(\frac{39.37 \text{ in.}}{1 \text{ m}}\right)\left(\frac{39.37 \text{ in.}}{1 \text{ m}}\right)}_{\substack{\text{to convert} \\ \text{m}^2 \text{ to in}^2}} \times \underbrace{\left(\frac{1 \text{ ft}}{12 \text{ in.}}\right)\left(\frac{1 \text{ ft}}{12 \text{ in.}}\right)}_{\substack{\text{to convert} \\ \text{in}^2 \text{ to ft}^2}}$$

This is the same as writing

$$? \text{ ft}^2 = 1.00 \text{ m}^2 \times \frac{(39.37)^2 \text{ in.}^2}{1 \text{ m}^2} \times \frac{1 \text{ ft}^2}{(12)^2 \text{ in.}^2} = 10.8 \text{ ft}^2$$

Another way to look at the problem is to convert the length 1.00 m to feet,

$$? \text{ ft} = 1.00 \text{ m} \times \frac{39.37 \text{ in.}}{1 \text{ m}} \times \frac{1 \text{ ft}}{12 \text{ in.}} = 3.28 \text{ ft}$$

and square the result

$$? \text{ ft}^2 = 3.28 \text{ ft} \times 3.28 \text{ ft} = 10.8 \text{ ft}^2$$

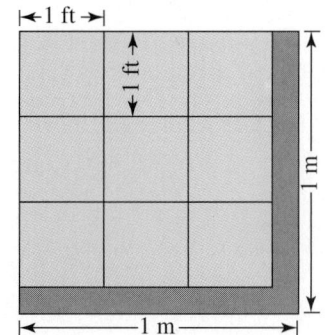

▲ FIGURE A-2
Comparison of one square foot and one square meter
One meter is slightly longer than 3 ft; 1 m² is somewhat larger than 9 ft².

Our last example incorporates several ideas discussed above. Here we will examine the situation in which the units in both the numerator and denominator must be converted. Consider the question, "how many meters per second (m/s) correspond to a speed of 63 mph, given that 1 mi = 5280 ft?"

We need to convert from miles to meters in the numerator and from hours to seconds in the denominator. We will need to use conversion factors from elsewhere in Section A-5 in addition to the given value. Also, we must be careful that our conversion factors produce the correct cancellation of units.

$$? \frac{m}{s} = \frac{63 \text{ mi}}{1 \text{ h}} \times \frac{1 \text{ h}}{60 \text{ min}} \times \frac{1 \text{ min}}{60 \text{ s}} \times \frac{5280 \text{ ft}}{1 \text{ mi}} \times \frac{12 \text{ in.}}{1 \text{ ft}} \times \frac{1 \text{ m}}{39.37 \text{ in.}}$$

$$= 28 \frac{m}{s}$$

In an alternative approach we break down the problem into three steps: (1) Convert 63 miles to a distance in meters; (2) convert 1 hour to a time in seconds; and (3) express the speed as a ratio of distance over time.

Step 1.

$$\text{distance} = 63 \text{ mi} \times \frac{5280 \text{ ft}}{1 \text{ mi}} \times \frac{12 \text{ in.}}{1 \text{ ft}} \times \frac{1 \text{ m}}{39.37 \text{ in.}} = 1.0 \times 10^5 \text{ m}$$

Step 2.

$$\text{time} = 1 \text{ h} \times \frac{60 \text{ min}}{1 \text{ h}} \times \frac{60 \text{ s}}{1 \text{ min}} = 3.6 \times 10^3 \text{ s}$$

Step 3.

$$\text{speed} = \frac{\text{distance}}{\text{time}} = \frac{1.0 \times 10^5 \text{ m}}{3.6 \times 10^3 \text{ s}} = 28 \frac{m}{s}$$

In conclusion, we have shown (1) how to make a conversion factor; (2) that a conversion factor may be inverted; (3) that a series of conversion factors may be used to make a conversion pathway; (4) that conversion factors may be raised to powers, if necessary; and (5) that conversions of values with units in both the numerator and the denominator (such as miles per hour or pounds per square inch) can be performed in one step or in several steps.

Some Basic Physical Concepts

Appendix

B-1   Velocity and Acceleration

Time elapses as an object moves from one point to another. The *velocity* of the object is defined as the distance traveled per unit of time. An automobile that travels a distance of 60.0 km in exactly one hour has a velocity of 60.0 km/h (or 16.7 m/s).

Table B.1 contains data on the velocity of a free-falling object. For this motion, velocity is not constant—it increases with time. The falling object "speeds up" continuously. The rate of change of velocity with time is called *acceleration*. Acceleration has the units of distance per unit time per unit time. With the methods of calculus, mathematical equations can be derived for the velocity (u) and distance (d) traveled in a time (t) by an object that has a constant acceleration (a).

$$u = at \qquad\qquad\qquad \textbf{(B.1)}$$
$$d = \tfrac{1}{2}at^2 \qquad\qquad\qquad \textbf{(B.2)}$$

For a free-falling object, the constant acceleration, called the *acceleration due to gravity*, is $a = g = 9.8 \text{ m/s}^2$. Equations (B.1) and (B.2) can be used to calculate the velocity and distance traveled by a free-falling object.

TABLE B.1 Velocity and Acceleration of a Free-Falling Body			
Time Elapsed, s	Total Distance, m	Velocity, m/s	Acceleration, m/s^2
0	0		
		4.9	
1	4.9		9.8
		14.7	
2	19.6		9.8
		24.5	
3	44.1		9.8
		34.3	
4	78.4		

B-2   Force and Work

Newton's *first law* of motion states that an object at rest remains at rest, and that an object in motion remains in uniform motion unless acted upon by an external force. The tendency for an object to remain at rest or in uniform motion is called *inertia; a force* is what is required to overcome inertia. Since the application of a force either gives an object motion or changes its motion, the actual effect of a force is to change the velocity of an object. Change in velocity is an *acceleration*, so force is what provides an object with acceleration.

Newton's *second law* of motion describes the force F required to produce an acceleration a in an object of mass m.

$$F = ma \tag{B.3}$$

The basic unit of force in the SI system is the *newton* (N). It is the force required to provide a one-kilogram mass with an acceleration of one meter per second per second.

$$1\,N = 1\,kg \times 1\,m\,s^{-2} \tag{B.4}$$

The force of gravity on an object (its weight) is the product of the mass of the object and the acceleration due to gravity, g.

$$F = mg \tag{B.5}$$

Work (w) is performed when a force acts through a distance.

$$\text{work } (w) = \text{force } (F) \times \text{distance } (d) \tag{B.6}$$

The *joule* (J) is the amount of work associated with a force of one newton (N) acting through a distance of one meter.

$$1\,J = 1\,N \times 1\,m \tag{B.7}$$

From the definition of the newton in expression (B.4), we can also write

$$1\,J = 1\,kg \times 1\,m\,s^{-2} \times 1\,m = 1\,kg\,m^2\,s^{-2} \tag{B.8}$$

B-3   Energy

Energy is defined as the capacity to do work, but there are other useful descriptions of energy as well. For example, a moving object possesses a kind of energy known as *kinetic energy*. We can obtain a useful equation for kinetic energy by combining some of the other simple equations in this appendix. Thus, because work is the product of a force and distance (equation B.6), and force is the product of a mass and acceleration (equation B.3), we can write

$$w \text{ (work)} = m \times a \times d \tag{B.9}$$

Now, if we substitute equation (B.2) relating acceleration (a), distance (d), and time (t), into equation (B.9), we obtain

$$w \text{ (work)} = m \times a \times \frac{1}{2}at^2 \tag{B.10}$$

Finally, let's substitute expression (B.1) relating acceleration (a) and velocity (u) into (B.10). That is, because $a = u/t$,

$$w \text{ (work)} = \frac{1}{2}m\left(\frac{u}{t}\right)^2 t^2 \tag{B.11}$$

Think of the work in (B.11) as the amount of work necessary to produce a velocity of u in an object of mass m. This amount of work is the energy that appears in the object as kinetic energy (e_k).

$$e_k \text{ (kinetic energy)} = \frac{1}{2}mu^2 \tag{B.12}$$

An object at rest may also have the capacity to do work by changing its position. The energy it possesses, which can be transformed into actual work, is called *potential energy*. Think of potential energy as energy "stored" within an object. Equations can be written for potential energy, but the exact forms of these equations depend on the manner in which the energy is "stored."

B-4   Magnetism

Attractive and repulsive forces associated with a magnet are centered at regions called *poles*. A magnet has a north and a south pole. If two magnets are aligned such that the north pole of one is directed toward the south pole of the second, an attractive force develops. If the alignment brings like poles into proximity, either both north or both south, a repulsive force develops. *Unlike poles attract; like poles repel.*

A *magnetic field* exists in that region surrounding a magnet in which the influence of the magnet can be felt. Internal changes produced within an iron object by a magnetic field, not produced in a field-free region, are responsible for the attractive force that the object experiences.

B-5   Static Electricity

Another property with which certain objects may be endowed is *electric charge*. Analogous to the case of magnetism, *unlike charges attract, and like charges repel* (recall Figure 2-4). The unit of charge is called a *coulomb*, C. In Coulomb's law, stated below, a *positive* force between electrically charged objects is *repulsive; a negative* force is *attractive.*

$$F = \frac{Q_1 Q_2}{4\pi \varepsilon r^2} \qquad \textbf{(B.13)}$$

where Q_1 is the charge on object 1,

$\quad\quad\;\; Q_2$ is the charge on object 2,

$\quad\quad\;\; r$ is the distance between the objects, and

$\quad\quad\;\; \varepsilon$ is a proportionality constant called the *dielectric constant*, whose numerical value reflects the effect that the medium separating two charged objects has on the force existing between them. For a vacuum, $\varepsilon = \varepsilon_0 = 8.85419 \times 10^{-12}\,\mathrm{C^2\,N^{-1}\,m^{-2}} = 8.85419 \times 10^{-12}\,\mathrm{C^2\,J^{-1}\,m^{-1}}$; for other media, ε_0 is greater than 1 (for example, for water $\varepsilon = 78.5\varepsilon_0$).

An *electric field* exists in that region surrounding an electrically charged object in which the influence of the electric charge is felt. If an uncharged object is brought into the field of a charged object, the uncharged object may undergo internal changes that it would not experience in a field-free region. These changes may lead to the production of electric charges in the formerly uncharged object, a phenomenon called *induction* (illustrated in Figure B-1).

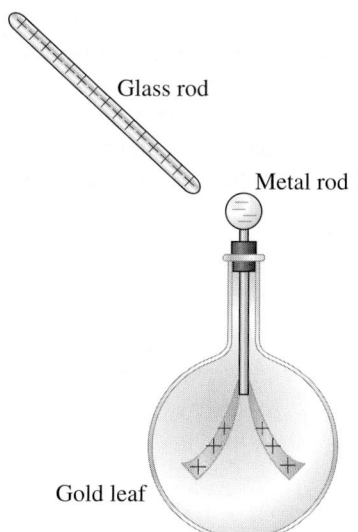

Glass rod

Metal rod

Gold leaf

◀ FIGURE B-1
Production of electric charges by induction in a gold-leaf electroscope
The glass rod acquires a positive electric charge by being rubbed with a silk cloth. As the rod is brought near the electroscope, a separation of charge occurs in the electroscope. The leaves become positively charged and repel one another. Negative charge is attracted to the spherical terminal at the end of the metal rod. If the glass rod is removed, the charges on the electroscope redistribute themselves, and the leaves collapse. If the spherical ball is touched by an electric conductor before the glass rod is removed, negative charge is removed from the ball. The electroscope retains a net positive charge, and the leaves remain outstretched.

The potential energy (*PE*) associated with the interaction of two charged objects is given by

$$PE = \frac{Q_1 Q_2}{4 \pi \varepsilon r} \tag{B.14}$$

PE is equal to the work done when the distance between the two objects is decreased from infinity to r.

B-6 Current Electricity

Current electricity is a flow of electrically charged particles. In electric currents in metallic conductors, the charged particles are electrons; in molten salts or in aqueous solutions, the particles are both negatively and positively charged ions.

As pointed out in Section B-5, the unit of electric charge is called a *coulomb* (C). The unit of electric current known as the *ampere* (A) is defined as a flow of one coulomb per second through an electrical conductor. Two variables determine the magnitude of the electric current I flowing through a conductor. These are the potential difference, or voltage drop, E, along the conductor, and the electrical resistance of the conductor, R. The units of voltage and resistance are the *volt* (V) and *ohm*, respectively. The relationship of electric current, voltage, and resistance is given by Ohm's law.

$$I = \frac{E}{R} \tag{B.15}$$

One joule of energy is associated with the passage of one coulomb of electric charge through a potential difference (voltage) of one volt. That is, one joule = one volt-coulomb. Electric *power* refers to the rate of production (or consumption) of electric energy. It has the unit *watt* (W).

$$1\,W = 1\,J\,s^{-1} = 1\,V\,C\,s^{-1}$$

Since one coulomb per second is a current of one ampere,

$$1\,W = 1\,V \times 1\,A \tag{B.16}$$

Thus, a 100-watt light bulb operating at 110 V draws a current of 100 W/110 V = 0.91 A.

B-7 Electromagnetism

The relationship between electricity and magnetism is an intimate one. Interactions of electric and magnetic fields result in (1) magnetic fields associated with the flow of electric current (as in electromagnets), (2) forces experienced by current-carrying conductors when placed in a magnetic field (as in electric motors), and (3) electric current being induced when an electric conductor is moved through a magnetic field (as in electric generators). Several observations described in this text can be understood in terms of electromagnetic phenomena.

Appendix

SI Units

The system of units that will in time be used universally for expressing all measured quantities is Le Système International d'Unités (The International System of Units), adopted in 1960 by the Conference Générale des Poids et Measures (General Conference of Weights and Measures). A summary of some of the provisions of the SI convention is provided here.

C-1   SI Base Units

A single unit has been established for each of the basic quantities involved in measurement. These are as follows:

Physical Quantity	Unit	Symbol
Length	Meter	m
Mass	Kilogram	kg
Time	Second	s
Electric current	Ampere	A
Temperature	Kelvin	K
Luminous intensity	Candela	cd
Amount of substance	Mole	mol
Plane angle	Radian	rad
Solid angle	Steradian	sr

C-2   SI Prefixes

Distinctive prefixes are attached to the base unit to express quantities that are *multiples* (greater than) or *submultiples* (less than) of the base unit. The multiples and submultiples are obtained by multiplying the base unit by powers of ten.

Multiple	Prefix	Symbol	Submultiple	Prefix	Symbol
10^{12}	tera	T	10^{-1}	deci	d
10^{9}	giga	G	10^{-2}	centi	c
10^{6}	mega	M	10^{-3}	milli	m
10^{3}	kilo	k	10^{-6}	micro	μ
10^{2}	hecto	h	10^{-9}	nano	n
10^{1}	deka	da	10^{-12}	pico	p
			10^{-15}	femto	f
			10^{-18}	atto	a

C-3   Derived SI Units

A number of quantities must be derived from measured values of the SI base quantities [for example, volume has the unit (length)^3]. Two sets of derived units are given, those whose names follow directly from the base units and those that are given special names. Notice that the units used in the text differ in some respects from those in the table. For example, for the most part, the text expresses density as $g\ cm^{-3}$, molar mass as $g\ mol^{-1}$, molar volume as $mL\ mol^{-1}$ or $L\ mol^{-1}$, and molar concentration (molarity) as $mol\ L^{-1}$, or M.

▶ Two other SI conventions are illustrated through this table: (a) Units are written in singular form—meter or m, *not* meters or ms; (b) negative exponents are preferred to the shilling bar or solidus (/), that is, $m\ s^{-1}$ and $m\ s^{-2}$, *not* m/s and m/s/s. In some chapters both ways of expressing units have been used as you need to be comfortable with both systems.

Physical Quantity	Unit	Symbol
Area	Square meter	m^2
Volume	Cubic meter	m^3
Velocity	Meter per second	$m\ s^{-1}$
Acceleration	Meter per second squared	$m\ s^{-2}$
Density	Kilogram per cubic meter	$kg\ m^{-3}$
Molar mass	Kilogram per mole	$kg\ mol^{-1}$
Molar volume	Cubic meter per mole	$m^3\ mol^{-1}$
Molar concentration	Mole per cubic meter	$mol\ m^{-3}$

Physical Quantity	Unit	Symbol	In Terms of SI Units
Frequency	hertz	Hz	s^{-1}
Force	newton	N	$kg\ m\ s^{-2}$
Pressure	pascal	Pa	$N\ m^{-2}$
Energy	joule	J	$kg\ m^2\ s^{-2}$
Power	watt	W	$J\ s^{-1}$
Electric charge	coulomb	C	$A\ s$
Electric potential difference	volt	V	$J\ A^{-1}\ s^{-1}$
Electric resistance	ohm	Ω	$V\ A^{-1}$

C-4   Units to Be Discouraged or Abandoned

There are several commonly used units whose use is to be discouraged and ultimately abandoned. Their gradual disappearance is to be expected, though each is used in this text. A few such units are listed.

▶ Another SI convention is implied here. No commas are used in expressing large numbers. Instead, spaces are left between groupings of three digits, that is, 101 325 rather than 101,325. Decimal points are written either as periods or commas. Numbers in this text retained the comma separators in numbers of at least five digits.

Physical Quantity	Unit	Symbol	Definition of SI Units
Length	ångstrom	Å	$1 \times 10^{-10}\ m$
Force	dyne	dyn	$1 \times 10^{-5}\ N$
Energy	erg	erg	$1 \times 10^{-7}\ J$
Energy	calorie	cal	4.184 J
Pressure	atmosphere	atm	101 325 Pa
Pressure	millimeter of mercury	mmHg	133.322 Pa
Pressure	torr	Torr	133.322 Pa

Appendix

Data Tables

TABLE D.1 Ground-State Electron Configurations

Z	Element	Configuration	Z	Element	Configuration	Z	Element	Configuration
1	H	$1s^1$	37	Rb	[Kr] $5s^1$	72	Hf	[Xe] $4f^{14}5d^26s^2$
2	He	$1s^2$	38	Sr	[Kr] $5s^2$	73	Ta	[Xe] $4f^{14}5d^36s^2$
3	Li	[He] $2s^1$	39	Y	[Kr] $4d^15s^2$	74	W	[Xe] $4f^{14}5d^46s^2$
4	Be	[He] $2s^2$	40	Zr	[Kr] $4d^25s^2$	75	Re	[Xe] $4f^{14}5d^56s^2$
5	B	[He] $2s^22p^1$	41	Nb	[Kr] $4d^45s^1$	76	Os	[Xe] $4f^{14}5d^66s^2$
6	C	[He] $2s^22p^2$	42	Mo	[Kr] $4d^55s^1$	77	Ir	[Xe] $4f^{14}5d^76s^2$
7	N	[He] $2s^22p^3$	43	Tc	[Kr] $4d^55s^2$	78	Pt	[Xe] $4f^{14}5d^96s^1$
8	O	[He] $2s^22p^4$	44	Ru	[Kr] $4d^75s^1$	79	Au	[Xe] $4f^{14}5d^{10}6s^1$
9	F	[He] $2s^22p^5$	45	Rh	[Kr] $4d^85s^1$	80	Hg	[Xe] $4f^{14}5d^{10}6s^2$
10	Ne	[He] $2s^22p^6$	46	Pd	[Kr] $4d^{10}$	81	Tl	[Xe] $4f^{14}5d^{10}6s^26p^1$
11	Na	[Ne] $3s^1$	47	Ag	[Kr] $4d^{10}5s^1$	82	Pb	[Xe] $4f^{14}5d^{10}6s^26p^2$
12	Mg	[Ne] $3s^2$	48	Cd	[Kr] $4d^{10}5s^2$	83	Bi	[Xe] $4f^{14}5d^{10}6s^26p^3$
13	Al	[Ne] $3s^23p^1$	49	In	[Kr] $4d^{10}5s^25p^1$	84	Po	[Xe] $4f^{14}5d^{10}6s^26p^4$
14	Si	[Ne] $3s^23p^2$	50	Sn	[Kr] $4d^{10}5s^25p^2$	85	At	[Xe] $4f^{14}5d^{10}6s^26p^5$
15	P	[Ne] $3s^23p^3$	51	Sb	[Kr] $4d^{10}5s^25p^3$	86	Rn	[Xe] $4f^{14}5d^{10}6s^26p^6$
16	S	[Ne] $3s^23p^4$	52	Te	[Kr] $4d^{10}5s^25p^4$	87	Fr	[Rn] $7s^1$
17	Cl	[Ne] $3s^23p^5$	53	I	[Kr] $4d^{10}5s^25p^5$	88	Ra	[Rn] $7s^2$
18	Ar	[Ne] $3s^23p^6$	54	Xe	[Kr] $4d^{10}5s^25p^6$	89	Ac	[Rn] $6d^17s^2$
19	K	[Ar] $4s^1$	55	Cs	[Xe] $6s^1$	90	Th	[Rn] $6d^27s^2$
20	Ca	[Ar] $4s^2$	56	Ba	[Xe] $6s^2$	91	Pa	[Rn] $5f^26d^17s^2$
21	Sc	[Ar] $3d^14s^2$	57	La	[Xe] $5d^16s^2$	92	U	[Rn] $5f^36d^17s^2$
22	Ti	[Ar] $3d^24s^2$	58	Ce	[Xe] $4f^26s^2$	93	Np	[Rn] $5f^46d^17s^2$
23	V	[Ar] $3d^34s^2$	59	Pr	[Xe] $4f^36s^2$	94	Pu	[Rn] $5f^67s^2$
24	Cr	[Ar] $3d^54s^1$	60	Nd	[Xe] $4f^46s^2$	95	Am	[Rn] $5f^77s^2$
25	Mn	[Ar] $3d^54s^2$	61	Pm	[Xe] $4f^56s^2$	96	Cm	[Rn] $5f^76d^17s^2$
26	Fe	[Ar] $3d^64s^2$	62	Sm	[Xe] $4f^66s^2$	97	Bk	[Rn] $5f^97s^2$
27	Co	[Ar] $3d^74s^2$	63	Eu	[Xe] $4f^76s^2$	98	Cf	[Rn] $5f^{10}7s^2$
28	Ni	[Ar] $3d^84s^2$	64	Gd	[Xe] $4f^75d^16s^2$	99	Es	[Rn] $5f^{11}7s^2$
29	Cu	[Ar] $3d^{10}4s^1$	65	Tb	[Xe] $4f^96s^2$	100	Fm	[Rn] $5f^{12}7s^2$
30	Zn	[Ar] $3d^{10}4s^2$	66	Dy	[Xe] $4f^{10}6s^2$	101	Md	[Rn] $5f^{13}7s^2$
31	Ga	[Ar] $3d^{10}4s^24p^1$	67	Ho	[Xe] $4f^{11}6s^2$	102	No	[Rn] $5f^{14}7s^2$
32	Ge	[Ar] $3d^{10}4s^24p^2$	68	Er	[Xe] $4f^{12}6s^2$	103	Lr	[Rn] $5f^{14}6d^17s^2$
33	As	[Ar] $3d^{10}4s^24p^3$	69	Tm	[Xe] $4f^{13}6s^2$	104	Rf	[Rn] $5f^{14}6d^27s^2$
34	Se	[Ar] $3d^{10}4s^24p^4$	70	Yb	[Xe] $4f^{14}6s^2$	105	Db	[Rn] $5f^{14}6d^37s^2$
35	Br	[Ar] $3d^{10}4s^24p^5$	71	Lu	[Xe] $4f^{14}5d^16s^2$	106	Sg	[Rn] $5f^{14}6d^47s^2$
36	Kr	[Ar] $3d^{10}4s^24p^6$						

The electron configurations printed in red are those of the noble gases. Each noble gas configuration serves as the core of the electron configurations of the elements that follow it, until the next noble gas is reached. Thus, [He] represents the core configuration of the second period elements; [Ne], the third period; [Ar], the fourth period; [Kr], the fifth period; [Xe], the sixth period; and [Rn], the seventh period.

TABLE D.2 Thermodynamic Properties of Substances at 298.15 K*. Substances are at 1 bar pressure. For aqueous solutions, solutes are at unit activity (roughly 1 M). Data for ions in aqueous solution are relative to values of *zero* for ΔH_f°, ΔG_f°, and S° for H^+

Inorganic Substances

	ΔH_f°, kJ mol^{-1}	ΔG_f°, kJ mol^{-1}	S°, J mol^{-1} K^{-1}	C_p, J mol^{-1} K^{-1}
Aluminum				
Al(s)	0	0	28.33	24.2
Al^{3+}(aq)	−531	−485	−321.7	—
AlCl$_3$(s)	−704.2	−628.8	110.7	91.1
Al$_2$Cl$_6$(g)	−1291	−1220.	490.	157.72
AlF$_3$(s)	−1504	−1425	66.44	75.1
Al$_2$O$_3$(α solid)	−1676	−1582	50.92	79.0
Al(OH)$_3$(s)	−1276	—	—	93.1
Al$_2$(SO$_4$)$_3$(s)	−3441	−3100.	239	259.4
Barium				
Ba(s)	0	0	62.8	28.1
Ba^{2+}(aq)	−537.6	−560.8	9.6	—
BaCO$_3$(s)	−1216	−1138	112.1	85.35
BaCl$_2$(s)	−858.6	−810.4	123.7	75.1
BaF$_2$(s)	−1207	−1157	96.36	71.2
BaO(s)	−553.5	−525.1	70.42	47.3
Ba(OH)$_2$(s)	−944.7	—	—	101.6
Ba(OH)$_2 \cdot$ 8 H$_2$O(s)	−3342	−2793	427	—
BaSO$_4$(s)	−1473	−1362	132.2	101.8
Beryllium				
Be(s)	0	0	9.50	16.4
BeCl$_2$(α solid)	−490.4	−445.6	82.68	62.4
BeF$_2$(α solid)	−1027	−979.4	53.35	51.8
BeO(s)	−609.6	−580.3	14.14	25.6
Bismuth				
Bi(s)	0	0	56.74	25.5
BiCl$_3$(s)	−379.1	−315.0	177.0	105.0
Bi$_2$O$_3$(s)	−573.9	−493.7	151.5	113.5
Boron				
B(s)	0	0	5.86	11.1
BCl$_3$(l)	−427.2	−387.4	206.3	106.7
BF$_3$(g)	−1137	−1120.	254.1	50.45
B$_2$H$_6$(g)	35.6	86.7	232.1	56.7
B$_2$O$_3$(s)	−1273	−1194	53.97	62.8
Bromine				
Br(g)	111.9	82.40	175.0	20.8
Br$^-$(aq)	−121.6	−104.0	82.4	−141.8
Br$_2$(g)	30.91	3.11	245.5	36.0
Br$_2$(l)	0	0	152.2	75.7
BrCl(g)	14.64	−0.98	240.1	35.0
BrF$_3$(g)	−255.6	−229.4	292.5	66.6
BrF$_3$(l)	−300.8	−240.5	178.2	124.6

*Data for inorganic substances and for organic compounds with up to two carbon atoms per molecule are adapted from D. D. Wagman, et al., "The NBS Tables of Chemical Thermodynamic Properties: Selected Values for Inorganic and C$_1$ and C$_2$ Organic Substances in SI Units" *Journal of Physical and Chemical Reference Data* 11 (1982) Supplement 2. Data for other organic compounds are from J. A. Dean, *Lange's Handbook of Chemistry*. 15th ed., McGraw-Hill, 1999, and other sources.

Inorganic Substances

	ΔH_f°, kJ mol^{-1}	ΔG_f°, kJ mol^{-1}	S°, J mol^{-1} K^{-1}	C_p, J mol^{-1} K^{-1}
Cadmium				
Cd(s)	0	0	51.76	26.0
Cd^{2+}(aq)	−75.90	−77.61	−73.2	—
CdCl$_2$(s)	−391.5	−343.9	115.3	74.7
CdO(s)	−258.2	−228.4	54.8	43.4
Calcium				
Ca(s)	0	0	41.42	25.9
Ca^{2+}(aq)	−542.8	−553.6	−53.1	—
CaCO$_3$(s)	−1207	−1129	92.9	80.6
CaCl$_2$(s)	−795.8	−748.1	104.6	72.9
CaF$_2$(s)	−1220.	−1167	68.87	67.0
CaH$_2$(s)	−186.2	−147.2	42	41.0
Ca(NO$_3$)$_2$(s)	−938.4	−743.1	193.3	149.4
CaO(s)	−635.1	−604.0	39.75	42.0
Ca(OH)$_2$(s)	−986.1	−898.5	83.39	87.5
Ca$_3$(PO$_4$)$_2$(s)	−4121	−3885	236.0	227.8
CaSO$_4$(s)	−1434	−1322	106.7	99.7
Carbon (See also the table of organic substances.)				
C(g)	716.7	671.3	158.0	20.8
C(diamond)	1.90	2.90	2.38	6.1
C(graphite)	0	0	5.74	8.5
CCl$_4$(g)	−102.9	−60.59	309.9	83.3
CCl$_4$(l)	−135.4	−65.21	216.4	130.7
C$_2$N$_2$(g)	309.0	297.4	241.9	56.8
CO(g)	−110.5	−137.2	197.7	29.1
CO$_2$(g)	−393.5	−394.4	213.7	37.1
CO$_3{}^{2-}$(aq)	−677.1	−527.8	−56.9	—
C$_3$O$_2$(g)	−93.72	−109.8	276.5	67.0
C$_3$O$_2$(l)	−117.3	−105.0	181.1	—
COCl$_2$(g)	−218.8	−204.6	283.5	57.7
COS(g)	−142.1	−169.3	231.6	41.5
CS$_2$(l)	89.70	65.27	151.3	76.4
Chlorine				
Cl(g)	121.7	105.7	165.2	21.8
Cl$^-$(aq)	−167.2	−131.2	56.5	−136.4
Cl$_2$(g)	0	0	223.1	33.9
ClF$_3$(g)	−163.2	−123.0	281.6	63.9
ClO$_2$(g)	102.5	120.5	256.8	42.0
Cl$_2$O(g)	80.3	97.9	266.2	45.4
Chromium				
Cr(s)	0	0	23.77	23.4
[Cr(H$_2$O)$_6$]$^{3+}$(aq)	−1999	—	—	—
Cr$_2$O$_3$(s)	−1140.	−1058	81.2	118.7
CrO$_4{}^{2-}$(aq)	−881.2	−727.8	50.21	—
Cr$_2$O$_7{}^{2-}$(aq)	−1490.	−1301	261.9	—
Cobalt				
Co(s)	0	0	30.04	24.8
CoO(s)	−237.9	−214.2	52.97	55.2
Co(OH)$_2$(pink solid)	−539.7	−454.3	79	68.8

(continued)

Inorganic Substances

	ΔH°_f, kJ mol^{-1}	ΔG°_f, kJ mol^{-1}	S°, J mol^{-1} K^{-1}	C_p, J mol^{-1} K^{-1}
Copper				
Cu(s)	0	0	33.15	24.4
Cu^{2+}(aq)	64.77	65.49	−99.6	—
CuCO$_3$·Cu(OH)$_2$(s)	−1051	−893.6	186.2	—
CuO(s)	−157.3	−129.7	42.63	42.3
Cu(OH)$_2$(s)	−449.8	—	—	95.19
CuSO$_4$·5 H$_2$O(s)	−2280.	−1880.	300.4	—
Fluorine				
F(g)	78.99	61.91	158.8	22.7
F$^-$(aq)	−332.6	−278.8	−13.8	−106.7
F$_2$(g)	0	0	202.8	31.3
Helium				
He(g)	0	0	126.2	20.8
Hydrogen				
H(g)	218.0	203.2	114.7	20.8
H$^+$(aq)	0	0	0	0
H$_2$(g)	0	0	130.7	28.8
HBr(g)	−36.40	−53.45	198.7	29.1
HCl(g)	−92.31	−95.30	186.9	29.1
HCl(aq)	−167.2	−131.2	56.5	−136.4
HClO$_2$(aq)	−51.9	5.9	188.3	—
HCN(g)	135.1	124.7	201.8	35.9
HF(g)	−271.1	−273.2	173.8	—
HI(g)	26.48	1.70	206.6	29.2
HNO$_3$(l)	−174.1	−80.71	155.6	109.9
HNO$_3$(aq)	−207.4	−111.3	146.4	−86.6
H$_2$O(g)	−241.8	−228.6	188.8	33.6
H$_2$O(l)	−285.8	−237.1	69.91	75.3
H$_2$O$_2$(g)	−136.3	−105.6	232.7	43.1
H$_2$O$_2$(l)	−187.8	−120.4	109.6	89.1
H$_2$S(g)	−20.63	−33.56	205.8	34.2
H$_2$SO$_4$(l)	−814.0	−690.0	156.9	138.9
H$_2$SO$_4$(aq)	−909.3	−744.5	20.1	−293.0
Iodine				
I(g)	106.8	70.25	180.8	20.8
I$^-$(aq)	−55.19	−51.57	111.3	−142.3
I$_2$(g)	62.44	19.33	260.7	36.9
I$_2$(s)	0	0	116.1	54.4
IBr(g)	40.84	3.69	258.8	36.4
ICl(g)	17.78	−5.46	247.6	35.6
ICl(l)	−23.89	−13.58	135.1	135.1
Iron				
Fe(s)	0	0	27.28	25.1
Fe^{2+}(aq)	−89.1	−78.90	−137.7	—
Fe^{3+}(aq)	−48.5	−4.7	−315.9	—
FeCO$_3$(s)	−740.6	−666.7	92.9	82.1
FeCl$_3$(s)	−399.5	−334.0	−142.3	96.7
FeO(s)	−272.0	—	—	49.91
Fe$_2$O$_3$(s)	−824.2	−742.2	87.40	103.9
Fe$_3$O$_4$(s)	−1118	−1015	146.4	143.4
Fe(OH)$_3$(s)	−823.0	−696.5	106.7	101.7

Inorganic Substances

	ΔH_f°, kJ mol^{-1}	ΔG_f°, kJ mol^{-1}	S°, J mol^{-1} K^{-1}	C_p, J mol^{-1} K^{-1}
Lead				
Pb(s)	0	0	64.81	26.4
Pb^{2+}(aq)	−1.7	−24.43	10.5	—
PbI$_2$(s)	−175.5	−173.6	174.9	77.4
PbO$_2$(s)	−277.4	−217.3	68.6	64.6
PbSO$_4$(s)	−919.9	−813.1	148.6	103.2
Lithium				
Li(g)	159.4	126.7	138.8	20.8
Li(s)	0	0	29.12	24.8
Li$^+$(aq)	−278.5	−293.3	13.4	68.6
LiCl(s)	−408.6	−384.4	59.33	48.0
LiOH(s)	−484.9	−439.0	42.80	49.6
LiNO$_3$(s)	−483.1	−381.1	90.0	—
Magnesium				
Mg(s)	0	0	32.68	24.9
Mg^{2+}(aq)	−466.9	−454.8	−138.1	—
MgCl$_2$(s)	−641.3	−591.8	89.62	71.4
MgCO$_3$(s)	−1096	−1012	65.7	75.5
MgF$_2$(s)	−1123	−1070	57.24	61.6
MgO(s)	−601.7	−569.4	26.94	37.2
Mg(OH)$_2$(s)	−924.5	−833.5	63.18	77.0
MgS(s)	−346.0	−341.8	50.33	45.6
MgSO$_4$(s)	−1285	−1171	91.6	96.5
Manganese				
Mn(s)	0	0	32.01	26.3
Mn^{2+}(aq)	−220.8	−228.1	−73.6	50.0
MnO$_2$(s)	−520.0	−465.1	53.05	54.1
MnO$_4^-$(aq)	−541.4	−447.2	191.2	−82.0
Mercury				
Hg(g)	61.32	31.82	175.0	20.8
Hg(l)	0	0	76.02	28.0
HgO(s)	−90.83	−58.54	70.29	44.1
Nitrogen				
N(g)	472.7	455.6	153.3	20.8
N$_2$(g)	0	0	191.6	29.1
NF$_3$(g)	−124.7	−83.2	260.7	53.4
NH$_3$(g)	−46.11	−16.45	192.5	35.1
NH$_3$(aq)	−80.29	−26.50	111.3	—
NH$_4^+$(aq)	−132.5	−79.31	113.4	79.9
NH$_4$Br(s)	−270.8	−175.2	113	96.0
NH$_4$Cl(s)	−314.4	−202.9	94.6	84.1
NH$_4$F(s)	−464.0	−348.7	71.96	65.3
NH$_4$HCO$_3$(s)	−849.4	−665.9	120.9	—
NH$_4$I(s)	−201.4	−112.5	117	—
NH$_4$NO$_3$(s)	−365.6	−183.9	151.1	139.3
NH$_4$NO$_3$(aq)	−339.9	−190.6	259.8	−6.7
(NH$_4$)$_2$SO$_4$(s)	−1181	−901.7	220.1	187.5
N$_2$H$_4$(g)	95.40	159.4	238.5	48.4
N$_2$H$_4$(l)	50.63	149.3	121.2	98.9
NO(g)	90.25	86.55	210.8	29.9
N$_2$O(g)	82.05	104.2	219.9	38.6

(continued)

Inorganic Substances

	ΔH_f°, kJ mol^{-1}	ΔG_f°, kJ mol^{-1}	S°, J mol^{-1} K^{-1}	C_p, J mol^{-1} K^{-1}
$NO_2(g)$	33.18	51.31	240.1	37.2
$N_2O_4(g)$	9.16	97.89	304.3	79.2
$N_2O_4(l)$	−19.50	97.54	209.2	142.7
$N_2O_5(g)$	11.3	115.1	355.7	95.3
$NO_3^-(aq)$	−205.0	−108.7	146.4	−86.6
$NOBr(g)$	82.17	82.42	273.7	45.5
$NOCl(g)$	51.71	66.08	261.7	44.7

Oxygen

$O(g)$	249.2	231.7	161.1	21.9
$O_2(g)$	0	0	205.1	29.4
$O_3(g)$	142.7	163.2	238.9	39.2
$OH^-(aq)$	−230.0	−157.2	−10.75	−148.5
$OF_2(g)$	24.7	41.9	247.4	43.3

Phosphorus

$P(\alpha \text{ white})$	0	0	41.09	23.8
$P(\text{red})$	−17.6	−12.1	22.80	21.2
$P_4(g)$	58.91	24.44	280.0	67.2
$PCl_3(g)$	−287.0	−267.8	311.8	71.8
$PCl_5(g)$	−374.9	−305.0	364.6	112.8
$PH_3(g)$	5.4	13.4	210.2	37.1
$P_4O_{10}(s)$	−2984	−2698	228.9	211.71
$PO_4^{3-}(aq)$	−1277	−1019	−222	—

Potassium

$K(g)$	89.24	60.59	160.3	20.8
$K(s)$	0	0	64.18	29.6
$K^+(aq)$	−252.4	−283.3	102.5	21.8
$KBr(s)$	−393.8	−380.7	95.90	52.4
$KCN(s)$	−113.0	−101.9	128.5	66.3
$KCl(s)$	−436.7	−409.1	82.59	51.3
$KClO_3(s)$	−397.7	−296.3	143.1	100.3
$KClO_4(s)$	−432.8	−303.1	151.0	112.4
$KF(s)$	−567.3	−537.8	66.57	49.0
$KI(s)$	−327.9	−324.9	106.3	52.9
$KNO_3(s)$	−494.6	−394.9	133.1	96.4
$KOH(s)$	−424.8	−379.1	78.9	68.9
$KOH(aq)$	−482.4	−440.5	91.6	−126.8
$K_2SO_4(s)$	−1438	−1321	175.6	131.5

Silicon

$Si(s)$	0	0	18.83	20.0
$SiH_4(g)$	34.3	56.9	204.6	42.8
$Si_2H_6(g)$	80.3	127.3	272.7	80.8
$SiO_2(\text{quartz})$	−910.9	−856.6	41.84	44.4

Silver

$Ag(s)$	0	0	42.55	25.4
$Ag^+(aq)$	105.6	77.11	72.68	21.8
$AgBr(s)$	−100.4	−96.90	107.1	52.4
$AgCl(s)$	−127.1	−109.8	96.2	50.8
$AgI(s)$	−61.84	−66.19	115.5	56.8
$AgNO_3(s)$	−124.4	−33.41	140.9	93.1
$Ag_2O(s)$	−31.05	−11.20	121.3	65.9
$Ag_2SO_4(s)$	−715.9	−618.4	200.4	131.4

Inorganic Substances

	ΔH_f°, kJ mol^{-1}	ΔG_f°, kJ mol^{-1}	S°, J mol^{-1} K^{-1}	C_p, J mol^{-1} K^{-1}
Sodium				
Na(g)	107.3	76.76	153.7	20.8
Na(s)	0	0	51.21	28.2
Na$^+$(aq)	−240.1	−261.9	59.0	46.4
Na$_2$(g)	142.1	103.9	230.2	37.6
NaBr(s)	−361.1	−349.0	86.82	51.4
Na$_2$CO$_3$(s)	−1131	−1044	135.0	112.3
NaHCO$_3$(s)	−950.8	−851.0	101.7	87.6
NaCl(s)	−411.2	−384.1	72.13	50.5
NaCl(aq)	−407.3	−393.1	115.5	−90.0
NaClO$_3$(s)	−365.8	−262.3	123.4	—
NaClO$_4$(s)	−383.3	−254.9	142.3	111.3
NaF(s)	−573.6	−543.5	51.46	46.9
NaH(s)	−56.28	−33.46	40.02	36.4
NaI(s)	−287.8	−286.1	98.53	52.1
NaNO$_3$(s)	−467.9	−367.0	116.5	92.9
NaNO$_3$(aq)	−447.5	−373.2	205.4	−40.2
Na$_2$O$_2$(s)	−510.9	−447.7	95.0	89.2
NaOH(s)	−425.6	−379.5	64.46	59.5
NaOH(aq)	−470.1	−419.2	48.1	−102.1
NaH$_2$PO$_4$(s)	−1537	−1386	127.5	−116.86
Na$_2$HPO$_4$(s)	−1748	−1608	150.5	135.3
Na$_3$PO$_4$(s)	−1917	−1789	173.8	153.47
NaHSO$_4$(s)	−1126	−992.8	113.0	—
Na$_2$SO$_4$(s)	−1387	−1270	149.6	128.2
Na$_2$SO$_4$(aq)	−1390.	−1268	138.1	−201.0
Na$_2$SO$_4 \cdot 10$ H$_2$O(s)	−4327	−3647	592.0	—
Na$_2$S$_2$O$_3$(s)	−1123	−1028	155	—
Sulfur				
S(g)	278.8	238.3	167.8	23.7
S(rhombic)	0	0	31.80	22.6
S$_8$(g)	102.3	49.63	431.0	156.06
S$_2$Cl$_2$(g)	−18.4	−31.8	331.5	124.3
SF$_6$(g)	−1209	−1105	291.8	97.0
SO$_2$(g)	−296.8	−300.2	248.2	39.9
SO$_3$(g)	−395.7	−371.1	256.8	50.7
SO$_4{}^{2-}$(aq)	−909.3	−744.5	20.1	−293.0
S$_2$O$_3{}^{2-}$(aq)	−648.5	−522.5	67	—
SO$_2$Cl$_2$(g)	−364.0	−320.0	311.9	77.0
SO$_2$Cl$_2$(l)	−394.1	—	—	−134.0
Tin				
Sn(white)	0	0	51.55	27.0
Sn(gray)	−2.09	0.13	44.14	25.8
SnCl$_4$(l)	−511.3	−440.1	258.6	165.3
SnO(s)	−285.8	−256.9	56.5	44.3
SnO$_2$(s)	−580.7	−519.6	52.3	52.6
Titanium				
Ti(s)	0	0	30.63	25.0
TiCl$_4$(g)	−763.2	−726.7	354.9	95.4
TiCl$_4$(l)	−804.2	−737.2	252.3	145.2
TiO$_2$(s)	−944.7	−889.5	50.33	55.0

(continued)

Inorganic Substances

	ΔH_f°, kJ mol^{-1}	ΔG_f°, kJ mol^{-1}	S°, J mol^{-1} K^{-1}	C_p, J mol^{-1} K^{-1}
Uranium				
U(s)	0	0	50.21	27.7
UF$_6$(g)	−2147	−2064	377.9	129.6
UF$_6$(s)	−2197	−2069	227.6	166.8
UO$_2$(s)	−1085	−1032	77.03	63.6
Zinc				
Zn(s)	0	0	41.63	25.4
Zn^{2+}(aq)	−153.9	−147.1	112.1	46.0
ZnO(s)	−138.3	−318.3	43.64	40.3

Organic Substances

	Name	ΔH_f°, kJ mol^{-1}	ΔG_f°, kJ mol^{-1}	S°, J mol^{-1} K^{-1}	C_p, J mol^{-1} K^{-1}
CH$_4$(g)	Methane(g)	−74.81	−50.72	186.3	35.7
C$_2$H$_2$(g)	Acetylene(g)	226.7	209.2	200.9	44.0
C$_2$H$_4$(g)	Ethylene(g)	52.26	68.15	219.6	42.9
C$_2$H$_6$(g)	Ethane(g)	−84.68	−32.82	229.6	52.5
C$_3$H$_8$(g)	Propane(g)	−103.8	−23.3	270.3	73.6
C$_4$H$_{10}$(g)	Butane(g)	−125.6	−17.1	310.2	97.5
C$_6$H$_6$(g)	Benzene(g)	82.6	129.8	269.3	82.4
C$_6$H$_6$(l)	Benzene(l)	49.0	124.5	173.4	136.0
C$_6$H$_{12}$(g)	Cyclohexane(g)	−123.4	32.0	298.4	106.3
C$_6$H$_{12}$(l)	Cyclohexane(l)	−156.4	26.9	204.4	154.9
C$_{10}$H$_8$(g)	Naphthalene(g)	150.6	224.2	333.2	131.9
C$_{10}$H$_8$(s)	Naphthalene(s)	77.9	201.7	167.5	165.7
CH$_2$O(g)	Formaldehyde(g)	−108.6	−102.5	218.8	35.4
CH$_3$CHO(g)	Acetaldehyde(g)	−166.2	−128.9	250.3	55.3
CH$_3$CHO(l)	Acetaldehyde(l)	−192.3	−128.1	160.2	89.0
CH$_3$OH(g)	Methanol(g)	−200.7	−162.0	239.8	44.1
CH$_3$OH(l)	Methanol(l)	−238.7	−166.3	126.8	81.1
CH$_3$CH$_2$OH(g)	Ethanol(g)	−235.1	−168.5	282.7	65.6
CH$_3$CH$_2$OH(l)	Ethanol(l)	−277.7	−174.8	160.7	112.3
C$_6$H$_5$OH(s)	Phenol(s)	−165.1	−50.4	144.0	127.4
(CH$_3$)$_2$CO(g)	Acetone(g)	−216.6	−153.0	295.0	74.5
(CH$_3$)$_2$CO(l)	Acetone(l)	−247.6	−155.6	200.5	126.3
CH$_3$COOH(g)	Acetic acid(g)	−432.3	−374.0	282.5	63.4
CH$_3$COOH(l)	Acetic acid(l)	−484.5	−389.9	159.8	123.3
CH$_3$COOH(aq)	Acetic acid(aq)	−485.8	−396.5	178.7	−6.3
C$_6$H$_5$COOH(s)	Benzoic acid(s)	−385.2	−245.3	167.6	146.8
CH$_3$NH$_2$(g)	Methylamine(g)	−22.97	32.16	243.4	50.1
C$_6$H$_5$NH$_2$(g)	Aniline(g)	86.86	166.8	319.3	107.9
C$_6$H$_5$NH$_2$(l)	Aniline(l)	31.6	149.2	191.3	191.9

TABLE D.3 Equilibrium Constants

A. Ionization Constants of Weak Acids at 25 °C

Name of acid	Formula	K_a	Name of acid	Formula	K_a
Acetic	$HC_2H_3O_2$	1.8×10^{-5}	Hyponitrous	$HON{=}NOH$	8.9×10^{-8}
Acrylic	$HC_3H_3O_2$	5.5×10^{-5}		$HON{=}NO^-$	4×10^{-12}
Arsenic	H_3AsO_4	6.0×10^{-3}	Iodic	HIO_3	1.6×10^{-1}
	$H_2AsO_4^-$	1.0×10^{-7}	Iodoacetic	$HC_2H_2IO_2$	6.7×10^{-4}
	$HAsO_4^{2-}$	3.2×10^{-12}	Malonic	$H_2C_3H_2O_4$	1.5×10^{-3}
Arsenous	H_3AsO_3	6.6×10^{-10}		$HC_3H_2O_4^-$	2.0×10^{-6}
Benzoic	$HC_7H_5O_2$	6.3×10^{-5}	Nitrous	HNO_2	7.2×10^{-4}
Bromoacetic	$HC_2H_2BrO_2$	1.3×10^{-3}	Oxalic	$H_2C_2O_4$	5.4×10^{-2}
Butyric	$HC_4H_7O_2$	1.5×10^{-5}		$HC_2O_4^-$	5.3×10^{-5}
Carbonic	H_2CO_3	4.4×10^{-7}	Phenol	HOC_6H_5	1.0×10^{-10}
	HCO_3^-	4.7×10^{-11}	Phenylacetic	$HC_8H_7O_2$	4.9×10^{-5}
Chloroacetic	$HC_2H_2ClO_2$	1.4×10^{-3}	Phosphoric	H_3PO_4	7.1×10^{-3}
Chlorous	$HClO_2$	1.1×10^{-2}		$H_2PO_4^-$	6.3×10^{-8}
Citric	$H_3C_6H_5O_7$	7.4×10^{-4}		HPO_4^{2-}	4.2×10^{-13}
	$H_2C_6H_5O_7^-$	1.7×10^{-5}	Phosphorous	H_3PO_3	3.7×10^{-2}
	$HC_6H_5O_7^{2-}$	4.0×10^{-7}		$H_2PO_3^-$	2.1×10^{-7}
Cyanic	$HOCN$	3.5×10^{-4}	Propionic	$HC_3H_5O_2$	1.3×10^{-5}
Dichloroacetic	$HC_2HCl_2O_2$	5.5×10^{-2}	Pyrophosphoric	$H_4P_2O_7$	3.0×10^{-2}
Fluoroacetic	$HC_2H_2FO_2$	2.6×10^{-3}		$H_3P_2O_7^-$	4.4×10^{-3}
Formic	$HCHO_2$	1.8×10^{-4}		$H_2P_2O_7^{2-}$	2.5×10^{-7}
Hydrazoic	HN_3	1.9×10^{-5}		$HP_2O_7^{3-}$	5.6×10^{-10}
Hydrocyanic	HCN	6.2×10^{-10}	Selenic	H_2SeO_4	strong acid
Hydrofluoric	HF	6.6×10^{-4}		$HSeO_4^-$	2.2×10^{-2}
Hydrogen peroxide	H_2O_2	2.2×10^{-12}	Selenous	H_2SeO_3	2.3×10^{-3}
Hydroselenic	H_2Se	1.3×10^{-4}		$HSeO_3^-$	5.4×10^{-9}
	HSe^-	1×10^{-11}	Succinic	$H_2C_4H_4O_4$	6.2×10^{-5}
Hydrosulfuric	H_2S	1.0×10^{-7}		$HC_4H_4O_4^-$	2.3×10^{-6}
	HS^-	1×10^{-19}	Sulfuric	H_2SO_4	strong acid
Hydrotelluric	H_2Te	2.3×10^{-3}		HSO_4^-	1.1×10^{-2}
	HTe^-	1.6×10^{-11}	Sulfurous	H_2SO_3	1.3×10^{-2}
Hypobromous	$HOBr$	2.5×10^{-9}		HSO_3^-	6.2×10^{-8}
Hypochlorous	$HOCl$	2.9×10^{-8}	Thiophenol	HSC_6H_5	3.2×10^{-7}
Hypoiodous	HOI	2.3×10^{-11}	Trichloroacetic	$HC_2Cl_3O_2$	3.0×10^{-1}

B. Ionization Constants of Weak Bases at 25 °C

Name of base	Formula	K_b	Name of base	Formula	K_b
Ammonia	NH_3	1.8×10^{-5}	Isoquinoline	C_9H_7N	2.5×10^{-9}
Aniline	$C_6H_5NH_2$	7.4×10^{-10}	Methylamine	CH_3NH_2	4.2×10^{-4}
Codeine	$C_{18}H_{21}O_3N$	8.9×10^{-7}	Morphine	$C_{17}H_{19}O_3N$	7.4×10^{-7}
Diethylamine	$(C_2H_5)_2NH$	6.9×10^{-4}	Piperdine	$C_5H_{11}N$	1.3×10^{-3}
Dimethylamine	$(CH_3)_2NH$	5.9×10^{-4}	Pyridine	C_5H_5N	1.5×10^{-9}
Ethylamine	$C_2H_5NH_2$	4.3×10^{-4}	Quinoline	C_9H_7N	6.3×10^{-10}
Hydrazine	NH_2NH_2	8.5×10^{-7}	Triethanolamine	$C_6H_{15}O_3N$	5.8×10^{-7}
	$NH_2NH_3^+$	8.9×10^{-16}	Triethylamine	$(C_2H_5)_3N$	5.2×10^{-4}
Hydroxylamine	NH_2OH	9.1×10^{-9}	Trimethylamine	$(CH_3)_3N$	6.3×10^{-5}

(continued)

C. Solubility Product Constants[a]

Name of solute	Formula	K_{sp}	Name of solute	Formula	K_{sp}
Aluminum hydroxide	$Al(OH)_3$	1.3×10^{-33}	Lead(II) hydroxide	$Pb(OH)_2$	1.2×10^{-15}
Aluminum phosphate	$AlPO_4$	6.3×10^{-19}	Lead(II) iodide	PbI_2	7.1×10^{-9}
Barium carbonate	$BaCO_3$	5.1×10^{-9}	Lead(II) sulfate	$PbSO_4$	1.6×10^{-8}
Barium chromate	$BaCrO_4$	1.2×10^{-10}	Lead(II) sulfide[b]	PbS	3×10^{-28}
Barium fluoride	BaF_2	1.0×10^{-6}	Lithium carbonate	Li_2CO_3	2.5×10^{-2}
Barium hydroxide	$Ba(OH)_2$	5×10^{-3}	Lithium fluoride	LiF	3.8×10^{-3}
Barium sulfate	$BaSO_4$	1.1×10^{-10}	Lithium phosphate	Li_3PO_4	3.2×10^{-9}
Barium sulfite	$BaSO_3$	8×10^{-7}	Magnesium ammonium phosphate	$MgNH_4PO_4$	2.5×10^{-13}
Barium thiosulfate	BaS_2O_3	1.6×10^{-5}			
Bismuthyl chloride	$BiOCl$	1.8×10^{-31}	Magnesium carbonate	$MgCO_3$	3.5×10^{-8}
Bismuthyl hydroxide	$BiOOH$	4×10^{-10}	Magnesium fluoride	MgF_2	3.7×10^{-8}
Cadmium carbonate	$CdCO_3$	5.2×10^{-12}	Magnesium hydroxide	$Mg(OH)_2$	1.8×10^{-11}
Cadmium hydroxide	$Cd(OH)_2$	2.5×10^{-14}	Magnesium phosphate	$Mg_3(PO_4)_2$	1×10^{-25}
Cadmium sulfide[b]	CdS	8×10^{-28}	Manganese(II) carbonate	$MnCO_3$	1.8×10^{-11}
Calcium carbonate	$CaCO_3$	2.8×10^{-9}	Manganese(II) hydroxide	$Mn(OH)_2$	1.9×10^{-13}
Calcium chromate	$CaCrO_4$	7.1×10^{-4}	Manganese(II) sulfide[b]	MnS	3×10^{-14}
Calcium fluoride	CaF_2	5.3×10^{-9}	Mercury(I) bromide	Hg_2Br_2	5.6×10^{-23}
Calcium hydroxide	$Ca(OH)_2$	5.5×10^{-6}	Mercury(I) chloride	Hg_2Cl_2	1.3×10^{-18}
Calcium hydrogen phosphate	$CaHPO_4$	1×10^{-7}	Mercury(I) iodide	Hg_2I_2	4.5×10^{-29}
			Mercury(II) sulfide[b]	HgS	2×10^{-53}
Calcium oxalate	CaC_2O_4	4×10^{-9}	Nickel(II) carbonate	$NiCO_3$	6.6×10^{-9}
Calcium phosphate	$Ca_3(PO_4)_2$	2.0×10^{-29}	Nickel(II) hydroxide	$Ni(OH)_2$	2.0×10^{-15}
Calcium sulfate	$CaSO_4$	9.1×10^{-6}	Scandium fluoride	ScF_3	4.2×10^{-18}
Calcium sulfite	$CaSO_3$	6.8×10^{-8}	Scandium hydroxide	$Sc(OH)_3$	8.0×10^{-31}
Chromium(II) hydroxide	$Cr(OH)_2$	2×10^{-16}	Silver arsenate	Ag_3AsO_4	1.0×10^{-22}
Chromium(III) hydroxide	$Cr(OH)_3$	6.3×10^{-31}	Silver azide	AgN_3	2.8×10^{-9}
Cobalt(II) carbonate	$CoCO_3$	1.4×10^{-13}	Silver bromide	$AgBr$	5.0×10^{-13}
Cobalt(II) hydroxide	$Co(OH)_2$	1.6×10^{-15}	Silver carbonate	Ag_2CO_3	8.5×10^{-12}
Cobalt(III) hydroxide	$Co(OH)_3$	1.6×10^{-44}	Silver chloride	$AgCl$	1.8×10^{-10}
Copper(I) chloride	$CuCl$	1.2×10^{-6}	Silver chromate	Ag_2CrO_4	1.1×10^{-12}
Copper(I) cyanide	$CuCN$	3.2×10^{-20}	Silver cyanide	$AgCN$	1.2×10^{-16}
Copper(I) iodide	CuI	1.1×10^{-12}	Silver iodate	$AgIO_3$	3.0×10^{-8}
Copper(II) arsenate	$Cu_3(AsO_4)_2$	7.6×10^{-36}	Silver iodide	AgI	8.5×10^{-17}
Copper(II) carbonate	$CuCO_3$	1.4×10^{-10}	Silver nitrite	$AgNO_2$	6.0×10^{-4}
Copper(II) chromate	$CuCrO_4$	3.6×10^{-6}	Silver sulfate	Ag_2SO_4	1.4×10^{-5}
Copper(II) ferrocyanide	$Cu_2[Fe(CN)_6]$	1.3×10^{-16}	Silver sulfide[b]	Ag_2S	6×10^{-51}
Copper(II) hydroxide	$Cu(OH)_2$	2.2×10^{-20}	Silver sulfite	Ag_2SO_3	1.5×10^{-14}
Copper(II) sulfide[b]	CuS	6×10^{-37}	Silver thiocyanate	$AgSCN$	1.0×10^{-12}
Iron(II) carbonate	$FeCO_3$	3.2×10^{-11}	Strontium carbonate	$SrCO_3$	1.1×10^{-10}
Iron(II) hydroxide	$Fe(OH)_2$	8.0×10^{-16}	Strontium chromate	$SrCrO_4$	2.2×10^{-5}
Iron(II) sulfide[b]	FeS	6×10^{-19}	Strontium fluoride	SrF_2	2.5×10^{-9}
Iron(III) arsenate	$FeAsO_4$	5.7×10^{-21}	Strontium sulfate	$SrSO_4$	3.2×10^{-7}
Iron(III) ferrocyanide	$Fe_4[Fe(CN)_6]_3$	3.3×10^{-41}	Thallium(I) bromide	$TlBr$	3.4×10^{-6}
Iron(III) hydroxide	$Fe(OH)_3$	4×10^{-38}	Thallium(I) chloride	$TlCl$	1.7×10^{-4}
Iron(III) phosphate	$FePO_4$	1.3×10^{-22}	Thallium(I) iodide	TlI	6.5×10^{-8}
Lead(II) arsenate	$Pb_3(AsO_4)_2$	4.0×10^{-36}	Thallium(III) hydroxide	$Tl(OH)_3$	6.3×10^{-46}
Lead(II) azide	$Pb(N_3)_2$	2.5×10^{-9}	Tin(II) hydroxide	$Sn(OH)_2$	1.4×10^{-28}
Lead(II) bromide	$PbBr_2$	4.0×10^{-5}	Tin(II) sulfide[b]	SnS	1×10^{-26}
Lead(II) carbonate	$PbCO_3$	7.4×10^{-14}	Zinc carbonate	$ZnCO_3$	1.4×10^{-11}
Lead(II) chloride	$PbCl_2$	1.6×10^{-5}	Zinc hydroxide	$Zn(OH)_2$	1.2×10^{-17}
Lead(II) chromate	$PbCrO_4$	2.8×10^{-13}	Zinc oxalate	ZnC_2O_4	2.7×10^{-8}
Lead(II) fluoride	PbF_2	2.7×10^{-8}	Zinc phosphate	$Zn_3(PO_4)_2$	9.0×10^{-33}
			Zinc sulfide[b]	ZnS	2×10^{-25}

D. Complex-Ion Formation Constants[c, d]

Formula	K_f	Formula	K_f	Formula	K_f
$[Ag(CN)_2]^-$	5.6×10^{18}	$[Co(ox)_3]^{3-}$	10^{20}	$[HgI_4]^{2-}$	6.8×10^{29}
$[Ag(EDTA)]^{3-}$	2.1×10^7	$[Cr(EDTA)]^-$	10^{23}	$[Hg(ox)_2]^{2-}$	9.5×10^6
$[Ag(en)_2]^+$	5.0×10^7	$[Cr(OH)_4]^-$	8×10^{29}	$[Ni(CN)_4]^{2-}$	2×10^{31}
$[Ag(NH_3)_2]^+$	1.6×10^7	$[CuCl_3]^{2-}$	5×10^5	$[Ni(EDTA)]^{2-}$	3.6×10^{18}
$[Ag(SCN)_4]^{3-}$	1.2×10^{10}	$[Cu(CN)_4]^{3-}$	2.0×10^{30}	$[Ni(en)_3]^{2+}$	2.1×10^{18}
$[Ag(S_2O_3)_2]^{3-}$	1.7×10^{13}	$[Cu(EDTA)]^{2-}$	5×10^{18}	$[Ni(NH_3)_6]^{2+}$	5.5×10^8
$[Al(EDTA)]^-$	1.3×10^{16}	$[Cu(en)_2]^{2+}$	1×10^{20}	$[Ni(ox)_3]^{4-}$	3×10^8
$[Al(OH)_4]^-$	1.1×10^{33}	$[Cu(NH_3)_4]^{2+}$	1.1×10^{13}	$[PbCl_3]^-$	2.4×10^1
$[Al(ox)_3]^{3-}$	2×10^{16}	$[Cu(ox)_2]^{2-}$	3×10^8	$[Pb(EDTA)]^{2-}$	2×10^{18}
$[CdCl_4]^{2-}$	6.3×10^2	$[Fe(CN)_6]^{4-}$	10^{37}	$[PbI_4]^{2-}$	3.0×10^4
$[Cd(CN)_4]^{2-}$	6.0×10^{18}	$[Fe(EDTA)]^{2-}$	2.1×10^{14}	$[Pb(OH)_3]^-$	3.8×10^{14}
$[Cd(en)_3]^{2+}$	1.2×10^{12}	$[Fe(en)_3]^{2+}$	5.0×10^9	$[Pb(ox)_2]^{2-}$	3.5×10^6
$[Cd(NH_3)_4]^{2+}$	1.3×10^7	$[Fe(ox)_3]^{4-}$	1.7×10^5	$[Pb(S_2O_3)_3]^{4-}$	2.2×10^6
$[Co(EDTA)]^{2-}$	2.0×10^{16}	$[Fe(CN)_6]^{3-}$	10^{42}	$[PtCl_4]^{2-}$	1×10^{16}
$[Co(en)_3]^{2+}$	8.7×10^{13}	$[Fe(EDTA)]^-$	1.7×10^{24}	$[Pt(NH_3)_6]^{2+}$	2×10^{35}
$[Co(NH_3)_6]^{2+}$	1.3×10^5	$[Fe(ox)_3]^{3-}$	2×10^{20}	$[Zn(CN)_4]^{2-}$	1×10^{18}
$[Co(ox)_3]^{4-}$	5×10^9	$[Fe(SCN)]^{2+}$	8.9×10^2	$[Zn(EDTA)]^{2-}$	3×10^{16}
$[Co(SCN)_4]^{2-}$	1.0×10^3	$[HgCl_4]^{2-}$	1.2×10^{15}	$[Zn(en)_3]^{2+}$	1.3×10^{14}
$[Co(EDTA)]^-$	10^{36}	$[Hg(CN)_4]^{2-}$	3×10^{41}	$[Zn(NH_3)_4]^{2+}$	4.1×10^8
$[Co(en)_3]^{3+}$	4.9×10^{48}	$[Hg(EDTA)]^{2-}$	6.3×10^{21}	$[Zn(OH)_4]^{2-}$	4.6×10^{17}
$[Co(NH_3)_6]^{3+}$	4.5×10^{33}	$[Hg(en)_2]^{2+}$	2×10^{23}	$[Zn(ox)_3]^{4-}$	1.4×10^8

[a]Data are at various temperatures around "room" temperature, from 18 to 25 °C.
[b]For a solubility equilibrium of the type $MS(s) + H_2O \rightleftharpoons M^{2+}(aq) + HS^-(aq) + OH^-(aq)$.
[c]The ligands referred to in this table are monodentate: Cl^-, CN^-, I^-, NH_3, OH^-, SCN^-, $S_2O_3^{2-}$; bidentate: ethylenediamine (en), oxalate ion (ox); tetradentate: ethylenediaminetetraacetato ion, $EDTA^{4-}$.
[d]The K_f values are cumulative or overall formation constants (see page 1094).

TABLE D.4 Standard Electrode (Reduction) Potentials at 25 °C

Reduction half-reaction	$E°$, V
$F_2(g) + 2\,e^- \longrightarrow 2\,F^-(aq)$	+2.866
$OF_2(g) + 2\,H^+(aq) + 4\,e^- \longrightarrow H_2O(l) + 2\,F^-(aq)$	+2.1
$O_3(g) + 2\,H^+(aq) + 2\,e^- \longrightarrow O_2(g) + H_2O(l)$	+2.075
$S_2O_8^{2-}(aq) + 2\,e^- \longrightarrow 2\,SO_4^{2-}(aq)$	+2.01
$Ag^{2+}(aq) + e^- \longrightarrow Ag^+(aq)$	+1.98
$H_2O_2(aq) + 2\,H^+(aq) + 2\,e^- \longrightarrow 2\,H_2O(l)$	+1.763
$MnO_4^-(aq) + 4\,H^+(aq) + 3\,e^- \longrightarrow MnO_2(s) + 2\,H_2O(l)$	+1.70
$PbO_2(s) + SO_4^{2-}(aq) + 4\,H^+(aq) + 2\,e^- \longrightarrow PbSO_4(s) + 2\,H_2O(l)$	+1.69
$Au^{3+}(aq) + 3\,e^- \longrightarrow Au(s)$	+1.52
$MnO_4^-(aq) + 8\,H^+(aq) + 5\,e^- \longrightarrow Mn^{2+}(aq) + 4\,H_2O(l)$	+1.51
$2\,BrO_3^-(aq) + 12\,H^+(aq) + 10\,e^- \longrightarrow Br_2(l) + 6\,H_2O(l)$	+1.478
$PbO_2(s) + 4\,H^+(aq) + 2\,e^- \longrightarrow Pb^{2+}(aq) + 2\,H_2O(l)$	+1.455
$ClO_3^-(aq) + 6\,H^+(aq) + 6\,e^- \longrightarrow Cl^-(aq) + 3\,H_2O(l)$	+1.450
$Au^{3+}(aq) + 2\,e^- \longrightarrow Au^+(aq)$	+1.36
$Cl_2(g) + 2\,e^- \longrightarrow 2\,Cl^-(aq)$	+1.358
$Cr_2O_7^{2-}(aq) + 14\,H^+(aq) + 6\,e^- \longrightarrow 2\,Cr^{3+}(aq) + 7\,H_2O(l)$	+1.33
$MnO_2(s) + 4\,H^+(aq) + 2\,e^- \longrightarrow Mn^{2+}(aq) + 2\,H_2O(l)$	+1.23
$O_2(g) + 4\,H^+(aq) + 4\,e^- \longrightarrow 2\,H_2O(l)$	+1.229

(continued)

Reduction half-reaction	$E°$, V
$2\,IO_3^-(aq) + 12\,H^+(aq) + 10\,e^- \longrightarrow I_2(s) + 6\,H_2O(l)$	$+1.20$
$ClO_4^-(aq) + 2\,H^+(aq) + 2\,e^- \longrightarrow ClO_3^-(aq) + H_2O(l)$	$+1.189$
$ClO_3^-(aq) + 2\,H^+(aq) + e^- \longrightarrow ClO_2(g) + H_2O(l)$	$+1.175$
$NO_2(g) + H^+(aq) + e^- \longrightarrow HNO_2(aq)$	$+1.07$
$Br_2(l) + 2\,e^- \longrightarrow 2\,Br^-(aq)$	$+1.065$
$NO_2(g) + 2\,H^+(aq) + 2\,e^- \longrightarrow NO(g) + H_2O(l)$	$+1.03$
$[AuCl_4]^-(aq) + 3\,e^- \longrightarrow Au(s) + 4\,Cl^-(aq)$	$+1.002$
$VO_2^+(aq) + 2\,H^+(aq) + e^- \longrightarrow VO^{2+}(aq) + H_2O(l)$	$+1.000$
$NO_3^-(aq) + 4\,H^+(aq) + 3\,e^- \longrightarrow NO(g) + 2\,H_2O(l)$	$+0.956$
$Cu^{2+}(aq) + I^-(aq) + e^- \longrightarrow CuI(s)$	$+0.86$
$Hg^{2+}(aq) + 2\,e^- \longrightarrow Hg(l)$	$+0.854$
$Ag^+(aq) + e^- \longrightarrow Ag(s)$	$+0.800$
$Fe^{3+}(aq) + e^- \longrightarrow Fe^{2+}(aq)$	$+0.771$
$O_2(g) + 2\,H^+(aq) + 2\,e^- \longrightarrow H_2O_2(aq)$	$+0.695$
$2\,HgCl_2(aq) + 2\,e^- \longrightarrow Hg_2Cl_2(s) + 2\,Cl^-(aq)$	$+0.63$
$MnO_4^-(aq) + e^- \longrightarrow MnO_4^{2-}(aq)$	$+0.56$
$I_2(s) + 2\,e^- \longrightarrow 2\,I^-(aq)$	$+0.535$
$Cu^+(aq) + e^- \longrightarrow Cu(s)$	$+0.520$
$H_2SO_3(aq) + 4\,H^+(aq) + 4\,e^- \longrightarrow S(s) + 3\,H_2O(l)$	$+0.449$
$C_2N_2(g) + 2\,H^+(aq) + 2\,e^- \longrightarrow 2\,HCN(aq)$	$+0.37$
$[Fe(CN)_6]^{3-}(aq) + e^- \longrightarrow [Fe(CN)_6]^{4-}(aq)$	$+0.361$
$VO^{2+}(aq) + 2\,H^+(aq) + e^- \longrightarrow V^{3+}(aq) + H_2O(l)$	$+0.337$
$Cu^{2+}(aq) + 2\,e^- \longrightarrow Cu(s)$	$+0.340$
$PbO_2(s) + 2\,H^+(aq) + 2\,e^- \longrightarrow PbO(s) + H_2O(l)$	$+0.28$
$Hg_2Cl_2(s) + 2\,e^- \longrightarrow 2\,Hg(l) + 2\,Cl^-(aq)$	$+0.2676$
$HAsO_2(aq) + 3\,H^+(aq) + 3\,e^- \longrightarrow As(s) + 2\,H_2O(l)$	$+0.240$
$AgCl(s) + e^- \longrightarrow Ag(s) + Cl^-(aq)$	$+0.2223$
$SO_4^{2-}(aq) + 4\,H^+(aq) + 2\,e^- \longrightarrow 2\,H_2O(l) + SO_2(g)$	$+0.17$
$Cu^{2+}(aq) + e^- \longrightarrow Cu^+(aq)$	$+0.159$
$Sn^{4+}(aq) + 2\,e^- \longrightarrow Sn^{2+}(aq)$	$+0.154$
$S(s) + 2\,H^+(aq) + 2\,e^- \longrightarrow H_2S(g)$	$+0.144$
$AgBr(s) + e^- \longrightarrow Ag(s) + Br^-(aq)$	$+0.071$
$2\,H^+(aq) + 2\,e^- \longrightarrow H_2(g)$	0
$Pb^{2+}(aq) + 2\,e^- \longrightarrow Pb(s)$	-0.125
$Sn^{2+}(aq) + 2\,e^- \longrightarrow Sn(s)$	-0.137
$AgI(s) + e^- \longrightarrow Ag(s) + I^-(aq)$	-0.152
$V^{3+}(aq) + e^- \longrightarrow V^{2+}(aq)$	-0.255
$Ni^{2+}(aq) + 2\,e^- \longrightarrow Ni(s)$	-0.257
$H_3PO_4(aq) + 2\,H^+(aq) + 2\,e^- \longrightarrow H_3PO_3(aq) + H_2O(l)$	-0.276
$Co^{2+}(aq) + 2\,e^- \longrightarrow Co(s)$	-0.277
$In^{3+}(aq) + 3\,e^- \longrightarrow In(s)$	-0.338
$PbSO_4(s) + 2\,e^- \longrightarrow Pb(s) + SO_4^{2-}(aq)$	-0.356
$Cd^{2+}(aq) + 2\,e^- \longrightarrow Cd(s)$	-0.403
$Cr^{3+}(aq) + e^- \longrightarrow Cr^{2+}(aq)$	-0.424
$Fe^{2+}(aq) + 2\,e^- \longrightarrow Fe(s)$	-0.440
$2\,CO_2(g) + 2\,H^+(aq) + 2\,e^- \longrightarrow H_2C_2O_4(aq)$	-0.49
$Zn^{2+}(aq) + 2\,e^- \longrightarrow Zn(s)$	-0.763

Reduction half-reaction	$E°$, V
$Cr^{2+}(aq) + 2\,e^- \longrightarrow Cr(s)$	-0.90
$Mn^{2+}(aq) + 2\,e^- \longrightarrow Mn(s)$	-1.18
$Ti^{2+}(aq) + 2\,e^- \longrightarrow Ti(s)$	-1.63
$U^{3+}(aq) + 3\,e^- \longrightarrow U(s)$	-1.66
$Al^{3+}(aq) + 3\,e^- \longrightarrow Al(s)$	-1.676
$Mg^{2+}(aq) + 2\,e^- \longrightarrow Mg(s)$	-2.356
$La^{3+}(aq) + 3\,e^- \longrightarrow La(s)$	-2.38
$Na^+(aq) + e^- \longrightarrow Na(s)$	-2.713
$Ca^{2+}(aq) + 2\,e^- \longrightarrow Ca(s)$	-2.84
$Sr^{2+}(aq) + 2\,e^- \longrightarrow Sr(s)$	-2.89
$Ba^{2+}(aq) + 2\,e^- \longrightarrow Ba(s)$	-2.92
$Cs^+(aq) + e^- \longrightarrow Cs(s)$	-2.923
$K^+(aq) + e^- \longrightarrow K(s)$	-2.924
$Rb^+(aq) + e^- \longrightarrow Rb(s)$	-2.924
$Li^+(aq) + e^- \longrightarrow Li(s)$	-3.040

Basic solution

Reduction half-reaction	$E°$, V
$O_3(g) + H_2O(l) + 2\,e^- \longrightarrow O_2(g) + 2\,OH^-(aq)$	$+1.246$
$ClO^-(aq) + H_2O(l) + 2\,e^- \longrightarrow Cl^-(aq) + 2\,OH^-(aq)$	$+0.890$
$H_2O_2(aq) + 2\,e^- \longrightarrow 2\,OH^-(aq)$	$+0.88$
$BrO^-(aq) + H_2O(l) + 2\,e^- \longrightarrow Br^-(aq) + 2\,OH^-(aq)$	$+0.766$
$ClO_3^-(aq) + 3\,H_2O(l) + 6\,e^- \longrightarrow Cl^-(aq) + 6\,OH^-(aq)$	$+0.622$
$2\,AgO(s) + H_2O(l) + 2\,e^- \longrightarrow Ag_2O(s) + 2\,OH^-(aq)$	$+0.604$
$MnO_4^-(aq) + 2\,H_2O(l) + 3\,e^- \longrightarrow MnO_2(s) + 4\,OH^-(aq)$	$+0.60$
$BrO_3^-(aq) + 3\,H_2O(l) + 6\,e^- \longrightarrow Br^-(aq) + 6\,OH^-(aq)$	$+0.584$
$2\,BrO^-(aq) + 2\,H_2O(l) + 2\,e^- \longrightarrow Br_2(l) + 4\,OH^-(aq)$	$+0.455$
$2\,IO^-(aq) + 2\,H_2O(l) + 2\,e^- \longrightarrow I_2(s) + 4\,OH^-(aq)$	$+0.42$
$O_2(g) + 2\,H_2O(l) + 4\,e^- \longrightarrow 4\,OH^-(aq)$	$+0.401$
$Ag_2O(s) + H_2O(l) + 2\,e^- \longrightarrow 2\,Ag(s) + 2\,OH^-(aq)$	$+0.342$
$Co(OH)_3(s) + e^- \longrightarrow Co(OH)_2(s) + OH^-(aq)$	$+0.17$
$2\,MnO_2(s) + H_2O(l) + 2\,e^- \longrightarrow Mn_2O_3(s) + 2\,OH^-(aq)$	$+0.118$
$NO_3^-(aq) + H_2O(l) + 2\,e^- \longrightarrow NO_2^-(aq) + 2\,OH^-(aq)$	$+0.01$
$CrO_4^{2-}(aq) + 4\,H_2O(l) + 3\,e^- \longrightarrow Cr(OH)_3(s) + 5\,OH^-(aq)$	-0.11
$S(s) + 2\,e^- \longrightarrow S^{2-}(aq)$	-0.48
$HPbO_2^-(aq) + H_2O(l) + 2\,e^- \longrightarrow Pb(s) + 3\,OH^-(aq)$	-0.54
$HCHO(aq) + 2\,H_2O(l) + 2\,e^- \longrightarrow CH_3OH(aq) + 2\,OH^-(aq)$	-0.59
$SO_3^{2-}(aq) + 3\,H_2O(l) + 4\,e^- \longrightarrow S(s) + 6\,OH^-(aq)$	-0.66
$AsO_4^{3-}(aq) + 2\,H_2O(l) + 2\,e^- \longrightarrow AsO_2^-(aq) + 4\,OH^-(aq)$	-0.67
$AsO_2^-(aq) + 2\,H_2O(l) + 3\,e^- \longrightarrow As(s) + 4\,OH^-(aq)$	-0.68
$Cd(OH)_2(s) + 2\,e^- \longrightarrow Cd(s) + 2\,OH^-(aq)$	-0.824
$2\,H_2O(l) + 2\,e^- \longrightarrow H_2(g) + 2\,OH^-(aq)$	-0.828
$OCN^-(aq) + H_2O(l) + 2\,e^- \longrightarrow CN^-(aq) + 2\,OH^-(aq)$	-0.97
$As(s) + 3\,H_2O(l) + 3\,e^- \longrightarrow AsH_3(g) + 3\,OH^-(aq)$	-1.21
$Zn(OH)_2(s) + 2\,e^- \longrightarrow Zn(s) + 2\,OH^-(aq)$	-1.246
$Sb(s) + 3\,H_2O(l) + 3\,e^- \longrightarrow SbH_3(g) + 3\,OH^-(aq)$	-1.338
$Al(OH)_4^-(aq) + 3\,e^- \longrightarrow Al(s) + 4\,OH^-(aq)$	-2.310
$Mg(OH)_2(s) + 2\,e^- \longrightarrow Mg(s) + 2\,OH^-(aq)$	-2.687

TABLE D.5 Isotopic Masses and Their Abundance*

Z	Name	Symbol	Mass of Atom, u	% Abundance
1	Hydrogen	^{1}H	1.007825	99.9885
	Deuterium	^{2}H	2.014102	0.0115
	Tritium	^{3}H	3.016049	—
2	Helium	^{3}He	3.016029	0.000137
		^{4}He	4.002603	99.999863
3	Lithium	^{6}Li	6.015122	7.59
		^{7}Li	7.016004	92.41
4	Beryllium	^{9}Be	9.012182	100
5	Boron	^{10}B	10.012937	19.9
		^{11}B	11.009305	80.1
6	Carbon	^{12}C	12.000000	98.93
		^{13}C	13.003355	1.07
		^{14}C	14.003242	—
7	Nitrogen	^{14}N	14.003074	99.632
		^{15}N	15.000109	0.368
8	Oxygen	^{16}O	15.994915	99.757
		^{17}O	16.999132	0.038
		^{18}O	17.999160	0.205
9	Fluorine	^{19}F	18.998403	100
10	Neon	^{20}Ne	19.992440	90.48
		^{21}Ne	20.993847	0.27
		^{22}Ne	21.991386	9.25
11	Sodium	^{23}Na	22.989770	100
12	Magnesium	^{24}Mg	23.985042	78.99
		^{25}Mg	24.985837	10.00
		^{26}Mg	25.982593	11.01
13	Aluminum	^{27}Al	26.981538	100
14	Silicon	^{28}Si	27.976927	92.2297
		^{29}Si	28.976495	4.6832
		^{30}Si	29.973770	3.0872
15	Phosphorus	^{31}P	30.973762	100
16	Sulfur	^{32}S	31.972071	94.93
		^{33}S	32.971458	0.76
		^{34}S	33.967867	4.29
		^{36}S	35.967081	0.02
17	Chlorine	^{35}Cl	34.968853	75.78
		^{37}Cl	36.965903	24.22
18	Argon	^{36}Ar	35.967546	0.3365
		^{38}Ar	37.962732	0.0632
		^{40}Ar	39.962383	99.6003
19	Potassium	^{39}K	38.963707	93.2581
		^{40}K	39.963999	0.0117
		^{41}K	40.961826	6.7302
20	Calcium	^{40}Ca	39.962591	96.941
		^{42}Ca	41.958618	0.647
		^{43}Ca	42.958767	0.135
		^{44}Ca	43.955481	2.086
		^{46}Ca	45.953693	0.004
		^{48}Ca	47.952534	0.187

*The isotopic mass data are from G. Audi and A. H. Wapstra, and M. Dedieu, *Nuclear Physics A*, volume 565, pages 1-65 (1993) and G. Audi and A. H. Wapstra, *Nuclear Physics A*, volume 595, pages 409-480 (1995). The percent natural abundance data are from K.J.R. Rosman and P.D.P. Taylor, *Pure and Applied Chemistry*, volume 70, pages 217-235 (1998).

Z	Name	Symbol	Mass of Atom, u	% Abundance
21	Scandium	^{45}Sc	44.955910	100
22	Titanium	^{46}Ti	45.952629	8.25
		^{47}Ti	46.951764	7.44
		^{48}Ti	47.947947	73.72
		^{49}Ti	48.947871	5.41
		^{50}Ti	49.944792	5.18
23	Vanadium	^{50}V	49.947163	0.250
		^{51}V	50.943964	99.750
24	Chromium	^{50}Cr	49.946050	4.345
		^{52}Cr	51.940512	83.789
		^{53}Cr	52.940654	9.501
		^{54}Cr	53.938885	2.365
25	Manganese	^{55}Mn	54.938050	100
26	Iron	^{54}Fe	53.939615	5.845
		^{56}Fe	55.934942	91.754
		^{57}Fe	56.935399	2.119
		^{58}Fe	57.933280	0.282
27	Cobalt	^{59}Co	58.933200	100
28	Nickel	^{58}Ni	57.935348	68.0769
		^{60}Ni	59.930791	26.2231
		^{61}Ni	60.931060	1.1399
		^{62}Ni	61.928349	3.6345
		^{64}Ni	63.927970	0.9256
29	Copper	^{63}Cu	62.929601	69.17
		^{65}Cu	64.927794	30.83
30	Zinc	^{64}Zn	63.929147	48.63
		^{66}Zn	65.926037	27.90
		^{67}Zn	66.927131	4.10
		^{68}Zn	67.924848	18.75
		^{70}Zn	69.925325	0.62
31	Gallium	^{69}Ga	68.925581	60.108
		^{71}Ga	70.924705	39.892
32	Germanium	^{70}Ge	69.924250	20.84
		^{72}Ge	71.922076	27.54
		^{73}Ge	72.923459	7.73
		^{74}Ge	73.921178	36.28
		^{76}Ge	75.921403	7.61
33	Arsenic	^{75}As	74.921596	100
34	Selenium	^{74}Se	73.922477	0.89
		^{76}Se	75.919214	9.37
		^{77}Se	76.919915	7.63
		^{78}Se	77.917310	23.77
		^{80}Se	79.916522	49.61
		^{82}Se	81.916700	8.73
35	Bromine	^{79}Br	78.918338	50.69
		^{81}Br	80.916291	49.31
36	Krypton	^{78}Kr	77.920386	0.35
		^{80}Kr	79.916378	2.28
		^{82}Kr	81.913485	11.58
		^{83}Kr	82.914136	11.49
		^{84}Kr	83.911507	57.00
		^{86}Kr	85.910610	17.30

(continued)

Z	Name	Symbol	Mass of Atom, u	% Abundance
37	Rubidium	^{85}Rb	84.911789	72.17
		^{87}Rb	86.909183	27.83
38	Strontium	^{84}Sr	83.913425	0.56
		^{86}Sr	85.909262	9.86
		^{87}Sr	86.908879	7.00
		^{88}Sr	87.905614	82.58
39	Yttrium	^{89}Y	88.905848	100
40	Zirconium	^{90}Zr	89.904704	51.45
		^{91}Zr	90.905645	11.22
		^{92}Zr	91.905040	17.15
		^{94}Zr	93.906316	17.38
		^{96}Zr	95.908276	2.80
41	Niobium	^{93}Nb	92.906378	100
42	Molybdenum	^{92}Mo	91.906810	14.84
		^{94}Mo	93.905088	9.25
		^{95}Mo	94.905841	15.92
		^{96}Mo	95.904679	16.68
		^{97}Mo	96.906021	9.55
		^{98}Mo	97.905408	24.13
		^{100}Mo	99.907477	9.63
43	Technetium	^{98}Tc	97.907216	—
44	Ruthenium	^{96}Ru	95.907598	5.54
		^{98}Ru	97.905287	1.87
		^{99}Ru	98.905939	12.76
		^{100}Ru	99.904220	12.60
		^{101}Ru	100.905582	17.06
		^{102}Ru	101.904350	31.55
		^{104}Ru	103.905430	18.62
45	Rhodium	^{103}Rh	102.905504	100
46	Palladium	^{102}Pd	101.905608	1.02
		^{104}Pd	103.904035	11.14
		^{105}Pd	104.905084	22.33
		^{106}Pd	105.903483	27.33
		^{108}Pd	107.903894	26.46
		^{110}Pd	109.905152	11.72
47	Silver	^{107}Ag	106.905093	51.839
		^{109}Ag	108.904756	48.161
48	Cadmium	^{106}Cd	105.906458	1.25
		^{108}Cd	107.904183	0.89
		^{110}Cd	109.903006	12.49
		^{111}Cd	110.904182	12.80
		^{112}Cd	111.902757	24.13
		^{113}Cd	112.904401	12.22
		^{114}Cd	113.903358	28.73
		^{116}Cd	115.904755	7.49
49	Indium	^{113}In	112.904061	4.29
		^{115}In	114.903878	95.71
50	Tin	^{112}Sn	111.904821	0.97
		^{114}Sn	113.902782	0.66
		^{115}Sn	114.903346	0.34

Z	Name	Symbol	Mass of Atom, u	% Abundance
50	Tin (continued)	^{116}Sn	115.901744	14.54
		^{117}Sn	116.902954	7.68
		^{118}Sn	117.901606	24.22
		^{119}Sn	118.903309	8.59
		^{120}Sn	119.902197	32.58
		^{122}Sn	121.903440	4.63
		^{124}Sn	123.905275	5.79
51	Antimony	^{121}Sb	120.903818	57.21
		^{123}Sb	122.904216	42.79
52	Tellurium	^{120}Te	119.904020	0.09
		^{122}Te	121.903047	2.55
		^{123}Te	122.904273	0.89
		^{124}Te	123.902819	4.74
		^{125}Te	124.904425	7.07
		^{126}Te	125.903306	18.84
		^{128}Te	127.904461	31.74
		^{130}Te	129.906223	34.08
53	Iodine	^{127}I	126.904468	100
54	Xenon	^{124}Xe	123.905896	0.09
		^{126}Xe	125.904269	0.09
		^{128}Xe	127.903530	1.92
		^{129}Xe	128.904779	26.44
		^{130}Xe	129.903508	4.08
		^{131}Xe	130.905082	21.18
		^{132}Xe	131.904154	26.89
		^{134}Xe	133.905395	10.44
		^{136}Xe	135.907220	8.87
55	Cesium	^{133}Cs	132.905447	100
56	Barium	^{130}Ba	129.906310	0.106
		^{132}Ba	131.905056	0.101
		^{134}Ba	133.904503	2.417
		^{135}Ba	134.905683	6.592
		^{136}Ba	135.904570	7.854
		^{137}Ba	136.905821	11.232
		^{138}Ba	137.905241	71.698
57	Lanthanum	^{138}La	137.907107	0.090
		^{139}La	138.906348	99.910
58	Cerium	^{136}Ce	135.907144	0.185
		^{138}Ce	137.905986	0.251
		^{140}Ce	139.905434	88.450
		^{142}Ce	141.909240	11.114
59	Praseodymium	^{141}Pr	140.907648	100
60	Neodymium	^{142}Nd	141.907719	27.2
		^{143}Nd	142.909810	12.2
		^{144}Nd	143.910083	23.8
		^{145}Nd	144.912569	8.3
		^{146}Nd	145.913112	17.2
		^{148}Nd	147.916889	5.7
		^{150}Nd	149.920887	5.6
61	Promethium	^{145}Pm	144.912744	—

(continued)

Z	Name	Symbol	Mass of Atom, u	% Abundance
62	Samarium	^{144}Sm	143.911995	3.07
		^{147}Sm	146.914893	14.99
		^{148}Sm	147.914818	11.24
		^{149}Sm	148.917180	13.82
		^{150}Sm	149.917271	7.38
		^{152}Sm	151.919728	26.75
		^{154}Sm	153.922205	22.75
63	Europium	^{151}Eu	150.919846	47.81
		^{153}Eu	152.921226	52.19
64	Gadolinium	^{152}Gd	151.919788	0.20
		^{154}Gd	153.920862	2.18
		^{155}Gd	154.922619	14.80
		^{156}Gd	155.922120	20.47
		^{157}Gd	156.923957	15.65
		^{158}Gd	157.924101	24.84
		^{160}Gd	159.927051	21.86
65	Terbium	^{159}Tb	158.925343	100
66	Dysprosium	^{156}Dy	155.924278	0.06
		^{158}Dy	157.924405	0.10
		^{160}Dy	159.925194	2.34
		^{161}Dy	160.926930	18.91
		^{162}Dy	161.926795	25.51
		^{163}Dy	162.928728	24.90
		^{164}Dy	163.929171	28.18
67	Holmium	^{165}Ho	164.930319	100
68	Erbium	^{162}Er	161.928775	0.14
		^{164}Er	163.929197	1.61
		^{166}Er	165.930290	33.61
		^{167}Er	166.932045	22.93
		^{168}Er	167.932368	26.78
		^{170}Er	169.935460	14.93
69	Thulium	^{169}Tm	168.934211	100
70	Ytterbium	^{168}Yb	167.933894	0.13
		^{170}Yb	169.934759	3.04
		^{171}Yb	170.936322	14.28
		^{172}Yb	171.936378	21.83
		^{173}Yb	172.938207	16.13
		^{174}Yb	173.938858	31.83
		^{176}Yb	175.942568	12.76
71	Lutetium	^{175}Lu	174.940768	97.41
		^{176}Lu	175.942682	2.59
72	Hafnium	^{174}Hf	173.940040	0.16
		^{176}Hf	175.941402	5.26
		^{177}Hf	176.943220	18.60
		^{178}Hf	177.943698	27.28
		^{179}Hf	178.945815	13.62
		^{180}Hf	179.946549	35.08
73	Tantalum	^{180}Ta	179.947466	0.012
		^{181}Ta	180.947996	99.988

Z	Name	Symbol	Mass of Atom, u	% Abundance
74	Tungsten	^{180}W	179.946706	0.12
		^{182}W	181.948206	26.50
		^{183}W	182.950224	14.31
		^{184}W	183.950933	30.64
		^{186}W	185.954362	28.43
75	Rhenium	^{185}Re	184.952956	37.40
		^{187}Re	186.955751	62.60
76	Osmium	^{184}Os	183.952491	0.02
		^{186}Os	185.953838	1.59
		^{187}Os	186.955748	1.96
		^{188}Os	187.955836	13.24
		^{189}Os	188.958145	16.15
		^{190}Os	189.958445	26.26
		^{192}Os	191.961479	40.78
77	Iridium	^{191}Ir	190.960591	37.3
		^{193}Ir	192.962924	62.7
78	Platinum	^{190}Pt	189.959930	0.014
		^{192}Pt	191.961035	0.782
		^{194}Pt	193.962664	32.967
		^{195}Pt	194.964774	33.832
		^{196}Pt	195.964935	25.242
		^{198}Pt	197.967876	7.163
79	Gold	^{197}Au	196.966552	100
80	Mercury	^{196}Hg	195.965815	0.15
		^{198}Hg	197.966752	9.97
		^{199}Hg	198.968262	16.87
		^{200}Hg	199.968309	23.10
		^{201}Hg	200.970285	13.18
		^{202}Hg	201.970626	29.86
		^{204}Hg	203.973476	6.87
81	Thallium	^{203}Tl	202.972329	29.524
		^{205}Tl	204.974412	70.476
82	Lead	^{204}Pb	203.973029	1.4
		^{206}Pb	205.974449	24.1
		^{207}Pb	206.975881	22.1
		^{208}Pb	207.976636	52.4
83	Bismuth	^{209}Bi	208.980383	100
84	Polonium	^{209}Po	208.982416	—
85	Astatine	^{210}At	209.987131	—
86	Radon	^{222}Rn	222.017570	—
87	Francium	^{223}Fr	223.019731	—
88	Radium	^{226}Ra	226.025403	—
89	Actinium	^{227}Ac	227.027747	—
90	Thorium	^{232}Th	232.038050	100
91	Protactinium	^{231}Pa	231.035879	100
92	Uranium	^{234}U	234.040946	0.0055
		^{235}U	235.043923	0.7200
		^{238}U	238.050783	99.2745
93	Neptunium	^{237}Np	237.048167	—
94	Plutonium	^{244}Pu	244.064198	—
95	Americium	^{243}Am	243.061373	—

(continued)

Z	Name	Symbol	Mass of Atom, u	% Abundance
96	Curium	^{247}Cm	247.070347	—
97	Berkelium	^{247}Bk	247.070299	—
98	Californium	^{251}Cf	251.079580	—
99	Einsteinium	^{252}Es	252.082972	—
100	Fermium	^{257}Fm	257.095099	—
101	Mendelevium	^{258}Md	258.098425	—
102	Nobelium	^{259}No	259.101024	—
103	Lawrencium	^{262}Lr	262.109692	—
104	Rutherfordium	^{263}Rf	263.118313	—
105	Dubnium	^{262}Db	262.011437	—
106	Seaborgium	^{266}Sg	266.012238	—
107	Bohrium	^{264}Bh	264.012496	—
108	Hassium	^{269}Hs	269.001341	—
109	Meitnerium	^{268}Mt	268.001388	—
110	Ununnilium	^{272}Uun	272.001463	—
111	Unununium	^{272}Uuu	272.001535	—
112	Ununbium	^{277}Uub	(277)	—
114	Ununquadium	^{289}Uuq	(289)	—
116	Ununhexium	^{289}Uuh	(289)	—
118	Ununoctium	^{293}Uuo	(293)	—

E Concept Maps

Appendix

As you study chemistry by reading this book and attending class, you will encounter many ideas and concepts. The task of linking them together can be quite daunting. An effective way to accomplish this task is through "concept mapping." A concept map is a visual map presenting the relationships among a set of connected concepts and ideas. It is a tangible way to display how your mind "perceives" a particular topic. By constructing a concept map, you reflect on what you understand and what you do not understand. In a concept map, each concept, usually represented by a word or two in a box, is connected to other concept boxes by lines or arrows. A word or brief phrase adjacent to the line or arrow defines the relationship between the connected concepts. Each major concept box has lines to and from several other concept boxes, thereby generating a network, or "map."

E-1   How to Construct a Concept Map

1. To create a concept map, construct a list of facts, terms, and ideas that you think are in any way associated with the topic, based on your reading and class attendance. Start by asking yourself, what was the class or reading assignment about? The answer to this question will provide the initial (most general) concepts. The list of concepts will grow as you think further about the answer to this question. You can review the chapter summaries, which emphasize the important points of the chapters, as well as the key terms of that chapter.

2. Review the concepts in your list, and categorize them from most general to most specific. Keep in mind that several of the concepts may have the same level of generality. At other times, it may be difficult to determine the relative importance of two related concepts; to get around this dilemma, try posing the following question: Which concept can be understood without reference to the other? The answer is likely the more general concept.

3. Once the categories have been decided, center the most general concept at the top of the page, and draw a box around it.

4. Arrange the next-most-general rank of concepts below the most general concept. Draw boxes around these concepts, and draw lines linking them to the most general concept. The links should have arrowheads to show the directions in which they should be read.

5. The next step is to label the linkages with short phrases, or even single words, which properly relate the linked concepts. When you place concept 1, a linkage phrase, and concept 2 in sequence, a sensible phrase should result. For example, measurements (concept 1) generate (linkage phrase) numbers (concept 2) that have (linkage phrase) uncertainty (concept 3). The inclusion of linkage labels is important. The appropriate linkage phrase shows that you understand the relationship between the concepts.

6. Proceed down the page, adding rows of ever-more-specific concepts. The most specific concepts should end up at the bottom of your map.

7. Throughout the map, search for cross-links between closely related concepts appearing on the same line. Use dashed lines with double arrowheads to indicate the cross-links.

8. As a last step, assess the map and redraw it if necessary to produce a more logical and neat map.

Once you have constructed the map, check that a concept appears only once and that you have labeled all linkages. Finally, remember that there is no one *correct* concept map for a collection of concepts. However, some concept maps are much more effective than others at displaying the relationships among a given set of concepts.

The diagram in this appendix represents a concept map for the scientific method and measurements. Notice that the concept of SI units could be further connected to such concepts as fundamental units and derived units.

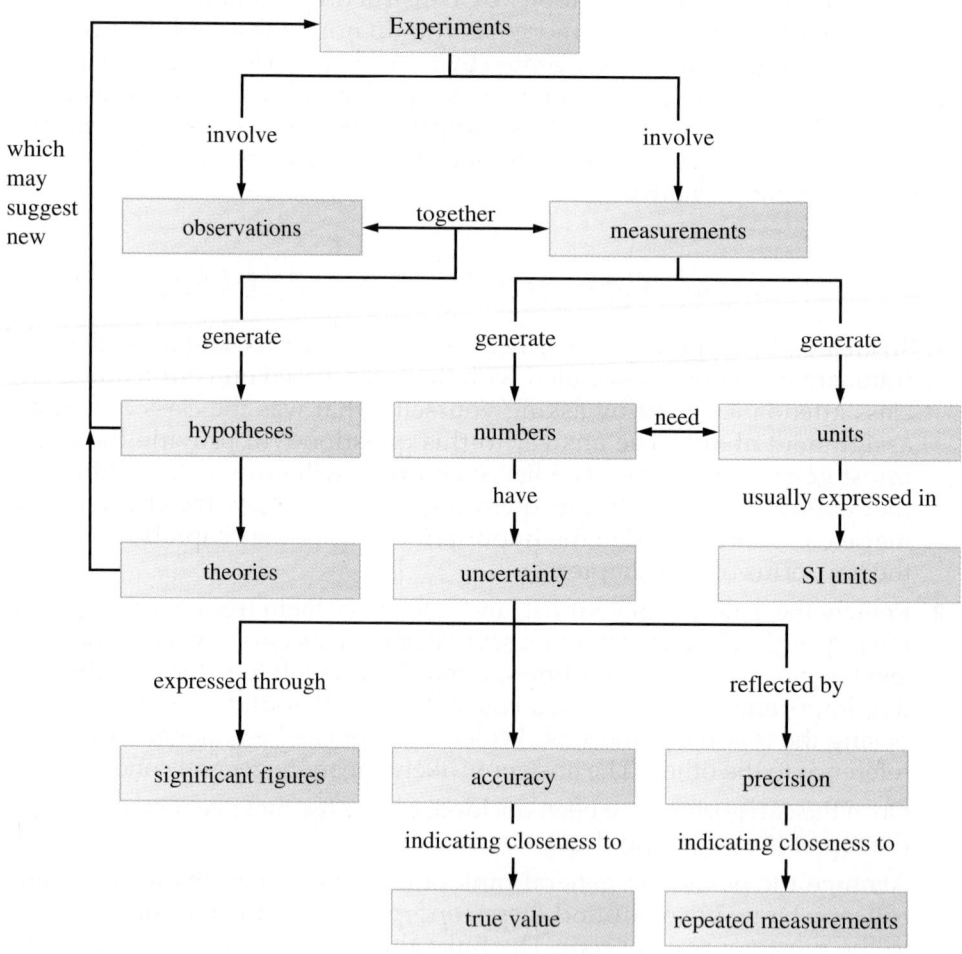

F

Appendix

Absolute configuration refers to the spatial arrangement of the groups attached to a chiral carbon atom. The two possibilities are D and L.

Accuracy is the "closeness" of a measured value to the true or accepted value of a quantity.

Acetyl group (See **acyl**.)

An **achiral** molecule has a structure that is superimposable on its mirror image. (See also **chiral**.)

An **acid** is (1) a hydrogen-containing compound that can produce hydrogen ions, H^+ (Arrhenius theory); (2) a proton donor (Brønsted–Lowry theory); (3) an atom, ion, or molecule that can accept a pair of electrons to form a covalent bond (Lewis theory).

An **acid–base indicator** is a substance used to measure the pH of a solution or to signal the equivalence point in an acid–base titration. The nonionized weak acid form has one color and the anionic form, a different color.

An **acid ionization constant, K_a,** is the equilibrium constant for the ionization reaction of a weak acid.

An **acid salt** contains an anion that can act as an acid (proton donor); examples are $NaHSO_4$ and NaH_2PO_4.

The **actinides** are a series of radioactive elements ($Z = 90 - 103$) characterized by partially filled $5f$ orbitals in their atoms.

An **activated complex** is an intermediate in a chemical reaction formed through collisions between energetic molecules. Once formed, it dissociates either into the products or back to the reactants.

Activation energy is the minimum total kinetic energy that molecules must bring to their collisions for a chemical reaction to occur.

Active sites are the locations at which catalysis occurs, whether on the surface of a heterogeneous catalyst or an enzyme.

Activity is the effective concentration of a species. It is obtained as the product of an activity coefficient and the ratio of the stoichiometric concentration or pressure to that of a reference state.

The **actual yield** is the *measured* quantity of a product obtained in a chemical reaction. (See also **theoretical yield** and **percent yield**.)

The **acyl group** is $-\overset{\overset{\displaystyle O}{\|}}{C}-R$. If $R = H$, this is called the **formyl** group; $R = CH_3$, **acetyl**; and $R = C_6H_5$, **benzoyl**.

An **addition–elimination reaction** is the overall reaction that occurs when compounds are interconverted. It involves (1) a nucleophilic addition to the carbonyl carbon to form a tetrahedral intermediate, followed by (2) an elimination reaction that regenerates the carbonyl group.

In an **addition reaction**, a molecule adds across a double or triple bond in another molecule.

An **adduct** is a compound formed by joining together two simpler molecules through a coordinate covalent bond, such as the adduct of $AlCl_3$ and $(C_2H_5)_2O$ pictured on page 948.

Adenosine diphosphate (ADP) and **adenosine triphosphate (ATP)** are agents involved in energy transfers during metabolism. The hydrolysis of ATP produces ADP, the ion HPO_4^{2-}, and a release of energy.

Adhesive forces are intermolecular forces between unlike molecules, such as molecules of a liquid and of a surface with which it is in contact.

ADP (See **adenosine diphosphate**.)

Alcohols contain the functional group $-OH$ and have the general formula ROH.

Aldehydes have the general formula $R-\overset{\overset{\displaystyle O}{\|}}{C}-H$.

Alicyclic hydrocarbon molecules have their carbon atom skeletons arranged in rings and resemble aliphatic (rather than aromatic) hydrocarbons.

Aliphatic hydrocarbon molecules have their carbon atom skeletons arranged in straight or branched chains.

Alkali metals is the family name for the group 1 elements of the periodic table.

Alkaline earth metals is the family name for the group 2 elements of the periodic table.

Alkane hydrocarbon molecules have only single covalent bonds between carbon atoms. In their chain structures alkanes have the general formula C_nH_{2n+2}.

Alkene hydrocarbons have one or more carbon-to-carbon double bonds in their molecules. The simple alkenes have the general formula C_nH_{2n}.

Alkyl groups are alkane hydrocarbon molecules from which one hydrogen atom has been extracted. For example, the group $-CH_3$ is the **methyl** group; $-CH_2CH_3$ is the **ethyl** group.

Alkyne hydrocarbons have one or more carbon-to-carbon triple bonds in their molecules. The simple alkynes have the general formula C_nH_{2n-2}.

An **alloy** is a mixture of two or more metals. Some alloys are solid solutions, some are heterogeneous mixtures, and some are intermetallic compounds.

An **alpha (α) particle** is a combination of two protons and two neutrons identical to the helium ion, that is, $^4He^{2+}$. Alpha particles are emitted in some radioactive decay processes.

Alums are sulfates of the general formula, $M(I)M(III)(SO_4)_2 \cdot 12\ H_2O$. M(I) is most commonly an alkali metal or ammonium ion, and M(III) is most commonly Al^{3+}, Fe^{3+}, or Cr^{3+}.

Amalgams are metal alloys containing mercury. Depending on their compositions, some are liquid and some are solid.

An **amide** is derived from the ammonium salt of a carboxylic acid and has the general formula $R-\overset{\overset{\displaystyle O}{\|}}{C}-NH_2$.

An **amine** is an organic base having the formula RNH_2 (primary), R_2NH (secondary), or R_3N (tertiary), depending on the number of hydrogen atoms of an NH_3 molecule that are replaced by R groups.

An **α-amino acid** is a carboxylic acid that has an amino group (—NH_2) attached to the carbon atom adjacent to the **carboxyl group** (—COOH).

Amplitude is the height of the crest of a wave above the center line of the wave.

Amphiprotic substances can act either as an acid or as a base.

Amphoteric is the term used to describe the ability of certain oxides and hydroxo compounds to act as either acids or bases.

An **angular wave function**, $Y(\theta, \phi)$, is the part of a wave function that depends on the angles θ and ϕ when the Schrüdinger wave equation is expressed in spherical polar coordinates. (See also **radial wave function**.)

Anhydride is a term meaning "without water." An acid anhydride is an element oxide that reacts with water to form an acid, and a base anhydride, to form a base.

An **anion** is a negatively charged ion. An anion migrates toward the anode in an electrochemical cell.

The **anode** is the electrode in an electrochemical cell at which an oxidation half-reaction occurs.

In the **anti** conformation, the methyl groups are diagonally opposite each other.

An **antibonding molecular orbital** describes regions in a molecule in which there is a low electron probability or charge density between two bonded atoms.

An **aprotic solvent** is a solvent whose molecules do not have a hydrogen atom bonded to an electronegative element.

An **arenium ion** is a cationic species with structural formula
An arenium ion is formed when an electrophile (E^+) accepts an electron pair from the π system of the benzene ring.

Aromatic hydrocarbons are organic substances whose carbon atom skeletons are arranged in hexagonal rings, based on benzene, C_6H_6.

Asymmetric is the term used to describe a C atom with four different substituent groups. A molecule with such a C atom is chiral.

One **atmosphere** (atm) is the pressure exerted by a column of mercury exactly 760 mm high when the density of mercury is 13.5951 g/cm^3 and the acceleration due to gravity is $g = 9.80665 \ m/s^2$.

The **atom** is the basic building block of matter. The number of different atoms currently known is 114. A chemical element consists of a single type of atom, and a chemical compound consists of two or more different kinds of atoms.

The **atomic mass (weight)** of an element is the average of the isotopic masses weighted according to the naturally occurring abundances of the isotopes of the element and relative to the value of exactly 12 u for a carbon-12 atom.

An **atomic mass unit, u,** is used to express the masses of individual atoms. One u is 1/12 the mass of a carbon-12 atom.

The **atomic number, Z,** is the number of protons in the nucleus of an atom. It is also the number of electrons outside the nucleus of an electrically neutral atom.

Atomic (line) spectra are produced by dispersing light emitted by excited gaseous atoms. Only a discrete set of wavelength components (seen as colored lines) is present in a line spectrum.

ATP (See **adenosine triphosphate**.)

The **aufbau process** is a method of writing electron configurations. Each element is described as differing from the preceding one in terms of the orbital to which the one additional electron is assigned.

An **average bond energy** is the average of bond-dissociation energies for a number of different species containing a particular covalent bond. (See also **bond dissociation energy**.)

The **Avogadro constant**, N_A, has a value of $6.02214 \times 10^{23} \ mol^{-1}$. It is the number of elementary units in one mole.

Avogadro's law (hypothesis) states that at a fixed temperature and pressure, the volume of a gas is directly proportional to the amount of gas and that equal volumes of different gases, compared under identical conditions of temperature and pressure, contain equal numbers of molecules.

An **azeotrope** is a solution that boils at a constant temperature, producing vapor of the same composition as the liquid. In some cases, the azeotrope boils at a lower temperature than the solution components, in other cases, at a higher temperature.

A **balanced equation** has the same number of atoms of each type on both sides. (See also **chemical equation**.)

Band theory is a form of molecular orbital theory to describe bonding in metals and semiconductors.

One **bar** is equal to 100 kilopascals (1 bar = 100 kPa).

A **barometer** is a device used to measure the pressure of the atmosphere.

Barometric pressure is the prevailing pressure of the atmosphere as indicated by a barometer.

A **base** is (1) a compound that produces hydroxide ions, OH^-, in water solution (Arrhenius theory); (2) a proton acceptor (Brønsted–Lowry theory); (3) an atom, ion, or molecule that can donate a pair of electrons to form a covalent bond (Lewis theory).

A **base ionization constant**, K_b, is the equilibrium constant for the ionization reaction of a weak base.

Basicity is a measure of the tendency of an electron pair donor to react with a proton.

The **basic oxygen process** is the principal process used to convert impure iron (pig iron) into steel.

A **battery** is a voltaic cell [or a group of voltaic cells connected in series (+ to −)] used to produce electricity from chemical change.

bcc (See **body-centered cubic**.)

A **benzyl** group is a methyl group with one hydrogen atom replaced by a phenyl group.

A **beta (β^-) particle** is an electron emitted as a result of the conversion of a neutron to a proton in certain atomic nuclei undergoing radioactive decay.

A **bidentate** ligand attaches itself to the central atom of a complex at two points in the coordination sphere.

A **bimolecular process** is an elementary process involving the collision of two molecules.

Binary compounds are compounds composed of *two* elements.

A **body-centered cubic (bcc)** crystal structure is one in which the unit cell has structural units at each corner and one in the center of the cube.

Boiling is a process in which vaporization occurs throughout a liquid. It occurs when the vapor pressure of a liquid is equal to barometric pressure.

A **bomb calorimeter** is a device used to measure the heat of a combustion reaction. The quantity measured is the heat of reaction at constant volume, $q_V = \Delta U$.

A **bond angle** is the angle between two covalent bonds. It is the angle between hypothetical lines joining the nuclei of two atoms to the nucleus of a third atom to which they are covalently bonded.

Bond-dissociation energy, D, is the quantity of energy required to break one mole of covalent bonds in a gaseous species, usually expressed in kJ mol^{-1}. (See also **average bond energy**.)

Bond length (bond distance) is the distance between the centers of two atoms joined by a covalent bond.

Bond order is one-half the difference between the numbers of electrons in bonding and in antibonding molecular orbitals in a covalent bond. A single bond has a bond order of 1; a double bond, 2; and a triple bond, 3.

A **bond pair** is a pair of electrons involved in covalent bond formation.

A **bonding molecular orbital** describes regions of high electron probability or charge density in the internuclear region between two bonded atoms.

Boyle's law states that the volume of a fixed amount of gas at a constant temperature is inversely proportional to the gas pressure.

In a **bridged halonium ion**, a halogen atom (X) is bonded to (bridges) two carbon atoms that are bonded to each

other: . The halogen atom has a complete octet, comprising two bonding pairs and two lone pairs, and it bears a formal charge of 1$^+$. If the halogen atom is chlorine or bromine, the ion is called a chloronium or bromonium ion.

Buffer capacity refers to the amount of acid and/or base that a buffer solution can neutralize while maintaining an essentially constant pH.

Buffer range is the range of pH values over which a buffer solution can maintain a fairly constant pH.

A **buffer solution** resists a change in its pH. It contains components capable of neutralizing small added amounts of acids and base.

By-products are substances produced along with the principal product in a chemical process, either through the main reaction or a side reaction.

Calcination refers to the decomposition of a solid by heating at temperatures below its melting point, such as the decomposition of calcium carbonate to calcium oxide and $CO_2(g)$.

The **calorie (cal)** is the quantity of heat required to change the temperature of one gram of water by one degree Celsius.

A **calorimeter** is a device (of which there are numerous types) used to measure a quantity of heat.

A **carbohydrate** is a polyhydroxy aldehyde, a polyhydroxy ketone, a derivative of these, or a substance that yields them upon hydrolysis. Carbohydrates can be viewed as "hydrates" of carbon, in the sense that their general formulas are $C_x(H_2O)_y$.

Carbon black is a finely divided amorphous form of carbon prepared by the incomplete combustion of hydrocarbons.

The carbonyl group is found in aldehydes, ketones, and carboxylic acids $\overset{}{\underset{}{>}}C{=}O$.

The **carboxyl group** is $-\overset{\displaystyle O}{\overset{\|}{C}}-OH$.

A **carboxylic acid** has one or more carboxyl groups attached to a hydrocarbon chain or ring structure.

A **catalyst** provides an alternative mechanism of lower activation energy for a chemical reaction. The reaction is speeded up, and the catalyst is regenerated.

The **cathode** is the electrode of an electrochemical cell where a reduction half-reaction occurs.

Cathode rays are negatively charged particles (electrons) emitted at the negative electrode (cathode) in the passage of electricity through gases at very low pressures.

Cathodic protection is a method of corrosion control in which the metal to be protected is joined to a more active metal that corrodes instead. The protected metal acts as the cathode of a voltaic cell.

A **cation** is a positively charged ion. A cation migrates toward the cathode in an electrochemical cell.

The **cell** is the fundamental unit of living organisms.

A **cell diagram** is a symbolic representation of an electrochemical cell that indicates the substances entering into the cell reaction, electrode materials, solution concentrations, etc.

The **cell voltage (potential)**, E_{cell}, is the potential difference (voltage) between the two electrodes of an electrochemical cell.

The **Celsius** temperature scale is based on a value of 0 °C for the normal melting point of ice and 100 °C for the normal boiling point of water.

A **central atom** in a structure is an atom that is bonded to two or more other atoms.

In **chain-reaction polymerization**, a reaction is initiated by "opening up" a carbon-to-carbon double bond. Monomer units add to free-radical intermediates to produce a long-chain polymer.

The **charge density**, ρ, is the charge per unit volume in a cation.

Charles's law states that the volume of a fixed amount of gas at a constant pressure is directly proportional to the Kelvin (absolute) temperature.

A **chelate** results from the attachment of polydentate ligands to the central atom of a complex ion. Chelates are five- or six-membered rings that include the central atom and atoms of the ligands.

A **chelating agent** is a polydentate ligand. It simultaneously attaches to two or more positions in the coordination sphere of the central atom of a complex ion.

Chelation is the process of chelate formation.

The **chelation effect** refers to an exceptional stability conferred to a complex ion when polydentate ligands are present.

Chemical change (See **chemical reaction**.)

Chemical energy is the energy associated with chemical bonds and intermolecular forces.

A **chemical equation** is a symbolic representation of a chemical reaction. Symbols and formulas are used to represent reactants and products, and stoichiometric coefficients are used to balance the equation. (See also **balanced equation**.)

A **chemical formula** represents the relative numbers of atoms of each kind in a substance through symbols and numerical subscripts.

A **chemical property** is the ability (or inability) of a sample of matter to undergo a particular chemical reaction.

A **chemical reaction** is a process in which one set of substances (reactants) is transformed into a new set of substances (products).

Chemical symbols are abbreviations of the names of the elements consisting of one or two letters (e.g., N = nitrogen and Ne = neon).

Chiral refers to a molecule with a structure that is not superimposable on its mirror image. (See also **enantiomers**.)

The term *cis* describes geometric isomers in which two groups are attached on the same side of a double bond in an organic molecule, or along the same edge of a square in a square-planar complex, or at two adjacent vertices of an octahedral complex. (See also **geometric isomerism**.)

cis–trans **isomerism** is a type of stereoisomerism.

A **closed system** is one that can exchange energy but not matter with its surroundings.

Cohesive forces are intermolecular forces between like molecules, such as within a drop of liquid.

Coke is a relatively pure form of carbon produced by heating coal out of contact with air (destructive distillation).

Colligative properties —vapor pressure lowering, freezing point depression, boiling point elevation, osmotic pressure—have values that depend on the number of solute particles in a solution but not on their identity.

A **colloid** is a mixture that contains particles that are larger than ions or molecules but are still submicroscopic.

The **common-ion effect** describes the effect on an equilibrium by a substance that furnishes ions that can participate in the equilibrium.

A **complementary color** is a secondary color that mixes with the opposite primary color on the color wheel to produce white light in additive color mixing or black in subtractive color mixing.

A **complex** is a polyatomic cation, anion, or neutral molecule in which groups (molecules or ions) called ligands are bonded to a central metal atom or ion.

A **complex ion** is a complex having a net electrical charge.

Composition refers to the components and their relative proportions in a sample of matter.

A **compound** is a substance made up of two or more elements. It does not change its identity in physical changes, but it can be broken down into its constituent elements by chemical changes.

Concentration (1) refers to the composition of a solution. (2) See **extractive metallurgy**.

In a **concentration cell** identical electrodes are immersed in solutions of different concentrations. The voltage (emf) of the cell is a function simply of the concentrations of the two solutions.

Condensation is the passage of molecules from the gaseous state to the liquid state.

A **condensed structural formula** is a simplified representation of a structural formula.

Conformations refer to the different spatial arrangements possible in a molecule. Examples are the "boat" and "chair" forms of cyclohexane.

A **conjugate acid** is formed when a Brønsted–Lowry base gains a proton. Every base has a conjugate acid.

A **conjugate acid–base pair** is pair of molecules or ions for which the chemical formulas differ by a single proton: H^+

(e.g., H_3O^+ and H_2O; H_2O and OH^-; NH_4^+ and NH_3; H_3PO_4 and $H_2PO_4^-$).

A **conjugate base** remains after a Brønsted–Lowry acid has lost a proton. Every acid has a conjugate base.

Consecutive reactions are two or more reactions carried out in sequence. A product of each reaction becomes a reactant in a following reaction until a final product is formed.

Constitutional isomers have different bond connectivities, and thus different skeletal structures.

The **contact process** is a process for the manufacture of sulfuric acid having as its key reaction the oxidation of $SO_2(g)$ to $SO_3(g)$ in contact with a catalyst.

Control rods are neutron-absorbing metal rods (e.g., Cd) that are used to control the neutron flux in a nuclear reactor and thereby control the rate of the fission reaction.

In a **coordinate covalent bond**, electrons shared between two atoms are contributed by just one of the atoms. As a result, the bonded atoms exhibit formal charges.

Coordination compounds are neutral complexes or compounds containing complex ions.

Coordination number is the number of positions around a central atom where ligands can be attached in the formation of a complex. Applied to a crystalline solid, coordination number signifies the number of nearest neighboring atoms (or ions of opposite charge) to any given atom (or ion) in a crystal.

Coupled reactions are sets of chemical reactions that occur together. One (or more) of the reactions taken alone is (are) *nonspontaneous* and other(s), *spontaneous*. The overall reaction is *spontaneous*.

A **covalent bond** is formed when electrons are shared between a pair of atoms. In valence bond theory, the sharing of the electrons is said to occur in the region in which atomic orbitals overlap.

Covalent radius is one-half the distance between the centers of two atoms that are bonded covalently. It is the atomic radius associated with an element in its covalent compounds.

The **critical point** refers to the temperature and pressure at which a liquid and its vapor become identical. It is the highest temperature point on the vapor pressure curve.

Crystal field theory describes bonding in complexes in terms of electrostatic attractions between ligands and the nucleus of the central metal. Particular attention is focused on the splitting of the d energy level of the central metal.

Cubic closest packed is one of the two ways in which spheres can be packed to minimize the amount of free space or voids among them.

Dalton's law of partial pressures states that in a mixture of gases, the total pressure is the sum of the partial pressures of the gases present. (See also **partial pressure**.)

The **dashed and solid wedge line notation** is a method of conveying a three-dimensional perspective to a structure plotted in a plane.

The *d* **block** refers to that section of the periodic table in which the process of orbital filling (aufbau process) involves a *d* subshell.

A **decay constant** is a first-order rate constant describing radioactive decay.

Degree of ionization refers to the extent to which molecules of a weak acid or weak base ionize. The degree of ionization increases as the weak electrolyte solution is diluted. (See also **percent ionization**.)

The **degree of unsaturation** is equal to the total number of π bonds and ring structures in a molecule.

Degenerate orbitals are orbitals that are at the same energy level.

A **delocalized molecular orbital** describes a region of high electron probability or charge density that extends over three or more atoms.

Denaturation refers to the loss of biological activity of a protein brought about by changes in its secondary and tertiary structures.

Density is a physical property obtained by dividing the mass of a material or object by its volume (i.e., mass per unit volume).

Deoxyribonucleic acid (DNA) is the substance that makes up the genes of the chromosomes in the nuclei of cells.

Deposition is the passage of molecules from the gaseous to the solid state.

Detergents are cleansing agents that act by emulsifying oils. Most common among synthetic detergents are the salts of organic sulfonic acids, $RSO_3^-Na^+$.

Dextrorotatory means the ability to rotate the plane of polarized light to the *right*, designated (+).

Diagonal relationships refer to similarities that exist between certain pairs of elements in different groups and periods of the periodic table, such as Li and Mg, Be and Al, and B and Si.

A **diamagnetic** substance has all its electrons paired and is slightly repelled by a magnetic field.

Diastereomers are optically active isomers of a compound, but their structures are *not* mirror images (as are enantiomers).

Diffraction is the dispersion of light into its different components as a result of the interference produced by the reflection of light from a grooved surface.

Diffusion refers to the spreading of a substance (usually a gas or liquid) into a region where it is not originally present as a result of random molecular motion.

A **dimer** is a molecule comprised of two simpler formula units, such as Al_2Cl_6, which is a dimer of $AlCl_3$.

Dipole moment, μ, is a measure of the extent to which a separation exists between the centers of positive and negative charge within a molecule. The unit used to measure dipole moment is the **debye**, 3.34×10^{-30} C m.

Dispersion (London) forces are intermolecular forces associated with instantaneous and induced dipoles.

In a **disproportionation reaction,** the same substance is both oxidized and reduced.

In a **double covalent bond,** *two pairs* of electrons are shared between bonded atoms. The bond is represented by a double-dash sign ($=$).

An **E1 reaction** is an elimination reaction in which the rate-determining step is unimolecular.

An **E2 reaction** is an elimination reaction in which the rate-determining step is bimolecular.

Effective nuclear charge, Z_{eff}, is the positive charge acting on a particular electron in an atom. Its value is the charge on the nucleus, reduced to the extent that other electrons screen the particular electron from the nucleus.

Effusion is the escape of a gas through a tiny hole in its container.

An **electrochemical cell** is a device in which the electrons transferred in an oxidation–reduction reaction are made to pass through an electrical circuit. (See also **electrolytic cell** and **voltaic cell**.)

An **electrode** is a metal surface on which an oxidation–reduction equilibrium is established between the metal and substances in solution.

Electrolysis is the decomposition of a substance, either in the molten state or in an electrolyte solution, by means of electric current.

An **electrolyte** is a substance that provides ions when dissolved in water.

An **electrolytic cell** is an electrochemical cell in which a nonspontaneous reaction is carried out by electrolysis.

Electromagnetic radiation is a form of energy propagated as mutually perpendicular electric and magnetic fields. It includes visible light, infrared, ultraviolet, X ray, and radio waves.

Electromotive force (emf) is the potential difference between two electrodes in a voltaic cell, expressed in volts.

Electron affinity (*EA*) is the energy change associated with the gain of an electron by a neutral gaseous atom.

Electron capture (EC) is a form of radioactive decay in which an electron from an inner electronic shell is absorbed by a nucleus. In the nucleus the electron is used to convert a proton to a neutron.

An **electron configuration** is a designation of how electrons are distributed among various orbitals in an atom.

Electronegativity (EN) is a measure of the electron-attracting power of a bonded atom; metals have low electronegativities, and nonmetals have high electronegativities.

The **electronegativity difference** between two atoms that are bonded together is used to assess the degree of polarity in the bond.

Electron-group geometry refers to the geometrical distribution about a central atom of the electron pairs in its valence shell.

Electron spin is a characteristic of electrons giving rise to the magnetic properties of atoms. The two possibilities for electron spin are $+\frac{1}{2}$ and $-\frac{1}{2}$.

Electrons are particles carrying the fundamental unit of negative electric charge. They are found outside the nuclei of all atoms.

Electron-withdrawing substituents are atoms or groups of atoms that draw electron density toward themselves. Highly electronegative atoms, such as F, O, N, and Cl, are examples.

An **electrophile** contains an electron-attracting region of positive charge (an electrophilic center) and is a reagent that forms a bond to its reaction partner (the nucleophile) by accepting both bonding electrons from that reaction partner.

An **electrophilic** center in a molecule is an electron-attracting region of positive charge.

In an **electrophilic substitution reaction**, an electrophile replaces another atom or group in a molecule. An example of an electrophilic substitution reaction is the replacement of an H atom in benzene with a nitro (NO_2) group.

An **electrostatic potential map** depicts the electron charge distribution in a molecule. The color red is used to represent the region with the most negative charge, and blue represents the most positive charge.

An **element** is a substance composed of a single type of atom. It cannot be broken down into simpler substances by chemical reactions.

An **elementary process** is an event that significantly alters a molecule's energy or geometry or produces a new molecule(s). It represents a single step in a reaction mechanism.

In an **elimination reaction**, atoms or groups that are bonded to adjacent atoms are eliminated as a small molecule (e.g., H_2O) and an additional bond is formed between carbon atoms.

An **empirical formula** is the simplest chemical formula that can be written for a compound, that is, having the smallest integral subscripts possible.

Enantiomers (optical isomers) are molecules whose structures are nonsuperimposable mirror images. The molecules are optically active, that is, able to rotate the plane of polarized light.

An **endothermic reaction** results in a lowering of the temperature of an isolated system or the absorption of heat by a system that interacts with its surroundings.

The **end point** is the point in a titration where the indicator used changes color. A properly chosen indicator has its end point coming as closely as possible to the equivalence point of the titration.

Energy is the capacity to do work. (See also **work**.)

An **energy-level diagram** is a representation of the allowed energy states for the electrons in atoms. The simplest energy-level diagram is that of the hydrogen atom.

The **English system** of measurement has the yard as its unit of length, the pound as its unit of mass, and the second as its unit of time.

Enthalpy, *H*, is a thermodynamic function used to describe constant-pressure processes: $H = U + PV$, and at constant pressure, $\Delta H = \Delta U + P\,\Delta V$.

Enthalpy change, ΔH, is the difference in enthalpy between two states of a system. For a chemical reaction carried out at constant temperature and pressure and with work limited to pressure–volume work, the enthalpy change is called the *heat of reaction at constant pressure*.

An **enthalpy diagram** is a diagrammatic representation of the enthalpy changes in a process.

Enthalpy (heat) of formation (See **standard enthalpy of formation**.)

Entropy, S, is a thermodynamic property related to the number of energy levels among which the energy of a system is spread. The greater the number of energy levels for a given total energy, the greater the entropy.

Entropy change, ΔS, is the difference in entropy between two states of a system.

An **enzyme** is a high molar mass protein that catalyzes biological reactions.

An **equation of state** is a mathematical expression relating the amount, volume, temperature, and pressure of a substance (usually applied to gases).

Equilibrium refers to a condition where forward and reverse processes proceed at equal rates and no further net change occurs. For example, amounts of reactants and products in a reversible reaction remain constant over time.

The **equilibrium constant** is the numerical value of the equilibrium constant expression.

An **equilibrium constant expression** describes the relationship among the concentrations (or partial pressures) of the substances present in a system at equilibrium.

The **equivalence point** of a titration is the condition in which the reactants are in stoichiometric proportions. They consume each other, and neither reactant is in excess.

An **ester** is the product of the elimination of H_2O from between an acid and an alcohol molecule. Esters have the

general formula R—C—O—R'.

An **ether** has the general formula R—O—R'.

Eutrophication is the deterioration of a freshwater body caused by nutrients such as nitrates and phosphates, which stimulate the growth of algae, oxygen depletion, and fish kills.

Evaporation is the physical process of a liquid changing to a vapor. (See also **vaporization**.)

In an **excited state** of an atom, one or more electrons are promoted to a higher energy level than in the ground state. (See also **ground state**.)

An **exothermic reaction** produces an increase in temperature in an isolated system or, for a system that interacts with its surroundings, the evolution of heat.

Expanded valence shell is a term used to describe Lewis structures in which certain atoms in the third or higher period of the periodic table appear to require 10 or 12 electrons in their valence shells.

An **extensive property** is one, like mass or volume, whose value depends on the quantity of matter observed.

Extractive metallurgy refers to the process of extracting a metal from its ores. Generally this occurs in four steps. **Concentration** separates the ore from waste rock (gangue). **Roasting** converts the ore to the metal oxide. **Reduction** (usually with carbon) converts the oxide to the metal. **Refining** removes impurities from the metal.

The **E, Z system** is a system of nomenclature used to describe the manner in which substituent groups are attached at a carbon-to-carbon double bond.

A **face-centered cubic (fcc)** crystal structure is one in which the unit cell has structural units at the eight corners and in the center of each face of the unit cell. It is derived from the cubic closest packed arrangement of spheres.

The **Fahrenheit** temperature scale is based on a value of 32 °F as the melting point of ice and 212 °F as the boiling point of water.

A **family** of elements is a numbered group from the periodic table, sometimes carrying a distinctive name. For example, group *17* is the halogen family.

The **Faraday constant, F,** is the charge associated with one mole of electrons, 96,485 C/mol e^-.

Fats are triglycerides in which saturated fatty acid components predominate. The *f* block is that portion of the periodic table where the process of filling of electron orbitals (aufbau process) involves *f* subshells. These are the lanthanide and actinide elements.

fcc (See **face-centered cubic**.)

Ferromagnetism is a property that permits certain materials (notably Fe, Co, and Ni) to be made into permanent magnets. The magnetic moments of individual atoms are aligned into domains. In the presence of a magnetic field, these domains orient themselves to produce a permanent magnetic moment.

A **Fischer projection formula** is a two-dimensional representation of a three-dimensional structural formula. It shows how the stereochemistry at a chiral carbon atom is represented in two dimensions, and how the carbon-chain backbone is arranged on the page.

The **first law of thermodynamics**, expressed as $\Delta U = q + w$, is an alternative statement of the law of conservation of energy. (See also **law of conservation of energy**.)

A **first-order reaction** is one for which the sum of the concentration-term exponents in the rate equation is 1.

Fission (See **nuclear fission**.)

A **flow battery** is a battery in which materials (reactants, products, electrolytes) pass continuously through the battery. The battery is simply a converter of chemical to electrical energy.

Formal charge is the number of outer-shell (valence) electrons in an isolated atom minus the number of electrons assigned to that atom in a Lewis structure.

The **formation constant, K_f,** describes equilibrium among a complex ion, the free metal ion, and ligands.

Formula mass is the mass of a formula unit of a compound, relative to a mass of exactly 12 u for carbon-12.

A **formula unit** is the smallest collection of atoms or ions from which the empirical formula of a compound can be established.

Fractional crystallization (recrystallization) is a method of purifying a substance by crystallizing the pure solid from a saturated solution while impurities remain in solution.

Fractional distillation (See **distillation**.)

Fractional precipitation is a technique in which two or more ions in solution, each capable of being precipitated by the same reagent, are separated by the use of that reagent.

The **Frasch process** is a method of extracting sulfur from underwater deposits. It is based on the use of superheated water to melt the sulfur.

Free radicals are highly reactive molecular fragments containing unpaired electrons.

Freezing is the conversion of a liquid to a solid that occurs at a fixed temperature known as the **freezing point**.

The **frequency** of a wave motion is the number of wave crests or troughs that pass through a given point in a unit of time. It is expressed by the unit $time^{-1}$ (e.g., s^{-1}, also called a hertz, Hz).

A **fuel cell** is a voltaic cell in which the cell reaction is the equivalent of the combustion of a fuel. Chemical energy of the fuel is converted to electricity.

A **function of state (state function)** is a property that assumes a unique value when the state or present condition of a system is defined. This value is *independent* of how the state is attained.

A **functional group** is an atom or grouping of atoms attached to a hydrocarbon residue, R. The functional group often confers specific properties to an organic molecule.

Fusion (See **nuclear fusion**.)

Galvanic cell (See **voltaic cell**.)

Gamma (γ) rays are a form of electromagnetic radiation of high penetrating power emitted by certain radioactive nuclei.

In a **gas,** atoms or molecules are generally much more widely separated than in liquids and solids. A gas assumes the shape of its container and expands to fill the container, thus having neither definite shape nor volume.

The **gas constant, R,** is the numerical constant appearing in the ideal gas equation ($PV = nRT$) and in several other equations as well.

In the **gauche** conformation, the methyl groups are to the left and right of each other.

In a **geminal dihalide**, a halogen adds across the double bond of an alkene, and the two halogen atoms are bonded to the same carbon.

The **general gas equation** is an expression based on the ideal gas equation and written in the form $P_1V_1/n_1T_1 = P_2V_2/n_2T_2$.

Geometric isomerism in organic compounds refers to the existence of nonequivalent structures (*cis* and *trans*) that differ in the positioning of substituent groups relative to a double bond. In complexes, the nonequivalent structures are based on the positions at which ligands are attached to the metal center.

Gibbs energy, G, is a thermodynamic function designed to produce a criterion for spontaneous change. It is defined through the equation $G = H - TS$.

Gibbs energy change, ΔG, is the change in Gibbs energy that accompanies a process and can be used to indicate the direction of spontaneous change. For a spontaneous process at constant temperature and pressure, $\Delta G < 0$. (See also **standard Gibbs energy change**.)

Glass is a transparent, amorphous solid consisting of Na^+ and Ca^{2+} ions in a network of SiO_4^{4-} anions. It is made by fusing together a mixture of sodium and calcium carbonates with sand.

Global warming refers to the warming of Earth that results from an accumulation in the atmosphere of gases such as CO_2 that absorb infrared radiation radiated from Earth's surface.

Graham's law states that the rates of effusion or diffusion of two different gases are inversely proportional to the square roots of their molar masses.

The **ground state** is the lowest energy state for the electrons in an atom or molecule.

A **group** is a vertical column of elements in the periodic table. Members of a group have similar properties.

A **half-cell** is a combination of an electrode and a solution. An oxidation–reduction equilibrium is established on the electrode. An electrochemical cell is a combination of two half-cells.

The **half-life** ($t_{1/2}$) of a reaction is the time required for one-half of a reactant to be consumed. In a nuclear decay process, it is the time required for one-half of the atoms present in a sample to undergo radioactive decay.

A **half-reaction** describes one portion of an overall oxidation–reduction reaction, either the oxidation or the reduction.

Halogens (group 17) are the most reactive nonmetals, having the electron configuration ns^2np^5 in the electronic shell of highest principal quantum number.

Hard water contains dissolved minerals in significant concentrations. If the hardness is primarily due to HCO_3^- and associated cations, the water has **temporary hardness**. If the hardness is due to anions other than HCO_3^- (e.g., SO_4^{2-}), the water has **permanent hardness**.

hcp (See **hexagonal closest packed**.)

Heat is a transfer of thermal energy as a result of a temperature difference.

Heat capacity is the quantity of heat required to change the temperature of an object or substance by one degree, usually expressed as $J\ °C^{-1}$ or $cal\ °C^{-1}$. **Specific heat capacity** is the heat capacity per gram of substance, i.e., $J\ °C^{-1}\ g^{-1}$, and **molar heat capacity** is the heat capacity per mole, i.e., $J\ °C^{-1}\ mol^{-1}$.

A **heat of reaction** is energy converted from chemical to thermal (or vice versa) in a reaction. In an isolated system, this energy conversion causes a temperature change, and in a system that interacts with its surroundings, heat (q) is either evolved to or absorbed from the surroundings.

The **Heisenberg uncertainty principle** states that, when measuring the position and momentum of fundamental particles of matter, uncertainties in measurement are inevitable.

The **Henderson–Hasselbalch equation** has the form, $pH = pK_a + \log$ [conjugate base]/[acid], in which stoichiometric concentrations of the weak acid and its conjugate base are used in place of the equilibrium concentrations. There are limitations on its validity.

Henry's law relates the solubility of a gas to the gas pressure maintained above a solution of the gaseous solute.

The solubility is directly proportional to the pressure of the gas above the solution.

The **hertz (Hz)** is the SI unit of frequency, equal to s^{-1}.

Hess's law states that the enthalpy change for an overall or net process is the sum of enthalpy changes for individual steps in the process.

Heterocyclic compounds are based on hydrocarbon ring structures in which one or more C atoms are replaced by atoms such as N, O, or S.

Heterogeneous catalysis is catalytic action that takes place on a surface separating two phases.

In a **heterogeneous mixture**, components separate into physically distinct regions of differing properties and often differing composition.

Hexagonal closest packed is one of the two ways in which spheres can be packed to minimize the amount of free space or voids among them. The crystal structure based on this type of packing is referred to as **hcp**.

In a **high-spin complex**, weak crystal field splitting leads to a maximum number of unpaired electrons in the *d* subshell of the central metal atom or ion.

Homogeneous catalysis refers to a catalytic reaction taking place in a single phase.

A **homogeneous mixture (solution)** is a mixture of elements and/or compounds that has a uniform composition and properties within a given sample. However, the composition and properties may vary from one sample to another.

A **homologous series** is a group of compounds that differ in composition by some constant unit ($-CH_2$ in the case of alkanes).

Hund's rule (rule of maximum multiplicity) states that whenever orbitals of equal energy are available, electrons occupy these orbitals singly before any pairing of electrons occurs.

A **hybrid orbital** is one of a set of identical orbitals reformulated from pure atomic orbitals and used to describe certain covalent bonds.

Hybridization refers to combining pure atomic orbitals to generate hybrid orbitals in the valence bond approach to covalent bonding.

A **hydrate** is a compound in which a fixed number of water molecules is associated with each formula unit, such as $CuSO_4 \cdot 5\ H_2O$.

Hydrides are compounds of hydrogen, usually divided into the categories of covalent (e.g., H_2O and HCl), ionic (e.g., LiH and CaH_2), and metallic (mostly nonstoichiometric compounds with the transition metals).

A **hydrocarbon** is a compound containing the two elements carbon and hydrogen. The C atoms are arranged in straight or branched chains or ring structures.

A **hydrogen bond** is an intermolecular force of attraction in which an H atom covalently bonded to one atom is attracted simultaneously to another highly nonmetallic atom of the same or a nearby molecule.

In a **hydrogenation reaction**, H atoms are added to multiple bonds between carbon atoms, converting carbon-to-carbon double bonds to single bonds and carbon-to-carbon triple bonds to double or single bonds. It is a reaction, for example, that converts an unsaturated to a saturated fatty acid.

Hydrolysis is a special name given to acid–base reactions in which ions act as acids or bases. As a result of hydrolysis, many salt solutions are not pH neutral, that is, pH $\neq$ 7.

Hydrometallurgy refers to metallurgical procedures where water and aqueous solutions are used to extract metals from their ores. In the first step, **leaching**, the target metal is obtained in soluble form in aqueous solution. Other steps include purifying the leached solution and depositing the metal from solution.

Hydronium ion, H_3O^+, is the form in which protons are found in aqueous solution. The terms "hydrogen ion" and "hydronium ion" are often used synonymously.

The **hydroxyl** group is —OH and is usually found attached to a straight or branched hydrocarbon chain (an alcohol) or a ring structure (a phenol).

A **hypothesis** is a tentative explanation of a series of observations or of a natural law.

An **ICE table** is a format for organizing the data in an equilibrium calculation. It is based on the **initial** concentrations of reactants and products, changes in concentrations to attain equilibrium, and equilibrium concentrations.

An **ideal (perfect) gas** is one whose behavior can be predicted by the ideal gas equation.

Ideal gas constant (See **gas constant**.)

The **ideal gas equation** relates the pressure, volume, temperature, and number of moles of ideal gas (n) through the expression $PV = nRT$.

An **ideal solution** has $\Delta H_{soln} = 0$ and certain properties (notably vapor pressure) that are predictable from the properties of the solution components.

An **indicator** is an added substance that changes color at the equivalence point in a titration.

The **inductive effect** refers to the shifting of electron density from one atom toward another through the chain of σ bonds that connects them.

Industrial smog is air pollution in which the chief pollutants are $SO_2(g)$, $SO_3(g)$, H_2SO_4 mist, and smoke.

Inert complex is the term used to describe a complex ion in which the exchange of ligands occurs very slowly.

The **inert pair effect** refers to the effects on the properties of certain post-transition elements that result from the presence of a pair of electrons in the s orbital of the valence shells of their atoms.

The **initial rate of a reaction** is the rate of a reaction immediately after the reactants are brought together.

An **inorganic compound** is any combination of elements that does not fit the category of organic compound. (See also **organic compound**.)

An **instantaneous rate of reaction** is the exact rate of a reaction at some precise point in the reaction. It is obtained from the slope of a tangent line to a concentration–time graph.

An **integrated rate law (equation)** is derived from a rate law (equation) by the calculus technique of integration. It relates the concentration of a reactant (or product) to elapsed time from the start of a reaction. The equation has different forms depending on the order of the reaction.

An **intensive property** is *independent* of the quantity of matter involved in the observation. Density and temperature are examples of intensive properties.

An **interhalogen** compound is a covalent compound between two or more halogen elements, such as ICl and BrF_3.

An **intermediate** is the product of one reaction that is consumed in a following reaction in a process that proceeds through several steps.

The **internal energy, U**, of a system is the total energy attributed to the particles of matter and their interactions within a system.

An **ion** is a charged species consisting of a single atom or a group of atoms. It is formed when a neutral atom or a covalently bonded group of atoms either gains or loses electrons.

Ion exchange is a process in which ions held to the surface of an ion exchange material are exchanged for other ions in solution. For example, Na^+ may be exchanged for Ca^{2+} and Mg^{2+}, or OH^- may be exchanged for SO_4^{2-}.

An **ion pair** is an association of a cation and an anion in solution. Such combinations, when they occur, can have a significant effect on solution equilibria.

An **ion product, Q_{sp}**, is formulated in the same manner as a solubility product constant, K_{sp}, but with *nonequilibrium* concentration terms. A comparison of Q_{sp} and K_{sp} provides a criterion for precipitation from solution.

The **ion product of water, K_w**, is the product of $[H_3O^+]$ and $[OH^-]$ in pure water or in an aqueous solution. This product has a unique value that depends only on temperature. At 25 °C, $K_w = 1.0 \times 10^{-14}$.

An **ionic bond** results from the transfer of electrons between metal and nonmetal atoms. Positive and negative ions are formed and held together by electrostatic attractions.

An **ionic compound** is a compound consisting of positive and negative ions that are held together by electrostatic forces of attraction.

Ionic radius is the radius of a spherical ion. It is the atomic radius associated with an element in its ionic compounds.

The **first ionization energy, I**, is the energy required to remove the most loosely held electron from a *gaseous* atom. The second ionization energy, I_2, is the energy required to remove an electron from a gaseous unipositive ion, and so on.

An **irreversible** process takes place in one or several finite steps such that the system is not in equilibrium with its surroundings.

The **isoelectric point, pI**, of an amino acid is the pH at which the dipolar structure or **zwitterion** predominates.

Isoelectronic species have the same number of electrons (usually in the same configuration). Na^+ and Ne are isoelectronic, as are CO and N_2.

An **isolated system** is one that exchanges neither energy nor matter with its surroundings.

Isomers are two or more compounds having the same formula but different structures and therefore different properties.

Isotopes of an element are atoms with different numbers of neutrons in their nuclei. That is, isotopes of an element have the same atomic numbers but different mass numbers.

Isotopic mass (See **nuclidic mass**.)

IUPAC (or **IUC**) refers to the International Union of Pure and Applied Chemistry.

K_c is the relationship among the *concentrations* of the reactants and products in a reversible reaction at equilibrium. Concentrations are expressed as molarities.

K_p, the partial pressure equilibrium constant, is the relationship that exists among the partial pressures of gaseous reactants and products in a reversible reaction at equilibrium. Partial pressures are expressed in atm.

The **Kelvin** temperature is an **absolute** temperature. That is, the lowest attainable temperature is 0 K = -273.15 °C (the temperature at which molecular motion ceases). Kelvin and Celsius temperatures are related through the expression T (K) = t(°C) + 273.15.

A **ketone** has the general formula

$$R-\overset{\overset{\text{O}}{\|}}{C}-R'.$$

A **kilopascal (kPa)** is a unit of pressure equal to 1000 pascals (Pa) or 1000 N/m^{-2}. The standard atmosphere of pressure is 101.325 kPa.

Kinetic energy is energy of motion. The kinetic energy of an object with mass m and velocity u is K.E. = $\frac{1}{2}mu^2$.

The **kinetic-molecular theory of gases** is a model for describing gas behavior. It is based on a set of assumptions and yields equations from which various properties of gases can be deduced.

Labile complex is the term used to describe a complex ion in which a rapid exchange of ligands occurs.

The **lanthanide contraction** refers to the decrease in atomic size in a series of elements in which an f subshell fills with electrons (an inner transition series). It results from the ineffectiveness of f electrons in shielding outer-shell electrons from the nuclear charge of an atom.

The **lanthanides** are the elements ($Z = 58 - 71$) characterized by a partially filled $4f$ subshell in their atoms. Because lanthanum resembles them, La ($Z = 57$) is generally considered together with them.

Lattice energy is the quantity of energy released in the formation of one mole of a crystalline ionic solid from its separated gaseous ions.

Gay-Lussac's law of combining volumes states that, when compared at the same temperature and pressure, the volumes of *gases* involved in a reaction are in the ratio of small whole numbers.

The **law of conservation of energy** states that energy can neither be created nor destroyed in ordinary processes.

The **law of conservation of mass** states that the total mass of the products of a chemical reaction is the same as the total mass of the reactants entering into the reaction.

The **law of constant composition (definite proportions)** states that all samples of a compound have the same composition, that is, the same proportions by mass of the constituent elements.

The **law of multiple proportions** states that if two elements form two or more compounds, the masses of one element combined with a fixed mass of the second are in the ratio of small whole numbers when the different compounds are compared.

The **leaving group** is the species expelled from an electrophilic molecule following attack by a nucleophile.

Le Châtelier's principle states that an action that tends to change the temperature, pressure, or concentrations of reactants in a system at equilibrium stimulates a response that partially offsets the change while a new equilibrium condition is established.

Levorotatory means the ability to rotate the plane of polarized light to the *left*, designated $(-)$.

Lewis acid (See **acid**.)

Lewis base (See **base**.)

A **Lewis structure** is a combination of Lewis symbols that depicts the transfer or sharing of electrons in a chemical bond.

In the **Lewis symbol** of an element, valence electrons are represented by dots placed around the chemical symbol of the element.

The **Lewis theory** refers to a description of chemical bonding through Lewis symbols and Lewis structures in accordance with a particular set of rules.

Ligands are the groups that are coordinated (bonded) to the central atom in a complex.

The **limiting reactant (reagent)** in a reaction is the reactant that is consumed completely. The quantity of product(s) formed depends on the quantity of the limiting reactant.

Line-angle formulas are shorthand representations of organic molecules in which bond lines are drawn, but chemical symbols are written only for elements other than carbon and hydrogen.

A **line spectrum** is produced from the emission of light produced from excited atom or ions. The spectrum contains lines at discrete wavelengths which arise from the transition of an electron from one energy level to another.

Lipids include a variety of naturally occurring substances (e.g., fats and oils) sharing the property of solubility in solvents of low polarity [such as in $CHCl_3$, CCl_4, C_6H_6, and $(C_2H_5)_2O$].

In a **liquid**, atoms or molecules are in close proximity (although generally not as close as in a solid). A liquid occupies a definite volume, but has the ability to flow and assume the shape of its container.

London forces (See **dispersion forces**.)

A **lone pair** is a pair of electrons found in the valence shell of an atom and *not* involved in bond formation.

In a **low-spin complex**, strong crystal field splitting leads to a minimum number of unpaired electrons in the d subshell of the central metal atom or ion.

Magic numbers is a term used to describe numbers of protons and neutrons that confer a special stability to an atomic nucleus.

The **main-group elements** are those in which s or p subshells are being filled in the aufbau process. They are also referred to as the s-block and p-block elements. They are found in groups 1, 2, and 13–18 in the periodic table (the A groups).

A **manometer** is a device used to measure the pressure of a gas, usually by comparing the gas pressure with barometric pressure.

Mass describes the quantity of matter in an object.

The **mass number**, A, is the total of the number of protons and neutrons in the nucleus of an atom.

A **mass spectrometer (mass spectrograph)** is a device used to separate and to measure the quantities and masses of different ions in a beam of positively charged gaseous ions.

Matter is anything that occupies space, has the property known as mass, and displays inertia.

Melting is the transition of a solid to a liquid and occurs at the **melting point**. The melting point and freezing point of a substance are identical.

A **meta (m-) isomer** has two substituents on a benzene ring separated by one C atom.

Metabolism refers to the totality of the chemical reactions occurring in living organisms.

A **metal** is an element whose atoms have small numbers of electrons in the outermost electronic shell. Removal of an electron(s) from a metal atom occurs without great difficulty, producing a positive ion (cation). Metals generally have a lustrous appearance, are malleable and ductile, and are able to conduct heat and electricity.

Metal carbonyls are complexes with d-block metals as central atoms and CO molecules as ligands, e.g., $Ni(CO)_4$.

Metallic radius is one-half the distance between the centers of adjacent atoms in a solid metal.

A **metalloid** is an element that may display both metallic and nonmetallic properties under the appropriate conditions.

A **millimeter of mercury (mmHg)** is a unit of pressure, usually applied to gases. For example, standard atmospheric pressure is equal to the pressure exerted by a 760-mm column of mercury.

A **millimole (mmol)** is one-thousandth of a mole (0.001 mol). It is especially useful in titration calculations.

A **mixture** is any sample of matter that is not pure, that is, not an element or compound. The composition of a mixture, unlike that of a substance, can be varied. Mixtures are either *homogeneous* or *heterogeneous*.

Moderator control slows down energetic neutrons from a fission process so that they are able to induce additional fission.

Molality, *m*, is a solution concentration expressed as the amount of solute, in moles, divided by the mass of solvent, in kg.

Molar mass, *M*, is the mass of one mole of atoms, formula units, or molecules of a substance.

A **mole** is an amount of substance containing Avogadro's number (6.02214×10^{23}) of atoms, formula units, or molecules.

Mole fraction describes a mixture in terms of the fraction of all the molecules that are of a particular type. It is the amount of one component, in moles, divided by the total amount of all the substances in the mixture.

A **mole percent** is a mole fraction expressed on a percentage basis, that is, mole fraction $\times$ 100%.

A **molecular compound** is a compound comprised of discrete molecules.

A **molecular formula** denotes the numbers of the different atoms present in a molecule. In some cases the molecular formula is the same as the empirical formula; in others it is an integral multiple of that formula.

Molecular geometry refers to the geometric shape of a molecule or polyatomic ion. In a species in which all electron pairs are bond pairs, the molecular geometry is the same as the electron-group geometry. In other cases, the two properties are related but not the same.

Molecular mass is the mass of a molecule relative to a mass of exactly 12 u for carbon-12.

Molecular orbital theory describes the covalent bonds in a molecule by replacing atomic orbitals of the component atoms by molecular orbitals belonging to the molecule as a whole. A set of rules is used to assign electrons to these molecular orbitals, thereby yielding the electronic structure of the molecule.

A **molecule** is a group of bonded atoms held together by covalent bonds and existing as a separate entity. A molecule is the smallest entity having the characteristic proportions of the constituent atoms present in a substance.

A **monodentate ligand** is a ligand that is able to attach to a metal center in a complex at only one position and using just one lone pair of electrons.

A **monosaccharide** is a single, simple molecule having the structural features of a carbohydrate. It can also be called a simple sugar.

A **multiple covalent bond** is a bond in which more than two electrons are shared between the bonded atoms.

A **natural law** is a concise statement, often in mathematical terms, that summarizes observations of certain natural phenomena.

The **Nernst equation** is used to relate E_{cell}, E°_{cell} and the activites of the reactants and products in a cell reaction.

A **net ionic equation** represents a reaction between ions in solution in such a way that all nonparticipant (spectator) ions are eliminated from the equation. The equation must be balanced both atomically and for net electric charge.

A **network covalent solid** is a substance in which covalent bonds extend throughout the crystal, making the covalent bond both an *intra*molecular and an *inter*molecular force.

In a **neutralization reaction**, an acid and a base react in stoichiometric proportions, so that there is no excess of either acid or base in the final solution. The products are water and a salt.

Neutrons are electrically neutral fundamental particles of matter found in all atomic nuclei except that of the simple hydrogen atom, protium, ^{1}H.

The **neutron number** is the number of neutrons in the nucleus of an atom. It is equal to the mass number (A) minus the atomic number (Z).

Noble gases are elements whose atoms have the electron configuration ns^2np^6 in the electronic shell of highest principal quantum number. (The noble gas helium has the configuration $1s^2$.)

A **nonelectrolyte** is a substance that is essentially non-ionized, both in the pure state and in solution.

A **nonmetal** is an element whose atoms tend to gain small numbers of electrons to form negative ions (anions) with the electron configuration of a noble gas. Nonmetal atoms may also alter their electron configurations by sharing electrons. Nonmetals are mostly gases, liquid (bromine), or low melting point solids and are very poor conductors of heat and electricity.

A **nonspontaneous process** is one that will not occur naturally. A nonspontaneous process can be brought about only by intervention from outside the system, as in the use of electricity to decompose a chemical compound (electrolysis).

The **normal boiling point** is the temperature at which the vapor pressure of a liquid is 1 atm. It is the temperature at which the liquid boils in a container open to the atmosphere at a pressure of 1 atm.

A **nuclear equation** represents the changes that occur during a nuclear process. The target nucleus and bombarding particle are represented on the left side of the equation, and the product nucleus and ejected particle on the right side.

Nuclear fission is a radioactive decay process in which a heavy nucleus breaks up into two lighter nuclei and several neutrons, accompanied by the release of energy.

In **nuclear fusion** small atomic nuclei are fused into larger ones, with some of their mass being converted to energy.

Nucleic acids are cell components comprised of purine and pyrimidine bases, pentose sugars, and phosphoric acid.

A **nucleophile** is a reactant that seeks out a center of positive charge as a point of attack in a chemical reaction.

A **nucleophilic substitution reaction** is a reaction between a nucleophile and an electrophile. The nucleophile attacks at a positive center on the electrophile, and the leaving group is ejected from another point.

Nucleophilicity is a measure of how readily (how fast) a nucleophile attacks an electrophilic carbon atom bearing a leaving group.

Nuclide is a term used to designate an atom with a specific atomic number and mass number. It is represented by the symbolism ^A_ZE.

An **octet** refers to *eight* electrons in the outermost (valence) electronic shell of an atom in a Lewis structure.

The **octet rule** states that the number of electrons associated with bond pairs and lone pairs of electrons for each of the Lewis symbols (except H) in a Lewis structure will be eight (an **octet**).

Oils are triglycerides in which unsaturated fatty acid components predominate.

Olefins are organic compounds that contain one or more carbon-to-carbon double bonds.

Oligosaccharides are carbohydrates consisting of two to ten monosaccharide units. (See also **sugar**.)

An **open system** is one that can exchange both matter and energy with its surroundings.

Optical isomerism results from the presence of a chiral atom in a structure, leading to a pair of optical isomers that differ only in the direction that they rotate the plane of polarized light. (See also **enantiomers**.)

Optical isomers, also called enantiomers (nonsuperimposable mirror images), are isomers that differ only in the direction they rotate the plane of polarized light.

An **orbital** is a mathematical function used to describe regions in an atom where the electron charge density or the probability of finding an electron is high. The several kinds of orbitals (s, p, d, f, ...) differ from one another in the shapes of the regions of high electron charge density they describe.

An **orbital diagram** is a representation of an electron configuration in which the most probable orbital designation and spin of each electron in the atom are indicated.

The **order of a reaction** relates to the exponents of the concentration terms in the rate law for a chemical reaction. The order can be stated with respect to a particular reactant (first order in A, second order in B, ...) or, more commonly, as the overall order. The overall order is the sum of the concentration-term exponents.

An **organic compound** is made up of carbon and hydrogen or carbon, hydrogen and a small number of other elements, such as oxygen, nitrogen, and sulfur.

An **ortho (o-) isomer** has two substituents attached to adjacent C atoms in a benzene ring.

Osmosis is the net flow of solvent molecules through a semipermeable membrane, from a more dilute solution (or from the pure solvent) into a more concentrated solution.

Osmotic pressure is the pressure that would have to be applied to a solution to stop the passage through a semipermeable membrane of solvent molecules from the pure solvent.

An **overall reaction** or **overall equation** is the overall or net change that occurs when a process is carried out in more than one step.

An **overpotential** is the voltage in excess of the theoretical value required to produce a particular electrode reaction in electrolysis.

Oxidation is a process in which electrons are "lost" and the oxidation state of some atom increases. (Oxidation can occur only in combination with reduction.)

In an **oxidation–reduction (redox)** reaction certain atoms undergo changes in oxidation state. The substance containing atoms whose oxidation states *increase* is **oxidized**. The substance containing atoms whose oxidation states *decrease* is **reduced**.

An **oxidation state** relates to the number of electrons an atom loses, gains, or shares in combining with other atoms to form molecules or polyatomic ions.

An **oxidizing agent (oxidant)** makes possible an oxidation process by itself being *reduced*.

An **oxoacid** is an acid in which an ionizable hydrogen atom(s) is bonded through an oxygen atom to a central atom, that is, E — O — H. Other groups bonded to the central atom are either additional — OH groups or O atoms (or in a few cases H atoms).

An **oxoanion** is a polyatomic anion containing a nonmetal, such as Cl, N, P, or S, in combination with some number of oxygen atoms.

Pairing energy is the energy requirement to force an electron into an orbital that is already occupied by one electron.

A **para (p-) isomer** has two substituents located opposite to one another on a benzene ring.

A **paramagnetic** substance has one or more unpaired electrons in its atoms or molecules. It is attracted into a magnetic field.

A **partial pressure** is the pressure exerted by an individual gas in a mixture, independently of other gases. Each gas in the mixture expands to fill the container and exerts its own partial pressure.

A **pascal (pa)** is a pressure of one N/m^2.

The **Pauli exclusion principle** states that no two electrons may have all four quantum numbers alike. This limits occupancy of an orbital to two electrons with opposing spins.

The **p block** is that portion of the periodic table in which the filling of electron orbitals (aufbau process) involves p subshells.

A **peptide bond** is formed by the elimination of a water molecule from between two amino acid molecules. The H atom comes from the — NH_2 group of one amino acid and the — OH group, from the — COOH group of the other acid.

The **percent ionization** of a weak acid or a weak base is the percent of its molecules that ionize in an aqueous solution.

Percent natural abundances refer to the relative proportions, expressed as percentages by number, in which the isotopes of an element are found in natural sources.

Percent yield is the percent of the theoretical yield of product that is actually obtained in a chemical reaction. (See also **actual yield** and **theoretical yield**.)

A **perfect gas** is one whose behavior can be predicted by the ideal gas equation. It is also used to describe a gas whose molecules are "point masses" that do not interact with one another. (See also **ideal gas**.)

A **period** is a horizontal row of the periodic table. All members of a period have atoms with the same highest principal quantum number.

The **periodic law** refers to the periodic recurrence of certain physical and chemical properties when the elements are considered in terms of increasing atomic number.

The **periodic table** is an arrangement of the elements, by atomic number, in which elements with similar physical and chemical properties are grouped together in vertical columns.

Permanent hard water (See **hard water**.)

The **peroxide** ion has the structure $\left[\,\ddot{\underset{\cdot\cdot}{O}} - \ddot{\underset{\cdot\cdot}{O}}\,\right]^{2-}$.

pH is a shorthand designation for $[H_3O^+]$ in a solution. It is defined as $pH = -\log[H_3O^+]$.

A **phase diagram** is a graphical representation of the conditions of temperature and pressure at which solids, liquids, and gases (vapors) exist, either as single phases or states of matter or as two or more phases in equilibrium.

A **phenol** has the functional group — OH as part of an aromatic hydrocarbon structure.

A **phenyl group** is a benzene ring from which one H atom has been removed: — C_6H_5.

The **photoelectric effect** is the emission of electrons by certain materials when their surfaces are struck by electromagnetic radiation of the appropriate frequency.

A **photon** is a "particle" of light. The energy of a beam of light is concentrated into these photons.

In a **physical change**, one or more physical properties of a sample of matter change, but the composition remains unchanged.

A **physical property** is a characteristic that a substance can display without undergoing a change in its composition.

A **pi (π) bond** results from the side-to-side overlap of p orbitals, producing a high electron charge density above and below the line joining the bonded atoms.

Pig iron is an impure form of iron (about 95% Fe and 3–4% C, together with small quantities of Mn, Si, and P) produced in a blast furnace.

pK is a shorthand designation for an ionization constant: $pK = -\log K$. pK values are useful when comparing the relative strengths of acids or bases.

Planck's constant, h, is the proportionality constant that relates the energy of a photon of light to its frequency. Its value is 6.626×10^{-34} J s.

Plaster of Paris, $CaSO_4 \cdot \frac{1}{2}H_2O$, is a hemihydrate of calcium sulfate obtained by heating gypsum, $CaSO_4 \cdot 2\,H_2O$. It is a widely used material in the construction industry.

pOH is a shorthand designation for $[OH^-]$ in a solution: $pOH = -\log[OH^-]$.

In a **polar covalent bond** a separation exists between the centers of positive and negative charge in the bond.

In a **polar molecule**, the presence of one or more polar covalent bonds leads to a separation of the positive and negative charge centers for the molecule as a whole. A polar molecule has a resultant dipole moment.

Polarizability describes the ease with which the electron cloud in an atom or molecule can be distorted in an electric field, that is, the ease with which a dipole can be induced.

A **polyatomic ion** is a combination of two or more covalently bonded atoms that exists as an ion.

A **polydentate ligand** is capable of donating more than a single electron pair to the metal center of a complex, from different atoms in the ligand and to different sites in the geometric structure.

In a **polyhalide ion** two or more halogen atoms are covalently bonded into a polyatomic anion, e.g., I_3^-.

Polymorphism refers to the existence of a solid substance in more than one crystalline form.

In a **polypeptide**, a large number of amino acid units join together through peptide bonds.

A **polyprotic acid** is capable of losing more than a single proton per molecule in acid–base reactions. Protons are lost in a stepwise fashion, with the first proton being the most readily lost.

A **polysaccharide** is a carbohydrate (such as starch or cellulose) consisting of more than ten monosaccharide units.

Positional isomers differ in the position on a hydrocarbon chain or ring where a functional group(s) is attached.

A **positron (β^+)** is a *positive* electron emitted as a result of the conversion of a proton to a neutron in a radioactive nucleus.

Potential energy is energy due to position or arrangement. It is the energy associated with forces of attraction and repulsion between objects.

The term **ppb** (parts per billion) refers to the number of parts of a component to one billion parts of the medium in which it is found.

The term **ppm** (parts per million) refers to the number of parts of a component to one million parts of the medium in which it is found.

The term **ppt** (parts per trillion) refers to the number of parts of a component to one trillion parts of the medium in which it is found.

A **precipitate** is an insoluble solid that deposits from a solution as a result of a chemical reaction.

Precision is the degree of reproducibility of a measured quantity—the closeness of agreement among repeated measurements.

Pressure is a force per unit area. Applied to gases, pressure is most easily understood in terms of the height of a liquid column that can be maintained by the gas.

Pressure–volume work is work associated with the expansion or compression of gases.

A **primary carbon** is attached to one other carbon atom.

A **primary battery** produces electricity from a chemical reaction that cannot be reversed. As a result the battery cannot be recharged.

A **primary color** is one of a set of colors that when added together as light produce white light. Subtractive mixing leads to an absence of color (black). Red, yellow, and blue are a set of primary colors.

Hydrogen atoms attached to a primary carbon atom are called **primary hydrogen atoms**.

Primary structure refers to the sequence of amino acids in the polypeptide chains that make up a protein.

A **principal electronic shell (level)** refers to the collection of all orbitals having the same value of the principal quantum number, n. For example, the $3s$, $3p$, and $3d$ orbitals comprise the third principal shell ($n = 3$).

The **products** are the substances formed in a chemical reaction.

Properties are qualities or attributes that can be used to distinguish one sample of matter from others.

A **protein** is a large polypeptide, that is, having a molecular mass of 10,000 u or more.

In a **protic solvent** the molecules have hydrogen atoms bonded to electronegative atoms, such as oxygen or nitrogen.

A **proton acceptor** is a base in the Brønsted–Lowry acid–base theory.

A **proton donor** is an acid in the Brønsted–Lowry acid–base theory.

Proton number (See **atomic number**.)

Protons are fundamental particles carrying the basic unit of positive electric charge and found in the nuclei of all atoms.

Pyrometallurgy is the traditional approach to extractive metallurgy that uses dry solid materials heated to high temperatures. (See also **extractive metallurgy** and **hydrometallurgy**.)

Qualitative cation analysis is a laboratory method, based on a variety of solution equilibrium concepts, for determining the presence or absence of certain cations in a sample.

A **quantum** refers to a discrete unit of energy that is the smallest quantity by which the energy of a system can change.

Quantum numbers are integral numbers whose values must be specified in order to solve the equations of wave mechanics. Three different quantum numbers are required: the *principal quantum number, n*; the *orbital angular momentum quantum number, l*; and the *magnetic quantum number, m_l*. The permitted values of these numbers are interrelated.

A **quaternary carbon** is attached to four carbon atoms.

Quaternary structure is the highest order structure that is found in some proteins. It describes how separate polypeptide chains may be assembled into a larger, more complex structure.

Quicklime is a common name for calcium oxide, CaO.

A **racemic mixture** is a mixture containing equal amounts of the enantiomers of an optically active substance.

A **rad** is a quantity of radiation able to deposit 1×10^{-2} J of energy per kilogram of matter.

A **radial wave function**, $R(r)$, is the part of a wave function that depends only on the distance r when the Schrödinger wave equation is expressed in spherical polar coordinates. (See also **angular wave function**.)

Radical (See **free radical**.)

The **radioactive decay law** states that the rate of decay of a radioactive material—the activity, A—is directly proportional to the number of atoms present.

A **radioactive decay series** is a succession of individual steps whereby an initial radioactive isotope (e.g., ^{238}U) is ultimately converted to a stable isotope (e.g., ^{206}Pb).

Radioactivity is a phenomenon in which small particles of matter (α or β particles) and/or electromagnetic radiation (γ rays) are emitted by unstable atomic nuclei.

A **random error** is an error made by the experimenter in performing an experimental technique or measurement, such as the error in estimating a temperature reading on a thermometer.

Raoult's law states that the vapor pressure of a solution component is equal to the product of the vapor pressure of the pure liquid and its mole fraction in solution: $P_A = x_A P_A^o$.

The **rate constant, k,** is the proportionality constant in a rate law that permits the rate of a reaction to be related to the concentrations of the reactants.

A **rate-determining step** in a reaction mechanism is an elementary process that is instrumental in establishing the rate of the overall reaction, usually because it is the slowest step in the mechanism.

The **rate law (rate equation)** for a reaction relates the reaction rate to the concentrations of the reactants. It has the form: rate $= k[A]^m[B]^n \ldots$.

The **rate of reaction** describes how fast reactants are consumed and products are formed, usually expressed as change of concentration per unit time.

Reactants are the substances that enter into a chemical reaction. This term is often applied to *all* the substances involved in a reversible reaction, but it can also be limited to the substances that appear on the *left* side of a chemical equation—the starting substances. (Substances on the *right* side of the equation are usually called products.)

A **reaction intermediate** is a species formed in one elementary reaction in a reaction mechanism and consumed in a subsequent one. As a result, the species does not appear in the equation for the overall reaction.

A **reaction mechanism** is a set of elementary steps or processes by which a reaction is proposed to occur. The mechanism must be consistent with the stoichiometry and rate law of the overall reaction.

A **reaction profile** is a graphical representation of a chemical reaction in terms of the energies of the reactants, activated complex(es), and products.

The **reaction quotient, Q,** is a ratio of concentration terms (or partial pressures) having the same form as an equilibrium constant expression, but usually applied to *nonequilibrium* conditions.

In a **rearrangement reaction**, a molecule is converted into another of its isomeric forms.

Recrystallization (See **fractional crystallization**.)

A **reducing agent (reductant)** makes possible a reduction process by itself becoming *oxidized*.

A **reducing sugar** is one that is able to reduce $Cu^{2+}(aq)$ to red, insoluble Cu_2O. The sugar must have available an aldehyde group, which is oxidized to an acid.

A **reduction** process is one in which electrons are "gained" and the oxidation state of some atom decreases. (Reduction can only occur in combination with oxidation.) (See also **extractive metallurgy**.)

Refining (See **extractive metallurgy**.)

A **rem** is a unit of radiation related to the rad, but taking into account the varying effects on biological matter of different types of radiation of the same energy.

Representative elements (See **main-group elements**.)

Resonance occurs when two or more plausible Lewis structures can be written for a species. The true structure is a composite or *hybrid* of these different contributing structures.

Reverse osmosis is the passage through a semipermeable membrane of solvent molecules *from a solution into a pure solvent*. It can be achieved by applying to the solution a pressure in excess of its osmotic pressure.

A **reversible process** is one that can be made to reverse direction by just an infinitesimal change in a system property.

Ribonucleic acid (RNA), through its **messenger RNA (mRNA)** and **transfer RNA (tRNA)** forms, is involved in the synthesis of proteins.

Roasting (See **extractive metallurgy**.)

The **root-mean-square speed** is the square root of the average of the squares of the speeds of all the gas molecules in a gaseous sample.

The **R, S system** is used to indicate the arrangement of the four groups bonded to a chiral center and to provide names that distinguish between optical isomers.

A **salt bridge** is a device (a U-tube filled with a salt solution) used to join two half-cells in an electrochemical cell. The salt bridge permits the flow of ions between the two half-cells.

The **salt effect** is that of ions *different* from those directly involved in a solution equilibrium. The salt effect is also known as the diverse or "uncommon" ion effect.

Salts are ionic compounds in which hydrogen atoms of acids are replaced by metal ions. Salts are produced by the neutralization of acids with bases.

Saponification is the hydrolysis of a triglyceride by a strong base. The products are glycerol and a soap.

Saturated hydrocarbon molecules contain only single bonds between carbon atoms.

A **saturated solution** is one that contains the maximum quantity of solute that is normally possible at the given temperature.

The **s block** refers to the portion of the periodic table in which the filling of electron orbitals (aufbau process) involves the s subshell of the electronic shell of highest principal quantum number.

The **Schrödinger equation** describes the electron in a hydrogen atom as a matter wave. Solutions to the Schrödinger equation are called wave functions.

The **scientific method** refers to the general sequence of activities—observation, experimentation, and the formulation of hypotheses, laws, and theories—that lead to the advancement of scientific knowledge.

The **second law of thermodynamics** relates to the direction of spontaneous change. One statement of the law is that all spontaneous processes produce an increase in the entropy of the universe.

A **secondary battery** produces electricity from a reversible chemical reaction. When electricity is passed through the battery in the reverse direction the battery is recharged.

A **secondary carbon** is attached to two other carbon atoms.

A **secondary color** is the complement of a **primary color**. When light of a primary color and its complement (secondary) color are added, the result is white light. When they are subtracted, the result is an absence of color (black).

The **secondary structure** of a protein describes the structure or shape of a polypeptide chain, for example, a coiled helix.

A **second-order reaction** is one for which the sum of the concentration-term exponents in the rate equation is 2.

Self-ionization is an acid–base reaction in which one molecule acts as an acid and donates a proton to another molecule of the same kind acting as a base.

The **shielding effect** refers to the effect of inner-shell electrons in shielding or screening outer-shell electrons from the full effects of the nuclear charge. In effect the inner electrons partially reduce the nuclear charge. (See also **effective nuclear charge**.)

A **side reaction** is a reaction that produces an undesired or unexpected product and accompanies a reaction intended to produce something else.

A **sigma** (σ) **bond** results from the end-to-end overlap of simple or hybridized atomic orbitals along the straight line joining the nuclei of the bonded atoms.

Significant figures are those digits in an experimentally measured quantity that establish the precision with which the quantity is known.

A **silicone** is an organosilicon polymer containing O—Si—O bonds.

Simultaneous reactions are two or more reactions that occur at the same time.

A **single covalent bond** results from the sharing of *one pair* of electrons between bonded atoms. It is represented by a single dash sign (—).

A **skeletal structure** is an arrangement of atoms in a Lewis structure to correspond to the actual arrangement found by experiment.

Slaked lime is a common name for calcium hydroxide, $Ca(OH)_2$.

Smog is the general term used to refer to a condition in which polluted air reduces visibility, causes stinging eyes and breathing difficulties, and produces additional minor and major health problems. (See also **industrial smog** and **photochemical smog**.)

S$_N$1 is the designation for a nucleophilic substitution reaction in which the rate-determining step is unimolecular.

S$_N$2 is the designation for a nucleophilic substitution reaction in which the rate-determining step is bimolecular.

Soaps are the salts of fatty acids, e.g., $RCOO^-Na^+$, where the R group is a hydrocarbon chain containing from 3 to 21 C atoms. Sodium and potassium soaps are the common soaps used as cleansing agents.

Solders are low-melting alloys used for joining wires or pieces of metal. They usually contain metals such as Sn, Pb, Bi, and Cd.

In a **solid**, atoms or molecules are in close contact, often in a highly organized arrangement. A solid has a definite shape and occupies a definite volume. (See also **crystal**.)

The **solubility** of a substance is the concentration of its saturated solution.

The **solubility product constant**, K_{sp}, is the equilibrium constant that describes the formation of a saturated solution of a slightly soluble ionic compound. It is the product of ionic concentration terms, with each term raised to an appropriate power.

A **solute** is a solution component that is dissolved in a solvent. A solution may have several solutes, with the solutes generally present in lesser amounts than is the solvent.

Solution (See **homogeneous mixture**.)

The **solvent** is the solution component in which one or more solutes are dissolved. Usually the solvent is present in greater amount than are the solutes and determines the state of matter in which the solution exists.

An *sp* **hybrid orbital** is one of the pair of orbitals formed by the hybridization of one s and one *p* orbital. The angle between the two orbitals is 180°.

An *sp*^{2} **hybrid orbital** is one of the three orbitals formed by the hybridization of one s and two *p* orbitals. The angle between any two of the orbitals is 120°.

An *sp*^{3} **hybrid orbital** is one of the four orbitals formed by the hybridization of one *s* and three *p* orbitals. The angle between any two of the orbitals is the tetrahedral angle—109.5°.

An *sp*^{3}*d* **hybrid orbital** is one of the five orbitals formed by the hybridization of one *s*, three *p*, and one *d* orbital. The five orbitals are directed to the corners of a trigonal bipyramid.

An *sp*^{3}*d*^{2} **hybrid orbital** is one of the six orbitals formed by the hybridization of one *s*, three *p*, and two *d* orbitals. The six orbitals are directed to the corners of a regular octahedron.

spdf **notation** is a method of describing electron configurations in which the numbers of electrons assigned to each orbital are denoted as superscripts. For example, the electron configuration of Cl is $1s^2 2s^2 2p^6 3s^2 3p^5$.

The **specific heat** of a substance is the quantity of heat required to change the temperature of one gram of the substance by one degree Celsius.

Spectator ions are ionic species that are present in a reaction mixture but do not take part in the reaction. They are usually eliminated from a chemical equation.

The **spectrochemical series** is a ranking of ligand abilities to produce a splitting of the *d* energy level of a central metal ion in a complex ion.

Speed of light, c, has a value of $2.99792458 \times 10^8 \text{ m s}^{-1}$.

A **spontaneous (natural) process** is one that is able to take place in a system left to itself. No external action is required to make the process go, although in some cases the process may take a very long time.

Stalactites and **stalagmites** are limestone ($CaCO_3$) formations in limestone caves produced by the slow decomposition of $Ca(HCO_3)_2(aq)$.

A **standard cell potential**, $E°$, is the voltage of an electrochemical cell in which all species are in their standard states. (See also **cell potential**.)

Standard conditions of temperature and pressure (STP) refers to a gas maintained at a temperature of exactly 0 °C (273.15 K) and 760 mmHg (1 atm).

A **standard electrode potential**, $E°$, is the electric potential that develops on an electrode when the oxidized and reduced forms of some substance are in their *standard* states. Tabulated data are expressed in terms of the reduction process, that is, standard electrode potentials are standard reduction potentials.

The **standard enthalpy of formation**, $\Delta H_f°$, of a substance is the enthalpy change that occurs in the formation of 1 mol of the substance in its standard state from the reference forms of its elements in their standard states. The reference forms of the elements are their most stable forms at the given temperature and 1 bar pressure.

The **standard enthalpy of reaction**, $\Delta H°$, is the enthalpy change of a reaction in which all reactants and products are in their standard states.

Standard Gibbs energy change, $\Delta G°$, is the Gibbs energy change of a process when the reactants and products are all in their standard states. The equation relating standard free energy change to the equilibrium constant is $\Delta G° = -RT \ln K$.

The **standard Gibbs energy of formation,** $\Delta G_f°$, is the standard free energy change associated with the formation of 1 mol of compound from its elements in their most stable forms at 1 bar pressure.

The **standard hydrogen electrode (SHE)** is an electrode at which equilibrium is established between H_3O^+ (a = 1) and H_2 (g, 1 bar) on an inert (Pt) surface. The standard hydrogen electrode is *arbitrarily* assigned an electrode potential of exactly 0 V.

The **standard molar entropy** ($S°$) is the absolute entropy evaluated when one mole of a substance is in its standard state at a particular temperature.

The **standard state** of a substance refers to that substance when it is maintained at 1 bar pressure and at the temperature of interest. For a gas it is the (hypothetical) pure gas behaving as an ideal gas at 1 bar pressure and the temperature of interest.

Standardization of a solution refers to establishing the exact concentration of the solution, usually through a titration.

A **standing wave** is a wave motion that reflects back on itself in such a way that the wave contains a certain number of points (nodes) that undergo no motion. A common example is the vibration of a plucked guitar string, and a related example is the description of electrons as matter waves.

Steel is a term used to describe iron alloys containing from 0 to 1.5% C together with other key elements, such as V, Cr, Mn, Ni, W, and Mo.

Step-reaction polymerization is a type of polymerization reaction in which monomers are joined together by the elimination of small molecules between them. For example, a H_2O molecule might be eliminated by the reaction of a H atom from one monomer with an —OH group from another.

A **stereocenter** is an asymmetric carbon atom.

In **stereoisomers**, the number and types of atoms and bonds in molecules are the same, but certain atoms are oriented differently in space. *Cis* and *trans* isomerism is one type of stereoisomerism; optical isomerism is another.

Stoichiometric coefficients are the coefficients used to balance an equation.

A **stoichiometric factor** is a conversion factor relating molar amounts of two species in a chemical reaction (i.e., a

reactant to a product, one reactant to another, etc.). The numbers used in formulating the factor are stoichiometric coefficients.

Stoichiometric proportions refer to relative amounts of reactants that are in the same mole ratio as implied by the balanced equation for a chemical reaction. For example, a mixture of 2 mol H_2 and 1 mol O_2 is in stoichiometric proportions, and a mixture of 1 mol H_2 and 1 mol O_2 is not, for the reaction $2\ H_2 + O_2 \longrightarrow 2\ H_2O$.

Stoichiometry refers to quantitative measurements and relationships involving substances and mixtures of chemical interest.

A **strong acid** is an acid that is completely ionized in aqueous solution.

A **strong base** is a base that is completely ionized in aqueous solution.

A **strong electrolyte** is a substance that is completely ionized in solution.

A **structural formula** for a compound indicates which atoms in a molecule are bonded together, and whether by single, double, or triple bonds.

Structural isomers have the same number and kinds of atoms, but they differ in their structural formulas.

Sublimation is the passage of molecules from the solid to the gaseous state.

A **subshell** refers to a collection of orbitals of the same type. For example, the three $2p$ orbitals constitute the $2p$ subshell.

A **substance** has a constant composition and properties throughout a given sample and from one sample to another. All substances are either elements or compounds.

In a **substitution reaction,** an atom, an ion, or a group in one molecule is replaced by (substituted with) another.

A **substrate** is the substance that is acted upon by an enzyme in an enzyme-catalyzed reaction. The substrate is converted to products, and the enzyme is regenerated.

A **sugar** is a monosaccharide (simple sugar), a disaccharide, or an oligosaccharide containing up to ten monosaccharide units.

The **superoxide** ion has the structure $\left[:\ddot{O} - \ddot{O}: \right]^-$.

Superphosphate is a mixture of $Ca(H_2PO_4)_2$ and $CaSO_4$ produced by the action of H_2SO_4 on phosphate rock.

A **supersaturated** solution contains more solute than normally expected for a saturated solution, usually prepared from a solution that is saturated at one temperature by changing its temperature to one where supersaturation can occur.

Surface tension is the energy or work required to extend the surface of a liquid.

The **surroundings** represent that portion of the universe with which a system interacts.

A **suspension** is a heterogeneous fluid containing solid particles that are sufficiently large for sedimentation and, unlike colloids, will settle.

Synthesis gas is a mixture of $CO(g)$ and $H_2(g)$, generally made from coal or natural gas, that can be used as a fuel or in the synthesis of organic compounds.

A **system** is the portion of the universe selected for a thermodynamic study. (See also **open, closed,** and **isolated** systems.)

A **systematic error** is one that recurs regularly in a series of measurements because of an inherent error in the measuring system (e.g., through faulty calibration of a measuring device).

Temporary hard water (See **hard water.**)

A **terminal atom** is any atom that is bonded to only one other atom in a molecule or polyatomic ion.

A **termolecular process** is an elementary process in a reaction mechanism in which three atoms or molecules must collide simultaneously.

A **ternary compound** is comprised of *three* elements.

A **tertiary carbon** is attached to three other carbon atoms.

The **tertiary structure** of a protein refers to its three-dimensional structure—for example, the twisting and folding of coils.

The **theoretical yield** is the quantity of product *calculated* to result from a chemical reaction. (See also **actual yield** and **percent yield.**)

A **theory** is a model or conceptual framework with which one is able to explain and make further predictions about natural phenomena.

Thermal energy is energy associated with random molecular motion.

The **thermite reaction** is an oxidation–reduction reaction that uses powdered aluminum metal as a reducing agent to reduce a metal oxide, such as Fe_2O_3, to the free metal.

The **thermodynamic equilibrium constant,** K, is an equilibrium constant expression based on activities. In dilute solutions activities can be replaced by molarities and in ideal gases, by partial pressures in atm. The activities of pure solids and liquids are 1.

The **third law of thermodynamics** states that the entropy of a pure perfect crystal is zero at the absolute zero of temperature, 0 K.

The **titrant** is the solution that is added in a controlled fashion through a buret in a titration reaction. (See also **titration**.)

Titration is a procedure for carrying out a chemical reaction between two solutions by the controlled addition (from a buret) of one solution to the other. In a titration a means must be found, as by the use of an indicator, to locate the equivalence point.

A **titration curve** is a graph of solution pH versus volume of titrant. It outlines how pH changes during an acid–base titration, and it can be used to establish such features as the equivalence point of the titration.

A **torr** is a unit of pressure equal to the unit millimeter of mercury.

The **torsional** energy is the energy difference between the eclipsed and staggered forms of ethane.

The term **trans** is used to describe geometric isomers in which two groups are attached on opposite sides of a double bond in an organic molecule, or at opposite corners of a square in a square-planar complex, or at positions above and below the central plane of an octahedral complex. (See also **geometric isomerism**.)

Transition elements or **transition metals** are those elements whose atoms feature the filling of a *d* or *f* subshell of an *inner* electronic shell. If the filling of an *f* subshell occurs, the elements are sometimes referred to as inner transition elements.

The **transition state** in a chemical reaction is an intermediate state between the reactants and products. (See also **activated complex** and **reaction profile**.)

Triglycerides are esters of glycerol (1,2,3-propanetriol) with long-chain monocarboxylic (fatty) acids.

In a **triple covalent bond**, *three pairs* of electrons are shared between the bonded atoms. It is represented by a triple-dash sign ($\equiv$).

A **triple point** is a condition of temperature and pressure at which three phases of a substance (usually solid, liquid, and vapor) coexist at equilibrium.

Trouton's rule states that at their normal boiling points the entropies of vaporization of many liquids have about the same value: 87 J mol^{-1} K^{-1}.

A **unimolecular** process is an elementary process in a reaction mechanism in which a single molecule, when sufficiently energetic, dissociates.

A **unit cell** is a small collection of atoms, ions, or molecules occupying positions in a crystalline lattice. An entire crystal can be generated by straight-line displacements of the unit cell in the three perpendicular directions.

Unsaturated hydrocarbon molecules contain one or more carbon-to-carbon multiple bonds.

An **unsaturated solution** contains less solute than the solvent is capable of dissolving under the given conditions.

The **valence bond method** treats a covalent bond in terms of the overlap of pure or hybridized atomic orbitals. Electron probability (or electron charge density) is concentrated in the region of overlap.

Valence electrons are electrons in the electronic shell of highest principal quantum number, that is, electrons in the outermost shell.

The **valence-shell electron-pair repulsion (VSEPR) theory** is a theory used to predict probable shapes of molecules and polyatomic ions based on the mutual repulsions of electron pairs found in the valence shell of the central atom in the structure.

The **van der Waals equation** is an equation of state for nonideal gases. It includes correction terms to account for intermolecular forces of attraction and for the volume occupied by the gas molecules themselves.

The term **van der Waals forces** is used to describe, collectively, intermolecular forces of the London type and interactions between permanent dipoles.

One type of measure of an atomic size are **van der Waals radii.** van der Waals radii are strictly hard sphere radii measured using atomic distances in closest packed crystals.

Vaporization is the passage of molecules from the liquid to the gaseous state.

Vapor pressure is the pressure exerted by a vapor when it is in dynamic equilibrium with its liquid at a fixed temperature.

A **vapor-pressure curve** is a graph of vapor pressure as a function of temperature.

In a **vicinal dihalide**, a halogen adds across the double bond of an alkene, and the halogen atoms are bonded to adjacent carbons.

Viscosity refers to a liquid's resistance to flow. Its magnitude depends on intermolecular forces of attraction and in some cases, on molecular sizes and shapes.

A **volt (V)** is the SI unit for cell voltage. It is defined as 1 joule per coulomb.

A **voltaic (galvanic) cell** is an electrochemical cell in which a *spontaneous* chemical reaction produces electricity.

Water gas is a mixture of CO(g) and H_2(g), together with some of the noncombustible gases CO_2 and N_2, produced by passing steam [H_2O(g)] over heated coke.

A **wave** is a disturbance that transmits energy through a medium.

The **wavelength** is the distance between successive crests or troughs of a wave motion.

Wave mechanics is a form of quantum theory based on the concepts of wave–particle duality, the Heisenberg uncertainty principle, and the treatment of electrons as matter waves. Mathematical solutions of the equations of wave mechanics are known as **wave functions** (ψ).

Wave–particle duality was postulated by de Broglie and states that at times particles of matter have wave-like properties and vice versa. This was demonstrated in the diffraction pattern observed when electrons were directed at a nickel crystal.

A **weak acid** is an acid that is only partially ionized in aqueous solution in a reversible reaction.

A **weak base** is a base that it only partially ionized in aqueous solution in a reversible reaction.

A **weak electrolyte** is a substance that is only partially ionized in solution in a reversible reaction.

Work is a form of energy transfer between a system and its surroundings that can be expressed as a force acting through a distance.

A **zero-order reaction** proceeds at a rate that is *independent* of reactant concentrations. The sum of the concentration-term exponent(s) in the rate equation is equal to *zero*.

The **zero-point energy** is the lowest possible energy in a quantum mechanical system, such as the "particle-in-a-box" energy corresponding to $n = 1$ (page 318).

Zone refining is a purification process in which a rod of material is subjected to successive melting and freezing cycles. Impurities are swept by a moving molten zone to the end of the rod, which is cut off.

A **zwitterion** is a compound (for example, an amino acid or polypeptide) containing both acid and base groups. Zwitterions, at neutral pH, typically have simultaneously positively charged groups (cations) and negatively charge groups (anions).

G

Appendix

Answers to Concept Assessment Questions

Note: Your answers may differ slightly from those given here, depending on the number of steps used to solve a problem and whether any intermediate results were rounded off.

CHAPTER 1

Concept Assessment 1-1. No, an experimental result contrary to that predicted by a hypothesis is reason to reject a hypothesis, not its proof. **1-2.** The product (mg) is the same for the unknown and the "weights" of known mass, regardless of the value of g. The measured mass with a two-pan balance is the same on the moon as on Earth. The single pan electronic balance measures *weight*, which is converted to a "mass" reading. With such a balance calibrated on Earth, the "mass" will appear less when measured on the moon. **1-3.** To find the *one* temperature, substitute $t(°C) = t(°F)$ into the equation $t(°F) = \frac{9}{5}t(°C) + 32$. Solve for $t(°F)$ to obtain the value $-40\ °F$; thus $-40\ °F = -40\ °C$. **1-4.** The volume of wood is $1000\ g/(0.68\ g\ cm^{-3}) = 1470\ cm^3$. The wood displaces its own mass of water—1000 g, which has a volume of $1000\ cm^3$. The fraction of wood under water is $1000/1470 = 0.68$. **1-5.** Yes to both questions. A measuring instrument might yield precise readings but be incorrectly calibrated—measurements might agree with one another but their average might not agree with the actual value. Measurements with an imprecise instrument might differ widely from the actual value, yet their average might, by chance, agree with the actual value. **1-6.** The relationship 1 in = 2.54 cm is a definition—an *exact* quantity. A more precise relationship between meters and inches is 1 m = $100\ cm \times (1\ in/2.54\ cm) = 39.370079\ in$.

CHAPTER 2

Concept Assessment 2-1. Helmont assumed, incorrectly, that the tree interacted with its surroundings only through the soil, sunlight, and watering. The tree also interacted with the atmosphere, that is, with carbon dioxide gas (see photosynthesis, page 273). **2-2.** The *final* mass, magnesium bromide plus unreacted bromine, equals the *initial* mass: 4.15 g + 82.6 g = 86.8 g, but without knowing the mass of unreacted bromine we cannot deduce the mass of magnesium bromide. **2-3.** The discovery of cathode rays (electrons) refuted the idea that atoms are indivisible.

The discovery of isotopes refuted the idea that all atoms of an element are alike in mass. The idea that atoms combine in simple numerical ratios remains valid. **2-4.** The exception is the protium atom, ^1_1H, the most abundant isotope of hydrogen. It consists of a lone proton as the atomic nucleus. **2-5.** The weighted average atomic mass of 51.9961 u, almost exactly 52 u, suggests that chromium might exist exclusively as $^{52}_{24}Cr$ atoms. Another possibility (which, in fact, is the case) is that chromium exists as a mixture of isotopes whose weighted average atomic mass is 51.9961 u. For zinc, we should conclude two or more isotopes. The weighted average atomic mass, 65.409 u, is too far from 65 u to suggest a single isotopic mass. **2-6.** If all naturally occurring Au atoms have the mass 196.967g Au/N_A, they must all be $^{197}_{79}Au$, with no isotopes. If no Ag atom has the mass 107.868g Ag/N_A, there must be two or more naturally occurring isotopes. (In fact there are two: $^{107}_{47}Ag$ and $^{109}_{47}Ag$.)

CHAPTER 3

Concept Assessment 3-1. For the *molecular* formula count the atoms in the condensed structural formula given: $C_4H_6O_2$. The *empirical* formula has the same ratio but with the smallest possible integers: $C_2H_3O_2$. Visualize the *structural* formula in terms of that of butane in Figure 3-2(a). Replace the —CH_3 groups at the two ends of the butane structure with —COOH groups (see structural formula of acetic acid in Figure 3-1). For the simplest *line-angle* formula, draw a line to represent H_2C—CH_2 and at an angle at each end of the line attach a —COOH group. **3-2.** Compare the other four quantities to **(d)**—the 20.000g brass weight. **(a)** 0.50 mol O_2 = 16.00 g; this might be the smallest mass but cannot be the largest. **(b)** 2.0×10^{23} Cu atoms = 1/3 mol Cu $\approx$ 21 g; this is now the largest mass. **(c)** 1.0×10^{24} H_2O molecules is more than 1.5 mol H_2O > 27 g; this is now the largest mass. **(e)** the mass of 1.0 mol Ne = 20. g. Conclusion: greatest mass, **(c)**; smallest mass, **(a)**. **3-3.** C contributes the greatest number of atoms—13—and Cl contributes the greatest mass. A Cl atom, of which there are six, has nearly three times the mass of a C atom. **3-4.** Obtain relative numbers of atoms in the formula by multiplying the molar mass by the mass fractions of the elements. In Example 3-5, for example, mol C = 0.6258 $\times$ 230 g $\times$ (1 mol C/12.011 g) = 11.98 mol C.

Similarly, obtain 21.97 mol H and 3.995 mol H. Thus, the molecular formula is $C_{12}H_{22}O_4$ and the empirical formula is $C_6H_{11}O_2$. **3-5** In each combustion, 1 mol CO_2 forms for every mole C and 1 mol H_2O for every 2 mol H. Determine these amounts for each combustion: **(a)** 5 mol CO_2 and 2 mol H_2O; **(b)** 1.25 mol CO_2 and 2.5 mol H_2O; **(c)** 1 mol CO_2 and 1.5 mol H_2O; **(d)** 6 mol CO_2 and 3 mol H_2O. Thus, C_6H_5OH [response **(d)**] produces a greatest number of moles (and thus mass) of both CO_2 and H_2O. **3-6.** The O.S. of N in NH_3 is -3. The O.S. of N is higher in H_2NNH_2; it is -2. Note also that the molar mass of N_2H_4 is 32 g/mol. **3-7.** No. The greatest mass ratio of H_2O to CO_2 is found in CH_4, which has the highest percent H of all hydrocarbons; and its combustion produces only 2 mol H_2O (36 g) to 1 mol CO_2 (44 g).

CHAPTER 4

Concept Assessment 4-1. (a) The product is $O_2(g)$, not O(g). **(b)** The product is $O_2(g)$ exclusively, not a mixture of O(g) and $O_2(g)$. **(c)** The product is KCl(s), not KClO(s). **4-2 (a)** incorrect (3 mol S per *2 mol* H_2S) **(b)** incorrect (stoichiometric coefficients refer to moles, not grams) **(c)** correct (1 mol H_2O/1 mol H_2S is the same as 2 mol H_2O/2 mol H_2S) **(d)** correct (2 of every 3 mole of S on the left are in 2 mol H_2S, yielding 3 mol S on the right) **(e)** incorrect (3 moles of reactants yield 5 moles of product) **(f)** correct (no atoms can be created or destroyed in the reaction) **4-3.** The reaction producing the greatest mass of $O_2(g)$ per gram of reactant is the one having the reactant of lowest molar mass. Clearly that reactant is $NH_4NO_3(s)$ in reaction **(a)**. **4-4 (a)** Tripling the solution volume reduces the molarity to 1/3 of its initial value: 0.050 M NaCl. **(b)** Reducing the volume from 250.0 mL to 200.0 mL increases the molarity by 5/4, that is, to 1.000 M $C_{12}H_{22}O_{11}$. **(c)** The molarity of the first solution is reduced to 1/3, to 0.0900 M KCl; that of the second is reduced to 2/3, to 0.0900 M KCl. The total molarity of the final solution is 0.180 M KCl. **4-5.** The balanced equation is 4 $NH_3(g)$ + 5 $O_2(g)$ $\longrightarrow$ 4 NO(g) + 6 $H_2O(l)$. Starting with 1.0 mol each of NH_3 and O_2, the limiting reactant is O_2. The amounts of products are 0.8 mol NO and 1.2 mol H_2O. The only true statement is **(d)** all the $O_2(g)$ is consumed. **4-6.** The answer must be consistent with the following facts.

(1) The factor 0.90 must appear twice in the setup (which can include $0.90 \times 0.90 = 0.81$); (2) CH_3Cl is an intermediate and does not enter in; (3) the only molar masses needed are those of CH_4 (16 g/mol) and CH_2Cl_2 (85 g/mol). The correct response is (a).

CHAPTER 5

Concept Assessment 5-1 (1). (e) 0.025 M $RbNO_3$, the only strong electrolyte of the group. **(2). (a)** a strong electrolyte with a total ion concentration of 0.024 M. **5-2.** Predicted as soluble based on guidelines in Table 5.1: **(a), (c), (g)**. Predicted as insoluble based on Table 5.1: **(b), (e), (h), (i)**. Inconclusive based on Table 5.1: **(d)** Li_2CO_3 might be soluble (Li^+ is a group 1 cation), but also might be one of the exceptions referred to in Table 5.1. **(f)** No data are available in Table 5.1 concerning permanganates, so the solubility of $Mg(MnO_4)_2$ is uncertain. **5-3.** $H_2O(l)$ is suitable for $K_2CO_3(s)$ and $ZnSO_4(s)$, and $HCl(aq)$ is suitable for $CaO(s)$ and $BaCO_3(s)$. $H_2SO_4(aq)$ would not be suitable for $CaO(s)$ and $BaCO_3(s)$ because $CaSO_4(s)$ and $BaSO_4(s)$ might precipitate. **5-4.** No reaction would occur in **(a)**. Two reduction half-reactions and no oxidation half-reaction. Reaction could occur under appropriate conditions in **(b)** because $Cl_2(g)$ undergoes both oxidation and reduction. **5-5.** This can occur as seen, for example, in the reverse of the disproportionation reaction directly above on this page. **5-6.** An inaccurate statement. An oxidizing agent is necessary to oxidize $Cl^-(aq)$ to $Cl_2(g)$, but neither $HCl(aq)$ or $NaOH(aq)$ is an oxidizing agent. **5-7.** An exactly neutral $NaCl(aq)$ formed when 3.11×10^{-3} mol each of H^+ and OH^- ions neutralize one another.

CHAPTER 6

Concept Assessment 6-1. A water siphon passes over a "hump" from a pool of water at a higher level to a receiver at a lower level. After the siphon is initially filled with water, the pressure of the atmosphere pushes water over the hump, beyond which the water flows freely. In a suction pump, air pressure pushes water up a partially evacuated pipe. **6-2. (b) 6-3. (a) 6-4.** The direct proportionality of V and T must be based on an absolute temperature scale. While a change from 100 K to 200 K produces a doubling of V, a change from 100 °C to 200 °C produces only a 27% increase: $[(200 + 273)/(100 + 273)] = 1.27$. **6-5.** Consider these facts. **(1)** V is directly proportional to the number of moles of $O_2(g)$, and this must be 60.0 g $O_2/32.00$ g O_2 mol^{-1}. Only responses **(c)** and **(d)** meet this requirement. **(2)** The effect of changing P and T on the STP-volume must be expressed through the product $[(760 \text{ mmHg}/825 \text{ mmHg}) \times (303 \text{ K}/273 \text{ K})]$. The only correct response is **(d)**. **6-6. (a)** 0.667 L $SO_2(g)/1.00$ L $O_2(g)$, because the actual P and T are immaterial as long as the two gases are compared at the same t and P. **(b)** The 0.667 L $SO_2(g)$ has to be adjusted for the an increase in T (by the factor 298 K/273 K) and a decrease in P (by the factor 760 mmHg/745 mmHg), leading to $V = 0.743$ L $SO_2(g)$. **6-7.** The correct

responses—**(b)** and **(e)**—follow from basic ideas. Dalton's law of partial pressures dictates that P_{He} is not affected by any other gases present, and addition of 0.50 mol H_2 (1.0 g) will increase the total mass of gas by 1.0 g, independent of anything else that may happen. By simple estimates the other three statements can be shown to be false. **6-8.** He(g) at 1000 K has twice the u_{rms} as at 250 K (change T to $4T$ in equation 6.20). At 250 K, u_{rms} of $H_2(g)$ exceeds that of He(g) by the factor $\sqrt{2}$ (change M to $\frac{1}{2} M$ in equation 6.20). The two-fold increase in the first case exceeds the $\sqrt{2}$-fold increase in the second. He(g) at 1000 K has a greater u_{rms} than $H_2(g)$ at 250 K. **6-9.** The correct responses are **(a)** and **(c)**. The average kinetic energy of gas molecules depends only on T, and to two significant figures the mass of 0.50 mol He is the same as that of 1.0 mol H_2. **6-10.** Rearrange equation (6.14) to the form $R = MP/dT$. Substitute molar masses, $P = 1$ atm (exactly), $T = 293.2$ K, and the density data. Solve for three values of R and see how closely they conform to the ideal gas constant $R = 0.08206$ L atm mol^{-1} K^{-1}. Increasing adherence to ideal gas behavior: $OF_2(R = 0.0724) < NO(R = 0.08194) < O_2(R = 0.08200)$.

CHAPTER 7

Concept Assessment 7-1. Dynamite exploding in an underground cavern is a close approximation to an isolated system. Titration of an acid with a base is an open system. A steam-filled cylinder in a steam engine with all valves closed constitutes a closed system. **7-2.** Basic principle: law of conservation of energy. Assumptions: no heat loss to surroundings, d and sp. ht. of $H_2O(l)$ independent of T. Because the mass of hot water is twice that of the colder water, the initial temperature difference of 60.00 °C is divided into a 40.00 °C warming of the cold water and a 20.00 °C cooling of the hot water; final $T = 50.00$ °C. **7-3.** The ΔT of a fixed mass of substance is inversely proportional to its specific heat; thus the object with the smaller ΔT has the greater specific heat. The second question requires us to recognize the difference in enthalpy of transition for the solid and liquid form of water. The enthalpy of fusion for ice is less than the enthalpy of vaporization for the liquid, meaning that the amount of heat required to vaporize water is greater than that for ice. **7-4.** This is accomplished by adding a measured amount of a substance in which the heat of reaction is known. **7-5.** This is a closed system. Since the pressure dropped while the volume remained constant then the temperature must have decreased. The internal energy of the system decreased. Therefore the energy transferred across the boundary was in the form of heat. The direction of energy transfer was from the system to the surroundings. **7-6.** The balloon feels warm because the dissolution of $NH_3(g)$ in $H_2O(l)$ is exothermic, $q < 0$. The balloon shrinks because the atmosphere (surroundings) does work on the system, $w > 0$. **7-7.** In the bottom row T is uniform throughout the object while in the top row the object is hotter

at the edges than in its interior. Heating in the top row is irreversible; the process is far from equilibrium. The bottom row represents reversible heating; removal of just a tiny amount of heat can change heating to cooling. **7-8.** Enthalpy is a function of state. When a process returns a system to its initial state H returns to its initial value, meaning that $\Delta H = 0$. **7-9.** The enthalpy change in forming 1 mol $C_2H_2(g)$ from its elements is represented by the top line; that for forming 1 mol $C_2H_4(g)$ is the next line down. Both of these lines are above the broken line representing $\Delta H = 0$. The formation of 1 mol C_2H_6 has $\Delta H < 0$ and is the first line below the broken line. $\Delta H°$ for the reaction of interest is represented by the distance between the first and third lines. **7-10.** Yes, this can be done. The additional data needed are heat capacities as a function of T. The procedure is outlined in Figure 7-16.

CHAPTER 8

Concept Assessment 8-1. The wavelength of red light is about 700 nm (see Fig. 8-3). Since frequency and wavelength are reciprocally related, doubling ν halves λ. The frequency-doubled light will have a wavelength about 350 nm—light in the near ultraviolet not visible to the human eye. **8-2.** By studying emission spectra from the collision, scientists hope to identify the elements present in the comet and also on Jupiter's surface. **8-3.** The threshold wavelength is 91.2 nm. Using 70.0 nm light as compared to 80.0 nm light produces more energetic electrons. Each photon produces one electron, and the number of electrons produced depends on the intensity (number of photons) of the light, provided its wavelength is less than the threshold wavelength. **8-4.** The transition $n = 1$ to $n = 4$ corresponds to the greatest ΔE, but it involves *absorption* of a photon not emission. Photons are emitted in the other two transitions, with the transition $n = 4 \longrightarrow n = 2$ corresponding to the greater ΔE and hence shortest wavelength. **8-5.** If the wavelengths are the same then the momenta are the same (equation 8.10). The speed of the proton will have to be 1/2000th of the speed of the electron (that is, $m_p \times u_p = m_e \times u_e$, and $u_p = u_e \times m_e/m_p = u_e \times 1/2000$). **8-6.** The state $n = 2$ has a peak $\frac{1}{4}$ of the length from either end of the box, corresponding to the greatest probability of the particle being at those points. **8-7.** An orbital with three angular nodes has $\ell = 3$; it is an f orbital. One radial node makes for a total of four nodes, and since the total number of nodes is $n - 1$, n must be 5. The orbital is a $5f$. **8-8.** The compound is arsenic. The ground state is $4s^2 3d^{10} 4p^3$ with all three arrows pointing in the same direction, one in each box. The anion is adding one more arrow to one of the $4p$ boxes pointing in the opposite direction.

CHAPTER 9

Concept Assessment 9-1. (a) Ne **(b)** N^{3-} or P^{3-} **(c)** Zn^{2+}, Cd^{2+}, or Hg^{2+} **9-2.** Z_{eff} increases and atomic radius decreases with increasing Z. The blue axis represents Z_{eff} and the blue line, Z_{eff} as a function of Z. The red axis represents atomic radius and the red line, atomic radius as a function of Z. **9-3. (a)** B (at the top of group 13) **(b)** Cl (at the right end of the third period) **(c)** P^{3-} in period 3, group 15 (strong electron repulsions in an anion of high negative charge) **(d)** Tl (at the bottom of group 13) **9-4. (a)** C (smallest group-14 atom, at the top of the group) **(b)** Kr (noble gas element in group 18) **(c)** Se (lower ionization energy than Br based on the expected trend; lower than As for the same reason as in the P/S comparison on page 377) **9-5. (a)** group 17 (the smallest atoms in their periods) **(b)** group 2 (a filled ns subshell and essentially no affinity for an additional electron) **(c)** group 18 (noble gases have all shells and subshells closed) **9-6 (a)** scandium (Sc^{3+} has noble gas electron configuration) **(b)** tellurium (Te^{2-} has a noble gas electron configuration) **(c)** manganese (Mn^{2+} has the electron configuration $[Ar]3d^5$) **9-7 (a)** thallium **(b)** nitrogen **(c)** rubidium **(d)** iodine **(e)** aluminum

CHAPTER 10

Concept Assessment 10-1. The first and last symbols are acceptable; each has six dots with two unpaired. The unacceptable symbols have seven and five dots. **10-2.** The bonds are all covalent, with one being coordinate covalent. **10-3.** groups 14, 15, and 16 (for example, the elements C, N, O, P, and S) **10-4. (a)** Br **(b)** Be **(c)** P **10-5.** If covalent bonds between atoms involve equal contributions from all the bonded atoms, there are no formal charges. Where coordinate covalent bonds are formed there will be formal charges. A polyatomic ion must have at least one atom with a formal charge, consistent with the charge on the ion. **10-6.** We can draw two possible Lewis structures. One structure has no formal charges. In the other structure, one of the O atoms (the one bonded to H) has a formal charge of +1 and the other has a formal charge of −1. The structure with formal charges is considered to be unimportant and so we never represent the structure of CH_3CO_2H as a resonance hybrid. **10-7.** The structure of the SO_2 molecule is

best represented as $\ddot{\text{O}}=\text{S}=\ddot{\text{O}}$, and

so the sulfur–oxygen bonds are best thought of as double bonds. **10-8.** ICl_2^- is a *linear* anion with five electrons pairs around the I atom (AX_2E_3). ICl_2^+ is a *bent* cation with four electron pairs around the I atom (AX_2E_2). The difference of one pair of electrons produces a completely different electron-group geometry and geometric shape. **10-9.**

$$H_3C-\ddot{\text{N}}=\text{C}=\ddot{\text{O}} \longleftrightarrow$$

Most satisfactory

$$H_3C-\text{N}\equiv\text{C}-\ddot{\text{O}}: \longleftrightarrow$$
$$\phantom{H_3C-\text{N}}{\scriptstyle(+1)}\phantom{\text{C}-\text{O}}{\scriptstyle(-1)}$$

$$H_3C-\ddot{\text{N}}-\text{C}\equiv\text{O}:$$
$${\scriptstyle(-1)}\phantom{-\text{C}\equiv}{\scriptstyle(+1)}$$

Least satisfactory

10-10. In NH_3 the lone pair of electrons on the N atom pulls electron density away from the H atoms, creating a large resultant dipole moment. In NF_3 the highly electronegative F atoms pull electron density away from the N atom, producing highly polar N—F bonds that counteract the effect of the nitrogen lone pair and a greatly reduced resultant dipole moment. **10-11.** Both the linear NO_2^+ cation and bent NO_2^- anion exhibit resonance that involves double bond character in the N—O bonds. However, electrons in the NO_2^+ cation will be more tightly held by the center of positive charge, leading to shorter N-to-O bond lengths in NO_2^+ than in NO_2^-.

CHAPTER 11

Concept Assessment 11-1. The cation CH_3^+ is isoelectronic with BH_3 and has three pairs of electrons around the C atom; we expect sp^2 hybridization. In the anion CH_3^- there are four electron pairs suggesting sp^3 hybridization, as in CH_4. **11-2.** The sp^3d^2 hybridization scheme corresponds to six electron pairs around a central atom. Similar to PF_5 in the period above it, we expect the compound AsF_5. Now imagine adding F^- to AsF_5 to create $[AsF_6]^-$, which has 6 electron pairs around the As atom and requires sp^3d^2 hybridization. **11-3.** Five bonding electron groups can be accommodated by sp^3d hybridization, but the distribution would be trigonal bipyramidal. What is needed is sp^3d^2 hybridization in the species AX_5E. The lone-pair electrons are directed to a corner of an octahedron, and the remaining five positions determine the molecular geometry—square pyramidal. **11-4.** To complete the octets of the N atoms, they must retain a lone pair of electrons, form a double bond between themselves, and a single bond to a H atom. The hybridization of the N atoms is sp^2. **11-5.** The H_2^+ ion is formed by removal of one electron from H_2—a larger energy requirement than promoting a σ_{1s} electron to the σ_{1s}^* MO in the excited state of H_2. On the other hand, the bond order in H_2^+ is 0.5 and 0 in the excited state of H_2. H_2^+ is a stable species and the excited state of H_2 is not. **11-6.** No. For example, the double bond in C_2 is made up of two π bonds and no σ bond (see Figure 11-26). **11-7.** The molecule NeO is isoelectronic with F_2 and should have a bond order of 1. We expect it to be stable, but it has never been observed. **11-8.** In HCO_2^- three atoms provide p orbitals for π bonding, just as in ozone. In the NO_3^- anion four atoms provide p orbitals. The delocalized π bonding in HCO_2^- is different than in NO_3^-. **11-9.** GaN, a combination of group 13 and group 15 elements, is a semiconductor; this combination is equivalent to a group 14 semiconductor like Si or Ge.

CHAPTER 12

Concept Assessment 12-1. The intermolecular interactions are both London dispersion forces and hydrogen bonds. In substances with small molecules, hydrogen bonding usually dominates. **12-2.** Because the ball drops faster through it, the 10W oil is less viscous than the 40W oil. Viscosity is inversely proportional to

T, and the lower weight oil (10W) is preferred for low-temperature use (where higher weight oils might solidify). In hot desert regions higher weight oils (40W) are preferred because lightweight oils might become so mobile as to lose their lubricating properties. The strengths of intermolecular forces are directly related to viscosity, and hence the higher viscosity 40W oil has the stronger intermolecular forces. **12-3.** Because of the different elevation (and barometric pressure) between landlocked, mountainous Switzerland and sea-level Manhattan Island, the lower boiling temperature results in a longer cooking time. **12-4.** Hydrogen bonding occurs in NH_3 but not in N_2, resulting in stronger intermolecular attractions and consequently lower vapor pressures, a higher boiling point, and a higher critical temperature in NH_3 than in N_2. **12-5.** The greater number of electrons (much larger molar mass) in CCl_4 causes the intermolecular attractions (London dispersion forces) to outweigh the effect of the polar bonds in CH_3Cl. **12-6.** Dew forms in the condensation of $H_2O(g)$ to $H_2O(l)$ and frost from the deposition of $H_2O(g)$ as $H_2O(s)$. Both processes are exothermic and give off heat to the surroundings—more heat in frost formation, because ΔH(deposition) = ΔH(condensation) + ΔH(freezing). **12-7.** Wet books are placed in a cold, evacuated chamber. Moisture in the books freezes and the ice that is formed sublimes to $H_2O(g)$. This process avoids heating and involves a minimum of handling of the damaged books. **12-8.** According to the Bragg equation, $n\lambda = 2d\sin\theta$, if the extra distance traveled by the diffracted wave ($2d\sin\theta$) is to remain the same when n is doubled, the wavelength of the wave must be halved, so that $2n(\lambda/2) = n\lambda$. The required multiple is $\frac{1}{2}$. **12-9.** The fcc unit cell contains four C_{60} molecules. The unit cell has four octahedral and eight tetrahedral holes occupied by 12 K atoms. The formula based on the unit cell is $K_{12}(C_{60})_4$, and the molecular formula is $K_3(C_{60})$.

CHAPTER 13

Concept Assessment 13-1. 3.011×10^{23}. **13-2.** Concentrations are independent of temperature if based solely on mass or temperature-independent properties related to mass, specifically, mass percent, molality, mole fraction, and mole percent. Concentrations based on volumes—volume percent and molarity—are temperature dependent. **13-3.** HCl is undissociated in $C_6H_6(l)$, and the concentration of HCl in $C_6H_6(l)$ should closely follow $P_{HCl(g)}$ above the solution. On the other hand, HCl(g) reacts with $H_2O(l)$ to produce $H_3O^+(aq)$ and $Cl^-(aq)$. The relationship between $P_{HCl(g)}$ and the aqueous concentrations of ions is more complex. **13-4.** Start with $(P_A^\circ - P_A)/P_A^\circ = x_B$. Note that $(P_A^\circ - P_A)/P_A^\circ = 1 - (P_A/P_A^\circ)$. According to Raoult's law, $P_A = x_A P_A^\circ$, which means that $P_A/P_A^\circ = x_A$. Thus, we arrive at the true statement that $1 - x_A = x_B$. **13-5.** This will happen if both components in an ideal solution have the same vapor pressure. The vapor pressure of the solution will be independent of the solution composition and the line will be parallel to the composition axis. The likelihood of this happening is not very great, although it might be found at one particular temperature where the vapor pressure curves of two liquids cross. **13-6.** The similarities are that two different solutions are involved, water is transported from the more dilute to the more concentrated solution, and the process continues until the two solutions have the same concentration. The chief difference is that water is transported via the vapor phase in Figure 13-16a and through a semipermeable membrane in Figure 13-17. **13-7.** If Figure 13-19 were based on water rather than some other solvent, the two fusion curves would have *negative* rather than positive slopes. However, there would still be a freezing-point depression and a boiling-point elevation. **13-8.** When enough NaCl is present to depress the freezing point to -21 °C, the NaCl(aq) is saturated. Any solute added beyond this point remains as undissolved NaCl(s) and can have no further effect on the freezing point of the solution.

CHAPTER 14

Concept Assessment 14-1. In the reaction $N_2(g) + 3 H_2(g) \longrightarrow 2 NH_3(g)$, compared to the rate of disappearance of N_2, the rate of disappearance of H_2 is three times as great and the rate of formation of NH_3 is twice as great. **14-2.** If the initial and instantaneous rates of reaction are initially equal and remain so throughout a reaction, the concentration-time graph must be a straight line with a negative slope, as seen in Figure 14-3. **14-3.** If the reaction were first order the initial rate would double and if second order, quadruple, thus $1 < \text{order } (m) < 2$. More precisely, $2^m = 2.83$ and $m = 1.50$ (solve this equation for m: $m \log 2 = \log 2.83$). **14-4.** All four graphs can be plotted on the same sheet of paper. For example, take $[A]_0 = 3.0$ M and $k = 0.20 \text{ s}^{-1}$ as the larger of two rate constants and $k = 0.10 \text{ s}^{-1}$ as the smaller. Both concentration vs. time graphs will resemble Figure 14-5, but after starting at the same point, $[A]_0 = 3.0$ M, the one with the larger k has a shorter half-life and falls off more rapidly than the other. The two plots of $\ln k$ vs. t are straight lines, both starting at $\ln [A]_0 = \ln 3.0 = 1.10$, and having negative slopes, the steeper slope for the larger value of k. **14-5. (a)** If the plot was linear, the reaction would be zero order; **(b)** Look at successive half-lives. If $t_{1/2}$ is constant, the reaction is first order; **(c)** Look at successive half-lives. If $t_{1/2}$ doubles each time, the reaction is second-order. **14-6 (a)** This condition can exist. The reaction is exothermic (similar to Figure 14-10). **(b)** This condition can exist, and the reaction is endothermic (imagine flipping the reaction profile in Figure 14-10 from left to right). **(c)** This condition cannot exist; E_a for an exothermic cannot be less than ΔH. **(d)** This condition can exist; its only distinction is that there is no heat of reaction. **(e)** This condition cannot exist; E_a cannot be negative. **14-7.** Consider this equation from Figure 14-12: $E_a = R \times (-\text{slope of } \ln k \text{ vs. } 1/T)$. The greater

the value of E_a, the greater the slope of the graph and the more rapidly the rate of reaction changes with temperature. **14-8.** The means are not the same. The increase in reaction rate caused by the presence of a catalyst is most likely because of a different reaction mechanism that lowers the reaction barrier. The increase in the rate of reaction caused by an increase in temperature is due to more molecules with kinetic energy greater than the barrier; more collisions occur per unit time.

CHAPTER 15

Concept Assessment 15-1. (a) represents solubility-phase equilibrium; **(b)** phase equilibrium; **(c)** chemical equilibrium.

15-2. $Q = \dfrac{a_{Cu^+(aq)} a_{H_2(g)}}{a_{Cu(s)} a_{H^+(aq)}^2} = \dfrac{[Cu^{2+}] P_{H_2}}{[H^+]^2}$.

15-3. Into the expression $K = [B]/[A]$, substitute $[B] = 54 - [A]$ and the given value of K; solve for $[A]$ and $[B]$. If $K = 0.02$, $[A]$ (open circles) $= 53$ and $[B]$ (filled circles) $= 1$. If $K = 0.5$, $[A] = 36$ and $[B] = 18$. If $K = 1$, $[A] = [B] = 27$. **15-4.** If $K > 1$ for the 2nd reaction, K for the 1st reaction will be the larger of the two, but if $K < 1$ for the 2nd reaction, K for the 1st reaction will be the smaller of the two. **15-5.** Reverse given equation (invert its K value). To that equation add $CH_4(g) + H_2O(g) \rightleftharpoons CO(g) + 3 H_2(g)$; $CO(g)$ cancels and the overall equation is the one we seek; its K is the ratio of the other two K values. **15-6.** The balanced equation is sufficient to determine the outcome of a reaction that goes to completion. If the reaction is reversible and reaches a state of equilibrium, the value of K is required as well. **15-7. (a)** incorrect: would require that $CO(g)$ and $H_2O(g)$ be completely consumed— impossible with $K_p = 10.0$. **(b)** incorrect: would be violation of the law of conservation of mass. **(c)** incorrect; would require the consumption of some $CO_2(g)$, but the direction of net change must be in the forward reaction. **(d)** correct: an outcome that would result from a net change in the forward direction. **(e)** incorrect: sufficient data are given to calculate the composition of the equilibrium mixture. **15-8.** Even though the pressure increases because of the addition of an inert gas, the reaction will shift to the right since the volume of the reaction vessel decreased. **15-9. (a)** True—more $H_2(g)$ will form at the expense of the $H_2S(g)$ and $CH_4(g)$. **(b)** False— an inert gas has no effect on a constant-volume equilibrium condition. **(c)** True—K changes with T and so does the composition of the equilibrium mixture. **(d)** Uncertain— the partial pressures of $H_2S(g)$ and $CH_4(g)$ will rise because a net reaction occurs to the left, but the increase in partial pressures of $CS_2(g)$ and $H_2(g)$ caused by forcing these two gases into a smaller volume will be at least partly offset by the equilibrium shift to the left. **15-10.** Equilibrium shifts in the forward direction, the endothermic reaction. Student B, by holding the beaker, stimulates heat flow into the reaction mixture, probably achieving a higher yield of product.

CHAPTER 16

Concept Assessment 16-1. (a) is a conjugate acid/base pair; HCO_3^- can transfer a proton to a base (e.g., OH^-) yielding CO_3^{2-}, and CO_3^{2-} can react with an acid (e.g., H_3O^+) to reform HCO_3^-. **(b)** is not a conjugate acid/base pair; SO_4^{2-} can be produced from HSO_3^- only through an oxidation process not in an acid–base reaction. **(c)** is not a conjugate acid/base pair; it is a pair of unrelated acids. **(d)** yes; **(e)** no. **16-2.** With pH $= \log[H_3O^+]$, the vast majority of solutions would have negative pH values. It is more convenient to incorporate the negative sign in the definition than to carry it in individual pH values. With pH $= -\ln[H_3O^+]$ the close relationship between pH and the powers of ten used in scientific notation would be completely lost. (The desire to establish this relationship is why the pH concept was devised in the first case.) **16-3.** A concentrated solution of a weak acid may often have a lower pH than a dilute solution of a strong one. For example, the pH of 0.10 M $HC_2H_3O_2$— calculated as pH $= 2.89$ in Example 16-6—is lower than the pH of 0.0010 M HCl. **16-4.** The bottle labeled $K_a = 7.2 \times 10^{-4}$ contains the more acidic solution. The bottle labeled $K_a = 1.9 \times 10^{-5}$ has the acid with the larger pK_a. The relevant equations are $^+NH_3CH_2CH_2NH_3^+(aq) + H_2O(l) \rightleftharpoons$ $H_3O^+(aq) + NH_2CH_2CH_2NH_3^+(aq) pK_1 = 6.85$ and $NH_2CH_2CH_2NH_3^+(aq) + H_2O(l) \rightleftharpoons H_2O(l) \rightleftharpoons H_3O^+(aq) +$ $NH_2CH_2CH_2NH_2(aq) pK_2 = 9.92$. From equation (16.18), the values of the base ionization constants are $pK_{b_1} = 14.00 - 9.92 = 4.08$ and $pK_{b_2} = 14.00 - 6.85 = 7.15$. The base ionization reactions are $NH_2CH_2CH_2NH_2(aq) + H_2O(l) \rightleftharpoons$ $NH_2CH_2CH_2NH_3^+(aq) + OH^-(aq) pK_{b_1} = 4.08$ and $NH_2CH_2CH_2NH_3^+(aq) + H_2O(l) \rightleftharpoons$ $^+NH_3CH_2CH_2NH_3^+(aq) + OH^-(aq) pK_{b_2} = 7.15$ $^+NH_3CH(CH_3)COOH + H_2O \rightleftharpoons$ $NH_3CH(CH_3)COO^- + H_3O^+ pK_a = 2.34$ $^+NH_3CH(CH_3)COO^- + H_2O \rightleftharpoons$ $NH_2CH(CH_3)COO^- + H_3O^+ pK_a = 9.87$ $pK_{b_1} = 14.00 - 2.63 = 11.37$ $pK_{b_2} = 14.00 - 9.87 = 4.13$ $NH_2CH(CH_3)COO^- + H_2O \rightleftharpoons$ $^+NH_2CH(CH_3)COO^- + OH^- \quad pK_{b_1} = 4.13$ $^+NH_3CH(CH_3)COO^- + H_2O \rightleftharpoons$ $^+NH_3CH(CH_3)COOH + OH^- \quad pK_{b_1} = 11.37$. **16-6.** Consider HPO_4^{2-}, which can act as an acid: $HPO_4^{2-}(aq) + H_2O(l) \rightleftharpoons H_3O^+(aq) +$ $PO_4^{3-}(aq)$, $K_a = 4.2 \times 10^{-13}$ or as a base: $HPO_4^{2-}(aq) + H_2O(l) \rightleftharpoons H_2PO_4^-(aq) +$ $OH^-(aq) K_b = K_w/K_a = 1.00 \times 10^{-14}/$ $4.2 \times 10^{-13} = 2.4 \times 10^{-2}$. Because K_b is much greater than K_a, $HPO_4^{2-}(aq)$ is basic. In a similar way, $H_2PO_4^-(aq)$ is seen to be acidic. Thus, depending on K values, ions in aqueous solutions may have pH values ranging from acidic to neutral to basic. **16-7.** We should expect pK_a for *ortho*-chlorophenol to be smaller than for phenol because of the electron-withdrawing effect of the Cl atom. (Its pK_a is 8.55 compared with 10.00 for phenol.) **16-8.** Picture three Br atoms joined by single bonds to a Fe(III) atom on which there is also a lone pair of electrons: Br_3Fe: Now imagine that a Br_2 molecule dissociates into Br^+ and Br^- ions. The electron-deficient Br^+, a Lewis acid, attaches to the lone-pair electrons of Fe(III), a Lewis base, forming $[FeBr_4]^+$. The final product is $[FeBr_4]^+Br^-$.

CHAPTER 17

Concept Assessment 17-1. (a) no; NH_4Cl lowers the pH through the common-ion effect **(b)** yes, but only slightly. Diethylamine is a somewhat stronger base than NH_3, but only a rather small amount is being added. **(c)** no; HCl is a strong acid that will neutralize some of the NH_3, producing an aqueous solution of NH_3 and NH_4Cl. **(d)** no; since the added $NH_3(aq)$ is more dilute than the 0.10 M $NH_3(aq)$ the overall solution will be 0.075 M $NH_3(aq)$. **(e)** yes; $Ca(OH)_2(s)$ is a strong base. **17-2.** benzoic acid/benzoate with a 1:2 ratio. **17-3. (a)** yellow; a low pH **(b)** yellow; a CH_3COOH/CH_3COO^- buffer solution is formed but its pH is about 5 **(c)** yellow; the buffer completely neutralizes the small amount of added OH^- **(d)** red; the buffer capacity is exceeded and the solution becomes basic. **17-4.** Choice **(c)** is the correct one; 0.60 mol $NaCH_3COO$, converts all the HCl to CH_3COOH, producing a $CH_3COOH/NaCH_3COO$ buffer solution of pH $\approx$ 4. Choices **(a)** and **(b)** have essentially no effect on the pH, and while choice **(d)** would neutralize 80% of the acid, the amount of strong acid remaining would still produce a pH $\approx$ 1. **17-5.** This is the titration of a weak base that ionizes in two stages. The titration curve would begin at a moderately high pH; the pH would drop during the titration, and there would be two equivalence points. In general, the curve would resemble that in Figure 17-13, but flipped from bottom to top. The two buffer regions and pH $= pK_b$ values would be in segments of the curve between the two equivalence points. **17-6. (a)** six species: $K^+, H_3O^+, I^-, CH_3COO^-, OH^-, CH_3COOH$ **(b)** most abundant, K^+ (the spectator ion in highest concentration); 2nd most abundant, CH_3COO^- (produced in the neutralization of 3/4 of the CH_3COOH) **(c)** least abundant, OH^- (the final solution is acidic, so $[OH^-] < 10^{-7}$ M); 2nd least abundant, H_3O^+ (final solution a buffer with pH $\approx$ 5)

CHAPTER 18

Concept Assessment 18-1. Because of the large excess of $MgF_2(s)$ the solution would remain saturated even when the volume of solution is doubled; $[Mg^{2+}]$ remains constant. **18-2.** CaF_2 and AgCl are insoluble with $K_{sp} = 5.3 \times 10^{-9}$ and $K_{sp} = 1.8 \times 10^{-10}$, respectively. Since K_{sp} values for $CaCl_2$ and AgF are not found in tables, they can be assumed to be soluble compounds. **18-3.** It is largely immaterial. The only requirement is that the $AgNO_3(aq)$ be concentrated enough to bring about the precipitation without unduly diluting the solution from which the precipitation occurs. **18-4.** It will affect the solubility of CaF_2 more. The formation of F^- is derived from the weak acid HF, and so it will undergo hydrolysis whose equilibrium can be changed on the addition of an acid or base. **18-5. (a)** At

pH = 10.00, $[OH^-]=1.0 \times 10^{-4}$ M and $[Mg^{2+}] = 1.8 \times 10^{-3}$ M, an entirely plausible result. **(b)** At pH = 5.00, $[OH^-] = 1 \times 10^{-9}$ M and $[Mg^{2+}] = 1.8 \times 10^7$ M, an impossible result. The situation here is that at pH = 5.00 the solution is not one of $Mg(OH)_2$. It is $MgCl_2(aq)$, and $[Mg^{2+}]$ in this solution depends on the solubility of $MgCl_2$. **18-6.** As expected, with $[Cl^-] = 0.0039$ M, the molar solubility of AgCl(s) is less than in pure water because of the common-ion effect. At higher concentrations of $Cl^-(aq)$, AgCl(s) becomes more soluble because of complex ion formation: $AgCl(s) + Cl^-(aq) \longrightarrow [AgCl_2]^-(aq)$. **18-7.** $Pb^{2+}(aq)$ is present because of the formation of the yellow $PbCrO_4(s)$; $Hg_2^{2+}(aq)$ is absent because of the negative test for that ion; the presence of $Ag^+(aq)$ is uncertain because it really wasn't tested for. **18-8.** We must consider the formation of $Pb(OH)_3^-$, $K_f = 3.8 \times 10^{14}$, in the solution. **18-9.** Will not work because both ions would precipitate: $(NH_4)_2CO_3, H_2S(aq)$, and $NaOH(aq)$; will not work because neither ion would precipitate: $HNO_3(aq)$ and $NH_3(aq)$. $HCl(aq)$ will work because $CuCl_2$ is water soluble and AgCl is not.

CHAPTER 19

Concept Assessment 19-1. No, spontaneous and nonspontaneous refer to the thermodynamics of a process, not the kinetics. A nonspontaneous process will not occur without external intervention, and a spontaneous reaction is not necessarily fast; it can occur very slowly. **19-2.** Doubling the volume available to the gas in Figure 19-1. is equivalent to doubling the length of the box from L to $2L$ in Figure 19-3a. The expansion of the gas in Figure 19-1 seems driven by a tendency to fill all the available volume. The gas expansion can also be explained, however, as the tendency of the system energy to be distributed among the greater number of available energy levels in the $2L$ box compared to the L box. **19-3.** ΔH as a function of T is a straight line with only a slight slope (positive or negative), in the negative energy region. The $T\Delta S$ line, in the same energy region, has a steep negative slope and intersects the ΔH line. The distance between the two lines ($\Delta H - T\Delta S$) represents ΔG. At the point of intersection, $\Delta G = 0$, at T below the intersection $\Delta G < 0$, and at T above the intersection $\Delta G > 0$. **19-4.** $G° = 326.4$ kJ mol^{-1} means that the Gibbs energy change for the system is 326.4 kJ when 3 mol O_2 is converted into 2 mol O_3. If 1.75 mol O_2 reacts, then the Gibbs energy change for the system is (326.4 kJ/3 mol O_2) × (1.75 mol O_2) = 190.4 kJ. **19-5.** $H_2O(l, 1 \text{ atm}) \rightleftharpoons$ $H_2O(g, 1 \text{ atm})$, is a process in which $\Delta H > 0$ and $\Delta S > 0$, as represented by Figure 19-9. Below 100 °C, condensation of $H_2O(g, 1 \text{ atm})$ is favored and at 100 °C (the normal boiling point) condensation and vaporization are at equilibrium and $\Delta G = 0$. At 120 °C, vaporization predominates, $\Delta G < 0$, and $T \Delta S > \Delta H$.

CHAPTER 20

Concept Assessment 20-1. At the anode, Zn(s) is oxidized to $Zn^{2+}(aq)$ and, to preserve

charge balance, $NO_3^-(aq)$ migrates in from the salt bridge. At the cathode, $Cu^{2+}(aq)$ is reduced to Cu(s) and, to preserve charge balance, $K^+(aq)$ migrates in from the salt bridge. **20-2.** No changes in mass at the inert Pt(s) electrodes; a gain in mass at the Cu(s) electrode through the half reaction $Cu^{2+}(aq) + 2e^- \longrightarrow$ Cu(s); and a loss in mass at the Zn(s) electrode through the half reaction $Zn(s) \longrightarrow$ $Zn^{2+}(aq) + 2e^-$. **20-3.** Standard-state conditions for $ClO_4^-(aq)$ and $H^+(aq)$ are $a \approx 1$ M; for $Cl_2(g)$, $a = 1$ bar ≈ 1 atm; for $H_2O(l)$, $a = 1$. **20-4.** The cell with $E° > 0$ proceeds toward the formation of more products. A net reaction also occurs in the case where $E° < 0$, but in the reverse direction; the concentrations on the left side of the equation increase and those on the right decrease, until equilibrium is reached. **20-5.** $E_{cell} = E°_{cell}$ if all reactants and products are in their standard states, but also for any set of concentrations where $Q = 1$ in equation (20.18). **20-6.** The precipitate is $PbSO_4(s)$, thereby reducing $[Pb^{2+}]$ in the anode compartment and increasing the value of E_{cell}, so that $E_{cell} > E°_{cell}$. **20-7.** The cell diagram $Pt(s)|Cl_2(g, 1 \text{ atm})|$ $Cl^-(0.50 \text{ M}) \| Cl^-(0.10 \text{ M})|Cl_2(g, 1 \text{ atm})|Pt(s)$ has the net cell reaction: $0.50 \text{ M } Cl^-(aq) \longrightarrow$ $0.10 \text{ M } Cl^-(aq)$ and $E_{cell} = -0.0592$ V × $\log(0.10/0.50) = 0.041$ V. **20-8.** In a calomel electrode, reduction potential depends on the chloride concentration. Therefore, the standard reduction potential for a calomel electrode has a different chloride concentration from the saturated calomel electrode. **20-9.** Dry cells and lead-acid cells "run down" as the concentrations of reactants and products eventually reach their equilibrium values, where ΔG and E_{cell} both become 0. This does not happen in a fuel cell because fuel is continuously added. **20-10.** Both Al and Zn can be used because they are more active than Fe; Ni and Cu are less active and cannot be used.

CHAPTER 21

Concept Assessment 21-1. AlF_3 will have the higher melting point. **21-2.** The Na^+, K^+, Rb^+ and Cs^+ ions have relatively low charge densities and are better able to stabilize large, polyatomic anions such as NO_2^-. Because the Li^+ ion has a very high charge density and high polarizing power, it may kinetically assist the decomposition of polyatomic anions, such as N_3^- and N_2^-, to smaller anions such as O_2^-. Balanced chemical equations for the reactions are: $MNO_3(s) \xrightarrow{\Delta} MNO_2(s) + \frac{1}{2} O_2(g)$ (M = Na, K, Rb, Cs); $2 LiNO_3(s) \xrightarrow{\Delta} Li_2O(s) +$ $2 NO_2(g) + \frac{1}{2} O_2(g)$. **21-3.** The hybridization of Be changes from sp to sp^2 to sp^3 on going from $BeCl_2$ to $(BeCl_2)_2$ and $(BeCl_2)_n$ and the geometry around the Be atom changes from linear to trigonal planar to tetrahedral. **21-4.** A plausible reaction is that Mg(s) is oxidized to MgO(s) and $CO_2(g)$ is reduced to C(s), that is, $2 Mg(s) + CO_2(g) \longrightarrow$ $2 MgO(s) + C(s)$. **21-5.** For all but beryllium, heating the group 2 carbonates (MCO_3) yields the corresponding oxide (MCO); see equation (21.11). One method for preparing BeO is to burn Be(s) in O_2. **21-6.** The small size of Be^{2+} precludes the possibility of six H_2O molecules coordinating to the central Be^{2+} ion, whereas this happens readily with the Mg^{2+} ion. **21-7.** Lithium and magnesium form ions, Li^+ and Mg^{2+} that have high charge densities and are strongly polarizing. Presumably, when these ions are formed, their charge densities are large enough to stabilize the N^{3-} anion. Balanced chemical equations for the reactions are: $3 Li(s) + \frac{1}{2} N_2(g) \rightarrow Li_3N(s)$; $3 Mg(s) + N_2(g) \rightarrow Mg_3N_2(s)$. **21-8.** The structure of $B_2H_2(CH_3)_4$ is analogous to diborane (Fig. 21-11a) with two bridging H atoms and four methyl groups ($-CH_3$) in the terminal positions. **21-9.** AlF_3 is an electron-deficient compound and forms $[AlF_4]^-$ in the presence of F^- from KF. BF_3 is a stronger Lewis acid than AlF_3 and abstracts a F^- ion from $[AlF_4]^-$. The result: $[AlF_4]^- + BF_3 \longrightarrow AlF_3(s) + [BF_4]^-$. **21-10.** Determine $\Delta H°$ for each reaction using equation (7.21), the only difference is replacing 1 mol CO_2 in equation (21.29) by 1 mol CO in equation (21.30). The difference in the heat of reaction is -110.5 kJ/mol $CO(g)$ − $[-393.5$ kJ/mol $CO_2(g)] = 283$ kJ. Reaction (21.30) liberates 238 kJ less heat than reaction (21.29). **21-11.** Refer to pages 958 and 959 and imagine a chain of tetrahedra starting with SiO_4^{4-} and increasing in increments of SiO_3^{2-}, yielding SiO_4^{4-}, $Si_2O_7^{6-}$, $Si_3O_{10}^{8-}$ $Si_4O_{13}^{10-}$, $Si_5O_{16}^{12-}$, $Si_6O_{19}^{14-}$. Now bend the six-unit silicate chain into a hexagonal ring (similar to that in the third structure on page 958) by eliminating one O^{2-} between the ends of the chain. The result is the anion $Si_6O_{18}^{12-}$. In beryl, 3 Be^{2+} and 2 Al^{3+} provide the necessary 12 units of positive charge. **21-12.** Dissolved O_2 (g) in $Sn^{2+}(aq)$ is able to oxidize Sn^{2+} to Sn^{4+}. The following spontaneous reaction reduces tin (IV) to tin (II): $Sn^{4+}(aq) + Sn(s) \longrightarrow 2 Sn^{2+}(aq)$; $E°_{cell} = 0.154 V - (-0.137 V) = 0.017 V$. Thus, as long as an excess of Sn(s) is present, the $Sn^{2+}(aq)$ can be maintained with little or no $Sn^{4+}(aq)$ present.

CHAPTER 22

Concept Assessment 22-1. The larger EN difference and shorter bond length between Xe and F, compared to Xe and Cl, makes XeF_2 a more stable molecule than $XeCl_2$. **22-2.** The ions do not have the same shape. ICl_2^+ (VSEPR notation, AX_2E_2) has a tetrahedral electron-group geometry and a bent molecular shape. ICl_2^- (VSEPR notation, AX_2E_3) has a trigonal-bipyramidal electron-group geometry, and a linear molecular shape. **22-3.** $CuSO_4(aq)$ yields Cu(s) at the Pt cathode and $O_2(g)$ at the Pt anode, and $NaI(aq)$ yields I_2 at the anode and $H_2(g)$ at the cathode. The other three solutions—$H_2SO_4(aq)$, $NaOH(aq)$, and $KNO_3(aq)$—all yield $H_2(g)$ at the cathode and $O_2(g)$ at the anode. **22-4.** O_3 and O_3^- are both V shaped. The Lewis structures for these species suggest that the central O atom is sp^2 hybridized in O_3 and sp^3 hybridized in O_3^-; thus, the ideal bond angles are 120° for O_3 and 109° in O_3^-. Experiment shows that the $O-O-O$ bond angles in these two

molecular species are much closer: 117° in $O_3(g)$ and 114° in $KO_3(s)$. The experimental results suggest that valence bond theory not entirely satisfactory for describing the bonding O_3 and O_3^-, and a molecular orbital approach is more appropriate. (According to molecular orbital theory, the extra electron in O_3^- occupies an antibonding orbital. As a result, the oxygen–oxygen bonds in O_3^- are slightly longer than in O_3, but the bond angle is not significantly affected. **22-5.** The Cl in OCl_2 shows more positive character (blue color) than does F in OF_2 because the electronegativity of Cl is considerably less than that of F. **22-6.** The condensed structural formulas for phosphoric acid and phosphorous acid are $OP(OH)_3$ and $HPO(OH)_2$, respectively. **22-7.** The Br atom is larger than the Cl atom. A central P atom can accommodate only four Br atoms, as in the tetrahedral PBr_4^+ ion. But the central P atom can accommodate either four Cl atoms, as in the tetrahedral PCl_4^+ ion, or six, as in the octahedral PCl_6^- ion.

CHAPTER 23

Concept Assessment 23-1. The electron configuration of Fe is $[Ar]3d^64s^2$ and for Fe^{3+}, $[Ar]3d^5$. The $3d^5$ subshell is half-filled and especially stable. Cobalt ($[Ar]3d^74s^2$) and nickel ($[Ar]3d^84s^2$) must lose four and five electrons, respectively, to achieve a half-filled $3d$ subshell. They simply lose the two $4s$ electrons instead. **23-2.** Both involve reducing metallic compounds to the free metal, often from the same ores. Pyrometallurgy employs high temperatures, yields impure metals that must be refined, and generates gaseous emissions and solid wastes. Hydrometallurgy involves leaching desired metal ions into an aqueous solution, followed by chemical or electrolytic reduction to the metal. Lower temperatures are used, gaseous emissions are largely eliminated, but liquid waste solutions are generated. **23-3.** $Fe_2O_3(s) + 3 H_2(g)$ $\xrightarrow{\Delta}$ $2 Fe(s) + 3 H_2O(g)$ **23-4.** The $[Cr_3O_{10}]^{2-}$ anion consists of three tetrahedral structures arranged around the backbone $Cr-O-Cr-O-Cr$. The central Cr atom of the backbone is bonded to two additional O atoms and the Cr atoms at the ends to three other O atoms for a total of 10 O atoms. The O. S. of Cr is +6 and that of O is −2 (accounting for the 2− charge on the anion). Similar anions are polysilicate (Fig. 21-31) and polyphosphate (Fig. 22-22). **23-5.** The CO groups in $Fe(CO)_5$ are neutral molecules; the sum of the O.S. of the C and O is 0, and the O.S. of Fe is also 0. **23-6.** Au ($Z = 79$, group 11) has the electron configuration $[Xe]4f^{14}5d^{10}6s^1$; the electrons lost in forming Au^{3+} are the 6s and two of the 5d, resulting in $[Xe]4f^{14}5d^8$. **23-7.** The five pairs of electrons around the Cd^{2+} central ion in $[CdCl_5]^{3-}$ is consistent with trigonal bipyramidal molecular geometry (see Table 10-1).

CHAPTER 24

Concept Assessment 24-1. Only one possibility for a six-coordinate complex—$[AlCl_3(H_2O)_3]$—a non-electrolyte that does not conduct

electricity and yields no precipitate with $AgNO_3(aq)$. Three possible four-coordinate complexes: $[AlCl(H_2O)_3]Cl_2$, $[AlCl_2(H_2O)_2]Cl \cdot H_2O$, and $[AlCl_3(H_2O)] \cdot 2 H_2O$. The first two of these three can be differentiated by conductivity measurements; the first is the better conductor. The third cannot be distinguished from the six-coordinate complex by Werner's method. **24-2.** Three N atoms can donate an electron pair, so the ligand is tridentate. **24-3.** The formula of the coordination compound is $K_3[FeBr_2Cl_2(OH)_2]$ and the name is potassium dibromodichlorodihydroxoferrate(III). **24-4.** As seen from the models in the margin, substitution of a fourth Cl^- in the *mer*-isomer leads to all four Cl^- ligands in the same plane, but this same isomer cannot be obtained by substituting a fourth Cl^- in the *fac*-isomer. **24-5.** Structures **(a)** and **(d)** are identical and are geometric isomers of **(b)**, **(c)**, and **(e)**. Structures **(e)** and **(c)** are enantiomers, and **(b)** and **(e)** are identical. **24-6 (i)** 4 **(ii)** 3 **(iii)** 5 **(iv)** 2 **(v)** 1 **24-7.** Both ligands form chelates, but $[Cr(EDTA)]^-$ does so in a single step, while $[Cr(en)_3]^{3+}$ requires a succession of three steps. The β_1 for $[Cr(EDTA)]^-$ is much larger than β_1 or β_2 of $[Cr(en)_3]^{3+}$, but the cumulative formation constant, β_3 (or K_f), of $[Cr(en)_3]^{3+}$ should be similar to β_1 (or K_f) of $[Cr(EDTA)]^-$.

CHAPTER 25

Concept Assessment 25-1. Radioactive decay that changes Z produces a different element. This includes α, β^-, β^+ emissions and electron capture. If there is no change in Z (emission of γ rays) the element remains the same. **25-2.** Francium, a radioactive element, is produced in the decay schemes of heavier elements and found only in conjunction with other decay products, not in natural sources of the alkali metals that Fr resembles. **25-3.** Radioactive nuclides with very long half-lives have a very low activity; those with very short half-lives have a high activity but they do not persist long. Those with intermediate half-lives may persist in the environment for a significant period of time at a high activity, making them potentially the most hazardous. **25-4.** Refer to Table 25.2 and Figure 25-7 and focus on magic numbers and the relative thickness and width of the belt of stability for a fixed proton number and a fixed neutron number. The greatest thickness comes at the magic number, $Z = 50$. (Tin has 10 stable isotopes.) The greatest width comes at the magic number, $N = 82$. (There are 7 stable nuclides with 82 n.) **25-5.** Refer to Figure 25-7. The point representing ^{44}Ca falls in the belt of stability; ^{44}Ca is a stable nuclide. ^{57}Cu falls below the $N = Z$ line and in the region of β^+ emission, while ^{100}Zr falls above the belt of stability and decays by β^- emission. ^{235}U falls above and beyond the belt of stability and decays by α-particle emission.

CHAPTER 26

Concept Assessment 26-1. All structures based on F atoms at two of the vertices of a tetrahedron and H atoms at the remaining two are superimposable. Only one molecule has the formula CH_2F_2. Two possibilities exist for four atoms at the corners of a square—two F atoms on one side (cis) or on opposite corners (trans). **26-2.** No. A quaternary carbon atom is bonded to four other carbon atoms.

26-3. 2-pentanol. **26-4.** The generic formula of an alkane is C_nH_{2n+2}, and for an alkyl halide where an X atom replaces one H atom, $C_nH_{2n+1}X$. If we limit the series to straight-chain alkanes with X as a terminal atom, we have $H(CH_2)_nX$. That is, $n = 1$, HCH_2X or CH_3X; $n = 2$, HCH_2CH_2X or CH_3CH_2X; $n = 3$, $HCH_2CH_2CH_2X$ or $CH_3CH_2CH_2X$; and so on. **26-5.** Yes. **26-6.**

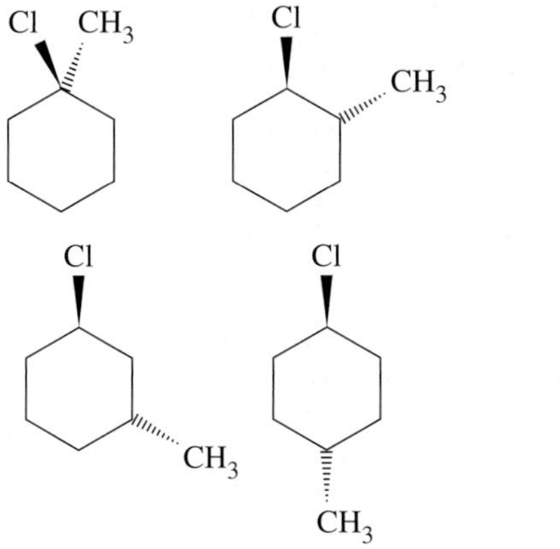

26-7. The conformer with the methyl group in the axial position has higher energy, and so it would release more energy, as heat, if burned. **26-8.** The lower energy conformation is:

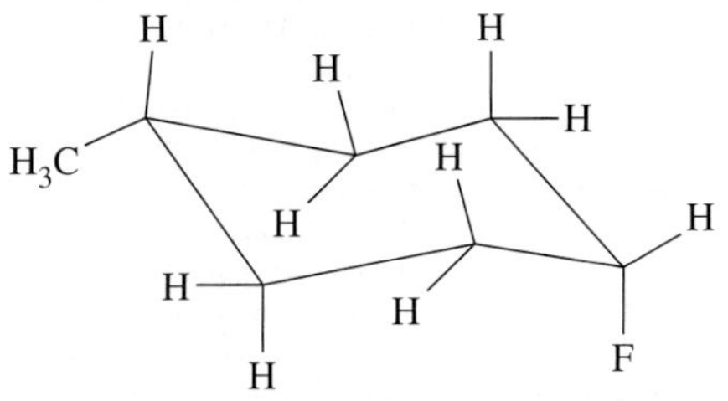

The larger group, $-CH_3$, is in the equatorial position. **26-9.** There are two chiral centers, at the 2nd and 3rd C atoms (to which Br atoms are attached), and four stereoisomers. To show this, sketch a dashed line-wedge structure with the two Br atoms on the same side of the molecule; next to it sketch its nonsuperimposable mirror image. Sketch another structure with the two Br atoms on opposite sides of the molecule and its mirror image, making a total of four stereoisomers. **26-10.**

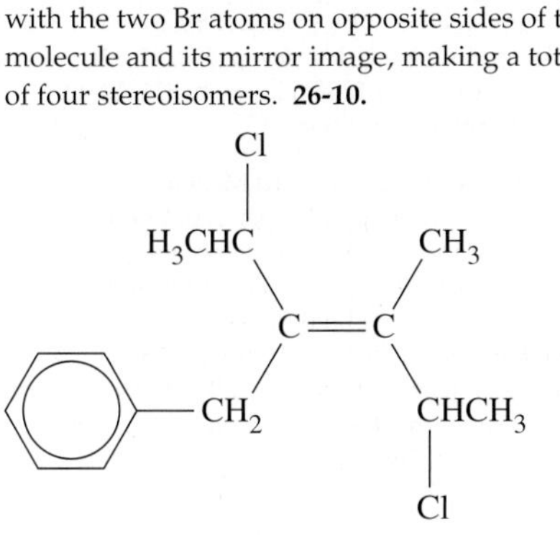

26-11. In butyl acohol $-OH$ attaches to either of the indicated C atoms in the skeleton

$C-C^*-C^*-C$; in both cases the molecule is the same and there is only one s-butyl alcohol. In pentyl alcohol, $-OH$ attaches to one of the indicated C atoms in $C-C^*-C^\#-C^*-C$. Here, the two structures are the same if attachment is at a C^* atom, but a different molecule results if attachment is to the $C^\#$ atom. The name "s-pentyl alcohol" is inadequate. **26-12.**

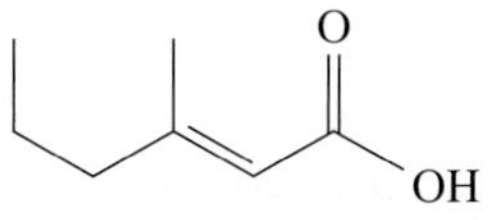

26-13. No, because $C_3H_6O_2$ has only one element of unsaturation. A dialdehyde has two π bonds, and thus two elements of unsaturation.

CHAPTER 27

27-1. CH_3CN is aprotic; NH_3 is protic; $(CH_3)_3N$ is aprotic; $HCONH_2$ is protic; CH_3COCH_3 is aprotic. **27-2.** The minor product is the substitution product, $(CH_3)_3C-OCH_2CH_3$, an ether. Because the substrate is a 3° haloalkane (which disfavors backside attack) and the solvent is polar protic (and stabilizes a carbocation), the ether is formed by the S_N1 mechanism.

27-3.

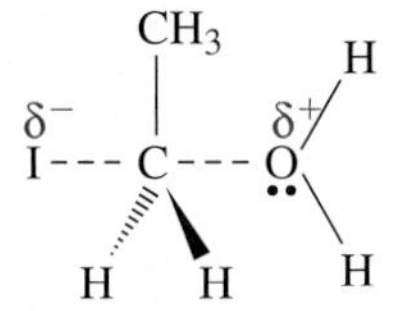

27-4.

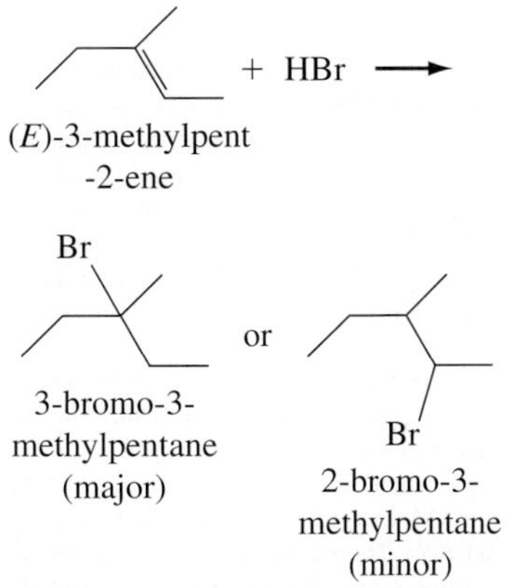

(E)-3-methylpent -2-ene

3-bromo-3-methylpentane (major)   or   2-bromo-3-methylpentane (minor)

27-5. A mixture of enantiomers would be obtained because hydration proceeds through a carbocation. The carbocation can be attacked from above or below by a water molecule, and so both the (R) and (S) configurations of 2-butanol will be produced. **27-6.** The product will be trans-1,2-dibromocyclopentane. If the bromonium ion is formed with the Br atom situated above the plane of the ring, then Br^- will attack from below the plane of the ring because it must attack from the backside. Therefore, the trans isomer is obtained:

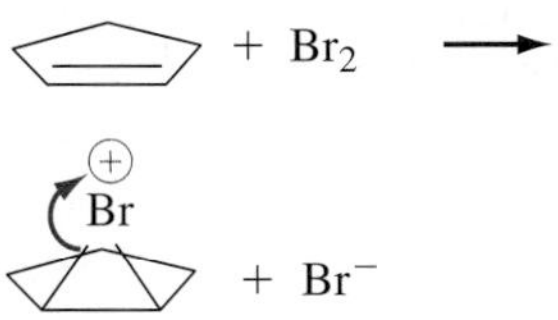

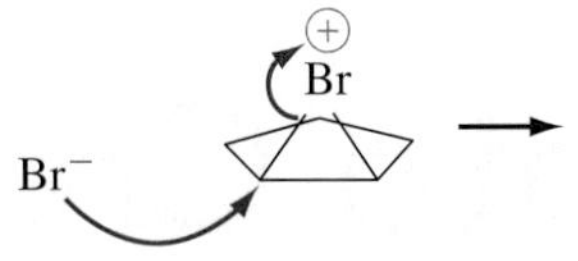

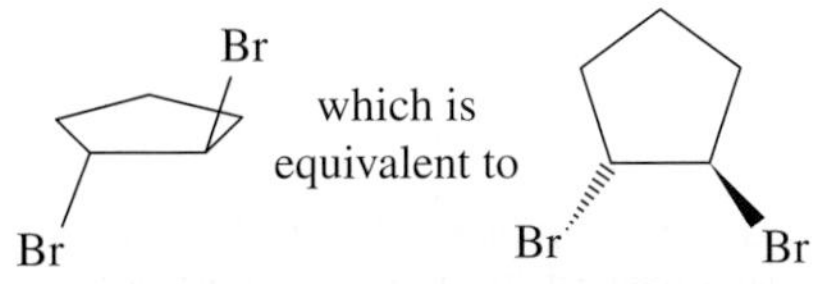

which is equivalent to

trans isomer

27-7. Once formed, CH_3Cl can react with a Cl to form a $H_2ClC\cdot$ radical. A $H_2ClC\cdot$ radical can then react with a $H_3C\cdot$ radical to form $ClCH_2CH_3$. **27-8.** $(CH_3)_2CBrCH_2CH_3$.

CHAPTER 28

Concept Assessment 28-1. The products are 1 mol glycerol and 1 mol each of sodium palmitate, sodium oleate, and sodium linoleate. **28-2.** The mirror image of a (+) enantiomer is the (−) enantiomer, so the mirror image of D-(+)-glucose is L-(−)-glucose. There can be no D-(−)-glucose. **28-3.** This polypeptide structure is shown in Example 28-1. Ionization occurs only at the N-terminal and C-terminal ends of the chain and nowhere else along the chain. Since the N-terminal and C-terminal amino acids are at a pH more than one unit above their isoelectric points, the only significant ionization is at the C-terminal amino acid, which would be present as a 1− anion, making the net charge on the tripeptide also 1−. **28-4.** Helix formation in a protein (Fig. 28-12) requires close proximity of carbonyl and amide groups and formation of hydrogen bonds between them, occurring regularly over an entire macromolecule. In the polysaccharides (Fig. 28-9), O atoms and $-OH$ groups could conceivably form hydrogen bonds between them, but more randomly and not in the very tight helical fashion seen in proteins.

Photo Credits

Index

Selected Physical Constants*

Acceleration due to gravity	g	9.80665 m s^{-2}
Speed of light (in vacuum)	c	$2.99792458 \times 10^8 \text{ m s}^{-1}$
Gas constant	R	$0.0820574 \text{ atm L mol}^{-1} \text{ K}^{-1}$
		$0.08314472 \text{ bar L mol}^{-1} \text{ K}^{-1}$
		$8.314472 \text{ J mol}^{-1} \text{ K}^{-1}$
Electron charge	e^-	$-1.602176462 \times 10^{-19} \text{ C}$
Electron rest mass	m_e	$9.10938188 \times 10^{-31} \text{ kg}$
Planck's constant	h	$6.62606876 \times 10^{-34} \text{ J s}$
Faraday constant	F	$9.64853415 \times 10^4 \text{ C mol}^{-1}$
Avogadro constant	N_A	$6.02214199 \times 10^{23} \text{ mol}^{-1}$

*Committee on Data for Science and Technology (CODATA) Recommended Values of the Fundamental Physics Constants: 2006 (http://physics.nist.gov/constants)

Some Common Conversion Factors

Length
1 meter (m) = 39.37007874 inches (in.)

1 in. = 2.54 centimeters (cm) (exact)

Mass
1 kilogram (kg) = 2.2046226 pounds (lb)

1 lb = 453.59237 grams (g)

Volume
1 liter (L) = 1000 mL = 1000 cm³ (exact)

    1 L = 1.056688 quart (qt)

1 gallon (gal) = 3.785412 L

Pressure
1 atmosphere (atm) = 101.325 kilopascals (kPa) (exact)

             = 1.01325 bar (exact)

             = 760 Torr (exact)

1 Torr ≈ 1 millimeter of mercury (mmHg)

Energy
1 joule (J) = 1 N m = $1 \text{ kg m}^2 \text{ s}^{-2}$

1 calorie (cal) = 4.184 J (exact)

1 kPa L = 1 J

1 bar L = 100 J

1 atm L = 101.325 J (exact)

1 electronvolt (eV) = $1.602176462 \times 10^{-19}$ J

1 eV/atom = $96.485 \text{ kJ mol}^{-1}$

1 kilowatt hour (kWh) = 3600 kJ (exact)

Mass-energy equivalence:

    1 unified atomic mass unit (u)

      = $1.66053873 \times 10^{-27} \text{ kg}$

      = 931.4866 MeV

Force
1 newton (N) = 1 kg m s^{-2}

Some Useful Geometric Formulas

Perimeter of a rectangle = $2l + 2w$

Circumference of a circle = $2\pi r$

Area of a rectangle = $l \times w$

Area of a triangle = $\frac{1}{2}(\text{base} \times \text{height})$

Area of a circle = πr^2

Area of a sphere = $4\pi r^2$

Volume of a parallelepiped = $l \times w \times h$

Volume of a sphere = $\frac{4}{3}\pi r^3$

Volume of a cylinder or prism = (area of base) × height

$\pi \approx 3.14159$